VOLUME TWO

MEDICAL
PHYSIOLOGY

VOLUME TWO

MEDICAL PHYSIOLOGY

Edited by

VERNON B. MOUNTCASTLE, M.D.

Professor and Director, Department of Physiology
The Johns Hopkins University
Baltimore, Maryland

FOURTEENTH EDITION

with 1668 illustrations

The C. V. Mosby Company

ST. LOUIS • TORONTO • LONDON   1980

FOURTEENTH EDITION

The C. V. Mosby Company
11830 Westline Industrial Drive, St. Louis, Missouri 63141

Library of Congress Cataloging in Publication Data

Mountcastle, Vernon B
   Medical physiology.

   Bibliography: p.
   Includes index.
   1. Human physiology. I. Title.
QP34.5.M76   1980   612   79-25943
ISBN 0-8016-3560-8

GW/VH/VH 9 8 7 6 5 4 3 2 1   01/B/006

Contributors

G. D. AURBACH

National Institutes of Health
Bethesda, Maryland

LLOYD M. BEIDLER

Florida State University
Tallahassee, Florida

JOHN D. BIGGERS

Harvard University
Boston, Massachusetts

F. J. BRINLEY, Jr.

National Institutes of Health
Bethesda, Maryland

JOHN R. BROBECK

University of Pennsylvania
Philadelphia, Pennsylvania

CHANDLER M. BROOKS

State University of New York
Brooklyn, New York

KENNETH T. BROWN

University of California
San Francisco, California

C. LOCKARD CONLEY

The Johns Hopkins University
Baltimore, Maryland

ROBERT D. DeVOE

The Johns Hopkins University
Baltimore, Maryland

ROBERT M. DOWBEN

The University of Texas
Dallas, Texas

ARTHUR B. DuBOIS

Yale University
New Haven, Connecticut

NORMAN GESCHWIND

Harvard University
Boston, Massachusetts

MOÏSE H. GOLDSTEIN, Jr.

The Johns Hopkins University
Baltimore, Maryland

H. MAURICE GOODMAN

University of Massachusetts
Worcester, Massachusetts

CARL W. GOTTSCHALK

University of North Carolina at Chapel Hill
Chapel Hill, North Carolina

JAMES D. HARDY

Yale University
New Haven, Connecticut

DALE A. HARRIS

Harvard University
Boston, Massachusetts

THOMAS R. HENDRIX

The Johns Hopkins University
Baltimore, Maryland

ELWOOD HENNEMAN

Harvard University
Boston, Massachusetts

JAMES C. HOUK

Northwestern University
Chicago, Illinois

KIYOMI KOIZUMI

State University of New York
Brooklyn, New York

CHRISTIAN J. LAMBERTSEN

University of Pennsylvania
Philadelphia, Pennsylvania

WILLIAM E. LASSITER

University of North Carolina at Chapel Hill
Chapel Hill, North Carolina

PETER C. MALONEY

The Johns Hopkins University
Baltimore, Maryland

JANICE W. MARAN

McNeil Laboratories
Fort Washington, Pennsylvania

THOMAS H. MAREN

University of Florida
Gainesville, Florida

DONALD J. MARSH

University of Southern California
Los Angeles, California

JEAN M. MARSHALL

Brown University
Providence, Rhode Island

LORNE M. MENDELL

Duke University
Durham, North Carolina

WILLIAM R. MILNOR

The Johns Hopkins University
Baltimore, Maryland

VERNON B. MOUNTCASTLE

The Johns Hopkins University
Baltimore, Maryland

WILLIAM L. NASTUK

Columbia University
New York, New York

BARRY W. PETERSON

Rockefeller University
New York, New York

JAMES M. PHANG

National Institutes of Health
Bethesda, Maryland

GIAN F. POGGIO

The Johns Hopkins University
Baltimore, Maryland

SID ROBINSON

Indiana University
Bloomington, Indiana

ANTONIO SASTRE

The Johns Hopkins University
Baltimore, Maryland

W. T. THACH, Jr.

Washington University
St. Louis, Missouri

PAOLA S. TIMIRAS

University of California at Berkeley
Berkeley, California

LESTER VAN MIDDLESWORTH

The University of Tennessee
Memphis, Tennessee

GERHARD WERNER

University of Pittsburgh
Pittsburgh, Pennsylvania

GERALD WESTHEIMER

University of California at Berkeley
Berkeley, California

VICTOR J. WILSON

Rockefeller University
New York, New York

F. EUGENE YATES

University of Southern California
Los Angeles, California

Preface

TO FOURTEENTH EDITION

The general principles by which this textbook is organized remain those described in the preface to its twelfth edition, namely, to present mammalian physiology as an independent biologic discipline as well as a basic medical science. Two new sections appear in this edition, the first on the principles of system theory as applied to physiology and the second on the physiology of development and aging. Sixty-five chapters of the present edition either are wholly new (ten) or have been extensively revised (fifty-five); twelve chapters remain essentially as they appeared in the thirteenth edition.

This edition has been written by forty-five authors, of whom twelve have joined this effort for the first time. Forty-one of these writers are continually engaged in research and teaching in physiology. Each has taken time from that dedicated life to summarize here the present state of knowledge in a particular field of interest. Whatever value this book possesses is wholly due to the contributors' depth of understanding, skill of exposition, and devotion to the task. For this I am indebted to each.

For them and for myself I wish to thank those authors and publishers who have allowed us to reproduce illustrations previously published elsewhere.

Vernon B. Mountcastle

Preface

TO TWELFTH EDITION

The twelfth edition of *Medical Physiology* presents a cross section of knowledge of the physiologic sciences, as viewed by a group of thirty-one individuals, twenty-three of whom are actively engaged in physiologic research and teaching. Each section of the book provides statements of the central core of information in a particular field of physiology, reflecting, by virtue of the daily occupations of its authors, the questioning and explorative attitude of the investigator and indeed some of the excitement of the search. These statements vary along a continuum from those with a high probability for continuing certainty to those that are speculative but, it is hoped, of heuristic value. An attempt has been made to maintain a balanced point of view. I hope this book will convey to the student who reads it the fact that physiology is a living and changing science, continuously perfecting its basic propositions and laws in the light of new discoveries that permit new conceptual advances. The student should retain for himself a questioning attitude toward all, for commonly the most important advances are made when young investigators doubt those statements others have come to regard as absolutely true. This is not a book that sets forth in stately order a series of facts which, if learned, will be considered adequate for success in a course in physiology. Many such "facts" are likely to be obsolete before the student of physiology reaches the research laboratory, or the student of medicine the bedside. Nor is it a book that provides ready-made correlations and integrations of the various fields of physiology necessary for a comprehensive understanding of bodily function. Those integrations are an essential part of scholarly endeavor not readily gained from books alone. It is my hope, however, that study of this book, combined with laboratory experience and scholarly reflection, will provide the student with a method and an attitude that will serve him long after the concepts presented here are replaced by new and more cogent ones.

The title *Medical Physiology* has been retained, for one of the purposes of this edition, in common with earlier ones, is "to present that part of physiology which is of special concern to the medical student, the practitioner of medicine, and the medical scientist in terms of the experimental inquiries that have led to our present state of knowledge." The scope of the book was and is still broader, however, and attempts to present mammalian physiology as an independent biologic discipline as well as a basic medical science. Mammalian physiology has its base in cellular physiology and biophysics, and it is from this point of view that many of the subjects treated here are approached. Above all, mammalian physiology must deal with problems of the interactions between large populations of cells, organs, and organ systems and, finally, the integrated function of an entire animal. Physiology thus must bridge the distance from cellular biology on the one hand to systems analysis and control theory on the other: each is important and any one is incomplete without the others. This approach to the problems of internal homeostasis, of reaction to the environment, and of action upon the environment is evidenced in several sections of this book.

Of the eighty chapters composing this book, twenty-nine are wholly new in this edition; forty-five from the last edition have been extensively revised either by their original authors or by new ones. Six have been allowed to stand substantially as previously written, for these seemed to comprise as balanced and modern a survey as any presently possible. The names and affiliations of my colleagues in this effort have been listed. They have taken time from busy lives to survey their fields of interest; for this I am greatly indebted to each. If this book possesses any worth it is in large part due to their continuing devotion to the task of its preparation.

For them and for myself I wish to thank those authors and publishers who have allowed us to reproduce illustrations previously published elsewhere.

Vernon B. Mountcastle

Contents

xi

VOLUME TWO

MEDICAL PHYSIOLOGY

IX

THE CIRCULATION

35 Cardiovascular system

WILLIAM R. MILNOR

The heart and blood vessels form a transportation system that delivers to all cells of the body the materials needed for their proper function and carries away the waste products of their metabolism. By bringing oxygen and nutritive materials to the cells and relieving them of carbon dioxide and other metabolites, the circulation provides a means of communication between the cells and the external environment. In the higher reaches of the evolutionary scale the circulation acquires additional functions (e.g., the dissemination of *hormones*) which are important to many cells of the body but manufactured only by specialized tissues or organs, and the distribution of fats and carbohydrates from storage depots to regions where they are utilized. A number of mechanisms for the defense of the organism are also transported by the vascular system and can be delivered by way of the bloodstream to regions of injury or invasion. These include leukocytes, certain proteins concerned with immunity (immune antibodies), and materials involved in blood coagulation, which not only guard the integrity of the vascular system but also provide in damaged areas a framework of fibrin on which new tissue can be built. Circulation also contributes to the control of body temperature by carrying heat generated by the processes of metabolism to areas where it can be dissipated.

The bloodstream communicates with the cells it supplies only indirectly, through the mediation of an *extracellular fluid* in which all cells are bathed. Substances that leave the vascular system through the walls of its smallest branches, the capillaries, enter the extracellular fluid and then penetrate the cell membrane to reach the interior of the cell. The composition of this pervasive extravascular and extracellular fluid, which Claude Bernard[1] termed the *milieu interieur,* is kept constant by a host of regulatory mechanisms, enabling the organism as a whole to adapt to a wide range of external conditions. This concept of a constant internal environment, communicating on the one hand with the circulating blood and on the other with the cells, is fundamental in modern physiology.

The cardiovascular system includes the heart, arteries, capillaries, and veins, all differing in function as well as structure. The heart provides the motive force to pump blood through the system, the arteries carry blood away from the heart to the tissues of the body, and the veins return blood to the heart again—a circular motion that escaped recognition for centuries. The arteries are a branching system of vessels that subdivide in a more or less orderly fashion into increasingly numerous branches of diminishing caliber and eventually lead into the capillaries. Capillaries are thin-walled vessels a few thousandths of a millimeter in diameter, extending from the smallest terminal branches of the arterial tree to the equally small branches that form the beginning of the venous system. They anastomose freely with their neighbors in a diffuse network quite different from the relatively systematic branching of the arterial and venous vessels. The capillary channels are in close contact with the cells they serve, although extracellular fluid always intervenes between the capillary wall and the cell membrane, and substances are delivered to or removed from the immediate environment of the cell by passage through the capillary wall. Blood travels from these capillary beds into the veins, which form a branching system like that of the arteries in reverse, joining together in fewer vessels of increasing diameter as they approach the heart.

The circulating blood traverses two such systems in succession: the *systemic circulation,* which supplies the tissues of the body, and the *pulmonary circulation,* which carries blood through the lungs. Blood is pumped into the systemic circulation by the left ventricle, returns to the right atrium, is pumped into the pulmonary circulation by the right ventricle, then flows back from the lungs to the left side of the heart where it beings its journey again. This continuous circulation of the blood out to the most distal parts of the body and back to the heart was first clearly described by William Harvey (1578-1657).[5] Fragments of evidence pointing to the circular motion of the blood had been encountered by earlier observers, but this evidence was ignored or misinterpreted because it conflicted with theories that had been accepted dogmatically since the writings of Galen in the second century. Harvey's careful experiments and observations, recorded in 1628, refuted this uncritical acceptance of ancient authority and established firmly the principle "that blood is pushed by the beat of the left ventricle and distributed through the arteries to the whole body, and back through the veins to the vena cava, and then returned to the right auricle, just as it is sent to the lungs through the pulmonary artery from the right ventricle and returned from the lungs through the pulmonary vein to the left ventricle. . . ."[5]

The vascular system properly includes not only the arteries, capillaries, and veins of the systemic and pulmonary circulations but another network called *lymphatic vessels* (Chapter 43). These begin in extracellular spaces as small, thin-walled vessels resembling the blood capillaries in size and permeability but closed at one end. The "lymph capillaries" lead into larger lymphatics, which converge into progressively wider channels until they drain into large veins near the heart. The fluid *lymph* contained in these channels carries solutes that diffuse into it from the extracellular fluid and returns them to the circulating blood. Many of the larger lymph channels are supplied with valves that direct the flow toward the veins and pass through one or more *lymph glands.*

The heart. In some biologic forms the heart—or at least the organ that moves blood—is a simple muscular tube. The evolutionary development of this uncomplicated structure into a heart with four chambers, valves, and independent outflows to the pulmonary and systemic circulation is retraced in the growth of the mammalian embryo. The sequence of development, which explains many features of the structure and function of the human heart, has been carefully summarized by Patten.[7]

As we ascend the evolutionary scale, this tube gradually becomes divided into specialized regions, so that the blood first enters a relatively thin-walled muscular sac (atrium) and then passes through a valve into a thick-walled muscular chamber (ventricle) that propels the blood onward by its contractions. Septa appear in the atrium and ventricle, and in birds and mammals these partitions become complete, producing in

effect two conjoined pumps, each with two chambers. The right atrium and ventricle receive blood from the systemic veins and send blood to the lungs, whereas the left atrium and ventricle receive blood from the pulmonic veins and eject it into the systemic arteries. This complete separation into "right heart" and "left heart" ensures that blood leaving the lungs, which is relatively rich in oxygen and poor in carbon dioxide, does not mix with blood returning from the body tissues through the systemic veins. The progress of blood through the heart is controlled at the entrance and exit of each ventricle by one-way valves that prevent retrograde flow from the outflow vessels (pulmonary artery or aorta) to the ventricles and from the ventricles to the atria. Atrial and ventricular contractions follow an orderly sequence determined by the pacemaker of the heart and by the conduction paths through myocardium and specialized tissues (Chapter 36). The strength of these contractions and the volume of blood ejected by the ventricles are controlled by the autonomic nervous system, by substances carried in the blood, and by physicochemical characteristics of the myocardial fibers and their contractile proteins (Chapter 37).

Arteries. As the arterial tree divides into successive branches, the individual vessels change gradually in physical dimensions and in the structure of their arterial wall, with important functional consequences. The data in Table 35-1, which represent estimates for a hypothetical dog weighing 13 kg, give some idea of the number and dimensions of arteries and other segments of the circulatory system. The values given are rough approximations based on a number of reasonable but unproved assumptions, and the classification into segments is somewhat arbitrary.[13,19]

Except for the major branches leaving the aortic arch and the bifurcation of the main pulmonary artery into right and left trunks, the total cross section of the arterial bed increases with each branching toward the periphery. In the simplest form of branching, where a single vessel divides in two, the ratio of the total cross section of the two branches beyond the bifurcation to that of the parent vessel usually ranges from 1.2 to 1.7. Subdivisions do not occur with perfect geometric regularity, of course, and multiple branching at some sites yields a more rapid increase in total cross section, whereas at others a single small lateral branch may issue from a much larger parent vessel. The net effect is a rapid increase of total cross section in the periphery of the arterial bed, so that total cross-sectional area increases more than 100-fold between aorta

Table 35-1. Changes that take place with branching of mesenteric artery of dog*

	Total number	Radius (cm)	Total cross-sectional area (cm²)	Length (cm)	Mean velocity (cm/sec)	Pressure fall (mm Hg)	Volumes (% of total)
Mesenteric artery	1	0.15	0.07	6.0	16.80	0.8	2.6
Main branches	15	0.05	0.12	4.5	10.10	3.2	3.4
End branches	45	0.03	0.13	3.91	9.30	7.4	3.2
Short and long intestinal branches	1,899	0.0068	0.20	1.42	5.80	23.5	1.7
Last branches	26,640	0.0025	0.57	0.11	2.10	7.2	0.4
Branches to villi	328,500	0.00155	2.48	0.15	0.48	5.4	2.3
Arteries of villi	1,051,000	0.00122	4.18	0.20	0.28	8.1	5.3
Capillaries of villi	47,300,000	0.00040	23.78	0.04	0.05	2.4	5.9
						58.0	24.8
Veins at base of villi	2,102,400	0.00132	11.59	0.10	0.10	1.03	7.2
Veins before entering submucosa	131,400	0.00375	5.80	0.10	0.20	0.29	3.6
Last branches of submucosa	18,000	0.0064	2.32	0.15	0.51	0.37	2.2
Last branches of short intestinal veins	28,800	0.0032	0.93	1.1	1.30	2.50	6.4
Long and short intestinal veins	1,899	0.0138	0.84	1.42	1.40	1.40	7.4
Last branches of mesenteric vein	45	0.075	0.79	3.91	1.50	0.22	19.2
Branches of mesenteric vein	15	0.12	0.67	4.5	1.70	0.07	18.7
Mesenteric vein	1	0.3	0.28	6.0	4.2	0.05	10.5
						5.93	75.2

*After Mall; modified from Schleier.[19]

and terminal systemic arterioles and is still greater in the systemic capillary bed (Table 35-1). Since the same average volume of blood passes through each cross section of the circulation per unit time, this overall widening of the bed results in a decrease in the velocity of blood flow as it moves through the arterial tree. Blood moves through the ascending aorta at an average velocity of about 20 cm/sec but slows with progressive branching of the arterial tree until its velocity in the capillaries is probably less than 0.2 cm/sec. Beyond the capillaries the process is reversed, and the flow of blood accelerates as the total cross-sectional area of the veins decreases.

The branching pattern and changing dimensions of the arteries are major factors in determining the resistance to flow in different parts of the bed. The concept of vascular resistance is described in detail in Chapter 39, but in general terms, it expresses the principle that the flow produced by a constant driving force is a function of the radius and length of the vessel and of the physical properties of blood. The smaller the radius and the greater the length, the higher the resistance to flow. The total resistance of a vascular bed depends on the number of parallel channels as well as on the dimensions of the individual vessels.

One manifestation of resistance to flow is the constant drop in pressure within the vascular tree from arteries to veins, which can be represented by the "pressure profile" shown in Fig. 35-1. As this diagram shows, the major part of the total fall in pressure from aorta to inferior vena cava occurs in the vessels smaller than 0.5 mm in diameter. The pressure drop across the capillary bed itself is difficult to measure exactly but is probably in the neighborhood of 20 mm Hg in the systemic capillaries—slightly greater than the drop in the venules but much less than that in the smallest arteries and arterioles, where almost half the total resistance to flow occurs.

The proximal and distal parts of the arterial tree differ not only in the dimensions of individual vessels but also in the structure of the vascular wall.[2] The walls of virtually all blood vessels, with the exception of capillaries, contain varying amounts of smooth muscle,[6] elastin, and collagen, as indicated diagrammatically in Fig. 35-2. The aorta and pulmonary artery, together with their initial branches, contain relatively large amounts of elastin and are sometimes called *elastic arteries*, whereas the more distal branches with their higher concentration of smooth muscle are termed *muscular arteries*. No sharp boundary marks the end of one type and the beginnings of the other, but the distinction serves to emphasize

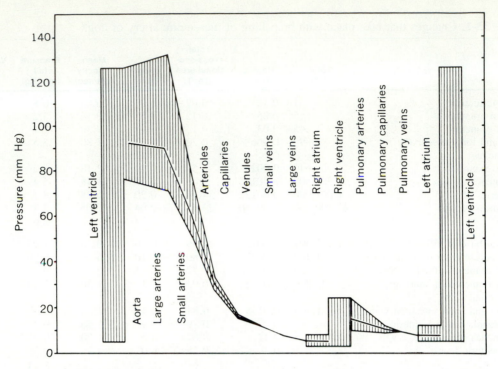

Fig. 35-1. Typical pressures in human cardiovascular system. Solid line indicates mean pressures, and shaded area indicates pulsations in systole and diastole.

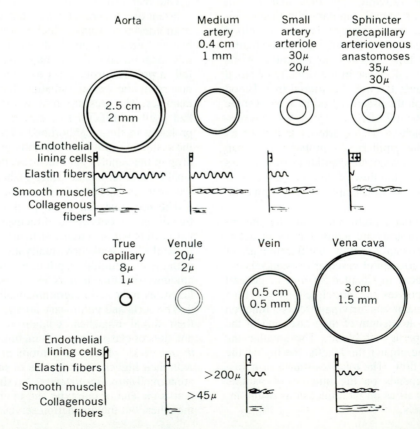

Fig. 35-2. Size, wall thickness, and admixture of four basic tissues in walls of different blood vessels. First figure below name of vessel represents diameter of lumen; figure below that indicates thickness of wall. (From Burton.[2])

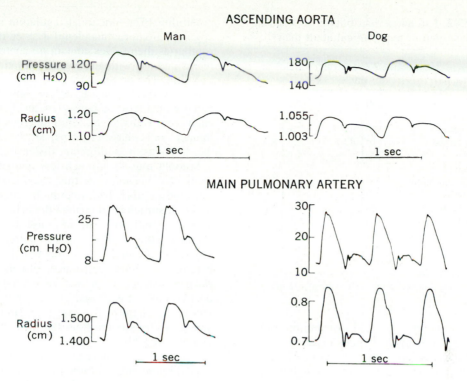

Fig. 35-3. Pulsatile pressures and changes in radius in ascending aorta and main pulmonary artery in man and in dog. (From Patel et al.[15])

a difference in function. The elastic arteries, by reason of their distensibility and proximity to the ventricles, are distended by each volume of blood discharged from the ventricles in systole. The energy thus stored in the arterial wall is returned to the bloodstream at the end of each ventricular contraction, after the aortic and pulmonic valves close, as the elastic recoil of the wall maintains a flow of blood from the large arteries into the peripheral bed throughout diastole. The increase in diameter of the aorta with ventricular systole is not very large—about 6% in the ascending aorta and 2% in the abdominal aorta[15]—but it is sufficient to transform the discontinuous ejections of blood from the heart into a pulsatile but continuous flow. The pulmonary artery is more distensible than the aorta, and its diameter normally increases about 10% with each stroke ejected by the ventricle (Fig. 35-3).

As the branches of the arterial tree extend peripherally, they contain increasing amounts of smooth muscle, and the ratio of wall thickness to lumen increases (Fig. 35-2). The small arteries and arterioles are richly supplied with smooth muscle, and the contraction or relaxation of this muscle can produce significant alterations in the diameter of the lumen. Since resistance to blood

flow depends in part on this diameter (Chapter 39), the activity of smooth muscle in the peripheral arteries is one of the factors controlling blood flow. Smooth muscle contraction in the arterioles and small arteries of any organ will tend to reduce the cross section of these vessels, raising their resistance to flow so that less of the total cardiac output flows through that organ. Contraction of smooth muscle in the larger arteries changes the elastic properties of the vascular wall, but it has little effect on their diameter. The metabolic needs of the cells that make up the arterial wall are supplied in two ways. The endothelial cells lining the vascular lumen are supplied directly by diffusion from the adjacent bloodstream, as are the innermost parts of the media. The midportion and outer layers of the media have their own blood supply, the *vasa vasorum,* tiny vessels that originate from branches of the artery itself and return to penetrate its adventitia.

Thus the arteries are not simply passive conduits but a system in which structure and function change gradually from the origin at the heart to the periphery. The large proximal vessels act as a reservoir to maintain flow during diastole of the heart, whereas the smaller peripheral ves-

Table 35-2. Estimated distribution of blood in vascular system of hypothetical adult man*

Region	Volume	
	ml	%
Heart (diastole)	360	7.2
Pulmonary		
Arteries	130 ⎫	2.6 ⎫
Capillaries	110 ⎬ 440	2.2 ⎬ 8.8
Veins	200 ⎭	4.0 ⎭
Systemic		
Aorta and large arteries	300 ⎫	6.0 ⎫
Small arteries	400 ⎪	8.0 ⎪
Capillaries	300 ⎬ 4,200	6.0 ⎬ 84.0
Small veins	2,300 ⎪	46.0 ⎪
Large veins	900 ⎭	18.0 ⎭
	5,000	100

*Age, 40 years; weight, 75 kg; surface area, 1.85 m².

sels control the pressure and distribution of flow in the capillary beds of various organs.

Microcirculation. The vascular bed lying between the arterioles and venules is complex and variable in pattern and includes not only the capillaries but also certain small vessels that differ from true capillaries in structure and function. This whole network of microscopic arteriovenous pathways is comprehended in the general term "microcirculation" (Chapter 43). *Capillaries* are thin-walled endothelial tubes that lack smooth muscle and adventitial layers. The vital functions of the capillary bed are carried on through the capillary walls, which transfer solutes in accordance with their permeability and the physicochemical forces acting the extracellular fluid and in the capillary lumen. The volume of blood in the capillaries is only a small part of the total blood volume (Table 35-2), but their functional surface area is enormous. Extrapolation from the meager data available suggests a surface area of perhaps 60 m² for the systemic capillaries and 40 m² for the capillaries of the lung in adult human subjects.

In many regions an intermediate type of vessel, or *metarteriole,* intervenes between the terminal arteriole and capillary. The metarteriole has a distinct muscular coat near its arteriolar origin but none in its more distal portion.

Arteriovenous anastomoses constitute a third class of vessels found in the microcirculation of many tissues. Through these anastomoses, which have a muscular coat and rich innervation, blood can be shunted directly from arteriole to venule, bypassing the capillary channels. The possibilities for functional regulation in the microcirculation are thus plentiful, even though the capillaries themselves contain no contractile elements and are without direct innervation.

Veins. The architecture of the venous system resembles that of the arterial tree; however, compared with arteries at equivalent levels in the vascular tree, veins have more numerous branches, thinner walls, larger lumens, less smooth muscle, less elastin, and greater distensibility.

Smooth muscle is much less abundant in the walls of the *venules,* or tiny veins leading away from the capillary bed, than in the arterioles, but it is still sufficient to change the caliber of these vessels significantly in response to neural or chemical stimuli. The venules control postcapillary resistance in the same way that the arterioles control precapillary resistance, although the arteriolar resistance is greater in magnitude. This dual control, at each end of the capillaries, provides a mechanism for adjusting capillary pressure as well as controlling the flow through the vascular bed.[9,14]

The size and number of venous branches is reflected in the large volume of blood contained in the systemic veins, which is more than half the volume of the entire cardiovascular system (Table 35-2). Relatively small changes in the diameter of the venous channels can therefore alter the volume, or capacity, of the venous system by several hundred milliliters. Since the total blood volume does not change rapidly under normal conditions, the blood vessels are in effect a closed system, and any change in the capacity of the veins must induce a redistribution of blood within the circulation. In this way the venous capacity, controlled by the smooth muscle in the walls of the veins, influences blood pressure, blood flow, and regional blood volumes throughout the cardiovascular system. The marked distensibility of veins is an additional factor in determining the volume distribution of blood, allowing fairly large changes in venous volume with only small alterations in venous pressure. Because of this ability to increase or diminish in volume to accommodate the needs of the circulation, the venous system is sometimes described as a blood reservoir.

There is little resistance to flow through venous channels beyond the venules, as the small pressure decrements shown in Fig. 35-1 indicate. The ability of the venous bed to alter its capacity by small changes in the dimensions of numerous individual vessels makes possible significant changes in venous capacity with negligible effects on this resistance. The contrast in this re-

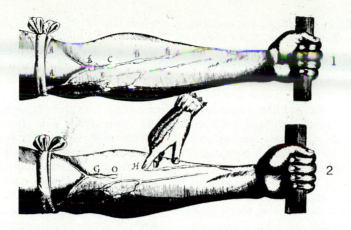

Fig. 35-4. From William Harvey's presentation of experiments on venous flow. See description in text.

spect between the veins, with their large capacity and low resistance, and the arterioles, in which marked changes in resistance are accomplished with relatively little change in volume, has led Folkow,[9] Folkow and Neil,[3] and others[14] to distinguish between *capacitance vessels* and *resistance vessels*. This dichotomy should be regarded as functional rather than anatomic, for vessels midway in size between the larger vessels and the arterioles—small arteries or veins from 2 to 0.5 mm in diameter—influence both capacitance and resistance to a measurable degree.

One further characteristic is the presence of valves at more or less regular intervals along the larger veins of the extremities, preventing retrograde flow of blood. These valves are readily demonstrated in superficial veins without dissection (Fig. 35-4), as William Harvey pointed out when he reasoned that blood flow in the veins was always toward the heart.

The valves are present solely that blood may not advance from the center of the body into the periphery through the veins, but rather from the extremities to the center. This fact may be more clearly shown by tying off an arm of a subject as if for blood letting. *(AA,* Fig. 35-4, *1.)* At intervals (especially in rustics) knots or swellings will appear *(B* to *F)* not only where there is branching but also where there is none. These are caused by the valves. If you will clear the blood away from a nodule or valve by pressing a finger below it *(H,* Fig. 35-4, *2),* you will see that nothing can flow back, being entirely prevented by the valve. If you press downward against valve *O,* you will note that nothing can be forced through it. From many such experiments it is evident that the function of the valves in the veins is the same as that of the three sigmoid valves placed at the opening of the aorta and pulmonary artery—to prevent, when they are tightly closed, the reflux of blood passing over them.*

These valves play a large role in returning blood to the heart from the extremities when the body is in the erect position, for the compression of deeper veins by contractions of the skeletal muscle through which they pass helps move the blood toward the venae cavae. This "muscle pumping" may contribute a significant amount of energy toward circulating the blood during running or other muscular exercise, although quantitative experimental evidence on this point is not available at present.

Blood. The volume and physical properties of the fluid that fills the blood's vascular system are as important as the characteristics of the heart and blood vessels. In man the total blood volume ranges from 60 to 80 ml/kg of body weight,[18] or about 5 L in an average adult male (Chapter 44). The cardiovascular system is in effect a closed, elastic container (assuming a state of equilibrium of fluid transfer at the capillary level), and the normal blood volume not only fills the system but distends it slightly. The blood vascular system is filled under pressure, so to speak, and the dynamic pressures generated in the system by the heart are superimposed on a static pressure determined by the total blood volume and the capacity of the system, which Guyton[12] has termed the *mean circulatory filling pressure.* This static filling pressure can be deter-

---

*From Harvey, W.: Exercitatio anatomica de motu cordis et sanguinis in animalibus, Frankfort, 1628, Guilielmi Fitzeri. (Translated by C. D. Leake, Springfield, Ill., 1928, Charles C Thomas, Publisher.)

mined experimentally by measuring the uniform pressure found in both arteries and veins after a short period of cardiac arrest. When the heart stops, arterial pressure falls, venous pressure rises, and a static equilibrium is reached throughout the system at a pressure of about 6 mm Hg.[12] As would be expected, the equilibrium pressure is higher when the total blood volume has been artificially increased or when vascular capacity has been diminished by widespread contraction of vascular smooth muscle.[12,17] The relation between vascular capacity and total blood volume is therefore one of the critical variables in circulatory function. When the heart is functioning and blood is circulating in the normal way, the blood volume of each part of the circulation depends on the vascular pressures and distensibility in that particular region, and a kind of dynamic equilibrium is maintained. The resulting volume of different segments of the circulation is summarized in Table 35-2. It is evident that a large part of the blood volume is accommodated in the most distensible vessels, the veins, and it is also evident that substantial shifts of volume from one region to another can be brought about by changes in cardiac function or vascular elasticity.

The physical property of blood that is of greatest interest to the circulatory physiologist is its viscosity, a property lucidly described by Newton as "a lack of slipperiness between the layers of a fluid." The viscosity of blood is intimately involved in the principles of hemodynamics, since the energy required to move a liquid through a tube increases with the viscosity of the liquid (Chapter 39). Blood is a suspension of cells rather than a simple fluid, and its viscosity varies with the concentration of these cells. The presence of cells and other factors discussed in Chapter 39 make it difficult to apply the physical laws that describe viscosity and fluid flow to the circulation with complete accuracy, and the relevant experimental data on blood usually refer to "apparent" viscosity. The "apparent" viscosity of blood varies directly with the concentration of red blood cells, although the relationship is not linear.

The *hematocrit,* or ratio of red blood cell volume to blood volume, normally averages about 0.4 for the total cells and plasma of the body, but the hematocrit of blood in the vascular beds of different organs varies considerably. In the spleen and vascular spaces of the bone marrow the hematocrit may be as high as 0.82,[10] while the circulation of the kidney tissues, excluding large vessels, is relatively poor in red blood cells, with a hematocrit in the region of 0.15.[10] These values refer to the ratio of cell volume to whole blood volume in the organ at any instant, and the rather surprising variations arise from differences in the velocity and vascular distribution of cells and plasma. The mechanism for this nonuniform velocity is still not entirely clear, but the structure of parts of the microcirculation may play a part, some channels diverting the flow of plasma and others of red blood cells. The size and shape of the erythrocytes obviously influence their motion through the smallest capillaries, some of which are considerably narrower than the diameter of the red blood cell. The hemodynamic consequences of this disparity and the deformation of erythrocytes in passing through small vessels have not yet been completely evaluated.[16]

In large vessels the erythrocytes are not evenly distributed across the lumen of the vessel but tend to travel with the axial portion of the stream more rapidly than the average velocity of the plasma. In the large peripheral veins, from which blood samples are usually drawn, the hematocrit is not identical with that of the body as a whole, the average venous hematocrit being about 0.45, whereas the ratio of total red blood cell volume to total blood volume is about 91% of that value.[8]

INTEGRATED CARDIOVASCULAR FUNCTION

The cardiovascular system functions as a coordinated unit, establishing the hemodynamic conditions appropriate to the needs of the moment in each part of the body. Muscular exercise, for example, initiates a set of responses that act together to ensure an adequate blood supply to the contracting skeletal muscles while maintaining the perfusion of other vital organs. The analytic method of studying one segment of the circulation independently of the others is unavoidable in the classroom and laboratory but should be employed as a means of understanding the integrated function of the system as a whole.

The classic experimental procedure in physiology aims at keeping constant all relevant variables save those of immediate interest. This is most readily accomplished in isolated vessels or organs, although it is never possible to control every variable or even to identify all those that are relevant. The disadvantage inherent in such preparations is their failure to reveal the normal interactions between the isolated component and the remainder of the circuit. On the other hand, experiments in vivo demonstrate the circulatory responses that emerge from the interplay of forces in different parts of the system, but the number of variables to be measured is great, and the results are often complicated by the effects of anesthetics or surgical manipulation. Both methods contribute useful information; historically, the method of isolation and strict control has played the greater role in cardiovascular physiology. The reasons for dominance of the analytic method lay partly in the lack of instruments capable of reliable measurements in vivo and partly in the unmanageable number of variables involved. Technical developments in electronics, in transducers that measure pressure, flow, force, and displacement with a high degree of fidelity, and in methods of data collection, storage, and analysis have gone far toward removing these obstacles. The increasing use of mathematics and computers in biology has opened new paths for the study of the circulation as a functional unit, not merely by easing the tasks of numeric and statistical analysis, but by giving the investigator free rein to express his data or hypotheses in the mathematical and logical form that seems most appropriate. The wealth of analytic techniques available in those branches of engineering and technology that afford analogies with the cardiovascular system is evident in the discussion of control systems in Chapter 42 and elsewhere.[4,20] Inanimate control systems, such as the familiar automatic controls for central heating, share with biologic systems the appearance of purposeful action, many simultaneous or consecutive processes cooperating to produce a specific result.

Mathematical models of the circulation[4,11,20] consist of equations that describe the characteristics of each part of the system—the dimensions and elasticity of vessels, the energy released by the myocardium per unit of initial fiber length, the total blood volume, etc.—and the relations between parts. From these formulas the dependent variables, such as blood pressure and flow in the various parts of the system, can be derived. In digital computers the computations are carried out numerically; in analog computers the parts of the circulation are represented by electrical components and results are presented as voltages. To embody the known responses of the circulation in even a rudimentary way requires a large number of mathematical expressions, usually in the form of differential equations.

The making of such models, obviously of little value as an end in itself, has a number of practical uses. First, it requires the investigator to be quite explicit in the interpretation of his experimental results and in the formation of hypotheses. Second, it allows him to compute the consequences of any hypothesis and compare them with the observed facts. Third, it identifies important questions on which present information is incomplete. This uncovering of gaps in our knowledge, one of the major benefits of circulatory models, has emphasized the central importance of ventricular length-tension relationships, for example, and the need for accurate measurements of vessel elasticity. Finally, models help to overcome some of the conceptual difficulties raised by the circular nature of the cardiovascular system. Will partial occlusion of the venae cavae, for example, raise arterial pressure by increasing peripheral resistance or lower arterial pressure by decreasing the filling pressure of the right and eventually the left ventricle? Either response is possible, depending on the initial state of the system and the degree of occlusion, and a mechanical or mathematical model of the circulation can help to explain why this is true. The relevant factors include total blood volume, compliances and resistances of the venous and arterial systems, and a number of other variables. An appropriate model would demonstrate how a change in any one of these variables would interact with the others to establish a new dynamic equilibrium.

REFERENCES
General reviews

1. Bernard, C.: Introduction à l'étude de la medecine expérimentale, Paris, 1865, J. B. Baillière et Fils. (Translated by H. C. Greene, New York, 1957, Dover Publications.)
2. Burton, A. C.: Relation of structure to function of the tissues of the wall of blood vessels, Physiol. Rev. **34:** 619, 1954.
3. Folkow, B., and Neil, E.: Circulation, New York, 1971, Oxford University Press.
4. Grodins, F. S.: Control theory and biological systems, New York, 1963, Columbia University Press.
5. Harvey, W.: Exercitatio anatomica de motu cordis et sanguinis in animalibus, Frankfort, 1628, Guilielmi Fitzeri. (Translated by C. D. Leake, Springfield, Ill., 1928, Charles C Thomas, Publisher.)
6. Mommaerts, W. F. H. M.: Perspectives in the study of arterial muscle. In Lansing, A. I., editor: The arterial wall, Baltimore, 1959, The Williams & Wilkins Co.
7. Patten, B. M.: The development of the heart. In Gould, S. E., editor: Pathology of the heart, Springfield, Ill., 1953, Charles C Thomas, Publisher.

Original papers

8. Chaplin, H., Jr., Mollison, P. L., and Vetter, H.: The body/venous hematocrit ratio: its constancy over a wide hematocrit range, J. Clin. Invest. **32:**1309, 1953.
9. Folkow, B.: The efferent innervation of the cardiovascular system, Verh. Dtsch. Ges. Kreislaufforsch. **24:** 84, 1959.
10. Gibson, J. G., et al.: The distribution of red cells and plasma in large and minute vessels of the normal dog, determined by radioactive isotopes of iron and iodine, J. Clin. Invest. **25:**848, 1946.

11. Grodins, F. S.: Integrative cardiovascular physiology: a mathematical synthesis of cardiac and blood vessel hemodynamics, Q. Rev. Biol. **34:**93, 1959.

12. Guyton, A. C., Polizo, D., and Armstrong, G. G.: Mean circulatory filling pressure measured immediately after cessation of heart pumping, Am. J. Physiol. **179:** 261, 1954.

13. Mall, J. P.: Die Blut und Lymphwege im Dünndarm des Hundes. Akademie der Wissenschaften, Mathematisch Classe, Abh. **14:**151, 1888.

14. Mellander, S.: Comparative studies on the adrenergic neuro-hormonal control of resistance and capacitance blood vessels in the cat, Acta Physiol. Scand. **50:**suppl. 176, 1960.

15. Patel, D. J., Greenfield, J. C., Jr., and Fry, D. L.: In vivo pressure-length-radius relationship of certain blood vessels in man and dog. In Attinger, E. O., editor: Pulsatile blood flow, New York, 1964, McGraw-Hill Book Co.

16. Prothero, J. W., and Burton, A. C.: The physics of blood flow in capillaries. III. The pressure required to deform erythrocytes in acid-citrate-dextrose, Biophys. J. **2:** 213, 1962.

17. Richardson, T. Q., Stallings, J. O., and Guyton, A. C.: Pressure-volume curves in live, intact dogs, Am. J. Physiol. **201:**471, 1961.

18. Samet, P., Fritts, H. W., Jr., Fishman, A. P., and Cournand, A.: The blood volume in heart disease, Medicine **36:**211, 1957.

19. Schleier, J.: Der Energieverbrauch in der Blutbahn, Arch. Gesamte Physiol. **173:**172, 1918.

20. Skalak, R.: Synthesis of a complete circulation. In Bergel, D. H., editor: Cardiovascular fluid dynamics, London, 1972, Academic Press, Ltd.

36

WILLIAM R. MILNOR

Properties of cardiac tissues

The effectiveness of the heart as a pump depends on an orderly spread of excitation through the cardiac muscle, or myocardium. To understand the complex sequence of events that occurs with each heartbeat (Chapter 37), we need to know something about the excitability of cardiac cells and the propagation of response through the tissues of the heart, as well as the fundamental mechanisms of the contractile process. The heart is primarily a muscular organ, made up predominantly of a form of striated muscle that is similar in many ways to skeletal muscle. Two other kinds of tissue that appear in certain regions of the heart are *nodes* and *conducting fibers*.

MORPHOLOGY

Nodal tissue is concentrated in two places (Fig. 36-1): (1) the *sinoatrial node* in the sulcus terminalis between the superior vena cava and the right atrium, and (2) the *atrioventricular node,* in the lower part of the interatrial septum, near the membranous portion of the interventricular septum. The cells in the sinoatrial node are spindle shaped, smaller than ordinary myocardial fibers, and possess relatively few myofibrils. In man, this node is about 2 mm thick and extends along the sulcus for a distance of 2 cm. The sinoatrial node is the normal pacemaker of the heart, and it is supplied with both sympathetic and parasympathetic fibers of the autonomic nervous system. Small strands of nodal fibers extend a short distance out from this region and become continuous with the typical muscle fibers of the atrium

The atrioventricular node was first identified by Kent in 1892. In 1893 His described a band of modified muscle fibers coursing from the atrium to the ventricle, and in 1906 Tawara[58] published a book containing a complete account of this nodal tissue in many species of animals. The atrioventricular node is located in the subendocardial region of the right atrium near the entry of the coronary sinus. It lies along the lower portion of the interatrial septum and contains fibers that closely resemble those of the sinoatrial node. The atrial border of the node is rather indistinct, as the nodal fibers merge with atrial muscle fibers. The precise morphologic characteristics of this junction are unknown at present. At the ventricular end the node merges with the atrioventricular bundle, sometimes called the bundle of His.

The specialized conducting tissue of the heart[60] begins with the bundle of His, separates into two main "bundle branches," subdivides into a network of Purkinje fibers, and terminates in fine ramifications in the subendocardial regions of both ventricles. The left main branch of the His bundle travels only a short distance before dividing into anterior and posterior subbranches. This specialized conducting tissue has been identified only in the ventricles of birds and mammals.[25] Fish, amphibia, and reptiles are usually said to contain no such specialized tissue and apparently rely on ordinary cardiac muscle fibers for impulse conduction. In birds and mammals the degree of differentiation of Purkinje tissue shows a marked species variation, being relatively poor in carnivores, rodents, and primates but highly differentiated in ungulates. Purkinje fibers have a large diameter, 50 to 70 μm, relative to that of ordinary cardiac muscle fibers, and their sarcoplasm contains much glycogen and relatively few myofibrils. These large fibers ramify beneath the endocardium and penetrate into the depths of the ventricular muscle, where they lose their specific structure and become more like ordinary cardiac muscle fibers. The exact nature of the junction between Purkinje tissue and cardiac muscle fibers has never been described.

The ordinary myocardium consists of elongated, multinucleated cells about 15 μm in diameter and 125 μm in length arranged in columns.

961

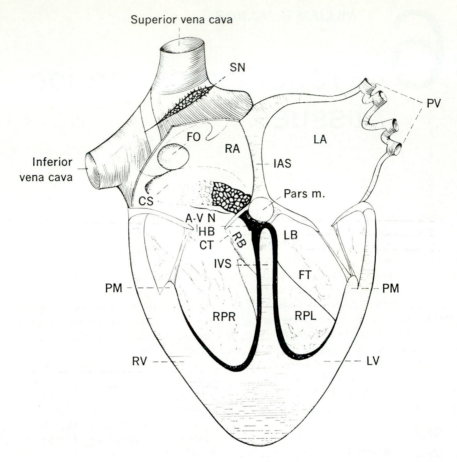

Fig. 36-1. Diagram of location and distribution of specialized nodal and conducting tissue in human heart. *SN*, Sinus node; *FO*, foramen ovale; *CS*, coronary sinus; *RA*, right atrium; *LA*, left atrium; *IAS*, interatrial septum; *PV*, pulmonary veins; *Pars m.*, pars membranacea; *A-V N*, atrioventricular node; *HB*, His bundle, *PM*, papillary muscle; *RB* and *LB*, right and left branches of His bundle; *IVS*, interventricular septum; *CT*, chorda tendinea; *FT*, "false tendon" of Purkinje network; *RV* and *LV*, right and left ventricles; *RPR* and *RPL*, ramifications of Purkinje network within right and left ventricles. (Redrawn from Schütz.[12])

These cells contain myofibrils aligned parallel to the long axis of the cell, organized in a way that produces the typical cross striations. The A, I, and Z bands (p. 84) of cardiac muscle are similar to those of skeletal muscle, with the exception that in some places the Z bands are replaced by thick, irregular boundaries. These landmarks, which were named *intercalated discs* because of their appearance under light microscopy, are actually continuous, stepwise boundaries between individual cells (Fig. 36-2). Electron microscopic studies[27,29,43] have shown that "intercalated discs" appear where two cells are opposed end to end and consist of the plasma membranes of the cells. They continue between cells that lie side by side, but in these longitudinal boundaries the cell membranes are usually very closely opposed in what is called a *close junction,* or *nexus* (Fig. 36-3). Such close junctions are sometimes also found within the transverse disc, although none is shown in Fig. 36-3, *B*).

This demonstration of cytoplasmic discontinuity at the level of the intercalated disc was disturbing to physiologists, because they were then confronted with the problem of explaining how the excitatory impulse was transmitted from one cell to another throughout the heart. Recent evidence suggests that transmission is electrical and occurs by means of local current spread from cell to cell via low-resistance pathways between the interiors of adjacent cells (probably the close junctions) in a manner analogous to that postulated for visceral smooth muscle[20,27,64]

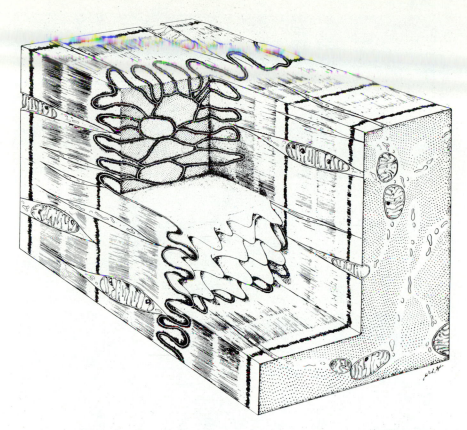

Fig. 36-2. Three-dimensional diagram of fine structure of small segment of junctional region between two cardiac muscle cells, showing intercalated disc. Disc consists of highly interdigitated plasma membranes of two closely apposed muscle cells. Dense material resembling substance of Z band is concentrated in cytoplasm adjacent to cell membranes. Myofilaments of I band insert into this dense material. Discs transect fibers in stepwise manner. Disc in upper portion of diagram is shown in cross and longitudinal sections, whereas in lower portion of diagram, cell surface of one fiber is shown in three dimensions. Mitochondria containing small dense granules and a few tubular and vesicular elements of endoplasmic reticulum appear in sarcoplasm between myofibrils. (Based on results of Fawcett and McNutt.[29])

(Chapter 4). Support for this suggestion comes from a variety of experiments. For example, when current is applied across the membrane of one cardiac muscle fiber, the membrane potentials of adjacent fibers are affected. Furthermore, the current spreads about twice as far in the fiber direction as it does at right angles to the fiber direction.[18] The space constant is about 1 mm in heart muscle.[64] Since the average cell length is about 125 μm, a space constant of 1 mm is possible only if cell-to-cell resistance is low, permitting electrical continuity between the individual cells. Finally, when cardiac muscle is exposed to hypertonic sucrose solutions, the cell-to-cell contacts are disrupted and conduction fails. When these structural changes are reversed by a return to normotonic solutions, conduction is reestablished.[20] The microscopic

transverse striations of myocardial tissue arise from the arrangement of the intracellular filaments that constitute the contractile apparatus, as they do in skeletal muscle (Chapter 3). The protein myosin is localized in thick filaments, which are 100 to 120 Å in diameter and produce the relatively high density of the A band. Actin is associated with the thin filaments, which are about 50 Å in diameter. The thin filaments are anchored at the Z-line, run through the I band, and interdigitate with the thick filaments in the A band.

The heart is innervated by branches of the autonomic nervous system, which travel to the nodal tissues, the atrial muscle, and throughout the ventricular muscle except for the most apical regions.[11] Some of the terminal fibers contain granular vesicles and are presumably adrenergic (p. 203). A smaller number of fibers contains

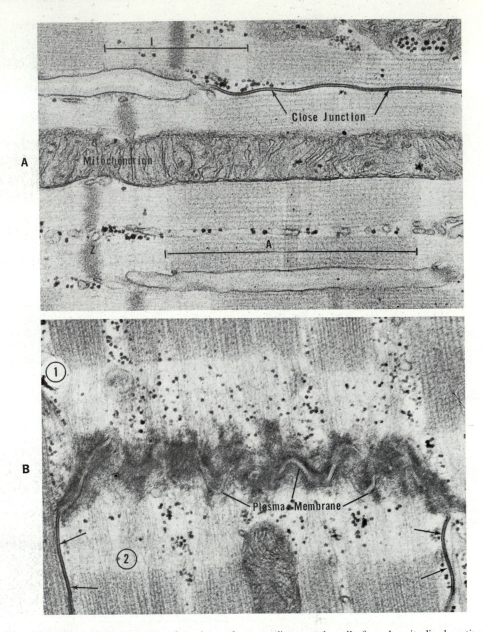

Fig. 36-3. Electron micrographs of portions of two cardiac muscle cells from longitudinal section of cat ventricle. **A,** Arrangement of myofilaments within myofibrils showing periodicity (A and I bands) characteristic of striated muscle. **B,** Intercalated disc, a structure representing part of boundary between two cardiac muscle cells. Disc is composed of plasma membranes of two adjacent muscle cells, *1* and *2*. Arrows indicate close junctions. (×41,000.) (Courtesy Dr. D. W. Fawcett.)

nongranular vesicles and are probably cholinergic. Other neural elements have been found that are believed to represent postganglionic parasympathetic fibers because they persist without structural change after complete denervation of the heart. Type C fibers that meet this criterion have been reported in the free walls of atria and ventricles, lying between myocardial elements.[48] The anatomic relation between myocardial cells and the terminations of these nerve fibers is not entirely clear, although the question has been studied by both light and electron microscopy.

Nothing resembling the neuromuscular junctions of skeletal muscle (p. 153) has been found in the heart. Some investigators have failed to find any direct terminations of nerve fibers on myocardial cell surfaces,[48] suggesting that released neurotransmitters must travel relatively large distances to affect their target cells. Others have demonstrated unmyelinated nerve fibers in immediate contact with atrial, ventricular, and nodal cells. In the atrioventricular node of the mouse heart, for example, Thaemert[59] has found vesiculated nerve processes that travel in sarcolemma-lined tunnels through nodal cells, providing convincing evidence of intimate contact. Afferent as well as efferent fibers travel in the cardiac nerves, but the origins of these afferents have not been identified. Although there is physiologic evidence that mechanical receptors exist in the heart, the morphology of these structures is unknown.

The heart receives its nerve supply from both the sympathetic and parasympathetic divisions of the autonomic nervous system. The parasympathetic supply is inhibitory in its action, whereas the sympathetics are stimulatory. Both sets of nerves exhibit tonic activity under normal conditions. The heart rate in the intact animal depends on the action of both the parasympathetic and sympathetic nerves. The tonic inhibitory activity of the parasympathetic supply is easily demonstrated by severing the vagus nerves. The heart rate immediately increases. A similar effect is produced by the administration of atropine, which blocks the action of acetylcholine, the chemical mediator liberated at the vagal nerve endings in the heart. That the sympathetic nerves also have a tonic action is suggested by similar experiments; sympathectomy slows the heart rate, although this slowing of the rate is not as marked as the increase in rate seen when the vagi are cut.

The parasympathetic preganglionic fibers arise from the vagal nuclei in the medulla and pass via the vagus nerves and their cardiac branches to the cardiac plexus. Here some fibers synapse with postganglionic fibers, whereas others proceed through the plexus and terminate in the intrinsic cardiac ganglia. From the cardiac ganglia, postganglionic fibers go out to cells in the sinoatrial node, interatrial septum, atrial muscle, atrioventricular node, bundle of His, Purkinje network, interventricular septum, and ventricular muscle.[24,42] The nodal regions are especially rich in parasympathetic innervation, whereas the atrial and ventricular muscle and the conducting network are less well endowed.[48,59]

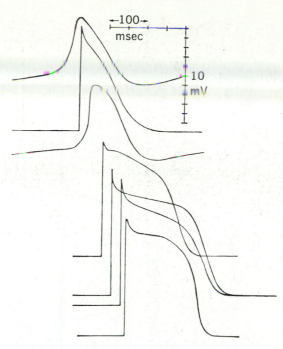

Fig. 36-4. Tracings of typical transmembrane action potentials recorded from the following regions (from above downward): sinoatrial node, atrial muscle, atrioventricular node, bundle of His, Purkinje fiber in false tendon, terminal Purkinje fiber, and ventricular muscle fiber. Note sequence of activation at various regions and also amplitude, configuration, and duration of action potentials. (From Hoffman and Cranefield.[6])

Transmembrane potentials[4,17,18]

The membranes of myocardial cells, like those of most other living cells, are "polarized," which is to say that a difference of electrical potential exists between the inside and outside of the cell (Chapter 1). In the resting state the potential difference between the tip of a microelectrode inserted into the cell and another electrode in the extracellular fluid is 80 to 95 mV, and the interior of the cell is negative with respect to the exterior. If this resting potential is reduced to a certain threshold level, either spontaneously or as a result of an external stimulus, an action potential develops. As in nerve cells (p. 54), the action potential consists of a rapid depolarization followed by a relatively slow repolarization. The membrane potential suddenly becomes less and less negative, briefly reverses so that the interior of the cell is positive with respect to the exterior, and then gradually returns to its initial level. The repolarization phase is much longer in myocardial cells than in nerve cells, and the time courses of the action potentials are not quite

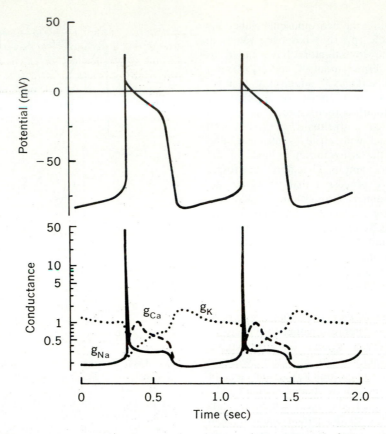

Fig. 36-5. Diagram of myocardial cell membrane conductances for sodium, potassium, and calcium (g_{Na}, g_K, g_{Ca}, below) in relation to action potentials (above). Conductance unit is millimhos per square centimeter. (Modified from Nobel.[10])

the same in the various kinds of cardiac tissue (Fig. 36-4). At normal heart rates the total duration of an action potential is in the neighborhood of 100 msec in the canine ventricular muscle and at least twice that long in the human ventricle. The duration is longer in ventricular than in atrial muscle cells and still more prolonged in Purkinje fibers. Cells in the sinoatrial and atrioventricular nodes depolarize more slowly than those in other parts of the heart and lack the early "spike" found in ordinary myocardium and conducting fibers. In all cardiac cells the early part of repolarization is slower than the later part, and in ventricular muscle this early phase of repolarization is an almost unvarying "plateau."

Myocardial resting and action potentials, like those in nerve cells, have their origin in the permeability of the cell membrane, which is highly selective and varies during the action potential, resulting in a characteristic pattern of ion movements through the membrane and concentration gradients between the inside and outside

of the cell (pp. 14 and 58). Modern theories of excitation began with discovery of the critical changes in permeability to sodium and potassium ions (Chapter 2). More recently, it has become apparent that the movement of calcium ions is equally important, at least in the cells of the heart.[7,10,16] Membrane conductances in a myocardial cell are shown schematically in Fig. 36-5, and the sequence of events is as follows: (1) A large and abrupt increase in sodium conductance, accompanied by a fast inward sodium current, is responsible for the rising (depolarization) phase of the action potential. The very high sodium conductance (g_{Na}) lasts only a short time, then falls to a lower level, which it maintains almost to the end of the action potential. (2) Potassium conductance *decreases* suddenly during depolarization (which is not seen in normal squid nerve), then rises slowly and continuously. (3) When the membrane has depolarized to about −30 mV, an inward current of Ca^{2+} begins, and calcium conductance follows the time course shown in Fig. 36-5. The relatively long repolarization phase in cardiac muscle is thus the result of the interaction of sodium, potassium, and calcium conductances. Changes in the intracellular concentration and distribution of Ca^{2+} are a crucial feature of the contractile process in heart muscle (p. 980 and Chapter 3). (4) The late, rapid phase of repolarization is attributable mainly to an outward current of potassium ions, whereas g_{Na} and $g_{Ca^{2+}}$ return to their levels in the resting state. The small changes in ionic distribution that have occurred during the action potential are then reversed by active metabolic processes, or "pumps," that extrude sodium and calcium from the cell and reaccumulate potassium. A more detailed discussion of these phenomena appears in a recent monograph by Noble.[10]

AUTORHYTHMICITY

The regular, rhythmic contractions of the normal mammalian heart originate in the heart itself. Impulses arriving through cardiac nerves can influence the heart rate, but the heart will continue to beat regularly after complete denervation. The tissue of the sinoatrial node, described anatomically by Keith and Flack in 1906 (Lewis[8]), is the normal cardiac pacemaker. This node represents the remnants of the embryonic sinus venosus, and the early development of cardiac rhythmicity in the embryo[52] is relevant to some aspects of pacemaker activity in the adult.

The primordia of the heart are contained in the mesoderm on either side of the foregut invagination of the mammalian embryo. The embryonic cardiac tube is formed by the progressive fusion of these paired primordia, the conoventricular region first, the atrial region next, and the region of the sinus venosus last. In the developing myocardium there is a succession of pacemaking areas. The first contractions in the embryonic rat heart occur in the ventricular regions at the somite 10 stage. At the time the ventricle is beginning its activity, the atrial tissue is forming behind it and soon begins to contract (somite 17 stage). The atria become dominant and initiate contractions that pass from the posterior to the anterior regions of the cardiac tube. The contractions gradually increase in magnitude and rate, blood is set in motion, and the circulation begins. Shortly thereafter the sinus venosus forms behind the atrium, starts to contract at a slightly faster rate than the atrium, and thereby assumes control of the cardiac rhythm. This is the final shift of the pacemaker in the developing heart. All these changes occur before any nerve fibers appear in the cardiac muscle and before any specialized conducting system can be identified, demonstrating that a rhythmic and effective heartbeat is possible in the absence of extrinsic control. The shift from ventricle to atrium to sinus venosus also shows that the fastest pacemaker cells tend to capture control of the cardiac rate, a phenomenon that occasionally appears in adult life when some region other than the sinoatrial node begins to fire spontaneously at a rapid rate.

The normal location of the pacemaker in the sinoatrial node was discovered experimentally in several different ways. Regional warming or cooling of the sinus venosus in amphibian hearts produced alterations in the heart rate, whereas local changes in ventricular temperature did not.[30,44,45] MacWilliam[41] experimented with temperature alterations in the mammalian heart and found that warming the terminal portions of the vena cava increased the heart rate. Other investigators excised or crushed this crucial area and obtained a temporary cessation of the heartbeat. With the advent of electrical recording methods a more precise localization of the pacemaker was possible. These techniques revealed that when an area of muscle becomes active, it becomes electrically negative with respect to the resting tissue. In 1910, with this information in mind, Lewis[8] placed one electrode on the region of the sinoatrial node of the dog's heart and a second electrode at successive points around the node. Relative negativity always appeared first at the electrode located on the sinoatrial node, demonstrating that the site of origin of the cardiac activity was the sinoatrial node and providing physiologic evidence to substantiate the anatomic findings of Keith and Flack.

Pacemaker potentials. The mechanism for spontaneous, regular autoexcitation in a cardiac cell resides in a slow depolarization of the membrane in the resting state (Figs. 36-4 and 36-6), which has been called the "pacemaker potential."[35] This unsteady state of the resting potential is normally present only in the cells of the sinoatrial and atrioventricular nodes, but it can appear in other cardiac tissues. Cells acting as pacemakers not only exhibit a slow depolarization between action potentials, but also have smaller resting and action potentials than other myocardial cells. Within the sinoatrial node the shape and magnitude of the pacemaker potentials vary to some extent in different cells.[65] The dominant cells have the lowest voltage at the end of the action potential and the most rapid depolarization thereafter.

The rate established by pacemaker cells depends on the time required for their membrane

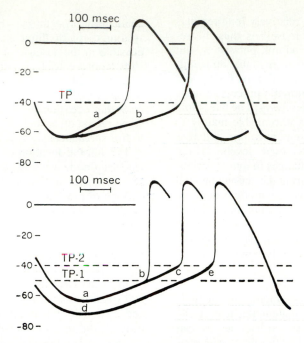

Fig. 36-6. Drawings of transmembrane action potentials from fiber located in mammalian sinoatrial node to show ways in which frequency of pacemaker discharge can be altered. Upper diagram illustrates decrease in frequency of discharge caused by slowing of rate of diastolic depolarization from rate shown in *a* to that in *b*, resulting in increase in length of time required to reach threshold potential, *TP*, for generation of action potential. Lower diagram shows effects of reduction in level of threshold potential from 50 mV, *TP-1*, to 40 mV, *TP-2*, which results in slowing of rate. Also evident in this diagram is effect of change in level of "resting" potential from *a* to *d*, resulting in slowing of rate of discharge (see *a-b, a-c, d-e*). (From Hoffman and Cranefield.[6])

potential to change from its initial resting value to the threshold for generation of the next action potential. Changes in heart rate are consequently mediated by alterations in (1) the magnitude of the initial resting potential, (2) the rate of depolarization, and (3) the threshold. If the initial resting potential becomes more negative and all other parameters are unchanged, the interval between action potentials will increase, and heart rate will be slower than before (Fig. 36-6). Alternatively, an increase in the rate of depolarization, other factors remaining constant, will increase heart rate. A shift of threshold to a more negative potential would also increase the heart rate. All these mechanisms, singly or in combination, are involved in the neural control of heart rate and in many responses to drugs.

Heart rate can also be changed by a shift in the pacemaker site. Marked slowing or complete suppression of normal sinoatrial pacemaker activity (by experimental stimulation of the vagus nerve, for example, or by disease) allows latent pacemakers elsewhere in the heart to assume control. Under such conditions the ectopic pacemaker cells are most frequently located in the atrioventricular node, His bundle, or Purkinje fibers, and their inherent rate is appreciably slower than that of the sinoatrial node. Given abnormal conditions, such as ischemia or injury, ordinary muscle cells in the atrial or ventricular muscle fibers can become the dominant pacemaker. Even under physiologic conditions a single ectopic beat originating in atrial muscle may, on rare occasions, interrupt a normal rhythm.

The slow resting depolarization characteristic of pacemaker cells is associated with time and voltage-dependent alterations in sodium and potassium conductances across the cell membrane (Fig. 36-5). According to the Hodgkin-Huxley theory described in Chapter 2, depolarization represents a movement of the membrane potential away from the potassium equilibrium potential and toward the sodium equilibrium potential.* For atrial fibers in the dog the equilib-

---

*For definition and discussion of equilibrium potentials, see p. 13.

rium potential for potassium has been estimated as −102 mV and that for sodium as about +70 mV.

Depolarization occurs if sodium conductance increases relative to potassium or if potassium conductance decreases. Indirect evidence shows that the pacemaker fibers in the sinoatrial node have a very high resting diastolic sodium conductance compared to that of atrial or ventricular fibers.[15] Furthermore, in the nodal cells, potassium conductance apparently declines slowly throughout diastole.[15] This decrease allows the sodium conductance to be more effective in determining the level of the membrane potential. The membrane slowly depolarizes, and, in so doing, the potential goes through threshold, where sodium conductance increases very rapidly and regeneratively and an action potential is produced. At the threshold potential, potassium conductance declines precipitately. This decline, combined with the regenerative increase in sodium conductance, allows the membrane potential to approach the sodium equilibrium potential, and the membrane reverses its polarity, producing the "overshoot" of the action potential. The potential never actually reaches the sodium equilibrium potential, however, because sodium conductance suddenly diminishes (Fig. 36-5). Accompanying the decline in sodium conductance is a rise in potassium conductance. These conductance changes bring the membrane potential toward the potassium equilibrium potential and produce the repolarization phase of the action potential. After repolarization, potassium conductance does not remain high but gradually declines throughout diastole, and the pacemaker potential again begins to develop.

Thus on the basis of evidence now available, the conditions for pacemaker activity seem to be (1) a relatively high resting diastolic sodium conductance and (2) a slow decline in potassium conductance during diastole. The selective ionic conductances of the membrane are probably maintained and governed by metabolic reactions within the cardiac muscle cell. Agents that alter the rate of development of the pacemaker potential or its magnitude may act either directly on the ionic conductances or indirectly on the metabolic reactions that maintain them.

CONDUCTIVITY

As indicated in an earlier section, the speed with which excitation is propagated varies greatly in different parts of the heart. The factors that determine propagation velocity are similar to those in nerve (Chapter 2) and include the rate of initial rapid depolarization, the magnitude of the action potential, and the diameter of the fibers.

Within the sinoatrial node itself, excitation spreads at about 0.05 m/sec, a relatively slow speed attributable to the slow rate of rise and overall amplitude of action potentials in this region (Fig. 36-4). Once excitation has reached the atrial cells adjoining the sinus node, however, it spreads rapidly over the atria at about 1 m/sec in a uniform radial pattern that Lewis[8] likened to "fluid poured over a flat surface." Although this wide, advancing front is the dominant pattern of atrial excitation, there are certain more or less direct conduction paths from the sinus to the atrioventricular node. The existence of such preferential pathways was long doubted because no specialized conducting tissue like that of the ventricular Purkinje system has been found in the atria. There is now convincing histologic and electrophysiologic evidence, however, that three internodal bundles of atrial muscle fibers exist in the rabbit, dog, and man.[36] The shortest and probably most important of these is the anterior internodal tract, which leaves the sinus node, curves anteriorly around the superior vena cava, enters the interatrial septum, and terminates at the upper margin of the atrioventricular node. Under normal conditions, conduction through this path brings excitation to the atrioventricular node slightly earlier than would otherwise be the case. The physiologic effects of the other internodal, and possibly interatrial, tracts is not known.

Atrioventricular conduction. The time required for excitation to travel from the atrial border of the atrioventricular node to the main branches of the His bundle is surprisingly long, extending to as much as 0.15 sec in the normal human heart. Almost all this delay occurs at the atrial end of the node, over a distance of about 1 mm.[28] Conduction velocity in the junctional area between the atria and the atrioventricular node may be as low as 0.05 m/sec. The resting potentials in this region are lower than those in atrial muscle fibers and sometimes show a slow diastolic depolarization, thus resembling the sinoatrial node cells. There is a gradual transition in the shape of the action potential from the atrial border of the atrioventricular node to the bundle of His, where the potentials resemble those in Purkinje fibers.

Hoffman and Cranefield[6] have proposed that the most probable explanation of the very slow conduction in the upper portion of the atrioventricular node is a gradual change of the properties of the tissue in that region. They found that as the action potential moves from atrium into the upper node, it diminishes in size and initial rate of depolarization, both of which would tend to produce slow conduction. Whether the resting potentials, membrane conductances, or fiber dimensions are the principal cause of this change is not yet known. In the lower portion of the node and in the bundle of His the diameter of muscle fibers becomes progressively larger, increasing from about 7 μm in the node to 50 μm in the Purkinje fiber system. The fibers of conducting tissue converge to form larger and larger bundles toward the lower end of the node, which may also tend to produce a faster conduction velocity.[6] The velocity of conduction through the atrioventricular node is influenced by neural impulses through the vagal and sympathetic nerves, as discussed in a later section. The atrioventricular node will transmit atrial impulses only up to a limiting frequency, about 200 impulses/min in man. If the atrial rate rises above that frequency, varying degrees of atrioventricular block occur. The site of this conduction block is in the junctional area between atrial tissue and upper atrioventricular nodal fibers.[6]

Conduction through the atrioventricular node is normally unidirectional from atrium to ventricle. Experimental stimulation of the ventricle produces a spread of action potentials through the ventricular myocardium, but they are not propagated in a retrograde fashion through the bundle of His. As excitation passes from muscle cells into the bundle, the action potentials begin to diminish in size and rate of depolarization, and they die out completely before reaching the lower portion of the atrioventricular node.[53] Electrophysiologic studies on experimental animals have been the primary source of information about atrioventricular conduction, but conduction velocity in different parts of the atrioventricular node and His bundle can now be measured in man by means of cardiac catheterization. Potentials recorded from multiple electrodes near the tip of a catheter placed in the right atrium indicate the conduction time from sinus node to atrioventricular node, His bundle, and ventricular muscle, so that abnormalities of conduction can be localized.[55]

Conduction through ventricles. Once the atrioventricular node and bundle of His are passed, the conduction velocity increases to 2 to 4 m/sec, and the impulse spreads rapidly along the right and left bundle branches and the Purkinje network. Because of the extensive ramifications of the specialized conducting system within the ventricles, the excitatory wave reaches all portions of the heart muscle rapidly. In the dog heart, no more than 22 msec elapse[6] between activation of the earliest and latest points of the ventricles reached by the impulse. In man the corresponding value is about 80 msec.

The sequence of ventricular activation has been studied in the dog heart, with external electrodes placed gently against the epicardium or with small plunge-type electrodes inserted directly into the ventricles of the exposed heart.[54] The impulse spreads from the atrioventricular septum via the right and left branches of the bundle of His. The right bundle branch runs as an individual unit to the anterior papillary muscle, where it divides into finer branches. These penetrate into the ventricular muscle and transmit the excitation into the muscle itself. The left bundle branch divides almost at once into anterior and posterior branches. On both the right and left sides of the septum there are numerous small Purkinje branches ("false tendons") that cross the cavity of the ventricles.

The endocardial surface of the ventricles of the dog heart is excited very rapidly by the Purkinje network, which conducts at a velocity of 2 to 4 m/sec. As the branching Purkinje fibers turn inward and terminate in the ventricular muscle proper, the conduction velocity slows from 2 to 4 m/sec to about 0.3 m/sec. Therefore excitation spreads at high speeds along the endocardial surface but at much lower speeds within the ventricular wall. The impulse appears first on the epicardial surface of the right ventricle near the thin midseptal region. In general terms, activation then spreads from left to right in the septum, from apex to base of the free ventricular walls, and from endocardial to epicardial regions. The locations of the moving boundary of excitation have been mapped in detail in the dog heart.[54] Robb and associates,[19,52] as a result of meticulous anatomic and electrophysiologic studies on the dog heart, concluded that the impulse follows the anatomic arrangements of the muscle strands. The impulse leaves the branches of the Purkinje network and follows the sinospiral and bulbospiral muscle bundles to the epicardial surface of the ventricles. Anatomic studies by these same investigators revealed that the Purkinje network actually follows these main muscle bundles in general. Each branch of the Purkinje system may activate a discrete bundle of muscle fibers within the ventricular wall. Such bundles collectively make up the main muscle strands just mentioned. In this manner, "innervation

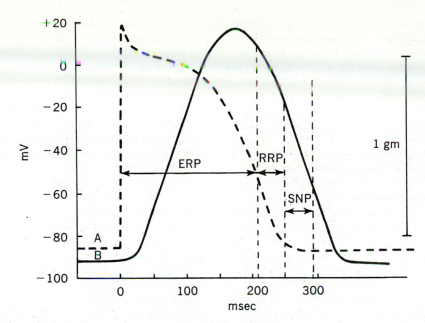

Fig. 36-7. Diagram showing relation between transmembrane potential, **A,** and isometric tension curve, **B,** as recorded from small segment of isolated cat papillary muscle. *ERP,* Effective refractory period; *RRP,* relative refractory period; *SNP,* supernormal period. Note that peak of tension curve is reached before end of effective refractory period. (Modified from Brooks et al.[2])

areas'' composed of one Purkinje strand leading into a group of muscle fibers might be built up.

Excitability and refractory period. The excitability of the myocardium, defined as its ability to respond to a normal or experimental stimulus, varies during the different phases of the cardiac cycle. This basic principle of cardiac action was recognized in 1785 by Fontana, who observed living hearts and concluded that a definite period of time must elapse after the heart contracts in order that normal excitability be restored.[32] In 1871 Bowditch[21] noted that a frog heart needed a certain period of time after a contraction in order to recover its excitability, but he failed to relate this finding to any definite interval of the cardiac cycle. Marey[9] demonstrated this fact experimentally. If the stimulus was applied early in the contraction or systolic period, it did not initiate an extra contraction. If the stimulus came later in systole, however, a contraction (extrasystole) could be elicited, but only after a considerable latent period. This latent period was shorter the later in systole the stimulus was applied. Finally, when the heart reached complete relaxation or diastole, an extrasystole could be evoked with a minimal latency. These experiments showed that during the initial phase of systole (contraction) the heart is refractory to stimulation. Later in the contraction phase,

extrasystoles do appear, that is, the heart has partially recovered its excitability. An extrasystole is often followed by a compensatory pause, a delay before the resumption of the normal rhythm. The pause can be explained in the following manner: As a result of the extrasystole, the heart is made temporarily inexcitable to the normal sinus node impulse, and therefore one beat is missed. By the time the next normal sinus impulse passes over the heart muscle the muscle has recovered its excitability, and the normal rhythm is reestablished.

Later investigations measured the strength of the electrical stimulus necessary to elicit a propagated action potential and determined the threshold of the heart muscle to electrical stimulation in various phases of the cardiac cycle (threshold is used here to mean the reciprocal of excitability). The various states of excitability measured in this manner during the cardiac cycle (Fig. 36-7) were (1) the effective refractory period, (2) the relative refractory period, (3) the period of supernormality, and (4) the period of normal excitability. The effective refractory period is defined as that period in the cardiac cycle during which a stimulus, no matter how strong, fails to produce a propagated electrical response. The term ''propagated'' is emphasized because it has recently been shown that there is no period in the cardiac cycle when a strong stimulus of sufficient duration will not either initiate a local membrane response or influence the recovery process in some manner. For this reason, it seems advisable to use the term "effective" rather than "absolute" for this portion of the refractory period.[2,6] The relative refractory period is defined as that period during which a propagated action potential can be elicited, but the stimulus required to do so is greater than that needed to elicit a similar response in diastole. For a short time after the relative refractory period the reverse situation holds, and the threshold is slightly lower than in diastole; this is the so-called supernormal period. The period of "normal" excitability extends from the end of the supernormal period, throughout diastole, until the beginning of the next action potential discharge. At any time during this period, stimuli of diastolic threshold strength will elicit fully propagated action potentials that possess normal latencies and conduction velocities. The relation of these electrical periods to the actual contraction of the cardiac musculature will be discussed more fully in the section on contractility of the heart.

Changes in excitability can be explained on the basis of the relationship between the resting membrane potential and the permeability of the membrane to the sodium ion. A lowering (depolarization) of the membrane potential results in an increase in the sodium permeability and an increase in the rate of inward movement of sodium ions across the membrane. The threshold potential is the level of membrane potential at which the rate of inward movement of sodium is sufficiently rapid to elicit a propagated action potential. The strength of the stimulus (amount of applied current) necessary to depolarize the membrane to the level at which the rate of inward sodium movement is sufficient to give rise to a propagated action potential is called the "threshold strength."[2] Weidmann[17] has shown that in Purkinje fibers stimulated with cathodal pulses there is an S-shaped relationship between the level of the membrane potential and the rate of inward movement of sodium ions. He used the rate of rise of the action potential resulting from stimulation as a measure of the inward movement of sodium. At membrane potentials of 90 mV, the inward movement of sodium was maximal, resulting in a large action potential with a rapid upstroke velocity. At a membrane potential of 70 mV, the inward movement of sodium was reduced to 50% of maximal; the action potential was much smaller in overall height and upstroke velocity. When the membrane potential was reduced to 50 mV or below, the inward movement

of sodium was so decreased that no propagated action potentials could be generated. Brooks et al.[2] have used these findings to explain the various periods of refractoriness in cardiac muscle. Fig. 36-7 is a schematic diagram of the relationship between the level of the membrane potential in a single ventricular fiber and the excitability of the fiber to cathodal stimulation.[6] The effective refractory period extends from the onset of depolarization until the membrane has repolarized to the level of 50 mV. During this period a stimulus, no matter how strong, cannot give rise to an action potential because there is virtually no inward sodium current. As repolarization proceeds and the membrane potential increases to 60 to 70 mV, a propagated action potential can be elicited. The strength of stimulus necessary to do this will be much larger than that required during diastole. At this level of membrane potential (60 to 70 mV), the rate of inward sodium current is so slow that a much greater amount of applied current is necessary to bring about an inward movement of sodium that is sufficiently rapid to generate an action potential. The action potential itself is small and has a slow-rising velocity characteristic of the low membrane potential from which it arises. When the membrane has returned to its resting level, stimuli of normal diastolic threshold intensity are sufficient to initiate a propagated action potential. At a resting potential of 90 mV, a slight depolarization of the membrane will cause a rapid inward movement of sodium, resulting in the production of a normal action potential.

On the basis of this analysis the recovery of excitability is related to the return of the membrane potential to its normal level, and the recovery parallels the repolarization of the fiber membrane. The rate at which the membrane repolarizes will determine the duration of the various phases of refractoriness. Since the time course of repolarization is much longer in cardiac muscle than in skeletal muscle or nerve fibers, the refractory period in heart muscle is correspondingly longer. The total refractory period in ventricular muscle of the dog is 250 to 300 msec at normal heart rates, as compared with 2 to 4 msec for mammalian skeletal muscle and less than 0.5 msec for mammalian nerve fibers. The duration of the refractory period also varies within different areas of the heart muscle; for example, ventricular fibers have a somewhat longer refractory period than do atrial fibers. As the heart rate increases, the muscle recovers its excitability more rapidly, and the total refractory period shortens.

Under certain abnormal conditions the recovery of excitability does not always parallel the repolarization of the cell membrane. Some metabolic inhibitors, such as dinitrophenol, hasten the repolarization phase of the action potential but increase the total refractory period. In these circumstances the myocardium remains inexcitable, although the membrane potential has returned to its normal level. Alterations in extracellular pH and certain drugs have similar effects.[4,6]

MYOCARDIAL CONTRACTION[7,14]

When a myocardial cell receives a suprathreshold stimulus, it is said to change from a "relaxed" to an "active" state. In the relaxed state the only tension in the muscle is the force that has been applied externally to stretch it to a particular resting length. This resting tension suggests the existence in muscle fibers of nonlinear elastic elements that are stretched as the muscle is passively extended. When the muscle is "activated," it can develop additional force, shorten its length, or both, depending on the external constraints imposed on it. If a muscle strip is anchored at both ends so that it cannot shorten, for example, it will nevertheless develop tension when effectively stimulated. The amount of force is influenced by many different factors, including the chemical environment, the physical state of the cell components, and the arrangement of the contractile protein filaments. The concentration of Ca^{2+} within and adjacent to the cell is one of the most important chemical factors, and the polarization of the cell membrane is one of the critical physical conditions. The crucial factor in the disposition of the contractile proteins is the amount of overlap between actin and myosin strands (Chapter 3). All these factors influence the performance of the muscle throughout its response, with the result that the "active state" is not a unique stable condition, but a dynamic process.

Cardiac muscle is very similar to skeletal muscle, but there are at least three important differences. First, the prolonged tetanic contraction that is possible in skeletal muscle cannot occur in the heart under normal conditions, and each myocardial contraction is analogous to a single skeletal muscle "twitch." Second, the mechanical response takes somewhat longer to develop and declines more slowly in cardiac than in skeletal muscle. Third, normal variations in the force of cardiac contraction come about through changes in the properties of muscle cells, and not, as in skeletal muscle, through changes in the total number of active fibers. The two kinds of muscle also differ quantitatively in their length-tension relationships.

Mechanics of isolated muscle

Much of our present knowledge of myocardial function has been gained from experiments in vitro on isolated papillary muscles. In the cat heart, which has been widely used, this muscle is very small, typically about 12 mm long and 0.7 mm wide, so that it can be kept in reasonably physiologic condition in a bath and can be activated as a unit by an electrical stimulus. The muscle fibers are parallel along its length. In such preparations, muscle tension and length can be measured as functions of time, and even transmembrane potentials can be recorded. The results are shown diagrammatically in Fig. 36-7. As in skeletal muscle, the onset of the action potential precedes the beginning of mechanical response. The peak of the tension curve occurs during electrical repolarization, near the end of the effective refractory period.

The phenomena observed in a papillary muscle preparation depend on the physical arrangements of the experiment. Many arrangements are possible, and four of the most important are shown in Fig. 36-8. In an *isometric* experiment (Fig. 36-8, *A*) the muscle is first stretched to some predetermined resting length, and both ends are then anchored so that the muscle cannot shorten. When stimulated, the muscle develops tension while maintaining a constant length. In an *isotonic* contraction (Fig. 36-8, *B*) the upper end of the muscle is anchored, and a weight (*preload*) is suspended from the lower end. The resting tension, or force, is then equal to the weight attached, and this force remains constant when the muscle is stimulated and shortens. To provide conditions that resemble ventricular contraction more closely, these two modes can be combined (Fig. 36-8, *C*) in a preparation that is both preloaded and afterloaded. First, the lower end of the muscle is anchored. Then the upper end is attached to a lever and the preload suspended from the opposite end of the lever. This stretches the muscle to a preselected resting length. A "stop" is next placed above the muscle end of the lever, so that it is impossible for the muscle to lengthen further, and an additional weight (the *afterload*) is added to the load, but the resting tension remains equal to the preload. When the muscle is stimulated, it will begin to develop more tension, but will not shorten until the force developed exactly equals preload plus afterload. Thereafter the muscle shortens, pulling down the lever, lifting the load, and maintaining a constant force. After reaching maximum shortening (which depends on the condition of the muscle, the chemical environment, and the load) the muscle begins to lengthen and eventually returns

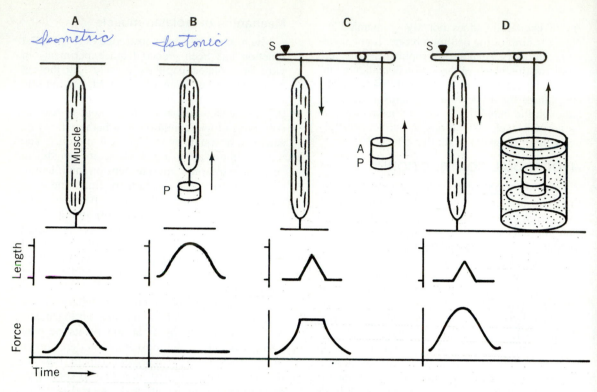

Fig. 36-8. Diagram showing four different experimental arrangements for studying papillary muscles in vitro (above) and the changes in length and force developed on stimulation of the muscle (below). **A,** Isometric contraction. Muscle is stretched to a preselected resting length and then anchored at both ends. On stimulation, muscle develops force but cannot shorten. **B,** Isotonic contraction. Upper end of muscle is anchored and selected weight ("preload", *P*) attached to lower end. Force remains constant (equal to weight of preload) during activation, and muscle shortens. **C,** Isotonic afterload. Lower end of muscle is anchored, and upper end is attached to a lever, with preload on opposite end of lever. A "stop" *(S)* is placed against lever, so that muscle cannot be stretched further, and an additional weight ("afterload", *A*) is then added. On stimulation, muscle maintains a constant length until it develops force equal to the combined preload and afterload, then begins to shorten. **D,** Viscous afterload. Preload, stop, and afterload are attached as for isotonic afterload, but a wide, thin disc of negligible weight is also added, and entire "load" is immersed in a container filled with viscous fluid. When muscle has developed force equal to preload and afterload and begins to shorten, it must develop additional force to pull plate through viscous fluid. Moving "load" against this "viscous resistance" is roughly analogous to ventricular ejection of blood into arterial system. **D** therefore resembles conditions of contraction of myocardial fibers in ventricle in vivo more closely than do other three arrangements.

to its resting length. Muscle tension is now less than the total load and gradually falls until it equals the preload. The response is thus isometric at the beginning and end of contraction and isotonic in the midportion.

Isometric contraction resembles the isometric periods of the intact ventricle in vivo (Chapter 37), which occur at the beginning and end of contraction. Isotonic contraction, however, is not an appropriate simulation of the ventricular ejection period. During muscle shortening (Fig. 36-8, *C*), force is constant, but in the ventricles, in vivo pressure varies with time as blood is

ejected (Fig. 37-1) and so does the force developed in the ventricular walls. The ventricle is not called on to lift a weight, but rather to move a viscous fluid into a viscoelastic arterial tree. The analogous conditions for a papillary muscle experiment would be those shown in Fig. 36-8, *D,* where a viscous afterload has been added. Preload and afterload weights are the same as in Fig. 36-8, *C,* but a circular plate submerged in a viscous fluid (a "dashpot") constitutes an additional afterload. As in the case of the isotonic afterload, the muscle develops force when activated, but it does not shorten until it develops

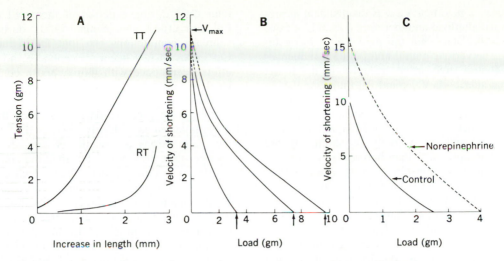

Fig. 36-9. Length-tension-velocity of shortening relationships in isolated papillary muscle of cat heart. **A,** Both resting tension *(RT)* and total developed tension *(TT)* under isometric conditions increase with increasing muscle length. **B,** Force-velocity (or tension-velocity) curves for three different initial muscle lengths. From left to right, first arrow indicates curve for initial muscle length of 11.7 mm; second arrow, 13.0 mm; third arrow, 13.4 mm. Velocity of shortening of contractile elements falls as tension (load) increases in all three instances. Load P_0, at which no shortening occurs (i.e., point where curves intercept absicca, indicating velocity = 0) becomes greater as initial length is increased, but maximum velocity (V_{max}), determined by extrapolating curves to intercept ordinate (load = 0), is virtually the same for all three initial lengths. **C,** Effect of norepinephrine on force-velocity curve. Both V_{max} and P_0 are greater after administration of norepinephrine than in control state. (Modified from Sonnenblick.[13])

enough force to lift the weights of preload and afterload. When it does begin to shorten, it must not only lift the weights, but also must pull the plate through the viscous medium, which requires additional work. The amount of extra work is a function of the fluid viscosity, the diameter of the plate, and the *velocity* with which the muscle shortens. The muscle must "choose," so to speak, whether to shorten rapidly, which requires relatively great force, or to shorten more slowly with less force. The particular time course of force and velocity of shortening that the muscle exhibits under these conditions depend on inherent properties of the muscle and its chemical environment in some way that is not yet clear. The viscous nature of the afterload in vivo has received little attention until recently,[46] and very few experiments of this kind have been reported. A large body of information about myocardial contraction has been obtained, however, by the other three kinds of experiment illustrated in Fig. 36-8.

Length-tension-velocity relationships. A relationship between length and tension is a fundamental property of both skeletal and cardiac muscle. If resting cardiac muscle is stretched to progressively greater lengths, the tension developed when it is stimulated under isometric

conditions becomes progressively greater. Fig. 36-9, *A*, shows this response in an isolated cat papillary muscle. The total tension developed is regarded as the sum of the resting tension induced by passive stretching at rest and an "active tension" developed by the muscle. The same relation can be demonstrated in intact ventricles and is the basis for what has come to be known as the "Frank-Starling law of the heart," which states that the work performed by the ventricle is a function of end-diastolic fiber length (Chapter 37).

An equally important relationship exists between tension and velocity of shortening of myocardial fibers. As Fig. 36-9, *B* indicates, in a constant physicochemical environment, shortening velocity of an isotonically contracting papillary muscle is inversely proportional to the weight the muscle is required to lift. If the weight is sufficiently large, the muscle is unable to shorten, but develops a *maximum isometric tension* (P_0). Extrapolation of the force-velocity curve to the velocity axis gives an estimate of the *maximum* (unloaded) *velocity of shortening* (V_{max}). An interaction between the force-length and force-velocity properties is also evident in Fig. 36-9, *B*. The greater the preload, and hence the initial muscle length, the greater the P_0. It has been argued that V_{max} is independent of preload and afterload and therefore a useful index of myocardial state, but the evidence is inconclusive. The various neural and humoral influences that control myocardial behavior in the living animal alter both V_{max} and P_0, thus establishing new force-length and force-velocity relationships. Norepinephrine, for example, increases both parameters (Fig. 36-9, *C*) so that the tension developed and the velocity of shortening are greater at any length than before.

All these relationships can be demonstrated in the heart in vivo (Chapter 37), although it is more difficult to measure the relevant variables in the intact cardiac chambers than in isolated muscle strips. The complex shape of the ventricle, the arrangement of its muscle fibers, and the nonhomogeneous distribution of tension within its walls all contribute to this difficulty. Nevertheless, the knowledge gained from observations of papillary muscles in vitro has proved to be a reliable foundation on which to build an understanding of ventricular function in vivo.

Frequency and force of contraction. Myocardial performance is also affected by the frequency and regularity of stimulation. Most of the experimental observations on the relationship between the frequency and the force of contraction in cardiac muscle have been made on isolated hearts or strips of heart muscle in which the frequency of stimulation was the primary variable. The observations can be classified into at least three categories: (1) the staircase or "treppe" phenomenon, (2) posttetanic potentiation, and (3) postextrasystolic potentiation.

The staircase phenomenon was first described by Bowditch,[21] who noted that when the heart was stimulated to beat after a long period of quiescence, the amplitude of each successive contraction increased until a maximum steady level was reached. The progressive increase in height of each contraction was a transitory phenomenon that occurred only at the beginning of the period of repetitive stimulation.

The phenomenon of treppe has subsequently been confirmed in the mammalian heart. Although the explanation of treppe is not entirely clear, it has been shown that an increase in the rate of influx of calcium into the muscle cell accompanies the increase in the tension during a staircase phenomenon.[66] Since the concentration of ionized calcium inside the cell is thought to reflect a dynamic balance among factors such as calcium influx, binding to intracellular storage sites, and active efflux (p. 979), a decrease in either binding or efflux would elevate intracellular Ca^{2+} during a period of increased influx. Calcium efflux lags behind influx during the staircase contractions, so that an increase in Ca^{2+} does in fact occur.[7] Eventually, however, efflux keeps pace with influx, and a consistent level of intracellular calcium accompanies each contraction.

Posttetanic potentiation occurs when the rate of stimulation is suddenly decreased. The first few contractions at the slower rate are much stronger than previously, but then the amplitude of the contractions progressively declines to a steady level as the slower stimulation continues. The mechanism involved in the production of this potentiation, as in the case of staircase, is thought to be related in some way to the internal Ca^{2+}. An attractive hypothesis that would account for a variety of frequency-force relationships in the heart has been advocated by Wood et al.[67] Their suggestion is based on the idea that, in addition to the influx of extracellular calcium, the amount of calcium bound to intracellular stores and released by the action potential (p. 979) is a major determinant of the contractile force. This hypothesis further assumes that the process in which calcium is rebound to intracellular sites occurs at an exponentially decreasing rate during and after repolarization of the action potential. Therefore the longer the interval between action potentials (i.e., stimulations), up to a period of several seconds, the greater the amount of calcium binding and hence the greater the "package" of calcium available for release by the next stimulus. When the interval between stimuli is suddenly lengthened (e.g., in going from a faster to a slower frequency), the contraction amplitude will become larger. Once the dynamic balance between the factors regulating calcium is achieved, as discussed, the force of contraction will return to a level characteristic of the slower frequency. Both the staircase phenomenon and posttetanic potentiation have been observed only in isolated hearts. According to Brooks et al.,[2] alterations of rate in the mammalian heart in situ have little effect on systolic tension over a wide range of frequencies.

Postextrasystolic potentiation, however, has been observed in intact hearts as well as in isolated preparations. The first few contractions following an extrasystole are much stronger than the normal ones, although the strength of the extrasystole itself is less than that of the normal contraction. For many years the phenomenon of postextrasystolic potentiation was explained on the basis of a change in initial fiber length. The premature contraction (extrasystole) occurs at a time when the heart is incompletely filled. Under these conditions the initial lengths of the muscle fibers are less than the lengths of the fibers in the normally filled heart. The Frank-Starling hypothesis would predict that under these conditions a weak contraction should occur. The increased tension of the following beat, coming after a compensatory pause, might be related to the greater initial fiber lengths that result from the additional filling of the ventricles during the long pause between contractions. However, evidence has shown that this explanation cannot be correct because postextrasystolic potentiation is also seen in isolated hearts maintained at constant initial lengths. Moreover, hearts in which right ventricular filling pressure is controlled show greater postextrasystolic potentiation in the absence of filling than under conditions of normal filling.[2] Finally, the potentiation resulting from an extrasystole is always greater than the small augmentation of contraction resulting from a single dropped beat, although the filling time for the ventricle is longer after a dropped beat than after an extrasystole.[2] It seems reasonable to conclude that the postextrasystolic potentiation is not due to a change in initial length of the muscle fibers. Both postextrasystolic potentiation and the decreased force of an extrasystolic contraction could be explained by the hypothesis offered by Wood et al.[67] and just described for posttetanic potentiation.

Myocardial contractility. The concept of "contractility" is one of the oldest notions in myocardial mechanics. The importance of the concept is in no way reduced by the fact that investigators cannot even agree on a definition of "contractility," much less measure it. The literal meaning "ability to contract" will not suffice as a definition, because the ability of muscle to develop force is as fundamental as its ability to shorten. Indeed, isometric and isotonic responses indicate that shortening and force development are in some sense equivalent expressions of muscle function.

The basic premise on which the concept is founded is that the response of cardiac muscle is controlled by the innate structure of the myocardial cell and by its physical and chemical environment. As long as the environmental conditions remain constant, the characteristics of myocardial contraction should remain constant. What troubles cardiac physiologists is an apparent exception to that rule, namely, that myocardial performance seems to change with the degree of external constraint on muscle contrac-

tion (the "load"), even when all other conditons are kept constant. The solution has been to assume that this apparent anomaly would vanish if the myocardial response were measured in an appropriate way. In other words, in a constant environment, myocardial response would remain constant no matter what the load. Myocardial contractility has thus been defined conceptually as the ability of cardiac muscle to respond mechanically to an adequate stimulus in a way that depends on environmental conditions but is independent of load. The relevant "environmental conditions," it must be remembered, include ion concentrations within as well as outside the cell, enzymes and energy substrates, polarization of cell membrane, and physical state of organelles, to list but a few. As always, our understanding of phenomena is limited by the nature of the variables we can measure—in this case, muscle length and force as functions of time. The myocardial "response" must somehow be expressed in these terms, but there is abundant experimental evidence to show that these particular variables are not independent of load. Muscle length, shortening velocity, and tension (or ventricular volume, outflow, and wall force) are all affected by preload and afterload. No satisfactory solution to this problem has yet been discovered. The difficulty in arriving at a rigorous operational definition of myocardial contractility that specifies how it should be measured may well arise from inconsistencies in the conceptual definition. Insisting that contractility be independent of load may be incompatible with the way the muscle cell actual works. For example, the assumption that contractility is not affected by load implies that it is not affected by fiber length. The intracellular distribution of Ca^{2+} probably depends to some extent on the overlap of actin and myosin filaments[3] and hence on fiber length. It is therefore paradoxical to assert that the concentration and distribution of calcium ions is one of the determinants of contractility but that fiber length is not.

All the reservations noted with respect to myocardial contractility apply with equal force to attempts to measure the "intensity of the active state."[1,3] This phrase was first defined by Hill[31] as the tension the contractile element can just bear without lengthening or shortening. Alternative definitions in terms of shortening velocity have been proposed,[3] but the abilities to shorten and to develop tension do not have the same time course in a single contraction.[3,37] The ability of cardiac (and skeletal) muscle to shorten develops very early in the response and reaches a maximum much earlier than does isometric tension. The difference may lie in the time required for cross-bridge formation as contrasted with calcium-troponin binding, but the important point is that contraction is a time-varying process that cannot be expressed in a single variable.

The lack of a firm theoretic basis for measuring muscle contractility does not eliminate the need to evaluate myocardial performance in the course of basic and clinical research. Many parameters have been proposed as "indices of contractility" in order to learn to what extent the heart is being influenced in a given situation by the cellular environment, by the load, or by myocardial disease. Some such indices appear to have at least an empirical value and are discussed in Chapter 37. The force-length-velocity relationships already discussed have also been used as the basis for determining myocardial contractility. These relationships (Fig. 36-9) reflect fundamental properties of cardiac muscle, but in some respects, they fall short as a description of myocardial function in vivo. Strictly speaking, they apply only to isometric or isotonic conditions. The fact that force, length, and shortening velocity all normally vary with time in each cardiac contraction is omitted from conventional force-length and force-velocity curves. The total tension plotted in Fig. 36-9, *A*, is peak tension, and the velocities in Fig. 36-9, *B* and *C*, are early velocities. This difficulty can be circumvented by assuming that the relationships vary with time only at the very beginning and end of contraction, but the evidence to support this assumption is equivocal at best. Three-dimensional graphs based on such a hypothesis (Fig. 37-5) can be constructed as a way of expressing the complicated force-length-velocity relationships, but they fail to portray the ejection period accurately.

All-or-none nature of contraction. In 1871 Bowditch[21] showed that a threshold stimulus gives rise to a maximal contraction of the heart muscle. If the stimulus is increased above the threshold strength, no further augmentation of the response can be elicited. From these observations, Bowditch concluded that once threshold was attained, the magnitude of the contraction was independent of the strength of the stimulus. The relationship between magnitude of the response and strength of the stimulus became known as the all-or-none law and was subsequently extended to include nerve and skeletal muscle.

It is important to emphasize that the "all," or in other words, the magnitude of the contraction in heart muscle, may not always be the same. The characteristics of the excitable membrane surrounding the muscle fiber, as well as the state of the contractile proteins within the fiber, can be altered by temperature, pH, and the ionic and chemical constituents of the extracellular fluid. These alterations will influence the magnitude of the contraction. Nevertheless, the magnitude of the contraction for any given set of conditions is independent of the strength of the stimulus, provided the stimulus is of threshold strength.

Contraction at the cellular level

Better understanding of myocardial function, and eventually a useful redefinition of contractility, must be founded on identification of the physical and chemical changes that accompany activation of cardiac muscle at the cellular level. Recent years have seen great progress in this field, and current investigation concentrates to a large extent on correlating these changes with the mechanical response. Many new techniques have contributed to this progress, including methods of measuring and manipulating membrane potentials[3,10,18] and histochemical procedures that can determine the amount and intracellular locations of particular substances. Instead of the classic measurements of overall muscle length, actual sarcomere lengths can now be measured by electron microscopy, and by methods based on diffraction of light or x rays.[3] Integration of the new morphologic, physicochemical, and physiologic data permits at least a tentative description of the events that underlie myocardial contraction at the cellular level. One crucial point, the way in which an effective stimulus elicits the characteristic response in a muscle cell, has still not been defined precisely, but many of the phenomena involved in this "excitation-contraction coupling" have been clearly delineated.

Excitation-contraction coupling.[5,7,16,47] The fundamentals of this subject have been considered in Chapter 3, but the main features of the process in myocardial cells will be indicated briefly here. The action potential signals a change in cell membrane conductance to Ca^{2+} as well as Na^+, K^+, and other ions and probably a change in the properties of the sarcotubular membrane and other organelles. With the onset of an action potential, an influx of Ca^{2+} across the cell membrane begins. Ionized (free) calcium is also released from the cisternae of sarcoplasmic reticulum and probably from mitochondria and other intracellular structures. The inward flow of Ca^{2+} continues during the plateau of the action poten-

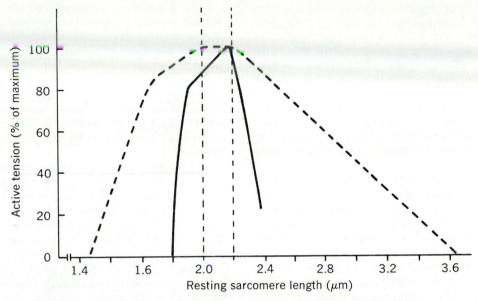

Fig. 36-10. Relation between resting sarcomere length and peak force developed during activation for cat papillary muscle (unbroken line) and for a single skeletal muscle fiber (broken line, see Fig. 3-11). Vertical broken lines indicate region of relatively constant force for skeletal muscle, which is absent in cardiac muscle. (Modified from Sonnenblick; from Ciba Foundation Symposium.[3])

tial so that the concentration of intracellular Ca^{2+} rises, even though some rebinding of free calcium begins almost immediately. By diffusion, the Ca^{2+} reaches the troponin molecules on the actin filaments and there initiates an actin-myosin interaction. The essential feature of this interaction is the formation of cross bridges linking thick and thin filaments, which generate a shearing force tending to increase the filament overlap. The basic mechanism of contraction is believed to be the asynchronous attachment, hingelike motion, and release of multiple cross bridges, just as in skeletal muscle (p. 85). When membrane repolarization begins, the calcium influx subsides, and intracellular Ca^{2+} is rapidly reduced as the ion is rebound to intracellular storage sites.

Excitation thus triggers contraction by giving the contractile proteins access to an appropriate concentration of free calcium. Intracellular concentration of Ca^{2+} in the resting state is below 1×10^{-7} M, and the force-generating interaction of actin and myosin does not appear until this concentration reaches a level above 5×10^{-7} M.[38] The relatively long plateau of the action potential in myocardial cells is largely attributable to calcium rather than to sodium influx, but membrane transports of these two ions are not totally independent. Increased extracellular Na^+ favors exit of Ca^{2+} from the cell, and increased intra-

cellular Na^+ inhibits it, but the exact nature of interdependent membrane transport has not been clarified.[5]

Calcium thus plays a central role in myocardial contraction, and the sequence of events just outlined accounts for the experimentally observed direct correlation between force of contraction and extracellular Ca^{2+} concentration. It also explains[7] the failure of membrane excitation to elicit a contraction at low levels of extracellular Ca^{2+}. The amount of force developed may be a function of the amount of calcium associated with troponin at every instant, which would account in part for the rise and fall of force in each contraction (Fig. 36-7). The time course of force development does not match that of the inward calcium current perfectly,[10,62] but intracellular calcium release must also be considered, and its characteristics are not known. Rebinding of calcium is not necessarily complete at the end of the action potential, and the force-frequency relation already described may depend on the time available for rebinding between stimuli.[5]

Sarcomere length and force development.[3,57] The force developed by myocardial fibers is also a function of sarcomere length. The relationship is qualitatively like that in skeletal muscle, but there are important quantitative differences (Fig. 36-10). First, maximum force is developed at a sarcomere length of about 2.2 μm, but the plateau of force observed in skeletal muscle from 2.0 to 2.2 μm is absent in cardiac muscle. Second, force declines much more steeply at sarcomere lengths above or below 2.2 μm than is the case in skeletal muscle. Third, resting myocardial sarcomere length is rarely less than 1.8 μm, whereas shorter resting lengths are common in skeletal muscle. Sarcomere lengths less than 1.8 μm are observed in actively contracting myocardial fibers, but there is apparently a restoring force that elongates them toward 1.8 μm during relaxation.

Interpretation of these observations must be tentative because the techniques for studying sarcomere length in cardiac muscle are still in the process of development, and the effects of tissue fixatives are uncertain. The relation between muscle length and sarcomere length is puzzling, for there appears to be considerable sarcomere motion in so-called isometric contractions.[3] In addition, the physical site of the "series elastic element" that must be assumed to explain the elastic behavior of muscle has not been identified. Nevertheless, most features of the myocardial force–sarcomere length curve in Fig. 36-10 can be explained in terms of the overlap of sliding thick and thin intracellular filaments.

As expected, maximum force is developed when sarcomere length is in a region of optimal overlap (p. 95). There is no obvious reason, however, why force falls appreciably from 2.2 μm to 2.0 μm, despite optimal overlap. At lengths below 2.0 μm the diminishing force can be attributed to an overlapping of thin fibers, inhibiting the formation of some cross bridges, and at very short lengths there may be buckling of the thick fibers at the Z band. At the other end of the curve the decrease in tension as sarcomeres are stretched beyond 2.2 μm is to be expected from the diminishing number of potential cross bridges. The spatial disposition of thick and thin filaments probably affects the amount of calcium bound to the contractile proteins as well as the number of physical sites available for bridges, but the force-length relationship almost certainly depends on other factors as well. The effects of fiber stretching on the sarcotubular system, on the intracellular release of Ca^{2+}, or on other processes as yet unknown may be involved.

CARDIOACTIVE AGENTS

Certain chemicals, which range from simple ions to complex organic drugs, can have profound effects on cardiac function and are consequently said to be "cardioactive." Many of the ions normally present in the body fluids come under this classification, as do the neurotrans-

mitters released by autonomic nerves and the adrenal medulla. Effects on the contraction of cardiac muscle have come to be called "inotropic," and those on pacemaker rates are designated "chronotropic." The latter term is also sometimes applied to alterations of myocardial excitability or conduction velocity. The cardiac effects of most of these agents were discovered empirically, but it is gradually becoming possible to explain their actions in terms of cellular mechanisms.

A "positive" inotropic effect, which is one in which the force of contraction is increased, can be brought about by any mechanism that makes a greater number of free intracellular calcium ions accessible to the troponin on actin filaments. An increase in pacemaker rate, a "positive" chronotropic action, may be associated with any of the three kinds of change in membrane potentials illustrated in Fig. 36-6. The decreased early resting potential that causes faster rates is usually attributable to an altered ratio of intracellular to external K^+. Pacemaker rates also increase if the membrane depolarizes more rapidly in diastole or if the excitation threshold is lowered, but the basic mechanism of these alterations is still not entirely clear. Conduction velocity depends predominantly on the initial rate of rise of the action potential and hence on the sudden early rise in the conductance of the cell membrane for sodium. The more rapid the rate of rise, the faster the velocity of conduction.

Ions

The maintenance of the normal activity of the heart depends on the presence of certain inorganic ions in the surrounding fluid environment. This fact has been recognized since the early studies of Ringer[51] in the 1880s. He showed that it was necessary to have the cations sodium, potassium, and calcium in proper proportions in any solution bathing the heart if its activity was to be preserved for periods of many hours. The oxygenated fluid he devised for the perfusion medium (which bears his name) of the frog heart contained the cations Na, K, and Ca and the anions Cl and HCO_3. Solutions for the perfusion of mammalian hearts also include dextrose and sometimes inorganic phosphate and $MgCl_2$. The composition of these various salt solutions is shown in Table 36-1.

Calcium. Calcium ions have a positive inotropic effect, which is to say that an increase in the extracellular concentration of Ca^{2+} produces an increase in maximum isometric force. The mechanism of this action was indicated in the

Table 36-1. Salt solutions—percentages of salts (anhydrous)

Author	Animal adapted to	NaCl	KCl	CaCl$_2$	NaHCO$_3$	Other constituents
Ringer	Frog	0.6	0.0075	0.01	0.01	
Howell	Frog	0.7	0.03	0.026	0.003	
Locke	Mammals	0.92	0.042	0.018	0.015	Dextrose 0.1
Tyrode	Mammals	0.8	0.02	0.01	0.1	MgCl$_2$ 0.01
						NaH$_2$PO$_4$ 0.005
						Dextrose 0.1

earlier discussion of the contractile process. Elevated extracellular Ca^{2+} tends to produce a greater calcium influx during the action potential, making more free calcium available to the troponin binding sites and thereby increasing the force of contraction. Very low extracellular calcium levels reduce myocardial force until there may be no contractile response at all, although action potentials still appear.[2] Cardiac muscle is much more sensitive in this respect than is skeletal muscle. Calcium is also a chronotropic agent,[4,6] and elevations of plasma calcium levels slow the heart rate, primarily by raising the excitation threshold (p. 968). Low calcium concentrations tend to increase the rate, although the effects are small unless Ca^{2+} is reduced to about one tenth the normal level. This effect is mediated by an increase in the rate of diastolic depolarization, the "pacemaker potential." Purkinje tissue is especially sensitive to low calcium concentrations and often develops spontaneously firing ectopic foci under such conditions.[2,17]

Sodium. The changes in the extracellular concentration of sodium ions that can occur in the intact animal have little effect on cardiac function. In experiments on the isolated heart, however, an apparent negative inotropic effect can be demonstrated, inasmuch as raising external sodium ion concentration reduces myocardial force, provided that calcium concentration is kept constant.[40] This response has its origin in a competition between calcium and sodium ions on a purely stoichiometric basis. The experimental data suggest that there is a membrane site that may bind either one sodium or two calcium ions. A decrease in external Na$^+$ permits an increase in the binding of Ca^{2+} and thus a greater influx of calcium in subsequent action potentials.[3,49] As a result, the tension developed depends on the calcium ion:sodium ion ratio.

Chronotropic effects of very low Na$^+$ concentrations include slowing of the heart rate and of conduction. The mechanism of the slowing is not clear, but it may involve the sodium-calcium interaction just mentioned. The decreased velocity of conduction is to be expected because the initial inward sodium current is responsible for the fast-rising phase of the action potential (p. 966). The lower the external sodium ion level, the slower the upstroke and the smaller the amplitude of the action potential, conditions that tend to slow propagation. Continued exposure to sodium concentrations less than 0.1N suppresses pacemaker activity completely, and the heart stops in a relaxed state.[6]

Potassium. The effects of changing potassium ion concentration in the extracellular fluid are predominantly chronotropic. A twofold increase above the normal serum potassium concentration slows heart rate and decreases conduction velocity. The impaired conduction is particularly marked in the atrioventricular node and may result in varying degrees of block. Higher levels of potassium can lead to arrhythmias, including fibrillation. Most, but not all, of these effects can be attributed to known actions of potassium on the cell membrane.

An increase in extracellular potassium alters membrane potentials in two ways: (1) it decreases the resting potential (partial depolarization)[4,6] and (2) it shortens the duration of the action potential.[2,17] The decrease in resting potential fits exactly the principles outlined in Chapter 1, whereby the membrane potential is directly related to external potassium concentration. The change is associated with a diminution in the rate of rise and overall amplitude of the propagated action potential, which slows the conduction velocity and may even produce block. As the resting potential becomes less negative, it approaches the threshold potential, which would tend to increase pacemaker rate. When the resting potential is very near or beyond the threshold level, excitability becomes depressed or is abolished. This is in accord with observations by Brooks et al.[2] showing that a fourfold increase in serum potassium level produces a transitory increase in cardiac excitability followed by an abrupt decrease.

The shortening of action potentials by high extracellular K$^+$ concentration is more difficult to explain, for one might expect that the repolarizing outward potassium current would be reduced in these circumstances, prolonging repolarization. The evidence for shortening of the action potential is unequivocal, however, and an hypothesis involving inward rectification of potassium currents has been devised to account for it.[3] The ratio of external K^+ to Ca^{2+} is also a critical factor in these effects. The influence of elevated K^+ on isolated ventricular muscle can be reduced if external Ca^{2+} is lowered concomitantly.[2,6] This interaction, unlike the Na^+/Ca^{2+} competition, probably does not depend on membrane binding sites, but on the duration of the action potential. The shorter the action potential, the less the net calcium influx, which may explain the negative inotropic action of potassium that was first observed by Ringer.[53]

Neural transmitters

Acetylcholine. The cardiac actions of acetylcholine are demonstrated by the effects of stimulating the parasympathetic (vagal) nerve fibers that supply the heart, which releases acetylcholine at the nerve terminations. Mild stimulation of the vagus nerve results in (1) slowing of the heart rate, (2) decreased strength of atrial contraction, and (3) marked reduction of conduction velocity through the atrioventricular node. If the frequency of stimulation is increased, heart rate slows even more, and 2:1 atrioventricular block may occur, that is, atrial excitations are alternately transmitted to the ventricles and blocked, so that only half the supraventricular impulses reach the ventricles. The refractory period of atrial muscle fibers is shortened and their conduction velocity slightly increased by vagal stimulation. Moreover, the atria become more susceptible to fibrillation, and direct electrical stimulation of the atria during vagal stimulation often elicits atrial fibrillation.[23] The ventricular myocardium also receives a parasympathetic nerve supply, and vagal stimulation has a negative inotropic action on ventricular contraction, but this effect is so weak that it can be demonstrated only under carefully controlled experimental conditions. Acetylcholine applied to the atria mimics the effects of vagal stimulation,[22,35,63] but the ventricles and Purkinje system are relatively insensitive to direct application of this substance, for reasons that are not known.

Almost all these effects can be understood in the light of one essential action of acetylcholine on cardiac cells, namely, an increase in permeability to potassium ion, which increases the resting membrane potential. Such "hyperpolarization" in response to vagal stimulation was demonstrated by Hutter and Trautwein,[35] who recorded potentials from a microelectrode inserted into a single muscle fiber in the sinus venosus of the frog heart. Their records, one of which appears in Fig. 36-11, showed that stimulation of the vagus nerve suppresses the development of pacemaker potentials and increases the diastolic membrane potential. If the frequency of vagal stimulation is high enough, the heart stops beating. If the vagus nerve is stimulated at lower frequencies, pacemaker discharge is slowed by virtue of a slowing of the diastolic depolarization rate. Evidence for increased permeability to potassium ions as the primary action of vagal stimulation or acetylcholine is provided by studies using radioactive potassium, which show that both inward and outward fluxes of potassium across the atrial cell membrane are increased, especially in the sinus region.[34] Furthermore, an increase of extracellular potassium concentration diminishes the inhibitory effects of acetylcholine on atrial muscle.[22] The large increase in membrane potassium conductance brought about by acetylcholine not only affects pacemaker rates, but also increases the net repolarizing current and thus reduces the duration of the action potential.[10,35] This reduction probably accounts in part for the negative inotropic effects. The atrioventricular block that can be produced by acetylcholine occurs at the atrial border of the node,[6] the same region in which most of the normal atrioventricular delay appears, but the cellular basis for the block is not clear.

It should be noted that the action of acetylcholine at the vagal endings in the heart differs from its action at the myoneural junction of skeletal muscle (Chapter 5). In both instances the permeability of the postjunctional muscle fiber membrane is increased, but in the myoneural junction, this increase is nonselective, and the membrane potential therefore tends to fall toward the low level that would be assumed if sodium and other ions were free to diffuse.[26] The result is a depolarizing (excitatory) action. In the sinus node cells, on the other hand, acetylcholine induces a highly selective increase in permeability, principally for potassium, resulting in hyperpolarization.

Norepinephrine. Norepinephrine has both inotropic and chronotropic effects. Stimulation of sympathetic nerves to the heart or administration of norepinephrine causes (1) a great increase in both rate and force of contraction; (2) an increase in conduction velocity through the atria, atrioventricular node, and ventricles; and (3) a transitory increase in excitability followed by de-

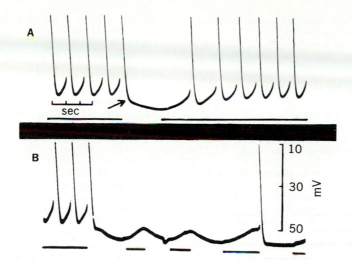

Fig. 36-11. Effects of vagal stimulation on transmembrane potential of pacemaker fiber in sinus venosus of spontaneously beating frog heart. Record in **A** shows vagal stimulation at rate of 20 pulses/sec, as indicated by break in solid line at bottom of tracing. At arrow, note beginning of hyperpolarization of membrane and inhibition of pacemaker action potential. Several seconds after vagal stimulation ceases, membrane slowly depolarizes to a level at which action potentials are again generated. Record in **B** shows effects of intermittent (20/sec) stimulation on same fiber as shown in **A.** Note hyperpolarization of membrane during vagal stimulation and slow return (depolarization) toward threshold when stimulation is discontinued. Note also escaped beat at beginning of fourth stimulation period. (From Hutter and Trautwein.[46])

creased excitability in the atria and ventricles. Purkinje fibers, on the other hand, become more excitable and tend to develop spontaneous pacemaker activity. The actions of epinephrine on the heart are essentially the same as those of norepinephrine.[2] The positive inotropic effect arises from an increase in calcium ion influx, there being a close correlation between inward calcium current and force developed.[62] The increased pacemaker rate is rather more complicated. It originates in a norepinephrine-induced increase in rate of depolarization during diastole, but in sinoatrial nodal and atrial fibers, this faster depolarization is thought to be caused primarily by increased calcium ion current, whereas in Purkinje fibers, it is related to the more rapid rate of decay of potassium conductance.[10] The changes in conduction velocity are apparently attributable to the increased amplitude of the initial action potential, although norepinephrine does not alter the excitatory sodium ion conductance.[10] Norepinephrine also shortens the duration of the action potential, probably by increasing the uptake of calcium ion at intracellular sites. It also acts to increase the activity of the sodium-potassium exchange "pump," a potentially important effect about which little is known. Recent evidence suggests that norepi-

nephrine stimulates adenylate cyclase activity, leading to increased intracellular levels of cyclic AMP, which could influence contractility by increasing calcium ion accumulation by the sarcoplasmic reticulum.[39,61]

Cardiac glycosides

Certain compounds that are the active components of such drugs as digitalis and oubain have long been known to exert a positive inotropic action and are called "cardiac glycosides." The cellular mechanism of their action has not been fully explained, but they apparently block the sodium-potassium pump of the cell membrane. The enzyme Na^+,K^+-ATPase may be a receptor for these agents, although this is only a hypothesis at present.[16,56]

REFERENCES
General reviews

1. Brady, A. J.: Active state in cardiac muscle, Physiol. Rev. **48:**570-600, 1968.
2. Brooks, C. M., et al.: Excitability of the heart, New York, 1955, Grune & Stratton, Inc.
3. CIBA Foundation Symposium: The physiological basis of Starling's law of the heart, Amsterdam, 1974, Associated Scientific Publishers.
4. Cranefield, P. F., and Hoffman, B. F.: Electrophysiology of single cardiac cells, Physiol. Rev. **38:**41, 1958.

5. Fozzard, H. A.: Heart: excitation-contraction coupling, Annu. Rev. Physiol. **39:**201-220, 1977.

6. Hoffman, B. F., and Cranefield, P. F.: Electrophysiology of the heart, New York, 1960, McGraw-Hill Book Co.

7. Langer, G. A.: Ionic movements and the control of contraction. In Langer, G. A., and Brady, A. L., editors: The mammalian myocardium, New York, 1974, John Wiley & Sons, Inc., pp. 193-217.

8. Lewis, T.: The mechanism and graphic registration of the heart beat, ed. 3, London, 1925, Shaw & Sons, Ltd.

9. Marey, E. J.: La méthode graphique, Paris, 1885, Masson et Cie Editeurs.

10. Noble, D.: The initiation of the heartbeat, Oxford, 1975, Clarendon Press.

11. Randall, W. C.: Neural regulation of the heart, New York, 1977, Oxford University Press.

12. Schutz, E.: Physiologie des Herzens, Berlin, 1958, Springer-Verlag.

13. Sonnenblick, E. H.: The mechanics of myocardial contraction. In Briller, S. A., and Conn, H. L., Jr., editors: The myocardial cell: structure, function, and modification by cardiac drugs, Philadelphia, 1966, University of Pennsylvania Press.

14. Tanz, R. D., Kavaler, F., and Roberts, J.: Factors influencing myocardial contractility, New York, 1967, Academic Press, Inc.

15. Trautwein, W.: Generation and conduction of impulses in the heart as affected by drugs, Pharmacol. Rev. **15:**277, 1963.

16. Van Winkel, W. B., and Schwartz, A.: Ions and inotropy, Annu. Rev. Physiol. **38:**247, 1976.

17. Weidmann, S.: Elektrophysiologie der Herzmuskelfaser, Bern, 1956, Hans Huber Medical Publisher.

18. Woodbury, J. W.: Cellular electrophysiology of the heart. In Hamilton, W. F., and Dow, P., editors: Handbook of physiology. Circulation section, Baltimore, 1962, The Williams & Wilkins Co., vol. 1.

Original papers

19. Baird, J. A., and Robb, J. S.: Study, reconstruction and gross dissection of the A-V conducting system in the dog heart, Anat. Rec. **108:**747, 1950.

20. Barr, L., Dewey, M. M., and Berger, W.: Propagation of action potentials and the structure of the nexus in cardiac muscle, J. Gen. Physiol. **48:**796, 1965.

21. Bowditch, H. P.: Ueber die Eigenthümlichkeiten der Reizbarkeit, welche die Muskelfasern des Herzens ziegen, Ber. Math. Phys. Sachs. Ges. Wissensch., p. 662, 1871.

22. Burgen, A. S. V., and Terroux, K. G.: On the negative inotropic effect in the cat's auricle, J. Physiol. **119:**449, 1953.

23. Burn, J. H., Vaughan-Williams, E. M., and Walker, M. J.: The effects of acetylcholine in the heart-lung preparation including the production of auricular fibrillation, J. Physiol. **128:**277, 1955.

24. Chiba, T., and Yamauchi, A.: On the fine structure of the nerve terminals in the human myocardium, Z. Zellforsch. Microsk. Anat. **108:**324, 1970.

25. Davies, F., and Francis, E. T. B.: The conduction of the impulse for cardiac contraction, J. Anat. **86:**302, 1952.

26. Del Castillo, J., and Katz, B.: Local activity at a depolarized nerve-muscle junction, J. Physiol. **128:**396, 1955.

27. Dewey, M. M., and Barr L.: A study of the structure and distribution of the nexus, J. Cell Biol. **23:**553, 1964.

28. Eyster, J. A. E., and Meek, W. J.: Experiments on the origin and conduction of the cardiac impulse; conduction of the excitation from the sino-auricular node to the right auricle and auriculoventricular node, Arch. Intern. Med. **18:**755, 1916.

29. Fawcett, D. W., and McNutt, N. S.: The ultrastructure of the cat myocardium. I. Ventricular papillary muscle, J. Cell Biol. **42:**1, 1969.

30. Gaskell, W. H.: The contraction of heart muscle. In Schafer, E. A., editor: Textbook of physiology, Pentland, England, 1900, The Macmillan Co., Ltd., vol. 2.

31. Hill, A. V., Abrupt transition from rest to activity in muscle. Proc. R. Soc. Lond. (Biol.) **136:**399, 1949.

32. Hoff, M. E.: The history of the refractory period, Yale J. Biol. Med. **14:**635, 1942.

33. Hoffman, B. F., and Suckling, E. E.: Cardiac cellular potentials; effect of vagal stimulation and acetylcholine, Am. J. Physiol. **173:**312, 1953.

34. Hutter, O. F.: Mode of action of autonomic transmitters on the heart, Br. Med. Bull. **13:**176, 1957.

35. Hutter, O. F., and Trautwein, W.: Vagal and sympathetic effects on the pacemaker fibers in the sinus venosus of the heart, J. Gen. Physiol. **39:**715, 1956.

36. James, T. N., and Sherf, L.: Specialized tissues and preferential conduction in the atria of the heart. Am. J. Cardiol. **28:**414, 1971.

37. Julian, F. J., and Moss, R. L.: Active state in striated muscle, Circ. Res. **38:**53, 1976.

38. Katz, A. M.: Regulation of cardiac muscle contractility, J. Gen. Physiol. **50:**185, 1967.

39. Katz, A. M., and Repke, D. I.: Calcium-membrane interactions in the myocardium: effects of oubain, epinephrine, and 3′,5′-cyclic adenosine monophosphate, Am. J. Cardiol. **31:**193, 1973.

40. Luttgau, H. C., and Niedergerke, R.: The antagonism between Ca and Na ions on the frog's heart, J. Physiol. **143:**486, 1958.

41. MacWilliam, M. A.: On the rhythm of the mammalian heart, J. Physiol. **9:**167, 1888.

42. McKibben, J. C., and Getty, R.: A comparative morphologic study of the cardiac innervation in domestic animals. I. The canine, Am. J. Anat. **122:**533, 1968.

43. McNutt, S. N.: Ultrastructure of intercellular junctions in adult and developing cardiac muscle, Am. J. Cardiol. **25:**169, 1970.

44. Meek, W. M., and Eyster, J. A. E.: Experiments on the origin and propagation of the impulse in the heart; effects of vagal stimulation and of cooling on the location of the pacemaker within the sinoauricular node, Am. J. Physiol. **34:**368, 1914.

45. Meek, W. J., and Eyster, J. A. E.: The origin of the cardiac impulse in the turtle's heart, Am. J. Physiol. **39:**291, 1916.

46. Milnor, W. R.: Arterial impedance as ventricular afterload, Circ. Res. **36:**565, 1975.

47. Morad, M., and Goldman, Y.: Excitation-contraction coupling in heart muscle; membrane control of development of tension, Prog. Biophys. Mol. Biol. **27:**257-313. 1973.

48. Napolitano, L. M., et al.: Intrinsic innervation of the heart, Am. J. Physiol. **208:**455, 1965.

49. Niedergerke, R.: Movements of Ca in frog heart ventricles at rest and during contractions, J. Physiol. **167:**515, 1963.

50. Patten, B. M.: Initiation and early changes in the character of the heart beat in vertebrate embryos, Physiol. Rev. **29:**31, 1949.

51. Ringer, S.: Concerning the influence exerted by each

of the constituents of the blood on the contraction of the ventricle, J. Physiol. **3**:380, 1880-1882.

52. Robb, J. S., and Robb, R. C.: The excitatory process in the mammalian ventricle, Am. J. Physiol. **115**:43, 1936.

53. Sano, T., Ohtsuka, E., and Shimamoto, T.: "Unidirectional" atrioventricular conduction studied by microelectrodes, Circ. Res. **8**:600, 1960.

54. Scher, A. M., and Young, A. C.: Ventricular depolarization and the genesis of the QRS, Ann. N.Y. Acad. Sci. **65**:768, 1957.

55. Scherlag, B. J., et al.: Catheter technique for recording His bundle activity in man, Circulation **39**:13, 1967.

56. Schwartz, A., Lindenmayer, G. E., and Allen, J. C.: The sodium potassium adenosine triphosphatase: pharmacological, physiological, and biochemical aspects, Pharmacol. Rev. **27**:3, 1975.

57. Sonnenblick, E. H., et al.: The ultrastructure of the heart in systole and diastole. Changes in sarcomere length, Circ. Res. **21**:423, 1967.

58. Tawara, S.: Das Reizleitungssystem des Saugetierherzen, Jena, 1906, Gustav Fisher Verlag.

59. Thaemert, J. C.: Atrioventricular node innervation in ultrastructural three dimensions, Am. J. Anat. **128**:239, 1970.

60. Todd, T. W.: The specialized systems of the heart. In Cowdry, E. V., editor: Special cytology, New York, 1932, Paul B. Hoeber, Inc., vol. 2.

61. Tsein, R. W., Giles, W. R., and Greengard, P.: Cyclic AMP mediates the action of adrenaline on the action potential plateau of cardiac Purkinje fibers, Nature N. Biol. **240**:181, 1972.

62. Vassort, G.: Existence of two components in frog cardiac mechanical activity, Eur. J. Cardiol. **1**:163, 1973.

63. Vaughan-Williams, E. M.: Some observations concerning the mode of action of acetylcholine on isolated rabbit atria, J. Physiol. **140**:327, 1957.

64. Weidmann, S.: Electrical coupling between myocardial cells, Prog. Brain Res. **31**:275, 1969.

65. West, T. C.: Ultramicroelectrode recording from the cardiac pacemaker, J. Pharmacol. Exp. Ther. **115**:283, 1955.

66. Winegrad, S., and Shanes, A.: Calcium flux and contractility in guinea pig atrium, J. Gen. Physiol. **45**:371, 1962.

67. Wood, E. H., Heppner, R. L., and Weidmann, S.: Inotropic effects of electric currents. I. Positive and negative effects of constant electric currents or current pulses applied during cardiac action potentials; II. Hypothesis: calcium movements, excitation-contraction coupling and inotropic effects, Circ. Res. **24**:409, 1969.

37

WILLIAM R. MILNOR

The heart as a pump

The physiologic characteristics of heart muscle are considered in Chapter 36. In this chapter, attention will be directed to the sequence of events that enables the myocardial contractions to pump blood around the circulation, the factors that influence ventricular function and cardiac output, and the methods employed to measure blood pressure and flow.

CARDIAC CYCLE

The events of the cardiac cycle can be understood best by studying the pressure changes in the atria and ventricles, the time relationship between these changes and the electrocardiogram, the action of the heart valves, and the blood flow into the pulmonary artery and aorta. These events are summarized diagrammatically in Fig. 37-1. The pressure ordinates in this diagram indicate pressure in relation to atmospheric pressure. Since the heart is contained in the thorax, where the pressure is subatmospheric, the actual pressure developed by the heart should be considered in relation to this intrathoracic pressure. The curves drawn in this figure represent values to be anticipated in a normal adult subject with a pulse rate of 75/min.

First, compare the pressure curves from the left atrium and ventricle. In this and other comparisons the electrocardiogram (ECG) may be used as the time standard. The start of the P wave indicates the beginning of excitation in the atria. A few hundredths of a second later a rise in pressure lasting about 0.1 sec develops in both atrium and ventricle as a result of the atrial contraction. The delay is due to the latent period between electrical and mechanical changes. Since excitation precedes contraction, this latent period is partly real, but mainly it is an artifact because the mechanical changes do not develop measurable pressures in the heart chambers until excitation has spread over a large part of the muscle. Since the atrioventricular valve is open and the two chambers are in open communication with one another, this rise in pressure occurs in both atrium and ventricle. Actually the total energy is

greater in the atrium, as flow occurs from the atrium to the ventricle, but the resistance in the valvular orifice is low and the difference in pressure is small. Atrial contraction reaches a peak and then begins to subside. At this time the atrial pressure tends to fall below that in the ventricle because, as the atrium relaxes, the mitral valve tends to close and to maintain the original pressure in the ventricle; the difference is small and not readily demonstrated, but a presystolic shutting of the mitral (or the tricuspid) valve may give rise to sound vibrations of low frequency that precede not only ventricular contraction but even the signal of ventricular excitation given by the onset of the QRS complex. This early valvular closing, of obvious functional value, is rendered possible by the delay of conduction at the atrioventricular node indicated in the ECG by the interval from the end of the P wave to the start of the QRS complex. The closing of these valves, however, most commonly occurs later with the onset of ventricular contraction.

Comparing intraventricular pressure with aortic pressure in diastole, the aortic pressure is seen to be very different from that in the ventricle, which at this time is in open communication with the left atrium and is being filled from the pulmonary veins discharging through the atrium. The aortic valve is closed. When the ventricle starts to contract, the pressure within it rises rapidly, but all the valves remain closed until the pressure is raised above that in the aorta, when the aortic valve opens. During this period of rising tension there are no changes in the volume of the contents of the ventricle, and the lengths of the fibers remain unchanged; only their tension is altered. It is therefore a period of isometric contraction. Under normal conditions in man, it lasts 0.02 to 0.06 sec, most commonly 0.04 to 0.05 sec.[14] Following the opening of the aortic valve, blood is forced out rapidly into the aorta. The pressure in that vessel is raised, and the period of ejection commences. Ejection is rapid at first and decreases somewhat in the later stages of systole. Pressure is slightly higher in

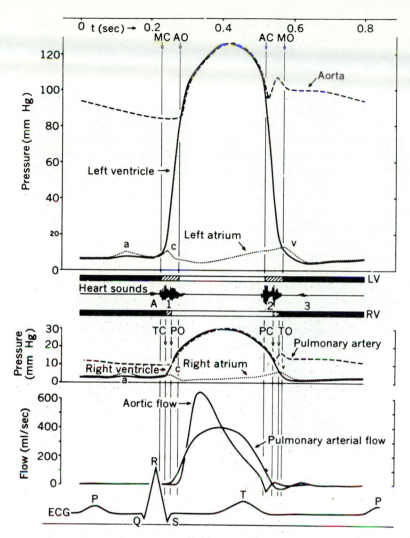

Fig. 37-1. Diagrammatic representation of events in cardiac cycle. From top downward: pressure in aorta, left ventricle, and left atrium; duration of left ventricular diastole (heavy shading), isometric periods (diagonal lines), and systole (unshaded); heart sounds; right ventricular diastole, isometric periods, and systole; pressure in pulmonary artery, right ventricle, and right atrium; blood flow in aorta and pulmonary artery; and electrocardiogram. Valvular opening and closure are indicated by *AO* and *AC,* respectively, for aortic valve; *MO* and *MC* for mitral valve; *PO* and *PC* for pulmonic valve; and *TO* and *TC* for tricuspid valve.

the ventricle than the aorta during the early part of ejection because of the small resistance offered by the aortic valve, but the difference is often undetectable with routine techniques. During the latter part of ejection, the movement of blood in the aorta is partly a result of the momentum it has already acquired, and aortic pressure may be slightly higher than ventricular pressure.

Eventually the gradual cessation of muscular activity in the ventricle supervenes; pressures in both the ventricle and aorta fall—the former faster than the latter—and the aortic valves close.

Tension in the ventricular muscle continues to fall, but no change in ventricular volume occurs until the pressure has fallen below that in the atrium, which causes the mitral valve to open. This is the period of isometric relaxation. During diastole, blood stored in the left atrium as well as that returning in the pulmonary veins enters the ventricle. The initial pressure in the left atrium is relatively high and inflow is rapid, but the blood that had been stored in the atrium is emptied into the ventricle, tending to lower slightly the pressure in the former and raise it in the latter.

Ventricular filling then becomes much slower and merely represents inflow from the pulmonary veins. If the ventricular pressure rises, even this inflow may be slowed. If the pulse rate is slow and the diastole long, inflow may be slowed in the latter part of diastole, and this is known as the period of diastasis. When the atrium contracts in the next cycle, it forces additional blood into the ventricle—not a large amount but a significant proportion of the total. The pressure changes in the right ventricle and pulmonary artery are similar to those on the left side but proceed at a much lower pressure level. The pulmonary circuit will be discussed separately in Chapter 45.

The cardiac cycle can be further subdivided into various phases. Ventricular systole may be divided into the isometric period of rising tension (about 0.05 sec) and a period of ejection, rapid at first and slower later, lasting through the rest of systole with a duration (at a pulse rate of 75) of about one third the cycle length. Ventricular diastole may be divided into an isometric relaxation phase, when the tension in the ventricle is falling but all valves are closed (about 0.05 sec), and a period of filling. The duration of this last period varies greatly with the duration of the cycle. It may be roughly divided into an initial period of rapid filling, a later period of diastasis, and a period of atrial systole, but the division into the first two periods is entirely arbitrary. The early rapid filling of the ventricles, dependent on the necessary damming back of blood in the atria during ventricular systole, is of considerable physiologic importance. As the pulse rate is increased and diastole shortened, the period of diastasis is reduced.

Atrial contraction follows the filling period and causes a small additional "boost" in the ventricular distention. Atrial relaxation and ventricular contraction are more or less coincident. Mechanical consequences of ventricular contraction are not demonstrated until about 0.05 sec after the beginning of the QRS complex, when ventricular pressure begins to rise. Estimates of this latency range from 0.02 to 0.08 sec, and the evidence suggests that apparent latency is least when the pulse rate is fast.

The sequence of events in the atria (limiting our account to the left atrium for the sake of simplicity) is as follows: As soon as intraventricular pressure rises, the mitral valve closes and the pressure curves of the atrium and ventricle become quite different. The rise in pressure within the ventricle causes the valve to bulge into the atrium, producing a short initial rise in the atrial pressure curve, but the main effect of the ventricular contraction is to cause pressure to fall in the atrium. When the ventricle contracts and thrusts blood into the aorta, the movement of blood causes a recoil reaction that forces the apex of the heart downward. Thus during ventricular shortening the base of the heart moves toward the apex, but the apex does not move toward the base. Since the venous orifices into the atria are relatively fixed, the atria are lengthened. A consequent increase in their capacity causes a fall of pressure that aids their filling. Later, as the inflowing blood, which is dammed back during ventricular systole, increases, intra-atrial pressure rises again. As the ejection of blood slackens, recoil weakens and the base of the heart begins to return upward to its more normal position. The resultant shortening of the atrium tends to produce a positive pressure wave. Thus the positive wave created by atrial filling during ventricular contraction may be increased as ventricular contraction begins to slacken and may be still more increased as ventricular contraction ceases. It is, however, reversed by a fall of pressure as soon as the atrioventricular valve opens and the ventricle fills. The greatest rise in intra-atrial pressure is seen during the protodiastolic or isometric relaxation phases of the cycle. When ventricular relaxation is complete, the intraventricular pressure falls below the atrial pressure level, and the mitral valve opens. The blood collected in the atrium as well as the blood returning from the lungs enters the ventricle, and again the pressure changes in the two connected chambers vary in unison.

The intra-atrial pressure curve recorded in the left atrium therefore shows three main positive waves. The first results from atrial contraction and the second from the initial bulging of the mitral valve; the third, which is more complicated, depends on the accumulation of blood in the atrium during ventricular contraction and on the impulse transmitted to this accumulated blood when the recoil reaction of the ventricle ceases and the base of the ventricle tends to return to its normal position. The initial rise of pressure resulting from atrial systole may cause some regurgitation of a small amount of blood into the veins. The pressure changes in the right atrium are similar to those in the left even though the pressures in the right ventricle do not rise nearly as high during systole as those in the left. The general relationship of the atrial and ventricular pressures on the right side, however, is the same as that described.

Action of valves

The heart valves—mitral, tricuspid, aortic, and pulmonic—are set in orifices. In each case the diameter of the chamber beyond the valve is greater than that of the orifice.

The relaxed ventricle is much wider than the aperture produced by the opening of the atrioventricular valve, and in the aorta the sinuses of Valsalva effect a widening of the vessel just above the semilunar valve. The functional significance of these anatomic arrangements is that eddy currents are generated when fluid is forced through an orifice into a broader vessel or chamber. Here the setting up of eddy currents is important because it keeps the valve cusps floating in the stream and prevents their being pressed against the side walls. Thus the valves are in a position to close rapidly when a critical change in pressure occurs. Also, this effect prevents the open aortic semilunar valves from obstructing the origin of the coronary arteries.

A valve opens when the pressure in the proximal chamber exceeds that in the distal chamber, whether this condition develops from a rise in pressure in the proximal chamber (as in the opening of the semilunar valve by the rise of pressure in the ventricle) or from the fall of pressure in the distal chamber (opening of mitral and tricuspid valves). A valve closes when the pressure in the distal chamber exceeds that in the proximal chamber, whatever the cause of the difference.

The mitral and tricuspid valves are supported by the chordae tendineae, which are attached to the papillary muscles. Consequently, the valves are prevented from bulging to any great extent in the atria as the ventricles shorten.

Ventricular outflow

The instantaneous rate of blood flow from the ventricles into the aorta and pulmonary artery is also shown in Fig. 37-1. Because of the slightly asynchronous activation of the right and left ventricles, outflow into the pulmonary artery leads outflow into the aorta by approximately 0.02 sec. The time course of the two flow curves is also quite different, the left ventricular discharge rising much more rapidly to a higher peak, although the mean output during each systole is ordinarily the same in both ventricles. The right ventricle has a slightly longer ejection period than the left, and its periods of isometric contraction are relaxation are shorter. The flow curves in Fig. 37-1 represent blood flow 1 or 2 cm beyond the aortic and pulmonic valves, and the transient negative or backward flow shown at the end of ejection, after the closure of these valves, indicates a small backflow that distends the aortic sinuses of Valsalva and the root of the pulmonary artery.

Heart sounds

Two distinct sounds may be heard by applying the ear or a mechanical sound transmitter (stethoscope) to the chest. The first occurs at the beginning of ventricular systole and is heard best over the site of the apex beat. The second occurs at the end of systole and is heard best in the sternal region near the second intercostal space. A third heart sound is occasionally detectable in diastole.

Auscultation of the heart with the aid of a stethoscope is an important diagnostic procedure in clinical medicine, a fact confirmed by a large body of empirical information relating the sounds generated by the heart to cardiovascular function. The vibrations generated in the heart are transmitted to the chest wall and are modified by the resonance of these structures; they tend to be conducted along bony structures as well as along columns of moving fluid in the direction of the movement. The sounds heard are modified by the resonance of the particular receiver used; the student should therefore become accustomed to the stethoscope of his choice.

Phonocardiography is the recording of heart sounds and murmurs, usually by means of sensitive microphones and electronic amplification. The timing and intensity of the sounds can be recorded with an oscillograph, as in Fig. 37-2, or special techniques can be applied to analyze their frequency content.

First heart sound. The first heart sound consists of irregular vibrations, mainly of a frequency of 30 to 45/sec. These may be grouped into preliminary vibrations, main vibrations, and final vibrations. They probably originate in three ways: (1) from the vibrations of the atrioventricular valves during and after closure, (2) from vibrations set up by eddies in the stream ejected through the aortic orifice into the broader sinuses of Valsalva, and (3) from vibrations in the muscle fibers themselves, generated by passage from relaxation to tension. In the first sound, as developed in the ventricle and heard at the apex, the closure of the atrioventricular valves probably gives rise to the one or two small, slow initial vibrations. The initial vibrations of the first sound sometimes develop before ventricular contraction and even before the QRS complex of the ECG. Under such circumstances the initial vibrations must depend partly on atrial relaxation. The development of atrial contraction also causes

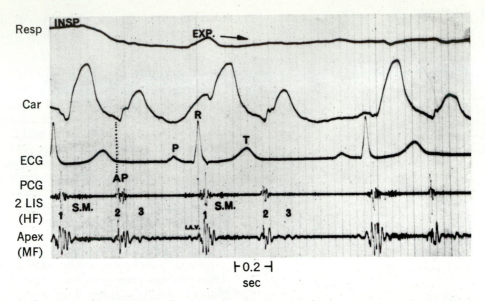

Fig. 37-2. Heart sounds in normal boy. *Resp.* Phase of respiration; *Car,* carotid pulsations in neck recorded externally; *ECG,* electrocardiogram; *PCG,* phonocardiogram. One phonocardiographic tracing was recorded from second intercostal space at left sternal border, *2 LIS,* and filtered electrically to accentuate high frequencies, *HF.* A second phonocardiogram was recorded simultaneously from apex of heart, *Apex,* in medium frequency range, *MF.* First, second, and third heart sounds are labeled *1, 2,* and *3,* respectively. Aortic component of second sound, *A,* precedes pulmonic component, *P,* and records demonstrate that interval between these two components is wider in first recorded cycle, near peak of inspiration, than in subsequent beats during expiration. Faint early systolic murmur, *SM,* is occasionally recorded in normal young people. (From Deuchar.[2])

sound vibrations, as in Fig. 37-2, which can sometimes be heard or recorded from the chest wall in a normal individual, and can be recorded in the normal subject with receivers introduced into the esophagus so that they are in close contact with the atria. These atrial sounds may occasionally be so exaggerated that they modify the first heart sounds as heard and recorded in a normal subject.

The main vibrations of the first sound develop usually toward the end of the QRS complex, at the beginning of the ventricular contraction. These vibrations probably depend on factors 2 and 3. That sounds develop in the muscle itself is definitely established, for sounds may be recorded from the beating excised and unfilled heart and they also may be recorded from contracting skeletal muscle.

Second heart sound. The second heart sound is developed from the closure of the aortic and pulmonary semilunar valves, and the vibrations that result are usually of a higher pitch than those of the first sound, about 50 to 70 Hz. The two valves do not close at exactly the same time (Fig.

37-1), and this may cause duplication of the second sound. The closure of the valves occurs at that point in the disappearance of ventricular systole when the pressures in the vessels exceed those in the ventricles; therefore closure varies with vascular conditions. The second sound normally occurs slightly after the end of the T wave, but this relationship varies considerably, and the T wave is not a reliable guide for the precise timing of any mechanical event.

Third heart sound. The third heart sound develops some 0.1 to 0.2 sec after the second sound and is particularly evident when the blood flow is rapid (e.g., after exercise). It probably develops in relation to the rapid inflow of blood from the atrium into the ventricle. A faint third heart sound is often audible in young adults, but it can rarely be heard in older persons except in pathologic conditions.

Intensity of heart sounds. The intensity of the heart sounds varies with the force of the heart's contraction and the tensions developed in the heart, aorta, and valves. The loudness of the second sound may be used to estimate changes in the tension of the large vessels, because increase in pulmonary or aortic pressures usually intensifies the sound.

Venous pulse

The pressure changes in the atrium are transmitted to the veins and produce a pulse recordable in the jugular vein. This may be recorded in human subjects by placing an appropriate transducer on the skin overlying the jugular vein. With the subject lying down so that the veins of the neck are clearly visible, a pulsation can be seen in this region, and the waves of this venous pulse correspond closely with those recorded within the atrium. There are three definite positive waves, conventionally labeled *a, c,* and *v.*

When the atrium contracts, the centripetal flow of blood in the large veins is temporarily impeded. The pressure in the large veins is raised, and the wave so produced is labeled *a.* When the tricuspid valve closes or bulges into the atrium at the start of ventricular contraction, a similar wave of raised pressure is produced. If the veins are congested, this wave may travel relatively rapidly up them and cause a second increase in pressure. Under more normal conditions of venous pressure, it travels more slowly, and the pressure wave in the carotid artery (starting later because of the isometric period) reaches the extrathoracic vessels at about the same time. This arterial wave is transmitted not only from the carotid artery but also from the aorta and other large arteries to the large veins, and a positive wave is thus created; because of its relation to the carotid pulse, it is called the *c* wave. This wave may therefore be derived from three sources: (1) from the carotid, (2) from the aortic pulse affecting the large veins at the heart, and (3) from a true venous wave. During the early part of ventricular expulsion the stretching action of the downward movement of the base of the heart lowers the pressures in the atrium and large veins. This gives a negative wave. As the incoming blood stagnates during ventricular systole, the atrium and veins fill and raise the pressure, whereas the return of the base of the heart as recoil ceases creates another positive wave. Together, these form the *v* wave. Finally, the tricuspid valve opens, and the venous pressure again falls as the ventricle fills. The waves of the venous pulse may be compared also in point of time with arterial pulse tracings from the brachial or radial artery. As a result of the time taken for the transmission of the arterial pulse to the arm, the part of the venous pulse that corresponds with it is recorded in the neck about 0.1 sec earlier.

VENTRICULAR FUNCTION

The amount of blood pumped into the circulation by the heart with each stroke and the mechanisms that regulate this cardiac output have engaged students of the circulation since the time of William Harvey. Although heart muscle responds to adequate stimuli with the maximal contraction of which it is capable at the time (the "all-or-none" law), this capability is determined by a host of factors that modify it continuously in the intact animal. The term "contractility" is often used to describe the myocardial property that enables the heart to perform its function as a pump, but complete agreement on an operational definition of contractility has yet to be reached (p. 977).

Because of the large number of interacting influences on myocardial contraction in the intact animal, much of the relevant experimental work has been done on strips of myocardium (Chapter 36 and Fig. 36-8) or the isolated heart-lung preparation. Early investigators identified many factors that altered the output of the heart, but one in particular emerged as the foundation for much of the subsequent research in this field—the myocardial fiber length prior to contraction.

In the hands of Frank,[20] Starling,[9] and others, the isolated heart-lung preparation provided evidence that the energy imparted to the blood by ventricular contraction was proportional to ventricular fiber length at the end of the preceding diastole. This relationship, later confirmed in intact animals[8] and in man,[7] has been widely adopted as a frame of reference for the investigation of other equally important factors.

Length-tension relations

Function in isolated heart. A mammalian heart can continue to beat for hours after its removal from the body if it is kept warm and supplied with an oxygenated salt solution containing the essential electrolytes in proper concentration. However, such a heart is far from normal and is incapable of obtaining from the solution enough oxygen to perform any extensive work. If the heart is supplied with blood, conditions are more nearly normal. The simplest method of oxygenating the blood is to utilize the animal's own lungs, although a mechanical oxygenator can be used[18] should the use of the lungs be undesirable.

Variations in the rate of venous filling and the pressures at which the atrium and ventricle are distended in diastole (and consequently their diastolic volumes) may be attained in the heart-lung preparation by varying the height of a reservoir. If this is raised, the pressure at which the right heart is filled is increased. With a greater diastolic volume, the heart contracts more forcibly and discharges more blood into the pulmonary artery, and the blood returned by the pulmonary veins is increased. A similar process then follows in the left heart chambers. Within limits, ventricles eject blood with a greater pressure as well as in a greater amount as the height of the

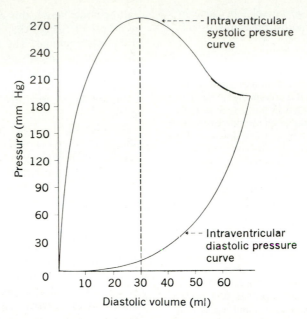

Fig. 37-3. Diagram of relationship between end-diastolic ventricular volume (index of length of ventricular muscle fibers just before contraction) and intraventricular end-diastolic and peak systolic pressures. Values, which apply to isolated dog ventricle, were estimated by Starling and his colleagues from data obtained by Frank on isolated frog heart. (Redrawn from Patterson et al.[32])

reservoir (the "filling pressure") is increased, and the external work done can be calculated from measurements of volume and pressure.

Starling[9] formulated his law of the heart on the basis of such experiments and also on the earlier investigations of Frank[20] concerning the relationship between diastolic volume of the heart and the magnitude of its developed tension. This law contains three propositions: (1) in cardiac muscle as in skeletal muscle the energy of contraction is a function of the length of the muscle fibers prior to contraction; (2) as the fiber length (diastolic volume) of the heart muscle increases, the energy of contraction also increases up to a certain optimal length of the muscle fiber, and then with further lengthening the energy of contraction diminishes; and (3) these relationships apply to the isolated mammalian heart devoid of neural or humoral control. Fiber length was assumed to be proportional to ventricular volume rather than measured directly, and the external work done by the heart was calculated from the ejected volume and pressure.

Fig. 37-3 illustrates the operation of this law in terms of end-diastolic ventricular volume and peak systolic pressure in the subsequent contraction. The effect of raising the filling pressure is to increase diastolic volume and presumably increase the length of the ventricular fibers. Diastolic volume can be increased up to approximately 30 ml in this example by relatively small increments in pressure, and the greater the end-diastolic volume, the greater the peak systolic pressure developed when the ventricle contracts. At higher end-diastolic volumes, however, this relation no longer holds, and the systolic pressures fall as diastolic volume increases further. These curves thus demonstrate a direct relation between end-diastolic fiber length and force of contraction up to a limit beyond which the ventricle becomes overdistended. Starling and his associates also showed that the energy as well as the pressure delivered by the contracting ventricle rose as filling pressure increased. A similar response can be demonstrated under proper conditions in vivo in animals and man, but the falling portion of the systolic pressure curve at very high diastolic volumes in Fig. 37-3 is outside the normal physiologic range.

In a comprehensive series of experiments, Sarnoff and Mitchell[8] confirmed the validity of Starling's law in vivo and introduced "ventricular function curves" as a graphic method of expressing other factors that influence cardiac function. A ventricular function curve (Fig. 37-4) relates stroke work to end-diastolic fiber length (or some indirect index of this length such as ventricular filling pressure) for one particular

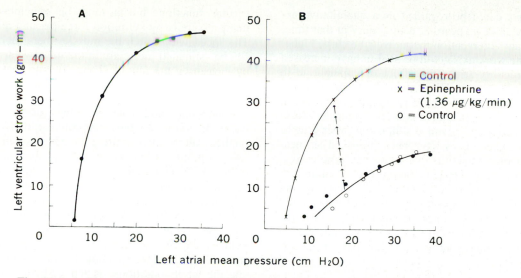

Fig. 37-4. A, Ventricular function curve obtained from open-chest dog preparation showing relation between left atrial mean pressure and left ventricular stroke work. **B,** Function curves from same heart as in **A,** but after left ventricle had begun to fail, as indicated by relatively low stroke work at each level of mean atrial pressure (lower curve in **B**). Upper curve shows effect of infusion of synthetic epinephrine. Note increase in ventricular contractility, as indicated by greater amount of stroke work at each level of mean atrial pressure. (From Sarnoff.[34])

set of conditions. If all conditions relevant to myocardial function remain constant, variations in stroke work move along this curve in response to variations in filling pressure, but when the conditions are changed, ventricular function moves to a new work–fiber length curve. The preparation used by Sarnoff allows measurement and control of most of the related variables in order to define these function curves.

The heart of a dog maintained on positive pressure breathing is exposed through the open chest. Both vagi are cut high in the cervical region. Cannulas are introduced into the right and left atria through small openings in the pericardium and are connected to reservoirs containing blood from a donor dog. Atrial pressures can be varied over a wide range by alterations in the heights of the blood reservoirs. The pressures within the right and left atria and in the aortic arch are measured with electromanometers. The mean atrial pressure is taken as an index of the ventricular filling pressure and also of ventricular end-diastolic pressure. The systemic blood flow (left ventricular output minus coronary flow) is measured continuously. From these measurements the ventricular stroke work is calculated in gram-meters.

The relationship between mean atrial pressure and ventricular stroke work at various levels of atrial pressure is expressed in a ventricular function curve. One such curve obtained from the left atrium and left ventricle in a normal dog is

shown in Fig. 37-4, *A*. It can be seen that there is an initial steep rise in ventricular stroke work at low atrial mean pressures (low ventricular filling pressures), and then the curve begins to level off as the pressures increase. Under normal conditions with the pericardium intact there is little or no tendency toward a descending limb on the function curve (i.e., a fall in stroke work at high mean atrial pressures). A descending limb does occur, however, when the myocardial metabolism is depressed, as, for example, during occlusion of the coronary circulation. The ventricular function curves shown in Fig. 37-4, *B*, were obtained from the same dog but at a time when there was evidence of myocardial failure, that is, a relatively small amount of stroke work at all levels of mean atrial pressure. A continuous intravenous infusion of synthetic epinephrine under these conditions causes an immediate increase in stroke work at all levels of mean atrial pressure; there is a shift upward in the ventricular function curve. Since the heart can shift from one function curve to another, the functional activity of cardiac muscle can be described by a family of curves, each corresponding to a specific set of circumstances.

In the search for a practical definition and measurement of contractility, the ventricular function curve thus offers one possible solution. The curve determined under any constant set of conditions could be regarded as a quantitative expression of myocardial contractility at that time, and any intervention that moves the curve higher on the work axis could be said to have increased contractility (e.g., the effect of epinephrine in Fig. 37-4, *B*). Displacement of the function curve downward would represent a decrease in contractility, in that less work would be delivered from any given end-diastolic fiber length. Many alternatives have been proposed as measures of contractility, including the time derivative of intraventricular pressure (maximum dP/dt), maximum acceleration of blood during ejection,[31] and specific functions of tension and time.[7,35] Each of these has its own practical and theoretical merits but each also limits description of myocardial contraction to one part of the cardiac cycle.

Length-tension-velocity relations*

The characteristic relationships between length, tension, and velocity of shortening that have been observed in the isolated papillary muscle of the heart (p. 975) can all be shown to exist in normally functioning intact ventricles. The anatomic arrangement of ventricular muscle into chambers that eject blood as they contract, as contrasted with the simple shortening of an isolated muscle strip, makes it necessary to measure these three variables indirectly. Intraventricular volume is usually measured rather than fiber length, and rough approximations of the length-volume relationship are sometimes made by assuming that the ventricle is spherical or ellipsoidal.[7] Intraventricular pressure takes the place of myocardial tension, although again the relation between the two can be estimated only by making simplifying assumptions about the shape and changes in wall thickness of the ventricle. At the end of diastole the intraventricular pressure is analogous to the ''preload'' in a simple muscle strip preparation (i.e., the weight that is suspended from such a strip to stretch it to the desired initial length). During ejection the aortic pressure is related to the ''afterload,'' or weight that the muscle is required to lift. Since the aortic pressure during systole is a function of arterial resistance as well as the rate of blood flow, the ventricular ''load'' depends in part on arterial resistance. Since the pressure and flow are pulsatile, aortic impedance (p. 1029) as well as resistance is involved in the ventricular load. The terms ''load,'' ''force,'' and ''pressure'' are often used interchangeably in discussions of ven-

tricular function, but it must be remembered that pressure is by definition force per unit area. The speed with which the muscle fibers shorten can be estimated indirectly in the intact ventricle by measuring the rate of change of intraventricular volume or the rate of blood flow into the aorta.

Sonnenblick and others[1,15,22] have proposed a useful definition of ventricular contractility based on the relation between these three key variables, using a graphic presentation such as that shown in Fig. 37-5. The three planes in the figure represent the relations between intraventricular volume and pressure (analogous to length-tension), volume and velocity of shortening of contractile elements (length-velocity), and pressure and velocity (force-velocity). The surface indicated in the diagram represents contractility under one specific set of conditions, as defined by the interrelation of the three variables. In this particular state of contractility (Fig. 37-5), ventricular contraction begins at point A, which represents the end-diastolic pressure, end-diastolic volume, and zero velocity. Contractile element velocity then rises to point B on the appropriate pressure-velocity curve. From B to C, during the remainder of the isometric contraction period, pressure rises and velocity slows accordingly. At C, intraventricular pressure reaches the level of diastolic aortic pressure, ejection of blood from the ventricle begins, and ventricular volume starts to diminish. During ejection, length-tension-velocity relationships depend in part on the aortic impedance (p. 1029), and the curve tends to rise above the surface shown in Fig. 37-5. Ejection ends at point D and isometric relaxation at point E.

The surface shown in Fig. 37-5 has been proposed as an operational definition of ventricular function. It is a useful way of visualizing the relations among three important variables under one particular set of conditions, but it does not constitute a satisfactory definition of ''myocardial contractility'' (p. 977) for at least three reasons. First, contractile element velocity (the vertical axis in Fig. 37-5) cannot be measured under anything approximating normal conditions in vivo. Shortening velocity of the muscle fiber as a whole is not an adequate substitute, because muscle behaves as though it contained elastic elements in series with the contractile elements (p. 975). During isometric contraction, for example, muscle length does not change, although it is assumed that contractile elements are shortening and thus elongating the elastic elements. Experiments designed especially to estimate

---

*See references 1, 7, 10, 15, 22, and 30.

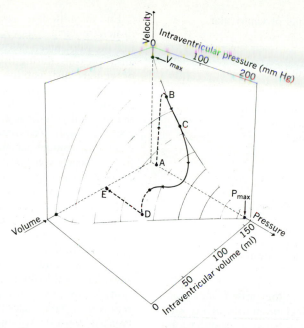

Fig. 37-5. Diagrammatic representation of ventricular contractility in terms of relations between pressure and velocity of shortening of muscle (right), volume and velocity (left), and volume and pressure (base). Unscaled vertical axis labeled "Velocity" represents instantaneous velocity of shortening of contractile elements in ventricular muscle. Intraventricular pressure and volume are scaled in accordance with typical conditions in human left ventricle. V_{max} indicates maximum velocity of shortening of unloaded muscle; P_{max} indicates maximum isometric tension at end-diastolic fiber length. Contraction begins at *A*, velocity reaches pressure-velocity curve appropriate to end-diastolic volume at *B*, and follows that curve to end of isometric contraction period at *C*. At *C*, intraventricular pressure reaches level of aortic pressure (initial load), aortic valve opens, and ejection begins. As intraventricular volume decreases during ejection period, pressure-volume-velocity relationships follow surface indicated until late in systole, when active state begins to decline. Ejection ends at *D*, followed by isometric relaxation period *(D to E)*. Three-dimensional surface shown in diagram in effect defines a particular contractile state (see text). (After Sonnenblick.[36])

length changes in each kind of element suggest that the series elastic component is stretched only slightly,[1,10] but enough to influence the shape of force-velocity curves.[15,30] Second, even under constant environmental conditions, ventricular pressure-volume-velocity relationships during the ejection period are not unique, but depend on the impedance of the aorta. Third, time-dependent properties are neglected in this approach, except for the assumption that the "active state" reaches a maximum early in systole (point B in Fig. 37-5) and remains constant throughout most of the contraction. The validity of this assumption is highly questionable, and it is much more likely that the "active state" is a time-varying process rather than a steady state (p. 977). For all these reasons, length-force-velocity relationships alone do not yield a complete description of ventricular function in vivo, nor do they define myocardial contractility in a wholly consis-

tent way. Numerous alternatives have been proposed, generated by the need to determine the functional ability of the myocardium in given experimental or clinical conditions. The maximum rate of change of pressure, dP/dt, during isometric contraction is frequently used as an "index" of contractility. Other parameters used for the same purpose include integrals of time and pressure,[35] maximum unloaded contractile element velocity (V_{max}),[36] maximum acceleration of blood during ejection,[31] and ventricular work. Each of these may be empirically useful under certain conditions, but an ideal measure of muscle contractility has yet to be discovered.

Determinants of ventricular function

The function of the heart as a pump is influenced by many factors, the most important of which are listed in Table 37-1. Some may be classed as *intrinsic* determinants of ventricular

Table 37-1. Determinants of ventricular function

Intrinsic (autoregulation)
  End-diastolic fiber length
    Filling pressure
    Duration of diastole
    Diastolic ventricular distensibility
  Myocardial tension (load)
  Frequency of contraction
  Temperature
Extrinsic
  Autonomic nervous system
  Humoral factors
    Oxygen and carbon dioxide
    Catecholamines
  Coordination of contractions
    Timing of atrial systole
    Sequence of ventricular activation

function, in the sense that they arise from certain inherent physicochemical properties of myocardial cells. The relation between end-diastolic fiber length and force of contraction that underlies Starling's law is an example. The cellular mechanisms responsible for these phenomena are not yet known, but they enable the myocardium to adapt to a variety of conditions without the intervention of external controls or, in other words, to exhibit *autoregulation.*

Myocardial autoregulation. To emphasize the difference between autoregulation that depends on diastolic fiber length and other varieties, Sarnoff[8] has suggested the terms "heterometric autoregulation" for the former type of response and "homeometric autoregulation" for the others. Some investigators have questioned whether heterometric responses play much part in the circulatory adjustments to ordinary activities in the dog or man, inasmuch as ventricular filling pressures and volumes in many situations are not closely correlated with ventricular work. The multitude of factors involved in vivo, however, makes it difficult to measure the contribution of any one factor. The Starling law, moreover, describes a kind of "feedback" mechanism (Chapter 42), which by its very nature should tend to prevent large changes in ventricular diastolic pressure or volume. The normal function of heterometric autoregulation appears to be the maintenance of balance between right and left ventricular outputs and the provision of a mechanism for relating cardiac function to total blood volume and vascular capacity. Homeometric autoregulation,[8] on the other hand, gives the circulation the ability to adapt to special demands, such as exercise, with little change in venous pressure or heart size.

End-diastolic fiber length, the crucial variable in heterometric responses, depends on both filling pressure and ventricular distensibility. The atrial pressure during filling is in turn a function of venous distensibility and pressure, whereas the amount of blood that passes from atrium to ventricle depends on the duration of diastole and the residual volume left in the ventricle from the previous beat. Ventricular distensibility, or the change in volume per unit change in pressure, is by no means constant, and experimental inference of fiber length from measurements of ventricular pressure or volume must therefore be made with caution.[16] The volume and stiffness of the distal parts of the coronary vascular tree, which act as a kind of skeletal structure within the myocardium, affect ventricular distensibility, as does the fluid content of the myocardial tissues. The pericardium also restricts the distensibility of the heart in the intact animal, and the removal of the pericardium is followed by an increased diastolic volume at normal filling pressures. The pericardial sac thus acts to prevent excessive dilatation of the ventricles, but the protection it affords seems to be nonessential, since the normal heart can function adequately in the absence of the pericardium. Pressure within the pericardial sac, when it is intact, varies to some extent with intraventricular pressure, a fact that must be taken into account in estimating the transmural pressure of the ventricles.[25]

A second type of autoregulation is described by the force-velocity curve, which arises from an intrinsic response of cardiac muscle to applied tension (p. 975). The instantaneous velocity of shortening of myocardial fibers is inversely proportional to the load they are required to work against, and the performance of the heart is consequently regulated in part by arterial pressure and impedance. If the load on the left ventricle is suddenly increased by partial occlusion of the aorta, the next beat produces a higher pressure but the ejection velocity and stroke volume fall.[38] In experiments in which the load is reduced the opposite effect occurs. Calculations of the external work of the heart in such experiments suggest that the normal aortic impedance is about optimal, if contractility is normal, for the production of maximum work.[38]

All other forms of regulation involve a change in contractility, as do the responses to extrinsic stimuli. The "staircase" phenomenon (p. 976) is an example of an intrinsic autoregulatory response, in which an increase in heart rate enhances contractility. The basic mechanism is unknown, but elevation of the frequency of contraction is accompanied by an increase in the velocity of shortening at all but the highest loads.

Tachycardia has another effect unrelated to this alteration of contractility, in that diastole is abbreviated more than systole, leaving a shorter time for ventricular filling. All other factors being equal, this means that an increase in heart rate tends to reduce end-diastolic volume, but the net effect on cardiac output is minimal, except with very fast rates.

Extrinsic control of myocardial function. The neural stimuli reaching the heart through efferents of the autonomic nervous system and the chemical content of the blood supplied to the myocardium are the two principal elements that may be classed as *extrinsic* controls of myocardial contractility, as indicated in Table 37-1. If heart rate is kept constant, stimulation of the stellate ganglia leads to an increase in ventricular contractility, a more rapid development of tension, a more rapid shortening of the muscle fibers, and a shorter period of ejection.[8] A similar myocardial response follows the peripheral activation of appropriate circulatory reflexes, as by altering pressure in the carotid sinuses.[17] The constant stream of impulses that travels over the sympathetic efferents to the ventricles does not maintain a constant level of "tone" in the myocardium as it does in vascular smooth muscle, but it does ensure that the force of each contraction can be made either smaller or greater by changing the frequency of these impulses.

Although it was long believed that the ventricles had no parasympathetic innervation, there is now convincing evidence that they do. When the heart rate and other relevant variables are strictly controlled, stimulation of the distal end of the cut vagus nerves produces a depression of ventricular contractility.[19] This direct effect on contractility is additional to the indirect vagal modifications of cardiac function in vivo by way of changes in heart rate or in the delay between atrial and ventricular contractions.

The oxygen, carbon dioxide, and catecholamines in the blood reaching the myocardium constitute a second group of extrinsic influences. Myocardial hypoxia and hypercapnia both depress contractility,[1] but this effect is not readily apparent in the intact animal because of the action of arterial chemoreceptors and respiratory mechanisms. In man the hypoxemia brought on by inhalation of 12% oxygen in nitrogen leads to an increase in cardiac output, as does inhalation of high concentrations of carbon dioxide.[5] An increased P_{CO_2} rather than a decreased pH appears to be the important factor in the latter response. As for epinephrine and norepinephrine, their direct effect on the myocardium is to in-

crease contractility, although the net effect on cardiac output and blood pressure in the intact animal depends in part on the reaction of the vascular system to these agents (p. 1055).

The sequence of activation in the heart is a third determinant of cardiac function that may arbitrarily be classed as extrinsic. Atrial contraction toward the end of diastole adds to the volume of blood distending the ventricles and in this way contributes to the filling pressure.[28] When the presystolic atrial contractions occur normally, the stroke work is greater than in their absence, as is apparent when the normal relationship is altered by atrial fibrillation or complete atrioventricular block. Stroke work also depends on the sequence of ventricular activation normally dictated by the conduction pathways through the bundle of His and Purkinje system (p. 969). When the usual sequence of contraction of myocardial fibers is altered by artificial stimuli applied directly to the ventricle, other factors remaining constant, stroke work decreases; in the ultimate disorganization of the contraction pattern represented by ventricular fibrillation, pumping by the ventricles ceases entirely.

Myocardial work and efficiency. The external work done by the heart with each stroke can be calculated from the volume, pressure, and velocity of the blood ejected by the ventricles. The component of this work that appears as potential energy equals the product of the volume discharged from the ventricle and the pressure imparted to it, integrated over the period of ejection (Chapter 39, equation 4). The kinetic energy component is relatively small (Table 40-1) and is frequently neglected in the calculation of cardiac work. Stroke work can be represented graphically by plotting the changes in ventricular volume and pressure with time,[26] as shown in Fig. 37-6. The area of the "work loop" constructed in this way is proportional to the stroke work, although the kinetic energy is omitted. The area between the abscissa and the bottom of the loop represents the energy remaining at the end of the previous heart cycle and that added by venous pressure as the ventricle fills.

The external work done by the heart is much less than the energy equivalent of the oxygen consumed by the myocardium. In other words, there appears to be a large expenditure of energy on "internal work" that is dissipated as heat. This is true of any machine that harnesses energy, and the ratio of useful work performed to energy required is referred to as the *efficiency* of the device. In this sense the efficiency of the heart is on the order of 10% to 20%.[13] This efficiency falls still lower in pathologic conditions,

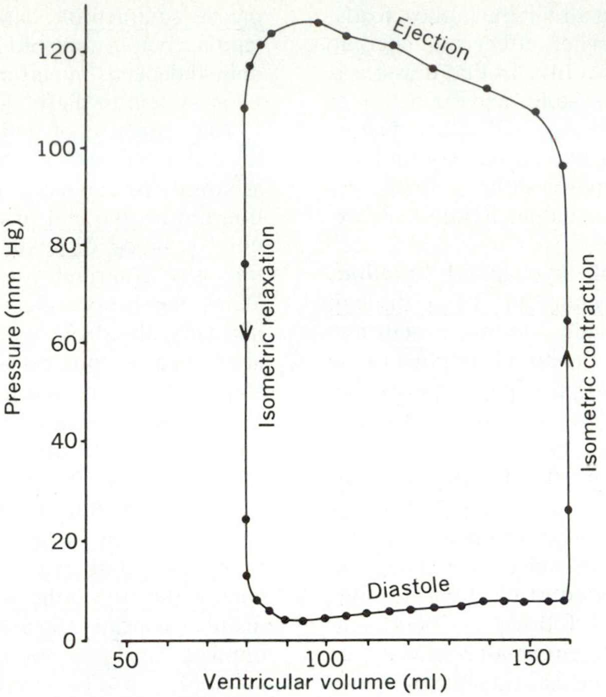

Fig. 37-6. Intraventricular volume and pressure changes in single heart cycle in man at rest. Intervals between points represent 0.02 sec. Area within loop equals potential energy component of stroke work (Chapter 39, equation 4).

but even under normal conditions, efficiency is far from constant. Myocardial oxygen consumption for a given stroke work is greater for a normal stroke volume at high pressure than for a large stroke volume at normal pressure.[35] This means that the efficiency of the heart decreases when it is made to pump against an increased resistance and that the heart can more efficiently increase its stroke volume than it can increase the pressure at which this volume is ejected.

Under controlled conditions, it is possible to demonstrate a significant correlation of end-diastolic fiber length with either oxygen consumption or external work, but it is clear from the inconstant efficiency with which oxygen is utilized by the heart that fiber length may determine one or the other of these variables, but not both. Data presented by Sarnoff and co-workers show that the variable that correlates most closely with myocardial oxygen consumption is not the work of the heart but the tension developed by the myocardium.[33]

Cardiac failure. The heart can ordinarily provide a cardiac output adequate to the needs of normal life even when handicapped by pathologic faults in its valves or other structures. When the demands on it are too great, however,

or large parts of the myocardium are deprived of an adequate blood supply, the heart can no longer compensate for these abnormalities and ventricular function deteriorates. The same phenomenon appears in isolated heart preparations after several hours of essentially normal function.

Ventricular failure in man and experimental animals is associated with depressed myocardial contractility, which would be even lower were it not for an augmentation of sympathetic adrenergic stimuli to the heart. In force-velocity terms the maximum isometric tension and V_{max} are both diminished. Ventricular end-diastolic volume and pressure are usually greater than normal, as though the Starling mechanism were being called on to compensate for the myocardial failure. In left ventricular failure in man the end-diastolic pressure may rise from about 10 mm Hg, which is a normal level, to 20 mm Hg or more, while stroke volume and cardiac output remain normal, at least at rest. Elevation of ventricular end-diastolic pressure is used in clinical physiology as a sign of ventricular failure, but unfortunately, it gives only limited information about the state of the myocardium. The clinical need to detect ventricular failure in its earliest stages is one reason for the long and intensive search for a sensitive index of myocardial contractility.

As failure progresses, the ventricles dilate, residual volume increases, and the fraction of end-diastolic volume that is ejected with each stroke may fall to one fourth or less. With prolonged failure or extreme overloading of the heart, resting cardiac output becomes subnormal.

The consequences of this deterioration include a rise in venous pressure that is eventually reflected in elevated capillary pressure and loss of fluid into the extracellular spaces, leading to edema of the extremities, accumulation of fluid within the abdomen, and exudation of fluid into the alveoli. The clinical state that accompanies these events is referred to as *cardiac decompensation* or simply *heart failure*. The syndrome of heart failure can assume a variety of different forms determined by the pathologic conditions that initiate it, and secondary responses of organ systems other than the circulation are involved. Efforts to identify the basic mechanisms of myocardial failure at the cellular level have as yet been unsuccessful, although biochemical explanations have been sought in the properties of the contractile proteins, the supply and utilization of energy, and the coupling between excitation and contraction.[1,5]

Measurement of cardiac output

Fick principle and direct Fick method. This principle, first set forth by Fick in 1870, is based on the obvious fact that the average output of the right ventricle must equal that of the left. Of course, brief periods of unbalanced flow do occur; for example, when the return of blood in the systemic veins suddenly increases, the output of the right ventricle will exceed that of the left, but only momentarily. In its passage through the lungs the blood put out by the right ventricle takes up O_2 and gives off CO_2. The extent of this exchange must depend on the volume of blood flowing through the lungs. Therefore, if the O_2 or CO_2 content of the mixed venous blood entering the pulmonary artery can be compared with that of arterial blood leaving the pulmonary vein, the volume of O_2 absorbed or of CO_2 given off by each liter of blood in its passage through the lungs can be stated. Then, if the volume of O_2 taken up by the lungs or the volume of CO_2 expelled from the lungs per minute is measured, the number of liters of blood flowing per minute through the lungs to allow this exchange can be readily calculated. In short, cardiac output (C.O.) in liters per minute is equal to the exchange of one of the respiratory gases per minute divided by the arteriovenous difference (A-VΔ) of that gas in milliliters per liter.

$$\text{C.O. (L/min)} = \frac{\text{ml/min of } O_2 \text{ used}}{\text{A-V } O_2 \ \Delta \text{ in ml/L}}$$

or

$$\frac{\text{ml/min } CO_2 \text{ produced}}{\text{A-V } CO_2 \ \Delta \text{ in ml/L}}$$

The consumption of O_2 and the production of CO_2 can be determined easily by analysis of inspired and expired air. The gaseous composition of blood in the pulmonary vein can be determined by a sample taken from any systemic artery, since the composition of blood does not change in its passage through the left heart and the systemic arterial tree. Obtaining a sample of mixed venous blood from the right ventricle or pulmonary artery can be accomplished in experimental animals with relative ease by direct puncture, which was used for this purpose by Barcroft et al.[11] as early as 1919. In man, such sampling became feasible only with the introduction of cardiac catheterization, which is now performed daily in many laboratories.

A small, suitably flexible catheter is introduced through an opening in an antecubital or other peripheral vein (made under local anesthesia) and passed through the axillary, subclavian, and innominate veins into the vena cava, from which its tip can be guided under fluoroscopic observation into the right heart or into the pulmonary artery. The catheter may be left in place for a sufficient time to obtain serial samples.

The following is a typical determination of cardiac output by the direct Fick method. The resting subject was found to have an O_2 intake of 250 ml/min. An analysis of blood samples taken simultaneously from the femoral artery by needle puncture and from the right ventricle through a venous catheter showed that his arterial blood contained 19 and his venous blood 14.5 ml O_2/100. The arteriovenous oxygen difference was therefore 45 ml/L. Accordingly, the cardiac output was estimated to be 250/45, or 5.5 L/min. Dividing the output per minute by the pulse rate gives the average stroke volume.

This method has definite limitations in its application to man, as have all other respiratory methods of measuring cardiac output. It is accurate only when the output and the pulmonary gas exchange remain stable during the period of measurement. It cannot be used to follow rapid changes in circulation rate such as occur at the beginning of muscular exercise. The method has nevertheless proved successful and yielded important information in many studies of patients with cardiovascular disease.

Indirect Fick methods. Long before it was realized that the direct Fick method was feasible for work on man, several indirect applications of the principle were developed and used extensively.

In one indirect method the O_2 or CO_2 contents of arterial and venous bloods are inferred from measurements of gas tensions in the lungs. For several reasons, CO_2 rather than O_2 is usually the gas dealt with. Advantage is taken of the fact that alveolar air is in equilibrium with arterial blood. A determination of the alveolar CO_2 tension and a reference to a CO_2 dissociation curve for arterial blood, which relates tension to content (Chapter 69), permit a calculation of the CO_2 content of arterial blood. The CO_2 content of mixed venous blood can be estimated in various ways by finding the composition of a gas mixture in the lungs that neither gives up nor acquires CO_2. The tension of the gas in such a mixture must equal that of blood entering the pulmonary capillaries from the right heart, and its CO_2 content is estimated from a dissociation curve for venous blood. Since the method is difficult in practice and subject to a variety of errors, it will not be described here in further detail.

Use of Fick principle with foreign gases. In these methods a foreign gas that is nontoxic is breathed and goes into solution in the plasma but does not combine with any constituent of blood. Its concentration in arterial blood is estimated from its tension in the alveolar air and its solubility in blood. When the experimental period is kept short enough (acetylene method) and the returning venous blood is considered to contain none of the foreign gas, the problem is to this extent simplified. When the period is long (ethyl iodide method), the venous tension is estimated by a rebreathing procedure, and the content is calculated.

Indicator-dilution method. The dilution method was first introduced by Stewart in 1897 and has subsequently been modified and used by many other investigators.[24,39]

The technique involves the injection of a known amount of material into the circulation and the measurement of the dilution of the material during a known period of time. The substance whose dilution is to be measured should remain within the vascular system during the period of analysis, and it should also lend itself to rapid and relatively easy analytic procedures. Dyes such as indocyanine green are the compounds most frequently used, although radioactive indicators are sometimes employed (tagged cells or plasma). The material whose dilution is to be measured is injected into a vein. A few seconds after injection the dye begins to appear in the arterial blood. The dye concentration gradually increases until it reaches a maximum; then it begins to decline until a second rise in concentration occurs as a result of recirculation. The downslope of the curve prior to recirculation approximates an exponential decay,[24] so that replotting the curve with the logarithm of dye concentration as the ordinate and time as the abscissa (a procedure simplified by using semilogarithmic graph paper) allows the time-concentration curve for the indicator on its first circula-

tion past the point of sampling to be defined by extrapolating the downslope (Fig. 37-7). The area under this curve then indicates the average concentration of indicator and the duration of this first passage. The volume of indicator-containing blood that must have passed the sampling site per unit time in order to produce the curve can be calculated by dividing the area of the curve into the amount of indicator injected. Dye-dilution curves are now commonly recorded by withdrawing a small sample of blood continuously from an artery and measuring the concentration of dye with an electrical photosensitive device. An example of such a record is shown at the bottom of Fig. 37-7 with the same curve replotted above on semilogarithmic coordinates. This curve was recorded from the femoral artery of a dog after injection of dye into the pulmonary artery, and the steps in calculating cardiac output are indicated.

The indicator-dilution method is now used even more widely than the direct Fick method for the determination of cardiac output, and a close agreement between the two techniques has been shown. The indicator-dilution method has an advantage in that a steady state is not needed except for the few seconds required for the first circulation of the indicator, and a disadvantage in that it becomes difficult to exclude recirculation by extrapolating the downslope when cardiac output is unusually low. In addition to providing the data needed to calculate cardiac output, indicator-dilution curves give a measurement of the *mean transit time* between injection and sampling sites, from which the volume of blood in that segment of the circulation can be estimated. This volume, which is calculated by multiplying mean transit time and cardiac output, includes not only the volume directly between the sites of injection and sampling but also the blood volume of temporally equidistant sites in other branches of the vascular tree.[24,39]

Circulation time. The average time required for a particle of blood to travel from one point in the circulation to another depends on the rate of blood flow and the regional blood volume. Rough measurements of circulation times have been found to have some diagnostic value in clinical medicine, and a number of test substances have been introduced for this purpose. These materials are usually injected into a vein in the arm, and the time that elapses before the substance appears in the systemic arterial circulation is measured. The substances employed are ones that can be detected by relatively simple means in the peripheral circulation: saccharin, which yields a sweet taste on reaching the vascular bed of the tongue; fluorescein, which can be detected in the skin by its fluorescence under ultraviolet light; or very small amounts of sodium cyanide, which induce a brief hyperpnea on reaching the carotid body. Gross slowing of the circulation, or enlargement

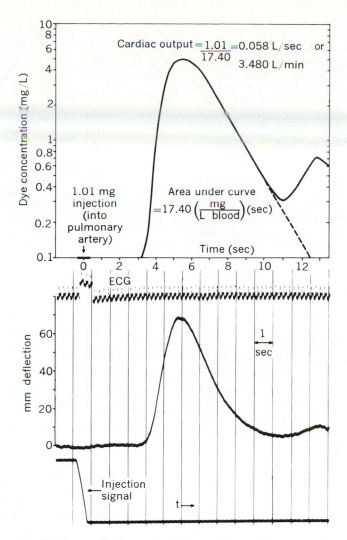

Fig. 37-7. Example of indicator-dilution method of measuring cardiac output. Curve in lower half of illustration is dilution curve recorded from femoral arterial blood by continuously recording densitometer after injection of 1.01 mg of indocyanine green dye into pulmonary artery of dog. Downward shift of lowest tracing corresponds to period of injection of dye. By means of calibration factor determined experimentally, deflections of dilution curve are translated into arterial dye concentration in milligrams per liter of blood, and curve is replotted above with concentration as ordinate on logarithmic scale. Downstroke of curve is extrapolated to eliminate portion due to recirculation of dye, and area of primary curve is measured. Cardiac output in liters of blood per second is calculated by dividing curve area into amount of dye injected. Mean transit time (not indicated) between injection and sampling sites can also be calculated from curve.

of the blood volume, prolongs such measurements of circulation time to an unmistakable degree, but the great variations in normal subjects make small deviations of no significance.

A refinement of this approach uses the injection of indicators into the circulation and the recording of dilution curves from various vessels or from the heart chambers themselves as a way of detecting abnormal circulatory shunts, such as those found in some forms of congenital heart disease.

Ballistocardiograph. This device records the thrusts transmitted to the body as the result of impacts and recoils produced by the ejection of blood from the ventricles and its movement through the two arterial trees. The forces thus developed are sufficient to move a slightly movable table on which the body lies. The table is suspended in such a way as to move in the direction of the long axis of the body, and the record of its movements is called a ballistocardiogram. The recoil phenomena thus revealed are complex. They arise not only from a footward recoil evoked when blood is ejected into the pulmonary artery and ascending aorta but also from the nearly simultaneous headward recoil produced when a large fraction of the aortic blood changes its direction as it rounds the aortic arch.

Empirical methods for estimating cardiac output from the waves of the ballistocardiogram that have been developed are of some value in detecting changes of output in normal supine subjects, but they do not estimate absolute values of output with any accuracy and are invalidated by the presence of myocardial disease.

Pulse-pressure methods. The pulse pressure is determined by the magnitude of the stroke volume and by various physical properties of the arterial tree. Attempts to estimate stroke volume from the arterial pulse pressure have been made by many investigators, but the assumptions that must be made about the other variables involved give little ground for confidence in the results. In general, pulse pressure varies directly with stroke volume as long as all other relevant variables remain constant, but in most circumstances, it is unwise to expect such constancy.

Measurement of blood flow

Numerous methods of measuring the flow of blood through vessels have been devised over the years, many of them applicable to the aorta or pulmonary artery and thus to the measurement of cardiac output, but the instruments that have been used for this purpose are too numerous to list here. A few of the techniques now employed to measure instantaneous blood velocity[6] may be considered briefly.

Electromagnetic flowmeters. Since blood is a conductor of electricity, the flow of blood within a vessel placed in a magnetic field generates an electrical potential across the vessel in accordance with the principles of electromagnetic induction. The practical application of this principle was first accomplished by Kolin,[27] who devised a probe incorporating a magnet and electrodes that could be placed around an exposed artery to detect the flow-generated signal. The formidable electronic problems associated with this technique have been gradually resolved, so that reliable flowmeters of this type are now widely available. In addition to the advantages inherent in an external device that leaves the vessel lumen intact and unobstructed, the electromagnetic flowmeter has a frequency response that is more than adequate for the pulsatile waveforms found in the circulation. Its response is linearly proportional to the velocity of blood flow; calibration either in vitro or in vivo is necessary to convert its electrical output into absolute units of velocity or volume flow per unit time. Typical electromagnetic flowmeter recordings of blood flow in the aorta and its branches are shown in Fig. 37-8.

Probes can be made in a wide range of diameters down to the dimensions of the coronary vessels in the

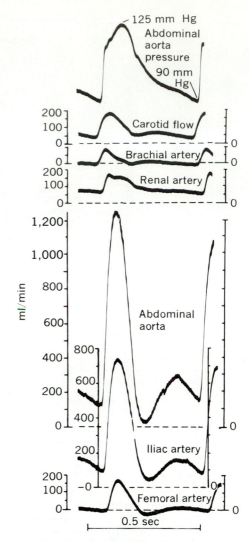

Fig. 37-8. Blood flow in aorta and its branches as recorded by electromagnetic flowmeter. (Uppermost tracing indicates abdominal aortic pressure.) All flow ordinates are scaled equally. Carotid and renal flow characteristically pulsate around mean value representing continuous forward velocity. Blood flow in brachial, iliac, and femoral arteries and in lower abdominal aorta may oscillate through zero in early diastole under resting conditions. (From Spencer and Denison.[37])

dog (Fig. 44-2), and they can be implanted chronically in experimental animals to allow measurements over periods of weeks to months. The recent development of miniature electromagnetic flowmeters fitted on the tip of a cardiac catheter makes it possible to measure pulsatile flow in many of the larger vessels in man.[6,29]

Ultrasonic flowmeters. The ultrasonic flowmeter is is also applied to the exterior of the blood vessel and shares many of the other advantages of the electromag-

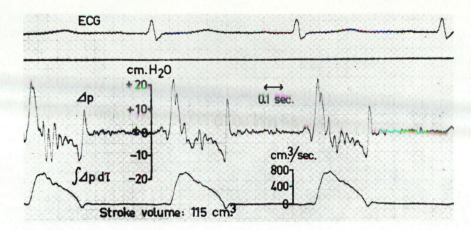

Fig. 37-9. Aortic blood flow computed from pressure difference between two sites in thoracic aorta. From above downward: electrocardiogram; differential pressure (Δp); and computed velocity of aortic blood flow. (From Rudewald.[33])

netic devices. The probe in this case is a small, light-weight sleeve containing ceramic crystals that transmit and receive ultrasound or vibrations of very high frequency. The transit time between upstream and downstream crystals is a function of the blood flow through which the ultrasound is transmitted, and from precise, rapidly repeated measurements of the transit time or the phase shift in the signal a virtually continuous record of flow can be obtained. The same principle has been applied to the continuous measurement of ventricular dimensions and has also been adapted to the telemetry of blood flow measurements.[21] Instruments that rely on Doppler phase shifts in ultrasonic signals have also been devised.[6]

Pressure gradient method. By relying on the hydrodynamic relationships between pressure and flow that have been derived analytically (Chapter 39, equation 15), pulsatile blood flow can be calculated from simultaneous measurements of pressure at two sites along the longitudinal axis of a blood vessel. The development of the theoretical background and practical application of this method is largely the work of Fry and collaborators,[23] and it has been applied successfully to the human[12,33] and animal aorta (Fig. 37-9). Double-bore catheters, with a distal opening of one lumen 2 to 10 cm beyond the other, are used for this purpose, and the size of the catheter precludes its application to vessels much smaller than the canine aorta without obstructing the lumen to an unacceptable degree. The electrical signals representing the two pressures are fed to a simple analog computing circuit, which solves the appropriate equation and gives a continuous record of blood velocity. A similar computation of flow can be based on the more complex pressure-flow equations of Womersley (p. 1028). The greatest weakness of the technique lies in the very small pressure differences that must be measured accurately. If the catheter openings are separated by a distance of 8 cm in the aorta of the dog, the peak pressure gradient is of the order of 3 to 20 cm of water; the mean gradient is at the limits of resolution of modern manometric techniques.

MEASUREMENT OF BLOOD PRESSURES

The first measurements of arterial and venous blood pressure were made by Stephen Hales (1677-1761), who reported, ''I caused a mare to be tied down alive on her back; . . . having laid open the left crural artery about 3 inches from her belly, I inserted into it a brass pipe whose bore was $\frac{1}{6}$ of an inch in diameter; and to that, by means of another brass pipe which was fitly adapted to it, I fixed a glass tube, of nearly the same diameter, which was 9 feet in length: then untying the ligature on the artery, the blood rose in the tube 8 feet 3 inches perpendicular above the level of the left ventricle of the heart. . . .''[4] The principle that intravascular pressures can be determined by connecting the vessel to a vertical tube and measuring the height of the fluid column that the pressure will support has been used to advantage by many subsequent investigators. The substitution of a U tube partially filled with mercury for the simple blood-filled tube of Hales makes the dimensions of the manometer more manageable when arterial pressures are being measured, and for venous pressures the tube may be filled with isotonic saline solution. Such direct manometry measures pressures as the height of the fluid column supported above a specific level of reference, usually the right atrium, and expresses them in centimeters of saline solution or water or millimeters of mercury. These units can be interconverted or expressed in standard physical units (dynes/cm²) by reference to the densities of saline solution, water, and mercury. The principal disadvantage of the method arises from the inertia of the fluid in the connecting tubes and manometer, which prevents it from responding instantaneously to the rapid fluctuations of the pressure-pulse, so that mean pressure is indicated accurately but not the systolic and diastolic extremes.

With the conversion of the manometer into a closed system by adding a small cup covered by a thin rubber membrane to the tube communicating with the vessel, it became possible to reduce the inertia of the system and record pressures continuously from a light lever attached to the membrane. The later replacement of the recording lever by a tiny mirror cemented to the membrane, which reflected a beam of light onto photographic paper or plates, gave a sensitive manometer that could record faithfully the excursions of pressure. In the hands of Wiggers and Hamilton, among others, these optical manometers served to advance the understanding of hemodynamics to a remarkable degree.

The development of electrical pressure transducers[3] provides readily available instruments that are less fragile and technically demanding than the older mechanical devices and that can reproduce circulatory pressures with high fidelity. The simplicity of operation of these transducers carries with it increased potentialities for unidentified sources of artifact or inaccuracy, however, and frequent testing of calibration and frequency response is essential. Faithful recording of the pressure-pulse contour depends on the frequency response of the pressure-sensing system (the relative magnitudes of the signals produced by sinusoidal pressures of different frequencies), which in turn depends on the catheter, tubing, or needle connecting the blood vessel to the transducer as well as the characteristics of the transducer itself.[3] Miniaturized transducers fitted to the tip of an intravascular catheter have been constructed to avoid some of the limitations imposed by fluid-filled connections.

Indirect measurement of arterial blood pressure in man

Since an indirect method without puncture of vessels is essential for routine clinical work, the principle first suggested by Riva Rocci[3] is of great importance. Measurements are made of the pressure that is required when applied to a limb to close the artery and obstruct the blood flow. By these methods the pressures may be measured with as much accuracy as can be attained with the cruder types of manometer and exposure of a vessel, but the reliability of the values is below that obtainable by direct puncture and modern transducers.

Technical procedures. The first step is the placing of an armlet or cuff around the upper arm or thigh. The vessels of the forearm and lower leg are sheltered by bones and are less easily used. The armlet must be broad enough to allow the pressure to be distributed over a large area: for an arm, a cuff 12 to 13 cm in breadth should be used; for a thigh, one 18 to 20 cm in breadth is used. For children with small arms a narrower cuff (7 cm) may be used. The cuff is distended with air by means of a pump, and the pressure is determined by a mercury manometer or aneroid pressure gauge. The latter should be checked at intervals against a mercury manometer. Usually the pressure is raised rapidly above the systolic level and then is allowed to fall slowly while the point of recurrence of the pulse below the cuff is determined.

Indices that may be used to detect the peripheral pulse may be grouped into several types. The palpatory is the simplest index and is used in the original method: when the cuff pressure exceeds the systolic pressure, the peripheral pulse in the radial or dorsalis pedis artery is obliterated, and as the pressure falls, the pulse returns, growing to a supernormal amplitude and then returning to normal. Only the initial return is clearly differentiated, and the systolic pressure so indicated is some 5 mm Hg below that indicated by auscultation; the discrepancy may be greater if the observer is inexperienced.

The auscultatory, because of its fair accuracy and simplicity, is the method of greatest clinical value. It was originally described by Korotkow. If a stethoscope is held over the brachial artery at the elbow, no sound is heard in the normal subject. But if the pressure is raised in a cuff around the upper arm, particularly if the lower limit of the cuff is close to the position of the stethoscope, sounds will be heard with each pulse, and these disappear as the circulation becomes occluded. If the cuff pressure then is allowed to fall slowly, the sounds heard may be described as falling into four phases, although any single subject often fails to provide clear examples of all of them. In the first phase, sounds become evident, and the pressure at the occurrence of the first sound is taken as the systolic pressure level; these sounds may be sharp or thudding and not unlike one or another of the heart sounds. If the systolic level is closely approximated and decompression is slow, the respiratory variations of blood pressure may cause a rhythmic disappearance of the sounds in late expiration or early inspiration. In the second phase the sounds become much louder and a hissing murmur may be added to them. In the third phase the sound consists again mainly of loud thuds. In the fourth phase these thudding sounds gradually fade and become muffled; the fading may be rapid or gradual, and in the latter case the effect may be compared with that of a band gradually marching into the distance. This fourth phase ends with a return to silence. Occasionally a silent phase may occur between the other phases. The development of the initial sounds indicates systolic pressure, probably a pressure that is intermediate between end systolic and lateral systolic pressure. The development of the second phase is associated with the transmission of a much larger pulsation; it is possibly correlated with a pressure within the cuff which has fallen just below the lateral systolic pressure. The development of the beginning of the fourth phase appears to indicate lateral diastolic pressure, but the values so determined are sometimes deceptive and often tend to be too low. Such errors are reduced if certain precautions are followed.

The auscultatory systolic criterion is always found at pressures some millimeters above those for the palpatory technique if the stethoscope is properly situated close to the point of compression of the brachial artery by the cuff; if this pressure difference is not found, it indicates that either the stethoscope is inaccurately placed or the brachial artery is abnormally situated so that the auscultatory readings are unreliable; occasionally, merely changing the cuff to the opposite arm will give quite different and more reliable pressure readings. In making readings by auscultation or any other method, several inflations and deflations should be made before readings are recorded, for even with experienced subjects an initial alteration of blood pressure because of excitement in response to the sensations of circulatory obstruction is the rule. Deflation should be carried out slowly, for rapid deflation alters the circulation by allowing the upper vessels to discharge suddenly into empty arteries below the cuff, and abnormally low diastolic readings result. Pressure within the cuff should be released between readings and not be maintained longer than necessary, since any congestion not only causes the subject discomfort but may induce reflex changes. The stethoscope bell should be applied to the elbow as lightly as possible; if it is applied forcibly, it may compress and distort the vessels and give rise to sounds. This is especially likely to happen when the diastolic pressure is low, for a pressure of 130 gm applied to a bell 2 cm in diameter is the equivalent of more than 30 mm Hg pressure. In the literature in many cases, sounds have been recorded as continuing without fading down to zero pressure, and diastolic pressure has been estimated as zero. It is certain that in animals the lowest diastolic pressures ever recorded are always well above the atmospheric level, so that these low estimates of diastolic pressure must be considered as erroneous. In some cases at least, they depend on failure to observe the aforementioned (particularly the last) precautions, although sounds may sometimes be generated in vessels in the absence of compression.

The indirect methods of measuring pressure are likely to remain important because of the simplicity of the equipment used. However, they can never be very precise. At best, they can only show that at one moment the falling pressure in the armlet is above and at the next moment below the pressure being measured, and the point of crossing is estimated approximately.

REFERENCES
General reviews

1. Braunwald, E., Ross, J., Jr., and Sonnenblick, E. H.: Mechanisms of contraction of the normal and failing heart, Boston, 1967, Little, Brown & Co.
2. Deuchar, D. C.: Clinical phonocardiography, London, 1965, English Universities Press, Ltd.
3. Geddes, L. A.: The direct and indirect measurement of blood pressure, Chicago, 1970, Year Book Medical Publishers, Inc.
4. Hales, S.: Statical essays: containing hemostatics; or, an account of some hydraulic and hydrostatical experiments made on the blood and blood-vessels of animals, ed. 3, London, 1769, W. Innys.
5. Marshall, R. J., and Shepherd, J. T.: Cardiac function in health and disease, Philadelphia, 1968, W. B. Saunders Co.
6. Mills, C. J.: Measurement of pulsatile flow and flow velocity. In Bergel, D. H., editor: Cardiovascular fluid dynamics, New York, 1972, Academic Press, Inc.
7. Mirsky, I., Ghista, D. N., and Sandler, H., editors: Cardiac mechanics: physiological, clinical, and mathematical considerations, New York, 1974, John Wiley & Sons, Inc.
8. Sarnoff, S. J., and Mitchell, J. H.: The control of the function of the heart. In Hamilton, W. F., and Dow, P., editors: Handbook of physiology. Circulation section, Baltimore, 1962, The Williams & Wilkins Co., vol. 1.
9. Starling, E. H.: Linacre lecture on the law of the heart, New York, 1918, Longmans, Green & Co.

Original papers

10. Abbot, B. C., and Mommaerts, W. F. H. M.: A study of inotropic mechanisms in the papillary muscle preparation, J. Gen. Physiol. **42:**533, 1959.
11. Barcroft, J., Boycott, A. E., Dunn, J. S., and Peters, R. A.: Observations on respiration and circulation in the goat, Q. J. Med. **13:**35, 1919.
12. Barnett, G. O., Greenfield, J. C., Jr., and Fox, S. M.: The technique of estimating the instantaneous aortic blood velocity in man from the pressure gradient, Am. Heart J. **62:**359, 1961.
13. Bing, R. J., et al.: The measurement of coronary blood flow, oxygen consumption, and efficiency of the left ventricle in man, Am. Heart J. **38:**1, 1949.
14. Braunwald, E., Fishman, A. P., and Cournand, A.: Time relationship of dynamic events in the cardiac chambers, pulmonary artery and aorta in man, Circ. Res. **4:**100, 1956.
15. Brutsaert, D. L., and Sonnenblick, E. H.: Force-velocity-length-time relations of the contractile elements in heart muscle of the cat, Circ. Res. **24:**137, 1969.
16. Buckley, N. M., Sidky, M., and Ogden, E.: Factors altering the filling of the isolated left ventricle of the dog heart, Circ. Res. **4:**148, 1956.
17. Cotten, M. deV., and Moran, N. C.: Effects of increased reflex sympathetic activity on contractile force of the heart, Am. J. Physiol. **191:**461, 1957.
18. Daly, I. deB., and Thorpe, W. V.: An isolated mammalian heart preparation capable of performing work for prolonged periods, J. Physiol. **79:**199, 1933.
19. DeGeest, H., Levy, M. N., Zieske, H., and Lipman,

R. I.: Depression of ventricular contractility by stimulation of vagus nerves, Circ. Res. **17:**222, 1965.

20. Frank, O.: Zur Dynamik des Herzmuskels, Z. Biol. **32:** 370, 1895.

21. Franklin, D. L., Watson, N. W., and Van Citters, R. L.: Blood velocity telemetered from untethered animals, Nature (Lond.) **203:**528, 1964.

22. Fry, D., Griggs, D., and Greenfield, J.: Myocardial mechanics: tension-velocity-length relationships of heart muscle, Circ. Res. **14:**73, 1964.

23. Greenfield, J. C., Patel, D. J., Mallos, A. J., and Fry, D. L.: Evaluation of Kolin type electromagnetic flowmeter and the pressure gradient technique, J. Appl. Physiol. **17:**372, 1962.

24. Hamilton, W. F., Moore, J. W., Kinsman, J. M., and Spurling, R. G.: Studies on the circulation. IV. Further analysis of the injection method, and of changes in hemodynamics under physiological and pathological conditions, Am. J. Physiol. **99:**534, 1932.

25. Holt, J. P., Rhode, E. A., and Kines, H.: Pericardial and ventricular pressure, Circ. Res. **8:**1171, 1960.

26. Katz, A. M., Katz, L. N., and Williams, F. L.: Registration of left ventricular volume curves in the dog with the systemic circulation intact, Circ. Res. **3:**588, 1955.

27. Kolin, A.: An electromagnetic flowmeter: principles of the method and its application to blood flow measurements, Proc. Soc. Exp. Biol. Med. **35:**53, 1936.

28. Linden, R. J., and Mitchell, J. H.: Relation between left ventricular diastolic pressure and myocardial segment length and observations on the contribution of atrial systole, Circ. Res. **8:**1092, 1960.

29. Mills, C. J., et al.: Pressure-flow relationships and vascular impedance in man, Cardiovasc. Res. **4:**405, 1970.

30. Noble, M. I. M., Bowen, T. E., and Hefner, L. L.: Force-velocity relationship of cat cardiac muscle, studied by isotonic and quick-release techniques, Circ. Res. **24:**821-833, 1969.

31. Noble, M. I. M., Trenchard, D., and Guz, A.: Left ventricular ejection in conscious dogs, I. Measurement and significance of the maximum acceleration of blood from the left ventricle, Circ. Res. **19:**139, 1966.

32. Patterson, S. W., Pieper, H., and Starling, H.: The regulation of the heart beat, J. Physiol. **48:**465, 1914.

33. Rudewald, B.: Hemodynamics of the human ascending aorta as studied by means of a differential pressure technique, Acta Physiol. Scand. **54**(suppl.):187, 1962.

34. Sarnoff, S. J.: Symposium on regulation of performance of heart; myocardial contractility as described by ventricular function curves; observations on Starling's law of heart, Physiol. Rev. **35:**107, 1955.

35. Sarnoff, S. J., et al.: Hemodynamic determinants of oxygen consumption of the heart with special reference to the tension-time index, Am. J. Physiol. **192:**148, 1958.

36. Sonnenblick, E. H.: Instantaneous force-velocity-length determinants in the contraction of heart muscle, Circ. Res. **16:**441, 1965.

37. Spencer, M. P., and Denison, A. B.: Pulsatile blood flow in the vascular system. In Hamilton, W. F., and Dow, P., editors: Handbook of physiology. Circulation section, Baltimore, 1962, The Williams & Wilkins Co., vol. 1.

38. Wilcken, D. E. L., Charlier, A. A., Hoffman, J. I. E., and Guz, A.: Effects of alterations in aortic impedance on the performance of the ventricles, Circ. Res. **14:**283, 1964.

39. Zierler, K. L.: A simplified explanation of the theory of indicator-dilution for measurement of fluid flow and volume and other distributive phenomena, Bull. Johns Hopkins Hosp. **103:**199, 1958.

38

WILLIAM R. MILNOR

The electrocardiogram

Myocardial action potentials described in Chapter 36 and the sequential development of these action potentials as excitation spreads over the myocardium generate an electrical field throughout the body. The electrocardiogram is a record of the difference of electrical potential, or voltage, between two points in this field, usually points on the body surface. There is a superficial resemblance between electrocardiographic tracings and records of the cellular action potential, but these two phenomena differ in one important respect: the cellular action potential represents differences between the inside and outside of a cell, whereas the electrocardiogram arises from potential differences between the excited and the unexcited portions of the heart. Both have their origin in the electrical events that accompany cellular excitation and recovery, but the electrocardiogram results from the *spread* of these events from one cell to another.

The first electrocardiograms recorded from the body surface were made by Waller[15] in 1887, using a capillary electrometer. This device responded too slowly for accurate reproduction of electrocardiographic signals, which are on the order of 1 or 2 mV and include components only a few milliseconds in duration; yet a remarkable amount of basic information was gained with it in the next few years by Waller, Burdon-Sanderson, Einthoven, and others.[1] The introduction of the string galvanometer by Einthoven[8] in 1903 made possible a high degree of accuracy in recording the electrocardiogram, although the apparatus was cumbersome and required some delicacy in operation. Current from the cardiac electrical field was led through a silver-plated quartz string suspended between the poles of an electromagnet, and the resulting deflections of the string in the magnetic field were recorded optically. In modern electrocardiographs the currents from the heart are amplified electronically, and the bulky machine of Einthoven's day has been replaced by compact, portable instruments that are widely used in medical practice.

Electrical field of the heart

As excitation spreads through the heart in the orderly fashion described in Chapter 36, the surface of cells that have just been excited, or *depolarized,* becomes electrically negative with respect to regions still in the resting state. It is this phenomenon that generates the cardiac electrical field and thus the electrocardiogram. The advancing *excitation front,* or line of demarcation between excited and as yet unexcited cells, is electrically equivalent to an array of batteries, each with a negative pole on the excited side of the front and a positive pole on the unexcited side. This metaphor is useful in correlating the spread of excitation with the deflections in the electrocardiogram, because the field produced in a conductive medium by such two-terminal electrical generators, or *dipoles,* can be expressed mathematically[18] and has been studied extensively in physical models. If the distance between the poles of the dipole is small in relation to the extent of the surrounding medium, the potential difference between two electrodes at a distance from the dipole will depend on three factors: the voltage of the dipolar generator itself, the distance of the electrodes from the dipole, and the orientation of the electrodes with respect to a line passing through the poles of the generator. The potential difference between the electrodes will be at a maximum when a line connecting the electrodes is parallel to the axis of the dipole, and it will be zero when this line is perpendicular to the dipole axis.

The relation between the spread of excitation in a small strip of cardiac muscle and the electrocardiogram recorded by electrodes placed in the saline solution surrounding the strip is illustrated in Fig. 38-1. The left end of the strip is stimulated so that excitation travels across the strip from left to right. The position of the excitation front at three successive instants in time is shown in Fig. 38-1, steps 2 to 4. The field generated by these dipoles makes electrode b positive with respect to electrode a, and the gal-

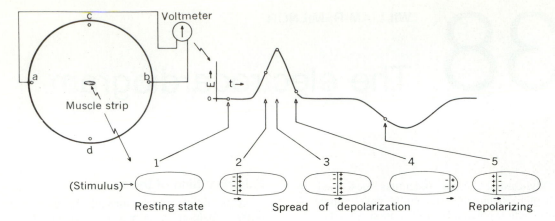

Fig. 38-1. Schematic representation of spread of activation and recovery in small strip of myocardial tissue placed in volume conductor and of resulting electrocardiogram recorded from two electrodes, *a* and *b,* at a distance from muscle strip. At beginning, muscle strip is in resting state and stimulus is applied to end on left, *1,* so that activation spreads from left to right. Advancing activation front as successive cells are depolarized, *2* to *4,* is electrically equivalent to array of dipoles with their positive poles in direction of spread (see text). Subsequent repolarization in same sequence, from left to right, is equivalent to dipole array with opposite orientation, *5.* Virtually no electrocardiographic deflections would be recorded from electrodes *c* and *d* because they are oriented at right angles to spread of depolarization and repolarization.

vanometer tracing has been connected to show an upward deflection. At the time represented by 5 in Fig. 38-1, excitation has traversed the entire length of the strip, and the cells at the left have *repolarized,* or recovered their normal resting state, so that a *wave of repolarization* is beginning to move from left to right. The dipoles in the repolarization wave are oriented with their negative poles in the direction of advance, so the voltage detected by the electrode pair a and b is now one in which the electrode b is negative with respect to a, and the deflection is downward. When the entire strip has repolarized, it has returned to the resting state, and no excitation fronts exist until the strip is stimulated again. Electrode pair c and d, being oriented at right angles to the dipoles of both excitation and repolarization, exhibits no potential difference during these events.

This example is a relatively simple one in that the dipoles lined up along the excitation front are very nearly parallel to each other. Even if they were diversely oriented along an irregular excitation front, however, their net effect on distant electrodes would be equivalent to that of a single dipole, the orientation and voltage of which could be determined by appropriate measurements from the surrounding medium. In the heart the excitation fronts that exist at any instant correspond to enormous numbers of dipoles distributed in an intricate pattern. Despite

this complexity and the limited extent and lack of electrical homogeneity of the body tissues that form the conducting medium around the heart, the electrocardiogram resembles to a marked degree the potential differences that would be generated by a single *equivalent dipole* situated anatomically near the center of the heart.

Einthoven hypothesis and limb leads

An electrocardiogram may be recorded by measuring the potential difference between any two points on the body, and any two sites selected for this purpose constitute an electrocardiographic *lead.* The number of possible leads is infinite, but the *limb leads*—those recorded from electrodes on the arms and legs—are the source of a large part of our present knowledge. The three limb leads introduced by Einthoven et al.[9] in what was called the "schema of the equilateral triangle" have been so extensively used that they are often referred to as the *standard limb leads.* Three electrodes are used to record these leads: one on the right arm, one on the left arm, and one on either the right or left leg. It makes no difference whether they are placed on the proximal or distal part of the extremity, since the arms and legs act as essentially linear conductors attached to the torso. A fourth electrode is frequently placed on one leg and connected to an electrical ground to minimize electrical interference.

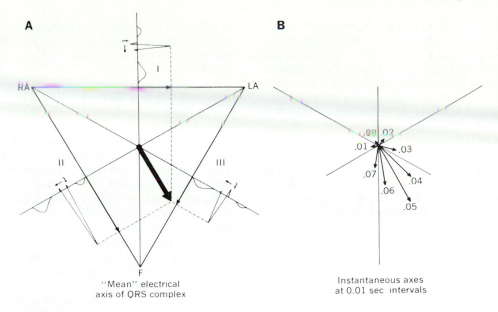

Fig. 38-2. Electrical axes of QRS complex derived from three standard limb leads represented by Einthoven triangle. **A,** Electrical QRS axis derived from peak of R waves in each lead. **B,** Instantaneous electrical axes at intervals of 0.01 sec throughout QRS complex showing their changing magnitude and orientation as activation spreads through ventricles. For further description see text.

Table 38-1. Standard limb leads formed by pairing electrodes

Lead	Positive terminal	Negative terminal
I	Left arm	Right arm
II	Leg	Right arm
III	Leg	Left arm

The standard limb leads formed by pairing these electrodes are designated by Roman numerals. In each lead, one electrode is designated as *positive* and the other as *negative,* as seen in Table 38-1. This arbitrary convention serves to indicate the polarity of the electrocardiographic signals, since the leads are connected to the electrocardiograph in such a way that an upward deflection of the tracing signifies electrical positivity of the positive electrode with respect to the negative one. To explain the relationship between these three leads, Einthoven et al.[9] assumed that the body was like an electrically homogeneous plate in which the right arm, left arm, and legs formed the corners of an equilateral triangle with the heart at its center. He then used the deflections observed in these leads at any instant to calculate what he called the *instantaneous electrical axis* and *manifest potential difference* of the heart or, in modern terms, the orientation and voltage of the single equivalent

dipole at that instant. The mathematical relationship implied by his basic assumption is that the potential difference observed in each lead will be proportional to the magnitude of the voltage generated by the heart and to cosine α, where α is an angle determined by the orientation of the lead and the orientation of the dipolar generator.

A geometric method for carrying out these calculations is illustrated in Fig. 38-2. The sides of the equilateral triangle correspond to the standard limb leads, and a set of coordinates is established by drawing perpendicular lines from the midpoint of each side so that they intersect in the center of the triangle. The potential difference observed in each lead under consideration at that instant is plotted on the line representing the lead, using an arbitrary scale for magnitude. The forces measured, having both magnitude and direction, may properly be regarded as vectors. In Fig. 38-2, *A,* they are symbolized by arrows that are superimposed on the appropriate side of the triangle, the head of the arrow being positive. Positive potentials are plotted from the zero coordinate toward the positive terminal of the lead; negative potentials are plotted in the opposite direction. Perpendiculars dropped from the head of each arrow should then intersect at a single point, and a vector drawn from the center of the triangle to this point of intersection represents the axis and magnitude of the equivalent dipole that would have generated the observed deflections. A check on the calculations is provided by the relationship: Lead I + Lead III = Lead II at any instant, which follows not from the Einthoven hypothesis but from the way in which electrodes are shared by these leads.

The measurement of instantaneous electrical axes and magnitudes is still used by investigators, but accurate determinations require the simultaneous recording of all three leads. An example of instantaneous electrical axes at eight successive instants in a normal QRS complex is shown in Fig. 38-2, *B*. As a compromise for practical purposes the peak of the R wave in each lead is often substituted for the instantaneous measurements by clinical electrocardiographers, and the direction of the vector calculated from these peaks is referred to as the "electrical axis of the QRS complex." It is also sometimes called the "mean" electrical axis, but that is a misnomer because it is in no sense an average. Moreover, the R wave peak usually does not occur at exactly the same instant in all leads.

The concept of the Einthoven triangle is an oversimplification, but it is a useful one. Investigation has shown that the "triangle" formed by the limb leads is skewed and nonequilateral and that it shows considerable individual variation.[6] Moreover, under some conditions the complexity of the heart as a generator is too great and the distance between heart and body surface is too small to allow a single equivalent dipole to represent accurately every detail of the electrocardiogram. Since such discrepancies are remarkably small in most subjects, the Einthoven hypothesis still serves as an approximation adequate for clinical work and a starting point for further investigation.

Sequence and origin of deflections

The successive deflections in the electrocardiogram reflect the sequence of excitation and repolarization in the myocardium. The normal cardiac cycle begins with an electrocardiographic wave produced by excitation of the atria, called the *P wave,* followed after a short interval by deflections corresponding to ventricular excitation, the QRS complex (Fig. 38-3). Excitation reaches the atrioventricular node before the end of the P wave and travels through the bundles of His and Purkinje system (Chapter 36) during the interval between the P wave and the QRS complex, but the electrical activity of these specialized tissues does not give rise to a widespread electrical field or to deflections in the electrocardiogram. Repolarization of the atria, which follows the P wave and continues during the QRS complex, produces an electrocardiographic wave so small in magnitude that it can rarely be detected in conventional leads.

The term "QRS complex" is used comprehensively to designate any deflections produced by ventricular excitation. The terminology for the different parts of this complex is based on the order and polarity of the deflections. A positive deflection is termed an *R wave,* and any subsequent negative deflection is termed an *S wave.* If the ventricular complex begins with a negative

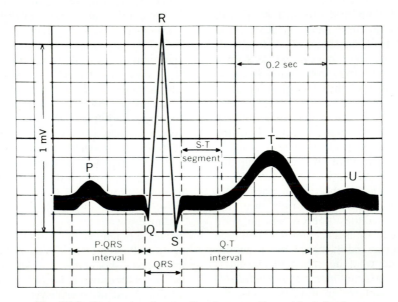

Fig. 38-3. Conventional terms for electrocardiographic deflections.

deflection, that deflection is called a *Q wave*. These terms are purely descriptive and not indicative of specific physiologic events in the heart; the same excitation front in the heart may be responsible for a Q wave in one lead and a simultaneous R wave in another. The QRS complex is followed by a slower wave representing ventricular repolarization, the *T wave*.

These waves provide a way of measuring the duration of the corresponding events in the heart. In adult man the time required for complete excitation of both atria, as indicated by the duration of the P wave, is approximately 0.1 sec. The *P-R interval,* which is the name given to the time interval from the beginning of the P wave to the onset of the QRS complex, indicates the time between excitation of the first atrial cell and excitation of the first ventricular cell and ranges from 0.12 to 0.2 sec. The duration of the QRS complex is about 0.08 sec and represents the period between depolarization of the first and the last ventricular cells to be excited. The interval from onset of the QRS complex to the end of the T wave is termed the *Q-T interval.* This period, which averages 0.35 sec, corresponds to the time required for complete electrical excitation and recovery of the ventricles. Each of the values given here is typical of those found in adult human subjects, but these intervals vary considerably with age and heart rate[3-5] and in other species.

Contour and polarity of the limb lead deflections

The contours and polarity of the waves in each lead of the electrocardiogram depend on the changing orientation of the excitation fronts in the heart. If we assume that the net effect of the many dipoles along excitation fronts at each instant is equivalent to that of a single dipole, the relationship between the spread of excitation through the heart and the contours of the electrocardiogram in each of the limb leads can be recognized. To picture this relationship the heart is visualized as a dipolar generator in the center of an equilateral triangle, as in Fig. 38-2.

Since the sinoatrial node lies in the dextrocephalad portion of the right atrium, the excitation front that spreads over the atria from this site is oriented with positive poles toward the left side of the thorax and somewhat caudad. A net or equivalent dipole to represent this front would have the same general orientation, although its direction would vary slightly from instant to instant until all atrial fibers were depolarized. The field generated by this dipole would make the left arm positive with respect to the right arm and

would thus produce a positive or upright P wave in standard limb lead I. It would also result in an upright P wave in lead II, but in lead III, which is almost perpendicular to the equivalent dipole, the P wave might be small or absent. This is in accord with the observation that the P wave in lead III in man is usually small and may be either positive or negative.

The spread of excitation through the ventricles is more complicated because of the specialized, rapidly conducting system that allows excitation to begin almost simultaneously in different parts of the ventricles. The first ventricular fibers to be activated electrically lie on the left side of the midportion of the interventricular septum, owing to the asymmetric distribution of Purkinje fibers from the bundles of His. The earliest ventricular excitation front therefore travels through the septal muscle from left to right (0.01 sec vector in Fig. 38-2, *B*). Within 10 msec, however, subendocardial fibers throughout most of the ventricular walls have been excited via the Purkinje system, and excitation fronts are moving through the muscle toward the epicardial surface. These fronts have been carefully plotted in some species,[14] and the resulting picture shows an orderly, although complex, progression from subendocardial to subepicardial regions and from the apex to the base of each ventricle.

During much of the QRS complex, parts of the excitation fronts in the right and left ventricle are diametrically opposite in orientation and partially cancel each other in their effects on the cardiac field. Under these conditions an equivalent dipole to represent these fronts at any instant can be estimated only approximately, but a dipole roughly parallel to lead II and with the same polarity (positive pole toward the left and caudad) is in accord with electrocardiographic observations during most of the QRS complex. The QRS complex is normally predominantly positive in lead II, positive but of lesser magnitude in lead I, and small and either positive or negative in lead III. This generalization does not apply to the very earliest and the latest portions of the QRS complex. The earliest part, arising from excitation of the septum, corresponds to an equivalent dipole of small magnitude with the positive pole toward the right. The last part of the ventricles to be excited is a region in the right ventricular outflow tract, and the excitation front that finally advances into this region corresponds to an equivalent dipole with the positive pole directed more or less cephalad. The beginning and end of the QRS complex may therefore consist of deflections opposite in direction to the major in-

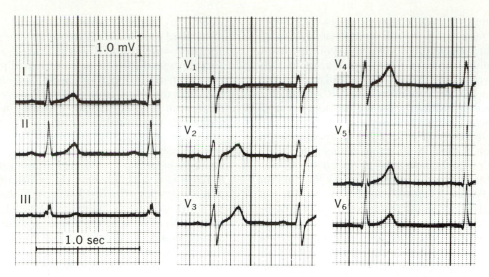

Fig. 38-4. Electrocardiogram from normal adult man showing three standard limb leads recorded simultaneously and six precordial leads.

tervening deflection in some leads, as for example in Fig. 38-4, lead I. The form of the QRS complex, as of all other waves in the electrocardiogram, varies appreciably from one individual to the next[5] because of anatomic differences in the position of the heart, the shape of the thorax, and probably small individual variations in the pattern of excitation. The duration of the QRS complex or of the P wave may not be exactly equal in all leads. This discrepancy appears when the earliest or latest excitation fronts are oriented perpendicularly to one lead for a few milliseconds and therefore yield no deflection in that particular lead during that time.

The sequence of repolarization in the ventricles is less well understood, and the relation between this sequence and the T wave is consequently not altogether clear. If the duration of the action potential, from onset to restoration of resting potential, were exactly equal in all ventricular fibers, repolarization would proceed through the myocardium in the same sequence as the preceding wave of excitation. Since the repolarization front would be negative on its advancing edge, the electrocardiographic wave caused by repolarization would be opposite in direction to the wave caused by depolarization, as in Fig. 38-1. In man, however, the T wave normally has the same polarity as the QRS complex in most leads (Fig. 38-4). This suggests that fibers are not repolarized in the same order in which they were depolarized and that some of the last fibers to be excited may be among the first

to recover. In some instances the T wave is followed by a relatively small, slow *U wave,* the origin of which is unknown.

The electrocardiographic tracing immediately following the QRS complex is called the *S-T segment* and usually consists of a short *isoelectrical period,* during which no difference in potential is recorded, followed by a gradual rise to the peak of the T wave. The S-T segment occurs at a time corresponding to the plateau of the action potential in many ventricular cells, and the T wave coincides with the return of membrane potentials toward the resting level. No sharp change marks the end of the S-T segment and beginning of the T wave in the electrocardiogram, nor should one be expected, since repolarization is gradual and not synchronous in all cells.

To summarize, the P, QRS, and T waves occur simultaneously in all leads but with differences in polarity and contour that depend on the orientation of each lead with respect to the heart. At any instant an equivalent dipole that would produce the observed deflections can be visualized, and from this the pattern of excitation and repolarization in the heart can be deduced to some extent. This forms the basis for many applications of electrocardiography in clinical diagnosis.[4]

Precordial and other unipolar leads

Although the standard limb leads introduced by Einthoven are still in general use today, they are often supplemented by other leads of a quite different kind. These leads measure the potential difference between a reference electrode with multiple connections to the body and an *exploring electrode* that may be placed anywhere on the body surface. The reference terminal most commonly used was devised by Wilson and consists of a *central terminal* connected through resistors in parallel to both arms and one leg.[19] This terminal, it was reasoned, would be analogous to an electrode very distant from the heart, since the potentials generated in electrodes at the apices of the Einthoven triangle would effectively cancel each other. The voltages recorded in a lead that paired the Wilson central terminal with an exploring electrode would therefore be dominated by the potential changes at the site of the exploring electrode, and these leads were accordingly named *unipolar*. The theoretical basis for this concept is debatable,[10] but the empirical value of unipolar leads from the chest wall has been firmly established.[20] Six specific unipolar precordial leads—leads in which the exploring electrode is placed on the thoracic wall near the heart and the Wilson central terminal serves as the reference electrode—have been adopted for routine clinical use,[4,5] These leads are designated by the letter V (for voltage) combined with a number indicating the specific position of the exploring electrode on the precordium.

Positions 1 and 2 are in the fourth intercostal space at the right and left sternal margins, respectively. Position 4 is the fifth interspace at its intersection with the midclavicular line, and position 3 lies midway between 2 and 4. Positions 4, 5, and 6 are all at the same transverse level, 5 being in the left anterior axillary line and 6 in the midaxillary line. Other sites are sometimes used for special purposes. The exploring electrode is defined as "positive" in each unipolar lead according to present conventions, although this practice was not consistently followed in records taken prior to 1938. In other words, relative positivity of the exploring electrode with respect to the central terminal should give an upward deflection.

The characteristics of the precordial electrocardiogram, like those of the standard limb leads, are in large measure consistent with the representation of the heart as a dipolar generator. In this respect, they supplement the limb leads by recording dorsoventral components of the three-dimensional electrical field of the heart, which the Einthoven leads fail to detect. The routine precordial positions are so close to the ventral surface of the heart, however, that the potential of the exploring electrode may in some circumstances be dominated by events in the nearby myocardium. When this happens, the hypothesis that the field sampled by the electrocardiogram is equivalent to that of a single dipole representing a summation of excitation fronts in all parts of the heart no longer applies. The extent to which such local events influence the precordial leads is still uncertain, but in the absence of pathologic conditions, their effects appear to be relatively small. The Wilson central terminal is also used to record *unipolar extremity leads* V_R, V_L, and V_F by placing the exploring electrode on the right arm, left arm, and leg, respectively. These leads contain the same information as the standard limb leads.

With suitable exploring electrodes, unipolar leads may be recorded from the body cavities as well as the body surface. Electrodes on the tip of a cardiac catheter have been used to record potentials from within the heart chambers, and special multielectrode catheters have been used in man to study conduction through the bundle of His.[7]

Clinical electrocardiography

Electrocardiography has become an important diagnostic technique in clinical medicine and is especially helpful in identifying disturbances of cardiac rhythm and certain specific abnormalities of ventricular function and structure. Clinical interpretation of the electrocardiogram involves a multitude of measurements in different leads and estimates of wave contours, a partly subjective process that may eventually be improved by the application of computer techniques.[2,13] The effects of various pathologic conditions are described in detail in textbooks of electrocardiography.[4] The electrocardiogram often provides significant diagnostic clues in conditions of abnormal ventricular function, including myocardial ischemia, myocardial infarction (localized destruction of part of the myocardium because of inadequate blood supply), and ventricular hypertrophy.

Many pathologic conditions produce electrocardiographic changes that are recognizably abnormal but not specific for a single disease entity. In some instances, these changes can be related logically to the underlying disturbances; in others, interpretation of the records is purely empirical. Although the potentials on the body surface allow us to predict the state of the heart with reasonable accuracy in many cases, potentially serious cardiac abnormalities sometimes exist in the presence of what must be regarded by present criteria as a normal electrocardiogram.

Arrhythmias.[3,4] Disturbances of the cardiac rhythm caused by abnormal pacemakers or by some failure in the normal conduction of excitation disrupt the usual P-QRS-T sequence and are ordinarily easily recognized in the electrocardiogram. Even under normal conditions the sinus rhythm is not always perfectly regular because of a slight waxing and waning of heart rate in phase with respiration. This *sinus arrhythmia* is a reflex response that originates in pulmonary stretch receptors that are stimulated by inflation of the lung (p. 1069). Under pathologic conditions, cells in some region other than the sinoatrial node may begin to fire intermittently, producing one or more *ectopic beats,* or they may capture control of the heartbeat by firing continuously at a rapid rate. The resulting ectopic rhythm is called atrial, atrioventricular nodal, or ventricular tachycardia, depending on the site of the abnormal pacemaker. "Ventricular" ectopic rhythms probably originate in Purkinje fibers more often than in ordinary muscle fibers.

Fibrillation is a completely uncoordinated, almost random, electrical and mechanical activity of the atria or ventricles. Atrial fibrillation is compatible with effective cardiac pumping of blood even though the ventricular rate becomes irregular. In ventricular fibrillation the heart ceases to eject blood because the rhythmic, coordinated contraction of the ventricles is lost. The circulation comes to a halt, and death ensues unless normal rhythm can be restored immediately. The mechanism of fibrillation is not entirely clear, but it is often preceded by a series of ectopic beats. Apparently, a single stimulus arriving at a time when various parts of the myocardium are in different stages of recovery can precipitate fibrillation, and any influence that disturbs the normal relations among excitability, conduction velocity, and refractory period predisposes the heart to this arrhythmia.

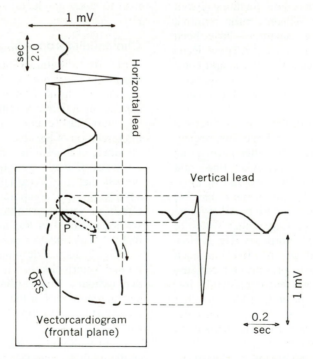

Fig. 38-5. Diagram of normal frontal plane vectorcardiogram. To record such a vectorcardiogram, horizontal or right-to-left electrocardiographic lead (such as standard limb lead I) is connected through amplifiers to horizontal deflection plates of cathode-ray oscilloscope and vertical or head-to-foot lead is connected to vertical deflection plates. Oscilloscope beam then traces out loops corresponding to P, QRS, and T waves, representing continuous measurements of instantaneous electrical axis in plane defined by these leads. Series of QRS vectors shown in Fig. 38-2, *B,* correspond to eight different points on such a QRS loop. To give indication of timing, oscilloscope trace is usually blanked out at regular intervals; in this vectorcardiogram, each dash begins 4 msec after previous one.

Several hypotheses concerning ventricular fibrillation have been proposed, none of which is entirely satisfactory. The "circus movement" theory postulates grossly irregular excitation spreading from one responsive cell to its neighbor and bypassing refractory cells, only to return to them later over a circuitous route. Another explanation involves multiple ectopic foci that divide the heart into several independently functioning units, leading to complete disorganization of contraction. Such ectopic sites can result from mechanical stimulation of the myocardium or from local application of certain drugs such as aconitine. An alternative theory invokes an electrotonic current flow, generated by injured cells in a region of anoxia, for example, as the agent that creates an inhomogeneous state in the myocardium. A strong electrical stimulus to the heart can stop ventricular fibrillation, and instruments that apply such a stimulus through the intact chest wall are now in clinical use. The electrical shock produces depolarization of all myocardial elements, and survival then depends on the reappearance of a rhythmic pacemaker and coordinated contractions. Defibrillation must be accomplished within a few minutes of the onset of the arrhythmia or else circulation must be maintained by cardiac massage or other artificial means.

Vectorcardiography

As we have already seen, any two of the Einthoven limb leads may be used to calculate instantaneous electrical axes. A series of such axes may be constructed for successive instants in the cardiac cycle, showing the changing direction and magnitude of the vectors corresponding to the P, QRS, and T waves, as in Fig. 38-2, *B*. Einthoven's original paper[9] on the electrical axis lists such a series of vectors in tabular form. In calculating these vectors from the electrocardiogram, it is customary to select a limited number of instants throughout the cycle for measurement, but it is evident that they actually form a continuum in time, rather than a discrete series. This continuum might be represented by connecting the heads of the arrows in Fig. 38-2, *B*, for example, so that the QRS complex would be represented by a loop, and similar loops for the P and T waves could be constructed. Mann[11] published the first diagrams of this kind in 1920, calling them *monocardiograms* and describing them as "a fusion of the three leads of the electrocardiogram into a single curve by an algebraic reversal of the process by which three leads are obtained from one heart." In later years, devel-

opment of the cathode-ray oscilloscope made possible the direct visualization of these loops, subsequently called vectorcardiograms, on the face of a cathode-ray tube.[16] From the recording of frontal plane vectorcardiograms (Fig. 38-5), it was a natural step to the sagittal and transverse plane projections of the loops, using appropriate leads, and to the concept of a three-dimensional spatial vectorcardiogram.[1,4,12,17] Investigators have used a variety of different leads to obtain spatial vectorcardiograms, and no single system has yet been universally adopted. Most lead systems are founded on an orthogonal arrangement of three leads, which record the right-to-left, head-to-foot, and ventrodorsal components of the cardiac electrical field.

The clinical usefulness of vectorcardiography is at present limited, although it is of great value as a device for teaching the concepts of electrocardiography. To the extent that the cardiac electrical field departs from the idealized form of Einthoven's hypothesis, the vectorcardiogram is an incomplete synthesis of the actual events. For this reason, and because it is inferior to conventional electrocardiography in depicting time relationships between successive heart cycles, it is probable that vectorcardiography will continue to supplement electrocardiography rather than replace it.

REFERENCES
General reviews

1. Burch, G. E., and DePasquale, N. P.: A history of electrocardiography, Chicago, 1964, Year Book Medical Publishers, Inc.
2. Caceres, C. A., and Dreifus, L. S., editors: Clinical electrocardiography and computers, London, 1970, Academic Press, Ltd.
3. Cassels, D. E., and Ziegler, R. F., editors: Electrocardiography in infants and children, New York, 1966, Grune & Stratton, Inc.
4. Lipman, B. S., and Massie, E.: Clinical scalar electrocardiography, ed. 6, Chicago, 1972, Year Book Medical Publishers, Inc.
5. New York Heart Association, Criteria Committee: Diseases of the heart and blood vessels: nomenclature and criteria for diagnosis, ed. 6, Boston, 1964, Little, Brown & Co.

Original papers

6. Burger, H. C., and van Milaan, J. B.: Heart vector and leads, Br. Heart J. **10:**229, 1948.
7. Damato, A. N., et al.: A study of heart block in man using His bundle recordings, Circulation **39:**397, 1969.
8. Einthoven, W.: Die galvanometrische Registrirung des menschlichen Elektrokardiogramms, zugleich eine Beurtheilung der Anwendung des Capillar-Elektrometers in der Physiologie, Arch. Gesamte Physiol. **99:**472, 1903.
9. Einthoven, W., Fahr, G., and deWaart, A.: Über die Richtung und die manifeste Grösse der Potential-

schwankungen im menschlichen Herzen und über den Einfluss der Herzlage auf die Form des Elektrokardiogramms, Arch. Gesamte Physiol. **150:**275, 1913.

10. Grant, R. P.: The relationship between the anatomic position of the heart and the electrocardiogram: a criticism of "unipolar" electrocardiography, Circulation **7:** 890, 1953.

11. Mann, H.: A method of analyzing the electrocardiogram, Arch. Intern. Med. **25:**283, 1920.

12. Milnor, W. R.: The normal vectorcardiogram and a system for the classification of vectorcardiographic abnormalities, Circulation **16:**95, 1957.

13. Pipberger, H. V., Arms, R. J., and Stallmann, F. W.: Automatic screening of normal and abnormal electrocardiograms by means of a digital electronic computer, Proc. Soc. Exp. Biol. Med. **106:**103, 1961.

14. Scher, A. M., and Young, A. C.: The pathway of ventricular depolarization in the dog, Circulation Res. **4:** 461, 1956.

15. Waller, A. D.: A demonstration on man of electromotive changes accompanying the heart's beat, J. Physiol. **8:**229, 1887.

16. Wilson, F. N., and Johnston, F. D.: The vectorcardiogram, Am. Heart J. **16:**14, 1938.

17. Wilson, F. N., Johnston, F. D., and Kossmann, C. E.: The substitution of a tetrahedron for the Einthoven triangle, Am. J. Heart J. **33:**594, 1947.

18. Wilson, F. N., Macleod, A. G., and Barker, P. S.: The distribution of the currents of action and of injury displayed by heart muscle and other excitable tissues, University of Michigan Studies, Scientific Series, Ann Arbor, 1933, University of Michigan Press, vol. 10.

19. Wilson, F. N., et al.: Electrocardiograms that represent the potential variations of a single electrode, Am. Heart J. **9:**447, 1934.

20. Wilson, F. N., et al.: On Einthoven's triangle, the theory of unipolar electrocardiographic leads, and the interpretation of the precordial electrocardiogram, Am. Heart J. **32:**277, 1946.

39

WILLIAM R. MILNOR

Principles of hemodynamics

The physical principles that apply to the motion of blood through the vascular system are derived from the general laws of hydrodynamics, with certain modifications imposed by the properties of blood and the vascular tree. In the study of hemodynamics, it is natural to begin with the laws that govern the flow of simple liquids through a rigid cylindric tube, since they are the foundation for the more elaborate laws needed to describe the flow of blood through the branching, distensible network of tubes that make up the cardiovascular system.

Viscosity. The flow of liquids through tubes is governed in part by a fundamental property of fluids called viscosity, which is a kind of internal friction between adjacent layers of a fluid. If one portion of a viscous fluid is set in motion by an applied force, this motion is communicated to the adjacent parts of the liquid. Isaac Newton described this phenomenon as a "lack of slipperiness" between the parts of a liquid and derived the earliest theoretical treatment of this general property of fluids from the hypothesis that the resultant internal resistance is proportional to the velocity with which the parts are separated from one another. Fluid moving through a tube may be thought of as a series of thin layers slipping against each other and proceeding at different velocities. Viscosity is defined as the ratio of stress to velocity gradient, where the stress is the force applied per unit area to produce motion of the fluid and the velocity gradient is the change in velocity from one layer to the next. Under ideal conditions the velocity gradient forms a symmetric pattern (Fig. 39-1). The standard unit of viscosity is the *poise*, 1 poise (1 P) being equal to 1 dyne sec/cm². *Relative* viscosity, relating the viscosity of a fluid to that of water (0.01 P at 20° C) is sometimes used instead of these absolute units. Viscosity varies with temperature, increasing as temperature falls.

This definition implies that viscosity is a property of the fluid itself, independent of the actual velocity gradient. Such independence can be demonstrated in water and a number of other liquids, which are therefore termed *newtonian* fluids. In blood and certain other substances, however, careful measurements reveal that Newton's fundamental assumption of a velocity gradient proportional to applied stress is not true at all velocities.[7,18,51,54] Nevertheless, the behavior of these nonnewtonian fluids can be described by their *apparent* viscosity under specified conditions. The anomalies in the viscous behavior of blood result partly from the presence and orientation of the cells suspended in plasma and partly from the accumulation of red blood cells in the axial portion of the bloodstream. The physical forces involved are not entirely clear, but the result is a decrease in apparent viscosity with increasing velocity of flow and with decreasing vessel radius. The magnitude of these effects is small in vessels more than 0.5 mm in diameter at flow rates in the physiologic range, and for most purposes the viscosity of blood at 37° C in

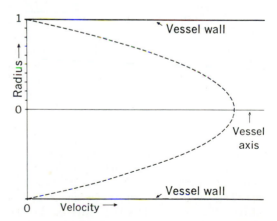

Fig. 39-1. Parabolic relationship between blood velocity and distance from vessel axis under conditions of laminar flow.

dog and man may be assumed to remain constant at 0.03 to 0.04 P.[4] This value applies to blood with normal red blood cell concentrations. Viscosity rises with increasing hematocrit.

LAMINAR FLOW IN RIGID TUBES

Newtonian liquids flow through a cylindric tube in a series of concentric cylindric laminae; within each lamina the fluid particles all move with the same velocity, but the velocity of each cylindric shell decreases with its distance from the axis of the stream. The relation between velocity and radial distance from the axis is parabolic (Fig. 39-1), and flow of this kind is therefore described as parabolic or alternatively as laminar or newtonian. The characteristics of laminar flow can be derived mathematically from the definition of viscosity if we assume that the infinitely thin outermost lamina in contact with the vessel wall is at rest. This derivation, which appears in the appendix to this chapter, demonstrates that the velocity at the axis ($r = 0$) is twice the average velocity across the stream. The rate of flow (Q), or volume of liquid passing a given cross section per unit of time, is inversely proportional to viscosity (η) and directly proportional to the drop in pressure ($P_1 - P_2$) per unit length of the vessel (L) and the fourth power of the radius (R).

$$Q = \frac{\pi(P_1 - P_2)\,R^4}{8\eta\,L} \tag{1}$$

In other words, under these conditions flow depends on four factors: (1) the constant ($\pi/8$), in which π relates to the circular cross section of the tube and 8 appears in the process of integration; (2) the reciprocal of the viscosity ($1/\eta$); (3) the dimensions of the tube, in the form R^4/L; and (4) the pressure exerted to produce flow ($P_1 - P_2$), in which P_1 is the pressure at the upstream end of the cylinder and P_2 is the pressure at the distal end. Since the radius of the tube appears to the fourth power, relatively small changes in radius produce large changes in flow. The fall in pressure per unit length, ($P_1 - P_2$)/L, is occasionally referred to as the *pressure gradient*. Strictly defined, the latter term means the rate of change of pressure along the longitudinal axis of a vessel, but it is sometimes loosely used as a synonym for the pressure head ($P_1 - P_2$), or difference in pressure between two intravascular sites. The relationship defined by equation 1 is usually called Poiseuille's law, after the French physician and physiologist Poiseuille (1799-1869), whose interest in flow through small blood vessels led him to measure the flow of

water in fine capillary tubes. His results, published in detail in 1846,[39] described the relations between pressure head, flow, and tubing dimensions with a high degree of accuracy, although it was not until some years later that the theoretical analysis of viscosity and laminar flow gave the law its present form.

Poiseuille's law describes the laminar, constant flow of newtonian liquids through rigid cylindric tubes, and its applicability to the circulation depends on the extent to which blood and the vascular system conform to these theoretical assumptions. Blood behaves as a newtonian fluid in vessels with an internal diameter greater than 0.5 mm, so that in larger vessels, at least, this requirement is satisfied. Flow in these vessels is also laminar as a rule, but flow is pulsatile rather than constant throughout most of the arterial tree. The acceleration and deceleration of blood in pulsatile flow introduce variations in kinetic energy not taken into account in the Poiseuille equation, which for this reason cannot be applied to oscillations of flow, although it gives a fair approximation of the relation between mean pressure and mean flow. A further departure from Poiseuille's law results from the distensibility of blood vessels, which makes vessel radius vary with the applied pressure.

Despite these discrepancies, Poiseuille's law has served as a useful concept in circulatory physiology, if only because its limitations were for a long time matched by experimental limitations imposed by the lack of a satisfactory theoretical treatment of pulsatile flow in distensible vessels and of reliable ways to measure pulsatile flow.

Vascular resistance

According to Poiseuille's law, the ratio of pressure drop to rate of flow is a function of all the forces that retard blood flow—viscosity, length, and radius. This ratio has been adopted as an expression of vascular resistance by analogy with Ohm's law for electrical circuits (voltage/current = resistance). The resistance (K) of a vascular bed is defined as the difference between the pressure at the inlet of the bed (P_1) and the pressure at the outlet (P_2) divided by the blood flow (Q):

$$K = \frac{P_1 - P_2}{Q} \tag{2}$$

For example, resistance of the whole systemic bed, sometimes called the *peripheral resistance*[25] in recognition of the fact that the greatest drop occurs in the small vessels, is calculated from equation 2, with mean systemic arterial pressure substituted for P_1, systemic venous (or else right atrial) pressure for P_2, and the cardiac output as Q. Pulmonary vascular resistance is calculated in the same way, with pulmonary arterial pressure for P_1 and left atrial pressure for P_2. The cgs unit for resistance is dyne sec/cm^5, pressure being expressed in dynes/cm^2 and flow in cm^3/sec. Because of the units in which pressures and cardiac output are customarily measured in the laboratory, it is sometimes convenient to express resistance in the arbitrary units that result from dividing millimeters of mercury by liters per minute. Since a pressure of 1 mm Hg is equivalent to 1,333 dynes/cm^2 and an output of 1 L/min equals 1,000/60 cm^3/sec, the factor for converting such arbitrary units into dyne sec/cm^5 is 80 ($= 1,333 \times 0.06$). As an example of the vascular resistances found in the human circulation, assume a cardiac output of 6 L/min, systemic arterial pressure of 90 mm Hg, right atrial pressure of 5 mm Hg, pulmonary arterial pressure of 15 mm Hg, and left atrial pressure of 6 mm Hg. The systemic resistance (K_s) and pulmonary resistance (K_p) would then be:

$$K_s = \frac{90 - 5}{6} \times 80 = 1,133 \text{ dyne sec/cm}^5$$

$$K_p = \frac{15 - 6}{6} \times 80 = 120 \text{ dyne sec/cm}^5$$

Since vascular length can be regarded as constant in most beds and blood viscosity is constant under most conditions, it is evident from Poiseuille's law that resistance depends principally on vessel radius. A change in vascular resistance therefore implies a change in radius of some vessels in the bed, although it does not indicate either the site or the mechanism of this change in caliber. Although the inference that a change in pressure:flow ratio in one part of the circulation is due to alteration of local arteriolar resistance is usually valid, it should be remembered that small arteries (diameter = 2 to 0.1 mm), venules, and even capillaries (Chapter 43) contribute a significant part of the total resistance in some circumstances.

In considering the mechanisms by which vessel radius and thus resistance can be altered, it is essential to distinguish between the fall in pressure along the length of a vessel, which is the factor related to resistance, and transmural pressure across the wall, which affects the radius of distensible vessels. The caliber of a blood vessel can change passively because of a change in transmural pressure, or it can be changed by activation of vascular smooth muscle. A decrease in resistance, for example, may arise from increased arterial pressure, which tends to raise transmural pressure and distend the vessels passively, or from relaxation of the smooth muscle fibers in the arteriolar walls, which are under active neural control. In other words, measurements of vascular resistance reflect changes in vessel caliber but tell nothing of the mechanism by which these changes are brought about. Moreover, because the vascular tree consists of many parallel channels, changes in resistance may be due to complete closure or reopening of some channels rather than to generalized changes in caliber.

The nonspecific nature of resistance measurements has important practical consequences. Consider, for example, some pharmacologic agent or neural reflex that is found to lower pulmonary vascular resistance, as evidenced by the data shown in Table 39-1. The pressure head across the lung vessels decreases by 33%, whereas the blood flow remains constant; thus it is clear that pulmonary vascular resistance falls. The absolute pressures in the bed rise, however, and if we assume that extravascular pressures remain constant, this agent must cause a considerable increase in transmural pressure, perhaps by effects on myocardial function. There is consequently no evidence in the data given that the agent has any direct or vasomotor effect on the pulmonary vessels; the fall in resistance may be simply a result of passive distention.

It is evident that measurements of vascular resistance must be interpreted with caution for a number of reasons. When resistance and trans-

Table 39-1. Example of decreasing pulmonary vascular resistance by passive distention*

	Before	After
Pressures (mm Hg)		
Pulmonary artery	20	30
Left atrium	5	20
Cardiac output (L/min)	5	5
Pulmonary vascular resistance (dyne sec/cm^5)	$\left(\dfrac{15}{5} \times 80\right) = 240$	$\left(\dfrac{10}{5} \times 80\right) = 160$

*Extravascular pressures assumed constant.

mural pressure change *in the same direction,* there has probably been an active change in the caliber of vessels somewhere in the vascular bed, attributable to changes in the tension of vascular smooth muscle. Under any other circumstances the mechanism of a change in resistance is less certain.

In the absence of laminar flow, even this limited applicability of the concept of resistance is altered. When flow becomes turbulent, the pressure:flow ratio is greater than that predicted by Poiseuille's law because energy is expended in random motion across the stream as well as in maintaining forward flow. Laminar flow obtains in the larger vessels of the circulation, but in very small collapsible vessels such as the capillaries a transition from laminar to turbulent flow may, under some conditions, make Poiseuille's law quite inapplicable to that particular segment. For similar reasons, equation 2 is invalid for partially or completely collapsed veins.[31] The flow of liquids at high velocity through a small orifice is a special instance of nonlaminar flow extensively studied by hydraulic engineers. The dynamic conditions at such an orifice are similar to those in heart valves constricted by disease, and equations have been developed to estimate the functional area in such valves from measurements of pressure and flow.[24]

ENERGETICS OF FLOW

Because of the assumptions on which it is based, Poiseuille's law is concerned with only part of the energy manifested in the motion of fluids—the portion associated with constant flow and pressure. A full accounting of hydraulic energy must also include the *kinetic energy* entailed in acceleration of fluid and the hydrostatic *energy of position* associated with gravitational force. The hydraulic energy (W) per unit volume of blood, which can also be regarded as the work done to endow a unit volume with this energy is as follows:

$$W = P + \tfrac{1}{2}\rho v^2 + \rho g h \qquad (3)$$

Here, P = pressure in dynes per square centimeter; ρ = density in grams per cubic centimeter; v = velocity in centimeters per second; h = height in centimeters above an arbitrary level to which all points in the circulation are referred; and g = gravity acceleration constant (980 cm/sec²). Each of these three terms expresses energy per cubic centimeter of blood in dynes per square centimeter. The first term represents a kind of potential energy, or *energy of pressure;* the second is kinetic energy, which can also be expressed as ½ mv², where m is the mass of 1 cm³

of blood; the third is gravitational potential energy, the energy associated with position above or below the arbitrary reference level.

Conversion of one type of energy to another can occur as blood moves through the circulation—in transitions from laminar to turbulent flow, for example—but the total fluid energy is conserved, the losses inherent in viscous flow, which are dissipated as heat, being balanced by the rhythmic additions of energy from the heart. This is a much more general law than the Poiseuille equation; in a freely communicating vascular bed, blood flow depends on differences in total fluid energy, flowing *downhill,* whether flow be laminar or turbulent. In this respect, Poiseuille's equation is misleading: restricted as it is to situations in which there is no change in kinetic energy, it suggests that liquids necessarily flow from a region of high pressure to one of low pressure. Empirically this is what happens in most parts of the circulation, but the driving force is the *total* fluid energy, not the energy of pressure alone. This is illustrated in the flow of blood through the aortic valve, for during the latter part of systole the pressure gradient across this valve is actually negative (i.e., pressure is higher in the aorta than in the ventricle), and the late systolic flow into the aorta is maintained only by the kinetic energy imparted to the blood by the left ventricle.[49] Equation 3 can be used to calculate the fluid energy per unit volume of blood at any point in the circulation and therefore offers a way of measuring the difference in energy between two points in the circulation. The work done by the heart with each stroke or the energy dissipated in moving blood through any vascular bed can thus be estimated. For such purposes, each term on the right side of equation 3 must be multiplied by the volume of blood being considered (V), and the energy introduced or dissipated over a period of time must be integrated, since the variables are functions of time unless flow is constant. In many applications no change in hydrostatic level is involved, and the gravitational term can be dropped, so that the total fluid energy (W_T) manifested at a specified point over a period of time (T) becomes:

$$W_T = \int_0^T P dV + \int_0^T \frac{v^2\, dm}{2} \qquad (4)$$

If pressure is expressed in dynes per square centimeter, volume in cubic centimeters, velocity in centimeters per second, mass in grams, and T in seconds, energy will be expressed in dyne centimeters or ergs (10^7 ergs/sec = 1 watt). In calculating the energy imparted to the blood with each contraction of the left ventricle, for example, T is the length of one heart cycle, P is left ventricular pressure, dV is the flow out of the ventricle (Q = dV/dt), and dm = ρ dV; the work done during systole minus the energy provided during diastole by filling pressure is the ventricular stroke work. Since only part of the total energy utilized in ventricular contraction appears as fluid energy, this portion associated with blood flow is referred to as *external work* (p. 997). Continuous measurements of the pulsations of pressure and flow are required in equation 4; the substitution of values averaged over the complete heart cycle gives errors of about 30% in the pulmonary artery and 10% in the aorta.[1] The order of magnitude of the external work of the right and left ventricles is indicated in Table 40-1, together with their average *power*, or rate of doing work, which is probably as important to ventricular function and metabolism as the absolute amount of work done. The hydraulic power represented by kinetic energy of flow is much less than that associated with potential or pressure energy under resting conditions, but as the cardiac output and blood velocity increase, the ratio of kinetic to potential energy rises. In the pulmonary artery with its relatively low pressure the kinetic component is always a greater fraction of the total fluid energy than in the aorta. Pathologic constriction of the pulmonic or aortic valve may produce a great increase in the kinetic energy of flow by virtue of the high velocity of flow through the narrowed orifice.

Conversion of kinetic energy into pressure, or the reverse, has a number of hemodynamic consequences, some of which have already been mentioned. At least two further instances of this interconvertibility are significant in the circulation. The first concerns the energy of liquid flowing through a tube in which the cross section varies along its length, as in Fig. 39-2. The Swiss physicist and mathematician Daniel Bernoulli (1700-1782) developed the theorem that applies to this situation in the course of his theoretical work on the behavior of "frictionless" or nonviscous fluids. If we ignore the effects of viscosity and assume that the tube represented in the lower part of Fig. 39-2 is horizontal, so that gravitational effects can be omitted, then the sum of pressure and kinetic energies at site b just before the expansion of the tube at *A* must equal the sum at site c.

$$W_T = P_b + \tfrac{1}{2}\rho v_b^2 = P_c + \tfrac{1}{2}\rho v_c^2 \qquad (5)$$

In a steady state of flow the volume rate of flow (Q) will be the same through all cross sections of the tube. Since the cross-sectional area increases, and velocity equals flow divided by cross-sectional area (v = Q/A), the velocity at site c (v_c) must be less than that at site b (v_b). It follows that the kinetic energies at these two sections are unequal, and that $\tfrac{1}{2}\rho v_b^2 > \tfrac{1}{2}\rho v_c^2$. For the conservation of energy expressed in equation 5 to be maintained, P_c must then be greater than P_b. In other words, as the tube widens, kinetic energy is con-

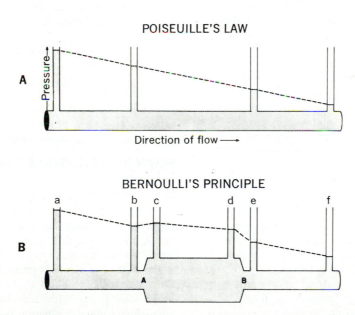

Fig. 39-2 Diagrammatic representation of the pressure-flow relationships described by Poiseuille's law and Bernouilli's principle. Constant flow through cylindric tube, **A,** is associated with linear fall in pressure. Abrupt change in cross section of tube, **B,** modifies fall in pressure. For description see text.

verted into pressure. The energy-dissipating effects of viscosity cannot, of course, be ignored in the circulation, but the operation of Bernoulli's principle in regions of increasing total vascular cross section should result in a pressure gradient smaller than that predicted by Poiseuille. The kinetic energy of blood flow is so small beyond the major aortic branches, however, that its influence on small vessel pressures is usually negligible.

Measurements of intravascular pressure through a catheter or needle can also create the conditions for conversion of kinetic energy to pressure. If the catheter opening faces directly upstream, the velocity and kinetic energy lost by the parts of the stream that impinge on it will be transformed into pressure and added to the intravascular pressure energy. The *end pressure* so measured will be higher than the *lateral pressure* detected by a catheter opening that lies parallel to the direction of flow. Conversely, a catheter with an opening facing downstream will record a pressure less than the lateral pressure. In both cases the pressure added to or subtracted from the true lateral pressure equals the kinetic energy transformed into pressure or derived from it. Whenever pressure measurements are reported, it should be made clear whether they are lateral or end pressures; for most purposes, lateral pressures are to be preferred.

Turbulent flow. At high velocities the pattern of laminar flow breaks down into turbulence, a fact known to Poiseuille but first treated quantitatively by Reynolds[43] in 1883. In turbulent flow the fluid particles move in irregular and constantly varying paths, forming eddies at some times and giving the appearance of random motion at others. The transition from laminar to turbulent flow depends on the density and viscosity of the liquid and on the diameter of the tube as well as on the average velocity, and the critical point at which this transition occurs can be defined by a dimensionless quantity known as the *Reynolds number*.

$$\text{Reynolds number} = \frac{\bar{v}D\rho}{\eta} \qquad (6)$$

Here $\bar{v}$ = average velocity in centimeters per second; D = diameter of the circular tube in centimeters; ρ = blood density, which is about 1.05 gm/cm³; and η = blood viscosity in dyne seconds per square centimeter. Increasing velocity, wider tube diameter, and decreasing viscosity all tend to encourage turbulence. Turbulence usually develops when the Reynolds number exceeds 2,000, but the critical value varies with the experimental conditions. The transition is not abrupt, and flow that is neither perfectly laminar nor fully turbulent may be seen at Reynolds numbers of 1,000 or even less. Pulsations also affect the stability of laminar flow and lead to the appearance of turbulence at lower Reynolds numbers.[4] Turbulent flow dissipates more energy than laminar flow at the same average velocity, so that the ratio of pressure head to flow through a tube changes more or less abruptly when turbulence develops. Under carefully controlled conditions, this phenomenon can be used to detect the appearance of turbulence. Applying this method to blood flowing through rigid tubes 0.126 to 0.340 cm in radius, Coulter and Pappenheimer[17] found the critical Reynolds number to be approximately 2,000, showing that blood in this respect behaves like a homogeneous fluid. From the limited data available in man, Reynolds numbers are probably well below 2,000 throughout the circulation except in the root of the aorta and the pulmonary artery, where values of 5,000 to 12,000 are reached during systole.[40,47] Laminar flow is the rule in most blood vessels, even though it may be modified by changes in hematocrit, unusually rapid flow, or the effects of disease on vascular structure.[47]

Turbulence plays a part in the generation of sounds and murmurs in the circulation, although it is probably less important in this regard than the vibrations induced in vessel walls, valve leaflets, or surrounding solid structures. Laminar flow is essentially noiseless, whereas turbulent flow, under proper conditions, can generate vibrations in the audible range; however, the energy of sound produced by turbulence per se is small and is further attenuated by conduction through the body tissues. Cavitation, the formation of gas bubbles by sharp, local pressure gradients within a liquid, has been suggested as the source of some cardiovascular sounds, although it seems unlikely that the large gradients needed to produce it occur in the circulation.

FLOW IN DISTENSIBLE TUBES

Blood vessels are not, of course, rigid tubes. Because of the elastic properties of the vascular wall, the circumference of a blood vessel is a function of the forces applied to it, and the principles that govern flow in rigid tubes cannot be applied without certain modifications.

Elasticity is a physical property manifested by the tendency of a solid to resume its original shape after being deformed. The force per unit area causing the deformation is called *stress* and the resulting deformation is termed *strain*. A substance that regains its initial form exactly when

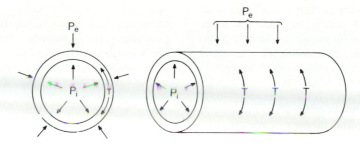

Transmural pressure $= P_i - P_e$

Fig. 39-3. Forces acting on distensible tube or blood vessel. P_i, Intravascular pressure; P_e, extravascular pressure; T, circumferential tension in wall.

forces acting to distort it are removed is said to be *perfectly elastic*. As Burton[2] has pointed out, this physical concept of elasticity differs from the popular one that labels substances as elastic only if they can undergo large deformations, or strains, without breaking, a property that might better be termed *extensibility*. To the physicist, steel and rubber are both equally good examples of elastic substances. This conflict with popular usage can be circumvented by speaking of the *distensibility* of vessels, or of their stiffness, rather than their elasticity. A consistent set of terms applying to biologic elasticity has been formulated by Landowne and Stacy.[34]

Within certain limits the deformation of an elastic material is proportional to the force applied, a law stated by Robert Hooke (1635-1703) with classic brevity: "Ut tensio, sic vis." The quantitative relation between applied stress and resulting strain is defined for any material by an *elastic modulus,* the quantity relevant to forces applied so as to stretch the material being *Young's modulus* (E).

$$E = \frac{\text{Applied force/unit area of cross section}}{\text{Increase in length/unit length}} \quad (7)$$

If F = force in dynes, A = cross section in square centimeters, L = initial length, and ΔL = increase in length in centimeters, then:

$$E = \frac{F/A}{\Delta L/L} = \frac{FL}{A\ \Delta L} \quad (8)$$

Young's modulus is therefore equivalent to the force per unit area (dynes/cm²) required to extend the material 100%. When applied to the circumferential distention of a tube, the force in equation 7 is the total circumferential tension, or tangential stress, over unit length of the tube (Fig. 39-3, *T*), L and ΔL refer to the circumference of the tube, and A is the product of wall thickness and unit length. Changes in dimension along the longitudinal axis of a blood vessel in vivo are usually assumed to be negligible, although in some vessels this is not true.

The relation between pressure and wall tension in an elastic cylinder was derived by Laplace:

$$T = Pr \quad (9)$$

In this instance, T = circumferential tension, P = pressure, and r = radius of the cylinder. The relevant pressure here is the force acting radially on each unit area of the vessel wall and includes both the intraluminal pressure (Fig. 39-3, P_i) and the pressure applied to the external surface of the cylinder (P_e). The *transmural pressure* (P_t) exerted on the wall is the difference between internal and external pressure.

$$P_t = P_i - P_e \quad (10)$$

Laplace's law, however, applies only to a cylinder with infinitely thin walls, as equation 9 implies by expressing tension in force per unit length (dynes/cm), rather than force per unit area (dynes/cm²). When wall thickness is taken into account, as it must be in blood vessels, the relation between tangential wall stress and transmural pressure becomes more complex,[52] but in tubes with a ratio of radius to wall thickness (r/δ) greater than 10 a good approximation is given by:

$$T = \frac{Pr}{\delta} \quad (11)$$

Vascular elasticity

The elastic behavior of blood vessels depends on the materials that make up the vascular wall and the arrangement of these materials. Elastin, collagen, and smooth muscle are the most important constituents. The extracellular fluid and mucopolysaccharide matrix that are the other major components of the tissues of the wall probably have relatively little influence on vessel elasticity under normal conditions, although this point merits further investigation.

Collagen and elastin together constitute about half the weight of the vessel wall in the arterial tree of the dog.[29] The thoracic aorta contains more elastin than collagen, but this proportion is reversed in the initial branches of the aorta and in the abdominal aorta, and collagen predominates throughout the distal arterial tree. Elastin is a truly rubberlike substance, highly extensible, with a Young's modulus of the order of 6×10^6 dynes/cm².[11,32] Collagen, which is the major component of tendons, is very inextensible and has a higher elastic modulus: 100×10^6 dynes/cm² or more.[11,32] The elastic modulus of vascular smooth muscle is on the order of 1×10^6 dynes/cm², although this varies widely with variations in the number of fibers in active contraction. These three major components thus have quite different elastic properties, and the elastic behavior of the vessel wall depends to a large extent on the histologic arrangement of these elements and the way in which they are linked.[10]

Collagen is arranged parallel to the other wall components but forms a loose interlacing network of fibers that becomes comparatively rigid when the vessel is distended. The collagenous

structure of the wall therefore sets an upper limit to the diameter of the vessel. In the aorta and the proximal parts of its major branches the elastic and smooth muscle fibers are arranged in a complex pattern that includes connections in series and in parallel between muscle and elastin; some smooth muscle fibers attach to elastic fibers, others to adjacent smooth muscle. The elastic behavior of these arteries is correspondingly complex. Smooth muscle and elastic fibers are generally unconnected in the remainder of the arterial bed, muscle attaching to muscle and elastin to elastin in separate parallel arrangements around the circumference of the artery. The properties of these vessels are dominated by the stiffest element, which varies with the radius of the vessel and the activity of smooth muscle.

With such inhomogeneity in its structure, it is not surprising that the vascular wall does not follow Hooke's law, which states that the ratio between stress and strain is constant. In all blood vessels, Young's modulus increases with increasing radius; the further the wall is stretched, the stiffer it becomes (Fig. 39-4).[11] This phenomenon probably has its origin in the physical properties of the different elements in the vascular wall and the way in which they are linked, and various models have been used to represent mechanical arrangements that would account for the experimental data on vascular elasticity.[1,2] At moderate degrees of distention the character-

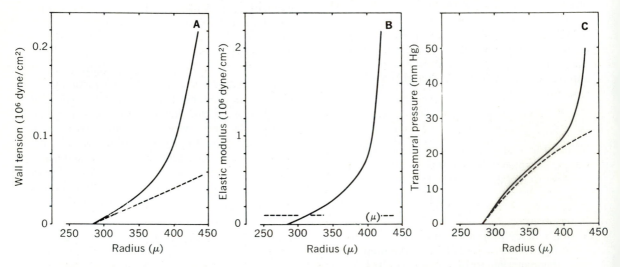

Fig. 39-4. Comparison of elastic behavior of small artery (continuous line, based on data from Hinke[30]) with that of thin-walled perfectly elastic tube having elastic modulus of 1×10^5 dynes/cm² (broken line). **A,** Relation between radius of vessel or tube and total tension in its wall. **B,** Radius and Young's modulus of elasticity showing increase in modulus (increasing "stiffness") with increasing radius in blood vessel.[11] **C,** Radius and transmural pressure showing relative rigidity of blood vessels at high transmural pressures (1 μm = 1×10^{-4} cm).

istics and arrangement of muscle fibers and elastin are probably dominant, whereas at maximum distention the very high moduli observed indicate that most of the stress has been transferred to collagen. An alternative hypothesis is that vessel walls resemble the class of substances known as *elastomers*. The behavior of rubber and many other elastomers can be related to their molecular structure, but the analysis of blood vessel walls in terms of elastomeric theory is far from complete.[1,4,5]

For a "perfectly elastic" material the ratio of stress to strain is independent of time, but the stretching of vessel walls, like that of rubber, depends on the rate and duration of the applied stress. In other words, the wall possesses viscosity as well as elasticity. The stress required to maintain a constant strain in a blood vessel, for example, decreases with time after the initial application of the stress until it reaches some equilibrium value, a phenomenon termed *stress relaxation*, or *delayed compliance*.[7] If the stress is kept constant, the strain gradually increases (creep). Such manifestations of viscoelastic behavior can have a significant influence on hemodynamics in the more distensible parts of the vascular bed.[45] The appropriate quantitative expression of viscoelasticity is a *dynamic elastic modulus* such as that derived by Hardung.[28]

Vascular smooth muscle. Smooth muscle is the one element in the vascular wall through which the caliber of small vessels can be modified quickly and precisely and is therefore the principal effector in control of the circulation (Chapter 42), yet its contractile process exhibits many puzzling aspects. Unlike striated muscle, vascular smooth muscle can change its length in response to neural impulses without significant changes in tension, and vice versa. Like other forms of smooth muscle,[13] it can maintain a constant tension for long periods of time without apparent fatigue, a phenomenon described as *tonic contraction*, or *maintenance of tone*. Whether this is accomplished by repetitive asynchronous activation of contractile elements or by some other mechanism is not known, but the smooth muscle of blood vessels appears to consist of structurally and functionally discrete cells[44] that would permit such multi-unit function. The elastic modulus of vascular smooth muscle has not been measured accurately, since it is difficult to apply the appropriate experimental procedure or to measure its cross-sectional area. The best evidence, however, indicates that its modulus is somewhat less than that of elastin, ranging from perhaps 0.1×10^6 to as much as 6×10^6 dynes/

cm² with different degrees of stimulation. Much lower values have been reported for some types of smooth muscle in a completely relaxed state, but these are probably of no relevance to blood vessels in vivo. Activation of smooth muscle produces little change in the radius of large vessels such as the aorta but can reduce the caliber of small vessels to a marked degree, as is apparent from the "spasm" produced experimentally by even slight mechanical manipulation of a small vein or artery. This difference is consistent with the estimates of smooth muscle elasticity and with the modified Laplace relationship (equation 11), since in a vessel of large radius the tension developed by active tonic contraction of smooth muscle would be a small part of the total wall tension, the major part arising from stretching of elastic elements. In vessels of small radius and lower transmural pressure the total wall tension would be smaller and predominantly due to smooth muscle contraction.

While the tension developed by smooth muscle contraction unquestionably influences the radius of small blood vessels, the extent to which it influences their elasticity is uncertain. Burton,[2,15] has emphasized the possibility that muscle generates an *active tension* that is independent of vessel circumference, contributing little to wall elasticity. The total wall tension would then be the sum of *active tension* and *passive tension*, the latter being determined by the elastic elements in the wall and the radius at which the total tension and transmural pressure reach equilibrium. The observation that drugs applied directly to the wall of small blood vessels produce marked changes in elastic modulus[30,38] strongly suggests, however, that physiologic stimulation of vascular smooth muscle also induces changes in wall elasticity.

In most vessels the degree of tonic contraction in smooth muscle not only affects the elastic modulus of the wall but also determines the unstressed radius, that is, the radius when transmural pressure is zero.[30] A comprehensive description of vessel elasticity and viscosity would therefore include the properties of the various tissues in the wall, the mechanical consequences of their linkage, and the unstressed radius at various levels of tone. The data available for this kind of description are as yet very limited in scope,* but some qualitative generalizations are possible: (1) the vascular wall does not obey Hooke's law but becomes stiffer with increasing diameter; (2) the structural arrangement of elements in the vascular wall is as important as their physical properties in determining vessel diameter; (3) viscous effects are prominent in vascular responses; (4) changes in the degree of tonic contraction of vascular smooth muscle can induce 10-fold alterations in the viscoelasticity of the wall; and (5)

---

*See references 11, 13, 19, 30, 38, and 55.

such alterations have little effect on the diameter of large vessels but marked effects on small vessels. In view of the paramount importance of these factors in circulatory control, further quantitative information is greatly to be desired.

Flow in distensible tubes

In distensible vessels, it is not only the pressure gradient along the tube but also the transmural pressure acting on its wall that is of hemodynamic significance. When an isolated organ is perfused, for example, keeping venous outflow pressure and extravascular pressure constant, an increase in arterial perfusion pressure will increase flow partly because of the increased pressure head (as in a rigid tube) and partly because the vessels are distended by the increased transmural pressure. Under these conditions the increment in flow for an increment of inflow pressure is greater for a distensible tube than for a rigid one. This distention, which causes a decrease in resistance with increasing inflow pressure, gives a curvilinear pressure-flow relationship, as shown in Fig. 39-5, *A*.

Raising venous outflow pressure while keeping inflow pressure constant will also raise transmural pressure and distend part of the vascular bed, but the pressure head will now diminish, and the net effect on flow will depend on the

quantitative relationship between the increased vessel radius and the decreased driving pressure. Any formal analysis of the pressure-flow-radius relationship in even a single distensible tube becomes extremely complex,[4,5,42,57] and in vascular beds, with their branching configuration and non-hookean elasticity, empirical curves from experimental data usually take the place of rigorous mathematical treatment. Such curves typically show the expected curvilinear pressure-flow relationship up to some characteristic pressure. Above this level (approximately 43 mm Hg in Fig. 39-5, *A*), pressure and flow are linearly proportional, as in a rigid tube. Once the vessels reach a certain degree of distention, the constant resistance presumably reflects the limit set by the relatively inextensible collagen fibers.

The positive intercept on the pressure axis in the Fig. 39-5, *A*, illustrates another characteristic of most vascular beds. As inflow pressure is raised from zero, no flow occurs until some critical level is reached, as though a small positive transmural pressure were required to open up some part of the bed. This *critical closing pressure* has been identified by Burton and his associates[36] as the minimum transmural pressure that can maintain equilibrium with mural tension at a finite radius, which is in accordance with equation 9. Any increase in smooth muscle tension,

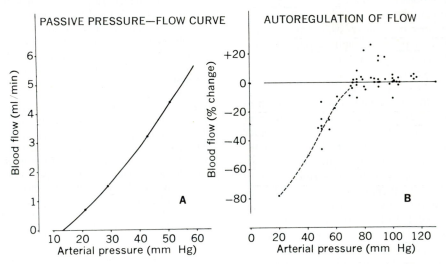

Fig. 39-5. Comparison of passive pressure-flow relationships in rabbit ear, **A,** with autoregulation of renal blood flow in anesthetized dog, **B. A,** Curve represents blood flow through vascular bed of rabbit ear at various arterial pressures. Slight convexity of curve toward pressure axis at low pressures is result of vascular distention by transmural pressures, as described in text. **B,** Renal arterial blood flow was measured with electromagnetic flowmeter, and renal arterial pressure was lowered acutely from control value by graded constriction of renal artery. Ordinate of this graph represents percentage change in flow from unconstricted control level. Renal blood flow did not diminish as arterial pressure was lowered until renal arterial pressure fell below approximately 70 mm Hg. (**A** based on data from Girling[23]; **B** based on data from Schmid and Spencer.[46])

and hence in total wall tension, should then raise the critical pressure at which the vessel closes completely—a prediction confirmed by experimental observation. The vessels ordinarily responsible for this phenomenon have not been identified, but they probably lie in the region of the terminal arterioles and are less than 50 μm in diameter. Total closure of metarteriolar capillary junctions (precapillary sphincters) with decreasing transmural pressure has been directly observed.[33] With normal vascular tone, critical closing pressure ranges between 10 and 25 mm Hg, but with intense stimulation of vascular smooth muscle, it may be much higher.[16] A critical pressure of 0.5 to 5 mm Hg exists even after all smooth muscle tone has been eliminated,[36] possibly because of a kind of interfacial tension between blood and the vessel lining.

Since transmural pressure in the small vessels that actually undergo closure cannot be measured in an intact vascular bed, all these estimates of critical closing pressure are inexact. The intravascular pressure in the relevant vessels is sometimes assumed to be midway between pressures at the inflow and outflow of the bed, a crude approximation at best, and no completely reliable technique exists for measuring extravascular pressure. The pressure outside a vessel is presumably that in the local interstitial fluid, but the distribution of this fluid along intercellular spaces perhaps 1 to 2 μm in width makes it difficult to measure local pressures directly without introducing artifacts. Values for normal *interstitial pressure*, obtained by inserting a needle directly into tissue, range from +0.5 to +7 mm Hg (Chapter 43),[53] whereas measurements with a long-term implanted pressure capsule indicate a negative pressure of approximately −6 mm Hg.[26] Until this conflict in evidence is resolved, estimates of critical closing pressure will remain largely speculative. It seems certain, nonetheless, that the phenomenon does occur in vivo and that the critical level of pressure varies in different organs.

Not all vascular beds exhibit the kind of pressure-flow curve shown in Fig. 39-5, *A*. In the kidney and in some other organs, blood flow is independent of perfusion pressure at pressures from about 70 mm Hg up to at least 150 mm Hg.[41,46,48] Below 70 mm Hg, as shown in Fig. 39-5, *B*, flow varies with pressure as expected, but at higher pressures some mechanism operates to alter resistance just enough to maintain constant flow. Since this same response can be demonstrated in the denervated kidney, it is appropriately described as *autoregulation*. The precision of regulation and the pressure range in which it is observed vary with experimental technique and are particularly sensitive to tissue damage. Phenomena that can properly be described as autoregulation have been described in the mesentery, skeletal muscle, brain, liver, and myocardium, testifying to its importance and generality in circulatory control.

The intrinsic mechanism responsible for autoregulation has not been identified, although it has been the subject of intensive study.[3] The earliest hypothesis, proposed by Bayliss[9] in 1902 to explain apparent autoregulation in the hind limbs of dogs and cats, is that vascular smooth muscle responds to increased stretch by increasing its tension (the *myogenic* hypothesis). An alternative explanation implicates tissue P_{O_2}, P_{CO_2}, or other metabolites as the agents responsible, whereas still a third hypothesis attributes autoregulation to changes in tissue pressure.[3] Each hypothesis has certain defects, and the evidence presented for each one warrants the verdict "not proved."

PULSATILE FLOW

Up to this point, we have been primarily concerned with steady flow and constant pressure gradients, yet the output of the heart is obviously pulsatile, and the pulsations of flow and pressure that it generates are transmitted far out into the vascular tree. Although the amplitude of these oscillations is greatly diminished by passage through the small arteries, pulsations of flow persist in many capillary beds and even in some venous channels. The part played by arterial distensibility in transforming the intermittent output of the ventricles into a relatively steady peripheral flow was described more than two centuries ago by Stephen Hales (1677-1761), who likened the arteries to a mechanical device familiar in his day: ". . . the blood is carried on in the finer capillaries, with an almost even tenor of velocity, in the same manner as the spouting water of some fire-engines is contrived to flow with a more even velocity, notwithstanding the alternate *systoles* and *diastoles* of the rising and falling *embolus* or force; and this by means of a large inverted globe, wherein the compressed air alternately dilating or contracting, in conformity to the workings to and fro of the *embolus*, and thereby impelling the water more equably than the *embolus* alone would do, pushes it out in a more nearly equal spout."[27]

The compression chamber of this analogy, *Windkessel* in its German translation, figured largely in the subsequent development of hemodynamics. The simile is well chosen; the volume ejected during each systole distends the proximal portion of the aorta and its branches, and the elastic energy thus stored in the walls of these vessels is then reconverted to energy of flow as the walls recoil during diastole. The gradual decline of arterial pressure during diastole, contrasting with the abrupt drop that would be found in a rigid system of tubes, is a manifestation of this "buffering" action. The large swings of pressure and flow in the proximal aorta are gradually damped as the blood moves peripherally, yielding an almost steady flow in the veins except for perturbations transmitted backward into the great veins by the action of the heart and respiratory movements. The mathematical treatment of these events by Otto Frank[20] is a landmark in the history of hemodynamics, although he treated the arteries as a single chamber in accordance with the Windkessel model, and the transmission of pressure and flow waves along the arteries was consequently not included in his original analysis.

In the search for analogs of Poiseuille's law appropriate to pulsatile flow a number of equations that have been derived describe the relation between pressure gradient, flow, and vascular dimensions when each of these variables is a function of time, some being modifications and extensions of Frank's work, whereas others abandon the Windkessel concept.* The formulation achieved by Womersley[56] and applied to the circulation by McDonald[4] and colleagues is in many ways the most promising because it rests on fairly realistic assumptions about vascular structures and function and is mathematically tractable. In this kind of analysis a first-order approximation is sought first by adopting a simplified model of the circulation, since the irregular branching of the real vascular tree, the nonlinear elasticity of its walls, the complex output function of the heart, and the anomalous viscosity of blood do not lend themselves to a direct approach. The accuracy with which the analysis describes the behavior of the circulation then depends to a large extent on the fidelity of the model. The Windkessel model, useful as it has been, treats the whole arterial tree as a single chamber and fails to represent the time required for pressure and flow pulses to move along the arteries. This is such an important feature of the circulation that it now seems wise to adopt a quite different model for describing transmission along the lon-

---

*See references 1, 4, 5, 50, and 56.

gitudinal axis of a vessel as well as elastic characteristics. Womersley approached this problem by considering the pulsations of pressure and flow as regular periodic waves, describing them quantitatively through harmonic analysis, and deriving the relationship between pressure and flow in a perfectly elastic tube by steps similar to those used to derive Poiseuille's law.[4,56] Since blood vessels are generally embedded in tissues that exert some restraint on the vessel wall and prevent much change in length, his equations take into account the effects of this "tethering" of the vessel to surrounding structures.

The use of harmonic analysis was a major step forward because it provided a relatively simple way to express numerically the asymmetric contours of the pressure and flow curves found in the circulation.[4,38]

Harmonic analysis. For many practical purposes a rough indication of the shape of a pressure pulse may be conveyed by the maximum systolic and minimum diastolic values and by estimates of its rate of rise and fall, but a more complete form of expression is required for any general statement of hemodynamic principles. The methods of harmonic analysis, which can be found in introductory form in the monograph by McDonald,[4] express any periodic waveform as the sum of a series of sinusoidal waves and a mean term. Pulsatile flow, for example, can be represented as a function of time, $Q(t)$, by the following series:

$$Q(t) = \overline{Q} + \sum_{n=1}^{N} M_n \sin(n\omega t + \phi_n) \qquad (12)$$

The set of terms on the right side of this equation is called a *Fourier series*. The first term ($\overline{Q}$) stands for the mean flow, and the subsequent terms represent a series of sinusoidal waves, or *harmonics*, in which n, the harmonic number, takes on successive integral values from 1 to N. For each harmonic, M_n is the amplitude of the wave, $n\omega$ its frequency in radians/sec, and ϕ_n its phase angle, which specifies its time relation to the other harmonics. The graphic significance of a Fourier series is illustrated in Fig. 39-6. Experimental records of pressure or flow can be transformed into such series by appropriate calculations, and the accuracy with which the analytic function $Q(t)$ coincides with the experimentally observed waveform increases with the number of harmonics included in a series; the first six harmonic (n = 1 to 6) suffice to reproduce very closely the pulses observed in the circulation.

Womersley's analysis of pressure and flow. The Womersley equations[4,56] relating oscillatory pressure and flow when both are expressed as Fourier series have the same general form as the Poiseuille equation. Mean pressure, flow, and radius in the presence of pulsations are still related as in equation 1, whereas each harmonic Q_n of the oscillations of flow around the mean is described by:

$$Q_n = \frac{\pi R^4 P_n}{\eta L} \Phi_n \cdot \sin(n\omega t + \phi_n + \beta_n) \qquad (13)$$

Each symbol has the same meaning as in equations 1 and 12 except that R is now a function of the oscillations of the vessel radius around its mean; P_n is the amplitude, or *modulus*, and ϕ_n the phase of the pressure gradient; Φ_n and β_n are functions of R, ω, η, blood density, wall thickness, and longitudinal tether-

ing. (Equation 13 is given in a form slightly different from that used by Womersley in order to show its resemblance to equation 1.)

The calculation of harmonic series and other operations required in this kind of analysis have been rendered much less formidable by the increasing availability of digital computers, and the experimental application of Womersley's equations has opened a number of new channels of investigation. The resistance defined by the ratio of mean pressure gradient to mean flow, for example, can be regarded as just one element in the more general concept of vascular impedance,[4,5,35] which includes the ratio of oscillatory pressure gradient to oscillatory flow over a spectrum of frequencies. The methods developed by Womersley,[56] McDonald,[4] and others not only define the theoretical relation of the impedance of a tube to its dimensions and elastic characteristics but also permit impedance to be determined experimentally. The impedance of both the systemic and pulmonic arteries has been found to vary significantly with frequency, suggesting that the energy needed to move blood through the circulation is also frequency dependent and hence influenced indirectly by the frequency of the heart rate.

Pulse wave velocity. The pulse wave velocity,[4,14] or speed with which a pressure wave travels along an elastic tube, is another variable that can be incorporated in this approach to hemodynamics. If the effects of viscosity are ignored, the velocity (c) of a pressure wave in a thin-walled elastic tube filled with fluid is given (in symbols defined earlier) by:

$$c = \sqrt{\frac{E\delta}{2R\rho}} \qquad (14)$$

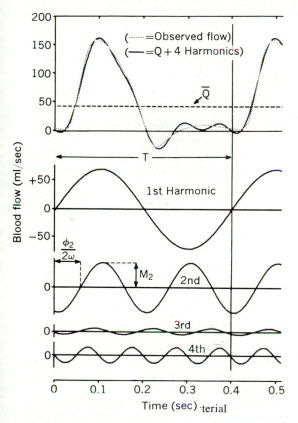

Fig. 39-6. Example of pulmonary arterial flow pulse measured in vivo (observed flow) and its representation by Fourier series (see equation 12). The Fourier series, which is derived mathematically from observed flow pulse, gives numeric description of shape and amplitude of pulsation. Each term, or harmonic, in the series is a sinusoidal wave of amplitude M_n, phase ϕ_n, and frequency $n\omega$. Sum of these harmonics and mean flow ($\overline{Q}$), shown as unbroken line in top tracing, can be made to approximate observed pulsation to any degree of exactness by increasing number of harmonics calculated, but for purpose of this illustration, only four harmonics have been used.

This expression, often referred to as the Moens-Korteweg equation, does not accurately describe pulse wave velocity in blood vessels because the vascular wall has finite and variable thickness, the vessels are restrained longitudinally by external attachments, and the blood viscosity cannot be neglected. When these factors are taken into account, the true pulse wave velocity in large arteries is estimated by Bergel[12] to be about 12% greater than the value given by equation 14. This velocity is also a function of frequency and is therefore different for each harmonic component of the wave. Pulse wave velocities in the systemic arteries increase as the vessels become smaller and stiffer, ranging from about 5 m/sec in the thoracic aorta to perhaps 20 m/sec in the terminal arterioles. These values represent the velocity of propagation of *pressure* waves through the circulation, which should not be confused with the much slower velocity of *blood flow*.

Reflected waves. All the aforementioned analyses of pulsatile phenomena must be modified to include one further characteristic of the circulation: the reflection of waves from branch points and other irregularities in the vascular tree. Such reflections are apparently the principal cause of the changing contour of the arterial pressure pulse as it travels toward the periphery. The behavior of similar reflections in electrical and hydraulic transmission lines has been thoroughly studied, and their influence on impedance and pulse wave velocity provides a consistent explanation for such experimental data as the well-documented observation that the arterial pulse pressure increases from the aortic root to the distal aorta, instead of decreasing as the effects of viscosity alone would predict.[4,50]

Pressure gradient and pulsatile flow. An alternative approach to pressure-flow relationships in large vessels, somewhat simpler than that adopted by Womersley, has been developed by Fry[21] and Fry et al.[22] Starting from the Navier-Stokes equation and the assumption that the inertia of flow is much larger than the effects of viscosity over a short longitudinal segment of a vessel, they arrived at an equation that relates longitudinal pressure gradient to velocity of flow:

$$P(t) = \rho \Delta X \frac{dv}{dt} = kv(t) \qquad (15)$$

$P(t)$ is the difference in pressure in dynes per square centimeter between two sites along the longitudinal axis of the vessel separated by ΔX cm, as a function of time; v is velocity in centimeters per second; ρ is blood density in grams per cubic centimeter; and k is a constant that depends on the relation of viscous pressure drop to velocity. Rigorous interpretation of this equation would require that $P(t)$ be the space derivative of the pressure at one point in the vessel, but for practical applications, it is necessary to assume that a satisfactory approximation of the space derivative is given by measurements of the pressure difference between two sites several centimeters apart. If the pressure difference is recorded from the distal openings of a double-lumen catheter in the aorta, for example, relatively simple electrical circuits[22] can be used to solve equation 15 continuously for velocity (p. 1003).

APPENDIX

Derivation of Poiseuille's law

The flow of newtonian liquids through rigid cylindric tubes may be pictured as a series of concentric cylindric laminae; the velocity of each cylindric shell is constant, but less that that the velocity of its more axial neighbor. From the definition of newtonian liquids and viscosity, it follows that the force acting to retard flow over unit area is proportional to the fluid viscosity and the velocity gradient. For a cylindric unit of liquid of length L and radius r the viscous retarding force F will then be the product of its surface area $(2\pi rL)$, its viscosity (η), and the velocity gradient across the width of the tube (dv/dr):

$$F \text{ (viscous)} = 2\pi rL\eta(dv/dr) \qquad (A-1)$$

The force exerted to move this unit of liquid is the pressure applied to the end of the cylinder (P_1) multiplied by its cross-sectional area, less the product of pressure (P_2) and area at the distal end:

$$F \text{ (pressure)} = \pi r^2(P_1 - P_2) \qquad (A-2)$$

Since these two forces are equal and opposite:

$$2\pi r \, L\eta(dv/dr) = -\pi r^2(P_1 - P_2) \qquad (A-3)$$

which may be arranged to give:

$$\frac{dv}{dr} = -\frac{r(P_1 - P_2)}{2L\eta} \qquad (A-4)$$

Assuming that viscosity (η) and pressure gradient $(P_1 - P_2)/L$ are constant, integration of equation A-4 to calculate velocity gives:

$$v = -\frac{r^2(P_1 - P_2)}{4L\eta} + C \qquad (A-5)$$

The integration constant (C) can be evaluated from the boundary conditions contained in the assumption that the lamina in contact with the vessel wall is stationary (i.e., $v = 0$ when $r = R$, the radius of the vessel lumen). Substituting these values in equation A-5 gives:

$$C = \frac{R^2(P_1 - P_2)}{4L\eta} \qquad (A-6)$$

Thus equation A-5 becomes:

$$v = -\frac{r^2(P_1 - P_2)}{4L\eta} + \frac{R^2(P_1 - P_2)}{4L\eta} \qquad (A-7)$$

$$= \frac{(P_1 - P_2)}{4L\eta}(R^2 - r^2) \qquad (A-8)$$

This is the equation for a parabola (Fig. 37-1), and it remains only to derive the volume flow represented by this distribution of velocities.

The volume (Q) of the solid of revolution of this parabola is:

$$Q = \int_0^R 2\pi v \, r \cdot dr \qquad (A-9)$$

Substituting the expression for v given by equation A-8 into equation A-9 gives:

$$Q = \frac{2\pi(P_1 - P_2)}{4L\eta} \int_0^R (R^2 - r^2) \cdot dr \qquad (A-10)$$

$$= \frac{\pi(P_1 - P_2)R^4}{8L\eta} \qquad (A-11)$$

Equation A-11 is the form in which Poiseuille's law is usually expressed.

REFERENCES
Reviews and monographs

1. Bergel, D. H., editor: Cardiovascular fluid dynamics, London, 1972, Academic Press, Ltd.
2. Burton, A. C.: Relation of structure to function of the tissues of the walls of blood vessels, Physiol. Rev. **34:** 619, 1954.
3. Johnson, P. C., editor: Autoregulation of blood flow, Circ. Res. **15**(suppl. 1):2, 1964.
4. McDonald, D. A.: Blood flow in arteries, ed. 2, Baltimore, 1974, The Williams & Wilkins Co.
5. Noordergraaf, A.: Hemodynamics. In Schwan, H. P., editor: Biological engineering, New York, 1969, McGraw-Hill Book Co., p. 391.
6. Patel, D. J., et al.: Hemodynamics, Ann. Rev. Physiol. **36:**125, 1974.
7. Whitmore, R. L.: Rheology of the circulation, Oxford, 1968, Pergamon Press, Inc.

Original papers

8. Alexander, R. S., Edwards, W. S., and Ankeney, J. L.: The distensibility characteristics of the portal vascular bed, Circ. Res. **1:**271, 1953.
9. Bayliss, W. M.: On the local reactions of the arterial wall to changes of internal pressure, J. Physiol. **28:**220, 1902.
10. Benninghoff, A.: Über die Beziehunger zwischen Elastischem Gerüst und glatter Muskulatur in der Arterienwand und ihre funktionelle Bedeutung, Z. Zellforsch. Mikrosk. Anat. **6:**348, 1927.
11. Bergel, D. H.: The static elastic properties of the arterial wall, J. Physiol. **156:**445, 1961.
12. Bergel, D. H.: The dynamic elastic properties of the arterial wall, J. Physiol. **156:**458, 1961.
13. Bozler, E.: Conduction, automaticity, and tonus of visceral muscles, Experientia **4:**213, 1948.
14. Bramwell, J. C., and Hill, A. V.: The velocity of the pulse wave in man, Proc. R. Soc. Lond. (Biol.) **93:** 298, 1922.
15. Burton, A. C., and Stinson, R. H.: The measurement of tension in vascular smooth muscle, J. Physiol. **153:** 290, 1960.
16. Burton, A. C., and Yamada, S.: Relation between blood pressure and flow in the human forearm, J. Appl. Physiol. **4:**329, 1951.
17. Coulter, N. A., Jr., and Pappenheimer, J. R.: Development of turbulence in flowing blood, Am. J. Physiol. **159:**401, 1949.
18. Fåhraeus, R., and Lindqvist, T.: The viscosity of the blood in narrow capillary tubes, Am. J. Physiol. **96:** 562, 1931.
19. Folkow, B., and Lofving, B.: The distensibility of the systemic resistance blood vessels, Acta Physiol. Scand. **38:**37, 1956.
20. Frank, O.: Die Theorie der Pulswellen, Z. Biol. **85:** 91, 1926.
21. Fry, D. L.: Certain aspects of hydrodynamics as applied to the living cardiovascular system, IRE Trans. Med. Electronics **ME-6:**252, 1959.
22. Fry, D. L., Noble, F. W., and Mallos, A. J.: An electric device for instantaneous and continuous computation of aortic blood velocity, Circ. Res. **5:**75, 1957.
23. Girling, F.: Vasomotor effects of electrical stimulation, Am. J. Physiol. **170:**131, 1952.
24. Gorlin, R., and Gorlin, S. G.: Hydraulic formula for calculation of the area of the stenotic mitral valve, other cardiac valves, and central circulatory shunts, Am. Heart J. **41:**1, 1951.
25. Green, H. D., et al.: Blood flow, peripheral resistance and vascular tonus, with observations on the relationship between blood flow and cutaneous temperature, Am. J. Physiol. **141:**518, 1944.
26. Guyton, A. C.: A concept of negative interstitial pressure based on pressures in implanted perforated capsules, Circ. Res. **12:**399, 1963.
27. Hales, S.: Statical essays: containing hemostatics, ed. 3, London, 1769, W. Innys, vol. 2.
28. Hardung, V.: Vergleichende Messungen der dynamischen Elastizität und Viskosität von Blutgefässen, Kautschuk. und synthetischen Elastomeren, Helv. Physiol. Acta **11:**194, 1953.
29. Harkness, M. L. R., Harkness, R. D., and McDonald, D. A.: The collagen and elastin content of the arterial wall in the dog, Proc. R. Soc. Lond. (Biol.) **146:**541, 1957.
30. Hinke, J. A. M., and Wilson, M. L.: A study of elastic properties of a 550-micron artery in vitro, Am. J. Physiol. **203:**1153, 1962.
31. Holt, J. P.: Flow of liquids through "collapsible" tubes, Circ. Res. **7:**342, 1959.
32. Krafka, J., Jr.: Comparative study of the histophysics of the aorta, Am. J. Physiol. **125:**1, 1939.
33. Lamport, H., and Baez, S.: Physical properties of small arterial vessels, Physiol. Rev. **42**(suppl. 5):328, 1962.
34. Landowne, M., and Stacy, R. W.: Glossary of terms. In Remington, J. W., editor: Tissue elasticity, Washington, D.C., 1957, American Physiological Society.
35. Milnor, W. R., et al.: Pulmonary arterial pulse wave velocity and impedance in man, Circ. Res. **25:**637, 1969.
36. Nichol, J., et al.: Fundamental instability of the small blood vessels and critical closing pressures in vascular beds, Am. J. Physiol. **164:**330, 1951.
37. Patel, D. J., et al.: Harmonic analysis of pressure pulses obtained from the heart and great vessels of man, Am. Heart J. **69:**785, 1965.
38. Peterson, L. H., Jensen, R. E., and Parnell, J.: Mechanical properties of arteries in vivo, Circ. Res. **8:**622, 1960.
39. Poiseuille, J. L. M.: Recherches expérimentales sur le mouvement des liquides dans les tubes de très-petits diamètres, Mémoires Présentes par divers savants, à l'Acad. Sci. de l'Institut de France **9:**433, 1846.
40. Prec, O., et al.: Determination of kinetic energy of the heart in man, Am. J. Physiol. **159:**483, 1949.
41. Rein, H.: Vasomotorische regulationen, Ergeb. Physiol. **32:**28, 1931.
42. Reshevsky, N.: A problem in the mathematical biophysics of blood circulation. II. Relation between pressure and flow of a viscous fluid in an elastic distensible tube, Bull. Math. Biophys. **7:**35, 1945.
43. Reynolds, O.: An experimental investigation of the circumstances which determine whether the motion of water shall be direct or sinuous, and of the law of resistance in parallel channels, Philos. Trans. **174:**935, 1883.
44. Rhodin, J. A.: Fine structure of vascular walls in mammals, Physiol. Rev. **42**(suppl. 5):48, 1962.
45. Sarnoff, S. J., and Berglund, E.: Pressure-volume characteristics and stress relaxation in the pulmonary vascular bed of the dog, Am. J. Physiol. **171:**238, 1952.
46. Schmid, H. E., Jr., and Spencer, M. P.: Characteristics of pressure-flow regulation by the kidney, J. Appl. Physiol. **17:**201, 1962.
47. Schultz, D. L., et al.: Velocity distribution and transition

in the arterial system. In Wolstenholme, G. E. W., and Knight, J., editors: Circulatory and respiratory mass transport, Boston, 1969, Little, Brown & Co.

48. Selkurt, E. E.: The relation of renal blood flow to effective arterial pressure in the intact kidney of the dog, Am. J. Physiol. **147:**537, 1946.

49. Spencer, M. P., and Greiss, F. C.: Dynamics of ventricular ejection, Circ. Res. **10:**274, 1962.

50. Taylor, M. G.: An approach to an analysis of the arterial pulse wave. I. Oscillations in an attenuating line, Phys. Med. Biol. **1:**258, 1957.

51. Taylor, M. G.: The influence of the anomalous viscosity of blood upon its oscillatory flow, Phys. Med. Biol. **3:**273, 1959.

52. Timoshenko, S.: Strength of materials, ed. 3, Princeton, 1955, D. Van Nostrand Co., Inc. pt. 2, p. 236.

53. Wells, H. S., Youmans, J. B., and Miller, D. G., Jr.: Tissue pressure (intracutaneous, subcutaneous, and intramuscular) as related to venous pressure, capillary filtration, and other factors, J. Clin. Invest. **17:**489, 1938.

54. Whittaker, S. R. F., and Winton, F. R.: The apparent viscosity of blood flowing in the isolated hind limb of the dog, and its variation with corpuscular concentration, J. Physiol. **78:**339, 1933.

55. Wiederhielm, C. A., Woodbury, J. W., Kirk, S., and Rushmer, R. F.: Pulsatile pressures in the microcirculation of frog's mesentery, Am. J. Physiol. **207:**173, 1964.

56. Womersley, J. R.: An elastic tube theory of pulse transmission and oscillatory flow in mammalian arteries, Wright Air Development Center Technical Report TR 56-614, 1957.

40

WILLIAM R. MILNOR

Normal circulatory function

BASAL CIRCULATORY FUNCTION

Cardiovascular functioning is adjusted to the metabolic needs of the tissues it serves and can adapt to a wide range of environmental conditions. Circulatory conditions when the organism is at rest constitute a basal state on which all such adjustments are superimposed. Hemodynamic studies in man are the source of much of our knowledge of the circulation in vivo, and the values cited throughout this chapter refer to normal adult human subjects unless otherwise indicated.

Basal state. Basal conditions may be defined as those that entail the smallest expenditure of energy compatible with normal metabolism and the conscious state. Such conditions are easier to define than to achieve, and it is difficult to do more than approach basal conditions when making physiologic measurements. Although some animals can be trained to cooperate to an extent that allows reliable measurements without anesthesia,[16,72] most data relating to the circulation in intact animals have been gained under the influence of anesthetic agents that modify normal reflex responses profoundly. Human subjects can be studied without the complications of general anesthesia, but the procedures used to measure circulatory function often produce a degree of apprehension and discomfort sufficient to elevate heart rate, blood pressure, and cardiac output above their basal levels. The wide range of values that must be considered "within normal limits" in circulatory physiology is due in part to this elusiveness of the basal state as well as to differences in techniques of measurement and true biologic variation. The result is a standard deviation of ±20% for many hemodynamic variables in presumably normal populations.

Cardiac output. The average cardiac output in healthy adult subjects under basal conditions is about 6 L of blood/min, but the individual measurements are widely distributed. Part of this variation is related to differences in body size, and this factor must be taken into account in comparing cardiac outputs from different individuals. No completely satisfactory way of expressing the relation between output and body size has been found, but the *cardiac index*, or ratio of output to body surface area, is widely used for this purpose. Since basal metabolic rate and oxygen consumption are directly correlated with body surface area,[18] it is reasonable to expect a similar relationship for the blood flow that serves these metabolic processes, and this expectation has been amply confirmed (Fig. 40-1). Cardiac output and surface area are not strictly proportional over the physiologic range, but the relationship is close enough to make the cardiac index an acceptable compromise for comparative measurements,[75] even though the surface area is commonly estimated from the Dubois height-weight equation rather than measured directly (Fig. 56-8).

As the nearly constant ratio of cardiac output to oxygen consumption implies (Fig. 40-1), the arteriovenous oxygen difference at rest is relatively constant. Reeves and colleagues found the mean arteriovenous oxygen difference in a series of normal subjects to be 3.98 vol%, with a variance much smaller than that of the cardiac index.[68] It can be argued that this parameter is a better comparative measure of cardiovascular function than is the cardiac index, since the arteriovenous oxygen difference is not only relatively independent of body size but reflects the relation between blood flow and metabolic demand.[68] Although this claim is valid and particularly relevant when metabolic rate is abnormal, the relation of blood flow to body size or total cell mass is also of physiologic significance. The use of total lean body mass instead of surface area would be a more rational way of expressing the relation between output and body size, but for practical purposes the cardiac index is a satisfactory alternative.

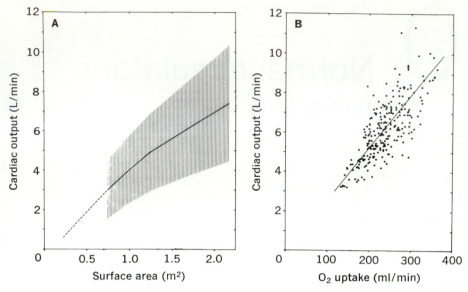

Fig. 40-1. A, Relation between cardiac output at rest and body surface area in human subjects. Shaded area includes 2 SD above and below mean. Body surface areas of less than 1 m² are found in children and infants, on whom relatively few observations have been made. Above 1 m², relationship is almost linear, lending validity to use of cardiac index (cardiac output/surface area) as a standard of comparison.[12,13,25,26,51,65] **B,** Relation between cardiac output and oxygen uptake showing linear relationship corresponding to arteriovenous oxygen difference of 38.4 ml/L. (**B** from Reeves et al.[68])

Apart from the problems introduced by differences in body size, attempts to define the normal range for cardiac output were long beset by discrepancies in the results given by the various techniques evolved for its measurement, a situation critically reviewed by Wade and Bishop.[12] Data obtained more recently by the direct Fick and indicator-dilution methods (Chapter 37), which are the most reliable methods now available for use in man, indicate a normal mean cardiac index of approximately 3.5 (L/min)/m², with 95% of healthy adults having indices in the range of 2.1 to 4.9 (L/min)/m².* These values do not apply to persons below the age of 18 years, since there is some evidence to suggest that the cardiac index is higher in childhood and declines during adolescence toward the adult range.[13,26,51] On the other hand, in the first 24 hr of infancy the ratio of cardiac output to surface area is lower than in adults.[13,65] After the first two decades the cardiac index falls gradually with increasing age.[23] There appears to be no significant difference in cardiac output between men and women apart from differences in body stature. The external work of the heart associ-

*See references 12, 30, 34, 38, and 68.

Table 40-1. Representative values of ventricular work and power in man at rest

	Right ventricle	Left ventricle
Heart rate (strokes/min)	75	75
Stroke volume (ml)	77	77
Intraventricular pressure (mm Hg systolic/ diastolic)	25/4	125/8
Stroke work (gm-m)	24.8	119
Average hydraulic power (milliwatts)		
Potential energy	290	1,430
Kinetic energy	15	29
	305	1,459

ated with the output of each ventricle under basal conditions is summarized in Table 40-1.

Comparison of different species throughout the animal kingdom reveals that cardiac output is more closely correlated with body weight than with surface area. Animals ranging from the anesthetized rabbit[35] and dog to man and domestic cattle[39] have resting outputs of about 100 (ml/min)/kg body weight. The influence of anesthetics is particularly evident in the dog, in which

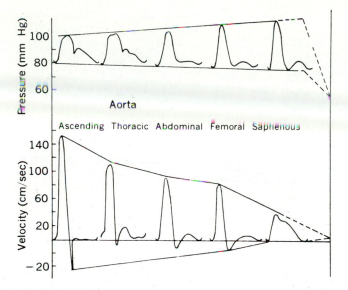

Fig. 40-2. Diagrammatic comparison of transformation of pressure and flow pulses in aorta and its branches. (From McDonald.[8])

cardiac output when unanesthetized and apparently relaxed is usually closer to 150 (ml/min)/kg body weight.

Cardiac output depends on heart rate and stroke volume, and the average volume normally ejected by each ventricle is in the neighborhood of 80 ml at rest, with a range of 60 to 100 ml/stroke at rates of 60 to 84 strokes/min. Ventricular contraction expels only a part of the blood contained in the chamber; from 40% to 60% of the end-diastolic volume remains in the ventricle at the end of systole. The roentgenographic and dilution methods that are used to measure this *residual volume* are not very accurate, but this same ratio of residual to end-diastolic volume under resting conditions has been reported for man and a number of other mammals for both right and left ventricles.* This residual blood serves as a small but effective reserve that can be drawn on for sudden changes in stroke volume or for temporary adjustments needed to balance the output of the two ventricles.[43]

Pressure and flow in the aorta and its branches

Blood pressure. The outflow from right and left ventricles was considered in Fig. 35-1, and the subsequent transformation of pressure and flow pulsations as they travel along the aorta is illustrated in Fig. 40-2. The increasing pulse

*See references 36, 41, 50, 66, and 67.

pressure, which is the most striking feature of this transformation, was reported by Frank[42] in 1905, when adequate manometers were coming into use, and has been a subject for intensive investigation ever since. In addition to the peripheral "peaking" of the pressure wave out as far as the femoral or saphenous arteries, with increasing systolic and decreasing diastolic pressures, the wave front rises more steeply, the sharp inflection after the peak disappears, and a secondary dip and rise develop in diastole. A quantitative description of these pressure waves can be substituted for these qualitative terms by the methods of harmonic analysis (Chapter 39).

Three factors involved in this metamorphosis have long been recognized: (1) damping of waves traveling in a viscous fluid, (2) differences in velocity of the harmonic components of the pulse, and (3) wave reflection. The best evidence at present assigns the dominant role to reflected waves. Damping has been demonstrated directly by the rapid attenuation of waves imposed artificially postmortem,[74] and the nature of viscous fluids requires that pulse wave velocity increase with frequency,[8] but in the large arteries both these factors are relatively small and their effects are virtually lost in the interaction between incident and reflected waves. In a simple elastic tube filled with fluid and completely closed at one end, this interaction would augment the pressure pulse at some sites in the tube and diminish it at others in a systematic fashion.[8] This orderly pattern is not found in its entirety in the circulation because the aortic tree does not conform to so simple a model, and the wavelength of the fun-

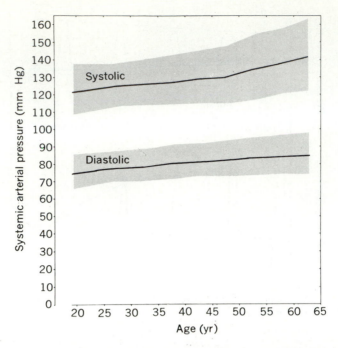

Fig. 40-3. Average systolic and diastolic blood pressures in a population of 7,722 healthy men between 16 and 65 years of age. Shaded area includes ±1 SD around mean averages. (Modified from Master et al.[7])

damental component is much longer than the aorta, but the augmentation of pulse pressure in the aorta is an analogous phenomenon.[8,70] Although the oscillations of pressure increase in amplitude toward the distal aorta, mean pressure necessarily declines steadily toward the periphery, the mean gradient along the aorta amounting to only a few millimeters of mercury because of its large diameter.

Arterial blood pressure varies with age, sex, weight, metabolic rate, emotional state, and countless other factors, so that "normal" values can be expressed properly only by specifying means and frequency distributions for homogeneous groups. To convey a general idea of the usual levels, however, it may be said that in the great majority of healthy adult subjects, under reasonably basal conditions, the systolic pressure in arteries the size of the brachial is between 100 and 150 mm Hg and the diastolic pressure is between 60 and 90 mm Hg. A representative study is illustrated in Fig. 40-3. The groups studied in such investigations can rarely be random samples, and the population studied in this instance consisted of industrial workers at a number of plants throughout the country. Bias in selection was carefully minimized, however, and the results are consistent with many other observations. It is evident that blood pressure

gradually increases with age, emphasizing the dangers of drawing conclusions from the young adults most readily available for physiologic investigations. The shaded areas in Fig. 40-3 include values that lie within ±1 SD from the mean, indicating that approximately two thirds of the observations fell within this range. As with all physiologic variables, "normal limits" should be indicated by a broad gray smudge rather than a sharp line. Blood pressures in a similar group of women examined in the same study were 5 to 10 mm Hg lower on the average up to the age of 50 years, after which there was no difference.

Most clinical studies of blood pressure, including that summarized in Fig. 40-3 are not carried out under strictly basal conditions, but this is not necessarily a disadvantage, since the conditions are usually similar to those in clinical practice. These clearly do not represent minimal activity, because pressures obtained during sleep are sometimes 10 to 20 mm Hg below those in the waking state. No great emphasis should be placed on any single reading of blood pressure, for there are considerable variations throughout the day in any individual. Systolic pressure rises moderately after meals, for example, and diastolic pressure often falls. Fuller discussion of

Table 40-2. Estimated distribution of cardiac output and oxygen consumption in normal human subject* at rest under usual indoor conditions†

Circulation	Blood flow		Arteriovenous oxygen difference (vol%)	Oxygen uptake	
	ml/min	%Total		ml/min	%Total
Splanchnic	1,400	24	4.1	58	25
Renal	1,100	19	1.3	16	7
Cerebral	750	13	6.3	46	20
Coronary	250	4	11.4	27	11
Skeletal muscle	1,200	21	8.0	70	30
Skin	500	9	1.0	5	2
Other organs	600	10	3.0	12	5
TOTAL	5,800	100	4.0	234	100

*Weight, 70 kg; surface area, 1.7 m².
†From Wade and Bishop.[12]

blood pressure in normal human subjects and in the presence of disease can be found in monographs by Master et al.[7] and Pickering.[9]

Arterial pressure in other mammalian species is fairly uniform; systolic, diastolic, and mean pressures in animals ranging in size from rats to cattle do not differ radically from those in man.[35] Higher pressures have been reported for the dog, but these apparently represent the effects of barbiturate or other anesthetic agents that elevate both heart rate and systemic pressure. The giraffe is an exception, with a mean aortic pressure of about 250 mm Hg,[45] a finding doubtless related to the position of the brain some 10 feet above heart level.

Blood flow in arteries. Pulsations of flow,[8,63] in contrast to pulsations of pressure, diminish steadily along the aorta, and the onset of the wave is perceptibly later at each successive stage. A rounded diastolic peak waxes and wanes in the descending aorta, becoming an almost steady diastolic flow in the branches. The transient backward flow found at the origin of the aorta usually persists out to the femoral artery (Fig. 40-3), although it varies in magnitude and may be absent beyond the midabdominal aorta. Vasoconstriction in the peripheral bed increases this backflow, but the effects of vasodilation in this regard are variable.[8] When cardiac output increases during exercise, the rise in mean flow through the abdominal aorta and its branches elevates the flow curves so that the oscillations no longer pass through zero.

Regional distribution of flow. The distribution of the cardiac output to the various organ systems of the body under resting conditions is indicated in Table 40-2, which gives estimated typical values derived by Wade and Bishop[12] from their own studies and a number of reports

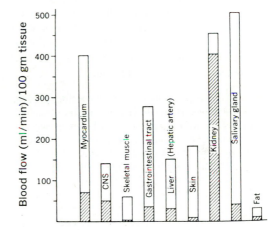

Fig. 40-4. Diagram of regional blood flows in basal state (shaded areas) and at maximum vasodilatation (unshaded areas), scaled for normal adult human subject weighing 70 gm. Ordinate is blood flow per 100 gm tissue, showing relative vascular resistance of each organ system. (From Mellander and Johansson.[60])

in the literature. The methods available for measuring regional blood flows in man are by no means as accurate as one could wish, and the data in this table are only approximate.

Some of the regional circulations are discussed more fully in Chapter 42, but several aspects of the partitioning of flow should be noted here. The division of the cardiac output among different organs depends on their relative resistance to blood flow (Fig. 40-4). The resistance of each organ system is a function of the local vasomotor tone and of the overall size of its vascular bed, in the sense that resistance falls as the number of parallel channels increases. Vasodilatation in one organ, for example, in the absence of any change in cardiac output or resistance of other organ systems, would increase the share of the total output flowing to the dilated bed; the absolute blood flow to other regions would necessarily decrease. In most situations the regulatory mechanisms of the cardiovascular system operate to maintain blood flow to critical regions at an optimum level by appropriate adjustments of cardiac output and regional resistances. Blood flow in the cerebral circulation, and to a limited extent in the kidney, remains almost constant when marked changes in arterial pressure and cardiac output occur, demonstrating the effectiveness of regulation by regional resistances and the priority accorded the central nervous system. Splanchnic and cutaneous blood flow, on the other hand, vary greatly according to the functional demands made on the liver and gastrointestinal tract in the one case and the needs of temperature regulation in the other. The same holds true for blood flow to skeletal muscles, which can increase 20-fold in response to muscular exertion.

Considerable uncertainty enters into estimates of the total blood flow to skeletal muscle, since it is necessary to extrapolate from measurements of flow to one segment of a limb and to make assumptions as to the fraction of this flow that reaches skin and bone instead of muscle. The circulation to muscle in the upper limbs, in very approximate terms, is about 300 ml/min, and to the lower limbs about 900 ml/min.[12] In the forearm the flow to resting muscle is estimated to be about 4 (ml/min)/100 gm of muscle.[12,15] No reliable method has been found for the measurement in vivo of the circulation to bone, which is a major system included among the "other organs" in Table 40-2. Reports of the total blood flow to the bone marrow range from 2% to 7% of the cardiac output.[82]

In the case of the cerebral and renal[33] circulations there is evidence to suggest that blood flow to these regions declines with age, although it is not certain that this exceeds the simultaneous decline in total cardiac output.

The regional arteriovenous oxygen differences and oxygen uptakes in Table 40-2 give some indication of the relation between blood flow and metabolic rate in different areas. The greatest arteriovenous difference is found in the coronary circuit, where 11% of the total oxygen consumed by the body is extracted from a flow that amounts to 4% of the total cardiac output.[71] The blood in the coronary sinus draining the myocardium has the lowest oxygen content in the circulation.

Peripheral circulation

Beyond the large arteries, blood pressure drops rapidly (Fig. 35-1) and the linear velocity of flow decreases. Pulsations are gradually transformed

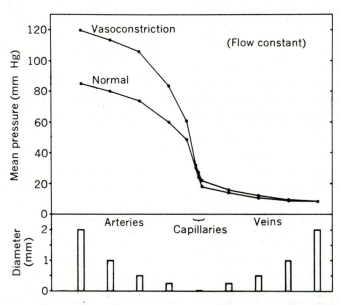

Fig. 40-5. Diagrammatic representation of pressure drop in small vessels of systemic circulation under normal conditions and after vasoconstriction by norepinephrine, assuming that flow remains constant. Vertical bars at bottom of graph represent normal diameter of vessels in which indicated pressures occur. Under ordinary conditions in vivo, capillary pressures often fall slightly with vasoconstriction, depending on changes in blood flow and on relative amounts of arteriolar and venular constriction. (Based on data from Haddy[47]; Landis[53]; and other sources.)

into a more nearly steady flow, although small oscillations of both pressure and flow persist in the capillaries and for an undetermined distance into the venous bed. Traditionally, the most marked drop in pressure is attributed to the arterioles, but it would be more exact to say that the greatest pressure decrement per unit vessel length occurs in the microscopic terminal branches of the arterial tree. For the histologist, an arteriole is a precapillary vessel with a prominent muscular coat and a diameter less than 200 μm; for the physiologist, arterioles are the tiny arteries that control local blood flow by altering their diameter. That these two definitions conflict to some extent is evident in the gradients of pressure and vessel dimensions shown in Fig. 40-5. The pressure drop as vessels narrow from 1,000 to 200 μm in diameter is not only subject to vasomotor control but also represents a large part of the total vascular resistance in most beds. Resistance is not sharply localized but is distributed in varying degrees along a fairly wide range of vessels, and it would be appropriate to label arteries of perhaps 0.5 mm or less in diameter as *functional arterioles*.

The pressure gradient of some 10 to 20 mm Hg across the capillaries is controlled by the activity of the muscular vessels at the entrances and exits of the capillary bed. Under the microscope, capillary flow presents a picture of almost random opening of some channels and closing of others from time to time, but this apparently patternless activity is presumably the result of changes in capillary transmural pressures, which in turn depend on the resistance of the terminal arterioles and venules. Mean capillary pressure is not the same in all parts of the circulation, being relatively high in the kidney[46] and low in the lung.

Veins[10]

In reclining subjects there is a further drop in pressure between the systemic venules and right atrium of about 20 mm Hg; pressure falls from 15 to 20 mm Hg in the venules to a level of 4 to 10 mm Hg in the veins of the extremities, and approximately atmospheric pressure in the right atrium. The very small gradient along the inferior vena cava, amounting to 3 or 4 mm Hg, has been measured in human subjects with a miniature intravascular manometer, but the decline in pressure along the intrathoracic portions of the venae cavae is too small to be measured accurately. The transmural pressure of the intrathoracic veins is increased by the subatmospheric pressure in the thoracic cavity; the effect of the respiratory excursions of this negative intratho-

racic pressure on blood flow through the central veins is considered in Chapter 45. Venous pressures in the erect position are modified by the weight of the intravascular columns of blood extending to the heart, the results of which will be considered in the discussion of the effects of posture.

None of the arterial pulsation is transmitted forward to the large systemic veins, but the great veins near the heart show small pressure waves transmitted in a retrograde direction from the right atrium. Pulsations 1 to 3 mm Hg in amplitude are found in the inferior vena cava down to the diaphragm and in the proximal branches of the superior vena cava. The resulting pulses in the jugular vein are often visible in supine subjects, and with suitable recording equipment, these venous pulse waves can be shown to resemble the characteristic *a, c,* and *v* waves of the atrium (Fig. 37-1).

NORMAL CIRCULATORY RESPONSES
Effects of posture

When man rises from a recumbent position and stands erect, the weight of intravascular columns of blood in vessels above and below the heart places stresses on the circulation that are absent in the supine position. These hydrostatic pressures do not *directly* influence the driving force, or gradient of pressure, from arteries to veins; however, they do alter local transmural pressures, so that the larger veins, being readily distended, tend to enlarge in the legs and collapse in the upper portions of the body. The magnitude of the hydrostatic force at any point can be calculated from the vertical distance between that point and the right atrium, where the systemic circulation terminates, taking into account the density of blood and the gravitational constant (see equation 3, Chapter 39).

The veins of the lower leg, for example, lie about 100 cm below the atrial reference level in the erect position in an adult of average stature, and the hydrostatic pressure in these veins, by virtue of the weight of this vertical column of blood (ignoring the interruption of the column by venous valves), is approximately 105 cm H_2O, or 77 mm Hg. To this is added the dynamic pressure remaining from the energy imparted to the blood by the left ventricle upstream. Assuming a dynamic venous pressure of 10 mm Hg in both supine and erect positions, the total intravascular pressure, or force acting to distend these veins, would increase from +10 mm Hg when supine to +87 mm Hg when standing. Transmural pressure does not increase to the same degree because of the venous valves and a simultaneous, lesser increment in extravascular pressure. Similar hydrostatic forces exist in the capillaries at the same level, where they influence fluid exchange through the capillary wall, and in the arteries, where their effect on the less distensible arterial wall is relatively small.

In the cerebral circulation and other vessels above the level of the heart the hydrostatic effects of standing are just the opposite. The column of blood extending downward from a point in the cerebral veins to the atrium reduces the intravascular pressure at that point by an amount proportional to its weight and may thereby decrease the intravascular component of the transmural pressure by as much as 30 mm Hg (assuming a vertical distance of 40 cm; estimates vary with the height of the author).

These radical changes in intravascular pressure would be intolerable were it not for compensating changes in extravascular pressure that result from the structure of the body and changes in venous tone that are mediated by reflex activity. The cerebral vessels are protected by their situation in virtually incompressible tissues rigidly enclosed by the skull. Under these conditions the hydraulic principle known as *Pascal's law* dictates that any fall in pressure within the vessels be communicated to all the intracranial tissues, so that transmural pressure is unchanged. The absence of this means of compensation in the extracranial veins above heart level causes visible collapse of the superficial veins, such as the jugular, in the erect position. Such collapse does not completely obliterate the lumen of the vein, however, and a limited flow through it persists.

For the deeper veins of the extremities a similar, although much less rigid, enclosure is formed by the limited confines of the connective tissue spaces, muscles, and skin. These structures reduce the effects of changes in hydrostatic pressure, but they do not prevent a moderate engorgement of the leg veins in the erect position and a net loss of fluid from the capillaries into the extracellular fluid of the legs.[82] Pressure in the subcutaneous tissues of the leg and in the gastrocnemius muscles rises about 5 cm H_2O within the first few minutes after standing as a result of venous congestion.[59] The contraction of skeletal muscle provides an additional counterforce to hydrostatic pressures; with passive tilting to a semivertical position on a tilt table, the intramuscular pressure in the lower leg often rises by 15 to 20 cm H_2O, and weight bearing on the toes or ball of the foot can increase this by 40 cm H_2O.[59] Muscle contraction serves an even more important function when alternate contraction and re-

laxation of muscle segments compresses the veins externally, tending to propel blood toward the heart. As a result of such "muscle pumping," pressure in the veins of the foot in a relaxed standing posture falls as soon as the subject begins walking.[64] The effectiveness of this mechanism arises from the presence of valves directed toward the heart at intervals along the course of the deep veins and from the low resistance to forward flow, as contrasted with the resistance to backflow through the small veins and venules. Competent valves prevent backflow in periods between muscle contractions, whereas the relation between upstream and downstream resistances favors flow toward the heart when vessels are externally compressed, even in those veins that lack valves.[14] By this process, skeletal muscle contributes toward the energy expended in circulating the blood, and it thus adds to the work done by the heart. The magnitude of this contribution in various forms of exertion merits further investigation.

Venoconstriction is another mechanism for adjusting to the hydrostatic effects of the erect posture, in addition to the "muscle pump" and the physical limitation of the extravascular spaces. Evidence of increased venomotor tone, and hence decreased venous distensibility, on standing or being tilted has been found in the forelimbs[84] as well as the legs.[62] In the limbs, at least, this is part of a generalized vasoconstriction, because the total vascular resistance of the legs and forearms[24] increases when standing quietly. The involvement of the sympathetic nervous system in this response is demonstrated by the failure of such vasoconstriction to occur with tilting in sympathectomized patients.[24]

Increased heart rate in the erect position is a further manifestation of increased sympathetic activity. Although the rate is usually 5 to 10 beats/min faster when standing quietly than when supine, stroke volume is 10% to 50% smaller. Cardiac output, as a consequence, is about 20% lower in the erect position.[30,69,78]

The volume of blood that shifts from the upper to the lower portions of the body when standing as a result of the force of gravity and of imperfect circulatory adjustments to it, has not been measured exactly, but it is probably about 500 ml.[11,44,73] Most of the vessels in the thorax and arms are diminished in volume to some extent by this shift, and the blood volume of the heart itself decreases approximately 100 ml.[58] The capillary bed of the lungs decreases in volume by some 30 ml, largely as the result of closure of many channels at the apices of the lungs; total pulmonary blood volume diminishes by an unknown amount, probably not more than 150 ml, or one third of its reclining value; and the vascular volume of the arms decreases significantly.[84] Transudation of fluid from the capillaries of the lower extremities leads to an increased extracellular fluid volume in that region, manifested by an increase of 60 to 120 ml in total volume of the lower legs after 40 min of standing,[81] together with a decrease of about 15% in total plasma volume.[80]

Postural hypotension and fainting. The effectiveness of all the compensatory mechanisms just discussed varies widely in different individuals. Passive tilting, head upward, on a tilt table is a severe stress that commonly leads to fainting in healthy persons, but some subjects will compensate adequately for long periods of time, whereas others lose consciousness within a few minutes. Standing motionless can have the same result, but the effectiveness of muscle pumping can be exploited by small movements of the foot and calf muscles, a practice learned by experienced soldiers. To some extent individual differences in tolerance can be traced to differences in strength of the vasomotor response, particularly venoconstriction,[62] sometimes accompanied by differences in intramuscular pressure.[59] Fainting brought on by tilting or quiet standing, like that induced by hemorrhage or emotional reactions, is characterized by dizziness, blurring or loss of vision, and unconsciousness. The arterial blood pressure and pulse rate fall suddenly, while the cardiac output may show little initial change.[5] Despite the fall in arterial pressure, blood flow to skeletal muscle increases, and Barcroft's observation that this flow is lessened by local nerve block was the primary evidence for the existence of sympathetic vasodilator fibers to muscle.[1] The normal vasoconstrictor response to standing is thus replaced by a marked vasodilatation in muscle as well as in the splanchnic and renal beds.[5] The resulting fall in arterial perfusion pressure in the absence of any increase in cardiac output reduced cerebral blood flow and thus accounts for the characteristic symptoms and loss of consciousness. Ordinarily, fainting of this kind is preceded by such signs of excessive autonomic activity as sweating and pallor due to cutaneous vasoconstriction. Because this sequence represents a failure of vasomotor compensation and the bradycardia suggests a dominance of vagal influence on the heart, it is sometimes termed *vasovagal* fainting. The response to pharmacologic agents that induce vasodilatation or block sympathetic activity is somewhat similar. Blood pressure can usually be maintained in the supine position after the administration of such drugs, but syncope occurs on standing[83] unless cuffs or tight bindings are applied around the legs.

Exercise

The cardiovascular response to the demands of muscular exertion is one of the most common and essential physiologic adaptations and as such has intrigued investigators for many years. Studies of the cardiac output in exercise, for example, stretch from the work of Zuntz and Hagemann in 1898 on the horse[85] to the present-day application of cardiac catheterization and indicator-dilution methods to exercising human subjects.* It has long been apparent that the output of the heart increases with exercise, but reliable quantitative information about this increment was delayed by the technical difficulty of making the necessary measurements, and many questions about the underlying mechanisms are still unanswered.

One factor complicating the cardiac response to exercise is the effect of posture on cardiac output. Stroke volume and cardiac output diminish on rising from a supine position and standing quietly, as discussed before. Moderate exercise in the erect position, as in walking on a treadmill, increases stroke volume to about the resting supine level.[19,30,78] When the effects of exercise in erect and supine positions are compared at equal levels of oxygen consumption, cardiac output is smaller in the erect position.[69] In either position, cardiac output increases almost linearly with oxygen consumption,[38] and a similar relationship obtains in the dog[16] and horse.[12]

Cardiac output can be increased by raising the heart rate or by increasing stroke volume. With mild exercise, increased rate is usually the dominant change; with increasingly severe exertion, tachycardia and increased stroke both contribute to the elevation of cardiac output. The evidence now available suggests that individuals differ, both naturally and as a result of training, in the way these two variables are utilized to increase total blood flow. Trained athletes, for example, show much greater increments in stroke than in rate at all levels of exercise. There may well be species differences in this respect, since increases in rate appear to play a more prominent part in dog than in man. In the exercising dog the increase in rate greatly exceeds that in stroke volume, but this relationship can be altered at will by atropinization or infusion of catecholamines.[77] Experiments in dogs in which the heart has been

---

*See references 3, 30, 34, 38, and 69.

totally denervated[32,37] and other experiments in which heart rate is controlled artificially[79] further suggest that stroke volume can be used to raise cardiac output during exercise when the mechanism for increasing rate is no longer available.

Although the operation of Starling's "law of the heart" (Chapter 37) in vivo has been convincingly demonstrated, there is no general agreement on the extent to which it participates in the exercise response. At the onset of exercise the blood squeezed from the peripheral veins by the pumping action of the skeletal muscles may increase the filling pressure of the right ventricle and thus bring the Starling mechanism into play. The changes in central venous pressure that occur with exercise, however, are small and highly variable; this fact, together with the observation that ventricular volume (and presumably myocardial fiber length) does not increase with exercise,[56,72] has led some investigators to conclude that Starling's law is of little significance in this circulatory adjustment, and they emphasize instead the importance of control mechanisms that operate through the central nervous system. The cardiovascular reaction to exercise is an integration of many kinds of response, however, and the evidence now at hand is not sufficient to rule out the intrinsic myocardial mechanism related to fiber length as one component among others.

In the first 1 to 2 min after the beginning of exercise, cardiac output and arteriovenous oxygen difference reach a constant level that is maintained for at least 10 min under ordinary laboratory conditions.[38,55] The output that can be attained with maximal exertion varies with different individuals, but most normal subjects can double their resting cardiac output, and many can reach levels of 15 to 25 L/min for short periods of time. Arterial oxygen saturation is little affected by moderate exercise, but the venous oxygen saturation falls in proportion to the oxygen uptake. Arteriovenous oxygen difference thereby increases with increasing oxygen consumption, rising from about 4 vol% at rest to the neighborhood of 12 vol% at an oxygen uptake of 1,200 (ml/min)/m². The increased demand of the tissues for oxygen is thus met by extracting a larger proportion of the oxygen from each unit volume of blood as well as by delivering a larger volume of blood per minute.

When exercise stops, the output, arteriovenous oxygen difference, oxygen uptake, and pulse rate do not immediately return to resting levels but decline gradually over a period of 2 to 5 min, depending in part on the severity of the exer-

Table 40-3. Distribution of cardiac output in normal human subject at rest and during light, strenuous, and maximal exercise that has continued for 10 min*

	Blood flow (ml/min)			
Circulation	Rest	Light exercise	Strenuous exercise	Maximal exercise
Splanchnic	1,400	1,100	600	300
Renal	1,100	900	600	250
Cerebral	750	750	750	750
Coronary	250	350	750	1,000
Skeletal muscle	1,200	4,500	12,500	22,000
Skin	500	1,500	1,900	600
Other organs	600	400	400	100
Cardiac output	5,800	9,500	17,500	25,000
Oxygen uptake (ml/min)/m²	140	400	1,200	2,000

*From Wade and Bishop.[12]

cise.[38] This recovery period represents repayment of an *oxygen debt* incurred during exercise, and it indicates that anaerobic sources of energy were mobilized as a temporary expedient. Ventilation, which is slower than cardiac output to reach a steady state after the beginning of exercise, also takes longer to return to the preexercise level.

The greatest part of the increased output of the heart during exercise goes to the working skeletal muscles, as would be expected, and the blood flow to muscles may increase 20-fold. In man, the distribution of flow to other regions during exercise follows the general patterns summarized in Table 40-3. Splanchnic[22] and renal[31] blood flows decrease in exercising human subjects, although not in trained dogs.[76] Coronary blood flow increases in accordance with the extra work demanded of the heart, whereas cerebral flow remains constant. Initial vasoconstriction in the cutaneous vessels is replaced after a few minutes by dilatation and increased circulation to the skin,[20] increasing the ability to dissipate heat.

Effects of training.[4] Trained athletes frequently have relatively slow pulse rates at rest (44 to 60 beats/min) and a large stroke volume.[21,48] These individuals increase their heart rate with exercise to a much smaller extent than untrained subjects, relying largely on increased stroke to raise cardiac output, and obtaining a significant part of this increased stroke by reducing the residual diastolic volume. Lower heart rates and larger stroke volumes for a given work load after a period of training have been demonstrated in dogs as well as in man.[49] The bradycardia is the result of vagal action on the sinus node, but its origin is otherwise unexplained. These differences in rate and stroke volume, together with a slight increase in total blood volume[49,61] and elevation of maximum oxygen uptake and cardiac output,[40] are the only differences in circulatory function between trained and untrained subjects that have been unequivocally established, despite numerous investigations over a long period.[21,48] Neither resting cardiac output nor the output reached at specific levels of oxygen consumption are altered by training,[40] although the trained subject can continue at a higher level for longer periods of time. Differences in metabolism have been suspected[21] but never proved. The individual differences in performance that are a matter of common experience in competitive athletics and physical work certainly suggest that some counterpart exists in the physiology of the circulation, but their source remains obscure.

Effects of temperature

Exposure to a hot environment has a number of effects on the circulation, most of them traceable to the cutaneous vasodilatation that constitutes one of the mechanisms for heat regulation. Along with a greater blood flow to the skin, the cardiac output and resting pulse rate increase, systolic arterial pressure rises moderately, and diastolic pressure falls. Cutaneous blood flow approximately doubles as temperature rises from 35° to 45° C. Cardiac output rises only moderately, perhaps 10% to 20%, as temperature rises from 20° to 37° C.[29] The intensity of the initial adjustment to high temperatures varies greatly in different individuals, as does the effectiveness of acclimitization with time. Burch and Hyman[28] observed increases in output of 5 to 20 L/min at a temperature of 43.9° C and at the uncomfortably high relative humidity of 86%, the smallest increment occurring in a man who was accustomed to working at this high temperature. Changes in arterial pressure may amount to only a few millimeters of mercury, or the diastolic level may fall by 40 mm Hg or more, and postural hypotension or syncope is common. All these hemodynamic changes are consistent with an enlargement of the capacity of the vascular system secondary to dilatation of the more superficial vessels. With continued exposure, as by residence in tropical climates, the tachycardia, elevation of output, and other circulatory responses become less marked, but remain at levels different from those found in the same individual in temperate regions. The physiologic basis

for such acclimatization has yet to be discovered; the small increase in blood volume that occurs with time[17] seems an inadequate explanation.

Cold environments have just the opposite effects on the circulation. With decreasing temperature, cardiac output falls, heart rate slows, and there is peripheral vasoconstriction, with diminished blood flow to the extremities. The acclimatization that comes with continued exposure is probably related to a lessening of the vasoconstriction, resulting in more blood flow to the limbs and to a small decrease in blood volume.[17] Eskimos and some other populations continually exposed to very low temperatures not only have greater blood flow to the extremities than do nonnatives at equivalent temperatures, but they function at higher basal metabolic rates.[27]

Acceleration and weightlessness

In human subjects carried by aircraft or other vehicles moving with a changing velocity the effects of acceleration on the cardiovascular system are closely related to those of changing from the supine to the erect posture. The gravitational force to which all objects at the earth's surface are exposed is proportional to their mass and the gravitational constant of acceleration for this planet; these conditions are conventionally adopted as the standard for comparison, and such objects are said to be exposed to an acceleration of 1 gravity *(g)* vectorially directed toward the center of the earth. The action of gravitational stress on the body depends on both the magnitude of the acceleration and its direction with respect to the long axis of the body. This fact is of practical importance in aviation, since the *positive g* accompanying a sharp pullout from a dive can have such profound effect in shifting blood away from the head that loss of vision and consciousness ("blackout") follow. Conversely, the oppositely directed *negative g* that lifts the pilot from his seat as he levels out from a steep climb may produce similar end results ("redout") by overcongestion of the cerebral circulation.[6] Rotation at 1 *g* in a human centrifuge, with the long axis of the body oriented radially and the feet toward the periphery, duplicates in essence the effects of the erect posture. With increasing acceleration the pulse rate rises, often reaching 130 to 160 beats/min at 3 to 4 *g*. Cardiac output tends to decrease, but the change is surprisingly small up to 4 *g*.[57] The baroreceptors are obviously responsible for at least some of these circulatory reactions; when the body is oriented so that acceleration tends to displace blood toward the head, the heart rate slows, and this response has been shown to depend on the integrity of the vagus and carotid sinus nerves.[52] The initial effects of positive gravity on arterial blood pressure differ in vessels above and below the heart; pressure rises in the iliac artery, for example, and falls in the carotid artery.[54]

The beginnings of space travel emphasize the vulnerability of the cardiovascular system to gravitational stress. In addition to the tremendous acceleration experienced by astronauts at the time of launching, the effects of which are minimized by orienting the long axis of the body at right angles to the direction of travel, orbital or space flights place the ship and its passengers in a state of "weightlessness." This does not relieve the heart of the work of moving blood through the circulation, since the blood still has inertia, but it does eliminate hydrostatic effects. The immediate effect of "weightlessness" on the circulation is therefore much like that of assuming the supine position. The long-term effects in manned orbital and lunar flights thus far have been remarkably small, although a moderate reduction in total blood volume and a measurable alteration in the cardiovascular responses to changes in posture have been observed.[2]

REFERENCES
General reviews

1. Barcroft, H., and Swan, H. J. C.: Sympathetic control of human blood vessels, London, 1953, Edward Arnold & Co.
2. Berry, C. A.: Medical experience in manned space flight. In Randel, H. W., editor: Aerospace medicine, ed. 2, Baltimore, 1970, The Williams & Wilkins Co.
3. Bevegård, B. S., and Shepherd, J. T.: Regulation of the circulation during exercise, Physiol. Rev. **47:**178, 1967.
4. Clausen, J. P.: Effect of physical training on cardiovascular adjustments to exercise in man, Physiol. Rev. **57:** 779, 1977.
5. Edholm, O. G.: Physiological changes during fainting. In Wolstenholme, G. E. W., editor: Visceral circulation, Boston, 1953, Little, Brown & Co.
6. Gauer, O. H., and Zuidema, G. D.: Gravitational stress in aerospace medicine, Boston, 1961, Little, Brown & Co.
7. Master, A. M., Garfield, C. I., and Walters, M. B.: Normal blood pressure and hypertension, Philadelphia, 1952, Lea & Febiger.
8. McDonald, D. A.: Blood flow in arteries, ed. 2, London, 1974, Edward Arnold, Ltd.
9. Pickering, G. W.: High blood pressure, ed. 2, London, 1968, J. & A. Churchill, Ltd.
10. Shepherd, J. T., and Vanhoutte, P. M.: Veins and their control, Philadelphia, 1975, W. B. Saunders Co.
11. Sjöstrand, T.: Volume and distribution of blood and their significance in regulating the circulation, Physiol. Rev. **33:**202, 1953.
12. Wade, O. L., and Bishop, J. M.: Cardiac output and regional blood flow, Oxford, 1962, Blackwell Scientific Publications.

Original papers

13. Agustsson, M. H., Bicoff, J. P., and Arcilla, R. A.: Hemodynamic studies in fifty-two normal infants and children, Circulation **28:**683, 1963.
14. Alexander, R. S.: Influence of the diaphragm upon portal blood flow and venous return, Am. J. Physiol. **167:** 738, 1951.
15. Andres, R., et al.: Measurement of blood flow and volume in the forearm of man; with notes on the theory of indicator-dilution and on production of turbulence, hemolysis, and vasodilation by intra-vascular injection, J. Clin. Invest. **33:**482, 1954.
16. Barger, A. C., et al.: Regulation of the circulation during exercise: cardiac output (direct Fick) and metabolic adjustments in the normal dog, Am. J. Physiol. **184:**613, 1956.
17. Bass, D. E., and Henschel, A.: Responses of body fluid compartments to heat and cold, Physiol. Rev. **36:**128, 1956.
18. Berkson, J., and Boothby, W. M.: Studies of the energy of metabolism of normal individuals: a comparison of the estimation of basal metabolism from a linear formula and "surface area," Am. J. Physiol. **116:**485, 1936.
19. Bevegård, S., Holmgren, A., and Jonsson, B.: The effect of body position on the circulation at rest and during exercise, with special reference to the influence on the stroke volume, Acta Physiol. Scand. **49:**279, 1960.
20. Bishop, J. M., et al.: The blood flow in the human arm during supine leg exercise, J. Physiol. **137:**294, 1957.
21. Bock, A. V., et al.: Studies in muscular activity. III. Dynamical changes occurring in man at work, J. Physiol. **66:**136, 1928.
22. Bradley, S. E.: Variations in hepatic blood flow in man during health and disease, N. Engl. J. Med. **240:**456, 1949.
23. Brandfonbrener, M., Landowne, M., and Shock, N. W.: Changes in cardiac output with age, Circulation **12:**557, 1955.
24. Brigden, W., Howarth, S., and Sharpey-Schafer, E. P.: Postural changes in the peripheral blood-flow of normal subjects with observations on vasovagal fainting reactions as a result of tilting, the lordotic posture, pregnancy and spinal anesthesia, Clin. Sci. **9:**79, 1950.
25. Brotmacher, L., and Deuchar, D. C.: The systemic blood flow in congenital heart disease, with an examination of the validity of the cardiac index, Clin. Sci. **15:** 441, 1956.
26. Brotmacher, L., and Fleming, P.: Cardiac output and vascular pressures in ten normal children and adolescents, Guy's Hosp. Rep. **106:**268, 1957.
27. Brown, G. M., et al.: The circulation in cold acclimatization, Circulation **9:**813, 1954.
28. Burch, G. E., and Hyman, A.: Influence of a hot and humid environment upon cardiac output and work in normal man and in patients with chronic congestive heart failure at rest, Am. Heart J. **53:**665, 1957.
29. Carlsten, A., Gustafson, A., and Werkö, L.: Hemodynamic influence of warm and dry environment in man with and without rheumatic heart disease, Acta Med. Scand. **169:**411, 1961.
30. Chapman, C. B., Fisher, J. N., and Sproule, B. J.: Behavior of stroke volume at rest and during exercise in human beings, J. Clin. Invest. **39:**1208, 1960.
31. Chapman, C. B., et al.: The effect of exercise on renal plasma flow in normal male subjects, J. Clin. Invest. **27:**639, 1948
32. Cooper, T., et al.: Chronic extrinsic cardiac denervation by regional neural ablation, Circ. Res. **9:**275, 1961.

33. Davies, D. F., and Shock, N. W.: Age changes in glomerular filtration rate, effective renal plasma flow, and tubular excretory capacity in adult males, J. Clin. Invest. **29:**496, 1950.

34. Dexter, L., et al.: Effect of exercise on circulatory dynamics of normal individuals, J. Appl. Physiol. **3:**439, 1951.

35. Dittmer, D. S., and Grebe, R. M., editors: Handbook of circulation, Philadelphia, 1959, W. B. Saunders Co.

36. Dodge, H. T., Hay, R. E., and Sandler, H.: An angiocardiographic method for directly determining left ventricular stroke volume in man, Circ. Res. **11:**739, 1962.

37. Donald, D. E., and Shepherd, J. T.: Responses to exercises in dogs with cardiac denervation, Am. J. Physiol. **205:**393, 1963.

38. Donald, D. W., et al.: The effect of exercise on the cardiac output and circulatory dynamics of normal subjects, Clin. Sci. **14:**37, 1955.

39. Doyle, J. T., Observations on the circulation of domestic cattle, Circ. Res. **8:**4, 1960.

40. Ekblom, B., et al.: Effect of training on circulatory response, J. Appl. Physiol. **24:**518, 1968.

41. Folse, R., and Braunwald, E.: Determination of fraction of left ventricular volume ejected per beat and of ventricular end-diastolic and residual volumes, Circulation **25:**674, 1962.

42. Frank, O.: Der Puls in den Arterien, Z. Biol. **46:**441, 1905.

43. Franklin, D. L., Van Citters, R. L., and Rushmer, R. F.: Balance between right and left ventricular output, Circ. Res. **10:**17, 1962.

44. Fries, E. D., et al.: The collapse produced by venous congestion of the extremities or by venesection following certain hypotensive agents, J. Clin. Invest. **30:**435, 1951.

45. Goetz, R. H., et al.: Circulation of the giraffe, Circ. Res. **8:**1049, 1960.

46. Gottschalk, C. W., and Mylle, M.: Micropuncture study of pressures in proximal tubules and peritubular capillaries of the rat kidney and their relation to ureteral and renal venous pressures, Am. J. Physiol. **185:**430, 1956.

47. Haddy, F. J.: Vasomotion in systemic arteries, small vessels, and veins determined by direct resistance measurements, Minn. Med. **41:**162, 1958.

48. Henderson, Y., Haggard, H. W., and Dolley, F. S.: The efficiency of the heart, and the significance of rapid and slow pulse rates, Am. J. Physiol. **82:**512, 1927.

49. Holmgren, A., et al.: Effect of training on work capacity, total hemoglobin, blood volume, heart volume and pulse rate in recumbent and upright positions, Acta Physiol. Scand. **50:**72, 1960.

50. Holt, J. P., et al.: Left ventricular function in mammals of greatly different size, Circ. Res. **10:**798, 1962.

51. Jegier, W., et al.: Cardiac output and related hemodynamic data in normal children and adults, Can. J. Biochem. Physiol. **39:**1747, 1961.

52. Jongbloed, J., and Noyons, A. K.: Der Einfluss von Bieschleunigungen auf den Kreislaufapparat, Arch. Gesamte Physiol. **233:**67, 1934.

53. Landis, E. M.: Microinjection studies of capillary blood pressure in human skin, Heart **15:**209, 1929.

54. Lawton, R. W., et al.: Arterial blood pressure responses to G forces in the monkey. I. Sinusoidal positive G, J. Aviation Med. **29:**97, 1958.

55. Levy, A. M., Tabakin, B. S., and Hanson, J. S.: Cardiac output in normal men during steady-state exercise utilizing dye-dilution technique, Br. Heart J. **23:**425, 1961.

56. Liljestrand, G., Lysholm, E., and Nylin, G.: The im-
mediate effects of muscular work on the stroke and heart volume in man, Skand. Arch. Physiol. **80:**265, 1938.

57. Lindberg, E. F., et al.: Measurement of cardiac output during headward acceleration using the dye-dilution technique, Aerospace Med. **31:**817, 1960.

58. Linderholm, H., and Strandell, T.: Heart volume in the prone and erect positions in certain heart cases, Acta Med. Scand. **162:**247, 1958.

59. Mayerson, H. S., and Burch, G. E.: Relationships of tissue (subcutaneous and intramuscular) and venous pressures to syncope induced in man by gravity, Am. J. Physiol. **128:**258, 1940.

60. Mellander, S., and Johannson, B.: Control of resistance, exchange, and capacitance functions in the peripheral circulation, Pharmacol. Rev. **20:**117, 1968.

61. Oscai, L. B., Williams, B. T., and Hertig, B. A.: Effect of exercise on blood volume, J. Appl. Physiol. **24:**622, 1968.

62. Page, E. B., et al.: Reflex venomotor activity in normal persons and in patients with postural hypotension, Circulation **11:**262, 1955.

63. Patel, D. J., et al.: Pressure-flow relationships in the ascending aorta and femoral artery of man, J. Appl. Physiol. **20:**459, 1965.

64. Pollack, A. A., and Wood, E. H.: Venous pressure in the saphenous vein at the ankle in man during exercise and changes in posture, J. Appl. Physiol. **1:**649, 1949.

65. Prec, K. J., and Cassels, D. E.: Dye dilution curves and cardiac output in newborn infants, Circulation **11:**789, 1955.

66. Rapaport, E., et al.: Right ventricular volumes in patients with and without heart failure, Circulation **31:**531, 1965.

67. Reedy, T., and Chapman, C. B.: Measurement of right ventricular volume by cineangiofluorography, Am. Heart J. **66:**221, 1963.

68. Reeves, J. T., et al.: Cardiac output in normal resting man, J. Appl. Physiol. **16:**276, 1961.

69. Reeves, J. T., et al.: Cardiac output response to standing and treadmill walking, J. Appl. Physiol. **16:**283, 1961.

70. Remington, J. W., and Wood, E. H.: Formation of peripheral pulse contour in man, J. Appl. Physiol. **9:**433, 1956.

71. Rowe, G. G.: The nitrous-oxide method for determining coronary blood flow in man, Am. Heart J. **58:**268, 1959.

72. Rushmer, R. F., et al.: Continuous measurements of left ventricular dimensions in intact, unanesthetized dogs, Circ. Res. **2:**14, 1954.

73. Sjöstrand, T.: The regulation of the blood distribution in man, Acta Physiol. Scand. **26:**312, 1952.

74. Starr, I.: Studies made by simulating systole at necropsy. X. State of peripheral circulation in cadaver preparations, J. Appl. Physiol. **11:**174, 1957.

75. Taylor, H. L., and Tiede, K.: A comparison of the estimation of the basal cardiac output from a linear formula and the cardiac index, J. Clin. Invest. **31:**209, 1952.

76. Van Citters, R. L., and Franklin, D.: Cardiovascular performance of Alaska sled dogs during exercise, Circ. Res. **24:**33, 1969.

77. Wang, Y., Marshall, R. J., and Shepherd, J. T.: Stroke volume in the dog during graded exercise, Circ. Res. **8:**558, 1960.

78. Wang, Y., Marshall, R. J., and Shepherd, J. T.: The effect of changes in posture and of graded exercises on stroke volume in man, J. Clin. Invest. **39:**1051, 1960.

79. Warner, H. R., and Tortono, A. F.: Regulation of car-

diac output through stroke volume, Circ. Res. **8:**549, 1960.

80. Waterfield, R. L.: The effects of posture on the circulating blood volume, J. Physiol. **72:**110, 1931.

81. Waterfield, R. L.: The effect of posture on the volume of the leg, J. Physiol. **72:**121, 1931.

82. Weinman, D. T., et al.: Skeletal clearance of Ca^{47} and Sr^{85} and skeletal blood flow in dogs, Proc. Staff Meet. Mayo Clin. **38:**559, 1963.

83. Wilkins, R. W., Haynes, F. W., and Weiss, S.: The role of the venous system in circulatory collapse induced by sodium nitrite, J. Clin. Invest. **16:**85, 1937.

84. Wood, J. E., and Eckstein, J. W.: A tandem forearm plethysmograph for study of acute responses of the peripheral veins of man: the effect of environmental and local temperature change, and the effect of pooling blood in the extremities, J. Clin. Invest. **37:**41, 1958.

85. Zuntz, N., and Hagemann, O.: Stoffwechsel des Pferdes bei Ruhe und Arbeit, Landwirtschaftliche Jahrbücher **27:** 371, 1898.

41

WILLIAM R. MILNOR

Autonomic and peripheral control mechanisms

The cardiovascular system maintains an adequate circulation of blood to all parts of the body under a variety of external conditions and can alter regional blood flows to meet a wide range of demands. During muscular exercise, for example, not only does the total cardiac output increase, but vascular adjustments direct most of this increment to the exercising muscles. The physiologic control system that accomplishes these adaptations to changing circumstances has three major subdivisions: (1) the sensors, or receptors, that monitor the state of the circulation; (2) the cells in the CNS that receive afferent signals from receptors and generate appropriate efferent impulses; (3) the effector tissues of the heart and blood vessels. Two important effectors, the myocardium and the cardiac pacemaker, have been considered in Chapters 37 and 38. This chapter is concerned with the efferent functions of the autonomic nerves to the cardiovascular system and with vascular smooth muscle, which is the effector in blood vessels. Cardiovascular receptors, the contributions of the CNS, and unified operation of the control system as a whole are treated in Chapter 42.

NEURAL CONTROL

Neural control of blood vessels is exerted through the autonomic nervous system, including both its sympathetic and parasympathetic divisions (Chapter 33). Sympathetic postganglionic fibers whose cell bodies lie in the thoracic paravertebral ganglia travel to the heart through a maze of "cardiac nerves," most of which also contain preganglionic parasympathetic fibers and sensory fibers. Impulses arriving through these nerves alter the rate of firing of pacemaker cells in the sinoatrial node, the velocity of conduction through the myocardium and specialized conducting tissues, and the contractility of the myocardium[10,51] (p. 997). Nerve fibers originating in the right and the left sides of the nervous system apparently have somewhat different distributions within the heart. Stimulation of the right stellate ganglion frequently causes an increase in heart rate with relatively little change in ventricular function, whereas increased contractility is the most striking effect of left stellate stimulation. Vagal stimulation slows the heart rate and conduction velocity in the atrioventricular node, but the effects on rate predominate with right vagal stimulation, and effects on conduction predominate with left vagal stimulation. These differences are usually quantitative rather than qualitative, indicating a considerable overlap in distribution of nerve fibers; there are also marked individual differences.

The nature of the terminal innervation of the heart is still uncertain, but there is strong evidence for direct motor innervation of mammalian cardiac cells.[10] The finding of significant amounts of catecholamines throughout the myocardium and their disappearance 3 to 4 days after total denervation of the heart provides neurochemical evidence for such innervation; there is morphologic evidence as well. Axons and very fine nerve terminals have been demonstrated on the myocardium[50] and also nerve cells thought to represent peripheral ganglia. This last finding is consistent with recent demonstrations that vagal impulses directly influence ventricular contractility,[24] although the ventricles were long believed to be devoid of parasympathetic innervation. The possibility that a limited number of adrenergic postganglionic cell bodies may lie within the myocardium has not yet been ruled out.

Branches of the autonomic nervous system innervate almost all blood vessels, but the density and function of this innervation varies greatly in different organs and in different types of vessels.[37] Norepinephrine is stored in the nerve endings, and its release is triggered by the arrival of a nerve impulse. The termination of adrenergic vasomotor nerves is a fine plexus that lies just outside the medial layers of the vascular wall, except in the case of relatively large arteries such as the aorta, pulmonary artery, and their major branches; in these the terminal fibers may extend into the outer half of the media. Most of the vascular smooth muscle cells are consequently a great distance (by neurophysiologic standards) from the nearest nerve endings. Even in the outer layers of the media the separation is rarely less than 1,000Å,[13] so that diffusion of the chemical transmitter must play a larger role in the activation of vascular smooth muscle than in other muscles. Considering the long diffusion distances and the prompt local uptake of much of the released transmitter, it is possible that neurally released chemicals never reach the inner layers of these muscle cells. Stimulation of a vasomotor nerve can nevertheless lead to contraction of smooth muscle cells a considerable distance from the nerve terminals, indicating that excitation can spread by cell-to-cell propagation. In at least some blood vessels a limited number of smooth muscle cells act as spontaneous "pacemakers," generating action potentials that can be modified by nerve impulses but that continue even in the absence of innervation.[13]

The great majority of sympathetic nerves act on the cardiovascular system by the release of norepinephrine at their postganglionic terminals. In all parasympathetic and in some sympathetic nerves, acetylcholine (ACh) is the chemical transmitter between postganglionic nerve endings and the heart or blood vessels. A few exceptions to the rules in the form of vasomotor fibers that are neither adrenergic nor cholinergic have been reported,[6] but these are still under investigation.

VASOMOTOR ACTIVITY

Efferent nerve impulses play an important part in vascular control, but there are other factors of equal importance. Bozler's "single-unit" and "multi-unit" types of smooth muscle (Chapter 4) are both represented in vascular walls, and some muscle cells may be able to assume the properties of either type, depending on local conditions. The relatively direct innervation of the outer layers of the vascular media suggests that muscle cells in that region are probably of the multi-unit type, whereas the absence of innervation elsewhere in the media points to single-unit properties. The spontaneous electrical and mechanical activity of smooth muscle in some vessels (Fig. 41-1) suggests that a few cells in these single-unit regions can act as pacemakers, generating a response that spreads from one cell to the next.[13] The frequency of discharge of these pacemakers can be altered by stimulation of local vasomotor nerves, by catecholamines, and by physical

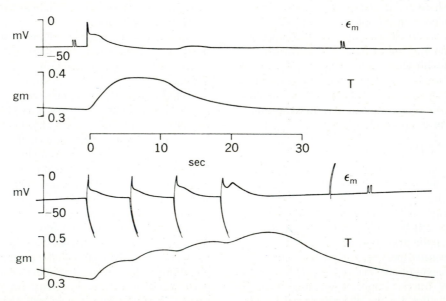

Fig. 41-1. Simultaneous records of transmembrane potential, ϵ_m, and tension, T, in turtle venous smooth muscle. Top: spontaneous activity. Bottom: summation of mechanical contractions when repetitive electrical stimuli were applied to tissue. Negative deflections preceding action potentials are shock artifacts. Short positive pulses are calibration signals. (From Roddie.[52])

stretching of the wall. As a result, each of these factors influences the net tension developed by the whole population of muscle cells in the vascular wall.[3,12,13]

Vascular tone. Under almost all in vivo conditions the contraction of vascular smooth muscle exerts some degree of tension on the vascular wall; this is called "active" tension. "Passive" tension, by way of contrast, is that resulting only from the stress of transmural pressure, which acts to distend vessels and thus stretch their walls. Vessels in which the active tension is comparatively small, as is usually the case in cerebral and pulmonary vessels, are said to have a low degree of vascular "tone." Strong active tension, on the other hand, is referred to as a high degree of vascular "tone." Vascular tone is the result of smooth muscle contraction, which may be triggered by the arrival of neuronally released transmitters, by cell-to-cell spread of excitation, or by chemical events at the cell membrane.[13,42,55] "Basal tone" of blood vessels refers to the tone when the animal is under basal conditions. "Intrinsic tone" is that generated by the muscle in the absence of neural impulses or vasoactive substances. In each smooth muscle fiber, excitation evokes a mechanical contraction followed by relaxation, and the tension developed is a function of initial fiber length, afterload, catecholamine levels, and the chemical environment.[3] The frequency with which contraction is repeated is determined by the frequency of incoming nerve impulses or by the rate of firing of "spontaneous" pacemakers. In the population of cells that make up the layers of smooth muscle in a blood vessel, contraction goes on asynchronously and at many different rates, leading to summation of the mechanical events or incomplete tetanus (p. 131). The larger the number of muscle cells contracting and the higher the frequency of contractions, the greater will be the resulting active tension in the vascular wall. As in other types of muscle, excitation leads to a shortening of the contractile apparatus, but because of stretching in the series-elastic elements, the muscle as a whole may shorten to a lesser degree or not at all (isometric contraction). In blood vessels the equivalent of isometric contraction occurs if simultaneous increases in active smooth muscle tension and in transmural pressure exactly balance each other, leaving the vessel diameter unchanged.

The hemodynamic effects of vasomotor activity arise from changes in diameter (or cross-sectional area) of the vascular lumen and from changes in the compliance of the vessel wall. An increase in active tension tends to reduce the circumference and hence the diameter of a blood vessel, although the quantitative relationship between muscle shortening and change in circumference is a function of the complex linkage between muscle fibers, elastin, and collagen in the wall (p. 1024). As one would expect from the laws that govern viscous flow in tubes (p. 1018), such "vasoconstriction" raises the vascular resistance, that is, increases the drop in pressure along the vessel per unit of blood flow. In many situations, vasoconstriction is also accompanied by an elevation of pressure proximal to the constricted vessels and a decreased blood flow through them, but these changes depend to a large extent on the performance of the heart and on the resistance of other vascular beds.

It is essential to keep in mind the influence of transmural pressure on vascular caliber when interpreting vasomotor responses. Suppose, for example, that activation of arteriolar smooth muscle and a marked increase in the external work of the heart occurred simultaneously. If the increased cardiac work produced an elevation of arterial and arteriolar pressure, the increase in transmural distending pressure in the arterioles might be high enough to counterbalance (or even outweigh) the efforts of arteriolar smooth muscle to shorten. In that case the arterioles would not actually constrict, although there would be an increase of active tension in their walls and thus a decrease in their compliance. Vascular resistance would ordinarily remain unchanged in this situation, but it would be a mistake to assume from this that no vasomotor response had occurred. In other situations in which smooth muscle activity is constant, moreover, changes in transmural pressure alone can alter vascular caliber. For this reason there are advantages in reserving the terms "vasoconstriction" and "vasodilatation" for changes in vascular caliber that result directly from the activity of vascular smooth muscle. Alterations in active smooth muscle tension can be described in terms of "tone," whether or not there are changes in vascular caliber. Other terms such as "distention" or "collapse" should be applied for passive responses to altered transmural pressure.

Resistance and capacitance vessels. Smooth muscle activation in relatively large vessels has effects qualitatively similar to those in small vessels, but the quantitative results differ radically. Increased active tension in the wall of the aorta or vena cava, for example, produces only a slight decrease in diameter. The resistance in such wide vessels is low in any case, and the change in resistance that follows activation of their smooth muscle is trivial. Even so, a small change in the cross-sectional area of large vessels can have important effects if it extends over a large part of the circulation and thus significantly alters the intravascular volume, or capacity. The hemodynamic consequences of vasomotor activity in the large veins arises from just such changes in capacity. Because a large part of the total blood volume resides in veins, a small but widely distributed decrease in venous diameters can displace several hundred milliliters of blood into other parts of the circulation. This affects the filling pressure of the ventricles, and a new dynamic equilibrium is established throughout the circulation.

Because of the different consequences of vasomotor activity in small and in large blood vessels, the former have been designated "resistance vessels" and the latter "capacitance vessels."[49] Arterioles are considered the principal resistance vessels and large veins the capacitance vessels. This is a useful functional distinction, even though the properties of resistance and capacity are distributed in varying degrees throughout the circulatory system. The resistance of large arteries is finite, although very small, and the volume of the microcirculation is not altogether negligible in hemodynamics. Nevertheless, the generalization that small-vessel resistance and large-vein capacity are the two major variables gives a broad and reasonably accurate picture of vascular control. Resistance vessels control the distribution of cardiac output among various organ systems, whereas capacitance vessels contribute to the overall regulation of cardiac output and blood pressure. A third functional class is that of "exchange vessels," in which the transfer of blood gases, ions, and other substances takes place between blood and extracellular fluid. The capillaries and a small part of the adjacent arterioles and venules belong in this category. Transmural pressure in the exchange vessels is one force that controls the passage of materials into and out of the vascular compartment, and this pressure is in turn controlled by vasomotor activity in the vessels upstream and downstream. Arteriolar vasoconstriction tends to reduce, and venular constriction to elevate, transmural pressure in exchange vessels.

This tripartite classification leaves out the factors that govern impedance and pulsatile flow (p. 1027); yet the compliance or "stiffness" of arteries, quite apart from their resistance or volume, is hemodynamically important because it determines the transmission of pulse waves from heart to capillaries (p. 1029). The amplitude and timing of the pressure and flow waves that reach the microcirculation thus depend in part on the compliance of the arterial tree, or what may be called "conduit vessels." Theoretically, increased distensibility of the arterial tree would tend to reduce the amplitude of pulses at the termination of the bed if other factors remained constant, but the presence of reflected waves in vivo makes transmission patterns somewhat complicated. In the pulmonary circulation, at least, stimulation of sympathetic nerves produces much greater changes in pressure and flow pulses than in resistance,[41] suggesting that wave transmission through the large arteries may be regulated independently of small-vessel constriction. How this fits into the general patterns of cardiovascular control has yet to be learned.

NEURAL VASOCONSTRICTION

All known vasoconstrictor nerves belong to the sympathetic division of the autonomic nervous system and act by releasing norepinephrine at their postganglionic terminals. Most of the postganglionic fibers are unmyelinated and belong to the C group of nerve fibers.

Efferent nerves in which an increased frequency of impulses augments active tension in vascular smooth muscle are usually referred to as constrictor nerves. Those in which a higher impulse rate is associated with relaxation of smooth muscle are termed dilator nerves. Strictly speaking, these designations are misnomers. As mentioned earlier, there are times when an increase in transmural pressure may prevent actual constriction of a vessel, so that stimulation of its "constrictor nerves" may simply stiffen the wall of the vessel without changing its caliber. The term "positive inotropic response" would perhaps be a more exact description of the effects of stimulating a constrictor nerve, but the imperfections of the present terminology are sanctioned by long usage, and in most circumstances the meaning is free of ambiguity.

The density of sympathetic constrictor innervation varies widely in different parts of the circulation, although virtually all arteries and veins, of whatever size, receive some fibers. Small arteries and arterioles in skin, skeletal muscle, and the splanchnic bed, for example, receive a relatively large number of sympathetic constrictor fibers, and the adrenergic nerve supply of veins is usually less dense than that of the corresponding arteries[13,28] (Fig. 41-2). Regional differences range from a very high concentration of fibers in the vessels of the skin to a very sparse innervation in the cerebral vessels. The range of vasomotor control (i.e., the difference between maximal

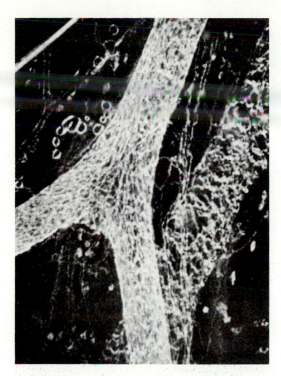

Fig. 41-2. Plexus of adrenergic nerve fibers around small arteries and vein in rat mesentery, as demonstrated by fluorescence method. Arteries are surrounded by dense plexus; network around vein has larger meshes. (From Falck.[29])

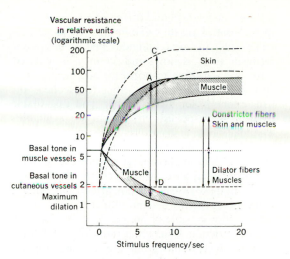

Fig. 41-3. Schematic illustration of range of vasomotor control of blood vessels in terms of frequency of efferent nerve impulses in cat. *A* and *B*, Leg muscles; *C* and *D*, skin of paw. Distal end of cut abdominal sympathetic chains stimulated electrically at frequencies indicated on abscissa. Shaded areas indicate range of results observed in series of experiments. In case of muscle vessels, atropine was given to block effects of cholinergic sympathetic vasodilator fibers; dihydroergotamine was given to block effects of adrenergic sympathetic vasoconstrictor fibers. Note that basal tone of muscle vessels is much greater than that of skin vessels and that stimulation at a given frequency induces greater vasoconstriction in cutaneous than in muscle vessels. (Based on data from Folkow[7] and Celander.[21])

constriction and maximal dilatation) is in most instances directly correlated with the density of innervation.

Vascular smooth muscle response is directly correlated with the frequency of efferent stimuli, and the full range of vasomotor activity can be reproduced by experimental stimulation of sympathetic constrictor nerves at rates from 0 to 10 impulses/sec (Fig. 41-3).

Under normal resting conditions the discharge of impulses through adrenergic constrictor fibers that contributes to the "basal tone" of most blood vessels is presumably very slow, perhaps 1 to 3 impulses/sec (Fig. 41-3). Central nervous modulation of the frequency of these impulses can raise or lower the resistance of small blood vessels, and this constitutes the principal central mechanism for control of the circulation. Lowering impulse frequency below the basal tonic level (release of constrictor tone) reduces the degree of constriction and in effect causes vasodilatation. The resistance of vascular beds can thus be either raised or lowered from the basal state by adjustments in the frequency of impulses delivered through sympathetic constrictor fibers.

The basal level of vasomotor tone is by no means the same in all vascular beds. Basal tone in cutaneous vessels, for example, is much lower than that in the vascular bed of skeletal muscle, as shown diagrammatically in Fig. 41-3. The experimental results in this figure were obtained in the paw of the cat,[21,30] and vascular resistance is used as a measure of relative vasomotor response. The relation between impulse frequency and resistance gives similar curves in both sets of vessels, but there are characteristic quantitative differences. First, the level of resistance in the absence of vasomotor impulses (frequency = 0), which may be attributed to intrinsic or nonneurogenic tone, is higher in the vascular bed of skeletal muscle than in skin. Second, the constrictor effect of any given frequency of stimulation is greater in cutaneous than in muscular vessels. The range of control exerted by sympathetic vasoconstrictors is consequently smaller in muscle than in skin, but the operation of vasodilator fibers in muscle effectively compensates for this difference. Two slightly different modes of vasomotor control are thus exemplified by these vascular beds. In one (cutaneous bed of the paw), intrinsic tone is low and control is exerted through a sensitive response to neurogenic constrictor impulses. In the other (muscular), intrinsic tone is somewhat higher, and vasodilator as well as vasoconstrictor fibers are available. The total range of control is essentially the same in both cases, but the mechanisms involved are different. A third class is typified by the cerebral and myocardial vascular beds (not shown in Fig. 41-3). In these organs the constrictor innervation is meager, remains inactive under resting conditions, and is capable of producing only moderate vasoconstriction at maximal rates of stimulation. Vascular beds in this group are regulated mainly by intrinsic vascular tone and the vasodilator effects secondary to tissue metabolism (p. 1058).

Regional differences in vascular control are even more complex than this threefold classification implies. Some degree of tonic vasoconstrictor discharge under resting conditions is found in the skeletal muscles, the gastrointestinal tract, and some cutaneous areas but probably not in the skin of the human forearm, calf, or forehead. Exposure to cold evokes a typical vasoconstrictor discharge in the latter regions, but vasodilatation in response to body heating is brought about by agents formed as a result of activation of the sweat glands. The renal circulation is another in which tonic vasoconstrictor tone is probably absent under basal conditions, although it may become marked in situations in which the proper distribution of an increased cardiac output must be ensured.

Effects of constrictor denervation. The early discovery that vasodilatation follows interruption of sympathetic constrictor nerves laid the foundation for vasomotor physiology (Barnard and Brown-Séquard, 1851 and 1852), but an observation of equal importance is the failure of this vasodilatation to persist. Vascular tone begins to return a few days after sympathetic denervation of the human hand, for example, and blood flow is often back to normal a few weeks later.[5] This is not due to regeneration of sympathetic nerves, and no completely satisfactory explanation has yet been put forward. Autonomic effectors in general become hypersensitive to catecholamines after denervation, particularly when it is the postganglionic fibers that are interrupted, and this could contribute to the reestablishment of vascular tone, just as it accounts for the intense vasoconstriction sometimes seen in the sympathectomized hand on exposure to cold or to other stimuli that provoke increased secretion from the adrenal medulla. Sensitivity to ACh and a number of other substances is increased at the same time, however, and it is probable that some more specific mechanism will eventually be identified.

The ability of blood vessels to establish some degree of partial constriction, or tone, after denervation may involve the noninnervated layers of vascular smooth muscle that exist in most (if not all) small blood vessels. Folkow et al.[34] have proposed an hypothesis that bears on this phenomenon and on vascular control in general, based on the contrast between more or less direct innervation of the outer medial layers of smooth muscle and the absence of nerve terminals in the inner layers. When regional sympathetic discharge is at a high level, they suggest, the innervated outer multi-unit muscles recruit the inner muscles by cell-to-cell propagation.[42] Alternatively, constriction of the outer sheath may simply "unload" the inner layers. In either case, the state of the vessel is then controlled predominantly by nervous impulses. On the other hand, when sympathetic discharge is relatively low or absent, relaxation of the outer layers would transfer control to the single-unit muscles of the inner layers. The spontaneous activity of these fibers, modulated by local concentrations of norepinephrine and other vasoactive materials, would then provide a certain degree of wall tension, even in a denervated vessel. This attractive theory is consistent with most observations, but its full confirmation awaits further experiment.

NEURAL VASODILATATION

True vasodilator fibers, in which nerve impulses release a transmitter that causes relaxation of vascular smooth muscle, are fewer in number and more limited in their distribution than are the constrictors. Nerves of this kind exist in both the sympathetic and parasympathetic systems, and transmission from postganglionic fiber to smooth muscle is cholinergic in both cases. The principal distribution of sympathetic vasodilator fibers is to the vessels of skeletal muscle. In addition, both sympathetic and parasympathetic dilator nerves supply the external genitalia. Parasympathetic stimulation leads indirectly to vasodilatation in the submandibular salivary gland[38] and tongue,[39] but this response is produced by glandular secretions, not by nerve-generated ACh.

Sympathetic vasodilators. Experimental studies of sympathetic vasodilatation are complicated by the intermixture of constrictor and dilator fibers in most nerve trunks. Stimulation of the cervical sympathetic trunk or the lumbar sympathetic chain activates both pathways, and vaso-constriction usually dominates the response that follows. This difficulty can be circumvented by selectively blocking one function or the other; ergotoxin, for example, blocks the constrictor but not the dilator activity, whereas reserpine depletes norepinephrine stores in nerve terminals so that only cholinergic terminals remain active. Stimulating the areas of the CNS in which sympathetic discharges originate is another useful technique. All these methods have been used, and they mutually confirm the existence of a sympathetic dilator innervation of skeletal muscles.[14] These fibers are not tonically active at rest, but they discharge profusely in at least the initial phase of what has come to be called the "defense reaction."[15] This is an organized response occasioned by situations that are threatening or that evoke intense emotion[18] and can be interpreted teleologically as a preparation for flight, attack, or strenuous effort. A marked dilatation of the muscular vessels is part of this reaction, together with increased cardiac rate and other adjustments. A *reduction* of oxygen uptake by skeletal muscle reported in experiments where the sympathetic vasodilator outflow from the hypothalamus is stimulated[14] seems an inappropriate part of the reaction, and this is surely not what happens when the muscles are contracting actively. During exercise in vivo the oxygen consumption of skeletal muscle increases enormously, and vaso-dilatation is a means of directing a large proportion of the increased cardiac output to the active muscles.

Activity of sympathetic cholinergic fibers is only one of the mechanisms that produce vaso-dilatation in these circumstances, and probably the least important. The direct local action of metabolic products is a much more potent stimulus to vasodilatation (p. 1058). Release of constrictor tone is a contributing factor, but its relatively minor role is indicated by the failure of intensive vasoconstrictor discharge to alter resistance in the muscular vessels during exercise.[45] All these vasodilator effects, whether neurogenic or locally mediated, are limited to the arterioles and do not involve postcapillary vessels.[33]

Sympathetic erector fibers. Sympathetic as well as parasympathetic cholinergic vasodilator fibers innervate the external genitalia.[17] The existence of the sympathetic pathway was indicated as long ago as 1901 when L. R. Muller reported that a dog from which the entire sacral portion of the cord had been removed developed penile erections whenever it became sexually excited by

the presence of a bitch in estrus. This observation was later confirmed in cats by Root and Bard.[53] These vasodilator fibers have their origin in lumbar spinal segments (chiefly L2 to L4), course through the lumbar sympathetic chains, and reach the pelvis by way of the hypogastric nerves. This innervation probably serves in man, as it does in the dog and cat, as a physiologically important adjunct to the parasympathetic erector fibers described in the following paragraphs. Its elimination by abdominal sympathectomy in no way alters overt sexual behavior, but this operation does abolish the emission of seminal fluid because it interrupts the parallel sympathetic motor pathways to the seminal vesicles.

Parasympathetic vasodilators. The parasympathetic fibers supplying the external genitalia are the one group of parasympathetic vasodilators whose existence and function are clearly established. Another set of vessels that receives parasympathetic vasodilator innervation is found in the small pial arteries of the cerebral circulation, but their extent and function are obscure. The pathway of the vasodilator innervation of these arteries begins in the medulla and travels in the facial nerve, so they apparently belong to the cranial division of the parasympathetic system. The pathway has been traced through the geniculate ganglion, the great superficial petrosal nerve, and then along a branch of the latter to the plexus surrounding the internal carotid artery.[23] In cat and monkey, stimulation of the facial nerve near the medulla or of the geniculate ganglion causes an ipsilateral dilatation of arterial vessels of the parietal cortex. Both the extent and functional significance of this vasodilator supply remain obscure; the existence of parasympathetic vasodilator fibers has been suspected in a number of other vascular beds, but the evidence adduced thus far is not convincing.

Secretomotor nerves. Vasodilatation can be produced by stimulation of the parasympathetic nerves that evoke secretion in the glands of the salivary system, tongue, and skin. These are not true vasodilator nerves, however, for the agent directly responsible for vascular dilatation is a substance produced by the secretory cells of the glands, not a neural transmitter. Investigation of this response has been in progress for more than a century, and its history is instructive. In 1858 Bernard found that stimulation of the chorda tympani nerve increased blood flow through the submandibular salivary gland. He noted that stimulation of the cervical sympathetic trunk reduced greatly the flow of blood issuing from the gland and caused it to darken in color, suggesting vasoconstriction, but stimulation of the chorda tympani caused the venous outflow to increase, become pulsatile, and turn arterial in color as if the resistance to flow had been greatly reduced.

Bernard's demonstration of the vasodilator effect of chorda stimulation has been confirmed many times over, and for nearly 100 years the chorda tympani was regarded as a typical example of a nerve containing vasodilator fibers. In 1872 Heidenhain found that the vasodilator action of chorda stimulation was not abolished by doses of atropine that prevented salivary secretion. Because of this difference in the effects of atropine, the concept was developed that the chorda tympani contains separate secretomotor and vasodilator fibers. That this may not be the case was strongly suggested by the observation of Joseph Barcroft (1914) that the oxygen consumption of the atropinized salivary gland is still increased by chorda stimulation, even when no secretion appears. Barcroft therefore rejected the idea of a special vasodilator innervation and attributed the vasodilatation to the local action on the blood vessels of the products of the metabolic activity of the gland cells. His views were not generally accepted, and it has been only recently that they have been shown to be quite correct in principle. The vasodilatation is actually not completely resistant to atropine, but only relatively so, and as the cholinergic nature of postganglionic parasympathetic fibers gradually became evident, this incomplete blockade became difficult to explain.

In 1955 and 1956 Hilton and Lewis[38] presented a large body of evidence to show that no fibers of the chorda tympani (more precisely the intraglandular postganglionic fibers of the chorda) are true vasodilator fibers to the vessels of the sub-mandibular salivary gland of the cat. They confirmed and extended Barcroft's conclusion that the vasodilatation is caused by a material produced as a result of the activation of the gland cells. This material appears whether the gland cells are caused to secrete by stimulation of the chorda tympani or by arterial injection of ACh. The possible existence of noncholinergic vasodilator fibers in this innervation of the gland was apparently ruled out by the finding that injection of botulinus toxin into the gland (which selectively inactivates cholinergic fibers) simultaneously abolishes the secretory and vasodilator effects of chorda stimulation. During brief periods of perfusion of the gland with Locke's solution, chorda stimulation caused the blood-free perfusate (venous effluent) to acquire vasodilating properties, which could be revealed when it was injected into the reestablished arterial inflow of the gland. Since the vasodilator response was not antagonized by atropine or by an antihistamine drug, it was not due to ACh or histamine.

As finally unraveled, the sequence of events initiated by stimulation of the chorda begins with glandular release of an enzyme *(kallikrein)* to the local extracellular spaces (Fig. 41-4). This enzyme acts on a normally occurring globulin *(kininogen)* to produce a polypeptide, *kallidin,* which is rapidly transformed to *bradykinin.* The latter two substances are potent vasodilators, and all the evidence now points to them as mediators of the local vasodilatation that follows neural stimulation of the salivary, glossal, and sweat glands. They also increase capillary permeability, a phenomenon that may well be related to the generation of fluid glandular secretions. To bring the story full circle, recent experiments in which the submandibular gland of the cat was perfused

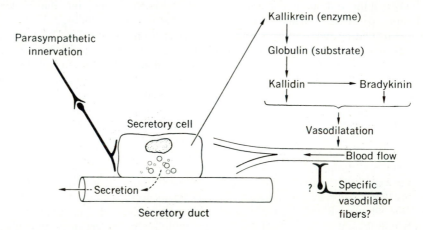

Fig. 41-4. Mechanism of vasodilatation in salivary gland on stimulation of parasympathetic nerve supply. See text for discussion. (After Hilton and Lewis[38]; Folkow and Neil[32] and others.)

with kininogen-free solutions suggest that some parasympathetic cholinergic vasodilator fibers do, after all, function in this region, *in addition* to the bradykinin-generating mechanism.[35]

Axon reflex. The terminal branches of some sensory nerves in the skin and external mucous membranes form the pathway for another localized vasodilatation. Mechanical or chemical irritation of the skin is the stimulus for this response, generating nerve impulses in local sensory fibers that not only travel in the normal direction back toward the spinal cord but also enter nearby branches of the same nerve and travel peripherally ("antidromic conduction") to local cutaneous blood vessels, causing them to dilate. The neurons involved are true afferents, for their cell bodies lie in ganglia of the somatic sensory system, and the response can be evoked by stimulation of the distal end of a cut dorsal spinal root. Pain fibers appear to be the only ones that act in this way.[22] The vasodilating transmitter presumably released from the sensory nerve terminals is unknown but adenosine triphosphate (ATP) has been suspected,[40] among other candidates (p. 401).

This short circuit, or axon reflex, in peripheral sensory branches is the source of one element of the so-called triple response of the skin, which can be evoked by firmly stroking the skin with a needle (p. 400). The reddening of the skin *adjacent* to the scratch (flare) represents arteriolar dilatation initiated by this mechanism. It can be demonstrated for a short time after all nerves to the skin have been cut, but not after the somatic innervation has degenerated (Chapter 13). The same is true of some vasodilator responses to cold.[47]

HUMORAL CONTROL
Epinephrine and norepinephrine

Epinephrine and norepinephrine are vasoactive catecholamines that occupy a central role in vascular reactions. Norepinephrine is one of the two principal neural transmitters, and epinephrine is the major secretion of the adrenal medulla in man and some other species. The secretory cells of the adrenal medulla are in some ways analogous to a sympathetic ganglion, being innervated by sympathetic preganglionic fibers (p. 898). As a result, the CNS controls delivery of catecholamines to vessels by way of the bloodstream as well as by way of vasomotor nerve terminals. The ways in which the two parts of this dual sympatho-adrenal system interact in reflex cardiovascular control have been intensively studied.[21,46]

Epinephrine secretion by the medulla appears to be a much less powerful vascular stimulant than are the impulses in sympathetic vasoconstrictor nerves when the two are compared at their respective physiologic levels. Experiments on the skin and leg muscles of the cat, for example,[21] showed that infusion of catecholamines at even the highest rates of reflex medullary secretion produced much less vasoconstriction than did stimulation of sympathetic constrictors (Fig. 41-5). Norepinephrine produces constriction in both of these vascular beds, as it does in virtually all blood vessels, with the possible exception of the coronaries, but the response to epinephrine in muscle is dose dependent, yielding vasodilatation at very low concentrations and constriction at higher levels (Fig. 41-5). It has been suggested that the dilatation results from the glycogenolytic action of epinephrine and the formation of lactic acid,[7] but this remains uncertain.

Adrenergic receptors. Blood vessels differ in their response to catecholamines, not merely in sensitivity but in whether they constrict or dilate. The reaction of smooth muscle in other organs is equally variable; epinephrine contracts the iris constrictors, for example, but relaxes the rat uterus, although the smooth muscle cells in these structures are anatomically similar. A semblance of order has been imposed on a chaos of such observations by Ahlquist's concept of smooth muscle receptors.[16] The relevant parts of this concept will be summarized only briefly here; further details can be found in recent reviews[1,9] (Chapters 4 and 33).

The contractile state of vascular smooth muscle, it is postulated, can be altered by the combination of agonists such as epinephrine and norepinephrine with receptors that form part of the muscle itself. Whether the agonist is released from nerve terminals or delivered by the bloodstream is irrelevant, as long as it has access to the muscle cell. Receptors that respond to epinephrine or norepinephrine are termed *adrenergic* and are of at least two different pharmacologic types, designated *alpha* and *beta.* This classification is based entirely on functional characteristics; no structures corresponding to these receptors have been identified morphologically, and the "reception" may be accomplished by some chemical molecule rather than a structural component of the cell. The receptor must be considered part of the effector cells (smooth muscle, in the case of blood vessels), however, since denervation does not prevent the response. Experimentally, the receptors are identified by the use of drugs that inhibit or activate one type but not the other.

Vascular smooth muscle that is endowed with

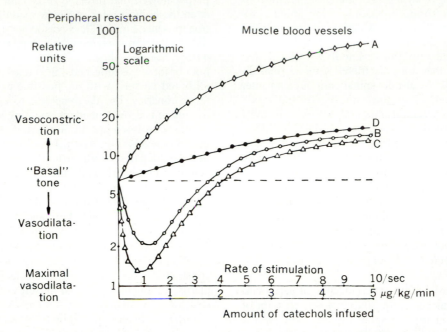

Fig. 41-5. Data summarizing control of decentralized blood vessels of leg muscles in cat. *A,* Stimulation of sympathetic vasoconstrictor fibers (sympathetic vasodilator fibers blocked by atropine); *B,* stimulation of splanchnic innervation of adrenal medulla; *C,* infusion of L-epinephrine; *D,* infusion of L-norepinephrine. Note marked basal tone, vasoconstriction by stimulation of sympathetic constrictors and infusion of L-norepinephrine, and vasodilatation from low concentrations of L-epinephrine or low-intensity stimulation of adrenal medulla. (From Celander.[21])

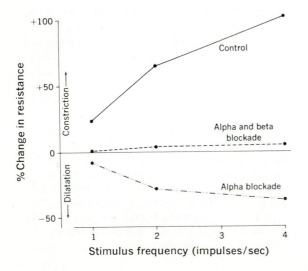

Fig. 41-6. Changes in resistance of vascular bed in perfused dog skeletal muscle in response to stimulation of its sympathetic nerves. Abscissa: Stimulus frequency, a measure of intensity. Ordinate: Change in resistance on stimulation, calculated from changes in perfusion pressure and rate of blood flow. Stimulation in absence of adrenergic blocking agents (control) caused vasoconstriction proportional to intensity of stimulus, presumably because of dominance of alpha receptors. Administration of drug that blocked alpha-adrenergic receptors reversed vascular response, in that sympathetic stimulation now produced vasodilatation (alpha blockade). Prevention of alpha-receptor response unmasked, so to speak, the beta-adrenergic dilatation. When both alpha- and beta-blocking agents were present (alpha and beta blockade), there was little or no response to nerve stimulation. (After Viveros et al.[57])

alpha receptors *contracts* when these receptors are activated. Smooth muscle cells that possess *beta* receptors *relax* on activation of their receptors. (This generalization does not extend to all varieties of nonvascular smooth muscle; relaxation of intestinal smooth muscle is its response to both alpha- and beta-receptor activation.) In many blood vessels (perhaps all), both types exist, and reaction to agents such as epinephrine, which activates both alpha and beta receptors, depends on the relative numbers of each type in the smooth muscle population, and perhaps on their accessibility. Vasoconstriction occurs if alpha receptors predominate, and dilatation occurs if beta receptors prevail. The most striking effect of norepinephrine infusion in vivo is vasoconstriction by virtue of its action on alpha receptors, but it also acts on beta receptors. Low concentrations of norepinephrine have been shown to dilate small coronary arteries,[19] presumably a beta-receptor effect, and vasodilatation has been produced by norepinephrine in skeletal muscle in dogs after alpha blockade.[36] Sympathetic control of veins in the systemic circulation probably involves both alpha and beta receptors.[27,44]

For some time, it was supposed that vascular beta receptors, unlike the alpha type, were not innervated or, to be more exact, were not accessible to neurally released transmitter. Recent experiments, however, have demonstrated sympathetic nerve–mediated beta-adrenergic vasodilatation in the gracilis muscle of the dog[57] (Fig. 41-6).

The differing responses to blood-borne and nerve-generated norepinephrine argue, nevertheless, that vascular beta receptors are normally less subject to nervous control than are alpha receptors. Braunwald and his associates,[36] for example, found that after alpha blockade, norepinephrine infusion of splanchnic vessels or of the perfused hind limb of the dog produced vasodilatation, whereas a reflexly induced sympathetic discharge did not. From this and other observations, they concluded that neuronally released and intravascularly administered norepinephrine did not act on the same population of adrenergic receptors. One may well ask what is meant by the "innervation" of an hypothetical "receptor" that may eventually prove to be an enzyme or substrate rather than a discrete structure. The answer, as indicated by the experiments cited, is that "innervation" in this context refers only to the accessibility of receptors to transmitter released by nerve endings.

Beta-adrenergic receptors are present in the heart as well as in blood vessels. Activation of these receptors in cardiac tissue increases the myocardial force of contraction, increases pacemaker rate, and shortens the duration of systole. These responses are properly called adrenergic because they are produced by norepinephrine. Their classification as beta adrenergic is based on studies with drugs that imitate or block alpha or beta actions on blood vessels. The cardiac effects of norepinephrine are prevented by administration of a beta-blocking agent, for example, but not by alpha-blocking agents.[9] Beta-adrenergic agonists (e.g., the drug isoproterenol) have qualitatively the same cardiac effects as norepinephrine. Pharmacologic classification thus dictates that beta-adrenergic receptors mediate increased contractile force in cardiac muscle, but relaxation in vascular smooth muscle. It is by no means certain, however, that the receptors are identical in the two cases. Some beta-adrenergic blockers, for example, act predominantly on the heart, and others act on the blood vessels. Whether or not alpha-adrenergic receptors are present in physiologically significant amounts in the heart is uncertain at present.

Vascular effects. The hypothesis of adrenergic receptors provides a consistent explanation for many of the vascular responses to epinephrine and norepinephrine, if we assume that both alpha and beta types exist in the smooth muscle layers of many blood vessels and that the relative numbers of each type differ in different vascular beds. This could account for the diversity of responses to physiologic concentrations of epinephrine, which produce vasodilatation in vessels of the heart, skeletal muscle, and liver but vasoconstriction in the kidney, spleen, and mesentery. This difference can be attributed to the predominance of beta receptors in the first group of vascular beds and of alpha receptors in the second. Numeric superiority of one type of receptor is not, however, enough to account for all experimental observations; in some cases there also appear to be differences in receptor *sensitivity*. Norepinephrine is more potent than epinephrine in activating beta receptors in coronary vessels, for example, but in the vessels of skeletal muscle, this relationship is reversed.[56] A further differentiating factor, in addition to numbers and sensitivity of receptors, arises from the influence of catecholamines on the spontaneous "pacemakers" found in the smooth muscles of some blood vessels.[13] For example, norepinephrine raises the frequency of smooth muscle action potentials at the same time that it increases the tension developed in canine mesenteric veins in vitro.[12] In the portal vein, on the other hand, a beta-receptor activator (isoproterenol) increases the frequency of spontaneous firing but *decreases* the mechanical tension in the vessel wall.[43] Clearly, knowledge of the exact ways in which vasoactive agents in general operate on smooth muscle cells to trigger a response is in a state of expansion and flux. Evidence is accumulating to indicate that the concept of adrenergic receptors, which served initially as a unifying hypothesis, needs to be modified substantially. As Ahlquist observes, "When better knowledge of a receptor is obtained, for example, the exact identification of the enzyme or enzyme system involved, the need for receptor vanishes."[1] In the interim the alpha-beta receptor scheme is a useful conceptual framework, but it is to be hoped that the multiplicity of new receptors that has been proposed to explain specific experimental findings will be only temporary expedients, leading eventually to more fundamental physicochemical explanations.[4]

Whatever the basic mechanism of catecholamine action may be, the net effect of intravenous infusion of norepinephrine in man and most laboratory animals is an increase in systemic vascular resistance. Epinephrine, in contrast, usually lowers the peripheral resistance, at the same time altering the relative resistances in different organs so as to redistribute blood flow. The overall response to epinephrine includes actions on the heart as well as on the blood vessels, and the cardiac effects are mediated entirely through beta receptors. Activation of these receptors enhances myocardial contractility as well as increasing heart rate and the velocity of impulse conduction.

Vasoactive agents

Metabolically active substances.[2] Oxygen and carbon dioxide are the most important of a number of materials involved in smooth muscle metabolism that have direct effects on vascular tone. In vitro studies have shown that in the physiologic range, low P_{O_2} is associated with diminished contractility of vascular smooth muscle and also with decreased responsiveness to epinephrine.[25] This finding is consistent with numerous in vivo observations of local vasodilatation in response to decreased arterial oxygen saturation. The reasonable assumption that this represents a means of local regulation of blood flow, ensuring increased flow to metabolically active regions, would be untenable if oxygen exchange took place only in the capillaries, for arteriolar smooth muscle could not then "be aware" of lowered P_{O_2} in the downstream ex-

change vessels. This difficulty has been removed by the demonstration that oxygen diffuses from arterioles as well as capillaries,[26] so that changes in the gradient of P_{O_2} along the length of an arteriole provide means of indirectly sensing the local tissue P_{O_2}. A closely analogous situation has been reported in the pulmonary microcirculation (p. 1115), but in that vascular bed, lowered P_{O_2} causes *constriction*.

Increased P_{CO_2} causes vasodilatation in most beds, although it is possible that this action may be mediated by the hydrogen ion. Hydrogen ions have a relaxing effect on vascular smooth muscle, and this is the source of the vasodilatation produced by lactic acid. Lactate ion itself is not vasoactive, but pyruvate has a mild intrinsic vasodilator action. ATP is a potent vasodilator as well as a key agent in metabolism,[31] and the mono- and diphosphates are of equal activity in most vascular beds. These and many other "metabolies"[4] as well as the local release of potassium ions have been investigated extensively in an effort to explain the hyperemia of active skeletal muscle, which is much greater than can be accounted for by neural mechanisms alone, and to account for some aspects of vascular autoregulation (p. 1027) in other beds. The rapid initial vasodilatation in contracting muscle appears to be the result of an increase in extracellular potassium concentration. This increase is only transient, however, and wanes within the first minute of exercise. The subsequent maintenance of vasodilatation is attributable to a more slowly developing fall in local oxygen tension and increase of hydrogen ion concentration.[2]

Other vasoactive agents

Kallidin and bradykinin are just two of a number of naturally occurring polypeptides that have vasoactive properties.[11] One of these, *angiotensin II,* is an extremely powerful vasoconstrictor. Although it plays an important part in salt and water balance, its normal plasma levels are probably too low to contribute to vascular control. The same is true of *vasopressin* (antidiuretic hormone), except that this posterior pituitary secretion may have a physiologic role in cutaneous vasoconstriction and in the renal vascular bed. *Histamine,* which is released from cells of the skin in response to injury or antibody-antigen reactions, not only dilates arterioles and venules but also increases their permeability. This localized response, which is probably its principal vasomotor role, is not accompanied by significant amounts of circulating histamine. The effects of experimental histamine infusion vary widely among different species, but in cat, dog, and man, it usually lowers the total systemic vascular resistance. *Prostaglandins* are a group of biologically active lipids that exert vasodilator effects,[8] and their role in normal physiology is under active investigation. Normal plasma contains other substances, as yet unidentified, that appear to have a considerable influence on locally generated vascular tone.[20]

The osmotic activity of fluid within smooth muscle cells or in extracellular fluid may in itself influence vasomotor activity, and local hyperosmolarity is the most recent candidate for the role of dilator in exercising skeletal muscle, although the evidence on its qualifications for the part is conflicting.[4,48,54] Finally, although studies of vasomotor tone and control properly concentrate on smooth muscle, the nonmuscular portions of the vascular wall should not be neglected. Collagen and elastin are presumably fairly stable components of the wall, but the connective tissue elements and the water content of the wall must contribute in at least a minor way to modifications of the viscoelastic properties of blood vessels.

REFERENCES
General reviews

1. Ahlquist, A. R. P.: Development of the concept of alpha and beta adrenotropic receptors, Ann. N.Y. Acad. Sci. **139**:549, 1967.
2. American Physiological Society Symposium: Local regulatory mechanisms in the microcirculation. Fed. Proc. **34**:2005, 1975.
3. Axelsson, J.: Mechanical properties of smooth muscle, and the relationship between mechanical and electrical activity. In Bulbring, E., et al., editors: Smooth muscle, Baltimore, 1970, The Williams & Wilkins Co.
4. Axelsson, J.: Catecholamine functions, Ann. Rev. Physiol. **33**:1, 1971.
5. Barcroft, H., and Swan, H. J. C.: Sympathetic control of human blood vessels, London, 1953, Edward Arnold & Co.
6. Campbell, G.: Autonomic nervous supply to effector tissues. In Bulbring, E., et al., editors: Smooth muscle, Baltimore, 1970, The Williams & Wilkins Co.
7. Folkow, B.: The nervous control of the blood vessels. In McDowall, R. J. S., editor: The control of the circulation of the blood (supplemental volume), London, 1956, Wm. Dawson & Sons.
8. Messina, E. J., Weiner, R., and Kaley, G.: Prostaglandins and local circulatory control, Fed. Proc. **35**:2367, 1976.
9. Moran, N. C.: Beta adrenergic blockade. An historical review and evaluation. In Kattus, A. A., Ross, G., and Hall, V. E., editors: Cardiovascular beta adrenergic responses, Los Angeles, 1970, University of California Press.
10. Randall, W. C.: Neural regulation of the heart, New York, 1977, Oxford University Press.
11. Schachter, M.: Vasoactive peptides, Fed. Proc. **27**:49, 1968.
12. Somlyo, A. P., and Somlyo, A. V.: Vascular smooth muscle. I. Normal structure, pathology, biochemistry and biophysics, Pharmacol. Rev. **20**:197, 1968.
13. Speden, R. N.: Excitation of vascular smooth muscle. In Bulbring, E., et al., editors: Smooth muscle, Baltimore, 1970, The Williams & Wilkins Co.
14. Uvnäs, B.: Sympathetic vasodilator system and blood flow, Physiol. Rev. **40**(suppl.4):69, 1960.

Original papers

15. Abrahams, V. C., Hilton, S. M., and Zbrozyn, A. W.: The role of active muscle vasodilatation in the alerting stage of the defense reaction, J. Physiol. **171**:189, 1964.
16. Ahlquist, R. P.: A study of the adrenotropic receptors, Am. J. Physiol. **153**:586, 1948.
17. Bacq, Z. M.: Nature cholinergique et adrénergique des diverses innervations vasomotrices du pénis chez le chien, Arch. Int. Physiol. **40**:311, 1935.
18. Blair, D. A., et al.: Excitation of cholinergic vasodilator nerves to human skeletal muscles during emotional stress, J. Physiol. **148**:633, 1959.
19. Bohr, D. F.: Adrenergic receptors in coronary arteries, Ann. N.Y. Acad. Sci. **139**:799, 1967.
20. Bohr, D. F., and Johansson, B.: Contraction of vascular smooth muscle in response to plasma, Circ. Res. **19**:593, 1966.
21. Celander, O.: The range of control exercised by the "sympatheticoadrenal system," Acta Physiol. Scand. **32**(suppl.):116, 1954.
22. Celander, O., and Folkow, B.: The nature and the distribution of afferent fibres provided with the axon reflex arrangement, Acta Physiol. Scand. **29**:359, 1953.
23. Chorobski, J., and Penfield, W.: Cerebral vasodilator nerves and their pathway from the medulla oblongata, Arch. Neurol. Psychiatry **28**:1257, 1932.
24. Degeest, H., et al.: Depression of ventricular contractility by stimulation of vagus nerves, Circ. Res. **17**:222, 1965.
25. Detar, R., and Bohr, D. F.: Oxygen and vascular smooth muscle contraction, Am. J. Physiol. **214**:241, 1968.
26. Duling, B. R., and Berne, R. M.: Longitudinal gradients in periarteriolar oxygen tension: a possible mechanism for the participation of oxygen in local regulation of blood flow, Circ. Res. **27**:669, 1970.
27. Eckstein, J. W., Wendling, M. G., and Abboud, F. M.: Forearm venous responses to stimulation of adrenergic receptors, J. Clin. Invest. **44**:1151, 1965.
28. Ehinger, B., Falck, B., and Sporong, D.: Adrenergic fibers to the heart and to peripheral vessels, Bibl. Anat. **8**:35, 1966.
29. Falck, B.: Observations on the possibilities of the cellular localization of monoamines by a fluorescence method, Acta Physiol. Scand. **56**(suppl. 197):1, 1962.
30. Folkow, B.: The vasodilator action of adenosine triphosphate, Acta Physiol. Scand. **17**:311, 1949.
31. Folkow, B.: Impulse frequency in sympathetic vasomotor fibres correlated to the release and elimination of the transmitter, Acta Physiol. Scand. **25**:49, 1952.
32. Folkow, B., and Neil, E.: Circulation, London, 1971, Oxford University Press.
33. Folkow, B., Melander, S., and Oberg, B.: The range of effect of the sympathetic vasodilator fibers with regard to consecutive sections of the muscle vessels, Acta Physiol. Scand. **53**:7, 1961.
34. Folkow, B., Oberg, B., and Rubenstein, E. H.: A proposed differentiated neuro-effector organization in muscle resistance vessels, Angiologica **1**:197, 1964.
35. Gautvik, K.: The interaction of two different vasodilator

mechanisms in the chorda tympani–activated submandibular salivary gland, Acta Physiol. Scand. **79:**199, 1970.

36. Glick, G., et al.: Physiological differences between the effects of neuronally released and bloodborne norepinephrine on beta adrenergic receptors in the arterial bed of the dog, Circ. Res. **21:**217, 1967.

37. Hillarp, N. A.: The construction and functional organization of the autonomic innervation apparatus, Acta Physiol. Scand. **46**(suppl. 157):1, 1959.

38. Hilton, S. M., and Lewis, G. P.: The relationship between glandular activity, bradykinin formation and functional vasodilatation in the submandibular salivary gland, J. Physiol. **134:**471, 1956.

39. Hilton, S. M., and Lewis, G. P.: Vasodilatation in the tongue and its relation to plasma kinin formation, J. Physiol. **144:**532, 1958.

40. Holton, P.: The liberation of adenosine triphosphate on antidromic stimulation of sensory nerves, J. Physiol. **145:**494, 1959.

41. Ingram, R. H., et al.: Effects of sympathetic nerve stimulation on the pulmonary arterial tree of the isolated lobe perfused in situ, Circ. Res. **22:**801, 1968.

42. Johansson, B., and Ljung, B.: Role of myogenic propagation in vascular smooth muscle response to vasomotor nerve stimulation, Acta Physiol. Scand. **73:**501, 1968.

43. Johansson, B., et al.: Electrical and mechanical characteristics of vascular smooth muscle response to norepinephrine and isoproterenol, Circ. Res. **21:**619, 1967.

44. Kaiser, G. A., Ross, J., Jr., and Braunwald, E.: Alpha and beta adrenergic receptor mechanisms in the systemic venous bed, J. Pharmacol. Exp. Ther. **144:**156, 1964.

45. Kjellmer, I.: Studies on exercise hyperemia, Acta Physiol. Scand. **64**(suppl. 244):1, 1965.

46. Korner, P. I., Chalmers, J. P., and White, S. W.: Some mechanisms of reflex control of the circulation by the sympatho-adrenal system, Circ. Res. **22**(suppl. 3):157, 1967.

47. Lewis, T.: Observations upon the reactions of the vessels of the human skin to cold, Heart **15:**177, 1930.

48. Lundvall, J., Mellander, S., and White, T.: Hyperosmolality and vasodilatation in human skeletal muscle, Acta Physiol. Scand. **77:**224, 1969.

49. Mellander, S.: Comparative studies of the adrenergic neuro-hormonal control of resistance and capacitance blood vessels in the cat, Acta Physiol. Scand. **50**(suppl. 176):1, 1960.

50. Napolitano, L. M., et al.: Fine structures of the heart after transplant; with special reference to the neural elements, Circulation **30**(suppl. 1):1, 1964.

51. Randall, W. C., and Rohse, W. G.: The augmentor action of the sympathetic cardiac nerves, Circ. Res. **4:**470, 1956.

52. Roddie, I. C.: The transmembrane potential changes associated with smooth muscle activity in turtle arteries and veins, J. Physiol. **163:**138, 1962.

53. Root, W. S., and Bard, P.: The mediation of feline erection through sympathetic pathways, with some remarks on sexual behavior after deafferentiation of the genitalia, Am. J. Physiol. **151:**80, 1947.

54. Scott, J. B., et al.: Role of osmolarity, K^+, H^+, Mg^{++}, and O_2 in local blood flow regulation, Am. J. Physiol. **218:**338, 1970.

55. Somlyo, A. V., and Somlyo, A. P.: Electromechanical and pharmacomechanical coupling in vascular smooth muscle, J. Pharmacol. Exp. Ther. **159:**129, 1968.

56. Uchida, E., and Bohr, D. F., and Hoobler, S. W.: A method for studying isolated resistance vessels from rabbit mesentery and brain and their responses to drugs, Circ. Res. **21:**525, 1967.

57. Viveros, O. H., Garlick, D. G., and Renkin, E. M.: Sympathetic beta adrenergic vasodilatation in skeletal muscle of the dog, Am. J. Physiol. **215:**1218, 1968.

42

WILLIAM R. MILNOR

The cardiovascular control system

The cardiovascular control system consists of the effectors and efferent pathways considered in Chapter 41, certain parts of the central nervous system (CNS), peripheral receptors, and afferent pathways. Afferent signals in this system are generated by specific receptors in the cardiovascular system called *baroreceptors* and *chemoreceptors* and also by the stimulation of somatic sensory and other nerve fibers extrinsic to the circulation. The central processing that intervenes between afferent signals and reflex efferent impulses in autonomic motoneurons to the effectors involves almost all major subdivisions of the CNS, but the most important centers for cardiovascular control lie in the medulla. The higher levels of the brain make their contributions principally by acting on these medullary cells.

Normal circulatory function, including both homeostasis in the basal state and the adjustments of pressure and flow appropriate to changing circumstances, depends on a repertoire of cardiovascular reflexes, many of which have been studied in great detail. Much of our knowledge of cardiovascular control has been gained from experiments in which a single reflex arc is artificially isolated and examined and by identifying each component of the reflex—the receptor properties, afferent signals, central processing, efferent signals, and effector activity. The reflex effects of stimulating a single class of receptors when the interacting effects of other stimuli are prevented are called the *primary* reflex response. On the other hand, physiologic control in the intact animal operates through the integration of multiple reflex responses. Here there are significant gaps in our information, although the cardiovascular responses to such stresses are exercise, anoxia, and intense emotion have been carefully investigated. The nature of central processing and integration is the area of greatest uncertainty.

RECEPTORS AND AFFERENT SIGNALS
Baroreceptors

Organs sensitive to mechanical deformation exist in the walls of some blood vessels and produce afferent nerve impulses when the wall is stretched. The term "baroreceptors" has been applied to these units because intravascular pressure is the most obvious force acting to stretch the vessel wall under physiologic conditions, but any stress that deforms the wall in which they lie can activate them, and "strain receptors" would be a more accurate name. If a plaster cast is applied around a baroreceptor-containing vessel so as to prevent its distention, for example, raising the intravascular pressure no longer produces the increased baroreceptor firing that is the ordinary response. Distention of a vessel is not, moreover, the only kind of deformation that stimulates baroreceptors. Longitudinal stretching is also an effective stimulus, as is the abnormal reduction in circumference caused by very low levels of pressure.[8]

The most concentrated and functionally important baroreceptive regions lie in the two carotid sinuses and in the arch of the aorta, although baroreceptors have been found in other segments of the common carotid arteries, as shown in Fig. 42-1. The carotid sinus is a dilatation of the internal carotid artery at its origin (Fig. 42-2) and is found in virtually all mammals except ruminants.[7] The arterial wall becomes somewhat thinner in this region, with more elastin and less smooth muscle than in more distal parts of the vessel. Sensory innervation of these areas is by way of branches of the glossopharyngeal (carotid sinus nerves), whereas the aortic baroreceptors are supplied by the left aortic nerve. The right aortic nerve innervates baroreceptors at the root of the right subclavian and common carotid arteries. Both aortic nerves supply the aortic (chemoreceptor) body. Most

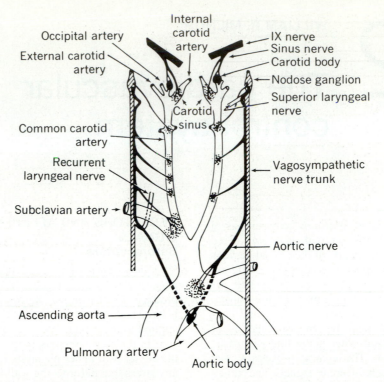

Fig. 42-1. Diagrammatic representation of location and innervation of baroreceptors and chemo-receptors of aortic and carotid regions in cat and dog.[4,7,8] Baroreceptor areas are indicated by stippling. Nerves and arteries are labeled.

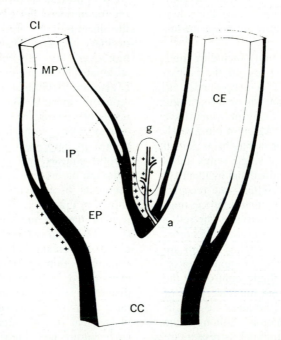

Fig. 42-2. Diagram of carotid bifurcation in man. Black areas indicate predominantly elastic structure and unshaded areas those portions with muscular structure. Small crosses indicate regions of baroreceptor endings. Common carotid, *CC*, internal carotid, *CI*, and external carotid, *CE*, arteries are shown, together with artery, *a*, of carotid body, *g*. "Carotid sinus" is dilatation of origin of internal carotid artery, in which three regions may be distinguished: a proximal elastic portion, *EP*, abundantly supplied with baroreceptors; an intermediate portion, *IP*; and a muscular distal portion, *MP*. (From Muratori.[80])

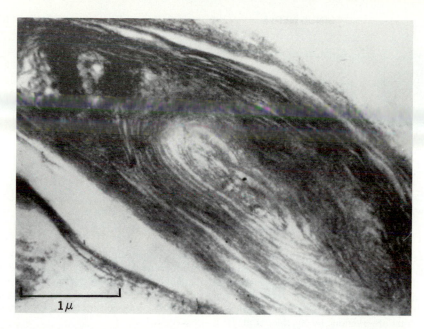

1μ

Fig. 42-3. Cross section of lamellar body in outer media of rabbit carotid sinus, as recorded by electron microscope. (From Dropmann.[50])

of the baroreceptor nerve fibers are of intermediate size (3 to 5 μm in diameter), but a few are considerably larger.[7,43] On the basis of this histologic difference in the nerve supply and the varying spike heights in recordings from multifiber preparations of sinus nerves, two types of baroreceptors have been distinguished, but the only functional difference appears to be a somewhat lower threshold for the receptors innervated by the larger fibers.[7,75] Histologically, the nerve endings in baroreceptor regions form a profuse arborization, limited almost exclusively to the adventitia of the arterial wall. With light microscopy, it is impossible to see the final termination of the nerve endings, but electron microscopy has revealed layered ovoid structures 4 μm or less in length (Fig. 42-3) in the outermost part of the media.[50]

Within limits, the more a baroreceptive vessel is distended, the greater the frequency of afferent impulses from the receptors. The curve relating mean distending pressure and the frequency of afferent impulses is sigmoid (Fig. 42-4), but the relationship is approximately linear for mean systemic arterial pressures from about 100 to 180 mm Hg. When the distending pressure is pulsatile, as it is in vivo, the afferent impulses recur in groups, the most rapid firing appearing during the rising phase and peak of the pressure pulse (Fig. 42-5). Not only are the impulses grouped phasically in accordance with the pulsating distention, but the average impulse rate is higher

with pulsatile pressure than with a constant pressure of the same mean value, as is evident on comparison of the two curves in Fig. 42-4. Baroreceptor response thus includes at least two different elements, one proportional to the instantaneous strain, the other sensitive to its rate of change. It is evident in records such as those shown in Figs. 42-5 and 42-6 that instantaneous pressure is not the sole determinant of afferent impulse frequency, for firing in a single fiber sometimes ceases on the downstroke of the pressure wave at a pressure that was accompanied by firing on the upstroke. The reaction to rates of change is thus asymmetric, the response being greater to rising than to falling pressures (Figs. 42-5 and 42-6). This "unidirectional rate sensitivity" accounts in part for the differences in response to pulsatile and steady distention of the carotid sinus at the same mean pressure.[51]

Records from single afferent fibers[7,8] show an average firing rate of between 10 and 30 impulses/sec at normal pressures. The total activity recorded from multifiber afferents represents the response of a large population of receptors, which differ in threshold and do not fire exactly in synchrony. The end result is a continuous, highly patterned stream of signals from the baroreceptive areas to the CNS. These signals play a major role in maintaining normal levels of vasomotor tone (p. 1049) and cardiac output, which in turn determine arterial pressure, so that the baroreceptor system constitutes a "feedback

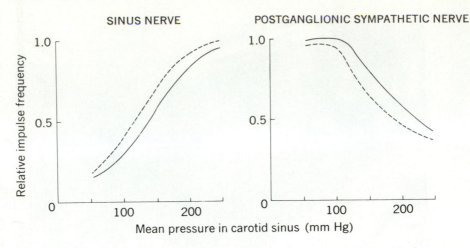

Fig. 42-4. Impulse frequencies in carotid sinus nerve and in postganglionic sympathetic nerve as function of pressure in dog carotid sinus. Unbroken line represents response to constant pressures; broken line represents response to pulsations around indicated mean pressure. As pressure is raised in sinus, there is increase in average frequency of afferent impulses from baroreceptors and decrease in frequency of efferent impulses in sympathetic vasoconstrictor nerve. (Based on experimental observations by Spickler et al.[91])

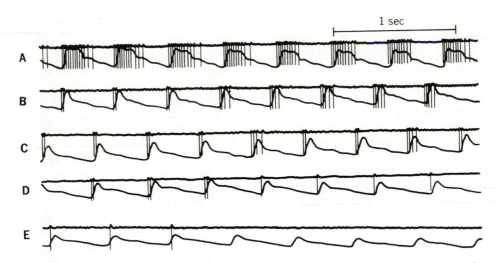

Fig. 42-5. Impulse activity in single fiber of left aortic nerve, and blood pressure recorded from left common carotid artery. Mean pressures in five records shown, beginning at top, are 125, 80, 62, 55, and 42 mm Hg. (From Neil.[81])

loop'' (Chapter 7). The existence of multiple baroreceptive regions means that selective stimulation or elimination of any one of them can be counteracted by the activity of the others. Distention of only one carotid sinus, for example, has only transient effects unless the other sinus nerve and aortic nerves have been cut.[7] Baroreceptors belong to the slowly adapting class of sensory transducers.[7,31] When pressure within the carotid sinus is gradually raised to a new level,

the afferent impulse rate increases proportionately and may remain constant at the new frequency (i.e., show no adaptation) for more than 1 hr.[46] If the increase in pressure is sudden, as in a ''stepwise'' increment, there is an initial burst of rapid firing because of the rate-sensitive component of baroreceptor response before the impulse rate settles to a new constant level.

Smooth muscle and sympathetic constrictor innervation exist in the vascular walls of barore-

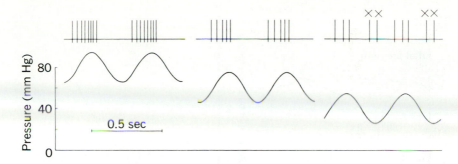

Fig. 42-6. Baroreceptor response to pulsatile pressure at three different levels of mean pressure. Upper records show action potentials in single fiber of right common carotid baroreceptor nerve in cat. Lower records represent pressures applied to lumen of artery. Amplitude and frequency of pressure pulsations are the same in each instance. Rate of firing in nerve depends on rate and direction of change of pressure as well as on mean pressure. Note paradoxical firing, *X*, when pressure falls to very low levels. (After Kezdi.[8])

ceptive regions, as in other blood vessels,[70] suggesting that sympathetic discharge may modify the baroreceptor response to intravascular pressure. Although stimulation of the sympathetic efferents to the carotid sinus does evoke an afferent response,[56,72] present evidence suggests that this is not because of neural activation of smooth muscle in the sinus wall.[59] The physiologic significance of the efferent innervation to baroreceptive regions is therefore unclear.

The cardiovascular response in baroreceptor reflexes, that is to say the *complete* response as contrasted with the afferent signals from receptors, is considered in detail in a later section (p. 1076). In brief, elevation of pressure in the baroreceptive areas leads to a reflex slowing of the heart rate, a decrease in myocardial contractility, arteriolar dilatation in most vascular beds, and decreased tone in large veins. The arteriolar and venous responses are caused by a reduction in the normal frequency of efferent sympathetic constrictor impulses. The cardiac response involves both sympathetic and parasympathetic efferents.

Pulmonary arterial baroreceptors. Strain receptors similar to those in the aortic arch and carotids have also been found in the pulmonary artery. They are concentrated near the bifurcation of the main pulmonary artery into the right and left branches. Their functions and reflex effects are essentially the same as those of the baroreceptors of the systemic circulation,[37] and a rise in pulmonary arterial pressure tends to produce a reflex decrease in systemic arterial pressure. A fall in pulmonary arterial pressure has the opposite effect and at the same time causes a striking increase in the rate and depth of respira-

tion. Occasional experiments[7] have raised the possibility that distention of the main pulmonary artery elicits a reflex response in the pulmonary as well as in the systemic vessels, a possibility that merits further investigation.

Cardiac stretch receptors.* The presence of stretch receptors in the chambers of the heart has been demonstrated by electrophysiologic as well as morphologic techniques. In the right and left atria, these receptors lie subendocardially in the region of the venoatrial junctions, and their afferent signals travel in the vagal nerves. There appear to be two types of atrial receptors. One group generates impulses during atrial contraction (type A), whereas the other (type B) fires late in ventricular systole, near the peak of the atrial "v" wave (Fig. 42-7). The two kinds of receptors exist in approximately equal numbers.[12] Both types can be activated by stretching the atrial wall, and their different patterns of firing are therefore somewhat puzzling. Type B receptors are slower adapting, and it may be that they are less sensitive to rate of stretch than type A receptors. The failure of type B receptors to fire during atrial contraction could be explained if they were arranged in parallel with the atrial muscle fibers, but there is no firm evidence that this is the case.[12] The reflex responses to stretching of type B receptors are qualitatively similar to those elicited by increased pressure in the arterial baroreceptors, but they are sensitive to much smaller changes in pressure, as is appropriate in this "low pressure" part of the circulation. The rate of firing of type B receptors is raised by increased atrial filling[12] and by increased blood

---

*See references 12, 36, 62, 82, and 84.

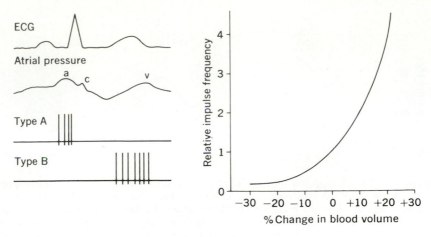

Fig. 42-7. Atrial baroreceptor activity. Left: relation between electrocardiogram (ECG), left atrial pressure, and impulses in afferent vagal fibers from type A and type B atrial receptors. Right: relative frequency of impulses (ordinate) in afferent fiber from type B receptor when total blood volume was changed from its normal level. Change in atrial pressure in these experiments was approximately ±5 cm H_2O for ±20% change in blood volume. (Based on data of Gupta et al.[61] and Paintal.[11a])

volume (Fig. 42-7). Type A receptors apparently stimulate sympathetic discharge whereas type B receptors inhibit it, and the pattern of the complete reflex response may be a function of the A/B impulse ratio.[62]

The so-called *Bainbridge reflex,* an increase in heart rate sometimes produced by large intravenous infusions of saline solution or blood, now appears to be the result of the activation of type A receptors in the left atrium.[76] The afferent pathway for this reflex is in the vagus nerve and its efferents predominantly in sympathetic nerves to the heart. The inconsistent results obtained by many investigators who have tried to study this reflex are probably explained by the mixture of A and B receptors that were stimulated in their particular experiments. In view of the high atrial pressures required to demonstrate the reflex (2 to 4 times the normal levels), its physiologic significance is dubious.

Atrial receptors are strategically located for monitoring terminal venous pressure and the dynamics of ventricular filling. Their afferent signals are transmitted to the hypothalamic centers that control antidiuretic hormone (ADH) secretion (p. 928) as well as to the medullary cardiovascular centers,[61] and they are consequently part of a feedback loop in the system that controls total blood volume (Chapter 46). A rise in atrial pressure leads to the inhibition of ADH secretion and thus promotes fluid loss. Moderate reduction of blood volume triggers a much more marked decrease in afferent signals from atrial than from arterial baroreceptors.[61] In addition, the atrial receptors are probably involved (together with those of the aorta and carotids) in the control of sodium excretion by the kidney by way of the renin-angiotensin-aldosterone mechanisms (p. 1193). Efferent sympathetic impulses to the renal arterioles modify the rate of renin release, but some form of intrinsic local response in the renal vessels appears to be of equal importance.[28]

Ventricular stretch receptors lie in the subendocardial layers of both ventricles, are relatively few in number, and normally fire a few impulses just after the QRS complex of the ECG at a time corresponding to the period of isometric contraction in the ventricles.[12] The afferent fibers of ventricular receptors, like those of the atria, are in the vagal trunks and cardiac nerves, where they travel beside efferent fibers. Extreme distention of the ventricles causes marked reflex bradycardia and hypotension,[7] but their contribution to reflex control under ordinary conditions is unknown.

At least two other sets of baroreceptors function in the heart. One group, located in the pericardium, discharges irregularly with no relation to the cardiac cycle.[36] The physiologic function of these receptors is not yet clearly understood. A second set in the coronary sinus functions like the atrial receptors. Mechanical receptors also exist in the mesentery in the form of pacinian corpuscles that lie in contact with the walls of arteries in this region.[29,58,63] No cardiovascular reflexes from these organs have been discovered.

Neither the atrial nor the ventricular stretch receptors are sensitive to changes in the normal chemical constituents of the blood, but they can be activated by certain substances foreign to the body, notably the veratrum alkaloids. Stimulation of the cardiac receptors by veratrine derivatives produces reflex bradycardia and vasodilatation, and the same drugs act on pulmonary receptors to induce a slowing or arrest of respiration.[5] This sequence of events is called the *Bezold-Jarisch reflex* and plays no part in normal cardiovascular control, although it provides an experimental tool for the investigator.

CHEMORECEPTORS

Arterial receptors that respond to changes in O_2, CO_2, and H ion concentrations are found in specialized tissues near the aortic arch and the carotid bifurcations (Fig. 42-1 and 42-2). Like the baroreceptors, these chemoreceptors are an important part of the cardiovascular control system, but their function is quite different, and not merely in the mode of stimulation. Under normal conditions the chemoreceptors send few impulses to the CNS and have little or no effect on the circulation. When they are stimulated by some deviation of the blood gases from normal levels, the respiratory response is even greater than that of the heart and blood vessels. As far as the circulation is concerned, the chemoreceptors act as a kind of emergency system, unlike baroreceptors, which perform a tonic regulatory function. Respiration, on the other hand, is constantly and delicately controlled by chemoreceptor reflexes. The physiology of chemoreceptors has been lucidly reviewed by Torrance[17] in an introduction to an informative series of papers on this subject.

The aortic and carotid "bodies" are highly vascular, as befits organs that "sample" the chemical content of the blood. The total blood flow to these tiny structures is very small, but the flow per gram of tissue is on the order of 20 (ml/gm)/min, by far the highest in the body. The finest branches of the microcirculation in these organs are wide sinusoids lined by chemosensitive cells.[17,43] Blood is separated from the most distant parts of the cells by less than 20 μm, providing excellent conditions for diffusion. The chemoreceptors are sensitive to the partial pressures of O_2 and CO_2 in the blood, and their response depends on the tensions of both gases (Fig. 42-8). Individual receptors differ widely in threshold, and at least some single afferent fibers respond to both O_2 and CO_2 tensions.[65] In general, afferent impulses increase as Po_2 falls or Pco_2 rises, but there is an interaction that makes the re-

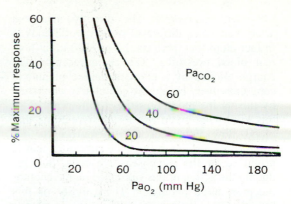

Fig. 42-8. Average chemoreceptor impulse frequency recorded from multifiber preparation, expressed as percent of maximum response (ordinate), in relation to changes in arterial Po_2 and Pco_2 (abscissa). (From Korner.[9])

flex response greatest when both changes occur together. High Pco_2, for example, causes little discharge when Po_2 is relatively high but has a marked effect when Po_2 is low. Whether Pco_2 has an effect per se, apart from its influence on pH, is still an unsettled question.

Chemoreceptor function is more complex than this straightforward response to gas tensions would suggest. A decrease in blood flow through the aortic or carotid bodies increases chemoreceptor discharge, and increasing blood flow has the opposite effect, even when the arterial O_2 content is kept constant.[53] This appears to be the explanation for the activation of chemoreceptors by arterial hypotension or by stimulation of the sympathetic constrictor fibers to the arterioles that supply chemosensitive areas.[27] One might suppose that decreased flow simply causes a lowering of Po_2 at the venous end of the sinusoids (a widening of the arteriovenous O_2 difference), which would be sensed by at least some chemoreceptor cells. This explanation is contradicted, however, by virtually all the experimental evidence. Observations made in experimental CO poisoning, which produces typical "anemic" anoxia (Chapter 00), indicate that arterial Po_2 rather than O_2 content is the critical factor. In such experiments, one might expect that the reduction of O_2 capacity of the blood by CO would lower the end-capillary Po_2; yet chemoreceptor discharge bears no relation to O_2 capacity of the incoming arterial blood as long as its Po_2 is held constant. Evidently the arteriovenous O_2 difference is very small, and the amount of O_2 carried in simple solution in the plasma is more than sufficient for the metabolic and sensory needs of the chemoreceptors.[38] The directly measured arteriovenous O_2 difference in the carotid body is in fact quite low, perhaps 0.5 ml/100 ml at normal blood pressures. The chemoreceptor discharge elicited by a constant degree of anoxia can, moreover, be almost completely abolished by raising the blood pressure.[17] Observations of this kind have given rise to the hypothesis that blood flow influences the chemoreceptor response by carrying away some substance that is produced by the chemosensitive cells, thus reducing the action of this substance on the chemoreceptor nerve endings.[68] The validity of this hypothesis and the nature of the "transmitter" are still in question.[17]

The primary cardiovascular response to activation of the carotid chemoreceptors consists of bradycardia, arteriolar constriction in all the principal vascular beds, and increased adrenal secretion of catecholamines. The cardiac effect is brought about by increased parasympathetic and decreased sympathetic activity, whereas the vasoconstriction results from an increase in the rate of efferent sympathetic constrictor impulses.[40] The cardiac output falls, but the total peripheral resistance increases to such an extent that arterial pressure becomes higher. These *primary* effects are seen in pure form, however, only in experiments where respiration is artificially kept constant, for chemoreceptor activation causes a marked increase in rate and depth of breathing. These respiratory effects are not only more striking than the effects on heart and blood vessels, but they also modify the circulatory response. When hypercapnia is allowed to occur, the heart rate and cardiac output usually increase.[41] In addition, the local vasodilatation of resistance vessels in response to hypoxia partially counteracts the reflex vasoconstriction, with results that differ in different regions. The cerebral blood flow usually increases, whereas the gastrointestinal and renal flows often fall.

EXTRINSIC SOURCES OF AFFERENT SIGNALS

Although the strain and chemosensitive tissues located within the cardiovascular system are the key receptors in its regulation, there are many situations in which afferent signals arising outside the heart and blood vessels evoke reflex changes in the circulation. Stimulation of somatic sensory nerves, inflation of the lungs, contraction of skeletal muscles, and cerebral ischemia are the major extrinsic sources of such reflexes.

Somatic and visceral sensory inputs. The un-physiologic procedure of stimulating the central stump of a cut somatic or visceral nerve may produce either a rise or a fall in blood pressure, depending on the experimental conditions. The "pressor" response is produced by relatively strong stimulation and includes tachycardia and peripheral vasoconstriction. It probably involves the activation of small myelinated A-delta fibers involved in the mediation of pain. Two different groups of sensory fibers appear to be capable of initiating the "depressor" response, which can be elicited either by rapid stimulation of high-threshold C fibers or by low-frequency stimulation of smaller myelinated fibers. The latter mediate the "deep pain" associated with manipulation of the viscera or puncture of an artery, procedures that are often accompanied by the "depressor" response of bradycardia, profound hypotension, and sweating. One source of reflex circulatory responses brought on by stimulation of a cutaneous sensory nerve may be activation of fibers from thermoreceptors (p. 422). Stimulation of afferent nerves from skeletal muscle can also elicit either vasoconstriction or dilatation, again depending on the particular fibers stimulated. The degree of response in different vascular beds depends on their initial tone, but there is no evidence that preferential dilatation of any one region is brought about by this means.[69] The extraordinary vasodilatation in exercising muscle is due predominantly to the local action of metabolites (p. 1058), but there is some evidence that these same metabolites can produce true *reflex* effects by their action on some kind of local receptors.[20] If the muscles of one limb are exercised while the circulation to and from the extremity is occluded, tachycardia and elevation of blood pressure result, even with degrees and durations of exercise far below those that produce pain. These reflex effects persist after exercise until the occlusion is released, but the nature of the receptors and their effective stimulus are unknown.

Submersion and diving. Somatic inputs through the trigeminal nerve serve a unique reflex function in diving animals.[1,25] In the duck, for example, submersion of the head or even the nostrils triggers an immediate bradycardia and intense peripheral vasoconstriction everywhere except in the heart, brain, and lungs. Cardiac output is greatly reduced, while arterial pressure is maintained and venous pressure rises slightly. This is not a response to asphyxia, because it develops instantaneously and cannot be duplicated by simple occlusion of the trachea. The overall result is an adaptation that minimizes oxygen usage by all but the most essential tissues, and the energy necessary for muscular activity during prolonged submersion is obtained largely from anaerobic metabolism. A similar reflex occurs in human subjects, although the vasoconstriction and bradycardia are much less marked.

Lung inflation. The most numerous vagal afferent fibers are those that carry signals from stretch receptors in the parenchyma (not the vessels) of the lung and respond to pulmonary inflation (the Hering-Breuer reflex). Although the most prominent reflex effects of activating these receptors are on the mechanics of respiration, there are cardiovascular effects as well. Lung inflation causes reflex vasodilatation in skin, muscle, and the splanchnic bed through a reduction of sympathetic constrictor impulses to these vessels.[42] The dilatation is greatest in skeletal muscle, and the cardiac output is consequently redistributed so as to increase muscle blood flow at the expense of other regions. This vascular response is insignificant with normal inspiration and expiration, but it presumably contributes to increased blood flow to muscle when hyperventilation accompanies muscular exercise.

Cerebral chemoreceptors. Cerebral ischemia has long been known to produce acute vasoconstriction elsewhere in the body and elevation of the arterial pressure. The fundamental stimulus is apparently not the cerebral blood pressure per se but the action of local hypoxia and hypercapnia on cells somewhere in the CNS, probably neurons of the medullary vasomotor centers. Tachycardia and a sharp rise in total peripheral resistance are the principal features of the response, which can be produced either by lowering the cerebral perfusion pressure[87] or by lowering the arterial oxygen saturation, but it does not appear at low perfusion pressures if blood oxygen is kept at a high level.[48] The efferent pathways of the reflex are the same as those activated by aortic and carotid chemoreceptor stimulation. In generalized anoxia the tachycardia produced by this cerebral component of the response, plus the effects of hyperventilation, often outweigh the primary bradycardia that the arterial chemoreceptors would otherwise produce. It is unlikely that specialized ''chemoreceptors'' in the cerebral circulation are responsible for the reaction, but the end result has all the characteristics of a chemoreceptor reflex. Because of the highly developed autoregulatory ability of the cerebral circulation, this is an emergency reflex called into play only in cases of extreme hypotension or when local mechanisms of vasomotor control can no longer cope with arterial hypoxia or hypercapnia. Increased intracranial pressure of sufficient mag-

nitude to reduce cerebral blood flow has very similar effects (p. 1097).

CENTRAL NERVOUS CONTROL SYSTEMS

All levels of the CNS from cortex to spinal cord contribute to the control of the mammalian circulation. The most essential control elements, however, are concentrated in the medulla oblongata and lower pons (the bulbar region). Bulbar centers generate tonic and reflex efferent impulses to all the cardiovascular effectors. Higher levels of the brain influence the circulation through their action on these medullary centers, with the single exception of the sympathetic vasodilator outflow. Although decerebrate animals survive and demonstrate very effective baroreceptor and chemoreceptor reflexes, these responses are modulated in the intact animal by the hypothalamus and cerebral cortex, and hypothalamic participation is essential for the cardiovascular components of certain generalized patterns of response such as thermoregulation and the defense reaction. The complex interaction of these different levels is by no means completely understood. Contributions to cardiovascular control from various levels have been studied principally by progressive ablation of parts of the brain, although mapping of the afferent projections from periphery to medulla by electrophysiologic techniques has been carried out in considerable detail. The efferent pathways have been determined largely by stimulation of specific points in the brain, naturally an imperfect imitation of physiologic events. The picture that emerges, however, is one of a central processing system through which a particular combination of afferent signals—the ''input profile''[9,92]—is converted into a qualitatively and quantitatively appropriate set of effector responses. In general, tonic and adaptive control of the blood vessels is exerted through medullary centers that increase or decrease the degree of sympathetic constrictor tone, while the heart is controlled by the interplay of parasympathetic impulses from the vagal centers and sympathetic impulses from the medulla.

Spinal level

The cells of the spinal cord alone can effect a surprising degree of vascular regulation under certain conditions.[55] Transection of the cervical spinal cord leads to a precipitous fall in blood pressure and other manifestations that have been termed ''spinal shock'' (p. 781). If the section is made between segments C6 and T1, preserving the respiratory function of the phrenic nerve and the sympathetic preganglionic outflow, the animal survives and, in about a week, is able not only to maintain normal arterial pressure but also to respond with an elevation of pressure and tachycardia to central stimulation of a cut sensory nerve. Such chronic spinal animals can even compensate almost as well as normal specimens for moderate degrees of blood loss (10% to 25% of total blood volume). These responses occur by virtue of sympathetic discharge to the heart and blood vessels, but they have their origin in hypoxia of spinal cells rather than true reflex loops.[21] Although it is conceivable that this mechanism may be called on in the intact animal under conditions of drastic hypotension or anoxia, it probably has little significance in ordinary cardiovascular control and regulation.

Although the spinal preganglionic cells can be considered the "final common pathway" of autonomic control,[13] the autonomic ganglion itself it one more link in the path to the effectors. Adrenal catecholamines are among the factors influencing transmission through these ganglia,[45] which suggests the possibility, not yet completely explored, that they should be regarded as another active component in the circulatory control system.

Medulla and pons (bulbar level)

What has come to be known as the *vasomotor center* of the medulla oblongata is the longest known and most thoroughly studied of the central nervous regions that influence the circulatory system. Its existence was discovered in the laboratory of Carl Ludwig at Leipzig during the early 1870s when two of Ludwig's students, Owsjannikow and Dittmar, showed in rabbits and cats that (1) central stimulation of the sciatic nerve induces a reflex rise in arterial pressure after separation of the medulla oblongata from the rest of the brain and (2) successive transections of the brainstem from above downward have no effect on the height of arterial pressure or on pressor reflexes until they reach the lower part of the pons; transection at still lower levels produces a greater and greater hypotension and reduction in pressor responses until, at a point a few millimeters above the calamus scriptorius, the arterial pressure falls to the low level (40 to 50 mm Hg) produced by acute section of the cervical cord, and pressor responses virtually disappear. Thus it was demonstrated that a central mechanism essential for the maintenance of normal arterial pressure and for reflex increases in pressure extends from about 3 mm above the point of the

calamus scriptorius to the fovea superior and includes the diffuse part of the superior olive. Subsequent studies have in general confirmed this first delimitation of the vasomotor center.

Further information regarding the position and nature of the bulbar centers has been secured by the method of localized stimulation. The first to do this were Ranson and Billingsley (1916) who, on exploring the floor of the fourth ventricle with a needle electrode, found two small reactive areas. Rises in arterial pressure were consistently obtained from one of these, whereas the other yielded depressor responses. Subsequent experiments have confirmed this rather distinct localization of "pressor" and "depressor" centers, although the latter is somewhat less sharply outlined than the former. The location of these two areas in the brain stem is shown in Fig. 42-9, in which it can be seen that transection at appropriate levels can eliminate the pressor region while leaving most of the depressor region intact. Such a procedure is followed by a fall in arterial pressure, a reduction of tonic discharge in the cardiac sympathetic nerves, and abolition of the response to somatic nerve stimulation. Transection below the depressor center restores some of the tonic nervous activity. The neurons in these two areas are essential, then, for reflexes that call on sympathetic constrictors and the heart, and they also act as sources of tonic impulses for these effectors. Neurons in the depressor area, moreover, deliver tonic inhibition of the sympathetic cardiovascular discharge directly, rather than by inhibition of the pressor area.

The terminology here, as in the case of vasomotor nerves (p. 1050), is slightly misleading. The terms "pressor" and "depressor" were originally adopted because the changes in blood pressure that could be induced experimentally were striking and relatively easy to observe. In the light of subsequent work, it is clear that stimulation of the pressor area can activate all the effectors that respond to sympathetic adrenergic impulses, increasing the heart rate, the force and velocity of myocardial contraction, the wall tension in capacitance vessels, the resistance of arterioles, and the adrenal secretion of catecholamines. Stimulation of the depressor region, on the other hand, *diminishes* sympathetic adrenergic discharge. No true vasodilator fibers are controlled by these centers, or called on by depressor reflexes. Central representation of sympathetic cholinergic vasodilator discharge extends from the motor cortex through the hypothalamus and mesencephalon, but the pathways bypass the medullary centers. Excitation and in-

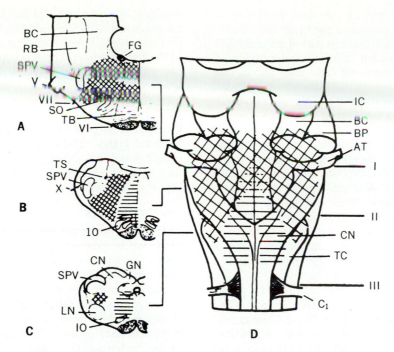

Fig. 42-9. Localization of pressor and depressor centers in brain stem of cat. Pressor regions indicated by cross-hatching, depression regions by horizontal ruling. **A** to **C,** Cross sections through medulla at levels indicated by guidelines to **D. D,** Semidiagrammatic projection of pressor and depressor regions onto dorsal surface of brain stem viewed with cerebellar peduncles cut across and cerebellum removed. *AT,* Auditory tubercle; *BC,* brachium conjunctiva; *BP,* brachium pontis; *C₁*, first cervical nerve; *CN,* cuneate nucleus; *FG,* facial genu; *GN,* gracile nucleus; *IC,* inferior colliculus; *IO,* inferior olivary nucleus; *LN,* lateral reticular nucleus; *RB,* restiform body; *SO,* superior olivary nucleus; *SPV,* spinal trigeminal tract; *TB,* trapezoid body; *TC,* tuberculum cinereum; *TS,* tractus solitarius; *V* to *VII,* corresponding cranial nerves; *I* to *III,* levels of transection discussed in text. (From Alexander.[22])

hibition of the parasympathetic cardiac innervation and cholinergic vasodilators take place in the dorsal vagal nuclei, but in view of the reciprocal changes in sympathetic and parasympathetic discharge by which reflexes alter heart rate, the activity of the vagal nuclei and the medullary centers must be closely coordinated. It is obvious now, as it was not when the study of cardiovascular reflexes began, that changes in blood pressure and heart rate are not always the most sensitive indicators of reflex activity. Alterations in ventricular function, for example, or redistribution of blood flow by changes in the relative resistances of different organ systems, can occur without a significant change in blood pressure.

Although many suprabulbar parts of the brain wave have an influence on the circulation, the only descending pathways that are definitely known to bypass the medullary centers are the sympathetic cholinergic vasodilators. This fact, together with the relatively small deviations from normality of the baroreceptor responses in ani-

mals that lack all centers rostral to the pons,[9,73] is the basis for the hypothesis that medullary centers are the principal control site for cardiovascular effectors. Most investigators would agree that this is a conceptually convenient but oversimplified view.[3,9,13] Although it is of some importance to know how the organism functions when certain parts of the nervous system have been destroyed, such information is only a partial guide to function in the intact animal. Ablation and local stimulation experiments have added enormously to our knowledge, but too little is known as yet to be certain that our standards for judging what constitutes a ''slight'' contribution from one part of the brain to cardiovascular control, as opposed to a ''dominant'' one, are the same as those of the CNS.

Hypothalamus

Stimulation at various points in the hypothalamus can produce pressor or depressor responses involving virtually all the cardiovascular effectors. Indeed, if nothing else were known about central nervous control of the heart and vessels, one might conclude that this part of the diencephalon was the principal "cardiovascular center."[13] A variety of sympathetic constrictor and cardio-acceleratory effects in particular can be elicited, and by careful localization and attention to stimulus intensity and frequency, it has been possible to influence selectively the renal, muscular, or splanchnic vascular beds,[54] or adrenal medullary secretion. High-frequency, low-intensity stimulation in the hypothalamus can produce characteristically slow tonic impulses in single efferent sympathetic fibers, and the frequency of these efferent signals then varies as would be expected with baroreceptor activation.[86] This and other experiments demonstrate that impulses of hypothalamic or higher origin pass through a synaptic break in the bulbar region.

The hypothalamus is involved in certain generalized responses such as the defense reaction and temperature regulation (p. 1043 and Chapter 59), and the cardiovascular element in these reactions is centered in the hypothalamus. Cutaneous vasodilatation and visceral vasoconstriction are an important part of the total thermoregulatory response to a hot environment, in addition to sweating and, in some animals, panting. The opposite stimulus of exposure to cold evokes cutaneous vasoconstriction as well as muscular contractions and shivering. In both cases the vascular and nonvascular components of the reaction depend on the same parts of the hypothalamus.

One of the most detailed studies of an integrated cardiovascular response to stimulation of selected telencephalic areas is that of Rushmer et al.[15] They found that stimuli applied in the fields of Forel produced changes in the dimensions, pressures, and external work of the left ventricle, as well as in heart rate, that closely simulated those seen during exercise. The "defense reaction," which is accompanied by discharge of the sympathetic cholinergic vasodilators to skeletal muscle and sympathetic adrenergic excitation of most other effectors, can also be elicited by localized hypothalamic stimulation.[57]

Hypothalamic influence on the cardiovascular system is not limited to these generalized reactions. The full primary response to chemoreceptor activation, for example, including bradycardia and the rise in total peripheral resistance, requires an intact hypothalamus. Although this appears to be less true for the baroreceptor reflex, it is probable that hypothalamic function modulates all reflex responses to some extent and determines in part the levels of tonic medullary discharge. The suprabulbar interneurons, interposed

between afferent fibers and the autonomic efferent mechanisms in the medulla, have been only partially mapped, but it is clear that projections of the baroreceptors, chemoreceptors, and vagal and glossopharyngeal afferents reach almost all levels of the brain, including the cerebral cortex.[9,39] The majority of the vagal pathways extend to the fronto-orbital cortex, whereas the principal glossopharyngeal projections reach the temporal cortex. The cingulate gyrus and connected rhinencephalic regions, together with the diencephalon, receive baroreceptor and chemoreceptor projections.[9]

Cerebral cortex

The influence of the highest levels of the brain on the circulation is a matter of common observation, as in the tachycardia and hypertension of emotional stress, or the now archaic vasomotor response known as blushing. Pertinent examples in animal experimentation are not lacking, for as Rushmer et al.[15] and others who have studied dogs exercising on a treadmill have noted, many of the cardiovascular responses begin when the animal sees the investigator reaching for the treadmill switch, anticipating by a few seconds the actual muscular exertion.

Quite marked cardiovascular effects can be produced by electrical stimulation of the cerebral cortex, even when precautions are taken to limit the physical effect of the stimulus to the superficial layers.[64] Responses that are predominantly of the depressor type can be elicited from the fronto-orbital cortex, the temporal cortex, piriform lobe, and parts of the amygdala.[9] The opposite effects (i.e., increased blood pressure, cardioacceleration, and increased adrenal catecholamine secretion) can be produced by stimulation of other parts of the amygdala and, to a lesser extent, from the motor cortex. Wall and Davis[94] concluded from experiments on the monkey that there are three cortical systems affecting autonomic function: the sensory-motor cortex, a temporal-cingulate system, and a posterior orbital-anterior insular system. The first two act through descending pathways that avoid the hypothalamus, but the cardiovascular portions of the orbital-insular group are eliminated by hypothalamic destruction. Included in the pathways from the motor cortex are neurons that subserve sympathetic cholinergic vasodilatation, which relay in the hypothalamus and tectum, but then descend without synaptic interruption to the cells of origin of the preganglionic spinal outflow.[78] Autonomic and somatic effects can be evoked independently by cortical stimulation, but they are usually functionally and anatomically related. The existence of pathways from the cortex to cardiovascular effectors is therefore not in question, but their internal organization and proper function remain a mystery. At least two firm conclusions are warranted, however, based in part on experiments in which parts of the cortex were removed. First, it seems safe to assume that parts of the cortex have a significant, although not essential, role in termperature regulation. Removal of the sensorimotor area of one hemisphere in the cat or dog produces vasodilatation in the opposite paws, and a limitation of the cutaneous vasoconstrictor response to cold.[3] Second, there is conclusive evidence that parts of the cortex and diencephalon are critical in the lung inflation reflex and thus have an indirect influence on the circulation. Korner and his colleagues[9,73,92] found that removal of the occipital lobe, parietal cortex, and midportion of the *area parecentralis* in the rabbit (*rhinencephalic* animals, in their terminology) produced few if any changes in the circulation apart from those secondary to the respiratory effects, the latter being manifested chiefly by an increase in respiratory minute volume. In this as in most normal responses, autonomic activity appears to be geared to respiratory performance. Whether the cyclic effects of normal respiration on central sensitivity to the baroreceptor input depend on cortical pathways has not yet been established.

Obviously a comprehensive understanding of cortical and diencephalic autonomic mechanisms is still far beyond our reach, but it is apparent that these higher subdivisions are not only engaged in affective or conditional responses but also act continuously in conjunction with the medullary centers. The ways in which the cortex and each of the lower centers work together to enforce the precise cardiovascular control achieved in the normal animal are still obscure. Certainly, it is not a matter of simple linear pathways in series, but rather, in Livingston's words, one "... of mutually interdependent loop circuits which stitch together the various parts of the brain into a functional whole."[13]

ANALYSIS OF CONTROL SYSTEM

The receptors, central neurons, and effectors discussed in this and the preceding chapter constitute an efficient control system by which the heart and blood vessels are made to function in accordance with the needs of the organism. In seeking to understand the operation of this system, the complementary techniques of analysis and synthesis are equally useful. Both the isolated parts of the control machinery and the total cardiovascular responses in vivo must be examined, in other words, the goal being an understanding of the interaction of the parts that accounts for the whole repertoire of natural responses. In this enterprise the physiologist and physician have found an ally in the control systems engineer. Concepts and analytic techniques developed for the design of mechanical and electrical control systems have proved to be especially valuable in experimental cardiovascular physiology.

Control theory

An introduction to control theory can be found in Chapter 7, and it will suffice here to identify it as a body of concepts, definitions, and mathematical relationships concerned with the transformation of input into output signals. The physical nature of the signals is irrelevant. Nerve impulses, blood pressures, chemical concentrations, and a host of other physiologic variables can be treated with equal facility. The principal values of this approach lie in its quantitative statements of relationships, in the recognition of crucial variables, and in the existing body of information about analogous control systems in engineering.

The first step in applying control theory is the identification of the components of the system and their interconnections, usually in the form of an explicit schema such as that shown in Fig. 7-5. In the baroreceptors, for example, the transmural pressure acting to distend the carotid sinus might be regarded as the "input signal." The system to be analyzed could be limited to the sinus wall and receptors, in which case the afferent impulses in the sinus nerve would be the "output." Alternatively, the complete baroreceptor reflex could be examined, using the systemic arterial pressure (or perhaps the impulses in some efferent cardiovascular nerve fiber) as the output. As this suggests, at least some prior information about the working of the system is needed to guide the tentative selection of a model. Next, experiments can be designed to get quantitative data relating output to input. This amounts to determining the "transfer function" of the system. The form of this function will depend initially on the assumptions made in the model, but the transfer function and the model itself are subject to change in the light of experimental results. The relation between frequency of impulses in the carotid sinus nerve and pressure in the sinus, for example, has been stated by Warner and his colleagues[35] as:

$$F(t) = A\frac{dP}{dt} + B\frac{dP'}{dt} + C[P(t) - P_{th}]$$

where F(t) is the firing rate of a single unit and P(t) the sinus pressure at time t, while dP/dt and dP′/dt are the time derivatives of pressure on the *rising* and *falling* parts of the pressure curve, respectively, and P_{th} is the pressure threshold for that unit, below which it stops firing. A, B, and C are sensitivity coefficients. Here P(t) is the input, F(t) is the output, and the equation expresses their relationship or "coupling function" (the term usually applied to a transfer function in which input and output dimensions are not identical). The third term on the right describes the relationship when pressure is kept constant at any level and would suffice in itself to express the response if the impulses generated by a series of steady-pressure levels were the only data available. Experiments show, however, that there is an additional effect when pressure is changing or is pulsatile, and that the receptor is not as sensitive to falling as to rising pressures (Fig. 42-5); the derivatives in the first and second terms describe these characteristics of the response. The component of the response related only to the absolute level of pressure would be referred to in control theory as proportional control, and the components that depend on rates of change would be termed rate control. Both are involved in almost all cardiovascular reflexes.

Analysis of circulatory control

Cardiovascular control operates through systems with feedback loops, which is to say that controlled variables—pressures, for example—have a voice in their own control. A rise in arterial pressure elicits through the baroreceptors a set of responses that tend to return the pressure to normal, which in turn is sensed and acknowledged by the baroreceptors. Such a closed-loop operation is typical of many efficient control systems, biologic and otherwise, but it makes detailed experimental analysis of the response difficult. For this reason, reflexes are often studied by eliminating the feedback (opening the loop). This can be done for the carotid baroreceptors, for example, by isolating the carotid sinuses from the rest of the circulation, preserving their innervation, and perfusing them at controlled pressures. A large body of information about cardiovascular control has been obtained from such open-loop preparations.* Because projections from many different centers converge on each autonomic effector, every effort is made in most such experiments to eliminate all potential variations of input other than that being studied. If

the primary baroreceptor response is to be tested, for example, chemoreceptor activity must be kept constant. Measurement of the interaction of reflexes, such as the alterations in baroreceptor response caused by various degrees of chemoreceptor activation,[9] requires a more complicated experimental design. In determining the output of the cardiovascular control system, it is essential to remember that there are outputs to each of the regional vascular beds and to all kinds of effectors. Operation of the control system under closed-loop conditions in vivo yields a more or less specific set of outputs for any given input profile, but reflex effects are usually not uniform in all parts of the circulation. A response is labeled "pressor" if vasoconstriction is sufficiently strong and widespread to raise systemic arterial pressure, but some regional beds may show little change and others may actually dilate. Investigation has consequently been focused on each part of the response as well as on the net effects represented by cardiac output and total peripheral vascular resistance.

Static properties: gain. The simplest experimental input to analyze is an unvarying stimulus of known intensity, such as a constant pressure in the carotid sinus. The outputs from a set of such constant stimuli at different levels of intensity constitute the "static response" of the system (Fig. 42-4). Its "dynamic response," on the other hand, is the relation between output and input for such signals as sinusoidal pressure waves or the normal pulsatile pressure. Certain parameters are conventionally used for quantitative description of both types of response.

The numeric value of the change in output per unit change of input in a control system (or one of its parts) is called "gain." The output/input relationship of all cardiovascular reflexes is nonlinear, often resembling the sigmoid curve plotted in Fig. 42-4, which represents the baroreceptor response. This response is minimal at low sinus pressures and reaches a maximum at high pressures (saturation), but for small segments in the middle of its range the relationship is approximately linear. The gain is thus minimal at the extremes of input, and maximal somewhere in the midregion. Values cited in the literature are usually average gains in the near-linear portion of the response.

The open-loop gain reported for complete reflex responses in the cardiovascular system is usually no greater than 2, although the range is about 0.5 to 15. Values of 1 to 2 are considered low in the general context of control systems, for they predict closed-loop operation that

*See references 24, 26, 31, 35, 44, 51, and 71.

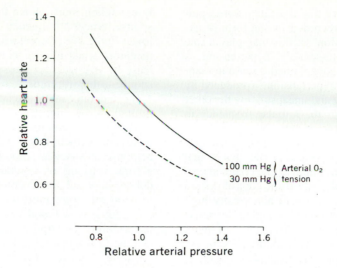

Fig. 42-10. Interaction of rabbit baroreceptor and chemoreceptor reflexes. Ordinate represents heart rate and abscissa represents systemic arterial pressure; both are scaled in relation to their resting value (1.0). Relationship is illustrated for two different arterial oxygen tensions: normal level of 100 mm Hg (unbroken line) and 30 mm Hg (broken line). (Modified from Korner.[9])

would correct only one half to two thirds of any disturbance. The homeostatic control of blood pressure seems to do much better than that, presumably because a large number of additive control loops are always at work.

The gain of any reflex, moreover, can be changed by other parts of the control system. This is illustrated by the increased gain in the arterial baroreceptor reflex from one isolated carotid sinus when all other baroreceptors are denervated.[24,88] Among other things, this finding demonstrates a significant conclusion arrived at by systematic analysis of control, for it suggests that denervation of the other receptors keeps them from occluding signals from the isolated sinus.[9] It supports, in other words, the idea that afferent-efferent pathways from many different baroreceptor units sometimes converge on the same autonomic effector.

Set point. The "set point," or level that a control system is designed to maintain for a given variable, is difficult to measure unequivocally in any cardiovascular response, but it is a useful operational concept. Response curves like those in Fig. 42-4 are conventionally assumed to represent operation around some "set point" on the ordinate, and a vertical shift in the experimentally measured points indicates a change in set point. Fig. 42-10 illustrates such a change, in this case displacement of the set point of the baroreceptor system by arterial hypoxia.[9] The measured output in this instance is heart rate, and the experimental input signal is the sinus pressure.

Lowering the arterial oxygen tension in unanesthetized rabbits with an intact CNS from 100 mm Hg (the level while breathing air) to 30 mm Hg displaced the relationship between sinus pressure and heart rate vertically downward (Fig. 42-10). This means that lowering the oxygen tension yielded a slower heart rate at any given sinus pressure. The set point for baroreceptor control of heart rate is thus altered by hypoxia, although the gain (indicated by the slope of the response) is unchanged.

Study of this phenomenon after ablation of successive parts of the brain shows that baroreceptor and chemoreceptor projections not only converge to some extent on the same efferent neurons but also travel through at least two functionally different sets of interneurons. Those in the diencephalon apparently render the chemoreceptor effects more susceptible to incoming baroreceptor signals, and those in the bulbar region render them less susceptible.[9] These interactions, which can be expressed at least crudely and in part by changes in set point and gain, make it clear that the convergence of virtually all inputs on every effector, the possibilities for occlusion or facilitation, and the variable size of the neuron pools involved combine to provide the diversity and selectivity that are characteristic of circulatory control.

Apart from this kind of "central resetting," it is possible for changes in the behavior of the receptors themselves to produce something resembling a change in set point. Electroneurograms of the carotid sinus nerve that indicate baroreceptor resetting have been recorded in one form of experimental hypertension[8,79]; this may play a role in the genesis of "essential" hypertension, a widespread and still unexplained disease in man. In dogs with "renal hypertension" (produced by removing one kidney and encapsulating the other in a plastic wrapping) the impulse frequency in the carotid sinus nerve at any given pressure is lower than normal. This would explain the hypertension if the converse were also true, that is, that the "normal" afferent impulse rate could be produced in these animals only if the control system pushed arterial pressure up to abnormally high levels. The central set point implied in this interpretation is an "expected" frequency of input from the baroreceptors. The abnormality in experimental renal hypertension appears to lie in the receptors themselves, however, not in the central set point. The alteration in the receptor, which has been called "peripheral resetting," consists of a change in the receptor coupling function. Whether it is a primary cause of hypertension or a kind of slow adaptation to elevations of blood pressure that have their origin elsewhere is not known. Changes in the physical properties of the sinus wall related to smooth muscle activity, water content, or aging have been considered as possible explanations but they remain unproved.

Dynamics. As we have already seen, the rate of change of input signals is as important as their absolute value in circulatory control. Time and frequency are therefore key variables in the operation of reflexes, and their influence can be investigated in several different ways. One of the simplest methods measures the time between input and resulting output of a complete reflex, expressed as the time constant of the recorded change from zero to maximal response. Sympathetic and parasympathetic responses differ considerably in this respect; the reflex changes in heart rate mediated by the vagus nerve have time constants of less than 1 sec, whereas those for sympathetic adrenergic effects are typically 15 sec or longer.[89]

Time constants can be measured experimentally by the application of a step function, but a more complete analysis of the dynamic behavior of a system can be made by examining its response to sinusoidal waves (Fig. 42-6). Comparison of the amplitudes and phase of input and output signals then gives the gain and phase shift of the system. The dynamic range of operation for complete responses of the cardiovascular system is restricted to very low frequencies, rarely extending much above 0.5 Hz. The static response, where "frequency" is zero, defines the lower limit. The afferent neural response of baroreceptors tends to rise with frequency up to at least 10 Hz, whereas the sympathetic nerve response to baroreceptor activation peaks at about 1 Hz and falls off at higher frequencies.[91] The limitations on high-frequency response of the cardiovascular control system as a whole evidently have their origin in the relatively slow response of vascular smooth muscle, which has a time constant of several seconds. The time delays imposed by receptors and neural transmission are comparatively very small. The *local* regulation of blood vessels (p. 1058), which plays almost as large a part in circulatory control as do the central reflex systems, shares the dynamic characteristics of smooth muscle. Experimental tests with sinusoidal inputs allow the time lag between input and output to be expressed as a *phase shift*, which generally increases with frequency. When examining phase shift in responses like those of the baroreceptors, the complete reflex response to a pressure elevation in the isolated sinus is represented by a *decrease* in systemic pressure, which is the equivalent of an inherent phase shift of 180 degrees.

One aspect of control theory that is potentially relevant to the circulation under abnormal conditions is the notion of instability. Engineers, understandably concerned about avoiding systems that may go "out of control," have devised mathematical methods of predicting this event, given certain basic properties of the system. The one convincing example of such behavior in the cardiovascular system is Sagawa's report on a reflex initiated by cerebral ischemia.[87] The data obtained in his analysis of this response indicated that an unstable condition would be reached as cerebral perfusion pressure was progressively lowered, and further experiments in animals with an isolated cerebral circulation confirmed this prediction. At very low cerebral perfusion pressures a cyclic rise and fall of systemic pressure amounting to some 80 mm Hg appeared and persisted as long as the cerebral pressure was kept at the same low level.

PATTERNS OF RESPONSE

Each complete reflex response has a distinctive pattern because of qualitative and quantitative differences in the effectors employed. The typical differentiation among efferent paths and effectors in the primary baroreceptor response to an increase in the pressure distending the baroreceptive vessels, for example, consists of:

1. Slowing of the heart rate, owing to a prominent increase in efferent vagal parasympathetic impulses and a lesser decrease in sympathetic discharge.
2. Slowing of conduction in myocardium and conducting tissues mediated in the same way.
3. Decreased myocardial contractility, produced by increased vagal tone and by the simultaneous decrease in sympathetic discharge.[44,77]
4. Decreased cardiac output, in which the slower heart rate plays a greater part than the depressed ventricular performance.
5. Systemic arteriolar vasodilatation, brought about by a lessening of sympathetic constrictor tone. All vascular beds share in this response to some extent, and total peripheral resistance falls, but the vasodilatation in skeletal muscle far exceeds that elsewhere; renal vasculature is the least affected and sometimes shows almost no response. Because of these differential changes in regional resistance, there is a redistribution of the cardiac output so that blood flow increases in skeletal muscle, whereas renal and splanchnic flow are almost unchanged.
6. Decreased mean systemic arterial pressure, owing more to the fall in vascular resistance than to the decreased cardiac output[90]
7. Systemic venodilatation[30,32] as a result of diminished sympathetic adrenergic impulses.

The effects of *decreasing* pressure in an isolated innervated carotid sinus are the reverse of those listed, and in addition there is sympathetic stimulation of adrenal catecholamine secretion. The differentiated character of the response arises in the CNS, not in the receptors or their afferent signals. The actions of baroreceptors are similar whatever their location, although the gain (Fig. 42-11) and the ratio of vagal to sympathetic activity called into play by baroreceptors vary with arterial pressure; the receptor population of the aortic arch is also slightly less sensitive and wider in its range of operation than that of the carotid sinuses[60,66]; and the aortic chemoreceptors seem to have somewhat less influence on heart rate than do those in the carotid bodies.

The cardiovascular control system is capable of many different patterns of response, but a few general statements can be made that apply to almost all of them. As a rule, reflex changes in cardiac output owe more to vagally induced changes in heart rate than to changes in the force

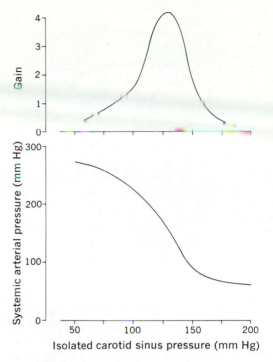

Fig. 42-11. Characteristics of baroreceptor reflex when carotid sinuses have been isolated from rest of circulation, while their innervation remains intact (''open-loop'' conditions, see text). When pressure is raised in isolated sinuses, reflex responses in heart and blood vessels lead to fall in systemic arterial pressure (bottom). Change in arterial pressure for a given change in sinus pressure can be expressed quantitatively as ''open-loop gain'' (top). (Modified from Korner.[9])

of contraction. This is true in the regulation of the basal state and in moderate activations of baroreceptors or chemoreceptors. With more severe demands on the circulation, changes in myocardial contractility as well as heart rate come into play. Under unusual circumstances, however, the cardiac output can be increased as needed with little or no increase in rate, as Donald and Shepherd[47] have shown in their work with exercising dogs whose ability to raise their heart rate was limited by cardiac denervation (Fig. 42-12).

Resistance versus output. Reflex elevations of arterial blood pressure are usually accomplished primarily by an increase in vascular resistance rather than cardiac output, although this depends in part on the starting point.[85] The control system behaves, in fact, as if it set the cardiac output at a level dictated by the metabolic needs of the body, without regard to baroreceptor activation. Reflexes obviously influence the heart,

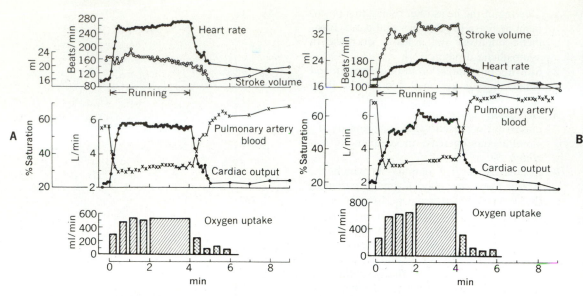

Fig. 42-12. Cardiorespiratory changes with exercise (5.5 km/hr, 21% grade). **A,** Normal dog. **B,** Dog with chronic cardiac denervation. (From Donald and Shepherd.[47])

but cardiac output rather than rate or contractility appears to be the critically controlled variable. During exercise, stimulation of the baroreceptors can radically alter ventricular work as well as peripheral resistance, but the net result is maintenance of the cardiac output at the required level and some elevation of the arterial pressure.[67,93,95]

Capacitance versus resistance. The extent to which resistance as compared with capacitance vessels is engaged by reflexes is doubtless an important factor in control, but one that is difficult to estimate quantitatively. Both are controlled through sympathetic adrenergic fibers. The potency of changes in sympathetic constrictor tone of the veins has been measured by experimental techniques, specifically designed for that purpose,[23,30,32,33] which show that reflex venoconstriction can readily displace 10% of the total blood volume from the veins into other parts of the circulation. Such alterations in venous capacity have a great effect on the rest of the circulation, although this is not obvious from the relatively small changes in venous and arterial pressures. The impact on cardiac output is apparent in the experiments of Braunwald et al.,[30] in which total vascular capacity in dogs was measured and controlled by an external reservoir and a pump was substituted for the heart. When venodilatation was produced in these animals by raising pressure in the isolated carotid sinuses, a decrease of 20% to 30% in the "cardiac output" of the pump was sometimes required to restore equilibrium and keep the total intravascular vol-

ume constant. Similar changes in cardiac output presumably accompany changes in venomotor tone in the intact animal. Stimulation of chemoreceptors by arterial hypoxemia and activation of the cardiac stretch receptors both reduce venous capacity.

Venoconstriction increases venous return, in the sense that contraction of the veins can redistribute the blood volume only by temporarily increasing venous flow into the heart. The period during which more blood may be returned to the heart and lungs than leaves them is necessarily brief, however, and the principal effect of increased venomotor tone is the new dynamic equilibrium established in the circulation, including an increased cardiac output. It is this effect that makes the reflex changes in venous capacity as important as those in arteriolar resistance.

Precapillary versus postcapillary effects. The action of reflexes on precapillary as contrasted with postcapillary parts of the microcirculation is another kind of differentiation, which allows the cardiovascular control system to take part in the regulation of plasma volume. In skeletal muscle, in particular, reflex sympathetic vasoconstriction increases precapillary more than postcapillary resistance and briefly constricts the precapillary sphincters.[10] The result in the capillaries, or exchange vessels, is a tendency toward increased resorption of fluid into the vascular system, hence an increase in plasma volume. Reflex sympathetic constriction in vessels of the intestine has similar effects, but only transiently, because local mechanisms act to restore the earlier state despite continued reflex stimulation (autoregulatory escape). The atrial receptors are the most important ones in governing the precapillary to postcapillary resistance ratio and are sensitive to quite small changes in blood volume.[10,61,83] Minimal changes in circulatory pressure thus suffice to regulate the intravascular volume.

Controlled responses to stress

Operation of the cardiovascular control system in vivo manifests itself in specific and appropriate patterns of response. Sympathetic adrenergic discharge is an efferent arm of control common to a great many responses, but it takes the form of highly selective, individualized reactions rather than a general response to a hypothetical "push button" in the bulbar centers. Cardiovascular behavior during exercise, hypoxia, and hemorrhage will serve to illustrate the diversity and specificity of reactions.

Exercise. The hemodynamic changes accompanying exercise are summarized in Chapter 58, and we will be concerned here only with the mechanisms that bring them about. The response begins immediately with the onset of exercise, although a steady state may not be reached for several minutes. The principal changes can be attributed to two kinds of response: sympathetic adrenergic discharge to the heart and vessels, and vasodilatation in the vascular beds of skeletal muscle produced by a local, presumably metabolic, stimulant (p. 1058). Tachycardia is one of the sympathetic effects, and at maximal work rates the heart rate is controlled solely by sympathetic stimuli. The alterations in regional resistance are graded so that the blood flow to the kidneys, brain, and mesentery changes little if at all, whereas most of the increment in blood flow goes to the exercising muscles. Despite vasoconstriction in nonmuscular beds, total peripheral vascular resistance falls, and there is some evidence that this fall actually precedes the changes in output and rate.[95] The muscular and abdominal "pump" mechanisms acting on deep veins (p. 957) also come into play immediately.

The increase in cardiac output is directly correlated with oxygen consumption and the work load, but how the control system achieves this adjustment is not clear. In the first place, experimental changes in heart rate alone do not change cardiac output, except with extreme bradycardia or tachycardia. Second, the denervated heart, in which the initial rapid development of tachycardia with exercise does not appear, neverthe-

less increases its stroke volume so as to reach an appropriate cardiac output[47] (Fig. 42-12). Venoconstriction, leading to redistribution of blood and an increased cardiac output, is probably a key part of the response. The system reacts as though vasomotor tone and the total blood volume, between them, determine the cardiac output, and the increased venomotor tone with exercise ensured that cardiac output increased. When the usual tachycardia is available and can increase the output sufficiently, there is little change in stroke volume, as in Fig. 42-12, *A*. When heart rate cannot be increased rapidly, however, as in the denervated heart (Fig. 42-12, *B*), the volume distribution and intravascular pressure relationships ensure that stroke volume increases enough to create the necessary output. The performance of the denervated heart must depend to a considerable degree on heterometric autoregulation (p. 996) and perhaps on circulating catecholamines.

The location of the receptors in the control loop, however, and what they monitor, are not readily apparent. There is normally no change in the arterial O_2 saturation, and baroreceptor signals from the raised arterial pressure would for the most part yield reflex responses opposite in direction to those that actually occur. The only "monitors" that are clearly hard at work in exercise are those nearest the exchange vessels of the contracting muscles, namely, the smooth muscle cells in terminal arterioles, whose purely local response to the levels of Po_2 and other metabolites is the major source of the vasodilatation in exercising muscle. If the blood flow to muscle were to become inadequate, the local accumulation of metabolites would increase local vasodilatation, and the exercising muscle would receive a correspondingly larger share of the cardiac output. The proportion of the cardiac output that goes to exercising muscle is thus locally controlled, but some additional mechanism is needed to sense the absolute output. At present, one can only guess at the factors involved, but cerebral blood flow is a possible candidate.

The higher level, or psychologic, contributions in exercise are evident in the anticipatory responses of animals trained to run on a treadmill or track runners at the starting line. In many ways, this part of the response resembles the so-called defense reaction,[18,19] which appears in animals when they become alert to any threat that may dictate aggressive behavior or escape. In man the same response is made to situations that evoke anxiety or intense emotion.[18] Like the exercise response the defense reaction can be simulated by stimulating selected areas in the hypothalamus, and the combination of cardiac excitation with general adrenergic vasoconstriction and cholinergic vasodilatation in muscle occurs in both.[52] Arterial pressure rises, but the part of the baroreceptor response that would otherwise tend to limit tachycardia and cardiac output is suppressed at the bulbar level,[74] an interaction also seen in exercise.[95] The centers involved are not limited to the hypothalamus but include parts of the central gray matter and midbrain tegmentum. This evidence suggests that the suprabulbar centers responsible for the defense reaction also play some part in the cardiovascular adaptations to exercise, but it seems unlikely that they can account for the exact matching of cardiac output to demand.

Hypoxia. The respiratory and cardiovascular responses to hypoxemia are quite different from those in exercise and are for the most part encountered only when the animal is exposed to an abnormal environment or is affected by pathologic processes. The response is initiated by activation of the chemoreceptors, but the secondary responses from stretch receptors in the lung and perhaps the atria, as well as the conventional baroreceptors, largely determine the cardiovascular effects. The threshold for the reflex corresponds to an arterial oxygen saturation of about 96%, and the response becomes more intense as the saturation falls. With moderate degrees of hypoxia the increase in rate and depth of breathing that is the most prominent effect is inversely proportional to the P_{O_2}, and the heart rate rises because of the secondary effects of hyperventilation. Sympathetic adrenergic discharge causes an increase in myocardial contracility,[49] venoconstriction, and a general increase in arteriolar resistance except in the skin. These effects combine to produce a rise in cardiac output and arterial pressure. At very low levels of P_{O_2} (less than 30 mm Hg) the respiratory effects begin to diminish, and the bradycardia characteristic of open-loop stimulation of the isolated chemoreceptors appears. Both the latter effects are diencephalically mediated.[9,92]

The interaction of different levels of the CNS has been defined more exactly in hypoxia than in any other reflex[9] (Fig. 42-13). In effect, suprabulbar centers act to suppress the primary bradycardia but enhance the arteriolar vasoconstric-

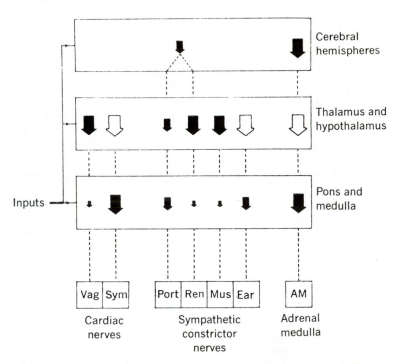

Fig. 42-13. Relative contribution of cardiovascular centers in pons and medulla, thalamus and hypothalamus, and cerebral hemispheres to autonomic effector response during severe arterial hypoxia (rabbit). Magnitude of afferent inputs assumed constant. Shaded arrows represent excitation of indicated effectors. Unshaded arrows indicate inhibitory effects. Size of arrow is roughly proportional to magnitude of effect. (From Korner.[9])

tion. At different levels of P_{O_2}, the magnitude of the autonomic effects depends on the ratio of chemoreceptor to lung inflation inputs, the latter acting primarily through suprabulbar centers. With severe hypoxia, there are regional differences in the degree of vasoconstriction, which depend in part on local vascular sensitivity to lowered O_2 tension.[34] At arterial tensions near 30 mm Hg, in unanesthetized rabbits, for example, the vasoconstriction in the portal circulation and ears is greater than elsewhere. Adrenal secretion of catecholamines is also increased, and this appears to play a more significant part in the total response of conscious animals than was suggested by earlier experiments using anesthetics.[9,92] In severe hypoxia the time course of the cardiovascular events reflects an initial primary chemoreceptor response followed by the slower development of secondary effects. When arterial P_{O_2} is dropped suddenly from normal to 30 mm Hg, bradycardia, decreased output, and a sharp rise in peripheral resistance appear very quickly, together with a transient rise in arterial pressure. In this earliest phase, input from the chemoreceptors is maximal, and the response is essentially that of open-loop experimental preparations. After a few minutes, all these effects diminish to some degree as baroreceptor and per-

haps atrial receptor input modify the response, and blood pressure falls as local vascular reactions to hypoxia partially counteract the reflex vasoconstriction. In control system terminology the early chemoreceptor activity provides rate sensitivity and the later baroreceptor response the proportional element of control.[9]

Teleologically, the purpose served by the response to hypoxia is only partly clear. The propriety of the increased respiratory effort is obvious, and with moderate degress of hypoxia the cardiovascular responses may do little more than meet the needs incident to the increased work of breathing. On the other hand, the bradycardia of severe anoxia, like the "diving response" described earlier (p. 1068), resembles a move toward a much lower metabolic state in which hypoxia could be better tolerated.

Hemorrhage. The circulatory reflexes triggered by acute hemorrhage are summarized in Fig. 42-14, together with an indication of the way in which they tend to compensate for blood loss. The cardiovascular control system in this situation responds primarily to the acute reduction in total blood volume, which is detected principally by atrial type B receptors. A reflex sympathetic adrenergic discharge increases heart rate and myocardial contractility, produces peripheral ar-

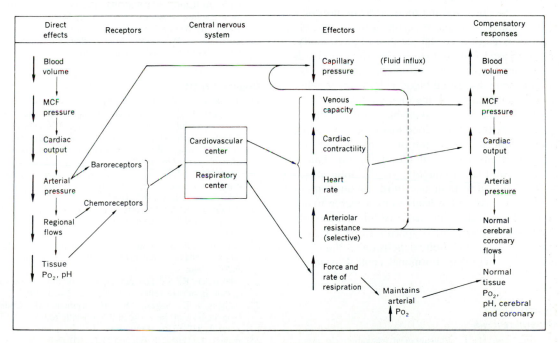

Fig. 42-14. Schematic summary of responses to hemorrhage. MCF pressure = mean circulatory filling pressure (p. 957). Compensatory responses tend to counteract direct effects of hemorrhage. Release of antidiuretic and adrenal cortical hormones is not included in diagram.

teriolar constriction and venoconstriction, and increases catecholamine secretion. As noted in other responses, the vasoconstriction is regionally selective, and in this case it is most marked in skin, muscle, splanchnic beds, slightly less in the renal vasculature, and absent in the coronary and cerebral circulations. These responses may or may not be sufficient to maintain cardiac output and arterial blood pressure at their normal levels, but even if output and pressure fall, the blood flow to heart and brain is preserved at the expense of other organs. Systemic arterial pressure usually remains normal during blood losses up to 10% of the total blood volume (nonhypotensive hemorrhage) but falls with greater amounts. Arterial pulse pressure is usually decreased by even small degrees of hemorrhage.[61] As blood flow to the limbs and splanchnic bed declines, the O_2 content of venous blood from these regions falls and the arteriovenous O_2 difference widens, reflecting a maintenance of tissue O_2 uptake in the face of diminished flow. When flow becomes too low to meet the O_2 needs directly, the tissues turn to anaerobic metabolism, with a concomitant fall in pH of the venous effluent. Acidosis then provides a chemoreceptor stimulus leading to hyperventilation. The by-products of anaerobic metabolism also act locally to produce vasodilatation and thus counteract the sympathetic constriction to some degree.

The venoconstriction produced by hemorrhage well illustrates the place occupied by venous capacity in cardiovascular control. Functionally, the increased venous tone in this situation tends to restore normal *mean circulatory pressure* (p. 957) despite decreased blood volume or, in other words, tends to match the total vascular capacity to the amount of blood available to fill it at physiologic pressure. If the blood loss is neither severe nor prolonged, restoration of normal blood volume by transfusion can return the circulatory state to normal. Reduction of the total blood volume by more than one third for an hour or more, however, often leads to irreversible hypotension and death. The cause of this irreversible state is unknown, but it appears to be a consequence of local metabolic deficits rather than failure of the neural or hormonal control systems.

REFERENCES
General reviews

1. Andersen, H. T.: Physiological adaptations in diving vertebrates, Physiol. Rev. **46**:212, 1966.
2. Aviado, D. M., and Schmidt, C. F.: Reflexes from stretch receptors in blood vessels, heart and lungs, Physiol. Rev. **35**:247, 1955.
3. Bard, P.: Anatomical organization of the central nervous system in relation to control of the heart and blood vessels, Physiol. Rev. **40**(suppl. 4):3, 1960.
4. Comroe, J. H., Jr.: Peripheral chemoreceptors. In Fenn, W. O., and Rahn, H., editors: Handbook of physiology. Respiration section, Baltimore, 1964, The Williams & Wilkins Co., vol. 1.
5. Dawes, G. S., and Comroe, J. H., Jr.: Chemoreflexes from the heart and lungs, Physiol. Rev. **34**:167, 1954.
6. Gauer, O. H., and Henry, J. P.: Circulatory basis of fluid volume control, Physiol. Rev. **43**:423, 1961.
7. Heymans, C., and Neil, E.: Reflexogenic areas of the cardiovascular system, Boston, 1958, Little, Brown, & Co.
8. Kezdi, P., editor: Baroreceptors and hypertension, Oxford, 1967, Pergamon Press, Ltd.
9. Korner, P. I.: Integrative neural cardiovascular control, Physiol. Rev. **51**:312, 1971.
10. Mellander, S., and Johansson, B.: Control of resistance, exchange and capacitance functions in the peripheral circulation, Pharmacol. Rev. **20**:117, 1968.
11. Oberholzer, R. J. H.: Circulatory centers in medulla and midbrain, Physiol. Rev. **40**(suppl. 4):179, 1960.
11a. Paintal, A. S.: Vagal afferent fibres, Ergeb. Physiol. **52**:75, 1963.
12. Paintal, A. S.: Vagal sensory receptors and their reflex effects, Physiol. Rev. **53**:159, 1973.
13. Randall, W. C., editor: Neural regulation of the heart, New York, 1977, Oxford University Press.
14. Reeve, E. B., and Guyton, A. C. editors: Physical bases of circulatory transport: regulation and exchange, Philadelphia, 1967, W. B. Saunders Co.
15. Rushmer, R. F., Smith, A. O., Jr., and Lasher, E. P.: Neural mechanisms of cardiac control during exertion, Physiol. Rev. **40**(suppl. 4):27, 1960.
16. Smith, A. O., Jr.: Anatomy of the central neural pathways mediating cardiovascular functions. In Randall, W. C., editor: Nervous control of the heart, Baltimore, 1965, The Williams & Wilkins Co.
17. Torrance, R. W., editor: Arterial chemoreceptors, Oxford, 1968, Blackwell Scientific Publications.

Original papers

18. Abrahams, V. C., Hilton, S. M., and Zbrozyua, A. W.: The role of active muscle vasodilatation in the alerting stage of the defense reaction, J. Physiol. **171**:189, 1964.
19. Adams, D. B., et al.: Cardiovascular changes during naturally elicited fighting behavior in the cat, Am. J. Physiol. **216**:1226, 1969.
20. Alam, M., and Smirk, F. H.: Observations in man upon a blood pressure raising reflex from the voluntary muscles, J. Physiol. **89**:372, 1937.
21. Alexander, R. S.: The effects of blood flow and anoxia on spinal cardiovascular centers, Am. J. Physiol. **143**:698, 1945.
22. Alexander, R. S.: Tonic and reflex functions of medullary sympathetic carciovascular centers, J. Neurophysiol. **9**:205, 1946.
23. Alexander, R. S.: The participation of the venomotor system in pressor reflexes, Circ. Res. **2**:405, 1954.
24. Allison, J. L., Sagawa, K., and Kumada, M.: Open-loop analysis of the aortic arch barostatic reflex, Am. J. Physiol. **217**:1576, 1969.
25. Angell, J., Daley, J. E., and Daly, M. DeB.: Nasal reflexes, Proc. R. Soc. Med. **62**:1287, 1969.
26. Angell, J., Daley, J. E., and Daly, M. DeB.: Comparison of reflex vasomotor responses to separate and combined stimulation of carotid sinus and aortic arch baro-

receptors by pulsatile and non-pulsatile pressure in the dog, J. Physiol. **209:**257, 1970.

27. Biscoe, T. J., and Purves, M. J.: Observations on carotid body chemoreceptor activity and cervical sympathetic discharge in the cat, J. Physiol. **190:**413, 1967.

28. Blaine, E. H., and Davis, J. O.: Evidence for a renal vascular mechanism in renin release: new observations with graded stimulation by aortic constriction, Circ. Res. **28**(suppl. 2):118, 1971.

29. Boyer, G. O., and Scher, A. M.: Significance of mesenteric arterial receptors in the reflex regulation of systemic blood pressure, Circ. Res. **8:**845, 1960.

30. Braunwald, E., et al.: Reflex control of the systemic venous bed, Circ. Res. **12:**534, 1963.

31. Bronk, D. W., and Stella, G.: The response to steady pressure of single end organs in the isolated carotid sinus, Am. J. Physiol. **110:**708, 1935.

32. Browse, N. L., and Shepherd, J. T.: Response of veins of canine limb to aortic and carotid chemoreceptor stimulation, Am. J. Physiol. **210:**1435, 1966.

33. Browse, N. L., Donald, D. E., and Shepherd, J. T.: Role of veins in the carotid sinus reflex, Am. J. Physiol. **210:**1424, 1966.

34. Chalmers, J. P., Korner, P. I., and White, S. W.: Local and reflex factors affecting the distribution of the peripheral blood flow during arterial hypoxia in the rabbit, J. Physiol. **192:**537, 1967.

35. Christensen, B. N., Warner, H. R., and Pryor, T. A.: A technique for quantitative study of carotid sinus behavior. In Kezdi, P., editor: Baroreceptors and hypertension, Oxford, 1967, Pergamon Press, Ltd.

36. Coleridge, H. M., Coleridge, J. C. G., and Kidd, C.: Cardiac receptors in the dog, with particular reference to two types of afferent ending in the ventricular wall, J. Physiol. **174:**323, 1964.

37. Coleridge, J. C. G., and Kidd, C.: Reflex effects of stimulating baroreceptors in the pulmonary artery, J. Physiol. **166:**197, 1963.

38. Comroe, J. H., Jr.: The location and function of the chemoreceptors of the aorta, Am. J. Physiol. **127:**176, 1939.

39. Crill, W. E., and Reis, D. J.: Distribution of carotid sinus and depressor nerves in the cat brain stem, Am. J. Physiol. **214:**269, 1968.

40. Daly, M. DeB., and Scott, M. J.: The cardiovascular responses to stimulation of the carotid body chemoreceptors in the dog, J. Physiol. **165:**179, 1963.

41. Daly, M. DeB., Hazzledine, J. L., and Howe, A.: Reflex respiratory and peripheral vascular responses to stimulation of the isolated perfused aortic arch chemoreceptors of the dog, J. Physiol. **177:**300, 1965.

42. Daly, M. DeB., Hazzledine, J. L., and Ungar, A.: The reflex effects of alterations in lung volume on systemic vascular resistance in the dog, J. Physiol. **188:**331, 1967.

43. De Castro, F.: Sur la structure de la synapse dans les chemorecepteurs: leur mecanism d'excitation et rôle dans la circulation sanguine locale, Acta Physiol. Scand. **22:**14, 1951.

44. DeGeest, H., Levy, M. N., and Zieske, H.: Carotid sinus baroreceptor reflex effects upon myocardial contractility, Circ. Res. **15:**327, 1964.

45. Degroat, W. C.: Actions of catecholamines on sympathetic ganglia, Circ. Res. **21**(suppl. 3):135, 1967.

46. Diamond, J.: Observations on the excitation by acetylcholine and by pressure of sensory receptors in the cat's carotid sinus, J. Physiol. **130:**513, 1955.

47. Donald, D. E., and Shepherd, J. T.: Initial cardiovascular adjustment to exercise in digs with chronic cardiac denervation, Am. J. Physiol. **207:**1325, 1964.

48. Downing, S. E., Mitchel, J. H., and Wallace, A. G.: Cardiovascular responses to ischemia, hypoxia, and hypercapnia of the central nervous system, Am. J. Physiol. **204:**881, 1963.

49. Downing, S. E., Talner, N. S., and Gardner, T. H.: Influences of hypoxemia and acidemia on left ventricular function, Am. J. Physiol. **210:**1327, 1966.

50. Dropmann, K.: Electron microscopic findings in the carotid sinus of rabbits. In Kezdi, P., editor: Baroreceptors and hypertension, Oxford, 1967, Pergamon Press, Ltd.

51. Ead, H. W., Green, J. H., and Neil, E.: A comparison of the effects of pulsatile and nonpulsatile blood flow through the carotid sinus on the reflexogenic activity of the sinus baroreceptors in the cat, J. Physiol. **118:**509, 1952.

52. Eliasson, S., et al.: Activation of sympathetic vasodilator fibers to the skeletal muscles in the cat by hypothalamic stimulation, Acta Physiol. Scand. **23:**333, 1951.

53. Eyzaguirre, C., and Lewin, J.: Effect of different oxygen tensions on the carotid body in vitro, J. Physiol. **159:**238, 1961.

54. Feigl, E. O.: Vasoconstriction resulting from diencephalic stimulation, Acta Physiol. Scand. **60:**372, 1964.

55. Fernandez de Molina, A., and Perl, E. R.: Sympathetic activity and the systemic circulation in the spinal cat, J. Physiol. **181:**82, 1965.

56. Floyd, W. F., and Neil, E.: The influence of the sympathetic innervation of the carotid bifurcation on chemoreceptor and baroreceptor activity of the cat, Arch. Int. Pharmacodyn. Therap. **91:**230, 1952.

57. Folkow, B., et al.: Changes in cardiac output upon stimulation of the hypothalamic defense area and the medullary depressor area in the cat, Acta Physiol. Scand. **72:**220, 1968.

58. Gammon, G. D., and Bronk, D. W.: Discharge of impulses from pacinian corpuscles in the mesentery and its relation to vascular changes, Am. J. Physiol. **114:**77, 1935.

59. Gero, J., and Gerova, M.: Dynamics of carotid sinus elasticity during pressor reaction, Circ. Res. **11:**1010, 1962.

60. Glick, G., and Covell, J. W.: Relative importance of the carotid and aortic baroreceptors in the reflex control of heart rate, Am. J. Physiol. **214:**955, 1967.

61. Gupta, P. D., et al.: Responses of atrial and aortic baroreceptors to nonhypotensive hemorrhage and to transfusion, Am. J. Physiol. **211:**1429, 1966.

62. Hakumaki, M. O. K.: Function of the left atrial receptors, Acta Physiol. Scand. **344**(suppl.):1, 1970.

63. Heymans, C., De Schaepdryver, A. F., and De Vleeschhouwer, G. R.: Abdominal baro- and chemosensitivity in dogs, Circ. Res. **8:**347, 1960.

64. Hoff, E. C., and Green, H. D.: Cardiovascular reactions induced by electrical stimulation of the cerebral cortex, Am. J. Physiol. **117:**411, 1936.

65. Hornbein, T. F., Griffo, Z. J., and Roos, A.: Quantitation of chemoreceptor activity: interrelation of hypoxia and hypercapnia, J. Neurophysiol. **24:**561, 1961.

66. Irisawa, H., and Ninomiya, I.: Comparison of the averaged nervous activities of aortic and carotid sinus nerves, Am. J. Physiol. **213:**504, 1967.

67. Iriuchijima, J., Soulsby, M. E., and Wilson, M. F.: Participation of cardiac sympathetics in carotid occlusion pressor reflex, Am. J. Physiol. **215:**1111, 1968.

68. Joels, N., and Neil, E.: The idea of a sensory transmitter. In Torrance, R. W., editor: Arterial chemoreceptors, Oxford, 1968, Blackwell Scientific Publications.

69. Johannson, B.: Circulatory responses to stimulation of somatic afferents, Acta Physiol. Scand. **57**(suppl. 198): 1, 1962.

70. Kezdi, P.: Control by the superior cervical ganglion of the state of contraction and pulsatile expansion of the carotid sinus arterial wall, Circ. Res. **2**:367, 1954.

71. Kezdi, P., and Geller, E.: Baroreceptor control of postganglionic sympathetic nerve discharge, Am. J. Physiol. **214**:427, 1968.

72. Koizumi, K., and Sato, A.: Influence of sympathetic innervation on carotid sinus baroreceptor activity, Am. J. Physiol. **216**:321, 1969.

73. Korner, P. I., Uther, J. B., and White, S. W.: Central nervous integration of the circulatory and respiratory responses to arterial hypoxia in the rabbit, Circ. Res. **24**:757, 1969.

74. Kylstra, P. H., and Lisander, B.: Differentiated interaction between the hypothalamic defense area and baroreceptor reflexes. II. Effects on aortic blood flow as related to work load on the left ventricle, Acta Physiol. Scand. **78**:386, 1970.

75. Landgren, S.: On the excitation mechanism of the carotid baroreceptors, Acta Physiol. Scand. **26**:1, 1952.

76. Ledsome, J. R., and Linden, R. J.: The effect of distending a pouch of the left atrium on the heart rate, J. Physiol. **193**:121, 1967.

77. Levy, M. N., et al.: Vagus nerves and baroreceptor control of ventricular performance, Circ. Res. **18**:101, 1966.

78. Lindgren, P.: The mesencephalon and the vasomotor system, Acta Physiol. Scand. **35**(suppl. 121):1, 1955.

79. McCubbin, J. W., Green, J. H., and Page, I. H.: Baroreceptor function in chronic renal hypertension, Circ. Res. **4**:205, 1956.

80. Muratori, G.: Histological observations on the structure of the carotid sinus in man and mammals. In Kezdi, P., editor: Baroreceptors and hypertension, Oxford, 1967, Pergamon Press, Ltd.

81. Neil, E.: Reflexogenic areas of the circulation, Arch. Middlesex Hosp. **4**:16, 1954.

82. Neil, E., and Joels, N.: Impulse activity in cardiac afferent vagal fibers, Arch. Exp. Pathol. Pharmakol. **240**: 453, 1961.

83. Oberg, B.: Effects of cardiovascular reflexes on net capillary fluid transfer, Acta Physiol. Scand. **62**(suppl. 229):1, 1964.

84. Oberg, B., and White, S.: Role of vagal cardiac nerves and arterial baroreceptors in the circulatory adjustments to hemorrhage in the cat, Acta Physiol. Scand. **80**:395, 1970.

85. Olmsted, F., McCubbin, J. W., and Page, I. H.: Hemodynamic causes of the pressor response to carotid occlusion, Am. J. Physiol. **210**:1342, 1966.

86. Pitts, R. F., Larrabee, M. G., and Bronk, D. W.: An analysis of hypothalamic cardiovascular control, Am. J. Physiol. **134**:359, 1941.

87. Sagawa, K.: Analysis of the CNS ischemic feedback regulation of the circulation. In Reeve, E. B., and Guyton, A. C., editors: Physiological bases of circulatory transport, Philadelphia, 1967, W. B. Saunders Co.

88. Sagawa, K., and Watanabe, K.: Summation of bilateral carotid sinus signals in the barostatic reflex, Am. J. Physiol. **209**:1278, 1965.

89. Scher, A. M., and Young, A. C.: Reflex control of heart rate in the unanesthetized dog, Am. J. Physiol. **218**: 780, 1970.

90. Shepherd, J. T., Corcondilas, A., and Donald, D. E.: Assessment by two independent methods of the role of cardiac output in the pressor response to carotid occlusion, J. Physiol. **170**:250, 1964.

91. Spickler, J. W., Kezdi, P., and Geller, E.: Transfer characteristics of the carotid sinus pressure control system. In Kezdi, P., editor: Baroreceptors and hypertension, Oxford, 1967, Pergamon Press, Ltd.

92. Uther, J. B., et al.: Bulbar and suprabulbar control of the cardiovascular autonomic effects during arterial hypoxia in the rabbit, Circ. Res. **26**:491, 1970.

93. Vatner, S. F., et al.: Effects of carotid sinus nerve stimulation of blood flow distribution in conscious dogs at rest and during exercise, Circ. Res. **27**:495, 1970.

94. Wall, P. D., and Davis, G. D.: Three cerebral cortical systems affecting autonomic function, J. Neurophysiol. **14**:507, 1951.

95. Warner, H. R., and Topham, W. S.: Regulation of cardiac output during transition from rest to exercise. In Kezdi, P., editor: Baroreceptors and hypertension, Oxford, 1967, Pergamon Press, Ltd.

43

WILLIAM R. MILNOR

Capillaries and lymphatic vessels

CAPILLARY CIRCULATION

The capillaries are the most important part of the cardiovascular system in one sense, for they are the only part of the system in which exchange between blood and tissue fluid takes place. The activities of the heart and all other parts of the system, together with their neural and humoral controls, are directed toward ensuring this essential exchange. The capillary circulation is remarkable for its division into numerous small channels, with a great broadening of the cross section of the bed. As a result, the velocity of blood flow in capillaries is relatively slow, and the total capillary surface to which blood is exposed is enormous (Table 35-1).

Architecture of microcirculation

Blood normally flows through only a small number of the capillary channels under resting conditions. In skeletal muscle, where the demand for oxygen varies widely in accordance with the degree of muscular activity, Krogh's quantitative anatomic studies indicate that the muscular capillaries in humans would be equivalent to 5% or more of the entire volume of muscle if all were open at once.[12] Since the muscles represent about half the total body weight, this implies a potential total capillary volume of about 1.5 L. In inactive muscles, Krogh found only about 1% of the capillaries patent; in humans, this would mean a muscle-capillary blood volume of about 150 ml. In the microcirculation of the skin, intestine, and kidneys, on the other hand, the number of capillaries open during activity may not be much greater than at rest. On the basis of such estimates the volume of capillary blood in the whole systemic circulation at rest has been estimated at 300 ml (Table 35-2). A muscle fiber probably receives oxygen by diffusion from capillaries across a distance that averages some 50 μm during rest and somewhere between 1 and 5 μm during work. The increase in the number of ac-

tive capillaries is of importance in decreasing the distance covered by the diffusion of oxygen molecules, a change that is necessary to speed the delivery of oxygen to the active fibers. Although Krogh estimated a 100-fold increase in the number of capillaries open in active muscle, others have concluded that the difference is much less, although still highly significant.[18]

The arrangement of the vessels interposed between the arterioles and the venules differs from tissue to tissue, but the pattern almost everywhere appears to be some variation of that described by Zweifach[31] and his co-workers (Fig. 43-1). It can be seen that the arrangement permits large changes not only in the total amount of blood flowing through the tissue but also in the course of the flow. The true capillaries, endothelial tubes devoid of perivascular contractile elements, are not in the direct path of flow from arterioles to venules. The main direct path is via thoroughfare or preferential channels, the metarterioles. These vessels are side branches of the well-muscled arterioles, which they leave almost at right angles. In its proximal portion the metarteriole is discontinuously coated with smooth muscle cells, but distally, these effector cells become fewer and fewer and finally disappear. The network of true capillaries carries blood between the proximal and distal portions of one or more metarterioles.

Just as it leaves the parent vessel, each capillary offshoot is embraced by one or two smooth muscle cells. This is the only muscular portion of a true capillary; it is called a precapillary sphincter. It is obvious that contraction of these strategically placed muscle cells will remove the true capillaries from the active circulation and confine the blood flow to the preferential channels, a change that can be observed under the microscope. Furthermore, the arrangement is such that the flow through the preferential channels (and through any capillaries issuing from them whose

1085

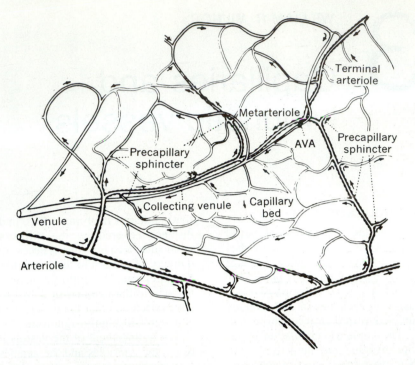

Fig. 43-1. Camera lucida drawing of microcirculation in mesentery of cat. Vascular smooth muscle is indicated by scalloped thickening of vessel wall. Several different types of preferential channels from arteriole to venule are seen. Capillary network is supplied by two separate arterioles. Note anastomosis, *AVA*, between metarteriole and collecting venule. (From Zweifach.[30])

sphincters are relaxed) can be regulated by the degree of contraction of the muscle of the arterioles.

It is evident that the microcirculation is so arranged that it can readily be adjusted to meet the varying metabolic requirements of the tissues. Exchange of water, respiratory gases, and other materials doubtless occurs through the walls of the preferential channels, especially in their distal segments, as well as through the bare endothelium of the true capillaries, but the total surface of the latter is much greater. When the tissue cells require a greater exchange, it is provided by an opening of the precapillary sphincters as well as by a dilatation of arterioles and metarterioles.

The relative number of metarterioles and true capillaries varies in different tissues. In tissues such as the mesentery, where metabolic activity does not vary greatly, there are 2 or 3 times as many true capillaries as preferential channels (Fig. 43-2). In skeletal muscle, which undergoes enormous fluctuations in its nutritional requirements, the ratio is 8 or 10 to 1.

In their careful studies of the circulation in the bat wing, Nicoll and Webb[21] found no thoroughfare or preferential channels; the arterioles were seen to branch into true capillaries after becoming

less and less muscular. Zweifach[30] has indicated that the capillary loops characteristic of the nail bed skin in humans are direct offshoots of the terminal arterioles, but each has a precapillary sphincter at its origin. Such may be the case in other cutaneous areas.

Although the metarterioles constitute the main direct path between the arterial and venous portions of the peripheral circulation, there are in addition arteriovenous anastomoses (AVA), or short connecting vessels, that have a strongly developed muscular coat and are under nervous control. These tend to be either widely dilated or fully constricted. These shunts are found not only between small arteries and small veins and between arterioles and venules but also, as shown in Fig. 43-1, between metarterioles and adjacent venules. They are found in many tissues and organs. In the skin, they are especially abundant in apical regions such as the fingers, toes, and earlobes.

The ultrastructure of the capillaries as revealed by electron microscopy is not the same in all parts of the circulation.[24] Three different kinds of capillary walls have been identified, for which the terms "continuous," "fenestrated," and "discontinuous" are convenient and descriptive.

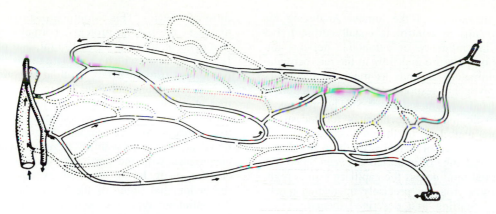

Fig. 43-2. Camera lucida drawing of capillary bed in mesentery of dog. Interlacing network of true capillaries (indicated by dotted lines) supplied by metarterioles derived from two arterioles and drained by two venules. (From Zweifach.[30])

In the *continuous* type the capillary wall is a continuous membrane of endothelial cells, 0.1 to 0.3 μm thick except at the sites of the nuclei. The wall is not literally continuous, for there are numerous intercellular channels, 40 to 50 Å wide, connecting the lumen of the capillary with the interstitial space around it.[11] The existence of "pores" of this magnitude was predicted by studies of capillary function long before supporting morphologic evidence was forthcoming.[14] This is perhaps the most common kind of capillary, being found in skeletal and smooth muscle, fat, connective tissue, and the pulmonary circulation. The cerebral capillaries are similar in appearance except for the absence of intercellular pores.

Fenestrated capillaries are so labeled because of intracellular fenestrations in the endothelial wall. These openings in the cell are 0.1 μm or less in diameter, often but not always closed by a very thin membrane (Fig. 43-3). Such vessels are characteristic of the renal glomeruli and intestinal mucosa. In the *discontinuous* type of capillary the wall is interrupted at intervals by relatively large intercellular gaps, through which formed elements of the blood as well as fluid can pass freely. Capillaries in the bone marrow, hepatic sinusoids, and spleen are of this type. To a large extent the transfer of substances between blood and interstitial fluid in the capillaries of different regions can be correlated with these differences in capillary structure.

Capillary pressure and flow

Although blood flow through a capillary segment is obviously a function of the pressure applied as a driving force and the viscous resistance of blood and the vascular tube, it is not possible to present a quantitative description of capillary

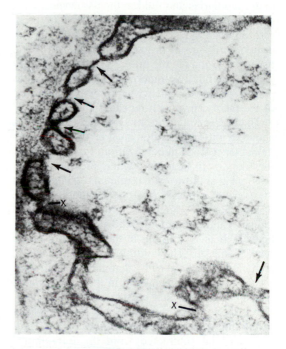

Fig. 43-3. Electron photomicrograph of section across venous end of capillary in mouse jejunum. Capillary lumen lies on right. Arrows indicate fenestrae in capillary wall. *X*, Fenestrae that lie partly out of section. (From Casley-Smith.[3])

resistance to flow similar to that in the larger vessels (p. 1018). The architecture of capillary networks makes physical analysis of pressure-flow relationships a difficult challenge for the engineer of fluid dynamics as well as the physiologist.[9,31]

No simple law applies, as might be expected in channels in which the erythrocytes and other formed elements of the blood are of roughly the same magnitude as the lumen. Analyzing the mechanisms of control is equally difficult because the influence of autonomic nervous impulses on pressure and flow in the microcirculation ultimately depends as much on secondary local myogenic responses as on the reaction of innervated smooth muscle.[26] Steady progress in the development of techniques for measuring pressure and flow in the microcirculation* is beginning to resolve many of these problems.

The exchange of materials between blood and interstitial fluid across the capillary wall is regulated by a group of precisely interacting mechanisms. Starling first described them in 1896, and subsequent investigation has modified his hypothesis in some details without altering the general conclusions. In essence, he proposed that the direction and rate of transfer between plasma in the capillaries and fluid in the tissue spaces depended on three things: (1) the hydrostatic pressure on each side of the capillary wall, (2) the osmotic pressure of protein in plasma and in the tissue fluid, and (3) the properties of the capillary wall.[25] The relative impermeability of the wall to plasma proteins, which was an essential feature of the theory, has been repeatedly confirmed.

The average pressure in systemic capillaries is about 25 mm Hg, but this must be regarded as no more than a broad generalization, for it represents the midpoint in a capillary where pressure may be 35 mm Hg at the arteriolar inlet and 15 mm Hg at the venular outlet.[14] Average capillary pressure is less than 10 mm Hg, moreover, in the pulmonary capillaries and hepatic sinusoids and may be as high as 70 mm Hg in the capillaries of the renal glomeruli. The continual opening, closing, and shifting of flow paths that can be seen when a capillary bed is visualized in the living animal also makes it clear that pressure in any one capillary is by no means constant with time. Pressure in the interstitial fluid environment of the capillary must be subtracted from the intraluminal pressure in the capillary to arrive at the effective transmural pressure, and here the experimental data are meager and conflicting. Some measurements suggest a negative or subatmospheric pressure in the tissue fluid spaces; others, made under somewhat more physiologic conditions but thus far only in the wing of the bat, indicate pressures between 0 (atmospheric pressure) and 2 mm Hg.[28]

The colloid osmotic pressure of plasma is in the neighborhood of 26 mm Hg in humans, but it varies considerably in different individuals and in different species. This osmotic pressure is attributable to the plasma proteins, normally present in a concentration of 7 gm/100 ml, and particularly to albumin, which exerts two thirds or more of the total osmotic effect. Colloid osmotic pressure in the interstitial fluid is much lower than that of plasma in most regions, so that the net osmotic force favors adsorption of fluid from the tissue spaces. Marked regional differences exist, however, since permeability to protein varies in accordance with the particular structure of the capillary wall. In the liver, for example, where the endothelial lining of the sinusoids is of the discontinuous type, the protein concentration of lymph is 80% to 90% of that in plasma.

The classic Starling explanation of fluid exchange in capillaries is shown diagrammatically in Fig. 43-4. In this hypothetical example the hydraulic pressure (derived from the driving force of the heart) is assumed to be 35 mm Hg near the entrance of the capillary and 15 mm Hg near its exit. Tissue fluid pressure is arbitrarily set at 5 mm Hg along the whole length of the capillary. Colloid osmotic pressures of 26 mm Hg in the plasma and 1 mm Hg in the tissue fluid are assumed, giving a net, or effective, osmotic pressure of 25 mm Hg. Near the entrance the forces tending toward fluid movement from capillary to tissue fluid consequently outweigh those acting in the opposite direction by 5 mm Hg. At the capillary midpoint the forces are about balanced, whereas near the venule they favor capillary resorption from the interstitial fluid.

The Starling hypothesis thus predicts outward flow at the arteriolar end and inward flow at the venular end. This is doubtless the case in some capillaries under some conditions, but several modifications of this simple scheme are required to make it generally applicable and consistent with experimental observations. First, it represents only one of many possible sets of conditions, and it is easy to conceive of changes in precapillary or postcapillary resistance that would produce resorption along the entire length of one capillary, while filtration occurred outward at all points along another. Furthermore, recent measurements indicate that venular hydraulic pressure may be well above the effective colloid osmotic pressure, at least in the mesentery.[10,32] Lymph drainage would necessarily play an important part in maintaining local fluid equilibrium under such conditions, and there is other evidence that this is the case. In resting skeletal muscle, on the other hand, in which lymph flow is minimal, filtration and resorption must balance out, and measurements in this bed show that

---

*See references 10, 28, 29, 31, and 32.

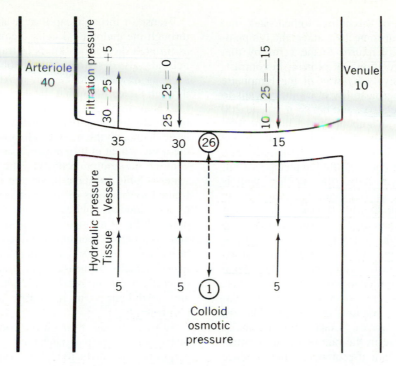

Fig. 43-4. Diagram of capillary carrying blood from arteriole to venule and forces that influence transfer across capillary wall. Numbers represent pressures (expressed in millimeters of mercury). Pressures in tissue fluid indicated at bottom, intravascular pressures shown within vessels, and net filtration pressure indicated above. Colloid osmotic pressures are circled. Osmotic pressure in tissue fluid is assumed to be 1 mm Hg, and colloid osmotic pressure in blood 26 mm Hg. These are opposed forces, and at all points along capillary, net resultant force of 25 mm Hg tends to cause movement of water and noncolloid solutes into capillary. Hydraulic pressure in tissue spaces is assumed to be 5 mm Hg at all locations. Hydraulic pressure in capillary falls from 35 mm Hg at arteriolar end to 15 mm Hg near venous end. Net hydraulic pressure, or difference between intravascular and tissue fluid pressures, therefore tends to move fluid out of capillary at arteriolar end and tends to resorb fluid from tissue spaces back into capillary near venous end. Filtration pressure includes both hydraulic and colloid osmotic pressures.

venular pressures are distinctly lower than the colloid osmotic pressure of plasma.[10]

Further modification of the Starling principle has been made necessary by the discovery that the interstitial space is not a homogeneous fluid compartment but an organized material resembling a gel, in which molecules of even medium size do not diffuse freely.[6,27] Transport within the tissue spaces themselves is consequently a problem of some complexity, and the solute concentration immediately adjacent to the capillary wall, which is the factor involved in capillary transfer, may differ appreciably from that at a greater distance. Some experimental observations lead one to suspect that tissue fluid pressures gradually decline from the arteriolar to the venular ends of the capillary,[28] but a firm conclusion on this point is not yet possible. Finally, the traditional view that true capillaries are the only site of exchange with interstitial fluid is no longer strictly tenable, particularly in the pulmonary microcirculation (p. 1109). A limited portion of the microcirculation apparently shares the exchange functions to some degree, and there is some evidence that the total surface of the venules may be greater than that of the capillaries.[26,27]

Transcapillary exchange

When all the capillaries in the body are taken into account, the fluid movement between capillaries, the extracellular space, and lymphatics constitute a kind of secondary "circulation,"[14] comparable in importance if not in magnitude to the cardiovascular system. Leaving aside the renal glomerular filtrate, an average of perhaps 14 ml of fluid/min is transferred from the capillary system to the extracellular compartment.

The greater part of this is reabsorbed back into the capillaries, but some 10% enters the lymphatics and eventually returns to the veins by that route. The "circulation" of protein, in contrast, is almost exclusively by way of the lymphatic system. The impermeability of capillary walls to protein is only relative, and between 80 and 200 gm protein/day (an amount approaching the total intravascular protein content) leaves the blood vascular system through the capillaries, all but a few percent of it entering the lymphatic system. Passage of protein across the capillary wall is often referred to as "leakage," as if it were in some way undesirable, but the steady transfer of relatively small amounts of plasma protein into the interstitial fluid is a perfectly normal physiologic process, and virtually all of it is returned to the bloodstream by lymph flow under normal conditions. In certain pathologic states, however, capillary permeability to protein is greatly increased and large amounts are irretrievably lost into the tissue spaces. Local anoxia usually causes an increase in the rate of capillary filtration (Fig. 43-5) and, if prolonged, may produce irreversible damage to the wall and high rates of protein extravasation.

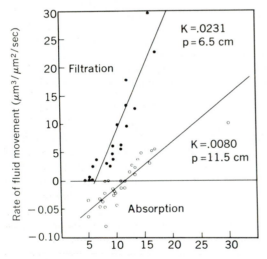

Fig. 43-5. Chart showing rate of filtration in frog's mesenteric capillary 1 min (dots) and 15 min (circles) after 3 min interruption of blood flow. Effect of oxygen lack is shown by greater steepness of upper curve. Ordinates indicate rate of fluid movement in cubic micrometers per square micrometer of capillary wall per second; abscissae, capillary pressure in centimeters of water. *K*, Filtration constant representing filtration rate; *p*, apparent effective osmotic pressure of plasma proteins or pressures at which no fluid transfer occurred. (From Landis.[13])

Transport through capillary walls takes place through the endothelial cells themselves as well as through the various kinds of transmural channels already described. Lipid-soluble materials such as oxygen and carbon dioxide can travel through the cells, so that the entire capillary surface is available for their transfer. The total area of functioning capillary surface depends to a large degree on the number of capillaries open at the time, which in turn depends on arteriolar and venular vasomotor activity. The exchange of water-soluble substances, on the other hand, is limited to the pores of the capillary wall. As far as such aqueous solutes are concerned, capillary permeability is a function of the size of the pores and their density per unit capillary surface. Water-soluble "tracers" are therefore often used experimentally to measure permeability. Certain naturally occurring substances, notably histamine and bradykinin, can reversibly alter capillary permeability.

Substances pass through the capillary wall by one of two principal mechanisms: filtration-absorption and diffusion.[6,14,17] Pinocytosis, or transport by vacuoles that traverse the endothelial cells, is a third possibility that has been little explored, but it is probably of minor significance in this context. *Filtration* through the pores of the capillary wall into the interstitial space, and the reverse process of absorption from the interstitial space, have already been discussed along with their relation to hydraulic and colloid osmotic pressures. *Diffusion* across the capillary membrane is of two kinds, one the conventional free diffusion through an available space and the other a restricted, or pore-determined, diffusion in which the relation between pore size and molecular size becomes an important element. Free diffusion is available for lipid-soluble materials, which can pass through the endothelial cells as well as through pores. Pore-restricted diffusion applies for water-soluble substances, which can traverse the pores but not the cell walls. Such substances can diffuse freely if their particle size is small in comparison with the pore openings, but not otherwise. Small molecules like glucose, for example, pass readily. The albumin molecule, which is almost as large as the average pore width, has a much lower diffusion coefficient.

Methods have been developed for measuring capillary transfer in preparations ranging from a single capillary to more or less intact regional vascular beds. Landis and Pappenheimer[14] devised one of the earliest techniques for measuring the filtration-absorption process, using a glass microtool to occlude a capillary segment, and the observed motion of red blood cells to estimate the subsequent rate of fluid movement into or out of the vessel. The method has been refined in diverse applications, including studies of diffusion coefficients in the interstitial space and of differences in permeability along the length of capillaries.[27] In isolated perfused hind limbs of the dog and cat, filtration-absorption has been examined by equating changes in weight of the preparation with changes in net fluid transfer across the capillary membrane. When the weight is constant, the rates of filtration and absorption presumably balance each other exactly, and the experiments are designed to estimate the capillary pressure needed to ensure this *isogravimetric* condition. Blood flow must be maintained through the vascular bed to preserve normal function of the microcirculation and capillary walls; therefore capillary pressure must be inferred from measurements of arterial and venous pressure. Despite this quite indirect way of estimating the level of capillary pressure that produces no net gain or loss in the intra- and extravascular compartments, this approach has produced much useful information.[14,22] For any given protein concentration of plasma, for example, there is a corresponding isogravimetric capillary pressure. Changes in arterial and particularly in venous pressures produce a gain or loss of limb weight in accordance with the displacement of capillary pressure from the isogravimetric level. The same method and principles can be applied to determine whether the predominant effect of vasomotor stimuli or drugs is on the arteriolar or venular segments of the microcirculation.

Quantitative analysis of diffusion requires knowledge of the available surface, the diffusion coefficient, and the concentration gradient. None of these can be measured with any precision under physiologic conditions, but several functional methods of estimating capillary diffusion are available. One is the measurement of capillary transfer of radioactive rubidium (^{86}Rb), which exhibits a pore-restricted diffusion similar to that of potassium.[23] Under proper conditions the clearance (p. 1174) of rubidium from the blood can be used to arrive at what is sometimes called a *capillary transport coefficient,* a function of diffusion rate and the capillary surface area. The method cannot separate these two factors but represents the product of permeability per unit capillary surface and total surface area. It has been useful, nevertheless, in detecting regional differences in capillary diffusion. The intestinal mucosa, for example, has been found to have a

transport coefficient many times higher than that of resting muscle. From histologic and other evidence, it is reasonable to assume that this reflects a difference in both density of capillary distribution and permeability of the individual capillaries. Indicator-dilution curves (p. 1000) have been used for the same purpose, one indicator being used as a tag for the diffusible substance under investigation and another strictly intravascular material as a reference indicator.

Capillaries of the skin

One vascular response in the cutaneous microcirculation is of particular interest because it involves an axon reflex and a local response to some vasoactive substance. If a blunt point is drawn lightly over the skin, a white line slowly develops along the line of stimulation. This appears to be a direct and sharply localized vascular response to mechanical stimulation or to some substance liberated by it; it does not depend on nerves and can be obtained in denervated areas. If the stroke across the skin is made with more force, three somewhat different effects follow, called the cutaneous *triple response.*[16] They have a latency of about 20 sec and reach a maximum in approximately 3 min. One component is a red line of vasodilatation along the course of the stroke, with a pale area of constriction on either side. The second is a reddened area of vasodilatation (flare) extending some centimeters around the first reaction. The mechanism of this second component of the response is an axon reflex, involving local nerves but not the CNS (p. 1055). The third component is formation of a wheal along the line initially scratched, due to extrusion of fluid into the tissue spaces. A similar triple response can be elicited by the intradermal injection of a minute quantity of histamine, and the local release of a histamine-like substance may account for the mechanically induced phenomena.

LYMPHATIC SYSTEM

The lymphatic system consists of a branched arrangement of vessels, and the finest branches of the system resemble capillaries, except that they are closed at one end.[1,4,8] This "blind" ending lies in the interstitial space, and the single layer of endothelial cells that makes up the wall of the "lymph capillary" is the site for absorption of substances from tissue fluid into the lymph. These small vessels join together to form larger trunks that eventually enter lymph glands, where they form a rich plexus of small vessels before converging again to form an efferent vessel. Lymph may pass through one or more such glands before it reaches the smaller collecting ducts, from whence it drains into the thoracic ducts and so into the veins. Lymphatics are found in almost all tissues and are almost as numerous as the blood capillaries.[7,8] The function of the lymphatic system appears to be that of a homeostatic mechanism, clearing the interstitial spaces of proteins, lipids, and even foreign materials.

Pressure, flow, and composition of lymph

Pressure in the peripheral lymph vessels is low, ordinarily 1 to 3 mm Hg.[20,28] Direct measurements with microtechniques indicate that intralymphatic pressure is very nearly the same as local interstitial pressure.[28] A small pressure gradient from the periphery to the major central trunks[7] favors flow in that direction, but lymph flow is largely attributable to external compression combined with the effect of centrally directed valves in the bigger vessels. External massage, for example, can raise peripheral lymphatic pressure in the limb of a dog to 50 mm Hg.[7] Contractions of the skeletal muscles, and in some locations the normal pulsation of arteries, promote lymph flow by similar compression. Lymph moves slowly, but the total lymph flow in humans is 2 to 4 L/day, an amount roughly equivalent to the plasma volume.[1,14] The rate of formation and flow of lymph is highly variable. The fluid in a lymph vessel may be quite stagnant for long periods of time,[5] but muscular activity, massage, hyperemia, or increased venous pressure will produce a relatively rapid flow.[1,7,8,15] Inflammatory reactions to chemical irritants have a similar effect.

The colloid osmotic pressure of lymph depends on its protein content, which varies in different regions of the body. At rest, the protein concentration in lymph from the skeletal muscles of the limbs is about 15% of that in plasma, whereas the corresponding figure for lymph from the intestinal tract is nearer 50%. The highest protein concentration is found in lymph from the liver, which contains almost as much protein as the blood plasma. Lymph will clot in vitro, although less readily than blood, and it contains fibrinogen in amounts approximating 50% of the serum level. The lymphatic system is the major route for transfer of protein from the tissue spaces back to the blood, as discussed under capillary function. Apart from protein, the concentrations of solutes in the lymph are close to those in plasma, although calcium is somewhat lower in lymph than in plasma, and chlorides and nonelectrolytes are slightly higher.

Permeability

The terminal lymphatics are highly permeable in the usual sense, but they also have the ability to absorb large particles with diameters of at least 25 μm. Lymph is in equilibrium with tissue fluid with respect to small molecules such as glucose or urea and allows two-way diffusion up to molecular weights of 2,000 to 6,000. Larger molecules, together with red blood cells, white blood cells, carbon, or other foreign particles, are admitted to the lymphatics but leave only by way of the thoracic ducts or by extraction in the lymph glands. The most striking example of particulate absorption is that of the lacteals in the intestines, and lipid globules, or *chylomicrons* (Chapter 53), appear in great abundance in the thoracic ducts after a meal. Ultrastructure of the wall in lymph capillaries has been studied in only a few species and regions, but open intercellular junctions similar to those in blood capillaries have been found in some places, as have quite large gaps in the endothelium. Some observations suggest that passage of vesicles through the endothelial cells may be an important mechanism of transport.[1]

One function of the lymphatic system is absorption from the pleural, peritoneal, pericardial, and joint spaces.[2,20] The fluid in these serous cavities is not in direct communication with lymph, but absorption takes place through lymphatic capillaries in the same way as in other tissue spaces. Dyes or particulate matter injected into these cavities appear fairly promptly in the thoracic duct lymph.

REFERENCES

1. Allen, L.: Lymphatics and lymphoid tissues, Ann. Rev. Physiol. **29:**197, 1967.
2. Allen, L., and Vogt, E.: A mechanism of lymphatic absorption from serous cavities, Am. J. Physiol. **119:**776, 1937.
3. Casley-Smith, J. R.: Endothelial fenestrae in intestinal villi: differences between the arterial and venous ends of the capillaries, Microvasc. Res. **3:**49, 1971.
4. Casley-Smith, J. R., and Florey, H. W.: The structure of normal small lymphatics, Q. J. Exp. Physiol. **46:**101, 1961.
5. Clark, E. R., and Clark, E. L.: Further observations on living lymphatic vessels in the transparent chamber in the rabbit's ear: their relation to the tissue spaces, Am. J. Anat. **52:**273, 1933.
6. Crone, C., and Lassen, N. A., editors: Capillary permeability, Copenhagen, 1970, Munksgaard International Booksellers & Publishers, Ltd.
7. Drinker, C. K., and Field, M. E.: Lymphatics, lymph and tissue fluid, Baltimore, 1933, The Williams & Wilkins Co.
8. Drinker, C. K., and Yoffey, J. M.: Lymphatics, lymph and lymphoid tissue, Cambridge, Mass., 1941, Harvard University Press.
9. Fung, Y. C., and Zweifach, B. W.: Microcirculation: mechanics of blood flow in capillaries, Ann. Rev. Fluid Mechanics **3:**189, 1971.
10. Intaglietta, M., Pawula, R. F., and Tompkins, W. R.: Pressure measurements in the mammalian microvasculature, Microvasc. Res. **2:**212, 1970.

11. Karnovsky, M. J.: The ultrastructural basis of transcapillary exchanges, J. Gen. Physiol. **52**(suppl.):64, 1968.

12. Krogh, A.: The anatomy and physiology of capillaries, New Haven, 1929, Yale University Press.

13. Landis, E. M.: Micro-injection studies of capillary permeability. III. The effect of lack of oxygen on the permeability of the capillary wall to fluid and to the plasma proteins, Am. J. Physiol. **83**:528, 1928.

14. Landis, E. M., and Pappenheimer, J. R.: Exchange of substances through the capillary walls. In Hamilton, W. F., and Dow, P., editors: Handbook of physiology. Circulation section, Baltimore, 1963, The Williams & Wilkins Co., vol. 2.

15. LeBrie, S. J., and Mayerson, H. S.: Influence of elevated venous pressure on flow and composition of renal lymph, Am. J. Physiol. **198**:1037, 1960.

16. Lewis, T.: Blood vessels of the human skin and their responses, London, 1927, Shaw & Sons, Ltd.

17. Lundgren, O., and Mellander, S.: Augmentation of tissue-blood transfer of solutes by transcapillary filtration and absorption, Acta Physiol. Scand. **70**:26, 1967.

18. Martin, E. G., Woolley, E. C., and Miller, M.: Capillary counts in resting and active muscles, Am. J. Physiol. **100**:407, 1932.

19. Maurer, F. W., Warren, M. F., and Drinker, C. K.: The composition of mammalian pericardial and peritoneal fluids, Am. J. Physiol. **129**:635, 1940.

20. McMaster, P. D.: Conditions in the skin influencing interstitial fluid movement, lymph formation and lymph flow, Ann. N.Y. Acad. Sci. **46**:743, 1946.

21. Nicoll, P. A., and Webb, R. L.: Blood circulation in the subcutaneous tissue of the living bat's wing, Ann. N.Y. Acad. Sci. **46**:687, 1946.

22. Pappenheimer, J. R., and Soto-Rivera, A.: Effective osmotic pressure of the plasma proteins and other quantities associated with the capillary circulation in the hindlimbs of cats and dogs, Am. J. Physiol. **152**:471, 1948.

23. Renkin, E. M.: Transport of potassium[42] from blood to tissue in isolated mammalian skeletal muscles, Am. J. Physiol. **197**:1205, 1959.

24. Rhodin, J. A. G.: Ultrastructure of mammalian venous capillaries, venules, and small collecting veins, J. Ultrastruct. Res. **25**:452, 1968.

25. Starling, E. H.: On the fluids of the body, Chicago, 1909, W. T. Renner & Co.

26. Wiedeman, M. P.: Blood flow through terminal arterial vessels after denervation of the bat wing, Circ. Res. **22**:83, 1968.

27. Wiederhielm, C. A.: Transcapillary and interstitial transport phenomena in the mesentery, Fed. Proc. **25**:1789, 1966.

28. Wiederhielm, C. A.: The interstitial space and lymphatic pressures in the bat wing. In Fishman, A. P., and Hecht, H. H., editors: The pulmonary circulation and interstitial space, Chicago, 1969, The University of Chicago Press.

29. Wiederhielm, C. A., et al.: Pulsatile pressures in the microcirculation of frog's mesentery, Am. J. Physiol. **207**:173, 1964.

30. Zweifach, B. W.: Basic mechanisms in peripheral vascular homeostasis. In Third conference on factors regulating blood pressure, New York, 1950, Josiah Macy, Jr., Foundation.

31. Zweifach, B. W.: Functional behavior of the microcirculation, Springfield, Ill., 1961, Charles C Thomas, Publisher.

32. Zweifach, B. W., and Intaglietta, M.: Mechanics of fluid movement across single capillaries in the rabbit, Microvasc. Res. **1**:83, 1968.

33. Zweifach, B. W., and Lipowsky, H. H.: Quantitative studies of microcirculatory structure and function. III. Microvascular hemodynamics of cat mesentery and rabbit omentum, Circ. Res. **41**:380, 1977.

WILLIAM R. MILNOR

44 Regional circulations

Each regional circulation has certain special characteristics, and in this chapter we will consider separately the blood supplies to the brain, myocardium, skin, skeletal muscles, kidneys, and splanchnic region. The distribution of total cardiac output among the various organ systems is summarized in Table 40-2.

CEREBRAL CIRCULATION

The circulation through the brain and spinal cord has certain peculiarities that are imposed by the rigid containers in which these structures are enclosed. Even if bone is removed, the dura mater still forms an inelastic envelope. The concept that the volume of the intracranial contents cannot change was formulated in 1783 by Alexander Monro (the second of the three Edinburgh anatomists of that name) and elaborated in 1824 by Kellie. It is now generally agreed that this concept is strictly applicable to the cranium of the normal adult human being.

The total volume of the contents of the cranial cavity is made up of nerve cells, nerve fibers, neuroglia, intercellular fluid, cerebrospinal fluid, meninges, vascular tissue, and blood. Since each of these is incompressible, the total volume of blood within the cranial cavity cannot be increased. Although no general dilatation of all the cerebral vessels at any one time is possible, reciprocal volume changes can and do occur between different parts of the intracranial vascular bed and between the vessels and the extravascular contents of the cranial cavity. For example, the extra space required when there is dilatation of arterioles can be provided by a reduction in the volume of the large intracranial veins in which pressure is low. The change in the venous cross-sectional area thus induced may be considerable without appreciably interfering with the overall reduction in flow resistance brought about by the arteriolar dilatation. In addition, the fraction of intracranial space available for blood can be modified by changes in the volume of cerebrospinal, intercellular, or intracellular fluid. It follows from the Monro-Kellie doctrine that any rise in intracranial pressure will be transmitted equally throughout the contents of the cavity, including the blood vessels. The cerebral veins will be primarily affected, since they have the lowest intravascular pressure, one that is normally no higher than that of the cerebrospinal fluid. But a rise in intracranial pressure resulting from a tumor or other extravascular encroachment on space will not seriously reduce blood flow until it exceeds the pressure within the arteries.

The great importance of the supply of blood to the central nervous system (CNS) and of the mechanisms concerned in its maintenance are too obvious to require special comment.

Anatomic complexities and methods of measuring cerebral blood flow

In terms of afferent and efferent vessels the cerebral circulation is the antithesis of that of an organ with a pedicle through which a single artery and vein enter and leave. This feature has made it impossible or at least extremely difficult to apply to this circuit many methods of measurement that can give quantitative results elsewhere.

In man and the macaque the arrangements of the vessels are more favorable for studies of the total cerebral flow than in the cat or dog. In these two primates the greater part of the inflow is through the internal carotid arteries. In the case of man, this is attested by the fact that occlusion of the common carotid arteries in the neck produces unconsciousness within a few seconds. The vertebral arteries, which are smaller than the internal carotids, carry the remainder of the blood that reaches the brain. Although the internal carotids do not branch before entering the skull, each gives off an ophthalmic artery before joining the circle of Willis, and this artery supplies not only the neural tissue of the orbit but also the ocular muscles, the lacrimal glands, and part of the nasal mucosa. It is not likely that this extracerebral distribution represents any large source of error in determinations of cerebral blood flow by measuring arterial inflow. Unlike the internal carotids the vertebral arteries give off numerous branches before entering the skull. These branches go to neck muscles; in monkeys and man, they are not significant, but in animals with heavy necks, they doubtless carry a considerable volume of blood destined for extracerebral delivery. Some animals, including the cat and dog, have a fine network of intracranial and extra-cranial anastomoses between cerebral arteries and the external carotids.

The venous outflow is even more complicated than the arterial inflow in most higher species, and it is im-possible to determine by direct experimental measure-ment the total venous flow from the brain. The verte-bral vein plexus forms a particularly important path-way, which in some individuals can accommodate the entire drainage after occlusion of the jugular veins.

Dumke and Schmidt,[22] taking advantage of the fact that in the monkey (as in man) there are no significant communications between the intracranial and extra-cranial portions of the cephalic arteries, made the first quantitative measurements of total cerebral blood flow under conditions that approach normal. In anesthetized monkeys, they inserted a bubble flowmeter in the course of the common carotid arteries and, with basilar and external carotids ligated, determined the carotid inflow to the brain. In three experiments, they suc-ceeded in measuring flow in the carotid and basilar system simultaneously. It was found that the vertebral-basilar system carried 26%, 43%, and 80% of the vol-ume flow contributed by the two carotids. These large individual variations may be characteristic of man.

In any application of the Fick principle to the mea-surement of cerebral blood flow the problem is to ob-tain a sample that is representative of mixed cerebral venous blood. Kety and Schmidt have shown that in the great majority of human beings "blood from one internal jugular vein at the level of the superior bulb is fairly representative of mixed cerebral venous blood not significantly contaminated with blood from extra-cerebral sources."[39] Inhalation of a low concentration of nitrous oxide (15%), measurement of arterial and jugular venous N_2O concentrations, and knowledge of the partition coefficient between brain and blood allow calculation of cerebral blood flow by the Fick principle.[39]

Other methods have also been devised,[6,10] including techniques for measuring regional cerebral blood flow by arterial injection of an inert, diffusible radioiso-tope and extracranial monitoring.[42] Studies with this last method show that blood flow to the gray matter of the brain is about 4 times that in the white matter.

Basal blood flow and oxygen consumption of the brain

The work of Kety and Schmidt[39,40] has shown that under resting conditions a normal young man of average size has a cerebral blood flow of about 750 ml/min (54 ml/100 gm brain/min) and a cerebral O_2 consumption of 46 ml/min (3.3 ml/100 gm brain). The average cerebral arteriove-nous O_2 difference is relatively high, 62 ml/L. Under essentially basal conditions the brain re-ceives approximately 14% of the output of the left ventricle and 18% of the O_2 absorbed by the lungs. Table 40-2 shows that in terms of resting blood flow and O_2 consumption per unit of mass the CNS is surpassed only by the kidneys and myocardium, among the larger single organs of the body. Until recent years, it has been assumed that the metabolic needs of the brain are small, since no effect of mental activity on the O_2 con-sumption of the body as a whole could be de-tected. It is now apparent that such efforts at de-tection failed because the resting value is so high and perhaps because any variations with mental activity cannot be readily distinguished from those caused by muscular contractions that may accompany mental activity.

Under resting conditions the cerebrovascular resistance is moderate in degree. The relative resistance is well above the resistance of the renal vessels but is less than that of the skin and much less than the resistance of resting muscles (Fig. 44-4). The flow resistance of the vessels of the brain seems to be normally much less subject to large changes than is that of the muscular, renal, splanchnic, or cutaneous circuits; yet an effective regulation is much in evidence. The vessels of the brain differ from those of other systemic cir-cuits in that many influences dilate them strongly, whereas only a few have a constrictor effect. An active contraction of the walls of the cerebral ves-sels is never a strong response. There is substan-tial evidence that they are not played on by a tonic sympathetic vasoconstrictor discharge. Ac-tually, their "basal" state seems to be the result of the operation of some factor or combination of factors that produces vasodilatation.

Control mechanisms

In the past, it has been supposed that under most circumstances an adequate supply of blood to all parts of the brain is mainly attained by sub-ordinating the level of arterial pressure to this end. There is no doubt that the maintenance of a normal mean arterial pressure ensures an ade-quate flow through the CNS. In this sense the circuit does have an extrinsic (extracerebral) con-trol that, as pointed out in the discussion of the renal circulation, is put to a test when arterial pressure is low, but this is not the sole regulatory mechanism. Quantitative studies have made it clear that within a considerable range of arterial pressures the cerebral blood flow is not affected by the perfusion (arterial) pressure to which it is exposed. It is only when arterial pressure is re-duced to about 50% or more of its normal value that cerebral vasodilatation of intrinsic origin fails to provide the blood flow necessary to main-tain an adequate cerebral metabolism.[6] Auto-nomic innervation appears to play only a small part in control of the cerebral vessels. There is no evidence of tonic sympathetic constrictor discharge, and very little increase in resistance can be produced by stimulation of vasoconstrictor fibers. Vasodilator nerves to cerebral vessels exist,[10,20] but their normal function is unknown.

In 1905 the probability of an intrinsic control of the cerebral circulation was suggested by Wiggers' demonstration that epinephrine reduces the flow through the excised perfused brain of the dog. After 1928 the studies of Forbes and Wolff[26] and his other collaborators, who especially perfected the technique of direct microscopic observation of pial vessels, supplied further evidence that the circulation through the brain is somewhat independent of extracerebral circulatory events. They found that electrical stimulation of the cervical sympathetic ganglion, carried out mainly in cats, produces constriction of vessels of the pia-arachnoid over the parietal cortex. The constriction thus obtained is wholly ipsilateral, takes place entirely independently of changes in systemic arterial pressure, and occurs consistently only in arteries larger than 50 μm in diameter. The average narrowing produced was only 8%, one tenth as great as that evoked in skin vessels by the same stimulus. Subsequent work by Schmidt, in which a thermoelectric flow recorder was used, also revealed the occurrence of sympathetic vasoconstriction in the parietal cortex of cat and rabbit and in the cat's hypothalamus, but it failed to give any indication of it in the occipital cortex, pons, or medulla of the cat. Again, evidence was obtained that cervical sympathetic stimulation evokes a far stronger and more widespread vasoconstriction in the extracranial than in the intracranial parts of the cephalic circulation. With weak stimulation the extracranial action may even shunt blood so that an increase of flow in the brain results. Further, there is agreement that, in cats, section of the cervical sympathetic ganglion does not affect cerebral blood flow, although it produces an increase in flow in the extracranial vessels innervated by that nerve. Neither in normal individuals nor in patients with hypertension does blocking of the stellate ganglia significantly affect cerebral blood flow.[35]

The dilator effect of the products of metabolism is infinitely more important in the control of the cerebral vessels than any other known factor. Rise in temperature, increase in CO_2 or hydrogen ion concentration, and decreased O_2 tension all produce dilatation of cerebral vessels. This means of regulation is intimately related to the fact that the blood supply of the brain as a whole or of any part of it must rise and fall with the metabolic activity of the neurons, and there is evidence that cerebral blood flow is more closely correlated with tissue than with arterial CO_2 tension.[49] The range of cerebral metabolic activity is quite wide. Reference to Table 44-1 will show that the basal O_2 consumption may be halved in man during Pentothal anesthesia, whereas it may be almost doubled in a monkey during a Metrazol convulsion. Therefore it is of obvious physiologic significance that changes in the arterial tensions of the respiratory gases and in arterial hydrogen ion concentration evoke marked alterations in cerebral vascular resistance and blood flow.

Table 44-1. Cerebral blood flow and metabolism in man under various conditions (average figures from studies on groups of various sizes)*

Condition	Mean arterial tension (mm Hg)	Cerebral blood flow		Cerebral O_2 consumption (ml/100 gm/min)	Cerebrovascular resistance (mm Hg) (ml/100 gm/min)
		Total (ml/min)	ml/100 gm/min		
Normal young men					
At rest breathing air	85	750	54	3.3	1.6
Voluntary hyperventilation (low CO_2)	98	473	34	3.7	2.9
Inhalation of:					
CO_2 5%-7% in air	93	1293	93	3.3	1.1
Low (10%) O_2 in N_2	78	1015	73	3.2	1.06
High (85%-100%) O_2	98	625	45	3.2	2.2
Pentothal anesthesia	71	834	60	2.1	1.2
Diabetic acidosis					
Severe, without coma	86	625	45	2.7	2.1
Diabetic coma	66	904	65	1.7	1.1
Hypertension (no cerebral change)	155	778	56	3.4	2.8
Cerebral arteriosclerosis	136	584	42	2.8	3.2
Brain tumor (comatose)	122	122	34	2.5	3.6

*Modified from Schmidt[10] and from Sokoloff and Kety.[13]

In the cat, it has been found that acids dilate and alkalies constrict cerebral vessels, but the results are less marked than those produced by changes in CO_2 tension. In studies on patients in diabetic acidosis, it was found that cerebral vascular resistance increases as arterial CO_2 tension decreases over the arterial pH range 7.4 to 7.1 but falls sharply when the acidosis becomes more severe. The constricting effect of decreased CO_2 and increased O_2 tension is noteworthy (Table 44-1). Hypocapnia has the more powerful action, and Schmidt points out that it is the "only agency by which a significant grade of cerebral vasoconstriction has ever been produced under physiological conditions." This response is presumably an expression of the withdrawal of the dilator effect of normal CO_2 tensions and is found in the vessels of no other region.

The function of exchange vessels in the cerebral circulation is distinctive in that transfer of substances from blood to the cerebral extracellular fluid is a highly selective process that does not depend entirely on molecular size and transmural pressure.[9] O_2 and CO_2 pass readily, as in other capillaries, but some molecules are transported actively in one direction or the other, whereas the passage of still others is completely blocked. The environment of the brain cells is thus critically controlled by a "blood-brain barrier."

Local responses to changes in functional activity. Since the brain, unlike such organs as the liver, is divisible into distinct functional areas, it is reasonable to suppose that its vasculature is subject to local adjustments related to localized changes in neuronal activity. Actually, several investigators[10,13,38] have secured evidence that blood vessels dilate in cerebral areas that are specifically active. Schmidt and Hendrix, for example, by using a thermoelectric probe in the anesthetized cat, found that illumination of one retina by a small flashlight caused an increase in blood flow that was restricted to the visual area of the cerebral cortex when there was no change in arterial pressure or in the tensions of the respiratory gases in arterial blood. Kety and Sokoloff[13] and their collaborators showed in animals that the sensorimotor, visual, and auditory areas of the cortex normally have relatively greater blood flow than other regions, and that thiopental anesthesia depressed this flow and abolished the regional differences. Regions of what might be called "functional hyperemia" have also been detected in conscious human subjects subjected to various physical or emotional stimuli.[38]

Effects of increased intracranial pressure. At the beginning of this century, Cushing[21] showed in animals that an acutely induced rise in intracranial pressure is accompanied by a rise in arterial pressure that reaches a level somewhat above that within the cranial cavity. This generalized vascular response, of obvious compensatory value, he found to be due to vasoconstrictor discharge from the bulbar centers. Little further information about this important phenomenon was obtained until Kety et al.,[41] using the N_2O method, analyzed the effects of increased intracranial pressure on cerebral circulatory functions in man. Their subjects were patients suffering from an elevated intracranial pressure produced by brain tumors. It was found that with a rise in cerebrospinal fluid pressure there was a proportionate rise in cerebrovascular resistance, but a definite decrease in cerebral blood flow did not occur until the pressure in the fluid reached a value of about 450 mm H_2O. Since at all levels of intracranial pressure above the normal (about 100 mm H_2O in the horizontal position in which the subjects were studied) a linear relationship between cerebrospinal fluid pressure and mean arterial pressure was found, it is probable that up to this point (450 mm H_2O) the increased cerebrovascular resistance was successfully counterbalanced by the rise in arterial pressure. Above that point there was a good correlation between intracranial pressure and a decrease in cerebral blood flow. All patients who failed to compensate and showed a flow less than 40 ml/100 gm brain/min were comatose. The cerebral O_2 utilization of the conscious patients was close to the normal level, but in those in coma, it was definitely depressed (Table 44-1). As Kety et al. point out, the data obtained do not reveal whether "consciousness failed because of the severe restriction in blood flow, or whether the compensation failed because of the generalized neuronal depression manifested in unconsciousness. . . ."[41]

The results of these quantitative studies of blood flow, O_2 use, and vascular resistance in the brain are in accord with Cushing's hypothesis that the compensatory rise in arterial pressure is due to ischemia of bulbar centers. The response is similar to that seen with decreased cerebral perfusion after severe hemorrhage (p. 1069).

CORONARY CIRCULATION
Anatomic considerations

The blood supply to the heart is delivered through the two coronary arteries that arise from the root of the aorta, travel out over the surface of the heart in numerous branches, and ultimately penetrate the muscle to supply the deeper layers. These deeper vessels are exposed to compression as the heart contracts, so that blood flow through the peripheral coronary bed is regularly impeded to some extent by ventricular systole.

Venous drainage of the heart is through a superficial system and a deep system of vessels. The most prominent parts of the superficial system are the large coronary veins that converge to form the coronary sinus. Together with some other vessels that open near the mouth of the sinus, they provide the main drainage for the subepicardial veins of the left heart. The other part of the superficial system is made up of vessels of the right heart that merge to form the anterior cardiac veins that empty directly into the right auricle quite separately from the coronary sinus.[4,31] The deep system of vessels for arterial drainage connects with the heart chambers. These vessels are of three kinds: (1) arterioluminal vessels, which are communications between a small coronary arterial branch and a heart cavity; (2) the more numerous arteriosinusoidal vessels, which, beginning as arterioles, break up into large, irregular, capillary-like structures emptying into an auricle or ventricle; and (3) thebesian vessels, which are channels between coronary veins or the distal ends of capillaries and the heart chambers. The complex arrangements and the probable directions of flow are expressed in Fig. 44-1. Only a very small part of the total venous flow travels through these last three groups of vessels, by far the greater part draining into the coronary sinus and anterior cardiac veins.

Coronary flow

A consideration of the anatomic features of the coronary vessels makes it easy to understand the difficulties that are encountered in attempts to determine the total coronary blood flow. In animal experiments, inflow through one artery or a branch may be measured and the total flow through the whole system estimated, or measurements may be made of the venous flow into the right auricle through the coronary sinus or from the anterior cardiac veins. In the dog, it has been found[4] that (1) of the total inflow about 85% enters the left coronary artery and about 15% the right; (2) although practically all the flow into the coronary sinus arises from the left coronary artery, this flow constitutes but 64% to 83% of the left coronary inflow; (3) a small fraction of the left coronary inflow and a large part of the right inflow are drained by the anterior cardiac veins; and (4) a small fraction of blood from the right coronary artery drains into the coronary sinus. The finer branches of the arterial bed include anastomotic channels through which blood can reach regions whose major artery has become occluded, but relatively little blood can be supplied by this route to an ischemic area if a major artery is occluded abruptly. If the occlusion is gradual over a period of months or years, enlargement of these anastomoses often compensates for the impairment of the normal flow pathway to a considerable degree.

Measurements of coronary flows have been made in a variety of animal preparations. In the older experiments the heart-lung preparation or the isolated perfused heart was generally employed. Later, left coronary inflow was determined, under less abnormal conditions, directly in anesthetized open-chest dogs by the rotameter, the bubble flowmeter, and other devices and indirectly by measuring coronary sinus outflow. More recently, left coronary inflow has been determined by applying to the hearts of intact, anesthetized or unanesthetized dogs and normal human beings the N_2O method developed by Kety and Schmidt[39] for the measurement of cerebral blood flow. In this application of the Fick principle the arteriovenous N_2O difference is determined by obtaining samples of blood from any systemic artery and from the coronary sinus, the latter vessel being reached by a catheter inserted into a peripheral vein and advanced under fluoroscopic guidance into the desired position. As in the case of

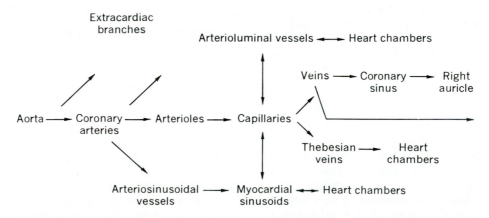

Fig. 44-1. Diagram of connections in coronary system. (Modified from Wearn et al.[51])

the brain, the quantity of gas absorbed by the tissue is calculated from its tension in the venous blood and its solubility in the myocardium. At the same time the O_2 uptake of the myocardium may be calculated from the coronary blood flow and the arteriovenous O_2 difference.

Coronary flow has been measured by very small electromagnetic flowmeter probes in conscious animals[29] and by intracoronary injection of radioisotopes and external monitoring in dogs[36] and in man. The highest figures reported for left ventricular blood flow and O_2 usage were secured in experiments on trained unanesthetized dogs in which the flows were determined by the N_2O method[50]; the mean flow was 133 ml and the mean O_2 consumption 15.57 ml/100 gm of left ventricle/min, but in each case the variations were great. On the other hand, the arteriovenous O_2 differences showed far less scattering and averaged 11.9 vol% (119 ml/L). In human subjects, Bing et al.[17] found "left ventricular" blood flows and O_2 usages that averaged, respectively, 65 and 7.8 ml/100 gm/ min (arteriovenous O_2 difference, 11.5 vol%), whereas in anesthetized dogs, this same method in the hands of several groups of investigators[4,23,28] has yielded values that range from 66 to 81 ml/100 gm/min for flow and that cluster around 9.7 ml for O_2 consumption.

Phasic variations. Phasic flow patterns in an extramural coronary artery have been recorded with various instruments, but the most reliable data relevant to coronary flow in vivo have been obtained by Gregg and Fisher[4] and colleagues[29] with permanently implanted electromagnetic flowmeters (Fig. 44-2). These records show that

the flow during systole is greater than had been supposed from earlier work on the open-chest dog, amounting to approximately 30% of that during diastole. This ratio is maintained even when the total coronary blood flow increases threefold with excitement or stimulation of the cardiac sympathetic nerves.

At the onset of isometric contraction of the ventricle, coronary flow falls abruptly and there may be a transient backflow. With opening of the aortic valve and the systolic increase in aortic pressure, coronary flow rises to a moderate peak and falls again. As soon as the aortic valve closes, coronary flow rises to a maximum, to decline slowly during diastole[4] (Fig. 44-2). The resistance to flow in the coronary bed obviously fluctuates considerably throughout the cardiac cycle, and the shape of the coronary flow curves indicates that left ventricular contraction is the principal cause of these fluctuations. As left coronary inflow is reduced during systole, venous flow in the coronary sinus is increased, demonstrating that ventricular contraction squeezes blood out of the intramural coronary vessels into the veins at the same time that it increases the inflow resistance of the bed. The flow pattern in the right coronary artery is in general similar to that in the left.

The phasic pattern imposed on coronary blood flow by the mechanical action of the ventricle has two important physiologic consequences. First, changes in the velocity and force of muscle con-

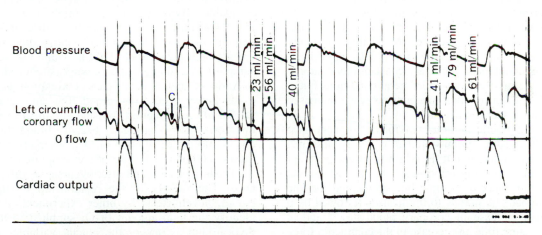

Fig. 44-2. Records of aortic blood pressure and phasic blood flow in left circumflex coronary artery and ascending aorta, obtained with strain-gauge and electromagnetic flowmeters in unanesthetized dog standing at rest. In middle of coronary flow record, artery is occluded temporarily to determine signal for zero flow. Values on coronary flow tracing indicate flows at about peak aortic pressure, early diastole, and late diastole. Vertical time lines at intervals of 0.1 sec. Cardiac output = 4,116 ml/min; mean aortic pressure = 133 mm Hg; mean left circumflex coronary flow = 36 ml/min. (From Gregg et al.[30])

traction can affect the pattern of flow and perhaps the mean coronary flow rate. Second, the relative shortening of diastole as heart rate rises tends to reduce net coronary flow. This tendency is more than offset in the intact animal, however, by the indirect metabolic effects of tachycardia.[2]

Control mechanisms[2,7]

The existence of highly effective mechanisms for regulating coronary blood flow is demonstrated by the prominent vasodilatation during exercise, when flow may increase threefold or more over the resting level. Myocardial metabolism is predominantly aerobic, so it is not surprising that cardiac muscle should need an increased blood supply when called on to deliver an increased cardiac output, but the basic cause of this vasodilatation is not clear. Most of the evidence points to the local action of some product of metabolism as the principal mechanism. Any procedure that increases the work or metabolic activity of the heart leads to an increase in coronary blood flow, and this indirect effect complicates investigation of the specific action of nerve impulses or other factors on the coronary vessels.

Aortic pressure, which is the driving force, is obviously one of the factors that determine coronary blood flow, becoming especially important when pressure falls to pathologically low levels. In the usual range of arterial pressures, however, the coronary circulation exhibits some degree of autoregulation (p. 1027). A sudden increase in coronary perfusion pressure in an experimental preparation produces an equally abrupt rise in blood flow, but, after a few minutes at the new pressure, blood flow gradually falls toward its previous level. This autoregulation maintains a relatively constant blood supply to the myocardium under basal conditions but is overridden by other mechanisms when increased cardiac work is needed.

Autonomic nervous control of coronary vessels can be demonstrated in carefully designed experiments, but it is relatively weak and apparently of little importance under most normal circumstances. Ordinarily, stimulation of the sympathetic nerves to the heart is followed by a great increase in coronary blood flow[29] and at the same time an increase in the contractile force of the heart. Coronary vasodilatation is thus the normal response to sympathetic discharge when the heart is functioning normally, but under these conditions, it is impossible to tell how much of this response is connected with myocardial performance and how much with nerve-mediated vasomotion. The same is true of the effects of circulating epinephrine and norepinephrine. If the ventricles are made to fibrillate, however, so that their external work is negligible and no longer influenced by autonomic impulses, sympathetic stimulation then causes a weak coronary vasoconstriction.[16] Direct sympathetic vasoconstrictor action can be demonstrated, in other words, but in the intact animal, sympathetic discharge to the heart leads indirectly to coronary vasodilatation because of increased myocardial work and metabolism. Surgical interruption[18] or chemical blockade of nerve impulses shows that the tonic state of the coronaries is a slight degree of constriction. The presence of both alpha- and beta-adrenergic receptors (p. 1055) has been reported, but the extent to which they function under normal conditions is unknown. A functional parasympathetic innervation of the coronary vessels also exists, and the direct effect of vagal stimulation is a faint coronary vasodilatation.[16,29]

The large and consistent reduction of coronary resistance whenever the heart is required to work harder stands in marked contrast to the almost trivial vasomotor phenomena that can be elicited in the laboratory by direct neural or neurohumoral actions. In preparations ranging from the isolated heart to conscious subjects, myocardial blood flow has been shown to be directly correlated with cardiac output, cardiac work, and to a lesser extent with heart rate.[2,4,25] The implication that flow in the coronary bed is somehow connected with myocardial energetics seems inescapable. Myocardial O_2 consumption appears to be the most important variable, since under certain conditions, sympathetic stimulation increases both myocardial blood flow and O_2 consumption even though cardiac work is reduced.[24]

A direct and almost linear correlation also exists between coronary blood flow and myocardial O_2 consumption, but the significance of this correlation is suspect[46] because blood flow enters into the calculation of O_2 consumption in most experiments, the rate of consumption being the product of flow and arteriovenous O_2 difference. The validity of the relationship is supported, nevertheless, by other experiments that identify the total amount of O_2 supplied to the myocardium as one of the crucial variables in determining coronary blood flow.[2] Hypoxemia produces coronary vasodilatation, for example, but arterial O_2 content, not tension, is the critical factor.[2,33]

If the coronary arteries are perfused with hemoglobin solutions of constant O_2 tension but differing O_2 content, the coronary resistance falls as arterial O_2 content is lowered.[33] Evidently a lowering of the O_2 tension in exchange vessels or their fluid environment is the primary stimulus for vasodilatation, a conclusion strengthened by the observation that myocardial blood flow correlates more consistently with coronary venous than with arterial O_2 content. Coronary flow increases with arterial hypoxemia only when the coronary venous O_2 content falls to 5 or 6 vol%, indicating a critical level of tissue hypoxia at which some local mechanism of vascular control comes into play. Conversely, oversupplying the needs of the myocardium in the intact animal by ventilation with 100% O_2 produces moderate coronary vasoconstriction.[2] Clearly, the resistance of the coronary bed is regulated in accordance with the metabolic activity of the heart, predominantly by vasodilatation as metabolic demands increase. The possibility that Po_2 in tissue and extracellular fluid is in itself the agent of dilatation has not been ruled out, but it is more likely that some product of metabolism—possibly adenosine—is responsible. Elevated blood CO_2 tension and lowered pH also act as coronary vasodilators, but the extent to which this is a direct action on the vessels is uncertain.

CUTANEOUS CIRCULATION

The circulation in the skin is convenient for the study of vascular reactions to injury and inflammation, and these have already been considered (pp. 1055 and 1068). The special function of the blood supply of the skin is the maintenance of heat balance, and its regulation is chiefly directed to that end (Chapter 59).

The measurement of blood flow through the skin is fraught with many difficulties, both practical and theoretical. In general, three methods have been used (see the account by Day in the monograph edited by Newburgh[8]). They are (1) the plethysmograph adapted for hand or arm, (2) the photoelectric plethysmograph of Hertzman, and (3) skin thermometry. A combination of calorimeter and plethysmograph has been used to measure skin temperature, heat loss, and blood flow through the hand.[27] Blood in passage through the skin cools over a range of 1 to 8 kcal/L. Under basal conditions, it loses heat to the extent of about 2.5 kcal/L.

The anatomic arrangements not only of the vessels in the skin itself but also of the arteries leading to the skin and of the venous channels for drainage are important factors in the heat-regulating function of the cutaneous circulation. The arteries that carry blood to the skin reach it by piercing the underlying muscles. They are branches of arteries that are accompanied by venae comites carrying blood from both muscles and skin. Thus the arterial blood on its way to the skin may be warmed or cooled to a temperature above or below that of deep body tissues. This has been demonstrated by means of thermocouples protected by fine plastic tubing, which can be threaded through a fine needle and passed for a considerable distance up a vessel.[15] With muscular exercise the warming effect predominates, and heat is carried directly to the surface of the skin.

Blood from the skin can return by two alternate channels. It may pass up the superficial veins, cooling in transit, or it may pass through communicating veins to the venae comites. In the latter case the returning blood is rewarmed while in passage by the adjacent arterial blood, which is thereby cooled. When the arterial blood is exposed to cold, the precooling is a matter of internal heat exchange, with relatively little heat loss to the environment. Under this circumstance, peripheral parts such as the hands are allowed to cool, thus minimizing the heat loss from their extensive surfaces. This process involves marked changes in arterial temperature. Temperatures below 30° C have been found in the blood flowing in the radial artery at the wrist when the subject has not been exposed to the point of vigorous shivering and has a nearly normal rectal temperature. Even the temperature of blood in the brachial artery may be significantly lowered. On exposure to warmth and during muscular exercise the cutaneous blood flow is much more rapid, and most of the blood returns through the superficial veins, which are dilated; the venae comites now play a minor role as a path for the returning blood. The arterial blood is cooled very little in transit, and it reaches the apical areas such as the hands and feet at temperatures of 35° to 37° C. The returning blood continues to cool in the superficial veins so that temperatures recorded in those of the hand exceed those at the level of the elbow.

The anatomic arrangement of vessels in the dermis is peculiar and complex. There are several layers of arterial and venous channels, and these determine the thermal gradients of the skin (Newburgh[8,p.171]). Between the surface and the subcutaneous tissue the highest temperature is found at a depth of about 1 mm. This corresponds to the position of an arteriolar plexus that receives branches from an arterial plexus beneath the dermis. The arteriolar plexus gives rise to branches that supply the capillary loops beneath the epidermis, but between the capillary loops and the arteriolar complex are two superficial venous plexuses. A third deeper venous plexus is found below the arteriolar plexus. Finally, at the junction of dermis with subcutaneous fat, at a depth of about 2 mm, veins draining the skin run beside the arteries as venae comites. From the arteriolar plexus to the surface of the epidermis (a distance of about 1 mm) the temperature falls precipitously; between the arteriolar plexus and subcutaneous tissue (a distance of about 1.5 mm), it shows a conspicuous dip that corresponds to the position of the deep dermal venous plexus. Beyond this the temperature again rises as the subcutaneous area is approached.

There are variations in the blood flow in different parts of the skin under constant conditions of thermal balance.[8] The regional differences in the rates of skin blood flow have been studied especially by Hertzman and colleagues, who found that the flows in various areas vary from greater to less in the following order: finger pad, earlobe, toe pad, palm of hand, forehead and face, forearm, knee, and tibial areas. It appears that the blood flow through the hands and feet is especially related to body temperature regulation. Scott et al.[47] observed that on changing from a cold to a hot environment the flow through the fingers rose from 1 ml to 80 or 90 ml/100 gm tissue/min, and Wilkins et al.[52] have reported a change from 0.2 to 120 ml/100 ml finger/min, a 600-fold increase. On the other hand, the vessels of the skin of the proximal parts of the limbs and trunk are much less responsive to temperature changes.

The regulation of the cutaneous blood vessels is chiefly nervous. There is a wide distribution of sympathetic adrenergic vasoconstrictor fibers to the skin, and in the case of apical regions (hands, feet, and ears), this innervation is tonically active under conditions of rest and thermal neutrality. Here the intrinsic myogenic activity of the vascular smooth muscle of the resistance vessels is slight, and marked cutaneous vasodilatation is produced by a centrally engendered release of vasoconstrictor tone or by sympathectomy. On exposure to heat the former accounts chiefly, if not entirely, for the increase in blood flow. In contrast to these cutaneous areas are those of forearm, calf, and forehead (and doubtless other regions of the skin that have been little studied) where under ordinary thermal conditions the resistance vessels possess a tone independent of their innervation. Yet on exposure to cold, they constrict because of a discharge of sympathetic vasoconstrictor impulses.

There is little convincing evidence that the smooth muscle of cutaneous vessels receives a direct inhibitory (vasodilator) innervation from the CNS. What little there is has been well set forth by Hertzman.[5] Cutaneous vasomotor responses of central origin in man appear to be mediated solely by sympathetic fibers, and these may be divided into two groups: the adrenergic constrictor fibers, which directly innervate the vessels, and the cholinergic sudomotor fibers, which indirectly cause vasodilatation through the formation of bradykinin (p. 1054). Barcroft has summarized the facts as follows: "The vasomotor center maintains vasoconstrictor tone which is strong in the hand and weak in the forearm skin; it is released during body warming. The sudomotor center causes the formation of bradykinin in the forearm skin during body warming: during emotional stress it causes large amounts of bradykinin to be formed in the palmar skin of hyperhidrotic subjects. Its action causes cutaneous vasodilatation."[1]

The important role played by the cutaneous veins and their innervation, which is entirely vasoconstrictor and sympathetic, should not be overlooked. These vessels account for the greater part of the capacity of the skin to hold blood, and according to Hertzman,[5] their innervation is much more potent than that of the cutaneous resistance vessels. As suggested earlier, they must play a large role in the adjustments of the cutaneous circulation.

In addition to the factors governing cutaneous blood flow so far discussed, there are the arteriovenous anastomoses that, when open, allow blood to flow directly from arterioles into venules or veins. This arrangement makes it possible to transfer heat rapidly to the superficial cutaneous veins. These anastomoses are subject to the same nervous control as are the arterioles. They open in response to extreme environmental heat and so further increase the volume flow of blood at the surface from which heat may be lost. They are also made use of on exposure to extreme cold, in which case the brisk flow raises local surface temperature at the expense of the rest of the body when injury from frostbite is imminent. It is significant that they are found in greatest numbers over exposed and poorly insulated areas such as the fingertips, nail beds, and thenar and hypothenar eminences. They are also abundant in the rabbit's ear and in the feet of birds, where insulation is minimal.

BLOOD SUPPLY TO SKELETAL MUSCLES

The O_2 consumption of muscles during work is enormous, but under basal conditions it is amazingly small. Here, as elsewhere, metabolic activity is paralleled by variations in blood flow.[11] The O_2 utilized by all the skeletal muscles during rest has been estimated by determining the effect of sudden occlusion of the arterial inflow to the legs on the total O_2 absorption by the lungs.[14] The results indicate that muscle accounts for about 20% to 30% of the basal O_2 consumption. This is a small fraction in view of the fact that the skeletal musculature makes up nearly half of the mass of the entire body. The integrative competence of the cardiovascular system is well illustrated by the fact that at rest the muscles receive only 15% of the output of the left ventricle and extract from the arterial blood less O_2 per unit of mass than any other major subdivision of the body. The low blood flow is due to the exceptionally high relative resistance of the vessels in resting skeletal muscle. It follows that vasodilatation in this circuit would be the most effective means of lowering the flow resistance of the systemic circulation as a whole.

The variation in the number of capillaries and metarterioles that are open in muscle according to the absence or presence of activity has already been described (p. 1085), as have the vasomotor changes during exercise (Chapters 40 and 42). Considerable uncertainty remains as to the relative importance of vasomotor nerves and local effects of metabolites in the development and maintenance of the vasodilatation. Since the vascular expansion develops rapidly with the onset of exercise, the initial phase is probably nervous in origin. The tremendous increase in blood flow through muscles that takes place with vigorous exercise can be appreciated best by reference to available figures for O_2 use. When the lungs of the resting subject are absorbing 200 to 250 ml O_2/min, each 100 gm of muscle uses, on the average, only 0.16 ml/min. When the vigorously exercising individual is taking in 4 L O_2/min, his muscles are extracting it from the arterial blood at a rate of about 12 ml/100 gm muscle/ min. This is a greater increment with activity than is shown by any other major subdivision of the body supplied by the systemic vessels. The rapid delivery of O_2 is, of course, the result of the increased cardiac output and the great fall in flow resistance in the muscles; these changes are supplemented by increased flow resistance in the renal vessels. A factor of some importance is that the blood in the arteries supplying the muscles is warmed by blood returning from the muscles in the venae comites. This facilitates the warming up of muscles during the early phases of work to a temperature at which hemoglobin more readily unloads O_2. Another important feature of the vascular arrangement is one to which attention has already been drawn—that the skin is supplied by arteries that arise from muscular branches and pierce the muscles to reach the skin. Thus in ex-

ercise the blood destined for the skin is warmed by the muscles and carries some of the heat generated in them directly to the surface, from which is may be eliminated by radiation, convection, and vaporization.

RENAL CIRCULATION

The flow of blood through the kidneys is a matter so intimately related to the formation of urine that a detailed consideration of this important portion of the systemic circulation will be deferred until the physiology of the kidney is presented (Chapter 50). Renal blood flow is a significant fraction of the total cardiac output, however, and the control of renal vascular resistance is important in the overall operation of the cardiovascular system.

Studies of renal blood flow in intact unanesthetized animals and human beings have become possible through an almost ideal application of the Fick principle in the estimation of peripheral flow. It has been found that in a single passage of blood, normal kidneys are capable of almost completely removing from the blood low concentrations of Diodrast or *p*-aminohippurate (PAH). PAH is preferred because its quantitative determination in blood and urine is more easily accomplished. The kidneys extract it from blood to the extent of 87% to 90% and excrete that quantity into the urine. Although PAH does enter erythrocytes to some extent, the Fick principle can be applied to determine the flow of plasma through the kidneys.[12,44] The amount appearing in the volume of urine put out in a reasonable unit of time divided by the renal arteriovenous hippurate difference gives a figure that, when multiplied by the reciprocal of the renal extraction of PAH in percent, is the renal plasma flow. The renal blood flow can then be readily calculated from the hematocrit. On the basis of a number of studies, it is justifiable to assume a renal extraction of about 87%. If, however, samples of blood from a renal vein can be obtained, the renal extraction of PAH and the arteriovenous difference can be directly determined.

It is now well established that under basal conditions the average human male has a renal blood flow of about 1.2 L/min. Since the two kidneys weigh about 300 gm, this means that the flow is in excess of 400 ml/100 gm renal tissue/min. In anesthetized dogs, Selkurt[48] found renal blood flows that averaged 426 ml/100 gm of kidney. From the figures given in Table 40-2, it is apparent that under resting conditions the kidneys receive, per unit of weight, a greater volume flow of blood than any other major region of the body supplied from the left ventricle. In this respect the kidneys easily outdistance the myocardium, which has approximately the same mass; every minute during bodily rest the renal vessels carry about 5 times the volume of blood that flows through the coronary vessels. This is due to the exceptionally low resistance of the kidney vessels, which apparently obtains not only under basal conditions but also with mild degrees of activity. At rest the vasoconstrictor nerves to the kidneys exhibit little or no tonic activity (p. 1052).

The large volume flow and the low flow resistance that characterize the normal renal circulation make this circuit an ideal site for the occurrence of compensatory vasoconstrictor responses. Constriction of renal vessels is an important factor in maintaining arterial pressure when there is a reduction in cardiac output. The renal vessels participate in the widespread vasoconstriction that occurs when blood volume is reduced; this reaction is of great importance in maintaining an adequate flow of blood to the brain and the myocardium under such conditions. The reduction in renal flow after a severe hemorrhage or following severe trauma of muscle and bone has been measured in dogs[44,48] and in patients[43]; it is too great to be explained by the fall in arterial pressure and is unquestionably the result of a vigorous vasoconstriction in the vessels of the kidneys. In a series of persons in shock from skeletal trauma, hemorrhage, or both, Lauson et al.[43] found that the renal vascular resistance was raised and the fraction of the total blood flow going through the kidneys (renal fraction) was decreased. As the cardiac output fell, blood was shunted away from the kidneys.

A similar compensatory shunting of blood occurs on exposure to heat and during muscular exercise. For example, Radigan and Robinson[45] measured renal blood flows in five normal and fully hydrated young men under these four conditions: rest in a cool environment, rest in a hot chamber, exercise in the cool environment, and exercise in the heat. The renal blood flows averaged, under these different conditions, 1,250, 775, 524, and 496 ml/min, respectively. In the case of exercise the cardiac output increases, but the fraction that passes through the kidneys is greatly decreased. It was estimated that renal fractions under the four conditions of this experiment were 26%, 17.2%, 4.3%, and 4.5%. On exposure to heat and during exercise the rate of heat loss from the body must be augmented by an increase in blood flow through the skin. During exercise there is of course an enormous increase in flow through the active muscles. The evidence is clear that these changes are facilitated by a rise in flow resistance in the kidneys because of active vasoconstriction.

In 1858 Claude Bernard noted that the kidneys remove less O_2 from the arterial blood than do most tissues. This observation has been confirmed by experiments on dogs and men. Although there has been considerable variation in the figures that have been obtained by different workers for the renal arteriovenous O_2 difference, it can be said that they are all low, far below the resting (basal) values for all other major organs or regions except perhaps the skin (Table 40-2). In the case of normal human subjects the reported differences range from 6 to 30 ml/L blood.[12]

HEPATIC-PORTAL CIRCULATION
The liver

The liver receives blood from the hepatic artery and the portal vein, the portal blood having already passed through capillaries in the gastrointestinal tract or spleen. Hepatic arterioles and portal venules both enter the peripheral part of the sinusoids of the liver, so that a mixture of arterial and venous blood flows through the sinusoids. The sinusoidal channels, which are the "exchange vessels" (p. 1050) of the liver, drain into the hepatic veins and thence into the inferior vena cava. Functional sphincters have been observed at the entrance of the hepatic arterioles and portal venules into the sinusoids and at the junction of the sinusoids with central veins of the liver lobules. The hemodynamic consequence of this dual blood supply are complex and not yet entirely clear in all respects.[3]

Hepatic blood flow has been measured in animals with the thermostromuhr[32] and more recently with electromagnetic flowmeters.[3,37] In addition, a large body of data has been collected in man as well as in animals by a method based on the Fick principle.[19] This technique employs the intravenous injection of a substance (e.g., Bromsulphalein) that is removed from the blood as it passes through the liver. The accuracy of the method depends in part on the validity of the assumption that extrahepatic removal of the injected material is negligible, an assumption that has been questioned by some investigators but is supported empirically by the close agreement between this method and such direct procedures as measurement of total hepatic venous outflow. The arteriovenous difference that is essential in any application of the Fick method is determined from measurements on samples of blood from a peripheral artery and hepatic venous blood obtained by venous catheterization.

The proportion of total hepatic blood flow delivered by each of the two routes, portal and arterial, is quite variable under physiologic conditions, but 25% to 35% is usually supplied by the hepatic artery. This highly saturated arterial blood provides the greater part of the O_2 supply to the liver, but the O_2 content of hepatic veins is lower than that of the portal vein, and calculations show that the portal inflow supplies perhaps one third of the total hepatic O_2 uptake.

Since the portal venous blood traverses two capillary beds, the gastrointestinal and the hepatic, it is evident that it cannot exert a pressure that fully balances the colloid osmotic pressure of plasma in both of them. Pressures are in the neighborhood of 26 mm Hg in the mesenteric capillaries, 6 to 12 mm Hg in the portal vein, and probably 2 to 6 mm Hg in the hepatic sinusoids, although measurements in this last region are difficult to obtain and not altogether reliable at present. The sinusoidal walls are highly permeable; yet the protein content of hepatic interstitial fluid is relatively low, less than 300 mg/100 ml. This apparent paradox has led to the suggestion that the interstitial fluid is not a homogenous compartment, but rather a gel-like matrix through which albumin diffuses only slowly,[3] whereas smaller molecules and ions are distributed more freely. Hepatic lymph, on the other hand, is abdundant and almost as high as blood plasma in protein content. From one fourth to one half of the thoracic duct lymph flow comes from the lymphatics of the liver.

A clear understanding of hemodynamics in the hepatic vascular bed is impeded by the technical difficulty of measuring pressures and flows at the necessary points, but a fairly consistent picture is beginning to emerge. Portal venular resistance is relatively low, as is the postsinusoidal resistance, whereas that in the hepatic arterioles is at least 50 times higher. The evidence at present indicates that hepatic arteriolar resistance is regulated predominantly by local myogenic activity (p. 1027). An increase in pressure in the portal vein, hepatic veins, or hepatic artery produces vasoconstriction in the hepatic arterioles, a response characteristic of "autoregulation" in many vascular beds. This arteriolar reaction is particularly sensitive to changes in the venous pressures, suggesting that pressure in the sinusoids or in the terminal part of the arterioles is the effective stimulus. Portal venules do not share this response. On the contrary, an increase in hepatic venous pressure tends to reduce portal resistance slightly, a phenomenon that could be explained by simple passive distention.[3,34] The apparent absence of myogenic reaction in the terminals of the portal vein is rather puzzling in view of the spontaneous contractions and active response to stretch that can be demonstrated in strips of portal vein in vitro.

This account of local control of the hepatic microcirculation is consistent with many experiments showing autoregulation of hepatic arterial flow, even after hepatic denervation. This phenomenon is often observed at arterial perfusion pressures above 80 mm Hg, although, like autoregulation under experimental conditions in other organs, it may be present in only two out of every three experiments. It also accounts for the interaction between arterial and portal inflows, whereby occlusion of the portal vein sometimes leads to a decrease in hepatic arterial resistance and increased hepatic arterial flow, if we assume that obstruction of portal inflow lowers the sinusoidal pressure slightly. Changes in hepatic arterial resistance and flow influence the amount of O_2 available for hepatic metabolism as well as the transmural pressure in the sinusoids.

Stimulation of sympathetic nerves to the liver increases hepatic arterial resistance, but this vasoconstriction gradually disappears over a period of minutes even though stimulation is continued, an example of "autoregulatory escape." In addition to the predominant alpha-adrenergic receptors responsible for this vasoconstriction, the presence of beta-receptors has been shown by stimulation after administration of an alpha blocking agent. A mild sympathetic vasoconstriction can also be demonstrated in the portal inflow circuit, portal venous pressure rising moderately with little or no change in flow, but these vessels do not "escape" on continued stimulation. Functional parasympathetic cholinergic fibers have not been found in the hepatic vascular bed. The greatest physiologic significance of neural vasomotor control in the liver probably lies in the effects on vascular capacity. Hepatic blood volume normally constitutes about 10% of the total blood volume, and as much as half of this amount can be expelled by stimulation of the hepatic nerves.[3] In this sense the vascular bed of the liver functions as a blood reservoir, and changes in its capacity affect the function of the cardiovascular system as a whole.

Much remains to be learned about the involvement of the hepatic vessels in reflex and other physiologic adjustments of the circulation. Activation of baroreceptors, like sympathetic stimulation, causes only temporary alterations in hepatic arterial resistance. Many investigators have reported a fall in splanchnic blood volume and hepatic blood flow in man during exercise,[3] but direct measurements with electromagnetic flowmeters in exercising dogs show no change in portal or hepatic arterial flow.[37] Whether this is a species difference or reflects some inaccuracy of the indirect methods of measurement in man remains to be seen. Changes in the hepatic circulation during hemorrhage have been intensively studied because pooling of blood in the splanchnic beds and hypoxic damage to the liver have been suspected as contributors to the clinical syndrome of circulatory "shock." Severe hemorrhage leads to a fall in hepatic blood flow, in which decreased cardiac output and increased intestinal vascular resistance play at least as great a part as hepatic arterial vasoconstriction. Prolonged depletion of the total blood volume leads to a stage of irreversible shock, in which death occurs even if the volume is restored to normal, but there is no clear evidence in man that the hepatic circulation is involved in this irreversibility.

Splenic circulation

In many species of animals the capsule and trabeculae of the spleen contain smooth muscle, contraction of which reduces the vascular capacity of that organ and expels blood into the general circulation. Stimulation of the sympathetic fibers to the spleen produces such contraction, as does epinephrine. In such species the spleen acts as a blood reservoir in addition to its functions as part of the reticuloendothelial and leukocyte-producing systems (p. 1133). Splenic contraction probably does not occur in man, where attempts to demonstrate it under physiologic conditions have uniformly failed, but it is conspicuous in the dog during exercise and after hemorrhage.

REFERENCES
General reviews

1. Barcroft, H.: Sympathetic control of vessels in the hand and forearm skin, Physiol. Rev. **40**(suppl. 4):81, 1960.
2. Berne, R. M.: Regulation of coronary flow, Physiol. Rev. **44**:1, 1964.
3. Greenway, C. V., and Stark, R. D.: Hepatic vascular bed, Physiol. Rev. **51**:23, 1971.
4. Gregg, D. E., and Fisher, L. C.: Blood supply to the heart. In Hamilton, W. R., and Dow, P., editors: Handbook of physiology. Circulation section, Baltimore, 1963, The Williams & Wilkins Co., vol. 2.
5. Hertzman, A. B.: Vasomotor regulation of cutaneous circulation, Physiol. Rev. **39**:280, 1959.
6. Lassen, N. A.: Cerebral blood flow and oxygen consumption in man, Physiol. Rev. **39**:183, 1959.
7. Marchetti, G., and Taccardi, B., editors: International symposium on the coronary circulation and energetics of the myocardium, Basel, 1967, S. Karger.
8. Newburgh, L. H., editor: Physiology of heat regulation and the science of clothing, Philadelphia, 1949, W. B. Saunders Co.
9. Pappenheimer, J. R.: The ionic composition of cerebral extracellular fluid and its relation to the control of breathing, Harvey Lect. **61**:71, 1967.
10. Schmidt, C. F.: The cerebral circulation in health and disease, Springfield, Ill., 1950, Charles C Thomas, Publisher.
11. Shepherd, J. T.: Physiology of the circulation in human limbs in health and disease, Philadelphia, 1963, W. B. Saunders Co.
12. Smith, H. W.: The kidney, New York, 1951, Oxford University Press, Inc.
13. Sokoloff, L., and Kety, S. S.: Regulation of cerebral circulation, Physiol. Rev. **40**(suppl. 4):38, 1960.

Original papers

14. Asmussen, E., Christensen, E. H., and Nielsen, M.: Die O$_2$-Aufnahme der ruhenden und der arbeitender Skelettmuskeln, Skand. Arch. Physiol. **82**:212, 1939.
15. Bazett, H. C., et al.: Temperature changes in blood flowing in arteries and veins in man, J. Appl. Physiol. **1**:3, 1948.
16. Berne, R. M., DeGeest, H., and Levy, M. N.: Influence of cardiac nerves on coronary resistance, Am. J. Physiol. **208**:763, 1965.
17. Bing, R. J., et al.: The measurement of coronary blood flow, oxygen consumption, and efficiency of the left ventricle in man, Am. Heart J. **38**:1, 1949.
18. Brachfeld, N., Monroe, R. G., and Gorlin, R.: Effect of pericoronary denervation on coronary hemodynamics, Am. J. Physiol. **199**:174, 1960.
19. Bradley, S. E., et al.: The estimation of hepatic blood flow in man, J. Clin. Invest. **24**:890, 1945.
20. Chorobski, J., and Penfield, W.: Cerebral vasodilator nerves and their pathway from the medulla oblongata, Arch. Neurol. Psychiatry **28**:1257, 1932.
21. Cushing, H.: Concerning a definite regulatory mechanism of the vaso-motor centre which controls blood pressure during cerebral compression, Bull. Johns Hopkins Hosp. **12**:290, 1901.
22. Dumke, P. R., and Schmidt, C. F.: Quantitative measurements of cerebral blood flow in the macaque monkey, Am. J. Physiol. **138**:421, 1943.
23. Eckenhoff, J. E., et al.: Measurement of coronary blood flow by the nitrous oxide method, Am. J. Physiol. **152**:356, 1948.
24. Eckstein, R. W., et al.: Effects of control of cardiac work upon coronary flow and O$_2$ consumption after sympathetic nerve stimulation, Am. J. Physiol. **163**:539, 1950.
25. Foltz, E. L., et al.: Factors in variation and regulation of coronary blood flow in intact anesthetized dogs, Am. J. Physiol. **162**:521, 1950.
26. Forbes, H. S., and Wolff, H. G.: Cerebral circulation. III. The vasomotor control of cerebral vessels. Arch. Neurol. Psychiatry **19**:1057, 1928.
27. Forster, R. E., II, Ferris, B. G., and Day, R.: The relationship between total heat exchange and blood flow in the hand at various ambient temperatures, Am. J. Physiol. **146**:600, 1946.
28. Goodale, W. T., et al.: Coronary sinus catheterization for studying coronary blood flow and myocardial metabolism, Am. J. Physiol. **152**:340, 1948.
29. Granata, L., et al.: Coronary inflow and oxygen usage following cardiac sympathetic nerve stimulation in unanesthetized dogs, Circ. Res. **16**:114, 1965.
30. Gregg, D. E., Khouri, E. M., and Rayford, C. R.: Systemic and coronary energetics in the resting unanesthetized dog, Circ. Res. **16**:102, 1965.

31. Gregg, D. E., Shipley, R. E., and Bidder, T. G.: The anterior cardiac veins; their functional importance in the venous drainage of the right heart, Am. J. Physiol. **139:** 732, 1943.

32. Grindlay, J. H., Herrick, J. F., and Mann, F. C.: Measurement of the blood flow of the liver, Am. J. Physiol. 132:189, 1941.

33. Guz, A., Kurland, G. S., and Freedberg, A. S.: Relation of coronary flow to oxygen supply, Am. J. Physiol. **199:**179, 1960.

34. Hanson, K. M., and Johnson, P. C.: Local control of hepatic arterial and portal venous flow in the dog, Am. J. Physiol. **211:**712, 1966.

35. Harmel, M. H., et al.: The effect of bilateral stellate ganglion block on the cerebral circulation in normotensive and hypertensive patients, J. Clin. Invest. **28:** 415, 1949.

36. Herd, J. A., et al.: Myocardial blood flow determined with krypton[85] in unanesthetized dogs, Am. J. Physiol. **203:**122, 1962.

37. Hopkinson, B. R., and Schenk, W. G.: The electromagnetic measurement of liver blood flow and cardiac output in conscious dogs during feeding and exercise, Surgery **63:**970, 1968.

38. Ingvar, D. H., and Risberg, J.: Increase of regional cerebral blood flow during mental effort in normals and in patients with focal brain disorders, Exp. Brain Res. **3:**195, 1967.

39. Kety, S. S., and Schmidt, C. F.: The nitrous oxide method for the quantitative determination of cerebral blood flow in man; theory, procedure and normal values, J. Clin. Invest. **27:**476, 1948.

40. Kety, S. S., and Schmidt, C. F.: The effects of altered arterial tensions of carbon dioxide and oxygen on cerebral blood flow and cerebral oxygen consumption of normal young men, J. Clin. Invest. **27:**484, 1948.

41. Kety, S. S., Shenkin, H. A., and Schmidt, C. F.: The effects of increased intracranial pressure on cerebral circulatory functions in man, J. Clin. Invest. **27:**493, 1948.

42. Lassen, N. A., and Hedt-Rasmussen, K.: Human cerebral blood flow measurement by two inert gas techniques: comparison of the Kety-Schmidt method and the intra-arterial method, Circ. Res. **19:**681, 1966.

43. Lauson, H. D., Bradley, S. E., and Cournand, A.: The renal circulation in shock, J. Clin. Invest. **23:**381, 1944.

44. Phillips, R. A., et al.: Effects of acute hemorrhagic and traumatic shock on renal function of dogs, Am. J. Physiol. **145:**314, 1946.

45. Radigan, L. R., and Robinson, S.: Effects of environmental heat stress and exercise on renal blood flow and filtration rate, J. Appl. Physiol. **2:**185, 1949

46. Scott, J. C., and Balourdas, T. A.: The interpretation of "spurious" correlations in coronary flow literature, Circ. Res. **7:**169, 1959.

47. Scott, J. C., Bazett, H. C., and Mackie, G. C.: Climatic effects of cardiac output and the circulation in man, Am. J. Physiol. **129:**102, 1940.

48. Selkurt, E. E.: Renal blood flow and renal clearance during hemorrhagic shock, Am. J. Physiol. **145:**699, 1946.

49. Shapiro, W., Wasserman, A. J., and Patterson, J. L., Jr.: Mechanism and pattern of human cerebrovascular regulation after rapid changes in blood CO_2 tension, J. Clin. Invest. **45:**913, 1966.

50. Spencer, F. C., et al.: Coronary blood flow and cardiac oxygen consumption in unanesthetized dogs, Am. J. Physiol. **160:**149, 1950.

51. Wearn, J. T., et al.: The nature of the vascular communications between the coronary arteries and the chambers of the heart, Am. Heart J. **9:**143, 1933.

52. Wilkins, R. W., Doupe, J., and Newman, H. W.: The rate of blood flow in normal fingers, Clin. Sci. **3:**403, 1938.

45

WILLIAM R. MILNOR

Pulmonary circulation

PULMONARY HEMODYNAMICS

The pulmonary circulation differs hemodynamically from the systemic circulation in that it is a low-pressure, low-resistance system, with corresponding differences in vascular architecture. The large arteries and veins are shorter than their systemic counterparts,[46] and the pulmonary vessels in general have thinner, more distensible walls containing less smooth muscle and elastin. All blood returning to the heart from the systemic circuit passes through the pulmonary bed, so that the outputs of the two ventricles must be equal except for brief periods of adjustment. Any temporary imbalance between right and left ventricular outputs[25] leads to an increase or decrease in the volume of blood contained in the pulmonary bed. The pulmonary artery and its initial branches distend with each systolic ejection from the heart, just as the aorta does. Pressure and flow in the pulmonary arteries during systole and diastole are determined by the elasticity of these vessels and by the vascular resistance and impedance of the bed (pp. 1018 and 1029).

The extravascular pressure of a large part of the pulmonary circulation is related to the subatmospheric pressure of the surrounding intrapleural space, whereas the pulmonary capillaries are exposed to the pressures within the alveoli. Because these pressures change systematically with inspiration and expiration, there is an intimate relationship between pulmonary hemodynamics and ventilation of the lungs. A consideration of the changes that take place at birth,[3] when the lungs are first filled with air, is an appropriate introduction to the pulmonary circulation.

Hemodynamic changes at birth

In the fetus the pulmonary alveoli are filled with amniotic fluid, and the respiratory exchange of oxygen and carbon dioxide is carried on entirely by the maternal pulmonary circulation. Studies of the fetal circulation in sheep[3] reveal that blood flow through the lungs during intrauterine life is much smaller than systemic blood flow. About one fourth of the systemic flow returns to the right ventricle, the remainder being diverted into the left heart through an opening in the interatrial septum, the *foramen ovale*. An even smaller amount flows through the lungs, since about two thirds of the right ventricular output of the fetus goes from the pulmonary artery to the aorta through the *ductus arteriosus*. Pulmonary and systemic arterial pressures are approximately equal at this time, and pulmonary vascular resistance in the fluid-filled fetal lung is of the same order as the systemic resistance.

With the first breath after birth, the pulmonary resistance falls and pulmonary blood flow increases, a response to which the mechanical expansion of the lungs and the change in alveolar tensions of oxygen and carbon dioxide both contribute. Within a few hours after delivery the pulmonary resistance decreases 10-fold, and by 6 months of age the pulmonary arterial pressure and resistance are down almost to adult levels.[28] Over the same period the heavy elastic layers of the fetal pulmonary artery largely disappear, and the arteriolar walls become thinner.[20]

Pressures in pulmonary vessels

Mean pressure in the main pulmonary artery of the adult is normally from 10 to 20 mm Hg. Representative values for this and other relevant hemodynamic variables are given in Table 45-1. The pulsations are about as large as the mean pressure, and diastolic pressures in the pulmonary artery may fall to only a few millimeters of mercury, particularly if the heart rate is slow and the diastolic interval correspondingly long. Since the magnitude of the pressure in the pulmonary artery is small compared with that in the aorta, whereas the velocity of flow is approximately the same, the kinetic energy of blood flow represents a much greater proportion of the total hydraulic energy in the pulmonary artery than in the aorta (see equation 3, Chapter 39, and Table 39-1). The relatively low pressures also make technical considerations in the measurement of pressure of greater importance; when the customary fluid-filled catheters are used, differences

Table 45-1. Representative values pertaining to pulmonary hemodynamics under basal conditions

	Dog (10 kg, unanesthetized)		Man (75 kg, 1.85 m², 40 years of age)	
	Mean	Systolic/diastolic	Mean	Systolic/diastolic
Pulmonary blood flow (L/min)	1.5		6.10	
Radius, pulmonary artery lumen (cm)	0.77	(±0.07)*	1.35	(±0.08)*
Average velocity of flow, pulmonary artery (cm/sec)	13.4	47/0	17.8	85/0
Pressure (mm Hg)				
Right atrium	2		4	
Right ventricle		27/2		25/4
Pulmonary artery	14	27/9	15	25/10
Pulmonary capillaries	8		10	
Left atrium	4		7	
Pulmonary vascular resistance (dyne sec/cm⁵)	534		105	
Pulmonary blood volume (ml)	110		440	
Pulmonary mean transit time (sec)	4.4		4.33	

*Maximal systolic/diastolic excursion around mean radius.

of a few centimeters in the selection of the "zero reference level" (the hydrostatic pressure on the gauge when determining the signal for atmospheric pressure) will give relatively large differences in the values recorded.

As in other vascular beds, the drop in pressure along the arteries is greatest in the smallest branches, and the greatest resistance to flow is in the terminal precapillary vessels. Under resting conditions, however, the total fall in mean pressure from the pulmonary artery to the capillaries is only about 4 or 5 mm Hg, and a significant part of the pressure gradient occurs across the capillary bed.[33] Although the pulsations of pressure are attenuated as they pass through the arterial bed, oscillations of perhaps 3 to 5 mm Hg are transmitted into the capillaries.

Pulmonary capillaries. Pressure and flow in the pulmonary microcirculation are inevitably complex, since the vessels are collapsible; the hydrodynamic, hydrostatic, and extravascular pressures are all of the same order of magnitude, and possibilities for interconversion of kinetic and potential energy may be significant. The capillaries of the lung form a diffuse network around the pulmonary alveoli, presenting a very large capillary surface for gas exchange through the walls of the air spaces. The total area of the capillary membrane that serves this function in the human lung, based on anatomic measurements, is estimated to be about 30 m².[8] It is misleading to specify a length for the pulmonary capillaries, since they form a cross-anastomosing

mesh around the alveoli, but their diameter is 5 to 8 μm, and the time spent by each particle of blood in passing through the capillary bed under resting conditions is around 0.5 to 1 sec. Traditionally the surface for pulmonary gas exchange has been considered to be limited to the capillaries, but some diffusion occurs in the terminal precapillary vessels as well.[47] About 1% of the cardiac output normally traverses the lung without taking part in the exchange of respiratory gases[29] through either unventilated regions or arteriovenous shunts.

Because of the low dynamic pressure in the lumen of the pulmonary capillaries, they are particularly susceptible to the influence of extravascular and hydrostatic pressures. The extracapillary pressure in the sparse tissue separating the capillary from the adjacent air space is very close to intra-alveolar pressure, which ordinarily swings from +3 mm Hg to −3 mm Hg with each respiration. The influence of alveolar pressure on the diameter of capillaries is demonstrated by the great increase in resistance to flow through the pulmonary capillary bed when large positive pressures are applied to inflate the lung, as in some devices for artificial respiration. The importance of hydrostatic pressure in the pulmonary vessels is evident from anatomic considerations; in the erect adult, vessels at the apex of the lung may be 15 cm above the pulmonary artery and those at the base an equal distance below it. Since the corresponding hydrostatic pressures are of the same order as the pulmonary arterial pressure, capillaries at the lung apices should be barely open (or perhaps entirely closed) in the erect position, whereas those at the bases are fully distended, and measurements of regional pulmonary blood flow in human subjects show this to be the case.[9] West[9] and others[5,37] have studied in detail the interaction of arteriolar, venular, and alveolar pressures on the patency or collapse of pulmonary capillaries, and the resulting regional differences in blood flow must be taken into account in virtually all respiratory and hemodynamic responses in the lung. Capillaries are not readily distensible, and it is probable that local capillary resistance depends more on the relative number of vessels opened and closed than on graded variations in individual diameter. This emphasizes once again the distinction between transmural pressures that act on the vessel wall and the intravascular pressure gradients that produce blood flow (p. 1018). Transmural pressure is also a factor in transcapillary exchange of water and solutes[16] (p. 1089), and here the oncotic pressure of plasma (some 25 mm Hg) encourages the movement of fluid into the capillaries, whereas intracapillary pressure acts in the opposite direction. Although microscopic examination reveals very little tissue between the alveolar walls and the capillaries, the total volume of interstitial fluid in the lungs is substantial,[40] and there is an active fluid exchange between the tissue spaces and the lumen of the capillaries.[16] When pressure is abnormally high or the endothelial wall is damaged, there is a net loss of fluid from the capillary, which leads to an increased interstitial volume and eventually to pulmonary edema.[7]

The extravascular, or tissue, pressure applied to pulmonary vessels other than capillaries is determined by the intrapleural pressure and the degree of inflation of the lungs.[5,17,38] Interstitial pressure around the vessels is probably not the same near the pleural surface as in the deeper portions of the lung, and there is some evidence to suggest that interstitial pressure in at least some parts of the lung is more negative than the intrapleural pressure under the conditions of inflation normally maintained in vivo.[38]

Pulmonary veins. The drop in pressure from the venous end of the pulmonary capillaries to the entrance of the pulmonary veins into the left atrium is probably 3 to 4 mm Hg. The muscular coat of the venules and their resistance to flow is much less marked than that of the arterioles. Nevertheless, the caliber of these small veins can be altered by vasomotor activity of their smooth muscle, and their resistance presumably influences pressure in the capillaries upstream. In the larger pulmonary veins, pressure gradients are so small that they are at the limits of resolution of present methods of measuring pressure.

When a catheter is advanced out the pulmonary artery until it "wedges" in a peripheral branch (a practice introduced in clinical investigation in an effort to estimate pulmonary capillary pressure), the pressure recorded probably approximates that in the small pulmonary veins. Under such conditions, blood stands stagnant in the vascular bed supplied by the occluded artery out to the point at which these vessels are joined by collaterals draining blood from unobstructed regions. Theoretically, it is the pressure at this junction—presumably somewhere in the pulmonary venous tree—that is recorded, but inasmuch as the site of the confluence is not known and the pressure and flow in adjacent unobstructed areas may well be affected by the procedure, such measurements are of questionable value.

Pulmonary blood flow

The pulsations of flow in the main pulmonary artery normally have the contour shown in Fig. 45-1. Pulmonary flow rises more slowly and falls less steeply than does flow in the aorta (Fig. 37-1). Flow recorded from the midportion of the main pulmonary trunk shows a brief reversal at the end of systole, occasioned by a small backflow of blood that distends the root of the artery. The intermittent systolic ejections of the right ventricle are converted by the distention and recoil of the pulmonary artery and its major branches into a pulsatile flow that continues during diastole, just as in the systemic circulation. Pulsatile flow in the pulmonary capillaries was first demonstrated by direct microscopic observation[49] and can now be measured in human subjects through an application of the principles of plethysmography.[30,32,36] The normal pattern of flow in the capillaries, shown in Fig. 45-2, indicates that some 25% of the total flow takes place during diastole. Flow pulsations gradually diminish as blood travels down the pulmonary veins, and very small oscillations persist up to the left atrium. Left atrial contraction produces a small transient backflow in the main pulmonary veins near the atriovenous junction. The magnitude and velocity of transmission of pressure and flow waves from pulmonary artery to capillaries and from capillaries to terminal pulmonary veins have been measured in animals[33] and in man.[15,30] The factors that influence this transition (p. 1038) in the pulmonary bed have been carefully analyzed,[6,50] and abnormalities of transmission have been noted in pulmonary vascular disease.

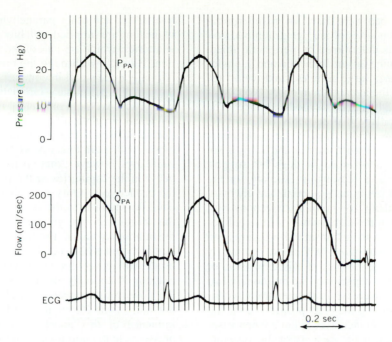

Fig. 45-1. Records from unanesthetized dog standing quietly. P_{PA}, Pressure recorded by catheter-tip manometer in pulmonary artery; $\dot{Q}_{PA}$, pulmonary arterial flow measured by electromagnetic flowmeter (small spikes in flow baseline are artifacts from P and QRS waves of the electrocardiogram detected by flowmeter electrodes); *ECG*, electrocardiogram. Time lines at intervals of 0.02 sec.

In man there is a small but consistent difference in blood flow to the lungs, about 55% of the cardiac output flowing through the right lung and 45% through the left.[4] The distribution of flow within each lung varies as the effects of hydrostatic pressure would predict; in the seated normal subject the lower lobes are much better perfused than the upper lobes.[9]

Pulmonary blood volume

The results of early animal experiments in which the lungs were exsanguinated[31] and more recent measurements using indicator-dilution methods[10,22,35] agree that approximately 9% of the total blood volume is contained in the pulmonary vessels. A number of reports in the literature that assign more than twice this volume to

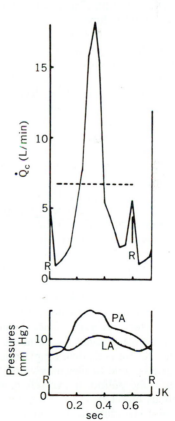

Fig. 45-2. Instantaneous pulmonary capillary blood flow, $\dot{Q}_c$, and pressures in pulmonary artery, *PA*, and left atrium, *LA*, in human subject with normal pulmonary circulation. Pressures and flow were not measured simultaneously, and length of cardiac cycle is not identical in upper and lower records. *R* indicates time corresponding to peak of electrocardiographic QRS complex in upper record. (From DuBois; in Adams and Vieth.[1])

the pulmonary bed were based on misinterpretations of the boundaries of the "central blood volume," which is determined by peripheral injection and sampling of an indicator and includes the lungs, heart, large arteries, and veins (p. 1000). In man the average pulmonary blood volume is about 240 ml/m² of body surface. The pulmonary capillary bed accounts for some 60 ml/m², or 25% of this total[10]; the partitioning of the remainder between arteries and veins is not certain, but data obtained in dogs[24,39] suggest that 70 ml/m² is a reasonable estimate for the volume of the pulmonary arterial bed in man, leaving 110 ml/m² for pulmonary venous volume. Changes in the distensibility of the pulmonary vessels or in their transmural pressures can raise or lower total pulmonary blood volume by 25% to 50%.[6,22] Examples of this expansion or contraction of the capacity of the pulmonary bed range from the decrease in pulmonary vascular volume on assuming the erect position to the accommodation of an increased volume when vasoconstriction reduces the capacity of the systemic bed.[45] The lung vessels constitute a very small blood reservoir in comparison with the systemic veins, however, since the total volume shifted into or out of the pulmonary bed in man probably rarely exceeds 200 ml. These alterations of the pulmonary blood volume imply concomitant alterations in the caliber of some of the pulmonary vessels, and in a general way the pulmonary vascular resistance is elevated when pulmonary blood volume falls, and vice versa. It would be illogical, however, to regard the change in vascular volume as a *cause* of changes in resistance, as is sometimes asserted. Regional blood volume and vascular resistance both depend on the dimensions of the vessels, which in turn depend on transmural pressure, smooth muscle tone, and a host of other factors.

The mean pulmonary circulation time, or average time required for a particle of blood to travel from the pulmonary valve to the left atrium, is a function of cardiac output and pulmonary blood volume. In normal adults at rest the pulmonary mean transit time is 4 to 5 sec. The circulation time through the most rapid path in the pulmonary circuit, indicated by the earliest appearance in the left atrium of a substance injected into the pulmonary artery, averages 2 sec in man. The mean transit time through the pulmonary bed is slightly shorter for cells than for plasma, and the ratio of cell volume to plasma volume is somewhat less in the pulmonary vessels than in systemic venous blood.[41]

Bronchial circulation. The bronchi, connective tissue, and other parenchymatous structures of the lung are supplied with blood by the bronchial arteries, which are part of the systemic circulation. Blood flow through the bronchial vessels is normally quite small in comparison with the flow through the pulmonary circulation, but the two circulations are not completely independent. The bronchial circulation has a complicated distribution[2,14] in which there are normally no anastomoses with pulmonary vessels proximal to the capillaries; however, there is free communication between the capillaries of the two systems along the bronchioles. Only a small part of the bronchial flow returns to the systemic circuit through the azygos vein, the greater portion draining into the pulmonary veins. The highly oxygenated pulmonary venous blood is thereby contaminated by bronchial venous blood of lower oxygen content, but this is normally of little significance, since the flow from bronchial to pulmonary circuit is not more than 1% of the cardiac output.[14] The influence of this anastomotic arrangement on pressures in the pulmonary microcirculation may well be significant, although it cannot be evaluated at present. Relevant in this regard are reports that bronchial arteries are affected by vasomotor nerves and drugs in a manner characteristic of other parts of the systemic circuit and that the pulmonary vasomotor response to neural stimuli is altered by interruption of the bronchial circulation.[2,14,19]

CONTROL OF PULMONARY CIRCULATION
Pulmonary vascular resistance

The distribution of the pulmonary vascular resistance between arterioles, capillaries, and venules is difficult to determine, but it seems clear that the capillary bed and venules contribute a larger share than is the case in the systemic circulation. The pressure drop across the capillaries is about one third of the total fall in pressure from pulmonary artery to left atrium, and that in the veins is not much less. These proportions show that pulmonary vascular resistance can be altered significantly by events in the capillaries and pulmonary venous bed, although it is controlled principally by the arterioles.

The caliber of the pulmonary vessels not only changes with active contraction or relaxation of the smooth muscle in their walls but also responds passively to changes in transmural pressure. This is true of all blood vessels, but in the thin-walled pulmonary vessels the effects of transmural pressures on vessel diameter are especially prominent.

The potentialities for partial or complete collapse of some pulmonary capillaries makes the relation between blood flow and pressure particularly complex in the lung.[5,9,37] Two different phenomena combine to govern hemodynamics in the pulmonary microcirculation. On the one hand an increase in capillary hydraulic transmural pressure tends to open up a greater number of capillaries and, to a lesser extent, distend those already open.[5] This response depends on critical closing pressures and on stress-strain relationships (p. 1026). Equally important, on the other hand, is the relationship between alveolar and venous pressures. If alveolar pressure is higher than intravascular pressure at the venous end of the pulmonary capillary, which is the normal condition in the upper parts of the human lung in the erect position, the capillary tends to collapse at its venous end. If pulmonary arterial and arteriolar pressures are higher than the intra-alveolar pressure, the collapse is not complete, and blood flow continues through the capillaries so affected. The capillary-to-venous pressure gradient under these circumstances is much higher per unit blood flow under these conditions that it is when the capillaries are widely open, as they normally are in the lower parts of the lung where venous is greater than alveolar pressure. Total pulmonary blood flow is thus the sum of regional flows that differ considerably in different parts of the lung.

The vascular resistance (p. 1018) of the pulmonary vascular bed is consequently susceptible to alteration by (1) the intravascular component of transmural pressure, which depends in part on hemodynamic forces generated by the heart and other structures beyond the pulmonary circulation, and (2) the extravascular component of transmural pressure, which is complexly related to pressures in the intrapleural, interstitial, and alveolar spaces. Because of these ample opportunities for passive response quite apart from vasomotor activity, measurements of pulmonary vascular resistance have sometimes been regarded as having no physiologic significance. This unduly pessimistic view is not necessary if the pressure-flow relationships expressed as vascular resistance are clearly defined (p. 1019). First, three variables must be measured to determine pulmonary vascular resistance: pulmonary arterial pressure, left atrial pressure, and pulmonary blood flow, or cardiac output. Inferences about resistance based on the inflow pressure at the pulmonary artery alone are unreliable, since this pressure may rise as a result of increased cardiac output or elevation of left atrial pres-

sure.[21] Second, the ratio of pressure head to blood flow varies with blood viscosity and vessel length as well as with vessel diameter. If viscosity and length remain constant, a change in resistance implies a change in the diameter of the lumen of vessels somewhere in the bed, but it does not reveal the location of the particular vessels involved. The change may be a uniform increase or decrease in the caliber of one particular class of vessels or it may be a collapse of capillaries in a limited region. Third, although resistance reflects the dimensions of vessels in the vascular bed, it tells nothing about vasomotor tone unless the passive determinants of vessel diameter can be measured simultaneously and taken into account. Active and passive vascular responses are integrated in vivo, and agents that act as pulmonary vasoconstrictors on the isolated perfused lung may entrain extrapulmonary effects in the intact animal that result in pulmonary vasodilatation (Fig. 45-3). Finally, the calculation of resistance from mean pressures and flows instead of from actual pulsations expresses only part of the opposition that must be overcome in moving blood through a vascular bed. A complete expression requires that pulmonary vascular impedance be computed from records of pulsatile flow and pressure.[6,12] Fortunately, the presence of pulsations has negligible effects on the relation between the mean terms and vessel diameter, so that conventional calculation of resistance gives an accurate estimate of the opposition to flow associated with mean terms, although it omits the additional effect of pulsations. When each of these limitations is strictly observed in the interpretation of vascular resistance, its measurements can serve a useful purpose in physiologic investigation.

An important example of the modifications of pulmonary vascular resistance that occur in vivo is the characteristic response to exercise. As the cardiac output increases, mean pulmonary arterial pressure rises only moderately,* reaching 3 to 10 mm Hg above resting levels when the pulmonary blood flow has increased two- or threefold. The concomitant alterations in left atrial pressure have been measured only rarely, but there appears to be little or no significant change. Pulmonary vascular resistance therefore falls during exercise, but it remains to be seen whether this is a passive response to the relatively small increase in transmural pressures or if it is an active vasomotor dilatation.

---

*See references 4, 6, 23, 42, and 44.

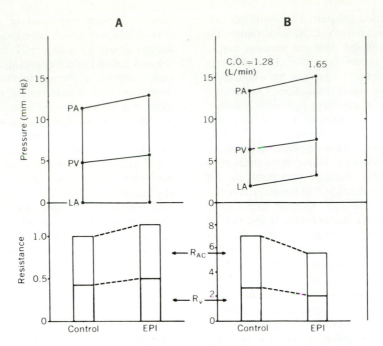

Fig. 45-3. Comparison of effects of epinephrine infusion on pulmonary vessels in **A,** lung perfused at constant flow, and in **B,** anesthetized open-chest dog. Mean pressures during control period and after infusion of drug are indicated for pulmonary artery, *PA,* small pulmonary vein 1 to 2 mm in diameter, *PV,* and left atrium, *LA,* the latter being open to atmospheric pressure in perfused preparation. Vertical bars at bottom of graph represent resistance between pulmonary artery and pulmonary vein, R_{AC}, and between vein and left atrium, R_V; resistances are scaled in arbitrary units for perfused lung and in 10^2 dyne sec/cm^5 for anesthetized dog.

In perfused lung, where flow is held constant and left atrial outflow pressure cannot change, infusion of 5 μg of epinephrine raises pressure gradient and hence resistance in both venous and arteriocapillary segments of pulmonary bed. In anesthetized dog, infusion of 1 μg/min/kg of the same drug affects heart and systemic circulation so as to elevate left atrial pressure and consequently raise transmural pressures in pulmonary vessels, while cardiac output increases from 1.28 to 1.65 L/min. Epinephrine presumably makes pulmonary vessels less distensible in both preparations; but when circulation is intact, this effect is outweighed by increased transmural pressure, and vascular resistance falls. (Based on data from Gilbert et al.[26] and Milnor.[6])

Vasomotor control

Because of varying interpretations of changes in pulmonary vascular resistance and the technical difficulty of measuring the small pressures in the pulmonary circulation, the existence of active vasomotion in the pulmonary vessels was long regarded as doubtful. The rich innervation that has been found in the vessels of the lung and the demonstration of a number of fairly unequivocal vasomotor responses under experimental conditions[2,19,27] have gone far to remove this doubt, although the extent to which vasomotor control operates in vivo is still not clear. In a series of experiments, Daly and Hebb[2] have shown the existence of sympathetic vasoconstrictor and vasodilator fibers to the pulmonary vessels, both groups being atropine resistant. They have also demonstrated cholinergic parasympathetic fibers in the vagus that cause pulmonary vasodilatation in their preparations, but the possibility that these act on the bronchial rather than the pulmonary vessels has not been ruled out. That stimulation of such fibers constricts or dilates the pulmonary vessels does not, of course, prove that they are part of the normal mechanism for circulatory control, but the evidence tends strongly in that direction. Additional clues that point to a constant active autonomic control of the pulmonary circulation are provided by the discovery of reflexes that originate in the systemic baroreceptors and moderate pulmonary vascular resistance. Stimulation of the carotid sinus baroreceptors leads to a reduction of pulmonary resistance, whereas stimulation of the carotid body chemoreceptors by hypoxic blood is followed by pulmonary vasoconstriction,[19] although the latter response is sometimes obscured by changes in the bronchial circulation. Apart from the efferent vasomotor supply that mediates these reactions, the neural connections of the pulmonary vascular bed also include afferent fibers from stretch receptors in the pulmonary artery and elsewhere[18] (p. 1065). In short, the pulmonary vessels now appear to be subject to autonomic control in much the same way as the systemic vessels, although under resting conditions the normal vasomotor tone is minimal and the vessels are almost maximally dilated.

Local vasomotor responses. In addition to the responses elicited through central nervous system pathways, the pulmonary vessels are capable of reacting directly to substances in contact with their walls. Perhaps the most important response of this kind is the pulmonary vasoconstriction brought on by low oxygen or high carbon dioxide tensions in the blood. This phenomenon was first described by von Euler and Liljestrand,[48] who postulated that the vasomotor response to respiratory gases was a mechanism for adjusting local alveolar perfusion to ventilation. The local vascular resistance in different parts of the lung, in other words, could be modified in this way to favor perfusion of the best-ventilated alveoli. Further experiments have confirmed this hypothesis and show that pulmonary vasoconstriction develops in man where the oxygen saturation of systemic arterial blood falls below about 80%.[4] The site of vasoconstriction is uncertain, but both pre- and postcapillary small vessels are probably involved. The effects of chronic hypoxia are similar, and residents at high altitudes develop chronic pulmonary hypertension and enlargement of the right ventricle as well as polycythemia.[43] The pulmonary vascular effects of hypercapnia are more variable and depend on the production of acidosis, a decrease in blood pH acting as a stimulus to vasoconstriction.[4] Recent evidence suggests that the mechanism of hypoxic constriction of small pulmonary vessels involves hyperpolarization of the membranes of vascular smooth muscle cells and increased calcium influx.[34]

The pulmonary vasculature can also be shown to respond directly to drugs introduced into the pulmonary circulation. In the isolated lung, epinephrine, norepinephrine, and histamine act as pulmonary vasoconstrictors; in the intact animal, the increased vascular stiffness they induce by stimulating smooth muscle is sometimes outweighed by passive changes in transmural pressure secondary to their action on the heart and systemic circulation.[4,26] In the case of histamine, modifications of ventilation and intra-alveolar pressure resulting from bronchoconstriction contribute to the vascular response. One form of pulmonary vasoconstriction that is still unexplained is the reaction to embolization of small pulmonary vessels, which causes intense constriction even in regions distant from the sites of occlusion.

CIRCULATORY EFFECTS OF RESPIRATION

The dimensions of the pulmonary vessels vary with the degree of inflation of the lungs. Pulmonary vascular resistance is at a minimum when the lungs are inflated to a volume approximating the normal state in vivo and increases when the lungs are inflated or deflated beyond this range.[5,38] In the intact animal, respiratory variations in the ventricular outputs, vena caval blood flow, and vascular transmural pressures are superimposed on these purely mechanical effects. On inspiration the intrapleural pressure falls from its resting expiratory level of about -3 mm Hg to -6 mm Hg, the intrathoracic portions of the venae cavae tend to be distended, blood flow into these veins and the right atrium increases, and the right ventricle ejects a larger stroke volume.[13,25] Elevation of intra-abdominal pressure by the respiratory movements of the diaphragm and abdominal muscles during inspiration may also help to increase flow into the intrathoracic portion of the inferior vena cava.

Up to this point, most of the experimental evidence is in agreement, but the net effect of these events on pulmonary vascular resistance is a subject of debate.[1] It is certain, however, that any respiratory change in pulmonary vascular resistance must be quite small, and it is probable that the resistance rises slightly with inspiration but does not prevent a simultaneous increase in pulmonary blood flow. The left ventricle increases its output almost in phase with the right ventricle[25] but to a lesser extent, so that pulmonary blood volume enlarges with inspiration. During deep breathing or chronic pulmonary disease, these respiratory swings of flow, resistance, and volume are greatly exaggerated.

Artificial respiration by mechanical devices affects the pulmonary circulation by altering the intra-alveolar and intrapleural pressures. Positive-pressure breathing, in which the lungs are inflated intermittently by applying positive pressure to the airway, compresses the pulmonary capillaries so that pulmonary vascular resistance rises and right ventricular output falls with each inflation.

When intra-alveolar and intrapleural pressures are raised by forced expiratory effort against a closed glottis (the Valsalva maneuver) or a column of water, venous return to the right heart is impeded and the output of first the right and then the left ventricle declines gradually. The mean and pulse pressures in systemic arteries such as the brachial artery fall, but they return abruptly to levels above the control values when the increased pressure is released. Intrapleural pressures of more than 100 mm Hg can be developed, and prolonged maintenance of such high levels may be terminated by fainting. Coughing or straining also caused a marked increase in intrathoracic pressure, which is reflected in systemic arterial pressures. Although pressure in the peripheral arteries is increased by the external pressure on their intrathoracic origins, the corresponding pressures in the arteries within the thorax or abdomen are counterbalanced by the rise in pressure within these cavities. The arteries of the central nervous system are protected in a similar way by transmission of the intrathoracic and intra-abdominal pressures to the cerebrospinal fluid through the intervertebral foramina.

Respiration and heart rate. Waxing and waning of the heart rate in phase with respiration is a common observation. Because it usually arises from variations in the firing rate of the normal pacemaker in the sinoatrial node, it is called *sinus arrhythmia*. The same phenomenon originates in some instances by a gradual wandering of the pacemaker from its normal site to fibers in the atrium or atrioventricular node and back again.

Sinus arrhythmia is evoked through the vagus nerves and has its origin in at least two different sources: the stimulation of pulmonary stretch receptors by inflation of the lung and an interaction between the respiratory centers and the vagal nucleus.

REFERENCES
General reviews
1. Adams, W. R., and Veith, I., editors: Pulmonary circulation, New York, 1959, Grune & Stratton, Inc.
2. Daly, I. deB., and Hebb, C.: Pulmonary and bronchial vascular systems, Baltimore, 1966, The Williams & Wilkins Co.
3. Dawes, G. S.: Foetal and neonatal physiology, Chicago, 1968, Year Book Medical Publishers, Inc.
4. Fishman, A. P.: Dynamics of the pulmonary circulation. In Hamilton, W. F., and Dow, P., editors: Handbook of physiology. Circulation section, Baltimore, 1963, The Williams & Wilkins Co., vol. 2.
5. Fishman, A. P., and Hecht, H. H., editors: The pulmonary circulation and interstitial space, Chicago, 1969, The University of Chicago Press.
6. Milnor, W. R.: Pulmonary hemodynamics. In Bergel, D. H., editor: Cardiovascular fluid dynamics, London, 1972, Academic Press, Ltd.
7. Staub, N. C.: Pulmonary edema, Physiol. Rev. **54:**678, 1974.
8. Weibel, E. R.: Morphometry of the human lung, New York, 1963, Academic Press, Inc.
9. West, J. B.: Ventilation/blood flow and gas exchange, Oxford, 1965, Blackwell Scientific Publications.
10. Yu, P. N.: Pulmonary blood volume in health and disease, Philadelphia, 1969, Lea & Febiger.

Original papers
11. Bates, D. V., et al.: Variations in the pulmonary capillary blood volume and membrane diffusion component in health and disease, J. Clin. Invest. **39:**401, 1960.
12. Bergel, D. H., and Milnor, W. R.: Pulmonary vascular impedance in the dog, Circ. Res. **16:**401, 1965.
13. Brecher, G. A., and Hubay, C. A.: Pulmonary blood flow and venous return during spontaneous respiration, Circ. Res. **3:**210, 1955.
14. Bruner, H. D., and Schmidt, C. F.: Bloodflow in the bronchial artery of the anesthetized dog, Am. J. Physiol. **148:**648, 1947.
15. Caro, C. G., Harrison, G. K., and Mognoni, P.: Pressure wave transmission in the human pulmonary circulation, Cardiovasc. Res. **1:**91, 1967.
16. Chinard, F. P., and Enns, T.: Transcapillary pulmonary exchange of water in the dog, Am. J. Physiol. **178:**197, 1954.
17. Clements, J. A., et al.: Pulmonary surface tension and alveolar stability, J. Appl. Physiol. **16:**444, 1961.
18. Coleridge, J. C. G., and Kidd, C.: Reflex effects of stimulating baroreceptors in the pulmonary artery, J. Physiol. **166:**197, 1963.
19. Daly, I. deB., and Daly, M. deB.: The effects of stimulation of the carotid body chemoreceptors on the pulmonary vascular bed in the dog: the "vasosensory controlled perfused living animal" preparation, J. Physiol. **148:**201, 1959.
20. Dammann, J. F., and Ferencz, C.: The significance of the pulmonary vascular bed in congenital heart disease, Am. Heart J. **52:**7, 1956.
21. Dexter, L., et al.: Studies of the pulmonary circulation in man at rest; normal variations and the interrelations between increased pulmonary blood flow, elevated pulmonary arterial pressure, and high pulmonary "capillary" pressures, J. Clin. Invest. **29:**602, 1950.
22. Dock, D. S., et al.: The pulmonary blood volume in man, J. Clin. Invest. **40:**317, 1961.
23. Donald, K. W., et al.: The effect of exercise on the cardiac output and circulatory dynamics of normal subjects, Clin. Sci. **14:**37, 1955.
24. Feisal, L. A., Soni, J., and DuBois, A. B.: Pulmonary arterial circulation time, pulmonary arterial blood volume, and the ratio of gas to tissue volume in the lungs of dogs, J. Clin. Invest. **41:**390, 1962.
25. Franklin, D. L., Van Citters, R. L., and Rushmer, R. F.: Balance between right and left ventricular output, Circ. Res. **10:**17, 1962.
26. Gilbert, R. P., et al.: Effects of histamine, 5-hydroxytryptamine and epinephrine on pulmonary hemodynamics with particular reference to arterial and venous segment resistances, Am. J. Physiol. **194:**165, 1958.
27. Ingram, R. H., Jr., Szidon, J. P., and Fishman, A. P.: Response of the main pulmonary artery of dogs to neu-

ronally released versus blood-borne norepinephrine, Circ. Res. **26:**249, 1970.

28. James, L. S., and Rowe, R. D.: The pattern of response of pulmonary and systemic arterial pressure in newborn and older infants to short periods of hypoxia, J. Pediatr. **51:**5, 1957.

29. Jose, A. D., and Milnor, W. R.: The demonstration of pulmonary arteriovenous shunts in normal human subjects, and their increase in certain disease states, J. Clin. Invest. **38:**1913, 1959.

30. Karatzas, N. B., and Lee, G. de J.: Propagation of blood flow pulse in the normal human pulmonary arterial tree, Circ. Res. **25:**11, 1969.

31. Kuno, Y.: On the amount of blood in the lungs, J. Physiol. **51:**154, 1917.

32. Linderholm, H., et al.: Pulmonary capillary blood flow during cardiac catheterization, J. Appl. Physiol. **17:**135, 1962.

33. Maloney, J. E., and Castle, B. L.: Dynamic intravascular pressures in the microvessels of the frog lung, Resp. Physiol. **10:**51, 1970.

34. McMurty, I. F., et al.: Inhibition of hypoxic vasoconstriction by calcium antagonists in isolated rat lungs, Circ. Res. **38:**99, 1976.

35. Milnor, W. R., Jose, A. D., and McGaff, C. J.: Pulmonary vascular volume, resistance, and compliance in man, Circulation **22:**130, 1960.

36. Morkin, E., et al.: The nature of pulmonary capillary blood flow and gas exchange, J. Clin. Invest. **41:**1386, 1962.

37. Permutt, S., Bromberger-Barnea, B., and Bane, H. N.: Alveolar pressure, pulmonary venous pressure, and the vascular waterfall, Med. Thorac. **19:**239, 1962.

38. Permutt, S., et al.: Effect of lung inflation on static pressure-volume characteristics of pulmonary vessels, J. Appl. Physiol. **16:**64, 1961.

39. Piiper, J.: Verhalten des Strömungswiderstandes und der Blutfullung am isolierten Lungenlappen des Hundes, Arch. Gesamte Physiol. **264:**596, 1957.

40. Ramsey, L. H., et al.: Pericapillary gas and water distribution volumes of the lung calculated from multiple indicator dilution nerves, Circ. Res. **15:**275, 1964.

41. Rapaport, E., et al.: Pulmonary red cell and plasma volumes and pulmonary hematocrit in the normal dog, Am. J. Physiol. **185:**127, 1956.

42. Riley, R. L., et al.: Studies of the pulmonary circulation at rest and during exercise in normal individuals and in patients with chronic pulmonary disease, Am. J. Physiol. **152:**372, 1948.

43. Rotta, A., et al.: Pulmonary circulation at sea level and at high altitudes, J. Appl. Physiol. **9:**328, 1956.

44. Sancetta, S. M., and Rakita, L.: Response of pulmonary artery pressure and total pulmonary resistance of untrained, convalescent man to prolonged mild steady state exercise, J. Clin. Invest. **36:**1138, 1957.

45. Sarnoff, S. J., Berglund, E., and Sarnoff, L. C.: Neurohemodynamics of pulmonary edema. III. Estimated changes in pulmonary blood volume accompanying systemic vasoconstriction and vasodilation, J. Appl. Physiol. **5:**367, 1953.

46. Schleier, J.: Der Energieverbrauch in der Blutbahn, Arch. Gesamte Physiol. **173:**172, 1919.

47. Staub, N. C.: Gas exchange vessels in the cat lung, Fed. Proc. **20:**107, 1961.

48. von Euler, U. S., and Liljestrand, G.: Observations on the pulmonary arterial blood pressure in the cat, Acta Physiol. Scand. **12:**301, 1946.

49. Wearn, J. T., et al.: The normal behavior of the pulmonary blood vessels with observations on the intermittence of the flow of blood in the arterioles and capillaries, Am. J. Physiol. **109:**236, 1934.

50. Wiener, F., et al.: Wave propagation in the pulmonary circulation, Circ. Res. **19:**834, 1966.

46 Blood volume

WILLIAM R. MILNOR

The volume of blood contained within the vascular system is remarkably constant under ordinary conditions because of the dynamic equilibrium maintained by capillary exchange and lymph flow (p. 1088). This is an important factor in circulatory homeostasis in the resting state, because intravascular pressures and cardiac output are determined in part by the magnitude of the blood volume in relation of vascular capacity (p. 957). Physical activity, extremes of environmental temperature, or other functional demands on the circulation can alter blood volume, but the changes take place relatively slowly.

MEASUREMENT OF BLOOD VOLUME

Most of the methods currently used to determine blood volume measure the dilution of some nontoxic test substance injected into the bloodstream. Blood volume is defined as the sum of the volumes of plasma and formed elements of the blood (chiefly erythrocytes) *within* the blood vascular system, and the ideal test material is one that mixes completely with either plasma or the cellular elements and does not leave the vascular compartment. Materials that combine with plasma proteins or with chemical components of the red blood cell fulfill this requirement adequately in practice, although virtually all substances that can be injected disappear from the blood sooner or later. Most of the constituents of blood—albumin, for example—are also present to some degree in extravascular compartments, and the rate at which the test substance moves from blood into these other areas of distribution must be taken into account.

A great many test substances are now available, and the extensive literature on the development of methods of measuring blood volume has been reviewed by Lawson.[4] Most of the commonly used procedures measure either plasma volume or cell volume and then calculate total blood volume from the hematocrit (the ratio of packed cells to total volume in a centrifuged blood sample). Alternatively, two different test materials may be used, one to "tag" red blood cells, and the other plasma. A detailed description of technical procedures and precautions for measuring blood volume and body fluid volumes in general has been published by Chien and Gregersen.[1]

Measurement of plasma volume with Evans blue dye (T-1824)[1,26,28] illustrates the principles common to most methods. This dye, which is bound by plasma albumin,[44] and radioiodine-(^{131}I) labeled albumin are the materials most often used in measuring plasma volume. The determination is usually made with the subject recumbent and at rest, and in the postabsorptive state to avoid lipemia, since lipemia and hemolysis interfere with the optical density methods used to measure plasma dye concentration.

After drawing a control sample of blood to serve as a "blank," an exactly measured amount of Evans blue dye is injected into a peripheral vein. After a period suitable to allow the dye to mix with circulating plasma, one or more blood samples are taken from another vein. A series of at least three samples taken 10, 20, and 30 min after injection is desirable in order to define the slope of concentration with time (Fig. 46-1). The samples are centrifuged, the supernatant plasma decanted, and dye concentrations determined by comparison with the blank in a spectrophotometer. Concentrations are plotted against time and then extrapolated back to the time of injection so as to estimate the plasma concentration that would have been attained had all the dye injected been instantaneously and uniformly mixed with the entire plasma volume. This extrapolation is necessary because the concentration oscillates during the first minute or so after injection while mixing takes place and then begins to fall at a rate determined by the loss of dye-tagged albumin from the bloodstream into the extravascular albumin pool (Fig. 46-1). The plasma volume (V_{plasma}) can be calculated in milliliters from M, the amount of dye injected (in milligrams), and C, the concentration of dye in plasma at time

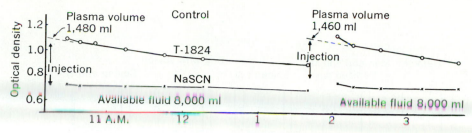

Fig. 46-1. Time-concentration curves of Evans blue dye (T-1824) and of sodium thiocyanate in a dog weighing 25 kg. Note agreement in values when determinations are repeated by second injection. During period required for uniform distribution of dye, some escapes from bloodstream. By projecting disappearance curve back to time of injection, one may correct for this loss and estimate concentration of dye (expressed in terms of optical density) that would exist if uniform distribution occurred immediately. Plasma concentration of sodium thiocyanate usually remains practically the same for several hours after uniform distribution has occurred. Therefore no correction need be made for disappearance, as in case of Evans blue dye. Volume of fluid available for distribution of thiocyanate is approximate measure of volume of extracelllular fluid (Chapter 49). (From Gregersen and Stewart.[29])

zero (expressed in milligrams per milliliter), by the following simple relation:

$$V_{plasma} = \frac{M}{C} \qquad (1)$$

Although the collection of multiple samples is theoretically necessary, the error involved in using the concentration in a single 10 min sample in place of the extrapolated value has been found to be extremely small and therefore acceptable for routine clinical measurements. Plasma volume and the hematocrit (H) can be used to calculate total blood volume (V_{blood}):

$$V_{blood} = \frac{V_{plasma}}{(1 - H)} \qquad (2)$$

In practice the hematocrit is usually determined on a sample of venous blood, and it is often assumed that the proportion of cells to plasma in the blood volume as a whole is the same as in this venous sample. This assumption is incorrect, but the quantitative relation between the hematocrit in large peripheral vessels and that in the circulation as a whole has been studied in great detail,[4,20,46] so that an appropriate correction can be made under most conditions. The "overall hematocrit" (H_o), or ratio of total cell volume to total blood volume, can be determined experimentally by simultaneous measurement of cell and plasma volumes with two different indicators. The ratio of H_o to the "venous hematocrit" (H_v) has been termed F_{cells}:

$$F_{cells} = \frac{H_o}{H_v} \qquad (3)$$

F_{cells} normally averages 0.91 in man,[45] and 0.88 in the splenectomized dog.[46] The overall hematocrit is thus lower than that in peripheral samples, and the difference arises largely from the comparative scarcity of cells in the smallest peripheral vessels.[45] The splenic pulp, with a local hematocrit of 0.7 to 0.8, is a notable exception to this rule. Contraction of the splenic capsule in the dog expels blood of high hematocrit into the general circulation and can raise the hematocrit in venous samples (Fig. 46-2). This does not occur in man, where F_{cells} is fairly constant under normal conditions and can be used to derive an appropriate correction factor in calculating total blood volume. Changes in the F_{cells} factor do occur under some conditions, including pregnancy[53] and congestive heart failure.[49] A curious observation in comparative physiology is an apparent decrease in F_{cells} along with body size in mammalian species.

NORMAL BLOOD VOLUME

A large body of information on the normal blood volume in man is available, and the data are in general agreement. In normal adult males the average plasma volume is 45 ml/kg body weight, and the cell volume 30 ml/kg. The range of variation within normal populations is such that about 95% of individuals fall between ±20% of the mean. Total blood volume varies with the size of the individual and shows correlations with body surface, height, and weight.[7,28] Body weight is the reference most commonly used. As might be expected, total blood volume correlates more closely with lean body mass than

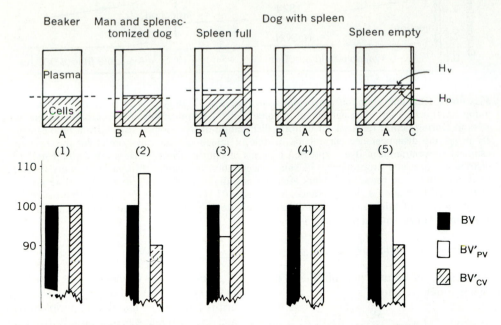

Fig. 46-2. Upper row: effects of cell distribution on relationship between large vessel cell percentage (H_v) and overall cell percentage $(H_o,$ broken lines). Compartment A indicates large vessels, B denotes (extrasplenic) minute vessels, and C represents spleen. Lower row: effects of cell distribution on relationships between "blood volumes" calculated from plasma volume (BV'_{PV}) or cell volume (BV'_{CV}) and the venous hematocrit, compared with true blood volume as determined by simultaneous measurement of both plasma and cell volume (BV), which is taken as 100%. (Modified from Gregersen.[27])

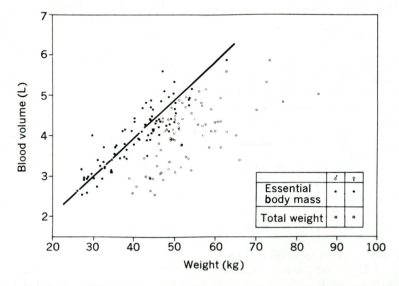

Fig. 46-3. Relationships of blood volume to total body weight (open symbols) and essential body mass (total weight minus adipose tissue, closed symbols) in Chinese subjects. Note that when essential body mass is used, scatter of blood volume data is greatly reduced, and results on males (squares) and females (circles) can be fit with straight line. (Based on data from Allen et al.[7])

with total body weight[1,7,41] (Fig. 46-3), but the former is not readily measured. The approximate distribution of volume in vessels of different size is indicated in Table 46-1.

Normal variations in blood volume

Age. Blood volume per unit weight is high in the newborn infant, averaging 100 ml/kg.[48,50] During the second and third years of life, this value declines, and thereafter total blood volume increases more or less linearly with weight.[40] No age-related change in blood volume has been found between the ages of 20 and 90 years,[10,16,54] although there is often an increase in weight and hence a slight decrease in milliliters per kilogram.[15]

Posture and activity. Plasma volume tends to be lower in the erect than in the recumbent position, and after a period of quiet standing, it may decrease by as much as 15%, presumably because of increased capillary filtration secondary to the increased venous and capillary pressures in the lower extremities. Physical and muscular activity must be taken into account in evaluating the effects of body position, however. Patients recumbent during long periods of bed rest also show a diminution of blood volume, and decreases of 10% in total blood volume over a period of 3 weeks have been reported in such subjects.[52] A decrease of similar magnitude has been observed in astronauts after periods of "weightlessness" (p. 1043).

Physical exercise decreases plasma volume, although the change depends on the type, degree, and duration of the exertion. Running or exercise on a bicycle ergometer can reduce plasma volume in man by several hundred milliliters.[35]

Physical training. Blood volume is increased by intensive physical training, although investigators disagree on the magnitude of the change. Kjellberg et al.[37] found values averaging 103 ml/kg in 23 trained men, as compared with 75 ml/kg in 174 untrained individuals. In eight college freshmen studied by Reuschlein et al.[47] during a 5-month period of rigorous training as a rowing crew, however, only a barely significant increase in plasma volume per unit body weight was found. Adjustments of blood volume can apparently take place in a relatively short time, for total hemoglobin and blood volume have been found to increase 10% to 19% over a 9-day period of ski training. Courtice[18] reports that trained greyhounds have red blood cells volumes higher than those in mongrel dogs, but whether this difference is a result of training is not known.

Seasonal variations. In man, at least, blood volume is higher in summer than in winter.[8] The relation between temperature and physical activity is an additional variable, but the seasonal variation has been confirmed in well-controlled studies.[55] Virtually all the difference is in plasma volume, which rises from 15% to 30% over a period of days after a move from a cool to a warm environment.[17]

Nutrition. The most conclusive information on the effect of nutrition has been given by Henschel et al.,[32] who conducted carefully controlled studies on 32 volunteers subjected to 24 weeks of semistarvation. Body weight fell from an average control of 69.5 to 53.5 kg and, although total blood volume fell by an average of 8.6%, plasma volume increased 8.3%, indicating a marked re-

Table 46-1. Estimates of relative volumes of blood contained in component parts of the circulatory system of a dog*

Structure	Diameter (mm)	Number	Total cross-sectional area (cm²)	Length (cm)	Total volume (ml)
Aorta	10.0	1	0.8	40.0	30
Large arteries	3.0	40	3.0	20.0	60
Main arterial branches	1.0	600	5.0	10.0	50
Terminal branches	0.6	1,800	5.0	1.0	5
Arterioles	0.02	40,000,000	125.0	0.2	25
Capillaries	0.008	1,200,000,000	600.0	0.1	60
Venules	0.03	80,000,000	570.0	0.2	110
Terminal veins	1.5	1,800	30.0	1.0	30
Main venous branches	2.4	600	27.0	10.0	270
Large veins	6.0	40	11.0	20.0	220
Vena cava	12.5	1	1.2	40.0	50
TOTAL					910

*From Green.[25]

duction in red blood cell volume (anemia). At the end of the period the average plasma volume was 67.5 ml/kg as compared with the control of 45.5 ml/kg. The shrinkage in cell volume and expansion of the plasma volume are consistent with observation made on ill-nourished inmates of concentration camps during World War II. Milder forms of nutritional disturbances that are not grossly apparent or are considered mere differences in dietary habits may well account for some of the reported variations in blood volume of "normal" populations in different countries.[36]

Pregnancy. The total blood volume rises gradually during pregnancy, reaching a peak during the ninth month, after which it falls off slightly until delivery. The change in volume is large, amounting to an increase of 30% to 50% or, in absolute values, 1,200 to 1,800 ml.[13,51] In the early stages the rise is mainly in the plasma volume. Cell volume may actually fall during the first month or two, thus producing an absolute as well as a relative anemia.[9] Subsequently, the red blood cell volume rises to a level of 300 to 500 ml above the control value. Immediately after delivery the blood volume returns to the nonpregnant level. It is noteworthy that only one fourth to one third of the decrement can be accounted for by the blood lost in the delivery.[51]

Abnormal changes in blood volume

Hypovolemia. Severe reduction of the total blood volume by hemorrhage or other pathologic events affects cardiovascular function adversely because of the decreased mean circulatory filling pressure (p. 957). This and other effects of hemorrhage have been considered in detail elsewhere (p. 1081). A decrease in total red blood cell volume, as in chronic anemia, reduces the oxygen supplied to tissues even though the total blood volume may remain normal. Red blood cell volumes of less than half the normal level are compatible with life, although not with health, provided the plasma volume enlarges to maintain a fairly normal total blood volume and hence an adequate circulatory state. Hemoglobin content rather than total volume of red blood cells is the critical factor in supplying oxygen to the tissues, and hemoglobin concentration in the cell falls in certain pathologic states. External disturbances of water balance (e.g., dehydration) and internal disturbances that accelerate salt and water excretion can also reduce the plasma volume.

Hypervolemia. Abnormally high blood volumes can be caused by faulty operation of the salt- and water-excreting mechanisms, by patho-logic overproduction of red blood cells, or by excessive infusion of blood or other fluids. Congestive heart failure is an example of the first situation.[51] When ventricular function "fails" to the degree that venous pressures begin to rise, renal excretion no longer keeps up with dietary intake of sodium and water, with the result that plasma and extracellular fluid volumes increase. Accumulation of fluid in the interstitial spaces, which the effect of gravity makes most striking in dependent parts of the body, is called *edema*. Renal function in this state has been exhaustively studied, but the connection between hemodynamic abnormalities and renal function remains elusive. In the absence of heart failure, atrial distention accompanying experimental increases in blood volume elicits reflex diuresis (p. 1066), making the fluid retention in congestive failure all the more puzzling. Increased blood volume, within limits, must have a beneficial effect on cardiovascular function by raising the mean circulatory filling pressure. A 10% increase in blood volume ordinarily raises right and left atrial pressures by approximately 3 cm H_2O,[2] which would tend to improve myocardial performance.

REGULATION OF BLOOD VOLUME

Total blood volume is remarkably constant in any one individual, and repeated measurements over a period of years show very little change.[15] The same is generally true of the red blood cell and plasma volumes that are its two major components, but any decrease in red blood cell volume, as in anemia, is quickly followed by an increase in plasma volume, returning total volume to its normal level. Red blood cell volume is determined by the balance between the rate at which erythrocytes are produced and the rate at which they age and are destroyed, a process described in Chapter 47. Plasma volume is thus the more readily controlled variable in acute regulation of total blood volume.

Adjustments in capillary transmural pressure are the mechanisms through which plasma volume is ultimately controlled (p. 1088). When this pressure rises, for example, fluid tends to move from the intravascular to the extracellular compartment. Vasomotor activity plays a part in regulation of plasma volume by virtue of alterations in precapillary/postcapillary resistance ratios, which influence the capillary hydraulic pressure. The relative compliances of the interstitial and intravascular spaces[2,3] as well as the local pressures determine the distribution of fluid between these compartments.

The renal circulation is the principal regulator

of plasma volume, although appreciable shifts of fluid into or out of the vascular compartment can also take place through the capillaries of skeletal muscle. In the renal capillaries, however, transmural pressures are important in controlling the volume and solute concentrations of the urine, and both plasma and extracellular fluid volumes are largely regulated by renal excretion of water and salt (Chapter 50).

The mechanisms involved in this regulation are not yet entirely clear, but one point does seem firmly established—that changes in blood volume are detected primarily by the atrial stretch receptors. Complete denervation of the heart, eliminating responses from atrial or other cardiac receptors, attenuates water and sodium diuresis.[23] The arterial baroreceptors may also have a small role but are not nearly as sensitive to minor alterations of blood volume as the atrial receptors. Structures sensitive to *osmotic* pressure of the blood ("osmoreceptors"), which are another conceivable mechanism for regulating blood volume, have been reported in the hepatic circulation,[31] but these observations have yet to be confirmed.[2]

The unresolved question of greatest importance is the precise way in which activation of atrial receptors influences renal function. Distention of the left atrium leads to a decrease in the rate of efferent impulses in the renal nerves, a decrease in the production of renin, angiotensin, and aldosterone, and diuresis of both water and sodium.[2,33] All these changes tend to reduce plasma volume, an appropriate response in view of the known activation of atrial receptors by an increase in blood volume. Although experiments with atrial distention sometimes increase total renal blood flow,[38] the diuretic response is the same when total flow remains unchanged.[2,11] In fact, neither renal blood flow nor perfusion pressure seems to occupy a central place in control of blood volume. Although it is true that urinary output in the experimentally isolated kidney is directly correlated with arterial pressure under some conditions, autoregulation prevents much variation in renal blood flow in the living animal (p. 1027).

If the reflex initiated by atrial distention is mediated through the renal nerves, the intrarenal mechanism must be vascular because these nerves, as far as is known, have no direct cellular effects on tubular resorption and excretion. The crucial phenomenon appears to be a redistribution of blood flow within the kidney. Barger and his colleagues[12,43] have shown that low-intensity stimulation of the renal nerves reduces flow in the cortical region and increases it in the renal medulla. Since adrenergic nerves innervate the afferent but not the efferent arterioles of the glomeruli,[39] the primary vasomotor effect appears to be constriction of preglomerular vessels in the cortex of the kidney. The cause of increased medullary flow is uncertain, since there is some disagreement about the existence of adrenergic vascular innervation in the outer medullary region, but cortical vasoconstriction in itself may be enough to account for the redistribution of flow. Whatever effect these vascular changes have on salt and water resorption, it is not in terms of the volume filtered, for the total glomerular filtration rate remains unchanged during mild stimulation, although it can be reduced by strong stimulation of the renal nerve. Impulses arriving through the renal nerves increase sodium resorption, however, by some as yet unidentified mechanism. Many hypotheses have been proposed, including the possibility that blood flow is shifted from one population of nephrons to another of different resorptive capacity.[43]

The evidence that renal control of blood volume is at least partly nerve mediated is quite convincing, but this is not the whole story, for a decrease in atrial volume produces sodium retention even after the renal nerves has been cut.[19] Some humoral agent is evidently involved, and a specific natriuretic hormone has been sought, so far without success.[2] Antidiuretic hormone (ADH) itself (p. 928) is certainly one element in the neuroendocrine control of blood volume, and the reflex stimulation of ADH secretion through the projections of atrial receptors onto the hypothalamus may be the most important control mechanism.[42] Distention of the left atrium causes a decrease in ADH secretion, and a sensitive inverse correlation between atrial pressure and plasma ADH concentration has been demonstrated.[34] The direct action of ADH on the functioning cells of the kidney is well known, and the possibility that it can also alter the intrarenal distribution of blood flow has not been ruled out.

REFERENCES
General reviews

1. Chien, S., and Gregersen, M. I.: Determination of body fluid volumes. In Nastuk, W. L., editor: Physical techniques in biological research, New York, 1962, Academic Press, Inc., vol. 4.
2. Gauer, O. H., Henry, J. P., and Behn, C.: The regulation of extracellular fluid volume, Ann. Rev. Physiol. **32:**547, 1970.
3. Guyton, A. C., and Coleman, T. G.: Long-term regulation of the circulation; interrelationships with body fluid volumes. In Reeve, E. B., and Guyton, A. C., editors: Physical bases of circulatory transport: regula-

tion and exchange, Philadelphia, 1967, W. B. Saunders Co.

4. Lawson, H. C.: The volume of blood—a critical examination of methods for its measurements. In Hamilton, W. F., and Dow, P., editors: Handbook of physiology. Circulation section, Baltimore, 1962, The Williams & Wilkins Co., vol. 1.

5. Mayerson, H. S.: Blood volume and its regulation, Ann. Rev. Physiol. **27:**307, 1965.

6. Share, L.: Extracellular fluid volume and vasopressin secretion. In Ganong, W. F., and Martini, L., editors: Frontiers in neuroendocrinology, London, 1969, Oxford University Press.

Original papers

7. Allen, T. H., et al.: Prediction of blood volume and adiposity in man from body weight and cube of height, Metabolism **5:**328, 1956.

8. Bazett, H. C., Sunderman, F. W., and Scott, J. C.: Climatic effects on the volume and composition of blood in man, Am. J. Physiol. **129:**69, 1940.

9. Berlin, N. I., et al.: The blood volume in pregnancy as determined by P^{32} labelled red blood cells, Surg. Gynecol. Obstet. **97:**173, 1953.

10. Brown, E., et al.: Red cell, plasma, and blood volume in healthy women measured by radiochromium cell labeling and hematocrit, J. Clin. Invest. **41:**2182, 1962.

11. Burag, R. D., Page, I. H., and McCubbin, J. W.: Neural stimulation of release of renin, Circ. Res. **19:**851, 1966.

12. Carriere, S., et al.: Intrarenal distribution of blood flow in dogs during hemorrhagic hypotension, Circ. Res. **19:**167, 1966.

13. Caton, W. L., et al.: The circulating red cell volume and body hematocrit in normal pregnancy and the puerperium by direct measurement using radioactive red cells, Am. J. Obstet. Gynecol. **61:**1207, 1951.

14. Chaplin, H., Mollison, P. L., and Vetter, H.: The blood/venous hematocrit ratio: its consistency over a wide hematocrit range, J. Clin. Invest. **32:**1309, 1953.

15. Chien, S., et al.: Blood volume and age: repeated measurements on normal men after 17 years, J. Appl. Physiol. **21:**583, 1966.

16. Cohn, J. E., and Shock, N. W.: Blood volume studies in middle-aged and elderly males, Am. J. Med. Sci. **217:**388, 1949.

17. Conley, C. L., and Nickerson, J. L.: Effects of temperature change on the water balance in man, Am. J. Physiol. **143:**373, 1945.

18. Courtice, F. C.: The blood volume of normal animals, J. Physiol. **102:**290, 1943.

19. Davis, J. O., et al.: An extra-adrenal factor essential for chronic renal sodium retention in presence of increased sodium-retaining hormone, Circ. Res. **14:**17, 1964.

20. Ebert, R. V., and Stead, E. A.: Demonstration that the cell plasma ratio of blood contained in minute vessels is lower than that of venous blood, J. Clin. Invest. **20:**317, 1941.

21. Farnsworth, P. N., Paulino-Gonzalez, C. M., and Gregersen, M. I.: F_{cells} values in the normal and splenectomized cat: relation of F_{cells} to body size, Proc. Soc. Exp. Biol. Med. **104:**729, 1960.

22. Gibson, J. G., II, and Evans, W. A., Jr.: clinical studies of the blood volume. III. Changes in blood volume, venous pressure and blood velocity rate in chronic congestive heart failure, J. Clin. Invest. **16:**851, 1937.

23. Gilmore, J. P.: Contribution of cardiac nerves to the control of body salt and water, Fed. Proc. **27:**1156, 1968.

24. Grant, W. C., and Root, W. S.: Fundamental stimulus for erythropoiesis, Physiol. Rev. **32:**448, 1952.

25. Green, H. D.: Circulatory system; physical principles. In Glasser, O., editor: Medical physics, Chicago, 1944, Year Book Medical Publishers, Inc., vol. 2.

26. Gregersen, M. I.: A practical method for determination of blood volume with the dye T-1824, J. Lab. Clin. Med. **29:**1266, 1944.

27. Gregersen, M. I.: Effects of circulatory states on determinations of blood volume, Am. J. Med. **15:**785, 1953.

28. Gregersen, M. I., and Nickerson, J. L.: Relation of blood volume and cardiac output to body type, J. Appl. Physiol. **3:**329, 1950.

29. Gregersen, M. I., and Stewart, J. D.: Simultaneous determinations of the plasma volume with T-1824 and the "available fluid" volume with sodium thiocyanate, Am. J. Physiol. **125:**142, 1939.

30. Gregersen, M. I., Gibson, J. G., II, and Stead, E. A.: Plasma volume determinations with dyes: errors in colorimetry: use of blue dye T-1824, Am. J. Physiol. **113:**54, 1935.

31. Haberich, F. J.: Osmoreception in the portal circulation, Fed. Proc. **27:**1137, 1968.

32. Henschel, A., et al.: Plasma volume and thiocyanate space in famine edema and recovery, Am. J. Physiol. **150:**170, 1947.

33. Hodge, R. L., et al.: Role of the vagus nerve in the control of the concentration of angiotensin II in the circulation, Nature **221:**177, 1969.

34. Johnson, J. A., Moore, W. W., and Segar, W. E.: Small changes in left atrial pressure and plasma anti-diuretic hormone titers in dogs, Am. J. Physiol. **217:**210, 1969.

35. Kaltreider, N. L., and Meneely, G. R.: The effect of exercise on the volume of the blood, J. Clin. Invest. **19:**627, 1940.

36. Keys, A., et al.: The biology of human starvation, Minneapolis, 1950, University of Minnesota Press, Vol. 1.

37. Kjellberg, S. R., Rudhe, U., and Sjöstrand, T.: Increase in the amount of hemoglobin and blood volume in connection with physical training, Acta Physiol. Scand. **19:**146, 1949.

38. Lyatin, H.: Untersuchungen über Mechansmen der Osmo- und Volumenregulation. I. Uber den Einfluss akutur Änderungen des Drucks im linken Vorhof auf die Diurese. Untersuchungen am wachen und narkotisierten Hund, Z. Gesamte Exp. Med. **149:**91, 1966.

39. McKenna, O. C., and Angelakos, E. T.: Adrenergic innervation of the canine kidney, Circ. Res. **22:**345, 1968.

40. Morse, M., Cassels, D. E., and Schlutz, F. W.: Blood volumes of normal children, Am. J. Physiol. **151:**448, 1947.

41. Muldowney, F. P.: The relationship of total red cell mass to lean body mass in man, Clin. Sci. **16:**163, 1957.

42. Perlmutt, J. H.: Contribution of carotid and vagal reflex mechanisms, Fed. Proc. **27:**1149, 1968.

43. Pomeranz, B. H., Birtch, A. G., and Barger, A. C.: Neural control of intrarenal blood flow, Am. J. Physiol. **215:**1067, 1968.

44. Rawson, R. A.: The binding of T-1824 and structurally related diazo dyes by the plasma proteins, Am. J. Physiol. **138:**708, 1943.

45. Reeve, E. B.: Use of radioactive phosphorus for the measurement of red-cell and blood volume, Br. Med. Bull. **8:**181, 1952.

46. Reeve, E. B., et al.: Distribution of cells and plasma in the normal and splenectomized dog and its influence on

blood volume estimates with P^{32} and T-1824, Am. J. Physiol. **175:**195, 1953.

47. Reuschlein, P. S., et al.: Effect of physical training on the pulmonary diffusing capacity during submaximal work, J. Appl. Physiol. **24:**152, 1968.

48. Robinow, M., and Hamilton, W. F.: Blood volume and extra-cellular fluid volume of infants and children; studies with improved dye micro-method, Am. J. Dis. Child **60:**827, 1940.

49. Schreiber, S. S., et al.: Blood volume alterations in congestive heart failure, J. Clin. Invest. **33:**578, 1954.

50. Smith, C. A.: The physiology of the new-born infant, ed. 2, Springfield, Ill., 1951 Charles C Thomas, Publisher.

51. Tatum, H. J.: Blood volume variation during labor and early puerperium, Am. J. Obstet. Gynecol. **66:**27, 1953.

52. Taylor, H. L., et al.: The effect of bed rest on the blood volume of normal young men, Am. J. Physiol. **144:**227, 1945.

53. Verel, D., Bury, J. D., and Hope, A.: Blood volume changes in pregnancy and the puerperium, Clin. Sci. **15:**1, 1956.

54. Wennesland, R., et al.: Red cell, plasma and blood volume in healthy men measured by radiochromium (Cr^{51}) cell tagging and hematocrit: influence of age, somatotype and habits of physical activity on the variance after regression of volumes to height and weight combined, J. Clin. Invest. **38:**1065, 1959.

55. Yoshimura, H.: Acclimatization to heat and cold. In Yoshimura, H., Ogata, K., and Itoh, S., editors: Essential problems in climatic physiology, Kyoto, 1960, Nakoda Publishing Co., Ltd.

47 The blood

C. LOCKARD CONLEY

The blood is composed of specialized cells suspended in an extracellular liquid medium. Unique among tissues because of its fluid state, the blood has as its essential role the *transport* of many agents to and from the organs and tissues. The circulatory system provides the mechanism by which the blood performs this vital function. The circulation of the blood ensures that there is little variation in the environment of individual cells throughout the body. Heat is distributed in such a manner as to provide uniformity and constancy of body temperature. The chemical composition of the fluid that bathes tissue cells is maintained in equilibrium with that of the circulating blood. Consequently, the concentrations of hydrogen ions, oxygen, electrolytes, nutrient materials, and other essential substances remain relatively fixed, whereas the products of cellular metabolism are not permitted to accumulate but are continuously removed. This almost unvarying cellular environment, *le milieu interieur* described by Claude Bernard,[32] is essential for normal function and for life. Cessation of blood circulation for only a few minutes may cause irreparable damage or death because of the resulting unfavorable alterations in the environment of highly susceptible cells. The concentration of each of the many components of the blood is kept within physiologically appropriate limits by effective regulatory devices. Information concerning many of these is summarized in other chapters. The maintenance of a constant state is an aspect of the precise biologic regulation to which the term ''homeostasis'' was applied by Cannon.[1,41]

The constancy of the composition of the blood is only relative. As the blood flows through the body, materials are added here and removed there, so that variations occur continuously. There are differences among normal persons in every measurable constituent. These facts must be borne in mind in attempting to establish ''normal'' values, which invariably include a range of values rather than a single figure. The composition of the blood is affected by age, sex, genetic factors, environmental conditions, physiologic state of the individual, and disease.

PROPERTIES

The constant motion of the blood ensures that the blood *cells* are dispersed throughout the liquid portion, the *plasma*. The specific gravity of the plasma is about 1.026, whereas that of the red blood cells is 1.093. When blood containing an anticoagulant is undisturbed, the rate at which the cells settle (erythrocyte sedimentation rate) is slow because of the small size and relatively large surface area of the red blood cells. Red blood cells tend to form cylindric aggregates resembling rolls of coins (rouleaux), which sediment more rapidly because of the decreased surface:volume ratio. Certain plasma proteins, fibrinogen in particular, enhance rouleaux formation and accelerate sedimentation by reducing the zeta potential of the suspended red blood cells.[65,75,76,109] Separation of plasma from the cellular elements of blood can be achieved quickly by centrifugation. Somewhat less than half of the volume of the blood is composed of cells, and these form two layers when sedimentation is complete. The red blood cells (*erythrocytes*), easily recognized by their color, are denser and form the bottom layer. The packed red blood cells constitute approximately 45% of the volume of the blood specimen (hematocrit value).[24] Above the red blood cells is a narrow white layer, the buffy coat, comprising only about 1% of the volume of the blood and consisting of the white blood cells (*leukocytes*) and the blood platelets (thrombocytes). The hematocrit of the circulating blood is slightly lower in capillaries than in larger vessels because of the axial flow of red blood cells.[46]

That the blood is a suspension has important effects on its properties as a flowing liquid (Chapter 39). The relatively high viscosity of blood as compared to water is largely attributable to the suspended cells and bears a nonlinear relationship to the hematocrit value; blood viscosity is also influenced by the plasticity of the cells and their degree of aggregation. Since its shear rate is not proportional to shear stress, blood is a non-newtonian fluid. Accordingly, its viscosity cannot be expressed in absolute terms apart from the conditions of measurement. Blood is more viscous at low than at high shear rates, a discrepancy that is exaggerated at high hematocrit values.[87,123,130] The relatively low viscosity of blood at high rates of flow and shear is in part attributable to the ease of deformability of the red blood cell.[59]

PLASMA

The blood plasma, a solution of extraordinary complexity, is a clear, slightly yellowish liquid that transiently may appear turbid after a fatty meal. Containing about 90% water, plasma has as its major solute a heterogeneous group of proteins that make up approximately 7% of its weight. Plasma albumin, the principal component, consists of relatively small molecules whose osmotic activity is essential to prevent loss of fluid through the semipermeable membranes of the capillary blood vessels. Albumin serves an important transport function through its capacity to bind reversibly many insoluble substances, including certain hormones and metabolic products (e.g., corticosteroids, bilirubin). The lipids in plasma are carried as lipoproteins. Specific metal-binding proteins are essential for transport of iron (transferrin) and copper (ceruloplasmin). Numerous other substances are transported by unique proteins. A heterogenous group of plasma proteins is required for coagulation of the blood and for other defense reactions. The immuno-globulins consist of antibodies, reacting with many different antigens. Essential nutrient substances in the plasma include glucose and amino acids, which are transported in low concentrations but at high turnover rates. The electrolytes are held within narrow concentration limits, and the hydrogen ion concentration is precisely regulated. Sodium is the major cation of plasma, which also contains potassium, calcium, and magnesium. Chloride, bicarbonate, and phosphate are the principal inorganic anions. Waste products of metabolism are transported through the plasma to excretory organs, including the lungs, liver, and kidneys. Although the plasma compartment is small in comparison with the extravascular fluid compartments of the body, it is the most dynamic in terms of flow and turnover rate of its many solutes. Individual constituents of the plasma are considered in relation to their metabolic roles in other chapters of this book.

ERYTHROCYTES

Approximately 5×10^{12} red blood cells are contained in 1 L of blood. Each cell is a biconcave disk with a mean diameter of slightly more than 8 μm and a depth of about 2 μm in the thickest part (Fig. 47-1). The shape of the erythrocyte provides it with a large surface:volume ratio, a feature that permits swelling and distortion of the cell without stretching of the membrane. Erythrocytes are much deformed during passage through small blood vessels but quickly revert to their usual shape.[72,120] The biconcave form is determined by the cell membrane[39,40,126] and may depend in part on a contractile protein.[100] The membrane is a triple-layered structure, the central stratum of which is a lipid bilayer (Chapter 1).[10,93] Virtually all the erythrocyte lipid is in the membrane; about 70% is phospholipid, with the remainder largely cholesterol. About half the dry weight of the membrane is protein, including a distinct filamentous contractile protein "spectrin."[92] Sialic acid, bound to glycoproteins, gives the red blood cell surface a negative charge. Glycoproteins and glycolipids of the membrane have antigenic properties and confer blood group specificity.[11,19] The mature red blood cell does not have a nucleus, nor does it contain mitochondria, ribosomes, or other organelles. It does have an active metabolism, on which its survival in the circulation depends. Potassium is retained within the erythrocyte against a large cell-plasma gradient. The principal solid component is hemoglobin, which makes up about 95% of the dry weight of the cell and gives the erythrocyte its primary attribute, the ability to transport oxygen.

Hemoglobin

Hemoglobin, a conjugated protein with a molecular weight of about 64,500, is composed of *heme,* a protoporphyrin to which an iron atom is linked, and *globin,* a protein. A molecule of adult hemoglobin (Hb A) contains two pairs of polypeptides, the alpha and beta chains, each containing more than 140 amino acid residues. The predominant hemoglobin of the fetus and newborn, fetal hemoglobin (Hb F), consists of alpha and gamma chains.[13] A heme group is attached to each chain. Each of the four iron atoms of the molecule combines reversibly with an atom of oxygen.[13,106] The complex structure of hemoglobin provides precisely the environment for the iron atom so that it binds and releases oxygen under conditions appropriate for oxygen transport without oxidative damage to the molecule. The affinity of hemoglobin for oxygen is so great that

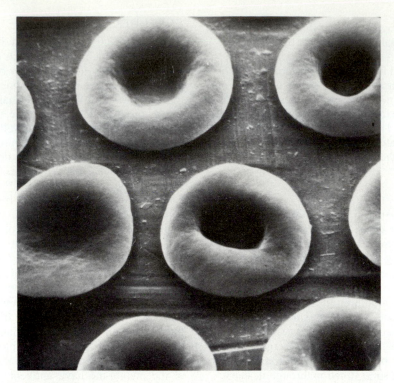

Fig. 47-1. Normal human red blood cells photographed by stereoscan electron microscopy. (Courtesy Dr. Wallace N. Jensen.)

in the lungs about 95% of the hemoglobin is saturated with oxygen. In tissues in which the oxygen tension is lower, oxygen dissociates from hemoglobin and diffuses through the red blood cell membrane and the plasma to the sites at which it is utilized. The relationship between partial pressure of oxygen and hemoglobin saturation is not linear but is described by a sigmoid curve; oxygenation of one iron atom increases the oxygen affinity of remaining iron atoms of the molecule by stereochemical effects involving the polypeptide subunits (heme-heme interactions.[105,113]) The oxygen affinity of hemoglobin lessens when pH is lowered (Bohr effect)[36,107]; it is also reduced by certain phosphate constituents of the red blood cell, in particular 2,3-diphosphoglycerate.[45] The role of hemoglobin in the transport of oxygen and carbon dioxide is considered in Chapter 69.

Red blood cell production (erythropoiesis)

In the fetus, primitive blood cells are formed first in the blood islands of the yolk sac and later in the liver and spleen, but by the time of birth the bone marrow is the sole site of hematopoiesis.[15,34] Formation of blood cells occurs princi-

pally in the red marrow of the axial skeleton, ribs, sternum, pelvis, and proximal bones of the extremities. When erythropoiesis is accelerated, there is extension of hematopoietically active red marrow to areas that were previously fatty. The erythrocyte precursor is derived from a stem cell of uncertain identity.[15] In the early stages of development the immature cell contains a large nucleus. Maturation proceeds by cell division and differentiation. During this process the nucleus becomes progressively smaller, hemoglobin appears within the cytoplasm, the amount of ribonucleic acid (RNA) lessens, and the cell decreases in size. At the time of entry into the circulation, newly formed erythrocytes do not have a nucleus but usually do contain mitochondria and ribosomes. In appropriately stained preparations a reticulum composed of RNA is seen within these young cells, which are therefore designated as *reticulocytes*. They constitute about 1% of the red blood cells of a normal person but increase when blood production is accelerated. Reticulocytes are less dense than mature cells and, on centrifugation of blood, are found to be concentrated in the upper portion of the sedimented red blood cells.

Erythrocytes are continuously produced at a

rate sufficient to replace the aged cells that are removed from the circulation. Almost 1% of the red blood cells are regenerated each day. The balance between red blood cell production and red blood cell destruction in normal humans is so precisely maintained that there is little variation in the number of circulating erythrocytes from week to week. When blood is lost, the erythropoietic activity of the marrow accelerates until the normal number of cells has been restored. If the volume of circulating red blood cells is increased, as by transfusion of blood to a normal person, erythropoiesis is thereupon retarded until the normal red blood cell volume has been regained. The rate of destruction of red blood cells depends on aging of the erythrocytes, a function not susceptible to physiologic control. Regulation of the volume of circulating red blood cells is achieved only by the control of the rate of production of erythrocytes.[12]

Materials required for erythropoiesis

Materials required for red blood cell production include the precursors of the structural components of the red blood cells as well as numerous accessory factors. Thus amino acids are essential for the synthesis of globin, and iron is indispensable for the formation of heme.[78,131] Erythropoiesis depends on the availability of relatively minute amounts of certain other substances, including vitamin B_{12}, folic acid, pyridoxine, and copper. In addition, production of red blood cells at the normal rate requires the hormones of several endocrine glands, the pituitary, thyroid, adrenal glands, and gonads.[6,97]

Iron. Iron is an integral part of the hemoglobin molecule. The body of a normal man contains about 4 gm of iron, of which more than half is in the hemoglobin of the red blood cells. A smaller fraction occurs as *storage iron* and is deposited in the form of ferritin or hemosiderin, chiefly in the liver, spleen, and bone marrow.[5,9] Storage iron is readily available for hemoglobin synthesis when needed. Iron is an essential component of myoglobin and of certain enzymes, but the amount combined in these substances is small in relation to that in hemoglobin. The concentration of iron in the plasma is low,[42] approximately 100 $\mu g/100$ ml, but the transfer of iron through the plasma is rapid.[50] Iron is carried by a specific binding protein, a globulin known as *transferrin* or siderophilin.[89] The concentration of this protein limits the concentration of iron in plasma. Normally only about one third is saturated with iron.[111] The iron-binding protein plays a specific role in the transfer of iron across the membrane of the immature erythrocyte.[86]

In prenatal life, iron is transported from the mother to the fetus against a concentration gradient.[37] At birth, most of the iron is in hemoglobin. The hemoglobin concentration of the blood decreases after birth, and the iron released by hemoglobin degradation serves as a small reserve.[121] As growth proceeds, iron must be provided in the diet. Iron is not readily absorbed from the gastrointestinal tract; therefore most of the iron ingested is excreted in the feces. Iron must be reduced to the ferrous state for absorption, a process that occurs principally in the proximal small intestine.[5] There is some degree of regulation of the amount of iron absorbed, for example, absorption is enhanced by iron deficiency.[73,108] The limited absorption of iron is physiologically advantageous, since the body has no mechanism for excreting iron after it has been absorbed.[95]

Within the body there is a continuous circulation of iron.[81] As aging erythrocytes are destroyed, the iron is salvaged, transported to the erythroid cells of the marrow by transferrin, and reutilized for new hemoglobin synthesis. The overall loss from the body of an adult man is less than 1 mg/day, including the tiny amounts contained in urine, bile, sweat, and desquamated cells.[62] Iron absorbed from the diet approximately equals this loss, and the iron content of the body is fairly stable. Larger amounts are lost from the mother to the fetus during pregnancy. Iron is lost whenever blood is shed, since 1 ml of blood contains about 0.5 mg iron. Menstrual blood loss has been estimated to double the iron requirement, which is then about 2 mg/day for the average woman. Iron deficiency is characterized by deficient hemoglobin synthesis manifested by *hypochromic anemia;* the red blood cells are smaller than normal and contain less than the usual concentration of hemoglobin.[24]

Copper. Deficiency of copper, experimentally induced in animals, causes an anemia resembling that of iron deficiency.[90] Iron absorption is impaired, but iron does not prevent or correct the anemia. Ceruloplasmin, the copper-containing protein of plasma, is a ferroxidase and appears to be necessary for transfer of iron from reticuloendothelial cells to plasma transferrin.[112]

Cobalt. Cobalt forms a part of the molecule of vitamin B_{12}, and there appears to be no other need for it. Sheep grazing in soil that is deficient in cobalt acquire an anemia that responds to oral but not to parenteral administration of cobalt salts. In sheep, vitamin B_{12} is synthetized by the bacteria of the rumen, and cobalt is necessary for the synthesis.[94] A comparable situation does not exist in man, for preformed vitamin B_{12} is obtained from dietary sources.

Vitamin B$_{12}$. Biologically active forms of vitamin B$_{12}$ are the cobamide coenzymes that are essential for several specific biochemical reactions.[119,128] Vitamin B$_{12}$ is required for DNA synthesis, and it is this function that appears to be its primary role in hematopoiesis.[101] Deficiency of the vitamin is manifested by abnormal maturation of proliferating cells; DNA synthesis is impaired, and cell division is retarded. As a result, erythrocyte precursors are larger than normal and display other characteristic abnormalities (megaloblasts).[2] Mature red blood cells also are large (macrocytes), but each cell contains hemoglobin in normal concentration. Vitamin B$_{12}$ is a large molecule that does not easily penetrate the mucosa of the gastrointestinal tract.[132] The daily requirement for the vitamin is only about 1 μg, but even this tiny amount cannot be absorbed from the normal diet except in the presence of a specific component of the gastric secretion. The gastric factor has the properties of a glycoprotein and is known as *intrinsic factor* according to the terminology of Castle,[43] whose investigations developed this field. Intrinsic factor binds and facilitates the absorption of vitamin B$_{12}$.[51,68] Because of the abundance of the vitamin in the diet, deficiency is rarely the result of dietary inadequacy. *Pernicious anemia* is a disease in which atrophy of the gastric mucosa leads to the absence of intrinsic factor and consequent malabsorption of vitamin B$_{12}$; the characteristic anemia and other manifestations are attributable to the resulting deficiency of the vitamin.[44]

Folic acid. Derivatives of tetrahydrofolate, coenzymes involved in transfers of methyl groups, are essential for hematopoiesis.[2] A deficiency of folic acid produces anemia, with abnormalities of the blood and bone marrow that are indistinguishable from those resulting from deficiency of vitamin B$_{12}$.[79] Links between the biochemical activities of these vitamins probably explain their similar hematopoietic effects.[101] Folic acid is required for synthesis of thymidylate, a DNA precursor. The megaloblastic anemia of folate deficiency is probably attributable to impaired DNA synthesis.[29]

Pyridoxine. Pyridoxal phosphate, the coenzyme derivative of pyridoxine (vitamin B$_6$), is required for the condensation of glycine and succinate in the formation of δ-aminolevulinic acid, a precursor of heme.[69] Deficiency of pyridoxine experimentally induced in animals causes a hypochromic anemia resembling that of iron deficiency, except that there is no deficiency of iron and the concentration of iron in the serum is high.[133] A similar anemia responsive to pyridox-ine has in rare instances been encountered in man.[80]

Other factors. All the nutrient substances and cofactors required for cell growth and division are needed by proliferating erythropoietic cells. Pantothenic acid is specifically required for heme synthesis,[116] but anemia attributable to deficiency of pantothenate has not been encountered in man. There is experimental evidence indicating that deficiency of riboflavin may cause anemia.[88]

Control of erythropoiesis

The capacity of the bone marrow to produce new red blood cells is so great that when maximally stimulated, its output may be increased about 6 times.[56] Reduction of the circulating red blood cell volume by hemorrhage or excessive red blood cell destruction provides a stimulus for enhanced marrow activity. When the oxygen tension of the arterial blood is reduced, erythropoiesis increases.[71] There is a direct relationship between the degree of oxygen lack (hypoxia) and the degree of marrow stimulation. Chronic severe oxygen deficiency is associated with a pronounced increase in the volume of circulating red blood cells (polycythemia). Persons living at high altitudes where the oxygen pressure of the atmosphere is low have higher red blood cell counts and circulating red blood cell volumes than are normally encountered at sea level.[82] Pathologic states that cause a reduction of the arterial oxygen tension are similarly associated with polycythemia.[12]

The rate of production of erythrocytes is controlled by a hormonelike substance (erythropoietin) that is transported through the plasma to the bone marrow.[6,12,64] The chemical identity of erythropoietin has not been established, but the substance appears to be a relatively heat-stable glycoprotein of small molecular size. The amount of hormone produced is governed by the relationship between oxygen supply and oxygen demand. When the circulating red blood cell volume declines or when hypoxia occurs, a receptor organ senses the change in oxygen delivery and the output of erythropoietin is increased. Production of erythropoietin occurs primarily in the kidneys, but there appear to be extrarenal sources. The mechanism by which production of the hormone is enhanced by diminished oxygen delivery is unknown. Erythropoietin appears to exert its major effect on a class of stem cells, causing them to differentiate into erythroblasts, but there is evidence that later stages of erythropoiesis also may be affected. Production of erythrocytes is increased by androgenic hormones causing the red blood cell count and blood volume to increase.[25] The higher red blood cell counts of men as compared with women can be accounted for by the higher rate of production of androgens in the male.

Metabolism of red blood cells

The nucleated erythrocyte of the bone marrow is comparable in its metabolic activity to cells of other tissues.[91] The early red blood cell precursor synthesizes deoxyribonucleic acid (DNA), RNA, proteins, lipids, and carbohydrates. By the time the reticulocyte stage has been reached, most of the hemoglobin has been formed. Reticulocytes contain RNA and mitochondria, and they have a functional Krebs tricarboxylic acid cycle as well as active glycolytic and hexose monophosphate shunt pathways. They are capable of synthesizing hemoglobin and certain other red blood cell components. Mature red blood cells lack mitochondria and RNA, and the Krebs cycle is inactive. These cells are incapable of protein synthesis.

Energy of the mature erythrocytes is derived from the metabolism of glucose via the glycolytic and hexose monophosphate shunt pathways.[38] Under physiologic conditions, approximately 95% of the glucose is converted to lactate. Both pathways are essential for maintaining the viability and function of the cell. Energy is needed in particular to maintain the cationic gradients across the red blood cell membrane and to sustain oxidation-reduction homeostasis within the cell.[84]

Two important compounds generated by anaerobic glycolysis are adenosine triphosphate (ATP) and 2,3-diphosphoglycerate (DPG). One of the physiologic roles of ATP is to preserve the shape of the cell by regulating the flow of potassium and sodium across the membrane. When the level of ATP is decreased, erythrocytes tend to become more spheric because of the inward movement of sodium and water.[22] Another function of ATP is control of the deformability of the membrane.[125] As concentration of ATP decreases, the membrane becomes more rigid. The amount of DPG present in the erythrocyte is much larger than in other cells. The binding of phosphorylated compounds, in particular DPG, to hemoglobin reduces the oxygen affinity of the hemoglobin.[45] This phenomenon facilitates the release of oxygen in the tissue capillaries.

During oxygen transport, iron is maintained in the ferrous state. Oxidation of the iron to the ferric state results in the formation of a brown pigment, *methemoglobin,* which is incapable of reversibly combining with oxygen. Two enzymatic systems are present in the red blood cell for rapidly reducing the iron of any methemoglobin that is formed. One is active in red blood cells of the newborn child and requires reduced nicotinamide-adenine dinucleotide phosphate (NADPH) as an essential cofactor. The other enzyme is present in red blood cells at all ages, and reduced nicotinamide-adenine dinucleotide (NADH) is its essential cofactor. NADH is generated by the glycolytic pathway and NADPH by the hexose monophosphate shunt, the only pathway available to the red blood cell for forming this compound.[117]

The erythrocyte has a high concentration of reduced glutathione (GSH). Most of this thiol-containing tripeptide exists in the reduced form, an enzyme-catalyzed reduction requiring NADPH. The redox state of sulfhydryl-containing proteins (enzymes, hemoglobin, and possibly membrane proteins) depends on the availability of GSH. Another metabolic function of GSH is to prevent oxidative injury to proteins by removing hydrogen peroxide. Conversion of low levels of hydrogen peroxide to water and oxygen is catalyzed by an enzyme requiring GSH.[33]

Hemoglobin synthesis

Production of hemoglobin by the immature red blood cell involves two parallel synthetic processes: one resulting in the formation of the iron-containing porphyrin structure *heme,* the other in the production of the polypeptide chains of *globin.* The porphyrin moiety is derived from glycine and succinate through δ-aminolevulinic acid, and mitochondria are required for its synthesis.[69] In the final stage of heme formation, iron is introduced into protoporphyrin.[70] Synthesis of globin subunits occurs on ribosomes; the amino acids are added sequentially from the amino terminal end under instruction of messenger RNA.[60] Synthesis of the alpha and beta chains is under the control of separate genes.[13]

Destruction of red blood cells

After their release into the circulation, red blood cells have a finite life span averaging about 120 days.[31] Synthesis of new protein is not accomplished by mature erythrocytes, and intracellular reparative and regenerative processes are not possible.[91] Some of the enzymes of aging erythrocytes lessen in activity. Continuing glycolysis is required for regeneration of the reduced pyridine nucleotides essential for maintenance of oxidation-reduction homeostasis and for production of the ATP needed to maintain the shape and volume of the red blood cells.[8] When these metabolic processes are impaired, red blood cells are no longer viable and are removed from the blood. The destruction of erythrocytes normally does not permit the release of hemoglobin into the plasma. Phagocytic cells that form a part of the lining of the blood vessels, particularly in the spleen, liver, and bone marrow, ingest aging red blood cells.[85,114,125] Within the phagocytes, erythrocytes are rapidly lysed, hemoglobin is degraded, protein is removed and conserved, iron is detached from the porphyrin structure, and iron-free porphyrin residue is converted to bilirubin.[53] Bilirubin, a yellow pigment, is transported to the liver, where it is combined with glucuronide and excreted into the biliary tract. Bilirubin unconjugated with glucuronide is not excreted by the kidney; therefore bilirubin does not appear in normal urine.[115] Because bilirubin is derived from hemoglobin, the quantity of the pigment produced is directly related to the amount of hemoglobin destroyed.[104] Excretion of bilirubin is not associated with the loss of either protein or iron.[61] The iron released by degradation of hemoglobin is available for hemoglobin synthesis and is used preferentially in the production of new erythrocytes.[57] Accordingly, there is a continuous recirculation of iron through the plasma from the sites of red blood cell destruction to the erythropoietic tissue of the bone marrow. If the amount of iron released exceeds that needed for hemoglobin synthesis, the excess is deposited as ferritin or hemosiderin, principally in the marrow, liver, and spleen.[66]

Because few red blood cells undergo lysis (hemolysis) within the circulation, normal plasma contains virtually no hemoglobin (less than 1 mg/100 ml). In some pathologic states, intravascular hemolysis does occur, the plasma then becoming discolored by the hemoglobin in solution.[74] Hemoglobin released into the plasma combines with a specific hemoglobin-binding protein, *haptoglobin*.[103] The hemoglobin-haptoglobin complex is removed from the bloodstream but, because of the large size of the molecule, is not excreted by the kidneys. There is sufficient haptoglobin in normal plasma to bind about 150 mg hemoglobin/100 ml. When the concentration of hemoglobin in plasma exceeds the binding capacity of haptoglobin, the excess or unbound hemoglobin passes through the kidneys and is reabsorbed by the renal tubules or excreted in the urine (hemoglobinuria).

There is no physiologic mechanism for altering the life span of the erythrocyte, but in many disease states the survival time is shortened. The abbreviated life span may result from intrinsic abnormalities of the red blood cell, from extrinsic agents that affect red blood cells adversely, or from a combination of both.[4] Accelerated destruction of red blood cells leads to a compensatory increase in erythrocyte production. Anemia occurs if the enhanced erythropoiesis is inadequate for complete compensation. An inevitable result of accelerated destruction of red blood cells is an increase in the rate of production of bilirubin, often with an increase in its concentration in the plasma. The resulting yellow color of the serum and often of the sclerae of the eyes, the skin, and the mucous membranes is known as *jaundice*.

Hemolysis and osmotic properties

Damage to the membrane of the red blood cell may cause the hemoglobin to be liberated in solution in the plasma; the remainder of the erythrocyte persists as a virtually hemoglobin-free membranous structure, or "ghost." This phenomenon, known as *hemolysis,* may be induced in a number of ways.[4,18] Mechanical injury, freezing and thawing, heat, and many chemical agents may cause hemolysis; bile salts, saponin, detergents, and lipid solvents in particular are hemolytic because of their injurious effects on the membrane of the red blood cell. Biologic products, including the venom of certain snakes and the toxins of some bacteria, have the property of lysing erythrocytes. The red blood cell surface contains antigenic substances capable of reacting with specific antibodies. In the presence of *complement,* a group of substances existing in normal plasma, the antigen-antibody reaction may result in hemolysis.[118]

The envelope of the red blood cell is a semipermeable membrane that does not readily permit the passage of cations.[18,22,83] When erythrocytes are suspended in a solution containing cations in a concentration lower than that of plasma, water passes into the cells, causing their volume to increase. Because of the biconcave shape of the red blood cell, pronounced swelling and sphering occur before disruption. Normal erythrocytes are hemolyzed by distilled water, but most erythrocytes resist hemolysis when exposed to a solution of 0.4% sodium chloride.

Agglutination and antigenic properties

Carbohydrate structures on the surface of the erythrocyte have the properties of antigens.[11,98,124] The existence of a large number of red blood cell antigens confers a high degree of immunologic specificity on the erythrocyte and makes it unlikely that the red blood cells of one person will be identical to those of any other except a uniovular twin.[19] Each of the antigenic groups is determined genetically, and several series of antigens are known to be inherited independently. Knowledge of the blood group antigens is of importance in genetic and anthropologic studies and is vital to the practical use of blood transfusion therapy.[17] When red blood cells having a certain antigen are suspended in plasma containing the reactive antibody, the antibody attaches to the antigenic sites and may damage the red blood cells and cause them to agglutinate. From the standpoint of blood transfusion the most important red blood cell antigens are those of the ABO system, which divide the population into four major groups. Red blood cell antigens unrelated genetically to the ABO system include the extremely complex Rh system and numerous others.[19]

LEUKOCYTES

The white blood cells (leukocytes) are nucleated motile cells, larger and much more heterogeneous than the erythrocytes.[3] Unlike red blood cells and platelets, leukocytes exist predominantly outside the circulation.[35,52] The relatively small number in the blood may be considered to be in transit. Normal blood contains three principle categories of white blood cells—granulocytes, lymphocytes, and monocytes. They have their origin in primitive stem cells of the bone marrow.

The principal functions of the leukocytes are related to the defense of the body against invasion by infectious agents, to the repair of damaged tissue, and to immune reactions. Granulocytes and monocytes are capable of engulfing bacteria and other particles, a phenomenon known as *phagocytosis*.[102,110,122] Intracellular mechanisms are available for killing bacteria.[30,48] The granules of these cells are lysosomes, containing lytic enzymes that facilitate the degradation of engulfed organic material.[26,49,58,129] Some granules contain histamine and other agents that may be involved in inflammatory and vascular reactions.[47] The lymphocytes are concerned with cellular immunity, hypersensitivity, and production of antibodies.[16,21,67,96]

There are physiologic control mechanisms that regulate the number of leukocytes in the circulating blood, but the mode of operation is unknown. Leukocytes may be attracted in large numbers to an area of local infection or inflammation or to a site of tissue death, a phenomenon known as *chemotaxis*.[3,77]

SPLEEN

The spleen is intimately concerned with the blood and circulation.[127] It is a large lymphoid organ engaged in the production of mononuclear leukocytes and in the formation of antibodies. As one of the most important areas of the reticuloendothelial system, it plays a significant role in the removal of cells and cellular detritus from the circulation. In this activity, it is favored by its peculiar circulation, which in some respects resembles a filter. The spleen is normally the major site of destruction of aging red blood cells, although it is not essential for this activity. Damaged red blood cells may be preferentially sequestered in the spleen.[85] The spleen has the ability to remove intracellular inclusion bodies from the red blood cells without causing destruction of the erythrocytes.[54] Presumably for this reason, splenectomy is often followed by the appearance in the blood of red cells containing nuclear remnants and other intracellular inclusion bodies.[55] Iron is stored in the spleen as ferritin and hemosiderin. In some mammals the spleen serves as a reservoir that stores concentrated red blood cells and releases them into the circulation by contraction.[28] In man the small size of the spleen excludes a significant reservoir function, although red blood cells and platelets are at a higher concentration in the spleen than in the bloodstream.[27]

The spleen is not an essential organ and, except in infants, can be removed without important effects. Its functions are readily assumed by other lymphoid and reticuloendothelial organs. In addition to the changes in the erythrocytes that follow splenectomy, there is a transient or persistent increase in the leukocyte and platelet counts.[55] The alterations noted in the blood probably reflect changes in the destruction rather than in the production of blood cells.[99] Splenectomy appears to enhance the susceptibility to serious infection in infants.[63]

REFERENCES
General reviews

1. Adolph, E. F.: Early concepts of physiological regulations, Physiol. Rev. **41:**737, 1961.
2. Chanarin, I.: The megaloblastic anemias, Philadelphia, 1969, F. A. Davis Co.
3. Cline, M. J.: The white cell, Cambridge, Mass., 1975, Harvard University Press.
4. Dacie, J. V.: The haemolytic anaemias, congenital and acquired, ed. 2, London, 1960 to 1967, J. & A. Churchill, parts 1 to 4.
5. Fairbanks, V. F., Fahey, J. L., and Beutler, E.: Clinical disorders of iron metabolism, ed. 2, New York, 1971, Grune & Stratton, Inc.
6. Gordon, A. S., editor: Regulation of hematopoiesis, New York, 1970, Appleton-Century-Crofts.
7. Greenwalt, T. J., and Jamieson, G. A., editors: For-

mation and destruction of blood cells, Philadelphia, 1970, J. B. Lippincott Co.

8. Harris, J. W., and Kellermeyer, R. W.: The red cell, ed. 2, Cambridge, Mass., 1970, Harvard University Press.

9. Jacobs, A., and Worwood, M., editors: Iron in biochemistry and medicine, London, 1974, Academic Press, Ltd.

10. Jamieson, G. A., and Greenwalt, T. J., editors: Red cell membranes, Philadelphia, 1969, J. B. Lippincott Co.

11. Kabat, E. A.: Blood group substances: their chemistry and immunochemistry, New York, 1956, Academic Press, Inc.

12. Krantz, S. B., and Jacobson, L. O.: Erythropoietin and the regulation of erythropoiesis, Chicago, 1970, University of Chicago Press.

13. Lehmann, H., and Huntsman, R. G.: Man's hemoglobins, ed. 2, New York, 1974, Elsevier North Holland Inc.

14. Macfarlane, R. G., and Robb-Smith, A. H. T., editors: Functions of the blood, New York, 1961, Academic Press, Inc.

15. Metcalf, D., and Moore, M. A. S.: Haematopoietic cells, New York, 1971, Elsevier North Holland Inc.

16. Mills, J. A., and Cooperband, S. R.: Lymphocyte physiology, Annu. Rev. Med. **22:**185, 1971.

17. Mollison, P. L.: Blood transfusion in clinical medicine, ed. 4, Philadelphia, 1967, F. A. Davis Co.

18. Ponder, E.: Hemolysis and related phenomena, New York, 1948, Grune & Stratton, Inc.

19. Race, R. R., and Sanger, R.: Blood groups in man, ed. 6, Oxford, 1975, Blackwell Scientific Publications.

20. Stohlman, F., Jr., editor: Hemopoietic cellular proliferation, New York, 1970, Grune & Stratton, Inc.

21. Weiss, L.: The cells and tissues of the immune system. Structure, functions, interactions, Englewood Cliffs, N.J., 1972, Prentice-Hall, Inc.

22. Whittam, R.: Transport and diffusion in red blood cells, Baltimore, 1964, The Williams & Wilkins Co.

23. Williams, W. J., et al., editors: Hematology, New York, 1972, McGraw-Hill Book Co.

24. Wintrobe, M. M., et al.: Clinical hematology, ed. 7, Philadelphia, 1974, Lea & Febiger.

Original papers

25. Alexanian, R.: Erythropoietin and erythropoiesis in anemic man following androgens, Blood **33:**564, 1969.

26. Archer, G. T., and Hirsch, J. G.: Isolation of granules from eosinophil leukocytes and study of their enzyme content, J. Exp. Med. **118:**277, 1963.

27. Aster, R. H.: Pooling of platelets in the spleen: role in the pathogenesis of "hypersplenic" thrombocytopenia, J. Clin. Invest. **45:**645, 1966.

28. Barcroft, J.: Alterations in the volume of the normal spleen and their significance, Am. J. Med. Sci. **179:**1, 1930.

29. Beck, W. S.: Abnormalities of nucleic acid metabolism and growth balance in the megaloblastic anemias, Medicine **43:**715, 1964.

30. Bellanti, J. A., and Dayton, D. H., editors: The phagocytic cell in host resistance, New York, 1975, Raven Press.

31. Berlin, N. I., Waldmann, T. A., and Weissman, S. M.: Life span of red blood cell, Physiol. Rev. **39:**577, 1959.

32. Bernard, C.: Lecons sur les phénomènes de la vie communes aux animaux et aux végétaux, Paris, 1879, B. Baillière et Fils.

33. Beutler, E.: Drug-induced hemolytic anemia, Pharmacol. Rev. **21:**73, 1969.

34. Bloom, W., and Bartelmez, G. W.: Hematopoiesis in young human embryos, Am. J. Anat. **67:**21, 1940.

35. Boggs, D. R.: The kinetics of neutrophilic leukocytes in health and in disease, Semin. Hematol. **4:**359, 1967.

36. Bohr, C., Hasselbalch, K., and Krogh, A.: Über einem in biologischer Beziehung wichtigen Einfluss, den die Kohlensaurespannung des Blutes auf dessen Sauerstoffbindung ubt, Scand. Arch. Physiol. **16:**402, 1904.

37. Bothwell, T. H., et al.: Iron metabolism in the pregnant rabbit. Iron transport acorss the placenta, Am. J. Physiol. **193:**615, 1958.

38. Brewer, G. J., and Eaton, J. W.: Erythrocyte metabolism: interaction with oxygen transport, Science **171:**1205, 1971.

39. Bull, B. S., and Brailsford, J. D.: The biconcavity of the red cell: an analysis of several hypotheses, Blood **41:**833, 1973.

40. Canham, P. B.: The minimum energy of binding as a possible explanation of the biconcave shape of the human red blood cell, J. Theor. Biol. **26:**61, 1970.

41. Cannon, W. B.: Organization for physiological homeostasis, Physiol. Rev. **9:**399, 1929.

42. Cartwright, G. E., et al.: Studies on free erythrocyte protoporphyrin, plasma iron and plasma copper in normal and anemic subjects, Blood **3:**501, 1948.

43. Castle, W. B.: Development of knowledge concerning the gastric intrinsic factor and its relation to pernicious anemia, N. Engl. J. Med. **249:**603, 1953.

44. Castle, W. B.: Current concepts of pernicious anemia, Am. J. Med. **48:**541, 1970.

45. Chanutin, A., and Curnish, R. R.: Effect of organic and inorganic phosphates on the oxygen equilibrium of human erythrocytes, Arch. Biochem. Biophys. **121:**96, 1967.

46. Chaplin, H., Jr., Mollison, P. L., and Vetter, H.: The body/venous hematocrit ratio: its constancy over a wide hematocrit range, J. Clin. Invest. **32:**1309, 1953.

47. Code, C. F., Hurn, M. M., and Mitchell, R. G.: Histamine in human disease, Proc. Staff Meet. Mayo Clin. **39:**715, 1964.

48. Cohn, Z. A.: The fate of bacteria within phagocytic cells. I. The degradation of isotopically labeled bacteria by polymorphonuclear leukocytes and macrophages, J. Exp. Med. **117:**27, 1963.

49. Cohn, Z. A., and Hirsch, J. G.: The isolation and properties of the specific cytoplasmic granules of rabbit polymorphonuclear leukocytes, J. Exp. Med. **112:**983, 1960.

50. Cook, J. D., et al.: Ferrokinetics: a biologic model for plasma iron exchange in man, J. Clin. Invest. **49:**197, 1970.

51. Corcino, J. J., Waxman, S., and Herbert, V.: Absorption and malabsorption of vitamin B_{12}, Am. J. Med. **48:**562, 1970.

52. Craddock, C. G.: Kinetics of lymphoreticular tissue, with particular emphasis on the lymphatic system, Semin. Hematol. **4:**387, 1967.

53. Crosby, W. H.: The metabolism of hemoglobin and bile pigment in hemolytic disease, Am. J. Med. **18:**112, 1955.

54. Crosby, W. H.: Siderocytes and the spleen, Blood **12:**165, 1957.

55. Crosby, W. H.: Hyposplenism: an inquiry into normal

functions of the spleen, Annu. Rev. Med. **14:**349, 1963.

56. Crosby, W. H., and Akeroyd, J. H.: The limit of hemoglobin synthesis in hereditary hemolytic anemia, its relation to the excretion of bile pigment, Am. J. Med. **13:**273, 1952.

57. Cruz, W. O., Hahn, P. F., and Bale, W. F.: Hemoglobin radioactive iron liberated by erythrocyte destruction (acetylphenylhydrazine) promptly reutilized to form new hemoglobin, Am. J. Physiol. **135:**595, 1942.

58. deDuve, C., and Wattiaux, R.: Functions of lysosomes, Annu. Rev. Physiol. **28:**435, 1966.

59. Dintenfass, L.: Molecular and rheological considerations of the red cell membrane in view of the internal fluidity of the red cell, Acta Haematol. **32:**299, 1964.

60. Dintzis, H. M.: Assembly of the peptide chains of hemoglobin, Proc. Natl. Acad. Sci. **47:**247, 1961.

61. Drabkin, D. L.: Metabolism of the hemin chromoproteins, Physiol. Rev. **31:**345, 1951.

62. Dubach, R., Moore, C. V., and Callender, S.: Studies in iron transportation and metabolism. IX. The excretion of iron as measured by the isotope technique, J. Lab. Clin. Med. **45:**599, 1955.

63. Erickson, W. D., Burgert, E. O., Jr., and Lynn, H. B.: The hazard of infection following splenectomy in children, Am. J. Dis. Child. **116:**1, 1968.

64. Erslev, A. J.: The role of erythropoietin in the control of red cell production, Medicine **43:**661, 1964.

65. Fåhraeus, R.: The suspension stability of the blood, Physiol. Rev. **9:**241, 1929.

66. Finch, C. A., et al.: Ferrokinetics in man, Medicine **49:**17, 1970.

67. Gatti, R. A., Stutman, O., and Good, R. A.: The lymphoid system, Annu. Rev. Physiol. **32:**529, 1970.

68. Glass, G. B. J.: Gastric intrinsic factor and its function in the metabolism of vitamin B_{12}, Physiol. Rev. **43:**529, 1963.

69. Granick, S.: Porphyrin biosynthesis in erythrocytes: I. Formation of δ-aminolevulinic acid in erythrocytes. J. Biol. Chem. **232:**1101, 1958.

70. Granick, S., and Levere, R. D.: Heme synthesis in erythroid cells, Prog. Hematol. **4:**1, 1964.

71. Grant, W. C., and Root, W. S.: Fundamental stimulus for erythropoiesis, Physiol. Rev. **32:**449, 1952.

72. Guest, M. M., et al.: Red cell deformation in capillaries, Science **142:**1319, 1963.

73. Hahn, P. F., et al.: Radioactive iron absorption by gastrointestinal tract; influence of anemia, anoxia, and antecedent feeding; distribution in growing dogs, J. Exp. Med. **78:**169, 1943.

74. Ham, T. H.: Hemoglobinuria, Am. J. Med. **18:**990, 1955.

75. Ham, T. H., and Curtis, F. C.: Sedimentation rate of erythrocytes. Influence of technical, erythrocyte and plasma factors and quantitative comparison of five commonly used sedimentation methods, Medicine **17:**447, 1938.

76. Hardwicke, J., and Squire, J. R.: The basis of the erythrocyte sedimentation rate, Clin. Sci. **11:**333, 1952.

77. Harris, H.: Role of chemotaxis in inflammation, Physiol. Rev. **34:**529, 1954.

78. Heath, C. W., Strauss, M. B., and Castle, W. B.: Quantitative aspects of iron deficiency in hypochromic anemia, J. Clin. Invest. **11:**1293, 1932.

79. Herbert, V.: Experimental nutritional folate deficiency in man, Trans. Assoc. Am. Physicians **75:**307, 1962.

80. Horrigan, D. L., and Harris, J. W.: Pyridoxine-responsive anemia: analysis of 62 cases, Adv. Intern. Med. **12:**103, 1964.

81. Huff, R. L., et al.: Ferrokinetics in normal persons and in patients having various erythropoietic disorders, J. Clin. Invest. **30:**1512, 1951.

82. Hurtado, A., Merino, C., and Delgado, E.: Influence of anoxemia on hemopoietic activity, Arch. Intern. Med. **75:**284, 1945.

83. Jacobs, M. H.: The quantitative measurement of the permeability of the erythrocyte to water and to solutes by the hemolysis method, J. Cell. Comp. Physiol. **4:**161, 1934.

84. Jandl, J. H.: Analytical review: leaky red cells, Blood **26:**367, 1965.

85. Jandl, J. H., et al.: Clinical determination of sites of red cell sequestration in hemolytic anemias, J. Clin. Invest. **35:**842, 1956.

86. Jandl, J. H., et al.: Transfer of iron from serum iron-binding protein to human reticulocytes, J. Clin. Invest. **38:**161, 1959.

87. LaCelle, P. L., and Weed, R. I.: The contribution of normal and pathologic erythrocytes to blood rheology, Prog. Hematol. **7:**1, 1971.

88. Lane, M., and Alfrey, C. P., Jr.: The anemia of human riboflavin deficiency, Blood **25:**432, 1965.

89. Laurell, C. B.: Studies on the transportation and metabolism of iron in the body with special reference to the iron-binding component in human plasma, Acta Physiol. Scand. **14**(suppl.):46, 1947.

90. Lee, G. R., et al.: Iron metabolism in copper-deficient swine, J. Clin. Invest. **47:**2058, 1968.

91. London, I. M.: The metabolism of the erythrocyte, Harvey Lect. **56:**151, 1961.

92. Marchesi, S. L., et al.: Physical and chemical properties of a protein isolated from red cell membranes, Biochemistry **9:**50, 1970.

93. Marchesi, V. T., and Furthmayr, H.: The red cell membrane, Annu. Rev. Biochem. **45:**667, 1976.

94. Marston, H. R.: The cobalt story, Med. J. Aust. **2:**105, 1959.

95. McCance, R. A., and Widdowson, E. M.: Absorption and excretion of iron following oral and intravenous administration, J. Physiol. **94:**148, 1938.

96. McMaster, P. D., and Hudack, S. S.: The formation of agglutinins within lymph nodes, J. Exp. Med. **61:**783, 1935.

97. Meineke, H. A., and Crafts, R. C.: Effect of combined thyroxin–cortisone–growth hormone therapy on hematopoiesis in hypophysectomized rats, Proc. Soc. Exp. Biol. Med. **96:**74, 1957.

98. Morgan, W. T. J.: A contribution to human biochemical genetics; the chemical basis of blood-group specificity, Proc. R. Soc. Lond. (Biol.) **151:**308, 1960.

99. Motulsky, A. G., et al.: Anemia and the spleen, N. Engl. J. Med. **259:**1164, 1958.

100. Nakao, M., et al.: Adenosine triphosphate and shape of erythrocytes, Biochem. J. **49:**487, 1961.

101. Nixon, P. F., and Bertino, J. R.: Interrelationships of vitamin B_{12} and folate in man, Am. J. Med. **48:**555, 1970.

102. North, R. J.: Endocytosis, Semin. Hematol. **7:**161, 1970.

103. Nyman, M.: Serum haptoglobin. Methodological and clinical studies, Scand. J. Clin. Lab. Invest. **11**(suppl.):39, 1959.

104. Ostrow, J. D., Jandl, J., and Schmid, R.: The formation of bilirubin from hemoglobin in vivo, J. Clin. Invest. **41:**1628, 1962.

105. Perutz, M. F.: Stereochemistry of cooperative effects in haemoglobin, Nature **228:**726, 1970.

106. Perutz, M. F., et al.: Three dimensional Fourier synthesis of horse oxyhemoglobin at 2.8 Å resolution: the atomic model, Nature **219:**131, 1968.

107. Perutz, M. F., et al.: Identification of residues responsible for the alkaline Bohr effect in haemoglobin, Nature **222:**1240, 1969.

108. Pirzio-Biroli, G., Finch, C. A., and Loden, B.: Iron absorption. III. The influence of iron stores on iron absorption in the normal subject, J. Lab. Clin. Med. **55:**216, 1960.

109. Pollack, W., et al.: A study of the forces involved in the second stage of hemagglutination, Transfusion **5:**158, 1965.

110. Rabinovitch, M.: Phagocytosis: the engulfment stage, Semin. Hematol. **5:**134, 1968.

111. Rath, C. E., and Finch, C. A.: Chemical, clinical, and immunological studies on the products of human plasma fractionation. XXXVIII. Serum iron transport: measurement of iron-binding capacity of serum in man, J. Clin. Invest. **28:**79, 1949.

112. Roeser, H. P., et al.: The role of ceruloplasmin in iron metabolism, J. Clin. Invest. **49:**2408, 1970.

113. Roughton, F. J. W., Otis, A. B., and Lyster, R. L. J.: The determination of the individual equilibrium constants of the four intermediate reactions between oxygen and sheep haemoglobin, Proc. R. Soc. Lond. (Biol.) **144:**29, 1955.

114. Rous, P.: Destruction of the red blood corpuscles in health and disease, Physiol. Rev. **3:**75, 1923.

115. Schmid, R.: Jaundice and bilirubin metabolism, Arch. Intern. Med. **101:**669, 1958.

116. Schulman, M. P., and Richert, D. A.: Heme synthesis in vitamin B_6 and pantothenic acid deficiencies, J. Biol. Chem. **226:**181, 1957.

117. Scott, E. M.: Congenital methemoglobinemia due to DPNH-diaphorase deficiency. In Beutler, E., editor: Hereditary disorders of erythrocyte metabolism, New York, 1968, Grune & Stratton, Inc.

118. Sears, D. A., Weed, R. I., and Swisher, S. N.: Differences in the mechanism of in vitro immune hemolysis related to antibody specificity, J. Clin. Invest. **43:**975, 1964.

119. Silber, R., and Moldow, C. F.: The biochemistry of B_{12}-mediated reactions in man, Am. J. Med. **48:**549, 1970.

120. Skalak, R., and Branemark, P.-I.: Deformation of red blood cells in capillaries, Science **164:**717, 1969.

121. Smith, C. A., et al.: Persistence and utilization of maternal iron for blood formation during infancy, J. Clin. Invest. **34:**1391, 1955.

122. Stossel, T. P.: The mechanism of phagocytosis, J. Reticuloendothel. Soc. **19:**237, 1976.

123. Usami, S., Chien, S., and Gregersen, M. I.: Viscometric characteristics of blood of the elephant, man, dog, sheep and goat, Am. J. Physiol. **217:**884, 1969.

124. Watkins, W. M.: Blood group substances, Science **152:**172, 1966.

125. Weed, R. I., and Reed, C. F.: Membrane alterations leading to red cell destruction, Am. J. Med. **41:**681, 1966.

126. Weed, R. I., Reed, C. F., and Berg, G.: Is hemoglobin an essential structural component of human erythrocyte membranes? J. Clin. Invest. **42:**581, 1963.

127. Weiss, L.: The structure of the normal spleen, Semin. Hematol. **2:**205, 1965.

128. Weissbach, H., and Taylor, R. T.: Metabolic role of vitamin B_{12}, Vitam. Horm. **26:**395, 1968.

129. Weissmann, G.: Lysosomes, Blood **24:**594, 1964.

130. Wells, R. E., Jr., and Merrill, E. W.: Influence of flow properties of blood upon viscosity-hematocrit relationships, J. Clin. Invest. **41:**1591, 1962.

131. Whipple, G. H., and Madden, S. C.: Hemoglobin, plasma protein and cell protein—their interchange and construction in emergencies, Medicine **23:**215, 1944.

132. Wilson, T. H.: Membrane transport of vitamin B_{12}, Medicine **43:**669, 1964.

133. Wintrobe, M. M., et al.: Pyridoxine deficiency in swine, with particular reference to anemia, epileptiform convulsions and fatty liver, Bull. Johns Hopkins Hosp. **72:**1, 1943.

48

C. LOCKARD CONLEY

Hemostasis

Blood in the normal circulatory system remains fluid. Pressure is higher within the vascular tree than in the surrounding tissues, and the vast area of delicate capillary membrane is susceptible to damage even by the events of physiologic activity. Accordingly, there is an ever-present hazard of loss of blood, a circumstance that may be fatal if more than one third is lost. *Hemostasis,* the prevention and control of hemorrhage, is achieved by several mechanisms functioning in concert.[14] Resistance to bleeding is provided by the integrity of the blood vessels and the turgor of the tissues. Injured blood vessels constrict, retarding the flow of blood. Blood platelets aggregate at sites of injury, tending to seal leaking vessels. Finally, blood coagulates within damaged vessels and wounds, forming plugs that prevent further bleeding. These events take place so rapidly that a normal person can withstand injury without grave danger of major blood loss.[24]

VASCULAR FACTORS

The resistance of delicate blood vessels to extravasation of blood can be overcome if the pressure within the channels is excessive. When capillary pressure is increased, disruption of the thin walls may occur, producing punctate hemorrhages known as *petechiae*. A tourniquet tightly applied to an extremity increases the capillary pressure distal to the constriction and often causes petechiae to appear in the skin, even in normal persons. Superficially placed vessels in susceptible areas, such as those in the mucous membrane of the nose, are readily traumatized and may bleed profusely. A blow to the surface of the body damages capillaries, and the resulting leakage of blood produces a black-and-blue spot, or *ecchymosis*. The cushioning effect of normal supporting tissues protects blood vessels against injury. Women tend to bruise more readily than men, in part because women have more delicate skin and subcutaneous tissue. In elderly persons, atrophy of the subcutaneous tissue often is associated with a tendency to bruise; ecchymoses

appear particularly over bony prominences. The extent to which blood spreads from a bleeding site is influenced by the character of the tissue. The lax tissue surrounding the eye offers little resistance to extension of bleeding, and spreading ecchymosis in the periorbital area often occurs after injury to this region (the "black eye"). Integrity of vascular walls depends on a healthy state of the blood vessels. Abnormal fragility of the capillaries, manifested by the appearance of ecchymoses and petechiae, is characteristic of vitamin C deficiency, indicating that the vitamin is essential for the normal resistance of capillary membranes.[19,35] A similar increase in vascular fragility occurs in the absence of the blood platelets, which are required to maintain the integrity of the capillaries.[123]

When blood vessels are severed, vascular constriction occurs as a reflex response, a step not adequate in itself to achieve hemostasis.[151] Pressure on a bleeding point or application of a tourniquet similarly retards blood flow until a coagulum has effectively blocked the leaking vessel. After an occlusive plug has been formed, release of external pressure or relaxation of constricted vessels is not followed by bleeding.

PLATELET FACTORS

Platelets are essential for hemostasis.[7,12,15] Within 1 or 2 sec of damage to the endothelium, platelets in large numbers adhere to the injured surface and form a dense aggregate.[65,79,123] This is the beginning of a hemostatic plug, which is completed by the formation of the blood clot. Platelets contribute substances that accelerate coagulation,[16,59] and they are required for the retraction of a clot after it has been formed.[3] In addition, they release substances with vasoconstrictor properties, in particular 5-hydroxytryptamine (serotonin).[70] Depletion of platelets, *thrombocytopenia*, causes a hemorrhagic disorder manifested by persistent bleeding from cuts and wounds and by petechiae, ecchymoses, and oozing of blood from vascular beds, attributable to increased fragility of the capillaries. The

1137

mechanism by which platelets participate in maintaining the integrity of capillary membranes is unknown. Thinning of the membranes with fenestrations and increased permeability to particulate matter has been demonstrated by electron microscopy after experimentally induced thrombocytopenia.[89] Transfused platelets correct these defects, apparently by contributing material to endothelial cells.[86]

The blood platelets are the smallest of the blood cells, discoid in shape and averaging 2 μm in diameter.[25] Normal blood contains 150,000 to 300,000 platelets/mm³, a number greatly in excess of that required for hemostasis. Derived from the megakaryocytes of bone marrow,[147] platelets are nonnucleated, but they have a more complex internal structure than the red blood cells.[12,15,145] The rate of platelet production is regulated by a feedback control mechanism mediated by an unidentified plasma factor, *thrombopoietin*.[95] Platelets occur in relatively high concentration in the spleen, but there is no other appreciable reserve.[31] Their life in the circulation, averaging about 10 days,[74] is terminated primarily by removal of senescent platelets by the macrophages of the reticuloendothelial system.[32] Protein synthesis is retarded in aging platelets,[136] and reparative processes presumably are impaired. Platelets do not circulate and do not exert hemostatic effects when their metabolic activity has been terminated.[84] They have a tricarboxylic acid cycle and utilize glucose via the glycolytic and hexose monophosphate shunt pathways.[7,12,15] Adenine nucleotides are present in high concentration. Platelets take up serotonin, epinephrine, and norepinephrine by means of an active transport mechanism. Serotonin in blood is found almost exclusively in the platelets, which concentrate this substance against a large cell-plasma gradient.[48] Although serotonin has vasoconstrictor properties, it does not appear to be essential for hemostasis.[77]

Platelets adhere to collagen,[152,153] to the basement membranes of capillaries,[142] and to subendothelial microfibrils.[137] They are aggregated by adenosine diphosphate (ADP) released from platelets on contact with these surfaces,[46,79,104] a reaction believed to be involved in the formation of the hemostatic plug in injured blood vessels. A unique and remarkable property of platelets is their extreme sensitivity to the action of the enzyme *thrombin,* a substance formed during the coagulation of blood.[34] In the presence of thrombin, platelets aggregate and rapidly undergo a series of morphologic changes leading to their ultimate dissolution, a phenomenon known as *viscous metamorphosis.*[148] Glucose and calcium

ions are required for this reaction,[56] in which adenosine triphosphate (ATP) is utilized as a source of energy.[45] Both ADP and thrombin are bound to the platelet surface.[47,141] The manner in which they exert their effects on platelets is unknown. Within seconds of the onset of the reaction, some of the intracellular components of platelets are extruded into the surrounding medium.[70,72,150] Potassium, ADP, and serotonin are released, along with platelet enzymes and other substances. Simultaneously there is rapid disappearance of ATP. Changes in the platelets induced by thrombin cause shrinking or *retraction* of the blood clot[3] by the action of a contractile protein of platelets, *thrombosthenin,* a substance similar to actomyosin.[7,38,44]

The lipids of platelets have important functional activities. Arachidonic acid, a prostaglandin precursor, appears to be especially significant. Arachidonic acid itself causes platelet aggregation, a phenomenon prevented by aspirin and indomethacin, inhibitors of prostaglandin synthesis.[41,135] Certain derivatives of prostaglandin synthesis, notably thromboxane A_2, are extremely potent agents in causing aggregation of platelets.[134] These substances may have a primary role in the hemostatic activity of platelets. Platelets release substances that are concerned with the coagulation of blood, the most important of which are certain lipoproteins and phospholipids (platelet factor 3), which participate in the reactions accelerating the conversion of prothrombin to thrombin.[15,59,144]

COAGULATION OF BLOOD

The final stage in the control of bleeding is the formation of a blood clot within the injured vessel and in the wound. The manner in which coagulation is initiated in this setting is uncertain. Materials liberated from the damaged tissues or from aggregated platelets and the activation of procoagulant substances of plasma by other mechanisms undoubtedly set in motion the reactions that lead to clot formation. Most of our information about coagulation has been derived from studies of clotting in vitro. Blood contains all the substances required for coagulation, which occurs within a few minutes when blood is removed from the body. The blood clot is the product of a complex series of reactions involving a number of components of the blood. The theories and the development of knowledge that have led to our present concepts of the mechanism of blood coagulation are described in a number of reviews.* There are close interrela-

---

*See references 2, 6, 10, 11, 16, 18, 21, and 23.

tionships among coagulation, lysis of blood clots, and the inflammatory response.[20,21,61]

Formation of fibrin

The formation of the clot is the result of the appearance of a network of insoluble filaments of fibrin. At first needlelike, the fibrin strands join to form a tough, resilient, durable mesh in which the fluid and cellular components of the blood are trapped. The clot subsequently retracts, extruding the liquid portion of the blood (serum). Serum differs from plasma in that it lacks fibrinogen and certain other proteins involved in blood coagulation (Table 48-1), and it contains substances released from platelets during coagulation. If the cellular elements are removed before the blood coagulates, fibrin forms the entire solid material of the clot and retraction does not occur.[75] fibrin is formed from a soluble precursor, *fibrinogen*, a protein produced in the liver and present in plasma in a concentration of about 0.3 gm/100 ml.[64,68,103] Fibrinogen is continuously removed from the circulation, with a half-life of about 4.5 days. The mechanism of its catabolism has not been established; a continuous process of intravascular coagulation and lysis has been suggested[149] but seems unlikely.[53] Fibrinogen is a stable substance that is converted to fibrin only by enzymatic action.[62,91] Fibrin formation in physiologic systems is produced by *thrombin*, an enzyme normally absent from the circulating blood.[133] Fibrinogen is a dimeric protein with a molecular weight of about 340,000; each half molecule is composed of three unlike polypeptide chains joined by disulfide bridges.[42,43,132] Thrombin is a highly specific hydrolytic enzyme that produces limited proteolysis of fibrinogen, splitting off two small peptides (fibrinopeptides A and B) by hydrolysis of arginyl-glycine bonds. Fibrinogen thus activated by thrombin contains about 1,500 amino acid residues and polymerizes to form the fibrin thread. Electron photomicrographs of fibrin in clots demonstrate the orderly arrangement of the polymerized components.[78] The action of factor XIII (*fibrin stabilizing factor*), a constituent of normal plasma, strengthens the clot and decreases its solubility.[92] In the presence of thrombin and calcium, factor XIII is converted to an enzxme that causes cross linking of the fibrin polypeptides by a transamidation reaction involving specific lysine and glutamine residues.[9,51,98,128]

Formation of thrombin

Thrombin, a serine protease, is derived solely from *prothrombin*, an inert precursor in normal plasma.[23] In the circulating blood, prothrombin is not converted to thrombin at an appreciable rate, if at all; however, it is continuously removed from the blood with a half-life of about 3 days.[129] The molecular weight of prothrombin is approximately 70,000, whereas that of thrombin is much lower.[6,94,127,140] Thrombin is formed by the splitting of the molecule as a result of an enzymatically accelerated reaction during which prothrombin is bound to phospholipid by calcium ions.[36,50,100] Thrombin appears soon after blood is shed, and it is readily demonstrated in fresh

Table 48-1. Substances participating in coagulation of blood

Numeric designation*	Common name†	Molecular weight (approximate)	Present in serum	Dependent on vitamin K	Pathway Intrinsic	Pathway Extrinsic
I	Fibrinogen	340,000			+	+
II	Prothrombin	68,700		+	+	+
III	Tissue factor	220,000-330,000				+
IV	Calcium		+		+	+
V	Proaccelerin[110]	290,000-400,000			+	+
VII	Proconvertin (SPCA)[28]	63,000	+	+		+
VIII	Antihemophilic factor (AHF)[111]	1,100,000			+	
IX	Christmas factor (PTC)[26,40]	55,400	+	+	+	
X	Stuart factor[81]	55,000	+	+	+	+
XI	Plasma thromboplastin antecedent (PTA)[122]	160,000			+	
XII	Hageman factor[116]	90,000	+		+	
XIII	Fibrin-stabilizing factor[92]	320,000	+		+	+

*Note that factor VI does not exist.
†Other designations for some of the clotting factors have been in common use. Proconvertin originally was designated "serum prothrombin conversion accelerator (SPCA)." Christmas factor was first designated "plasma thromboplastin component (PTC)."

serum. The rate of its formation is the principal determinant of the time required for clotting. The concentration of thrombin potentially available in normal blood is sufficient to cause coagulation to occur within a few seconds. Within minutes of the time thrombin activity reaches a maximum, the enzyme is progressively inactivated by antithrombin III, a constituent of serum with which thrombin reacts stochiometrically.[80,115,121]

The enzymatic activity responsible for thrombin formation evolves during the interaction of a number of constituents of the blood. *Calcium ions* are essential for conversion of prothrombin to thrombin under physiologic conditions, participating in several steps of the reaction.[8,30] Substances that precipitate or bind calcium are anticoagulants. Blood that has been rendered incoagulable by decalcification clots when a calcium salt is added. The concentration in which calcium occurs in normal plasma is approximately optimal, and higher as well as lower concentrations are inhibitory.[18] Lipoproteins and phospholipids play an essential role in physiologic thrombin formation.[82] The procoagulant lipids derived from platelets are known collectively as "platelet factor 3."[15] Purified phospholipids, notably phosphatidylserine and phosphatidylethanolamine, are active, but less so than the naturally occurring lipids, which are mixtures containing lipoproteins and phospholipids.[144] The clot-promoting activity of an extract of many tissues (tissue factor) is associated with subcellular particles containing phospholipid and protein.[106,146] The tissue factor (tissue thromboplastin) differs from the platelet lipid in its effect on the coagulation process. Phospholipids appear to exert their effects in part because of phenomena related to their surface activity and electrical charge.[73]

Formation of thrombin from prothrombin is the end result of sequential reactions involving several plasma components along either of two converging pathways.[6] Blood drawn without admixture with any other substance clots on contact with glass or other surfaces as a result of activation of the *intrinsic pathway.* Blood mixed with material derived from damaged tissue clots because of reactions initiated by *tissue factor* via the *extrinsic pathway.* Interruption of either sequence of reactions may cause a hemorrhagic disorder.

Intrinsic pathway. When blood is drawn in a manner that precludes contamination with tissue juice, the rate of thrombin formation (and of clotting) is largely a function of the character and extent of the surface.[60,97] Coagulation occurs slowly when blood is exposed to oiled or sili-

cone-coated surfaces and much more rapidly on contact with glass. Plasma containing no extraneous material, platelets, or other cells coagulates on contact with glass and other "inert" surfaces.[55] The reaction initiated by plasma-surface contact is the activation of factor XII (Hageman factor) as a result of adsorption of an inhibitor.[116,119] Factor XII can be activated also by collagen and by the basement membrane of blood vessels. The activated protein (factor XII$_a$) reacts with factor XI, converting it to an enzyme, factor XI$_a$, which has endopeptidase activity toward factor IX.[69,120] The reaction of factor XI$_a$ with factor IX requires calcium and involves the cleavage of factor IX to form factor IX$_a$, a serine protease.[6,118] In the presence of calcium and phospholipid (most of which is derived from platelets), factor IX$_a$ interacts with factor VIII, apparently by forming a complex that acts as a catalyst in the activation of factor X.[6] The activated form (X$_a$) interacts with factor V, calcium, and phospholipid to cause the formation of thrombin from prothrombin. Factor X$_a$, a serine protease, appears to be the enzyme of the reaction, and the specific role of factor V is unknown[6] (Fig. 48-1).

There are additional complexities. Plasma contains inhibitors of many of the reactions, and activated clotting factors are inactivated or removed from the circulation. The coagulant activity of factors V and VIII is enhanced by traces of thrombin.[54,94] There is evidence that substances derived from platelets may partially substitute for factors XI and XII.[126] These reactions may be important in accelerating coagulation after it has been initiated. In addition, relationships have been shown to exist between coagulation and other systems of plasma proteins involved in the defenses of the body.[4,88] Fletcher factor, a prekallikrein, and Fitzgerald factor, a kininogen, are involved with Hageman factor in the activation of factor XI.* Activated Hageman factor, in addition to its role in initiating the coagulation sequence, is involved in activation of the plasmin, kinin, and complement systems.[20]

Extrinsic pathway. Even when exposed to an extensive glass surface, blood does not clot as rapidly as when it is mixed with tissue extracts.[49] An aqueous extract of many tissues, if injected intravenously, causes intravascular coagulation.[117] The reactions initiated by tissue factor bypass most of the sequences of the intrinsic pathway and lead more directly to thrombin formation. Tissue factor contains phospholipid and

---

*See references, 21, 61, 76, 124, and 125.

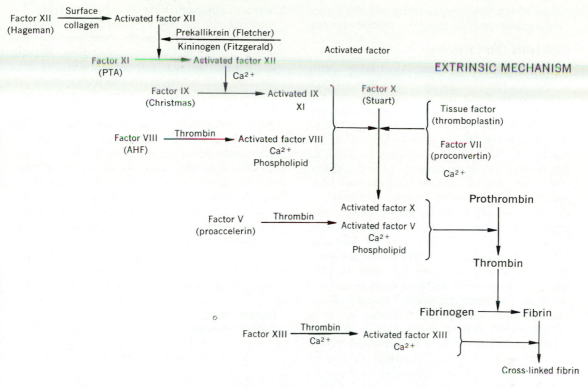

Fig. 48-1. Mechanisms of blood coagulation. Many proteins involved are proenzymes, activated during sequence of reactions. Autocatalytic phenomena greatly accelerate coagulation process. Alteration of factor V and factor VIII by thrombin enhances their potency but does not produce enzymatic activity. Action of thrombin on platelets releases phospholipid, important in intrinsic pathway. (Modified from Davie and Ratnoff[58] and Macfarlane.[99])

protein, both of which are necessary for its activity.[106,107] In the presence of calcium, tissue factor reacts with factor VII to form an enzymatic complex that activates factor X.[107] The extrinsic and intrinsic pathways converge with the activation of factor X.

Vitamin K–dependent clotting factors. Naturally occurring substances with vitamin K activity are fat-soluble derivatives of naphthoquinone.[29,57] Deficiency of vitamin K causes depletion of four of the procoagulant proteins of plasma: prothrombin, factor VII, factor IX, and factor X.[33,114] These glycoproteins have similarities of physical properties and of some amino acid sequences, suggesting that they are derived from reduplicated genes. Vitamin K is not required for synthesis of the proteins in the liver[138] but is necessary for enzymatic carboxylation of glutamyl residues in the glycoprotein precursors.[66] Gamma-carboxyglutamic acid moieties appear to be essential for calcium binding, by which the proteins are linked to phospholipids.

Attachment to a phospholipid surface apparently is necessary to accelerate activation of these clotting factors.[105]

Certain derivatives of coumarin that are structurally related to vitamin K impede the action of the vitamin; on administration, they impair utilization of vitamin K and reduce the plasma concentration of the vitamin K–dependent clotting factors. These compounds have been extensively used in medicine to retard thrombin formation and thus to prevent thrombosis (the formation of an abnormal occlusive intravascular clot).[17,96]

Deficiency of clotting factors. Severe deficiency of one or more of the plasma procoagulant proteins usually leads to a hemorrhagic disorder. Factor VIII deficiency is the cause of hemophilia,[111] a heritable sex-linked disease in which an abnormal protein is produced that lacks normal factor VIII coagulant activity.[37] Persons with hereditary deficiency of factor XII, Fletcher factor, or Fitzgerald factor have impaired blood coagulation in vitro, but hemostasis appears to be normal. In contrast, a hemophilia-like disorder results from a deficiency of factor VII, the unique procoagulant protein of the extrinsic system.

Clot lysis (fibrinolysis)

If blood is collected and stored under sterile conditions, the clot that is formed may undergo complete liquefaction.[113] For normal blood, this process requires many days. In certain abnormal states, clot lysis may be greatly accelerated, so that dissolution of the clot occurs in a few hours.[71] Study of this phenomenon has revealed that normal plasma contains *plasminogen,* the inactive precursor of a proteolytic enzyme, *plasmin.*[67,131] The activated enzyme is capable of digesting not only fibrin but also numerous other proteins, including fibrinogen and prothrombin, by splitting arginyl-lysine bonds.[143] Hageman factor is involved in the activation of plasminogen.[109] Naturally occurring activators of plasminogen can be extracted from many tissues and are found in plasma and urine.[27] Filtrates of cultures of certain streptococci contain an agent (streptokinase) that is a potent activator.[52,139] Plasma and platelets have pronounced antiplasmin activity.[87,108] Enhanced proteolytic activity of plasma in vivo can be produced by the injection of streptokinase or urokinase.[112,130] Activation of plasminogen causes rapid dissolution of recently formed intravascular clots. Presumably, plasmin plays a role in the removal of fibrin deposits occurring as a result of injury or disease. Proteolysis of fibrinogen and fibrin by plasmin leads to the production of split products, some of which inhibit the thrombin-fibrinogen reaction.[90,101]

ANTICOAGULANTS

Many substances, when added to blood, have the property of retarding or inhibiting coagulation. Only by the use of anticoagulants can blood be maintained in a fluid state outside of the body. Simple neutral salts in relatively high concentration inhibit clotting.[18] Compounds that precipitate or bind calcium are effective anticoagulants and are the agents most commonly employed in the laboratory. The conversion of fibrinogen to fibrin by thrombin is inhibited by a number of substances; of these, heparin is the most important. Thrombin formation is prevented by the decalcifying anticoagulants, by heparin, and by numerous other chemical agents. Specific inhibitors of procoagulant proteins have been described.[39,102]

Heparin. Among substances with anticoagulant properties, heparin has attracted the greatest interest because it is produced in the body of the normal individual. First isolated from liver,[83] it subsequently has been obtained from other organs, particularly the lung and intestinal mucosa, which are the chief commercial sources. Heparin is a polysulfuric acid ester of mucoitin.[13] For its anticoagulant action, it requires antithrombin III, a protein component of plasma. In the presence of the plasma cofactor, heparin inhibits both the production of thrombin and the effect of thrombin on fibrinogen.[63] The physiologic significance of the occurrence of heparin in the body is unknown. Only trace amounts are present in plasma, too little to account for the fluidity of blood. In anaphylactic shock in dogs, heparin may appear in sufficient concentration to have a pronounced anticoagulant effect,[85] but a comparable occurrence has not been observed in man. Injection of heparin inhibits the coagulation of blood in vivo and has been extensively employed in medicine to prevent intravascular clotting.

REFERENCES
General reviews

1. Baldini, M. G., and Ebbe, S., editors: Platelets: production, function, transfusion and storage, New York, 1974, Grune & Stratton, Inc.
2. Biggs, R., editor: Human blood coagulation, haemostasis and thrombosis, Oxford, 1972, Blackwell Scientific Publications, Ltd.
3. Budtz-Olsen, O. E.: Clot retraction, Springfield, Ill., 1951, Charles C Thomas, Publisher.
4. Colman, R. W., et al.: The human plasma kallikrein-kinin system. In Brown, E. B., and Moore, C. V., editors: Progress in hematology, New York, 1971, Grune & Stratton, Inc. vol. 7.
5. Cooper, H. A., Mason, R. G., and Brinkhous, K. M.: The platelet: membrane and surface reactions, Annu. Rev. Physiol. **38:**501, 1976.
6. Davie, E. W., and Fujikawa, K.: Basic mechanisms in blood coagulation, Annu. Rev. Biochem. **44:**799, 1975.
7. Elliott, K., and Knight, J., editors: Biochemistry and pharmacology of platelets, Ciba Foundation Symposium 35, New York, 1975, Elsevier North Holland, Inc.
8. Ferguson, J. H.: The blood calcium and the calcium factor in blood coagulation, Physiol. Rev. **16:**640, 1936.
9. Folk, J. E., and Chung, S. I.: Molecular and catalytic properties of transglutaminases, Adv. Enzymol. **38:** 109, 1973.
10. Hemker, H. C., Loeliger, E. A., and Veltkamp, J. J., editors: Human blood coagulation, biochemistry, clinical investigation and therapy, Leiden, Netherlands, 1969, Leiden University Press.
11. Howell, W. H.: Theories of blood coagulation, Physiol. Rev. **15:**435, 1935.
12. Johnson, S. A., editor: The circulating platelet, New York, 1971, Academic Press, Inc.
13. Jorpes, J. E.: Heparin: its chemistry, physiology and application in medicine, London, 1939, Humphrey Milford.
14. Macfarlane, R. G., editor: The haemostatic mechanism

in man and other animals, London, 1970, Academic Press, Ltd.

15. Marcus, A. J., and Zucker, M. B.: The physiology of blood platelets, New York, 1965, Grune & Stratton, Inc.

16. Morawitz, P.: Die Chemie der Blutgerinnung, Ergeb. Physiol. **4:**307, 1905.

17. O'Reilly, R. A.: Vitamin K and the oral anticoagulant drugs, Annu. Rev. Med. **27:**245, 1976.

18. Quick, A. J.: The hemorrhagic diseases and the physiology of hemostasis, Springfield, Ill., 1942, Charles C Thomas, Publisher.

19. Ralli, E. P., and Sherry, S.: Adult scurvy and the metabolism of vitamin C, Medicine **20:**251, 1941.

20. Ratnoff, O. D.: Some relationships among hemostasis, fibrinolytic phenomena, immunity and the inflammatory response, Adv. Immunol. **10:**145, 1969.

21. Ratnoff, O. D.: Some recent advances in the study of hemostasis, Circulation Res. **35:**1, 1974.

22. Schmidt, A.: Zur Blutlehre, Leipzig, Germany, 1892, F. C. W. Vogel.

23. Seegers, W. H.: Prothrombin, Cambridge, Mass., 1962, Harvard University Press.

24. Stormorken, H., and Owren, P. A.: Physiopathology of hemostasis, Semin. Hematol. **8:**3, 1971.

25. Tocantins, L. M.: The mammalian blood platelet in health and disease, Medicine **17:**155, 1938.

Original papers

26. Aggeler, P. M., et al.: Plasma thromboplastic component (PTC) deficiency; a new disease resembling hemophilia, Proc. Soc. Exp. Biol. Med. **79:**692, 1952.

27. Albrechtsen, O. K.: Fibrinolytic activity in the organism, Acta Physiol. Scand. **47**(suppl.):165, 1959.

28. Alexander, B., Goldstein, R., and Landwehr, G.: The prothrombin conversion accelerator of serum (SPCA): its partial purification and its properties compared with serum AC-globulin, J. Clin. Invest. **29:**881, 1950.

29. Almquist, H. J.: Vitamin K, Physiol. Rev. **21:**194, 1941.

30. Arthus, M., and Pagès, C.: Nouvelle théorie chimique de la coagulation du sang, Arch. Physiol. Norm. Pathol. **2:**739, 1890.

31. Aster, R. H.: Pooling of platelets in the spleen: role in the pathogenesis of "hypersplenic" thrombocytopenia, J. Clin. Invest. **45:**645, 1966.

32. Aster, R. H.: Studies of the fate of platelets in rats and man, Blood **34:**117, 1969.

33. Babior, B. M., and Kipnes, R. S.: Vitamin K dependent formation of factor VII by a cell-free system from rat liver, Biochemistry **9:**2564, 1970.

34. Baranofsky, I. D., and Quick, A. J.: Heparin and the agglutination of platelets in vitro, Proc. Soc. Exp. Biol. Med. **53:**173, 1943.

35. Barnes, M. J., and Kodicek, E.: Biological hydroxylations and ascorbic acid with special regard to collagen metabolism, Vitam. Horm. **30:**1, 1972.

36. Barton, P. G., and Hanahan, D. J.: Some lipid-protein interactions involved in prothrombin activation, Biochim. Biophys. Acta **187:**319, 1969.

37. Bennett, E., and Huehns, E. R.: Immunologic differentiation of three types of haemophilia and identification of some female carriers, Lancet **2:**956, 1970.

38. Bettex-Galland, M., and Lüscher, E. F.: Extraction of an actomyosin-like protein from human thrombocytes, Nature **184:**276, 1959.

39. Bidwell, E., et al.: Antibody nature of the inhibitor to

40. Biggs, R., et al.: Christmas disease: a condition previously mistaken for haemophilia, Br. Med. J. **2:**1378, 1952.

41. Bills, T. K., Smith, J. B., and Silver, M. J.: Metabolism of (^{14}C) arachidonic acid by human platelets, Biochim. Biophys. Acta **424:**303, 1976.

42. Blombäck, B.: Studies on fibrinogen: its purification and conversion into fibrin, Acta Physiol. Scand. **43** (suppl.):148, 1958.

43. Blombäck, B.: A story about a fiber, Scand. J. Clin. Lab. Invest. **24**(suppl. 107):179, 1969.

44. Booyse, F. M., et al.: Human platelet myosin, J. Biol. Chem. **246:**4291, 1971.

45. Born, G. V. R.: Changes in the distribution of phosphorus in platelet-rich plasma during clotting, Biochem. J. **68:**695, 1958.

46. Born, G. V. R., and Cross, M. J.: The aggregation of blood platelets, J. Physiol. **168:**178, 1963.

47. Born, G. V. R., and Feinberg, H.: Binding of adenosine diphosphate to intact human platelets, J. Physiol. **251:**803, 1975.

48. Born, G. V. R., and Gillson, R. E.: Studies on the uptake of 5-hydroxytryptamine by blood platelets, J. Physiol. **146:**472, 1959.

49. Buchanan, A.: On the coagulation of the blood and other fibriniferous liquids, Glasgow, Scotland, 1845, Bell & Bail.

50. Bull, R. K., Jevons, S., and Barton, P. G.: Complexes of prothrombin with calcium ions and phospholipids, J. Biol. Chem. **247:**2747, 1972.

51. Chen, R., and Doolittle, R. F.: Isolation, characterization, and location of a donor-acceptor unit from cross-linked fibrin, Proc. Natl. Acad. Sci. **66:**472, 1970.

52. Christensen, L. R.: Streptococcal fibrinolysis: a proteolytic reaction due to a serum enzyme activated by streptococcal fibrinolysin, J. Gen. Physiol. **28:**363, 1945.

53. Collen, D., et al.: Metabolism and distribution of fibrinogen. I. Fibrinogen turnover in physiological conditions in humans, Br. J. Haematol. **22:**681, 1972.

54. Colman, R. W.: The effect of proteolytic enzymes on bovine factor V. I. Kinetics of activation and inactivation by bovine thrombin, Biochemistry **8:**1438, 1969.

55. Conley, C. L., Hartmann, R. C., and Morse, W. I., II: The clotting behavior of human "platelet-free" plasma: evidence for the existence of a "plasma thromboplastin," J. Clin. Invest. **28:**340, 1949.

56. Corn, M., Jackson, D. P., and Conley, C. L.: Components of blood necessary for clot retraction, Bull. Johns Hopkins Hosp. **107:**90, 1960.

57. Dam, H., and Schönheyder, F.: The occurrence and chemical nature of vitamin K, Biochem. J. **30:**897, 1936.

58. Davie, E. W., and Ratnoff, O. D.: Waterfall sequence for intrinsic blood clotting, Science **145:**1310, 1964.

59. Deutsch, E., and Lechner, K.: Platelet clotting factors. In Kowalski, E., and Niewiarowski, S., editors: The biochemistry of blood platelets, New York, 1967, Academic Press, Inc.

60. Dick, F. W., Jackson, D. P., and Conley, C. L.: Surface as a quantitative factor in prothrombin utilization, J. Clin. Invest. **33:**1423, 1954.

61. Donaldson, V. H.: Kininogen deficiency in Fitzgerald trait: role of high molecular weight kininogen in clotting and fibrinolysis, J. Lab. Clin. Med. **87:**327, 1976.

62. Doolittle, R. F.: Structural aspects of the fibrinogen

to fibrin conversion, Adv. Protein Chem. **27:**1, 1973.

63. Douglas, A. S.: The action of heparin in the prevention of prothrombin conversion, J. Clin. Invest. **35:**533, 1956.

64. Drury, D. R., and McMaster, P. D.: The liver as the source of fibrinogen, J. Exp. Med. **50:**569, 1929.

65. Eberth, J. C., and Schimmelbusch, C.: Experimentalle Untersuchungen über Thrombose, Virchows Arch. Pathol. Anat. **103:**39, 1886.

66. Esmon, C. T., and Suttie, J. W.: Vitamin K–dependent carboxylase. Solubilization and properties, J. Biol. Chem. **251:**6238, 1976.

67. Fearnley, G. R.: Fibrinolysis. In Poller, L., editor: Recent advances in blood coagulation, London, 1969, J. & A. Churchill.

68. Forman, W. B., and Barnhart, M.: Cellular site for fibrinogen synthesis, J.A.M.A. **187:**128, 1964.

69. Fujikawa, K., et al.: The mechanism of activation of bovine factor IX (Christmas factor) by bovine factor XI$_a$ (activated plasma thromboplastin antecedent), Biochemistry **13:**4508, 1974.

70. Gaintner, J. R., Jackson, D. P., and Maynert, E. W.: The action of thrombin on platelet 5-hydroxytryptamine, Bull. Johns Hopkins Hosp. **111:**185, 1962.

71. Goodpasture, E. W.: Fibrinolysis in chronic hepatic insufficiency, Bull. Johns Hopkins Hosp. **25:**330, 1914.

72. Grette, K.: Studies on the mechanism of thrombin-catalyzed hemostatic reactions in blood platelets, Acta Physiol. Scand. **56**(suppl.):195, 1962.

73. Hanahan, J., Barton, P. G., and Cox, A.: The interactions of prothrombin and factors V and X with phospholipids and calcium ions. In Hemker, H. C., Loeliger, E. A., and Veltkamp, J. J., editors: Human blood coagulation, biochemistry, clinical investigation and therapy, New York, 1969, Springer-Verlag New York, Inc.

74. Harker, L. A., and Finch, C. A.: Thrombokinetics in man, J. Clin. Invest. **48:**963, 1969.

75. Hartmann, R. C., and Conley, C. L.: Clot retraction as a measure of platelet function. I. Effects of certain experimental conditions on platelets in vitro, Bull. Johns Hopkins Hosp. **93:**355, 1953.

76. Hathaway, W. E., Belhasen, L. P., and Hathaway, H. S.: Evidence for a new plasma thromboplastin factor. I. Case report, coagulation studies, and physicochemical properties, Blood **26:**521, 1965.

77. Haverbach, B. J., et al.: Serotonin changes in platelets and brain induced by small daily doses of reserpine; lack of effect of depletion of platelet serotonin on hemostatic mechanisms, N. Engl. J. Med. **256:**343, 1957.

78. Hawn, C. V. Z., and Porter, K. R.: The fine structure of clots formed from purified bovine fibrinogen and thrombin: a study with the electron microscope, J. Exp. Med. **86:**285, 1947.

79. Hellem, A. J.: The adhesiveness of human blood platelets in vitro, Scand. J. Clin. Invest. **12**(suppl.):51, 1960.

80. Hensen, A., and Loeliger, E. A.: Antithrombin III. Its metabolism and its function in blood coagulation, Thromb. Diath. Haemorrh. **9**(suppl.):1, 1963.

81. Hougie, C., Barrow, E. M., and Graham, J. B.: Stuart clotting defect. I. Segregation of an hereditary hemorrhagic state from the heterogeneous group heretofore called "stable factor" (SPCA, proconvertin, factor VII) deficiency, J. Clin. Invest. **36:**485, 1957.

82. Howell, W. H.: The nature and action of the thromboplastic (zymoplastic) substance of the tissues, Am. J. Physiol. **31:**1, 1912.

83. Howell, W. H., and Holt, E.: Two new factors in blood coagulation—heparin and proantithrombin, Am. J. Physiol. **47:**328, 1918.

84. Jackson, D. P., et al.: Effectiveness of transfusions of fresh and lyophilized platelets in controlling bleeding due to thrombocytopenia, J. Clin. Invest. **38:**1689, 1959.

85. Jaques, L. B., and Waters, E. T.: The identity and origin of the anticoagulant of anaphylactic shock in the dog, J. Physiol. **99:**454, 1941.

86. Johnson, S. A.: Endothelial supporting function of platelets. In Johnson, S. A., editor: The circulating platelet, New York, 1971, Academic Press, Inc.

87. Johnson, S. A., and Schneider, C. L.: The existence of antifibrinolysin activity in platelets, Science **117:**229, 1953.

88. Kaplan, A. P., Goetzl, E. J., and Austen, K. F.: The fibrinolytic pathway of human plasma. II. The generation of chemotactic activity by activation of plasminogen proactivator, J. Clin. Invest. **52:**2591, 1973.

89. Kitchens, C. S., and Weiss, L.: Ultrastructural changes of endothelium associated with thrombocytopenia, Blood **46:**567, 1975.

90. Kowalski, E.: Fibrinogen derivatives and their biologic activities, Semin. Hematol. **5:**45, 1968.

91. Laki, K., and Gladner, J. A.: Chemistry and physiology of the fibrinogen-fibrin transition, Physiol. Rev. **44:**127, 1964.

92. Laki, K., and Lóránd, L.: On the solubility of fibrin clots, Science **108:**280, 1948.

93. Lamy, F., and Waugh, D. F.: Certain physical properties of bovine prothrombin, J. Biol. Chem. **203:**489, 1953.

94. Legaz, M. E., et al.: Isolation, subunit structure, and proteolytic modifications of bovine factor VIII, Ann. N.Y. Acad. Sci. **240:**43, 1975.

95. Levin, J., Evatt, B. V., and Schreiner, D. P.: Measurement of plasma thrombopoiesis stimulating activity using selenomethione-[75]Se: studies in rabbits and mice. In Baldini, M. G., and Ebbe, S., editors: Platelets: production, function, transfusion and storage, New York, 1974, Grune & Stratton, Inc.

96. Link, K. P.: The anticoagulant from spoiled sweet clover hay, Harvey Lect. **39:**162, 1943.

97. Lister, J.: On the coagulation of the blood, Proc. R. Soc. **12:**580, 1862-1863.

98. Lorand, L., and Chenoweth, D.: Intramolecular localization of the acceptor cross-linking sites in fibrin, Proc. Natl. Acad. Sci. USA **63:**1247, 1969.

99. Macfarlane, R. G.: An enzyme cascade in the blood clotting mechanism, and its function as a biochemical amplifier, Nature **202:**498, 1964.

100. Magnusson, S.: Thrombin and prothrombin. In Boyer, P. D., editor: The enzymes, New York, 1971, Academic Press, Inc., vol. 3.

101. Marder, V. J., and Budzynski, A. Z.: The structure of the fibrinogen degradation products, Prog. Hemostasis Thromb. **2:**141, 1974.

102. Margolius, A., Jr., Jackson, D. P., and Ratnoff, O. D.: Circulating anticoagulants: a study of 40 cases and a review of the literature, Medicine **40:**145, 1961.

103. Miller, L. L., et al.: The dominant role of the liver in plasma protein synthesis. A direct study of the isolated perfused rat liver with the aid of lysine-ε-C[14], J. Exp. Med. **94:**431, 1951.

104. Mills, D. C. B., Robb, I. A., and Roberts, G. C. K.: The release of nucleotides, 5-hydroxytryptamine and enzymes from human blood platelets during aggregation, J. Physiol. **195:**715, 1968.

105. Nelsestuen, G. L., Broderuis, M., and Martin, G.:

Role of γ-carboxyglutamic acid. Cation specificity of prothrombin and factor X–phospholipid binding, J. Biol. Chem. **251**:6886, 1976.

106. Nemerson, Y., and Pitlick, F. A.: Purification and characterization of the protein component of tissue factor, Biochemistry **9**:5100, 1970.

107. Nemerson, Y., and Pitlick, F. A.: The tissue factor pathway of blood coagulation, Prog. Hemostasis Thromb. **1**:1, 1972.

108. Norman, P. S., and Hill, B. M.: Studies of the plasmin system. III. Physical properties of the two plasmin inhibitors in plasma, J. Exp. Med. **108**:639, 1958.

109. Ogston, D., et al.: Studies on a complex mechanism for the activation of plasminogen by kaolin and by chloroform: the participation of Hageman factor and additional cofactors, J. Clin. Invest. **48**:1786, 1969.

110. Owren, P. A.: The coagulation of blood: investigations on new clotting factor, Acta Med. Scand. **128**(suppl.): 194, 1947.

111. Patek, A. J., and Taylor, F. H. L.: Hemophilia. II. Some properties of a substance obtained from normal human plasma effective in accelerating the coagulation of hemophilic blood, J. Clin. Invest. **16**:113, 1937.

112. Pechet, L.: Fibrinolysis, N. Engl. J. Med. **273**:966, 1965.

113. Pinkus, S. N.: On fibrinolysis, J. Physiol. **35**:xiii, 1906.

114. Pool, J., and Robinson, J.: In vitro synthesis of coagulation factors by rat liver slices, Am. J. Physiol. **196**: 423, 1959.

115. Quick, A. J.: The normal antithrombin of the blood and its relation to heparin, Am. J. Physiol. **123**:712, 1938.

116. Ratnoff, O. D., and Colopy, J. E.: A familial hemorrhagic trait associated with a deficiency of a clot-promoting fraction of plasma, J. Clin. Invest. **34**:602, 1955.

117. Ratnoff, O. D., and Conley, C. L.: Studies on afibrinogenemia. II. The defibrinating effect on dog blood of intravenous injection of thromboplastic material, Bull. Johns Hopkins Hosp. **88**:414, 1951.

118. Ratnoff, O. D., and Davie, E. W.: The activation of Christmas factor (factor IX) by activated plasma thromboplastin antecedent (activated factor XI), Biochemistry **1**:677, 1962.

119. Ratnoff, O. D., and Rosenblum, J. M.: Role of Hageman factor in the initiation of clotting by glass. Evidence that glass frees Hageman factor from inhibition, Am. J. Med. **25**:160, 1958.

120. Ratnoff, O. D., Davie, E. W., and Mallett, D. L.: Studies on the action of Hageman factor: evidence that activated Hageman factor in turn activates plasma thromboplastin antecedent, J. Clin. Invest. **40**:803, 1961.

121. Rosenberg, R. D., and Damus, P. S.: The purification and mechanism of action of human antithrombin-heparin cofactor, J. Biol. Chem. **248**:6490, 1973.

122. Rosenthal, R. L., Dreskin, O. H., and Rosenthal, N.: New hemophilia-like disease caused by deficiency of a third plasma thromboplastin factor, Proc. Soc. Exp. Biol. Med. **82**:171, 1953.

123. Roskam, J.: L' hémostase spontanée, Paris, 1951, Masson & Cie Editeurs.

124. Saito, H., Goldsmith, G., and Waldmann, R.: Fitzgerald factor (high molecular weight kininogen) clotting activity in human plasma in health and disease in various animal plasmas, Blood **48**:941, 1976.

125. Saito, H., et al.: Fitzgerald trait. Deficiency of a hitherto unrecognized agent, Fitzgerald factor, par-

ticipating in surface-mediated reactions of clotting, fibrinolysis, generation of kinins, and the property of diluted plasma enhancing vascular permeability (PF/ Dil), J. Clin. Invest. **55**:1082, 1975.

126. Schiffman, S., Rapaport, S. I., and Chong, M. M. Y.: Platelets and initiation of intrinsic clotting, Br. J. Haematol. **24**:633, 1973.

127. Schrier, E. E., Broomfield, C. A., and Scheraga, H. A.: Molecular weight of bovine thrombin, Arch. Biochem. Biophys. **1**(suppl.):309, 1962.

128. Schwartz, M. L., et al.: The effect of fibrin-stabilizing factor on the subunit structure of human fibrin, J. Clin. Invest. **50**:1506, 1971.

129. Shapiro, S. S., and Martinez, J.: Human prothrombin metabolism in normal man and in hypocoagulable subjects, J. Clin. Invest. **48**:1292, 1969.

130. Sherry, S.: Urokinase, Ann. Intern. Med. **69**:415, 1968.

131. Sherry, S., Fletcher, A. P., and Alkjaersig, N.: Fibrinolysis and fibrinolytic activity in man, Physiol. Rev. **39**:343, 1959.

132. Shulman, S.: The size and shape of bovine fibrinogen. Studies of sedimentation, diffusion and viscosity, J. Am. Chem. Soc. **75**:5846, 1953.

133. Shuman, M. A., and Majerus, P. W.: The measurement of thrombin in clotting blood by radioimmunoassay, J. Clin. Invest. **58**:1249, 1976.

134. Smith, J. B., Ingerman, C., and Silver, M. J.: Persistence of thromboxane A_2–like material and platelet release–inducing activity in plasma, J. Clin. Invest. **58**:1119, 1976.

135. Smith, J. B., Ingerman, M., and Silver, M. J.: Prostaglandins and precursors in platelet function. In Biochemistry and pharmacology of platelets, Ciba Foundation Symposium 35, New York, 1974, Elsevier North Holland, Inc.

136. Steiner, M., and Baldini, M.: Protein synthesis in aging blood platelets, Blood **33**:628, 1969.

137. Stemerman, M. B., Baumgartner, H. R., and Spaet, T. H.: The subendothelial microfibril and platelet adhesion, Lab. Invest. **24**:179, 1971.

138. Suttie, J. W.: The effect of cycloheximide administration on vitamin K–stimulated prothrombin formation, Arch. Biochem. Biophys. **141**:571, 1970.

139. Tillett, W. S., and Garner, R. L.: The fibrinolytic activity of hemolytic streptococci, J. Exp. Med. **58**: 485, 1933.

140. Tishkoff, G. H., Williams, L. C., and Brown, D. M.: Preparation and properties of human prothrombin complex, Thromb. Diath. Haemorrh. **24**:325, 1971.

141. Tollefsen, D. M., Feagler, J. R., and Majerus, P. W.: The binding of thrombin to the surface of human platelets, J. Biol. Chem. **249**:2646, 1974.

142. Tranzer, J. P., and Baumgartner, H. R.: Filling gaps in the vascular endothelium with blood platelets, Nature **216**:1126, 1967.

143. Troll, W., Sherry, S., and Wachman, J.: The action of plasmin on synthetic substrates, J. Biol. Chem. **208**: 85, 1954.

144. Troup, S. B., et al.: Thromboplastic factors in platelets and red blood cells: observations on their chemical nature and function in in vitro coagulation, J. Clin. Invest. **39**:342, 1960.

145. White, J. B.: Platelet morphology. In Johnson, S. A., editor: The circulating platelet, New York, 1971, Academic Press, Inc.

146. Williams, W. J.: The activity of human placenta microsomes and brain particles in blood coagulation, J. Biol. Chem. **241**:1840, 1966.

147. Wright, J. H.: The origin and nature of the blood plates, Boston Med. Surg. J. **154:**643, 1906.

148. Wright, J. H., and Minot, G. R.: The viscous metamorphosis of the blood platelets, J. Exp. Med. **26:** 395, 1917.

149. Zetterquist, E.: Elimination of iodine-labelled fibrinogen in intravascular coagulation and in the normal state. Experimental and clinical studies, Acta Physiol. Scand. **325**(suppl.):1, 1969.

150. Zieve, P. D., Gamble, J. L., Jr., and Jackson, D. P.: Effects of thrombin on the potassium and ATP content of platelets, J. Clin. Invest. **43:**2063, 1964.

151. Zucker, M. B.: Platelet agglutination and vasoconstriction as factors in spontaneous hemostasis in normal, thrombocytopenic, heparinized and hypoprothrombinemic rats, Am. J. Physiol. **148:**275, 1947.

152. Zucker, M. B., and Borrelli, J.: Platelet clumping produced by connective tissue suspensions and by collagen, Proc. Soc. Exp. Biol. Med. **109:**779, 1962.

153. Zucker, W. H., and Mason, R. G.: Ultrastructural aspects of interactions of platelets with microcrystalline collagen, Am. J. Pathol. **82:**129, 1976.

X

THE KIDNEY AND
BODY FLUIDS

49

WILLIAM E. LASSITER and **CARL W. GOTTSCHALK**

Volume and composition of the body fluids

Water is by far the most abundant single constituent of the human body. Estimates of the water content of various individuals range from 45% to 75% of total body weight, the variations being largely a function of the amount of adipose tissue present. Because the water content of fat is relatively low (approximately 10%), the fraction of body mass that is water varies inversely with the degree of obesity, and individual variations in total body water are much smaller if it is computed as a fraction of *lean* body mass.

The importance of water to the organism derives not only from its abundance but also from the fact that it is the medium in which the body solutes are dissolved and in which all metabolic reactions take place. Metabolic processes may be profoundly influenced by small changes in the composition of the medium in which they occur; hence precise regulation of the volume and composition of body fluids is essential. Body water remains remarkably constant in the healthy individual, but acute changes in its volume or distribution may occur in disease states, with extremely damaging physiologic consequences. It should not be assumed from the constancy of its volume and composition that body water is a static system. On the contrary, a dynamic steady state exists, both between the individual and his external environment, and internally, between the various body fluid compartments.[19]

It is important in the clinical setting to be able to estimate changes in fluid balance quickly and accurately. The volume of total body water is often difficult to determine with precision, but acute *changes* in the water content of the body can be readily detected. A most sensitive indicator is found in the daily measurement of body weight. The distribution of body water between its various compartments may also be altered with or without a change in total body water. Recognition and estimation of these changes require an understanding of the physical factors governing the distribution of water in the body.

BODY FLUID COMPARTMENTS

Body water is distributed chiefly between two phases or compartments differing markedly in composition; these are the intracellular and extracellular fluid compartments. The intracellular fluid compartment includes the water and solutes contained within cells of the various tissues of the body and constitutes the medium in which chemical reactions of cell metabolism occur. The water contained within erythrocytes and other formed elements of the blood is included in this compartment. Extracellular fluid surrounds the cells and is the medium through which all internal metabolic exchanges (i.e., between cells and tissues) and exchanges with the external environment are made. The extracellular fluid compartment may be divided into several subcompartments, generally similar in electrolyte composition: plasma, interstitial fluid, lymph, and the extracellular portions of dense connective tissue, cartilage, and bone. The term ''transcellular fluid'' denotes fluids of varying composition that are separated from the remainder of the extracellular fluid by a layer of epithelial cells. This fluid is considered by some to be a part of the extracellular fluid compartment, but since the composition of various components of transcellular fluid may differ profoundly from that of the remainder of the extracellular fluid, it is perhaps best considered as being contained in a separate compartment or compartments. Water and solutes in the renal tubules and urinary bladder, cerebrospinal and intraocular fluids, and digestive secretions are included in this compartment, the total volume of which is usually small, constituting no more than 1% to 3% of body weight. Most techniques for estimating the volume of extracellular water inevitably measure, in addition, an indeterminate volume of transcellular water.

The relative volumes of the various body fluid compartments change with age and are different in the two sexes. Typical volumes of the various

1149

compartments, expressed as percent of body weight, are shown in Table 49-1. This table is intended to illustrate only the general magnitude of the body water compartments, and the reader is cautioned not to attach too much significance to the absolute numeric values tabulated. Wide variations may be observed in water content between individuals of differing habitus. In addition, as noted below, the estimation of water distribution is subject to numerous uncertainties, and measurements of the same compartment by different techniques may disagree. However, it is apparent that infants and very young children contain relatively more water per unit body weight than do adults, and the difference is most marked in the volume of extracellular fluid.[8] Children not yet in puberty have values intermediate between those tabulated for infants and for adult males, but the greatest changes in percent of water content occur in the first 2 years of life. The relatively lower water content of adult females reflects the greater development of subcutaneous adipose tissue that is characteristic of women.

Indicator dilution principle. Many problems have been encountered in attempts to estimate the volume of body water and its partition into various compartments. Only total body water and blood or plasma volume are even theoretically accessible to direct measurement. The water content of small animals has been estimated from the change in weight after desiccation, and blood volume has been determined by exsanguination. These procedures require destruction of the animal and are obviously proscribed in human studies. Water content of the body and the volumes of the various fluid compartments can be measured in humans and in experimental animals with varying degrees of precision by indicator dilution methods. The principle on which these measurements is based is simple, but difficulties are encountered in practical application. Suppose it is desired to measure the volume of water in an odd-shaped container of indeterminate dimensions. If a known quantity (Q) of an indicator substance is added to the water and sufficient time is allowed for it to become uniformly distributed, the volume (V) of water can be estimated by determining the concentration (C) of the indicator in a small aliquot of the solution, since $Q = VC$, and therefore:

$$V = \frac{Q}{C}$$

An indicator dilution measurement provides an accurate estimate of the volume of a body fluid compartment only if certain conditions are met:

1. The indicator must be uniformly distributed throughout the compartment in question and must not enter any other compartment, where it might be present in different concentration.
2. The concentration of the indicator in plasma, the fluid actually sampled, must be representative of the concentration in the entire compartment, and it must be possible to determine its concentration in plasma with a high degree of analytic precision.
3. If excretion or metabolic degradation of the reference substance occurs during the interval in which the measurement is made, it must be possible to estimate this accurately, so that an appropriate correction can be applied.

Obviously the indicator substance must also be nontoxic and must not itself induce changes in water distribution.

Correction for the loss of indicator from the body is often accomplished most conveniently by obtaining several plasma samples at timed intervals after injection of the indicator. Disappearance of the indicator from the circulation usually follows first-order kinetics, that is, its rate of disappearance is proportional to concentration, and a semilogarithmic plot of concentration against time will yield a straight line after the indicator is uniformly dispersed in its volume of distribution. This relationship is illustrated in Fig. 49-1. The beginning, nonlinear portion of the curve represents initial mixing and dispersal of the indicator after injection, and the linear portion shows the exponential decrease in

Table 49-1. Typical volumes of body fluid compartments (percent of body weight)

	Total body water	Intracellular fluid	Extracellular fluid		
			Plasma	Interstitial fluid	Total
Infants	75	45	4	26	30
Adult males	60	40	5	15	20
Adult females	50	35	4	11	15

plasma concentration as the result of excretion and/or metabolism. Extrapolation of this linear portion of the curve to zero time yields the plasma concentration of indicator that would have been found had there been instant and uniform mixing and no excretion or destruction. This method of correction is convenient and useful when the rate of disappearance of the indicator from the body is slow, but it is impractical if the indicator is a substance such as inulin that is rapidly excreted.

An alternative procedure may be employed with indicators that are rapidly excreted in the urine. After a suitable priming dose is given, a continuous intravenous infusion of the indicator is maintained until the plasma concentration reaches a constant level. This may take several hours. The infusion is then stopped, the bladder is evacuated, and all subsequently formed urine is collected over a period of time sufficiently long to ensure complete recovery of all the indicator. The amount excreted then equals the amount present in the body at the time the infusion was discontinued, and this quantity, divided by the plasma concentration at the moment the urine collection was started, is the volume of distribution. This method is useful only if the reference substance is excreted rapidly enough to permit total recovery, and only if the sole pathway of loss from the body is by urinary excretion.

Most substances used for estimation of body fluid compartments are water soluble and are distributed only in the aqueous phase of plasma; thus calculations must be based on the concentration of indicator in plasma water, not in whole plasma. If 7% of plasma volume is occupied by protein, the concentration of indicator in plasma water equals its plasma concentration divided by

0.93. If, in addition, the indicator is ionic (e.g., radiosulfate), the effects of the Donnan equilibrium on its distribution must also be considered. The Donnan effect is discussed in detail in a later section of this chapter.

Total body water. Substances commonly employed to estimate the volume of total body water include antipyrine, a compound that appears to be distributed rapidly and evenly throughout all the water of the body, and the two forms of heavy water, deuterium oxide (D_2O)[16] and tritiated water (HTO). D_2O is quantified by its effect on the specific gravity of water, and the concentration of HTO, which is radioactive, is usually estimated in a liquid scintillation counter. Both water isotopes distribute rapidly and uniformly in the total body water, and an equilibrium time of 2 hr or less is usually sufficient for a reasonably accurate determination of body water. Strictly speaking, these isotopes are indicators, not of the total volume of water, but of the apparent volume of distribution of all exchangeable hydrogen, including labile protons on the amino, carboxyl, hydroxyl, and phenolic groups of proteins and other organic molecules. One might thus expect the actual volume of water to be overestimated by as much as 3% to 4%. However, other sources of error exist that cannot be readily quantified but that limit the overall accuracy of the determination, and the correction is usually ignored.

Plasma and blood volumes. Two substances commonly employed for the determination of plasma volume are radioiodinated human serum albumin (Risa) and Evans blue dye.[17] Evans blue dye binds tightly to plasma albumin and assumes a distribution identical to that of the carrier molecule. The indicators are rapidly mixed in the circulation, and their distribution throughout the vascular compartment is uniform within 10 to 15 min. Actually the tagged albumin (as well as natural albumin) slowly escapes from the circulation, the loss approximating 3% to 4%/hr, but a correction for this loss can be readily made from a semilogarithmic plot of the plasma concentration against time, as previously described (Fig. 49-1). Plasma volume measured in this fashion is typically 4% to 5% of body weight.

Since the volume of circulating blood equals plasma volume plus red blood cell volume, blood volume may be estimated as follows:

$$\text{Blood volume} = \frac{\text{plasma volume}}{1 - \text{PCV}}$$

where PCV is the fractional packed cell volume, or hematocrit.

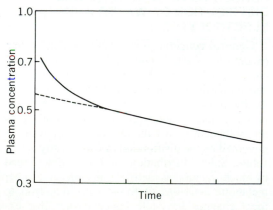

Fig. 49-1. Plasma concentration of indicator as function of time.

Although this formula gives an estimate that is accurate enough for most purposes, blood volume is actually somewhat overestimated. In laminar blood flow through minute peripheral vessels the red blood cells tend to be segregated in the more rapidly moving central stream, whereas plasma moves more sluggishly near the vessel wall. As a result, the mean velocity of transit of red blood cells through these small vessels is more rapid than that of plasma, and the volume of cells relative to plasma is smaller than in the larger arteries and veins. The average hematocrit of the entire blood volume is therefore overestimated when blood from a large vessel such as the antecubital vein is sampled. A more accurate estimate of blood volume is given by the sum of independent measurements of plasma and red blood cell volumes. Total red blood cell volume may be estimated by indicator dilution, using erythrocytes tagged with radioactive phosphorus, chromium, or iron.

Extracellular fluid volume. The extracellular compartment is conventionally defined as that volume of fluid which is external to the cells and within which there is (except for the colloids) diffusion equilibrium. This definition partially excludes transcellular fluid, and although it is probable that most dense connective tissue fluid is in diffusion equilibrium with the remainder of the extracellular compartment, the rate of equilibration is very slow. Penetration of indicator substances into such tissues may therefore be incomplete. Since there is not even uniform agreement as to the boundaries of the extracellular compartment, it should not be surprising that its volume cannot be measured with as much precision as total body water or plasma. The measured volume depends on the nature of the reference substance employed, and it is more precise, although physiologically less meaningful, to speak of the volume of distribution of a specific indicator rather than of the extracellular volume per se. Many substances have been used for the estimation of extracellular volume, including inulin, sucrose, mannitol, radiosulfate, thiosulfate, thiocyanate, bromide, radiochloride, and radiosodium. The large size of the inulin molecule limits its rate of diffusion through the interstitial fluid, and the apparent volume of extracellular fluid may be underestimated if insufficient time is allowed for uniform distribution of the indicator. The large inulin molecule may also be excluded from some of the smaller interstices, as, for example, in dense connective tissue. The distribution of sodium and chloride ions is not totally extracellular; therefore the use of radioisotopes of these ions leads to overestimation of the extracellular volume. Radiosulfate is considered to be the most satisfactory indicator for extracellular fluid volume[7,22] because it diffuses more rapidly and distributes more completely through tissue interstices than does inulin, is slowly excreted, and only very slowly enters cells. The volume of distribution of radiosulfate is believed to equal plasma volume plus that fraction of the interstitial fluid that is in ready diffusion equilibrium with plasma. Whether this is the same as the true anatomic extracellular space is not known. Although the magnitude of the extracellular volume cannot be determined with great accuracy, it may be assumed, as a reasonable and useful rule of thumb, that the extracellular fluid compartment includes somewhat more than one third, but distinctly less than half, of the total body water.[2]

INTERSTITIAL FLUID VOLUME. The interstitial fluid is defined as that portion of the extracellular fluid which is extravascular. Its determination is subject to the same uncertainties as that of intracellular fluid. A decrease in the relative volume of interstitial fluid is primarily responsible for the decrease with age in the fraction of body mass that is water. Interstitial fluid amounts to 20% to 25% of total body water in a typical adult.

Intracellular fluid volume. The volume of intracellular fluid cannot be measured directly but must be calculated as the difference between total body water and the extracellular volume. Its estimation is therefore limited by the uncertainties of both measurements. Underestimation of extracellular fluid volume will lead to a corresponding overestimation of intracellular water. Somewhat more than half and up to two thirds of the body water is within the cells.

FORCES GOVERNING INTERNAL DISTRIBUTION OF WATER AND ELECTROLYTES

General considerations. With certain limited exceptions, water is considered to be freely exchangeable between all body fluid compartments, and its partition is therefore determined primarily by physical forces, particularly hydrostatic pressure and osmotic pressure. A corollary to this statement is that the distribution of water is governed by the distribution of osmotically active solute. Solute distribution, in turn, is determined by such factors as active transport and other metabolic activities of living cells that lead to local changes in solute concentration, for example, the synthesis of large molecules or the metabolic degradation of large molecules into a larger number of small molecules. The passive distribution of ions across selectively permeable membranes must, in addition, obey the requirements of Donnan equilibrium.

Donnan equilibrium. Let us consider two compartments (I and II) separated by a membrane impermeable to protein but freely permeable to water and to smaller molecules and ions such as sodium and chloride (Fig. 49-2). If aqueous solutions of sodium chloride of differing concentrations are placed on either side of the membrane, the ions will exchange rapidly in both directions between the two compartments. The constraints of electrical neutrality dictate that there can be no *net* movement of cation from one compartment to the other unless there is a comparable movement of anion. If both solutions contain only ions that diffuse freely across the membrane (Fig. 49-2, *A*), the composition of the two solutions must ultimately become the same.

Now let us alter the system by adding protein to compartment I (Fig. 49-2, *B*). At the pH of body fluids, most protein molecules carry a net negative charge and are therefore anions. Because the membrane is impermeable to protein, these protein molecules are not free to diffuse into compartment II, and electrical neutrality requires the presence of sufficient excess of cations in compartment I to balance the charge on the protein molecules. Diffusible ions (Na^+ and Cl^- in this model) are still able to exchange freely across the membrane and will distribute themselves passively between the two compartments, but the electrical constraint imposed by the negative charge on the protein molecules ensures that at equilibrium the concentration of diffusible cation (Na^+) must always be greater, and the concentration of diffusible anion (Cl^-) must always be less on the side containing the nondiffusible anion (protein).

Donnan[9] was able to show from thermodynamic considerations that the activity product for each passively distributed diffusible cation-anion pair must be the same on either side of the membrane at equilibrium. (See Chapter 1 for derivation.) Thus in our example:

$$[Na^+]_I \cdot [Cl^-]_I = [Na^+]_{II} \cdot [Cl^-]_{II}$$

Furthermore, even though no active transport is involved, a stable electrical potential (E) exists across the membrane, negative on the side containing the impermeant anions. The magnitude of this potential is related to the passive distribution of diffusible ions and may be predicted by the Nernst equation. At 37° C the potential in millivolts is as follows:

$$E = -61 \log \frac{[Na^+]_I}{[Na^+]_{II}} = -61 \log \frac{[Cl^-]_{II}}{[Cl^-]_I}$$

Although the *product* of the concentrations of any pair of diffusible ions must be the same on either side of the membrane at equilibrium, it can be shown mathematically that if impermeant ions are present, the *sum* of the concentrations of *diffusible* ions must be greater on the side containing the impermeant species. Thus the osmotic pressure of solution I exceeds that of II in our example, not only because of the presence of the nondiffusible protein molecules, but also because of a greater concentration of diffusible ions ($[Na^+] + [Cl^-]$) on that side. Net water movement must necessarily occur from compartment II to compartment I, unless there is an effective counterforce (hydrostatic pressure) balancing the difference in osmotic pressure. As we shall see, the nonuniform distribution of diffusible ions across the capillary wall contributes importantly to the effective colloid osmotic pressure of plasma.

Osmotic pressure. If two aqueous solutions are separated by a membrane that is permeable to water, an exchange of water molecules will occur between them. The tendency of water to pass from one solution to the other is designated its "escaping tendency" and is a function of its vapor pressure, or chemical potential. If the chemical potential of the water of the two solutions is different, water will move from the solu-

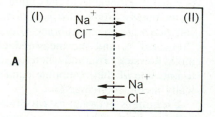

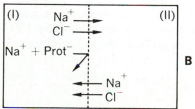

Fig. 49-2. Donnan equilibrium. (See text.)

tion of higher to that of lower potential until equilibrium is reached. The chemical potential of water is increased when the hydrostatic pressure is elevated or the temperature rises, and it is reduced by the presence of solutes. Therefore if two solutions of differing solute concentration are separated by a membrane that is permeable to water but that does not permit passage of the solute, water will move from the solution of lower solute concentration (or osmolality) to that of higher osmolality until the solute concentrations on the two sides of the membrane are equal, or until the hydrostatic pressure rises on the side of higher osmolality to a level that exactly balances the difference in osmotic pressure. At this point the chemical potential of water on either side of the membrane is the same, and the net water flux across the membrane is zero. This leads to the following operational definition of osmotic pressure: The osmotic pressure of an aqueous solution is equal in magnitude to the hydrostatic pressure that must be applied to the solution to restore the chemical potential of the solvent (water) to that of pure water at the same temperature. It is important to understand that the osmotic pressure of a solution is a function of the chemical potential of the solvent, and that this in turn depends on the concentration, but not the nature, of the dissolved solute(s).

Substances in true solution are uniformly distributed throughout the solution and behave in many respects as if they were molecules of a gas, and the magnitude of the osmotic effect of dissolved solutes can be calculated from a modification of the gas law.[21] For an ideal gas:

$$PV = nRT$$

where P is the pressure in atmospheres, V the volume of gas in liters, n the number of moles, and T the absolute temperature in degrees Kelvin. The gas constant, R, has the value of 0.082 L atm/mole degree.

Another property of gases that is applicable to solutions is described by Dalton's law of partial pressures, which states that in a mixture of gases the total pressure of the gas mixture equals the sum of the partial pressures of each of its components, provided the components do not interact. The partial pressure of any component (X) of a gas mixture is given by the following equation:

$$P_X = \frac{N_X RT}{V} = C_X RT$$

where C_X is the concentration of gas X in moles per liter. The equivalent equation for osmotic pressure (π) is:

$$\pi = CRT$$

where C is the concentration of dissolved solute.

At body temperature (37° C or 310° K) and at a pressure of 1 atm, the volume occupied by 1 mole of an ideal gas is 25.4 L. By analogy, in an ideal solution 1 mole of solute dissolved in 25.4 L (or more precisely, 25.4 kg) of water will exert a partial or osmotic pressure of 1 atm, and 1 mole of solute dissolved in 1 kg of water will exert an osmotic pressure of 25.4 atm. The chemical potential of the solvent (in this case water) is reduced by an equivalent amount, and a hydrostatic pressure equal to this must be applied to the solution to restore the chemical potential of the solvent to that of pure water.

It is apparent that the osmotic pressure of biologic solutions must be enormous. A solute concentration of 1 mOsm/kg water is equivalent to a pressure of 19.3 mm Hg at 37° C, and the absolute osmotic pressure of a solution with a solute concentration of 300 mOsm/kg (typical of body fluids) is therefore nearly 6,000 mm Hg.

In practice, we are not often concerned with the absolute level of osmotic pressure but rather with changes in osmolality or with differences in osmotic pressure between two solutions. No net water transfer will occur across a selectively permeable membrane placed between two solutions of identical total solute concentration, even if the nature of the solutes is different. Biologic membranes are rarely completely selective in their permeability properties, however, and to the extent that a solute is able to cross the membrane, its contribution to an observed osmotic pressure difference (P*) will be less than that predicted from its chemical activity in the solution. A quantitative measure of this is given by the reflection coefficient, σ,[20] and the observed osmotic pressure difference (P*) is related to the theoretical difference (π) as follows:

$$P^* = \sigma \pi$$

For most solutes and membranes of biologic interest, σ has a value between 0 and 1. A reflection coefficient of 0 indicates that the membrane is equally permeable to the solute and to water. At the other extreme, if σ is 1, none of the solute can penetrate the membrane (i.e., it is totally "reflected"), and the measured change in osmotic pressure observed when the solute is added to one side of the membrane equals the theoretically predicted change (π).

Exchanges between extracellular and intracellular fluid. The ionic composition of cell water differs markedly from that of extracellular fluid, but the overall concentration of osmotically active solute is essentially the same in the two fluids. Most animal cells are more or less freely permeable to water, and since no appreciable hydrostatic pressure gradients can be maintained across animal cell membranes, it follows that the osmotic pressure, and hence the total concentration of osmotically active solutes, must be the same within the cells as in the surrounding extracellular fluid. The distribution of water between the extracellular and intracellular compartments therefore corresponds to their content of osmotically active solute. Because the concentration of protein and other nondiffusible solutes is so much greater in cells than in extracellular fluid, the cells would inevitably swell until they burst if there were not a mechanism for limiting their solute content via the active extrusion of ions.

Although most animal cells exist in osmotic equilibrium with their extracellular fluid environment, there are exceptions. Many secretory cells, for example, elaborate a secretion that is not precisely isosmotic; in the kidney the cells of the ascending limb of the loop of Henle and, in the absence of antidiuretic hormone, the distal tubule and collecting ducts are relatively impermeable to water and are bathed on opposite surfaces by fluids of differing osmolality. The special permeability characteristics of these cells are essential to the elaboration of urine that has a higher or lower total solute concentration than extracellular fluid (Chapter 50).

The amount of osmotically active solute in cells is not normally subject to rapid alteration, and changes in extracellular osmolality ordinarily induce the net transfer of water from one compartment to the other until osmotic uniformity is restored. This is illustrated by the simplified system depicted in Fig. 49-3. Let us assume that compartment I is enclosed by a collapsible membrane that is completely impermeable to glucose,

sodium, and chloride, but is freely permeable to urea and, of course, water. If the fluid enclosed by the membrane (I) contains glucose at a concentration of 300 mOsm/kg and the outer chamber (E) is filled with a solution of sodium chloride at the same osmolality (Fig. 49-3, *A*), the two compartments will be in osmotic equilibrium. Although water molecules may pass rapidly in both directions across the membrane, unidirectional fluxes in each direction will be equal, and there will be no change in the volume of either compartment. Now let us suppose that additional crystalline sodium chloride is added to the outer compartment and is brought into solution (Fig. 49-3, *B*). Because of the higher solute concentration, the chemical potential of the water in compartment E is reduced, and a net flow of water from compartment I to compartment E occurs. This water movement causes the concentration of glucose in the inner compartment to increase and that of sodium and chloride in the outer compartment to decrease; this will continue until the total solute concentrations in the two compartments are again equal (Fig. 49-3, *C*). Note that the final concentration of sodium chloride in compartment E is less than it would have been had not the volume of this compartment been increased by the translocation of water from the inner compartment. Addition of nondiffusible solute to the outer compartment therefore produced an increase in the size of the outer compartment at the expense of the inner compartment, and the solute concentration increased equally in *both* compartments.

The sequence of events would have been different had the added solute been urea. Since in this model the membrane was defined as being freely permeable to urea (i.e., reflection coefficient equal to 0), the added urea would ultimately become distributed uniformly between the two compartments, and at equilibrium the

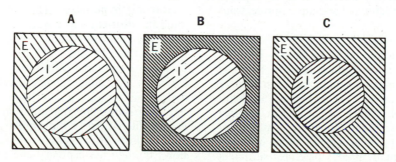

Fig. 49-3. Principle of osmotic uniformity. Compartments *E* and *I* are separated by collapsible membrane that is freely permeable to water and urea, but completely impermeable to Na^+, Cl^-, and glucose. (See text.)

volumes of the compartments would be unchanged.

The requirement of osmotic uniformity between intracellular and extracellular fluid has important implications in the clinical management of disturbances of hydration. Since most of the solute of extracellular fluid is sodium chloride, the concentration of sodium in plasma is a reasonably accurate index of extracellular fluid osmolality. If sodium chloride and water are lost from the body in isotonic proportions, the extracellular fluid osmolality is unchanged, there is no net transfer of water between the extracellular and intracellular compartments, and the entire volume of the fluid loss is at the expense of the extracellular fluid volume. On the other hand, if water is lost without solute, as in evaporation, the increase in extracellular fluid osmolality causes water to move out of the intracellular compartment, and the fluid loss is distributed uniformly throughout the total body water. Loss of extracellular solute without a proportionate loss in water decreases the solute concentration of extracellular fluid, and water must move from the extracellular compartment into the cells until osmotic uniformity is restored; thus the extracellular fluid volume may be decreased while at the same time there is intracellular overhydration.

A short clinical example will be instructive. The plasma sodium concentration is closely regulated and seldom deviates far from normal, even in disease, but occasionally a dangerous level of hyponatremia requiring prompt correction may appear. One of a number of conditions in which this may occur is Addison's disease, or primary adrenal insufficiency. This is a complex syndrome, one feature of which is failure to produce adequate amounts of the hormone aldosterone. This in turn leads to decreased ability of the kidney to conserve sodium and, if salt intake is inadequate, to a decrease in the plasma sodium concentration. This hyponatremia may be corrected if necessary by infusion of an appropriate amount of a concentrated sodium chloride solution.

Suppose it is required to raise the plasma sodium concentration of an 80 kg man by 10 mM/kg water. How much sodium must be infused? An increase in plasma sodium concentration of 10 mM/kg is equivalent to an increase in extracellular fluid osmolality of 20 mOsm/kg. An increase in extracellular osmolality of this magnitude requires a similar increase in the osmolality of intracellular fluid; therefore sufficient solute must be given to raise the solute concentration of the entire body water by 20 mOsm/kg.

Let us assume that the total body water of this patient (60% of body weight) is 48 kg. Then $48 \times 20 = 960$ mM of solute are required. This could be provided by the administration of 480 mM of sodium chloride. Although the administered solute is largely confined to the extracellular space, its effect on osmolality is the same as if it had been distributed in the entire body water. Of course, the intracellular solute concentration increased as water moved out of the cells into the extracellular compartment.

Exchanges between plasma and interstitial fluid. The distribution of extracellular fluid between the vascular and extravascular subcompartments is determined primarily by the balance between hydrostatic and osmotic forces at the level of the blood capillaries. The most important of these forces are the capillary hydrostatic pressure, which promotes filtration of water across the capillary wall, and the plasma colloid osmotic pressure, which opposes it. Most ions and other small molecules are more or less freely diffusible across the capillary wall and contribute little to the effective osmotic pressure of plasma. On the other hand, the capillaries are relatively impermeable to protein, and the concentration of protein in the interstitial fluid of most organs is presumed to be normally quite low; hence virtually the entire colloid osmotic pressure of the plasma proteins is effective in opposing loss of fluid from the vascular compartment into the interstitium. At body temperature (37° C) a solute concentration of 1 mOsm/kg exerts an osmotic pressure of 19.3 mm Hg. Since the concentration of plasma proteins is about 0.9 mM/kg, the osmotic pressure of the proteins alone is approximately 17 mm Hg. In addition, because of the Donnan equilibrium, the concentration of diffusible ions in plasma water exceeds that in interstitial fluid by about 0.5 mM/kg, contributing an additional osmotic pressure of 9 or 10 mm Hg and bringing the effective colloid osmotic pressure of plasma to the range of 25 to 30 mm Hg.

Starling hypothesis. The forces that determine the distribution of water between the vascular and interstitial fluid compartments were first described by Starling[18] in 1896 and are illustrated diagrammatically in Fig. 49-4. Starling assumed that blood enters capillaries with a hydrostatic pressure greater than the colloid osmotic pressure opposing filtration and that fluid is thus filtered out of the capillary at its arterial end. Hydrostatic pressure decreases as the blood flows through the capillary and is less than the colloid osmotic pressure at the venous end; hence the balance of forces favors reabsorption at that end of the capillary. Although Starling was probably correct in his basic premise, it is quite certain that the pressure relationships in all capillaries are not at all times precisely as he described. It is more likely that in capillaries in which the hydrostatic pressure is relatively high, filtration occurs along the entire length of the capillary, whereas in other capillaries, or at other times in the same capillary, the pressure is lower and the capillary functions as a reabsorptive channel.[12,13,24] Pulsatile variations in capillary hydrostatic pressure, furthermore, may produce oscillatory flow back and forth across the capillary wall.

It should not be assumed that bulk filtration and reabsorption of fluid by capillaries is the means by which oxygen and other nutrients are delivered to cells and waste products removed. On the contrary, Pappenheimer[14] has shown that the rate at which fluid circulates in this fashion is quite small. Bidirectional diffusion of water and solutes across the capillary wall, in contrast, occurs at phenomenally high rates, and it is chiefly through this means that exchange of nutrients and waste products between cells and capillaries is accomplished. Although bulk filtration and reabsorption are small in relation to total water and solute exchange across the capillaries, a sustained distortion in the balance of Starling forces can nonetheless lead in time to a marked distortion in the distribution of the extracellular fluid.

Smaller, but at times significant, forces influencing transcapillary exchange are the colloid osmotic pressure of the interstitial fluid, which favors filtration, and tissue hydrostatic pressure. This tissue pressure opposing filtration is generally thought to be small, only 2 to 5 mm Hg, and Guyton et al.[11] have even suggested that the tissue pressure is negative (i.e., less than atmo-

spheric pressure) and hence tends to facilitate filtration.

Although all would agree that tissue pressure is ordinarily small, it occasionally becomes an important factor in determining the distribution of edema in pathologic states associated with the abnormal accumulation and retention of extracellular fluid. The skin of the palm of the hand, for example, is very tightly bound down by dense bands of fibrous tissue, and a very small increase in the volume of interstitial fluid in this area is accompanied by a marked increase in tissue pressure that limits the further extravasation of fluid; thus clinically detectable edema is seldom seen here. However, the tissues on the dorsum of the hand are quite loose, and a large volume of fluid may accumulate here before further extravasation is limited by the rise in tissue pressure.

Lymphatics. Blood capillaries are relatively impermeable to protein, but small amounts do leak across, and if there were no mechanism for returning this protein to the plasma, its concentration in interstitial fluid would eventually equal that in plasma, reducing the effective colloid osmotic pressure of the plasma to 0. This does not occur because protein that has leaked into the interstitial fluid is able to gain access to lymphatic channels. The volume of fluid passing through intercellular spaces and lymphatic channels is always significant but is greatly increased when interstitial fluid is expanded. Movement of fluid along the lymph channels is promoted by muscular contraction and tissue movements. Smaller channels progressively coalesce into larger vessels, and the lymph passes ultimately into the thoracic duct, which empties into the venous circulation.

This mechanism for returning protein from interstitial fluid to plasma is disrupted when the lymphatic channels are interrupted surgically or

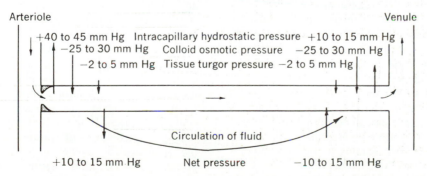

Fig. 49-4. Starling hypothesis of fluid distribution between blood plasma and interstitial fluid compartments. (From Pitts.[15])

obstructed by inflammatory disease, and massive edema of the affected part of the body may result. A dramatic example of the latter is the tropical parasitic disease elephantiasis. A common example of surgically induced lymphedema is that seen in women after radical surgery for breast cancer. In this procedure, lymph nodes in the axilla, as well as the primary tumor, are removed, and the attendant damage to the lymph vessels in the region may result in chronic edema of the arm.

Mechanisms of edema formation. In its most general sense the term "edema" refers to the abnormal swelling of a tissue without regard to whether the excess fluid is anatomically extracellular or intracellular. The most commonly encountered forms of edema, however, involve expansion of the extracellular volume, and edema is defined operationally as an abnormal accumulation of extravascular interstitial fluid of sufficient magnitude to be clinically detectable. It may occur in a wide variety of unrelated pathologic states, but in every instance the proximate cause is a disturbance in the balance of the Starling forces governing the distribution of fluid across the capillaries. Thus increased capillary hydrostatic pressure, decreased colloid osmotic pressure, increased permeability of capillaries to protein, and obstruction to lymphatic flow may all be associated with edema formation. Edema may be sharply circumscribed in its distribution, as, for example, that associated with local inflammation, or it may be more widespread. Generalized edema is invariably associated with the abnormal retention of sodium and water and expansion of the extracellular fluid volume.

Perhaps the simplest example of edema formation is that seen in the lower extremities after prolonged periods of quiet standing or the edema associated with varicose veins in the legs. In both these circumstances the return of blood through the veins in the legs is impaired and the hydrostatic pressure at the venous end of the capillaries is abnormally high. A similar situation exists in congestive heart failure, in which inability of the heart to pump all the blood presented to it leads to pooling of blood in the venous circulation and an increase in peripheral venous pressure. The pathogenesis of cardiac edema is considerably more complicated than this, however, since as a homeostatic response to decreased cardiac output, mechanisms are activated that promote salt and water retention, leading to a generalized increase in extracellular fluid volume. This increased volume is shared by both the vascular and the interstitial fluid compartments.

Localized edema associated with inflammation or with obstruction or destruction of lymphatic channels can also be understood in terms of the disturbance in the balance of Starling forces. In local inflammation the permeability of capillary walls to protein is increased, and more rapid leakage of protein into the interstitial space is permitted. Accumulation of protein in the interstitial space provides an osmotic force that opposes the colloid osmotic pressure of plasma, and edema fluid accumulates in the interstitium. A similar mechanism is responsible for edema formation when lymphatic channels are blocked, since in the absence of lymph flow, no mechanism exists for the removal of the small amounts of protein that normally leak out of the capillaries into the interstitial space.

Nephrotic syndrome. Renal disease is commonly associated with leakage of protein into the urine. In certain forms of glomerular disease the loss of protein by this route may be so great that the concentration of protein, especially albumin, in the plasma falls to subnormal levels, resulting in a condition known as the nephrotic syndrome. This is a symptom complex characterized by massive proteinuria, hypoalbuminemia, increased plasma cholesterol concentration, and edema. Reasons for the increase in plasma cholesterol concentration are not fully understood, but edema formation is clearly a consequence of the altered balance of Starling forces. The following sequence of events is thought to occur. Massive loss of protein in the urine leads to a reduction in plasma albumin concentration and hence in the effective colloid osmotic pressure of plasma. (It should be apparent to the reader that a reduction in the rate of albumin synthesis, because of poor nutrition or liver disease, might have a similar effect, and indeed hypoalbuminemia and edema are frequently seen in such circumstances.) Redistribution of extracellular fluid follows, with a decrease in vascular volume and an increase in interstitial fluid volume. Homeostatic mechanisms are then called into play that tend to restore vascular volume toward normal; thus retention of sodium and water is promoted, and the total volume of extracellular fluid is increased. Because of the low plasma colloid osmotic pressure, a disproportionate share of the retained fluid enters the interstitial space, and edema increases. If the underlying renal defect that permitted loss of protein is corrected, the sequence of events is reversed. Plasma protein concentration returns to normal, and a diuresis of the retained salt and water ensues.

ELECTROLYTE COMPOSITION OF BODY FLUIDS

Plasma. The composition of blood plasma varies only slightly from individual to individual or from day to day in a healthy individual. Average normal values for the several ionic components of plasma and other body fluids are summarized in Table 49-2. The major cation in plasma is sodium, and the major anions are chloride, bicarbonate, and protein. Certain ions (Ca^{2+} or Mg^{2+}, for example) may carry more than one charge, and whereas 1 mM of a univalent cation such as sodium or potassium equals 1 mEq, 1 mM of a divalent cation equals 2 mEq. Because each protein molecule contains multiple charges, the contribution of proteins to electrical charge balance is much greater than their contribution to the plasma osmolality. Electrical neutrality requires only that the number of positive charges equals the number of negative charges, not that the number of particles bearing positive charge (cations) equals the number of anions. Since each protein molecule carries several negative charges, electrical neutrality obligates the presence of several positive ions for each protein molecule, and the molar concentration of cations in plasma is greater than the concentration of anions.

Protein molecules, because of their large size, occupy a volume that is far out of proportion to their molar concentration, and at normal protein concentrations, water represents about 93% of the total volume of the plasma. Many ions and other small molecules are distributed only in the aqueous phase of the plasma, and their chemical activities are a function of their concentrations in the plasma water. The concentrations in plasma water of several of the ionic species, including sodium, potassium, and chloride, may be estimated by dividing plasma concentration by 0.93. Certain ions, of which calcium is an example, are reversibly bound to plasma proteins, and the concentration of free ions in the aqueous phase may be lower than the total concentration in plasma as determined by chemical analysis.

Interstitial fluid. Interstitial fluid is an ultrafiltrate of plasma, and as such its composition is similar. One might expect the concentration of freely diffusible solutes in interstitial fluid to be identical to that in plasma water, but this is true only for uncharged particles, since the distribution of ions across the capillary wall is subject to the constraints of the Donnan equilibrium. The concentrations of diffusible cations are somewhat lower, and of anions higher, in interstitial fluid than in plasma water because of the Donnan effect (Table 49-2). Interstitial fluid cannot normally be sampled in amounts sufficient for extensive chemical analysis. However, edema fluid, lymph, and glomerular filtrate are thought to closely resemble normal interstitial fluid in composition, and the detailed analysis of these fluids agrees well with the predictions of the Donnan equilibrium.

Intracellular fluid. The intracellular compartment is not homogeneous, and no description of its composition can be given that is generally applicable to all cells. Important differences

Table 49-2. Electrolyte composition of body fluids*

	Plasma		Interstitial fluid (mEq/kg H_2O)	Cell water of muscle (mEq/kg H_2O)
	mEq/L	mEq/kg H_2O		
Cations				
Na^+	138	148	141	10
K^+	4	4.3	4	150
$Ca^{2+} + Mg^{2+}$	7	7.5	(6)	40
Total	149	160	151	200
Anions				
Cl^-	102	110	115	—
HCO_3^-	26	28	29	10
Protein	15	16	—	40
Others	6	6.5	(7)	150
Total	149	160	151	200

*Plasma water content of 93% is assumed. Donnan ratio (interstitial fluid:plasma water) is 0.95 for monovalent cations and 1.05 for monovalent anions. Donnan ratio for Ca^{2+} and Mg^{2+} is 0.90, but computation of concentration of these ions in interstitial fluid is inexact because of protein binding. Average net charge per ion, and therefore mean Donnan ratio, is uncertain in the case of undetermined anions.

exist, for example, in the composition of red blood cells, which contain appreciable amounts of chloride, and muscle cells, which are virtually chloride free. Moreover, cells exhibit a high level of structural organization, and the distribution of ions is not uniform even within individual cells. The gross characteristics of intracellular fluid composition can be derived only indirectly, since intracellular fluid is not available for direct analysis.

As an example of the sort of analytic approach that must be employed, suppose one wishes to estimate the average sodium concentration in intracellular fluid. The volumes of total body water and of extracellular fluid can be estimated with conventional indicator dilution techniques, although, as we have seen, these measurements, especially the estimation of extracellular volume, are subject to considerable uncertainty. The total amount of sodium in the body that is freely exchangeable (this excludes sodium tightly bound in the crystal structure of bone) can be estimated by administering a known quantity of $^{22}Na^+$ or $^{24}Na^+$ and measuring the specific activity (radioactivity per millimole) and plasma sodium concentration after a suitable interval of time. Total exchangeable sodium (Na_t) equals the intracellular (Na_i) plus extracellular (Na_e) sodium concentrations, and the latter can be calculated from the volume of extracellular water (ECW) and plasma sodium concentration, after appropriate corrections for plasma water and the Donnan effect. Intracellular sodium concentration is then calculated as follows:

$$[Na]_i = \frac{Na_t - ECW[Na]_e}{TBW - ECW}$$

Because intracellular sodium, a small quantity, is estimated as the difference between two much larger quantities, total sodium and extracellular sodium, the calculation is subject to large uncertainties.

The average composition of cell fluid in small samples of tissue, such as pieces of muscle, can be determined somewhat more accurately, provided an appropriate correction can be made for the amount of extracellular water in the specimen. A typical analysis of the composition of muscle cell water is given in Table 49-2.

Certain features of the composition of all cells are qualitatively similar and serve to distinguish cell water from extracellular fluid. The major cations of intracellular fluid are potassium and magnesium, and the concentration of sodium is always relatively low. Sodium, by contrast, is the chief cation of extracellular fluid. Chloride,

the most abundant anion of extracellular fluid, is present in much lower concentration in cell water and is thought to be virtually excluded from muscle cells. Bicarbonate also is usually present in much lower concentration in intracellular fluid than in extracellular fluid, reflecting the lower intracellular pH. The chief anions of cell fluids are proteins and organic phosphates and sulfates.

Cells contain high concentrations of nondiffusible anions, and, as should be expected, the Donnan equilibrium influences the composition of intracellular water. But the distribution of ions between intracellular and extracellular fluid is not governed solely by the Donnan equilibrium. The Donnan effect, if unopposed, would require a much greater concentration of diffusible ions inside cells than in the extracellular water, and osmotic forces should then favor the continuous entry of water into the cells. Because the animal cell membrane is incapable of sustaining a large opposing hydrostatic gradient, cells would inevitably swell up until they burst. This does not occur because of the active extrusion of solute, chiefly sodium, by transport processes that require the continuous expenditure of energy. In most cells, this active outward transport of sodium appears to be linked (although not necessarily in a 1:1 fashion) to inward transport of potassium, leading to the accumulation of potassium in intracellular fluid. Both of these active processes are opposed by the tendency for sodium to move passively down its concentration gradient from the extracellular fluid into the cells and for potassium to leak out of the cells along its concentration gradient. The cell membrane is far more permeable to potassium than to sodium, however, and the tendency of positively charged potassium ions to leak out of the cell, leaving behind negatively charged nondiffusible anions, is considered to be largely responsible for the electrical potential across the cell membrane, which can reach values as high as 80 to 90 mV (inside negative). This large potential difference serves to virtually exclude diffusible anions such as chloride from the cell's interior. A more detailed discussion of the relationship between membrane potentials and ion distribution is found in Chapter 1.

EXTERNAL EXCHANGES OF WATER AND ELECTROLYTES

The remarkable day-to-day constancy of the volume of body water is the consequence of a dynamic balance between intake and output. The major routes of water intake and excretion and the contribution of each to water balance in a

Table 49-3. Normal routes of water intake and output in adults (in milliliters/day)*

Intake		Output	
Water and beverages	500-1,600	Urine	600-1,600
Food water	800-1,000	Evaporation	850-1,200
Water of oxidation	200- 400	Feces	50- 200
TOTAL	1,500-3,000	TOTAL	1,500-3,000

*Adapted from Muntwyler.[5]

typical normal adult are summarized in Table 49-3. Under ordinary circumstances the loss of water in the feces and by evaporation is very nearly equal to intake derived from the water of oxidation and the water content of solid foods. Evaporative and fecal water losses are obligatory and cannot be directly altered to maintain water balance, but the voluntary intake of water can be varied within relatively wide limits, and urine volume is subject to homeostatic regulation.

Insensible perspiration—evaporation. An appreciable volume of solute-free water is lost via the lungs and respiratory passages and by evaporative loss through the skin. The expired air is saturated with water vapor at the temperature of the respiratory passages with which it last comes in contact, and the magnitude of water loss by this route depends on the temperature and humidity of the inspired air, the body temperature, and the volume of respiratory exchange. Additional water is lost by simple evaporation through the skin. The rate of evaporative loss by these routes is a function of the chemical potential, or vapor pressure, of extracellular water. Its magnitude is therefore increased by an increase in body temperature and decreased when the osmolality of body fluids rises, as in progressive dehydration. It should be understood that this "insensible perspiration" represents a loss of pure water, without solute, from the body fluids. It is not to be confused with sweat, which contains solutes.

Evaporative losses in normal adults average close to 1 L/day, but the loss may be greatly augmented by fever or vigorous exercise. Water loss in the adult is increased by 100 to 150 ml/day for each degree Celsius rise in the body temperature above normal. Evaporation also varies under extreme environmental conditions. Insensible water loss in the normal adult approximates 42 ml for each 100 calories metabolized, and since the heat of vaporization of water at 37° C is 575 calories/kg, it can be demonstrated that dissipation of heat through evaporative water loss accounts for approximately one fourth of the total heat production of the body. Evaporation is

Table 49-4. Composition of thermal sweat in 19 healthy male subjects*

Solute	Range (mM)	Average (mM)
Na	9.8-77.2	47.9
K	3.9- 9.2	5.9
Cl	5.2-65.1	40.4
NH_3	1.7- 5.6	3.5
Urea	6.2-12.1	8.6

*From Amatruda and Welt.[6]

therefore an important factor in the regulation of body temperature. The rate of evaporation is proportional to body surface area, and insensible losses are thus relatively greater in infants and young children, in whom the ratio of body surface to volume is large, than in adults. This is a factor in the more rapid development of severe dehydration in infants and children when water intake is curtailed.

Sweat. Sweat, or sensible perspiration, is secreted by the sweat glands of the skin and differs from insensible perspiration in that it always contains solute, both electrolytes and urea, in significant concentrations. The magnitude of fluid loss by this route is negligible in an individual who is resting quietly in a cool environment, but the losses may become quite large when problems of heat dissipation are augmented, as with increased environmental temperature, fever, or vigorous exercise. Sweat losses as great as 8 to 10 L/day have been recorded in outdoor workers in the summertime.[23] The volume of sweat is determined primarily, but not exclusively, by requirements for heat regulation. The induction of hypertonicity of body fluids, whether by dehydration or by infusion of hypertonic salt solution, is accompanied by a decrease in the rate of sweat secretion.

Sweat qualitatively resembles extracellular fluid in its composition, with sodium and chloride the major ionic components, but it is almost invariably hyposmotic in comparison to plasma, and its composition can be quite variable, as seen in Table 49-4. The relative concentrations of sodium and potassium appear to be partially under the control of adrenal cortical hormones, particularly aldosterone.

The importance of adequate and appropriate fluid replacement under conditions of active sweating must be emphasized. Since sweat is hypotonic, failure to provide replacement leads to hypertonicity of body fluids and a decrease in both extracellular and intracellular fluid volumes. However, because the solute that is lost is primarily from the extracellular fluid, this compartment bears a disproportionate share of the loss in volume; if sufficiently severe, this loss may lead to vascular collapse. Adequate water replacement (driven by thirst) without replacement of the lost solute will result in a reduction in the total solute concentration of body fluids and a shift of water from the extracellular to the intracellular compartment, leaving the individual with a contracted extracellular volume but an expanded intracellular volume, or cellular edema.

Gastrointestinal secretions. Net water loss via the gastrointestinal tract is ordinarily small, averaging only 100 to 150 ml/day, but the daily unidirectional exchanges of water and electrolytes between the extracellular compartment and gastrointestinal tract are quite large. The volumes of saliva and other gastrointestinal secretions in an average adult are shown in Table 49-5. Ordinarily, most of this fluid is reabsorbed and there is little or no net loss, but obviously the gastrointestinal tract may become an important avenue of water and electrolyte loss if reabsorption is impaired. Except for saliva, which is hypotonic, the total solute concentration in most gastrointestinal secretions is similar to that of extracellular fluid, but marked differences exist in the specific composition of secretions from various parts of the gastrointestinal tract, and the consequences of fluid loss via the gastrointestinal tract depend on the composition of the fluid that is lost. The approximate compositions of various gastrointestinal secretions are shown in Fig.

Table 49-5. Daily volume of gastrointestinal secretions in a typical adult*

Secretion	Volume (ml)
Saliva	1,500
Gastric secretions	2,500
Bile	500
Pancreatic juice	700
Intestinal secretions	3,000
TOTAL	8,200

*From Gamble.[10]

49-5. It is obvious that large losses from the gastrointestinal tract, which may occur from vomiting, diarrhea, or drainage from an intestinal fistula, unless adequately replaced with fluids of appropriate composition, will lead rapidly not only to profound contraction of the extracellular fluid volume but also to marked distortion of its electrolyte composition.

Renal excretion of water. The daily losses of water via the gastrointestinal tract and through evaporation are largely unavoidable and uncontrollable. Furthermore, sweating occurs, not in response to the demands of water balance, but chiefly as a mechanism of temperature control. The kidneys therefore are the principal organs of regulation of the volume and solute composition of the body fluids. Under normal circumstances the volume of urine can be varied over a wide range, and the rate of solute excretion regulated independently in response to the requirements of water and electrolyte balance. Conditions of stress may be associated with an increase in the obligatory urinary solute load, however, and the ability of the kidneys to limit water excretion under these circumstances may be impaired. The ability to vary urine volume and composition in response to need is likewise diminished in advanced renal disease, and under these conditions the control of water and solute intake becomes more critical if balance is to be maintained. Regulation of the renal excretion of water and solute is considered in detail in Chapter 50.

Water intake. The amount of water ingested daily by humans is highly variable. Approximately half the intake of the typical adult results from the voluntary drinking of water and other beverages; the remainder is derived from the

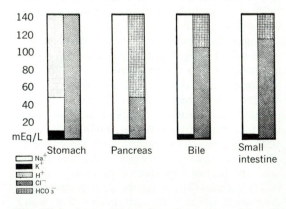

Fig. 49-5. Approximate electrolyte composition of gastrointestinal secretions. (From Welt.[23])

water content of food and from metabolic water formed in the process of oxidation of foods (Table 49-3). The water content of solid foods is seldom less than 40% and may be more than 90% in certain vegetables and fruits. In a typical diet, as much as 1 L of water is ingested daily as a component of "dry" food.

The amount of water formed as a result of metabolic processes depends on the nature of the substrate being oxidized. Complete oxidation of 100 gm of fat yields about 100 ml of water; oxidation of a similar quantity of carbohydrate or protein yields 60 or 45 ml water, respectively. If the availability of water is limited, it is obvious that water balance can best be maintained with a diet high in fat and carbohydrate and low in protein. In addition to the differences in the yield of water of oxidation, the chief end product of fat and carbohydrate metabolism other than water is carbon dioxide, which can be excreted by the lungs without the obligation of additional water, whereas the nonvolatile end products of protein metabolism, of which urea is the most abundant, require water for their excretion in the urine. Under starvation conditions, catabolism of tissue proteins is accelerated, with increased urea production, and the resulting increase in urinary solute excretion may decrease the ability of the individual to conserve water.

Urine output usually varies with the amount of water ingested. Some desert rodents that do not have ready access to drinking water have a highly developed urinary concentrating mechanism and are able to excrete very small volumes of extremely concentrated urine. These animals never drink, and their water intake is derived entirely from food water and water of oxidation. In humans the urinary concentrating ability is not nearly as great, and some daily intake of water as such is obligatory if water balance is to be maintained.

CONTROL OF WATER INTAKE—THIRST. Thirst is the subjective sensory impression that activates a desire to drink water. Water deficiency usually manifests itself promptly by thirst, the satisfaction of which relieves the deficiency. The mechanism of thirst is complex and still not completely understood, and it is beyond the scope of this discussion to attempt a detailed consideration of all the factors involved. Thirst may be induced in a variety of ways, but regardless of the mechanism of induction a common feature appears to be a diminished output of saliva, and the resulting dryness of the mouth and throat is apparently the immediate cause of the thirst sensation.

Thirst is related to the concentration of solutes in the body fluids as these reflect an absolute or relative deficit of water. An increase in the effective osmolality of extracellular fluids may be produced either by water deprivation or by the administration of hypertonic sodium chloride solutions. In both cases the requirements of osmotic uniformity lead to cellular dehydration, and a sensation of thirst is experienced. Although administration of a hypertonic sodium chloride solution leads to thirst, administration of a urea solution of similar osmolality does not. Since urea freely traverses cell membranes, it does not alter the effective osmolality of extracellular fluid and does not produce cellular dehydration. It is attractive to speculate, therefore, that dehydration of thirst receptors is a factor leading to their stimulation. However, thirst may also be stimulated in the absence of cellular dehydration. Reduction in vascular volume, which may occur as the result of acute hemorrhage, may be accompanied by profound thirst, although there has been no change in the effective osmolality of body fluids. Thirst may also occur in the absence of a significant change in total water balance with salt depletion. In this circumstance, body fluids are hypotonic and cells are actually overhydrated but, in common with the effects of hemorrhage, the extracellular fluid volume is contracted, and this, too, appears to be an effective stimulus to thirst.

Understanding of the sensation of thirst is complicated by the fact that in humans water ingestion is seldom determined solely by requirements for water balance. Thirst may be stimulated, for example, by emotional stress, and the fluid intake of most individuals is controlled less by true thirst than by cultural and social factors. A typical individual allowed free access to water and other beverages will ordinarily ingest more water than he needs to stay in balance, and his urinary concentrating ability is rarely taxed to its capacity.

REFERENCES
General reviews

1. Berl, T., et al.: Clinical disorders of water metabolism, Kidney Int. **10:**117, 1976.
2. Edelman, I. S., and Leibman, J.: Anatomy of body water and electrolytes, Am. J. Med. **27:**256, 1959.
3. Fitzsimons, J. T.: The physiological basis of thirst, Kidney Int. **10:**3, 1976.
4. MacKnight, A. D. C., and Leaf, A.: Regulation of cellular volume, Physiol. Rev. **57:**510, 1977.
5. Muntwyler, E.: Water and electrolyte metabolism and acid-base balance, St. Louis, 1968, The C. V. Mosby Co.

Original papers

6. Amatruda, T. T., and Welt, L. G.: Secretion of electrolytes in thermal sweat, J. Appl. Physiol. **5:**759, 1953.
7. Barratt, T. M., and Walser, M.: Extracellular fluid in individual tissues and in whole animals: the distribution of radiosulfate and radiobromide, J. Clin. Invest. **48:**56, 1969.
8. Christian, J. R., et al.: Total body water and exchangeable sodium in normal full-term newborn infants, J. Dis. Child. **92:**325, 1956.
9. Donnan, F. G.: The theory of membrane equilibria, Chem. Rev. **1:**73, 1924.
10. Gamble, J. L.: Chemical anatomy, physiology and pathology of extracellular fluid. A lecture syllabus, ed. 6, Cambridge, Mass., 1954, Harvard University Press.
11. Guyton, A. C., Granger, H. J., and Taylor, A. E.: Interstitial fluid pressure, Physiol. Rev. **51:**527, 1971.
12. Intaglietta, M., and Zweifach, B. W.: Indirect method for measurement of pressure in blood capillaries, Circ. Res. **19:**199, 1966.
13. Landis, E. M.: The capillary pressure in frog mesentery as determined by microinjection methods, Am. J. Physiol. **75:**548, 1926.
14. Pappenheimer, J. R.: Passage of molecules through capillary walls, Physiol. Rev. **33:**387, 1953.
15. Pitts, R. F.: Physiology of the kidney and body fluids, ed. 3, Chicago, 1974, Year Book Medical Publishers, Inc.
16. Schloerb, P. R., et al.: The measurement of total body water in the human subject by deuterium oxide dilution, J. Clin. Invest. **29:**1296, 1950.
17. Schultz, A. L., et al.: A critical comparison of the T-1824 dye and iodinated albumin methods for plasma volume measurement, J. Clin. Invest. **32:**107, 1953.
18. Starling, E. H.: Physiological factors involved in the causation of dropsy, Lancet **1:**1407, 1896.
19. Strauss, M. B.: Body water in man. The acquisition and maintenance of the body fluids, Boston, 1957, Little, Brown & Co.
20. Talen, J. L., and Staverman, A. J.: Osmometry with membranes permeable to solvent and solute, Trans. Faraday Soc. **61:**2794, 1965.
21. Van t'Hoff, J. H.: Die Rolle des osmotischen Druckes in der Analogie zwischen Loesungen und Gasen, Z. Physikal. Chem. **1:**481, 1887.
22. Walser, M., Seldin, D. W., and Grollman, A.: An evaluation of radiosulfate for the determination of the volume of extracellular fluid in man and dogs, J. Clin. Invest. **32:**299, 1953.
23. Welt, L. G.: Clinical disorders of hydration and acid-base equilibrium, ed. 2, Boston, 1959, Little, Brown & Co.
24. Zweifach, B. W.: Functional behavior of the microcirculation, Springfield, Ill., 1961, Charles C Thomas, Publisher.

50

CARL W. GOTTSCHALK and WILLIAM E. LASSITER

Mechanisms of urine formation

The kidney has the primary responsibility for maintaining constant the composition and volume of the body fluids. Other organs participate in this process, notably the lungs in respect to acid-base regulation, but the kidney, through its excretory activity, provides the major mechanisms for maintaining homeostasis. In order to maintain a constant internal environment the kidney must respond appropriately to variations in dietary intake and in extrarenal losses of solutes and water, which may result from environmental stresses and disturbances in function of other organs. Many of these excretory processes are under some type of regulatory control. The kidney's excretory function must also be maintained in the presence of renal disease if the patient is to survive. This chapter will be concerned only with the excretory function of the normal mammalian kidney. The renal production of certain humoral substances such as erythropoietin and the role of the kidney in the pathogenesis of hypertension, etc., are not discussed here.

The process of urine formation is a complicated one that begins with the formation at a rapid rate of an ultrafiltrate of plasma in the glomerulus, with subsequent reabsorption by the uriniferous tubules of most of the filtered water and electrolytes, and with secretion of certain substances into the tubular urine. Reabsorption is defined as the transport across the tubular epithelium of a substance from the tubular lumen to blood in the surrounding vessels. Secretion is transtubular transport in the opposite direction, from blood to lumen. In this context the terms "reabsorption" and "secretion" refer only to the direction of the tubular transport process without any implication as to mechanism. Either process can be active or passive. Excretion is defined as the elimination of a substance from the body, in this case in the urine.

FUNCTIONAL ANATOMY OF THE KIDNEY

Nephron. Because the structure and function of the kidney are so closely related, it is necessary to have some knowledge of its structure before one can understand its function. The basic unit of structure and function of the kidney is the nephron (Fig. 50-1). The nephron includes the glomerulus and its attached tubule. The tubular epithelium is composed of a single layer of cells surrounded by a supporting basement membrane. The tubules empty into collecting ducts that coalesce into a decreasing number as they course through the medulla before emptying into the pelvis as the papillary ducts of Bellini. The collecting ducts are generally considered an integral part of the uriniferous tubules, although they have an embryologic origin separate from that of the nephrons. A normal human kidney contains approximately 1,200,000 nephrons. This estimate has been arrived at both from direct counts of the number of glomeruli in portions of kidney tissue and from consideration of the anatomic structure of the collecting duct system and the number of nephrons that drain into individual collecting ducts.[80] The number of nephrons does not increase after birth, and the increase in size of the kidney with growth results from an increase in size of individual nephrons, particularly their length, without change in their numbers. The response to reduction in the number of nephrons, as from removal of one kidney or destruction by disease, is an increase in the size of the remaining nephrons. New nephrons cannot develop.

The major parts of the nephron are the glomerulus, proximal convoluted tubule, loop of Henle, and distal convoluted tubule. The tubule begins with an invaginated sphere, Bowman's capsule, that envelops the glomerulus, an anastomotic network of freely branching capillaries (Fig.

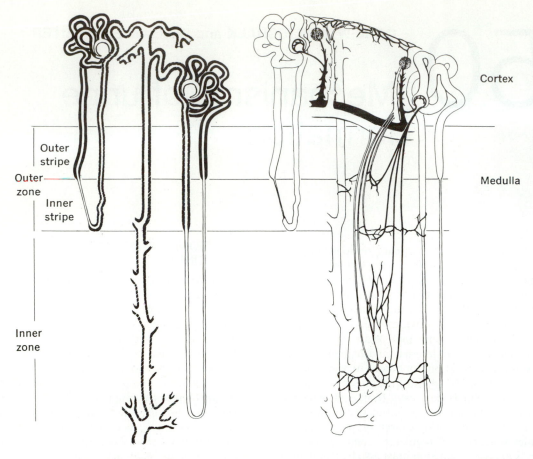

Fig. 50-1. Schematic diagram of two nephrons, one with short and other with long loop, and their associated blood vessels. (From Gottschalk.[44])

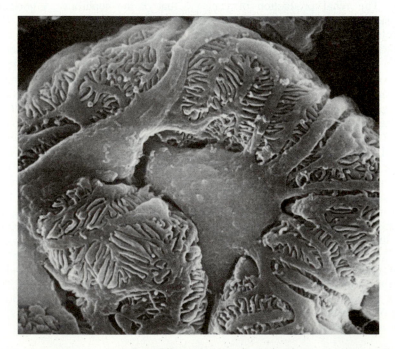

Fig. 50-2. Scanning electron micrograph of capillary loop in normal rat glomerulus. Two podocytes with their processes are shown magnified ×4,200. (From Arakawa.[10])

50-2). The entire structure is properly known as a malpighian corpuscle but is generally called the glomerulus. Glomeruli are found only in the cortex of the kidney. Each tuft of glomerular capillaries originates from an afferent arteriole, and the capillaries reunite before leaving the glomerulus as a single efferent arteriole. This arteriole in turn breaks up to form the numerous peritubular capillaries that surround the cortical tubules. The efferent arterioles of the juxtamedullary glomeruli drain into the specialized blood vessels of the medulla, the vasa recta. Mesangial cells, which resemble pericytes, are found in the axial or stalk region of the capillary tufts. These cells increase in number in proliferative glomerulonephritis.

Glomerular capillaries are unique in that they are interposed between two arterioles. This arrangement serves as a means for maintaining within them the high hydrostatic pressure necessary for an efficient filtration mechanism and at the same time provides a large surface area for filtration.

The endothelium lining the glomerular capillaries is of the fenestrated type and has a higher permeability than capillaries of this type elsewhere in the body. This may be because the fenestrae of the glomerular capillaries are larger and generally lack the thin diaphragm seen across the pores of other fenestrated capillaries. The glomerular filtrate must pass not only through the endothelial pores, but also across the basement membrane, an extracellular layer located between the endothelium and the inner layer of Bowman's capsule, the podocytes. The filtrate then traverses the numerous intercellular channels formed by the complicated branched interdigitating processes of the glomerular podocytes. Spanning from one process to another is a thin layer of material called the filtration slit membrane. Each of these components of the filtration barrier (endothelial pores, basement membrane, filtration slit membrane) seems to restrict the passage of substances above a certain size (Fig. 50-3).

The proximal tubule begins with Bowman's capsule and ends at the abrupt transition into the thin descending limb of the loop of Henle. In man, it varies from 12 to 24 mm in length and 50 to 65 μm in outside diameter. Proximal tubules form the major part of the tubular mass of the kidney and the bulk of the renal parenchyma. The proximal tubule consists of two parts, the pars convoluta, which is highly convoluted and lies close to its own glomerulus, and the pars recta, the terminal straighter portion, which plunges through the cortex toward the medulla. The tubule is lined by a single layer of epithelial cells that exhibit numerous grooves and ridges that interlock with adjacent cells. The luminal

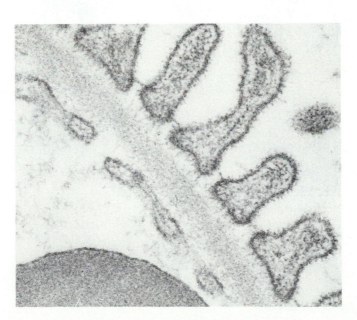

Fig. 50-3. High-power electron micrograph of filtration membrane in normal rat glomerulus showing endothelial pores, basement membrane, and filtration slit membrane magnified ×95,000. (Courtesy Dr. R. Bulger.)

surface of these cells has a prominent brush border that greatly increases the surface area of the membrane in contact with the tubular fluid (Fig. 50-4).

The loop of Henle consists of several parts: the pars recta of the proximal tubule, the thin descending limb, the thick ascending limb, and, if the loop is a long one and dips into the inner medulla, a thin ascending limb. The loops penetrate into the medulla to varying depths before returning to the cortex. Loops arising from glomeruli situated in the juxtamedullary area of the cortex descend most deeply. The thin, descending limbs begin close to the corticomedullary boundary. Their cells are squamous in type, with few microvilli and few mitochrondria. The epithelium of the short loops changes just before or at the bend to that of the thick ascending limbs of the distal tubules. The thin ascending limbs of the longest loops transform into thick ascending limbs as they return to the outer medulla. The cells of the thick ascending limbs are elaborately interdigitated with their neighbors and contain many mitochondria. Each thick ascending limb returns to the cortex and for a short distance is in contact with its own glomerulus. The point of contact is termed the macula densa, since the cells are narrower and their nuclei are densely crowded together.

The distal convoluted tubule (distal convolution) begins at the macula densa and joins with several other distal convoluted tubules to form a collecting duct. It is about one third the length of the proximal tubule. The microvilli of the luminal surface of the cells are much fewer in number, but there are many mitochondria and basal interdigitating processes, as in the proximal tubule.

The collecting ducts descend through the cortex and medulla, and in the latter are close to the loops of Henle. In the human kidney, they undergo eight or nine series of anastomoses before they terminate as the ducts of Bellini. This markedly reduces their total luminal cross-sectional area even though the diameter of successive individual ducts increases. The cells are relatively undifferentiated, but become more columnar as the pelvis is approached.

Blood vessels. The four or five large branches of the renal artery arch around the renal pelvis before they enter the kidney parenchyma as interlobar vessels. At the corticomedullary junction, these vessels branch to form arcuate arteries, which course along the boundary zone between the cortex and medulla. Interlobular arteries radiate from the arcuate arteries toward the surface of the kidney and give off the afferent arterioles. Each of the latter enters a glomerulus and breaks

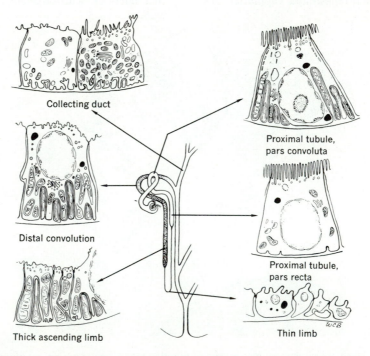

Fig. 50-4. Diagram of nephron with long loop showing structural features of its several major portions. (Courtesy Dr. R. Bulger.)

up into the glomerular capillaries, which in turn coalesce to form an efferent arteriole. The efferent arterioles quickly divide into a network of anastomotic capillaries that surround the proximal and distal convoluted tubules in the cortex. The peritubular capillaries drain into the interlobular veins that enter the arcuate veins. The arcuate veins drain into the interlobar veins, which form the renal vein leaving the kidney.

The efferent arterioles of the juxtamedullary glomeruli do not drain into the cortical capillary network but into the specialized blood vessels of the medulla, the vasa recta. The vasa recta are relatively straight vessels that descend for varying distances into the medulla before they return to the corticomedullary boundary and empty into the arcuate veins. There is a rich anastomotic network between the descending and ascending vasa recta.

Blood (about 25% of the cardiac output) is thus supplied to the glomerular tufts through short muscular vessels at a high but controlled pressure. The peritubular network, on the other hand, is a low-pressure system, facilitating its function as a reabsorptive network.

Juxtamedullary nephrons. Those nephrons originating most deeply in the cortex are specialized in several respects. Their glomeruli have a larger diameter than those located more superficially, their efferent arterioles have a larger diameter than their afferent arterioles (the converse of the relation in the more superficial glomeruli), they supply the blood that perfuses the medulla, and they have the longest loops of Henle. The proportion of nephrons with long loops is species dependent. In man and some other mammals, long-looped nephrons originate only from juxtamedullary glomeruli. In certain other species, including the dog and the cat, all nephrons have long loops that enter the inner medulla and have a thin ascending as well as thin descending limb. In the rat, only about 30% of nephrons have long loops. The percent of nephrons with long loops in man is not certain, but it is estimated to be about 15%. Also in man, certain nephrons are entirely restricted to the cortex. In these nephrons the thin limb of the loop of Henle is absent or poorly developed.

It appears likely that juxtamedullary nephrons have certain functional specializations. They may have larger filtration rates than smaller glomeruli nearer the surface of the kidney, and they may respond in a different fashion to certain stresses.

Juxtaglomerular apparatus. The juxtaglomerular apparatus (JGA) is morphologically a highly specialized structure with both vascular and tubular components. It probably has an important, but at present poorly understood, function in a feedback mechanism controlling glomerular filtration rate and/or renal blood flow. It consists of several parts: granulated cells in the afferent and efferent arterioles, the macula densa, and the extraglomerular mesangium (polar cushion, lacis cells). Numerous nonmyelinated nerve fibers are associated with the vessels in the JGA. Some of the cells in the media of the afferent and probably the efferent arteriole close to the glomerulus are specialized myoepithelioid cells and contain granules that appear to be composed of the enzyme renin. Their granularity (i.e. renin content) is a function of the physiologic state of the animal.

The macula densa cells are ideally situated to monitor some aspect of distal tubular function, possibly hydrostatic pressure or the concentration of one or more solutes, and in some fashion to transfer this information to the epithelioid cells, thereby regulating the release of renin and perhaps controlling some aspect of nephron function. Renin is a proteolytic enzyme that acts on the plasma α_2-globulin angiotensinogen to release the physiologically inactive decapeptide angiotensin I. Converting enzyme splits off the two terminal amino acids to form the octapeptide angiotensin II, which is a powerful vasoconstrictor. Most converting enzyme activity is found in the lung, but apparently, angiotensin II may be formed locally in the glomerular region. The latter must occur if some aspect of nephron function, either filtration rate and/or sodium reabsorption, is regulated by angiotensin II on an individual nephron basis. The renin-angiotensin system is involved in the regulation of aldosterone secretion and in at least some forms of hypertension.

GLOMERULAR FILTRATION

The initial step in the formation of urine is ultrafiltration at the glomerulus, with separation of a fraction of the water and dissolved solutes from the formed elements and macromolecules of the blood flowing through the glomerular capillaries. This process was first postulated by Bowman[19] in 1842 when he described the structure of the glomerulus, but somewhat later he and Heidenhain[51] both thought that filtration mainly provided the water in the urine and that the urinary solutes were secreted by the tubular epithelium. In 1844 Ludwig[69] proposed a mechanical theory, which held that all the urinary solutes and the water were first filtered at the glomerulus and that the solutes were concentrated as water diffused out of the tubules under the influence of physical forces.

Experimental proof of the existence of glomerular filtration was obtained in a series of elegant studies performed in the laboratory of A. N. Richards at the University of Pennsylvania in the 1920s and 1930s.[90,112] Richards and his colleagues were the first to study kidney function by micropuncture techniques that involved collection in a micropipet of minute quantities of fluid for chemical analysis. Because of technical considerations, they worked primarily with amphibians. They demonstrated that the total molecular concentration and the concentrations of a variety of solutes were the same in fluid collected from Bowman's capsule as in plasma water when corrections were made for the differences in concentration that arise from the absence of proteins (i.e., the Donnan effect). The process of tubular reabsorption was proved at the same time when two substances that are normal constituents of blood plasma, sodium chloride and glucose, were found to be present in fluid collected from Bowman's space but absent from the final urine under the experimental conditions. The occurrence of tubular secretion in the kidneys of higher vertebrates was more difficult to prove, and this concept was resisted by Richards until Marshall of The Johns Hopkins University provided incontrovertible evidence that the rate of excretion of phenol red in the urine was greater than could possibly occur as the result of glomerular filtration alone.[77] Certain fish are aglomerular, and the presence of tubular secretory mechanisms has never been questioned in them.

Rate of glomerular filtration. The rate of filtration across the glomerular filtration membrane into Bowman's space is determined by physical forces and the permeability of the membrane, as is the case with other capillary beds. The hydrostatic pressure in the glomerular capillaries (P_{gc}) is the driving force for filtration. Opposing filtration are the hydrostatic pressure in the Bowman's space (P_{bs}) and the colloid osmotic pressure (COP) exerted by the proteins of the capillary plasma. Because the filtrate in Bowman's space is essentially protein free, its COP is negligible and need not be considered. The glomerular filtration rate (GFR) is directly proportional to the net balance of these forces averaged over the length of the capillary bed, the mean effective filtration pressure (EFP), and to the permeability (Kf) of the filtration membrane. The latter has two components: (1) The surface area and (2) the specific hydraulic permeability of the membrane. Thus:

$$GFR = Kf(P_{gc} - [P_{bs} + COP])$$

or

$$GFR = Kf \cdot EFP$$

Understanding of the dynamics of glomerular ultrafiltration has been impaired by the unknown value of P_{gc}. Indirect evidence has suggested that it is about two thirds of normal mean arterial pressure, approximately 60 mm Hg. It is known that P_{bs} is approximately 10 mm Hg; COP is about 25 mm Hg in blood entering and 35 mm Hg in blood leaving the glomerular capillaries, the difference being due to the increase in plasma protein concentration during the filtration process. Thus the net balance of forces has been thought to result in filtration along the entire length of the glomerular capillaries. Filtration pressure equilibrium, the condition in which the balance of forces favoring and opposing filtration are equal, would be reached in the efferent arteriole. Further along, in the peritubular capillaries, the net pressure favors reabsorption from the interstitial spaces, since plasma COP plus interstitial fluid hydrostatic pressure exceeds the sum of the peritubular capillary hydrostatic and interstitial fluid COPs. Thus, as in capillaries elsewhere in the body, the transfer of fluid is determined by the balance of forces but, in contrast to other organs, filtration occurs in one capillary bed and reabsorption in another capillary bed.

Variations in GFR are due to changes in EFP or glomerular permeability. According to the conventional view (i.e., filtration pressure disequilibrium), they would generally result from changes in glomerular capillary pressure or permeability; changes in Bowman's space pressure or in arterial COP would be less frequent causes. A change in renal blood flow (RBF) alone would have little effect on GFR, since it would produce only a minimal change in the EFP. This is because the increase in COP is small along the glomerular capillaries if their Kf is relatively low, and the balance of forces favors filtration even at their efferent end (Fig. 50-5). If Kf is sufficiently high, filtration pressure equilibrium is reached in the glomerular capillaries, and an increase in RBF increases GFR, as EFP is increased by distal displacement of the point along the glomerular capillaries at which equilibrium is achieved. If RBF is sufficiently great, disequilibrium will occur even with a high Kf. If equilibrium is achieved, a further increase in Kf alone will not change GFR. In this circumstance the point of equilibrium would move toward the afferent end of the glomerular capillaries and EFP

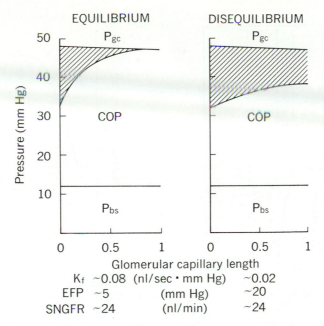

Fig. 50-5. Idealized representation of forces producing (P_{gc}) and opposing (P_{bs} + COP) ultra-filtration in an individual glomerulus. Shaded area divided by capillary length is mean effective filtration pressure. With high Kf filtration pressure, equilibrium is reached; with low Kf, disequilibrium persists at end of glomerular capillary. With same P_{gc}, P_{bs}, and arterial COP in two cases, single nephron GFR would be identical.

would diminish. At all Kf values an increase in P_{gc} will increase GFR and the filtration fraction.

Recent studies by Brenner et al.[23] and Blantz et al.[18] indicate that filtration equilibrium is normally achieved in the glomerular capillaries. Utilizing a strain of rats with an unusually large number of surface glomeruli, Brenner measured by micropuncture the P_{gc}, P_{bs}, and COP in small samples of plasma from surface efferent arterioles. The value for P_{gc} was 45 mm Hg, lower than has been generally predicted, and this result was the same as the sum of P_{bs} and efferent arteriolar COP, demonstrating the presence of filtration pressure equilibrium. Moreover, filtration equilibrium was present with moderate increases in RBF, but was not achieved at very high RBF values, permitting the calculation of the specific value for Kf. That value, 0.08 ml/sec/mm Hg, is much higher than earlier estimates. Thus in these studies the GFR was highly plasma flow dependent and was not independent of glomerular plasma flow, as has been usually postulated. It is not certain at this time whether these findings have general applicability to all glomeruli (superficial vs deep) in all species of mammals. Filtration equilibrium apparently is present in the squirrel monkey[70], but recent less direct measure-ments by others in dogs and rats suggest that Kf is lower and that filtration equilibrium is not usually achieved.[11,82]

In rats with acute experimental glomerulonephritis, Kf was much less and filtration equilibrium was not achieved.[17,71] In rats with chronic experimental glomerulonephritis, single-nephron GFR, glomerular capillary pressure, and proximal intratubular pressure were quite variable from nephron to nephron in the same kidney, and it is likely that single-nephron plasma flow, filtration fraction, and Kf are also variable.[9] Heterogeneity among nephrons makes a complete experimental analysis exceedingly difficult, since all determinations must then be performed on the same nephron.

A glomerular capillary pressure of 45 to 50 mm Hg indicates that approximately 50% of the renal vascular resistance is located in preglomerular vessels, presumably largely in the afferent arterioles (Fig. 50-6). Another large fall in hydrostatic pressure occurs in the efferent arterioles, since the pressure in the peritubular capillaries is of the order of 10 mm Hg. A smaller fall in pressure occurs in the intrarenal venous system.

In experimental animals the GFR and RBF are

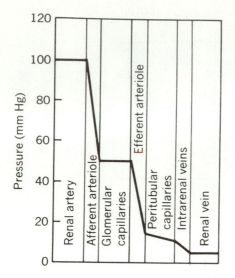

Fig. 50-6. Pressure profile across renal vasculature.

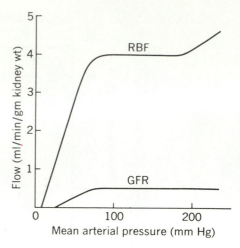

Fig. 50-7. Pressure-flow relations in mammalian kidney showing autoregulation of renal blood flow and glomerular filtration rate over range of 80 to 180 mm Hg perfusion pressure.

usually independent of the arterial pressure over a range of pressures from 80 to 180 mm Hg (Fig. 50-7).[99] The glomerular capillary pressure presumably remains unchanged over this range of arterial pressure, an indication that under these circumstances the resistance of the preglomerular vessels varies directly with the arterial pressure. Changes in diameter of the afferent and efferent arterioles will have opposite effects on glomerular capillary pressure, which will decrease with afferent arteriolar constriction and increase with efferent constriction.

The relative constancy of GFR and RBF, despite wide variations in arterial pressure, is termed autoregulation, since it has also been observed in isolated perfused kidneys. Autoregulation does not always occur and may disappear after administration of vasoactive drugs, in shock, on stimulation of the afferent renal nerves, etc.[107]

Gomerular membrane. Pappenheimer[83] has estimated that the permeability of glomerular capillaries is approximately 100 times that of capillaries in skeletal muscle, and the recent studies by Brenner and others indicate that they are even more permeable. Some years ago Bayliss et al.[14] studied in dogs the excretion of intravenously administered proteins as a function of their molecular size. Proteins with molecular weights of approximately 35,000 (gelatin, egg albumen, Bence-Jones protein) were readily excreted, whereas hemoglobin, with a molecular weight of 68,000, was excreted only when present in high

concentration in the plasma. Proteins with a higher molecular weight were not excreted. The molecular weight of hemoglobin is similar to that of human serum albumin, which is not normally excreted in significant amounts.

Passage of molecules across the glomerular filter is not determined solely by molecular size, however, since macromolecules of comparable size (i.e., albumin and dextran) may have differing rates of filtration. Recent electron microscopic as well as physiologic observations indicate that molecular charge is also important. A positive charge facilitates and a negative charge impedes filtration, indicating that there are fixed negative charges in the glomerular filter. The enhanced filtration of the polyanion albumin in disease states may be explained, at least in part, by the loss of fixed negative charges.[21]

In normal mammals, all glomeruli are believed to be functional at all times, and the total glomerular capillary surface area in humans has been estimated to be more than 1.5 m². In contrast, in the frog the number of filtering glomeruli is variable and is determined by the animal's state of hydration. It is not known whether the total glomerular capillary surface area available for filtration is constant in mammals under all circumstances. The marked increase in GFR that can be produced in mammals by saline infusion has been thought to result from an increase in EFP and/or a change in membrane characteristics. However, if filtration equilibrium is achieved, it must result directly from the increased glomerular blood flow that occurs under these circumstances.

Hydrostatic pressure in Bowman's space. It appears likely that in humans the hydrostatic pressure in Bowman's space remains relatively constant under normal circumstances. Micropuncture determinations in rats indicate that the pressure in Bowman's space is normally 10 to 13 mm Hg.[43] The pressure here and in the tubules rises with increased urine flow because the collecting ducts provide a limiting resistance to the outflow of urine from the kidney. In the animal given antidiuretics there is only a low rate of flow through the collecting ducts, and only a small pressure difference between the distal tubule and renal pelvis is required to produce the flow. An increased urine flow requires a greater pressure fall across the collecting duct system. At low rates of urine flow there is also a significant fall in intratubular pressure as fluid flows through the thin Henle's limbs. When intratubular pressure is elevated, as during diuresis, the thin limb is dilated and resistance to flow is markedly decreased. With ureteral obstruction in experimental animals, the pressure in the uriniferous tubules increases in a retrograde fashion back to Bowman's space and filtration is reduced. This probably happens in humans under such abnormal circumstances as obstruction to the urinary outflow by an enlarged prostate or stones.

Colloid osmotic pressure of plasma proteins. The COP exerted by the plasma proteins rises in the glomerular capillaries from approximately 25 to 35 mm Hg as water is removed by filtration. This is the major force opposing filtration and sets a minimal limit on the hydrostatic pressure required for filtration. As far as is known, variations in the COP are not major determinants of the rate of filtration in humans. Changes in COP could be compensated for by changes in the resistance of the afferent or efferent arterioles.

Measurement of glomerular filtration rate. The rate of filtration of individual glomeruli may be measured directly by micropuncture techniques in some experimental animals, but it is obviously impossible to measure directly the filtration rate of all the glomeruli in either animals or humans. Indirect methods must therefore be employed to estimate the GFR of the whole kidney. An extensive search was made in the 1930s for a chemical marker with the necessary characteristics to permit its use as a measure of the rate of glomerular filtration. It would be possible to measure the rate of glomerular filtration if a substance in the blood were freely filterable at the

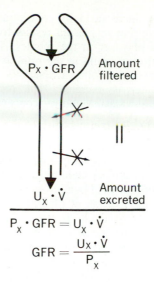

$$P_x \cdot GFR = U_x \cdot \dot{V}$$

$$GFR = \frac{U_x \cdot \dot{V}}{P_x}$$

Fig. 50-8. Diagrammatic representation of principles underlying measurement of GFR with substance that is freely filtered at glomerulus and neither reabsorbed nor secreted by tubule. (See text.)

glomerulus and neither reabsorbed from nor secreted into the tubule (Fig. 50-8). In this case the amount of the substance filtered would equal the amount excreted, and dividing the rate of excretion by the concentration of the substance in plasma would provide a measure of the GFR. For a substance (x) that has these characteristics:

$$GFR = \frac{U_x \cdot V}{P_x}$$

where GFR equals the glomerular filtration rate, V the rate of urine flow, P the concentration of x in plasma, and U the concentration in the urine. It is obvious that to be useful the material must be nontoxic, must not induce variations in the rate of filtration, and must be subject to precise analytic determination. Smith[101] and Richards[90] independently identified inulin as a substance that had the particular characteristics that make it a satisfactory marker for measuring the GFR. Inulin is a polymer of fructose with a nominal molecular rate of 5,400. It is used extensively to measure the GFR in humans and experimental animals.

In normal adult males the GFR averages 125 (ml/min)/1.73 m^2 and in normal females about 10% less. In order to compare the GFRs in individuals of various sizes, it is customary to normalize the values to those of an "ideal" person with a body surface of 1.73 m^2.

Because inulin gains access to the urine only by filtration and is neither reabsorbed nor secreted, any increase in its concentration in the urine above that in plasma results from the reabsorption of water in the tubules. Thus if the concentration of inulin in the urine is 100 times that in plasma, it follows that only 1% of the filtered water is excreted. The reciprocal of the urine to plasma inulin concentration ratio $\left(\text{i.e., } 1/\dfrac{U_{in}}{P_{in}}\right)$ is a measure of the fraction of filtered water excreted. Inulin can be used as a measure of water movement all along the nephron, and the tubular fluid:plasma (F:P) inulin ratio is routinely used in micropuncture studies to quantitate the fraction of filtered water reaching some point in the tubule. If, for example, the F:P inulin ratio at the end of the proximal pars convoluta were 3, then one third of the water filtered in that nephron reached that point and two thirds were reabsorbed. By comparing the urine:plasma (U:P) or F:P ratio of another substance with that of inulin, one obtains a measure of the fraction of that substance arriving at that point relative to the amount filtered. If the other substance is freely filtered and neither reabsorbed nor secreted, its F:P ratio will be the same as that for inulin. If the substance is reabsorbed, its F:P concentration ratio will be less than that of inulin; conversely, it will be higher if the substance is secreted. For example, if the F:P sodium ratio is 1 and the F:P inulin ratio is 3, then one third of the filtered sodium reached the point in the tubule at which the fluid was collected.

Other markers used for measuring glomerular filtration rate. Inulin is the most satisfactory substance yet found for determining the GFR, and there is now direct experimental evidence that it is freely filterable at the glomerulus and is neither reabsorbed nor secreted by the tubules. Since inulin does not normally occur in man, it must be infused intravenously at such a rate that an accurately measurable concentration is produced in the plasma and maintained at a relatively constant level during the period of urine collection. Obviously, it would be much more convenient to use a normally occurring substance with the same renal properties as inulin. For this reason the clearance of endogenous creatine is often used in studies on humans as a measure of GFR, although its clearance rate does not exactly equal the GFR. There are two inadequacies in the use of creatinine for this purpose, but fortunately these tend to cancel each other out. There is probably some creatinine secreted by the tubules, thereby increasing the amount of creatinine excreted, but its chromogenic determination in plasma probably includes substances other than creatinine.

Since the rate of creatinine production in the body is relatively constant, its plasma concentration should vary inversely with the rate of glomerular filtration. This is a necessary mathematical consequence of the fact that the plasma concentration must rise or fall until the product of concentration and clearance (i.e., the excretion rate) equals the amount being produced. If the latter were constant and creatinine were excreted only by filtration, its plasma concentration would necessarily bear a precise inverse relationship to the rate of filtration. Even though neither condition holds exactly, serial determinations of the plasma creatinine concentration are useful clinically in following the course of a patient with renal disease. In many experimental animals the clearance of exogenously infused creatinine is a useful measure of the GFR, but there is significant tubular secretion in the male dog and rat that limits its usefulness in these animals.

Other substances, particularly mannitol, sucrose, allantoin, and thiosulfate, have been used as measures of the GFR, but they are limited in their application either because their clearance is not precisely the same as that of inulin or because of problems in their analytic determination.

CLEARANCE CONCEPT

The rate of urinary excretion of a substance ($U_x \times V$) divided by its plasma concentration is a measure of the minimal volume of plasma required to supply the amount of the substance excreted in the urine in a given period of time. This is termed the clearance of a substance (C_x) and is expressed mathematically as follows:

$$C_x = \frac{U_x \cdot V}{P_x}$$

The clearance of any solute appearing in the urine can be calculated from this formula. Because of the special properties of inulin, its clearance equals the rate of glomerular filtration. As an example of the utility of the clearance concept in studying the excretion of other solutes, sodium for example, assume that $V = 1$ ml/min, $U_{Na} = 280$ mEq/L, and $P_{Na} = 140$ mEq/L; then $C_{Na} = 2$ ml/min. In other words, an amount of sodium equal to that contained in 2 ml of plasma is excreted in the urine each minute. If GFR (C inulin) equals 125 ml/min, an amount of sodium equal to that contained in 125 ml plasma is filtered each minute, and the amount of sodium contained in 123 ml of plasma is reabsorbed each minute by the tubule. Thus if the clearance of any freely filtrable substance is less than that of inulin, the substance undergoes a process involving net reabsorption in the tubule. If, on the other hand, the clearance of a substance, freely filtrable or not, is greater than that of inulin, its excretion must involve net tubular secretion. The amount of a substance actually secreted or reabsorbed in the tubule cannot always be calculated in this fashion, since some substances, such as potassium and urate, undergo bidirectional transport in the tubule. Comparison of the clearance of a substance with the simultaneously determined inulin clearance provides quantitation only of *net* transtubular transport.

The volume of plasma "cleared" of a substance is a virtual, not a real, volume, and one should not infer that all the substance has been removed from that volume of plasma. On the contrary, only some of the substance is removed from a much larger volume of blood.

Calculation of rate of tubular transport. The rate of net tubular transport of a substance is calculated from the filtered load and its rate of excretion. The filtered load equals the rate of filtration multiplied by the filtrable plasma concentration:

$$\text{Filtered x} = \text{GFR} \cdot P_x$$

and

$$\text{GFR} \cdot P_x + T_x = U_x \cdot V$$

where T_x has a negative value if x is reabsorbed from the tubule and a positive value if it is secreted.

Determination of renal blood flow by clearance methods. The clearance technique permits the estimation of RBF and renal plasma flow (RPF) in man or animal with minimal interference or trauma to the subject. According to the Fick principle, the blood flow through an organ can be calculated from the rate of extraction of some substance in the blood entering and leaving the organ.

The rate of extraction of x by the kidneys equals blood flow times the difference between A_x and V_x, where A_x and V_x are the concentrations in arterial and renal venous blood, respectively. In a steady state the rate of extraction of x by the kidney will equal the rate of urinary excretion of x if the substance is not manufactured, destroyed, or stored in the kidney, and:

$$\text{RBF} = \frac{U_x \cdot V}{A_x - V_x}$$

In principle the RBF can be calculated from any excreted solute for which these restrictions hold, but the estimate will be accurate only if there is a reasonably large difference between the arterial and renal venous concentrations. However, if the equation is to be applied in this form, it is neces-

sary to obtain a sample of renal venous blood. This necessitates either operative exposure of the kidney and its vessels or passing a catheter from a superficial vein into the renal vein. Obviously the determination would be much simpler if a compound could be used whose rate of extraction by the kidney and excretion in the urine was so high that its concentration in renal venous blood approached zero. Certain substances, most notably *p*-aminohippurate (PAH) and iodopyracet (Diodrast), approach this limiting condition when present in the blood in low concentrations. The calculation then involves only the determination of concentration in renal arterial blood, which is the same as in systemic venous blood, and the amount of the substance excreted per unit time in the urine; thus the "effective" renal plasma flow (ERPF) is as follows:

$$\text{ERPF} = C_{PAH} = \frac{U_{PAH} \cdot V}{P_{PAH}}$$

The clearance of PAH will equal the renal plasma flow only if it is extracted completely in each circulation through the kidney. In fact, its extraction is not complete even at low plasma concentration, but averages approximately 85% in normal individuals under ordinary circumstances of study. Therefore the clearance of PAH is termed the "effective" renal plasma flow (ERPF). The rate of filtration of PAH is not as great as that of inulin, because some of the plasma PAH is protein bound and is not filtrable. The high renal clearance of PAH results from the secretion of PAH by the cells of the proximal tubule and removal for this reason from the blood flowing through the peritubular capillaries. It is believed by some that the reason for lack of complete extraction is that some of the blood perfusing the kidney goes through areas such as the fibrous capsule that are incapable of secreting PAH.

The true RPF can be calculated from C_{PAH} if the extraction (E) of PAH is determined by measuring the PAH concentration in arterial and renal venous plasma:

$$E = \frac{A_{PAH} - RV_{PAH}}{A_{PAH}}$$

and

$$\text{RPF} = \frac{C_{PAH}}{E}$$

The effective renal blood flow (ERBF) is calculated from the ERPF as follows:

$$\text{ERPF} = \frac{\text{ERPF}}{1 - \text{Hematocrit}}$$

This method of determining the RPF was devised and extensively employed by Smith.[101] Normal values of ERPF are 650 (ml/min)/1.73 m² in males and 600 (ml/min)/1.73 m² in females, corresponding to an ERBF of approximately 1,200 ml/min, or approximately one fourth of the cardiac output under basal conditions.

The term "filtration fraction" is used to refer to that fraction of the plasma perfusing the kidney that is filtered and is determined as C_{inulin}/C_{PAH}. In humans, this ratio is normally 0.2.

TUBULAR REABSORPTION

Simultaneous measurement of the GFR and the rate of solute and water excretion demonstrates the quantitative importance of tubular reabsorption in urine formation. Most of the filtered water and the major filtered solute, sodium chloride, are reabsorbed in the tubule. The mechanisms of reabsorption may be passive, as with water and urea, or active, as with sodium and glucose.

Active reabsorption is defined as reabsorption that cannot be accounted for by the existing electrochemical concentration gradients and that requires expenditure of metabolic energy at the site of transport. Membrane carriers are involved, and, as in other active transport systems, evidence of saturation (i.e., a transport maximum) and competition between various solutes for transport have been found in at least some of the transport systems.

Transport of a substance is considered passive if it occurs along an existing electrochemical gradient. The active transport of sodium plays a dominant role in passive reabsorption, since reabsorption of sodium with chloride and bicarbonate sets up a driving force for water reabsorption. This in turn results in an increased concentration in the tubular fluid of certain solutes, such as urea, leading to their reabsorption.

The suprabasal oxygen consumption of the kidney is directly correlated with the sodium transport and is supported by oxidative metabolism in the cortex. Fatty acids appear to be the major and glucose a minor substrate for metabolism in the cortex. In the medulla, in contrast, glucose is preferentially utilized for a much higher rate of aerobic and anaerobic glycolysis.

Glucose reabsorption and the transfer maximum concept

Glucose is generally found only in trace amounts in the urine of normal individuals. This indicates that in the fasting state about 100 mg of glucose is reabsorbed each minute, since approximately 125 ml of plasma with a glucose concentration of 80 mg/100 ml are filtered each minute. The process is by definition an active transport mechanism, since glucose is an uncharged molecule and its reabsorption occurs against a chemical concentration gradient. The transport mechanism is poorly understood but is undoubtedly complicated and related in some fashion to sodium transport. Phlorhizin completely blocks glucose reabsorption, presumably by binding strongly to membrane carriers and preventing their attachment to glucose molecules. Micropuncture experiments have demonstrated that reabsorption of glucose is restricted to the proximal tubule, primarily the convoluted portion.[111]

Under certain physiologic conditions the maximum amount of glucose that can be reabsorbed per minute is more or less fixed, hence the term "transfer maximum" (Tm). Similar considerations hold for phosphate, sulfate, certain organic acids and amino acids, etc.[85] More recent work indicates that the maximum rate of reabsorption of glucose and the other solutes mentioned is not the same under all physiologic conditions. The rate of glucose reabsorption, for example, is a function of the rate of proximal salt and water reabsorption; phosphate reabsorption is influenced by parathormone, etc. Nevertheless, the concept of a Tm-limited reabsorptive mechanism is well established in renal physiology and a careful consideration of the concept is helpful in understanding certain reabsorptive mechanisms considered to represent active transport processes.

The concept is illustrated by quantitative consideration of the relations observed when the filtered load of glucose is increased as a result of increased plasma glucose concentration (P_G) due to intravenous infusion of glucose without change in GFR (Fig. 50-9). The amount of glucose filtered per unit time equals the GFR $\times$ P_G; the amount excreted is $U_G \times V$, and the amount reabsorbed by the tubules, T_G, is the difference between these two quantities:

$$T_G = GFR \cdot P_G - U_G \cdot V$$

As the filtered load increases with increased plasma concentration, the urine at first remains essentially glucose free. When the plasma concentration is raised above some critical level, the "renal plasma threshold," an increasing amount of glucose appears in the urine even though the amount reabsorbed is still increasing. Eventually, however, a plasma glucose level is reached above

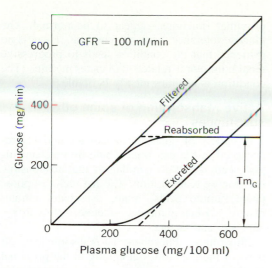

Fig. 50-9. Idealized representation of relationship between plasma concentration and filtration, reabsorption, and excretion of glucose in man. GFR is assumed to be constant at 100 ml/min.

which increasing amounts of filtered glucose are quantitatively excreted in the urine, an indication that the rate of glucose reabsorption has reached a maximum for the circumstances (i.e., Tm_G). In normal individuals, Tm_G averages approximately 375 and 300 (mg/min)/1.73 m² of body surface area in men and women, respectively. The demonstration of saturation kinetics, that is, the existence of a limited capacity to transport a substance despite an increasing load available for transport, is a characteristic feature of many active transport processes and is often used as one of the criteria for identification of carrier-mediated transport systems.

The line showing the increased rate of reabsorption with increasing P_G is curved at its upper end and does not break sharply as Tm_G is approached, since some glucose is excreted in the urine at a time when the amount reabsorbed by the tubule is still increasing. This is termed the "splay" of the titration curve. The cause of the "splay" in a glucose titration curve has been the object of much discussion and investigation. Two explanations that are not necessarily mutually exclusive have received serious consideration. One relates to a kinetic description of the transport mechanisms. The other has a morphologic basis and relates to differences in the rate of glomerular filtration and tubular transport of individual nephrons as a function of their morphologic characteristics.

The kinetic description has been developed from the views of Shannon.[98] The limitation in

transport is believed to result from the presence of a carrier in the system, which is necessary for transmembrane transport of the substance under consideration. The transported substance, glucose (G) in this instance, is believed to be able to penetrate and cross the luminal membrane of the tubular cell by virtue of combination with a carrier (C). After traversing the membrane the carrier-glucose complex dissociates. The free glucose is then presumably able to diffuse across the cell and into the interstitial fluid, and the membrane carrier returns to the luminal surface to accept another glucose molecule.

$$G + \underset{\text{(Lumen)}}{C} \rightarrow \underset{\text{(Membrane)}}{GC} \rightarrow \underset{\text{(Cytoplasm)}}{C + G}$$

Two reactions are thus involved. Although operation of the system requires the expenditure of energy, there is no evidence that the rate of transport is limited by the amount of energy available. The first reaction is written as a mass action expression:

$$K = \frac{(G) \cdot (C)}{(GC)}$$

where K is the dissociation constant for the glucose-carrier complex. The greater the affinity of the carrier for glucose, the smaller the value of K. The smaller the value of K, the more completely the glucose will be removed from the tubular fluid up to the point of complete saturation of the carrier. This is the point at which Tm is reached. K undoubtedly has a finite value, however, and an appreciable concentration of glucose in the tubular fluid is probably needed to saturate the carrier. Thus the concentration of glucose in the tubular fluid needed to saturate the carrier may be higher than that at which the filtered load equals Tm, and some glucose is excreted in the urine before the carrier is completely saturated.

The alternative, but not mutually exclusive, explanation of the splay has to do with variations in the filtering and reabsorptive capacities of individual nephrons. The term "glomerulotubular balance"* has been used to describe this relationship. Morphologic observations of the size of the glomerular tuft and of the cellular mass of the proximal convoluted tubule support the physiologic concept that variations are to be expected

---

*The term "glomerulotubular balance" was introduced to describe the relations between glomerular and tubular activities in general, but more recently it has been used specifically in respect to the relationship between glomerular filtration and the reabsorption of salt and water in the proximal tubule.

in the relationship between filtering and reabsorptive capacities of individual nephrons. A nephron with relatively high filtering capacity and low reabsorptive capacity would be expected to saturate and spill glucose in the urine at a lower plasma glucose concentration than a nephron with relatively low filtering capacity and high reabsorptive capacity. Smith[100] has considered the implications of glucose titration curves, assuming that there is no kinetic contribution to the splay (K approaching zero) and that the splay therefore results solely from variations in glomerulotubular balance among individual nephrons. His analysis indicates that glomerulotubular balance of glucose transport in individual nephrons of normal kidneys deviates by no more than 50% from the average of all nephrons in the kidney. The population is relatively uniform in normal kidneys. The nature of the morphologic changes seen in patients with chronic renal disease would lead one to predict a greater degree of splay under these circumstances. The extent of the splay, however, appears to remain unchanged in moderate kidney disease and increases only with marked renal insufficiency. The explanation for this is not clear at this time, but it suggests that the splay has primarily a kinetic basis.

In contrast to the results obtained in experiments with unchanging GFR and increasing P_G, a Tm_G is usually not demonstrable when the filtered load of glucose is altered by changing the glomerular filtration rate without altering plasma glucose concentration. The rate of salt and water reabsorption also changes under these circumstances, and a considerable body of evidence relating both to the kidney and to other tissues indicates that glucose reabsorption is in some fashion linked to salt and water reabsorption. As long as the latter is constant, glucose reabsorptive capacity may also be constant, but variations in proximal salt and water reabsorption appear to be associated with parallel variations in Tm.[13,91] Presumably the correlation with salt and water reabsorption is restricted to that in the proximal tubule, since glucose is not reabsorbed from the more distal parts of the nephron.

Renal glycosuria. Renal glycosuria is a congenital defect characterized by the excretion of moderate amounts of glucose in the urine at normal blood glucose levels. This may occur as a consequence of a reduction in Tm_G. More commonly, however, Tm_G is normal and glycosuria results from an increase in the splay of the reabsorption curve, perhaps reflecting an abnormal degree of nephron heterogeneity. All other aspects of renal function, including GFR and other proximal functions, appear to be normal. The glucose tolerance test is also normal. There is no evidence that the condition leads to progressive renal failure or is related to diabetes mellitus. The condition is at times clearly heritable.

Active reabsorption of some other substances

A variety of other substances are actively reabsorbed in the proximal tubule by mechanisms that seem generally similar to that responsible for glucose reabsorption. The list includes phosphate, amino acids, urate, ascorbate, citrate, malate, α-ketoglutarate, acetate, β-hydroxybutyrate, and lactate.

Phosphate. Phosphate is reabsorbed in all parts of the nephron, but chiefly in the proximal tubule by an active transport mechanism that is Tm limited. In contrast to glucose, however, appreciable quantities of phosphate are normally present in the urine. Thus the kidney participates in the regulation of the plasma phosphate concentration. Parathormone promptly depresses Tm_{PO_4} and increases phosphate excretion when given intravenously to normal and hypoparathyroid patients. The variability of Tm_{PO_4} seen in normal individuals apparently results from differences in rate of secretion of parathormone.

Amino acids. Amino acids are present in significant concentrations in plasma, but in only trace amounts in normal urine. A Tm has been demonstrated for some amino acids. A number of mechanisms with varying degrees of specificity are responsible for reabsorption of the various amino acids. One mechanism is responsible for reabsorption of the dibasic amino acids and cystine, another for the neutral amino acids, a third for the dicarboxylic amino acids, a fourth for the iminoacids and glycine, and a fifth for the beta aminoacids.

The term "Fanconi syndrome" has been applied to a wide variety of clinical conditions characterized by multiple proximal tubular transport defects leading to generalized aminoaciduria, glycosuria, phosphaturia, and disturbances of the acidification mechanism and urate excretion. The syndrome may appear in early childhood as a generally fatal condition or later in life in a less serious form. In recent years a number of rare hereditary and acquired diseases have been described in which there is disturbance of the reabsorption of one or more amino acids.

Urate. Urate is subject to filtration, reabsorption, and secretion, but the relative importance of these three mechanisms differs in various species of mammals. Thus it has been difficult to extrapolate from experiments on animals to urate excretion in humans. Ordinarily in humans the clearance of urate approximates 5% to 10% of the GFR, indicating that most of the filtered urate is reabsorbed. An active reabsorptive mechanism with Tm characteristics, presumably located in the proximal tubule, has been demonstrated. It is not certain whether the urate excreted in the urine represents a fraction of the filtered urate that escapes reabsorption or whether it is urate that has been secreted at some point along the tubule. Urate secretion regularly occurs in some mammals and in humans under certain experimental conditions. The urate clearance may exceed the inulin clearance, demonstrating tubular secretion. The uricosuric action of several drugs is utilized in the treatment of gout. Interestingly, some drugs may either increase or decrease urate excretion, depending on the dosage employed. For example, salicylates at low-dosage levels decrease urate excretion, but at high-dosage levels, urate excretion increases. Since urate is subject to bidirectional tubular transport, it is difficult to determine whether the effect of the drug is on the reabsorptive or the secretory component of the transport mechanism.[113,121]

Sulfate. Sulfate is ordinarily almost completely reabsorbed from the urine, and a Tm can be demonstrated. There is some evidence of competition between sulfate, phosphate, and glucose for reabsorption.

TUBULAR SECRETION

Tubular secretion, transtubular transport from blood to lumen, may be by either an active or passive mechanism. The dye phenol red (phenolsulfonphthalein, commonly known as PSP in clinical studies) was the first substance that was proved to be excreted as a result of tubular secretion in the mammalian kidney. The same mechanism that actively transports phenol red seems to transport a variety of organic acids. Another mechanism is capable of transporting several organic bases. As with active reabsorptive mechanisms, there is evidence for saturation and competition for the carrier in both secretory systems. *Passive* mechanisms for secretion are highly specialized and include the influx of cations along a transtubular electrical potential gradient and nonionic diffusion of uncharged lipoid-soluble molecules. In the distal convolution, reabsorption of sodium ion results in the establishment of a transtubular potential gradient with the lumen negatively charged. This provides a driving force that results in the inward movement of potassium and hydrogen ions, as well as the outward movement of chloride ions. This

mechanism is described in greater detail in the sections on potassium and hydrogen ion excretion. Nonionic diffusion is discussed at the end of this section.

Active secretion. Studies by Marshall and Vickers[77] provided the first definitive evidence for a process of tubular secretion in the mammalian kidney, a topic that had been warmly debated for many years. Several lines of evidence were obtained, the most convincing of which was the fact that following administration of phenol red, as much as 70% of the dye could be removed from plasma during a single circulation through the kidney. Since only 25% of the dye could have been filtered (75% is protein bound) even if all the plasma were filtered, which was clearly impossible, it was quite obvious that the rate of excretion was many times greater than could be explained by filtration alone. Marshall also found that the rate of excretion of phenol red did not increase in direct proportion to the plasma concentration, as is required for a substance excreted only by filtration, but approached a maximal value at high plasma concentrations typical for active transport processes. Following the introduction of the technique for measuring glomerular filtration rate a number of organic molecules such as *p*-aminohippurate (PAH), penicillin, some sulfonamides, such x-ray contrast materials as iodopyracet (Diodrast), sodium methiodal (Skiodan), etc., have been shown to have a clearance exceeding that for inulin, demonstrating the process of tubular secretion. It has often been noted that none of these substances occur normally in plasma and the purpose of the secretory system has been questioned. Pitts speculates that perhaps the mechanism was evolved to secrete some unknown substance that, because of high toxicity, must be maintained in very low concentration in the body fluids.[85] In any event, it serves for the rapid excretion of ingested foreign substances as conjugates in the form of hippurates, glucuronides, and sulfates. As is the case with many of the active reabsorptive mechanisms, the active secretory mechanisms show evidence of Tm limitation and are localized to the proximal tubule.

Again, quantitative consideration of the relations between rates of filtration, secretion, and excretion as functions of the plasma concentration are helpful in understanding the mechanism. These relations for PAH are shown in Fig. 50-10 and are determined as follows:

$$T_{PAH} = U_{PAH} V - GFR \cdot P_{PAH} \cdot f$$

A portion of the PAH in plasma is bound to the plasma proteins, as is the case with most foreign

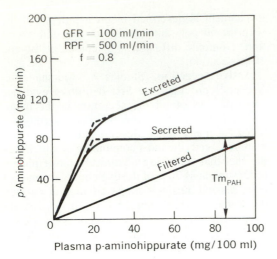

Fig. 50-10. Idealized relationship between plasma concentration and filtered, secreted, and excreted *p*-aminohippurate in humans. GFR and renal plasma flow are assumed to be constant at 100 and 500 ml/min, respectively.

substances secreted by the tubule. About 20% is protein bound and 80% is filtrable (f = 0.8). At constant GFR the rate of filtration of PAH is, of course, proportional to the filtrable plasma concentration. At low plasma concentrations, almost all the plasma PAH that escapes filtration in the glomerular tufts is removed from the peritubular capillary blood by the secretory activity of the proximal tubular cells. It is this high rate of extraction at low plasma concentrations, 85% on the average in normal humans, that is the basis for the use of PAH for measurement of renal plasma flow, as discussed earlier. As the plasma concentration is increased, the rate of secretion also increases, but, as with glucose reabsorption, a plasma concentration is reached above which there is no further increase in the amount of PAH secreted. The amount of PAH excreted continues to increase with increasing plasma concentration as more is filtered. Also, as is the case with glucose reabsorption, there is splay in the titration curve that can be explained in a similar fashion. The magnitudes of the Tm_{PAH} and Tm_G are often used as measures of the functional mass of the proximal convoluted tubules.

Nonionic diffusion. Certain ionic substances may at times appear in the urine in concentrations that suggest active secretion or reabsorption, but it should not automatically be assumed that an active transport mechanism is indeed involved. These ions are generally the anions of weak organic acids or cations of weak bases, and on careful examination, it will be observed that their excretion rates are strongly influenced by changes in urinary pH and to a lesser extent by changes in the rate of urine flow. Many weak acids and bases in their nonionized form are relatively lipid soluble and diffuse readily across cell membranes. If the pH of the tubular urine is appropriate, these substances, on diffusing into the tubular lumen, may be converted into an ionized form that, because of its low lipid solubility, is unable to diffuse out of the tubule and is thus trapped in the lumen and excreted. The excretion of the anions of weak acids is thus maximal when the urine is alkaline, and the excretion of weak bases increases when the urinary pH falls. The excretion of the weak base ammonia occurs by nonionic diffusion, the NH_3 becoming trapped in an acid urine in the form NH_4^+ (see later discussion). The reabsorption of certain ionic substances filtered at the glomerulus may also occur through a process of nonionic diffusion, and the reabsorption of bicarbonate after conversion to carbon dioxide, described in greater detail later, is an example.

The phenomenon of nonionic diffusion is at times employed to clinical advantage in the treatment of barbiturate intoxication, particularly when phenobarbital is the offending drug. Phenobarbital is a weak organic acid that diffuses across biologic membranes in its un-ionized form. A patient suffering from phenobarbital intoxication is customarily given a large intravenous infusion of $NaHCO_3$ to produce an alkaline urine. The phenobarbital thus becomes trapped in the urine in its anionic form, and its elimination from the body is hastened.

WATER EXCRETION: THE COUNTERCURRENT MECHANISM

An essential function of the kidney is control of the water content of the body so that the total solute concentration (osmolality) of the body fluids in normal humans is maintained within very narrow limits despite wide variations in fluid intake and extrarenal water losses. The basic regulatory mechanism involves the excretion of urine with high osmotic concentration when the osmolality of the body fluids increases and the excretion of urine osmotically less concentrated than the body fluids when their osmolality falls. In this fashion the osmolality of the body fluids is closely regulated and normally varies little from the average value of 285 mOsm/kg of water.

It is important to differentiate between the

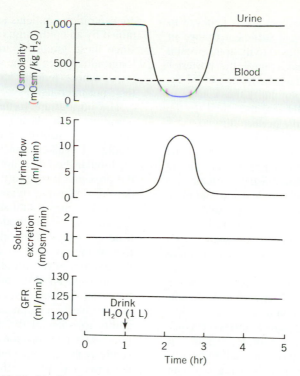

Fig. 50-11. Idealized representation of findings during water diuresis.

osmolality of the body fluids and the COP exerted by the plasma proteins. Osmolality is determined by the total number of osmotically active particles in solution, and proteins contribute less than one mOsm/kg to the plasma osmolality because of their low molar concentration. They play an important role in fluid transfers across the capillary wall, however, because their large molecular size restricts their transcapillary movement (Chapter 49). One mole of an "ideal" solute (i.e., all the particles are osmotically active) in 1 kg of water produces a solution with an osmolality of 1 Osm/kg. One mole of sodium chloride in 1 kg of water would produce a solution with an osmolality of 2 Osm/kg, if all the sodium and chloride ions were dissociated and osmotically active. Since the body fluids are dilute solutions, it is more convenient to express their osmolality in terms of milliosmoles (1 Osm = 1,000 mOsm). The osmolality of a solution is usually determined by measuring its freezing point depression. One mole of "ideal" solute in 1 kg of water depresses the freezing point of the solution by 1.86° C. Thus normal plasma with an osmolality of 285 mOsm/kg of water has a freezing point of $-0.53°$ C.

The osmolality of human urine can vary between extremes of 85 and 1,400 mOsm/kg of water. In the basal state the rate of urine flow is low, perhaps 1 ml/min, and the urine is moderately concentrated. If under these circumstances 1 L of water is ingested, a series of events occurs, as shown in Fig. 50-11. Urine flow begins to increase within 15 to 20 min, and in about 1 hr, it reaches a maximum rate that does not exceed 12% to 15% of the GFR. If no more water is ingested, the urine flow soon decreases and returns to the previous rate in approximately 1 hr. The urine osmolality is inversely related to the rate of urine flow, and the rate of solute excretion remains constant. Since the GFR also remains the same, the change in water excretion results from differences in tubular reabsorption of water. The decrease in water reabsorption under these circumstances results from a decreased secretion of antidiuretic hormone (ADH), or vasopressin, which is caused by the fall in plasma osmolality following water ingestion. A water diuresis can be interrupted by administration of ADH.

The rate of release of ADH is determined by the osmolality of the blood perfusing the brain in the area of the supraoptic nuclei. Certain cells in this area are believed to act as osmoreceptors, swelling when the osmolality of the blood to which they are exposed is diminished and shrinking when the osmolality is increased. A change in osmolality of no more than 2% is believed to be effective in producing a sufficient change in osmoreceptor cell size to lead to the transmission of nervous impulses that control the posterior pituitary gland and regulate the release of ADH. An increase in plasma osmolality thus causes an increased release of ADH into the blood. In the kidney, ADH increases the permeability to water of the cells of the distal convolution and collecting ducts, and less water is excreted. The increase in water reabsorption results in a fall in plasma osmolality toward normal. Similarly, a decrease in plasma osmolality leads to inhibition of ADH release, and the cells of the distal convolution and collecting ducts become less permeable to water. More water is excreted, resulting in an increase in osmolality of the body fluids towards normal.

The classic work on the control of ADH secretion was performed by Verney.[110] He demonstrated that antidiuresis could be evoked in dogs by the injection into the carotid artery of hypertonic solutions of sodium chloride, sodium sulfate, or sucrose in amounts too small to have an effect when given intravenously. Urea was without effect on ADH release, suggesting that the osmoreceptors are freely permeable to urea and that they do not change in volume when exposed to urea. More recent studies demonstrate that the rate of ADH release is also determined by the volume of the blood as well as its tonicity. "Volume receptors," probably stimulated by changes in stretch, are located in the left atrium, and perhaps elsewhere, and modulate the rate of ADH secretion (Chapter 61).

Countercurrent mechanism. One of the remarkable achievements in mammalian physiology in recent years has been the increase in understanding of the mechanism of water reabsorption from the nephron. The countercurrent hypothesis for urinary concentration and dilution has provided great insight, and the renewed application of micropuncture technology has provided much new information concerning the location and nature of the processes involved. The countercurrent hypothesis was originally proposed by the Swiss physical chemist Werner Kuhn[50]; his associate Wirz[117] was the first to provide experimental data in its support. The great theoretical advantage of the countercurrent mechanism is that it becomes possible to explain all water movement as passive and secondary to solute movement, and it is no longer necessary to postulate active transport of water to explain the process of urine concentration. Also, at no location in the kidney is it necessary for large osmotic

concentration gradients to be produced and maintained by an epithelium only one cell layer thick, since these gradients are established along the longitudinal axes of the straight medullary structures, whose lengths are measured in centimeters, not microns.

The ultrafiltrate formed at the glomerulus is isosmotic with plasma, and it has been known since the pioneering mammalian micropuncture studies of Walker et al.[111] that the osmolality of the tubular fluid does not change as it flows along the proximal tubule in the concentrating kidney, despite extensive reabsorption of solute and water. It is now established that within the limits of the analytic methodology, proximal tubular fluid is always isosmotic with plasma irrespective of the presence or absence of ADH and of the fraction of the filtrate reabsorbed in this part of the nephron (Fig. 50-12). The driving force for water reabsorption is the reabsorption of solute, primarily sodium salts, and the high permeability to water of the proximal epithelium permits water reabsorption to occur pari passu with solute reabsorption. Only the first two thirds of the proximal tubule, the pars convoluta, is accessible to micropuncture, but other studies indicate that the tubular fluid remains isosmotic as it flows through the cortical portion of the pars recta. The tubular fluid first becomes hyperosmotic to arterial plasma as it flows down the medullary portion of the descending limb of the loop of Henle, and it reaches its maximum concentration at the bend of the loop. The osmolality of fluid at the bend of the loop is higher in the presence of ADH than in its absence, but in all conditions, it is hyperosmotic to arterial plasma (Fig. 50-13). The increase in osmolality of fluid flowing along the thin descending limb of the loop results from the outward diffusion of water and inward diffusion of solutes, primarily sodium, chloride, and urea. The relative contribution of these two processes to the increase in osmotic concentration is not clear at this time.

Beyond the bend of the loop the tubular fluid becomes progressively more dilute as it returns toward the cortex. This results from the reabsorption of sodium chloride by an epithelium that has a relatively low permeability to water. The mechanism for reabsorption of sodium chloride probably differs in the thick and thin ascending limbs of the loop. In the thick ascending limb, chloride is actively reabsorbed and sodium follows passively. There is uncertainty about the mechanism in the thin ascending limb and, as discussed subsequently, both active and passive mechanisms have been proposed.

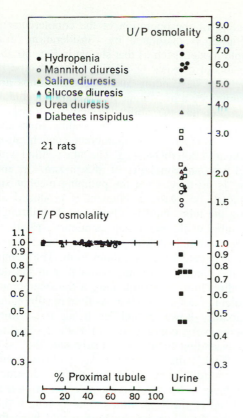

Fig. 50-12. Osmolality ratios of proximal tubular fluid and urine in rats. (From Gottschalk and Mylle.[46])

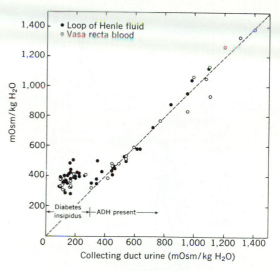

Fig. 50-13. Relation between osmolality of collecting duct urine and fluid from loops of Henle and vasa recta blood in various normal desert rodents and hamsters with experimental diabetes insipidus. Some values obtained in presence of ADH were from DI hamsters following administration of exogenous vasopressin. (From Gottschalk.[43])

Thus solute, primarily sodium chloride, is transported transtubularly with minimal movement of water. This increases the solute concentration in the interstitium of the medulla and decreases the osmolality of the fluid remaining in the tubule. It is the basic mechanism for producing hypertonicity of the medulla, and the fluid arriving in the early part of the distal convolution is always hypoosmotic. In the absence of ADH the epithelium of the distal convolution and collecting ducts has a very low permeability to water, the tubular urine flows through the remainder of the tubule without achieving osmotic equilibrium with the surrounding interstitial fluid, and a hypoosmotic urine is excreted. ADH causes the permeability to water of these parts of the nephron to increase[44,116]; thus in the presence of ADH the tubular fluid becomes isosmotic with interstitial fluid in the distal convolutions and/or the cortical collecting ducts by the outward diffusion of water (Fig. 50-14). In rats, tubular fluid becomes approximately isosmotic with plasma by the end of the distal convolution, whereas in dogs[29] and perhaps in man the fluid does not become isosmotic until it reaches the cortical collecting ducts. In the presence of ADH, water diffuses freely across the collecting duct epithelium into the medullary interstitium, and the urine emerging from the ducts of Bellini has the same osmotic concentration as the surrounding interstitium. Thus a hyperosmotic urine is formed without requiring the active transport of water. The water reabsorbed from the cortical collecting ducts and distal convolutions enters the peritubular capillaries of the cortex and is removed from the kidney without interfering with the concentrating mechanism in the medulla. Water reabsorbed from the collecting ducts is carried away from the medulla in the vasa recta.

The mechanism by which ADH increases the permeability to water of some epithelial membranes is poorly understood, and it will be mentioned here only in passing. A more detailed discussion is found in Chapter 61. Koefoed-Johnsen and Ussing[59] proposed that vasopressin accelerates osmotic flow of water across epithelial membranes by increasing the diameter and/or the number of aqueous channels (pores) in a

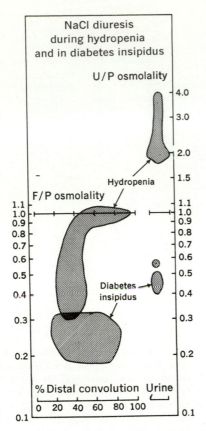

NaCl diuresis
during hydropenia
and in diabetes insipidus

U/P osmolality

Hydropenia

F/P osmolality

Diabetes
insipidus

% Distal convolution   Urine

Fig. 50-14. Comparison of osmolality ratios of fluid from distal convolution and of urine in presence and absence of ADH. (From Gottschalk.[43])

limiting cell membrane. The validity of this interpretation has been tested by comparing diffusional and osmotic water permeabilities, and it is now realized that these calculations are subject to error due to the presence of an unstirred layer of fluid adjacent to the membrane. Alternative hypotheses have been proposed that postulate the existence of two barriers in series with different properties, including differences in response to ADH. Although vasopressin stimulates sodium transport by the frog skin and toad bladder, it does not produce changes in renal salt excretion when administered in physiologic amounts. The action of vasopressin on the renal tubule and isolated epithelial membranes is mimicked by cyclic 3',5'-AMP, and the urinary excretion of this substance varies in parallel with ADH activity in the kidney. Whatever the mechanism of action of ADH, it appears likely that cyclic 3',5'-AMP plays an integral role.[48,81]

Countercurrent multiplication. An under-standing of the operation of the countercurrent multiplier is facilitated by consideration of a simple model, such as that shown in Fig. 50-15. The model consists of a hairpin-shaped tube in which the descending and ascending limbs are separated by a membrane that is impermeable to water but has the ability to actively transport solute from ascending to descending limb and thereby to create a transmembrane concentration gradient of 200 mOsm/kg. The tube is filled with fluid with an osmolality of 285 mOsm/kg, and there is no flow when the pumping mechanism is turned on. After a gradient is produced all along the tube, fluid is allowed to flow through the tube while the pump continues. Fluid of higher osmolality flows around the bend of the loop and enters the ascending limb. The pumping mechanism raises the osmolality in the descending limb by maintaining a transmembrane gradient of 200 mOsm/kg. As fluid of still higher osmolality flows around the bend, it becomes progressively concentrated in both descending and ascending limbs. In a steady state the "single effect," the 200 mOsm/kg gradient at any level, is multiplied in the longitudinal direction to achieve a high but transient concentration at the bend of the loop. In a simple system of this sort the fluid leaving the loop would have the same osmolality as that which enters, and in the build-up phase, solute is trapped and recycled in the loop.

In the kidney (Fig. 50-16) the limbs of the loop are not in direct contact, and the exchanges occur between the limbs of the loop and the surrounding interstitial fluid. Also there is net transport of sodium chloride out of the water-impermeable ascending limb into the interstitium, and the fluid leaving the loop of Henle is hypoosmotic. It is apparent that the magnitude of the longitudinal gradient established will depend on (1) the magnitude of the single effect, (2) the length of the loop along which it is multiplied, and (3) the volume flow through the loop, unless the amount of solute transported is in some fashion adjusted to the volume flow. Longitudinal diffusion in the two limbs will limit the magnitude of the total gradient that can be established.

Although the overall operations of the countercurrent mechanism are clear, important details of the process remain to be established, particularly the function of the thin ascending limb of the loop of Henle in the inner medulla. Transport of solute with restricted water movement must occur out of some tubular structure in each portion of the kidney in which a longitudinal interstitial osmotic gradient develops. The osmolality

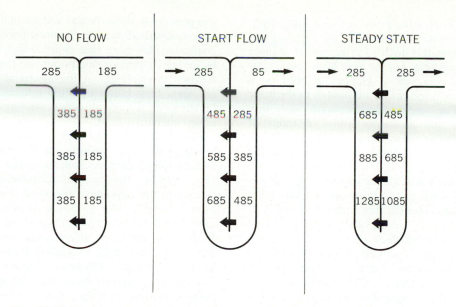

Fig. 50-15. Model to demonstrate principle of countercurrent multiplication. (See text.)

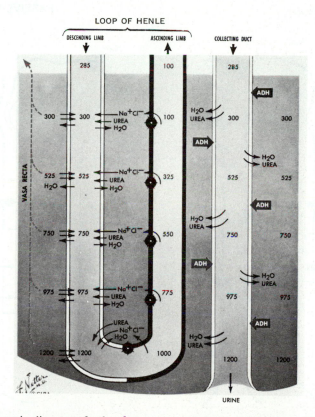

Fig. 50-16. Schematic diagram of role of countercurrent multiplier in loop of Henle in creating high interstitial gradient and interactions between loop of Henle and collecting duct in presence of ADH. (© Copyright 1973 CIBA Pharmaceutical Company, Division of CIBA-GEIGY Corporation. Reproduced, with permission, from THE CIBA COLLECTION OF MEDICAL IL-LUSTRATIONS by Frank H. Netter, M.D. All rights reserved.)

increases progressively but not uniformly from the corticomedullary junction to the papillary tip. It is approximately doubled at the border of inner and outer medulla, but the very high solute concentrations achieved in kidneys elaborating a highly concentrated urine are limited to the inner medulla. The inner medulla contains only thin descending and ascending limbs of loops of Henle, collecting ducts, and vasa recta. Thick ascending limbs of Henle are absent from the inner medulla.

The epithelium of the thick ascending limb has a very low water permeability and actively reabsorbs sodium chloride; this is the mechanism of the "single effect" in the outer medulla. Recent studies on isolated thick segments from rabbit kidneys by Burg and Green[26] and Rocha and Kokko[92] demonstrate that it is chloride and not sodium that is actively transported by the thick segments. The transtubular potential difference

was positive in the lumen as sodium and chloride were reabsorbed against chemical concentration gradients, indicating that chloride is reabsorbed by an active electrogenic pump and that sodium follows passively down the electrical potential gradient. This dilutes the fluid remaining in the lumen while osmotically concentrating the medullary interstitial fluid.

The function of the thin ascending limb of Henle in the inner medulla and its participation in the concentrating mechanism remain uncertain (Fig. 50-17). The original formulation of the countercurrent hypothesis proposed that the thin ascending limb, like the thick, is water impermeable and actively reabsorbs sodium chloride. In this view the single effect is the same or similar in thin and thick ascending limbs, despite marked structural differences. Many in vivo micropuncture results have been interpreted as being consistent with this view, but the data are

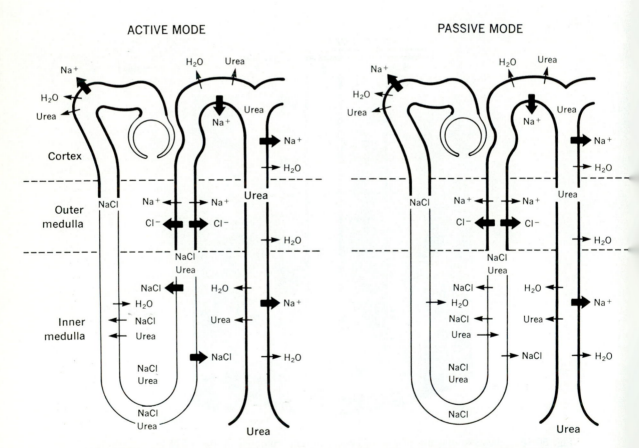

Fig. 50-17. Diagram of countercurrent mechanism with active or passive mode of operation of thin ascending limb of Henle's loop. (See text.) Narrow and heavy lines outlining nephron indicate thin and thick epithelium. Heavy and thin arrows indicate active and passive transport, respectively.

not conclusive.[34,75,76,84] Isolated thin ascending limbs have a very low water permeability and high passive sodium and chloride permeability, but, contrary to the original hypothesis, there is no definitive evidence of active transport of either sodium or chloride in this segment.[53,55,56,60]

An alternative hypothesis concerning the function of the thin ascending limbs and the generation of the osmotic gradients in the inner medulla has recently been independently proposed by Stephenson[105] and by Kokko and Rector.[61] In their models, osmotic equilibrium in the thin descending limb occurs predominantly through water abstraction, creating a very high concentration of sodium chloride at the bend of the loop (Fig. 50-17). Although the total solute concentration at the bend of the loop is equal to that of the surrounding interstitial fluid, its sodium chloride concentration is higher, since the latter contains both sodium chloride and a high concentration of urea, which has diffused out of collecting ducts in the inner medulla. Thus there is a concentration gradient down which sodium chloride can passively diffuse out of the water-impermeable, highly sodium chloride–permeable thin ascending limbs in amounts sufficient to produce an osmotic "single effect."

Urea plays an essential role in the passive model. It is the "messenger" through which energy generated by active chloride transport in the thick ascending limb is transmitted to the inner medulla. To ensure delivery of large amounts of urea into the interstitium of the inner medulla, the thin descending and thick ascending limbs, distal convolution, and cortical collecting ducts all must have restricted urea permeability, whereas the medullary collecting ducts must have a high urea permeability. The necessary permeability to urea, water, sodium, and chloride have been described by Kokko and Rector[61] for isolated segments of rabbit nephrons. Although usually discussed as alternative and incompatible views, active and passive modes of operation of the thin ascending limbs are not mutually exclusive. Even in the same species, it is conceivable that both systems are operative, with variable contributions to the concentrating process under various conditions.

The countercurrent hypothesis, which postulates that the loop of Henle functions as a source of sodium chloride for the medulla, provides an explanation for certain puzzling observations made in the past. Only those animals that have loops of Henle are able to elaborate a hyperosmotic urine. Phylogenetically, loops of Henle first appear in certain birds, but they are best developed in mammals. Vertebrates without loops of Henle usually produce a hypoosmotic urine, and the maximum urine concentration equals that of arterial plasma. Certain desert rodents have very long renal papillae and a remarkable urinary concentrating ability. The maximal concentrating ability is related to the length of the longest loops and not to the percent of nephrons possessing long loops.

Countercurrent diffusion. Specialized blood vessels in the medulla, the vasa recta, also participate in the countercurrent mechanism. Water and solutes removed from the medullary portions of the loop of Henle and the collecting ducts enter the vasa recta under the influence of the COP. These vessels, because of their anatomic configuration, act as countercurrent diffusion exchangers, minimizing the loss of solute in the blood flowing out of the medulla. Freely diffusible substances such as sodium chloride or urea, which are added to the medullary interstitium by reabsorption from the tubules, tend to be trapped there because of countercurrent diffusion in the vasa recta. Freely diffusible substances brought to the medulla by the vasa recta tend to be excluded, and some of the plasma water short-circuits across the tops of the vasa recta.

Blood flowing down the descending vasa recta becomes osmotically more concentrated because of the inward diffusion of solutes from the interstitium and the outward diffusion of water.[115] The reverse occurs in the ascending vasa recta. Since all water and solute movements across the walls of the vasa recta are believed to be passive, osmotic equilibration is not complete at any level, and the blood leaving the medulla must be somewhat hyperosmotic to that which enters. The osmolality of the blood leaving the medulla is unknown, but the loss of solutes from the medulla in this fashion is certainly far less than if the blood left the medulla while maximally concentrated (Fig. 50-18). The low volume and velocity of medullary blood flow are important factors in efficient countercurrent diffusion exchange. Increased medullary blood flow would be expected to result in increased "wash out" of medullary solutes and decreased urine concentration.

Countercurrent diffusion has implications with respect to oxygen utilization and carbon dioxide production in the medulla. Carbon dioxide is highly diffusible, and if it were produced in the medulla, it would be trapped there, and high concentrations of carbon dioxide in the medullary interstitium and in the urine would result. Also,

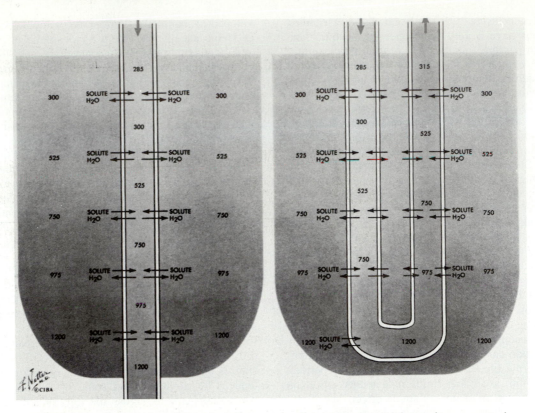

Fig. 50-18. Diagram of model to demonstrate role of countercurrent exchange in vasa recta in minimizing dissipation of medullary osmotic gradient. In both models, vasa recta are assumed to be freely permeable to water and solutes. If blood exits while highly concentrated, as in model on left, medullary solute concentration will be quickly dissipated. This does not occur in kidney, as is shown on right, since vasa recta return to cortex and blood leaves medulla only slightly more concentrated than on entry, thereby minimizing loss of solutes from medulla. (© Copyright 1973 CIBA Pharmaceutical Company, Division of CIBA-GEIGY Corporation. Reproduced, with permission, from THE CIBA COLLECTION OF MEDICAL ILLUSTRATIONS by Frank H. Netter, M.D. All rights reserved.)

if the energy requirements of the medulla were met by oxidative metabolism, the P_{O_2} of the blood leaving the medulla would be less than that on entry, and the gradient would cause a short-circuit of the oxygen across the tops of the vasa recta, with the unfortunate consequence that the greater the utilization of oxygen in the medulla, the less would be the delivery of oxygen to this region. Medullary energy requirements, not surprisingly, are met in large part by anaerobic glycolysis.

Osmotic diuresis. An osmotic diuresis is a diuresis that results from an increased rate of solute excretion. The solute excretion rate and maximal urine osmolality are inversely related, and the latter falls toward the isosmotic value as solute excretion increases. Regularly in dogs and

rarely in man the urine may actually become hypoosmotic at very high excretion rates, even in the presence of maximum ADH activity. This reflects a failure to achieve osmotic equilibration in the collecting ducts because of the very high volume flow through them. In osmotic diuresis a greatly increased volume of fluid is presented to the collecting ducts, leading to an increased rate of water reabsorption from them and a decrease in medullary tonicity.

An osmotic diuresis can result from the infusion of a poorly reabsorbable solute such as mannitol or glucose in excess of its transfer maximum, increased excretion of sodium salts following increased dietary intake, intravenous saline solution infusion or administration of a diuretic drug to block sodium reabsorption,

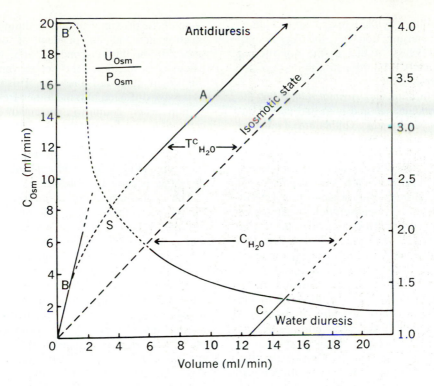

Fig. 50-19. Concentrating and diluting operations of human kidney. (See text.) (From Smith.[102])

increased amounts of urea for excretion, etc.

The osmolar clearance is readily calculated as follows:

$$C_{Osm} = \frac{U_{Osm}\ V}{P_{Osm}}$$

Smith[101] pointed out that C_{Osm} is greater than the rate of urine flow when the urine is hyperosmotic and less than the urine flow when the urine is hypoosmotic. If one considers the volume of water required to contain the excreted solute in isosmotic concentration (C_{Osm}), an additional volume of water (C_{H_2O}) must be added when the urine is hypoosmotic and subtracted when the urine is hyperosmotic to equal the actual volume of urine excreted. Thus:

$$V = C_{Osm} + C_{H_2O}$$

C_{H_2O}, termed the free water clearance, has a positive value when the urine is hypoosmotic and a negative value when the urine is hyperosmotic. A negative free water clearance is designated as $T^C_{H_2O}$, and:

$$T^C_{H_2O} = C_{Osm} - V$$

The usual relation between these variables is shown in Fig. 50-19.

Smith postulated that $T^C_{H_2O}$ represented the rate of water abstraction from the collecting ducts in the concentrating kidney and C_{H_2O} the volume of isosmotic tubular fluid from which salt was reabsorbed in the diluting kidney. Although $T^C_{H_2O}$ and C_{H_2O} give some estimate of the magnitude of these operations, they can by no means be considered exact measures. This is because solute reabsorption continues in the collecting ducts and is accompanied by unimpeded water reabsorption in the presence of ADH. Furthermore, some water reabsorption occurs across the distal convolution and collecting duct epithelium, even in the absence of ADH. Thus although $T^C_{H_2O}$ and C_{H_2O} may be precisely measured, they are of questionable physiologic significance.

SODIUM EXCRETION

Sodium salts constitute 90% or more of the solutes of extracellular fluid and are thus the major solutes filtered as well as reabsorbed by the nephrons. Their rate of reabsorption must be regulated to maintain the proper sodium content in the body. Water reabsorption is regulated in turn to maintain a proper osmolality of the body fluids. The active transport of sodium appears to represent the greatest requirement for energy expenditure by the kidney, and the supra-basal oxygen consumption (i.e., the oxygen consumption in excess of that in the nonfiltering kidney) is directly proportional to the rate of sodium reabsorption.[62,106]

The epithelium of the *proximal tubule* has the high ionic conductance and high hydraulic conductivity typical of a leaky epithelium. Transport of solute and water is isotonic in such epithelia and only relatively small ionic concentration gradients are generated. Sodium reabsorption here is characterized by (1) a large capacity, (2) the ability to produce only a small transtubular sodium gradient, (3) reabsorption of an equal amount of anions, and (4) a small transtubular electrical potential difference (PD).* The last appears to vary from 1 to 2 mV (lumen negative) in the first part to 1 to 2 mV (lumen positive) in the latter part of the tubule. The fraction of filtered sodium reabsorbed in the proximal tubule is variable, but probably amounts to about two thirds of the filtered load under conditions of normal hydration. The concentration of sodium in glomerular filtrate is essentially the same as that in plasma, and ordinarily it remains unchanged along the proximal tubule despite a high rate of reabsorption. This is because there is equal reabsorption of anions and rapid reabsorption of water so that the total solute concentration of the tubular fluid remains unchanged. In the presence of a nonreabsorbable solute such as mannitol, a transtubular sodium gradient of as much as 40 mM can be produced in the proximal tubule of the rat kidney. Under these circumstances the nonreabsorbable solute remains in the tubular fluid, osmotically obligating water, and the sodium concentration falls. The minimum tubular fluid sodium concentration is achieved when the rate of reabsorption equals the net passive back flux into the tubule, which in turn is determined by the transtubular sodium gradient and the permeability of the epithelium. These observations provide evidence that sodium reabsorption involves active transport, since the sodium is being moved against its electrochemical gradient. Although total anion reabsorption must equal total cation reabsorption, the proximal tubular fluid chloride concentration rises because of the preferential reabsorption of bicarbonate. This provides a gradient favoring chloride reabsorption, and there is no need to postulate an active reabsorptive mechanism for chloride in this part of the tubule.

The *loop of Henle* has different sodium transport properties in its various portions. In the pars recta of the proximal tubule, sodium transport is electrogenic and independent of organic solutes.[57] In the thin descending limb, which is located in the medulla, the net movement of sodium is in an inward direction. This is the result of passive diffusion into the lumen from the interstitium, where sodium is present in higher concentration. As discussed earlier, the mechanism of reabsorption of sodium chloride in the thin ascending loop remains uncertain, and in the thick ascending loop, it is chloride and not sodium that is actively reabsorbed.

The epithelium of the *distal convolution* has the characteristics of a "tight" epithelium. Sodium reabsorption is characterized by (1) a small capacity, (2) the ability to produce large transtubular concentration gradients, (3) the establishment of a large transepithelial electrical potential difference, and (4) association with either anion reabsorption or cation secretion, that is, exchange of sodium ion for hydrogen or potassium ion.[39,41,109,114] Under conditions of normal hydration, approximately 10% of the filtered sodium is reabsorbed in the distal convolution and collecting ducts, but the magnitude of reabsorption is variable, depending on the physiologic state of the individual, and is precisely regulated in some fashion. The "fine tuning" of sodium excretion probably occurs in the collecting ducts. There are significant but variable transtubular sodium concentration and electrical potential gradients all along the distal convolution, indicating that sodium reabsorption occurs by active transport. The sodium concentration of distal tubular fluid may be as low as 20 mM (transtubular gradient of 120 mM) in rats. The difference in maximum transtubular gradients across proximal and distal tubules may relate more to differences in passive permeability and back leak of sodium, probably via the intercellular junctional pathways, than to differences in the sodium pump itself. The transtubular electrical potential difference along the rat distal convolution varies with the location and averages about −10 mV, lumen negative, in the first third, and −45 mV or more in the latter part of the convolution.[118] The potential difference is a function of the anions available for reabsorption and becomes greater if poorly reabsorbable anion is present in the tubular fluid in high concentration. The chloride concentration may decrease to very low levels in the distal tubule and collecting ducts. It appears that active transport must be involved to account for chloride reabsorption from these parts of the nephron under these circumstances. When the chloride concentration does not fall so low, its reabsorption can be explained

---

*See references 1, 38, 41, 54, 109, and 114.

as passive and resulting from the negative intratubular potential.

There is little information concerning the molecular aspects of the mechanism(s) for transtubular sodium movement, but presumably it has features in common with the ubiquitous sodium pump in animal cells and is related in some poorly understood fashion to sodium-potassium–activated ATPase. The entry of sodium into the tubular epithelial cell across the luminal membrane is presumably a passive process, and the pump is thought to be located in the membrane along the lateral intercellular spaces and/or at the base of the cell. There is evidence for a shunt pathway for sodium reabsorption, presumably involving the intercellular junctions, and there are probably several mechanisms of transport.[41]

The possibility that salt and water reabsorption from the proximal tubule is passive and is due to the COP of the plasma proteins has been proposed periodically for many years, only to be rejected. The permeability of the nephron does not appear to be great enough for the plasma proteins to provide the primary driving force for tubular reabsorption of water. This view is given further support by the finding that protein placed in the fluid in the proximal tubule in a concentration similar to that of plasma exerts little influence on reabsorption. Nor could the COP lead to the formation of transtubular gradients for sodium and electrical potentials that have already been mentioned as evidence for active transport. However, recent work by Brenner and Troy,[20] Earley,[37] Lewy and Windhager,[67] and Martino and Earley,[78] indicates that the COP and other physical factors, including hydrostatic pressure, may play a role in regulating the rate of proximal salt and water transport, perhaps by some effect on flow in the intercellular labyrinths and absorption into the peritubular capillaries. This is discussed later in more detail.

Factors regulating sodium excretion

For the body to remain in sodium balance the loss of sodium must equal the dietary intake. Except for variable and largely uncontrollable but usually small losses in sweat, the loss of sodium from the body occurs almost exclusively via the urine, and the rate of sodium excretion is appropriate for the rate of salt ingestion even in advanced renal disease. Thus it is apparent that the renal excretion of sodium is precisely regulated in some fashion. The nature of this regulation is not well understood, but almost certainly there are several control mechanisms that participate in the process, and their relative importance probably varies with the circumstances. The control factors that are currently recognized are discussed here.

Glomerular filtration rate. The rates of filtration and reabsorption of salt and water are both much greater than the rate of excretion; this has important implications. Consider the consequences if the reabsorptive rate were unchanged while the GFR varied. For purposes of exposition, it is simpler to consider rates of filtration, reabsorption, and excretion of water, assuming that all water reabsorption is isosmotic and secondary to sodium transport. Under these circumstances the following relationships would hold:

Condition	GFR (ml/min)	Reabsorption (ml/min)	Excretion (ml/min)
Normal	125	124	1
0.8% decrease in GFR	124	124	0
20% increase in GFR	150	124	26

Experimental data demonstrate that these relationships do not exist. A priori one would doubt that such an important physiologic function as sodium excretion would result from the interaction of two operations of such large magnitude without a close coupling mechanism. If GFR and tubular reabsorption were not closely coupled, small variations in one function or the other would result in large percentage variations in rate of excretion, with disastrous consequences. Studies in which acute changes in filtration rate are experimentally induced demonstrate that the type of relationship just given does not hold. Instead there is a close coupling between the rate of filtration and rate of reabsorption, so that a change in one is usually associated with a proportional change in the other. Generally an increase in GFR is accompanied by an increased rate of reabsorption and a considerably smaller increase in the rate of excretion; with decrease in GFR there is decreased absolute reabsorption and decreased excretion. The following values are more illustrative of actual events, but the figures shown have meaning only in demonstrating the general nature of the relationship.

Condition	GFR (ml/min)	Reabsorption (ml/min)	Excretion (ml/min)	% Excreted	% Reabsorbed
Normal	125	124	1.0	0.8	99.2
Decrease in GFR	62.5	62.25	0.25	0.4	99.6
Increase in GFR	150	145	5	3.3	96.7

At all levels of GFR the rate of sodium reabsorption closely parallels its rate of filtration, a relationship included in the concept of glomerulotubular balance. The relationship holds for the entire uriniferous tubule, but in the contemporary micropuncture literature the use of the term is usually restricted to indicate "unchanged" fractional reabsorption in the proximal tubule. This is a much narrower definition of "glomerulotubular balance" than originally used by Smith,[101] who had in mind a broader concept that included filtration and reabsorption of all solutes throughout the kidney. Glomerulotubular balance for salt and water apparently can be maintained by an entirely intrarenal mechanism, since changes similar to those shown can be produced by decreasing or increasing the perfusion pressure to a single kidney without perturbations in extracellular volume, systemic arterial pressure, or other known changes outside the kidney. The mechanism whereby the coupling takes place is unknown, but it has been generally assumed that the GFR is the independent variable and that the appropriate adjustments in the reabsorptive rate occur in response to primary changes in GFR.

One hypothesis in vogue in recent years proposed that the rate of salt and water reabsorption out of the proximal tubule was proportional to the cross-sectional area of the tubule.[24,40] If this were true, increases or decreases in rates of filtration and reabsorption would be accompanied by proportional changes in tubular cross-sectional area. Although this relationship may exist under certain restricted circumstances, there have been convincing demonstrations from several laboratories that this is not generally true, and the hypothesis has now been discarded.*

Physical factors—the hydrostatic pressures and COP, hematocrit, blood viscosity, etc.—are considered by some investigators to be of primary importance in the regulatory mechanism.[20,37,67] It is accepted that hydrostatic and osmotic pressures are responsible for reabsorption from interstitium into peritubular capillaries in the kidney, as proposed by Starling for capillaries elsewhere in the body. For example, if renal plasma flow is unchanged, an increased rate of filtration and hence increased filtration fraction will result in an increase in colloid osmotic pressure in the blood leaving the glomeruli, and this presumably induces increased reabsorption of fluid from the interstitium into the peritubular capillaries surrounding the tubules. Whether this in turn in some fashion, perhaps indirectly, influences the rate of sodium reabsorption out of the tubule is not known at this time. It is possible that increased flow out of the lateral intercellular spaces permits less backflow of sodium into the tubule or that the pumping mechanism is directly affected by the hydrostatic pressure or other physical factors. With increased hydrostatic pressure the nephron becomes leaky to certain molecules to which it is normally impermeant, and this may affect transport through shunt pathways.

Another type of control that has been proposed in several forms is that a feedback mechanism operates, perhaps at the individual nephron level, through the juxtaglomerular apparatus (JGA).[49,68,95,108] It is assumed that some aspect of distal tubular function is monitored and appropriate changes induced in the rate of filtration and/or reabsorption. Formation of angiotensin II in the region of the afferent arteriole could produce arteriolar constriction and result in a decrease in GFR; on the other hand the result would be an increase in GFR if the efferent arteriole were constricted. Various feedback mechanisms in response to change in distal tubular fluid osmolality, sodium concentration, hydrostatic pressure, etc., have been proposed, but there is no definitive evidence to support any of these proposals. Nor is there unequivocal evidence for the presence in the region of the glomerulus of converting enzyme, which is necessary to convert angiotensin I to the active form, angiotensin II. Yet the unique anatomic locus of the JGA, with its physiologic implications, begs for an explanation.

Most investigators have assumed that appropriate adjustments in reabsorptive rate occur in response to primary changes in GFR, but a totally different regulatory mechanism, also involving the JGA, has been proposed by Leyssac.[68] Although recognizing that changes in glomerular capillary pressure may affect filtration pressure and thereby filtration rate, he believes that the independent variable is generally the rate of reabsorption. Changes in the rate of reabsorption would produce changes in proximal intratubular pressure and thereby EFP and GFR. Leyssac postulates that variations in some function monitored at the macula densa lead to changes in the rate of renin release and angiotensin formation. He suggests that angiotensin constricts the efferent arteriole, increasing EFP, and also has a direct inhibitory effect on active sodium transport in the proximal tubule. In this view, decreased delivery of sodium to the distal convolution (which might be reflected in a lowered hydrostatic pressure at the macula densa) brings

---

*See references 12, 22, 93, 94, and 103.

about an increased release of angiotensin, which in turn restores to normal the delivery of sodium to the distal convolution. Other investigators report that they can find no direct effect of angiotensin on proximal sodium transport and for this reason discard the hypothesis. The nature of the coupling between filtration and reabsorption thus remains obscure and controversial.

Effect of adrenocortical steroids. It has long been known that salt excretion is importantly affected by the steroid hormones secreted by the adrenal cortex. When secretion of these hormones is deficient (the clinical syndrome known as Addison's disease), tubular reabsorption of sodium is decreased, leading to excessive loss of salt and water from the body and extracellular volume depletion. Hypersecretion results in increased reabsorption and retention of excess quantities of salt and water and volume expansion. This is one feature of Cushing's syndrome, or hyperactivity of the adrenal cortex. Although all corticosteroids possess some sodium-retaining activity, aldosterone is the most physiologically important mineralocorticoid in humans. Micropuncture investigations have demonstrated that aldosterone increases sodium reabsorption in all parts of the tubule, in the proximal as well as the distal convolution, and presumably also in the collecting ducts, and that its effect is not restricted to the distal nephron as was previously supposed.[52]

The mechanism of action of aldosterone involves the synthesis of a new protein that plays a poorly understood role in the sodium transport process. The new protein may act as a permease and make more sodium available to the transport process by facilitating the entry of sodium ions into the cells. Alternatively, stimulation of sodium transport by aldosterone may result from an effect of the new protein to provide more energy for the active transport system (Chapter 64).

The effect of aldosterone on salt excretion appears relatively slowly after its administration and requires some hours for its maximal development. This is in contrast to the coupling discussed earlier between filtration rate and rate of reabsorption, which is rapidly responsive to changes. Aldosterone is generally considered to exert fine control on a process of prodigious magnitude, and other control mechanisms can override the aldosterone effect. For example, continued administration of aldosterone leads to only a relatively brief period of salt retention in normal humans. The individual then comes into salt balance and salt excretion again equals salt

ingestion, but the extracellular volume remains expanded.

The renin-angiotensin system participates in the regulation of aldosterone secretion. This and the mechanism of action and effects of aldosterone are discussed more fully in Chapter 64.

Other factors. It has long been recognized that extracellular fluid (ECF) volume expansion and contraction are associated with increased and decreased sodium excretion, respectively. ECF volume changes of considerable magnitude are generally associated with parallel changes in filtration rate, and until recently, changes in salt excretion under these circumstances were thought to result entirely from changes in the rate of filtration. It has been shown, however, in experiments in which GFR was meticulously controlled, that an increase in ECF volume can be associated with increased salt and water excretion and decreased volume with decreased salt and water excretion, even when GFR does not change.[35,36,66] The following values assume isosmotic water movement secondary to sodium transport and demonstrate the general relationship:

Condition	GFR (ml/min)	Reabsorption (ml/min)	Excretion (ml/min)
Normal	125	124	1
Volume expansion	125	120	5
Volume contraction	125	124.5	0.5

When expansion or contraction are marked, changes in filtration rate may also occur and will accentuate the changes in excretion of salt and water. The essential feature, however, is that with alterations in ECF volume there are parallel changes in the rate of salt excretion that result directly from changes in tubular sodium transport and do not depend on altered GFR. Micropuncture studies demonstrate that these changes in sodium reabsorption occur primarily in the proximal tubule.[33] Changes in the more distal parts of the tubule are of lesser magnitude and are somewhat variable. Fig. 50-20 shows diagrammatically how glomerulotubular balance might be maintained with changes in GFR at various levels of proximal fractional reabsorption in different states of hydration. In a given state of hydration, acute changes in GFR would result in proportional changes in reabsorption if glomerulotubular balance were perfect. Balance is not perfect, however, and there is a tendency for fractional reabsorption to increase at very low filtration rates. At any given level of GFR, fractional reabsorption in the proximal tubule is greater

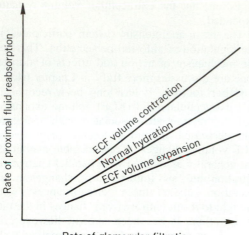

Fig. 50-20. Diagram showing how glomerulotubular balance could theoretically be maintained at three different levels with varying states of body hydration. Fractional reabsorption in proximal tubule would be highest during volume contraction and lowest during volume expansion.

when ECF is contracted and is reduced when ECF is expanded.*

Two general mechanisms have been suggested to explain this phenomenon: (1) A natriuretic hormone is released during volume expansion,[35,36] or (2) the changes in rate of reabsorption result from changes in physical factors subsequent to the change in ECF volume.[37,67] The effect of physical factors on salt and water reabsorption may well be restricted to the leaky epithelium of the proximal tubule. Conversely, any effect of a natriuretic hormone may be restricted to the tight epithelium of more distal parts of the nephron. It is not certain at this time whether either, both, or neither of these explanations is correct. Several groups of investigators have reported inhibition of sodium transport in the kidney or by isolated epithelial membranes on exposure to blood or to a substance or substances extracted from blood obtained from volume-expanded animals.[25,32,89,96] Others have been unable to verify these observations,[120] and the subject remains under active investigation in many laboratories.

There has been a revival in recent years of interest in neural control of salt excretion. Stimulation of the renal sympathetic nerves increases

and denervation (transection) decreases salt and water reabsorption in the proximal tubule.[31] It is not known whether the results are due to an effect on the passive backflow of salt and water across the epithelium or to an effect on active sodium transport. It is of interest in this regard that norepinephrine has been shown to stimulate active sodium transport across several tight and leaky epithelia. Nor is it known whether the neural effect is direct, an effect of the neurotransmitter or electrical events at the nerve endings, or indirect, involving the renin-angiotensin, prostaglandin, kallikrein, or other humoral systems. The neural effect, however, is independent of any change in hemodynamics and does not result from changes in GFR, renal blood flow, or physical factors that affect reabsorption.

It is apparent from this discussion that sodium excretion is subject to the direction of multiple control systems. Several mechanisms, possibly including some not yet discovered, might be simultaneously involved in its regulation. This is not unreasonable, since the maintenance of sodium balance is so critical to survival.

POTASSIUM EXCRETION

The magnitude of potassium excretion depends, of course, on dietary intake, and in a normal human on an average diet, it approximates 10% to 20% of the rate at which it is filtered at the glomerulus. Under special circumstances, however, such as following administration of potassium salts and inhibition of hydrogen ion excretion, the rate of potassium excretion may exceed the rate at which it is filtered, indicating tubular secretion.[16,79] The nephron must therefore be capable of both reabsorbing and secreting potassium, and this has been verified in recent years by micropuncture studies.[41]

Most of the filtered potassium is reabsorbed in the proximal tubule under the various experimental conditions that have been studied. An amount of potassium equal to 5% to 20% of the filtered load is present in the fluid emerging from the loop of Henle and entering the distal convolution. Differences in the rate of excretion of potassium under varying conditions depend on secretory and reabsorptive activity in the more distal parts of the tubule.

There is uncertainty concerning the nature of the proximal reabsorptive mechanism. In the monkey the tubular fluid potassium concentration remains the same as in glomerular filtrate,[15] whereas in the rat, it may decrease slightly but significantly below the concentration in plasma.[72] In the presence of a small transtubular

---

*The effect of the state of hydration on fractional sodium reabsorption has been termed "third factor" because its existence was recognized after the effects of GFR and aldosterone (factors 1 and 2) were known.

electrical potential gradient, lumen negative, a tubular fluid:plasma (F:P) concentration ratio for a positively charged ion that falls below unity suggests reabsorption by an active mechanism, but an unchanging F:P concentration ratio despite extensive reabsorption of water is most consistent with passive reabsorption and suggests that the proximal tubular epithelium is highly permeable to potassium.

In both dogs and rats, most of the potassium appearing in the urine originates from secretion in the more distal parts of the nephron, primarily the late distal convolution. Distal transtubular influx of potassium occurs down an electrical potential gradient established by the prior active reabsorption of sodium; thus the addition of potassium to tubular fluid is presumed to be passive. In recent experiments in rats, however, the observed luminal potassium concentration has exceeded that predicted from the transtubular potential difference.[119] These results bring into question the role of the potential difference as a driving force for potassium secretion. Although the distal tubule is short (about 3 mm in rats), it is morphologically and physiologically diverse, and it is likely that potassium transport varies in several portions.

In potassium-depleted animals there is continued net reabsorption in the distal tubule and/or collecting ducts. This must involve active transport, since potassium is being transported against its electrochemical potential gradient.

To summarize, filtered potassium is largely reabsorbed in the proximal tubule, and the amount of potassium presented to the early distal convolution varies only slightly despite great variations in potassium excretion. Ordinarily there is net potassium secretion in the distal tubule but, under conditions of potassium depletion, further net reabsorption by an active transport mechanism occurs in the distal tubule and/or collecting ducts.

Control of potassium excretion. The major determinants of the rate of potassium excretion are the relative concentrations of potassium and hydrogen in the distal tubular cells and the rate of delivery of sodium to this part of the nephron. When the concentration of potassium in the distal tubular cells is relatively high the rate of potassium excretion is increased, but when hydrogen ions are relatively more available than potassium ions in the cells, the rate of excretion of hydrogen ions increases and potassium excretion decreases. This may occur during systemic acidosis or potassium deficiency.

Since the secretion of both potassium and hydrogen ions depends on the availability of sodium ions for reabsorption in the distal convolution, the excretion of these ions and the excretion of sodium frequently vary in parallel rather than inversely, as might at first be anticipated. Under conditions of dehydration or other circumstances in which the rates of filtration and delivery of sodium to the distal convolution are decreased, the excretion of potassium is also reduced. Osmotic diuresis increases the excretion of potassium, but a water diuresis does not. This is not surprising, since delivery of sodium to the distal convolution is increased during osmotic diuresis but is unchanged in water diuresis.

Patients with Addison's disease have an impaired ability to reabsorb sodium in the distal convolution because of decreased aldosterone secretion. As a result, potassium secretion is also impaired, and this might well be a major cause of the striking hyperkalemia frequently observed in this condition.

The conservation of potassium by the kidney is more efficient than was once thought, but it is less efficient than the conservation of sodium. Some potassium continues to be excreted in the urine even in the face of a significant deficiency of body potassium and hypokalemia, possibly because distal tubular reabsorption of sodium continues, with obligatory cation exchange.

Potassium deficiency is characteristically associated with impairment of maximal urinary osmotic concentrating ability. Despite many studies in man and animals, the pathogenesis of this abnormality is poorly understood.

UREA EXCRETION

Urea is the chief end product of nitrogen metabolism in the mammal and is the major urinary solute under ordinary circumstances. It has a high lipid solubility and is freely diffusible across most cell membranes. Urea undergoes two net operations in the kidney, glomerular filtration and tubular reabsorption. Urea reabsorption is passive and occurs because of the increase in its concentration in the tubular urine as water is reabsorbed.

The urea clearance is not constant but varies directly with the rate of urine flow. Quantitative aspects of the relationship between urea excretion and urine flow in the dog were investigated in detail over 40 years ago by Shannon.[97] The open symbols in Fig. 50-21 represent observations made at normal urine flows and during water diuresis, and the solid points represent values found during profuse osmotic diuresis. Because of the difference in slopes of the lines relating fractional

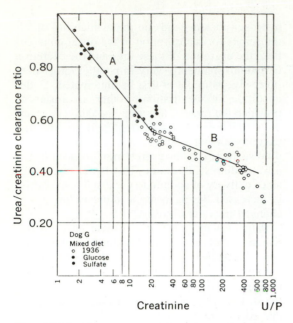

Fig. 50-21. The urea:creatinine clearance ratio in typical dog in relation to reabsorption of water as indicated by U/P ratio of creatinine. Open circles represent observations at normal urine flows and during water diuresis and solid circles during osmotic diuresis. (From Shannon.[97])

Table 50-1. Fraction of filtered urea remaining within nephron

Location	Nondiuretic	NaCl diuresis
Glomerulus	1.0	1.0
Late proximal	0.5	0.6
Early distal	1.1	0.6
Late distal	0.7	—
Ureteral urine	0.1	0.6

reabsorption of water and urea with varying increases in urine flow, Shannon postulated that urea reabsorption occurred as separate processes in two locations in the tubule. Since he correctly believed water diuresis to result from a change in water permeability in the distal convolution, he localized the process indicated by line B to the distal part of the nephron. The process illustrated by line A he localized to the proximal tubule. Since line A extrapolates to a urea:creatinine clearance ratio of 1 (i.e., no urea reabsorption), at a creatinine urine to plasma (U:P) ratio of 1 (i.e., no water reabsorption), Shannon concluded that urea reabsorption is due entirely to passive diffusion out of the tubule as a consequence of the increase in urea concentration resulting from water reabsorption.

Although the pattern of urea movement in the kidney is far more complex[63] than that originally proposed by Shannon, micropuncture studies have provided remarkable confirmation for his conclusions and support the hypothesis that tubular reabsorption of urea is passive and depends on concentration gradients established by water reabsorption.[64,65] In nondiuretic rats approximately 50% of the filtered urea is reabsorbed from the proximal convolution, but an amount of urea approximately equal to the filtered load reaches the beginning of the distal convolution (Table 50-1). This indicates addition of urea to the tubular fluid in the loop of Henle, and micropuncture studies indicate that this occurs primarily in the thin descending limb of the loop. Some urea is lost from the distal convolution, and a comparison of the urea content of fluid in the late distal convolution and ureteral urine indicates that under nondiuretic conditions a large amount of urea is reabsorbed by the collecting ducts. This has been confirmed by microcatheterization of the collecting ducts in nondiuretic hamsters.[58] Much of the urea that diffuses out of the collecting ducts diffuses into the loops of Henle and is recirculated through the distal convolution back to the collecting ducts. This mechanism, along with efficient countercurrent diffusion in the vasa recta, leads to a high concentration of urea in all the medullary structures under nondiuretic conditions. Because of its high solubility, ease of diffusion, and great abundance, urea plays a very important role in the concentrating mechanism, especially in obtaining urine of maximal concentration. All these characteristics are optimal for the accumulation of urea in the medulla by countercurrent diffusion.

It is noteworthy that a high rate of urea reabsorption out of the proximal convolution occurs despite a relatively small concentration gradient favoring diffusion out of this portion of the nephron (maximum urea F:P = 1.5). Urea loss from the distal convolution is much less, although the concentration of urea in the distal tubular fluid is considerably greater. These observations indicate that the permeability of the proximal convolution to urea is much higher than that of the distal convolution. This has been directly confirmed in microperfusion studies indicating that the permeability to urea of the rat distal convolution is only one twentieth of that of the proximal convolution.[27] Interestingly, there is no evidence in either in situ or in vitro microperfusion studies of distal convolutions and/or collecting ducts that ADH has an effect on urea permeability.[27,48] These observations suggest that the alteration in urea excretion noted by Shannon and other workers in graded water diuresis is secondary to change in urea concentration and transit time through distal convolutions and collecting ducts and not to any direct effect of ADH on urea permeability per se.

The passive nature of urea reabsorption leads to predictable changes when osmotic diuresis is induced. During massive saline diuresis in rats, 40% of the filtered urea was reabsorbed in the proximal tubule, but there was no measurable reabsorption beyond that point (Table 50-1). The large decreases in fractional water reabsorption during osmotic diuresis occur in the distal convolution and collecting ducts, where the tubular fluid urea concentration is much less and the transit time much more rapid than during antidiuresis. Therefore a much smaller urea loss is to be expected.

Micropuncture studies of urea transport in the dog are more limited, but the available data confirm the findings in rats.[28] Urea transport in the human kidney has not been studied by micropuncture, but it is probably qualitatively similar to that in other mammals. Fractional excretion of filtered urea averages about 40% in the nondiuretic state and increases rapidly with increasing rates of urine flow up to about 2 ml/min, at which level the urea clearance averages about 60% to 70% of the inulin clearance. With further increases in urine flow there is only a slight further increase in urea clearance. In certain mammals, especially under unusual circumstances, there is evidence of a small component of active reabsorption of urea. In all likelihood, this plays a minimal role, if any, in urea transport in humans.

REGULATION OF ACID-BASE BALANCE

The activity of cellular enzymes is profoundly influenced by small changes in pH, and maintenance of normal metabolic activity therefore requires that the pH of body fluids be closely regulated. The pH of arterial blood is ordinarily maintained within the narrow range of 7.35 to 7.45, and deviation much outside this range is indicative of a profound and potentially life-threatening metabolic disturbance. The utilization of foodstuffs leads to the continuous production of acidic metabolites that must be excreted if a constant pH is to be maintained. The most abundant of these is carbonic acid, which is excreted as carbon dioxide by the lungs. In addition, there is a variable but continuous production of nonvola-

tile acids such as phosphoric, sulfuric, and various organic acids, which cannot be excreted by the lungs. Relative constancy of pH in the face of the continuous production of these acids is obtained by the action of various intracellular and extracellular buffer systems, most notably the carbon dioxide–bicarbonate system, but the continuous titration of these buffers would lead to their eventual depletion if no means were available for excretion of the anions of nonvolatile acids ingested in food or produced in metabolism, at the same time conserving cation and conserving and regenerating buffer.

Bicarbonate buffer system. The carbon dioxide–bicarbonate system is described by the following equation:

$$H_2O + CO_2 \rightleftharpoons H_2CO_3 \rightleftharpoons H^+ + HCO_3^-$$

Carbonic acid (H_2CO_3) is a relatively strong acid, but the equilibrium concentration of H_2CO_3 is much smaller than the concentration of dissolved anhydrous CO_2, the dissolved CO_2: H_2CO_3 ratio being approximately 400:1 at body temperature.[73] Since at equilibrium the ratio of H_2CO_3 to dissolved CO_2 is constant, pH may be expressed in terms of bicarbonate concentration and partial pressure of CO_2, as follows:

$$pH = pK' + \log \frac{[HCO_3^-]}{\alpha P_{CO_2}}$$

This is the Henderson-Hasselbalch equation applied to the bicarbonate buffer system. pK' has the value 6.10, and α, the solubility coefficient for CO_2, is 0.0301 mM/mm Hg at a temperature of 37° C.

The concentration of dissolved CO_2 in plasma at a normal pH of 7.4 is only one twentieth the concentration of bicarbonate, which for a conventional buffer pair would indicate that it should be relatively ineffective in controlling pH. For most buffer pairs the total amount of buffer, that is, HA + A⁻, is fixed, and as the concentration of A⁻ decreases, HA must increase. This is fortunately not true for the bicarbonate system, because CO_2 is manufactured continuously in the tissues, and P_{CO_2} can be maintained at a constant level or adjusted according to need through respiratory control. Likewise, the concentration of bicarbonate may be varied over a wide range without the need for a reciprocal change in the concentration of dissolved CO_2. As shown in Fig. 50-22, the buffer capacity of the bicarbonate system increases with rising pH, and the system is in fact a much more efficient buffer at a pH of 7.4 than it would be at a lower pH near its pK'.

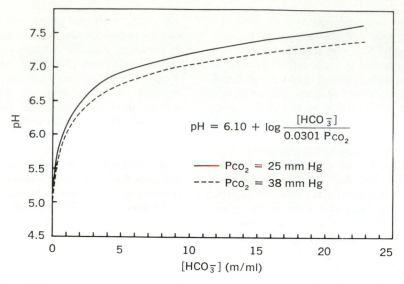

$$pH = 6.10 + \log \frac{[HCO_3^-]}{0.0301 \, P_{CO_2}}$$

— P_{CO_2} = 25 mm Hg

---- P_{CO_2} = 38 mm Hg

Fig. 50-22. Titration curve of HCO_3^- at constant P_{CO_2}. (From Gottschalk et al.[47])

A strong acid added to this system is buffered as follows:

$$HA + Na^+ + HCO_3^- \rightarrow Na^+ + A^- + H_2CO_3 \rightarrow Na^+ + A^- + H_2O + CO_2 \uparrow$$

The CO_2 thus produced is excreted by the lungs, preventing a rise in P_{CO_2} and minimizing the change in pH.

Excretion of anion A^- as such by the kidney obligates the excretion of an equivalent amount of cation. If the cation is sodium, 1 mM of sodium bicarbonate or other buffer will have disappeared from the circulation for each millimole of acid added to the extracellular fluid and subsequently excreted; the body would rapidly become depleted of cation and bicarbonate if there were no mechanism for conservation of cation and regeneration of bicarbonate. For weak acids, this is accomplished in the kidney by exchange of hydrogen ion for sodium, and the acid is excreted in its un-ionized form. The capacity of the kidney to secrete hydrogen ion is limited, however, by the gradient against which secretion must occur. The lowest pH that can normally be achieved in the urine is approximately 4.5, equivalent to a 1,000-fold increase in $[H^+]$ over the concentration in plasma. Strong acids such as sulfuric acid remain almost completely dissociated even at this low pH, and conservation of sodium by excretion of these acids in the un-ionized form is not possible. In this circumstance, sodium conservation is accomplished through the intrarenal synthesis and excretion of ammonia, as described in the following section.

Mechanisms of urine acidification. Acidification of the urine could in theory be accomplished by any of several mechanisms. The pH of the urine is determined by the state of its buffers. The major buffer pairs in glomerular filtrate are $HCO_3^- : CO_2$ and $HPO_4^{2-} : H_2PO_4^-$, and selective reabsorption of the basic form of either or both of these buffers, HCO_3^- and HPO_4^{2-}, while the acidic form is retained in the urine, would cause the urine pH to fall, with net acid excretion. Selective reabsorption of either HCO_3^- or HPO_4^{2-} would be adequate to account for the magnitude of urinary acidification observed normally, but acid excretion may be greatly enhanced in metabolic acidosis. Pitts and his associates[86,87] demonstrated that the selective reabsorption of these buffers, either alone or in combination, is inadequate to account for the magnitude of acidification under these circumstances. They found that the amount of phosphate excreted as $H_2PO_4^-$ was greater than the amount filtered in that form; thus some of the filtered HPO_4^{2-} must have been converted to $H_2PO_4^-$ in the tubular lumen, that is, hydrogen ion secretion must have occurred.

It is now generally agreed that renal tubular cells are capable of secreting hydrogen ion by a mechanism that requires the enzyme carbonic anhydrase. In simplified form, the process may be visualized as follows (Fig. 50-23): CO_2, a product of cellular metabolism, readily permeates cell membranes and, in the presence of carbonic anhydrase, exists within the cell in equilibrium with H_2CO_3. The latter, a relatively strong acid,

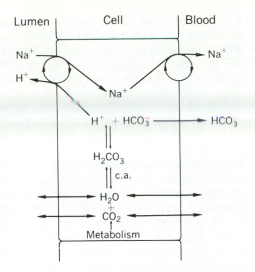

Fig. 50-23. Schematic representation of postulated mechanism of H$^+$ secretion. (See text.)

dissociates into H$^+$ and HCO$_3^-$, providing H$^+$ for secretion and HCO$_3^-$ for return to the blood in company with reabsorbed cation, chiefly sodium. Thus as H$^+$ is secreted, an equivalent quantity of sodium or other cations is absorbed and returned to the circulation with HCO$_3^-$. Hydrogen ion secretion is thought not to be directly coupled to sodium reabsorption, but rather is considered to be an active electrogenic process at the luminal membrane, requiring the expenditure of energy.[104] The simultaneous movement of other ions (by cation reabsorption and/or anion secretion) limits the membrane potential and permits the process to continue—thus the apparent coupling.

It should not be assumed that H$_2$CO$_3$ itself is necessarily the proximate source of hydrogen ion for secretion. Indeed, by far the most abundant reservoir of protons in any aqueous medium is water itself, and most of the hydrogen ion is almost surely derived from this source. Furthermore, there is evidence that the reaction catalyzed by carbonic anhydrase may not be that between CO$_2$ and water, but between CO$_2$ and OH$^-$.[73] Production of the hydrogen ion for secretion involves two reactions:

$$H_2O \rightleftharpoons H^+ + OH^-$$

and

$$OH^- + CO_2 \rightleftharpoons HCO_3^-$$

The immediate source of hydrogen ion is thus ionic dissociation of water, and the role of CO$_2$ is to remove the OH$^-$. In a mathematical sense,

the sum of these two reactions is formally identical to the scheme illustrated in Fig. 50-23.

It is assumed in this discussion that the continuous supply of hydrogen ion for secretion is provided by the intracellular hydration of CO$_2$. The evidence for this is indirect and is derived from observation of the effects of carbonic anhydrase inhibitors on the urinary acidification process. As noted previously, the equilibrium between CO$_2$ and H$_2$CO$_3$ strongly favors the anhydrous form. The concentration of H$_2$CO$_3$ in the cell is therefore at all times very small, and it must be rapidly replenished if it is to serve as a continuing source of hydrogen ion. However, in the absence of the enzyme carbonic anhydrase the hydration of CO$_2$ proceeds toward equilibrium very slowly. Administration of carbonic anhydrase inhibitors markedly depresses bicarbonate reabsorption, initially producing bicarbonaturia and an increased urine pH, and it is thought that this occurs because hydrogen ion secretion is limited by the slow rate of hydration of CO$_2$ to H$_2$CO$_3$.[88]

The fate of hydrogen ion after its secretion into the tubular lumen depends on the composition of the urine at that point, since hydrogen ion will react with whatever buffers happen to be present. In the proximal tubule the major buffer is HCO$_3^-$, which combines with the secreted hydrogen ion to form H$_2$CO$_3$. Both H$_2$CO$_3$ and its dehydration product, CO$_2$, are apparently able to diffuse back across the luminal membrane of proximal tubular cells, providing a continuing source of hydrogen ion for secretion. The net result is the reabsorption of one Na$^+$ and one HCO$_3^-$ from the tubular lumen for each hydrogen ion secreted. Note, however, that the HCO$_3^-$ ion delivered to the blood is not necessarily the same as the one that disappeared from the tubular lumen.

As bicarbonate reabsorption proceeds, the HCO$_3^-$ concentration and pH of the tubular fluid decrease, and phosphate becomes increasingly important as a buffer. The cellular mechanism of hydrogen ion secretion is unchanged, but the intraluminal effect is conversion of HPO$_4^{2-}$ to H$_2$PO$_4^-$. Now, as before, the anion returned to the blood is HCO$_3^-$; thus the net effect of the conversion of one HPO$_4^{2-}$ to H$_2$PO$_4^-$ is reabsorption of one Na$^+$ and generation of one HCO$_3^-$.

When large amounts of strong acids must be excreted, the buffer capacity of the urine may be exhausted and a limiting urinary pH (4.5 to 5.0) is reached. Under these circumstances, no more hydrogen ion can be secreted unless another mechanism is brought into play to generate new buffer. Renal tubular cells are rich in the enzyme glutaminase and have the capacity to synthesize ammonia, utilizing glutamine and to a lesser extent certain other amino acids as substrate.[85] Prolonged acidosis stimulates ammonia synthesis, apparently through a process of enzyme induction. Ammonia in the un-ionized form (NH_3) is lipid soluble and diffuses freely across cell membranes into both peritubular capillaries and the tubular urine; it will become distributed more or less uniformly in the three compartments. To the extent that the pH of the tubular fluid is more acidic than blood, ammonia diffusing in this direction will tend to combine more avidly with hydrogen ion and become "trapped," in a sense, in the tubule in the impermeant NH_4^+ form. Further diffusion of NH_3 into the tubule is thus favored, reducing the hydrogen ion concentration and permitting additional hydrogen ion secretion. Micropuncture studies in rats and dogs have shown that, whereas the entire nephron participates in ammonia secretion, a large fraction, possibly more than half, of the excreted ammonia is synthesized in the cells of the proximal tubules.[30,42] The pH of proximal tubular fluid is only slightly lower than that of plasma; hence conditions are less favorable for ammonia trapping here than at more distal sites in the nephron, where the pH is lower. However, ammonia synthesized in proximal tubules may be carried by either tubular fluid or peritubular blood into the renal medulla, where it may diffuse into the highly acidic collecting duct urine and be excreted.

The foregoing discussion suggests that carbonic anhydrase–dependent hydrogen ion secretion is responsible both for the conservation of filtered HCO_3^- and for the excretion of acid loads. To a large extent, this is true, but HCO_3^- reabsorption and acid excretion do not cease entirely when carbonic anhydrase activity is blocked. Although patients may temporarily excrete an alkaline urine when inhibitors are administered, a new steady state of mild stable acidosis is soon reached, and net urinary acid excretion returns to normal. Although HCO_3^- reabsorption is diminished, the amount filtered is also decreased, and HCO_3^- is no longer lost in the urine. Maren administered potent carbonic anhydrase inhibitors to dogs and observed continued HCO_3^- reabsorption at rates as great as 60% of normal.[74] The uncatalyzed rate of hydration of CO_2 is insufficient to maintain hydrogen ion secretion of this magnitude, but two other possibilities exist to account for the findings. The bulk of HCO_3^- reabsorption occurs in the proximal tubule and, as previously noted, the cells of this tubular segment appear to be permeable to H_2CO_3 and CO_2.

H_2CO_3 formed in the tubular lumen might therefore diffuse into cells and serve directly as a source of hydrogen ion without the need for carbonic anhydrase. Alternatively, a fraction of filtered HCO_3^- may be reabsorbed directly in anionic form in a manner analogous to that of the reabsorption of chloride. In either case the net result is the translocation of HCO_3^- from tubular lumen to blood, without the intervention of carbonic anhydrase.

Localization of urine acidification. Available evidence supports the view that all parts of the mammalian nephron participate in the process of hydrogen ion secretion. Qualitatively, the mechanism of secretion appears to be the same in all nephron segments, but important quantitative differences are apparent. Results of micropuncture studies in nondiuretic rats are shown in Fig. 50-24. The pH of tubular fluid falls approximately 0.5 pH units along the proximal tubule, equivalent to a threefold rise in hydrogen ion concentration, effecting a decrease in HCO_3^- concentration from a normal value of approximately 25 mEq/L to 10 mEq/L. Since two thirds to three fourths of the glomerular filtrate is reabsorbed in this segment, it follows that about 90% of the filtered bicarbonate is reabsorbed in the proximal tubule. Reabsorption in this segment appears to be limited by the rate of hydrogen ion secretion and not by the pH gradient, since an even greater decrease in pH along the proximal tubule may be observed if the animals are first made acidotic, reducing bicarbonate concentration and hence the buffering capacity of the glomerular filtrate.

It is probable that the process of hydrogen ion secretion continues in the distal convolution in a manner qualitatively similar to that occurring in the proximal tubule. However, the contribution of this portion of the nephron to total hydrogen ion secretion is much less than that of the proximal tubule because of the smaller buffer load presented to it. Similarly, although in the nondiuretic state the greatest fall in pH occurs in the collecting ducts, the contribution of this segment to HCO_3^- reabsorption and to total hydrogen ion secretion may be even less than that of the distal convolution. The volume of fluid presented to the collecting ducts is small, and a relatively low rate of secretion could produce a large fall in pH. The presence of phosphate and other buffers in high concentration in collecting ducts may increase hydrogen ion secretion somewhat, but this contribution is still small compared to that of the HCO_3^- buffer reabsorbed in more proximal portions of the nephron. The collecting duct, how-

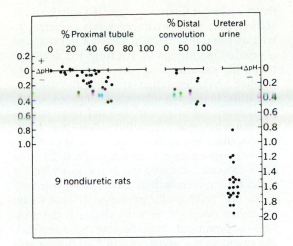

Fig. 50-24. Difference in pH of tubular fluid and ureteral urine from that of arterial blood in rats. Minus ΔpH values represent acidification relative to arterial pH. (From Gottschalk et al.[47])

ever, does have the capacity to secrete hydrogen ion against a much steeper concentration gradient than other parts of the nephron, a gradient of about 1,000:1 existing when the urine pH is 4.5.

It should be apparent that the terms "hydrogen ion secretion" and "hydrogen ion excretion" are not synonymous. To the extent that hydrogen ion secretion is buffered by HCO_3^- in the tubular fluid, secretion will be reflected only in the conservation and return to body fluids of filtered bicarbonate and not in the elimination of acid from the body. Indeed, if bicarbonate escapes reabsorption and is excreted in the urine, net acid excretion is reduced, since loss of bicarbonate is equivalent to the retention of a similar quantity of nonvolatile acid in the body. On the other hand, to the extent that secreted hydrogen ion is buffered by phosphate and other nonvolatile buffers or by ammonia, net excretion of hydrogen ion (i.e., acid excretion) is accomplished. That portion of acid excretion not buffered by ammonia is known as titratable acidity. The titratable acidity of urine is equal to the amount of NaOH that must be added to raise the pH of the urine to that of plasma (i.e., 7.4). The magnitude of the titratable acidity depends to a certain extent on the urinary pH but more importantly on the concentration and nature of the urinary buffers. Normally more than 5,000 mEq of hydrogen ion is secreted by the renal tubules daily, which is reflected in the excretion of some 10 to 30 mEq of titratable acidity and 30 to 50 mEq of

NH_4^+, and the reabsorption of about 5,000 mEq of HCO_3^-.

Factors affecting hydrogen ion secretion. The dependence of the acidifying mechanism on the enzyme carbonic anhydrase has already been mentioned. Renal tubular cells contain a generous excess of the enzyme, however, and its activity is never a limiting factor except when markedly reduced, as after the administration of powerful inhibitors. However, since CO_2 is postulated to be the ultimate source of the secreted hydrogen ion, it might be predicted that the secretory rate should be influenced by variations in P_{CO_2}, and this is indeed the case. In general, in the absence of other limiting factors, bicarbonate reabsorption (i.e., hydrogen ion secretion) is increased when P_{CO_2} is elevated and is decreased when P_{CO_2} is lowered, as in hyperventilation. Since there is a limiting gradient of approximately 1,000:1 against which hydrogen ion can be secreted in the collecting ducts, the pH of the urine is also a factor limiting secretion under certain circumstances. The fall in pH with a given acid load is in turn determined by the nature and concentration of buffers in the urine. Hydrogen ion secretion may also be modified indirectly by factors influencing sodium reabsorption, since the processes of sodium ion reabsorption and hydrogen ion secretion are linked. The availability of potassium ion also influences hydrogen ion secretion, particularly in the distal portion of the nephron where, as noted previously, potassium competes with hydrogen in the cation exchange process.

Renal response to acid-base disturbances. It is not within the scope of this chapter to consider in detail the alterations of renal function that occur in response to disturbances in acid-base homeostasis, but a brief review of the renal responses to certain uncomplicated perturbations in acid-base balance will serve to illustrate the principles governing the regulation of urinary acidification. Usually, but not always, the response of the kidney to acid-base disturbances serves to correct the disturbance and to minimize changes in the ratio $[HCO_3^-]:P_{CO_2}$, and therefore in the pH, of the plasma.

Respiratory acidosis. Respiratory acidosis is seen in individuals in whom pulmonary ventilation is reduced, as in pulmonary emphysema, and may be produced experimentally by inhaling gas mixtures containing an increased concentration of CO_2. P_{CO_2} is elevated, and as a result there is increased availability of hydrogen ion for secretion. Bicarbonate reabsorption is therefore increased, and plasma HCO_3^- concentration rises, minimizing the decrease in pH resulting from CO_2 retention. Concurrently with increased $Na^+:H^+$ exchange, chloride ion reabsorption is decreased and excretion increased. Also since urinary pH is lower, NH_4^+ excretion increases.

Respiratory alkalosis. Respiratory alkalosis may be produced by hyperventilation and is seen occasionally in patients with head injuries or after ingestion of drugs such as aspirin that stimulate the respiratory center. As a result of the fall in P_{CO_2} there is a decrease in the availability of hydrogen ion for secretion and therefore decreased HCO_3^- reabsorption and a fall in plasma HCO_3^- concentration, minimizing the change in plasma pH. In addition, since less hydrogen ion is available for exchange with sodium, the reabsorption of chloride may be increased, and potassium excretion is also increased. Since the urine is more alkaline, NH_4^+ excretion is diminished.

Metabolic acidosis. Metabolic acidosis is commonly encountered in renal failure and in a variety of metabolic disorders, most notably uncontrolled diabetes mellitus. In the latter disorder, because of the inability to utilize glucose as a source of energy, there is accelerated breakdown of fatty acid molecules to 4-carbon ketoacids. The abnormal accumulation of nonvolatile acids in the body fluids is buffered by bicarbonate and other buffers, and plasma HCO_3^- concentration is reduced. The resulting fall in plasma pH leads to a stimulation of respiration, with hyperventilation and decreased P_{CO_2}. Because of the lowered P_{CO_2}, hydrogen ion secretion by the renal tubule may actually be reduced, and a larger fraction of HCO_3^- reabsorption occurs by mechanisms independent of carbonic anhydrase. Since the filtered HCO_3^- load is reduced, a larger share of the total hydrogen ion secretion appears as hydrogen ion excretion, which, reflected in titratable acidity and NH_4^+ excretion, is usually greater than normal. In prolonged acidosis, renal NH_3 synthesis is increased and NH_4^+ excretion may be markedly elevated.

Metabolic alkalosis. An increase in plasma HCO_3^- concentration and pH may be induced by ingestion of large amounts of sodium bicarbonate or other alkali and may also occur as a consequence of intractable vomiting, with loss from the body of a large volume of highly acidic gastric contents. The increased pH may reduce the stimulus to respiration and produce a slight rise in P_{CO_2}, but this effect is limited because with hypoventilation, hypoxia quickly assumes a larger role in the control of respiration. The slight elevation in P_{CO_2} may induce increased hydrogen ion secretion by the renal tubule, but because of the much larger filtered load of HCO_3^- to be reabsorbed, little or no hydrogen ion is excreted, so long as sodium deficiency does not develop and adequate chloride ion is available for reabsorption with sodium. Since the urine is relatively alkaline, NH_4^+ excretion is markedly reduced. The urine may contain large amounts of sodium, potassium, and other cations, and there is danger that a cation deficit will develop. If a significant sodium ion deficit occurs, the stimulus to sodium conservation is increased, and if, as is often the case, chloride ion is not available for reabsorption, there will be an obligatory increase in $Na^+:H^+$ and $Na^+:K^+$ exchange. If a potassium ion deficit develops, as is inevitable if the alkalosis is prolonged and adequate potassium replacement is not given, $Na^+:K^+$ exchange will decrease, with a further increase in $Na^+:H^+$ exchange. In this latter circumstance, hydrogen ion excretion increases and an acid urine is formed despite the presence of metabolic alkalosis, and the alkalosis is made worse. Thus in contrast to the renal response in most other acid-base disturbances, in metabolic alkalosis the response does not invariably serve to automatically correct the disturbance.

REFERENCES
General reviews

1. Boulpaep, E. L.: Electrical phenomena in the nephron, Kidney Int. **9:**88, 1976.
2. Brenner, B. M., Deen, W. M., and Robertson, C. R.: Determinants of glomerular filtration rate, Annu. Rev. Physiol. **38:**7, 1976.
3. Burg, M., and Stoner, L.: Renal tubular chloride transport and the mode of action of some diuretics, Annu. Rev. Physiol. **38:**37, 1976.
4. Dousa, T. P., and Valtin, H.: Cellular actions of vasopressin in the mammalian kidney, Kidney Int. **10:**46, 1976.
5. Hierholzer, K., and Wiederholt, M.: Some aspects of distal tubular solute and water transport, Kidney Int. **9:**198, 1976.
6. Kokko, J. P., and Tisher, C. C.: Water movement across nephron segments involved with the countercurrent multiplication system, Kidney Int. **10:**64, 1976.
7. Malnic, G., and Steinmetz, P. R.: Transport processes in urinary acidification, Kidney Int. **9:**172, 1976.
8. Windhager, E. E., and Giebisch, G.: Proximal sodium and fluid transport, Kidney Int. **9:**121, 1976.

Original papers

9. Allison, M. E. M., Wilson, C. B., and Gottschalk, C. W.: Pathophysiology of experimental glomerulonephritis in rats, J. Clin. Invest. **53:**1402, 1974.
10. Arakawa, M.: A scanning electron microscopy of the glomerulus of normal and nephrotic rats, Lab. Invest. **23:**489, 1970.
11. Arendshorst, W. J., and Gottschalk, C. W.: Apparent filtration pressure disequilibrium and filtration coefficient in the rat kidney. Seventh Annual Meeting of the

American Society of Nephrology, November 25-26, 1974, Washington, D.C. (Abstr.)

12. Arrizurieta-Muchnik, E. E., et al.: Micropuncture study of glomerulotubular balance in the rat kidney, Nephron **6:**418, 1969.

13. Baines, A. D.: Effect of extracellular fluid volume expansion on maximum glucose reabsorption rate and glomerular tubular balance in single rat nephrons, J. Clin. Invest. **50:**2414, 1971.

14. Bayliss, L. E., Kerridge, P. N. P., and Russell, C. W.: The excretion of protein by the mammalian kidney, J. Physiol. **82:**62, 1934.

15. Bennett, C. M., Brenner, B. M., and Berliner, R. W.: Micropuncture study of nephron function in the Rhesus monkey, J. Clin. Invest. **47:**203, 1968.

16. Berliner, R. W.: Renal mechanisms for potassium excretion, Harvey Lect. **55:**141, 1961.

17. Blantz, R. C., and Wilson, C. B.: Acute effects of antiglomerular basement membrane antibody on the process of glomerular filtration in the rat, J. Clin. Invest. **58:**899, 1976.

18. Blantz, R. C., Rector, F. C., and Seldin, D. W.: Effect of hyperoncotic albumin expansion upon glomerular ultrafiltration in the rat, Kidney Int. **6:**209, 1974.

19. Bowman, W.: On the structure and use of the malpighian bodies of the kidney with observations on the circulation through that gland, Philos. Trans. R. Soc. Lond. (Biol.) **1:**57, 1842.

20. Brenner, B. M., and Troy, J. L.: Postglomerular vascular protein concentration: evidence for a causal role in governing fluid reabsorption and glomerulotubular balance by the renal proximal tubule, J. Clin. Invest. **50:**336, 1971.

21. Brenner, B. M., Baylis, C., and Deen, W. M.: Transport of molecules across renal glomerular capillaries, Physiol. Rev. **56:**502, 1976.

22. Brenner, B. M., Bennett, C. M., and Berliner, R. W.: The relationship between glomerular filtration rate and sodium reabsorption by the proximal tubule of the rat nephron, J. Clin. Invest. **47:**1358, 1968.

23. Brenner, B. M., Troy, J. L., and Daugharty, T. M.: The dynamics of glomerular ultrafiltration in the rat, J. Clin. Invest. **50:**1776, 1971.

24. Brunner, F. P., Rector, F. C., Jr., and Seldin, D. W.: Mechanism of glomerulotubular balance. II. Regulation of proximal tubular reabsorption by tubular volume, as studied by stopped-flow microperfusion, J. Clin. Invest. **45:**603, 1966.

25. Buckalew, V. M., Jr., Martinez, F. J., and Green, W. E.: The effect of dialysates and ultrafiltrates of plasma of saline-loaded dogs on toad bladder sodium transport, J. Clin. Invest. **49:**926, 1970.

26. Burg, M. B., and Green, N.: Function of the thick ascending limb of Henle's loop, Am. J. Physiol. **224:**655, 1973.

27. Capek, K., et al.: Harnstoffpermeabilität der corticalen Tubulusabschnitte, Pflüegers Arch. **290:**237, 1966.

28. Clapp, J. R.: Urea reabsorption by the proximal tubule of the dog, Proc. Soc. Exp. Biol. Med. **120:**521, 1965.

29. Clapp, J. R., and Robinson, R. R.: Osmolality of distal tubular fluid in the dog, J. Clin. Invest. **45:**1847, 1966.

30. Clapp, J. R., Owen, E. E., and Robinson, R. R.: Contribution of the proximal tubule to urinary ammonia excretion by the dog, Am. J. Physiol. **209:**269, 1965.

31. Colindres, R. E., and Gottschalk, C. W.: Neural control of renal tubular sodium reabsorption in the rat: single nephron analysis, Fed. Proc. **37:**1218, 1978.

32. Cort, J. H., Pliška, V., and Douša, T.: The chemical nature and tissue source of natriuretic hormone, Lancet **1:**230, 1968.

33. Cortney, M. A., et al.: Renal tubular transport of water, solute, and PAH in rats loaded with isotonic saline, Am. J. Physiol. **209:**1199, 1965.

34. de Rouffignac, C., and Morel, F.: Micropuncture study of water, electrolytes, and urea movements along the loops of Henle in Psammomys, J. Clin. Invest. **48:**474, 1969.

35. de Wardener, H. E.: Control of sodium reabsorption, Br. Med. J. **3:**611, 1969.

36. de Wardener, H. E., et al.: Studies on the efferent mechanism of the sodium diuresis which follows the administration of intravenous saline in the dog, Clin. Sci. **21:**249, 1961.

37. Earley, L. E.: Influence of hemodynamic factors on sodium reabsorption, Ann. N.Y. Acad. Sci. **139:**312, 1966.

38. Frömter, E., and Gessner, K.: Free-flow potential profile along rat kidney proximal tubule, Pfluegers Arch. **351:**69, 1974.

39. Frömter, E., and Hegel, U.: Transtubulare Potentialdifferenzen an proximalen und distalen Tubuli der Rattenniere, Pfluegers Arch. **291:**107, 1966.

40. Gertz, K. H., et al.: On the glomerular tubular balance in the rat kidney, Pfluegers Arch. **285:**360, 1965.

41. Giebisch, G., and Windhager, E. E.: Electrolyte transport across renal tubular membranes, In Orloff, J., and Berliner, R. W., editors: Handbook of physiology. Section 8: renal physiology, Washington, D.C., 1973, American Physiological Society, p. 315.

42. Glabman, S., Klose, R. M., and Giebisch, G.: Micropuncture study of ammonia excretion in the rat, Am. J. Physiol. **205:**127, 1963.

43. Gottschalk, C. W.: Micropuncture studies of tubular function in the mammalian kidney, Physiologist **4:**35, 1961.

44. Gottschalk, C. W.: Osmotic concentration and dilution of urine, Am. J. Med. **36:**670, 1964.

45. Gottschalk, C. W., and Mylle, M.: Micropuncture study of pressures in proximal tubules and peritubular capillaries of the rat kidney and their relation to ureteral and renal venous pressures, Am. J. Physiol. **185:**430, 1956.

46. Gottschalk, C. W., and Mylle, M.: Micropuncture study of the mammalian urinary concentrating mechanism: evidence for the countercurrent hypothesis, Am. J. Physiol. **196:**927, 1959.

47. Gottschalk, C. W., Lassiter, W. E., and Mylle, M.: Localization of urine acidification in the mammalian kidney, Am. J. Physiol. **198:**581, 1960.

48. Grantham, J. J., and Burg, M. B.: Effect of vasopressin and cyclic AMP on permeability of isolated collecting tubules, Am. J. Physiol. **211:**255, 1965.

49. Guyton, A. C., Langston, J. B., and Navar, G.: Theory for renal autoregulation by feedback of the juxtaglomerular apparatus, Circ. Res. **15**(suppl. I):1, 1964.

50. Hargitay, B., and Kuhn, W.: Das Multiplikationsprinzip als Grundlage der Harnkonzentrierung in der Niere, Z. Elektrochemie **55:**539, 1951.

51. Heidenhain, R.: Hermanns Handb. Physiol. **5:**279, 1883.

52. Hierholzer, K., and Stolte, H.: The proximal and distal tubular action of adrenal steroids on Na reabsorption, Nephron **6:**188, 1969.

53. Imai, M., and Kokko, J. P.: Mechanism of sodium and chloride transport in the thin ascending limb of Henle, J. Clin. Invest. **58:**1054, 1976.

54. Jacobson, H. R., and Kokko, J. P.: Intrinsic differences in various segments of the proximal convoluted tubule, J. Clin. Invest. **57**:818, 1976.

55. Jamison, R. L.: The urinary concentrating mechanism, N. Engl. J. Med. **295**:1059, 1976.

56. Johnston, P. A., et al.: Evidence for a concentration gradient favoring outward movement of sodium from the thin loop of Henle, J. Clin. Invest. **59**:234, 1977.

57. Kawamura, S., et al.: Characteristics of salt and water transport in superficial and juxtamedullary straight segments of proximal tubules, J. Clin. Invest. **55**:1269, 1975.

58. Klümper, J. D., Ullrich, K. J., and Hilger, H. H.: Das Verhalten des Harnstoffs in den Sammelrohren der Saugetierniere, Pfluegers Arch. **267**:238, 1958.

59. Koefoed-Johnsen, V., and Ussing, H. H.: The contributions of diffusion and flow to the passage of D_2O through living membranes, Acta Physiol. Scand. **28**:60, 1953.

60. Kokko, J. P.: Membrane characteristics governing salt and water transport in the loop of Henle, Fed. Proc. **33**:25, 1974.

61. Kokko, J. P., and Rector, F. C., Jr.: Countercurrent multiplication system without active transport in inner medulla, Kidney Int. **2**:214, 1972.

62. Lassen, N. A., Munck, O., and Thaysen, J. H.: Oxygen consumption and sodium reabsorption in the kidney, Acta Physiol. Scand. **51**:371, 1961.

63. Lassiter, W. E.: Urea transport in the mammalian nephron, In Schmidt-Nielsen, B., editor: Urea and the kidney, Excerpta Med. Int. Cong. series, No. 195, Princeton, N.J., 1968, Excerpta Medica, p. 206.

64. Lassiter, W. E., Gottschalk, C. W., and Mylle, M.: Micropuncture study of net transtubular movement of water and urea in nondiuretic mammalian kidney, Am. J. Physiol. **200**:1139, 1961.

65. Lassiter, W. E., Mylle, M., and Gottschalk, C. W.: Net transtubular movement of water and urea in saline diuresis, Am. J. Physiol. **206**:669, 1964.

66. Levinsky, N. G., and Lalone, R. C.: The mechanism of sodium diuresis after saline infusion in the dog, J. Clin. Invest. **42**:1261, 1963.

67. Lewy, J. E., and Windhager, E. E.: Peritubular control of proximal tubular fluid reabsorption in the rat kidney, Am. J. Physiol. **214**:943, 1968.

68. Leyssac, P. P.: Dependence of glomerular filtration rate on proximal tubular reabsorption of salt, Acta Physiol. Scand. **58**:236, 1963.

69. Ludwig, K.: Wagners Handworterb, Physiologie **2**:637, 1844.

70. Maddox, D. A., Deen, W. M., and Brenner, B. M.: Dynamics of glomerular ultrafiltration. VI. Studies in the primate, Kidney Int. **5**:271, 1974.

71. Maddox, D. A., et al.: Determinants of glomerular filtration in experimental glomerulonephritis in the rat, J. Clin. Invest. **55**:305, 1975.

72. Malnic, G., Klose, R. M., and Giebisch, G.: Micropuncture study of renal potassium excretion in the rat, Am. J. Physiol. **206**:674, 1964.

73. Maren, T. H.: Carbonic anhydrase: chemistry, physiology, and inhibition, Physiol. Rev. **47**:595, 1967.

74. Maren, T. H.: Chemistry of the renal reabsorption of bicarbonate, Can. J. Physiol. Pharmacol. **52**:1041, 1974.

75. Marsh, D. J., and Azen, S. P.: Mechanism of NaCl reabsorption by hamster thin ascending limbs of Henle's loop, Am. J. Physiol. **228**:71, 1975.

76. Marsh, D. J., and Martin, C. M.: Origin of electrical

PD's in hamster thin ascending limbs of Henle's loop, Am. J. Physiol. **232**:F348, 1977.

77. Marshall, E. K., and Vickers, J. L.: The mechanism of the elimination of phenolsulphonphthalein by the kidney: a proof of secretion by the convoluted tubules, Bull. Johns Hopkins Hosp. **34**:1, 1923.

78. Martino, J. A., and Earley, L. E.: Relationship between intrarenal hydrostatic pressure and hemodynamically induced changes in sodium excretion, Circ. Res. **23**:371, 1968.

79. Mudge, G. H., Foulks, J., and Gilman, A.: The renal excretion of potassium, Proc. Soc. Exp. Biol. Med. **67**:545, 1948.

80. Oliver, J.: Nephrons and kidneys. A quantitative study of developmental and evolutionary mammalian renal architectonics, New York, 1968, Harper & Row, Publishers, Inc.

81. Orloff, J., and Handler, J.: The role of adenosine $3',5'$-phosphate in the action of antidiuretic hormone, Am. J. Med. **42**:757, 1967.

82. Ott, C. W., et al.: Determinants of glomerular filtration rate in the dog, Am. J. Physiol. **231**:235, 1976.

83. Pappenheimer, J. R.: Passage of molecules through capillary walls, Physiol. Rev. **33**:387, 1953.

84. Pennell, J. P., Lacy, F. B., and Jamison, R. L.: An in vivo study of the concentrating process in the descending limb of Henle's loop, Kidney Int. **5**:337, 1974.

85. Pitts, R. F.: Physiology of the kidney and body fluids, ed. 3, Chicago, 1974, Year Book Medical Publishers, Inc.

86. Pitts, R. F., and Alexander, R. S.: The nature of the renal tubular mechanism for acidifying the urine, Am. J. Physiol. **144**:239, 1945.

87. Pitts, R. F., et al.: The renal regulation of acid base balance in man. I. The nature of the mechanism for acidifying the urine, J. Clin. Invest. **27**:48, 1948.

88. Rector, F. C., Jr., Carter, N. W., and Seldin, D. W.: The mechanism of bicarbonate reabsorption in the proximal and distal tubules of the kidney, J. Clin. Invest. **44**:278, 1965.

89. Rector, F. C., Jr., et al.: Demonstration of a hormonal inhibitor of proximal tubular reabsorption during expansion of extracellular volume with isotonic saline, J. Clin. Invest. **47**:761, 1968.

90. Richards, A. N.: The Croonian lecture. Processes of urine formation, Proc. R. Soc. Lond. (Biol.) **126**:398, 1938.

91. Robson, A. M., Srivastava, P. L., and Bricker, N. S.: The influence of saline loading on renal glucose reabsorption in the rat, J. Clin. Invest. **47**:329, 1968.

92. Rocha, A. S., and Kokko, J. P.: Sodium chloride and water transport in the medullary thick ascending limb of Henle. Evidence for active chloride transport. J. Clin. Invest. **52**:612, 1973.

93. Rodicio, J., et al.: Studies on glomerulotubular balance during aortic constriction, ureteral obstruction and venous occlusion in hydropenic and saline-loaded rats, Nephron **6**:437, 1969.

94. Schnermann, J., Levine, D. Z., and Horster, M.: A direct evaluation of the Gertz hypothesis on single rat proximal tubules in vivo: failure of the tubular volume to be the sole determinant of the reabsorptive rate, Pfluegers Arch. **308**:149, 1969.

95. Schnermann, J., et al.: Regulation of superficial nephron filtration rate by tubuloglomerular feedback, Pfluegers Arch. **318**:147, 1970.

96. Sealey, J. E., Kirshman, J. D., and Laragh, J. H.:

Natriuretic activity in plasma and urine of salt-loaded man and sheep, J. Clin. Invest. **48:**2210, 1969.

97. Shannon, J. A.: Urea excretion in the normal dog during forced diuresis, Am. J. Physiol. **122:**782, 1938.

98. Shannon, J. A.: Renal tubular excretion, Physiol. Rev. **19:**63, 1939.

99. Shipley, R. E., and Study, R. S.: Changes in renal blood flow, extraction of inulin, glomerular filtration rate, tissue pressure and urine flow with acute alterations of renal artery blood pressure, Am. J. Physiol. **167:**676, 1951.

100. Smith, H. W.: Lectures on the kidney, Lawrence, Kan., 1943, University Extension Division, University of Kansas.

101. Smith, H. W.: The kidney: structure and function in health and disease, London, 1951, Oxford University Press, Inc.

102. Smith, H. W.: Principles of renal physiology, New York, 1956, Oxford University Press.

103. Steinhausen, M.: Messungen des tubulären Harnstromes und der tubularen Reabsorption unter erhöhtem Ureterdruck, Pfluegers Arch. **298:**105, 1967.

104. Steinmetz, P. R.: Cellular mechanisms of urinary acidification, Physiol. Rev. **54:**890, 1974.

105. Stephenson, J. L.: Concentration of urine in a central core model of the renal counterflow system, Kidney Int. **2:**85, 1972.

106. Thurau, K.: Renal Na-reabsorption and O_2-uptake in dogs during hypoxia and hydrochlorothiazide infusion, Proc. Soc. Exp. Biol. Med. **106:**714, 1961.

107. Thurau, K.: Renal hemodynamics, Am. J. Med. **36:** 698, 1964.

108. Thurau, K., and Schnermann, J.: Die Natriumkonzentration an den Macula densa-Zellen als regulierender Faktor fur das Glomerulumfiltrat (Mikropunktionsversuche), Klin. Wochenschr. **43:**410, 1965.

109. Ullrich, K. J.: Renal transport of sodium. In Proceedings of the Third International Congress of Nephrology, Washington, D.C., 1966, Basel, Switzerland, 1967, Karger, vol. 1, p. 48.

110. Verney, E. B.: Antidiuretic hormone and the factors which determine its release, Proc. R. Soc. Lond. (Biol.) **135:**25, 1947.

111. Walker, A. M., et al.: The collection and analysis of fluid from single nephrons of the mammalian kidney, Am. J. Physiol. **134:**580, 1941.

112. Wearn, J. T., and Richards, A. N.: Observations on the composition of glomerular urine with particular reference to the problem of reabsorption in the renal tubules, Am. J. Physiol. **71:**209, 1924.

113. Weiner, I. M., and Mudge, G. H.: Renal tubular mechanisms for excretion of organic acids and bases, Am. J. Med. **36:**743, 1964.

114. Windhager, E. E., and Giebisch, G.: Electrophysiology of the nephron, Physiol. Rev. **45:**214, 1965.

115. Wirz, H.: Der osmotische Druck des Blutes in der Nierenpapille, Helv. Physiol. Acta **11:**20, 1953.

116. Wirz, H.: Der osmotische Druck in den corticalen Tubuli der Rattenniere, Helv. Physiol. Acta **14:**353, 1956.

117. Wirz, H., Hargitay, H. B., and Kuhn, W.: Lokalisation des Konzentrierungsprozesses in der Niere durch direkte Kryoskopie, Helv. Physiol. Acta **9:**196, 1951.

118. Wright, F. S.: Increasing magnitude of electrical potential along the renal distal tubule, Am. J. Physiol. **220:**624, 1971.

119. Wright, F. S.: Sites and mechanisms of potassium transport along the renal tubule, Kidney Int. **11:**415, 1977.

120. Wright, F. S., et al.: Failure to demonstrate a hormonal inhibitor of proximal sodium reabsorption, J. Clin. Invest. **48:**1107, 1969.

121. Zins, G. R., and Weiner, I. M.: Bidirectional urate transport limited to the proximal tubule in dogs, Am. J. Physiol. **215:**411, 1968.

51

WILLIAM E. LASSITER and CARL W. GOTTSCHALK

Urine formation in the diseased kidney

Renal insufficiency is accompanied by numerous physiologic and biochemical disturbances, including loss of urinary concentrating ability, acidosis, and an increase in the plasma concentration of urea and creatinine. Nevertheless, many patients with mild or moderate degrees of renal insufficiency remain essentially asymptomatic except in times of stress, such as during acute infection or when dietary intake is markedly altered, when the decreased ability of the kidney to respond to changing homeostatic requirements may become manifest. Advanced renal insufficiency, on the other hand, is accompanied by a complex of signs and symptoms known as uremia. The clinical picture of uremia is variable and is influenced by such factors as the age and nutritional status of the patient, the rapidity with which renal failure has developed, and the presence or absence of hypertension. The manifestations of the uremic syndrome result from disturbances in both the excretory and the endocrine functions of the kidney.

Uremic syndrome. Most patients with chronic renal failure remain in fluid and electrolyte balance, but the ability of their kidneys to respond quickly and appropriately to changing excretory demands is limited, and at times of stress, they may experience wider fluctuations than normal in volume and electrolyte composition of their body fluids. In far-advanced renal failure, if intake is not carefully controlled, retention of salt and water due to absolute limits on excretory capacity may result in overexpansion of the extracellular compartment to the point of central venous congestion, heart failure, and pulmonary edema. Conversely, in certain forms of chronic renal disease the ability to conserve sodium may be impaired, and contraction of the extracellular space, hypotension, and decreased renal blood flow and intensification of the underlying renal insufficiency may occur when dietary intake of salt is inappropriately restricted.

Hypertension is commonly associated with chronic renal failure. In many instances the hypertension itself is the primary disorder, the microvascular disease associated with sustained blood pressure elevation leading in time to destruction of the kidneys. Often, however, renal disease occurs first. Usually in such circumstances the hypertension is volume dependent, blood pressure rising when extracellular volume is expanded and falling with volume contraction. Occasionally a particularly severe, accelerated variety of hypertension occurs that is not volume dependent. This form of hypertension, which may be immediately life threatening, is associated with increased synthesis and release of renin by the kidney, and it is occasionally necessary to remove the kidneys surgically in order to achieve blood pressure control.

Metabolic acidosis is the rule in chronic renal failure. The capacity to secrete hydrogen ion is related to the functional mass of renal tubular cells, which is reduced. Although most patients with chronic renal disease are able to excrete nearly normal acid loads, they can do this only at the expense of lower than normal plasma pH and bicarbonate concentration. If the plasma concentration and filtered load of bicarbonate are increased by the administration of alkali, the diminished capacity to secrete hydrogen ion is uncovered, and bicarbonate wasting in the urine ensues. The capacity to excrete ammonia, which, too, is related to the functional renal mass, is also inappropriately low for the level of acidosis.

Metabolic bone disease (renal osteodystrophy) is not seen in acute renal failure but is a common finding in chronic uremia. The origins of the bone disease are complex. On pathologic and radiologic examination the bones show features of both hyperparathyroidism (osteitis fibrosa) and osteomalacia. Slatopolsky et al.[34] have suggested that secondary hyperparathyroidism may result in the following manner: Decreased glomerular filtration leads to a slight rise in the plasma phosphate level, which in turn causes the plasma calcium concentration to decrease. This stimulates release of parathyroid hormone, which acts on bone to accelerate the release of calcium and phosphate and on the kidney to depress tubular reabsorption of phosphate, thus increasing phosphate excretion and causing the plasma phosphate and calcium concentrations to return toward normal. In the extreme, this process can lead to the characteristic bony changes associated with hyperparathyroidism. The mechanism whereby phosphate retention induces hypocalcemia and thus initiates this chain of events is not fully understood. It is probably not a direct solubility product effect. Rather, it seems that patients with renal failure may exhibit skeletal resistance to the calcium-mobilizing action of parathyroid hormone, partly because of vitamin D lack, and hyperphosphatemia appears to interfere with the metabolism of vitamin D.[12,26]

Absorption of calcium from the gastrointestinal tract is decreased in renal failure, and the resulting calcium deficit, in addition to stimulating parathyroid gland activity, leads to defective calcification of bony matrix, or osteomalacia. The defect in calcium absorption is a consequence of disordered vitamin D metabolism.[12] Vitamin D itself is physiologically inactive and must first be converted in the liver to 25-OH-D. This metabolite undergoes further hydroxylation to 1,25-$(OH)_2$-D. This second hydroxylation, at the 1 position, is impaired in renal failure, since it requires an enzyme, 1,α-hydroxylase, that is found only in the kidney. Administration of 1,25-$(OH)_2$-D to uremic patients restores intestinal calcium absorption to normal[11] and may prove to be of value in the treatment of renal osteodystrophy.

Neurologic abnormalities are commonly observed in patients with chronic uremia, particularly lassitude and mental depression, weakness, twitching, and, in terminal stages, convulsions and coma. Peripheral neuropathies may be troublesome in patients with long-standing disease. Almost all patients have some degree of anemia, and this is a consequence of both decreased red blood cell production and shortened red blood cell survival time. The kidney is the chief source of erythropoietin, a hormone that stimulates red blood cell production, and synthesis of the hormone is decreased in renal failure.[13,21] Gastrointestinal symptoms are almost always present, particularly anorexia. Vomiting and diarrhea are also common, and in the terminal stages of the disease, gastrointestinal bleeding may occur. Purpuric lesions of the skin are commonly observed, as well as bleeding from the gums or nasal mucosa, apparently because of defective platelet function.

Although many of the symptoms of uremia may be related to acidosis or to disturbances in water and electrolyte balance, these electrolyte abnormalities cannot explain all the features of the syndrome. Many symptoms are thought to be caused by accumulation of a toxic substance or substances, produced most likely in the course of protein metabolism and normally excreted by the kidney.[36] The nature of these substances remains undefined. Clearly the toxin is not urea, which, although present in high concentration in the uremic patient, is relatively nontoxic. Many of the signs and symptoms of uremia may be reversed by chronic hemodialysis, and this is true even if urea is placed in the dialysis bath so that the plasma urea concentration does not decrease.

ACUTE RENAL FAILURE

An acute deterioration of renal function may occur under a variety of circumstances, resulting in oliguria or anuria and progressive alterations in the composition of the body fluids.* Acute renal failure may occur without the appearance of primary morphologic changes in the kidneys themselves as a result of disturbed renal hemodynamics in dehydration and peripheral vascular collapse, heart failure, liver disease, etc. or because of temporary lower urinary tract obstruction. In these circumstances, if the disturbance is of short duration, necrosis of renal tubular cells may not occur, and kidney function returns promptly to normal if the underlying disorder is corrected. Acute renal failure may also appear as a consequence of acute inflammatory disease of the glomeruli (acute glomerulonephritis), interstitial inflammation resulting from a fulminating kidney infection, occlusion of the renal artery, or, most commonly, acute tubular necrosis.

Acute tubular necrosis. Necrosis of the renal tubular epithelium may occur as a result of prolonged renal ischemia, as in shock, after exposure to substances that are directly toxic to renal tubular cells, or, most frequently, as the result of a combination of these factors. The kidney is more susceptible to damage by nephrotoxic substances when its circulation is impaired, and it is for this reason that acute tubular necrosis is most often observed in individuals who are already severely ill for other reasons. A wide variety of

---

*Oliguria is conventionally defined by clinicians as a urine volume of 100 to 500 ml/1.73 m²/day, and anuria is defined as less than 100 ml/1.73m²/day.

drugs and poisons have been shown to be capable of producing acute tubular necrosis, including heavy metals, a number of organic solvents in common industrial and household use, the products of hemolysis and tissue trauma, and many of the drugs commonly employed in the treatment of severely ill patients.

Dr. Jean Oliver and his co-workers,[28] utilizing the technique of microdissection to obtain entire nephrons for examination, have provided a classic description of the pathology of this disorder. Two types of lesion are described. The first, called the nephrotoxic lesion, is limited to the proximal tubule and consists of destruction of epithelial cells, but with preservation of the tubular basement membrane. Generally, similar lesions are observed in all proximal tubules. The second type, designated tubulorhexis, results from renal ischemia, which produces much more intense destruction of the tubular architecture and disruption of the basement membrane. These ischemic lesions usually appear in patchy fashion throughout the kidney cortex and involve both proximal and distal tubules in the involved areas. Their extent depends on the size of the vessels involved.

The distribution of nephrotoxic lesions in the proximal tubule depends on the poison to which the individual was exposed. For example, it has been shown in animal studies, in which the dosage of the toxic substance was carefully controlled and not excessive, that dichromate produces damage primarily in the convoluted portion of the proximal tubule, whereas compounds containing mercury preferentially damage the distal third, or pars recta, of the tubule. Interestingly, uranium damages the middle third of the proximal tubule, leaving intact a short segment next to the glomerulus as well as the pars recta. The reason for the anatomic specificity of various nephrotoxins is unknown. Large doses of any of these poisons produce more extensive damage.

The mechanism of oliguria in acute tubular necrosis has been a source of controversy. Experimental evidence has been presented in support of each of the following as causes: diminished or absent glomerular filtration,[15] blockage of tubules by casts or cellular debris,[6] and excessive reabsorption or gross leakage of glomerular filtrate across a damaged and abnormally permeable tubular epithelium.[5] Renal blood flow is usually greatly reduced,[20] but comparable reductions in blood flow in chronic renal disease may be associated with the production of large volumes of urine, and anatomic evidence of glomerular damage is usually absent in acute tubular necrosis.

It is probable that there is no single cause for the oliguria and that all three suggested mechanisms may be involved to varying degrees in its genesis, their relative importance depending on the situation and on the stage of evolution of the disease process.[6] Recent observations in an experimental model of ischemic tubular injury lend support to this conclusion.[4] Anoxic injury was induced in rat kidneys by occluding the renal artery for a period of 1 hr. In the hours immediately following release of the occlusion, renal blood flow returned to about half its normal value, but despite the reduced blood flow, glomerular capillary pressure was high, and proximal intratubular pressure was greatly elevated. Urine flow was nonetheless markedly depressed, suggesting tubular obstruction, and this was confirmed on histologic study. Twenty-four hours later, glomerular capillary and proximal intratubular pressures were both depressed, a consequence of marked afferent arteriolar vasoconstriction, and there was also evidence that glomerular filtrate had leaked back across the damaged tubular wall. At this point, preglomerular vasoconstriction, continued tubular obstruction, and passive backflow of tubular fluid all appeared to be involved in the maintenance of oliguria. After acute volume expansion the glomerular capillary and proximal tubular pressures were restored to normal, but oliguria continued. Other studies showed that afferent arteriolar vasoconstriction may occur as a delayed response to tubular obstruction, even in the absence of ischemia or tubular necrosis.[3] It is thus clear that tubular obstruction is a significant factor in both the genesis and the maintenance of oliguria in ischemia-induced acute renal failure in rats. Preglomerular vasoconstriction and tubular backflow appear to be secondary phenomena that may be of increasing importance as the disease evolves.

The physiologic effects of acute renal failure are a direct consequence of loss of the homeostatic functions the kidney serves. Since control of solute and water excretion is lost, profound disturbances in water and electrolyte balance are inevitable unless intake is carefully controlled. Nonvolatile end products of cell metabolism that are normally excreted by the kidneys also accumulate in the body fluids, notably the anions of nonvolatile acids such as sulfate and phosphate and nitrogenous products of protein catabolism. Among the latter, urea is by far the most abundant, and the rate of rise of the blood urea concentration is a convenient indicator of the rate of protein breakdown in anuric patients. Urea is relatively nontoxic, however, and other products of protein metabolism, present in far smaller concentration, may be more dangerous. The earliest life-threatening consequence of acute renal failure may be an increase in the plasma potassium concentration, which, if not recognized and controlled, may lead to cardiac arrest.

The nephrotoxic lesions of acute tubular necrosis are reversible, and eventual recovery of renal function is the rule if the patient can be kept alive for the duration of the period of kidney failure. The onset of returning renal function may be apparent in a matter of a few hours in mild degrees of acute tubular necrosis, but in more severe instances the recovery process may take several weeks to a month or longer. The lesions due to ischemia produce more extensive structural damage than the toxic lesions, and where these predominate, complete functional recovery may not occur. An extreme example of this is known as renal cortical necrosis. In this condition there is generalized or widespread patchy ischemic necrosis of the entire renal cortex, except that in some instances nephrons lying immediately adjacent to the corticomedullary border or just below the renal capsule are spared. The condition apparently results from severe vasospasm or widespread thrombosis in the small intralobular arteries and afferent arterioles, and recovery of renal function is rare.

Since acute tubular necrosis is potentially reversible, treatment of the patient is directed toward careful maintenance of fluid and electrolyte balance and minimizing the rate of accumulation of toxic products of cell metabolism until recovery occurs. Salt and water intake must be carefully regulated to replace losses by evaporation, in sweat, feces, vomitus, etc., and protein and usually potassium intake must be curtailed. However, adequate carbohydrate, at least 100 gm daily in the adult, must be supplied to meet minimum caloric requirements, and it is desirable to keep caloric intake as near normal as possible. Otherwise there will be accelerated utilization of tissue fat stores, with the development of ketoacidosis, as well as a more rapid breakdown of cell proteins, with the consequent release into the extracellular fluid of potassium and toxic end products of protein metabolism. Although dietary protein intake must be severely restricted, the judicious administration of synthetic mixtures of essential amino acids or their α-keto analogs may be of value in reducing the catabolism of cellular protein. If the anuria is prolonged, or in severely ill patients, hemodialysis or peritoneal dialysis may be employed for removal of nonvolatile waste products and correction of disturbances in fluid balance. Patients with no renal function at all have been kept alive for months or years with these techniques.

CHRONIC RENAL INSUFFICIENCY

The fundamental pathophysiologic defect in the usual types of chronic renal insufficiency is the loss of nephrons and not damage to some one or another of their specific parts. The physiologic results are similar regardless of whether the loss occurs as the result of glomerulonephritis, pyelonephritis, or renal vascular disease. Destructive morphologic changes leading to disappearance of nephrons result in a fall in the glomerular filtration rate and tubular transport capacities and evoke compensatory morphologic and functional responses. The compensatory morphologic changes are hypertrophy and hyperplasia in remaining nephrons, particularly of the glomerulus and the proximal convoluted tubule. These changes are similar to those seen in the remaining kidney after unilateral nephrectomy. According to Oliver,[27] who made extensive microdissection studies of nephrons of diseased kidneys, *all* remaining nephrons in the end-stage kidney exhibit structural alterations. No morphologically unaltered nephrons are left to produce urine. The changes in the surviving nephrons are of two basic types: either an increase or a decrease in size. Oliver states, "Two such nephrons, the one large (hypertrophic) and the other small (atrophic), can therefore be considered the basic architectural units from which the kidney of chronic Bright's disease is built. It is true . . . that an infinite number of modifications of these types are found in the abnormal kidney, but always there is present the fundamental variation of increase or decrease of at least some part of the altered structure." Included among the possible types of nephron changes are hypertrophy and/or atrophy of the glomerulus and proximal tubule of the same nephron, loss of the glomerulus with persistence of the tubule forming an aglomerular nephron, alternating areas of atrophy, hypertrophy, hyperplasia, or dilatation in individual tubules, and varying degrees of damage to and/or obliteration of the glomerulus.

We believe in the general correlation of structure and function and therefore anticipate that such marked changes in structural configuration must be associated with changes in function. These presumed alterations in function are changes in rates of transport and perhaps in some of the permeability properties of the tubule. This should result in greater heterogeneity in function among the surviving nephrons, just as there is

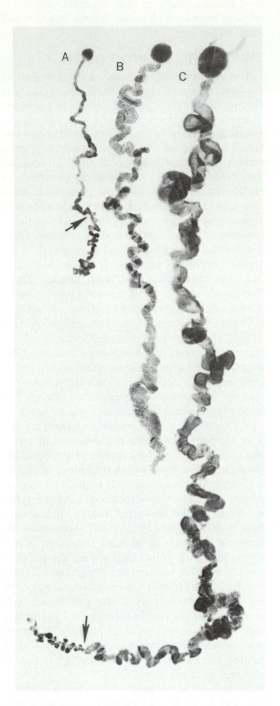

Fig. 51-1. Mosaic photomicrograph of three proximal tubules and attached glomeruli isolated by microdissection from kidney of patient with juvenile familial nephronophthisis. Arrows mark ends of proximal tubules of nephrons *A* and *C*. Note marked disparity in size of nephrons and that, in these three nephrons, size of glomerulus and attached tubule vary in parallel. (From Fetterman et al.[14])

greater structural heterogeneity (Fig. 51-1). We consider these compensatory morphologic and functional changes to represent a special example of adaptive growth.[18] Adaptive morphologic changes with parallel changes in function are essential for the survival of the patient. This view is somewhat different from the "intact nephron hypothesis" evolved by Bricker,[10] who has had difficulty resolving what he considers to be a discrepancy between the functional and structural changes in the chronically diseased kidney. Bricker stresses that the function of the chronically diseased kidney is organized and highly appropriate in a regulatory context despite a diversity of morphologic changes, a point of view with which we agree. Clearly the diseased kidney must be able to function appropriately and respond to changing needs and stresses; when it is no longer able to do this, the patient dies. Recognition that the correlation of structure and function relates to quantitative rather than qualitative changes in the function of individual surviving nephrons resolves this apparent discrepancy.[24,31]

An understanding of the pathophysiology of the chronically diseased kidney is facilitated by recognition of the fact that the primary function of the kidney is to regulate the volume and composition of the body fluids.[29] Excretory function is determined accordingly. This perhaps can be best illustrated by comparing the urine excreted by a person with chronic kidney disease with that excreted by one with normal kidney function. Given two such individuals who ingest an identical diet containing the same amount of protein, carbohydrate, fat, salt, etc. and the same relatively generous fluid intake, the volume and composition of their urine will be identical. Each person will excrete the same amount of nitrogen, phosphate, sulfate, salt, water, etc. Such a relationship must hold for one to be in a steady state, and the sustained absence of a steady state is incompatible with prolonged continuation of life. The urine of the patient with chronic renal disease may, of course, also contain an additional amount of protein, red blood cells, white blood cells, and casts, but the major excretory products will be the same.

To understand how this happens it is helpful to review as prototypes the mechanisms of excretion of creatinine, urea, sodium, and potassium in the presence of a reduced mass of nephrons. Since urine formation is the net result of two processes of great magnitude, glomerular filtration and tubular reabsorption and/or secretion, a decrease in glomerular filtration rate (GFR) has

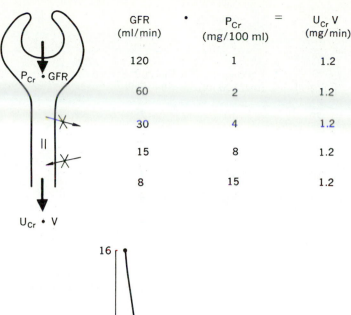

GFR (ml/min)	•	P_{Cr} (mg/100 ml)	=	$U_{Cr}V$ (mg/min)
120		1		1.2
60		2		1.2
30		4		1.2
15		8		1.2
8		15		1.2

Fig. 51-2. Relationships between glomerular filtration rate, plasma creatinine concentration, and rate of creatinine excretion at normal and reduced filtration rates. (See text.)

predictable consequences relating to tubular function.[16]

Creatinine clearance. The simplest relationship holds for a substance that is excreted only as a result of glomerular filtration, without tubular reabsorption or secretion. Let us assume that this is true in humans for creatinine and that the amount of creatinine produced each day is the same in a person with serious renal disease as it was during health. The relationships that must exist if the individual is in a steady state are shown in Fig. 51-2. If the normal GFR is 120 ml/min and the normal plasma creatinine concentration is 1.0 mg/dl, then 1.2 mg/min of creatinine are excreted in the urine. For these to be steady-state values the rate of excretion must equal the rate of production. If either rate changes, the plasma creatinine concentration will change until the amount of creatinine excreted again equals that produced. If the filtration rate is reduced by disease to half of normal, the plasma creatinine concentration will rise to 2 mg/dl; if the filtration rate is a fourth of normal, to 4 mg/dl, etc. As shown in Fig. 51-2, the GFR and plasma creatinine concentration must be reciprocally related in order for the rate of creatinine excretion to remain unchanged. Because of slight deviations from the simplifying assumptions made previously (creatinine production is not necessarily precisely constant and there is probably some tubular transport of creatinine), plasma endogenous creatinine concentrations in patients should not be interpreted in these precise terms. However, the relationship holds qualitatively, and serial determinations of the plasma creatinine concentration are very useful in following the clinical course of a patient with chronic renal disease.

Urea clearance. The relationships for urea are quite similar, although the urea clearance is always less than the GFR because of tubular reabsorption (Fig. 51-3). In a steady state the product of the urea clearance and plasma concentration

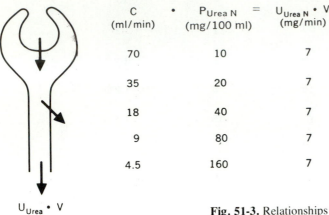

$$C = \frac{[U]V}{[P]}$$

C (ml/min)	•	$P_{Urea\,N}$ (mg/100 ml)	=	$U_{Urea\,N}$ • V (mg/min)
70		10		7
35		20		7
18		40		7
9		80		7
4.5		160		7

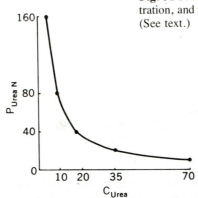

Fig. 51-3. Relationships between clearance of urea, plasma urea concentration, and rate of urea excretion at normal and reduced urea clearances. (See text.)

(i.e., urea excretion) must equal the rate of urea production. The latter will be constant only if protein intake and metabolism are constant. When the GFR is reduced, as by the destruction of glomeruli, the clearance of urea is also reduced, and the plasma urea concentration increases until the product of clearance and plasma concentration again equals the rate of urea production. As was true for creatinine, urea clearance and plasma concentration must be related in the form of a rectangular hyperbola in order for the rate of urea excretion to remain constant.

Not unexpectedly the experimental findings conform to the predicted results. Thirty years ago Goldring and Chasis[17] demonstrated that these relationships hold in patients with chronic glomerulonephritis and hypertension. In Fig. 51-4, their data are plotted on a logarithmic scale to form a straight line. On a linear plot, these points would fall along a hyperbola, as in Fig.

51-3. The upper diagonal line describes the relation when the rate of urea excretion equals 24 mg and the lower line when it equals 8 mg/min, which would result from daily protein intakes of approximately 100 and 24 gm/day, respectively.

The tubular reabsorption of urea is largely but perhaps not exclusively passive, and when the rate of filtration is decreased, the plasma urea concentration must rise until the amount of urea that is filtered and escapes reabsorption equals the amount produced. Thus with both urea and creatinine there is *no regulation* of plasma concentration and there is *no adaptation* of the mechanisms of excretion in response to changes in the internal environment.

Sodium excretion. The situation with regard to sodium excretion is very different. The body cannot tolerate large changes in either total body sodium content or plasma sodium concentration; the tubular reabsorption of sodium must therefore be regulated. Again, what happens in the chronically diseased kidney is predictable, since the amount of sodium excreted must equal that ingested; otherwise, one would either swell up and burst or wither away and disappear. The basic relationships are shown in Fig. 51-5. In order for the amount of sodium excreted to remain the same when the GFR decreases, the fraction of the filtered sodium that is excreted must increase. If, for example, the dietary intake demands that

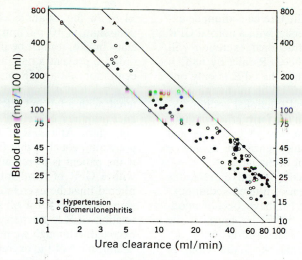

Fig. 51-4. Relationship of blood urea concentration to urea clearance in 103 observations in patients with hypertension and glomerulonephritis (urine flow above 1.5 ml/min). (From Goldring and Chasis.[17])

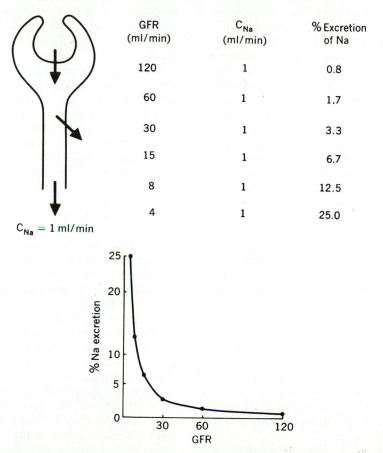

GFR (ml/min)	C_{Na} (ml/min)	% Excretion of Na
120	1	0.8
60	1	1.7
30	1	3.3
15	1	6.7
8	1	12.5
4	1	25.0

$C_{Na} = 1$ ml/min

Fig. 51-5. Relationships between filtration rate, clearance of sodium, and percent of filtered sodium excreted at normal and reduced filtration rates with unchanging plasma sodium concentration. (See text.)

approximately 0.8% of the filtered sodium be excreted in order for the person with a normal GFR to remain in balance, fractional excretion will double to 1.7% when the GFR falls by half; it will increase fourfold when the GFR is only one fourth the normal rate, etc. With unchanging sodium intake the GFR and fractional sodium excretion must be related in the form of a rectangular hyperbola if both the rate of excretion and plasma concentration, and hence the clearance of sodium, are to remain unchanged.

Although the predicted adaptive relations between GFR and percent of filtered sodium excreted shown in Fig. 51-5 have been demonstrated experimentally,[7,23,30,33] the cause of the decrease in fractional reabsorption with decreased GFR remains unknown. Alterations in several of the normal control systems have been suggested. It has been proposed that the decreased fractional sodium and water reabsorption is a consequence of the osmotic diuresis per nephron caused by the high plasma urea concentration found in such patients.[19] A priori, it seems highly unlikely that so vital a function as sodium balance should be determined in such an indirect fashion. Furthermore, there was no natriuresis and only slight increase in water excretion when an acute elevation of the plasma urea concentration was produced in rats without volume expansion.[22] Various clinical observations lead to a similar conclusion. There is no evidence that the change in sodium reabsorption results from changes in either mineralocorticoid secretion or physical factors. Bricker et al.[9] have suggested that it results from a circulating natriuretic factor found in the serum of uremic patients. It is not known whether the proposed natriuretic substance is the same as the factor responsible for the natriuretic response of normal subjects to extracellular volume expansion. If so, the implication is that there is either an undetected element of volume expansion in many patients with chronic renal disease, or the sensitivity of the system regulating the production of natriuretic hormone is increased.

Fractional excretion of filtered water is also increased in the presence of a reduced mass of nephrons and normal water intake. The antidiuretic hormone mechanism controls water reabsorption in the distal convolution and collecting ducts, so that the sodium concentration of plasma is usually normal.

Potassium excretion. The relationships between GFR, percent of filtered potassium excreted, and clearance of potassium with advancing renal insufficiency are diagrammed in Fig.

51-6. In the example chosen the clearance of potassium is 10 ml/min, and with constant dietary intake, it must be maintained at that rate in order to prevent a rise in plasma potassium concentration. At a GFR of 120 ml/min an amount of potassium is excreted equal to 8% of that filtered; with a fall in GFR to half, the same clearance requires excretion of 16% of the filtered potassium. At a GFR of 10 ml/min an amount of potassium equal to that filtered must be excreted if the patient is to remain in a steady state, and with a GFR of 5 ml/min, twice as much as is filtered must be excreted. Again, these relationships take the form of a rectangular hyperbola relating GFR to the percent of filtered potassium excreted. Although we now know that, at least in health, most of the excreted potassium is derived from a process of potassium secretion in the distal convolutions and collecting ducts, the process of potassium secretion in man was initially demonstrated by Leaf and Camara[25] and independently by Platt[30] in clearance studies on patients with severe renal insufficiency. When such patients are loaded with potassium, the clearance of potassium regularly exceeds the amount filtered.

The control system responsible for these adaptive changes in potassium excretion remains unidentified. The appropriate rate of potassium excretion with a decrease in GFR does not depend on the rate of sodium excretion.[23,32] It cannot be fully explained by an increase in nonpermeant ion excretion per nephron, nor does it depend on changes in mineralocorticoid activity or in the rate of hydrogen ion secretion per nephron. Whether an undefined control factor contributes to the kaliuresis per nephron remains unknown. In some patients with severe renal insufficiency the control of potassium excretion is inadequate, and the plasma concentration may rise to levels that are life threatening.

"Trade-off" hypothesis. Not all the consequences of functional adaptation in the surviving nephrons of diseased kidneys are necessarily salutary. We have previously noted that as renal disease progresses, phosphate balance is maintained at the expense of rising levels of parathyroid hormone, with the unfortunate consequences of hyperparathyroidism appearing as a trade-off. Bricker[8] has suggested that many other such trade-offs may be necessary in progressive renal disease and that many of the signs and symptoms of uremia may be unfortunate side effects of required homeostatic adaptations. For example, the adaptative increase in sodium excretion per nephron that must occur as the num-

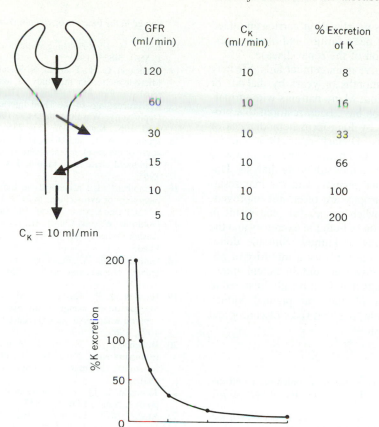

GFR (ml/min)	C_K (ml/min)	% Excretion of K
120	10	8
60	10	16
30	10	33
15	10	66
10	10	100
5	10	200

$C_K = 10$ ml/min

Fig. 51-6. Relationships between filtration rate, clearance of potassium, and percent of filtered potassium excreted at normal and reduced filtration rates with unchanging plasma potassium concentration. (See text.)

ber of functioning nephrons diminishes might involve the intervention of a natriuretic substance. Such a substance, present in high levels in far-advanced renal disease, might interfere with sodium transport in cells throughout the body and thus contribute to the signs and symptoms of the uremic state.[9] Conceivably, certain other so-called uremic toxins might be hormonal substances normally involved in the homeostatic regulation of renal function, but necessarily present in abnormal concentrations in renal failure.

Therapeutic considerations. Conservative management of the patient with chronic renal failure is based on recognition of the fact that the capacity of the diseased kidney to respond appropriately to rapidly changing homeostatic demands is compromised. The dietary intake of water and electrolytes must therefore be more carefully controlled, and any unusual losses, such as excessive sweating or diarrhea, promptly re-

placed. Acidosis, if severe, may be partially corrected by the administration of sodium bicarbonate or sodium citrate. Many uremic patients experience striking symptomatic improvement when dietary protein is restricted. Adequate amounts of the essential amino acids must be provided if accelerated metabolism of cellular protein is to be avoided, however, and this is most conveniently accomplished by restricting protein intake to those sources rich in essential amino acids, such as eggs and milk. The same end may be accomplished more efficiently, although far more expensively, with synthetic mixtures of amino acids or their α-keto analogs.[35] Although life-threatening hyperkalemia is unusual except in the terminal stages of chronic renal insufficiency, it is generally wise to avoid excessive potassium intake, and this is most readily accomplished on a protein-restricted diet. Hypertension, if present, should be controlled, since sustained blood pressure elevation carries an added risk of stroke and heart failure and also accelerates the progression of the renal disease.

When conservative management fails, life may be sustained for months or years by dialysis or renal transplantation. Many patients with virtually no renal function have been restored to a productive existence by long-term hemodialysis or peritoneal dialysis. Dietary restrictions can be relaxed somewhat, and many of the symptoms of uremia are ameliorated by adequate dialysis. Unfortunately, anemia persists, and the manifestations of osteodystrophy are often not improved. Transplantation, when successful, can result in complete restoration of both the excretory and the endocrine functions of kidney. Suitable donor kidneys are available for only a minority of potential recipients, however, and the potent immunosuppressive drugs that must be administered to prevent rejection of the transplanted kidney expose the recipient to added risks of cancer and fulminant infection.

REFERENCES
General reviews

1. Giordano, C.: Proceedings of a conference on uremia, Capri, Italy, Sept. 1974, Kidney Int. **7**(suppl. 3):267, 1975.
2. Levinsky, N. G.: Pathophysiology of acute renal failure, N. Engl. J. Med. **296**:1453, 1977.

Original papers

3. Arendshorst, W. J., Finn, W. F., and Gottschalk, C. W.: Nephron stop-flow pressure response to obstruction for 24 hours in the rat kidney, J. Clin. Invest. **53**:1497, 1974.
4. Arendshorst, W. J., Finn, W. F., and Gottschalk, C. W.: Pathogenesis of acute renal failure following temporary renal ischemia in the rat, Circ. Res. **37**:558, 1975.
5. Bank, N., Mutz, B. F., and Aynedjian, H. S.: The role of "leakage" of tubular fluid in anuria due to mercury poisoning, J. Clin. Invest. **46**:695, 1967.
6. Biber, T. U. L., et al.: A study by micropuncture and microdissection of acute renal damage in rats, Am. J. Med. **44**:664, 1968.
7. Black, D. A. K., Platt, R., and Stanbury, S. W.: Regulation of sodium excretion in normal and salt-depleted subjects, Clin. Sci. **9**:205, 1950.
8. Bricker, N. S.: On the pathogenesis of the uremic state. An exposition of the "trade-off hypothesis," N. Engl. J. Med. **286**:1093, 1972.
9. Bricker, N. S., Bourgoignie, J. J., and Klahr, S.: A humoral inhibitor of sodium transport in uremic serum. A potential toxin? Arch. Intern. Med. **126**:860, 1970.
10. Bricker, N. S., et al.: The pathophysiology of renal insufficiency. On the functional transformations in the residual nephrons with advancing disease, Pediatr. Clin. North Am. **18**:595, 1971.
11. Brickman, A. S., et al.: 1,25 dihydroxy-vitamin D in normal man and patients with renal failure, Ann. Intern. Med. **80**:161, 1974.
12. DeLuca, H. F.: The kidney as an endocrine organ involved in the function of vitamin D, Am. J. Med. **58**:39, 1975.
13. Erslev, A. J.: Renal biogenesis of erythropoietin, Am. J. Med. **58**:25, 1975.
14. Fetterman, G. H., Fabrizio, N. S., and Studnicki, F. M.: Microdissection in the study of normal and abnormal renal structure and function. In Heptinstall, R. H., editor: Proceedings of the Third International Congress of Nephrology, Washington, D.C., 1966, Basel, Switzerland, 1967, Karger, vol. 2, p. 235.
15. Flanigan, W. J., and Oken, D. E.: Renal micropuncture study of the development of anuria in the rat with mercury-induced acute renal failure, J. Clin. Invest. **44**:449, 1965.
16. Gamble, J. L.: Chemical anatomy, physiology and pathology of extracellular fluid. A lecture syllabus, ed. 6, Cambridge, Mass., 1954, Harvard University Press.
17. Goldring, W., and Chasis, H.: Hypertension and hypertensive disease, New York, 1944, The Commonwealth Fund.
18. Gottschalk, C. W.: Function of the chronically diseased kidney. The adaptive nephron, Circ. Res. **29**(suppl. 2):1, 1971.
19. Hayslett, J. P., Kashgarian, M., and Epstein, F. H.: Mechanism of change in the excretion of sodium per nephron when renal mass is reduced, J. Clin. Invest. **48**:1002, 1969.
20. Hollenberg, N. K., et al.: Acute renal failure due to nephrotoxins. Renal hemodynamic and angiographic studies in man, N. Engl. J. Med. **282**:1329, 1970.
21. Jacobson, L. O., et al.: Role of the kidney in erythropoiesis, Nature **179**:633, 1957.
22. Kauker, M. L., Lassiter, W. E., and Gottschalk, C. W.: Micropuncture study of effects of urea infusion on tubular reabsorption in the rat, Am. J. Physiol. **219**:45, 1970.
23. Kleeman, C. R., Okun, R., and Heller, R. J.: The renal regulation of sodium and potassium in patients with chronic renal failure (CRF) and the effect of diuretics on the excretion of these ions, Ann. N.Y. Acad. Sci. **139**:520, 1966.
24. Kramp, R. A., et al.: A study by microdissection and micropuncture of the structure and the function of the kidneys and the nephrons of rats with chronic renal damage, Kidney Int. **5**:147, 1974.
25. Leaf, A., and Camara, A. A.: Renal tubular secretion of potassium in man, J. Clin. Invest. **28**:1526, 1949.
26. Massry, S. G., Ritz, E., and Verberckmoes, R.: Role of phosphate in the genesis of secondary hyperparathyroidism of renal failure, Nephron **18**:77, 1977.
27. Oliver, J.: Architecture of the kidney in chronic Bright's disease, New York, 1939, Paul B. Hoeber, Inc.
28. Oliver, J., MacDowell, M., and Tracy, A.: The pathogenesis of acute renal failure associated with traumatic and toxic injury. Renal ischemia, nephrotoxic damage and the ischemuric episode, J. Clin. Invest. **30**:1305, 1951.
29. Pitts, R. F.: Physiology of the kidney and body fluids, ed. 3, Chicago, 1974, Year Book Medical Publishers, Inc.
30. Platt, R.: Sodium and potassium excretion in chronic renal failure, Clin. Sci. **9**:367, 1950.
31. Platt, R.: Structural and functional adaptation in renal failure, Br. Med. J. **1**:1313, 1372, 1952.
32. Schultze, R. G., et al.: On the adaptation in potassium excretion associated with nephron reduction in the dog, J. Clin. Invest. **50**:1061, 1971.

33. Slatopolsky, E., et al.: Studies on the characteristics of the control system governing sodium excretion in uremic man, J. Clin. Invest. **47:**521, 1968.

34. Slatopolsky, E., et al.: On the pathogenesis of hyperparathyroidism in chronic experimental renal insufficiency in the dog, J. Clin. Invest. **50:**492, 1971.

35. Walser, M., et al.: The effect of keto-analogues of essential amino acids in severe chronic uremia, J. Clin. Invest. **52:**678, 1973.

36. Welt, L. G., Black, H. R., and Krueger, K. K.: Symposium on uremic toxins, Arch. Intern. Med. **126:**773, 1970.

THOMAS H. MAREN

52 Cerebrospinal fluid, aqueous humor, and endolymph

The normal physiology and biochemistry of the cerebrospinal fluid (CSF) stand at the crossroads of three vital subjects in medical biology: active transport, respiration, and neurology. The concepts and information in these subjects are changing and developing rapidly. Not long ago the CSF was regarded as a filtrate of plasma, whose role, if any, was the mechanical one of keeping the central nervous system (CNS) afloat. Now, as we shall see, it is clear that the composition of CSF is actively and closely regulated at secretory sites; that the respiratory center is sensitive to this composition; and that the continuity between CSF and fluids of the brain makes inevitable some relationship between fluid composition and neuronal activity.

The delicate pathways of the CSF within the brain are subject to aberrations during development, trauma, and disease. The life-threatening nature of these changes makes it essential for the physician to understand the normal physiology and the areas susceptible to pathologic change.

There is a remarkable similarity between the CSF and aqueous humor (AH), including the histology of the secretory and absorbing tissue, the protein-free fluid formed, the rates of ionic movement, the active nature of ionic entry, the passive nature of ionic and fluid reabsorption, and disease due to obstruction of flow. A further remarkable point, which has received insufficient attention, is that the CSF, like AH, might be regarded as a two-compartment system: ventricular space analogous to posterior chamber and subarachnoid space analogous to anterior chamber.

The endolymphatic fluid (EL) is relatively neglected in general physiology. Recent work shows basic similarities to CSF and AH in the chemistry of its formation. As with CSF and AH, there are serious disease states associated with impairment of flow of EL.

CEREBROSPINAL FLUID

PHYSIOLOGIC ANATOMY

Thorough descriptions of the morphology of CSF pathways, including the comparative and embryologic aspects, are available.[6,21,30] Fig. 52-1 shows two three-dimensional models of the ventricular system; Fig. 52-2 is a diagram of the choroid plexuses, ventricles, and pathways of the subarachnoid space. The physiology of these systems, including brain fluid, is fully described in four excellent monographs.[2a,8,10,27] Other reviews emphasize the acid-base physiology of CSF, certain clinical problems, and the role of CSF in the regulation of respiration.[15,25] An analysis of CSF formation based largely on observations of the frog choroid plexus in vitro has just appeared.[30a]

CSF is formed predominantly in the choroid plexuses, which are secretory modifications of the pia mater projecting into the lateral ventricles and the roofs of the third and fourth ventricles.

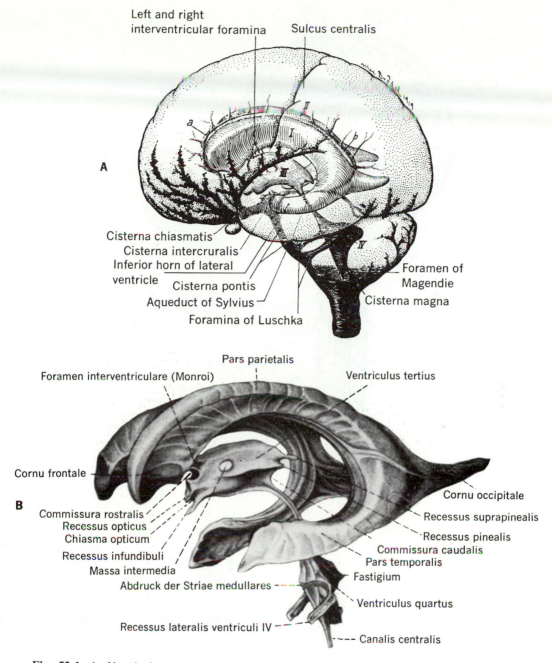

Fig. 52-1. A, Ventricular system and cisterns of human brain. *a,* Cistern of corpus callosum; *b,* superior cistern. **B,** Cast of human brain. (From Dandy.[45])

The detailed structure of this tissue, including histochemistry and pathology, is the subject of a recent monograph.[22] Of particular importance is characterization of choroid plexus with respect to permeability. Structural and electrical studies show this tissue to be intermediate in "leakiness" between the extremes of open proximal tubule and tight frog skin.[27] Histology of this tissue is remarkably like that of the proximal tubule of the kidney (Fig. 52-3); there are striking physiologic similarities as well. Attention has been given to the structure of microvilli and blebs in the choroid plexus following the use of drugs that reduce secretion.[43] Blood flow through the choroid plexus is remarkably rapid (3 ml/gm/min), and CSF secretion is 25% of this rate.[109] CSF is

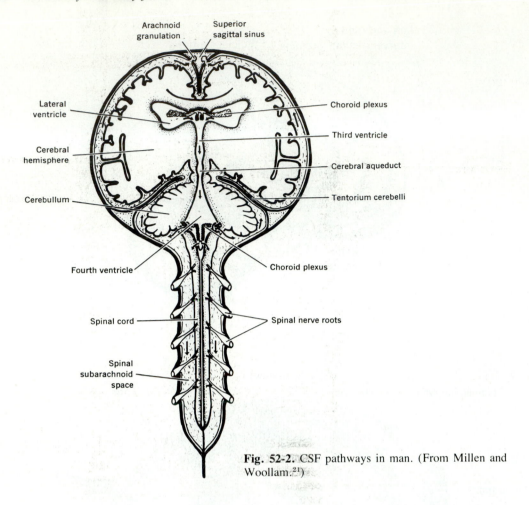

Arachnoid granulation

Superior sagittal sinus

Lateral ventricle

Cerebral hemisphere

Cerebellum

Fourth ventricle

Spinal cord

Spinal subarachnoid space

Choroid plexus

Third ventricle

Cerebral aqueduct

Tentorium cerebelli

Choroid plexus

Spinal nerve roots

Fig. 52-2. CSF pathways in man. (From Millen and Woollam.[21])

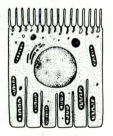
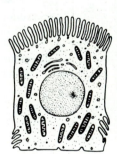

Proximal convoluted tubule

Adult choroid plexus

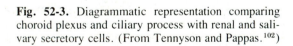

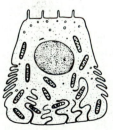
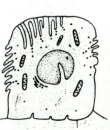

Striated salivary duct

Ciliary epithelium

Fig. 52-3. Diagrammatic representation comparing choroid plexus and ciliary process with renal and salivary secretory cells. (From Tennyson and Pappas.[102])

formed by extrachoroidal tissue as well; it appears likely that glial elements in the ependymal wall of the brain are responsible. Perhaps 35% of CSF is formed here.[84]

In man and other mammals, CSF flows in the direction shown by the arrows in Fig. 52-2. The volume within the ventricular system of man (approximately 23 ml) communicates with the volume in the subarachnoid space of brain and spinal cord (approximately 117 ml). CSF escapes from the roof of the fourth ventricle (foramen of Magendie) into the subarachnoid cisterna magna and from the lateral recesses of the fourth ventricle (foramina of Luschka) to the ventral aspect of the brain, spreading to the basal cisterns. The entire subarachnoid space of the brain and spinal cord is thus permeated with fluid that ultimately exits from that space via the veins (see subsequent discussion). CSF flow, in relation to its volume, is rather constant in mammals.[6] In man, flow/volume[98] yields the turnover rate constant for formation as follows:

$$k_{in} = \frac{0.37 \text{ ml/min}}{140 \text{ ml}} = 0.0026/\text{min}$$

The half-time ($t_{1/2}$) for CSF renewal is ln 2/rate constant, or 270 min. Fluid formation is rarely susceptible to increase but can be decreased by drugs that lower the rate of ion secretion (see discussion of mechanisms underlying ion movement later in this chapter). CSF formation also is decreased by the elevation of intraventricular pressure, although the mechanism is not clear.[63]

The measurement of CSF formation and bulk fluid absorption has been performed in a variety of ways; perhaps the most effective is the system using perfusion from lateral ventricle to cisterna magna. Inulin (or dextran) is added to the perfusate; its dilution measures the addition of nascent CSF, and its clearance measures absorption or drainage of fluid distal to the cisterna.[61,98]

Bulk CSF exits from the subarachnoid space, which contains projections of microscopic (villi) or macroscopic (granulations) size, into the venous sinuses. The CSF pressure is about 150 mm H_2O, almost twice that in the veins. CSF pressure is little influenced by arterial blood pressure but is affected by venous pressure and by the volume of blood in the head. The villi are valve-like, requiring a gradient of 10 mm H_2O to open and allowing particles as large as 7 μm in diameter to pass through passively.[110] It follows that the rate of loss of tagged albumin or colloid dyes from CSF will give a measure of the outflow of bulk fluid, which in turn must equal the formation rate. However, it will be shown that individual ions and drugs, particularly those carrying a charge, can be secreted outward into the blood by choroidal tissue at a rate considerably faster than that provided by bulk flow.

The rabbit choroid plexus receives well-developed cholinergic and adrenergic nerves.[15a] Agonists of both systems can increase CSF flow, but adrenergic effects are complicated by multiple receptor systems.[60a] Norepinephrine and stimulation of the superior cervical ganglion reduce CSF flow.[15a] On the other hand, the increase of cyclic AMP by activating adenyl cyclase (using cholera toxin[52a]) or inhibiting its breakdown (using theophylline[60a]) augments CSF flow. These effects appear to be mediated by β-adrenergic agonism.[86a] None of the changes cited exceed twofold in either direction. The specific pathways involved and their connections to or control of ion transport remain to be discovered.

RELATION BETWEEN CSF AND FLUID COMPARTMENTS OF THE BRAIN: DISPOSITION OF CHEMICALS IN COMPARTMENTS

Fig. 52-4 shows fluid exchange within the brain; the subject is a difficult one, due in part to the facts that secretory elements are distributed through the brain, probably in glia, and that neither extracellular fluid (ECF) nor intracellular fluid (ICF) can be sampled directly. It is convenient and a reasonably true estimate to regard the ECF as similar to CSF in composition and as continuous with it through intercellular channels (Fig. 52-4). However, there is at least one distinction: sodium entry into CSF is determined by unidirectional active flux at the choroid plexus, whereas at the blood-brain barrier there is a major diffusional component for this ion.[8] This will be evident from the pharmacologic data presented in a later section of this chapter.

Fluid compartments of the brain may be defined by the disposition of certain drugs or markers of known physicochemical properties, whose movement and concentration can be measured readily. Lipid-insoluble substances (excluding small ions), for example, inulin and sucrose, are distributed in the ECF but do not enter the ICF. Furthermore, blood-CSF and blood-brain barriers prevent their direct entry into CSF and brain. Their distribution in ECF is measured by adding them to CSF or to brain tissue in vitro. Analysis of brain in either of these circumstances yields a volume of distribution of 10% to 15% of tissue weight, which generally is accepted as the ECF space.[97]

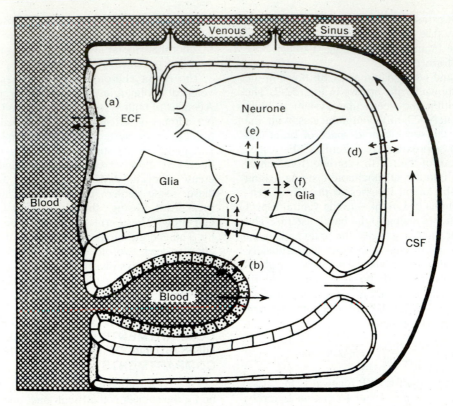

Fig. 52-4. Scheme illustrating exchanges within CNS. Dashed lines show diffusion; solid arrows show proved directions of fluid flow. (From Davson and Bradbury.[47]) See also Bradbury.[2a]

Table 52-1. Rate constants for entry of certain drugs from plasma to cisternal CSF and anterior chamber AH*

	CSF (k_{in}, min^{-1})	AH (k_{in}, min^{-1})	$CHCl_3$ Buffer (pH 7) partition
Thiopental	0.6	—	102
Aniline	0.5	—	17
D_2O	0.3	0.13	—
Aminopyrine	0.3	—	73
Ethanol	0.25	0.05	—
Pentobarbital	0.17	—	—
Antipyrine	0.12	—	28
Acetanilide	0.04	—	3
Barbital	0.03	—	2
Sulfanilamide	0.02	0.03	0.014
Sulfadiazine	0.01	0.016	0.004
Thiourea	0.006	0.008	—
Urea	0.004	0.005	—
Sulfanilic acid	0.005	—	0.0005

*Data chiefly from Davson[6,7] and Rall[96]; various species.

Lipid-soluble substances, on the other hand, enter CSF and brain rapidly, with a fair relation between solubility and rate. An example is shown in Fig. 52-5, in which a comparison is made between the kinetics of entry of two drugs, closely related but differing 10- to 45-fold in lipid solubility.[80] Antipyrine, the more lipid-soluble substance, enters the brain and CSF far more rapidly than *N*-acetyl-4-aminoantipyrine. It will also be noted that for antipyrine, accession to brain precedes that of CSF; this is generally true for highly lipid-soluble substances such as thiourea and ethanol.[8] Flowing CSF serves as a "sink" into which the drug diffuses down its concentration gradient from the relatively stagnant brain fluid. In such cases (see the lipid-soluble drugs of Table 52-1) the entry to CSF reflects direct passage from blood and diffusion from brain. The rate constant, k_{in}, is then a composite term. *N*-Acetyl-4-aminoantipyrine, by contrast, enters brain and CSF slowly (compared to liver) and at the same rate, suggesting parallel

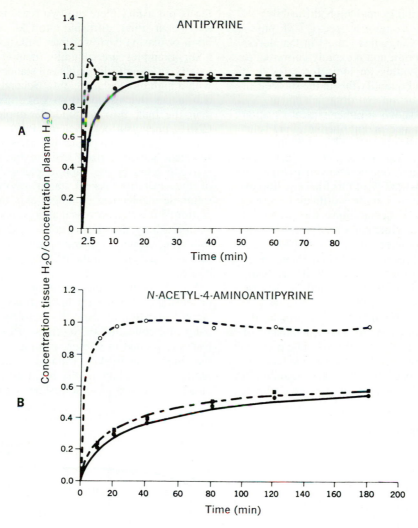

Fig. 52-5. Rates of entry of two drugs into liver (o), whole brain (■), and CSF (●) of rabbit. Constant plasma levels were maintained. Both drugs are negligibly bound to plasma proteins and are weak bases un-ionized at body pH. Organic solvent–aqueous buffer pH 7.4 fractions are as follows:

	Olive oil or heptane	**Benzene**	**CHCl$_3$**
Antipyrine	0.04	0.09	28.0
N-Acetyl-4-aminoantipyrine	0.004	0.002	1.5

(From Mayer et al.[80])

blood-CSF and blood-brain barriers. There are no analogous structures in liver or muscle; in these somatic tissues the ECF may be regarded generally as a continuum of plasma. In the brain and CSF, as we have seen, there are barriers at the plasma interface. Anatomic and kinetic evidence for these has been reviewed.[2a,6,8]

Table 52-1 shows rates of entry of some dozen drugs to CSF and AH. The most rapid, thiopental, has a CHCl$_3$:water partition ratio of about 100; this property, reflecting high lipid solubility,

is the basis of its use as an immediate-acting intravenous anesthetic. The slowest listed, sulfadiazine, has a partition ratio of 0.04. Highly polar compounds, such as *d*-tubocurarine, have a ratio on the order of 10^{-5} and cannot even be measured in CSF. This is due also to active removal of such charged substances, which is discussed subsequently. Table 52-1 also shows that highly diffusible substances gain access to CSF at much higher rates than bulk flow of fluid, the k_{in} of which (see previously) is 0.0026/min.

The lipid solubility and high diffusibility of CO_2 are keys to a major aspect of CSF physiology, whereby CO_2 gains almost instant access to the CSF and brain and is rapidly converted to HCO_3^-, thus providing an ionic force for fluid movement. I shall quantify this in the subsequent section on rates of ion and drug movement.

Fig. 52-5, *B*, shows a rather typical situation in which a substance of intermediate lipid solubility slowly penetrates the CSF and brain, gradually approaching (after due correction for the water content) the concentration in plasma. In the example shown, 60% of this final equilibrium value is achieved in 3 hr. It is difficult to sustain constant plasma concentrations for prolonged periods in an experimental situation; we might guess that, if this were done, *N*-acetyl-4-aminoantipyrine would reach equilibrium among brain water, CSF, and plasma in about 10 hr. Many common drugs such as aspirin and the several types of sulfonamides follow this general pattern. If the drug is bound to plasma protein, the reference concentration (as the denominator of the ordinate in Fig. 52-5) is the unbound portion, since only it is free to diffuse across the barriers.

Certain substances fail to reach equilibrium between CSF and plasma. If plasma protein bind-ing is put aside, there are two reasons why this may occur. First, diffusion into CSF from either brain or plasma is slower than bulk flow leaving CSF; examples are albumin, mannitol, and a variety of slowly diffusible substances or those with a high molecular weight.[5] Second, there are substances that escape from CSF more rapidly than the bulk flow of fluid. Examples are I^-, CNS^-, phenolsulfonphthalein (PSP), iodopyracet (Diodrast), and penicillin. These drugs are secreted outward (from CSF to blood) by an active process at the choroid plexus.[5] Kinetics of the operation reveal an energy-dependent, saturable mechanism strikingly akin to renal secretion.[92] It is this process that prevents penicillin from reaching an effective concentration in brain following ordinary systemic doses, an important factor to recognize in the treatment of infectious disease.

Fig. 52-6 shows the access of a wide variety of substances to the rat brain, following intracarotid injection of their ^{14}C-radiolabeled molecules and collection of the tissue 15 sec later.[89] Data are plotted in reference to the accession of the highly diffusible tritiated water, which reaches equilibrium in the brain after a single passage of blood. The solid line is drawn with

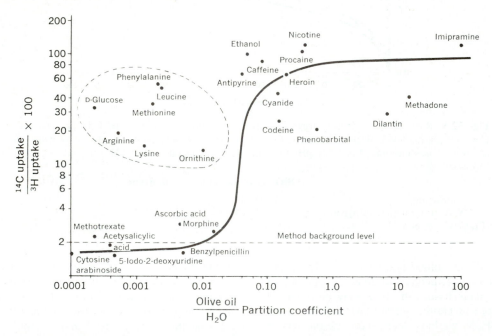

Fig. 52-6. Lipid solubility and blood-brain barrier penetration. Plot of percentage of clearance of radiolabeled substances vs lipid/water partition coefficient during course of single brain passage following carotid arterial injection. Drugs with partition coefficient greater than about 0.03 show nearly complete clearance. Encircled substances on left have minimal lipid affinity, yet show appreciable clearances. These brain metabolites penetrate blood-brain barrier by virtue of specific transport systems. (From Oldendorf.[89])

points for 20 drugs, and shows, once again, the classic relation between lipid solubility and brain uptake. When the olive oil:buffer partition ratio exceeds 0.03, brain uptake of the drug is complete. (Note from the legend of Fig. 52-5 that olive oil partitions are some 100-fold lower than those for $CHCl_3$.) Plasma protein binding is not considered in Fig. 52-6 but probably accounts for numbers lower than 100 for phenobarbital and phenytoin (Dilantin). The pharmacologic basis for a major socioeconomic problem is seen in Fig. 52-6; heroin gains access to the brain much faster than morphine, due to the relative lipid solubility conferred by its acetyl groups. This is the basis for the rush or orgasm-like sensation following intravenous heroin and contributes greatly to its continued position as an abused street drug. Fig. 52-6 (dotted enclosure) also shows that amino acids and glucose have the special feature of rapid access to brain despite minimal lipid affinity; this is shown to be due to carrier-mediated penetration, which exhibits properties of stereospecificity and saturation.[23]

COMPOSITION OF CSF: POTENTIALS

The ionic composition of CSF in man is given in Table 52-2. If CSF were a filtrate of plasma, that is, if only the laws of diffusion dictated the CSF concentrations, ion distribution would be governed first by the Gibbs-Donnan relation for chemical equilibrium and second by the Nernst relation for electrochemical equilibrium. The Gibbs-Donnan relation gives the ion ratios across a membrane, one side of which contains small

Table 52-2. Concentration of ions in human lumbar CSF expressed in milliequivalents per kilogram of water*

Substance	Arterial plasma	CSF	R_{CSF}†	R_{Dial}‡
Na^+	150	147	0.98	0.95
K^+	4.6	2.9	0.62	0.96
Mg^{2+}	1.6	2.2	1.4	0.80
Ca^{2+}	4.7	2.3	0.49	0.65
HCO_3^-	27	24	0.90	1.04
Cl^-	112	123	1.10	1.04
PO_4^{2-}	1.5	1.1	0.73	—
Glucose	4.0	2.2	0.55	0.97
Osmolality	289	289	1.0	1.0
pH	7.40	7.32	—	—
P_{CO_2} (mm Hg)	41	50	—	—

*From Davson[6]; Plum and Seisjo[25]; and Katzman and Pappius.[10]
†Ratio of CSF to plasma concentration.
‡Ratio of concentration in plasma dialysate to that in plasma.

ions only (CSF); the other (plasma) has, in addition, a large molecule carrying a charge, in this case chiefly albumin. It may be calculated that in this situation the theoretical ratio across the membrane would be 0.96 for monovalent cations and 1.04 for monovalent anions. The equilibriums may also be approached experimentally by preparing and analyzing a dialysate of plasma, as shown in Table 52-2.

The data show that none of the ions gives the distribution demanded by the Gibbs-Donnan equilibrium; furthermore, each ion has a unique value for the ratio of CSF to plasma concentration.

We must now inquire into the potential across the CSF-blood barrier. In most mammals, this is about 5 mV, CSF positive.[29] For a passive distribution of monovalent ions, the Nernst equation would yield the following:

$$+5 \text{ mV} = 61 \log \frac{[\text{ion}]_1}{[\text{ion}]_2}$$

Subscripts 1 and 2 refer to the concentration in CSF and plasma water, respectively, for anions and to plasma and CSF for cations. The ratio of monovalent ions in CSF to plasma (R_{CSF}) for anions would be 1.2, and for cations 0.8, if they were distributed according to the electrochemical gradient. Table 52-2 shows that this is not the case. Corresponding calculations for $[Ca^{2+}]$ and $[Mg^{2+}]$ yield 0.64 as the ratio for passive distribution; again, this is not the observed value for either ion.

The electrochemical gradient does not appear to determine the distribution of the ions in CSF. They are intrinsically regulated and usually quite independent of plasma composition.[5] Cases in point are K^+, HCO_3^-, Mg^{2+}, and Ca^{2+}. A very large reduction in plasma $[Cl^-]$ by dialysis leaves CSF $[Cl^-]$ relatively unchanged and the potential still about +5 mV.[31]

The source of the potential remains unknown, but an important finding is that its magnitude depends on plasma pH and K^+. For each 0.1 unit of pH rise the potential falls about 3 mV; at pH 7.55 the potential is zero.[62] For each millimolar rise in plasma K^+ the positive CSF-blood potential difference rises 1.5 mV. This effect was used to show that when the potential difference was increased sevenfold, there was no change in distribution of other ions between CSF and blood. From the impedance of the blood-CSF barrier, it was calculated that a relatively large potential difference change should not alter ion distribution.[36]

The composition of CSF must be considered

not only as it flows through the cavities of the ventricular-subarachnoid pathway, but also as the fluid is freshly formed. This topic has been approached in two ways: by direct measurement of the microvolume secreted by the choroid plexus[32] and by calculation from the net influx rates.[73] The latter will be discussed in the next section.

Freshly formed choroid plexus fluid in the cat has been collected under oil in situ from the exposed lateral ventricle[32] and also from a small chamber set over the plexus in situ.[66,85] Under these conditions the formation rate of fluid is the same as that in vivo. The data from these two experiments agree fairly well, and the composition of choroid plexus fluid (CPF) is summarized as follows, in reference to Table 52-2: $[Na^+]$, CPF about the same as CSF and about 10 mM higher than an ultrafiltrate (UF) of plasma; $[Cl^-]$, CPF 8 mM lower than CSF and like UF; $[HCO_3^-]$, by difference of measured cations minus $[Cl^-]$, about 8 mM higher than plasma (measurement of HCO_3^- in freshly formed "chamber fluid" shows it to be regulated independently of its plasma concentration[66]); $[K^+]$ in CPF, as for CSF, lower than in plasma; and $[Ca^{2+}]$ and $[Mg^{2+}]$, about 30% higher in CPF than CSF. These data are important in showing that deductions about mechanisms cannot be made from the composition of lumbar (or cisternal) CSF and that significant exchanges take place between newly formed fluid and brain tissue, influencing the composition of measured CSF. I shall return to this point in connection with rate measurements.

The protein concentration of CSF is less than 0.1% that of plasma, having an approximate value of 25 mg/100 ml. In the main the proteins are those of plasma; the albumin:globulins ratio is between 1 and 2. Essentially all substances with a molecular weight greater than about 500 are "excluded" from CSF and brain fluid, but the data for protein show that exclusion is only relative. Elevation of CSF protein levels is an important sign in neurologic disease. The mechanism is multifaceted; certainly the "leak" is not confined to the choroid plexus, but may reflect lesions at many sites, including blood vessels and meninges. The distribution of proteins, amino acids, lipids, and other metabolites in CSF and brain is beyond the scope of this chapter but admirably reviewed elsewhere.[6,8]

As Table 52-2 shows, measured lumbar CSF is isosmotic with plasma. Formation and pressure of CSF are altered by changes in the osmolality of the plasma; fluid is drawn out of the brain and CSF by the introduction of hypertonic urea, mannitol, or glycerol to the bloodstream.

This may control cases of increased intracranial pressure or cerebral edema, but the effect is short-lived due to the gradual equilibration of the solute between plasma and brain fluids. The relation between the osmolality of artificial, perfused ventricular fluid and the bulk flow of nascent CSF has been demonstrated in the cat. When osmolality was varied between 6 and 780 mOsm, bulk flow ranged from 0 to 88 $\mu l/min$ (normal = 25 $\mu l/min$). In these studies, it was also possible to show the effect of acetazolamide (see subsequent discussion) on bulk flow and to suggest that, under conditions of low CSF osmolality, a fluid of very high osmolar concentration may be secreted.[107]

RATES OF ION AND DRUG MOVEMENT: CONCENTRATION OF NEWLY FORMED FLUID

The rates of movement of ions from plasma to CSF (or CSF to plasma) are best measured with isotopes, since it is not usually feasible to alter "cold" plasma or CSF composition enough to change the steady-state relations between the fluids, and such a change imposes an unphysiologic situation. Fig. 52-7 shows experiments in which $^{22}Na^+$, $^{36}Cl^-$, and $H^{14}CO_3^-$ were injected intravenously and their concentration in plasma sustained by continuous infusion.[106] The data yield turnover rate constants for the ions, k_{in}, from which the percent equilibrium (X, the ordinate) may be computed as a function of time (the abscissa) from the following equation:

$$\ln \frac{100}{100 - X} = \frac{k_{in} \cdot Time}{R}$$

R is the equilibrium ratio from Table 52-2. Fig. 52-7 yields visual evidence of the relations among the three rate constants and shows the effect of carbonic anhydrase inhibition on these constants. However, for a true analysis of the chemistry of fluid formation, it is necessary to introduce the term *"accession rate."* This is $k_{in} \times$ plasma concentration, and its use depends on two assumptions: that the ion in CSF is derived from plasma and that k_{in} does not include diffusional exchange between labeled and unlabeled ions.[106] Accession rates show the stoichiometry among the ions, and when these rates are multiplied by the turnover time (reciprocal of formation rate constant) for bulk fluid, they yield the concentration in newly formed fluid. Table 52-3 shows these relations, which will be discussed in terms of the individual ions.

Ion movement

Sodium ion. The turnover rate constant $k_{in} = 0.016/\text{min}$ is very close to that for fluid turnover. This value in cation = formation rate/ventilation volume = 20 μl per min/1.4 ml = 0.014/min. Thus there is no backflow for Na^+ at the formation site (see arrow at bottom of Fig. 52-4). The concentration of Na^+ in newly formed fluid is slightly higher than that of plasma, suggesting an osmotic gradient for flow. Fig. 52-8 shows a remarkable experiment in man, yielding the accession rate of $^{24}Na^+$ to three regions of the CSF; rates differ sevenfold between lateral ventricle and lumbar space.[101] The $^{24}Na^+$ rate to the lateral ventricle is taken to be that at the formation site; from these and other data,[104] we may compute the rate constant, the accession, and the concentration in new fluid. From open circles in Fig.

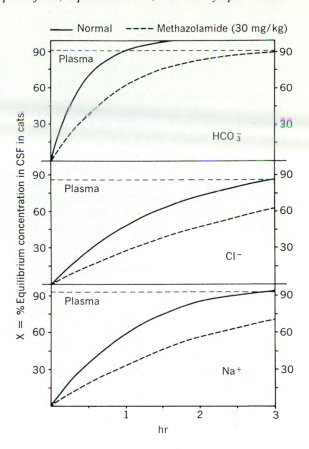

Fig. 52-7. Accession of isotopes from plasma to CSF as a function of their first-order rate constant, k_{in}. Ordinate, X, is related to time *(t)* by expression $\ln \frac{100}{100 - X} = \frac{k_{in} \, t}{r}$. r is equilibrium ratio $\frac{\text{CSF}}{\text{Pl}}$ for individual ion. Horizontal dashed line gives plasma concentration relative to CSF as 100 for equilibrium state, yielding r. Methazolamide given 1 to 2 hr before injection of isotope at zero time.[106]

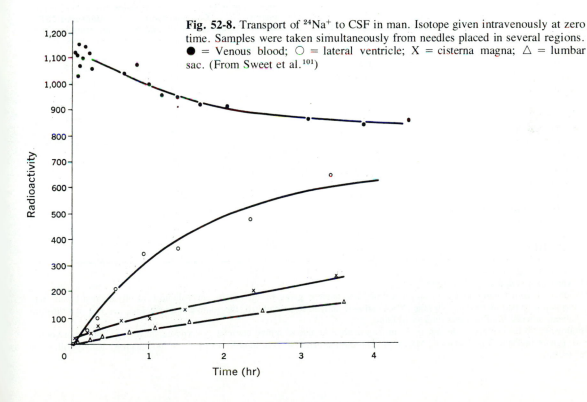

Fig. 52-8. Transport of $^{24}Na^+$ to CSF in man. Isotope given intravenously at zero time. Samples were taken simultaneously from needles placed in several regions. ● = Venous blood; ○ = lateral ventricle; X = cisterna magna; △ = lumbar sac. (From Sweet et al.[101])

52-8 the $t_{1/2}$ for sodium accession is 45 min, yielding the following:

$$k_{in} = \frac{\ln 2}{t_{\frac{1}{2}}} = 0.015/min$$

Plasma concentration is 150 mM, and the rate of accession is 2.25 mM/min. Thus to the 23 ml of ventricular volume, 52 μmoles of Na^+ is added each minute. Since 0.37 ml CSF is formed each minute, Na^+ concentration in newly formed fluid is 142 mM, in fair agreement with data of Table 52-3. Thus Na^+ movement from plasma to new fluid to ventricle is attended by relatively little change in concentration. This is also evident by comparing the fluid turnover constant for the ventricle:

$$\frac{370 \ \mu l/min}{23 \ ml} = 0.016/min$$

with that given for Na^+. They are virtually the same, indicating that Na^+ entry is a valid measure of flow. As will be seen later, Na^+ entry is an active process, allied chemically to its transport at other sites in the body.

Chloride ion. Cl^- accession is shown in Table 52-3. The k_{in} is nearly 20% less than that for Na^+; this is shown to be a significant difference.[106] The difference in accession rates is even greater, showing 37% less Cl^-. Newly formed CSF is thus 60 mM lower in Cl^- than in Na^+. This important result agrees qualitatively with direct determinations of nascent fluid,[32,85] although the isotopic experiments yield a larger $Na^+ - Cl^-$ difference. Recognition of this fact leads to quite a different concept of the chemistry of CSF formation than older, ideas based on the chemistry of bulk CSF, which shows a Cl^- excess (Table 52-2). The mechanism of transfer of Cl^- from plasma to CSF is undecided. When plasma $[Cl^-]$ is reduced by dialysis against isothionite (2-hydroxymethanesulfonate), CSF $[Cl^-]$ is relatively unaffected, compared to reduction in concentration in brain, liver, and muscle.[31,39] To use this in favor of active Cl^- transport, it would be necessary to show that newly formed CSF is being measured. More compelling is the finding that in such experiments the accession of ^{36}Cl from plasma to CSF is unaffected by plasma $[Cl^-]$, with the putative transport system saturated at plasma $[Cl^-] < 60$ mM.[39] As Fig. 52-7 and Table 52-3 show, carbonic anhydrase inhibition reduces Cl^- entry to CSF (also in monkey[38]); this is best explained on the basis of a lowered Cl^- gradient from plasma to nascent fluid secondary to reduction of HCO_3^- formation. Thus in Table 52-3, this gradient is normally 11

Table 52-3. Accessions of ions to CSF in cat*

	Plasma (mM)	k_{in} (min⁻¹)	Accession rate (mM/min)†	Concentration in new fluid (mM)‡	Measured CSF concentration (mM)
Na^+					
Normal	147	0.016	2.35	164	158
Carbonic anhydrase inhibition§		0.0076	1.12	157	
Cl^-					
Normal	115	0.013	1.49	104	134
Carbonic anhydrase inhibition		0.0073	0.84	118	
HCO_3^-					
Normal					
As CO_2	0.95	0.91			
As HCO_3^-	20	0.043	.86	60	22
Carbonic anhydrase inhibition					
As CO_2	0.95	0.29			
As HCO_3^-	20	0.014	.28	39	

*Data from Vogh and Maren.[106]

†Product of first and second columns.

‡Product of third column and fluid turnover time. The latter is given by the ventricular volume (1.4 ml) divided by the rate of formation (0.020 ml/min) = 70 min. When carbonic anhydrase is inhibited, the volume is unchanged, and rate of formation is halved, whence fluid turnover time = 140 min.

§Following 50 mg/kg acetazolamide or 30 mg/kg methazolamide to inhibit completely carbonic anhydrase. Data from normal P_{CO_2} (given in Table 2 of Vogh and Maren[106]). The plasma and CSF concentrations of ions are not changed acutely following this treatment.

mM but is abolished following acetazolamide administration.

Bicarbonate ion. Studies of the third major ion, HCO_3^-, are complicated by its equilibrium with gaseous CO_2. However, this may be taken into account and the accumulation of the ionic species $H^{14}CO_3^-$ measured in CSF. When this is done in the dogfish, *Squalus acanthias,* the calculated HCO_3^- concentration in new fluid is 10 times that of plasma, and this concentration (some 85 mM) agrees with that calculated from nascent $[Na^+] - [Cl^-]$. The accession of $H^{14}CO_3^-$ to CSF as well as that of "cold HCO_3^-" is reduced some 50% by carbonic anhydrase inhibition.[73] Taken as a whole, the data indicate that HCO_3^- arrives at the CSF by formation from gaseous CO_2 at the luminal side of the choroid plexus, as shown in Fig. 52-9. This model is consistent with classic experiments in general physiology that show that CO_2 (as well as undissociated acids) traverses cell membranes far more rapidly than HCO_3^- (or other conjugate bases). Here also is the basis for the observation that when a compensated metabolic acidosis is induced in the blood (low $[HCO_3^-]$ and low Pco_2), the CSF immediately becomes alkaline, since the low CO_2 equilibrates rapidly with the normal $[HCO_3^-]$ of the CSF. A metabolic alkalosis in blood, generating high Pco_2, produces the same paradoxical effect—acidosis in CSF.[6]

In mammals the rate of accumulation of $H^{14}CO_3^-$ in CSF is so rapid that isotope exchange must be involved, and its net rate is calculated as the difference between rates of Na^+ and Cl^- entry. Fig. 52-7 and Table 52-3 show the data. If we wish to calculate the rate of formation in the process $CO_2 + OH^- \rightarrow HCO_3^-$, the appropriate

rate constant (k_{in}) is based on the CO_2 of plasma (or tissues) as the substrate and is 0.91/min, indicating a $t_{1/2}$ of 46 sec (Table 52-3). If we think of the process (quite apart from mechanism) in terms of plasma HCO_3^- and approaching the equilibrium ratio (R) between CSF and plasma for HCO_3^-, k_{in} may be based on the plasma $[HCO_3^-]$, as indicated in Table 52-3. The latter convention is followed in Fig. 52-7 for comparison to rates of approach to equilibrium of $[Na^+]$ and $[Cl^-]$.

This process for HCO_3^- formation is faster than that of the ionic transport of Na^+ and Cl^- (Fig. 52-7). HCO_3^- accumulation matches 37% of the Na^+ transport, and its movement in the three-compartment system described previously leads to establishment of the Cl^- gradients (Table 52-3). It has been shown that Na^+ movement in isolated frog choroid plexus critically depends on the presence of HCO_3^-[112] and that in this preparation (as in vivo), some 40% of Na^+ transport is matched by HCO_3^-.[30a]

Table 52-3 shows that the newly formed fluid, conceived as a microvolume at or immediately distal to the secretion site, has a composition quite different from that of CSF sampled in the ventricular cavity. Nascent fluid appears changed by the addition of a relatively acidic solution of NaCl and by some element of Cl^-/HCO_3^- exchange with brain.

It is of interest to compare the basal rates of HCO_3^- formation (about 0.9 mM/min or 53 mM/hr, Table 52-3) to the increments induced by the elevation of plasma Pco_2 or plasma HCO_3^-. Such an analysis reveals that doubling of either of these modalities increases HCO_3^- accession only about 5 mM/hr—a rise of 10%. Two important conclusions are reached: (1) that HCO_3^- for-

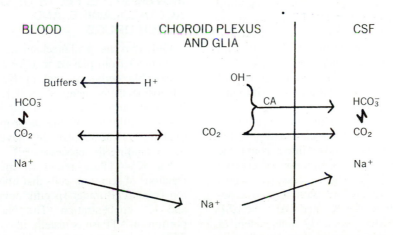

Fig. 52-9. Primary process is protolysis of water within cell to yield OH^- at CSF surface. *CA,* Carbonic anhydrase. (From Maren.[73])

mation from CO_2 is normally running near maximal rate and (2) that normally the contribution of ionic HCO_3^- passage "as such" is negligible.[77]

Other ions. The movements of K^+, Ca^{2+}, Mg^{2+}, and PO_4^{2-} into CSF are not involved with fluid formation, since they occur in low concentrations (Table 52-2). Their accession, however, is of intrinsic importance and will be outlined.

K^+ in all mammals and many "lower" vertebrates is about 35% lower in CSF than in plasma. Most significantly, CSF $[K^+]$ is maintained at its normal level of about 3 mM despite wide variations in plasma $[K^+]$ due to experimental manipulation or disease. This strict homeostasis appears to be maintained by two mechanisms: a carrier-mediated system from plasma to CSF across the choroid plexus that is saturated at plasma $[K^+]$ of about 3.5 mM and an efflux system into brain tissue. The latter appears to depend on an Na^+, K^+ pump that can be inhibited by ouabain. The extracellular fluid of brain is about 2.5 mM in K^+. Since the movement of K^+ is independent of its plasma concentration, a first-order rate constant (as assigned for the ions of Table 52-3) is not applicable. The overall clearance rate of K^+ is about fourfold greater than fluid turnover.[5,6,10]

$[Ca^{2+}]$ in CSF is normally about half that in plasma. Since plasma protein binding of the ion is one half to one third, it might appear that CSF $[Ca^{2+}]$ is in diffusional equilibrium with plasma $[Ca^{2+}]$. However, there is clear evidence against this, the most compelling being the near-stability of CSF $[Ca^{2+}]$ at about 2.8 mEq/L whereas plasma $[Ca^{2+}]$ varies from 2.9 to 7.3 mEq/L. As for K^+, there appears to be an active or carrier-mediated system for Ca^{2+} movement to CSF, localized at the choroid plexus. About one third of Ca^{2+} efflux from CSF is to brain, and it appears likely that this is also a transport mechanism with regulatory properties.[10] Thus, like K^+, Ca^{2+} movement appears independent of its concentration in plasma or CSF. Turnover rate of Ca^{2+} is slow, the $t_{1/2}$ for equilibrium with plasma being several hours.[6]

$[Mg^{2+}]$ is nearly 40% higher in human CSF than in plasma, and when the electrical potential and binding to plasma are taken into account, the gradient is even steeper. Collection of choroid plexus fluid also reveals Mg^{2+} excess, so it seems certain that we are dealing with an active secretory process.[32] As for K^+ and Ca^{2+}, the CSF concentration of Mg^{2+} is nearly independent of that in plasma. Kinetic studies show that there is little or no effect of plasma $[Mg^{2+}]$ on influx to

CSF, and that the outflow of Mg^{2+} from CSF perfusates is regulated by the level of the ion in the perfusate. At normal levels the $t_{1/2}$ for accession to CSF is 140 min. The ECF of brain has about the same $[Mg^{2+}]$ as CSF.[8]

In summary, K^+, Ca^{2+}, and Mg^{2+} concentrations in CSF and ECF of brain are regulated by both influx and efflux mechanisms. The terms "k_{in}" and "k_{out}" in such cases are not true constants but vary with the plasma and CSF concentrations of the ions to preserve strict homeostasis in brain fluids.

Phosphate appears in deficit in CSF compared to plasma (Table 52-2). Uptake from plasma is a relatively slow process, with approach to equilibrium taking at least several hours. The kinetics are complex due to continuous uptake by brain, and simple rate constants cannot be assigned.[6]

Drug movement

Table 52-1 lists rate constants for the entry of drugs to CSF and AH. Many lipid-soluble substances enter more rapidly than fluid is formed (formation rate constant in the various species is in the range 0.004 to 0.016/min), since they diffuse quickly across both blood-CSF and blood-brain barriers. Fig. 52-6 also shows this phenomenon. As noted previously, this principle underlies the high formation rate of HCO_3^-, since it is formed from the lipid-soluble CO_2. Urea, a small water-soluble molecule that enters muscle rapidly, enters CSF very slowly, since it is lipid-insoluble. Many sulfonamides and antibiotics enter CSF even more slowly than does sulfadiazine; again, it is a question of lipid solubility, which has been treated in detail elsewhere.[6,27,96]

MECHANISMS UNDERLYING ION MOVEMENT: EFFECTS OF OUABAIN, ACETAZOLAMIDE, AND OTHER DRUGS

Only two chemical mechanisms underlying ion movement from plasma to CSF are known with reasonable certainty: that of Na^+,K^+-activated adenosine triphosphatase (Na^+,K^+-ATPase) for Na^+ and carbonic anhydrase for HCO_3^-. In both cases, inhibitors of the appropriate enzyme, ouabain and acetazolamide, have been keys to understanding the processes.[9,18]

Na^+,K^+-ATPase refers to membrane-bound fractions of secretory cells that liberate phosphate ion from ATP under specific conditions of Na^+ and K^+ concentration. The Na^+,K^+-activated portion of ATPase is usually about half the total ATPase activity, and this moiety can be inhibited by 10^{-5} M of the cardiac glycoside ouabain. In most in vivo transport systems (e.g., kidney, pancreas) for Na^+, ouabain can be shown to reduce Na^+ transfer; because of the toxicity of ouabain, special means must be taken to deliver it to the organ site.[88] In the case of CSF, this may be done by introducing it into a ventriculocisternal perfusion system, thus obviating the cardiac effects. Under these circumstances, $^{22}Na^+$ turnover from plasma to CSF is reduced, and the secretion rate is reduced by 55%.[105]

These results suggest an important association between Na^+,K^+-ATPase and sodium transport that involves energy for the process through the familiar $ATP \rightleftharpoons ADP$ cycle, rather than the actual transport step. The chemistry of this transport step remains a mystery, as it is in all tissues. This enzyme is also involved in the efflux of K^+ from CSF, since 10^{-5} M ouabain inhibits the process.[5,10]

The accumulation of HCO_3^- in CSF is similar to that elsewhere in the body.[15] Fig. 52-7 shows the rapid transfer of total CO_2 (gaseous CO_2 and HCO_3^-) from plasma to CSF; at the pH of the fluids, 95% is HCO_3^-. The process follows the steps shown in Fig. 52-9. Here we deal not with transport of an ion but with its formation in secretory cells from water and CO_2. The primary event may be regarded as the protolysis of water to yield OH^- at the secretory surface and H^+ at the blood surface. *This is a fundamental property of the cell leading to secretion.* CO_2 attacks OH^- at the secretory side to yield HCO_3^-, which then is secreted or diffuses into CSF; H^+ crosses into the blood, where it is buffered by hemoglobin.[74] Choroid plexus and glia, like most (but not all) cells that carry out this process, contain carbonic anhydrase that speeds the reaction $CO_2 + OH^- \rightarrow HCO_3^-$. Drugs with the general formula aryl-SO_2NH_2 inhibit the enzyme; studies with such compounds, notably acetazolamide, have been instructive in elucidating these processes. Unlike ouabain, acetazolamide is nontoxic, so that doses that yield complete inhibition may be given safely by mouth or by injection.[18]

Thus Fig. 52-7 and Table 52-3 show that when carbonic anhydrase is inhibited, the rate of entry of HCO_3^- is reduced. Na^+ entry also is reduced, since it is the counterion to HCO_3^-. The effect on Cl^- is explained by the data of Table 52-3, which show that when carbonic anhydrase is inhibited, the Cl^- gradient from plasma to nascent fluid is abolished. The result of lowered ion entry is the reduction of fluid formation by half in animal species (Fig. 52-10) and in man.[98] The effects on pressure depend on dose, route of administration, and the resting pressure.[18]

When ouabain and acetazolamide are given together, the reduction of sodium transport and CSF flow is no greater than that for each drug alone (Fig. 52-10).[48] This fact suggests that the

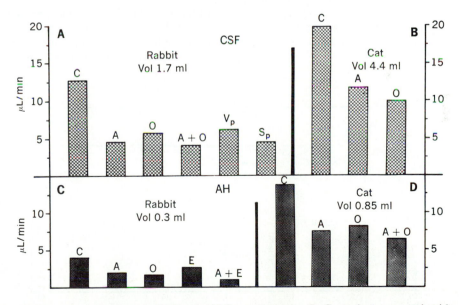

Fig. 52-10. Effect of drugs on CSF and AH formation rates. *C,* Control; *A,* acetazolamide; *O,* ouabain; V_p, vasopressin; S_p, spironolactone; *E,* epinephrine. (**A** based on data from Davson and Segal[48]; **B** based on data from Vates et al.[105]; **C** based on data from Kolker and Hetherington[14]; **D** based on data from Garg and Oppelt.[57])

coupling between Na^+ transport and HCO_3^- formation is tight; Na^+ may be affected either through reduction of energy liberated by Na^+,K^+-ATPase or by reduction of the carbonic anhydrase–catalyzed rate of formation of its counterion HCO_3^-.

Drugs of varied type that increase renal sodium excretion are also reported to decrease CSF formation by about 50% in the cat.[51] They are hydrochlorthiazide (2.5 mg/kg), chlorothiazide (20 mg/kg), triamterene (5 mg/kg), and ethacrynic acid (10 mg/kg). The effect of acetazolamide (10 mg/kg) was confirmed. Furosemide (20 mg/kg) was without effect despite its high activity on renal Cl^- transport.[3] The reduction in CSF flow shown for a 50 mg/kg dose of furosemide[81] could be due to its carbonic anhydrase inhibitory effect (it is one tenth as active as acetazolamide), although this was not the conclusion of the study. These varied findings require confirmation, and drug action in CSF formation needs further and vigorous work.

In contrast to their effects on CSF, neither ouabain nor acetazolamide reduces turnover of $^{22}Na^+$ into brain tissue. These facts have a bearing on the relation between the formation of CSF and that of fluid of the brain. They suggest that whereas there is a unidirectional flow of ions from plasma into CSF, the blood-brain barrier has a large two-way diffusional component. This is the meaning of the double arrows between blood or glia and ECF in Fig. 52-4. Thus brain ECF does not appear to be formed by net transport; the existence of active pumps is at least doubtful (for Na^+), since none of some dozen drugs alter $^{22}Na^+$ entry.[48]

CSF AND CONTROL OF RESPIRATION

Until recent times, chemical control of the medullary respiratory centers was thought to be mediated through CO_2 balance in the plasma. Early in the 1950s this concept was overthrown,[72] and it appeared that the CO_2 equilibriums of CSF and interstitial fluid were the final controlling factors. The pioneer worker in this field has provided an admirable recent review.[15] Neural control of respiration is treated elsewhere in this text (Chapter 70).

When CO_2 gas level (P_{CO_2}) is elevated in the blood, the measured pH of CSF drops within 3 min and may be presumed to drop much sooner at interfaces between blood and CSF or brain. In this small interval of time, respiration increases (Fig. 52-11).[70] The same effect on respiration can be obtained when the P_{CO_2} (as well as $[H^+]$) is increased only in the perfused ventricles (Fig. 52-12). Lowering P_{CO_2} (or $[H^+]$) in blood or perfused ventricles decreases respiration (Fig. 52-13). Both rate and amplitude of respiration are changed in the same direction.

When P_{CO_2} is held constant in the ventricular perfusate, respiration increases as $[HCO_3^-]$ drops and decreases as $[HCO_3^-]$ rises.[24] Fig. 52-14 illustrates these relationships in terms of $[H^+]$, since:

$$K = \frac{[H^+][HCO_3^-]}{P_{CO_2} \cdot 0.03 \text{ mM/mm Hg}} \tag{1}$$

Calculating the P_{CO_2} of the perfusate from the $[HCO_3^-]$ (top) and pH (abscissa), the P_{CO_2} at the upper right is equal to that at the lower left, but there is a fivefold difference in alveolar ventila-

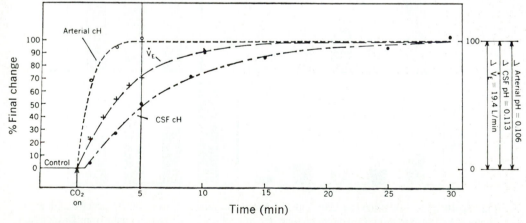

Fig. 52-11. Respiratory response to 7% CO_2 in anesthetized dogs correlated with changes in arterial and CSF $[H^+]$. cH, $-pH$; $\dot{V}_E$, minute ventilation. (From Lambertsen et al.[70])

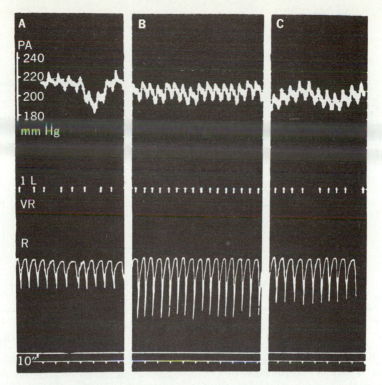

Fig. 52-12. Effect on respiration of perfusing CSF with solution approximately 2 times normal P_{CO_2}. Dog under morphine-chloralose anesthesia, vagotomized, and with carotid sinus denervation. *PA*, Arterial blood pressures; *VR*, respiratory volume, 500 ml between marks; *R*, respiratory movements. **A,** Perfusion of lateral ventricles with mock CSF at pH 7.45 [HCO_3^-] = 21 mM. **B,** Perfusion with same fluid altered by CO_2 enrichment to yield pH 7.15. **C,** Perfusion with original (pH 7.45) solution. (From Leusen.[72])

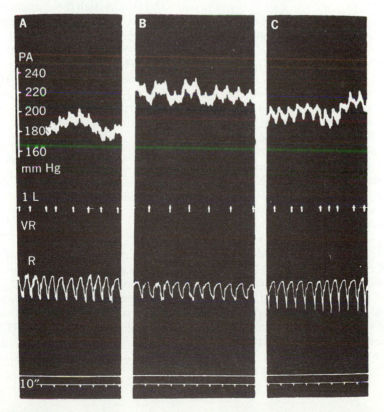

Fig. 52-13. Experimental design as in Fig. 52-12, except that in **B,** solution is CO_2 poor to yield pH of 7.66. (From Leusen.[72])

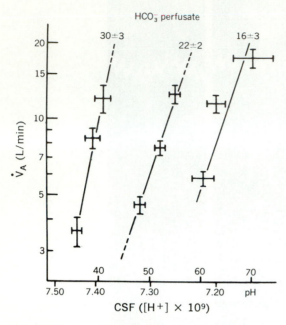

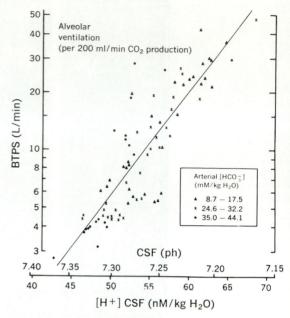

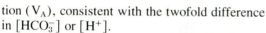

Fig. 52-14. Ventilatory response of unanesthetized goats to inhalation of various concentrations of CO_2 during ventriculocisternal perfusion at three levels of HCO_3^- concentration in perfusate. $\dot{V}_A$, Alveolar ventilation. Crosses give means $\pm$ SE from 75 periods in six goats. (From Pappenheimer et al.[93])

Fig. 52-15. Alveolar ventilation vs CSF $[H^+]$ in chronic alterations of acid-base balance. Inset shows range of metabolic changes (HCO_3^- concentrations). Respiratory changes induced by 0% to 10% CO_2. (From Fencl et al.[56])

tion (V_A), consistent with the twofold difference in $[HCO_3^-]$ or $[H^+]$.

A significant clue to the underlying mechanism is the classic observation that the ventilatory response to metabolic acidosis or alkalosis in the blood (i.e., primary changes in $[H^+]$) is small or absent compared with $[H^+]$ changes induced by primary alterations in CO_2. The blood-CSF and blood-brain barrier to H^+ or HCO_3^- is far greater than to CO_2; thus it appears that the significant acid-base changes occur in CSF and interstitial fluid but originate from the CO_2 changes in the blood. This is evident in all vertebrate species that have been examined, including shark,[73] rat,[82] dog,[111] and man.[25] Fig. 52-16 shows representative changes in CSF and brain HCO_3^- and compares these to lack of change in plasma and muscle.

The chemical modality in brain fluids that appears to control resting ventilation is $[H^+]$. Fig. 52-15 suggests that alveolar ventilation is a single function of CSF $[H^+]$ during normal conditions of acid-base balance and during acidosis and alkalosis.[56] The entire range of ventilation (20-fold) is associated with a range of 0.2 pH of CSF. In acute changes the intracellular brain pH and

HCO_3^- move in the same direction as CSF (Fig. 52-16), although these recent data now show significant quantitative differences in their responses.

Consider the chemistry of the regulation of CSF $[H^+]$ in terms of the normal equilibrium of equation 1, in which CSF $[H^+]$ is 50 nM (pH 7.32), $[HCO_3^-]$ is 20 mM, and P_{CO_2} is 40 mm Hg. K is the equilibrium constant, $10^{-6.1}$ M. In the normal situation:

$$K = \frac{(50 \text{ nM}) (20 \text{ mM})}{1.2 \text{ mM}} \qquad (2)$$

It is essential to realize that the gaseous CO_2 is in equilibrium with H_2CO_3:

$$\begin{array}{ccc} CO_2 & \rightleftarrows & H_2CO_3 \\ (1.2 \text{ mM}) & & (3.0 \text{ nM}) \end{array} \qquad (3)$$

In the normal organism, carbonic anhydrase in red blood cells and lung ensures instant equilibration between CO_2 and H_2CO_3.

If arterial P_{CO_2} is increased to 80 mm Hg by administering 10% CO_2 gas to a dog, the molar concentrations of CO_2 and H_2CO_3 roughly double, demanding (since K is constant) that $[H^+]$ increase proportionately. Assuming perfect mix-

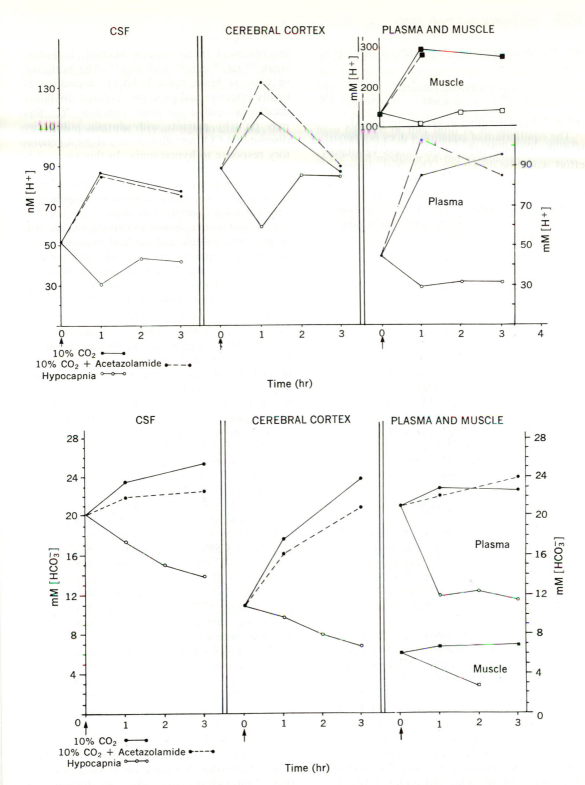

Fig. 52-16. Effects of hypercapnia (10% CO_2, arterial P_{CO_2} = 85 mm Hg) shown in solid points and of hypocapnia (mechanical ventilation adjusted to arterial P_{CO_2} = 15 mm Hg) shown in open points, on HCO_3^- and H^+ concentration of CSF, cerebral cortex, plasma, and muscle of dogs. Abscissae show time from zero of altered P_{CO_2}. Dotted lines show experiments in which dogs were pretreated with intravenous acetazolamide (100 mg/kg at 0 and 2 hr) to inhibit completely carbonic anhydrase. (Based on data from Arieff et al.[34])

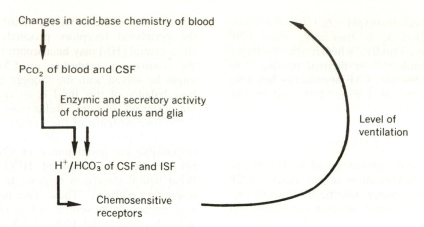

Fig. 52-18. Simplified view of chemical control of respiration. High Pco_2 initially favors high H^+/HCO_3^- ratio, stimulating respiration. Ultimately high Pco_2 generates HCO_3^-, so that H^+/HCO_3^- returns to near normal. (See example in text.)

respiratory centers on the dorsal aspect (Fig. 52-17, right). Central nervous chemoreceptors have been reviewed in an essay that marshalls evidence from anatomy, physiology, and pharmacology.[16] CSF flows from the fourth ventricle into its lateral recesses, which wind around the medulla and enter the subarachnoid space ventrally just below the level of the pons (Fig. 52-1). It will be recalled that there is choroid plexus in the roof and recesses of the fourth ventricle, so that newly formed CSF is close to the chemosensitive areas. As shown previously, the final chemical stimuli may be mediated by the intracellular fluid of the brain formed in glia by reactions similar to those that form CSF at the choroid plexuses (Fig. 52-9).

In summary, the chemical control of respiration is mediated by the CO_2 equilibria of brain ICF in a servomechanism that keeps ventilation close to normal or adjusts it to special needs during wide swings in the acid-base balance of the blood. A simplified scheme is given in Fig. 52-18. Other aspects of respiratory control, notably the roles of oxygen and neural stimuli, are treated elsewhere (Chapters 63 to 65).

PATHOLOGY OF CSF PATHWAYS: HYDROCEPHALUS

A brief description of the pathophysiology and treatment of hydrocephalus is presented to show how strongly rooted the knowledge of disease must be in physiology and pharmacology. Sixty years ago America's first neurosurgeon made notable contributions to the understanding of this disease, observing that, for the neurosurgeon, the CSF "may prove his most stubborn enemy or his most valued ally."[44]

Hydrocephalus (Greek, water-head) describes a disease state in which there is enlargement of the ventricles and usually (but not always) an increase in intracranial pressure from the normal of about 100 to 160 mm H_2O. In its obvious and most common form, it is a disease of the perinatal period, due to malformation, injury, or infection. The incidence is 1 to 3 cases/1,000 live births.[20] Since the fontanels of the skull are not yet closed, the increased pressure expands the head. Pressure on the brain may be slowly or rapidly destructive, depending on its degree. In adults, lack of expansion of the skull in the face of continued production of CSF results in marked dilation of the ventricles, usually with increased intracranial pressure. In some cases, ventricular dilatation is not accompanied by increased pressure but is explained on the basis of increased force due to certain pressure-area relationships of the ventricles following injury.[58]

The pathways followed in the production, flow, and absorption of CSF (Figs. 52-1 and 52-2) are susceptible to pathologic obstruction at virtually every point. These will be briefly traced; a full description is found in recent monographs that emphasize relations among the pathophysiology, clinical course, and possible means of treatment of the disease.[20,22]

1. The choroid plexus generates a tumor that secretes CSF, leading to overproduction. To accommodate the higher flow and absorption rate, a new steady state at higher than normal pressure develops. Although a few cases of choroid plexus

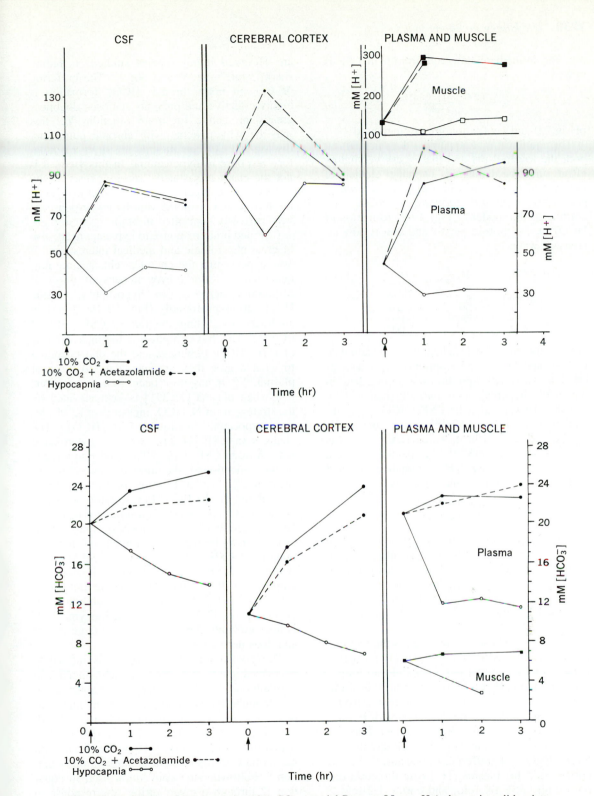

Fig. 52-16. Effects of hypercapnia (10% CO_2, arterial P_{CO_2} = 85 mm Hg) shown in solid points and of hypocapnia (mechanical ventilation adjusted to arterial P_{CO_2} = 15 mm Hg) shown in open points, on HCO_3^- and H^+ concentration of CSF, cerebral cortex, plasma, and muscle of dogs. Abscissae show time from zero of altered P_{CO_2}. Dotted lines show experiments in which dogs were pretreated with intravenous acetazolamide (100 mg/kg at 0 and 2 hr) to inhibit completely carbonic anhydrase. (Based on data from Arieff et al.[34])

ing, the immediate new equilibrium in CSF is now:

$$K = \frac{(100 \text{ nM}) (20 \text{ mM})}{2.4 \text{ mM}} \qquad (4)$$

pH is 7.0 and ventilation greatly increased.

The equilibrium of equation 4 does not persist; both the lowered pH and the excess ventilatory effort would be damaging. The chief element of the system now comes into play, *regulation of pH and respiration by formation of HCO_3^-*. According to the model of Fig. 49-9, formation of HCO_3^- at the choroid plexus and glia is now increased, whence:

$$CO_2 \ + \ OH^- \ \xrightarrow[\text{anhydrase}]{\text{Carbonic}} \ HCO_3^- \qquad (5)$$

(Plasma and tissue fluid)	(Secretory cell)	(CSF and brain fluid)

Within minutes, HCO_3^- begins to accumulate in CSF and brain, and respiration decreases as $[H^+]$ drops. A new equilibrium is approached, in which the hypercapnia is partially matched by elevated HCO_3^-. In 3 hr CSF $[HCO_3^-]$ has increased 6 mM (Fig. 52-16), and a plateau is reached in 12 to 24 hr in which CSF $[HCO_3^-]$ is elevated some 12 mM.[82] This persists into the chronic phase. CSF $[H^+]$ remains elevated (pH = 7.2). The new equilibrium is:

$$K = \frac{(62 \text{ nM}) (32 \text{ mM})}{2.4 \text{ mM}} \qquad (6)$$

This is the common clinical situation of chronic lung disease.[25] Thus HCO_3^- formation in CSF partially but not completely regulates $[H^+]$ toward normal. Note from Fig. 52-16 that such regulation does not appear to be passive movement, since it does not occur in plasma or muscle.

Of great significance, however, is the fact that in hypercapnia, cerebral cortex HCO_3^- formation is more rapid than that in CSF and does result in bringing brain $[H^+]$ to normal. This is shown in Fig. 52-16 in the doubling of cortex $[HCO_3^-]$ within 3 hr when P_{CO_2} has been doubled. Further experiments with very high ambient P_{CO_2} (3 times normal, 114 mm Hg) show clearly the different degrees of buffering in plasma, CSF, and cortex. In 3 hr, plasma $[H^+]$ rose threefold (no buffering), CSF twofold, and cortex 1.3-fold.[34] This "buffering," as indicated earlier (Fig. 52-9), is actually the formation of HCO_3^- catalyzed by carbonic anhydrase. Fig. 52-16 shows this (dotted line), in that specific inhibition of this enzyme reduces the rate of HCO_3^- formation in CSF and cortex. These effects of acetazolamide

are observed in all species studied, including shark,[73] rat,[68] cat,[106] and dog.[111] The buffering of CO_2 in brain through HCO_3^- formation is highly effective and prevents acidosis, doubtless greatly protective for brain function. Yet this may be why patients with chronic pulmonary disease and CO_2 retention do not show a ventilatory response to hypercapnia. In this view, it is not the classically stated "receptor insensitivity to CO_2" in such patients, but the fact that $[H^+]$, the true mediator of the ventilatory response, is very narrowly controlled in brain.

CSF and brain response to hypocapnia are also of great physiologic and medical interest, and it now appears that the chemical changes are not, quantitatively, the mirror images of those of respiratory acidosis. In hypocapnia, plasma HCO_3^- drops profoundly (Fig. 52-16; 9 mM in an hour), exceeding the fall in CSF $[HCO_3^-]$ (3 mM). This has complicated the interpretation of CSF $[HCO_3^-]$ changes and the understanding to what degree they are secondary to those of plasma.[60,94] It has now been shown, however, that when plasma $[HCO_3^-]$ is kept elevated by the strategems of $NaHCO_3$ ingestion or K^+ depletion, hypocapnia still lowers CSF $[HCO_3^-]$. The changes in CSF $[HCO_3^-]$ were best correlated with those in CSF P_{CO_2}.[50] Thus, despite the differences mentioned, the ultimate controlling element of CSF $[HCO_3^-]$ in both respiratory alterations is indeed the P_{CO_2}. In respiratory acidosis, elevated CO_2 appears to set a signal for increased HCO_3^- production; in alkalosis decreased CSF $[HCO_3^-]$ may be mediated through lowered production of CSF.[91] The decline in CSF $[HCO_3^-]$ is not affected by acetazolamide.[111] It should be emphasized that although carbonic anhydrase catalyzes HCO_3^- formation and inhibition reduces it (Fig. 52-7 and 52-16), the enzyme itself is not rate controlling, since it is in large excess and does not vary.[18]

The physiology of hypocapnia is intimately connected to that of ventilatory acclimatization at high altitude. Until very recently (Chapter 73), it was thought that acclimatization in the high-altitude sojourner was due to normalization of CSF pH by the appropriate fall in CSF $[HCO_3^-]$ to match the lowered P_{CO_2}.[99] However, it now appears that CSF pH remains alkaline in this situation.[49] Following the convention used in equations 2 through 6 given earlier, the response to low P_{CO_2} in CSF may be set down. Before 1 hr (Fig. 52-16) there is little or no change in $[HCO_3^-]$, whence:

$$K = \frac{(25 \text{ nM}) (20 \text{ mM})}{0.6 \text{ mM}} \qquad (7)$$

This extreme alkalosis (pH 7.62), like the acidosis of equation 4, is transitory, since CSF [HCO$_3^-$] drops rapidly, whether due to hypocapnia,[34] altitude,[49,99] or thermal panting.[59] In the final equilibrium, CSF approaches but does not reach normal [H$^+$] (Fig. 52-16) and may be shown as follows:

$$K = \frac{(40 \text{ nM}) (12.5 \text{ mM})}{0.6 \text{ mM}}$$

With respect to the altitude problem, it then appears that acclimatization occurs despite CSF alkalosis. This matter has been critically reviewed, with emphasis on need for further explanation.[11]

Such explanations may be at hand, when we study Fig. 52-16. For although CSF [HCO$_3^-$] does not rise or fall enough to normalize CSF [H$^+$], there is the strong suggestion that intracellular brain [H$^+$] may indeed be normalized by the sharp changes in brain [HCO$_3^-$]. At

higher altitudes the anoxemic ventilatory drive at the peripheral receptors proceeds unchecked, since central [H$^+$] may be at control levels. Thus the controversy centering on CSF regulation might be solved with the thought that the acid-base balance of this fluid is an approximation, but not exact, of intracellular brain fluid, which is the true mediator of the ventilatory response.

Choroid plexus and glia are the structures responsible for maintenance of CSF and brain pH by their regulation of HCO$_3^-$ formation. What neural structures respond to [H$^+$] in the respiratory pathway? These have been identified as dominant on the ventral surface of the medulla at levels of the eighth nerve (CSA in Fig. 52-17) and caudal at an area just medial to the root of the twelfth nerve (L in Fig. 52-17). The chemosensitive points lie about 0.2 mm deep. Impulses from CSF and L appear to converge at an area intermediate between them. Note that these chemosensitive areas are quite separate from the

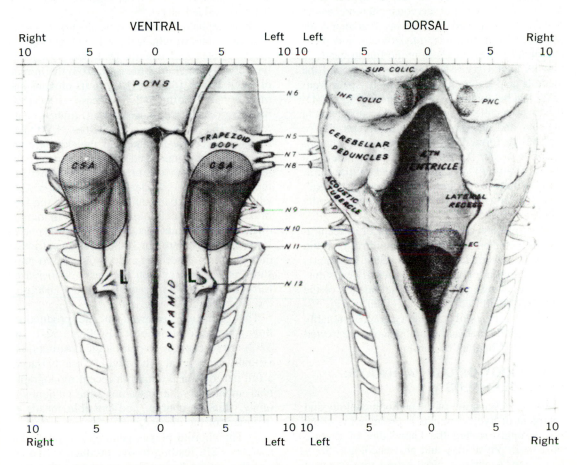

Fig. 52-17. Medulla and pons of cat. PNC, Pneuomtaxic center in region of locus ceruleus; *EC* and *IC,* expiratory and inspiratory centers in reticular formation; *CSA,* region first shown sensitive to CO$_2$, H$^+$, nicotine, and acetylcholine[86]; *L,* region of greatest sensitivity to H$^+$, described in text.[16] Coordinates are in millimeters. (From Mitchell.[86])

1238 *The kidney and body fluids*

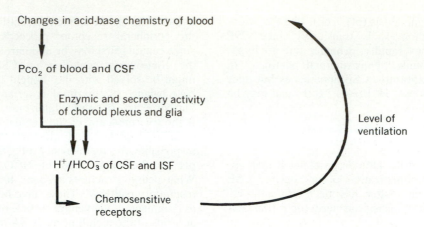

Fig. 52-18. Simplified view of chemical control of respiration. High P_{CO_2} initially favors high H^+/HCO_3^- ratio, stimulating respiration. Ultimately high P_{CO_2} generates HCO_3^-, so that H^+/HCO_3^- returns to near normal. (See example in text.)

respiratory centers on the dorsal aspect (Fig. 52-17, right). Central nervous chemoreceptors have been reviewed in an essay that marshalls evidence from anatomy, physiology, and pharmacology.[16] CSF flows from the fourth ventricle into its lateral recesses, which wind around the medulla and enter the subarachnoid space ventrally just below the level of the pons (Fig. 52-1). It will be recalled that there is choroid plexus in the roof and recesses of the fourth ventricle, so that newly formed CSF is close to the chemosensitive areas. As shown previously, the final chemical stimuli may be mediated by the intracellular fluid of the brain formed in glia by reactions similar to those that form CSF at the choroid plexuses (Fig. 52-9).

In summary, the chemical control of respiration is mediated by the CO_2 equilibria of brain ICF in a servomechanism that keeps ventilation close to normal or adjusts it to special needs during wide swings in the acid-base balance of the blood. A simplified scheme is given in Fig. 52-18. Other aspects of respiratory control, notably the roles of oxygen and neural stimuli, are treated elsewhere (Chapters 63 to 65).

PATHOLOGY OF CSF PATHWAYS: HYDROCEPHALUS

A brief description of the pathophysiology and treatment of hydrocephalus is presented to show how strongly rooted the knowledge of disease must be in physiology and pharmacology. Sixty years ago America's first neurosurgeon made notable contributions to the understanding of this disease, observing that, for the neurosurgeon, the

CSF "may prove his most stubborn enemy or his most valued ally."[44]

Hydrocephalus (Greek, water-head) describes a disease state in which there is enlargement of the ventricles and usually (but not always) an increase in intracranial pressure from the normal of about 100 to 160 mm H_2O. In its obvious and most common form, it is a disease of the perinatal period, due to malformation, injury, or infection. The incidence is 1 to 3 cases/1,000 live births.[20] Since the fontanels of the skull are not yet closed, the increased pressure expands the head. Pressure on the brain may be slowly or rapidly destructive, depending on its degree. In adults, lack of expansion of the skull in the face of continued production of CSF results in marked dilation of the ventricles, usually with increased intracranial pressure. In some cases, ventricular dilatation is not accompanied by increased pressure but is explained on the basis of increased force due to certain pressure-area relationships of the ventricles following injury.[58]

The pathways followed in the production, flow, and absorption of CSF (Figs. 52-1 and 52-2) are susceptible to pathologic obstruction at virtually every point. These will be briefly traced; a full description is found in recent monographs that emphasize relations among the pathophysiology, clinical course, and possible means of treatment of the disease.[20,22]

1. The choroid plexus generates a tumor that secretes CSF, leading to overproduction. To accommodate the higher flow and absorption rate, a new steady state at higher than normal pressure develops. Although a few cases of choroid plexus

papilloma with hydrocephalus have been described, rigorous proof of hypersecretion has been lacking[20] until recently. Using the ventriculocisternal perfusion technique, CSF formation rate of 1.4 μl/min (4 times normal) was measured in a 5-month-old boy with papilloma contiguous with normal choroid plexus.[52]

2. There is stenosis or obstruction of the aqueduct of Sylvius. This may be due to tumor, congenital anomaly, or inflammation, usually as a result of infection. This is the most common cause of the disease.

3. There is an obstruction of the outflow of the fourth ventricle at the foramen of Luschka or of Magendie. Such obstructions usually develop as the result of infection but may be due to tumor or injury. In the latter instance, traumatic hemorrhage is the principal etiologic factor.

In the second and third situations just cited, the disease is termed noncommunicating (or, in the older literature, obstructive) because the connection between the ventricular and subarachnoid spaces has been reduced or obliterated. An interesting but unexplained point is that in infant aqueductal stenosis, $^{22}Na^+$ cannot escape from the ventricles, but in older patients (9 to 20 years of age) with similar lesions, escape is almost normal.[83] Presumably, $^{22}Na^+$ (and fluid) can leave by the ependymal route in the older patient. This observation suggests that, following aqueductal closure, ventricular dilatation in the adult will be less severe than that in children.

4. There is obstruction in the subarachnoid space. This is most commonly in the region of the basal cisterns and is due to infection, intracranial bleeding, or tumor. The subarachnoid space dorsally may be distorted and blocked, as in a congenital defect called the Arnold-Chiari malformation, in which the inferior portions of cerebellum and medulla are found in the cervical spinal canal.

Failure of adequate CSF absorption has been an elusive but important issue in the study of hydrocephalus. A defect of outflow from subarachnoid space to venous sinus is implied by the absence of any overt lesion, as just described in the fourth point. In every series of hydrocephalus there is a fairly large group, at least 20%, in which no anatomic cause can be found. This may also be the cause of so-called pseudotumor cerebri, a self-limiting disease that usually affects young women. This type of elevated CSF pressure is analogous to chronic simple glaucoma (see subsequent discussion).

The classic treatment for hydrocephalus is surgical. Of a variety of procedures (i.e., choroid plexectomy, third ventriculostomy), a few are notably successful, except where removal of tumor or obstruction can restore flow. Shunt operations, in which CSF is passed from lateral ventricle to cisterna magna (Torkildsen procedure) or to blood or cavities outside the brain,[20] are most in favor. The numerous complications have made a pharmacologic approach to the disease appear a most desirable goal.

Reduction of CSF flow by acetazolamide (Fig. 52-10) offered the possibility of treatment with this agent, particularly since it is reasonably well tolerated in large doses. Some half-dozen studies now show that the drug is effective in only some cases, perhaps those that are partially arrested.[20] The physiologic basis for this is quite clear; maximal inhibition of carbonic anhydrase only reduces flow and ion transport 50%. Even if the entire secretory process is linked to the formation of HCO_3^-, the uncatalyzed rate of formation could account for the half not affected by the drug.[106] The pathology of hydrocephalus is such that a 50% reduction in flow cannot reduce pressure in all patients; yet trials remain to be done under carefully controlled conditions in order to select patients in whom surgery may be avoided. This will surely be facilitated by the new noninvasive method of measuring brain size by the scanning procedure of computerized axial tomography.

As noted previously, a number of "renal drugs" appear to reduce CSF flow by a direct action and may deserve trial in hydrocephalus. Digoxin, which inhibits Na^+, K^+-ATPase and lowers flow, as described earlier for ouabain, has been tried clinically and appeared to reduce CSF production about 70% in three cases of hydrocephalus. Acetazolamide was somewhat less effective.[87] Long-term controlled studies of these drugs, alone or in combination with acetazolamide, remain to be done.

NOTE ON BEHAVIORAL CORRELATES OF CSF PHYSIOLOGY

Despite its obvious implications and the relative ease of making measurements, reliable data regarding the behavioral correlates of CSF physiology are almost nonexistent; there is no reference to behavior in the leading monographs. In the literature of neurology and psychiatry there are numerous claims that abnormalities of secretion, composition, and circulation of CSF were responsible for a variety of disorders, but these have been largely discredited. Reference may also be made to a review on the biochemistry of affective disorders, in which some interesting work on brain and CSF electrolytes and metabolites is cited.[4] In the context of the present chapter, one challenging finding is that in depression the rate of $^{24}Na^+$ entering the lumbar CSF is about one third normal. Confirmation of this finding for the ventricles would be of importance.

Particular attention has been given in the past few years to the measurement of catecholamines, their metabolites, and their precursors in the CSF of patients with psychiatric disease. Resulting data thus far are conflicting. Such studies ideally should be carried out so that the turnover of these substances is measured under controlled conditions. Data presented in this chapter show that CSF has an important protective function in brain; all the ions are regulated in CSF when they fluctuate in plasma. Whether such protection extends to brain metabolites is not known with certainty, although the capacity of the choroid plexus to secrete polar substances into the blood suggests it. CSF may function as a sink for such metabolites, with the capacity to deliver them rapidly to the blood for excretion.

Of further importance in psychiatry is the accession of the newer drugs to CSF, many of which have not been studied. Enough is known of older and model drugs (Fig. 52-6 and Table 52-1) to show patterns that may be expected, based on their physical and chemical properties, and to suggest that access to CSF is a proper measure of access to brain.

Great interest has been aroused in the finding of two peptides in CSF that affect behavior. Reviving earlier findings that had not captured attention, a series of elegant and difficult experiments showed that CSF taken from sleep-deprived goats induces sleep when placed in the ventricle of rats. The substance may be secreted from sleep-regulating centers in the hypothalamus or thalamus that are stimulated during sleep deprivation. The target for the factor is not known, but it could be in the reticular activating system. The precise structure continues to be elusive.[55] The second factor from human CSF (also found in brain) appears to be the normal physiologic substance that acts on morphine receptors in brain, having the same relations as acetylcholine to the cholinergic receptor.[103] Several pentapeptides of defined sequence show this activity and are termed enkephalins.[66] Since opiate receptors involved in analgesia are situated in the periaqueductal gray matter, the levels of enkephalin in CSF may be critical in the pathophysiology of pain (Chapter 12).

AQUEOUS HUMOR

A remarkable early paper emphasizes the close correspondence between the physiologic anatomy and pathology of the CSF and aqueous humor (AH) pathways; even included is the prophecy that the treatment of hydrocephalus and glaucoma "will eventually be along the same lines."[108]

Figs. 16-1 and 16-2 and Fig. 52-19 depict the anatomy of the eye and the pathway of AH flow. Fluid is secreted by epithelial cells of the ciliary process, whose histology is quite similar to that shown for choroid plexus and renal tubule in Fig. 52-3. The fluid enters the posterior chamber (volume about 50 μl in rabbit and man) and then flows forward between lens and iris to enter the anterior chamber (about 250 μl in rabbit and man). Exit is at the corneoscleral junction, where a "trabecular meshwork" of collagen and endothelial cells leads into the drainage canal of Schlemm and thence to the aqueous veins.

The turnover rate of fluid is reasonably constant among mammals, being about 1% of the volume (total of both chambers) per minute. In man the formation rate is 2 μl/min and plasma flow is 7.5 μl/min.[1] The perfusion system described for CSF, using the dilution of inulin, has been adapted to the measurement of AH formation.[90] Fluid dynamics in the eye, including the important matter of blood circulation, has been well reviewed.[1]

The subject of intraocular pressure is a large one, intimately connected with the understanding and treatment of glaucoma (Greek, green), a disease that affects 1% to 2% of adults over the age of 40 years.[14] (Glaucoma will be discussed at the end of this section.) The normal human eye has a pressure of about 16 mm Hg, almost twice that of CSF. The total pressure (at constant volume) may be expressed as follows:

$$P = P_{ev} + F \cdot R \qquad (7)$$

where P_{ev} is the venous pressure as transmitted to the sclera (episcleral venous pressure), F is the

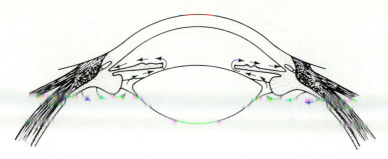

Fig. 52-19. Direction of aqueous humor flow. See Fig. 16-2 to identify structures. (From Davson.[7])

Table 52-4. Entry of ions to posterior aqueous of dog: normal values and those following carbonic anhydrase inhibition (CAI)*†

| | Plasma (mM) | k_{in} (min⁻¹) | Accession rate (mM/min)‡ | New fluid calculation (mM)§ | Measured fluid concentration (mM) | |
					Posterior	Anterior
Na⁺						
Control	152	0.044	6.7	149	153	153
CAI	156	0.031	4.8	149	152	151
Cl⁻						
Control	117	0.028	3.3	73	132	128
CAI	117	0.027	3.2	100	130	129
HCO₃⁻						
Control	22	0.109	2.4	53	26	25
CAI	21	0.039	0.8	25	24	24

*Data from Maren.[19]
†Aqueous flow is 9 µl/min in control and 6.4 µl/min during inhibition, as calculated from sodium accession rates.
‡Product of first and second columns.
§Product of third column and $\dfrac{\text{Volume posterior chamber } (=0.2 \text{ ml})}{\text{Aqueous flow}}$.

aqueous flow, and R is outflow resistance in millimeters of mercury per unit of flow.[7] It is evident that a variety of factors can influence P. In man the P_{ev} is about 8 mm Hg; this means (equation 7) that a relatively large change in F or R is necessary to alter P.

Unlike CSF the composition of AH varies among species, and this has led to special problems in understanding the mechanism of formation.[74] Fortunately the animal most widely used in this research, the rabbit, has a large (16 mM) excess of HCO₃⁻ in the posterior chamber over that in plasma.[7,35] This, together with the finding of carbonic anhydrase in the ciliary process and the effect of inhibition,[18] leads to the scheme of Fig. 52-9, analogous to that for CSF described earlier.

Table 52-4 shows the ionic composition of ocular fluid in dog, the rate constants for ion accession to the posterior chamber, and the effect of inhibiting carbonic anhydrase. This species shows about 15% excess of both HCO₃⁻ and Cl⁻

in the posterior chamber, and preliminary data suggest that even smaller gradients occur in a primate.[75] The data of Table 52-4 and Fig. 52-20 show HCO₃⁻ formation in dog and monkey, just as earlier data do in rabbit[67]; clearly the mechanism is not related to or revealed by the measured concentration of ions. About 36% of Na⁺ entry is matched by HCO₃⁻ formation in the normal state; accession of both ions is reduced by carbonic anhydrase inhibitors. When this is done, only 17% of Na⁺ entry is matched by HCO₃⁻, and flow is reduced by about half. (See also Fig. 52-10.) As in other physiologic systems, the formation of HCO₃⁻ is intimately connected to fluid flow in a way not understood.

These data give a view of the kinetics of entry and formation of AH; the turnover of ions is more complex because of exchange with lens and vitreous and diffusion across the iris.[12]

The calculated concentrations of ions in new fluid (Table 52-4), with its special characteristics of low [Cl⁻] and high [HCO₃⁻], agree with ex-

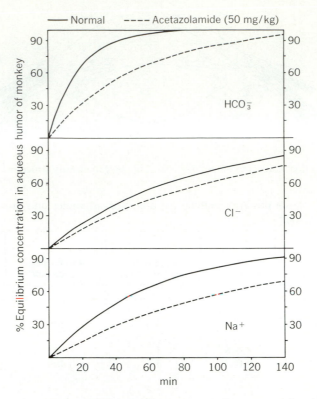

Fig. 52-20. Accession of isotopes injected in plasma at zero time, to posterior chamber AH of the monkey, *Macaca irus*.[75] See legend for Fig. 52-7.

periments in which the outflow channels of rabbit eyes were blocked and the chambers perfused with mannitol. Entering solutes were $[Na^+]$, 172 mM; $[Cl^-]$, 106 mM; and $[K^+]$, 4.6 mM. By difference, $[HCO_3^-]$ would be 70 mM. When flow was reduced with dinitrophenol or ouabain, the ionic composition approached that of NaCl, that is, the HCO_3^- component was reduced.[42] The potential difference between AH and plasma in rabbit is about +8 mV, and, as for CSF, none of the ions appear to be passively distributed.[7]

Table 52-4 and data in the monkey[75] show that the rate of fluid production is virtually the same as that of the entrance of Na^+. This is relevant to the questions of whether there is ion exchange at the secretory site and whether a considerable component of AH is formed by ultrafiltration-diffusion. The answer to both appears to be no, with evidence also from flow dynamics and anatomy.[1] Fig. 52-21 gives a synthesis of data and ideas on this important point, taking the position that all AH formation is by secretion.

In this context, it is significant that the ciliary process contains Na^+,K^+-ATPase[7] as well as

carbonic anhydrase and that ouabain reduces AH flow (Fig. 52-10) and the transport of Na^+ and Cl^- from plasma to AH.[57] The data suggest that ouabain works on the "sodium pump," whereas acetazolamide lowers HCO_3^- formation. The nature of the linkage between these two processes is such that the two drugs do not have additive effects.

As for Na^+, Cl^- carries membrane current, and these two processes are linked.[41,64] Active Cl^- transport is maximal at pH 8.5 and falls off rapidly as pH drops. This suggests, as described previously for CSF (data and discussion of Table 52-3), that the formation of HCO_3^- is critical in establishing a Cl^- gradient.

The ciliary processes are also capable of transporting anions and amino acids into and out of the eye. Of particular interest is ascorbate, which is found in the AH of most species at levels 20 times the concentration in plasma. The system for transporting p-aminohippurate and iodide out of CSF also operates at the ciliary process.[7,12] The rate of entry of various drugs into AH is given in Table 52-1, and it is seen that the same principles described for CSF are at work.

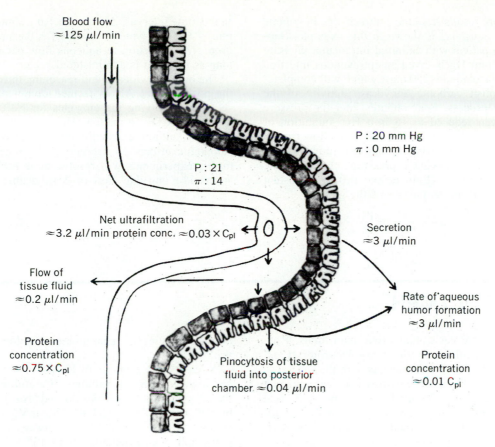

Blood flow
≈125 µl/min

P : 20 mm Hg
π : 0 mm Hg

P : 21
π : 14

Net ultrafiltration
≈3.2 µl/min protein conc. ≈0.03 × C_{pl}

Secretion
≈3 µl/min

Flow of
tissue fluid
≈0.2 µl/min

Rate of aqueous
humor formation
≈3 µl/min

Protein
concentration
≈0.75 × C_{pl}

Pinocytosis of tissue
fluid into posterior
chamber ≈0.04 µl/min

Protein
concentration
≈0.01 C_{pl}

Fig. 52-21. Formation of AH in rabbits. *P*, hydrostatic pressure; *π*, oncotic pressure; C_{pl}, plasma concentration. (From Bill.[1])

Glaucoma, like hydrocephalus, may result from overproduction of fluid, tumor, injury, or obstruction of outflow. The last is the most common (chronic simple glaucoma) and involves an impairment of drainage (increase of R in equation 7) at the trabecular meshwork in the corneoscleral angle (Figs. 16-1 and 16-2). The increase in pressure is transmitted to the optic nerve head, and blindness results if the disease is untreated. Treatment is either surgical (enlargement of the drainage angle) or medical.[14] The latter therapy will be briefly described, since it results directly from the principles given and illuminates them.

Drugs that mimic cholinergic nerve stimulation (e.g., acetylcholine and pilocarpine) or inhibit cholinesterase (e.g., eserine and diisopropyl fluorophosphate) reduce intraocular pressure, particularly when it has been pathologically elevated. The effect is to lower R in equation 7, increasing what ophthalmologists call C, the facility of outflow, which is equal to 1/R. The mechanism is that of opening channels in the trabecular meshwork by inducing contraction of the ciliary muscles.

The complex effects of adrenergic agents on the eye differ among drugs and species.[1] Of particular interest is epinephrine, which is topically effective in the treatment of glaucoma. This action is shared by isoproterenol; thus it appears that this is beta-adrenergic stimulation, which seems to reduce AH secretion and increase facility of outflow (F and 1/R, respectively, in equation 7).[14] A seeming but important paradox is the effect of a new beta-adrenergic antagonist, timolol, in reducing both AH formation and outflow resistance.[60b] It must be recognized that drugs of this type might reduce AH formation either by effecting blood flow or acting directly on secretion.

Carbonic anhydrase inhibitors, as shown in Fig. 52-20, decrease the formation of HCO_3^- and the transport of Na^+ and Cl^- into AH. The effect on AH fluid production is a reduction of 50% to 60% (Fig. 52-11). An immediate decrease in pressure results from the acute change in volume of the aqueous, as shown in this example. Suppose a patient with an initial intraocular pressure of 26 mm Hg is given an intravenous injection of acetazolamide (250 mg), which will completely abolish carbonic anhydrase activity in the ciliary process. Four minutes later the pressure has been reduced to 15 mm Hg. Inhibition of carbonic anhydrase reduces aqueous flow from 3 to 1.4 μl/min, whereas outflow remains constant. The following expression relates volume (V) and pressure (P) in the eye; 0.02 is a constant obtained from the physics of the globe[14]:

$$\Delta V = \frac{1}{0.02} \log \frac{P_{initial}}{P_{final}} \qquad (8)$$

In this time interval, ΔV will be 1.6 μl/min $\times$ 4 min = 6.4 μl, which accounts for the pressure drop. The reduction in aqueous flow persists as long as the drug is administered.

The bulk of evidence and reasoning thus favor the position that AH is formed dominantly by secretion involving the enzymes for Na^+ transport and HCO_3^- formation. However, there are complex, although usually minor, central and autonomic effects on secretion.[1,60a] Hopefully, further experiments will resolve these problems and yield a unifying view of AH dynamics.

ENDOLYMPH

The inner ear contains two main fluids. Perilymph circulates in the bony labyrinth (Figs. 52-22 and 14-2) and connects with the subarachnoid space through the narrow cochlear aqueduct (Fig. 52-23). Perilymph has the same electrolyte composition as CSF and may reasonably be regarded as the same fluid. The volume of perilymph is of the order of 1 ml, so it can be regarded as an "infinite" source of endolymph in connection with the discussion to follow. It is important to note that the fluid bathing the organ of Corti is perilymph.

Endolymph (EL) flows in all the membranous channels of the labyrinth, which communicate only with the endolymphatic sac (Fig. 52-22). The relations among the various fluids are shown in model form in Fig. 52-23. The source of EL has been difficult to determine; all investigation in this field is conditioned by the relative inaccessibility of the fluid and the fact that its volume in laboratory mammals and man is about 3 μl. Fig. 14-9 shows the tissues of the cochlea; the stria vascularis contains the type of epithelial cells associated with secretion (Fig. 52-3), and Reissner's membrane provides a possible site for interchange between perilymph and EL. Discussion of these points is found in several reviews.[17,26,28] As will be shown, EL appears to be formed at the stria vascularis, at least in part from perilymph. The site of absorption is probably the endolymphatic duct and sac, which invade the dura and at the terminus have epithelial folds.[26] Reissner's membrane may be involved in fluid transport, and fluid may pass in each direction at the stria (akin to the renal tubules of aglomerular fish).

A remarkable property of EL is the high $[K^+]$ and low $[Na^+]$ in all mammals. Careful work in the rat yields $[K^+]$ of 154 mM and $[Na^+]$ of 0.9 mM.[37] The EL potential is +92 mV, but the passive distribution of $[Na^+]$ between EL and perilymph would demand about 135 mV. The Cl^- content of EL is about 110 mM, suggesting the possibility of high $[HCO_3^-]$ from the $[K^+] + [Na^+] - [Cl^-]$ value. Electrolyte data from various species are rather fragmentary and summarized in a monograph on the biochemistry of the ear.[28]

Fig. 52-24 shows overall accession of K^+ from plasma to EL in the cat. The 48 hr value corresponds to the equilibrium ratio of the ion between the fluids, about 35. The half-time for entry is about 32 hr. The actual secretory step, however, now appears to be from perilymph to EL. In an admirable recent study, $^{42}K^+$ injected into the perilymph of the scala tympani and scala vestibuli was shown to appear immediately in EL of the scala media or cochlear duct (see Figs. 14-9 and 52-22) and to have a half-time (to equilibrium) of about 1 hr (Fig. 52-25).[68a] The turnover constant for K^+ is 0.013 min^{-1}; interestingly, this is close to that for Na^+ to CSF (see earlier discussion). If this value is tentatively regarded as representing fluid transport (as for Na^+ in CSF and AH), it also yields for the first time the formation rate of EL. The $^{42}K^+$ accession was reduced about 50% by 10^{-4}M ouabain (Fig. 52-25). Authors concluded that the site of K^+ secre-

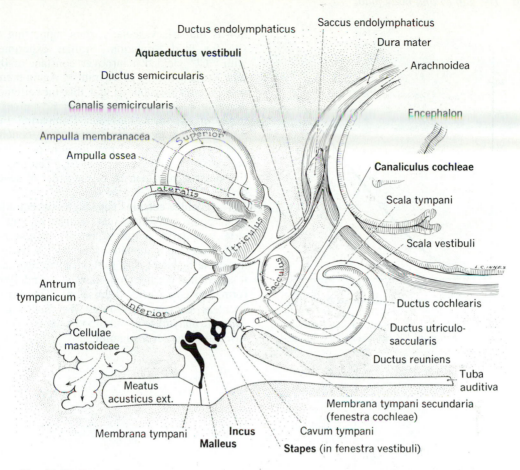

Fig. 52-22. Parts of ear, emphasizing cochlear canaliculus (or aqueduct) and vestibular aqueduct, which transmits endolymphatic duct and sac. See also Fig. 14-2. (From Anson et al.[33])

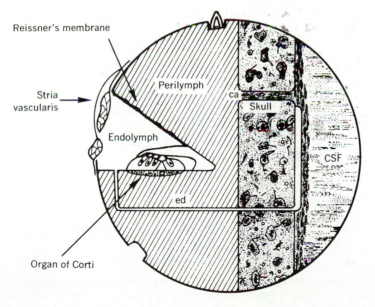

Fig. 52-23. Interrelationship of fluid systems in ear. Endolymphatic system is closed, subject to CSF pressure through endolymphatic sac and duct, *ed*. Perilymphatic system connects with CSF through tissue-filled cochlear aqueduct, *ca*. Mechanical blockage of either duct could produce fluid imbalance. See also Fig. 14-6. (From Lawrence.[71])

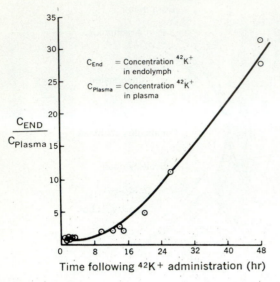

$$\frac{C_{END}}{C_{Plasma}}$$

C_{End} = Concentration $^{42}K^+$ in endolymph

C_{Plasma} = Concentration $^{42}K^+$ in plasma

Time following $^{42}K^+$ administration (hr)

Fig. 52-24. Uptake of $^{42}K^+$ in EL of cat following intraperitoneal injection of $^{42}K_2CO_3$. (From Choo and Tabowitz.[41])

tion is the stria vascularis, thus supporting this site for fluid formation. Similar experiments with $^{36}Cl^-$ yielded a turnover constant of 0.01 min^{-1} or a half-time of 69 min.[68b] Again there is good reason to suppose that this measurement, agreeing as it does with that for K^+, gives an index of EL formation rate. Comparing Figs. 52-24 and 52-25, we may tentatively conclude that K^+ passes slowly from blood to perilymph, but rapidly from perilymph to EL. The time for EL renewal may then be of the order of several hours.

The cochlea contains very high concentrations of carbonic anhydrase (cat, guinea pig) in the stria vascularis, spiral ligament, membranous wall, and endolymphatic sac.[51a,53] Activity in the spiral ligament is of particular interest, since it is here that perilymph has access to endolymph formation (see earlier discussion). The purified cochlear enzyme resembles the high-activity carbonic anhydrase in red blood cells and is fully susceptible to acetazolamide.[51a]

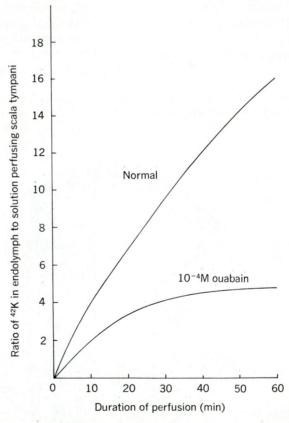

Ratio of ^{42}K in endolymph to solution perfusing scala tympani

Normal

$10^{-4}M$ ouabain

Duration of perfusion (min)

Fig. 52-25. Accession of ^{42}K to EL in guinea pigs during perfusion of scala tympani with modified Ringer's solution containing K label in 5 mM KCl. Ouabain was added to perfusate. (Modified from data of Konishi et al.[68a])

In experiments analogous to those of Fig. 52-20, HCO_3^- in EL of the guinea pig was found to reach equilibrium with plasma, at 1.3 times plasma concentration, in 30 min. The rate was diminished about 60% by acetazolamide.[39a] These rapid rates may reflect fast diffusion of gaseous CO_2 from plasma to stria vascularis or initially into perilymph, with catalytic formation of HCO_3^- as a counterion for K^+ transport to EL. What appears to be a related finding is that carbonic anhydrase inhibition abolishes the effect of 10% CO_2 in increasing endocochlear potential.[94a]

The stria vascularis also contains high concentrations of Na^+,K^+-ATPase; it is supposed that this plays a primary role in maintenance of the cation gradients and in generation of cochlear potentials.[2] Glycolytic intermediates have been found in several tissues of the cochlea, again with particularly high concentrations in the stria vascularis.[69,79] The excised inner ear of the frog continues to secrete K^+ into the EL for several days, and the process is abolished by 10^{-3} M ouabain, presumably through inhibition of Na^+,K^+-ATPase.[100]

The only study of the overall pattern of electrolyte transport to EL has been done in the dogfish, *Squalus acanthias*, using the appropriate radioactive labels.[78] Fig. 52-26 shows the data. Rb^+ is used as a marker for K^+. Of significance are the findings that, although K^+ and CO_2 (or HCO_3^-) are in low concentration in plasma, their

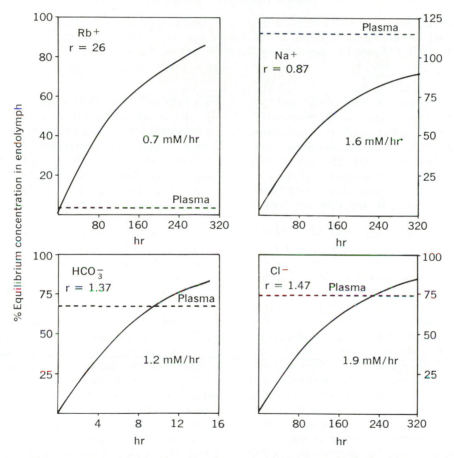

Fig. 52-26. Accession of ions to EL in dogfish, *Squalus acanthias*.[78] Solid lines show movement of isotope into EL, relative to its concentration in plasma (dotted lines). Small numbers at upper left *(r)* give equilibrium ratio of EL:plasma. Larger numbers at right give actual accession rates to EL. Electrolyte concentrations (mM) in this species are as follows:

	Na^+	K^+	Cl^-	CO_2	Urea
Plasma	256	4	230	7.6	350
EL	223	107	314	11.2	282

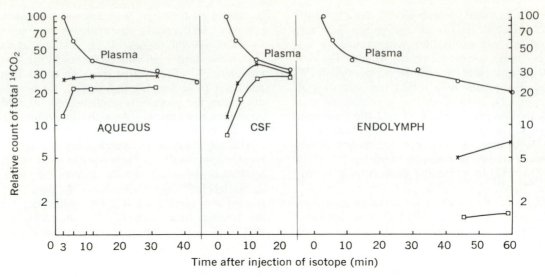

Fig. 52-27. Accession of total $^{14}CO_2$ to fluids in dogfish, *Squalus acanthias*. Control (X) and following carbonic anhydrase inhibition (□). Isotope injected intravenously at zero time; each sample of AH, CSF, and EL taken from a different fish. (From Maren.[76])

accession rates approach those of Na^+ and Cl^-. The same is probably true for the mammal.[68a] Fig. 52-27 shows reduction of HCO_3^- accession to EL by inhibition of carbonic anhydrase, similar to effects in CSF and AH. The much slower observed rates for EL compared to AH and CSF may be due to the fact that EL is formed from perilymph, whereas AH and CSF are formed from blood.

EL [K^+] in dog is reduced from 148 to 24 mM in 10 to 40 min following the injection of 1 to 2 mg/kg of ethacrynic acid. The EL potential was unaffected.[40] A lesser and more delayed decline in EL [K^+] occurred in the rat following a much larger dose, but here the electrical potential (EP) dropped profoundly, from +90 to −40 mV in 15 min.[37] In a general way, these findings give the basis for temporary deafness and tinnitus following this drug and other "high ceiling" diuretics (i.e., furosemide and bumetanide) in man. However, the relation between changes in EP and K^+ are not direct, and the electrolyte changes are so rapid that they seem to reflect disruption of Reissner's membranes rather than inhibition of secretion. The changes suggest that in the cochlea there are high-energy systems that are readily compromised by drugs affecting oxidizing pathways; indeed, high Po_2 protects against the effects of ethacrynic acid on EP and possibly on electrolyte balance in EL.[95]

These data suggest secretion of K^+-rich, pos-

sibly high HCO_3^-, EL fluid at the stria vascularis. The special ionic content of EL must be essential to the physiology of equilibration; high [K^+] is found in the EL of fish, which lack the organ of Corti, and as noted above it is the low [K^+] perilymph (cortilymph) which bathes the hearing organ in the mammal. Indeed, the lateral line canals of certain fish do have K^+ concentrations some 30 times that of plasma.[54]

As with CSF and AH, EL physiology has an associated disorder, Ménière's disease,[26] which is often referred to as inner ear or endolymphatic hydrops. Symptoms are vertigo, hearing loss, and tinnitus. The usual cause is dilatation and pressure within the cochlear duct (scala media, Figs. 14-9 and 52-22) due to failure of absorption of EL. Just as the optic nerve head suffers injury from elevated pressure in glaucoma, so in Ménière's disease the hair cells of the organ of Corti are the target. As for glaucoma, acetazolamide has been used in Ménière's disease, for which the classic treatment has been surgical. Results are difficult to evaluate, but relief of symptoms has been reported. The data reviewed above give a clear basis for such trials, and it may be emphasized that the unusually high concentration of carbonic anhydrase in the inner ear[53] dictates high doses of the drug. From the study of the physiology, anatomy, and biochemistry of EL (and also CSF and AH), it is evident that a sound pharmacologic approach to the associated diseases is possible.

SUMMARY

Fig. 52-27 is taken from a recent study in which the transport of ions to CSF, AH, and EL are compared in a single species. The rate constants for formation of fluid are in the order AH > CSF > EL. The figure shows this also to be the case with HCO_3^- accession as measured by $^{14}CO_2$, lending credence to the theme that has run through this essay, that HCO_3^- and fluid formation are linked. Finally, among the three fluids there is the unifying feature of similarity of response to carbonic anhydrase and Na^+,K^+-ATPase inhibition.[76] The study of these fluids reveals a basic pattern of nature in which there is diversity for specific means overlaid on a single fundamental plan.

REFERENCES
General reviews

1. Bill, A.: Blood circulation and fluid dynamics in the eye, Physiol. Rev. **55:**383, 1975.
2. Bonting, S. L.: Sodium-potassium activated adenosine-triphosphatase and cation transport. In Bittar, E. E., editor: Membranes and ion transport, London, 1970, John Wiley & Sons, Ltd., vol. 1.
2a. Bradbury, M.: The concept of a blood-brain barrier, New York, 1979, John Wiley & Sons, Inc.
3. Burg, M. B.: Mechanisms of action of diuretic drugs. In Brenner, B. M., and Rector, F. C., editors: The kidney, Philadelphia, 1976, W. B. Saunders Co., vol. 1.
4. Coppen, A.: The biochemistry of affective disorders, Br. J. Psychiatry **113:**1237, 1967.
5. Cserr, H. F.: Physiology of the choroid plexus, Physiol. Rev. **51:**273, 1971.
6. Davson, H.: Physiology of the cerebrospinal fluid, London, 1967, J. & A. Churchill, Ltd.
7. Davson, H.: The intraocular fluids. In Davson, H., editor: The eye, ed. 2, New York, 1969, Academic Press, Inc., vol. 1.
8. Davson, H.: The blood-brain barrier, J. Physiol. **255:**1, 1976.
9. Glynn, I. M.: The action of cardiac glycosides on ion movements, Pharmacol. Rev. **16:**381, 1964.
10. Katzman, R., and Pappius, H. M.: Brain electrolytes and fluid metabolism, Baltimore, 1973, The Williams & Wilkins Co.
11. Kellogg, R. H.: Oxygen and carbon dioxide in the regulation of respiration, Fed. Proc. **36:**1658, 1977.
12. Kinsey, V. E.: Ion movement in ciliary processes. In Bittar, E. E., editor: Membranes and ion transport, New York, 1971, John Wiley & Sons, Inc.
13. Kinsey, V. E., and Reddy, D. V. N.: Chemistry and dynamics of aqueous humor. In Prince, J. H., editor: The rabbit in eye research, Springfield, Ill., 1964, Charles C Thomas, Publisher.
14. Kolker, A. E., and Hetherington, J., Jr.: Becker-Schaffer's diagnosis and therapy of the glaucomas, ed. 4, St. Louis, 1976, The C. V. Mosby Co.
15. Leusen, I.: Regulation of cerebrospinal fluid composition with reference to breathing, Physiol. Rev. **52:**1, 1972.
15a. Lindvall, M., Edvinsson, L., and Owman, C.: Histochemical, ultrastructural, and functional evidence for a neurogenic control of CSF production from the choroid plexus. In Cervos-Navarro, J., et al., editor: Advances in neurology, New York, 1978, Raven Press, vol. 20.
16. Loeschcke, H. H.: Central nervous chemoreceptors. In Guyton, A. C., and Widdicombe, J. G., editors: MTP International Review of Science physiology series. 1. Respiratory physiology, London, 1974, Butterworth & Co., Ltd., vol. 2.
17. Maggio, E.: The humoral system of the labyrinth, Acta Otolaryngol. **218**(suppl.):9, 1966.
18. Maren, T. H.: Carbonic anhydrase: chemistry, physiology and inhibition, Physiol. Rev. **47:**594, 1967.
19. Maren, T. H.: The rates of movement of Na^+, Cl^-, and HCO_3^- from plasma to posterior chamber: effect of acetazolamide and relation to the treatment of glaucoma, Invest. Ophthalmol. **15:**356, 1976.
20. Milhorat, T. H.: Hydrocephalus and the cerebrospinal fluid, Baltimore, 1972, The Williams & Wilkins Co.
21. Millen, J. W., and Woollam, D. H. M.: The anatomy of the cerebrospinal fluid, London, 1962, Oxford University Press.
22. Netsky, M. G., and Shuangshoti, S.: The choroid plexus in health and disease, Charlottesville, Va., 1975, University Press of Virginia.
23. Oldendorf, W. H.: Permeability of the blood-brain barrier. In Tower, D. B., editor: The nervous system, New York, 1975, Raven Press, vol. 1.
24. Pappenheimer, J. R.: The ionic composition of cerebral extracellular fluid and its relation to control of breathing, Harvey Lect. **61:**71, 1965-1966.
25. Plum, F., and Siesjo, B. K.: Recent advances in CSF physiology, Anesthesiology **42:**708, 1975.
26. Pulec, J.: Ménière's disease, Philadelphia, 1968, W. B. Saunders Co.
27. Rapoport, S. I.: Blood-brain barrier in physiology and medicine, New York, 1976, Raven Press.
28. Rauch, S.: Biochemie des Hörorgans, Stuttgart, Germany, 1964, George Thieme Verlag.
29. Siesjö, B. K., and Sørensen, S. C., editors: Ion homeostasis of the brain, Alfred Benzon Symposium III, Copenhagen, 1971, Munksgaard, International Booksellers & Publishers, Ltd.
30. Weed, L. H.: The cerebrospinal fluid, Physiol. Rev. **2:**171, 1922.
30a. Wright, E. M.: Transport processes in the formation of cerebrospinal fluid, Rev. Physiol. Biochem. Pharmacol. **83:**1, 1978.

Original papers

31. Abbott, J., et al.: Chloride transport and potential across the blood-CSF barrier, Brain Res. **29:**185, 1971.
32. Ames, A., Sakanoue, M., and Endo, S.: Na, K, Ca, Mg and Cl concentrations in choroid plexus fluid and cisternal fluid compared with plasma ultrafiltrate, J. Neurophysiol. **27:**672, 1964.
33. Anson, B. J., et al.: The developmental and adult anatomy of the membranous and osseous labyrinths and of the otic capsule. In Pulec, J. L., editor: Meniere's disease, Philadelphia, 1968, W. B. Saunders Co.
34. Arieff, A. I., et al.: Intracellular pH of brain: alterations

in acute respiratory acidosis and alkalosis, Am. J. Physiol. **239:**804, 1976.

35. Becker, B.: Carbonic anhydrase and the formation of aqueous humor, the Friedenwald Memorial Lecture, Am. J. Ophthalmol. **47:**342, 1959.

36. Bledsoe, S. W., and Mines, A. H.: Effect of plasma (K^+) on the DC potential and on ion distributions between CSF and blood, J. Appl. Physiol. **39:**1012, 1975.

37. Bosher, S. K., Smith, C., and Warren, R. L.: The effects of ethacrynic acid upon the cochlear endolymph and stria vascularis, Acta Otolaryngol. **75:**184, 1973.

38. Bourke, R. S., and Nelson, K. M.: Studies on the site of mediated transport of chloride from blood into cerebrospinal fluid: effects of acetazolamide, J. Neurochem. **19:**1225, 1972.

39. Bourke, R. S., Gabelnick, H., and Young, O.: Mediated transport of chloride from blood into cerebrospinal fluid, Exp. Brain Res. **10:**17, 1970.

39a. Brusilow, S.: Unpublished data.

40. Brusilow, S. W., and Gordes, E.: The mutual independence of the endolymphatic potential and the concentrations of sodium and potassium in endolymph, J. Clin. Invest. **52:**2517, 1973.

41. Choo, Y. B., and Tabowitz, D.: The formation and flow of the cochlear fluids, Ann. Otol. Rhinol. Laryngol. **74:**140, 1965.

42. Cole, D. F.: Aqueous humor formation, Doc. Ophthalmol. **21:**116, 1966.

43. Collins, P., and Morriss, G. M.: Changes in the surface features of choroid plexus of the rat following the administration of acetazolamide and other drugs which affect CSF secretion, J. Anat. **120:**571, 1975.

44. Cushing, H.: Studies on the cerebrospinal fluid, J. Med. Res. **31:**1, 1914.

45. Dandy, W. E.: Der Liquor cerebrospinalis. In Clara, H., editor: Das Nervensystem den Menschen, ed. 2, Leipzig, Germany, 1953, Johann Ambrosius Barth Verlagsbuchhandlung.

46. Davson, H.: Comparative study of aqueous humor and cerebrospinal fluid in rabbit, J. Physiol. **129:**111, 1955.

47. Davson, H., and Bradbury, M. W.: The fluid exchange of the central nervous system, Symp. Soc. Exp. Biol. **19:**349, 1964.

48. Davson, H., and Segal, M. B.: The effects of some inhibitors and accelerators of sodium transport on the turnover of ^{22}Na in the cerebrospinal fluid and the brain, J. Physiol. **209:**131, 1970.

49. Dempsey, J. A., et al.: Effects of moderate hypoxemia and hypocapnia on CSF [H^+] and ventilation in man, J. Appl. Physiol. **38:**665, 1975.

50. Dempsey, J. A., et al.: Regulation of CSF [HCO_3^-] during long-term hypoxic hypocapnia in man, J. Appl. Physiol. **44:**175, 1978.

51. Domer, F. R.: Effect of diuretics on cerebrospinal fluid formation and potassium movement, Exp. Neurol. **24:** 54, 1969.

51a. Drescher, D. G.: Purification of a carbonic anhydrase from the inner ear of the guinea pig, Proc. Natl. Acad. Sci. USA **74:**892, 1977.

52. Eisenberg, H. M., McComb, J. G., and Lorenzo, A. V.: Cerebrospinal fluid overproduction and hydrocephalus associated with choroid plexus papilloma, J. Neurosurg. **40:**381, 1974.

52a. Epstein, M. H., Feldman, A. M., and Brusilow, S. W.: Cerebrospinal fluid production: stimulation by cholera toxin, Science **196:**1012, 1977.

53. Erulkar, S. D., and Maren, T. H.: Carbonic anhydrase and the inner ear, Nature **189:**459, 1961.

54. Fänge, R., Larsson, Å., and Lidman, U.: Fluids and jellies of the acousticolateralis system in relation to body fluids in *Coryphaenoides rupestris* and other fishes, Marine Biol. **17:**180, 1972.

55. Fencl, V., Koski, G., and Pappenheimer, J. R.: Factors in cerebrospinal fluid from goats that affect sleep and activity in rats, J. Physiol. **216:**565, 1971.

56. Fencl, V., Miller, T. B., and Pappenheimer, J. R.: Studies on the respiratory response to disturbances of acid-base balance, with deductions concerning the ionic composition of cerebral interstitial fluid, Am. J. Physiol. **210:**459, 1966.

57. Garg, L. C., and Oppelt, W. W.: The effect of ouabain and acetazolamide on the transport of sodium and chloride from plasma to aqueous humor, J. Pharmacol. Exp. Ther. **175:**237, 1970.

58. Hakim, S., and Adams, R. D.: The special clinical problem of symptomatic hydrocephalus with normal CSF pressure, J. Neurol. Sci. **2:**307, 1965.

59. Hales, J. R. S., Bligh, J., and Maskrey, M.: Cerebrospinal fluid acid-base balance during respiratory alkalosis in the panting animal, Am. J. Physiol. **219:** 469, 1970.

60. Hasan, F. M., and Kazemi, H.: Dual contribution theory of regulation of CSF HCO_3^- in respiratory acidosis, J. Appl. Physiol. **40:**559, 1976.

60a. Haywood, J. R., and Vogh, B. P.: Some measurements of autonomic nervous system influence on production of cerebrospinal fluid in the cat, J. Pharmacol. Exp. Ther. **208:**341, 1979.

60b. Heel, R. C., Brogden, R. N., Speight, T. M., and Avery, G. S.: Timolol: a review of its therapeutic efficacy in the topical treatment of glaucoma, Drugs **17:**38, 1979.

61. Heisy, S. R., Held, D., and Pappenheimer, J. R.: Bulk flow and diffusion in the cerebrospinal fluid system of the goat, Am. J. Physiol. **203:**775, 1962.

62. Held, D., Fencl, V., and Pappenheimer, J. R.: Electrical potential of cerebrospinal fluid, J. Neurophysiol. **27:**942, 1964.

63. Hochwald, G. M., and Sahar, A.: Effect of spinal fluid pressure on cerebrospinal fluid formation, Exp. Neurol. **32:**30, 1971.

64. Holland, M. G.: Chloride ion transport in the isolated ciliary body. II. Ion substitution experiments, Invest. Ophthalmol. **9:**30, 1970.

65. Hughes, J., et al.: Identification of two related pentapeptides from the brain with potent opiate agonist activity, Nature **258:**577, 1975.

66. Husted, R. F., and Reed, D. J.: Regulation of cerebrospinal fluid bicarbonate by the cat choroid plexus, J. Physiol. **267:**411, 1977.

67. Kinsey, V. E., and Reddy, D. V. N.: Turnover of carbon dioxide in the aqueous humor and the effect thereon of acetazolamide, Arch. Ophthalmol. **62:**78, 1959.

68. Kjallquist, A., Messeter, K., and Siesjö, B. K.: The in vivo buffer capacity of rat brain tissue under carbonic anhydrase inhibition, Acta Physiol. Scand. **78:** 94, 1970.

68a. Konishi, T., Hamrick, P. E., and Walsh, P. J.: Ion transport in guinea pig chochlea. I. Potassium and sodium transport, Acta Otolaryngol. **86:**22, 1978.

68b. Konishi, T., and Hamrick, P. E.: Ion transport in the cochlea of guinea pig. II. Chloride transport, Acta Otolaryngol. **86:**176, 1978.

69. Kuijpers, W., and Bonting, S. L.: Studies on (Na^+-K^+)-activated ATPase. XXIV. Localization and properties of ATPase in the inner ear of the guinea pig, Biochim Biophys. Acta **173:**477, 1969.

70. Lambertson, C. J., Gelfand, R., and Kemp, R. A.: Dynamic response characteristics of several CO_2-reactive components of the respiratory control system. In Brooks, C. M., Kao, F. F., and Lloyd, B. B., editors: Cerebrospinal fluid and the regulation of ventilation, Philadelphia, 1965, F. A. Davis Co.

71. Lawrence, M.: Theories of the cause of hydrops. In Pulec, J. L., editor: Ménière's disease, Philadelphia, 1968, W. B. Saunders Co.

72. Leusen, I.: Chemosensitivity of respiratory center; influence of CO_2 in cerebral ventricles on respiration, Am. J. Physiol. **176:**39, 1954.

73. Maren, T. H.: Bicarbonate formation in the cerebrospinal fluid. Its role in sodium transport and pH regulation, Am. J. Physiol. **222:**885, 1972.

74. Maren, T. H.: HCO_3^- formation in aqueous humor: mechanism and relation to the treatment of glaucoma, Invest. Ophthalmol. **13:**479, 1974.

75. Maren, T. H.: Ion secretion into the posterior aqueous humor of dogs and monkeys, Exp. Eye Res. (suppl.) **25:**245, 1977.

76. Maren, T. H.: Physiology and chemistry of cerebrospinal fluid, aqueous humor and endolymph in *Squalus acanthias,* J. Exp. Zool. **199:**317, 1977.

77. Maren, T. H.: The effect of varying CO_2 equilibria upon rates of HCO_3^- formation in cerebrospinal fluid, J. Appl. Physiol. **47:**471, 1979.

78. Maren, T. H., Swenson, E. R., and Addink, A. D. F.: The rates of ion movement from plasma to endolymph in the dogfish, *Squalus acanthias,* Ann. Otol. Rhinol. Laryngol. **84:**847, 1975.

79. Matschinsky, F. M., and Thalmann, R.: Quantitative histochemistry of microscopic structures of the cochlea. II. Ischemic alterations of levels of glycolytic intermediates and cofactors in the organ of Corti and stria vascularis, Ann. Otol. Rhinol. Laryngol. **76:**638, 1967.

80. Mayer, S., Maickel, R. P., and Brodie, B. B.: Kinetics of penetration of drugs and other foreign compounds into cerebrospinal fluid and brain, J. Pharmacol. Exp. Ther. **127:**205, 1959.

81. McCarthy, K. D., and Reed, D. J.: The effect of acetazolamide and furosemide on cerebrospinal fluid production and choroid plexus carbonic anhydrase activity, J. Pharmacol. Exp. Ther. **189:**194, 1974.

82. Messeter, K., and Siesjö, B. K.: Electrochemical gradients for H^+ and HCO_3^- between blood and CSF during sustained acid-base changes. In Siesjö, B. K., and Sørensen, S. C., editors: Ion homeostasis of the brain, Alfred Benzon Symposium III, Copenhagen, 1971, Munksgaard, International Booksellers & Publishers, Ltd.

83. Migliore, A., Paoletti, P., and Villani, R.: Studies on the rate of disappearance of Na^{24} from the ventricular cavities in obstructive hydrocephalus, J. Neurosurg. **21:**521, 1964.

84. Milhorat, T. H., et al.: Cerebrospinal fluid production by the choroid plexus and brain, Science **173:**330, 1971.

85. Miner, L. C., and Reed, D. J.: Composition of fluid obtained from choroid plexus tissue isolated in a chamber in situ, J. Physiol. **227:**127, 1972.

86. Mitchell, R. A.: The regulation of respiration in metabolic acidosis and alkalosis. In Brooks, C. M., Kao, F. F., and Lloyd, B. B., editors: Cerebrospinal fluid and the regulation of ventilation, Philadelphia, 1965, F. A. Davis Co.

86a. Nathanson, J. A.: β-Adrenergic-sensitive adenylate cyclase in secretory cells of choroid plexus, Science **204:**843, 1979.

87. Neblett, C. R., et al.: Effect of cardiac glycosides on human cerebrospinal-fluid production, Lancet **2:**1008, 1972.

88. Nechay, B. R., et al.: The problem of $Na^+ + K^+$ adenosine triphosphatase as the receptor for diuretic action of mercurials and ethacrynic acid, J. Pharmacol. Exp. Ther. **157:**599, 1967.

89. Oldendorf, W. H.: Lipid solubility and drug penetration of the blood brain barrier, Proc. Soc. Exp. Biol. Med. **147:**813, 1974.

90. Oppelt, W. W.: Measurement of aqueous humor formation rates by posterior-anterior chamber perfusion with inulin: normal values and the effect of carbonic anhydrase inhibition, Invest. Ophthalmol. **6:**76, 1967.

91. Oppelt, W. W., et al.: Effects of acid-base alterations on cerebrospinal fluid production, Proc. Soc. Exp. Biol. Med. **114:**86, 1963.

92. Pappenheimer, J. R., Heisey, S. R., and Jordan, E. F.: Active transport of Diodrast and phenolsulfonphthalein from cerebrospinal fluid to blood, Am. J. Physiol. **200:**1, 1961.

93. Pappenheimer, J. R., et al.: Role of cerebral fluids in control of respiration as studied in unanesthetized goats, Am. J. Physiol. **208:**436, 1965.

94. Pelligrino, D. A., and Dempsey, J. A.: Dependence of CSF on plasma bicarbonate during hypocapnia and hypoxemic hypocapnia, Resp. Physiol. **26:**11, 1976.

94a. Prazma, J.: Carbonic anhydrase in the generation of cochlear potentials, Am. J. Physiol. **235:**F317, 1978.

95. Prazma, J., and Pecorak, J. B.: Ethacrynic acid: effects on the cochlear potentials in normal and high blood oxygen, J. Clin. Invest. **55:**840, 1975.

96. Rall, D. P.: Drug entry into brain and CSF. In Brodie, B. B., and Gillette, J. R., editors: Concepts in biochemical pharmacology, Heffter's handbook of experimental pharmacology, Berlin, 1971, Springer-Verlag, vol. 28.

97. Rall, D. P., Oppelt, W. W., and Patlak, C. S.: Extracellular space of brain as determined by diffusion of inulin from the ventricular system, Life Sci. **1:**43, 1962.

98. Rubin, R. C., et al.: The production of cerebrospinal fluid in man and its modification by acetazolamide, J. Neurosurg. **25:**430, 1966.

99. Severinghaus, J. W., et al.: Respiratory control at high altitude suggesting active transport regulation of CSF pH, J. Appl. Physiol. **18:**1155, 1963.

100. Simon, E. J., Hilding, D. A., and Kashgarian, M.: Micropuncture studies of the mechanism of endolymph production in the frog, Am. J. Physiol. **225:**114, 1973.

101. Sweet, W. H., et al.: Studies of formation, flow and absorption of cerebrospinal fluid. II. Studies with heavy water in the normal man, Surg. Forum **1:**376, 1950.

102. Tennyson, V. M., and Pappas, G. D.: Electron-microscope studies of the developing telencephalic choroid plexus in normal and hydrocephalic rabbits. In Fields, W. S., and Desmond, M. M., editors: Disorders of the developing nervous system, Springfield, Ill., 1961, Charles C Thomas, Publisher.

103. Terenius, L., and Wahlström, A.: Morphine-like ligand for opiate receptors in human CSF, Life Sci. **16:**1759, 1975.

104. Tubiana, M., Benda, P., and Constans, J.: Sodium radioactif ^{22}Na et liquide céphalorachidien, Rev. Neurol. **85:**17, 1951.

105. Vates, T. S., Bonting, S. L., and Oppelt, W. W.: Na-K activated adenosine triphosphatase formation of cerebrospinal fluid in the cat, Am. J. Physiol. **206:**1165, 1964.

106. Vogh, B. P., and Maren, T. H.: Sodium, chloride, and bicarbonate movement from plasma to cerebrospinal fluid in cats, Am. J. Physiol. **228:**673, 1975.
107. Wald, A., Hochwald, G. M., and Malhan, C.: The effects of ventricular fluid osmolality on bulk flow of nascent fluid into the cerebral ventricles of cats, Exp. Brain Res. **25:**157, 1976.
108. Wegefarth, P., and Weed, L. H.: The analogous processes of the cerebral and ocular fluids, J. Med. Res. **31:**167, 1914.
109. Welch, K.: Secretion of cerebrospinal fluid by choroid plexus of the rabbit, Am. J. Physiol. **205:**617, 1963.
110. Welch, K., and Friedman, V.: The cerebrospinal fluid valves, Brain **83:**454, 1960.
111. Wichser, J., and Kazemi, H.: CSF bicarbonate regulation in respiratory acidosis and alkalosis, J. Appl. Physiol. **38:**504, 1975.
112. Wright, E. M.: Mechanisms of ion transport across the choroid plexus, J. Physiol. **226:**545, 1972.

XI

PHYSIOLOGY OF THE DIGESTIVE SYSTEM

53

THOMAS R. HENDRIX

The absorptive function of the alimentary canal

Absorption of fluid and nutrients is the raison d'être of the alimentary canal. Absorption is accomplished in unicellular organisms by transfer from the surrounding media through the cell membrane into the cell. In higher organisms, absorbed materials must cross the cells lining the alimentary canal. Hence the transit of two membranes, one at the mucosal surface of the absorptive cells and the other at the serosal surface, must be explained.

ABSORPTIVE SURFACE

The cell that performs the function of absorption is the columnar epithelial cell of the intestinal villi (Fig. 53-1). The surface of the cell facing the intestinal lumen is composed of a brush border of slender, closely packed, finger-shaped projections.[118] It has been estimated that there are 1,700 of these microvilli per cell in man and that they increase the surface area facing the luminal content by 15- to 40-fold. It seems likely that this complex membrane serves a more complex function than that of merely increasing the surface area of the cell. Since it contains many digestive enzymes and transport sites, it is not unreasonable to suggest that this structure provides important spatial arrangements of these functional sites.[15] The plasma membrane of microvilli is wider than that of the lateral and basal portions of the cell and is covered with a glycocalyx, or fuzzy coat. The outer leaflets of the plasma membranes of adjacent cells are fused for a distance of 0.1 to 0.2 μm on the lateral surfaces just below the luminal surface to form the tight junction that obliterates the intercellular space.[100,118]

The absorptive columnar cell has a limited life span. Cells extruded from the tips of the villi are replaced by cells migrating up the villi from the crypts of Lieberkühn, where cell division normally occurs. The travel from crypt to villus tip requires 2 to 5 days, depending on the species studied. As the cell moves out of the crypt onto the villus, it develops brush border enzymes characteristic of a mature absorbing cell.

In recent years the crypts of Lieberkühn have been considered primarily as a nursery for absorptive cells. Physiologists of an earlier generation, however, ascribed a glandular or secretory function to them.[31] This concept has been reactivated to explain the massive fluid production by the intestine in cholera.[49]

Not only is the surface for absorption increased by the brush border of the absorbing surface of the columnar cell, but the epithelium is also thrown up into finger- and tonguelike projections, the villi, and the mucosa in turn is thrown up into the valvulae conniventes (Fig. 53-2). It has been estimated that these folds increase the absorbing surface 600 times over that of a simple cylinder of similar dimensions and provide a total surface area of 2,000,000 cm².

Most nutrients are absorbed in the duodenum and jejunum. In man it has been shown that a meal of protein, carbohydrate, and fat is almost completely absorbed in the first 100 cm of the jejunum.[9] The remainder of the small intestine is thought to provide a reserve absorptive function. Although this is probably true, the ileum has some special absorptive functions. The best defined are the active transport of vitamin B_{12} and bile acids, functions unique to the ileum (Fig. 53-3). In addition, there are differences in the handling of water and electrolytes between the upper and lower small bowel. In the duodenum and upper jejunum there is a rapid and voluminous fluid circuit, that is, water and electrolytes enter the lumen about as rapidly as they leave. As a consequence, there is little decrease in the volume of the luminal content. On the other hand, the exchange is less voluminous in the ileum and, because movement out of the lu-

1255

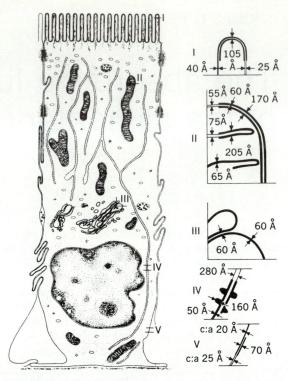

Fig. 53-1. Schematic drawing of columnar absorbing cell. Dimensions at right are for, *I,* plasma membrane of microvilli; *II,* mitochondrial membranes; *III,* Golgi membranes; *IV,* endoplasmic reticulum membranes; and *V,* lateral plasma membrane. (From Zetterqvist.[118])

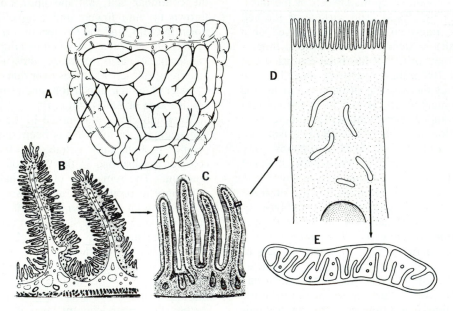

Fig. 53-2. Diagram illustrating arrangements at various levels of morphologic organization in the alimentary tract that increase efficiency by amplifying area of physiologically significant interfaces. First, hollow viscus is greatly elongated, **A.** Its internal surface is further increased by circumferential folds, or plications, of mucosa visible with naked eye, **B.** These in turn are covered with microscopic villi, **C.** Individual cells on villi are covered with myriad microvilli, **D.** Membranes of intracellular organelles, for example, mitochondria, are arranged in folds, **E.** See Fig. 53-1. (From Fawcett.[30])

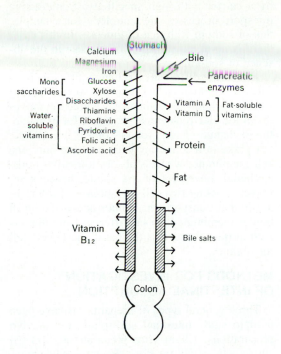

Fig. 53-3. Sites of absorption in small intestine. Absorption of primary nutrients (carbohydrates, proteins, and fat) commences in duodenum and is usually completed in upper one third to one half of small intestine. (From Booth.[7])

men exceeds movement in, the volume of the luminal contents is greatly decreased.[32,107]

MECHANISMS OF ABSORPTION

The membranes of cells behave as though they were lipoid, and hence they impede the flow of water and charged solutes in and out of cells. Since water and charged solutes do move in and out of cells in measurable quantities, a model has been suggested to account for this movement in which the lipoid component of the membrane is interrupted at intervals. The sizes of these interruptions in the lipoid membrane have been calculated after measuring the movement of uncharged solutes of varying size. The smaller the solute, the less its movement down a concentration gradient is impaired by the cell membrane. Most solutes with a molecular weight greater than 180 cannot pass through the cell membrane; for example, urea (mol wt 60) equilibrates across cell membranes, but mannitol (mol wt 182) does not.

In addition to the transcellular route of absorption, which involves transit of two lipoidal membranes, water and small solutes are absorbed by an intercellular route through the tight junctions in response to electrochemical, osmotic, and hydrostatic gradients (Fig. 53-4).[91] Resistance to flow via the intercellular route is low in the jejunum, moderate in the ileum, and high in the

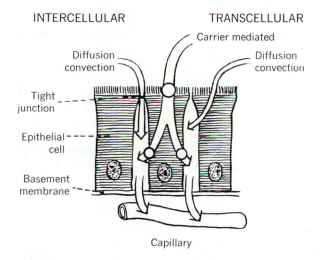

Fig. 53-4. Small solutes and water are absorbed from intestinal lumen via two routes. First, transcellular route, is path taken by solutes that require membrane carriers, both active transport and carrier-mediated diffusion. This transport route involves passage through two lipoidal membranes: brush border and lateral cell membrane. In addition, limited but undefined proportion of water and solutes absorbed by diffusion and convection (solvent drag) absorbed by this route. Second, intercellular route, is probably pathway for majority of absorption driven by diffusion and convection. It does not involve passage through intestinal epithelial cells but rather through tight junction into lateral intercellular space.

colon. To explain the movements of materials across cell membranes, a summary of the mechanisms involved is useful (Fig. 53-4).[32,79]

Simple passive diffusion. The driving force for passive diffusion across a membrane is the electrochemical gradient of the material under study. There is no net movement of a material across a membrane by passive diffusion when its concentration is equal on both sides of the membrane, the potential difference (electrical gradient) across the membrane is zero, and the net flow of water (solvent drag) is also zero. The diffusion of small, charged solutes (passive ionic diffusion) occurs transcellularly through the nonlipoid sites in the membrane and intercellularly through the tight junctions. Nonionized solutes, on the other hand, diffuse through the lipoid membrane. The relative solubilities of the solute in aqueous extracellular fluid and in the lipoid membrane and the proportion in the nonionized form determine the rate of diffusion, which is related linearly to the magnitude of the gradients.

Solvent drag or convection. Solvent drag and convection describe the movement of solutes swept along by the bulk flow of water driven by osmotic or hydrostatic gradients across the cell membrane and the tight junction. The smaller the solute size, the less its movement will be impeded by "channels" through which the water is flowing and the more closely it will follow the water movement. Charges on the solute and in the "channel" also impose a constraint on movement by solvent drag.

Facilitated or carrier-mediated diffusion. Many solutes enter the cell readily even though they are too large to enter by simple diffusion or solvent drag and are not sufficiently lipid soluble to enter by nonionic passive diffusion. Membrane carrier mechanisms are proposed to explain the translocation of these solutes. This type of transport mechanism shows (1) substrate and stereo specificity, (2) competition between related substances, and (3) saturability of the mechanism with increasing concentration (see the section on carbohydrates later in this chapter). Facilitated diffusion is similar to simple passive diffusion in that the substance only moves down its electrochemical gradient.

Active or uphill transport. The movement of a solute against its electrochemical gradient and in the absence of bulk flow is active or uphill transport. It is characterized by (1) efficiency at low luminal concentration, so that sizable transmucosal concentration gradients are achieved, (2) Michaelis-Menton (saturation) kinetics, since a maximal transport rate is reached as luminal concentration is increased, (3) requirement for expenditure of cellular energy, and (4) mediation by a carrier with high specificity. Some active transport processes are coupled, for example, that of sodium with that of glucose. In this case, there is no uphill transport of glucose in the absence of sodium.

Exchange diffusion. Exchange diffusion indicates transport in which one ion moves against its electrochemical gradient as a paired ion moves down the electrochemical gradient, for example, the exchange of Na^+ for H^+ or Cl^- for HCO_3^-.

Uptake of a solute involves crossing only one cell membrane, whereas absorption is the transepithelial movement of the solute; hence it involves crossing two cell membranes. The mechanism that transports the solutes across the brush border membrane may be different from that effecting transport across the basal and lateral cell membranes.

METHODS FOR INVESTIGATION OF INTESTINAL ABSORPTION

Three general types of preparations have been used to study intestinal absorption: (1) in vivo preparations, (2) in vitro preparations, and (3) single cells or cell fractions. Each has advantages and disadvantages, hence several approaches must be used to gain an understanding of how the absorption of any substance occurs in the intact organism.[60,82]

In vivo techniques

Conceptually, a balance study is the simplest of the in vivo techniques; the difference between the amount ingested and the amount passed in the stools is the amount absorbed. Such a technique, however, gives no insight into the mechanism involved, the section of the gut performing the absorption, how much of the disappearance measured is due to absorption, how much is due to the metabolism of colonic bacteria, and how much has been added during the passage through the gut.

Tolerance tests. Tolerance tests measure the appearance of the test substance in the venous blood. The concentration appearing in the blood is the consequence of several factors other than absorption, for example, volume of distribution in the body, rate and site of metabolism, and renal excretion of the test material. Moreover, most absorbed substances, except lipids, enter the circulation via the portal vein so that the appearance curve in peripheral venous blood is the algebraic sum of intestinal, hepatic, and peripheral tissue handling of the absorbed substance.

Intestinal loops. The use of in vivo intestinal loops makes it possible to measure directly the appearance and disappearance of material from the intestinal lumen. The addition of isotopes to the luminal fluid or to the blood makes it possible to calculate bidirectional fluxes as well as net flux. Transmucosal concentration and electrical gradients can also be measured and interferences drawn as to the processes involved. The in vivo loop allows the intestinal epithelium to function under optimal conditions of oxygenation, nutrition, and efficient removal of absorbed material from the subepithelial space. The use of nonabsorbable volume markers makes it possible to obtain identical information from the intact in vivo intestine by intubation.[34,47]

In vitro techniques

Present-day interest and understanding of intestinal absorption is the direct result of the introduction of the everted gut sac, a simple but powerful tool for the study of intestinal absorption.[82] With the same solution inside and outside the sac, it is a simple matter to determine whether a substance is transported against a concentration gradient. *Sheets of intestinal wall* or mucosa stripped of the muscularis propria may be clamped as a diaphragm between two chambers. In such a preparation volume flow, concentration gradients, and electrical gradients can be manipulated at will, and unidirectional fluxes can be measured by isotopes. This rigorously controlled situation makes it possible to identify the component of total transport due to separate factors such as diffusion, convection, and active transport and to identify the elements of coupled transport.[104] *Individual villi*, for example, have been isolated for measurement of the mucosal accumulation of sugar. *Individual cells* and *cell fragment preparations* can be used to measure uptake rather than transport. Isolation of the brush border membranes makes it possible to localize enzymes to the digestive absorptive surface. It is from such preparations that isolation of the carriers that mediate active transport and carrier-mediated diffusion will be accomplished.

WATER AND ELECTROLYTE ABSORPTION

In addition to 1.5 to 2.5 L of water ingested daily, 8.5 to 10 L of water are added to the intestinal fluid in the form of alimentary tract secretions. Most of this fluid is absorbed in the small intestine with only 1.0 to 1.5 L remaining to enter the colon; of this volume, less than 150 ml

normally appears in the feces[12,32] (Table 53-1 and Fig. 53-5).

In the intestine, as in all other tissues, water absorption is passive and depends on solute absorption. All cells transport sodium out of the cell against an electrochemical gradient. This extrusion of sodium that "leaks" into the cell down an electrochemical gradient is accomplished by a "sodium pump," which utilizes membrane-bound ATPase and depends on the aerobic metabolism of the cell. This does not, however, lead

Table 53-1. Alimentary tract fluid balance, expressed in milliliters per day

	Into lumen (ingested and secreted)*	Out of lumen (absorbed)
Food and drink	1,500	—
Saliva	1,500	—
Gastric juice	3,000	—
Pancreatic juice	2,000	—
Bile	500	—
Small intestine†	—	5,850
Colon‡	—	2,500
Feces	—	150
TOTAL	8,500	8,500

*From Carter et al.[12]
†Figure for net small intestinal absorption is minimal and is calculated as the difference between fluid entering the small intestine (8,500 ml) and that absorbed by the colon (2,500 ml) and lost in the feces (150 ml). This figure ignores the fluid that is secreted into the lumen by the small intestine, the amount of which has been estimated to be as great as 48 L/24 hr by Watten et al.[109]
‡Calculated from Levitan et al.[61] The net absorptive capacity of colon was 2,500 ml/day, flux into colon was 8,640 ml/day, and flux out was 11,200 ml/day.

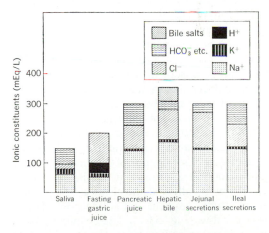

Fig. 53-5. Ionic constituents of digestive and intestinal secretions. (From Fordtran and Ingelfinger.[33])

to net transport of solute and water. Transport requires an asymmetry in the rates of transfer across different areas of the cell membrane. Anatomically, the intestinal cell exhibits polarity. Functionally, it is useful to consider the brush border as the specialized membrane conferring polarity on the cell, with the cell membrane below the tight junction having properties more or less similar to cell membranes of symmetric cells such as muscle cells and erythrocytes.[19,89]

Sodium transport and water flow

The most likely as well as the most abundant solute to account for water transport in the intestine is sodium. Although there is little or no concentration gradient for sodium across the intestinal epithelium, sodium absorption is against an electrochemical gradient since there is a potential difference of 5 to 10 mV across the intestinal mucosa, with serosal surface positive to luminal surface. In in vitro experiments, it is possible to short-circuit the mucosa so that there is neither a concentration nor electrical gradient across the mucosa; in this circumstance, sodium absorption continues. The translocation of sodium is diagrammed in Fig. 53-6. Sodium enters the brush border of the cell down both a chemical and an electrical gradient. It is not known whether this entry is by simple diffusion, carrier-mediated diffusion, or both. The entry of sodium into the cell stimulates the sodium pump operating along the lateral and basal margins of the cell; sodium is then extruded, resulting in net transport of sodium and maintenance of a low intracellular concentration of sodium. When sodium transport is increased by adding glucose or an amino acid, the transmucosal potential difference increases.[89] This often is attributed directly to transport of sodium ions; however, the phenomenon should be considered as the algebraic sum of the potential differences across two membranes (i.e., the brush border and the basilar membranes in series). Changes in the overall potential difference are the consequence of polarization of these membranes rather than transport of cations in excess of anions.

Although water movement can be demonstrated against an activity gradient, it should not be interpreted as evidence of a primary transport system for water. Indeed, it can be shown that water flow parallels solute flow; in the absence of an activity gradient and solute flow, there is no water flow (Fig. 53-7). The coupling of sodium movement and water movement has been explained by a double membrane model (Fig.

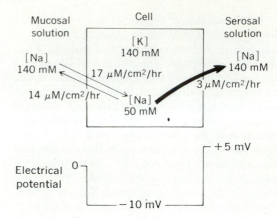

Fig. 53-6. Model for Na transport across intestinal cell. Intracellular Na and K concentrations were obtained on mucosal strips incubated in the presence of 15 mM K and 140 mM Na. Heavy arrow represents steady-state net flux across tissue in absence of actively transported nonelectrolytes. Unidirectional efflux from cell across mucosal membrane is difference between influx and steady-state net flux. If mucosal solution is taken as reference, cell interior has potential difference of −10 mV and serosal surface of cell has potential difference of +5 mV. Sodium enters cell down electrochemical gradient but its transport out the lateral and basal membrane is against a gradient. (From Schultz and Curran.[89])

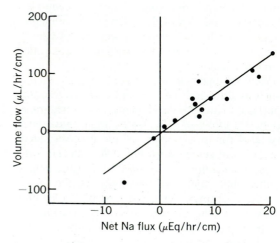

Fig. 53-7. Relations between volume flow and Na⁺ transfer in rat ileum in vitro. At zero solute flux there is zero water flow.

53-8). If the membrane is relatively impermeable to a solute and offers little or no interference to solvent flow, there will be net fluid movement from compartment 1 to compartment 3 when the concentration of the solute is greater in compartment 2 than in compartment 1. The coupling of

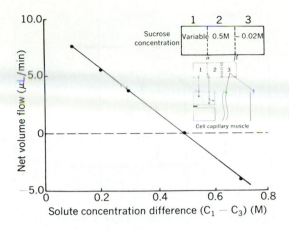

Fig. 53-8. Double membrane model to explain coupling of water flow to solute flux. Positive net volume flow indicates volume increase in compartment *3* and corresponding decrease in compartment *1*. If membrane α is permeable to water but not solute and membrane β is permeable to both but retards diffusion of solute from compartment *2* to *3*, increased concentration of solute in compartment *2* will be associated with volume flow from compartment *1*, through *2*, to *3*. At upper right, possible analogs between system and intestine are suggested. (From Curran.[19])

Table 53-2. Intestinal fluid composition, expressed in milliequivalents per liter

	Jejunum				Ileum			
	Na+	**K+**	**Cl−**	**HCO3**	**Na+**	**K+**	**Cl−**	**HCO3**
Man	148	5.6	138	15	146	5.7	121	42
Dog	159	6.9	122	26	145	9.6	68	76
Rabbit	153	4.3	45	92	148	4.4	48	91

sodium and water movement in the intestine has been explained by this model.

To fit the model to the intestine and to provide an explanation for the coupling of water and sodium movement, it only need be assumed that the brush border is relatively permeable to sodium and water, the tight junction is more permeable to water than sodium, and there is an active transport system for sodium in the lateral membrane of the cell. The osmotic gradient developed by active sodium transport into the intracellular space will draw water into this space by both the transcellular and intracellular routes. Because flow is restricted by the tight junction, hydrostatic pressure develops and fluid flows into the subepithelial space where the lymphatics and capillaries course (Fig. 53-4).

Although this model contributes to our understanding of the coupling of sodium absorption to absorption of monosaccharides and amino acids, it does not completely explain the movement of the other ions that are absorbed in significant amounts from the intestine nor their differing concentration in jejunum and ileum (Table 53-2).[49]

Coupled transport and water flow

The most widely held view, in its simplest form, is that fluid and electrolyte absorption is due to active sodium transport and that the electrical potential difference generated in the process (serosal surface positive to mucosal surface) is the main force for anion absorption.

Jejunum. Data obtained from in vivo studies in the jejunum of man by Fordtran and co-workers[101] were difficult to fit with the concept that active sodium transport, by establishing osmotic and electrochemical gradients, determined the rate and direction of water and ion movements. Specific points of concern were the following: (1) No sodium movement occurred without water movement (this is the reverse of the argument presented previously in support of the concept that fluid movement is coupled with active sodium movement). (2) Sodium absorption against an electrochemical gradient required the presence of bicarbonate in the jejunal fluid. (3) Bicarbonate appeared to be absorbed against a steep electrochemical gradient. (4) Jejunal fluid had a bicarbonate concentration one third that of plasma (8 mEq/L), a slightly acidic pH (6.5), and a partial pressure of CO_2 twice that of plasma (P_{CO_2} = 100 mm Hg). Such values could be explained by the addition of 15 mM HCl to 1 L of an ultrafiltrate of plasma moving into the jejunum. It was found that increasing the concentration of bicarbonate in the lumen to 40 mM/L was associated with an increasing P_{CO_2}, an increased HCO_3^- absorption as CO_2, and sodium absorption against an electrochemical gradient without any change in transmural potential difference. These observations led to the suggestion that the primary event is an H^+ for Na^+ exchange, which leads to sodium absorption without an alteration in potential difference across the mucosa. In addition, the secreted hydrogen ion lowers jejunal pH by neutralizing intraluminal bicarbonate and producing the elevated P_{CO_2} with increased diffusion of CO_2 out of the jejunum, thus producing apparent bicarbonate absorption (Fig. 53-9).

These studies revealed no evidence of sodium absorption against an electrochemical gradient in the absence of bicarbonate absorption (in the form of CO_2). On the other hand, only net flux, which is the difference between the unidirectional flux from lumen to blood and the flux from blood to lumen, was measured. As can be seen from Fig. 53-10, net flux can be quite small despite sizable unidirectional fluxes. In both in vitro and in vivo studies, it has been assumed that the movement of sodium from blood to lumen is solely passive. Recently, evidence from the study of disease states (e.g., cholera) suggests that there is a secretory mechanism in the crypts of Lieberkühn.[52] If such a mechanism is more active in vivo than in vitro and more active in jejunum than ileum, the sum of the two sodium transport processes might very well equal zero and lead to the assumption that there is no uncoupled active transport mechanism for sodium.

Ileum. Several considerations led Fordtran and co-workers to consider other explanations for water and electrolyte movement in the ileum. (1) Electrolyte absorption from the ileum in vivo has not been associated with an increased transmural potential difference; this discrepancy from in vitro studies is attributed to coupled active transport of sodium and chloride, so that the generated potential differences cancel each other. (2) In vivo concentrations of chloride and bicarbonate in ileal fluid vary reciprocally, their sum approximating 135 mEq/L; generally the bicarbonate concentration is found to be higher and chloride lower than in plasma. (3) The chloride absorption from fluid with ion concentrations within physiologic ranges is always much greater than is bicarbonate secretion.[102]

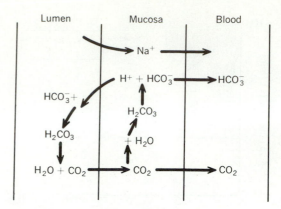

Fig. 53-9. Model to explain dependence of "uphill" sodium absorption on presence of bicarbonate in jejunal fluid and concentration differences between jejunal fluid and plasma of bicarbonate and P_{CO_2}. (Modified from Turnberg et al.[101])

It was suggested that a double-exchange model provided the most complete explanation for observations on ileal electrolyte transport (Fig. 53-11). It provides an explanation of how Na^+, Cl^-, and HCO_3^- can be transported against electrochemical gradients without generating a transmural potential difference, how the ileum can alkalinize its contents during absorption, and how the contents become acidified in the absence of chloride in the ileal fluid. Many differences among results obtained with various experimental techniques remain to be explained, but those models that attempt to interrelate the several ion movements, net water movement, and transmural potential differences will undoubtedly stimulate studies to bridge these gaps.

Fluid circuit—intestinal secretion

It has generally been held that movement of water and electrolytes into the intestinal lumen is a passive process; that is, all movement is down electrochemical gradients. On the other hand, the intestine has a large mass of tubular glands, the crypts of Lieberkühn, to which physiologists of an earlier generation ascribed a secretory function.[52] Florey et al. concluded their review of intestinal secretion with the suggestion that "one may envisage a circulation of fluid during active digestion, the secretion passing out from the crypts of Lieberkühn into the lumen and back into the villi."[31] Indeed, a number of authors have invoked the concept of a fluid circuit to explain their findings.[54,105,107] This hypothesis has largely been discredited, in part because some of its proponents suggested that the fluid circuit accounted for the active transport of water.

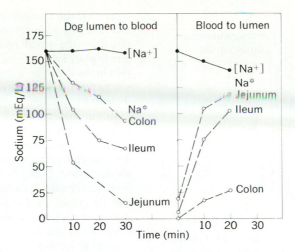

Fig. 53-10. Movement of sodium across intestinal mucosa of unanesthetized dogs with surgically prepared, chronic loops of jejunum, ileum, and colon. Left: intestinal loops were filled with isotonic NaCl containing trace amount of ^{24}Na. Total sodium concentration, [Na$^+$], remained constant, but labeled sodium, *Na**, originally present in loop, rapidly disappeared. Right: intestinal loops were filled with approximately isotonic Na$_2$SO$_4$ solutions. Trace amount of ^{24}Na was injected intravenously and its appearance in contents of lumen was observed. Total sodium concentration fell, but sodium in lumen of loop was rapidly replaced by labeled sodium from blood. Movement of labeled sodium in both directions is in the following order: jejunum → ileum → colon. (Modified from Visscher et al.[106]; from Davenport.[24])

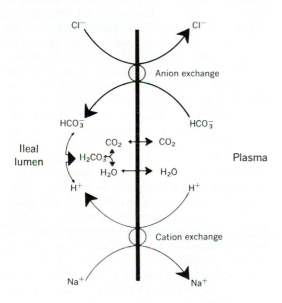

Fig. 53-11. Double exchange model that best explains in vivo observations on relation of different ion movements in ileum of man. If anion exchange is more active than cation exchange, ileal content becomes alkalinized. However, if ileal content contains nonabsorbable anion, content becomes acidified. (From Turnberg et al.[102])

As noted previously, water does move against its activity gradient, but all evidence indicates that this movement is in response to solute transport and concentration gradients. It is fair to say that the suggestion of Florey et al. has not been tested in the context of present-day concepts of water and solute transport. Indeed, even the most sophisticated in vitro techniques are unlikely to give a clear answer because crypt epithelium does not receive a constant supply of fluid and

nutrients via its capillary network. On the other hand, evidence in support of secretion from the crypts has been obtained from studies of experimental cholera.[50] It seems likely that some of the differences between in vitro and in vivo studies of water and electrolytes may be resolved by a consideration of the contribution of the secretion of the crypts of Lieberkühn to the unidirectional and net fluxes (Fig. 53-12).

Colon and gallbladder. In addition to the

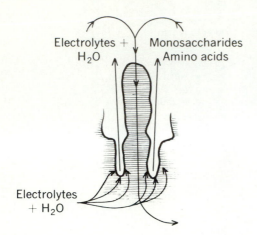

Fig. 53-12. Fluid circuit as proposed by Florey et al.[31] Water and electrolytes are secreted by crypts and are reabsorbed with monosaccharides and amino acids by villi. Net fluid and electrolyte movement is algebraic sum of fluxes into lumen from crypts of Lieberkühn and fluxes out of lumen into villi.

small intestine, two other alimentary canal organs, the colon and gallbladder, play an important role in water and electrolyte absorption.

COLON. On the average 500 to 700 ml of fluid is delivered to the colon per day. This fluid contains 50 to 100 mEq of sodium and chloride. In passage through the colon, water and electrolytes are absorbed so that only 100 to 150 ml of water is lost in the feces. Over 90% of the sodium and chloride is absorbed. The daily electrolyte losses in the feces are approximately 3 mEq of sodium, 8 mEq of potassium, 2 mEq of chloride, and 4 mEq of bicarbonate.[24,61,83] Organic anions derived primarily from bacterial hydrolysis of unabsorbed carbohydrates and cellulose account for 50% to 75% of the anions in stool water. At the pH of stool they are ionized and hence are not appreciably absorbed. Since stool water is essentially isosmotic with plasma, any increase in carbohydrate reaching the colon will increase the amount of unabsorbable anions and as a consequence increase stool water. By infusing isotonic electrolyte solution into the cecum and collecting the stool it was found that the colon has the capacity to absorb up to 5.0 L of fluid/day.[23]

The fluid absorbed by the colon is isosmotic and is driven by electrogenic sodium absorption with the generation of a transepithelial potential difference (PD) of 40 to 50 mV (serosa positive to mucosa). The PD generated in the colon is much greater than in the jejunum or ileum. This difference has been attributed to the higher resistance of the tight junctions in the colon. Further-

more, sodium absorption and PD in the colon are increased by aldosterone, which has little or no effect on the small intestine.[36,117] On the other hand, glucose does not enhance sodium absorption in the colon as it does in the intestine. Potassium enters the colon by passive movement down its activity gradient.

In the passage of luminal content through the colon the concentrations of sodium and chloride in the fecal water are greatly decreased, whereas the concentrations of potassium and bicarbonate are increased. It has been postulated that a chloride-bicarbonate exchange mechanism similar to that described in the ileum is operative in the colon because both ions can move against electrochemical gradients.

GALLBLADDER. The gallbladder concentrates the bile produced by the liver to 10% to 20% of its original volume. Although the sum of the concentrations of the solutes in gallbladder bile may be twice that of plasma, the osmolalities are equal because macromolecular aggregates are formed. Because of the relative simplicity of its structure and its high rate of fluid transport, study of the gallbladder has provided important insight into water and electrolyte transport. It has been shown that (1) in the absence of an activity gradient between the mucosal and serosal surface of the gallbladder, water transport is directly proportional to NaCl transport, (2) sodium and chloride can be transported simultaneously against their electrochemical gradients and against the direction of water flow, (3) the solute transport mechanism requires sodium and chloride or bicarbonate, and (4) this active salt transport is accomplished without generating a transmucosal potential difference, an indication that sodium and chloride transports are tightly coupled.[26,27] These considerations correlate with the observation that when the gallbladder is absorbing salt and water, the lateral intracellular spaces are distended (in the absence of transport they are empty) and lead to the description of a model called the standing gradient osmotic flow (Fig. 53-13). It provides a convincing explanation for solute-linked water transport in the gallbladder and for other absorbing or secreting epithelia as well.

CARBOHYDRATE ABSORPTION

Carbohydrates account for approximately 50% of the calories in the diets of residents of Western Europe and the United States. In many tropical regions the percentage of carbohydrates is considerably higher. Starch makes up about 60% of the carbohydrate calories, sucrose 30%, and lactose (milk sugar) 10%.[45] Continued drinking of

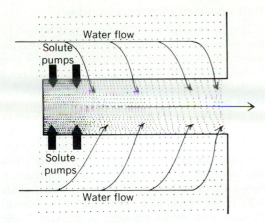

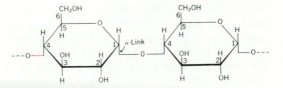

Fig. 53-14. Structure of 1,4-α-glucose links in amylose and amylopectin. Two adjacent glucose molecules are shown. Numbers refer to carbon atoms. (From Gray.[39])

Fig. 53-13. Standing gradient osmotic flow: model for fluid transport across epithelia. Solute pumped into closed end of long and narrow channel makes it hypertonic and pulls in water osmotically. Solute moves toward open end of channel as result of water flow and of diffusion down its concentration gradient. Osmotic equilibrium is progressively approached as water enters along length of channel until emergent solution is isotonic. Standing gradient in this flow system is continually maintained by active solute transport. For diagrammatic purposes, solute pump is indicated only at base of channel, but it may operate over a greater fraction of channel length. Given appropriate values for channel length and water permeability, model can also yield hypertonic solution of fixed osmolarity. Epithelial structures to which model may find application in understanding solute-linked water transport are lateral intercellular spaces (intestine and gallbladder), basal membrane infoldings (kidney, salt gland, ciliary body, etc.), and intracellular canaliculi (stomach). (From Diamond and Tormey.[26])

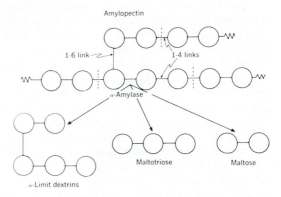

Fig. 53-15. Final products of amylopectin hydrolysis. Segment of amylopectin is shown. Each circle represents a glucose unit. Sites of α-amylase action are shown by dotted lines. (From Gray.[39])

milk, hence intake of lactose, in adult life is limited for the most part to people of European origin, for only they continue to have the enzyme necessary for the hydrolysis of lactose in their intestinal mucosa. The rest of the world's population loses the ability to hydrolyze lactose in childhood, as do all other mammals after weaning.

Hydrolysis of polysaccharides

Starches are polysaccharides (polyglucose) that serve as nutritional stores in plants; there are two types of compounds, amylose and amylopectin. Amylose accounts for 20% of the starch and is a straight chain of 25 to 2,000 glucose molecules linked by an oxygen bridge between carbon number 1 of one glucose molecule and carbon number 4 of the next (Fig. 53-14). Salivary and pancreatic α-amylases hydrolyze the interior 1,4-α-linkages; the final product is a mixture of maltose (two glucose molecules joined by a 1,4-α-link) and maltotriose (three glucose molecules similarly joined).[39]

Amylopectin is a very large branched polysaccharide with 6,000 or more glucose molecules and accounts for 80% of starch. The glucose molecules in the straight chains are joined by 1,4-α-links such as those found in amylose. The branch points that occur every 25 to 30 glucose molecules, however, are 1,6-α-linkages (Fig. 53-15). The 1,6-α-linkages as well as the adjacent 1,4-α-linkages resist hydrolysis by salivary and pancreatic amylases. The final products of amylase hydrolysis of amylopectin are maltose, maltotriose, and α-dextrins; the smallest is a pentaglucose, but on the average a dextrin contains 8 glucose molecules.

There are other polysaccharides in our diets, but they account for a very small percentage of our carbohydrate calories. Glycogen (animal starch) is a polyglucose resembling amylopectin, containing 1,700 to 22,000 glucose molecules.

Inulin, contained in artichokes, is a fructose polysaccharide. Agar is sulfated galactose polysaccharide, and other types of polysaccharides are contained in vegetable gums and fruit pectins. The most abundant vegetable polysaccharide is cellulose, which contains up to 2,500 glucose molecules joined by 1,4-β-linkage. Mammals have no enzyme that can hydrolyze this bond. However, cellulose is hydrolyzed by bacteria in the stomachs of ruminnants and in the cecum and colon of herbivores, making its glucose available for absorption.

Hydrolysis of starch is begun in the mouth by the α-amylase ptyalin contained in saliva. This enzyme is active between pH 4.0 and 11.0, with the optimum at pH 6.9. Its activation requires the presence of chloride ions. Hydrolysis of only 0.1% of the glucoside linkages in starch produces particles $1/100$ the original size, with a consequent decrease in viscosity. The extent to which salivary amylase hydrolyses starch depends on a number of factors: (1) degree of mixing with saliva in the mouth, (2) duration of contact in the mouth, and (3) time required for gastric juice to acidify the swallowed bolus and to inactivate ptyalin. As much as 50% of the starch ingested may be hydrolyzed by the action of salivary amylase before it is inactivated in the stomach.[84]

In the duodenum, pancreatic α-amylase attacks unhydrolyzed starch. The enzyme has the same pH optimum as salivary amylase and differs in action only in that it is active against raw as well as cooked starch. During a meal, 1 ml of duodenal juice contains sufficient amylase to hydrolyze 1 to 9 gm starch/hr.[22] Starch hydrolysis is very rapid in the duodenum; within 10 min of contact, hydrolysis is practically complete. This rapid hydrolysis converts the large starch molecules with trivial osmotic activity into a large number of small molecules; for example, if 10 gm of starch is hydrolyzed completely to glucose, the osmolality will increase by 60 mOsm and, if none of the glucose is absorbed, 200 ml of water will have to be added to restore the duodenal contents to isotonicity.[24] In addition to the high amylytic activity in the fluid in the lumen, amylase is found to be bound to the intestinal mucosal surface.[24] This bound amylase appears to have two origins; (1) that elaborated by the intestinal epithelium and (2) pancreatic amylase adsorbed onto the mucosa. Although there is sufficient amylytic activity in the luminal fluid to account for starch hydrolysis, it has been suggested that the physiologic function of pancreatic α-amylase is carried out on the surface of the intestinal epithelium, a process called "membrane digestion."[103]

Hydrolysis of oligosaccharides

It was once believed that the final hydrolytic step to monosaccharides also occurred within the intestinal lumen prior to absorption, and it was postulated that the disaccharidases were secreted by the intestinal glands, the crypts of Lieberkühn. It has now been demonstrated that these enzymes are part of the membrane of the brush border of the intestinal epithelium,[70] and that the disaccharidase found in the intestinal fluid (succus entericus) is derived from shed epithelial cells and has little or no functional role.[31] Disaccharidase activity is absent in the brush border of the cells in the crypts of Lieberkühn but appears as the cells migrate up the sides of the villi.[21] The disaccharidases (more exactly, oligosaccharidases) of the human intestine are listed in Table 53-3. Although several species of maltase, sucrose, and lactose have been described, it is not certain whether those of the first two enzymes exist in vivo as distinct proteins or whether their chemical differences are produced during isolation. Two lactases (β-galactosidases) with different pH optima have been separated from intestinal mucosa, but only the one with a pH optimum of 6.0 plays a role in the digestion of dietary lactose.

Evidence that the hydrolysis of the disaccharides occurs at the surface of the brush border rather than within the cell is of three types. (1) Molecules of the size and nature of disaccharides do not enter cells readily. (2) In lactase-deficient individuals, less than 1% of ingested lactose appears in urine. Since lactose given intravenously has been shown to be totally excreted in the urine, absorption of lactose must be very limited.[44,110] (3) In experiments in which the intestine is perfused with a disaccharide, considerable quantities of the hydrolysis products are found in the luminal fluid (Fig. 53-16).[41] The activity of disaccharidases requires sodium ions in much the same pattern as is required for active transport

Table 53-3. Intestinal mucosal oligosaccharidase

Enzyme	Substrate	Form in diet
α-Dextrinase*	α-Dextrins	Starch (amylopectin)
Maltase	Maltose, maltotriose	Starch (amylopectin, amylose)
Sucrase	Sucrose	Sucrose
Lactase	Lactose	Lactose
Trehalase	Trehalose	Trehalose

*This is an oligo-1,6-α-glucosidase, more commonly called isomaltose but, since isomaltose is not a physiologic substrate in the intestine, one of the other terms is preferred.

of monosaccharides. It has been suggested that the processes of hydrolysis and absorption may be mediated at a single site.[93] Recently feedback control of disaccharidase activity by the hydrolysis products has been suggested. Such a system limits the liberation of osmotically active molecules to the capacity of the transport system to remove them from the lumen.

The levels of oligosaccharidase activity are low in the first part of the duodenum, peak in the jejunum, and decrease in the ileum.[77] Two enzymes, lactase and trehalase, warrant special comment. Lactase is normally present in the intestine of human infants as well as other suckling mammals (a notable exception is the California sea lion, which has no lactose in its milk). As in other mammalian species, most humans lose intestinal lactase (the enzyme with pH 6.0 optimum) between infancy and adulthood and, as a consequence, can no longer hydrolyze lactose to the absorbable monosaccharides glucose and galactose. Lactase activity cannot be made to persist or to return by feeding lactose.[92] In a minority of the world's population, those individuals whose family origins are in Europe, lactase activity and the ability to hydrolyze lactose persists throughout life.[6] On the other hand, the activities of other oligosaccharidases can be altered by diets; changing from a sucrose-free diet

to one high in sucrose will lead to a doubling of sucrase and maltase activity.[86] The role of trehalase in the human mucosa is a mystery since its substrate, trehalose, is found primarily in insects and some mushrooms, and thus contributes little to our caloric intake.[92] Is it a relic that has persisted without purpose, or is its physiologic role yet to be discovered? Just as α-dextrinase (isomaltase) can hydrolyse isomaltose, although its real physiologic role is the hydrolysis of α-dextrins.

Absorption of monosaccharides

The primary monosaccharides presented to the intestinal mucosa are glucose, fructose, and galactose, with glucose accounting for 80% of the total load (Fig. 53-17). The absorption of monosaccharides presents two problems; first, the transport of hydrophilic monosaccharides is through the lipid cell membrane, which restricts the flow of water and hydrophilic solutes, and second, the monosaccharides are too large to enter the cells by simple diffusion through the "aqueous channels" in the membrane. To account for this absorption of hydrophilic solutes such as glucose, a carrier hypothesis has been developed.

Carrier hypothesis. The carrier hypothesis suggests that the molecule to be transported binds with a specific site on the carrier and that this complex then diffuses through the lipid cell membrane and dissociates at the inner border of the membrane, releasing the transported molecule into the aqueous interior of the cell. It is as though the carrier had covered up the hydrophilic groups of the transported molecule, making it temporarily lipid soluble. The extent of the combination between carrier and transported molecule (substrate) is proportional to the concentration of the substrate in the aqueous phase in contact with the membrane. If the concentration is greater on the outside than on the inside of the cell, the carrier will be more saturated on the outside surface than on the inner. Since both free and combined carrier are free to move back and forth through the membrane, the substrate will move into the cell by what has been called facilitated or carrier-mediated diffusion. The use of Michaelis-Menten kinetics has ordered the analysis of such transfer by the intestine:

1. $E + S \rightleftharpoons$ Enzyme-substrate complex $\rightarrow E + P$ (products)

2. $C + S \rightleftharpoons$ Carrier-substrate complex $\rightleftharpoons C + S$

where 1 represents the Michaelis-Menten hypothesis of enzyme kinetics and 2 represents its appli-

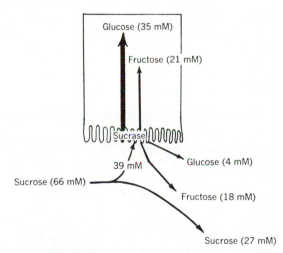

Fig. 53-16. Sucrose absorption in human intestinal perfusion experiments. Diagrammatic representation of what occurs over 30 cm segment of intestine when sucrose (73 mM) is infused. Values are expressed in millimoles per hour. Of glucose released by hydrolysis, 90% is captured at site of hydrolysis and transported into cell, whereas only 54% of fructose is absorbed at site of hydrolysis. (From Gray and Ingelfinger.[41])

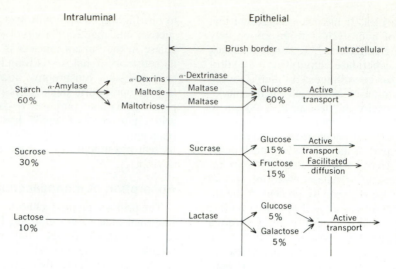

Fig. 53-17. Carbohydrate digestion and absorption. Percentages refer to fraction of total dietary carbohydrate. (Modified from Gray.[39])

cation to the carrier hypothesis. It should not be assumed that the carrier is necessarily an enzyme, since such kinetics require only that there is a rate-limiting step in the process under study. Although there is an intermediate carrier-substrate complex, the product of the reaction is the substrate delivered to the other side of the membrane.

MICHAELIS-MENTEN KINETICS. If Θ is the fraction of the carrier saturated and [S] is the concentration of the substrate in contact with the carrier, then:

$$\Theta = \frac{[S]}{[S] + K_t} \tag{1}$$

where K_t, the transport constant, is equal to the concentration of [S] at which the velocity of transport is one half the maximum velocity. It is a measure of the affinity of the carrier for the substrate. The smaller the constant, the greater the velocity at any given concentration. This relation holds for the outer and inner surfaces of the membrane where the concentrations of substrate are S_1 and S_2. If it is assumed that the rate of transfer depends on the difference between the fraction of carrier saturated at the two surfaces of the membrane, then the rate of transport (V) is expressed by the equation:

$$V = V_{max} \left(\frac{[S_1]}{[S_1] + K_t} - \frac{[S_2]}{[S_2] + K_t} \right) \tag{2}$$

where V_{max} is a constant equal to the maximal transport rate when $[S_2]$ is zero. It also is related to the amount of carrier and its rate of movement

through the membrane. If the concentration on one side is zero ($S_2 = 0$), then the equation takes the form:

$$V = V_{max} \left(\frac{[S_1]}{[S_1] + K_t} \right) \tag{3}$$

which states in effect that the rate of transport is proportional to the saturation of the carrier. When the reciprocal of the initial substrate concentration is plotted against the reciprocal of the uptake measured in a series of experiments (a Lineweaver-Burk plot), the values of V_{max} and K_t can be derived (Fig. 53-18). This provides a useful way to describe the effect of one substrate on the absorption of another and to determine whether they share the same carrier.

Rate of absorption of monosaccharides

It has been known for a long time that monosaccharides differ greatly in the rates at which they are absorbed from the intestine (Figs. 53-19 and 53-20). It had once been thought that monosaccharides could be divided into two classes, those that entered the cell by simple diffusion (passive absorption), for example, mannose, xylose, and arabinose, and those whose absorption was rapid, required energy, and could move against a concentration gradient (active absorption), for example, glucose and galactose. When the effect of one sugar on the absorption of another is measured, it appears that differing affinities for the carrier explains the differing rates of uptake better than the two separate mechanism hypothesis for monosaccharide absorp-

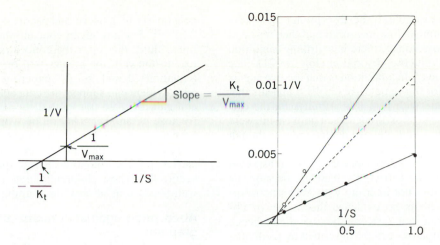

Fig. 53-18. Lineweaver-Burk plot. S, Substrate concentration; V, rate of transport; V_{max}, maximal velocity when carrier is saturated; K_t, transport constant (Michaelis constant, K_m), for example, substrate concentration at which $v = \frac{1}{2} V_{max}$. Use of Lineweaver-Burk plot is shown at right, illustrating inhibition of L-alanine transport by 2 mM L-leucine. ● = L-Alanine alone; ○ = L-alanine in presence of 2 mM L-leucine. V_{max} is same in presence and absence of inhibitor (leucine) of alanine transport. If concentration of inhibited amino acid is increased sufficiently, inhibition of its transport can be overcome. Hence L-alanine and L-leucine are competing for same transport mechanism. (From Matthews and Laster.[67])

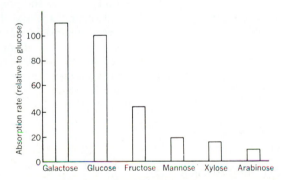

Fig. 53-19. Selective sugar absorption by unanesthetized rats. Sugar was fed by stomach tube, and absorption rate was determined as disappearance of sugar from gastrointestinal tract in a given period of time. (Based on data from Cori[14]; from Wilson.[112])

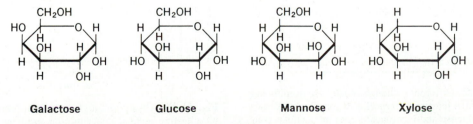

Fig. 53-20. Structural formulas of several sugars shown in Fig. 53-19. Seemingly relatively minor changes in configuration result in greatly altered affinity for carrier and hence in greatly altered transport rates.

tion.[3] For example, if the absorption of 6-deoxy-glucose alone is first measured and then repeated with glucose added, there is a striking inhibition of 6-deoxyglucose absorption (Fig. 53-21). The explanation of the inhibition is that the two sugars compete for the same carrier.[17,113] It appears that those sugars with very low affinities for the carrier can behave as though their movement were directed by simple passive diffusion, even though it is carrier mediated.[79]

Fructose. Fructose appears to be transported by a different carrier mechanism, because its presence does not interfere with glucose absorption.[35] There is no evidence that it can be absorbed against a concentration gradient; hence its absorption can best be described by facilitated or carrier-mediated diffusion. The rapidity of fructose absorption (Fig. 53-20) is explained in part by the maintenance of a low intracellular concentration of fructose, which results from its

conversion to glucose and lactic acid (Fig. 53-22).[81,112] For any given concentration in the luminal fluid, the lower the concentration of fructose in the cell, the greater will be the transport (equations 2 and 3). An experiment of nature adds evidence to support the concept that glucose and galactose are transported by one carrier and fructose by another. On rare occasions a child is born with an inability to absorb glucose or galactose, presumably because of the absence of the carrier protein (monosaccharide malabsorption syndrome). These children tolerate fructose quite well because its absorption is unimpaired.[63]

Absorption against concentration gradient

Although carrier-mediated diffusion can account for specificity and for differing rates of transfer, it cannot alone account for absorption against a gradient, as demonstrated most strikingly by glucose and galactose but also by a variety of other sugars (Fig. 53-23). The process has two characteristic features: (1) it is energy dependent, as demonstrated by its inhibition by dinitrophenol or the absence of oxygen, and (2) it depends on the presence of sodium in the mucosal fluid.[17,78,90] In the process, not only can glucose and other "actively absorbed sugars" be transported against a concentration gradient, but the absorption of sodium is increased as well. To explain these observations Crane[16] suggested the "ion gradient model." In essence, it proposes that sodium and glucose absorption are coupled

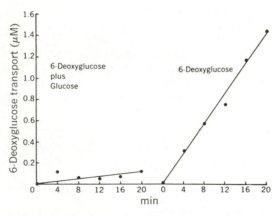

Fig. 53-21. Effect of glucose (12 mM) on transport of 6-deoxyglucose (1 mM). (From Wilson et al.[114])

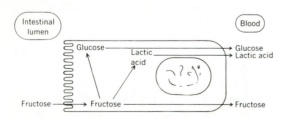

Fig. 53-22. Different pathways of fructose in epithelial cell of small intestine. Intracellular conversion of fructose to glucose and lactic acid keeps intracellular concentration of fructose low, maintaining favorable gradient across brush border for rapid transport. (From Wilson.[112])

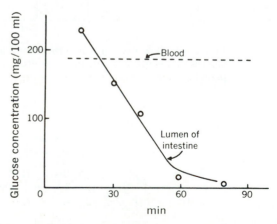

Fig. 53-23. Absorption of glucose from lumen of rabbit intestine against concentration gradient. Glucose solution containing sodium sulfate was placed in loop of intestine of rabbit anesthetized with ether. (Based on data from Bárány and Sperber[5]; from Wilson.[112])

as diagrammed in Fig. 53-24. In the brush border there is a carrier that must bind with sodium as well as with a transportable monosaccharide in order to be activated. Since the concentration of sodium within the cell is low, the concentration gradient for sodium may provide the driving force for monosaccharide absorption. Also, the low intracellular concentration limits the amount of sodium that is available to bind with the carrier when it is empty at the inner surface of the brush border membrane. This limits the transport of sodium and monosaccharide from the cell back into the lumen. It has been shown both by radioautography[57] and chemical analysis that the highest concentration of sugar is in the cell. Therefore the active concentrating mechanism is best attributed to the brush border carrier. In the absence of a monosaccharide (e.g., glucose), sodium enters the cell from the lumen down a concentration gradient either by diffusion, via another carrier, or by the sugar-sodium carrier that operates less efficiently in the absence of glucose. When glucose is added, the carrier is fully activated, leading to increased sodium absorption. The glycoside phlorizin has a very high affinity for the carrier but is not transported.[4] In low concentration, 10^{-4} or 10^{-5}M, it blocks glucose absorption and the associated increase in sodium absorption.

For the process of glucose absorption to continue, the sodium entering the intestinal cell with the glucose must be pumped out. This requires energy and is apparently mediated through Na^+, K^+-ATPase in the lateral and basal cell membrane. If this process that keeps intracellular sodium low is inhibited by the cardiac glycoside ouabain, then glucose and the associated sodium absorption cease.[18] The active transport of sodi-

um through the lateral cell membrane establishes a standing gradient that provides the osmotic force to move water and solutes from the cell into the lamina propria. How glucose gets out of the cell is not known, but since it is too large and hydrophilic to get in through the brush border by simple diffusion, this process cannot be invoked with confidence to explain its exit. It probably leaves via some type of carrier-mediated diffusion.[91]

PROTEIN ABSORPTION

Protein, unlike carbohydrate, is essential for growth of the young as well as maintenance of health in adults. The protein content of the average American diet is 75 gm/day or more. This is in excess of the minimum of 30 to 40 gm necessary to replace daily losses in an adult. Growing children and pregnant and lactating women, of course, have greater requirements; young children require as much as 4 gm/kg body weight for adequate growth. The proteins of the body are made up of 20 different amino acids, and only half of them can be synthesized by animal species; the remainder, the essential amino acids, must be provided in the diet. The relative proportion of amino acids in different proteins varies greatly. In general, animal protein, milk, and eggs have an optimal distribution of amino acids, whereas many vegetable proteins are deficient in one or another essential amino acid; for example, gliadin, the protein of wheat, is deficient in lysine and zein, or corn protein, is deficient in lysine and tryptophan. Even if the protein in the diet is adequate in amount, normal nutrition may not be maintained if it comes from a single vegetable source (Fig. 53-25).

Protein presented to the gut for hydrolysis and

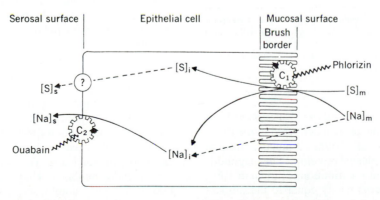

Fig. 53-24. Hypothetical model of Na-sugar interaction in distal rabbit ileum. See text. (From Schultz and Zalusky.[90])

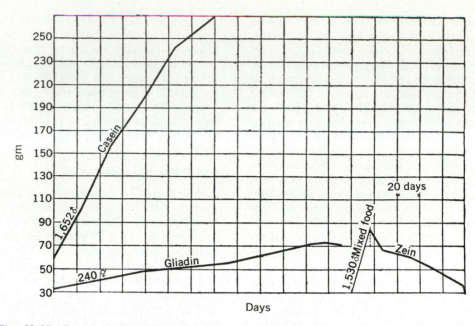

Fig. 53-25. Growth curves of rats maintained on diets containing single protein. Satisfactory growth is obtained on casein, a milk protein; little more than maintenance of body weight is possible on gliadin, a wheat protein (deficient in lysine); even maintenance of body weight is impossible on zein, a corn protein (deficient in lysine and tryptophan). (From Mendel.[69])

Table 53-4. Daily protein intake and turnover for adult man, expressed in grams per day

Daily dietary intake	60-120
Protein secreted into gastrointestinal tract	
Saliva	3
Gastric juice	5
Bile	<1
Pancreatic juice	8
Mucosal shedding*	50
TOTAL	67
Absorption—total presented for absorption	130-190
Fecal loss†	5-15
Total body protein synthesis	160

*Leblond and Walker[59] estimated that 250 gm of gastrointestinal epithelium is shed per day; 50 gm protein will be shed if assumed cells are 20% protein.
†Predominantly in the form of bacteria and desquamated cells and mucoproteins from distal ileum and colon. Many bacteria are not dependent on amino acids for their protein synthesis and can use urea or ammonia as nitrogen source.

absorption consists not only of that derived from food but also of the protein delivered to the intestine in the form of enzymes, mucus, desquamated cells, and plasma protein.[75] The magnitude of this endogenous contribution is shown in Table 53-4[74]; it is approximately equal to the amount derived from the diet and is approximately one third of the total daily synthesis of protein.

Hydrolysis of proteins

The hydrolysis of protein begins in the stomach with the action of pepsin. This enzyme is an endopeptidase, which means that it attacks the interior peptide bonds of protein. It is active in an acid pH, ranging from 1.0 to 4.0, depending on the substrate, and is most active against peptide bonds adjacent to aromatic amino acids (phenylalanine and tyrosine). During this process some free amino acids and polypeptides are produced.[99] The extent of the gastric digestion of protein is determined by the physical state of the ingested protein, the length of time that it stays in the stomach, and the activity of the gastric juice. At best, peptic hydrolysis is incomplete and ceases when the pH of the chyme is raised in the duodenum. Gastric hydrolysis is not an essential step in protein digestion, since individuals with achlorhydria have no impairment of protein digestion and absorption.

In the duodenum the proteins in the chyme are mixed with the proteolytic enzymes of the pancreas, trypsin, chymotrypsin, carboxypeptidases A and B, and elastase, as well as ribonuclease and deoxyribonuclease. Trypsin and chymotrypsin are endopeptidases. They break specific peptide bonds; for example, trypsin hydrolyses only the carbonyl bonds of amino acids with a positively charged side chain (e.g., lysine or argi-

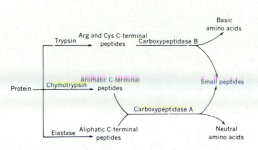

Fig. 53-26. Intraduodenal sequential action of pancreatic endopeptidases and exopeptidases on dietary protein. Final products at right are substrates that must be handled by intestinal cell. *Arg*, Arginine; *Cys*, cysteine. (From Gray and Cooper.[40])

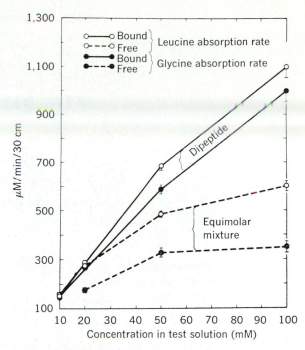

Fig. 53-27. Rates of intestinal absorption of leucine and glycine from solutions of increasing concentrations of the dipeptide Gly-Leu (bound) or from equimolar solutions of the constituent amino acids (free). Unequal absorption of leucine and glycine from dipeptide solution would not be expected if uptake was entirely in unhydrolyzed form via peptide carrier pathway. It appears that there are two pathways available. Major one is uptake of intact dipeptides with intracellular hydrolysis; secondary pathway involves surface hydrolysis at brush border and subsequent absorption via amino acid pathway. (From Adibi.[1])

nine) and chymotrypsin attacks only the carbonyl bonds of aromatic amino acids (e.g., phenylalanine and tyrosine). Carboxypeptidase A, on the other hand, is an exopeptidase that hydrolyses in sequence the terminal peptide bonds at the carboxyl end of the peptide chains. The combined action of the endopeptidases and exopeptidases results in a mixture of peptides two to six amino acids in length and free amino acids (Fig. 53-26).[40] The proportion of amino acids in the peptide form after a meal has been observed to be 4 times as great as in the free amino acid form.[2] However, the proportions of three amino acids—methionine, tyrosine, and arginine—were found to be greater in the free amino acid form. The peptidases of the brush border further hydrolyse the peptides to di- and tripeptides and to a much smaller extent to amino acids.[55,56]

Peptide absorption

Until recently, it was generally held that protein absorption was analogous to carbohydrate absorption in that the combination of intraluminal and brush border hydrolysis reduced the absorbate to its primary units, monosaccharides in the case of carbohydrates and amino acids from protein. Absorption was accomplished by specific carriers for amino acids as well as for monosaccharides. This view had to be modified when it was shown that amino acids are absorbed more rapidly from a solution of di- and tripeptides than from an equimolar solution of their constituent amino acids[65,66] (Fig. 53-27). Most evidence suggests that there is a single transport system for a wide variety of di- and tripeptides. Di- and tripeptides compete for this carrier, but tetrapep-

tides do not.[96] Also, amino acids do not interfere with peptide transport by this mechanism.

This peptide transport mechanism has characteristics of an "active carrier-mediated" process: (1) transport against an electrochemical gradient can be shown by using peptides that can be transported but are resistant to hydrolysis in the cell, for example, carnosine (β-alanylhistidine), glycylsarcosine (Gly-Sar), and Gly-Sar-Sar; (2) the peptide transport process is sodium dependent; and (3) peptide transport is inhibited by metabolic inhibitors. The current view is that peptide uptake is coupled to sodium transport. In addition, the process is facilitated because most peptides, once in the cell, are rapidly hydrolyzed to their constituent amino acid, thus maintaining a favorable gradient for uptake.

The localization and activities of peptide hydrolases (peptidases) in the intestinal cell is of

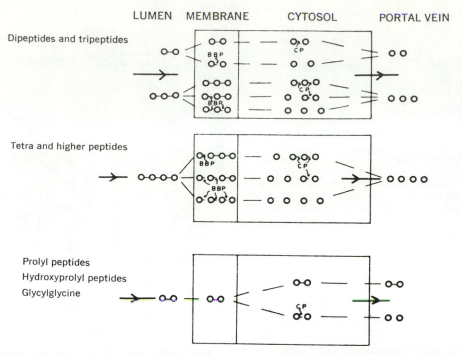

Fig. 53-28. Schematic representation of peptide absorption. In upper figure, di- and tripeptides enter intestinal epithelial cell intact via peptide carrier, hydrolyzed by cytosal peptide hydrolases (CP), and enter portal blood as amino acids. Small portion of di- and tripeptides is hydrolyzed by brush border peptide hydrolases (BBP) to amino acids, which then enter cell via amino acid carriers. Middle figure illustrates that tetra- and higher peptides cannot enter cell intact but enter as amino acids, di- and tripeptides after hydrolysis by brush border peptide hydrolases. Lower figure shows that prolyl, hydroxylprolyl, and glycine dipeptides are absorbed intact, but a portion of these dipeptides escapes intracellular hydrolysis and appears in portal blood as peptides. *BBP,* Brush border peptide hydrolase; *CP,* cytosol peptide hydrolase. (From Kim et al.[56])

importance for an understanding of protein hydrolysis and absorption. These enzymes are in the cytosol and the brush border. The majority of intestinal di- and tripeptidase activity is found in the cytosol, whereas this fraction has little or no activity against tetra- and higher peptides[55,56] (Fig. 53-28). These observations provide support for the view that intraluminal hydrolysis of proteins produces a mixture of small peptides and amino acids. Tetra- and higher peptides are hydrolyzed by brush border peptide hydrolases to di- and tripeptides and, to a lesser extent, to amino acids. The di- and tripeptides are transported intact into the cytosol, where the hydrolysis to amino acids is completed. The end products of protein hydrolysis that appear in portal blood are almost entirely amino acids. Hydroxyproline peptides are the exception, for it has been observed that as much as 8% of hydroxyproline peptides ingested as gelatin, a protein with a high hydroxyproline content, appears as hydroxypro-line peptides in the blood and subsequently in the urine.

Amino acid absorption

Absorption of amino acids had been attributed in the past to simple diffusion because it was observed that their rates of absorption correlated with the reciprocal of their molecular size (Fig. 53-29).[58] This notion had to be abandoned, however, when it was shown that the absorption of amino acids showed striking stereospecificity for the L-stereoisomers, the naturally occurring amino acids, being absorbed at a much greater rate than the D-isomers (Fig. 53-30).[112] In addition, the L-amino acids can be transported against striking gradients, whereas the D forms cannot (Fig. 53-31); finally, amino acid transport exhibits saturation kinetics, implying a rate-limiting or carrier-mediated step (Fig. 53-32).[68] The structural features that determine the affinity for the transport site are (1) the α-amino group—

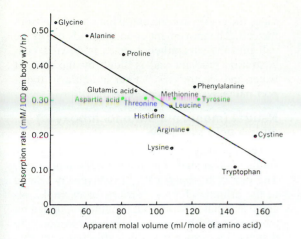

Fig. 53-29. Relation between rate of absorption of amino acids from chick intestine and apparent molal volume of amino acid. (From Kratzer.[58])

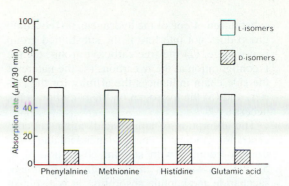

Fig. 53-30. Selective absorption of stereoisomers of amino acids from loops of rat intestine in vivo; 1 ml of racemic mixture of amino acid (0.2 to 0.4 M) was introduced into loop of intestine of anesthetized rat. After 0.5 to 1 hr, loss of amino acid from loop was determined by enzymatic methods. (Based on data from Gibson and Wiseman[37]; from Wilson.[112])

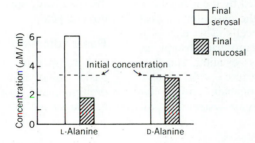

Fig. 53-31. Active transport of L-alanine across rat intestine in vitro. Racemic mixture of alanine was placed at equal concentration on both sides of rat intestine. Following incubation for 1 hr, amino acids on the two sides were determined by specific L-amino acid oxidase and D-amino acid oxidase.[105] (Based on data from Wiseman[115]; from Wilson.[112])

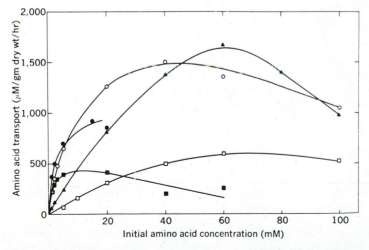

Fig. 53-32. Relationship between initial amino acid concentration in mucosal and serosal fluids and transport rate into serosal fluid. ▲ = Glycine; ○ = L-alanine; ● = L-valine; ■ = L-leucine; □ = AIB. (From Matthews and Laster.[68])

substitution of one of the hydrogens on the amino group does not interfere if the substituted group is not large, (2) the free carboxyl group, (3) the L configuration at the α-carbon, (4) the nature of the side chain influences the affinity for the carrier, and (5) the hydrogen on the α-carbon is not necessary.[67]

The absorption of amino acids against a concentration gradient is energy dependent and functionally coupled with sodium absorption (Fig. 53-33). As was seen in the examples of monosaccharide and peptide absorption, it is also necessary for the amino acid carriers to be activated by combination with sodium to form a ternary complex. Intracellular sodium concentration is lower than is the extracellular concentration, and therefore the complex will be less stable on the inside surface of the membrane; hence dissociation and uptake will occur. The energy required for transport from lumen to subepithelial space and into the blood is provided by the mechanism that pumps sodium out of the basolateral aspect of the intestinal absorptive cells. This later process, and the associated amino acid absorption, is inhibited by anaerobiasis, cyanide, and 2,4-dinitrophenol, which interfere with aerobic energy production, and by ouabain, an inhibitor of Na^+,K^+-ATPase, the enzyme system that pro-

vides the energy for the extrusion of sodium from the cell. The movement of amino acids out of the cell into the subepithelial space is down a concentration gradient and is aided by the flow of water and electrolytes.[19] Four amino acid transport pathways have been described.[116]

Neutral (monoamino-monocarboxylic) amino acid pathway

The pathway that has been most extensively studied is that shared by the neutral amino acids. The rate of transport (V_{max}) is inversely related to the affinities for the carrier; for example, alanine associates poorly but is quickly released by the carrier and hence has a higher transport rate than leucine, which has a high affinity for the carrier and is released much less rapidly. As expected, leucine is a more potent inhibitor of the absorption of other amino acids than is alanine. On the other hand, amino acids that do not share this pathway, such as dibasic amino acids, do not compete.

Basic (diamino) amino acid pathway

A transport mechanism has been shown for the basic amino acids (e.g., lysine, arginine and ornithine, and cystine; the latter also uses the neutral amino acid pathway).[45] In addition to

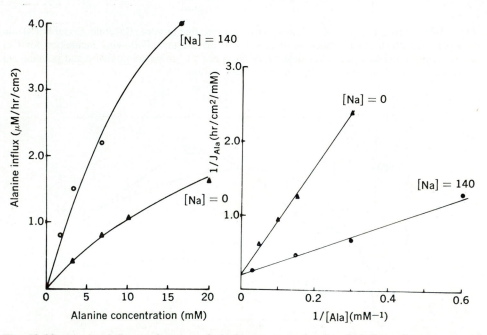

Fig. 53-33. Alanine influx as function of alanine concentration in rabbit ileum in presence and absence of sodium. At all intraluminal concentrations, alanine absorption is enhanced by presence of sodium. At right, if reciprocal of influx is plotted against reciprocal of alanine concentration, straight lines are obtained. In this way, alanine influx can be described as function of concentration. (From Curran.[19])

competitive inhibition within this group, the pathway is inhibited by several neutral amino acids, most notably methionine. The basic amino acid pathway operates at a rate one tenth to one twentieth of that of the most rapidly absorbed neutral amino acids (e.g., glycine and alanine). This difference may be explained in part by virtue of their charge, basic amino acids being transported against an electrical gradient.

Proline and other pathways

A third transport mechanism is shared by proline, hydroxyproline, and the N-methyl–substituted derivates of glycine, sarcosine, N-dimethyl glycine, and betaine.[62] Proline and hydroxyproline are absorbed quite well by the neutral amino pathway but have a higher affinity for the "betaine carrier."

Two other pathways have been suggested, but opinion is divided concerning the interpretation of the data on which they are based.[117] It has been suggested that one pathway transports glycine and thereby provides an explanation of the apparent anomalies of competitive inhibition between proline, methionine, and glycine. The other pathway has been proposed to provide an additional path for the neutral amino acids valine, leucine, and isoleucine.

Acidic (dicarboxylic) amino acids

The preceding discussion leaves the "acidic" amino acids, that is, glutamic and aspartic acids, unaccounted for. It is not possible to demonstrate active transport for these two amino acids by the technique employed with the others, since no concentration gradient develops across the membrane.[116] On the other hand, during the absorption of glutamic acid an increase in alanine is found on the serosal side (Fig. 53-34). It appears that on entering the cell, most of the dicarboxylic amino acids exchange their amino group by transamination with pyruvate to form alanine and a keto acid. Presumably, the keto acids formed are metabolized within the intestinal cell, since their concentration in mesenteric blood does not rise during absorption of dicarboxylic amino acids.

Amino acids, as well as monosaccharides, enter the circulation via the portal blood, primarily because the volume flowing through the intestinal capillary bed is so much greater than the volume of lymph flow.[46]

Absorption of intact protein

In many species there is little or no transfer of immune globulin across the placenta. "Pas-

sive immunization" of the newborn occurs by means of the absorption of intact immunoglobulins contained in colostrum. The absorption of antibodies and other proteins is accomplished by pinocytosis, in which microglobules of protein are engulfed by invaginations of the plasma cell membrane, which in turn pinch off and move into the cytoplasm.[72] The globules coalesce and move to the base of the cell, where the ingested protein is transferred to the subepithelial space. This process terminates from 1 to 18 days after birth, varying with the species. In man and other primates, this process does not provide antibody protection for the newborn to any appreciable extent. Small amounts of proteins, however, must get across the mucosa intact, as evidenced by the existence of antibodies to various food proteins. Recently, it has been shown that soluble antigens are absorbed intact by adult animals. Specific secretory immunoglobulin A (IgA) in the lumen and mucus coat overlying the intestinal epithelium prevents attachment of the antigen to the epithelial cells and subsequent pinocytosis.[108] In addition, the binding of IgA to intraluminal antigen appears to permit more complete degradation of the antigen by pancreatic enzymes.

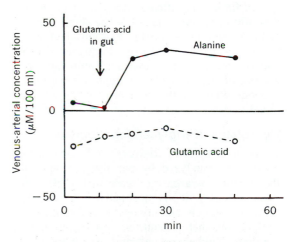

Fig. 53-34. Production of alanine during glutamic acid absorption. Amino acid concentration was determined in an artery and mesenteric vein of an anesthetized dog. Broken line indicates difference in concentration of glutamic acid; solid line indicates difference in concentration of alanine; arrow shows time at which 10 ml of 0.15% glutamic acid solution was introduced into lumen of intestine.[105] (Redrawn from Neame and Wiseman[76]; from Wilson.[112])

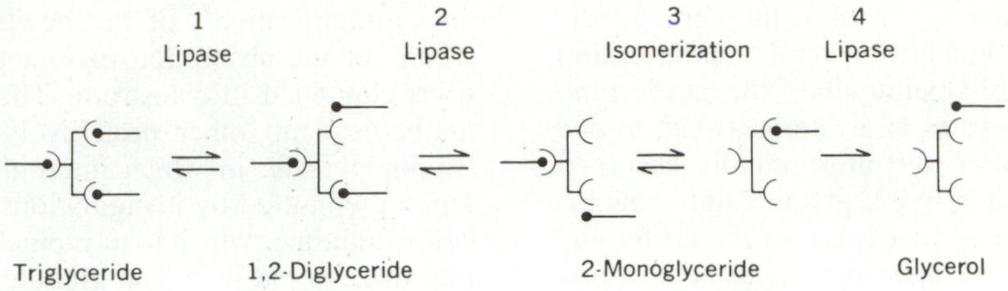

Fig. 53-35. Steps in lipolysis. See text.

LIPID ABSORPTION

Lipids include a variety of compounds grouped together on the basis of their solubilities. Dietary lipids of importance include triglycerides, which account for 90% of dietary lipid, cholesterol esters, phospholipids, and the fat soluble vitamins A, D, E, and K. Lipids are the most concentrated source of calories in the diet, yielding more than twice as many calories per gram as other nutrients (9 calories/gm as compared to 4 calories/gm for carbohydrates). However, if small amounts of polyunsaturated fatty acids and fat-soluble vitamins are provided, normal nutrition can be maintained on a lipid-free diet, since the necessary lipids can be synthesized from other nutrients. Such a diet, however, is unpalatable. On the average, 30% to 40% of calories in the diet are lipid, but fat intake varies widely, ranging from 25 to 160 gm/day in different ethnic and socioeconomic groups. The absorption of fat is very efficient; it is completed in the first 100 cm of the jejunum.[8] The small amount of lipid found in the stool (2 to 4 gm/day) originates to a major extent from desquamated cells and colonic bacteria. Lipid absorption involves a series of steps to make the lipids soluble in the aqueous phase of the luminal contents and the blood.

Hydrolysis of triglycerides (lipolysis)

The hydrolysis of triglycerides begins in the duodenum, catalyzed by pancreatic lipase, although there is a gastric lipase with a pH optimum of 5.5, which is most active against tributyrin and without activity against the long-chain triglycerides that account for most of the lipid in our diets. Unlike most enzymes, which are active in the aqueous phase, pancreatic lipase acts at the lipid-water interface and is inactive if the triglyceride substrate is in solution.[11,25] Lipase is protected from inactivation by bile salts by a pancreatic protein colipase.[11]

Bile salts and calcium maintain lipase activity by carrying the products of hydrolysis away into the aqueous phase. The rate of hydrolysis is de-termined by the amount of lipase present in the area of water-lipid interface and the concentration of the hydrolysis products remaining at the interface. The interface available for lipase action is increased by emulsification of the ingested lipids produced by the churning action of the gastric antrum and duodenum and by the emulsifying agents such as proteins, lecithin, and lysolecithin.

Lipolysis occurs in four steps (Fig. 53-35). The first two are rapid, but the third and fourth are very slow, proceeding only at 10% of the rate of the first two steps; hence they occur to a very limited extent in vivo.

Lipase demonstrates substrate specificity: the rate of hydrolysis is low for triglycerides with acetyl (C_2) chains in the 1 positions, is maximal for proprionyl (C_3) and butyryl (C_4), and decreases to a constant plateau for lauryl (C_{12}) and longer chains. If the rate of hydrolysis of triglyceride with C_{12} and longer chains is taken as 1.0, the rate for triglyceride with acetyl chains is 0.2, and for proprionyl and butyryl chains is 2.5.[25] Two additional pancreatic enzymes are involved in lipolysis: phospholipase, which hydrolyzes lecithin at the 2 position, yielding 1-lysolecithin and fatty acid, and an ester hydrolase that hydrolyzes cholesterol esters and other lipid esters.

Solubilization of lipolytic products

Bile salts are necessary for the efficient transport of the water-insoluble products of lipolysis from the lipid droplets in the lumen to the absorbing epithelial cells of the intestine. Two bile acids are synthesized from cholesterol in the liver of man, cholic and chenodeoxycholic acids. They account for 80% of the bile acids in bile; the remaining 20% is in the form of deoxycholic acid, a secondary bile acid formed by bacterial removal of the 7-α-OH group of cholic acid. The secondary bile acid is absorbed in the distal bowel and resecreted in the bile (Fig. 53-36). The bile acids are secreted as conjugates of glycine or taurine, with glycine conjugates accounting for 75% of the bile acids. Since the taurine

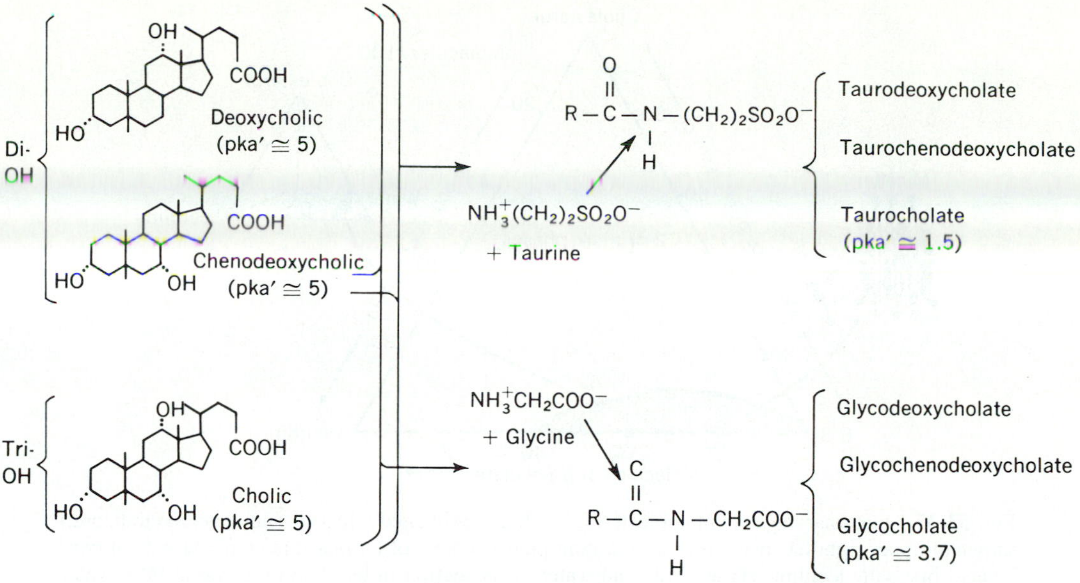

Fig. 53-36. Predominant bile salts of human bile. Two primary bile acids, cholic and chenodeoxycholic, and the secondary bile acid, deoxycholic, are conjugated with glycine or taurine in peptide bond resistant to all digestive enzymes. Conjugation alters solubility properties: taurine conjugates are soluble at all pH values, glycine conjugates precipitate from solution below pH 4 to 5, and unconjugated bile salts precipitate from solution below pH 7. pK_a values given are those of bile salts below their critical micellar concentration. When bile salts are in micellar form, pK_a values will be about 1 unit higher. Mean bile salt composition of bile in healthy man has been reported to be (moles%): glycocholate, 30%; glycochenodeoxycholate, 30%; glycodeoxycholate, 15%; taurocholate, 10%; taurochenodeoxycholate, 10%; taurodeoxycholate, 5%. (From Hofmann.[51])

conjugates have low pK_a values, they will be in the form of ionized salts at intestinal pH and therefore will not be absorbed until they reach the level of the active transport mechanism in the ileum.[98] A significant amount of the glycine conjugates are absorbed in the jejunum by passive nonionic diffusion; because they have a higher pK_a, a larger fraction is un-ionized at intestinal pH.

Bile salts function as ionic detergents. At low concentrations, conjugated bile salts form molecular solutions and, as the concentration rises, they form micelles or aggregates with their polar groups facing the aqueous solution; as a consequence, a nonpolar core is formed.[52] In bile the micelles contain not only bile salts but also lecithin, a polar lipid, and cholesterol, a nonpolar lipid—hence the term "mixed micelles." In the intestinal lumen the micelle adds other polar lipids, fatty acids, and monoglycerides. The resulting mixed micelle has a nonpolar hydrocarbon center in which nonpolar lipids such as cholesterol and the fat-soluble vitamins are effectively solubilized (Fig. 53-37). Mixed mi-

celles form when the bile salt concentration reaches about 2 mM, a level that is called the critical micellar concentration. Micellar solubilization increases the concentration of the lipolytic products in the aqueous phase by a factor of 1,000. The micelle, being larger than a fatty acid molecule, diffuses somewhat more slowly but nevertheless increases the rate of delivery of fatty acids to the absorptive epithelium by 100 to 200 times. Such a micellar solution is water clear and stable. When the micellar and lipid phases are separated by ultracentrifugation, it is found that the lipid in the micellar solution is predominantly fatty acids and monoglyceride, whereas practically all the triglycerides and diglycerides are in the oil phase (Fig. 53-38).[51] The lipolytic products of medium-chain triglycerides (MCT) are water soluble and hence do not require micellar solubilization for absorption. The fatty acids of MCT are 8 to 12 carbon atoms in length as compared to fatty acids with chain length of 14 atoms or greater found in long-chain triglycerides. Almost all the triglycerides in the diet are long-chain triglycerides.[43]

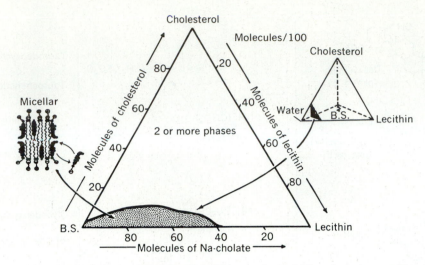

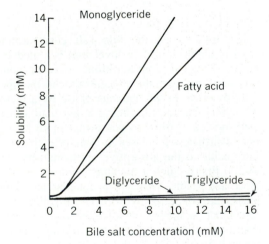

Fig. 53-37. Quaternary system of cholesterol, lecithin, conjugated bile salt, and water. Tetrahedron shown at upper right has been used to represent physical state of all possible combinations of conjugated bile salt, lecithin, cholesterol, and water. This section in tetrahedron taken at 90% water results in triangular phase diagram that has been enlarged and is shown at left. Diagram shows physical state of all possible combinations of bile salt, lecithin, and cholesterol in aqueous solutions containing a total of 10% solids and 90% water. Small zone at bottom of triangular diagram represents mixtures that form single micellar phase. Structure of these mixed micelles of bile salt, lecithin, and cholesterol is given in inset at left. It is very similar to structure of lecithin bile salt micelles except that cholesterol is interdigitated between lecithin molecules. Any mixtures having composition falling above this line separate into two or more phases. Phases are continuous micellar phase, cholesterol crystals, and mixed liquid crystals of lecithin and cholesterol. (From Small.[97])

Fig. 53-38. Solubility in bile salt solution of four classes of lipids present in lumen of small intestine during digestion and absorption of fat. Values were obtained with triolein, diolein, mono-olein, and oleic acid under conditions similar to those present in jejunal lumen during digestion (pH 6.3; 37° C; Na+, 0.15M). When present in excess, diglyceride and triglyceride form separate oil phase that appears as crude emulsion. Monoglyceride and fatty acid (if partially ionized) form liquid crystalline state, together with bile salts. Solubility curves for monoglyceride and fatty acid are edges of micellar region if phase diagram is used to depict phase equilibriums. (From Hofmann.[51])

As the micelles approach the brush border of the absorptive cells, the lipid products are thought to enter molecular solution, and dissolve in the lipid plasma cell membrane of the cell, and thus enter the cell by passive nonionic diffusion. This can be very efficient, for it is thought that the exchange of monoglycerides and fatty acids between the micelles and molecular solution is very rapid, any lipid molecule spending only 10 msec in a particular micelle. The rate of transfer depends on the solubility of the lipid in the cell membrane and the concentration gradient across the membrane. The emptied bile salt micelle then picks up another load of lipid hydrolysis products for delivery to the absorbing cells.

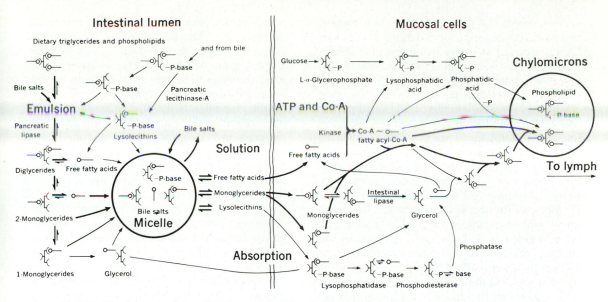

Fig. 53-39. Scheme of fat digestion, absorption, and resynthesis. Heavy arrows indicate more important pathways. (From Davenport.[24])

Intracellular phase

A favorable concentration gradient for diffusion of fatty acids and monoglycerides through the cell membrane is maintained by their rapid resynthesis into triglycerides within the cell (Fig. 53-39). Uptake and transfer of fatty acids to the site of resynthesis is believed to be facilitated by a fatty acid–binding protein. The endoplasmic reticulum contains the enzymes and cofactors necessary for the conversion of fatty acids to their CoA derivatives, acyl CoA synthetase, ATP, coA, and Mg^{++}. Mediated by monoglyceride transacylase, the fatty acyl CoA reacts with 2-monoglyceride to give first the diglyceride. The third fatty acid is added by diglyceride transacylase to complete the resynthesis of triglyceride. There is no exchange of the fatty acid in the 2 position in 75% to 80% of the monoglyceride absorbed and resynthesized into triglyceride; on the other hand, there is striking randomization of the fatty acids in the 1 positions of triglycerides and phospholipids.

In addition to the monoglyceride acylation pathway outlined previously, the intestinal cell can convert glucose to glycerol phosphate, which in turn is esterified via CoA with fatty acid to form phosphatidic acid, which may be converted to triglyceride. This phosphatidic acid pathway normally plays a minor role (Fig. 53-39). Phospholipids may be synthesized from glycerol or from absorbed lysolecithin.

Chylomicron formation—transport into lymph

The final step in lipid transport by the intestine is to convert the absorbed, resynthesized lipids into a water-soluble form that can be transported in the lymph and ultimately in the blood. Chylomicrons perform this function. Their chemical composition is approximately 80% triglycerides, 9% phosphatides, 3% cholesterol and cholesterol esters, and 2% protein. The protein is an essential component in the "packaging" of the lipid for transport. In its absence, as observed in patients with a rare congenital disorder (a-β-lipoproteinemia) and in animals given inhibitors of protein synthesis, the intestinal epithelial cells become engorged with absorbed fat that cannot be transported across the lateral and basal membrane of the cell.[87] The monoglycerides and fatty acids produced by lipolysis of MCTs are not resynthesized in the intestine and, because they are water soluble, enter the circulation directly via the portal blood rather than via the lymphatics. Just as with the other nutrients discussed, little is known of the process by which lipid leaves the cell. Once in the extracellular space, it gains entrance to the lacteals through fenestra and flows to the thoracic duct and into the venous return to the heart.

Absorption of steroids

Cholesterol. Cholesterol is presented for absorption from three sources: bile, diet, and desquamated cells. Of these, bile is the most important source, accounting for two thirds of the total. Cholesterol and its esters are almost insoluble in water and are found in the lipid emulsion phase. Esterified cholesterol is hydrolyzed by an ester hydrolase secreted in the pancreatic juice. Free cholesterol is taken into mixed bile salt micelles and delivered to the absorbing cells of the intestine. Cholesterol leaves the micelles to enter the intestinal cells less readily than do the monoglycerides and fatty acids. With no fat in the diet, little cholesterol is absorbed because it is not very soluble in simple bile salt micelles.

Once within the epithelial cell, cholesterol mixes with the intracellular pool of cholesterol, is reesterified with fatty acids, and enters the chylomicrons, which also contain small amounts of free cholesterol. The intestinal mucosa of the ileum is active in the de novo synthesis of cholesterol. Cholesterol synthesis in the intestine is not altered by fasting or cholesterol feeding as it is in the liver, but it is inhibited by feeding bile salts and is increased greatly when the bile is diverted from the intestinal tract.[28,111]

Bile salts. Absorption of bile salts during fat digestion and absorption is very limited in the upper intestine. However, in the ileum there is an efficient active transport mechanism for the reabsorption of bile salts. Less than 5% of the bile salts entering the ileum are lost into the colon. The absorbed bile salts are carried to the liver via the portal blood where they are totally removed and reexcreted into the bile. In the course of the digestion and absorption of a meal the entire bile salt pool may make two or more passes through the enterohepatic circulation.[29] The liver maintains the pool size by synthesizing new bile salts to replace those that are lost. This synthetic capacity of the liver is limited, however, and with failure of ileal reabsorption, the bile salt pool may decrease so that the concentration of bile salts in the jejunum falls below the critical micellar concentration, with resultant impairment of lipid absorption.

Vitamin D and other fat-soluble vitamins. β-Carotene, the dietary precursor of vitamin A, and viatmins D, E, and K are quite insoluble in aqueous solutions, so that their absorption depends on the presence of bile salts. Their solubility, as is the case with cholesterol, is greater in mixed than pure bile salt micelles. Fat-soluble vitamins, with the exception of vitamin E, function in the synthesis of specific proteins.

Drugs

The alimentary canal receives a number of diverse chemical compounds ingested as therapeutic agents, sedatives, analgesics, antibiotics, diuretics, etc. The absorption of most of these substances from the stomach, small intestine, and colon can be accounted for by passive, nonionic diffusion. The rate of absorption by nonionic diffusion is determined by three factors: (1) the proportion in the un-ionized form, (2) the lipid solubility of the un-ionized form and (3) the pH gradient between the lumen and the interior of the cell. Drugs having pK_as between 3 and 8 are effectively absorbed by passive nonionic diffusion if the un-ionized form has appreciable lipid solubility. It has been estimated that if as much as $1/100$ to $1/1,000$ of the drug is in the un-ionized form, effective absorption will take place, for as the un-ionized drug is absorbed, equilibrium will be shifted and more drug will enter the un-ionized form.

Strongly ionized substances are not lipid soluble, so if they do not have a small molecular size or a specific transport mechanism, they will not be absorbed and will hold water in the intestinal lumen, thus being effective cathartics.[95]

Absorption of divalent ions

Iron. Iron is an essential element in the oxygen-transporting pigments. The total body iron in an adult is about 4 gm, and daily losses are of the order of 0.5 to 1.0 mg in the form of desquamated cells, minor blood loss into the gastrointestinal tract, and menstrual blood loss. These losses are balanced by absorption of the 10 to 15 mg of iron that is ingested daily, primarily in the form of iron complexes.

The availability of iron varies widely with the source; for example, iron in liver and meat is better absorbed than that in eggs and green vegetables, whereas iron in wheat, corn, and beans is relatively unavailable for absorption. Iron from most dietary sources is thought to be split from its organic complexes prior to entry into the mucosal cell. Hemoglobin and myoglobin iron, however, enters the cell as heme, and iron is then split off by heme oxygenase and enters the intracellular iron pool. In the fasting state, inorganic and heme iron are absorbed with equal efficiency. Heme iron, however, is much more effectively absorbed from a meal. The intraluminal environment determines the availability of iron for absorption. Anything that promotes solubility of iron in the alkaline environment of the duodenum and upper jejunum will promote iron absorption; for example, ascorbic acid and fructose form soluble chelates, whereas formation of insoluble iron salts or chelates will interfere with absorption (e.g., phosphates, oxalate, and phytates). Ferrous iron is absorbed 3 times

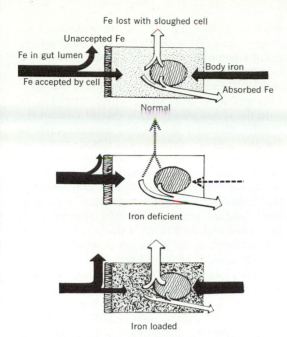

Fig. 53-40. Concept of control of iron absorption by intestinal mucosa. It is predicted that iron absorption is regulated primarily through columnar epithelium of small intestine. In normal, iron-replete subjects, mucosal cells may contain variable amount of iron supplied from body store. This deposit regulates—within limits—quantity of iron that can enter cell from gut lumen. After dietary iron has entered cell, it may proceed into body to fulfill a requirement. Alternatively, a portion of iron may become fixed in epithelial cytoplasm to be lost when cell is sloughed at end of its lifespan. In iron-deficient subjects there appears to be little inhibition to entrance of iron into villous epithelial cells and no capacity to retain it. Thus dietary iron readily proceeds into body. In iron-loaded subjects, body iron incorporated in epithelial cells is eventually lost, but, during life-span of cells, its presence inhibits entrance of dietary iron into cells. (From Conrad and Crosby.[13])

as efficiently as ferric iron, due in part to the lower solubility of ferric salts at duodenal pH.

Iron absorption occurs primarily in the duodenum and upper jejunum and takes place in two steps, mucosal uptake and transfer to plasma. Both steps depend on oxidative glycolysis. Uptake appears to have rate-limiting features, as it is greater in iron-deficient individuals and less in individuals with iron overload and is presumably mediated by a membrane carrier. The barrier to excess iron absorption is not at the brush border or in the uptake step but rather in the cell itself.

The iron content of the epithelial cells is determined by the level of circulating plasma iron at the time the cells divide in the crypts. There appear to be two pools of iron in the epithelial cell, one as ferritin and the other as tri- and divalent iron that has been absorbed from the lumen. The amount of divalent iron transferred to transferrin, the iron-carrying protein in the blood, is determined by the amount of ferritin in the cell. The mucosal cells of an iron-deficient individual will not have received much iron when the cell was formed, hence uptake is greater than normal, as is transfer to the plasma. Since little iron is retained in the cell, little is lost as the cell sloughs off the tip of the villus into the lumen. On the other hand, if the individual is iron overloaded, the intestinal cells will have more iron than normal. As a result, iron uptake will be decreased, less iron is transferred to the plasma, and more is lost into the lumen when the cell is shed at the end of its functional life[10,13] (Fig. 53-40).

Calcium. Milk is the richest source of dietary calcium, 1 pint containing about 500 mg, an amount that equals or exceeds the calcium in the remainder of the daily food intake. To the dietary intake of calcium, about 200 mg is added in the digestive juices. Calcium absorption increases with dietary intake to reach a plateau of 10 to 15 mg/kg/day when the dietary intake reaches 50 mg/kg/day.[80]

In order that it may be absorbed, calcium must be in solution and ionized; this means that bound calcium such as calcium caseinate (from milk) must be freed, and insoluble calcium salts such as phosphate and carbonates must be solubilized. A maximum amount of calcium is in the absorbable form at pH 5.0 to 7.0. Lactose in some unknown way promotes calcium absorption.

Calcium absorption occurs in the duodenum by an active process that can transport it against a concentration gradient; it is transported preferentially over other divalent ions such as magnesium and strontium. The entry into the cell appears to be carrier mediated and occurs down an electrochemical gradient. The transport from the cell to the subepithelial space is "uphill" against an electrochemical gradient. Both steps require vitamin D, but to be effective the vitamin must be given in vivo and its effect becomes manifest only after several hours.[88] Parathyroid hormone is also necessary for optimal function of the calcium transport system. The more distal intestine also absorbs calcium by a low-affinity, low-capacity mechanism. Although less specific, it is important because of its greater surface area and contact time.

Absorption of water-soluble vitamins

The water-soluble vitamin group of essential nutrients ranges from relatively simple molecules such as ascorbic acid (vitamin C, mol wt 176) to cyanocobalamin (vitamin B_{12}, mol wt more than 1,300). As might be anticipated, their absorption is also variable but not in a manner that can be predicted from molecular size. Except for ascorbic acid, the water-soluble vitamins function as essential cofactors for enzymatic reactions.

Vitamin B_{12}. Vitamin B_{12} is synthesized by bacteria, but our primary dietary sources are meat and seafood. This vitamin, which is essential for deoxyribonucleic acid (DNA) synthesis, is stored in the liver at concentrations ranging from 50 to 100 $\mu g/100$ gm. It is bound to protein by peptide bonds that are broken by proteases in the stomach and intestinal lumen. Preparation for absorption depends on the binding of vitamin B_{12} to intrinsic factor, a glycoprotein (mol wt $\sim$ 115,000) synthesized by the parietal cells in man.[38] One molecule of intrinsic factor tightly binds two molecules of vitamin B_{12}. It has been suggested that this complexing of vitamin B_{12} protects it from hydrolytic enzymes during its transit to the ileum. The ileum, where vitamin B_{12} is absorbed, has receptor sites on the microvilli for the intrinsic factor–vitamin B_{12} complex that depend on calcium ions for normal function. The translocation of vitamin B_{12} across the epithelium is not understood. In the process, it becomes dissociated from intrinsic factor and appears in the portal blood attached to B_{12}-binding α- and β-globulins (Fig. 53-41).[53,83a]

A normal diet contains 5 to 15 μg vitamin B_{12}/day, and it has been estimated that the daily requirement is 1 μg/day. In the presence of intrinsic factor, 60% to 80% of a small dose (2 μg of vitamin B_{12}) will be absorbed. An increase in the intake of the vitamin is associated with a striking decrease in the proportion absorbed.[71] This appears to be a carrier-mediated process.

In the absence of intrinsic factor, vitamin B_{12} can be absorbed if massive amounts (100 to 100,000 μg) are given. In this dosage range, no more than 1% is absorbed. The mechanism for the absorption in the absence of intrinsic factor is not understood. However, it is not limited to the ileum, is not calcium dependent, is temperature related, and, as with iron absorption, is associated with an 8 to 12 hr delay before a peak concentration in the plasma is reached.

Folic acid. Folates are found in a wide variety of foods, including vegetables and nuts, but they are found in especially high levels in yeast and liver, which contain 2 mg and 300 $\mu g/100$ gm. Of the dietary folates, 70% to 80% are polyglutamate conjugates linked by a γ- rather than the more common α-peptide bond. (Folic acid is unreduced pteroyl monoglutamic acid, mol wt 441). The principal form in yeast is pteroyl heptaglutamate. About 90% of folate in vegetables is in the 5- or 10-formyl form, with less than 10% existing as unreduced folate. The folate in liver is primarily 5-methyltetrahydrofolate. The daily intake of folate is estimated to be between 700 to 1,500 μg/day, whereas the daily requirement is only 50 μg. How much of the dietary intake of folate reaches the intestine in an active form is difficult to estimate, for it is sensitive to extremes of pH and boiling.

Folates are absorbed preferentially in the upper

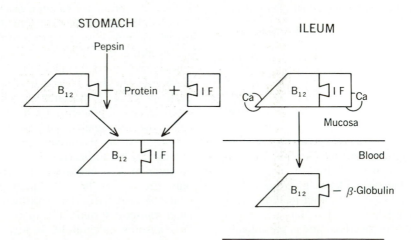

Fig. 53-41. Vitamin B_{12} absorption. See text. (Modified from Herbert.[50])

jejunum. A prerequisite for absorption is the removal of the highly charged polyglutamate chain. None of the gastric or pancreatic enzymes is able to cleave the γ-peptide bond. Although it has been clear that folates are hydrolyzed to the monoglutamate form in the process of absorption, the exact site of this important step remained to be determined (Fig. 53-42). Since γ-glutamyl carboxypeptidase activity could only be identified in lysosomes, the intracellular site was favored. Recently, kinetic studies and isolation of the enzyme from the brush border indicate that the brush border site is the one of physiologic importance in folate absorption.[53,83a]

Evidence has been presented that folates can be transported against a concentration gradient by a saturable mechanism. The interpretation of this fact is complicated by observations indicating that different forms of folate are handled differently by the intestine; that is, folic acid (unreduced pteroylglutamic acid) is only partially methylated during passage through the intestinal epithelium, whereas the partially reduced dietary folates appear to be quantitatively reduced to methyltetrahydrofolate. It is not possible to adequately analyze the absorptive mechanism if the material changes form in its passage through the intestinal epithelium (see discussion of fructose and acidic amino acid absorption).[85]

Other B vitamins and ascorbic acid. The

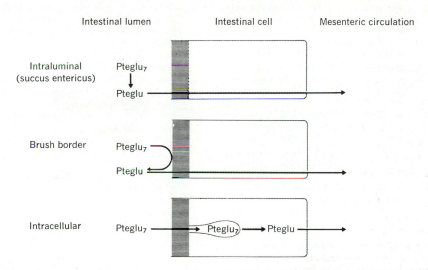

Fig. 53-42. Schematic demonstration of alternate hypotheses for site of deconjugation of conjugated folate. (From Rosenberg and Godwin.[85])

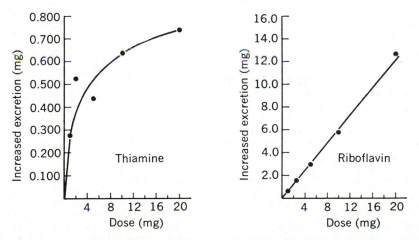

Fig. 53-43. Effect of oral dose level on urinary excretion of thiamine and riboflavin in human subjects. (From Morrison and Campbell.[73])

details of the absorption of other B vitamins and ascorbic acid have not been determined, but it is generally believed that absorption of these vitamins occurs by simple diffusion.[64,94] Thiamine, however, appears to be handled differently, because its absorption does not appear to be linear with intraluminal concentration, as is characteristic of passive diffusion; it appears to be absorbed by a mechanism that can be saturated, that is, a carrier-mediated diffusion (Fig. 53-43).[73]

REFERENCES

1. Adibi, S. A.: Intestinal transport of dipeptides in man: relative importance of hydrolysis and intact absorption, J. Clin. Invest. **50:**2266, 1971.
2. Adibi, S. A., and Mercer, D. W.: Protein digestion in the human intestine as reflected in luminal, mucosal and plasma amino acid concentrations after meals, J. Clin. Invest. **52:**1586, 1973.
3. Alvarado, F.: D-Xylose transport in the chicken small intestine, Comp. Biochem. Physiol. **20:**461, 1967.
4. Alvarado, F., and Crane, R. K.: Phlorizin as a competitive inhibitor of the active transport of sugars by hamster small intestine in vitro, Biochim. Biophys. Acta **56:**170, 1962.
5. Bárány, E., and Sperber, E.: Absorption of glucose against concentration gradient by small intestine of rabbit, Skand. Arch. Physiol. **81:**290, 1939.
6. Bayless, T. M., Paige, D. M., and Ferry, G. D.: Lactose intolerance and milk drinking habits, Gastroenterology **60:**605, 1971.
7. Booth, C. C.: Effect of location along the small intestine on absorption of nutrients. In Code, C. F., editor: Handbook of physiology. Alimentary canal section, Baltimore, 1968, The Williams & Wilkins Co., vol. 3.
8. Borgström, B., and Erlanson, C.: Pancreatic juice colipase: physiological importance, Biochim. Biophys. Acta **242:**509, 1971.
9. Borgström, B., et al.: Studies of intestinal digestion and absorption in the human, J. Clin. Invest. **36:**1521, 1957.
10. Bothwell, T. H., and Charlton, R. W.: Absorption of iron, Annu. Rev. Med. **21:**145, 1970.
11. Brocherhoff, H.: A model of pancreatic lipase and the orientation of enzymes at interfaces, Chem. Phys. Lipids **10:**215, 1973.
12. Carter, C. W., et al.: Biochemistry in relation to medicine, London, 1959, Longmans, Green & Co.
13. Conrad, M. E., and Crosby, W. H.: Intestinal mucosal mechanisms controlling iron absorption, Blood **22:**406, 1963.
14. Cori, C. F.: Fate of sugar in animal body; rate of absorption of hexoses and pentoses from intestinal tract, J. Biol. Chem. **66:**691, 1925.
15. Crane, R. K.: Enzymes and malabsorption: a concept of a brush border membrane disease, Gastroenterology **50:**254, 1960.
16. Crane, R. K.: Hypothesis for mechanisms of intestinal active transport of sugars, Fed. Proc. **21:**891, 1962.
17. Crane, R. K.: Absorption of sugars. In Code, C. F., editor: Handbook of physiology. Alimentary canal section, Baltimore, 1968, The Williams & Wilkins Co., vol. 3.

18. Csaky, T. Z., Hartzog, H. C., III, and Fernald, G. W.: Effect of digitalis on active intestinal sugar transport, Am. J. Physiol. **200:**459, 1961.
19. Curran, P. F.: Coupling between transport processes in the intestine, Physiologist **11:**3, 1968.
20. Dahlqvist, A., and Borgström, B.: Digestion and absorption of carbohydrates in man, Biochem. J. **81:**411, 1961.
21. Dahlqvist, A., and Nordström, D.: The distribution of disaccharidase activities in the villi and crypts of the small intestinal mucosa, Biochim. Biophys. Acta **113:**624, 1966.
22. Dahlqvist, A., and Thompson, D. L.: Separation and characterization of two rat intestinal amylases, Biochem. J. **89:**272, 1963.
23. Das, M., and Radhakushnan, A. N.: Studies on a widespectrum dipeptide uptake system in the monkey and the human, Biochem. J. **146:**133, 1975.
24. Davenport, H. W.: Physiology of the digestive tract, Chicago, 1971, Year Book Medical Publishers, Inc.
25. Desnuelle, P.: Pancreatic lipase. In Code, C. F., editor: Handbook of physiology. Alimentary canal section, Baltimore, 1968, The Williams & Wilkins Co., vol. 5.
26. Diamond, J. M., and Tormey, J. M.: Studies on the structural basis of water transport across epithelial membranes, Fed. Proc. **25:**1458, 1966.
27. Dietschy, J. M.: Recent developments in solute and water transport across the gallbladder epithelium, Gastroenterology **50:**395, 1966.
28. Dietschy, J. M., and Siperstein, M. D.: Cholesterol synthesis by the gastrointestinal tract: localization and methods of control, J. Clin. Invest. **44:**1311, 1965.
29. Dowling, R. H.: Enterohepatic circulation, Gastroenterology **62:**122, 1972.
30. Fawcett, D. W.: Physiologically significant specializations of the cell surface, Circulation **26:**1105, 1962.
31. Florey, H. W., Wright, R. D., and Jennings, M. A.: The secretions of the intestine, Physiol. Rev. **21:**36, 1941.
32. Fordtran, J. S., and Dietschy, J. M.: Water and electrolyte movement in the intestine. Gastroenterology **50:**263, 1966.
33. Fordtran, J. S., and Ingelfinger, F. J.: Absorption of water, electrolytes, and sugar from the human gut. In Code, C. F., editor: Handbook of physiology. Alimentary canal section, Baltimore, 1968, The Williams & Wilkins Co., vol. 3.
34. Fordtran, J. S., et al.: The kinetics of water absorption in the human intestine, Trans. Assoc. Am. Physicians **74:**195, 1961.
35. Fridhandler, L., and Quastel, J. H.: Absorption of sugars from isolated surviving intestine, Arch. Biochem. **56:**412, 1955.
36. Gallagher, N. D., Harrison, D. D., and Skyring, A. P.: Fluid and electrolyte disturbances in patients with long-established ileostomies, Gut **3:**219, 1962.
37. Gibson, Q. H., and Wiseman, G.: Selective absorption of stereo-isomers of amino-cells from loops of small intestine of rat, Biochem. J. **48:**426, 1951.
38. Glass, G. B. J.: Gastric intrinsic factor and its function in the metabolism of vitamin B_{12}. Physiol. Rev. **43:**529, 1963.
39. Gray, G. M.: Carbohydrate digestion and absorption, Gastroenterology **58:**96, 1970.
40. Gray, G. M., and Cooper, H. L.: Protein digestion and absorption, Gastroenterology **61:**535, 1971.
41. Gray, G. M., and Ingelfinger, F. J.: Intestinal absorption of sucrose in man: interrelation of hydrolysis and

monosaccharide product absorption, J. Clin. Invest. **45**:388, 1966.

42. Greaves, J. P., and Hollingsworth, D. F.: Changes in the pattern of carbohydrate consumption in Britain, Proc. Nutr. Soc. **23**:136, 1964.

43. Greenberger, N. J., and Skillman, T. G.: Medium chain triglycerides. Physiologic considerations and clinical implications, N. Engl. J. Med. **280**:1045, 1969.

44. Haemmerli, U. P., et al.: Acquired milk intolerance in the adult caused by lactose malabsorption and to a selective deficiency of intestinal lactase deficiency, Am. J. Med. **38**:7, 1965.

45. Hagihira, H., Lin, E. C. C., and Wilson, T. H.: Active transport of lysine, ornithine, arginine, and cystine by the intestine, Biochem. Biophys. Res. Commun. **4**:478, 1961.

46. Hendrix, B. M., and Sweet, J. E.: A study of amino nitrogen and glucose in lymph and blood before and after injection of nutrient solutions in the intestine. J. Biol. Chem. **32**:299, 1917.

47. Hendrix, T. R.: Effect of a hypertonic solution on intestinal absorption, Am. J. Dig. Dis. **2**:643, 1957.

48. Hendrix, T. R.: Cholera toxin and intestinal transport. In Csaky, T. Z., editor: Intestinal absorption and malabsorption, New York, 1975, Raven Press.

49. Hendrix, T. R., and Bayless, T. M.: Intestinal secretion, Annu. Rev. Physiol. **32**:139, 1970.

50. Herbert, V.: Absorption of vitamin B_{12} and folic acid, Gastroenterology **54**:110, 1968.

51. Hofmann, A. F.: Function of bile in the alimentary canal. In Code, C. F., editor: Handbook of physiology. Alimentary canal section, Baltimore, 1968, The Williams & Wilkins Co., vol. 5.

52. Hofmann, A. F., and Small, D. M.: Detergent properties of bile salts: correlation with physiological function. Annu. Rev. Med. **18**:333, 1967.

53. Icelani, Dhar, G., et al.: Direct in vivo demonstration of the sequence of events in intestinal polyglutamyl folate absorption, Clin. Res. **25**:309A, 1977.

54. Ingram, R. C., Peters, H. C., and Visscher, M. B.: On movement of materials across living membranes against concentration gradients, J. Phys. Chem. **42**:141, 1938.

55. Kim, Y. S., Bertwhistle, W., and Kim, Y. W.: Peptide hydrolases in the brush border and soluble fractions of small intestinal mucosa of rat and man, J. Clin. Invest. **51**:1419, 1972.

56. Kim, Y. S., Nicholson, J. A., and Curtis, K. J.: Intestinal peptide hydrolases: peptide and amino acid absorption, Med. Clin. North Am. **58**:1397, 1974.

57. Kinter, W. G., and Wilson, T. H.: Autoradiographic studies of sugar and amino acid absorption by everted sacs of hamster intestine, J. Cell. Biol. **25**:19, 1965.

58. Kratzer, F. H.: Amino acid absorption and utilization in the chick, J. Biol. Chem. **153**:237, 1944.

59. Leblond, C. P., and Walker, B. E.: Renewal of cell populations, Physiol. Rev. **36**:255, 1956.

60. Levin, R. J.: Techniques, terminology and parameters in intestinal absorption, Br. Med. Bull. **23**:209, 1967.

61. Levitan, R., et al.: Water and salt absorption in the human colon, J. Clin. Invest. **41**:1754, 1962.

62. Lin, E. C. C., Hagihira, H., and Wilson, T. H.: Specificity of the transport system for neutral amino acids in the hamster intestine, Am. J. Physiol. **202**:919, 1962.

63. Marks, J. F., Norton, J. B., and Fordtran, J. S.: Glucose-galactose malabsorption syndrome, J. Pediatr. **69**:255, 1966.

64. Matthews, D. M.: Absorption of water soluble vitamins, Br. Med. Bull. **23**:258, 1967.

65. Matthews, D. M.: Intestinal absorption of peptides, Physiol. Rev. **55**:537, 1975.

66. Matthews, D. M., and Adibi, S. A.: Peptide absorption. Gastroenterology **71**:151, 1976.

67. Matthews, D. M., and Laster, L.: Absorption of protein digestion products: a review, Gut **6**:411, 1965.

68. Matthews, D. M., and Laster, L.: The kinetics of intestinal active transport of five neutral amino acids, Am. J. Physiol. **208**:593, 1965.

69. Mendel, L. B.: Nutrition and growth, Harvey Lect. **10**:111, 1914-1915.

70. Miller, D., and Crane, R. K.: The digestive function of the epithelium of the small intestine. II. Localization of disaccharidase hydrolysis in the isolated brush border portion of intestinal epithelial cells, Biochim. Biophys. Acta **52**:319, 1961.

71. Molin, D. L.: Radioactive vitamin B_{12} in the study of blood diseases, Br. Med. Bull. **15**:8, 1959.

72. Morris, I. G.: Gamma globulin absorption in the newborn. In Code, C. F., editor: Handbook of physiology. Alimentary canal section, Baltimore, 1968, The Williams & Wilkins Co., vol. 5.

73. Morrison, A. B., and Campbell, J. A.: Vitamin absorption studies. I. Factors influencing the excretion of oral test doses of thiamine and riboflavin by human subjects. J. Nutr. **72**:435, 1960.

74. Munro, H. N.: Protein secretion into the gastrointestinal tract. In Thomson, T. J., and Gillepsie, I. E., editors: Postgraduate gastroenterology, London, 1966, Bailliere, Tindall & Cassell, Ltd.

75. Nasset, E. S., Schwartz, P., and Weiss, H. V.: The digestion of proteins in vivo, J. Nutr. **56**:83, 1955.

76. Neame, K. D., and Wiseman, G.: The transamination of glutamic and aspartic acids during absorption by the small intestine of the dog in vitro, J. Physiol. **135**:442, 1957.

77. Newcomer, A. D., and McGill, D. B.: Distribution of disaccharidase activity in the small bowel of normal and lactase deficient subjects, Gastroenterology **51**:481, 1966.

78. Newey, H.: Absorption of carbohydrates, Br. Med. Bull. **23**:236, 1967.

79. Newey, H., and Smyth, D. H.: Basic concepts in intestinal absorption. In Bittar, E. E., and Bittar, N., editors: The biological basis of medicine, New York, 1969, Academic Press, Inc., vol. 5.

80. Nordin, B. E. C.: Measurement and meaning of calcium absorption, Gastroenterology **54**:294, 1968.

81. Öckerman, P. A., and Lundborg, H.: Conversion of fructose to glucose by human jejunum. Absence of galactose to glucose conversion, Biochim. Biophys. Acta **105**:34, 1965.

82. Parsons, D. S.: Methods for investigation of intestinal absorption. In Code, C. F., editor: Handbook of physiology. Alimentary canal section, Baltimore, 1968, The Williams & Wilkins Co., vol. 3.

83. Phillips, S. F.: Absorption and secretion by the colon, Gastroenterology **56**:966, 1969.

83a. Reisenauer, A. M., Krumdieck, C. L., and Halstead, C. H.: Folate conjugase: two separate activities in human jejuneum, Science **198**:196, 1977.

84. Roberts, P. J. P., and Whelan, W. J.: The mechanism of carbohydrase action. V. Action of human salivary

α-amylase on anylopectin and glycogen, Biochem. J. **76:**246, 1960.

85. Rosenberg, I. H., and Godwin, H. A.: The digestion and absorption of dietary folate, Gastroenterology **60:**445, 1971.

86. Rosensweig, N. S., and Herman, R. H.: Control of jejunal sucrase and maltase activity by dietary sucrose or fructose in man, J. Clin. Invest. **47:**2253, 1968.

87. Sabesin, S. M., and Isselbacher, K. J.: Protein synthesis inhibition: mechanism for impaired fat absorption, Science **147:**1149, 1965.

88. Schachter, D., et al.: Tissue concentration differences during active transport of calcium by the intestine, Am. J. Physiol. **211:**1131, 1966.

89. Schultz, S. G., and Curran, P. F.: Intestinal absorption of sodium chloride and water. In Code, C. F., editor: Handbook of physiology, Alimentary canal section, Baltimore, 1968, The Williams & Wilkins Co., vol. 3.

90. Schultz, S. G., and Zalusky, R.: Transport of sodium and sugar in rabbit ileum, J. Gen. Physiol. **47:**1043, 1964.

91. Schultz, S. G., Frizzell, R. A., and Nellans, H. N.: Ion transport by mammalian small intestine, Annu. Rev. Physiol. **36:**51, 1974.

92. Semenza, G.: Intestinal oligosaccharidases and disaccharidases. In Code, C. F., editor: Handbook of physiology. Alimentary canal section, Baltimore, 1968, The Williams & Wilkins Co., vol. 5.

93. Semenza, G., et al.: Sodium activation of human intestinal sucrase and its possible significance in the enzymatic organization of brush borders, Biochim. Biophys. Acta **89:**109, 1964.

94. Serebro, H. A., et al.: The intestinal absorption of vitamin B_6 compounds by the rat and hamster, Bull. Johns Hopkins Hosp. **119:**166, 1966.

95. Shanker, L. S.: Mechanisms of drug absorption and distribution, Annu. Rev. Pharmacol. **1:**29, 1961.

96. Sleisenger, M. H., et al.: Evidence for a single common carrier for uptake of a dipeptide and a tripeptide by hamster jejunum in vitro, Gastroenterology **71:**76, 1976.

97. Small, D. M.: A physical chemical approach to bile. In Schiff, L., Carey, J. B., Jr., and Dietschy, J. M., editors: Bile salt metabolism, Springfield, Ill., 1969, Charles C Thomas, Publisher.

98. Small, D. M.: The enterohepatic circulation of bile salts, Arch. Intern. Med. **130:**552, 1972.

99. Taylor, W. H.: Biochemistry of pepsins. In Code, C. F., editor: Handbook of physiology. Alimentary canal section, Baltimore, 1968, The Williams & Wilkins Co., vol. 5.

100. Trier, J. S.: Morphology of the epithelium of the small intestine. In Code, C. F., editor: Handbook of physiology. Alimentary canal section, Baltimore, 1968, The Williams & Wilkins Co., vol. 3.

101. Turnberg, L. A., et al.: Mechanism of bicarbonate absorption and its relationship to sodium transport in the human jejunum, J. Clin. Invest. **49:**548, 1970.

102. Turnberg, L. A., et al.: Interrelationships of chloride, bicarbonate, sodium, and hydrogen transport in human ileum, J. Clin. Invest. **49:**557, 1970.

103. Ugolev, A. M.: Membrane (contact) digestion, Physiol. Rev. **45:**555, 1965.

104. Ussing, H. H., and Zerahn, K.: Active transport of sodium as the source of electrical current in the short circuited frog skin, Acta Physiol. Scand. **23:**110, 1951.

105. Vaughan, B. E.: Intestinal electrolyte absorption by parallel determination of unidirectional sodium and water transfers, Am. J. Physiol. **198:**1235, 1960.

106. Visscher, M. B., et al.: Sodium ion movement between intestinal lumen and blood, Am. J. Physiol. **141:**488, 1944.

107. Visscher, M. B., et al.: Isotopic tracer studies on the movement of water and ions between intestinal lumen and blood, Am. J. Physiol. **142:**550, 1944.

108. Walker, W. A., Wu, M., and Isselbacher, K. J.: Intestinal uptake of macromolecules. III. Studies on the mechanism by which immunization interferes with antigen uptake, J. Immunol. **115:**854, 1975.

109. Watten, R. H., et al.: Water and electrolyte studies in cholera, J. Clin. Invest. **38:**1879, 1959.

110. Weser, E., and Sleisenger, M. H.: Metabolism of circulating disaccharides in man and the rat, J. Clin. Invest. **46:**499, 1967.

111. Wilson, J. D.: Biosynthetic origin of serum cholesterol in the squirrel monkey: evidence for a contribution by the intestinal wall, J. Clin. Invest. **47:**175, 1968.

112. Wilson, T. H.: Intestinal absorption, Philadelphia, 1962, W. B. Saunders Co.

113. Wilson, T. H., and Wiseman, G.: The use of sacs of inverted small intestine for the study of transference of substances from the serosal to the mucosal surface, J. Physiol. **123:**116, 1954.

114. Wilson, T. H., et al.: Intestinal transport of sugars and amino acids, Fed. Proc. **19:**870, 1960.

115. Wiseman, G.: Absorption of amino-acids using in vitro technique, J. Physiol. **120:**63, 1953.

116. Wiseman, G.: Absorption of amino acids. In Code, C. F., editor: Handbook of physiology. Alimentary canal section, Baltimore, 1968, The Williams & Wilkins Co., vol. 3.

117. Wrong, W., and Metcalfe-Gibson, A.: Electrolyte content of faeces, Proc. R. Soc. Med. **58:**1007, 1965.

118. Zetterqvist, H.: The ultrastructural organization of the columnar absorbing cells of the mouse jejunum, Stockholm, 1956, Stockholm Akademisk Avhandling Karolinska Institutet.

54

THOMAS R. HENDRIX

The secretory function of the alimentary canal

The alimentary tract produces both exocrine and endocrine secretions. Exocrine secretions prepare the food for absorption by diluting it to the osmolality of plasma, altering the pH for optimal hydrolysis, and hydrolyzing complex foodstuffs in preparation for the final mucosal hydrolysis and absorption; finally, they protect the mucosa from physical and chemical trauma by their content of mucus, antibodies, and copious amounts of fluid. Endocrine secretions, on the other hand, play a major role in the control and coordination of the secretory and motor activities involved in the assimilation of foodstuffs. In addition, gastrointestinal hormones interact with other hormones such as insulin to modify the way in which absorbed nutrients are handled.

SALIVARY SECRETION

The processing of food for absorption begins in the mouth, where the food is mixed with saliva. Saliva moistens and lubricates food so that it may be swallowed easily; it dissolves food so that the taste buds may be activated, thus stimulating the appetite; it begins the process of diluting the food to a level isotonic with plasma; and it begins the digestion of starch and glycogen. In addition, saliva is necessary for the health of the oral mucosa and to retard the development of dental caries. Its flow increases to dilute and wash out unpleasant or damaging substances from the mouth. Despite these important functions, the salivary secretion is not essential for normal digestion and absorption of nutrients. The salivary glands, however, have been extensively studied as relatively accessible models of an exocrine secretory gland, and from these studies a great deal has been learned relating structure to secretory function.

Salivary glands

The major salivary glands in man are the paired parotid, submaxillary, and sublingual glands. Based on the type of secretory product and types of secretory cells, they have been classified as mucous, serous, or mixed salivary glands.[3] In addition, the oral mucosa contains many small glands, the buccal glands, which are purely mucous glands. The parotids are serous glands that produce a watery secretion containing proteins, including salivary amylase, an enzyme that hydrolyzes starch. The submaxillary and sublingual glands are mixed, containing both mucin- and serous-producing cells. These secretory cells are arranged in acini that connect to intercalated ducts lined by small cuboid cells. These ducts in turn join intralobular ducts that are also called striated ducts because the lining cells have pronounced striations in their basal third caused by extensive infolding of the basal plasma membrane. The parotid and submaxillary glands have extensive "striated" ducts, whereas these are not so extensively developed in the sublingual glands. In turn, the striated ducts empty into the excretory ducts (Fig. 54-1).

Secretory process

The salivary glands are capable of secreting at very high rates; for example, the human parotid gland can secrete its weight in saliva during a 20 min period.[79]

Water and electrolytes. The composition of saliva changes with increasing secretory rates. At rates between 5% and 10% of maximum, the osmolality of saliva is proportional to flow. At low rates of flow, salivary osmolality is as low as 50 mOsm/L, whereas at the maximal secretory rate, it approaches the plasma level, 300 mOsm/L.[8] The electrolyte composition also changes with flow rate, and it differs considerably from that of plasma in that the concentrations of sodium and chloride are less than plasma, whereas potassium and bicarbonate are present in greater concentrations (Fig. 54-2). The explanation proposed for these findings is that an isotonic

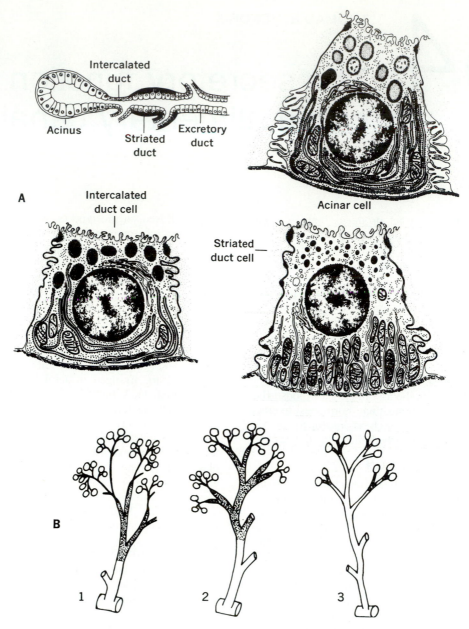

Fig. 54-1. A, Diagrammatic representation of locations and fine structural characteristics of various cell types in mouse submandibular gland. **B,** Duct systems of salivary glands of man. *1,* Parotid; *2,* submaxillary; *3,* sublingual. Solid black areas: intercalated ducts; stippled areas: striated ducts; unshaded areas: interlobular and interlobar excretory ducts. (**A** from Bloom and Fawcett[6]; **B** from Leeson.[54])

fluid with an electrolyte composition similar to that of plasma is produced in the acini and subsequently modified by absorption of electrolytes in excess of water during passage through the ducts.

Micropuncture studies of salivary glands have provided direct evidence in support of this the-

ory. Fluid obtained from the junction of acini and intercalated ducts resembles an ultrafiltrate of plasma. However, this primary secretion is not produced by ultrafiltration, but rather as the result of active ion transport, because (1) it is associated with a striking transmembrane potential dif-

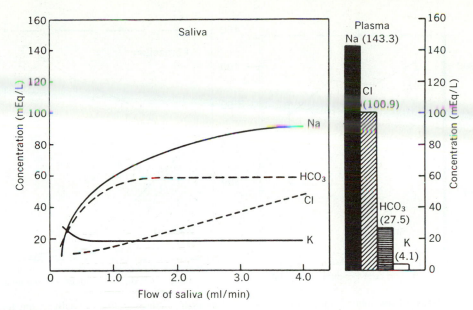

Fig. 54-2. Electrolyte composition of parotid saliva as function of rate of secretion compared to electrolyte composition of plasma. (From Thaysen et al.[85])

ference, (2) it can be secreted against a pressure greater than arterial pressure or even in the absence of circulation, and (3) it stops and the transmembrane secretory potential difference is abolished when sodium is removed from the perfusing fluid or after application or ouabain, which inhibits Na^+, K^+-activated ATPase, the enzyme believed to be responsible for providing energy for active sodium transport.[96] The acinar, or primary secretory fluid, however, has not been obtained for analysis. Since it would be remarkable if an ion pump or pumps produced a secretion equivalent to an ultrafiltrate of plasma over a wide range of secretory rates, it has been suggested that there is equilibration of ionic constituents without volume transfer in the thin intercalated ducts. Support for this notion is provided by the observation that it is in this segment that urea enters saliva most readily. In the striated lobar ducts, sodium is absorbed rapidly and efficiently against a steep electrochemical gradient, leading to a sodium concentration in the salivary fluid as low as 2 mEq/L. Potassium enters the duct but not simply on the basis of a 1:1 exchange for sodium (Fig. 54-3). The transfers of sodium out of and potassium into the saliva are active, energy-requiring processes but are nonelectrogenic, since the potential difference across the duct epithelium is not altered by these processes.[78,79] In addition, aldosterone, an adrenal hormone promoting increased reabsorption of

sodium by the renal tubules, influences the sodium:potassium ratio in saliva. Raising the level of aldosterone increases absorption of sodium and secretion of potassium. The sites of transfer of a variety of substances have been identified by injecting two isotopes into the arterial blood supply of a salivary gland and analyzing timed fractions of saliva. The more distal the transfer site for a particular isotope, the earlier it will appear in the serial specimens collected from the excretory duct (Fig. 54-4). The transport of iodide is very efficient, 88% being cleared from the blood in one pass through the gland. Since such a large percentage of the salivary blood flow passes the salivary ducts, it is believed that arterial blood entering the gland flows through capillaries surrounding the ducts before it reaches the acini.

Protein and mucosubstances. Saliva contains two types of protein: serum proteins and intrinsic secretory substances synthesized in the salivary glands. The concentration of salivary protein is 0.1 to 0.5 gm/dl compared to 6 gm/dl in plasma. Immunoglobulins of the IgA class are joined to a secretory piece synthesized in the salivary epithelium and transferred to the saliva.[86] Of the proteins synthesized by the salivary gland, ptyalin or salivary amylase, the most extensively studied, accounts for 25% of salivary protein. Salivary amylase exists as several isoenzymes and is stored in the resting gland in the secretory

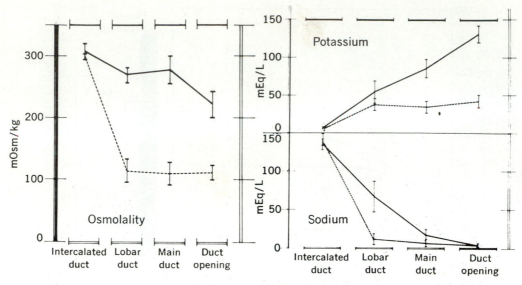

Fig. 54-3. Left, osmolality and, right, sodium and potassium concentrations in saliva from four levels of submaxillary duct system. Vertical bars represent ±2 SE of mean. Unbroken lines indicate samples collected at rest (flow rate less than 3 μL/gm/min); broken lines indicate samples collected after stimulation with pilocarpine (flow rates 20 to 40 μL/gm/min). Data for graphs were derived from micropuncture studies of rat submaxillary gland. Fluid collected from intercalated ducts close to acini is similar to plasma. Sodium is reabsorbed preferentially in lobular duct. (From Young et al.[96])

or zymogen granules of serous acinar cells. Of the protein in the secretory granules, 30% to 40% is amylase. When the gland is stimulated, the serous cells discharge the contents of the secretory granules into the lumen, and synthesis of new protein is initiated at once. If the stimulus for secretion continues, the newly synthesized amylase is discharged directly into the lumen without first being accumulated in secretory granules.

The lubricant characteristics of saliva are due to glycoproteins and mucopolysaccharides. Although the parotid is characterized as a serous gland and its secretion is watery, parotid saliva contains considerable glycoprotein with properties different from the mucosubstances of the other salivary glands. Some individuals secrete their blood group–specific mucopolysaccharide in their saliva. This blood group secretor characteristic is genetically determined.[18]

Control of salivary secretion

The control of secretion of the salivary glands, unlike that of other glands of the alimentary tract, is exclusively neural. The salivary glands receive their primary secretory innervation from the parasympathetic system; hence the neurochemical mediator of salivary secretion is acetylcholine (ACh). Stimulation of the parasympathetic nerves to the salivary glands also causes vasodilatation in the glands. It has been suggested that the stimulation of the gland activates an enzyme, kallikrein, which releases bradykinin, a potent octapeptide vasodilator. The relative importance of these two mechanisms, direct neural stimulation and increased blood flow, is still unsettled.[75] The bulk of the available evidence favors the first mechanism because salivary secretion can be blocked by atropine without blocking vasodilatation, and merely increasing blood flow does not increase salivary secretion. In addition to its primary role in the control of salivary secretion, parasympathetic innervation has a trophic effect necessary for the structural and functional integrity of the salivary glands.[79]

The salivary glands are also innervated by sympathetic fibers from the superior cervical ganglion, and the primary neurochemical mediator of these fibers is norepinephrine. Stimulation of these nerves leads to vasoconstriction and, in some circumstances, to a salivary secretion that is neither as voluminous nor as well maintained as that elicited by parasympathetic stimulation. Neither the role nor the significance of the sympathetic nerves in normal salivary gland function is clear.[20] What is clear, however, is that the sympathetic and parasympathetic nervous systems cannot be thought of as balanced opposing systems controlling salivary secretion.

The cell bodies of preganglionic parasympathetic nerves supplying the salivary glands are in the medulla in the superior salivary nucleus (submaxillary and sublingual) and in the inferior salivary nucleus (parotid). Influences of many sorts—taste, smell, the thought of food, and nausea—lead to salivary stimulation. Pavlov, in his famous experiments on conditioned reflexes, used salivation conditioned to the ringing of a bell to demonstrate the phenomenon.

GASTRIC SECRETION

In the stomach the ingested food, drink, and saliva are mixed with gastric juice. Here the first step in the hydrolysis of proteins is accomplished by pepsin, the main digestive enzyme of the stomach. Peptic hydrolysis of protein, however, is unnecessary for normal absorption of protein, as demonstrated by the absence of impaired protein absorption in individuals having no acid or pepsin in their gastric juice. On the other hand, intrinsic factor contained in normal gastric secretion is essential for the absorption of vitamin B_{12}. In addition to their roles in protein hydrolysis and vitamin B_{12} absorption, gastric secretions contribute to the dilution of ingested nutrients to the isotonicity required for normal absorption, and their low pH destroys many, but not all, bacteria ingested.

Gastric mucosa

The mucosa of the stomach is arranged in folds, or rugae, with the surface covered by simple columnar epithelium, the surface mucous cells. Opening onto the surface are gastric pits, about 100/cm², which are also lined by surface mucous cells. Three to seven tubular glands empty into each gastric pit. In addition to the surface mucous cell, four other cell types make up the gastric mucosa: (1) the oxyntic or parietal cells, which are responsible for hydrochloric acid and intrinsic factor secretion, (2) the chief or zymogen cells, responsible for pepsinogen secretion, (3) mucous neck cells, and (4) argentaffin cells, the endocrine cells that produce alimentary tract hormones. Some of these cells have argentaffin and argyrophil staining reactions[82] (Fig. 54-5). The stomach is divided into three areas, cardiac, oxyntic or fundic, and pyloric or antral, based on histologic features of the mucosa. The cardiac glands occupy a narrow rim around the entrance

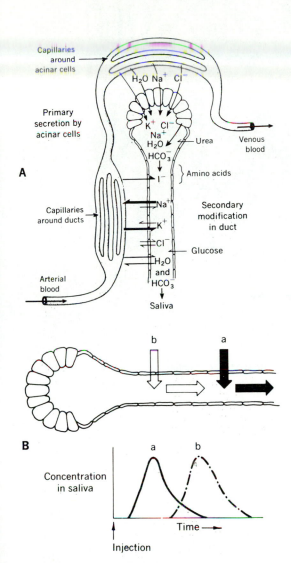

Fig. 54-4. A, Diagrammatic representation of electrolyte exchanges during secretion by parotid gland. **B,** Method of locating point at which substance crosses salivary tubules to enter saliva. Gland is stimulated to secrete at constant rate, and saliva is collected in small serial samples. Small volume of solution containing two isotopically labeled substances, *a* and *b*, is quickly injected into artery close to gland. Substance *a,* because it crosses at more distal portion of tubule, appears in saliva and reaches peak concentration earlier than substance *b,* which enters at more proximal part of tubule. Relative positions of entry of iodide, sodium, potassium, chloride, water, and bicarbonate, shown in **A,** were determined in this manner. (Modified from Burgen and Emmelin[8]; from Davenport.[13])

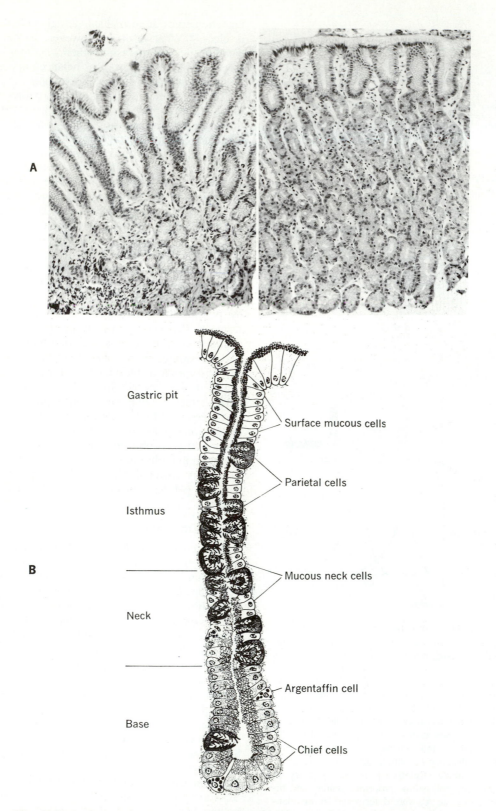

Gastric pit

Surface mucous cells

Parietal cells

Isthmus

Mucous neck cells

Neck

Argentaffin cell

Base

Chief cells

Fig. 54-5. A, Human gastric mucosa. At left, pyloric gland with deep gastric pits and shallow glands. At right, fundic gland area with shallow gastric pits and deep glands. **B,** Diagram of tubular gland from fundic area of stomach. (Modified from Stevens and Leblond[84]; from Ito and Winchester.[45])

of the esophagus. These glands are composed almost entirely of mucus-secreting cells.

The oxyntic or fundic gland area occupies the proximal four fifths of the stomach. These glands contain all four types of gastric epithelial cells: mucous neck, argentaffin, oxyntic, and chief cells. Because of the abundance of the latter two cell types, this region is the major source of hydrochloric acid and pepsin in the gastric juice. The pyloric gland area or antrum constitutes the distal one fifth of the stomach, extending a greater distance proximally from the pylorus along the lesser than along the greater curvature.[57] In this area the gastric pits penetrate half the thickness of the mucosa, and the glands are composed primarily of cells resembling the mucous neck cells of the oxyntic glands and G cells, the endocrine cells that produce the hormone gastrin, which stimulates gastric secretion as its major function.[60]

Gastric juice

Gastric juice is composed of the secretion of the mucous cells, oxyntic cells, and chief cells. It is slightly hypotonic compared to plasma. The electrolyte content changes with the rate of secretion (Fig. 54-6). As secretion increases, the

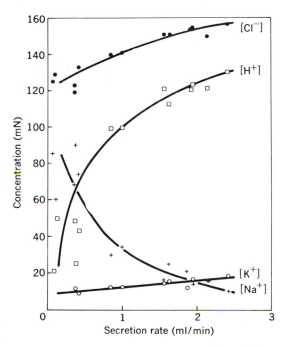

Fig. 54-6. Relation between concentration of electrolytes and rate of secretion of gastric juice in man. Secretion was stimulated by intravenous histamine at increasing rates. (From Nordgren.[63])

sodium ion concentration falls and the hydrogen ion concentration rises to levels as high as 150 mEq/L. The chloride and potassium concentrations, however, remain relatively unchanged over a wide range of secretory rates.

Several hypotheses have been advanced to explain the changing ionic composition of gastric juice.[44] The two-component hypotheses, suggested by Pavlov and developed by Hollander, hold that the ionic composition of gastric juice depends on the proportion of two secretions: the first is produced by the parietal cells and contains hydrochloric acid and small amounts of potassium; the second, produced by the other cells of the gastric mucosa, is an alkaline secretion containing sodium, potassium, chloride, and bicarbonate as well as mucus and enzymes. The ionic composition of the parietal component has been estimated to be H^+, 149 mEq/L; K^+, 17 mEq/L; and Cl^-, 166 mEq/L; and the composition of the nonparietal component Na^+, 140 mEq/L; K^+, 5 mEq/L; Cl^- 120 mEq/L; and HCO_3^-, 25 mEq/L.[44] Since the concentration of chloride has been estimated to be about 166 mEq/L in parietal cell secretion and 120 mEq/L in nonparietal secretion, chloride concentration will rise slightly as the hydrogen ion concentration rises steeply with increased parietal cell secretion.

Because of the relative constancy of the chloride ion concentration in gastric juice over a wide range of secretory rates, a second hypothesis for the formation of gastric juice has been postulated as follows. The gastric glands secrete a sodium chloride solution with a concentration 155 to 160 mEq/L, and with increasing secretion, there is increased exchange of sodium ion for hydrogen ion by the gastric mucosa. This hypothesis has been put on an anatomic basis by the suggestion that gastric juice, like saliva, is produced by the modification of a precursor fluid produced by the chief cells located in the lower third of the gastric glands. As the fluid passes the parietal cells, which are located in the middle and upper part of the gastric glands, hydrogen ion is exchanged for sodium ion.[41]

The third hypothesis is based on the assumption that the primary secretion is hydrochloric acid and that gastric juice as it is collected from the stomach is the consequence of back diffusion of the secreted hydrogen ion into the mucosa in exchange for sodium ion.

No one hypothesis completely explains all the observations. The bulk of the evidence, however, suggests that the parietal cells secrete hydrochloric acid into their intracellular canaliculi, a secretion that is then mixed with the secretions of the other cells of the gastric mucosa. To maintain the striking gradients of hydrogen ion and sodium ion between plasma and gastric juice, the gastric epithelium is practically impermeable to the passive diffusion of these ions and maintains a transepithelial potential difference of 40 mV.

Hydrochloric acid secretion. One of the most striking biophysical phenomena is the concentration gradient of hydrogen ions that is developed across the gastric epithelium. The parietal cells, which constitute about one tenth of the gastric mucosal volume, increase the concentration of hydrogen ions 1 million–fold in the fluid extracted from the plasma and secreted as gastric juice. The energy for this work is provided by increased mucosal blood flow and oxygen uptake. In the secreting stomach, four hydrogen ions are delivered into the gastric juice for every oxygen atom used. This is the average for the whole stomach and does not give precise information about parietal cell metabolism and oxygen consumption involved in acid secretion.[14] In addition, carbon dioxide is taken up from arterial blood entering the secreting stomach, and bicarbonate is added to the venous blood draining it. Fig. 54-7 diagrams the steps involved in one hypothesis of hydrochloric acid secretion by the parietal cell. It is believed that water is hydrolyzed at the canalicular membrane, with the hydrogen ion being secreted into the lumen and the resultant hydroxyl ion neutralized by a hydrogen ion from the hydrolysis of carbonic acid. The carbonic acid, which serves as a hydrogen ion donor, is formed by the hydration of carbon dioxide through the action of carbonic anhydrase, and the bicarbonate formed is removed in the venous blood. Acetazolamide, an inhibitor of carbonic anhydrase, also inhibits gastric acid secretion. The energy for secretion is derived from adenosine triphosphate (ATP). Intracellular control of hydrogen ion production appears to be mediated by the "second messenger," adenosine 3′,5′-monophosphate (cyclic AMP).

Chloride ions are also actively secreted into the gastric juice, although they are not transported against as great a concentration gradient as are the hydrogen ions. (The plasma to gastric juice concentration gradient of hydrogen ions is 5×10^{-5} mEq/L to 150 to 170 mEq/L, whereas the concentration gradient of chloride is only 100 to 105 mEq/L in plasma to 155 to 160 mEq/L in gastric juice.) Chloride, however, moves against a steep electrical gradient (a potential difference of 44 mV, lumen negative to blood).[17] If gastric mucosa is mounted in a chamber, as mentioned in the discussion of water and electrolyte transport by the intestine (Chapter 53) so that there is neither an electrical nor a chemical gradient across the mucosa, chloride will be secreted in the direction of serosa to mucosa. On the other hand, when hydrochloric acid secretion is stimulated, there is little change in transmural potential, presumably because bicarbonate leaves at

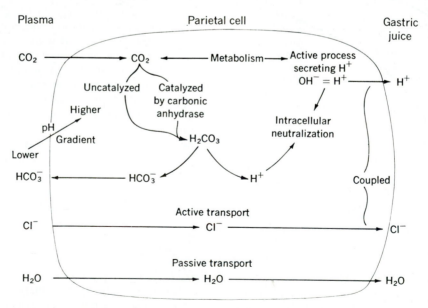

Fig. 54-7. One hypothesis for generation of hydrochloric acid by parietal cells. (From Davenport.[13])

the serosal pole as chloride is extruded at the mucosal pole of the cell. In in vitro experiments the hydrogen ion and chloride ion pumps can be made to operate independently. In vivo, however, they appear to be coupled.

Water movement into gastric juice is the consequence of solute transport. Gastric juice, especially at low secretory rates, is hypotonic to plasma due to the mixing of parietal secretion containing hydrochloric acid and the nonparietal secretion containing bicarbonate with the production of water and carbon dioxide. At high secretory rates the hydrochloric acid output increases markedly, with little increase in HCO_3^- output; hence gastric juice approaches isotonicity.

The maximal output of hydrochloric acid by the stomach shows a positive correlation with the number of oxyntic cells in the mucosa.[58]

Nonelectrolyte constituents of gastric juice. The important nonelectrolytes of gastric juice—pepsinogen, intrinsic factor, and mucoproteins—account for less than one thirtieth of the osmoles of gastric juice.

PEPSINOGEN SECRETION. Pepsinogen is the inactive precursor of pepsin, the major enzyme of gastric juice. Seven different pepsinogens have been described in man on the basis of electrophoretic mobility, pH optima, and immunologic reactivity. They have been separated into two groups. Group I pepsinogens are found in chief cells and mucus neck cells of the oxyntic gland area, whereas group II pepsinogens are distributed in oxyntic, pyloric, and duodenal mucosa. At present the functional importance of these findings is not known.[73] The chief cells store pepsinogen in encapsulated secretory granules. When the cell is full, the synthetic process is inhibited, although not completely, for there is a steady resting secretory rate of pepsinogen that is about one fifth the maximal stimulated rate.[42] A secretory stimulus such as vagal stimulation, histamine, or gastrin, causes the preformed granules of pepsinogen to be discharged into the lumen of the gastric glands. Pepsinogen secretion continues at a high and sustained rate with continued stimulation, but under these conditions the proenzyme leaves the cell without going through the granule stage. When the secretory stimulus ceases, the chief cell continues to synthesize pepsinogen until the granules are restored.

Pepsinogen is converted by acid and pepsin to active pepsin by the cleavage of a peptide from the molecule. (The role of pepsin in protein digestion is discussed in Chapter 53.)

INTRINSIC FACTOR SECRETION. The secretion of intrinsic factor is the only essential function of the stomach. In its absence, vitamin B_{12} is not absorbed, and when the body stores are exhausted, as in untreated pernicious anemia, death occurs. Intrinsic factor has not been completely characterized, but it appears to be a mucoprotein, (mol wt approximately 100,000). In man and most other animals, intrinsic factor is secreted by the oxyntic cells but is synthesized by the chief cells in the rat and mouse and by pyloric and Brunner's glands in the pig. Intrinsic factor is continuously secreted in amounts exceeding minimal needs for the absorption of vitamin B_{12}. The secretion produced in 1 hr is normally sufficient to bind with the minimal daily requirement of vitamin B_{12}. When gastric secretion is stimulated, the output of intrinsic factor increases, the peak output preceding peak acid output.[49]

MUCOPROTEIN SECRETION. The application of irritants to the gastric mucosa is followed by the copious outpouring of mucus; hence it is not surprising that mucus has been assigned a barrier or protective role. Mucus is secreted as a gel, and in this form it is thought to control the microenvironment at the cell surface. At a low pH, molecules are detached from the gel to produce "soluble mucus," but as long as production keeps pace with the solubilization of mucus, the "mucus barrier" will be maintained.[79] The "mucus barrier" is no impediment to hydrogen ion diffusion, however. One of the stimuli for increasing mucus secretion is a drop in the pH of gastric contents. Vagal stimulation also leads to increased mucus production; whether mucus production is the consequence solely of increased vagal activity or whether an increased hydrogen ion concentration can stimulate mucus production directly has not been resolved.[61]

Control of gastric secretion

Study of the control of gastric secretion has been concerned almost exclusively with the control of hydrogen ion secretion by the stomach. Acid secretion has been divided into basal, cephalic, gastric, and intestinal phases. Although useful in formulating concepts, it is an oversimplification of interacting factors.

Much of our knowledge of the control of gastric secretion has been derived from experimental surgical preparations in dogs that permit the study of neural, humoral, and circulatory factors determining the rate of secretion. In order to determine the effect of the central nervous system (CNS) on gastric secretion, uncomplicated by effects of food in the stomach, sham feeding techniques that divert the food to the outside

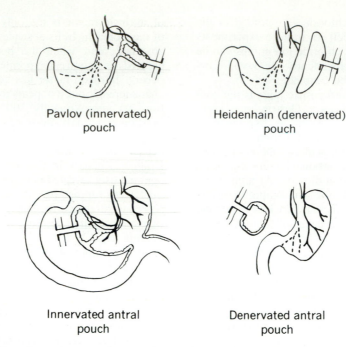

Pavlov (innervated)
pouch

Heidenhain (denervated)
pouch

Innervated antral
pouch

Denervated antral
pouch

Fig. 54-8. Experimental preparations used to study neural and humoral control of gastric secretion. Preparations combining antral and fundic pouches are sometimes used.

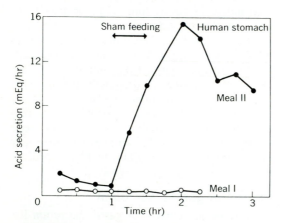

Fig. 54-9. Rate of acid secretion by stomach of 24-year-old woman with complete stenosis of esophagus. In her usual manner of eating, food was tasted, chewed, partly swallowed, regurgitated, expectorated, and then placed in stomach through a gastrostomy. In two instances recorded here, food was not placed in stomach after it had been tasted and chewed. Meal I (mean of 4 experiments) consisted of 8 oz cereal gruel, eaten dutifully but with obvious distaste. Meal II (mean of 12 experiments), subject's unrestricted choice, included fresh vegetables, salad with dressing, three slices of white bread with butter, two glasses of milk, potatoes, half a fried chicken or two broiled lamb chops or fried ham steak or two fried eggs, and ice cream and cake. (From Janowitz et al.[47])

through an esophageal fistula are employed. The acid output can be conveniently measured from fundic pouches that contain the oxyntic cells. Such preparations make it possible to sample gastric juice without the complicating factor of food contamination while maintaining alimentary tract continuity and the animal's nutrition. A Pavlov or innervated pouch is responsive to both humoral and neural stimulation, whereas a Heidenhain pouch or vagally denervated pouch responds to humoral but not to vagal stimuli.[21] Since sympathetic fibers accompany the blood vessels, the sympathetic innervation is intact in both. These preparations may be combined with antral pouches having a vagal innervation that is either intact or separated (Fig. 54-8).

Basal secretion. The stomach of man secretes acid at about 15% of maximal acid output during fasting. The regular fluctuations observed in basal secretion, peak in the evening and low in the morning, are not associated with changes in circulating gastrin.[62] The stimulus for basal gastric secretion is not known, since removal of neural and humoral stimuli by vagotomy and antrectomy decrease but do not abolish basal secretion. Histamine may play a role, as indicated by the finding in the rat that depletion of mast cells of histamine was followed by decreased

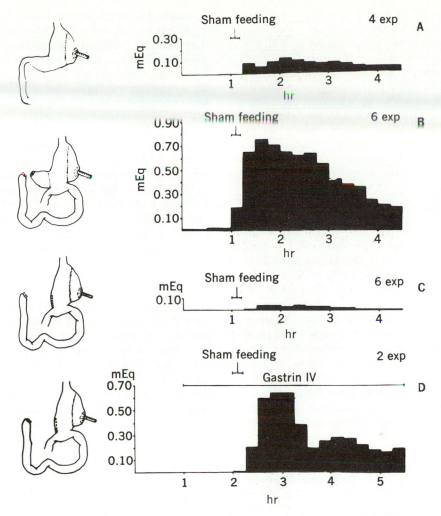

Fig. 54-10. Vagal release of gastrin. Sham feeding response in dog demonstrating importance of gastrin in cephalic phase of gastric acid secretion as measured by secretion from Pavlov (innervated) fundic pouch. **A,** Pavlov pouch preparation. Acid in stomach limits gastrin release. **B,** Isolation of innervated antrum. In absence of acid in antrum, vagal stimulation increases gastrin secretion, which greatly augments acid output by Pavlov pouch. **C,** Resection of antrum removes source of gastrin and limits response to sham feeding. **D,** Intravenous infusion of gastrin at one third threshold dose greatly augments secretory response to sham feeding. (From Olbe.[65])

basal secretion but no decrease in response to vagal and gastrin stimulation.[36]

Cephalic phase. The thought, smell, or taste of food stimulates acid secretion by activation of vagal efferent pathways. Stimulation of these pathways by sham feeding (Fig. 54-9), direct electrical stimulation, or in response to hypoglycemia produced by insulin or 2-deoxy-D-glucose also produces acid secretion. This vagus-induced secretion is produced by the combined effect of (1) ACh released at the parietal cells by stimulated postganglionic parasympathetic (vagal) nerves and (2) gastrin, the primary hormonal mediator of gastric acid secretion, released from antral G cells in response to vagal stimulation. In man the first pathway is the predominate one, whereas the second is the most important in the dog (Fig. 54-10).[27,34,91]

Gastric phase. Although in 1879 Heidenbain recognized that denervated fundic pouches secreted acid when food was in the main stomach, the explanation for this humorally mediated gastric secretion has been unraveled only recently (see discussion of endocrine secretion, p. 1314).

The components of food responsible for gastrin release are peptides and amino acids that stimulate chemoreceptors to release ACh, which in turn stimulates the release of gastrin from the G cells in the antrum. In addition, chemoreceptors in the oxyntic gland area directly stimulate the parietal cells through the release of ACh from postganglionic parasympathetic neurones. Distention of the stomach also contributes to the gastric phase of acid secretion. Stretch receptors in the gastric wall stimulate gastric secretion directly by stimulating the release of ACh at the oxyntic cells (these impulses reach the oxyntic cells through short (intramural) and long (vagovagal) reflexes and indirectly by stimulating gastrin release. The secretory response to stimulation of the chemoreceptors is blocked by local anesthetics and by anticholinergic drugs. In addition, acid in the antrum, probably by direct action on the G cells, inhibits the release of gastrin by all stimuli. Thus stimulation of chemoreceptors by protein digestion products and stretch receptors by distention leads to acid secretion through ACh-mediated release of gastrin and direct stimulation of parietal cells by ACh released by activation of short and long reflex arcs (Fig. 54-11).

Intestinal phase. Denervated fundic pouches are stimulated to secrete when food enters the intestine or the intestine is distended. Since the increased gastric stimulation cannot be explained by neural stimuli or by the absorbed products of digestion, an intestinal hormone has been suggested as the responsible agent.[29] Although cells containing immunoreactive gastrin are found in the duodenum and upper jejunum,[91] recent studies suggest that neither gastrin of intestinal origin nor cholecystokinin-pancreozymin, which has weak acid stimulatory properties, is the hormone responsible for the intestinal phase of gastric secretion.[15] "Entero-oxyntin" is used to designate this hormone, which has not yet been isolated.

In summary, the current view is that three interacting chemical mediators, ACh, gastrin, and histamine, are responsible for stimulating the parietal cell to secrete hydrochloric acid. The physiologic roles of the first two mediators have been well characterized, whereas histamine's physiologic role in gastric secretion is yet to be resolved. A role for histamine is supported by the following observations: (1) release of histamine from gastric stores by feeding, gastrin stimulation, and hypoglycemia (vagal stimulation); (2) activation of histidine decarboxylase (an enzyme that converts histidine to histamine) by gastrin stimulation, ingestion of food, and gastric distention; and (3) inhibition of the gastric secretion induced by food, gastrin stimulation, and vagal stimulation, as well as histamine stimulation

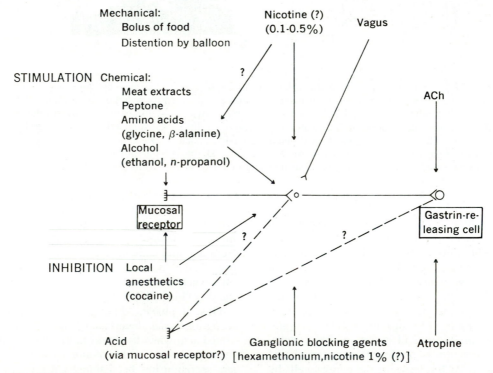

Fig. 54-11. Hypothetical mechanisms for release of gastrin from antral mucosa. (From Elwin and Uvnäs.[19])

by H$_2$-receptor antagonists. (Histamine has two types of receptors: H$_1$ receptors, which mediate histamine's effect on the smooth muscle of the bronchi, blood vessels, and gut and are blocked by antihistaminic drugs, and H$_2$ receptors, which mediate histamine's effect on gastric secretion, heart, and uterine muscle and are not blocked by antihistaminic agents.[55]) In addition, histamine, cholinergic agents, and gastrin potentiate the action of each other, and blockade of cholinergic receptors inhibits not only cholinergic stimulation of acid secretion but also the response to histamine and gastrin. These observations have led to the suggestion that the parietal cell has separate but interacting receptors for the three mediators of gastric secretion. Finally, there is suggestive evidence that cyclic AMP is the intracellular "messenger" for acid secretion.[16] Since caffeine is an inhibitor of phosphodiesterase, the enzyme that breaks down cyclic AMP, elevation of parietal cell levels of cyclic AMP may provide the explanation for caffeine stimulation of gastric secretion.

Inhibition of gastric secretion

Inhibition of acid secretion is effected by inhibiting the release of gastrin and decreasing the responsiveness of the parietal cell. Low pH in

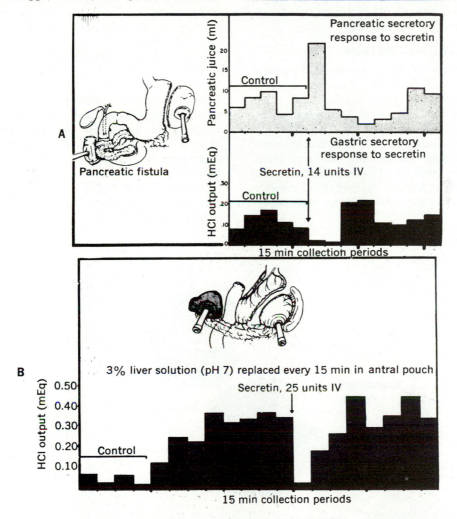

Fig. 54-12. Inhibition of gastric secretion by secretin. **A,** Effect of intravenous injection of secretin on secretion of gastric and pancreatic juice in dog provided with total pancreatic fistula and Heidenhain pouch. Simultaneous stimulation of pancreatic secretion and inhibition of gastric secretion was produced. **B,** Effect of intravenous injection of secretin on gastric secretion in dog with Heidenhain pouch. This animal was provided with an isolated pouch of antrum, which was stimulated continuously during course of experiment by instillation of 3% liver solution every 15 min. Injection of secretin produced marked temporary depression in gastric secretion. (From Greenlee et al.[26])

the antrum inhibits gastrin release regardless of the stimulus. The effect appears at pH 3 and reaches a maximum at pH 1, thus providing a negative feedback control of gastric acid secretion. This negative feedback has been attributed in part to a hormone isolated from the duodenal bulb (bulbogastrone).[33] A variety of hormones (gastric inhibitory peptide, enteroglucagon, vasoactive intestinal peptide, and cholecystokinin) released by glucose, amino acids, fatty acids, and hyperosmotic fluid, respectively, are capable of inhibiting parietal cell response, but their physiologic significance in this role is not established. Hormones from the intestine that inhibit acid secretion have in the past been called "enterogastrones"[50] (Fig. 54-12).

PANCREATIC SECRETION

The pancreas is the primary source of digestive enzymes active in the lumen of the small intestine and of bicarbonate required to raise the pH of the acid gastric chyme to a level compatible with pancreatic enzyme activity. The pancreas synthesizes and secretes more protein than any other gland, except for the lactating mammary gland. The enzymes synthesized by the pancreas include amylase, lipase, trypsin, chymotrypsin, carboxypeptidase, elastase, ribonuclease, and deoxyribonuclease.

Exocrine pancreas

The exocrine pancreas is composed of acini containing a single type of zymogen-containing cell. Extending into the acini is a very small duct lined with a flattened epithelium, the centroacinar cells, which connects to the intercalated ducts. The latter are interposed between the acini and the intralobular ducts, which are lined by columnar epithelium (Fig. 54-13). The anatomic arrangement resembles that of the parotid gland.

Pancreatic juice

Water and electrolyte secretion. Pancreatic juice is isosmolar with plasma. The ionic content of pancreatic juice varies with the secretory rate, as does that of the salivary and gastric glands (Fig. 54-14). The concentrations of sodium and potassium do not change with increasing secretory rates and equal plasma levels. On the other hand, the concentrations of chloride and bicarbonate change in a reciprocal fashion, so that the sum of their concentration equals the sum of sodium and potassium ion concentrations.[10] With an increasing rate of secretion the chloride concentration falls as the bicarbonate concentration rises.

A number of hypotheses have been advanced to explain the electrolyte composition of pancreatic juice.[10,54,57] The admixture theory, like

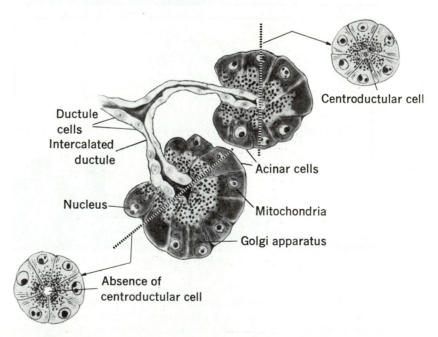

Fig. 54-13. Diagrammatic representation of pancreatic acini and terminal ductules. Upper acinus shows centroductular cell extending into acinus. Zymogen granules occupy luminal pole of acinar cells. (From Herman et al.[40])

that advanced to explain the composition of gastric juice, proposes the secretion of three fluids: a neutral chloride solution and a bicarbonate solution, both with an anionic concentration of 174 mEq/L, and a third solution containing enzymes, mucus, and chloride. It has been suggested that the centroacinar cells are analogous in location and function to the parietal cells in the gastric glands, secreting bicarbonate rather than hydrogen ions.[35]

The exchange theory suggests that the primary secretion originates from the centroacinar cells and the intercalated ducts and is an isosmotic bicarbonate solution. As it passes down the duct, there is a passive exchange of the bicarbonate in the lumen for chloride in the interstitial fluid. At low flow rates there is sufficient time for the exchange to come to equilibrium; hence the concentrations of the anions as well as the cations approach levels found in plasma. At high secretory rates, however, there is little time for chloride-bicarbonate exchange across the duct epithelium, and the ionic concentration of pancreatic juice approaches that of the primary secretion. The strongest support for this hypothesis is provided by the results of experiments in which a bicarbonate solution isosmotic with plasma was perfused at varying rates through the main pancreatic duct. For each flow rate the ionic composition of the fluid flowing from the duct was almost identical with the composition of normally produced pancreatic juice.[9]

Studies using micropuncture techniques have

led to the suggestion that the primary pancreatic secretion is a fluid with an electrolyte composition similar to that of plasma. As the primary fluid passes down the duct, bicarbonate is secreted against a concentration gradient.[81] This suggests a similarity between the secretory mechanisms of the parotid gland and the pancreas that parallels their anatomic similarities. It should be noted, however, that there is no morphologically specialized duct epithelium in the pancreas that is analogous to the striated intralobular duct epithelium of salivary glands.

It is not possible to present a clear picture of how bicarbonate is secreted against an electrochemical gradient, since the cells responsible have not been clearly identified. Several points, however, can be made: (1) The acinar cells are responsible for the production of the enzymes of pancreatic juice. (2) They do not appear to be primarily involved in the volume and bicarbonate production of pancreatic juice because ethionine, which damages the acinar cells, does not affect the volume or bicarbonate content of pancreatic juice.[52] (3) The ductal epithelium, or some part of it, is responsible for the volume and electrolyte composition of pancreatic juice because alloxan, which produces histologic damage to the pancreatic ducts, decreases the volume and bicarbonate response to secretin stimulation[32]; carbonic anhydrase, which is thought to play a role in bicarbonate secretion, is concentrated in ductal epithelium.[5] (4) Electrolyte transport must be by a neutral pump that presumably transfers sodium

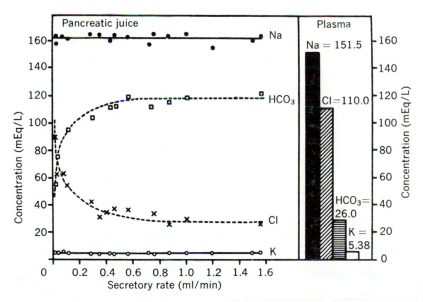

Fig. 54-14. Relation between rate of secretion and concentration of electrolytes in pancreatic juice compared with electrolyte composition of plasma. (From Bro-Rasmussen et al.[7])

bicarbonate to the lumen, since the potential difference across the ductal epithelium is small and is not changed by secretin stimulation or carbonic anhydrase inhibition.[81] Carbonic anhydrase plays a role, but probably not a central role, in the pancreatic secretion of bicarbonate because inhibition of this enzyme associated with a 50% decrease in volume response causes a much smaller decrease in bicarbonate concentration.[9] (5) The bicarbonate of pancreatic juice is formed primarily from carbon dioxide derived from the blood. In vitro fluid and electrolyte secretion by the pancreas depends on aerobic metabolism and the presence of chloride, sodium, and carbon dioxide in the bathing medium.[72]

Enzyme secretion. The pancreatic enzymes are synthesized by the acinar cells and perform three major functions: proteolytic, lipolytic, and amyolytic. The proteolytic enzymes are secreted as inactive zymogens or proenzymes (e.g., trypsinogen, chymotrypsinogen, proelastase). When trypsinogen enters the duodenum, it is converted to active trypsin by the splitting off of a peptide by enterokinase, an intestinal enzyme.[59] In turn, the trypsin activates more trypsinogen as well as the other proenzymes. Pancreatic lipase and amylase are secreted in an active form. Earlier it had been believed that the pancreas secreted its enzymes in a constant proportion, but recent studies indicate that the enzyme proportions can be modified.[74] The protein content of pancreatic juice varies from 0.1% to 10%, depending on the secretory state. On the other hand, the enzyme secretion adds very little to the volume of pancreatic juice. In some species, probably including man, there is a continuous low basal secretion by the acinar cells that is strikingly increased when stimulated. (For a discussion of pancreatic enzyme function, see the sections on protein, carbohydrate, and lipid absorption in Chapter 53.)

The enzymes are synthesized in the rough endoplasmic reticulum. In the Golgi complex, they are condensed and encapsulated to form zymogen granules. In the resting stage the zymogen granules accumulate in the apical portion of the cell. When stimulated to secrete by cholecystokinin-pancreozymin or ACh released from vagal fibers, the membrane of the zymogen granule fuses with the apical cell membrane and spills the zymogen into the lumen.[46]

Control of pancreatic secretion

Traditionally the control of pancreatic secretion has been thought of as the sum of the action of two independent hormonal stimulants, se-cretin controlling fluid and electrolyte secretion and cholecystokinin-pancreozymin controlling enzyme secretion. Although the concept is less completely developed than in the case of gastric secretion, it seems likely that pancreatic secretion is the consequence of interacting neural and humoral stimuli.

Gastric phase. Distention of the fundus of the stomach increases the enzyme, volume, and bicarbonate output of the pancreas. This response cannot be elicited after section of the vagus nerve. It is not known whether this phenomenon is produced by a direct action of the vagus on the secretory cells of the pancreas or whether it is mediated through the neural release of secretin and pancreozymin. Distention of the antrum, on the other hand, is followed by increased enzyme output by the pancreas, which is not blocked by vagotomy. This response is most probably elicited by gastrin released from the antrum which, having the same C-terminal tetrapeptide as cholecystokinin-pancreozymin, causes enzyme secretion by the pancreas (Fig. 54-15).

Intestinal phase. Two intestinal hormones, secretin and cholecystokinin-pancreozymin, play the major role in the control of pancreatic exocrine secretion.

SECRETIN. Secretin is a polypeptide hormone composed of 27 amino acids that is released from the mucosa of the duodenum and upper intestine. It stimulates fluid and bicarbonate secretion by the pancreas, presumably by acting on the centroacinar and ductal epithelial cells.[43] Acid in the duodenum is the most potent stimulus for secretin release, but other materials such as protein hydrolysates, amino acids, and fatty acids also stimulate its release (Fig. 54-16). The cells responsible for secretin production and release have been located by the immunofluorescence techniques in the duodenum and upper jejunum. The secretin-containing S cells belong to the series of endocrine-polypeptide cells that produce hormones such as gastrin and serotonin; they are located in the transitional zone between crypts and villi and have contact with the lumen.[67] It has not been established how secretin is released, but it is reasonable to suppose that, as with the gastrin cell, both a local neuron arc and secretagogues from the lumen that act directly on the secretin cell are involved.

The intravenous injection of secretin produces secretion that rises to a maximum rate within 1 min, reaches a plateau, and returns to basal level within 20 to 30 min. Enzyme content in the fluid collected immediately after secretin stimulation is high due to the "washout" of the enzyme that

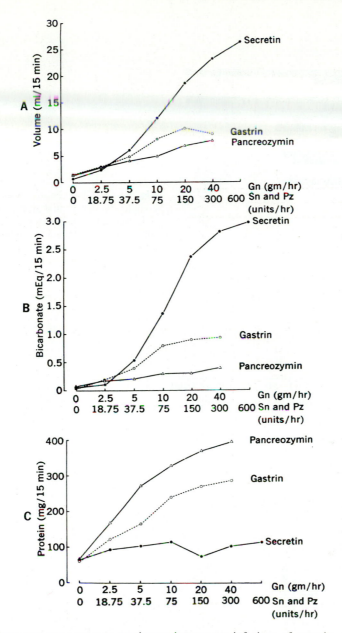

Fig. 54-15. Dose-response curves to continuous intravenous infusions of secretin, pancreozymin, and gastrin (mean results on four dogs with chronic pancreatic fistulas), **A,** Volume response. **B,** Bicarbonate output. **C,** Enzyme (protein) output. (From Harper.[35])

has accumulated in the ducts during the basal period from the resting secretion by acinar cells, but it quickly falls to low levels (Fig. 54-17).

CHOLECYSTOKININ-PANCREOZYMIN. When investigators found that extracts of duodenal mucosa produced such different effects as stimulation of enzyme secretion by the pancreas and contraction of the gallbladder, it was natural to suppose that the effects were due to separate hor-

mones that therefore deserved separate designations, that is, pancreozymin and cholecystokinin. When pancreozymin was purified, it was found that making a preparation free of secretin activity was relatively simple but that cholecystokinin activity increased in parallel with pancreozymin activity. Subsequently, it has been shown that cholecystokinin-pancreozymin (CCK-PZ) is a 33 amino acid peptide with the C-terminal penta-

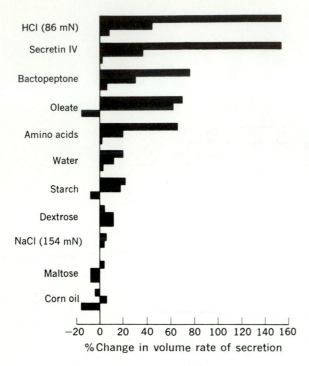

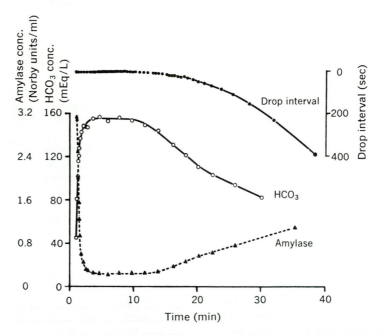

Fig. 54-16. Changes in volume rate of secretion by completely denervated transplanted pancreas of dog in response to substances placed in duodenum. (Secretin was injected to test magnitude of response to maximal stimulation.) Groups of bars represent three consecutive 20 min periods of collection. (Modified from Wang and Grossman[92]; from Davenport.[13])

Fig. 54-17. Pancreatic response to single intravenous injection of 0.35 mg secretin in anesthetized cat. Secretion began in 1 min and lasted 38 min, during which 3.26 gm of juice was secreted. Flow rate is indicated by interval between drops. Contour of alterations in bicarbonate and enzyme concentrations was demonstrated by analysis of successive small fractions of juice. (From Harper.[35])

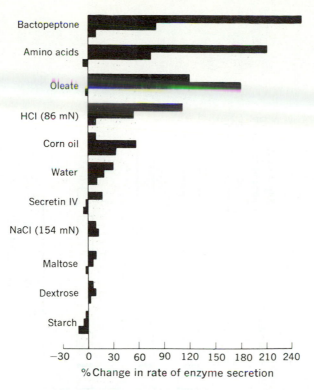

Fig. 54-18. Changes in rate of enzyme secretion by completely denervated, transplanted pancreas of dog in response to substances placed in duodenum. (Secretin was injected.) Groups of bars represent three consecutive 20 min periods of collection. (Modified from Wang and Grossman[92]; from Davenport.[13])

peptide identical to that of gastrin.[51] CCK-PZ causes discharge of zymogen granules from the acinar cells of the pancreas. With continuing stimulation, enzymes continue to be secreted but do not pass through the zymogen granule stage. Since very little fluid volume accompanies zymogen secretion, pancreatic enzyme secretion is best studied with background secretin stimulation, so that the secreted enzymes will be washed out of the gland.

CCK-PZ is released from the mucosa of the duodenum and upper intestine by protein hydrolysates, amino acids, fatty acids, and, to a lesser extent, acid (Fig. 54-18). A mixture of amino acids within the duodenum at physiologic concentrations is as potent a stimulus for pancreatic enzyme secretion as maximally tolerated doses of CCK-PZ. This ability to stimulate the release of CCK-PZ is attributed to four essential amino acids: phenylalanine, valine, methionine, and tryptophan.[25] The origin of CCK-PZ has been localized to endocrine cells (I cells) in the mucosa of the duodenum and jejunum.[68]

Vagal stimulation. Vagal stimulation causes the release of enzymes from the pancreas in amounts similar to those produced by CCK-PZ. This effect can be blocked by atropine, which interferes with transmission from the postganglionic parasympathetic neuron to the effector cell, and by hexamethonium, which blocks transmission from the pre- to the postganglionic neuron. CCK-PZ stimulation, on the other hand, is not altered by either of these two drugs, which suggests that the neural and humoral stimuli work through separate receptor sites on the acinar cells (Fig. 54-19). In addition, vagal stimulation augments flow and bicarbonate secretion, an effect not altered by atropine.[35] The present views on the stimulation of pancreatic secretion are diagrammed in Fig. 54-20.

Effect of diet. The enzyme content of pancreatic juice can be altered in experimental animals by changes in the proportion of carbohydrate, protein, and fat in the diet. An increase in amylase is evident after 3 to 5 days on a high-starch diet, and chymotrypsin increases on a high-casein diet. Lipase, however, was not altered by a high-fat diet.[70,74]

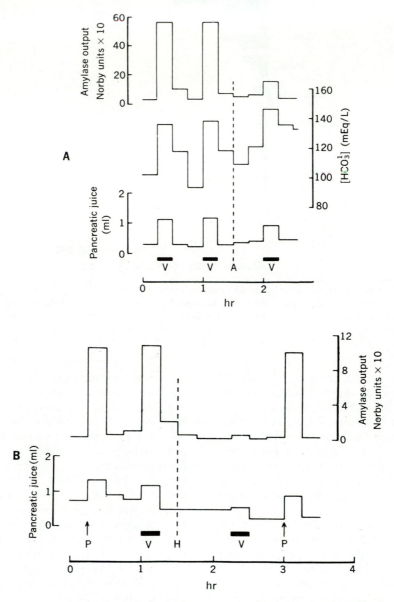

Fig. 54-19. A, Potentiating effect of stimulation of dorsal vagus trunk, *V*, on response to continuous intravenous infusion of secretin in anesthetized cat. Vagal stimulation increased volume, bicarbonate concentration, and enzyme content of juice. After atropine, *A*, effects of vagal stimulation on volume and bicarbonate concentration persisted. (Small increases in enzyme output on vagal stimulation after atropine can be attributed to "washing out" of enzymes by increased flow of juice.) **B,** Potentiating effect of 2 mg pancreozymin intravenously, *P*, and vagal stimulation, *V*, on response to continuous intravenous infusion of secretin in anesthetized cat. After intravenous injection of 3 mg/kg hexamethonium, *H*, effect of vagal stimulation on volume and enzyme content of juice was abolished, but response to pancreozymin persisted. (From Harper.[35])

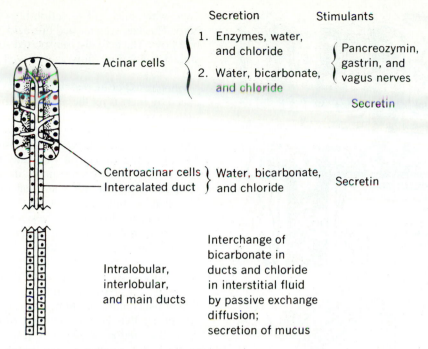

Fig. 54-20. Site of action and secretory effects of secretin, pancreozymin, gastrin, and vagus nerves on pancreas. (From Harper.[35])

BILE SECRETION

The secretion of bile serves two important physiologic functions: (1) it provides the bile salts necessary for the efficient absorption of lipids (Chapter 53) and (2) it is the excretory pathway for water-insoluble substances such as cholesterol, steroid hormones, drugs, and bilirubin, the breakdown product of hemoglobin and other heme proteins.[2] Bile is secreted continuously, but in species having a gallbladder, it is delivered only intermittently to the intestine.

Intrahepatic biliary tract

The parenchymal cells of the liver are arranged in fenestrated, interconnecting plates one cell thick that radiate from the central hepatic vein to the periphery of the hepatic lobule. Blood from the portal vein and hepatic artery enters the sinusoids at the periphery of the lobule, bathing both sides of the plates as it flows centrally to enter the central hepatic vein. Within the plates, each parenchymal cell is ringed by a bile capillary of canaliculus that is interconnected with those around adjacent cells in a "chicken-wire" pattern. The wall of the canaliculus is composed of specialized areas of the cell membrane of two adjacent hepatic parenchymal cells

and is separated from the rest of the cell membrane by a tight junction similar to that found in intestinal cells adjacent to the brush border (Fig. 54-21). The membrane of the canaliculus forms microvilli, structures generally associated with transport function. At the periphery of the lobule the bile canaliculi empty into cholangioles, or bile ductules, that are lined by squamous epithelium; these in turn connect with the intralobular ducts that are lined by cuboidal epithelium and main ducts lined by columnar epithelium.

Composition of bile

In the fasting state, bile coming from the liver has an electrolyte composition similar to that of plasma, with added bile salts and pigments.[88] In addition, bile contains cholesterol and lecithin; the latter with bile salts forms micelles that normally keep cholesterol in solution. The solubilization of cholesterol in bile is precarious at best, so that minor alterations in bile salt or phospholipid concentrations may lead to cholesterol precipitation and the formation of gallstones. Macromolecular aggregates of bilirubin in bile have also been described. These aggregates allow greater amounts of materials to be

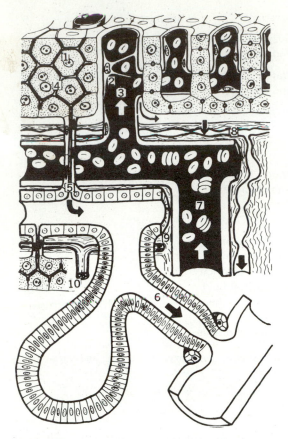

Fig. 54-21. Basic structures and currents of flow in liver. *1,* Hepatic cells; *2,* Küpffer cell; *3,* sinusoid; *4,* bile canaliculus; *5,* bile ductule; *6,* bile duct; *7,* portal vein; *8,* lymphatic vessel; *9,* stroma; *10,* nerve. (From Popper and Schaffner.[69])

carried in bile without exceeding the osmotic pressure of plasma.

Formation of bile

Although the electrolyte composition of bile is similar to that of plasma, bile is not formed from an ultrafiltrate of plasma. The strongest support for this statement is the observation that bile can be secreted against a pressure in the biliary tree that is 4 to 5 times the pressure in the sinusoids. The observation of an excellent correlation between canalicular bile flow and bile acid excretion has lead to the concept that the active secretion of bile acids into the canaliculus provides the osmotic gradient that brings water and electrolytes into the lumen. This has been called bile acid–dependent bile flow. At low rates of bile flow (not shown in Fig. 54-22), bile acid excretion does not account for all the bile flow. There is still bile flow (a positive inter-

cept) when bile acid secretion is extrapolated to zero. This has been termed bile acid–independent bile flow. The actively transported solute responsible for this bile flow is believed to be sodium.[48,83]

Choleresis, or increased bile flow, can be produced by increasing the amount of bile salts presented to the liver, either by intravenous injection or by feeding bile salts that then enter the enterohepatic circulation through ileal absorption (Fig. 54-22). If the amount of bile salts reaching the liver is decreased by interrupting the enterohepatic circulation with either an ileal resection or a bile fistula, bile flow decreases, because the capacity of the liver to increase synthesis of bile salts from cholesterol is not adequate to compensate for the loss of bile salts.

Not all bile flow can be explained as the consequence of active secretion of bile salts into the canaliculi. Secretin, in addition to its stimulation of pancreatic secretion, has a striking choleretic effect that is not mediated through an increased secretion of bile salts (Fig. 54-23). The response with an increased volume and bicarbonate content and reciprocal decrease in chloride concentration resembles the response of the pancreas to secretin. The volume response in bile secretion to secretin is, however, only one fifth to one tenth the pancreatic response.[94] Present evidence indicates that secretin induces choleresis by stimulation of secretion by the bile ductules rather than by the parenchymal cells at the level of the canaliculi. Although the question of whether secretin produces choleresis by stimulating secretion or by inhibiting reabsorption from the bile ducts is unresolved, the generally held view is that secretin stimulates the secretion of a second fluid that is added to the primary canalicular bile. In addition to secretin, gastrin and CCK-PZ stimulate volume and bicarbonate output in the bile, although to a lesser degree than does secretin. The extent of neural regulation of bile secretion is uncertain. It is not clear whether vagal stimulation increases bile flow by direct stimulation or by stimulating the release of gastrin and other hormones.

A great number of organic compounds in addition to conjugated bile salts and conjugated bilirubin are excreted in the bile, presumably by means of active transport mechanisms.[76] The list includes a wide variety of drugs and dyes. Within the liver cell, these substances are made more water soluble, usually by conjugation to a highly polar group, such as sulfate, amino acids, or glucuronic acid, so that they may be secreted into the bile (Fig. 54-24).[71,94]

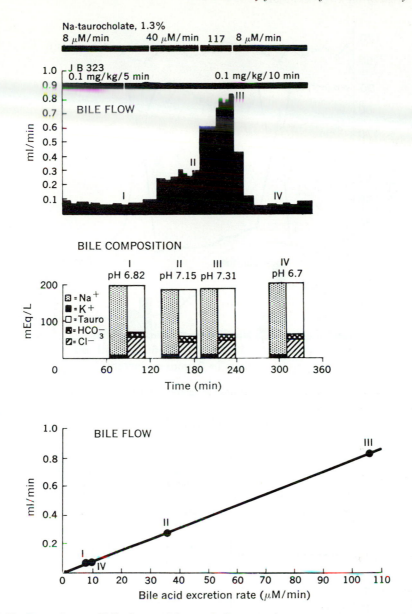

Fig. 54-22. Dependence of bile flow and inorganic ion excretion on taurocholate secretion rate. Anticholinergic drug pipenzolate methylbromide (JB 323) was administered, and sodium taurocholate infusion rate was varied in four successive stages. Bile flow exhibited linear dependence on taurocholate secretion rate. Bile composition changed very little, indicating that output of chloride and bicarbonate also depend on taurocholate secretion. (From Wheeler.[93])

A great deal is known about the transport of one endogenous organic compound, bilirubin. It is very insoluble in aqueous solution, but in the plasma, it is kept in solution by attachment to albumin. It is transferred to the liver cells because they contain an acceptor protein with a very high affinity for bilirubin. Once within the

liver cell, bilirubin is converted to bilirubin diglucuronide by a microsomal enzyme, glucuronyl transferase, and thus the gradient between sinusoidal blood and liver cytoplasm for unconjugated bilirubin is maintained.[2,77] The water-soluble bilirubin glucuronide is actively transferred into the canaliculus.

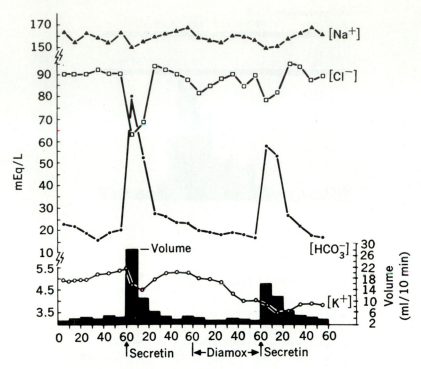

Fig. 54-23. Bile flow and electrolyte response to secretin and acetazolamide (Diamox), a carbonic anhydrase inhibitor. (From Waitman et al.[88])

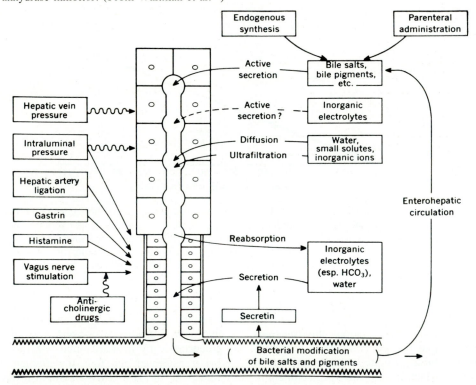

Fig. 54-24. Diagrammatic summary of some factors known to affect bile flow and electrolyte composition. Mechanisms shown at right are believed to act at cannalicular or duct level as shown. Locus and mode of action of factors shown at left are speculative. (From Wheeler.[94])

INTESTINAL SECRETION

Intestinal secretion was at one time considered to be an important source of digestive enzymes responsible for the final hydrolytic step preceding absorption. It has been shown, however, that these intestinal enzymes play their physiologic role in the epithelium rather than in the lumen.[95] A proposed but largely unexplored function of intestinal secretion is that of providing fluid to dilute the products of digestion to isotonicity so that they may be efficiently absorbed.[23] Finally, intestinal secretions perform a protective function, since they increase in response to chemical irritants and enterotoxins produced by bacteria such as *Vibrio cholerae*, staphylococci, and *Escherichia coli*.

There are two types of glands in the intestine: (1) Brunner's glands, which are submucosal and are found in the duodenum and extend into the upper jejunum in some individuals, and (2) intestinal glands or crypts of Lieberkühn, which are mucosal in location and found throughout the intestine.

Brunner's glands

Brunner's glands lie below the muscularis mucosae in the submucosa and are composed of branching ducts ending in simple acini; the ducts pass through the muscularis mucosae and empty into the crypts of Lieberkühn. Their secretion is mucoid and alkaline. It is stimulated by both vagal and humoral stimuli, but the latter are the more potent. Secretin probably is the major humoral agent,[56] although some investigators have been unable to produce Brunner's gland secretion with purified secretin and have suggested an unidentified humoral agent, "duocrinin," as that responsible for the observed secretion.[12,33]

Since the duodenum is more resistant to the damaging effects of acid gastric juice than is the small intestine below the Brunner gland area, it seems reasonable to attribute this resistance to a protective property of Brunner's gland secretion.

Intestinal glands

The intestinal glands are simple glands, densely packed, with their bases just above the muscularis mucosae. As they approach the mucosal surface, four glands join to form a vestibule that interconnects in turn with other vestibules to form the intervellous space.[11] There are four cell types in the crypts. (1) Undifferentiated crypt cells move up onto the villi to become absorptive cells, replacing the cells shed from the tips of the villi. New cells are formed by mitosis in the crypts. While in the crypts, these cells have secretory granules and enzymes of nucleic acid synthesis but none of the brush border hydrolytic enzymes such as the disaccharidases. As these cells migrate out of the crypts onto the villi, they lose functions such as nucleic acid synthesis and develop digestive and absorptive functions. (2) Paneth cells are highly organized secretory cells that occupy the base of the crypts and do not migrate. Their granules are released by feeding and by parasympathomimetic drugs, but the nature of the secretory product is unknown. (3) Goblet cells, which secrete mucus, are generally considered to perform a protective and lubricating function. (4) The argentaffin cells are endocrine in function.[66,87] Recent electron microscopic and immunofluorescence studies suggest that there are at least eight different types of endocrine cells in the intestine. Most types of endocrine cells are concentrated in the duodenum and upper jejunum.[67,82]

Composition of intestinal secretion

There are no direct measurements of the composition of intestinal secretion. What has been measured is the composition of intestinal juice, which is determined by what is absorbed from the lumen as well as what is secreted into it. Intestinal fluid is isosmotic to plasma, and the concentrations of sodium, potassium, calcium, and total anions are relatively constant and differ little from those of plasma. On the other hand, chloride and bicarbonate concentrations show a reciprocal relation. In the jejunum, bicarbonate concentration is lower than in plasma, whereas in the ileum, it is higher.[38]

In addition to water and electrolytes, mucus is added to intestinal juice by goblet cells. All serum proteins enter the intestinal lumen in small amounts, but IgA immunoglobulins have a special secretory form, in that a secretory or transfer piece that is synthesized in the epithelial cells is added to the immunoglobulin, thus facilitating its selective secretion.[86] Secretory IgA plays a role in protection against potentially harmful antigens in the intestinal lumen.[89,90]

Only two of the enzymes found in intestinal juice, enterokinase and intestinal amylase, are thought to have any physiologic role within the lumen; the remainder perform their physiologic roles in situ in the epithelial cells of the mucosa. Intestinal amylase appears to be a specific intestinal enzyme and not merely pancreatic amylase adsorbed onto the intestinal mucosa.[23]

Enterokinase splits a peptide from trypsinogen, thus forming the active enzyme trypsin, which in turn activates additional trypsinogen and other proteolytic pancreatic enzymes. Although it is generally believed that the crypts of Lieberkühn secrete enterokinase, recent evidence suggests that enterokinase may be a brush border enzyme. If true, this hypothesis greatly strengthens the notion of "contact digestion" and would suggest that the processes of digestion and absorption are truly one, located at the brush border and with interdependent rates determined in the final analysis by the rate of transfer across the luminal membrane of the intestinal absorptive cell (Chapter 53).

Control of intestinal secretion

Since little attention has been devoted to intestinal secretion in the last three decades, it is not surprising that there is little recent work on its control. *Local stimulation* by chemical agents such as mustard oil, 0.5% hydrochloride, soap, peptone, and sugar solutions leads to increased intestinal secretion, but the character of the fluid and the pathways involved have not been defined. *Distention* of the intestine increases intestinal secretion. *Neural stimulation* through the vagus has been said by some but not all investigators to produce intestinal secretion. There is agreement, however, that parasympathomimetic drugs such as pilocarpine and physostigmine produce copious intestinal secretion, which is blocked by atropine.[95] It has been suggested in the past that there is a *humoral* stimulant of intestinal secretion derived from the intestinal mucosa that is not one of the known intestinal hormones and has been called enterocrinin.[33,37] Recently, several hormones, including gastrin, secretin, CCK-PZ, and vasoactive intestinal peptide (VIP), have been reported to produce intestinal secretion. Of these, only VIP produced an associated elevation of cyclic AMP, a mechanism believed to mediate the intestinal secretion of cholera.[4,22,39,53] Gastric inhibitory peptide (GIP), calcitonin, and glucagon also have been reported to stimulate intestinal secretion. If the concept that intestinal secretion provides one limb of a fluid circuit to facilitate intestinal absorption is validated, then proof that intestinal secretion is stimulated by gastrointestinal hormones is to be expected.

ENDOCRINE SECRETION

Although endocrinology had its origin in the gastrointestinal tract at the turn of the century with the description of secretion by Bayless and Starling, alimentary tract endocrinology was dormant until the last decade. Now there are, in addition to the established hormones gastrin, secretin, and CCK-PZ, at least 18 "candidate hormones."[33] Criteria to establish a hormone include the following: (1) demonstration that a stimulus applied to one part of the gastrointestinal tract alters the activity of a distant part of the gastrointestinal tract, (2) demonstration that this effect persists when all neural connections between the stimulated and responding part have been severed, (3) isolation from the site of stimulation of a substance that mimics the physiologic action, and (4) chemical identification of the substance and proof of its structure by syntheses.[28] At this time, these criteria can be met only for gastrin, secretin, and CCK-PZ. Physiologic evidence suggests the following hormone actions, but since chemical confirmation of the existence of a specific hormone for each action is lacking, they are termed candidate hormones: bulbogastrone, a hormone released by acidification of the duodenal bulb that inhibits gastric acid secretion; duocrinin, a hormone that stimulates Brunner gland secretion; enterocrinin, a hormone that stimulates intestinal secretion; enterogastrone, an intestinal hormone that inhibits gastric acid secretion; enteroglucagon, a glycogenolytic intestinal hormone; incretin, an insulin-releasing hormone; and villikinin, a hormone that stimulates contraction and relaxation of intestinal villi. Some of these actions are possessed by established hormones, for example, gastrin, secretin, CCK-PZ, and glucagon all stimulate Brunner's gland, but it is not established that this is a physiologic role of any of these hormones, and gastrin, secretin, CCK-PZ, GIP, and VIP all release insulin, hence possess incretin activity. In addition, four bioactive pure peptides have been isolated from the intestinal mucosa, but the physiologic stimuli for their release, hence their hormone status, has not been established. Chymodenin elicits the secretion of chymotrypsin; GIP inhibits gastric secretion (enterogastrone activity) and is insulinotropic (incretin activity); motilin stimulates gastric motility when the duodenum is alkalinized; and VIP is a vasodilator that also inhibits gastric secretion and stimulates electrolyte and water secretion by the pancreas, liver, and intestine (enterocrinin activity).[33] Undoubtedly, some on this list will turn out to be minor manifestations of the activities of established hormones, just as pancreozymin and cholecystokinin have been shown to be a single hormone that influences different functions.

Human gastrin II

```
1   2   3   4   5   6   7   8   9   10  11  12  13  14  15  16   17
Pyr-Gly-Pro-Trp-Leu-Glu-Glu-Glu-Glu-Glu-Ala-Tyr-Gly-Trp-Met-Asp-Phe-NH2
                                              |
                                              O
                                              |
                                            HSO3
```

Porcine cholecystokinin-pancreozymin

```
1    2    3    4    5    6    7   8    9    10    11  12  13   14    15 16-17 18   19-20
Lys-(Ala, Gly, Pro, Ser)-Arg-Val-(Ile, Met, Ser)-Lys-Asn-(Asn, Gln, His, Leu2, Pro, Ser2)-

21 22 23  24   25 26 27  28 29 30 31 32  33
Arg-Ile-(Asp, Ser)-Arg-Asp-Tyr-Met-Gly-Trp-Met-Asp-Phe-NH2
                            |
                            O
                            |
                          HSO3
```

Porcine secretin

```
1   2   3   4   5   6   7   8   9   10  11  12  13  14  15  16  17  18  19  20
His-Ser-Asp-Gly-Thr-Phe-Thr-Ser-Glu-Leu-Ser-Arg-Leu-Arg-Asp-Ser-Ala-Arg-Leu-Gln-

21  22  23  24  25  26  27
Arg-Leu-Leu-Gln-Gly-Leu-Val-NH2
```

Glucagon

```
1   2   3   4   5   6   7   8   9   10  11  12  13  14  15  16  17  18  19  20
His-Ser-Gln-Gly-Thr-Phe-Thr-Ser-Asp-Tyr-Ser-Lys-Tyr-Leu-Asp-Ser-Arg-Arg-Ala-Gln-

21  22  23  24  25  26  27  28  29
Asp-Phe-Val-Gln-Try-Leu-Met-Asn-Thr
```

Fig. 54-25. Structure of gastrointestinal hormones. Gastrin I is identical to gastrin II except that tyrosine in position 12 is not sulfated. Identical sequences in two hormone pairs are underlined.

It has been suggested that the endocrine cells of the alimentary tract are members of a larger class of polypeptide hormone–producing cells. These cells share many biochemical, histologic, and ultrastructural characteristics, including an origin from the alimentary canal. In addition to nine types of polypeptide endocrine cells in the gut, the list includes the several types of pancreatic islet cells, the calcitonin-producing cell of the thyroid gland, the chief cell of the parathyroid gland, and the corticotropin and melanotropin cells of the pituitary gland.[24,67,82] With the use of immunofluorescent microscopy, nine types of the gastrointestinal polypeptide hormone–producing cells have been related to specific hormones.[68,82]

The chemical structures of gastrin, CCK-PZ, secretin, and glucagon are shown in Fig. 54-25. Gastrin and CCK-PZ share a similar structure at the carboxyl terminus and, as shown in Table 54-1, they produce similar physiologic effects. Secretin and glucagon likewise show striking similarities in structure and function. From these considerations, it has been suggested that these hormones act on one receptor that has two interacting sites, one for gastrin and CCK-PZ, and the other for secretin and other secretin-like hormones (e.g., glucagon, VIP, and GIP). The physiologic response will be the algebraic sum of the degree of activation of the two receptor sites (Fig. 54-26).[30,31] Such a system provides for an enormous number of interactions and possibilities for modulating and

Table 54-1. Tentative list of effects of gastrointestinal hormones

Site of action	Gastrin	Cholecysto-kinin-pancreozymin	Secretin	Glucagon
Lower esophageal sphincter	+	+	−	−
Stomach				
Acid secretion	⊕	+	⊖	⊖
Pepsinogen secretion	⊕	+	+	+
Antral motility	⊕	⊖	⊖	⊖
Pyloric sphincter	−	⊕	⊕	+
Pancreas				
HCO_3^- secretion			⊕	+
Enzyme secretion	+	⊕		
Liver				
HCO_3^- secretion			⊕	
Gallbladder				
Contraction	+	⊕		
Duodenum and small intestine				
Brunner's glands	+	+	⊕	+
Intestinal secretion	+	+	+	+
Motility	+	+	−	−
Ileocecal sphincter	−			
Metabolic effects				
Insulin release	+	+	+	+

+ = stimulation; − = inhibition; ○ = predominant effect.

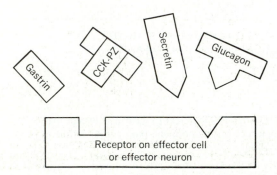

Fig. 54-26. Proposed mechanism for mediation of gastrointestinal hormones. Response of effector cell is determined by (1) relative concentration of competing hormone pairs (gastrin–CCK-PZ and secretin-glucagon); (2) affinity of hormone for receptor site; (3) efficacy of hormone in activating effector cell; and (4) interaction of two receptor sites. For example, if affinity of receptor is equal for gastrin and CCK-PZ and gastrin:CCK-PZ ratio is 4:1, response of effector cell will be predominantly what is expected of gastrin. If CCK-PZ is much less effective in activating effector cell than gastrin, CCK-PZ will appear to inhibit effect of gastrin, since response will be less than that seen with gastrin alone. If the two sites (gastrin–CCK-PZ and secretin-glucagon) have similar effect on effector cell, effect of hormone pairs will be augmented. On the other hand, if they have opposite effects, result will be noncompetitive inhibition.

coordinating functions of the elements of the alimentary tract.

REFERENCES

1. Andersson, S.: Gastric and duodenal mechanisms inhibiting gastric secretion of acid. In Code, C. F., editor: Handbook of physiology. Alimentary canal section, Baltimore, 1968, The Williams & Wilkins Co., vol. 2.
2. Arias, I. M.: Formation of bile pigment. In Code, C. F., editor: Handbook of physiology, Alimentary canal section, Baltimore, 1968, The Williams & Wilkins Co., vol. 5.
3. Babkin, B. P.: Secretory mechanism of digestive glands, ed. 2, New York, 1950, Paul B. Hoeber, Inc.
4. Barbezat, G. O., and Grossman, M. I.: Intestinal secretion: stimulation by peptides, Science **174:**422, 1971.
5. Becker, V.: Histochemistry of the exocrine pancreas. In de Reuck, A. V. S., and Cameron, M. P., editors: Ciba Foundation symposium on the exocrine pancreas: normal and abnormal functions, Boston, 1961, Little, Brown & Co.
6. Bloom, W., and Fawcett, D. W.: Textbook of histology, Philadelphia, 1968, W. B. Saunders Co.
7. Bro-Rasmussen, F., Kilman, S. A., and Thaysen, J. H.: The composition of pancreatic juice as compared to sweat, parotid saliva and tears, Acta Physiol. Scand. **37:** 97, 1956.
8. Burgen, A. S. V., and Emmelin, N. G.: Physiology of the salivary glands, Baltimore, 1961, The Williams & Wilkins Co.
9. Case, R. M., Harper, A. A., and Scratcherd, T.: The secretion of electrolytes and enzymes by the pancreas of the anaesthetized cat, J. Physiol. **201:**335, 1969.

10. Case, R. M., Harper, A. A., and Scratcherd, T.: Water and electrolyte secretion by the pancreas. In Botelho, S. Y., Brooks, F. P., and Shelley, W. B., editors: Exocrine glands, Philadelphia, 1969, University of Pennsylvania Press.

11. Cocco, A. E., Dohrmann, M. J., and Hendrix, T. R.: Reconstruction of normal jejunal biopsies: three-dimensional histology, Gastroenterology **51**:24, 1966.

12. Cooke, A. R.: The glands of Brunner. In Code, C. F., editor: Handbook of physiology, Alimentary canal section, Baltimore, 1968, The Williams & Wilkins Co., vol. 2.

13. Davenport, H. W.: Physiology of the digestive tract, ed. 3, Chicago, 1971, Year Book Medical Publishers, Inc.

14. Davies, R. E.: The metabolism of gastric mucosa during secretion of hydrochloric acid. In Shnitka, T. K., Gilbert, J. A. L., and Harrison, R. C., editors: Gastric secretion, London, 1967, Pergamon Press, Ltd.

15. Debos, H. T., Slaff, G. F., and Grossman, M. I.: Intestinal phase of acid secretion: augmentation of maximal response of Heidenhain pouch to gastrin and histamine, Gastroenterology **68**:691, 1975.

16. Donsa, T. P., and Code, C. F.: Effect of histamine and its methyl derivative on cAMP metabolism in gastric mucosa and its blockade by an H_2 receptor antagonist, J. Clin. Invest. **53**:334, 1974.

17. Durbin, R. P.: Electrical potential difference of the gastric mucosa. In Code, C. R., editor: Handbook of physiology, Baltimore, 1968, The Williams & Wilkins Co., vol. 2.

18. Ellison, S. A.: Proteins and glycoproteins of saliva. In Code, C. F., editor: Handbook of physiology. Alimentary canal section, Baltimore, 1968, The Williams & Wilkins Co., vol. 2.

19. Elwin, C. E., and Uvnäs, B.: Distribution and local release of gastrin. In Grossman, M. I., editor: Gastrin, UCLA forum in medical sciences No. 5, Berkeley, Calif., 1966, University of California Press.

20. Emmelin, N.: Nervous control fo salivary glands. In Code, C. F., editor: Handbook of physiology. Alimentary canal section, Baltimore, 1968, The Williams & Wilkins Co., vol. 2.

21. Emos, S., Swan, K. G., and Jacobsen, E. D.: Methods of studying gastric secretion. In Code, C. F., editor: Handbook of physiology, Alimentary canal section, Baltimore, 1968, The Williams & Wilkins Co., vol. 2.

22. Field, M.: Intestinal secretion, Gastroenterology **66**:1063, 1974.

23. Florey, H. W., Wright, R. D., and Jennings, M. A.: The secretions of the intestine, Physiol. Rev. **21**:36, 1941.

24. Forssman, W. G.: Ultrastructure of hormone-producing cells of the upper gastrointestinal tract. In Creutzfeld, W., editor: Origin, chemistry, physiology and pathophysiology of the gastrointestinal hormones, Stuttgart, Germany, 1969, F. K. Schattauer Verlag.

25. Go, V. L. W., Hofmann, A. F., and Summerskill, W. H. J.: Pancreozymin bioassay in man based on pancreatic enzyme secretion: potency of specific amino acids and other digestive products, J. Clin. Invest. **49**:1558, 1970.

26. Greenlee, H. B., et al.: Inhibitory effort of pancreatic secretin on gastric secretion, Am. J. Physiol. **190**:396,- 1957.

27. Grossman, M. I., editor: Gastrin, UCLA forum in medical sciences No. 5, Berkeley, Calif., 1966, University of California Press.

28. Grossman, M. I.: Gastrointestinal hormones, Med. Clin. North Am. **52**:1297, 1968.

29. Grossman, M. I.: Neural and humoral stimulation of gastric secretion of acid. In Code, C. F., editor: Handbook of physiology, Alimentary canal section, Baltimore, 1968, The Williams & Wilkins Co., vol. 2.

30. Grossman, M. I.: Effect of gastrin, cholecystokinin and secretin on gastric and pancreatic secretion: a theory of interaction of hormones. In Cruetzfeld, W., editor: Origin, chemistry, physiology and pathophysiology of the gastrointestinal hormones, Stuttgart, Germany, 1969, F. K. Schattauer Verlag.

31. Grossman, M. I.: Gastrin, cholecystokinin and secretin act on one receptor, Lancelot **1**:1088, 1970.

32. Grossman, M. I., and Joy, A. C.: Effect of alloxan upon external secretion of the pancreas, Proc. Soc. Exp. Biol. Med. **63**:62, 1946.

33. Grossman, M. I. et al.: Candidate hormones of the gut, Gastroenterology **67**:730, 1974.

34. Hanksy, J.: Clinical aspects of gastrin physiology, Med. Clin. North Am. **58**:1217, 1974.

35. Harper, A. A.: Hormonal control of pancreatic secretion. In Code, C. F., editor: Handbook of physiology. Alimentary canal section, Baltimore, 1968, The Williams & Wilkins Co., vol. 2.

36. Heap, B. J., and Kiernan, J. A.: Mast cells and gastric secretion in the rat, J. Exp. Physiol. **60**:307, 1975.

37. Heggeness, F. W., and Nasset, E. S.: Purification of enterocrinin, Am. J. Physiol. **167**:149, 1951.

38. Hendrix, T. R., and Bayless, T. M.: Digestion: intestinal secretion, Annu. Rev. Physiol. **32**:139, 1970.

39. Hendrix, T. R., and Paulk, H. T.: Intestinal secretion, Int. Rev. Physiol. **12**:257, 1977.

40. Herman, L., Sato, T., and Fitzgerald, P. J.: The pancreas. In Kurtz, S. M., editor: Electron microscopic anatomy, New York, 1964, Academic Press, Inc.

41. Hirschowitz, B. I.: Gastric osmolar clearance and ionic barrier. Two mechanisms of electrolyte secretion by the human stomach, J. Appl. Physiol. **15**:933, 1960.

42. Hirschowitz, B. I.: Secretion of pepsinogen. In Code, C. F., editor: Handbook of physiology. Alimentary canal section, Baltimore, 1968, The Williams & Wilkins Co., vol. 2.

43. Hubel, K. A.: Secretin: a long progress note, Gastroenterology **62**:318, 1972.

44. Hunt, J. N., and Wan, B.: Electrolytes of mammalian gastric juice. In Code, C. F., editor: Handbook of physiology, alimentary canal section, Baltimore, 1968, The Williams & Wilkins Co., vol. 2.

45. Ito, S., and Winchester, R. J.: The fine structure of the gastric mucosa in the bat, J. Cell Biol. **16**:541, 1963.

46. Jamieson, J. D., and Palade, G. E.: Intracellular transport of secretory proteins in the pancreatic acinar cell. I. Role of the peripheral elements of the Golgi apparatus. II. Transport of condensing vacuoles and zymogen granules, J. Cell. Biol. **34**:577, 1967.

47. Janowitz, H. D., et al: A quantitative study of the gastric secretory response to sham feeding in a human subject, Gastroenterology **16**:104, 1950.

48. Javitt, N. B.: Hepatic bile formation, N. Engl. J. Med. **295**:1464, 1976.

49. Jeffries, G. H.: Gastric secretion of intrinsic factor. In Code, C. F., editor: Handbook of physology, Alimentary canal section, Baltimore, 1968, The Williams & Wilkins Co., vol. 2.

50. Johnson, L. R., and Grossman, M. I.: Intestinal hor-

mones as inhibitors of gastric secretion, Gastroenterology **60**:120, 1971.

51. Jorpes, J. E.: The isolation and chemistry of secretin and cholecystokinin, Gastroenterology **55**:157, 1968.

52. Kalser, M. H., and Grossman, M. I.: Pancreatic secretion in dogs with ethionine-induced pancreatitis, Gastroenterology **26**:189, 1954.

53. Klaeveman, H. L., et al.: Effects of gastrointestinal hormones on adenylate cyclase activity in human jejunal mucosa, Gastroenterology **68**:667, 1975.

54. Leeson, C. R.: Structure of salivary glands. In Code, C. F., editor: Handbook of physiology. Alimentary canal section, Baltimore, 1968, The Williams & Wilkins Co., vol. 2.

55. Lin, T. M.: Possible relation of gastrin and histamine receptors in gastric hydrochloric acid secretion, Med. Clin. North Am. **58**:1247, 1974.

56. Love, J. W., et al.: The hormonal control of Brunner's gland secretion and duodenal motility in the dog, J. Surg. Res. **10**:395, 1970.

57. Makhlouf, G. M., and Blum, A. L.: Assessment of models for pancreatic secretion, Gastroenterology **59**:896, 1970.

58. Marks, I. N., Komarov, S. A., and Shay, H.: Acid secretory responses to histamine and the parietal cell mass in the dog, Am. J. Physiol. **195**:528, 1958.

59. McCarthy, D. M., Nicholson, J. A., and Kim, Y. S.: Alteration in enterokinase trypsin, and alkaline phosphatase in response to variation in dietary protein content in the rat, J. Lab. Clin. Med. **89**:72, 1977.

60. McGuigan, J. E.: Gastric mucosal intracellular localization of gastrin by immunofluorescence, Gastroenterology **55**:315, 1968.

61. Menguy, R.: Regulation of gastric mucus secretion. In Shnitka, T, K., Gilbert, J. A. L., and Harrison, R. C., editors: Gastric secretion, mechanisms and control, London, 1967, Pergamon Press, Ltd.

62. Moore, J. G., and Englert, E.: Circadian rhythm of gastric acid secretion in man, Nature **226**:1261, 1970.

63. Nordgren, B.: The rate of secretion and electrolyte content of normal gastric juice, Acta Physiol. Scand. **58** (suppl. 202):1, 1963.

64. Oi, M., Oshida, K., and Sugimura, S.: The location of gastric ulcer, Gastroenterology **36**:45, 1959.

65. Olbe, L.: Vagal release of gastrin. In Grossman, M. I., editor: Gastrin, UCLA forum in medical sciences No. 5, Berkeley, Calif., 1966, University of California Press.

66. Pearse, A. G. E., et al.: The endocrine polypeptide cells of the human stomach, duodenum and jejunum, Gut **11**:649, 1970.

67. Polak, J. M., et al.: Immunofluorescent localization of secretin and enteroglucagon in human intestinal mucosa, Scand. J. Gastroenterol. **6**:739, 1971.

68. Polak, J. M., et al.: Cellular localization of cholecystokinin, Lancet **2**:1016, 1975.

69. Popper, H., and Schaffner, F.: Liver: structure and function, New York, 1957, McGraw-Hill Book Co.

70. Reboud, J. P., et al.: Adaptation of the rate of biosyntheses of pancreatic amylase and chymotrypsinogen to starch-rich and protein-rich diets, Biochim. Biophys. Acta **117**:351, 1966.

71. Remmer, H.: The role of the liver in drug metabolism Am. J. Med. **49**:617, 1970.

72. Rothman, S. S., and Brooks, F. P.: Pancreatic secretion in vitro in "Cl-free," "CO_2-free" and low-Na^+ environment, Am. J. Physiol. **209**:790, 1965.

73. Samloff, I. M.: Pepsinogens, pepsin and pepsin inhibitors, Gastroenterology **60**:586, 1971.

74. Sarles, H., et al.: Comportement different de la lipase, de l'amylase et des enzymes proteolytiques pancréatiques après differents modes d'excitation du pancreas humain, Bull. Soc. Chim. Biol. **48**:951, 1966.

75. Schachter, M.: Control of blood flow in the salivary glands. In Schneyer, L. H., and Schneyer, C. A., editors: Secretory mechanisms of salivary glands, New York, 1967, Academic Press, Inc.

76. Schanker, L. S.: Secretion of organic compounds in the bile. In Code, C. F., editor: Handbook of physiology. Alimentary canal section, Baltimore, 1968, The Williams & Wilkins Co., vol. 5.

77. Schmid, R.: Bilirubin metabolism in man. N. Engl. J. Med. **287**:703, 1972.

78. Schneyer, L. H. Secretion of potassium by perfumed excretory duct of rat submaxillary gland, Am. J. Physiol. **217**:1324, 1969.

79. Schneyer, L. H., and Emmelin, N.: Salivary secretion of electrolytes, Physiol. Rev. **52**:720, 1972.

80. Schrager, J.: The chemical composition and function of gastrointestinal mucus, Gut **11**:450, 1970.

81. Schultz, I.: Micropuncture studies of the pancreas of the rabbit. In Botelho, S. Y., Brooks, F. P., and Shelley, W. B., editors: Exocrine glands, Philadelphia, 1969, University of Pennsylvania Press.

82. Solcia, E., et al.: Endocrine cells of the intestinal mucosa. In Thompson, J. C., editor: Proceedings of Symposium on Gastrointestinal Hormones, Austin, Texas, 1975, University of Texas Press, p. 155.

83. Sperber, I.: Biliary secretion of anions and its influence on bile flow. In Taylor, W., editor: The biliary system, Oxford, 1965, Blackwell Scientific Publications, Ltd.

84. Stevens, C. E., and Leblond, C. P.: Renewal of the mucosal cells in the gastric mucosa of the rat, Anat. Rec. **115**:231, 1953.

85. Thaysen, J. H., Thorn, N. A., and Schwartz, I. L.: Excretion of sodium, potassium, chloride and carbon dioxide in human parotid saliva, Am. J. Physiol. **178**:155, 1954.

86. Tomasi, T. B., Jr., et al.: Characteristics of an immune system common to certain external secretions, J. Exp. Med. **121**:101, 1965.

87. Trier, J. S.: Morphology of the epithelium of the small intestine. In Code, C. F., editor: Handbook of physiology. Alimentary canal section, Baltimore, 1968, The Williams & Wilkins Co., vol. 3.

88. Waitman, A. M., Dyck, W. P., and Janowitz, H. D.: Effect of secretin and acetazolamide on the volume and electrolyte composition of hepatic bile in man, Gastroenterology **56**:286, 1969.

89. Waldman, R. H.: Local mucosal immunity, Am. J. Med. Sci. **260**:255, 1970.

90. Walker, W. A.: Host defense mechanism in the gastrointestinal tract, Pediatrics **57**:901, 1975.

91. Walsh, J. H.: Control of gastric secretion. In Sleisenger, M. H., and Fordtran, J. S., editors: Gastrointestinal disease, Philadelphia, 1973, W. B. Saunders Co., p. 144.

92. Wang, C. C., and Grossman, M. I.: Physiological determination of release of secretin and pancreozymin from intestine of dogs with transplanted pancreas, Am. J. Physiol. **164**:527, 1951.

93. Wheeler, H. O.: Inorganic ions in bile. In Taylor, W., editor: The biliary system, Oxford, 1965, Blackwell Scientific Publications, Ltd.

94. Wheeler, H. O.: Water and electrolytes in bile. In Code, C. F., editor: Handbook of physiology. Alimentary canal section, Baltimore, 1968, The Williams & Wilkins Co., vol. 5.

95. Wright, R. D., et al.: Influence of nerves and drugs on secretion by the small intestine and an investigation of the enzymes in intestinal juice, Q. J. Exp. Physiol. **30:**72, 1940.

96. Young, J. A., et al.: Micropuncture and perfusion studies of fluid and electrolyte transport in the rat submaxillary gland. In Schneyer, L. H., and Schneyer, C. A., editors: Secretory mechanisms of salivary glands, New York, 1967, Academic Press, Inc.

55

THOMAS R. HENDRIX

The motility of the alimentary canal

The motility or muscular activity of the alimentary canal serves three functions: (1) transport of ingested food and fluid through the gut in an orderly fashion and finally the discharge or defecation of the unabsorbed residua, (2) conversion of ingested food to a soluble, absorbable form by grinding and kneading it with the digestive juices, and (3) mixing the luminal contents so that the material in contact with the absorbing surface is constantly renewed, thus increasing the efficiency of absorption by minimizing the distance across which materials prepared for absorption must diffuse. In general the smooth muscle of the gut is arranged in three coats: an inner muscularis mucosa, a thin coat that serves as a base for the mucosa; the circular muscle coat with the fibers arranged approximately at right angles to the axis of the gut; and the outer longitudinal coat with the fibers paralleling the axis of the gut. The musculature of the alimentary canal is smooth muscle, except for that of the mouth, pharynx, proximal esophagus, and external anal sphincter (Chapter 4).

The alimentary tract is divided into functional compartments by a series of sphincters; the upper esophageal sphincter (the circopharyngeus), the lower esophageal sphincter, the pyloric sphincter, the ileocecal sphincter, and the internal and external anal sphincters. A sphincter is characterized by two properties: (1) is has a resting tone greater than the adjacent segments of the alimentary canal, which results in an intraluminal, high-pressure zone that separates the lumen into two compartments, and (2) it relaxes in response to the appropriate stimulus so that flow may occur from one compartment to the next (Fig. 55-1). In simplest terms the upper esophageal sphincter prevents air from being drawn into the esophagus during respiration; the lower esophageal sphincter prevents irritant gastric juice from entering the esophagus; the pyloric sphincter separates the acid environment of the stomach from the alkaline environment of the duodenum, a separation necessary if the respective digestive enzymes are to function properly; the ileal-cecal sphincter limits the reflux of bacteria-laden colonic contents into the small intestine; and the anal sphincters are necessary for continence.

Innervation of the muscle of the gut is provided by efferents from both the sympathetic and parasympathetic portions of the autonomic nervous system by a complex and as yet poorly understood intrinsic nerve plexus with myenteric ganglia that lie between the circular and longitudinal muscle coats, and by a submucosal nerve plexus with its ganglia lying between the muscularis mucosae and the circular muscle coat. Except for the muscles of mastication, the tongue and the oropharynx at the entrance, and the external anal sphincter at the exit of the alimentary canal, these motor functions are not under conscious control and usually proceed without awareness. The esophagus, although not under conscious control, does not function in the absence of its central nervous system (CNS) connections, whereas the functions of the stomach, small intestine, gallbladder, and colon appear to be but little affected by the interruption of their external nerve supplies.

The motility of the distal stomach, small intestine, and colon is regulated through a combination of myogenic, neural, and hormonal factors. Of these three the intrinsic myogenic control system is fundamental.[10,14,19] This control system is based on intrinsic, rhythmic fluctuations of membrane potential of the muscle cells (Fig. 55-2). These rhythmic depolarizations have been called slow waves, control activity, basic electrical rhythm (BER), and pacesetter potentials (PSP). The configuration, frequency, and distance of propagation of these waves of depolarization vary from site to site and from species to species. The slow waves do not cause muscle contraction because they are present

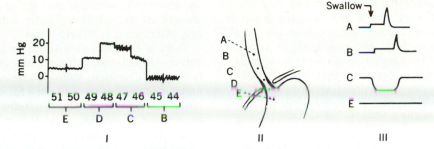

Fig. 55-1. Characteristics of sphincter as demonstrated by lower esophageal sphincter. **I,** Intraluminal pressure is recorded as perfused catheter is withdrawn in 1 cm steps from stomach into esophagus. Numbers along bottom indicate distance in centimeters between recording orifice and anterior nares; letters indicate relative position of recording orifice as shown in **II.** Zone of high intraluminal pressure extending from 49 through 46 cm separates stomach from esophagus. **III,** High-pressure zone, *C,* behaves like a sphincter by relaxing to appropriate stimulus, a swallow in this case. *A* and *B* demonstrate first simultaneous pressure rise generated by pharynx projecting bolus into esophagus, followed by progressive wave of contraction, the peristaltic wave. (From Skinner et al.[78])

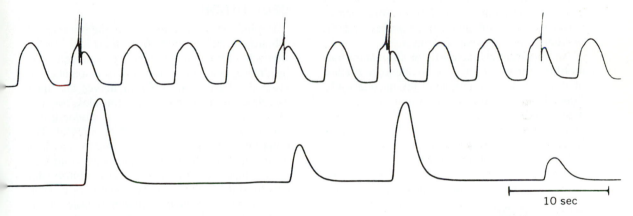

Fig. 55-2. Upper trace: intestinal slow waves (electrical activity) recorded from cat jejunum, and lower trace: contractions (mechanical activity). Contractions follow those slow waves that have superimposed spike potentials. Amplitude of contraction correlates with number of spike potentials. (From Bortoff.[8])

when the muscle is inactive as well as active. Rather, they determine when the muscle can be stimulated to contract; they determine the frequency, duration, speed, and direction of muscle contractions.[10,14,19] The slow waves originate in the longitudinal muscle of the stomach, small intestine, and probably colon and spread to involve the circular muscle.[7] When a contraction follows a slow wave, it is associated with a second potential, or burst of action potentials. Whether or not an action potential is superimposed on a slow wave to produce muscle contraction is determined by the algebraic sum of all excitatory and inhibitory factors of muscular, neural, and humoral origin.

Although intestinal motility was known to the ancients, and despite the fact that it has been studied since the beginning of modern physiology, its mechanisms are still poorly understood.

Methods of study. A number of techniques have been employed, both singly and in combination, in the study of alimentary tract motor function.[36] The simplest is to follow the movement of the luminal contents by radiographic methods. This method provides information about the movement of the bolus but gives no insight into the organization of the muscle contraction and relaxation that determines the speed and direction of propulsion.

The measurement of intraluminal pressures has provided information about the timing of muscular contraction and the propulsive forces generated. These measurement do not, however, indicate whether the muscle activity has produced useful work, that is, the transport or mixing of intestinal contents. When radiographic and manometric methods are combined, it is observed that the greatest movement of a bolus occurs when the pressure gradients are low. Higher pressures are recorded when the segment of bowel under observation is empty or its contents are not being moved. Intraluminal pressures are measured most often by catheters that are perfused with liquid, so that accurate intraluminal pressures may be recorded when the segment is empty as well as when it contains a bolus. These pressures also may be measured by intraluminal balloons that induce motor activity by distending the wall and acting on stretch receptors. Finally, intraluminal pressure has been measured by pressure-sensitive telemetering capsules that are swallowed.

Electrical activity originating in the alimentary canal can be recorded in vivo and thus can be correlated with intraluminal pressures and the movement of intestinal contents.

An important characteristic of alimentary tract motility, for which there is no adequate method of study, is the resting tone of the muscle walls. Tone is the resistance to distention. When techniques are developed to measure time-pressure-volume relations, the understanding of motor function will be greatly advanced.

Finally, preparations of gut muscle may be studied in vitro to determine certain neurohumoromyogenic interrelations (Chapter 4).

MASTICATION

The first step in the preparation of food for absorption occurs in the mouth. In addition to reducing the particle size of the food and mixing it with saliva in preparation for deglutition, mastication provides for stimulation of the rich sensory supply of the mouth and tongue that is so necessary for the full enjoyment of eating. These pleasant sensations arising from the sight, smell, taste, and chewing of food lead to parasympathetic stimulation that institutes salivary, gastric, and pancreatic secretion.

Mastication involves opening and closing the jaws, in addition to small forward, backward, and lateral movements before, during, and after tooth contact. During mastication, considerable force (as high as 15 to 20 kg) is applied to food caught between the teeth.[2] The tongue and cheeks play an essential role in keeping the food between the teeth. Coordination of a great volume of sensory information is required to use such forces without injuring the participating tongue, cheeks,

and teeth. Although the muscles of mastication are striated and are hence under voluntary control, mastication is largely reflex.

There is little quantitative information on the importance of mastication for the efficiency of subsequent digestion, although edentia or poorly fitting dentures are often blamed for malnutrition in the elderly. In one study, 29 different foods were tested by placing a small portion of each food, chewed and unchewed, into separate small cotton bags. The bags were swallowed and their contents were examined when recovered in the feces. There was more residual food material in the unchewed samples for 18 of the 29 foods, including meat, potatoes, and carrots. Fish, egg, and bread were well digested whether chewed or not. There seemed to be little correlation between masticatory efficiency as measured by other means and the completeness of digestion.[37]

DEGLUTITION

Deglutition, or swallowing, is the transport of a bolus, liquid or solid, from the mouth to the stomach. Although swallowing is a complex, integrated, continuous act involving somatic and visceral, afferent and efferent nerves and their associated striated and smooth muscles, for simplicity of description it is useful to divide it into two phases, oropharyngeal and esophageal. The initiation of the first phase is subject to conscious control and involves striated muscles controlled by a complex of stimulatory and inhibitory signals from the brain stem. The second or esophageal phase involves the smooth muscle of the esophageal wall and depends on both central coordination and local, intramural neural arcs. Swallowing has been studied from two points of view: (1) the mechanical, in which the movement of the bolus is correlated with the relaxation and contraction of the muscles involved in the act of swallowing, (cineradiography combined with intraluminal manometry provides graphic documentation[20,71]) and (2) the neurophysiologic, in which the neural impulses involved in swallowing are traced from afferent signals through the swallowing center to the motor nerves in an attempt to understand the "wiring diagram" that programs this complex, stereotyped, synergistic muscular response.[32]

Oropharyngeal phase

Swallowing begins with the segregation of the bolus to be swallowed into a depression or groove in the middle of the tongue. As the anterior portion of the tongue is pressed up against the hard palate, the bolus is forced posteriorly.

Afferent impulses arising from contact of the bolus with the anterior pillars of the pharynx, base of the tongue, and soft palate then take subsequent events out of conscious control by initiating the swallowing reflexes. The upper esophageal sphincter relaxes, respiration is inhibited, and the glottis closes as the larynx is drawn forward and upward. The nasopharynx is closed by a combination of an elevation of the soft palate and a contraction of the superior pharyngeal constrictor. The epiglottis tips down, deflecting the oncoming bolus laterally and posteriorly away from the larynx. The combined action of the tongue and the superior constrictor of the pharynx empty the oropharynx and strip the mucosa clean. The pharyngeal constrictors contract sequentially, propelling the bolus into the esophagus. The muscles of the oropharynx relax and respiration resumes after the upper esophageal sphincter contracts to separate the bolus in the esophagus from the pharynx and the airway (Fig. 55-3).[20,71]

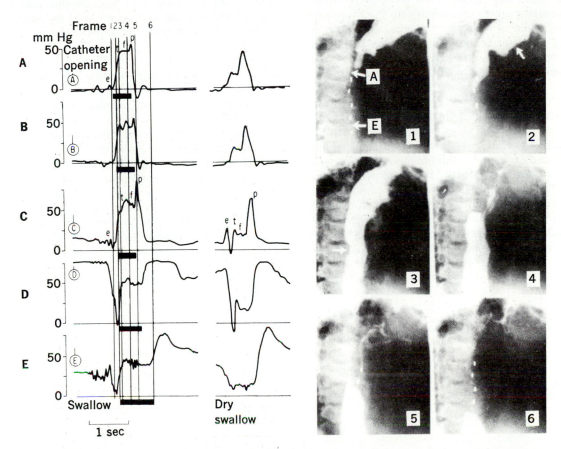

Fig. 55-3. Simultaneous cinemanometric study with catheter openings spanning upper esophageal sphincter. Subsequent swallow with small bolus (dry swallow) is shown for comparison. Maximum resting pressure is recorded from **D**. Lesser elevations of resting pressure at **C** and **E**. Atmospheric pressure at **A** and **B**. Onset of relaxation (tracings **D** and **E**) also occurs prior to frame *1*. Backward movement of tongue (arrow in frame *2* points to root of tongue) begins between frames *1* and *2* with onset of t wave. Frame *3* shows prominent posterior cricopharyngeal lip (arrow) indenting barium column during continuous barium flow. Frames *4, 5,* and *6* show sequential emptying coincident with onset of p wave (**A** to **E**). Catheter assembly and posterior cricopharyngeal lip descend during emptying. Posterior lip is located immediately above flattened tail of barium column in frame *6*, that is, at level of **D** opening. Bars below manometric tracings indicate period of time during which barium is present at level of indicated catheter opening. The e wave corresponds to elevation of hyoid bone and larynx; t wave is associated with vigorous posterior movement of tongue; and p wave is peak of peristaltic wave. (From Cohen and Wolf.[20])

Upper esophageal sphincter

During quiet respiration, withdrawal of a pressure-recording catheter from the esophagus into the oropharynx reveals a segment 2.5 to 4.5 cm in length with increased intraluminal pressure. This high-pressure zone corresponds in location to the cricopharyngeus muscle (Fig. 55-3, *D* and *E*). Yield pressures as high as 75 mm Hg have been recorded in this segment. The upper esophageal sphincter serves two major functions: (1) it prevents air from entering the esophagus during inspiration, thus minimizing the respiratory dead space, and (2) it prevents reflux of material from the esophagus into the hypopharynx, where it might be aspirated. As shown in Fig. 55-3, relaxation of the sphincter is the first recorded manometric event occurring while the bolus is still in the mouth.[20]

Neural organization of deglutition

The oropharyngeal phase of swallowing, completed in less than 1 sec, is an intricate, stereotyped, bilaterally symmetric sequence of inhibition and excitation involving more than 25 muscle groups controlled by the swallowing center of the brain stem. This complex neuromuscular reaction pattern can be initiated by stimulation of a single afferent nerve, the superior laryngeal (Fig. 55-4).[32] Although the initiation of swallowing is under voluntary control, like other voluntary movements, it is often effected without conscious effort. It appears in the fetus as early as 12 weeks of gestation, long before suckling and respiratory movements occur. Normal adults on the average swallow 600 times/24 hr, 50 swallows occurring during sleep and only 200 during eating. While drinking, swallows can be repeated as frequently as 1/sec, but in the absence of a bolus the frequency is greatly reduced, an indication that swallowing depends on peripheral stimuli from the oropharynx as well as messages from higher centers. In fact, it appears that swallowing cannot be initiated voluntarily if afferents from the pharynx are blocked by administration of a local anesthetic. In this situation, swallowing can still be initiated by direct electrical stimulation of the superior laryngeal nerve.

In addition to the superior laryngeal nerves, other afferents stimulating swallowing converge on the nucleus solitarius via the maxillary branch of the trigeminal nerve and the glossopharyngeal nerve. Afferent stimuli over these same nerves also initiate other patterned responses such as gagging, coughing, and chewing. The temporal and spatial patterns of the afferent stimuli determine the patterned response that is evoked. Appropriate afferent stimuli are relayed to the swallowing center in the reticular formation of the rostral medulla. The center is actually two paired half centers that continue to initiate swallowing responses from the ipsilateral muscles when the connections between them are severed. Impulses from the swallowing center then activate the ap-

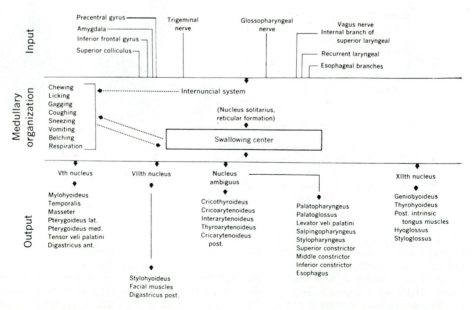

Fig. 55-4. Outline of afferent and efferent systems involved in swallowing, and requirement of interaction between swallowing and other synergies. (From Doty.[32])

propriate motoneurons in the nucleus ambiguus and the fifth, seventh, and twelfth cranial nerve nuclei. The resultant motor activity can be analyzed from electromyographic tracings (Fig. 55-5). Once the swallowing center is activated, there is little evidence that afferent impulses play any role in modifying the oropharyngeal patterned muscular response. Activation of the swallowing center influences the activities of other centers, most notably the respiratory center, as evidenced by an apneic pause of 0.5 to 3.5 sec that accompanies every swallow.[32] The continuous train of

action potentials that maintains resting tone of the upper esophageal sphincter is interrupted, with resultant sphincter relaxation before the onset of pharyngeal peristalsis (Figs. 55-3 and 55-5). Closure is effected as the neural discharge propagating the peristaltic wave passes through the sphincter and resting tone is reestablished.[46]

Esophageal phase

The esophagus is functionally the simplest segment of the alimentary canal. Its only purpose is the transport of materials from the pharynx to

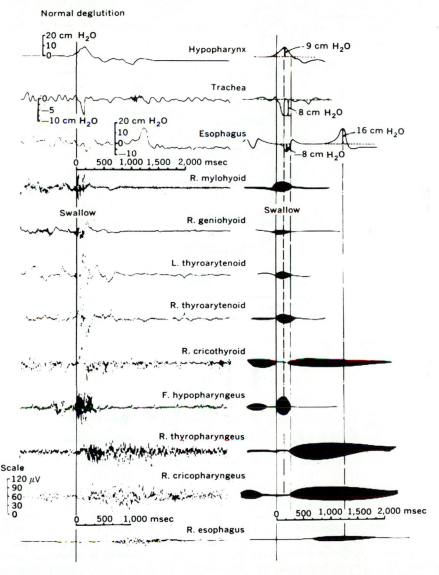

Fig. 55-5. Electromyographic and pressure recordings during swallowing in alert dog 12 days after implantation of electrodes. Left: actual records. Right: idealized summary. Note inhibition in inferior constrictor (thyropharyngeus and cricopharyngeus) and cricothyroideus during first 200 msec of swallowing and delay of over 60 msec before onset of esophageal activity. (From Kawasaki et al.[60])

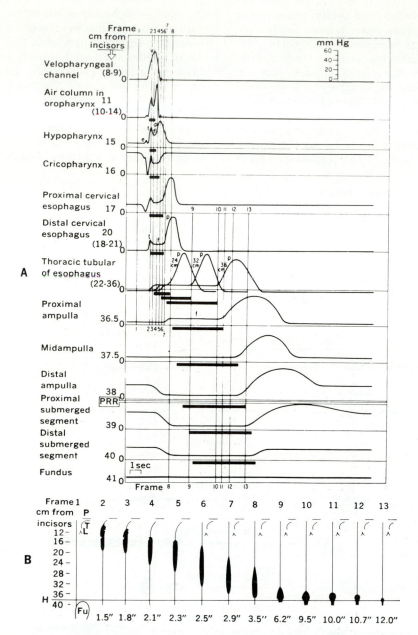

Fig. 55-6. A, Prototype resting pressures and swallowing complexes from velopharyngeal channel to fundus of stomach. Atmospheric pressure is taken as zero pressure. Vertical lines through curves indicate times selected to demonstrate bolus transport and roentgen configurations in **B.** Period during which barium is present at each level is indicated by bar below each pressure curve. *PRR* is level at which respiratory excursions change from thoracic to abdominal. Positive-pressure curve from velopharyngeal channel is indicated by *V.* **B,** Prototype "frames" corresponding to schematic curves in **A.** Barium bolus is shaded. Phrenic ampulla in emptying phases has been exaggerated. Sequence demonstrates perfectly orderly swallowing in which one continuous peristaltic wave sweeps barium through pharynx and esophagus into stomach. Approximate times of each frame are indicated in seconds below each frame. Frame *1,* the empty state, shows no barium. *P* indicates level of hard palate; *T,* tongue; and *L,* entrance to larynx. Larynx is closed in frames *2* to *5. H* indicates level of top of hiatus. Small posterior cricopharyngeal lip is shown in frames *3* and *4.* (From Cohen and Wolf.[20])

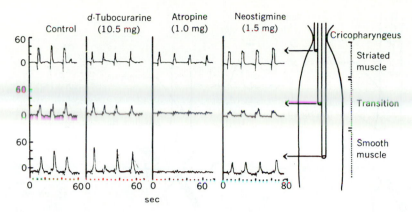

Fig. 55-7. Portions of manometric recordings obtained during control period and after administration of *d*-tubocurarine, atropine, and neostigmine. Amplitude of peristaltic wave is decreased in striated muscle by *d*-tubocurarine but not in smooth muscle. Atropine abolishes peristalsis in smooth muscle but produces no further change in striated muscle. Neostigmine reverses both effects. (From Kantrowitz et al.[59])

the stomach. As the esophagus has no normal spontaneous activity, its motility can be initiated and studied at will. In man the upper quarter of the esophagus is striated muscle, the inner circularly arranged coat being continuous with the muscle of the hypopharynx and upper esophageal sphincter. The lower third of the esophagus is smooth muscle, with the longitudinal and circular layers of the muscle coat being continuous with those of the stomach. In the segment between the upper striated and the lower smooth muscle segments there is a variable mixture of the two types of muscle. In some species such as dogs, rats, and cattle the esophagus is composed wholly of striated muscle.[74] Irrespective of the composition of the muscular coat, its activity, the peristaltic wave, is remarkable uniform.[56] The peristaltic wave, which begins in the superior constrictor of the pharynx, sweeps through striated and smooth muscle to the cardia without interruption (Fig. 55-6). The wave becomes broader and its advance slows as it moves aborally. Differences between the striated and smooth muscle portions can be identified pharmacologically; atropine blocks the peristaltic response to swallowing in the smooth muscle segment, whereas curare blocks peristalsis in the striated muscle segment (Fig. 55-7).[59]

Esophageal peristalsis

A distinction has been made between primary and secondary peristalsis (Fig. 55-8). Originally, it was believed that primary peristalsis, that initiated by a swallow, was mediated by impulses originating from the swallowing center, and sec-

ondary peristalsis, that initiated by a bolus in the esophagus, was mediated by local intramural neural pathways. Secondary peristalsis serves to return material to the stomach that refluxes into the esophagus or to carry material to the stomach that has been left behind by an ineffectual peristaltic wave. Secondary peristalsis can be produced experimentally by inflation and deflation of a balloon in the esophagus. It now appears that the designations primary and secondary peristalsis should be used only to indicate how peristalsis is initiated, since both central and local neural pathways appear to be involved regardless of the means of initiation.[7,40,77] The initial event in esophageal peristalsis is stimulation of the longitudinal muscle of the esophagus followed by segmental activation of circular muscle and inhibition of distal esophagus.[31,72] The stimulation of the longitudinal muscle is cholinergic (atropine sensitive). The circular muscle displays a different response in that it does not contract until after the end of the electrical stimulus (off-response) and is not inhibited by antagonists to any of the known autonomic neurotransmitters.[16] It has been suggested that vagal noncholinergic nonadrenergic inhibitory nerves hyperpolarize the circular muscle by an as yet unidentified neurotransmitter. When the stimulation is turned off, rebound depolarization leads to contraction.[28] Esophageal peristalsis will not be understood until the relation between longitudinal and circular muscle activity is clarified and the nature of the mediator of the vagal inhibitory nerves is identified. The ability of anticholinergic drugs to block peristalsis in the esophagus (Fig. 55-7)

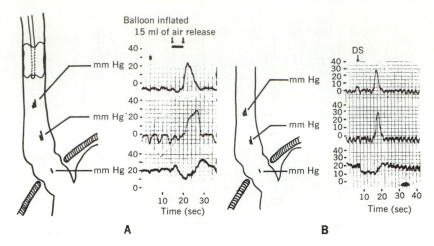

Fig. 55-8. Comparison of secondary and primary esophageal peristaltic waves. **A,** Momentary distention with air-filled balloon fixed in midesophagus produced drop in pressure in lower esophageal sphincter (bottom tracing), followed by progressive contraction recorded at two esophageal sites distal to balloon (upper two tracings). These pressure phenomena, elicited without swallow or moving bolus, are comparable to those characterizing primary esophageal peristalsis initiated by swallowing, as shown in **B.** Drop in pressure in sphincter (bottom tracing) and progressive contraction in esophagus (upper two tracings) are released by dry swallow, *DS* (one containing only air and a little saliva). (From Fleshler et al.[40])

may be through their action on the cholinergically innervated longitudinal muscle.

In most young adults a swallow uniformly triggers the swallowing sequence shown in Fig. 55-6. With increasing age the frequency of "misfires" and abnormal, segmental, nonprogressive contractions in the smooth muscle segment increases. If, instead of a single swallow, the subject takes a series of swallows, as in drinking a glass of water, each swallow initiates the complete oropharyngeal sequence, but peristalsis in the esophagus is inhibited until the last swallow of the series. The upper esophageal sphincter relaxes and closes with each swallow, whereas the lower esophageal sphincter opens with the first swallow and only closes when the peristaltic wave enters the sphincter (Fig. 55-9).[40] Secondary peristalsis initiated by balloon distention of the body of the esophagus can be inhibited by a swallow.[55] These observations are best interpreted as an indication of the inhibition of distal neuromuscular activity during the activity of proximal segments, presumably mediated through the swallowing center. In addition, local afferents from the esophagus alter esophageal peristalsis. For example, the force of the peristaltic contraction is proportional to the size of the bolus. In addition, swallows of hot liquid increase the speed and amplitude of esophageal peristalsis, whereas cold swallows decrease speed and amplitude and finally abolish esophageal

peristalsis.[84] Acid swallows disorder and delay esophageal peristalsis.[47] Recent studies suggest that afferent impulses of local origin may not only modify the pattern of esophageal peristalsis but may, indeed, be essential for its normal propagation. In conscious dogs with a cervical esophageal cannula in place, it was found that with the cannula closed, esophageal peristalsis occurred after 90% of water swallows. With the cannula open and the water swallows diverted to the outside, no peristalsis was recorded in the lower esophagus.[65] These findings may be relevant to esophageal function in man, for transection of the smooth muscle portion of the esophagus in the monkey interferes with primary peristalsis.[82] It appears that primary and secondary peristalsis both depend on descending neural activity originating in the swallowing center as well as on sequential afferent input from the esophagus.

Innervation of the esophagus

Although knowledge of the afferent innervation of the esophagus is of great importance to an understanding of its motor function, very little is presently known. Neither the location and characteristics of the sensory receptors nor the functional significance of the afferents that are found in the vagus and sympathetic nerves have been defined. Electrical records from the nodose ganglion in the cat show two kinds of afferent signals arising from the esophagus in response

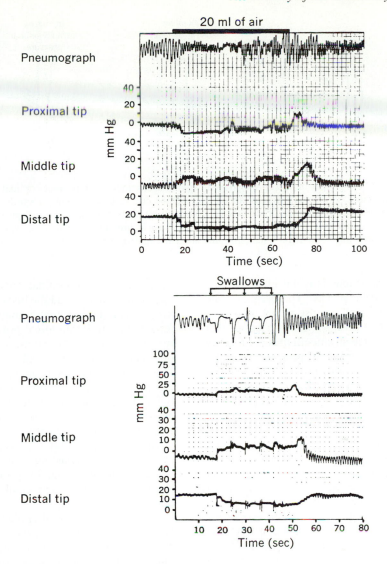

Fig. 55-9. Pressure tracings during prolonged distention compared with frequent swallowings. Prolonged distention (solid bar at top of record) of balloon fixed in midesophagus was attended by no changes in body of esophagus distal to balloon (proximal and middle tips) and produced sustained drop in intrasphincteric pressure (distal tip) that was maintained until balloon was deflated. Thereafter, pressure wave passed through lower esophagus and sphincter. Overall effect on esophageal motility is similar to that produced by rapidly repeated swallows (lower graphs). Liquid swallows taken at approximately 7 sec intervals produced sustained drop in pressure in sphincter (distal tip), and no peristaltic wave appeared in body of esophagus (proximal and middle tips) during this time. Sustained rise and occasional spikes of pressure recorded are ascribable to fluid introduced into esophagus and to transmitted oropharyngeal pressures. After swallowing efforts ceased, peristaltic contraction traveled down esophagus. (From Fleshler et al.[40])

to balloon distention. One is a continuous discharge for the duration of the distention and the other is an "on-off" signal firing one burst after inflation and another after deflation of the balloon.[14]

Although the striated portion of the esophagus appears to have the same innervation as does other striated muscle, it contracts grossly like smooth muscle, that is, it contracts slowly, remains contracted for 1 to 2 sec, and then relaxes slowly.

The motor innervation of the smooth muscle

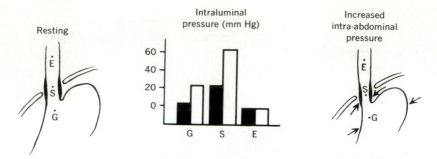

Fig. 55-10. Response of lower esophageal sphincter to increased intra-abdominal and intragastric pressure. *G, S,* and *E* refer to gastric, sphincteric, and esophageal recording sites and pressures. Resting intraluminal pressures are indicated by black bars. Intraluminal pressure within sphincter exceeds intragastric pressure. When intragastric pressure is increased, sphincter pressure is greatly augmented, far exceeding increased intragastric pressure. (Modified from Skinner et al.[78])

portion of the esophagus is believed to be from the vagal esophageal plexus. The preganglionic parasympathetic fibers synapse in the myenteric plexuses with the postganglionic neurons that innervate the esophageal smooth muscle. Exactly how the peristaltic wave is propagated is not understood, but it is unlikely that it is mediated by serial excitation of cells in the motor nuclei of the vagus. Distention of isolated segments of the opossum esophagus, which like human esophagus is smooth muscle in its distal half, produces peristalsis. This peristalsis is neural rather than myogenic in origin because it is blocked by tetrodotoxin, a pharmacologic agent that selectively depresses transmission along nerve fibers.[13] Studies of this preparation suggest that there are motor nerves in the esophagus that are neither cholinergic nor adrenergic, the so-called nonadrenergic inhibitory nerves.[14]

The contribution of the adrenergic nerves to esophageal motor function is still undefined. It seems clear, however, that the classic view that cholinergic impulses are excitatory and adrenergic impulses are inhibitory is an oversimplification.

Lower esophageal sphincter

It has long been recognized that there is a barrier in the lower esophagus that prevents the reflux of gastric contents into the esophagus. In the past, it was thought to be a passive mechanical barrier, but more recently, it has been recognized that there is a functional sphincter in the distal esophagus, although there are no morphologic features that explain any specialized function of this segment of esophagus.

Withdrawal of pressure-sensing catheters from the stomach into the esophagus identifies a segment with an intraluminal pressure greater than the stomach below and the body of the esophagus above. This high-pressure zone is 2 to 4 cm in length, has an intraluminal pressure of 10 to 40 mm Hg, and usually extends into the intra-abdominal portion of the esophagus (Figs. 55-1 and 55-6). The lower esophageal sphincter relaxes 2.5 sec after the initiation of swallowing, well before the arrival of the bolus in the lower esophagus, and remains open for 10 to 12 sec until the peristaltic wave enters the sphincter and closes it. Sphincter relaxation is prolonged by repetitive swallows, by a persisting distention of the body of the esophagus, and by prolonged spasm or abnormal repetitive segmental esophageal contraction (Fig. 55-9).

The lower esophageal sphincter is not a static barrier but responds to a variety of stimuli with changes in the force of its contraction. Increased intra-abdominal pressure is associated with increased sphincter pressure; the increase in sphincter pressure may be more than 2 to 4 times as great as the increased intragastric pressure (Fig. 55-10). On the other hand, gastric distention that is not associated with an increase in intragastric pressure is accompanied by a fall in sphincter pressure, thus facilitating eructation.[69] It is believed that there responses are mediated by vagovagal reflexes.[24] Sphincteric pressure also is modified by meals of differing composition; sphincter pressure is augmented by a protein meal and decreased by a fat meal (Fig. 55-11).[70] In addition, sphincter pressure is increased by alkalinization of stomach and decreased by acidification.[49] Because of the relation between intragastric pH and gastrin release and the effect of protein meals on gastrin levels, it was assumed that these changes in sphincter pressure were

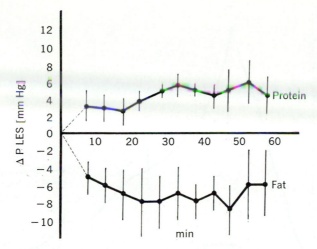

Fig. 55-11. Change in lower esophageal sphincter *(LES)* pressure from basal after ingestion of protein and fat meals (mean ± SEM). Test meals consisted of 150 kcal in a volume of 150 ml. Protein was in the form of cooked minced beef, and fat meal was emulsified corn oil. Protein meal was associated with 31% increase in LES pressure over basal, whereas after the fat meal, LES pressure fell 40%. When a combined protein and fat meal was ingested, LES pressure fell 19%. (From Nebel and Castell.[70])

caused by gastrin action on the sphincter. Although gastrin injection can strikingly augment lower esophageal sphincter pressure, gastrin's physiologic role of maintaining resting sphincter pressure and modulating postprandial pressure is unsettled.[39,44]

Basal sphincter tone has been attributed to special characteristics of the muscle,[17,42] gastrin,[21] and alpha adrenergic activity,[30] but more investigation will have to be done to determine the relative importance of these factors and to determine which, if any, of these explain the special functional characteristics of this sphincter segment. Sphincter tone is modified by a variety of gastrointestinal hormones: (1) Sphincter pressure is increased by gastrin, and this effect is inhibited by secretin, but secretin by itself does not alter sphincter pressure (Fig. 55-12); (2) cholecystokinin, which is structurally similar to gastrin, competitively inhibits gastrin effect; and (3) glucagon, which is in the secretin family of hormones, also blocks the gastrin effect but, unlike secretin, is able to decrease basal sphincter pressure.[39] The physiologic role of these interactions is not known, but it is clear that the measured sphincter pressure at any time may be the algebraic sum of many stimulatory and inhibitory influences. The site of gastrin action has been placed on postganglionic cholinergic nerves by one set of experiments[64] and on the sphincteric muscle itself by other experiments.[86]

Relaxation of the lower esophageal sphincter

is mediated by nonadrenergic, noncholinergic vagal nerves.[80] The mediator of these inhibitory nerves has not been identified, but several candidates, including adenosine triphosphate and dopamine, have been suggested. Also prostaglandin E_1, which stimulates the production of cyclic AMP, has been suggested for a role in sphincter relaxation.[39]

In summary, deglutition, or swallowing, is a complex neuromuscular operation initiated consciously but carried to completion by an integration of afferent impulses and CNS efferent impulses organized both in the swallowing center and in local intramural arcs; these act on a muscular tube made up of both striated and smooth muscle. The esophagus is separated from the pharynx above and the stomach below by a pair of sphincters, the upper and lower esophageal sphincters, which maintain a resting intraluminal pressure greater than the adjacent segments. Swallowing may be viewed as a relaxation of the swallowing tube to receive the swallowed bolus, followed by caudally progressing muscular contracture, the peristaltic wave, sweeping the bolus before it. Relaxation is most obvious in these segments with a high resting tone, the sphincters, whereas the progressive nature of the peristaltic wave is most obvious in the remaining segments that exhibit a low resting tone. It may be that the sphincters are differentiated from adjacent segments not much by special neural arrangements but rather by an increased sensitivity of the neu-

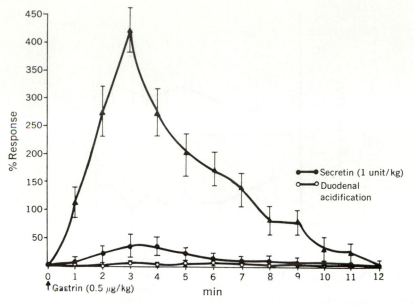

Fig. 55-12. Lower esophageal sphincter response in seven patients given 0.5 μg/kg of gastrin I alone as compared to identical dose of gastrin I given during duodenal acidification at 18 mEq/15 min and directly after intravenous secretin (1 U/kg). Secretin, given parenterally or released by duodenal acidification, does not decrease sphincter tone but rather inhibits augmentation of sphincter tone induced by exogenous gastrin. (From Cohen and Lipshutz.[21])

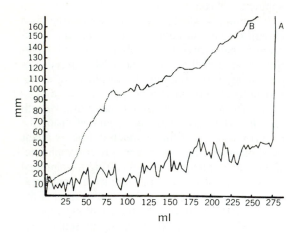

Fig. 55-13. Pattern of accommodation to added volume in vivo and in vitro. *A,* Curve of intragastric pressure in living rabbit; 2.5 ml added every 2½ min. Abdomen closed. *B,* Curve of intragastric pressure in excised rabbit stomach immersed in physiologic saline solution at 38° C; 2.5 ml added every 2½ min. (From Grey.[43])

romuscular elements of the sphincters to the neurohumoral determinants of motor function.

GASTRIC MOTILITY

The stomach retains an ingested meal while its physical state is altered to facilitate intestinal digestion and absorption. The storage and preparative activities of the stomach are accomplished by two functional regions: (1) the proximal fundus and body that function as a reservoir and (2) the antrum that churns the gastric contents and delivers it in a metered fashion to the duodenum.[23]

Gastric handling of a meal

Receptive relaxation of the stomach. The resting pressure within the body and fundus of the stomach equals intra-abdominal pressure. Only a slight rise in intraluminal pressure is recorded when the fasting gastric volume of less than 50 ml is increased to greater than 1.5 L. This accommodation is not due solely to passive stretching of the stomach (Fig. 55-13). The relaxation induced by swallowing precedes the bolus, allows it to pass through the esophagus and sphincter, and also spreads to the fundus and

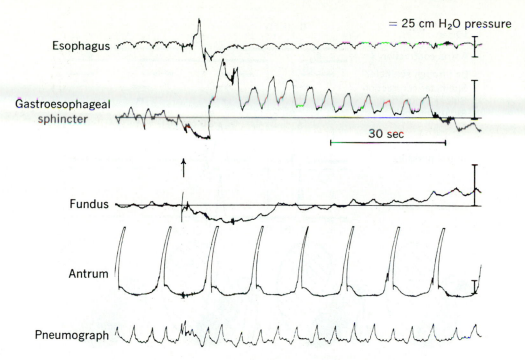

Fig. 55-14. Receptive relaxation in dog gastric fundus during deglutition (indicated by ↑). Peristaltic contraction of esophagus is depicted (top tracing) with relaxation and contraction of lower esophageal sphincter as well as no change in character of contractions in antrum of stomach during deglutition response. (From Lind et al.[63])

body of the stomach, thus permitting the bolus to be accommodated without an increase in intragastric pressure (Fig. 55-14).[63] Afferent vagal impulses, presumably from stretch receptors in the pharynx and esophagus, as well as in the stomach, initiate a reflex inhibition of the muscle of the stomach wall. The vagal efferents of this arc are nonadrenergic inhibitory fibers.[42] Repeated swallows produce a summation of the inhibition of tone in the upper half of the stomach.[57] This receptive relaxation of the stomach can occur in the absence of swallowing, however, for the stomach can be filled by tube without an increase in intragastric pressure.[69]

Gastric peristalsis. Gastric peristalsis serves two functions. First, it provides the force that transports gastric contents into the duodenum; second, it emulsifies and macerates the gastric contents in preparation for intestinal digestion and absorption. The gastric peristaltic wave originates in the distal half of the stomach. This ring of contraction begins as a feeble indentation of the gastric wall but, as it sweeps toward the pylorus, the contraction becomes more powerful and increases in speed so that the terminal antrum

appears to contract as a unit, the so-called antral systole. The pylorus narrows as the peristaltic wave approaches the antrum, so that only 5 ml or so of liquefied and emulsified chyme is discharged into the duodenum. As pressure rises in the distal antrum as a consequence of the closure of the pylorus and advancing peristaltic wave, the trapped material squirts back into the body of the stomach through the advancing ring of contraction (Fig. 55-15). This antral mechanism permits the discharge of gastric chyme into the duodenum but returns solid, unmacerated food to the stomach for further processing.

CONTROL OF GASTRIC PERISTALSIS. The frequency and strength of gastric peristalsis are determined by the interaction of three control mechanisms: (1) myogenic (gastric slow waves or pacesetter potentials), (2) neurogenic, and (3) hormonal.

Gastric slow waves. Slow waves set the basic pattern of frequency, velocity, and direction of gastric peristalsis. These rhythmic depolarizations of the smooth muscle originate every 20 sec in the longitudinal muscle of the greater curvature of the stomach, approximately halfway

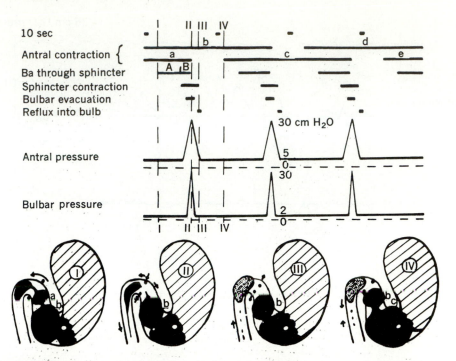

Fig. 55-15. Evacuation-pressure cycle in antrum and duodenal bulb, schematized in four phases. Sketches depict events at phase of cycle designated by vertical lines and Roman numerals. (From Werle et al.[83])

between the pylorus and fundus, and spread toward the pylorus with increasing speed (Fig. 55-16). Slow waves can be recorded from the inactive as well as the contracting stomach. This wave of depolarization of gastric musculature is myogenic, for it continues in muscle preparations separated from extrinsic nerves and with intrinsic nerves blocked by tetrodotoxin. The muscle cells in the region where the gastric slow waves originate have a higher intrinsic frequency of depolarization than do the other cells of the stomach; hence they serve as the pacemaker for the stomach. The pattern of the gastric slow waves can be modified, however, by vagotomy, acetylcholine (ACh), and norepinephrine. Given at the appropriate time in the cycle, ACh can cause a premature slow wave that is analogous to a premature heartbeat.[20]

Although gastric peristalsis is paced by the slow waves, muscle contraction occurs only when spike potentials are superimposed on a slow wave. The appearance of spike potentials and subsequent gastric peristalsis is determined by the sum of stimulatory and inhibitory neural and hormonal stimuli. The depolarization associated with gastric peristalsis is produced by ACh and blocked by atropine. ACh is released by postganglionic parasympathetic neurons as a con-

sequence of their stimulation by impulses from vagal fibers, intramural reflex arcs, or hormones such as gastrin.[64] It is believed that both alpha and beta adrenergic receptors act to prevent the release of ACh rather than acting directly on the gastric musculature.[25] Secretin decreases the strength of gastric peristalsis, presumably in a manner similar to its interference with gastrin-stimulated contraction of the lower esophageal sphincter, as discussed earlier in this chapter.

In summary, the gastric slow wave sweeps down the antrum every 20 sec. Whether spike potentials and consequent gastric peristalsis occur is determined by the algebraic sum of stimuli augmenting and inhibiting ACh release.

Control of gastric emptying. The rate of gastric emptying is determined by the balance between the force and frequency of gastric peristalsis and the resistance to flow offered by the pylorus. The determinant of the rate of gastric emptying that has been most thoroughly studied is gastric volume. These studies have shown that the rapidity of gastric emptying is proportional to the volume introduced, that is, the distention of the stomach (Fig. 55-17).[54] Gastric distention increases afferent vagal discharges from stretch receptors. The resulting efferent vagal discharge is responsible in part for increased gastric peri-

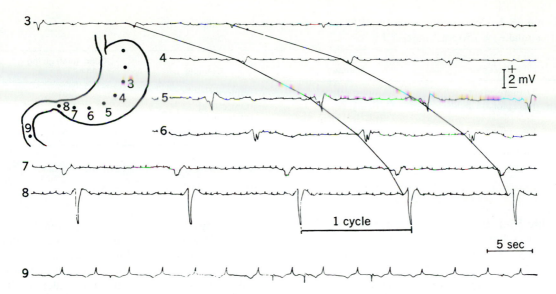

Fig. 55-16. Caudal propagation of gastric slow waves in fasting dog. Velocity increases as they approach pylorus. Slow-wave activity in duodenum is more rapid. (From Code et al.[19])

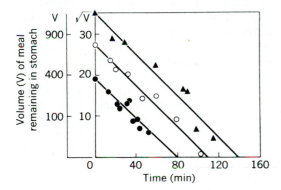

Fig. 55-17. Gastric emptying of 1,250, 750, and 330 ml of sucrose solution (35 gm/L) plotted as square roots of volume remaining in stomach against time. The larger the gastric volume, the greater the rate of gastric emptying. (From Hopkins.[52])

stalsis and increased gastric emptying. The release of gastrin by antral distention also augments gastric peristalsis.

The duodenum, on the other hand, acts to decrease gastric peristalsis and hence limits gastric emptying to a volume that can be efficiently and optimally mixed with bile and pancreatic juice. There is evidence that receptors for fats and acids and osmoreceptors exist in the duodenum. The relative potency of these inhibitory stimuli is shown in Table 55-1. Recently, it has been suggested that the variable most closely controlled is caloric density, so that re-

gardless of the type of meal ingested the number of calories presented to the intestine per unit time is maintained within narrow limits. The relative contribution of neural and hormonal stimuli in duodenal inhibition of gastric emptying has not been resolved. The inhibition produced by acid is thought, however, to be primarily nerve mediated, whereas inhibition by fat is hormonal. Cholecystokinin-pancreozymin (CCK-PZ) has been shown to be a potent inhibitor of gastric emptying at physiologic concentrations.[27] The effects of other gastrointestinal hormones on gastric emptying are listed in Table 55-2. In addition, glucagon, vasoactive intestinal peptide (VIP), and gastric inhibitory peptide (GIP) inhibit, whereas motilin stimulates gastric motility, but as yet these responses have not been integrated into a model describing the control of gastric motility and gastric emptying.[23,33]

The conventional view is that the pylorus is not a sphincter because the methods employed in earlier studies did not reveal a high-pressure zone between the stomach and duodenum.[35] Recent studies, however, have detected pressures in the resting pylorus that were 5.5 mm Hg greater on the average than pressures in the resting antrum and duodenum. More striking were the increased pressures recorded in this high-pressure zone after placing acid, fat, or amino acids in the duodenum. Duodenal acidification increased pressure within the pylorus by 23%. In addition, secretin and CCK-PZ were additive in increasing pyloric pressure.[38] It ap-

Table 55-1. Concentrations giving 500 ml recovery from stomach 20 min after intake of 750 ml "meal"*†

Receptor	Stimulus	Example	Conc. (mM) giving 500 ml recovery
Osmoreceptor	Osmotic effect	Glucose	500
Receptor for acids	Ordinary acids	Citric acid	120
	(pK less than 5)	Hydrochloric acid	40
Receptor for fats	Salts of fatty acids	Sodium myristate	8

*Modified from Hunt and Knox.[54]

†By comparison, less than 100 ml of a 750 ml "meal" of isotonic sodium chloride is recovered 20 min after ingestion.

Table 55-2. Effects of three hormones on pyloric sphincter tension, antral motility, and emptying*

Agent	Pyloric sphincter tension	Antral motility	Emptying
Gastrin	No effect	Increases	Delays
Secretin	Increases, no effect	Decreases	Delays
CCK-PZ†	Increases	Decreases	Delays

*From Cooke.[23]

†CCK-PZ = cholecystokinin-pancreozymin.

pears that the pyloric sphincter serves three functions: (1) it limits the rate of flow (primarily of solids) into the duodenum, (2) it increases the effectiveness of antral mixing and churning by closing before the peristaltic wave has passed completely over the antrum, and (3) it prevents the reflux of duodenal contents into the stomach.[23] Many more studies will be required to define the complex interactions between the antrum, pyloric sphincter, and duodenum that control gastric emptying.

Eructation. As the stomach is distended by food and swallowed air, tone in the lower esophageal sphincter is decreased until it ceases to provide a barrier between stomach and esophagus, and gas refluxes into the esophagus. The distention of the esophagus triggers secondary peristalsis, the refluxed material is returned to the stomach, and sphincter tone is restored. Then the sequence repeats. When the individual becomes aware of distention, he inspires against a closed glottis, thus increasing the gradient from stomach to esophagus, and increases intragastric pressure by contracting the abdominal muscles. This forces gas up through the esophagus and upper esophageal sphincter, producing an eructation.[69]

Vomiting. Vomiting involves a complex reflux and is associated with a widespread autonomic discharge, including salivation, pallor, sweating, and tachycardia. The antrum and duodenum contract, and the remainder of the stomach, the esophagus, and its sphincters relax. The glottis closes off the pulmonary airway and the soft palate closes the nasopharynx; then the diaphragm and the abdominal muscles contract, raising intra-abdominal pressure sharply and propelling the gastric contents into the esophagus and out the mouth. Several cycles of reflux into the esophagus usually precede the actual vomiting.[66]

The coordination of this complex reflex is effected by the vomiting center in the dorsolateral reticular formation of the medulla. Afferent impulses from the olfactory epithelium, labyrinths, pharynx, gastrointestinal tract, and testis may converge on the vomiting center to initiate vomiting. Gastric irritants such as copper sulfate initiate vomiting through action on the gastric afferent nerves. If the stomach is denervated, copper sulfate will still initiate vomiting, but only when introduced in much higher concentrations. Vomiting after gastric denervation is produced by absorption of the toxic material, which is carried by the blood to the chemoreceptor trigger zone located on the floor of the fourth venticle, which in turn stimulates the vomiting center. There are some emetics such as apomorphine that exert their entire emetic effect by stimulating the trigger zone.[6]

SMALL INTESTINE MOTILITY

The motor activity of the small intestine mixes the chyme with bile and pancreatic juice and is continually bringing newly prepared material into contact with the absorbing mucosa. As absorption progresses, the contents move into the ileum and then into the colon. The motor pattern of the small intestine is very complex and may be

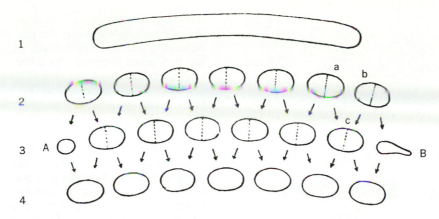

Fig. 55-18. Diagrammatic representation of process of rhythmic segmentation of intestinal contents derived from Cannon's original radiographic observation. Lines *1, 2, 3,* and *4* indicate sequence of appearances in loop. Dotted lines mark regions of division. Arrows show relation of particles to segments they subsequently form. (From Cannon.[11])

considered to be of three types: tone, mixing waves, and peristaltic waves. *Tone* provides the basis on which the other motor activities act. If the intestinal tone is low, that is, offers little resistance to distention, mixing of intestinal contents is feeble and transit is slow, whereas if the tone is high, transit and mixing are rapid. *Mixing waves* were originally described by Cannon: "A small mass of food is seen lying quietly in one of the intestinal loops. Suddenly an unidentified activity appears in the mass and a moment later constriction at regular intervals along its length cut it into little ovoid pieces. . . . A moment later each of these segments is divided into two particles, and immediately after the division neighboring particles rush together. . . . The next moment these new segments are divided, and neighboring particles unite to make a third series, and so on . . . ." (Fig. 55-18).[12] Little has been added to this early account. When present, these mixing or segmenting contractions occur at the rate of 11/min in the upper jejunum of man and decrease in frequency to 8/min in the terminal ileum.[50] *Peristaltic waves* travel for only short distances in the intestine, except in the fasting state, when every 90 to 120 min a peristaltic wave may sweep the entire length of the small intestine.[18] Bayliss and Starling, from their studies of intestinal peristalsis concluded, "Since the whole act is evoked by the presence of a bolus in the gut, we must say that the irritation of the mucous membrane and the stretching of the walls of the gut at any point set up impulses which are transmitted both up and down the intestine, and cause excitation above, and inhibition below

. . . more generally, if cerebrospinal reflexes be excluded, excitation at any point of the gut excites contraction above, inhibition below. This is the law of the intestine."[5] Mixing and peristaltic waves probably have the same origin, differing only in their distance of propagation.[8]

Part of the difficulty in making meaningful correlations in the study of intestinal motility lies in the fact that in the fasting state the intestine is minimally active, and most studies have been conducted in this state. Records taken with balloons that provide the intestine with a bolus on which to work show the most activity. Such studies reveal the response of the intestine to stretch but reflect little of the hormonal effect on motility, which is a prominent if not the dominant physiologic determinant of muscle activity in response to a meal.

Control of intestinal motility

The circular muscle coat of the stomach is separated from that of the duodenum by a connective tissue septum. Three quarters of the gastric longitudinal muscle bundles dip into the circular muscle just proximal to the septum, and the remaining quarter of the bundles, particularly those on the lesser curve, pass on to the duodenal cap.[35] Although the duodenal cap displays slow waves of greater frequency (11/min) than those of the antrum (3/min), most of the contractions follow a gastric peristaltic wave. This coordination is affected by two factors: first, distention of the duodenal cap by gastric chyme discharged into it by the gastric peristaltic wave increases the probability that spike potentials

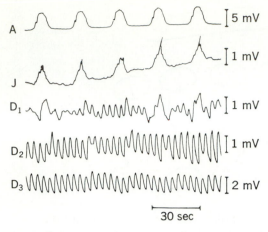

Fig. 55-19. Slow wave potentials recorded simultaneously from points on antrum, pylorus, and proximal duodenum of rhesus monkey in situ. Antral electrode, A, was 14 mm oral to pylorus; J was on pylorus; D_1, D_2, and D_3 were 4, 10, and 17 mm aboral to pylorus, respectively. Antral augmentation of duodenal slow waves is evident as far as 17 mm from pylorus. (From Bortoff and Davis.[9])

leading to contraction will occur on a slow wave; second, gastric slow waves are transmitted to the duodenum via the longitudinal muscle bundles and thus serve to couple the contractions of the antrum and duodenal cap (Fig. 55-19).

In the small intestine, as in the stomach, bioelectrical slow waves passing aborally determine when spike discharges that cause muscle contraction can occur. The intestinal waves also arise in the longitudinal muscle and pass into the circular muscle, probably through low-resistance muscle connections between the two layers. The pacemaker for small intestinal slow waves is in the region of the entrance of the bile ducts and in man has a frequency of 11/min. Unlike gastric slow waves, those of the small intestine do not sweep in an uninterrupted fashion down the gut. Only short segments are coupled with the pacemaker. With increasing distance from the pacemaker the muscle fibers are less likely to follow the pacemaker frequency, and a more distal pacemaker with a slower frequency takes over near the point of failure to follow. By the time the ileum is reached in man the fre-

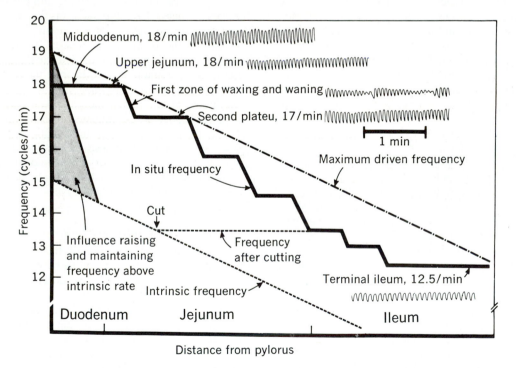

Fig. 55-20. Frequency of slow waves in small intestine of anesthetized cat. Stepwise heavy line shows that observed frequency is 18/min in duodenum and upper jejunum and descending to 12.5/min in terminal ileum. Segments of electrical record from which frequency was measured are shown. Lower slanting dotted line shows intrinsic frequency displayed by short, isolated segments of intestine. If intestine is transected at point labeled *Cut*, frequency distal to cut falls to that of intrinsic frequency at point of transection. (Modified from Diamant and Bortoff[29]; from Davenport.[26])

quency of the slow waves has decreased from the rate of 11/min in the duodenum to 8/min (Fig. 55-20). It is as though the muscle cells were oscillators with intrinsic frequencies, highest in the duodenum and lowest in the ileum. If the intestine is transected, the muscle distal to the cut is no longer driven by the proximal high-frequency muscle pacemaker and generates slow waves at its own slower, intrinsic frequency, which in turn drives the more distal, lower frequency muscle segments. The propagated slow waves of the intestinal smooth muscle determine when it can contract. Whether it does so or not is determined by the imposition of action potentials on the slow wave, thought to be triggered by ACh released by the combined action of hormones such as gastrin and cholecystokinin and extrinsic and intrinsic nerve impulses.

Intrinsic nervous control. The primary intrinsic stimulus for intestinal muscle contraction is radial stretching. The receptors are located in the mucosa, but their precise nature has not been established. The threshold of these receptors is decreased by 5-hydroxytryptamine (5-HT), or serotonin, but the physiologic significance of this observation is not established.[41] With distention of the intestinal lumen, contrac-

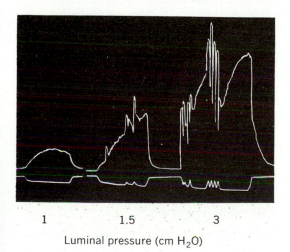

1   1.5   3

Luminal pressure (cm H₂O)

Fig. 55-21. Guinea pig isolated ileum. Modified Trendelenburg method. Upper trace: isotonic record of longitudinal muscle, contraction upward. Lower trace: record of filling of lumen, increased filling downward. At 1 cm H_2O intraluminal pressure, only graded reflex contraction of longitudinal muscle; at 1.5 and 3 cm H_2O intraluminal pressure, fast contractions of longitudinal muscle superimposed on graded reflex contraction. Fast contractions occur only when circular muscle contracts and expels fluid from lumen; in this preparation, peristaltic reflex is elicited intermittently only. (From Kosterlitz.[61])

tion is first elicited in the longitudinal muscle; the strength of contraction is graded in proportion to distention (Fig. 55-21). This contraction is only partially blocked by anticholinergic agents; hence it is both myogenic and neurogenic. With greater distention the threshold for the peristaltic reflex is reached, and both longitudinal and circular muscles contract. This response is cholinergic (Fig. 55-22).

The peristaltic reflex is inhibited by adrenergic agents. It appears that the alpha receptors are on the postganglionic cholinergic neurons and the beta receptors are on the muscle cells. In addition, evidence for a nonadrenergic, noncholinergic (''purinergic'') inhibitory mechanism can be found in the intestine.[61] How these four pathways interact in vivo to determine the complex motor activity of the intestine is not known.

Recent studies employing direct recording of myenteric ganglia neurons have provided new insight into their integrative function in the control of motility. These studies have lead to the classification of myenteric neurons into three categories on the basis of their pattern of discharge: (1) mechanosensitive, (2) single spike, and (3) burst-type neurons.[85] From these studies a model has been suggested that envisions the role of the myenteric ganglia in the programming of motility to involve inhibition of myogenic mechanisms leading to contraction of the circular muscle. The neurotransmitter released from the tonically active inhibitory neurons appears not to be a catecholamine; hence by exclusion these neurons may be ''purinergic.''[10] Propagation of muscle contraction results from the release of the circular muscle from neural inhibition (Fig. 55-23). The control of longitudinal muscle appears to differ from the control of circular muscle in that the activity of the myenteric neurons increases the likelihood that a slow wave will be associated with contraction of longitudinal muscle. This action is mediated by ACh.

Extrinsic innervation. The influence of the vagus on the intestine is diffuse rather than localized. It is greater in the jejunum than in the ileum. It has been estimated that the ratio of myenteric ganglion cells to preganglionic neurons may be as high as 6,000:1. Vagal stimulation increases the likelihood and the strength of intestinal contractions. Adrenergic nerves inhibit motor activity of the intestine. The intestino-intestinal inhibitory reflex seen in anesthetized animals is mediated through a spinal reflex arc and is eliminated by interrupting the splanchnic (adrenergic) nerve supply to the in-

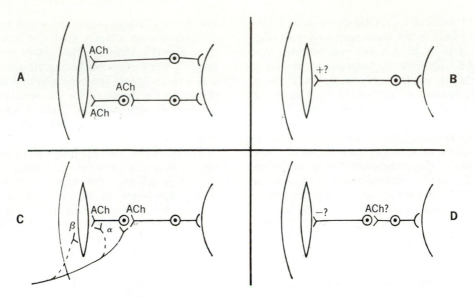

Fig. 55-22. Diagrams of some possible circuits of contractor and relaxant responses of longitudinal muscle of guinea pig ileum. Serosal surface on left and mucosal surface on right. **A,** At top, postulated pathway of graded reflex contraction. Sensory receptor that responds to distention is assumed to be near mucosal surface. Since reflex is not blocked by hexamethonium, there is no synapse or synapse with noncholinergic transmission. At bottom, since contraction during peristaltic reflex is blocked by hexamethonium, there is at least one cholinergic synapse involved. **B,** Graded reflex contraction is resistant to combination of morphine and hyoscine and hence may be mediated by unidentified neuroeffector transmitter. **C,** Possible sites of adrenergic inhibition. Adrenergic nerves terminate around ganglion cells. α-Receptors on nerve terminals cause depression of ACh release. It appears that these α-receptors as well as β-receptors on smooth muscle are activated by circulating catecholamines. **D,** Speculative organization of nonadrenergic relaxant mechanism. (From Kosterlitz.[61])

testine. What role this and other intestinal reflexes such as the gastroileal and ileogastric reflexes play in normal function, and how much of the observed effect is neural or humoral, is not known.

Hormonal control. The interaction of hormonal and neural control of motility is just beginning to be studied. In general, gastrin and CCK-PZ stimulate intestinal motility and glucagon, and secretin and epinephrine inhibit it. A number of substances capable of modifying intestinal motility have been isolated from the intestine, such as serotonin, prostaglandins, darmstoff, and substance P, but how or whether they play a role in the normal neurohumoral control of intestinal motor function has not been clarified.[4,48]

Terminal ileal sphincter. In the past the prevention of the reflux of cecal contents into the distal ileum has been attributed to a mechanical action of a valve at the ileocecal junction. Recently a high-pressure zone analogous to the lower esophageal sphincter has been identified in the ileum just proximal to the ileal-cecal junc-

tion. It is approximately 4 cm in length and exerts a resting pressure of 20 mm Hg. Distention in the ileum above the sphincter leads to relaxation, whereas distention of the cecum leads to increased pressure within the sphincter. Gastrin inhibits tone in the terminal ileal sphincter, an effect opposite that on the lower esophageal sphincter but similar to the effect on the pyloric sphincter (Fig. 55-24).[22,58]

COLON MOTILITY

The colon reduces the volume of the enteric chyme delivered to it by absorbing water and electrolytes and collects the remainder for elimination by defecation. It appears that the colon, like the stomach, is made up of functionally distinct segments. The predominate motor activity in the ascending colon is retrograde contractions. Stationary segmental contractions predominate the motor pattern of the transverse colon, whereas mass movement produced by a strong contraction moving slowly caudally is seen in transverse and descending colon.[15] Taking a meal stimulates gastrointestinal motility,

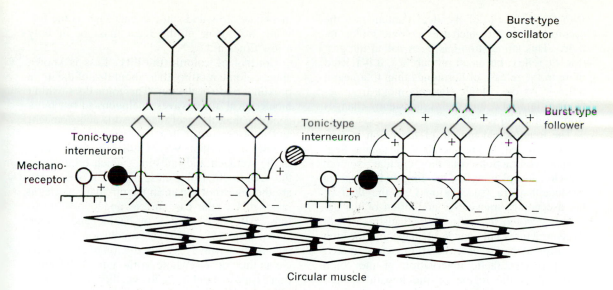

Fig. 55-23. Proposed model for inhibitory nervous control of intestinal circular muscle by myenteric (Auerbach's) plexus. Circular muscle behaves like a functional syncytium that is activated by myogenic electrical activity originating in longitudinal muscle. Burst-type oscillator cells discharge continuously and drive follower neurones that have no intrinsic spontaneous discharge activity. Burst-type followers are intrinsic inhibitory neurones that release a nonadrenergic transmitter substance at neuromuscular appositions. Mechanoreceptors activate tonic-type interneurones. One kind of tonic-type cell (dark circle) forms inhibitory synapses with burst-type followers. These tonic-type cells function to stop discharge of burst-type followers and release muscle from inhibition. Sequential activation of mechanoreceptors and consequent release of inhibition account for propagation of contractile response from left to right on diagram. Second kind of tonic-type neurone (cross-hatched circle) has excitatory synapses on burst-type follower cells, which account for descending inhibition that proceeds peristaltic wave. − = Inhibitory synapses; + = excitatory synapses. (From Wood.[85])

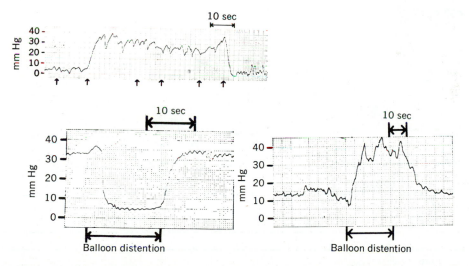

Fig. 55-24. Ileocecal sphincter. Upper tracing demonstrates high-pressure zone in terminal ileum that separates ileum from cecum. Each arrow indicates 1 cm movement of recording tip from colon (left) to ileum (right). Lower tracings indicate that sphincter relaxes in response to balloon distention in ileum above sphincter (left) and contracts in response to distention of balloon in colon (right). (From Cohen et al.[22])

with the discharge of the ileal content into the cecum, and in turn increases colonic motor activity. This stimulation has been called the gastrocolic reflex, but in all probability, it is related more to the release of hormones than to a neural reflex.[51] As the colonic contents move from cecum to sigmoid colon, their consistency changes from semiliquid to solid.

The longitudinal muscle of the colon in man is concentrated in three bundles, or taenia. Segmental contractions of the circular smooth muscles and pleating by the longitudinal taenia produce haustra (sacculations) of the wall of the ascending, transverse, and often of the descending colon as well. These haustra are not fixed structures but change in shape and location as different bands of circular muscle contract. The parasympathetic innervation of the colon proximal to the region of the hepatic flexure reaches it via the vagus, whereas the more distal colon is supplied by the pelvic nerves from the sacral segments of the spinal cord.[79]

The most prominent motor acticity is segmentation, which kneads the colonic contents to facilitate absorption. Periodically, after minutes or hours, segmentation ceases, the haustra are obliterated, and the contents are moved en masse to a more distal segment of the colon. This mass movement is thought to be produced by colonic peristalsis (Fig. 55-25).[45] The study of colonic motility is difficult because the move-ments are slow and some events such as the forward movement of the fecal mass occur only a few times a day.

Control of colonic motility. Less is known about colonic motility than about that of the more proximal parts of the gut. The colon does exhibit bioelectrical slow waves analogous to but more complex than those of the intestine and stomach, and it is presumed that they control the motor activity in a similar way. Unlike the slow waves in the stomach and intestine, which originate in the longitudinal muscles, those of the colon are thought to originate in the circular muscle; at least, this is the case in cat and opossum colons.[15] In vitro studies of canine and human colons lead to a different conclusion—that the colonic slow waves arise in the longitudinal muscle as they do elsewhere in the gut.[7,81] Information is not available as to whether this discrepancy is to be attributed to species or methodologic differences. In all species the frequency of the slow waves is greater in the distal than in the proximal colon (Fig. 55-26). It is assumed that this rising frequency of the slow waves serves to retard the progress of the fecal mass in its journey to the rectum.

INTRINSIC AND EXTRINSIC NERVOUS CONTROL. Nervous control of colonic motor function appears to be qualitatively similar to that of the small intestine. Contractions in the colon are stronger, less frequent, and more prolonged than

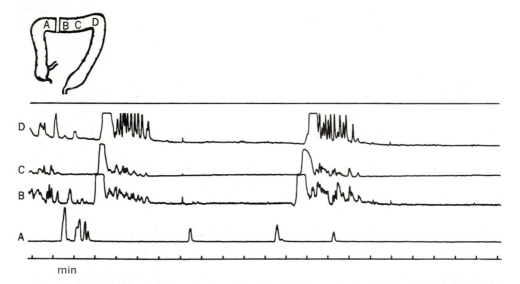

Fig. 55-25. Spontaneous peristalsis in colon of individual with colostomy between points *A* and *B*. Two progressive waves are seen passing from recording sites *B* through *D*. Spontaneous peristaltic waves are infrequently recorded in colon. Their frequency is increased by a variety of laxative preparations. (From Hardcastle and Mann.[45])

in the small intestine. In addition, the intraluminal pressure required to elicit a peristaltic reflex is greater in the large than in the small intestine. Inhibitory mechanisms appear to be more dominant in the colon, especially in the left colon, than in the small intestine. Both adrenergic and nonadrenergic inhibitory fibers have been described.[62]

DEFECATION

Defecation is the temporary reflex interruption of anal continence. Although authorities disagree on their relative importance, continence is maintained by the tonic contraction of the internal and sphincter and the angle between the axis of the rectum and the anal canal (Fig. 55-27).[34]

The rectum is usually empty. When material enters from the sigmoid colon, stretching the wall of the rectum, it produces an awareness

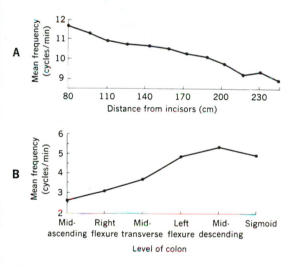

Fig. 55-26. Comparison of gradients in mean slow-wave frequency in, **A,** small bowel, and **B,** colon. **A,** Mean frequency of slow waves was determined in unanesthetized human subjects with intraluminal electrode. Vertical axis shows frequency, and horizontal axis shows distance below incisor teeth. Graph begins at 80 cm, approximately the level of ligament of Treitz. Frequency does not decline along duodenum. Graph does not show stepwise gradient described in animals, perhaps because technique does not permit sufficient accuracy in definition of loci of electrode along bowel. **B,** Mean frequency of slow waves was determined in isolated cat colon. Vertical axis shows frequency, and horizontal axis shows six recording points from proximal colon to distal colon. In contrast to small bowel, frequency seems to rise slightly along colon. No corresponding data are available from man or from colon in situ in any species. (From Christensen.[14])

of the necessity to defecate, and the anal-rectal reflex is initiated (Fig. 55-28). This spinal reflex produces relaxation of the smooth muscle of the internal anal sphincter and contraction of the striated muscle of the external anal sphincter. The latter allows the individual the choice of whether to defecate or suppress the urge. The defecation reflex is reinforced by voluntary effort, for example, increasing intra-abdominal pressure. If the urge is suppressed, external and sphincter contraction is reinforced by voluntary effort and contraction of the pubo rectalis bundles of the levator ani, which increases the angle between rectum and the anus. The rectum accommodates to the new volume, the internal and sphincter regains resting tone, and the urge to defecate passes, only to return with the next distention of the rectum.[75] The rich sensory supply in the anal canal is believed to play an important role in continence and in the control of defecation. Sensations from the upper anal canal that arise during brief internal anal sphincter relaxation appear to play a role in making the important distinction between flatus and feces.

GALLBLADDER EMPTYING

Bile is secreted continuously by the liver but is discharged into the duodenum only at the time of feeding. During the interdigestive periods the bile duct, as it enters the duodenal wall, is closed by the sphincter of Oddi, and the flow is diverted through the cystic duct to the gallbladder. Although egress of bile from the biliary tract is prevented, pressure does not rise because the volume of the bile is decreased by the absorption of water and electrolytes and receptive relaxation by the gallbladder.[3]

Food entering the duodenum provides the normal stimulus for gallbladder emptying. Both vagal impulses and hormone stimulation produce gallbladder contraction and play a role in the control of emptying, although their interrelation is not clear. CCK-PZ is the principal hormone controlling gallbladder emptying (Fig. 55-29). In addition, gallbladder emptying occurs in the absence of vagal innervation. CCK-PZ release increases with increasing load of fatty acids presented to the upper intestine. An excess of bile acids, on the other hand, decreases CCK-PZ release.[68] Secretin inhibits CCK-PZ stimulation of gallbladder contraction, although secretin alone has no effect.[73] Gastrin, which has a structure similar to CCK-PZ, is only one thirtieth as potent as CCK-PZ, hence it is unlikely that it plays any role in the physiologic control of

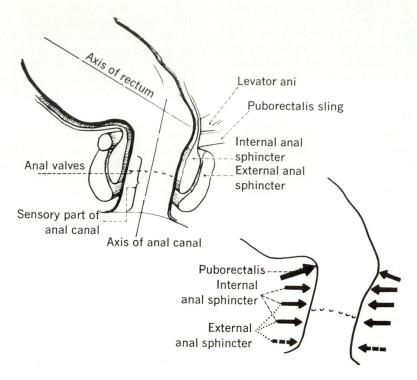

Fig. 55-27. Diagram of sagittal section through anal canal and lower rectum to demonstrate factors involved in anal continence. Arrows in outline diagrams at right represent forces involved. (From Duthie.[34])

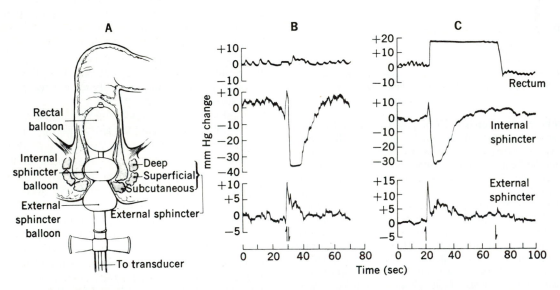

Fig. 55-28. Responses of internal and external anal sphincters of normal human subject. **A,** Schematic diagram of apparatus for distending rectum and recording pressure changes within sphincters. **B,** Relaxation of internal sphincter and contraction of external sphincter following brief distention (arrows) of rectal balloon. **C,** Relaxation of internal sphincter and contraction of external sphincter followed by return to baseline pressures during continued distention of rectum. (Modified from Schuster et al.[76])

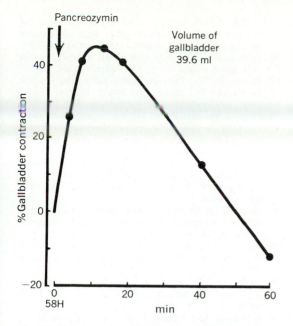

Fig. 55-29. Humoral stimulation of gallbladder contraction in man. CCK-PZ, 1.7 units/kg body weight, given at zero time. Reciprocal relaxation of sphincter of Oddi is associated with gallbladder contraction. (From Howat.[53])

gallbladder function. CCK-PZ causes the gallbladder to contract and the sphincter of Oddi to relax so that bile is discharged into the duodenum.[67] Duodenal contractions are superimposed on the tone of the sphincter, so that when the muscle of the duodenal wall surrounding the sphincter contracts, bile flow is interrupted. Stimulation of the sympathetic nerves to the gallbladder inhibits the effect of CCK-PZ and vagal stimulation.[67] The gallbladder empties about 50% to 70% of its contents over a period ranging from 15 to 60 min. Then, after a quiescent period, the gallbladder completes its emptying.

REFERENCES

1. Abrahamson, H.: Studies on the inhibitory nervous control of gastric motility, Acta Physiol. Scand. **390** (suppl.):1, 1973.
2. Anderson, D. J.: Mastication. In Code, C. F., editor: Handbook of physiology. Alimentary canal section, Baltimore, 1968, The Williams & Wilkins Co., vol. 4.
3. Banfield, W. J.: Physiology of the gallbladder, Gastroenterology **69**:770, 1975.
4. Bass, P., and Bennett, D. R.: Local chemical regulation of motor action of the bowel, substance P and lipid-soluble acids. In Code, C. F., editor: Handbook of physiology. Alimentary canal section, Baltimore, 1968, The Williams & Wilkins Co., vol. 4.
5. Bayliss, W. M., and Starling, E. H.: The movements and innervation of the small intestine, J. Physiol. **24**:99, 1899.
6. Borrison, H. L., and Wang, S. C.: Physiology and pharmacology of vomiting, Pharmacol. Rev. **5**:193, 1953.
7. Bortoff, A.: Digestion: motility, Ann. Rev. Physiol. **34**:261, 1972.
8. Bortoff, A.: Myogenic control of intestinal motility, Physiol. Rev. **56**:418, 1976.
9. Bortoff, A., and Davis, R. S.: Myogenic transmission of antral slow waves across the gastroduodenal junction in situ, Am. J. Physiol. **215**:889, 1968.
10. Burnstock, G.: Purinergic nerves, Pharmacol. Rev. **24**:509, 1972.
11. Cannon, W. B.: The movements of the intestines studied by means of the Röntgen rays, Am. J. Physiol. **6**:251, 1902.
12. Cannon, W. B.: The mechanical factors of digestion, New York, 1911, Longmans, Green & Co.
13. Christensen, J.: Patterns and origins of some esophageal responses to stretch and electrical stimulation, Gastroenterology **59**:909, 1970.
14. Christensen, J.: The controls of gastrointestinal movements: some old and new views, N. Engl. J. Med. **285**:85, 1971.
15. Christensen, J.: Myoelectric control of the colon, Gastroenterology **68**:601, 1975.
16. Christensen, J.: Pharmacology of the esophageal motor function, Anu. Rev. Pharmacol. **15**:243, 1975.
17. Christensen, J.: The controls of oesophageal movement, Clin. Gastroenterol. **5**:15, 1976.
18. Code, C. F., and Schlegel, J. F.: The gastrointestinal interdigestive housekeeper: Motor correlates of the interdigestive myoelectric complex of the dog. In Proceeding of the Fourth International Symposium on Gastrointestinal Motility, Vancouver, B. C., 1974, Mitchell Press, p. 631.
19. Code, C. F., et al.: A concept of control of gastrointestinal motility. In Code, C. F., editor: Handbook of physiology. Alimentary canal section, Baltimore, 1968, The Williams & Wilkins Co., vol. 5.
20. Cohen, B. R., and Wolf, B. S.: Cine-radiographic and intraluminal pressure correlations in the pharynx and esophagus. In Code, C. F., editor: Handbook of physiology. Alimentary canal section, Baltimore, 1968, The Williams & Wilkins Co., vol. 5.
21. Cohen, S., and Lipschutz, W. H.: Hormonal regulation of human lower esophageal sphincter competence: interaction of gastrin and secretin, J. Clin. Invest. **50**:449, 1971.
22. Cohen, S., Harris, L. D., and Levitan, R.: Manometric characteristics of the human ileocecal junctional zone, Gastroenterology **54**:72, 1968.
23. Cooke, A. R.: Control of gastric emptying and function, Gastroenterology **68**:804, 1975.
24. Crispin, J. S., McIver, D. K. and Lind, J. F.: Manometric study of the effect of vagotomy on the gastroesophageal sphincter, Can. J. Surg. **10**:299, 1966.
25. Daniel, E. E., and Irwin, J.: Electrical activity of the gastric musculature. In Code, C. F., editor: Handbook of physiology. Alimentary canal section, Baltimore, 1968, The Williams & Wilkins Co., vol. 4.
26. Davenport, H. W.: Physiology of the digestive tract, ed. 3, Chicago, 1971, Year Book Medical Publishers, Inc.
27. Debas, H. T., Farooq, O., and Grossman, M. I.: Inhibition of gastric emptying is a physiologic action of cholecystokinin, Gastroenterology **68**:1211, 1975.

28. Diamant, N. E.: Electrical activity of the cat smooth muscle esophagus: A study of hyperpolarizing responses. In Proceedings of the Fourth International Symposium on Gastrointestinal Motility, Vancouver, B.C., 1974, Mitchell Press, p. 593.

29. Diamant, N. E., and Bortoff, A.: Effects of transection on the intestinal slow-wave frequency gradient. Am. J. Physiol. **216:**734, 1969.

30. DiMarino, A. J., and Cohen, S.: The adrenergic control of lower esophageal sphincter function, J. Clin. Invest. **52:**2264. 1973.

31. Dodds, W. J., et al.: Movement of the feline associated with respiration and peristalsis. An evaluation using tantalum markers, J. Clin. Invest. **52:**1, 1972.

32. Doty, R. W.: Neural organization of deglutition, In Code, C. F., editor: Handbook of physiology. Alimentary canal section, Baltimore, 1968, The Williams & Wilkins Co., vol. 4.

33. Dryburgh, J. R., and Brown, J. C.: Radioimmunoassay for motilin, Gastroenterology **68:**1169, 1975.

34. Duthie, H. L.: Anal continence, Gut **12:**844, 1971.

35. Edwards, D. A. W., and Rowlands, E. N.: Physiology of the gastroduodenal junction. In Code, C. F., editor: Handbook of physiology. Alimentary canal section, Baltimore, 1968, The Williams & Wilkins Co., vol. 4.

36. Farrar, J. T.: Gastrointestinal smooth muscle function, Am. J. Dig. Dis. **8:**103, 1963.

37. Farrell, J. H.: The effect of mastication on the digestion of food, Br. Dent. J. **100:**149, 1956.

38. Fisher, R. S., and Cohen, S.: Physiological characteristics of the pyloric sphincter, Gastroenterology **64:**67, 1973.

39. Fisher, R. S., and Cohen, S.: The influence of gastrointestinal hormones and prostaglandins on the lower oesophageal sphincter, Clin. Gastroenterol. **5:**29, 1976.

40. Fleshler, B., et al.: Characteristics and similarity of primary and secondary peristalsis, J. Clin. Invest. **38:**110, 1959.

41. Gershon, M. D.: Serotonin and the motility of the gastrointestinal tract, Gastroenterology **54:**453, 1968.

42. Goyal, R. K., and Rattan, S.: Genesis of basal sphincter pressure: effect of tetradotoxin on lower esophageal sphincter pressure in opposum in vivo, Gastroenterology **71:**62, 1976.

43. Grey, E. G.: Observations on the postural activity of the stomach, Am. J. Physiol. **45:**272, 1918.

44. Grossman, M. I.: What is physiological: round 2, Gastroenterology **67:**766, 1974.

45. Hardcastle, J. D., and Mann, C. V.: Study of large bowel peristalsis, Gut **9:**512, 1968.

46. Hellemans, J., Vantrappen, G., and Vanderbrouche, J.: Electrical activity of the human esophagus, Gastroenterology **58:**959, 1970.

47. Hendrix, T. R., and Yardley, J. H.: Consequences of gastro-oesophageal reflux, Clin. Gastroenterol. **5:**155, 1976.

48. Hendrix, T. R., et al.: The effect of 5-hydroxytryptamine on intestinal motor function in man, Am. J. Med. **23:**886, 1957.

49. Higgs, R. H., Smyth, R. D., and Castell, D. O.: Gastric alkalinization. Effect on lower esophageal sphincter pressure and serum gastrin, N. Engl. J. Med. **29:**486, 1974.

50. Hightower, N. C., Jr.: Motor action of the small bowel. In Code, C. F., editor: Handbook of physiology. Alimentary canal section, Baltimore, 1968, The Williams & Wilkins Co., vol. 4.

51. Holdstock, D. J., and Misiewicz, J. J.: Factors controlling colonic motility: colonic pressures and transit after meals in patients with total gastrectomy, pernicious anemia, or duodenal ulcer, Gut **11:**100, 1970.

52. Hopkins, A.: The pattern of gastric emptying: a new view of old results, J. Physiol. **182:**144, 1966.

53. Howat, H. W.: Tests on human gallbladder function. In Taylor, W., editor: The biliary system, Oxford, 1965, Blackwell Scientific Publications, Ltd.

54. Hunt, J. N., and Knox, M. T.: Regulation of gastric emptying. In Code, C. F., editor: Handbook of physiology. Alimentary canal section, Baltimore, 1968, The Williams & Wilkins Co., vol. 4.

55. Hwang, K.: Mechanism of transport of the content of the esophagus, J. Appl. Physiol. **6:**781, 1954.

56. Ingelfinger, F. J.: Esophageal motility, Physiol. Rev. **38:**533, 1961.

57. Jansson, G.: Extrinsic nervous control of gastric motility: an experimental study in the cat, Acta Physiol. Scand. **326**(suppl.):1, 1969.

58. Jarrett, R. J., and Gazet, J. C.: Studies in vivo of the ileosaeco-colic sphincter in the cat and dog, Gut **7:**271, 1966.

59. Kantrowitz, P. A., et al.: Response of the human esophagus to d-tubocurarine and atropine, Gut **11:**47, 1970.

60. Kawasaki, M., Ogura, J. H., and Takenouchi, S.: Neurophysiologic observations of normal deglutition. I. Its relationship to the respiratory cycle. II. Its relationship to allied phenomena, Laryngoscope **74:**1747 and 1766, 1964.

61. Kosterlitz, H. W.: Intrinsic intestinal reflexes, Am. J. Dig. Dis. **12:**245, 1967.

62. Kosterlitz, H. W.: Intrinsic and extrinsic nervous control of motility of the stomach and intestine. In Code, C. F., editor: Handbook of physiology. Alimentary canal section, Baltimore, 1968, The Williams & Wilkins Co., vol. 4.

63. Lind, J. F., et al.: Motility of the gastric fundus, Am. J. Physiol. **201:**197, 1961.

64. Lipshutz, W., Tuch, A. F., and Cohen, S.: A comparison of the site of action of gastrin I on lower esophageal sphincter and antra circular muscle, Gastroenterology **61:**454, 1971.

65. Longhi, E. H., and Jordan, P. H., Jr.: Necessity of a bolus for propagation of primary preistalsis in the canine esophagus, Am. J. Physiol. **220:**609, 1971.

66. Lumsden, K., and Holden, W. S.: The act of vomiting in man, Gut **10:**173, 1969.

67. Magee, D. F.: Physiology of gallbladder emptying. In Taylor, W., editor: The biliary system, Oxford, 1965, Blackwell Scientific Publications, Ltd.

68. Malagelada, J. R., et al.: Regulation of pancreatic and gallbladder functions by intraluminal fatty acids and bile acids in man, J. Clin. Invest. **58:**493, 1976.

69. McNally, E. F., Kelly, J. E., Jr., and Ingelfinger, F. J.: Mechanism of belching: effects of gastric distention with air, Gastroenterology **46:**254, 1964.

70. Nebel, O. T., and Castell, D. O.: Inhibition of the lower oesophageal sphincter by fat–a mechanism for fatty food intolerance, Gut **14:**270, 1973.

71. Ramsey, G. H., et al.: Cinefluorographic analysis of the mechanism of swallowing, Radiology **64:**498, 1955.

72. Roman, C., and Tieffenbach, L.: Motricite de l'oesophage a musculeuse lisse apres vivagotomie: etude electromyographique (EMG), J. Physiol. (Paris) **63:**733, 1971.

73. Ryan, J., and Cohen, S.: Interaction of gastrin I, secretin and cholecystokinin on gallbladder smooth muscle, Am. J. Physiol. **230:**553, 1976.

74. Schofield, G. C.: Anatomy of muscular and neural tissues in the alimentary canal. In Code, C. F., editor: Handbook of physiology. Alimentary canal section, Baltimore, 1968, The Williams & Wilkins Co., vol. 4.

75. Schuster, M. M: Motor action of rectum and anal sphincters in continence and defecation. In Code, C. F., editor: Handbook of physiology. Alimentary canal section, Baltimore, 1968, The Williams & Wilkins Co., vol. 4.

76. Schuster, M. M., et al.: Simultaneous manometric recording of internal and external anal sphincteric reflexes, Bull. Johns Hopkins Hosp. **116:**79, 1965.

77. Siegel, C. I., and Hendrix, T. R.: Evidence for central mediation of secondary peristalsis in the esophagus, Bull. Johns Hopkins Hosp. **108:**297, 1961.

78. Skinner, D., et al.: Gastroesophageal reflux and hiatus hernia, Boston, 1972, Little, Brown & Co.

79. Truelove, S. C.: Movements of the large intestine, Physiol. Rev. **46:**457, 1966.

80. Tuck, A., and Cohen, S.: Neurogenic basis of lower esophageal sphincter relaxation, J. Clin. Invest. **52:**14, 1973.

81. Vanasin, B., Ustach, T. J., and Schuster, M. M.: Electrical and motor activity of human and dog colon in vitro, Johns Hopkins Med. J. **134:**201, 1974.

82. Vantrappen, G., and Hellemans, J.: Esophageal motility, Rendic. R. Gastroenterol. **2:**7, 1970.

83. Werle, J. M., et al.: The mechanics of gastric evacuation, Am. J. Physiol. **131:**606, 1941.

84. Winship, D. H., Viegas de Andrade, S. R., and Zboralshe, F. F.: Influence of bolus temperature on human esophageal motor function, J. Clin. Invest. **49:**243, 1970.

85. Wood, J. D.: Neurophysiology of Auerbach's plexus and control of intestinal motility, Physiol. Rev. **55:**307, 1975.

86. Zwich, R., et al.: Mechanism of action of pentagastrin on the lower esophageal sphincter, J. Clin. Invest. **57:**1644, 1976.

XII

METABOLISM

56 Energy exchange

JOHN R. BROBECK and ARTHUR B. DuBOIS

PRINCIPLES

Between its body and its environment a living animal or plant carries on a continual exchange of both materials and energy (Table 56-1). The exchange follows the conservation principle in that none of the material or energy is lost in the exchange. The amount retained by the body is the difference between the intake and the output. Energy exchange is also in accord with the principles of thermodynamics that apply to nonliving physical and chemical systems.

There is a fundamental difference between plants and animals in the nature of the energy they utilize for their life processes. Plants convert the energy of solar photons into the energy of chemical bonds; thus they synthesize those compounds that compose their several structures, including the leaves, fruit, tubers, or seeds that serve as food for animals. Animals cannot do this to any significant degree. During photosynthesis, plants give off oxygen. Plants are interposed, therefore, between the sun and all animals in the energy flow sheet of nature; all the energy utilized by animals first passes through the photosynthetic processes of plants.

Of all the variety of chemical bondings found in organic and inorganic compounds, only one type is significant as a source of energy for animals. It is the bonding between hydrogen and carbon in compounds known as carbohydrate, edible fat, or the carbohydrate-like carbon-hydrogen chains belonging to certain amino acids. In the release of energy in biologic reactions the C and H are oxidized to CO_2 and H_2O. Other types of bondings may serve certain specialized functions of the body. One example is the inorganic bonds of the mineral salts giving rigidity to bone. Another example is the bonding between a complex organic compound and a phosphate radical, which cells utilize for storage or for transport of energy. This particular bonding is so important in biochemistry that it has been given a name—the "high-energy bond" or the "high-group potential." Its energy is made available to cells on hydrolysis. Yet this particular bonding is likewise derived ultimately from the oxidation of the C—H bonds of carbohydrates and lipids. One can say that the energy comes from the bonding potential of molecular oxygen, provided that one remembers that out of all the different atoms capable of being oxidized by oxygen, only C—H bonds of a few types of compounds are important in physiologic energy supply.

It is apparent that versatility of reactions, behavior, and achievement shown by animals lies not in any corresponding variety of energy-yielding processes but rather in the diversity of uses animals make of the C—H bonding energy. It is used for beating of cilia, for shortening of muscles, for conduction of nerve impulses, for secretion by glands, and for synthesis of compounds that become a part of the body's substance. It energizes growth, repair, and reproduction as well as the commercial production of meat, milk, and eggs. The sum total of the processes in which this energy is made available and utilized is the *metabolism* of the body. The word is from a Greek term meaning "change," and it should not be used merely in the sense of exchange. Thus it is better to speak of water exchange than of water metabolism and of energy exchange in preference to energy metabolism. It is correct to say that the metabolism of carbohydrate and fat permits the body to carry on its energy exchange.

Table 56-1. Physiologic exchanges*

Variable	Body content	Daily exchange
Oxygen	1.5 L	360 L
Water	42 L	3 L
Heat	990 kcal	2,800 kcal
Energy	146,500 kcal	2,800 kcal

*Estimated for a man weighing 70 kg at room temperature of 20° C. The daily exchange of heat, 2,800 kcal, includes the amount of heat equal to the mechanical work done by the person on the environment.

The chemical changes by which energy is made available to the body are known as *catabolism;* usually they are oxidative or hydrolytic, and they result in the conversion of either food or body tissue into carbon dioxide and water. The converse of these reactions is the synthesis of materials stored as a part of the body, and it is known as *anabolism*. In animals, it cannot occur alone because the energy for synthetic reactions, including formation of protein or body fat, comes only from catabolic reactions. This means that anabolism for the body as a whole can occur only after feeding has made available both the molecules to be stored and the energy for their synthesis into tissue constituents. In the absence of food supply, anabolism can occur in a given tissue only at the expense of molecules and energy derived from some other location in the body. Exchange of this type between one part of the body and another takes place with astonishing ease. The compounds and molecules of many tissues are in a state of more or less continual interchange with other compounds and molecules of the same and other tissues, so much so that it is very often impossible to say just where a given molecule or atom is or belongs at any one instant. The discovery of this lability of the body's composition[35] led to the concept that the compounds in question exist as components of a metabolic "pool"—a volume of fluid including many constituents of the plasma and interstitial fluid but also including the labile components of all cells. Substances are continually entering the pool while other compounds are constantly leaving it, either to be metabolized for liberation of energy as work or heat or to be converted into other molecules.

MEASUREMENT OF ENERGY EXCHANGE

The study of the body's input and output of energy, the energy exchange, is based on the first law of thermodynamics.

First law of thermodynamics. Count Rumford at the end of the eighteenth century noticed that the amount of heat evolved during the boring of a cannon was roughly proportional to the mechanical energy expended in the process. During the fifth decade of the last century, Mayer, Joule, and Helmholtz independently stated that a definite amount of mechanical energy produces a definite amount of heat. Their observations and discussions led to the general conclusion that energy is neither gained nor lost when it is converted from one form to another. This "law of the conservation of energy" has been applied to all forms of energy—kinetic, thermal, electrical, and chemical. According to this law, any system in a given condition contains a definite quantity of energy, and when the system undergoes change, any gain or loss of its internal energy is equal to the loss or gain in the energy of surrounding systems. The application of this law to biologic systems was one of the outstanding events in the development of our knowledge of metabolism. It involved a comparison of the actual energy expended by an animal with the energy supplied to the animal either from ingested foodstuffs or from the reserve stores of the *energy balance of metabolism*. It was applied later to the energy expenditure in muscular exercise, in hyperthyroidism and hypothyroidism, and in many other conditions.

Thermal physiologists use the International System of Units (SI), which is the meter (m), kilogram (kg), second (s), ampere (A), degree kelvin (°K), and candela (cd). In SI the unit of force is the newton (N), which is in kg · m/sec² or kg × 9.807. Consequently, the unit of work, energy, or a quantity of heat is the joule (J), in N · m or kg · m²/sec². Finally, the unit of power is the watt (W), in J/s or N · m/s.

The unit of energy used in studies of energy balance in biologic systems is the large or kilogram calorie. One kilocalorie (kcal) is the amount of heat required to raise the temperature of 1 kg of water 1° C of temperature from 15° to 16° C. To convert kilocalories to joules, multiply by 4,187. To convert kilocalories per square meter per hour to watts per square meter, multiply by 1.163. The value of the energy supplied in the food is determined by burning a measured quantity of material in a bomb calorimeter. This method is called the *direct* method. The apparatus is placed in a known amount of water at a certain temperature. After combustion of the material inside the bomb the resulting rise in the temperature of the water is multiplied by the specific heat of water and by the volume of the water in liters to obtain the caloric value. Many different types of calorimeters have been designed for the direct measurement of the caloric value of foodstuffs. Benedict and Fox[8] designed an *indirect* method in which the amount of oxygen used in burning the food material is measured. This value is converted into calories. The caloric values obtained for commonly ingested materials and some intermediates are given in Table 56-2. The values that have been obtained for complex substances such as glycogen or starch or compounds containing nitrogen or sulfur have not been in precise agreement. Although the differ-

Table 56-2. Gaseous exchange, respiratory quotient, heat production, and caloric value of carbon dioxide and oxygen of various commonly metabolized compounds*

| Material | O_2 required to oxidize 1 gm (ml) | Produced in oxidation of 1 gm | | | Heat/L | |
		CO_2 (ml)	Heat (kcal)	RQ	O_2 (kcal)	CO_2 (kcal)
Starch	829.3	829.3	4.20	1.00	5.06	5.06
Cane sugar	785.5	785.5	3.96	1.00	5.04	5.04
Dextrose	746.2	746.2	3.74	1.00	5.01	5.01
Lactic acid	745.9	745.9	3.62	1.00	4.85	4.85
Animal fat	2,013.2	1,431.1	9.50	0.71	4.72	6.64
Protein	956.9	773.8	4.40	0.81	4.60	5.69
Acetone	1,542.9	1,157.2	7.43	0.75	4.82	6.42
Ethyl alcohol	1,459.5	972.9	7.08	0.67	4.85	7.28

*From Carpenter.[12]

ences are small, they become significant when used in attempts to balance the energy exchange of tissues.

Essentially the same methods are used to measure the heat liberated by the animal body. Here again it may be determined *directly* by placing the animal in a calorimeter or *indirectly* by measuring the amount of oxygen used or the amount of carbon dioxide given off and then converting these values into calories. From the *respiratory quotient,* which is the ratio of the volume of carbon dioxide given off to the volume of oxygen taken in during the same interval of time, a rough indication of the type of food material metabolized is obtained. It will be advantageous to give an outline of the techniques of the two methods.

Direct method of calorimetry

The direct method has the advantage that it measures in calories the output of heat. When small animals are studied, the simplest form of apparatus is a Dewar flask, in which the animal is placed, and the resulting temperature change is recorded by a sensitive thermometer. Hill[25] constructed a differential calorimeter using two flasks, in each of which was placed one junction of a thermocouple. With such an apparatus the animal or tissue is placed in one flask and the other flask is used as a control, whereby the effects of changes in external temperature on the recording mechanism are excluded. This type of apparatus has been used to study the metabolism of frogs and other small animals. Chemists have used apparatus of similar but more elaborate design for the study of heats of chemical reactions catalyzed by enzymes.

Adiabatic calorimetry. Calorimeters have been constructed for the purpose of measuring, by the direct method, the heat production of larger animals and man.[6] The subject occupies a thermally insulated chamber in which heat lost from the body by radiation and conduction (about three fourths of the total lost) is removed by a stream of water flowing through tubes attached to the inner walls. The number of calories thus lost by the subject can easily be calculated from the volume of water passing through, the difference in temperature of the ingoing and outgoing water, and the specific heat of water. The calories carried away from the apparatus during the period plus the heat value of the water vaporized from the body of the subject give the total heat elimination. Any other loss of heat from the calorimeter is prevented by a double copper wall surrounding the chamber. These copper walls are connected with thermocouples so that an electrical current is generated when any difference in temperature occurs between the walls. If there is a change, temperature equilibrium is restored by passing cold water through coils or by electrical heating. In this manner the two walls are kept at the same temperature, there can be no flow of heat between them, and hence the calorimeter is adiabatic.

A more convenient method of calorimetry has been devised based on the measurement of heat transfer through a layer of insulating material having constant size, shape, and thermal properties. This insulation is the gradient layer; the temperatures of its two surfaces are functions of the rate of heat production to which it is exposed. The principles of this technique have been described by Benzinger and Kitzinger as follows (see also Fig. 56-1):

When a source of heat is accommodated in a cavity . . . and the surface of that cavity is completely lined with a layer of insulating material, uniform in thickness and in specific thermal conductivity, the integrated average difference, or gradient, of temperatures between the inner and outer surfaces of this layer is proportional to the rate of loss or gain of heat from the source inside, once a steady state of heat flow has been established. In response to a sudden transition from one constant rate of heat flow to another, the temperature gradient will show an exponential rise or fall to a new level which represents the new steady rate of heat flow. The time constant of this exponential response decreases with the first power of the specific thermal diffusivity of the layer. It increases

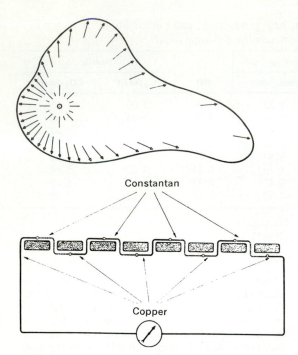

Fig. 56-1. Gradient principle of calorimetry. Scheme showing cross sections through cavity with source of heat and through gradient layer (enlarged). "Gradient layer" shown below completely and uniformly lines entire inner surface of cavity (above) and records correctly total heat from source within, regardless of its location and size or shape of cavity. (From Benzinger and Kitzinger.[10])

with the square of the thickness. Thin layers respond more swiftly, thick layers produce higher gradients.*

It is easier to measure accurately such an average temperature gradient with thermocouples than to measure temperature change and volume of water circulated through an adiabatic calorimeter. This makes the gradient layer method not only more convenient but considerably more accurate, especially when a short response time is needed for the physiologic study. Fig. 56-2 shows a human subject reclining in the gradient layer calorimeter at the Naval Medical Research Institute in Bethesda, Maryland.

Indirect methods of calorimetry

There are two general indirect methods for the study of metabolism—the closed-circuit and the open-circuit types.

Closed-circuit method. In the closed-circuit method the expired air is passed through bottles containing suitable reagents for the absorption of water and CO_2,

---

and the amount of O_2 used is determined from the amount added during the experiment to keep the total volume of the system unchanged. This method may be used with a calorimeter (Fig. 56-3), in which case the results of the indirect method may be compared with those of the direct. The chamber is connected with a closed system of tubes through which air is passed by a pump. Outside the chamber the air passes through a spirometer, a bottle of concentrated sulfuric acid, a bottle of soda lime, and a second bottle of sulfuric acid. The first sulfuric acid absorbs water vapor in the air coming from the chamber, the soda lime takes out the CO_2, and the second sulfuric acid takes up the water produced by the reaction between the soda lime and the CO_2. The amounts of water and CO_2 given off by the subject are found by weighing these bottles before and after the subject has been placed in the chamber. During the experiment, O_2 is admitted to the system from a side tube at such a rate that the amount used by the subject is exactly replaced. The quantity of O_2 taken up by the subject may be determined by weighing the cylinder from which it comes before and after the experiment or by passing the gas through a sensitive meter placed between the O_2 tank and the tubing of the system.

To avoid the necessity of building a closed cabinet for these determinations, numerous devices such as the Benedict helmet, masks, mouthpieces, and nosepieces have been designed for connection of the subject with the apparatus. These contrivances are also used with the open circuit method.

Open-circuit method. In the open-circuit method the subject breathes through a valve system so arranged that it separates the inspired from the expired air. The volume of expired air is measured by collecting it in a spirometer (Tissot method) or a rubber bag (Douglas method[18]), or it may be measured directly by passage through a gas meter. After use of the gas laws for correction of gas volumes the CO_2 production is taken to be equal to the amount of this gas present in the expired mixture per unit time. O_2 *consumption* may be calculated if the O_2 and nitrogen contents of the expired mixture are determined. Inasmuch as nitrogen is a "passive" gas in respiratory exchange, the total amount of expired nitrogen must equal the total amount of inspired nitrogen. If the inspired mixture is dry air, its nitrogen concentration is 79.02%; the volume of inspired gas therefore is equal to the amount of nitrogen present in the expired mixture multiplied by $\frac{100}{79.02}$. As atmospheric air, the O_2 concentration of the inspired mixture is 20.95%. Hence O_2 consumption is the difference between the amount of O_2 calculated to be present in the inspired mixture and the amount of O_2 found by analysis to be present in the expired mixture. O_2 consumption is usually expressed as milliliters per minute standard temperature and pressure (STP). Although the inert gases nitrogen and argon are loosely referred to as "nitrogen," actual atmospheric nitrogen is 78.084% and argon 0.934%. Adding them and rounding off to two decimal places yields 79.02%. Atmospheric O_2, 20.946% (or 20.95%), and CO_2,

Fig. 56-2. Human subject in gradient layer calorimeter. Man is suspended on frame with wire nets held by nylon cords. Gradient layers at sides are covered with aluminum foil. (From Benzinger and Kitzinger.[10])

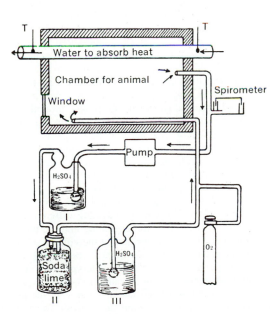

Fig. 56-3. Diagram of Atwater-Benedict respiration calorimeter. As animal uses up O_2, total volume of air decreases. This change is indicated by spirometer, and corresponding amount of O_2 is delivered from weighed O_2 cylinder. Increase in weight of bottles II and III gives amount of CO_2; that of I, amount of water vapor.

0.033% (or 0.03), make up the remainder. The amount of CO_2 is often taken as zero, since it is less than 0.05%.

Gas analysis

DIRECT READING. After a period of collection a sample of expired air is analyzed for CO_2 and O_2 concentration. This analysis is usually done now by some instrument giving results in the form of an electrical signal. All such instruments must be calibrated, however, against a more direct method of analysis based on changes in volume or pressure of the gas mixture as either CO_2 or O_2 is removed by absorption in an appropriate solution.

INDIRECT METHODS. Gases that are the constituents of the gas mixtures significant in energy exchange differ from one another in a number of physical properties, or for a given type of property they exhibit different numeric values when the property is measured. The indirect methods for their estimation take advantage of these differences. In one class of methods a characteristic property of a given gas is measured as an index of the concentration of that gas. In the other class of methods the gases are separated by some physical process, and then the concentration of each one is determined by measurement of a property they all share. The classic example of the use of inherent differences in properties is the mass spectrometer, in which gases are ionized and then separated according to mass. Another method widely used is based on the fact that O_2 has paramagnetic properties exceeding those of the other respiratory gases. Using a magnetic torsion balance, Pauling et al.[31] developed a technique in which O_2 concentration in a gas mixture can be measured without separating the components of the mixture. In a sense the gases are inherently separated because of the difference in paramagnetic property. Yet a third technique is applied in the same fashion to measurement of CO_2 in gas mixtures. It is based on infrared absorption where, in a suitably designed instrument, the degree of absorption is proportional to the CO_2 concentration. Here again the gas mixture does not need to be separated physically and is not altered by the analysis.

If a gas mixture can be separated into its constituent gases, other types of measurement are possible, two of which are thermal conductivity[27] and electrical conductivity. The former is measured as a resistance change in a heated platinum wire, where resistance varies with temperature, and temperature is a function of the amount of heat taken from the wire by the gas mixture flowing over it. The wire is "bathed" continuously in a flowing stream of some inert gas, for example, helium. When another gas such as O_2 is present in the helium, the thermal conductivity of the gas mixture is altered to a degree proportional to the amount of O_2 present. Similar principles are used in measuring electrical conductivity. The measurements are easy to make and accurate; they do require, as noted earlier, the calibration of the instrument by passage of gas mixtures of known composition. Also, this standard composition must be determined by using a Haldane or a Scholander analyzer.[17]

Finally, one must ask just how the gases of a mixture can be separated so as to determine their thermal conductivity one at a time. The separation is made using a technique known as gas chromatography, but one must understand that the term is a misnomer, since no color reaction nor process is involved. The name was given by analogy to other types of chromatography, which were originally used to separate colored compounds. In brief, the gases of a mixture become separated because on any substance that adsorbs gases the several types of molecules will have varied affinities for the adsorbent. A gas column filled with an adsorbent (a solid, not a liquid) is perfused by a stream of the carrier or inert gas. When the gas mixture to be analyzed is added to the carrier stream, the molecules of the several constituent gases will tend to be adsorbed at different locations in the column because of their different affinities for the adsorbent. But when all the unknown mixture has passed into the column, the carrier stream will begin to carry out of the column those molecules least tightly bound to the adsorbent. Later the stream carries out the more tightly bound molecules. By suitable choice of length of adsorbent column, of nature of adsorbing material, and of nature and rate of flow of carrier gas, it is possible to separate, for example, CO_2 from the rest of the mixture so that its conductivity can be measured. It can then be removed from the mixture by a Haldane-type adsorbent, so that the remaining O_2 and nitrogen can be separated and also measured one at a time.

Clinical methods. The indirect measurements just outlined are now extensively used in clinics and hospitals as well as in research laboratories. Nevertheless, there are many occasions when the precision of such specialized instruments is not needed and when less expensive and more simple techniques are good enough for the purpose at hand. This is particularly true when the only data needed are overall measurements of energy exchange in a fasting subject, or when the O_2 consumption alone can serve as an index of the rate of heat production or energy exchange. For many years, O_2 consumption was so measured by physicians in the diagnosis of diseases of the thyroid gland. (More reliable diagnostic methods are now available.) The apparatus used consisted of a closed system filled with O_2. It was made up of a spirometer with tubes and valves arranged so that the O_2 flowed around a circuit where some of the O_2 was removed in the subject's lungs and some CO_2 added. The CO_2 was then taken up by soda lime adsorbent inside the spirometer bell (Fig. 56-4). The position of the bell changed in proportion to the amount of O_2 taken up in the lungs, whereas the inspirations and expirations of the subject were reflected in alternate downward and upward movements of the spirometer bell. The difference in the height of the bell at the beginning and the end of the run (usually 6 min) represented the volume of O_2 used in the interval. This was then corrected for aqueous tension and reduced to standard temperature and pressure. A record of the breathing movements, which slanted because of the fall in volume in the system, was usually made on graph paper, from which the rate

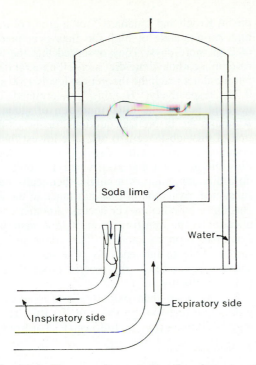

Fig. 56-4. Diagram of connections in spirometer of Benedict-Roth apparatus.

Table 56-3. Respiratory quotients calculated from chemical equations for oxidation*

Substance	RQ
Glycerol	0.86
Glyceric aldehyde	1.00
β-Hydroxybutyric acid	0.89
Acetoacetic acid	1.00
Acetone	0.75
Lactic acid	1.00
Pyruvic acid	1.20
Alcohol	0.667

*From Richardson.[32]

of O_2 consumption could be converted to calories per hour.

RESPIRATORY QUOTIENT

The respiratory quotient (RQ) is defined as the ratio of the *volume* of CO_2 expired to the *volume* of O_2 consumed during the same period of time. It has long been known that among animals this ratio varies to some extent with food. The quantity of O_2 required and the amount of CO_2 produced in the combustion of food, either within or outside the body, depend on the chemical composition of the material. Therefore the RQ for any individual foodstuff may be determined from the equation representing its oxidative breakdown to the final products. Since equal volumes of gases at the same temperature and pressure contain the same number of molecules (law of Avogadro), it is not necessary in making this determination to convert the volumes of the two gases into number of molecules. The following examples illustrating the method of calculating the RQ are taken from Lusk.[30]

Carbohydrate. The general formula for carbohydrate foodstuffs may be written $(CH_2O)_n$; it is plain that the RQ will be unity from the following equation:

$$C_6H_{12}O_6 + 6O_2 = 6CO_2 + 6H_2O$$
$$\text{(glucose)}$$

$$RQ = \frac{6 \text{ vol } CO_2}{6 \text{ vol } O_2} = 1.00$$

Fat. In the oxidation of fat, extramolecular O_2 is required not only for the oxidation of carbon but also for the oxidation of hydrogen. For this reason the RQ will be less than unity. If we take as an example tripalmitin, $C_{51}H_{98}O_6$, this fact becomes evident from the following equation:

$$2C_{51}H_{98}O_6 + 145O_2 = 102CO_2 + 98H_2O$$

$$RQ = \frac{102 \text{ vol } CO_2}{145 \text{ vol } O_2} = 0.703$$

Food fats, which are composed chiefly of the glycerides of palmitic, stearic, and oleic acids, give a quotient of 0.706.

Protein. It is more difficult to calculate the RQ of the combustion of protein in the body because some of the O_2 and carbon of the constituent amino acids remains combined with nitrogen and is excreted as nitrogenous compounds of the urine and feces. The quantities of carbon and hydrogen involved in oxidative processes in the body require 138.18 gm of O_2 for the production of 152.17 gm of CO_2. Converting these quantities into volumes,* we find that the ratio for protein is as follows:

$$RQ = \frac{77.52 \text{ L } CO_2}{96.70 \text{ L } O_2} = 0.802$$

Other substances. In a similar manner, RQs have been calculated for the oxidation of a number of compounds of biologic significance (Table 56-3). It may be noted that oxidation of aceto-

---

*Conversion is carried out on the basis of the fact that at STP 1 gm of O_2 occupies 0.6998 L and 1 gm of CO_2, 0.5094 L.

acetic acid, a substance derived from fatty acid oxidation, gives an RQ of 1.00, whereas pyruvic acid, derived from carbohydrate, gives a quotient over 1 (1.20).

Nonprotein respiratory quotient. The protein metabolized, in grams per hour, can be calculated from the amount of nitrogen excreted in the urine per hour. Each gram of nitrogen represents the catabolism of 6.25 gm of protein. Twelve hours after the last meal, protein contributes about 15% of the total calories. The caloric energy, O_2 consumption, and CO_2 output derived from protein catabolism can be calculated from the measured or assumed hourly rate of its utilization and subtracted from the total caloric output and O_2 and CO_2 exchanges to leave the nonprotein components and nonprotein RQ. Assuming that carbohydrate has an RQ of 1.00 and fat 0.71, the nonprotein RQ can be divided, by simple proportion, into the fractions attributable to carbohydrate and fat. From this the O_2 consumption and CO_2 output of these are calculated individually. Thus the caloric contribution and amount of each consumed can be calculated for carbohydrate, protein, and fat individually.

Significance of respiratory quotient. From the foregoing discussion, it might seem that the determination of the RQ furnishes exact information concerning the kind of foodstuff being metabolized. Such indeed is the general impression. However, experimental evidence has not completely supported this idea, especially when the metabolic changes have been examined over short periods of time. The difficulty arises from the fact that the cells of the body are able not only to oxidize various foodstuffs concurrently but also to convert one food material into another. The general idea of a "metabolic pool" has replaced the idea of the metabolism of a single foodstuff. In fact, when a single food material is supplied to the animal body, it might be expected that this food material would be oxidized exclusively. In this case the RQ as determined after the ingestion of this food material should indicate its metabolism. For example, the ingestion of glucose should raise the RQ to 1.00 if glucose alone is oxidized. Benedict and Higgins[9] fed various amounts of sugar to fasting subjects. After the subject had ingested 600 gm of glucose in 1 day, the highest RQ they observed was 0.87.

Cathcart and Markowitz[15] studied the change in the RQ and the increase in metabolism after the ingestion of various sugars. Glucose and maltose raised the metabolism, but the RQ did not reach unity. On the other hand, cane sugar raised the quotient to 1.00 but did not change the metabolism. Krogh and Lindhard[28] fed a group of men high-carbohydrate or high-fat diets over periods of 2 or more days. They observed that the RQ rose on a carbohydrate diet and fell on a fat diet, but it did not reach the theoretical levels 1.00 and 0.71, respectively. Higgins[24] determined the quotient before and after ingestion of alcohol. Although the theoretical quotient for the oxidation of this substance is 0.667, the lowest figure obtained experimentally was 0.775. These facts illustrate the point that even when the subject is on a diet consisting of a single chemically pure substance, oxidation of that substance to the exclusion of all others never occurs. Another complication is the transformation of food materials from one form to another. The conversion of carbohydrate to fat tends to raise the quotient above unity, whereas the reverse process tends to lower it below 0.71.

Factors other than oxidative processes exert profound effects on the value of the RQ. These may be summarized as follows (Richardson[32]):

A. Increases RQ
   1. Hyperventilation
   2. Acid formation
      a. Lactic acid (muscular exercise, convulsions, asphyxia, epinephrine)
      b. Keto acids (diabetes, carbohydrate starvation)
   3. Acid retention (as in nephritis)
B. Decreases RQ
   1. Hypoventilation
   2. Acid removal (by oxidation, excretion of free acid, synthesis, loss of HCl from stomach by vomiting)
   3. Alkalosis produced by alkali therapy

In view of the evidence presented the RQ should not be regarded as an exact indicator of specific metabolic processes but as a resultant of all the changes in the body that require O_2 and lead to pulmonary excretion of CO_2.

In metabolism experiments, however, it is necessary to convert liters of O_2 into calories. The assumption underlying most calculations is that when the RQ has a value of 1.00, it indicates an exclusive oxidation of carbohydrate, whereas one of 0.71 signifies an oxidation of fat alone; any quotient between these two values is the resultant of the burning of a mixture of fat and carbohydrate in definite proportions. The situation has been shown to be more complex than this simple theory demands. The question arises as to what should be used for the conversion of liters of O_2 to calories. The answer is that it does not make much difference (Table 56-4). At an RQ of 0.71 the amount of heat produced per liter of

O_2 used is only about 6% less than when the RQ is 1.00.

CALCULATION OF BASAL METABOLIC RATE

Metabolic rate is usually calculated from O_2 consumption by assuming that under standard conditions (see below) the RQ is 0.82 and the caloric value of O_2 is 4.825 kcal/L. For reasons to be discussed, it is customary to calculate the metabolism on the basis of kilocalories per square meter of body surface per hour and to compare this value with a normal standard. Assuming that a subject weighs 80 kg and is 180 cm in height, his surface area, as obtained from the chart constructed by DuBois (Fig. 56-5), is 2 m². From the table of standard metabolic rates (Table 56-5), it is found that a man between 20 and 30 years of age should have a metabolism of 39.5 kcal/m² body surface/hr. The observed value is then divided by the standard value and expressed as a percentage, from which 100 is subtracted algebraically to give the metabolic rate. Since normal individuals vary ±15% from the standard, the variation due to the method of calculation is small, and it is safe to assume an average caloric value for this determination.

Comparison of direct and indirect methods of calorimetry

It was necessary to discuss methods and the RQ before continuing the consideration of the application of the first law of thermodynamics to biologic systems. The first law of thermodynamics states that energy is neither created nor destroyed. Thus when mechanical work is transformed into heat or heat into work, the amount of work is always equivalent to the quantity of heat. According to this law, the heat produced by the combustion of food material should be equivalent to the heat liberated by the subject at rest plus the external work done by the subject on the surroundings. Rubner[33] in 1894 determined by means of a calorimeter the heat produced by dogs at rest and at the same time measured the CO_2 expired and the nitrogen excreted

Table 56-4. Caloric value of 1 L of O_2 at various RQs*

RQ	0.707	0.75	0.80	0.85	0.90	0.95	1.00
Kilocalories	4.686	4.739	4.801	4.862	4.924	4.985	5.047

*Kilocalories $= 4{,}686 + \dfrac{(RQ - 0.707) \times 0.361}{0.293}$ (Lusk[30]).

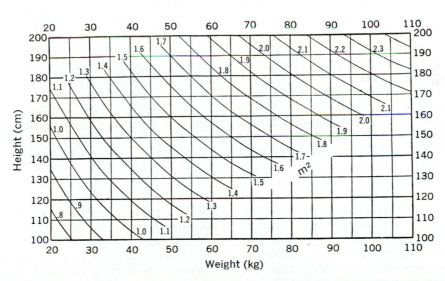

Fig. 56-5. Chart for determining surface area of man in square meters from weight in kilograms and height in centimeters according to formula: Area (m²) $= W^{0.425} \times H^{0.725} \times 71.84$. (From DuBois and DuBois.[21])

Table 56-5. Kilocalories per square meter of body surface (height-weight formula) per hour (Aub and DuBois standards)

Age (yr)	Males	Females
14 to 16	46.0	43.0
16 to 18	43.0	40.0
18 to 20	41.0	38.0
20 to 30	39.5	37.0
30 to 40	39.5	36.5
40 to 50	38.5	36.0
50 to 60	37.5	35.0
60 to 70	36.5	34.0
70 to 80	35.5	33.0

Table 56-6. Comparison of heat calculated from gaseous exchange with heat actually produced*

Food	Days	Heat calculated (kcal)	Heat determined (kcal)
Starvation	5	1,296.3	1,305.2
	2	1,091.2	1,056.6
Fat	5	1,510.1	1,495.3
	6	2,249.8	2,276.9
Meat	7	4,780.8	4,769.3

*From Rubner.[33]

in the urine. Assuming that under fasting conditions a dog oxidizes only protein and fat, he was able to calculate the amount of food material burned in 24 hr. By comparing the heat available from the oxidation of this material with the actual amount of heat liberated by the animal during the same period of time, he found close agreement between these two observations on fasting dogs and also on dogs that were fed meat and fat (Table 56-6). Rubner's work was continued by Atwater and Benedict in this country. With the aid of Rosa, Atwater[3] constructed a calorimeter in which both the heat production and the respiratory metabolism of a human being could be measured. Some of the results are given in Table 56-7.

From the previous examples, it is apparent that the first law of thermodynamics holds for such a complex biologic system as the human body. The slight deviations between the two quantities are probably due to the use of the tables for the calculation of the caloric value of O_2 and the assumption that in feeding one type of food material only that type is oxidized.

Second law of thermodynamics. Heats of combustion that are used at the present time in metabolic experiments do not give any information concerning that part of the energy, the free energy, that is available for work. Free energy content of a given compound may be greater or less than its heat of combustion. If greater than the heat of combustion, the process may be improved so that more work may be obtained from the system. If the free energy is less, even if all this energy may be converted into work, there is still additional energy that is unavailable for work. These facts have been incorporated into the second law of thermodynamics, which deals with this loss of availability of energy for external work and with entropy in a quantitative

fashion.[27] The application of this law enables the observer to decide whether a certain reaction can take place and to determine the maximal amount of work that may be obtained from the system. It tells nothing of the speed of the reaction or of the mechanisms involved. At the present time, free energy data obtained from simple systems such as enzyme reactions or bacterial metabolism can be employed cautiously to predict the amount of useful energy available as adenosine triphosphate (ATP) equivalent values, which can be used to perform catabolic or anabolic functions, carry ions or amino acids across cell membranes, deposit calcium in bone, aid muscular contraction and relaxation,[1] and form new tissue during a period of growth.[26] Thus although the heat of combustion is determined by the substance burned and not by its metabolic pathway, the amount of ATP produced or utilized in the intermediate reactions is subject to phenotypic metabolic control. Also, the difference between ATP equivalents generated or used up in a process helps to govern the direction the reaction will proceed. If ATP is used in muscular work performed on external objects, as in lifting a weight, the gain or loss of energy by the external object must be counted in balancing the total energy equation stated as follows: Food energy equals body heat loss plus external work performed (expressed as heat). For a laborer the physiologic efficiency in the performance of a task is the external work done divided by the metabolic energy used in performing this work. This efficiency usually is less than 20%. Therefore the free energy change expressed as ATP equivalents plays a role in regulating the distribution and nature of the work of the parts of the body, and this in turn is regulated through the enzymes and hormones that facilitate or impede the different anabolic and catabolic pathways.

Table 56-7. Comparison of indirect and direct calorimetry applied to human subjects at rest and at work*

Type of experiment	Number of experimental days	Average kcal/24 hr	
		Indirect calorimetry	Direct calorimetry
Rest	41	2,246	2,246
Work	66	4,682	4,676

*From Atwater.[2]

Table 56-8. Relationship between heat production and body surface*

Dogs		Various mammals		
Weight (kg)	Kcal/m² body surface/ 24 hr	Animal	Weight (kg)	Kcal/m² body surface/ 24 hr
31.20	1,036	Hog	128.00	1,074
24.00	1,112	Man	64.00	1,042
19.80	1,207	Dog	15.00	1,039
18.20	1,097	Guinea pig	0.50	1,246
9.61	1,183	Mouse	0.018	1,185
6.50	1,153			
3.19	1,212			
AVERAGES 1,142				1,145

*From Rubner.[33]

It may be that in this century the use of the second law of thermodynamics will lead to an advance in our knowledge of metabolic processes comparable to that which resulted in the application of the first law of thermodynamics to biologic problems in the last century.

BASAL METABOLISM

In order that a *comparison* of the metabolic activity of individuals and species may be made, certain conditions have been selected for its determination. DuBois[19] lists the many factors that influence metabolism as follows:

A. Factors excluded in all basal tests on normal controls
  1. Muscular movements during test
  2. Recent muscular exertion (within ½ to 1 hr)
  3. Food within 12 or 14 hr
  4. Strong emotions, noises, discomforts
  5. Extremes of environmental temperature
  6. Disease
B. Factors taken into account in interpretation of determinations of basal metabolism
  1. Age
  2. Sex
  3. Weight and height
  4. Surface area
  5. Undernutrition or overnutrition
  6. Athletic training
  7. Climate
  8. Altitude
  9. Sleep
  10. Body temperature
C. Additional factors that may or may not be of importance
  1. Occupation
  2. Race
  3. Previous diet
  4. Menstruation
  5. Vigorous exercise or emotion on day before test
  6. Novelty of situation
  7. Time of year

The three factors that have the most to do with determining the level of metabolism are muscular activity, external temperature if the subject is lightly clothed or nude, and digestion of food. If a subject is removed from the influence of these factors, the metabolism corresponds to the energy exchange for the minimal functional activity of the body. The term "basal metabolism" has been applied to the energy changes of the body under these conditions. It is usually determined 12 to 14 hr after the last meal, with the subject at rest and in a comfortable environmental temperature. Krogh suggested the term "standard metabolism" for the heat production under these conditions, since the metabolism can be lowered from the "basal" level by sleep or drugs. Benedict used the term "postabsorptive" for the same reasons. Under basal conditions, there is no food intake; therefore a person cannot remain at basal level indefinitely. The term "basal metabolism," however, has taken a definite meaning and is commonly used in this country.

Surface area. Basal metabolism is reported in terms of kilocalories per square meter of body surface. Rubner established this empirical relationship by measuring the metabolism of dogs of various sizes and of other mammals varying in weight from 18 gm to 128 kg. His results are given in Table 56-8. It was later shown that Rubner's figures for the number of kilocalories per square meter of body surface were too high as compared to modern standards. The same relationship, however, holds for the revised values (Table 56-9). It is of interest to note that vital capacity, cardiac output, and cross section of the aorta and trachea are also related to the surface area of the body.

In 1879 Meeh marked out in geometric designs the bodies of 6 adults and 10 children. From these measurements he derived the formula $S = KW^{2/3}$, in which K is a constant for a given species. Rubner used this formula in his calculation

Table 56-9. Relationship between heat production and body surface in men*

No. of individuals	Average weight (kg)	Variation in weights (kg)	Kcal/ m² body surface/24 hr
6	48.7	40-50	922
41	53.4	50-60	914
164	64.5	60-70	927
24	74.7	70-80	924
8	83.7	80-90	924
AVERAGE			922

*Analysis of material taken from Benedict; from Rubner.[34]

of the surface areas of animals. Years later DuBois and DuBois[20,21] also determined the surface area of a small group of individuals by measuring paper casts of their bodies. They derived a formula that was based on height and weight:

$$SA = W^{0.425} \times H^{0.725} \times 71.84$$

in which SA is the surface area in square meters, W is weight in kilograms, and H is height in centimeters. This formula has been used to construct graphs from which the surface area of any individual may be determined when the height and weight are known (Fig. 56-5). Although they derived this general formula from a study of a small group of individuals, other observers have determined the surface area of man and have found that this mathematical relationship does hold for a large number of individuals.

Kleiber, however, in a critical review of the evidence for and against the principle that metabolic rate is proportional to surface area,[27] agreed with Benedict,[5] saying that this so-called law is not confirmed by recent findings. On the contrary, his data in *The Fire of Life* (p. 212) show the following relationship: "For all practical purposes, one may assume that the mean standard metabolic rate of mammals is seventy times the three-fourth power of their body weight (in kg) per day, or about three times the three-fourth power of their body weight (in kg) per hour."[27] This generalization is attractive because it is so easy to apply to animals of any size and because it is supported by the data Kleiber reviewed.

There are therefore two variables in the conventional forms of expression of metabolic rates in man and animals. The metabolism itself can be determined accurately by a large number of methods, whereas the surface area is more difficult to measure. The variability in surface area measurements may account for the fact that the

basal metabolism of some individuals differs widely from the accepted standards. Boothby and Sandiford analyzed the results of determinations of basal metabolic rates of 8,614 patients and found that 77% of this group fell within ±10% of the normal standards and 90% fell within ±15%. Thus 10% of this group fell outside the wider limits. It is customary, however, to consider ±15% the normal variation from the accepted standards of basal metabolism. Since these standards were derived from tests on patients, it is not surprising that individuals accustomed to metabolic tests should give results that are below these values.

The various factors that affect basal metabolism were listed at the beginning of this section. Carpenter[14] has reviewed the magnitude of the changes due to several of these and a discussion of these factors follows.

Minor muscular movements. The subject must remain perfectly quiet during the determination. Benedict[5] and Carpenter[13] have determined the effect of slight arm and leg movements such as raising the hand to the forehead every few seconds or crossing and uncrossing the legs every 20 sec. In one individual the arm movements changed the metabolism from a basal value of 200 ml O_2 to 210 ml/min, whereas the leg movements raised it to 222 ml. Therefore movements such as these should be avoided. The subject should be placed in a comfortable position so that there will be no desire to change position during the determination. The usual motions of rising, dressing, and walking to the laboratory do not affect the basal values, provided that the subject rests for 30 min before the test. On the other hand, after strenuous muscular exercise the metabolic rate may not return to the basal level for several hours.

Temperature. The subject must be in a room at a comfortable environmental temperature. In the course of his studies of cardiac output, Grollman[23] measured the metabolism of a subject exposed to temperatures ranging from 0° to 45° C. The metabolism at 0° was 330 ml O_2; it decreased to 240 ml at 30° C and then rose with increasing temperatures, reaching 260 ml at 40° C. The rise at the lower temperature is attributable to shivering and increased muscular tone. The gradual decline from 20° to 30° C is due to greater muscular relaxation. The increase from 30° to 45° C may be due to an increased reaction velocity of the chemical processes in the cells. It may also result, in part at least, from the increased activity of the sweating, respiratory, and cardiac mechanisms. DuBois has calculated that

Table 56-10. Specific dynamic action of glucose*

Variable measured	Dog (Lusk)		Man (Gephart and DuBois)	Man (Deuel)	
Weight (kg)	12.5	12.2	78.4	81.8	73.9
Basal rate (kcal/hr)	16.7	16.7	78.6	62.4	74.0
Glucose (gm)	70.0	70.0	100.0	75.0	75.0
Glucose (kcal)	287.0	287.0	410.0	307.0	307.0
Duration (hr)	6.0	6.0	3.0	4.75	4.75
Specific dynamic action					
Total kilocalories	28.13	31.06	25.94	31.90	13.90
Kilocalories/gram	0.40	0.44	0.26	0.43	0.19

*After Deuel; from Ball.[4]

the metabolism increases about 13% for each degree rise of body temperature. It is therefore important in metabolic experiments to record the temperature of the subject and of the room.

Food. One of the most important conditions listed by DuBois that must be satisfied to determine a true basal metabolic rate is food. The subject must have had no food for at least 12 hr.[7] This usually means that the subject is instructed to have his usual evening meal around 6 PM and then to eat no food or nutritious beverage from that time until the testing is completed sometime the following morning. As shown in Table 56-10, a considerable increase in metabolism and in heat production occurs after ingestion of food. In 1885 Rubner found that a characteristic amount of extra heat is produced for each class of foodstuffs—protein, carbohydrate, or fat. After a meal exclusively of protein the resting heat production may increase by an amount equal to 25% to 30% of the energy value of the protein. Carbohydrate and fat evoked less marked but appreciable rises. The significance of this heat in the energy balance of the body is sometimes a confusing subject in calculations. An example may clarify the situation. If the heat production of a fasting, resting subject is known to be 75 kcal/hr, or approximately 375 kcal in 5 hr, and the subject is given a protein meal with a caloric value of 375 kcal, his heat production during the ensuing 5 hr will not remain at 375 kcal total but will increase to a total of perhaps 450 kcal (90 kcal/hr). This means that his body did not use the protein completely to supply the energy for basal energy exchange and that in the utilization of the protein, it was necessary to draw on other energy stores of the body. On first inspection, one wonders, therefore, how energy balance can ever be attained if only protein is given as food. The answer is that to attain a bal-

ance or to induce actual storage of energy from protein, the amount of protein given must exceed the basal metabolism by at least 25%. In the case of carbohydrate or fat the margin is smaller because their effect on metabolism is less (Table 56-10).

Rubner gave to this increase in metabolism a name translated into English as *specific dynamic action* (abbreviated SDA). If this term implies that any foodstuff acts as a stimulant to metabolic processes in general, the term is misleading. Food is not a metabolic stimulant as is the thyroid hormone. More correctly, certain tissues of the body—principally the liver, although other parts of the digestive system may be involved to a small degree—spend extra energy in preparing the products of protein digestion for their addition to the metabolic or energy pool of the body as a whole. Rubner guessed, and later work has confirmed his hypothesis, that most of the extra energy is required for oxidative deamination of amino acids in the liver.

The significance of the SDA and much of the confusion in the science of nutrition as to how it should be calculated and expressed have been reviewed critically by Kleiber,[27] who has also given an explanation of how the confusion was initiated by statements and in translation of statements made by Rubner and other investigators of his period. For translation into English of the word used by Rubner, Kleiber suggests that *wirkung* is represented better by *effect* rather than by *action*. According to his proposal, the increase in heat production after ingestion of food may be called the *specific dynamic effect* (SDE).

Other factors

AGE. The standards given in Table 56-5 are divided into various age groups. Few persons, however, have been followed over long periods of time. Carpenter gives data on two individuals

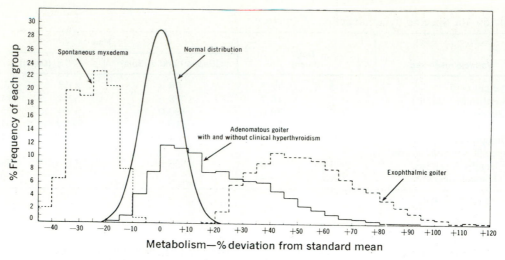

Fig. 56-6. Basal metabolism in three thyroid diseases compared with group of normal individuals. (From Boothby et al.[11])

whose metabolism was charted over a period of 20 years. In one person the metabolic rate fell from 27.8 kcal/kg/day to 23.5 kcal, whereas in the other subject the change was less, from 22.5 to 20.6 kcal/kg/day. The metabolic rate of many more persons should be studied over long periods of time in order to obtain sufficient data on the effect of age.

RACIAL VARIATIONS. Compared with American standards, natives in Yucatan have a higher metabolic rate (+8.4%), whereas a group of female students at Madras, India, were found to have a much lower one (−17.4%). A group of Chinese women living in the United States had an average metabolic rate 10.4% below American standards. These few examples illustrate that there may be significant deviations in metabolism according to race.

CLIMATE. Galvão,[22] studying a group of men in Brazil, found the basal metabolism of men in that climate to be less than in corresponding groups in the United States. The metabolic activity of his group did not follow the body surface law but was proportional to $W^{0.83}$. The reason given for the difference between the standards in the United States and those obtained in Brazil is that there is a much greater difference in temperature between the body surface and the external environment in the United States. In the colder climate, heat loss is the dominant factor, whereas in the warmer climate, heat loss is much smaller. Therefore heat production becomes more proportional to the active weight of the body and is independent of surface area.

ALTITUDE. Many studies have shown that man's exposure to changes in barometric pressure in low pressure chambers, in aircraft, or on mountains does not change the basal metabolism. This statement also applies to the acclimatized subjects. Careful experiments on mice[16] and rats[29] have also shown no changes in the basal metabolic level during acclimatization to low barometric pressure.

DRUGS. Numerous drugs affect the basal metabolism. Caffeine, theophylline, and epinephrine cause an increase in metabolic rate. Also, thyroxine will raise metabolic rate, and 3,3',5-triiodothyronine has a similar effect. Thiouracil and similar compounds will block the production of thyroxine and thus decrease the metabolic rate. Metabolic antagonists to thyroxine block the action of thyroxine at the cellular level and lower the metabolic rate. Obviously there are many ways of increasing and decreasing the metabolic activity of the body by drug action.

Application of basal metabolic studies to clinical conditions. Since the introduction of standard methods of determination of basal metabolism and of adequate standards for comparison, many studies have been made on patients. The most practical result of this work was the use of basal metabolic studies in aiding the diagnosis of doubtful cases of hypothyroidism or hyperthyroidism. Boothby et al.[11] have analyzed the percentage frequency distribution for thyroid diseases (Fig. 56-6). It may be seen from the figure that patients suffering from myxedema and exophthalmic goiter fall, with few exceptions, outside the normal group, but those with goiter are distributed between the normal and exophthalmic groups. In addition to an aid to diagnosis, basal metabolic tests were used to obtain evidence concerning the value of medication in thyroid diseases and to determine the time for operation in hyperthyroid patients.

REFERENCES

1. Atkinson, D. E.: Adenine nucleotides as universal stoichiometric metabolic coupling agents, Adv. Enzyme Regul. **9:**207, 1971.
2. Atwater, W. O.: Neue Versuche über Stoff- und Kraftwechsel in menschlichen Körper, Ergeb. Physiol. **3:**497, 1904.
3. Atwater, W. O., and Rosa, E. B.: A new respiration calorimeter, Bull. No. 63, Washington, 1899, U.S. Department of Agriculture.
4. Ball, E. G.: Some energy relationships in adipose tissue, Ann. N.Y. Acad. Sci. **131:**225, 1965.
5. Benedict, F. G.: Vital energetics; a study in comparative basal metabolism, Publ. No. 503, 1938, Carnegie Institution of Washington.
6. Benedict, F. G., and Carpenter, T. M.: Respiration calorimeters for studying respiratory exchange and energy transformations of man, Publ. No. 123, 1910, Carnegie Institution of Washington.
7. Benedict, F. G., and Carpenter, T. M.: Food ingestion and energy transformations with special reference to the stimulating effect of nutrients, Publ. No. 261, 1918, Carnegie Institution of Washington.
8. Benedict, F. G., and Fox, F. L.: A method for the determination of the energy values of foods and excreta, J. Biol. Chem. **66:**783, 1925.
9. Benedict, F. G., and Higgins, H. L.: The influence on the respiratory exchange of varying amounts of carbohydrate in the diet, Am. J. Physiol. **30:**217, 1912.
10. Benzinger, T. H., and Kitzinger, C.: Gradient layer calorimetry and human calorimetry. In Hardy, J. D., editor: Temperature—its measurement and control, New York, 1963, Reinhold Publishing Corp., vol. 3, pt. 3.
11. Boothby, W. M., Berkson, J., and Plummer, W. A.: The variability of basal metabolism: some observations concerning its application in conditions of health and disease, Ann. Intern. Med. **11:**1014, 1937.
12. Carpenter, T. M.: A comparison of methods for determining the respiratory exchange of man, Publ. No. 216, 1915, Carnegie Institution of Washington.
13. Carpenter, T. M.: Tables, factors, and formulas for computing respiratory exchanges and biological transformations of energy, ed. 2, Publ. No. 303A, 1924, Carnegie Institution of Washington.
14. Carpenter, T. M.: Problems in the determination of the basal metabolism of man and factors affecting it, Ohio J. Sci. **33:**315, 1933.
15. Cathcart, E. P., and Markowitz, J.: The influence of various sugars on the respiratory quotient; a contribution to the significance of the R.Q., J. Physiol. **63:**309, 1927.
16. Clark, R. T., Jr., and Otis, A. B.: Comparative studies on acclimatization of mice to carbon monoxide and to low oxygen, Am. J. Physiol. **169:**285, 1952.
17. Consolazio, C. F., Johnson, R. E., and Pecora, L. J.: Physiological measurements of metabolic functions in man, New York, 1963, McGraw-Hill Book Co.
18. Douglas, C. G.: A method for determining the total respiratory exchange in man, J. Physiol. **42:**17, 1911.
19. DuBois, E. F.: Basal metabolism in health and disease, ed. 3, Philadelphia, 1936, Lea & Febiger.
20. DuBois, D., and DuBois, E. F.: The measurement of the surface area of man, Arch. Intern. Med. **15:**868, 1915.
21. DuBois, D., and DuBois, E. F.: Clinical calorimetry. X. A formula to estimate the approximate surface area if height and weight be known, Arch. Intern. Med. **17:**863, 1916.
22. Galvão, P. E.: Human heat production in relation to body weight and body surface, J. Appl. Physiol. **1:**395, 1948.
23. Grollman, A.: The effect of variations in the environmental temperature on the pulse rate, blood pressure, oxygen consumption, arteriovenous oxygen difference and cardiac output of normal individuals, Am. J. Physiol. **95:**263, 1930.
24. Higgins, H. L.: Effect of alcohol on the respiration and the gaseous metabolism in man, J. Pharmacol. Exp. Ther. **9:**441, 1917.
25. Hill, A. V.: The heat-production of surviving amphibian muscles, during rest, activity and rigor, J. Physiol. **44:**466, 1912.
26. Hommes, F. A., et al.: The energy for growth: an application of Atkinson's metabolic price system, Pediatr. Res. **9:**51, 1975.
27. Kleiber, M.: The fire of life, New York, 1961, John Wiley & Sons, Inc.
28. Krogh, A., and Lindhard, J.: The relative value of fat and carbohydrate as sources of muscular energy, Biochem. J. **14:**290, 1920.
29. Lipin, J. L., and Whitehorn, W. V.: Role of metabolism in the acclimatization of albino rats to reduced barometric pressure, J. Aviation Med. **21:**405, 1950.
30. Lusk, G.: The elements of the science of nutrition, ed. 4, Philadelphia, 1928, W. B. Saunders Co.
31. Pauling, L., Wood, R. E., and Sturdivant, J. H.: An instrument for determining the partial pressure of oxygen in a gas, J. Am. Chem. Soc. **68:**795, 1946.
32. Richardson, H. B.: The respiratory quotient, Physiol. Rev. **9:**61, 1929.
33. Rubner, M.: Die Quelle der thierischen Wärme, Z. Biol. **30:**73, 1894.
34. Rubner, M.: Geschichte der Entwicklung des Energieverbrauches bei den Wirbeltieren, Sitzungsb. Preuss. Akad. Wissenschaft, Berlin **17:**313, 1931.
35. Schoenheimer, R.: The dynamic state of body constituents, Cambridge, Mass., 1942, Harvard University Press.

57

JOHN R. BROBECK

Energy balance and food intake

COMPONENTS OF ENERGY EXCHANGE AND THEIR MEASUREMENT

Because there is conservation of material in the metabolism of the body, every atom of carbon, hydrogen, oxygen, or nitrogen can be accounted for as *intake* in food and then as *output* (in urine, feces, sweat, or expired gas) or as *materials stored* within the body (e.g., water, carbohydrate, protein, fat, or minerals). By arithmetic the amount stored can be calculated if intake and output are known, or output can be calculated if input and storage can be measured. The amount of a given compound can be expressed as weight, volume, or moles. A similar calculation can be made for the energy exchange, where the principle of conservation also applies because none of the energy disappears in metabolic processes. The energy value of food is converted in part into the energy value of certain excretory products (principally urea and other constituents of the urine), in part into physical work performed by the body, and in large part into heat—plus or minus the energy equivalent of gain or loss of materials stored within the body (Table 57-1). For energy, as for substance, the relationships among the several factors are expressed by simple arithmetic after they have been converted into any given system of units.

As noted in the preceding chapter, the common denominator of units for energy exchange is the heat equivalent—the large calorie or kilocalorie. It is used to express the energy value of food as determined from analysis of chemical composition or by bomb calorimetry. A value obtained in this way is not the "total energy" of any compound but rather the energy equivalent of its heat of oxidation. That is to say, it is an energy difference between the unoxidized and the oxidized state of the carbon and hydrogen, with proper accounting for any change in entropy accompanying the oxidation.

For tissues of the body the methods for measuring energy value are the same as those for food, and they have the same significance (Table 57-2). They require the combustion of a sample (from biopsy or autopsy) of a given tissue or organ in a bomb calorimeter, or they may be based on a complete chemical analysis and the use of known heats of oxidation for the constituents found to be present. As ordinarily used in studies of metabolism, these values are not very accurate, yet they serve the purpose for which they are needed. When the principles of thermodynamics in biologic systems were shown to be the same as in physical and chemical systems, it was no longer necessary to measure the factors of energy exchange precisely. They vary so much under even standard conditions that the precautions needed to secure accurate measurement are difficult beyond the significance of the data. Consequently, approximate values are widely used, and authors are satisfied when they can arrive at even approximate figures for a balance, as seen in Passmore.[49]

Measurement of heat exchange is done by techniques given in the preceding chapter, using either direct or indirect methods. The possibility that heat might be stored within the body must be considered; it is usually estimated from the mass of the tissue storing heat, its temperature change, and its specific heat. Control mechanisms that determine how much heat is stored are described in Chapter 59.

Finally, there is the energetic equivalent of physical work to be considered. Its measurement is a matter of some confusion in physiology, although in a strict sense there is no ambiguity. Work is done when a mass is moved, and it is measured as force × distance, the units being kilogram-meters, foot-pounds, or equivalent expressions such as ergs or joules. These units may be interconverted and may be expressed also as calories. A kilogram-meter is approximately

Table 57-1. Relationships of factors in energy exchange*

Category	Energy added	Energy lost
Potential chemical	Food (including body tissues)	Excreted in sweat, urine, feces, and miscellaneous excreta
	Synthesis of high-energy bonds	Degradation of high-energy bonds
Potential physical	Potential energy stored in body	Potential energy lost from body
Kinetic	Metabolic heat	
	Motion imparted to body	Motion lost as heat
	Heat added from environment	Heat lost to environment
Work (physical sense)		Effective work done
Miscellaneous	For example, static electricity added to body	For example, bioelectrical potentials lost as electrical energy or heat

Calorie balance = kcal in food − (kcal in excreta + kcal of metabolism)

*From Consolazio et al.[17]

Table 57-2. Composition of adult male human body*

Parts of body	Percentage of total body	Composition of parts (%)				
		H_2O	Ash	Crude fat (ether extract)	Crude protein (N × 6.25)	Heat of combustion (kcal/gm)
Skin	7.81	64.68	0.68	13.00	22.19	2.29
Skeleton	14.84	31.81	28.91	17.18	18.93	2.50
Teeth	0.06	5.00	70.90	—	23	—
Striated muscle	31.56	79.52	0.93	3.35	16.50	1.24
Brain and cord	2.52	73.33	1.37	12.68	12.06	1.90
Liver	3.41	71.46	0.88	10.35	16.19	2.20
Heart	0.69	73.69	0.80	9.26	15.88	1.82
Lungs	4.15	83.74	0.95	1.54	13.38	0.98
Spleen	0.19	78.69	1.13	1.19	17.81	1.19
Kidneys	0.51	79.47	0.96	4.01	14.69	1.33
Pancreas	0.16	73.08	0.93	13.08	12.69	1.98
Alimentary tract	2.07	79.07	0.86	6.24	13.19	1.34
Adipose tissue	13.63	50.09	0.51	42.44	7.06	4.16
Remaining tissue:						
Liquid	3.79	93.33	0.94	0.17	5.68	0.38
Solid	13.63	70.40	1.01	12.39	16.06	2.04
Contents of alimentary tract	0.80	—	—	—	—	—
Bile	0.15	—	—	—	—	—
Hair	0.03	—	—	—	—	—
Total body weight = 70.6 kg	100.00	67.85	4.84	12.51	14.39	1.93

*From Mitchell et al.[47]

equal to 9.81 joules (absolute), 9.81×10^7 ergs, 7.233 foot-pounds, or 2.34 gram-calories. In physiologic experiments for measuring work, special machines are used, for example, a treadmill (for animals or man) or bicycle ergometer (for man). If the machine is regarded as being external to the physiologic system under study, the heat produced by the machine may be equated with the "work" output of the body. But if a treadmill or bicycle is used for exercise within a calorimeter, the heat from the machine becomes a part of the total heat production and in a physical sense, no work appears. This illustrates that special attention is usually needed to measure true work and that in physiologic literature one must look carefully to learn whether work is meant in a physical sense or if the term is used merely to identify what muscles do when they contract. Strictly, there is no work unless the muscle shortens against a resistance—no shortening, no work; no resistance, no work. Exercise, therefore, is a general and nonspecific term, whereas work has a unique physical definition. The physiology of exercise is discussed in Chapter 58, and the energy costs of various kinds of exercise are listed in Table 57-3.

ENERGY EQUIVALENT OF BODILY STORES

A more detailed account of measurement is needed for one factor of energy exchange, the energy stores of the body. Since their energy value is calculated from standard tables if their composition is known (Table 57-2), the fundamental problem is deciding the chemical nature of the material being put into or taken from the energy depots. In laboratory animals the question is usually answered by analysis of the carcass. Water content is determined by drying the carcass to a constant weight. Lipid content is measured by extracting the carcass with lipid solvents (e.g., petroleum ether or benzine, chloroform, diethyl ether, or acetone); the amount of fat is the weight of the extracted material after the solvent has been evaporated off, or it is the loss in weight of the dried carcass after the extraction—the two calculations should yield the same value. Glycogen is determined by direct chemical analysis, usually of muscle or liver samples. Nitrogen and protein are estimated by a technique such as the Kjeldahl method, using either the entire carcass or some known fraction of it (an aliquot). Finally, the minerals may be estimated together by combustion and weighing of ash, or they may be determined one at a time by an appropriate chemical or physical method applied to an aliquot. For example, sodium and potassium are measured by flame photometry.

Although techniques like these can be applied to organs removed at autopsy or to tissues taken at biopsy of living human subjects, they are not suitable for the measurements of greatest interest in studying energy balance, which are the changes in composition of a living body as a result of experimental procedures. For this reason, indirect methods for estimating body composition have been invented, based on certain facts about bodily constituents, as follows:

1. In general the size of the skeleton changes little in short periods of observation; when it does change, the magnitude of the change may be detected by experiments on calcium or phosphate balance.

2. The amount of water in the several compartments of the body (outside or within cells) may be estimated by the conservation principle, by injecting into the body a material that will distribute itself through a given compartment. *Plasma volume* is measured by injection of a dye that becomes attached to plasma proteins; *blood volume* by determining both plasma volume and the volume of distribution of red blood cells labeled with an isotope such as radioactive chromium (^{51}Cr); *extracellular volume* by use of ions such as thiosulfate, thiocyanate, or isotopic chloride, any one of which tends to be distributed throughout the water of plasma plus the interstitial fluid; and *total body water* by injection of deuterium oxide (heavy water) followed by analysis with the aid of mass spectrometry.[2] *Intracellular water* is the difference between total and extracellular water volumes. These measurements give reproducible values that seem to be precise. The anatomic nature of the compartments they measure, however, is not always clear, especially in the case of the extracellular determinations.

3. Nitrogenous and lipid components of the body differ in specific gravity and specific vol-

Table 57-3. Metabolic costs of various activities*

Activity	Body weight (kg)	Kcal/ min	Kcal/ kg/10 min
Sleeping	68.1	1.17	0.172
Resting in bed	73.2	1.26	0.174
Sitting, normally	73.2	1.29	0.176
Sitting reading	73.2	1.29	0.176
Lying, quietly	68.1	1.33	0.195
Sitting, eating	73.2	1.49	0.204
Sitting, playing cards	73.2	1.53	0.210
Standing, normally	73.2	1.50	0.206
Classwork, lecture	68.1	1.67	0.245
Conversing	68.1	1.83	0.269
Personal toilet	73.2	2.02	0.278
Sitting, writing	82.0	2.20	0.268
Standing, light activity	73.2	2.60	0.356
Washing and dressing	68.0	2.60	0.382
Washing and shaving	62.0	2.60	0.419
Driving a car	64.0	2.80	0.438
Washing clothes		3.13	
Walking indoors		3.11	
Shining shoes	73.2	3.20	0.437
Making bed	59.1	3.38	0.572
Dressing	73.2	3.40	0.466
Showering	73.2	3.40	0.466
Driving motorcycle	64.0	3.40	0.531
Cleaning windows	61.0	3.70	0.607
Sweeping floors	73.2	3.91	0.535
Ironing clothes	67.0	4.20	0.627
Mopping floors	73.2	4.86	0.665
Walking downstairs	73.2	7.14	0.976
Walking upstairs	73.2	18.58	2.540

*From Consolazio et al.[17]

ume. Whereas the lipid of the body has a specific gravity of about 0.90 gm/ml and a specific volume of about 1.10 ml/gm (i.e., 1.10 L/kg), the remainder of the body has a specific gravity of about 1.10 gm/ml and a specific volume of about 0.91 ml/gm. This nonfat portion of the body has been called the lean body mass. From these figures, it is possible to estimate the proportions of fat and nonfat material if the body's weight and volume are known or if its specific gravity can be calculated.

Specific gravity of body

Two measurements are required for this determination: body weight and body volume. The principle used is attributed to Archimedes, the Greek mathematician who discovered it while trying to learn whether a crown made for the king of Syracuse contained all the gold it was supposed to have. The principle is that an object immersed in water displaces water equal to the volume submersed and is buoyed up by a force equal to the weight of the water displaced. To utilize the principle of Archimedes, one determines, using a sensitive balance, the weight of the object in air and in water. For the human body the second weighing is not easy to obtain because air in the respiratory passages (and any large amount of flatus in the intestine) gives the body a buoyancy that obscures the relationship between fat and nonfat tissue as just outlined. Hence another measurement is needed, the volume of air in the respiratory tract, determined preferably while the underwater weighing is being done. For a more detailed account of how these values are obtained, one should refer either to the original papers or to the summaries of Kleiber,[36] Consolazio et al.,[17] and Pearson.[2] As illustrated by Behnke in Table 57-4, density may

Table 57-4. Measuring density of men by immersion*

Weight in air (kg)		83.20
Weight immersed in water (kg)	10.55	
Weight of belt in water (kg)	6.25	
Weight of body in water without belt (kg)		4.30
Weight loss by immersion (kg)		78.90
Volume (L)		78.90
Estimated residual air in lung (L)		1.20
Net volume of body (L)		77.70

$$\text{Density} = \frac{83.20}{77.70} = 1.071 \text{ kg/L}$$

*Modified from Behnke et al.[7]; from Kleiber.[36]

be calculated at least to hundredths of a unit by these methods.

The volume of the body is used to determine the amount of fat and nonfat material it contains by the following relationship:

Volume of the body =
(Weight of fat × Specific volume of fat) +
(Weight of nonfat × Specific volume of nonfat)

In the illustration given in Table 57-4 the following values may be utilized:

$$77.70 \text{ L} = (X \times 1.10) + (83.20 - X) \times 0.91$$
$$77.70 = 1.10 \text{ X} + (83.20 \times 0.91) - 0.91X$$
$$= 1.10 \text{ X} - 0.91 \text{ X} + 75.71$$
$$0.19 \text{ X} = 1.99$$
$$X = 10.47 \text{ kg of fat}$$
$$83.20 - 10.47 = 72.73 \text{ kg of nonfat material}$$

Pneumatic measurement of body volume

Kleiber[36] has given a good account and an amusing illustration (Fig. 57-1) of a method in which the volume of the body is measured in a device known as a body plethysmograph. This is an airtight chamber large enough to contain a person and having a known volume. With the subject inside, a measured amount of air is injected into the chamber, and the resulting change in pressure is measured. The change in pressure is a function of the relationship between the two volumes, that of the air originally surrounding the subject in the chamber and that of the air injected. (The equations are on the figure.) With this method, no correction is necessary for air in respiratory passages, since it is a part of the air originally in the chamber, or it may be excluded from the calculations by using a tubing and mouthpiece to connect the respiratory system to the air outside.

Thickness of skin folds

To enable investigators to estimate the fatness of the body more quickly and more easily than these methods will permit, another technique has become widely used, the measurement of the thickness of skin folds. A fold of skin is elevated by gentle pinching, and its thickness is measured using calipers specially designed to close on the skin with a standard force controlled by a spring (Fig. 57-2). Measurements are made at certain standard locations over the body surface; the validity of the method is based on the assumption that fat is stored within the skin in proportion to its storage elsewhere in the body. Keys and

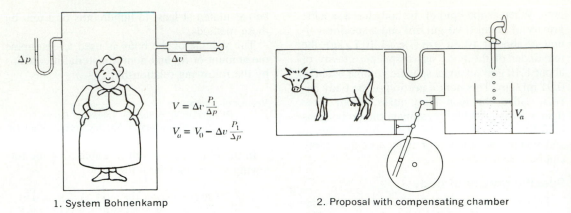

1. System Bohnenkamp   2. Proposal with compensating chamber

Fig. 57-1. Pneumatic measurement of animal volume. (From Kleiber.[36])

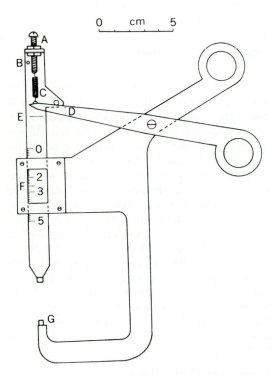

Fig. 57-2. Diagram of USAMRNL calipers for measuring skinfold thickness. (Design by Best; from Consolazio et al.[17])

Brožek[34] concluded that the method gives consistent results and that they correlate well with specific gravity (Tables 57-5 and 57-6). These judgments are based on averaged data and do not mean that in any given subject the skinfold thickness will reveal subtle changes in bodily fat content. For further information about the subject, consult Brožek.[11]

Sensitivity of measurement of energy stores

More complete instructions for measuring energy exchange are given in the monographs by Kleiber,[36] by Consolazio et al.,[17] and by Albanese.[2] In some instances, both the reproducibility and the accuracy of a method can be stated. This is not necessary here because an exact accounting is not intended. Nevertheless, a few relevant comments may stimulate thinking in a quantitative way.

First, the precision of measurement of carbohydrate stores may be considered. Most of the carbohydrate in the body is stored as glycogen in liver or muscle. The amount in *muscle* of fasting human subjects varies from about 1% to perhaps 4%[31] and has been estimated to increase by as much as 3% after prolonged overfeeding.[49] The method by which the values of Hildes et al.[31] were obtained is precise; hence the variability must be of biologic origin, and it illustrates why metabolic data often cannot be used with any great accuracy. If one assumes that muscle makes up about half the weight of the body, a man weighing 70 kg with 35 kg of muscle will have during fasting somewhere between 350 and 1,400 gm of glycogen in muscles. With feeding, this may rise, although the data now available do not show what the minimal and maximal levels may be in a given muscle or person. The calculations of Passmore and associates, made indirectly from balance data and measurements of metabolic rate, suggest that one of their obese subjects stored an excess of 1,010 gm of glycogen during the overfeeding period.[49]

Glycogen in the *liver* is perhaps even more variable, with concentrations from 0.95 to 4.1% in fasting human subjects.[31] In experimental

Table 57-5. Consistency of measurement of skinfold thickness*†

Skin folds	n	Observer 1		Observer 2		r
		M	SD	M	SD	
Abdomen	76	18.6	7.9	16.6	7.0	0.941
Chest	83	14.5	6.9	14.4	6.7	0.955
Arm	83	12.5	5.0	13.0	5.1	0.916
Calculated fat content (% of weight)	76	12.4	5.6	12.4	5.7	0.996

*From Keys and Brožek.[34]
†n, Size of sample; M, mean (mm); SD, standard deviation; r, coefficient of correlation between the values obtained by each of the two observers.

Table 57-6. Correlation of absolute values of skin folds with specific gravity as found in 20 normal men (ages 48 to 58 years)*

Site	Absolute skin folds
Abdomen	0.641
Chest	0.808
Back	0.648
Arm	0.649
Thigh	0.608
Waist	0.678
MEAN	0.775

*Based on data from the Laboratory of Physiological Hygiene, University of Minnesota[37]; from Keys and Brožek.[34]

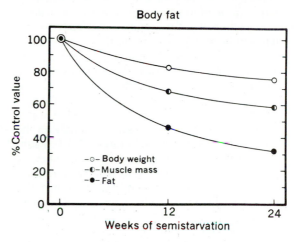

Fig. 57-3. Percentage decrements of body weight, muscle mass, and body fat. ($N = 32$.) (From Keys et al.[35])

animals, where estimation of liver glycogen and total liver weight is fairly routine, values range from 5% in fed animals to less than 1% during fasting. With similar changes in concentration a human liver weighing 1,700 gm would contain a maximum of 85 gm and a minimum of less than 17 gm of glycogen. This does not represent a very large reserve of energy, since 70 gm of glycogen has a caloric value of less than 300 kcal. Just how much energy the muscle glycogen might contribute under similar conditions is uncertain, but even if 1,000 gm or 4,000 kcal of energy could be drawn from this reserve, it would amount to only about twice the daily energy exchange of an adult subject. If fasting continues for longer than 48 hr, one can assume that there is no further net loss of glycogen stores and that any carbohydrate oxidized by the tissues must have been derived from other sources via glyconeogenesis.

Changes in protein storage can be estimated fairly accurately, even in living human subjects. The conventional method for doing this is by nitrogen balance procedures in which the nitrogen of food intake and of urine and feces is measured over a period of time long enough to pro-

vide a reliable index of how much protein has been stored or broken down. If urine and feces can be collected without loss for several days, and if the composition of the diet is known, this is one of the most accurate of the metabolic balance procedures.

The estimate that cannot be made accurately is for body fat. When the total amount of lipid in the body is more than 10 kg, this represents more than 95,000 kcal of potential heat (at 9.5 kcal/gm). Methods of its estimation probably do well if they are accurate to ±5%, and thus they cannot reveal how much energy is taken from fat in any period shorter than several days. One can see, therefore, that, taken altogether, the techniques for studying the overall energy balance and storage are useful only for long-term experiments in which changes are relatively large, as in the comparison of obese with normal subjects or in studying starvation and recovery as Keys et al.[35] did in their experiments at the University of Minnesota (Fig. 57-3).

CONSTANCY OF ENERGY STORES

In human patients and in most laboratory animals there is no conclusive evidence that the body's energy stores are constant in amount; if they are not constant, the range of their variability is likewise unknown. The only way to resolve these questions would be to determine the body's volume or specific gravity day by day and week after week in a group of subjects that would provide a sufficiently large sample to establish that the composition is or is not "constant" within certain limits. Yet despite the lack of direct evidence, a constancy of composition is often inferred from information that is indirect in nature. It is well known, for example, that body weight tends to be almost constant in most persons; this is believed to reflect a constancy of body stores. This may not be the case, because a record of weight does not reveal the composition and caloric value of the tissues of the body. There is a marked difference in the energy level of stored lipid (more than 9.0 kcal/gm as stored) on the one hand and stored protein (about 1.4 kcal/gm of tissue as stored with water) on the other. A person who gains 5 kg of weight by storing fat has an energy reserve far different from that of a person in whom the added 5 kg was 1 part protein and 2 parts water.

Another type of indirect evidence for constancy is the fact that experimental animals of a given strain, age, sex, and nutritional history have almost the same body weight and composition. The distribution of body weight or composition follows the frequency distribution curve known as bell shaped or gaussian (Fig. 57-4). Data in a curve of this type can be treated statistically, so that any difference in mean values or in variance can be analyzed for its significance. These analyses, however, do not deal with the fundamental problem of constancy. If an animal has a body weight outside the 95% range of values, is it in that position because it is an organism somehow different from the other animals, or because any animal might occupy such a position at some stage of its inherent variations in body weight? In this particular instance the question can be answered, because weighing can be repeated, many times if necessary. Usually the lighter animals and the heavier animals maintain their relative positions on the curve of weight distribution. But in the case of frequency distribution of body composition or of total fat content, repeated analysis is not possible, and one cannot find out how widely a given animal may change its rank in a total population. Perhaps it is true to say that lean animals remain lean and fat animals remain fat, or perhaps the statement is only "relatively" true.

In any event, physiologists generally assume that the processes of alimentation and nutrition proceed so as to regularize the energy stores of the body, at least when average values are considered. This leads to the question of how the energy stores are controlled. The mechanisms of

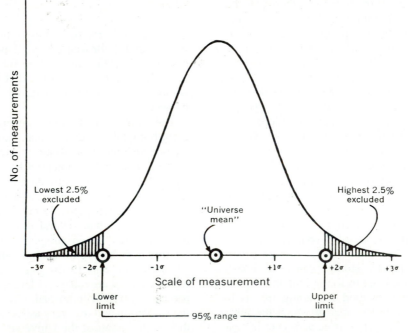

Fig. 57-4. Normal frequency distribution curve. (From Altman and Dittmer.[3])

storage are biochemical and involve synthesis of glycogen, storage of lipid, synthesis and storage of protein, and interconversions of carbohydrate, lipid, and protein. They take place in more than one location in the body—the digestive tract, liver, muscle, and adipose tissue. Moreover, they are under the influence of many of the hormones of the body. In the synthesis of protein, for example, insulin, growth hormone, thyroid hormone, and gonadal steroids are involved. The amount of carbohydrate stored as glycogen is influenced by insulin, glucagon, growth hormone, glucocorticoids, and catecholamines. These same hormones act on adipose tissue to determine the rates at which fat is stored or broken down. One sees, therefore, that the mechanisms of control of energy storage are so complex that here they can only be summarized, not delineated. They function in normal persons so as to bring about the following conditions (Fig. 57-5):

1. Young adult subjects have a relatively constant body weight and bodily composition.
2. Growth to adulthood involves actively the synthesis and storage of added protein and passively the accumulation of water and of electrolytes. There is also an increase in percentage of lipid.
3. With advancing age, lipid storage exceeds protein storage.

Control of energy storage

There is a sense in which energy stores depend completely on the other factors of energy exchange, since they must be equal to the difference between intake and output of energy. Food is the only source of energy, and food energy not converted into work or heat (nor excreted in urine and feces) must be retained in storage depots. Similarly, when food is not available, the energy for basal heat production, motion, and work must come from the reserves. Thus the stores of energy, particularly in adipose tissue, seem to function as a great warehouse into which metabolic processes pack away energy unneeded at the moment and from which they requisition energy when it is required. In this respect, both protein and carbohydrate differ from fat in that when their content in the body approaches a certain level, they tend to be converted into fat and stored in that form. In contrast, there seems to be almost no limit to the ability to store fat.

One of the most important facts about energy balance is that neither heat loss nor exercise is utilized by the body so as to preserve energy balance. Body temperature is not turned down when food is not available, and an animal without food does not immediately stop moving. On the contrary, deprivation of food seems to stimulate the locomotor activity of many animals and no doubt

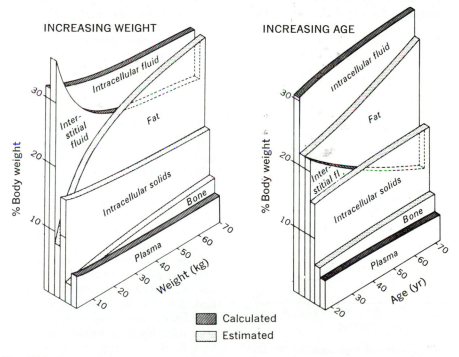

Calculated
Estimated

Fig. 57-5. Body composition with increasing weight and age. (From Altman and Dittmer.[3])

increases the probability of their finding food and being able to replenish their energy stores (Fig. 57-6). Consequently, if the body cannot protect its energy reserves by cutting off energy expenditure, and if the reserves are in fact to be maintained, this can be done only by adjusting food intake to energy requirements. This means that the control of energy storage is virtually the same as the control of food intake.

Before this latter topic is considered, there is one other possibility that must be discussed. The influence of the several hormones on lipid synthesis has already been mentioned. One may adopt the tentative hypothesis that these hormones work together to promote lipid storage when energy derived from food is available to the body, and they function to draw energy out of the adipose tissue when energy sources are deficient. According to this hypothesis the actions of the hormones are appropriate to the mass action direction of energy flow. The hormones merely speed up what would normally happen at a much slower rate. The opposite of this conclusion, however, cannot be excluded as a theoretical possibility; it is conceivable that the hormones actually control the direction of energy flow. They might stimulate the deposition of fat even if food energy was not available at the moment; this would then compel the animal to find food for nourishment of the rest of the body. For almost 100 years, this concept has been offered as a possible explanation for certain cases of obesity. Whether or not it occurs in human patients is still uncertain; yet it apparently does occur in laboratory mice of certain strains, where it is known as a "metabolic" obesity.[42] The exact nature of the metabolic deficit in these animals is not clear; they do show hyperphagia, but it is regarded as secondary to the metabolic change. The fact that the obesity is transmitted by genetic mechanisms has suggested to a number of authors that if genetic factors can be identified in human obesity, they may be found to operate in a similar fashion.[43]

One of the reasons why the ability of adipose tissue actively to withdraw energy from the metabolic pool seems plausible is that the protein synthetic mechanisms seem to be able to accomplish this same feat. In the spurt of growth associated with puberty, for example, or during recovery from starvation the enhanced protein synthesis and deposition may prevent the accumulation of fat and at the same time lead to increased food intake. For example, the enormous food intake and relative leanness of rapidly growing youngsters is known everywhere. The experiments of Keys et al.[35] on human volunteers subjected to partial starvation indicate that an actual protein retention is not a required part of the overall picture. Their subjects used up their fat and were ravenously hungry under conditions in which they could not gain in protein stores and were perceptibly losing nitrogen from the body because of the caloric deficit (Fig. 57-3). The men seemed to be more hungry than they would have been during complete starvation. Thus their appetite seemed to be related to what the body would have done with protein were it available, rather than to what was actually happening in the tissues.

CONTROL OF FOOD INTAKE

If evidence is lacking that energy content of the body remains constant, one must next inquire as to the evidence that food intake is controlled in a fashion that might be expected to create such a constancy. The evidence is of three types, of which the first is the correlation of food intake with exercise level. As estimated in tables of nutrition, a man doing heavy work will need and will eat an amount of food double that of his sedentary brother. Levels as high as 6,000 kcal/day are described, in contrast to perhaps 3,000 kcal/day in more conventional living. If a man does heavy work more or less continuously day after day, his food intake matches output at a high level of energy exchange. Mayer et al.[44] described an equivalent situation in exercised

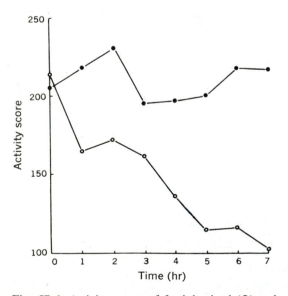

Fig. 57-6. Activity scores of food-deprived (●) and control groups (○) of *Drosophila*. (From Connolly.[16])

rats. When the exercise is not performed daily but is episodic in nature, a different relationship is observed. Edholm et al.,[21] studying the possibility of day-to-day correlations between exercise and food intake in military cadets, found that the compensatory increase in food intake came 2 days after a day of exercise (Fig. 57-7).

A second type of observation was made originally by Cowgill,[18] using dogs, when he found that in a 3 hr period of feeding the animals ate approximately the same quantity of food in calories when fed either of two diets that differed in their caloric density. These experiments have been repeated and extended by Adolph,[1] Strominger et al.,[54] and other authors. Their data may be summarized by the statement that in a long-term experiment when the nutrients of the diet are diluted with inert material such as cellulose, clay, mineral oil, or a large amount of water, the animal will increase the total food intake so that the caloric intake remains almost constant. The ability of animals to make this adjustment has been used by other authors as a test for "regulation" of caloric intake.[55]

A third type of evidence favoring the idea that energy balance occurs through change in food intake is the data of Cohn and Joseph,[15] who performed experiments in which they arbitrarily overfed rats by forcing a liquid diet through a stomach tube day after day until the animals had become grossly overweight (Fig. 57-8). When the forced feeding was discontinued, the animals failed to eat a normal amount of food until they had lost enough weight to return to the weight level of control animals. Although these experiments clearly revealed that animals tend to preserve certain relationships between food intake, energy expenditure, and energy reserves, no one knows by what mechanisms the body measures the size of the depots and the energy they contain. No kind of sensory cell is known to respond to body weight, body size, or body composition per se.

Mechanisms of control of feeding

In contemporary research on feeding mechanisms the hypothalamus of the diencephalon is the focus of interest because dramatic changes in feeding follow injury to this region. The observations that began this type of study were made over a period of about 10 years[5,10,59] after Hetherington and Ranson[29] had shown that obesity can be produced in rats by lesions restricted to the hypothalamus without injury to the pituitary gland. In a later paper, Hetherington and Ranson[30] reported that hypothalamic obesity could be produced even in hypophysectomized dwarf rats; thus they clearly disproved the older

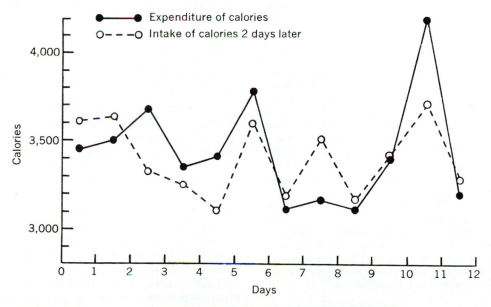

Fig. 57-7. Relationship between calorie expenditure and intake in cadets. Close agreement between the two plots indicates that daily intake of food is affected by degree of activity 2 days previously. (From Edholm et al.[21])

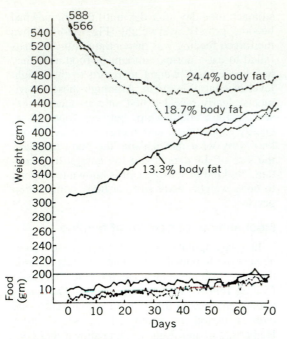

Fig. 57-8. Body weights, food intakes, and final body fat contents of rats that had been force-fed and were here fed ad libitum. Curves show low food intake until body fat had approached control level. (From Cohn and Joseph.[15])

idea that this type of obesity arose from pituitary disturbances. Subsequent experiments by other investigators showed that the rats become obese primarily because of their overeating or hyperphagia.[59,10] A description of these animals follows:

> Hypothalamic lesions which eventually induced adiposity were found to produce an increase in the amount of food eaten by the rat—an increase which was usually evident even before the animal had completely recovered from the operation. As the effects of the Evipal (a short-acting barbiturate) disappeared many of the animals showed locomotor hyperactivity as described by Hetherington and Ranson (1942), but when food was given to these rats they substituted ravenous eating behavior for the locomotion. They voraciously gnawed and ate chow pellets before their pharyngeal reflexes were sufficiently reestablished to maintain an adequate airway. . . . Because of their voracity the rats increased their body weight by as much as from 20 to 23 grams within the 18 hours following operation. Most of this weight gain undoubtedly represented food in the gut with the water mixed with it.*

---

*From Brobeck, J. R., Tepperman, J., and Long, C. N. H.: Experimental hypothalamic hyperphagia in the albino rat, Yale J. Biol. Med. **15**:831, 1943.

In 1951 Anand and Brobeck described the converse of this condition—an aphagia in rats having lesions of the hypothalamus just lateral to the other operation.[5] Their animals died without eating, although food was continuously present in the cage. Other authors have confirmed their observations and have added the point that animals with lateral hypothalamic lesions may recover their feeding behavior if they are kept alive by tube feeding for a few days or weeks,[56] whereas some animals recover spontaneously if they are given certain "preferred" diets.[55]

Techniques other than making lesions have been used to analyze these mechanisms further. One is the technique of electrical stimulation used by Delgado and Anand,[19] by Larsson,[38] and by other authors; another is the injection of chemicals directly into the brain[26,46] (Figs. 57-9 and 57-10). All these experiments suggest the same conclusion: by injury or by stimulation within the hypothalamus, it is possible to induce quantitative changes in food intake. This conclusion is often expressed as a statement that the hypothalamus controls feeding and food intake. There is a difference, however, between these two statements. It is not certain even now that mechanisms for feeding originate in the regions where they can be destroyed or stimulated. One of the interesting problems of this field of study is to identify just which neurons are responsible. There seems to be no question, however, that if one wishes to interrupt or stimulate these neurons, the most convenient place to approach them or their processes is in the tuberal portion of the hypothalamus.

The attention given to this portion of the brain and its functions has turned up a number of interesting and still puzzling phenomena. One is the recovery of feeding in animals with lateral lesions, as previously noted and as observed in monkeys by Anand[4] and associates. This recovery, together with the observation by Woods[63] that decerebrated rats show feeding behavior (although they cannot control their food intake), emphasizes the fact that feeding is ultimately accomplished by reflex and other mechanisms that lie well below the hypothalamus in the spinal cord and brain stem. When food is present in the environment as a sensory stimulus and when the sensory and reflex mechanisms are intact, feeding may be simply a reflex response (Table 57-7). Neural systems of the hypothalamus or those passing through it act by enhancing or suppressing the reflex centers, just as facilitation and inhibition occur in other reflex pathways. It may be that recovery of feeding in animals with lateral

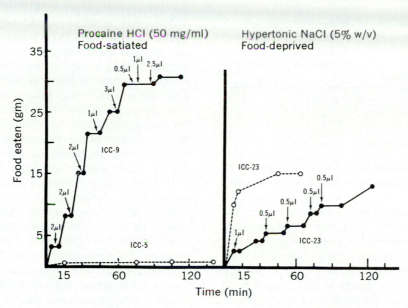

Fig. 57-9. Summary of experiments illustrating two major effects of chemical injections into medial hypothalamus, that is, elicitation of eating by procaine HCl and suppression of eating by hypertonic saline solution. (From Epstein.[22])

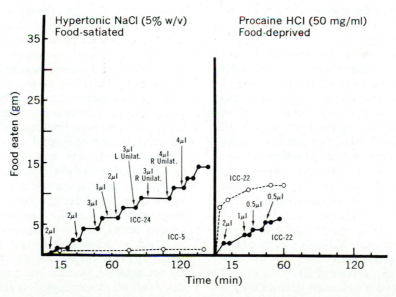

Fig. 57-10. Summary of experiments illustrating two major effects of chemical injections into lateral hypothalamus, that is, elicitation of eating by hypertonic saline solution and suppression of eating by procaine HCl. (From Epstein.[22])

Table 57-7. Classification of feeding reflexes, on sensory (left-hand column) and behavioral bases (right-hand column)*

Sensory basis	Behavioral basis
Visual reflexes	Reflexes of attention
Olfactory reflexes	Reflexes of approach
Auditory reflexes	Reflexes of examination
Tactile reflexes	Reflexes of incorporation
Gustatory reflexes	Reflexes of rejection
Enteroceptive reflexes	

*From Brobeck.[8]

hypothalamic lesions is analogous to recovery of spinal reflexes after the shock of spinal transection.

A second unexplained phenomenon is the somewhat transitory character of hyperphagia following ventromedial lesions. After hypothalamic lesions, some rats regain a normal food intake when they have become obese; if they are fasted so that they return to the control body weight, on refeeding they show a second period of hyperphagia and become obese a second time (Fig. 57-11).[10] Kennedy[33] commented on this phenomenon and proposed that it has the same explanation as the one later given to the experiments by Cohn and Joseph[15] in which rats were made fat by artificial feeding (see earlier section). Kennedy suggested that there is a regulation of the total amount of fat stored in the body and that following ventromedial lesions in the hypothalamus this regulation is "set" at a higher level. Conceptually, this is similar to the idea that in fever the body's temperature is set at a level higher than normal. According to Kennedy's view the hypothalamic lesions do not remove any of the control systems, they merely alter the amount of fat required in the body to shut off feeding. According to this interpretation, in the Cohn and Joseph experiments in which the control system was not altered but the amount of fat was increased by the tube feeding, food intake was naturally depressed until the added fat had been disposed of by oxidation.

To explain the ability of the body to determine how much fat is present, Kennedy suggested that some metabolic product circulates in body fluids in proportion to the amount of fat in the depots. What this might be is unknown. Unesterified or free fatty acids have been considered, but it seems doubtful whether they can serve this function because they rise under a variety of condi-

tions and thus do not precisely reflect the amount of stored fat. As Tepperman[57] has noted, the correlation is with what is happening to fatty acids rather than with their levels in the blood. Satiety seems to be correlated with fatty acid uptake by the depots and hunger with fatty acid release.

One other problem should be mentioned—whether animals with hypothalamic obesity have any primary abnormality other than the hyperphagia. That is to say, is there some metabolic change, a metabolic lesion in which the tissues reach out into the energy pool of body fluids and store energy in excess of the normal storage, so that hyperphagia follows? As just noted, this seems to be true of the obese-hyperglycemic syndrome in mice. In the obesity that follows hypothalamic operations, however, Tepperman et al.[60] decided that changes in metabolism were secondary in nature and were caused by the changes in feeding patterns. This conclusion opened the way to the investigation of "adaptation" of enzyme activities in various states of nutrition.[58] Nevertheless, Han et al.[28] discovered that certain rats with hypothalamic obesity do have another abnormality in addition to hyperphagia. They have shown that in this condition two animals may weigh the same but one may have more fat in its body than the other. Animals with the greater fat content were shorter than control animals, and their long bones and skulls had undergone either slowing or failure of growth. The affected animals were believed to have a deficiency of growth hormone–releasing factor. Hypothalamic lesions, therefore, may lead to both obesity and dwarfing. As noted earlier, however, hypothalamic obesity can be produced in rats that have been previously hypophysectomized and thus does not require any change in the level of pituitary function.

Han's observation that hypothalamic lesions may cause two disturbances is not unique; investigators have been troubled for many years by the intermingling of control systems within the hypothalamus. Lesions that abolish drinking are in regions close to those that prevent feeding,[48] whereas the wide spectrum of controls of body temperature, reproductive functions, diuresis, water intake, food intake, and the neurohumoral systems acting on the pituitary gland is sometimes almost impossible to sort into its several parts.[53] Perhaps under natural conditions they are not separate; perhaps the neurons and their processes are intermingled rather than segregated by function. If so, we are fortunate in our study to be able to distinguish them as well as we can.

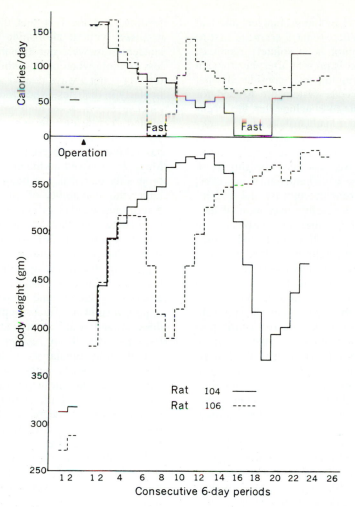

Fig. 57-11. Initial hyperphagia leading to obesity, followed by return of food intake to normal levels in two rats with ventromedial lesions in hypothalamus. After prolonged fasting the hyperphagia reappeared. (From Brobeck et al.[10])

NATURE OF SIGNALS

If mechanisms in (or running through) the hypothalamus control food intake, they must respond to changes in the economy of the body brought about by the presence or absence of food. These changes might have their influence on the hypothalamus either through neural pathways reaching the hypothalamus from other parts of the nervous system or from other parts of the body or through more generalized reactions that affect, for example, the chemical composition of blood perfusing the brain, its temperature, or some other physical or chemical quality. To understand the control of food intake, it is necessary to know both the changes and how they are impressed on the hypothalamus.

Gastrointestinal factors

More than 50 years ago, Cannon and Washburn[13] described a correlation between contractions of the stomach and sensations of hunger referred to the epigastrium. Carlson[14] and associates confirmed their observation and made a systemic study of gastric contractions, hunger sensations, or both as they are influenced by procedures such as exposure to a warm or cold temperature, smoking, or drinking water. The studies did not show, however, the effect of these same factors on food intake, and there is as yet no quantitative correlation between feeding and gastric contractions. Even before the appearance of Cannon and Washburn's paper, Sherrington[50] had considered the possible mechanisms for "general sensation," including hunger, and had noted that although the stomach seems at first to be involved, it cannot be absolutely required, since desire for food is present after the stomach is removed. (Sherrington also considered what portions of the nervous system are needed for feeding and was able to conclude that the mechanisms must be largely subcortical.)

In addition to the mechanisms suggested in these and other studies of gastric *hunger*, a gastric *satiety* mechanism has been the subject of many investigations. Taken altogether, however, these experiments remain controversial and inconclusive, with the possible exception of data collected by McHugh and his associates in studies on monkeys.[45] They implanted a tube into the stomach of the animals, trained the monkeys to eat ad libitum for only a 4-hr period each day, and then measured food intake on days when either saline or a nutrient solution was administered via the tube directly into the stomach. They found that food intake was reduced in proportion to the caloric value of the material given by the tube, whether this was a mixture of nutrients or glucose, starch, triglycerides, or hydrolyzed casein. Their experiments suggest not only a gastric satiety mechanism, but also one sensitive to ingested (or administered) caloric equivalents.

In related experiments, other investigators have made injections into the duodenum and then measured the food intake of animals. Beginning with the paper of Hill et al.[32] and including, among others, the work of Snowdon[52] and Liebling et al.,[39] these studies nearly all agree that satiety can be produced rather easily by any of a variety of materials in the duodenum. Substances used include glucose solution, liquid diets, suspensions of cellulose, urea, and hypertonic saline solution. The mechanism for this inhibition of feeding may be neural or hormonal. Smith and his associates have emphasized the importance of the latter. They have reported that in rats[24] or in monkeys[23] satiety follows the intraperitoneal injection of the duodenal hormone cholecystokinin (CCK). They believe this to be at least one component of the satiety that normally follows food ingestion.

Blood glucose

At a time when gastric contractions were prominent in the analysis of hunger, Bulatao and Carlson[12] proposed that they were initiated by low blood sugar levels. If this were true, it might provide an almost direct coupling of blood glucose, gastric contractions, hunger sensation, and

feeding. But the fact that the stomach can be removed without abolishing feeding made the concept somewhat questionable. It became still less credible when other authors pointed out that feeding is not always related to blood sugar, particularly in diabetes, when food intake and blood glucose are both at high levels, not inversely related. Yet despite these problems the hypothesis that blood glucose somehow controls feeding has retained its interest for many investigators. Mayer[42] improved on the original formulation by proposing that it is not blood glucose level but rather the availability of glucose to tissues that determines whether or not feeding is stimulated. In diabetes the lack of insulin makes glucose relatively unavailable, so that the animal responds as though glucose were deficient in supply. In the Mayer hypothesis the neurons of the hypothalamus respond directly to the availability of glucose and must be presumed to require insulin for glucose uptake just as muscle does. Anand et al.[6] attempted to study the problem more directly by recording from the medial and lateral hypothalamus. Using microelectrodes near spontaneously active single neurons, they reported that the intravenous infusion of glucose solution increased the activity of cells in the medial region and decreased the activity of those in the lateral region (Fig. 57-12), whereas insulin had the opposite effect.

Two other types of evidence offer support to the "glucostatic" hypothesis. First, insulin administration is a well-established means of increasing food intake. If given with the proper dosage and scheduling, it will increase food intake even to the point of inducing a mild obesity.[41] Second, on administration of a chemical known as gold thioglucose to mice, an obesity practically indistinguishable from the obesity that follows hypothalamic operations is induced. The animals show hyperphagia and resulting metabolic changes that are similar to those of hyperphagic rats.[20] On examination of their brains, lesions in the ventromedial region are found, and neurons of the ventromedial nucleus are injured and eventually destroyed. Gold thiocarbohydrates other than glucose do not have these actions on the hypothalamus and do not cause obesity.

Taken together, these data seem to give the glucostatic hypothesis strong support; most authors now accept it in whole or in part.[27] Nevertheless, it has a few conspicuous deficiencies. Perhaps the first is that whereas everyone agrees that insulin hypoglycemia provokes feeding, no one has presented conclusive evidence that hy-

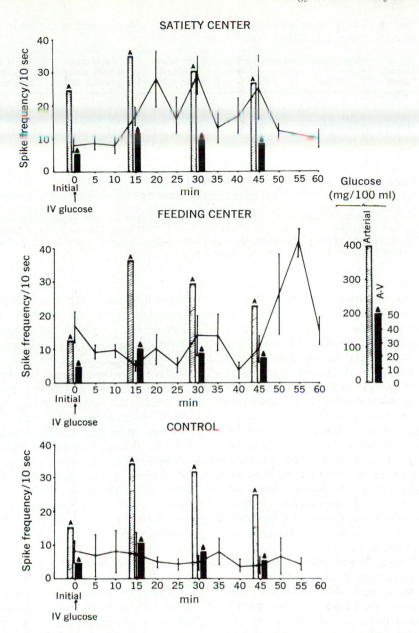

Fig. 57-12. Mean frequencies of unit discharges, with their standard errors, recorded from satiety center (22 units), feeding center (9 units), and control hypothalamic regions (6 units) from dogs anesthetized with Dial. Hyperglycemia is followed by changes in frequency of unit activity except in control records. (From Anand et al.[6])

perglycemia suppresses food intake. In fact, a number of authors have tried to demonstrate such a suppression without notable success.[25,51] If there were a simple relationship between blood glucose and feeding, it should be easy to design experiments to reveal it. Second, feeding cycles occur apparently normally in animals given diets free of carbohydrate and in human subjects on a

low carbohydrate diet[61] in whom swings of blood glucose levels are minimal or absent. Third, Epstein[22] (Figs. 57-9 and 57-10) and Wagner and de Groot[62] were unable to inhibit feeding by injecting glucose directly into the hypothalamus of unanesthetized animals. Their results seem to be contrary to those of Anand et al., but Epstein and Wagner and de Groot made their injections

Table 57-8. Body weight, food intake, fat content, and water compartment in the fat-free body weight of normal and pancreatic diabetic rats*

Groups†	No. of animals used	Body weight (gm) (mean ± SE)		Food intake (gm/day) (mean ± SE)	Fat content (% body weight) (mean ± SE)	Water compartment (% fat-free body weight) (mean ± SE)
		Initial	Final			
CC	7	209 ± 7.2	260 ± 8.7	18.4 ± 1.2	9.3 ± 1.1	70.9 ± 0.68
CH	3	204 ± 6.5	329 ± 9.8	23.9 ± 0.6	20.8 ± 3.0	73.6 ± 0.33
DC	7	188 ± 6.4	150 ± 4.8	27.9 ± 2.8	0.7 ± 0.3	70.5 ± 0.76
DH	6	187 ± 6.8	156 ± 5.2	27.0 ± 2.5	0.4 ± 0.1	71.9 ± 0.43
DIC	4	186 ± 6.2	238 ± 6.2	19.0 ± 1.6	9.0 ± 1.9	71.0 ± 1.07
DIH	4	188 ± 5.4	339 ± 14.6	34.2 ± 4.1	21.9 ± 4.7	72.7 ± 0.32
DNC	4	192 ± 5.7	218 ± 6.9	15.4 ± 0.8	9.7 ± 1.8	70.0 ± 0.28

*From Young and Liu.[64]
†CC, Control rats without lesions; CH, control rats with lesions; DC and DH, diabetic rats without and with lesions; DIC and DIH, diabetic rats treated with protamine zinc insulin without and with lesions; DNC, pancreatectomized rats without diabetes and without lesions. Significant difference (p < 0.05) of food intake and fat content exist between CC and CH and between DIC and DIH but not between DC and DH.

into the brain, whereas Anand injected either glucose or insulin into the systemic circulation. It is not known in the Anand experiments whether there were one or more reactions interposed between blood sugar levels and the hypothalamic activity. The timing of the recordings is compatible with the concept that other changes had occurred in the body before the hypothalamic cells altered their rate of firing.

Although they do not reveal the nature of the control mechanisms, experiments described by Young and Liu[64] show the relationship of glucose utilization to food intake more clearly than any other data (Table 57-8). The study utilized normal animals, rats with hypothalamic hyperphagia, and animals made diabetic by partial pancreatectomy. As the table indicates, diabetic rats manifested a hyperphagia relative to control animals but ate the same amount of food as diabetic rats with hypothalamic lesions. If rats of this latter type were treated with insulin, they ate even more food, gained weight, and gained in fat content similar to the gain of rats having only the hypothalamic hyperphagia. The data illustrate, therefore, the high food intake characteristic of insulin deficiency in animals that could not deposit fat and become obese. They also illustrate the role of insulin in promoting fat deposition and show that in diabetic animals in which the hypothalamus was intact, the insulin actually reduced food intake.

Thermal effects

Another type of signal that might influence hypothalamic feeding systems is the temperature of the blood perfusing this region. Body temperature is regulated in part through thermal de-

tection by cells in the rostral hypothalamus or the preoptic region just ahead of it (Chapter 59). Since a rapid increase in heat production follows the ingestion of food (p. 1363, specific dynamic effect [SDE]), and since this part of the brain is sensitive to thermal change, it would not be surprising if the added heat were utilized as a signal that feeding had occurred. When exposed to high or low temperatures, the higher animals (mammals and birds) alter their food intake in the appropriate direction. They eat more in the cold and less in the heat (Fig. 57-13); these results prove that control of temperature and of feeding are interrelated. Moreover, during undernutrition, human subjects complain that they feel cold; Keys et al.[35] reviewed the evidence for such an association from the experience of earlier authors, in addition to making these comments about their own subjects: "They asked that their food, coffee, and tea be served unusually hot . . . cold temperatures were poorly tolerated. Complaints of being cold or having cold hands and feet were frequent and persistent. In hot summer weather many of the subjects slept under heavy blankets and wore extra clothing during the day."

One of the possible advantages to the body of having the feeding signal related to the SDE is that the magnitude of the SDE is not constant.[40] Kleiber[36] points out that the SDE is a function of environmental temperature; it is constant above the fasting neutral or "critical" temperature of the animal and then decreases as the temperature falls, until at about 12° C, it disappears. The SDE also is related to the composition of the diet and is high when the diet contains either a deficiency or an excess of protein. In all these

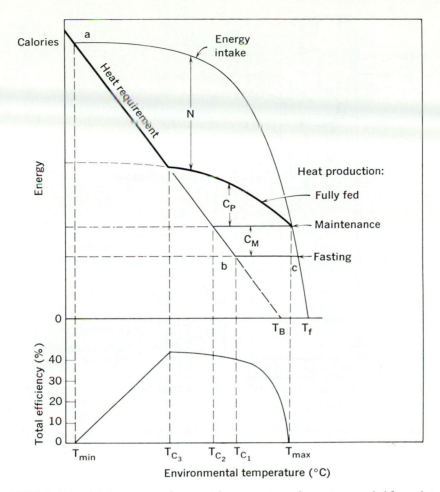

Fig. 57-13. Relationship between environmental temperature and energy expended for maintenance of body temperature *(heat requirement),* in comparison with total *energy intake. Total efficiency* is low at low temperatures because of high requirement for heating of body; efficiency is also low at high temperatures because intake is low and little energy is available for any productive process. In middle range of temperature, however, efficiency is relatively high; in this range, production of eggs, milk, or meat is efficient, being optimal at temperature where food intake is falling while efficiency remains high. (From Kleiber.[36])

conditions the food intake follows the SDE inversely. Kleiber has shown this effect in a diagram that summarizes the metabolic processes related to food intake (Fig. 57-14).

CONCLUSION

Control of food intake cannot be explained by any single physiologic mechanism. Intake tends to be suppressed when the body is well supplied with energy reserves and enhanced when fat stores are depleted, when blood sugar levels are low, or when amino acids are being diverted into the pathways of protein synthesis. These observations suggest the presence of sensors that re-

spond to concentrations, or to changes in concentrations, of materials in the metabolic pool (Fig. 57-15). A sensor in the fat depots is not needed because any signal present in the metabolic pool will be affected by whether material is being taken from the pool for lipid deposition or is being added to the pool by lipolysis. In addition to this type of monitoring of energy sources, there are other mechanisms that signal information about gastrointestinal factors, for example, hunger contractions, distention, or release of gastrointestinal hormones. Finally, there are more general factors that may limit food intake, such as the availability of water for digestive and

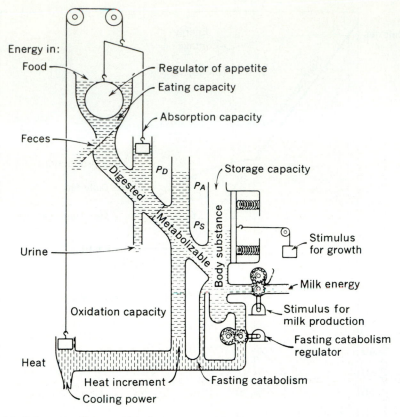

Fig. 57-14. Schematic diagram showing energy utilization as related to control of food intake. (From Kleiber.[36])

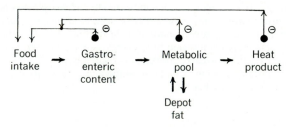

Fig. 57-15. Diagram to illustrate three types of negative feedback loops controlling food intake. Dark circle designates site of detection of some change induced by presence of food, for example, in gastrointestinal tract, in metabolic pool, or as change in rate of heat production.

metabolic processes, the conditions of environmental temperature stress in relation to the bodily heat content, and the rate of change of heat production. In its simplest form, therefore, a diagram of the mechanisms that limit food intake (i.e., create satiety) must have at least three types of interdependent loops producing negative feedback.[9]

REFERENCES

1. Adolph, E. F.: Urges to eat and drink in rats, Am. J. Physiol. **151:**110, 1947.
2. Albanese, A. A., editor: Newer methods of nutritional biochemistry, New York, 1965, Academic Press, Inc.
3. Altman, P. L., and Dittmer, D. S., editors: Biology data book, Washington, D.C., 1964, Federation of American Societies for Experimental Biology.

4. Anand, B. K.: Nervous regulation of food intake, Physiol. Rev. **41:**677, 1961.

5. Anand, B. K., and Brobeck, J. R.: Hypothalamic control of food intake in rats and cats, Yale J. Biol. Med. **24:**123, 1951.

6. Anand, B. K., et al.: Activity of single neurons in the hypothalamic feeding centers: effect of glucose, Am. J. Physiol. **207:**1146, 1964.

7. Behnke, A. R., Jr., Feen, B. G., and Welham, W. C.: The specific gravity of healthy men, J.A.M.A. **118:**495, 1942.

8. Brobeck, J. R.: Neural basis of hunger, appetite, and satiety, Gastroenterology **32:**169, 1957.

9. Brobeck, J. R.: Nature of satiety signals, Am. J. Clin. Nutr. **28:**806, 1975.

10. Brobeck, J. R., Tepperman, J., and Long, C. N. H.: Experimental hypothalamic hyperphagia in the albino rat, Yale J. Biol. Med. **15:**831, 1943.

11. Brožek, J., editor: Human body composition: approaches and applications, New York, 1965, Pergamon Press, Inc.

12. Bulatao, E., and Carlson, A. J.: Influence of experimental changes in blood sugar level on gastric hunger contractions, Am. J. Physiol. **69:**107, 1924.

13. Cannon, W. B., and Washburn, A. L.: An explanation of hunger, Am. J. Physiol. **29:**441, 1912.

14. Carlson, A. J.: The control of hunger in health and disease, Chicago, 1916, University of Chicago Press.

15. Cohn, C., and Joseph, D.: Influence of body weight and body fat on appetite of "normal" lean and obese rats, Yale J. Biol. Med. **34:**598, 1962.

16. Connolly, K. J.: Locomotor activity in Drosophila as a function of food deprivation, Nature **209:**224, 1966.

17. Consolazio, C. F., Johnson, R. E., and Pecora, L. J.: Physiological measurements of metabolic functions in man, New York, 1963, McGraw-Hill Book Co.

18. Cowgill, G. R.: The energy factor in relation to food intake: experiments on the dog, Am. J. Physiol. **85:**45, 1928.

19. Delgado, J. M. R., and Anand, B. K.: Increase of food intake induced by electrical stimulation of the lateral hypothalamus, Am. J. Physiol. **172:**162, 1953.

20. Drachman, R. H., and Tepperman, J.: Aurothioglucose obesity in the mouse, Yale J. Biol. Med. **26:**394, 1954.

21. Edholm, O. G., et al.: The energy expenditure and food intake of individual men, Br. J. Nutr. **9:**286, 1955.

22. Epstein, A. N.: Reciprocal changes in feeding behavior produced by intrahypothalamic chemical injections, Am. J. Physiol. **199:**969, 1960.

23. Gibbs, J., Falasco, J. D., and McHugh, P. R.: Cholecystokinin decreases food intake in rhesus monkeys, Am. J. Physiol. **230:**15, 1976.

24. Gibbs, J., Young, R. C., and Smith, G. P.: Cholecystokinin decreases food intake in rats, J. Comp. Physiol. Psychol. **84:**488, 1973.

25. Grossman, M. I.: Integration of current views on the regulation of hunger and appetite, Ann. N.Y. Acad. Sci. **63:**76, 1955.

26. Grossman, S. P.: Direct adrenergic and cholinergic stimulation of hypothalamic mechanisms, Am. J. Physiol. **202:**872, 1962.

27. Hamilton, C. L.: Control of food intake. In Yamamoto, W. S., and Brobeck, J. R., editors: Physiological controls and regulations, Philadelphia, 1965, W. B. Saunders Co.

28. Han, P. W., et al.: Hypothalamic obesity in weanling rats, Am. J. Physiol. **209:**627, 1965.

29. Hetherington, A. W., and Ranson, S. W.: Hypothalamic lesions and adiposity in the rat, Anat. Rec. **78:**149, 1940.

30. Hetherington, A. W., and Ranson, S. W.: Effect of early hypophysectomy on hypothalamic obesity, Endocrinology **31:**30, 1942.

31. Hildes, J. A., Sherlock, S., and Walshe, V.: Liver and muscle glycogen in normal subjects, in diabetes mellitus and in acute hepatitis. I. Under basal conditions, Clin. Sci. **7:**287, 1949.

32. Hill, R. G., et al.: The small intestine as a factor in regulation of eating, Am. J. Physiol. **170:**201, 1952.

33. Kennedy, G. C.: The role of depot fat in the hypothalamic control of food intake in the rat, Proc. R. Soc. Lond. (Biol.) **140:**578, 1953.

34. Keys, A., and Brožek, J.: Body fat in adult man, Physiol. Rev. **33:**245, 1953.

35. Keys, A., et al.: The biology of human starvation, Minneapolis, 1950, University of Minnesota Press.

36. Kleiber, M.: The fire of life, New York, 1961, John Wiley & Sons, Inc.

37. Laboratory of Physiological Hygiene, University of Minnesota: Unpublished data.

38. Larsson, S.: On the hypothalamic organization of the nervous mechanism regulating food intake, Acta Physiol. Scand. **32**(suppl. 115):entire issue, 1954.

39. Liebling, D. S., et al.: Intestinal satiety in rats, J. Comp. Physiol. Psychol. **89:**955, 1975.

40. Lusk, G.: The elements of the science of nutrition, ed. 4, Philadelphia, 1928, W. B. Saunders Co.

41. MacKay, E. M., Callaway, J. W., and Barnes, R. H.: Hyperalimentation in normal animals produced by protamine insulin, J. Nutr. **20:**59, 1940.

42. Mayer, J.: The obese hyperglycemic syndrome of mice as an example of "metabolic" obesity, Am. J. Clin. Nutr. **8:**712, 1960.

43. Mayer, J.: Genetic factors in human obesity, Ann. N.Y. Acad. Sci. **131:**412, 1965.

44. Mayer, J., et al.: Exercise, food intake and body weight in normal rats and genetically obese adult mice, Am. J. Physiol. **177:**544, 1954.

45. McHugh, P. R., Moran, T. H., and Barton, G. N.: Satiety: A graded behavioral phenomenon regulating caloric intake, Science **190:**167, 1975.

46. Miller, N. E.: Chemical coding of behavior in the brain, Science **148:**328, 1965.

47. Mitchell, H. H., et al.: Chemical composition of the adult human body and its bearing on the biochemistry of growth, J. Biol. Chem. **158:**625, 1945.

48. Montemurro, D. G., and Stevenson, J. A. F.: The localization of hypothalamic structures in the rat influencing water consumption, Yale J. Biol. Med. **28:**396, 1955/56.

49. Passmore, R.: Carbohydrates, the cinderella of nutrition. In Wolstenholme, G. E. W., editor: Diet and bodily constitution, Boston, 1964, Little, Brown & Co.

50. Sherrington, C. S.: Cutaneous sensations. In Schafer, E. A., editor: Textbook of physiology, London, 1900, Young J. Penland.

51. Smith, M., and Duffy, M.: Some physiological factors that regulate eating behavior, J. Comp. Physiol. Psychol. **50:**601, 1957.

52. Snowdon, C. T.: Production of satiety with small intraduodenal infusions in the rat, J. Comp. Physiol. Psychol. **88:**231, 1975.

53. Stevenson, J. A. F.: The hypothalamus in the regulation of energy and water balance, Physiologist **7:**305, 1964.

54. Strominger, J. L., Brobeck, J. R., and Cort, R. L.: Regulation of food intake in normal rats and in rats

with hypothalamic hyperphagia, Yale J. Biol. Med. **26:** 55, 1953.

55. Teitelbaum, P., and Epstein, A. N.: The lateral hypothalamic syndrome: recovery of feeding and drinking after lateral hypothalamic lesions, Psychol. Rev. **69:** 74, 1962.

56. Teitelbaum, P., and Stellar, E.: Recovery from the failure to eat produced by hypothalamic lesions, Science **120:**894, 1954.

57. Tepperman, J.: Metabolic and endocrine physiology, Chicago, 1962, Year Book Medical Publishers, Inc.

58. Tepperman, J., and Tepperman, H. M.: Adaptive hyperlipogenesis: late 1964 model, Ann. N.Y. Acad. Sci. **131:**404, 1965.

59. Tepperman, J., Brobeck, J. R., and Long, C. N. H.: A study of experimental hypothalamic obesity in the rat, Am. J. Physiol. **133:**468, 1941.

60. Tepperman, J., Brobeck, J. R., and Long, C. N. H.: The effects of hypothalamic hyperphagia and of alterations in feeding habits on the metabolism of the albino rat, Yale J. Biol. Med. **15:**855, 1943.

61. Van Itallie, T. B., and Hashim, S. A.: Biochemical concomitants of hunger and satiety in man, Am. J. Clin. Nutr. **8:**587, 1960.

62. Wagner, J. W., and de Groot, J.: Changes in feeding behavior after intracerebral injections in the rat, Am. J. Physiol. **204:**483, 1963.

63. Woods, J. W.: Behavior of chronic decerebrate rats, J. Neurophysiol. **27:**635, 1964.

64. Young, T. K., and Liu, A. C.: Hyperphagia, insulin and obesity, Chinese J. Physiol. **19**(3):247, 1965.

58

SID ROBINSON

Physiology of muscular exercise

Since muscular exercise involves not only the intricate neuromuscular coordination of bodily movements but also many complex adjustments of metabolism, respiration, and circulation, practically the entire organism may be involved in the adjustment of a man to work. The actual coordination of movement depends on the nervous system. The release of energy by the contracting muscles and the processes of excitation and conduction require many complex physical changes as well as both anaerobic and oxidative chemical reactions. Since the ultimate source of the energy involved in all these metabolic processes is the oxidation of fuel, the tissues must be supplied with O_2, and CO_2 must be removed in amounts proportional to the man's energy requirements for the work. To supply the need for O_2 and CO_2 exchange in work the respiratory and circulatory systems are automatically and precisely adjusted. To ensure rapidity and precision in these adjustments to the requirements of the body, a multiplicity of regulatory mechanisms for each system may be involved. These regulations of respiration and circulation involve both the nervous and endocrine systems. They may be set into action by a number of chemical, thermal, and mechanical stimuli associated with the neuromuscular activity itself. Thus an exercising human must supply the muscles with metabolic material from lungs, liver, and intestine by way of the circulatory system and maintain chemical homeostasis by transporting the metabolites of work to lungs, kidneys, and skin for excretion.

A large proportion of the energy released by the working muscles takes the form of heat, and this must be dissipated through the skin. The conduction of heat at high rates from the muscles to the skin requires a large increase of blood flow to the skin, and this puts another burden on the circulation, particularly if the individual is working in a warm environment. As an example of the complex interdependence of these processes, the circulatory system of a person working in a warm environment must meet a threefold demand in addition to the requirements for the usual metabolic processes: (1) a large blood flow through the working muscles provides for respiratory exchange and conducts away the excess heat produced there; (2) blood flow through the skin cools the blood and supplies the sweat glands with water at high rates; (3) flow through the alimentary tract and liver transfers nutrients and water to muscles and sweat glands. Humans are able to achieve these remarkable adjustments in circulation by increasing heart output to several times its resting level and by very fine vasomotor adjustments of blood flow operating on a blood volume that may be considerably increased over the volume at rest in a cool environment.

Dill[26] has classified work on the basis of the rate of energy expenditure or O_2 consumption of the persons performing the work. *Moderate work,* which includes most of the everyday jobs of workers, requires average hourly energy expenditures up to about 3 times the resting metabolism of the individual. For an average man weighing 70 kg, this would include O_2 consumption up to about 0.8 L/min and energy metabolism of 4 kcal/min or 1,900 kcal in 8 hr of work. Assuming 500 kcal expended during 8 hr of sleep and 1,400 kcal expended for the 8 hr spent off work, this man would expend a total of 3,800 kcal during the 24 hr day. Müller[83] and Lehmann[72] consider an energy expenditure of 5 kcal/min (2,400 kcal in 8 hr) as the maximal rate at which average laborers can continue working 8 hr/day for years. This energy expenditure is about equal to that required to walk at 3.8 mph on the level or 30 miles in 8 hr. Dill's category of moderate work would appear to include a more realistic range of the energy expenditures that laborers could sustain for years.

Hard work, according to Dill, includes work rates with energy expenditure of 4 to 8 times the basal metabolism and O_2 consumption of 1 to 2 L/min. This range includes most of the manual jobs of heavy industry, building, mining, agriculture, and war. A 70 kg man consuming 1.84 L O_2/min (9 kcal/min) in walking at 3.5 mph up a 10% grade would transform a total of about 4,300 kcal of energy in 8 hr. This is probably as hard as a well-conditioned man of average size could work for that length of time. The man performing this work would transform a total of about 6,200 kcal in 24 hr, including the basal metabolism, the specific dynamic action of foods, and the energy of activities other than the work. This is probably too hard for even highly fit men to continue day after day, although timber workers and men on certain jobs in other heavy industries have been reported to expend energy at even higher rates than this.[40,75] Such hard work could be endured only by unusually strong men on jobs in which the work is seasonal rather than continuous. The relationship of work rate to energy expenditure in the ranges of moderate and hard work is well illustrated by the data of Margaria[77] in Fig. 58-1.

Dill's category of *maximal work* includes work rates in which men cannot maintain a steady state of energy output for long and even greater rates in which the steady state is never reached. Exhaustion may result from (1) hyperpyrexia due to inability to dissipate heat fast enough in warm environments, (2) hypoglycemia due to depletion of the body's carbohydrate reserves, (3) circulatory strain with high heart rate, and (4) inadequate O_2 supply in the most intense work, such as competitive races, in which the anaerobic reserves of the muscles are called on. In the last category, exhaustion is associated with the accumulation of metabolites of anaerobic metabolism, mainly lactic acid, in the muscles, which diffuses into the blood with significant increases in hydrogen ion concentration.

METABOLISM IN EXERCISE

The energy of muscular work is derived from the oxidation of fuels, predominantly carbohydrate and fat. Most of the work performed by humans is at moderate rates in which the energy required is provided during the work by oxidative mechanisms in the muscles, but even in exhausting work of short duration where energy may be

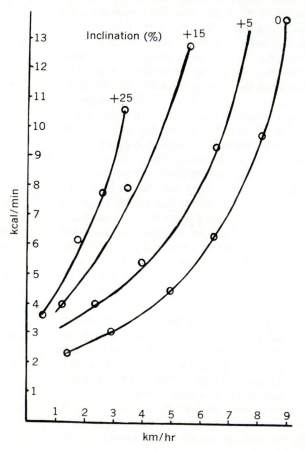

Fig. 58-1. Effects of speed and grade on energy cost of walking on motor-driven treadmill. (Based on data by Margaria[77]; from Passmore and Durnin.[88])

released at rates far in excess of the capacity of the individual to supply and utilize O_2 during the work, the ultimate source of all energy expended is the oxidation of fuel. In the latter case the work is performed predominantly with energy from anaerobic chemical reactions in the muscles, and extra oxidations are carried on in recovery to restore the energy to this system. Measurements of the O_2 consumption of humans in relation to various types and intensities of work provide a relatively simple basis for calculating the total energy expended in the work. The relationships of O_2 consumption to the energy of general metabolism and to the types of fuels being metabolized are given in Chapter 56.

The three principal methods for determining the energy metabolism as outlined in Chapters 56 and 57 may all be used to determine the energy cost of work by humans: (1) direct calorimetry in which the subject's heat production is measured directly in a human calorimeter, (2) indirect calorimetry in which the subject's respiratory exchange (O_2 consumption and CO_2 production) is determined and used as a basis for calculating energy metabolism, and (3) indirect calorimetry in which the balance between the subject's calorie intake in food and measured changes in the body's energy stores as glycogen, fat, and protein is followed over a period of several weeks or months of work on the job. Under laboratory conditions the work done by a person is most easily controlled and measured if it is performed on a treadmill or stationary bicycle ergometer; the total energy expenditure in the work can be measured by one of the methods just given.

Of these methods the indirect method, based on calculation of energy expenditure from O_2 consumption and CO_2 output, has the advantages of being rapid, accurate, and relatively simple in technique. With this method, it is possible to follow the course of oxidative metabolism in humans during the rapid transitions from rest to work, in recovery after work, or during more gradual changes associated with fatigue during a prolonged period of hard work at constant rate.

Oxygen consumption

When an individual starts walking or running on the treadmill, he immediately begins to expend energy at a rate proportional to the rate of movement. However, the O_2 consumption ($\dot{V}O_2$) does not rise instantly to the level required to supply by oxidation all the energy being expended, even in light work, nor does it return instantly to the resting level when the work is stopped. O_2 consumption, beginning at the resting level, is rapidly accelerated in the first 2 min of work and levels off at a steady rate after the second minute (Fig. 58-2). As the work is continued at a constant rate, the O_2 consumption remains constant.

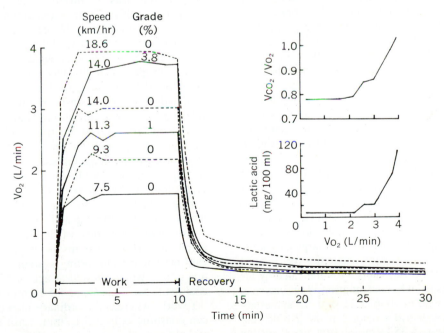

Fig. 58-2. Effects of increasing intensities of work on treadmill on man's O_2 consumption, respiratory exchange ratio ($\dot{V}CO_2/\dot{V}O_2$), and blood lactate level. (Based on data from Margaria et al.[78])

During recovery after work, O_2 consumption declines as a logarithmic function of time.

This steady state of O_2 consumption during constant submaximal work is just sufficient to provide by oxidation the energy of basal metabolism plus all the energy required to carry on the work. In submaximal work a person's steady state of O_2 consumption is proportional to the rate of work (Fig. 58-2). The only deficiency in the O_2 supply for this type of work is the "O_2 deficit" that occurs in the first few minutes.[112,126] These relationships of O_2 consumption to work rate are illustrated in Fig. 58-2 for a subject working on the treadmill at six different rates.[78] It will be noted that the time required for acceleration of O_2 consumption from 0.24 L/min in rest to a steady 1.6 L/min in the lightest work was about 2 min, which was the same length of time required to accelerate to 3.9 L/min in the most severe work.

The *maximal O_2 consumption* of a man refers to his maximal rate of aerobic energy release for work; it is attained in the second or third minute of any exhausting work in which the energy requirement of the work exceeds his maximal capacity for aerobic metabolism. The highest O_2 consumption of the subject represented in Fig. 58-2, 3.9 L/min, was maintained from the second minute through the tenth minute of a run on the treadmill at 18 km/hr. This probably represents the maximal O_2 consumption of this subject, since the run exhausted him and he was unable to exceed this rate of O_2 consumption even in faster runs with higher energy expenditures in which he was exhausted in less time than in this run. The maximal O_2 consumption is probably the best single physiologic indicator of a man's capacity for maintaining extremely heavy work.[82,92,119] For a given man, it is remarkably constant, although it may be decreased by long-continued inactivity[52,120] and significantly increased by a few months of intensive athletic training[66,97] (Fig. 58-3).

The maximal O_2 consumption declines with age in men, from an average of 50 ml/kg · min at 18 years of age to 26 ml/kg · min at 75 years of age.[4,92] Highly trained distance runners,[101] swimmers,[62] bicyclists,[18] and skiers[20] reach championship performances to a large extent because of exceedingly high capacities for carrying on aerobic work; this capacity is indicated by the man's maximal O_2 consumption. For instance, Don Lash, who in 1936 at 21 years of age established a new world's record in the 2-mile run, could consume 5.35 L/min during an exhausting run on the treadmill.[101] In relation to body

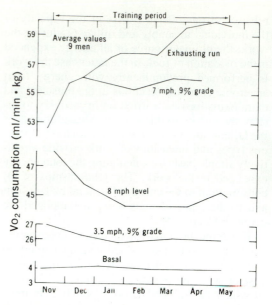

Fig. 58-3. Effect of strenuous athletic training on O_2 consumption of men, ages 18-22 years. In exhausting runs, grade and speed of running were increased as training period progressed in order to exhaust men in 5 min and raise O_2 consumption to maximum. Other grades of work were the same in all experiments, and rates of O_2 consumption represent steady-state values during 10 to 15 min periods of work. Averages of determinations on nine men are given. (Based on data from Robinson and Harmon.[97])

weight, this was 81 ml/min · kg for Lash, as compared with an average maximum of 50 ml/min · kg for untrained college men of this age. Asmussen et al.[3] have found that the resting muscles of a man consume about 20% of the basal O_2 consumption. On this basis, if one half of the body weight is muscle, Lash's muscular system weighed 33 kg and at rest consumed O_2 at 0.16 ml/min · 100 gm of muscle and in maximal work consumed O_2 at 98 times this rate. The average young man in maximal work would show a corresponding increase to 60 times the resting level of muscle O_2 consumption. The great ability of Lash to release energy oxidatively during exhausting work was due in part to intensive athletic training and in part to inherent physiologic properties of his respiratory and metabolic enzyme systems as well as to a superb cardiovascular transporting system. Evidence of this is that when retested at age of 43 years, he could still consume O_2 at 4.5 L/min in exhausting work, whereas the ordinary man of 43 years can consume only 2.5 L/min. Lash has a non-athletic son who at 15 years of age consumed 4 L/min in a run on the treadmill and another son who at age 17 years could consume 4.5 L/min and was state high school champion in the mile run and in cross-country running.

The principal physiologic factors that may limit the maximal O_2 consumption of humans are (1) rate of O_2 transport by the circulation, (2) O_2 utilization by the tissues, and (3) O_2-diffusing capacity of the lungs. O_2 transport by the blood depends on cardiac output and O_2 capacity of the blood. It is possible that the O_2 utilization by the working muscles may be more limited than the capacity of circulation and respiration to supply O_2. There is good evidence, to be discussed in this chapter under respiration, that neither the capacity to ventilate the lungs nor the O_2-diffusing capacity of the lungs is a limiting factor in the maximum O_2 consumption of a normal healthy person.

Oxygen debt

After a period of work an individual's O_2 consumption does not immediately return to the pre-work level. It declines as a logarithmic function of time, and thus during recovery the person consumes more O_2 than the resting requirement (Fig. 58-2). This excess O_2 consumed during recovery was termed the "O_2 debt" by Hill et al.[58] The extra oxidation involved in the O_2 debt produces energy required for restoring the high-energy systems in the muscles, which have released energy anaerobically during the work, and therefore the phenomenon might more properly be called an "energy debt." The total O_2 debt measured in recovery is considered to have three components: (1) a small net diminution in O_2 stored in venous blood and muscle hemoglobin (100 ml in moderate work to 250 ml in exhausting work), which is not really a part of the metabolic energy debt and is paid off in the first minute of recovery; (2) the rapid resynthesis of high-energy phosphate bonds in adenosine triphosphate, phosphocreatine, etc.; and (3) the slow removal of lactic acid formed from pyruvate in the anaerobic breakdown of glycogen in the working muscles (Chapter 3). Hill originally ascribed the entire metabolic O_2 debt to the oxidation of about one fifth of the anaerobically accumulated lactate and the resynthesis of the remainder to glycogen. Since then, Fiske and Subbarow,[34] Eggleton and Eggleton,[30] Lundsgaard,[76] Lohmann,[74] and a long series of other physiologists and biochemists have shown the role of the high-energy phosphate bonds of phosphocreatine and adenosine triphosphate in the anaerobic release of energy in muscular work. After hard and prolonged work the O_2 consump-

tion of an individual may not return to the pre-work level for several hours. Hill et al.[58] and Margaria et al.[78] did not consider this prolonged elevation of metabolism after exhausting work a part of the metabolic O_2 debt.

The most abundant metabolite of anaerobic muscular metabolism is lactate, which in the intact human accumulates in the working muscles and diffuses into the blood. Hill[56] and others[60,78] have found the blood lactate of men after heavy work to vary with the O_2 debt (Fig. 58-4). This has been disputed by Sacks and Sacks,[107] who questioned the uniformity of lactate distribution between the working muscles and blood on the basis of observations on the local blood flow from a cat's muscle working in response to electrical stimulation. On the other hand, Newman[85] found equal increments in blood and muscle lactates in rats working in a steady state. Whether or not the concentration of lactate in the blood provides an accurate estimate of total lactate in the human body, there is a high correlation between the O_2 debt and the lactate concentration of blood samples taken 5 min after the end of heavy work. Huckabee[60] has reported that significant increases of blood lactate brought about by hyperventilation or by intravenous infusion of glucose, pyruvate, or $NaHCO_3$ occur without evidence of tissue hypoxia or an O_2 debt. He concludes therefore that under these circumstances changes in blood lactate are not related to tissue hypoxia.

Margaria et al.[78] recognized an alactacid mechanism in the O_2 debt on the grounds that there is usually no elevation of blood lactate during the steady state of moderate work and yet O_2 debts up to 2 L are developed at these work levels (Figs. 58-4 and 58-5). Furthermore, payment of the alactacid debt is 30 times as fast as that associated with removal of lactate. Huckabee[60] reported that the O_2 debt of exercise was closely related to "excess lactate," which is a function of both lactate and pyruvate concentrations in arterial blood during work. In more recent studies by Margaria et al.,[79] Knuttgen,[67] and Schneider et al.,[112] this relationship has not been found to hold at all levels of exercise. The latter authors found no rise in "excess lactate" in moderate work involving O_2 debts up to 2 L.

After submaximal work, both the time required for recovery and the amount of the O_2 debt are related directly to the intensity of the work, and as indicated previously, there is little or no elevation of blood lactate (Figs. 58-2 and 58-4). In maximal or exhausting work the time of recovery and the O_2 debt are related to the degree

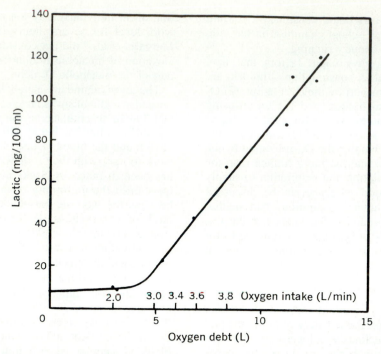

Fig. 58-4. Relation between lactic acid concentration in blood and O_2 debt as calculated by Hill's method, and O_2 consumption in exercise. Duration of exercise 10 min in each case. (From Margaria et al.[78])

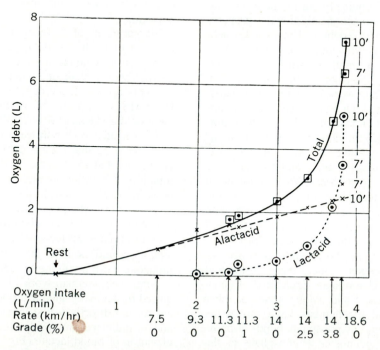

Fig. 58-5. Amounts of alactacid and lactacid O_2 debts as a function of O_2 intake and work rate on treadmill. (From Margaria et al.[78])

of exhaustion reached and therefore to the intensity of the work in relation to the length of time the work is continued. It is significant to note in Fig. 58-2 that the O_2 deficit incurred in the first 2 min of exercise was not repaid before work was stopped, even in the lighter rates of work where the O_2 requirement for maintaining the work oxidatively was far below the subject's capacity for consuming O_2. This indicates that the level of O_2 consumption in submaximal exercise depends on the degree of deficiency or need for O_2 in the tissues and is not determined by the supply. The lag in O_2 consumption in the early minutes of exercise appears to depend more directly on the time required to develop an O_2 need and thus to prepare the tissues to accept O_2 than on a lag in the O_2 supplied by the respiratory and circulatory systems. When the O_2 need is developed, the rate of O_2 consumption exactly satisfies this need and no more, thus leaving an O_2 debt to be paid after exercise.[112,126] Fig. 58-2 also indicates that in exhausting work the maximal O_2 debt need not be incurred in order that the tissues utilize O_2 at the maximum rate; the subject's O_2 debt continued to increase throughout the run at 18 km/hr, during which time O_2 intake was constant at the maximum.

From a practical viewpoint the anaerobic mechanisms of energy release in the muscle are extremely important in maximal work, where they permit an individual to expend energy far in excess of the capacity for carrying on oxidative metabolism. This is illustrated in Fig. 58-6, which shows the changes of blood lactate in two champion runners during track meets lasting 2 hr. One man (C. K.) ran three races and the other (R. C.) ran four races and performed in the broad jump. During the warm-up, lactate level in C. K. was raised to 30 mg/100 ml; he ran the mile race and it rose to 175 mg/100 ml; in 40 min of recovery it dropped to about the prerace level. He then ran the half-mile and again lactate level was elevated to 175 mg/100 ml; in another 40 min of rest it returned to 30 mg/100; and a fast 440 yd run in the relay raised it again to 180 mg/100 ml. All three of his performances resulted in victories in the "Big Ten" track championship meet. The other man (R. C.) competing in a dual track meet warmed up and ran the 100 yd dash before the first blood sample was drawn (lactate level, 70 mg/100 ml). He then ran a fast 440 yd race that elevated his lactate level to 200 mg/100 ml. His recovery times between races were brief, and he was competing in the broad jump. Therefore recovery was incomplete between races, and his performances in the later races were not up to his capacity.

Even in moderate work the anaerobic reactions

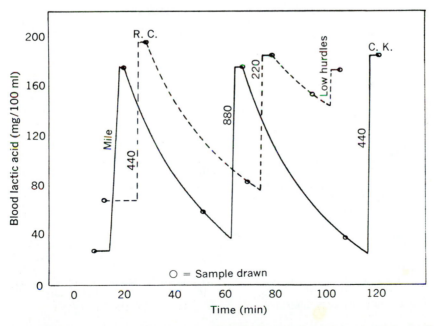

Fig. 58-6. Blood lactate changes in two champion athletes competing in track meet. C. K. ran three races as indicated, and R. C. ran 100 yd dash before competing in three races indicated for him. See text for interpretation. (Based on unpublished data of Robinson and Turrell.[100])

in muscle metabolism are important in that they allow the individual to start releasing the required energy instantly without the delay involved in acceleration of O_2 utilization. The amount of energy that a person may borrow from these anaerobic mechanisms is probably limited by his tolerance for acidosis resulting from the accumulation of lactic acid in the muscles.[121] Lactic acid formed in the working muscles during strenuous exercise is accompanied by a decrease in base-bound bicarbonate in the body fluids. The reaction $BHCO_3$ + Lactic acid $\rightleftharpoons$ B (lactate) + H_2CO_3 results in loss of CO_2 through the lungs in excess of that being formed metabolically. As a result the expiratory exchange ratio ($\dot{V}_{CO_2}/\dot{V}_{O_2}$) may exceed unity during the work (Fig. 58-2) and fall below the range of metabolic respiratory quotient as the lactate is being removed during recovery. The changes in CO_2-combining capacity and pH of the arterial blood of athletic young men in relation to increments of lactic acid accumulated in work of varying intensities are given in Fig. 58-7. After the most exhausting work of 2 to 5 min duration, the blood lactate level had increased to 22 mEq/L and the arterial pH had fallen to 7.00, as compared with resting values of 1 mEq/L and 7.40, respectively. Recovery of athletes from such fatigue is rapid enough that they may perform well again in less than an hour, although complete recovery may require 2 hr or more. Muscle lactate at the moment work ceased in these experiments was higher than the concentration of 22 mEq/L or 0.2% observed in the blood. In this connection, it is interesting to recall that, when an isolated frog muscle is stimulated to work anaerobically, it is completely fatigued and in a state of contracture when the muscle lactate concentration increases to about 0.45%.

Strenuous athletic training reduces the extent to which humans are required to utilize the lact-acid O_2 debt in performing a fixed rate of submaximal work and greatly increases the ability to borrow energy from the anaerobic mechanisms in exhausting work.[96] This is shown in Fig. 58-8 for men during a 6-month training period. Both running at 7 mph up a 9% grade, which originally exhausted all the men in 5 min or less, and running at 8 mph on the level for 10 min became easier for the men, and the elevations of blood lactate in the runs became progressively lower during the training period. This improvement was associated with greater efficiency or skill in running, as shown by reduced O_2 requirement in running at 8 mph and by an increase in maximal O_2 consumption in exhausting work (Fig. 58-3). The mean values of blood lactate after exhausting runs increased progressively from 13 mEq/L before training to 18 mEq/L at the end of the training period (Fig.

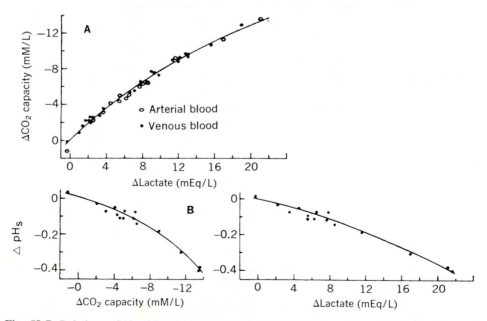

Fig. 58-7. Relations of blood lactate increase (Δlactate) of men in various intensities of exercise to, **A,** decreases of CO_2-combining capacity (ΔCO_2 capacity) below basal level and, **B,** decreases of the plasma pH of arterial blood (ΔpH_s). (From Turrell and Robinson.[121])

58-8). The increase in tolerance for lactic acid was not due to increased alkali reserve of the blood because the latter did not change during the training period. With training, the men acquired the ability to expend 18% more energy from the O_2 debt mechanism than before training.[96] Associated with the increased ability of the men to expend energy anaerobically was an increase of 18% in their maximal O_2 consumption (Fig. 58-3) and a marked improvement in their ability to run the mile race—average time, 6.3 min before training and 5.3 min after 6 months of training. On the treadmill, it required progressively harder runs to exhaust the subjects.

The reader should remember that both anaerobic and aerobic mechanisms are involved in all intensities of muscular work by intact humans. The relative contribution of each to the work depends on the intensity and duration of the work. Actually the greatest utilization of energy from the O_2 debt mechanism occurs when the O_2 utilization is also greatest, as it is in highly trained athletes.[96,121] On the other hand, it is surprising to learn that when the O_2 supply is limited, as in cardiac insufficiency[125] and in old age (Fig. 58-9),[92] persons are unable to utilize the O_2 debt mechanism as fully as when the O_2 supply is

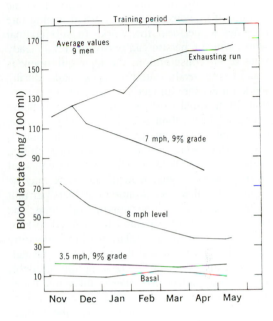

Fig. 58-8. Influence of strenuous athletic training on elevation of blood lactate in exercise. In exhausting runs, speed and grade were increased progressively during training period in order to exhaust men in 5 min. Other grades of work were the same in all experiments. Conditions were same as those in Fig. 58-3. (Based on data from Robinson and Harmon.[96])

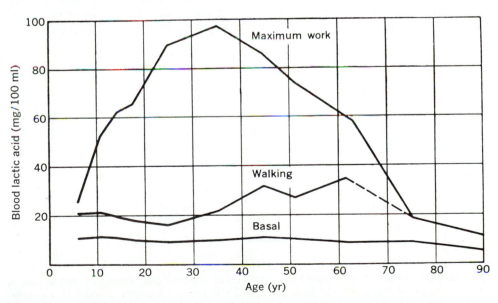

Fig. 58-9. Blood lactate at rest, during moderate work, and after maximal work in relation to age of human subjects. Curves represent average values observed on 10 to 12 subjects in each age group. (From Robinson.[92])

high. It seems that there may be some protective factor or factors involved that limits the extent to which individuals under conditions of restricted O_2 supply can draw on their anaerobic energy sources.

Oxygen requirement of work

The "O_2 requirement of work" is a useful indicator of energy expenditure in work, since the energy cost of work is proportional to the amount of O_2 consumed above usual resting rate in order to perform the work. In submaximal work, which a person can perform with a steady state of O_2 consumption, the O_2 requirement is equal to the steady state rate of O_2 intake during work minus the subject's resting O_2 consumption. In maximal work, in which energy is being expended faster than it can be supplied by the individual's maximal rate of oxidation, the O_2 debt must be added to the excess O_2 consumption during work in calculating the O_2 requirement of work. The O_2 requirement of the work may be expressed in liters per minute or as liters per unit of distance traveled or per unit of work done. The O_2 requirement of a person for a fixed task of work shows an inverse relation to the economy or efficiency with which the work is performed.

The relationships of O_2 consumption, O_2 debt, and O_2 requirement of a man in three intensities of work are illustrated diagrammatically in Fig. 58-10. In the sprint lasting 30 sec the subject acquired a high O_2 debt but had little time to utilize O_2 during the run, and therefore about 90% of the energy expenditure was anaerobic. In the 2 min run to exhaustion he acquired about

the same O_2 debt as in the shorter run, but in 2 min he was able to consume much more O_2, so that anaerobic energy sources contributed only about two thirds of the total requirement for the run. In the third bout of work, O_2 requirement was only 2.5 L/min, which was considerably less than the maximal O_2 consumption of the man. Therefore he attained a steady balance between oxidative metabolism and energy expenditure and could have continued the work for an hour or more without exhaustion.

The O_2 requirement of a given man for work is affected by a number of factors, including speed, load, fatigue, training, and type of work. Sargent[108] found that the energy or O_2 required to run a given distance (120 yd) increased as the 2.8th power of the speed, whereas that required to run a given time increased as the 3.8th power of the speed (Fig. 58-11, *A*).

Margaria[77] found a similar relation between speed and energy cost of walking (Fig. 58-1). More recently, Margaria et al.[80] found in champion runners a linear relation between energy cost and speed of running at steady rates between 3 and 6 yd/sec, speeds that the athletes could maintain aerobically. Aerobic capacity was smaller in an untrained man in whom the linear relationship was found to range from 2 to 4.2 yd/sec. It will be noted in Fig. 58-11 that Sargent's data are for speeds of running between 5.6 and 9.3 yd/sec, the latter being the maximum speed of the subject. Karpovich and Millman[62] and more recently Bullock[14] have found similar relationships between speed and energy requirements of swimming (Fig. 58-11, *B*). Be-

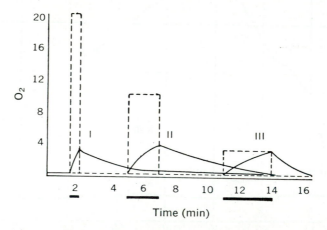

Fig. 58-10. O_2 requirement (broken lines) and O_2 consumption (solid lines) in l/min of man in three intensities of exercise: sprint lasting ½ min, 2 min run to exhaustion in which O_2 consumption was maximal, and less intense work in which O_2 consumption was equal to O_2 requirement. Area of O_2 requirement rectangle is equal to sum of O_2 consumed plus O_2 debt. (From Furusawa et al.[35])

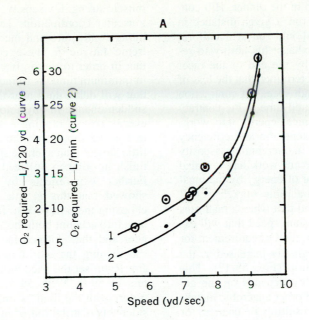

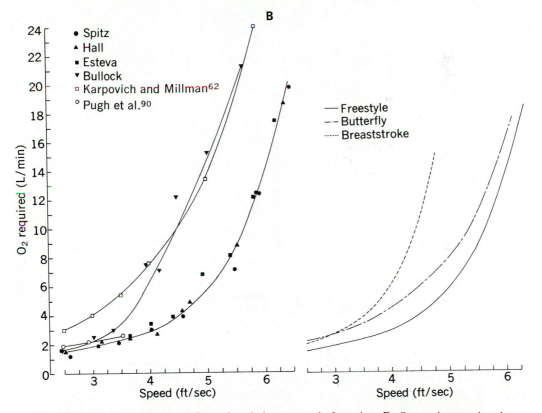

Fig. 58-11. A, O_2 requirement of man in relation to speed of running. **B,** O_2 requirement in relation to speed of swimming. Left: data of Bullock on recent champions Spitz, Hall, and Esteva compared with data on former swimmer Bullock, and with data of Karpovich[62] on good swimmers in 1944 and data of Pugh[90] on channel swimmers in 1960. Right: comparison of O_2 requirements of recent champion swimmers performing three competive strokes. (**A** from Sargent[108]; **B** from Bullock.[14])

cause of this relationship in the runner, Hill concluded that in order to run a given distance in minimum time a man should run at a uniform speed that would just exhaust his ability to release anaerobic energy by the end of the race. Fenn[32] and Hill[57] have determined that the loss of muscular efficiency at high speeds of contraction is due largely to the limited time in the contraction for converting chemical energy into work, and to a much lesser extent, the low efficiency is due to waste of energy in overcoming viscosity in the muscles. More recent work has revealed that the anaerobic release of energy by the muscles is less efficient than aerobic work.[1,19,114]

Robinson et al.[104] found that when a man runs on the treadmill at a constant speed that will exhaust him in about 3 min, the O_2 requirement for maintaining the pace is greatly increased as the man approaches exhaustion (Fig. 58-12). This loss of efficiency probably depends on the accumulation of lactate and other anaerobic metabolites in the muscles, resulting in changes of muscle strength, viscosity, relaxation, and neuromuscular coordination. The isolated muscle exhibits contracture and increased viscosity in extreme fatigue.[37] These considerations indicate that in order to run a given middle distance race in minimum time the runner should follow a pace that will delay until near the end of the race the sudden change in physiologic state in which the energy cost of running is so greatly accelerated; that is, he should run the first part of the race a little slower than average speed and make a faster finish in order to utilize the maximum O_2 debt. Further experiments by Robinson et al.[104] showed that this is true. Thus the runner in choosing the best pace to follow in his race must take into account both the "speed factor" of Sargent and the "fatigue factor" of Robinson. In establishing the world record in the mile run, Jim Ryun set the pace himself and ran the quarters in 59.0, 59.9, 58.5, and 53.7 sec in that order, with the final 1/8 mile in 26.3 sec.[84] Consider this distribution of his energy expenditure,

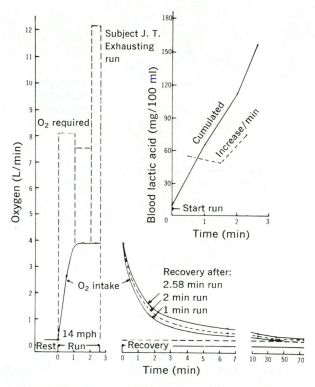

Fig. 58-12. Blood lactic acid (4 min after work) and O_2 requirement of man running at a constant speed of 6.84 yd/sec (14 mph). For each period of the run:

$$O_2 \text{ req/min} = \frac{O_2 \text{ req for run} - O_2 \text{ req for next shorter run}}{\text{Time of run} - \text{time of next shorter run}}$$

(From Robinson et al.[104])

taking into account the effects of both fatigue and speed on the energy cost of running.

Efficiency of work

Under favorable conditions the human body may perform work with a mechanical efficiency of 20% to 30%. Thus a man is an efficient machine as compared with a steam engine with 10% efficiency and a gasoline engine with an efficiency of 20% to 30%. The efficiency of work by a man may be calculated according to the following formula:

$$\text{Efficiency} = \frac{\text{Work} \times 100}{\text{Energy expended}}$$

The total energy expended is calculated from the O_2 consumption during the work. The net energy expended in the work equals the total energy in work minus the basal metabolism of the man. Measurement of work accomplished in climbing stairs, walking up a grade on a treadmill, carrying a load up a grade, working on a bicycle ergometer, lifting a weight, shoveling, etc. may be readily taken. It is more difficult to estimate the work accomplished in activities such as swimming and running on the level.

Fenn,[32] with the aid of moving pictures taken of men running on a level track, has calculated the total external work accomplished in the form of movements of the limbs in relation to the center of gravity of the body and the horizontal and vertical movements of the center of gravity of the whole body. Calculations based on the external work and energy expenditure determined from O_2 consumption showed the runners' efficiencies above 20%. In swimming the work accomplished in overcoming the resistance of the water to

movement of the body has been measured by Goff et al.,[41,42] by Karpovich and Millman,[62] and by Karpovich and Pestrecov,[63] and the efficiency of swimming was found to be only 2% to 8%. Efficiencies of men performing other common types of work have been measured: walking upgrade on the treadmill, 18%[31]; climbing uphill, 21%[29]; and riding a bicycle ergometer, 23%.[25] Efficiencies in climbing a grade appear to be lower than they actually are because the work is calculated as the vertical lift of the body without accounting for the horizontal component of the movement.

The efficiency of work may vary with speed, load, fatigue, training, and fuel being used by the body in performing the work. As is true with machines, there is an optimal speed of work for humans. This is shown for stair climbing in Fig. 58-13. At very low speeds, gross efficiency is low because little work is accomplished relative to the energy cost of muscle tension exerted in supporting the body weight. At high speeds a greater proportion of the energy of contraction takes the form of heat for reasons previously discussed under O_2 requirement. The loss of efficiency characterized by increased O_2 requirement caused by fatigue associated with anaerobic work and lactate accumulation in fast running is illustrated in Fig. 58-12. In long-continued, hard but aerobic work a fasting person shows a gradual increase in the energy cost of maintaining the work (Fig. 58-14). This loss of efficiency is associated with fatigue, which develops progressively without a corresponding rise in blood lactate or a rise in body temperature. A gradual decline of the respiratory quotient indicates an increasing proportion of fat in the fuel of metabolism; this may be partially responsible for the increased energy cost of the work. The mechanism of this effect will be discussed under fuel of muscular work. The data in Fig. 58-14 are also interesting in showing the loss of efficiency of a man walking on the treadmill in heavy shoes. In these experiments in which the man wore light shoes and heavy shoes during alternate hours, it was observed that it cost him 4 times as much energy to carry an extra kilogram of weight on his feet as it would cost him to carry it in a pack on his back. The loss of efficiency depended on the fact that the weight on the feet was being carried at the end of a long lever and it had to be started and stopped with each step.

Training has little effect on the efficiency of walking, but it does increase efficiency in some other types of exercise. It was found that 3 months of practice in running improved the ef-

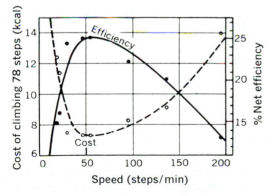

Fig. 58-13. Influence of speed of climbing stairs on energy cost (broken curve) and on efficiency (smooth curve). Cost was lowest when the 78 steps were climbed at about 50 steps/min. (Based on data from Lupton; from Brody and Trowbridge.[12])

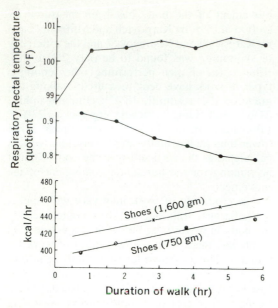

Fig. 58-14. Increase in metabolic energy expenditure of man during 6 hr of walking on treadmill at constant rate (5.6 km/hr up 5.7% grade). Heavier shoes worn in third and fifth hours increased energy cost by 5% or about 4 times as much as if extra weight had been carried in pack on subject's back. (Based on data from Turrell and Robinson.[122])

ficiencies of men running at 8 mph, as indicated by an 8% reduction in their O_2 requirements for performing the work in a steady state (Fig. 58-3). Much greater improvements in the efficiency of swimming may result from training.

Fuel of muscular work

Both carbohydrate and fat are utilized by the working muscles, whereas little protein is used directly as fuel. Protein metabolism, as represented by the rate of nitrogen excretion in the urine, is not significantly increased in exercise.[16,27,28,33] Dill et al.[27] found that a fasting man excreted 4.6 gm of nitrogen in 24 hr of rest and 7.4 gm during a similar 24 hr period in which he expended 4,000 kcal in hard work. Protein metabolism associated with the work was only 70 kcal, which is negligible compared with the total of 4,000 kcal of energy expended in the work. Aerobic work must therefore be performed by oxidation of fuel made up of fat, carbohydrate, or a mixture of the two.

The respiratory quotient (RQ) is the ratio of CO_2 produced to the O_2 consumed in metabolism and may be determined by collection and analysis of expired air of a man at rest or working at a steady rate (Chapter 56). The RQ of carbohydrate metabolism is 1.0, of fat 0.7, and of protein 0.8. The nonprotein RQ is determined by subtracting from the total respiratory exchange that of the protein metabolism, which is calculated from the nitrogen excretion in the urine. Evidence from the nonprotein RQs of working men indicates that in moderate work the fuel is made up of mixtures of carbohydrate and fat, the proportions of the two varying with their proportions in the diet and the duration and intensity of the work.[28,70]

If a man is in energy balance so that his daily energy output is equal to the energy content of his diet, the mean RQ of his total metabolism over a period of days will be the same as the average RQ of the foods in his diet. In an average diet, nonprotein RQ is about 0.8, indicating that about two thirds of the nonprotein respiratory exchange is used in the oxidation of fat and one third in carbohydrate metabolism. In light to moderate work the proportions of carbohydrate and fat are about the same as in rest. In more strenuous steady-state work the RQ increases toward unity, indicating an increase in the proportion of carbohydrate being utilized (Fig. 58-2). Thus carbohydrate appears to be the preferred fuel for strenuous work.

If a fasting man continues to perform hard work for several hours, the RQ will gradually decline, indicating an increase in the proportions of fat in the fuel (Figs. 58-14 and 58-15). Further evidence that fat is being used for the muscular work in such an experiment is that ketosis develops in the man when the RQ has dropped to about 0.75. Furthermore, in a prolonged bout of work a fasting man may expend several times as much energy as the total energy stored in his carbohydrate reserve. Dill et al.[28] found that a fasting man expended 5,000 kcal during 24 hr in which he alternately rested and worked (500 kcal/hr). The subject developed ketosis and during the last 3 hr had a blood sugar level of 80 mg/100 ml and an RQ of 0.73, indicating that fat was 90% of the fuel being utilized (Fig. 58-15). Thus a fasting man carries on hard work by expending his fat reserves—he conserves his more limited carbohydrate reserve and maintains a normal blood sugar level for many hours. The investigators found that injections of adrenaline (a solution containing epinephrine and norepinephrine in unknown proportions) at times during the 24 hr experiments increased the fasting subject's utilization of carbohydrate and decreased ketosis (Fig. 58-15).

In very strenuous work, such as marathon running lasting 2.5 to 3 hr, exhaustion may result

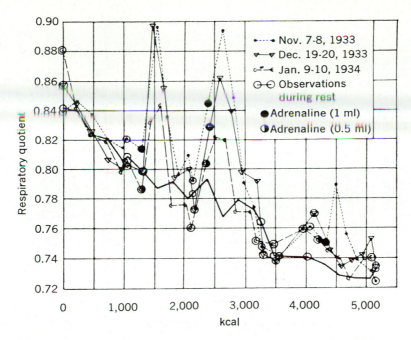

Fig. 58-15. RQ of fasting man during 24 hr in which he expended a total of 5,090 kcal of energy by alternate periods of hard work (O_2 consumption, 1.75 L/min) and rest. Administration of adrenaline, as indicated, temporarily raised RQ, indicating an increase in carbohydrate utilization. (From Dill et al.[27])

from serious depletion of carbohydrate reserves.[9,43] In this case, it appears that fat cannot be mobilized or utilized rapidly enough so that the organism is forced to expend its carbohydrate reserves, and exhaustion is associated with a serious drop in blood sugar.[9,43]

Krogh and Lindhard,[70] Christensen,[18] Henschel et al.,[53] and Benedict and Cathcart[8] have all reported that the efficiency of work is greater when carbohydrate rather than fat is utilized as fuel. This is due only partly to the increased energy used in the oxidation of fat. Henschel et al.[53] suggest that in a fasting man the reduced efficiency of work performed with fat as a fuel may be due in part to reduction of blood sugar. The work of Krogh and Lindhard,[70] Christensen,[18] Karpovich,[61] Henderson and Haggard,[51] and others indicates that the capacity of men for strenuous work is greater on a high-carbohydrate than on a rich fat diet.

Under normal working conditions, men eat often enough, and they digest and assimilate most of their food during actual work periods and therefore never experience such depletions of carbohydrate stores as those just described.[50,117,118,124] On the other hand, in intensive work of short duration in which the energy is derived largely from the anaerobic mecha-

nisms, digestive functions are markedly reduced.[117]

RESPIRATION

When a person exercises, pulmonary ventilation (volume of expired air in liters per minute, V_E) is increased in proportion to the intensity of the work. This response is brought about by increasing both the tidal air and the frequency of respiration, as shown in Fig. 58-16 constructed from data obtained on a man who performed three different intensities of work on a treadmill. At submaximal work rates (running at 5.8 and 10 mph) the subject's ventilation was accelerated in the first 3 min and thereafter continued in a steady state, with the ventilatory rate being precisely adjusted to meet the O_2 requirement of the work and the need for CO_2 excretion. In work severe enough to exhaust the man in 5 min or less (running 13 mph up a 2% grade), his O_2 consumption reached a maximum at the end of the second minute when ventilation was about 100 L/min. As the work was continued through the third to fifth minutes, O_2 consumption remained at the maximum, and the ventilation continued to rise gradually due to the progressively increasing stimulus of anaerobic metabolites, principally lactic acid with increased [H] (Fig. 58-7). The

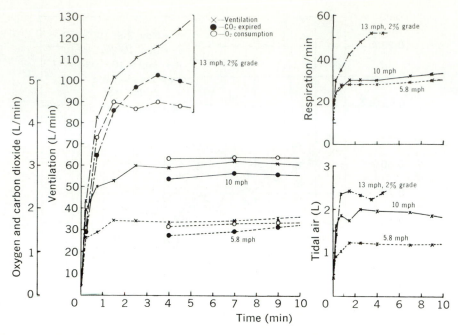

Fig. 58-16. Respiratory and metabolic responses of man during three different intensities of exercise on treadmill (gas volumes at STP). Run at 13 mph up 2% grade was exhausting; maximal O_2 consumption was reached in second minute, ventilation continued to increase throughout 5 min, and blood lactate level rose to 148 mg/100 ml at end. (Based on data from Robinson and Turrell.[99])

subject's blood lactate level had risen to 148 mg/100 ml at the end of the run. Thus for the subject illustrated in Fig. 58-16, a ventilation of 100 L/min was sufficient for maximal O_2 consumption, and, since ventilation later in the run rose far above this value, we must conclude that the pulmonary ventilation did not limit the maximal O_2 consumption of the man in this exhausting work.

The regulation of these respiratory adjustments in exercise is complex and not entirely understood. An analysis of this problem is presented in Chapters 72 to 74, and therefore it will be discussed only briefly here. The entire picture of the response does not depend on any one chemical stimulus acting via the arterial blood either directly on the respiratory center or through the chemoreceptors of the carotid and aortic bodies. For that matter, no combination of changes among three chemical stimuli to respiration (reduced O_2 tension, increased CO_2 tension, and increased hydrogen ion concentration of arterial blood) has been observed in exercise that could be responsible for the respiratory adjustments. Direct evidence bearing on the question is that when men perform moderate rates of aerobic work requiring ventilation rates up to 35 L/min,

there is little or no elevation of blood lactate level (Figs. 58-2 and 58-17), and arterial pH is essentially unchanged from rest[46,86]; arterial (alveolar) P_{CO_2} may rise in some men and fall in others, with the average man (Fig. 58-17) showing no significant change[46,86]; and alveolar P_{O_2} tends to rise (Fig. 58-17), with arterial P_{O_2} and O_2 saturation remaining about the same as at rest.[73] Thus in moderate activity, respiration is not driven by chemical changes in the arterial blood; rather it is regulated so as to minimize chemical change. In this range of work rates, ventilation is directly proportional to O_2 consumption and CO_2 production (Figs. 58-16 and 58-17).

In more severe work rates of from 5 to 10 min duration, in which lactic acid accumulates in the blood, the arterial pH may drop (Fig. 58-7), and ventilation is excited out of proportion to O_2 consumption (Figs. 58-16 and 58-17). The relative hyperventilation observed in severe work causes both alveolar and arterial P_{CO_2} to fall significantly below resting values and to a considerable extent compensates for the accumulated lactate (Fig. 58-17). This drop in P_{CO_2} plus the increase in lactic acid, which combines with base from the bicarbonate of the body fluids, results in the pulmonary release of CO_2 in excess of that being

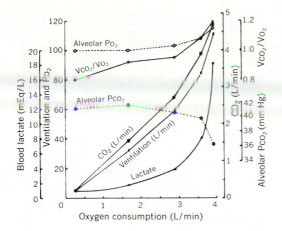

Fig. 58-17. Respiratory responses and blood lactate level in relation to O_2 consumption of men at rest and during four intensities of work on treadmill (3.5 mph, 9% grade, 15 min; 8 mph, level, 10 min; 7 mph, 9%, 5 min; and exhausting runs of 5 min duration). Values plotted represent averages of determinations on nine men. (Based on data from Robinson and Harmon.[96])

produced metabolically. Hence $\dot{V}_{CO_2}/\dot{V}_{O_2}$ may exceed unity during anaerobic work (Figs. 58-16 and 58-17).

There are other mechanisms that may participate in this regulation. There is good evidence of reflexes originating from receptors in the joints that might well contribute to respiratory control in direct proportion to the intensity of the muscular activity.[23,36,46] Krogh and Lindhard[69] suggested that irradiation of impulses over collateral fibers from the motor tracts might excite the respiratory center in direct proportion to the intensity of the work. Such mechanisms explain the instantaneous respiratory response at the start of exercise and the precipitous decline in the first few seconds of recovery, before chemical or temperature changes occur. The Hering-Breuer reflex originating in the stretch receptors of the lungs may contribute to ventilation in exercise by augmenting the force of the expiratory movements.

Although a rise in body temperature excites respiration, the changes in core temperature of the body that occur in exercise could contribute to respiratory regulation only in a secondary manner. The major respiratory responses to moderate exercise occur within 2 min (Fig. 58-16), and it requires a half hour or more for the rectal temperature to rise to the steady-state level characteristic of the work (Fig. 58-24).

It should be stressed that the regulation of respiration in exercise must depend not on any

single stimulus but on the combined effects of a number of the factors previously discussed and probably other factors that have not been identified. Excellent reviews and interpretations of the literature on these relationships have been written by Comroe,[22] Schmidt,[109,110] Gray,[44] Grodins,[46] and Lambertsen (Chapter 72).

The respiratory adjustments in exercise are affected by such factors as age, sex, barometric pressure, environmental temperature, and physical fitness. The lung volume, the maximal capacity for lung ventilation in exercise,[92] and the O_2-diffusing capacity of the lungs decline with age in men.[21] Older men, even at rest, may have moderately reduced O_2 saturation of the arterial blood as compared with young men.[92] This is probably due largely to uneven alveolar ventilation in older persons. Although the ventilatory capacity of adults in exercise declines with age, when normal individuals of all ages perform exhausting work of 5 to 10 min duration, they hyperventilate relative to the maximal O_2 consumption, indicating that normally ventilation is not a limiting factor in determining the maximal capacity for aerobic work. This relative hyperventilation of humans performing anaerobic work is shown for a young man in Fig. 58-16.

Hard physical training does not alter the pulmonary ventilation of humans in the resting state, but it produces profound changes in their respiratory adjustments in exercise.[66,96,111] The pulmonary ventilations of men in performing submaximal work (running at 8 mph) declined markedly as the work became easier for them during a period of vigorous training (Fig. 58-18). This adaptation is related directly to the fact that the elevation of blood lactate levels in the work experiments became progressively lower during the training period, thus reducing an important stimulus to respiration (Fig. 58-8). Improvement in the efficiency of the work, as indicated by reductions in both O_2 consumption (Fig. 58-3) and lactacid O_2 debt, also contributed to the decrease in ventilation of the men in running at 8 mph. The work rate (7 mph, 9% grade) that originally exhausted all the men in 5 min or less became progressively easier for them during the training period. As their fitness improved, there was a gradual reduction in the men's ventilation and in the concentration of lactate accumulated during repetitions of this run for 5 min. The training produced a substantial increase in the maximal ventilation of the men in exhausting work (Fig. 58-18). This increase in the capacity of the men to ventilate depended directly on the fact that their tolerance for lactic acid increased with

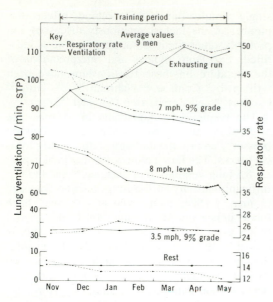

Fig. 58-18. Influence of strenuous athletic training on respiratory response of men at rest and during four intensities of exercise on treadmill. Conditions same as those in Fig. 58-3. (Based on data from Robinson and Harmon.[98])

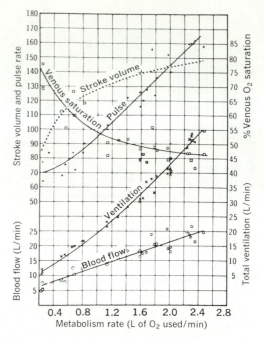

Fig. 58-19. Circulatory and respiratory changes of man in relation to increments in O_2 consumption produced in experiments at different intensities of work on bicycle ergometer. (Modified from Bock et al.[10])

the improvement in their ability to endure progressively harder bouts of work on the treadmill (Fig. 58-8). Other contributing factors include neuromuscular reflexes in relation to the increasing rates of work and improvement in the strength and endurance of the respiratory muscles. Neither the vital capacities nor the total lung volumes of the men increased during the 6-month training period.

CIRCULATORY ADJUSTMENTS IN EXERCISE

The basic function of the human circulation is to maintain a relatively constant environment for the body tissues. The actual exchange of materials such as nutrients, metabolites, respiratory gases, and hormones between the circulating blood and the tissues takes place through the thin-walled capillaries and the interstitial fluid. The larger blood vessels serve only to conduct the blood. The heart provides the energy and pumping action necessary to move the blood through the transportation system.

Transportation of O_2 and CO_2 by the blood is the immediate and principal function of the circulation in a person performing exercise under ordinary environmental conditions. This is true because there is practically no storage of O_2 in

the tissues and therefore the increased O_2 requirement of the muscles in work must be supplied currently by the circulation. The nutrients required are stored in the muscles in sufficient amounts to maintain the increased metabolism in exercise until reserves can be mobilized. Need for increasing the transport of metabolic heat to the surface is delayed for many minutes after the onset of work as the body accumulates heat and body temperature rises. When the core temperature of the body reaches a critical level, the demand for heat dissipation predominates, cutaneous vasodilatation occurs, and blood flow to the skin is greatly increased.

The increased O_2 supply to the muscles during work depends on (1) increasing the rate of blood flow through the muscles and (2) increasing the utilization of O_2 from each unit of blood as it passes through the muscles (Fig. 58-19). The increased volume of blood flowing through the working muscles depends on (1) increases in the heart's output of blood up to 5 or 6 times the output at rest, (2) capillary and arteriolar dilatation in the active muscles, lowering the resistance to flow therein, and (3) compensatory vasoconstriction in large inactive vascular beds such as the skin and viscera and the resultant shunting of the principal blood flow through the muscles. The utilization of O_2 from each unit of blood flowing through the working muscles may increase by a factor of 3 or more above the average resting utilization. Evidence of this is that the mean arteriovenous O_2 difference (A-V $O_2\Delta$) may be 2 or 3 times as great during strenuous exercise as at rest. The increased utilization of O_2 in work is a result of lowered O_2 tension in the active muscles and a shift of the O_2 dissociation curve of the blood to the right by increases of CO_2 tension, hydrogen ion concentration, and temperature in the tissues.

The heart output (minute volume) of men resting in the reclining position is about 6 L/min. With this value and an average heart rate of 65 beats/min the output per beat (stroke volume) would be 92 ml. In the standing position the heart rate is usually increased and the minute volume and stroke volume reduced because of the gravitational resistance to venous return to the heart from the legs and abdomen. At rates of exercise that can be performed in a steady state the minute volume of the heart increases in direct porportion to the O_2 consumption and thus also to the work rate. This relationship is well illustrated by the data in Fig. 58-19. When an individual performs heavy work that will lead to exhaustion in 5 to 20 min, both heart output and O_2 consumption are elevated to maximal values.

The increases in heart output in work depend on raising both heart rate and stroke volume, with heart rate increasing relatively more than stroke volume, particularly in untrained men. An average young man in severe work has a maximal heart output of about 23 L/min, heart rate 190, stroke volume 120 ml, and maximal O_2 consumption of about 3.5 L/min. The runner Lash, who could consume O_2 at 5.35 L/min during exhausting runs on the treadmill,[101] was found in later experiments, using the acetylene method, to have a maximal heart output of 32 L/min. In this experiment, his heart rate was 180, stroke volume 178, and A-V $O_2\Delta$ 165 ml/L of blood. The high A-V $O_2\Delta$ depended in part on a remarkably great capacity for hemoconcentration; O_2 capacity of his blood was 21.3 vol% in rest and 24.5 vol% in exhausting work. Assuming 93% O_2 saturation of the arterial blood in the run, his mixed venous blood would have been 24.5% saturated. Lash's O_2 consumption and heart output when at rest and in performing moderate work were about the same as those for other young men. In walking at 3.5 mph up a 9% grade, his heart rate was 100, minute volume 15 L, and stroke volume 150 ml as compared with an average college student with averages of 140,

15.2, and 109, respectively. The great difference in the capacities of these men for transporting and utilizing O_2 in severe work is responsible for the tremendous difference in the rates at which they could work—the athlete could run 2 miles in 8 min, 58 sec; the untrained man required 10 min to run 1½ miles. These differences are due in part to strenuous training and in part to inherent physiologic qualities. In long, sustained aerobic work of moderate intensity the efficiencies and endurance of these two men would be equal after a few weeks of work and training. The experiences of coaches show that no amount of training would develop in the average young man the physiologic capacities required to run 2 miles in 9 min.

The mechanism of bringing about an increase in stroke volume in exercise is a subject of considerable controversy among circulation physiologists at this time.* For 40 years, physiologists believed the increase to be largely the result of increased diastolic volume in exercise, acting according to the principle of Starling's law of the heart[89,116]; that is, within physiologic limits the force and amplitude of the heart's contraction are proportional to the diastolic volume. This principle was established on the heart-lung preparation in which the heart operates without its normal chemical and nervous regulation. Although the normal heart undoubtedly adheres to the principle, the extent to which normal cardiac function in exercise depends on it is debatable. X-ray evidence has been accumulating for more than 40 years that the diastolic volume of the heart does not increase significantly in exercise. Some workers have found a slight increase, others no change, and some even a slight decrease of diastolic volume in exercise.[106] This indicates that Starling's law does not make an important contribution to increasing mean stroke volume in exercise. Hamilton[47] believed that in the normal person the Starling phenomenon functions principally to keep the output of the two ventricles in pace with each other under conditions of changing venous return and arterial resistance.

The stroke volume of the heart may increase normally by more complete systolic emptying. A number of investigators[59,65,115] have found that in rest a rather large volume of blood remains in the heart at the end of systole. Holt[59] found in resting dogs that the systolic volume of blood in the heart averaged 55% of the diastolic volume. Thus, without increasing diastolic volume, stroke

---

*See references 2, 13, 38, 45, 47, and 107.

volume could be doubled by ejecting in systole 90% instead of 45% of its diastolic content. There are at least two important factors indicating that this must be the principal mechanism that operates normally to increase stroke volume in exercise: (1) increased force of ventricular contraction resulting from the stimulating effects of sympathetic nervous action and epinephrine and (2) a decreased resistance to the ejection of blood into the arterial system. Direct recordings of arterial pressure in men performing strenuous exercise show a substantial fall in diastolic pressure, a rise in systolic pressure, and usually a small reduction in mean pressure (Fig. 58-20). The reduction in diastolic arterial pressure under conditions of rapid heart rate in exercise indicates a large reduction in the arterial resistance to flow. The increase in systolic and pulse pressures that is associated with the reduced resistance indicates an increase in stroke volume, although the increase may be due in part to a more rapid ventricular contraction.

In the final analysis the heart output depends on the rate of venous return; the heart cannot put out more blood than it receives; normally it adjusts its output to the inflow so that blood is not dammed up in the venous system. Thus the venous return to the heart determines its output. This means that in exercise the venous return

may be as great as 32 L/min in an athlete, since this is the maximal heart output. In a resting man, such a high rate of venous inflow and heart output would be impossible except momentarily as a result of rapid vasomotor adjustments. The "effective filling pressure" determines the return of blood to the heart from the systemic circulation and is proportional to the rate of muscular activity. This relationship depends on two factors: (1) the massaging action on the veins in the rhythmically contracting skeletal muscles and (2) the mechanical pumping action of the respiratory movements on the venous cistern of chest and abdomen. Thus during the deep inspirations of exercise in which the intrathoracic pressure is more negative, the filling pressure of the right ventricle may be high without an absolute rise in auricular pressure.

The veins are thin walled and collapsible, and in the limbs they have valves that permit blood to flow only toward the heart. During muscular contraction the veins in the muscles are compressed and the contained blood is forced toward the heart. When the muscles relax, blood from the capillaries quickly fills the veins again, and in turn the next contraction empties them. Thus in exercises such as walking or running, in which the muscles alternately contract and relax in a rapid rhythm, they add a powerful pumping

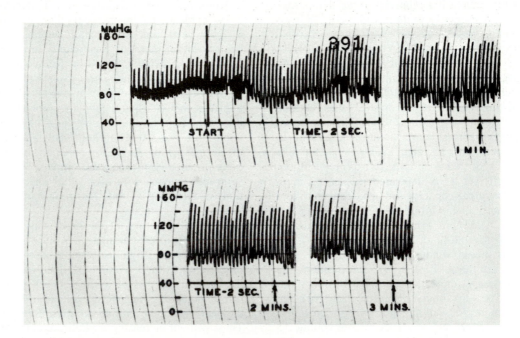

Fig. 58-20. Recording of arterial pressure of man before and during walk on treadmill at 5.6 km/hr up 9% grade; O_2 consumption 1.8 L/min. (Based on data from Robinson.[94])

action in promoting return to the heart. This effect is proportional to the rhythm and force of the movements and to the total mass of muscles entering into the activity. Powerful static contractions such as weight lifting may actually hinder venous return and heart output and therefore lead to rapid fatigue.

The respiratory movements promote venous return to the heart. The large veins of the thorax and abdomen constitute a considerable venous cistern. Blood can flow out of this cistern only into the heart because of the valves in the veins leading into the cistern from the limbs and neck. During inspiration a considerable negative pressure is created in the thoracic cavity; at the same time this tends to aspirate blood into the veins in the chest and thus into the heart during diastole. There is considerable pooling of blood in the pulmonary circulation during inspiration. The lowering of the diaphragm in inspiration tends to increase intra-abdominal pressure and empty the abdominal veins toward the heart. During expiration the pressure effects are reversed and the venous cistern tends to fill again with blood from the working limbs. This respiratory pumping action on venous return to the heart increases with the depth and rate of respiration in exercise.

It is obvious from the previous statements that changes in heart rate contribute greatly to the changes in heart output that occur in exercise (Fig. 58-21). The resting heart rate of human beings is affected by many factors, including age, sex, posture, emotional excitement, food intake, temperature, and physical fitness. The basal heart rate declines with age from 130 at birth to one half of this at maturity. The maximal heart rate declines with age from 200 in youth to 150 at age 70 years.[92] The basal rate, the steady-state rate of moderate work, and the maximal heart rate are all reduced by strenuous athletic training (Fig. 58-22). The mechanism of these reductions of heart rate in training is not known, but they are accompanied by improvement in the strength and efficiency of the myocardium and corresponding increases in stroke volume. An increase in vagal cardioinhibitory tone has been suggested as a possible factor in reducing the basal heart rate in training. However, this is not responsible for the reduction in maximal heart rate in training, since blocking the vagi with atropine does not increase the maximal rate.[103] This indicates that in strenuous work the heart is normally released entirely from vagal inhibition.

The acceleration of the heart in exercise is illustrated in Fig. 58-21. There is often an anticipatory acceleration just before work due to nervous influences from the cerebral cortex acting on the cardiac centers in the medulla. A rapid

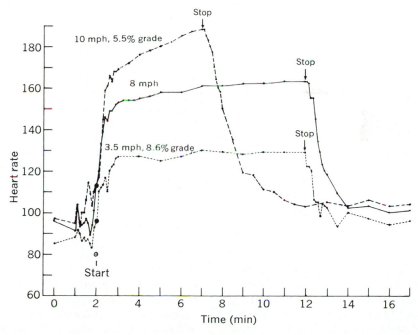

Fig. 58-21. Heart rate of man in response to three different intensities of exercise on treadmill. Run at 10 mph up 5.5% grade exhausted subject in 5 min, and his heart rate rose to 191 at the end. (Based on data from Robinson and Turrell.[100])

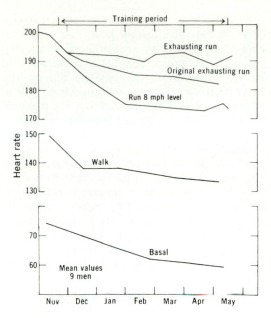

Fig. 58-22. Influence of strenuous athletic training on heart rates of men in basal state and in response to four different intensities of work. Conditions of experiments same as those in Fig. 58-3. (Based on data from Harmon and Robinson.[49])

acceleration during the first 30 sec of work occurs as the heart is reflexly released from the tonic cardioinhibitory restraint that is characteristic of the resting state.[37a] Evidence of continuous cardioinhibitory tone in rest is that when the vagus endings are blocked with atropine, a resting man's heart rate increases about 40 beats/min.[103] The initial rise in exercise is due principally to a reduction in diastasis.[11,69] Diastole of the first cardiac cycle at the start of work is shortened, indicating that the effect is of nervous rather than humoral origin. The suddenness of this response indicates that it must be due to inhibition of the cardioinhibitory center by impulses irradiating from the motor tracts and/or proprioceptor impulses from the muscles and joints. In light work the heart rate may decline somewhat after the first minute and attain a steady state at a level proportional to the work rate and O_2 consumption. In harder work the rate usually drops slightly after the rapid acceleration in the first 30 or 40 sec and then begins a more gradual rise to a higher steady-state level that is also proportional to the work rate. In exhausting work of about 5 min the initial rapid rise is followed by continued acceleration, and the rate reaches a maximum of about 190 as the man approaches exhaustion. Further acceleration of the heart and the

continuance of a high rate after the initial rise in exercise depend on increases of cardioaccelerator tone and adrenal activity. Thus the acceleration of the heart during exercise takes place in two stages: an initial rise due largely to reflex release from cardioinhibitory tone and the delayed rise and maintenance of a high rate due primarily to increased accelerator tone. In severe anaerobic work, changes in arterial pH and Po_2 may contribute to the delayed phase of cardiac acceleration by reflexes initiated from the chemoreceptors of carotid and aortic bodies. In prolonged work a rise in body temperature may contribute secondarily to cardiac acceleration. Humoral influences from the thyroid gland and liver are known to affect heart rate, but their significance in exercise is not known.

The nervous factors in the regulation of heart rate in exercise depend on afferent impulses from a number of sources acting reciprocally through the cardiac centers: (1) pressor reflexes originating in the working muscles and joints and acting in proportion to the rate of work, (2) acceleration due to nervous influences from the higher centers and possible irradiation from the motor tracts, (3) pressor reflexes from the chemoreceptors of the carotid and aortic bodies that may exert a delayed effect in anaerobic work, and (4) depressor reflexes from the pressure receptors in the carotid sinus and aortic arch. Reflexes from pressure receptors in the great veins and right auricle, the Bainbridge reflex, are of doubtful significance in exercise.

Recovery is characterized by a precipitous drop in heart rate, followed by a more gradual decline to the prework level (Fig. 58-21). This early drop-off after work is due largely to the sudden cessation of accelerator reflexes. The time required for the heart to complete its return to the prework level is only a few minutes after light to moderate work performed in a comfortable environment. After exhausting work, 2 or 3 hr of rest may be required for complete recovery. Recovery is delayed in proportion to the degree of fatigue developed during the work, and this is affected by the rate and duration of the work and the physical fitness of the person for the work. Physiologically the recovery involves instantaneous cessation of the reflexes associated with neuromuscular work, a more gradual removal of epinephrine, the slow removal of lactic acid and excess body heat accumulated during the work, and vasomotor adjustments to compensate for the sudden drop in heart output and the greatly dilated vascular beds in the muscles at the cessation of work.

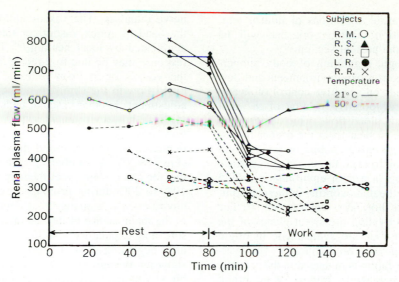

Fig. 58-23. Effects of exercise and environmental heat stress on effective renal plasma flow in five subjects. Work included walking at 5.6 km/hr up 5% grade. (From Radigan and Robinson.[91])

Peripheral vascular adjustments are as important in increasing blood flow to the muscles in exercise as the changes in cardiac output. An analysis of the mechanisms of these changes is given in Chapter 41; only a summary statement will be given here. In severe work, vasodilatation in the active muscles themselves involves up to a 100-fold increase in the number of open capillaries between resting and working muscle.[68,81] Associated with this great increase in surface area available for respiratory exchange is a reduction in the average distance that the materials have to diffuse between capillaries and the muscle fibers from 50 μm in rest to 1 to 5 μm in exercise. This tremendous opening up of the vascular beds of the muscles is made possible by compensatory vasoconstriction in the skin and abdominal viscera. Direct evidence that visceral vasoconstriction occurs in men during work is provided by the fact that renal plasma flow is significantly reduced during exercise (Fig. 58-23).[17,91] The vascular dilatation in the working muscles is much greater than the compensatory constriction in other vascular beds. The net effect of these integrated vascular adjustments in exercise is a marked reduction in the overall arterial resistance to flow. Evidence for this is found by direct recordings of arterial pressure of men in exercise that show a significant fall in diastolic pressure (Fig. 55-20). The reduction in diastolic pressure in exercise occurs even though heart rate and heart output are both greatly increased. This indicates a lowered peripheral resistance. The lowered resistance makes the high minute output of the heart in exercise possible without heart strain. It has been estimated that the resistance in the muscle circuits is reduced from 4.9 resistance units in rest to 0.31 unit in exercise. The rapid drop in diastolic arterial pressure at the beginning of exercise indicates that the initial vascular dilatation in the muscles is probably of nervous origin, but undoubtedly metabolites (CO_2, lactic acid, adenylic acid, etc.) and elevations of temperature enter into the adjustments as work continues. The vasodilator fibers to the muscles are of sympathetic origin but cholinergic in their action, and compensatory vasoconstriction in skin and viscera is by sympathetic fibers acting adrenergically.[15] These responses are maximal in exhausting work of short duration, but in prolonged moderate work, more of the blood being circulated must be distributed to such organs as the skin, the gastrointestinal tract, and the liver, as discussed in the following section.

TEMPERATURE REGULATION IN EXERCISE

The basic mechanisms of physical heat exchange between a human and the environment and the physiologic regulation of body temperature are analyzed in Chapter 59. The special autonomic and vasomotor factors concerned with adjustments of blood flow to the skin in temperature regulation are also discussed in Chapter 59. The reader should review these sections in studying the special conditions of human temperature regulation during work that will be emphasized in the present section.

The physiologic regulation of body temperature in humans in warm environments and during work depends on sensitive control of sweating to provide evaporative cooling of the skin and on cutaneous blood flow, which determines the conductance of heat from the deeper tissues to the skin. Evaporative cooling may vary from insensible loss of 10 kcal/hr · m² of skin surface in a man resting in a cool environment to 375 kcal/hr · m² in the man working in thermal equilibrium with a hot, dry environment.[99] The motor nerves activating the sweat glands of the general body surface are sympathetic in origin and yet they function cholinergically. The overall control of sweating is managed in the preoptic and anterior hypothalamic regions of the brain; this center may be activated by afferent impulses from the thermal receptors in the skin and directly by temperature changes within the center itself (Chapter 59). The nature of this control and other factors affecting it in work will be discussed later.

Heat conductance to the skin, which is proportional to the rate of cutaneous blood flow, varies from a minimum of about 10 kcal/hr/ m²/°C difference between central body temperature and mean surface temperature in a man resting in a cold environment to 78 kcal when he is working in thermal equilibrium in a hot environment.[93] Blood flow through the fingers has been found to vary from 0.5 to 90 ml/100 gm tissue · min when a man changes from a cold to a hot environment.[113,127] The fingers, with their high ratio of surface area to mass, are especially adapted for heat exchange, and they are capable of making unusually great adjustments of blood flow. Hertzman[54] and Hertzman and Dillon[55] have found the increase in blood flow to different skin areas of the body in response to heat to vary in decreasing order as follows: finger pad, earlobe, hand, forehead, forearm, knee, and tibial area. These cutaneous vascular adjustments that determine heat transfers to the surface are regulated reflexly through the vasomotor and thermoregulatory centers. It is now known that maximal rates of blood flow and the vasomotor regulation of flow to the hands and feet are radically different from those of the forearms and legs. Flow to the hands and feet is increased in heat largely by graded release of these vascular beds from vasoconstrictor tone, whereas cutaneous flow to the forearm and legs in response to heat is regulated largely by active vasodilator nerve impulses. The vasodilatation produced in the hand by reflex vasomotor inhibition can be augmented by local heating.[105] The augmented response may be due to direct effects of heat on the vessels or to axon reflexes. Variations of blood flow to the skin depend on compensatory adjustments in other vascular beds. A man working in a hot environment requiring large blood flow to the skin and working muscles may have a reduction of 60% in blood flow to his kidneys (Fig. 58-23). A number of authors have found that the blood volumes of men are increased from 10% to 20% during prolonged exposures to hot environments, a change that contributes greatly to the maintenance of normal arterial blood pressure and circulatory stability.[5,7,24] Within 2 hr after beginning work in a hot environment, men show an average of 11% increase in circulating blood volume as compared with values observed when they are at rest in a cool room.[102]

The central or core temperature (usually measured as the rectal temperature) rises gradually during the first hour of work to a new level that is proportional to the rate of work. This is illustrated by the data of Nielsen[87] shown in Fig. 58-24, *A*, representing 1 hr experiments on a man performing different rates of work in a cool room (22° to 23° C). The rise of core temperature in work occurs even in cold environments in which heat dissipation is not a limiting factor. The man's thermostat, represented by the heat-regulatory center in the hypothalamus, is apparently set at a higher temperature in work, and it serves to maintain his core temperature at the higher level until the work stops (Fig. 58-24, *D*). Under ordinary environmental conditions a man's core temperature is more precisely regulated at the elevated level of steady-state work than is the core temperature of the same man at rest. During recovery after work the core temperature slowly returns to the resting level (Fig. 58-24, *D*). This adjustment of core temperature in the working man occurs independently of changes in the environmental temperature ranging from cold to moderately warm. This is illustrated in Fig. 58-25 by the elevations of rectal temperature of a working man during 90 min exposures to "effective temperatures" of 9.5° and 31.1° C. The effective temperature scale of Yaglou and Miller[128] is a physiologic temperature scale based on direct observations of the effects of variations of air temperature, humidity, and air movement on the comfort of human subjects. The effective temperature of an environment is the temperature of a still, saturated atmosphere that would produce the same physiologic effect and sensation

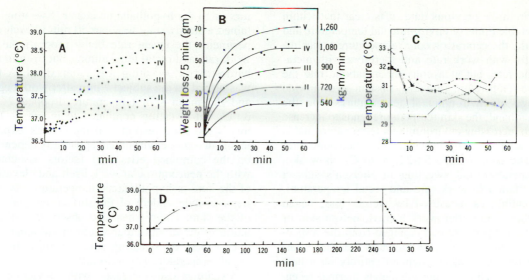

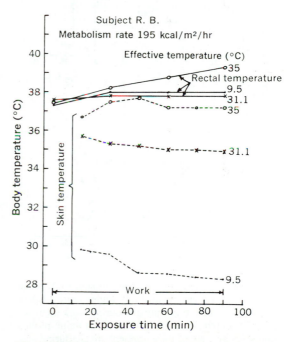

Fig. 58-24. A, Rectal temperature of man during 1 hr of work at each of five different rates: *I,* 360 kg-m/min; *II,* 540 kg-m/min; *III,* 900 kg-m/min; *IV,* 1,080 kg-m/min; *V,* 1,260 kg-m/min. Room temperature 22° to 23° C in all experiments. **B,** Weight loss of subject in five work experiments described in **A.** Weight loss indicates rates of sweating and evaporation. **C,** Mean skin temperature of man in four work experiments performed in cool room (23° C): -○-, 360 kg-m/min; -Δ-, 540 kg-m/min; -●-, 1,080 kg-m/min; -x-, 1,260 kg-m/min. Fan was turned on subject beginning at 7 min in each experiment. **D,** Rectal temperature during 250 min of work at 1,080 kg-m/min. (From Nielsen.[87])

of temperature as the environment being evaluated. In extremely hot environments the resistance to heat dissipation may be so great that core temperature will rise above the thermostatic temperature setting of the hypothalamic center. This is shown in Fig. 58-25 by the continued elevation of the working subject's rectal temperature during a 90 min exposure to an effective temperature of 35° C.

The average human skin temperature increases with increments in the environmental temperature (Fig. 58-25). This relation of skin temperature to environmental temperature is the result of three factors: (1) direct exchange of heat by radiation and convection between the skin and environment, (2) evaporative cooling of the skin, and (3) variations of cutaneous blood flow, which, when increased, tends to warm the skin. In a given environment, skin temperature may fall moderately during work due to increased sweating and evaporative cooling. This is illustrated in Fig. 58-25 for the subject in 90 min work experiments at effective temperatures of 9.5° and 31.1° C. This illustration shows that as the effective temperature of the environment increased, the skin temperature of the man increased, and the gradient between core and mean surface temperature decreased.

Fig. 58-25. Rectal and mean skin temperatures of man during exposures to three different effective temperatures. In each experiment he walked on treadmill for 90 min at 5.6 km/hr up 2.5% grade. (From Robinson.[95])

From the previous data, it is clear that within tolerable limits of environmental temperature and work, the central or core temperature varies directly with work rate and therefore with metabolic rate, but at a constant metabolic rate the core temperature remains constant over a wide range of environmental temperatures. On the other hand, the skin temperature tends to increase with increasing environmental temperature and is not altered greatly by changes in metabolic rate.

The data in Fig. 58-24, *A* to *C,* show skin temperature and sweating in Nielsen's subject in relation to work rate and rectal temperature. Sweating, as measured by weight loss, began within 5 min after work started, before skin or rectal temperature had changed from the respective resting levels. This indicates that the sweating response did not depend entirely on a heat stimulus, either to the cutaneous thermal recep-

tors or to the hypothalamic center. Sweating attained a steady state in about 30 min in each experiment, the sweat rate being proportional to work rate and rectal temperature, but even in the hardest work, sweating was submaximal. A fan was turned on the subject 7 min after work began and the mean skin temperature dropped 2° to 3° C, but sweating continued to increase. The man's sweating response in these work experiments in a cool environment apparently depended on the combined effects of factors associated with the neuromuscular work itself and elevation of the core or hypothalamic temperature. Sweating was independent of the surface temperature of the skin. Kuno[71] has also observed a stimulating effect of muscular work on sweating and a cessation of sweating after exercise before rectal temperature began to fall.

Sweating also increases with elevations of mean skin temperature and with reductions in the gradient between rectal and skin temperature under steady-state conditions in which rectal temperature and metabolic rate are constant.[95] This is also true of tissue heat conductance, which is directly proportional to the rate of blood flow to the skin. These relationships are shown in Figs. 58-26 and 58-27 in a series of 2 hr exposures of a

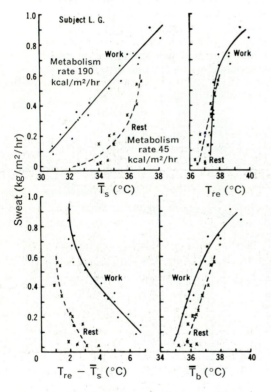

Fig. 58-26. Relations of sweating to skin ($\overline{T}_s$), rectal (T_{re}), $T_{re} - \overline{T}_s$, and mean body temperature ($\overline{T}_b$) of subject L. G. in resting state and during work on treadmill (5.6 km/hr up 2.5% grade). Each value plotted represents average of measurements made during second hour of 2 hr exposure to constant environment. Environment was varied (effective temperature, 15° to 35° C) from experiment to experiment to produce variations in body temperature and sweating. (From Robinson.[95])

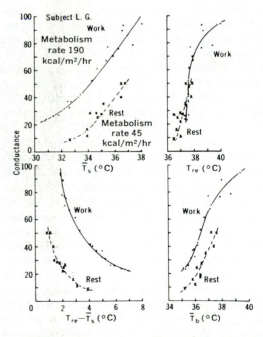

Fig. 58-27. Tissue heat conductance ($kcal/m^2/°C$ $T_{re} - \overline{T}_s/hr$) of subject L. G. in same series of experiments described in Fig. 58-26. Conductance was similar to sweating in its relations to each of the four measures of body temperature ($\overline{T}_s$, T_{re}, $T_{re} - \overline{T}_s$, and $\overline{T}_b$). (From Robinson.[95])

man to effective temperatures ranging from cool (15° C) to extremely hot (35° C). In the work experiments performed in different environments the subject's sweat rate varied from 0.2 to 0.7 kg/m² · hr, and the coefficient of heat conductance varied from 22 to 70, with no change in rectal temperature. In the same experiments there were close relations of sweating and conductance to average surface temperature ($\overline{T}_s$), mean body temperature ($\overline{T}_b$), and the gradient between central and surface temperature ($T_{re} - \overline{T}_s$). These facts indicate that the stimulus for the increments of sweating and conduction in the constant work experiments were of peripheral rather than central origin. The responses of the subject were much greater in work than in rest at any given thermal stimulus, as represented by $\overline{T}_s$, $\overline{T}_b$, or $T_{re} - \overline{T}_s$.

The evidence just presented serves to emphasize the complexity of the mechanism of eliciting the temperature-regulatory responses. The responses are not due entirely to reflexes originating in the thermoreceptors of the skin, because in a cool environment a working man will sweat profusely even with a cool skin (Figs. 58-24 and 58-26). Hardy,[48] in discussing temperature regulation of men in exercise, states that in work, reflex regulation originating from the cutaneous thermoreceptors is overridden by powerful effects from the hypothalamic centers. The late H. C. Bazett[6] believed that the elevation of central temperature in exercise increases the excitability of the center to sensory effects originating in the skin and proposed that thermal receptors located in the deeper cutaneous venous plexuses draining the muscles may serve to elicit the heat-regulatory responses during work when the temperature of the blood from the muscles is elevated. The sweating of a working man with his elevated core temperature may be due in part to the increased

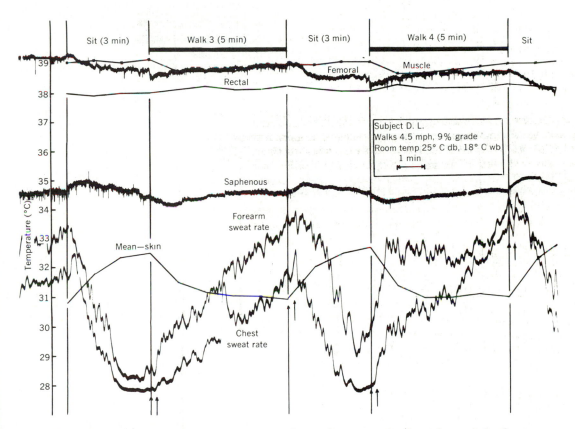

Fig. 58-28. Changes of sweating recorded by resistance hygrometry in alternating periods of rest and hard treadmill work. Double arrows show 12 sec lag time in hygrometer record of sweating. Note extremely rapid change of sweat rate with each start or cessation of work. Central temperatures (rectal, gastrocnemius muscle, and femoral vein blood) were elevated, and subject was sweating as result of previous periods of work. Skin and saphenous vein blood were relatively cool. (From Gisolfi and Robinson.[39])

temperature of the heat-sensitive hypothalamic center; but stimuli dependent on the neuromuscular activity must facilitate the response, since at any given mean body or skin temperature a man sweats much more and circulates more blood to the skin during work than in rest; after work, if the skin is cool, he ceases to sweat before his rectal temperature begins to fall. Keller[64] has also found evidence of a neuromuscular factor in temperature regulation of dogs in which panting was elicited by body movements in anticipation of increased heat production and elevated body temperature.

Further evidence that a neuromuscular factor participates in the regulation of sweating in exercise is provided in data by Van Beaumont and Bullard[123] and Gisolfi and Robinson.[39] These investigators have shown in men who are warm and already sweating at low rates in rest that the rate of sweating will begin to increase within 1 to 2 sec after the man begins vigorous work (Fig. 58-28). If the man is not excessively hot, his sweating will begin to decrease within 1 to 2 sec after the cessation of the work. Gisolfi and Robinson found that these rapid sweating responses associated with work occur on a background of a warm skin (35° to 36° C) with a cool core (T_{re} 37° C), on a background of an elevated core temperature (38° C) with a cool skin (31° to 32° C), or when both $\overline{T}_s$ and T_{re} are elevated. These abrupt responses of the sweat glands probably depend on proprioceptor reflexes, or possibly irradiation of impulses from the motor tracts in the brain stem, acting reflexly through the hypothalamic thermoregulatory center. They occur so rapidly that they precede temperature changes even in the working leg muscles (Fig. 58-28).

REFERENCES

1. Asmussen, E.: Aerobic recovery after anaerobiosis in rest and work, Acta Physiol. Scand. **11:**197, 1946.
2. Asmussen, E., and Nielsen, M.: Cardiac output during muscular work and its regulation, Physiol. Rev. **35:**778, 1955.
3. Asmussen, E., Christensen, E. H., and Nielsen, M.: Die O_2-Aufnahme der ruhenden und der arbeitenden Skelettmuskeln, Skand. Arch. Physiol. **82:**212, 1939.
4. Astrand, I.: Aerobic work capacity in men and women with special reference to age, Acta Physiol. Scand. **49:**suppl. 169, 1960.
5. Bass, D. E., and Henschel, A.: Responses of body fluid compartments to heat and cold, Physiol. Rev. **36:**128, 1956.
6. Bazett, H. C.: The regulation of body temperatures. In Newburgh, L. H., editor: Physiology of heat regulation and the science of clothing, Philadelphia, 1949, W. B. Saunders Co.
7. Bazett, H. C., Sunderman, F. W., and Scott, J. C.: Climatic effects on the volume and composition of blood in man, Am. J. Physiol. **129:**69, 1940.
8. Benedict, F. G., and Cathcart, E. P.: Muscular work, Publ. No. 187, 1913, Carnegie Institution of Washington.
9. Best, C. H., and Partridge, R. C.: Observations on Olympic athletes, Proc. R. Soc. Lond. (Biol.) **105:**323, 1929.
10. Bock, A. V., et al.: Studies in muscular activity, J. Physiol. **66:**136, 1928.
11. Bowen, W. P.: Changes in heart rate, blood pressure and duration of systole resulting from bicycling, Am. J. Physiol. **11:**59, 1904.
12. Brody, S., and Trowbridge, E. A.: Efficiency of horses, men and motors, Bull. No. 383, 1937, Missouri Agriculture Experiment Station.
13. Brouha, L., and Radford, E. P., Jr.: Cardiovascular system in muscular activity. In Johnson, W. R., editor: Science and medicine of exercise and sports, New York, 1960, Harper & Brothers.
14. Bullock, C. L.: Relations of O_2 requirement with speed of swimming in men of championship ability. M.A. dissertation, Bloomington, Ind., 1970, Indiana University.
15. Burn, J. H.: Sympathetic vasodilators, Physiol. Rev. **18:**137, 1938.
16. Cathcart, E. P.: The influence of muscle work on protein metabolism, Physiol. Rev. **5:**225, 1925.
17. Chapman, C. B., et al.: Effect of exercise on renal plasma flow in normal male subjects, J. Clin. Invest. **27:**639, 1948.
18. Christensen, E. H.: Beiträge zur Physiologie schwerer körperlicher Arbeit. V. Minuten Volume und Schlagvolumen des Herzens während schwerer körperlicher Arbeit, Arbeitsphysiol. **4:**470, 1931.
19. Christensen, E. H., and Hogberg, P.: The efficiency of anaerobical work, Arbeitsphysiol. **14:**249, 1950.
20. Christensen, E. H., and Hogberg, P.: The physiology of skiing, Arbeitsphysiol. **14:**292, 1950.
21. Cohn, J. E., et al.: Maximal diffusing capacity of the lung in normal male subjects of different ages, J. Appl. Physiol. **6:**588, 1954.
22. Comroe, J. H.: The hyperpnea of muscular exercise, Physiol. Rev. **24:**319, 1944.
23. Comroe, J. H., and Schmidt, C. F.: Reflexes from the limbs as a factor in the hyperpnea of muscular exercise, Am. J. Physiol. **138:**536, 1943.
24. Conley, C. L., and Nickerson, J. L.: Effects of temperature change on the water balance in man, Am. J. Physiol. **143:**373, 1945.
25. Dickinson, S.: Efficiency of bicycle pedalling as affected by speed and load, J. Physiol. **67:**242, 1929.
26. Dill, D. B.: The economy of exercise, Physiol. Rev. **16:**263, 1936.
27. Dill, D. B., Edwards, H. T., and de Meio, R. H.: Effects of adrenalin injection in moderate work, Am. J. Physiol. **3:**9, 1935.
28. Dill, D. B., Jones, B. F., and Edwards, H. T.: Les combustibles de l'activité musculaire, Le Travail Humain **2:**1, 1934.
29. Durnin, J. V. G. A.: Oxygen consumption, energy expenditure, and efficiency of climbing with loads at low altitudes, J. Physiol. **128:**294, 1955.
30. Eggleton, P., and Eggleton, G. P.: The inorganic phosphate and a labile form of organic phosphate in the gastrocnemius of the frog, Biochem. J. **21:**190, 1927.
31. Erickson, L., et al.: The energy cost of horizontal and

grade walking on the motor driven treadmill, Am. J. Physiol. **145:**391, 1946.

32. Fenn, W. O.: Work against gravity and work due to velocity in running, Am. J. Physiol. **93:**433, 1930.

33. Fick, A., and Wislicenus, J.: On the origin of muscular power, Philos. Mag. **31:**485, 1866.

34. Fiske, C. H., and Subbarow, Y.: The nature of the inorganic phosphate in voluntary muscle, Science **65:**401, 1927.

35. Furusawa, K., et al.: Muscular exercise, lactic acid, and the supply and utilization of oxygen, Proc. R. Soc. Lond. (Biol.) **97:**155, 1924.

36. Gardner, E., and Jacobs, J.: Joint reflexes in the regulation of respiration in exercise, Am. J. Physiol. **153:**567, 1948.

37. Gasser, H. S.: Contractures of skeletal muscle, Physiol. Rev. **10:**35, 1930.

37a. Gasser, H. S., and Meek, W. J.: A study of the mechanism by which muscular exercise produces acceleration of the heart, Am. J. Physiol. **35:**48, 1914.

38. Gauer, O. H.: Volume changes of the left ventricle during blood pooling and exercise in the intact animal, Physiol. Rev. **35:**143, 1955.

39. Gisolfi, C., and Robinson, S.: Central and peripheral stimuli regulating sweating in intermittent work in men, J. Appl. Physiol. **29:**761, 1970.

40. Gläser, H.: Untersuchungen über die Seklagarbeit mit Hämmern oder Axten, Arbeitsphysiol. **14:**448, 1952.

41. Goff, L. G., Bruback, H. F., and Specht, H.: Measurements of respiratory responses and work efficiency of underwater swimmers utilizing improved instrumentation, J. Appl. Physiol. **10:**197, 1957.

42. Goff, L. G., et al.: Work efficiency and respiratory response of trained underwater swimmers, J. Appl. Physiol. **10:**376, 1957.

43. Gordon, B., et al.: Sugar content of the blood in runners following a marathon race, J.A.M.A. **85:**508, 1925.

44. Gray, J. S.: The multiple factory theory of the control of respiratory ventilation, Science **103:**739, 1946.

45. Gregg, D. E., Sabiston, D. C., and Theilen, E. O.: Changes in left ventricular end-diastolic pressure and stroke work during infusion and following exercise, Physiol. Rev. **35:**130, 1955.

46. Grodins, F. S.: Regulation of breathing in exercise, Physiol. Rev. **30:**220, 1950.

47. Hamilton, W. F.: Role of the Starling concept in regulation of normal circulation, Physiol. Rev. **35:**161, 1955.

48. Hardy, J. D.: Control of heat loss and heat production in physiological temperature regulation, Harvey Lect. **49:**242, 1954.

49. Harmon, P. M., and Robinson, S.: Unpublished data, 1941.

50. Hellebrandt, F. A., and Hoopes, S. L.: Studies on the influence of exercise on the digestive work of the stomach. I. Its effect on the secretory cycle, Am. J. Physiol. **107:**348, 1934.

51. Henderson, Y., and Haggard, H. W.: The maximum power and its fuel, Am. J. Physiol. **72:**264, 1925.

52. Henschel, A., Taylor, H. L., and Keys, A.: Experimental malaria in man; physical deterioration and recovery, J. Clin. Invest. **29:**52, 1950.

53. Henschel, A., Taylor, H. L., and Keys, A.: Performance capacity in acute starvation with hard work, J. Appl. Physiol. **6:**624, 1954.

54. Hertzman, A. B.: The blood supply of various skin areas as estimated by the photoelectric plethysmograph, Am. J. Physiol. **124:**328, 1938.

55. Hertzman, A. B., and Dillon, J. B.: Selective vascular reaction patterns in nasal septum and skin of the extremities and head, Am. J. Physiol. **127:**671, 1939.

56. Hill, A. V.: Muscular movement in man; the factors governing speed and recovery from fatigue, New York, 1927, McGraw-Hill Book Co.

57. Hill, A. V.: The dynamic constants of human muscle, Proc. R. Soc. Lond. (Biol.) **128:**263, 1940.

58. Hill, A. V., Long, C. N. H., and Lupton, H.: Muscular exercise, lactic acid and the supply and utilization of oxygen, Proc. R. Soc. Lond. (Biol.) **96:**438, 1924.

59. Holt, J. P.: Estimation of the residual volume of the ventricle of the dog's heart by two indicator dilution techniques, Circ. Res. **4:**187, 1956.

60. Huckabee, W. A.: Relationship of pyruvate and lactate during anaerobic metabolism. II. Exercise and formation of O_2 debt, J. Clin. Invest. **37:**255, 1958.

61. Karpovich, P. R.: Ergogenic acids in work and sport, Res. Q. **12:**432, 1941.

62. Karpovich, P. V., and Millman, N.: Energy expenditure in swimming, Am. J. Physiol. **142:**140, 1944.

63. Karpovich, P. V., and Pestrecov, K.: Mechanical work done and efficiency in swimming the crawl and back strokes, Arbeitsphysiol. **10:**504, 1939.

64. Keller, A. D.: The role of the circulation in the physiology of heat regulation, Physiotherapy Rev. **30:**1, 1950.

65. Kjellberg, S. R., Rudhe, U., and Sjöstrand, T.: The amount of hemoglobin (blood volume) in relation to the pulse rate and heart volume during work, Acta Physiol. Scand. **19:**152, 1950.

66. Knehr, C. A., Dill, D. B., and Neufeld, W.: Training and its effects on man at rest and work, Am. J. Physiol. **136:**148, 1942.

67. Knuttgen, H. G.: Oxygen debt, lactate pyruvate and "excess lactate" after muscular work, J. Appl. Physiol. **17:**639, 1962.

68. Krogh, A.: The anatomy and physiology of the capillaries, New Haven, Conn., 1929, Yale University Press.

69. Krogh, A., and Lindhard, J.: The regulation of respiration and circulation during the initial stages of exercise, Skand. Arch. Physiol. **47:**112, 1913.

70. Krogh, A., and Lindhard, J.: The relative value of fats and carbohydrates as source of muscular energy, Biochem. J. **14:**290, 1920.

71. Kuno, Y.: Human perspiration, Springfield, Ill., 1956, Charles C Thomas, Publisher.

72. Lehmann, G.: Praktische Arbeitsphysiologie, Stuttgart, Germany, 1953, Georg Thieme.

73. Lilienthal, J. L., Jr., et al.: An experimental analysis in man of the oxygen pressure gradient from alveolar air to arterial blood during rest and exercise at sea level and at altitude, Am. J. Physiol. **147:**199, 1946.

74. Lohmann, K.: Konstitution der Adenylpyrophosphorsäure und Adenosindiphosphorsäure, Biochem. Z. **282:**120, 1935.

75. Lundgren, N. P. V.: Physiological effects of time schedule work on lumber-workers, Acta Physiol. Scand. **13**(suppl. 41):entire issue, 1946.

76. Lundsgaard, E.: Weitere Untersuchungen über Muskelkontraktionen ohne Milchsäurebildung, Biochem. Z. **227:**51, 1930.

77. Margaria, R.: Sulla fisiologia, e specialmente sul consumo energetico della Marcia e della corsa a varie velocita ed inclinazioni del terreno, Atti dei Lencei **7:**299, 1938.

78. Margaria, R., Edwards, H. T., and Dill, D. B.: The possible mechanisms of contracting and paying the

oxygen debt and the role of lactic acid in muscular contraction, Am. J. Physiol. **106**:689, 1933.

79. Margaria, R., et al.: Energy cost of running, J. Appl. Physiol. **18**:367, 1963.

80. Margaria, R., et al.: Kinetics and mechanism of oxygen debt contraction in man, J. Appl. Physiol. **18**:371, 1963.

81. Martin, E. G., Wooley, E. C., and Miller, M.: Capillary counts in resting and active muscles, Am. J. Physiol. **100**:407, 1932.

82. Mitchell, J. H., Sproule, B. J., and Chapman, C. B.: The physiological meaning of the maximal oxygen intake test, J. Clin. Invest. **37**:538, 1958.

83. Müller, E. A.: Physiological basis of rest pauses in heavy work, Q. J. Exp. Physiol. **38**:205, 1953.

84. Nelson, C.: The A.A.U. championships, Track and Field News **20**:12, 1967.

85. Newman, E. V.: Distribution of lactic acid between blood and muscle of rats, Am. J. Physiol. **122**:359, 1938.

86. Nielsen, M.: Untersuchungen über Atemregulation beim Menschen, Skand. Arch. Physiol. **74**(suppl. 10): 87, 1936.

87. Nielsen, M.: Die Regulation der Körpertemperatur bei Muskelarbeit, Skand. Arch. Physiol. **79**:193, 1938.

88. Passmore, R., and Durnin, J. V. G. A.: Human energy expenditure, Physiol. Rev. **35**:808, 1955.

89. Patterson, S. W., Piper, N., and Starling, E. H.: The regulation of the heart beat, J. Physiol. **48**:465, 1914.

90. Pugh, L. G. C., et al.: A physiological study of channel swimmers, Clin. Sci. **19**:257, 1960.

91. Radigan, L. R., and Robinson, S.: Effects of environmental heat stress and exercise on renal blood flow and filtration rate, J. Appl. Physiol. **2**:185, 1949.

92. Robinson, S.: Experimental studies of physical fitness in relation to age, Arbeitsphysiol. **10**:251, 1938.

93. Robinson, S.: Physiological adjustments to heat. In Newburgh, L. H., editor: Physiology of heat regulation and the science of clothing, Philadelphia, 1949, W. B. Saunders Co.

94. Robinson, S.: Unpublished data, 1960.

95. Robinson, S.: Temperature regulation in exercise, Pediatrics **32**:691, 1963.

96. Robinson, S., and Harmon, P. M.: The lactic acid mechanism in relation to training, Am. J. Physiol. **132**:757, 1941.

97. Robinson, S., and Harmon, P. M.: The effects of training and of gelatin upon certain factors which limit muscular work, Am. J. Physiol. **133**:161, 1941.

98. Robinson, S., and Harmon, P. M.: Unpublished data, 1941.

99. Robinson, S., and Turrell, E. S.: Unpublished data, 1939.

100. Robinson, S., and Turrell, E. S.: Unpublished data, 1941.

101. Robinson, S., Edwards, H. T., and Dill, D. B.: New records in human power, Science **85**:409, 1937.

102. Robinson, S., Kincaid, R. K., and Rhamy, R. K.: Effects of desoxycorticosterone acetate on acclimatization of men to heat, J. Appl. Physiol. **2**:399, 1950.

103. Robinson, S., et al.: Effects of atropine on heart rate and oxygen intake in working man, J. Appl. Physiol. **5**:508, 1953.

104. Robinson, S., et al.: Fatigue and efficiency of men during exhausting runs, J. Appl. Physiol. **12**:197, 1958.

105. Roddie, I. C., and Shepherd, J. T.: Blood flow through the hand during local heating, release of sympathetic vasomotor tone by indirect heating and a combination of both, J. Physiol. **131**:657, 1956.

106. Rushmer, R. F., and Smith, O. A.: Cardiac control, Physiol. Rev. **39**:41, 1959.

107. Sacks, J., and Sacks, W. C.: Blood and muscle lactic acid in the steady state, Am. J. Physiol. **118**:697, 1937.

108. Sargent, R. M.: The relation between oxygen requirement and speed in running, Proc. R. Soc. Lond. (Biol.) **100**:10, 1926.

109. Schmidt, C. F.: The revolution in respiratory physiology, Anesthesiology **5**:77, 1944.

110. Schmidt, C. F.: Respiration, Annu. Rev. Physiol. **7**: 23, 1945.

111. Schneider, E. C., and Ring, G. C.: The influence of a moderate amount of physical training on the respiratory exchange and breathing during physical exercise, Am. J. Physiol. **91**:103, 1929.

112. Schneider, E. G., Robinson, S., and Newton, J.: The oxygen debt in aerobic work, Physiologist **7**:247, 1964.

113. Scott, J. C., Bazett, H. C., and Mackie, L. C.: Climatic effects on cardiac output and the circulation in man, Am. J. Physiol. **129**:102, 1940.

114. Simonson, E., and Sirkina, G.: Wirkungsgrad und Arbeitsmaximum, Arbeitsphysiol. **7**:457, 1934.

115. Sjöstrand, T.: The volume and distribution of blood and their significance in regulating the circulation, Physiol. Rev. **33**:202, 1953.

116. Starling, E. H.: Linacre lecture on the law of the heart, London, 1918, Longmans, Green & Co.

117. Stickney, J. C., and Van Liere, E. J.: The effects of exercise upon the function of the gastrointestinal tract. In Johnson, W. R., editor: Science and medicine of exercise and sports, New York, 1960, Harper & Row, Publishers.

118. Stickney, J. C., Northrup, D. W., and Van Liere, E. J.: Resistance of the small intestine (motility) against stress, J. Appl. Physiol. **9**:484, 1956.

119. Taylor, H. L., Buskirk, E., and Henschel, A.: Maximal oxygen intake as an objective measure of cardiorespiratory performance, J. Appl. Physiol. **8**:73, 1955.

120. Taylor, H. L., et al.: The effects of bed rest on cardiovascular function and work performance, J. Appl. Physiol. **2**:223, 1949.

121. Turrell, E. S., and Robinson, S.: Acid-base equilibrium of blood in exercise, Am. J. Physiol. **137**:742, 1942.

122. Turrell, E. S., and Robinson, S.: Report No. 3 C.M.R. of the O.S.R.D., 1943.

123. Van Beaumont, W., and Bullard, R. W.: Sweating: its rapid response to muscular work, Science **141**:643, 1963.

124. Van Liere, E. J., Hess, H. H., and Edwards, J. E.: Effect of physical training on the propulsive motility of the small intestine, J. Appl. Physiol. **7**:186, 1954.

125. Weiss, S., and Ellis, L. B.: Oxygen utilization and lactic acid production in the extremities during rest and exercise, Arch. Intern. Med. **55**:665, 1935.

126. Whipp, J. B., Seard, C., and Wasserman, K.: O_2 deficit–O_2 debt relationships and efficiency of anaerobic work, J. Appl. Physiol. **28**:452, 1970.

127. Wilkins, R. W., Doupe, J., and Newman, H. W.: The rate of blood flow in normal fingers, Clin. Sci. **3**:403, 1938.

128. Yaglou, C. P., and Miller, W. E.: Equivalent conditions of temperature, humidity and air movements, J. Am. Soc. Heat. Vent. Eng. **31**:59, 1925.

59

JAMES D. HARDY

Body temperature regulation

GENERAL EFFECTS OF TEMPERATURE

All living forms are limited by their abilities to survive extremes of temperature; in general, the more complex the organisms, the more severe the limitations.[4] Under conditions that prevent dehydration and excessive crystal formation, single cells may be cooled to temperatures near absolute zero ($-273°$ C) and maintain their viability for some time. However, all metabolic activities are completely inhibited at these temperatures. Multicellular organisms are usually unable to endure sustained subfreezing temperatures because of the large crystal formation that injures cell membranes and the associated dehydration, both of which accompany the slow transmission of temperature changes through tissues. For most endotherms, or warm-blooded organisms, low temperatures, even above freezing, will interfere with metabolic processes and thus result in injury or death. The effect of temperature on rates of chemical reactions underlying much biologic activity can be expressed in the form of an equation first suggested by Arrehnius (1889):

$$k = Ce^{-\Delta E/RT} \qquad (1)$$

where k is the reaction rate, C is a constant, e is the base of naperian logarithms (2.718), ΔE is the activation energy of the reaction, R is the gas constant, and T is the absolute temperature. This equation has been found to apply to contraction rates of smooth muscle, respiratory movements, etc. as they are affected by temperature. A convenient form of the equation can be obtained by selecting two temperatures $10°$ C apart and determining the activity of the organism at these temperatures. The ratio of these activities[4] is called the "Q_{10}" or

$$Q_{10} = \left(\frac{k_1}{k_2}\right)^{\frac{10}{T_1 - T_2}} \qquad (2)$$

The Q_{10} for the human metabolic rate has been estimated for patients with fever to be about 2.5; for many biologic activities the Q_{10} lies between 2 and 3. Since equation 1 indicates that the dependence of k on temperature is determined to a great extent by the value of ΔE, and since ΔE for many enzymatic reactions lies between 5,000 and 20,000 calories/mole, giving a Q_{10} near 2, it is not surprising that the biologists consider a Q_{10} of 2 to 3 as "normal." Thus if the average body temperature of an individual is lowered by $10°$ C, the metabolic rate would be reduced by a factor of 2.5. Advantage is taken of this fact by inducing local or general hypothermia in some types of surgery when the blood flow must be occluded; lowering the metabolic needs of the tissues by hypothermia tends to prevent excessive hypoxia.

At high tissue temperature, danger arises from another area. Some cellular proteins are inactivated rapidly with a Q_{10} that is greater than 2.5 (about 200) because ΔE in this case equals 100,000 to 200,000 calories/mole.[7] Thus, as the temperature is increased, tissues are destroyed by heat far more rapidly than they can be repaired by metabolic activity, and tissue damage or burns result.

Although it is true that human tissues are burned at rather low temperatures ($45°$ to $50°$ C), other forms with more thermally resistant tissues can live and develop in hot springs near $100°$ C. However, even for thermally stable organisms continued existence is impossible at extremely high or low temperatures; these forms of life are also temperature limited.

1417

BODY TEMPERATURES OF MAN AND THEIR NORMAL VARIATIONS
Internal temperatures

Most tissues and organs of the human body function best when they are maintained at a relatively constant temperature near 37° C. Significant departures from this "normal range" of tissue temperatures are associated with illness and death.[25] The temperatures of the peripheral tissues (e.g., skin, muscles, and subcutaneous tissues) are generally cooler than the visceral temperatures and are subject to much wider fluctuations. Skin temperature, for example, may fluctuate between 20° and 40° C without damage. However, prolonged exposures to cold or hot environments, causing local skin temperatures as low as 18° C or as high as 45° C, are usually associated with both pain and tissue injury.

For clinical or experimental purposes the temperatures most often recorded are those of the rectum, mouth, and axilla.[25] The rectal temperatures are usually about 0.65° C (1.2° F) higher than those of the mouth or of a well-closed axilla, but, of course, orally (sublingually) recorded temperatures will be too low if the subject talks or breathes through the mouth or if the thermometer is not in the proper place for a sufficient length of time.[5] Even in the rectum there are variations of 0.1° to 0.9° C, depending on the position of the recording instrument.[37] The lowest rectal temperatures are found in those parts of the rectal wall closest to the veins carrying blood from the buttocks and legs. The temperature of blood flowing in the arteries situated on the pelvic wall can exert a similar effect if the legs are cold, for it has been demonstrated that during cooling of a part of the body, arterial blood is cooled as it passes close to veins draining blood from the cooled region. Therefore any rectal temperature must represent that of the deep abdominal and pelvic viscera modified by the temperature and volume flow of blood in adjacent veins and arteries. It can be said that in any internal organ or mass of tissue the temperature depends on (1) the metabolic activity of the region, (2) the temperature and amount of blood flowing through the area, and (3) the steepness of the temperature gradients to surrounding tissues.

Oral temperatures are the ones most often taken. The clinician or investigator who relies on them would do well to bear in mind the following statement by DuBois: "Oral temperatures, if taken with the proper precautions, are of great value and as a rule are the only ones needed except in the case of children or adults who are seriously ill. It must always be recognized that there are many parts of the body warmer and many parts cooler than the mouth."[25]

The old assumption that in homoiothermic animals the blood has a constant temperature has been shown to be entirely erroneous. Although many earlier studies had suggested the occurrence of considerable variations in the temperature of blood in peripheral vessels, the magnitude and cause of the changes that can and do occur in blood during transit through the arteries was first demonstrated by Bazett et al.[18] Needle thermocouples were introduced into the brachial and radial arteries, or thermocouples protected by a fine plastic tube and threaded through a needle were inserted far up the radial, brachial, and femoral arteries and into various veins of the subject (usually Professor Bazett himself). When the environment was thermally neutral (i.e., a comfortable room temperature), the temperatures were found to be 36° to 37° C in the brachial and 35.5° to 36.5° C in the radial artery. On exposure to cold these fell to 31.1° and 21.5° C, respectively, but "without the subject's being unduly cold, or the rectal temperature particularly low." This means that the blood leaving the left ventricle (at approximately 37° C) was cooled 15.5° C in its progress toward the hand! It was made clear that this lowering of the temperature of arterial blood by distal cooling is achieved by heat transfer from the artery to the much cooler blood flowing in adjacent veins. Much of the blood from the distal portions of the extremities is carried by the *venae comites,* which are closely applied to the arteries, an arrangement admirably suited for interchange of heat between the two streams. A significant feature of this interchange is the fact that it involves warming of the cooled venous blood; this heat is not lost to the environment but to a medium that returns it to the core of the body.

When temperatures are taken from the axilla, mouth, or rectum in a group or series of normal adult subjects, individual differences are invariably found. From both a clinical and an experimental point of view the question is, "What is the normal range and when is a temperature febrile or subnormal?" The search for an accurate answer to this question has demonstrated clearly that the little red arrow at 37° C (98.6° F) on clinical thermometers should be replaced by a fairly wide band denoting a range within which the temperatures of normal, healthy, resting adults may be recorded. DuBois[5] has listed the results of a number of studies on the range of normal human temperatures, and from this list it is apparent that the range is greater than is generally realized.

Fig. 59-1 indicates the ranges of internal body temperature at rest and in many conditions of life that are compatible with health. At rest during the day, internal temperature changes from a minimum 0.5° C below the mean occurring during the early morning hours to a maximum about 0.5° C above the mean level occurring in the afternoon (Fig. 59-2). This circadian rhythm is closely associated with the level of metabolic heat production but has a component that is "free running" and independent of activity.[13] The cause of the free-running rhythm is not understood but is suspected to be due to hormonal influences. The monthly temperature rhythm observed in nonpregnant women, a low temperature level after menses and a high temperature level after ovulation, has been correlated with shifts in the blood concentrations of estrogen and other hormones.

In Fig. 59-1, it is seen that temperatures above the normal range are potentially more dangerous than those below it, and that man lives only a few degrees below his death point. During vigorous exercise or work the internal body temperature may increase to levels of 39° to 40° C. It should be noted that this shift in temperature occurs even though the work may be done in cold surroundings and depends only on the level of activity as long as the thermoregulatory system can function effectively. Fig. 59-3 shows the linear relationship that exists between the metabolic rate and rectal temperature in young, healthy men; it is clear that these temperatures would be considered fever levels if the men were resting. Many children are continually active and their body temperatures will therefore appear elevated most of the time; only after a half hour of rest or sleep can their resting temperature be measured.

Skin temperatures

Although the skin temperatures over the body surface are highly variable as compared to the internal temperatures, both the mean skin temperature and the distribution of skin temperatures are important to health. For example, hot hands and cold feet do not combine their effects to pro-

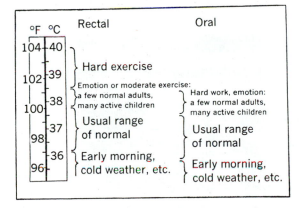

Fig. 59-1. Ranges in rectal and oral temperatures found in normal persons. (From DuBois.[5])

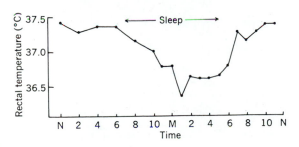

Fig. 59-2. Twenty-four–hour temperature cycle (circadian rhythm). Temperature chart of normal man on bed rest (average of 7 days).

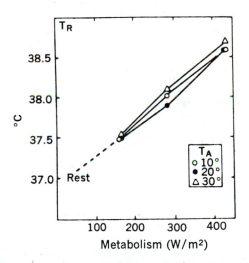

Fig. 59-3. Effect on body temperature of healthy young men working at different rates at 10°, 20°, and 30° C. Note that cold exposure has little effect on internal temperature, although subjects worked in shorts for 45 min. (From Stolwijk et al.[53])

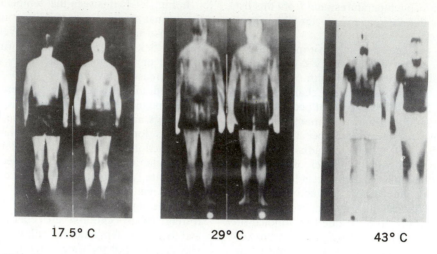

17.5° C   29° C   43° C

Fig. 59-4. Thermogram showing variations in skin temperature over body surface after 1 hr exposure in cold, neutral, and hot environments. Dark areas are relatively cool; light areas are warm. (From Stolwijk, quoted by Hardy.[8])

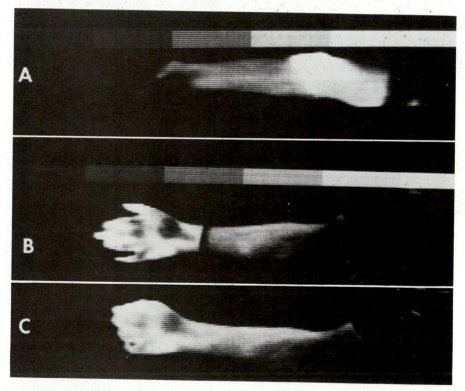

Fig. 59-5. Thermogram showing abnormal and normal blood flow. **A,** Inflammation due to bruise. **B,** Circulation to hand partially occluded. **C,** Normal circulation restored. (Modified from Barnes.[17])

duce a comfortable condition, and a hot flushed face is often symptomatic of disease.

Fig. 59-4 is an infrared thermogram of a man, a picture taken with the infrared radiation from the skin.[8] On exposure to cold (left panel) the hands, feet, ears, nose, and fat pads are cold, whereas the face, trunk, arms, and legs are relatively warm. In heat (right panel) the reverse is the case (note pattern of sweating on top of head and around neck serving to cool cerebral tissues). The normal, comfortable individual is shown in the center panel with indicated skin temperatures in the narrow range between 32° and 34° C. The thermogram images of the hands and feet indicate good circulation in that they are warmer than the arms and legs. Diseased states are quickly evident in alterations of peripheral circulation resulting in either hot spots or cold areas as shown in Fig. 59-5.

The human skin is exquisitely sensitive to its own temperature, being able to sense changes as small as 0.01° C in mean skin temperature. However, the temperature sensations adapt quickly if the changes are not great. Warmth will not adapt to extinction if the skin temperature of the arm, for example, is above 37° C. Temperature sensations have the property of spatial summation so that small areas of skin (1 cm²) are a thousand times less sensitive than the whole body surface.

REGULATION OF BODY TEMPERATURE

In view of the vulnerability of the tissues to temperatures that differ very much from 37° C, it is not surprising that the human has evolved an elaborate mechanism for regulating body temperatures. As stated by Claude Bernard in 1865, "La fixité du milieu intérieur est la condition de la vie libre." For example, the ability to preserve body temperatures near their normal levels enables man to brave the cold depths of outer space or explore the furnace-hot surface of the moon. To accomplish these regulations, two distinct control systems are employed—behavioral regulation, involving the conscious, voluntary use of all available means, and physiologic regulation, employing the involuntary responses of the body that tend to maintain constant temperatures. The relationships between these systems are indicated in Fig. 59-6 as they provide for the control of body temperatures in the environmental extremes of terrestrial and near-space environments. By means of protective clothing and air conditioning, man can live for extended periods in the coldest or hottest terrestrial climates and make excursions into space. Physiologic regulation in this context provides fine control of body temperature for the normal resting man and the principal control during exercise. Thus, although behavioral actions such as adjusting the thermostat of an efficient air-conditioning system may be the only means of providing a habitable thermal environment, these adjustments alone may not be necessarily effective in eliminating heat from the deep body structures of the exercising man.

Phylogenetically, behavioral regulation is the older system. Many so-called cold-blooded or exothermic species actually control body temperature within rough limits by moving their bodies into more favorable thermal environments. An example of such an animal is the rather large lizard *Tiliqua scincoides* (Australian blue-tongued skink), which has no significant physiologic regulation of body temperature. In an experiment in which the animal was forced to choose between a hot or cold environment

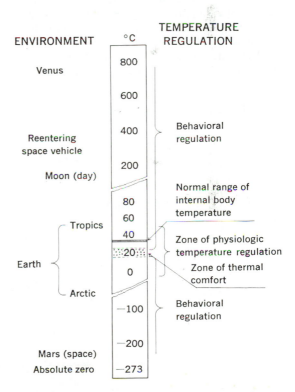

Fig. 59-6. Environmental temperatures on earth and in near space as related to physiologic and behavioral regulations of body temperature. (From Hardy.[8])

(45° and 15° C), the lizard moved back and forth between these environments and maintained its body temperature between 28° and 38° C (Fig. 59-7). Hammel et al.[31] also found that warming or cooling the hypothalamus in these animals caused marked disturbances in their responses to heat and cold. As will be noted later, such experiments imply that behavioral and physiologic temperature regulations have a common basis in neural activity. This impression is further strengthened by the discovery of neurons in the hypothalamus of *Tiliqua* that respond to temperature in a manner similar to neurons in the anterior hypothalamus of the dog, cat, and rabbit.[21] Generally it can be thought that the thermoregulatory response to temperature is the sum of behavioral and physiologic activity. A major factor determining the combination of these two regulations in a particular situation is the presence or absence of the means for conscious, voluntary behavior. In the physiology laboratory with the animal restrained or anesthetized, behavioral activity may be reduced to insignificance so that the physiologic reactions can be studied more or less in isolation. With the unrestrained animal in its natural habitat, physiologic responses may be reduced to such low levels that they cannot be measured. It is desirable therefore to separate the two activities for individual measurement and then to observe them together as they complement each other in natural situations. It can be noted in Fig. 59-7 that behavioral regulation, although effective, is rough; that is, the capacity of the regulator is adequate but the permitted deviations in the value of the controlled temperature are great. The addition of physiologic regulation to behavioral regulation provides both the precise control of the former and the great power of the latter. Fig. 59-6 indicates that in man both regulations are highly developed.

The block diagram of a regulator as shown in Fig. 59-8 provides an outline for the analysis of the large body of physiologic data now available. To the right in the figure is the regulated or passive system, which is the body of the man in whom the temperature is controlled. Complete characterization of the passive system for temperature regulation involves descriptions of the cardiovascular, respiratory, endocrine, metabolic, and other systems that are treated elsewhere in this text; attention here will be limited to heat transfer for the passive system and the detection, transfer, and integration of information concerning body temperatures in the controlling system. In Fig. 59-8 the heavy lines indicate energy transfer and the thin lines represent information transfer. The body produces heat and exchanges thermal energy within itself and with the environment. The temperature receptors in the skin and body core detect the level and rate of change of temperature and transmit this information

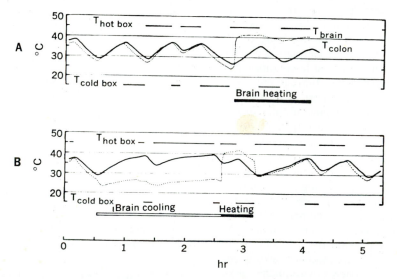

Fig. 59-7. Behavioral temperature regulation in cold-blooded lizard *Tiliqua scincoides,* which moved back and forth between hot and cold environments. Note that brain heating or cooling altered behavior pattern. (From Hammel et al.[31])

through nerve channels to the hypothalamus in which much of the integration occurs. When deviations are noted, effector signals are generated to modify the heat flow in the body by changing the distribution of the circulation, by increasing the heat production through shivering, or by altering the rate of heat removal from the skin by sweating.

Heat balance in the body

The overall energy exchange of humans may be roughly characterized by the balance diagram in Fig. 59-9. In the steady state of heat flow the heat produced in the body is balanced by the heat loss to the environment, so that over an extended period the internal body temperature tends to remain near 37° C. The action of the thermoregulator rather than the properties of the passive system maintains the internal temperature near 37° C, since in theory the balance between heat production and heat loss could occur equally well at any body temperature. Thus the chemical energy made available by the combustion of carbohydrate, fat, and protein in body tissues is

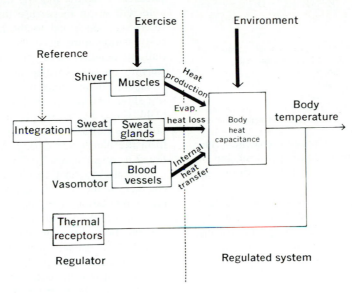

Fig. 59-8. Simplified block diagram of human thermoregulatory system. (Modified from Stolwijk.[51])

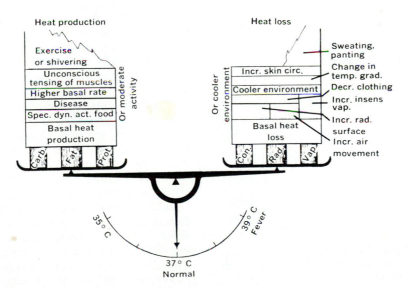

Fig. 59-9. Balance between factors increasing heat production and heat loss. (From DuBois.[5])

converted into heat or external work, and the heat is dissipated into the environment by convection, radiation, and the evaporation of water from the skin and respiratory tract. In this process the internal body temperature may vary from 35° to 41° C, depending on the circumstances, but when normal conditions prevail, the temperature will return to its resting level near 37° C by thermoregulator action. Thus, since the body heat storage is equal to the heat produced minus the heat lost, the heat balance equation may be written as follows[28]:

$$S = M - W - [E - (R + C + D)] \qquad (3)$$

where:

> S = Rate of body heat storage (considered positive if body temperature rises)
> M = Total metabolic rate
> W = Useful physical work rate
> E = Evaporative heat loss rate
> R + C + D = Rates of heat loss (or gain) by radiation (R), convection (C), and conduction (D)

In cool environments, radiation, convection, and conduction are all channels of heat loss, but in hot environments, they may help to store heat in the body.

Body heat storage. In the resting or sedentary individual the changes in body heat content normally take place entirely in the peripheral tissues, as shown in Fig. 59-10. With cold exposure,

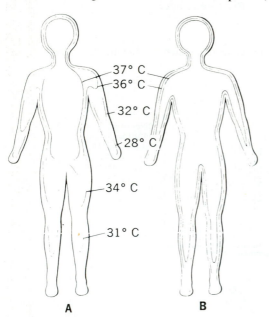

Fig. 59-10. Distribution of internal body temperatures and thermal gradients after exposure to cold, **A,** and heat, **B.** (From Aschoff.[13])

selective vasoconstriction reduces the supply of blood to the skin and peripheral tissues and diverts the blood to the visceral organs. This increase in central blood volume evokes a temporary "cold" diuresis. Under the conditions of mild cold exposure the mean body temperature will decrease, although the rectal temperature may increase or remain near 37° C. However, the cold, relatively bloodless peripheral tissues provide excellent insulation to vital internal structures. On exposure of a chilled individual to a warm environment, vasoconstriction may be released so rapidly that the warm blood from the core will begin to perfuse the cold tissues and return as cold blood to the heart. The internal body temperature may then decrease substantially (depending on the previous cold exposure), causing nausea, cardiac arrhythmia, and even collapse. For this reason a severely chilled person should be placed in a hot tub of water (40° to 44° C) so that the supply of heat to the internal organs is rapid. Aviators and mariners rescued from the North Sea during World War II were found to respond well under this regimen.

Exposure to heat results in relatively small changes in mean body temperature as long as thermoregulation is effective. Even in an environment of 38° to 40° C (100° to 104° F) the positive body heat storage may represent less than 0.5° C. As shown in Fig. 59-10, *B,* under these conditions the isotherms for 36° and 37° C move out into the extremities. The blood rapidly perfuses the skin and peripheral tissues so that the metabolic heat is efficiently transferred to the sweating skin. Since some of the values in equation 3 can be expressed most conveniently in terms of heat flow per unit of body area (watts per square meter),[28] the body heat storage will also be expressed in these terms (to convert from kilocalories per hour to watts multiply by 1.163. For example, 40 kcal/hr × m² = 46.5 W/m²).

Metabolic rate. In the quiet, resting person, most heat is produced in the body core—the trunk, viscera, and brain—even though this constitutes only a little over one third of the body mass (Table 59-1). Thus heat is supplied to the tissues of the resting individual from the central organs, and the escape of this heat from the body is easily controlled by the vasomotor system. During work the principal site of heat production shifts to the musculature. The actual values for metabolic rates (M) and their dependence on the work done and various physiologic factors such as age, sex, and size are presented in detail in

Chapters 56 and 57; suffice it to say here that the metabolic rate may range from 40 W/m² to 800 W/m², depending on the activity level and condition of physical fitness.

Work rate. The work rate (W) represents the rate of doing useful work such as lifting a weight or winding a clock spring. This energy would not be measured by a calorimeter that is collecting all the heat lost from the man. For example, if a man pedals a stationary bicycle that is connected to an electrical generator, which in turn charges a battery, the energy stored in the battery will not appear in the measurement of heat loss. Thus equation 3 cannot be balanced until the work energy has been measured and put into the equation. The useful work performed by the body is termed *positive work*. Reversing the experiment just described (i.e., using the electrical generator as a motor to drive the bicycle backward) the man has to oppose this effect by keeping the bicycle speed constant. In this case, work will be done on the man, and he will absorb the motor's mechanical energy into his body. As he will actually be performing work, his metabolic rate will increase at the same time. The same effect is seen when a person descends the stairs or walks downhill. The work is then called *negative work,* and the sign of W must be changed in equation 3. A major difference between positive work and negative work is the great efficiency that can be estimated for negative work.[41] It has been proposed that the energy absorbed during negative work can be used to reconstitute adenosine triphosphate (ATP) or other high-energy compounds and thus "run the metabolic machine backward"; however, study has indicated that such an effect is unlikely and that all negative work appears as heat in man.

Table 59-1. Relative masses and rates of metabolic heat production of various body compartments at rest and during exercise*

	Body mass (%)	Heat production (%)	
		Rest	**Work**
Brain	2	16	1
Trunk viscera	34	56	8
Muscle and skin	56	18	90
Other	8	10	1

*Modified from Stolwijk and Hardy.[52]

Heat transfer

Thermal radiation. Radiation means the exchange of thermal energy between objects in space through a process that depends only on the absolute temperature and nature of the radiating surfaces. The flow of radiant heat does not require the presence of an intervening medium; energy will pass from a hot object to a cooler one through a vacuum. Radiation in the general sense includes all energy transfers of this type—ultraviolet, visible light, infrared, etc. Fig. 59-11 shows the spectral distributions of the skin reflectance and the thermal radiation (R) emitted from the human body. The atmospheric gases (water vapor, CO_2, O_3, etc.) absorb infrared radiation, so that the thermal radiation from the skin and from the earth itself is absorbed in the air layers near the ground. The sun's infrared radiation also is almost entirely absorbed by the earth's atmosphere. The solar visible and near-infrared radiation is transmitted to the earth's surface and is trapped because the associated increase in ground temperature results in thermal radiation from the trees, etc., which cannot pass back to the sky. This selective absorption of the atmospheric gases is called the "greenhouse" effect of the atmosphere, and it provides an important moderating effect on the daily temperature swings at the earth's surface. However, on the mountaintops or in the desert the clear dry atmosphere allows rapid radiant cooling at night and (sometimes unexpected) temperature changes between midday and midnight. The low-temperature infrared radiation does not penetrate water or tissues containing water and thus thermal radiation does not partake in the heat transfer within the human body.

The net transfer of heat is the difference between the radiation emitted by a surface and that which it receives. The Stefan-Boltzmann law expresses this difference as follows[8]:

$$R = \sigma \epsilon_1 \epsilon_2 (T^4 - T_w^4) \tag{4}$$

where:

R = Radiant heat transfer in W/m²

σ = 5.75×10^{-8} W/m² °K⁴ (Stefan-Boltzmann constant)

T, T_w = Temperatures of hot object and surfaces of environmental objects in degrees absolute (°K)

ϵ_1, ϵ_2 = Emissivities of radiator surface and environmental surfaces with maximum values of unity

In this equation the surface *quality* or *emissivity* enters as an important factor. By definition, a *black body* absorbs completely all types of radiation and reflects nothing; the term "black body"

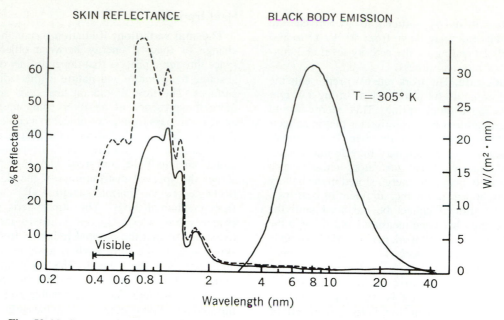

SKIN REFLECTANCE   BLACK BODY EMISSION

Fig. 59-11. Spectral distribution of reflectance and radiation emittance of human skin. Note that both white skin (dashed line) and pigmented skin (solid line) are nonreflecting surfaces (black bodies) in infrared spectral region.

is thus appropriate. If one imagines a hypothetical surface that absorbs all incident radiation on one side and reflects nothing (e.g., an open window), it will emit all the radiation from the other side if there is equilibrium. Thus good radiators, like good absorbers, are black bodies for which the emissivity is by definition equal to unity. No physical object is completely black because every object reflects some light even though the amount may be small. In contrast to the perfect absorber is the perfect reflector with an emissivity of zero. This type of surface is approximated in nature by highly polished metallic surfaces. If an object reflects some light of all wavelengths it is termed a "gray" body. Many surfaces are almost black for some radiations but reflect others quite well. These are colored objects; it is to this class of surfaces that the human skin belongs. The reflecting power of human skin varies with wavelength, but (Fig. 59-11) the reflectivity in the infrared regions is the same for both skin colors; that is, all skin is black in the dark! In the visible spectrum there is a difference in reflectance that presumably is in some way related to solar radiation, although a completely satisfactory explanation for different skin coloration is not available. From the point of view of energy balance, it has been shown that dark skin absorbs more solar

Table 59-2. Low-temperature (300° to 400° K) emissivities

Hole in black enclosure	0.998
Human skin	0.95
Matte black (optical black)	0.97
Glossy painted surfaces (regardless of color)	0.95
Water	0.95
Wood or plastic surfaces	0.95
Copper (polished)	0.10 to 0.20
Silver (highly polished)	0.02

heat than white, a fact that may have some importance for behavioral temperature regulation in hot climates. All human skins are excellent radiators in the infrared (Table 59-2), and for this reason skin is often called a "black body radiator."

In nearly all circumstances the surfaces in the environment are highly absorbing for infrared radiation; thus the radiation exchange of human skin (and clothing) can be estimated with little error from equation 4, assuming that all surfaces are black (i.e., $\epsilon_1 = \epsilon_2 = 1$). If the skin temperature and environmental temperature are not very

Table 59-3. Specific thermal conductivities*

Substance	Conductivity (W/m · °C)
Silver	1.16
Glass	0.0028
Water	0.0014
Fat	0.0005
Muscle	0.0011
Bone	0.0011
Air	0.00006

*From Newberg.[10]

different (i.e., within 20° C), equation 4 can be further simplified to the following:

$$R = k_r(\overline{T}_s - T_w) \qquad (5)$$

$$k_r = 4\sigma \overline{T}_s^3$$

T_w and $\overline{T}_s$ can be measured with radiometers. For a man dressed in shorts and sitting quietly in an environment at 25° C, R equals about 50% to 70% of the heat lost from the body or about 30 W/m². Not all the body surface is effective in radiation exchange with the environment because some surfaces exchange energy with other skin areas. For example, between the legs, under the arms, and between fingers the radiant heat lost from one area is absorbed by the opposite skin surface and there is no net loss to the environment. A standing man with arms at his side has an effective radiating area of about 75% of the total; with arms and legs extended the value may go as high as 85%; in the tightly curled-up position the radiating area can be reduced to 50% of the total body area calculated from the DuBois formula.[5] In unusual environments such as those in outer space or on the moon, radiation may be the only channel of energy loss or gain from the environment.

Conduction. The flow of heat from one object to another with which it is in contact is called conduction (D); the process involves no transfer of material between the two objects. Heat is conducted from the warmer tissues within the body to the skin surface and from the skin through the thin layer of air with which it is in contact or into cooler objects such as the floor or clothing that may touch the skin. Hot objects conduct heat into the skin; the direction of heat flow is always toward the lower temperature. Thus cold or hot objects are identified as describing flow of heat out of or into the skin. The rate at which heat is

transferred by conduction is calculated in watts per square meter using the following equation:

$$D = K(T_1 - T_2) \qquad (6)$$

in which:

K = Conductance = Specific thermal conductivity divided by length of conducting path and multiplied by area of contact

T_1, T_2 = Temperatures of warm and cool surfaces

For the resting individual, heat exchange by conduction of heat from solid objects is usually small, and this factor is often neglected in the heat balance.

Metals generally have high thermal conductivities and gases low conductivities (Table 59-3). Tissues have values generally near that of water, although fat is lower. Often it is desirable to think of *insulation* rather than *conduction* because insulation represents the resistance to heat flow. The insulation is the reciprocal of conductance or:

$$I = \frac{1}{K}$$

If there are a series of insulators such as several layers of clothing, the total insulation is the sum of the resistance of each layer or:

$$I_{total} = I_1 + I_2 + I_3 \ldots$$

The insulation value of an ordinary suit of men's winter clothing is called 1 "clo unit"[27] and the usual spring suit is about 0.6 "clo." The insulating value of 1 clo is:

$$I_{clo} = 0.155 \ °C/m^2 \cdot W \qquad (7)$$

Table 59-4 gives some idea of the magnitude of the insulation that can be obtained with clothing or wraps.

The greatest insulation is provided by the natural insulators such as feathers and fur, particularly winter fur. These coats have soft down hair that traps a highly insulating layer of air and long guard hairs that overlay the down, thus preventing the disturbance of the warm air layer by wind. The major contributing factors to heat loss control are the physiologic change of insulation that occurs spring and fall in furred animals and the human behavioral regulation of clothing.

The resistance to conduction of heat from the internal organs to the skin is important physiologically, since fat is a good insulator. Fat people can generally prevent deep chilling in cold water; a layer of grease applied to the skin of swimmers can have the same effect, giving some protection

Table 59-4. Data for various insulators*

	Insulation (clo)	Permeance to water vapor
Nude	0	1
Shorts	0.1	0.97
Typical tropical clothing ensemble; shorts, open-neck shirt with short sleeves, light socks, and sandals	0.3 to 0.4	0.90
Light summer clothing: long light-weight trousers, open-neck shirt with short sleeves	0.5	0.85
Light working ensemble: athletic shorts, woolen socks, cotton work shirt (open-neck), and work trousers	0.6	0.82
Typical American business suit without vest	0.6	0.82
Typical business suit and vest	1.0	0.75
Heavy traditional European business suit: includes cotton underwear with long legs and sleeves, shirt, woolen socks	1.5	0.65
Heavy wool pile ensemble (Polar weather suit)	3.4	0.48
Raccoon fur	3.9	—
Husky dog	4.1	—
Lynx fur	4.1	—
Gray wolf fur	6.6	—
Red fox fur	7.8	—

*Normal air movement (10 m/min).

against the cold water. Women have a thicker layer of subcutaneous fat than men, providing a greater insulation between the skin and deeper tissues and thus reducing heat loss. Advantage is taken of this fact by the amas (female pearl divers) of Japan, who can dive in the cold coastal waters without suffering as severe a reduction in internal temperature as men would. Aquatic mammals such as the seal, whale, and walrus rely on the 10 cm thick layer of blubber fat that surrounds their internal organs to prevent loss of body heat in arctic waters.

Convection. The type of heat transfer referred to as convection (C) depends on the existence of a fluid or gaseous medium between hot and cold objects and on the exchange of warm and cold molecules between the warm and cold surfaces.

The heat loss due to streams of warm air rising from the skin and passing into the cooler environment is called *natural convection,* whereas the cooling of the skin with an electrical fan is an example of *forced convection.* The heat that is convected from the skin must first be conducted through a layer of air that sticks tightly to the skin surface. The thickness of the film depends on the skin temperature and the speed with which air is blown over the surface, but even in high winds a very thin air layer will cling to the skin. Fig. 59-12 illustrates convective heat loss from the skin for both types of convection. The skin at temperature T is in contact with the air at temperature T_a and there is air movement in the environment due to both natural and forced convection. The air layers very near the skin do not move and heat is conducted through this boundary layer; thus the thermal gradient through this layer will be linear. As the distance from the skin increases, the air molecules participate more and more in the general movement of the air over the surface, and finally the heat will be transported away by streams of faster moving air. This layer of air has been called the ''private climate,'' and for conditions of natural convection, it may have a thickness of 2 to 4 mm for exposed skin surfaces; in a strong wind or other condition of high forced convection the private climate layer may be reduced to a few microns in thickness.

The rate of heat loss from the skin by convection will depend on the skin and air temperatures and the air velocity (V); the relationship between these quantities has been shown to be as follows:

$$C = h_c(\overline{T}_s - T_a) \tag{8}$$

in which h_c equals the convective heat transfer coefficient and depends on the air velocity.

Equation 8 can be rewritten as:

$$C = 10\sqrt{V}(\overline{T}_s - T_a)$$

to express C in watts per square meter (velocity in meters per second).[10] The convective heat loss depends on the thermal gradient between skin and air and roughly on the square root of the air velocity; conditions for still air are represented by a velocity of 0.1 m/sec, an amount that takes natural convection into account. For weak air currents such as occur with the movements of the head, arms, or legs, the forced convection will be small and the larger convective heat loss will be due to natural convection. However, with walking or running the forced convection will play the major role, and in a wind, most of the

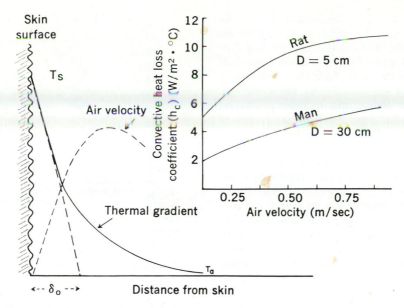

Fig. 59-12. Heat loss by convection (diagrammatic) from warm skin (T_s) to ambient air (T_a); δ_0 = boundary layer or Grenzschicht.[8] Inset: convective heat loss coefficient as function of air velocity. D = diameter; 5 cm = rat-equivalent cylinder; 30 cm = man-equivalent cylinder.

body heat loss will be due to forced convection. The relationship between the convective heat loss (per degree thermal gradient) and air velocity is shown in the top panel of Fig. 59-12. It is seen that for low velocities there is a relatively rapid increase in heat loss; with higher velocities the increase becomes smaller and at very high air velocities (60 to 80 km/hr), changes in velocity have little effect on convective heat loss. At extremely high speeds (~1,000 km/hr) the heat developed by air friction may reverse the heat flow and cause a marked increase in surface temperature. However, air pressures that may distort and even tear the skin prevent human exposures to such high velocities. Under outdoor conditions, persons are exposed to the chilling effects of winter winds, a fact that gives rise to the term "wind chill factor,"[10] often reported by weather stations to indicate how "cold it feels." This factor can be calculated from equation 8 by assuming a reasonable value for skin temperature and comparing the value of C for no wind with that for the measured wind velocity at the weather station. Naturally, it will "feel colder" in the wind, since C will have a much larger value, and it thus may be reported that "the temperature is 20° F but the wind chill factor causes it to feel like 4° above zero." Physiologically, it can be seen that reducing the value of C may be impor-

tant to maintaining body heat in the cold. Shivering and muscular work increase metabolic heat, but they also increase C; brisk walking, stamping feet, etc. may actually increase heat loss more than the increase in metabolic heat production. However, shivering is an efficient way of increasing metabolic heat (unpleasant as it is), inasmuch as up to 50% of the increased heat production due to shivering can be retained in the body, thus permitting sleep under cold conditions. For example, after a couple of sleepless nights, a healthy young man can sleep unclothed reasonably well under one wool blanket with temperatures below freezing.

Convective heat transfer by the bloodstream to the skin is of vital physiologic importance and is under control of the sympathetic nervous system. For each liter of blood at 37° C that flows to the skin and returns to the body core at 36° C the body loses roughly 1 kcal or 1.16 W · hr of heat. Thus, in the cold, vigorous vasoconstriction nearly shuts off blood flow to the arms, legs, and skin of the trunk. This action ensures that body heat from the core can be lost only by conduction through the muscle and fatty tissues, which are reasonably good insulators. The heat flow cannot be stopped but it can be reduced to 5 to 9 W/m² for each degree of thermal gradient between core and skin (the actual value will depend to a large extent on the amount of fat on the body). In the heat, during vigorous exercise the blood flow to the skin can increase the heat flow almost 10-fold, providing an efficient channel of heat transfer in the body. Indeed, the exercising muscles cannot get rid of their metabolic heat rapidly enough without this increase in convective heat transfer. Fig. 59-13 shows results of studies on heat balance in men and women exposed nude to temperatures of 22° to 36° C. At the bottom are plotted the changes in tissue "conductance." Conductance is the combined effect of the two heat transfer channels, conduction of heat through the muscle and fat layers and convective heat transfer by the blood. This conductance can be estimated on the assumption that all the body heat is produced in the body core, conducted to the skin, and then transferred into the environment, that is:

$$H_L = K(T_{re} - T_s)$$

where

H_L = Heat loss (except for the small amount from the respiratory tract)
K = Effective conductance
T_{re} = Rectal or core temperature
T_s = Mean skin temperature

In the cold (22° to 28° C) the conductance is not greatly affected by ambient temperature and represents the fully vasoconstricted state in which peripheral blood flow is minimal. The effect of the layer of subcutaneous fat in women is clearly seen to reduce their conductance below that of men. In the heat (30° to 36° C) the conductance increases progressively, with women having a steeper increase, indicating their greater control of the peripheral blood flow that carries heat to the skin. The rate of the "effective" blood flow can be estimated using the value of conductance in the cold as the condition for no blood flow and assuming that any increase above this minimum is due to increased flow of blood from core to skin. The warm blood, as it leaves the body core via the arteries, soon comes into contact with the cooler venous blood returning from the skin and other peripheral tissues: heat exchange begins so that some of the heat in the arterial stream is returned to the body core. This process is known as "countercurrent exchange," and its effect is to reduce the convective heat transfer or to require more blood to produce the same heat transfer that would be accomplished without countercurrent exchange. In general, measurements of heat transfer can give values only for

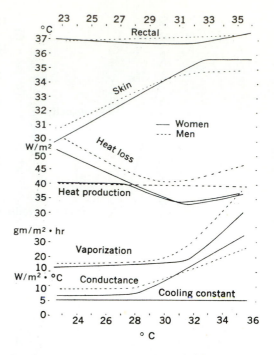

Fig. 59-13. Average values of thermal factors for men and women in near steady state at different calorimeter temperatures. (From Hardy et al.[32])

the *minimum* blood flow rate that can affect the observed heat flow. The actual blood flow may be and usually is greater because heat transfer calculation do not allow for the exchange of heat that takes place in the networks of arteries and veins supplying the skin or other area.

As will be noted below in discussing the passive system, the control of blood flow rate and distribution is a major function of the vasomotor system. However, temperature changes affect the state of vasodilatation directly by action on the blood vessels. Using an isolated vein of a dog, measurements have been made of the pressure required to keep blood flow constant while the temperature of the vessel was being altered by changing the perfusate temperature. The results of one study of this phenomenon are shown in Fig. 59-14. As the perfusate temperature was lowered, it was necessary to increase the driving pressure, and, of course, this pressure is proportional to the peripheral resistance or degree of apparent vasoconstriction. It can be seen also in Fig. 59-14 that the increase in the esophageal temperature of the animal caused a decrease in

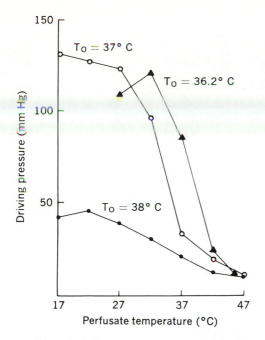

Fig. 59-14. Response of isolated cutaneous vein to temperature changes (results from three dogs). Driving pressure is saphenous perfusion pressure minus femoral vein pressure at equilibrium following each change in perfusate temperature. Note that response is least in dog with highest esophageal temperature. $T_0 = 38°$ C. (From Webb-Peploe and Shepherd.[56])

response of the vessels. The overall effect of temperature on mean thermal conductance of men and women on exposure to heat (Fig. 59-13) can thus be attributed in part to action of the sympathetic innervation of the vasomotor system and in part to the effect of warming the vessels of the skin.

Evaporative heat loss. When water is converted into vapor at body temperature, 0.7 W/hr (0.6 kcal) of heat per gram of water is absorbed in the process. Thus evaporating water (sweat) from the skin is an efficient way of losing heat even in an environment hotter than the skin. In this case the evaporative heat loss (E) must take care of both the metabolic heat and the heat absorbed by radiation and convection from the hot environment. The body is always losing some heat by evaporation from the skin and respiratory tract, even in very cold weather when the physiologic problem is to retain body heat. As indicated in Fig. 59-13, at ambient temperatures below 30° C the evaporation is fairly constant at 12 to 15 gm/m² · hr, about half of which is due to the loss of moisture in breathing and half to

the slow transudation of water through the almost dry skin. Even though the skin appears dry, moisture is diffusing through the outer skin layers and evaporating. This drying effect can be troublesome in a dry environment, causing skin scaling and itching as well as nasal and pharyngeal discomfort. At 30° C the evaporative heat loss amounts to about 25% of the total heat loss, which, at this neutral temperature, equals the basal heat production. Clinical use has been made of this fact by employing a method of careful weighing of the patient to evaluate the daily energy balance.

Above 30° C the evaporative heat loss increases linearly with temperature as active sweating is stimulated to take care of the decreased heat loss due to radiation and convection. When the environmental and skin temperatures are equal there can be no heat loss by radiation and convection, since the thermal gradient is zero; in this case, usually at an ambient temperature of 35° to 36° C, all the metabolic heat must be lost by evaporation.

Sweat glands. The overriding importance of normal sweat gland function in warm climates is clearly evident in preventing dangerous hyperthermia. The secretory activity of these glands, which in the average man number something like 2.5 million, is under control of the central nervous system (CNS) via sympathetic innervation. When skin disease—ichthyosis and hyperkeratosis of the skin, for example—interferes with sweating, exposure to heat, and especially exercise in the heat, may raise the internal body temperature to lethal levels. Such individuals depend completely on behavioral temperature regulation in stressfully warm conditions and must repeatedly dampen clothing with water even during normal sedentary activity. Sweating rate also depends on the skin temperature in its own locality.[19,39] Fig. 59-15 shows the result of warming the entire skin surface and independently controlling the skin temperature in a small area from which sweat was being collected and measured. This effect has been attributed to the dependence of sweat gland activity on the temperature-sensitive transmitter action of the sympathetic nerves stimulating the sweat gland.

For the exercising man, sweating is even more important than for the resting or sedentary individual because the rate of heat production can reach high values during heavy work and heat must be eliminated rapidly. To provide for this high heat flux, it is first necessary to stimulate the vasomotor system to transfer the heat from

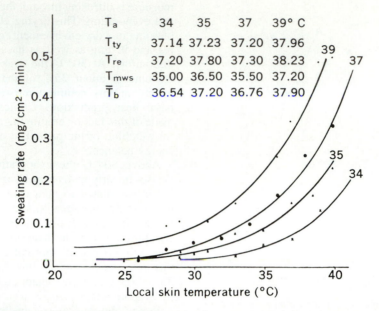

T_a	34	35	37	39° C
T_{ty}	37.14	37.23	37.20	37.96
T_{re}	37.20	37.80	37.30	38.23
T_{mws}	35.00	36.50	35.50	37.20
$\overline{T}_b$	36.54	37.20	36.76	37.90

Fig. 59-15. Effect of local skin temperature on local sweat rate. Thermal status of subject is indicated in Table. T_{ty} = tympanic temperature; T_{re} = rectal temperature; T_{mws} = mean weighted skin temperature; $\overline{T}_b$ = calculated mean body temperature; T_a = ambient temperature. (From Bullard et al.[19])

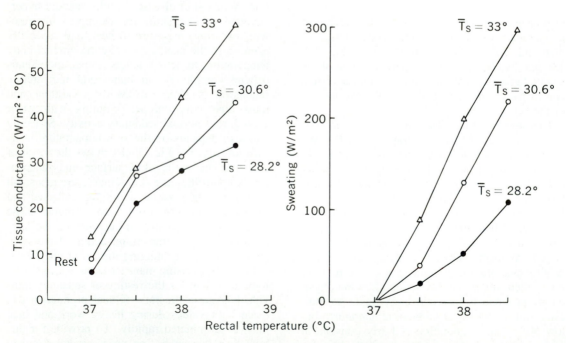

Fig. 59-16. Tissue conductance and sweating heat loss as affected by mean skin temperature ($\overline{T}_s$) and rectal temperature. Experiments were performed on young men working at 25, 50, and 75% of maximum $\dot{V}o_2$ at ambient temperatures 10°, 20°, and 30° C. (Redrawn from Stolwijk et al.[53])

the muscles and body core to the skin; second, in all except very cold environments, the sweat glands must be activated so as to eliminate the heat from the skin surface. The dependence of both sweating and vasodilatation on internal body temperature (esophageal) is illustrated in Fig. 59-16 for exercising men at three rates of work. During exercise the internal body temperature *must* increase even though the environment is cold, because it is necessary to activate vaso-dilatation to transfer the heat to the skin. As the internal temperature rises in proportion to the work rate (watts per square meter) regardless of the external temperatures (within ranges of possible heat balance), we can assume that the temperature controller action is of the "proportional control" type.[7]

The heat loss by sweating is often termed "mass transfer" because water is lost from the body and the transfer of heat by evaporation can occur even though the skin is much cooler than the environment; in other words, the heat can be transferred *against* the thermal gradient. Thus the term "*mass transfer*" is used as distinct from sensible heat transfer (via radiation, conduction, or convection), which is always in the direction of the thermal gradient. The total evaporative heat loss from man can be expressed as the sum of the water losses from the respiratory tract, from transudation, and via active sweating. The first two avenues are not affected by thermoregulation in the resting or sedentary man and are generally small; these will be combined as E_{ex}, and the sweating loss will be represented as E_{sw}. Therefore the total evaporative heat loss, E, in watts per square meter is[28]:

$$E = E_{ex} + E_{sw} = m\lambda/A_D \qquad (9)$$

where:

    m = Rate of moisture loss from the body, excluding urine and fecal losses
    A_D = Body area according to DuBois
    λ = Heat required to vaporize 1 gm of water

The evaporative loss from the membranes of the respiratory tract and transudation, E_{ex}, is:

$$E_{ex} = \mathring{V}(\rho_{ex} - \phi_a\rho_{in})\lambda/A_D + E_{tn}\lambda/A_D \qquad (10)$$

where:

    $\mathring{V}$ = Ventilation rate due to breathing
    ρ_{ex} = Saturation density of expired air
    ϕ_a = Relative humidity of ambient air (RH)
    ρ_{in} = Saturation density of inhaled air
    E_{tn} = Moisture transudation rate of dry skin (Transudation decreases, of course, as the skin becomes wet with sweat.)

In cold environments, $\phi_a\rho_{in}$ is always quite small because cold air, even when saturated, will hold little water vapor. Thus in the winter the weather service may report a relative humidity of 90%, but since the air is cold, ρ_{in} will be small and the air will be quite dry relative to the respiratory tract. Also, at a high altitude (in an aircraft or in a city on a high plateau), $\mathring{V}$ may be increased and $\phi_a\rho_{in}$ will be small due to the reduced atmospheric pressure. The result of these conditions is the drying of the nasal mucosa. Effective relief from such conditions can be provided by the "croup-kettle," which is a teakettle attached to some sort of breathing mask. With the kettle boiling at a low rate, steam at roughly body temperature can be provided to a face mask at a rate that raises the value of ϕ_a and ρ_{in}; as a result, the evaporation from the respiratory tract is greatly decreased and there is at least temporary relief. In general, little can be done to humidify the ordinary home or office during cold winter months because the rate of condensation of moisture against cold window panes and other surfaces and the infiltration of dry air from outside will defeat most efforts.

The sweating term, E_{sw}, depends on the air motion over the wet skin surface, the amount of skin that is actually wet with sweat, and the water vapor pressure gradient between the skin and the environment. That is E_{sw} in watts per square meter may be expressed as:

$$E_{sw} = h_\epsilon(P_{ws} - \phi_a P_{wa})A_w/A_D \qquad (11)$$

where:

    h_ϵ = Vaporization heat transfer coefficient that depends on the air velocity; its value is about twice the convection coefficient, or $h_\epsilon = 2.2\ h_c$ in $W/m^2 \cdot$ mm Hg
    P_{ws} = Water vapor pressure at skin temperature
    P_{wa} = Water vapor pressure saturated at air temperature
    A_w = Area of skin wet with sweat; depends on rate of sweating up to the point of completely covering the skin surface (sweat that drips from the body does not help with evaporative cooling)

The importance of air motion to the sweating man is seen to be greater than for the man with a dry skin because the vaporization coefficient, h_ϵ, is more than twice as large as the convective coefficient, h_c. Thus in warm climates, it is desirable to maintain some moisture on the skin and to provide air movement over the wet surfaces. The dark patches seen in the thermogram in Fig. 59-4 represent cool skin wet with sweat. These surfaces include the head (except the face), neck, trunk, and to a lesser extent the areas of the arms and legs, but not the hands and feet. For the heat-acclimatized man, more of the skin would have been wet with sweat and there would have been a more even distribution of cool skin.

The importance of the vapor pressure in the ambient air is apparent from equation 11, because if the value of the product $\phi_a P_{wa}$ becomes equal to P_{ws}, there can be no evaporative cooling. The weather service usually reports the value of the relative humidity, ϕ_a, but it is the product of ϕ_a and P_{wa}, the saturated vapor pressure at the ambient temperature, that is the important environmental limitation to the evaporative heat loss. The values of ϕ_a and P_{wa} can be obtained by measurement of the ambient dry- and wet-bulb temperatures. However, since the value of P_{wa} is usually contained in the "steam tables" and is not easily described in terms of everyday life, several substitutes have been proposed over the years to indicate "how hot it feels." One of the best of these is the "effective temperature," which is an empirical combination of dry- and wet-bulb thermometer readings based on experi-mental comparisons of warmth sensation in conditions of high and low vapor pressure. Another commonly reported factor is the THI, or the "temperature-humidity index," which is calculated from an equation that sums the wet- and dry-bulb readings as follows[20]:

$$\text{THI} = 0.4(T_{db} + T_{wb}) + 4.8$$

There is little guidance to be received from the THI, and its basis is not secured by physiologic and psychologic data, as is the scale of effective temperature. Keeping in mind that it is the ability of the environment to take up moisture from the skin that is the important factor, use has been made of the wet-bulb thermometer reading alone. Maximum values can be given for certain activities, and its reading has been used successfully to prevent occurrences of heat exhaustion and heat stroke during training and industrial activities in hot weather (Fig. 59-17).

It should be noted that to achieve heat balance in equation 3 the value of E_{sw} must be maintained despite increasing humidity. For example, if half

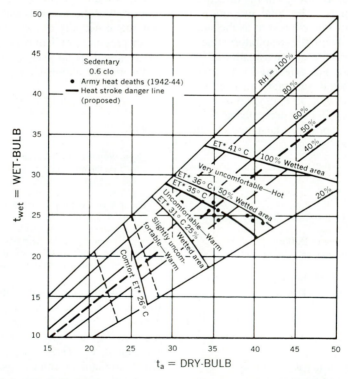

Fig. 59-17. Wet- and dry-bulb temperature diagram indicating comfort and discomfort zones and lines of equal wetted area. Dashed lines are constant effective temperature. Note that zone for heat stroke danger begins near 30% wetted area and much below zone of thermoregulatory failure. (Redrawn from Hardy.[8])

the metabolic heat must be eliminated from the body by evaporative heat loss, then the man must sweat enough and increase the value of A_w to make up for any loss of evaporative power due to an increase in $\phi_a P_{wa}$. There is little significant change that can be achieved by altering the value of P_{sw} because the skin temperature cannot rise unless the body temperature goes up also (hyperthermia). Of course, if $\phi_a P_{wa}$ is low, then the sweating can be reduced and people can live in comfort in hot dry climates by sweating gently all the time. Thus the desert areas of New Mexico, Arizona, and California are often used as resort areas, and their climate is considered salubrious even though hot.

The influence of clothing on sweating rate should be noted. As with convective and conductive heat losses, clothing represents a resistance to the passage of water vapor from the skin to the ambient air. The value of this "permeance" to water vapor has been measured by observing the evaporative heat losses from wet surfaces covered over with various combinations of layers of cloth. Fortunately much clothing allows free diffusion of moisture even while providing satisfactory insulation. For example, if water vapor permeance has a maximum value of 1.0 for the nude man (Table 59-4), light summer clothing has a value of 0.90, and even a winter suit has a permeance of 0.75. So-called "wind resistant" clothing has a permeance of about 0.60. These high values of permeance mean that when work is done in heavy outdoor clothing the evaporation of sweat is permitted. Only for truly vapor barrier materials—plastic or rubberized—is the permeance reduced to very low values.

TEMPERATURE-REGULATING SYSTEM

With the information concerning the production of heat in the body, its transfer to the skin by conduction through the tissues and convection via the blood, and the further transfer of the heat from the skin by radiation, conduction, convection, and evaporation, the detailed discussion of temperature regulation in man can proceed with presentation of the passive system. As illustrated in Fig. 59-8, the environment is shown as causing disturbances in heat loss, and exercise is indicated as producing internal disturbances in heat production. Taken together, these are the principal challenges to the regulating system, which strives to maintain body temperatures in the normal ranges indicated in Fig. 59-1. With the present state of knowledge, it will be necessary

to make many simplifications. For example, it will be assumed, at least initially, that the cardiovascular, respiratory, electrolyte, endocrine, and other regulated systems that directly affect temperature regulation are all functioning normally within known physiologic limits. Ideally, all systems should be included with temperature regulation and studied simultaneously, but such an approach is impractical at the present time due to lack of data. Study of a limited system is sometimes called "modeling." The principal value of models is to outline what is currently known and to indicate some of the specific areas that can be profitably studied. Thus a model should not be considered as providing new truth or new data. Modeling has only recently become useful to physiologists, although astronomers, physicists, engineers, etc. have used the method for many years. Increased detailed physiologic knowledge and the advent of electrical calculating machines (analog and digital computers) have made modeling a valuable analytic procedure for the physiologist and many physicians. The rapid increase in biologic data on man will unquestionably make these approaches even more important.

Description of passive system

The model is based on the 71 kg man with a body surface area of 1.83 m². The complex human form is represented by three cylinders, as shown in Fig. 59-18. The ratios of the surface area to the volume of the model correspond to the anatomic values for the head and trunk. The four extremities are combined into a single cylinder representing their total surface areas and volumes. The choice of lengths and radii of the cylinders was made on the assumption that the ends of the cylinders could be neglected by considering them perfectly insulated. The outer 2 mm of each cylinder is considered to represent the skin compartment, and the inner layers are divided according to anatomic data among muscle, viscera, brain, and central blood compartments; for all compartments the available data on mass, specific heat of tissues, and thermal conductivities were used. In the model, heat can only flow radially, the exchange of heat between cylinders is by blood convection, and the temperature within each compartment is assumed to be uniform. Physiologic data do not completely support the model described and a more complex model with more compartments would provide a greater precision. (To follow the argument, exact

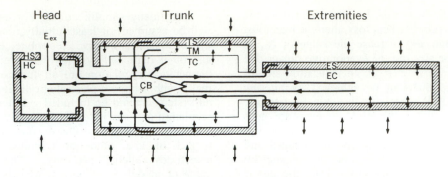

Head Trunk Extremities

Environmental heat exchange

Fig. 59-18. Three-cylinder diagram of thermal model of man. E_{ex} = respiratory evaporative heat loss; *HS* = head skin; *HC* = head core; *CB* = central blood; *TC* = trunk core; *TM* = trunk muscle; *TS* = trunk skin; *ES* = extremity skin; *EC* = extremity core. (From Stolwijk and Hardy.[52])

numeric values are not necessary. However, they are required to carry through the detailed analysis; they are available together with the data sources in Stolwijk and Hardy.[52])

The blood performs so important a function in the heat transfer in the body that the model contains a separate central blood compartment. This compartment represents the volume of blood in the heart and large vessels; this blood flows so rapidly that it does not come into thermal equilibrium with any compartment and so exchanges heat with all. This convective heat flow is by far the most important method of heat transport in the passive system and the controlling or regulating system must exercise a careful control over both the rate and distribution of blood flow. The tissue compartments that exhibit the greatest variability in their blood flow are the skin, muscles, and the splanchnic area (trunk core); of these, the flow through the skin and muscles is of special importance for thermoregulation.

The block diagram in Fig. 59-19 summarizes the heat flow exchanges between the eight compartments of the model: these exchanges need to be expressed by eight heat flow equations, one for each compartment. The equations are all of the same form and state that the change in the heat content of a compartment equals the heat produced by metabolism plus the heat inflow minus the heat outflow. A basal heat production, M_o, is assigned to all compartments except the central blood compartment for which it is assumed that its metabolism contributes insignificantly to heat production. The muscles of the trunk and extremities are assigned additional heat production, ΔM, due to shivering or to physical

work. The dashed lines for skin blood flow (SBF) and muscle blood flow (MBF) are indicated as being under vasomotor control. The solid lines represent passive heat conduction, or a blood flow that is not under vasomotor control (e.g., to the head core and trunk core). The control of E_{sw}, ΔM, MBF, and SBF is the function of the regulating system to be described later as algebraic equations that depend on the temperature signals from the head core, skin, etc. These controller equations close the loop and describe the mechanism of the regulating system.

As an example of the procedure, the heat balance equation for one compartment is presented here. The head skin (HS) will serve the purpose, since it is in thermal contact with the head core, the central blood compartment, and the environment. The equation for this compartment in algebraic terms is as follows:

$$m_{HS} \cdot S \frac{dT_{HS}}{dt} = K_{HCHS}(T_{HC} - T_{HS}) + M_{0HS} + \quad (12)$$
$$\alpha_{HS}\rho c \cdot 0.138 SBF(T_{CB} - T_{HS}) - 0.09(E_{ex})_{HS} -$$
$$0.09(E_{sw})_{HS} - A_{HS} \cdot h_0(T_{HS} - T_a)/A_D$$

where:

$m_{HS} \cdot S$ = Thermal capacitance (product of mass and specific heat) of HS = 1.13 W · hr/°C

K_{HCHS} = Thermal conductance (passive) through tissues from head core, HC, to HS = 3.06 W/°C

T_{HC}, T_{HS} = Temperatures of HC and HS

M_{0HS} = Basal heat production in HS = 0.14 W

α_{HS} = Countercurrent heat exchange factor for HS = 0.92

0.138 SBF = Fraction of total SBF assigned to HS

ρc = Density times specific heat of blood = Thermal capacitance of 1 L of blood

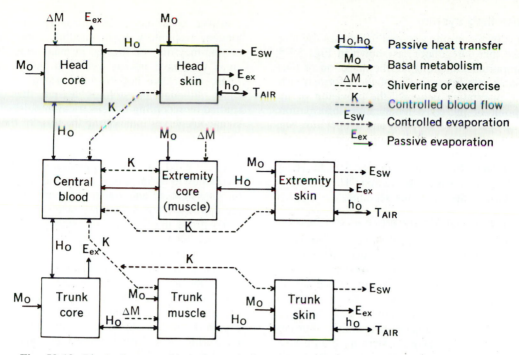

Fig. 59-19. Block diagram of heat flow relationship between various compartments of thermal model of man. Solid lines are channels of passive heat exchange, dashed lines indicate heat exchange under controller action via blood flow (K) and sweating (E_{sw}). (From Stolwijk.[51])

T_{CB} = Temperature of central blood
0.09 $(E_{ex})_{HS}$ = Fraction of insensible evaporative heat loss asigned to HS = 0.94 W
0.09 $(E_{sw})_{HS}$ = Fraction of sweating evaporative heat loss assigned to HS
A_{HS} = Area of HS
h_0 = Combined heat transfer coefficient for radiation and convective heat loss of HS to environment = 7.0 W/m² · °C
T_a = Ambient air and surface temperatures (assumed to be equal)
A_D = Total area of man = 1.83 m²

The first three terms of the equation represent heat put into the head skin compartment, and the remaining three terms are the heat dissipation into the environment. The difference, of course, is the rate of heat gain or loss by the compartment. The equation cannot be solved by itself, since its solution depends on such quantities as the head core and central blood temperatures and the skin blood flow and sweat rate; these in turn depend on the heat balances in the other seven compartments. Therefore the equations for all compartments must be solved at the same time and the interdependent values substituted continually in all equations. Solutions can be obtained by hand calculation, but the task is slow and laborious. Analog or digital computers are well suited to the problem once the actions of the controller or thermoregulator have been determined.

The large variety of anatomic and physiologic data required for the formulation of the passive system is evident from the equation for the head skin compartment and the way in which the other compartments are established. Even so simple a model as just described can be assembled only for man. Physiologic and anatomic data are missing for other species, even the dog and cat, which have served for so many years as subjects for physiologic studies of thermoregulation. Also, the precision of the model presented leaves much to be desired. Even so, it represents much that is known of the human body as it applies to temperature regulation. With the use of more compartments (e.g., more layers to the cylinders and more cylinders so that the legs, arms, hands, and feet can be represented) and better descriptions of regional blood flow, effects of respiration, water balance, etc., the precision and understanding of the model can be enhanced. However, the present model is quite good enough for the study of short-term (a few hours) exposure to heat, cold, work, fever. etc.

Controlling system

It is the controlling system that closes the loop of the thermoregulator, as shown in Fig. 59-8; its important functions are the detection of temperature, the transmission of the information to some integrating network, the provision of the set-point or preferred temperature information, the comparison of set-point and temperature signals, the production of deviation or error signals, and the transmission of the error signals to the appropriate effector mechanisms for shivering, sweating, and vasomotor action. These will be discussed in the following sections.

Detection of temperature in the body

Skin. As discussed in the opening paragraph of this chapter, all biologic tissues have a sensitivity to temperature because the rates of the biochemical reactions underlying life processes are temperature dependent. The activity of many cells in the body changes between two- and three-fold for a 10° C change in temperature, but certain cells of the nervous system have an unusually high temperature sensitivity; it is these cells that are suspected of serving the function of temperature detectors. Their locations have been established by physiologic and psychologic studies to be in the skin (giving rise to sensations of heat and cold), the deep viscera (Auerbach's plexus), the spinal cord, the mesencephalic reticular formation, the hypothalamus, and the preoptic region of the brain stem. Temperature-sensitive cells have also been found in the tongue, respiratory tract, medulla, and motor cortex; they may be located in many other body areas that have not as yet been investigated. These cells have not been described anatomically, although physiologic activity has been studied by recording the electrical impulses from cutaneous nerves innervating the nose of the cat. The average steady-state response of "warm" and "cold" fibers is illustrated in Fig. 59-20 for different fixed levels of temperature. Cold fibers are identified by increased activity as the nose of the cat is cooled and warm fibers are those with the opposite temperature response.

To prove that heating the skin produces not only sensations of warmth but also thermoregulatory signals for the central controller, it was necessary to observe a physiologic response (e.g., the sweating from an area of unheated skin) while the general level of skin temperature was raised. By exposing the ventral surface of a man dressed in shorts to a radiant heater and measuring the sweat rate from areas on the right and left thighs, which were screened from the radiation, it was found (Fig. 59-21) that the sweating rate was closely associated with the level of mean skin temperature.[39] The internal body temperature and the skin temperature of the two areas of the thighs were observed to be constant during the three radiation exposures shown. Thus mean skin temperature has a central action in the control of body sweating by the contribu-

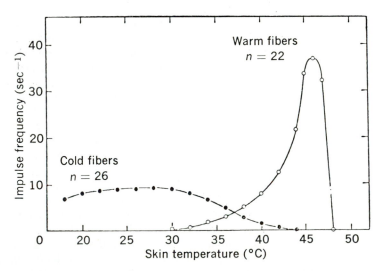

Fig. 59-20. Static discharge rate of cutaneous warm and cold fiber populations as function of temperature. (From Hensel and Wurster.[34])

tion of temperature information to the central regulator.

Viscera. The abdominal viscera have also been shown to be sensitive to temperature, and the central thermoregulatory system is affected when the visceral temperature is selectively altered. Rawson and Quick[45] observed that when electrical heaters were implanted in the ventral and lateral abdominal cavity of sheep and heated

to 43° to 44° C, a rapid increase in respiratory frequency and evaporative water loss occurred, as shown in Fig. 59-22. After 3 to 5 min of heating the animal began to pant, and at the same time there was a marked fall in the temperature of the hypothalamus, although the skin directly over the heaters was raised about 1° C. Other temperatures were unchanged or decreased slightly. The animals' response to heat must have originated

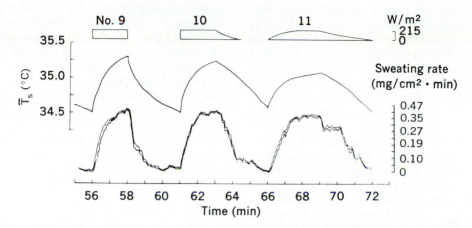

Fig. 59-21. Effect of changing mean skin temperature on sweat rate from unheated skin area. Top line: pattern of irradiation of ventral skin surface; middle line: mean skin temperature; bottom line: sweat rate from unheated areas of both thighs. (From Nadel et al.[39])

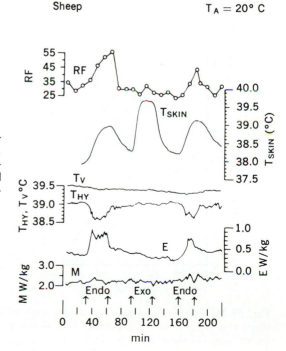

Fig. 59-22. Change in rate of breathing *(RF)*, evaporative heat loss *(E)*, hypothalamic temperature *(T_{HY})*, vaginal temperature *(T_V)*, and overlying skin temperature *(T_{SKIN})* during endogenous heating of ventral abdominal viscera *(Endo)* and external heating of overlying skin *(Exo)*. (From Rawson et al.[45])

either from the visceral tissues near the heaters or from the locally heated skin. To evaluate the effect of the warm skin alone, the internal heaters were turned off and external heat applied to the same skin areas (Exo in Fig. 59-22). Although the skin temperature was increased more than before, there was no effect on the breathing rate or the evaporative water loss, indicating the visceral origin of the thermoregulatory effect. Little is known at present of the neural connections and pathways involved aside from the possible relation of the observed effect to the temperature-sensitive neurons recently reported by Wood.[57] Recording from single units of the enteric nervous system in Auerbach's plexus of the jejunum of the cat, Wood observed units that discharged in bursts and were spontaneously active at temperatures 32° to 42° C. As shown in Fig. 59-23, both the number of spikes per burst and the frequency of the bursts were increased at high temperature and decreased at low temperature. Their sensitivity to temperature was high, thus providing, at least hypothetically, a good temperature transducer in the viscera.

Spinal cord. Thauer[55] and his colleagues, in an extensive series of experiments on several species, have shown that selective heating or cooling of the spinal cord evokes thermoregulatory activity. For example, Fig. 59-24 illustrates an experiment in which a small plastic tube, through which cold water could be circulated, was passed into the peridural space of the spinal canal. The metabolic rate and several body temperatures were measured; the presence of shivering was noted by eye or electromyograph. During the experiment, periods of local cooling of the cord between lower thoracic and seventh lumbar segments evoked marked shivering with increased metabolic rate and vasoconstriction. In other experiments, warming the spinal canal was associated with panting and vasodilatation. From these and other experiments, which excluded effects from the dura, the presence in the spinal cord of temperature-sensitive neurons must be inferred. If the spinal cord is sectioned above the site of temperature change, some thermoregulatory action still can be observed. Thauer suggests that at least part of this effect may be due to the direct action of temperature on the membrane

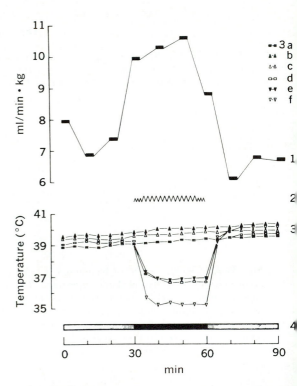

Fig. 59-24. Changes in O_2 consumption and body temperatures during cooling of vertebral canal (C2 to L7) of anesthetized dog. *1*, Metabolic rate; *2*, shivering; *3*, temperatures (*a*, aorta; *b*, hypothalamus; *c*, cisternal occipitalis; *d*, cervical canal; *e*, thoracic canal; *f*, lumbar canal); *4*, cooling period (black space). (Modified from Thauer.[55])

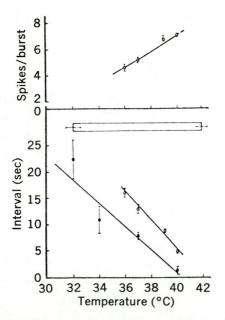

Fig. 59-23. Effect of temperature on discharge of single unit firing in bursts from Auerbach's plexus in cat jejunum. (Modified from Wood.[57])

resistance of the motoneurons and on the mono- and polysynaptic excitatory postsynaptic potentials.

Preoptic-anterior hypothalamic area. The small volume of tissue lying ventral to the anterior commissure and near the walls of the third ventricle (preoptic-anterior hypothalamus, POA) is especially important as a temperature sensing area. If small steel thermodes are stereotaxically located bilaterally to the area and perfused with warm or cool water, marked thermoregulatory responses are evoked. Fig. 59-25 illustrates this effect.[33]

This experiment involved temperature change in only a small tissue volume and suggests the presence in the area of temperature-sensitive neurons that respond to temperatures above and below the normal range. Using microelectrodes, such neurons have been located and their response to temperature measured, as shown in Fig. 59-26 for an anesthetized cat.[40] The re-corded unit activity and respiration frequency are to the left in the figure; it is of note that these neurons are active even at normal body temperature. When the heating with weak diathermy radiofrequency current began (see lower right), and the local temperature rose from 38° to 39.8° C, the unit firing rate increased immediately, followed about 90 sec later by the increase in breathing rate, usually associated with panting. Only about 20% of the units studied in this manner respond to temperature change; the unresponsive units presumably have functions unrelated to temperature regulation. Of the thermally sensitive units, 10% were stimulated by local temperature *only,* and these cells may serve as the POA temperature sensors. Some units responded to a variety of stimuli, including mechanical and thermal stimulation of the skin, and visual, auditory, or noxious senses, etc.; other units were sensitive to local temperature and to temperature changes in the midbrain, spinal cord, and viscera.

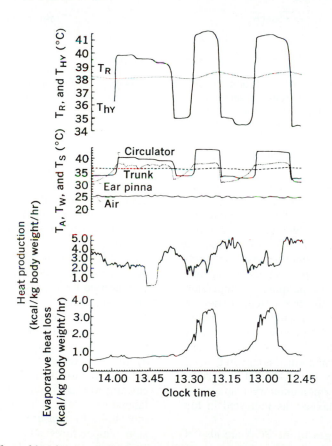

Fig. 59-25. Effect of heating or cooling of preoptic region of dog on metabolic rate and panting (evaporative heat loss). *Tr, Thy* = rectal and skin temperatures. (Modified from Hellstrøm and Hammel.[33])

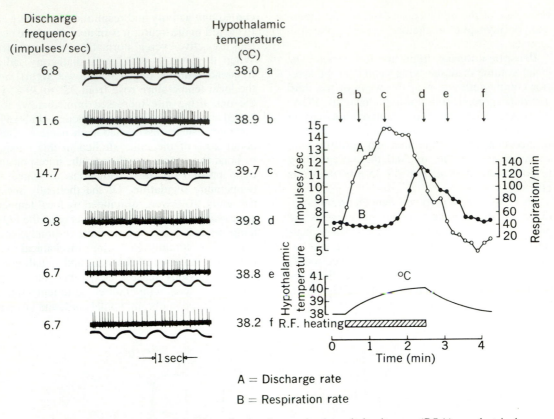

Fig. 59-26. Effects of local heating of preoptic anterior hypothalamic area (POA) on electrical activity of temperature-sensitive unit in POA of anesthetized dog. At right, time course of experiment showing effect on breathing rate. (From Nakayama et al.[40])

The responses of these cells suggest an integrating network for temperature information from anywhere in the body. Fig. 59-27 illustrates the response of such a unit, which responded to rubbing the fur and to warming (but not cooling) the thoracic spinal cord of a lightly anesthetized rabbit.[30] Similar experiments have identified units sensitive to local and peripheral temperature changes scattered in the brain stem tissues extending from the septum caudal to the rostral reticular formation. Thermally sensitive units have also been located in the thalamus, pons, medulla, and in several cortical areas. In general, however, it is mainly the POA that is effective in producing the thermoregulatory responses to local temperature changes, as indicated in Fig. 59-26.

Temperature changes in the POA are also effective in evoking behavioral temperature regulatory activity, as has been shown by the quantitative experiments of Adair et al.[11] in the sub-

human primate *Saimiri sciureus* (squirrel monkey). These animals were trained to select appropriate amounts of hot (50° C) and cold (10° C) air to provide a preferred ambient temperature of about 35° C. The animals were then implanted stereotaxically with a thermode in the medial preoptic region 0.5 mm lateral to the midline; a reentrant tube for thermocouples was centered 2.0 mm further laterally on the same side. After complete recovery the animals were again placed in their environmental chamber and allowed to select an environmental temperature. Fig. 59-28, *A*, illustrates the effect. During the control period the animal turned on the heat, as was its usual custom, to warm itself after being kept in a cool laboratory while being instrumented with thermocouples to measure hypothalamic, rectal, and skin temperatures and to make connections to perfuse the thermode. The usual heating process was interrupted by suddenly increasing the hypothalamic temperature (T_H); the animal responded

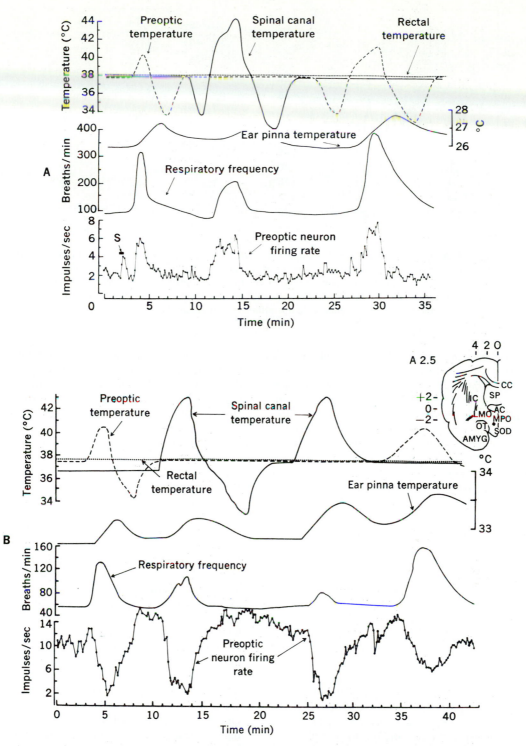

Fig. 59-27. Electrical activity of two units in POA of rabbit that responded to heating of POA or thoracic spinal cord. **A,** Unit excited by heating unresponsive to cooling. **B,** Unit inhibited by heating but unresponsive to cooling. Thermoregulatory activity, for example, vasomotor responses and breathing rate, were evoked by heating both regions. Approximate location of units indicated at bottom. Note that unit at top responded to rubbing skin of forepaw. (From Guieu and Hardy.[30])

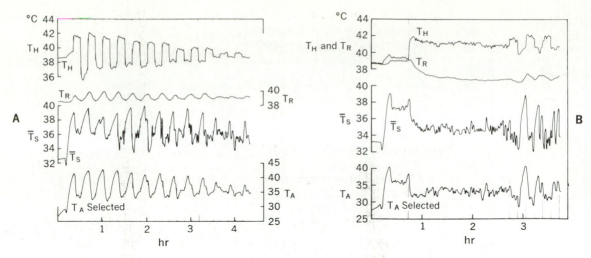

Fig. 59-28. Effect of heating and cooling hypothalamus (T_H) of squirrel monkey on behavioral temperature regulation. **A,** 10 min cycles of decreasing amplitude to obtain threshold (less than $\pm 0.3°$ C). **B,** 90 min thermal clamp of (T_H) at 41° to 42° C during which animal "regulated" its body temperature about 3° C lower than its usual normal level. T_R = Rectal temperature; T_s = mean skin temperature; T_A = ambient temperature. (From Adair.[11])

at once by cooling the air in the chamber. A series of 10 min changes of T_H of decreasing amplitude stimulated the animal to respond in each instance (even to $\pm 0.3°$ C) with an *opposite* change in the chamber, skin, and rectal temperatures. The very high sensitivity of the preoptic temperature sensors is clearly seen from this experiment. In Fig. 59-28, *B,* is another experiment in which the preoptic temperature was raised and held at 41° to 42° C for 90 min. During the initial phase of the preoptic heating the animal cooled its environment, thereby losing body heat until its rectal temperature reached 36.5° C and its skin temperature fell to 35° C. At these levels the animal came into a purposeful thermal balance and "regulated" its temperature at these new levels. This is indicated by the fact that when the preoptic temperature was lowered from 41.5° to 40° C the animal's immediate response was to heat itself even though the 40° C temperature was *higher* than the animal's usual normal level of 39.5° C. This behavior indicates that the output from central receptors stimulated by heat from the thermode was balanced by an output from peripheral receptors responding to body cooling. The converse experiment has shown that if an animal is given the opportunity to alter its hypothalamic temperature, it will cool its brain stem if the environmental chamber is heated, thus suggesting a conscious sensory response to the POA receptors.[23]

The thermal receptor system for temperature regulation appears to be widespread throughout the body, and the information on the thermal state of the body converges in the preoptic region, hypothalamus, and reticular formation. The manner in which the integration of the signals is accomplished is unknown, but recordings from units with multiple signal inputs indicate summation, inhibition, and facilitation (multiplication). Some sort of "set point" or preferred level of body temperature appears to be a feature of the integration because many units change firing rates only for temperatures *above* normal levels (Fig. 59-27). The pathways by which this widespread input reaches the brain stem have yet to be worked out, although the close relationship of the pain and temperature pathways in the cord, as described in Chapter 13, would suggest that the visceral and somatic afferents for temperature have connections in the reticular formation and the thalamus that finally lead to the hypothalamus.

The special importance of the internal thermal receptor system is its function during exercise when the heat of the working muscles must be eliminated even through a skin cooled by sweating. It is clear that under these circumstances the cutaneous receptor system must be reporting cool conditions even though the internal body temperature is rising. The visceral, spinal, and POA sensitive units, however, are giving ample afferent input to the regulator. Data from neurophysiologic, behavioral, and physiologic experiments indicate that the afferent inputs to the regulator are summed in weighted sequence, depending on the number and sensitivity of the units in many body areas with an added function that indicates a multiplication of inputs from skin and POA receptors. The functions of the POA, posterior hypothalamus, and rostral reticular formation include the integration of the summed inputs and provide set points and the development of the efferent outputs. The neuronal network involved in the integration action begins in the septal-preoptic region and extends caudally to involve the hypothalamus and reticular formation. Progressive ablations of the brain stem, beginning with the POA and proceeding caudally, produce preparations with more and more impairment of thermoregulatory ability, even though temperature information from many body areas is still reaching the remaining brain stem.[14,35,36] The central integration of the temperature information from the POA and other areas appears to be an essential thermoregulatory function of the hypothalamus.

Central nervous control of body temperature

All the changes in the body that normally act in the face of thermal stresses to preserve a relatively constant temperature are due to an increased or decreased number of impulses in cerebrospinal and autonomic nerves. Since these alterations in impulse discharge are of central origin, we must search in the CNS for the mechanisms that execute the delicate control outlined in the preceding paragraphs of this chapter.

Spinal cord. When the spinal cord is transected in the lower cervical region (just below the origins of the phrenic nerves), body temperature regulation is greatly impaired. After cord section the extent to which the animal or man comes to behave as if cold-blooded varies with the level at which the cut is made. If it is in the upper thoracic region, for example, the regulation against cold, although distinctly less effective than normal, is far better than when the section is made through the lower cervical cord. The difference depends largely on the fact that after the lower section, more muscle groups are left innervated from the brain, but preservation of the cerebral connections of the uppermost portion of the sympathetic vasoconstrictor outflow is also a factor.

In dogs surviving spinal transection for an extended period of time and entirely free from symptoms of spinal shock, Sherrington[49] showed that there is, in the region innervated from behind the lesion, marked failure of vasomotor responses to even great changes in environmental temperature. Shivering does not occur as a spinal reflex no matter how considerable the exposure to cold, and Sherrington was unable to induce sweating in the hind paws of "spinal" dogs by exposure to heat.

The dependence of normal thermal responses on prespinal central mechanisms is strikingly illustrated by the appearance of shivering in the muscles innervated anterior to the level of a spinal section when the parts below are immersed in cold water. Sherrington found that "the nonappearance of cold shivering in the paraplegic region of the dog can reveal with great clearness the exact segmental level of the spinal lesion by demarcating the muscle-field innervated behind the lesion from that in front. The boundary so shown will actually subdivide an individual muscle; for example, the anterior part of the deltoid will shiver violently, whereas in its posterior part, no shivering can be evoked." If the transection is above the first thoracic segment and the room temperature is kept at its usual level, the shivering in the anterior parts of the body is not accompanied by any signs of cutaneous vasoconstriction. The forepaws and ears remain warm to the touch because of the separation of the sympathetic outflow from prespinal centers. With a midthoracic transection the response to cooling the hindparts includes vasoconstriction. These reactions in regions not directly exposed to environmental cold and yet separated from the cooled parts by cord transection may be reflexes evoked from the sentient parts by cold blood from the chilled region or they may be the results of a direct effect of cooled blood on the thermoregulatory mechanisms of the brain.

It seems quite clear that the isolated human spinal cord, like that of the carnivore, is unable to mediate effective thermal responses. Yet, as Seckendorf and Randall[48] have shown, slight thermal sweating may occur as a spinal reflex in a paraplegic man with complete transection of the cord at levels between T3 and T8. On exposure to quite high ambient temperatures, this response began, as does thermal sweating in normal subjects, on the dorsum of the foot and was then progressively recruited in a cranial direction to and across the cutaneous area corresponding to the level of the spinal lesion. The sweating on the foot, calf, and thigh associated with spinal control was far less intense and the corresponding skin temperatures were higher than on these parts in normal control subjects or on the skin of the paraplegics innervated from the upper thoracic cord. In the normal subjects, central (rectal) temperature did not rise during the exposure to heat; in the "spinal" subjects, this temperature (oral) rose but reached a plateau because of the profuse sweating over the normally innervated areas of skin. In a more recent series of experiments Randall et al.[44] obtained the same meager thermal sweating in five patients with neurologic signs of complete transections between C4 and C6 and in one with a surgically verified transection at T1. It appears that in each the entire sympathetic outflow had been disconnected from the thermoregulatory centers of the brain. On exposure to heat, all six patients became hyperthermic but showed a progressive increase in a sparse form of sweating and signs of cutaneous vasodilatation. Apparently there is within the thoracolumbar segments of the cord a specific mechanism for sweating organized to effect its sequential recruitment but whose activation normally depends on descending impulses from the cerebral thermoregulatory centers.

Hindbrain and mesencephalon. The cat or dog maintained for many weeks or months after truncation or transection of the brain stem at an upper pontile or a mesencephalic level (decerebration) is unable to regulate its body temperature when exposed to cold. Indeed, the core temperature of such an animal falls rapidly when the ambient temperature is that of a moderately warm room. There is no difference in this respect between cats with truncation in the upper third of the pons and cats with truncation at the junction of the diencephalon and mesencephalon. Bard and Macht[14] and later Bard et al.[16] studied many such animals whose long survivals (up to 11.5 months) were made possible by careful nursing, control of ambient temperature, and leaving of a neurally isolated "island" of hypothalamus and attached pituitary gland rostral to the truncated brain stem. The "island," the only part of the forebrain remaining, served to maintain a normal water balance and provided a basal release of tropic hormones from the pars distalis of the hypophysis. Since its only connection with the rest of the body was vascular, it could not influence the remaining brain stem, cerebellum, and spinal cord through any direct nervous channel.

When the truncation of the brain stem is between the level of exit of the third pair of cranial nerves and the caudal extremity of the diencephalon, the animal is able to right itself, walk, sleep, and, in the case of the cat, climb.[14] But these preparations, which have shown so many unexpected modes of behavior, cannot control their body temperature. As just indicated, on exposure to cold the "pontile" or "mesencephalic" cat or dog shows a fall in core temperature. There are no signs of cutaneous vasoconstriction or piloerection. If the body temperature falls rapidly from a normal level (38.0° C), a gross form of shivering, actually a jerking or twitching of large muscle masses, begins when the colonic temperature reaches a point between 34° and 33° C. The high threshold of this thermal response renders it utterly ineffective, for it does not occur until the animal is in a state of hypothermia and it fails to check the rate of fall of core temperature. If the body temperature is allowed to fall slowly, no muscular twitching is seen, no matter how great the fall. All available evidence is explicit in showing that any effective defense against body cooling depends on cerebral mechanisms situated rostral to the mesencephalon.

One of the chief difficulties in maintaining pontile or mesencephalic cats is their tendency to develop a fatal hyperthermia when a rise in ambient temperature occurs, especially if it happens to coincide with a period of increased locomotor activity. When the colonic temperature reaches a level (about 43.5° C) that, if maintained for long, is incompatible with life, respiration increases to a rate of 90 to 100/min (contrasted with a rate of 250 to 300/min in normal cats exposed to the same degree of warmth), and true panting occurs spontaneously or can be evoked by opening the mouth. The effectiveness of this thermolytic response is limited and cannot prevent death.

Hypothalamus. Isaac Ott of Philadelphia was the first to indicate that this ventral division of the diencephalon is the cerebral region primarily concerned in body temperature regulation, but his observations, first published in 1887, seem to have had little influence on contemporary opinion. It was not until 1912 and 1914 that Isenschmid and his collaborators, Krehl and Schnitzler, who used rabbits in their experiments, provided clear evidence that after removal of most of the forebrain lying above the hypothalamus, a mammal retains a nearly normal ability to regulate its body temperature, whereas removal of the entire forebrain (producing a mesencephalic animal) resulted in a poikilothermic state. Much later, other investigators,[14] who also studied animals (cats) for rather brief postoperative periods, obtained further evidence that the hypothalamus is prepotent in the central control of temperature. Essentially the same conclusion was derived from experiments by Pinkston et al.,[49] who studied cats and dogs during much longer survivals after removal of all parts of the forebrain except the hypothalamus and the most medial portions of the thalamus. These animals maintained their core temperatures within the normal range when exposed to extreme cold and heat. Fig. 59-29 illustrates the ability of such a hypothalamic cat to withstand an ambient temperature that averaged 12° C over a period of more than 46 hr. This test was carried out 64 days after removal of all forebrain except the hypothalamus and very small fragments of amygdala and globus pallidus on one side and a very little of the hippocampus on both sides. This same animal on its two hundred and twenty-second postoperative day was exposed to an ambient temperature of 1.5 to 5.0° C for 4.1 hr without any fall in colonic temperature; in fact, it rose 0.3° C, a response that is not uncommon in normal cats. A comparison of the response of the hypothalamic animal (Fig. 59-29, *A*) with the record of the temperature fall in the mesencephalic cat as shown in Fig. 59-29, *B*, will illustrate clearly the importance of

the hypothalamus in the maintenance of a normal core temperature. When exposed to heat (39° to 41° C) the hypothalamic cat, not unlike many normal cats, showed a rise in colonic temperature to 41.3° C and remained at that level as a result of its capacity to increase heat loss by panting. In view of the fact that no part of the diencephalon except the hypothalamus remained in this animal, any influence of cutaneous or other peripheral thermoreceptors must have been mediated by ascending fibers that bypass the somatic sensory nuclei of the thalamus.

Combined thermoregulatory system

Because much of the data have been obtained from animal studies and their application to man is presumptive, it must be said at this time that the control equations that close the loop for the model indicated in Fig. 59-19 can be written only in general form. The relationship that best fits both the human and the animal data states that the effects of the central signal are facilitated by the direct action of temperature on the effector organs. For example, the sweating response may be expressed as follows:

Sweat is equal to the central warm signal, S_w, multiplied by local temperature effect on sweat glands, L_{sw}, or:

$$E_{sw} = S_w \cdot L_{sw}$$

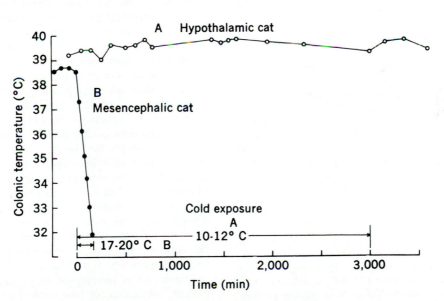

Fig. 59-29. Rectal temperatures of hypothalamic cat, *A*, during cold exposure (12° C) and mesencephalic cat, *B*, exposed to temperatures of 17° to 20° C. (From Bard and Woods.[15])

Skin vasodilatation is equal to the central warm signal multiplied by local temperature effect on skin vessels, L_{sbf}, or:

$$S(DILAT) = S_w \times L_{sbf}$$

Shivering is equal to the central cold signal multiplied by muscle sensitivity per unit mass, M_c, or:

$$\Delta M = S_c \times M_c$$

Skin vasoconstriction is equal to the central cold signal multiplied by local temperature effect on skin vessels, L_{sbf}, or:

$$S(STRIC) = S_c \cdot L_{sbf}$$

These four control equations can be expressed in several forms, depending on the detail with which the physiologic responses are described and the relative importance attached to the various factors. As an example, let us consider an equation that describes much of the data for sedentary man. In this model[51] the central signal was simplified to consider only the head core temperature (T_{hc}), the mean muscle temperature ($\overline{T}_m$), and the mean skin temperature ($\overline{T}_s$). Local temperature effects were assumed to be essentially constant. Combining the control equations into the equations for the passive model provides the feedback that "closes the loop" and completes the model of physiologic temperature regulation. Tests were conducted to evaluate the model against physiologic experiments on man. As an example of such a test an experimental fever was induced.

Healthy young males were studied by Palmes and Park[42] after intravenous injection of immune tyhpoid vaccine. Measurements were made of rectal, mean skin, and ambient temperatures, metabolic rate, evaporative heat loss, and effective skin blood flow as estimated from skin conductance values. Cooper et al.[22] have shown that the site of action of pyrogens in producing fever is principally in the POA, an area we have shown to be concerned with the detection of temperature and the integration of temperature information. Neurophysiologic data indicate that pyrogens affect the central neurons by decreasing the warm signals and increasing the cold signals. Essentially this action is equivalent to moving the set points to a higher temperature level so that the subject "feels cold," even though he may be in a neutral or even a warm environment. Thus the thermoregulatory set points were considered to be slowly elevated by the pyrogens so that the central signal became a "cold" signal, activating shivering and vasoconstriction. Neurophysiologic data indicate that the upward shift of the set point begins 7 to 10 min after the intravenous injection of an endoge-

nous pyrogen, reaches a maximum after a period of 1 to 2 hr, and slowly decreases in 2 to 4 hr. We will assume these conditions pertain to the man in the experiment and that the maximum shift in central set-point temperature was 2.5° C.

In the actual experiment, performed at a neutral ambient temperature of 30° C and low wet-bulb temperature, the pyrogen injections occurred about 30 min prior to the beginning of the records for heat balance studies. The progress of the experiment is illustrated in Fig. 59-30. The rectal (T_{Tc}) and skin ($\overline{T}_s$) temperatures were normal at the start, as were the metabolic rate (M), evaporative heat loss (sweating), and skin blood flow. In the figure the actual data are indicated by symbols and the predictions of the model by solid lines. About 30 min after the start of recording the subject broke into a violent chill and increased his metabolic rate some fivefold; the model also indicated a chill, but at a somewhat lower level, and an accompanying vasoconstriction not observable from the skin conductance measurements. However, vasoconstriction is a usual characteristic of the prodromal phase of a chill and was often reported in other subjects studied by Palmes and Park. As the defervescence of the fever proceeded, there was a vasodilatation and sweating, which were predicted by the model.

For so simple a model, agreement with experimental findings as indicated in Fig. 59-30 is as good as can be expected. A more detailed description of both the passive and controlling systems is of course possible, with better agreement resulting between prediction and physiologic data. However, such models are more complex with many more equations to be solved. To completely represent the human physiologic response the model would have to be at least as complex as the man himself! For practical purposes, models are always simplifications of an actual system rather than reproductions of it. In many cases, models can make predictions that are well within the limits of the biologic data.

A point of importance in connection with the fever experiment is that the *only* way in which the model can predict the course of the fever is by an upward shift in set-point temperature for the regulator. The set point not only must be displaced upward, but it also must be maintained at appropriate levels during the entire fever episode. Thus in a patient who is "running a fever" or "spiking a fever" the central set-point shifts may be approximated by corresponding levels of rectal temperature. For situations in which thermoregulation is interrupted (e.g., death of the patient, anesthesia, or heat stroke) a subsequent hypothermia or hyperthermia is not, of course, due to a thermoregulation relatable to any devi-

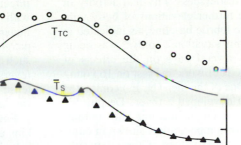

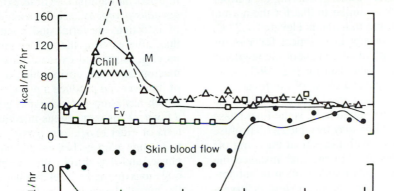

Fig. 59-30. Experimental and theoretical study of heat balance during fever induced experimentally by intravenous injection of immune typhoid vaccine into healthy young subject (30° C). Experimental data shown as symbols; model predictions indicated by solid lines. (Data from Palmes and Park[42]; modified from Stolwijk and Hardy.[52])

ations of set point. However, in fever, it appears that the patient's body temperature is physiologically regulated but at a higher set-point level.

Fever[1,5,58]

Elevations of internal body temperature associated with infections and other disease states are generally classified as fevers. Although often the fever may develop slowly over a period of several hours, in some infections there may be a rapid rise in body temperature, such as that noted in the experimental fever discussed in the preceding section of this chapter. In any case, there must be a storage of heat in the body either by means of a temporary increase in heat production over heat loss, a decrease in heat loss below the level of heat production, or both. A febrile episode can often be divided into three recognizable periods. The chill phases correlating with the fever development are characterized by a sensation of intense cold, during which the patient may break into a violent shaking chill with high metabolic rate or seek to cover himself with extra blankets and heating pads without the appearance of an actual chill. If the body temperature rise is rapid, the chill may last 20 to 40 min, but in other instances the elevation of temperature may be so slow that only a general restlessness lasting some hours can be observed. The second phase is the fever plateau, during which the body temperature is maintained at an elevated level; the chill has subsided and the metabolic rate returned to near resting level, with the body coming into a thermal equilibrium 1° to 4° C above the afebrile level. This phase may last for a few hours or for as long as several weeks, depending on the course of the infection. The patient usually has hot dry skin during this period and experiences discomfort and restlessness. Physical activity or work will cause further elevations of body temperature over the febrile baseline, as indicated in Fig. 59-31. In this situation a study[26] was being conducted over several days on the effects of intermittent periods of work and rest on 16 young male subjects. One of the men (N. E.) contracted an infection and developed a fever. Despite his illness, he insisted on maintaining the routine, and a record of his temperature is shown in comparison with the mean values of the other subjects for 1 day during his febrile episode. The pattern of temperature changes during the periods of rest and work is similar to that for the normal men except that the baseline is elevated 1.2° C, supporting the concept that clinical fever is the result of an upward displacement of the regulated body temperature. The data for Fig. 59-31 were taken during the plateau phase of the fever.

Defervescence, a period during which the body loses its extra stored heat, is the final phase of the fever. Heat loss may be rapid, with profuse sweating, or slow, with the patient throwing off blankets and sleeping. In the first instance the body temperature falls rapidly, often to subnormal levels, 35.5° to 36.5° C; in slow defervescence, some light sweating may occur, but decreased restlessness and sleep accompanied by lower metabolic rate are characteristic. Either type of defervescence may be brought about by the administration of an antipyretic such as aspirin. Depending on the underlying disease, vari-

ations in the pattern of fever will occur. In some infections, it is common to have a series of chills with "spike" fevers. Brucellosis, typhus fever, malaria, and many acute viral disease present patterns of repeated febrile episodes. Diseases such as pneumonia, typhoid fever, and enteric fevers are generally characterized by a single intense chill phase with an extended plateau that can last up to several weeks if not treated.

Pathogenesis of fever. Observations by Atkins and Bodel[1] and Wood[58] in the 1950s established the fact that substances extracted from inflammatory exudates could produce fever when injected into animals. These extracts, called *exogenous pyrogens,* are associated with a number of infectious bacterial and nonmicrobial agents that, after entering the body, interact with granulocytes, blood and exudate monocytes, alveolar macrophages and eosinophils, and Kupffer cells of the liver and produce a pyrogen (endogenous), which has a much more rapid action in evoking a fever. After intravenous injection of killed bacteria or other exogenous pyrogen a period of 30 to 60 min elapses before the onset of the chill phase, and it is during this interval that the endogenous pyrogen is elaborated and carried to the hypothalamus via the bloodstream. The preoptic–anterior hypothalamic area is by far the most sensitive part of the brain to pyrogens, and fevers can be stimulated following microinjection of nanograms of endogenous pyrogen of prostaglandins of the E series into the tissues of this region.[38,47,50] The onset of thermoregulatory activity causing the fever in these experiments on rabbits may follow within 2 min of the injection.

Among the microbial agents that have been shown to produce endogenous pyrogen in vivo and in vitro are the lipopolysaccharides that form part of the cell wall of gram-negative bacteria, various gram-positive bacteria, pathogenic fungi, and viruses. The mechanisms by which these agents activate the release of endogenous pyrogen are multiple and probably include particulate stimulation, including phagocytosis, and chemical interaction with the cell membrane. Endogenous pyrogen has now been purified to a protein with a molecular weight of about 14,000, which is clearly different from all known exogenous pyrogens, including endotoxins, which are lipopolysaccharides with molecular weights of several million.

The ultimate action of pyrogens on the hypothalamic thermoregulatory center is not cearly understood, but recent investigations have suggested the possibility that prostaglandins may be

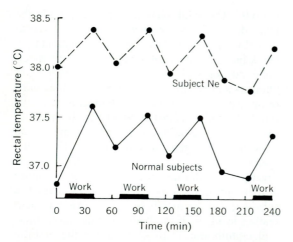

Fig. 59-31. Effect of fever on body temperature during work-rest cycles. N. E. with fever shows a febrile level about 1° C higher than normal subjects. (Modified from Fox and MacPherson.[26])

involved in the final "common pathway" for fever production. Since prostaglandin E_1 is so potent in evoking fever, and aspirin and its congener antipyretics inhibit prostaglandin synthetase but do not prevent a prostaglandin fever, the possibility is suggested that the antipyretics may block the production of prostaglandin. Fig. 59-32 shows a hypothetical flow chart of the events that recent work suggests occur between the introduction of an infectious agent into the host and the production of fever. The possible action of endopyrogen through as yet unknown pathways on areas other than the preoptic–anterior hypothalamus is included because such effects cannot be completely excluded at present.

Anesthetic hyperthermia[29,46]

A lethal condition associated with the administration of general anesthesia is termed "anesthetic hyperthermia." The condition is rare, with an incidence of two or three cases a year in a major medical center, and is characterized by a sudden, rapid, and uncontrolled rise in body temperature accompanied by tachycardia and cardiac arrhythmias. There is a documented instance in which the rectal temperature rose 4.4° C in 1 hr.

To account for this rise the metabolic rate must have risen more than fourfold, and the thermoregulatory reflexes for hyperthermia must have been completely suppressed. The most likely source of the heat production is an intense muscular activity, perhaps associated with the clinical observation of muscular spasticity and rigidity reported in many cases. The failure of the thermoregulatory system may be attributed in large measure to the effects of the general anesthesia. The cause of the syndrome is unknown, but several hypotheses have been offered. One is that the condition results from an abnormal response to succinylcholine because muscular rigidity and the onset of anesthetic hyperthermia have occurred soon after the injection of commonly administered amounts of this agent. However, instances of anesthetic hyperthermia have occurred without administration of succinylcholine or other muscle relaxant. Another suggestion has proposed the uncoupling of the intracellular oxidative phosphorylation by some unknown anesthetic agent, resulting in greatly increased heat production. Two further observations may be of significance: (1) Termination of anesthesia seems to have no effect on the devel-

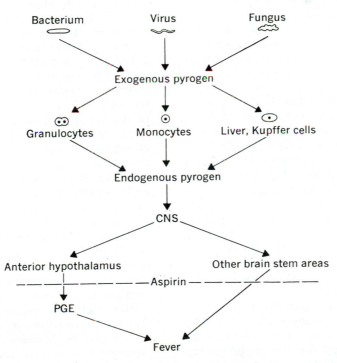

Fig. 59-32. Postulated pathways in pathogenesis of fever. (From Atkins and Bodel[1]; reprinted by permission from the New England Journal of Medicine **286:**27, 1972.)

opment of the hyperthermia, and (2) when it is possible by drastic means to lower the body temperature to normal, there is frequently no further tendency for hyperthermia to develop.

Acclimatization[3,4,6,12]

Among the seasonal changes observable in man and many vertebrates is a pattern of physiologic adjustments to temperature. In some instances these changes are profound, involving long periods of hibernation or torpidation in response to winter cold, and in others the physiologic changes can be seen only with careful testing.[3,24] For example, rats exposed for a few days to mildly cold temperatures of 15° to 20° C can survive continued exposure at 2° to 4° C (an otherwise lethal temperature) due to the development of a nonshivering thermogenesis that enables the animals to decrease their shivering and yet maintain body temperature. The extra metabolic heat results from the metabolism of brown fat from the tissue pads usually located in the interscapular region of the back; some days of cold exposure are required for the mobilization of this heat source by norepinephrine.[3]

Physiologic temperature regulation of humans in a cold environment depends on vasoconstriction and the increased heat production due to shivering or exercise. These latter are exhausting and can be maintained for prolonged periods only by the physically fit; adult humans do not have the capability of developing nonshivering thermogenesis so far as is known, although newborns are said to possess depots of brown fat that disappear in a few weeks. Vasoconstriction is an effective means of preventing heat loss from internal organs, as indicated in Fig. 59-13, but the process is limited by the vulnerability of the tissues to low temperature. Qualitatively, vasoconstriction is associated with the following:

1. Large axial thermal gradients in the extremities, ears, and nose (Fig. 59-4) with intense cold sensation and pain
2. Hypoxia in the hands, feet, legs, etc. with development of numbness, cramps, and tissue damage, particularly in the elderly
3. Skeletal muscle and joint stiffness
4. Dry, irritated skin (chapping)
5. Irreversible tissue damage due to local freezing (frostbite)

Since prolonged or repeated exposure to intense cold can result in any or all of these concomitants of vasoconstriction, it is obvious that humans cannot depend alone on physiologic regulation of body temperature under very cold con-

ditions. However, the mean conductance of peripheral tissues during cold exposure has been shown to be closely related to feelings of discomfort—the motivating principle for behavioral temperature regulation. Fig. 59-33 illustrates this effect for male and female subjects exposed in swimsuits to a hot environment (48° C) for 1 hr followed by exposure to a mildly cold environment (15° C). Reports of discomfort are plotted on a 4-point scale as ordinates, and the tissue conductance measured with a heat flow disk taped to the palm of the hand. During the heat exposure, vasodilatation was greater for the women than the men, as expected from the data shown in Fig. 59-13, and the women felt relatively less uncomfortably hot. On exposure to cold, vasoconstriction developed rapidly and the heat flow to the hand became progressively smaller until at the end of the cold exposure, almost no heat was being lost from the palmar surface. The initial vasoconstriction that reduced tissue conductance (and blood flow) by 40% was associated with sensations of thermal comfort; however, further vasoconstriction was felt at the uncomfortable level in the men and at the "unbearable" level in the women, who, despite their greater capacity for vasoconstriction (see conductance values in Fig. 59-13), were obviously more sensitive to these changes as they affected their extremities. The discomfort and pain associated with vasoconstriction produce behavioral activity by which humans can regulate body temperature in the cold (i.e., modifying their microclimate). It is not surprising, therefore, that human acclimatization to cold involves only the adoption of appropriate behavior to ensure that the body remains warm. The utilization of the means available in a particular environment is sometimes called "habituation" and is the basis of much that has been referred to as human acclimatization to cold.

Failure of the sense of cold discomfort to evoke activity can be serious, particularly in the elderly person living alone. These individuals sometimes develop spontaneous hypothermia while sitting or walking out-of-doors on a mild day. When these patients are admitted to the hospital, an initial medical examination reveals an internal body temperature of 30° C or lower. Revival is often difficult.

Physioiogic acclimatization to heat provides man with thermoregulatory capacities in addition to sweat and vasodilatation that can be lifesaving. Acute exposures to heat, especially when accompanied by exercise or work, place an unusual

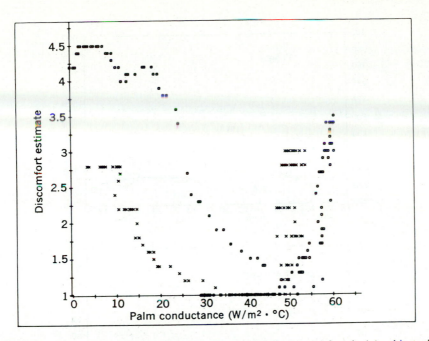

Fig. 59-33. Palmar conductance (blood flow to hand) of male (**x**) and female (□) subjects during exposure to heat (48° C) followed by cold (15° C) as related to reports of discomfort. Scale is as follows: 1 = comfortable; 2 = slightly uncomfortable; 3 = uncomfortable; 4 = very uncomfortable. (Modified from Hardy.[8])

strain on the cardiovascular system and on sweat gland function. Failure of the former results in hypotension, dyspnea, and fainting, which are symptomatic of heat syncope and *heat exhaustion.* The skin is usually damp and cool and the internal body temperature within normal ranges. Placing the individual in a supine position in the shade results in return of consciousness and disappearance of symptoms. Elderly individuals or those with cardiovascular impairment may not react as well to the strains of heat exhaustion. *Heat stroke* is characterized by failure of the sweating mechanism for reasons not understood at this time. Hot, dry skin and rising internal body temperature accompanied by periods of delirium and unconsciousness are signs of this dangerous syndrome. Cooling the body and immediate sustaining measures will lower internal body temperature and support the patient, but brain damage may cause death a day or so later. These "thermal disorders" and the difficulties of diagnosis and classification have been noted by Leithead and Lind[9]; the physiologic pathology of heat stroke is not well understood. Bridger and Helfand[20] made an epidemiologic study of the 1966 heat wave in southern Illinois, July 1 to 31, and related the death rates of various age groups

to the weather data. Current disease state, if any, reported cause of death, and sex were noted also for the three heat spells—July 1 to 7, July 9 to 15, and the last transient spell that had the highest temperatures on July 18. In analyzing their data the authors used July 1965 as a "normal" period. Fig. 59-34 shows the relative effect of the 1966 heat wave on the weekly deaths in St. Louis and the percentage of deaths of those 65 years and older. In their analysis the authors noted a 33% increase in mortality during the first spell, a 36% increase during the second, and a 10% increase in the third; the "sliding" averaging method resulted in the curve shown at the top of Fig. 59-34. As expected, of the population at risk, people 65 years of age and over suffered the greatest mortality, which was attributed largely to lesions of the vascular system of the brain and other diseases of the circulatory system (excluding heart diseases). Circulatory impairment could lead rapidly to failure of the physiologic thermoregulation, on which these individuals were entirely dependent. The authors also noted that the heat wave was sudden in onset, that it was the first truly hot spell of the year, and that the effect of such heat exposure on an essentially unacclimatized population might thus be severe.

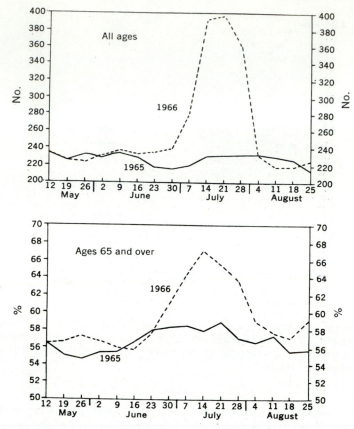

Fig. 59-34. Three-week average of number of deaths per week in St. Louis, Mo., with percentages for age 65 and over, summers of 1965 and 1966. (From Bridger and Helfand.[20])

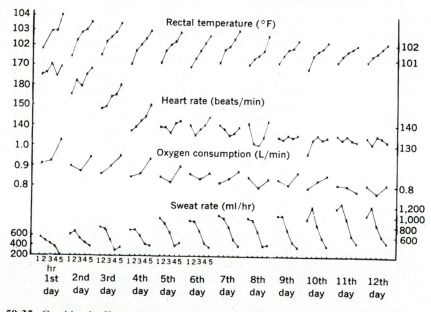

Fig. 59-35. Combined effects of heat acclimatization and physical training on five young men during 4 hr heat-work exposure. Dry- and wet-bulb temperatures were 36° and 34° C, respectively; wind velocity was 40 m/min. (Modified from Strydom et al.[54])

Acclimatization to heat consists of two parts—the development of additional capacity for cutaneous vasodilation and the capacity for an earlier onset and maintenance of sweating. When these physiologic developments are added to acquiring additional physical fitness, the results can be dramatic. Fig. 59-35 illustrates the combined effects of developing heat acclimatization and physical fitness in young men who were tested during a programmed period of work of 4 hr heat exposure. None of the four subjects tested could complete the test on the first day. Internal body temperature rose rapidly to unacceptable levels, pulse rate was high, and sweat rate declined before reaching high levels. At the end of 2 weeks, all subjects completed the work period with apparent ease, without undue elevations of body temperature or pulse rates, and with sweat rates at higher levels. Metabolic rate was also decreased, perhaps partly because of increased physical fitness. The requirement for producing heat acclimatization is repeated exposures to heat, which result in some elevation of internal body temperature (most easily accomplished by exercise) and in moderate to profuse sweating. Exposures of 1 to 4 hr for 5 to 7 days are usually sufficient. Failure to develop an acclimatization to heat may underlie a number of problems encountered by youthful athletes during the football season, particularly in the northern United States. Heat acclimatization is transient, usually disappearing in 3 to 4 weeks, if not maintained by repeated exposures to heat.

REFERENCES
General reviews

1. Atkins, E., and Bodel, P.: Fever, Physiology in Medicine, N. Engl. J. Med. **286:**27, 1972.
2. Bligh, J., and Moore, R. E.: Essays in temperature regulation, New York, 1972, Elsevier North Holland, Inc.
3. Carlson, L. D., and Hsieh, A. C. L.: Control of energy exchange, New York, 1970, Macmillan Publishing Co., Inc.
4. Dill, D. B., Adolph, E. F., and Wilber, C. G., editors: Handbook of physiology. Adaptation to the environment section, Baltimore, 1964, The Williams & Wilkins Co.
5. DuBois, E. F.: Fever and the regulation of body temperature, American lecture series, Publ. No. 13, Springfield, Ill., 1948, Charles C Thomas, Publisher.
6. Hammel, H. T.: Regulation of internal body temperature, Annu. Rev. Physiol. **30:**641, 1968.
7. Hardy, J. D.: Physiology of temperature regulation, Physiol. Rev. **41:**521, 1961.
8. Hardy, J. D.: Thermal comfort and health, ASHRAE J. **13:**43, 1971.
9. Leithead, C. S., and Lind, A. R.: Heat stress and heat disorders, Philadelphia, 1964, F. A. Davis Co.
10. Newberg, L. H., editor: Physiology of heat regulation and the science of clothing, Philadelphia, 1949, W. B. Saunders Co.

Original papers

11. Adair, E. R., Casby, J. U., and Stolwijk, J. A. J.: Behavioral temperature regulation in the squirrel monkey, J. Comp. Physiol. Psychol. **72:**17, 1970.
12. Andersson, B., et al.: Control of thyrotropic hormone (TSH) secretion by the "heat loss center," Acta Physiol. Scand. **59:**12, 1963.
13. Aschoff, J., and Wever, R.: Kern und Schale im Wärmehaushalt des Menschen, Naturwissenschaften **20:**477, 1958.
14. Bard, P., and Macht, M. B.: The behaviour of chronically decerebrate cats. In Ciba Foundation symposium on the neurological basis of behaviour, Boston, 1958, Little, Brown & Co.
15. Bard, P., and Woods, J. W.: Unpublished paper, 1962.
16. Bard, P., Woods, J. W., and Bleier, R.: The effects of cooling, heating and pyrogen on chronically decerebrate cats, Commun. Behav. Biol. **5:**31, 1970.
17. Barnes, R. B.: Thermography of the human body, Science **140:**870, 1963.
18. Bazett, H. C., et al.: Temperature changes in blood flowing in arteries and veins in man, J. Appl. Physiol. **2:**3, 1948.
19. Bullard, R. W., et al.: Skin temperature and thermoregulatory sweating: a control system approach. In Hardy, J. D., Gagge, A. P., and Stolwijk, J. A. J., editors: Physiological and behavioral temperature regulation, Springfield, Ill., 1970, Charles C Thomas, Publisher.
20. Bridger, C. A., and Helfand, L. A.: Mortality from heat during July 1966 in Illinois, Int. J. Biometeorol. **12:**51, 1968.
21. Cabanac, M., Hammel, T., and Hardy, J. D.: Tiliqua scincoides: temperature-sensitive units in lizard brain, Science **158:**1050, 1967.
22. Cooper, K. E., Cranston, W. I., and Honour, A. J.: Observations on the site and mode of action of pyrogens in the rabbit brain, J. Physiol. **191:**325, 1967.
23. Corbit, J.: Behavioral regulation of hypothalamic temperature, Science **166:**256, 1969.
24. Cottle, W. H., and Carlson, L. D.: The regulation of heat production in cold-adapted rats, Proc. Soc. Exp. Biol. Med. **92:**845, 1956.
25. DuBois, E. F.: The many different temperatures of the human body and its parts, West. J. Surg. **59:**476, 1951.
26. Fox, R. H., and MacPherson, R. K.: The regulation of body temperature during fever, J. Physiol. **125:**21P, 1954.
27. Gagge, A. P., Burton, A. C., and Bazett, H. C.: A practical system of units for description of the heat exchange of man with his environment, Science **94:**428, 1941.
28. Gagge, A. P., Hardy, J. D., and Stolwijk, J. A. J.: Proposed standard system of symbols for thermal physiology, J. Appl. Physiol. **27:**439, 1969.
29. Gordon, R. A., Britt, B. A., and Kalow, W., editors: International symposium on malignant hyperthermia, Springfield, Ill., 1973, Charles C Thomas, Publisher.
30. Guieu, J. D., and Hardy, J. D.: Effects of heating and cooling of the spinal cord on preoptic unit activity, J. Appl. Physiol. **29:**675, 1970.
31. Hammel, H. T., Caldwell, F. T., Jr., and Abrams, R. M.: Regulation of body temperature in the blue-tongued lizard, Science **156:**1260, 1967.
32. Hardy, J. D., Milhorat, A. T., and DuBois, E. F.: Basal metabolism and heat loss of young women at temperatures from 22° C to 35° C, J. Nutr. **21:**383, 1941.
33. Hellström, B., and Hammel, H. T.: Some characteristics

of temperaure regulation in the unanesthetized dog, Am. J. Physiol. **213**:547, 1967.

34. Hensel, H., and Wurster, R. D.: Static behavior of cold receptors in the trigeminal area, Pfluegers Arch. **313**: 153, 1969.

35. Keller, A. D., and Hare, W. K.: The hypothalamus and heat regulation, Proc. Soc. Exp. Biol. Med. **29**:1069, 1932.

36. Keller, A. D., and McClaskey, E. B.: Localization, by the brain slicing method, of the level or levels of the cephalic brainstem upon which effective heat dissipation is dependent, Am. J. Phys. Med. **43**:181, 1964.

37. Mead, J., and Bonarito, C. L.: Reliability of rectal temperatures as an index of internal body temperature, J. Appl. Physiol. **2**:97, 1949.

38. Milton, A. S., and Wendlandt, S.: A possible role for prostaglandin E_1 as a modulator for temperature regulation in the central nervous system of the cat, J. Physiol. **207**:76P, 1970.

39. Nadel, E. K., Bullard, R. W., and Stolwijk, J. A. J.: The importance of skin temperature in regulation of sweating, J. Appl. Physiol. **31**:80, 1971.

40. Nakayama, T., Eisenman, J. S., and Hardy, J. D.: Single unit activity of anterior hypothalamus during local heating, Science **134**:560, 1961.

41. Nielsen, M.: Heat production and body temperature during rest and work. In Hardy, J. D., Gagge, A. P., and Stolwijk, J. A. J., editors: Physiological and behavioral temperature regulation, Springfield, Ill., 1970, Charles C Thomas, Publisher.

42. Palmes, E. D., and Park, C. R.: The regulation of body temperature during fever, Arch. Environ. Health **11**: 749, 1965.

43. Pinkston, J. O., Bard, P., and Rioch, D. M.: The responses to changes in environmental temperature after removal of portions of the forebrain, Am. J. Physiol. **109**:515, 1934.

44. Randall, W. C., Wurster, R. B., and Lewin, R. J.: Responses of patients with high spinal transection to high ambient temperatures, J. Appl. Physiol. **21**:985, 1966.

45. Rawson, R. O., and Quick, K. P.: Evidence of deep-body thermoreceptor response to intraabdominal heating of the ewe, J. Appl. Physiol. **28**:813, 1970.

46. Roe, C. F.: Temperature in anesthesia. In Hardy, J. D., Gagge, A. P., and Stolwijk, J. A. J., editors: Physiological and behavioral temperature regulation, Springfield, Ill., 1970, Charles C Thomas, Publisher.

47. Rosendorff, C., and Mooney, J. J.: Central nervous system sites of action of a purified leucocyte pyrogen, Am. J. Physiol. **220**:597, 1971.

48. Seckendorf, R., and Randall, W. C.: Thermal reflex sweating in normal and paraplegic man, J. Appl. Physiol. **16**:796, 1961.

49. Sherrington, C. S.: Notes on temperature after spinal transection with some observations on shivering, J. Physiol. **58**:405, 1924.

50. Stitt, J. T., Hardy, J. D., and Stolwijk, J. A. J.: PGE_1 fever: its effect on thermoregulation at different low ambient temperatures, Am. J. Physiol. **227**:622, 1974.

51. Stolwijk, J. A. J.: Mathematical model of thermoregulation. In Hardy, J. D., Gagge, A. P., and Stolwijk, J. A. J., editors: Physiological and behavioral temperature regulation, Springfield, Ill., 1970 Charles C Thomas, Publisher.

52. Stolwijk, J. A. J., and Hardy, J. D.: Temperature regulation in man—a theoretical study, Pfluegers Arch. **291**:129, 1966.

53. Stolwijk, J. A. J., Saltin, B., and Gagge, A. P.: Physiological factors associated with sweating during exercise, Aerosp. Med. **39**:1101, 1968.

54. Strydom, N. B., et al.: Acclimatization to humid heat and the role of physical conditioning, J. Appl. Physiol. **21**:636, 1966.

55. Thauer, R.: Thermosensitivity of spinal cord. In Hardy, J. D., Gagge, A. P., and Stolwijk, J. A. J., editors: Physiological and behavioral temperature regulation, Springfield, Ill., 1970, Charles C Thomas, Publisher.

56. Webb-Peploe, M. M., and Shepherd, J. T.: Response of dogs cutaneous veins to local and central temperature changes, Circ. Res. **23**:693, 1968.

57. Wood, J. D.: Electrical activity from single neurons in Auerbach's plexus, Am. J. Physiol. **219**:159, 1970.

58. Wood, W. B., Jr.: The pathogenesis of fever, infectious agents and host reactions, Mudd, S., editor, Philadelphia, W. B. Saunders Co., 1970, p. 146.

XIII

ENDOCRINE GLANDS

60

H. MAURICE GOODMAN

Introduction to endocrinology

The endocrine system comprises a series of glands scattered throughout the body. Phylogenetically, it is an ancient system, with all but the parathyroid glands and placenta present and presumably functional in all classes of vertebrates. Superficially, at least, the endocrine glands appear quite unrelated. They may be paired or unpaired and derived from all three embryonic germ layers. While all the endocrine glands consist of secretory cells usually arranged in cords, or acini, the cells may or may not contain secretory granules and may have either a rough or a predominantly smooth endoplasmic reticulum. Unlike other secretory glands, which release their products into special ducts that lead to the body surface or to a hollow viscus, all the endocrine glands are ductless. Their secretions are delivered *internally* into the bloodstream—hence the name "endocrine" (from the Greek *endon*, within; *krinein*, to separate) and the term "internal secretion." The endocrine glands, together with the carotid bodies, are among the most highly vascularized tissues in the body. They frequently contain blood sinusoids, and their capillary endothelium is typically of the fenestrated type.

The secretory product of an endocrine gland is a *hormone* (from the Greek *hormaein*, to excite), a term first applied by Bayliss and Starling in the early 1900s to describe secretin, which is released into the blood by the duodenal mucosa and stimulates the flow of pancreatic juice (Chapter 54). We may define a hormone as *a chemical substance that is released into the blood in small amounts and that, on delivery by the circulation, elicits a typical physiologic response in other cells.* The restriction "in small amounts" excludes such substances as glucose, free fatty acids, various ions, and carbon dioxide, which might otherwise fit the definition. The term has been extended by some to include the "local hormones" such as acetylcholine (ACh), norepinephrine, or histamine, which reach their target cells by diffusion through the extracellular fluid rather than by transport through the general circulation. Although the distinction is somewhat arbitrary, some neurosecretory products such as the hypothalamic releasing factors are considered true hormones because they are transported in blood, albeit only for the short distance between the hypothalamus and the pituitary gland (Chapter 61). In recent years, some authors have also considered an analogous group of substances called pheromones to be yet another type of hormonelike agent. Pheromones such as the various sexual attractants in the invertebrates are released by one individual into the environment; on reaching another individual of the same species the attractants elicit behavioral or physiologic responses. To date, pheromones have not been extensively studied in mammals, although their existence has been well documented.[15]

Although the list is growing, the generally recognized endocrine glands are the pituitary, thyroid, parathyroid, pancreas, intestinal mucosa, adrenal, gonads, and placenta. Over the years a variety of other structures such as the pineal and thymus glands have tentatively been assigned an endocrine role, although supporting evidence is incomplete and agreement is not general.* The endocrine role of the kidney in volume regulation is discussed in Chapter 46. Its putative regulatory role in erythropoiesis is discussed by Stohlman.[19]

FUNCTIONS OF THE ENDOCRINE SYSTEM

Endocrinology has traditionally been organized in textbooks along anatomic lines, with

---

*Because their role in human physiology has not yet been elucidated and because space is limited, these glands will not be considered further in this section. The interested student can find detailed discussions of these glands and their possible physiologic significance in Goldstein et al.,[5] Kitay and Altschule,[9] and Wurtman et al.[24]

each gland and its hormones discussed separately. Although this approach tends to obscure the physiologic interrelationships among the hormones, it increases the usefulness of the text for reference purposes. Consequently, the anatomic format will be followed in the ensuing chapters. However, it is emphasized that, in life, hormones from a variety of endocrine glands work in concert to achieve physiologic control of the fundamental vital processes. It is this interrelationship among the hormones, be it in a cooperative or antagonistic manner, that provides both flexibility and subtle gradations in response that are the hallmark of homeostatic control. Furthermore, it is these functional interactions among the diverse glands that justifies our consideration of them as a physiologic system. We can define at least six major areas in which the hormones play a crucial integrative or regulatory role.

Control of digestive tract and its accessories. This function has already been discussed in Chapters 53 to 55.

Control of energy production. This complex area involves the storage, mobilization, interconversion, and utilization of the various metabolic fuels to permit continued survival despite intermittent food consumption. At least six hormones (insulin, glucagon, epinephrine, growth hormone, cortisol, and thyroxine) from four different glands are directly involved (Chapter 66).

Control of composition and volume of extracellular fluid. Seven hormones—antidiuretic hormone (ADH), adrenocorticotropic hormone (ACTH), aldosterone, cortisol, renin, thyrocalcitonin, and parathyroid hormone—acting in harmony, regulate both the ionic composition of the body fluids and the volume of the vascular compartment to maintain an optimal cellular environment and ensure its efficient renewal.

Adaptation to hostile environment. This area encompasses a wide range of responses from simple adjustment of skin coloration for camouflage in lower vertebrates to such complex adjustments in mammals as acclimatization to cold and resistance to infection, noxious agents, and stress. Pituitary, adrenal, and thyroid hormones are involved.

Growth and development. Complex interactions among virtually all the hormones govern normal growth and maturation.

Reproduction. This function, in contrast to all of those just cited, is not required for the survival of the individual but rather is concerned with the survival of the species. The endocrine system is involved in most phases, from production and maturation of the gametes through gestation and lactation (Chapter 65).

REGULATION OF HORMONE SECRETION

To regulate the foregoing processes, endocrine glands must be able to turn their secretions on and off at appropriate times. In maintaining the internal environment, control of hormone secretion is achieved by a self-regulating, circular series of events known as *negative feedback;* that is, a hormone produces a biologic effect that, on attaining sufficient magnitude, inhibits further secretion. Such a system contains the following components: (1) a detector of an actual or threatened homeostatic imbalance, (2) a means of signaling secretion, (3) the secretory cells, (4) an end-organ or organs capable of an appropriate response to the hormone, and (5) a means of promptly shutting off secretion when the response becomes adequate to restore homeostatic balance. Since control of the secretion of the individual glands will be considered in detail in the subsequent chapters, it is sufficient to note here that there is a wide range in the complexity of the regulatory systems. In the simplest case (e.g., the parathyroids or the alpha cells of the pancreas), components 1, 2, 3, and 5 reside in the same secretory cells. In other systems such as the control of thyroid secretion, complex interactions between the central nervous system (CNS), anterior pituitary gland, and thyroid gland are involved (Chapter 61).

Increased secretion of hormones can also be elicited in response to signals from the external environment. Oxytocin, which is necessary for milk release from the mammary glands (Chapter 61), is secreted in response to the young suckling at the nipples. Threatening changes in the environment evoke the release of pituitary and adrenal hormones. In such responses to external stimuli a sensory component of the nervous system detects the stimulus, and secretion is effected in a manner analogous to a simple reflex. There is no closed feedback loop as such, but hormone secretion is shut off with the cessation of the external stimulus, after satisfying the hungry infant or successfully coping with the threatening situation.

Hormones are not "used up" in exerting their biologic effects. They must therefore be inactivated or excreted if they are to remain effective signals. Inactivation occurs enzymatically in blood or intercellular spaces, in liver or kidney, or in the target tissues themselves. Inactivation may consist of cleavage, reduction, oxidation, or addition of some component such as a methyl group or glucuronic acid. Excretion is by way of the urine and, to a lesser extent, the bile.

Most hormones are destroyed rapidly and have

a half-life in blood of 10 to 30 min. However, some, such as the catecholamines, have half-lives on the order of seconds, while the half-lives of the thyroid hormones are measured in days. It should be emphasized here, however, that knowing the half-life of a hormone in blood does not necessarily provide insight into the time course of hormonal action. Some hormones produce their physiologic effects almost instantaneously, while others require a lag period of minutes or hours before the events we recognize as physiologic responses can occur. Some responses disappear immediately after the hormone disappears from the circulation, while others may persist for hours after hormone concentrations have returned to basal levels. A wide range in time constants for both onset and duration of hormone action presumably enhances the flexibility of hormonal responses.

MEASUREMENT OF HORMONES

Measurement of the concentrations of hormones in plasma is a formidable task. All the recognized hormones are proteins, polypeptides, amino acid derivatives, or steroids. With the exception of the thyroid hormones, which contain large amounts of iodine, there is no unique chemistry that sets them apart from other bodily constituents. Consequently, it has been difficult to devise chemical methods to measure hormone concentrations in biologic fluids. Furthermore, most hormones normally circulate at concentrations of 10^{-7} M or less. Measuring blood levels of a protein hormone such as growth hormone therefore requires a method capable of discerning a few thousandths of a microgram of hormone in the presence of 1 million times as much plasma protein. Several general methods have been developed.

Bioassay

A biologic assay utilizes some physiologic response to a hormone as a means of detecting its presence. Thus the induction of ovulation in the rabbit in response to the injection of urine from a pregnant woman is an indication of the presence of the placental hormone chorionic gonadotropin and is the basis for the familiar "rabbit test." In most bioassays the magnitude of the response measured is related to the quantity of hormone present. By comparing the response of test animals or tissues to the unknown material with a series of responses to known amounts of purified hormones, an estimate can be made of the amount of hormone present in the unknown sample. Thus if 1 ml of plasma produced the same degree of a biologic response as 10 μg of purified standard, it can be inferred that the milliliter of plasma contained about 10 μg of that hormone. Because of the small amounts of hormone in blood, it is often necessary to extract and concentrate the sample before assay. Some other difficulties of bioassays include variability in biologic responses, the presence in the unknown sample of substances that either enhance or inhibit the response to the hormone, and the presence of other substances that might exert a biologic effect similar to that of the hormone in question.

Before chemically pure hormones were available, hormones were quantitated biologically. Consequently, standardization of hormone activity was in units of biologic activity rather than units of weight. The amount of hormone that when administered under a particular set of circumstances produced a given response was defined as a unit. To this day, many hormones are standardized in terms of biologic units. For example, although there are approximately 25 units/mg, insulin is still packaged as 40 or 80 units/ml rather than 1.6 or 3.2 mg/ml. (A unit of insulin is defined as one third of the amount of insulin that will lower the blood sugar of a rabbit weighing 2 kg to convulsive levels in 3 hr.) The biologic responses on which standard unitages are based often require amounts of hormone that are tens of thousands of times greater than the concentrations of the hormones in blood. Consequently, hormone levels in blood are sometimes expressed as milliunits (thousandths of units) or microunits (millionths of units).

Radioimmunoassay

An ingenious technique devised by Berson and Yalow[2] has made it possible to measure minute concentrations of hormones in blood plasma. The method exploits the finding that protein hormones prepared from one species of animals may induce the formation of antibodies when injected into individuals of another species. These antibodies bind firmly and specifically to the hormone that induced their formation. To perform the assay, radioactive iodine, which can readily be measured in minute amounts, is first attached chemically to a pure sample of the hormone. A constant amount of the labeled hormone is then added to a test tube, together with a constant amount of antibody and the plasma sample, which contains some unknown amount of unlabeled hormone. The labeled and unlabeled hormones compete for binding sites on the antibody; the more unlabeled hormone present in the unknown sample, the less labeled hormone will be bound. Bound and unbound radioactivity can readily be separated by any of a number of physicochemical means, including electrophoresis, adsorption onto charcoal, etc. The ratio of bound to free labeled hormone is related in a precise manner to the amount of native, unlabeled hormone present in the unknown sample.

It is also possible to apply this technique for the assay of nonprotein hormones and other substances that ordinarily do not elicit antibody formation. Production of specific antibodies can be induced by first attaching these compounds to some protein such as serum albumin. Radioimmunoassays are now available for ADH, which is an octapeptide, for estrogen, which is a steroid, and even for thyroxine and its metabolites, which are amino acids.

The major limitation of radioimmunoassay is that *immunologic* rather than *biologic* activity is measured. The structural determinants of immunoreactivity are not necessarily the same as those of biologic activity. In fact, it is likely that variations in amino acid composition may be limited to nonessential positions of the hormone molecule. Consequently, a partially degraded, biologically inert hormone might retain immunoreactivity and thus might be indistinguishable from physiologically active hormone. Furthermore, an inborn error might lead to the production of a biologically inert hormone analog, which might well retain normal immunologic reactivity, or vice versa. Another important limitation of radioimmunoassay is that only hormones that are available in absolutely pure form can be studied, for even small amounts of contaminating protein can result in an assay of the contaminant rather than the hormone. Despite these theoretical objections, the technique of radioimmunoassay has made possible the most significant advances in our knowledge of the physiology of the protein hormones since their discovery and isolation. The importance of this technique was recognized by the awarding of a Nobel prize to Dr. Rosalind Yalow.

Competition for binding between the natural hormone and a radioactively labeled hormone has also been used in other nonimmunologic assays such as that devised by Murphy[13] for some of the steroid hormones. In this assay the corticosteroid-binding globulin of blood plasma is used instead of the antibody. Theoretically, any substance that binds specifically and reversibly to a hormone can be used for such an assay. The finding that many hormones bind with high affinity and great specificity to certain "hormone receptors" (see following section) on the surface of target cells or within their cytoplasm has led to the development of "radioreceptor assays," in which purified preparations of cell membranes or cytoplasmic binding proteins are used as ligands in place of the antibodies in radioimmunoassays.[17] Theoretically, at least, radioreceptor assays are more specific than radioimmunoassays in that biologic activity and the ability to bind to the receptor are presumably closely related.

HOW HORMONES WORK

Before embarking on a detailed survey of the hormones, it will be useful to consider briefly some current ideas of how they work. New information (and unfortunately, misinformation) is accumulating rapidly in this area of very active investigation. Consequently, the following paragraphs will doubtlessly require modification as more data are obtained.

Since all hormones travel in the blood from their glands of origin to their "target" tissues, all cells must be exposed to all hormones. Yet under normal circumstances, tissues respond only to their appropriate hormones. Such *specificity* of hormone action appears to reside in the capacity of receptors in the target tissue to recognize only their own signal. We may define a hormone *receptor* as a hypothetical, unique molecular grouping in or on a cell that interacts with a hormone in a highly specific manner so that a characteristic response or group of responses is initiated.

In producing their biologic effects, hormones appear to commandeer and redirect some of the normal regulatory processes that control day-to-day cellular activities. They may do so by (1) controlling the formation or destruction of some intracellular regulator such as cyclic adenosine monophosphate (cyclic AMP), (2) controlling the protein synthetic apparatus, thus causing cells to make new kinds of enzymes or increased amounts of some enzymes already in production, or (3) controlling the movements of molecules across the plasma membrane, thus regulating the availability of substrates, ions, and cofactors. Clearly, these mechanisms are not mutually exclusive, and no one of them alone can account for all of the actions of certain hormones such as insulin. Indeed, the question of whether a hormone has just one primary action or a multiplicity of primary actions has yet to be resolved. Some investigators believe that there is a single unique interaction between a hormone and a receptive cell and that all subsequent events follow as secondary and tertiary consequences of that interaction. Other investigators suggest that a single hormone may interact with more than one receptor in a target cell and that each interaction produces its own initial event that may in turn entrain secondary and tertiary events. The idea of a single primary action adheres more closely to the "principle of parsimony" and therefore seems more intellectually satisfying to many investigators. In an effort to salvage this explanation for the action of such hormones as insulin, Hechter and Halkerston[7] offered the "cytoskeleton" hypothesis, which suggests that cellular function depends on the orderly arrangement of cellular constituents on some framework, the cytoskeleton. Interaction between cell and hormone may result in a disturbance or rearrangement of the cytoskeleton that produces characteristic changes in various cellular functions. To date, it has not been possible to obtain the data necessary to either refute or support this interesting idea.

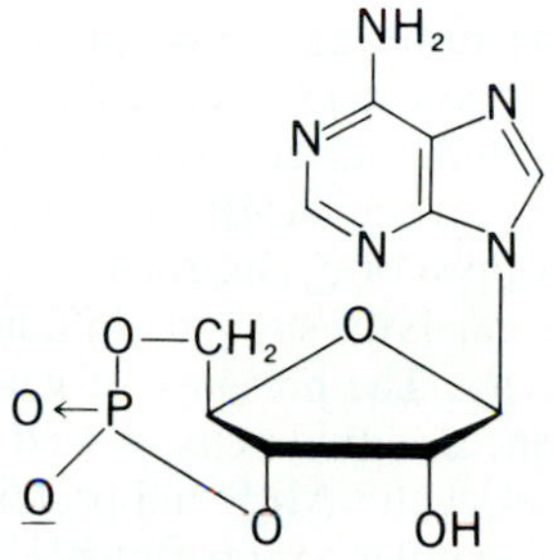

Cyclic adenosine monophosphate and the second messenger hypothesis

While studying the hormonal control of glycogen breakdown, Sutherland and Rall[20] and associates discovered the previously unknown adenine nucleotide 3',5'-adenosine monophosphate, or cyclic AMP (Fig. 60-1). The series of reactions that is initiated by the glycogenolytic hormones epinephrine and glucagon and mediated by cyclic AMP are shown in Fig. 60-2. Glycogenolysis is catalyzed by the enzyme phosphorylase, which exists in both an active and a relatively inactive form. The "inactive" form is activated by the addition of two phosphate groups to two of its serine residues. This activation in turn is catalyzed by the enzyme phosphorylase kinase, which also exists in an inactive dephospho- form and an active phosphorylated form. Cyclic AMP activates the enzyme phosphorylase kinase kinase, which converts phosphorylase kinase to its active form. The enzyme adenylate

Fig. 60-1. Adenosine 3',5'-monophosphate (cyclic AMP).

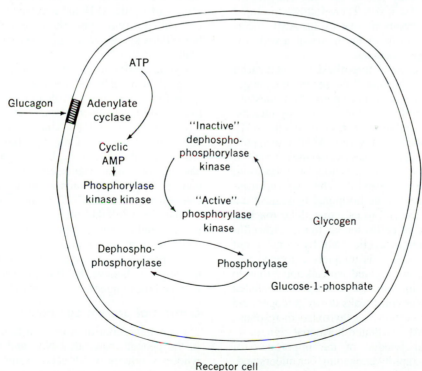

Fig. 60-2. Schematic representation of enzymatic steps involved in stimulation of glycogen breakdown by glucagon.

cyclase, located in the plasma membrane of hepatic cells, catalyzes the formation of cyclic AMP from adenosine triphosphate (ATP). Epinephrine and glucagon trigger the whole process by interacting with receptor molecules that are closely associated with or somehow linked to adenylate cyclase.

Subsequent observations established that cyclic AMP and the enzymes involved in its formation and destruction are present in virtually all nucleated mammalian cells as well as in a wide variety of other organisms, including even bacteria. Shortly after its discovery in liver, Haynes and Berthet[6] proposed that cyclic AMP might also mediate the effects of ACTH in adrenal cortical cells. They suggested that increased hormone production depends on increased breakdown of glycogen to glucose, whose oxidation provides the reduced pyridine nucleotide (NADPH) needed for hormone synthesis. Although subsequent studies revealed that activation of phosphorylase is probably not causally involved in stimulation of adrenal steroid production and secretion, the Haynes hypothesis implicated cyclic AMP as an intermediate in the actions of a hormone other than epinephrine and glucagon and led ultimately to the realization that phosphorylase may not be the only enzyme responsive to cyclic AMP. There followed in rapid succession observations that cyclic AMP mediates the actions of many hormones on a wide variety of cellular processes.

Second messenger hypothesis. As advanced by the Sutherland group,[21] the second messenger hypothesis states that hormones (first messengers) may exert their biologic effects by increasing the intracellular formation of cyclic AMP (the second messenger). Cyclic AMP in turn governs such diverse processes as enzyme activation, secretion, muscular contraction or relaxation, and membrane permeability. The "second messenger" thus relays the hormonal message from the surface of the cell to its intracellular machinery. Each kind of cell responds to the intracellular accumulation of cyclic AMP by carrying out whatever processes it is uniquely equipped for. Thus, just as a sweat gland secretes and a muscle cell contracts in response to an ACh-mediated signal, so a liver cell breaks down glycogen and a thyroid cell synthesizes thyroxine in response to a cyclic AMP–mediated hormonal signal.

Although knowledge of the importance of cyclic AMP is rapidly expanding our understanding of an ever-widening range of biochemical and physiologic processes, the nature of its actions at the molecular level are still under intensive investigation. Protein phosphorylation seems to be related to many of its effects, and cyclic AMP–dependent "protein kinases" have been found in a variety of tissues. Phosphorylation of a protein is thought to change its configuration and hence its ability to interact with other substances. Cyclic AMP–dependent protein kinases (two are known) are tetramers consisting of two "catalytic subunits" and two "regulatory subunits."[1] The catalytic subunits are responsible for catalyzing the transfer of the terminal phosphate group of ATP to the hydroxyl group of a serine residue in a receptive protein. The regulatory subunit has the dual capability of binding cyclic AMP and inhibiting catalytic activity. In binding to the regulatory subunits, cyclic AMP lowers their affinity for the catalytic subunits. The resulting dissociation of the catalytic subunits unmasks their catalytic activity. When the hormone effect wanes and local concentrations of cyclic AMP fall, the regulatory subunit is stripped of cyclic AMP and once again binds to the catalytic subunit, inhibiting its enzymatic activity. The presence of various modulating substances within cells as well as possible segregation of cyclic AMP and protein kinase in localized subcellular compartments have made this system quite difficult to study, and hence all the details of these reactions are not known. The effects of cyclic AMP are reversed by the action of another enzyme, phosphoprotein phosphatase, but little is known about this enzyme and its regulation.

Cyclic AMP may not be the only second messenger. Many laboratories are engaged in the quest for the physiologic role of cyclic guanosine monophosphate, which in many respects is analogous to cyclic AMP. The prostaglandins, a series of derivatives of the essential fatty acids, which are pharmacologically very potent, have also been nominated as second messengers, but definite evidence for such a role is lacking. The calcium ion also acts as a second messenger to mediate such cellular processes as muscular contraction and neurosecretion and thus conveys the message from an event at the cell surface, the action potential, to the intracellular effects. Many investigators believe that the calcium ion may be a second messenger for hormonal signals as well.

Control of protein synthesis

Ultimately, control of cellular function resides in the genetic material, DNA, and the proteins whose synthesis it directs. Some hormones, notably the steroids, affect DNA and evoke the synthesis of messenger RNA, which carries the information for the new proteins to be synthesized from the chromosomes to the protein factories, the ribosomes. The end result is a modification in the physiologic behavior of the target cells.

Steroid hormones penetrate the cell membranes and accumulate within the nucleus. It is still not certain whether they enter cells by simple diffusion or by a carrier-mediated process. Because of their high lipid solubility, they may enter many cells, but only target cells contain the protein receptors that tightly bind the steroid hormone molecules and prevent their escape. In the most thoroughly studied system to date, the chick oviduct, each target cell contains about 10,000 receptors, which, in the absence of hormone stimulation, are found exclusively in the cytoplasm. The receptors are dimers of two nonidentical subunits, each of which has a molecular weight of about 100,000. One molecule of hormone binds to each subunit. Once binding has occurred, the receptor-hormone complex migrates to the nucleus by some still unknown process. Within the nucleus the hormone-laden receptors bind to specific sites in the chromatin. Target cells of the chick oviduct contain about 5,000 such binding sites. In addition to the DNA, chromatin also contains basic proteins or histones and perhaps as many as 500 different species of acidic proteins associated with DNA in some highly organized fashion. Only the chromatin from target cells contains the specific acidic proteins capable of recognizing and binding to the hormone receptor complexes. When nuclei or chromatin isolated from nontarget cells is added to purified hormone receptor complexes, little or no binding occurs. It appears that one of the subunits of the receptor recognizes and binds to a specific acidic protein in the chromatin. The other subunit of the receptor then dissociates and binds to the adjacent portion of the DNA strand. In this way the hormone and receptor subunit are directed to the appropriate locus on the DNA. Binding of the receptor subunit somehow allows the enzyme RNA polymerase to bind to a specific initiation site on DNA. A segment of DNA is then transcribed as a strand of messenger RNA. The new messenger RNA directs the synthesis of new "induced proteins," which are thought to be responsible for the various hormone effects. These reactions are illustrated schematically in Fig. 60-3. Although not all these steps have been

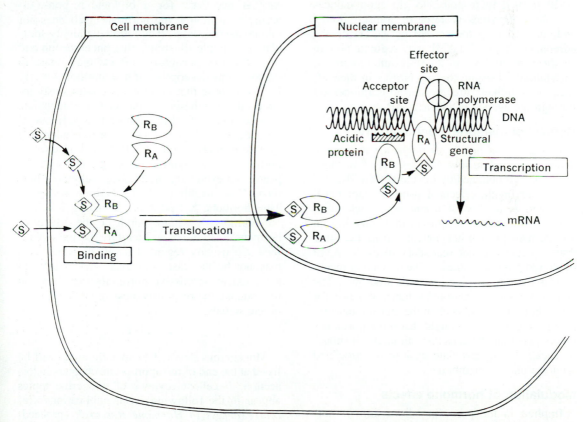

Fig. 60-3. Schematic representation of intracellular events involved in the molecular action of steroid hormones. S, Steroid; R_A, receptor subunit A; R_B, receptor subunit B. (See text.) (From Buller et al.[3])

demonstrated for each of the steroid hormones, the great similarities in their behavior appear to justify consideration of a single mode of action for all. A delightfully readable account of the brilliant research that uncovered the preceding steps of steroid hormone action can be found in O'Malley and Schrader.[14] Still to be defined are the processes that account for turning off the hormone effect and restoring responsiveness of the cells to subsequent stimulation by hormone.

The steroid hormones appear to control cellular function at the level of RNA synthesis (transcription). Hormones can also exert control at the level of the ribosome (translation) by controlling protein synthesis by ribosomes already charged with messenger RNA. Growth hormone[10] and insulin[23] are thought to work in this way, but the molecular mechanisms are not understood. These hormones probably do not interact directly with the ribosomes.

Cyclic AMP can also induce protein synthesis in some cells. Observations with liver cells suggest an interaction between cyclic AMP and nuclear histones, which are thought to have a role in the regulation of genetic expression.[11] Cyclic AMP formed in response to glucagon induces new liver enzymes (control of transcription)[22] and, in mediating the actions of ACTH on the adrenal cortex, cyclic AMP may regulate protein synthesis at the level of the ribosome (control of translation).[4] Despite the similarities in their effects, it is unlikely that the steroid hormones act through the agency of cyclic AMP.

Control of membrane permeability

Evidence that hormones might act at the cell surface to regulate inward transport of metabolites was first obtained by Levine and Goldstein[12] while studying the effects of insulin (Chapter 66). Although a number of hormones affect membrane permeability, it appears that only insulin and perhaps growth hormone do so in a manner that apparently is not secondary to cyclic AMP or altered protein synthesis. The molecular mechanisms that are involved are still unknown. The cell membrane is a reasonable locus of action for many hormones. Other than the steroid and thyroid hormones, it is thought that most hormones do not penetrate the cells they stimulate. It should be recalled that adenylate cyclase is associated with the plasma membrane.

Modulation of hormone effects

Implicit in the preceding discussion of hormone action is the idea that hormones act as signals that turn specific responses in receptive tissues on or off. This is not always true. Hormones may also affect the responsiveness of receptive tissues to physiologic signals. One way this may be achieved is through regulating the number or perhaps the accessibility of hormone receptors in target tissues. One hormone may modify the sensitivity of a tissue to a second hormone by increasing or decreasing the number of hormone receptors available to interact with the second hormone. For example, the ovarian hormone estrogen increases the number of binding sites for the pituitary hormones luteotropic hormone and follicle-stimulating hormone in granulosa cells in the ovary and thereby increases their responsiveness.[16] Hormones may even regulate their own receptors in target tissues. Exposure to high concentrations of insulin decreases both the number of insulin receptors on target cells and the responsiveness of these cells to insulin.[18] The mechanisms involved in regulation of hormone receptors are not known.

Another way that hormones may affect the responsiveness of tissues has been called *permissive action,* a term that describes without explaining. A hormone acts permissively when its presence is necessary for a biologic response to occur, even though the hormone itself does not initiate the response.[8] The following highly speculative example illustrates this phenomenon and indicates how permissive effects may relate to the foregoing description of how hormones work. Let us suppose that hormone X continuously induces the synthesis of the enzyme adenylate cyclase. Let us further suppose that hormone Y produces its biologic effects by causing adenylate cyclase to make cyclic AMP. In the absence of hormone X, adenylate cyclase gradually disappears, and even very high concentrations of hormone Y are unable to elicit a biologic response. Thus hormone X "permits" hormone Y to act, but hormone X cannot duplicate the effects of hormone Y. In the older literature, permissive effects are usually regarded as all or none. This may not be the case, however, and at least to some extent permissive hormones may function to modulate the responsiveness of cells to physiologic signals.

• • •

Monographs devoted to specific areas will be listed at the end of the appropriate chapters in this section. Excellent reviews of endocrine topics appear in the following serial publications: *Recent Progress in Hormone Research* (proceedings of the annual Laurentian Hormone Conferences); *Vitamins and Hormones; Annual Review of Physiology, Annual Review of Pharmacology,* and *Annual Review of Biochemistry; Physiological Reviews;* and *Pharmacological Reviews.* A comprehensive multivolume treatise published recently by the American Physiological Society in the *Handbook of Physiology* series offers in-depth coverage of the most important topics in endocrinology.

REFERENCES

1. Beavo, J. A., Bechtel, P. J., and Krebs, E. G.: Mechanisms of control for cAMP-dependent protein kinase from skeletal muscle, Adv. Cyclic Nucleotide Res. **5:** 241, 1975.
2. Berson, S. A., and Yalow, R.: Radioimmunoassays of peptide hormones in plasma, N. Engl. J. Med. **277:**640, 1967.
3. Buller, R. E., et al.: Progesterone binding components of chick oviduct: in vitro effect of receptor subunits on gene transcription in vitro, J. Biol. Chem. **251:**5166, 1976.
4. Garren, L. D.: On the mechanism of action of ACTH, Recent Prog. Horm. Res. **27:**433, 1971.
5. Goldstein, A. L., Asanuma, Y., and White, A.: The thymus as an endocrine gland: properties of thymosin, a new thymus hormone, Recent Prog. Horm. Res. **26:** 505, 1970.
6. Haynes, R. C., Jr., and Berthet, L.: Studies on the mechanism of action of adrenocorticotropic hormone, J. Biol. Chem. **225:**115, 1957.
7. Hechter, O., and Halkerston, I. D. K.: On the action of mammalian hormones. In Pincus, G., Thimann, K. V., and Astwood, E. B., editors: The hormones, New York, 1964, Academic Press, Inc., vol. 5.
8. Ingle, D. J.: Permissibility of hormone action. A review, Acta Endocrinol. **17:**172, 1954.
9. Kitay, J. I., and Altschule, M. D.: The pineal gland, Cambridge, Mass., 1954, Harvard University Press.
10. Korner, A.: Anabolic action of growth hormones, Ann. N.Y. Acad. Sci. **148:**408, 1968.
11. Langan, T. A.: Action of adenosine 3′,5′-monophosphate-dependent histone kinase in vivo, J. Biol. Chem. **244:**5763, 1969.
12. Levine, R., and Goldstein, M. S.: On the mechanism of action of insulin, Recent Prog. Horm. Res. **11:**343, 1969.
13. Murphy, B. E. P.: Protein binding and the assay of nonantigenic hormones, Recent Prog. Horm. Res. **25:**563, 1955.
14. O'Malley, B. W., and Schrader, W. T.: The receptors of steroid hormones, Sci. Am. **234:**32, 1976.
15. Parkes, A. S., and Brice, H. M.: Olfactory stimuli in mammalian reproduction, Science **134:**1049, 1961.
16. Richards, J. S., et al.: Ovarian follicular development in the rat: hormone receptor regulation by estradiol, follicle stimulating hormone and leutinizing hormone, Endocrinology **99:**1562, 1976.
17. Roth, J.: Peptide hormone binding to receptors: a review of direct studies in vitro, Metabolism **22:**1059, 1973.
18. Roth, J., et al.: Receptors for insulin, NSILA-S and growth hormone. Applications to disease states in man, Recent Prog. Horm. Res. **31:**95, 1975.
19. Stohlman, F., Jr.: The kidney and erythropoiesis, N. Engl. J. Med. **279:**1437, 1968.
20. Sutherland, E. W., and Rall, T. W.: The relation of adenosine-3′,5′-phosphate and phosphorylase to the action of catecholamines and other hormones, Pharmacol. Rev. **12:**265, 1960.
21. Sutherland, E. W., Øye, I., and Butcher, R. W.: The action of epinephrine and the role of the adenyl cyclase system in hormone action, Recent Prog. Horm. Res. **21:**623, 1965.
22. Wicks, W. D.: Induction of hepatic enzymes by adenosine 3′,5′-monophosphate in organ culture, J. Biol. Chem. **244:**3941, 1969.
23. Wool, I. G., et al.: Mode of action of insulin in the regulation of protein biosynthesis in muscle, Recent Prog. Horm. Res. **24:**139, 1968.
24. Wurtman, R. J., Axelrod, J., and Kelly, D. E.: The pineal, New York, 1968, Academic Press, Inc.

61

H. MAURICE GOODMAN

The pituitary gland

The pituitary gland was so named by the great Renaissance anatomist Vesalius, who thought its function was to collect the *pituita,* or mucus, produced by the brain and to discharge it as a lubricant of the nose and throat. Because of its position under the brain the pituitary gland is also called the hypophysis (from the Greek *hypon,* under; and *phyein,* to grow). Over the years, various functions have been ascribed to it, including the role of center of intellectuality, center of piety, and center of the emotions. Willis (1664) thought the pituitary produced the cerebrospinal fluid (CSF), while Magendie (1847) thought it absorbed the CSF and discharged it into the blood. The first inkling of one of its real functions came in 1871, when Lorain noted that damage to the pituitary was associated with failure of growth. The first demonstration that it contained biologically active material came in 1895 When Oliver and Shäfer[117] found that aqueous extracts of the pituitary gland caused intense vasoconstriction and hypertension. Harvey Cushing (1905) performed one of the first complete hypophysectomies (removal of the pituitary gland) on a puppy and noted the ensuing failure of growth, but the dog survived for only 3 weeks. Aschner (1912) was more successful and clearly established that hypophysectomized animals fail to grow. Real appreciation of the importance of the pituitary gland did not come until 1926, when P. E. Smith developed a simple, relatively atraumatic surgical technique for removing the pituitary glands from rats. He found that in addition to preventing growth, hypophysectomy also produced atrophy of the adrenal and thyroid glands, the gonads, and the accessory reproductive structures. Administration of aqueous pituitary extracts reduced or abolished these effects.[148,149] These observations, coupled with the findings of Evans and his colleagues and of Aschheim and Zondek that pituitary extracts given to normal animals augmented growth and produced gonadal hypertrophy, ultimately led to the idea that the pituitary is the ''master gland'' that regulates

the function of many of the other endocrine glands (Fig. 61-1).

Morphology

The human pituitary gland is a round or ovoid structure about the size of a sweet pea; it weighs about 0.5 gm. It sits in a small depression in the sphenoid bone called the *sella turcica,* located just beneath the hypothalamus, and is connected to the hypothalamus by a thin stalk. Grossly and microscopically, the pituitary can be divided into two regions (Fig. 61-2). The anterior portion, which appears quite glandular histologically, is called the *adenohypophysis.* It is composed of three regions: (1) the *pars distalis,* which forms the bulk of the *anterior lobe,* (2) the thin collar of cells that surrounds the stalk, the *pars tuberalis,* and (3) the thin rim of cells located at the junction with the neural lobe, the *pars intermedia,* or intermediate lobe. The posterior portion, or *neurohypophysis,* also consists of three regions: (1) the neural lobe, (2) the stalk or *infundibulum,* and (3) the *median eminence* of the *tuber cinereum,* which forms the attachment with the hypothalamus. Embryologically, the adenohypophysis develops from an evagination of the primitive gut called Rathke's pouch. The neurohypophysis arises from an outgrowth of the primitive brain stem.

Blood supply

The anterior pituitary gland receives nearly all its blood supply from the superior and inferior hypophyseal arteries, which arise from the internal carotids. The superior hypophyseal arteries enter the median eminence and break up into numerous branches and capillary loops (the primary plexus). These capillaries reconverge to form a system of parallel venules (the long portal vessels) that descend the stalk, penetrate the adenohypophysis, and empty into capillary sinusoids bathing the secretory cells of the anterior lobe. The inferior hypophyseal arteries supply a similar capillary plexus in the lower portion of

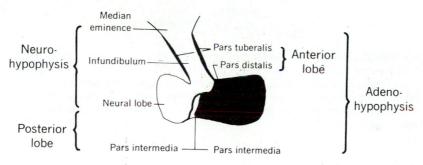

Fig. 61-2. Diagram of human pituitary gland in midsagittal section. (Redrawn from Xuereb et al.[162])

the infundibular stem. These vessels reconverge to form the "short portal vessels," which break up into a second sinusoidal capillary bed within the adenohypophysis.[37] This vascular arrangement is called the *hypophyseal portal system* and is analogous to the portal blood supply of the liver. In addition, the anterior pituitary may receive some small portion of its blood supply through the paired trabecular arteries that also course down the stalk (Figs. 61-3 to 61-5). The neurohypophysis receives its blood supply through the inferior hypophyseal arteries. The intermediate lobe is virtually avascular. Venous drainage is by way of short hypophyseal veins that empty into nearby venous sinuses.

Microscopic anatomy

Histologically, the adenohypophysis consists of large polygonal cells arranged in cords surrounded by a sinusoidal capillary system. Most of the cells contain secretory granules, although some are only sparsely granulated or agranular. The secretory granules take up either acidic or basic stains and are thus *acidophilic* or *basophilic*. The cells that are agranular or sparsely granulated take up little stain and are hence *chromophobic*. Acidophilic and basophilic cells can be further subdivided on the basis of histochemical reactions, shape, and location. An experienced cytologist can identify at least as many cell types as there are known anterior pituitary hormones and, thanks in part to immunofluorescent techniques, can assign a particular cell type to each hormone. Although certain cells may be found more commonly in either the peripheral or central areas, the various cell types are not segregated but are freely interspersed throughout the anterior lobe.

The neural lobe contains small cells, the pituicytes, that are transformed neuroglia cells. It also

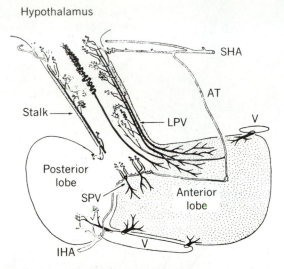

Fig. 61-3. Vascular supply of human pituitary gland. Diagram shows origin of long portal vessels, *LPV*, from primary capillary bed and origin of short portal vessels, *SPV*, from capillary bed in lower part of stalk. Both sets of portal vessels break up into sinusoidal capillaries in anterior lobe. *SHA* and *IHA*, Superior and inferior hypophyseal arteries, respectively. *AT* is artery of trabecula that forms an anastomotic pathway between *SHA* and *IHA*. *V* indicates venous sinuses. (Redrawn from Daniel and Prichard.[38])

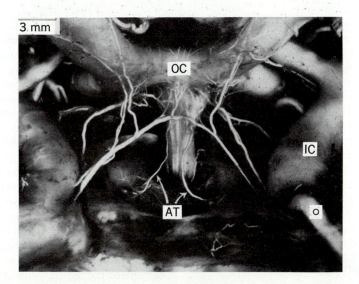

Fig. 61-4. Neoprene latex injection of vasculature of human pituitary stalk viewed from front. *OC,* Optic chiasm; *IC,* internal carotid artery; *O,* ophthalmic artery; *AT,* trabecular artery. (From Xuereb et al.[162])

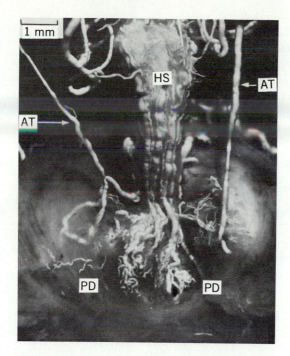

Fig. 61-5. Anterior aspect of human pituitary stalk, *HS*, with vasculature injected with neoprene latex. *AT* indicates paired trabecular arteries and *PD* pars distalis. (From Xuereb et al.[162])

contains connective tissue elements and a profusion of nonmedullated nerve fibers whose cell bodies lie in the supraoptic and paraventricular nuclei of the hypothalamus.

Adenohypophysis

PHYSIOLOGY OF THE ADENOHYPOPHYSEAL HORMONES

The adenohypophysis secretes at least seven recognized hormones, four of which directly control the functioning of their particular "target" glands, the adrenals, thyroid, or gonads. These hormones were dubbed "trophic" or "tropic" (from the Greek *trophos,* to nourish, or *tropos,* to turn toward). Actually, they neither nourish nor turn toward their target glands, but the terms have remained in common usage. For no stronger reason than euphonics, the suffix *tropic* will be used in this discussion. A variety of other pharmacologically active peptides have also been isolated from the pituitary, but since their role in normal physiology is still uncertain, they will not be discussed here.

Adrenocorticotropic hormone (corticotropin)

The primary physiologic role of the adreno-corticotropic hormone (ACTH) is to evoke the secretion and synthesis of cortisol and corticosterone by the adrenal cortex. Probably as an indirect consequence of these actions, ACTH also maintains the size and blood flow of the adrenal cortex. When present in excessive amounts, it causes enlargement of the two inner zones of the adrenal cortex, the *zona fasiculata* and the *zona reticularis*. In its absence, adrenal cortical atrophy occurs, particularly in these zones.

The effects of ACTH on the adrenal cortex are probably mediated by cyclic adenosine monophosphate (cyclic AMP).[132] A number of extra-adrenal effects of ACTH have also been observed after administration to experimental animals or addition to tissues in vitro.[50] The physiologic importance of these extra-adrenal effects of ACTH are open to serious question. It is of interest to note, however, that most if not all of these extra-adrenal effects of ACTH are on cyclic AMP–mediated systems. The human diseases that stem from derangements in ACTH secretion are the same as those caused by over- or under-production of cortisol (Chapter 64).

ACTH was the first of the anterior pituitary hormones to have its structure completely defined[17,23] and verified by artificial synthesis.[76,100,141] It is an unbranched peptide chain of 39 amino acids and has a molecular weight of 4,567 (Fig. 61-6). Some biologic activity is present in a molecule consisting of only the first 13 amino acids (starting at the amino terminal), and elongation of the chain leads to progressively greater activity. Full activity is attained with the first 20 amino acids in some systems. The remaining amino acids are responsible for the immunologic activity of the molecule. Species differences in amino acid sequence occur only in this region (residues 25 to 32).[79] Physiologically, this portion of the molecule may play some role in hormone transport or in protection from enzymatic degradation.

When extracts of human pituitary glands were placed into reaction with antibodies to ACTH, two forms of immunologically reactive ACTH were found.[163] One, which had a molecular weight of about 5,000, was biologically active; the other form had a higher molecular weight and no biologic activity. Controlled digestion of the so-called "big ACTH" with the proteolytic enzyme trypsin converted it to a form that had full biologic activity and was indistinguishable from authentic ACTH. Evidence is mounting that

1	2	3	4	5	6	7	8	9	10	11	12	13	14	15	16	17	18	19	20	21	22	23	24
Ser·	Tyr·	Ser·	Met·	Glu·	His·	Phe·	Arg·	Try·	Gly·	Lys·	Pro·	Val·	Gly·	Lys·	Lys·	Arg·	Arg·	Pro·	Val·	Lys·	Val·	Tyr·	Pro·

Species

25	26	27	28	29	30	31	32	33	34	35	36	37	38	39	
Asp·	Gly·	Glu·	Ala·	Glu	·Asp	·Ser·	Ala·	Gluta·	Ala·	Phe·	Pro·	Leu·	Glu·	Phe	Cow
Ala·	Gly·	Glu·	Asp·	Asp	·Glu	·Ala·	Ser·	Gluta·	Ala·	Phe·	Pro·	Leu·	Glu·	Phe	Sheep
Gly·	Ala·	Asn·	Glu·	Asp	·Glu	·Ser·	Ala ·	Glu	·Ala·	Phe·	Pro·	Leu·	Glu·	Phe	Man
Asn·	Gly·	Ala·	Glu·	Asp	·Glu	·Leu·	Ala ·	Glu	·Ala·	Phe·	Pro·	Leu·	Glu·	Phe	Pig
Asp·	Gly·	Ala·	Glu·	Asp·	Gluta·	Leu·	Ala·	Glu	·Ala·	Phe·	Pro·	Leu·	Glu·	Phe	Synthetic

Fig. 61-6. Amino acid sequence of ACTH showing species variations that occur only in positions 25 to 39. (Redrawn from Ney[113] to include corrections and structure described by Riniker et al.[129])

many of the protein and peptide hormones are synthesized as larger molecules or *prohormones,* which are enzymatically clipped to release the biologically active hormone. Big ACTH may be an example of such a prohormone. Normal blood plasma seldom contains much big ACTH, but in disease states characterized by the hypersecretion of ACTH, appreciable amounts of big ACTH may be seen. It is of interest that big ACTH is often produced and secreted by malignant tumors of nonpituitary origin, particularly those of the lung.

The parent molecule from which ACTH is derived appears to be a glycoprotein with a molecular weight of about 30,000 to 40,000. This large precursor molecule also contains other biologically active peptides, including α- and β-MSH (see p. 1478) and α- and β-lipotropin, which stimulate fatty mobilization in some bioassays.[78a] A physiologic role for α- and β-lipotropin as metabolic regulators is doubtful, but these peptides contain the amino acid sequences of two new classes of peptides, which may prove to be of great physiologic importance. These are the so-called enkephalins and endorphins. The enkephalins are pentapeptides that bind to morphine receptors in the CNS and gastrointestinal tract and mimic the actions of morphine and other opiate alkaloids. The endorphins (α, β, and γ) are larger peptides, up to 30 amino acid residues, that include the enkephalins and also bind to opiate receptors. The name endorphin derives from *endogenous morphine.* Little is known of the physiologic role of these pituitary compounds or the processes that regulate their enzymatic formation from larger precursor peptides. It is sig-

nificant that they are also found in nerve endings in the CNS, where they may function as neurotransmitters that modulate perception of pain.

The human pituitary contains about 250 μg of ACTH (approximately 50 units), probably stored in granules of chromophobe cells. This is enough for almost 1 month under nonstressful conditions, assuming an average daily secretion rate of 8 to 10 μg. With stressful situations, however, the secretion of ACTH may increase as much as 10-fold. Thus to withstand prolonged stress, hormone synthesis must be coordinated with secretion. ACTH secretion follows a diurnal pattern, with the highest rates occurring in the early morning. In the absence of stress, plasma levels of ACTH vary from less than 1 mU/L in the evening to 3 or 4 mU/L in the early morning hours. In stressful situations, ACTH concentrations in plasma may reach 20 mU/L and in some disease states may reach as high as 2,000 mU/L. These normal plasma levels are in the range of 10^{-11} to 10^{-12} M.

ACTH in plasma can now be measured by radioimmunoassay, but it is still quantitated biologically and standardized with an assay based on its ability to deplete the rat adrenal cortex of ascorbic acid.[139] The physiologic significance of this response to ACTH is unknown. In man, ACTH disappears rapidly from the blood, with a half-life of about 4 to 18 min.[107] The biologic response of the adrenal cortex disappears more slowly, with a half-life of about 1 hr. Little is known of the metabolic fate of ACTH, except that it is avidly taken up by the kidney as well as the adrenal cortex. Little or none appears in the urine.

Thyroid-stimulating hormone (thyrotropin)

Although extrathyroidal effects of the thyroid-stimulating hormone (TSH) have been suggested, its only known physiologic role is to stimulate the synthesis and secretion of the thyroid hormones. In so doing, it maintains the size of the thyroid and its rate of blood flow. Excessive secretion of TSH leads to enlargement of the thyroid gland (goiter). The only known ailments that follow derangements in TSH secretion are the same as those caused by too much or too little secretion of thyroid hormones (Chapter 62). Hyperthyroidism in man is often accompanied by *exophthalmos* (bulging eyes) caused by the deposition of fluid rich in mucopolysaccharides in the retro-orbital tissues. Strong evidence suggests that the circulating factor that causes exophthalmos is not identical with TSH.

The human pituitary gland contains about 300 μg of TSH stored in the granules of basophils. As judged by chemical analyses, by variations in potency in different biologic assay systems, and by immunologic criteria, TSH from different species varies in molecular structure. Human TSH is a glycoprotein and has a molecular weight of about 30,000. Its chemistry is remarkably similar to that of the gonadotropins (see following discussion).

Although a variety of biologic assays for TSH are in use, the radioimmunoassay is the most specific and sensitive method available. Average plasma levels are of the order of 1 to 3 μU/ml (about 10^{-10}M) but may increase 50-fold or more in thyroid insufficiency.[116] TSH has a half-life in normal plasma of about 77 min. It distributes in a theoretical volume that is only slightly larger than the plasma volume. The mean production rate of TSH is about 100 mU/day.[91] The kidney appears to be a major site of TSH degradation. Little TSH is recoverable in the urine.

Gonadotropins

The pioneering studies of Zondek[166] and Aschheim in the 1920s suggested that the pituitary glands of females must contain not just one but two gonadotropic principles: one to stimulate ripening of the ovarian follicles and a second to induce the formation of corpora lutea from the collapsed, ruptured follicles. The first successful separation of gonadotropin into two fractions was achieved by Fevold et al.[52] in 1931. The gonadotropins are now called *follicle-stimulating hormone (FSH)* and *luteinizing hormone (LH)*, which is also called *interstitial cell–stimulating hormone (ICSH)* for its action on the interstitial cells of both the ovary and the testis. In the female, FSH promotes the development of the graafian follicles. In a manner not yet definable

in molecular terms, some LH is also required to bring the follicles to full maturity and to stimulate estrogen secretion. At the appropriate moment a sudden, massive release of LH triggers ovulation (i.e., the rupture of the mature follicle and the escape of the ovum with its attendant cells). This is followed by the transformation and inward growth of granulosa cells to form the *corpus luteum* in the crater left by the rupture of the follicle. The corpus luteum is itself an endocrine gland that under the influence of LH secretes both estrogen and progesterone. In the male, FSH stimulates the formation and maturation of sperm in the seminiferous tubules of the testis. LH stimulates the synthesis and secretion of testosterone by the interstitial cells of Leydig in the testis. These complicated relationships and activities are discussed in detail in Chapter 65.

FSH, LH, and TSH are all glycoproteins with molecular weights of about 30,000. They each contain two biologically inactive subunits and are produced, probably separately, in pituitary basophil cells. Some evidence suggests that FSH and LH may be produced by the same cell. The subunits, designated alpha and beta, are not covalently linked and are readily dissociable. The alpha subunit appears to be identical in its amino acid sequence in all three hormones.[16,146] It comprises a single chain of 89 amino acid residues cross-connected by five disulfide bridges. Two carbohydrate moieties are bound to asparagine residues at positions 49 and 75. The beta subunit, which provides the hormonal specificity, is somewhat larger than the alpha subunit and consists of 115 amino acid residues for FSHβ and LHβ and 113 residues for TSHβ. The beta subunits also contain carbohydrate. Although free alpha and beta subunits of all three hormones can be found in blood plasma, particularly during periods of rapid secretion, only the alpha-beta dimers have biologic activity. Another glycoprotein hormone, human chorionic gonadotropin (HCG), produced by the placenta in early pregnancy, it is also quite similar in structure to the glycoprotein hormones produced by the pituitary. HCG is also made up of alpha and beta subunits that are virtually identical with those of LH except that the subunits of HCG contain some extra amino acids at the amino terminus of the alpha chain and at the carboxyl terminus of the beta chain. Biologic assays, particularly for FSH, are complicated because of synergistic interaction with LH. FSH is best quantitated by a radioreceptor assay using testicular tissue.[126] Specific biologic assays for LH are available based on growth of the ventral lobe of the prostate (caused by secretion of testosterone elicited by LH from the Leydig cells) or depletion of ascorbic acid from the extensively luteinized ovary. Radioimmunoassays for LH and FSH are routine. Although subject to wide variations, both FSH and LH circulate at concentrations of about 10^{-11}M. LH is cleared from plasma more rapidly than FSH. Less than 5% of the gonadotropin secreted by the pituitary is excreted in the urine. The metabolic fate of the remainder is unknown.

Growth hormone

Although Aschner (1912) found that total hypophysectomy and not just ablation of the neural lobe led to arrest of growth in pups,[19] the idea that the anterior pituitary might produce growth-promoting substances was not seriously regarded until Evans and Long (1921) produced giant rats by giving repeated daily injections of homogenized pituitary glands.[51] The question of whether the pituitary produced a specific growth-promoting hormone or whether it affected growth only indirectly through the interactions of its various tropic hormones was finally resolved in the mid-1940s with the isolation of highly purified and crystalline preparations of growth hormone (GH) that were devoid of thyroid-, adrenal-, and gonad-stimulating activity.[102,161] Subsequent studies indicated that GH produces a variety of metabolic effects that are not readily related to growth and gave rise to the idea that GH may also regulate energy metabolism. Because its effects are widespread and are not directed at specific target tissues or limited to enhancement of growth, it was proposed that GH be renamed somatotropin (STH) from the Greek *soma,* meaning body. Both names are now in general usage.

Chemistry

GH isolated from human pituitary glands (HGH) is a protein with a molecular weight of 21,000 and consists of an unbranched chain of 191 amino acids containing two internal disulfide bridges. The amino acid sequence is known.[101,114] It is chemically and immunologically distinct from GH produced in other species. The GH molecules isolated from pituitaries of animals of different species vary in amino acid composition. GH preparations from the pituitaries of rats, horses, cows, sheep, pigs, whales, monkeys, and man are all equally effective in rats and produce a characteristic pattern of physiologic responses. However, only the human and monkey GH preparations are effective in man and other primates. Until this unique "species specificity" was discovered,[84] GH was prepared only from the pituitary glands of slaughterhouse animals and was totally ineffective in human patients. Now human pituitary glands collected at autopsy are funneled through a national agency for production of the HGH that is used clinically. Fortunately the human pituitary stores very large amounts of GH; as much as 10% of the dry weight of the anterior lobe has been recovered as highly purified HGH. Li[99] has proposed that the various species of GH have a common "active core" wrapped in a mantle of species-specific peptide. According to this hypothesis the rat, unlike the primate, can respond to the various mammalian GH molecules, presumably because it has the enzymatic capacity to unmask the active cores of heterologous GH molecules. Attempts at partial degradation of bovine GH into a molecule that is effective in primates have met with occasional success, but the problem is complex and the results highly variable. The question of a common active core will only be resolved when the amino acid sequences of the various GH molecules are determined and their three-dimensional structures revealed.

Virtually all the information available on the chemistry of GH is based on the molecule as it exists in storage granules within the pituitary gland. Some studies suggest that the GH molecule may be chemically altered sometime during the secretory process and that the circulating form may be different than the storage form. Some investigators find that the biologic activity of GH isolated from blood plasma may be as much as 200 times greater than the activity estimated by radioimmunoassay,[48] which uses the hormone isolated from pituitaries as a standard reference. Other investigators have observed that mild proteolytic digestion of GH may lead to the deletion of the 12 amino acids between 137 and 149 and produce a 10-fold activation of GH in some biologic assays.[98] Removal of such a fragment converts GH from a single-chain peptide to a two-strand molecular structure held together by a disulfide bridge. It is of interest that the N (amino) terminal two thirds of the molecule (amino acids 1 to 134) can account for all the different biologic activities of GH,[89] even though this fragment is less potent than the intact molecule. The nature of the biologically important circulating form of GH remains a subject for future research. Other forms of GH (e.g., "big growth hormone") are also found in the blood, but these appear to be dimeric and polymeric forms that have reduced biologic activity.

Physiologic actions

Growth. The simple word "growth" describes a variety of living and nonliving processes that are associated with an increase in mass but are otherwise quite unrelated. Simple increase in mass such as occurs with edema formation, for example, is not true growth. By growth we mean the organized addition of new tissue such as occurs normally in development from infancy to adulthood.

The effects of GH on this process are not straightforward. Although overall body growth (i.e., the attainment of adult stature) is absolutely dependent on GH, many bona fide forms of growth are independent of it. Some of these are worthy of note: The increase in mass (hypertrophy) that occurs in the biceps of a weight lifter is a form of growth that occurs without cell division and without GH. The regenerative growth of wound healing or following partial hepatectomy requires increased cell division but also occurs whether or not GH is present. Normal replacement of cells such as erythrocytes or the cells lining the intestinal tract, as well as the growth of most tumors, results from rapid cell division and can also occur in the absence of GH. The growth of hair, which is acellular, similarly occurs without GH. Increases in mass of the target glands (adrenals, thyroids, gonads) and the accessory organs of reproduction (uterus, prostate, seminal vesicles) are true growth processes that do not require GH. Bodily growth during fetal and early neonatal life is possibly the most rapid form of normal growth, yet this too is independent of GH. The question of why some tissues have an absolute requirement for GH while others are independent or only partially dependent on GH is still unanswered. It should be emphasized that some of the foregoing processes that can continue without GH may nevertheless proceed more rapidly when GH is present.

To complicate matters even further the concentration of GH in plasma, which presumably reflects the rate of GH secretion, is not noticeably higher in the rapidly growing child than in the adult whose growth has ceased.[59] Paradoxically, in the immediate neonatal period, GH levels in plasma may be at their highest,[59] but growth at this time is independent of GH.[20] Other clinical situations have been described in which growth does not always correlate with blood levels of GH.[59,77]

In man there are several disorders of growth that can be ascribed to faulty production of GH. Overproduction of GH in the immature individual results in excessive growth of the long bones and gigantism. Afflicted individuals may attain heights of as much as 8 feet. When excessive production of GH occurs in the adult, after the epiphyseal plates of the long bones have fused, growth is only possible in those areas in which responsive cartilage persists. Thus there is elongation of the jaw bone and characteristic deform-

ities of the bones of the face, hands, and feet. Growth in these "acral" parts gives rise to the name *acromegaly* (from the Greek *akros*, highest or outermost; *megas* or *megalou*, big). In such acromegalic patients there is thickening and coarsening of the skin, overgrowth of the underlying tissues, and hypertrophy of muscles and nonendocrine organs. Deficiency of GH in the immature individual leads to a severe stunting of growth, or dwarfism. Such deficiency may be part of an overall lack of anterior pituitary hormones (panhypopituitarism) or may result from an isolated genetic deficiency in GH (sexual ateliotic dwarfism). In the former case, dwarfism is accompanied by sexual immaturity, hypothyroidism, and adrenal insufficiency. Dwarfs suffering from isolated deficiency of GH mature sexually and may marry, reproduce, and show no overt abnormalities other than shortness of stature.[128] Shortness of stature can also result from genetic unresponsiveness to GH, as may occur in the African pygmy.[77] There are no known diseases that result from deficiency of GH that may occur in the adult.

From the foregoing, we may infer that GH has an important, although poorly understood, role in growth and development. It should be emphasized that other genetic and nutritional as well as hormonal factors also govern growth. Thyroxine, the gonadal hormones, and insulin are indispensable for the attainment of normal adult size and form.

EFFECTS ON BONE. The effects of GH on the growth of bone are manifest principally on chondrogenesis and calcification. After hypophysectomy, cartilage formation ceases and the epiphyseal plates in the long bones atrophy. Administration of GH renews cartilage formation and brings about a widening of the epiphyseal plates. This effect of GH on the epiphyseal plate of the rat tibia is the basis for a sensitive biologic assay for GH.[57] Continued treatment with GH for prolonged periods produces sustained growth of the long bones without hastening epiphyseal fusion or bone maturation. In contrast, the gonadal hormones and thyroxine may also stimulate bone growth, but their effects are self-limiting; unlike GH, they hasten bone maturation and fusion of the epiphyseal plates.

GH increases the incorporation of radioactive sulfate into condroitin sulfate of epiphyseal cartilage. When added to excised cartilage in vitro, however, GH itself has little or no effect. In contrast, serum from GH-treated animals markedly increases incorporation of sulfate into mucopolysaccharides and incorporation of amino acids into collagen. The effective substance present in the serum of GH-treated animals is called the "sulfation factor,"[39,137] or "somatomedin." Somatomedin,[40,156] is found in blood plasma of normal human subjects and experimental animals. It is thought to have a molecular weight in the range of 6,000 to 9,000 and to circulate tightly bound to a large molecular weight complex. The liver appears to be the principal site of somatomedin formation. Somatomedin concentrations in blood generally show good correspondence with GH status. After hypophysectomy, somatomedin concentrations in rat blood plasma fall to about 30% of normal, with a half-life of 3 to 4 hr. Administration of GH to humans with hypopituitarism or to experimental animals restores somatomedin levels to normal in 1 to 2 days. One form of familial dwarfism caused by an inability to respond to GH is accompanied by low plasma levels of somatomedin.[41] Despite all this a clear case cannot be made for somatomedin as the mediator of the growth-promoting effects of GH. Somatomedin generation does not always correlate with and precede responses to GH. Thus far, biologic responses to somatomedin have been produced mainly in isolated tissues, and these more closely resembled the actions of insulin than of GH. Most disturbing of all, somatomedin failed to produce growth when injected into hypophysectomized rats[154] as would be expected if it were truly the mediator of GH action. Although all these discrepancies may prove to result from methodologic problems, it is possible that somatomedin is the by-product rather than the mediator of GH action.

EFFECTS ON SOFT TISSUES. GH also promotes symmetric enlargement of soft tissue mass in proportion to skeletal growth. The kidneys seem particularly sensitive, but whether their enlargement reflects a direct response to GH or a compensatory hypertrophy in response to increased glomerular filtration and hemodynamic factors is unknown. Connective tissue and skin are also markedly affected by GH. With prolonged treatment there is a thickening and overgrowth of the skin and increased collagen formation. The nonendocrine viscera also appear to increase in mass (splanchnomegaly). The thymus and lymphoid tissue may hypertrophy.

Metabolism. After prolonged treatment with GH, the body composition resembles that of the very young animal; the proportions of protein and water are high, whereas that of fat is low.[95] There is increased retention of dietary nitrogen, reflecting protein accumulation. Such individuals are said to be in positive nitrogen balance. Positive balances of calcium, sodium, potassium, and phosphorus also occur.

PROTEIN METABOLISM. Administration of GH leads to a decrease in the concentration of amino acids and urea in blood. When given along with an amino acid load, GH decreases the degradation of amino acids to urea.[134] This effect is indirect and probably results from an accelerated uptake (active transport) of amino acids by tissues, particularly muscle. The effects of GH on the transport of amino acids across cell boundaries have been studied with the aid of an amino acid analog, α-amino isobutyric acid (AIB), which, although transported by normal cellular mechanisms, can be neither incorporated into protein nor degraded. Within a few minutes after administration, GH accelerates the accumulation of radioactive AIB within muscle cells,[88] but this effect is transient and disappears in less than 3 hr.[17]

GH also increases protein synthesis, as evidenced by the incorporation of radioactive amino acids into protein. The mechanisms for this action are not understood, but it is likely that GH accelerates some stage of protein synthesis that occurs at the level of the ribosome.[87] Within the first few hours after administration, GH augments protein synthesis even when RNA synthesis and amino acid transport are inhibited. Prolonged treatment with GH eventually results in increased RNA synthesis in muscle and liver.

CARBOHYDRATE METABOLISM. Although early studies with various crude extracts of pituitary tissue provided some hints that it might influence the metabolism of carbohydrate and lipid, the nature of the active material in the extracts was unknown. Subsequent studies revealed that all the metabolic activity present in these extracts could probably be accounted for by ACTH and GH.[20]

Diabetogenic effects. Houssay[78] and his collaborators found that hypophysectomy alleviated much of the impairment of glucose metabolism of diabetic animals (Chapter 66). This effect is due, at least in part, to the removal of GH. Repeated administration of pituitary extracts that were rich in growth-promoting activity impaired glucose metabolism in dogs, and continued treatment resulted in permanent diabetes.[164] For reasons still unknown, rapidly growing pups and lactating dogs were immune to this diabetogenic effect of GH. Related observations indicate that hypophysectomized animals are hypersensitive to the effects of insulin in lowering blood sugar.[78] Conversely, GH decreases sensitivity to insulin partly by antagonizing its effects on the uptake and utilization of glucose in muscle and adipose tissue and partly by stimulating glucose production by the liver.[43] It is therefore not surprising that the hypophysectomized subject is particularly susceptible to hypoglycemia (low blood sugar) and cannot withstand long periods of fasting.

Glycostatic effect. During fasting the hypophysectomized animal cannot maintain glycogen reserves in skeletal or cardiac muscle. In contrast, glycogen in skeletal muscle of normal animals remains unchanged during fasting and cardiac glycogen actually increases. GH prevents the loss of glycogen from skeletal muscle in fasting hypophysectomized animals and increases their cardiac glycogen.[133] This has been called the "glycostatic" effect and probably results, at least in part, from an inhibition of glucose metabolism in the muscle consequent to accelerated lipid oxidation. The diabetogenic and glycostatic effects are reflections of the overall tendency of GH to minimize glucose consumption by muscle and adipose tissue.

Hypoglycemic effect. Immediately after administration to hypophysectomized subjects, GH causes a paradoxical increase in glucose uptake and hypoglycemia.[42] A similar but less pronounced effect is sometimes seen in normal individuals and can also be demonstrated by adding GH to muscle or adipose tissue in vitro.[61,120] This effect lasts for less than 1 hr, and after 3 hr may be superseded by a prolonged decrease in glucose utilization.[42,62] The physiologic significance of the hypoglycemic effect is unknown. The role of GH in the metabolism of carbohydrates will be discussed further in Chapter 66.

LIPID METABOLISM. The reduction in body fat seen after chronic administration of GH results from decreased synthesis of fatty acids from glucose[62] and, more importantly, increased mobilization of free fatty acids from adipose tissue depots.[63,125] Increased fat mobilization results in increased fatty acid utilization, a concomitant fall in the respiratory quotient, and increased production of ketone bodies (β-hydroxybutyric acid and acetoacetic acid) by the liver. The role of GH in lipid metabolism will be discussed more fully in Chapter 66. For the present, it will suffice to say that GH tends to limit the synthesis of triglycerides from glucose and to accelerate the breakdown of stored fat.

Regulation of growth hormone secretion

Since growth is a slow, prolonged process, it was anticipated that plasma levels of GH might remain fairly constant, high in growing children and low or even absent in the adult. Startling results were obtained, however, with the introduction of the radioimmunoassay, which was sufficiently sensitive to measure minute-to-minute changes in plasma GH concentrations. First, as already mentioned, plasma concentrations of GH are not particularly different in children and adults. More startling, perhaps, were the observations that GH has a half-life of only 10 to 20 min and that its concentrations in plasma fluctuate widely in response to a variety of stimuli.[59] GH concentrations in plasma may increase as much as 50-fold within a few minutes and then plummet rapidly to low baseline levels. In a pioneering series of investigations, Glick et al.[60] showed that stimuli that seemingly signal a change from glucose to fat metabolism (e.g., hypoglycemia, exercise, and fasting) all evoked GH secretion. Arginine is also a potent stimulus.[85] In man and the rhesus monkey, bursts of growth hormone secretion also occur in response to nonspecific stressful stimuli[108] and during certain periods of sleep.[151] Females and children are more susceptible than men to fluctuations in circulating concentrations of GH.[59] The role of the GH secreted in response to any of these stimuli is unknown, particularly since its lipid-mobilizing action has a lag time of at least 2 hr.

Prolactin

Hypophysectomy of the lactating animal terminates lactation and causes involution of mammary tissue.[54] This result is due, at least in part, to deficiency of an anterior pituitary hormone that was appropriately named prolactin (also called lactogenic hormone and mammotropin). In the presence of ovarian and adrenal hormones, prolactin promotes the growth of the mammary glands and is required to initiate and maintain milk production.

Prolactin, or some similar material, is present in the pituitary glands of most, if not all, vertebrates. In its fascinating evolutionary history, prolactin has played a variety of physiologic roles, including osmoregulation in fish, stimulation of the drive to return to water in salamanders, stimulation of growth and antagonism of metamorphosis in tadpoles, and stimulation of growth and secretion of the crop gland in pigeons. In addition, in some rodents, it prolongs the functional life of the corpus luteum and hence has been called luteotropin in these species. A detailed account of these various effects is found in Bern and Nicholl.[25] In human beings, the only reasonably well-established physiologic role for prolactin is in breast development and lactation. A role in water and electrolyte balance has been suggested, but definitive evidence is lacking.

Although the presence of prolactin in pituitaries of lower animals has been known for decades, the existence of prolactin in the primate pituitary gland was not definitely established until 1971.[69] Several factors conspired to complicate the isolation of prolactin from human pituitary glands: GH and prolactin have similar molecular properties; GH has intrinsic lactogenic activity; and GH may be as much as 100 times more abundant than prolactin in acetone-dried pituitaries, which are the starting material for the isolation procedure. The complete amino acid sequence of human prolactin has not yet been determined, but those portions that are known are nearly identical with corresponding portions of sheep prolactin whose total structure has been worked out.[115] Prolactin has a molecular weight of about 20,000 and consists of a single chain of 198 amino acid residues cross-linked internally by three disulfide bridges. Because of the great similarity in their primary structures, it has been suggested that GH and prolactin arose from a single ancestral molecule.[115] Some pituitary tumor lines cloned from a single cell secrete both GH and prolactin, although the hormones are normally produced in separate cells. There is a third hormone, human placental lactogen (Chapter 65), which has lactogenic activity, but little growth-promoting activity, even though its structure is more like GH than prolactin. Although of placental rather than pituitary origin, human placental lactogen may also have arisen from the same ancestral gene as GH and prolactin.

The mean concentration of prolactin in random samples of human blood plasma is about 10 mμg/ml (5×10^{-10}M) Slightly higher values are found in women than men. It has a half-life in human plasma of 15 to 30 min. Prolactin concentrations vary widely and exhibit diurnal variation in both males and females. From 60 to 90 min after the onset of sleep, prolactin levels rise in episodic fashion and reach their highest values (25 to 50 mμg/ml) shortly before awakening. Estrogen therapy increases prolactin levels in plasma and the number of prolactin secreting cells in the pituitary. During pregnancy, prolactin concentrations in plasma increase progressively and average about 200 mμg/ml at term. They decline slowly after parturition, and in nonlactating women, they return to basal levels within 2 to 3 weeks. Prolactin, probably originating in the fetal hypophysis, is found in high concentration in amniotic fluid and increases in concentration as pregnancy progresses. Prolactin concentrations are high in the plasma of the newborn and gradually fall to adult levels in about 6 weeks. In the lactating woman, plasma prolactin may increase 25-fold within 30 min of the start of nursing. Suckling at the nipple or tactile stimulation of the breast produces a striking increase in plasma prolactin even in the nonlactating woman, and in some male subjects. A variety of stressful stimuli also evoke the secretion of prolactin, but the physiologic implications of this are not yet understood.

Melanocyte-stimulating hormone (intermedin)

Hypophysectomy of amphibia causes blanching of the skin,[18,147] and administration of pituitary extracts produces darkening.[21] These observations led to the discovery that the pituitary gland contains a substance capable of dispersing melanin granules within the melanocytes of the skin. Two different substances with melanocyte-stimulating hormone (MSH) activity were isolated from porcine pituitaries almost simultaneously.[24,25] They were dubbed α-MSH and β-MSH.[96] Both are polypeptides closely related to ACTH. In fact, α-MSH is made up of the first 13 amino acids of ACTH. β-MSH varies somewhat from species to species. In man, it consists of 22 amino acids with the heptapeptide 11 to 17 having an amino acid sequence identical with that of residues 4 to 10 of corticotropin and α-MSH (Fig. 61-7). This heptapeptide is thought to represent the active region.[75] α- and β-MSH are allegedly produced in the intermediate lobe of the pituitary but actually can be extracted from all parts of the pituitary. Although the physiologic role of this hormone in mammals is unknown, darkening of the skin can be produced in man by the daily injection of 8 mg of either type of MSH for 7 to 12 days.[97] ACTH also darkens the skin, but larger amounts are needed. MSH activity may be an evolutionary vestige of an important camouflaging reaction in lower species.

REGULATION OF ANTERIOR PITUITARY FUNCTION

The early endocrinologists had to account for their findings that (1) hypophysectomy results in profound atrophy of the target glands (adrenals, thyroid, and gonads), (2) administration of pituitary extracts causes hypertrophy of these glands, and (3) prolonged administration of a hormone or extract of a target gland produces atrophy of that gland. These and other observations led to the conclusion that the pituitary and its target glands

ACTH
(pig, sheep, cow)
Ser·Tyr·Ser·Met·Glu·His·Phe·Arg·Try·Gly·Lys·Pro·Val·Gly·Lys·Lys·Arg·Arg·Pro
1   2   3   4   5   6   7   8   9   10 11 12 13 14 15 16 17 18 19

α·MSH
(pig, cow, horse)
CH₃CO–Ser·Tyr·Ser·Met·Glu·His·Phe·Arg·Try·Gly·Lys·Pro·Val·NH₂
1   2   3   4   5   6   7   8   9   10 11 12 13

β·MSH
(pig)
Asp·Glu·Gly·Pro·Tyr·Lys·Met·Glu·His·Phe·Arg·Try·Gly·Ser·Pro·Pro·Lys·Asp
1   2   3   4   5   6   7   8   9   10 11 12 13 14 15 16 17 18

β·MSH
(beef)
Asp·Ser·Gly·Pro·Tyr·Lys·Met·Glu·His·Phe·Arg·Try·Gly·Ser·Pro·Pro·Lys·Asp
1   2   3   4   5   6   7   8   9   10 11 12 13 14 15 16 17 18

β·MSH
(horse)
Asp·Glu·Gly·Pro·Tyr·Lys·Met·Glu·His·Phe·Arg·Try·Gly·Ser·Pro·Arg·Lys·Asp
1   2   3   4   5   6   7   8   9   10 11 12 13 14 15 16 17 18

β·MSH
(man)
Ala·Glu·Lys·Lys·Asp·Glu·Gly·Pro·Tyr·Arg·Met·Glu·His·Phe·Arg·Try·Gly·Ser·Pro·Pro·Lys·Asp
1   2   3   4   5   6   7   8   9   10 11 12 13 14 15 16 17 18 19 20 21 22

Fig. 61-7. Relationship between structure of MSH and ACTH. (From Dixon and Li.[44])

are coupled in a *negative feedback* relationship wherein a tropic hormone of the pituitary stimulates its target gland to secrete a hormone that in turn inhibits further secretion of the tropic hormone by the pituitary (Fig. 61-8). When the plasma concentration of the target gland hormone falls below some critical level, inhibition of the pituitary is relieved and the tropic hormone is again secreted until the target gland produces enough hormone to raise its concentration in plasma high enough to shut off the pituitary. Such an arrangement can fully account for the foregoing observations. The concept of a negative feedback relationship between the pituitary and its target glands remains a valid cornerstone of modern endocrinology. It has widespread applications both diagnostically and therapeutically and was the rationale behind the development of the contraceptive pill. It is a highly effective mechanism for maintaining constant levels of target gland hormones in blood. However, the negative feedback concept alone cannot explain the changes in plasma levels of target gland hormones that result from changing demands of the internal and external environment. Some examples may be cited. (1) ACTH secretion is increased in response to such internal threats as bacterial endotoxins and hypoglycemia as well as to a variety of real or potential hazards in the external environment. This increase occurs even in adrenalectomized animals in which the pituitary is free of feedback inhibition by adrenal cortical hormones.[150] (2) In the ferret, for example, the gonads atrophy between breeding seasons, but as the days grow longer in springtime, the gonads develop under the influence of renewed gonadotropin secretion. By gradually increasing

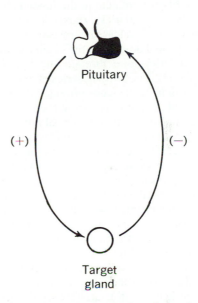

Fig. 61-8. Schematic representation of negative feedback relation between pituitary and target glands.

the hours of exposure to artificial light each day, increased secretion of gonadotropins and gonadal maturation can be induced in the dead of winter.[69] In this case, too, increased secretion of gonadotropin cannot result from release of feedback inhibition by gonadal hormones, because initially gonadal secretion is virtually nil. (3) In the rabbit, sexual stimulation induces the release of sufficient LH to cause ovulation.

The foregoing examples suggest strongly that the central nervous system (CNS) plays a decisive role in adjusting pituitary function to fit environmental demands. Direct experimental evidence supports this concept and makes inescapable the conclusion that the anterior pituitary gland must be under the control of the CNS.[71]

The elucidation of how the CNS controls the anterior pituitary gland has formed an exciting chapter in contemporary endocrinology. Careful anatomic studies in many laboratories have made it certain that there are no secretomotor nerve fibers in the anterior pituitary gland. The few nerve fibers that are present serve vasomotor functions.[65] Green and Harris[66] were among the first to note that the peculiar blood supply to the anterior pituitary gland is ideally suited for the transmission of chemical signals from the CNS to the anterior pituitary gland. Before reaching the secretory cells of the adenohypophysis, arterial blood passes through the primary capillary plexus in the median eminence. A vast number of nerve fibers terminate in the immediate vicinity of these capillaries. Harris postulated that these nerve fibers release neurosecretory products that are washed down into the capillary sinusoids of the anterior lobe through the hypophyseal portal vessels. In a brilliant series of experiments, Harris[4] and his colleagues provided indirect support for this hypothesis. They found that if the pituitary stalk is sectioned, normal pituitary function is lost and returns only after the hypophyseal portal vessels regenerate. Placing an impermeable barrier between the pituitary and the hypothalamus prevented both regeneration of the hypophyseal portal vessels and the restoration of normal pituitary function. When the pituitary was transplanted to some site away from the *sella turcica*, it failed to function normally, despite adequate blood flow, but when it was transplanted back to the *sella turcica*, normal function returned as soon as the portal vessels regenerated. Although it flows at a rate of only 0.4 ml/hr in the rat, Porter and his colleagues have collected hypophyseal portal blood and found that it does indeed contain factors that evoke the release of anterior pituitary hormones.[124,159]

PHYSIOLOGY OF THE HYPOTHALAMIC REGULATORY HORMONES

The neurosecretory products that regulate the activity of the anterior pituitary gland have been called "releasing factors" or "release-inhibiting factors." The term "hypophysiotropic hormones" has been coined by analogy with the terminology used for the anterior pituitary hormones. There are at least nine hypothalamic factors that regulate anterior pituitary function, but only three have been isolated and chemically defined. Their isolation and characterization have

been the object of a long and tedious quest that began in the early 1950s and did not reach fruition until 1970 when the first of these hormones, the TSH-releasing hormone, was isolated and defined chemically.

Thyrotropin-releasing hormone

Thyrotropin-releasing hormone (TRH), whose structure is shown in Fig. 61-9, is a tripeptide[27,29] that is now readily synthesized artificially. Its original isolation was the culmination of heroic efforts by two laboratories. In one, 265,000 porcine hypothalami were extracted to obtain just 8.2 mg of TRH; in the other, 25 kg of sheep hypothalami yielded only 1 mg of TRH. TRH has now been found in neural tissue of all classes of vertebrates, in primitive chordates, and even in snails.[127] In its evolutionary history, it apparently predates the appearance not only of TSH, but of the pituitary gland itself. Although it is found in highest concentration in the median eminence of the hypothalamus, almost 80% of the TRH in the brain is extrahypothalamic.[127] Its ubiquitous distribution, coupled with reports of its behavioral effects in rodents, and its ability to suppress action potentials in some brain cells after application by microiontophoresis suggest that TRH may be a neurotransmitter as well as a releasing hormone.[160] There can be little doubt that TRH is indeed a releasing hormone, since in addition to being found in the hypothalamus it has been detected in hypophyseal-portal blood, where its concentration was increased by electrical stimulation of the hypothalamus.[159] When injected intravenously to intact subjects or added to pituitary tissue in culture, TRH increases the secretion of TSH. Paradoxically, TRH is at least as effective in provoking the secretion of prolactin.[80] The biological significance of this dual effect is not understood, especially since TSH and prolactin are usually secreted independently. Exactly how such independent secretion can be achieved has not been fully resolved, but it is

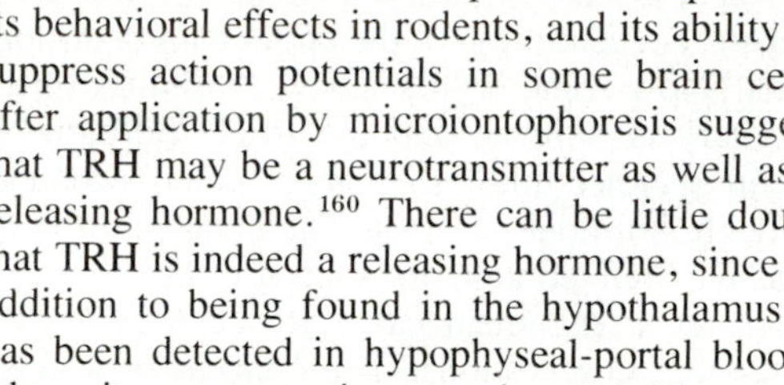

Fig. 61-9. Molecular structure of thyrotropin-releasing hormone (TRH): L-2 pyrrolidone-5-carboxyl-L-histidyl-L-proline.

likely that other factors modify the sensitivity of pituitary cells to TRH.[26] For example, the classic negative feedback effects of thyroid hormones on TSH secretion can be at least partially explained by the ability of thyroid hormones to decrease the sensitivity of the TSH-producing cells to TRH. In addition, the secretion of prolactin appears to be governed by a release-inhibiting factor that may block the response of prolactin-producing cells to TRH. The GH release–inhibiting hormone (see p. 000) blocks the effects of TRH on TSH secretion but not on prolactin secretion. Clinically, TRH is now used to discriminate between diseases of thyroid function that are ascribable to either pituitary or hypothalamic disorders.

Luteinizing hormone–releasing hormone

Luteinizing hormone–releasing hormone (LRH) is a decapeptide (Fig. 61-10) that was isolated from both porcine and ovine hypothalami in 1971.[30,106] Although originally sought after and assayed for its ability to provoke LH release from the pituitary, it is now apparent that LRH elicits the secretion of both LH and FSH. Hence, its more appropriate name should be gonadotropin-releasing hormone (GnRH). It is still possible that a separate releasing hormone for FSH exists, but most investigators now favor the idea of a single GnRH. Independent secretion of LH and FSH is possible with only one releasing hormone, for the gonadal target-gland hormones appear to modulate the sensitivities of the FSH- and LH-producing cells to GnRH. GnRH is found principally in the median eminence, the arcuate nucleus, and the organum vasculosum of the lamina terminalis.[28,165] It is probably synthesized in nerve cells and concentrated in secretory granules in the nerve terminals that impinge on the capillaries of the primary plexus of the hypophyseal portal system. GnRH, like other releasing hormones, is also found in the cerebrospinal fluid. It has been suggested that the cerebrospinal fluid may play a role in transporting GnRH and other releasing hormones to the pituitary[82] or other parts of the CNS. It may be recalled that the third ventricle dips down into the hypothalamus in the vicinity of the median eminence. Specialized ependymal cells, or *tanycytes*, which extend from the floor of the third ventricle to the capillaries of the primary plexus, have been shown by immunocytochemical means to be rich in GnRH and to have structural features consistent with a transport role.[165] Establishment of the relative importance of this route of trans-

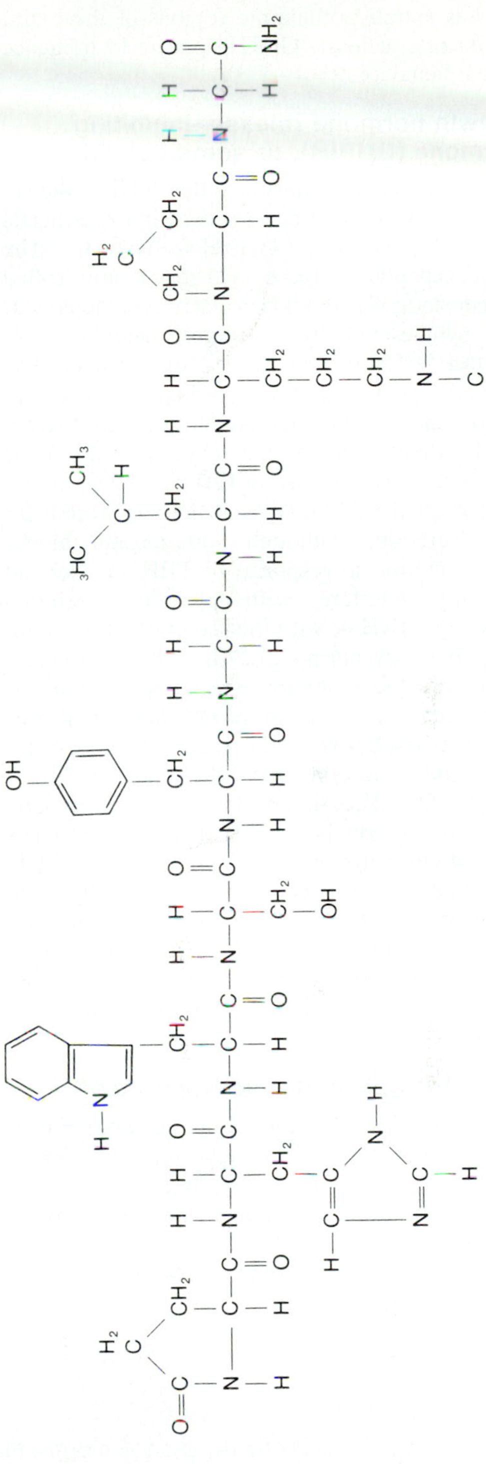

Fig. 61-10. Molecular structure of LH-releasing hormone (LRH), also called gonadotropin-releasing hormone (GnRH). (From Guillemin.[67])

port of hypophysiotropic hormones and of the direction of that transport awaits further investigation. It is reasonably certain that GnRH reaches extrahypothalamic regions of the brain. In rodents, at least, GnRH appears to influence sexual behavior.[111]

Growth hormone release–inhibiting hormone (GHRIH), or somatostatin

A hypothalamic substance that inhibits the secretion of GH was first observed in experiments designed to study the GH-releasing factor. The tetradecapeptide (Fig. 61-11) that is now called somatostatin was quickly isolated, characterized, and synthesized by Guillemin and his colleagues.[31,68] Somatostatin is found in nerve terminals[121] in the anterior hypothalamus and median eminence and is secreted in response to electrical stimulation of the hypothalamus.[105] It blocks the basal release of GH as well as the responses of the pituitary to all known stimuli for GH secretion. Although somatostatin blocks TSH secretion in response to TRH, it does not normally interfere with prolactin secretion evoked by TRH or with the secretion of other anterior pituitary hormones.[68] Infusion of somatostatin into the systemic circulation of baboons produced a decrease in blood sugar that was traced to inhibition of the secretion of both insulin and glucagon from the pancreas[68] (see Chapter 66). This finding led to the observation that somatostatin is present in pancreatic islets and various parts of the gastrointestinal tract in cells that perhaps derive embryologically from the neural crest.[127] Like acetylcholine and epinephrine, somatostatin may prove to be a neurohumor whose physiologic role extends far beyond the function that led to its original discovery.

Other hypophysiotropic hormones[90,127]

Although they are not yet chemically defined, there is good reason to believe that other hypophysiotropic hormones exist. Because secretion of prolactin increased when contact between the pituitary and hypothalamus was interrupted, it was suggested that prolactin secretion is controlled by a prolactin-inhibiting factor (PIF). Hypothalamic extracts are rich in a prolactin release–inhibiting activity that meets the criteria for PIF. Some investigators believe that PIF may prove to be dopamine, a precursor in the biosynthesis of norepinephrine. In addition, experimental evidence suggests there is a prolactin-releasing hormone distinct from TRH. A hypothalamic factor that triggers the release of ACTH in response to stress undoubtedly exists, but it has been particularly elusive. There is good experimental evidence for a GH-releasing hormone, and MSH may be under the control of both a releasing and a release-inhibiting factor. With the possible exception of PIF, it is likely that the hypothalamic regulatory hormones are all peptides.

Isolation of the hypophysiotropic hormones provides unequivocal proof of the neurohumoral link between the CNS and the anterior pituitary gland and shifts the focal point of investigation from the brain-pituitary interface to the brain itself. A detailed account of the distribution of the releasing hormones in the brain and what is known of the pathways involved in their secretion is beyond the scope of this text. Most of the available information has been derived from the study of rodents, and even within this group, species differences are evident. The morphologic, physiologic, and pharmacologic aspects of what controls the release of the releasing factors have been the subjects of intense research in many laboratories and will undoubtedly consume the energies of many investigators for some time to come. The tools available for such study include (1) production of lesions electrolytically or by knife cuts in various areas of the hypothalamus or brain stem and determination of which physiologic responses are lost or impaired; (2) electrical stimulation of certain areas of the brain and determination of what physiologic responses are thus artificially produced; (3) assay of specific hypophysiotropic activities in homogenates of carefully microdissected regions of the hypothalamus or other areas; and (4) immunohistochemical identification of hypophysiotropic hormones within specific cells. This latter approach, of course, requires the isolation and chemical

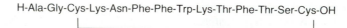

H-Ala-Gly-Cys-Lys-Asn-Phe-Phe-Trp-Lys-Thr-Phe-Thr-Ser-Cys-OH

Fig. 61-11. Amino acid sequence of growth hormone–release inhibiting hormone, or somatostatin.

characterization of hypophysiotropic hormones so that specific antibodies can be raised. It is reassuring that inferences of the involvement of particular areas of the hypothalamus derived from one type of study are usually confirmed using other techniques. Another approach that is becoming increasingly prominent is the pharmacologic characterization of the neural elements that influence the release of the hypophysiotropic factors according to their neurotransmitters.[90,159] Thus stimulation of prolactin release, for example, is thought to involve at least one serotonergic synapse, and inhibition of its release at least one dopaminergic synapse. In addition to contributing to understanding of the control of hypophyseal function, knowledge of the transmitters involved in triggering or inhibiting their release permits specific pharmacologic intervention to correct oversecretion or undersecretion of hypophysiotropic hormones and the hormones of the pituitary itself.

Regulation of anterior pituitary function is achieved through integration of several neural and hormonal inputs (Fig. 61-12). The neural components may arise from external stimuli such as light or temperature as well as from internal stimuli that may be either emotional or of the housekeeping, homeostatic variety. In addition, there are the as yet poorly understood daily rhythms and sleep-related signals for hormone secretion.[158] These neural signals are transmitted to the anterior pituitary gland by the hypophysiotropic hormones that may, as we have seen, exert either stimulatory or inhibitory effects. One component of hormonal input is feedback inhibition exerted by the target gland hormones. These hormones may inhibit the secretion of tropic hormones by blocking secretion of hypothalamic releasing factors, by decreasing the sensitivity of pituitary cells to releasing factors, or both. The possibility has also been raised that some of the anterior pituitary hormones may influence their own secretion through ''short-loop'' feedback inhibition of their own releasing factors.[112]

Little can be said about the molecular events that occur in the anterior pituitary cells after stimulation with releasing factors. Anterior pituitary hormones are stored in granules. It is likely that the secretory process involves a calcium-dependent mechanism in which the secretory granules are translocated to the cell surface where they are extruded. The details of the process are not known, nor is the coupling between synthesis and secretion yet understood. Available data suggest that GnRH and TRH bind to receptors on the surface of the cells they stimulate and activate the enzyme adenylate cyclase.[92] The cyclic AMP thus produced is thought in some way to activate the secretory process. There is evidence to suggest that somatostatin may interfere with cyclic AMP formation and also with some action of the cyclic nucleotide.

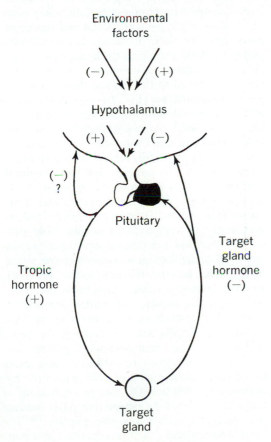

Fig. 61-12. Diagrammatic representation of regulation of anterior pituitary secretion.

Neurohypophysis

PHYSIOLOGY OF THE NEUROHYPOPHYSEAL HORMONES

Oxytocin and *vasopressin,* which is also called the antidiuretic hormone (ADH), are the two hormones secreted by the posterior pituitary. They were the first peptide hormones to be completely characterized chemically and the first to be synthesized artificially.[46] Each is composed of nine amino acids in peptide linkage and formed into a ring structure by the disulfide bridge connecting the cysteine residues at positions 1 and 6. Many people refer to these compounds as octapeptides and consider the two cysteine residues as halves of the amino acid cystine. Vasopressin and oxytocin differ only in the amino acid residues present in positions 3 and 8 (Fig. 61-13) and may have arisen phylogenetically from a common precursor.[138] Most mammalian species produce vasopressin molecules containing arginine in position 8 (arginine vasopressin). Members of the pig family, however, produce a vasopressin that has lysine substituted for the arginine at position 8 (lysine vasopressin). Lysine vasopressin is somewhat less potent than arginine vasopressin. The name "vasopressin" derives historically from the observation that extracts of the posterior pituitary gland increase blood pressure and constrict vascular smooth muscle.[117] To call this hormone vasopressin is misleading, however, for this name describes what is now believed to be a pharmacologic rather than a physiologic action of the hormone. ADH is a more appropriate name, since it describes the only recognized physiologic function of this hormone, namely to decrease the loss of free water (Chapter 49). The vasopressor effect requires a concentration of hormone that is about 1,000 times higher than that needed for antidiuretic activity and considerably more than that normally found in plasma.

The term "oxytocin," meaning rapid birth, derives from the ability of this hormone to cause contraction of uterine muscle. This property of posterior pituitary extracts was first observed by Dale in 1906.[36] Oxytocin also contracts the myoepithelial cells of the mammary gland to cause the "let-down" reflex or ejection of milk into the mammary ducts. This effect was first described by Ott and Scott in 1911.[119] Because of the great similarity in their structures, vasopressin has some intrinsic oxytocic activity and vice versa.

Source of neurohypophyseal hormones: neurosecretion

Unlike the adenohypophysis the neurohypophysis contains no large epithelial cells full of secretory granules. It consists primarily of unmyelinated nerve fibers and modified glial cells, the *pituicytes*. Although it was thought for many years that the pituicytes were the source of the posterior pituitary hormones, it is now recognized that these hormones are neurosecretory products released from nerve cells whose axons terminate in bulbous swellings on the basement membrane of neurohypophyseal capillaries. This concept developed from an observation made originally by Herring (1908) that the nerve fibers in the pituitary stalk contain droplets of darkly staining colloid.[73] By means of special staining procedures, it was demonstrated that these so-called Herring bodies are present throughout the hypothalamic-hypophyseal tract, including nerve cell bodies of the paraventricular and supraoptic nuclei and the nerve terminals in the neurohypophysis. After sectioning the hypophyseal stalk, stainable material accumulates in the axon stumps proximal to the cut and disappears from the neurohypophysis (Fig. 61-14), an indication that the colloid flows from the hypothalamus into the neural lobe.[74] A relationship between stainable colloid and ADH was strongly suggested by findings that physiologic manipulations known to release ADH produce a coincident loss of colloid. Furthermore, destruction of the supraoptic nuclei through disease processes or experimental means causes loss of the "Herring substance" and the appearance of symptoms of severe ADH deficiency, even when the neural lobe remains uninjured. Most authors now agree that the supraoptic nuclei are the primary source of ADH, whereas oxytocin comes principally from the paraventricular nuclei, but neither hormone is produced exclusively in one or the other nucleus.

Although details are still lacking, the broad outlines of the neurosecretory process are known.[22,140] Hormone biosynthesis takes place in the perinuclear region of cells within the hypothalamus. Studies of the incorporation of radioactively labeled cysteine into ADH indicate that biosynthesis is a two-step process.[135] The labeled amino acid is first incorporated into some hormonally inactive precursor, with a molecular weight of about 20,000 daltons. The appearance

Arginine vasopressin

Cys-Tyr-Phe-Gln-Asn-Cys-Pro-Arg-Gly-NH$_2$
1   2   3   4   5   6   7   8   9

Lysine vasopressin

Cys-Tyr-Phe-Gln-Asn-Cys-Pro-Lys-Gly-NH$_2$

Oxytocin

Cys-Tyr-Ile-Gln-Asn-Cys-Pro-Leu-Gly-NH$_2$

Fig. 61-13. Amino acid sequence of vasopressin (ADH) and oxytocin.

of the label in biologically active hormone requires several hours and is achieved by a process that is insensitive to inhibitors of protein synthesis. It is likely that this process involves proteolytic cleavage of the active peptide from a larger protein molecule. The hormone precursor is probably packaged with a proteolytic enzyme in the Golgi apparatus into dense membrane-enclosed secretory granules 0.1 to 0.3 μm in diameter. The active hormone is thought to be cleaved from the precursor within the secretory granule during the 1 to 2 hr period in which the granules are transported down the axons to the posterior pituitary gland.[55,122] Another product of this cleavage is thought to be the binding protein, neurophysin, which is probably the darkly staining protein observed by Herring. The posterior hormones are bound noncovalently to neurophysin in the secretory granules, probably in a molar ratio of 1. The human posterior pituitary gland contains two distinct but structurally closely related neurophysins, one apparently associated with ADH and the other with oxytocin. Neurophysin accounts for about 5% of the weight of the posterior pituitary gland and is concentrated along with the hormones in the neurosecretory granules in axon terminals. Despite their specialized function, neurosecretory cells retain their capacity to conduct electrical impulses.[34] Action potentials that arise from stimulation of the cell bodies in the hypothalamus are conducted down the axons of the neurosecretory fibers and trigger hormone release. It has been suggested that depolarization of the neuronal membrane is followed by an influx of calcium ions that are thought to mediate this secretory process.[45] Energy input in the form of ATP may be required.[123] Neurophysin is released along with the hormone in the process of exocytosis. Although the foregoing description was derived from studies on ADH, similar mechanisms of synthesis and neurosecretion probably occur for oxytocin.

The neurophysins that are released along with the posterior pituitary hormones circulate in the plasma at a concentration of about 1 mμg/ml.[130] They are cleared rapidly with a half-life of 1 to 2 min. The kidney appears to be the principal site of inactivation. Blood levels of each of the neurophysins usually correlate quite well with their related hormones.[131] Thus nicotine, which is a specific stimulus for vasopressin secretion,

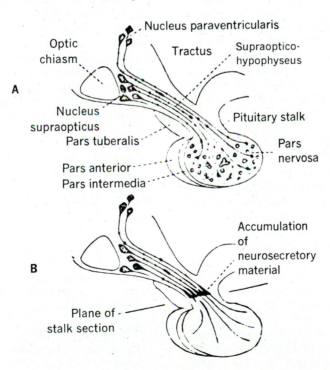

Endocrine glands

Fig. 61-14. Diagram of hypothalamiconeurohypophyseal relationships showing distribution of secretory granules, **A,** before and **B,** after stalk section. (See text.) (From Scharrer and Scharrer.[140])

increases the concentration of one of the neurophysins (nicotine-sensitive neurophysin). Smoking two cigarettes in 15 min triples the concentration of the neurophysin. The other neurophysin thought to be related to oxytocin secretion is increased in blood by estrogen (estrogen-sensitive neurophysin). This neurophysin circulates in slightly higher concentrations in women and is elevated in the second half of pregnancy. Despite their appearance in plasma in response to specific stimuli, no hormonal role has yet been established for the neurophysins.

Since the neurohypophyseal hormones are produced in the hypothalamus and not in the neurohypophysis, what then is the role of the neurohypophysis? Clearly, one function is storage. The posterior pituitary contains about 15 units of ADH, which is enough to maintain a man in maximum antidiuresis for more than 1 week. By contrast, the hypothalamus contains very little hormone—less than 5% of the total amount stored in the posterior pituitary. Thus while the hypothalamus may be capable of meeting some demands for ADH, it may not be capable of supporting intense or sustained antidiuresis. A second factor worthy of consideration is that the blood supply of the posterior pituitary is separate from that of the hypothalamus and lies outside of the blood-brain barrier. Locating a large depot of hormone outside of the blood-brain barrier may facilitate release of hormone into the circulation.

ADH and oxytocin in blood are probably not bound to plasma proteins. Both are cleared rapidly from the plasma and have half-lives of less than 5 min. Both are inactivated principally in the kidney and liver, but about 10% is excreted in the urine in active form.

Antidiuretic hormone
Physiologic actions

Just a few hundred microunits of ADH administered to well-hydrated subjects elicits a prompt decrease in urine production. This decrease can be accounted for almost exclusively by a decrease in "free water" excretion. Solute excretion remains virtually unchanged, with the result that the urine becomes hypertonic with respect to plasma. A detailed explanation of the effects of ADH on the kidney is given in Chapter 50. For the present, it may be recalled that the interstitial fluid in the renal medulla is hypertonic due to the action of the countercurrent mechanism in the loop of Henle, whereas the intratubular fluid that emerges from the loop is hypotonic. In the absence of ADH the distal tubule and the collecting ducts in the well-hydrated subject are quite impermeable to water. Consequently, the hypotonic urine passes out into the bladder unchanged in its osmolarity or made even more hypotonic by the reabsorption of solutes by the distal convoluted tubules. ADH in some way increases permeability to water in the collecting ducts and probably the distal convoluted tubules. Thus water flows out through the tubular wall down its osmotic gradient into the hypertonic interstitium of the renal medulla, from whence it is carried off in the blood flowing through the vasa recta.

The molecular events that accompany the increased permeability of the collecting ducts to water in response to ADH are not yet understood. What little is known of this process has been learned using model tissues from amphibia. Toads, which are constantly threatened with dehydration, utilize the urinary bladder as a reservoir from which water can be reabsorbed as needed. The toad bladder wall is only one or two cell layers thick and becomes permeable to water in the presence of ADH. The ventral skin of frogs and toads is also sensitive to ADH. ADH increases the net flow of water across these membranes, but only when there is an osmotic gradient.[86,93,94] Kinetic studies with heavy water indicate that ADH increases the movement of water by "bulk flow" without greatly increasing the rate of diffusion across the tissue. These results have been interpreted to imply that ADH increases the size of pores in some layer of the tissue without increasing the total area available for diffusion. To date, such pores have not been seen on electron micrographs. ADH produces virtually identical effects in isolated perfused collecting ducts obtained from rabbit kidneys.[64]

ADH also stimulates the unidirectional active transport of sodium across the toad bladder from the urinary to the blood side and increases the passive diffusion of urea. Similar effects of ADH on solute movement in the mammalian kidney have not been established. Many of the molecular events that lead to increased permeability are not yet known. It is almost certain that ADH increases the formation of cyclic AMP within the nephron[143] as well as in the toad bladder.[118] Cyclic AMP mimics the effects of ADH, but how it produces the subsequent changes in tissue permeability is unknown.

Physiologic roles

The principal physiologic effect of ADH is to increase the retention of "free water" by the kidney (Chapters 49 and 50). Water thus conserved maintains or expands the volume of plasma and reduces its osmolarity by diluting circulating solutes. In a water-deprived individual, for example, increased secretion of ADH offsets tendencies toward increased plasma osmolarity and decreased volume. Increased osmolarity of the plasma and decreased effective vascular volume are now recognized as the principal physiologic stimuli for ADH secretion. ADH is also secreted in response to a variety of nonspecific emotional or stressful stimuli such as pain or fainting. Secretion of ADH is inhibited by epinephrine and ethyl alcohol. The physiologic significance of ADH release under such circumstances is unclear at this time.

ADH and regulation of osmolarity. Most of what is now understood of the role of ADH in the regulation of plasma osmolarity was elucidated by the experiments of Verney.[157] To test the idea that increased osmolarity of the plasma evokes ADH secretion, he injected small volumes of hypertonic solution directly into the carotid artery of well-hydrated dogs and observed a prompt decrease in urine flow. This effect was not seen when the same volume of hypertonic solution was injected into a systemic vein or after removal of the neurohypophysis. From the concentration and volume of the injected material and the rate of blood flow in the internal carotid artery, Verney calculated that a sustained increase of as little as 2% in the osmolarity of the plasma bathing the anterior hypothalamus was sufficient to induce the release of ADH. Solutions containing electrolytes or sucrose were particularly effective in stimulating the hypothetical osmoreceptors, but a readily diffusible substance such as urea was ineffective. These observations led to the idea that the osmoreceptors might consist of small vesicles surrounded by a semipermeable membrane within which nerve endings are embedded. Increased osmolarity of the plasma would cause the outward movement of water from these tiny osmometers, thereby decreasing their volume and diminishing the degree of stretch sensed by the neurons. By ligating the vessels supplying various portions of the bed served by the internal carotid artery, the locus of the osmoreceptors was narrowed down to the anterior hypothalamus, possibly even to the region of the supraoptic nuclei.[81] Although Verney described small vesicles near the supraoptic nuclei, the exact location of the osmoreceptors and their nature have not yet been established definitively. Intracarotid infusions of hypertonic solutions that cause an increase in ADH in jugular blood[47] also cause an increase in unit electrical activity in specific neurons within the supraoptic nucleus and the perinuclear region. The neurons of the perinuclear region that exhibit monophasic responses have been nominated as the osmosensitive cells. The neurons of the nucleus itself exhibit biphasic responses and are thought to be secretory cells.[72]

ADH and control of plasma volume. Verney also demonstrated that removing as little as 6% of the blood volume was sufficient to decrease urine flow. Subsequent findings revealed an increase in ADH concentration in the jugular vein under these conditions.[58] The extracellular fluid volume can also be reduced by the process of peritoneal lavage, in which a solution of electrolytes identical in composition with plasma but made hypertonic with glucose can be introduced into the peritoneal cavity. Since water and electrolytes are drawn out of the circulation faster than the glucose is absorbed, there is net movement of fluid into the peritoneal cavity. This procedure produced a sixfold increase in the concentration of ADH in jugular blood.[144] Apparently it is not the total extracellular fluid nor the absolute vascular volume that is monitored in regulating ADH release. Rather it is the *effective* vascular volume as sensed in the thoracic region. Internal shifts in blood volume, such as pooling of blood in the extremities after prolonged standing, results in ADH release. Positive-pressure respiration that decreases the volume of blood in the major vessels of the thorax results in an apparent release of ADH, whereas negative pressure respiration, which engorges the thoracic vessels with blood, results in diuresis.[56,145] Using a sensitive bioassay for ADH, it was found that simply changing from a supine to a sitting position results in a threefold increase in blood levels of ADH and nearly an eightfold increase in switching from a supine to a standing position.[142] Exposure of human subjects to high ambient temperatures, which causes increased blood flow in the skin and a concomitant shift of blood from deeper to more superficial regions, resulted in a threefold increase in the concentration of ADH in plasma. Conversely, exposure to cold, which shunts blood away from the skin to the deeper regions of the body, halved the circulating concentration.[142] These findings indicate the exquisite sensitivity of the volume receptors and suggest a role for ADH in the minute-to-minute regulation of blood volume.

The location of the volume receptors and their afferent pathways are not known with certainty. Unlike the osmoreceptors, the volume receptors involved in ADH release cannot be located in the CNS, since the various vascular reflexes ensure constant blood flow to the brain. Sectioning of the vagus nerves in the neck prevents the release of ADH in response to hemorrhage.[145] Since the vagus carries afferent signals from the aortic arch, the atrial and ventricular stretch receptors, and receptors that might be located in the great veins of the chest and possibly from the pulmonary circulation, any or all of these structures might house volume receptors. The carotid sinus pressor receptors have also been implicated.[145] Occlusion of the common carotid arteries below the bifurcation causes an increase in ADH release, but only when this manipulation is performed after the vagus nerves have been sectioned. It appears that in the normal animal the first receptors to sense a change in volume are located on the low pressure side of the vascular system (i.e., the great veins of the chest, the pulmonary circulation, and the atria).

A hypothetical feedback scheme that summarizes the possible regulation of ADH secretion is illustrated in Fig. 61-15. Increased concentration of solute in the plasma evokes the release of ADH, which travels in the blood to the kidney, where it increases water reabsorption. The free water thus reabsorbed dilutes the solutes in the plasma and thereby shuts off the signal for ADH secretion. When the solute concentration of plasma is too low, as might occur after drinking large volumes of water, ADH secretion is inhibited and free water is excreted until the solute concentration in plasma reattains some optimal level that again initiates basal secretion of ADH. When the volume of blood decreases following hemorrhage, dehydration, or sodium depletion, signals originating from receptors that monitor either the central venous pressure or the arterial pressure are transmitted to the CNS and initiate the release of ADH. In addition, renin released from the kidney may trigger increased ADH secretion through the intermediate angiotensin II (Chapter 46).[103] Increased water reabsorption reduces further loss of volume in the urine, and in the presence of water intake increases plasma volume, thereby shutting off the signal for ADH secretion.

An additional complexity must be considered. ADH secretion is responsive to two different inputs, osmolarity and volume, that may either reinforce or contradict each other. In dehydration the call for more ADH signaled by hypertonicity is reinforced by receptors monitoring decreased volume. In severe sodium depletion, however, there is a conflict between osmotic and volume receptors, for if osmolarity is to be maintained, the volume must be diminished by inhibiting ADH secretion. Conversely, if volume is to be

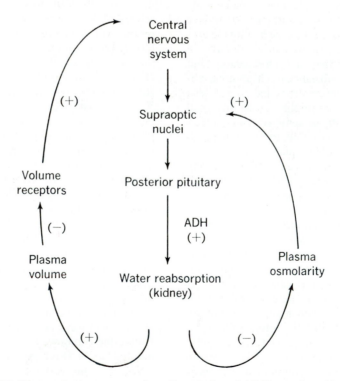

Fig. 61-15. Diagrammatic representation of regulation of ADH secretion. (See text.)

maintained by water retention, hypotonicity must result. Presumably because hyponatremia (low sodium) is a lesser evil than circulatory collapse, isotonicity is sacrificed in favor of isovolemia.[110] A "happy medium" may be struck by a mechanism in which signals from the volume receptors modulate the sensitivity of the osmoreceptors.[109]

Deficiency of ADH results in a continuous, copious flow of as much as 15 L of dilute urine/day. This disease state is known as *diabetes insipidus* (diabetes, meaning to flow through, as a siphon; insipidus meaning flat or tasteless in contrast to the sweet urine of *diabetes mellitus,* which results from insufficiency of insulin) (Chapter 66). It is accompanied by a profound thirst to compensate for the water loss in the urine. Just a few units of ADH per day are usually sufficient to control this disease. Diabetes insipidus may result from any injury to or destruction of the supraoptic nucleus of the hypothalamus. A rarer form of the disease (*nephrogenic* diabetes insipidus) is characterized by an abundance of ADH but insensitivity of the nephron to its actions. Overproduction of ADH produces a disease state that is characterized by abnormal retention of water and dilution of plasma sodium. Renal compensation for the change in volume results in increased sodium loss, which aggravates the hyponatremia.

Oxytocin
Physiologic actions

Mammary effects. The lactating mammary gland consists of an intricate system of branching tubules and ducts that end in bulbous evaginations of secretory epithelial cells called alveoli (Fig. 61-16). Milk is continuously secreted by these cells and stored in the lumina of the alveoli. Simply applying negative pressure at the nipple is not sufficient to draw milk from the alveoli into the ducts and then to the exterior. The secretory cells of the alveoli are encased in a meshwork of myoepithelial cells that are endowed with contractile properties similar to those of smooth muscle. Contraction of the myoepithelial cells generates a positive pressure of the order of 10 to 15 mm Hg and forces the milk out of the alveoli into

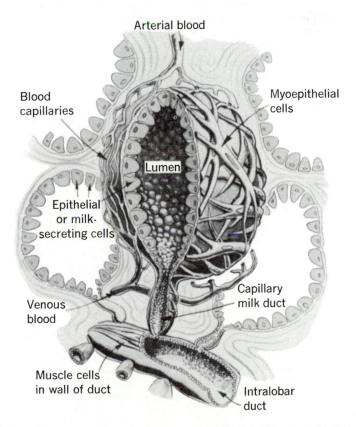

Fig. 61-16. Schematic representation of mammary alveolus encased in myoepithelial cells. (From Turner.[153])

the ducts and the lacteal sinuses opening to the nipple. This phenomenon is known as milk let-down or milk ejection. In the absence of suckling the pressurized milk is retained in the mammary gland by sphincterlike elements in the nipple. Distortion of the nipple and the application of slight negative pressure overcomes the resistance of the nipple sphincters and allows the milk to escape.

It is now recognized that milk ejection is a reflex phenomenon that occurs in response to tactile stimulation of the nipple and has a lag time of 30 to 60 sec.[33,35,49] The reflex has both neural and hormonal components. Blocking the sensory input by denervation or local or general anesthesia blocks the reflex. The finding that interruption of the motor nerves to the mammary gland did not prevent milk ejection led to the realization that the motor component of the reflex is humorally mediated. Oxytocin was implicated because of its known ability to cause milk let-down. Strong support was lent to this idea by the finding that removal of the posterior pituitary gland abolished milk ejection in response to tactile stimulation of the mammary gland. Furthermore, electrical stimulation of neural pathways to the posterior pituitary caused milk ejection, and lesions in these pathways block the reflex. Emotional states such as fear, anxiety, and pain can also inhibit the release of oxytocin and thereby prevent the suckling young from receiving milk. As might be expected, epinephrine is a potent inhibitor of oxytocin secretion. Auditory or visual stimuli related to suckling or milking cause release of oxytocin by a ''conditioned reflex.'' Simply seeing their infants or hearing them cry is sufficient to initiate milk ejection in many lactating women. Dairy cows often let down their milk on hearing the clatter of milking pails.

Contraction of myoepithelial cells is so specific and sensitive a response to oxytocin that it has been used as the basis of a bioassay. Increased pressure recorded from polyethylene tubes inserted in the mammary ducts is presumptive evidence of endogenous oxytocin secretion in response to physiologic stimuli. Understanding of the cellular events triggered by oxytocin to cause milk ejection is still lacking. It is of interest to note, however, that sensitivity of the mammary myoepithelial cells to oxytocin increases 1,000-fold as pregnancy progresses.[136]

Uterine effects. There is little doubt that oxytocin causes uterine contraction. The cellular and molecular events that mediate this response are described by Marshall[104] and in Chapter 65. The role of oxytocin in parturition has not been fully defined.[53,152,155] Oxytocin can initiate the birth process and is used clinically to induce delivery in women entering the final days of their pregnancy. However, because even radioimmunoassay is not sensitive enough to detect the presence of oxytocin in normal unextracted plasma, it has not been possible to determine if small increases in oxytocin secretion precede and possibly even trigger parturition. The large amounts of oxytocin in jugular blood found during the expulsive phase of parturition are thought to be released reflexly in response to stretching of the uterine cervix and vagina, in much the same way that suckling triggers milk ejection. It is well known that genital stimulation can result in oxytocin release and milk ejection. During parturition, oxytocin secretion may be evoked by a positive feedback mechanism to hasten expulsion of the infant; that is, more forceful contractions of uterine muscle result in greater stretching of the cervix and vagina, which in turn results in more oxytocin release. Measurements of intraductal pressure in the mammary gland during parturition indicate that oxytocin indeed is secreted at this time but reveal that not all the contractions of the uterus can be attributed to oxytocin, since the fluctuations in intraductal pressure and intrauterine pressure do not always coincide. There are conflicting reports concerning the ability of hypophysectomized animals or women suffering from hypothalamic damage to deliver their offspring normally. Most of the available data support the idea that oxytocin is essential for rapid expulsion of the infant, but it appears unlikely that increased secretion of oxytocin is the event that initiates parturition.[32]

Although the nonpregnant uterus is relatively insensitive to oxytocin, it has been suggested that oxytocin released during coitus may facilitate the transport of sperm from the vagina to the fallopian tubes where fertilization occurs.[70] Definitive evidence in support of this idea is lacking. Similarly, there is little to indicate that oxytocin plays any important physiologic role in the male.

REFERENCES
General reviews

1. Ganong, W. F., and Martini, L.: Frontiers in neuroendocrinology 1969, New York, 1969, Oxford University Press, Inc.
2. Ganong, W. F., and Martini, L.: Frontiers in neuroendocrinology 1971, New York, 1971, Academic Press, Inc.
3. Ganong, W. F., and Martini, L.: Frontiers in neuroendocrinology 1973, New York, 1973, Oxford University Press, Inc.
4. Harris, G. W.: Neural control of the pituitary gland, London, 1955, Edward Arnold, Ltd.

5. Harris, G. W., and Donovan, B. T.: The pituitary gland, Berkeley, Calif., 1966, University of California Press, vols. 1-3.
6. Heller, H., and Pickering, B. T.: Pharmacology of the endocrine systems and related drugs: the neurohypophysis. In International encyclopedia of pharmacology and therapeutics, London, 1970, Pergamon Press, Ltd., vol. 1.
7. Knobil, E., and Sawyer, W. H., editors: Handbook of physiology. Section 7: endocrinology, the pituitary gland and its neuroendocrine control, Washington, D.C., 1974, The American Physiological Society, vol. 4.
8. Martini, L., and Ganong, W. F.: Neuroendocrinology, New York, 1966, Academic Press, Inc., vols. 1 and 2.
9. Martini, L., and Ganong, W. F.: Frontiers in neuroendocrinology 1976, New York, 1976, Raven Press, vol. 4.
10. Meites, J., ed.: Hypophysiotropic hormones of the hypothalamus: assay and chemistry, Baltimore, 1970, The Williams & Wilkins Co.
11. Pecile, A., and Müller, E. E.: Growth hormone. In Excerpta Medica Foundation International Congress Series, No. 158, Princeton, N.J., 1968, Excerpta Medica Foundation.
12. Pecile, A., and Müller, E. E.: Growth and growth hormone. In Excerpta Medica Foundation International Congress Series, No. 244, Princeton, N.J., 1972, Excerpta Medica Foundation.
13. Pecile, A., and Müller, E. E.: Growth hormone and related peptides. In Excerpta Medica Foundation International Congress Series, No. 381, Princeton, N.J., 1976, Excerpta Medica Foundation.
14. Raiti, S.: Advances in human growth hormone research, Publ. No. 74-612, Washington, D.C., 1973, Department of Health, Education, and Welfare.
15. Schwartz, I. L., and Schwartz, W. B., editors: Symposium on antidiuretic hormones, Am. J. Med. **42:**651, 1967.

Original papers

16. Acher, R.: Molecular evolution of the polypeptide hormones. In Polypeptide hormones: molecular and cellular aspects, Ciba Foundation Symposium No. 41 (new series), New York, 1976, Elsevier North Holland, Inc., p. 31.
17. Ahrén, K., and Hjalmarson, A.: Early and late effects of growth hormone on transport of amino acids and monosaccharides in the isolated rat diaphragm. In Pecile, A., and Müller, E. E., editors: Growth hormone, Excerpta Medica Foundation International Congress Series, No. 158, Princeton, N.J., 1968, Excerpta Medica Foundation.
18. Allen, B. M.: The results of extirpation of the anterior lobe of the hypophysis and of the thyroid of Rana pipiens larvae, Science **44:**755, 1916.
19. Aschner, B.: Über die Funktion der Hypophyse, Pflüegers Arch. **146:**1, 1912.
20. Astwood, E. B.: Growth hormone and corticotropin. In Pincus, G., and Thimann, K. V., editors: The hormones, New York, 1955, Academic Press, Inc., vol. 3.
21. Atwell, W. J.: On the nature of the pigmentation changes following hypophysectomy in the frog larva, Science **49:**48, 1919.
22. Bargmann, W.: Neurosecretion, Int. Rev. Cytol. 1966.
23. Bell, P. H.: Purification and structure of β-corticotropin, J. Am. Chem. Soc. **76:**5565, 1954.
24. Benfey, B. J., and Purvis, J. L.: Purification and amino acid composition of melanophore expanding hormone from hog pituitary gland, J. Am. Chem. Soc. **77:**5167, 1955.
25. Bern, H. A., and Nicoll, C. S.: The comparative endocrinology of prolactin, Recent Prog. Horm. Res. **24:**681, 1968.
26. Bowers, C. Y., et al.: Interactions of the L-thyroxine or L-triiodothyronine and thyrotropin-releasing factor on the release and synthesis of thyrotropin from the anterior pituitary gland of mice, Endocrinology **81:**741, 1967.
27. Bowers, C. Y., et al.: Porcine thyrotropin releasing hormone is (Pyro) Glu-His-Pro(NH$_2$), Endocrinology **86:**1143, 1970.
28. Brownstein, M. G., et al.: Distribution of hypothalamic hormones and neurotransmitters within the diencephalon. In Martini, L., and Ganong, W. F., editors: Frontiers in neuroendocrinology, New York, 1976, Raven Press, vol. 1, p. 1.
29. Burgus, R., et al.: Characterization of ovine hypothalamic hypophysiotropic TSH-releasing factor, Nature **226:**321, 1970.
30. Burgus, R., et al.: Structure moléculaire du facteur hypothalamique (LRF) d'origine ovine contrôlant la sécrétion de luteinisation (LH), C.R. Acad. Sci. [D] (Paris) **273:**1611, 1971.
31. Burgus, R., et al.: Primary structure of somatostatin, a hypothalamic peptide that inhibits the secretion of pituitary growth hormone, Proc. Natl. Acad. Sci. USA **70:**684, 1973.
32. Chard, T.: Neurohypophyseal hormones. In Fuchs, F., and Klopper, A., editors: Endocrinology of pregnancy, New York, 1974, Harper & Row, Publishers, p. 271.
33. Cross, B. A.: Neural control of oxytocin secretion. In Martini, L., and Ganong, F. W., editors: Neuroendocrinology, New York, 1966, Academic Press, Inc.
34. Cross, B. A., and Green, J. D.: Activity of single neurones in the hypothalamus: effect of osmotic and other stimuli, J. Physiol. **148:**554, 1959.
35. Cross, B. A., and Harris, G. W.: The role of the neurohypophysis in the milk ejection reflex, J. Endocrinol. **8:**148, 1952.
36. Dale, H. H.: On some physiological actions of ergot, J. Physiol. **34:**163, 1906.
37. Daniel, P. M.: The blood supply of the hypothalamus and pituitary gland, Br. Med. Bull. **22:**202, 1966.
38. Daniel, P. M., and Prichard, M. M. L.: Observations on the vascular anatomy of the pituitary gland and its importance in pituitary function, Am. Heart J. **72:**147 1966.
39. Daughaday, W. H.: Regulation of skeletal growth by sulfation factor, Adv. Intern. Med. **17:**237, 1971.
40. Daughaday, W. H., Herington, A. C., and Phillips, L. S.: The regulation of growth by endocrines, Annu. Rev. Physiol. **37:**211, 1975.
41. Daughaday, W. H., et al.: Defective sulfation factor generation: a possible etiologic link in dwarfism, Trans. Assoc. Am. Physicians **82:**129, 1969.
42. deBodo, R. C., and Altszuler, N.: Insulin hypersensitivity and physiological insulin antagonists, Physiol. Rev. **38:**389, 1958.
43. deBodo, R. C., et al.: On the hormonal regulation of carbohydrate metabolism; studies with C^{14} glucose, Recent Prog. Horm. Res. **19:**445, 1963.
44. Dixon, J. S., and Li, C. H.: The isolation and structure of beta-melanocyte-stimulating hormone from horse pituitary glands, Gen. Comp. Endocrinol. **1:**167, 1961.
45. Douglas, W. W.: Mechanism of release of neuro-

hypophysial hormones: Stimulus secretion coupling. In Knobil, E., and Sawyer, W. H., editors: Handbook of physiology. Section 7: endocrinology, the pituitary gland and its neuroendocrine control, Washington, D.C., 1974, American Physiological Society, vol. 4, pt. 1, p. 191.

46. DuVigneaud, V.: Hormones of the posterior pituitary gland: oxytocin and vasopressin, Harvey Lect. **50:** 1, 1954-1955.

47. Dyball, R. E. J., and Koizumi, K.: Electrical activity in the supraoptic and paraventricular nuclei associated with neurohypophysial hormone release, J. Physiol. **201:**711, 1969.

48. Ellis, S., et al.: Studies on the nature of plasma growth hormone. In Pecile, A., and Müller, E. E., editors: Growth hormone and related peptides, Princeton, N.J., 1976, Excerpta Medica Foundation, p. 75.

49. Ely, F., and Petersen, W. E.: Factors involved in the ejection of milk, J. Dairy Sci. **24:**211, 1941.

50. Engel, F. L.: Extra-adrenal actions of adrenocorticotropin, Vitam. Horm. **19:**189, 1961.

51. Evans, H. M., and Long, J. A.: The effect of the anterior lobe administered intraperitoneally upon growth, maturity and oestrus cycles of the rat, Anat. Rec. **21:** 62, 1921.

52. Fevold, H. L., Hisaw, F. L., and Leonard, S. L.: The gonad stimulating and the luteinizing hormones of the anterior lobe of the hypophysis, Am. J. Physiol. **97:** 291, 1931.

53. Fitzpatrick, R. J.: The posterior pituitary gland and the female reproductive tract. In Harris, G. W., and Donovan, B. T., editors: The pituitary gland, Berkeley, Calif., 1966, University of California Press, vol. 3.

54. Folley, S. J.: Lactation. In Parkes, A. S., editor: Marshall's physiology of reproduction, ed. 3, London, 1952, Longmans, Green & Co., Ltd., vol. 2.

55. Gainer, H., Sarne, Y., and Brownstein, M. J.: Neurophysin biosynthesis: conversion of a putative precursor during axonal transport, Science **195:**1354, 1977.

56. Gauer, O. H., and Henry, J. P.: Circulatory basis of fluid volume control, Physiol. Rev. **43:**423, 1963.

57. Geschwind, I. I., and Li, C. H.: The tibia test for growth hormone. In Smith, R. W., Jr., Gaebler, O. H., and Long, N. H., editors: Hypophyseal growth hormone, nature and actions, New York, 1955, McGraw-Hill Book Co.

58. Ginsburg, M., and Heller, H. H.: Antidiuretic activity in blood obtained from various parts of the cardiovascular system, J. Endocrinol. **9:**274, 1953.

59. Glick, S. M.: The regulation of growth hormone secretion. In Ganong, W. F., and Martini, L., editors: Frontiers in neuroendocrinology 1969, New York, 1969, Oxford University Press, Inc.

60. Glick, S. M., et al.: The regulation of growth hormone secretion, Recent Prog. Horm. Res. **21:**241, 1965.

61. Goodman, H. M.: In vitro actions of growth hormone on glucose metabolism in adipose tissue, Endocrinology **76:**216, 1965.

62. Goodman, H. M.: Growth hormone and the metabolism of carbohydrate and lipid in adipose tissue, Ann. N.Y. Acad. Sci. **148:**419, 1968.

63. Goodman, H. M., and Knobil, E.: The effects of fasting and of growth hormone administration on plasma fatty acid concentration in normal and hypophysectomized rhesus monkeys, Endocrinology **65:**451, 1959.

64. Grantham, J. J., and Burg, M. B.: Effect of vasopressin and cyclic AMP on permeability of isolated collecting tubules, Am. J. Physiol. **211:**255, 1966.

65. Green, J. D.: The comparative anatomy of the portal vascular system and of the innervation of the hypophysis. In Harris, W. G., and Donovan, B. T., editors: The pituitary gland, Berkeley, Calif., 1966, University of California Press, vol. 1.

66. Green, J. D., and Harris, G. W.: The neurovascular link between the neurohypophysis and adenohypophysis, J. Endocrinol. **5:**136, 1947.

67. Guillemin, R.: Physiology and chemistry of the hypothalamic releasing factors for gonadotropins: a new approach to fertility control, Contraception **5:**1, 1972.

68. Guillemin, R., and Gerich, J. R.: Somatostatin: physiological and clinical significance, Annu. Rev. Med. **27:**379, 1976.

69. Guyda, H. J., and Freisen, H. G.: The separation of monkey prolactin from monkey growth hormone by affinity chromatography, Biochem. Biophys. Res. Comm. **42:**1068, 1971.

70. Hammond, J., Jr.: Light regulation of hormone secretion, Vitam. Horm. **12:**157, 1954.

71. Harris, G. W.: The innervation and actions of the neurohypophysis; an investigation using the method of remote control stimulation, Philos. Trans. R. Soc. Lond. (Biol.) **232:**385, 1947.

72. Hayward, J. N., and Vincent, J. D.. Osmosensitive single neurons in the hypothalamus of unanesthetized monkeys, J. Physiol. **210:**947, 1970.

73. Herring, P. T.: The histological appearance of the mammalian pituitary body, Q. J. Exp. Physiol. **1:**121, 1908.

74. Hild, W., and Zetler, G.: Experimenteller Beweis für die Entstehung der sog. Hypophysenhinterlappenwirkstoffe in Hypothalamus, Pfluegers Arch. **257:**169, 1953.

75. Hofmann, K.: Chemistry and function of polypeptide hormones, Annu. Rev. Biochem. **31:**213, 1962.

76. Hofmann, K., and Yajima, H.: Synthetic pituitary hormones, Recent Prog. Horm. Res. **18:**41, 1962.

77. Holmes, L. B., et al.: Normal growth with subnormal growth hormone levels, N. Engl. J. Med. **279:**359, 1968.

78. Houssay, B. A.: The hypophysis and metabolism, N. Engl. J. Med. **214:**961, 1936.

78a. Hughes, J.: Opioid peptides and their relatives, Nature **278:**394, 1979.

79. Imura, H., et al.: Immunologic studies of adrenocorticotropic hormone (ACTH): dissociation of biologic and immunologic activities, J. Clin. Endocrinol. Metab. **25:**1361, 1965.

80. Jacobs, L. S., et al.: Increased serum prolactin after administration of synthetic thyrotropin releasing hormone (TRH) in man, J. Clin. Endocrinol. Metab. **33:**996, 1971.

81. Jewell, P. A., and Verney, E. B.: Experimental attempt to determine the site of neurohypophysial osmoreceptors in the dog, Philos. Trans. R. Soc. Lond. (Biol.) **240:**197, 1957.

82. Knigge, K. M., and Silverman, A. J.: Transport capacity of the median eminence. In Knigge, K. M., Scott, D. E., and Weindl, A., editors: Brain-endocrine interaction. Median eminence: structure and function Symposium, March 1971, Basel, Switzerland, 1972, Karger, p. 350.

83. Knobil, E.: The pituitary growth hormone: some physiological considerations, In Zarrow, M. X., editor: Growth in living systems, New York, 1961, Basic Books, Inc., Publishers.

84. Knobil, E., and Greep, R. O.: The physiology of growth hormone with particular reference to its action in the rhesus monkey and the "species specificity"

problem, Recent Prog. Horm. Res. **15**:1, 1959.

85. Knopf, R. F., et al.: Plasma growth hormone response to intravenous administration of amino acids, J. Clin. Endocrinol. Metab. **25**:11, 1965.

86. Koefed-Johnsen, V., and Ussing, H. H.: The contributions of diffusion and flow to the passage of D_2O through living membranes, Acta Physiol. Scand. **28**:60, 1953.

87. Korner, A.: Anabolic action of growth hormone, Ann. N.Y. Acad. Sci. **148**:408, 1968.

88. Kostyo, J. L.: Rapid effects of growth hormone on amino acid transport and protein synthesis, Ann. N.Y. Acad. Sci. **148**:389, 1968.

89. Kostyo, J. L., et al.: The nature of fragments of human growth hormone produced by plasmin digestion. In Pecile, A., and Müller, E. E., editors: Growth hormone and related peptides, Princeton, N.J., 1976, Excerpta Medica Foundation, p. 33.

90. Krieger, D. T.: Neuroendocrinology. In Ingbar, S. H., editor: The year in endocrinology 1975-76, Plenum Medical Book Co., New York, 1976, p. 1.

91. Kuku, S. F., et al.: Human thyrotrophic hormone kinetics and effects in euthyroid males, Horm. Metab. Res. **7**:54, 1975.

92. Labrie, F., et al.: Mode of action of hypothalamic regulatory hormones in the adenohypophysis. In Martini, L., and Ganong, W. F., editors: Frontiers in neuroendocrinology, New York, 1976, Raven Press, vol. 4, p. 63.

93. Leaf, A.: Membrane effects of antidiuretic hormone, Am. J. Med. **42**:745, 1967.

94. Leaf, A., and Frazier, H.: Some recent studies on the actions of neurohypophyseal hormones, Prog. Cardiovasc. Dis. **4**:47, 1961.

95. Lee, M. O., and Schaffer, N. K.: Anterior pituitary growth hormone and the composition of growth, J. Nutr. **7**:337, 1934.

96. Lerner, A. B., and Lee, T. H.: Isolation of a homogenous melanocyte stimulating hormone from hog pituitary glands, J. Am. Chem. Soc. **77**:1066, 1955.

97. Lerner, A. B., and McGuire, J. S.: Effect of α- and β-melanocyte stimulating hormones on the skin colour of man, Nature **189**:176, 1961.

98. Lewis, V. J., et al.: Human growth hormone: a family of proteins. In Pecile, A., and Müller, E. E., editors: Growth hormone and related peptides, Princeton, N.J., 1976, Excerpta Medica Foundation, p. 64.

99. Li, C. H.: Properties of and structural investigations on growth hormones isolated from bovine, monkey and human pituitary glands, Fed. Proc. **16**:775, 1957.

100. Li, C. H.: Synthesis and biological properties of ACTH peptides, Recent Prog. Horm. Res. **18**:1, 1962.

101. Li, C. H.: The chemistry of human pituitary growth hormone: 1956-1966. In Pecile, A., and Müller, E. E., editors: Growth hormone, Excerpta Medica Foundation International Congress Series, No. 158, Princeton, N.J., 1968, Excerpta Medica Foundation.

102. Li, C. H., Evans, H. M., and Simpson, M. E.: Isolation and properties of the anterior hypophyseal growth hormone, J. Biol. Chem. **159**:353, 1945.

103. Malayan, S. A., and Reid, I. A.: Antidiuresis produced by injection of renin into the third cerebral ventricle of the dog, Endocrinology **98**:329, 1976.

104. Marshall, J. M.: Effects of neurohypophyseal hormones on the myometrium. In Knobil, E., and Sawyer, W., editors: Handbook of physiology. Section 7: endocrinology, Washington, D.C., 1974, American Physiological Society, vol. 4.

105. Martin, J. B., Kontor, J., and Mead, P.: Plasma growth hormone responses to hypothalamic hippocampal and amygdaloid stimulation: effects of variation in stimulus parameters and treatment with α methyl-tyrosine (αMT), Endocrinology **92**:354, 1973.

106. Matsuo, H., et al.: Structure of the porcine LH and FSH releasing hormone in the proposed amino acid sequence, Biochem. Biophys. Res. Comm. **43**:1334, 1971.

107. Meakin, J. W., et al.: The rate of disappearance of ACTH activity from the blood of humans, J. Clin. Endocrinol. Metab. **19**:1491, 1959.

108. Meyer, V., and Knobil, E.: Growth hormone secretion in the unanesthetized rhesus monkey in response to noxious stimuli, Endocrinology **80**:163, 1967.

109. Moses, A. M., and Miller, M.: Osmotic threshold for vasopressin release as determined by saline infusion and by dehydration, Neuroendocrinology **7**:219, 1971.

110. Moses, A. M., and Miller, M.: Osmotic influences on the release of vasopressin. In Knobil, E., and Sawyer, W. H., editors: Handbook of physiology. Section 7: endocrinology, Washington, D.C., 1974, American Physiological Society, vol. 4, pt. 1, p. 225.

111. Moses, R. L., McCann, S. M., and Dudley, C. A.: Releasing hormones and sexual behavior, Prog. Brain Res. **42**:37, 1975.

112. Motta, M., Fraschini, F., and Martini, L.: Short feedback mechanisms in the control of anterior pituitary function. In Ganong, F. W., and Martini, L., editors: Frontiers in neuroendocrinology 1969, New York, 1969, Academic Press, Inc.

113. Ney, R. L.: The anterior pituitary gland. In Bondy, P. K., and Rosenberg, L. E., editors: Duncan's diseases of metabolism, ed. 6, Philadelphia, 1969, W. B. Saunders Co.

114. Niall, H. D.: Revised primary structure for human growth hormone, Nature N. Biol. **230**:90, 1971.

115. Niall, H. D., et al.: The chemistry of growth hormone and lactogenic hormones, Recent Prog. Horm. Res. **29**:387, 1973.

116. Odell, W. D., Wilber, J. F., and Utiger, R. D.: Studies of thyrotropin physiology by means of radioimmunoassay, Recent Prog. Horm. Res. **23**:47, 1967.

117. Oliver, G., and Shäfer, E. A.: On the physiological action of extracts of pituitary body and certain other glandular organs, J. Physiol. **18**:277, 1895.

118. Orloff, J., and Handler, J. S.: The role of adenosine 3',5'-phosphate in the action of antidiuretic hormone, Am. J. Med. **42**:757, 1967.

119. Ott, I., and Scott, J. C.: The action of infundibulin upon the mammary secretion, Proc. Soc. Exp. Biol. Med. **8**:48, 1910.

120. Park, C. R., et al.: The effect of growth hormone on glucose uptake by isolated rat diaphragm, J. Biol. Chem. **197**:151, 1952.

121. Pelletier, G., et al.: Electron microscopic immunochemical localization of growth hormone–release inhibiting hormone (somatostatin) in the rat median eminence, Am. J. Anat. **140**:445, 1974.

122. Pickering, B. T., et al.: The role of neurophysin proteins: suggestions from the study of their transport and turnover, Ann. N.Y. Acad. Sci. **284**:15, 1975.

123. Poisner, A. M., and Douglas, W. W.: A possible mechanism of release of posterior pituitary hormones involving adenosine triphosphate in the neurosecretory granules, Mol. Pharmacol. **4**:531, 1968.

124. Porter, J. C., et al.: Neurovascular regulation of the anterior hypophysis, Recent Prog. Horm. Res. **29**:161, 1973.

125. Raben, M. S., and Hollenberg, C. H.: Effect of growth

hormone on plasma fatty acids, J. Clin. Invest. **38:**484, 1959.

126. Reichert, L. E., Jr., Ramsey, R. B., and Carter, E. B.: Application of a tissue receptor assay to measurement of serum follitropin (FSH), J. Clin. Endocrinol. Metab. **41:**634, 1975.

127. Reichlin, S., et al.: Hypothalamic hormones, Annu. Rev. Physiol. **38:**389, 1976.

128. Rimoin, D. L., et al.: Genetic aspects of clinical endocrinology, Recent Prog. Horm. Res. **24:**365, 1968.

129. Riniker, B. P., Sieber, P., and Rittel, W.: Revised amino acid sequences for porcine and human adrenocorticotropic hormones, Nature N. Biol. **235:**114, 1972.

130. Robinson, A. G.: Isolation, assay and secretion of individual human neurophysins, J. Clin. Invest. **55:**360, 1975.

131. Robinson, A. G., et al.: Physiological control of two neurophysins in humans, J. Clin. Endocrinol. Metab. **44:**330, 1977.

132. Robison, G. A., Butcher, R. W., and Sutherland, E. W.: Cyclic AMP, New York, 1971, Academic Press, Inc.

133. Russell, J. A.: Hormonal control of glycogen storage, Ciba Found. Coll. Endocrinol. **6:**193, 1954.

134. Russell, J. A.: Effects of growth hormone on the metabolism of amino acids. In Smith, R. W., Jr., Gaebler, O. H., and Long, C. N. H., editors: Hypophyseal growth hormone, nature and actions, New York, 1955, McGraw-Hill Book Co.

135. Sachs, H.: Biosynthesis of the neurohypophysial hormones. In Heller, H., and Pickering, B. T., editors: International encyclopedia of pharmacology and therapeutics. Section 41: pharmacology of the endocrine system and related drugs; the neurohypophysis, Oxford, 1970, Pergamon Press, Ltd.

136. Sala, N. L.: The milk ejecting effect induced by oxytocin and vasopressin during human pregnancy, Am. J. Obstet. Gynecol. **89:**626, 1964.

137. Salmon, W. D., Jr., and Daughaday, W. H.: A hormonally controlled serum factor which stimulates sulfate incorporation by cartilage in vitro, J. Lab. Clin. Med. **49:**825, 1958.

138. Sawyer, W. H.: Neurohypophysial principles of vertebrates. In Harris, G. W., and Donovan, B. T., editors: The pituitary gland, Berkeley, Calif., 1966, University of California Press, vol. 3.

139. Sayers, M. A., Sayers, G., and Woodbury, L. A.: The assay of adrenocorticotropic hormone by the adrenal ascorbic acid depletion method, Endocrinology **42:**379, 1948.

140. Scharrer, E., and Scharrer, B.: Hormones produced by neurosecretory cells, Recent Prog. Horm. Res. **10:**193, 1954.

141. Schwyze, R., and Sieber, P.: Total synthesis of adrenocorticotropic hormone, Nature **199:**172, 1963.

142. Segar, W. E., and Moore, W. W.: The regulation of antidiuretic hormone release in man. I. Effects of change in position and ambient temperature on blood ADH levels, J. Clin. Invest. **47:**143, 1968.

143. Senft, G., et al.: Effects of hydration and dehydration on cyclic adenosine 3′,5′-monophosphate concentration in the rat kidney, Pfluegers Arch. **298:**348, 1968.

144. Share, L.: Acute reduction in extracellular fluid volume and the concentration of antidiuretic hormone in blood, Endocrinology **69:**925, 1961.

145. Share, L.: Extracellular fluid volume and vasopressin secretion. In Ganong, W. F., and Martini, L., editors: Frontiers in neuroendocrinology 1969, New York, 1969, Oxford University Press, Inc.

146. Shome, B., and Parlow, A. F.: Human follicle stimulating hormone (hFSH): first proposal for the amino acid sequence of the α subunit (hFSHα) and first demonstration of its identity with the α subunit of human leutinizing hormone (hLHα), J. Clin. Endocrinol. Metab. **39:**199, 1974.

147. Smith, P. E.: Experimental ablation of the hypophysis in the frog embryo, Science **44:**280, 1916.

148. Smith, P. E.: The disabilities caused by hypophysectomy and their repair, J.A.M.A. **88:**158, 1927.

149. Smith, P. E.: Hypophysectomy and replacement therapy in the rat, Am. J. Anat. **45:**205, 1930.

150. Syndor, K. L., and Sayers, G.: Blood and pituitary ACTH in intact and adrenalectomized rats after stress, Endocrinology **55:**621, 1954.

151. Takahashi, V., et al.: Growth hormone, glucose and insulin in human plasma during sleep, J. Lab. Clin. Med. **70:**1021, 1967.

152. Theobald, G. W.: Clinical pharmacology: oxytocin. In Heller, H., and Pickering, B. T., editors: International encyclopedia of pharmacology and therapeutics. Section 41: pharmacology of the endocrine system and related drugs, Oxford, 1970, Pergamon Press, Ltd.

153. Turner, C. W.: Harvesting your milk crop, Oak Brook, Ill., 1969, Babson Bros. Co.

154. Uthne, K.: Preliminary studies of somatomedin in vitro and in vivo in rats. Adv. Metab. Disord. **8:**115, 1975.

155. Uttenthal, L. O., and Hope, D. B.: The isolation of three neurophysins from porcine posterior pituitary lobes, Biochem. J. **116:**899, 1970.

156. Van Wyk, J. J., et al.: The somatomedins: a family of insulin-like hormones under growth hormone control, Recent Prog. Horm. Res. **30:**259, 1974.

157. Verney, E. B.: The antidiuretic hormone and the factors which determine its release, Proc. R. Soc. Lond. (Biol.) **135:**25, 1947.

158. Weitzman, E. N.: Circadian rhythms and episodic hormone secretion in man, Annu. Rev. Med. **27:**225, 1976.

159. Wilber, J. F., and Porter, J. C.: Thyrotropin and growth hormone releasing activity in hypophysial portal blood, Endocrinology **87:**807, 1970.

160. Wilber, J. F., et al.: Gonadotropin-releasing hormone: distribution and effects in the central nervous system, Recent Prog. Horm. Res. **32:**117, 1976.

161. Wilhelmi, A. E., Fishman, J. B., and Russell, J. A.: A new preparation of crystallin anterior pituitary growth hormone, J. Biol. Chem. **176:**735, 1948.

162. Xuereb, G. P., Prichard, M. M. L., and Daniel, P. M.: The arterial supply and venous drainage of the human hypophysis cerebri, Q. J. Exp. Physiol. **39:**199, 1954.

163. Yalow, R. S.: Heterogeneity of peptide hormones, Recent Prog. Horm. Res. **30:**597, 1974.

164. Young, F. G.: Permanent experimental diabetes produced by pituitary (anterior lobe) injections, Lancet **2:**372, 1937.

165. Zimmerman, E. A.: Localization of hypothalamic hormones by immunocytochemical techniques. In Martini, L., and Ganong, W. F., editors: Frontiers in neuroendocrinology, New York, 1976, Raven Press, vol. 4, p. 25.

166. Zondek, B.: Über die hormon des hypophysenvorder lappen. I. Wachtsumhormon, Follikelreifungshormon (Prolan A), Luteinisnerungshormon (Prolan B), Stoffwechselhormon, Klin. Wochenschr. **8:**245, 1930.

62

H. MAURICE GOODMAN and LESTER VAN MIDDLESWORTH

The thyroid gland

The thyroid gland was first described in the second century A.D. by Galen, who thought its function was to moisten and lubricate the pharynx. In 1656 Wharton[120a] named it after the adjacent thyroid cartilage. The name "thyroid" derives from the Greek *tiros,* meaning shield-shaped, which describes the cartilage. Wharton assigned four functions for the thyroid gland: (1) to "take up certain superfluous fluids emanating from the recurrent nerve," (2) to "warm the cartilage of the trachea," (3) to "lubricate the larynx and so to render the voice smoother, more melodious and sweeter," and (4) to "contribute to the contour and beauty of the neck . . . especially in the female sex to whom on this account a larger gland has been assigned." Other functions were also ascribed to the thyroid, particularly in reference to its inordinately rich supply of blood. It was thought to divert the flow of blood from the brain during periods of quiescence or to buffer the large changes in blood flow to the brain that might occur in sudden periods of excitement.

As in many other areas in endocrinology, clues to the role of the thyroid and its physiology were derived from clinical observations.[100a] In 1802 Flajani reported some of the clinical signs of what is now recognized as hyperthyroidism. Subsequent reports by Parry (1815) and Graves (1842) described a disease entity characterized by sweating, tremor, and tachycardia. Von Basedow (1840) described the triad of exophthalmos* (bulging eyes), goiter, and palpitation associated with hyperactivity of the thyroid. Even now the condition of hyperthyroidism is called Graves' disease or von Basedow's disease.

Goiter (from the Latin *guttur,* meaning throat), or enlarged thyroid, is the most common consequence of deranged thyroid physiology and has been observed since antiquity. Ancient and modern observations show a majority of some populations to have enlarged thyroid glands. In some areas of the world where iodine is in short supply, goiter is endemic, but goiter exists sporadically in all populations. Goiter may result from either hypo- or hyperthyroidism, and it may even occur in the euthyroid state. Ancient therapy for goiter has included sheep thyroids (627 AD) and ashes of sponge or seaweed, which we now know are rich in iodine, an essential constituent in thyroid hormone biosynthesis. For cosmetic reasons the surgeons Kocher and the Reverdins (1883) independently devised an operation for removing the thyroid gland (thyroidectomy) from persons living in the endemic goiter region of Switzerland. Several months postoperatively the symptoms of somnolence, impaired orientation, intolerance to cold, coarseness and dryness of skin, and accumulation of mucoproteins in subcutaneous areas were noted. These symptoms resembled those of another disease that Ord called myxedema (from the Greek *myxa,* mucus; *oidema,* swelling) because of the extensive deposit of mucin under the skin. Among the early descriptions of myxedema or hypothyroidism was that of Gull (1874), who recognized its relationship to cretinism, a disease characterized by dwarfism and mental retardation and caused by congenital hypothyroidism. Murray[75] in 1891 established that the thyroid might actually secrete some material that protected against thyroid insufficiency and successfully treated hypothyroid patients with extracts of sheep thyroid glands. In 1895 Bauman[18] found that the thyroid gland not only contained iodine but that some 25% of the total body iodine store is located in the thyroid colloid. On Christmas Day of 1914 Kendall, who later won the Nobel Prize for his work on adrenal hormones, isolated thyroid hormone in crystalline form.[62] He erroneously thought the hormone was an iodinated oxyindole and hence proposed the name "thyroxin," derived from *thyroid oxyindole.* Although this structure was incorrect, the name thyroxine has been retained.

---

*Exophthalmos often accompanies hyperthyroidism and appears to result from infiltration of mucopolysaccharides in the retro-orbital space and damage to the extraocular muscles. Some investigators attribute these effects to an exophthalmos-producing substance (EPS) released from the pituitary gland, whereas other attribute them to an immune reaction involving the retro-orbital tissues.[123]

The correct structure was proposed by Harington and was proved by synthesis by Harington and Barger in 1927.[54] Although thyroxine appeared to possess all the physiologic activity of material extracted from thyroid tissue, a second, even more active thyroid hormone, triiodothyronine (Fig. 62-1), was isolated in 1952 by Gross and Pitt-Rivers.[51] Triiodothyronine differs from thyroxine only in the absence of one iodine molecule from the B ring.

The thyroid-stimulating effect of the pituitary was suggested by P. E. Smith in early experiments on amphibia[98] and subsequently confirmed in studies on hypophysectomized rats.[99] This discovery led ultimately to the discovery and isolation of the pituitary thyroid-stimulating hormone (TSH) thyrotropin. Recently TSH secretion has been shown to be partially controlled by a neurohormone thyrotropin-releasing hormone (TRH).

Development of thyroid gland

Phylogenetically, the thyroid is an ancient gland that is derived from the endostyle in the floor of the pharynx of even the most primitive chordates. In primitive larval forms the endostyle is an exocrine gland that secretes its iodinated mucuslike product into the pharynx to entrap particles of food. During metamorphosis from larval to adult form, some of the epithelial cells of the endostyle transform into thyroid follicles. The iodine-rich secretion of the endostyle appears to be the evolutionary forerunner of the thyroid colloid.[49]

In mammals the thyroid gland develops embryologically as an evagination from the floor of the embryonic pharynx. It enlarges and migrates posteriorly. Initially, it is connected to the pharynx by the thyroglossal duct, which attenuates and eventually is obliterated. In most species the thyroglossal duct atrophies early, but occasionally isolated portions persist and give rise to thyroglossal duct cysts or even to ectopic bits of thyroid tissue, particularly in the lingual and sublingual regions. In man a recognizable thyroid organized in follicles appears 74 days after fertilization. Even at this early date, iodine accumulation and colloid formation[96] indicate that the gland is functional. Early stages of thyroid development appear to be independent of the pituitary and its TSH.

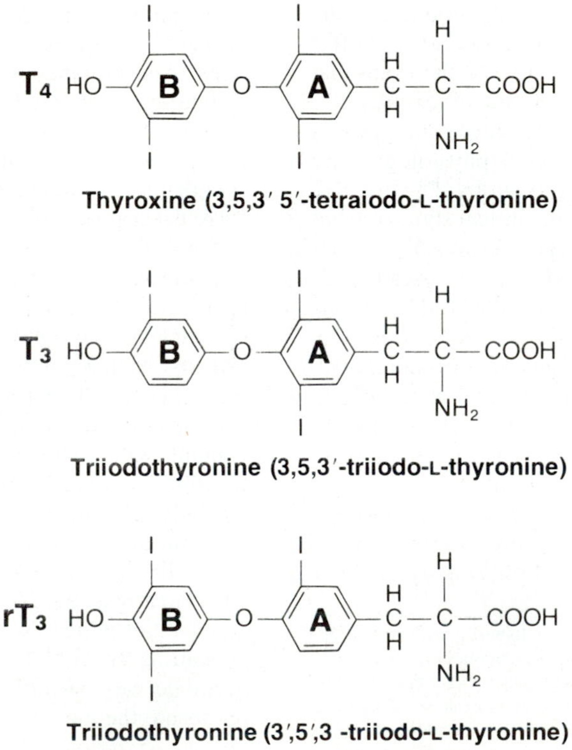

Fig. 62-1. Structural formulas of thyroxine, triiodothyronine, and "reverse" triiodothyronine.

Gross anatomy

The human thyroid gland is a dumbbell-shaped organ consisting of two lateral lobes connected by a thin isthmus. It wraps around the trachea at the level of the thyroid cartilage and is closely bound to it by loose connective tissue. Lateral to each lobe are the carotid sheaths. The recurrent laryngeal nerves lie on the medial borders of the lobes in the grooves formed between them and the trachea. The thyroid of normal adult men weighs 15 to 25 gm. It is often asymmetric, with the right lobe being slightly larger. The thyroid is capable of enormous growth and may attain a weight of several hundred grams in conditions of severe goiter. In addition to the lateral lobes and the isthmus, a third structure, or pyramidal lobe, is sometimes seen as a fingerlike projection extending headward from the isthmus; it apparently represents a remnant of the thyroglossal duct.

The thyroid gland is among the most richly vascularized tissues of the body; only the lungs and the carotid body receive a greater blood supply.[100] Normal blood flow to the thyroid is approximately 5 ml/min/gm, as compared, for example, to roughly 3 ml/min/gm for the kidney. Blood flow may become as high as 1 L/min after extreme stimulation. The thyroid receives its blood supply from the superior and inferior thyroid arteries that arise from the external carotids and subclavian arteries. Venous drainage is by way of the superior and middle thyroid veins into the internal jugular veins and by way of the inferior thyroid vein into the innominate veins. The thyroid also is endowed with a rich lymphatic system that may play an important role in the delivery of hormone to the general circulation.

The thyroid gland is innervated by both sympathetic and parasympathetic fibers that arise from the cervical sympathetic ganglia and the vagus nerves, respectively. The physiologic importance of the rich nerve supply to the gland is not understood. These nerves may regulate blood flow to the gland, especially when their effect is combined with TSH.[74] There is no evidence to indicate that changing blood flow significantly alters either the rate of hormone release or the delivery of iodine and substrates. Denervation of the thyroid or transplantation to a distant site leads to no obvious impairment of function.

Microscopic anatomy[124]

The functional unit of the thyroid gland is the follicle (Fig. 62-2), which is composed of closely packed vesicles of various shapes and sizes (from 0.02 to 0.30 mm in diameter) and filled with a proteinaceous colloid, thyroglobulin. The epithelial cells lining the follicles vary in configuration from low cuboidal to columnar, depending on their functional state; the height of the epithelium is greatest when the gland is most active. Each follicle is invested with a dense capillary network separated from the epithelial cells by a well-defined basement membrane. The capillary

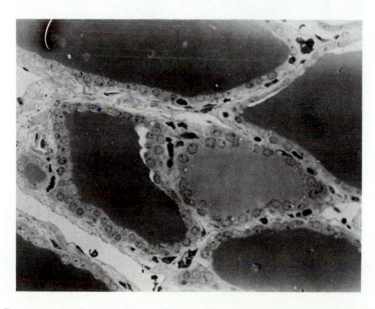

Fig. 62-2. Cross section of normal human thyroid. Thyroid follicular cells surround "lakes" of stored thyroglobulin. (Magnification ×150.) (From Matthews and Martin.[70])

endothelium is fenestrated and contains pores that range from 400 to 600 Å in diameter. Groups of follicles are set off by connective tissue septa to form lobules that receive their blood supply from a single small artery. The functional state of one lobule may differ widely from that of an adjacent lobule. The large cells between the follicles are the parafollicular or "C" cells, which produce calcitonin (Chapter 63).

Ultrastructure[124]

Unstimulated thyroid follicular cells are unremarkable in their ultrastructure. The cells (Fig. 62-3) are clearly polarized; their apical surfaces, which face the follicular lumen, have many microvilli that are not prominent in the normal human thyroid gland. Terminal bars are prominent on the lateral borders at the apical end of the cells. The endoplasmic reticulum appears to be composed of irregular tubules of large caliber and greatly expanded cisternae. The Golgi complex surrounds the apical side of the nucleus. Numerous vesicles are found throughout the cytoplasm. Lysosomes are abundant. The secretory product of the thyroid follicular cells is stored in the follicular lumen, and no storage granules are seen within the cells.

THYROID HORMONES

The mammalian thyroid gland produces three recognized hormones: thyroxine, triiodothyronine, and thyrocalcitonin. This chapter will deal solely with thyroxine and triiodothyronine, which are classically regarded as the thyroid hormones. Calcitonin, which arises from the parafollicular cells rather than the follicles, may serve in the regulation of calcium metabolism and is discussed in Chapter 63. Thyroxine and triiodothyronine are amino acid derivatives of the L configuration. The thyroxine nucleus consists of two benzene rings connected in ether linkage with an alanine side chain in the para position on the A ring (Fig. 62-4). Although represented in the same plane on the printed page, the A and B rings lie perpendicular to each other in three-dimensional configuration. Since the thyroid hormone molecules (thyroxine and triiodothyronine) contain 4 and 3 atoms of iodine, respectively,

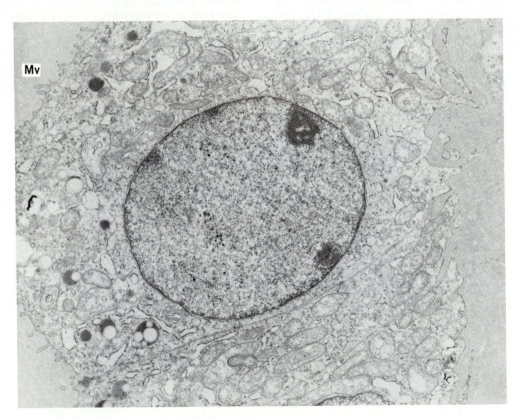

Fig. 62-3. Electron micrograph of human thyroid follicular cell. Microvilli, *Mv,* are shown on apical edge projecting into stored thyroglobulin. (From Matthews and Martin.[70])

iodine comprises more than half their molecular weight. Because of their iodine content, thyroxine is abbreviated as T_4 and triiodothyronine as T_3. This abundance of iodine in the thyroid hormones has proved to be a great boon to endocrinologists in their quest to understand the physiology of the thyroid. The availability of radioactive iodine of high specific activity has been exploited for the study of hormone biosynthesis, storage, degradation, and excretion. Consequently, a fairly detailed knowledge of thyroidal iodine metabolism has been acquired.

Biosynthesis

Biosynthesis of thyroxine and triiodothyronine requires synthesis of thyroglobulin and accumulation and oxidation of iodide. Thyroxine and triiodothyronine are produced by iodinating thyroglobulin. The peptide portion of the thyroglobulin is believed to be synthesized, as are all other proteins, on ribosomes throughout the endoplasmic reticulum of the follicular cells. The protein is then transported to the Golgi apparatus, where it is thought that the carbohydrate component is added. Vesicles containing thyroglobulin then migrate to the apical surface of the follicular cells, where they fuse with the apical membrane and deliver their contents into the follicular lumen.[77] Addition of the iodine component may occur extracellularly within the follicular lumen at the apical surface of the follicular cells.[76] The synthesis of the thyroglobulin is independent of iodination, but the protein is unstable until iodinated. Incorporation of iodine into thyroid hormones can be conveniently divided into a four-step process.

Iodine trapping. The thyroid avidly takes up iodide from the extracellular fluid and concentrates it to many times the amount found in plasma. The ratio of the iodide ion in the thyroid gland to that in the serum, commonly called the T/S ratio, under normal circumstances is 25:1.

Following hypophysectomy or other conditions in which the thyroid is inactive, the T/S ratio falls to about 10:1; conversely, under intense TSH stimulation, the T/S ratio may approach 250:1. Iodide trapping results from a highly efficient process that involves the participation of an active transport system located in the basal portion of the follicular cells. Active transport is believed to depend on the characteristic size, shape, and charge of the iodide ion.[127] This "iodide pump" requires energy supplied from oxidative metabolism and can be blocked by a variety of metabolic poisons, including cyanide, azide, and dinitrophenol. A number of anions inhibit iodide trapping, presumably because they compete for binding sites on the carrier. These include perchlorate, pertechnetate, perrhenate and perborate.[127,130,131] Normally, none of these ions exists in plasma in sufficient concentration to influence iodide trapping. However, another ion, thiocyanate, competitively blocks iodide concentration by the thyroid, and it is continually produced within the mammalian organism under physiologic conditions. The normal concentration of thiocyanate in human serum is 50 to 100 times greater than that of iodide, but no role for thiocyanate in normal thyroid physiology has been described.

Oxidation of iodide. It is generally believed that the iodide anion cannot itself be incorporated into organic molecules but must first be oxidized to a higher oxidation state such as atomic iodine I^0, or perhaps even I_3^-, although neither of these intermediates has been isolated from the thyroid. Such oxidation is accomplished through the action of a peroxidase enzyme system[112] that transfers an electron from an iodide anion to hydrogen peroxide to yield water and free oxygen. The source of the peroxide has not been established. Ninety percent of the peroxidase activity is present in a particulate fraction of thyroid homogenates. In life the peroxidase activity is probably

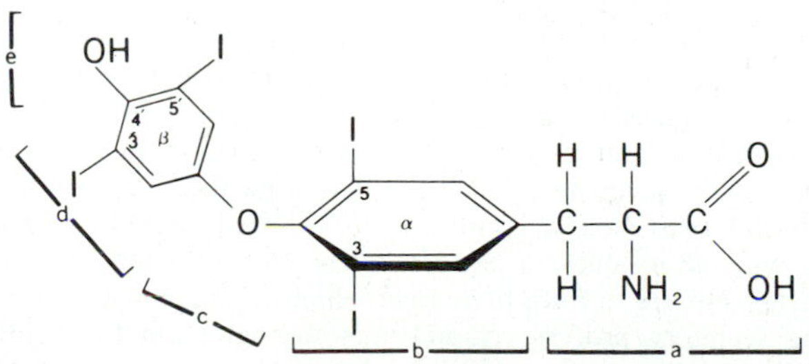

Fig. 62-4. Three-dimensional representation of thyroxine molecule. *a*, Alanine side chain; *b*, α- or A ring lies in plane perpendicular to page; *c*, oxygen linkage forms 110-degree angle; *d*, β- or B ring lies in plane of page; *e*, the phenolic hydroxyl group. (From Barker and Klitgaard.[17])

localized at the apical membrane of the thyroid follicle cells.

Iodination. The addition of iodine to tyrosine residues in the thyroglobulin molecule is thought to occur within the lumen of the follicle, probably at the luminal surfaces of the apical microvilli. Iodination appears to be coupled with peroxidase activity and may occur spontaneously as soon as I^0 has been formed.[112] The initial product of iodination appears to be monoiodotyrosine (MIT), with the tyrosine receptor residue present in peptide linkage within the thyroglobulin molecule. Biochemical studies indicate that any number of large molecular weight proteins in addition to thyroglobulin can be iodinated in this way. Abundance of iodine generated through the peroxidase reaction leads to the addition of a second iodine to form diiodotyrosine (DIT).

Coupling. The final stage of synthesis requires the coupling of two DIT molecules to form thyroxine or one DIT and one MIT to form triiodothyronine. Triiodothyronine may also be formed by removal of one iodine from the thyroxine molecule. Such coupling involves two iodinated tyrosines present in peptide linkage within the same or adjacent thyroglobulin molecules. It has been suggested that free radicals of diiodotyrosine are generated in some way that is associated with the action of the thyroid peroxidase and that two such free radicals react to form an ether linkage, leaving a serine or alanine behind in the peptide chain. Thus the synthesis of thyroxine depends on the action of the peroxidase, which, in addition to iodine oxidation, triggers both iodination of tyrosine and coupling of the two tyrosine precursor residues.[112,115] These reactions are summarized in Fig. 62-5.

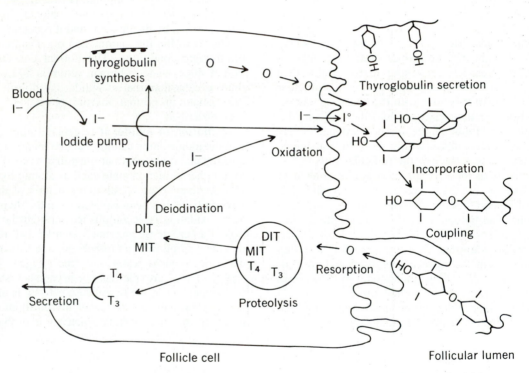

Fig. 62-5. Schematic representation of thyroid hormone biosynthesis and secretion. Protein portion of thyroglobulin is synthesized in rough endoplasmic reticulum. It then travels to Golgi apparatus, where carbohydrate moieties are added, and proceeds to apical surface in secretory vesicles, which fuse with apical membrane and discharge their contents into lumen. Iodide is pumped into cell from blood. At apical surface, it is oxidized through the action of a peroxidase. Iodine attaches to tyrosine residues in peptide linkage in thyroglobulin. Two iodinated tyrosyl groups couple in ether linkage to form thyroxine, which is still trapped in peptide linkage within thyroglobulin. The secretory process requires that thyroglobulin be engulfed by pseudopods thrown out into follicular lumen to resorb thyroglobulin into vesicles that fuse with lysosomes. Lysosomal protease breaks thyroglobulin down to amino acids T_4, T_3, MIT, and DIT. T_4 and T_3 are released from cell. DIT and MIT are deiodinated to free tyrosine and iodide, both of which are recycled back to iodinated thyroglobulin.

Hormone storage

The thyroid is unique among endocrine glands in that it stores its product extracellularly. Normally, approximately 30% of the mass of the thyroid comprises stored thyroglobulin, which contains enough thyroid hormone to meet normal human requirements for 2 to 3 months, even if no further synthesis were to occur during that time. Thyroxine and triiodothyronine are stored in peptide linkage within the thyroglobulin molecule in the lumen of the thyroid follicle. Thyroglobulin is probably a dimer[116a] and has a molecular weight of 670,000[39]; it may comprise as much as 80% of the total protein of the thyroid. About 10% is carbohydrate, and about 0.5% is iodine. It contains 5,650 amino acid residues. Of these, 110 are tyrosyl residues, of which only 7 to 20 are normally iodinated.[7,100b] Only a few of the iodinated tyrosine molecules are converted to active hormone; most remain as the biologically inactive MIT and DIT. Each molecule of thyroglobulin contains an average of 6 molecules of MIT, 5 molecules of DIT, and 1 to 3 molecules of T_4. A molecule of T_3 is found only in 20% to 30% of the thyroglobulin molecules. Therefore only 1 to 3 molecules of thyroid hormone are present in each thyroglobulin molecule. If iodination occurs in the presence of a relatively subnormal amount of iodide, the total iodination is reduced and the proportion of MIT is increased relative to DIT. This results in much less T_4 being formed and a slight increase in the formation of T_3. The result is that the T_3/T_4 ratio in thyroglobulin is increased when the total iodination is subnormal.[109]

Hormone secretion

Because of the unusual mode of storage of thyroid hormones, the secretory process is also unique. Under normal circumstances, very little thyroglobulin escapes from the thyroid follicle into the circulation.[25] Thyroglobulin is broken down to its amino acid components, including the thyroid hormones T_3 and T_4, by a proteolytic enzyme. Some proteolytic digestion of thyroglobulin may occur extracellularly within the follicular lumen where proteolytic enzymes are known to be present,[36] but electron microscopic evidence supports an alternate hypothesis. Small droplets of colloid appear to be taken up into the follicular cells by pinocytosis; these droplets appear to fuse with lysosomes within the follicular cells. Proteolytic digestion within the vesicles results in the release of the thyroid hormones.[16,34] Intense TSH stimulation causes the appearance of long strands of protoplasm, which reach out from the apical surface of the follicular cells into the lumen, where they engulf chunks of thyroglobulin (Fig. 62-6).[124]

The synthesis of thyroglobulin and its export in vesicles into the follicular lumen is an ongoing process that takes place simultaneously with uptake of iodinated thyroglobulin and digestion to release the thyroid hormones. These opposite processes, involving vesicles laden with thyroglobulin moving both into and out of the cells, are somehow regulated and coordinated so that under normal circumstances thyroglobulin neither accumulates nor is depleted. In the early stages of increased secretory activity the T/S

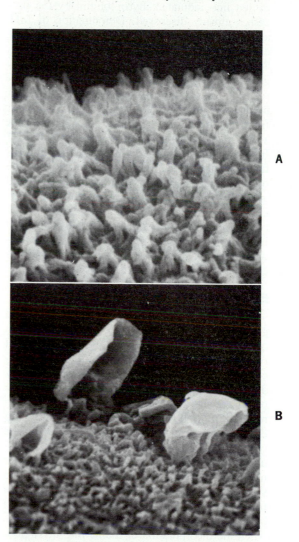

Fig. 62-6. Scanning electron micrograph of microvilli lining apical border of dog thyroid follicular cells. **A,** TSH secretion suppressed by feeding thyroid hormone. (×36,000.) **B,** One hour after dog was injected with TSH. (×16,500.) (From Balasse et al.[15])

ratio falls somewhat, an indication that secretion exceeds formation.

Digestion of the iodinated thyroglobulin releases not only thyroxine and triiodothyronine but also relatively large amounts of MIT and DIT as well as the other amino acid components of the thyroglobulin. Only thyroxine and triiodothyronine are secreted into the circulation.[113] MIT and DIT can neither escape from the cell into the circulation to any significant degree nor be directly resynthesized into thyroglobulin. The iodine is stripped off the iodotyrosines by a deiodinase enzyme,[103] and the tyrosine can be recycled into thyroglobulin. The iodide so produced is also recycled into iodinated thyroglobulin by the processes already discussed. Deiodination of MIT and DIT is of great importance in the iodine economy of the thyroid cell. In rats, as much as 70% of the iodide incorporated into the thyroglobulin may derive from deiodination of iodotyrosine.[97] Individuals suffering from genetic deficiency of thyroid deiodinase[103] readily suffer from iodine deficiency because the iodotyrosines are excreted in the urine and the iodine is lost.[102] It should be reemphasized that thyroxine and triiodothyronine in a ratio of about 20:1 are the only iodinated products released by the thyroid gland in significant quantities under normal circumstances. The normal daily utilization of thyroxine and triiodothyronine in man is approximately 90 and 30 μg, respectively[78]; this utilization ratio of thyroxine to triiodothyronine is clearly less than that stored in the thyroid.

Effects of thyroid-stimulating hormone

Autonomous function of the normal thyroid gland is too sluggish to supply normal bodily requirements for hormone. Stimulation by TSH from the anterior lobe of the pituitary gland accelerates every function of the thyroid. TSH independently promotes iodine trapping, iodination of thyroglobulin, and the elaboration of the thyroglobulin itself. Inhibition of iodine trapping does not interfere with the action of TSH on the other synthetic phases. Consequently, thyroglobulin synthesis and secretion into the follicular lumen proceed at a rate commensurate with the stimulation caused by TSH in iodine-deficient individuals or when the iodine pump is inhibited by an ion such as perchlorate.[114] In addition, when the incorporation of iodine into thyroglobulin is inhibited by drugs such as propylthiouracil, TSH still accelerates the pumping of iodine into the thyroid and stimulates the synthesis and proteolysis of thyroglobulin, whether or not it contains thyroid hormone.[91]

Stimulating multiple, separate, but complementary reactions does not necessarily imply that TSH has a multiplicity of actions. A single, unique interaction between TSH and follicular cells may still produce some intermediate product that independently accelerates a series of separate but functionally linked biochemical reactions. A preponderance of evidence indicates that the effects of TSH on the thyroid gland are mediated through cyclic AMP. Within 1 min after the addition of TSH, cyclic AMP concentration within the thyroid increases sharply.[41a,46] Virtually all effects of TSH on the thyroid can be duplicated by inhibition of cyclic AMP degradation or by appropriate administration of cyclic AMP or one of its analogs.[90] Cyclic AMP enhances the release of thyroid hormone as well as the incorporation of radioactive iodine into thyroglobulin and the synthesis of thyroglobulin.

Within minutes after injection, TSH also produces a striking increase in blood flow to the thyroid gland. This effect may be due to the release of serotonin from mast cells within the thyroid gland.[32]

Since TSH produces growth of the thyroid gland as well as enhanced hormone synthesis and secretion, it is not surprising that it increases the synthesis of RNA, protein, and phospholipid. Stimulation of the protein synthetic machinery of thyroid cells and of cellular division requires a considerably longer time lag than hormone synthesis and secretion. The immediate effects do not appear to require the formation of new RNA. After many hours of continuous stimulation with TSH, however, inhibition of RNA synthesis with actinomycin D prevents further response of the thyroid cells to TSH.[73]

Other agents that affect the thyroid gland

The plasma of most patients suffering from hyperthyroidism contains only normal or subnormal concentrations of TSH,[117] but it also contains certain IgG gamma globulins that stimulate the thyroid to produce its hormones.[71,72,78a] These stimulators are antibodies and have been termed "thyroid-stimulating antibodies" (TSAb).[10a] They are species specific with variable cross-reactions,[72] and therefore when they are injected into laboratory animals, the effects may be different from those in humans. TSAbs include substances sometimes referred to as "long-acting thyroid stimulator (LATS)" or "LATS-protector"[10a] and, like TSH, their effects on human thyroids are mediated by cyclic AMP. Possible antigens for TSAb are TSH receptor sites of the thyroid gland;[97a] the probability of a person producing TSAb may depend partly on his genetic background.[31a]

A TSH-like material that may arise from the placenta is present in the plasma of pregnant women.[55a,56] Although not yet characterized chemically, placental TSH appears to differ only to a minor degree from pituitary TSH. Some trophoblastic tumors such as the hydatidiform mole produce high concentrations of chorionic gonadotropin (Chapter 65), which may stimulate the thyroid.[53a]

Another category of compounds that has found widespread clinical use both stimulates and inhibits the thyroid gland. These compounds are called goitrogens because they lead to the formation of goiter. They do so indirectly by blocking the incorporation of iodine into tyrosine residues of thyroglobulin.[87] This blocks hormone formation even though the thyroid continues to make uniodinated thyroglobulin. When the hormone reserves become exhausted, T_4 concentration in the blood falls, causing increased secretion of TSH by the pituitary. In the presence of the goitrogen, TSH cannot increase hormone production but does stimulate formation of uniodinated thyroglobulin and growth of the thyroid cells. Small quantities of natural goitrogens are found throughout most of the *Brassica* genus of plants, which includes cabbage. Synthetic goitrogens, principally propylthiouracil (PTU), are used therapeutically to treat hyperthyroidism. The common structural features of these goitrogens is a thioamide grouping (Fig. 62-7).

It has already been mentioned that a variety of inorganic anions can competitively inhibit iodide transport and consequently interfere with the synthesis of thyroxine. The iodine ion itself has interesting paradoxical effects on the thyroid gland. Obviously too little iodine interferes with the formation of thyroid hormone. When the concentration of iodide in plasma exceeds 35 $\mu g/100$ ml (the normal concentration is less than 0.5 $\mu g/100$ ml), the intracellular concentration of iodide increases and the concentrating mechanism for iodide is overwhelmed. When the concentration of iodide within the follicular cells is greatly increased, the oxidation of iodide is abruptly blocked, and the synthesis of T_4 and T_3 can no longer occur.[128,129] Excessive iodide also inhibits the release of previously formed hormone. These effects are independent of TSH and are seen even in thyroid glands of hypophysectomized animals. The thyroid cells thus appear to autoregulate their function according to their intracellular iodide content.[59,107] Autoregulation is thought to be achieved through the action of one or more iodinated inhibitory substances whose formation is dependent on iodide concentration.

Transport of thyroid hormones in blood

Of the circulating iodine in plasma, 90% is in compounds that are neither ultrafiltrable nor dialyzable. The predominant portion (more than 99%) of thyroid hormone in the blood is firmly bound to plasma proteins. The term "protein-bound iodine" (PBI) has been used to describe

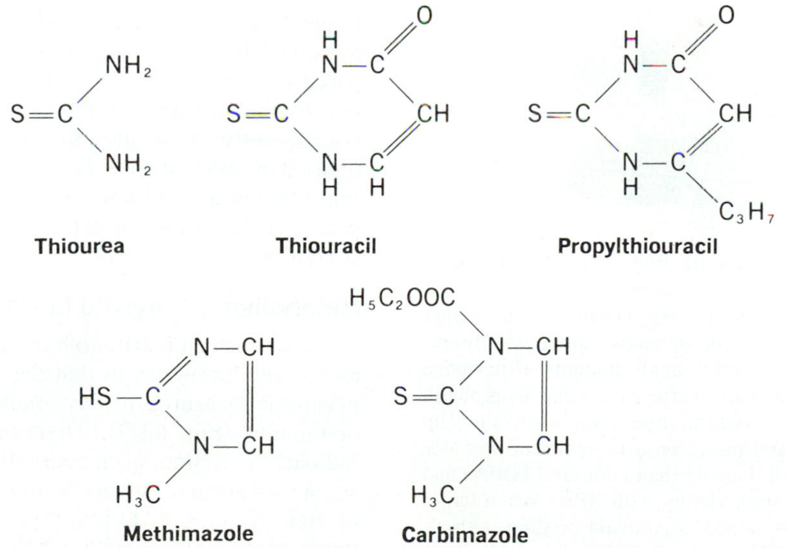

Fig. 62-7. Some common goitrogens.

circulating levels of the thyroid hormone expressed as iodide. In normal man the PBI ranges from 4 to 8 μg/100 ml of plasma. Thyroid hormones do not bind indiscriminately to the plasma proteins but are linked to particular proteins (Fig. 62-8). The first of these to be discovered is called thyroxine-binding globulin (TBG).[50,64,89] It is probably a glycoprotein and has a molecular weight of 63,000. In human beings the blood contains approximately 1.5 mg TBG/100 ml of serum. Each molecule of TBG can bind one molecule of T_4. TBG is potentially capable of binding 99.9% of the plasma thyroxine, but normally it accounts for only about 70% of the protein-bound thyroxine. TBG also binds about 70% of the circulating T_3, but T_3 is bound less firmly. Not all the binding capacity of TBG is utilized because some of the hormone is also transported by a fraction of protein, which migrates more rapidly to the anode than albumin during electro-phoresis.[58] For this reason it has been called the thyroxine-binding prealbumin (TBPA). There is enough TBPA in normal human serum to bind 200 to 340 μg T_4/100 ml, as compared to only 16 to 24 μg T_4/100 ml for TBG, but the binding affinity for TBPA is weaker than TBG.[79] TBPA is responsible for binding about 15% of the plasma thyroxine[125] but does not bind T_3. TBPA has a molecular weight of 73,000 and is present in normal human serum at a concentration of 30 mg/100 ml. Each molecule of TBPA can bind one molecule of T_4. The remaining T_3 and T_4 are bound to albumin, which, although more abundant than TBG and TBPA, has a low affinity for thyroid hormones. The amounts of the thyroid-binding proteins present in plasma can vary with different physiologic and pathologic states. Pregnancy or estrogen therapy increases TBG[88] without changing the concentration of free T_3 or T_4 in the serum.[105]

Protein-bound thyroxine and triiodothyronine are in equilibrium with unbound or free T_3 and T_4 in plasma. The equilibrium heavily favors binding so that only 0.3% of the T_3 and 0.03% of the T_4 are free. Because T_3 is much less firmly bound than T_4, it can diffuse out of the vascular compartment into the extracellular fluid some 5 times more rapidly than T_4. In the rat, perhaps 20 times more T_3 is bound to peripheral cells than to circulating plasma proteins.[81] Of the 6 to 12 μg of thyroid hormone present/100 ml of serum, only about 0.10 μg/100 ml is triiodothyronine. The serum therefore contains only 10^{-7} moles of T_4 and 10^{-9} moles of T_3/L. Although the total amount of T_3 in plasma is only 1% of that of T_4, it is bound less firmly than T_4. Consequently, free T_3 reaches 10% of the concentration of the free T_4 (0.0003 μg T_3/100 ml compared to 0.003 μg T_4/100 ml). Since only the free thyroid hormone is available to the cells and consequently constitutes the biologically active pool, it is evident that T_3 is considerably more important at the cellular level than might be suspected from its rate of secretion or plasma concentration.[78]

Metabolism of thyroid hormones

Thyroxine and triiodothyronine are unique among the hormones in that their half-life in the plasma is measured in days rather than seconds or minutes (Fig. 62-9). Thyroxine labeled with radioactive iodine disappears from the bloodstream of normal human beings with a half-life of 6 to 7 days.[78] Triiodothyronine turns over much more rapidly, with a half-life of approximately 1 day.[78] Interestingly, the thyroid status

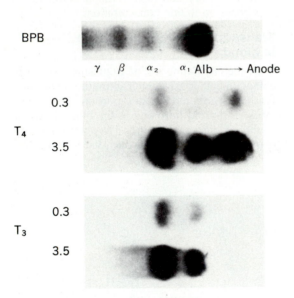

Fig. 62-8. Electrophoretic migration of ^{131}I-labeled T_4 and T_3 added to normal human serum (0.3 or 3.5 μg/100 ml of serum) in tris-maleate buffer. Upper portion shows major protein bands stained with bromphenol blue (BPB). With small amounts of hormone present, T_4 travels partly between α_2- and α_1-globulin on a protein whose concentration is too low to show up with BPB stain and partly on a protein that migrates more rapidly toward anode than albumin (TBPA) and that also fails to stain visibly with BPB. When larger amounts of T_4 are added, significant binding with albumin is seen. T_3 binds only to TBG and albumin. (From Ingbar.[58])

of the individual determines the rate of disappearance of labeled hormone from his plasma. In hyperthyroid states, thyroxine disappears from plasma with a half-life of only 3 or 4 days and, in severe hypothyroidism, 9 or 10 days may be required for half the thyroxine to leave the plasma.[108] In children the biologic half-life of thyroxine is only about 5 days.[53] The turnover rate of thyroid hormone in blood can be calculated from the half-life and the measured size of the extrathyroidal pool of hormone. Approximately half the total extrathyroidal pool of triiodothyronine, or about 30 to 60 μg, is replaced daily. This compares with approximately 80 to 90 μg of thyroxine removed and replaced daily, despite the fact that thyroxine is so much more abundant in plasma.[104] About 80% of the thyroxine produced each day is deiodinated[45,83a] in the extrathyroidal tissues in an enzymatic reaction[119a] that can remove an iodine molecule from either the B ring or the A ring of the thyronine nucleus (Fig. 62-1). Deiodination of the B ring (the outer ring) of thyroxine produces 3,5,3'-triiodothyronine, which has greater biologic potency than thyroxine, and which probably accounts for most of the biologic activity of the thyroid hormones.[25,80] Removal of an iodine molecule from the A ring (the inner ring) produces 3,3',5'-triiodothyronine, which is also called reverse T_3 (rT_3). As yet, no physiologic role of rT_3 has been shown, although experimentally it interferes with deiodination of thyroxine and hence could decrease T_3 formation.[30] Normally, about 60% of the thyroxine

deiodinated gives rise to rT_3 and about 40% to T_3.[45] Therefore much of the T_3 utilized each day originates from peripheral deiodination of thyroxine, which is now viewed as a circulating prohormone.[60] Because rT_3 is degraded more rapidly than T_3, its plasma concentration is normally only about one third as high as that of T_3.[31,83a,107] In starvation and in chronic liver disease the concentration of rT_3 in the blood increases with respect to T_3. This has led to the speculation that under some circumstances the production of rT_3 from thyroxine may be selectively increased at the expense of T_3.[118] In the newborn human being the serum concentration of rT_3 is 4 to 6 times greater than the concentration of T_3, but by 9 to 11 days after birth the concentrations of these two compounds are similar to those found in the adult.[31] Fluctuations in the ratios of T_3 and rT_3 may result from changes in the rates of destruction or production of these compounds and, because of their vastly different biologic activities, may be indicative of another level of physiologic control.

Within 4 hr after intravenous administration of a dose of labeled T_4, about one third can be found in the liver.[29] From this and other observations, it has been concluded that the liver is probably the principal site of thyroid hormone inactivation. Thyroxine and triiodothyronine can be inactivated by conjugation through the phenolic hydroxyl group with glucuronic acid or, to a lesser extent, with sulfuric acid. Conjugated thyroid hormone along with some free thyroxine and

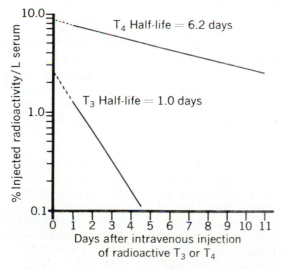

Fig. 62-9. Rate of loss of serum radioactivity after injection of labeled thyroxine or triiodothyronine into human subjects. (Plotted from data of Nicoloff et al.[78])

triiodothyronine are then excreted by the liver into the bile and thence into the intestine. Intestinal bacteria can split the ester bonds, releasing thyroid hormone that can then be reabsorbed from the intestine to travel once again to the liver. This excretion through the bile and reabsorption from the intestine has been called the enterohepatic circulation of thyroxine.[24] The portion of hormone that is not reabsorbed is excreted in the feces. In man, fecal excretion accounts for some 20% to 40% of the thyroxine degraded. Conjugation is not confined to the liver but can also occur in the kidneys and to a small extent in other tissues.

The other major route of thyroid hormone degradation is through deiodination.[83a,101] Virtually all tissues can remove iodine from the thyroid hormones. There has even been some evidence to suggest that deiodination of thyroxine may be linked to its action within the tissues. In some cases, drugs that inhibit deiodination also inhibit the action of thyroxine.[119] Ultimately, T_3 and T_4 are either deiodinated to diiodothyronine[45a] or deaminated or the side chain is oxidized.[83a] Administered T_3 appears to follow the same fate of degradation as T_4. The iodine removed from thyroxine and triiodothyronine may be excreted in the urine or may find its way back to the thyroid and be reconverted to hormone (Fig. 62-10).

Nonthyroidal metabolism of iodine

Iodine is a ubiquitous trace element in nature. All its natural inorganic salts are soluble. All vertebrates require it for synthesis of thyroid hormone, although most natural food products contain very little iodine. Seafood and seaweed are particularly rich in iodine. Epithelial cells of many organs, gastric mucosa, the ducts of the salivary glands, and parts of the small intestine have a capacity to concentrate iodide as much as 30 to 40 times the concentration found in the extracellular fluid.[26] No biologic importance has yet been attached to these processes. Iodide is also concentrated by the placenta, the mammary gland, and the skin of the newborn of some species, which increases the supply of iodide for the newborn where thyroid hormone is of critical importance (p. 1507 and Fig. 62-11). The choroid plexus actively transports iodide out of the cerebrospinal fluid (CSF) into the blood, so that the concentration of iodide in the CSF is maintained at only one third to one fifth that of the serum. All these iodide-concentrating mechanisms consume adenosine triphosphate (ATP) and oxygen and compete with the thyroid for ingested iodide ion. Iodide is excreted passively

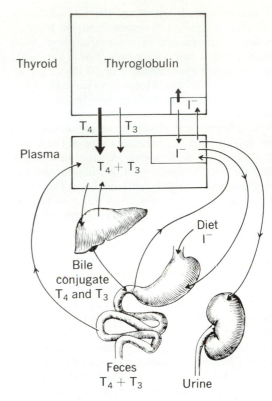

Fig. 62-10. Schematic representation of human iodine metabolism.

by the human kidney under normal circumstances, but when its concentration in plasma becomes extremely low, active renal tubular reabsorption can occur in some species.

PHYSIOLOGIC EFFECTS OF THYROID HORMONES

Although a vast number of observations have been made concerning the effects of too much or too little thyroid hormone in clinical conditions, a precise delineation of the physiologic role of the thyroid hormones is not yet possible. Hypo- and hyperthyroidism produce signs and symptoms that are not easily related to what might be regarded as a physiologic function of the hormone. For example, in hypothyroidism there is an accumulation of a mucuslike material in the subcutaneous and interstitial spaces. Are we to conclude that the normal physiologic role of thyroid hormones is to prevent such accumulation? Conversely, one of the striking aspects of thyrotoxicosis is severe protein catabolism. Are we to conclude that the normal physiologic role of the thyroid hormone is to promote protein

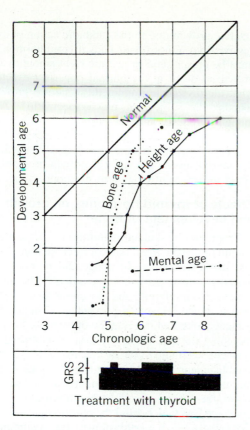

Fig. 62-11. Effects of thyroid therapy on growth and development of child with no functional thyroid tissue. Treatment with 1½ grains (100 mg) of thyroid daily began at 4½ years of age. Bone age rapidly returned toward normal, and rate of growth (height age) paralleled normal curve. Mental development, however, remained infantile. (From Wilkins.[121])

pothyroid individual may be noticeable for as long as 2 or 3 weeks. This delay in the onset of action, coupled with the prolonged action of a single dose of thyroid hormones, tends to rule them out as minute-to-minute regulators of physiologic function. Rather, their role appears to be more in establishing or permitting long-term functions such as growth, maturation, and some types of adaptation. Because it takes so long before any of the typical responses are seen, it is reasonable to infer that these typical responses lie many steps away from the initial interaction of hormone and tissue.

Thyroxine and triiodothyronine have thus far been considered together, since at present there appears to be no qualitative difference between the physiologic actions of triiodothyronine and thyroxine. On a molar basis, triiodothyronine is 3 to 5 times as potent as thyroxine in most biologic assays.[51] T_3 has a shorter latency and shorter duration of action.[5] These observations are all consistent with the fact discussed previously that a significant fraction of thyroxine is converted to triiodothyronine extrathyroidally and that triiodothyronine is the major physiologically active hormone.[104]

Finally, although fatal hypothyroid coma may occur on occasion, it should be noted that in general the thyroid hormones are not necessary for the maintenance of life. Experimental animals can survive for prolonged periods of time after thyroidectomy. Similarly, hypothyroidism in human beings resulting either from thyroid disease or thyroid ablation is compatible with prolonged survival, even if exogenous therapy is not administered. Clearly, the quality of life of these individuals is diminished.

Effects of thyroid hormones on growth and maturation

One of the most dramatic of all hormonal effects is that seen following the treatment of some amphibian larvae with thyroid hormones. In 1912 Gudernatsch[52] fed slices of bovine thyroid glands to frog tadpoles. Within the next few days the tadpoles underwent premature metamorphosis and became adult frogs that were normal in every respect except for their small size. The tadpole stage of the bullfrog lasts for about 2 years. If these tadpoles are fed dessicated thyroid, or if the purified hormones are added to the water in which they swim, full metamorphosis to adult form occurs within just a few days. Some of these adult frogs may be no bigger than houseflies. The morphologic changes that occur include loss of the tail, development of the hind

catabolism? Even observations of effects of thyroid hormones made in thyroidectomized or normal experimental animals are not always easily related to the maintenance of homeostasis.

Before discussing the various actions of thyroid hormones, some generalizations might be appropriate. First, the thyroid hormones do not seem to have any discrete target organ or organ systems. Their effects are generally diffuse and are manifest throughout the body. No special tissue or cell type can be identified as being particularly affected by the thyroid hormones. Second, thyroxine and, to a lesser extent, triiodothyronine are unique among the hormones in that there is a considerable lag period or latency that may last for several days before some of their typical effects are noticed. This is particularly true in man but is less so in other animals. Furthermore, the effects of a single injection of thyroxine to a hy-

limbs for hopping, the appearance of forelimbs, and replacement of the gills with lungs. Biochemically, changes can be observed in hemoglobin[43] and in the appearance of the urea cycle enzymes in the liver. In amphibia, triggering of metamorphosis is a normal physiologic function of the thyroid gland. In its absence, metamorphosis does not occur. Hypophysectomy also prevents metamorphosis. It was this observation by P. E. Smith[98] that provided the first insight into the role of the pituitary in thyroid function and ultimately led to the discovery of TSH. Examination of the tadpole thyroid just prior to metamorphosis reveals all the signs of increasing endocrine activity, including increased cell height and increased iodine trapping.

The mechanisms by which the thyroid hormone produces metamorphic changes is not known. Presumably the hormone simply acts as a trigger for a genetically determined sequence of changes in various cells. It is difficult to conceive of a mechanism that allows thyroxine to increase the synthesis of protein in some cells (e.g., those of the limb buds) while it hastens the degradation of protein in other cells, particularly the tail, which is totally reabsorbed. In producing these changes, it appears that thyroxine (or T_3) acts directly and locally on the cells whose developmental processes it directs. Implantation of small crystals of thyroxine in the tail of a tadpole produces local reabsorption in the vicinity of the implant without producing systemic metamorphic changes.[61]

In mammals, thyroxine and triiodothyronine have a less dramatic effect but are nevertheless indispensable for normal growth and development. Although the thyroid develops early in embryonic life, it is not certain that its hormones are required for growth and development of the embryo. Postnatally, however, in all warm-blooded animals, including man, the presence of a functional thyroid is a prerequisite for normal growth. Congenital thyroid deficiency results in cretinism, a common characteristic of which is dwarfed stature (Fig. 62-11). Thyroidectomy of rats in the early weeks of life leads to retardation of growth, which is almost as severe as that seen following hypophysectomy. Administration of T_3 or T_4 to thyroidectomized rats rapidly reinitiates growth. This response to thyroid hormone in thyroidectomized animals is one of the most sensitive recorded.[41] As little as 0.10 μg of thyroxine/100 gm thyroidectomized rat given daily is sufficient to restore the normal growth rate. Interestingly enough, giving supranormal amounts of hormone may depress rather than accelerate growth due to the catabolic action of excessive thyroid hormone. Thus thyroid hormones must be present and in just the right amounts for growth to occur at the normal rate. Just exactly what the thyroid hormones do to permit growth remains the subject of much speculation. Although synthesis and secretion of growth hormone by the pituitary gland are deficient in the absence of thyroid hormone,[67,94] it is likely that the thyroid hormones have a growth-promoting effect independent of growth hormone.

Effects of thyroid hormones on growth and development of bone

Promotion of skeletal maturation is a specific, cumulative, and irreversible effect of thyroxine and triiodothyronine. In congenital or neonatal hypothyroidism growth is retarded[92]; the long bones remain infantile and closure of their epiphyses is either delayed or prevented.[121] In normal subjects, ossification centers of various bones, including the ankle and wrist, are established at very characteristic times of life. So regular is the establishment of these ossification centers that radiologic examination of patients can be used to determine "bone age." In hypothyroidism the bone age lags well behind the chronologic age.[121] Conversely, in hyperthyroidism the maturation of bone may be accelerated and the bone age will exceed the chronologic age. This can lead to premature closure of the epiphyses and consequent stunting of growth. Not all bones are equally susceptible to the effects of thyroid hormone or have an equal requirement for it. Consequently, bone formation, particularly in the skull, may be anomalous in experimental animals subjected to thyroidectomy at an early age. In untreated cretins, failure of proper formation of the nasal and orbital bones gives a characteristic facial appearance. Thus, in the immature individual, it may be concluded that the thyroid gland plays an indispensable role in normal bone growth and maturation. The underlying mechanisms responsible for these effects are wholly unknown.

Excess thyroid hormone in the adult causes excessive demineralization of bone and a resultant increase in urinary and fecal loss of calcium and phosphate.[14] This effect is not mediated through the parathyroid glands and parathormone. Once again the mechanism is unknown but may represent an exaggeration of the normal catabolic processes that go on constantly in bone. In what appears to be a related phenomenon, the hormones of the thyroid gland also seem to control the rate of eruption of the teeth.[19] Increased secretion of thyroxine hastens tooth eruption in newborn animals and, conversely, thyroidectomy delays both the formation and eruption of teeth. This latter effect is manifest by a retardation in both dentine formation and root development.

Effects of thyroid hormones on the nervous system

In addition to shortness of stature, the congenitally hypothyroid individual suffers from severe mental retardation.[121] Not only is there a need for thyroid hormone for normal development of the CNS, but the hormone must be present at just the right time. In both clinical studies and animal experiments there is a critical period during which thyroid hormone must be present for normal development of the cerebrum. After this period has passed, even excessive amounts of thyroxine cannot restore function to normal. In man the critical period centers around the time of birth and may extend through the first year of life. Delayed recognition of hypothyroidism leads tragically to irreversible mental retardation despite subsequent hormone replacement (Fig. 62-11).

The neonatal rat appears to be at a relatively less mature stage than the human neonate. If rats are made hypothyroid at birth, either by surgical thyroidectomy or feeding of propylthiouracil to the mother during pregnancy, cerebral growth and nerve myelination are severely delayed.[4,37] There is delayed appearance of many of the classic fiber tracts and maturation of the nerve cells themselves is retarded. Development of cell processes and the density of axon and dendrite networks are diminished. Overall size of the brain is reduced and vascularity, particularly at the capillary level, is markedly reduced. As might be expected, hypothyroidism occurring before the critical period in rats significantly impairs their learning performance. Congenitally hypothyroid rats have impaired ability to negotiate a maze and tend to repeat mistakes over and over again. No such obvious learning impairments are observed if hypothyroidism occurs after the first few weeks of life. Similarly, treatment with thyroid hormone in the early neonatal days improves performance, but treatment beyond the critical period has little effect.[38]

In the adult, too, thyroid hormones have a profound influence on the function of the CNS. Hyperthyroidism in man results in hyperexcitability, irritability, restlessness, exaggerated responses to environmental stimuli, and emotional instability that can lead to full-blown psychosis.[68] Conversely, when hypothyroidism occurs in the adult, there is a characteristic listlessness, lack of energy, slowness of speech, decreased sensory capacity, impairment of memory, and somnolence. There is a general dulling of mental capacity and, in untreated cases, overt psychosis may occur (myxedematous madness).[13] In hypothyroidism there are characteristic changes in brainwave or electroencephalographic patterns.[27] Both the amplitude and frequency of the alpha rhythm may be diminished, and in some cases the alpha rhythm may disappear entirely.

Attempts at characterization of the effects of thyroid hormone on the CNS in adults leads to the frustrating conclusion that they are nonspecific and difficult to pinpoint. Speed of nerve conduction appears to be diminished in hypothyroidism, and reaction times for complex responses are increased.[37] Simple reflexes such as the myotonic reflex are slowed, but this can be explained, at least in part, by a slowing of both contraction and relaxation of skeletal muscle in the hypothyroid state.[63] As with the effects of thyroid hormone on growth and skeletal development, no information is available concerning the molecular actions of the thyroid hormone that produce the critical changes that permit normal development and function of nervous tissue.

Effects of thyroid hormones on oxidative metabolism

One of the most familiar actions of thyroid hormones is on oxidative metabolism. The relationship between thyroid function and oxygen consumption was first observed by Magnus-Levy in 1895.[69] Measurement of oxygen consumption or the so-called basal metabolic rate (BMR) became a popular diagnostic tool for assessment of thyroid status for the subsequent 50 years until more sophisticated biochemical measurements such as that of T_4 became available. To determine BMR, the oxygen consumption at rest is measured after an overnight fast. Deviations from normal are expressed as a percent of the normal. Both thyroxine and triiodothyronine increase BMR, but this effect is limited to warm-blooded animals. Although increased oxygen consumption has perhaps the longest latency of all effects of thyroid hormones, it has become commonly regarded as one of the principal effects of thyroid hormones. The tissues principally affected include liver, heart, kidney, skeletal muscle, pancreas, salivary glands, epidermis, and anterior pituitary gland[8,17]; increased oxygen consumption can be demonstrated in isolated tissues removed after the administration of thyroid hormone to intact animals. The oxygen consumption of the adult brain is not appreciably affected by thyroid hormone. Similarly, the gonads and accessory sex organs, lungs, spleen, and gastric smooth muscle are unaffected by thyroid status (Fig. 62-12). Increased oxygen consumption appears only after a lag period of many hours after the injection of a single dose of thyroid hormone. In man the peak effect of thyroxine is reached only after about 9 days, and the subsequent exponential fall of metabolic rate has a half-time of about 11 to 15 days.[84] Since thyroxine disappears from the blood with a half-time of about 6 days,[108] it is apparent that the metabolic effect persists after the hormone leaves the blood. The effects of triiodothyronine are more rapid; increased oxygen consumption occurs within hours and reaches a maximum in 2 days or less, and the effect on oxygen consumption disappears with a half-time of about 8 days, whereas the half-time of the hormone in the blood is approximately 1 day.

It now appears[40] that the increase in oxygen consumption caused by thyroid hormones may be primarily attributable to increased activity of the sodium pump, which normally extrudes sodium from cells. Blocking the pump with ouabain in isolated tissues eliminates the calorigenic effect of thyroid hormone. Why thyroid hormone increases the amount of the so-called sodium-potassium ATPase is not known.

Effects of thyroid hormones on carbohydrate and lipid metabolism

In line with its effects on oxygen consumption, the hormones of the thyroid promote the catabolism of both glucose and fat. The effects on carbohydrate metabolism do not appear to be specific or limited to any particular enzymatic reaction. Rather, the thyroid hormones act simply to accelerate glucose catabolism. The effects on lipid metabolism are more complex.

One of the typical findings in hypothyroidism is an elevation of circulating levels of cholesterol. Administration of thyroid hormone, after a characteristic lag period, reduces plasma cholesterol levels in normal subjects and restores to normal the elevated cholesterol levels of hypothyroid individuals.[83] Plasma cholesterol level is elevated even though the ability to synthesize cholesterol from precursors is impaired in thyroid deficiency. Administration of thyroid hormone increases cholesterol synthesis, but more importantly, it also enhances the ability of the liver to excrete cholesterol in the bile.[44] The effects of thyroid hormones on cholesterol excretion outweigh the effects on synthesis and thus produce a net decrease in plasma concentrations.

Thyroid hormones also affect both the synthesis and oxidation of fatty acids but, despite the popular belief, hypothyroidism seldom leads to severe obesity. In hypothyroid rats the synthesis of long-chain fatty acids from glucose and other carbohydrates precursors is diminished.[23] Similarly, oxidation of long-chain fatty acids is reduced.[23] Since both of these processes require movements of metabolites across the mitochondrial membrane, it has been suggested that the metabolic defect in hypothyroidism may reside in the mitochondrial membrane.[23] The mo-

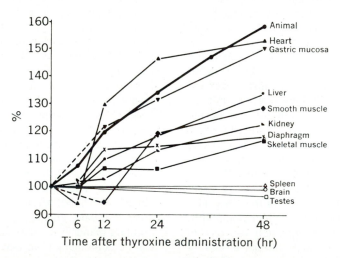

Fig. 62-12. Changes in oxygen consumption of whole animal and excised tissues after administration of large dose of thyroxine to thyroidectomized rat. Note lag between hormone administration and onset of response. Lag period is much longer in man. (From Barker and Klitgaard.[17])

bilization of free fatty acids (FFA) (Chapter 66) from adipose tissue is also highly sensitive to thyroid hormone. The rate of FFA release in response to epinephrine is increased in hyperthyroidism and reduced in thyroid deficiency.[48] FFA mobilization in response to epinephrine depends on cyclic AMP. Although no precise relationship has yet been uncovered, it appears the effects of thyroid hormones on fat mobilization are intimately related to the metabolism of cyclic AMP in adipose tissue.

Effects on nitrogen metabolism

As might be expected from the role of thyroid hormones in growth, normal accretion of protein mass cannot occur in the absence of the thyroid gland. In hypothyroidism, both the synthesis and degradation of protein are diminished, but the effects on synthesis are more profound.[33] When thyroid hormones are given to hypothyroid individuals in amounts that approximate levels secreted at the normal rate, net synthesis of protein is increased. The individual excretes less nitrogen in his urine and feces than he consumes in his diet and hence is in positive nitrogen balance. Within 4 to 6 hr after the administration of a single injection of triiodothyronine to thyroidectomized rats, hepatic synthesis of RNA was increased.[110] Stimulation of protein synthesis was seen about 12 hr later.[111]

Although low or physiologic amounts of thyroid hormones are anabolic, excessive doses or overactivity of the thyroid gland leads to net catabolism of protein.[7] This is reflected in increased nitrogen excretion. One of the characteristic changes seen in hyperthyroidism is increased creatine excretion (creatinuria). Excess thyroid hormone increases the concentration of creatine in blood and decreases the creatinine levels of both blood and urine. This phenomenon is probably secondary to the excessive breakdown of muscle mass seen in thyrotoxicosis. The molecular mechanisms responsible for this phenomenon are still unknown.

Another abnormality of nitrogen metabolism related to thyroid function is the deposition of mucoprotein in the subcutaneous and extracellular spaces in hypothyroid individuals. This material is osmotically active and causes retention of water in the subcutaneous spaces. It is this watery, mucuslike deposit that causes the characteristic puffiness of myxedema. Administration of thyroid hormone causes absorption of the mucoprotein and the subsequent loss of the water in the urine. The mechanisms involved in the deposition and mobilization of the mucoprotein and the role of thyroid hormones in this process are not understood.

Other effects of thyroid hormones

Normal function of a number of organ systems depends on optimal amounts of thyroid hormone. These will be mentioned only briefly here. Detailed discussion of these effects can be found in other sources.[3,5-9] Any attempts at defining the precise role of the thyroid hormones in these systems are again frustrated by the scarcity of any but descriptive data. These other effects will therefore be mentioned only in the most cursory terms.

1. *Digestion.* Thyroidectomy decreases absorption of sugar from the gut and decreases the content of various enzymes in the pancreas. These changes are reversed by treatment with thyroid hormone.

2. *Water and electrolytes.* In deficiency states (myxedema), water and electrolytes accumulate in the subcutaneous spaces. Administration of T_3 or T_4 causes a loss of fluid from these spaces along with an appropriate amount of electrolytes. Consequently, an acute diuretic effect is seen following the administration of T_3 or T_4 to a hypothyroid individual.

3. *Cardiovascular system.* Thyroid hormones cause tachycardia, increased pulse pressure, and increased cardiac output, effects that resemble increased activity of the sympathetic nervous system. There is recent evidence that T_3 causes an increased number of beta adrenergic receptors in the myocardium.[1,116,122] In the rat, these are among the earliest detectable responses to a single dose of hormone.[22] The opposite effects are seen in myxedema. On a longer time scale, cardiac output is further increased by thyroid hormones secondary to increased oxygen consumption, and this can lead to cardiac hypertrophy and even high-output failure.

4. *Lactation.* Thyroxine is necessary for normal lactation. In its absence the fat content of milk and the total milk yield are markedly reduced.

5. *Reproduction.* Normal rhythmicity in the reproductive cycle requires thyroid hormones. The mechanism for this effect is not known.

Thyroid hormones and temperature regulation

The effects of thyroid hormone on oxidative metabolism have been termed "calorigenic" and appear to be related to thermoregulation, the maintenance of normal body temperature. One of the most prominent symptoms of myxedema is a disturbance in heat production with a concomitant decrease in tolerance to cold. Thyroidectomized animals have a severely reduced survival when exposed to cold temperatures.[65,95] Survival of normal adult rats was unimpaired when they were transferred from an ambient temperature of 30° to 1.5° C; they promptly doubled their metabolic rate and increased heat production. Rats thyroidectomized several weeks earlier survived for only a few days at 1.5° C. It may be asked how the thyroid hormones can be of use in acute exposure to cold, since the onset of their calorigenic action requires many hours or days. When the thyroidectomized animals were given a constant daily injection of thyroxine in an amount that was probably somewhat smaller than their normal daily secretory rate, exposure to the cold temperature had no apparent deleterious effects on survival and their metabolic rates rose normally.[95] Cold adaptation caused an increased conversion of thyroxine to triiodothyronine in most tissues of the rat.[11] These observations suggest that increased secretion of thyroid hormone is not required for survival after cold exposure. Catecholamines and increased muscular activity are directly responsible for the rapid increase in metabolic rate. Indeed, anesthesia completely abolished the hypermetabolism of rats exposed to cold and led to a rapid fall in body temperature.[95] From these experiments, it is suggested that the thyroid hormones act in a permissive sense, that is, they must be present for the normal increase in heat production to occur, although they themselves may not initiate this response. To increase heat production there must be increased mobilization of metabolic fuel. In this connection the dependence on thyroid hormone for a normal fat-mobilizing response to epinephrine comes to mind, especially since fat is the fuel chiefly responsible for increased heat production. Another component of resistance to cold resides in the ability of animals with intact thyroids to decrease heat loss. This involves circulatory adjustments and piloerection to decrease circulation to the skin and increase the thickness of the insulating layers. These responses are mediated by catecholamines and the sympathetic nervous system. Many responses to catecholamines are magnified by thyroid hormones.[120] Thus acute adaptive responses to lowered ambient temperature may depend on the interplay of the thyroid hormones and the sympathetic nervous system. In rats, exposure to cold is an important stimulus for increased thyroid secretion[35] secondary to increased TSH secretion.[74a] In humans, however, it appears that TSH secretion is not increased in response to cold exposure,[20,57] except perhaps in the newborn.[42]

Mechanism of action of thyroid hormones

A detailed understanding of the biochemical mechanisms that underlie the physiologic actions of the thyroid hormones is still lacking. Because many aspects of thyroid hormone action are related to development and cellular differentiation, it is logical to assume that altered expression of the genome must be involved. The observation that DNA-directed RNA synthesis was increased in rat liver 4 to 6 hr after a single dose of thyroid hormone was given to hypothyroid rats[110,111] is in accord with this idea, but whether these were direct or indirect effects of thyroid hormone was uncertain. Recent evidence favors a direct interaction between thyroid hormones and the cell nucleus.[82,93] Several groups of investigators have now shown that T_3 binds specifically to "receptor sites" within the nuclei of liver, kidney, pituitary, and white blood cells. The binding proteins have a high affinity and limited capacity for T_3. They appear to be acidic proteins with a molecular weight of 60,000 to 70,000 daltons and are thought to bind to chromatin. Substantial binding of labeled hormone to nuclear receptor sites occurs within 30 min of injection. In contrast to steroid hormones, interaction of thyroid hormones with cytoplasmic binding sites does not appear to be important and cytoplasmic proteins probably are not needed to translocate the hormone to the nucleus. In addition to the observations that inhibitors of RNA and protein synthesis block the expression of many of the effects of T_3, there is good evidence to support the idea that nuclear binding is relevant to the expression of biologic responses to the hormone; for instance: (1) Tissue responsiveness correlates with binding capacity; spleen and testis, which do not increase their oxygen consumption in response to T_3 (Fig. 62-12), have few binding sites relative to more responsive tissues. (2) The degree of saturation of binding sites correlates with the degree of hormonal stimulation of function. (3) The affinity of the binding sites for chemically modified thyroid hormone analogs corresponds with their biologic activity.

Little is known about events that follow nuclear binding. How T_3 ultimately interacts with the genome and how it prolongs its effects well beyond its disappearance from the nucleus will be exciting topics for future research. Since T_3 stimulates the synthesis of constitutive proteins as well as induces the synthesis of new proteins, it is possible that it affects the process of translation as well as transcription. It is also possible that some of the responses to thyroid hormones are accomplished by other mechanisms.

Because thyroid hormones increase oxidative metabolism, much attention has been devoted to the mitrochondrion as the site of hormone action.[104,108] A great deal of evidence suggests that the mitochondrion may be specifically affected by thyroid hormone, and, whereas the precise interaction awaits definition, some evidence suggests that T_3 may bind specifically to mitochondria as well as nuclei.[107] However, the recent finding that increased oxygen consumption is secondary to increased activity of the sodium pump[40] suggests that the mitochondria are responding to increased cellular demand for ATP and hence may argue against a primary effect on mitochondria. Earlier notions that thyroid hormones might increase oxidative metabolism by uncoupling oxidation from phosphorylation have largely been abandoned, since this effect is observed only in severely hyperthyroid states. Furthermore, other agents such as dinitrophenol that can uncouple oxidation from phosphorylation do not produce any of the other effects of the thyroid hormones. Because the effects of the thyroid hormones on oxidative metabolism appear rather late, it may be that increased oxidative metabolism is a secondary rather than a primary response to thyroid hormone.

Yet another aspect of the action of thyroid hormones is the interaction with catecholamines and the cyclic AMP system. Thyroid hormones increase myocardial adenyl cyclase, the enzyme that generates cyclic AMP,[66] and may decrease the activity of cyclic AMP phosphodiesterase, the enzyme that degrades cyclic AMP.[12] Although a mechanism has not yet been elucidated, these are among the earliest effects of thyroid hormones and consequently should lie nearer, at least in time, to the initial, primary action of thyroid hormones.[22] Increased sensitivity to the fat-mobilizing action of epinephrine does not appear to require either RNA or protein synthesis.[47]

Regulation of thyroid hormone secretion

Early studies of hypophysectomized tadpoles indicated that the pituitary and thyroid glands were somehow related.[98] When these studies were extended to hypophysectomized rats, it was noted that their thyroid glands atrophied after removal of the pituitary.[99] Administration of crude extracts of the pituitary gland partially restored the size of these glands. Subsequent stud-

ies revealed that pituitary extracts cause hypertrophy of the thyroid gland and can produce goiter. Conversely, administration of thyroid hormones, like hypophysectomy, produced atrophy of the thyroid gland. These effects can best be explained by the negative feedback hypothesis, which contends that secretion of TSH by the pituitary gland is inhibited when the circulating concentration of thyroid hormone reaches some critical level (Fig. 62-13). This notion had to be expanded to include the CNS when it was recognized that the hypothalamus undoubtedly controls the function of the anterior pituitary gland. Indeed, a hypothalamic hormone, TRH, has now been isolated, characterized, and synthesized[21,28] (Chapter 61, p. 1480). Inclusion of the CNS in the negative feedback loop complicates the control system but also provides a mechanism for altering the level of thyroid hormone in the plasma in response to changing physiologic demands. Inclusion of the hypothalamus in the feedback loop, however, raises the question of whether the thyroid hormone acts back in a negative feedback way to inhibit the release of TRH by the hypothalamus or TSH by the pituitary. Several observations suggest that the pituitary is probably the major site of negative feedback inhibition.[85] Implantation of small crystals of thyroxine in the pituitary itself is more effective in decreasing TSH secretion than similar implantation in the hypothalamus.[85] Isolating the pituitary from the hypothalamus by cutting the infundibular stalk[85] or transplanting the pituitary to a site away from the hypothalamus does not abolish the ability of thyroid hormone to regulate TSH secretion.[85]

It appears that both TRH and the thyroid hormones are necessary for normal secretion of TSH

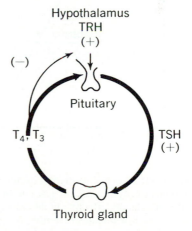

Fig. 62-13. Feedback regulation of thyroid function.

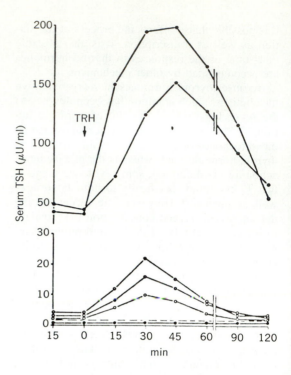

Fig. 62-14. Changes in plasma concentrations of TSH in response to intravenous administration of TRH in two hypothyroid patients (solid circles), three normal subjects (open circles), and one hyperthyroid patient with undetectable serum TSH (below dashed line). (From Hershman[55]; reprinted by permission from the New England Journal of Medicine **290:**876, 1974.)

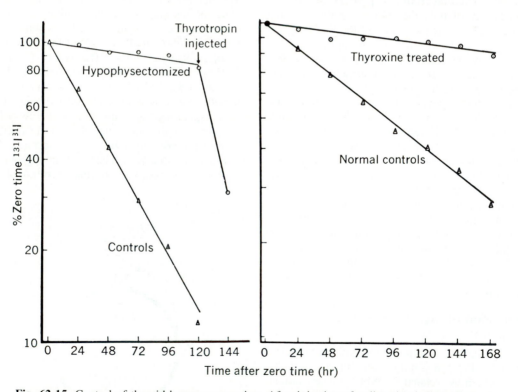

Fig. 62-15. Control of thyroid hormone secretion. After injection of radioactive iodine, decrease in radioactivity in neck reflects rate of thyroxine secretion. Hypophysectomy and thyroxine treatment are equally effective in decreasing hormone release, reflecting absence of TSH. Administration of TSH stimulates secretion. Steeper slope of control group in left-hand panel resulted from administration of propylthiouracil to this group to prevent recycling of iodine within gland. (From Wolff.[126])

by the pituitary gland. Without TRH, the pituitary has only a limited ability to secrete TSH. The thyroid hormones decrease the sensitivity of the TSH-secreting cells of the pituitary to TRH, and when their circulating levels become sufficiently high, as in hyperthyroidism, TSH levels in blood may fall below the limits of detectability. Such modulation of the sensitivity of the pituitary to TRH can be seen in Fig. 62-14. Even though their circulating concentrations of TSH were already high, a test dose of TRH elicited a profound increase in serum TSH levels in hypothyroid patients. The same dose of TRH elicited only a modest increase in serum TSH in normal individuals and had no effect in the hyperthyroid patient whose TSH remained below the limits of detectability.[55] This is not to say that the hypothalamus is not involved at all, for clearly the pituitary isolated from the hypothalamus is not normal with regard to TSH secretion.[85] Environmental influences can change the rate of release of TSH by the pituitary gland. In many species, cold increases TSH secretion, an effect probably mediated through the CNS and TRH. In the rat a variety of noxious stressful situations appear to decrease TSH secretion. These effects, too, are mediated centrally. Furthermore, the thyroid hormones may be required for normal synthesis of TRH in the hypothalamus, a finding that has led to the suggestion that under some conditions, T_3 and T_4 may regulate hypothalamic production of TRH by a positive rather than a negative feedback influence.[86]

Because thyroid hormone is rich in iodine, many aspects of negative feedback control of TSH secretion can readily be followed after the administration of a dose of radioactive iodine to an experimental subject. ^{131}I is a gamma emitter of sufficient energy that it can be detected outside the body. Following oral administration, radioactive iodine is taken up by the thyroid gland and incorporated into thyroid hormones. The radioactivity is gradually released from the gland in relation to the degree of secretory stimulus imparted by TSH. Hypophysectomy slows the release of radioactive iodine from the thyroid gland, whereas TSH accelerates it.[126] Thyroxine treatment, on the other hand, inhibits the pituitary secretion of TSH and slows the rate of release of thyroxine from the thyroid gland. These effects are illustrated in Fig. 62-15.

REFERENCES
General reviews

1. DeGroot, L. J., et al., editors: Endocrinology, vol. 1, New York, 1979, Grune & Stratton, Inc.
2. Grave, G. D., editor: Thyroid hormones and brain development, New York, 1977, Raven Press.
3. Greer, M. A., and Solomon, D. H., editors: Handbook of physiology, thyroid section, Baltimore, 1974, The Williams & Wilkins Co., vol. 3.
4. Hamburgh, M., and Barrington, E. J. W., editors: Hormones in development, New York, 1971, Appleton-Century-Crofts.
5. Pitt-Rivers, R., and Tata, J. R.: The thyroid hormones, New York, 1959, Pergamon Press, Inc.
6. Pitt-Rivers, R., and Trotter, W. R., editors: The thyroid gland, London, 1964, Butterworth & Co., Ltd.
7. Rall, J. E., Robbins, J., and Lewallen, C. G.: The thyroid. In Pincus, G., Thimann, K. V., and Astwood, E. B., editors: The hormones, New York, 1964, Academic Press, Inc.
8. Werner, S. C., and Ingbar, S. H., editors: The thyroid, ed. 3, New York, 1971, Harper & Row, Publishers.
9. Werner, S. C., and Ingbar, S. H., editors: The thyroid, a fundamental and clinical text, ed. 4, New York, 1978, Harper & Row, Publishers.

Original papers

10. Adams, D. D., and Purves, H. D.: The role of thyrotropin in hyperthyroidism and exophthalmos, Metabolism **6:**26, 1957.
10a. Adams, D. D., et al.: Nomenclature of thyroid-stimulating antibodies Lancet **1:**1201, 1975.
11. Albright, E. C., Heninger, R. W., and Larson, F. C.: Effect of cold-induced hyperthyroidism on iodine-containing compounds of extrathyroidal tissues. In Cassano, C., and Andreoli, M., editors: Current topics in thyroid research, New York, 1965, Academic Press, Inc.
12. Armstrong, K. J., et al.: Effects of thyroid hormone deficiency on cyclic adenosine 3':5'-monophosphate and control of lipolysis in fat cells, J. Biol. Chem. **249:** 4226, 1974.
13. Asher, R.: Myxoedematous madness, Br. Med. J. **2:** 555, 1949.
14. Aub, J. C., et al.: Studies of calcium and phosphorus metabolism. III. The effects of the thyroid hormone and thyroid disease, J. Clin. Invest. **7:**97, 1929.
15. Balasse, P. D., et al.: Observation en microscopie à balayage de la surface apicale et des cellules folliculaires thyroidiennes chez le chien, C. R. Acad. Sci. [D] (Paris) **274:**2332, 1972.
16. Balasubramaniam, K., et al.: Effect of thyrotropin on iodoprotein of thyroid cell fractions, Endocrinology **77:** 54, 1965.
17. Barker, S. B., and Klitgaard, H. M.: Metabolism of tissues excised from thyroxine-injected rats, Am. J. Physiol. **170:**81, 1952.
18. Baumann, E.: Über das normale Vorkommen von jod im Thierkörper, Z. Physiol. Chem. **21:**319, 1895.
19. Baume, L. J., Becks, H., and Evans, H. M.: Hormonal control of tooth eruption. I. The effect of thyroidectomy on the upper rat incisor and the response to growth hormone, thyroxin, or the combination of both, J. Dent. Res. **33:**80, 1954.
20. Berg, G. R., et al.: Effect of central cooling in man on pituitary-thyroid function and growth hormone secretion, J. Appl. Physiol. **21:**1791, 1966.
21. Bowers, C. Y., et al.: Porcine thyrotropin releasing hormone is (Pyro) Glu-His-Pro (NH$_2$), Endocrinology **86:**1143, 1970.
22. Bray, G. A., and Goodman, H. M.: Studies on the early effects of thyroid hormones, Endocrinology **76:**323, 1965.
23. Bray, G. A., and Goodman, H. M.: Metabolism of adi-

pose tissue from normal and hypothyroid rats, Endocrinology **82:**860, 1968.

24. Briggs, F. N., Taurog, A., and Chaikoff, I. L.: The enterohepatic circulation of thyroxine in the rat, Endocrinology **52:**559, 1953.
25. Brown, J., et al.: Thyroid physiology in health and disease, Ann. Intern. Med. **81:**68, 1974.
26. Brown-Grant, K.: Extrathyroidal iodide concentrating mechanisms, Physiol. Rev. **41:**189, 1961.
27. Browning, T. B., Atkins, R. W., and Weiner, H.: Cerebral metabolic disturbances in hypothyroidism, Arch. Intern. Med. **93:**938, 1954.
28. Burgus, R., et al.: Characterization of ovine hypothalamic hypophysiotropic TSH-releasing factor, Nature **226:**321, 1970.
29. Cavalieri, R. R., and Searle, G. L.: The kinetics of distribution between plasma and liver of ^{131}I-labeled L-thyroxine in man: observations of subjects with normal and decreased serum thyroxine-binding globulin, J. Clin. Invest. **45:**939, 1966.
30. Chopra, I. J.: Extrathyroidal conversion of T_4 to T_3 in vitro: evidence that reverse T_3 is a potent inhibitor of T_3 production, Clin. Res. **249:**426, 1976.
30a. Chopra, I. J.: Nature, source and biologic significance of thyroid hormones in blood. In Werner, S. C., and Ingbar, S. H., editors: The thyroid, a fundamental and clinical text, ed. 4, New York, 1978, Harper & Row, Publishers.
31. Chopra, I. J., Sack, J., and Fisher, D. A.: Circulating 3,3',5'-triiodothyronine (Reverse T_3) in the human newborn, J. Clin. Invest. **55:**1137, 1975.
31a. Chopra, I. J., et al.: Abnormalities in thyroid function in relatives of patients with Graves' disease and Hashimoto's thyroiditis: lack of correlation with inheritance of HLA-B8, J. Clin. Endocrinol. Metab. **45:**45, 1977.
32. Clayton, J. A., and Szego, C. M.: Depletion of rat thyroid serotonin accompanied by increased blood flow as an acute response to thyroid-stimulating hormone, Endocrinology **80:**689, 1967.
33. Crispell, K. R., Parson, W., and Hollifield, G.: A study of the rate of protein synthesis before and during the administration of L-triiodothyronine to patients with myxedema and healthy volunteers using N-15 glycine, J. Clin. Invest. **35:**164, 1956.
34. Deiss, W. P., Jr., et al.: Stimulation of proteolysis in thyroid particles by thyrotropin, Endocrinology **79:**19, 1966.
35. Dempsey, E. W., and Astwood, E. B.: Determination of the rate of thyroid hormone secretion at various environmental temperatures, Endocrinology **32:**509, 1943.
36. DeRobertis, E.: Proteolytic enzyme activity of colloid extracted from single follicles of the rat thyroid, Anat. Rec. **80:**219, 1941.
37. Eayrs, J. T.: Influence of the thyroid on the central nervous system, Br. Med. Bull. **16:**122, 1960.
38. Eayrs, J. T.: Age as a factor determining the severity and reversibility of the effects of thyroid deprivation in the rat, J. Endocrinol. **22:**409, 1961.
39. Edelhoch, H.: The properties of thyroglobulin. I. The effects of alkali, J. Biol. Chem. **235:**1326, 1960.
40. Edelman, I. S., and Ismail-Beigi, F.: Thyroid thermogenesis and active sodium transport. Recent Prog. Horm. Res. **30:**235, 1974.
41. Evans, E. S., Rosenberg, L. L., and Simpson, M. E.: Relative sensitivity of different biological responses to thyroxine, Endocrinology **66:**433, 1960.
41a. Field, J. B., et al.: Activation of protein kinase in thy-

roid slices by thyroid-stimulating hormone, J. Biol. Chem. **250:**4903, 1975.
42. Fisher, D. A., and Odell, W. D.: Acute release of thyrotropin in the newborn, J. Clin. Invest. **48:**1670, 1969.
43. Frieden, E.: Thyroid hormones and biochemistry of amphibian metamorphosis, Recent Prog. Horm. Res. **23:**139, 1967.
44. Friedman, M., Byers, S. O., and Rosenman, R. H.: Changes in excretion of intestinal cholesterol and sterol digitonides in hyper- and hypothyroidism, Circulation **5:**657, 1952.
45. Gavin, L., et al.: Extrathyroidal conversion of thyroxine to 3,3',5'-triiodothyronine (reverse T_3) and to 3,5,3'-triiodothyronine (T_3) in humans, J. Clin. Endocrinol. Metab. **44:**733, 1977.
45a. Gavin, L. A., et al.: 3,3'-Diiodothyronine production, a major pathway of peripheral iodothyronine metabolism in man, J. Clin. Invest. **61:**1276, 1978.
46. Gilman, A. G., and Rall, T. W.: Studies on the relation of cyclic 3',5'AMP (CA) to TSH action in beef thyroid slices, Fed. Proc. **25:**617, 1966.
47. Goodman, H. M.: Permissive effects of hormones on lipolysis, Endocrinology **86:**1064, 1970.
48. Goodman, H. M., and Bray, G. A.: Role of thyroid hormones in lipolysis, Am. J. Physiol. **210:**1053, 1966.
49. Gorbman, A.: Comparative anatomy and phylogeny of the thyroid. In Werner, S. C., and Ingbar, S. H., editors: The thyroid, ed. 3, New York, 1971, Harper & Row, Publishers.
50. Gordon, A. H., et al.: Nature of the circulating thyroid hormone-plasma protein complex, Nature **169:**19, 1952.
51. Gross, J., and Pitt-Rivers, R.: Identification of 3:5:3'-triiodothyronine. I. Isolation from thyroid gland and synthesis, Biochem. J. **53:**645, 1953.
52. Guternatsch, J. F.: Feeding experiments on tadpoles. The influence of specific organs given as food on growth and differentiation. A contribution to the knowledge of organs with internal secretions, Arch. Entwicklungsmech. Org. **35:**457, 1913.
53. Haddad, H. M.: Studies on thyroid hormone metabolism in children, J. Pediatr. **57:**391, 1960.
53a. Harada, A., et al.: Comparison of thyroid stimulators and thyroid hormone concentrations in the sera of pregnant women, J. Clin. Endocrinol. Metab. **48:**793, 1979.
54. Harington, C. R., and Barger, G.: Chemistry of thyroxine. III. Constitution and synthesis of thyroxine, Biochem. J. **21:**169, 1927.
55. Hershman, J. M.: Clinical application of thyrotropin releasing hormone, N. Engl. J. Med. **290:**886, 1974.
56. Hershman, J. M., and Starnes, W. R.: Placental content and characterization of human chorionic thyrotropin, J. Clin. Endocrinol. Metab. **32:**52, 1971.
57. Hershman, J. M., et al.: Effects of cold exposure on serum thyrotropin, J. Clin. Endocrinol. Metab. **30:**430, 1970.
58. Ingbar, S. H.: Pre-albumin: a thyroxine-binding protein of human plasma, Endocrinology **63:**256, 1958.
59. Ingbar, S. H.: Autoregulation of the thyroid: response to iodide excess and depletion, Mayo Clin. Proc. **47:**814, 1972.
60. Ingbar, S. H., and Braverman, L. E.: Active form of the thyroid hormone, Annu. Rev. Med. **26:**443, 1975.
61. Kaltenbach, J. C.: Local action of thyroxine analogues on amphibian metamorphosis Gen. Comp. Endocrinol. **7:**329, 1966.
62. Kendall, E. C.: The isolation in crystalline form of the

compound containing iodin which occurs in the thyroid; its chemical nature and physiological activity, Trans. Assoc. Am. Physicians **30:**420, 1915.

63. Lambert, E. H., et al.: A study of the ankle jerk in myxedema, J. Clin. Endocrinol. Metab. **11:**1186, 1951.

64. Larson, F., Deiss, W. P., and Albright, E. C.: Localization of protein-bound radioactive iodine by filter paper electrophoresis, Science **115:**626, 1952.

65. Leblond, C. P., and Gross, J.: Effect of thyroidectomy on resistance to low environmental temperature, Endocrinology **33:**155, 1943.

66. Levey, G. S., Skelton, C. L., and Epstein, S. E.: Decreased myocardial adenyl cyclase activity in hypothyroidism, J. Clin. Invest. **48:**2244, 1969.

67. Lewis, U. J., Cheever, E. V., and Vanderlaan, W. P.: Alteration of the proteins of the pituitary gland of the rat by estradiol and cortisol, Endocrinology **76:**362, 1965.

68. Lidz, T.: Emotions and mentation. In Werner, S. C., and Ingbar, S. H., editors: The thyroid, ed. 3, New York, 1971, Harper & Row, Publishers.

69. Magnus-Levy, A.: Über den respiratorischen Gaswechsel unter dem Einfluss der Thyreoidea sowie unter verschiedenen pathologischen Zuständen, Berl. Klin. Wochenschr. **32:**650, 1895.

70. Matthews, J. L., and Martin, J. H.: Atlas of human histology and ultrastructure, Philadelphia, 1971, Lea & Febiger.

71. McKenzie, J. M.: Humoral factors in the pathogenesis of Graves' disease, Physiol. Rev. **48:**252, 1968.

72. McKenzie, J. M., and Zakarija, M.: A reconsideration of a thyroid-stimulating immunoglobulin as the cause of hyperthyroidism in Graves' disease, J. Clin. Endocrinol. Metab. **42:**778, 1976.

73. McKenzie, J. M., Adiga, P. R., and Murthy, P. V. N.: Effects of actinomycin D, cycloheximide and puromycin on thyroid stimulation, Endocrinology **83:**1132, 1968.

74. Melander, A., et al.: Intrathyroidal amines in the regulation of thyroid activity, Rev. Physiol. Biochem. Pharmacol. **73:**39, 1975.

74a. Montoya, E., Seibel, M. J., and Wilber, J. F.: Thyrotropin-releasing hormone secretory physiology: studies by radioimmunoassay and affinity chromatography, Endocrinology **96:**1413, 1975.

75. Murray, G. R.: Note on the treatment of myxoedema by hypodermic injections of an extract of the thyroid gland of a sheep, Br. Med. J. **2:**796, 1891.

76. Nadler, N. J.: Anatomy and histochemistry of the thyroid: comparative anatomy. In Werner, S. C., and Ingbar, S. H., editors: The thyroid, ed. 3, New York, 1971, Harper & Row, Publishers.

77. Nadler, N. J., et al.: Elaboration of thyroglobulin in the thyroid follicle, Endocrinology **74:**333, 1964.

78. Nicoloff, J. T., et al.: Simultaneous measurement of thyroxine and triiodothyronine peripheral turnover kinetics in man, J. Clin. Invest. **51:**473, 1972.

78a. Onaya, T., Masanobu, K., Yamada, T., and Ochi, Y.: New *in vitro* tests to detect the thyroid stimulator in sera from hyperthyroid patients by measuring colloid droplet formation and cyclic AMP in human thyroid slices, J. Clin. Endocrinol. Metab. **36:**859, 1973.

79. Oppenheimer, J. H.: Role of plasma proteins in the binding, distribution and metabolism of the thyroid hormones, N. Engl. J. Med. **278:**1153, 1968.

80. Oppenheimer, J. H., Schwartz, H. L., and Surks, M. I.: Propylthiouracil inhibits the conversion of L-thy-roxine to L-triiodothyronine, J. Clin. Invest. **51:**2493, 1972.

81. Oppenheimer, J. H., et al.: Differences in primary cellular factors influencing the metabolism and distribution of 3,5,3'-L-triiodothyronine and L-thyroxine, J. Clin. Invest. **49:**1016, 1970.

82. Oppenheimer, J. H., et al.: Nuclear receptors and the initiation of thyroid hormone action, Recent Prog. Horm. Res. **32:**529, 1976.

83. Peters, J. P., and Man, E. B.: The significance of serum cholesterol in thyroid disease, J. Clin. Invest. **29:**1, 1950.

83a. Pittman, C. S.: Hormone metabolism. In DeGroot, L. J., et al., editors: Endocrinology, New York, 1979, Grune & Stratton, Inc., vol. 1.

84. Rawson, R. W., et al.: L-Triiodothyronine versus L-thyroxine; a comparison of their metabolic effects in human myxedema, Trans. Assoc. Am. Phys. **66:**86, 1953.

85. Reichlin, S.: Control of thyrotropic hormone secretion. In Martini, L., and Ganong, W. F., editors: Neuroendocrinology, New York, 1966, Academic Press, Inc., vol. 1.

86. Reichlin, S., et al.: The hypothalamus in pituitary-thyroid regulation, Recent Prog. Horm. Res. **28:**229, 1972.

87. Richards, J. B., and Ingbar, S. H.: The effects of propylthiouracil and perchlorate on the biogenesis of thyroid hormone, Endocrinology **65:**198, 1959.

88. Robbins, J., and Nelson, J. H.: Thyroxine-binding by serum protein in pregnancy and in the newborn, J. Clin. Invest. **37:**153, 1958.

89. Robbins, J., and Rall, J. E.: Zone electrophoresis in filter paper of serum 131I after radioiodide administration, Proc. Soc. Exp. Biol. Med. **81:**530, 1952.

90. Robison, G. A., Butcher, R. W., and Sutherland, E. W.: Cyclic AMP, New York, 1971, Academic Press, Inc.

91. Rosenberg, I. N., and Bastomsky, C. H.: The thyroid, Annu. Rev. Physiol. **27:**71, 1965.

92. Salmon, T. N.: The effect on the growth rate of thyroparathyroidectomy in newborn rats of the subsequent administration of thyroid, parathyroid and anterior hypophysis, Endocrinology **23:**446, 1938.

93. Samuels, H. H., et al.: Thyroid hormone action. In vitro characterization of solubolized nuclear receptors from rat liver and culture GH cells, J. Clin. Invest. **54:**853, 1974.

94. Schooley, R. A., Friedkin, S., and Evans, E. S.: Reexamination of the discrepancy between acidophil numbers and growth hormone concentration in the anterior pituitary gland following thyroidectomy, Endocrinology **79:**1053, 1966.

95. Sellers, E. A., and You, S. S.: Role of the thyroid in metabolic responses to a cold environment, Am. J. Physiol. **163:**81, 1950.

96. Shepard, T. H.: Onset of function in the human fetal thyroid: biochemical and radioautographic studies from organ culture, J. Clin. Endocrinol. Metab. **27:**945, 1967.

97. Simon, C.: The iodide pool of the thyroid studied by means of the isotopic equilibrium method and double labelling with 125I and 131I, Biochim. Biophys. Acta **724:**565, 1963.

97a. Smith, B. R., and Hall, R.: Thyroid-stimulating immunoglobulins in Graves' disease, Lancet, **2:**427, 1974.

98. Smith, P. E.: The effect of hypophysectomy in the early

embryo upon growth and development of the frog, Anat. Rec. **11:**57, 1916.

99. Smith, P. E.: The disabilities caused by hypophysectomy and their repair. The tuberal (hypothalamic) syndrome in the rat, J.A.M.A. **88:**158, 1927.

100. Söderberg, U.: Short-term reactions in the thyroid gland, Acta Physiol. Scand. **42**(suppl. 147):1, 1958.

100a. Solomon, D. H., and Kleenman, K. E.: Concepts of pathogenesis of Graves' disease, Adv. Intern. Med. **22:**273, 1976.

100b. Spiro, M. J.: Studies on protein portion of thyroglobulin, J. Biol. Chem. **245:**5820, 1970.

101. Stanbury, J. B.: Deiodination of the iodinated amino acids, Ann. N. Y. Acad. Sci. **86:**417, 1960.

102. Stanbury, J. B.: The metabolic errors in certain types of familial goiter, Recent Prog. Horm. Res. **19:**547, 1963.

103. Stanbury, J. B.: Familial goiter. In Stanbury, J. B., Wyngaarden, J. B., and Fredrickson, D. S., editors: The metabolic basis of inherited disease, ed. 3, New York, 1972, McGraw-Hill Book Co.

104. Sterling, K.: The significance of circulating triiodothyronine, Recent Prog. Horm. Res. **26:**249, 1970.

104a. Sterling, K.: Thyroid hormone action at the cell level, N. Engl. J. Med. **300:**117, 173, 1979.

105. Sterling, K., and Brenner, M. A.: Free thyroxine in human serum: simplified measurement with the aid of magnesium precipitation, J. Clin. Invest. **45:**153, 1966.

106. Sterling, K., and Chodos, R. B.: Radiothyroxine turnover studies in myxedema, thyrotoxicosis and hypermetabolism without endocrine disease, J. Clin. Invest. **35:**806, 1956.

107. Sterling, K., and Lazarus, J. H.: The thyroid and its control, Annu. Rev. Physiol. **39:**349, 1977.

108. Sterling, K., and Milch, P. O.: The mitochondria as a site of thyroid hormone action. In Robbins, J., and Braverman, L. E., editors: Thyroid research, Excerpta Medica International Congress Series, No. 378, New York, 1976, Elsevier North Holland, Inc., p. 342.

109. Studer, H., and Greer, M. A.: The regulation of thyroid function in iodine deficiency, Berne, Switzerland, 1966, Hans Huber Medical Publisher.

110. Tata, J. R., and Widnell, C. C.: Ribonucleic acid synthesis during the early action of thyroid hormones, Biochem. J. **98:**604, 1966.

111. Tata, J. R., et al.: The actions of thyroid hormone at the cell level, Biochem. J. **86:**408, 1963.

112. Taurog, A.: Thyroid peroxidase and thyroid biosynthesis, Recent Prog. Horm. Res. **26:**189, 1970.

113. Taurog, A., Porter, J. C., and Thio, D. T.: Nature of the ^{131}I compounds released into the thyroid veins of rabbits, dogs and cats, before and after TSH administration. Endocrinology **74:**902, 1964.

114. Tong, W.: Thyrotropin stimulation of thyroxine synthesis in isolated thyroid cells treated with perchlorate, Endocrinology **75:**968, 1964.

115. Tong, W.: Thyroid hormone synthesis and release. In Werner, S. C., and Ingbar, S. H., editors: The thyroid, ed. 3, New York, 1971, Harper & Row, Publishers.

116. Tsai, J. S., and Chen, A.: L-T$_3$ increases the level of β-adrenergic receptor in cultured myocardial cells, Clin. Res. **25:**303A, 1977.

116a. Ui, N.: Synthesis and chemistry of iodoproteins. In Greer, M. A., and Solomon, D. H., editors: Handbook of physiology, Thyroid section 1, Baltimore, 1974, The Williams & Wilkins Co., vol. 3.

117. Utiger, R. D.: Radioimmunoassay of human plasma thyrotropin, J. Clin. Invest. **44:**1277, 1965.

118. Vagenakis, A. G., et al.: Diversion of peripheral thyroxine metabolism from activating to inactivating pathways during complete fasting, J. Clin. Endocrinol. Metab. **41:**191, 1975.

119. Van Middlesworth, L.: Metabolism and excretion of thyroid hormones. In Greer, M. A., and Solomon, D. H., editors: Handbook of physiology, Endocrinology section, Baltimore, 1974, The Williams & Wilkins Co., vol. 3.

120. Waldstein, S. S.: Thyroid-catecholamine interrelations, Annu. Rev. Med. **17:**123, 1966.

120a. Wharton, T.: Adenographia: sive, glandularum totius corporis descriptio. De glandulis thyreoidis, London, 1656.

121. Wilkins, L.: The diagnosis and treatment of endocrine disorders in childhood and adolescence, Springfield, Ill., 1965, Charles C Thomas, Publisher.

122. Williams, L. T., et al.: Thyroid hormone regulation of beta-adrenergic receptor number: possible biochemical basis for the hyperadrenergic state in hyperthyroidism, Clin. Res. **25:**458(A), 1977.

123. Winand, R. J., Salmon, J., and Lambert, P. H.: Characterization of the exophthalmogenic factor isolated from the serum of patients with malignant exophthalmos. In Fellinger, K., and Hofer, R., editors: Further advances in thyroid research, Vienna, 1971, Verlag der Wiener Medizinischer. Akademie, p. 583.

124. Wissig, S. L.: Morphology and cytology. In Pitt-Rivers, R., and Trotter, W. R., editors: The thyroid gland, London, 1964, Butterworth & Co., Ltd.

125. Woeber, K. A., and Ingbar, S. H.: The contribution of thyroxine-binding prealbumin to the binding of thyroxine in human serum, as assessed by immunoadsorption, J. Clin. Invest. **47:**1710, 1968.

126. Wolff, J.: Some factors that influence the release of iodine from the thyroid gland, Endocrinology **48:**284, 1951.

127. Wolff, J.: Transport of iodide and other anions in the thyroid gland, Physiol. Rev. **44:**45, 1964.

128. Wolff, J., and Chaikoff, I. L.: The inhibitory action of excessive iodide upon the synthesis of diiodotyrosine and of thyroxine in the thyroid gland of the normal rat, Endocrinology **43:**174, 1948.

129. Wolff, J., et al.: The temporary nature of the inhibitory action of excess iodide on organic iodine synthesis in the normal thyroid, Endocrinology **45:**504, 1949.

130. Wyngaarden, J. B., Stanbury, J. B., and Rapp, B.: The effects of iodide, perchlorate, thiocyanate and nitrate administration upon the iodide concentrating mechanism of the rat thyroid, Endocrinology **52:**568, 1953.

131. Wyngaarden, J. B., Wright, B. M., and Ways, P.: The effect of certain anions upon the accumulation and retention of iodide by the thyroid gland, Endocrinology **50:**537, 1952.

63

G. D. AURBACH and JAMES M. PHANG

Vitamin D, parathyroid hormone, and calcitonin

Hormonal regulation of calcium, within narrow limits of concentration, is essential for normal activity of many divergent physiologic and biochemical functions. Studies in this area arose in response to gross clinical and pathologic observations of metabolic bone diseases. Rickets was known in the sixteenth century as a disease of bone, but the underlying disorder of calcium metabolism was not recognized until modern times. Similarly, a description of the pathology of osteitis fibrosa cystica (the bone disease of hyperparathyroidism) seems to have been made in the eighteenth century, although the etiology of the disease was not appreciated until Mandl seized on the significance of a parathyroid tumor in his patient, Albert. Research into these abnormalities of bone eventually led to the discoveries that calcium metabolism is regulated by endocrine function and that this function is effected through the secretion of the polypeptide hormones, parathyroid hormone and calcitonin, and the action of vitamin D.

The parathyroid glands were recognized first in the nineteenth century as anatomic entities distinct from the thyroid gland. Although described as separate bodies, they were not assigned any particular function, and tetany following thyroidectomy was not properly attributed to parathyroidectomy until the twentieth century. Vassale and Generali[160] in 1900 selectively removed the parathyroid glands and proved thereby that the latter, not the thyroid, were the structures required to prevent tetany. Thereafter, MacCallum and Voegtlin discovered that tetany following parathyroidectomy was associated with hypocalcemia and was correctable by infusing calcium salts. Ultimately, it was Collip who proved the endocrine function of the parathyroids by preparing an active extract of the glands that corrected hypocalcemic tetany.

Interest in the endocrine control of calcium has intensified during the past 15 years, and studies on the regulation of parathyroid secretion led to discovery of calcitonin, another polypeptide hormone important in regulating calcium metabolism and bone. The latter hormone is elaborated from the parafollicular cells found in the mammalian thyroid but related embryologically to the ultimobranchial body of lower vertebrate species.

Rickets, an important public health problem to urban man, was the impetus for research that led to the discovery of vitamin D, the fat-soluble substance that can correct rickets or osteomalacia. Subsequent investigations into the actions of vitamin D have led to the discovery that it must be metabolized further in vivo to be active and that this metabolic conversion itself is regulated by intricate control mechanisms.

Recently the parathyroid and calcitonin polypeptide hormones have been isolated in pure form, analyzed for amino acid sequence, and synthesized chemically. Further, the key metabolites of vitamin D that represent the metabolically active forms in vivo have been isolated, characterized, and synthesized. It is expected that the enzymes required for these metabolic conversions soon will be purified and characterized.

The regulation of calcium metabolism, the chemical characteristics of the hormones involved, and their secretory control, mechanisms of action, and metabolic fates are discussed in this chapter. Several reviews[2-35] as well as the *Handbook of Physiology*[1] provide extensive detailed accounts of this field.

CALCIUM METABOLISM

Calcium is important in many physiologic processes. At the molecular level, calcium interacts with a large number of proteins to alter their conformation and function. This interaction may be characterized by (1) an absolute requirement for calcium in low concentrations, (2) sensitivity to changes in calcium ion activity in intracellular fluid (10^{-6} M), and (3) sensitivity to changes in calcium ion activity in extracellular fluid (10^{-3} M).

The first group includes enzymes that specifically require calcium for activation. A high affinity between the ion and these enzymes allows maximal activation at physiologic concentrations of calcium.[22] Other divalent cations, Mg^{2+}, Mn^{2+}, etc., are not substitutes for calcium. Examples of such enzymes or enzyme systems include amylase, lipase, and thromboplastin. Since the concentrations of calcium found in vivo greatly exceed the Michaelis constant (K_m) of these enzymes for calcium, it is doubtful that the metal is a physiologically important regulator of these enzymes.

On the other hand, intracellular calcium and its translocation across cellular and subcellular membranes is important in other cellular processes. In muscle, changes in intracellular concentration of calcium ion appear to trigger mechanical events.[75] Intracellular calcium ion also participates in regulating the activity of enzymes and the function of membranes and cell organelles. Mitochondria in vitro translocate calcium and phosphate coupled to electron transfer and adenosine triphosphate (ATP) hydrolysis. Several secretory processes and tissue responses to hormones depend on extracellular calcium. Intracellular translocation of calcium may be a concomitant event associated with interaction of a hormone with the cell membrane and/or activation of adenyl cyclase.[30]

The concentration of calcium in the extracellular fluid is closely regulated. Both parathyroid hormone and calcitonin are coupled through feedback control to plasma calcium ion activity to maintain plasma calcium homeostasis. This biologic regulation is particularly important for electrically excitable tissue. Within the range compatible with life, changes in the concentration of calcium ion activity greatly affect the excitability of neural tissue.[82] The central nervous system (CNS) is particularly sensitive to such changes in extracellular calcium ion activity. In cardiac muscle, calcium affects the mechanical properties of muscle as well as its excitability. The maximum rate of tension development and the peak isometric tension are all directly proportional to the concentration of extracellular calcium within the physiologically permissible range.[154] Thus two vital physiologic processes, nerve conduction and cardiac function, are each sensitive to small changes in extracellular calcium ion activity.

Extracellular calcium is also important in regulating hormone release, the secretory activity of exocrine glands, and synaptic transmission at nerve endings. Secretion of insulin in basal as well as stimulated states increases in direct proportion to calcium concentration.[71] Salivary protein release is similarly related to calcium concentration, as is the release of acetylcholine (ACh) at nerve endings. All these effects appear to be phenomena of calcium-stimulated release of vesicle contents.

Calcium in the body serves a dual role. The importance of calcium in the regulation of cellular function was emphasized in the previous section. Calcium in the form of crystalline structures is vital in the maintenance of skeletal integrity. Thus calcium metabolism may be conceptualized as two physiologic systems, each sensitive to particular homeostatic mechanisms. Bone serves as a major interface between the two systems, since the calcium in the skeleton not only contributes its mechanical properties to bone but also provides the reservoir for maintenance of plasma calcium homeostasis. The physiologic system for calcium will be described first, and the mechanical function for calcium will be discussed in connection with the properties of bone.

General control system

Calcium ion activity ($A_{Ca^{2+}}$) in the extracellular fluid is regulated within a narrow range. The control system that regulates $A_{Ca^{2+}}$ contains negative feedback loops that alter the properties of system components in the face of external perturbations (Fig. 63-1). The control system includes a controlled system and a controlling system. The controlled system comprises the plasma pool, the extravascular-extracellular fluid, and three organ subcomponents: the gastrointestinal tract, kidneys, and skeleton. The controlling system includes the parathyroid glands and the "C cells" of the thyroid that secrete calcitonin. When an external perturbation or a change in system properties causes levels of $A_{Ca^{2+}}$ to deviate from the desired value, changes in secretion of the hormones of the controlling system produce alterations in the system properties of the controlled system to compensate for the error.

The controlled system

Plasma pool. The plasma pool is the central component of the controlled system. The extravascular-extracellular fluid is in thermodynamic equilibrium with plasma, and both of these fluid compartments show nearly identical calcium ion activity at all times. The plasma calcium ion activity is the object of the entire control system and forms the input to the controlled system as well as the controlling system.

Ionized calcium represents a fraction of the

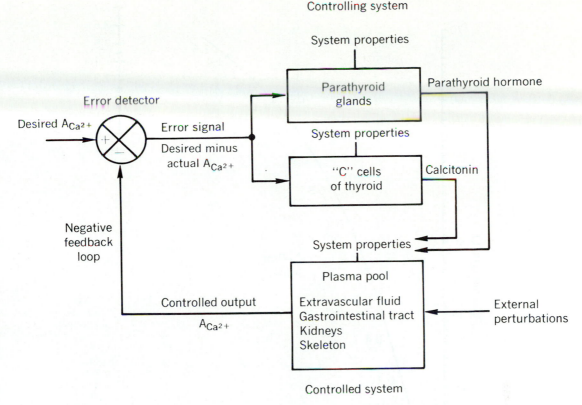

Fig. 63-1. "Block diagram" of closed-loop control system for calcium. External perturbations produce an error signal that causes controlling system to secrete parathyroid hormone or calcitonin. These controlling signals in turn lead to changes in properties of controlled system so as to minimize error. A positive error signal, Desired $A_{Ca^{2+}}$ > Actual $A_{Ca^{2+}}$, increases parathyroid hormone secretion and a negative error signal, Desired $A_{Ca^{2+}}$ < Actual $A_{Ca^{2+}}$, increases calcitonin secretion.

total calcium in plasma. At any instant the magnitude of this fraction depends on the concentration of a number of plasma constituents, but primarily albumin.[120] Albumin binds calcium and, as albumin concentration increases, the concentration of bound calcium increases (Fig. 63-2). At "normal" plasma albumin concentrations, about 50% to 60% of the total calcium concentration is in the bound form.

The plasma calcium ion activity is a function of ionized calcium concentration. The function depends on pH and on the concentration of anions with which calcium ion may interact (especially phosphate ion). As phosphate concentration increases, calcium ion activity decreases. With increases in hydronium ion concentration (i.e., a decrease in plasma pH), calcium ion activity increases. At any given concentration of calcium and phosphate ion an increase in plasma pH will increase the association of calcium and phosphate.

Gastrointestinal tract. The gastrointestinal

tract represents the interface between the calcium metabolic system and the external environment. Under physiologic conditions, perturbations of the system arise from changes in the amount of calcium entering the plasma pool via the gastrointestinal tract as dietary calcium. The processing of dietary calcium depends on (1) the intrinsic properties of the calcium absorption mechanism, (2) vitamin D and its metabolites, and (3) hormonal control mediated through the feedback loop.

Transmural absorption of calcium appears to be an active transport process saturable relative to concentration of calcium in the intestinal lumen.[148] The daily intake of dietary calcium,[135] the frequency of calcium ingestion during the day,[105] and the secretion of calcium into the gastrointestinal tract influence the concentration of calcium to which the duodenal and jejunal mucosa are exposed. In addition, hydronium ion increases and complexing anions decrease the effective concentration of calcium. Since trans-

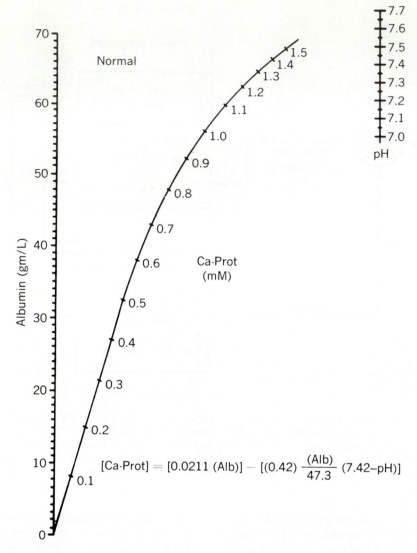

Fig. 63-2. Nomogram for estimating concentrations of calcium bound to protein, *Ca-Prot,* at given albumin concentrations and pH. Straight line connecting albumin and pH values will intersect curve at *Ca-Prot* value. (From Moore.[120])

mural calcium absorption is a saturable function of calcium ion concentration in the lumen, calcium fluxes reach a plateau value as lumenal concentrations increase. Therefore all the factors that alter lumenal concentration will influence calcium absorption.

Recent investigations have clarified the mechanism by which vitamin D promotes gastrointestinal absorption of calcium.[10] Vitamin D, its biosynthesis, and its metabolic actions are discussed in a separate section of this chapter. Its action on the gastrointestinal tract seems to be mediated by

a calcium-binding protein[35] and possibly a calcium-sensitive ATPase.[119] In hypovitaminosis D, calcium absorption is greatly impaired and rickets or osteomalacia may develop.[39] Conversely, excessive amounts of the vitamin can increase calcium absorption, so that toxicity with hypercalcemia occurs.

The intestinal transport of calcium also is influenced by parathyroid hormone but most probably in a secondary manner through a mechanism involving metabolic conversion of vitamin D (p. 1533). Unlike kidney and bone, the gastrointestinal tract does not appear to respond directly to parathyroid hormone.*

The animal can adapt to changes in dietary intake of calcium by altering the rate of intestinal absorption of calcium.[25] Calcium transport in vitro is high in specimens of intestine taken from animals fed a low-calcium diet.[107] Recent studies suggest that the mechanisms of this adaptation to diet involve an increase in the quantity of calcium-binding protein as well as an increase in the affinity of the protein for calcium.[161] These changes are regulated through increased in vivo production of 1,25-dihydroxy-vitamin D (p. 1534).

Kidneys. The kidneys of a normal adult filter from 5,000 to 8,500 mg of calcium/day. Urinary calcium usually ranges from 100 to 300 mg/day, so that 90% to 95% of the filtered calcium is reabsorbed by renal tubules. Changes in either the filtered load of calcium or in the efficiency with which calcium is reabsorbed in the renal tubule markedly influence urinary calcium excretion and plasma calcium as well. A number of studies suggest that renal tubular reabsorption of calcium involves an active transport process, although a true transport maximum has not been found.

Calcium complexes are not reabsorbed by renal tubules. Thus the tubular fluid concentration of complexing anions such as citrate and phosphate, as well as tubular fluid pH, will affect the rate of calcium reabsorption. Under physiologic conditions, however, the acidity of tubular fluid will limit calcium-anion interactions.

Sodium, magnesium, and glucose interfere with the renal tubular reabsorption of calcium. At any given filtered load of calcium an increase in the filtered load of sodium,[162] magnesium,[146] or glucose[111] lowers the rate of renal tubular calcium reabsorption and increases calcium clearance. The interference of magnesium with calcium reabsorption probably is due to competition between the two divalent cations for a common transport mechanism. The mechanism of the sodium and glucose effects on calcium tubular reabsorption remains unknown.

The renal tubular effect of parathyroid hormone is often obscured by the hypercalcemic action of the hormone, with a concomitant increase in the calcium-filtered load. However, at any given filtered load of calcium, parathyroid hormone produces an increase in calcium reabsorption in renal tubules[133] and a consequent reduction in urinary excretion of calcium.

Parathyroid hormone also causes inhibition of phosphate reabsorption by the renal tubules.[97] At a constant phosphate-filtered load an increase in the plasma concentration of parathyroid hormone reduces the tubular reabsorption of phosphate, causing increased urinary excretion of phosphate and reduction of the concentration of phosphate in plasma.

Bone

The primary function of bone is to maintain skeletal integrity. Bone, since it contains over 99% of body calcium, serves also as the major reservoir in the maintenance of calcium homeostasis. Both of these vital functions depend on bone turnover, that is, the continuous resorption and formation that constitute remodeling of bone. The shape and contour of bone respond to local mechanical stress.[102] The greater the strain placed on bone, the greater the amount of bone formed to support the stress. Although the mechanism of this response is not well understood, hydroxyapatite, through its piezoelectrical properites, may function as the transducer, converting mechanical into electrical and biologic signals.[43] Hormonal influences on bone turnover appear to be important for these signals to produce a tissue response.[27] The maintenance of calcium homeostasis requires rapid, sustained, and integrated mobilization or sequestration of bone-calcium. These mechanisms are regulated through the feedback loops, already discussed, involving parathyroid hormone and calcitonin. Although these two regulatory processes, remodeling and maintenance of calcium homeostasis, respond to different signals for different physiologic purposes, they share cellular and hormonal mechanisms.

The deposition of mineral in bone occurs within the framework of a well-organized organic matrix. This matrix, about 35% of tissue weight, contains collagen, polysaccharides, lipids, and sialoproteins. Collagen, however, is the main constituent and accounts for about 90% of the total organic matrix.[16]

Collagen is unique among body proteins in that it contains all the hydroxyproline found in the body. The source of this imino acid in collagen is not from exogenously derived hydroxyproline. Instead, enzymatic hydroxylation of proline occurs after the collagen molecule has been

*Many studies show that calcium absorption from the gut is increased in chronic hyperparathyroidism. The converse is true in hypoparathyroidism, but there is little convincing evidence that these changes reflect a direct action of the hormone on the gut (see the discussion of the mechanisms of action of vitamin D and parathyroid hormone). A receptor for parathyroid hormone in the gut has been sought by direct experiment but has not been found.

assembled. The hydroxylation depends on the (1) existence of the substrate within a polypeptide of requisite length and amino acid sequence, (2) oxygen, (3) ascorbic acid, and (4) iron in the ferrous form.

After proline has been hydroxylated, collagen is processed further by a number of enzyme-mediated steps. These include cross-linking, cleavage of an N-terminal peptide, and glycosylation. In mature collagen, that present in calcifying sites in bone, three polypeptide chains are wound about each other in a helical structure. The helices in turn are organized into collagen fibrils.

The mechanisms by which collagen synthesis is regulated remain largely unknown. Regulation may occur at the level of amino acids. Rates of collagen synthesis can be affected in vitro by experimentally manipulating the cellular uptake of proline.[78,93,136] Recent studies suggest that hormones may affect collagen synthesis selectively without affecting the synthesis of other proteins. In studies using fetal rat calvaria in 2- to 3-day cultures, proline incorporation into collagen is stimulated by insulin and inhibited by parathyroid hormone.[56a,73a] The intracellular mechanisms mediating these observations, however, remain unexplained.

Three inorganic phases containing calcium and phosphate have been identified in bone: (1) an amorphous calcium phosphate that is poorly crystallized but chemically well defined, (2) octacalcium phosphate, a crystalline structure with calcium:phosphorus molar ratios of 8:6 and with an x-ray diffraction pattern distinct from hydroxyapatite, and (3) hydroxyapatite, a highly crystalline structure found in enamel, dentin, and bone.[16] The sequence of events from calcium and phosphate ions in solution to cyrstalline hydroxyapatite is not completely understood. It has been suggested that amorphous calcium phosphate forms first, is converted to octacalcium phosphate, and finally to hydroxyapatite. Alternatively hydroxyapatite may form directly from amorphous calcium phosphate.

The deposition of calcium and phosphate occurs in a well-ordered spatial relationship within the collagen fibril. In the packing of collagen molecules within the fibril, repeating "holes" occur due to the systematic linear array of collagen molecules. These appear as bands on electron micrographs. In collagen that has been denatured and reconstituted, a number of configurational states may be identified, including that with the recurring bands resembling native collagen. Only the form of collagen, which has the recurring bands, will allow nucleation of hy-droxyapatite in vitro. These findings suggest that these "holes" within the collagen fibril serve as the sites for calcification.

Recent investigations have elucidated some of the cellular and biochemical events in the initiation of collagen calcification.[16] Calcium-phosphorus complexes (Ca-P) are accumulated in mitochondria by energy-dependent processes. Extracellular organelles derived from cells (vesicles) containing Ca-P appear prior to the deposition of Ca-P in the noncellular matrix. Although the temporal relationship of these events suggest a causal relationship, firm evidence for such a relationship is still lacking. The mechanism of nucleation on specific sites within the collagen fibril may involve enzymatic processes. Phosphorylation of specific amino acid residues (e.g., serine) is an attractive hypothesis. O-Phospho-serine has been found in calcifying collagen, and in the poorly calcified enamel of rachitic animals contains reduced amounts of organic phosphorus. Direct evidence, however, relating phosphorylation to the nucleation process is not yet available. Nevertheless, once nucleation has occurred, the concentration of ions in the extracellular fluid will sustain crystal growth.

There are two distinct types of bone formation: endochondral and intramembranous. Endochondral bone formation occurs at the ends of long bones behind the epiphyseal growth plate and produces the linear growth in long bones like the femur (Fig. 63-3). In this process the epiphyseal cartilage plate migrates away from the original ossification center, and the cartilage undergoes calcification behind the growing edge. The bone formed by this process appears spongy and porous and is therefore called cancellous or trabecular bone.

Intramembranous bone formation occurs in the flat bone (e.g., skull) during growth and in all bone after skeletal maturation. In this process, new bone is laid down by osteoblasts on the surfaces of existing bone. Osteoclasts may resorb bone at adjacent surfaces so that remodeling occurs simultaneously with new bone synthesis. Some of the osteoblasts are trapped in an area of bone deposition and become incorporated into lacunar spaces within bone. These cells, osteocytes, are interconnected by an extensive network of canaliculi that radiate from a central haversian canal. Concentric rings of bone are laid down and appear as lamellae on micrographs. These lamellae, together with haversian canal, comprise a haversian system or osteon, the basic unit of bone (Fig. 63-4). Remodeling processes are constantly altering the size, shape, and num-

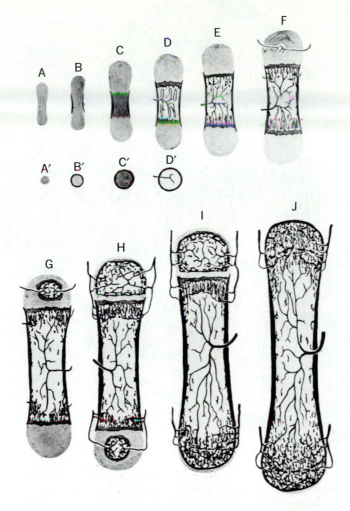

Fig. 63-3. Formation of bone in phalanx as example of endochondral bone formation. Ossification begins in area of preformed cartilage, *C*. Distal ends also calcify, leaving epiphyseal plate, *H*. Bone growth occurs in wake of growing cartilage at epiphyseal plate. (From Bloom and Fawcett.[6])

ber of osteons as well as the relationship of osteons to those adjacent to it in an area of compact bone.

Three types of bone cells can be distinguished by morphologic and functional characteristics. These are the osteoblast, osteoclast, and osteocyte.

Osteoblasts, located on the surfaces of forming bone, synthesize and secrete organic matrix and the enzymes mediating calcification of matrix (Fig. 63-5). Protein synthesis is a major function of osteoblasts, which contain a large number of ribosomes arranged in an endoplasmic reticulum. The Golgi apparatus also is prominent in these cells and may represent sites of mucopolysaccharide synthesis. A number of enzymes are located on the external surface of the osteoblast cell membrane. These include neutral, alkaline, and acid phosphatases that may function in nucleation and calcification or organic matrix.

Osteoclasts are multinucleated giant cells that resorb and remodel bone (Fig. 63-6). Since proteolytic enzymes are involved in this process, a large number of lysosomes presumably containing these enzymes are present in the cytoplasm of these cells. Adjacent to the surface of resorbing bone, osteoclasts show numerous infoldings of the cell membrane that appear to engulf bone crystals. Unlike the osteoblast, alkaline phosphatases do not appear to be associated with the

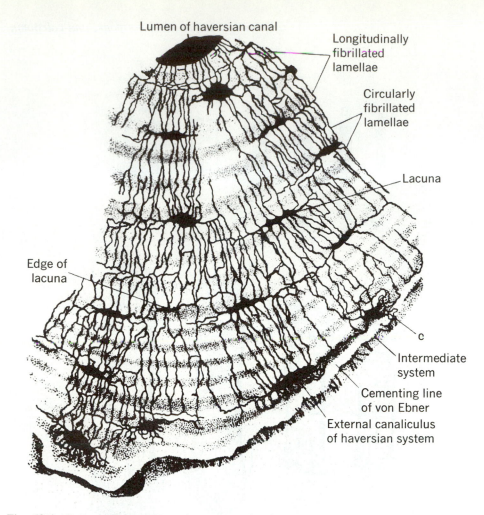

Fig. 63-4. Cross section of haversian system showing network of canaliculi between lacunae containing osteocytes. These canaliculi are filled with protoplasmic extensions of osteocytes. Dye has been used to demonstrate these features. (After Maximow; from Bloom and Fawcett.[6])

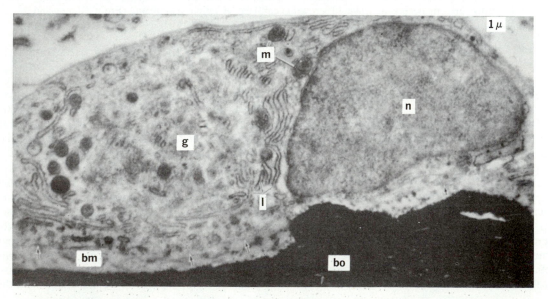

Fig. 63-5. Electron micrograph of section of undecalcified preparation of humerus showing contact of osteoblast with surface of bone matrix (arrows). *bm*, Uncalcified collagenous bone matrix; *bo*, calcified bone; *g*, Golgi apparatus; *l*, lamellae of endoplasmic reticulum; *m*, mitochondria; *n*, nucleus of osteoblast. (From Scott and Pease.[149])

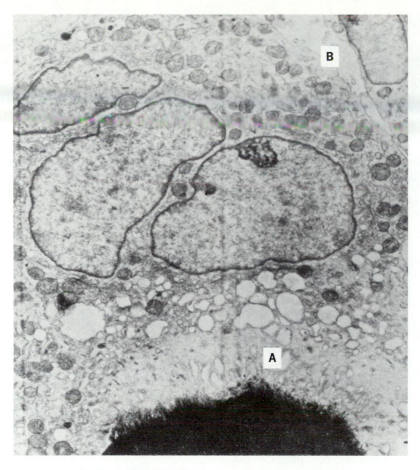

Fig. 63-6. Part of multinucleated osteoclast and matrix that is being resorbed. *A,* Irregular plasma membrane in resorption area; *B,* smoother plasma membrane not in contact with protein being resorbed. (Magnification ×6,500.) (From Robinson and Cameron.[145])

cell membrane. These enzymes are present, however, in the lysosomes, together with acid phosphatases.

Osteocytes are entrapped within compact bone (Fig. 63-7). The endoplasmic reticulum is less prominent than in osteoblasts, and Golgi apparatus and mitochondria are fewer in number. Although the osteocytes are located in lacunar spaces within compact bone, communication between adjacent cells takes place through cytoplasmic extensions coursing through the canalicular network. Because of these cytoplasmic processes, the osteocyte has a large surface area exposed to the fluid that bathes bone surfaces. These recent morphologic findings have altered the concept of osteocytic function. Previously considered a "dead-end" cell, the osteocyte is now thought to be an important component in the cellular response to parathyroid hormone and maintenance of plasma calcium homeostasis, since its surface area in contact with bone could facilitate rapid calcium mobilization or sequestration.[44] These cells, as well as osteoblasts and osteoclasts, participate in the remodeling discussed in connection with bone formation.

Traditionally, the osteoclast has been considered to be the main cellular mediator of calcium mobilization from bone.[85] However, the parathyroid hormone–induced changes in osteoclastic resorption are slow and require at least 6 hr of exposure to the hormone. Recent studies with isolated bone preparations show that parathyroid hormone–induced mobilization of bone calcium is very rapid. A significant increase is seen within minutes. The osteocyte has now been implicated in the rapid mobilization of calcium with parathyroid hormone. The osteocyte, through its cytoplasmic processes in the canalicular system, has a large surface area in contact with fluids that bathe calcium crystals, and it may function as a

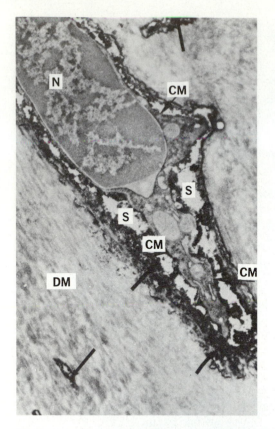

Fig. 63-7. Electron micrograph of osteocyte and canaliculi (arrows). *DM,* Dense matrix; *CM,* osteocyte cytoplasmic membranes; *S,* canalicular space; *N,* nucleus. (From Doty et al.[13])

calcium pump, transporting calcium into blood in response to parathyroid hormone. Although the exact mechanism for this transport has not been clarified, the hypothesis for "osteocytic osteolysis" is an attractive one. It may be that osteocytes produce the rapid, transient mobilization of calcium from bone. The capacity of this mobilization, however, may be limited. Osteoclasts, on the other hand, may mediate the slower but sustained mobilization of calcium through active resorption. The hormonally mediated integration of these two processes is undoubtedly important in the close regulation of plasma calcium concentration.

In the controlled system, bone is affected by several hormones in addition to parathyroid hormone and calcitonin. These hormones may alter osteoblast function and collagen synthesis or osteoclastic function and bone resorption.

Insulin,[93] growth hormone (GH), and thyroid hormone increase collagen synthesis, whereas adrenal glucocorticoids decrease it. The mechanism of these effects on collagen is probably related to the effects of these hormones on protein synthesis in general. The sex steroids as well as pharmacologic doses of glucocorticoids inhibit osteoclastic bone resorption.[95] The mechanism of this inhibition is not well understood. Many of the reported studies involved the use of pharmacologic amounts of the steroid. The calcium controlling system may also be affected by other hormones. For example, gut hormones can act as secretogogues for calcitonin, thereby leading to a decrease in $A_{Ca^{2+}}$. Prostaglandins also may be important regulators of skeletal metabolism. Prostaglandins E_1 and E_2, added to bone in vitro, cause increased concentrations of cyclic 3′,5′,-adenosine monophosphate (cyclic 3′,5′-AMP) and bone resorption.[62,108]

Calcium kinetics

The relationships among individual components of the controlled system as well as the coupling between the controlled and controlling systems may be investigated with radioactive calcium kinetics and metabolic balance studies. Observations made on patients with calcium or bone disorders and on normal subjects with dietary and hormone perturbations provide information concerning the coupling between controlled and controlling systems.

The specific activity of plasma or urine as well as cumulative excretion of radioactivity into urine and feces can be monitored after an intravenous injection of radiocalcium. A semilogarithmic plot of plasma-specific activity against time shows a curvilinear fall-off in specific activity; cumulative excretion of radioactivity in urine and feces is an integral of plasma-specific activity multiplied by a proportionality factor. The rate of disappearance of radioactivity from plasma is greater than can be accounted for by urine and fecal excretion. This difference is due predominantly to uptake and accretion of calcium into bone. Since the calcium metabolic balance (i.e., the difference between the amounts of calcium entering and leaving the body) must be the difference between calcium accreted into bone and resorbed from bone, one can derive estimates of bone turnover. By combining kinetic data with measured dietary intake and excretion via urine and feces, one can calculate rates of calcium metabolism in all the organ components of the controlled system.

Several mathematical approaches yield solutions for the desired metabolic parameters. In recent years, investigators have utilized the com-

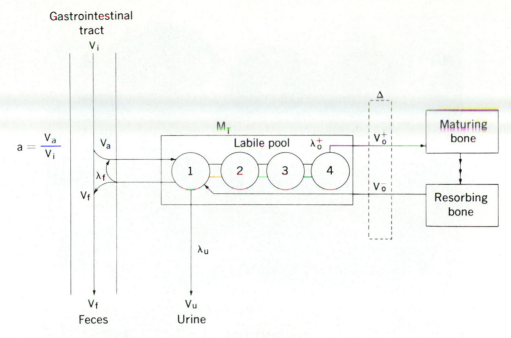

Fig. 63-8. Schematic representation of calcium metabolism based on model of four compartments in series. Gastrointestinal tract, labile (exchangeable) calcium pool, and nonexchanging calcium in bone are shown. Transfer rate (Vs) expressed in grams of calcium per day. Estimates are from studies of normal subjects.[135] V_i (ingested) = 0.80 gm/day; V_a (absorbed) = 0.35 gm/day; V_f (endogenous fecal) = 0.15 gm/day; V_F (total fecal) = 0.60 gm/day; V_u (urinary excretion) = 0.20 gm/day; V_0^+ (bone accretion) = 0.55 gm/day; V_0^- (bone resorption) = 0.55 gm/day. λs represent rate constant from compartment expressed as d^{-1}. λ_f (gastrointestinal rate constant from compartment *1* into feces) = 0.12 d^{-1}; λ_u (rate constant from compartment *1* into urine) = 0.16 d^{-1}; λ_0^+ (rate constant from compartment *4* into nonexchanging bone) = 0.10 d^{-1}. M_T (labile calcium pool) = 10 gm; a (fractional absorption of calcium) = 0.45; and Δ (calcium balance) = 0 gm/day. (From Phang.[135])

partmental model approach with digital computer analysis.[95,124] In this approach the metabolic system "seen" by the tracer is considered to be composed of discrete compartments. The transitional probabilities between compartments, λs, are calculated from the kinetics data. Clearly, one cannot assign an anatomic or physiologic identity for these mathematical compartments. On the other hand, by comparing compartment sizes and λs from studies performed in a variety of metabolic states, one can obtain insight into the physiologic processes embedded in the model.

A multicompartmental model using plasma, urine, fecal, and surface counting data has been developed, and subsequent studies have provided insight into the functions of the metabolic system as described by this model[124] (Fig. 63-8). The exchangeable calcium pool is divided into four compartments connected in series. Additional compartments were beyond the resolution of the

data, and fewer compartments were inadequate to fit the plasma-specific activity disappearance curve. Each compartment does not represent any one anatomic site, nor does each rate constant represent any one physiologic process. Compartment one is the site of tracer injection but is larger than the plasma pool. The subsequent compartments represent those with progressively slower turnover; compartments three and four contain predominantly rapidly exchangeable calcium in bone. Surface-exchangeable bone calcium is probably included in all the compartments. Urine and fecal excretion derives from the plasma pool, and bone accretion occurs from the pool most distal to the plasma pool. Bone resorption returns calcium directly into the plasma pool.

Studies in the steady state using this compartmental model have shown that parathyroid hormone and dietary perturbation produce changes in urine and fecal excretion and calcium absorp-

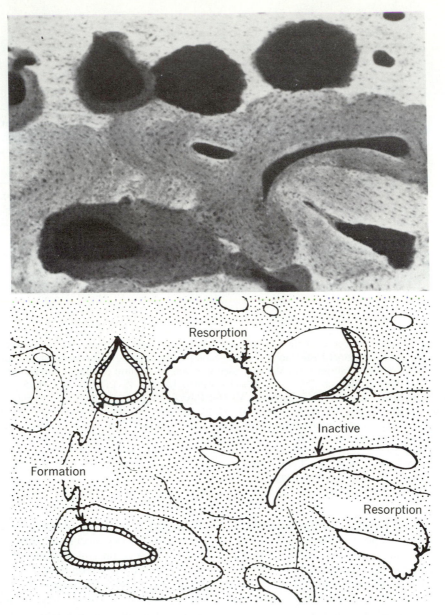

Fig. 63-9. Micrograph and schematic representation of bone specimen from iliac crest biopsy showing resorption and formation surfaces. Area of these surfaces is index for estimating bone resorption and formation rates. (From Jowsey.[103])

tion from the gastrointestinal tract.[47,135] Prolonged lack of parathyroid hormone in hypoparathyroidism is associated with decreased bone accretion rates as well as diminished bone resorption rates. In the homeostatic adjustments produced by dietary perturbation, however, bone accretion rate does not participate in the compensatory changes.[104,135] The rate constants describing exchange within the exchangeable calcium pool remained unaltered. Thus it appears that in the steady state, regulation does not occur at the intercompartmental level. However, with rapid perturbations in the nonsteady state, these rapidly exchangeable compartments may play a role in the transient behavior of the system.

Many other methods are available for estimating rates of bone metabolism.[95] These include morphometric (Fig. 63-9) and intravital staining techniques (Fig. 63-10). Although these methods allow for direct estimates of bone deposition and resorption, it is dangerous to extrapolate from a single biopsy site to overall calcium me-

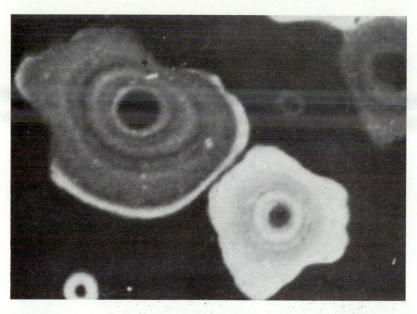

Fig. 63-10. Estimates of bone formation using intravital staining techniques. Tetracycline can be administered over defined period of time. Volume of bone deposited containing this fluorescent label may be estimated from biopsy specimen. (From Harris and Heaney.[95])

tabolism. On the other hand, analysis of radiocalcium kinetics should not be overinterpreted either. It is difficult to correlate calculated parameters with actual physiologic phenomena. Relatively slowly exchanging calcium in bone may be included in the estimates of bone accretion rate. Dystrophic calcifications, especially a problem for the elderly, may produce irreversible trapping of calcium that would tend to produce overestimates of bone accretion rates. The best estimate of overall osseous calcium metabolism is obtained by a combination of available techniques. Histologic methods provide a direct, albeit localized, measure of bone metabolism, whereas calcium kinetics provide an indirect estimate of calcium metabolism in the entire skeleton.

VITAMIN D

It was not until this century that vitamin D, the antirachitic vitamin, was discovered, although rickets had been recognized as a disease in the sixteenth century. The problem of rickets seems to have developed as man migrated from rural agricultural life to the crowded quarters of urban living[21] at northern latitudes. Physicians in the late nineteenth century recognized these factors, and early in the twentieth century the great discoveries were made that rickets could be cured by administration of a vitamin found in fish oils

and by exposure to ultraviolet light. The history of rickets and development of knowledge leading to the discovery of vitamin D is elegantly reviewed by Fourman et al.[15] DeLuca and Schnoes[12] have provided an important recent review of the chemistry, biosynthesis, actions, and clinical significance of the D vitamins and metabolites.

Chemistry

During the early 1930s, it was discovered that rickets could be corrected by feeding foods irradiated with ultraviolet light. It is now known that vitamin D can be formed from precursors by photochemical reactions in vitro or in vivo. On irradiation the plant precursor ergosterol gives rise to vitamin D. The animal precursor 7-dehydrocholesterol gives rise in an analogous manner to vitamin D_3. The structures of these several compounds are shown in Fig. 63-11. Irradiation of ergosterol gives rise also to a vitamin D–like compound, dihydrotachysterol. This compound recently has been crystallized and shows biologic properties similar to those of the D vitamins.

Biologic actions and assay

Vitamin D is one of the factors essential for normal calcium homeostasis. In adults the deficiency state is manifested as osteomalacia (defective mineralization of growing bone), and in

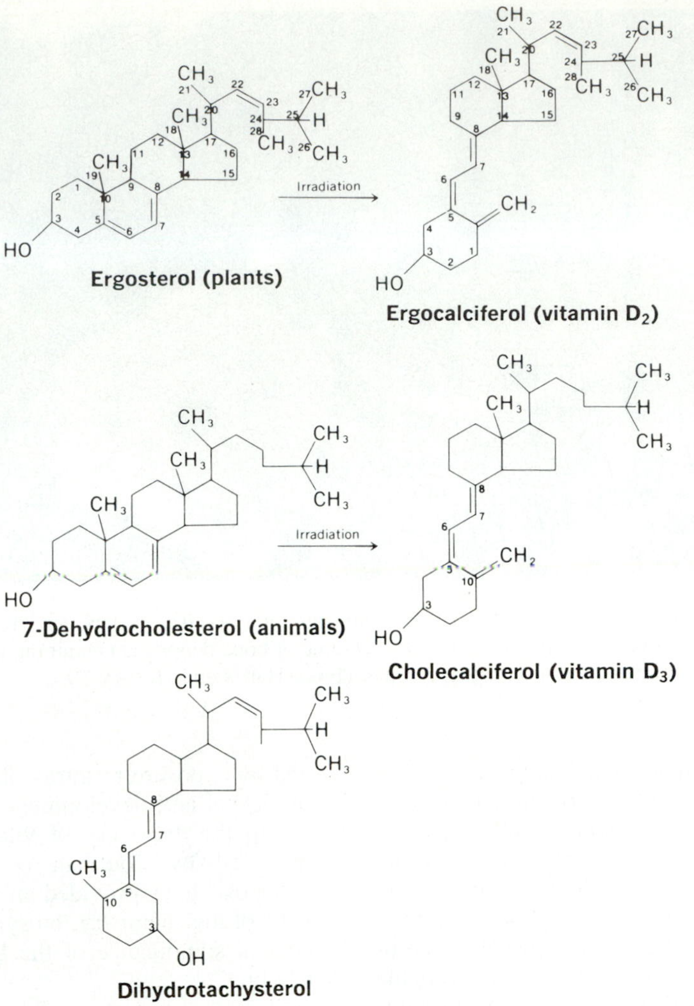

Fig. 63-11. Structures of vitamins D_2 and D_3 and their precursors. Structure of dihydrotachysterol is also shown. The latter is another compound with vitamin D–like activity found after irradiation of plant sterols.

children it is expressed as rickets. In the latter instance the bones become bowed and there is widening and cupping of the epiphyses. In either instance, serum calcium and phosphate levels are reduced. Microscopically, the bones in rickets and osteomalacia show wide areas of uncalcified matrix. There has been a controversy as to whether the influence of vitamin D in correcting rickets can be accounted for solely by its action on the gastrointestinal tract to increase absorption of calcium as well as phosphate or whether there is also a direct action on bone. A direct action on bone is apparent when the vitamin is administered in large doses, as in the treatment of hypoparathyroidism. The rise in serum calcium under these conditions must be explained at least partly

by increased resorption of mineral from bone. Recent investigations show that vitamin D metabolites definitely cause bone resorption (see the following discussion of metabolism and mechanism of action). On the other hand, the importance of vitamin D for gastrointestinal absorption of calcium has been known for years. Early clinical studies utilizing the relatively gross technique of calculating calcium balance from total intake and excretion had indicated that the improvement in blood calcium concentration was associated with a fall in fecal calcium output. More recent tests employing radioactive calcium show increased intestinal absorption of calcium in response to vitamin D. Vitamin D given in vivo leads to enhanced calcium transport detected in vitro.[74,147] Vitamin D increases the absorption of phosphate as well as calcium; indeed, certain forms of osteomalacia show apparent cure with the administration of phosphate alone. The influence of vitamin D on phosphate absorption, however, appears secondary to increased transport of calcium from the gut. The vitamin may influence the absorption of magnesium, strontium, or other divalent cations as well.

The physiologic importance of vitamin D for the renal transport of phosphate is still unresolved. It is well known that in rickets and osteomalacia the renal clearance of phosphate is abnormally high. However, this is attributable to compensatory hypersecretion of parathyroid hormone, which is the most important hormone influencing the renal clearance of phosphate. It is also known (p. 1537) that hypocalcemia, which occurs in severe rickets or osteomalacia, is the principal stimulus to increased secretion of parathyroid hormone. Thus merely infusing calcium into subjects with rickets or osteomalacia produces a fall in urinary excretion of phosphate that can be attributed to inhibition of secretion of parathyroid hormone. On the other hand, when vitamin D is used in pharmacologic doses to treat hypoparathyroidism, phosphate excretion increases. Large doses then are phosphaturic in the absence of the parathyroid glands. Reduction of phosphaturia, the converse effect of vitamin D that is observed in rickets and osteomalacia, can be attributed to suppressed secretion of parathyroid hormone as the blood calcium is raised in response to the vitamin.

Classic bioassays for vitamin D are based on the prevention or correction of rickets in chicks or young rats. The oldest form of the assay is the line test,[117] which is based on calcification of the epiphyseal cartilage of the tibia of the rachitic rat. Competitive binding assays[46] for vitamin D metabolites are being developed[56,77,92] and should be of great value in further studies in physiology as well as clinical medicine.

Mechanism of action and metabolism

It has been recognized for years that the gross physiologic effects of vitamin D do not develop immediately on administration of the vitamin.[10-12,26] This lag phase, shortened but not obliterated with large doses of the vitamin, provided a significant lead toward deciphering the mechanism of action of the vitamin. The delay might be accounted for by slow absorption from a storage pool for the vitamin, by the gradual conversion of the vitamin to an active form, or by the intermediation of secondary biologic

events, protein synthesis for example. It is now clear that at least two of these mechanisms are involved in the lag phase for the action of the vitamin. An experiment with actinomycin D[76] provided one important clue. This antibiotic, an inhibitor of the transcriptional process of protein biosynthesis, prevents the hypercalcemia caused by pharmacologic doses of vitamin D. Indeed, several inhibitors of protein synthesis, including puromycin and 5-fluorouracil, in addition to actinomycin D, prevent the action of vitamin D on intestinal calcium transport as well as induction of hypercalcemia.[11,35] Thus at least part of the delay in vitamin D action is accounted for by the interval required for synthesis of new protein. This thesis is further supported by tests for a calcium-binding protein in the intestine. Vitamin D, after a delay of several hours, induces the production of a calcium-binding protein in the intestinal mucosa of test animals, chicks or rats.[35] The rate of appearance of this binding protein parallels the increase in calcium transport activity of the intestine. Vitamin D_3 is biologically more active than vitamin D_2 in chicks, and, commensurate with higher biopotency, it causes more rapid accumulation of the calcium-binding protein.[35] Further, it has been shown that administration of actinomycin D, which, as noted before, inhibits vitamin D–stimulated absorption of calcium, causes comparable depression in the rate of formation of the binding protein.[35] Vitamin D induces synthesis of an ATPase in the intestinal brush border following either oral or parenteral administration of vitamin D_3 in chicks.[119] This ATPase seems to be specifically stimulated by calcium. However, the K_m for the binding of calcium to the ATPase is in the range of 2 mM. The K_m for the binding of calcium to the intestinal protein of Wasserman is in the range of 10M to 4M. It remains to be seen what the relationship between the calcium-activated ATPase and binding proteins might be and how each is specifically involved in the vitamin D–mediated intestinal transport of calcium.

The other factor in the lag phase is related to the metabolism of the vitamin in vivo. It long had been suspected that "activation" of vitamin D in vivo (or, more properly, conversion to a biologically active metabolite) was essential in the physiologic actions of the vitamin. The first clues in support of this hypothesis came from the early experiments of DeLuca,[12] his student Norman,[26] and their collaborators, who found metabolites in kidney, liver, intestinal mucosa, and serum after injecting tritiated or vitamin D_3 into rats. At least one of the metabolites appeared to be biologically active and showed chromatographic behavior, indicating that it was more polar than the parent vitamin D_3 compound. DeLuca and his associates showed that vitamin D_3 is converted by the liver to 25-hydroxycholecalciferol, which is then a key intermediate in the action of vitamin D in several species, including chicks, rats, and man. The metabolite is more active than the parent compound in curing rickets and produces an increase in calcium transport in vivo at a time earlier than that produced by the native compound. Further, 25-hydroxycholecalciferol in vitro causes bone resorption[158]; this had never been clearly observed with physiologic doses of vitamin D_3 itself. 25-Hydroxydihydrotachysterol and 25-hydroxyergocalciferol also have been synthesized and similarly found to be more ac-

tive than the parent compounds. Thus hydroxylation at the 25 position is a general requisite for the action of compounds related to vitamin D.

Further metabolites of vitamin D have been discovered, and it is evident that 25-hydroxycholecalciferol itself is converted in specific receptor tissues to compounds still more polar than 25-hydroxy vitamin D (Fig. 63-12). Most significant is the identification of 1,25-dihydroxycholecalciferol (Fig. 63-12, *C*) formed from 25-hydroxycholecalciferol by an enzyme in the kidney.* This renal enzyme system may be the most important control point of all in the regulation of

*See references 83, 100, 101, 110, 129.

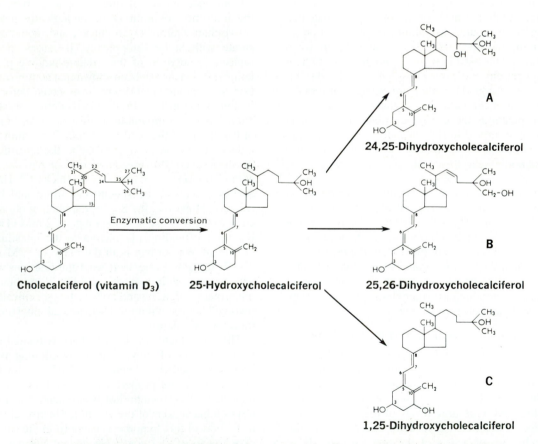

Fig. 63-12. Vitamin D_3 (cholecalciferol) is converted by liver to 25-hydroxycholecalciferol, which then is transported to kidney to be converted to 1,25-dihydroxycholecalciferol, **C,** the active metabolite for induction of calcium transport by gut. Other metabolites, 24,25-dihydroxycholecalciferol, **A,** and 25,26-dihydroxycholecalciferol, **B,** have been identified, but their physiologic significance has not been established.

vitamin D action.* After enzymatic formation in the kidney the dihydroxymetabolite is presumably transported to the intestine, where it enters the intestinal cell (possibly acting at the nucleus, as suggested by experiments with radioautography to localize radioactive vitamin D) and causes synthesis or activation of the intestinal calcium transport system responsible for the physiologic response to vitamin D.[11] 1-α-25-Dihydroxy vitamin D is 3 to 5 times as active as vitamin D_3 or 25-hydroxycholecalciferol.

PARATHYROID HORMONE

The parathyroid glands are small bodies, usually four in number, located near the thyroid and sometimes embedded in it. They are derived from the third and fourth pharyngeal pouches, the inferior glands coming from the upper or third pouch and the superior glands from the lower or fourth pouch. The parathyroids are not found in fish but exist in amphibia and all higher vertebrates. The glands are composed of two general cell types, the chief cells and the oxyphil cells. The chief cells are the usual active secretory cells, but occasional cases of hyperparathyroidism clearly are attributable to adenomas of oxyphil cells, thus proving the potential of the latter for hormone secretion. The chief cell is polyhedral, 6 to 8 μm in diameter, with the nucleus occupying half or more of the apparent cell volume. The cytoplasm is acidophilic. The oxyphil cell is larger, 11 to 14 μm in diameter, and has a nucleus about the same size as that of the chief cell but less dense. The cytoplasm is more eosinophilic than that of the chief cell. Oxyphil cells are rarely observed in children, become more prominent after puberty, and increase in numbers with advancing age. They are frequently replete with mitochondria.

Electron microscopy has shown that actively secreting chief cells can be distinguished from inactive cells. The latter show little development of subcellular organelles and few secretory granules.[57] The plasma membranes show sparse interdigitations between contiguous cells, and the intracellular spaces are narrow (Fig. 63-13, *A*). The cytoplasm contains lipid droplets (Fig. 63-13, *A*). Conversely, actively secreting chief cells contain well-developed organelles and numbers of secretory granules and are replete with endoplasmic reticulum and Golgi apparatus (Fig. 63-13, *B*). Induction of hypercalcemia by administration of vitamin D, for example, inhibits secretion; microscopic examination shows inactive cells. Glands taken from chronically hypocalcemic animals, on the other hand, show the histologic characteristics of active secretion. Studies with fluorescent antibodies appear to localize the hormone to chief cells. It is presumed that the "secretory granules" of the chief cells contain parathyroid hormone, but this has not yet been proved.

Bioassay

The classic bioassay for parathyroid hormone was developed by Collip and Clark[65] in 1925. It has been used with little modification as the standard USP bioassay and is based on determination of blood calcium in normal dogs. However, it is a cumbersome test, and more facile assays have been developed based on the concentration of calcium in the serum of young calcium-depleted rats injected with the hormone.[72,121] Calcium-depleted rats are more sensitive to parathyroidectomy; parathyroid hormone given to these animals prevents hypocalcemia (Fig. 63-14). Graded doses of hormone injected at the time of parathyroidectomy produce a log-linear increase in serum calcium level that can be determined 4 to 6 hr later. This is the basis of a simple and rapid bioassay. Assays have also been developed using the phosphaturic effect of the hormone as a parameter.[121]

The finding that parathyroid hormone directly activates adenyl cyclase in the renal cortex[3] led to the development of an in vitro bioassay for the hormone. The

---

*The mitochondrial 1-α-hydroxylase system of the kidney appears to be regulated in vivo by phosphate as well as parathyroid hormone itself. (In vitro experiments with isolated mitochondria suggest that calcium may also have a direct regulatory effect.) Low phosphate concentration stimulates synthesis of 1-α-25-dihydroxy vitamin D even in the absence of the parathyroid glands. Parathyroid hormone through its hypophosphatemic action causes enhanced 1-α-hydroxylase activity. Moreover, parathyroid hormone itself acts on the kidney through a mechanism mediated by cyclic 3',5'-AMP (see the discussion of the mechanism of action of parathyroid hormone). Hypocalcemia, the recognized stimulus for parathyroid hormone secretion, causes increased activity of the 25-hydroxy vitamin D hydroxylase that catalyzes formation of 1,25-dihydroxy vitamin D_3. This control mechanism may be closely analogous to the 3',5'-AMP–mediated adrenocorticotropic hormone (ACTH) regulation of steroid hydroxylation in the adrenal gland. Hydroxylase of kidney thus may represent the site at which parathyroid hormone influences intestinal absorption of calcium. 1,25-Hydroxycholecalciferol, the compound formed in the kidney from 25-hydroxy vitamin D, represents the active form of the vitamin that stimulates intestinal absorption of calcium.

The kidney hydroxylase mechanism, like that of the adrenal gland, also seems to involve cytochrome P-450.[12] A 24-hydroxylase found in kidney as well as other tissues converts 25-hydroxy vitamin D to 24,25-dihydroxy vitamin D. Production of the latter compound is favored when there is decreased need for and suppressed synthesis of the 1-α-25-hydroxy metabolite.

There is already evidence for abnormal metabolism of vitamin D in chronic renal failure.[42] It seems likely that the malabsorption of calcium in chronic renal disease as well as the clinical syndrome of vitamin D–dependent rickets will be explainable by defective biosynthesis or genetic lack of the enzyme needed for formation of 1,25-dihydroxycholecalciferol. This key renal enzyme must turn over at a high rate in vivo in view of its rapid depletion following exposure to inhibitors of protein synthesis.

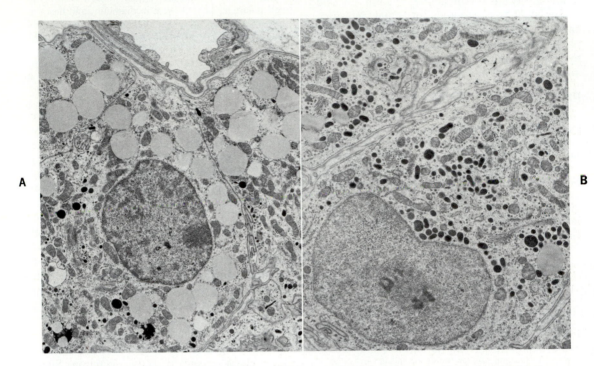

Fig. 63-13. A, Chief cells from bovine parathyroid gland in inactive state. **B,** Active chief cells showing numerous secretory granules, arrays of endoplasmic reticulum, and prominent Golgi apparatus. In contrast to inactive cell shown in **A,** there are few lipid bodies. (From Capen.[57])

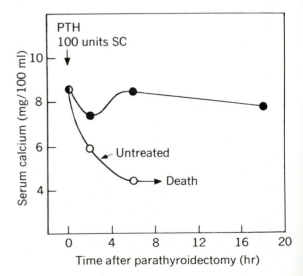

Fig. 63-14. Changes in serum calcium of calcium-deficient rats parathyroidectomized at time zero. Injection of parathyroid extract prevents hypocalcemia and death due to tetany. (Modified from Munson.[121])

assay is sensitive to as little as 0.14 USP units of parathyroid hormone, is more precise than bioassays in vivo, and shows a log-linear dose response over the range of 0.14 to 1.1 units.[113]

Standards

The USP unit is defined as 1/100 the amount of hormone required to raise the serum calcium 1 mg/100 ml in dogs. Obviously this is a crude definition of activity, based on a rough, imprecise assay. It is presumed that a new unit will soon be defined now that pure hormone and much more precise assays are available. Some progress has been made toward preparing a uniform preparation of hormone that could be adopted as a universal standard.[32]

Chemistry

Collip's preparation[64] of the first active parathyroid extract was a singularly important contribution to knowledge in the area of calcium metabolism.* His studies established that calcium metabolism was under hormonal control and proved the endocrine nature of the parathyroid glands. It is of interest to point out that his method for extraction, which yielded the first biologically effective preparation, involved boiling parathyroid tissue in 0.1N hydrochloric acid. Such extracts were difficult to purify because, as we now know, hot dilute hydrochloric acid generates a series of partially hydrolyzed hormone molecules. On the other hand, studies of peptides hydrolyzed from parathyroid hormone by this method have recently yielded highly significant information about the parathyroid hormone molecule. The availability of a crude extract fostered a series of investigations into the physiologic actions of the hormone. There was, however, a hiatus of more than a quarter century before methods were developed that allowed isolation of parathyroid hormone in purified form. In 1959 a method was devised with which parathyroid hormone could be extracted as a unique polypeptide molecule.[41] Isolation of the hormone as a virtually homogeneous substance (85% to 90% purity) was then possible at a 25,000-fold purification from bovine parathyroid glands.[41,141] Isolation of porcine, bovine, and human parathy-

---

*Hanson[94] reported also that an active extract could be prepared using a hot-acid technique similar to Collip's. But Hanson's report, a few short paragraphs, was bare of actual serum calcium measurements (he cited a range for percent increases in calcium content). Collip, on the other hand, took a meticulous approach to the problem, developed a bioassay based on determining serum calcium with a method he adapted for the purpose, and used this novel assay to document, with tables of data, the rise in serum calcium produced by injecting his extract into parathyroidectomized dogs.[64]

roid hormone has now been accomplished.[106] The purified hormones show both of the classically recognized biologic activities, mobilization of calcium from skeleton and enhanced renal excretion of phosphate.

The entire amino acid sequences of human, porcine, and bovine parathyroid hormone have been analyzed (Fig. 63-15).[3,17] Early in the course of the analysis it became apparent that a biologically active fragment might be prepared from the amino terminus of the molecule by hydrolysis in dilute acid. Dilute acid hydrolysis (under conditions similar to those used by Collip in the preparation of crude gland extracts) characteristically causes cleavage of polypeptide chains at aspartate residues. Since aspartate is not found within the first 30 amino-terminal residues of bovine parathyroid hormone, it seemed likely that this portion of the sequence contained the biologically active region of the hormone. This projection proved correct and led to synthesis of the tetratriacontapeptide consisting of residues 1 to 34 (Fig. 63-15); the synthetic peptides for bovine as well as human hormone show all the biologic properties characteristic of native parathyroid hormone.[3,123]

Biosynthesis and secretion

Shortly after Collip prepared an active extract the thesis was formulated that secretion of the hormone was regulated through a feedback control mechanism. It was presumed that hypocalcemia would stimulate and hypercalcemia would inhibit secretion. Patt and Luckhardt[132] were the first to obtain direct evidence in support of this thesis. They perfused the isolated thyroid-parathyroid apparatus of the dog with calcium-deficient blood. The perfusate, when given intravenously to a second parathyroidectomized dog, caused hypercalcemia and phosphaturia. Interest in the perfusion technique was revived by Copp and Davidson[66]; they also found that perfusion of the isolated glands with blood low in calcium (actually they perfused with the calcium chelator, ethylenediaminetetra-acetate, EDTA) caused systemic hypercalcemia. Their further experiments with this system led to the discovery of calcitonin (p. 1547).

Calcium may be a regulator of biosynthesis as well as secretion of hormone. Incubations of parathyroid tissue in vitro show that in low-calcium media the rate for incorporation of amino acids into hormone increases. Incorporation is partially inhibited by calcium. Calcium also inhibits transport of amino acids into parathyroid tissue. Biosynthesis of parathyroid hormone pro-

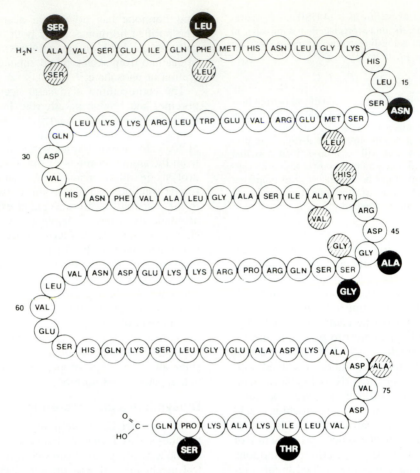

Fig. 63-15. Complete amino acid sequence of major form of bovine parathyroid hormone (open symbols). Porcine parathyroid hormone differs only in residues represented by shaded symbols. The human hormone differs from the bovine hormone in the residues shown by heavy shading. Highly active fragments (first 34 residues) of both human and bovine hormones have been synthesized. (Modified from Habener et al.[17])

ceeds through several precursor forms before storage and release of the major secretory form (84 amino acids in length; Fig. 63-15). Studies with biosynthesis in cell-free systems indicate that the initial product translated by parathyroid hormone in RNA is "pre-ProPTH," a polypeptide 115 amino acids in length.[17] As this polypeptide emerges from the ribosome, 25 amino acids are removed from the amino terminus, yielding "ProPTH." Six further amino acids are then cleaved from the amino terminus to yield PTH, the 84-amino acid polypeptide hormone (Fig. 63-15) that is stored in the gland and is secreted therefrom.[17,91]

The development of the radioimmunoassay for bovine parathyroid hormone[49] made it feasible to carry out extensive physiologic studies on factors regulating secretion of the hormone. Studies utilizing this assay confirmed the earlier thesis that hypercalcemia causes a fall and hypocalcemia causes a rise in the concentration of hormone in peripheral plasma (Fig. 63-16).

There have been suggestions that phosphate as well as calcium might influence secretion of the hormone, and this hypothesis has been tested by giving infusions of phosphate. In general, two types of response are observed when cows are given intravenous phosphate.[153] In some instances (Fig. 63-17, *A* and *C*) the rise in blood phosphate leads to a gradual rise in rate of hormone secretion; however, the increase in circulating hormone occurs later than the maximum

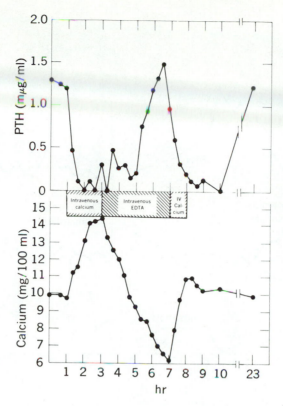

Fig. 63-16. Regulation of parathyroid hormone secretion by blood calcium level. Induction of hypercalcemia or hypocalcemia with infusions of calcium or EDTA, respectively, causes fall or rise in concentration of parathyroid hormone in circulation. (From Sherwood et al.[152])

for phosphate and corresponds to the slight degree of hypocalcemia that occurs in some of these experiments. In other instances, blood phosphate does not cause hypocalcemia; then there is no increase in secretory rate (Fig. 63-17, *B*). Conversely, in animals that show a slight rise in secretory rate in response to phosphate, this response is inhibited by giving a supplemental infusion to maintain normal calcium (Fig. 63-17, *D*). Again, calcium seems to be the primary determinant for parathyroid secretion. On the other hand, chronic unremitting hyperphosphatemia causes a slight but persistent hypocalcemia that may represent an important factor in chronic renal disease. Thus the persistent hypocalcemia attendant on chronically high concentrations of phosphate may well be significant in the development of secondary hyperparathyroidism.[143]

The hormone is continuously secreted in normal animals. Results shown in Fig. 63-18 represent radioimmunoassay determinations on a large

number of cows in various physiologic states. Note that secretion ceases as serum calcium level approaches 12 mg/100 ml and that there is a highly significant inverse correlation (Fig. 63-18) between the concentration of calcium and parathyroid hormone in plasma. This indicates that under steady-state conditions, secretion of parathyroid hormone is regulated by calcium through a proportional mechanism, that is, for every small change in calcium concentration there is a linear inverse change of hormonal secretory rate. The apparent half-life of the hormone in the circulation is about 20 min, and the hormone is distributed into a volume equivalent to approximately 30% of the body weight. These results allow calculation of the secretory rate at steady-state conditions with average hormone concentrations of 1 ng/ml of plasma. Such calculations yield a secretory rate of 1 to 3 USP units of hormone/hr/kg body weight. This result approximates closely the amount of hormone

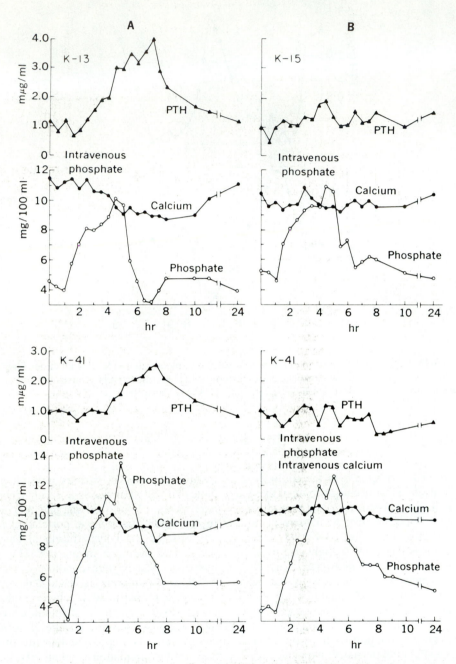

Fig. 63-17. Response of cows to infusion of phosphate. **A,** Results observed for two animals that developed delayed hypocalcemia; rise in hormone concentration in plasma coincided with slight fall in calcium that appeared after maximum in phosphate concentration. **B,** Calcium remained normal or was maintained normal with a supplemental infusion. (From Sherwood et al.[151])

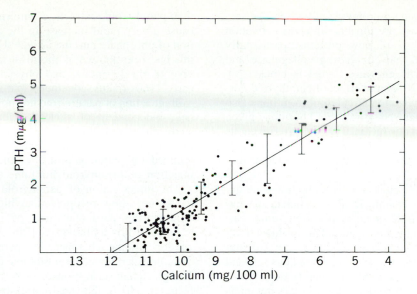

Fig. 63-18. Linear inverse relationship between concentration of calcium and parathyroid hormone in plasma. Vertical bars define standard deviation of line at several points. (From Sherwood et al.[151])

required by constant intravenous infusion to maintain normal calcium concentrations in parathyroidectomized animals.

Magnesium also may be a factor in regulating hormone secretion. Studies with direct perfusion or tissue culture of parathyroid glands show that high concentration of either divalent ion, calcium or magnesium, inhibits secretion. As the concentration of calcium approaches 3 mM, hormone secretion ceases. Substitution of magnesium for any or all of this amount of calcium is equally effective in suppressing secretion. Under normal physiologic conditions, however, magnesium in plasma does not rise to nearly the concentrations normally found for calcium. Thus one cannot determine the degree to which magnesium is important for regulation of secretion under normal physiologic conditions. Indeed, the modest changes in the normally occurring concentration of magnesium would not be expected to be of major physiologic significance for hormone secretion. Very marked reduction in magnesium concentrations also interferes with hormone secretion.

Physiologic actions

The principal function of parathyroid hormone is regulation of the concentration of calcium in the extracellular fluid. This regulatory influence is accomplished by actions of the hormone on the kidney and bone to produce an increase in the rate of transfer of calcium from the glomerular filtrate and skeletal tissue into the extracellular fluid. In addition, the hormone influences gastrointestinal absorption of calcium, probably through an indirect mechanism.* Under normal physiologic conditions, calcium transport mechanisms in the three separate tissues act in concert to provide close regulation of transfer of calcium into the extracellular fluid compartment. On the other hand, the action of the hormone on each tissue, bone, or kidney can each be effected, at least experimentally, independently of its influence on the other. Removal of the parathyroid glands causes a 30% to 40% drop in the concentration of calcium in the extracellular fluid. In the parathyroidectomized animal, calcium is maintained at this reduced concentration so that a new steady state is established. A further fall in calcium concentration can be produced (e.g., by infusing EDTA) in the parathyroidectomized animal, but the concentration returns to the hypoparathyroid steady state when the hypocalcemic challenge is removed.

Bone has been considered the principal tissue influenced by parathyroid hormone in maintaining blood calcium, and it has been believed that the influence of the hormone on this tissue is relatively slow. Only recently have the relatively rapid effects of the hormone on bone been recognized. Indeed, the effect of parathyroid hormone on bone in tissue culture develops in two phases. The earlier phase is more rapid, is manifested by the increased release of calcium into the medium within the first 2 to 3 hr of hormone action, and

---

*See footnotes on pp. 1523 and 1535.

does not depend on protein synthesis.[29] The second, later phase presumably involves activation and biosynthesis of new proteins, particularly lysozomal enzymes, including collagenase and other hydrolytic enzymes,[29] and is blocked by actinomycin D. A rapid effect of parathyroid hormone on mobilization of calcium from bone also has been recognized. Parathyroid hormone given to cats causes an increase in the arteriovenous difference of the calcium content in plasma perfusing the isolated tibia; this is detected within 10 to 15 min.

The hypercalcemic action of the hormone depends also on vitamin D. Rats deficient in vitamin D show little or no response in serum calcium levels to the hormone.[38] Part of this defect, however, can be corrected by giving calcium alone, and it has been observed that dogs respond to the hormone even when severely depleted of the vitamin.[126] Hypovitaminosis D does not interfere with the renal responsiveness to the hormone in parathyroidectomized animals. Even parathyroid-intact animals respond normally as assessed by cyclic 3',5'-AMP determinations. On the other hand, the high rate of phosphaturia due to secondary hyperparathyroidism in hypovitaminosis D may obscure a further phosphaturic response to exogenous hormone. The latter response can be readily elicited by raising blood calcium level, thereby inhibiting endogenous secretion of hormone, before a test dose of exogenous parathyroid hormone is infused.

Hydroxyproline content increases in urine or in the media after the hormone is introduced into the intact animals or into incubations of bone fragments in vitro. This phenomenon represents a slowly developing effect of the hormone.[29] Induction of collagenolytic activity in bone is similarly a late (12 to 24 hr) effect of parathyroid hormone. Thus it appears that the hormone promotes release of calcium from bone before any apparent resorption of bone matrix (the latter effect requiring hydrolytic enzymes). Indeed, there is some evidence that resorption of bone mineral occurs before resorption of matrix. Thus it is reasonable to propose that the early phase of parathyroid hormone action on bone represents an active transport process for calcium from the periosteocytic spaces of bone. The latter action of the hormone then involves further actions of these cells, increased communicative channels between osteocytes, increased numbers of osteoclasts, and the synthesis of hydrolytic enzymes.

Actions on kidney

It was discovered in 1925, immediately after the development of an active extract by Collip, that parathyroid extracts administered in vivo cause a very rapid increase in the rate of excretion of phosphate into the urine.[89] Until recently, this had been known as the most rapid effect in vivo of the hormone, and physiologists in the early 1960s finally concluded that it represented a direct action of parathyroid hormone on the kidney, independent of skeletal effects.[50,138] One study illustrating this direct action shows a rise in phosphate excretion by the parathyroidectomized rat within 8 min after the intravenous injection of parathyroid hormone (Fig. 63-19).

The direct action of parathyroid hormone on the kidney, now recognized as highly specific, requires the particular area of the molecule (represented by amino acids 1 to 29; see the discussion on the chemistry of parathyroid hormone) needed for general biologic effectiveness. The phosphaturic response, then, represents effects on cells in the renal cortex by the specific polypeptide structure that characterizes parathyroid hormone. Nevertheless, the mechanism of the renal phosphaturic effect remains incompletely understood. Many investigators have concluded that the hormone inhibits reabsorption of phosphate from the glomerular filtrate; evidence for active hormone-induced secretion of phosphate from the renal tubule has not been conclusive. On the other hand, phosphaturic response to the hormone may represent merely a secondary event consequent to a general increase in metabolic activity of the renal cell (see the discussion on the mechanism of action of parathyroid hormone). This leaves open the possibility that the initial increase in phosphate excretion represents phosphate elaborated from the proximal tubular cell into the tubular luminal fluid.

The site of action of the hormone in the kidney has been localized to the cells of the cortical tubules, proximal as well as distal tubular cells, and is distinct from regions containing receptors for calcitonin or vasopressin.[59,60] It is assumed that these are the same cells that are responsible for the phosphaturic effect, but this has not been proved conclusively. Studies utilizing micropuncture techniques support the conclusion that hormone-induced phosphaturia represents inhibition of a proximal tubular function.[36] Indeed, it is the transport of sodium that seems to be under the influence of the hormone[36]; phosphate may simply follow sodium as the accompanying anion. Even these more refined physiologic studies, however, do not exclude the possibility that cells in the same region, perhaps even the same cells, elaborate phosphate back into the lumen as well as reabsorb phosphate from the tubular fluid.

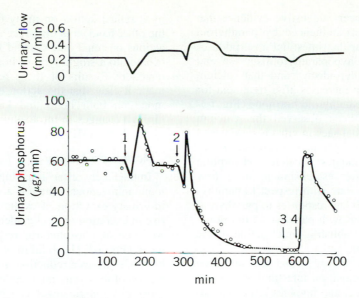

Fig. 63-19. Changes in rate of phosphate excretion by rat immediately following parathyroidectomy, *2*, or injection of parathyroid extract, *4*. Saline solution was injected at *3*. Small fluctuation in phosphate excretion following *1* is due to manipulations for anesthesia (sham parathyroidectomy as control). (From Beutner and Munson.[50])

Calcium excretion. It is widely known that prolonged administration of parathyroid hormone gives rise to increased calcium excretion. The latter is a consequence of hypercalcemia produced by the hormone. On the other hand, suggestions of hormonal effects on tubular reabsorption of calcium have been relatively new. In the first few hours after parathyroidectomy in the rat the urinary excretion of calcium increases[156] as phosphate excretion falls. It is not until later, with hypocalcemia established, that calcium excretion decreases. These findings led to the discovery that parathyroid hormone causes decreased renal clearance of calcium. At any given concentration of calcium in plasma, parathyroid hormone reduces markedly the amount of calcium excreted per 100 ml of glomerular filtrate. This renal effect is readily apparent in studies on the golden hamster, which develops a marked loss of calcium into the urine following parathyroidectomy. It was calculated that hypocalcemia following parathyroidectomy in this species can be totally accounted for by increased renal loss of calcium during the immediate postoperative period. One can conclude, then, that the hormonal influence on the renal reabsorption of calcium is important for the minute-to-minute regulation of calcium in the extracellular fluid.

Other metabolic effects. It had been known for

years that parathyroid extracts cause increased urine flow, and this observation was confirmed with the first preparations of purified parathyroid hormone. This finding has been explained recently by the observation that parathyroid hormone, as well as isoproterenol or dibutyryl cyclic AMP, causes inhibition of sodium reabsorption in the proximal tubule.[36,86] The sodium thus excluded from proximal reabsorption passes to the distal nephron, where most of it is reabsorbed. The water associated with the sodium is incompletely resorbed, giving rise to a net increase in free water clearance. In addition to the increased clearance of water, there is also a rise in urinary excretion of potassium, sodium, chloride, and bicarbonate with an increase in urine pH. It has been proposed[36] that the phosphaturic response is secondary to this hormone-induced, cyclic $3',5'$-AMP–mediated inhibition of proximal sodium reabsorption. Other functions influenced by the hormone in the kidney include enhanced reabsorption of hexoses in vivo and increased transport of amino acids as well as increased gluconeogenesis in vitro. It is likely that all these effects on solute transport can be attributed to the specific effect of parathyroid hormone in causing a rise in cyclic AMP in the proximal renal tubular cell. This subject is discussed further in the section on the mechanism of action of parathyroid hormone.

Effects on intestine

The intestine represents one of the three organs important in mediating the transfer of calcium into the extracellular fluid, and in a number of studies there has been suggestive evidence that this function also is influenced by parathyroid hormone. It is clear that intestinal absorption of calcium is low in hypoparathyroidism[53,150] and high in hyperparathyroidism[53] and that calcium transport by the gut increases after treatment for several days with parathyroid hormone (Fig. 63-20). Also, it has been observed that calcium transport by intestinal sacs in vitro is diminished in tissue taken from parathyroidectomized animals.[140,150] Thus changes in calcium absorption clearly develop after establishment of a new physiologic steady state with respect to parathyroid secretion. Such late responses to parathyroid ablation or treatment with parathyroid hormone, however, do not establish a direct action of the hormone on the gut.

Recent evidence indicates that the influence of parathyroid hormone on intestinal absorption of calcium is indirect and involves a mechanism regulating synthesis of 1,25-dihydroxycholecalciferol in the kidney.[11] It is the latter metabolite of vitamin D that then causes enhanced absorption of calcium from the gastrointestinal tract. This system of control for calcium absorption is discussed in the section on vitamin D.*

Mechanism of action

The physiologist has long recognized that the diverse effects produced by parathyroid hormone

---

*See also footnotes on pp. 1523 and 1535.

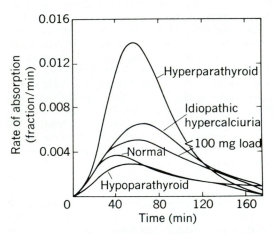

Fig. 63-20. Fractional rate of absorption of calcium from human gastrointestinal tract in normal and abnormal physiologic states. Absorption determined by giving radioactive calcium by mouth and determining isotope by external radiation monitoring equipment. (From Birge et al.[53])

must reflect action on many organ systems. On the other hand, it is only recently that the several actions on renal and skeletal tissue could be attributed to a single type of hormone-receptor interaction. Results of these recent studies have made it clear that the action of parathyroid hormone on both bone and kidney is mediated through changes in intracellular concentration of cyclic 3',5'-AMP.

Action on kidney

Immediately after injecting parathyroid hormone intravenously there is a very rapid increase in urinary excretion of cyclic 3',5'-AMP.[61] The rise in excretion of cyclic adenylate precedes the well-known phosphaturic response to the hormone (Fig. 63-21). Inhibition of the secretion of hormone causes a reduction in rate of urinary excretion of the cyclic nucleotide. This effect at the gross physiologic level can be traced to direct activation of renal adenyl cyclase (the enzyme catalyzing formation of cyclic 3',5'-AMP from ATP) by the hormone. Activation of the enzyme leads to a rise in tissue concentration of 3',5'-AMP; within 1 min following intravenous injection of parathyroid hormone one can detect a rise in concentration of 3',5'-AMP within the renal cortex.*

One consequence of increased concentration of 3',5'-AMP within proximal tubular cells of the renal cortex is increased elaboration of cyclic 3',5'-AMP into the luminal fluid. This is manifested by the urinary excretion of 3',5'-AMP observed in rats as well as in man.[61,63] Glucagon also influences the urinary excretion of 3',5'-AMP,[55,63] but in this instance the phenomenon is related to the elaboration of the cyclic nucleotide into the plasma from hepatic cells. Cyclic 3',5'-AMP in plasma is then cleared by glomerular filtration through the kidney. This difference in the action of parathyroid hormone and glucagon and action on urinary 3',5'-AMP can be recognized by changes in 3',5'-AMP to inulin clearance ratios. The clearance ratio does not change with glucagon administration, whereas

---

*Vasopressin is another hormone that acts on the kidney through an increase in concentration of 3',5'-AMP within the cell. It is clearly established, however, that the action of vasopressin within the kidney is localized to a site anatomically distinct from that influenced by parathyroid hormone. Vasopressin activates adenyl cyclase within the renal medulla, whereas parathyroid hormone activates the enzyme only in the cortex. These findings fit the current concepts of the renal physiologists, who localize the action of vasopressin to the collecting tubules. The tubules of the renal cortex are the sites of action for parathyroid hormone.

it rises 15-fold or more after injection of parathyroid hormone.

The infusion of cyclic 3',5'-AMP or dibutyryl cyclic 3',5'-AMP intravenously produces changes in the renal clearance of phosphate analogous to those caused by parathyroid hormone itself. Thus the biochemical sequence of events in the cells of the proximal renal tubule can be represented as follows: (1) Parathyroid hormone activates the enzyme adenyl cyclase in the plasma membrane of the cell. (2) The concentration of cyclic 3',5'-AMP increases intracellularly. (3) At very high concentrations, some of the 3',5'-AMP diffuses (or may even be transported) out of the cell into the luminal fluid to be excreted in the urine. (4) The rise in intracellular 3',5'-AMP causes activation of one or more cell processes. (5) Inorganic phosphate appears in the urine as a consequence of cell activation, part of which is manifested by inhibition of sodium reabsorption from the glomerular filtrate.

Action on bone

The action of the hormone on bone is of major physiologic importance for the maintenance of normal concentrations of calcium in the extracellular fluids. The cell types involved in these actions of the hormone are the osteocytes and osteoclasts. The initial events in the action of the hormone on these cells involves binding to the plasma membrane of the cell and then activation of adenyl cyclase. An interaction of the hormone with the cell membrane is evident from changes in cell membrane potential as well as activation of the enzyme. Transmembrane potentials of osteoclasts are found in a bimodal distribution among the cell population. Treatment of animals with parathyroid hormone causes a shift in distribution of cells to those with a more positive voltage distribution (decrease in negative voltage).[118] Calcitonin seems to inhibit this effect of parathyroid hormone. It is possible that interaction of parathyroid hormone with cells in vivo is associated with the intracellular entry of calcium. Kidney cells exposed to parathyroid hormone in vitro show an increased rate of radiocalcium exchange. In vivo, parathyroid hormone causes a slight but rapid and transient fall in blood calcium,[8] a phenomenon that has been attributed to the rapid entry of calcium into bone cells.

The interaction of parathyroid hormone with its receptor tissues has been studied indirectly by determining adenyl cyclase activity after addition of the hormone in vitro. The enzyme from bone or kidney is half-maximally activated at hormone concentrations of 5×10^{-7} M.[3] Presumably there are other receptor sites on the cells with greater affinities, since hormone concentrations of 0.1 nM are sufficient for biologic activity in vivo.

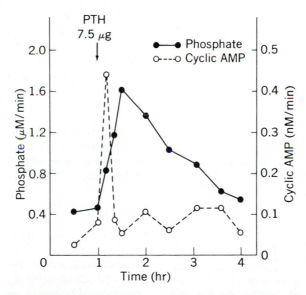

Fig. 63-21. Effect of parathyroid hormone on excretion of phosphate and cyclic AMP by parathyroidectomized rat. Parathyroid hormone (arrow) was infused over 2 min period. Note that phosphate is excreted in amounts on the order of micromoles per minute and 3',5'-AMP in nanomoles per minute. This difference in quantity of three orders of magnitude shows that phosphate excreted cannot be derived from cyclic AMP itself. (From Chase and Aurbach.[61])

Specific activation of adenyl cyclase in receptor tissues for parathyroid hormone occurs immediately after interaction of the hormone with the plasma membrane of the cell and is a key process in ultimate expression of the physiologic actions of the hormone. The attendant rapid increase in concentration of 3',5'-AMP in bone cells can account for the actions of the hormone on bone.* Theophylline, or dibutytyl cyclic 3',5'-AMP in vivo[163] or dibutyryl 3',5'-AMP in vitro, mimics the action of the hormone on bone, producing resorption of bone mineral[29] and increased formation and release of skeletal acid hydrolases.[159] The actions of theophylline, or dibutyryl cyclic 3',5'-AMP, are brought about through an increase in intracellular concentration of 3',5'-AMP. The rise in intracellular concentration of endogenous 3',5'-AMP can be accounted for by inhibition of cyclic nucleotide phosphodiesterase by dibutyryl cyclic AMP.

Role of cyclic 3',5'-AMP within cells or receptor tissues

As noted, it is the rise in the concentration of cyclic AMP within cells of the specific receptor tissues that causes the ultimate expression of the physiologic response to parathyroid hormone. A number of studies indicate that 3',5'-AMP activates protein kinases within cells responding to hormones through the intermediation of cyclic 3',5'-AMP.[3,30,109]

Phosphorylation of a particular substrate, catalyzed intracellularly by protein kinases, might be an intermediate step in the action of parathyroid hormone on bone cells. Several functions mediated in receptor tissues by cyclic AMP, particularly secretory processes, appear to involve microtubules of cells. For example, melanocyte-stimulating hormone (MSH) causes dispersion of melanin in melanophores. The granules of melanin appear to disperse along lines demarcated by microtubules.[52] Microtubular systems bind colchine and, in several instances, colchicine has been shown to disrupt microtubules or block se-

cretory processes in cells in which microtubules have been identified.[112] Further, it has been suggested that microtubular protein is the substrate for phosphorylation catalyzed by protein kinases that in turn are activated, as just noted, by cyclic 3',5'-AMP.[87] It is possible that bone resorption under the influence of parathyroid hormone is another process involving microtubules. Microtubules have been identified in bone cells,[32] and recently it has been shown that administration of colchicine causes hypocalcemia in vivo and inhibits bone resorption in vivo and in vitro. Colchicine also prevents the rise in serum calcium produced by injection of parathyroid hormone into parathyroidectomized rats.

Nature and metabolic fate in circulation

The evolution of radioimmunoassays for parathyroid hormone has allowed direct determination of hormone concentration in the plasma. Initial studies[152] indicated that circulating hormone was immunologically identical to that purified from glands. The first contrary evidence was reported by Berson and Yalow[48] concerning the hormone circulating in man. They had developed two different types of antisera to bovine parathyroid hormone. With one type of antiserum the apparent half-life of the hormone in human plasma appeared to be inordinately long. Results with another type of antiserum showed a relatively rapid disappearance of hormone from plasma after parathyroidectomy or induction of hypercalcemia. In uremia the half-life for either form of the hormone (i.e., detected by either of the two types of antisera) was markedly prolonged. These results indicated that there is more than one form of parathyroid hormone in the circulation of humans. A further deduction was that human parathyroid hormone is secreted from the gland as one species of polypeptide and is then converted into another form that disappears more slowly from the circulation. The more slowly disappearing fraction circulating in cows or man is lower in molecular weight as well as immunologically different from the hormone in the gland.*

Recent experiments show that parathyroid hormone is synthesized in the gland first as a larger molecule that is cleaved to a smaller polypeptide before secretion (see the discussion of biosynthesis and secretion). Further recent experiments (Fig. 63-22) utilizing direct venous sampling of endogenous hormone secreted into the inferior

---

*It has been found that, in addition to parathyroid hormone, prostaglandins, calcitonin, and epinephrine can influence the concentration of cyclic 3',5'-AMP in skeletal tissue in vitro.[62] The finding that prostaglandin causes an increase in cyclic AMP content led to the discovery that prostaglandin E_1 or E_2 can actually produce bone resorption in vitro.[108] Epinephrine in high doses also produces a rise in serum calcium levels in parathyroidectomized rats,[128] but this might reflect other phenomena (e.g., parathyroid secretion or hemodynamics) rather than calcium transports. The catecholamines do not produce bone resorption in vitro, presumably because the half-life under these conditions is short. Studies on the influence of calcitonin on cyclic AMP in bone are discussed in relationship to the mechanism of action of calcitonin. It is probable that calcitonin acts on cells that are distinct from those responding to parathyroid hormone.

---

*Some have suggested that the hormone must be converted to a smaller form in order to be active at the peripheral target tissues. This conclusion does not seem justified by current evidence.

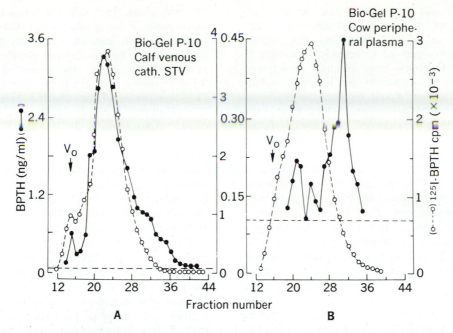

Fig. 63-22. Immunologically reactive parathyroid polypeptides in circulation. Samples were analyzed for molecular size by gel filtration on Bio-Gel P-10. Sample in **A** was obtained by directly catheterizing inferior thyroid vein of a cow; the sample in **B** was obtained from peripheral vein. Immunoreactive hormone in **A** fractionates similarly to native bovine parathyroid hormone (mol wt 9,800). In peripheral plasma, **B**, immunologically reactive hormone fractionates in a manner characteristic of smaller polypeptides. (Courtesy Dr. J. Habener.)

thyroid veins of cattle or man show that it is the native hormone that is released into the circulation. The relatively quick reductions in blood calcium, urinary excretion of phosphate, and 3',5'-AMP excretion following parathyroidectomy suggest a short biologic half-life compatible with that reported initially for immunologically reactive or radioiodinated hormone.[153] Thus the slowly disappearing fraction is probably of little physiologic significance and represents immunologically reactive, biologically inert peptide fragments of the native molecule. The half-life of some of these immunologically reactive fragments markedly exceeds that of the biologically effective molecule.

Site of inactivation

Early experiments indicated that the liver was a major site of inactivation of parathyroid hormone, but the kidney now seems to be most important in metabolism of the hormone. Microsomes obtained from rat kidney cause very rapid degradation of iodine-labeled parathyroid hormone in vitro,[115] and it has been found that the plasma membrane fraction of rat kidney (which also contains the receptor site for parathyroid

hormone molecule and adenyl cyclase) contains an enzyme capable of destroying the biologic activity of the hormone.[3]

Both the microsomal system and the plasma membrane system appear to inactivate the hormone at a rate commensurate with the rapid half-life in vivo. It is possible that the systems are similar; indeed, the microsomal system might represent the same enzyme found on the plasma membrane. Partial disruption of the system from the plasma membrane during homogenization could account for its appearance in the "microsomal fraction." The final identification of the physiologic site of inactivation in vivo and characterization of the hormone and metabolites in the circulation await results of further investigation.

CALCITONIN (THYROCALCITONIN)

To the physiologist in the first half of this century it seemed that parathyroid hormone must be the sole major humoral regulator of calcium metabolism. The essential function of vitamin D was recognized, but humoral control of calcium on a minute-to-minute basis appeared to be a function regulated solely by parathyroid hormone. Classic experiments had revealed that secretion of parathyroid hormone was regulated by calcium,[132] indicating that calcium metabolism could be effectively controlled by a simple feedback loop; clacium would regulate secretion of parathyroid hormone and the hormone in turn would regulate the concentration of calcium in the extracellular fluid. Copp and Davidson[69] tested this hypothesis, further utilizing the isolated perfused thyroparathyroid apparatus of the dog. Calcium concentration in the peripheral circulation was taken as the response index. Certain results of their experiments could not be explained by the action of parathyroid hormone alone. They found that hypercalcemic perfusion of the thyroparathyroid apparatus produced systemic hypocalcemia more rapidly than did simple parathyroidectomy (Fig. 63-23). This prompted Copp to propose that a heretofore unrecognized hormone that he called calcitonin must be elaborated to correct hypercalcemia.[69] At that time it was believed that the parathyroid glands were the source of calcitonin. Further information of major significance evolved from the experiments of Hirsch and Munson, who were studying the effects of surgical excision or cauterization of the parathyroid glands of rats. Hirsch et al.[98] discovered that cauterization caused the more rapid and profound hypocalcemia because concurrent damage to the thyroid caused release of a hypocalcemic factor

from the gland. They called this factor, found in the thyroid of many mammals, thyrocalcitonin to distinguish it from the calcitonin presumably elaborated by the parathyroid glands. Later investigations in many laboratories showed that thyrocalcitonin and calcitonin are indeed the same substance and that in mammals it is the thyroid gland that elaborates the hypocalcemic hormone.* Subsequently, the remarkable discovery was made that calcitonin does not exist in the thyroid of submammalian vertebrates but is synthesized and secreted by the ultimobranchial body, a distinct anatomic entity in these species.[67,68,157] Of further interest is the fact that extracts of ultimobranchial glands are 10 to 20 times or more as biologically active as extracts of mammalian thyroids. The high activity of ultimobranchial glands is explained not merely by high concentrations of hormone in the tissue but also by the much greater biologic potency of the hormone molecules of submammalian origin. This topic is discussed further in the section on the chemistry of these hormones.

Cells of origin

The identification of the thyroid as a gland of origin for calcitonin in mammals led to an effort to characterize the type of cells that secrete the hormone. The

---

*See references 8, 9, 24, 31, 81, and 99.

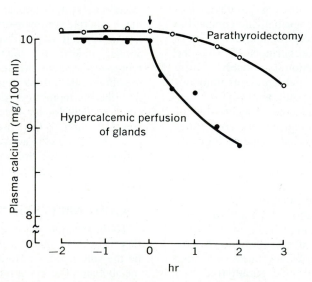

Fig. 63-23. Change in plasma calcium of dogs after parathyroidectomy or perfusion of thyroparathyroid apparatus with hypercalcemic blood. Note that the latter caused more rapid and greater hypocalcemia, presumably because it induced secretion of calcitonin.

first suggestion that the parafollicular cells (Fig. 63-24) might be the calcitonin secretors came from experiments of Foster et al.[81] Later, immunofluorescent techniques were utilized in an elegant study for direct identification of calcitonin in these cells ("C" cells). The morphology of these cells had been described in detail years earlier by Nonidez,[128] who had guessed that these cells might have some endocrine function yet to be discovered. A particularly useful histochemical marker for C cell types has been discovered.[134] These cells take up dihydroxyphenylalanine (dopa) and 5-hydroxytryptophan; intracellularly these precursors are decarboxylated to form dopamine and serotonin, which are detectable by fluorescence. Embryologic studies utilizing this marker indicate that the C cells in mammalian thyroids are derived from the last pharyngeal pouch, or ultimobranchial body.* In submammalian species, this ultimobranchial anlage does not fuse with the thyroid but remains as a separate organ, the ultimobranchial gland. Copp[67,68] found that extracts of ultimobranchial glands of elasmobranchs, fish, amphibia, reptiles, and birds contain active calcitonin peptides. Almost simultaneously, Tauber[157] reported also that the ultimobranchial glands of chicks were high in calcitonin content. Both Copp and Tauber noted that ultimobranchial glands contain more calcitonin than mammalian thyroids.

Chemistry

The observation that active extracts of thyroid tissue were readily obtainable[99] provided the impetus to purify large amounts of porcine calcitonin for structural analysis and possible synthesis. Within 4 years of its discovery the hormone had been isolated[139] and structural determination had been reported by three different groups.[45,125,137] Shortly thereafter the polypeptide was synthesized and the product found to be fully active biologically.[90,144] Investigators then turned their attention to analyzing the structure and synthesizing calcitonin obtained from human tissue[34] (medullary carcinoma of the thyroid*) as well as ultimobranchial calcitonin obtained from salmon.[34] The structures of these several molecules are illustrated in Fig. 63-25. Note that the structure in the region of the seven-membered ring is virtually identical (except glycine at position 2 in the human hormone) in all species of the hormone, and all contain prolineamide at the C terminus. There are, however, extensive differences in sequence throughout the remainder of the molecule. These intramolecular differences are very important in terms of biologic properties. Of particular note is the fact that salmon calcitonin shows 20 times the biologic potency of the porcine hormone. Chicken calcitonin also is similar to salmon calcitonin in structure and in its high potency.[1]

---

*Medullary carcinoma of the thyroid gland is derived from ultimobranchial-type cells (C cells).[155] The tumors contain high concentrations of active calcitonin. Calcitonin is secreted in this disease at abnormally high rates,[70] particularly when infusions of calcium or glucagon are given.[23] The radioimmunoassay for human calcitonin can be utilized to detect this syndrome before it is overtly recognized clinically.

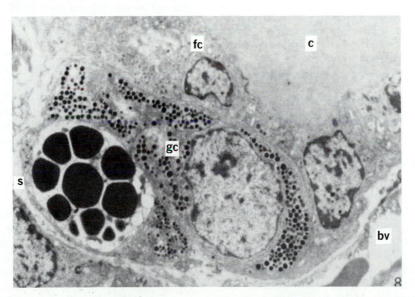

Fig. 63-24. Electron micrograph of section of bat thyroid. Parafollicular cells, *gc,* are replete with secretory granules and separated from colloid follicles, *c,* by follicular cells, *fc.* (From Nunez et al.[130])

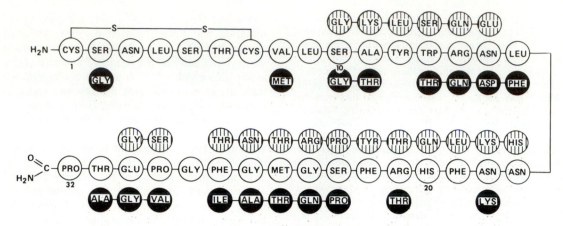

Fig. 63-25. Structure of calcitonin molecules. Open symbols represent structure of porcine calcitonin. Positions at which human calcitonin (darkened symbols) and salmon calcitonin (vertically hatched symbols) differ in sequence are shown. All molecules with calcitonin activity contain the 7-membered disulfide ring. The only substitution known within the ring is the glycine at position 2 of the human hormone.

Bioassays

In general, the bioassays for calcitonin have been based on reduction in serum calcium levels of the rat and have proved quite satisfactory in work on extraction and purification. The first practicable bioassay[99] allowed detection of approximately 10 MRC (Medical Research Council) milliunits. (The unit of calcitonin activity is currently defined in terms of a standard prepared by the Medical Research Council of the United Kingdom. Each vial of the standard contains 0.25 MRC units.)

Radioimmunoassays have been developed for calcitonin from several species.[73] The first useful radioimmunoassay for thyrocalcitonin was applied to determine calcitonin in the plasma of rabbits. The hormone in the latter species apparently shows a high degree of cross-reactivity with the porcine molecule. The method also has been applied to determination of calcitonin in the plasma of pigs. More recent assays have been developed successfully for human and ovine calcitonin.

Control of secretion

Early experiments showed that a hypocalcemic factor detected in the thyroid venous effluent was released from the thyroid in response to perfusion with fluids high in calcium content. Further, the thyroid gland of the rat was found to be important for calcium metabolism even at normal or subnormal concentrations of calcium in blood. With the advent of a radioimmunoassay method for calcitonin, it was possible through direct analysis of peripheral plasma to confirm that calcium was the stimulus for secretion of calcitonin from the

thyroid (Fig. 63-26). Calcitonin is undetectable in the peripheral plasma of thyroidectomized rabbits. Further studies with this technique indicated that calcitonin is secreted continuously at normal concentrations of blood calcium and that secretion is probably not suppressed until blood calcium level falls below normal.

Calcitonin secretion is regulated by mechanisms more intricate than feedback control through blood calcium level alone. Pentagastrin, glucagon, pancreozymin, or any of several polypeptides containing in common the tetrapeptide C terminus of pentagastrin, can stimulate secretion of calcitonin.[58] The possibility that a gastrointestinal hormone functions as a secretogogue for calcitonin has been raised also by other physiologic experiments. Normal rats challenged with 1% of calcium chloride instilled through a stomach tube show little change in plasma calcium. Thyroidectomized animals, on the other hand, develop a marked degree of hypercalcemia (Table 63-1). These several observations suggest that a gastrointestinal hormone (pancreozymin, gastrin, intestinal glucagon, or another) may be secreted as a physiologic response to a calcium-rich meal. The intestinal hormone thus elaborated into the circulation would act as a secretogogue for calcitonin, which in turn would partially or completely overcome the potential hypercalcemia attendant on absorption of calcium from the gastrointestinal tract. Whether this is a physiologically significant control mechanism remains

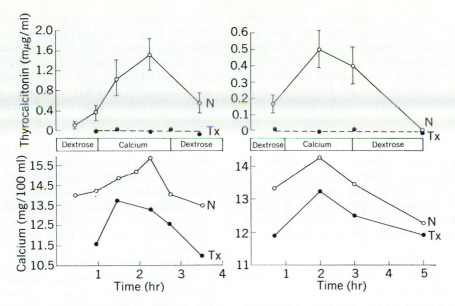

Fig. 63-26. Changes in concentration of calcium and thyrocalcitonin (calcitonin) in plasma during infusion of calcium into normal, *N*, or surgically thyroidectomized, *TX*, rabbits. (From Deftos et al.[73])

Table 63-1. Concentration of calcium in serum of rats with thyroids intact (sham-operated) or thyroids ablated (TPTX) 60 min after operation and intragastric administration of calcium chloride*†

Intragastric calcium (mg/100 gm)	Serum calcium (mg/100 ml)	
	Sham-operated	**TPTX**
1	10.60	11.40
2	10.42	11.40
4	10.70	12.12
8	10.92	13.52

*Data from Gray and Munson.[88]
†The difference between sham-operated and thyroparathyroidectomized animals was significant at each dose of intragastric calcium given.

to be proved, and this question must be answered by isolation and identification of the putative hormone involved.

Biologic effects, physiologic significance, and mechanism of action

The hypocalcemic effect of calcitonin is readily apparent at physiologic and pharmacologic levels. The earlier studies on calcitonin indicated that this hypocalcemia was accompanied by hypophosphatemia,[69,99] and the occurrence of these two actions suggested to investigators that the effects of calcitonin were brought about through a reduction in mobilization of mineral from bone. There is in addition a modest influence of the hormone on the kidney in some species, but there is little to suggest that there is an important influence on gastrointestinal absorption of calcium.

Effects on bone

Calcitonin directly inhibits resorption of bone mineral in vitro as well as in vivo. This influence of calcitonin on the skeleton can be observed histologically or can be determined by radiocalcium release from bones incubated in vitro. Analysis of radiocalcium kinetics in vivo provides further evidence for this thesis.

One of these approaches[40,84] utilizes in vitro cultures of bones of fetal rats taken from mothers previously given radioactive calcium in vivo. Parathyroid hormone added to this preparation causes an increased rate of release of radiocalcium from the bones. Calcitonin inhibits the parathyroid-induced release of radioactive calcium[84] and the hormone-induced resorption of mouse calvaria in tissue culture.[37]

The overall effect of calcitonin in reducing bone resorption is similarly apparent in vivo. It decreases the rate of removal of calcium from bone perfused in situ.[33] Studies using radiocalcium kinetics in vivo provide further evidence that calcitonin produces hypocalcemia by causing diminished resorption of bone, the principal source of calcium supplying the extracellular fluid. These experiments show that the rate of disappearance of calcium from plasma is not influenced by calcitonin. However, the action of the hormone can be explained by a reduction in rate of entry of calcium into the plasma (indeed, entire extracellular compartment) from bone. Inhibition of bone resorption accounts also for calcitonin-produced reduction in excretion of hydroxyproline, an amino acid that appears in the urine, in part as a consequence of bone catabolism.[34,142] The potent inhibition of bone resorption by salmon calcitonin has made it efficacious in the control of Paget's disease of bone.[34]

Effects on kidney

It was clear from the earliest experiments with calcitonin that the kidney was not necessary for elaboration of the gross physiologic effects of the hormone. Nevertheless, the hormone can influence renal function. Phosphate excretion increases following injection of calcitonin[2,34] but at least part of this effect is explained by the secondary secretion of parathyroid hormone attendant on the hypocalcemia produced by calcitonin. Occasional reports indicate that phosphaturia is produced in thyroparathyroidectomized animals, but consistent effects on phosphate excretion have not been achieved with highly purified preparations of calcitonin. Calcitonin also influences renal clearance of calcium and sodium[34,51] as well as phosphate, but one cannot yet conclude that calcitonin is an important physiologic influence on the kidney. On the other hand, it is important to note that calcitonin (as discussed in the section on mechanism of action) does act on the kidney in vitro. Whether this in vitro action is of physiologic significance remains to be established.

Biochemical mechanism of action

The preceding section describing the mechanism of action of parathyroid hormone referred to observations that dibutyryl cyclic 3′,5′-AMP caused resorption of bone in tissue culture. In the same experiments, dibutyryl cyclic AMP added at still higher concentrations caused inhibition of bone resorption and inhibition of the effects of parathyroid hormone.[29] These observations suggested that 3′,5′-AMP is involved in the action of calcitonin. It is now clear that the initial event in calcitonin action involves specific binding to high-affinity receptors identified in kidney as well as bone.[116] Apparent affinities for receptor

of the several calcitonins from different species parallel their potencies in vivo. Interaction of calcitonin peptides with these specific tissue receptors causes activation of adenyl cyclase generation of cyclic 3′,5′-AMP. Salmon calcitonin in doses as low as 1 ng/ml causes a significant increase in the concentration of cyclic AMP in bone and direct activation in vitro of adenyl cyclase from kidney as well as bone.[96] Mixtures of parathyroid hormone and salmon calcitonin at wide ranges of concentration consistently produce additive effects even at maximally effective concentrations of each hormone.[32] These findings indicate that calcitonin and parathyroid hormone receptors must be distinct in bone. Presumably, different cell types respond to each hormone.

Clear experimental evidence shows that calcitonin stimulates in vitro adenyl cyclase prepared from the renal cortex[122] and injections of calcitonin produce natriuresis in several species.[31] These observations suggest that this presumed inhibition of sodium reabsorption from the glomerular filtrate is also a 3′,5′-AMP–mediated phenomenon, analogous to the effect of parathyroid hormone on the kidney.

Metabolism

We have already pointed out that salmon calcitonin is 20 times as potent on a weight basis when tested in rats and 100 times more potent than porcine calcitonin in mice. The striking differences in biologic activity are related to a longer half-life for salmon calcitonin in the circulation and to a higher affinity for the receptor site itself. The high ratio of potency of salmon calcitonin to porcine calcitonin also is observed with skeletal tissue in vitro.[96] Calcitonin may be inactivated by an enzyme system in the plasma membranes of the receptor tissues themselves. Incubation of calcitonin with a system in the plasma membrane fraction of rat kidneys causes rapid loss of biologic activity.[96] Salmon calcitonin is relatively resistant to inactivation by this enzyme. The relatively slow rate of disappearance of salmon calcitonin from the circulation thus reflects resistance of the latter molecule to inactivation by enzymes in the plasma membrane of the kidney and possibly in other tissues as well. In addition to inactivation by the plasma membrane itself, there appear to be heat-labile factors in serum that are capable of destroying porcine calcitonin[33]; rabbit, rat and human sera cause inactivation of calcitonin at 37° C. Whether the latter factors are important in vivo is uncertain.

Physiologic significance

Experiments already cited have shown that calcitonin is important, at least in certain mammals, in the regulation of calcium homeostasis. Protection against the hypercalcemic challenge of calcium given intravenously or enterically to rats or rabbits depends on the thyroid and thus presumably on an intact secretory mechanism for calcitonin. Indeed, the relatively rapid hypocalcemic action of calcitonin makes it attractive to propose the latter hormone as an important factor in the minute-to-minute regulation of calcium in the extracellular fluid. This thesis is supported further by observations suggesting that, at least in the rabbit, rat, and pig, there appears to be a continuous secretion of calcitonin even at normal concentrations of blood calcium. On the other hand, man is relatively insensitive to the action of calcitonin. Massive doses are required to produce significant hypocalcemia, and there is little evidence that a syndrome of calcitonin deficiency is clinically significant. These arguments must be tempered, however, with the recognition that throughout the mammalian world it is only young, rapidly growing animals that show significant sensitivity to calcitonin or lack thereof. It is possible, then, that in fetal or neonatal life all mammals may depend on the influence of calcitonin; indeed, a role in hibernation has been postulated.[130] There is even greater uncertainty concerning the physiologic function of calcitonin in lower vertebrate species. Although calcitonins obtained from lower classes are more potent than mammalian calcitonins in the rat, no clear-cut role has been established for calcitonins in fish, birds, or other submammalian species. Perhaps at some earlier time in evolution, calcitonin had a function other than regulating calcium concentration. The natriuretic effect of the hormone may represent a hint that this is so.

REFERENCES
General reviews

1. Aurbach, G. D., editor: Parathyroid gland. In Greep, R. O., and Astwood, E. B., editors: Handbook of physiology. Endocrinology section, Baltimore, 1976, The Williams & Wilkins Co., vol. 7.
2. Aurbach, G. D., et al.: Polypeptide hormones and calcium metabolism, Ann. Intern. Med. **70:**1243, 1969.
3. Aurbach, G. D., et al.: Structure, synthesis and mechanism of action of parathyroid hormone, Recent Prog. Horm. Res. **28:**353, 1972.
4. Becker, R. O., Bassett, C. A., and Machman, C. H.: Bioelectrical factors controlling bone structure. In Frost, H. M., editor: Bone biodynamics, Boston, 1964, Little, Brown & Co.
5. Bloodworth, J. M. B., editor: Endocrine pathology, Baltimore, 1968, The Williams & Wilkins Co.
6. Bloom, W., and Fawcett, D. W.: A textbook of histology, ed. 9, Philadelphia, 1968, W. B. Saunders Co.
7. Bronner, F.: Dynamics and function of calcium. In Comar, C. L., and Bronner, F., editors: Mineral metabolism, New York, 1964, Academic Press, Inc., vol. 2.
8. Copp, D. H.: Parathyroids, calcitonin and control of plasma calcium, Recent Prog. Horm. Res. **20:**59, 1964.
9. Copp, D. H.: Parathyroid hormone, calcitonin, and calcium homeostasis. In Comar, C. L., and Bronner, F., editors: Mineral metabolism, New York, 1969, Academic Press, Inc., vol. 3.
10. DeLuca, H. F.: The role of vitamin D and its relationship to parathyroid hormone and calcitonin, Recent Prog. Horm. Res. **27:**479, 1971.
11. DeLuca, H. F.: Metabolism and function of vitamin D. In Astwood, E. B., and Greep, R. O., editors: Handbook of physiology. Endocrinology section, Baltimore, 1976, The Williams & Wilkins Co.
12. DeLuca, H. F., and Schnoes, H. K.: Metabolism and mechanism of action of vitamin D, Annu. Rev. Biochem. **45:**631, 1976.
13. Doty, S. B., Robinson, R. A., and Schofield, B.: Morphology of bone and histochemical staining characteristics of bone cells. In Greep, R. O. and Astwood, E. B., editors: Handbook of physiology. Endocrinology section, Baltimore, 1976, The Williams & Wilkins Co., p. 3.
14. Foster, G. V.: Calcitonin (thyrocalcitonin), N. Engl. J. Med. **279:**349, 1968.
15. Fourman, P., et al., editors: Calcium metabolism and the bone, Oxford, 1968, Blackwell Scientific Publications, Ltd.
16. Glimcher, M. K., and Krane, S. M.: Organization and structure of bone and the mechanism of calcification. In Gould, B. S., and Ramachandran, G. N., editors: Treatise on collagen, New York, 1968, Academic Press, Inc., part B.
17. Habener, J. B., et al.: Biosynthesis of parathyroid hormone, Recent Prog. Horm. Res. **33:**249, 1977.
18. Hirsch, P. L., and Munson, P. L.: Thyrocalcitonin, Physiol. Rev. **49:**548, 1969.
19. Johnson, L. C.: Morphologic analysis in pathology: kinetics of disease and general biology of bone. In Frost, H. M., editor: Bone biodynamics, Boston, 1964, Little, Brown & Co.
20. Kurty, S. M., editor: Electron microscopic anatomy, New York, 1964, Academic Press, Inc.
21. Loomis, W. F.: Rickets, Sci. Am. **223:**76, 1970.
22. Mahler, H. R.: Interrelationships with enzymes. In Comar, C. L., and Bronner, F., editors: Mineral metabolism, New York, 1961, Academic Press, Inc., vol. 1.
23. Melvin, K. E. W., Tashjian, A. H., Jr., and Miller, H. H.: Studies in familial thyroid carcinoma, Recent Prog. Horm. Res. **28:**399, 1972.
24. Munson, P. L., et al.: Thyrocalcitonin, Recent Prog. Horm. Res. **24:**589, 1968.
25. Nicolaysen, R., Eeg-Larson, N., and Malm, O. J.: Physiology of calcium metabolism, Physiol. Rev. **33:**424, 1953.
26. Norman, W. W.: The mode of action of vitamin D, Biol. Rev. **43:**97, 1968.
27. Potts, J. T., Jr., and Deftos, L. J.: Parathyroid hormone, thyrocalcitonin, vitamin D, bone and bone mineral metabolism. In Bondy, P. K., editor: Duncan's diseases of metabolism, ed. 7, Philadelphia, 1974, W. B. Saunders Co.
28. Rahamimoff, R.: Role of calcium ions in neuromus-

cular transmission. In Cuthbert, A. W., editor: Calcium and cellular function, 1970, New York, St. Martin's Press, Inc.

29. Raisz, L. G.: Mechanisms of bone resorption. In Astwood, E. B., and Greep, R. O., editors: Handbook of physiology. Endocrinology section, Baltimore, 1976, The Williams & Wilkins Co., p. 117.

30. Rasmussen, H.: Cell communication, calcium ion, and cyclic adenosine monophosphate, Science **170:**404, 1970.

31. Talmage, R. V., and Belanger, L. F., editors: Parathyroid hormone and thyrocalcitonin (calcitonin), Princeton, N.J., 1968, Excerpta Medica Foundation.

32. Talmage, R. V., and Munson, P. L, editors: Parathyroid hormone and the calcitonins, Princeton, N.J., 1972, Excerpta Medica Foundation.

33. Taylor, S., editor: Calcitonin. Proceedings of the symposium on thyrocalcitonin and the C cells, London, 1968, William Heinemann Medical Books Ltd.

34. Taylor, S., and Foster, G., editors: Calcitonin, 1969. Proceedings of the second international symposium, London, 1970, William Heinemann Medical Books Ltd.

35. Wasserman, R. H., and Taylor, A. N.: Some aspects of the intestinal absorption of calcium with special reference to vitamin D. In Comar, C. L., and Bronner, F., editors: Mineral metabolism, New York, 1969, Academic Press, Inc., vol. 3.

Original papers

36. Agus, Z. S., et al.: Mode of action of parathyroid hormone and cyclic adenosine 3′,5′-monophosphate on renal tubular phosphate reabsorption in the dog, J. Clin. Invest. **50:**617, 1971.

37. Aliapoulios, M. A., Goldhaber, P., and Munson, P. L.: Thyrocalcitonin inhibition of bone resorption induced by parathyroid hormone in tissue culture, Science **151:**330, 1966.

38. Arnaud, C., Rasmussen, H., and Anast, C.: Further studies on the interrelationship between parathyroid hormone and vitamin D, J. Clin. Invest. **45:**1955, 1966.

39. Arnstein, A. R., Frame, B., and Frost, H. M.: Recent progress in osteomalacia and rickets, Ann. Intern. Med. **67:**1296, 1967.

40. Au, W. Y. W., and Raisz, L. G.: Detection of thyrocalcitonin in rat serum by direct tissue culture bioassay, Endocrinology **86:**1231, 1970.

41. Aurbach, G. D.: Isolation of parathyroid hormone after extraction with phenol, J. Biol. Chem. **234:**3179, 1959.

42. Avioli, L. V., et al.: The metabolic fate of vitamin D3-3H in chronic renal failure, J. Clin. Invest. **47:**2239, 1968.

43. Bassett, C. A. L.: Biologic significance of piezoelectricity, Calcif. Tissue Res. **1:**252, 1968.

44. Belanger, L. F.: Osteolysis: an outlook on its mechanism and causation. In Gaillard, P. J., Talmage, R. V., and Budy, A. M., editors: The parathyroid glands, ultrastructure, secretion and function, Chicago, 1965, University of Chicago Press.

45. Bell, P. H., et al.: Purification and structure of porcine calcitonin-1, J. Am. Chem. Soc. **90:**2704, 1968.

46. Belsey, R., DeLuca, H. F., and Potts, J. T., Jr.: Competitive binding assay for vitamin D and 25-hydroxy vitamin D, J. Clin. Endocrinol. Metab. **33:**554, 1971.

47. Berlin, N. I., et al.: The application of multicompartmental analysis to problems of clinical medicine, Ann. Intern. Med. **68:**423, 1968.

48. Berson, S. A., and Yalow, R. S.: Immunochemical heterogeneity of parathyroid hormone in plasma, J. Clin. Endocrinol. Metab. **28:**1037, 1968.

49. Berson, S. A., et al.: Immunoassay of bovine and human parathyroid hormone, Proc. Natl. Acad. Sci. USA **49:**613, 1963.

50. Beutner, E. H., and Munson, P. L.: Time course of urinary excretion of inorganic phosphate by rats after parathyroidectomy and after injection of parathyroid extract, Endocrinology **66:**610, 1960.

51. Bijvoet, O. L. M., et al.: Natriuretic effect of calcitonin in man, N. Engl. J. Med. **284:**681, 1971.

52. Bikle, D., Tilney, L. G., and Porter, K. R.: Microtubules and pigment migration in the melanophores of Fundus heteroclitus L, Protoplasma **61:**322, 1966.

53. Birge, S. J., et al.: Study of calcium absorption in man: a kinetic analysis and physiologic model, J. Clin. Invest. **48:**1705, 1969.

54. Brewer, H. B., Jr., and Ronan, R.: Bovine parathyroid hormone: amino acid sequence, Proc. Natl. Acad. Sci. USA **67:**1862, 1970.

55. Broadus, A. E., et al.: Effects of glucagon on adenosine 3′,5′-monophosphate and guanosine 3′,5′-monophosphate in human plasma and urine, J. Clin. Invest. **49:**2337, 1970.

56. Brumbaugh, P. F., et al.: Filter assay for 1-alpha-25-dihydroxyvitamin D₃. Utilization of the hormone's target tissue chromatin receptor, Biochemistry **13:**4091, 1974.

56a. Canalis, E. M., et al.: Hormonal control of bone collagen synthesis in vitro: effect of insulin and glucagon, Endocrinology **100:**668, 1977.

57. Capen, C. C.: Fine structural alterations of parathyroid glands in response to experimental and spontaneous changes of calcium in extracellular fluids, Am. J. Med. **50:**598, 1971.

58. Care, A. D., et al.: The role of pancreozymincholecystokinin and structurally related compounds as calcitonin secretogogues, Endocrinology **89:**262, 1971.

59. Chabardes, D., et al.: PTH sensitive adenyl cyclase activity in different segments of the rabbit nephron, Pfluegers Arch. **354:**220, 1975.

60. Chabardes, D., et al.: Distribution of calcitonin-sensitive adenylate cyclase activity along the rabbit kidney tubule, Proc. Natl. Acad. Sci. USA **73:**3608, 1976.

61. Chase, L. R., and Aurbach, G. D.: Parathyroid function and the renal excretion of 3′,5′-adenylic acid, Proc. Natl. Acad. Sci. USA **58:**518, 1967.

62. Chase, L. R., and Aurbach, G. D.: The effect of parathyroid hormone on the concentration of adenosine 3′,5′-monophosphate in skeletal tissue in vitro, J. Biol. Chem. **245:**1520, 1970.

63. Chase, L. R., Melson, G. L., and Aurbach, G. D.: Pseudohypoparathyroidism: defective excretion of 3′,5′-AMP in response to parathyroid hormone, J. Clin. Invest. **48:**1832, 1969.

64. Collip, J. B.: The extraction of a parathyroid hormone which will prevent or control parathyroid tetany and which regulates the level of blood calcium, J. Biol. Chem. **63:**395, 1925.

65. Collip, J. B., and Clark, E. P.: Further studies on the physiological action of a parathyroid hormone, J. Biol. Chem. **64:**485, 1925.

66. Copp, D. H., and Davidson, A. G. F.: Direct humoral control of parathyroid function in the dog, Proc. Soc. Exp. Biol. Med. **107:**342, 1961.

67. Copp, D. H., Cockroft, D. W., and Kueh, H.: Calcitonin from ultimobranchial glands of dogfish and chickens, Science **158:**924, 1967.

68. Copp, D. H., Cockroft, D. W., and Kueh, Y.: Ultimo-

branchial origin of calcitonin. Hypocalcemic effect of extracts from chicken glands, Can. J. Physiol. Pharmacol. **45:**1095, 1967.

69. Copp, D. H., et al.: Evidence for calcitonin. A new hormone from the parathyroid that lowers blood calcium, Endocrinology **70:**638, 1962.

70. Cunliffe, W. J., et al.: A calcitonin-secreting thyroid carcinoma, Lancet **2:**63, 1968.

71. Curry, D. L., Bennett, L. L., and Grodsky, G. M.: Requirement for calcium ion in insulin secretion by the perfused rat pancreas, Am. J. Physiol. **214:**174, 1968.

72. Davies, B. M. A., Gordon, A. H., and Mussett, M. V.: A plasma calcium assay for parathyroid hormone, using parathyroidectomized rats, J. Physiol. **125:**383, 1954.

73. Deftos, L. J., Lee, M. R., and Potts, J. T., Jr.: A radioimmunoassay for thyrocalcitonin, Proc. Natl. Acad. Sci. USA **60:**293, 1968.

73a. Dietrich, J. W., et al.: Hormonal control of bone collagen synthesis in vitro: effects of parathyroid hormone and calcitonin, Endocrinology **98:**943, 1976.

74. Dowdle, E. B., Schachter, D., and Schenker, H.: Requirement for vitamin D for the active transport of calcium by the intestine, Am. J. Physiol. **198:**269, 1960.

75. Ebashi, S., and Endo, M.: Calcium ion and muscle contraction, Prog. Biophys. Mol. Biol. **18:**123, 1968.

76. Eisenstein, R., and Passavoy, M.: Actinomycin D inhibits parathyroid hormone and vitamin D activity, Proc. Soc. Exp. Biol. Med. **117:**77, 1964.

77. Eisman, J. A., et al.: A sensitive, precise, and convenient method for determination of 1,25-dihydroxyvitamin D in human plasma, Arch. Biochem. Biophys. **176:**235, 1976.

78. Finerman, G. A. M., Downing, S., and Rosenberg, L. E.: Amino acid transport in bone. II. Regulation of collagen synthesis by perturbation of proline transport, Biochim. Biophys. Acta **135:**1008, 1967.

79. Flanagan, B., and Nichols, G., Jr.: Metabolic studies of bone in vitro. V. Glucose metabolism and collagen biosynthesis, J. Biol. Chem. **239:**1261, 1964.

80. Foster, G. V., MacIntyre, I., and Pearse, A. G. E.: Calcitonin production and the mitochondrion-rich cells of the dog thyroid, Nature **203:**1029, 1964.

81. Foster, G. V., et al.: Thyroid origin of calcitonin, Nature **202:**1303, 1964.

82. Frankenhaeuser, B.: The effect of calcium on the myelinated nerve fibre, J. Physiol. **137:**245, 1957.

83. Fraser, D. R., and Kodicek, E.: Unique biosynthesis by kidney of a biologically active vitamin D metabolite, Nature **228:**764, 1970.

84. Friedman, J., and Raisz, L. G.: Thyrocalcitonin: inhibitor of bone resorption in tissue culture, Science **150:**1465, 1965.

85. Gaillard, P. J.: Observations on the effect of parathyroid products on explanted mouse limb bone rudiments. In Gaillard, P. J., Talmage, R. V., and Budy, A. M., editors: The parathyroid glands, ultrastructure, secretion, and function, Chicago, 1965, University of Chicago Press.

86. Gill, J. R., Jr., and Casper, A. G. T.: Depression of proximal tubular sodium reabsorption in the dog in response to renal beta adrenergic stimulation by isoproterenol, J. Clin. Invest. **50:**112, 1971.

87. Goodman, D. G. P., et al.: Cyclic adenosine 3′,5′-monophosphate-stimulated phosphorylation of isolated neurotubule subunits, Proc. Natl. Acad. Sci. USA **67:**652, 1970.

88. Gray, T. K., and Munson, P. L.: Thyrocalcitonin: evidence for physiological function, Science **166:**512, 1969.

89. Greenwald, I., and Gross, J.: The effect of the administration of a potent parathyroid extract upon the excretion of nitrogen, phosphorus, calcium, and magnesium with some remarks on the solubility of calcium phosphate in serum and on the pathogenesis of tetany, J. Biol. Chem. **66:**217, 1925.

90. Guttmann, S. T., et al.: Synthese des thyreocalcitonins, Helv. Chim. Acta **51:**1155, 1968.

91. Habener, J. F., et al.: Parathyroid hormone: secretion and metabolism in vivo, Proc. Natl. Acad. Sci. USA **68:**2986, 1971.

92. Haddad, J. G., and Stamp, T. C. B.: Circulating 25-hydroxyvitamin D in man, Am. J. Med. **57:**57, 1974.

93. Hahn, T. J., Downing, S. J., and Phang, J. M.: Insulin effect on amino acid transport in bone: dependence on protein synthesis and Na^+, Am. J. Physiol. **220:**1717, 1971.

94. Hanson, A. M.: The hormone of the parathyroid gland, Proc. Soc. Exp. Biol. Med. **22:**560, 1925.

95. Harris, W. H., and Heaney, R. P.: Skeletal renewal and metabolic bone disease, N. Engl. J. Med. **280:**193, 253, 303, 1969.

96. Heersche, J. N. M., Marcus, R., and Aurbach, G. D.: Calcitonin and the formation of 3′,5′-AMP in bone and kidney, Endocrinology **94:**241, 1974.

97. Hiatt, H. H., and Thompson, D. D.: The effects of parathyroid extract on renal function in man, J. Clin. Invest. **36:**557, 1957.

98. Hirsch, P. F., Gauthier, G. F., and Munson, P. L.: Thyroid hypocalcemic principle and recurrent laryngeal nerve injury as factors affecting the response to parathyroidectomy in rats, Endocrinology **73:**244, 1963.

99. Hirsch, P. F., Voelkel, E. F., and Munson, P. L.: Thyrocalcitonin: hypocalcemic hypophosphatemic principle of the thyroid gland, Science **146:**412, 1964.

100. Holick, M. F., Schnoes, H. K., and DeLuca, H. F.: Identification of 1,25-dihydroxycholecalciferol, a form of vitamin D_3 metabolically active in the intestine, Proc. Natl. Acad. Sci. USA **68:**803, 1971.

101. Holick, M. F., et al.: Isolation and identification of 1,25-dihydroxycholecalciferol. A metabolite of vitamin D active in intestine, Biochemistry **10:**2799, 1971.

102. Johnson, L. C.: Kinetics of skeletal remodeling. In Bergsma, D., editor: Structural organization of the skeleton: a symposium. Birth defects. Original article series, New York, 1966. The National Foundation, vol. 2.

103. Jowsey, J.: Quantitative microradiography, Am. J. Med. **40:**485, 1966.

104. Kales, A. N., and Phang, J. M.: Dietary calcium perturbation in patients with abnormal calcium deposition, J. Clin. Endocrinol. Metab. **31:**204, 1970.

105. Kales, A. N., and Phang, J. M.: Effect of divided calcium intake on calcium metabolism, J. Clin. Endocrinol. Metab. **32:**83, 1971.

106. Keutmann, H. T., et al.: Isolation and characterization of the bovine parathyroid isohormones, Biochemistry **10:**2779, 1971.

107. Kimberg, D. V., Schachter, D., and Schenker, H.: Active transport of calcium by intestine: effects of dietary calcium, Am. J. Physiol. **200:**1256, 1961.

108. Klein, D. C., and Raisz, L. G.: Prostaglandins: stimulation of bone resorption in tissue culture, Endocrinology **86:**1436, 1970.

109. Langan, T. A.: Histone phosphorylation: stimulation by adenosine 3′,5′-monophosphate, Science **162:**579, 1968.

110. Lawson, D. E. M., et al.: Identification of 1,25-dihydroxycholecalciferol, a new kidney hormone—con-

trolling calcium metabolism, Nature **230:**228, 1971.

111. Lemann, J., Jr., Piering, W. F., and Lennon, E. J.: Possible role of carbohydrate-induced calciuria in calcium oxalate kidney-stone formation, N. Engl. J. Med. **280:**232, 1969.

112. Malawista, S. E.: On the action of colchicine, J. Exp. Med. **122:**361, 1965.

113. Marcus, R., and Aurbach, G. D.: Bioassay of parathyroid hormone in vitro with a stable preparation of adenyl cyclase from rat kidney, Endocrinology **85:** 801, 1969.

114. Marcus, R., Heersche, J. N. M., and Aurbach, G. D.: Effects of calcitonin on formation of 3′,5′-AMP in bone and kidney, Program of the fifty-third meeting of the Endocrine Society, Philadelphia, 1971, J. B. Lippincott Co. (abstract 52).

115. Martin, T. J., Melick, R. A., and deLuise, M.: Metabolism of parathyroid hormone: degradation of ^{125}I-labeled hormone by the kidney enzyme, Biochem. J. **111:**509, 1969.

116. Marx, S. J., Woodard, C. J., and Aurbach, G. D.: Calcitonin receptors of kidney and bone, Science **178:** 999, 1972.

117. McCollum, E. V., et al.: Studies on experimental rickets, J. Biol. Chem. **51:**5, 1922.

118. Mears, D. C.: Effects of parathyroid hormone and thyrocalcitonin on the membrane potential of osteoclasts, Endocrinology **88:**1021, 1971.

119. Melancon, M. J., Jr., and DeLuca, H. F.: Vitamin D stimulation of calcium-dependent adenosine triphosphatase in chick intestinal brush borders, Biochemistry **9:**1658, 1970.

120. Moore, E. W.: Ionized calcium in normal serum, ultrafiltrates and whole blood determined by ion-exchange electrodes, J. Clin. Invest. **49:**318, 1970.

121. Munson, P. L.: Studies on the role of the parathyroid in calcium and phosphorus metabolism, Ann. N.Y. Acad. Sci. **60:**776, 1955.

122. Murad, F., Brewer, H. B., Jr., and Vaughan, M.: Effect of thyrocalcitonin on adenosine 3′,5′-cyclic phosphate formation by rat kidney and bone, Proc. Natl. Acad. Sci. USA **65:**446, 1970.

123. Neer, R. M., Tregear, G. W., and Potts, J. T., Jr.: Renal effects of native parathyroid hormone and synthetic biologically active fragments in pseudohypoparathyroidism and hypoparathyroidism, J. Clin. Endocrinol. Metab. **38:**420, 1977.

124. Neer, R., et al.: Multicompartmental analysis of calcium kinetics in normal adult males, J. Clin. Invest. **46:**1364, 1967.

125. Neher, R., et al.: Thyrocalcitonin. II. Struktur von alpha-Thyrocalcitonin, Helv. Chim. Acta **51:**917, 1968.

126. Ney, R. L., et al.: Actions of parathyroid hormone in the vitamin D-deficient dog, J. Clin. Invest. **44:**2003, 1965.

127. Niall, H., et al.: The amino acid sequence of bovine parathyroid hormone, Hoppe-Seyler's Z. Physiol. Chem. **351:**1586, 1970.

128. Nonidez, J. F.: The origin of the "parafollicular" cell, a second epithelial component of the thyroid gland of the dog, Am. J. Anat. **49:**479, 1931.

129. Norman, A. W., et al.: 1,25-dihydroxycholecalciferol: identification of the proposed active form of vitamin D_3 in the intestine, Science **173:**51, 1971.

130. Nunez, E. A., et al.: Seasonal changes in the fine structure of the basal granular cells of the bat thyroid, J. Cell Sci. **2:**401, 1967.

131. Owen, M.: The origin of bone cells, Int. Rev. Cytol. **28:**213, 1970.

132. Patt, H. M., and Luckhardt, A. B.: Relationship of a low blood calcium to parathyroid secretion, Endocrinology **31:**384, 1942.

133. Peacock, M., Robertson, W. G., and Nordin, B. E. C.: Relation between serum and urinary calcium with particular reference to parathyroid activity, Lancet **1:**384, 1969.

134. Pearse, A. G. E., and Carvalheira, A. F.: Cytochemical evidence for an ultimobranchial origin of rodent thyroid C cells, Nature **214:**929, 1967.

135. Phang, J. M., et al.: Dietary perturbation of calcium metabolism in normal man: compartmental analysis, J. Clin. Invest. **48:**67, 1969.

136. Phang, J. M., et al.: Compartmental analysis of collagen synthesis in fetal rat calvaria. I. Perturbations of proline transport, Biochim. Biophys. Acta **230:**146, 1971.

137. Potts, J. T., Jr., et al.: The amino acid sequence of porcine thyrocalcitonin, Proc. Natl. Acad. Sci. USA **59:**1321, 1968.

138. Pullman, T. N., et al.: Direct renal action of a purified parathyroid extract, Endocrinology **67:**570, 1960.

139. Putter, I., et al.: The isolation and properties of thyrocalcitonin, J. Am. Chem. Soc. **89:**5301, 1967.

140. Rasmussen, H.: The influence of parathyroid function upon the transport of calcium in isolated sacs of rat small intestine, Endocrinology **65:**517, 1959.

141. Rasmussen, H., and Craig, L. C.: Isolation and characterization of bovine parathyroid hormone, J. Biol. Chem. **236:**759, 1961.

142. Rasmussen, H., Anast, C., and Arnaud, C.: Thyrocalcitonin, EGTA, and urinary electrolyte excretion, J. Clin. Invest. **46:**746, 1967.

143. Reiss, E., et al.: The role of phosphate in the secretion of parathyroid hormone in man, J. Clin. Invest. **49:** 2146, 1970.

144. Rittel, W., et al.: Thyrocalcitonin III. Die Synthese des alpha-Thyrocalcitonins, Helv. Chim. Acta **51:** 924, 1968.

145. Robinson, R. A., and Cameron, D. A.: Bone. In Kurtz, S. M., editor: Electron microscopic anatomy, New York, 1964, Academic Press, Inc.

146. Samiy, A. H. E., et al.: Interrelation between renal transport systems of magnesium and calcium, Am. J. Physiol. **198:**599, 1960.

147. Schachter, D., and Rosen, S. M.: Active transport of ^{45}Ca by the small intestine and its dependence on vitamin D, Am. J. Physiol. **196:**357, 1959.

148. Schachter, D., Dowdle, E. B., and Schenker, H.: Active transport of calcium by the small intestine of the rat, Am. J. Physiol. **198:**263, 1960.

149. Scott, B. D., and Pease, D. C.: Electron microscopy of the epiphyseal apparatus, Anat. Rec. **126:**465, 1956.

150. Shah, B. G., and Draper, H. H.: Depression of calcium absorption in parathyroidectomized rats, Am. J. Physiol. **211:**963, 1966.

151. Sherwood, L. M., Rodman, J. S., and Lundberg, W. B.: Evidence for a precursor to circulating parathyroid hormone, Proc. Natl. Acad. Sci. USA **67:**1631, 1970.

152. Sherwood, L. M., et al.: Evaluation by radioimmunoassay of factors controlling the secretion of parathyroid hormone. I. Intravenous infusions of calcium and ethylenediamine tetraacetic acid in the cow and goat, Nature **209:**52, 1966.

153. Sherwood, L. M., et al.: Regulation of parathyroid hor-

mone secretion: proportional control by calcium, lack of effect of phosphate, Endocrinology **83:**1043, 1968.

154. Sonnenblick, E. H.: Force-velocity relations in mammalian heart muscle, Am. J. Physiol. **202:**931, 1962.

155. Steiner, A. L., Goodman, A. D., and Powers, S. R.: Study of a kindred with pheochromocytoma, medullary thyroid carcinoma, hyperparathyroidism and Cushing's disease: multiple endocrine neoplasia, type 2, Medicine **47:**371, 1968.

156. Talmage, R. V., and Kraintz, F. W.: Progressive changes in renal phosphate and calcium excretion in rats following parathyroidectomy or parathyroid administration, Proc. Soc. Exp. Biol. Med. **87:**263, 1954.

157. Tauber, S. D.: The ultimobranchial origin of thyrocalcitonin, Proc. Natl. Acad. Sci. USA **58:**1684, 1967.

158. Trummel, C. L., et al.: 25-Hydroxycholecalciferol: stimulation of bone resorption in tissue culture, Science **163:**1450, 1969.

159. Vaes, G.: Parathyroid hormonelike action of N6-2′-0-dibutyryl-adenosine-3′5′ (cyclic) monophosphate on bone explants in tissue culture, Nature **219:**939, 1968.

160. Vassale, G., and Generali, F.: Fonction parathyroidienne et fonction thyroidienne, Arch. Ital. Biol. **33:**154, 1900.

161. Walling, M. W., and Rothman, S. S.: Apparent increase in carrier affinity for intestinal calcium transport following dietary calcium restriction, J. Biol. Chem. **245:**5007, 1970.

162. Walser, M.: Calcium clearance as a function of sodium clearance in the dog, Am. J. Physiol. **200:**1099, 1961.

163. Wells, H., and Lloyd, W.: Hypercalcemic and hypophosphatemic effects of dibutyryl cyclic AMP in rats after parathyroidectomy, Endocrinology **84:**861, 1969.

64

F. EUGENE YATES, DONALD J. MARSH, and JANICE W. MARAN

The adrenal cortex

Without our adrenal cortices, we die; blood volume, blood pressure, and blood sugar levels fall, muscles weaken, and resistance to effects of noxious stimuli collapses. This chapter will discuss the adrenal hormones that ordinarily prevent such disasters, and how they act.

The adrenal cortex functions as two different endocrine glands. One secretes aldosterone, which acts on salt-transporting epithelia such as those of the kidney and the sweat glands. This hormone leads to sodium chloride retention, potassium loss, hydrogen ion excretion, and, indirectly, to water retention. It helps maintain extracellular fluid volume. The other part of the adrenal cortex secretes chiefly cortisol (and/or corticosterone, depending on the species). Cortisol helps to sustain blood sugar and blood pressure, and it protects against "stresses," as explained subsequently. Aldosterone is informally called a "mineralocorticoid"; cortisol is called a "glucocorticoid." These terms are not precise, but they are convenient, and we shall accept them here. A general term for both kinds of hormones is "corticosteroid."

MORPHOLOGY OF THE ADRENAL CORTEX

In most vertebrates, including man, the adrenal glands are small, paired structures lying on either side of the midline of the abdominal cavity above the kidneys. Each gland consists of an outer cortex of mesodermal origin and an inner medulla of ectodermal origin. The cortex secretes substances moderately soluble in lipids, of a chemical class known as steroids, and is not controlled by secretomotor nerves. The medulla secretes the water-soluble substances epinpehrine and norepinephrine and is innervated by preganglionic, cholinergic, sympathetic nerve fibers. The medullary cells are properly considered to be part of the postganglionic sympathetic nervous system and are discussed in detail in Chapter 33. In the normal adult human being a single adrenal gland (including the medulla) weighs about 4 gm. The medulla represents approximately 10% of the weight.

The mature adrenal cortex consists of three morphologically distinct zones[108] (Fig. 64-1). These zones differ in enzymatic content, histologic and ultrastructural morphology,[56,141] mitotic rates, lipid distribution, and functional activity. The outer layer is the zona glomerulosa, the middle layer is the zona fasciculata, and the inner layer, next to the medullary tissue, is the zona reticularis. The cells of all three layers are arranged in cordlike arrays surrounded by basement membranes. The cords are especially prominent in the fasciculata. Capillaries run between neighboring basement membranes. The capillary endothelium is discontinuous, and there is a well defined subendothelial space.

SECRETORY PRODUCTS OF THE ADRENAL CORTEX

The first truly effective crude cortical extracts were prepared in the years 1928 to 1930.[72,142,164] By 1948 several corticosteroids were available for clinical use. The structures of the three chief secretory products of the adrenal cortex are shown in Fig. 64-2. Their informal names and the correct chemical name for cortisol are shown in the figure. At physiologic concentrations in human beings, cortisol and corticosterone are glucocorticoids, and aldosterone is the only effective mineralocorticoid. The glucocorticoids are secreted by the zona fasciculata and zona reticularis; aldosterone is secreted by the zona glomerulosa.

In addition to secreting two glucocorticoids and one mineralocorticoid, the human adrenal cortex also leaks variable amounts of the biosynthetic precursors of the three end products (especially progesterone, 11-deoxycorticosterone, and 11-deoxycortisol). The adrenal cortex also secretes significant amounts of androgenic substances (but not testosterone), especially dehydroepiandrosterone and its sulfate ester. These hormones are not strongly androgenic in men,

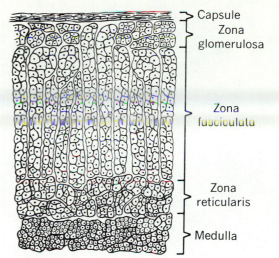

Fig. 64-1. Diagrammatic representation of histologic appearance of human adrenal cortex.

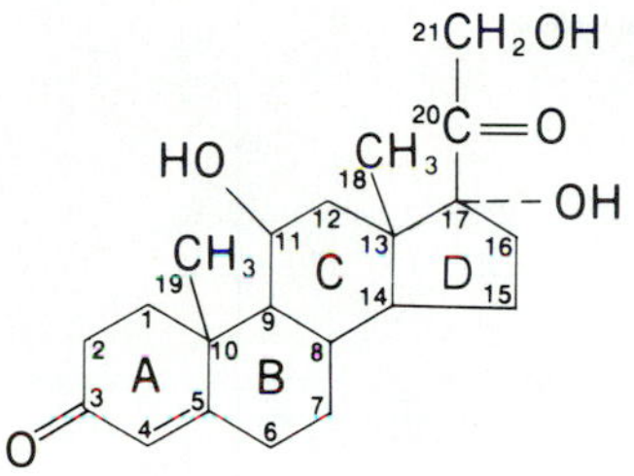

Cortisol
Hydrocortisone
Compound F (Kendall)
Δ^4-Pregnene-11β, 17α, 21-triol-3,20-dione

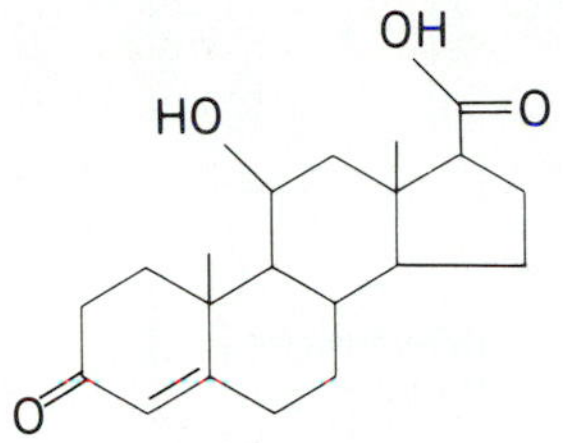

Corticosterone
Compound B (Reichstein)

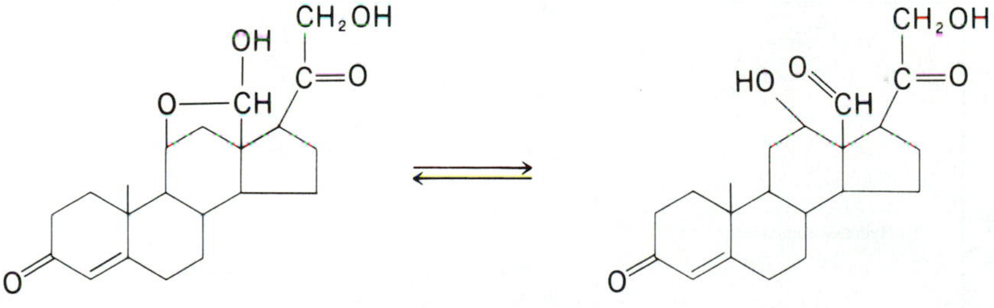

Aldosterone

Fig. 64-2. Structure of three chief secretory products of human adrenal cortex. All adrenal steroids are derivatives of the 21-carbon, fully saturated compound pregnane. Numbering of carbons is indicated in structure shown for cortisol. Methyl groups designated as carbons 18 and 19 and hydroxyl group attached at carbon 11 would project toward reader from plane of paper. The four rings lie roughly in plane of paper, and substituents that project away from reader are indicated with dashed lines, as is shown for hydroxyl group at carbon 17. The 2-carbon side chain attached at carbon 17 projects toward reader. All substituents attached to molecule on side toward reader are designated beta, and those that project away are designated alpha. Four names are given for cortisol molecule. In structure shown for corticosterone, methyl groups at positions 18 and 19 are indicated merely by lines showing bonds, as is customary. Sometimes the carbons at positions 20 and 21 are also suppressed. (See notation in Fig. 64-3.) Aldosterone in solution exists in equilibrium between 18-aldehyde and 11,18-hemiacetal forms.

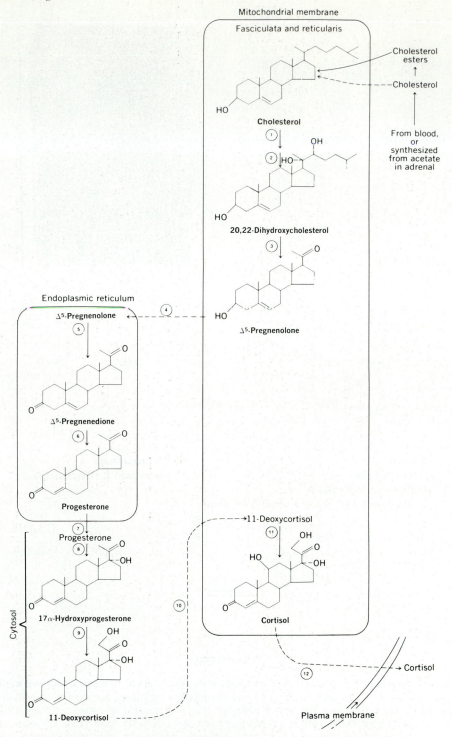

Fig. 64-3. Main sequence of cortisol biosynthesis. Starting compound, cholesterol, can be accumulated from blood or synthesized in adrenal cortex from acetate. Double hydroxylation of cholesterol and its side-chain cleavage to produce pregnenolone, as well as hydroxylation of 11 position, all occur within mitochondrial membranes. Oxidation at 3 position and A-B ring isomerization occur in microsomes. Hydroxylations at 21 and 17 positions occur in cytosol. Sequence of reactions is not obligatory, and hydroxylations may occur in order different from that shown. Corticosteroidogenesis involves at least 12 transfer and transformation steps, as numbered.

but they cause slight masculinization of post-menopausal women whose ovaries have stopped secreting estrogens, and they may support the libido of normal women. It has become apparent that the weak androgens secreted from the adrenal support female sexual behavior in the sub-human primates.[53,170] Replacement therapy in adrenalectomized human beings requires administration of a glucocorticoid and a mineralocorticoid in men and both of these hormones plus small amounts of androgens in women if well-being is to be fully restored.

The normal human adrenal cortex does not secrete physiologically effective amounts of estrogenic substances, but after chronic over-stimulation by stress, or in the case of certain tumors, the adrenal gland can produce estrogens or precursors that are converted to estrogens elsewhere. Feminization of males can then occur.

Biosynthesis of corticosteroids (corticosteroidogenesis)

The zona glomerulosa of the human adrenal cortex synthesizes aldosterone and corticosterone, whereas the zona fasciculata and zona reticularis synthesize mainly cortisol, but also some corticosterone. The starting material for the biosynthesis of corticosteroids is cholesterol. Cholesterol is stored in the adrenal cortex in both free and an esterified form. The adrenal cortex can synthesize cholesterol from acetate. It can also acquire cholesterol from blood, where it appears from dietary sources, or from the liver, in which cholesterol may be synthesized from acetate.

The main sequence of reactions leading from cholesterol to cortisol is shown in Fig. 64-3. The corticosterone pathway is the same up to progesterone, but then the reaction sequence is 9 to 12, omitting reaction 8. The various reactions include hydroxylations, side-chain cleavage, ring isomerization (migration of the double bond from ring B to ring A), and dehydrogenation of the C_3 hydroxyl group. The sequence of hydroxylations at C_{17}, C_{21}, and C_{11} usually goes in that order, as shown, but the order is not obligatory. Consequently, the chain branches and is more complex than is shown. It will be noted that during biosynthesis the steroid skeleton may be shuttled from mitochondrion to endoplasmic reticulum to cytosol and back to the mitochondrion. The exact subcellular location of the various reactions is not certain because of the problem of cross contamination of components during differential ultracentrifugation. It is possible that all the reactions can occur in the mitochondrion.

The biosynthesis of corticosterone and aldosterone is shown in Fig. 64-4. The pathway is similar to that for cortisol, except that 17-hydroxylation does not take place. The C_{18} position of corticosterone is oxidized into an aldehyde that then forms a hemiacetal by reacting with the C_{11} hydroxyl group, creating aldosterone.

Stimulation of corticosteroidogenesis

Synthesis and secretion of aldosterone from the zona glomerulosa are stimulated by multiple factors, including angiotensin II, potassium ion, and (transiently) adrenocorticotropic hormone (corticotropin, ACTH) from the adenohypophysis (anterior pituitary gland). Sodium ions are inhibitory, and therefore low sodium levels are often thought of as being a stimulus (despite the illogicalness of regarding the absence or decrease of an agent as a "stimulus"). Control of aldosterone production is discussed in detail in the latter part of this chapter.

Synthesis and secretion of cortisol and corticosterone are stimulated by ACTH, and the stimulation persists as long as ACTH is present. In the absence of ACTH, the adrenal cortex shows extreme atrophy of the zona fasciculata and zona reticularis, and cortisol and corticosterone secretion decreases to 0. (If there is any residual secretion of these hormones in the absence of ACTH, it is too low to produce physiologic effects.) After more than 30 years of intense study, it still is not clear how ACTH stimulates production and secretion of glucocorticoids. The biochemical effects of ACTH on the adrenal cortex are numerous and difficult to interpret. A labile protein is formed rapidly during ACTH stimulation, even though there is new mRNA production. Adenylate cyclase, located on adrenal cell membranes, is activated. Glucose metabolism is enhanced, cytochrome P-450 levels increase, ascorbic acid levels decrease, and cholesterol levels decrease. Over a longer period the adrenal cortex grows (hypertrophies) in response to ACTH, so DNA synthesis is increased, too.

The great problem in analyzing the effects just listed, and the many others not mentioned, is to decide which ones are involved in minute-to-minute regulation of rates of glucocorticoid secretion. Because the answer is not yet clear, we shall not elaborate on the biochemical coupling of ACTH to corticosteroidogenesis, even though the phenomenon is an essential component in the operation of the whole adrenal glucocorticoid system. The reader looking for information and theories about actions of ACTH on the adrenal

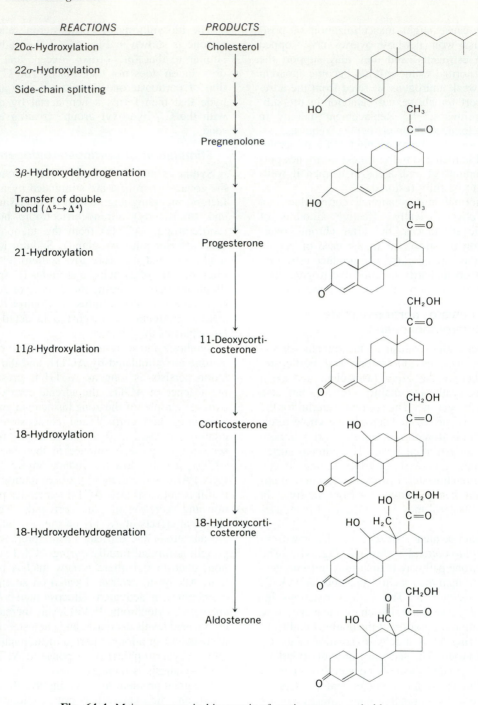

REACTIONS	PRODUCTS
20α-Hydroxylation	Cholesterol
22σ-Hydroxylation	
Side-chain splitting	
	Pregnenolone
3β-Hydroxydehydrogenation	
Transfer of double bond (Δ⁵→Δ⁴)	
	Progesterone
21-Hydroxylation	
	11-Deoxycorti-costerone
11β-Hydroxylation	
	Corticosterone
18-Hydroxylation	
	18-Hydroxycorti-costerone
18-Hydroxydehydrogenation	
	Aldosterone

Fig. 64-4. Main sequence in biogenesis of corticosterone and aldosterone.

fasciculata and reticularis should consult the review by Haynes.[73]

The adrenal cortex secretes some progesterone and contains prolactin and luteinizing hormone–releasing hormone (LH-RH). In female rats there is a coupling between the reproductive system and the adrenal production of progesterone,[134] but it is not known whether adrenal secretion of progesterone is physiologically variable and significant in women. The functions of prolactin and LH-RH in the adrenal cortex are unknown.

Table 64-1. Steady-state distribution and metabolism of corticosteroids in humans (virtual single-compartment model without binding)

	Cortisol	Corticosterone	Aldosterone
Total plasma concentration (c) (μg/L)			
AM	300	30	
PM	30	3	
mean ($\bar{c}$)	150	12	0.15
Mean extra-adrenal pool (μg)—$V_D \times \bar{c}$	2,250	100	7
Half-life ($t_{1/2}$) (min)	90	30	30
Turnover time (time constant, τ) (min)—$\tau = \dfrac{t_{1/2}}{0.693}$	130	43	43
Virtual volume of distribution (at plasma concentration) (V_D) (L)	15	9	47
Metabolic clearance rate (MCR) (L/min)—MCR = V_D/τ	0.1	0.2	1.1
24 hr secretion rate (μg/day)—MCR $\times \bar{c} \times$ 1440	21,600	3,500	250

Dynamics of the cortisol secretory response to adrenocorticotropic hormone

ACTH stimulates corticosteroidogenesis and cortisol and corticosterone secretion rates within 1 or 2 min after arrival at the adrenal gland. If the ACTH concentration in adrenal arterial blood is suddenly increased and then held constant at sub-maximal levels, the glucocorticoid secretion rates overshoot to reach a peak within 15 min and then settle to a lower plateau value that lasts as long as the ACTH stimulus is maintained.[22,117]

When ACTH stimulation of the adrenal cortex is diminished or stopped, glucocorticoid secretion declines approximately exponentially over a period of time that depends on the extent of prior stimulation of the gland. If the glucocorticoid secretion rate has been stimulated to maximal levels, the start of the decline may be delayed many minutes, but if the stimulation has been mild, glucocorticoid secretion rate falls almost immediately. The time required to reach new steady-state levels for both "on" and "off" glucocorticoid secretory responses to steady changes in ACTH levels is in the range of 40 min or less.

DISTRIBUTION, BINDING, AND DISPOSAL

Any substance secreted or infused into blood may undergo the four processes of (1) distribution (intravascular and extravascular), (2) excretion (in feces, urine, sweat, tears, saliva, or milk), (3) metabolism (chemical transformation), and (4) binding to plasma macromolecules or at cellular sites. For the purposes of a rough characterization of the kinetics of corticosteroid pro-

cessing after the hormones are secreted by the adrenal gland, it is convenient to consider distribution as if it occurred into a single, homogenous body fluid compartment and removal as if it were accomplished by a single inactivating-excretion process. In this approximation, protein binding of the steroids is not explicitly considered. Because none of these assumptions are valid the resulting information is inaccurate, but it does give a compact "order-of-magnitude" description. Table 64-1 shows the results of applying these assumptions, plus the additional assumption of a steady state between secretion and removal, to data from experiments involving injections of radioactively labeled corticosteroids into human beings.[131] Similar data can be obtained from infusions of the natural corticosteroids, but it is necessary to confine the infusions to rates within the physiologic range of adrenocortical secretion.[128]

Units of concentrations

It should be noted that the data in Table 64-1 present corticosteroid concentrations in the unit micrograms per liter (μg/L), but in current literature the trend is to give the concentrations as nanograms (ng) or picograms (pg) per milliliter. In the older literature and in the clinic the common unit is micrograms per 100 ml (μg/dl or μg%), although this is not the most convenient chemical or thermodynamic unit. We shall accept this common unit because it is firmly entrenched in the habits of physicians.

Daily turnover rates

In normal human beings the two adrenal cortices have a combined total turnover rate of approximately 20 mg of cortisol, 2 mg of corticosterone, and 0.2 mg of aldosterone each day. The figures vary widely from person to person, and all the numbers shown in Table 64-1 are merely representative of what normal values might be. As the data indicate, cortisol and corticosterone concentrations in human plasma are higher in the morning than in the evening. This circadian variation is discussed later. The 24 hr average concentration of cortisol is about 10-fold higher than that of corticosterone and 1,000-fold higher than that of aldosterone.

Details of the computations necessary to create Table 64-1 from primary data and a more elaborate mathematical treatment of the general problem of characterizing steroid dynamics can be found elsewhere.[26,165]

Some of the considerations involved in the more realistic models are given here.

Metabolism and excretion of corticosteroids

The loss of chemically unaltered corticosteroids through bile (feces), urine, sweat, or other bodily secretions is extremely low. A slight enterohepatic recirculation of the trace amounts of corticosteroids that escape metabolism in the liver and gut and appear in the bile does occur. The removal of these hormones consists primarily of chemical transformation followed by biliary or urinary excretion of the metabolites. The largest fraction, perhaps 70% or more, of the chemical transformation and inactivation of these substances is carried out in the liver, but other organs of the splanchnic bed that drain into the portal vein, and the kidney may also metabolize appreciable amounts of cortisol.[32,114]

The main sequence of hepatic reactions involved in the transformations is shown in Fig. 64-5. The double bond in ring A is reduced first and then the C_3 ketone follows. C_{20} reduction occurs, and the 2-carbon side chain attached to C_{17} can then be removed. Reduced corticosteroids may be conjugated through etherification or esterification with glucuronic or sulfuric acid. The usual process is conjugation with glucuronic acid at the C_3 hydroxyl. The order of the reactions is not obligatory, except that ring A reduction must precede C_3 ketone reduction. The total number of possible metabolites of cortisol is very high (Fig. 64-5).

Corticosteroids can be reduced by extrahepatic tissues, especially at the C_{20} carbonyl group. The various reduction and conjugation reactions all serve to increase the water solubility of corticosteroids and to facilitate their excretion in the urine. Any one of the listed reactions is sufficient to reduce or abolish the glucocorticoid activity of these hormones.

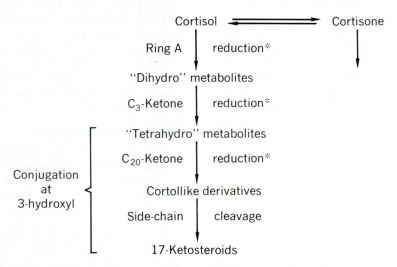

Fig. 64-5. Main sequence of inactivation of cortisol. Reductions of double bond in ring A and of ketones at C_3 and C_{20} each create an asymmetric carbon atom (i.e., carbon atoms with four different substituents on four covalent bonds) indicated by asterisk in figure. Such asymmetric carbon atoms manifest stereoisomerism; consequently, the number of metabolites of cortisol is very large. Cortisol can also be converted to cortisone, which in turn can undergo same sequence of reactions. Thus the total number of cortisol metabolites possible before conjugation is at least 37. Conjugation usually occurs by esterification with either glucuronic or sulfuric acid.

Protein binding of corticosteroids

The plasma macromolecule albumin has a single, high-affinity binding site for corticosteroids and a larger number of secondary sites, perhaps 20, with only weak affinity. All three corticosteroids shown in Table 64-1 are bound to albumin. The association constant for cortisol at 37° C is approximately 10^3 L/mole. In addition to binding on albumin, cortisol and corticosterone are also bound to a globulin that is a specific transport protein for these hormones. The globulin has been called transcortin, or corticosteroid-binding globulin (CBG), and it has a single, strong binding site for corticosteroids. The association constant at 37° C for cortisol is approximately 10^7 to 10^8 L/mole. Since the affinity of transcortin for cortisol is so much greater than that of albumin, at normal plasma concentrations, cortisol is largely adsorbed on transcortin. Cortisol and corticosterone compete for the strong binding sites on both albumin and transcortin.

The relationship between the concentration of unbound cortisol in plasma and its total concentration, as affected by its binding to the two plasma macromolecules, is shown in Fig. 64-6. The binding sites on transcortin saturate at plasma concentrations of approximately 20 μg/100 ml. Beyond that concentration, additional binding is accomplished only by albumin, which has a capacity that cannot be saturated within physiologic levels of plasma cortisol.

It can be seen from Fig. 64-6 that less than 10% of plasma cortisol is unbound in the basal,

resting physiologic range of variation of cortisol concentration, between 0 and 30 μg/100 ml. Even at maximum physiologic concentration (about 70 μg/100 ml), reached during a stress response, cortisol is still less than 20% unbound. As expected, cortisol and corticosterone binding in plasma greatly influences the transfer and distribution of the hormones from the circulation into extravascular fluid compartments.

Effects of protein binding on biologic activity and distribution dynamics

At most target tissues, including those that inactivate corticosteroids, only the unbound corticosteroids are recognized at receptors. However, at some receptors in the liver, binding may not interfere with the activity,[92,143] and this fact has been interpreted as representing an important aspect of information flow in the glucocorticoid system.[92] Because the plasma concentration of transcortin is itself a physiologic variable (that of albumin is more closely regulated and less variable), it follows that the protein binding of cortisol has clinical implications of some importance. The various clinical methods for measuring cortisol in plasma yield the total concentration of the hormone. The measurement of the unbound fraction is still a research technique. Yet it can be seen that two different patients with the same total concentration of cortisol could have very different levels of unbound cortisol in their plasmas and therefore different degrees of stimulation at most corticosteroid receptors. Conversely, in pregnant women transcortin levels are greatly increased by estrogens,[26] and total cortisol levels are higher than normal, but the concentration of unbound cortisol increases relatively less than does the total level, and so pregnancy is not a general hyperadrenocortical state, despite the high total concentration of cortisol. It is an interesting possibility that the function of the increased transcortin levels in pregnancy is to provide intensified, selective glucocorticoid stimulation of the hepatic receptors that read total concentrations, without overstimulating all other cortisol-sensitive tissues that read only unbound cortisol.[26,92]

Compartment models of cortisol dynamics

The effects of binding of corticosteroids are so striking that a single-compartment model fails to give an adequate description of the distribution processes. For a precise description, at least a four-compartment model is necessary. The compartments are (1) plasma water, (2) interstitial

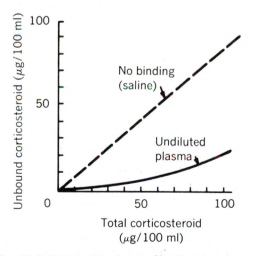

Fig. 64-6. Relationship between unbound and total cortisol levels in human plasma. Data are obtained by limited ultrafiltration of undiluted plasma. In control experiments, physiologic amounts of corticosteroids are added to protein-free medium such as saline solution, in which no binding occurs.

fluid, (3) transcortin-bound component in plasma, and (4) albumin-bound component in plasma. Computer simulations of complete cortisol distribution, binding, and disposal processes have been reported.[9,66,181]

GLUCOCORTICOIDS

In 1949 Hench and his colleagues made the astonishing discovery that after treatment with a glucocorticoid, patients severely crippled with rheumatoid arthritis underwent a remission of the disease and were able to move joints that had previously been stiff for many years.[77] The amelioration of this disease, previously almost untreatable, led to exploration of the pharmacologic, physiologic, and biochemical aspects of adrenocortical hormone actions. It soon became clear that patients or animals with low glucocorticoid levels were extremely vulnerable to complications in surgical procedures, infections, trauma, and other "stresses" and that many inflammatory and immunologic responses could be suppressed by giving glucocorticoids.

STRESS

The concept of stress is as elusive as it is important. In 1936, Selye noted that animals exposed to a wide variety of seemingly nonspecific but noxious agents or treatments underwent a somewhat stereotypic pattern of physiologic changes that included adrenocortical hypertrophy.[155] He designated the response pattern as the "general adaptation syndrome,"[156] and the stimuli that could provoke the syndrome became familiarly known as "stresses" or "stressors." In the 1950s it was discovered that plasma levels of glucocorticoids rise significantly during many stress responses. For a full account of the stress concept the works of Selye[157,158] and Mason[112,113] should be consulted. The strongest stresses (i.e., those that provoke the highest levels of glucocorticoids in blood) are of two kinds: (1) certain emotional states (particularly those involving a "failure of expectancy"[104]; and (2) very severe trauma, such as extensive burns with infection of the wounds. There is no close correlation, however, between emotional state and adrenal state.[125,162] Patients who have inadequate adrenocortical function and who are receiving baseline corticosteroid replacement therapy need increased doses of glucocorticoids when they encounter other diseases, accidental trauma, emotional distress, or surgical procedures.

GLUCOCORTICOID RECEPTORS

To produce their effects at target cells, hormones must first be recognized. Recognition occurs through binding of hormones to protein receptor macromolecules that may be located on the cell surface, in the cytosol, in membranes of the cell interior, or in the nucleus. In Fig. 64-7, we present the standard view of receptor identification of steroid hormones generally. The schema does not specify the nature of the process by which corticosteroids, specifically, are transferred through cell membranes. Because corticosteroids are lipid-soluble, and because cell membranes have a lipid phase, simple solution and passive diffusion appear to be the mechanisms of penetration. Although active transport of cortisol out of mouse fibroblasts growing in culture has been suggested,[70] no other instance of active transport of corticosteroid hormones has been claimed. Once inside the cell, glucocorticoids are bound to a specific receptor in the cytosol, and after binding of receptor and hormone the resulting complex enters the nucleus, where it is recognized by other receptors (of unknown structure). Nuclear receptors are not able to recognize either the native hormone or the unoccupied cytosol receptor, but only the complex.

Cytosol receptors for glucocorticoids have been found in every tissue that has been examined for them,* including parts of the brain, the pituitary gland, and even the adrenal cortex.[146] In most tissues the concentration of cytosol glucocorticoid receptors (expressed as the number of high-affinity sites per unit weight of cytosol protein) decreases with age.[144] The nuclear receptors for the complex formed by binding of glucocorticoids to their cytosol receptors have also been found in all tissues in which the search has been made. There may be two forms of the nuclear receptor.[116] In general it is thought that peptide and protein hormones are recognized at cell surfaces and steroid hormones in the cytosol. However, rat liver cells have been reported to have a corticosteroid receptor in their plasma membranes, as well as in the cytosol.[163] There is no a priori physicochemical argument against cell surface recognition of steroids, and recently other instances have been described.

Regulation of receptor number

An important discovery about receptors for polypeptide signals (both hormones and immunologic agents) has shown that receptor number, or more rarely, the binding affinity, may change in either direction as a function of the concentration of the agent being detected.[35] The change is usually a decrease in number of binding sites. Acetylcholine receptor concentration on skeletal

---

*See references 33, 51, 58, 116, and 139.

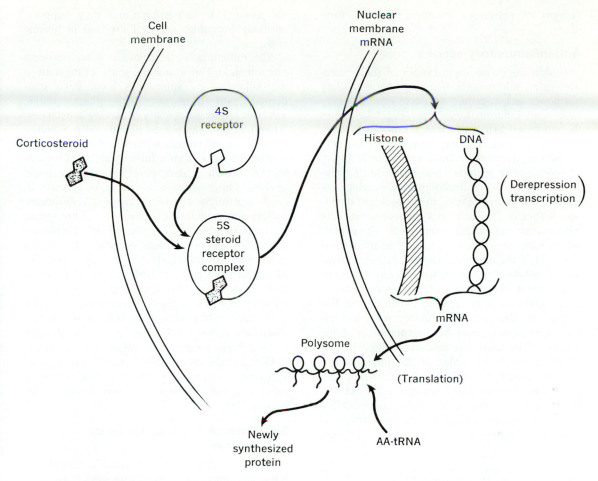

Fig. 64-7. Diagrammatic representation of proposed mechanism of action through which cortico-steroids may affect protein synthesis. Suggested mechanism has been documented, at least in part, for selected actions of steroid hormones in specific tissues. Other actions are also possible and are discussed in text. Symbols *4S* and *5S* indicate that macromolecular steroid receptor is altered after complexing with steroid hormones so that sedimentation constant (S) is increased. Complex then penetrates nuclear membrane and is "recognized" in nucleus, after which dere-pression occurs.

muscle membranes may also vary inversely with the flow of the transmitters, and similarly, eryth-rocyte membranes decrease their catecholamine receptor concentration when stimulated by adren-ergic agents. Thus the phenomenon of regula-tion by changing receptor concentration is not restricted to polypeptide signals; however, it may be restricted to plasma membrane receptors. It is not yet clear whether or not the cytosol receptors for glucocorticoids change in concentration or affinity as the flow of steroid signals changes, but estrogens appear to increase the amount of cy-tosol receptor specific for them in the rat uterus,[42] and so the phenomenon of adjustment of receptor level by hormone level may be widespread, or even universal. In any case, where it exists, this

phenomenon introduces a striking nonlinearity in pharmacokinetic analyses because it means that cell receptor responsiveness to a hormone may change in either direction while the hormone signals are flowing.

PHYSIOLOGIC EFFECTS OF GLUCOCORTICOIDS

Striking actions of glucocorticoids observable at the whole animal level of organization include (1) inhibition of the inflammatory response, (2) maintenance of normal arterial systemic blood pressure and volume, (3) effects on metabolism, including maintenance of blood sugar levels dur-ing starvation, (4) thymolysis, lympholysis, eosinopenia, and immunosuppression, and (5) support of resistance to many noxious stimuli.*

Anti-inflammatory actions

Within the physiologic ranges of their secretion rates, adrenocortical and adrenomedullary hormones act together to accelerate recovery from many inflammatory agents. Fig. 64-8 shows the combined effect of these hormones in ameliorating the inflammatory response of the skin of the ear of unanesthetized rats to mild injury by xylene. The recovery was supported by glucocorticoids and catecholamines, acting together, but neither alone, in physiologic doses, was sufficient.[159] The two systems are partially independent and partially coupled.[177] The experiments were performed on fasting animals. Surprisingly, if the adrenalectomized rats simply ate a pure carbohydrate meal, without any hormonal replacement, the defect in the recovery process could again be fully repaired. Therefore in this preparation the anti-inflammatory hormonal effects may merely have been expressions of the actions of these agents on carbohydrate metabolism. In other examples of the inflammatory process, glucocorticoids can be shown to decrease histamine release from mast cells and thus help to limit the enhanced capillary permeability of the inflammatory response[152]; glucocorticoids decrease phagocytosis and stickiness of poly-

morphonuclear leukocytes, and they suppress antibody formation and cell division in plasma cells.

The inflammatory response is a multiprocess phenomenon. Unfortunately, many of the data on anti-inflammatory substances have been concerned with drugs or with pharmacologic amounts of physiologically occurring substances. Attention has been placed heavily on prevention of onset of inflammation rather than on facilitation of recovery from a fully established inflammation, which is also an important physiologic process. Other subcellular actions of glucocorticoids that might contribute to their anti-inflammatory action include (1) inhibition of the spread of hyaluronidase, (2) sphering of fibroblasts (fibroblasts do not divide or produce collagen when they are spherical), (3) increased porosity of hyaluronate gels, and (4) increased granulation of connective tissue ground substance. The capacity of glucocorticoids to potentiate the vasoconstrictor effect of catecholamine hormones, discussed subsequently, has also been thought to be a basis for their anti-inflammatory action.

Glucocorticoids are able to stabilize lysosomal membranes and thereby inhibit release of proteolytic enzymes during the inflammatory response.[49,176]

Maintenance of blood pressure and volume

Glucocorticoids are not usually thought to have any important vasoactive effects of their

---

*See references 10, 11, 106, 123, and 124.

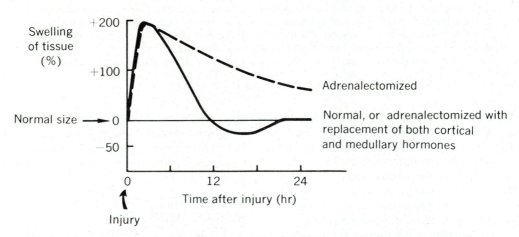

Fig. 64-8. Example of anti-inflammatory action of corticosteroids. Diagrammatic representation of course of swelling and recovery of skin of ear of unanesthetized rats following xylene-induced injury. In normal animal, injury produces prompt swelling, with recovery occurring in about 12 hr. In adrenalectomized animal, recovery is delayed. However, if adrenalectomized animal is treated with physiologic amounts of adrenocortical and adrenomedullary hormones, recovery is normal. Data were obtained from studies in fasted animals. (Redrawn from Sendelbeck and Yates.[159])

own, except possibly in the kidney, where they increase glomerular filtration rate by an unknown mechanism, but they are required for the vaso-constrictor action of norepinephrine. In the absence of glucocorticoids the vasopressor action of catecholamine hormones is diminished or lost and blood pressure falls. The subcellular basis for this supportive effect of glucocorticoids has been explained[88]: glucocorticoids inhibit the enzyme catecholamine-*O*-methyl transferase (COMT) that inactivates catecholamine hormones and transmitters at their sites of action. If this enzyme is not inhibited by corticosteroids, catecholamines are metabolized so rapidly that they cannot accumulate in amounts sufficient to cause vasoconstriction.

The mineralocorticoids play a major role in the regulation of extracellular fluid volume, as will be described later. Less well recognized is the fact that, by a different mechanism, the glucocorticoids also help to restore blood volume after a hemorrhage. Cardiovascular receptors lying in the right atrium and carotid arteries detect relative changes in blood volume.[31] Fibers from these receptors project through to the central nervous system (CNS) and converge in the hypothalamus to control the release of ACTH and, ultimately, cortisol[63] by pathways to be described later. Increased cortisol secretion, along with one or more other adrenal factors and one or more pituitary factors still unknown, leads to increased extracellular osmolality and a shift of fluid from the intracellular to the interstitial space, thereby increasing interstitial volume and pressure.[61,132,133] The increased interstitial pressure leads to an increase in capillary and lymphatic movement of fluid and to an increase in lymphatic return of interstitial protein to the vascular system. This combination of events is necessary for the support of cardiovascular function, for the subsequent cessation of the initial neuroendocrine response to hemorrhage, and for the eventual full recovery of blood volume and protein mass. It is important to note here that basal levels of cortisol are not sufficient to repair the blood loss; an increased secretion rate (the stress response) is required.

Effects on metabolism

Carbohydrate. Glucocorticoids maintain blood sugar[105,107] and liver, kidney, and muscle glycogen content. They do this through the process of gluconeogenesis in liver and kidney. In this process, extrahepatic tissue proteins, especially muscle proteins, are broken down into their component amino acids, and the amino acids

(especially alanine) are then delivered to the blood, taken up by the liver and kidney, and synthesized into glucose, which then enters the blood and the glycogen stores. Glucocorticoids facilitate this process at several points. They augment protein catabolism in muscle and other extrahepatic tissues. As a result, plasma amino acid levels increase. Glucocorticoids also increase trapping of amino acids by liver and kidney (but inhibit the uptake by muscle), and they stimulate the enzymatic activity of hepatic amino transferases (transaminases). Glucocorticoids stimulate conversion of approximately 10 of the 20 amino acids into carbohydrate precursors, but alanine is the chief amino acid precursor of glucose by gluconeogenesis. A preparation reaction for this conversion is transamination. Finally, glucocorticoids augment the activity of those enzymes that favor synthesis of glucose by gluconeogenesis, and they inhibit certain of the glycolytic enzymes.[175]

The enzymes favoring glycolysis that may be inhibited by glucocorticoid action are glucokinase, phosphofructokinase, and pyruvate kinase. The enzymes favoring production of glucose from gluconeogenesic precursors include pyruvate carboxylase, phosphoenolpyruvate carboxykinase, fructose-1,6-diphosphatase, and glucose-6-phosphatase, and these are stimulated by glucocorticoid action.

An important general effect of glucocorticoids on carbohydrate metabolism is to interfere with glucose uptake and utilization in lymphoid tissues, fibroblasts, and fat, but apparently not in muscle.[106] Thus glucocorticoids may appear to produce insulin resistance, and in so doing, they reroute carbohydrate fuel fluxes toward the brain. It should noted, however, that the effects of glucocorticoids on glucose uptake and utilization are conditioned by so many nutritional, hormonal and physiologic factors that it is hard to predict which effects, increased, decreased, or unchanged uptake of glucose, will be observed during corticosteroid treatment.

Protein and amino acids. The prominent effects of glucocorticoids on amino acid metabolism are those involved in transamination in support of gluconeogenesis.[175] The stimulation of the breakdown of extrahepatic tissue protein to provide an amino acid flow for gluconeogenesis is not a direct catabolic action, but represents inhibition of protein synthesis by glucocorticoids through decreased synthesis of mRNA. In fibroblasts, DNA synthesis is also impaired rather directly by corticosteroids.[138] The normal processes of nucleic acid and protein breakdown and turnover become unbalanced, and a net breakdown occurs. In liver, corticosteroids stimulate synthesis of some species of nucleic acids and proteins, but inhibit the synthesis of other species in the same organ.[78]

Fat. Glucocorticoids favor the mobilization of fatty acids from adipose tissue to liver, where the metabolism of free fatty acids may lead to the appearance of products that inhibit glycolytic enzymes and favor gluconeogenesis. Glucocorticoids also indirectly stimulate fatty acid oxidation by interfering with peripheral glucose utilization.[48,54,123] These peripheral effects of corticosteroids may require high levels, near the upper limit of the physiologic range. As a result of increased fatty acid oxidation, mainly in muscle, glucocorticoids may lead to mild ketosis, especially in diabetic animals. The steroids also potentiate the lipolytic actions of growth hormone,[54,55] and ACTH,[36] which involve stimulation of the hormonally sensitive lipase. They inhibit reesterification of fatty acids within adipocytes by inhibiting glucose utilization by these cells.[54]

Glucocorticoids and starvation. The brain uses carbohydrate preferentially after feeding and in the early stages of starvation. Stimulation of gluconeogenesis supports the metabolism of the brain during starvation, fasting, or limited caloric intake. Enhanced flow of free fatty acids during starvation and inhibition of peripheral utilization of glucose also spare carbohydrate for utilization by the brain. Therefore the major metabolic effects of glucocorticoids are such that these hormones might be thought of as "hormones of starvation," which permit a continuous supply of energy to the brain in animals without food who are hunting for it. Unfortunately for this view, all the processes described occur within a few hours after the last meal has been taken, and at that time, plasma concentrations of glucocorticoid hormones are not elevated, although they may rise later during a prolonged starvation. The acute switching signals that convert animals rapidly from "feasted state" metabolism to "fasted state" metabolism probably do not involve glucocorticoids, even though the actions of these hormones give support during more prolonged starvation. In any case, adrenalectomized animals being kept alive only by mineralocorticoid therapy die very quickly if their food intake is restricted, so glucocorticoids seem to have a "permissive" action during inanition.

Thymolytic, lympholytic, and eosinopenic actions, immunosuppression, and fibroblast inhibition

Glucocorticoids rapidly cause extreme atrophy of the thymus and a decrease in numbers of lymphocytes, plasma cells, and eosinophils in blood. Glucocorticoids depress the rate of growth and cell division of mouse fibroblasts in vitro by decreasing the rate of thymidine and uridine incorporation into macromolecular material.[138] DNA synthesis is depressed more rapidly and more extensively than is RNA synthesis. Protein synthesis is depressed in these cells later, and they then become spherical and quiescent. The initial locus of this biochemical effect of the steroids appears to be at DNA-dependent polymerization of deoxyribonucleotide triphosphates into DNA. Similar actions may be a basis of thymolysis and lympholysis; there is high mitotic activity in thymus and other lymphoid tissues, and these tissues cannot be maintained if mitosis is inhibited. A more general cytotoxic action of glucocorticoids on lymphoid tissue and plasma cells may also be involved.[124]

By decreasing the population of lymphocytes and plasma cells and interfering with the conversion of lymphocytes into antibody-producing plasma cells by phytohemagglutinin, glucocorticoids ultimately interfere with the immune response[68] and are therefore immunosuppressants useful in delaying rejection in patients with organ or tissue transplants.

Psychoneural effects

During therapy with glucocorticoids, patients may become initially euphoric and then psychotic, paranoid, or depressed if the dosage is very high and prolonged. Corticosteroids are bound in parts of the brain, presumably by neurons rather than glia, and the binding is very specific.[38,51,115] Receptors have been found in the forebrain limbic structures that are involved in the genesis of emotions. The psychogenic effects of glucocorticoids are sufficiently noteworthy that it has even been suggested that during prolonged therapy with these hormones, patients may become addicted to them.[40] Whether glucocorticoids physiologically affect mood is not clear, but it has been proposed that the major function of the stress-response increase in glucocorticoid levels is to switch the brain from a "dysphoric" (stressed) state to a "euphoric" state in which stable cognitive powers are restored.[25]

Corticosteroids affect several sensory processes by decreasing our ability to detect signals; thresholds for taste, smell, and hearing are higher in normal subjects than in patients with low levels of corticosteroids.[79] There seems to be little doubt that many CNS information-processing capabilities are glucocorticoid dependent.

SIGNAL PATHWAYS FROM BRAIN TO ADRENAL CORTEX

The plasma levels of glucocorticoids are ultimately determined by special neurons of the medial basal hypothalamus (also called the median eminence) that act as neuroendocrine transducers and signal integrators. These neurons receive and integrate neural inputs from other regions of the CNS and chemical inputs from blood and cerebrospinal fluid (the "blood-brain barrier" is incomplete in the midline hypothalamus and third cerebral ventricle). The output of these neurons is a hormone called corticotropin-releasing hormone (CRH). CRH is carried from the medial basal hypothalamus, via the infundibular stalk, to the corticotropic cells of the adrenohypophysis (in pars distalis and pars tuberalis) through the hypothalamohypophysial portal capillary system. Page and Bergland have prepared excellent illustrations of this vascular system in different species.[130] After stimulation by CRH, corticotropic cells release (and synthesize) ACTH, which then enters the general circulation and finally stimulates corticosteroidogenesis and increased glucocorticoid secretion at the adrenal cortex.

The following section is an account of ACTH, CRH, and the neural inputs to the CRH neurons.

Adrenocorticotropic hormone

In the late 1920s and early 1930s, it was discovered that the normal morphology of the adrenal cortex depended on the presence of an intact pituitary gland and that substances extracted from the pituitary could stimulate and maintain a remaining adrenal gland in unilaterally adrenalectomized, hypophysectomized animals. Additional lines of evidence that adrenocortical function depended on the pituitary gland came from studies of unilateral adrenalectomy in animals with intact pituitaries. It was found that removal of one adrenal cortex was followed by enlargement (compensatory hypertrophy) of the remaining gland. This phenomenon was impaired in hypophysectomized animals. In hypophysectomized animals (or in pituitary-deficient man), adrenal atrophy and failure of the synthesis and release of cortisol and corticosterone occur. All these influences of the pituitary gland on adrenocortical function are exerted through the adrenocorticotropic hormone (ACTH).

The physiologic range of ACTH in the blood of normal human beings and experimental animals lies between 0 and 100 μU/ml of plasma.[22] A microunit of ACTH is defined by bioassay and is equivalent to approximately 7 pg (7×10^{-12} gm). The relation between the unit value and weight of ACTH is inexact because ACTH potency determined by bioassay varies according to the route of administration of the hormone.

The structure, half-life in human beings, daily secretion rate, plasma concentration ranges and some actions of ACTH have been presented in Chapter 61.

ACTH is the major, but not the only, physiologic signal that determines the rates of glucocorticoid synthesis and release (secretory effects) and the mass of adrenal cortical tissue (trophic effects). Secretory effects and trophic effects are usually coupled, but there are cases in which one can be augmented without changing the other in the same direction. Besides ACTH, growth hormone and sympathetic nerves participate in trophic and secretory effects, including compensatory adrenal hypertrophy.[52,96] Whether or not sympathetic nerves or growth hormone have true regulatory functions in the adrenal cortex, as opposed to "permissive actions" that keep the machinery in operable condition but do not precisely determine what the operating point will be, remains to be seen.

ACTH negative feedback

A possibly important extra-adrenal action of ACTH is its capacity to inhibit its own release,[121] apparently by an action on the brain that leads to diminution in secretion of corticotropin-releasing hormone from the hypothalamus. The negative feedback inhibition of ACTH release by ACTH itself does not depend on its adrenal actions; the effect can be shown in adrenalectomized animals. It should be emphasized that the capacity of ACTH to inhibit its own release has been demonstrated only in relatively long-term experiments. There is no evidence yet that ACTH acts rapidly on a moment-to-moment basis to help set its own secretion rate.

ACTH is normally present in cerebrospinal fluid in concentrations that may be as high as those in plasma,[29] even though ACTH does not readily transfer from blood to cerebrospinal fluid. Either ACTH is synthesized by some regions of the brain or it can be transferred from the pituitary gland to the brain without first entering the general circulation. (Transfer of corticosteroids in the reverse direction, from cerebrospinal fluid to pituitary gland, can be demonstrated experimentally.[93] There may be a two-way hemodynamic exchange between these regions, as J. C. Porter has suggested.)

ACTH is produced by certain kinds of extrapituitary tumors, especially those in the lung, as well as by tumors involving corticotropic cells. In either case the patient will have severe hyperadrenocortical disorders, in addition to the risk from the neoplasm itself.

ACTH in the intermediate and posterior lobes of the pituitary

ACTH is present in all parts of the pituitary gland, although it is most concentrated in the corticotropic cells of the pars distalis. There has been much conjecture about the conditions (if any) under which ACTH might be preferentially released from the intermediate or neural lobes instead of from the anterior lobe, as is usual.[59,69,119,120] At present we believe that only the adenohypophysis secretes functionally significant amounts of ACTH.[69] The ACTH molecule is derived from a large precursor molecule that also produces endorphins. Enkephalins are sequences within endorphins, and αMSH is a sequence at the *N*-terminal end of ACTH. There is a close biochemical "familial" relationship among these hormones.

Vasopressin and ACTH release

Vasopressin administered intravenously strongly provokes the release of ACTH. Electrical stimulation of the paraventricular nucleus, a locus of neurons that synthesize vasopressin and send projections to the medial basal hypothalamus[30] and posterior pituitary gland, increases the release of ACTH.[110] Vasopressin has been found in the basal hypothalamus and in blood of the hypothalamohypophysial portal system.[183]

The mechanism by which vasopressin increases the release of ACTH is still being debated. Three possibilities have been considered. They are that vasopressin (1) is an independent corticotropin-releasing factor (CRF), (2) potentiates the action of corticotropin-releasing hormone (CRH),[182] or (3) causes release of CRH from neural structures in the medial basal hypothalamus, infundibular stalk, or posterior pituitary gland.[19,24,26] This problem is not solved, but newer approaches using normal rat anterior pituitary cells either in culture or superfused after dispersion suggest that only possibility 3 is correct.[137,178] The rightness of this conclusion depends on the degree of normality of the corticotropic cells in these preparations. In conscious dogs with small chronic cannulae in their anterior pituitaries, vasopressin microinfused into the adenohypophysis causes ACTH release.[67] If the vasopressin stayed in the anterior pituitary, a point never absolutely certain, then possibilities 1 and 2 continue to deserve consideration. A possibility excluded by the chemical work on CRH done thus far is that vasopressin and CRH are identical.

Corticotropin-releasing hormone

Although corticotropin-releasing factors (CRFs) have been known since 1955, and many attempts have been made to purify and characterize the primary hypothalamic corticotropin-releasing hormone (CRH),[2,26] CRH still has not been identified or synthesized. The problems in developing a suitable assay for CRH have been discussed.[2,85] Recently, several new approaches have been used to assay CRH. One method utilizes the release of ACTH from cultured anterior pituitary cells.[167,171] Another method uses various preparations of combined, dispersed anterior pituitary cells and dispersed adrenal cells[122,136,153] and so eliminates the need for a separate ACTH assay. A third method developed by Jones and his associates[85] involves stimulation of an in vitro explant of rat hypothalamus with the neurotransmitter serotonin and then examination of extracts of the medium for substances with the properties of CRH. Two molecules with CRH properties having molecular weights of 1,300 and 2,500 have been found.

The region of the brain in which CRH is synthesized is not known. It is stored in and released from the hypothalamic median eminence region in particular[100] and the anterior medial basal hypothalamus more generally in all species so far studied. CRH is probably synthesized in the same region, but it could be transferred from other regions through cerebrospinal fluid to the neural transducers in the medial basal hypothalamus.

Brodish and colleagues, using an in vivo CRH assay, proposed the existence of an extrahypothalamic CRF termed "tissue CRF."[37,109] They found that tissue CRF provoked a more prolonged increase in corticosteroids than did hypothalamic CRH in rats with extensive ventral hypothalamic lesions subjected to a severe stress. They suggested that tissue CRF may be released in situations of extreme stress when hypothalamic CRH is not adequate to maintain sustained high levels of corticosteroids. This extrahypothalamic CRF does not appear to be the same substance as CRH.[179] These claims about tissue CRF are interesting, but not yet confirmed.

Neural inputs to the hypothalamic CRH-containing transducer cells

Neural pathways converging on the CRH-releasing neurons have been summarized recently.[63] They may be roughly categorized by location as anterior or posterior or by function as stimulatory or inhibitory. They may also be categorized as having a role either in the "basal" state or in the "stressed" state of the adrenal glucocorticoid system. The basal state is oscillatory (discussed later), and its rhythmic inputs enter the hypothalamus anteriorly.[71,160]

The stress inputs to the adrenal glucocorticoid system arise through both anterior[135,140] and posterior[63,65,140] pathways. The posterior pathways entering from the brain stem sometimes provoke a slower response of the system than do anterior inputs.[140] At least some of the anterior stimulatory pathways arise in the amygdala; stimulation of the amygdala is very effective in causing CRH and ACTH release,[111] whereas lesions interfere.[95]

Some of the neural pathways that inhibit CRH release originate in the hippocampus.[95]

Fig. 64-9 summarizes neural inputs to the adrenal glucocorticoid system. Both anterior and posterior pathways are drawn and are indicated as stimulatory or inhibitory for the release of ACTH.

Most of the evidence for the posterior pathways to the adrenal glucocorticoid system shown in Fig. 64-9 comes from the work designed to elucidate the central neural pathways that mediate the release of ACTH following hemodynamic changes.[63] Atrial and carotid receptors detecting changes in blood volume project to the medulla in the region of the lateral solitary nucleus and then to the lateral nucleus intercalatus. Projections from these nuclei then extend to the dorsal rostral pons in the region of the locus subceruleus and also the principal locus ceruleus, dorsal and ventral tegmental nuclei, and dorsal raphe nucleus. Both stimulatory and inhibitory pathways for the release of ACTH ascend dorsally via the dorsal longitudinal fasciculus to the hypothalamus. The dorsal stimulatory pathway projects to the posterior medial hypothalamus and more anteriorly to the region of the parvocellular nucleus and to the paraventricular nucleus; this pathway probably mediates release of ACTH at least in part through release of vaso-

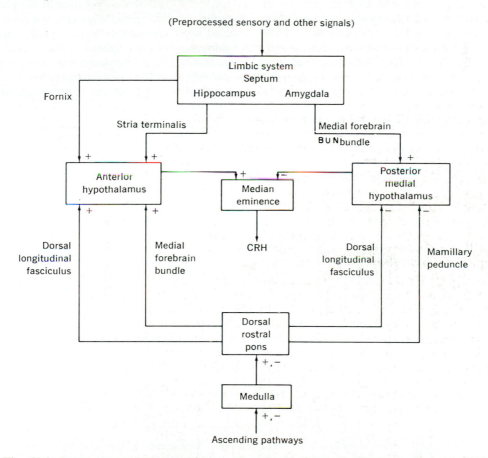

Fig. 64-9. Proposed neural input pathways to the adrenal glucocorticoid system. Both stimulatory (+) and inhibitory (−) pathways are shown.

pressin. The dorsal inhibitory pathway terminates in a region extending from the posterior medial hypothalamus to the anterodorsal hypothalamic area. Similarly, there are both stimulatory and inhibitory pathways ascending ventrally to the hypothalamus. The ventral stimulatory pathway projects via the medial forebrain bundle and turns medially at the region of the supraoptic decussation to the suprachiasmatic nucleus and ultimately the median eminence; this pathway may mediate release of ACTH through release of CRH. The ventral inhibitory pathway projects via the mamillary peduncle and converges with the dorsal inhibitory path in the posterior medial hypothalamus. Thus there appear to be three principal pathways in the hypothalamus: two stimulatory paths, one anterodorsal and mediated by vasopressin and the other anteroventral and mediated by CRH, and one posterior inhibitory path.

Neurotransmitters

Various neurotransmitters have been implicated in the neural control of the release of ACTH. The neurotransmitters in the stress stimulatory pathways include acetylcholine,[34,74,81,86] and those in the inhibitory pathways include norepinephrine.[60,64,86,172] The basal circadian rhythm pathways may involve cholinergic[101] or serotonergic[98] neurons.

Jones and associates, using an in vitro hypothalamic preparation, found that norepinephrine, dopamine, and histamine did not affect the basal secretion of CRH,[81] but that norepinephrine (via a receptor mechanism) and gamma aminobutyric acid (GABA) were able to inhibit the enhanced release of CRH following acetylcholine[81] or serotonin.[85] There is considerable controversy as to whether serotonin is stimulatory* or inhibitory.[168,169,172] The various input pathways converging on CRH-containing neurons are so multisynaptic and interconnected within the limbic system that we should hardly expect to find simple generalizations concerning the roles of various neurotransmitters in the operation of this system.

NEGATIVE FEEDBACK REGULATION OF PLASMA GLUCOCORTICOID LEVELS

In 1947 Sayers proposed that adrenocortical function must be under negative feedback control.[151] This hypothesis was later extended[149] and reformulated.[26] Since then the exact relationship between glucocorticoids and release of ACTH and CRH has been the subject of many studies.[14,24,26] Usually the operation of the glucocorticoid negative feedback process has been examined by observing the degree of inhibition of a subsequent stress response of the adrenal glucocorticoid system when corticosteroids are administered before the stressor. Data from an experiment of this type in rats[45] are diagrammed in Fig. 64-10.

---

*See references 28, 39, 82, 85, 98, and 126.

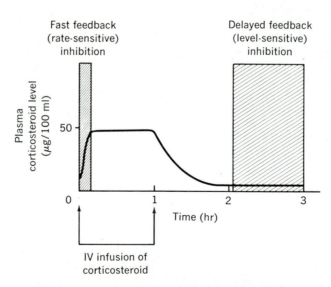

Fig. 64-10. Duration of inhibition of stress response. Rate-sensitive (fast) and level-sensitive (delayed) feedback inhibition demonstrated by an infusion of corticosteroid intravenously for 1 hr. Cross-hatched areas indicate times when stress-induced release of ACTH is prevented by the corticosteroid infusion.

Corticosteroids steadily infused intravenously at rates within the physiologic range of adrenocortical secretion cause a rapid increase in plasma corticosteroid levels to a new elevated steady-state value. When the infusion is stopped, the levels decline slowly to the initial low values. This dynamic asymmetry arises from the corticosteroid distribution, binding, and metabolism component of the system.[46] When attempts are made to stimulate a stress response of the adrenal glucocorticoid system while such infusions are being conducted, it is found that during the early part of the infusion, when corticosteroid levels are low but rising rapidly, a striking inhibition of ACTH release is present. This initial inhibition appears to depend on the first derivative with respect to time, rather than the absolute value of the corticosteroid levels.[45,84,89] This rate-sensitive inhibition is of very brief duration, and it is followed by a period lasting at least 1 hr, during which the stress response is uninhibited, even though plasma corticosteroid levels are elevated. Finally, a delayed inhibition, dependent on the earlier plateau levels of plasma corticosteroids, appears and persists for an appreciable period of time, even after the corticosteroid levels may have decayed back to resting levels after the cessation of the infusion.

The two components of the inhibition of glucocorticoid secretion by glucocorticoids have been designated as (1) fast rate-sensitive derivative feedback and (2) delayed level-sensitive proportional feedback. For convenience the terms "fast" and "delayed" are sufficient. Before the two components of glucocorticoid negative feedback were discovered there was considerable confusion generated by investigations carried out during the delay period, when no feedback effects were observable, even though corticosteroid levels were elevated. The erroneous conclusion then was that the feedback inhibition was trivial or nonexistent as a physiologic process. Unfortunately, even though this error has been eliminated by the recognition of the two types of feedback actions and the silent, delay period separating them, there is a residual problem, as described in the following section.

The problem of noncausal experiments

The term "physiologic" and its negation appear frequently in reports of experiments on whole animals. Usually the terms refer to whether a dose of a substance is natural to the animal. If the dose is larger than the maximal endogenous production rate of the substance by a normal animal, the effects produced are said to be "unphys-

iologic," or "pharmacologic." However, it is important to note that an inappropriate *sequence* of signals, each of which is of "physiologic" amplitude or form, also creates unphysiologic conditions. For example, plasma corticosteroids in a normal animal do not rise until *after* arrival of an input signal through one of the pathways shown in Fig. 64-9. But in experiments of the kind shown in Fig. 64-10 the corticosteroid levels are raised *before* the stressor input by an experimenter acting as a deus ex machina. This sequence of events violates physiologic causality within the glucocorticoid system.

Fast rate-sensitive feedback

The existence of fast inhibition of ACTH release by corticosteroids was first shown in 1947.[151] That this fast feedback is unidirectionally rate sensitive was first discovered in 1969.[45] Since then the details of operation of this fast feedback have been developed by several groups.*

The fast feedback receptors appear to recognize both cortisol and corticosterone, but their 11-deoxygenated derivatives, although recognized, are inactive.[87] These receptors are present in the hypothalamus[90,148] and perhaps also in the pituitary.[150] A critical rate of rise of corticosteroid levels (a rate threshold) must be reached before the rate-sensitive receptors are activated. The processes of activation by the rate of rise have been successfully modeled by applying the rate-of-occupancy principle of Paton to the binding of corticosteroids by the receptors.[89] Fast inhibition has recently been demonstrated with dispersed anterior pituitary cells activated in vitro by a CRH preparation after various periods of exposure to corticosteroids in vivo or in vitro.[150]

In engineering practice, rate-sensitive (derivative) control is used whenever an anticipatory effect is needed. Using rate of change rather than level as a signal allows a prediction of where the system is going, and powerful corrections for disturbances can be initiated before the perturbations have severely challenged the system.

Delayed level-sensitive feedback

The level-sensitive feedback receptors are apparently present in both hypothalamus and pituitary,† and they are sensitive to the concentration of unbound corticosteroids.[91] Cortisol and corticosterone are active, as are their 11-deoxyge-

---

*See references 47, 83, 84, 87, 89, 90, and 148.

†See references 44, 50, 67, 75, 80, 122, 145, 161, 173, 174, 178, and 180.

nated derivatives.[87] This specificity is different from that given for fast feedback, and so, presumably, different receptors are involved for the two types of feedback.

Both the duration of elevation of corticosteroid levels and the intensity of the ACTH-releasing stimulus are important determinants of activation of the hypothalamus-pituitary-adrenal system.[41,94] The level-sensitive corticosteroid negative feedback process can be saturated readily, with the result that strong stresses can provoke CRH and ACTH releases that cannot be inhibited by additional corticosteroids. These consequences of saturation of feedback sites are illustrated in the two panels of Fig. 64-11. Failure to recognize the ease with which level-sensitive corticosteroid feedback sites can be saturated within the physiologic range of plasma corticosteroid concentrations has led to great confusion about the importance of corticosteroids in the modulation of ACTH release. Many experiments have been carried out with stressors so strong that the inhibitory effect of corticosteroids even at saturating levels is inadequate to diminish the response. The erroneous conclusion has then been drawn that corticosteroids have no important effect on the ACTH release provoked by stressors.

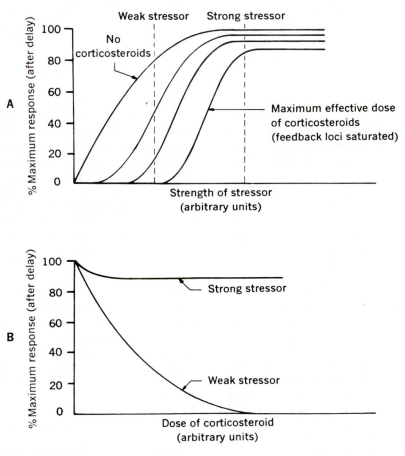

Fig. 64-11. Relationship between strength of stressor, corticosteroid levels, and ACTH release. **A** shows that at a given time (after delay in level-sensitive feedback), inhibition caused by corticosteroids is a function of dose of corticosteroid and strength of stressor. Eventually, maximum effect of corticosteroid is reached (maximum dose curve), and further increase in strength of stressor will continue to provoke substantial response of system because corticosteroid feedback loci saturate before stimulation loci. **B** shows how relationships presented in **A** appear when response to graded levels of stressor ("weak" or "strong") are examined after different, graded doses of corticosteroids have been given. Corticosteroids can completely inhibit response to "weak" stressor but have little effect on response to "strong" stressor because of saturation effects shown in **A**. (Modified from Yates and Maran.[24])

Response of the glucocorticoid system when plasma corticosteroid levels decrease

The discussion of feedback is based on results obtained when corticosteroid levels are elevated. We shall now examine what happens when they are decreased. Drugs are available that interfere with the enzyme 11-β-hydroxylase in the adrenal cortex and impair the biosynthesis of cortisol. Plasma cortisol levels then decrease. Decreases in cortisol levels relieve the level-sensitive inhibition of ACTH secretion, and if the hypothalamic and pituitary components of the system are intact, ACTH release then increases, and the adrenal gland is overstimulated in the production of cortisol precursors, metabolites of which can be collected in the urine. This chain of events has been used clinically to test for adequacy of glucocorticoid system function, but the drugs are not specific in their effects, and they make patients very sick.

Bilateral adrenalectomy leads to a prompt decrease in plasma corticosteroid levels and an increase in ACTH levels. The contributions of both the fast and the delayed feedback pathways to the shaping of the ACTH patterns in blood normally are revealed by adrenalectomy.[41]

A new experimental preparation[103,103a] that permits the investigator to open the adrenal glucocorticoid feedback loop at the throw of a switch, so that feedback signals diminish or disappear while all system components are in normal condition, is shown in Fig. 64-12. Adrenal venous blood is collected continuously from both adrenal cortices of a conscious, normal dog into a small reservoir. It is then pumped promptly

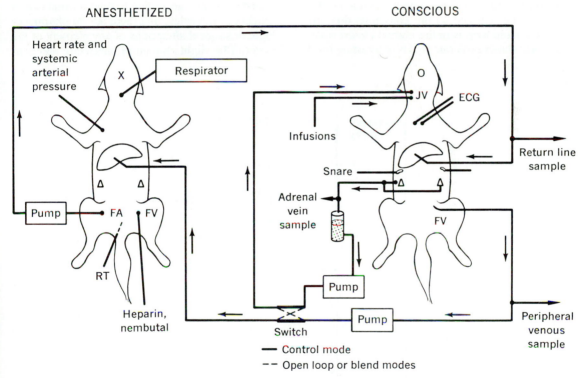

Fig. 64-12. New preparation for studying adrenal glucocorticoid system under both closed- and open-loop conditions in conscious animal with all parts of system intact. (See text.) *RT*, Rectal temperature; *0*, pituitary gland; *FA*, femoral artery; *FV*, femoral vein; *X*, site of hypophysectomy; Δ, adrenal glands; *ECG*, electrocardiogram leads. Control mode: adrenal venous effluent returns promptly to the conscious dog (closed-loop). Open-loop mode: glucocorticoid secretions are inactivated by anesthetized dog so conscious dog has no glucocorticoid signals. Blend mode: partial opening of switch to achieve partial opening of loop, if desired. All pumps have identical, matched rates set by spontaneous blood flow of two adrenal cortices: computer monitors flow into adrenal vein sample reservoir and drive pumps accordingly. Adrenal venous blood is heparinized as it leaves glands by small heparin catheters (not shown). Large elliptic form in middle of each dog represents liver. Snares cause adrenal venous blood to take external path to reservoir, instead of returning to interior vena cava in normal fashion. (From Lefcourt and Yates.[103a])

either back to the dog's jugular vein (closed-loop, control mode) or into the liver of an hypophysectomized, anesthetized dog, where the glucocorticoids are rapidly inactivated (open-loop, experimental mode). In the open-loop mode, blood is returned from the circulation of the anesthetized dog through the portal system of the conscious dog to permit additional hepatic removal of any remaining glucocorticoids. By the time the returning blood reaches the general circulation of the conscious dog, all corticosteroid feedback signals are removed; the adrenal loop is opened chemically, but closed hemodynamically. There is no perturbation of the blood volume of either animal because all pump flows are exactly matched.

The data diagrammed in Fig. 64-13 show a representative result of such experiments. During the control period the switch in Fig. 64-12 is in the control mode position, and the adrenal effluent is being returned to the conscious dog. Its glucocorticoid loop is in the normal closed state, and cortisol secretion rate (F_c) is at a resting level

(e.g., 4 μg/min). After 1 hr the switch is thrown to the open-loop mode. Plasma cortisol levels start to fall immediately because liver, gut, and kidney continue to metabolize the hormone, but there is no influx. As cortisol levels drop, feedback inhibition of CRH and ACTH releases is diminished, and their secretion rates increase. Cortisol secretion rate therefore rises to a new higher steady-state level ($F_x = 24$ μg/min). The absolute value of the ratio $(F_x - F_c)/F_c$ defines an engineering figure of merit for a feedback system—the dimensionless, steady-state, open-loop gain. For the dog, this gain is about 5, a figure that indicates the presence of strong continuous negative feedback action in the normal resting state.

The appeal of this new experimental demonstration lies in the fact that during an experiment all components of the system under study are allowed to operate according to their normal causal sequences and rules, and there are no pharmacologic or surgical alterations of any element of the system (the slight amount of anesthetic and hep-

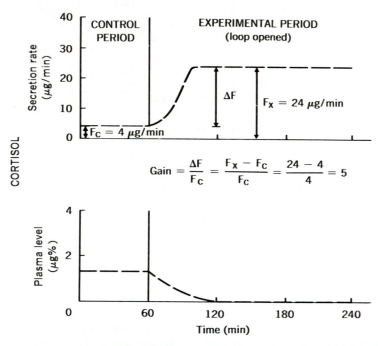

Fig. 64-13. Results of open-loop experiment. Data are from the conscious dog in Fig. 64-12. Switch was in control mode for 60 min and then was thrown to open-loop mode. As plasma glucocorticoid levels of conscious dog fell, cortisol secretion rate promptly increased to new, steady (submaximal) level. Maximum secretory response to ACTH was 40 μg/min (not shown). F_c, closed loop, control cortisol secretion rate (basal or resting conditions); F_x, steady-state cortisol secretion rate after feedback loop was opened; *gain*, dimensionless, steady-state, open-loop gain of whole adrenal glucocorticoid system with all components present and operating.

arin returning from the anesthetized dog have no effect on the conscious dog; they are immediately removed by the liver). The conscious dog is unrestrained, free to change posture during the experiment.

The data in Fig. 64-13 raise a new question. Which feedback pathway has the gain of 5? Is it the delayed feedback only that is seen in these experiments? The answer is not known, partly because the dog has not previously been used in studies on the two feedback types. The data show little or no delay in onset of increased cortisol secretion when the loop is opened, but the effect persists; cortisol secretion rates reach a new high plateau that lasts as long as the loop is opened, suggesting that the operation of level-sensitive feedback is being seen.

Repeated or chronic stresses

The adrenal gland can respond readily to repeated stimulation with pulses of ACTH, and it can sustain a high corticosteroid output if ACTH levels are chronically high. Similarly, the adenohypophysis can respond repeatedly to pulses of CRH,[122] and after adrenalectomy it can sustain high ACTH output indefinitely. However, it takes about 24 hr to develop this capacity, and in the short term, after an initial burst of ACTH, the adrenalectomized animal briefly shows a subsequent decline in levels before the final high sustained output is achieved.[46] Both repeated stresses and chronic stress create a hyperexcitable or facilitated state of the brain or pituitary components of the glucocorticoid system, so that for a given stress input the ACTH release tends to increase on subsequent or continuing exposure.[43,62,147]

At no level in the adrenal glucocorticoid system is "fatigue" or "exhaustion" observed. In rare instances the stress of a severe bacteremia in human beings can lead to hemorrhagic destruction of the adrenal glands, causing them to fail, but this is exceptional.

TOTAL ADRENAL GLUCOCORTICOID SYSTEM

The total adrenal glucocorticoid system is represented in Fig. 64-14. This figure shows all the components and processes discussed in this chapter: input pathways, hypothalamus, pituitary, adrenal glands, cortisol distribution, binding, metabolism, and receptors. Negative feedback inhibition of the system by corticosteroids acting at both the brain and the pituitary is shown, as well as inhibition by ACTH acting at the brain (the indicated stimulatory action of vasopressin

on the anterior pituitary gland is disputed). The system arrangements shown in the figure have served as the basis for computer simulations of the global system.[9,181] The simulations can perform a useful bookkeeping function, codifying fact and hypothesis concerning the adrenal glucocorticoid system. They also can predict values of inaccessible variables within the system and aid in the diagnosis of difficult cases of adrenocortical malfunction.

In order to describe the operation of a multiloop system such as that shown in Fig. 64-14, it is helpful to identify the many time domains in which the components and loops can process signals. Some of the actions are fast, occurring in less than 1 sec, and others are very slow, requiring at least 1 week for completion. Table 64-2 presents estimates of the dominant time constants of the adrenal glucocorticoid system in human beings. The usefulness of the time estimates lies in the help they give the physician in deciding which adrenocortical process will have reached equilibrium and which will still be active during any period of observation of a patient. If a patient is receiving a stress test of the competence of his adrenal glucocorticoid system and if he is being observed for several hours thereafter, with data sampled every 30 min, then processes 7 through 13 in the table will be at equilibrium at each sampling point, whereas processes 3 through 6 will be determining the changes observed. Processes 1 and 2 will be too slow to affect the results. In contrast, if the tests consist of administration of a slowly acting inhibitor of 11-β-hydroxylase to lower plasma cortisol levels, processes 1 and 3 are likely to be revealed, and the observations would have to be continued for 1 or 2 days, with samples obtained once every 4 to 8 hr.

The time domain estimates are merely guidelines, and it should be noted that even the processes that have reached equilibrium in the course of a clinical study may affect the results obtained if equilibrium operating points are abnormal. However, dynamics of the various processes will be inconsequential in shaping the results of clinical testing unless the period of observation fits the time domain appropriate for those dynamics. The concept of dynamic testing of endocrine and metabolic processes has not yet achieved prominence in clinical medicine, but it promises to offer much more powerful descriptions of diseases than can be achieved by static tests alone. Proper dynamic tests cannot be designed without consideration of information such as that shown in Table 64-2.

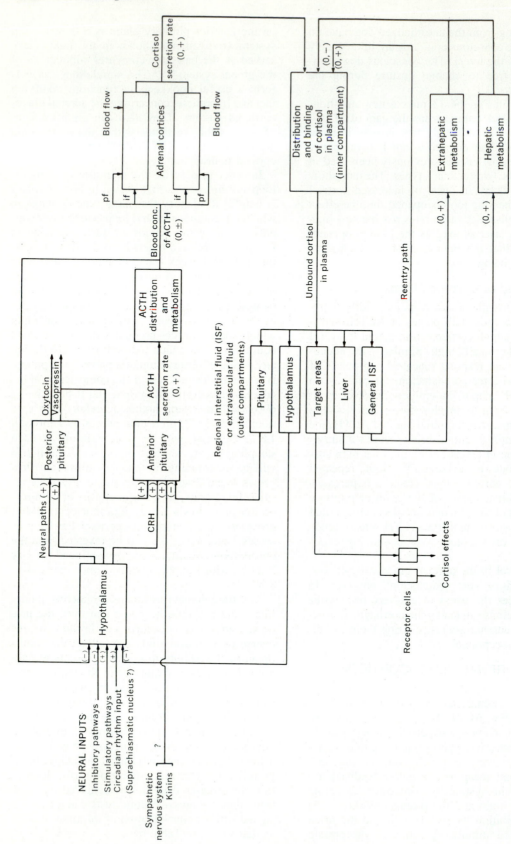

Fig. 64-14. Complete block diagram of adrenal glucocorticoid system. Term *if* indicates input forcing of adrenal by ACTH; pathway labeled *pf* indicates parametric forcing of adrenal (hypertrophic effect) caused by ACTH over longer time period. Parametric effect of changes in adrenal blood flow is also indicated. Symbols *0*, + and *0*, − indicate that signals in pathways are restricted in value (e.g., there are no negative masses or frequencies, and removal processes or inhibitors are negative in effects). Importance of these sign restrictions has been discussed elsewhere.[45] (Modified from Yates and Maran.[21])

Table 64-2. Time domains in adrenal glucocorticoid system

Process	Estimated longest (dominant) time constant*
1. ACTH feedforward (hypertrophic effect)	1 week
2. ACTH negative feedback	3 days
3. Onset of loss of adrenocortical function after withdrawal of ACTH	6 hr
4. Delay in level-sensitive feedback	2 hr
5. Cortisol unbinding, metabolism, and removal	1.5 hr
6. Settling time for ACTH-stimulated change in rate of corticosteroidogenesis	40 min
7. Distribution of cortisol	40 min
8. Rate-sensitive corticosteroid feedback	15 min
9. ACTH metabolism	<10 min
10. Delay in adrenal response to ACTH	<1 min
11. CRH stimulation of ACTH release	<1 min
12. Neural stimulation or inhibition of CRH release	<1 min
13. Circulation time for hormones in blood	15 sec
14. Binding of corticosteroids by plasma proteins	<1 sec

*A time constant is the time required for a process to reach 62.3% completion, a value that is equal to the fraction $1 - 1/e$: values will depend on species and sometimes on the sex of the subject. The numbers chosen are our best estimate for the human being.

GLUCOCORTICOID RHYTHMS

The adrenocortical system shown in Fig. 64-14 does not oscillate intrinsically. However, it does receive periodic inputs arriving through neural pathways to the medial basal hypothalamus. Two rhythms, the circadian and the ultradian, appear to be present. There is evidence that these inputs enter anteriorly through the retinohypothalamic tract. Surgical deafferentation of the anterior hypothalamus[71,160] and lesions of the suprachiasmatic nucleus, an area in the anterior hypothalamus receiving projections from the retinohypothalamic tract,[118] abolish adrenocortical periodicity. An apparent discrepancy showing that the periodic input was abolished in monkeys with complete disconnections of the medial basal hypothalamus, but not with anterior disconnections,[97] can be explained by the fact that the anterior cut spared pathways from the suprachiasmatic nucleus to the medial basal hypothalamus.

Circadian rhythm

The circadian (daily) rhythm can be observed in blind subjects and in persons isolated from geophysical rhythms, as well as in people leading normal lives. The circadian rhythm is entrained by the activity cycle. Food and water intake, the onset of light after a period of darkness, and the sleep/wakefulness cycle act to shape the waveform of plasma cortisol concentration. Ordinarily, the dark/light and sleep/wakefulness cycles are in phase, but experiments have been contrived so that they are out of phase, and these experiments have revealed the separateness of the two influences on the cortisol rhythm.[99,129]

In normal subjects sleeping in the dark from 11 PM to 7 AM, plasma cortisol levels begin to rise in the early hours of the morning and reach a peak after the subjects awaken. The levels then fall very slowly from the peak, reaching very low values, even zero, in the evening and during the early part of sleep. Thus the waveform is more sawtooth than sinusoid in appearance.

When humans travel from one time zone to another, the adrenocortical circadian rhythm requires a period of at least a week to adopt the proper phase relationship to the new local time. During that period the traveler often feels somnolent and inattentive. It is not yet certain that the distorted phase relationships between sleep/wakefulness cycles and adrenocortical cycles are responsible for these effects, but the possibility deserves serious consideration.

Although the sleep/wakefulness, eating and drinking, and the dark/light cycles all entrain the adrenocortical rhythm, it should be reemphasized that the brain rhythms that are inputs to the adrenal glucocorticoid system are endogenous and can be observed even when the entraining stimuli are absent, as in the case of constant wakefulness for several days under conditions of constant light and in the case of fasting. The entraining stimuli help to time the rhythm, but they do not create it. Circadian rhythms have been shown for hypothalamic CRH, pituitary ACTH, and adrenal corticosteroids, as would be expected. Because they are driven by an oscillating input to the median eminence, they can persist in the absence of negative feedback signals.[154,166]

An important aspect of the circadian variations in the adrenocortical system of human beings is that corticosteroid therapy is much more potent when it is given at midnight, than when it is given at noon. At midnight, corticosteroids suppress ACTH release for many hours, but at noon the effect may be only very brief.[127]

Ultradian rhythm

In addition to the circadian rhythm, the human adrenal glucocorticoid system shows faster periodicities in cortisol secretion rate and in plasma cortisol levels. Periodic functions with frequencies higher than once per 24 hr have been referred to as ultradian rhythms. About four to eight adrenal glucocorticoid secretory bursts occur every 24 hr.[76,102] following similar bursts in CRH and ACTH releases. These bursts are clustered in such a way as to be closer together and more prominent during the circadian rise in plasma glucocorticoid levels in the early hours of the morning than at other times, when they may be spaced so widely that much of the time adrenal secretion rate is zero. As a result, the adrenal cortex is secreting glucocorticoids only about 25% of the time in unstressed humans. It is possible that the bursts of adrenocortical secretion should be described as being intermittent rather than periodic, because they tend to cluster as described and give shape to the circadian rhythm. In fact, the ultradian rhythm may merely be the mechanism whereby an intermittent system creates a circadian rhythm; it may not be a separate rhythm at all.

ALDOSTERONE AND MINERALOCORTICOIDS

Aldosterone, whose structure is shown in Fig. 64-2, is the only physiologically important mineralocorticoid synthesized in the adrenal cortex. Aldosterone is the principal product of the cells of the zona glomerulosa. Deoxycorticosterone, one of the intermediates in the synthesis of corticosteroids, also has mineralocorticoid activity, but it has less potency than aldosterone and is released at much lower rates.

The only known physiologic action of aldosterone is to stimulate transport of sodium and potassium across epithelia. Every sodium reabsorbing epithelium studied thus far appears to respond to the hormone; the list includes sweat[195] and salivary gland ducts,[208] intestine,[186] and most notably the kidney.[250] Because aldosterone stimulates reabsorption of sodium into the extracellular fluid, the result of its action is to restrict the loss of sodium and its accompanying anions that would otherwise occur as various fluids (sweat, saliva, feces, and urine) are excreted. Sodium and its accompanying anions, chloride and bicarbonate, are the solutes that provide the bulk of extracellular fluid osmolality. Aldosterone, by minimizing the loss of these ions, plays a key role in maintaining the organism's extracellular fluid volume despite any vagaries of the dietary salt load.

Aldosterone also stimulates renal potassium secretion and thus plays an important role in the regulation of body potassium stores. Because potassium exchanges with intracellular hydrogen ions, any increase of plasma potassium promotes a shift of protons from cells to extracellular fluid. This effect of potassium on cell pH occurs in all tissues, including the kidneys, where the intracellular pH is an important determinant of the rate of urinary acidification. High plasma potassium levels inhibit acid excretion and contribute to extracellular acidosis.[226] Low plasma potassium levels have the opposite effects, both in tissues generally and in the kidney as well. Thus through its control of potassium distribution, aldosterone also affects acid-base balance.[217]

DISTRIBUTION, BINDING, AND METABOLISM
Protein binding and distribution

Aldosterone released into blood perfusing the adrenal gland binds to plasma albumin and to transcortin.[243] In this respect, its behavior is similar to that of the glucocorticoids, although the binding affinities are not as great for aldosterone as they are for the glucocorticoids. Under physiologic conditions, approximately half the plasma aldosterone is bound to these two proteins, whereas the other half is free. When labeled aldosterone is injected into the human circulation, its rate of disappearance follows simple first-order kinetics, with a half-life of about 25 to 40 min.[265] This rate of disappearance depends on the thyroid state of the organism; hyperthyroidism accelerates removal, hypothyroidism slows it.[235]

Metabolism

Under normal circumstances, processes occurring in the liver account for about 85% of all aldosterone removal from plasma.[266] The principal product of liver activity is tetrahydroaldosterone.[236] In addition, the liver and the kidney form a glucuronide of aldosterone itself by transferring glucuronic acid from uridine diphosphoglucuronide (UDPG). This glucuronide is sometimes referred to as acid-labile conjugate because free aldosterone can be liberated simply by treating the conjugate with a strong acid. The renal and hepatic pathways account completely for aldosterone disappearance from plasma.[237]

Nearly complete extraction of aldosterone by the liver is found over a large range of concentrations, so that the rate of aldosterone removal by the liver is simply the product of the plasma aldosterone concentration and hepatic blood flow rate.[236] Therefore the plasma aldosterone concentration should depend rather strongly on the rate of hepatic blood flow. This prediction has been confirmed by studying states in which the hepatic blood flow is reduced, either by disease or by experimental design. In hemorrhage or in acute, low-output congestive heart failure produced by partial constriction of the inferior vena cava lying within the thorax, decreased hepatic blood flow contributes significantly to the high plasma aldosterone concentrations.[199] In chronic human congestive heart failure and in chronic liver disease, reduced hepatic extraction combines with low blood flow rate to raise plasma aldosterone levels.[234]

Synthesis

Synthetic pathways. The main elements of aldosterone synthesis are shown in Fig. 64-4. The pathways for glucocorticoids and aldosterone are identical until progesterone is reached. The 17-hydroxylation does not occur in aldosterone synthesis, so that corticosterone is formed as an intermediate. Corticosterone is hydroxylated and then oxidized at the 18 position to form aldosterone.[256]

Control. Just as in the case of the glucocorticoids, there is no appreciable storage of aldosterone within the adrenal cortex, and the rate of synthesis is essentially equal to the rate of release. All the extra-adrenal factors that control the rate of aldosterone release exert their effects on various steps in the synthesis. Several signals are known to stimulate hydroxylation and side-chain cleavage of cholesterol to pregnenolone in the zona glomerulosa. These are ACTH,[212] high plasma potassium and low plasma sodium levels,[199] and angiotensin II,[227] a peptide hormone. The most widely accepted hypothesis is that ACTH, angiotensin II, and the cations act by increasing the production of cyclic AMP.[191,224] In this respect the stimulation is analogous to the response of zona fasciculata cells to ACTH stimulation, except that only the glomerulosa cells respond to angiotensin II and the cations.

In addition to these signals, chronic dietary salt restriction stimulates synthesis at later stages by an effect on the conversion of corticosterone to aldosterone.[242] This stimulus does not involve cyclic AMP or any of the other stimuli known to affect cholesterol hydroxylation. The nature of the signaling in response to manipulations of dietary salt intake is something of a mystery, although the effect is powerful. The response of the zona glomerulosa to angiotensin II or high plasma potassium, for example, is potentiated in animals with dietary salt restriction and attenuated in animals fed a surfeit of salt.[191]

CELLULAR MODE OF ACTION
Protein synthesis

Much of our understanding of how aldosterone stimulates sodium transport comes from experiments performed on the urinary bladder of the toad. The urinary bladder of mammals serves only to store urine so that it may be voided intermittently, but in amphibians such as toads, salt or water or both may be reabsorbed from the urine back into the blood, depending on the needs of the organism. Aldosterone is a normal secretory product of the amphibian adrenal gland, and it exerts a strong regulatory effect on salt transport.

Toad bladders mounted as membranes separating baths of Ringer's solution generate electrical potential differences between the mucosal and serosal surfaces. Leaf et al.[229] showed that this potential difference could be reduced to zero by the application of an externally generated electrical current, called the short-circuit current. They also found that the magnitude of this short-circuit current was accounted for entirely by the charge carried by the net flux of sodium ions. The magnitude of the short-circuit current is therefore a direct measure of the rate of active sodium transport.[272] Crabbé[197] found that aldosterone stimulated the short-circuit current in the toad bladder. The result of such measurements is shown in Fig. 64-15.

Crabbé also noted a latent period of 60 to 90 min between the time of administration of aldosterone and the onset of the short-circuit current increase. The duration of the latent period was independent of the concentration of aldosterone, and Crabbé concluded that the latency was not a function of slow penetration of the hormone to its site of action.

Edelman et al.[201] were the first to suggest that

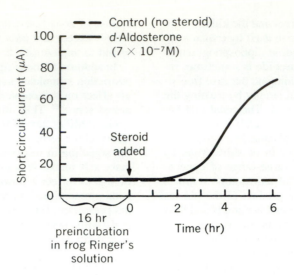

Fig. 64-15. Effect of aldosterone on active sodium transport in toad urinary bladder. Short-circuit current, defined as electrical current needed to bring spontaneously generated electrical potential difference to zero, equals rate of active sodium transport as explained in text. Preincubation period in aldosterone-free medium permits any residual aldosterone effects to disappear. Delayed onset of action following administration of hormone reflects time needed for synthesis of proteins that carry out action of aldosterone.

intermediate steps in aldosterone action involved the synthesis of proteins. They found that radioactive aldosterone accumulated in the nucleus of cells responding to the hormone. They suggested that aldosterone stimulates transcription of mRNA and that the newly synthesized mRNA leads in turn to the de novo synthesis of proteins at polysomes. The chief pieces of evidence that support this hypothesis are the findings that aldosterone stimulates production of mRNA in the toad bladder[249] and rat kidney[206] and that inhibitors of protein synthesis that act either on transcription in the nucleus or on translation at the ribosome can completely block the response to aldosterone.[204] Subsequently, it was shown that recognition of the aldosterone signal involves formation of a complex between the hormone and a 4-S-cytosol receptor protein. The resultant 5-S complex moves into the cell nucleus, where it derepresses a specific operon and enables the transcription process to begin.[263] This process is similar to that shown in Fig. 64-7.

Identity of newly synthesized proteins

The function of the newly synthesized proteins is not known, but three hypotheses have attracted attention. These hypotheses take as their starting point the classic model of transport across amphibian epithelia first advanced by Koefoed-Johnson and Ussing.[225] This model and the various hypotheses of aldosterone action are shown in Fig. 64-16. Sodium enters the cell across its luminal surface by moving down an electrochemical potential gradient, diffuses across the cell to the side facing the interstitial fluid or the lateral intercellular space, and is pumped out of the cell against an electrochemical potential gradient by a sodium pump. The sodium pump derives its energy from the hydrolysis of adenosine triphosphate (ATP), so that the action of an ATPase is an integral part of the transport mechanism. If one argues from this model, there appear to be three sites where aldosterone-induced proteins might act: (1) They could facilitate the entry of sodium into the cell at its luminal surface.[255] If this mechanism, the so-called permease hypothesis, is correct, the intracellular sodium concentration would rise in response to aldosterone administration and the rate of sodium pumping would increase in response to the higher concentrations. (2) The newly synthesized proteins could increase the activity of the sodium pump either by activating preexisting transport sites or by making new ones. (3) The aldosterone-induced proteins could act by increasing the rate of synthesis of ATP, which would raise the local concentration of the energy donor at the active site.[207]

Evidence is available to support each of these hypotheses.[198,252] None of the three excludes any

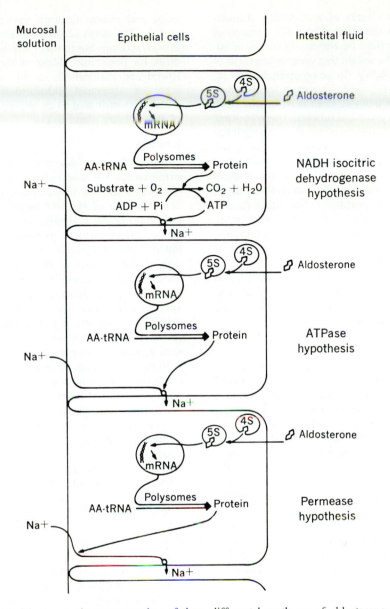

Fig. 64-16. Diagrammatic representation of three different hypotheses of aldosterone action on active sodium transport. In all three cases, it is assumed that sodium enters cell by diffusion along electrochemical potential gradient and is pumped out of cell by carrier associated with membrane-bound Na_+,K^+-dependent ATPase. In all cases, aldosterone forms a complex with a cytosol receptor, complex causes derepression at a genome, mRNA synthesis is stimulated, and de novo protein synthesis begins. Three hypotheses differ as to function of aldosterone-induced proteins.

of the others, so they need not be viewed as alternatives. Moreover, a single operon can control the synthesis of more than one protein. Thus aldosterone could stimulate the de novo formation of proteins acting at some or all of the sites shown in Fig. 64-16, and the increased pumping rate, supported by an increased flow of ATP and by an increased flow of sodium chloride across

the apical membrane into the cell, could generate the observed response.

ACTIONS ON SODIUM TRANSPORT

Aldosterone stimulates sodium reabsorption in epithelia, but the effects of this stimulation vary from tissue to tissue because the details of sodium pumping are not everywhere the same. There are two general effects of aldosterone stimulation. First, aldosterone regulates the volume of the extracellular fluid by restricting losses of sodium chloride, the solute that contributes most of the fluid's osmolality. In some epithelia the reabsorbed sodium chloride may exert local osmotic effects so that isotonic fluid is reabsorbed (intestine and renal proximal tubule). These isotonic reabsorbates are returned to the extracellular fluid directly. In other epithelia (sweat gland duct) the sodium chloride may be reabsorbed hypertonically, which leads to higher sodium chloride concentrations in the extracellular fluid. However, as this hypertonic extracellular fluid is distributed by the circulation, water will quickly move from cells in response to osmotic gradients, and the effect, again, is to retain the volume of the extracellular fluids.

The second general effect of aldosterone is on acid-base balance and potassium metabolism and is largely due to the action of the hormone on the distal tubule of the kidney.

Sweat glands

Sweat is formed in a two-stage process. The nascent fluid is secreted into the coils of the gland under cholinergic stimulation. Primary sweat is an isotonic fluid. This fluid then moves through the gland's duct toward the surface of the skin. In the duct, sodium chloride is reabsorbed hypertonically, leaving the emergent sweat hypotonic. This sweat gland duct reabsorption is strongly aldosterone dependent.[195] Patients with adrenal insufficiency characteristically have abnormally high sodium chloride concentrations in their sweat. They show poor tolerance to hot climates because their inability to restrict salt losses in sweat threatens them with the loss of extracellular fluid volume and, in turn, vascular collapse.

Intestine

Intestinal sodium chloride reabsorption is essentially an isotonic process and is subject to aldosterone stimulation.[202] Patients with adrenal insufficiency may suffer from diarrhea because they cannot reabsorb fluid from the intestinal lumen at normal rates. These fecal fluid losses occur at the expense of the extracellular fluid volume and contribute to the hypovolemia and hypotension characteristic of these patients.

Salivary glands

Saliva is also formed in a two-stage process. The primary secretion is formed in the gland's acini as an isotonic salt solution that contains the various enzymes of saliva. The fluid leaves the acini and moves through various ducts to the mouth. The ducts of the salivary glands also reabsorb sodium, but in this case they exchange sodium for potassium. Most of the saliva is swallowed and reabsorbed, so that the systemic effects of abnormal aldosterone levels cannot usually be attributed to an altered state of activity of the salivary gland ducts.

Kidney

Renal effects of aldosterone have been studied extensively. The consensus is that the hormone stimulates sodium reabsorption in the thick ascending limb of Henle's loop, in the distal tubule, and in the collecting ducts. The existence of an effect in the proximal tubule is disputed.[221,238] A complication preventing resolution of this conflict is that evidence favoring a proximal tubule action is entirely from studies on the rat, whereas that against it is from the dog. Species differences sometimes do occur and could be responsible for this disagreement. Evidence on whether or not aldosterone affects human proximal tubule function is lacking.

In the rat and dog, adrenalectomy reduces both glomerular filtration rate and proximal tubule resorption. The fall in glomerular filtration rate is due to the lack of glucocorticoids.[210] Administration of glucocorticoids restores the filtration rate to normal, but not proximal tubule function; aldosterone is required to correct this deficit.[221] When rats are fed a diet rich in NaCl, aldosterone secretion is inhibited, filtration rate rises, and proximal tubule reabsorption falls; aldosterone administration restores proximal reabsorption to normal.[262] Microperfusion experiments also support a proximal tubule action of aldosterone.[221]

Although the evidence supporting a locus of action of aldosterone in proximal tubules of the rat is convincing, the quantitative significance of this effect is unclear. When the excretion of sodium chloride is studied for the whole organ and not just the proximal tubule, the full range of aldosterone action appears to control no more than 3% or 4% of the filtered load of sodium. These are amounts that are consistent with the effects of the hormone's known actions on more distal regions of the nephron. The finding of a proximal tubule locus of action for aldosterone is therefore something of a quantitative embarrassment, since it is not necessary to explain the whole organ response of the kidney and, if present, must be compensated by other adaptations whose identity and nature are unknown.

The effects of aldosterone in more distal regions—the thick ascending limb,[220] the distal tubule,[220] and the collecting duct,[271]—are unmistakable. The ascending limb reabsorbs sodium chloride hypertonically, leaving the tubule fluid in a hypotonic state as it enters the distal tubule.[279] (Chapter 50). The early distal tubule fluid of adrenalectomized animals is not as dilute[220] as that of normal animals because the ascending limb requires aldosterone to maintain normal reabsorptive rates. The late distal tubule is the principal site of potassium secretion in the nephron. As shown by Giebisch and his colleagues, the secretion is a passive process in which the movement of potassium into the tubular fluid depends on the magnitude of the electrochemical potential gradient.[239] However, the magnitude of the driving force for potassium depends in part on the rate of sodium reabsorption. Aldosterone controls sodium reabsorption; in the absence of the hormone, reabsorption is diminished and potassium secretion is also reduced.[220] When the hormone is present in large amounts, distal tubule sodium reabsorption is stimulated and potassium secretion also rises. Because aldosterone has opposite effects on sodium and potassium excretion, the urinary sodium/potassium concentration ratio falls when aldosterone acts. Changes in this ratio are often used as an indication that the hormone's level has changed.

It should be emphasized that sodium reabsorption and potassium secretion in the distal tubule are not tightly coupled in the operation of a single carrier mechanism. In fact, some observations suggest that aldosterone actions on the transport of the two ions may involve more than one hormone receptor. Restoration of sodium balance in adrenalectomized animals requires only small doses of aldosterone that do not alter potassium excretion,[264] and actinomycin D, an inhibitor of protein synthesis, blocks the action of aldosterone on urinary sodium excretion but not the action on potassium secretion.[206] In the rat, aldosterone increases potassium permeability of the luminal membrane of the distal tubule, an action that would enhance potassium secretion.[278] Finally, a mouse strain has been found that responds to aldosterone only with increased potassium secretion and not with sodium retention.[261] Aldosterone binds to kidney cell nuclei in this strain as well as in normal mice, but some of the characteristics of the binding differ. Taken together, these findings again tend to suggest that aldosterone-induced proteins have more than one action and that the effect of aldosterone is the result of changes at more than one site in the responding cells.

Aldosterone has also been shown to stimulate sodium reabsorption in collecting ducts of the inner medulla.[271] This effect probably accounts for the final adjustments of sodium excretion made under aldosterone control.

SYSTEMIC EFFECTS

As we have already noted, aldosterone is involved in two tasks—the regulation of extracellular fluid volume and the maintenance of acid-base and potassium balance.

Effects on extracellular fluid volume

Removal of aldosterone by adrenalectomy leads to an overall reduction of sodium reabsorption by the various organs that normally respond to the hormone. The result is a powerful and uncontrolled loss of extracellular fluid volume. Because the plasma is a part of the extracellular compartment, its volume drops, too, until the subject enters into a state of hypovolemic shock. This shock is a frequent cause of death in adrenally insufficient animals, and only access to an exogenous supply of a mineralocorticoid or to large amounts of salt and water can prevent the development of shock.

Inappropriately high levels of aldosterone stimulate reabsorption of sodium from luminal fluids and lead to isotonic expansion of the extracellular fluid volume. However, this expansion of the extracellular fluid volume does not continue indefinitely, even in the face of continuing excesses of aldosterone. After about 3 days of excessive mineralocorticoid administration a phenomenon known as "escape" occurs.[194] This response is a diuresis similar to that found when sodium chloride solutions are infused intravenously.[280] Under these circumstances, proximal tubule reabsorption of sodium chloride is inhibited, the glomerular filtration rate is increased, and the combination of these circumstances produces an increased excretion of sodium chloride and water in the urine. Isotonic expansion of extracellular fluids dilutes plasma proteins.[241] These proteins exert important osmotic effects across capillary membranes; when their concentration falls, glomerular filtration rate increases and proximal tubular reabsorption of fluid declines.[192,260] Thus escape is a response to an effect produced by excess aldosterone activity and does not result directly from the action of the hormone on the proximal tubule. Moreover, escape ameliorates the extracellular fluid volume expansion caused by aldosterone excess but has no effect on distal tubular actions of the hormone on potassium excretion.[186]

Effects on acid-base and potassium balance

The systemic effects of aldosterone on acid-base balance can be understood primarily as a result of its effects on the distal tubule. The aldosterone deficit of adrenal insufficiency inhibits sodium reabsorption, and potassium secretion falls as a result. Urinary potassium excretion decreases, and in the face of normal dietary intake, plasma potassium level rises. The mild acidosis that develops in adrenal insufficiency can probably be attributed to the hyperkalemia, as discussed earlier.

When aldosterone is administered in excess, potassium secretion in the distal tubule is stimulated and, unless dietary intake keeps pace, body potassium stores are depleted and hypokalemia results. For reasons already detailed, hypokalemia stimulates increased acidification of the urine and leads to metabolic alkalosis.

Aldosterone probably has a direct effect on urinary acidification in addition to the action mediated by changing plasma potassium levels. For example, the alkalosis produced by a given degree of negative potassium balance is greater if that negative potassium balance is caused by aldosterone excess than if it is caused by dietary potassium restriction.[254] Administration of sodium chloride can completely correct the alkalosis of dietary potassium restriction but has no effect on the alkalosis of aldosterone excess. These results suggest that aldosterone has a specific effect on tubular acidification mechanisms, but no studies have yet been performed to confirm this conclusion, except in the toad bladder, where aldosterone stimulates acidification of the mucosal solution, even when sodium transport is blocked.[233]

CONTROL OF ALDOSTERONE RELEASE
Renin-angiotensin system

Tigerstedt and Bergman[269] were the first to report that crude saline extracts of rabbit kidney, when injected into an anesthetized rabbit, produced generalized vasoconstriction and a rise in blood pressure. They named the active substance in their extracts renin. In 1934 Goldblatt et al.[213] found that a persistent rise in blood pressure could be produced in the dog by constriction of a renal artery. This work stimulated several groups of investigators to reexamine the nature of the substance that might cause renal hypertension,

and renin was implicated. Renin is an enzyme that hydrolyzes a glycoprotein of the alpha-2-globulin fraction of the plasma called renin substrate.[246] to liberate a decapeptide called angiotensin I.[259] This decapeptide is physiologically inert except as the substrate for still another enzyme, the converting enzyme, which hydrolyzes two amino acids and yields an octapeptide called angiotensin II.[258] This final product is one of the most potent vasoconstrictors known, and for many years has been the focus of considerable interest in research in high blood pressure. In 1960 it was found that angiotensin II could stimulate the zona glomerulosa to synthesize and release aldosterone.[227] Subsequent work has confirmed that it is one of the most important signals controlling aldosterone release.

Renin. Goormaghtigh[214] was the first to suggest that the cells of the renal afferent arteriole, now known to contain granules, were the site of renin formation. His suggestion was entirely speculative and based on morphologic arguments. Subsequently, it was shown that the density of granules in the granular cells depended on the rate of renin release, high rates of release being accompanied by increased granularity.[270] Finally, Cook succeeded in removing the granules from individual cells with micropipets and showed that the granules acted on plasma to form angiotensin I and subsequently angiotensin II,[196] confirming the origin of renin in granules of the afferent arteriolar cells.

Renin can be released either into the blood flowing through the afferent arteriole or into the interstitial space, whence it must leave the kidney in the lymphatics. Lever and Peart[231] have shown that the renal lymph concentration of renin is much higher than the renal venous concentration. The renal blood flow is, of course, considerably greater than the renal lymphatic flow, so that the flow of renin leaving in the blood may still be preponderant; nevertheless, renal lymph drains a substantial fraction of renin from the kidney.

Renin injected into a normal dog disappears with a half-life of 15 to 20 min.[248] The liver appears to be responsible for about 90% of the total plasma clearance of renin. Hepatic clearance is accomplished by extracting about 20% to 30% of the renin in blood flowing through the liver. When hepatic blood flow falls, as in experimentally produced thoracic inferior vena caval constriction, the hepatic clearance of renin does not change significantly because the extraction of renin increases to compensate for the decreased blood flow rate.[253]

Reninlike activity has also been found in organs other than the kidney. These include the pregnant uterus, brain, vascular walls, spleen, lung, and liver.[215] No activity has been detected in muscle, heart, or the nonpregnant uterus. A currently prevalent notion is that reninlike enzymes are probably isoenzymes of kidney renin. The function of the extrarenal isoenzymes remains obscure, although it has been suggested that uterine renin plays a role in parturition, since angiotensin is a potent stimulus to uterine contraction.

Renin substrate. The source of renin substrate appears to be the liver.[230] The normal plasma substrate concentration is usually about one fourth the level needed to achieve maximal reaction velocity, and so the rate of angiotensin I production follows first-order kinetics in vivo.[248] Thus the control of angiotensin production depends as much on the supply of substrate as it does on the supply of renin. Where both liver and kidneys are intact, renin substrate concentrations vary little; apparently the release of renin and its substrate are under common or mutually interdependent control.

Estrogens stimulate liver production of renin substrate. High substrate concentrations are found in pregnancy and also in women taking contraceptive steroids,[228] presumably because of the high levels of estrogenic steroids.

Converting enzyme. Converting enzyme activity was first shown in the plasma,[259] but it is now known that the major site of angiotensin I conversion to angiotensin II is in the lung.[244] The pulmonary enzyme binds angiotensin I 10 times more avidly than does the plasma enzyme and appears to be responsible for substantially more than half of the conversion in vivo.[223] Converting enzyme activity has also been found to be associated with the juxtaglomeular apparatus itself.[216] This observation has considerable significance for the current theories of the control of renin release because it implies that angiotensin II may be formed locally at the site of renin release and exert direct effects on the kidney.

Angiotensin I and II. The major factors regulating the production of angiotensin I are the concentrations of renin and renin substrate. Angiotensin I appears to be consumed mainly in the production of angiotensin II by the action of the converting enzyme. The vasopressor response to a single injected dose of angiotensin II lasts only 4 to 6 min.[245] The evanescence of angiotensin action arises from the fact that it is quickly removed from plasma.[193] Injected angiotensin is found to accumulate rapidly in the kidney, the adrenal gland, and the uterus. The binding of angiotensin II by these tissues appears to be related at least in part to the physiologic ac-

tion of the octapeptide. In addition to the binding related to physiologic activity, angiotensin II is hydrolyzed by enzymes in plasma. Although these enzymes are collectively called angiotensinase, the action is attributable to several nonspecific peptidases rather than to the action of a single specific enzyme.[219] Despite the lack of specificity, the destruction of angiotensin II by these enzymes is a quantitatively significant phenomenon and contributes to the short duration of angiotensin action. The short lifetime permits the renin-angiotensin system to participate in relatively high-frequency control mechanisms.

Structure of the macula densa–juxtaglomerular apparatus. Fig. 64-17 shows in some detail the juxtaglomerular apparatus, the site of renin formation. An invariant feature of renal morphology is the contact made between the vascular pole of the nephron at the glomerulus and the loop of Henle ascending from the kidney medulla. The point of contact on the tubular side is a specialized group of cells referred to collectively as the macula densa. The macula densa cells are in intimate contact with another group of specialized cells known as the mesangium. The mesangium lies between the afferent and efferent arterioles. Embedded in the wall of the afferent arteriole are modified vascular smooth muscle cells containing prominent granules that store renin.[214,270] Together the mesangium and the granular cells comprise the juxtaglomerular apparatus. There is some disagreement about the arterioles (afferent or efferent) with which the macula densa cells make direct contact; however, the macula densa and mesangium are always in contact,[187] and the mesangium is in contact with both arterioles. The interest in the exact nature of the cell contacts arises in the attempt to define a signal flow path.

Function of macula densa–juxtaglomerular apparatus. There are three currently accepted ideas of how renin release is regulated. None explains all the data, and none excludes the others. It is possible and even likely that all three are to some extent correct. The basic observation each of these theories attempts to explain is that reduction of arterial blood pressure[23] or reduction of extracellular fluid volume,[205] which may result in arterial hypotension, leads to increased release of renin from the juxtaglomerular apparatus.

RENAL BARORECEPTOR THEORY. A potent stimulus for renin release is a fall in arterial blood pressure.[23] This response occurs even when the kidney is denervated and when glomerular filtration is abolished by acute renal fail-

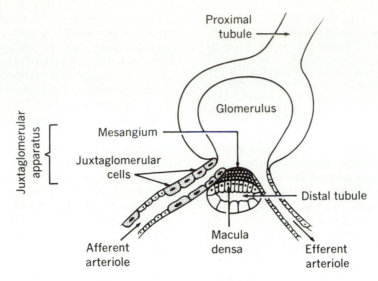

Fig. 64-17. Macula densa–juxtaglomerular apparatus. Reader is warned that terminology varies from one author to the next, and none is regarded as a final authority. Collection of mesangial cells is sometimes referred to as Pollkiasen, or lacis, and juxtaglomerular apparatus is a term that sometimes refers only to granular cells, sometimes to granular cells and mesangium (as in this figure), and sometimes to granular cells, mesangium, and macula densa. Macula densa is specialized collection of tubular cells that are always in direct contact with mesangium and in variable contact with afferent and efferent arterioles: Macula densa is demarcation between thick ascending limb of Henle's loop and distal tubule. Granular cells are modified vascular smooth muscle cells whose granules store renin.

ure.[190] These findings suggest that elements associated with the juxtaglomerular apparatus sense the arterial or arteriolar blood pressure and adjust renin release accordingly. The location of the sensor and the details of the signaling are unknown, although it has been assumed that afferent arteriolar wall tension is the critical parameter.[203,270]

RENAL NERVES. The kidneys are innervated by sympathetic postganglionic fibers; stimulation of these renal nerves leads to a prompt increase of renin release together with renal vasoconstriction. Norepinephrine infused directly into the renal artery provokes the same response,[275] and propanolol, a beta adrenergic antagonist, inhibits renin release.[185] At present it is not possible to decide whether renal sympathetic nerves excite a mechanism for renin release that is different from the others or whether neurally induced vasoconstriction at arteriolar levels anterior to the juxtaglomerular apparatus produces the same sequence of events as any other cause of renal artery hypotension. For example, mild hemorrhage can trigger cardiovascular reflexes from the venous circulation while leaving arterial blood pressure essentially unaltered. Under these circumstances, renin release increases despite the persistence of normal arterial pressure.[222] This response could be accounted for by increased stimulation through sympathetic nerves, leading to vasoconstriction in the renal artery, so that the arterial pressure at the renal baroreceptor site is reduced.

MACULA DENSA THEORY. Goormaghtigh, who suggested that granular cells might main renin, also proposed that the close contact between the macula densa and the remainder of the juxtaglomerular apparatus provided a pathway for information flow between the distal nephron and the arterioles whose resistances control glomerular filtration. If the concentration of some component of distal tubular fluid varied with flow rate, a change in filtration rate could alter the concentration of this substance, and the juxtaglomerular apparatus could use the news of this event either for adjusting arteriolar resistances until filtration rate returned to normal, for modifying the rate of renin release, or for both purposes. As evidence accumulated to support both functions for the macula densa signal pathway, it was perhaps natural to assume that the two functions were interrelated and that angiotensin II was the transmitter for the glomerular filtration feedback response, but this final supposition has not been supported.

The material whose concentration varies with flow is sodium chloride. The flow rate dependence arises because the thick ascending limb reabsorbs NaCl hypertonically; in any given element of tubular volume the extent to which the transport process lowers the NaCl concentration varies inversely with volume flow rate. As flow rate through the thick ascending limb falls, so too does the NaCl concentration in tubular fluid at the macula densa. The fall in NaCl concentration signals the juxtaglomerular apparatus to increase renin secretion. Flow through the loop of Henle varies with glomerular filtration rate, and a reduction of arterial blood pressure, which activates the baroreceptor mechanism, would also reduce filtration rate. It has been difficult, therefore, to establish that two distinct mechanisms exist, but two studies support the existence of a separate macula densa mechanism mediating renin release. Thurau, et al.[268] perfused single loops of Henle in a retrograde direction from the distal tubule. The perfusate NaCl concentration was varied, and after a 20 min perfusion the juxtaglomerular apparatus was removed and the renin activity analyzed. There was a direct linear relationship between NaCl concentration and renin activity. In the second study, solutions containing high concentrations of protein were infused into the renal artery to reduce golmerular filtration rate without changing glomerular capillary hydrostatic pressure. Renin secretion increased.[218] Although the macula densa and baroreceptor mechanisms can be differentiated experimentally, the two will act in concert in most circumstances.

When renin secretion is increased by the macula densa or by any other mechanism aldosterone secretion increases, which in turn stimulates retention of NaCl. Because aldosterone acts on the thick ascending limb, the primary result would seem to be still further reduction of tubular fluid NaCl concentration and a further increase of renin secretion. This potentially unstable scenario does not occur, however, because the salt retention raises arterial blood pressure; glomerular filtration rate rises with it, so that flow of tubular fluid and NaCl concentration in the tubular fluid increase.

Finally, most workers have abandoned the view that angiotensin is the transmitter in the autoregulatory response because experiments such as those in which hypotension stimulates renin secretion generally provoke dilatation of the afferent arteriole as part of the renal attempt to autoregulate; those experiments that inhibit renin secretion cause afferent arterioles to constrict. It is hard to imagine how angiotensin II, a potent vasoconstrictor, could produce vasodilatation in the afferent arteriole.

Another potent stimulus to renin release is chronic dietary salt deprivation.[205] Restriction of salt intake will lead initially to reduction of the extracellular fluid volume until salt excretion can be reduced to match the new intake level. The signal that provokes the renin response is unknown. Arterial blood pressure is normal,[189] as is glomerular filtration rate, during states of salt deprivation sufficient to cause severe elevation of plasma renin activity. The search for an extracellular fluid volume receptor has been long and not very fruitful. Whatever that receptor may be, and it may be several different things, one of its most important agents is the renin-angiotensin-aldosterone system.

Renin-angiotensin-aldosterone negative feedback. Whatever the mechanism causing it, any increase in renin release will lead to an increased rate of angiotensin II production and will thus stimulate the adrenal zona glomerulosa to increase aldosterone secretion. Since aldosterone acts to increase extracellular fluid volume, which in turn inhibits renin release, a powerful negative feedback loop is closed.

Angiotensin II also inhibits renin release from the juxtaglomerular apparatus.[276] This inhibitory feedback is similar to that seen with numerous other hormone systems; its physiologic significance is not known.

Other actions of angiotensin II. Renin and angiotensin II were first discovered because of the vasoconstrictor action of angiotensin, and many workers have wondered if this system participates in blood pressure regulation. Recent studies with competitive inhibitors of angiotensin I and angiotensin II have shown that it does, in some circumstances. In subjects on a normal salt intake, administration of these angiotensin inhibitors had no effect on arterial blood pressure whether the subjects were recumbent or upright. Chronically salt-deprived subjects with elevated plasma renin levels given an inhibitor suffer a fall in blood pressure when they stand upright, but not when they are recumbent. Animals that have received an inhibitor and are then subjected to hemorrhage become more hypotensive and remain so longer than untreated controls.[251] Thus when renin secretion is stimulated, the renin angiotensin system plays a significant role in blood pressure regulation.

Angiotensin II is also active in the brain when it is injected into the third ventricle, where it produces four responses: thirst, increased vasopressin secretion, increased blood pressure, and increased ACTH secretion.[211,240] There have been reports that renin is made in the brain and that there is a converting enzyme present, so that angiotensin II could act as a neurotransmitter. Whether it does and whether the responses it causes are physiologically significant remain questions to be answered. Angiotensin II may also be a CRF; it can release ACTH from the pituitary gland.[240]

Plasma potassium concentration

An elevation of potassium concentration by as little as 0.5 mEq/L is a potent stimulus to aldosterone production but has no effect on corticosterone or cortisol output.[209] High plasma potassium levels stimulate aldosterone release; aldosterone stimulates potassium excretion and lowers plasma potassium level, which diminishes the stimulus intensity acting on aldosterone. The potassium-aldosterone system is a simple yet important negative feedback loop.

A high potassium intake maintained in chronic feeding experiments produces a widening of the zona glomerulosa[200] and a sustained increase of aldosterone output. This increased aldosterone production appears in part and possibly in full to be responsible for renal adaptations to high dietary potassium intake.[184]

Large amounts of potassium salts administered suddenly to normal animals can produce death by ventricular fibrillation. Comparable doses given to animals who have been chronically fed high-potassium diets are not fatal but lead instead to a brisk secretion of potassium into the urine.[267] This secretion occurs in the distal tubule and collecting ducts and is simply a quantitatively more vigorous response than is possible in unconditioned animals.[281] The difference in behavior of these distal sgements is at least in part attributable to the high levels of aldosterone found in the potassium-fed animals, but definitive studies in which the hormone level was both measured and controlled have not been done. Potassium stimulation of aldosterone secretion is mediated by cyclic AMP, as discussed earlier.

Adrenocorticotropic hormone

ACTH also stimulates aldosterone secretion.[257] Its importance in this regard is highly species specific. In the rat, sodium deprivation stimulates aldosterone secretion only when the pituitary gland is present, and angiotensin II appears to be of less importance in this species than in man and dog.[247] In man and dog, species for which the angiotensin II system is a potent stimulus to aldosterone secretion, chronic stimulation by ACTH may have little or no effect on aldosterone production.[188] However, even in these species aldosterone secretion may fall considerably in the absence of the pituitary gland and may respond less vigorously to stimuli that normally cause increased release of the mineralocorticoid.[232] Because hypophysectomy removes more than just ACTH, the sensitivity to anterior pituitary removal may reflect the absence of other hormones. Growth hormone has been implicated.[247] However, when negative feedback loops, such as those previously described involving plasma potassium and angiotensin II, act simultaneously with an ACTH stimulus, chronically elevated levels of ACTH may be compensated for by a reduction in the stimulus intensity from the other two factors. If such compensations are permitted to occur, the sum of all stimuli can remain unchanged so that the rate of aldosterone secretion will remain normal. Apparent unresponsiveness to ACTH does not necessarily imply that ACTH is without effect but may simply reflect the presence of several degrees of freedom in the control of aldosterone secretion.

Sodium balance

Yet another very potent stimulus to aldosterone secretion is dietary salt restriction. There is little doubt that, at least in man and dog, this dietary restriction leads to a contraction of the extracellular fluid and blood volume and to increased renin and angiotensin II production and, further, that high angiotensin II levels can produce aldosterone secretion. What is not clear, however, is that angiotensin II or any of the other factors mentioned, either singly or in combination, can account entirely for all the changes in aldosterone secretion seen in response to dietary salt restriction. Angiotensin II, high plasma potassium level, low plasma sodium level (which rarely occurs in sodium restriction), and ACTH all appear to have their action early in the biosynthetic pathway, probably in the step from cholesterol to pregnenolone. A number of studies now indicate that dietary salt restriction stimulates not only this early step, but also the conversion of corticosterone to aldosterone.[191,242] None of the other signals known to influence aldosterone production appears to be capable of stimulation at this step. Moreover, as discussed, even the release of renin in response to salt deprivation lacks an adequate explanation.

TOTAL ALDOSTERONE SYSTEM

The total adrenal mineralocorticoid system is shown in Fig. 64-18. The figure shows many of the components and processes that have been discussed and includes some of the components in the nervous system that will be involved in the system response to perturbations.

For the sake of clarity the connections between the aldosterone and water systems have been omitted, but it should be remembered that the two systems interact at several points. Atrial stretch receptors, when stimulated, act to inhibit the release of antidiuretic hormone (ADH) from the posterior pituitary gland. Circumstances that increase central blood volume inhibit release of ADH as well as aldosterone, and reduced blood volume stimulates release of both hormones. Mild hemorrhage, for example, provokes the release of renin[222] and of ADH by activation of neural pathways; the interaction is reinforced because angiotensin II also stimulates the release of ADH directly.[23] Aldosterone and ADH both act to restrict fluid losses, ADH by increasing water reabsorption in the distal tubule and collecting duct, aldosterone by stimulating sodium reabsorption.

Although a block diagram such as the one shown in Fig. 64-18 is helpful in showing how the components are connected, it offers no help in evaluating the quantitative significance of the parts. This problem becomes particularly important when there is redundancy. For example,

because there are four inputs to zona glomerulosa cells, it is not unreasonable to ask if all are equally important. Plasma potassium and angiotensin II concentrations have received the most attention and are probably more important than the others. We might still wish to know which of these is more important, but there is no simple answer to this question. The context of the experiments used to answer it is all important. Suffice it to say that under appropriate circumstances, loss of either volume regulation or potassium regulation can be life threatening.

ALDOSTERONE RHYTHMS

As is true of glucocorticoids, aldosterone secretion undergoes a circadian rhythm. Maximum secretion occurs in the morning after arising; minimum secretion occurs after retiring.[277] In contrast to the glucocorticoid rhythm, the aldosterone pattern depends on the activity cycle because it is abolished promptly by enforced recumbency and reversed within 1 day after reversal of the day/night activity pattern.[273] When subjects stand upright, renin secretion increases promptly, so that activation of the renin-angiotensin system is the most likely cause of the aldosterone rhythm.

Urinary sodium and potassium excretion also have circadian rhythms. The potassium excretion rhythm coincides with the aldosterone rhythm during normal activity patterns, but the potassium rhythm persists with no change when the aldosterone rhythm is abolished or phase shifted. Although aldosterone is probably not immediate-

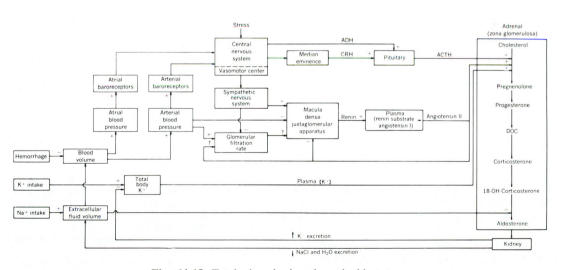

Fig. 64-18. Total adrenal mineralocorticoid system.

ly responsible for the potassium rhythm in normal subjects, the potassium rhythm is abolished in adrenal insufficiency[101,277] and distorted in primary aldosteronism.[274] It is not understood how aldosterone excess or deficit interferes with the potassium rhythm.

The sodium rhythm has a prominent midday maximum in recumbent subjects. Because bed rest abolishes the aldosterone rhythm, this persistent sodium rhythm connot be attributed to aldosterone.[273] The glomerular filtration rate also has a circadian rhythm with a maximum and a minimum at the same times as the recumbent sodium excretion maximum and minimum. The recumbent sodium rhythm is probably entrained in the glomerular filtration rate rhythm. The filtration rate rhythm is unaffected by postural variations as long as the basic day/night pattern is maintained, and, since glucocorticoids increase glomerular filtration rate, the recumbent sodium rhythm is possibly related to the glucocorticoid rhythm.

Normal activity cycles with upright posture shift the sodium rhythms. The circadian pattern persists, but maximum sodium excretion now occurs at night, after retiring. The time of peak sodium excretion then corresponds to the time of minimal aldosterone secretion, and maximal aldosterone secretion corresponds to that of minimal sodium excretion. In this case the aldosterone rhythm appears to drive the sodium rhythm.

DISORDERS OF ADRENOCORTICAL FUNCTION

Clinically important disorders of adrenocortical function comprise five categories, some of which represent composite diseases with multiple causes. The categories are (1) Cushing's syndrome (hyperglucocorticalism), (2) Addison's disease (generalized hypoadrenocorticalism), (3) iatrogenic hypoglucorticalism, (4) primary aldosteronism (Conn's syndrome), and (5) the adrenogenital syndrome.

Cushing's syndrome can be caused by lesions in the hypothalamus, by ACTH-secreting tumors located in the pituitary gland or elsewhere, including the lung; by tumors of the adrenal cortex; by adrenal hyperplasia and hypertrophy of unknown origin; or by overexuberant corticosteroid therapy. Since the appropriate treatments of the various forms of Cushing's disease are therefore necessarily very different, exact diagnosis is essential.

Adrenocortical hypofunction can be caused by destruction of the adrenal glands by tuberculosis (now rare), by surgical removal of the adrenal glands in the treatment of Cushing's syndrome, or as a consequence of certain disorders of the pituitary gland and probably also of the brain, including head injury. Iatrogenic hypoglucocorticalism occurs in patients who have received long-term anti-inflammatory or immunosuppressive therapy with glucocorticoids and have developed atrophy of the adrenal fasciculata and reticularis layers because of chronic feedback inhibition of ACTH release by the exogenous corticosteroids.[40] If the physician stops the therapy abruptly, the patients have low or zero glucocorticoid levels and may die.

Primary aldosteronism is usually caused by a tumor of the adrenal cortex. The tumors are benign, but the hyperaldosteronism is serious, and removal of the tumor-bearing adrenal gland is necessary.

The adrenogenital syndrome is caused by excessive production of androgens by the adrenal cortex. In otherwise normal men the disease may have almost no clinical manifestations except for increased secretion of 17-ketosteroids in the urine. In women, however, and in prepuberal children, striking masculinization can occur. One form of the disease is congenital, but other forms involve tumors.

The relatively elaborate and highly branched pathways involved in steroid biosynthesis have the potential for manifesting many inborn errors of metabolism associated with defects in one or more enzymes. Precise biochemical studies of steroid metabolites in urine often uncover exotic patterns, sometimes not associated with clinically detectable illness, that suggest the existence of a striking degree of biochemical individuality with respect to these processes in human beings.

REFERENCES
General reviews

1. Applezweig, N.: Steroid drugs, New York, 1962, McGraw-Hill Book Co.
2. Burgus, R., and Guillemin, R.: Hypothalamic releasing factors, Annu. Rev. Biochem. **39:**499, 1970.
3. Christy, N. P. editor: The human adrenal cortex, New York, 1971, Harper & Row, Publishers.
4. Cope, C. L.: Adrenal steroids and disease, Philadelphia, 1965, J. B. Lippencott Co.
5. de Weid, D., and Weijnen, J. A. W. M., editors: Pituitary, adrenal and the brain, Prog. Brain Res. **32:**entire issue, 1970.
6. Edelman, I. S., and Fimognari, G. M.: On the biochemical mechanism of action of aldosterone, Recent Prog. Horm. Res. **24:**1, 1968.
7. Eisenstein, A. B., editor: The adrenal cortex, Boston, 1967, Little, Brown & Co.
8. Fisher, J. W.: Kidney hormones, New York, 1971, Academic Press, Inc.
9. Gann, D. S., and Cryer, G. L.: Models of adrenal cortical control, Adv. Biomed. Engin. **2:**1, 1972.

10. Greep, R. O., and Astwood, E. B., editors: Handbook of physiology. Section 7, endocrinology, the adrenal gland, Baltimore, 1975, The Williams & Wilkins Co., vol. 6.

11. Grieco, M. H., and Cushman, P.: Adrenal glucocorticoids after twenty years—a review of their clinically relevant consequences, J. Chronic Dis. **22:**637, 1970.

12. Harris, G. W., and Donovan, B. T.: The pituitary gland, Berkeley, Calif., 1966, University of California Press, vols. 1 to 3.

13. Jones, I. C.: The adrenal cortex, Cambridge, Mass., 1957, Cambridge University Press.

14. Kendall, J. W.: Feedback control of adrenocorticotropic hormone secretion. In Martini, L., and Ganong, W. F.: Frontiers in neuroendocrinology, New York, 1971, Oxford University Press, Inc., p. 177.

15. Lee, M. R.: Renin and hypertension, London, 1969, Lloyd-Luke, Ltd.

16. McKerns, K. W., editor: Functions of the adrenal cortex, New York, 1968, Appleton-Century-Crofts, vols. 1 and 2.

17. Meites, J., editor: Hypophysiotropic hormones of the hypothalamus: assay and chemistry, Baltimore, 1970, The Williams & Wilkins Co.

18. Müller, J.: Regulation of aldosterone biosynthesis, New York, 1971, Springer Verlag New York, Inc.

19. Sayers, G., and Portanova, R.: Regulation of the secretory activity of the adrenal cortex: cortisol and corticosterone. In Greep, R. O., and Astwood, E. B., editors: Handbook of physiology. Section 7, endocrinology, Baltimore, 1975, The Williams & Wilkins Co., vol. 6, p. 41.

20. Schneeberg, N. G.: Essentials of clinical endocrinology, St. Louis, 1970, The C. V. Mosby Co.

21. Soffer, L. J., Dorfman, R. I., and Gabrilove, J. L.: The human adrenal gland, Philadelphia, 1961, Lea & Febiger.

22. Urquhart, J.: Physiological actions of adrenocorticotropic hormone. In Greep, R. O., and Astwood, E. B., editors: Handbook of physiology. Section 7, endocrinology, Baltimore, 1974, The Williams & Wilkins Co., vol. 6, pt. 2, p. 133.

23. Vander, A. J.: Control of renin release, Physiol. Rev. **47:**359, 1967.

24. Yates, F. E., and Maran, J. W.: Stimulation and inhibition of adrenocorticotropin release. In Greep, R. O., and Astwood, E. B., editors: Handbook of physiology. Section 7, endocrinology, Baltimore, 1974, The Williams & Wilkins Co., p. 367.

25. Yates, F. E., and Maran, J. W.: The physiology of the mammalian hypothalamo-adenohypophysial-adrenal glucocorticoid system—a new hypothesis. Chronobiologia **1**(suppl. 1):195, 1975.

26. Yates, F. E., and Urquhart, J.: Control of plasma concentrations of adrenocortical hormones, Physiol. Rev. **42:**359, 1962.

27. Yates, F. E., Russell, S. M., and Maran, J. W.: Brain adenohypophysial communication in mammals, Annu. Rev. Physiol. **33:**393, 1971.

Original papers
Glucocorticoids

28. Abe, K., and Hiroshige, T.: Changes in plasma corticosterone and hypothalamic CRF levels following intraventricular injection or drug-induced changes of brain biogenic amines in the rat, Neuroendocrinology **14:**195, 1974.

29. Allen, J. P., et al.: Immunoreactive ACTH in cerebrospinal fluid, J. Clin. Endocrinol. Metab. **38:**586, 1974.

30. Antunes, J. L., et al.: Paraventricular nucleus: the origin of vasopressin secreting terminals on hypophysial portal vessels in the monkey, Endocrinology **98:** A324, 1976.

31. Baertschi, A. J., Ward, D. G., and Gann, D. S.: Role of atrial receptors in the control of ACTH, Am. J. Physiol. **213:**692, 1976.

32. Bailey, E., and West, H. F.: The secretion, interconversion and catabolism of cortisol, cortisone and some of their metabolites in man, Acta Endocrinol. **62:**339, 1969.

33. Ballard, P. L., et al.: General presence of glucocorticoid receptors in mammalian tissues, Endocrinology **94:**998, 1974.

34. Bradbury, M. W. B., et al.: Stimulation electrically and by acetylcholine of the rat hypothalamus in vitro, J. Physiol. **239:**269, 1974.

35. Bradshaw, R. A., and Frazier, W. A.: Hormone receptors as regulators of hormone action, Curr. Top. Cell. Regul. **12:**1, 1977.

36. Braun, T., and Hechter, O.: Glucocorticoid regulation of ACTH sensitivity of adenyl cyclase in rat cell membranes, Proc. Natl. Acad. Sci. USA **66:**995, 1970.

37. Brodish, A.: Tissue corticotropin releasing factors, Fed. Proc. **36:**2088, 1977.

38. Carroll, B. J., Heath, B., and Jarrett, D. B.: Corticosteroids in brain tissue, Endocrinology **97:**290, 1975.

39. Chambers, J. W., and Brown, G. M.: Neurotransmitter regulation of growth hormone and ACTH in the Rhesus monkey: effects of biogenic amines, Endocrinology **98:**420, 1976.

40. Christy, N. P.: Iatrogenic Cushing's syndrome. In Christy, N. P., editor: The human adrenal cortex, New York, 1971, Harper & Row, Publisher, p. 395.

41. Cook, D. M., et al.: The effect of acute or chronic ether stress on plasma ACTH concentration in the rat, Endocrinology **93:**1019, 1973.

42. Coulson, P. B., and Pavlik, E. J.: Effects of estrogen and progesterone on cytoplasmic estrogen receptor and rates of protein synthesis in rat uterus, J. Steroid Biochem. **8:**205, 1977.

43. Dallman, M. F., and Jones, M. T.: Corticosteroid feedback control of ACTH secretion; effect of stress-induced corticosterone secretion on subsequent stress responses in the rat, Endocrinology **92:**1367, 1973.

44. Dallman, M. F., and Yates, F. E.: Anatomical and functional mapping of central neural input and feedback pathways of the adrenocortical system, Mem. Soc. Endocrinol. **17:**39, 1968.

45. Dallman, M. F., and Yates, F. E.: Dynamic asymmetries in the corticosteroid feedback path and distribution-metabolism-binding elements of the adrenocortical system, Ann. N. Y. Acad. Sci. **156:**696, 1969.

46. Dallman, M. F., DeManincor, D., and Shinsako, J.: Diminishing corticotrope capacity to release ACTH during sustained stimulation: the twenty-four hours after bilateral adrenalectomy in the rat, Endocrinology **95:**65, 1974.

47. Dallman, M. F., et al.: Corticosteroid feedback control of ACTH secretion: rapid effects of bilateral adrenalectomy on plasma ACTH in the rat, Endocrinology **91:** 961, 1972.

48. deBodo, R. C., and Altsuler, N.: Insulin hypersensitivity and physiological insulin antagonists, Physiol. Rev. **38:**389, 1958.

49. DeDuve, C.: Effects of fat-soluble compounds on lysosomes in vitro, Biochem. Pharmacol. **9**:97, 1962.

50. DeKloet, E. R., Van der Vies, J., and deWied, D.: The site of the suppressive action of dexamethasone on pituitary-adrenal activity, Endocrinology **94**:61, 1974.

51. DeKloet, E. R., Wallach, G., and McEwen, B. S.: Differences in corticosterone and dexamethasone binding to rat brain and pituitary, Endocrinology **96**:598, 1975.

52. Engeland, W. C., and Dallman, M. F.: Neural mediation of compensatory adrenal growth, Endocrinology **99**:1659, 1976.

53. Everitt, B., and Herbert, J.: Adrenal glands and sexual receptivity in female rhesus monkeys, Nature **222**:1065, 1969.

54. Fain, J. N., and Czech, M. P.: Glucocorticoid effects on lipid mobilization and adipose tissue metabolism. In Greep, R. O., and Astwood, E. B., editors: Handbook of physiology. Section 7, endocrinology, Baltimore, 1975, The Williams & Wilkins Co., vol. 6, p. 169.

55. Fain, J. N., and Saperstein, R.: The involvement of RNA synthesis and cyclic AMP in the activation of fat cell lipolysis by growth hormone and glucocorticoids. In Jeanrenaud, B., and Hepp, B., editors: Adipose tissue: regulation and metabolic functions, New York, 1970, Academic Press, Inc.

56. Fawcett, D. W., Long, J. A., and Jones, A. L.: The ultrastructure of endocrine glands, Recent Prog. Horm. Res. **25**:315, 1969.

57. Feldman, D., and Loose, D.: Glucocorticoid receptors in adipose tissue, Endocrinology **100**:398, 1977.

58. Feldman, D., et al.: Cytoplasmic glucocorticoid binding proteins in bone cells, Endocrinology **96**:29, 1975.

59. Fischer, J. L., and Moriarty, C. M.: Control of bioactive corticotropin release from the neuro-intermediate lobe of the rat pituitary in vitro, Endocrinology **100**:1047, 1977.

60. Fuxe, K., et al.: Brain and pituitary-adrenal interactions—studies on central monoamine neurons. In Brodish, A., and Redgate, E. S., editors: Brain-pituitary-adrenal interrelationships, Basel, Switzerland, 1973, S. Karger AG.

61. Gann, D. S.: Endocrine control of plasma protein and volume, Surg. Clin. North Am. **56**:1135, 1976.

62. Gann, D. S., Cryer, G. L., and Pirkle, J. C., Jr.: Physiological inhibition and facilitation of adrenocortical response to hemorrhage, Am. J. Physiol. **232**:R5, 1977.

63. Gann, D. S., et al.: Neural control of ACTH release in response to hemorrhage, Ann. N.Y. Acad. Sci. **297**:477, 1977.

64. Ganong, W. F.: Evidence for a central noradrenergic system that inhibits ACTH secretion. In Knigge, K. M., Scott, D. E., and Weindl, A., editors: Brain-endocrine interaction, median eminence: structure and function, Basel, Switzerland, 1972, S. Karger AG.

65. Gibbs, F. P.: Area of pons necessary for traumatic stress-induced ACTH release under pentobarbital anesthesia, Am. J. Physiol. **217**:84, 1969.

66. Glantz, S. A., et al.: A compartment description for cortisol secretion, distribution, binding and metabolism in man, IEEE Trans. Biomed. Eng. **BME-23**(1):36, 1976.

67. Gonzalez-Luque, A., et al.: Stimulation of corticotropin release by corticotropin-releasing factor (CRF) or by vasopressin following intrapituitary infusions in un-anesthetized dogs: inhibition of the responses by dexamethasone, Endocrinology **86**:1134, 1970.

68. Grant, N.: Metabolic effects of adrenal glucocorticoid hormones—other effects. In Eisenstein, A. B., editor: The adrenal cortex, Boston, 1967, Little, Brown & Co.

69. Greer, M. A., et al.: Evidence that the pars intermedia and pars nervosa of the pituitary do not secrete functionally significant quantities of ACTH, Endocrinology **96**:718, 1975.

70. Gross, S. R., Aronow, L., and Pratt, W. B.: The active transport of cortisol by mouse fibroblasts growing in vitro, Biochem. Biophys. Res. Comm. **32**:66, 1968.

71. Halász, B., Slusher, M. A., and Gorski, R. A.: Adrenocorticotrophic hormone secretion in rats after partial or total deafferentation of the medial basal hypothalamus, Neuroendocrinology **2**:43, 1967.

72. Hartman, F. A.: et al.: The hormone of the adrenal cortex, Am. J. Physiol. **86**:353, 1928.

73. Haynes, R. C., Jr.: Theories on the mode of action of ACTH in stimulating secretory activity of the adrenal cortex. In Greep, R. O., and Astwood, E. B., editors: Handbook of physiology. Section 7, endocrinology, Baltimore, 1975, The Williams & Wilkins Co., vol. 6, p. 69.

74. Hedge, G. A., and DeWied, D.: Corticotropin and vasopressin secretion after hypothalamic implantation of atropine, Endocrinology **88**:1257, 1971.

75. Hedge, G. A., and Smelik, P. G.: The action of dexamethasone and vasopressin on hypothalamic CRF production and release, Neuroendocrinology **4**:242, 1969.

76. Hellman, L., et al.: Cortisol is secreted episodically by normal man, J. Clin. Endocrinol. Metab. **30**:411, 1970.

77. Hench, P. S., Kendall, E. C., Slocumb, C. H., and Polley, H. F.: The effect of a hormone of the adrenal cortex (17-hydroxy-ll-dehydrocorticosterone: compound E) and of pituitary adrenocorticotropic hormone on rheumatoid arthritis. Preliminary report, Proc. Mayo Clin. **24**:181, 1949.

78. Henderson, I. C., Fischel, R. E., and Loeb, J. N.: Suppression of liver DNA synthesis by cortisone, Endocrinol. **88**:1471, 1971.

79. Henkin, R. I.: The role of adrenal corticosteroids in sensory processes. In Greep, R. O., and Astwood, E. B., editors: Handbook of physiology. Section 7, endocrinology, Baltimore, 1975, The Williams & Wilkins Co., vol. 6, p. 209.

80. Hillhouse, E. W., and Jones, M. T.: Effect of bilateral adrenalectomy and corticosteroid therapy on the secretion of corticotrophin-releasing factor activity from the hypothalamus of the rat in vitro, J. Endocrinol. **71**:1, 1976.

81. Hillhouse, E. W., Burden, J., and Jones, M. T.: The effect of various putative neurotransmitters on the release of corticotropin releasing hormone from the hypothalamus of the rat in vitro. I. The effect of acetylcholine and noradrenaline, Neuroendocrinology **17**:1, 1975.

82. Imura, H., Nakai, Y., and Yoshimi, T.: Effect of 5-hydroxytryptophan (5-HTP) on growth hormone and ACTH release in man, J. Clin. Endocrinol. Metab. **36**:204, 1973.

83. Jones, M. T., and Hillhouse, E. W.: Structure-activity relationship and the mode of action of corticosteroid feedback on the secretion of corticotrophin-releasing

factor (Corticoliberin), J. Steroid Biochem. **7:**1189, 1976.

84. Jones, M. T., Brush, F. R., and Neame, R. L. B.: Characteristics of fast feedback control of corticotrophin release by corticosteroids, J. Endocrinol. **55:**489, 1972.

85. Jones, M. T., Gillham, B., and Hillhouse, E. W.: The nature of corticotropin releasing factor from rat hypothalamus in vitro, Fed. Proc. **36:**2104, 1977.

86. Jones, M. T., Hillhouse, E., and Burden, J.: Secretion of corticotropin-releasing hormone in vitro. In Martini, L., and Ganong, W. F., editors: Frontiers in neuroendocrinology, New York, 1976, Raven Press, vol. 4, p. 195.

87. Jones, M. T., et al.: Evidence for dual corticosteroid-receptor mechanisms in the feedback control of adrenocorticotrophin secretion, J. Endocrinol. **60:**223, 1974.

88. Kalsner, S.: Mechanism of hydrocortisone potentiation of responses to epinephrine and norepinephrine in rabbit aorta, Circ. Res. **24:**383, 1969.

89. Kaneko, M., and Hiroshige, T.: Fast, rate-sensitive corticosteroid negative feedback during stress, Am. J. Physiol. **234**(1):R39, 1978.

90. Kaneko, M., and Hiroshige, T.: Site of fast, rate-sensitive feedback inhibition of corticotropin secretion during stress. Am. J. Physiol. **234**(1):R46, 1978.

91. Kawai, A., and Yates, F. E.: Interference with feedback inhibition of adrenocorticotropin release by protein binding of corticosterone, Endocrinology **79:**1040, 1966.

92. Keller, N., Richardson, U. I., and Yates, F. E.: Protein binding and the biological activity of corticosteroids: in vivo induction of hepatic and pancreatic alanine aminotransferases by corticosteroids in normal and estrogen-treated rats, Endocrinology **84:**49, 1969.

93. Kendall, J. W., Grimm, Y., and Shimshak, G.: Relation of cerebrospinal fluid circulation to the ACTH-suppressing effects of corticosteroid implants in the rat brain, Endocrinology **85:**200, 1969.

94. Kendall, J. W., et al.: The importance of stimulus intensity and duration of steroid administration in suppression of stress-induced ACTH secretion, Endocrinology **90:**525, 1972.

95. Knigge, K. M.: Adrenocortical response to stress in rats with lesions in hippocampus or amygdala, Proc. Soc. Exp. Biol. Med. **108:**18, 1961.

96. Kramer, R. E., Greiner, J. W., and Colby, H. D.: Site of action of growth hormone on adrenocortical steroidogenesis in rats, Endocrinology **101:**297, 1977.

97. Krey, L. C., et al.: Surgical disconnection of the medial basal hypothalamus and pituitary function in the rhesus monkey. II. GH and cortisol secretion, Endocrinology **96:**1088, 1975.

98. Krieger, D. T., and Rizzo, F.: Serotonin mediation of circadian periodicity of plasma 17-hydroxycorticosteroids, Am. J. Physiol. **217:**1703, 1969.

99. Krieger, D. T., Kreuzer, J., and Rizzo, F. A.: Constant light: effect on circadian pattern and phase reversal of steroid and electrolyte levels in man, J. Clin. Endocrinol. Metab. **29:**1634, 1969.

100. Krieger, D. T., Liotta, A., and Brownstein, M. J.: Corticotropin releasing factor distribution in normal and Brattleboro rat brain, and effect of deafferentation, hypophysectomy and steroid treatment in normal animals, Endocrinology **100:**227, 1977.

101. Krieger, D. T., et al.: Abolition of circadian periodicity of plasma 17-OHCS levels in the cat, Am. J. Physiol. **215:**959, 1968.

102. Krieger, D. T., et al.: Characterization of the normal temporal pattern of plasma corticosteroid levels, J. Clin. Endocrinol. Metab. **32:**266, 1971.

103. Lefcourt, A. M.: The open loop gain of cortisol feedback system in dogs. Ph.D. dissertation, Los Angeles, 1977, University of Southern California.

103a. Lefcourt, A. M., and Yates, F. E.: Feedback in the adrenocortical system is very significant in the basal state, Fed. Proc. **36:**275, 1977.

104. Levine, S., Goldman, L., and Conyer, G. D.: Expectancy and the pituitary-adrenal system. In Physiology, emotion and psychosomatic illness, Ciba Foundation Symposium No. 8, Amsterdam, 1972, Elsevier, p. 281.

105. Long, C. N. H., and Lukens, F. D. W.: The effects of adrenalectomy and hypophysectomy upon experimental diabetes in the cat, J. Exp. Med. **63:**465, 1936.

106. Lueng, K., and Munck, A.: Peripheral actions of glucocorticoids, Ann. Rev. Physiol. **37:**245, 1975.

107. Lumley, F. H., and Nice, L. B.: Blood sugar of adrenalectomized rats, Am. J. Physiol. **93:**152, 1930.

108. Luse, S.: Fine structure of adrenal cortex. In Eisenstein, A. B., editor: The adrenal cortex, Boston, 1967, Little, Brown & Co.

109. Lymangrover, J. R., and Brodish, A.: Tissue CRF: an extra-hypothalamic corticotrophin releasing factor (CRF) in the peripheral blood of stressed rats, Neuroendocrinology **12:**225, 1973.

110. Maran, J. W., et al.: Organization of the medial hypothalamus for control of adrenocorticotropin in the cat, Endocrinology **103:**957, 1978.

111. Mason, J. W.: Plasma 17-hydroxycorticosteroid levels during electrical stimulation of the amygdaloid complex in conscious monkeys, Am. J. Physiol. **196:**44, 1959.

112. Mason, J. W., A historical review of the stress field, J. Hum. Stress, **1:**6, 1975.

113. Mason, J. W., et al.: Organization of psychoendocrine mechanisms, (15 articles with co-workers), Psychosom. Med. **30**(5):565, 1968, pt. 2.

114. McCormick, J. R., et al.: Hydrocortisone metabolism in the adrenalectomized dog: the quantitative significance of each organ system in the total metabolic clearance of hydrocortisone, Endocrinology **94:**17, 1974.

115. McEwen, B. S., Weiss, J. M., and Schwartz, L. S.: Uptake of corticosterone by rat brain and its concentration by certain limbic structures, Brain Res. **16:**227, 1969.

116. Middlebrook, J. L., and Aronow, L.: Physicochemical properties of glucocorticoid receptors, Endocrinology **100:**271, 1977.

117. Miller, R. E., et al.: Static gain and dynamics of cortisol secretory response to intravenous corticotropin (ACTH) in unanesthetized dogs—analysis by a linear mean-squared estimator, Ann. Biomed. Eng. **4:**364, 1976.

118. Moore, R. Y., and Eichler, V. B.: Loss of a circadian adrenal corticosterone rhythm following suprachiasmatic lesions in the rat, Brain Res. **42:**201, 1972.

119. Moriarty, C. M., and Moriarty, G. C.: Bioactive and immunoactive ACTH in the rat pituitary: influence of stress and adrenalectomy, Endocrinology **96:**1419, 1975.

120. Moriarty, G. C., Halmi, N. S., and Moriarty, C. M.: The effect of stress on the cytology and immunocytochemistry of pars intermedian cells in the rat pituitary, Endocrinology **96:**1426, 1975.

121. Motta, M., Mangili, G., and Martini, L.: A "short" feedback loop in the control of ACTH secretion, Endocrinology **77**:392, 1965.
122. Mulder, G. H., and Smelik, P. G.: A superfusion system technique for the study of the site of action of glucocorticoids in the rat hypothalamus-pituitary-adrenal system in vitro. I. Pituitary cell superfusion, Endocrinology **100**:1143, 1977.
123. Munck, A., and Brinck-Johnsen, T.: Specific metabolic and physiochemical interactions of glucocorticoids in vivo and in vitro with rat adipose tissue and thymus cells. In Martini, L., Fraschini, F., and Motta, M., editors: International Congress on Hormonal Steroids, Proceedings of the Second Congress, Princeton, N.J., 1967, Excerpta Medica Foundation.
124. Munck, A., and Young, D. A.: Corticosteroids and lymphoid tissue. In Greep, R. O., and Astwood, E. B., editors: Handbook of physiology. Section 7, endocrinology, Baltimore, 1975, The Williams & Wilkins Co., vol. 6, p. 231.
125. Natelson, B. H., Krasnegor, N., and Holaday, J. W.: Relationship between behavioral arousal and plasma cortisol levels in monkeys performing repeated free-operant avoidance sessions, J. Comp. Physiol. Psychol. **90**:958, 1976.
126. Naumenko, E. V.: Role of adrenergic and cholinergic structures in the control of the pituitary-adrenal system, Endorcinology **80**:69, 1967.
127. Nichols, T., Nugent, C. A., and Tyler, F. H.: Diurnal variation in suppression of adrenal function by glucocorticoids, J. Clin. Endocrinol. Metab. **25**:343, 1965.
128. Nugent, C. A., Eik-Nes, K., and Tyler, F. H.: The disposal of plasma 17-hydroxycorticosteroids. I. Exponential disposal from a single compartment, J. Clin. Endocrinol. **21**:1106, 1961.
129. Orth, D. N., and Island, D. P.: Light synchronization of the circadian rhythm in plasma cortisol (17-OHCS) concentration in man, J. Clin. Endocrinol. Metab. **29**:479, 1969.
130. Page, R. B., and Bergland, R. M.: The neurohypophyseal capillary bed, Am. J. Anat. **148**:345, 1977.
131. Peterson, R. E.: Miscible pool and turnover rate of adrenal cortical steroids in man, Recent Prog. Horm. Res. **15**:231, 1959.
132. Pirkle, J. C., Jr., and Gann, D. S.: Restitution of blood volume after hemorrhage: mathematical description, Am. J. Physiol. **228**:821, 1975.
133. Pirkle, J. C., Jr., and Gann, D. S.: Restitution of blood volume after hemorrhage: role of the adrenal cortex, Am. J. Physiol. **230**:1683, 1976.
134. Piva, F., et al.: Adrenal progesterone: factors controlling its secretion, Endocrinology **93**:1178, 1973.
135. Porter, J. C.: Secretion of corticosterone in rats with anterior hypothalamic lesions, Am. J. Physiol. **204**:715, 1963.
136. Portonova, R., and Sayers, G.: An *in vitro* assay for corticotropin releasing factor(s) using suspensions of isolated pituitary cells. Neuroendocrinology **12**:236, 1973.
137. Portanova, R., and Sayers, G.: Isolated pituitary cells: CRF-like activity of neurohypophysial and related polypeptides, Proc. Soc. Exp. Biol. Med. **143**:661, 1973.
138. Pratt, W. B., and Aronow, L.: The effect of glucocorticoids on protein and nucleic acid synthesis in mouse fibroblasts growing in vitro, J. Biol. Chem. **241**:5244, 1966.
139. Raisz, L. G., et al.: Effect of glucocorticoids on bone resorption in tissue culture, Endocrinology **90**:961, 1972.
140. Redgate, E. S.: ACTH release evoked by electrical stimulation of brain stem and limbic system sites in the cat: the absence of ACTH release upon infundibular area stimulation, Endocrinology **86**:806, 1970.
141. Rhodin, J. A. G.: The ultrastructure of the adrenal cortex of the rat, under normal and experimental conditions, J. Ultrastruct. Res. **34**:23, 1971.
142. Rogoff, J. M., and Stewart, G. N.: Studies on adrenal insufficiency. V. The influence of adrenal extracts on the survival period of adrenalectomized dogs, Am. J. Physiol. **84**:660, 1928.
143. Rosner, W., and Hochberg, R.: Corticosteroid-binding globulin in the rat: isolation and studies on its influence on cortisol action in vivo, Endocrinology **91**:626, 1972.
144. Roth, G. S.: Age-related changes in specific glucocorticoid binding by steroid-responsive tissues of rats, Endocrinology **94**:82, 1974.
145. Russell, S. M., et al.: Inhibition by dexamethasone of the in vivo pituitary response to corticotropin-releasing factor (CRF), Endocrinology **85**:512, 1969.
146. Saez, J. M., Morera, A. M., and Gallet, D.: Opposite effects of ACTH and glucocorticoids on adrenal DNA synthesis in vivo, Endocrinology **100**:1268, 1977.
147. Sakellaris, P. C., and Vernikos-Danellis, J.: Increased rate of response of the pituitary-adrenal system in rats adapted to chronic stress, Endocrinology **97**:597, 1975.
148. Sato, T., et al.: Corticosterone-induced changes in hypothalamic corticotropin-releasing factor (CRF) content after stress, Endocrinology **97**:265, 1975.
149. Sayers, G.: The adrenal cortex and homeostasis, Physiol. Rev. **30**:241, 1950.
150. Sayers, G., and Portanova, R.: Secretion of ACTH by isolated anterior pituitary cells: kinetics of stimulation by corticotropin-releasing factor and of inhibition by corticosterone, Endocrinology **94**:1723, 1974.
151. Sayers, G., and Sayers, M. A.: Regulation of pituitary adrenocorticotrophic activity during the response of the rat to acute stress, Endocrinology **40**:265, 1947.
152. Schayer, R. W.: A unified theory of glucocorticoid action. II. On a circulatory basis for the metabolic effects of glucocorticoids, Perspect. Biol. Med. **10**:409, 1967.
153. Seeling, S., and Sayers, G.: Bovine hypothalamic corticotropin releasing factor: chemical and biological characteristics, Fed. Proc. **36**:2100, 1977.
154. Seiden, G., and Brodish, A.: Persistence of a diurnal rhythm in hypothalamic corticotropin-releasing factor (CRF) in the absence of hormonal feedback, Endocrinology **90**:1401, 1972.
155. Selye, H.: A syndrome produced by diverse nocuous agents, Nature **138**:32, 1936.
156. Selye, H.: The general adaptation syndrome and the diseases of adaptation, J. Clin. Endocrinol. Metab. **6**:117, 1946.
157. Selye, H.: The stress of life, New York, 1956, McGraw-Hill Book Co.
158. Selye, H.: Confusion and controversy in the stress field, J. Hum. Stress, **1**:37, 1975.
159. Sendelbeck, L. R., and Yates, F. E.: Adrenal cortical and medullary hormones in recovery of tissues from local injury, Am. J. Physiol. **219**:845, 1970.
160. Slusher, M. A.: Effects of chronic hypothalamic lesions on diurnal and stress corticosteroid levels, Am. J. Physiol. **206**:1161, 1964.
161. Smelik, P. G.: The effect of a CRF preparation on ACTH release in rats bearing hypothalamic dexametha-

sone implants. A study on the "implantation paradox," Neuroendocrinology **5:**193, 1969.

162. Smith, G. P.: Adrenal hormones and emotional behavior, Prog. Physiol. Psychol. 5:299, 1973.
163. Suyemitsu, T., and Terayama, H.: Specific binding sites for natural glucocorticoids in plasma membranes of rat liver, Endocrinology **96:**1499, 1975.
164. Swingle, W. W., and Pfiffner, J. J.: An aqueous extract of the suprarenal cortex which maintains the life of bilaterally adrenalectomized cats, Science **71:**321, 1930.
165. Tait, J. F., and Burstein, S.: In vivo studies of steroid dynamics in man. In Pincus, G., Thimann, K. V., and Astwood, E. B., editors: The hormones, New York, 1964, Academic Press, Inc., vol. 5.
166. Takebe, K., Sakakura, M., and Mashimo, K.: Continuance of diurnal rhythmicity of CRF activity in hypophysectomized rats, Endocrinology **90:**1515, 1972.
167. Takebe, K., Yasuda, N., and Greer, M. A.: A sensitive and simple in vitro assay for corticotropin-releasing substances utilizing ACTH release from cultured anterior pituitary cells, Endocrinology **97:**1248, 1975.
168. Telegdy, G., and Vermes, I.: The role of serotonin in the regulation of the hypophysis-adrenal system. In Brodish, A., and Redgate, E. S., editors: Brain-pituitary-adrenal interrelationships, Basel, Switzerland, 1973, S. Karger AG.
169. Telegdy, G., and Vermes, I.: Changes induced by stress in the activity of the serotoninergic system in limbic brain structures. In Usdin, E., Kvetnansky, R., and Kopin, I. J., editors: Catecholamines and stress, Oxford, 1976, Pergamon Press.
170. Trimble, M., and Herbert, J.: The effect of testosterone or estradiol upon the sexual and associated behavior of the adult female rhesus monkey, J. Endocrinol. 42:171, 1968.
171. Vale, W., and Rivier, C.: Substances modifying the secretion of ACTH by cultured anterior pituitary cells, Fed. Proc. **36:**2094, 1977.
172. VanLoon, G. R.: Brain catecholamines and ACTH secretion. In Ganong, W. F., and Martini, L., Frontiers in neuroendocrinology, New York, 1973, Oxford University Press, Inc.
173. Vermes, I., Mulder, G. H., and Smelik, P. G.: A superfusion system technique for the study of the site of action of glucocorticoids in the rat hypothalamus-pituitary-adrenal system in vitro. II. Hypothalamus pituitary cell-adrenal cell superfusion, Endocrinology **100:**1153, 1977.
174. Watanabe, H., Nicholson, W. E., and Orth, D. N.: Inhibition of adrenocorticotrophic hormone production by glucocorticoids in mouse pituitary tumor cells, Endocrinology 93:411, 1973.
175. Weber, G.: Hormonal control of gluconeogenesis. In Bittar, E. E., and Bittar, A., editors: The biological basis of medicine, London, 1968, Academic Press, Ltd., vol. 2, p. 263.
176. Weissman, G., and Thomas, L.: The effects of corticosteroids upon connective tissue and lysosomes, Recent Prog. Horm. Res. 20:215, 1964.
177. Wurtman, R. J., and Axelrod, J.: Control of enzymatic synthesis of adrenaline in the adrenal medulla by adrenal cortical steroids, J. Biol. Chem. **241:**2301, 1966.
178. Yasuda, N., and Greer, M. A.: Studies on the corticotrophin-releasing activity of vasopressin, using ACTH

secretion by cultured rat adenohypophyseal cells, Endocrinology **98:**936, 1976.
179. Yasuda, N., and Greer, M. A.: Distribution of corticotropin releasing factor(s) activity in neural and extraneural tissues of the rat, Endocrinology 99:944, 1976.
180. Yasuda, N., Takebe, K., and Greer, M. A.: Studies on ACTH dynamics in cultured adenohypophyseal cells: effect of adrenalectomy or dexamethasone in vivo, Endocrinology **98:**717, 1976.
181. Yates, F. E., and Brennan, R. D.: Study of the mammalian adrenal glucocorticoid system by computer simulation. In Stear, E. B., and Kadish, A. H., editors: Hormonal control systems, New York, 1969, Elsevier North Holland, p. 20.
182. Yates, F. E., et al.: Potentiation by vasopressin of corticotropin release induced by corticotropin-releasing factor, Endocrinology **88:**3, 1971.
183. Zimmerman, E. A.: Localization of hypothalamic hormones by immunocytochemical techniques. In Martini, L., and Ganong, W. F., editors: Frontiers in neuroendocrinology, New York, 1976, Raven Press, vol. 4.

Aldosterone and mineralocorticoids

184. Alexander, E. A., and Levinsky, N. G.: An extrarenal mechanism of potassium adaptation, J. Clin. Invest. 47:740, 1968.
185. Assaykeen, T. A., Clayton, A., and Goldfien, A.: Effect of α- and β- adrenergic blocking agents on the renin response to hypoglycemia and epinephrine in dog, Endocrinology **87:**1318, 1970.
186. August, J. T., Nelson, D. H., and Thorn, G. W.: Response of normal subjects to large amounts of aldosterone, J. Clin. Invest. 37:1549, 1958.
187. Barajas, L.: Renin secretion: an anatomical basis for tubular control, Science **172:**485, 1971.
188. Biglieri, E., et al.: Hypermineralocorticoidism, Am. J. Med. **45:**170, 1968.
189. Binnion, P. B., et al.: Mechanisms regulating aldosterone secretion during sodium depletion, Am. J. Physiol. **208:**665, 1965.
190. Blaine, E. H., Davis, J. O., and Prewitt, R. L.: Evidence for a renal vascular receptor in control of renin secretion, Am. J. Physiol. **220:**1593, 1971.
191. Blair-West, J., et al.: Effect of changes in sodium balance on the corticosteroid response to angiotensin II, Aust. J. Exp. Biol. Med. Sci. **48:**253, 1970.
192. Brenner, B. M., et al.: Dynamics of glomerular filtrations in the rat. II. Plasma-flow dependence of GFR, Am. J. Physiol. **223:**1191, 1972.
193. Bumpus, F. M., et al.: Distribution and metabolic fate of angiotensin II and various derivatives, Can. Med. Assoc. J. **90:**190, 1964.
194. Clinton, M., and Thorn, G. W.: Effect of desoxycorticosterone acetate administration on plasma volume and electrolyte balance of normal human subjects, Bull. Johns Hopkins Hosp. **72:**255, 1943.
195. Conn, J. W.: Electrolyte composition of sweat. Clinical implications as an indication of adrenocortical function, Arch. Intern. Med. **83:**416, 1949.
196. Cook, W. F.: Cellular localization of renin. In Fisher, J. W., editor: Kidney hormones, London, 1971, Academic Press, Ltd.
197. Crabbé, J.: Stimulation of active sodium transport by the isolated toad bladder with aldosterone in vitro, J. Clin. Invest. **40:**2103, 1961.
198. Cuthbert, A. W., and Shum, W. K.: Effects of vasopressin and aldosterone on amiloride binding in toad

bladder epithelial cells, Proc. R. Soc. Lond. (Biol.) **189:**543, 1975.

199. Davis, J. O., et al.: Metabolism of aldosterone in several experimental situations with altered aldosterone secretion, J. Clin. Invest. **44:**1433, 1965.

200. Deane, H. W., Shaw, J. H., and Greep, R. O.: The effect of altered sodium or potassium intake on the width and cytochemistry of the zona glomerulosa of the rat's adrenal cortex, Endocrinology **43:**133, 1948.

201. Edelman, I. S., Bogoroch, R., and Porter, G. A.: On the mechanism of action of aldosterone on sodium transport: the role of protein synthesis, Proc. Natl. Acad. Sci. USA **50:**1169, 1963.

202. Edmonds, C. J., and Marriott, J. C.: Effect of aldosterone and adrenalectomy on the electrical potential difference of rat colon and on the transfer of sodium, potassium, chloride and bicarbonate, J. Endocrinol. **93:**517, 1967.

203. Eide, I., Loyning, E., and Kiil, F.: Evidence for a hemodynamic autoregulation of renin release, Circ. Res. **32:**237, 1973.

204. Fanestil, D. D., and Edelman, I. S.: On the mechanism of action of aldosterone on sodium transport: effects of inhibitors of RNA and protein synthesis, Fed. Proc. **25:**912, 1966.

205. Fasciolo, J. C., et al.: The renin content of the blood of humans and dogs under several conditions, Can. Med. Assoc. J. **90:**206, 1964.

206. Fimognari, G. M., Fanestil, D. D., and Edelman, I. S.: Induction of RNA and protein synthesis in the action of aldosterone in the rat, Am. J. Physiol. **213:**954, 1967.

207. Fimognari, G. M., Porter, G., and Edelman, I. S.: The role of the citric acid cycle in the action of aldosterone on sodium transport, Biochim. Biophys. Acta **135:**89, 1969.

208. Frawley, T. F., and Forsham, P. H.: Salivary Na/K ratio and adrenal salt regulating factors: prolonged salt retention with desoxycorticosterone trimethylacetate, J. Clin. Invest. **11:**772, 1951.

209. Funder, J. W., et al.: Effect of plasma (K^+) on the secretion of aldosterone, Endocrinology **85:**381, 1969.

210. Garrod, O., Davies, S. A., and Cahill, G., Jr.: The action of cortisone and desoxycorticosterone acetate in glomerular filtration rate and sodium and water exchange in the adrenalectomized dog, J. Clin. Invest. **34:**761, 1955.

211. Ganong, W. F. The renin-angiotensin system and the central nervous system, Fed. Proc. **36:**1771, 1977.

212. Giroud, C. J. P., et al.: Production of aldosterone by rat adrenal glands in vitro, Proc. Soc. Exp. Biol. Med. **92:**855, 1956.

213. Goldblatt, H., et al.: Studies on experimental hypertension. I. The production of persistent elevation of systolic blood pressure by means of renal ischaemia, J. Exp. Med. **59:**347, 1934.

214. Goormaghtigh, N.: La fonction endocrine des arterioles rénales, Louvain, Belgium, 1944, Fonteyn Medical Books.

215. Gould, A. B., Skeggs, L. T., and Kahn, J. R.: The presence of renin activity in blood vessel walls, J. Exp. Med. **119:**389, 1964.

216. Granger, P., Dahlheim, H., and Thurau, K.: Determination of renin, angiotensinase, and converting enzyme activity in microdissected juxtaglomerular apparatus and parts of the nephron, Pfluegers Arch. **312:**R87, 1969.

217. Grollman, A. P., and Gamble, J. L., Jr.: Metabolic alkalosis, a specific effect of adrenocortical hormones, Am. J. Physiol. **196:**135, 1959.

218. Hall, J. E., and Guyton, A. C.: Changes in renal hemodynamics and renin release caused by increased plasma oncotic pressure, Am. J. Physiol. **231:**1550, 1976.

219. Helmer, O. M., Kohlstaedt, K. G., and Page, I. H.: Destruction of angiotonin by extracts of various tissues, Fed. Proc. **1:**114, 1942.

220. Hierholzer, K., Wiederholt, M., and Stolte, H.: Hemmung der Natriumresorption im proximalen und distalen konvolut adrenalektomierter Ratten, Pfluegers Arch. **291:**43, 1966.

221. Hierholzer, K., et al.: Micropuncture study of renal transtubular concentration gradients of sodium and potassium in adrenalectomized rats, Pfluegers Arch. **285:**193, 1965.

222. Hodge, R. L., Lowe, R. D., and Vane, J. R.: The effects of alteration of blood-volume on the concentration of circulating angiotensin in anesthetized dogs, J. Physiol. **185:**613, 1966.

223. Huggins, C. G., et al.: Kinetics of the plasma and lung angiotensin I converting enzymes, Circ. Res. **27**(suppl. 1):93, 1970.

224. Kaplan, N, M.: The biosynthesis of adrenal steroids: effect of angiotensin II, adrenocorticotropin, and potassium, J. Clin. Invest. **44:**2029, 1965.

225. Koefoed-Johnson, V., and Ussing, H. H.: The nature of the frog skin potential, Acta Physiol. Scand. **42:**298, 1958.

226. Kunau, R. T., Jr., et al.: Micropuncture study of the proximal tubular factors responsible for the maintenance of alkalosis during potassium deficiency in the rat, Clin. Sci. **34:**223, 1968.

227. Laragh, J. H., et al.: Hypotensive agents and pressor substances. The effect of epinephrine, norepinephrine, angiotensin II and others on the secretory rate of aldosterone in man, J.A.M.A. **174:**234, 1960.

228. Laragh, J. H., et al.: Oral contraceptives: renin, aldosterone and high blood pressure, J.A.M.A. **201:**918, 1967.

229. Leaf, A., Anderson, J., and Page, L. P.: Active sodium transport by the isolated toad bladder, J. Gen. Physiol. **41:**657, 1958.

230. Leloir, L. F., et al.: Secrétion de rénine et formation d'hypertensine, C. R. Soc. Biol. **134:**487, 1940.

231. Lever, A. F., and Peart, W. S.: Renin and angiotensin-like activity in renal lymph, J. Physiol. **160:**549, 1962.

232. Lieberman, A. H., and Luetscher, J. A., Jr.: Some effects of pituitary, adrenal or thyroid function on excretion of aldosterone and the response to corticotropin or sodium deprivation, J. Clin. Endocrinol. Metab. **20:**1004, 1960.

233. Ludens, J. H., and Fanestil, D. D.: Aldosterone stimulation of acidification of urine by isolated urinary bladder of the Columbian toad, Am. J. Physiol. **226:**1321, 1974.

234. Luetscher, J. A., et al.: Observations on metabolism of aldosterone in man, Ann. Intern. Med. **59:**1, 1963.

235. Luetscher, J. A., et al.: Aldosterone secretion and metabolism in hyperthyroidism and myxedema, J. Clin. Endocrinol. Metab. **23:**873, 1963.

236. Luetscher, J. A., et al.: Conjugation of 1,2-H^3-aldosterone in human liver and kidneys and renal extraction of aldosterone and labelled conjugates from plasma, J. Clin. Endocrinol. Metab. **25:**628, 1965.

237. Luetscher, J. A., et al.: Conjugation and excretion of aldosterone: testing of models with an analog computer. In Pincus, G., Nakao, T., and Tait, J. F., editors: Steroid dynamics, New York, 1966, Academic Press, Inc.

238. Lynch, R. E., et al.: Absence of mineralocorticoid-

dependent sodium reabsorption in dog proximal tubule, Am. J. Physiol. **223:**40, 1972.

239. Malnic, G. R., Klose, M., and Giebisch, G.: Micropuncture study of distal tubular potassium and sodium transfer in rat nephron, Am. J. Physiol. **211:**529, 1966.

240. Maran, J. W., and Yates, F. E.: Cortisol secretion during intrapituitary infusion of angiotensin II in conscious dogs, Am. J. Physiol. **233**(4):E273, 1977.

241. Martino, J. A., and Earley, L. E.: Determination of a role of physical factors as determinants of the natriuretic response to volume expansion, J. Clin. Invest. **46:**1963, 1967.

242. Marusic, E. T., and Mulrow, P. J.: Stimulation of aldosterone biosynthesis in adrenal mitochondria by sodium depletion, J. Clin. Invest. **46:**2101, 1967.

243. Meyer, C. J., et al.: The binding of aldosterone to plasma proteins in normal, pregnant and steroid-treated women, J. Clin. Invest. **40:**1663, 1961.

244. Ng, K. K. F., and Vane, J. R.: Conversion of angiotensin I to angiotensin II, Nature **216:**762, 1967.

245. Page, I. H., et al.: Angiotonin activator, renin, and angiotonin inhibitor, and the mechanism of angiotonin tachyphylaxis in normal, hypertensive and nephrectomized animals, J. Exp. Med. **71:**495, 1940.

246. Page, I. H., et al.: Pharmacologic aspects of synthetic angiotonin, Circ. Res. **5:**552, 1957.

247. Palmore, W. P., Anderson, R., and Mulrow, P. J.: Role of the pituitary in controlling aldosterone production in sodium-depleted rats, Endocrinology **86:**728, 1970.

248. Plentl, A. A., and Page, I. H.: A kinetic analysis of the reninangiotonin pressor system and the standardisation of the enzymes renin and angiotonase, J. Exp. Med. **78:**367, 1943.

249. Porter, G. A., Bogoroch, R., and Edelman, I. S.: On the mechanism of action of aldosterone on sodium transport: role of RNA synthesis, Proc. Natl. Acad. Sci. USA **52:**1326, 1964.

250. Prunty, F. T. G., et al.: The effects of aldosterone in Addison's disease and adrenal pseudohermaphroditism, Lancet **2:**620, 1954.

251. Samuels, A. I., et al.: Renin-angiotensin antagonists and the regulation of blood pressure, Fed. Proc. **35:**2512, 1976.

252. Schmidt, U., et al.: Sodium-and-potassium-activated ATPase, J. Clin. Invest. **55:**655, 1975.

253. Schneider, E. G., et al.: The hepatic metabolism of renin and aldosterone, Circ. Res. **27**(suppl. 1):175, 1970.

254. Seldin, D. W., Welt, L. G., and Cort, J. H.: The role of sodium salts and adrenal steroids in the production of hypokalemic alkalosis, Yale J. Biol. Med. **29:**229, 1956.

255. Sharp, G. W. G., and Leaf, A.: Biological action of aldosterone in vitro, Nature **202:**1185, 1964.

256. Sheppard, H., Swenson, R., and Mowles, T. F.: Steroid biosynthesis by rat adrenal: functional zonation, Endocrinology **73:**819, 1963.

257. Singer, B., and Stack-Dunne, M. P.: The secretion of aldosterone and corticosterone by the rat adrenal, J. Endocrinol. **12:**130, 1955.

258. Skeggs, L. T., Kahn, J. R., and Shumway, N. P.: The preparation and function of hypertensin converting enzyme, J. Exp. Med. **103:**295, 1956.

259. Skeggs, L. T., et al.: The existence of two forms of hypertensin, J. Exp. Med. **99:**275, 1954.

260. Spitzer, A., and Windhager, E. E.: Effect of peritubular oncotic pressure changes on proximal tubular fluid reabsorption, Am. J. Physiol. **218:**1188, 1970.

261. Stewart, J.: Genetic studies on the mechanism of action of aldosterone in mice, Endocrinology **96:**711, 1975.

262. Stumpe, K. O., and Ochwadt, B.: Wirkung von Aldosteron auf die Natrium-und Wasserresorption im proximalen Tubulus bei chronischer Kochsalzbelastung, Pfluegers Arch. **300:**148, 1968.

263. Swaneck, G. E., Highland, E., and Edelman, I. S.: Stereospecific nuclear and cytosol aldosterone-binding proteins of various tissues, Nephron **6:**297, 1969.

264. Swingle, W. W., et al.: A comparative study of aldosterone and other adrenal steroids in adrenalectomized dogs, Endocrinology **55:**813, 1954.

265. Tait, J. F., et al.: The disappearance of 7-H³-*d*-aldosterone in the plasma of normal subjects, J. Clin. Invest. **40:**72, 1961.

266. Tait, J. F., et al.: Splanchnic extraction and clearance of aldosterone in subjects with minimal and marked cardiac dysfunction, J. Clin. Endocrinol. Metab. **25:**219, 1965.

267. Thatcher, J. S., and Radike, A. W.: Tolerance to potassium intoxication in the albino rat, Am. J. Physiol. **151:**138, 1947.

268. Thurau, K., et al.: Activation of renin in the single juxtaglomerular apparatus by NaCl in the tubular fluid at the macula densa, Circ. Res. **30/31**(suppl.):182, 1972.

269. Tigerstedt, R., and Bergman, P. G.: Niere und Kreislauf, Skand. Arch. Physiol. **8:**223, 1898.

270. Tobian, L.: Interrelationship of electrolytes, juxtaglomerular cells, and hypertension, Physiol. Rev. **40:**280, 1960.

271. Uhlich, E., Baldamus, C. A., and Ullrich, K. J.: The effect of aldosterone on sodium transport in the collecting ducts of the mammalian kidney, Pfluegers Arch. **308:**111, 1969.

272. Ussing, H. H., and Zerahn, K.: Active transport of sodium as the source of electric current in the short-circuited isolated frog skin, Acta Physiol. Scand. **23:**110, 1951.

273. Vagnucci, A. H., Shapiro, A. P., and McDonald, R. H., Jr.: Effect of upright posture on renal electrolyte cycles, J. Appl. Physiol. **26:**720, 1969.

274. Vagnucci, A. H., Shapiro, A. P., and McDonald, R. H., Jr.: Renal electrolyte cycles and mechanism of nocturia in primary aldosteronism, Metabolism **19:**952, 1970.

275. Vander, A. J.: Effect of catecholamines and the renal nerves on renin secretion in anesthetized dogs, Am. J. Physiol. **209:**659, 1965.

276. Vander, A. J., and Geelhoed, G. W.: Inhibition of renin secretion by angiotensin II, Proc. Soc. Exp. Biol. Med. **120:**399, 1965.

277. Wesson, L. G.: Electrolyte excretion in relation to diurnal cycles of renal function, Medicine **43:**547, 1964.

278. Wiederholt, M., et al.: Mechanism of action of aldosterone on potassium transfer in the rat kidney, Pfluegers Arch. **345:**159, 1973.

279. Wirz, H.: Der osmotische Druck in der corticalen Tubuli der Rattenniere, Helv. Physiol. Pharmacol. Acta **14:**353, 1956.

280. Wright, F. S., et al.: Reduced sodium resorption by the proximal tubule of Doca-escaped dogs, Am. J. Physiol. **216:**869, 1969.

281. Wright, F. S., et al.: Potassium secretion by distal tubule after potassium adaptation, Am. J. Physiol. **221:**437, 1971.

65

H. MAURICE GOODMAN

Reproduction

Fundamental to any definition of life is the ability of a species to reproduce. Reproduction in mammals is an enormously complex process, involving in one way or another each of the physiologic systems discussed in this text. No single chapter can deal justly with the physiology of reproduction and all its ramifications. Consequently, the emphasis in this chapter will be placed on hormones and hormonal mechanisms, with the realization that detailed coverage of a wide range of physiologically relevant topics can be found elsewhere.[1-19] The enormity of the literature devoted to the endocrine regulation of reproduction must be narrowed still further by limiting the discussion to those hormonal processes that are most pertinent to reproduction in man and other primates. In no other area of physiology is there such a diversity of control mechanisms as exists for the regulation of reproduction in the mammalian female. Although studies in a variety of species were historically important in the development of current understanding of human reproductive physiology, presentation of the broad array of these mechanisms is often more confusing than illuminating to the student of human biology.

The hormones that control reproduction differ from all others because they function not to ensure survival of the individual but rather survival of the species. Therefore the factors that control their function must be coordinated with important environmental events that ensure successful mingling of the gametes and nurture of the offspring. In fact, in many species (e.g., the salmon) the individual is sacrificed in the interest of survival of the species.

The endocrine glands that are chiefly involved in reproductive processes are the gonads and the pituitary, although optimal functioning of all the other endocrine organs is obviously necessary as well. The gonads have the dual function of producing both the gametes and the sex hormones that prepare the reproductive tract for successful mating and nurture of the young. The pituitary, by virtue of its special links with the central nervous system (CNS) (Chapter 61), regulates the function of the gonads and coordinates the timing of the various events of the reproductive process in response to signals in both the internal and external environment. Schematically, the reproductive system of both males and females consists of paired gonads and the tubular structures that are the avenue for delivery of the gametes or offspring to the outside. These tubes are endowed with special properties to nurture the gametes and fetus during their sojourn in the parent, as well as with appropriate muscular devices for expelling them at the appropriate time.

HORMONAL CONTROL OF REPRODUCTION IN THE FEMALE
THE OVARY
Development

Morphologically indifferent gonads are recognizable by about the sixth week of fetal life. They consist of the primitive germ cells embedded in a layer of cortical epithelium surrounding a core of medullary mesenchymal tissue. Differentiation into the male gonad involves development of the medullary portion of the gonad into seminiferous tubules, whereas predominant differentiation of the cortical region signals ovarian development. By about the twentieth week of gestation, primordial follicles are recognizable. At birth the human ovary contains about 1 million germ cells; most are arranged in primordial follicles. Between birth and puberty the continuing process of partial maturation and regression decreases the number of primordial follicles by about one third.

Morphology

The adult human ovaries are paired, flattened, ellipsoid structures measuring up to 5 cm in their longest dimension. They float free in the peritoneal cavity except for their attachments via the mesovaria to the broad ligaments that extend from either side of the uterus. The outer or cortical portion contains the primordial follicles and completely surrounds the ovary except for the

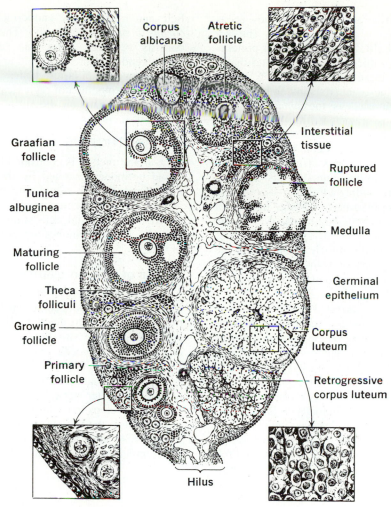

Fig. 65-1. Schematic representation of mammalian ovary showing different stages of follicular and luteal development. Obviously, events depicted occur sequentially and would not all be present in any section of a human ovary. (From Turner and Bagnara.[16])

point of attachment with the mesovarium at the region called the *hilus*. The inner portion, or medulla, consists chiefly of vascular elements that arise from anastomoses of the uterine and ovarian arteries and that penetrate the ovary at the hilus. Lymphatic drainage is abundant. A rich supply of unmyelinated nerve fibers also enters the medulla, together with the blood vessels.

The ovarian follicles are the source both of the gametes and of the ovarian hormones and hence constitute the functional unit of the ovary (Fig. 65-1). The follicles are surrounded by the *theca folliculi* (from the Greek *theke,* a container), which arises from stromal elements. The theca contains an inner rim of secretory cells, the *theca interna,* and an outer rim of connective tissue, the *theca externa*. Within the theca, and sepa-

rated from it by a thin basement membrane, are the granulosa cells, which in turn surround the ovum but are separated from it by an acellular layer of protein and mucopolysaccharide, the *zona pellucida*. Neither the ovum nor the granulosa cells are in direct contact with any capillaries, but the theca interna is richly vascularized. Most follicles remain quiescent in the immature state. Follicular development and maturation consists of proliferation of granulosa cells and the gradual elaboration of fluid within the follicle. Accumulation of this liquor folliculi is associated with enlargement of the follicle and the formation of a central cavity called the antrum. With follicular maturation the fluid content in the antrum rapidly increases and the follicle begins to bulge into the peritoneal cavity. The follicular fluid is derived from plasma that somehow seeps through the avascular granulosa region. The fact that gamma globulin and other plasma proteins are found in this follicular liquor indicates that it is not a simple ultrafiltrate of plasma.[58] As the follicle swells, the ovum remains embedded in a hillock of granulosa cells, the *cumulus oophorus*, which remains in contact with the theca. As more and more fluid accumulates, the cumulus attenuates until just a narrow column of cells connects the ovum with the rim of the follicle. The ripe preovulatory follicle thus consists of a ring of theca cells surrounding a thin rim of granulosa cells that connect by a narrow bridge of cells to the ovum and its surrounding layers of cells, the *corona radiata*, floating in a sea of follicular fluid. At ovulation the follicle ruptures and the ovum with its corona of granulosa cells is extruded into the peritoneal cavity in a bolus of follicular fluid (Fig. 65-2).

Any follicle can be arrested at any stage of its development and undergo degenerative changes known as *atresia*. Most of the primary follicles suffer this fate. In the reproductive lifetime of the normal woman, only some 400 follicles ovulate; the remaining hundreds of thousands undergo atresia. The physiologic mechanisms that control this process are poorly understood.

Following ovulation, there is ingrowth and differentiation of the remaining granulosa cells, which go on to fill the cavity of the collapsed follicle and form a new endocrine structure, the *corpus luteum* (yellow body). The corpus luteum consists of large polygonal cells containing smooth endoplasmic reticulum and a rich supply of fenestrated capillaries. Unless pregnancy ensues, the corpus luteum regresses after 2 weeks, leaving a remnant on the surface of the ovary known as a *corpus albicans* (Fig. 65-1).

GONADOTROPINS

Ovarian function is controlled by two gonadotropic hormones secreted by the anterior pituitary gland: follicle-stimulating hormone (FSH) and luteinizing hormone (LH), which is also called the interstitial cell–stimulating hormone (ICSH). These hormones are discussed in detail in Chapter 61.

Effects of FSH on ovaries

Little is understood of the mechanisms whereby primary follicles become sensitive to gonadotropins and begin to develop. In the absence of any stimulation by the pituitary gland, follicular development proceeds only until the stage at which the antrum is formed[135] in rats and rabbits, and in man gonadotropins are probably required at an even earlier stage.[3] A group, or cohort, of follicles develops in each ovarian cycle, but only one ovulates; the remainder become atretic. Although FSH may have a role in promoting the differentiation of the theca interna, its principal

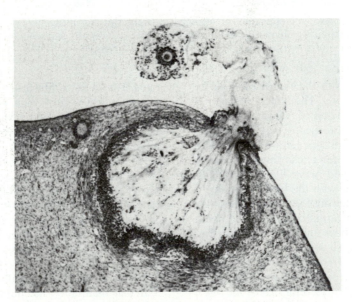

Fig. 65-2. Ovulation in rabbit. Follicular fluid, granulosa cells, some blood, and cellular debris continue to ooze out of follicle even after egg mass has been extruded. (From Blandau.[37])

and perhaps only target in the ovary is the granulosa cell.[48,122,123] Radioactively labeled FSH given to experimental animals or added to ovarian tissue in culture binds virtually exclusively to the granulosa cells, which proliferate and differentiate under its influence. FSH is also thought to increase the elaboration of the follicular liquor and thus cause swelling of the follicle. It may be noted that in controlling follicular size and development, FSH also governs the size of the whole ovary. Compensatory hypertrophy of the ovary, which occurs after unilateral ovariectomy in rodents, thus depends on FSH. Excessive administration of FSH results in the development of large numbers of vesicular follicles and pronounced ovarian enlargement. Increasing concentrations of FSH do not hasten the rate of follicular growth but rather increase the number of follicles that develop beyond the antrum stage. FSH alone apparently does not promote the synthesis or secretion of the ovarian steroid hormones,[114] but it undoubtedly collaborates with LH in this regard.[1] There appear to be no direct effects of FSH on the maturation of the oocyte.[78] The molecular events initiated by FSH to account for its effects are not understood except that increased generation of cyclic AMP in the granulosa cells seems to be involved.

Effects of LH on ovaries

LH is the more thoroughly studied of the two gonadotropic hormones. It synergizes with FSH in promoting follicular development and is required for the production and secretion of the follicular hormones.[81] Small amounts of LH are also required, at least in a permissive sense, for the corpus luteum to secrete its hormones, and large doses of LH sharply increase steroid hormone secretion by the corpus luteum.[25]

Effects on ovarian hormone production

The principal hormones secreted by the ovary are the estrogens and the progestins. These hormones are steroids and are derived from cholesterol by the series of reactions depicted in Fig. 65-3. Unlike man, of whom it has been said, "eats when he is not hungry, drinks when he is not thirsty, and makes love at all seasons of the year," most mammals mate only at certain times that coincide with maximum fertility in the female. This period of sexual receptivity is called estrus (from the Greek *oistros,* mad or vehement desire, frenzy). Estrogens are compounds that promote estrus in experimental animals. Pregnancy, or gestation, requires the presence of ovarian progestins, which promote gestation. Androgens (from the Greek *andros,* man), which are intermediates in the biosynthetic pathway for estrogen, are also produced and secreted by the ovary. The ovary also secretes a protein hormone with a molecular weight of about 9000 daltons called relaxin for its ability to relax the pubic ligament of the pregnant guinea pig. Since the function of this hormone in man and other primates has not been established, it will not be discussed further in this text.

The follicle destined to ovulate is the principal source of estrogen during the preovulatory period,[47] but the smaller follicles, which later become atretic, also participate in estrogen production. Although both the cells of the theca internal and the granulosa cells can synthesize estrogen, the weight of available evidence indicates that the thecal cells are the principal source of the estrogen that reaches the blood.[47,79,138] The biosynthetic activity of the granulosa cells probably accounts for the high concentrations of estrogen found in the follicular fluid. LH is usually credited with stimulating estrogen secretion, but this action depends on the presence of FSH and, oddly enough, estrogen. LH alone does not stimulate estrogen production.[79] Studies with radioactively labeled hormone indicate that LH binds to specific, high-affinity receptors on the thecal cells,[138] in which it provokes the formation of cyclic AMP.[116] Increased concentrations of cyclic AMP in the thecal cell cytoplasm trigger the conversion of cholesterol to pregnenolone, a complex reaction thought to be rate-limiting for steroid hormone biosynthesis.[26] LH also binds to granulosa cells,[138] but it is unlikely that they respond to LH until the immediate preovulatory period.

Estrogen production and the maturation of the follicle depend on complex interactions between the two gonadotropins and between the theca and the granulosa cells. A unique feature of this system is that estrogen plays a pivotal role in the entire maturation process, which includes increased capacity to produce estrogen. Estrogen causes proliferation of the granulosa cells,[75,87,138] and perhaps of the theca interna cells as well, and along with FSH enhances their ability to respond to stimulation by increasing their capacity to bind LH and FSH. Estrogen thus increases its own production, and FSH induces its own binding sites on the granulosa cells. This positive feedback system gives the follicle greater and greater capacity to produce estrogen and makes it more and more sensitive to stimulation as it matures. In the small immature follicle, LH appears to act

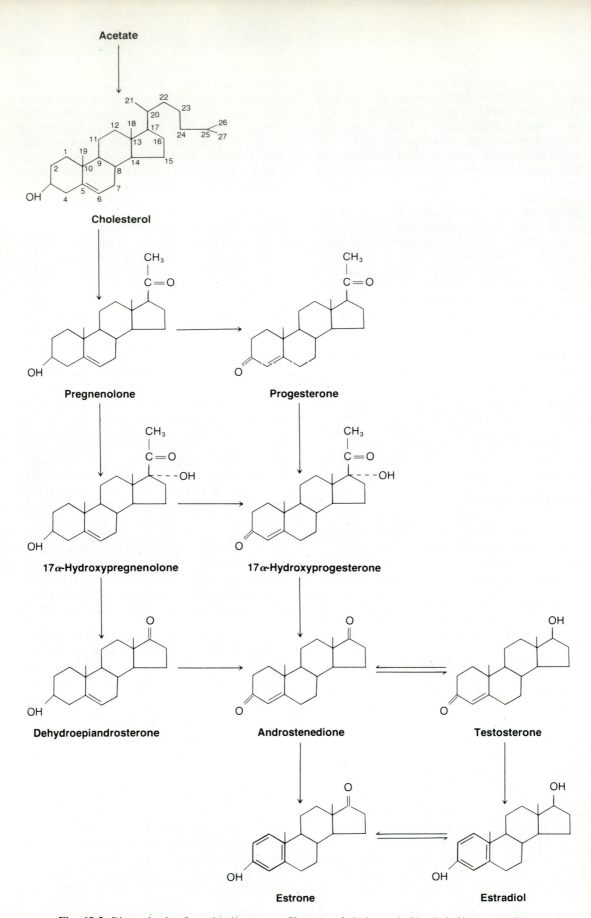

Fig. 65-3. Biosynthesis of ovarian hormones. Cleavage of cholesterol side chain between carbons 21 and 22 gives rise to 21-carbon progestins. Removal of carbons 20 and 21 produces 19-carbon androgen series. Aromatization of ring on far left (A ring) results in loss of angular methyl group (carbon-19) and in formation of 18-carbon estrogens.

only on the theca cells, which respond by producing estrogen and substantial amounts of its androgenic precursors. The granulosa cells lack the entire complement of enzymes needed to synthesize estrogens from cholesterol and are unable to convert 17-α-hydroxyprogesterone to androgen. They can readily accomplish the transformation of androgen to estrogen and synthesize estrogen by using the androgen produced by the theca interna as substrate.[1] FSH stimulates the conversion of androgen to estrogen in the granulosa cells,[1,60] probably through the agency of cyclic AMP.[1] The increased number of LH receptor sites appearing on the granulosa cells as the follicle matures prepares them to respond to LH shortly before ovulation.

Effects on ovulation

LH is usually credited with being the ovulatory hormone although, at least in rodents, both FSH and LH can produce ovulation in suitably prepared ovaries.[76,88] The manner in which LH produces ovulation and indeed the process of ovulation itself are not fully understood. Although the follicles undergo a preovulatory increase in follicular fluid, distensibility of the follicular wall increases, and no change in pressure results.[140] Rupture therefore is not due to increased follicular pressure but, from Laplace's law, increasing the radius at constant pressure increases the tension on the follicular wall. The increase in distensibility is probably caused by LH and may result from the formation and release of an ovulatory enzyme that destroys the collagen framework of the follicular wall. There is some evidence that the ovarian steroid hormone progesterone, which is formed in response to LH, triggers the formation and release of the ovulatory enzyme. Blockade of progesterone synthesis with cyanoketone also blocks ovulation in response to LH. Excellent discussions of this problem can be found in Rodbard,[139] Rondell,[140] and Lipner.[107]

The ability of the follicle to produce progesterone in response to LH derives from the events already described. Late in the preovulatory period the granulosa cells respond to LH because they have accumulated sufficient LH receptors on their surfaces and because circulating levels of LH rise sharply. LH presumably initiates the same events in the luteinized granulosa cells as in the theca cells; namely, conversion of cholesterol to pregnenolone. Presumably because of their inability to synthesize androgens, however, steroid biosynthesis in the granulosa cells can only proceed as far as progesterone and 17-α-hydroxyprogesterone, which therefore accumulate. Sufficient 17-α-hydroxyprogesterone is produced at this time to cause a significant increase in blood concentrations of this steroid. The corpus luteum does, however, produce estrogen, possibly from androgen precursors formed in the luteinized thecal cells.

In most primates a single ovum is shed at each ovulation, although many follicles in each ovary appear to be capable of responding to stimulation by gonadotropin. The factors that determine which follicle or even which ovary is selected to ovulate in a given cycle are uknown. Presumably, both ovaries and all follicles are exposed to the same concentrations of gonadotropin in the plasma. No hard evidence has yet been brought forth to indicate either that a dominant follicle suppresses development of other follicles or that ovulation makes an ovary less sensitive to gonadotropin during the next cycle so that alternation of ovarian function occurs.

Effects on corpus luteum formation

LH was named for its ability to induce the formation of the corpus luteum. However, luteinization is not an immediate response of the granulosa cells to LH.[46] Granulosa cells removed from mature follicles luteinize spontaneously in culture, even in the absence of gonadotropin. Less mature granulosa cells fail to do so unless FSH and LH are added. It appears that both gonadotropins are required at some point in the history of the granulosa cells for luteinization to occur, but neither gonadotropin can be considered as a direct trigger. Some observations suggest that simple separation of the granulosa cells from the influence of the ovum may be sufficient to induce luteinization. Removal of the ovum from the follicle (ovectomy)[63] or culture of isolated granulosa cells, as already mentioned, can induce luteinization. The ovum or its corona of cells may release an inhibitor of luteinization into the follicular fluid. When the follicle ruptures, the inhibitor is presumably removed in the escaping follicular fluid. It should be pointed out, however, that luteinization can occur even in the absence of ovulation in abnormal cycles. The appearance of luteinized follicles with the ovum entrapped is commonly encountered after massive doses of gonadotropin. Similar luteinization was seen when whole follicles from the rat ovary were transplanted to the kidney capsule.[64] Death or damage to the ovum cannot be ruled out in those cases in which luteinization occurred in the presence of the ovum.

Effects on oocyte maturation[3]

Primary oocytes begin the process of meiosis to reduce their chromosome number during late fetal life. Shortly before birth they reach the diplotene stage of the first meiotic division where further maturation is arrested until just prior to ovulation. The oocyte remains quiescent until the follicle begins to mature. It then increases in size from about 15 to 20 μm to about 150 μm at the time the antrum is formed, after which little further growth occurs. The oocyte is totally dependent on the granulosa cells for the nutrients required to achieve this nearly 10-fold increase in diameter or 1,000-fold increase in mass, since it has no direct vascular supply. Meiosis resumes at about the time of ovulation. LH is probably involved in triggering the resumption of meiosis, but its effects are poorly understood. It is likely that meiosis in the ripened oocyte is prevented by some inhibitory substance released by the granulosa cells into the follicular liquor. LH may block the production of the inhibitor or induce the formation of a substance that neutralizes its action. This effect of LH is probably not mediated by steroid hormones.

Effects on blood flow

LH also increases blood flow[134] to the ovary and produces ovarian hyperemia.[163] This effect may be secondary to the release of ovarian histamine[151] or perhaps prostaglandin.[9] In any event, increased blood flow to the ovary provides an increased opportunity for the distribution of steroid hormones in the general circulation. Increased blood flow to the ovary may also be important for preovulatory swelling of the follicles, which depends on increased elaboration of follicular liquor from plasma.

OVARIAN HORMONES
Estrogens

The estrogens, chiefly estradiol-17-beta and estrone, were originally isolated from the follicular liquor of sow ovaries by Allen and Doisy in 1923.[21] Both estradiol and estrone have also been isolated from whole human ovarian tissue but, characteristic of steroid-secreting tissues, very little hormone is stored. Only about 0.5 μg of estrone or estradiol can be isolated per gram of mature human ovary.[100] Pooled human follicular liquor contains about 0.06 μg of estrone/ml and about 5 times as much estradiol.[146] Although estrogen is most abundant in the preovulatory period, some estrogen is also produced after ovulation by the corpus luteum.[168]

Estrogens that circulate in blood are loosely bound to albumin and to the so-called sex hormone–binding globulin (p. 1629).[134] Plasma concentrations are of the order of 10^{-9}M and range from 30 to 500 pg/ml, depending on the point in the menstrual cycle in which the measurement was made. The daily secretory rate for estrogens range from 100 to 500 μg/day in normal women.[82]

The liver is the principal site of metabolic destruction; estrogen is completely cleared from the blood by a single passage through the liver. Estrogens are inactivated by enzymatic modification of the ring structure or by conjugation with sulfate and glucuronide. About half the protein-bound estrogen in blood is conjugated with sulfate and glucuronide. Although the liver may excrete some conjugated estrogens in the bile, these are reabsorbed in the lower gut and returned to the liver in portal blood in a typical enterohepatic circulatory pattern. The kidney remains the chief route of excretion of estrogen metabolites, but less than 50% of an administered radioactive dose of estrogen can be recovered from the urine.

Until recently, blood levels of estrogen were too low to be measured by available techniques. Therefore in the older literature, heavy reliance was placed on urinary estrogens as an index for ovarian function. Total urinary estrogens range from 12 to 75 μg excreted/24 hr[41] and can be readily detected by bioassay of suitably extracted urine specimens. Bioassays for estrogens make use of physiologic end points in young rats, including weight gain or water imbibition of the uterus and vaginal cornification. The development of radioimmunoassay for estrogen[93] has overcome this difficulty.

Physiologic actions

The principal effects of the estrogens are exerted on the organs of the reproductive tract. Estrogen is responsible for the growth and development of the fallopian tubes, uterus, vagina, and external genitalia that characteristically occur in the transition from sexual immaturity to the sexually mature state following puberty. Estrogen promotes cellular proliferation in the mucosal linings as well as in the muscular coats of these structures. Once maturation has occurred, however, these accessory organs of reproduction do not remain static; deprivation of ovarian hormones by castration results in severe involution and atrophy of these organs. Further, the mucosal lining of the uterus and vagina in particular are subject to constant turnover and change with waxing and waning of circulating levels of estrogen during the reproductive cycle.

Ovaries.[142] Although a gland of origin is seldom regarded as a target for its own hormone, estrogen appears to play a critical role in ovarian function. Some of its interactions with gonadotropins in controlling follicular development have already been discussed. In addition, because of its ability to stimulate proliferation of granulosa cells, estrogen produced by the corpus luteum in one cycle may provide the initial stimulus for growth of the cohort of follicles that develop in the next cycle. Local intrafollicular concentrations of estrogen or the ratio of estrogen and androgen may determine which one of these follicles will escape atresia and go on to ovulate.[87] The antiatretic effect of estrogen appears to be distinct from its ability to induce proliferation of the granulosa cells. Pituitary and ovarian hormone concentrations vary widely in follicular fluid obtained from different follicles in the same human ovary. The factors that determine these differences are not understood.

Fallopian tubes. The paired fallopian tubes, or oviducts, are responsible for the transfer of the ovum from the periovarial space to the cavity of the uterus. They are about 12 cm in length. The ovarian end has a funnel-shaped opening (the infundibulum) surrounded by fingerlike projections (the fimbriae), which come in close contact with the ovary. The newly shed egg mass is swept into the infundibular opening by the synchronous action of the cilia in the mucosal lining. The cilia beat at a rate of approximately 1,200/min[39] and thus set up a current of fluid flowing from the peritoneal cavity toward the uterine cavity. The lining of the oviduct also contains secretory cells, whose products may provide nourishment for the ovum in its 3 to 4 day journey to the uterine cavity as well as for the sperm swimming to meet the ovum. Estrogen increases the activity of both the secretory and ciliated cells that line the fallopian tubes. In the monkey the presence of cilia on the epithelial cells lining the fallopian tubes depends on estrogen.[40] Castration (surgical removal of the gonads) causes atrophy of the ciliated epithelium with a loss of cilia. Within 3 days of estrogen treatment the epithelial cells hypertrophy and regenerate their cilia.

The walls of the fallopian tubes contain layers of smooth muscle cells oriented both longitudinally and circularly. Rhythmic contraction of these smooth muscles accelerates the transport of the ovum toward the uterine cavity in much the same way that contraction of intestinal smooth muscle moves the chyme along the intestinal tract. Estrogen increases the contractility of fallopian smooth muscle, but its effects are complex and vary with the species studied, the dose of hormone administered, and the time elapsed between the injection of estrogen and the measurement of contractility. Estrogen withdrawal, rather than its administration, may be important in increasing contraction.[38] In some species, large doses of estrogen prevent egg transport,[42] a phenomenon called tubal locking.

Uterus.[12] In a normal reproductive cycle, estrogen increases the thickness of the endometrium two- to threefold. This thickening is accounted for by an increase in both the number and height of the endometrial cells. Estrogen also increases the number and length of the straight tubular endometrial glands. Not surprisingly, endometrial growth is accompanied by a marked increase in water content and blood flow. Part of this response to estrogen may be indirectly mediated through the release of histamine.[153] The endometrium receives its blood supply through two types of arteries. The so-called straight arteries supply the basal region and are unaffected by estrogen. The coiled or spiral arteries, on the other hand, are highly sensitive to estrogen and grow rapidly under its influence. The effects of estrogen on the muscular portion, or myometrium, are twofold. Estrogens increase the amounts of the contractile proteins, actin and myosin, in the myometrium and increase spontaneous muscular activity. Further, estrogen increases the irritability of uterine smooth muscle and, in particular, increases its sensitivity to oxytocin. This latter phenomenon may be of great significance in parturition. The effects of estrogen on uterine smooth muscle are discussed in greater detail in Chapter 4.

Under the influence of estrogen, the glands of the uterine cervix secrete an abundance of thin, watery mucus[17] that, when spread on a glass slide and allowed to dry, forms a characteristic typical fernlike pattern of sodium chloride cystals. Ferning of the uterine mucus is used diagnostically as an index of endogenous estrogen secretion.

Vagina. Estrogen causes thickening of the mucosa and cornification of the superficial vaginal epithelium. Increased mitotic figures appear in the basal layer of the vaginal mucosa, and growth and differentiation are evident in the intermediate and superficial layers. Under the influence of estrogen there is a heavy deposition of glycogen in the intermediate and superficial layers. The glycogen content of these cells may be as high as 3%, which is higher than in any other tissue except the liver.[136] The glycogen is thought to be important in the maintenance of the low pH (4 to 5) of the vaginal secretions. The products of glycogen breakdown are acted on by the vaginal bacteria to produce lactic acid. Vaginal acidity is thought to protect against infection. The appearance of a predominance of cornified cells in the vaginal smear of primates or rodents is a rough index of estrogenic activity.

External genitalia. At puberty, estrogen stimulates the growth of the labiae and, to a small extent, the clitoris. It probably also stimulates the development and function of Bartholin's glands, whose secretion provides some of the vaginal lubricant for intercourse.

Mammary glands. Growth and development of the mammary glands at sexual maturation is due primarily to estrogen. Estrogen alone promotes the development of the tubular duct system. In cooperation with progesterone, estrogen may also increase the lobuloalveolar portions of the glands, but alveolar development also requires the pituitary hormone prolactin. Development and pigmentation of the nipples and areolar portions of the breast are also attributable to estrogen.

Systemic effects. In addition to the foregoing, which are clearly and directly related to the reproductive process, estrogens also exert effects on the body as a whole. Some of these fall in the category of secondary sexual characteristics, which may be interpreted as those characteristics that make the female attractive to the male, and hence remain within the realm of sexually related phenomena. Estrogens hasten the maturation of bone and promote the fusion of the epiphyseal plates in the long bones and thus inhibit body growth. Estrogens also antagonize the action of growth hormone (GH).[144] They also produce changes in bone structure, particularly of the pelvis and the shoulder girdle. The softness and the curvaceous nature of the female body results from an effect of estrogen on the deposition of subcutaneous fat. The appearance of pubic and axillary hair at puberty apparently is due primarily to adrenal secretions, although estrogen may also play some role. Estrogens in large doses promote retention of sodium and water, possibly by an action on the renal tubules. Hepatic formation of the transport globulins for the thyroid and adrenal hormones is also stimulated by estrogen. Estrogen also acts on the CNS and is probably responsible for some behavioral patterns, particularly those related to mating in laboratory animals. Estrogen concentrations in blood indicate the level of ovarian activity and thus serve as an important link in the neurohumoral control of ovarian activity.

Mechanism of action

The molecular events set in motion by estrogen in target tissues have been the subject of intensive study in recent years. As described in Chapter 60, estrogen, like the other steroid hormones, appears to bind firmly to a cytoplasmic receptor and is translocated to the nucleus where it evokes the synthesis of new ribosomal and messenger RNA (mRNA). In the cytoplasm, this mRNA directs the synthesis of new proteins that presumably are responsible for the change in cellular activity that we regard as hormone action. Since estrogen promotes growth in most of its target tissues, it is not surprising that increased synthesis of RNA and protein occurs. Some observations have suggested that cyclic AMP may be involved in certain aspects of estrogen action on the uterus,[152] but there is doubt that the changes in cyclic AMP levels are directly produced by estrogen.

Progestins

Progesterone is the major progestational hormone in primates. It, too, is a steroid derived from cholesterol (Fig. 65-3). Although almost as much 17-OH-progesterone is secreted, it has less than 1% of the biologic activity of progesterone[28] and hence appears to be of little importance in normal reproductive function. Progesterone was originally isolated from swine ovaries in 1928 by Corner and Allen.[52] It is formed principally by the cells of the corpus luteum, but since it is a key intermediate in the biosynthesis of all steroid hormones, some progesterone may also be released from the adrenal cortex and the testis. Small amounts of progesterone are also produced by the granulosa cells during preovulatory period of the reproductive cycle. 17-OH-progesterone, which is intermediate between progesterone and estrogen in the biosynthetic pathway, is secreted in large amounts by these cells just prior to ovulation and has been used as an index of follicular maturity.[143]

The rate of progesterone production varies widely throughout the reproductive cycle and ranges from about 3 to 22 mg/day.[109] This is reflected in blood levels, which range from virtually nil in the early preovulatory part of the cycle to as high as 2 μg/100 ml after the corpus luteum has formed.[130] The corresponding molar concentrations are 10^{-8} to 10^{-7}M, levels considerably higher than those of estrogen. Progesterone travels in blood in association with the plasma proteins and has a high affinity for the corticosteroid-binding globulin (CBG).

As with estrogen, the liver is the principal site

of inactivation, which is achieved by A ring reduction and glucuronide formation. Considerable degradation also occurs in the uterus. The principal urinary metabolite is pregnanediol glucuronide. About 3 to 6 mg of this compound are excreted per day in the postovulatory portion of the cycle. Some pregnanetriol is also excreted, and this presumably arises from 17-OH-progesterone.

Progesterone has been difficult to quantitate by bioassay. Taking advantage of its high affinity for CBG, a competitive binding assay that has been developed[130] now makes it possible to monitor blood levels. More recently, a radioimmunoassay has been devised[71] and is in common use.

Physiologic actions

Progesterone is the hormone of pregnancy and is indispensable for implantation and retention of the conceptus. Removal of the ovaries or corpora lutea from pregnant rabbits interrupts pregnancy in 24 to 48 hr and causes the uterus to expel the products of conception unless exogenous progesterone is administered.[51] Despite these crucial functions, however, progesterone has little biologic effect when administered alone. To elicit most of the actions of progesterone on the female reproductive tract, previous or simultaneous treatment with estrogen is required, probably because it induces the formation of cytoplasmic receptors for progesterone.[67]

Uterus. Preparation of the estrogen-primed uterine lining for the nurture and implantation of the newly fertilized ovum is an essential action of progesterone. The endometrial lining, which has proliferated under the influence of estrogen, takes on a secretory function under the influence of progesterone. The endometrium proliferates further and becomes somewhat edematous; its surface is thrown into multiple folds and outpocketings, and blood flow to superficial areas is increased. The endometrial glands, which grew as straight tubes under the influence of estrogen, now become more tortuous as they elongate and branch and begin to produce their carbohydrate-laden secretions. This so-called uterine milk is thought to nourish the blastula until implantation and placentation are accomplished. Although the foregoing effects of progesterone can only be elicited in an estrogen-primed uterus, too much estrogen can antagonize the effects of progesterone on the endometrium and block the secretory changes. It is noteworthy that one of the effects of progesterone on the uterus is to suppress the formation of cytoplasmic binding proteins for estrogen.[94]

It has already been noted that under the influence of estrogen the myometrium grows and increases both its contractile apparatus and its spontaneous motility. Progesterone decreases the frequency and amplitude of spontaneous muscular contractions of the estrogen-primed myometrium.[12] Further, and perhaps more importantly, it decreases the sensitivity of uterine smooth muscle cells to the contractile stimulus of oxytocin. The ionic changes that accompany these effects of progesterone on uterine smooth muscle are discussed in detail in Chapter 4. Essentially similar effects of progesterone are seen in the fallopian tubes, where spontaneous motility is suppressed and secretory and ciliary activity of the fallopian mucosa are increased.[38]

Other effects

Under the influence of progesterone the volume of mucus secreted by the cervical glands is reduced, and it becomes thick and viscous.[17] In this way, progesterone prevents the "ferning" of the cervical mucus seen after estrogen stimulation. Absence of ferning of the cervical mucus provides presumptive evidence that ovulation and luteinization have occurred. As already mentioned, progesterone may mediate the effects of LH on ovulation.[140] In rodents, progesterone increases vaginal mucus formation and produces cytologic changes and leukocyte infiltration. In humans the progesterone-induced changes in the vaginal cytology are less striking but nevertheless evident. In the mammary gland, progesterone acts in concert with estrogen to promote lobulo-alveolar growth.[49] In discussing the biologic action of progesterone, it is of interest to note that cessation of progesterone secretion produces perhaps more dramatic effects than initiation of its secretion. Thus withdrawal of progesterone may trigger menstruation (p. 1612), lactation (p. 1623), parturition (p. 1622), and even the postpartum psychic depression experienced by many women.

One other physiologic effect of progesterone is worthy of note here, although it is only peripherally related to reproduction. Progesterone has a mild thermogenic effect and, when administered in amounts comparable to those secreted by the corpus luteum, increases basal body temperature by as much as 1° F. Increased basal body temperature can be observed in careful studies during the second half of the reproductive cycle when the corpus luteum is secreting progesterone. Because a sudden rise in progesterone secretion occurs immediately after ovulation with the formation of the corpus luteum or even 1 day or so before, the appearance of the increase in basal body temperature retrospectively provides an indication that ovulation has occurred and also indicates the approximate time of this event.

Mechanism of action

Like the other steroid hormones, progesterone appears to bind to specific cytoplasmic receptors that translocate it to the nucleus. Expression of progesterone effects is achieved through induction of DNA-directed RNA synthesis and the subsequent production of new proteins.

MENSTRUAL CYCLE

The overt indication of the cyclic nature of reproductive function in the mature human female is the periodic discharge of blood from the vagina, a result of the sloughing of the fully developed progestational endometrium. Although wide variations are seen, vaginal bleeding occurs on the average every 28 days, a time coinciding with the duration of the lunar month (hence the term "menstruation" from the Latin *mensis,* month). Menstruation occurs only in primates and is in no way analogous to estrus or heat observed in most other mammals. Estrus occurs just prior to ovulation at a time when conditions are optimal for fertilization and implantation. Menstruation occurs when there are no ripe follicles ready to ovulate and when the endometrium provides the least suitable conditions for implantation. The menstrual period typically lasts 3 to 5 days and the total flow of blood seldom exceeds about 50 ml. The first menstrual bleeding usually occurs at around the thirteenth year of life and continues until late in the fifth decade, interrupted only by periods of pregnancy. The initial menstrual period is called the *menarche* and the termination of reproductive function, preceded by irregular periods occurring at progressively longer and more erratic intervals, is called the *menopause.* A physiologic explanation for the termination of reproductive function in otherwise healthy women is not yet forthcoming, but depletion of follicles is a major factor. However, some follicles are sometimes present in the ovaries of postmenopausal women, an indication that not all the follicles have been used up by atretic processes and ovulation. Decreased sensitivity of the ovaries to gonadotropin is seen well before loss of cyclic function. Nonhuman primates apparently do not undergo menopause, but this is not firmly established.

The monthly flow of blood is but a marker for the cyclic pattern of events occurring in the ovary and reflected by changes in the endometrium; nonetheless, let us designate the onset of vaginal bleeding as the beginning of a reproductive cycle and trace the events that make up that cycle.

Several follicles may begin to develop, but only one reaches maturity and ovulates under usual circumstances. As it develops, the follicle secretes increasing amounts of estrogen and small amounts of progesterone. Under the influence of estrogen, the endometrium proliferates and increases in thickness. Comcomitant estrogenic changes are detectable in the secretions of the cervical glands and in the vaginal lining. Because this portion of the menstrual cycle is dominated by the developing follicle whose secretion of estrogen causes proliferation of the endometrium, it is referred to as the *follicular phase* with respect to the ovary and the *proliferative phase* with respect to the uterus. The follicular phase usually lasts about 2 weeks, but its duration can be quite variable; virtually all the variation seen in the duration of the menstrual cycle can be attributed to variability in the length of the follicular phase. Ovulation occurs at about the midpoint in the cycle. With the resulting development of the corpus luteum, progesterone production increases and evokes secretory changes in the endometrium. The second half of the reproductive cycle, which is dominated by the corpus luteum, is therefore called the *luteal phase* with respect to the ovary and the *secretory phase* with respect to the endometrium. Unless pregnancy occurs, the corpus luteum involutes within 2 weeks. With the involution of the corpus luteum, circulating levels of progesterone and estrogen decline sharply. In response to the loss of these hormones the endometrium gradually decreases in height, and the outer portion of the endometrium is shed and washed out with the menstrual flow. These events are summarized schematically in Fig. 65-4. All secretory portions of the endometrium are lost as the uterine lining is stripped down to the basalis.

The underlying mechanisms that precipitate menstrual bleeding are still not fully understood. One widely accepted view[92] holds that as the endometrium diminishes in height due to a decrease of luteal hormones, the spiral arteries form additional and even tighter coils. Increased coiling impairs blood flow, and stasis has been observed 1 to 3 days preceding menstruation in bits of monkey endometrial tissue grafted to the anterior chamber of the eye to permit direct observation of the menstrual process. A few hours before the onset of flow, vasoconstriction occurs in the portion of the spiral arteries adjacent to the myometrium and persists throughout the menstrual flow

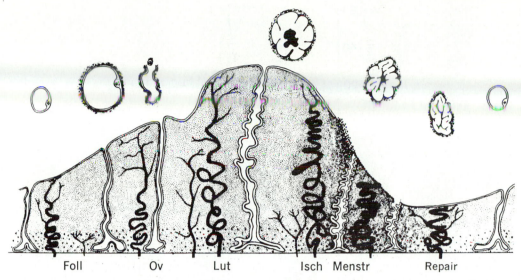

| Foll | Ov | Lut | Isch | Menstr | Repair |

Fig. 65-4. Schematic representation of endometrial changes during typical menstrual cycle. Simultaneous events in ovary are also indicated. Endometrium thickens during follicular phase. Uterine glands elongate, and spiral arteries grow to supply thickened endometrium. In early luteal phase there is further thickening of endometrium, marked growth of coiled arteries, and increased complexity of uterine glands. As corpus luteum wanes, there is reduction of endometrial thickness caused by loss of ground substance. Increased coiling of spiral arteries causes ischemia and finally sloughing of endometrium. (From Bartelmez.[31])

except for brief periods of relaxation of individual arteries. In this view, endometrial ischemia is the critical event that precipitates bleeding. Some objections raised to this idea include the observations that some monkeys that lack coiled arteries nevertheless menstruate. Further, ischemia produced by placing a tourniquet around the body of the uterus did not precipitate menstruation. More recently, it has been suggested[90] that the onset of menstruation may follow the release of lysosomal enzymes, which hydrolyze the intercellular mucopolysaccharides that hold the endometrial cells together.

REGULATION OF THE REPRODUCTIVE CYCLE
Pattern of hormones in blood during ovarian cycle

Thanks to the introduction of radioimmunoassay, it is now possible to monitor the concentration of pituitary and ovarian hormones in peripheral blood. Because the duration of the menstrual cycle varies both for a single individual and from one individual to another, some reference point other than the onset of vaginal bleeding is desirable to correlate observations made in different cycles. The central event of each ovarian cycle is ovulation, for which unfor-

tunately there is no simple means of detection short of direct observation of the ovary. Alternatively, the characteristic dramatic peak in LH concentration that precedes ovulation by 1 or 2 days can be used as the reference point to synchronize observations made in different individuals. Thus, if we call the peak of LH in blood day zero of the cycle, all the hormonal measurements can be normalized to this reference point (Fig. 65-5). Throughout both the follicular and luteal phase of the menstrual cycle (i.e., in the 2 weeks preceding and the 2 weeks following the peak), LH in plasma remains at a constant low level. The surge in LH secretion lasts for 2 to 3 days. A corresponding peak in FSH is also seen at this time, but it is considerably less pronounced. FSH concentrations tend to be higher during the follicular than the luteal phase.[143] Of interest is the observation that blood levels of FSH begin to rise 1 or 2 days before the onset of menstruation.[143] Thus the next ovarian cycle may begin even before the previous uterine cycle has come to an end. The steroid hormones show a somewhat different pattern. Progesterone is barely or not at all detectable during the follicular phase and then rises precipitously just at or immediately following the LH peak. During the luteal phase, progesterone levels remain at a high

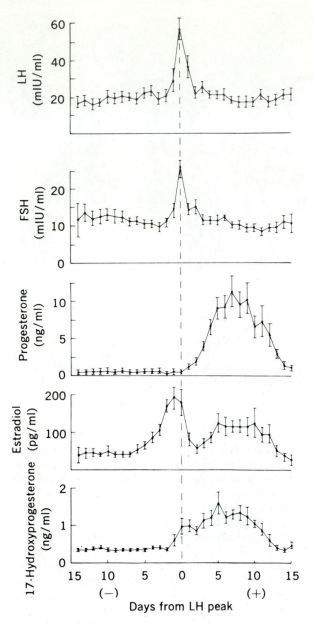

Fig. 65-5. Mean values of LH, FSH, progesterone, estradiol, and 17-hydroxyprogesterone in daily serum samples of nine women during ovulatory menstrual cycles. Data from different cycles are combined, using midcycle peak of LH as reference point (day 0). Vertical bars indicate standard error of mean (From Thorneycroft et al.[155])

plateau for about 7 days and then gradually fall. About 3 days before the onset of menstruation, progesterone levels have fallen to the very low levels seen during the follicular phase. At the beginning of the follicular phase, plasma concentrations of estradiol are low and gradually increase at an increasing rate, culminated by a surge of secretion that reaches its zenith about 12 hr before the peak in LH. Thereafter, estradiol levels fall abruptly and reach a nadir on the day of ovulation. During the luteal phase in women there is a secondary rise in estradiol levels, which then decline to low levels a few days before the beginning of the next cycle. In the rhesus monkey there is no secondary rise of estrogen, which remains at low levels throughout the luteal phase. The onset of menstruation, at least in the monkey, is not consistently associated with a decline in estrogen concentration.[103] Although little is known of its physiology, dramatic changes in circulating concentrations of 17-α-hydroxyprogesterone also occur during the cycle. The concentration of this steroid increases dramatically at the time of the LH surge and assumes a pattern during the luteal phase that is very similar to that of progesterone (Fig. 65-5).

Regulation of hormone secretion during menstrual cycle

The foregoing pattern of ovarian and pituitary hormones in blood results from intricate interactions between the pituitary, the ovaries, and the CNS. It has been known for many years through experiments in animals that some sort of negative feedback relationship between the ovary and the pituitary must exist.[80] The older observations include the findings that removal of one ovary led to hypertrophy of the other, but only when the pituitary was intact. While administration of pituitary extracts to intact animals caused hypertrophy of the ovaries, administration of ovarian hormones, either estrogen or progesterone or both, produced ovarian atrophy and failure of follicular development. In fact, it is this observation that led to the development of the contraceptive pill. Finally, the urine of postmenopausal women or castrated subjects contains very much more gonadotropin than that of individuals with normal functioning ovaries. These findings can all be explained by the hypothesis that the pituitary and ovaries are related in a typical negative feedback manner, with the pituitary hormones stimulating the secretion of estrogen and progesterone, which in turn act back on the pituitary to inhibit further secretion of gonadotropin. This simple scheme, however, cannot explain the pattern of hormone concentrations in the plasma described in the previous section. Further, simple negative feedback cannot explain the known influences of environmental factors on reproductive function. To explain the regulation of hormone secretion during the normal menstrual cycle, it appears that both positive and negative feedback control must be invoked, together with superimposition of control by the CNS through neurotransmitters released into the hypophyseal portal circulation.

Negative feedback aspects. LH concentrations in the blood of rhesus monkeys rise rapidly within the first day or two after ovariectomy but do not remain static at this high level.[164] Instead, the concentration of LH oscillates with a periodicity of about 1 hr as though LH were secreted intermittently in discrete bursts. Between these bursts of secretory activity, LH concentrations fall rapidly. Injection of estrogen, but not progesterone, prevents the appearance of the bursts of secretion and gradually brings the concentration of LH down to that seen during the follicular phase (Fig. 65-6). This appears to exemplify the classic negative feedback relationship, except that intermittent rather than constant pituitary secretion appears to be unique for the gonadotropins. Intermittent secretion of LH also occurs in normal men.[127] Progesterone alone, unlike estrogen, was ineffective in lowering the high levels of FSH and LH in the blood of postmenopausal women.[133] Although progesterone can synergize with estrogen to suppress gonadotropin secretion under experimental conditions, it is unlikely that progesterone plays a significant role in negative feedback control of gonadotropin secretion under normal circumstances. Whenever progesterone is present in the blood in significant concentrations, the concentration of estrogen is more than sufficient to account for the inhibition of gonadotropin secretion without synergism. It may be noted that the rise in FSH concentration, which presumably initiates the next ovarian cycle, follows the fall in both estrogen and progesterone at the end of the luteal phase.

Positive feedback aspects. The rising tide of estrogen during the late follicular phase appears to trigger the massive burst of LH secretion that just precedes ovulation. This can be considered positive feedback, because LH stimulates estrogen secretion, which in turn stimulates more LH secretion in a self-generating explosive pattern. The surge of LH secretion can be duplicated experimentally in monkeys and women given proper doses of estrogen.[103,165] This compelling evidence strongly implicates the increase in estrogen secretion by the ripening follicle, which begins several days before, as the causal event in triggering the surge of LH secretion by the pituitary. The less dramatic surge in FSH is controlled in the same way. Although progesterone can evoke LH secretion in certain experimental conditions, the scarcity of progesterone in the preovulatory period argues against it being the physiologic signal for the midcycle peak of gonadotropin. It is significant that progesterone blocks the estrogen-induced surge of LH. This probably accounts for the absence of repeated LH surges during the luteal phase, when the high concentrations of es-

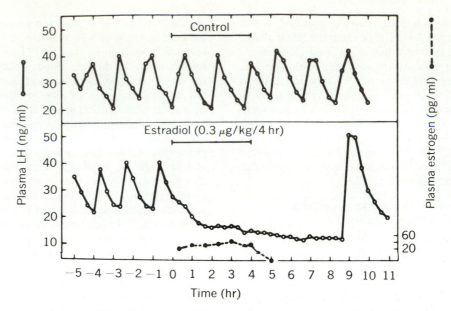

Fig. 65-6. Influence of physiologic levels of estrogen maintained by constant infusion on plasma LH in ovariectomized monkey. Horizontal line indicates duration of infusion beginning at 0 time. Plasma estrogen concentrations were undetectable prior to infusion. (From Yamaji et al.[164])

trogen might otherwise trigger the positive feedback effect.

Neural control of gonadotropin secretion

Although there is no doubt that the secretion of gonadotropins by the pituitary gland is influenced to a large measure by the ovarian steroid hormones, there is also little doubt that the secretion of pituitary hormones is under the control of the CNS. A wealth of experimental data collected in animals makes this conclusion inescapable. Some of these data are presented in the following examples:

1. Many animals are seasonal breeders; that is, gonadotropins secreted only at the appropriate season of the year cause maturation of the gonads and shedding of the gametes. In the nonbreeding season, gonadotropin secretion essentially ceases and the gonads and accessory organs of reproduction atrophy. In animals that normally breed in late spring, gonadal maturation can be brought about prematurely by exposure to increasing periods of light and corresponding decreasing periods of darkness. In response to this change in photoperiod the pituitary begins to secrete gonadotropin and the gonads develop prematurely in the dead of winter.[34] Thus the change in the ex-

ternal environment, mediated by the visual system, stimulates the pituitary to secrete gonadotropin.

2. Unlike primates that ovulate spontaneously, rabbits, cats, ferrets, and others only ovulate after mating.[6] The sensory stimuli associated with the act of mating reflexly evoke an ovulatory surge of LH. Electrical stimulation of the uterine cervix or mechanical stimulation with a glass rod are also effective stimuli for LH release. Electrical stimulation of the hypothalamus similarly causes LH secretion, whereas a variety of pharmacologic agents known to inhibit or depress neural function can block the secretion of LH induced by any of these stimuli. In the laboratory rat, which ovulates spontaneously every 4 or 5 days, the ovulatory surge of LH can also be inhibited by administration of a variety of pharmacologic agents that block autonomic function. Electrolytic lesions of the hypothalamus similarly block ovulation in these animals, as can such changes in the external environment as exposure to constant light.

The foregoing examples and many others brilliantly set forth by Harris[89] can only be explained by neural intervention in the control of gonadotropin secretion. Because the anterior pituitary gland is innervated only by vasomotor nerve fibers, it was postulated that chemical mediators must be released from hypothalamic nerve endings and travel by way of the hypophyseal portal vessels to the anterior pituitary, where gonadotropin secretion is elicited. Intensive efforts have been made over the past decades to isolate the neurohumoral factors responsible for eliciting gonadotropin secretion by the anterior pituitary gland. Two laboratories[22,117] have independently isolated from the hypothalami of sheep and swine a decapeptide that triggers the release of LH from the pituitary. This decapeptide has also been synthesized.[24] In addition to its effects on LH secretion, this material also evokes the release of FSH. Hence it is now called the gonadotropin-releasing hormone (GnRH). Most investigators now agree that a single releasing factor is probably responsible for the secretion of both FSH and LH, even though the gonadotropins are not secreted in a constant ratio to each other. Presumably, differences in the circulating levels of the ovarian hormones can selectively increase or decrease FSH or LH release.

Observations made in humans are also consistent with the idea that gonadotropin secretion must be controlled by the CNS. It is well known that the menstrual rhythm is sensitive to variations in the environment. The transition from day to night duty in nurses, fear of pregnancy, rapid transition across time zones by hostesses on jet planes, etc. are all phenomena that can disturb the menstrual rhythm. Further, it has commonly been observed that women who have been unable to conceive for many years become pregnant within a few months of adopting a child. These and numerous other examples that can be culled from the gynecologic literature leave little doubt that neural inputs play a role in regulation of the menstrual cycle.

Site of feedback control

The ovarian steroids can potentially influence the secretion of FSH and LH by acting at the level of the CNS, the anterior pituitary gland, or both. A variety of observations suggests that they act directly at the level of the pituitary gland, where they might either antagonize or enhance the activity of the GnRH. An equally impressive series of observations made principally in rodents and rabbits suggest that there are two hypothalamic centers concerned with regulation of gonadotropin secretion. One center, located in the anterior or preoptic region of the hypothalamus, appears to govern the ovulatory surge of gonadotropins and presumably is the site of positive feedback regulation. The center controlling ''tonic'' secretion appears to reside in the medial

basal region of the hypothalamus, which includes among others the arcuate nucleus. This region appears to be the site of negative feedback control.

By means of knife cuts, it is now possible to isolate surgically that area of the hypothalamus that sits directly above the infundibular stalk (the medial basal hypothalamus) from the rest of the brain. Rhesus monkeys subjected to this operation responded to increased circulating concentrations of estradiol with a normal surge of gonadotropin secretion, even though all connections between the preoptic area of the hypothalamus and the pituitary gland were severed.[104] Ovariectomy of these animals produced an increase in gonadotropin secretion characteristic of castrated animals, and estradiol treatment lowered gonadotropin levels to the normal range seen during the follicular phase. Thus the sites of both positive and negative feedback control in the primate must lie within the confines of the medial basal hypothalamus and the pituitary gland. If the preoptic area contributes at all to the ovulatory surge of gonadotropin secretion, its input must not be essential.

Further studies in ovariectomized monkeys indicated that destruction of the arcuate nucleus within the medial hypothalamus resulted in a total cessation of gonadotropin secretion.[126] Surges of FSH and LH could not be evoked by the same regimen of estrogen administration that evoked gonadotropin secretion in animals with intact hypothalami. Infusion of GnRH restored FSH and LH to the high levels normally seen in castrated animals. Administration of estradiol initially lowered blood levels of gonadotropins (negative feedback), and when estrogen concentrations were maintained at sufficiently high levels in blood, the typical surge of gonadotropin was seen (positive feedback). Both positive and negative feedback effects were seen even though the pituitary gland was exposed to an unchanging amount of GnRH. Interestingly, when GnRH was administered as a constant infusion, it was effective for only a limited time. The sustained responses to GnRH just described were obtained only when the hypothalamic hormone was administered intermittently in hourly pulses.

These results indicate that although the hypothalamic-releasing hormone is absolutely required for gonadotropin secretion by the primate pituitary gland, both positive and negative feedback can be accomplished by estrogen acting only at the level of the pituitary. This is consistent with observations in humans[166] and experimental animals that the responsiveness of the pituitary gland to a test dose of GnRH is greatest at midcycle. A contribution to regulation of gonadotropin secretion through variation in the secretion of GnRH by hypothalamic neurons is not ruled out by these observations. In fact, direct sampling of hypophyseal-portal blood in monkeys indicates that the concentration of GnRH is increased just prior to the anticipated ovulation.[131] These findings are not contradictory, but rather provide another example of redundant physiologic control of a critical function. Yet another variable in the regulation of gonadotropin secretion may be the pattern of GnRH secretion, that is, the frequency and duration of the secretory pulses.

Control of luteal function

It appears that once formed by the ovulatory surge of LH, the corpus luteum has a built-in life span of about 12 days. Although LH can increase the secretion of progesterone by the corpus luteum and maintain its continued function beyond this period of time, it is evident from Fig. 65-5 that no significant changes in LH concentration, other than the ovulatory surge, occur throughout the cycle. Changes in progesterone secretion during the luteal phase thus cannot be a reflection of changes in LH concentrations in plasma; the corpus luteum diminishes and finally stops its secretion of progesterone, even though it is exposed to constant low levels of LH. Furthermore, induction of ovulation in hypophysectomized women with exogenous gonadotropin is followed by the formation of a corpus luteum that secretes progesterone for the normal duration, although the amount is lower than normal.[157] Injection of LH in small, constant "permissive" doses restores the secretory rate to normal. These data suggest that some built-in property of the corpus luteum limits its functional life.

Before ascribing autonomous function to the corpus luteum, however, some mention should be made of the possibility that other factors might determine its functional life span. In rodents a third pituitary hormone, prolactin, stimulates the corpus luteum to secrete progesterone. No evidence exists that prolactin or any pituitary hormone other than LH has luteotropic effects in the primate. Almost a half century ago, Loeb[112] showed that hysterectomy (removal of the uterus) of the adult guinea pig interrupts normal cyclic function for a period of time that resembles the duration of pregnancy in this species. A large number of observations made in a variety of species suggest that the uterus produces some luteolytic substance that causes the corpus luteum to

involute. Studies in sheep and guinea pigs revealed that this substance, which may be a prostaglandin, is not carried in the general circulation from the uterus to the ovary, but appears to move by local diffusion. Interruption of the anatomic connection between an ovary and the adjacent portion of the uterus prevents luteolysis on the operated side, while the corpora involute normally on the side left intact. This now appears due to the direct diffusion of prostaglandin in a countercurrent manner from the uterine vein into the ovarian artery, which follows a highly convoluted and tortuous pathway on the surface of the uterine vein.[118] It is unlikely that a uterine luteolytic factor is of much importance in regulating the life span of the corpus luteum in primates, for hysterectomy of women or monkeys does not prolong luteal function or the duration of progesterone secretion.[129] Studies in rodents suggest the further possibility that some pituitary hormone or hormones may have a luteolytic function. In normal cycling rats, corpora lutea involute within just a few days, but hypophysectomy markedly prolongs their life span. Administration of crude pituitary extracts or of purified prolactin causes these corpora lutea to involute.[115] Available data provide no evidence for the existence of a pituitary luteolytic hormone in primates.

In summary, despite the presence of pituitary and uterine regulatory factors in some mammals, in the primate it appears that the life span of the corpus luteum is fixed, or preprogrammed, and that, during the normal menstrual cycle, neither luteotropic nor extraovarian luteolytic factors account for the duration of progesterone secretion. The single event that prolongs the life of the corpus luteum is pregnancy.

Timing of reproductive cycle

With the realization that the reproductive cycle in the primate female lasts 28 days and has regulatory components located in the ovary, the pituitary gland, and the CNS, it is tempting to jump to the conclusion that the timing mechanism that gives rhythmicity to the reproductive events is located within the CNS. This appears not to be the case. It has already been mentioned that irregularity in the timing of menstrual cycles stems primarily from variability in the duration of the follicular phase; the luteal phase is quite constant in length, and this appears to result from an inherent property of the corpus luteum. Thus the timing mechanism for at least half the productive cycle appears to reside in the ovary and not in the CNS.

The principal event around which the menstrual cycle revolves is ovulation, which depends on an ovulatory surge of LH. The timing of the ovulatory surge of LH also resides in the ovary. It is only when the developing follicle signals its readiness to ovulate by raising the blood levels of estrogen that the pituitary secretes the ovulatory spike of gonadotropin. Thus the ovary, and not the CNS, is the timekeeper and is responsible for the timing of two major events in the menstrual cycle: ovulation and the duration of steroid production by the corpus luteum, which determines when the next cycle begins.

FERTILIZATION, IMPLANTATION, AND PREGNANCY

At ovulation the ovum, together with its surrounding granulosa cells (the cumulus oophorus), is released into the peritoneal cavity from whence it is swept into the ostium of the fallopian tube by the beating of the cilia. A good account of fallopian tube activity at this time can be found in Blandau.[37] The high levels of estrogen that were present just prior to ovulation condition the fallopian tubes so that motility of the muscularis is maximal at this time and cause maximal development of cilia on the epithelial lining. The tube is thus prepared for speedy transport of the egg mass toward the uterus. The high preovulatory concentration of estrogen further optimizes conditions for fertilization. Converting the cervical mucus to a copious watery secretion facilitates the entry of sperm into the uterine cavity. Further, increased uterine motility may speed the transport of sperm from the cervical os to the fallopian tubes. Finally the high glycogen content of the vaginal mucosa contributes to the acidity of the vaginal secretions, heightening motility of the sperm.

Fertilization usually occurs in the ampullar portion of the fallopian tube. The period of fertility is short; from the time the ovum is shed until it must be fertilized is only on the order of 6 to 24 hr.[37] Sperm usually survive within the human female reproductive tract for 1 or 2 days, but a survival of 6 days has been reported. Since both the ciliary and muscular activity of the fallopian tube facilitate movement of the ovum away from the ovary toward the uterus, it is apparent that the sperm that meet the egg in the oviduct have had to move "upstream." Little is understood concerning the overall process of sperm transport within the female genital tract. Contractions of the vagina and uterus doubtlessly help propel the sperm toward the fallopian tubes, where they appear within 30 min after coitus.[37] There is a surprising lack of specific data on how the sperm traverse the muscular tubouteral junction and advance through the fallopian tubes. Suffice it to say that conditions within the female reproductive tract are sufficiently adverse that relatively few of the millions of sperm deposited in the vagina reach the upper levels of the fallopian tube.

As soon as a sperm penetrates the ovum, the second polar body is extruded and the fertilized ovum begins to divide, even while proceeding through the fallopian tubes toward the uterine cavity, which usually takes about 3 days.[36] By the time the fertilized egg reaches the uterine cavity, it has reached the 12-cell stage. In the rhesus monkey, and possibly in humans, the blastocyst does not implant for 4 to 6 days after reaching the uterine cavity; in some species (e.g., the armadillo), implantation is normally delayed for as long as 3 or 4 months. In rodents, at least, implantation depends on an appropriate balance between estrogen and progesterone. From the time the ovum is shed until the blastocyst implants, its energy needs are met through metabolism of the secretions of the fallopian tube and the endometrium.

Placental hormones

Within about 10 days after fertilization, the chorionic cells of the developing placenta begin to secrete a new hormone, human chorionic gonadotropin (HCG). HCG is luteotropic; that is, it rescues the waning corpus luteum and stimulates it to secrete progesterone and estrogen. This renewed surge of progesterone secretion is short-lived, lasting only about 10 days in the monkey despite continued stimulation with HCG.[102] By this time, however, the placental production of progesterone is adequate to maintain pregnancy. Measurements of progesterone secretion in human ovarian vein blood suggest that the corpus luteum remains functional throughout most of the first trimester of pregnancy.[124] Thereafter the corpus luteum becomes nonfunctional and involutes. Prolongation of the functional life of the corpus luteum, with the attendant production of steroids, prevents the onset of menstruation and shedding of the endometrium. During the major part of pregnancy the secretion of the pituitary gonadotropins is suppressed.

HCG production increases markedly in the early weeks of pregnancy. It is the appearance of this hormone in the urine in large amounts that is used as a test for pregnancy. HCG has biologic activity like that of LH; urine containing HCG induces ovulation in the rabbit or the release of sperm in frogs. It also gives positive results in any number of other biologic tests for LH activity. HCG can now be measured with a simple immunologic test[77] in the doctor's office, and early diagnoses of pregnancy can be made within just a few minutes.

HCG resembles pituitary LH, not only in its biologic activity but also in its chemical structure.[150] Like pituitary LH, it is a glycoprotein that can be dissociated into two subunits. The alpha subunit of HCG bears a striking resemblance to the alpha subunit of LH. HCG and human LH also cross-react immunologically. This property of HCG has been used to produce the antibodies against primate LH needed in the performance of the radioimmunoassay for LH. Needless to say, HCG that is purified from urine is available in abundance, whereas human pituitary LH is quite scarce.

The placenta is a multipotent endocrine organ. It secretes pituitary-like hormones as well as the gonadal steroids. A second placental protein hormone secreted in large amounts is the so-called human placental lactogen (HPL), also called placental sommatomammotropin. It is immunologically similar to human growth hormone, but it does not promote body growth.[98] The physiologic role of HPL is not understood. Although it induced lactation in test animals, lactation normally does not occur until after par-

turition. HPL may promote mammary growth in preparation for lactation. The placenta also produces large amounts of both progesterone and estrogen. Some data also suggest that it can secrete thyrotropin (TSH), adrenocorticotropin hormone (ACTH), and a variety of other substances. In experimental animals such as the rat, pregnancy is terminated if the pituitary gland is removed before the twelfth day of gestation. Pregnancy in the rat continues for the full 21 days if hypophysectomy is performed any time after day 12. In rats, pregnancy is also interrupted if at any time the ovaries, and consequently the corpora lutea, are removed. In primates, removal of either the ovaries or the pituitary gland after the first month does not interrupt the pregnancy or interfere with the delivery of normal offspring. That is because the placenta can substitute for the function of both the pituitary gland and the ovary and can produce adequate amounts of progesterone to maintain pregnancy.

The pattern of hormone secretion by the placenta changes with the duration of pregnancy. HCG peaks at about the sixth week and by the end of the first trimester falls to rather low levels, well beyond the time when the placenta can fully take over the secretion of progesterone (Fig. 65-7). Progesterone concentrations in plasma and the amount of pregnanediol in urine (pregnanediol is the principal metabolite of progesterone)

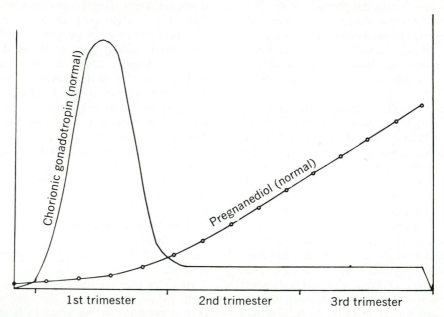

Fig. 65-7. Schematic representation of excretory pattern of chorionic gonadotropin and pregnanediol during pregnancy. (Modified from Lloyd.[111])

increase gradually throughout the first trimester and continue increasing at more rapid rates until parturition. Total estrogen secretion also increases during pregnancy, but the increase is most pronounced in estriol, especially as pregnancy progresses into the third trimester (Fig.

65-8). HPL becomes detectable in plasma about the second week in pregnancy. Thereafter, it gradually increases until it reaches very high concentrations (about 20 μg/ml in plasma) just before parturition (Fig. 65-9). This is about 100 times as much as any other protein hormone. It

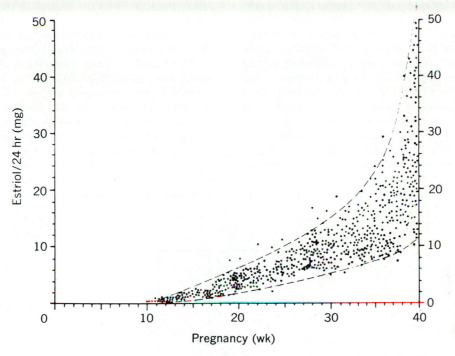

Fig. 65-8. Normal values for estriol excretion during pregnancy. (From Beling.[33])

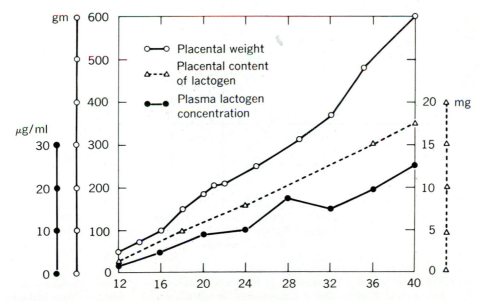

Fig. 65-9. Changes in placental weight, placental content of lactogen, and placental lactogen concentration in blood during pregnancy. (From Josimovich.[97])

has a short half-life and disappears from plasma within 1 or 2 days after birth.

PARTURITION

No one knows for sure what the exact triggering mechanisms are that induce parturition (labor). As has been pointed out, progesterone causes relaxation of the uterine musculature and suppresses spontaneous contractions. Direct measurements of circulating levels of progesterone, however, reveal no fall in progesterone at the onset of labor[70] and consequently appear to rule out any decrease in progesterone as the precipitating event (Fig. 65-10). Similarly, there are no changes in circulating concentrations of estrogen seen at this time. However, it is possible that a fall in progesterone is still involved, for there may be a decline in the local concentration of progesterone that diffuses into the myometrium, which lies directly behind the placenta.[55] Throughout most of pregnancy, a high con-

centration of progesterone in the immediate vicinity of the placenta may be required to overcome the local irritating effect of the placenta. As the placenta ages, however, or perhaps when it reaches some critical age, its ability to produce progesterone at a rate adequate to maintain high concentrations within the adjacent myometrium may diminish. With the decline in local progesterone concentration, the underlying myometrium might increase in excitability and, through its spontaneous contractions, initiate parturition.

Alternatively, some extrauterine event may precipitate parturition. Oxytocin, which is formed in the supraoptic nucleus of the hypothalamus and stored in the posterior pituitary gland, is known to increase uterine contractions (Chapter 61), yet in women with hypothalamic damage who therefore lack oxytocin, parturition can occur normally. Similarly, although removal of the posterior pituitary gland or placement of hypothalamic lesions in animals may prolong la-

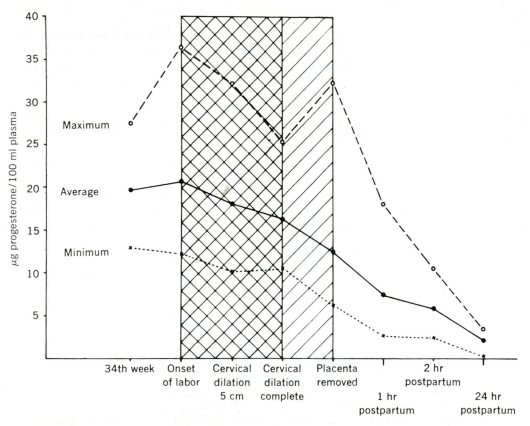

Fig. 65-10. Progesterone concentrations in peripheral blood in late pregnancy (normal duration 40 weeks), at onset of labor, during various stages of labor, and after parturition. Note that onset of labor is not preceded by change in progesterone, the concentrations of which decline only after onset of labor. (From Llauro et al.[110])

bor, pregnancy terminates on schedule. It is unlikely, therefore, that oxytocin acts as the trigger
for parturition. Stimulation of the uterine cervix,
particularly by stretching, is a potent stimulus
for the release of oxytocin in experimental animals. Therefore oxytocin may only reinforce a
process that has already begun. Whether or not
this is true of humans is unknown. It has also
been suggested that prostaglandin $F_2\alpha$ may be the
precipitating agent, but definitive evidence is
lacking. Whatever causes its release is unknown.[105] It is also possible that hormones of fetal origin may play some role in precipitating delivery. In sheep, destruction of the fetal pituitary
gland by electrocautery prevented the onset of
labor at the expected time.[106] Similarly, adrenalectomy of fetal lambs also prolonged gestation,
whereas infusion of the fetus with ACTH or glucocorticoid, but not estrogen, induced premature
delivery.[61] The relevance of these findings to
parturition in primates remains to be established.

In summary, it is still not known whether the
timing of parturition is governed by the fetoplacental unit or the CNS of the mother, nor is it
clear how the process of parturition is initiated.

LACTATION[9,54,137]

The appearance of special adaptations of the
sweat glands in the skin of the ventral body wall
to provide nourishment for newborn offspring is
the characteristic which gives the class Mam-
malia its name. The mammary glands furnish
maternal nourishment and antibodies for the
young for a varying period of postnatal life. The
process of lactation is also of enormous interest
economically, since it is the basis of the dairy industry and is of major importance in human nutrition. Consequently, this process has been
widely studied, particulatly in ruminants. Lactation is a complex process that is largely beyond
the scope of this chapter. A constellation of hormones is involved in growth and differentiation
of the mammary glands (mammogenesis), in initiation of lactation (lactogenesis), and in the
maintenance of milk production (galactopoiesis).
The hormonal requirements for each of these processes vary from species to species, and the exact
requirements for primates are still not known.
However, with few exceptions, prolactin, which
has been regarded as a third gonadotropin by
some, is uniquely required in all aspects of lactation. Since current understanding of the physiology of prolactin in primates is rudimentary at
best, only a cursory discussion based on animal
experiments follows.

Growth of mammary glands

Growth of the breast tissue in the nonpregnant
woman is primarily under the control of estrogen,
which promotes the growth of the ducts, nipples, and areolae. Progesterone, in combination
with estrogen, promotes growth of the glandular
or lobuloalveolar portion of the mammary
glands. These steroids are not very effective in
producing the growth of mammary tissue in hypophysectomized animals; prolactin, growth hormone, and insulin are also needed. Direct injection of prolactin into the mammary ducts in rabbits increases alveolar growth as well as milk formation, even after the ovaries are removed.
Breast tissues grow significantly during pregnancy, when estrogen and progesterone levels
are high and pituitary prolactin and placental
somatomammatropin levels are also elevated.[69]

Milk production

Once mammary tissue has been developed,
formation of milk appears to be primarily the result of increased prolactin levels, but the adrenal
glucocorticoids are also of critical importance.
Insulin is also required both in vivo and in explants of mammary tissue studied in culture.[154]
All these hormones and the placental lactogen are
present in the late stages of pregnancy, yet lactation does not begin until after parturition. The
reason for this is not entirely clear. It may be
that the high circulating levels of progesterone
interfere with the action of prolactin on the mammary epithelium[156] or that some component necessary for lactation, such as glucocorticoids, is
present in too low an effective concentration to
permit lactation. With parturition, the precipitous
fall in estrogen and progesterone may relieve the
inhibitory effect on the mammary epithelium.
Furthermore, the decline in corticosteroid-binding globulin may increase the unbound and physiologically effective form of glucocorticoid circulating in the plasma.[72,73]

Control of prolactin secretion

Continuation or maintenance of lactation requires not just the right complement of hormones
but also suckling and removal of milk by the
young. The phenomenon of milk let-down or
milk ejection and the role of oxytocin have already been discussed in Chapter 61. Suckling is
also an important stimulus for the secretion of
prolactin, which is an absolute requirement for
continued lactation. Observations in experimental animals, particularly rodents, indicate
that the control of prolactin secretion is exerted
through a hypothalamic prolactin inhibitory fac-

tor (PIF), which blocks spontaneous secretion of the hormone.[121] Interference with the connections between the anterior pituitary gland and the hypothalamus, either through sectioning of the pituitary stalk or transplantation of the pituitary to a distant site, causes marked autonomous secretion of prolactin. Hypothalamic extracts inhibit such secretion. Similar control of prolactin secretion probably exists for human pituitary glands as well.[128] Cultures of human pituitary glands secrete large amounts of prolactin, and this secretion can be inhibited by the addition of rat hypothalamic extracts. Surgical section of the human pituitary stalk increases plasma prolactin concentrations and leads to the onset of lactation. It is likely that suckling, which causes a 10- to 20-fold increase in plasma prolactin within 30 min, blocks the release of PIF, although the role of a prolactin-releasing factor cannot be excluded.

Estrogen is also an important stimulus for prolactin secretion in experimental animals. It, too, probably acts somewhere in the hypothalamus to inhibit the release of PIF and may further act directly on the pituitary gland to increase prolactin secretion. In addition, a variety of exteroceptive stimuli (e.g., sight or odor of the young), trigger prolactin secretion by what appears to be a conditioned reflex.

HORMONAL CONTROL OF REPRODUCTION IN THE MALE
TESTES
Development

As already mentioned, the testes and ovaries develop from sexually indifferent presumptive gonads during the sixth week of fetal life. The primordial germ cells become incorporated into the testicular cords that differentiate from the medulla of the primitive gonad. These cords are solidly packed with primordial germ cells and are destined to become the seminiferous tubules. By about the sixtieth day of fetal life, cells appear in the connective tissue between the seminiferous cords and give rise to the fetal interstitial or Leydig cells, which presumably secrete the androgens required for development of the male genitalia. Their differentiation and function may be under the influence of gonadotropin secreted by the fetal pituitary gland or the placenta.

The primitive gonads develop midabdominally in association with the mesonephric ridges. They descend into the scrotum through the inguinal canals. The testes pass through the inguinal ring separating the abdominal cavity from the scrotal cavity in about the seventh month of fetal life. At birth the human testis contains considerable numbers of interstitial cells that regress during the first 6 months of postnatal life.

Morphology

The testes, like the ovaries, are paired ovoid organs that serve the dual function of producing both gametes and steroid hormones. Unlike the ovary, in which these activities occur together in the follicles, the testis performs these functions in morphologically separate regions.

Seminiferous tubules. Formation of the sperm occurs throughout the length of the seminiferous tubules, which make up about 80% of the mass of the adult testis (Figs. 65-11 and 65-12). The seminiferous tubules are about 120 to 300 μm in diameter and 30 to 70 cm in length; they consist of multiple branching, highly convoluted loops. The overall length of all the tubules is estimated to be about 250 m. Several seminiferous tubules are present in each testicular lobule, and there are about 250 lobules in each testis. The seminiferous tubules connect by way of the *tubuli recti* with the *rete testis*, which is composed of a network of thin tubules that are contiguous with the seminiferous tubules and that collect the fluid and gametes produced therein. The seminiferous tubules are surrounded by a thin coating of smooth muscle whose contraction moves the nonmotile sperm through the seminiferous tubules and into the rete testis. The entire testis is encased in a dense, inelastic fibrous capsule, the *tunica albuginea,* which thickens at the posterior surface of the testis and projects inward to form the *mediastinum testes,* which houses the rete.

The seminiferous tubules are lined with a complex epithelium that produces the male gametes. The overall process of sperm production, or spermatogenesis, is too complex to be covered in detail in this chapter; excellent accounts can be found in Courot et al.[53] and Steinberger and Steinberger.[149] Spermatogenesis occurs continuously throughout the length of the seminiferous tubules from puberty to senescence. No abrupt loss of fertility with advancing age analogous to menopause is encountered in the male. Whereas the female sheds only a single gamete each month, the seminiferous tubules produce hundreds of millions of sperm each day. The epithelium consists of two types of cells; those in the processes of becoming sperm and the supporting or Sertoli cells. Spermatogenesis can be divided into three principal phases: *spermatocytogenesis,* in which the precursor spermatogonia undergo mitotic division to keep replenishing themselves and produce the spermatocytes;

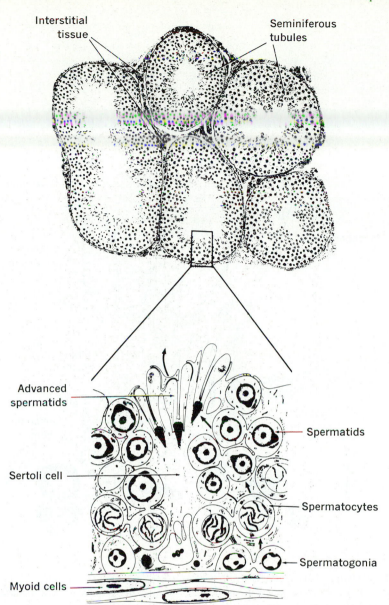

Interstitial tissue

Seminiferous tubules

Advanced spermatids

Spermatids

Sertoli cell

Spermatocytes

Spermatogonia

Myoid cells

Fig. 65-11. Histologic section of testis showing relation of various cell types. Upper figure shows cords of interstitial tissue interspersed between seminiferous tubules. Lower portion shows schematically a high-power magnification of seminiferous epithelium. Arrows indicate migration of germ cells through Sertoli cells as spermatogenesis progresses. (From Fawcett, D. W.: The male reproductive system. In Greep, R. O., Koblinsky, M. A., and Jaffe, F. S., editors: Reproduction and human welfare: a challenge to research. Reprinted by permission of the MIT Press, Cambridge, Massachusetts.)

meiosis, in which the spermatocytes undergo two maturational divisions that reduce the chromosome number and produce a cluster of spermatids; and finally *spermiogenesis,* in which the spermatids lose most of their cytoplasm and are transformed into sperm.

Sertoli cells. The Sertoli cells are remarkable polyfunctional cells whose activities are inti-

mately related to many aspects of the formation and maturation of the spermatozoa. In addition, during fetal life, they are credited with producing a protein, the *müllerian regression factor* (MRF), which is at least partially responsible for sexual differentiation.[35] Normal development of the male reproductive tract during fetal life depends on the diffusion of MRF from the fetal testis to

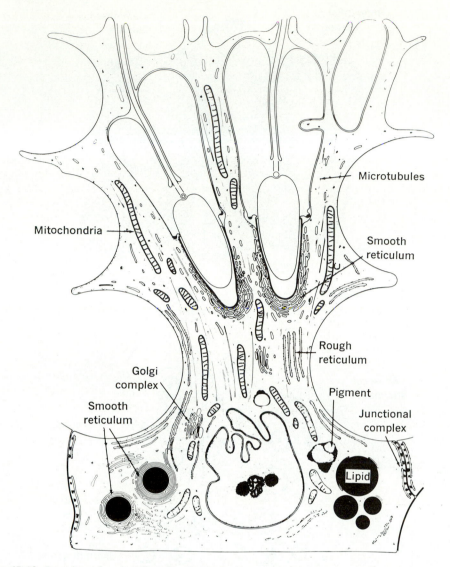

Fig. 65-12. Drawing of typical Sertoli cell. Two advanced spermatids are embedded in cytoplasm near apical surface. Irregular indentations on lateral borders indicate location of maturing spermatids. Various intracellular organelles are indicated. (From Fawcett.[65])

the primordia of the müllerian ducts, which develop into uterus and fallopian tubes unless inhibited by MRF. The Sertoli cells line the seminiferous tubules and extend through the entire thickness of the germinal epithelium from the basement membrane to the lumen. In the adult, these large, nonproliferating cells take on exceedingly irregular shapes determined by the changing conformation of the developing spermatogenic cells that are embedded in the sertolian cytoplasm (Fig. 65-12). Clusters of spermatogonia nestle between the Sertoli cells at the level of the basement membrane. Arching just

above the spermatogonia, adjacent Sertoli cells form occluding junctions in which the intercellular space is obliterated at multiple points to produce a series of tight junctions.[65] The occluding junctions between the Sertoli cells are thought to be responsible for the so-called *blood-testis barrier,* which, like its counterpart in the brain, prevents the passage of vital dyes and some physiologically relevant molecules into the seminiferous tubules. The blood-testis barrier actually has a wide range of permeabilities and allows rapid entry of testosterone, for example, while virtually completely excluding cholesterol.[145] The physiologic significance of the blood-testis barrier has not been established, nor has it been demonstrated that its role is to keep substances from entering the seminiferous tubules rather than to keep the tubular contents from reaching the blood. It may be of some importance that the spermatogonia are located on the blood side of the barrier, whereas the developing spermatids are restricted to the luminal side. At the start of their 64-day period of development into spermatozoa the developing spermatocytes cross the blood-testis barrier, presumably a result of the rupture and reformation of occluding junctions by the Sertoli cells. Throughout all the subsequent cell divisions the differentiating sperm cells are isolated from the bloodstream and must rely on the Sertoli cells for their sustenance and the removal of their metabolic waste products and even their excess cytoplasm, which is shed as the spermatozoa are released into the lumen. Little is known of the molecular processes in the Sertoli cells that gradually move the developing spermatids upward toward the lumen and ultimately expel them.

In addition to harboring the developing sperm, the Sertoli cells secrete a watery fluid that carries the nonmotile spermatozoa from the seminiferous tubules into the epididymis, where 99% of the fluid is reabsorbed. In the ram, flow is appreciable; each testis secretes about 40 ml of fluid/day.[160] Secretion of the fluid seems to be an active, energy-dependent process. The ionic composition of the fluid differs from blood plasma, and although all the plasma proteins are present, their concentration is quite low. The fluid elaborated by the seminiferous tubules of several species of experimental animals also contains a protein with a molecular weight of about 90,000 and has a high affinity for androgens.[86] Not surprisingly, it has been named the *andro-gen-binding protein* (ABP). ABP is distinct from both the cytoplasmic binding proteins involved in the intracellular actions of testosterone and from the sex hormone–binding globulin of the blood. Although the physiologic role of ABP is unknown, it may be of importance in delivery of androgen in high concentration to the epididymis, whose epithelial lining is androgen dependent. Synthesis and secretion of ABP by the Sertoli cells is under the control of both androgens and FSH.[85,119]

Although the Sertoli cells have a fairly extensive agranular cytoplasmic reticulum reminiscent of steroid-producing cells, a steroidogenic function for these cells has not been established. Available evidence indicates that the vast ma-jority, if not all, of testicular androgen production is attributable to the Leydig cells. Some evidence suggests, however, that the Sertoli cells, under the influence of FSH, may convert androgens to estrogens.[59]

Leydig cells. Embedded in the loose connective tissue that fills the spaces between the seminiferous tubules are clumps of epithelioid cells, the interstitial cells of Leydig, which are responsible for the production of the male sex hormone testosterone. They are large (20 μm in diameter) polyhedral cells with an extensive smooth endoplasmic reticulum characteristic of steroid-secreting cells. Secretion of testosterone by these cells during fetal life is required for the retention of the wolffian duct system and its differentiation into the internal male accessory organs and for the development of the external genitalia.[99] Although extensive at birth, these cells virtually disappear within the first 6 months of postnatal life. The reappearance of the interstitial cells heralds the onset of puberty.

Control of testicular function

The importance of the pituitary gonadotropins for normal testicular function is well established.[83] Hypophysectomy results in an almost immediate cessation of spermatogenesis and in atrophy of the testes and the accessory organs of reproduction. Treatment of hypophysectomized male rats with the same pituitary extracts that restore ovarian function in females also restores testicular function. Although the gonadotropins were originally named for their function in females, early investigators soon established the now-classic view that FSH controls spermatogenesis and LH controls androgen secretion by the Leydig cells. However, the observation that treatment with androgens alone can maintain spermatogenesis in hypophysectomized rats[136] and monkeys,[147] coupled with the lack of pure gonadotropins that contained only FSH or only LH activity, introduced discordant notes that have still not been entirely resolved.

Control of Leydig cells. It is now widely accepted that LH acts on the Leydig cells to increase their production of testosterone. LH[45] binds specifically to receptors on the Leydig cells and activates adenylate cyclase.[50] The resulting increase in cyclic AMP is presumed to trigger testosterone synthesis by increasing the rate of conversion of cholesterol to pregnenolone[84] (Fig. 65-13). FSH is thought to synergize with LH in evoking testosterone production, but how it does so is uncertain because the Leydig cells appear to lack receptors for FSH.[119] The Leydig cells atrophy after hypophysectomy and lose their lipid components. LH restores the Leydig cells to normal in hypophysectomized rats and can produce frank hypertrophy with continued administration. Together with these changes in histologic appearance of the Leydig cells, hypertrophy of the seminal vesicles and prostate gland attest to increased secretion of androgen. In fact, hypertrophy of the ventral lobe of the prostate after the administration of LH to hypophysectomized rats is the basis for a sensitive bioassay procedure for LH.

Control of spermatogenesis.[15] The idea that FSH is the pituitary hormone that controls spermatogenesis arose from the observation that administration of FSH alone to immature or hypophysectomized animals for a period of several days produced testicular enlargement accompanied by increased cell division in the seminiferous epithelium. Four decades later the precise role of FSH in this regard is still poorly understood, and what little is known derives largely from studies of rodents. FSH appears to be indispensable for initiating spermatogenesis at puberty or for restoring spermatogenesis after hypophysectomy, but once the process has been initiated, sperm formation can be maintained with testosterone alone or with LH alone. This observation has led to the speculation that FSH may not be involved in sustaining spermatogenesis in the mature individual. The principal, and perhaps only, site of action of FSH in the testis is the Sertoli cell.[120] FSH binds with high affinity to specific receptors and activates adenylate cyclase in Sertoli cells of very young or hypophysectomized rats. FSH also increases the synthesis of RNA and proteins in these cells. No such increase in cyclic AMP generation or protein synthesis is seen when FSH is given to more mature normal rats, even though the hormone still binds to receptors in these cells. At present there is no explanation for the loss of sensitivity of the Sertoli cells to FSH, and it is unknown whether these findings are applicable to man or even bear on the role of FSH in sustaining spermatogenesis.

Testosterone or one of its metabolites is now recognized as essential for the formation and maturation of sperm.[108] Appreciation of the physiologic importance of testosterone in this regard was hampered by the observation that low doses inhibited spermatogenesis and caused atrophy of the Leydig cells, whereas high doses supported spermatogenesis even in the absence of the pituitary gland. We now know that very high concentrations of androgens must be present in the seminiferous tubules. Attainment of such high concentrations depends on diffusion of testosterone from the Leydig cells to the adjacent tubules, which, as already indicated, allow free passage of testosterone across the blood-testis barrier. Administration of low doses of testosterone decreases the effective concentration of testosterone in the tubules by inhibiting LH secretion and hence androgen production. The concentration of testosterone in testicular venous blood, which presumably reflects its concentration in testicular interstitial fluid, is 40 to 50 times that found in peripheral blood. Hence extraordinarily high doses of testosterone must be given to duplicate this local concentration of the androgen. It is still not certain whether the site of action of testosterone in promoting spermatogenesis is at the level of the germ cells, the Sertoli cells that envelop them, or both. The Sertoli cells have cytoplasmic receptors for testosterone that can translocate the hormone to the nucleus, suggesting that they are targets for testosterone.[120]

EFFECT OF TEMPERATURE. In most mammals the testes descend from the abdominal cavity into the scrotum, where they are maintained at a few degrees below body temperature. Failure of testicular descent, either unilateral or bilateral, results in cryptorchidism (from the Greek *kryptos,* secret or hidden, and *orchis,* testicle), which is probably the most common defect in genital development. In the cryptorchid testis, spermatogenesis is defective or absent and androgen production is decreased.[20,160] The adverse effects of maintaining a testicle within the abdominal cavity appear to be due entirely to the increased temperature and can be duplicated by warming the normally descended testicle by scrotal insulation or high ambient temperature, or by transferring it surgically from the scrotum to the abdominal cavity. An intriguing series of experiments demonstrated the importance of temperature for steroid production.[91] When mouse ovaries were transplanted to the pinna of the ear of castrated male mice, the grafts produced sufficient androgen to maintain the accessory glands of reproduction. Ovaries transplanted to intra-abdominal sites, where they were maintained at higher temperatures, were unable to produce sufficient androgen to maintain the accessory glands. Similarly, when mice bearing the grafts in their ears were maintained at ambient temperature of 33° C rather than 22° C used in the previous experiment, androgen production was no longer in evidence. Thus the cooler temperature favored androgen production whereas the warmer temperature presumably favored estrogen production. The molecular basis for these findings has not been established.

As might be expected from the foregoing, a variety of morphologic modifications have evolved that maintain the testis at a reasonably constant temperature.[158] Prominent among these modifications is the vascular countercurrent heat exchanger formed by the vessels supplying and draining the testes. The testicular artery is not only highly tortuous and convoluted, but its convolutions are intimately intermingled with those of the testicular veins in the spermatic cord. The multiple branching of the spermatic artery allows for many points for counterflow of arterial and venous blood separated only by the thin vascular walls. This vascular arrangement, unique to mammals, is called the *pampiniform plexus*. In the ram, the species in which this heat-exchange mechanism has been studied most extensively, the internal spermatic artery is often more than 300 cm in length, and its tortuous gyrations permit the temperature of the blood that enters the testis to cool as much as 5.2° C below that of aortic blood.[159]

The scrotum is also endowed with features that facilitate temperature regulation.[158] Its skin is thin, virtually hairless in most species, and lacks subcutaneous fat. Cutaneous blood flow varies widely with temperature. The underlying sheet of smooth muscle (the tunica dartos) contracts in the cold, effectively thickening the scrotal wall while drawing the testes up closer to the body. Conversely, relaxation of the dartos muscle in response to heat lowers the testes away from the body and thins the scrotal wall. In the rat at least, these actions of the dartos muscle depend on androgen.[23] The scrotum also is richly endowed with abundant nerve endings that are sensitive to temperature and with sweat glands whose activity can be increased reflexly by warming various parts of the body. The importance of scrotal temperature receptors for overall body temperature regulation, however, is not known.

TESTOSTERONE

The first published evidence for the existence of hormones is credited to Berthold who, in 1849, transplanted testes into capons (castrated male chicks) and found that all the secondary sexual characteristics of roosters appeared. The quest for the male sex hormone was given impetus by the extravagant claims of rejuvenation by the eminent physiologist Brown-Séquard, who injected himself with aqueous testicular extracts that we now know probably were devoid of androgenic activity. Androsterone, now known to be a metabolite of testosterone, was isolated in crystalline form in 1932 by Butenandt, who extracted 15,000 L of male urine to obtain a yield of only 15 mg.[44] Crystalline testosterone, the physiologically important androgen secreted by the human testis, was isolated in 1935.[56] Other androgens, androstenedione, and dehydroepiandrosterone are also secreted but only at about 10% of the rate of testosterone; they have less than one tenth its biologic activity in most assays. The biosynthesis of these compounds is shown in Fig. 65-13. Testosterone is present in spermatic venous blood at concentrations between 10 and 50 $\mu g/100$ ml[74] and in peripheral blood at about 0.35 to 1.1 $\mu g/100$ ml.[30] At least 95% of the testosterone in plasma is secreted by the testis; the remainder is derived from adrenal secretion and peripheral conversion of other steroids to testosterone. As with the other steroid hormones, testosterone circulates bound to plasma protein. It is of interest that appreciable amounts of estrogens (20 pg/ml estradiol and 60 pg/ml estrone) are found in the plasma of normal men. This is comparable to the amounts found in the plasma of women in the early follicular phase. Two thirds of the estrogen produced each day in the normal man arises from extratesticular conversion of testosterone and androstenedione.[113] The Leydig cell is probably the chief source of the testicular estrogens although, as already mentioned, the Sertoli cells have the capacity to convert testosterone to estradiol.[27] The normal young man produces some 7 mg of testosterone/day.[95] Both the concentration in plasma and the estimated daily production rate decrease somewhat with age, so that by the seventh decade and beyond, testosterone production may decline to 4 mg/day.[101] In the absence of illness or injury, there is no sharp drop in testosterone production that is analogous to the abrupt cessation of estrogen and progesterone production in the menopausal woman.

Only about 2% of the testosterone circulates as the free hormone. The rest is bound to albumin (about 40%) and to the sex hormone–binding globulin (SHBG), which is also called testosterone-binding globulin (TEBG). This protein binds both estrogen and testosterone, but its single binding site has a higher affinity for testosterone than estrogen. Its concentration in plasma is increased by estrogens and decreased by androgens. There is about two to three times as much circulating SHBG in women as men, and in pregnancy the concentration increases even further. The concentrations of SHBG and testosterone in blood are quite similar, so that the protein is nearly saturated with testosterone. Thus fluctuations in estrogen can profoundly affect the amount of testosterone that is free in plasma and hence available to interact with target tissues.[43] Conversely, by lowering the amount of SHBG, androgens amplify their own unbound concentrations.

The liver is the principal site of degradation of testosterone, although other tissues also metabolize it. Metabolic breakdown consists of reduction of the A ring with or without previous oxidation to androstenedione. Following these degradative changes, the androgens are conjugated either with sulfate or glucuronide and released into the blood from whence they are excreted in the urine.

Physiologic actions

Testosterone acts principally to promote the growth, differentiation, and function of the accessory organs of reproduction, but significant effects are also exerted on a variety of other body tissues.

Prenatal development.[99] The seminal vesicles, epididymes, and vasa deferentia arise from the primitive mesonephric or wolffian ducts. In the female or androgen-deficient male fetus the wolffian ducts regress and are resorbed. Testosterone secreted by the fetal testis stabilizes the wolffian ducts and causes them to differentiate into the internal reproductive structures. The urogenital sinus and the genital tubercle are the primitive structures that give rise to the external genitalia in both sexes. Masculinization of these primitive structures to form the penis, scrotum, and prostate gland also depends on secretion of fetal testicular androgen. In the absence of androgen, these structures go on to form the female external genitalia; insufficient androgen in males or too much androgen in females results in incomplete differentiation.

Genitalia. Testosterone promotes the growth of the penis and scrotum at puberty. With the advent of puberty, testosterone increases the pigmentation of the genitalia as well as the depth of the rugal folds in the scrotal skin. The development of the dartos muscle and its sensitivity to temperature depend on testosterone. Testosterone also acts directly on the seminiferous tubules to promote spermatogenesis.

Accessory sexual organs. In addition to the external genitalia, the male reproductive tract consists of a series of excretory ducts and glands that function to transport, nurture, and mature the sperm and to provide the seminal fluid neces-

sary for their survival and motility after ejaculation. Testosterone controls the growth, development, and function of the accessory organs of reproduction.

Epididymis. After the completion of spermatogenesis in the seminiferous tubules the sperm are collected in the *rete testis,* from whence they enter the converging *ductuli efferentes,* whose ciliated epithelium facilitates passage into the highly convoluted and tortuous duct of the epididymis. In man, each epididymal duct would extend some 4 to 6 m in length if stretched out. The epididymis is the primary area for storage of sperm, and viable sperm may remain within its confines for months. Here maturation is completed, as evidenced by a time-dependent progressive increase in fertilizing capacity. The secretions of the epithelium are thought to nurture the sperm during their sojourn in the epididymis. This secretion presumably is under the control of testosterone, which has been shown to prolong the life of epididymal sperm in castrated guinea pigs.[167] The sperm are advanced through the epididymis, particularly during sexual arousal, by rhythmic contractions of the circular smooth muscle surrounding the duct. At ejaculation, sperm are expelled into the vas deferens and ultimately through the urethra. An accessory storage area lies in the ampulla of the vas deferens, posterior to the seminal vesicles. The combined secretions of the prostate gland, seminal vesicles, and Cowper's gland, together with the sperm, comprise the semen. The average volume of the ejaculate is of the order of 3 ml, of which less than 10% (50 to 150 million/ml) consists of sperm.

Seminal vesicles. The paired seminal vesicles are elongated, hollow evaginations of the ductus deferens. They have irregular lumina lined with secretory epithelium. The smooth muscle coat of the walls is richly innervated. Testosterone promotes both the prenatal development of the seminal vesicles and their growth at puberty. Normal function in the adult depends on testosterone; the secretory epithelium atrophies following castration and can be restored with injections of androgen. The secretion of the seminal vesicles is rich in both citric acid and fructose, both of which provide nutriment for the sperm after ejaculation. The content of fructose in the secretion of the seminal vesicle may be as high as 300 mg/100 ml and accounts almost entirely for the fructose content of the semen. Metabolism of fructose provides the energy for sperm motility.

Prostate. The prostate gland is about 4 cm in

diameter and completely surrounds the urethra just below the urinary bladder at the level where the two ejaculatory ducts join it. The prostate consists of compound tubuloalveolar glands, whose multiple excretory ducts open independently into the urethra. Growth of the prostate depends on testosterone and is one of the most sensitive indicators of androgen action in mammals. The prostatic secretion is thin and watery and contains a variety of enzymes, including acid phosphatase and fibrinolysin.

Bulbourethral glands (Cowper's glands). Testosterone also promotes the growth, development, and secretory activity of the bulbourethral glands, which are responsible for secretion of mucoprotein.

Secondary sexual characteristics. In addition to its effect on those organs that are directly related to the transport and delivery of sperm, testosterone also affects a variety of tissues and thus contributes to the morphologic and psychologic components of masculinity. These characteristics are clearly an integral part of reproduction, for they are related to the attractiveness of the male to the female and to his ability to compete with other males for her favors. Testosterone stimulates the growth of pubic hair in a typical masculine pattern, with hair extending upward to the umbilicus and beyond. Growth of chest, axillary, and facial hair is also stimulated, but scalp hair is affected in the opposite manner. Recession of the hair at the temples is a typical response to androgen, and adequate amounts of this hormone allow the expression of the genes for baldness. In early adolescence, testosterone causes the enlargement of the larynx and the thickening of the vocal chords, phenomena that lower the pitch of the male voice. At this time also the characteristic adolescent growth spurt results from an action of testosterone to promote growth of the long bones. This growth is self-limiting, since testosterone also accelerates maturation of bone and epiphyseal closure. Skeletal development, particularly of the shoulder girdle, is pronounced. Androgens markedly promote nitrogen retention and the formation of muscle, especially in the upper torso. In some rodents the temporal and masseter muscles are particularly sensitive to androgenic stimulation. This anabolic action of androgens has led to their use in combating effects of prolonged bed rest, disease, and surgical trauma. Androgen therapy can convert a mild negative nitrogen balance to net retention of nitrogen. This effect, which is exerted directly on the protein-synthetic apparatus of muscle, of course, is also related to stimulation of appetite

and increased food intake. Accordingly, androgens bring about increased physical vigor and a feeling of well-being. Growth and secretion of the sebaceous glands of the skin are also stimulated, a phenomenon undoubtedly related to the acne of adolescence.

Effects on the central nervous system. Neonatal androgen administration produces irreversible developmental changes in the CNS of rats. This finding has been offered to explain the effects of early androgen treatment on subsequent aggressive fighting behavior in adult female mice and in comparison on behavior of neonatally castrated males with animals castrated postpuberally.[62] Finally, androgens also affect the CNS by increasing sexual drive and libido. Androgens do not, however, affect the direction of such sex drive and appear unrelated to predilections for hetero- or homosexual activity.

Mechanism of action

Like the other steroid hormones, testosterone penetrates the target cells whose growth and function it stimulates. Labeled steroid is bound first to a cytoplasmic binding protein that is then transferred to the nucleus, where interaction with the genetic material occurs.[32] Some verification of this sequence of molecular events in human beings has been provided from a study of the syndrome of testicular feminization. Afflicted individuals are outwardly feminine and have normal female external genitalia and breast development. They are, however, genetic males and have intra-abdominal testes and circulating concentrations of testosterone and estrogen that are characteristic of normal males. These individuals are totally unresponsive to androgens. They develop as females because, as already mentioned, in the absence of androgen stimulation the presumptive genital tissue develops in the female pattern. Their absolute end-organ insensitivity has been traced to a lack of intracellular androgen receptor proteins.[162]

A unique feature of androgen action is that testosterone is converted within target cells to dihydrotestosterone by reduction of the double bond in ring A (Fig. 65-13).[161] The reductase enzyme

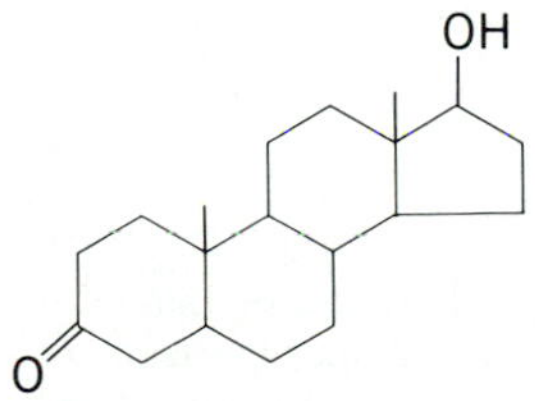

Fig. 65-13. Dihydrotestosterone.

required for the conversion is present in both the nuclei and cytoplasm in all the tissues of the accessory organs of reproduction. Dihydrotestosterone appears to be the predominant intracellular androgen that binds to the androgen receptors and ultimately interacts with chromatin. This places testosterone in the role of a prohormone. Related studies reveal that dihydrotestosterone is a more potent androgen than testosterone in most bioassay systems. Some androgen-sensitive cells can respond only to the dihydro form of testosterone. This is evident in individuals who are genetically deficient in the reductase enzyme. They have feminine-appearing external genitalia at birth, although all the structures that derive from the wolffian duct develop normally. Thus differentiation of the external genitalia requires dihydrotestosterone, whereas testosterone is adequate for development of the wolffian derivatives.[96]

REGULATION OF TESTICULAR FUNCTION[57,80]

Although regulation of reproductive function in the male has been regarded as less complex than in the female, it is perhaps even less well understood. The two pituitary hormones, FSH and LH, each appear to be primarily responsible for either spermatogenesis or steroidogenesis, while each apparently supports the other in carrying out these functions. It seems reasonably clear that the secretion of LH and testosterone are linked in a typical negative feedback way. LH, but not FSH, increases testosterone secretion. Conversely, physiologic doses of testosterone reduce LH secretion without decreasing the secretion of FSH[68] (Fig. 65-14), whereas castration results in a prompt increase in the release of both FSH and LH by the pituitary gland. A variety of injuries that lead to degeneration of the seminiferous tubules without inter-

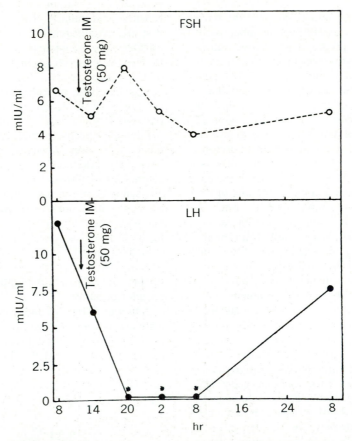

Fig. 65-14. Effect of intramuscular injection of testosterone propionate on serum FSH and LH levels. Each point represents mean value obtained from five individual determinations. Significant diminution indicated by asterisks was seen only in LH concentrations. (From Franchimont and Legros.[68])

fering with Leydig cell function similarly produce high circulating levels of FSH in the presence of normal concentrations of testosterone and LH. These observations have led to the suggestion that the seminiferous tubules produce a substance called inhibin that feeds back to the pituitary gland to diminish FSH secretion. Indeed, a water-soluble substance capable of selectively inhibiting FSH secretion has been found in fluid collected from the rete testis of rams and in seminal plasma.[29] Sertoli cells grown in tissue culture release a substance with similar properties, suggesting that they are the source of inhibin.[148] Several experiments suggest that inhibin acts at the level of the pituitary gland to decrease the sensitivity of the FSH-secreting cells to the gonadotropin-releasing hormone of the hypothalamus.[29,148]

The control of LH secretion in the male is probably analogous to the negative feedback control of LH secretion in the female (p. 1615). The finding that hypothalamic tissue can convert testosterone to estrogen[125] raises the possibility that estrogen rather than testosterone is monitored in the feedback control of LH secretion in the male as well as the female. On a molar basis, estrogen is considerably more effective an inhibitor of LH secretion than is testosterone. Furthermore, dihydrotestosterone, which has greater potency than testosterone on most target tissues, probably cannot be aromatized to estrogen and appears to exert no feedback suppression of LH secretion in primates, although it is quite potent in this regard in rodents.

FERTILITY CONTROL[83,141]

No aspect of contemporary physiology is as directly and immediately applicable to pressing contemporary social and economic problems as the study of reproduction. The oral contraceptive pills, which comprise synthetic estrogens and progestins administered alone, in combination, or in sequence, were developed because of an understanding, although rudimentary, of negative feedback relationships between the ovary and the pituitary gland. These drugs work principally through interference with gonadotropin secretion, although direct effects of the steroid hormones on the accessory organs of the female reproductive tract may also play a role in their effectiveness. For example, progesterone increases the viscosity of the cervical mucus, making it difficult for the sperm to penetrate the uterine cervix. Similarly effects of steroid hormones on motility of the fallopian tube and hence on transport of the ovum may interfere with either fertilization or implantation. Identification of the putative tubular compound that inhibits FSH secretion in the male may provide the basis for the development of a male contraceptive pill. Similarly, understanding of the mechanism for the relative decrease in fertility during lactation may provide another avenue of approach to fertility control. As an intellectual exercise, the interested student may find it of value to test his understanding of the material presented in this chapter by noting points of vulnerability in the reproductive process and theorizing on how they may be exploited for effective control of fertility.[7]

REFERENCES
General reviews

1. Armstrong, D. T., and Dorrington, J. H.: Estrogen biosynthesis in the ovaries and testes. In Thomas, J. A., and Singhal, R. L., editors: Advances in sex hormone research, Baltimore, 1977, University Park Press, p. 217.
2. Brenner, R. M., and West, N. B.: Hormonal regulation of the reproductive tract in female mammals, Annu. Rev. Physiol. **37:**273, 1975.
3. Channing, C. P., and Tsafriri, A.: Mechanism of action of luteinizing hormone and follicle stimulating hormone on the ovary in vitro, Metabolism **26:**413, 1977.
4. Everett, J. W.: Central neural control of reproductive functions of the adenohypophysis, Physiol. Rev. **44:** 373, 1964.
5. Grady, H. G., and Smith, D. E., editors: The ovary, Baltimore, 1963, The Williams & Wilkins Co.
6. Greep, R. O., and Astwood, E. B., editors: Handbook of physiology. Section 7: endocrinology, Female reproductive system and Male reproductive system, Washington, D.C., 1975, American Physiological Society, vols. 2 and 5.
7. Greep, R. O., Koblinsky, M. A., and Jaffe, T. S.: Reproduction and human welfare: a challenge to research. Cambridge, Mass., 1976, The M.I.T. Press.
8. Hisaw, F. L.: Development of the graafian follicle and ovulation, Physiol. Rev. **27:**95, 1947.
9. Johnson, A. D., Gomes, W. R., and Vandermakr, N. L., editors: The testis, New York, 1970, Academic Press, Inc.
10. Knobil, E.: On the control of gonadotropin secretion in the rhesus monkey, Recent Prog. Horm. Res. **30:** 1, 1974.
11. Odell, W. D., and Moyer, D. L.: Physiology of reproduction, St. Louis, 1971, The C. V. Mosby Co.
12. Reynolds, S. R. M.: Physiology of the uterus, New York, 1949, Paul B. Hoeber, Inc.
13. Rosemberg, E., and Paulsen, C. A., editors: The human testis, New York, 1970, Plenum Press, Inc.
14. Schwartz, N. B., and McCormack, C. E.: Reproduction: gonadal function and its regulation, Annu. Rev. Physiol. **34:**425, 1972.
15. Steinberger, E.: Hormonal control of mammalian spermatogenesis, Physiol. Rev. **51:**1, 1971.
16. Turner, C. D., and Bagnara, J. T., editors: General endocrinology, ed. 6, Philadelphia, 1976, W. B. Saunders Co.
17. Vickery, B. H., and Bennett, J. P.: The cervix and its secretion in mammals, Physiol. Rev. **48:**135, 1968.

18. Young, W. C., editor: Sex and internal secretions, Baltimore, 1961, The Williams & Wilkins Co.
19. Zuckerman, S., editor: The ovary, New York, 1962, Academic Press, Inc., vols. 1 and 2.

Original papers

20. Albert, A.: The mammalian testis. In Young, W. C., editor: Sex and internal secretions, ed. 3, Baltimore, 1961, The Williams & Wilkins Co., vol. 1.
21. Allen, E., and Doisy, E. A.: An ovarian hormone: a preliminary report on its localization, extraction and partial purification, and action in test animals, J.A.M.A. **81:**819, 1923.
22. Amoss, M., et al.: Purification, amino acid composition and N-terminus of the hypothalamic luteinizing hormone releasing factor (LRF) of ovine origin, Biochem. Biophys. Res. Commun. **44:**205, 1971.
23. Andrews, F. N.: Thermo-regulatory function of rat scrotum. I. Normal development and effect of castration, Proc. Soc. Exp. Biol. Med. **45:**867, 1940.
24. Arimura, A., et al.: Stimulation of release of LH by synthetic LH-RH in vivo: comparative study of natural and synthetic hormones, Endocrinology **90:**163, 1972.
25. Armstrong, D. T.: Gonadotropins, ovarian metabolism, and steroid biosynthesis, Recent Prog. Horm. Res. **24:**255, 1968.
26. Armstrong, D. T., Lee, T. P., and Miller, L. S.: Stimulation of progesterone biosynthesis in bovine corpora lutea by luteinizing hormone in the presence of an inhibitor of cholesterol synthesis, Biol. Reprod. **2:**29, 1970.
27. Armstrong, D. T., et al.: Synthesis of estradiol 17β by Sertoli cells in culture: stimulation of FSH and dibutyl cyclic AMP. In French, F. S., et al., editors: Hormonal regulation of spermatogenesis, New York, 1975, Plenum Press, p. 85.
28. Astwood, E. B.: Estrogens and progestins. In Goodman, L. S., and Gilman, A., editors: The pharmacological basis of therapeutics, New York, 1970, Macmillan Publishing Co., Inc.
29. Balcer, H. W. G., et al.: Testicular control of follicle-stimulating hormone secretion, Recent Prog. Horm. Res. **32:**429, 1976.
30. Bardin, C. W., and Lipsett, M. B.: Estimation of testosterone and androstenedione in peripheral plasma, Steroids **9:**71, 1967.
31. Bartelmez, G. W.: The phases of the menstrual cycle and their interpretation in terms of the pregnancy cycle, Am. J. Obstet. Gynecol. **74:**931, 1957.
32. Baulieu, E. E., et al.: Metabolism and protein binding of sex steroids in target organs: an approach to the mechanism of hormone action, Recent Prog. Horm. Res. **27:**351, 1971.
33. Beling, C. G.: Gel filtration of conjugated urinary oestrogens and its application in clinical assays, Acta Endocrinol. **43**(suppl. 79):80, 1963.
34. Bissonnette, T. H.: Modification of mammalian sexual cycles. III. Reversal of the cycle in male ferrets (Putorius vulgaris) by increasing periods of exposure to light between October second and March thirtieth, J. Exp. Zool. **71:**341, 1935.
35. Blanchard, M. G., and Josso, N.: Source of the anti-müllerian hormone synthesized by the fetal testis: müllerian-inhibiting activity of fetal bovine sertoli cells in tissue culture, Pediatr. Res. **8:**968, 1974.
36. Blandau, R. J.: Biology of eggs and implantation. In Young, W. C., editor: Sex and internal secretions, ed. 3, Baltimore, 1961, The Williams & Wilkins Co., vol. 2.
37. Blandau, R. J.: Gamete transport—comparative aspects. In Hafez, E. S. E., and Blandau, R. J., editor: The mammalian oviduct, Chicago, 1969, University of Chicago Press.
38. Boling, J. L.: Endocrinology of oviductal musculature. In Hafez, E. S. E., and Blandau, R. J., editors: The mammalian oviduct, Chicago, 1969, University of Chicago Press.
39. Borell, U., Nilsson, O., and Westman, A.: Ciliary activity in the rabbit fallopian tubes during oestrus and after copulation, Acta Obstet. Gynecol. Scand. **36:** 22, 1957.
40. Brenner, R. M.: The biology of oviductal cilia. In Hafez, E. S. E., and Blandau, R. J., editors: The mammalian oviduct, Chicago, 1969, University of Chicago Press.
41. Brown, J. B., and Matthew, G. D.: Application of urinary estrogen methods to problems in gynecology, Recent Prog. Horm. Res. **18:**337, 1962.
42. Burdick, H. O., and Pincus, G.: The effect of oestrin injections upon the developing ova of mice and rabbits, Am. J. Physiol. **111:**201, 1935.
43. Burke, C. W., and Anderson, D. C.: Sex hormone-binding globulin is an oestrogen amplifier, Nature **240:** 38, 1972.
44. Butenandt, A.: Über die chemische Untersuchung der sexual Hormons, Z. Angew. Chem. **44:**905, 1931.
45. Catt, K., and Dufau, M. L.: Interactions of LH and HCG with testicular gonadotropin receptors. In Means, A. B., and O'Malley, B. W., editors: Receptors for reproductive hormones, New York, 1973, Plenum Press, p. 279.
46. Channing, C. P.: Influences of the in vivo and in vitro hormonal environment upon luteinization of granulosa cells in tissue culture, Recent Prog. Horm. Res. **26:** 589, 1970.
47. Channing, C. P., and Coudert, S. P.: The role of granulosa cells and follicular fluid in estrogen secretion by the monkey ovary in vivo, Endocrinology **98:**590, 1976.
48. Channing, C. P., and Kammerman, S.: Binding of gonadotropins to ovarian cells, Biol. Reprod. **10:**179, 1974.
49. Chatterton, R. T., Jr.: Progesterone and mammary gland development. In McKerns, K. W., editor: The sex steroids, New York, 1971, Appleton-Century-Crofts.
50. Cooke, B. A., et al.: Effect of trophic hormones on 3′,5′-cyclic AMP levels in rat testis interstitial tissue and seminiferous tubules, FEBS Lett. **25:**83, 1972.
51. Corner, G. W.: Physiology of the corpus luteum. The effect of very early ablation of the corpus luteum upon embryos and uterus, Am. J. Physiol. **86:**74, 1928.
52. Corner, G. W., and Allen, W. M.: Physiology of the corpus luteum. Production of a special uterine reaction (progestational proliferation) by extracts of the corpus luteum, Am. J. Physiol. **88:**326, 1929.
53. Courot, M., Hochereau-deReviers, M. T., and Ortavant, R.: Spermatogenesis. In Johnson, A. D., Gomes, W. R., and Vandemark, N. L., editors: The testis, New York, 1970, Academic Press, Inc., vol. 1.
54. Cowie, A. T., and Folley, S. J.: The mammary gland and lactation. In Young, W. C., editor: Sex and internal secretions, ed. 3, Baltimore, 1961, The Williams & Wilkins Co., vol. 1.
55. Csapo, A. I.: Function and regulation of the myometrium, Ann. N.Y. Acad. Sci. **75:**790, 1959.
56. David, K., et al.: Uber krystallinisches mannliches

Hormon aus Hoden (Testosteron), wirksamer als aus Harn oder aus cholesterin bereitetes Androsteron, Hoppe-Seyler's Z, Physiol. Chem. **233:**281, 1935.

57. Davidson, J. M.: Control of gonadotropin secretion in the male. In Martini, L., and Ganong, W. F., editors: Neuroendocrinology, New York, 1966, Academic Press, Inc., vol. 1.

58. Davidson, O. W., and Alvarez, B.: Transference of circulating labeled serum proteins to follicle of rat ovary, J. Histochem. Cytochem. **11:**80, 1963.

59. Dorrington, J. H., and Armstrong, D. T.: Follicle stimulating hormone stimulates estradiol-17β synthesis in cultured sertoli cells, Proc. Natl. Acad. Sci. USA **72:**2677, 1975.

60. Dorrington, J. H., Moon, Y. S., and Armstrong, D. T.: Estradiol 17 β biosynthesis in cultured granulosa cells from hypophysectomized immature rats: stimulation by follicle-stimulating hormone, Endocrinology **97:**1328, 1975.

61. Drost, M., and Holm, L. W.: Prolonged gestation in ewes after foetal adrenalectomy, J. Endocrinol. **40:** 293, 1968.

62. Edwards, D. A.: Mice: fighting by neonatally androgenized females, Science **161:**1027, 1968.

63. El-Fouly, M. A., et al.: Role of the ovum in follicular luteinization, Endocrinology **87:**288, 1970.

64. Ellsworth, L. R., and Armstrong, D. T.: Effect of LH on luteinization of ovarian follicles transplanted under the kidney capsule in rats, Endocrinology **88:**755, 1971.

65. Fawcett, D. W.: Ultrastructure and function of the Sertoli cell. In Hamilton, D. W., and Greep, R. O., editors: Handbook of physiology. Section 7: Male reproductive system, Washington, D.C., 1975, American Physiological Society, p. 21.

66. Fawcett, D. W.: The male reproductive system. In Greep, R. O., Koblinsky, M. A., and Jaffe, F. S., editors: Reproduction and human welfare: a challenge to research, Cambridge, Mass., 1976, The M.I.T. Press, p. 165.

67. Fell, P. D., et al.: Progesterone binding in the mouse and rat uterus, Endocrinology **91:**738, 1972.

68. Franchimont, P., and Legros, J. J.: The control of gonadotropin secretion in the human. In Martini, L., Motta, M., and Fraschini, F., editors: The hypothalamus, New York, 1970, Academic Press, Inc.

69. Friesen, H. G., et al.: The synthesis and secretion of placental lactogen and pituitary prolactin. In Wolstenholme, G. E. W., and Knight, J., editors: Lactogenic hormones, London, 1972, J. & A. Churchill, Ltd.

70. Fuchs, F.: Endocrinology of labor. In Fuchs, F., and Klopper, A., editors: Endocrinology of pregnancy, New York, 1971, Harper & Row, Publishers.

71. Furuyama, S., and Nugent, C. A.: A radioimmunoassay for plasma progesterone, Steroids **17:**663, 1971.

72. Gala, R. R., and Westphal, U.: Corticosteroid-binding globulin in the rat: possible role in the initiation of lactation, Endocrinology **76:**1079, 1965.

73. Gala, R. R., and Westphal, U.: Corticosteroid-binding in serum of mouse, rabbit and guinea pig during pregnancy and lactation: possible involvement in the initiation of lactation, Acta Endocrinol. **55:**47, 1967.

74. Gandy, H., and Peterson, R. E.: Measurement of testosterone and 17-keto steroids in plasma by the double isotope dilution derivative technique, J. Clin. Endocrinol. Metab. **28:**949, 1968.

75. Goldenberg, R. L., Reiter, F. O., and Ross, G. T.: Follicle response to exogenous gonadotropins: an estro-

gen mediated phenomenon, Fertil. Steril. **24:**121, 1973.

76. Goldman, B. D., and Mahesh, V. B.: Fluctuations in pituitary FSH during the ovulatory cycle in the rat and a possible role of FSH in the induction of ovulation, Endocrinology **83:**97, 1968.

77. Goss, D. A., and Taymor, M. L.: A rapid immunologic method for the semiquantitative determination of human chorionic gonadotropin in urine, Fertil. Steril. **16:**151, 1965.

78. Green, S. H., and Zuckerman, S.: A comparison of growth of the ovum and follicle in normal rhesus monkeys and in monkeys treated with oestrogens and androgens, J. Endocrinol. **5:**207, 1947.

79. Greenwald, G. S.: Role of follicle stimulating hormone and luteinizing hormone in follicular development and ovulation. In Knobil, E., and Sawyer, W., editors: Handbook of physiology. Section 7, Washington, D.C., 1974, American Physiological Society, p. 293.

80. Greep, R. O.: Physiology of the hypophysis in relation to reproduction. In Young, W. C., editor: Sex and internal secretions, ed. 3, Baltimore, 1961, The Williams & Wilkins Co.

81. Greep, R. O., Van Dyke, H. B., and Chow, B. F.: Gonadotropins of the swine pituitary. I. Various biological effects of purified thylakentrin (FSH) and pure metakentrio (ICSH), Endocrinology **30:**635, 1942.

82. Gurpide, E., et al.: Determination of secretory rates of estrogen in pregnant women from specific activities of urinary metabolites, J. Clin. Endocrinol. Metab. **22:**935, 1962.

83. Hafez, E. S. E., and Evans, T. N., editors: Human reproduction and contraception, New York, 1973, Harper & Row, Publishers.

84. Hall, P. F., and Koritz, S. B.: The conversion of cholesterol and 20α hydroxycholesterol to steroids by acetone powder of particles from bovine corpus luteum, Biochemistry **3:**129, 1964.

85. Hansson, R., et al.: The hormonal specificity of the production and secretion of testicular androgen binding protein (ABP) from Sertoli cells. In James, V. H. T., editor: Endocrinology. Proceedings of the Fifth International Congress of Endocrinology, Excerpta Medica International Congress Series, No. 403, Princeton, N.J., 1977, Excerpta Medica Inc., p. 410.

86. Hansson, V. M., et al.: Androgen transport and receptor mechanisms in testis and epididymis. In Greep, R. O., and Hamilton, D. W., editors: Handbook of physiology. Section 7, Washington, D.C., 1975, American Physiological Society, vol. 5, p. 173.

87. Harmon, S. M., Louvert, J. P., and Ross, G. T.: Interaction of estrogen and gonadotropins on follicular atresia, Endocrinology **96:**1145, 1975.

88. Harrington, F. E., et al.: The ovulatory effects of follicle-stimulating hormone treated with chymotrypsin in chlorpromazine blocked rats, Endocrinology **65:** 222, 1970.

89. Harris, G. W.: Neural control of the pituitary gland, London, 1955, Edward Arnold, Ltd.

90. Henzyl, M. R., et al.: Lysosomal concept of menstrual bleeding in humans, J. Clin. Endocrinol. Metab. **34:** 860, 1972.

91. Hill, R. T.: Ovaries secrete male hormone. III. Temperature control of male hormone output by grafted ovaries, Endocrinology **21:**633, 1937.

92. Hisaw, F. L., and Hisaw, F. L., Jr.: Action of estrogen and progesterone on the reproductive tract of lower primates. In Young, W. C., editor: Sex and internal

secretions, ed. 3, Baltimore, 1961, The Williams & Wilkins Co.

93. Hotchkiss, J., Atkinson, L. E., and Knobil, E.: Time course of serum estrogen and luteinizing hormone (LH) concentrations during the menstrual cycle of the rhesus monkey, Endocrinology **89:**177, 1971.

94. Hsueh, A. J. W., Peck, E. J., Jr., and Clark, J. H.: Control of uterine estrogen receptor levels by progesterone, Endocrinology **98:**438, 1976.

95. Hudson, B., and Coghlan, J. P.: Abnormalities of testosterone secretion in the male. In Astwood, E. B., and Cassidy, C. E., editors: Clinical endocrinology, New York, 1968, Grune & Stratton, Inc., vol. 2.

96. Imperato-McGinley, J., et al.: 5α-reductase deficiency in man: an inherited form of male pseudohermaphroditism. In James, V. H. T., editor: Endocrinology. Proceedings of the Fifth International Congress of Endocrinology, Excerpta Medica International Congress Series, No. 403, Princeton, N.J., 1977, Excerpta Medica Inc., p. 312.

97. Josimovich, J. B.: Human placental lactogen. In Wynn, R. D., editor: Fetal homeostasis, New York, 1969, Appleton-Century-Crofts, vol. 4.

98. Josimovich, J. B.: Placental lactogenic hormone. In Fuchs, F., and Klopper, A., editors: Endocrinology of pregnancy, New York, 1971, Harper & Row, Publishers.

99. Jost, A.: Hormonal and genetic factors affecting the development of the male genital system, Andrologia **8**(suppl. 1):17, 1976.

100. Keksces, L., et al.: Paperchromatographische Isolation des Oestrons, 17β-Oestradiols und Oestriols aus menschlichen Ovarien, Acta Endocrinol. **39:**483, 1962.

101. Kent, J. R., and Acone, A. B.: Plasma testosterone levels in aging males. In Vermeulen, A., and Exley, D., editors: Androgens in normal and pathological conditions, Princeton, N.J., 1966, Excerpta Medica Inc.

102. Knobil, E.: Hormonal control of the menstrual cycle and ovulation in the rhesus monkey, Acta Endocrinol. **166**(suppl.):137, 1972.

103. Knobil, E., et al.: Role of estrogen in the positive and negative feedback control of luteinizing hormone secretions during the menstrual cycle of the rhesus monkey. In Saxena, B. B., editor: Gonadotropins, New York, 1972, John Wiley & Sons, Inc.

104. Krey, L. C., Butler, W. R., and Knobil, E.: Surgical disconnection of the medical basal hypothalamus and pituitary function in the rhesus monkey. I. Gonadotropin secretion, Endocrinology **96:**1073, 1975.

105. Liggins, G. C., et al.: Control of parturition in man, Biol. Reprod. **16:**39, 1977.

106. Liggins, G. C., Kennedy, P. C., and Holm, L. W.: Failure of initiation of parturition after electrocoagulation of the pituitary of the fetal lamb, Am. J. Obstet. Gynecol. **98:**1080, 1967.

107. Lipner, H.: Mechanism of mammalian ovulation. In Greep, R. O., editor: Handbook of physiology. Section 7: Female reproductive system, Washington, D.C., 1973, American Physiological Society, vol. 2, pt. 1, p. 409.

108. Lipsett, M. B.: Regulation of testicular functions, Andrologia **8**(suppl. 1):43, 1976.

109. Little, B., et al.: The metabolic clearance rate of progesterone in males and oophorectomized females, J. Clin. Invest. **45:**901, 1961.

110. Llauro, J. L., Runnebaum, B., and Zander, J.: Progesterone in human peripheral blood before, during, and after labor, Am. J. Obstet. Gynecol. **101:**867, 1968.

111. Lloyd, C. W.: The ovaries. In Williams, R. H., editor: Textbook of endocrinology, Philadelphia, 1968, W. B. Saunders Co.

112. Loeb, L.: Effects of hysterectomy on the system of sex organs and on periodicity of the sexual cycle in the guinea pig, Am. J. Physiol. **83:**202, 1927.

113. Longcope, C., Kato, T., and Horton, R.: Conversion of blood androgen to estrogens in normal men and women, J. Clin. Invest. **48:**2191, 1969.

114. Lostroh, A. J., and Johnson, R. E.: Amounts of interstitial cell-stimulating hormone and follicle stimulating hormone required for follicular development, uterine growth and ovulation in the hypophysectomized rat. Endocrinology **79:**991, 1961.

115. Malven, P. V.: Hypophysial regulation of luteolysis in the rat. In McKerns, K. W., editor: The gonads, New York, 1969, Appleton-Century-Crofts.

116. Marsh, J. M.: The role of cyclic AMP in gonadal steroidogenesis, Biol. Reprod. **14:**30, 1976.

117. Matsuo, H., et al.: Structure of the porcine LH- and FSH-releasing hormone. I. Proposed amino acid sequence, Biochem. Biophys. Res. Commun. **43:**1334, 1971.

118. McCracken, J. A., Baird, D. T., and Goding, J. R.: Factors affecting the secretion of steroids from the transplanted ovary in the sheep, Recent Prog. Horm. Res. **27:**531, 1971.

119. Means, A. R., and Vaitukaitis, J.: Peptide hormone receptors; specific binding of ³H-FSH to testis, Endocrinology **90:**39, 1972.

120. Means, A. R., et al.: Follicle stimulating hormone, the sertoli cell and spermatogenesis, Recent Prog. Horm. Res. **32:**477, 1976.

121. Meites, J.: Hypothalamic control of prolactin secretion. In Wolstenholme, G. E. W., and Knight, J., editors: Lactogenic hormone, London, 1972, J. & A. Churchill, Ltd.

122. Midgely, A. R., Jr.: Gonadotropin binding to frozen sections of ovarian tissue. In Saxena, B., Belmia, C., and Gandy, H., editors: Gonadotropins, New York, 1972, Wiley-Interscience, p. 248.

123. Midgely, A. R., Jr., et al.: Gonadotropin and estradiol receptors and follicular development. In Sadler, W. A., and Segal, S., editors: Advances in fertility regulation through basic research, New York, 1976, Plenum Press.

124. Mikhail, G., and Allen, W. M.: Ovarian function in human pregnancy, Am. J. Obstet. Gynecol. **99:**308, 1967.

125. Naftolin, F., Ryan, K. J., and Petro, Z.: Aromatization of androstenedione by the anterior hypothalamus of adult male and female rats, Endocrinology **90:**295, 1972.

126. Nakai, Y., et al.: On the sites of the negative and positive feedback actions of estradiol in the control of gonadotropin secretion in the rhesus monkey, Endocrinology **102:**1008, 1978.

127. Nankin, H. R., and Troen, P.: Repetitive luteinizing hormone elevations in serum of normal men, J. Clin. Endocrinol. Metab. **33:**558, 1971.

128. Neill, J. D.: Prolactin: its secretion and control. In Knobil, E., and Sawyer, W., editors: Handbook of physiology. Endocrinology section, Baltimore, 1974, The Williams & Wilkins Co.

129. Neill, J. D., Johansson, E. D. B., and Knobil, E.: Failure of hysterectomy to influence the normal pattern of cyclic progesterone secretion in the rhesus monkey, Endocrinology **84:**464, 1969.

130. Neill, J. D., et al.: Relationship between plasma levels of luteinizing hormone and progesterone during the normal menstrual cycle, J. Clin. Endocrinol. Metab. **27:**1167, 1967.

131. Neill, J. D., et al.: Leutinizing hormone releasing hormone (LHRH) in pituitary stalk blood of rhesus monkeys: relationship to level of LH Release. Endocrinology **101:**430, 1977.

132. Nelson, W. D., and Merckel, C.: Maintenance of spermatogenesis in the testis of the hypophysectomized rat with sterol derivatives, Proc. Soc. Exp. Biol. Med. **36:**825, 1937.

133. Nillius, S. J., and Wide, L.: Effects of progesterone on the serum levels of FSH and LH in postmenopausal women treated with oestrogen, Acta Endocrinol. **67:**362, 1971.

134. Niswender, G. D., et al.: Blood flow: a mediator of ovarian function, Biol. Reprod. **14:**64, 1976.

135. Paesi, F. J. A.: The influence of hypophysectomy and of subsequent treatment with chorionic gonadotropin on follicles of different size in the ovary of the rat, Acta Endocrinol. **3:**89, 1949.

136. Paschkis, K. E., et al.: Clinical endocrinology, ed. 3, New York, 1967, Harper & Row, Publishers.

137. Reynolds, M., and Folley, S. J., editors: Lactogenesis: the initiation of milk secretion at parturition, Philadelphia, 1969, The University of Pennsylvania Press.

138. Richards, J. S., and Midgely, A. R., Jr.: Protein hormone action: a key to understanding ovarian follicular and luteal cell development, Biol. Reprod. **14:**82, 1976.

139. Rodbard, D.: Mechanics of ovulation, J. Clin. Endocrinol. Metab. **28:**849, 1968.

140. Rondell, P.: Role of steroid synthesis in the process of ovulation, Biol. Reprod. **10:**199, 1974.

141. Rosemberg, E.: Newer approaches to fertility regulation. In Marcus, S. L., and Marcus, C. C., editors: Advances in obstetrics and gynecology, Baltimore, 1974, The Williams & Wilkins Co., vol. 2.

142. Ross, G. T.: An intraovarian control of oegenesis in the human. In Crosignani, P. G., and Mishell, D. R., editors: Ovulation in the human, London, 1976, Academic Press Ltd., p. 127.

143. Ross, G. T., et al.: Pituitary and gonadal hormones in women during spontaneous and induced ovulatory cycles, Recent Prog. Horm. Res. **26:**1, 1970.

144. Schwartz, E., et al.: Estrogenic antagonism of metabolic effects of administered growth hormone, J. Clin. Endocrinol. Metab. **29:**1176, 1969.

145. Setchwell, B. P., and Waites, G. M. H.: The blood testis barrier. In Greep, R. O. and Hamilton, D. W., editors: Handbook of physiology. Section 7: Male reproductive system, Washington, D.C., 1973, American Physiological Society, vol. 5, pt. 1, p. 143.

146. Smith, O. W.: Estrogens in the ovarian fluids of normally menstruating women, Endocrinology **67:**698, 1960.

147. Smith, P. E.: Maintenance and restoration of spermatogenesis in hypophysectomized rhesus monkeys by androgen administration, Yale J. Biol. Med. **17:**281, 1944.

148. Steinberger, A., and Steinberger, E.: Secretion of an FSH-inhibiting factor by cultured Sertoli cells, Endocrinology **99:**918, 1976.

149. Steinberger, E., and Steinberger, A.: Spermatogenic function of the testis. In Greep, R. O., and Hamilton, D. W., editors: Handbook of physiology. Section 7: Male reproductive system, Washington, D.C., 1973,

American Physiological Society, vol. 5, pt. 1, p. 1.

150. Swaminathan, N., and Bahl, O. P.: Dissociation and recombination of the subunits of human chorionic gonadotropin, Biochem. Biophys. Res. Commun. **40:**422, 1970.

151. Szego, C. M.: Role of histamine in mediation of hormone action, Fed. Proc. **24:**1343, 1965.

152. Szego, C. M., and Davis, J. S.: Adenosine 3'5' monophosphate in rat uterus: acute elevation by estrogen, Proc. Natl. Acad. Sci. USA **58:**1171, 1967.

153. Szego, C. M., and Sloan, S. H.: The influence of histamine and serotonin in promoting early uterine growth in the rat, Gen. Comp. Endocrinol. **1:**295, 1961.

154. Topper, Y. J.: Multiple hormone interactions in the development of mammary gland in vitro, Recent Prog. Horm. Res. **26:**287, 1970.

155. Thorneycroft, I. A., et al.: The relation of serum 17-hydroxyprogesterone and estradiol 17-beta levels during the human menstrual cycle, Am. J. Obstet. Gynecol. **111:**947, 1971.

156. Turkington, R. W., and Hill, R. L.: Lactose synthetase: progesterone inhibition of the induction of α-1-lactalbumin, Science **163:**1458, 1969.

157. Vande Wiele, R. L., et al.: Mechanisms regulating the menstrual cycle in women, Recent Prog. Horm. Res. **26:**63, 1970.

158. Waites, G. M. H.: Temperature regulation and the testis. In Johnson, A. D., Gomes, W. R., and Vandemark, N. L., editors: The testis, New York, 1970, Academic Press, Inc.

159. Waites, G. M. H., and Moule, G. R.: Relation of vascular heat exchange to temperature regulation in the testis of the ram, J. Reprod. Fertil. **2:**213, 1961.

160. Waites, G. M. H., and Setchell, B. P.: Some physiological aspects of the function of the testis. In McKerns, K. V., editor: The gonads, New York, 1969, Appleton-Century-Crofts.

161. Wilson, J. D.: Metabolism of testicular androgens. In Greep, R. O., and Hamilton, D. W., editors: Handbook of physiology. Section 7: male reproduction system, Washington, D.C., 1975, American Physiological Society, vol. 5, p. 491.

162. Wilson, J. D.: The use of genetic disorders for the analysis of sexual development, Andrologia **8**(suppl. 1):35, 1976.

163. Wurtman, R. J.: An effect of luteinizing hormone on the fractional perfusion of the rat ovary, Endocrinology **75:**927, 1964.

164. Yamaji, T., et al.: The negative feedback control by estradiol and progesterone of LH secretion in the ovariectomized rhesus monkey, Endocrinology **90:**771, 1972.

165. Yen, S. S. C., and Tsai, C. C.: Acute gonadotropin release induced by exogenous estradiol during the midfollicular phase of the menstrual cycle, J. Clin. Endocrinol. Metab. **34:**298, 1972.

166. Yen, S. S. C., et al.: The operating characteristics of the hypothalamic-pituitary system during the menstrual cycle and observations of biological action of somatostatin, Recent Prog. Horm. Res. **31:**321, 1975.

167. Young, W. C.: A study of the function of the epididymis. II. The importance of an aging process in sperm for the length of the period during which fertilizing capacity is retained by sperm isolated in the epididymis of the guinea pig, J. Morphol. **48:**475, 1929.

168. Zander, J., et al.: Identification and estimation of oestradiol-17β and oestrone in human ovaries, Acta Obstet. Gynecol. Scand. **38:**724, 1959.

H. MAURICE GOODMAN

66 The pancreas and regulation of metabolism

The disease state that results from a deficiency of the pancreatic hormone insulin has afflicted mankind since ancient times and even today remains a major health problem that affects almost 3% of the population. The earliest description of its symptoms are found in an Egyptian papyrus dating back to 1500 BC. In the second century, Aretaeus named this disease *diabetes,* meaning "to flow through, as a siphon." "Diabetes," he said, "is a strange disease which consists of the flesh and bones running together into the urine." This aptly describes the cardinal symptoms of copious urine flow accompanied by severe weight loss and wastage of both muscle and fat. In 1674 Sir Thomas Willis observed that the urine of diabetics is sweet to the taste; hence the descriptive term "mellitus," or honeyed. Pancreatic involvement was not suspected until 1889, when legend has it that an alert animal caretaker in the laboratory of Von Mering and Minkowski noted an unusual collection of flies hovering around the urine of pancreatectomized dogs. Analysis of the urine revealed the presence of glucose and alerted Von Mering and Minkowski to the similarity between the symptoms of pancreatectomy in the dog and human diabetes.[190] Attempts to alleviate diabetes by feeding raw pancreas to these dogs or injecting them with watery pancreatic extracts were fruitless. It is now recognized that insulin is a protein hormone that is destroyed by proteolytic enzymes in the gastrointestinal tract when given orally as well as by the proteolytic enzymes present in crude pancreatic extracts. Arnozant and Vaillard made the important observation that following ligation of the pancreatic duct the acinar or exocrine portion of the gland degenerated but the islets of Langerhans, discovered 20 years earlier by Paul Langerhans while he was still a medical student, remained unaffected. Since these animals did not develop the symptoms of diabetes, it was suggested that the islets of Langerhans rather than the acinar tissue protected against diabetes. The hypothetical antidiabetic substance thought to be secreted by the islets was named "insulin" in 1909 by de Meyer, but definitive proof of its existence was not obtained until 1921 when Frederick Banting and Charles Best (another medical student) successfully extracted insulin from dog pancreata depleted of proteolytic enzymes by prior ligation of their ducts.[30] This monumental achievement in endocrinology earned Banting and MacLeod, a pioneer investigator in cardohydrate metabolism, the Nobel Prize in medicine. As a weird sidelight, it may be noted that Eugene Gley, who is credited with recognizing the importance of the parathyroid glands, had performed essentially the same experiment and obtained an active preparation of insulin prior to 1905. For reasons still unknown, he described these results in a sealed letter deposited with the Société de Biologie de Paris and allowed it to be opened only after the experiments of Banting and Best were published.[36]

Morphology

The endocrine portion of the human pancreas consists of some 1 to 2 million islets, each about 300 μm in diameter, that comprise about 1% to 2% of the pancreatic mass. Although they appear throughout the gland, the islets are concentrated in the head of the pancreas. They are highly vascular, with each cell seemingly in direct contact with a capillary. The islets are also richly innervated with both vagal and sympathetic fibers. Embryologically, the pancreas arises as two outpocketings of the primitive gut that meet and fuse to give rise to a series of ramifying tubules of a

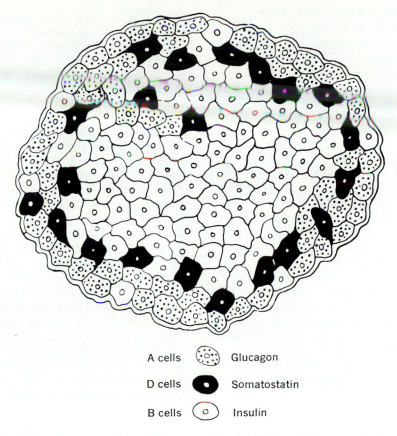

A cells — Glucagon

D cells — Somatostatin

B cells — Insulin

Fig. 66-1. Schematic representation of arrangement of cells in typical pancreatic islet. (From Orci and Unger.[136])

single layer of undifferentiated cells. These cells, in turn, give rise to the acinar cells, the duct cells, and the islets. Some evidence suggests that the cells of the pancreatic islets may derive from neuroectoderm cells that migrate to the primitive gut early in embryologic life[192] and hence may be closely related to the endocrine cells of the gut mucosa.

Histologically, the islets consist of a mixture of three cell types that can be distinguished by differential staining procedures. The alpha cells, which comprise about 20% of a typical islet, are the largest and contain large alcohol-insoluble granules. They are thought to be the source of *glucagon* (p. 1660). The beta cells, which make up about 75% of the islet, are somewhat smaller and more compact. Their granules are alcohol soluble and are thought to consist of insulin. The delta cells are a third granulated type and make up the remaining 5%. Their granules are considerably smaller than the alpha or beta granules and contain somatostatin[138,148] (Chapter 61, p.

1482). In some species (e.g., the guinea pig) there are also gamma cells or C cells that are ungranulated. The alpha, beta, and delta cells are not randomly distributed throughout the islets. The beta cells occupy the central region of the islet, whereas the alpha cells occupy the outer rim. The delta cells seem to be interposed between the alpha and beta cells and are thus in contact with both types (Fig. 66-1).[13] Ultrastructural studies reveal the existence of gap junctions linking alpha cells to each other, beta cells to each other, and alpha cells to beta cells.[137] Gap junctions are thought to represent sites of both electrical and metabolic coupling between cells. There are also tight junctions between the various islet cells. These junctions, which represent sites of close apposition or actual fusion of plasma membranes of adjacent cells, may affect the diffusion of substances into or out of the intercellular spaces. The physiologic consequences of these complex anatomic specializations are not understood.

BIOSYNTHESIS, SECRETION, AND METABOLISM OF INSULIN

Insulin consists of two unbranched peptide chains joined together at two points with disulfide bridges of cystine (Fig. 66-2). The smaller chain (the A chain) is composed of 21 amino acid residues and has an internal disulfide bridge composed of two half-cystines at positions 6 and 11. The B chain has 30 amino acid residues. Rupture of the disulfide bridges destroys biologic activity. The complete amino acid sequence of insulin was worked out by Sanger[161] and his colleagues and has been confirmed by artificial synthesis.[90] The question of how the beta cell assembles two separate peptides into the final hormone molecule was neatly resolved by Steiner et al.[173] who found that the A and B chains of insulin are formed as two ends of the same single-chain *proinsulin* molecule. The "connecting peptide" is cleaved out by a trypsinlike enzyme in the beta cells. Most of the proinsulin-insulin that is extractable from the pancreas is in the form of insulin. The connecting peptides vary in different species, in both composition and length. In contrast, the amino acid sequence of the insulin portion is constant in mammals except for the three amino acids at positions 8, 9, and 10 in the in-

ternal ring in the A chain and the C terminal amino acid in the B chain.[166]

Proinsulin is not formed directly by the ribosomes of the rough endoplasmic reticulum, but is produced as part of a larger "preproinsulin"[16] molecule that contains an additional segment with a molecular weight of 2,500 daltons attached at the amino terminus. "Preproinsulin" is short-lived and is rapidly converted to proinsulin by a protease presumably associated with the rough endoplasmic reticulum. On release from the rough endoplasmic reticulum the proinsulin molecule is transferred to the Golgi apparatus, where it is packaged in membrane-bounded storage granules. Conversion of proinsulin to insulin takes place slowly within the granules, which also contain the necessary peptidases.[175] The connecting peptide, therefore, accumulates within the granule in equimolar amounts with insulin. A small percentage of the proinsulin escapes cleavage and coprecipitates with insulin. Release from the cell occurs by the fusion of the membrane of the storage granule with the plasma membrane and a concomitant disgorgement and dissolution of the hormone into the perivascular space. The process, which is the reverse of pinocytosis, is called emiocytosis.[98]

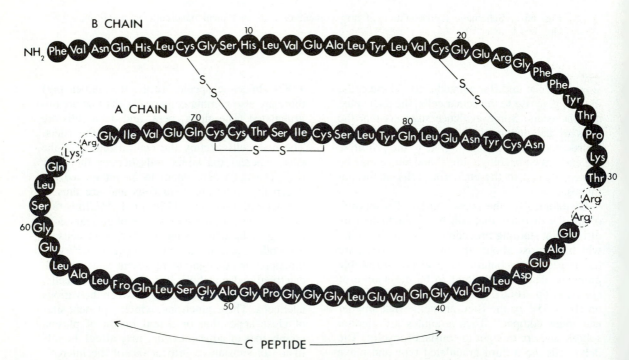

Fig. 66-2. Amino acid sequence of human proinsulin. Pair of basic amino acids shown as open circles at both ends of connecting peptide are removed during cleavage process. (From Oyer et al.[140])

Proinsulin and the connecting peptide are released along with insulin and enter the circulation.[155] When secretion is rapid, proinsulin may comprise as much as 20% of the circulating insulin immunoactivity. Proinsulin has only about 3% to 5% of the biologic activity of insulin when assayed in vitro. The connecting peptide has no known biologic activity. Insulin is cleared rapidly from the circulation with a half-life of less than 10 min. The half-life of proinsulin is at least twice as long. Proinsulin apparently is not converted to insulin outside of the pancreas. Insulin is destroyed by a specific enzyme system, *insulinase,* that is present in the liver, muscle, kidney, and other tissues.[127] The initial inactivating step appears to be reduction of the disulfide bonds linking the chains. Because of the peculiarities of the visceral circulation, insulin secreted into the pancreatic vein must pass through the liver by way of the hepatic portal vein before it enters the systemic circulation. The liver may inactivate as much as 40% of the insulin reaching it by this route[159] and is thus potentially able to regulate the amount of insulin that enters the systemic circulation.[3] Normally, little or no insulin is found in the urine.

Assay procedures

Although insulin varies little in amino acid sequence from species to species, it is nevertheless antigenic and induces the formation of antibodies in heterologous species. The small amounts of proinsulin that inevitably contaminate commercial preparations of insulin also contribute to the overall antigenicity. This can be a clinically important factor, since patients may develop antibodies to insulins derived from porcine or bovine sources. The antigenicity of insulin was put to good use by Yalow and Berson[201] in the development of a radioimmunoassay procedure that is now the method of choice for measuring circulating levels of insulin. Insulin activity of the plasma can also be assayed biologically using glucose uptake or oxidation by isolated adipose tissue[153] or diaphragm muscle[77] as indices. The radioimmunoassay and the in vitro tissue assays, however, give different estimates of the insulin content of plasma. Consequently, the term *"immunoreactive insulin"* (IRI) has been coined to describe the insulin content of plasma as determined by radioimmunoassay, whereas the term *"insulin-like activity"* (ILA) describes the activity detectable by biologic assay. IRI probably gives a true reflection of insulin in the plasma. Some ILA remains even after IRI is removed by adsorption onto insulin antibodies; this is called "nonsuppressible" ILA. The nature and physiologic importance of nonsuppressible ILA are not understood, especially since it may persist in the plasma long after pancreatectomy.[63] Nonsuppressible ILA is related to or is perhaps identical with somatomedin (Chapter 61, p. 1476).

Effects of insulin deficiency

In many areas of endocrinology, basic insights into the physiologic role of a hormone can be gained by studying the consequences of removal of its source of production. In the case of insulin, extirpation of the pancreas is not always feasible, since in rats and some other species the pancreas is quite diffuse and difficult (although not impossible) to remove entirely. Furthermore, pancreatectomy causes derangements in digestion that complicate interpretation of the effects of insulin deficiency. Two important pharmacologic tools, however, are available for the study of experimental diabetes. One of these is the drug *alloxan,* which is a cytotoxic agent with a high degree of specificity for the beta cells of the pancreas.[107] Another drug, streptozotocin, an antibiotic developed in recent years, also has a highly specific cytotoxic effect on the beta cells and produces fewer side effects than alloxan.[89] Both compounds produce a permanent diabetes in experimental animals. (It is doubtful that either of these drugs play any role in the etiology of human diabetes.)

Hyperglycemia (high blood sugar level)

In the normal individual the concentration of glucose in blood is maintained at around 80 mg/100 ml of plasma. Blood sugar concentrations in diabetics may be 300 to 400 mg/100 ml and even reach 1,000 mg/100 ml in extreme cases. Diabetics have a particularly difficult time in removing excess glucose from their blood. After a carbohydrate meal there is a transient increase in the concentration of blood glucose in normal individuals, but the excess sugar disappears rapidly from the plasma, and normal values are quickly regained. The diabetic, however, is "intolerant" of glucose from the plasma. Oral glucose tolerance tests are used diagnostically to evaluate existing or impending diabetic conditions. A standard load of glucose is given by mouth and the blood sugar concentration measured periodically over the course of the subsequent 4 hr. In the normal subject, blood sugar returns to baseline values within 2 hr and the peak of the blood sugar curve does not rise above 180 mg/100 ml. In the diabetic or "predia-

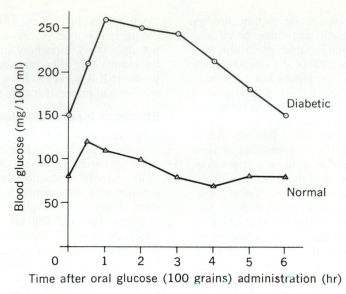

Fig. 66-3. Glucose tolerance tests in normal and diabetic subjects (idealized).

betic,'' blood sugar values rise much higher and take a longer time to return to basal levels (Fig. 66-3).

Glycosuria (excretion of glucose in urine)

Normally the capacity of the renal tubule to transport glucose is adequate to reabsorb all the glucose in the glomerular filtrate, and little or none escapes in the urine. Because of the hyperglycemia of diabetes, however, the concentration of glucose in the glomerular filtrate is so high (greater than 180 mg/100 ml) that the transport maximum is exceeded and glucose ''spills'' into the urine.

Polyuria (excess production of urine)

Because some of the glucose present in the glomerular filtrate cannot be reabsorbed (by the proximal tubule), it remains in the lumen of the proximal tubule and exerts an osmotic hindrance to water reabsorbtion in this portion of the nephron. The increased volume of glomerular filtrate that passes through the proximal tubule cannot be reabsorbed by the more distal portions of the nephron, with the result that water excretion is increased (osmotic diuresis). The increased flow through the nephron produces increased urinary loss of sodium and potassium.

Polydipsia (excessive drinking)

Dehydration results from the copious flow of urine and stimulates thirst. The untreated diabetic is characteristically thirsty and consumes large volumes of water to compensate for water lost in the urine.

Polyphagia (excessive food consumption)

In an effort to compensate for the loss of glucose in the urine, there is an increase in appetite.

Weight loss

Despite the increase in appetite, however, the deficiency of insulin reduces all anabolic processes and accelerates catabolic processes. There is increased mobilization and utilization of stored fats. This indirectly leads to increased triglyceride concentration in plasma and often results in *lipemia* (high concentration of lipids in blood). Fatty acid consumption by the liver results in *ketosis,* increased production of the ketone bodies (acetoacetic acid, acetone, and beta hydroxybutyric acid), which are released into the blood and cause *ketonemia.* Since ketone bodies are small, readily filtrable molecules that are actively reabsorbed by a renal mechanism of limited capacity, high blood levels may result in *ketonuria.* The ketone bodies are organic acids that can markedly lower blood pH and thus produce acidosis, which may be aggravated by excessive washout of sodium and potassium in the osmotic diuresis. Plasma pH may become sufficiently low to produce a comatose state (diabetic coma), which may terminate in death unless insulin therapy is instituted.

The hyperglycemia that causes this whole sequence of events arises from both an "underutilization" of glucose by peripheral tissues and an "overproduction" of glucose by gluconeogenesis in the liver. The increase in gluconeogenesis is accomplished at the expense of muscle protein, which is the chief source of the amino acid substrate. Consequently, there is a marked wasting of the muscle along with fat. *The net effect of insulin lack is a severe impairment of the ability to store foodstuffs as glycogen, fat, and protein*. In many respects, insulin lack resembles starvation. The starved individual, like the diabetic, decreases his utilization of glucose and increases mobilization and utilization of fat and eventually muscle protein. Diabetes appears to be a bizarre, inappropriate exaggeration of the normal physiologic response to starvation. Since it occurs in the presence of high blood glucose levels, diabetes has been called "starvation in the midst of plenty." In order to understand the physiologic role of insulin and how its absence brings all this about, it is necessary first to consider some of the general features of energy metabolism and their underlying biochemical regulatory mechanisms.

GENERAL FEATURES OF ENERGY METABOLISM

Constant availability of energy-yielding substrates is required for muscular activity, maintenance of body temperature and the host of other energy-demanding processes that constitute life. Evolution of the capacity for storage of energy-rich substrates in times of plenty, and the ability to draw on these stores in times of want, has liberated us from the necessity of constant feedings. However, certain biochemical limitations and restrictions have also evolved. Most important among these is that glucose must be constantly available in the blood, for the red blood cells, which lack mitochondria, and the renal medullae, which function under low oxygen tension, require glucose to meet their energy needs. Furthermore, the CNS, under normal circumstances, relies almost exclusively on glucose to meet its energy demands. Only after a total fast of 4 or 5 days do the plasma ketone bodies, the only alternative substrates, reach a sufficiently high concentration in blood to meet a significant fraction of the brain's energy needs, but even after several weeks of total starvation, the brain continues to derive about half its energy requirement from glucose.[39,139] As will be discussed, availability of glucose is ensured by (1) *gluconeogenesis*, the formation of glucose from nonglucose precursors, and (2) *inhibition of glucose utilization* in those tissues that can satisfy their energy needs by consuming other substrates, notably fatty acids and ketone bodies.

We may next ask how energy-rich substrates are stored.

Glucose. One gram of glucose yields about 4 calories. Since an average man weighing 70 kg requires approximately 2,000 calories/day, he would have to carry a reserve supply of approximately 500 gm of glucose to ensure sufficient substrate to survive 1 day of food deprivation. If this glucose were stored in isotonic solution, approximately 10 L of water (10 kg) would have to be carried around. The 70 kg man would have to carry around a storage depot equal to his own weight if he were to survive only 1 week of starvation. In reality, the 70 kg man carries only about 20 gm of free glucose dissolved in his extracellular fluids, or enough to supply energy for about 1 hr.

Glycogen. Polymerizing glucose to glycogen eliminates the osmotic requirement for large volumes of water. To meet a single day's energy needs, only about 1 kg of "wet" glycogen is required; that is, 500 gm of glycogen obligates only about 0.5 L of water. In a well-fed 70 kg man, enough glycogen is carried to meet less than 1 day's energy needs—about 100 gm in the liver and about 200 gm in muscle.

Fat. Triglycerides are by far the most concentrated storage form for high-energy fuel (9 calories/gm). One day's energy needs can be met by less than 250 gm of triglyceride, which is stored without the accompaniment of water. Thus a 70 kg man carrying 10 kg of fat maintains an adequate depot of fuel to supply caloric needs for more than 40 days. Most fat is stored in adipose tissue, but other tissues such as muscle also contain small reserves of triglycerides.

Protein. Calories can also be stored in the form of protein. The normal 70 kg man in nitrogen balance carries around 2 to 3 kg of expendable protein; that is, about 16% of his total body protein can be mobilized for energy. Unlike fat there is no special storage organ for protein. Rather the labile protein pool is distributed in various tissues, including muscle, liver, and lymphoid tissues.

As already indicated, fat is the most abundant and most efficient energy reserve. Efficiency has its price, however. (1) In converting dietary carbohydrate to fat, about 25% of the calories are dissipated as heat. (2) More importantly, *synthesis of fatty acids from glucose is a one-way street! Once the carbons of glucose are converted to fatty acids, there can be no net reconversion to glucose.* The glycerol portion of triglyceride is reconvertible to glucose, but glycerol represents only about 10% of the mass triglyceride.

Given these conditions of energy storage and the necessity to provide some 200 gm/day of glucose whether or not food is available, the need for mechanisms to control carbohydrate utilization becomes apparent. In times of plenty the excess carbohydrate beyond that which is set aside for storage must be utilized directly to prevent wasteful consumption of metabolically expensive stores. In times of want, carbohydrate utilization must be minimized in all but those tissues (CNS, red blood cells, and the renal medullae) that have an absolute requirement for glucose. Regulation of storage, mobilization, and interconversion of metabolic fuels can be simplified somewhat by focusing attention on (1) adipose tissue, the chief storage organ; (2) muscle, the chief consumer; and (3) the liver, the chief producer of glucose.

Adipose tissue

White adipose tissue is the principal repository for triglycerides and normally comprises about 15% of the body weight. This amount is highly variable, however, and may range from as low as 2% to more than 50% in obesity. As much as 80% to 90% of the wet weight of adipose tissue is triglyceride. Each adipose cell is composed of a single, large, lipid-filled vacuole surrounded by a thin rim of cytoplasm. The size of the fat vacuole varies widely with the nutritional state. In the adult, changes in the total amount of adipose tissue largely reflect changes in the size of the fat vacuoles rather than changes in the number of fat-storing cells.[82]

Fat storage

Uptake. Fat that arises from dietary sources or from de novo synthesis in the liver is transported to the adipose tissue complexed with proteins in the form of chylomicrons or low-density lipoproteins. The uptake of fat is complex and involves several processes. Some triglycerides may be taken up intact, but it is thought that most are first cleaved into fatty acids and glycerol by the enzyme lipoprotein lipase.[68,79] The site of this cleavage is not known with certainty and may occur within the capillaries, the endothelial cells, or even pinocytotic vacuoles within the fat cells. The fatty acids thus released are subsequently taken up by the fat cells and stored as triglycerides. Although the exact details of this process are still not known, there is general agreement that lipoprotein lipase plays an important role in the uptake of fat by adipose tissue. The uptake of the fatty acids derived from lipoproteins depends on (1) the availability of lipoprotein lipase and (2) the availability of α-glycerophosphate for esterification to triglyceride. The activity of lipoprotein lipase is high in adipose tissue obtained from fed individuals and depressed in adipose tissue of fasting subjects. There is some evidence that insulin may control the synthesis and activity of this enzyme.[196]

Once they enter the fat cell, the fatty acids released from lipoprotein complexes by lipoprotein lipase must be reesterified with a new glycerol molecule. The original glycerol liberated from triglycerides cannot be reutilized to esterify fatty acids in the adipocyte because, to be reincorporated into triglycerides, glycerol must first be converted to α-glycerol phosphate. Adipose tissue is almost totally devoid of the required enzyme glycerol kinase.[115] To store fatty acids as triglycerides, adipose tissue must therefore provide α-glycerol phosphate from some other source. That other source is glucose (Fig. 66-4). Thus the *availability of glucose determines the amount of fat that can be taken up and stored in adipose tissue.*

Although glucose is always present in the blood, it is not necessarily available as a source of α-glycerol phosphate, because adipose cells are not freely permeable to glucose. Glucose enters adipose cells by a process of facilitated diffusion (carrier-mediated transport down a concentration gradient).[49,85] The rate at which it penetrates adipose cells depends on its concentration in plasma and on the hormonal environment; insulin increases permeability of adipose cells to glucose; and growth hormone,[72] and possibly adrenal glucocorticoids have the opposite effect.

Fatty acid synthesis. Adipose tissue can also synthesize fatty acids from nonlipid precursors such as glucose and amino acids. Fatty acid synthesis, like the formation of triglycerides, depends on the availability of intracellular glucose. Although this process may go on at a substantial rate in adipose tissue of rats and mice, fatty acid synthesis in human adipose tissue is probably not of major importance.[7]

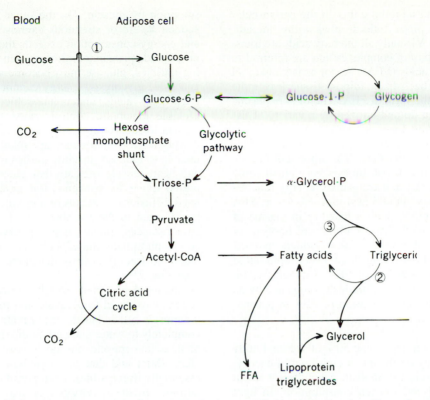

Fig. 66-4. Schematic representation of carbohydrate and lipid metabolism in adipose tissue. Important rate-limiting reactions controlling fatty acid production are, *1*, transport of glucose into adipose cell, *2*, cleavage of triglyceride into fatty acids and glycerol, and *3*, fatty acid esterification.

Fat mobilization

Even in the well-fed subject, triglycerides in adipose tissue are constantly subjected to the action of lipolytic enzymes (not to be confused with lipoprotein lipase), which split them into fatty acids and glycerol. In the fed subject, most of the fatty acids formed in this continuing process of lipolysis are reesterified back to triglycerides. As already indicated, reesterification requires the availability of α-glycerol phosphate and hence of glucose. Fatty acids that are not reesterified escape from the adipose cells into the perivascular space and eventually into the plasma. It is evident, then, that when glucose is abundant, little fat is permitted to leave the cells; when glucose is scarce, fewer of the fatty acids released by lipolysis can be reesterified and therefore more escape from the fat cell. Thus the *availability of glucose determines the amount of fatty acid that is mobilized.*

Although, under usual conditions of feeding and fasting, regulation of fat mobilization is largely achieved through the regulation of re-esterification (via regulation of glucose uptake), in emergencies the rate of lipolysis can also be accelerated by hormonal and possibly neural signals. Activation of the so-called hormone-sensitive lipase is doubtlessly mediated by cyclic adenosine monophosphate (cyclic AMP).[13] It appears that cyclic AMP formed in response to hormone stimulation of adenylate cyclase activates protein kinase, which in turn catalyzes the transfer of the terminal phosphate group of ATP to the relatively inactive form of the lipase, thereby activating it.[93] Inactivation of the lipase appears to depend on the action of a phosphatase that removes the phosphate group.[165] Under favorable conditions a wide variety of hormones and hormonelike agents increase lipolysis in adipose tissue, presumably by activating the lipase. It is doubtful, however, that any but epinephrine and norepinephrine are physiologically important. Norepinephrine released from sympathetic nerve endings in fat may be of particular importance.

The fatty acids that leave the adipose cells

complex with albumin either in the perivascular spaces or in pinocytotic droplets within the adipocytes.[50,51] Virtually all the fatty acids are transported as albumin complexes and are referred to as free fatty acids (FFA). As far as is known, all the fat mobilized from adipose tissue leaves the fat depot in the form of FFA. The FFA in plasma may be taken up and utilized by a variety of tissues, including liver, kidney, and muscle. The uptake of FFA is directly proportional to their concentration in plasma.[27] The amount of FFA in the plasma at any one time is very small (only about 1 mM/L) and can satisfy energy requirements for only about 3 or 4 min. However, FFA turn over rapidly, with a half-life in plasma of less than 3 min.[32] Thus inhibition of lipolysis or acceleration of reesterification, which shut off the influx of FFA, bring about a very rapid fall in plasma concentrations of FFA. Conversely, increased mobilization of fat is accompanied by an increase in the concentration of FFA in plasma.

Muscle

Muscle comprises some 50% of the total body mass and is by far the major consumer of energy. At rest, muscle metabolism accounts for about 30% of the total oxygen consumption. In light exercise, this value increases to 70%. Muscles can derive their energy from the metabolism of glucose, fatty acids, or ketone bodies, but fatty acids and ketone bodies are "preferred substrates." In the resting subject, more than half the oxygen consumed by muscle is utilized in the oxidation of fat.[26] Although muscles contain various transaminases and can completely oxidize some amino acids, little is known of the quantitative importance of amino acids as energy-yielding substrates.

In addition to the substrates they obtain through the circulation, muscles contain endogenous stores of fat, carbohydrates, and proteins. Skeletal muscles contain about 5 mg/gm of glycogen and about 10 to 30 mg/gm of fatty acids. This may be enough to supply normal metabolic needs for 1 or 2 days. Muscle glycogen, unlike its counterpart in liver, remains quite stable and usually does not decline in fasting. In cardiac muscle, glycogen may actually increase during fasting.[157]

Pathways of glucose metabolism

The utilization of glucose by muscle is controlled by both the metabolite and hormonal milieu. Generally speaking, physiologic situations in which the mobilization of fat is increased are accompanied by a decrease in glucose consumption by muscle. On the other hand, when sudden severe or strenuous exercise is required and energy consumption exceeds the capacity to provide adequate amounts of oxygen to the working muscles, glucose must become available for anaerobic glycolysis. Energy cannot be obtained from fatty acids under anaerobic conditions. Flexible metabolic controls must therefore be present to satisfy these conflicting requirements.

Muscle cell membranes are relatively impermeable to glucose. In vitro studies of diaphragm and heart muscle indicate that glucose is transported down its concentration gradient (facilitated diffusion) by means of a mobile carrier system located in the membrane.[143] Under normal circumstances, glucose is phosphorylated to glucose-6-phosphate immediately on entry, and virtually no free glucose is detectable in intracellular fluid.[142]

Glucose-6-phosphate may be converted to glycogen or oxidized to pyruvate and released from the muscle cell as lactate or pyruvate or oxidized completely to water and carbon dioxide (Fig. 66-5; note that muscle has no hexose monophosphate shunt and that the glycolytic pathway is essentially irreversible). Each possible fate is determined by the activity of a critical enzyme that acts like a valve and is located at or just beyond a branch point in the metabolic sequence. The activities of these enzyme "valves" are controlled by metabolites.

Control of glycogen formation and breakdown. Glycogen synthesis from glucose-1-phosphate is catalyzed by an enzyme (transferase), which transfers glucose from uridine diphosphoglucose to the growing glycogen molecule. Glycogen is broken down to glucose-1-phosphate by phosphorylase. Under physiologic circumstances, both of these are one-way reactions. Both the transferase and phosphorylase exist in active and inactive (less active) forms that can be interconverted by the addition or removal of a phosphate group.[99] Addition of phosphate to phosphorylase activates it, whereas addition of phosphate to the transferase "inactivates" it. Cyclic AMP activates phosphorylation of both these enzymes and thus accelerates the breakdown of glycogen while inhibiting its synthesis. Phosphorylation of these enzymes is also stimulated by electrical activity and by Ca^{2+}.[37,175] Excessive deposition of glycogen is normally prevented by glycogen itself, which in high concentrations inhibits the activation of the transferase.[99]

Under conditions of anaerobic muscular activity, the breakdown of glycogen may be accelerated by the accumulating AMP (not to be con-

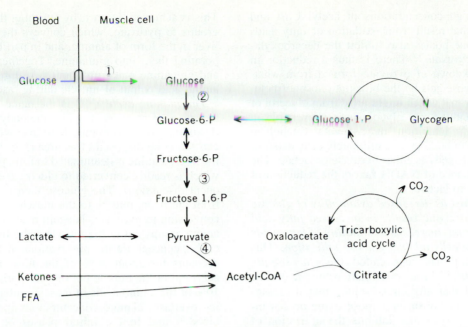

Fig. 66-5. Schematic representation of carbohydrate utilization in muscle. Important rate-limiting steps are, *1*, transport of glucose into muscle cell; *2*, phosphorylation of glucose by hexokinase; *3*, addition of second phosphate group by phosphofructokinase; and *4*, decarboxylation of pyruvate.

fused with cyclic AMP). In the presence of elevated concentrations of AMP the "inactive" (nonphosphorylated) phosphorylase has considerable glycogenolytic activity. This mechanism is probably operative only in extreme conditions of anoxia.

Control of glycolytic pathway. Whether glucose-6-phosphate goes down the glycogenic or the glycolytic pathway is determined by the relative activities of the transferase on the one hand and phosphofructokinase on the other. In general, phosphofructokinase is more abundant than transferase. Phosphofructokinase is inhibited by ATP[145] and citrate[68] and activated by AMP.[145] Cyclic AMP can overcome the inhibitory effects of ATP.[145]

Control of pyruvate metabolism. Pyruvate formed by glycolysis may either be converted to lactate and released from the muscle cell or it may be decarboxylated to form acetyl coenzyme A (acetyl CoA), which is then further broken down to carbon dioxide and water in the tricarboxylic acid cycle. Decarboxylation requires nicotinamide adenine dinucleotide (NAD) and coenzyme A and is accelerated when these cofactors are present in abundance. High concentrations of acetyl CoA or reduced NAD (NADH) inhibit pyruvate decarboxylation, and abundance of NADH promotes its conversion to lactate.

Utilization of fatty acids and ketone bodies

It has already been mentioned that fatty acids and ketone bodies are utilized by muscles in preference to glucose. Such preferential utilization is achieved in the following manner: Ketone bodies and fatty acids that enter muscle cells from plasma or that are produced by lipolysis from endogenous triglycerides are rapidly broken down to acetyl CoA. The ensuing increase in acetyl CoA activates *citrate synthetase*,[67] the enzyme that condenses acetate with oxaloacetate. The concentration of citrate therefore increases (as much as a 60-fold increase in citrate concentration has been found in heart muscle). Citrate diffuses out of the mitochondria and inhibits *phosphofructokinase*, thereby decreasing the flow of glucose-6-phosphate into the glycolytic pathway. Since muscle lacks *glucose-6-phosphatase* and the hexose monophosphate shunt, the only other pathway for glucose metabolism is conversion to glycogen. However, conversion of glucose to glycogen is self-limiting, since, as already mentioned, high concentrations of glycogen inhibit transferase. Consequently, glucose-6-phosphate accumulates. Since glucose-6-phosphate is a potent inhibitor of hexokinase,[169] further phosphorylation of glucose is depressed. This in turn ultimately leads to an inhibition of glucose uptake by muscle.

The high concentrations of acetyl CoA and NADH that result from oxidation of fatty acids and ketone bodies also inhibit the decarboxylation of pyruvate.[67] There is thus a reduction in the breakdown of pyruvate formed from whatever glucose gets by the block at phosphofructokinase or that which might arise from glycerol or amino acids. Since glycolysis is irreversible in muscle,[135] inhibition of pyruvate catabolism causes it to build up and ultimately diffuse out of the muscle cells as either pyruvate or lactate. The predominance of NADH favors the reduction of pyruvate to lactate.

Thus *just as decreased availability of glucose in adipose tissue results in increased fatty acid mobilization, increased availability of fatty acids in muscle cells inhibits glucose utilization.* This relationship has been called the *glucose-fatty acid cycle.*[152] From this relationship, it can be predicted that any circumstance that decreases glucose metabolism in adipose tissue or that increases lipolysis also inhibits the utilization of glucose in muscle and leads to elevated or at least stable levels of muscle glycogen.

Amino acid production and utilization

Because of their sheer bulk, muscles are the major repository for protein stores. The various proteins in muscle, both structural and enzymatic, are constantly breaking down to amino acids and being replaced by de novo synthesis. Both the synthesis and breakdown of muscle protein appear to be under hormonal and perhaps metabolite control. During fasting, protein synthesis is reduced; the continuing process of protein breakdown proceeds unabated or may even be accelerated. Most of the amino acids thus produced escape from muscle and serve as substrate for gluconeogenesis in the liver.

Although the quantitative importance of amino acid oxidation in muscle has not been established, it is clear that muscles oxidize amino acids for energy. In fact, muscle is the principal site of degradation of the branched chain amino acids, leucine, isoleucine, and valine.[122] Liver is deficient in the transaminase enzymes needed for the first reaction in the degradative pathway for the branched chain amino acids.[22] The output of leucine, isoleucine, and valine increases during fasting, indicating that muscle does not oxidize all the branched chain amino acids liberated from protein. Although the carbon skeletons of amino acids can be completely oxidized to carbon dioxide and water, some means is needed to protect the muscle cells from the buildup of ammonia.

This is achieved, in part, by attaching the amino groups to pyruvate, which conveys them to the liver in the form of alanine, and in part by incorporating them into glutamine. Together alanine and glutamine make up about two thirds of the amino acid output of muscle during fasting.[2]

The alanine cycle[112,133] is the name given to the series of reactions used by the body to shuttle amino groups from muscle to liver where they can safely be disposed of as urea (Fig. 66-6). In the liver, alanine is deaminated to form pyruvate, which is readily converted to glucose (see subsequent discussion). The glucose then is released from the liver, returns to the muscle, and, after conversion to pyruvate, is again ready to shuttle amino groups out of the muscle cells. In support of this concept are the observations that alanine accounts for about 30% of the flow of amino acids out of the muscle but only comprises some 10% of the amino acids in muscle protein.[2] During exercise, alanine concentrations increase in blood,[59] and in the initial phases of fasting, alanine release from muscle is quite marked.[152] Adding branched chain amino acids to muscle incubated in vitro increases in alanine production.[135] It should be noted that the carbon skeletons of about 70% of the alanine released from muscle derive from glucose,[2] and hence this fraction does not contribute to net glucose production.

The role of glutamine is less well understood. Unlike alanine, the amount of glutamine released by muscle does not fluctuate with nutritional state.[14] Further, glutamine is extracted from the blood by the gut and kidneys rather than by the liver. The gut converts the glutamine to alanine, which it then exports to the liver. Glutamine serves as an important substrate for the ammonia produced by the kidney in acidosis. The carbon

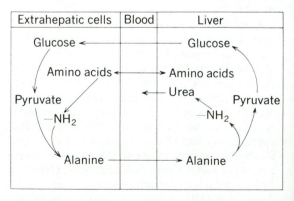

Fig. 66-6. Alanine cycle. (From Mallette et al.[113])

skeleton of glutamine serves as substrate for renal gluconeogenesis during fasting.

Liver

Although it comprises only about 3% of the body mass, the liver is the major chemical plant and site of substrate interconversion. Because of its unique vascular supply, the liver is the first organ to receive dietary substrates and hence is the first organ to be apprised of the availability of food in the gut. In receiving the entire venous drainage of the gut and pancreas, the liver is also the first organ to "see" (and degrade) gastrointestinal and pancreatic hormones. Even in the absence of hormones, the liver functions to maintain constant blood levels of glucose.[169] In hyperglycemia, it takes up glucose; in hypoglycemia, it releases glucose. The liver is the organ principally responsible for clearing dietary glucose from the blood.[60] Most of an ingested glucose load is promptly converted to liver glycogen. Glucose is produced both by breakdown of glycogen and by gluconeogenesis. Although the metabolic activity of the liver is normally subject to a wide variety of hormonal influences, substantial regulation is achieved by intrinsic mechanisms governed by metabolites themselves.

Pathways of glucose metabolism

When the concentration of glucose in portal venous blood is high, there is a net uptake of glucose by the hepatic cells. Unlike muscle and fat cells, hepatocytes are freely permeable to glucose,[40] and the concentration of glucose within the hepatocyte promptly equalizes with that of the plasma. Net uptake depends on phosphorylation and subsequent conversion to glycogen. In the fed subject, liver contains two enzymes capable of converting glucose to glucose-6-phosphate: hexokinase and glucokinase.[189,191] Hexokinase is present in constant amounts regardless of the nutritional state and also catalyzes the phosphorylation of fructose, galactose, mannose, and other hexoses. It has a high affinity for sugars but a relatively low capacity. Glucokinase is quite specific for glucose and has a high capacity for phosphorylation, but because its affinity for glucose is low, it catalyzes phosphorylation only when glucose concentrations are high. Glucokinase is part of the mechanism responsible for the net uptake of glucose by liver cells. The concentration of glucokinase falls dramatically during fasting and, after a lag period of several hours, increases on refeeding.[158]

Once formed, glucose-6-phosphate may experience any of four possible fates (Fig. 66-7): (1) it may be converted to glycogen, (2) it may be broken down in the hexose monophosphate shunt, (3) it may be broken down in the Embden-Meyerhof pathway, or (4) it may be dephosphorylated by glucose-6-phosphatase and released from the hepatic cell as free glucose. Glucose-6-phosphate cannot cross cell membranes.

The control of glycogen metabolism in the liver is analogous to that already described for muscle, but because of its special role in glucose production, the system may have an added complexity in the liver.[6] Activation of glycogen synthetase permits the liver to rapidly take up and store a sudden influx of glucose from the gut, and, conversely, activation of phosphorylase allows the liver to release glucose when the blood sugar is low. As in muscle, phosphorylase is present in both an active form (phosphorylase a) and a form that is virtually inactive under most physiologic circumstances (phosphorylase b). Conversion of phosphorylase b to phosphorylase a is catalyzed by a kinase that transfers the terminal phosphate group of ATP to the enzyme. Conversion of phosphorylase a to b is catalyzed by a phosphatase, which removes the phosphate group. As in muscle, activation of phosphorylase is sensitive to cyclic AMP, but can also be achieved by other means. Glycogen synthetase also exists in both an active and an inactive form. Interconversion of these forms also involves addition or removal of a phosphate group, but in this case, it is the phosphorylated form that is inactive. Synthetase phosphatase, the activating enzyme, is powerfully inhibited by phosphorylase a.[170] Thus activation of phosphorylase blocks glycogen synthesis while it accelerates glycogen mobilization. Glycogen synthetase cannot be activated by its phosphatase unless more than 90% of the phosphorylase is in the inactive form. Glucose, which freely enters liver cells from the blood, promotes its own storage in the liver. It binds to phosphorylase a and thereby inhibits the breakdown of glycogen. More importantly, in binding to phosphorylase a, glucose promotes its conversion to phosphorylase b by making it more susceptible to the phosphatase.[171] By decreasing phosphorylase a, the inhibition on synthetase phosphatase is removed, and glycogen synthetase becomes active. These complex interrelations are illustrated in Fig. 66-8.

The utilization of glucose-6-phosphate in the hexose monophosphate shunt is particularly brisk during periods of active lipogenesis (i.e., when glucose is plentiful). Similarly, utilization of glucose in the glycolytic pathway proceeds at a brisk pace when glucose is plentiful. Both these path-

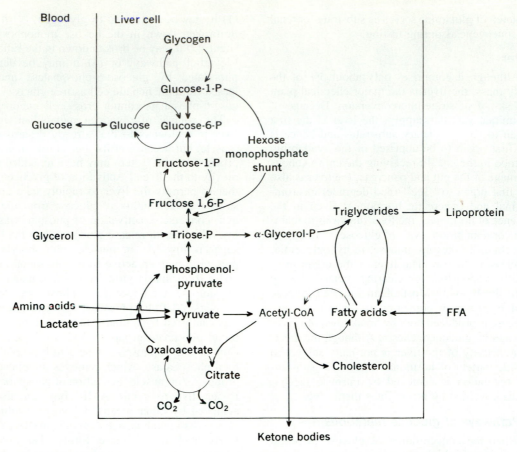

Fig. 66-7. Schematic representation of carbohydrate and lipid metabolism in liver. (See text.)

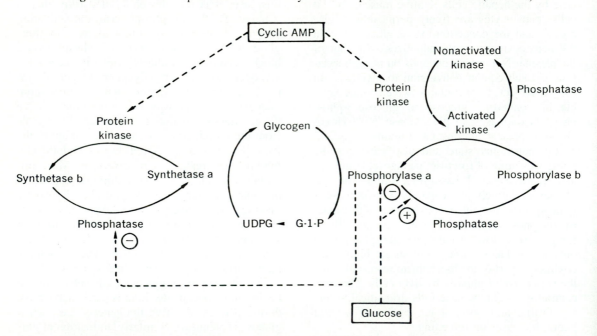

Fig. 66-8. Schematic representation of glycogen metabolism in liver. (See text.) (From Hers.[6] Reproduced, with permission, from the Annual Review of Biochemistry, Volume 45, © 1976 by Annual Reviews Inc.)

ways of glucose breakdown are severely curtailed when glucose is in short supply.

The pyruvate that is formed from glucose, alanine, and other amino acids or lactate may be decarboxylated to form acetyl CoA or be carboxylated to form oxaloacetate. Acetyl CoA can be condensed with oxaloacetate to form citrate and oxidized to carbon dioxide and water in the tricarboxylic acid cycle, it may be utilized in the formation of long-chain fatty acids and cholesterol, or it may be condensed with another molecule of acetyl CoA to form a "ketone body." (For the purpose of this discussion, note that the various synthetic reactions to form amino acids, purines, etc. will be ignored.)

Gluconeogenesis

In times of scarcity there is net production of glucose from various gluconeogenic precursors: lactate, "glucogenic" amino acids, and glycerol. Conversely, gluconeogenesis is inhibited by glucose.[156] The substrates for gluconeogenesis are produced principally by muscle and adipose tissue as already described, but the breakdown of endogenous triglyceride and protein in the liver itself also provides glycerol and amino acids. Red blood cells are an important source of lactate. Lymphoid tissue may provide an additional source of amino acids.

Gluconeogenesis cannot proceed by simple reversal of the reactions of the Embden-Meyerhof pathway. All of the steps in glucose metabolism involving either the *addition or removal of a phosphate group* are catalyzed by enzymes that react in only one direction under the conditions that prevail in hepatic cells. To reverse direction at points involving the addition or removal of phosphate, different enzymes must be used. Phosphorylation-dephosphorylation reactions occur between glucose and glucose-6-phosphate, between glycogen and glucose-1-phosphate, between fructose-6-phosphate and fructose-1,6-diphosphate, and between phosphoenol pyruvate and pyruvate. Generally speaking, the one-way enzymes involved in glucose breakdown are high in amount and/or activity when glucose is abundant and are low when it is scarce.[40] The reverse is true for the enzymes of gluconeogenesis. The amounts of these one-way enzymes may be influenced by hormones. The so-called bifunctional enzymes retain quite stable levels of activity[192] (Fig. 66-9).

Of all the one-way reactions in gluconeogenesis the most complex (and perhaps most crucial) is the conversion of pyruvate to phosphoenol pyruvate (PEP). Direct phosphorylation of pyruvate by pyruvate kinase is not thermodynamically feasible.[97] (The enzyme pyruvate kinase was named before this was appreciated. Actually, it catalyzes the reaction in the opposite direction and might better be called "PEP phosphatase.") To convert pyruvate to PEP, carbon dioxide must first be added by the enzyme pyruvate carboxylase to form oxaloacetate. Oxaloacetate, in turn, is converted to PEP by the action of the enzyme PEP carboxykinase.

Relationship between amino acid metabolism and glucose. Amino acids are the most important source of carbons for gluconeogenesis. Presentation of an increased load of amino acids to liver (particularly alanine) markedly increases glucose production.[112] This is accompanied by an increase in urea formation. Stimulation of gluconeogenesis by amino acids probably results simply from increased availability of substrate. Hence increased amino acid release from peripheral tissues or increased dietary amino acids can drive the liver to produce increased amounts of glucose. Three carbon amino acids enter the metabolic sequence by way of pyruvate; aspartate and glutamate enter at oxaloacetate and α-ketoglutarate, respectively. Other amino acids require some degradation before entering the gluconeogenic sequence.

Relationship between fatty acid metabolism and gluconeogenesis. As with the regulation of glucose breakdown in muscle, gluconeogenesis may be sensitive to the presence of fatty acids and their metabolites. Fatty acid breakdown to two-carbon fragments leads to accumulation of reduced pyridine nucleotides and ties up coenzyme A in the form of acetyl CoA. Both these consequences retard the conversion of pyruvate to acetyl CoA, thus preventing the loss of three-carbon fragments from the precursor pool for gluconeogenesis. The accumulation of long-chain fatty acyl CoA also inhibits the citrate synthetase reaction[195] and thus blocks the entry of acetyl CoA into the tricarboxylic acid cycle. The bulk of the energy required by the liver can be met by partial oxidation of long-chain fatty acids to acetyl CoA. The accumulating acetyl CoA may increase the activity of pyruvate carboxylase,[183] thus promoting formation of oxaloacetate, which is channeled into PEP formation, since citrate synthetase is inhibited. Thus lactate and all the amino acids that give rise to pyruvate are funneled into PEP formation. Cyclic AMP also promotes the formation of PEP from pyruvate, but its site of action is not known.[57] In this way, fatty acid metabolism may favor formation of glucose.

The quantitative importance of changes in enzyme activity in the acute regulation of gluconeo-

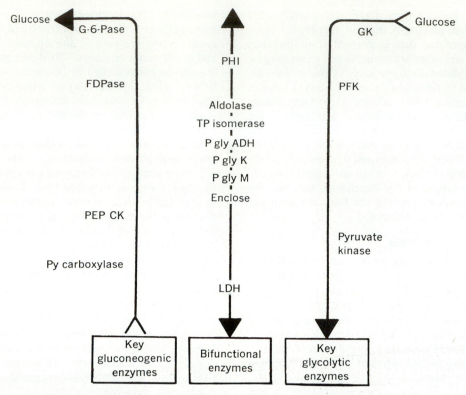

Fig. 66-9. Schematic division of enzymes of glucose metabolism into those that function in the direction of gluconeogenesis, glycolysis, or act in both directions. Gluconeogenic and glycolytic enzymes appear to be under separate control. *G-6-Pase,* Glucose-6-phosphatase; *FDFase,* fructose diphosphatase; *PEP CK,* phosphoenol pyruvate carboxykinase; *Py carboxylase,* pyruvate carboxylase; *PHI,* phosphohexose isomerase; *TP isomerase,* triose phosphate isomerase; *P Gly ADH,* phosphoglyceraldehydedehydrogenase; *P Gly K,* phosphoglyceric acid phosphokinase; *P Gly M,* phosphoglyceromutase; *LDH,* lactic dehydrogenase; *GK,* glucokinase; *PFK,* phosphofructokinase. (From Weber et al.[192])

genesis is questionable. Although high concentrations of unbound fatty acids do increase glucose formation in the isolated perfused rat liver, physiologic amounts presented in a form bound to albumin had little or no acute effect on glucose production.[57,163] In livers exposed to high levels of fatty acids for prolonged periods, however, the effects on enzyme activity might yet prove significant. The cyclic AMP–stimulated conversion of pyruvate to PEP remains the most likely site of acute stimulation of gluconeogenesis from three-carbon precursors.

Ketogenesis

Rapid oxidation of fatty acids leads to the formation of ketone bodies by the liver. Although the uptake of fatty acids by the liver is proportional to their concentration in the plasma, simply supplying the liver cells with increased amounts of fatty acids does not ensure that rapid oxidation will occur. After entering the hepatocyte the fatty acid is linked with CoA and then may either be oxidized or converted to triglyceride. Oxidation occurs entirely within the mitochondria, which are not freely permeable to long-chain acyl CoA, whereas triglyceride formation takes place in the cytoplasm. The rate at which the fatty acids can penetrate the mitochondrial membrane, therefore, determines the relative rates of fatty acid oxidation and esterification. To enter the mitochondrion the fatty acid must be transferred to a molecule of carnitine, for it is the fatty acyl-carnitine that is transported across the mitochondrial membrane. Availability of carnitine thus limits the rate at which fatty acids can be oxidized. Recent evidence indicates that hepatic concentrations of carnitine vary widely and are subject to nutritional and hormonal control[117];

increased ketogenesis coincides with increased hepatic levels of carnitine. The mechanism for increased ketogenesis once the fatty acids enter the mitochondria is inherent in what has already been discussed for the stimulation of gluconeogenesis. The breakdown of long-chain fatty acids results in increased production of acetyl CoA. Long-chain fatty acid CoA inhibits the entry of acetyl CoA into the tricarboxylic acid cycle (by inhibiting citrate synthetase) and also inhibits acetyl CoA carboxylase,[34] the first step in the synthesis of long-chain fatty acids from acetyl CoA. Thus with increased entry of two-carbon fragments into the acetate pool and inhibition of two of the important exits from the pool, acetyl CoA accumulates. This in turn results in a decrease in the amount of free CoA, which shifts the equilibrium of the reaction toward the formation of acetoacetate from two molecules of acetyl CoA.[161]

General comments

In most of the metabolic controls already discussed, whether they be metabolites or hormones via cyclic AMP, *it is changes in enzyme activity that produce regulation*. Changes in enzyme activity can result from changes in the concentration of substrates, products, or cofactors. Changes in enzyme activity can also occur via allosteric changes in enzyme configuration. *These are rapid responses that are rapidly reversible*. Control of metabolic processes can also be achieved through changes in the *amounts of enzymes*. Since all enzymes turn over, this can be achieved through stimulation or inhibition of the synthesis of enzymes either at the level of the gene or the ribosome. Conversely, rates of degradation may also be subject to regulation. Changes in enzyme amount are usually produced more slowly (measured in hours or days) and are reversed slowly. Thus in prolonged fasting, for example, enzymes responsible for the synthesis of long-chain fatty acids in both liver and adipose tissue decrease. Reconstitution of normal lipogenic function requires a matter of days, and until such reconstitution occurs, abnormal responses to feeding might be encountered as, for example, the accumulation of glycogen in adipose tissue of the refed rat. Although this may seem to be disadvantageous, it should be pointed out that unused enzymes constitute dispensable protein whose amino acids can yield life-prolonging substrate for gluconeogenesis.

From this brief discussion, it should be apparent that even in the absence of hormones, some regulation of substrate utilization can be achieved. Availability of glucose regulates fatty acid mobilization, and fatty acids regulate the utilization and perhaps the production of glucose. Against this background, we can now examine the physiologic effects of insulin.

PHYSIOLOGIC EFFECTS OF INSULIN

Insulin is the single most important hormone coordinating the utilization of substrate with the feeding-fasting cycle. *Its overwhelming effect is to promote storage of potential fuel; it favors buildup of triglyceride, glycogen, and protein reserves*. Other hormones, growth hormone, glucocorticoid, glucagon, and epinephrine are also involved, but their actions largely promote the breakdown of stored fuels.

Effects on carbohydrate and fat metabolism

Within a few minutes after the intravenous administration of insulin there is a striking decrease in the concentration of blood glucose. If the dose of insulin is great enough, the fall in glucose concentration may be so profound as to produce *hypoglycemic coma* (i.e., the concentration of glucose becomes too low to satisfy the fuel needs of nerve cells in the brain). Insulin lowers the blood glucose concentration in two ways: (1) it increases the uptake of glucose by muscle and adipose tissue and (2) it reduces the rate of glucose production by the liver.

Effects on glucose uptake

Insulin promotes the uptake of glucose from blood by increasing the movement of glucose across cell membranes. Not all cells are sensitive to this action of insulin. Notable exceptions are nerve cells, the intestinal mucosa, renal tubules, liver, and red blood cells. The most important targets for this action of insulin are skeletal muscle and adipose tissue.

The effects of insulin on glucose transport were first demonstrated by Levine and Goldstein,[102,103] who studied the distribution of galactose in eviscerated nephrectomized dogs. In the absence of the liver, galactose cannot be metabolized, and without the kidneys, it cannot be excreted. An administered dose of galactose therefore eventually distributes itself throughout the body water in all those compartments to which it is permeable. After intravenous injection the galactose concentration in plasma gradually falls as it diffuses out of the vascular compartment and equilibrates with body fluids. Its volume of distribution can be calculated from the amount of galactose injected and the equilibrium concentration reached in blood, since galactose passes freely through capillary membranes. In the absence of insulin, this volume very nearly approximates the total extracellular space. After insulin administration the concentration of galactose in the blood falls to a significantly lower level, indicating the galactose must have become diluted in a greater volume, that is, that some of the galactose must have entered intracellular spaces (Fig. 66-10).

Studies with both isolated diaphragm muscle[94] and perfused heart muscle[129,130,150] have established that glucose uptake can be considered as a two-step process. The rate-limiting step under normal physiologic circumstances is penetration of the cell membranes. Insulin increases the permeability to glucose not by simply increasing aqueous channels for free diffusion but by stimulating a process that has the characteristics of carrier-mediated transport: stereospecificity, saturability and competition between sugars of similar conformation, and counterflow (Chapter 1). Similar conclusions can be drawn for adipose cells.[47,48] The effects of insulin on glucose transport occur almost instantaneously and are thought to be direct.

While membrane transport is the rate-limiting step for glucose metabolism under normal circumstances, phosphorylation of glucose by the hexokinase reaction becomes rate limiting in tissues exposed to insulin or high concentrations of glucose.[130] Normally no free glucose accumulates intracellularly because of the efficiency of the hexokinase reaction. Significant amounts of free glucose may be found in the intracellular water after the addition of insulin to tissues obtained from insulin-deprived animals; in these animals, hexokinase can no longer phosphorylate glucose as rapidly as it enters the cell. This effect is understandable in the light of the foregoing discussion when it is recognized that deprivation of insulin leads to accelerated fat mobilization and utilization by muscle. This in turn leads to inhibition of phosphofructokinase and the ensuing accumulation of glucose-6-phosphate, which is a potent inhibitor of hexokinase. Administration of insulin eventually corrects the decreased hexokinase activity, but several hours are required. In addition, in muscle and adipose tissue there are several isozymes of hexokinase, at least one of which depends on insulin for its synthesis.[91] Thus insulin affects both the activity and the amount of hexokinase. Its effects on glucose phosphorylation are delayed.

Effects on free fatty acid production

Adipose tissue responds to insulin administration with an almost instantaneous curtailment of FFA release.[32] Increased entry of glucose into the fat cell permits rapid formation of α-glycerol phosphate that permits the fatty acids produced by lipolysis to be rapidly reesterified to triglyceride and thus prevented from escaping into the

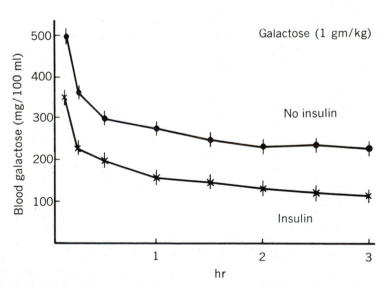

Fig. 66-10. Effect of insulin on distribution of galactose in eviscerated, nephrectomized dogs. (From Levine et al.[102])

bloodstream. Insulin also has an antilipolytic effect and appears to decrease the activity of the hormone-sensitive lipase. Just how this is achieved is still unknown. Insulin increases the activity of the enzyme cyclic nucleotide phosphodiesterase,[106] which degrades cyclic AMP, and it may also inhibit adenylate cyclase.[85] Other evidence suggests that insulin may also accelerate the rate of lipase inactivation, possibly by increasing the activity of the lipase phosphatase.[96] Any of these effects might slow the rate of lipolysis by decreasing the activation of the hormone-sensitive lipase. This latter effect is probably of secondary importance in shutting down the outflow of fatty acids from adipose tissue after a period of insulin deprivation.

Effects on ketone body production

In view of the regulatory effects of FFA on glucose utilization in muscle, it is worth noting that adipose tissue is more sensitive to insulin than is muscle[128]; that is, it takes less insulin to decrease plasma FFA than to decrease plasma glucose. Because fatty acids are withdrawn from the plasma so rapidly, the decrease in fatty acid production by adipose tissue results in a rapid fall in the concentration of FFA in blood.[32] In the face of decreased fatty acid input to the liver, the production of ketone bodies falls, resulting in a decrease in the concentration of ketone bodies in the blood. Insulin may also have an antiketogenic action at the level of the liver, exerted perhaps on the acylcarnitine system.[118]

Effects on glycogen metabolism

Insulin also increases the formation of glycogen in muscle and adipose tissue. In fact, a disproportionate amount of glucose taken up is converted to glycogen.[101,188] Insulin promotes the conversion of glycogen synthetase to its active form. In the liver, this may result in part from increased activity of cyclic nucleotide phosphodiesterase and hence a decrease in the concentration of cyclic AMP. It may be recalled that cyclic AMP not only promotes the formation of phosphorylase a, but it also inhibits glycogen formation by triggering the conversion of the synthetase to its relatively inactive form. By decreasing phosphorylase a, insulin may also remove the inhibition on the activation of glycogen synthetase.[6] The ability of glucose to activate glycogen synthetase is severely impaired in the livers of diabetic rats.[123] Insulin treatment rapidly corrects this defect. In muscle, insulin also promotes the conversion of synthetase to its active form, but no measurable decrease in cyclic AMP

is seen. The mechanism for this effect in muscle is unknown.

Effects on glucose production

Insulin decreases the production and secretion of glucose by the liver.[172] This effect occurs rapidly and probably results from the decrease in cellular levels of cyclic AMP[87] brought about by activation of cyclic nucleotide phosphodiesterase. The decrease in cyclic AMP favors the formation of glycogen over its breakdown and decreases hepatic gluconeogenesis. This acute effect on glucose production is often difficult to demonstrate in intact animals because homeostatic mechanisms set in motion by the fall in blood glucose concentrations increase glycogenolysis and mask the effect. However, when insulin is added to the isolated perfused liver or when an infusion of glucose accompanies the administration of insulin so that hypoglycemia does not occur, the inhibitory effect of insulin on hepatic glucose output is demonstrable.[172]

On a slower time scale, insulin evokes other changes that tend to reduce the production of glucose. By inhibiting the release of amino acids from muscle (p. 1656) and glycerol from adipose tissue, insulin deprives the liver of substrate for gluconeogenesis and thus decreases glucose production. Insulin may also regulate the synthesis of hepatic enzymes. For example, the enzyme glucokinase (p. 1649) virtually disappears from the liver in diabetes or prolonged fasting.[158] Both these conditions are characterized by a scarcity of insulin. Administration of insulin to diabetics or refeeding, which results in insulin secretion, promotes the reappearance of this enzyme after several hours. The key one-way enzymes of gluconeogenesis (glucose-6-phosphatase, fructose-1,6-diphosphatase, pyruvate carboxylase, and pyruvate carboxykinase) all decline when insulin becomes abundant, presumably because their synthesis has been shut off. A concomitant increase occurs in the amounts of enzymes catalyzing reactions in the direction of glycolysis. It has been suggested that each of these groups of enzymes may be controlled by a single operon and that insulin may control these operons. Insulin administration also tends to increase the enzymes of glycogen synthesis in a similar manner.[192]

In summary, insulin has both short-term and long-term effects on glucose production. The immediate effects are exerted through control of enzyme activity principally mediated through the cyclic AMP system. Long-term effects are achieved through control of the amounts of enzymes available for glucose production.

Effects on fatty acid synthesis

In the absence of insulin, hepatic lipogenesis virtually ceases.[44] Administration of insulin corrects this defect in fatty acid synthesis, but several hours are required.[42] The defect in fatty acid synthesis undoubtedly is related to fatty acid oxidation and the accumulation of fatty acyl CoA, which inhibits acetyl CoA carboxylase and citrate synthetase. As hepatic oxidation of fatty acids declines, inhibition of these enzymes is removed and fatty acid synthesis recommences. The renewed utilization of glucose by the glycolytic pathway provides the alpha glycerol phosphate for triglyceride synthesis. Furthermore, restoration of normal amounts of insulin increases the synthesis of the various enzymes of lipogenesis.

In adipose tissue, insulin increases fatty acid synthesis primarily by increasing the entry of glucose into the adipocytes, thus making substrate available. Decreases in cyclic AMP may also remove a stimulus for lipolysis.[162] Insulin also promotes the conversion of pyruvate to acetyl CoA, thereby making substrate for fatty acid synthesis more readily available.[88] This effect of insulin is of particular interest because it occurs entirely within the mitochondria and because it involves the conversion of the responsible enzyme, pyruvate dehydrogenase, from an inactive phosphorylated form to an active dephosphorylated form. Although analogous to the activation of glycogen synthetase, these reactions do not involve cyclic AMP.[8]

Effects on nitrogen metabolism

In the absence of insulin there is net catabolism of body proteins. Circulating concentrations of amino acids and urea are elevated, and there is increased loss of nitrogen in the urine. Administration of insulin causes a prompt decrease in plasma concentrations of amino acids and in blood urea nitrogen (BUN). Urinary nitrogen loss is subsequently diminished.

Under normal conditions, even in the nongrowing individual, all body constituents, including proteins, are constantly broken down and replaced by the balanced processes of synthesis and degradation. Insulin is required for protein synthesis in most, if not all, tissues. Insulin also retards protein degradation.[86] In its absence, protein degradation exceeds protein synthesis, and there is net loss of protein in muscle and other tissues. Amino acids are released into the blood and thus become available for gluconeogenesis.

When given to experimental animals or added to isolated tissues in vitro, insulin promotes the transport of amino acids across cell membranes.[95] This phenomenon has been studied in isolated rat diaphragm muscle with the aid of nonmetabolizable amino acid analogs such as alpha aminoisobutyric acid (AIB). Since AIB can neither be broken down nor incorporated into protein, any that is transported simply accumulates within cells. Stimulation of amino acid transport is demonstrable in the absence of glucose and is thus independent of and possibly unrelated to the effects of insulin on glucose transport. Transport of amino acids, unlike glucose, occurs against a concentration gradient and therefore requires energy. The presence of sodium in the extracellular fluid is essential.

Although acceleration of amino acid transport across cell membranes is obviously related to increased synthesis of proteins, it is not the underlying mechanism whereby insulin increases protein synthesis.[198] Even when the effects of insulin on amino acid transport are blocked by omitting sodium from the incubation medium, insulin promotes the incorporation of radioactive amino acids into protein. In the absence of extracellular amino acids, insulin increases the incorporation of radioactive carbon from pyruvate into proteins. Stimulation of protein synthesis is also independent of the effects of insulin on glucose transport, for it occurs even when there is no extracellular glucose.

Insulin also increases the synthesis of RNA in muscle, but the enhancement of protein synthesis apparently does not depend on formation of new RNA. Insulin accelerates RNA translation at the level of the ribosomes, but the specific mechanism for this is unknown.[199] Thus insulin appears to exert multiple, separate, but complementary effects on protein synthesis.

Insulin and growth

Since protein synthesis is reduced in the absence of insulin, it is not surprising that without insulin normal growth is markedly impaired. This is evident with regard to both developmental growth and the regenerative growth of wound healing. Characteristically, diabetics have great difficulty in healing even minor wounds. When insulin is deficient, growth hormone is ineffective in promoting growth and nitrogen retention.[124] Administration of insulin to young diabetic animals enables growth to recommence, but insulin cannot substitute for the pituitary growth hormone in causing normal growth of hypophysectomized animals.

Effects on electrolytes

It has long been known that insulin decreases plasma concentrations of potassium. This appears to result from increased entry of potassium into muscle and liver cells. Insulin increases the resting membrane potential of skeletal muscle and adipose tissue, and it is this hyperpolarization that accounts for the redistribution of potassium. The effect of insulin on the distribution of monovalent cations can be demonstrated in isolated diaphragm muscle in the absence of glucose, indicating that this effect is not secondary to the effects of insulin on glucose transport.[202]

MECHANISM OF INSULIN ACTION

It is evident that insulin can exert a variety of complementary biologic effects, all aimed toward building up body stores. Many of these effects appear to be independent or at least do not develop one from another. Although it is difficult at this time to envisage a single unique action of insulin that might give rise to all these diverse effects, such a unified view of insulin action remains an attractive possibility. Although some hormones may produce their characteristic actions in cell-free homogenates, consistent effects of insulin are only seen with intact cells or tissues. This suggests that the fundamental primary actions of insulin depend on normal cellular architecture. The initial interaction between insulin and responsive cells appears to be at the cell surface, where insulin binds reversibly to a limited number of specific, high-affinity receptor sites.[154] It is not known how binding at the cell surface is translated into the constellation of effects already described. We may speculate that some of the bound insulin may translocate to intracellular receptor sites,[69] that some second messenger (Ca^{2+}?) is released into the cytoplasm, or that conformational changes induced by the hormone-receptor interaction are propagated along the cell membrane and reverberate throughout the cell. Protein phosphatase activity may somehow be increased. It is significant that the number of insulin receptors on the surfaces of responsive cells is not constant and that their concentration may determine insulin sensitivity. States of insulin resistance are accompanied by high plasma concentrations of isulin and low concentrations of receptors on target cells. Conversely, insulin deficiency states may be accompanied by increased concentrations of receptors.[80] Roth and his colleagues[154] have obtained convincing evidence that insulin regulates the concentration of its own receptor binding sites on target cells. The molecular mechanisms responsible for these effects are not known.

REGULATION OF INSULIN SECRETION[4,6]

As might be expected of a hormone whose physiologic role appears to be promotion of fuel storage, insulin secretion is greatest immediately after eating and declines during fasting or between meal periods. Coordination of insulin secretion with nutritional state as well as with fluctuating environmental demands for energy production is achieved through stimulation by both metabolites and hormones. Secretion in response to stimuli is modulated or tempered by hormonal and neural influences.

Two technological achievements, development of the radioimmunoassay for insulin[201] and development of the methodology for preparing isolated pancreatic islets for in vitro incubations,[98] have permitted great advances in our understanding of insulin secretion. As might have been predicted, the regulation of insulin secretion is indeed complicated; numerous factors must be considered. Because insulin plays the primary role in regulating storage and mobilization of substrate, the beta cells of the pancreas must be constantly apprised of bodily needs, not only with regard to feeding and fasting, but also with regard to demands of the environment, for energy needs differ widely in individuals at peace with their surroundings and those fighting for survival. Maintaining the constancy of the internal environment is achieved through direct monitoring of circulating concentrations of energy-yielding metabolites by the beta cells themselves. This principal governor of insulin secretion can be overridden by hormonal mechanisms that prime the individual either for rapid storage of an influx of foodstuffs or for massive mobilization of bodily fuels to be consumed in responding to some environmental demand.

Circulating metabolites

Glucose is the most important signal for insulin secretion received by beta cells. Secretion of insulin is proportional to the concentration of glucose in the extracellular fluid bathing the islets, and this proportionality is maintained over a wide range of glucose concentrations extending well above and below the normal physiologic level.[120] The rat pancreas increases its secretion of insulin within 30 sec of exposure to increased concentration of glucose and can shut down secretion as rapidly.[75] When the beta cells are exposed to high concentrations of glucose for a prolonged period, insulin secretion follows a biphasic pattern.[76] There is an initial burst of secretion that declines after a few minutes and is then followed by a phase of gradually increasing secretion that reaches and maintains a level that is related to the concentration of glucose. The first phase appears to involve the release only of stored insulin, whereas the second phase appears to involve a combination of stored and newly synthesized hormone.[76] Just how the concentration of glucose is monitored by the beta cell and how this information is coupled to the secretory process are subjects still under intensive investigation. Glucose may play several roles in triggering insulin release.[115] Some evidence favors the existence of "glucoreceptors" located on the surface of the beta cells. The degree of interaction of the glucoreceptors with glucose is somehow transmitted intracellularly to trigger insulin release. Other evidence indicates that glucose must be metabolized to effect insulin secretion and suggests that the flux of glucose metabolites through some critical reaction may govern insulin secretion. Interference with glucose metabolism in the islet tissue by mannoheptulose or 2-deoxyglucose inhibits insulin secretion. It appears that metabolism of glucose by the glycolytic pathway and by the hexose monophosphate shunt is important for normal responsiveness of the beta cells, the former in the production of the required ATP and the latter as a supplier of reduced pyridine nucleotide. The intimate molecular mechanisms involved in coupling glucose transport and metabolism with the secretory process are not known.

Amino acids also are important stimuli for insulin secretion. Following the ingestion of a protein meal, there is a transient increase in plasma amino acids, and this is accompanied by an increase in insulin secretion.[58] The most potent amino acids in their ability to stimulate insulin secretion are arginine, lysine, and leucine. Since nonmetabolizable analogs of leucine and arginine also stimulate insulin secretion, it is apparent that the amino acids themselves rather than their metabolites are recognized as stimuli by the beta cells. Stimulation of insulin release may be related to the amino acid transport mechanisms in the beta cell membrane. Pharmacologic studies suggest that arginine and leucine operate by different mechanisms to evoke insulin secretion.[125] It should be noted that amino acids are effective signals for insulin release only in the presence of adequate levels of blood glucose. When glucose levels are too low, amino acids fail to evoke insulin secretion. This may act as a safety factor to prevent hypoglycemia after a protein meal that contains little carbohydrate.

Fatty acids and ketone bodies may also increase insulin secretion, but only when they are present in rather high amounts.[73,108] It may be recalled that fatty acid mobilization and hence ketone body formation depend on a low circulating level of insulin. The ability of these metabolites to induce insulin secretion when their concentrations become too high in plasma may be a feedback mechanism to protect against excessive mobilization of fatty acids and ketosis.

Hormonal control

The glucose tolerance test is a standard means for assessing the functional status of the beta cells. Both the oral and intravenous glucose tolerance tests have been used, and careful comparisons between them led to the discovery of an important hormonal mechanism for the control of insulin secretion.[119] Administration of the test dose of glucose by vein produces a higher level of blood glucose than administration of the same amount by mouth, but the oral route evokes a greater release of insulin (Fig. 66-11). This observation suggested that some factor released by the gastrointestinal tract in response to glucose stimulates the pancreas to secrete insulin. This is analogous to the stimulation of the exocrine portion of the pancreas by secretin in response to acid in the duodenum. Such an anticipatory secretion of insulin would prepare the individual for the rapid handling of the glucose about to be absorbed from the gastrointestinal tract. Examination of a variety of known intestinal hormones revealed that gastrin, secretin, and gastric inhibitory peptide (GIP), and possibly cholecystokinin-pancreozymin can evoke insulin release. Of these, GIP appears to be the physiologically important mediator of insulin secretion in response to carbohydrate ingestion.[38]

Insulin secretion by the human pancreas is virtually shut off during an infusion of epinephrine.[149] This effect cannot be explained on the basis of altered blood flow but rather is the result of a direct inhibitory action of epinephrine on the beta cells. Inhibition of insulin secretion by epinephrine is a beautiful example of how multiple actions of a hormone can reinforce each other and cooperate to enhance the overall effect of the hormone. At the level of the end-organ, insulin and the catecholamines have antagonistic effects on gluconeogenesis, glycogenolysis in both liver and muscle, and lipolysis in adipose tissue. The

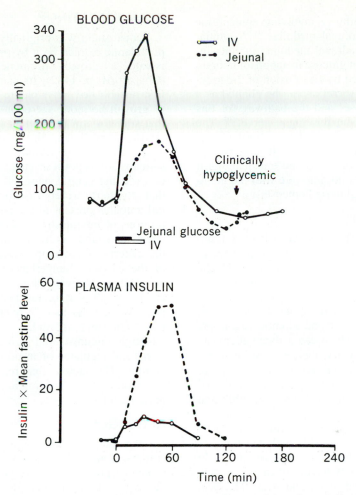

Fig. 66-11. Blood glucose and insulin concentrations after administration of 60 gm of glucose intrajejunally or intravenously. Glucose absorbed from gut produced much smaller rise in blood glucose but much greater increase in plasma insulin—so great, in fact, that frank hypoglycemia was observed. (From McIntyre et al.[119])

catecholamines are secreted in emergency situations in which survival is threatened and in which there may be a need for the rapid mobilization of energy-yielding fuels to provide the substrate for the "flight, fright, or fight" response (Chapter 64). Thus in responding to stress the mammalian organism can not only turn on those processes that produce abundant supplies of fuel, but in the same stroke it can also turn off the inhibition normally exerted on these processes by insulin.

Modulation of beta cell activity

The quantitative response of the beta cells to a stimulus for insulin secretion can be modified by both hormonal and neural influences. Growth hormone and adrenal glucocorticoids do not nec-

essarily increase the basal rate of insulin secretion, but these hormones increase the magnitude of the beta cell response to hyperglycemia and other stimuli for insulin release. The physiologic significance of this effect is unclear. On the one hand, these hormones appear to antagonize the action of insulin on glucose utilization, and on the other, they appear to enhance its secretion in such a manner that glucose utilization may be minimally affected.[52,146] Thyroxine similarly modulates the sensitivity of the beta cells to glucose.[111] Electrical stimulation of the vagus nerve tends to increase insulin secretion,[65] whereas sympathetic stimulation tends to inhibit it. Electrical stimulation of the ventromedial nucleus of the hypothalamus decreases insulin secretion by a neurohumoral pathway involving epinephrine release from the adrenal medullae.[64] A similar stimulation of the adrenal medullae inhibits insulin secretion when glucose metabolism in brain cells is compromised by an infusion of 2-deoxyglucose.[133] Insulin secretion can be elicited as a conditioned response, and the pathway for this response seems to be the vagus nerve.[197] The physiologic importance of these neural effects on insulin secretion is not well understood. Denervation of the pancreas and transplanting it to a distant site does not interfere with the maintenance of normal glucose homeostasis.

Cellular events

A great deal of effort has been directed at unraveling the processes that control insulin secretion at the cellular and subcellular level. Cyclic AMP is involved,[110] but the nature of its involvement is not understood. Glucose must be present for cyclic AMP to increase insulin secretion. Agents such as glucagon and adrenocorticotropic hormone (ACTH), which can activate adenylate cyclase in the beta cells, increase insulin secretion under experimental conditions[178] but are probably unimportant in this regard under normal physiologic circumstances. Similarly, inhibition of cyclic AMP degradation by drugs such as theophylline also increases insulin secretion.[181] These results are not at odds with the findings that epinephrine inhibits insulin secretion. In most tissues there are two types of catecholamine receptors, designated alpha and beta. The beta receptors appear to be related to increases in cyclic AMP, whereas the alpha receptors may lower cyclic AMP. When the effects of epinephrine on insulin secretion are examined in the presence of drugs that block the alpha receptors, epinephrine increases cyclic AMP concentrations in the beta cells and evokes insulin secretion.[180]

As with other secretory phenomena, calcium is required for insulin release.[49] Low concentrations of calcium in plasma found in some disease states can interfere with insulin secretion. Raising the potassium concentration in the solution perfusing the isolated rat pancreas increases insulin release.[74] Similarly, inhibition of the sodium:potassium–dependent ATPase with ouabain increases insulin release.[126] Both of the foregoing tend to depolarize the beta cell membrane. Depolarization also occurs during release of insulin in response to a variety of agents.[53] It is not known where these ionic and membrane events fit into the sequence of events leading to insulin secretion. Participation of the microtubular apparatus is suspected but not yet proved.

SOMATOSTATIN[5,184]

Somatostatin was originally isolated from hypothalamic extracts as a by-product of efforts to isolate a growth hormone–releasing factor (Chapter 61, p. 1482). Its name derives from its first established physiologic effect—inhibition of growth hormone secretion. It is now evident that somatostatin is a potent inhibitor of the secretion of insulin and glucagon as well. Immunoassay of extracts of various tissues indicated that somatostatin is not restricted to the hypothalamus or even the central nervous system, but rather that it is fairly widely distributed, with substantial amounts present in the pancreas and certain portions of the intestinal mucosa. Immunohistochemical studies indicated that somatostatin is the material stored in the granules of the D cells of the islets of Langerhans.[137,148] Somatostatin is found in relatively large amounts in the pancreata of all vertebrates tested, including the primitive jawless fish.[185] A physiologic role for pancreatic somatostatin has not been established, although circumstantial evidence implicates it as a local regulator of insulin and glucagon secretion. The factors that regulate somatostatin secretion from the pancreatic delta cells have not been elucidated.

GLUCAGON

Early preparations of insulin, including the crude acid-alcohol extracts of Banting and Best,[30] produced a transient hyperglycemia that preceded the expected fall in blood glucose concentration. Since some preparations, particularly the crystalline insulin prepared by Abel,[21] were devoid of hyperglycemic effects, it was apparent that transient hyperglycemia was not a property of insulin but rather of some contaminant. In 1923 the hyperglycemia-producing substance was separated from insulin and named glucagon.[131] Later workers referred to it as the hyperglycemic-glycogenolytic factor (HGF), since many years were to pass before the hormonal nature of glucagon was established. It is now generally accepted that glucagon is a bonafide hormone that exerts powerful effects on glucose production by the liver. Early indirect evidence indicated that the islets of Langerhans are the source of glucagon. The availability of antibodies to glucagon made possible an immunofluorescent examination of pancreas and confirmed that glucagon is localized in the alpha cells.[31]

Chemistry

Glucagon is an unremarkable single unbranched peptide chain with a molecular weight

$$NH_2 \qquad\qquad\qquad NH_2 \quad\; NH_2 \qquad\quad NH_2$$

H- HIS- SER- GLU-GLY-THR- PHE-THR- SER- ASP- TYR- SER- LYS-TYR- LEU-ASP-SER- ARG- ARG- ALA-GLU-ASP- PHE- VAL-GLU-TRY- LEU- MET- ASP- THR- OH

1　2　3　4　5　6　7　8　9　10　11　12　13　14　15　16　17　18　19　20　21　22　23　24　25　26　27　28　29

Fig. 66-12. Amino acid sequence of glucagon.

of 3,485; it consists of 29 amino acids (Fig. 66-12). Its complete amino acid sequence was worked out by Behrens and Bromer[35] in 1957 and has subsequently been verified through artificial synthesis.[200] The amino acid sequences of glucagon prepared from pigs, cows, rabbits, rats, and humans are identical.[20] The findings of a glucagon-like immunoreactivity in insects[179] attests to the fact that glucagon is a phylogenetically ancient molecule whose structure has been remarkably preserved through evolution.[176] Shark glucagon has the same number of amino acids and cross-reacts immunologically with the mammalian hormone. It is probable that all or nearly all the molecule is required for biologic potency, for glucagon is very sensitive to enzymatic cleavage and quickly loses its biologic activity after relatively mild treatment. Glucagon is synthesized as a large precursor protein that rapidly passes through several intermediate-sized molecules before its ultimate storage and secretion as the 29–amino acid "true" glucagon.[20] The most prominent precursor has a molecular weight of about 9,000 and is referred to as proglucagon. It is biologically inactive.

Glucagon is thought to be synthesized in the endoplasmic reticulum and thence transferred to the Golgi complex, where the alpha cell granules are formed. The granules migrate to the membrane of the alpha cell and are extruded in a process analagous to that described for the beta cell. The granules dissolve in the perivascular space and then enter the bloodstream. Glucagon is thought to circulate in the blood without binding to carrier proteins. It has a half-life of only about 3 to 4 min. The principal site of glucagon clearance from the blood is the liver, which degrades it enzymatically and also excretes about 0.5 mg into the bile each day.[29] The kidney is the other major site of degradation. Some glucagon is also inactivated in the blood.

Assay

Biologic assay is usually based on the hyperglycemic response to glucagon following injection into an animal such as the rabbit. The isolated perfused rat liver has also been used for this purpose. Since many agents produce hyperglycemia, however, such an assay is of limited usefulness. Activation of adenylate cyclase in homogenates of rat liver has proved effective, sensitive, and considerably more specific.[109] Radioimmunoassay of glucagon is now possible,[19] although development of a reliable assay specific for glucagon presented several difficulties. Glucagon immunoreactivity in plasma is heterogeneous[20] and may appear in as many as four separate fractions, ranging in molecular weight from about 2,000 to about 60,000. True glucagon appears to account for less than half the plasma immunoreactivity, partly because the half-life of true glucagon is considerably shorter than that of the other fractions of immunoreactive material. In addition, the biologic activity of true glucagon is rapidly destroyed by enzymes within the blood, but such destruction does not necessarily lead to a loss of immunologic activity. The concentration of true glucagon in hepatic portal blood ranges from 300 to 5,000 pg/ml, depending on environmental conditions[20] (1 pg = 10^{-12} gm). In the dog the concentration of glucagon in portal venous blood is 2 to 3 times as high as in peripheral blood. This difference reflects not only its greater dilution in peripheral blood, but also the fact that a considerable amount of glucagon is destroyed in passage through the liver.

The islets of Langerhans are not the only source of glucagon. Immunoreactive glucagon is found in the plasma of totally pancreatectomized dogs[186] and human subjects.[141] By immunologic, biologic, and physiochemical criteria, this extra-pancreatic material appears to be identical with pancreatic glucagon. It appears to be produced in the mucosa of the stomach and duodenum in cells that are histologically very similar to pancreatic alpha cells. Some immunoreactive glucagon may also be produced in the salivary glands.[100] Although some of the physiologic signals for pancreatic glucagon secretion also elicit the secretion of gastrointestinal glucagon, the physiologic role, if any, of extrapancreatic glucagon is not understood.

Biologic actions

As already indicated, the chief biologic response to an infusion of glucagon is an increase in blood glucose level. The physiologic importance of this action was suggested by elegant cross-circulation experiments.[62] Two anesthetized dogs were placed side by side so that the venous blood draining the pancreas of one dog (the donor) could be circulated into the femoral vein of the second dog (the recipient). Injection of sufficient insulin into the donor dog to produce hypoglycemia caused hyperglycemia in the recipient, suggesting that the donor pancreas released some material (glucagon) that increased the blood glucose in the recipient (Fig. 66-13). Despite these findings the physiologic importance of glucagon was not widely accepted, particularly since no disease state had been described that was attributable to either overproduction or underproduction of the hormone. Glucagon gained in respectability and ultimately acceptance as a hormone partly as a result of the studies of Sutherland and Rall[178] and colleagues. These studies culminated in the discovery of cyclic AMP and the cascade of biochemical reactions that lead to activation of phosphorylase (Chapter 60). Radioimmunoassay later revealed that glucagon is secreted into the bloodstream in varying amounts appropriate to changing physiologic conditions[19] and firmly established the hormonal role of glucagon.

The principal target organ for glucagon is the liver. Glucagon markedly increases the output

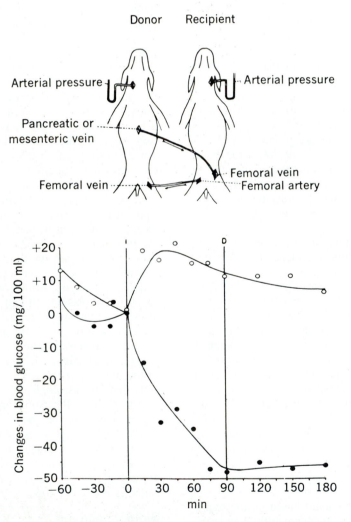

Fig. 66-13. Cross-circulation experiment demonstrating presence of hyperglycemic factor released into pancreatic venous blood during hypoglycemia. Donor dog was injected with insulin, which lowered its blood sugar level (closed circles) by nearly 50 mg/100 ml. Blood from pancreatic vein was allowed to flow into femoral vein of recipient, in whom it produced a mild hyperglycemia (open circles). (From Foa et al.[62])

of glucose in two ways, by increasing the breakdown of glycogen and by stimulating gluconeogenesis. Within 20 min after the intravenous injection of glucagon to appropriate subjects the concentration of glucose in the blood increases by about 50%. This effect is of short duration and disappears within 1 hr. The biochemical steps by which glucagon activates the breakdown of glycogen have been outlined in Chapter 60.

Glucagon may act at more than one biochemical locus in promoting gluconeogenesis. Some evidence favors an effect exerted between pyruvate and PEP,[56,147] whereas other data suggest an action on the phosphofructokinase/fructose diphosphatase enzymes.[45] In any case, cyclic AMP almost certainly mediates this action of glucagon, whether the gluconeogenic substrate is pyruvate or fructose and their respective precursors. Glucagon also increases the transport of amino acids across the hepatic cell membrane[112,113] and accelerates their degradation to gluconeogenic precursors and urea.[112,121] Increased urea formation can be demonstrated in the isolated perfused liver even in the absence of exogenous amino acid substrates, for glucagon also promotes the breakdown of endogenous hepatic protein to amino acids.[121] Glucagon also induces the synthesis of some enzymes of amino acid degradation.[194] When large amounts of glucagon are added to the perfusion mixture, a marked increase in lysosomal activity and the appearance of autophagic vacuoles can be demonstrated within hepatic cells.[28] Finally, glucagon stimulates lipolysis of endogenous hepatic triglycerides[46] and formation of ketone bodies from fatty acids. The glucagon-sensitive step in ketogenesis seems to be related to the acyl carnitine system which facilitates the entry of fatty acids into the mitochondria. In rats, glucagon increases the concentration of carnitine in the liver, possibly by increasing its uptake from the circulation,[117] and also increases the activity of the mitochondrial enzyme acyl transferase II, which catalyzes the conversion of acyl carnitine back to acyl CoA after the fatty acid has traversed the mitochondrial membrane.[118] It is likely that most, if not all, of the foregoing effects of glucagon are secondary to increased formation of cyclic AMP.

Other effects

Glucagon is also a potent stimulator of the hormone-sensitive lipase in adipose tissue.[78] It is unlikely, however, that this effect is physiologically important, since the concentration of glucagon needed to activate lipase is about a 1,000 times higher than that found in peripheral blood. Adipose tissue is notoriously nonselective and responds to virtually any hormone that can increase adenylate cyclase activity in any tissue.

In large doses, glucagon also produces interesting pharmacologic actions on the cardiovascular system. In cardiac but not in skeletal muscle, it activates phosphorylase and thereby increases glycogenolysis.[13] Glucagon also increases heart rate (chronotropic effect), force of contraction (inotropic effect),[144] and coronary blood flow[70] in a manner analogous to the catecholamines. All these effects appear to be due to the ability of glucagon to increase cyclic AMP concentrations in cardiac muscle.[116] These effects of glucagon on the heart are seen only when the hormone is added at a concentration at least 10,000 times that normally found in peripheral blood.

Glucagon exerts a variety of other effects that also appear to stem from its ability to increase cyclic AMP formation. Thus it increases bile secretion in man[55] and inhibits the effects of gastrin on gastric acid secretion in dogs and rats.[104] Infusion of glucagon into normal human subjects induces hypocalcemia,[33] and in the dog probably increases thyrocalcitonin production.[25] In large amounts, glucagon also increases insulin secretion.[92] All the foregoing effects—enhancement of bile secretion, inhibition of gastric acid secretion, and enhancement of thyrocalcitonin secretion, and enhancement of insulin secretion—are physiologic actions of specific enteric hormones. Because portions of the glucagon molecule share homologous sequences of amino acids with many of the enteric hormones, the foregoing effects may represent pharmacologic manifestations of these overlapping structures.

Regulation of secretion

The concentration of glucose in blood is the most important signal for glucagon secretion[187] as originally shown by cross-perfusion studies[62]; a decrease in blood sugar level signals increased glucagon secretion, whereas increased blood sugar level inhibits glucagon secretion. Similar results are obtained when isolated pancreatic islets are studied in vitro.[187] The precise molecular signal for glucagon release is not known. Insulin is required for normal function of the alpha cells; without it, glucose cannot suppress glucagon secretion.[20] From the foregoing, it is not surprising that the secretion of glucagon decreases after a carbohydrate meal and increases during periods of fasting[23] and in diabetes.[18] In this respect, glucagon secretion follows a pattern opposite that of insulin (Fig. 66-14). Since the effects of glucagon on the liver are opposite those of insulin, the balance between these hormones or their ratio may be of physiologic importance in the control of gluconeogenesis and glucose release from the liver. The secretion of glucagon is also enhanced by certain amino acids, particularly arginine.[182] In this respect, glucagon and insulin secretion increase in parallel. Once again, the balance reached between insulin and glucagon may determine whether the dietary amino acids are directed toward gluconeogenesis or spared for protein synthesis. The secretion of glucagon, in contrast to insulin, is inhibited by high concentrations of FFA and ketone bodies.[4] The influence of any of these circulating metabolites can be overridden by neurohumoral mechanisms involving principally the sympathetic nervous system, which stimulates the release of glucagon through the release of epinephrine from the adrenal medulla, and through apparently direct innervation of the pancreas.[4] Thus in stress,[20] hemorrhage, or exercise,[66] plasma concentrations of glucagon are elevated at the same time that insulin levels are depressed. Somatostatin, as already mentioned, inhibits glucagon secretion, but the physiologic relevance of this phenomenon is unknown.

From the foregoing discussion, it is evident that insulin and glucagon are physiologic antagonists whose secretory patterns often appear as mirror images of each other. Unger and his colleagues have suggested that the ratio of their concentrations, rather than the absolute amount of either, is the physiologic variable.[20] Since the numerator and denominator change in opposite directions in response to the same signals, this arrangement greatly increases both the overall sensitivity of the system and the precision with which it can be adjusted. In this light, diabetes mellitus is not a disease only of insulin deficiency, but one of glucagon excess as well.

HORMONE INTERACTION AND REGULATION OF METABOLISM

The findings that both hypophysectomy[83] and adrenalectomy[105] reduced the severity of the symptoms of diabetes in experimental animals raised the possibility that hormones of the pituitary gland and the adrenal cortex may counterbalance the actions of insulin. Hypophysectomized or adrenalectomized animals are hypersensitive to the hypoglycemic effects of insulin.[54] Furthermore, large doses of growth hormone or adrenal glucocorticoids given chronically can produce either temporary or permanent dia-

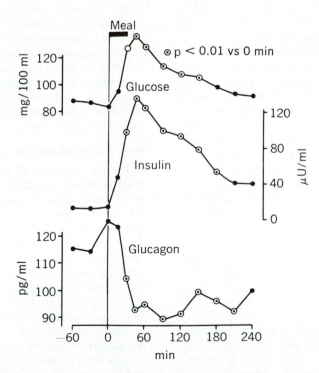

Fig. 66-14. Changes in blood levels of glucose, insulin, and glucagon after carbohydrate meal. (From Unger.[18])

betes.[81] These phenomena have been widely studied in many laboratories for more than a quarter of a century. The bulk of the evidence indicates that both growth hormone and the glucocorticoids tend to limit the uptake and utilization of glucose by muscle and adipose tissue and tend to increase glucose production by the liver. In this regard the actions of growth hormone and glucocorticoids set in slowly and tend to persist over a long period of time.

The effects of glucagon and epinephrine also oppose some of the actions of insulin. These hormones increase cyclic AMP production and consequently accelerate those metabolic processes that depend on cyclic AMP: glycogen breakdown, lipolysis, and gluconeogenesis. Epinephrine also accelerates hepatic glycogenolysis by mechanisms that are independent of cyclic AMP.[43] Glucagon and epinephrine are the rapid adjusters of metabolism; their effects are seen within just a few seconds or minutes and are countered within a few seconds or minutes by the action of insulin. Therefore minute-to-minute regulation of substrate availability is achieved through fluctuations in glucagon, epinephrine, and insulin secretion.

It is an oversimplification to regard insulin simply as an antagonist of epinephrine, glucagon, glucocorticoids, and growth hormone. The actions of thyroid hormone must also be taken into account, although by and large its effects are nonspecific and tend simply to accelerate ongoing metabolic processes. Before discussing the interaction of these hormones in the intact individual, it would be of value to take another look at adipose tissue, muscle, and liver and to examine the interactions of these hormones on metabolic processes in these representative tissues.

Adipose tissue

On the whole, insulin activates all those processes in adipose tissue that lead to the accumula-

tion of the triglyceride storage depot. It promotes the anabolic processes of fatty acid synthesis, triglyceride formation, protein synthesis, and glycogen synthesis. At the same time, insulin antagonizes the effects of the lipolytic hormones and reduces FFA production. Epinephrine accelerates FFA release by promoting lipolysis secondary to activation of the hormone sensitive lipase. Paradoxically, perhaps, it also increases fatty acid reesterification to triglyceride, although reesterification does not keep up with lipolysis and net output of FFA is increased.[12] Growth hormone and glucocorticoids behave in a similar and complementary way and reinforce each other's actions on adipose tissue. They inhibit synthesis of fatty acids and tend to increase lipolysis, but there is some question as to whether they do so directly or indirectly.[72] These hormones increase the ability of epinephrine to activate the hormone-sensitive lipase. Presumably their action is exerted on some aspect of the cyclic AMP system, but the exact details are still unknown. By inhibiting glucose utilization in adipose tissue, growth hormone and glucocorticoids block reesterification and de novo synthesis of fatty acids. None of these effects of growth hormone and glucocorticoids are seen for at least 2 hr after they are administered. The effects of growth hormone and glucocorticoids are probably permissive. Although large doses of growth hormone will increase the concentration of FFA in plasma, the amount of growth hormone needed to produce such an effect is about 50 times more than the pituitary gland secretes in response to physiologic stimuli. These effects are summarized in Table 66-1.

Muscle

Insulin favors the storage of both carbohydrate and protein in muscle and promotes the utilization of glucose when it is in abundant supply. Growth hormone and insulin work together to increase protein anabolism (Chapter

Table 66-1. Hormonal effects on metabolism of adipose tissue

	Fatty acid synthesis	Lipolysis	FFA reesterification	Protein synthesis	Glucose uptake
Insulin	↑	↓	↑	↑	↑
Growth hormone	↓	↑	↓	↑	↓
Glucocorticoids	↓	↑	↓	↑	↓
Epinephrine	↓	↑	↑		↑
Glucagon	↓	↑	↑		↑
Thyroxine	↑	↑	↓		↑

61). Both increase the transport of amino acids and their incorporation into proteins. This decreases the release of amino acids from muscle. Both may also inhibit the degradation of muscle protein, but little supporting data are yet available for this effect of growth hormone. The glucocorticoids, on the other hand, are catabolic insofar as muscle protein is concerned (Chapter 64). They accelerate protein degradation and amino acid release and inhibit amino acid transport and conversion to protein. The thyroid hormones accelerate both protein synthesis and protein degradation, but, when they are present in excess amounts, accelerated protein degradation predominates (Chapter 62).

All the effects of insulin on glucose metabolism in muscle are not antagonized by growth hormone and glucocorticoids. All three hormones favor the accumulation of glycogen in muscle, although by different mechanisms. Insulin appears to do so directly and rapidly by accelerating glucose transport and conversion to glycogen and by inhibiting glycogenolysis. Growth hormone and glucocorticoids promote glycogen synthesis and inhibit its breakdown indirectly through the glucose fatty acid cycle; that is, under the predominant influence of growth hormone or glucocorticoids, greater amounts of fatty acids are mobilized from adipose tissue and hence are available for metabolism in muscle, with the result that glucose utilization is hindered at the level of phosphofructokinase. The resulting accumulation of glucose-6-phosphate impedes glycogen breakdown and provides substrate for glycogen formation. At the same time, phosphorylation is reduced. Insulin promotes the phosphorylation of glucose, but this is one of the effects of insulin that requires a period of hours, since it probably occurs indirectly, requiring first an action on adipose tissue to reduce FFA production. Epinephrine acts in muscle as an immediate mobilizer of

energy by promoting the breakdown of glycogen, much of which is metabolized only as far as lactic acid, which is released from muscle and reconverted to glucose in the liver. Epinephrine may also decrease the phosphorylation of glucose, but this effect is transient and probably secondary to the increase in intracellular glucose-6-phosphate produced from glycogen; hormonal effects on muscle metabolism are summarized in Table 66-2.

Liver

In the liver, insulin and growth hormone exert complementary effects only insofar as hepatic growth is concerned. Insulin decreases all aspects of glucose production, decreases ketogenesis, and increases lipogenesis. Growth hormone increases glucose production, presumably through an effect on gluconeogenesis,[24] and increases ketone body formation indirectly through its effects on adipose tissue. Growth hormone also inhibits hepatic synthesis of fatty acids.[73] Although the glucocorticoids are severely catabolic insofar as muscle protein is concerned, they are anabolic insofar as liver is concerned. Glucocorticoids also act in a cooperative manner with insulin to promote the accumulation of liver glycogen; they increase the breakdown of muscle protein and the release of amino acids, which provide both substrate and stimulus for gluconeogenesis.[57] Glucocorticoids are required in a permissive way for the gluconeogenic action of glucagon.[58] They also promote glycogen formation in a third way. It may be recalled that the transferase required for glycogen synthesis exists in both an active, dephosphorylated form and an inactive, phosphorylated form. Glucocorticoids appear to increase the synthesis of a phosphatase whose action maintains the transferase in its active form.[81]

As already mentioned, the actions of glucagon are directly opposite those of insulin. This is also

Table 66-2. Hormonal effects on metabolism of muscle

	Glucose transport	Glucose phosphorylation	Glycogen synthesis	Glyco-genolysis	Amino acid transport	Protein synthesis	Protein degradation	Amino acid release
Insulin	↑	↑*	↑	↓	↑	↑	↓	↓
Growth hormone	↓	↓*	↑*	↓*	↑	↑	↓	↓
Glucocorticoids		↓*	↑*	↓*	↓	↓	↑	↑
Epinephrine	↓	↓		↑				
Glucagon								
Thyroxine						↑	↑	

*Delayed indirect effect by way of the glucose-fatty acid cycle.

the case for epinephrine, but there is some doubt that epinephrine circulates in sufficiently high concentrations in plasma to be physiologically important in glycogenolysis or gluconeogenesis,[167] but norepinephrine released from sympathetic nerve endings within the liver may be quite important in this regard. Norepinephrine activates these processes by a cyclic AMP—independent mechanism.[43] Once again, it should be pointed out that the actions of glucagon and insulin on glucose production are rapid in onset and of short duration. The effects of growth hormone and glucocorticoid require a lag period and tend to be of prolonged duration. These effects are summarized in Table 66-3. It cannot be overemphasized that the arrows depict relative tendencies and not absolute effects.

Hormonal interaction in normal individual—feeding and fasting

Considering the enormity in the fluctuations in the ingestion of carbohydrates, fat, and protein and in energy expenditure, the blood sugar concentration remains remarkably constant. The important factors that contribute to this constancy are illustrated in Fig. 66-15. Experimental per-

Table 66-3. Hormonal effects on metabolism of liver

	Glycogenolysis	Glycogenesis	Gluconeogenesis	Glucose production	Ketogenesis	Lipogenesis
Insulin	↓	↑	↓	↓	↓	↑
Growth hormone			↑	↑	↑	↓
Glucocorticoids		↑	↑	↑	↑	↓
Epinephrine	↑	↓	↑	↑	↑	
Glucagon	↑	↓	↑	↑	↑	
Thyroxine	↑	↓?				

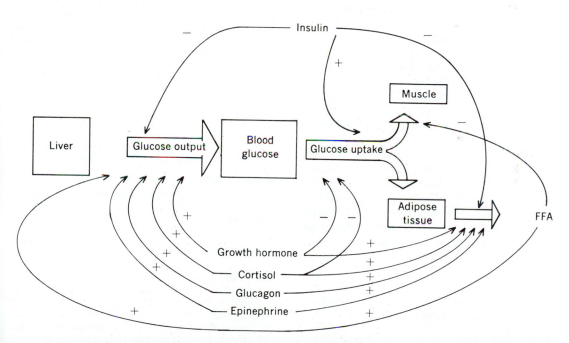

Fig. 66-15. Hormonal regulation of blood glucose concentrations. Decrease in blood glucose level triggers secretion of growth hormone, cortisol (secondary to ACTH secretion), glucagon, and epinephrine and reduces secretion of insulin. Rise in blood glucose triggers increased insulin secretion and decreases secretion of glucagon and perhaps growth hormone. + = stimulation; − = inhibition.

turbation of glucose concentration in either direction evokes an immediate counterregulatory secretion of hormone. Thus raising blood sugar level stimulates insulin secretion, whereas lowering it evokes glucagon, epinpehrine, growth hormone, and glucocorticoid (via ACTH). Obviously, however, glucose concentrations do not remain absolutely constant under normal circumstances. Immediately after a meal there is a transient increase in blood glucose; after 24 hr of fasting, blood glucose level may fall by as much as 15 or 20 mg/100 ml. Within the limits of these fluctuations, however, blood glucose level is kept remarkably constant.

Insulin and glucagon are the hormones that fluctuate most in reflection of feeding and fasting. The presence of food in the intestine signals insulin secretion in anticipation of absorption into the bloodstream. Insulin causes an almost immediate decline in FFA and amino acid concentrations in blood. Simultaneously, the mild hyperglycemia brought on by carbohydrate ingestion inhibits the secretion of glucagon; the predominance of insulin over glucagon inhibits gluconeogenesis. In the between-meal period, blood sugar level gradually falls, leading to a decrease in insulin and a rise in glucagon secretion. The braking effect of insulin on gluconeogenesis is diminished, and the stimulatory effect of glucagon prevails. As insulin concentrations diminish, the breakdown of muscle protein exceeds synthesis and amino acids are released into the blood to serve as a substrate for gluconeogenesis. The decline in insulin is also associated with an increase in FFA mobilization, although FFA concentrations in plasma increase before insulin falls.[164] There do not appear to be consistent changes in glucocorticoid or growth hormone concentrations in plasma in response to fasting.[39,41] Increased secretion would not be necessary for the effects of these hormones to prevail, however, for even if their average concentrations in plasma remain constant, they are increased relative to insulin and their inhibitory effects on glucose utilization become evident. It is of interest that with prolonged fasting, amino acid output by muscle diminishes and the brain adapts somewhat to consume ketone bodies as an energy source. The mechanisms for these responses to prolonged starvation are unknown.

REFERENCES
General reviews

1. deBodo, R. C., and Altszuler, N.: Insulin hypersensitivity and physiological insulin antagonists, Physiol. Rev. **38**:389, 1958.

2. Felig, P.: Amino acid metabolism in man, Annu. Rev. Biochem. **44**:933, 1975.
3. Frohman, L. A.: The endocrine function of the pancreas, Annu. Rev. Physiol. **31**:353, 1969.
4. Gerich, J. E., Charles, M. A., and Grodsky, G. M.: Regulation of pancreatic insulin and glucagon secretion, Annu. Rev. Physiol. **38**:353, 1976.
5. Guillemin, R., and Gerich, J. E.: Somatostatin: physiological and clinical significance, Annu. Rev. Med. **27**:379, 1976.
6. Hers, H. G.: The control of glycogen metabolism in the liver, Annu. Rev. Biochem. **45**:167, 1976.
7. Jeanrenaud, B.: Adipose tissue dynamics and regulation, revisited, Ergeb. Physiol. **60**:57, 1968.
8. Jungas, R. L.: Metabolic effects on adipose tissue in vitro. In Hasselblatt, A., and Bruchhausen, F. V., editors: Handbook of experimental pharmacology new series, Berlin, 1975, Springer-Verlag, vol. 32, pt. 2.
9. Lacy, P. E.: The pancreatic beta cell, N. Engl. J. Med. **276**:187, 1967.
10. Mayhew, D. A., Wright, P. H., and Ashmore, J.: Regulation of insulin secretion, Pharmacol. Rev. **21**:185, 1969.
11. Randle, P. J., and Morgan, H. E.: Regulation of glucose uptake in muscle, Vitam. Horm. **20**:199, 1962.
12. Renold, A. E., and Cahill, G. F., Jr., editors: Handbook of physiology. Adipose tissue section, Baltimore, 1965, The Williams & Wilkins Co.
13. Robison, G. A., Butcher, R. W., and Sutherland, E. W., Jr., editors: Cyclic AMP, New York, 1971, Academic Press, Inc.
14. Ruderman, N. B.: Muscle amino acid metabolism and gluconeogenesis, Annu. Rev. Med. **26**:245, 1975.
15. Steele, R.: The influences of insulin on the hepatic metabolism of glucose, Ergeb. Physiol. **57**:92, 1966.
16. Steiner, D. F., Insulin today, Diabetes **26**:322, 1976.
17. Steiner, D. F., and Freinkel, N.: The endocrine pancreas. In Knobil, E., and Sawyer, W., editors: Handbook of physiology. Endocrinology section, Baltimore, 1972, The Williams & Wilkins Co., vol. 1.
18. Unger, R. H.: Glucagon physiology and pathophysiology, N. Engl. J. Med. **285**:443, 1971.
19. Unger, R. H., and Eisentraut, A. M.: Glucagon. In Gray, C. H., and Bacharach, A. L., editors: Hormones in blood, New York, 1967, Academic Press, Inc., vol. 1.
20. Unger, R. H., and Orci, L.: Physiology and pathophysiology of glucagon, Physiol. Rev. **56**:778, 1976.

Original papers

21. Abel, J. J.: Crystaline insulin, Proc. Natl. Acad. Sci. USA **12**:132, 1926.
22. Adibi, S. A.: Metabolism of branched-chain amino acids in altered nutrition, Metabolism **25**:1287, 1976.
23. Aguilar, E., et al.: Effects of starvation on plasma pancreatic glucagon in normal man, Diabetes **18**:717, 1969.
24. Altszuler, N., et al.: Influence of growth hormone on glucose metabolism and plasma insulin levels in the dog. In Pecile, A., and Müller, E. E., editors: Growth hormone. Proceedings of the First International Symposium, Princeton, N.J., 1968, Excerpta Medica, Inc.
25. Alvioli, L. V., et al.: Role of the thyroid gland during glucagon-induced hypocalcemia, Am. J. Physiol. **216**:939, 1969.
26. Andres, R., Cader, G., and Zierler, K. L.: Quantitatively minor role of carbohydrate in oxidative metabolism by skeletal muscle in intact man in the basal state.

Measurements of oxygen and glucose uptake and carbon dioxide and lactate production in the forearm, J. Clin. Invest. **35**:671, 1956.

27. Armstrong, D. T., et al.: Regulation of plasma free fatty acid turnover, Am. J. Physiol. **201**:9, 1961.

28. Ashford, T. P., and Porter, K. R.: Cytoplasmic components in hepatic cell lysosomes, J. Cell Biol. **12**:198, 1962.

29. Assan, R.: In vivo metabolism of glucagon. In Lefebvre, P. J., and Unger, R. H., editors: Glucagon molecular physiology. Clinical and therapeutic implications, Oxford, 1972, Pergamon Press Ltd., p. 47.

30. Banting, F. G., and Best, C. H.: The internal secretion of the pancreas, J. Lab. Clin. Med. **7**:251, 1922.

31. Baum, J., et al.: Localization of glucagon in the alpha cells in the pancreatic islet by immunofluorescent technics, Diabetes **11**:371, 1962.

32. Bierman, E. L., Schwartz, I. L., and Dole, V. P.: Action of insulin on release of fatty acids from tissue stores, Am. J. Physiol. **191**:359, 1957.

33. Birge, S. J., and Alvioli, L. V.: Glucagon-induced hypoclacemia in man, J. Clin. Endocrinol. Metab. **29**:213, 1969.

34. Bortz, W. M., and Lynen, F.: The inhibition of acetyl CoA carboxylase by long chain acyl CoA derivatives, Biochem. Z. **337**:505, 1963.

35. Bromer, W. W., et al.: The amino acid sequence of glucagon, J. Am. Chem. Soc. **78**:3858, 1956.

36. Brooks, C. McC., et. al.: Humors, hormones, and neurosecretions, New York, 1962, University Publishers, Inc.

37. Brostrom, C. O., Hunkeler, F. L., and Krebs, E. G.: The regulation of skeletal muscle phosphorylase kinase by Ca^{2+}, J. Biol. Chem. **246**:1961, 1971.

38. Brown, J. C., and R. A. Pederson: GI hormones and insulin secretion. In James, V. H. T., editor: Endocrinology. Proceedings of the Fifth International Congress of Endocrinology, Princeton, N.J., 1977, Excerpta Medica Inc., vol. 2, p. 568.

39. Cahill, G. F., Jr.: Starvation in man, N. Engl. J. Med. **282**:668, 1970.

40. Cahill, G. F., Jr., et al.: Blood glucose and the liver, Am. J. Med. **26**:264, 1959.

41. Cahill, G. F., Jr., et al.: Hormone-fuel interrelationships during fasting, J. Clin. Invest. **45**:1751, 1966.

42. Chernick, S. S., and Chaikoff, I. L.: Insulin and hepatic utilization of glucose for lipogenesis, J. Biol. Chem. **186**:535, 1950.

43. Cherrington, A. D., et al.: Studies on the α-adrenergic activation of hepatic glucose output. II. Investigation of the roles of adenosine 3′:5′-monophosphate and adenosine 3′:5′-monophosphate-dependent protein kinase in the actions of phenylephrine in isolated hepatocytes, J. Biol. Chem. **251**:5209, 1976.

44. Christensen, H. N., et al.: In vitro stimulation of insulin release by nonmetabolizable, transport-specific amino acids, Biochim. Biophys. Acta **241**:341, 1971.

45. Clark, M. G., et al.: The fructose 1,6-diphosphatase-phosphofructokinase substrate cycle. A site of regulation of hepatic gluconeogenesis by glucagon, J. Biol. Chem. **249**:5695, 1974.

46. Claycomb, W. C., and Kilsheimer, G. S.: Effect of glucagon, adenosine 3′,5′-monophosphate and theophylline on free fatty acid release by rat liver slices and on tissue levels of coenzyme A esters, Endocrinology **84**:1179, 1969.

47. Crofford, O. B., and Renold, A. E.: Glucose uptake by incubated rat epididymal adipose tissue. Rate limiting steps and the site of insulin action, J. Biol. Chem. **240**:14, 1965.

48. Crofford, O. B., and Renold, A. E.: Glucose uptake by incubated rat epididymal adipose tissue. Characteristics of the glucose transport system and action of insulin, J. Biol. Chem. **240**:3237, 1965.

49. Curry, D. L., Bennett, L. L., and Grodsky, G. M.: Requirement for calcium ion in insulin secretion by the perfused rat pancreas, Am. J. Physiol. **214**:174, 1968.

50. Cushman, S. W.: Structure-function relationships in the adipose cell. II. Pinocytosis and factors influencing its activity in the isolated adipose cell, J. Cell. Biol. **46**:342, 1970.

51. Cushman, S. W., and Rizack, M. A.: Structure-function relationships in the adipose cell. III. Effects of bovine serum albumin on the metabolism of glucose and the release of nonesterified fatty acids and glycerol in the isolated adipose cell, J. Cell. Biol. **46**:354, 1970.

52. Daughaday, W. H., and Kipnis, D. M.: Growth promoting and anti-insulin actions of somatotropin, Recent Prog. Horm. Res. **22**:49, 1966.

53. Dean, P. M., and Matthews, E. K.: Electrical activity of pancreatic islet cells, Nature **219**:389, 1968.

54. deBodo, R. C., et al.: On the hormonal regulation of carbohydrate metabolism; studies with C^{14} glucose, Recent Prog. Horm. Res. **19**:445, 1963.

55. Dick, W. P., and Janowitz, H. D.: Effect of glucagon on hepatic bile secretion in man, Gastroenterology **60**:400, 1971.

56. Exton, J. H., and Park, C. R.: Control of gluconeogenesis in the liver. III. Effects of L-lactate, pyruvate, fructose, glucagon, epinephrine and adenosine 3′,5′-monophosphate on gluconeogenic intermediates in the perfused rat liver, J. Biol. Chem. **244**:1424, 1969.

57. Exton, J. H., et al.: The hormonal control of hepatic gluconeogenesis, Recent Prog. Horm. Res. **26**:411, 1970.

58. Fajans, S. A., et al.: Effect of amino acids and proteins on insulin secretion in man, Recent Prog. Horm. Res. **23**:617, 1967.

59. Felig, P., and Wahren, J.: Amino acid metabolism in exercising man, J. Clin. Invest. **50**:2703, 1971.

60. Felig, P., Wahren, J., and Hendler, R.: Influence of oral glucose ingestion on splanchnic glucose and gluconeogenic substrate metabolism in man, Diabetes **24**:468, 1975.

61. Field, J. B.: Insulin extraction by the liver. In Steiner, D. F., and Freinkel, N., editors: Handbook of physiology. Section 7, Washington, D.C., 1972, American Physiological Society, vol. 1, p. 175.

62. Foa, P. P., Galansino, G., and Pozza, G.: Glucagon, a second pancreatic hormone, Recent Prog. Horm. Res. **13**:473, 1957.

63. Froesch, E. R., et al.: Nonsuppressible insulinlike activity of human serum. Purification, physiochemical and biological properties and its relation to total serum ILA, Recent Prog. Horm. Res. **23**:565, 1967.

64. Frohman, L. A., and Bernardis, L. L.: Effect of hypothalamic stimulation on plasma glucose, insulin and glucagon levels, Am. J. Physiol. **221**:1596, 1971.

65. Frohman, L. A., Ezdinli, E. Z., and Javid, R.: Effect of vagotomy and vagal stimulation on insulin secretion, Diabetes **16**:443, 1967.

66. Galbo, H., Holst, J. J., and Christensen, N. J.: Glucagon and plasma catecholamine responses to graded and prolonged exercise in man, J. Appl. Physiol. **38**:70, 1975.

67. Garland, P. B., Randle, P. J., and Newsholme, E. A.: Citrate as an intermediary in the inhibition of phosphofructokinase in rat heart muscle by fatty acids, ketone bodies, pyruvate, diabetes and starvation, Nature **200:** 169, 1963.

68. Garfinkel, A. S., Baker, N., and Schotz, M. C.: Relationship of lipoprotein lipase activity to triglyceride uptake in adipose tissue, J. Lipid Res. **8:**274, 1967.

69. Goldfine, I. D.: Does insulin need a second messenger? Diabetes **26:**148, 1977.

70. Goldschlager, N., et al.: The effect of glucagon on the coronary circulation in man, Circulation **9:**829, 1969.

71. Goodman, H. M.: Effects of growth hormone on the penetration of L-arabinose in adipose tissue, Endocrinology **78:**891, 1966.

72. Goodman, H. M., and Schwartz, J.: Growth hormone and lipid metabolism. In Knobil, E., and Sawyer, W., editors: Handbook of physiology, Section 7: the pituitary gland and its neuroendocrine control, Baltimore, 1972, The Williams & Wilkins Co., vol. 2, p. 211.

73. Greenough, W. B., III, Crespin, S. P., and Steinberg, D.: Hypoglycemia and hyperinsulinaemia in response to raised free-fatty-acid levels, Lancet **2:**1334, 1967.

74. Grodsky, G. M., and Bennett, L. L.: Cation requirements for insulin secretion in the isolated perfused pancreas, Diabetes **15:**910, 1966.

75. Grodsky, G. M., et al.: Effect of pulse administration of glucose or glucagon on insulin secretion in vitro, Metabolism **16:**222, 1967.

76. Grodsky, G. M., et al.: In vitro studies suggesting a two-compartment model for insulin secretion. In Falkmer, S., Hellman, B., and Taljedal, L. B., editors: Structure and metabolism of the pancreatic islets, Oxford, 1970, Pergamon Press Ltd., p. 409.

77. Groen, J., et al.: Evidence for the presence of insulin in blood serum. A method for an approximate determination of the insulin content of blood, J. Clin. Invest. **31:**97, 1952.

78. Hagen, J.: Effect of glucagon on the metabolism of adipose tissue, J. Biol. Chem. **236:**1023, 1961.

79. Havel, R. J.: Metabolism of lipids in chylomicrons and very low density lipoproteins. In Renold, A. E., and Cahill, G. F., editors: Handbook of physiology. Adipose tissue section, Baltimore, 1965, The Williams & Wilkins Co.

80. Hepp, K. D., et al.: Increased insulin binding capacity of liver membranes from diabetic Chinese hamsters, Nature **258:**154, 1975.

81. Hers, H. G., deWolf, H., and Stalmans, W.: The control of glycogen metabolism in the liver, FEBS Lett. **12:**73, 1970.

82. Hirsch, J., and Han, P. W.: Cellularity of rat adipose tissue: effects of growth, starvation and obesity, J. Lipid Res. **10:**77, 1969.

83. Houssay, B. A., and Biassotti, A.: The hypophysis, carbohydrate metabolism, and diabetes, Endocrinology **15:**511, 1931.

84. Illiano, G., and Cuatrecasas, P.: Glucose transport in fat cell membranes, J. Biol. Chem. **246:**2472, 1971.

85. Illiano, G., and Cuatrecasas, P.: Modulation of adenylate cyclase activity in liver and fat cell membranes by insulin, Science **175:**906, 1972.

86. Jefferson, L. S., Li, J. B., and Rannels, S. R.: Regulation by insulin of amino acid release and protein turnover in the perfused rat hemicorpus, J. Biol. Chem. **252:**1476, 1977.

87. Jefferson, L. S., et al.: Role of adenosine 3′,5′-monophosphate in the effects of insulin and antiinsulin serum on liver metabolism, J. Biol. Chem. **243:**1031, 1968.

88. Jungas, R. L.: Effect of insulin on fatty acid synthesis from pyruvate, lactate, or endogenous sources in adipose tissue: evidence for the hormonal regulation of pyruvate dehydrogenase, Endocrinology **86:**1368, 1970.

89. Junod, A., et al.: Studies of the diabetogenic action of streptozotocin, Proc. Soc. Exp. Biol. Med. **126:**201, 1967.

90. Katsoyannis, P. G.: Synthetic insulins, Recent Prog. Horm. Res. **23:**505, 1967.

91. Katzen, H. M., and Schimke, R. T.: Multiple forms of hexokinase in the rat: tissue distribution, age dependency, properties, Proc. Natl. Acad. Sci. USA **54:** 1218, 1965.

92. Ketterer, H., Eisentraut, A. M., and Unger, R. H.: Effect upon insulin secretion of physiologic doses of glucagon administered via the portal vein, Diabetes **16:**283, 1967.

93. Khoo, J. C., Aquino, A. A., and Steinberg, D.: The mechanism of activation of hormone sensitive lipase in human adipose tissue, J. Clin. Invest. **53:**1124, 1974.

94. Kipnis, D. M.: Regulation of glucose uptake by muscle. Functional significance of permeability and phosphorylating activity, Ann. N.Y. Acad. Sci. **82:**354, 1959.

95. Kipnis, D. M., and Parrish, J. E.: Role of Na⁺ and K⁺ on sugar (2-deoxyglucose) and amino acid (2-amino isobutyric acid) transport in skeletal muscle, Fed. Proc. **24:**1051, 1965.

96. Kissebah, A. H., et al.: Role of calcium in insulin action. II. Effects of insulin and procaine hydrochloride on the enzymes regulating lipolysis, Horm. Metab. Res. **6:**357, 1974.

97. Krebs, H. A.: Considerations concerning the pathways of synthesis in living matter, Bull. Johns Hopkins Hosp. **95:**19, 1954.

98. Lacy, P. E., and Kostianovsky, M.: Method for the isolation of intact islets from the rat pancreas, Diabetes **16:**35, 1967.

99. Larner, J., et al.: Hormonal and nonhormonal control of glycogen synthesis-control of transferase phosphatase and transferase I kinase. In Whelan, W. J., editor: Control of glycogen metabolism, New York, 1968, Academic Press, Inc.

100. Lawrence, A. M., et al.: Salivary gland glucagon: a potent extrapancreatic hyperglycemic factor, Clin. Res. **23:**536, 1975.

101. Leonards, J. R., and Landau, B. R.: A study on the equivalent of metabolic patterns in rat adipose tissue: insulin versus glucose concentration, Arch. Biochem. Biophys. **91:**194, 1960.

102. Levine, R., et al.: The action of insulin on distribution of galactose in eviscerated nephrectomized dogs, J. Biol. Chem. **179:**985, 1949.

103. Levine, R., et al.: Action of insulin on the "permeability" of cells to free hexoses as studied by its effect on the distribution of galactose, Am. J. Physiol. **163:** 70, 1950.

104. Lin, T. M., and Spray, G. F.: Inhibitory effect of glucagon on gastric acid secretion induced by gastrin and its derivatives in dogs and rats, Arch. Int. Pharmacodyn. Ther. **191:**88, 1971.

105. Long, C. N. H., and Lukens, F. D. F.: The effects of adrenalectomy and hypophysectomy upon experimental diabetes in the cat, J. Exp. Med. **63:**465, 1936.

106. Loten, E. G., and Sneyd, J. G. T.: An effect of insulin

on adipose tissue adenosine 3′,5′-cyclic monophosphate phosphodiesterase, Biochem. J. **120:**187, 1970.

107. Lukens, F. D. W.: Alloxan diabetes, Physiol. Rev. **28:** 304, 1948.

108. Madison, L. L., et al.: The hypoglycemic action of ketones. II. Evidence for a stimulatory feedback of ketones on the pancreatic beta cells, J. Clin. Invest. **43:**408, 1964.

109. Makman, M. H., and Sutherland, E. W., Jr.: Use of liver adenyl cyclase for assay of glucagon in human gastrointestinal tract and pancreas, Endocrinology **75:** 127, 1964.

110. Malaisse, W. J., Malaisse-Lagae, F., and Mayhew, D. A.: A possible role for the adenyl cyclase system in insulin secretion, J. Clin. Invest. **46:**1724, 1967.

111. Malaisse, W. J., Malaisse-Lagae, F., and McCrawi, E. F.: Effects of thyroid function upon insulin secretion, Diabetes **16:**643, 1967.

112. Mallette, L. E., Exton, J. H., and Park, C. R.: Control of gluconeogenesis from amino acids in the perfused rat liver, J. Biol. Chem. **244:**5713, 1969.

113. Mallette, L. E., Exton, J. H., and Park, C. R.: Effects of glucagon on amino acid transport and utilization in the perfused rat liver, J. Biol. Chem. **244:**5724, 1969.

114. Margolis, S., and Vaughn, M.: α-Glycerophosphate synthesis and breakdown in homogenates of adipose tissue, J. Biol. Chem. **237:**44, 1962.

115. Matschinsky, F. M.: Enzymes, metabolites and cofactors involved in intermediatry metabolism of islets of Langerhans. In Steiner, D. F., and Freinkel, N., editors: Handbook of physiology. Section 7: endocrine pancreas, Washington, D.C., 1972, American Physiological Society, vol. 1, p. 199.

116. Mayer, S. E., Namm, D. H., and Rice, L.: Effect of glucagon on cyclic 3′5′-AMP, phosphorylase activity and contractility of heart muscle of the rat, Circ. Res. **26:**225, 1970.

117. McGarry, J. D., Robles-Valdes, C., and Foster, D. W.: Role of carnitine in hepatic ketogenesis, Proc. Natl. Acad. Sci. USA **72:**4385, 1975.

118. McGarry, J. D., Wright, P. H., and Foster, D. W.: Hormonal control of ketogenesis. Rapid activation of hepatic ketogenic capacity in fed rats by anti-insulin serum and glucagon, J. Clin. Invest. **55:**1202, 1975.

119. McIntyre, N., Holdsworth, C. C., and Turner, D. S.: Intestinal factors in the control of insulin secretion, J. Clin. Endocrinol. Metab. **25:**1317, 1965.

120. Metz, R.: The effect of blood glucose concentration on insulin output, Diabetes **9:**89, 1960.

121. Miller, L. L.: Glucagon: a protein catabolic hormone in the isolated perfused liver, Nature **185:**248, 1960.

122. Miller, L. L.: The role of the liver and the nonhepatic tissue in the regulation of the blood free amino acid levels. In Holden, J. T., editor: Amino acid pools, New York, 1962, Elsevier North Holland, Inc., p. 708.

123. Miller, T. B., Jr., Hazen, R., and Larner, J.: An absolute requirement for insulin in the control of hepatic glycogenesis by glucose, Biochem. Biophys. Res. Commun. **53:**466, 1973.

124. Milman, A. E., deMoor, P., and Lukens, F. D. W.: Relation of purified pituitary growth hormone and insulin in regulation of nitrogen balance, Am. J. Physiol. **166:**354, 1951.

125. Milner, R. D. G.: The mechanism by which leucine and arginine stimulate insulin release in vitro, Biochim. Biophys. Acta **192:**154, 1969.

126. Milner, R. D. G., and Hales, C. N.: The sodium pump and insulin secretion, Biochim. Biophys. Acta **135:** 375, 1967.

127. Mirsky, I. A.: The metabolism of insulin, Diabetes, **13:**225, 1964.

128. Mirsky, I. A.: Effect of biologically active peptides on adipose tissue. In Renold, A. E., and Cahill, E. F., Jr., editors: Handbook of physiology. Adipose tissue section, Baltimore, 1965, The Williams & Wilkins Co.

129. Morgan, H. E., et al.: Regulation of glucose uptake in muscle. I. The effects of insulin and anoxia on glucose transport and phosphorylation in the isolated, perfused heart of normal rats, J. Biol. Chem. **236:**253, 1961.

130. Morgan, H. E., et al.: Regulation of glucose uptake in muscle. II. Rate-limiting steps and effects of insulin and anoxia in heart muscle from diabetic rats, J. Biol. Chem. **236:**262, 1961.

131. Murlin, J. R., et al.: Aqueous extracts of pancreas. Influence on the carbohydrate metabolism of depancreatized animals, J. Biol. Chem. **56:**253, 1923.

132. Mortimore, G. E., and Mondon, C. E.: Inhibition by insulin of valine turnover in liver. Evidence for a general control of proteolysis, J. Biol. Chem. **245:**2375, 1970.

133. Müller, E. E., Frohman, L., and Cocchi, I. D.: Drug control of hyperglycemia and inhibition of insulin secretion due to centrally administered 2-deoxy-D-glucose, Am. J. Physiol. **224:**1210, 1973.

134. Newsholme, E. A., and Randle, P. J.: Regulation of glucose uptake by muscle, Biochem. J. **83:**387, 1962.

135. Odessey, R., Khairallah, E., and Goldberg, A. L.: Origin and possible significance of alanine production by skeletal muscle, J. Biol. Chem. **249:**7623, 1975.

136. Orci, L., and Unger, R. H.: Hypothesis: functional subdivisions of the islets of Langerhans and the possible role of the insular D-cell, Lancet **2:**1243, 1975.

137. Orci, L., et al.: Cell contacts in human islets of Langerhans, J. Clin. Endocrinol. Metab. **41:**841, 1975.

138. Orci, L., et al.: Evidence for the D-cell of the pancreas secreting somatostatin, Horm. Metab. Res. **7:**400, 1975.

139. Owen, O. E., et al.: Brain metabolism during fasting, J. Clin. Invest. **46:**1589, 1967.

140. Oyer, P. E., et al.: Studies on human proinsulin. Isolation and amino acid sequence of the human pancreatic C-peptide, J. Biol. Chem. **246:**1375, 1971.

141. Palmer, J. P., et al.: Immunoreactive glucagon responses to arginine in three pancreatectomized humans, Metabolism **25:**1483, 1976.

142. Park, C. R., Bornstein, J., and Post, R. L.: Effect of insulin on free glucose content of rat diaphragm in vitro, Am. J. Physiol. **182:**12, 1955.

143. Park, C. R., et al.: The action of insulin on the transport of glucose through the cell membrane, Am. J. Med. **26:**674, 1959.

144. Parmley, W. W., Glick, G., and Sonnenblick, E. H.: Cardiovascular effects of glucagon in man, N. Engl. J. Med. **279:**12, 1968.

145. Passoneau, J. V., and Lowry, O. H.: Phosphofructokinase and the Pasteur effect, Biochem. Biophys. Res. Commun. **7:**10, 1962.

146. Perley, M., and Kipnis, D. M.: Effect of glucocorticoids on plasma insulin, N. Engl. J. Med. **274:**1237, 1966.

147. Pilkis, S. J., et al.: Possible role of pyruvate kinase in the hormonal control of dihydroxyacetone gluconeogenesis in isolated hepatocytes, Metabolism **25:**1355, 1976.

148. Polak, J. M., et al.: Growth hormone release-inhibiting hormone in gastrointestinal and pancreatic D-cells, Lancet **1:**1220, 1975.

149. Porte, D., Jr., and Williams, R. H.: Inhibition of insulin release by norepinephrine in man, Science **152:** 1248, 1966.

150. Post, R. L., Morgan, H. E., and Park, C. R.: Regulation of glucose uptake in muscle. III. The interaction of membrane transport and phosphorylation in the control of glucose uptake, J. Biol. Chem. **236:**269, 1961.

151. Pozefsky, T., et al.: Amino acid balance across tissues of the forearm in postabsorptive man. Effects of insulin at two dose levels, J. Clin. Invest. **48:**2273, 1969.

152. Randle, P. J., et al.: The glucose fatty acid cycle. Its role in insulin sensitivity and the metabolic disturbances of diabetes mellitus, Lancet **1:**785, 1963.

153. Renold, A. E., et al.: Measurement of small quantities of insulin-like activity using rat adipose tissue. I. A proposed procedure. II. Evaluation of performance, J. Clin. Invest. **39:**1487, 1960.

154. Roth, J., et al.: Receptors for insulin, NSILA-S and growth hormone: Applications to disease states in man, Recent Prog. Horm. Res. **31:**95, 1975.

155. Rubinstein, A. H., Melani, F., and Steiner, D. F.: Circulating proinsulin: immunology, measurement and biological activity. In Steiner, D. F., and Freinkel, N., editors: Handbook of physiology. Section 7, Washington, D.C., 1972, American Physiological Society, vol. 1, p. 175.

156. Ruderman, N. B., and M. G. Herrera: Glucose regulation of hepatic gluconeogenesis, Am. J. Physiol. **214:**1346, 1968.

157. Russell, J. A., and Bloom, W.: Hormonal control of glycogen in the heart and other tissues in rats, Endocrinology **58:**83, 1956.

158. Salas, M., Vinuela, E., and Sols, A.: Insulin-dependent synthesis of liver glucokinase in the rat, J. Biol. Chem. **238:**3535, 1963.

159. Samols, E., and Ryder, J. A.: Studies on the tissue uptake of insulin in man a differential immunoassay for endogenous and exogenous insulin, J. Clin. Invest. **49:**837, 1970.

160. Sanger, F.: The chemical structure of insulin, Ciba Found. Colloq. Endocrinol. **9:**110, 1956.

161. Sauer, F., and Erfle, J. O.: On the mechanism of acetoacetate synthesis by guinea pig liver fractions, J. Biol. Chem. **241:**30, 1966.

162. Schimmel, R. J., and Goodman, H. M.: Effects of dibutyryl cyclic adenosine 3′,5′-monophosphate on glucose transport and metabolism in rat adipose tissue, Biochim. Biophys. Acta **239:**9, 1971.

163. Schimmel, R. J., and Knobil, E.: Role of free fatty acids in stimulation of gluconeogenesis during fasting, Am. J. Physiol. **217:**1803, 1969.

164. Schimmel, R. J., and Knobil, E.: Insulin, free fatty acids and stimulation of hepatic gluconeogenesis during fasting, Am. J. Physiol. **218:**1540, 1970.

165. Severson, D. L., Khoo, J. C., and Steinberg, D.: Role of phosphorprotein phosphatases in reversible deactivation of chicken adipose tissue hormone-sensitive lipase, J. Biol. Chem. **252:**1484, 1977.

166. Smith, L. F.: Species variation in the amino acid sequence of insulin, Am. J. Med. **40:**662, 1966.

167. Sokal, J. E., Sarcione, E. J., and Henderson, A. M.: Relative potency of glucagon and epinephrine as hepatic glycogenolytic agents: studies with the isolated perfused rat liver, Endocrinology **74:**930, 1964.

168. Sols, A., and Crane, R. K.: Heart muscle hexokinase, Fed. Proc. **13:**301, 1954.

169. Soskin, S., et al.: The mechanism of regulation of the blood sugar by the liver, Am. J. Physiol. **124:**558, 1938.

170. Stalmans, W., De Wulf, H., and Hers, H. G.: The control of liver glycogen synthetase phosphatase by phosphorylase, Eur. J. Biochem. **18:**582, 1971.

171. Stalmans, W., Laloux, M., and Hers, H. G.: The interaction of liver phosphorylase a with glucose and AMP, Eur. J. Biochem. **49:**415, 1974.

172. Steele, R., et al.: Inhibition by insulin of hepatic glucose production in the normal dog, Am. J. Physiol. **205:**301, 1965.

173. Steiner, D. F., et al.: Proinsulin and the biosynthesis of insulin, Recent Prog. Horm. Res. **25:**207, 1969.

174. Steiner, D. F., et al.: The biosynthesis of insulin. In Steiner, D. F., and Freinkel, N., editors: Handbook of physiology. Section 7, Washington, D.C., 1972, American Physiological Society, vol. 1, p. 175.

175. Stull, J. T., and Mayer, S. E.: Regulation of phosphorylase activation in skeletal muscle in vivo, J. Biol. Chem. **246:**5716, 1971.

176. Sundby, F., Species variations in the primary structure of glucagon, Metabolism **25:**1319, 1976.

177. Sussman, K. E., and Vaughan, G. D.: Insulin release after ACTH, glucagon and adenosine 3′,5′-phosphate (cyclic AMP) in the perfused isolated rat pancreas, Diabetes **16:**449, 1967.

178. Sutherland, E. W., and Rall, T. W.: The relation of adenosine-3′,5′-phosphate and phosphorylase to the actions of catecholamines and other hormones, Pharmacol. Rev. **12:**265, 1960.

179. Tager, H. S., et al.: Glucagon-like immunoreactivity in insect corpus cardiacum, Nature **254:**707, 1975.

180. Turtle, J. R., and Kipnis, D. M.: An adrenergic receptor mechanism for the control of cyclic 3′,5′ adenosine monophosphate synthesis in tissues, Biochem. Biophys. Res. Commun. **28:**797, 1967.

181. Turtle, J. R., Littleton, G. K., and Kipnis, D. M.: Stimulation of insulin secretion by theophylline, Nature **213:**727, 1967.

182. Unger, R. H., et al.: Studies of pancreatic alpha cell function in normal and diabetic subjects, J. Clin. Invest. **49:**837, 1970.

183. Utter, M. F., Keech, D. B., and Scrutton, M. C.: A possible role for acetyl CoA in the control of gluconeogenesis, Adv. Enzyme Reg. **2:**49, 1964.

184. Vale, W., et al.: Somatostatin, Recent Prog. Horm. Res. **31:**365, 1975.

185. Vale, W., et al.: Anatomic and phylogenetic distribution of somatostatin, Metabolism **25:**1491, 1976.

186. Valverde, I., Dobbs, R. E., and Unger, R. H.: Homogeneity of plasma glucagon immunoreactivity in normal, depancreatized and alloxan diabetic dogs, Metabolism **24:**1021, 1975.

187. Vance, J. E., et al.: Effect of glucose concentration on insulin and glucagon release from isolated islets of Langerhans of the rat, Diabetes **17:**187, 1968.

188. Villar-Pilasi, C., and Larner, J.: Insulin-mediated effect on the activity of UDPG-glycogen transglucosylase of muscle, Biochim. Biophys. Acta **39:**171, 1960.

189. Vinuela, E. L., Salas, M., and Sols, A.: Glucokinase and hexokinase in liver in relationship to glycogen synthesis, J. Biol. Chem. **238:**1175, PC, 1963.

190. Von Mering, J., and Minkowski, O.: Diabetes Mellitus nach Pankreasextirpation, Naunyn-Schmiedebergs Arch. Exp. Pathol. Pharmacol. **26:**371, 1889.

191. Walker, D. G.: On the presence of two soluble glucose phosphorylating enzymes in adult liver and the development of one of these after birth, Biochim. Biophys. Acta **77:**209, 1963.

192. Weber, G., Singhal, R. L., and Srivastava, S. K.: Ac-

tion of glucocorticoid as inducer and insulin as suppressor of biosynthesis of hepatic gluconeogenic enzymes, Adv. Enzyme Reg. **3:**43, 1965.

193. Welbourne, R. B., et al.: The APUD cells of the alimentary tract in health and disease, Surg. Clin. North Am. **58:**1359, 1974.

194. Wicks, W. D.: Induction of tyrosine-alphaketoglutarate transaminase in fetal rat liver, J. Biol. Chem. **243:**900, 1968.

195. Wieland, O., Weiss, L., and Eger-Neufeldt, I.: Enzymatic regulation of liver acetyl CoA metabolism in relation to ketogenesis, Adv. Enzyme Reg. **2:**85, 1964.

196. Wing, D. R., and Robinson, D. S.: Clearing factor lipase in adipose tissue, Biochem. J. **109:**841, 1968.

197. Woods, S. C.: Conditioned hypoglycemia: Effect of vagotomy and pharmacologic blockade, Am. J. Physiol. **223:**1424, 1972.

198. Wool, I. G.: Relation of effects of insulin on amino acid transport and on protein synthesis, Fed. Proc. **24:**1060, 1965.

199. Wool, L. G., et al.: Mode of action of insulin in the regulation of protein biosynthesis in muscle, Recent. Prog. Horm. Res. **24:**139, 1968.

200. Wunsch, E.: Die Totalsynthese des Pankreas-Hormons Glucagon, Z. Naturforsch. **22B:**1269, 1967.

201. Yalow, R. S., and Berson, S. A.: Immunoassay of endogenous plasma insulin in man, J. Clin. Invest. **39:**1157, 1960.

202. Zierler, K. L.: Possible mechanisms of insulin action membrane potential and ion fluxes, Am. J. Med. **40:**735, 1966.

XIV

RESPIRATION

67

CHRISTIAN J. LAMBERTSEN

The lung: physical aspects of respiration

In complete contrast with fish, humans and other air-breathing mammals carry the membrane for gas exchange between blood and air deep within the thorax, divided into millions of subsections at the ends of minute branches of a hollow tracheobronchial tree. Through this involuted design the pulmonary membrane, an incredibly fragile barrier between the external environment and the body fluids, is protected against trauma, dehydration, and even freezing. The delicate membrane imposes an immeasurably small barrier to gas exchange. But efficiency in gas exchange is not in all ways desirable, and the deeply involuted structure provides an important *in*efficiency of ventilatory exchange between the atmosphere and the terminal units of the lung, the alveoli. Respiratory movements are required to deliver oxygen and remove carbon dioxide. However, the respiratory movements do not clear the lung of its gas with each breath; a semirigid bellowslike system intermittently replaces only a small portion of the residual alveolar gas with fresh new air from the atmosphere. Some "dead space" air enters and is expelled without participating in gas exchange. The presence of a large residual volume of gas minimizes the degree to which each new breath can affect the composition of gas in diffusion exchange with the blood.

An important consequence of this design is the retention of metabolically produced carbon dioxide, resulting in the maintenance of a stable blood and tissue carbon dioxide pressure, grossly higher than in the ambient atmosphere. Here the contrived inefficiency, of extreme importance at rest and at work, is a vital asset. Maintenance of critical homeostasis for carbon dioxide pressure is of greater importance than precise control of oxygen, because so many essential electrical and metabolic functions themselves depend on the hydrogen ion concentration $[H^+]$ of the fluids in which these functions are carried on.

Thus the membrane has an extremely high capacity for gas exchange, which is damped by a ventilation system designed to protect against overventilation. The desirable level of retained carbon dioxide in the lung, about 40 mm Hg above the atmospheric level, is provided at the expense of an alveolar oxygen pressure approximately 50 mm Hg lower than that inspired.

With these considerations in mind, it will be important to examine respiration as a function that is not only for "uptake of oxygen and elimination of carbon dioxide," but which also in many ways participates actively in regulating the internal cellular environment.

Chapters 67 to 75 will emphasize the important combination of biophysical, chemical, physical, and physiologic factors that together comprise respiratory exchange with the environment.

DEVELOPMENT OF THE LUNG

Early in embryonic life, less than a month after fertilization and with the embryo only about 5 mm long, the lung appears as a groove located in the ventral wall of the gut. Development continues before and after birth, during the entire period of growth.[66]

The lung begins as a sheet of epithelium, becomes a simple sac by ventral fusion, then a solid mass of cells, and then develops into the deeply involuted structure of tubules described earlier. The initially cuboidal lining cells become the thin layer of alveolar epithelium prior to birth, with functions of the small airways in mucous secretion and surfactant secretion also developed by that time. At birth the lung weighs only about 60 gm, as compared with 700 gm in the adult.[67]

Prior to birth the lungs contain no air, the alveoli are collapsed, and the conducting passages are filled with fluid "lung liquid."[66] This fluid is formed in the lungs at a rate of about 3 ml/hr and exits via the trachea to the amniotic fluid or is swallowed.[65] In the human, surfactant secretion becomes evident in the amniotic fluid between the thirty-second and thirty-sixth week, then increasing until birth.[66] After lung expansion at birth the liquid is absorbed over the first 5 to 6 hours.[66]

Respiratory movements occur throughout much of gestation, beginning in lambs as early as 40 days.[66] These movements do not open the alveoli and cause essentially no movement of the fluid. During the period before birth, the central and peripheral reflex control mechanisms develop and integrate to the extent that the newborn infant, promptly after birth, shows the qualitative responses of an adult to chemical respiratory stimuli.[66]

In the fetus, from earliest development to birth, the system for respiration resembles that of a gilled marine organism. The fetal placenta serves as a gill, and its external environment is the placental blood of the mother. The placenta is the gas exchange organ, and the lungs are a ready reserve. Lungs exist, but most blood is shunted past them, with only about 7% to 10% passing through the pulmonary artery. This is because of the extremely high resistance of the heavily muscled pulmonary arteries in the otherwise well-developed pulmonary circulatory bed. The remainder of the blood from the right heart passes as a shunt to the left via the foramen ovale and ductus arteriosus.

The fetus lives and develops, adapted to what for an adult would be an intolerable hypoxic state. The umbilical vein delivers a Po_2 of about 32 mm Hg for distribution by the fetus to all its tissues. This is lower than the mixed venous Po_2 of the mother and is equivalent to residence at an altitude of about 25,000 ft, which is not humanly possible.

The changes and adaptions at birth are extreme and rapid. With the abrupt cessation of umbilical blood flow the cumulative stimuli and specific bombardments of the respiratory centers lead to a first breath. The collapsed, fluid-filled lung is expanded for the first time only by an extreme inspiratory force, generating a subatmospheric intrathoracic pressure of as much as 80 mm Hg. Expansion and filling of the surfactant-coated alveoli occurs, despite the requirement for moving fluid as well as air in this respiration.

At the beginning of breathing the increased oxygen pressure in the blood somehow affects the tone of the pulmonary vessels, and the gross shunting of blood past the lungs is overcome. By expansion and relaxation of the previously high-resistance pulmonary arteries the pulmonary circulation becomes and remains a low-resistance bed.[67] By this lowering of pressure in the right heart the foramen ovale valve is closed, evidently within minutes. The increased oxygen tension of arterial blood, through metabolic or related mechanisms not yet known, initiates contraction and closure of the ductus arteriosus. This is speeded by the fall in pressure in the pulmonary circulation that, by reversing the flow that had existed in the fetus, carries arterial rather than venous blood through the ductus and its own vessels.

In the normal infant, development of the lungs and respiratory function proceeds. Fetal hemoglobin, which had sustained the fetus through prenatal months of extreme hypoxia, is replaced over a period of about 4 months following birth. The number of alveoli increases from about 10 million at birth to about 300 million,[67] providing the area required for exercising states.

STRUCTURE OF THE AIRWAYS
Pulmonary air-conduction system

The trunk of the "pulmonary tree" is the *trachea,* which divides into two main *bronchi.* These in turn give rise to two smaller branches on the left and three on the right, corresponding to the lobes of the lung.

Within these major lobes the bronchi divide dichotomously,[51,52] again and again into successively smaller branches, finally into several million *terminal bronchioles.* Each of these continues into one, two, or more *respiratory bronchioles,* and each of these branches into 2 to 11 *alveolar ducts.* From these ducts the terminal, dominant gas exchange structures, the *alveolar sacs,* extend as simple groups of two to four or more *alveoli.*

The arrangement of the finer air passages and gas exchange surfaces is shown diagrammatically in Fig. 67-1. By electron microscopy and systematic statistical study of the human, it has been possible to estimate the dimensional characteristics and the number of these microscopic respiratory units.*

Up to and including the terminal bronchioles, the air-conducting system serves no important gas exchange function. The respiratory bronchioles have a few alveoli and can be

---

*See references 18, 37, 66, 67, 69, 70, and 71.

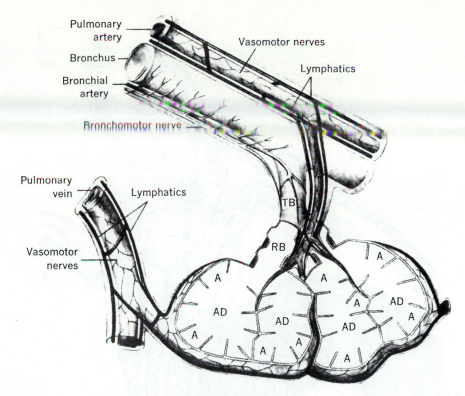

Fig. 67-1. Lung model showing simplified but anatomically accurate interrelations among structural elements of lung. Structures include alveolar gas space (*A,* alveoli; *AD,* alveolar ducts), alveolar walls, and loose collagen connective tissue spaces (gray overlays) containing pulmonary lymphatics. Terminal bronchiole *(TB)* and respiratory bronchiole *(RB)* are indicated. (From Staub.[65])

considered to have only a limited gas exchange function, mainly conducting air. From these to the alveoli themselves the gas conduction function is increasingly combined with respiratory gas exchange with the blood. The alveoli are purely respiratory, of such small size that they are transited by the diffusion of gas in hundredths of a second.

Structure of air passages

The larger units of the air-conducting passages are held open despite exposure to occasional increases in external pressure (forceful cough). Protection against collapse is provided by cartilage in the form of rings in the trachea and primary bronchi and of overlapping plates in the intrapulmonary bronchi. Bronchioles with a diameter of 1 mm or less do not have cartilage, and these finer passages are therefore easily collapsed in functional states. Smooth muscle is found in the walls of all the air passages, including the alveolar ducts. It is most abundant in the terminal bronchioles, where a complete obliteration

of the lumen can be produced by contraction of the muscle.[51,52] Peripheral to this point the muscle becomes more delicate, and the walls of the alveoli contain no functional muscle at all.[51,52,71]

The epithelium of the rigid tubes is columnar and bears cilia that beat toward the mouth; the surface is bathed with mucus secreted by goblet cells of the mucosa and by submucosal glands. No appreciable gas exchange can occur down to this point. In the respiratory bronchioles, which begin with a diameter of about 0.5 mm, the columnar epithelium of the air passages undergoes a transition to a flattened cuboidal type. A continuous, thin, single layer of flattened epithelium is present throughout the alveolar surfaces, covering the endothelium of the pulmonary capillaries.[49,67]

Alveoli

The number of alveoli increases from probably greater than 10 million at birth to

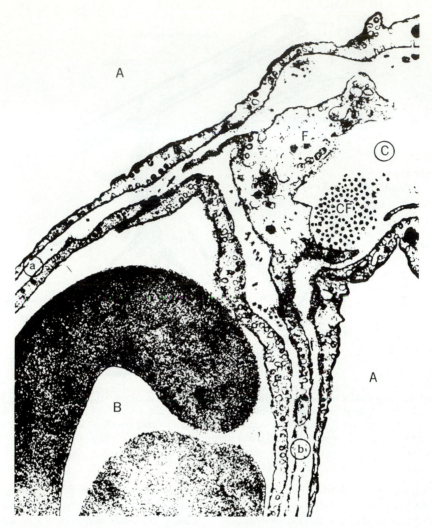

Fig. 67-2. Interalveolar septum showing alveolar-capillary membrane. *A,* Alveolar gas space; *B,* capillary; and *C,* space between capillaries containing fibroblast, *F,* and collagenous fiber, *CF.* The regions of interstitium shown include *a,* basement membranes of epithelium and endothelium, closely apposed, and *b,* fibroblast extension interposed between basement membranes. (Magnification ×28,000.) (From Weibel.[70])

about 300 million in the adult.[66,71] This more than 10-fold increase occurs mainly in the first 8 years.[31,66] Thereafter the increase in lung volume takes place largely by increases in the dimensions of existing alveoli. An average diameter for alveoli of about 250 μm in the adult combined with 8 μm for pulmonary capillaries provides between 60 and 80 m² of surface for gas exchange.[67,71]

The alveolocapillary membrane is now known to be composed of layers representing the involuted epithelium of the lung surface and the endothelium of the circulatory bed. Since

the thickness of this membrane, which separates gas from blood, is only about 0.2 μm, it was clearly visualized only by means of electron microscopy (Fig. 67-2). Gas molecules pass across a composite membrane of air-blood interface that consists of a single layer of alveolar epithelial cells, the basement membrane of these cells, a space containing elastic and collagenous connective tissue fibers, and the basement membrane and endothelial cells of the capillary.* This space may become distended by fluid in pa-

*See references 10, 11, 47, 67, and 71.

tients with pulmonary edema. Such fluid can be reabsorbed by the pulmonary capillaries or removed via the lymphatics, with which the lung is more richly supplied than almost any other organ.[11,68]

Pulmonary expansion

Increase in the volume of the lung on inspiration is predominantly due to a change in the alveolar volume with an insignificant volume change of the gas-conducting system. The increased volume of the lung occurs by expansion of individual patent alveoli and recruitment of atelectatic alveoli.[67]

Uniformity of expansion

The degree of expansion in various parts of the lung is not uniform, and "interregional differences in ventilation" result.[32,46,56] These are in part due to local differences in thoracic expansion and in part to position-related, gravity-dependent vertical gradients in pleural pressure.[46,56] Since inequalities of ventilation represent differences among alveoli, functional influence on gas exchange with pulmonary capillary blood can result. The significance of inequalities of alveolar ventilation and capillary perfusion are discussed in Chapter 68.

Pulmonary surfactant and surface tension of alveolar membrane

Any air-liquid interface, including biologic interfaces, has a surface tension that resists an increase in the dimensions of the surface and exerts force toward a decrease in surface area. Since alveoli resemble minute, communicating bubbles of gas in the lung fluid, they should have a powerful collapsing tendency that depends largely on the surface tension of the aqueous alveolar lining film and is countered by the subatmospheric intrapleural pressure.

The pressure (P) exerted by the surface tension of a bubble in a liquid is, from Laplace's law:

$$P = \frac{2T}{r} \qquad (1)$$

where T is the surface tension and r is the bubble radius. This indicates that as exhalation occurs and the alveolar radius decreases, the tendency of the alveolus to completely collapse should increase. Uncountered, this is an unstable condition of force in which atelectasis is inevitable and respiration could not continue; in fact, initial expansion of the lungs at birth could not occur.

Alveolar spaces persist because, beginning prior to birth, cells in the lung form and secrete a surface-active phospholipid, now designated surfactant,[25,26,61,66] that effectively reverses the aforementioned relationships of P, T, and r.

The surfactant is a complex structure, not yet fully identified in composition,[46,66] that reduces the surface tension of water in proportion to its concentration at the surface.

As the volume of an alveolus increases and decreases, its surface area does also. With a fixed *amount* of surfactant in the alveolar film the *concentration* of surfactant must increase on exhalation. The fortunate result is a *decrease* in collapsing tendency as the alveolar "bubble" becomes smaller, stabilizing the alveolus and overcoming the tendency for small gas spaces to disappear. On expansion at birth and thereafter, increase in volume of an alveolus decreases the concentration of surfactant on the membrane, allowing an increase in surface tension to contribute to alveolar stabilization.[27]

Surfactant is considered to be synthesized in the type II (compact) alveolar cells of the lung, which contain the inclusion bodies that led to their designation as granular pneumonocytes.[53] The inclusion bodies are presumed to contain the surface-active material, which must be continuously formed to replace that lost from the surfaces.

The contributions of surfactant are vital and stem from its single influence on alveolar surface tension. It promotes the stabilization of alveoli against collapse. By lowering alveolar surface tension, it simultaneously diminishes the force required for expansion (increases compliance) of the lung, thus reducing the work of breathing.

RESPIRATORY MOVEMENTS

The respiratory act of moving gas in and out of the lungs is fundamentally a neural and voluntary neuromuscular function, with the lungs and airways almost passively responding to these extrapulmonary influences. The neural drives are in turn controlled in response to intrinsic and whole-body metabolic functions, providing the feedback necessary to match ventilatory functions with requirements.

In some animals such as the frog, air is actually forced into the lungs by a process related to swallowing. In mammals, as a result of the force exerted by the respiratory muscles, the pressure exerted by the thoracic walls on the pleural surfaces of the lungs becomes alternately lower and higher. Since the air spaces of the lungs are normally in free communication with the outside air through the respiratory passages and since the pressure of the outside air remains constant, air enters the lungs when the intrathoracic pressure is lowered and is forced out when the muscular action raises pressure within the chest. When all the respiratory muscles are at rest, the thorax assumes the position of passive expiration. Deviations from this resting position involve the active contraction of inspiratory or expiratory muscles.

Expansion and contraction of thoracic cavity

Thoracic enlargement on inspiration is produced both by diaphragmatic contraction and by movements of the ribs, sternum, and vertebrae. The expansion is mainly downward because of contraction of the diaphragm, laterally via elevation of the ribs, and anteriorly because of the upward and anterior movement of the sternum. No expansion occurs at the posterior aspect of the lung apices or at the vertebral attachments of the ribs.[45,51,52]

Diaphragmatic action

The domed shape of the diaphragm exists despite the weight of the liver suspended from it. This shape is due partly to the elastic properties of the lungs, which, by reducing intrathoracic pressure at rest by approximately 3 mm Hg/cm² over the more than 250 cm² area of the diaphragm, cause a total upward force of about 1 kg to mold the diaphragm into the thorax. In its contraction the diaphragm behaves like a piston that moves downward and forward, enlarging the thoracic volume to be filled by expansion of the lungs. If the 250 cm² diaphragm descends 1 cm, there will be an increase in thoracic and lung volume of 250 ml. The excursion of the dome with resting inspiration varies greatly among normal individuals, with 12.5 mm being the average inspiratory descent of the whole diaphragm.[45]

The innervation of the diaphragm (the two phrenic nerves) comes from the third and fourth cervical segments of the spinal cord, whereas the intercostal, expiratory, and majority of the accessory inspiratory muscles are innervated from the thoracic and lumbar segments. The diaphragm therefore may remain active when the latter muscles have been completely paralyzed by a transverse lesion of the spinal cord in the upper thoracic or lower cervical region or by spinal anesthesia rising to these levels. Each phrenic nerve innervates its half of the diaphragm; therefore a phrenic nerve may be purposely divided to reduce the movements of a lung, as in pulmonary tuberculosis.

Movements of the ribs

Because of the angle at which the ribs are attached to the vertebral column, the progressive increase in the length of the ribs from above downward, and the outward rotation associated with an upward pull, any upward movement of the ribs means an increase (inspiration) and any depression means a diminution (expiration) in the size of the thoracic cavity. The *inspiratory muscles* therefore are mainly those that run from the head and neck (scaleni and sternocleidomastoideus) or from the arms to the ribs (pectoralis minor) and those that run from rib to rib in a downward and inward direction (intercostales externi).

The *expiratory muscles* are those that run from the pelvis to the sternum (rectus abdominis), to the ribs posteriorly (iliocostalis lumborum and quadratus lumborum), or from the vertebrae upward to the ribs (serratus posterior inferior); the triangularis sterni runs upward and outward on the inner surface of the thorax from the sternum to the third to sixth ribs and therefore depresses the ribs; the intercostales interni run downward and outward almost perpendicularly to the intercostales externi, and contraction of their fibers most probably depresses the ribs. The external and internal intercostal muscles are neurally activated alternately, the external intercostals with inspiration and the internal intercostals with expiration.[21] The strongest expiratory influence, however, is the contraction of the muscles of the abdominal wall, which presses the abdominal contents upward against the diaphragm, thus decreasing the vertical diameter of the thoracic cavity.

Eupnea and hyperventilation

Eupnea, the normal, quiet respiration at rest, is ordinarily carried out without the individual being aware of it. Approximately a third of the air movement in a single inspiration is accomplished by diaphragmatic descent, the rest by active elevation of the rib cage. Inspiration at rest is accomplished by contraction of the diaphragm and the external intercostals; no other muscles participate actively, although there are some associated movements such as expansion of the nostrils and abduction of the vocal cords. The expiratory muscles are inhibited automatically when the inspiratory muscles contract. The expiratory phase of eupnea involves no active muscular contraction; there are changes in the tone of the expiratory muscles—a decrease during inspiration and an increase during expiration—but there are none of the active expiratory movements that characterize increased respiration.

Hyperventilation, whether voluntary or the result of moderate muscular exercise, is not necessarily an uncomfortable experience. Abnormal respiratory discomfort is termed *dyspnea.* (For the causes of difficult and uncomfortable breathing, see Chapter 73.)

When it becomes necessary for the individual to move more air in and out of the lungs, the contractions of the diaphragm and external intercostals increase in strength, and the muscles that run from the ribs and sternum upward—the accessory muscles of inspiration—come into play. The expiratory muscles also begin to contract actively, so as to aid respiration and make it more complete. When breathing has been increased enough to bring the expiratory muscles into action, the accessory muscles of inspiration are also active.

PRESSURE CHANGES IN THE LUNGS AND THORAX

Under normal conditions, from intrauterine life to death, the outer surfaces of the lungs are always in contact with the inner surfaces of the thoracic cavity. The lungs of the fetus contain no air, and their volume is small; nevertheless, they fill the retracted chest completely. The first successful inspiratory effort after birth is due to an extremely powerful contraction of the diaphragm. As a result of the approximately 80 mm Hg subatmospheric pressure generated by this first inspiration, the lungs are stretched and the elastic tissue that they contain never again returns to its original length.

The muscular force required to accomplish the first inhalation is 15 to 20 times that required for a normal inhalation.[28] The continuous counterforce toward lung collapse accounts for the *subatmospheric pressure* that normally exists between the inner surfaces of the thorax and the outer surfaces of the lungs. It is due in part to the tension of the stretched elastic fibers in the lung parenchyma and in part to the surface tension of the alveolar surface lining film of phospholipid surfactant.[25,26,61]

Intrathoracic pressure can be measured by means of a manometer connected to a needle inserted through an intercostal space or by measuring the pressure within a balloon in the lower third of the esophagus. It is not the same at all locations, but during quiet inspiration it amounts to about −5 mm Hg.[29]

Intrapulmonary pressure is the pressure existing in the air passages and alveoli within the lungs. In the absence of air movement and with the glottis open the alveolar pressure should equal atmospheric pressure. During normal breathing, intrapulmonary pressure ranges from about −3 mm Hg during inspiration, through atmospheric pressure at the end of inspiration, to about +3 mm Hg at the beginning of expiration and again returns to atmospheric pressure at the end of expiration (Fig. 67-3). These fluctuations are due in part to the fact that the air passages interpose some resistance to the flow of air associated with inspiration and expiration. The limiting location for flow is apparently in the smallest bronchioles between the segmental bronchi and alveoli, but upper airway resistance may be responsible for a considerable proportion of the total airway resistance.[40,50] The degree of fluctuation of intrapulmonary pressure will vary more in the alveoli than in the bronchi or trachea.

The normal intrapulmonic pressure fluctuations can be greatly enhanced by vigorous contraction of the inspiratory and expiratory muscles (Fig. 67-3), with the greatest changes being elicited by forced inspiratory or expiratory effort while breath holding. The values obtainable by such effort in normal subjects approximate −90 and +110 mm Hg, respectively.[13,42,62] Pressures of this order will produce acute circulatory depression if applied for more than brief periods; the safe range for any but brief exposures is from about −20 to +30 mm Hg.[62]

The reason for the circulatory effects is that changes in *intrapulmonary* pressure are usually associated with corresponding changes in *intrathoracic* pressure, and these are exerted in turn on the thin-walled veins in the thoracic cavity. When, as in the Valsalva maneuver or positive-pressure artificial ventilation, the intrathoracic pressure exceeds the blood pressure inside the veins, the veins will collapse, and the flow within them will stop until enough blood accumulates distal to the obstruction (for venae cavae) to raise the venous pressure sufficiently to overcome the abnormal extravascular pressure. Meanwhile, right heart filling is impeded, cardiac output and blood pressure fall sharply, and unconsciousness may supervene. The ability of the venous pressure to rise, therefore, is the main factor determining the amount of positive intrapulmonary pressure that can be tolerated.

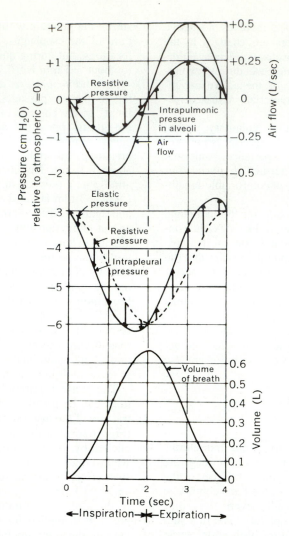

Fig. 67-3. Manner in which intrapulmonic (alveolar) pressure, intrapleural (intrathoracic) pressure, and rate of air flow change during cycle of inspiration and expiration at rest. During exercise, cycle duration would be shortened and magnitudes of all parameters grossly increased. Reduction in intrapulmonic pressure during inspiration is related to resistance to air flow through respiratory passages (resistive pressure). An additional factor (elastic pressure, due to increased stretch of elastic tissues of lungs and thorax) causes inspiratory change of intrapleural pressure to exceed change of intrapulmonic pressure. (From Perkins.[60])

STATIC VOLUMES AND DYNAMIC VOLUME CHANGES IN THE LUNGS

Pulmonary ventilation is a dynamic process best described in terms of the rate of movement of gas volume (V/t). However, certain additional designations, including the static parameters, are necessary for convenience in describing the characteristics of the lung and pulmonary function[1] (Fig. 67-4).

Designations of lung volume components

1 *Tidal air* (V_T) is the volume of air that enters and leaves the lungs at each natural respiratory effort at rest or in other states.
2. *Inspiratory reserve volume* (IRV) is the volume of air that can be taken in by a maximal inspiratory effort over and above the inspired tidal air.

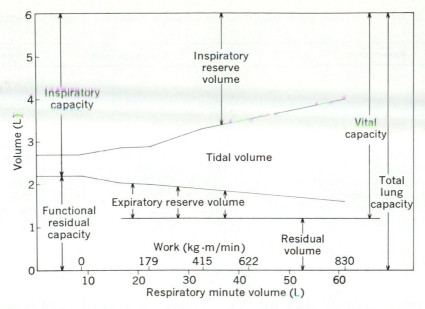

Fig. 67-4. Subdivisions of lung air in healthy young males in sitting or standing position and in work. When such subjects change to recumbent (supine) position, the following approximate corrections must be made in resting values: expiratory reserve decreases 14% of vital capacity; inspiratory reserve increases 11% of vital capacity; vital capacity decreases 3%; total capacity decreases 3%. (From Bates and Christie.[3])

3. *Expiratory reserve volume* (ERV) is the volume of air that can be expelled by the most powerful additional expiratory effort after the tidal air has been allowed to escape naturally.
4. *Vital capacity* (VC) is the sum of the tidal air and the inspiratory and expiratory reserves (i.e., the volume of air that can be expelled by the most vigorous possible expiratory effort after the deepest possible inspiration).[6]
5. *Residual volume* (RV) is the volume of air remaining in the lungs after the strongest possible expiration. It cannot be measured directly but can be determined by indirect methods.[5]
6. *Total lung capacity* (TLC) is the sum of the vital capacity and the residual air.
7. *Functional residual capacity* (FRC) is the volume of air left in the lungs at the end of a natural, unforced expiration.
8. *Pulmonary midcapacity* is the functional residual volume together with half the unforced tidal volume.
9. *Inspiratory capacity* (IC) is the volume that can be inspired from the end-expiratory level.

The magnitudes of the static lung volumes vary with body size and age[1] and closely correlate with the cube of body height.[43] Table 67-1 summarizes approximate normal values, which, of course, differ greatly from individual to individual.

Dynamic functions—changes in volume per unit time

The *respiratory minute volume* is the product of *tidal volume* and *frequency* of respiration per minute. In normal individuals, this value is representative of the overall output of the neural respiratory control mechanisms. The respiratory minute volume is composed of the *alveolar ventilation* per minute and the *dead space ventilation* per minute. The greatest rate of pulmonary ventilation that can be sustained for about 20 sec (a period short enough to obviate factors of respiratory fatigue and alkalosis) is termed the *maximal breathing capacity* or *maximum ventilatory volume*. It provides a useful overall measure of the ability to increase ventilation. *Timed* or *forced expiratory and inspiratory vital capacity* measurements of the rate at which a maximal inspiration can be completely exhaled offer a means of evaluating muscle power, pulmonary/thoracic compliance, and overall resistance to air movement.

The relationships among lung volumes are shown diagrammatically in Fig. 67-4, which also

Table 67-1. Static and dynamic pulmonary functions (average values in recumbent normal young men)

Measurement	Symbol	Average value	Reference
Total lung capacity	TLC	5.97 L	Kaltreider et al.[44]
Tidal volume	V_T	0.50 L	Comroe et al.[6]
Vital capacity	VC	4.78 L	Kaltreider et al.[44]
Inspiratory capacity	IC	3.79 L	Kaltreider et al.[44]
Expiratory reserve volume	ERV	0.98 L	Kaltreider et al.[44]
Functional residual volume	FRV	2.18 L	Kaltreider et al.[44]
Residual volume	RV	1.19 L	Kaltreider et al.[44]
Peak expiratory flow	PEF	500 L/min	Hyatt[39]
Maximal inspiratory flow rate (ATPS)		300 L/min	Comroe et al.[6]
Maximum breathing capacity	MBC	130 L/min	Baldwin et al.[20]
Pulmonary washout (time constant)		38 sec	Unpublished data

indicates the increase in minute volume and tidal air during exercise. The tidal air increases from about 12% of the vital capacity at rest to about 50% during heavy work. The respiratory minute volume during exercise can change as much as 8- to 10-fold from the resting value. The difference is accounted for by an increase in respiratory rate as well as depth during exercise. The increase in tidal air is derived by using portions of both the inspiratory and expiratory reserves. True residual air cannot change, but the functional residual air decreases as tidal volume rises. Tidal air greatly in excess of 50% of the vital capacity call for such vigorous contractions of the respiratory muscles that they cannot be continued without respiratory exhaustion. Further increments of pulmonary ventilation above this point are usually secured by progressive increases in respiratory rate.

RELATIONSHIPS OF PRESSURE, FLOW, AND VOLUME CHANGES

Gas flow through the airways is produced by the pressure gradients that can be generated there. Maximal rates of flow attainable and the degree of hyperventilation that can be sustained depend further on the work involved in the overall respiratory act. Respiratory work, in turn, is determined by the characteristics of the respired gas, the airways themselves, and the mechanical characteristics of the thorax.

Resistance to breathing

The total force exerted by respiratory muscles overcomes several types of resistance in accomplishing ventilation of the pulmonary alveoli.

Airway resistance is modified not only by the diameter and length of the respiratory airways but also by the nature of the gas breathed (i.e., helium-oxygen versus nitrogen-oxygen mixtures), factors affecting the viscosity and density of the gas (i.e., the deep-diving versus the high-altitude environment), and turbulence factors associated with high velocities of gas movement. In a situation involving a strictly laminar gas flow, pressure and flow are linearly related according to Poiseuille's law, which relates flow to the viscosity of the gas. Volume flow per unit time (cm^3/sec) through a passage permitting laminar flow is shown as follows:

$$\dot{V} = \frac{P}{(8 \times 1/\pi r^4)\mu} = \frac{P}{K_1} \tag{2}$$

in which P is the pressure gradient in dynes/cm², l the length, and r the radius of the passage in centimeters; the term μ describes the viscosity of the gas itself in poises.

Since flow through most of the respiratory passages is now believed to be turbulent rather than laminar,[36] an additional equation is required to predict resistance under unusual environmental circumstances. For turbulent flow the volume flow rate, again in cubic centimeters per second, is related to pressure as follows:

$$\dot{V} = \frac{P}{(f \times 1/4\pi^2 r^5)} = \frac{P}{K_2} \tag{3}$$

in which l, r, and π describe the dimensions of the passage and the new factor f relates to the friction between the gas and the wall of the passage.

Rohrer's equation[63] combines these separate

influences of purely laminar and turbulent flow as follows:

$$P = (K_1 \times \dot{V}) + (K_2 \times \dot{V}^2) \qquad (4)$$

to represent the "impure" situation that exists in normal physiologic states. K_1 is the constant for laminar flow from equation 2 and *depends on the viscosity of the gas*. K_2 is the constant for turbulent flow and *depends on gas density*.

Resistance to air flow through the respiratory passages during the act of breathing *(airway resistance)* is measured by simultaneous determinations of the pressure change between the alveoli and ambient air and the volume rate of gas flow between the lungs and the external environment.[7] The mean resistance of all air passages within the normal lung, expressed as *pressure/flow* or $P/\dot{V}$, approximates 2 cm H_2O/L/sec.[60]

A second component of resistance to respiration is the frictional and elastic resistance offered to movements or distortions of the lung tissue *(lung tissue resistance)* and the chest wall *(chest wall resistance)*. Airway and lung tissue resistance together are termed *pulmonary resistance,* whereas the term "*thoracic resistance*" refers to the sum of the three subdivisions cited.[7]

During respiration at sea level, gas density and viscosity are constant, and the resistance of the respiratory passages to the flow of air is unimportant under normal conditions of rest or exercise. This is probably because the smooth muscle in the bronchial walls is subject to autonomic nerve control and can therefore relax progressively as the demand for pulmonary ventilation increases. Any impediment such as contraction of the bronchial muscle, congestion of the bronchial mucous membrane, accumulation of secretions, foreign bodies, or disease processes impinging on the air passages has the effect of increasing the fluctuations in intrapulmonary and intrathoracic pressure associated with the same rate of alveolar ventilation. Changes in flow resistance accompany these sometimes severe and lethal obstructive changes in the respiratory passages (as in asthma). Normal individuals exposed to extreme gas density (as in deep diving) experience gross increases in resistance within otherwise normal airways.

Static pulmonary pressure-volume relationships

Compliance and its reciprocal, *elastance,* are terms used to designate the degree to which lung volume can be changed by imposed alterations of intrapulmonary pressure. Compliance is a measure of the ability of the lungs and thoracic cage to be expanded and is defined as $\Delta V/\Delta P$ in liters per centimeter of H_2O and normally has a value of about 0.23 L/cm H_2O[38,58]; elastance would then be 1/0.23 or about 4.5 cm H_2O/L. These are static relationships not affected by the velocity of air movement. Interest in assessing changes in compliance led to the evolution of pressure-volume diagrams, in which measured values pertaining to the distensibility of the respiratory structures are graphically illustrated.[62]

Respiratory gas-flow rates

In normal respiration at rest the mass flow of a tidal inspiration through the approximately 14 million alveolar ducts should not be rapid and should approach zero in the alveoli. However, in smaller bronchi and during a violent expiratory effort such as a cough, gas velocity equal to the speed of sound (Mach 1) is apparently achieved.[64] Under this circumstance of "critical flow," gas molecules all travel at the same speed, and the flow rate cannot rise even if the pressure gradient is grossly increased.

Maximal attainable flow rates for the entire lung can be directly measured, and in normal young men range from 300 to 800 L/min, with an average of about 684 L/min.[41,64] Such flow rates are achieved in performing the *timed vital capacity* and are approached in measuring the *maximal breathing capacity*. They are well above the maximum flows for normal exercise or respiratory stimulation by CO_2.

Work of breathing

Breathing involves application of *force* over *distance*. The respiratory muscles do *work* in producing the required changes in intrapulmonary pressure. This work is the result of changes in the size and configuration of the thorax produced against forces that resist the change.

The work of breathing can also be expressed in terms of *pressure* times *volume*, using an intraesophageal balloon or body plethysmograph to provide an estimate of the transpulmonary pressure throughout the respiratory cycle.[13]

The overall mechanical work accomplished by the respiratory muscles is made up of the following components:

Elastic work is related to the force required to stretch elastic tissues of the lungs and thorax. In the normal individual at rest the force required for inspiration is almost entirely related to overcoming the elastic resistance of the lungs and chest wall.

Inertial work is related to the force required to set tissues (e.g., liver and chest wall) in motion. In its extreme, it is exemplified by the situation in which a self-contained underwater swimmer or a liquid-breathing patient must, with each breath, displace a volume of water equal to his tidal volume. Under normal circumstances, inertial force and work are negligible.[30,55] Work is influenced by and is greatest at high rates of ventilation.

Work due to airway resistance is related to the force required to move respiratory gases through the airways. This is termed *flow-resistive work*[13] and is negligible at rest. During exercise or in other situations of vigorous alveolar ventilation when the velocity of air movement is high, flow-resistive work increases due to an exaggeration of the influences of air turbulence within the respiratory passages. This form of work can also become extreme, even at rest, in bronchial asthma or laryngeal edema due to narrowed air passages. At great diving depths such as 1,200 to 1,600 ft of seawater, the increased density of compressed

gas introduces an additional cause of flow-resistive work of breathing (Chapter 75). Finally, when the fluid breathed is a water solution used for lung lavage,[48] both flow-resistive and inertial work should be grossly increased.

Overall work of quiet breathing at rest in normal subjects ranges from about 0.3 to 0.6 kgm/min.[13,23,35,57] There is apparently a linear relationship between the increase in work of the inspiratory but not the expiratory muscles, and the magnitude of an experimentally induced rise in level of the alveolar P_{CO_2} (CO_2 administration).[57]

Oxygen (metabolic) cost of breathing

The work of breathing can also be expressed as "O_2 cost" in normal individuals and in patients with pulmonary disability.[13,23,35] In normal subjects the metabolic cost of pulmonary ventilation is approximately 0.5 to 1 ml O_2/L of pulmonary ventilation.[13]

This aspect of ventilatory efficiency is illustrated in Fig. 67-5, which shows that in the normal subject a large increase in ventilation can be accomplished at a low expenditure of energy for

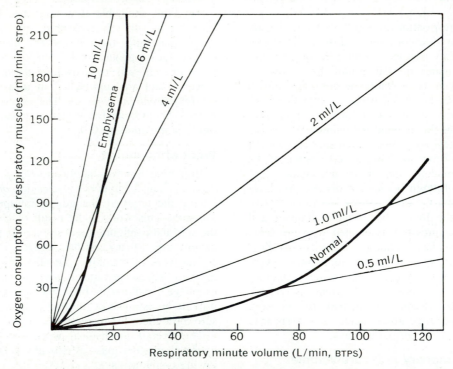

Fig. 67-5. O_2 cost of breathing in normal subject and in subject with emphysema, measured as increase in whole-body O_2 consumption during respiratory response to added dead space. Isopleths indicate ranges of O_2 cost per liter of respiratory volume and thus indicate efficiency of respiratory work. Up to about 40 L/min. efficiency of respiratory effort in patient with emphysema was less than one tenth that of normal subject. Cost of breathing in normal men over usual range of activity is less than 0.5 ml O_2/L air breathed. (Modified from Campbell et al.[23])

respiration. However, the patient with obstructive emphysema, limited in ability to increase his ventilation, is handicapped by the greater mechanical work that he must do to produce even a slight increase in breathing.[33-35] Such disturbances in the mechanics of breathing also produce alterations in the efficiency of pulmonary gas exchange, in the transport of respiratory gases, and even in the control of respiration.

Efficiency of respiration

The mechanical efficiency of breathing is defined as the respiratory work performed per unit O_2 required for the work, or work per ventilation liter/O_2 cost per liter of gas moved. In normal subjects the efficiency of the respiratory act has been estimated to average about 8%.[22,24,59] It undoubtedly depends on the frequency and depth of breathing and hence differs between rest and severe exercise. Nevertheless, in exercise the energy cost of breathing has been found to represent no more than about 3% of the total energy work rate of the subject.[54]

Respiratory decompensation

From the foregoing descriptions of forces, resistance, and work and the incessant neuromuscular function involved in respiration, it is evident that circumstances can occur in which the muscles of respiration can, like other voluntary muscles, become fatigued. Sustained strenuous respiratory work against sustained severe respiratory resistance must lead to respiratory decompensation. This is expected in certain pathologic states. It is conceivable also in normal, healthy individuals exposed for many days to working environments of extreme density (deep undersea).

ASSESSMENT OF PULMONARY FUNCTION

Although this chapter is concerned primarily with the normal mechanical functions of the lung, many types of derangements of pulmonary ventilation occur in disease. Quantitative assessments of these derangements require methods of measurement comparable to those used in the study of normal physiology. They may be found in several excellent sources.[5,9,14,17]

REFERENCES
General reviews

1. Agostoni, E., and Mead, J.: Statistics of the respiratory system. In Fenn, W. O., and Rahn, H., editors: Handbook of physiology. Respiration section, Baltimore, 1964, The Williams & Wilkins Co., vol. 1.

2. Altman, P. L., Gibson, J. F., and Wang, C. C.: Handbook of respiration, Philadelphia, 1958, W. B. Saunders Co.
3. Bates, D. V., and Christie, R. V., editors: Respiratory function in disease, Philadelphia, 1964, W. B. Saunders Co.
4. Campbell, E. J. M.: The respiratory muscles and the mechanics of breathing, Chicago, 1958, Year Book Medical Publishers, Inc.
5. Comroe, J. H., Jr.: Physiology of respiration, Chicago, ed. 2, 1975, Year Book Medical Publishers, Inc.
6. Comroe, J. H., Jr., et al.: The lung, ed. 2, Chicago, 1962, Year Book Medical Publishers, Inc.
7. DuBois, A. B.: Resistance to breathing. In Fenn, W. O., and Rahn, H., editors: Handbook of physiology. Respiration section, Baltimore, 1964, The Williams & Wilkins Co., vol. 1.
8. Engel, S.: Lung structure, Springfield, Ill., 1962, Charles C Thomas, Publisher.
9. Fenn, W. O., and Rahn, H., editors: Handbook of physiology. Respiration section, Baltimore, 1964, The Williams & Wilkins Co., vol. 1.
10. von Hayck, H.: The human lung, New York, 1960, Hafner Publishing Co., Inc. (Translated by V. E. Krahl.)
11. Krahl, V. E.: Anatomy of the mammalian lung. In Fenn, W. O., and Rahn, H., editors: Handbook of physiology. Respiration section, Baltimore, 1964, The Williams & Wilkins Co., vol. 1.
12. Macklem, P. T.: Respiratory mechanics, Annu. Rev. Physiol. **40:**157, 1978.
13. Otis, A. B.: The work of breathing. In Fenn, W. O., and Rahn, H., editors: Handbook of physiology. Respiration section, Baltimore, 1964, The Williams & Wilkins Co., vol. 1.
14. de Reuck, A. V. S., and O'Connor, M., editors: Ciba Foundation Symposium on pulmonary structure and function, Boston, 1962, Little, Brown, & Co.
15. Robin, E. D.: Overview: humoral agents processed in lung with extrapulmonary effects. In Robin, E. D., editor: Extrapulmonary manifestations of respiratory disease, New York, 1978, Marcel Dekker.
16. Shephard, R. J.: Some observations on peak expiratory flow, Thorax **17:**39, 1962.
17. Symposium: Entretiens de physiopathologie respiratoire, Paris, 1962, Vigot Frères, Series 5.
18. Weibel, E. R.: Morphometrics of the lung. In Fenn, W. O., and Rahn, H., editors: Handbook of physiology. Respiration section, Baltimore, 1964, The Williams & Wilkins Co., vol. 1.
19. Willson, H. G.: The terminals of the human bronchiole, Am. J. Anat. **30:**267, 1922.

Original papers

20. Baldwin, E. deF., Cournand, A., and Richards, D. W., Jr.: Pulmonary insufficiency; physiological classification, Clinical methods of analysis, standard values in normal subjects, Medicine **27:**243, 1948.
21. Bronk, D. W., and Ferguson, L. K.: The nervous control of intercostal respiration, Am. J. Physiol. **110:**700, 1935.
22. Cain, C. C., and Otis, A. B.: Some physiological effects resulting from added resistance to respiration, J. Aviation Med. **20:**149, 1949.
23. Campbell, E. J. M., Westlake, E. K., and Cherniack, R. M.: Simple methods of estimating oxygen consumption and efficiency of the muscles of breathing, J. Appl. Physiol. **11:**303, 1957.

24. Campbell, E. J. M., Westlake, E. K., and Cherniack, R. M.: The oxygen consumption and efficiency of the respiratory muscles of young male subjects, Clin. Sci. **18**:55, 1959.

25. Clements, J. A.: Surface phenomena in relation to pulmonary function, Physiologist **5**:11, 1962.

26. Clements, J. A.: Surface tension in the lungs, Sci. Am. **207**:120, 1962.

27. Clements, J. A.: Pulmonary surfactant, Am. Rev. Respir. Dis. **101**:984, 1970.

28. Cook, C. D., et al.: Studies of respiratory physiology in the newborn infant. III. Measurements of mechanics of respiration, J. Clin. Invest. **36**:440, 1957.

29. Daly, W. J., and Bondurant, S.: Direct measurement of respiratory pleural pressure changes in normal man, J. Appl. Physiol. **18**:513, 1963.

30. DuBois, A. B., et al.: Oscillation mechanics of lungs and chest in men, J. Appl. Physiol. **8**:587, 1956.

31. Dunnill, M. S.: Postnatal growth of the lung, Thorax **17**:329, 1962.

32. Engel, L. A., and Macklem, P. T.: Gas mixing and distribution in the lung. In Widdicombe, J. G., editor: International review of physiology. Respiratory physiology II, Baltimore, 1977, University Park Press.

33. Filley, G. F., Bower, G. C., and Mitchell, R. S.: The morphologic basis of pulmonary mechanics, Am. Rev. Respir. Dis. **81**:734, 1960.

34. Fowler, W. S.: Mechanics of ventilation in emphysema, Am. Rev. Respir. Dis. **80**:118, 1959.

35. Fritts, H. W., Jr., et al.: The efficiency of ventilation during voluntary hyperpnea: studies in normal subjects and in dyspneic patients with either chronic pulmonary emphysema or obesity, J. Clin. Invest. **38**:1339, 1959.

36. Gaensler, E. A., Maloney, J. V., Jr., and Björk, V. O.: Bronchospirometry. II. Experimental observations and theoretical considerations of resistance breathing, J. Lab. Clin. Med. **39**:935, 1952.

37. Gehr, P., Bachofen, M., and Weibel, E. R.: The normal human lung: ultrastructure and morphometric estimation of diffusion capacity, Respir. Physiol. **32**:121, 1978.

38. Granath, A., Horie, E., and Linderholm, H.: Compliance and resistance of the lungs in the sitting and supine positions at rest and during work, Scand. J. Clin. Lab. Invest. **11**:226, 1959.

39. Hyatt, R. E.: Dynamic lung volumes. In Fenn, W. O., and Rahn, H., editors: Handbook of physiology. Respiration section, Baltimore, 1964, The Williams & Wilkins Co., vol. 1.

40. Hyatt, R. E., and Wilcox, R. E.: Extrathoracic airway resistance in man, J. Appl. Physiol. **16**:326, 1961.

41. Higgins, I. T. T.: Respiratory symptoms, bronchitis, and ventilatory capacity in random sample of an agricultural population, Br. Med. J. **2**:1198, 1957.

42. Johnson, L. F.: Maximum pressure-volume relationships of the human respiratory system, Technical Documentary Report No. SAM-TDR-64-21, Washington, D.C., 1964, USAF School of Aerospace Medicine.

43. Jouasset, D.: Normalisation des épreuves fonctionnelles respiratoires, dans les pays de la communauté européenne du charbon et de l'acier, Poumon Coeur **16**:1145, 1960.

44. Kaltreider, N. L., Fray, W. W., and Phillips, E. W.: Effect of age on total pulmonary capacity and its subdivisions, Am. Rev. Tuberc. **37**:662, 1938.

45. Keith, A.: The mechanism of respiration in man: further advances in physiology, London, 1909, Edward Arnold & Co.

46. King, R. J., and Clements, J. A.: Surface active materials from dog lung. II. Composition and physiological correlations, Am. J. Physiol. **223**:715, 1972.

47. Krahl, V. E.: Observations upon the pulmonary alveolus and its capillary circulation in the living rabbit, Anat. Rec. **142**:350, 1962.

48. Kylstra, J. A.: Survival of submerged mammals, N. Engl. J. Med. **272**:198, 1965.

49. Low F. N.: The pulmonary alveolar epithelium of laboratory mammals and man, Anat. Rec. **117**:241, 1953.

50. Macklem, P. T., Fraser, R. G., and Bates, D. V.: Bronchial pressures and dimensions in health and obstructive airway disease, J. Appl. Physiol. **18**:669, 1963.

51. Macklin, C. C.: The musculature of the bronchi and lungs, Physiol. Rev. **9**:1, 1929.

52. Macklin, C. C.: Pulmonic alveolar epithelium, J. Thorac. Surg. **6**:82, 1936.

53. Macklin, C. C.: The pulmonary alveolar mucoid film and the pneumocytes, Lancet **1**:1099, 1954.

54. Margaria, R., et al.: Mechanical work of breathing during muscular exercise, J. Appl. Physiol. **15**:354, 1960.

55. Mead, J.: Measurement of inertia of the lungs at increased ambient pressure, J. Appl. Physiol. **9**:208, 1956.

56. Milic Emili, J.: Pulmonary statistics. In Widdicombe, J. G., editor: MTP international review of science, physiology series. I. Respiratory physiology I, Baltimore, 1974, University Park Press, vol. 2, p. 105.

57. Milic-Emili, J., and Tyler, J. M.: Relation between work output of respiratory muscles and end-tidal CO_2 tension, J. Appl. Physiol. **18**:497, 1963.

58. Naimark, A., and Cherniack, R. M.: Compliance of the respiratory system and its components in health and obesity, J. Appl. Physiol. **15**:377, 1960.

59. Otis, A. B., Fenn, W. O., and Rahn, H.: The mechanics of breathing in man, J. Appl. Physiol. **2**:592, 1950.

60. Perkins, J. F.: Respiration. In Encyclopaedia Brittanica, Chicago, 1961, Encyclopaedia Brittanica, Inc.

61. Polgar, G.: The first breath: a turbulent period of physiologic adjustment, Clin. Pediatr. **2**:562, 1963.

62. Rahn, H., Otis, A. B., Chadwick, L. E., and Fenn, W. O.: The pressure-volume diagram of the thorax and lung, Am. J. Physiol. **146**:161, 1946.

63. Rohrer, F.: Der Stromungswiderstand in den menschlichen Alemwegen und der Einfluss der unregelmassigen Verzweigung des Bronchialsystems auf den atmungsverlauf verschiedenen Lungenbezirken, Arch. Gesamte Physiol. **162**:225, 1915.

64. Ross, B. B., Gramiak, R., and Rahn, H.: Physical dynamics of the cough mechanism, J. Appl. Physiol. **8**:264, 1955.

65. Staub, N. C.: The pathophysiology of pulmonary edema, Hum. Pathol. **1**:419, 1970.

66. Strang, L. B.: Growth and development of the lung: fetal and post-natal, Annu. Rev. Physiol. **39**:253, 1977.

67. Thurbeck, W. M.: Structure of the lungs. In Widdicombe, J. G., editor: International review of physiology. Respiratory physiology II, Baltimore, 1977, University Park Press, vol. 14.

68. Tobin, C. E.: Lymphatics of the pulmonary alveoli, Anat. Rec. **120**:625, 1954.

69. Weibel, E. R.: Morphometrische Analyse von Zahl, Volumen und Oberflache der Alveolen und Kapillaren der Menschlichen Lunge, Z. Zellforsch. **57**:648, 1962.

70. Weibel, E. R.: Morphological basis of alveolar-capillary gas exchange, Physiol. Rev. **53**:419, 1973.

71. Weibel, E. R., and Gomez, D. M.: Architecture of the human lung, Science **137**:577, 1962.

68

CHRISTIAN J. LAMBERTSEN

Gas exchanges of the atmosphere with the lungs and blood

EVOLUTION OF OUR ATMOSPHERE

Four billion years ago, during the formation of our solar system, the gaseous composition of the nebular embryonic stage of this planet probably resembled that of the sun, where hydrogen and helium are predominant[43] as the precursor and product of the fusion reactions that are the ultimate source of all our energy. Conceivably the gas mantle that initially evolved around the solidifying earth contained largely hydrogen, helium, oxygen, and nitrogen, with oxygen amounting only to about 0.1%, nitrogen still less, and the sum of hydrogen and helium providing about 99.86% of the total.[43,63] As the earth cooled from solar temperatures, most of the original small concentration of atomic oxygen combined with hydrogen to form water molecules so that free oxygen was unavailable in the earth's envelope. It is presumed that the biologically adequate concentration of oxygen in the present atmosphere or "breathing sphere" of earth developed in two stages. In the first eons, ultraviolet energy from the sun continuously and transiently disassociated small amounts of gaseous water into oxygen and hydrogen atoms to reestablish a supply of molecular oxygen. In the continued presence of a great excess of free hydrogen, this oxygen could only form stable water again. However, as more and more of the low-mass hydrogen atoms escaped from earth, small amounts of free oxygen could begin to accumulate in the atmosphere. The second (and present) stage of atmospheric evolution is presumed to have followed the origins of life and the development of photosynthesis,[63] probably beginning about 400 million years ago. Presumably initial life was supported by anaerobic metabolic processes only. The photosynthetic process, by producing molecular oxygen and by fixing hydrogen

to carbon, helped to establish the characteristics of our present atmosphere. The progressive increase in oxygen pressure provided a gradient for oxygenation of multicellular plants and animals. However, as the oxygen pressure rose, the oxidative potential of our atmosphere increased above the oxygen pressure safe for the biologic processes that produced it, and life has since then been sustained by the development of antioxidant mechanisms.[64] It has been proposed that the oxygen tension in the atmosphere now is high enough to exert adverse biologic effects through oxygen toxicity (Chapter 75).[62,63]

A final stage of atmospheric evolution, influenced by man himself, is now in process. It involves not only the production of toxic atmospheric contaminants but also an increase in animal and industrial production of carbon dioxide, generating a heat-retaining mantle in the upper reaches of our atmosphere that traps infrared energy on earth to produce a "greenhouse effect."[34] This warming of the atmosphere of earth[90] introduces the possibility of a cycle of further gross climatic alteration coupled with the melting of polar ice, a rise in ocean level, and exaggerated botanical activity.[65]

GASES OF THE NORMAL ATMOSPHERE

The metabolically active and inert gases most involved in the respiratory exchange are the normal components of the earth's atmosphere, that is, O_2, CO_2, and nitrogen. The group of gases designated as chemically "inert" consists primarily of N_2 but also contains argon, xenon, krypton, hydrogen, and helium, which behave physiologically like N_2[91] and together are present only in extremely minute amounts. The inert gases are only passively involved in respiration

Table 68-1. Gaseous composition of the earth's atmosphere

Element	Percent in dry air
Nitrogen	78.09
Oxygen	20.94
Argon	0.93
Carbon dioxide	0.03
Neon	0.002
Helium	0.0005
Krypton	0.0001
Hydrogen	0.00005
Xenon	0.000008

and serve primarily as diluents and vehicles for oxygen. The proportions of atmospheric gases are shown in Table 68-1.

The normal gaseous environment of man is constant in percentage composition from sea level to extreme altitude. However, the surface of the earth is at the same time the bottom of the deep atmospheric "sea," and the total pressure of the air at any point of the earth is determined by the height (and hence the weight) of the air column above it. Differences in altitude result in gross alterations of inspired gas pressures (P_{O_2} and P_{N_2}) and thus in the tensions of these gases in alveolar air and body fluids.

Components of atmospheric-respiratory exchange

The respiratory function of air-breathing animals involves intimately interlocked processes, including (1) the regulation of respiration by spontaneous action of specialized neuronal structures and their reactivity to chemical or physical changes, (2) the influence of muscle activity on the chest to produce controllable alterations of pressure within the lungs, (3) the effect of these pressure changes in causing movement of air between the lungs and the external atmosphere, (4) the effect of metabolism and pulmonary ventilation on the composition of lung gases, (5) the exchanges of gases between the blood and the lungs, (6) the characteristics of the transport of O_2 and CO_2 between the lungs and the tissues, and (7) the effects of the overall sequence of events on the local cellular chemical environment and hence on the activity of the chemosensitive neurons regulating respiration.

These processes are all part of a coordinated system having a single fundamental purpose—to supply the cells of the body with the atmospheric O_2 they require and to remove the CO_2 they produce. This purpose can be frustrated by interference with any of the steps involved. Whereas each process has a considerable reserve, a decrease in the capacity of the tissues to function, culminating in irreversible changes, is inevitable if the interference is sufficiently severe and prolonged.

The portable gaseous environment

The life processes of humans do not occur in gaseous equilibrium with the atmosphere. Only a few of the cells of human tissues could survive extended, direct exposure to the partial pressures of O_2 or CO_2 present in the atmosphere of earth. O_2 at pressures found in air at sea level is distinctly toxic, and in homogenized preparations in vitro, enzyme systems of vital organs can be inhibited by the pressure of O_2 found in air at sea level. In addition, the pressure of CO_2 is so low in atmospheric air that severe alkalosis of cells would result from direct exposure to air, again interfering with metabolic processes. Therefore, while certain unicellular life forms must withstand direct exposure to the gas tensions of the atmosphere or water, the pulmonary structure evolved by man and other warm-blooded animals has provided a portable gaseous environment considerably different in composition from that of the ambient atmosphere. This, the *alveolar gas,* is the actual gaseous environment to which man presents an interface for gas exchange with his body fluids.

Alveolar gas composition is affected by two interacting factors. One is the influence of metabolism, which, by way of gas transport in blood flowing through pulmonary capillaries, causes the *continuous delivery of CO_2* to the alveolar gas and the *continuous removal of O_2* from it. The second factor is the *periodic partial replacement and dilution* of alveolar gas with fresh ambient air via the respiratory act. Close balance between these influences is attained not only at rest, when the normal man consumes about 250 ml O_2 and produces about 200 ml CO_2/min, but also during strenuous muscular exercise in which gas exchange may increase more than 10-fold. The design of lungs capable of maintaining a relatively constant composition of alveolar gas at the extremes of rest and severe exercise has occurred and involves an amazing composite of *extreme efficiency* on demand and a protective, *gross inefficiency* when metabolic demands are low. The factors inherent in this dual function will be the concern of this chapter.

Gradients between atmosphere and cells

The important gradients of gas tensions in a normal man at rest at sea level are given in Table 68-2 and Fig. 68-1. Since movement of gas molecules in air or within a liquid occurs by diffusion from a region of higher concentration of gas molecules to a region of low molecular concentration, these tension gradients (Fig. 68-2) are responsible for the entire gas exchange between atmosphere, alveoli, and blood and between blood and tissues.

Table 68-2 implies that, after gas exchange between pulmonary arterial (mixed venous) blood entering the lungs and the alveolar air, there is no significant alveolar-arterial gradient with respect to CO_2 and only a negligible one for O_2. The occurrence of a slightly lower O_2 tension in peripheral arterial blood than in alveolar gas is actually due to admixture of poorly oxygenated blood with the blood that leaves the alveoli at essentially full equilibrium with the prevailing alveolar P_{O_2}, as well as to differences in ven-

Table 68-2. Partial pressures of respiratory gases, expressed in millimeters of mercury

Gas	Ambient air	Tracheal air	Expired air	Alveolar air	Arterial blood	Mixed venous blood	Tissues
O_2	158.0	149.0	116	100	95	40	40 or less
CO_2	0.3	0.3	32	40	40	46	46 or more
H_2O	5.7	47.0	47	47	47	47	47
N_2	596.0	563.7	565	573	573	573	573
TOTALS	760	760	760	760	755	706	706 or less

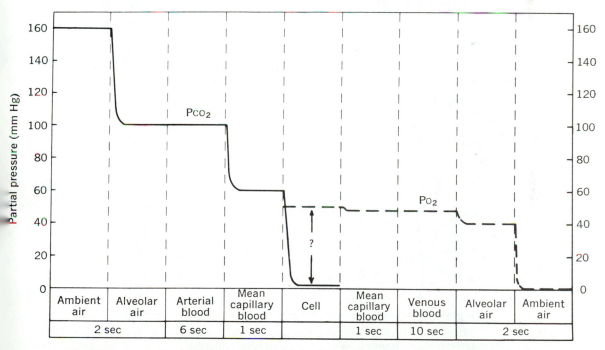

Fig. 68-1. Changes in P_{O_2} and P_{CO_2} in overall gas exchanges between tissues and ambient air. Gas tension gradients are arranged to indicate, from left, decreasing levels of P_{O_2} from inspired air to tissue and, to right, decreasing levels of P_{CO_2} from tissue to ambient air. At the cell a *?* is used to denote that gas tensions within any cell in a tissue must vary and may be as low as 1 mm Hg. Actually, tensions of O_2 and CO_2 change progressively between sites of uptake or production and sites of consumption or elimination (Fig. 68-2).[56]

tilation and perfusion among alveoli (p. 0000). No active gas transfer mechanisms ("secretions") are involved anywhere in these processes. The effectiveness of simple diffusion is greatly enhanced by (1) the rapid circulatory renewal of the blood in contact with the gas-permeable capillaries of the pulmonary alveoli and the metabolizing tissues, (2) the rapid chemical reactions into which the respiratory gases enter when they reach the blood, (3) specific enzymes that accelerate the uptake of O_2 in the metabolizing cells and the combination of CO_2 with water in the blood, (4) adjustment of circulation of the blood and the volume of alveolar ventilation to existing metabolic requirements.

In each of these physiologic processes the changes in composition, volume, and pressure of gases as well as the solution and diffusion of gases in body fluids are governed by a number of well-known *physical* principles.

GAS LAWS AND RESPIRATORY EXCHANGE

A characteristic property of gases is their ability to diffuse—to pass from a point of high concentration or pressure to one of lower concentration and to fill the entire available space. This diffusion towards equilibrium states occurs in gas mixtures or with gases dissolved in liquids.

The exchange of the respiratory gases between the blood and the lungs and the blood and tissues can be explained on the ground that gases are composed of discrete particles (molecules) constantly in motion in straight lines and with great velocity. The velocity of motion of a gas molecule such as O_2 is close to 500 mph, but the mean free path is short. The pressure or tension exerted by a gas is due to the sum of the bombardments of the walls of the containing vessel by the constantly moving molecules. Anything that increases the bombardment must therefore increase

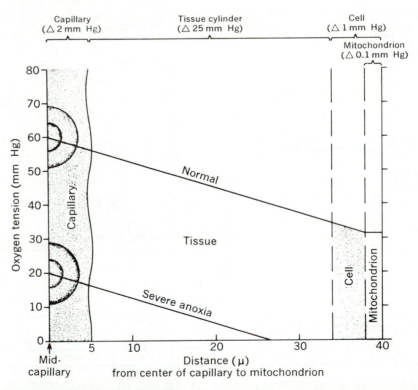

Fig. 68-2. Fall in O_2 pressure during passage from red blood cell to cell mitochondrion. Scales of distance, dimensions of diffusion compartments, and degree of fall in P_{O_2} are a composite of many individual studies or measurements. When pressure head available in the capillary for diffusion across these distances becomes too low, a physiologically effective P_{O_2} cannot be attained at the site of metabolism. (From Lambertsen.[73])

the pressure. The following gas laws express the factors of greatest physiologic importance.

Behavior of gases as gases

In the respiratory exchange between the ambient air atmosphere and the alveoli the separate and combined behavior of N_2, O_2, CO_2, and H_2O must be considered. The gas laws, disregarding the physiologically insignificant deviations of real gases from the behavior of the "ideal" gases the laws describe, are as follows:

For all gases an equal number of molecules in the same space and at the same temperature will exert the same pressure (Avogadro's hypothesis). In gases whose molecules are of different weights, equal numbers of molecules will be obtained by measuring the gases in quantities proportional to their molecular weights. The molecular weight of a gas in grams (called a gram-molecule, or mole) will contain 6×10^{23} molecules and will occupy a volume of 22.4 L at a temperature of 0° C and a pressure of 760 mm Hg.

In a gas mixture the pressure exerted by each individual gas in a space is independent of the pressures of other gases in the mixture (Dalton's law). Since each gas behaves as though it were the only gas present in the space, the total gas pressure is equal to the sum of the individual partial pressures (e.g., in the alveoli):

$$\text{Total pressure} = P_{H_2O} + P_{O_2} + P_{CO_2} + P_{N_2}$$

The partial pressure of a gas, or the ratio of the pressure of the gas to the total gas pressure, is of extreme importance both in the physical dynamics of pulmonary-blood-tissue gas exchange and in the participation of gases in the chemical processes of gas transport and metabolism.

As a gas is compressed, its volume decreases in exactly the same proportion as its pressure increases (Boyle's law). Thus 1 mole of any gas, when compressed at 0° C to a volume of 1 L, will exert a pressure of 22.4 atmospheres (atm). Because more molecules are present per unit volume, the impacts of gas molecules on the walls of the container are more numerous when the gas is compressed; the pressure is increased proportionately. The change from one pressure and volume to a second pressure and volume is expressed as:

$$P_1 V_1 = P_2 V_2$$

As an example, the 4 L volume of gas within the lungs of a breath-holding subject (body temperature constant) who dives from the water surface (1.0 atm) to a depth of 33 ft (2.0 atm) would be changed to the following:

$$V_2 = \frac{1.0 \text{ atm} \times 4.0 \text{ L}}{2.0 \text{ atm}} = 2.0 \text{ L}$$

Any gas expands when heated and contracts when cooled. Taking the volume at 0° C as the standard, the gas increases in volume by 1/273 for every degree C rise in temperature. This means that if the volume is kept constant, the pressure of a given gas is proportional to the temperature *(law of Gay-Lussac* or, independently, *Charles' law)*. At a temperature of $-273°$ C (absolute zero), gas molecules are motionless and exert no pressure. Increasing the temperature increases the speed at which the molecules move, hence the rise in pressure.

The relationships between temperature and volume and between temperature and pressure are written:

$$\frac{V_1}{V_2} = \frac{T_1}{T_2} \text{ and } \frac{P_1}{P_2} = \frac{T_1}{T_2}$$

Exhalation of 1 L of warm air from the lungs (body temperature, 37° C = 273° + 37° absolute) into a breathing bag at an ambient air temperature of 0° C (273° absolute) will result in a reduction of volume:

$$V_2 = V_1 \times \frac{T_2}{T_1} = 1.0 \text{ L} \times \frac{273°}{310°} = 0.88 \text{ L}$$

Since in physiologic situations temperature, volume, and pressure must be considered concurrently, the most useful relationship of these principles is expressed as the *ideal gas law:*

$$PV = nRT$$

in which P is the gas pressure expressed in atmospheres, V is its volume in liters, n is the number of moles of the contained gas, R is a constant, and T is the absolute temperature of the gas. As normally employed, R = 0.082 (L-atm/mole/°C). The ideal gas relationship is employed in calculating the volume of gas moved by the lungs at body temperature and atmospheric pressure, saturated with water vapor (BTPS) when the actual measurements have been made under a variety of laboratory conditions of ambient temperature and pressure, saturated with water vapor (ATPS). Thus at 760 mm Hg barometric pressure:

Respiratory minute volume (BTPS) =
$$\text{Observed volume (ATPS)} \times \frac{273 + 37°}{273 + t} \times \frac{760 - P_{H_2O_t}}{760 - 47}$$

in which t is the gasometer temperature and $P_{H_2O_t}$ is water vapor pressure at the gasometer temperature. This important conversion of experimental measurements to the equivalent intrapulmonary volume at BTPS can also be carried out automatically by maintaining the temperature of the water in a wet gasometer or spirometer at 37° C or at the subject's body temperature.[75]

Behavior of gases dissolved in body fluids

When gases are taken up by liquids, the situation is modified by the ability of the liquid to dissolve or to combine chemically with each gas. These factors now determine the volume of the gases that can be taken up by the liquid at a given temperature and pressure.

The law of solution of gases in liquids (Henry's law) states that the weight (i.e., the number of molecules) of a gas absorbed by a given liquid with which it does not combine chemically is directly proportional to the pressure of the gas to which the liquid is exposed. The volumes of O_2, CO_2, and N_2 that can be physically dissolved by water, blood plasma, and whole blood, when exposed at 37° C to 1 atm of each gas (760 mm Hg pressure), are given in Table 68-3.

The volumes of gas dissolved by the blood at the partial pressures existing in the alveolar air and at 37° C are approximately as follows:

O_2 0.30 ml/100 ml (vol%)
CO_2 3.0 ml/100 ml (vol%)
N_2 0.8 ml/100 ml (vol%)

These are the volumes of the gases taken up by the blood in purely physical solution; the amounts taken up in chemical combination are not included. The latter are much greater than this except for N_2, which does not combine chemically with the blood, does not participate in the metabolic gas exchange, and is carried entirely in physical solution. Note that CO_2 is by far the most soluble of the respired gases; this is partly responsible for the ease with which CO_2 diffuses in the body.

The *law of partial pressures of gases dissolved in liquids* has to do with the pressure exerted by each individual component of a mixture of gases. If the gases do not react chemically with one another (and none of the respiratory gases do), the partial pressure or tension of each constituent of the mixture is that which it would exert if no other gas were present. The tension of each gas in the liquid is the same as its partial pressure in the gaseous phase with which it was equilibrated. In a dynamic exchange the molecules of each gas move into the liquid until the pressure is equal in the gas and the liquid phases. At the stage of true equilibrium, gas molecules continue to move rapidly between the liquid and the gas phases, but the exchange is equal between the two phases. If the pressure in the gas phase is increased, more gas molecules pass into the liquid until the pressure in the two phases becomes equal at the higher level, and if it is decreased, gas molecules escape from the liquid until the pressure in the two phases is again equalized at a lower level. *The tension of a gas in a liquid can be regarded as the force with which the gas molecules are constantly striving to escape.* The tensions of the respiratory gases in the blood at any instant are due to that portion of each gas which is physically dissolved in the plasma; the larger amounts carried in chemical combination serve as reservoirs from which additional gas molecules are liberated to pass into physical solution when the tension of the gas is reduced or as reservoirs in which additional molecules are stored when the tension is increased.

Water vapor pressure of an aqueous fluid such as blood is practically dependent only on the temperature except at high ambient pressures. In extremes of body temperature change such as induced hypothermia, P_{H_2O} is considerably altered (Table 68-4). P_{H_2O} represents the tendency of H_2O molecules to escape into a gas phase. With adequate water present a gas phase will become saturated with water vapor at the P_{H_2O} corresponding to the existing temperature; at a body temperature of 37° C the P_{H_2O} is 47

Table 68-3. Physical solubility of gases in physiologic fluids (ml gas/100 ml fluid/atm at 37° C)

	Water*	Plasma	Blood
O_2	2.386	2.14†	2.36†
CO_2	56.7	51.5‡	—
N_2§	1.227	1.18	1.30

*Data from Altman et al.[31]
†Data from Fasciolo and Chiodi.[52]
‡Data from Bartels and Wrbitzky.[37]
§Interpolated from data at 38° C by Van Slyke et al.[110]

Table 68-4. Influence of body temperature on P_{H_2O}

Body temperature (°C)		Water vapor pressure (P_{H_2O} in mm Hg)
40	Fever	55.3
37	Normal	47.1
35		42.2
30		31.8
25		23.8
20	Hypothermia	17.5
15		12.8
10		9.2

mm Hg. This pressure constitutes a fixed part of the total pressure in a mixture of respiratory gases. For this reason the partial pressure of a gas such as O_2, present as 10% of a "dry" mixture at 37° C and a barometric pressure of 760 mm Hg, is calculated:

$$Po_2 = \%O_2 \text{ in dry gas} \times \text{Total pressure of dry gas}$$
$$= \%O_2 \times (\text{Barometric pressure} - P_{H_2O} \, 37° \text{C})$$
$$= \%O_2 \times (B - 47) = 71.3 \text{ mm Hg}$$

Of the purely physical factors that influence the rate of diffusion of gases the ones of greatest importance to the respiratory gas exchange are *(1) the difference in partial pressure, (2) the solubility of the gas in the fluid media involved, (3) the temperature, and (4) the molecular weight of the gas.* Under normal conditions, all these may be regarded as constant for a particular gas except the difference in partial pressure, which, along with mechanical ventilation of the alveoli and circulation of the blood, thus becomes responsible for the direction and degree of the gas exchange.[21]

INTRAPULMONARY GAS MIXING AND ALVEOLAR VENTILATION

The pulmonary ventilation produced by an inhalation and exhalation is not uniform, and not all air inhaled participates in exchange with the blood. The functional *unit* of the lung for gas exchange is a respiratory bronchiole, with the alveolar ducts and alveoli arising from it together with the corresponding blood vessels, lymphatics, nerves, and connective tissue.[83,87,109] This is the distensible or "bellows" part involved in ventilatory function.

Ventilation of the alveoli during lung expansion probably occurs in several ways, rather than only through passive expansion of elastic alveolar walls.[107] The alveolar surface area may in fact change very little during normal respiration, with any alveolar volume increase or decrease involving some recruitment of alveoli, but generally resembling the opening and closing of a paper bag.[107] Although the lung enlarges in all directions during inspiration, most of the change in volume appears to be due to the lengthening and expansion of the alveolar ducts accompanied by increases in the size of the openings into the air sacs.[81,82]

If the alveoli themselves do not greatly expand and the incoming gas can be drawn only as far as the alveolar ducts, diffusion and the convection currents set up in the passages by the inflowing stream of air, aided by the inspiratory enlargement of the openings into the air sacs, make possible a free interchange of gases between the ducts and the alveoli. With an alveolar diameter of only about 100 to 200 μm, the time required for diffusion across the alveolus is extremely short (2 to 4 msec).

Collapse of air passages on exhalation can occur in unusual conditions when the inherent rigidity of the small bronchi is overcome by a transmural pressure. At the beginning of an active and forceful exhalation, alveolar pressure is higher than that in the small bronchi. Since the bronchi are surrounded by lung parenchyma containing the alveoli, the pressure on the exterior wall of a bronchus is transiently greater than that within its lumen. In diseases such as emphysema or asthma there is a tendency for some pulmonary units to collapse during coughing. If collapse occurs, further expiratory force aggravates the bronchial blockage.

Alveolar air

In gas exchange between the blood and the external environment, the important volume of gas is the *alveolar "air."* An additional volume of inspired air ventilates the respiratory passages, but the relatively thick walls of these allow no significant gas exchange with the blood: this is the *dead space air.*

An excessively simplified concept of alveolar air is that it comprises all the air in the lungs except that in the dead space, has a uniform composition, can be collected at the mouth at the end of an exhalation large enough to displace with alveolar air the air previously contained in the dead space, and has tensions of O_2 and CO_2 extremely close to those of the arterial blood. The limitations of this simplified concept, not serious when discussing normal individuals at rest, become considerable in cases of acute and chronic pulmonary disease. Dead space air enters the respiratory passages not as a sharp "front," but must mix considerably with gas left in the respiratory passages on inspiration.[51] This makes it difficult to determine where dead space air ends and alveolar air begins. The venous blood entering the pulmonary capillaries continuously takes O_2 out of the alveolar air and gives off CO_2 into it, but the replenishment of alveolar air with freshly inhaled air occurs only intermittently. Consequently, the composition of the air expelled from the lungs by a forced exhalation inevitably differs with the phase of the respiratory cycle at which the forced exhalation is initiated (Table 68-5, Fig. 68-3). Furthermore, the composition in different localized groups of alveoli at any given instant depends on the relative

Table 68-5. Gas tensions in alveolar air (simultaneously estimated by five different procedures) and in arterial blood (directly determined): mean values of 33 measurements in 11 normal young men at rest at sea level*

Direct method	CO_2 tension (P_{CO_2} in mm Hg)	O_2 tension (P_{O_2} in mm Hg)	Calculation method	CO_2 tension (P_{CO_2} in mm Hg)	O_2 tension (P_{O_2} in mm Hg)
Haldane-Priestley (H-P)[67]			Effective alveolar air calculated from existing arterial P_{CO_2} and N_2 content of inspired and expired air (effective) (method of Riley et al.[96])	40.5†	100.3‡
End-expiratory sample	40.7	97.9			
End-inspiratory sample	38.6	104.2			
Mean of expiration and inspiration	39.7	101.1			
End of series of normal expirations automatically collected at the mouth by method of Rahn et al.[92]	38.6	103.1	Alveolar P_{O_2} estimated by subtracting sum of existing arterial tensions of CO_2, nitrogen, and water vapor from prevailing barometric pressure (method of Dill and Penrod[50])	40.5†	100.7
Air drawn at end of several successive normal expirations through catheter at the tracheal bifurcation[35] (bronchial)	40.5	99.7	Arterial blood-gas tension by microtonometric procedure[77]	40.5	96.6

*From Barker et al.[35]

†Assumed equal to P_{CO_2} of arterial blood.

‡The formula for calculating "effective alveolar air" is given by Riley et al.[96] as follows:

$$\text{Effective alveolar } P_{O_2} = \text{Tracheal } P_{O_2} \times \frac{\%N_2 \text{ in expired air}}{\%N_2 \text{ in inspired air}} - \frac{\text{Arterial } P_{CO_2}}{\text{Expired air R}}$$

The necessary data are (1) analysis of expired air (collected over a period of about 5 min) for O_2 and CO_2 content from which the values for N_2 percentages and expired air (R) are derived (tracheal P_{O_2} is the P_{O_2} of inspired air saturated with water vapor at body temperature) and (2) direct determination (by the microtonometric method) of the arterial P_{CO_2}.

The Dill-Penrod procedure[50] is based on the assumption that any difference between the total tensions of gases in the alveolar air and in the arterial blood is most probably due to an alveolar-arterial P_{O_2} gradient. Thus by determining the P_{O_2} and P_{CO_2} of arterial blood and by inserting an estimated value for P_{H_2O} in subjects breathing O_2 and therefore with negligible amounts of N_2 in alveoli and blood, these authors concluded that the alveolar-arterial P_{O_2} gradient increased with the volume of alveolar ventilation. Ferris[53] has applied the same principle to subjects breathing air at sea level, the arterial P_{N_2} being assumed to be equal to the alveolar P_{N_2}, which was assigned a standard value.

An example of this procedure is as follows:

	Arterial tensions	*Alveolar tensions*
P_{CO_2}	40 mm Hg (by analysis)	40 mm Hg (assumed equal to arterial)
P_{O_2}	95 mm Hg (by analysis)	Unknown
P_{H_2O}	47 mm Hg (assumed)	47 mm Hg (assumed)
P_{N_2}	573 mm Hg (assumed equal to alveolar)	573 mm Hg (assumed)
TOTAL	755 mm Hg	660 mm Hg

At a barometric pressure of 760 mm Hg the alveolar P_{O_2} must therefore have been $760 - 660 = 100$ mm Hg. The alveolar-arterial gradient here would be 5 mm Hg, representing a total gas pressure in arterial blood 5 mm Hg below that in the alveolar air. This conforms with the expectations of the venous admixture concept of the alveolar-arterial P_{O_2} gradient.

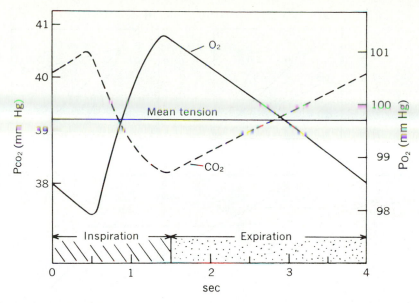

Fig. 68-3. Predicted time course of alveolar P_{CO_2} and P_{O_2} fluctuation during respiratory cycle in normal subject at rest. Addition to volume of residual air of another volume equal to tidal volume–dead space volume lowers the alveolar P_{CO_2}, causing fluctuations with respiration. Damping of fluctuation occurs due to continued arrival of CO_2 in mixed venous blood, to variations in blood flow, and to capacity of tissue buffers for CO_2. Calculation suggests that mean levels of P_{CO_2} and P_{O_2} are reached just after midexhalation. During air breathing at rest, washout of dead space air may not be complete with normal tidal exhalation, and a dilution error occurs in determinations of alveolar gas tensions. During respiratory stimulation, as by CO_2, it is practical to obtain samples of exhaled gas from mouth or nose (end-tidal samples) that closely match arterial P_{CO_2} and provide index of mean alveolar gas tensions useful for many types of respiratory study.[25,61]

degree of ventilation of the alveoli in different parts of the lungs.*

The composition of alveolar gas therefore is not a physiologic constant, but is variable both in time and space.[101] Fig. 68-4 describes the alveolar gas compositions at various gas exchange ratios. A calculated pattern of the nature and degree of variability of P_{O_2} and P_{CO_2} throughout a respiratory cycle at rest is shown in Fig. 68-3. On the basis of direct, practical comparisons of alveolar gas and arterial blood, there now is general agreement that the mean P_{CO_2} of alveolar air in normal men is nearly identical to that of arterial blood. During normal, shallow, unforced breathing the P_{CO_2} of the gas collected at the mouth at the end of expiration (end-tidal gas) will reflect the mixing with dead space air and will be lower than that of arterial blood. The P_{CO_2} and P_{O_2} of pulmonary capillary blood must be considered as coming within a minute fraction of a millimeter of mercury of the alveolar gas

to which it is exposed. Subsequently, changes occur between the lung, the peripheral artery, and the P_{O_2} of the peripheral arterial blood, which is normally several millimeters of mercury lower than that of alveolar gas, while its P_{CO_2} is not detectably different[35,36] (Table 68-5). The existence of a measured alveolar-arterial O_2 difference does not mean that blood and gas in the alveoli failed to approximate complete equality with respect to P_{O_2}. The alveolar-arterial P_{O_2} difference is largely due to contamination of fully oxygenated blood from well-ventilated alveoli with less oxygenated blood from (1) poorly ventilated alveoli and (2) venous blood from the bronchial circulation and arteriovenous shunts.[15,25,93] Due to the brief time spent in the pulmonary capillary, it is conceivable that essentially perfect equilibrium for CO_2 may be minutely upset by continued influence of carbonic anhydrase after the blood leaves the capillary* (Fig. 68-11).

---

*See references 18, 25, 51, 93, 101, and 116.

---

*See references 58, 66, 68, 69, 97, and 111.

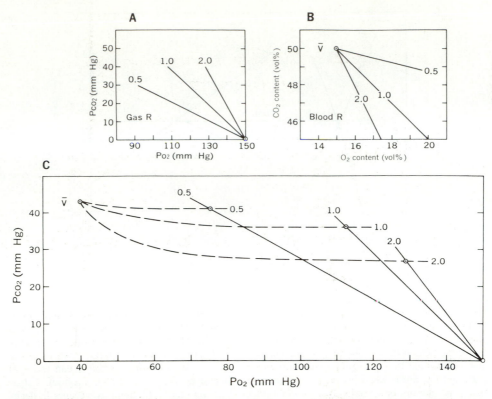

Fig. 68-4. Forms of P_{O_2}-P_{CO_2} diagram used to depict alveolar gas composition at different gas exchange ratios. **A,** Relation between alveolar partial pressures and respiratory exchange ratio *(R)* of alveolar or expired gas. For each value of R there is an infinite number of possible combinations of alveolar P_{O_2} and P_{CO_2} values. These are graphically represented by straight lines radiating from inspired gas point of 150 mm P_{O_2}. R lines for values of 0.5, 1.0, and 2.0 are drawn. **B,** Pulmonary capillary blood gas content. Mixed venous blood is indicated by point $\overline{V}$, with appropriate O_2 and CO_2 content coordinates. This blood can exchange in lung at different R values, here defined as CO_2 loss divided by O_2 gain. Hence any given slope represents all points for given R value of blood. Blood R lines for 0.5, 1.0, and 2.0 are shown. **C,** Composite of alveolar gas and capillary blood relationships, showing intersection of gas and blood R lines. Gas R lines were plotted as in **A.** Next the blood R lines from **B** were changed into partial pressure dimensions with aid of nomogram or appropriate blood gas dissociation curves, providing dashed lines that show transformed values. Intersection of each gas and blood R line represents only possible O_2 and CO_2 tension where gas can exchange between alveoli and blood at this R value. (After DuBois; modified from Rahn and Farhi.[25])

Definitions of standard symbols and abbreviations

In deriving and describing mathematical relationships among respiratory variables, some standardization and abbreviation of terminology is required. The system that has been developed[24,49,89] is shown in Table 68-6 and the following summary. These symbols are useful to the student and experienced pulmonary physiologist, since they provide a shorthand language. However, since they do not encompass all aspects of respiratory function and since they may tend to slow the introduction of the concepts of respiratory control to the new student, the symbols will not be used here exclusively.

Quantitative relationships in pulmonary ventilation and gas exchange

The relationships that describe the physiologic events involved in pulmonary gas exchange and the control of respiration can be expressed mathematically in convenient form using the symbols from Table 68-6.

Expired volume per breath:

$$V_T = V_A + V_D$$

Expired respiratory minute volume:

$$\dot{V}_E = V_T \times f$$

Inspired respiratory minute volume:

$$\dot{V}_I = \frac{F_{E_{N_2}}}{F_{I_{N_2}}} \dot{V}_E$$

Dead space per breath:

$$V_D = V_T - V_A$$

also

$$V_D = \frac{F_{A_{CO_2}} - F_{E_{CO_2}}}{F_{A_{CO_2}} - F_{I_{CO_2}}} V_E$$

Dead space ventilation per minute:

$$\dot{V}_D = \dot{V}_E - \dot{V}_A$$

Alveolar ventilation per breath:

$$V_A = V_E - V_D$$

Alveolar ventilation per minute:

$$\dot{V}_A = \dot{V}_E - (V_D \times f)$$

Table 68-6. Standard symbols in pulmonary gas exchange*

	Symbol	Definition	Example
Quantitative variables	P	Gas pressure in general	$P_{O_2} = 100$ mm Hg
	V	Gas volume	$V_{Tidal} = 500$ ml
	$\dot{V}$	Gas volume (flow) per unit time	$\dot{V}_{Expired} = 6$ L/min
	f	Frequency of respiration	$f = 10$ breaths/min
	F	Fractional concentration of gas in dry gas phase	F_{O_2} in air $= 0.2094$
	C	Concentration in blood or other aqueous phase	$C_{O_2\,Arterial} = 0.3$ vol%
	$\dot{Q}$	Blood volume flow per unit time	$\dot{Q}_{Bronchial} = 100$ ml/min
	R	Respiratory exchange ratio (RQ)	$\dot{V}_{CO_2}/\dot{V}_{O_2} = 0.80$
	D	Diffusing capacity	$D_{L_{O_2}} = 50$ ml O_2/min/mm Hg ΔP_{O_2}
Qualifying terms			
Gas	B	Barometric	$B = 760$ mm Hg
	A	Alveolar	$P_{O_{2_A}} = 100$ mm Hg
	D	Dead space	$V_D = 150$ ml
	T	Tidal	$V_T = 500$ ml
	I	Inspired	$F_{I_{O_2}} = 0.2094$
	E	Expired	$F_{E_{CO_2}} = 0.045$
Blood	a	Arterial	$P_{O_{2_a}} = 95$
	v	Venous	$P_{O_{2_v}} = 40$
	$\bar{v}$	Mixed venous	$P_{O_{2_{\bar{v}}}} = 40$
	c	Capillary	$P_{O_{2_c}} = 60$
	b	Unspecified site—blood in general	
Pulmonary	L	Lung	
Other terms describing gases	atm	Atmospheres	
	ata	Atmospheres absolute	
	ATPD	Ambient temperature and pressure, dry	
	ATPS	Ambient temperature and pressure, saturated with H_2O	
	BTPS	Body temperature and ambient pressure, saturated with H_2O	
	STPD	Standard temperature (0° C) and pressure (760 mm Hg), dry	
General qualifiers	$\overline{X}$	Indicates mean of any value	

*Dash above symbol denotes a mean value.

CO_2 production per minute:

$$\dot{V}_{CO_2} = \dot{V}_E F_{E_{CO_2}} - \dot{V}_I F_{I_{CO_2}}$$

$$\dot{V}_{CO_2} = \left(F_{E_{CO_2}} - \frac{F_{E_{N_2}}}{F_{I_{N_2}}} F_{I_{CO_2}}\right) \dot{V}_E$$

also

$$\dot{V}_{CO_2} = \dot{V}_A \times F_{A_{CO_2}}$$

O_2 consumption per minute:

$$\dot{V}_{O_2} = \dot{V}_I F_{I_{O_2}} - \dot{V}_E F_{E_{O_2}}$$

$$\dot{V}_{O_2} = \left(F_{I_{O_2}} \frac{F_{E_{N_2}}}{F_{I_{N_2}}} - F_{E_{O_2}}\right) \dot{V}_E$$

Respiratory exchange ratio:

$$R = \dot{V}_{CO_2} / \dot{V}_{O_2}$$

$$R = \frac{F_{E_{CO_2}} - F_{I_{CO_2}} \dfrac{F_{E_{N_2}}}{F_{I_{N_2}}}}{F_{I_{O_2}} \dfrac{F_{E_{N_2}}}{F_{I_{N_2}}} - F_{E_{O_2}}}$$

Percentage CO_2 expired:

$$F_{E_{CO_2}} = (F_{I_{CO_2}} V_D + F_{A_{CO_2}} V_A) / V_E$$

Alveolar CO_2 pressure:

$$P_{A_{CO_2}} = F_{A_{CO_2}}(P_B - P_{A_{H_2O}})$$

$$P_{A_{CO_2}} = P_{I_{CO_2}} + \frac{\dot{V}_{CO_2}}{\dot{V}_A}(P_B - 47) + P_{I_{CO_2}}\left(\frac{1}{R} - 1\right)\frac{\dot{V}_{CO_2}}{\dot{V}_A}$$

Alveolar O_2 pressure:

$$P_{A_{O_2}} = F_{BA_{O_2}}(P_A - P_{H_2O})$$

$$P_{A_{O_2}} = P_{Tracheal_{O_2}} \frac{F_{E_{N_2}}}{F_{I_{N_2}}} - \frac{Pa_{CO_2}}{R_E}$$

Ventilation-perfusion ratio:

$$\dot{V}_A/\dot{Q}_O = \frac{C\bar{v}_{CO_2} - Cc_{CO_2}}{0.83\, F_{A_{CO_2}}}$$

Percentage venous admixture:

$$\% \text{ Venous admixture} = \frac{Ca_{O_2} - Cc_{O_2}}{C\bar{v}_{O_2} - Cc_{O_2}}$$

Pulmonary diffusion capacity for O_2:

$$D_{L_{O_2}} = \frac{\dot{V}_{O_2}}{P_{A_{O_2}} - P_{C_{O_2}}}$$

Dead space and alveolar ventilation

Advantageously, pulmonary ventilation at rest is only about 70% efficient. The larger respiratory passages are filled with water-saturated air at inspiration and with alveolar air at expiration. Since no significant diffusion of gases occurs through the walls of these passages, the movement of air in and out within them plays no direct part in the pulmonary gas exchange. The functional volume of inspired gas that does take part in exchange with the blood is called the *alveolar*

ventilation. The remaining fraction of an inspiration that does not reach functioning alveolar units is termed *dead space ventilation*, and comprises about 30% of the tidal volume at rest and in exercise.

Single breath alveolar ventilation is expressed quantitatively as the amount of air remaining after the dead space volume is subtracted from the tidal volume. Therefore for a single breath:

$$V_A = V_T - V_D$$

Alveolar ventilation per minute is therefore the difference between total respiratory minute volume and dead space minute volume or:

$$\dot{V}_A = \dot{V}_E - \dot{V}_D = \dot{V}_E - (V_D \times f)$$

With a resting tidal volume of 500 ml and a normal dead space of approximately 150 ml the gas volume that carries out the function of alveolar ventilation in each breath is only 350 ml. At a respiratory rate of 12/min, total pulmonary ventilation will be 6 L/min, of which 4.2 L/min is the alveolar ventilation responsible for the functional gas exchange in the alveoli. The rest, 1.8 L/min, is dead space ventilation.

If the dead space per breath were increased, as by pulmonary disease, to 300 ml, $\dot{V}_A$ would be reduced to 2.4 L/min. This is what happens in pulmonary emphysema, a disease in which the effectiveness of alveolar ventilation is greatly reduced; the result is as though the anatomic dead space were increased to an equivalent degree.

If V_T is reduced while V_D is of a size usually associated with deeper inhalations, the result is to diminish $\dot{V}_A$ (if V_D were equal to V_T, there would be no alveolar ventilation at all). For example, if V_D remained unchanged at 150 ml, while V_T fell from 500 to 350 ml at a respiratory rate of 12/min, $\dot{V}_A$ would be reduced from 4.2 to 2.4 L/min. This is what happens in severe respiratory depression. Thus decreased respiratory depth has the same effect on the pulmonary gas exchange as increased volume of the dead space.

If both decreased tidal volume and increased dead space developed concurrently, the situation would of course be aggravated. An enlarged dead space can be compensated for by an increased tidal volume, but capacity for further enlargement of tidal volume as required for exercise would be diminished by this compensation.

Definitions and determination of dead space

Anatomic and physiologic (functional) dead space. The proportions of dead space to alveolar ventilation in a single breath and over a period of time are indices of the efficiency with which breathing affects the composition of alveolar air. The quantitative measurement of that efficiency depends on the ability to accomplish accurate separation of total pulmonary ventilation into alveolar and dead space components. Estimation of pulmonary dead space has been by three main approaches: (1) The use of plaster or plastic cast techniques to measure the volume of larger air passages in the human cadaver; this *anatomic dead space* is approximately 150 ml. (2) The use of a rapid gas analyzer such as the nitrogen meter, together with continuous measurement of exhaled volume, to measure the portion of a single breath of pure O_2 that is not appreciably mixed with N_2 in the lungs.[7,46,51,59] This method, which in effect uses O_2 to make a cast of the living lung, also measures the anatomic dead space and can be employed to study the influence of various conditions on this dead space. (3) Calculation of anatomic and physiologic dead space by means of the respiratory equation devised by Bohr in 1891.[39] When used in conjunction with N_2 measurements made in the above-mentioned single-breath, O_2-cast method, the Bohr equation serves to estimate anatomic dead space.[59] For measuring the somewhat larger *physiologic dead space*, this equation requires accurate values for the mean tension of CO_2 or O_2 in alveolar air. Due to variations in alveolar gas tension measurement the behavior of the physiologic dead space has been the subject of much controversy and interest over the years, ever since Bohr introduced the theoretical basis for its estimation.[*]

Bohr's important equation simply states that the total amount of any gas exhaled is the sum of that from the alveoli and that from the dead space. It assumes that the concentration of a gas in the dead space is the same as that inspired. Therefore for CO_2:

%CO_2 expired × Tidal vol =

(%CO_2 alveolar × Alveolar ventilation) +

(%CO_2 inspired × Dead space vol)

Since:

Alveolar ventilation = Tidal volume − Dead space volume

the equation can be rewritten as:

Dead space volume =

$$\frac{(\%CO_2 \text{ alveolar} - \%CO_2 \text{ expired}) \text{ (Tidal volume)}}{(\%CO_2 \text{ alveolar} - \%CO_2 \text{ inspired})}$$

During air breathing the percentage of inspired CO_2 is essentially zero, and this term of the Bohr equation can be eliminated. It is required during CO_2 breathing.

In the practical employment of this valuable equation, accurate measurement of three of the four values required is easy. Difficulties in direct measurement of the fourth, mean composition of alveolar gas, have been discussed previously. Because of these difficul-

ties, the most reliable estimations of physiologic dead space will be made when measurements of *arterial* P_{CO_2} are employed to indicate mean alveolar P_{CO_2}. By multiplying both the numerator and denominator of the equation by (barometric pressure − P_{H_2O}), Bohr's expression is converted to:

Dead space =

$$\frac{(\text{Arterial } P_{CO_2} - \text{Expired } P_{CO_2}) \text{ (Tidal volume)}}{(\text{Arterial } P_{CO_2} - \text{Inspired } P_{CO_2})}$$

Influence of tidal volume and P_{CO_2} on dead space. Fig. 68-5, *A*, in which dead space estimations were based on direct analyses of P_{CO_2} in arterial blood, shows the manner in which physiologic dead space is increased by the stretch effect of increased tidal volume and by other effects of CO_2 breathing.[31,47,48] The figure indicates that the effects of elevated CO_2 tensions are not entirely due to mechanical effects associated with the increase in tidal volume but are related in part to pharmacodynamic or other influences of CO_2.[15,48] On this basis the dead space increase observed in exercise[78] is most probably due to the larger tidal volume. Considering such changes, it appears that the classic designation "dead" space is misleading and that the volume in question is subject to dynamic alteration by mechanical, chemical, and probably even nervous influences.

Alveolar ventilation and whole-body metabolism

A fundamental relationship in breathing is that the amount of CO_2 exhaled per minute is equal to the alveolar ventilation times the percentage of CO_2 in the alveolar air. In a stable state, exhaled CO_2 equals metabolic CO_2 production and:

CO_2 production (ml/min) =

Alveolar ventilation (ml/min) × %CO_2 alveolar

EXAMPLE: 220 ml/min = 4,000 ml/min × 5.5%

These fixed relationships[*] for a constant resting rate of CO_2 production are shown in Fig. 68-6. The formula and the illustration show that at any particular level of CO_2 production, doubling of alveolar ventilation must eventually result in a reduction of alveolar P_{CO_2} to about one half its original value. This can occur in voluntary or

---

*See references 7, 48, 59, 102, and 103.

---

*CO_2 production is usually expressed as STPD, alveolar ventilation as BTPS. With the use of such values the formula shown above should be rewritten as CO_2 production (ml/min STPD) = 0.83 × Alveolar ventilation (ml/min BTSP) × %CO_2 alveolar, in which the standard conversion factor at 760 mm Hg is:

$$0.83 = \frac{P_B - 47}{P_B} \times \frac{273°}{273° + 37°}$$

To convert measurements at ATPS to BTPS the factor used is 1.21.

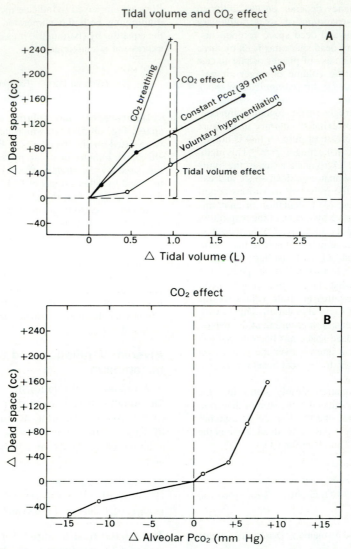

Fig. 68-5. Dynamic character of "dead space." **A,** Relation of change in tidal volume and P_{CO_2} to physiologic dead space in normal men at various levels of alveolar P_{CO_2}. Increase in tidal volume results in increase in physiologic dead space regardless of whether greater tidal volume results from CO_2 administration, from voluntarily controlled breathing pattern by which tidal volume is progressively raised while slowing respiratory rate to hold alveolar P_{CO_2} constant, or from voluntary hyperventilation with decreased alveolar P_{CO_2}. Middle curve, representing influence of tidal volume uncomplicated by altered P_{CO_2}, suggests increase in dead space of about 10 to 12 ml/ 100 ml elevation of tidal volume. (Data from Cooper and Lambertsen[47]; physiologic dead space calculated by Bohr equation, using arterial P_{CO_2} in six subjects to determine influence of CO_2 administration, and end-tidal P_{CO_2} in eight subjects for both voluntary hyperventilation and iso-P_{CO_2}.) **B,** Effect of change in alveolar P_{CO_2} on physiologic dead space. Data in **A** extracted to determine pharmacodynamic and other effects on physiologic dead space of change in alveolar P_{CO_2} independent of physical factors associated with increased tidal volume. It appears that influences of tidal volume on dead space can be increased about 30 ml for each mm Hg elevation of P_{CO_2} and decreased about 3 ml/mm Hg fall in P_{CO_2}.[47]

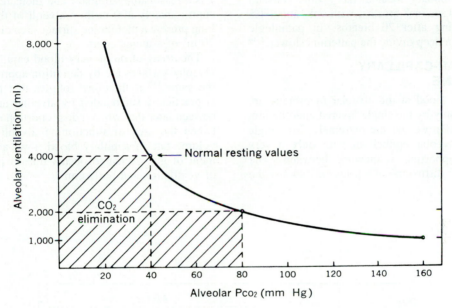

Fig. 68-6. Relationship between alveolar ventilation and alveolar P_{CO_2} at constant level of metabolic CO_2 production. Halving alveolar ventilation will eventually double alveolar P_{CO_2}, and doubling alveolar ventilation as in respiratory failure must halve alveolar P_{CO_2}. Considerable time is required to achieve new equilibrium state due to buffering of retained CO_2 or to release of CO_2 from body buffers.[74]

hysterical hyperventilation. On the other hand, halving of alveolar ventilation will eventually cause alveolar P_{CO_2} to double. Although the powerful respiratory stimulant effects of increased P_{CO_2} make the occurrence of the latter extreme inconceivable in the normal individual, subjects depressed by drugs and patients with pulmonary dysfunction do indeed show such degrees of hypercapnia.

Pulmonary "washout time"

One measure of the efficiency of overall pulmonary ventilation and the evenness of alveolar ventilation throughout the lung is the lung washout curve. This is produced by abrupt administration of 100% O_2 to an air-breathing individual and measurement of the time course of N_2 washout from the lungs (e.g., with a breath-by-breath nitrogen analyzer).[6,60] Such a curve is diagrammed in Fig. 68-7. When measured over a prolonged period, the experimentally derived curve is a complex function, consisting of an initial rapid exponential component representing N_2 washout from well-ventilated lung units, followed by a series of overlapping components representing N_2 elimination from less well-ventilated alveoli as well as rapidly and slowly perfused body tissues.[46,71] Fig. 68-7 shows that for the resting state the curve representing the pattern of pulmonary washout has a time constant of about 40 sec or 7 breaths, with only about 4% N_2 remaining after 20 breaths. In pathologic states and in deep diving the pattern is slower.[5,46]

ALVEOLAR-CAPILLARY MEMBRANE

Gas and blood at the alveolar membrane are separated only by the single-layered endothelium of the capillaries and the extremely thin single layer of alveolar epithelium with only a very tenuous supporting framework between. The barrier to diffusion of gases is about the smallest possible, hardly more than a gas-blood interface, and the surfaces are enormous. This area available for diffusion is closely related to the O_2 requirement in mammals from the tiny bat to the whale (Fig. 68-8).[106]

Pulmonary morphometric studies made by many investigators over the past 100 years have been reviewed and supplemented by Weibel.[113,115] Although some variation in the results is related to the methods used, Table 68-7 is a reasonable summary and indicates that the approximately 300 million alveoli of the human lung present a membrane surface area close to the 70 m² of a tennis court.

The *areas* of pulmonary blood capillaries and alveolar surfaces are by definition approximately the same.[113] It therefore follows that the blood is practically surrounded by alveolar air. It will be seen later that when a disproportion exists between the area of functioning alveoli and the volume rate of capillary blood flow, an impairment of gas exchange occurs due to inequality of ventilation and perfusion.

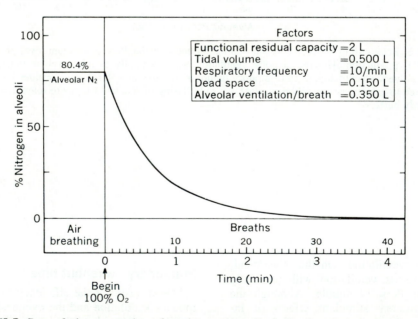

Fig. 68-7. Rate of nitrogen washout from human lung. Insert in diagram indicates factors that affect pulmonary washout time. Hypothetical washout curve for these values is shown. Initial phase of pulmonary nitrogen washout of normal lung is single exponential function. Subsequent elimination of nitrogen from blood and tissues follows complex composite of exponential rates of elimination.

Table 68-7. Estimates of dimensional features of pulmonary membrane

Membrane	Dimensions
Number of alveoli[112]	300×10^6
Alveolar diameter[112]	280 μm
Total alveolar area[115,117]	70 m²
Membrane thickness[85,98]	0.36 μm − 2.5 μm
Capillary length[112]	10.3×10^{-4} cm
Capillary volume[98,112]	140 ml
Blood transit time	
Rest[98]	0.75 sec
Exercise[98]	0.34 sec

PULMONARY BLOOD FLOW AND VASCULAR RESISTANCE

Whereas total *pulmonary blood flow* in normal individuals (without congenital shunts) is essentially identical to cardiac output, the *resistance of the pulmonary vessels* to the flow of blood can be increased by a number of factors. Pulmonary vascular resistance in the adult is low as compared with the mean for the many systemic organ vascular beds (1.7 as compared with 17 mm Hg/L of blood flow/min). In the fetus the overall vascular resistance in the uninflated lung is high, and most of the right heart output passes selectively through the specialized fetal shunts to the systemic circulation.

The low resistance of the pulmonary circulation and the desirably small degree of its alteration in normal states has led to long debate over its regulation. As an organ with no isolated requirement for change in organ work or energy metabolism, there is little need for intricate blood flow regulation in the lung, and the effects that

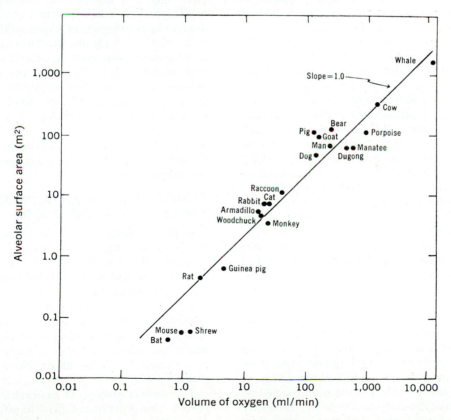

Fig. 68-8. Surface area of alveolar-capillary membrane in relation to O₂ requirements in mammals. (From Tenney and Remmers.[106])

do occur appear to be slight changes in the tone of the precapillary pulmonary arterioles.

Since tone is normally low, few physiologic substances or factors (except powerful vasodilators like acetylcholine) dilate the pulmonary bed. For the same reason, many direct or indirect influences can increase the tone. Among these is carotid chemoreceptor stimulation, which reflexly activates sympathetic nerves to the lung. Histamine, norepinephrine, serotonin, and angiotensin induce arteriolar constriction,[2,3,55,104] but it is not known whether these are functions or incidental effects.

A special feature of the pulmonary circulation is its local response to changes in overall circulatory and respiratory function. An increase in cardiac output during exercise, fever, or extrapulmonary disease is readily adapted to, evidently by simple passive dilatation of the distensible pulmonary vascular bed. Carotid body stimulation appears to induce small reflex increases in pulmonary tone via its activation of sympathetic nerves to the lung.[3]

It is the local effects of the respiratory gases, O_2 and CO_2, that have most confused respiratory physiologists, because they appear opposite to local effects of O_2 and CO_2 on vascular beds of systemic organs such as the brain and heart.

In the brain, increases in P_{CO_2} or $[H^+]$ or decreases in P_{O_2} dilate vessels, while increased P_{O_2} has little effect. Whereas increased P_{O_2} has a powerful constrictor influence on the ductus arteriosus,[41] it has little on normal pulmonary circulation,[3,104] and decreased P_{O_2} or increased P_{CO_2} or $[H^+]$ constrict local pulmonary precapillary vessels.[55,104] This inconsistency of response to fundamental substances such as O_2 and H^+ has led to speculations that these pulmonary vasoconstrictor effects are indirect, related to release of substances (e.g., histamine, serotonin) from the lung tissue itself[3,54,55,104] or to change in calcium ion concentration.[84]

The questions have special importance in two main directions: one is the abrupt redirection of blood flow and shunt closure at birth, the other is the conceived possibility that minute, localized influences of hypoxia in poorly ventilated alveoli may, in the mature animal, shunt blood to better ventilated sites for improved oxygenation.

Table 68-8 summarizes major factors known to modify pulmonary blood flow and relates them to states important in respiratory physiology or disease. Clearly the lung circulation is acted on simultaneously in each state by many, essentially all, of the factors cited, and the ultimate effect at any moment is therefore a composite of all

Table 68-8. Influences on pulmonary circulation in physiologic states

Physiologic state	Factors affecting pulmonary circulation
Exercise	Increased respiratory activity
Environmental or respiratory hypoxia	Increased cardiac output
	Increased sympathetic activity
Environmental or respiratory hypercapnia	Increased adrenal catecholamine release
Hyperthermia	Acidemia (increased $[H^+]$)
	Hypoxemia (arterial and venous)
	Carotid body reflex activation

influences. In the normal lung, it is the *absence* (not the presence) of prominent active influences on circulation that is the key to successful passive adaptation of the pulmonary circulation to the wide range of blood flow from rest to strenuous exercise. In disease, this absence of active regulation may indeed permit expression by a discrete local influence of hypoxia. If so, and it is the *absence* of O_2 that mediates constriction, then the mechanism must be indirect and not due to an action of O_2 itself. For reasons elaborated, pulmonary vascular tonus should be considered at least passively variable, not fixed, but *in the absence of pulmonary pathology there is no indication that either passive or active changes in pulmonary circulation are responsible for interference with pulmonary gas exchange in normal physiologic extremes.*

TRANSFER OF GASES BETWEEN THE LUNGS AND THE BLOOD

In the continuous stream of blood perfusing the pulmonary capillary bed a red blood cell requires only about 0.75 sec for its transit.[111] Gas diffusion within the alveolus to the alveolar wall is presumed to have a half-time of only about 0.001 sec, diffusion between alveoli and blood is rapid but not instantaneous, and the rate of transfer of each involves both physical diffusion through the interposed membranes and the chemical reactions with the blood gas transport system.[111]

Two distinct concepts of gas exchange are important. One is the overall mass transfer capacity of the entire lung and its circulation. The other is the complex of kinetic processes of gas movement toward equilibrium in the pulmonary capillary in the conversion of mixed venous to arterial blood. Each involves diffusion and chemical reaction.[58,111] Ultimately, it will be seen that, due to extensive and intimate association of blood and gas spaces, neither presents a functional limitation to respiratory gas exchange in the normal individual at rest or during strenuous exercise.

In the lungs of a normal man at rest, about 140 ml of pulmonary capillary blood is surrounded by approximately 2 L of alveolar air (functional residual capacity) in a capillary network having an area of about 70 m² and a length of about 1,000 miles, with only a freely permeable membrane averaging less than 1 μm in thickness to separate the alveolar gas from the blood.

Actually, in addition to diffusing through the alveolar and capillary membranes, a molecule of O_2 entering the blood must pass through the blood plasma in the pulmonary capillary and the wall and part of the intracellular fluid of the erythrocyte before entering into chemical union with hemoglobin. The sum of all these components represents the functional total "membrane" through which O_2 and CO_2 diffusion occurs.

Diffusion capacity of total lung system

A lung system could be 100% *efficient* in providing for specific gas pressure equilibrium, yet not be adequate in *capacity* to accomplish the gas exchange demands of metabolism. The measure of the adequacy of the overall transpulmonary gas exchange is the lung (pulmonary) diffusion capacity (D_L).[16,58,111] It differs for each respirable gas and is defined* as the *total amount of O_2 or other gas that passes across the alveolar membrane per minute per millimeter difference in the tension of the gas on the two sides of the membrane (i.e., between alveolar air and pulmonary capillary blood).*

$D_{L_{O_2}}$

The capacity of the lung for O_2 uptake is so great that in the 20-fold increase in O_2 demand associated with exercise, no significant failure of arterial oxygenation occurs. Diffusion capacity is therefore nonlimiting in the normal state. Expressed as an equation:

Pulmonary diffusion capacity for O_2 =

$$\frac{\text{Total } O_2 \text{ uptake (ml/min)}}{\text{Mean alveolar } P_{O_2} - \text{Mean pulmonary capillary } P_{O_2}}$$

---

*See references 40, 57, 58, 72, and 111.

or

$$D_{L_{O_2}} = \frac{\dot{V}_{O_2}}{P - p'}$$

Only one of these factors can be directly measured with definitive accuracy for O_2 during air breathing at sea level. This is $\dot{V}_{O_2}$, which is simply the existing whole-body O_2 consumption or metabolic rate. P, or mean alveolar P_{O_2}, can be estimated by a method such as the one devised for "effective" alveolar P_{O_2}.[96] The variable p' is the mean tension of O_2 in the pulmonary capillaries and thus is not directly measurable. By definition, p' is such that the difference between it and the mean alveolar tension would bring about exactly the same diffusion that actually takes place in the lungs. It lies somewhere between the P_{O_2} in the mixed venous blood entering the pulmonary capillaries and the P_{O_2} in the blood leaving them. It is not the simple arithmetic mean of these two tensions, but under resting conditions is much closer to the end-capillary value than to the venous value[16,40] (Fig. 68-9).

Mean pulmonary capillary P_{O_2}. Determination of p' theoretically can be accomplished by an integration procedure similar to that proposed many years ago by Bohr[40] and reviewed repeatedly.[16,58,111] Practically, this has not been accomplished during either rest or exercise while breathing air at sea level and probably will never be. The difficulty is related to (1) the existence of transpulmonary shunts that cause *peripheral arterial blood* to differ from the blood leaving functioning pulmonary capillaries and (2) the great efficiency of gas diffusion across the alveolar membrane, as a result of which a near-equilibrium for P_{O_2} occurs early in the passage of blood through the pulmonary capillary. Because of this efficiency, the final P_{O_2} difference between alveolar air and pulmonary capillary blood is small. Although the integration procedure for estimating mean P − p' requires precise knowledge of the final alveolar-capillary P_{O_2} difference, this value is estimated to be so small during air breathing (less than 0.01 mm Hg) that, even if end-pulmonary capillary blood was obtainable, no existing method for P_{O_2} estimation in blood or in alveolar air is accurate enough to be useful in determining (P − p').

Practical approaches to measurement of $D_{L_{O_2}}$. The magnitude of the aforementioned technical problems was appreciably reduced by the proposal of Lilienthal et al.[80] that during hypoxia the equilibrium between alveolar gas and pulmonary capillary blood would be less complete than in normal air breathing, and the final value of alveolar-pulmonary capillary ΔP_{O_2} should therefore be increased and the influence of shunted venous blood from peripheral arterial P_{O_2} minimized. This proposal permitted measurements of the P_{O_2} of peripheral arterial blood to be used along with "effective" alveolar P_{O_2} in estimating the final alveolar-blood O_2 pressure difference (membrane com-

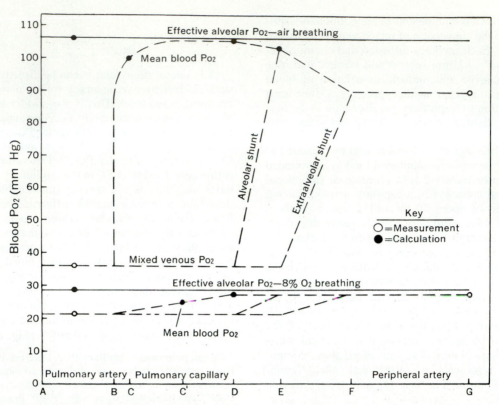

Fig. 68-9. Change in blood P_{O_2} during passage through lungs of six normal subjects (from data in Table 68-9).[76] Curves show influence of venous admixture on final alveolar-arterial P_{O_2} difference during air breathing (upper curves) and during anoxia (lower curves). Horizontal coordinates *A-B* represent mixed venous blood; *B-D,* length of pulmonary capillaries of optimally ventilated alveoli. *C* and *C'* indicate points in passage through capillary at which *mean* P_{O_2} is reached. This value cannot be calculated for air-breathing subjects because of inability to estimate final value for P_{O_2} in pulmonary capillary.[80] However, mean gradient for P_{O_2} between alveolar air and pulmonary capillary blood is, for purposes of illustration, considered to be the same as found during anoxia (about 3.5 mm Hg). This explains plotted value of over 100 mm Hg for mean pulmonary capillary P_{O_2} in subjects breathing air at sea level. At *E* is represented entrance of blood from poorly ventilated alveoli and at *F* from anatomic transpulmonary shunts. *F-G* indicates resultant peripheral arterial blood on which measurements are made. *During air breathing,* anatomic shunts add equivalent of mixed venous blood to blood that has come to nearly perfect equilibrium with alveolar air. Possibly this effect is slightly exaggerated by "alveolar" shunt, comprising blood from poorly ventilated alveoli (see text). Result is alveolar-arterial P_{O_2} difference that is not due to limitations of diffusion. *During anoxia,* in this case inhalation of 8% O_2 at 1.0 atm, influence of shunted mixed venous blood on already low P_{O_2} of arterial blood is minimized by steepness of Hb dissociation curve. Here alveolar-arterial P_{O_2} difference, although small, is almost entirely related to limitations of the low gradient available for diffusion. (Data from Lambertsen.[16,76])

ponent). Unfortunately the membrane (diffusion) component of the alveolar-arterial P_{O_2} gradient is small even during hypoxia (as low as 1 to 2 mm Hg) and therefore approximately equal to the error of an individual alveolar or arterial P_{O_2} measurement.[45,77,95] Review of a number of recent applications of this direct approach discusses its relation to indirect methods.[58,111]

The findings in two series of estimations of this type in normal men are summarized in Table 68-9. In the study of hypoxia at rest the data for blood P_{O_2} were obtained by microtonometric measurements on mixed venous blood (obtained by catheterization of the pulmonary artery) and peripheral arterial blood (collected by direct arterial puncture).[16,76] In the study of hypoxic

Table 68-9. Pulmonary diffusion constant for O_2 (DL_{O_2}) measured in hypoxia and in exercise: average Po_2 in mixed venous, arterial, and mean pulmonary capillary blood and alveolar air in normal young men at rest while breathing 8% O_2 in N_2[16,76] and during several levels of exercise while breathing 12% O_2 in N_2[108]

		Oxygen tension (mm Hg)						
Situation	Gas breathed	Mixed venous blood	Mean pulmonary capillary blood	Arterial blood	Alveolar air ("effective" formula)	P − p′	Oxygen consumption (ml/min)	DL_{O_2} (ml/min/ mm Hg)
Rest, normal oxygenation[16,76]	Air	36.0	?*	91.7	105.3	?*	292	?*
Rest, hypoxia[16,76]	8% O_2 in N_2	21.3	25.0	27.2	28.5	3.5	321	92
Rest, hypoxia[108]	12% O_2 in N_2	32.0†	38.5	44.0	48.5	10.0	261	26
Mild exercise, hypoxia[108]	12% O_2 in N_2	23.0	35.5	44.0	50.0	14.5	861	57
Moderate exercise, hypoxia[108]	12% O_2 in N_2	21.0	35.0	46.0	54.0	19.0	1,298	69
Severe exercise, hypoxia[108]	12% O_2 in N_2	19.0	35.0	47.0	55.0	20.0	1,820	91

*Mean capillary Po_2 and hence DL_{O_2} cannot be estimated from the data obtained during air breathing because of the influence of "venous admixture" on the value obtained for arterial Po_2[80] (Fig. 68-9) and must be derived from measurement of DL_{CO}.
†Mixed venous Po_2 estimated from measurements of arterial Po_2 and O_2 consumption together with assumed values for cardiac output and percent venous admixture.[108] Additional studies are cited in other reviews.[16,58,111]

exercising subjects, values for the Po_2 of mixed venous blood were indirectly estimated from measurements of arterial Po_2 and whole-body O_2 consumption; it was also necessary to assume values for cardiac output and percentage venous admixture.[80,108] It is evident that measurement of DL_{O_2} using O_2 itself is not practical for normal air breathing at rest and is a procedure applicable only to the hypoxic states used to make measurement possible. Although use of this procedure to study the effects of exercise adds the additional complication of hypoxia to the many other factors involved in the hyperpnea of exertion, the combined stress should indicate the true *maximum* diffusing capacity of the lungs for O_2 in physiologic extremes.

Table 68-9 shows that the value of about 90 ml O_2/min/mm Hg gradient obtained for DL_{O_2} in severe hypoxia (breathing 8% O_2 in N_2) is higher than that found with the lesser degree of hypoxia associated with breathing 12% O_2. It also indicates that increasingly strenuous exercise produces a progressive and prominent increase in the DL_{O_2}, again to about 90 ml/min/ mm Hg.

The bases for these physiologically desirable increases in DL_{O_2} are uncertain but may be related to (1) an increase in area of the pulmonary membrane, (2) a decrease in the effective diffusion distance, (3) a change in the initial Po_2 gradient, (4) an increase in the rate of pulmonary blood flow, and (5) gas exchange in the steep portion of the hemoglobin dissociation curve.

Since neither pulmonary blood flow nor alveolar ventilation were maximal at the exercise levels in Table 68-9, it is presumed that the value of 90 ml/min/mm Hg still does not represent a maximal level of DL_{O_2} in exercise or other states.[108] The correspondingly high DL_{O_2} in severe hypoxia indicates that in these two respiratory stresses O_2 uptake is not limited by diffusing capacity.[108] It is also important to realize that no real plateau or maximum possible value for pulmonary diffusion capacity has been measured.[111] It has been repeatedly emphasized that cardiac function rather than diffusion of O_2 across the alveolar membrane limits the capacity for sustained work.[108] These considerations, recognized many years ago, have been extensively reviewed.[16,17,58,111]

DL_{CO}

The impossibility of using O_2 itself for measurement of the pulmonary diffusion coefficient for O_2 during normal oxygenation led to employment of indirect procedures using CO as the test gas.

The CO method for estimating DL was first

suggested by Bohr[40] and first successfully employed by Krogh.[72] The principle of the method is the same as that underlying the O_2 method except that $\dot{V}_{CO}$ is used instead of $\dot{V}_{O_2}$ in the numerator of the diffusion capacity equation. The difficulty over determination of mean gas pressure in the pulmonary capillary is circumvented by assuming that CO in the low concentrations used combines so rapidly with hemoglobin after it enters the blood that p' can be assumed to be zero. The denominator then becomes simply P, the alveolar tension of CO, and the method should be applicable to any level of oxygenation. The calculation yields a diffusion coefficient for CO ($D_{L_{CO}}$), from which the corresponding coefficients for other gases may be computed on the basis that the diffusion of gases through membranes is directly proportional to their solubilities in water and inversely proportional to the square roots of their molecular weights. On this basis the $D_{L_{O_2}}$ would be 1.23 times the $D_{L_{CO}}$, and the $D_{L_{O_2}}$ would be 20.7 times the $D_{L_{CO_2}}$.[72] These calculations exclude all but physical diffusion factors.

Descriptions of the CO method, including practical applications, are available in Krogh's classic paper[72] and in a series of reviews on this topic.[16,17,58,111]

$D_{L_{CO_2}}$

CO_2, with its higher mass, diffuses less rapidly than O_2 in the gas phase. Since diffusion in water (and hence through alveolar membrane, plasma, and red blood cell water) is heavily influenced by solubility, diffusivity of CO_2 is about 20.7 times that for O_2. If the gradients for O_2 and CO_2 were the same and no chemical processes were involved, the $D_{L_{CO_2}}$ would be 20.7 times that for O_2.[45,111] However, the movement of molecular CO_2 from alveolus through its complex physicochemical exchanges with blood (Chapter 69) is not entirely governed by gas diffusion, and the rate-limiting physicochemical processes reduce the probable (but unmeasurable) diffusion capacity for CO_2 to a value close to that calculated for O_2. The influences of extreme diffusivity are counteracted primarily by the steep slope and linear characteristic of the CO_2 dissociation curve and to a smaller degree by its hydration reactions.[69,111]

$D_{L_{Inert}}$ gas

The lungs also have a diffusion capacity for inert respiratory gases such as nitrogen, helium, neon, and anesthetics.[70,111] Since inert gases are dissolved according to partial pressure and do not enter into rate-limiting chemical reactions in pulmonary capillary blood, their initial overall uptake by the perfused lung is diffusion related. However, once equilibrium has been reached with uptake and elimination in balance, the inert gas exchange has no influence on capacity for exercise or tolerance to hypoxia. It is pertinent to rates of development of anesthesia and rates of elimination of inert gases in diving (Chapter 75).

Membrane and red blood cell components of $D_{L_{O_2}}$

The overall passage of O_2 from alveolus to pulmonary capillary blood ($D_{L_{O_2}}$) involves two distinct steps in a continuous process.[105,111] Physical diffusion through the membrane and plasma into the red blood cell (Fig. 68-10) is followed by chemical combination of O_2 with hemoglobin.

The term D_M is used for the physical phase of diffusion through the composite membrane. Vc represents capillary blood volume, and, since the amount of O_2 or CO that can be taken up per minute is in part a function of the available hemoglobin, a large pulmonary capillary blood volume leads to a large $D_{L_{O_2}}$ or $D_{L_{CO}}$. This relates to the high $D_{L_{CO}}$ in athletes[88] and in altitude acclimatization.[58,111]

The symbol θ designates the velocity of O_2 in combination with hemoglobin per milliliter of blood, and the product of θ and Vc describes the overall rate for the entire lung (corresponding to D_M and also expressed in milliliters per minute per millimeters of mercury).[111] Therefore $D_L = D_M + \theta \times Vc$.

Influence of oxygen on diffusion capacity

In the measurement of the pulmonary diffusion capacity with CO, a peculiar defect in the basic method can be utilized to obtain information concerning the size of the membrane component of diffusion. The CO method of studying D_L depends on the occurrence of such a rapid combination of CO with hemoglobin that the mean pulmonary capillary P_{CO} is ordinarily assumed to be essentially zero and hence need not be measured.[16,57,58] However, it is now known that the $D_{L_{CO}}$ in man and the rate of combination of CO with hemoglobin are considerably affected by the level of oxygenation. The effect of P_{O_2} on the uptake of CO is apparently due to a dynamic, competitive interference by O_2 with the combination of CO and reduced hemoglobin. As a result, P_{CO} in the pulmonary capillary blood is not zero, and an error is introduced into the calcula-

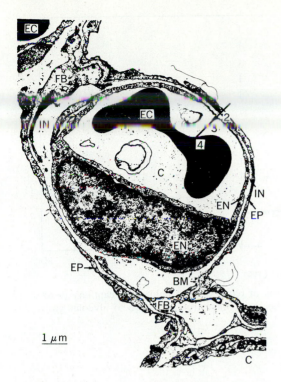

Fig. 68-10. Electron micrograph of alveolar capillary *(C)* from monkey lung with erythrocyte *(EC)*. Note endothelial cell lining of capillary *(EN)* and thin extensions of squamous alveolar epithelial cells *(EP)* covering alveolar surface. Interstitial space *(IN)* is bounded by two basement membranes *(BM)* and contains some fibroblast processes *(FB)* as well as a few connective tissue fibrils. (Magnification ×11,000.) (From Weibel.[114])

tion of $D_{L_{CO}}$. By performing the measurement both at a low alveolar P_{O_2} (where the reaction rate with hemoglobin is rapid) and during O_2 breathing, the "membrane component" of diffusion resistance can be estimated. It is high, approaching values measured with O_2 itself in hypoxia (Table 68-9). $D_{L_{CO}} \times 1.23$ even at rest is about 75 ml/min/mm Hg.[17,100] The red blood cell component of diffusion resistance, affected by the O_2-hemoglobin reaction rate, has been estimated to be less than the membrane resistance when hemoglobin saturation is less than 95%.[111] On the average the two components are of comparable importance.[111] During pure O_2 breathing at several atmospheres pressure, no red blood cell or hemoglobin function exists and only a membrane component is involved (Chapter 75).

Kinetics of equilibrium for oxygen and carbon dioxide

With a transit time of only 0.75 sec for blood in the pulmonary capillary, rates of O_2 reaction with hemoglobin and the related release of CO_2 from bicarbonate are important features in the adequacy of total pulmonary gas exchange.

Measurements in reaction chambers[58,111] and detailed computations[69,111] indicate that P_{O_2} rises rapidly on entry of mixed venous blood into the pulmonary capillary (Fig. 68-9).[58,111] At 37° C the half-time for O_2 uptake by red blood cells is about 0.07 sec.[99] Since the rise is not a simple exponential one, the pulmonary capillary P_{O_2} reaches essential equilibrium within about the first third of the capillary length. Therefore, while O_2-Hb reactions are considered rate-limiting in terms of overall pulmonary mass O_2 uptake, they do not prevent the P_{O_2} of end-pulmonary capillary blood from reaching within hundredths of a millimeter of the P_{O_2} of alveolar gas to which the blood was exposed. Even in exercise, with manifold increases in O_2 requirement, increased rate of pulmonary capillary blood flow, and reduced P_{O_2} of mixed venous blood entering the lung, arterial P_{O_2} attained is not importantly less than alveolar P_{O_2}.[78]

The situation for CO_2 is similar. Computation (not measurement) indicates that the rate of the composite CO_2 reactions and equilibrium may, despite a 20 times greater diffusivity from alveolus into plasma, be less rapid than for O_2.[58,111] The basis for rate limitation is that carbonic anhydrase does not exist in the plasma. Therefore CO_2 exchange requires action of carbonic anhydrase on CO_2 in the red blood cell and diffusion of bicarbonate ion into and chloride ion out of the red blood cell. It is presumed that because of time requirements for such exchanges, they are not complete as the blood leaves contact with the alveolar air.[58,111] Continued internal adjustment would then occur in the arterial blood, with a small (0.5 to 1.0 mm Hg) rise in P_{CO_2} (Fig. 68-11).[111] Whereas this change and the rate limitations that generate it have great significance to the understanding of pulmonary capillary gas exchange, the *degree* of effect on arterial blood–alveolar composition is not measurable (Table 68-5), and influence on overall respiratory control and capacity is negligible.

Alveolar-capillary diffusion block

Normal individuals breathing air at sea level have no significant limitation in pulmonary-blood exchange of either O_2 or CO_2 at rest or in exercise.[16,17,108] Physical diffusion is especially more

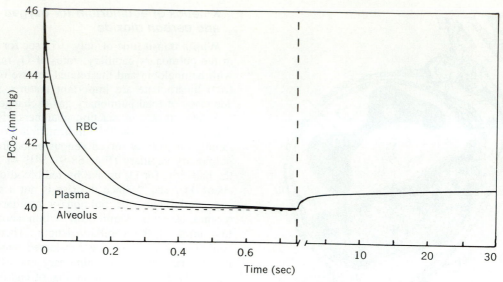

Fig. 68-11. Predicted changes in P_{CO_2} during transit of blood through pulmonary capillary (0 to 0.75 sec) and during continued CO_2 reactions after emergence from exposure to alveolar gas. (From Hill et al.[68])

than adequate in the normal state. However, when the alveolar–pulmonary capillary membrane becomes thickened (as by escape of fluid into the interstitial space to produce beginning pulmonary edema), an interference with O_2 diffusion across the membrane can occur even though alveolar ventilation is normal. This interference is termed *alveolar–pulmonary capillary diffusion block*.[32,116] As edema increases, fluid loss into the alveolar space occurs. If this is a film, it represents an added diffusion barrier; if it occupies the alveolus or becomes a froth of bubbles, ventilation of the alveoli involved is also abolished. Under such circumstances, positive-pressure breathing may aid in driving fluid back into the lungs, and high inspired O_2 pressures (even in a chamber at more than 1 atm) may provide the necessary increase in P_{O_2} gradient.

Interference with O_2 diffusion also occurs in chronic conditions such as interstitial pulmonary fibrosis. The resultant decrease in arterial P_{O_2} leads to chemoreflex respiratory drive. If this can increase alveolar ventilation, an actual lowering of arterial P_{CO_2} may occur despite the intrapulmonary defect. This is possible because CO_2 diffusivity is greater than that of O_2. Unless normal alveolar ventilation can be accomplished, interference with CO_2 elimination will occur and a rise in arterial P_{CO_2} will result.

BALANCE OF ALVEOLAR VENTILATION AND CAPILLARY PERFUSION

All alveoli are not ventilated equally during respiration. Neither are all alveoli perfused with blood at the same rate. A normal distribution pattern undoubtedly exists both for degrees of ventilation and for degrees of perfusion. This means that several possible functional states exist in diffusely scattered or regional groups of alveoli, including (1) well-ventilated alveoli, well perfused with blood; (2) poorly ventilated alveoli, well perfused with blood; (3) well-ventilated alveoli, poorly perfused with blood; and (4) poorly ventilated alveoli, poorly perfused with blood. The overall gas exchange, reflected by alveolar air and arterial blood, is the composite of all these, and even in the normal state, such factors as postural gravitational influences modify ventilation-perfusion relationships.[51]

It is possible, using measured values for alveolar ventilation and total pulmonary blood flow (cardiac output), to calculate an overall ventilation–blood flow ratio, which is a highly useful index of pulmonary gas exchange efficiency. In normal men at rest, this ratio is about 0.85. However, this is not a physiologic constant. In several normal situations such as exercise, hypoxia, and CO_2 breathing at rest the alveolar ventilation is increased out of proportion to changes in pulmonary blood flow. These conditions thus have a gross increase in composite ventilation-perfusion (V/P) ratio. In the severe alveolar hypoventilation of respiratory depression the overall V/P ratio becomes smaller. During breath-holding, it approaches zero. In peculiar physiologic stresses such as acceleration, different regions of the lung may experience V/P changes in opposite directions.[4]

There are two pathologic extremes in the concepts of alveolar V/P imbalance. One is that poor or zero ventilation of perfused lung (atelectasis) produces the same end result and influence on the composition of peripheral arterial blood as an anatomic vascular right-to-left shunt would produce. The opposite extreme is the infinitely ventilated alveolus with poor or zero perfusion; this is equivalent to an added component of physiologic dead space. Between these extremes the spectrum of ventilation and perfusion relationships influences the alveolar and arterial gas tensions and is subject to study and modification during unusual physiologic states and in pathologic conditions.[79,116]

ALVEOLAR-ARTERIAL GAS TENSION DIFFERENCES (A-a ΔP)

There is no reason to expect the normal individual to show appreciable A-a partial pressure differences for respiratory gases such as H_2O (dependent only on temperature), the inert gas N_2 (which has been in physiologic equilibrium in most individuals for generations), or CO_2.

The metabolically active CO_2 and O_2 differ considerably. Whereas both diffuse rapidly through normal alveolar membranes, fully arterialized blood from well-ventilated alveoli, when mixed with poorly arterialized or venous blood, will show a greater fall in Po_2 than rise in Pco_2 (this is because the difference in Pco_2 between arterial and mixed venous blood is only about 6 mm Hg, whereas the difference in Po_2 is 55 mm Hg). Moreover, the shapes of the dissociation curves of O_2 and CO_2 are such that an increase in the ventilation of alveoli that are already well ventilated increases the elimination of CO_2 more than the absorption of O_2. The Pco_2 of blood leaving such alveoli may be lowered sufficiently to compensate for subsequent admixture of blood from poorly ventilated alveoli or shunted venous blood. The result will then be a normal arterial Pco_2 in association with a subnormal arterial Po_2. These factors, now an integral part of the understanding of pulmonary respiratory functions, were first pointed out more than 50 years ago.[18]

They are described in detail in reviews of alveolar gas exchange.[25,51,116]

Alveolar-arterial ΔPo_2

Despite the near-perfect equilibrium between alveolar and pulmonary capillary gases, there is no doubt about the existence in normal persons of an alveolar-arterial difference for O_2.[37,86] There are three factors responsible for the magnitude of the A-a ΔPo_2 in normal and pathologic states:

1. *Limitation of diffusion.* Although this is undetectable in air breathing, it becomes evident in hypoxia[80] and in pulmonary disease.[46] Existence of a membrane (or diffusion) factor under resting conditions would imply an inefficiency of the alveolar gas exchange mechanisms that would be inconsistent with the highly effective oxygenation known to occur in exercise. During work an O_2 uptake of as much as 10 times the resting value can be accomplished, while the time spent by the blood in the pulmonary capillary is reduced by half.[98] A pathologic interference with diffusion therefore has an exaggerated effect in exercise.

2. *The effectiveness of alveolar ventilation in relation to the alveolar circulation.* This is significantly different in different alveoli, but because the top of the O_2 dissociation curve is flat, relative hyperventilation of some alveoli cannot raise the Po_2 of the mixed arterial blood enough to compensate for relative underventilation of others.

3. *True shunts.* The arterialized blood leaving optimally ventilated alveoli normally becomes contaminated to a small degree with venous blood from the bronchial circulation or other forms of anatomic shunts, including anatomic pulmonary arteriovenous communications.*

The two identifiable factors (uneven alveolar ventilation and contamination with venous blood from the bronchial venous and pulmonary arterial systems) functionally act as one and are designated *venous admixture*.[80] Both of these exist even in normal persons, the combination representing an effective shunt of about 2% of cardiac output at rest.[16,25,38] Fig. 68-9, based on measurements of normally oxygenated and hypoxic men, indicates the patterns and quantitative influences of oxygenation and subsequent contamination of pulmonary capillary blood.[76] It shows that, in alveoli in which both ventilation and circulation are optimal, the venous blood entering the alveolar capillaries takes up O_2

---

*See references 16, 25, 33, 80, and 94.

rapidly at first and then more slowly, but before it leaves the air-surrounded capillaries, all but a small portion of its hemoglobin molecules are oxygenated and its Po_2 becomes essentially equal to that of alveolar air. When it reaches the pulmonary venules, however, it mixes with blood from numbers of alveolus-capillary units in which the ventilation-perfusion relationships were less favorable and from which the blood emerged at less than full O_2 saturation and with a Po_2 less than that in the blood from optimally ventilated alveoli. Rediffusion of O_2 thereupon occurs *within* the pulmonary venous system from the streams of blood in which the Po_2 is high to those in which it is low. This is the functional component of unequal ventilation-perfusion labeled *alveolar shunt* in Fig. 68-9. The contributions from the bronchial circulation and from anatomic short circuits (normal or abnormal) between the pulmonary arterial and venous systems have a similar effect (the *extra-alveolar shunt*). Since the Po_2 cannot be higher than the alveolar level in any of the blood, it must be below the alveolar level in the mixed blood reaching the left side of the heart. The sum of the gas tensions of the mixed arterial blood also must be lower than the alveolar level by the difference between the fall in Po_2 and any rise in Pco_2.

Administration of 100% O_2 should equalize the Po_2 in all ventilated alveoli, thus wiping out the alveolar shunt factor and providing a means for separating the overall venous admixture into its components, inequalities of ventilation/blood flow and anatomic shunts.[44,58,111] Actually the A-a difference is markedly increased by O_2 inhalation, presumably because the anatomic or extra-alveolar shunts still operate or have a magnified influence in lowering arterial Po_2. The O_2 required to saturate the unoxygenated hemoglobin molecules now is provided from the increased amounts of the gas in physical solution in the plasma; the Po_2 must fall almost to 100 mm Hg before significant deoxygenation of the hemoglobin occurs (p. 1725), and until then any venous admixture will produce an extremely sharp decrease in the arterial Po_2. This fall in Po_2, due to shunted blood, is estimated to be about 14 mm Hg / 1% of cardiac output during O_2 breathing.[16]

Inhalation of mixtures low in O_2, on the other hand, diminishes the influence of both types of shunt and causes a decrease of the A-a difference. The contribution of inequalities of ventilation/perfusion is minimized primarily because blood leaving even the best ventilated alveoli is at a level of O_2 saturation where hemoglobin yields larger amounts of O_2 for a given drop of Po_2, so

that the sharing of some of its O_2 with less well-oxygenated blood produces a smaller drop in Po_2 of blood after it leaves the alveoli. The anatomic extra-alveolar shunt may become less effective in lowering arterial Po_2 because the contribution from pulmonary arteriovenous communications has a Po_2 closer to the arterial level than before. The same is true of the blood from the bronchial vessels, which are markedly dilated by hypoxemia.[8]

Quantitation of transpulmonary shunt (venous admixture)

It is possible to determine the approximate proportion of total cardiac output that does not become oxygenated in the lungs; measurements on peripheral arterial and mixed venous blood and expired air are employed, as follows:

$$\text{Venous admixture or physiologic shunt} =$$
$$(\% \text{ of cardiac output})$$
$$\frac{\text{Pul. cap. } O_2 \text{ cont.} - \text{Arterial } O_2 \text{ cont.}}{\text{Pul. cap. } O_2 \text{ cont.} - \text{Mixed venous } O_2 \text{ cont.}} =$$
$$\frac{\text{Pul. cap. \%Hb sat.} - \text{Arterial \%Hb sat.}}{\text{Pul. cap. \%Hb sat.} - \text{Mixed venous \%Hb sat.}}$$

Although arterial and mixed venous O_2 content or percent saturation can be measured, these values cannot be directly determined in pulmonary capillary blood. However, under normal conditions, it is possible to estimate the latter on the assumption that, with its normal near equilibrium, final pulmonary capillary Po_2 is essentially equal to the mean (effective) alveolar Po_2.[80] Since the approach to alveolar-pulmonary capillary Po_2 equilibrium is estimated as less than 0.01 mm Hg,[40] this is a practical assumption. An assumed normal value for mixed venous oxygenation is sometimes employed in the formula for estimating percent transpulmonary shunt.[16,80]

The magnitude of right-to-left shunt in certain pathologic states can be so large as to cause death. Normal values for venous admixture, including the components of true anatomic shunt and the influence of disproportionate alveolar ventilation and blood flow, have been reported as being approximately 2% of cardiac output.[16,38] Variations can be expected to be small. The effect of a relatively constant degree of venous admixture on the A-a ΔPo_2 in normal subjects in different ventilatory states leading to several different levels of alveolar Po_2 is shown in Fig. 68-12. The higher the alveolar (and end-pulmonary capillary) Po_2, the less steep is the hemoglobin dissociation curve and the greater is the lowering of pulmonary capillary Po_2 as it becomes contaminated with mixed venous blood.

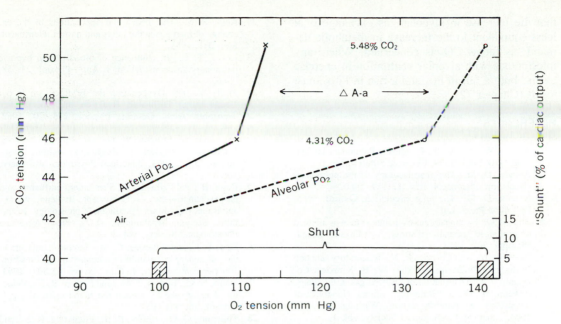

Fig. 68-12. Influence of increased ventilation at rest on alveolar-arterial Po_2 difference (average values in eight subjects[48]). During hyperventilation produced by CO_2 inhalation at rest, both arterial and effective alveolar Po_2 are increased. Since alveolar Po_2 is elevated out of proportion to rise in arterial Po_2, A-a ΔPo_2 increases from air-breathing value of 9.5 mm Hg to 24 and 27 mm Hg with hyperventilation. This gross change represents not change in ventilation/perfusion or interference with diffusion, but changing degree of influence of essentially constant degree of "venous admixture" (here about 4% of cardiac output) at different parts of hemoglobin dissociation curve. This effect has been demonstrated elsewhere.[42]

ADEQUACY OF PULMONARY GAS EXCHANGE

In respiration at rest the blood approaches its final equilibrium with the alveolar air early in its course through the alveolar capillaries (Fig. 68-9). When the exercising muscles remove more oxygen from the blood, the Po_2 of mixed venous blood is caused to be lower than it is at rest, and this increases the Po_2 gradient at the entrance to the alveolar capillary network. In exercise the rate of blood flow through this network also increases, which tends to push the point at which O_2 tension approaches equilibrium between pulmonary capillary blood and alveolar air farther toward the efferent end of the capillaries. Both factors increase the mean ΔPo_2, and therefore the amount of O_2 taken up at a given overall DL_{O_2}, but they would not increase the DL_{O_2} itself. This is determined by the characteristics of the membrane and the blood characteristics, not by the gradient across it. The DL_{O_2} increases during exercise, however, and the value must be large to provide the O_2 transfer that occurs in exercise.

According to the DL_{O_2} equation, a diffusion capacity of 80 ml/min/mm Hg during severe work would support an O_2 uptake of 4 L/min only at a mean ΔPo_2 of $4,000 \div 80 = 50$ mm Hg and a mean pulmonary capillary Po_2 of $100 - 50 = 50$ mm Hg. O_2 uptakes of this magnitude are close to the limit of capacity of men in excellent physical condition. The limit of oxygenation is set by the ability to achieve near equilibrium between alveolar and pulmonary capillary Po_2 at the existing DL_{O_2}, when pulmonary ventilation and pulmonary circulation are already maximally increased and the time spent in the alveolar capillaries greatly reduced. A further increase in the O_2 requirement above this during air breathing must necessarily result in a lower arterial O_2 saturation and tension, which could be obviated only by increasing the inspired Po_2.

In exercise, as under all physiologic conditions, the *elimination of CO$_2$* from blood is more readily accomplished than the uptake of O_2. The efficiency of CO_2 elimination is evidenced by the fact that alveolar Pco_2 tends to decrease even during mild and moderate exercise, indicating that the increase in alveolar gas exchange is at least equivalent to the increase in metabolic demand, as far as CO_2 is concerned. Where any interference with alveolar ventilation in exercise exists, both a fall in Po_2 and a rise in Pco_2 must result (Chapter 72).

REFERENCES
General reviews

1. Adams, W., and Veith, I., editors: Pulmonary circulation, New York, 1959, Grune & Stratton, Inc.
2. Aviado, D. M.: The pharmacology of the pulmonary circulation, Pharmacol. Rev. **12**:159, 1960.
3. Aviado, D. M.: The lung circulation, Oxford, 1965, Pergamon Press, Ltd.
4. Barr, P. O.: Pulmonary gas exchange in man as affected by prolonged gravitational stress, Acta Physiol. Scand. **58**:suppl. 207, 1963.
5. Bates, D. V., and Christie, R. V.: Respiratory function in disease, Philadelphia, 1964, W. B. Saunders Co.
6. Bouhuys, A.: Distribution of inspired gas in the lungs. In Fenn, W. O., and Rahn, H., editors: Handbook of physiology. Respiration section, Washington, D.C., 1964, American Physiological Society, vol. 1.
7. Bouhuys, A.: Respiratory dead space. In Fenn, W. O., and Rahn, H., editors: Handbook of physiology. Respiration section, Washington, D.C., 1964, American Physiological Society, vol. 1.
8. Bruner, H. D., and Schmidt, C. F.: Blood flow in the bronchial artery of the anesthetized dog, Am. J. Physiol. **148**:648, 1947.
9. Caro, C. G.: Physics of blood flow in the lung, Br. Med. Bull. **19**:66, 1963.
10. Comroe, J. H., Jr., and Dripps, R. D.: Artificial respiration, J.A.M.A. **130**:381, 1946.
11. Courtice, F. C.: Lymph flow in the lungs, Br. Med. Bull. **19**:76, 1963.
12. Draper, W. B., and Whitehead, R. W.: Diffusion respiration in the dog anesthetized with Pentothal sodium, Anesthesiology **5**:262, 1944.
13. Dripps, R. D., and Comroe, J. H., Jr.: The respiratory and circulatory response of normal man to inhalation of 7.6 and 10.4 per cent CO_2 with a comparison of the maximal ventilation produced by severe muscular exercise, inhalation of CO_2 and maximal voluntary hyperventilation, Am. J. Physiol. **149**:43, 1947.
14. Eger, E. I., and Severinghaus, J. W.: The rate of rise of Pa_{CO_2} in the apneic anesthetized patient, Anesthesiology **22**:419, 1961.
15. Fishman, A. P.: Respiratory gases in the regulation of the pulmonary circulation, Physiol. Rev. **41**:214, 1961.
16. Forster, R. E.: Exchange of gases between alveolar air and pulmonary capillary blood: pulmonary diffusing capacity, Physiol. Rev. **37**:391, 1957.
17. Forster, R. E.: Diffusion of gases. In Fenn, W. O., and Rahn, H., editors: Handbook of Physiology. Respiration section, Washington, D.C., 1964, American Physiological Society, vol. 1.
18. Haldane, J. S., and Priestley, J. G.: Respiration. London, 1935, Oxford University Press.
19. Heller, M. L., Watson, R. T., Jr., and Imredy, D. S.: Apneic oxygenation in man: polarographic arterial oxygen tension study, Anesthesiology **25**:25, 1964.
20. Hyde, R. W., Lawson, W. H., and Forster, R. E.: Influence of carbon dioxide on pulmonary vasculature, J. Appl. Physiol. **19**:734, 1964.

21. Kety, S. S.: The theory and applications of the exchange of inert gas at the lungs and tissues, Pharmacol. Rev. **3**:1, 1951.
22. Lloyd, T. C., Jr.: Influence of blood pH on hypoxic pulmonary vasoconstriction, J. Appl. Physiol. **21**:358, 1966.
23. Moll, W.: Die Oxygenation der Erythrocyten in der Lunge durch Diffusion, Reaktion und spezifischen Transport, Arch. Gesamte Physiol. **275**:420, 1962.
24. Otis, A. B.: Quantitative relationships in steady-state gas exchange. In Fenn, W. O., and Rahn, H., editors: Handbook of physiology. Respiration section, Washington, D.C., 1964, American Physiological Society, vol. 1.
25. Rahn, H., and Farhi, L. E.: Ventilation, perfusion, and gas exchange—the $\dot{V}_A/\dot{Q}$ concept. In Fenn, W. O., and Rahn, H., editors: Handbook of physiology. Respiration section, Washington, D.C., 1964, American Physiological Society, vol. 1.
26. Said, S. I., and Banerjee, C. M.: Venous admixture to the pulmonary circulation in human subjects breathing 100 per cent oxygen, J. Clin. Invest. **42**:507, 1963.
27. Staub, N. C., Bishop, J. M., and Forster, R. E.: Velocity of O_2 uptake by human red blood cells, J. Appl. Physiol. **16**:511, 1961.
28. Thilenius, O. G., Hoffer, P. B., Fitzgerald, R. S., and Perkins, J. F., Jr.: Response of pulmonary circulation of resting, unanesthetized dogs to acute hypoxia, Am. J. Physiol. **206**:867, 1964.
29. Viles, P. H., and Shepherd, J. T.: Relationship between pH, Po_2, and Pco_2 on the pulmonary vascular bed of the cat, Am. J. Physiol. **215**:1170, 1968.
30. West, J. B., and Hugh-Jones, P.: Pulsatile gas flow in bronchi caused by the heart beat, J. Appl. Physiol. **16**:697, 1961.

Original papers

31. Altman, P. L., Gibson, J. F., and Wang, C. C.: Handbook of respiration, Philadelphia, 1958, W. B. Saunders Co.
32. Austrian, R., et al.: Clinical and physiologic features of some types of diseases with impairment of alveolar-capillary diffusion: the syndrome of "alveolar-capillary block," Am. J. Med. **11**:667, 1951.
33. Ayres, S. M., Criscitiello, A., and Grabovsky, E.: Components of alveolar-arterial O_2 difference in normal man, J. Appl. Physiol. **19**:43, 1964.
34. Baes, C. F., Jr., et al.: Carbon dioxide and climate: the uncontrolled experiment, Am. Sci. **65**:310, 1977.
35. Barker, E. S., et al.: Comparative evaluation of several methods for determining alveolar gas tensions in man, Fed. Proc. **8**:7, 1949.
36. Bartels, H., and Rodewald, G.: The arterial oxygen tension, the alveolar-arterial oxygen tension difference and further respiratory physiological data in healthy men, Arch. Gesamte Physiol. **256**:113, 1952.
37. Bartels, H., and Wrbitzky, R.: Bestimmung des CO_2-Absorptionskoeffizienten Zwischen 15 und 38° C. in Wasser und Plasma, Arch. Gesamte Physiol. **271**:162, 1960.
38. Bartels, H., et al.: Bestimmung von Kurzschlussdurchblutung und Diffusionskapazitat der Lunge bei Gesunden und Lungenkranken, Arch. Gesamte Physiol. **261**:99, 1955.
39. Bohr, C.: Pulmonary ventilation, Scand. Arch. Physiol. **2**:236, 1891.
40. Bohr, C.: Über die spezifische Tätigkeit der Lungen bei der respiratorischem Gessaufnahme und ihre Ver-

halten zu der durch die Alveolarwand stattfindenden Gasdiffusion, Scand. Arch. Physiol. **22**:221, 1909.

41. Born, G. V. R., et al.: The constriction of the ductus arteriosus caused by oxygen and by asphyxia in newborn lambs, J. Physiol. **132**:304, 1956.

42. Briscoe, W. A.: Comparison between alveolar arterial gradient predicted from mixing studies and the observed gradient, J. Appl. Physiol. **14**:299, 1959.

43. Brown, H.: Rare gases and the formation of the earth's atmosphere. In Kuiper, G. P., editor: The atmosphere of the earth and planets, ed. 2, Chicago, 1952, University of Chicago Press.

44. Cohen, R., and Overfield, E. M.: The diffusion component of arterial hypoxemia, Am. Rev. Respir. Dis. **105**:532, 1972.

45. Comroe, J. H., Jr.: Physiology of respiration, Chicago, ed. 2, 1975, Year Book Medical Publishers, Inc.

46. Comroe, J. H., Jr., et al.: The lung, ed. 2, Chicago, 1962, Year Book Medical Publishers, Inc.

47. Cooper, D. Y., and Lambertsen, C. J.: Effect of changes in tidal volume and alveolar P_{CO_2} on physiological dead space, Fed. Proc. **15**:39, 1956.

48. Cooper, D. Y., et al.: Effects of CO_2 induced hyperventilation upon the alveolar-arterial P_{O_2} gradient and the functional respiratory dead space in normal men, Fed. Proc. **12**:28, 1953.

49. Dejours, P., editor: Note to authors, Respir. Physiol. **25**:inside cover, 1975.

50. Dill, D. B., and Penrod, K. E.: Man's ceiling as determined in the altitude chamber, J. Appl. Physiol. **1**:409, 1948.

51. Engel, L. A., and Macklem, P. T.: Gas mixing and distribution in the lung. In Widdicombe, J. G., editor: International review of physiology. Respiration physiology, Baltimore, 1977, University Park Press, vol. 14.

52. Fasciolo, J. C., and Chiodi, H.: Arterial oxygen pressure during pure O_2 breathing, Am. J. Physiol. **147**:54, 1946.

53. Ferris, B. G.: Comparison of the A-A gradient determined by different methods, Fed. Proc. **9**:40, 1950.

54. Fishman, A. P.: Hypoxia on the pulmonary circulation, Circ. Res. **38**:221, 1976.

55. Fishman, A. P.: The sensing of oxygen tension in the pulmonary circulation. In Reivich, M., et al., editors: Advances in experimental medicine and biology, New York, 1977, Plenum Publishing Co., vol. 78.

56. Forster, R. E.: Factors affecting the rate of exchange of O_2 between blood and tissues. In Dickens, F., and Neil, E., editors: Oxygen in the animal organism, Oxford, 1964, Pergamon Press, Ltd.

57. Forster, R. E.: Rate of gas uptake by red cells. In Fenn, W. O., and Rahn, H., editors: Handbook of physiology. Respiration section, Washington, D.C., 1964, American Physiological Society, vol. 1.

58. Forster, R. E., and Crandall, E. D.: Pulmonary gas exchange, Annu. Rev. Physiol. **38**:69, 1976.

59. Fowler, W. S.: Lung function studies. II. The respiratory dead space, Am. J. Physiol. **154**:405, 1948.

60. Fowler, W. S., Cornish, E. R., Jr., and Kety, S. S.: Lung function studies. VIII. Analysis of alveolar ventilation by pulmonary N_2 clearance curves, J. Clin. Invest. **31**:40, 1952.

61. Gelfand, R.: Electronic instrument for breath-by-breath end-expiratory CO_2 sampling. In Frommer, P. L., editor: Digest of the 1961 International Conference on Medical Electronics, Princeton, N.J., 1961, RCA Laboratories.

62. Gerschman, R.: Biological effects of oxygen. In Dickens, F., and Neil, E., editors: Oxygen in the animal organism, Oxford, 1964, Pergamon Press, Ltd.

63. Gilbert, D. L.: Atmosphere and oxygen, Physiologist **8**(1):9, 1965.

64. Gilbert, D. L.: The interdependence between the biosphere and the atmosphere, Respir. Physiol. **5**:68, 1968.

65. Goddard, D. R.: The biological role of carbon dioxide, Anesthesiology **21**:587, 1960.

66. Gurtner, G. H., and Forster, R. E.: Can alveolar P_{CO_2} exceed pulmonary end-capillary CO_2? J. Appl. Physiol. **42**:323, 1977.

67. Haldane, J. S., and Priestley, J. G.: The regulation of the lung ventilation, J. Physiol. **32**:225, 1905.

68. Hill, E. P., Power, G. G., and Longo, L. D.: Mathematical simulation of pulmonary O_2 and CO_2 exchange, Am. J. Physiol. **224**:904, 1973.

69. Hill, E. P., Power, G. G., and Longo, L. D.: Kinetics of O_2 and CO_2 exchange. In West, J. B., editor: Bioengineering aspects of lung biology, New York, 1975, Marcel Dekker, Inc.

70. Hills, B. X.: Gas transfer in the lung, London, 1974, Cambridge University Press.

71. Jones, H. B.: Respiratory system: nitrogen elimination. In Glasser, O., editor: Medical physics, Chicago, 1950, Year Book Medical Publishers, Inc., vol. 2

72. Krogh, M.: The diffusion of gases through the lungs of man, J. Physiol. **49**:271, 1914-15.

73. Lambertsen, C. J.: Medical implications of high oxygen pressures, Trans. Coll. Physicians Phila. **33**:1, 1965.

74. Lambertsen, C. J.: Therapeutic gases: oxygen, carbon dioxide, and helium. In DiPalma, J. R., editor: Drill's pharmacology in medicine, ed. 3, New York, 1965, McGraw-Hill Book Co.

75. Lambertsen, C. J., and Gelfand, R.: Breath-by-breath measurement of respiratory function. Instrumentation and application, J. Appl. Physiol. **21**:282, 1966.

76. Lambertsen, C. J., et al.: Attempt at direct measurement of values required for calculating the pulmonary diffusion coefficient for oxygen, Fed. Proc. **8**:90, 1949.

77. Lambertsen, C. J., et al.: Relationship of oxygen tension to hemoglobin oxygen saturation in the arterial blood of normal men, J. Appl. Physiol. **4**:873, 1952.

78. Lambertsen, C. J., et al.: Respiratory and cerebral circulatory control during exercise at .21 and 2.0 atmospheres inspired P_{O_2}, J. Appl. Physiol. **14**:966, 1959.

79. Lenfant, C., and Okuba, T.: Distribution function of pulmonary blood flow and ventilation-perfusion ratio in man, J. Appl. Physiol. **24**:688, 1968.

80. Lilienthal, J. L., Jr., et al.: An experimental analysis in man of the oxygen pressure gradient from alveolar air to arterial blood during rest and exercise at sea level and at altitude, Am. J. Physiol. **147**:199, 1946.

81. Macklin, C. C.: The musculature of the bronchi and lungs, Physiol. Rev. **9**:1, 1929.

82. Macklin, C. C.: Pulmonic alveolar epithelium, J. Thorac. Surg. **6**:82, 1936.

83. Maximow, A. A., and Bloom, W.: Textbook of histology, ed. 2, Philadelphia, 1934, W. B. Saunders Co.

84. McMurtry, I. F., Reeves, J. T., and Grover, R. F.: Inhibition by verapamil of the pulmonary vasoconstrictive response to alveolar hypoxia, Physiologist **17**:285, 1974.

85. Meessen, H.: Die Pathomorphologie der Diffusion und Perfusion, Verh. Dtsch. Ges. Pathol. **44**:98, 1960.

86. Mellemgaard, K.: The alveolar-arterial oxygen differ-

ence: its size and components in normal man, Acta Physiol. Scand. **67**:10, 1966.

87. Miller, W. S.: The lung, Springfield, Ill., 1943, Charles C Thomas, Publisher.

88. Mostyn, E. M., et al.: Pulmonary diffusing capacity of athletes, J. Appl. Physiol. **18**:687, 1963.

89. Pappenheimer, J. R.: Standardization of definitions and symbols in respiratory physiology, Fed. Proc. **9**:602, 1950.

90. Plass, G. N.: The carbon dioxide theory of climatic change, Tellus **8**:140, 1956.

91. Rahn, H.: The sampling of alveolar gas. In Boothby, W. M., editor: Handbook of respiratory physiology, Randolph AFB, Tex., 1954, USAF School of Aviation Medicine.

92. Rahn, H., et al.: A method for the continuous analysis of alveolar air, J. Aviation Med. **17**:173, 1946.

93. Riley, R. L., and Cournand, A.: "Ideal" alveolar air and the analysis of ventilation-perfusion relationships in the lungs, J. Appl. Physiol. **1**:825, 1949.

94. Riley, R. L., and Cournand, A.: Analysis of factors affecting partial pressures of oxygen and carbon dioxide in gas and blood of lungs: theory, J. Appl. Physiol. **4**:77, 1951.

95. Riley, R. L., Campbell, E. J. M., and Shepard, R. H.: A bubble method for estimation of P_{CO_2} and P_{O_2} in whole blood, J. Appl. Physiol. **11**:245, 1957.

96. Riley, R. L., et al.: On the determination of the physiologically effective pressures of oxygen and carbon dioxide in alveolar air, Am. J. Physiol. **147**:191, 1946.

97. Robertson, H. T., and Hlastala, M. P.: Elevated arterial P_{CO_2} relative to predicted values during normal gas exchange, J. Appl. Physiol. **43**:357, 1977.

98. Roughton, F. J. W.: The average time spent by the blood in the human lung capillary and its relation to the rates of CO uptake and elimination in man, Am. J. Physiol. **143**:621, 1945.

99. Roughton, F. J. W.: Transport of oxygen and carbon dioxide. In Fenn, W. O., and Rahn, H., editors, Respiration section: Handbook of physiology, Washington, D.C., 1964, American Physiological Society, vol. 1.

100. Roughton, F. J. W., and Forster, R. E.: Relative importance of diffusion and chemical reaction rates in determining rate of exchange of gases in the human lung, with special reference to true diffusing capacity of pulmonary membrane and volume of blood in the lung capillaries, J. Appl. Physiol. **11**:290, 1957.

101. Schoedel, W.: Alveolarluft, Ergeb. Physiol. **39**:450, 1937.

102. Severinghaus, J. W., and Stupfel, M. A.: Alveolar dead space as an index of distribution of blood flow in pulmonary capillaries, J. Appl. Physiol. **10**:335, 1957.

103. Shepard, R. H., et al.: Factors affecting the pulmonary dead space as determined by single breath analysis, J. Appl. Physiol. **11**:241, 1957.

104. Staub, N. C.: The sensing of oxygen tension in the pulmonary circulation — discussion. In Reivich, M., et al., editors: Advances in experimental medicine and biology, New York, 1977, Plenum Publishing Co., vol. 78.

105. Staub, N. C., Bishop, J. M., and Forster, R. E.: Importance of diffusion and chemical reactions in O_2 uptake in the lung, J. Appl. Physiol. **17**:21, 1962.

106. Tenney, S. M., and Remmers, J. E.: Comparative quantitative morphology of the mammalian lung: diffusing area, Nature **197**:54, 1963.

107. Thurlbeck, W. M.: Structure of the lungs. In Widdicombe, J. G., editor: International review of physiology. Respiratory physiology II, Baltimore, 1977, University Park Press, vol. 14.

108. Turino, G. M., Bergofsky, E. H., Goldring, R. M., and Fishman, A. P.: Effect of exercise on pulmonary diffusing capacity, J. Appl. Physiol. **18**:447, 1963.

109. Vandam, L. D.: The functional anatomy of the lung, Anesthesiology **13**:130, 1952.

110. Van Slyke, D. D., Dillon, R. T., and Margaria, R.: Studies of gas and electrolyte equilibria in blood. XVIII. Solubility and physical state of atmospheric nitrogen in blood cells and plasma, J. Biol. Chem. **105**:571, 1934.

111. Wagner, P. D.: Diffusion and chemical reaction in pulmonary gas exchange, Physiol. Rev. **57**:257, 1977.

112. Weibel, E. R.: Morphometry of the human lung, Berlin, 1963, Springer-Verlag.

113. Weibel, E. R.: Morphometrics of the lung. In Fenn, W. O., and Rahn, H., editors: Handbook of physiology, Respiration section, Washington, D.C., 1964, American Physiological Society, vol. 1.

114. Weibel, E. R.: Morphometric estimation of pulmonary diffusion capacity, Respir. Physiol. **11**:54, 1970.

115. Weibel, E. R., and Gomez, D. M.: Architecture of the human lung, Science **137**:577, 1962.

116. West, J. B.: Pulmonary gas exchange. In Widdicombe, J. G., editor: International review of physiology. Respiratory physiology II, Baltimore, 1977, University Park Press, vol. 14.

117. Willson, H. G.: The terminals of the human bronchiole, Am. J. Anat. **30**:267, 1922.

CHRISTIAN J. LAMBERTSEN

69

Transport of oxygen, carbon dioxide, and inert gases by the blood

The transport of metabolically active gases involves both physical and chemical processes. The limited exchange of inert respiratory "vehicle" gases, such as nitrogen and helium, occurs by physical processes alone. Because of the massive requirements for O_2 uptake and for CO_2 elimination (Table 69-1), physical solution of O_2 and CO_2 in the water of blood is in itself physiologically inadequate as a means of transport of these gases.

At a temperature of 37° C and a normal arterial P_{O_2} of 95 mm Hg, O_2 can be physically dissolved in the water of whole blood only to about 0.30 ml/100 ml. With a normal cardiac output of about 5 L/min, this concentration of physically dissolved O_2 would provide a total tissue O_2 flow (O_2 content times blood flow) of only 15 ml O_2/min in contrast to the approximately 250 ml/min required even for rest. Therefore if there were no other mechanism for meeting the tissue requirements, the resting O_2 consumption of 250 ml/min would demand a pulmonary blood flow (cardiac output) of at least:

$$\frac{0.250}{0.30} \times 100 = 83.3 \text{ L/min}$$

Only under one condition can the O_2 requirements of the body be met entirely by O_2 in physical solution. This is when the alveolar P_{O_2} is raised sufficiently by administration of O_2 at artificially high ambient pressures (Chapter 75). To provide physically dissolved O_2 in adequate supply for most organs at rest, O_2 must be breathed at a partial pressure considerably in excess of 3 atm (more than 2,000 mm Hg), and at such pressures O_2 soon becomes toxic. Moreover, no method exists for increasing the gradient of CO_2 *from* the body tissues. Therefore an animal supplied by physically dissolved O_2 and devoid of

hemoglobin and related buffer functions would still require a means of transporting and eliminating metabolically produced CO_2.

The importance of specific chemical transport mechanisms for metabolic gases is most evident from analysis of muscular exercise. An O_2 consumption of 4 L/min, even assuming the possibility of extracting all 0.30 ml of O_2 physically dissolved in each 100 ml of blood, would entail the following cardiac output:

$$\frac{4.0}{0.30} \times 100 = 1,333 \text{ L/min}$$

Table 69-1 shows for both O_2 and CO_2 the magnitude of the daily gas exchange of a resting individual and the blood flow required to transport it. A mass of approximately 1 kg each of these gases is exchanged daily between the atmosphere and the tissues. Although all transport at both the lung and the metabolizing tissue requires transfer through a phase of physical solution and diffusion,[13] the rounded normal values in Table 69-2 show that dependence on the transport of O_2 and CO_2 in physical solution becomes important only in abnormal circumstances.

The transport of both O_2 and CO_2 as respiratory gases is accomplished with the aid of the peculiar properties of hemoglobin; because of these the blood is able to carry more than 20 ml O_2 and 60 ml CO_2/100 ml blood in chemical combinations that are readily dissociated as a result of small changes in the gas tensions to which the blood is exposed.

TRANSPORT AND RELEASE OF OXYGEN

A continuous and massive flow of O_2 from the atmosphere to the mitochondrial sites in metabolizing tissue makes possible the existence of

Table 69-1. Twenty-four hr logistics of O_2 and CO_2

Structure	Mass (kg)	O_2 consumed/ day (L)	CO_2 produced/ day (L)	R	Blood flow/ day (L)
Brain	1.4	67	67	1.00	1,080
Heart	0.3	42	36	0.85	324
Kidney	0.3	25	20	0.80	1,814
Digestive organs	2.6	73	51	0.70	2,160
Skeletal muscle	31.0	71	57	0.85	1,205
Skin	3.6	17	14	0.85	663
Residual	24.0	72	58	0.85	547
Whole body	63.0	368	294	0.85	7,800

Table 69-2. Amounts of respiratory gases in blood (ml dry gas/100 ml blood)

Gas	Chemically combined		Physically dissolved	
	Arterial	Mixed venous	Arterial	Mixed venous
O_2	20.0	15.2	0.30	0.12
CO_2	46.4	50.0	2.62	3.00
N_2	0.0	0.0	0.98	0.98

multicellular animal life. In mammals and other warm- or cold-blooded animals, this *O_2 flow* [16,59] begins (in the lungs) and ends (in the tissues) with diffusion over very short distances. The beginning and end of the transfer process concern infinite numbers of discrete sites, each with different but short diffusion paths, blood perfusion patterns, and metabolic requirements.

The important, limiting intermediate phase of mass transport by the blood involves (1) blood distribution by *cardiac work,* (2) characteristics of hemoglobin as a chemical carrier of O_2, and (3) the mechanisms setting the concentration of hemoglobin in the blood.

The success of the chemical magnification of the O_2 transport function has depended on the many related accidents of natural selection that have evolved hemoglobin. This evolution provided a renewable, soluble substance (hemoglobin) capable of selectively binding molecular O_2 at a location of relatively high Po_2 (in the lung or gill) and releasing it as unchanged molecular O_2 at regions of low Po_2 (the tissues). In the process the hemoglobin itself remains unmodified and able to repeat the process thousands of times a day, for months of life on end. Without the aid of this or other O_2 transport pigments the body size of mobile land and marine animals would not exceed that of small insects.

The chemical characteristics and structural forms of hemoglobin as well as factors influencing its functions are described in detail in several excellent reviews.*

---

*See references 1, 3, 6, 8, 13, 17, and 44.

Structure of hemoglobins

Evolution has produced the tetrameric structure of a protein (globin) having four long and interlinked polypeptide chains of precisely sequenced amino acids (Fig. 69-1).[21,36,66,67] Each chain is attached to a protoporphyrin (heme) group with a ferrous (Fe^{2+}) atom centered among the symmetrically arranged pyrrols of heme. Fig. 69-2 diagrams this attachment, with the ends of the four chains binding to an Fe^{2+}, and by way of this to each of the four pyrrol nitrogen atoms of the protoporphyrin. The iron binds by a fifth linkage to one of the globin chains. The sixth site for ferrous iron is shown as freely available to bind with ligands (e.g., O_2, CO).

Each of the four heme groups with its iron atom lies in what appears to be a hydrophobic pocket of the porphyrin-globin complex between helices of its associated polypeptide chain, accounting in part for its protection against destructive ionic reactions.[1,3,6]

Globin chains

Hemoglobin (Hb) is not a single substance, but has evolved in many variations of the basic functional form. The modifications relate primarily to the structure of its four globin chains, with the porphyrin-Fe component being constant. The normal adult hemoglobin (HbA), representing more than 95% of Hb in the adult, has two α and two β chains ($\alpha_2\beta_2$). Embryonic Hb of several forms begins to be produced even prior to establishment of a circulation and is found within 7 to 12 weeks of life in the human embryo.[43] Fetal Hb (HbF) has predominantly two α and two γ chains ($\alpha_2\gamma_2$) and higher O_2 affinity than HbA. Synthesis of β chains begins about 6 weeks before birth, and by 4 months postpartum substitution by the HbA is almost complete. It is not surprising that at no stage is any form of Hb pure, since traces of different forms are synthesized in both immature and adult animals.

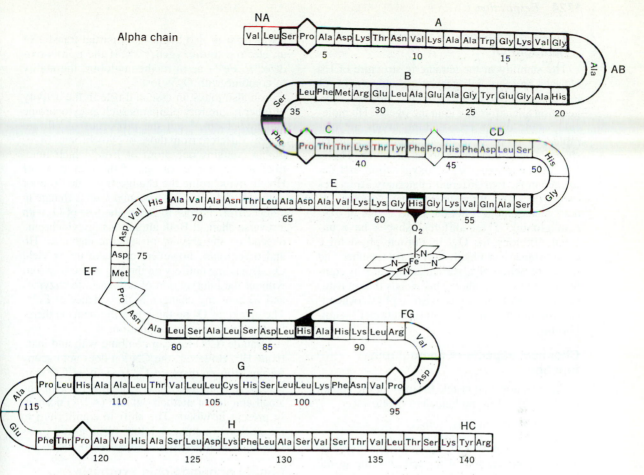

Fig. 69-1. The α globin chain of human hemoglobin, showing sequencing of amino acids and relation to prophyrin-Fe. (From Bartels and Baumann.[3])

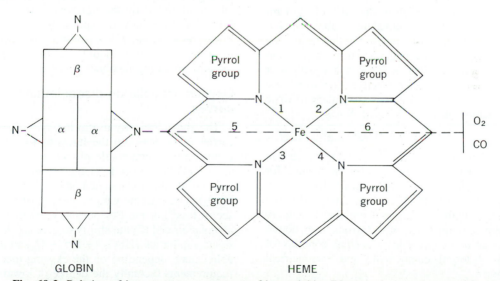

Fig. 69-2. Relation of iron atom to components of hemoglobin. Diagram shows only one of four heme groups that are attached to each globin molecule. Each heme contains one iron atom, joined by valency bonds 1, 2, 3, and 4 to the N atoms of the four pyrrol groups. By linkage 5 the iron atom is connected to an imidazole N in one chain of the globin. Actually, it is considered that heme is also attached to globin by two additional linkages.[35,45] Linkage to imidazole N may be responsible for maintaining iron in ferrous state and for ability to reversibly combine with O_2 or CO.[45] Sixth valency bond is available for combination with O_2, CO, and other substances. (Redrawn from Roughton; from Boothby.[34])

Three-dimensional structure of hemoglobin

The stability of the tetrameric structure of Hb is based on cross-linkages among amino acids of the α and β polypeptide chains.[1] Because oxygenation alters forces within the overall Hb molecule, some of the cross-linking bonds are disrupted by progressive oxygenation, altering the shape of the three-dimensional structure. Therefore in stable extremes of oxygenation and deoxygenation, two different quaternary structures exist (i.e., the three-dimensional structure of deoxyhemoglobin is not the same as that of oxyhemoglobin). These different shapes have unequal affinities for O_2. In normal physiologic circumstances in blood, they exist together and in the presence of a particular P_{O_2} are in equilibrium with each other. This means that in natural situations there is a coexisting physicochemical equilibrium and a "conformational" equilibrium.

Chemical aspects of hemoglobin function

The reaction responsible for most of the uptake of O_2 by blood in the lungs is as follows:

$$Hb + O_2 \rightleftarrows HbO_2$$

Hb combined with O_2 is the brilliant red pigment responsible for all the red coloring in the body. In the deoxygenated form, it is dark blue, and mixtures of oxygenated and deoxygenated (reduced) Hb show gradations of color between these extremes.

O_2 combines with Hb in the ratio of 1.34 ml O_2/gm Hb.[20] This is equivalent to about 22.4 L or 1 mole O_2/mole iron in Hb. On this basis, one molecule of O_2 reacts with one atom of iron, and one molecule of Hb reacts reversibly with four O_2 molecules. Since the reactive sites in Hb are the four iron atoms, it is now customary to write the stoichiometric reaction as follows:

$$Hb_4 + 4 O_2 \rightleftarrows Hb_4(O_2)_4$$

Identical relationships exist for CO, which can compete with O_2, molecule for molecule, for reversible attachment to the iron of Hb.

The ability of Hb to combine chemically and reversibly with O_2 derives from the amazingly complex circumstance of a ferrous iron bound within an *activating* and *protective* complex of protein (globin) components and symmetrical porphyrin.

Regardless of the chain involved, the Fe within the heme is bonded covalently to a terminal imidazole of the polypeptide chain (Fig. 69-2).

The bonding of O_2 to the iron of Hb is partially covalent; it is also bound by a partial transfer of an electron to dioxygen.[6,38,65] If the iron is oxidized to Fe^{3+}, as in methemoglobin, the ability to combine with O_2 is lost.[5]

Iron also plays an essential role in the activity of other respiratory pigments such as cytochrome and myoglobin,[19] but the properties of Hb are grossly different from those of cytochrome. Myoglobin contains one atom of iron, which functions in the manner of one of the iron atoms of Hb. In cytochrome the change from the O_2-poor to the O_2-rich state is associated with a change in valence from ferrous to ferric, the loss of O_2 with a reverse change. Both alterations of cytochrome depend on the activity of specific enzymes. Hb and myoglobin, however, can take up or yield O_2, depending entirely on the P_{O_2} of the solution, without the intervention of any specific enzymes and without any changes in the valence of Fe^{2+}. The uptake of O_2 by these two pigments is therefore oxygenation, not oxidation.

Cyanide (CN) ion can combine with and inactivate Hb. However, the CN ion does not occupy ferrous iron or displace O_2 from it as CO can. Rather it combines only with the Fe^{3+} of the small amount of methemoglobin (<1%) normally present in blood. The shift in equilibrium to the *oxidized* (methemoglobin) form allows gross formation of the stable cyanmethemoglobin.

Control of hemoglobin synthesis

Red blood cells and Hb are continuously destroyed and new cells and Hb formed. Evidently the equilibrium of formation and concentration of Hb is controlled by the hormone erythropoietin formed in the kidney in response to a lowering of P_{O_2} in renal tissue.[48,51] A resulting improvement in renal oxygenation should then inhibit Hb formation, providing feedback control.

Oxygen association-dissociation curves

The physiologically useful function of Hb as a carrier of O_2 between the lungs and tissues depends on two remarkable properties of this substance: (1) the ability to change from the deoxygenated to the oxygenated state (Hb + O_2 → HbO_2) at an extremely rapid rate (half-time, 0.01 sec or less[73]) at the P_{O_2} normally existing in the alveolar air and (2) the ability to give up O_2 in the tissue capillaries (HbO_2 → Hb + O_2) to a variable extent, depending on the existing metabolic requirements (actually the degree of lowering of capillary P_{O_2}). An effective release of O_2 at the tissue capillary depends on the fact that the amount of O_2 combined with Hb varies with all the major accompaniments of tissue metabolism, that is, P_{O_2}, acidity, and temperature. The relation of P_{O_2} to the degree of Hb oxygenation and the influences of metabolic factors are conveniently shown in the form of graphic *dissociation curves,* in which the proportion of HbO_2 to total Hb (percentage saturation) of the blood is plotted against P_{O_2}. Relationships of this type are shown in Figs. 69-3 and 69-4.

A maximal "capacity" of Hb for O_2 exists. However, the combination of O_2 with Hb is not an "all-or-none" reaction but involves reproducible patterns of equilibrium between the O_2 tension to which Hb is exposed and the degree of Hb oxygenation. This degree of oxygenation, designated as percentage Hb saturation, is measured in blood as shown in the following expression:

Basis for dissociation curve characteristics

The four iron atoms in one molecule of Hb do not become oxygenated simultaneously. As chance collisions with O_2 molecules occur, an iron will be occupied, adding to those already existing in blood as oxygenated iron. Evidently, HbA molecules are identical and have nearly equivalent affinity for O_2. Therefore 75% saturation of Hb with O_2 does not mean that three fourths of the Hb molecules in a blood sample are 100% oxygenated. It means instead that, on the average, three of the four iron atoms in each Hb molecule have become occupied by O_2. On either side of such a clear-cut ratio the degree of oxygenation of individual Hb molecules must vary.

$$\%\text{Hb saturation} = \frac{\text{Total blood } O_2 \text{ content (ml/100 ml)} - \text{Physically dissolved } O_2}{O_2 \text{ capacity of blood (ml/100 ml)} - \text{Physically dissolved } O_2}$$

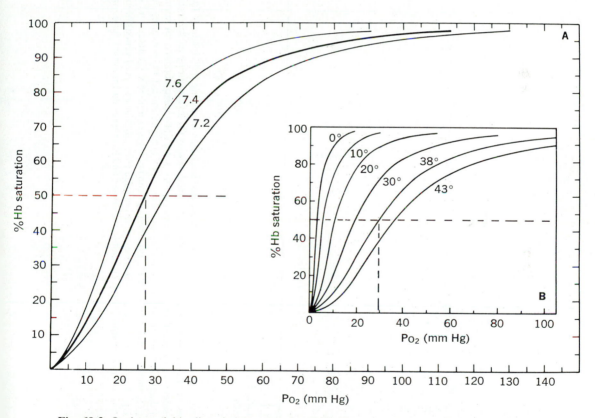

Fig. 69-3. Oxyhemoglobin dissociation curves for whole blood. Large diagram indicates influence of change in acidity of blood on affinity of blood for O_2. Curves are based on studies by Dill[41] and by Bock et al.[32] on blood of one man (A. V. Bock). At a particular P_{O_2} (e.g., 40 mm Hg), acidification of blood results in release of O_2. Action of changes in P_{CO_2} appear due in part to their effect on pH and in part to formation of carbamino compounds, displacing 2,3-DPG from Hb. Inset shows, for blood of sheep,[37] influence of temperature change on P_{O_2}-% HbO_2 relationships; increase in temperature (as in working muscle) aids in "unloading" O_2 from HbO_2, during hypothermia, hemoglobin has increased affinity for O_2.

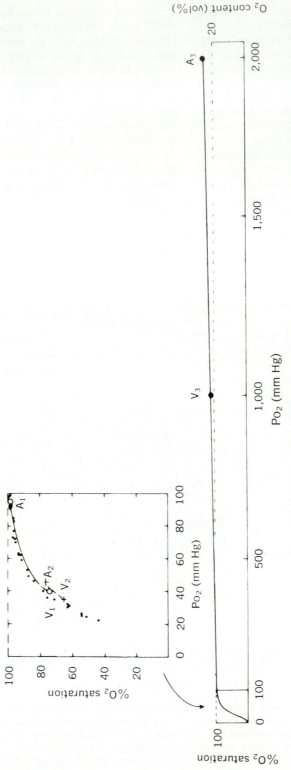

Fig. 69-4. Effect of environmental extremes on O_2 uptake and liberation in resting individuals. [16] In upper diagram are two pathways for air breathing in natural environments. A_1-V_1 shows transition from arterial to mixed venous O_2 levels during air breathing at sea level. [26] A_2-V_2 indicates same changes, in this case in 2 Andean residents at 14,900-foot altitude. [47] Numerous points (solid dots) were obtained on normal subjects breathing air and gas mixtures low in O_2 content. [53] In lower diagram, extended curve shows O_2 uptake by blood at high inspired O_2 pressures encountered in hyperbaric O_2 therapy and O_2 decompression after diving. Somewhat above 100 mm Hg Po_2, hemoglobin becomes completely saturated and slope of O_2 uptake curve then represents physical solubility of O_2/mm Hg. A_3-V_3 in lower curve shows expected change from arterial to mixed venous blood O_2 tension in individuals breathing pure O_2 at depth of 100 feet (about 4 atm). In this situation, HbO_2 circulates unchanged and Hb serves no important function. [16,54,56]

Adair concept

As proposed by Adair,[25] Hb combines with O_2 in four steps, each involving a different iron atom. The equilibrium constant for each reaction is different, since the combination of O_2 with one iron atom facilitates the reaction of O_2 with the remaining iron atoms.[21,62,66,67] This means that the affinity of Hb for O_2 increases with oxygenation. This change is so great that the oxy- form has 300 times the affinity for O_2 as the deoxy- form.[67] The result was conceived by Adair as having the following sequence of steps:

Reaction	Equilibrium constant
$Hb_4 + O_2 \rightleftarrows Hb_4O_2$	K_1
$Hb_4O_2 + O_2 \rightleftarrows Hb_4O_4$	K_2
$Hb_4O_4 + O_2 \rightleftarrows Hb_4O_6$	K_3
$Hb_4O_6 + O_2 \rightleftarrows Hb_4O_8$	K_4

In terms of mass action the equilibriums between oxyhemoglobin concentration and P_{O_2} for each oxygenation stage are shown as follows:

$$[Hb_4O_2] = K_1 P_{O_2}$$
$$[Hb_4O_4] = K_2 P_{O_2}$$
$$[Hb_4O_6] = K_3 P_{O_2}$$
$$[Hb_4O_8] = K_4 P_{O_2}$$

Adair expressed the relationships of these interacting steps to percentage Hb saturation and P_{O_2} mathematically in the equation below.

Monod-Wyman-Changeux (MWC) concept

Discovery that Hb exists in a three-dimensional, conformational equilibrium of deoxy- and oxyhemoglobin has provided a structural basis for the shape of the classic Hb association-dissociation curves.[6,64] The two main quaternary forms differ in affinity for O_2 as a result of differences in the degree of tight binding of the two α and two β chains by the salt bridges connecting oppositely charged amino acid side groups of these polypeptide chains.[3,6,64]

One physical configuration, occurring with deoxygenation, involves tight bridging, and this form is labeled "T," for *tense*. Its affinity for O_2 is the lowest. On full oxygenation the salt bridges are broken and the chains are less rigidly held; this is designated the "R," or *relaxed*, form. It has the highest affinity for O_2. These two forms are the extremes, and in transition from one to the other the irregular stepwise making or breaking of salt bridges in a population of millions of Hb molecules leads to the smooth rela-

tionship between O_2 *partial pressure* and percentage Hb saturation. This differs from the Adair concept in that not rate-constant, but rather absolute stable-state affinity is involved.

Influences of physiologic factors on hemoglobin oxygenation
Influence of P_{O_2}

As with many "dose-response" functions in biology, the relation between P_{O_2} and percentage Hb saturation is S shaped. Clearly a limit of 100% must be reached at some P_{O_2}. However, the steep middle portion of the curve and the characteristic early flattening of the upper portion are largely due to the increasing affinity for O_2 induced by preceding steps in oxygenation. A doubly fortunate situation exists: in the lungs, oxygenation of Hb is high or essentially complete over a wide range of the natural environmental conditions on earth. Then, as O_2 is released from Hb for use in the tissues, the resultant decrease in affinity for O_2 facilitates further release.

The flattened character of the upper portion of the Hb association curve means that over a considerable range of P_{O_2} at the upper extreme of the curve a change in arterial P_{O_2} does not grossly affect the volume of O_2 held by Hb. Therefore residents at moderate altitudes and patients with mild respiratory insufficiency suffer no important impairment in O_2 uptake under resting conditions.

The slopes of the curves indicate that in the tissue capillaries the amount of O_2 liberated per unit decrease in P_{O_2} is rather small until the P_{O_2} falls to about 60 mm Hg. Below this level, large amounts of O_2 can be liberated with only a slight further lowering of P_{O_2}. This is an important factor in the maintenance of an adequate diffusion gradient for O_2 from capillary to cell. At rest, even the mixed venous blood returning from the metabolizing tissues has a P_{O_2} of approximately 40 mm Hg and a Hb saturation of about 75%. This means that at rest near sea level most of the Hb iron of normal individuals remains oxygenated and does not truly participate in functional O_2 transport. Exercise, disease, and environmental change bring it into play.

Influence of exercise

To meet the needs of a living tissue in vivo for more O_2 requires an increase in the volume flow of blood. The circulation would be subjected to

$$\% \text{ Hb saturation} = \frac{K_1(P_{O_2}) + 2K_1K_2(P_{O_2})^2 + 3K_1K_2K_3(P_{O_2})^3 + 4K_1K_2K_3K_4(P_{O_2})^4}{4[1 + K_1(P_{O_2}) + K_1K_2(P_{O_2})^2 + K_1K_2K_3(P_{O_2})^3 + K_1K_2K_3K_4(P_{O_2})^4]}$$

less strain if each unit of blood yielded a higher proportion of its load of O_2, but this advantage to the circulation is purchased at the price of a lower tissue Po_2 and a higher tissue Pco_2, acidity, and temperature.

The most marked increases in total metabolic rate of O_2 consumption are due to increased activity of the skeletal muscles, which are probably less injured than most other body tissues by a low Po_2. The increased cardiac output of muscular exercise, although allocated mainly to meet the increased requirements of contracting muscles, is not enough to provide O_2 in adequate amounts for aerobic metabolism. The O_2 content and pressure in the blood leaving the exercising muscle are extremely low, and strenuous work ultimately depends on anaerobic processes. Thus a value as low as 10 mm Hg for Po_2, close to the lower physiologic limit, may be found in the venous blood of exercising muscle; this corresponds to a blood O_2 content of little more than 10% of the HbO_2 capacity. This means that Hb has released 90% of its transported O_2 and strikingly indicates the effectiveness of the physiologic reversibility of the O_2-Hb combination. If myoglobin were the O_2-carrying pigment, it would fail to release 50% of its O_2 even at 10 mm Hg Po_2 (Fig. 69-5).

It is important to recognize that the larger curves of Fig. 69-3, representing what is still probably the best available illustration of the influences of Po_2 on Hb saturation at various levels of pH, were actually constructed from data on one man, A. V. Bock.[32,46,53] Their usefulness depends on the apparent absence of important variation in the dissociation characteristics of Hb among different individuals.[53,70]

Influence of acidity and carbon dioxide

Change in pH, whether by metabolic acids or by CO_2, alters the affinity of Hb for O_2. The position of the dissociation curve is therefore shifted—to the right by an increase in $[H^+]$ and to the left by a decrease. The pH within the red blood cell is about 0.2 unit more acid than plasma,[41] but changes in plasma pH induce equivalent changes in the red blood cell interior. This means that, at a given Po_2, oxyhemoglobin will yield progressively larger amounts of its O_2 as the acidity of blood is increased in the tissues. It is therefore evident that tissue oxygenation can be attained at a higher mean tissue Po_2 as the acidity is increased both by entry of CO_2 into the tissue capillary and by any preexisting arterial acidosis. Thus the degree of hypoxia to which the tissues are exposed depends in part on the existing $[H^+]$.

The influence of $[H^+]$ on the dissociation curve of oxyhemoglobin is called the "Bohr ef-

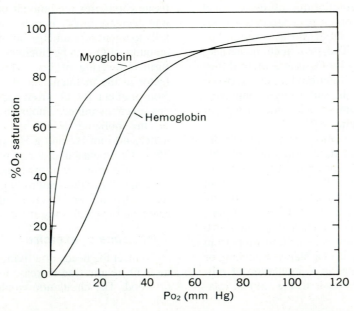

Fig. 69-5. Oxyhemoglobin and myoglobin dissociation curves of human blood at 38° C and pH 7.40.

fect'' after its discoverer.[33] Most of the shift produced by CO_2 is the result of P_{CO_2}-induced change in $[H^+]$ within the erythrocyte rather than the result of molecular CO_2, since similar effects are produced when the $[H^+]$ of the blood is increased without a concomitant rise in P_{CO_2}.[14,20]

The effect of changing pH appears related to change in the conformational equilibrium between the different three-dimensional forms of Hb, the change being brought about in increased $[H^+]$ by a stripping of protons from positively charged amino-acid residues, which provide the stabilizing salt bridges between globin chains in the tense, or T, form of Hb.[6] The affinity of iron for O_2 in the R conformation is high and essentially equal to that of separated α and β chains.[6]

CO_2 itself also has an effect by way of its formation of carbamino compounds with terminal amino acids of the globin chains; this CO_2 effect is extremely small.

The influence of $[H^+]$ on HbO_2 affinity is negligible or absent at the uppermost portion of the dissociation curve, where Hb is in the R form of mixed Hb_4O_6 and Hb_4O_8.[21] This is because the T to R form change induced in Hb by oxygenation occurs predominantly by bridge rupture on uptake of the third O_2 molecule; attachment of the fourth O_2 molecule does not induce major further proton release.[6] The upper portion of the curve is the important region in the final stages of O_2 uptake in the lungs. Therefore pH change due to loss of CO_2 aids only the initial phase of Hb oxygenation in the pulmonary capillaries.

Influence of temperature

Classic information regarding the influence of temperature on hemoglobin dissociation (Fig. 69-3, *B*) is largely based on the blood of one man.[37] The curves shown represent an application to measurements at 38° C on the blood of C. G. Douglas of relationships observed in pig and ox blood. The predicted curves have also been shown to represent rather closely the situation in vivo at low body temperature in the dog.

The effect of an increase in temperature resembles that of elevation of acidity and may actually be related to influences of temperature on H^+ activity. In the working muscle the rise in temperature should favor the release of O_2. In hypothermia, low tissue temperatures decrease cellular O_2 requirements. The lowering of blood temperature will at the same time increase the affinity of Hb for O_2 and require a lower capillary P_{O_2} for release of O_2. An imbalance in these changes can result in hypoxia and cell damage.

Influence of 2,3-diphosphoglycerate

The affinity of Hb for O_2 is prominently modified by organic phosphate concentration, as well as by H^+, P_{CO_2}, temperature, and P_{O_2} itself.[6] Actually each of these factors interacts with each of the others in physiologic states.[14,79,80] The physiologically significant phosphate in human and most animal red blood cells is 2,3-diphosphoglycerate (DPG).[69] It is normally present in adult red blood cells as one molecule per molecule of Hb and forms ionic bonds allosterically and reversibly with amino acid residues of the α or β chains.[6] The DPG alters release of O_2 from blood only as it can bind to the Hb molecule. It is probable that a resulting change in Hb conformation alters O_2 affinity.

The binding of DPG is selective for the deoxyform of Hb. This means that, in the tissue capillaries, a fall in P_{O_2} (and Hb saturation) leads to DPG combination with deoxyhemoglobin, diminishing the O_2 affinity of the mixed Hb[6,30,31] and facilitating further release of O_2.

The source of DPG in red blood cells appears to be the nucleoside inosine, along with pyruvate and inorganic phosphate. Synthesis of DPG occurs within the red blood cell, aided by red blood cell nucleoside phosphorylase. A higher concentration of DPG shifts the Hb dissociation curve to the right. A primary precipitating cause of increase in DPG in red blood cells is acute or chronic exposure to low O_2 pressure.[80] Change in concentration occurs in hours and evidently contributes to the initial adaptation to hypoxia at high altitude.[6] This increase in total DPG concentration is presumed to be triggered both by alkalosis and by the binding to the greater numbers of deoxyhemoglobin molecules, with an initial lowering of effective concentration.

Interaction among $[H^+]$, DPG, and CO_2 is at least in part due to a common site of binding for CO_2 and DPG at the NH_2 terminals of the α and β chains of Hb.[79] The DPG effect lessens with decreased $[H^+]$ as the binding sites lose protons and charge.[6,60]

Other influences on hemoglobin oxygenation and dissociation

Distinct differences in the shape of the relationship between P_{O_2} and percent HbO_2 exist among the various animal species that have been studied, from the tiny shrew to the whale.[28] Since the heme is identical in all species, the differences are due to specific globulins,[21] now largely identified.

Abnormal human hemoglobins are mostly the result of a substitution of one chain for another or of one amino acid for another in an α or β chain.

Human fetal oxyhemoglobin dissociation curves are to the left of those for the adult, indicating a higher affinity of HbF for O_2.[27,29] The HbF molecule is composed of two α globin chains with two γ chains in place of the adult β chains. With synthesis of the adult form there occurs a greater facility of O_2 release in the tissues. Since it coincides with a physiologic anemia during the postnatal period, the shift in position of the curve compensates for what would otherwise require a 50% increase in blood flow.[29]

Carbon monoxide competes with O_2 for the same bond with ferrous iron in Hb and, by its attachment, modifies the characteristics of the Hb reaction with O_2 (Chapter 75). The extremely high affinity of CO for Hb (about 200 times that of O_2) results in the firm occupation of one or more of the most reactive iron atoms, thus modifying the affinity of the others for O_2 (Fig. 75-11).

Methemoglobin is formed by oxidation of an iron atom to the ferric state (p. 1724). This form is incapable of oxygenation. The effect on the dissociation characteristics of the remaining oxyhemoglobin resembles that of CO.[21]

Range of blood oxygen uptake in extreme environments

Only a small range of the overall Hb dissociation curve is involved in most tissues, either at rest or in exercise. The portion involved is a function not only of metabolic demand, but also of the respiratory environment.

Besides exposure to air at or near sea level, the human can survive the low inspired P_{O_2} of high altitude and, for short periods of time, the high inspired P_{O_2} of O_2 or air breathing beneath the sea surface. Fig. 69-4 shows data obtained in a normal man from measurements of arterial and brain venous oxygenation near these extremes, together with values measured in arterial blood at sea level when air and low O_2 mixtures are breathed.[47,53,54]

The figure therefore illustrates the great range of the physiologic relationship between O_2 content and O_2 tension of arterial blood. The curve also indicates the approximate physiologic pathways for these relationships as blood becomes deoxygenated in the capillaries of a tissue and reoxygenated in the lung.

TRANSPORT OF CARBON DIOXIDE

CO_2 transport begins at the site of its formation within the metabolizing cell or mitochondrion. Here there is no flow of fluid to carry the metabolite away, and the interposed cell membrane and capillary endothelium prevent the ready passage of bicarbonate ions. Hence the entire amount of CO_2 produced must leave the cell by diffusion of dissolved, uncharged gas molecules that move from the regions of high P_{CO_2} within the cells to the lower partial pressure levels within the tissue capillaries. The distances involved in this diffusion are finite, but because the diffusion coefficient for CO_2 in the water of the tissues is very high (more than 20 times that for O_2), the P_{CO_2} gradient from cell to capillary probably does not exceed 1 to 2 mm Hg.[55]

Once the CO_2 molecule enters a blood capillary, the work of its transport to the lungs is accomplished by the heart. However, the efficiency of CO_2 transport depends on a number of important physicochemical characteristics of the blood itself and is intimately related to the function of O_2 transport.[15] The *dynamic events* that occur within the capillary and the resultant *distribution of CO_2 at equilibrium* in the venous blood can be pictured from the changes that occur in 1 L of arterial blood as it flows through the tissues. About 90% of the molecular CO_2 diffusing into tissue capillary blood is hydrated to become HCO_3^-. Table 69-3 shows such changes in the composition of blood in the tissue capillaries of normal men; the actual values appear in *Handbook of Respiration*.[26]

Carbon dioxide forms in arterial blood

Arterial blood, equilibrated with alveolar gas and on the way to metabolizing tissues, contains more CO_2 than it does O_2. Each liter of arterial blood (pH 7.40) that enters the tissues already carries about 22 mM of CO_2. This large amount is dictated by the total base available in the blood for the buffering of H_2CO_3 and by the CO_2 partial pressure imposed on the arterial blood by the balance between alveolar ventilation and the rate of whole body CO_2 production. In the subjects represented in Table 69-3, arterial P_{CO_2} averaged 41 mm Hg. The CO_2 carried to the tissues by arterial blood, in one sense an indication of inefficiency of the total mechanism for CO_2 removal, is important as a means of preventing excessive lowering of cell acidity as blood perfuses the tissues.

CO_2 in the arterial blood is distributed unevenly between the plasma and erythrocytes (Table 69-3). In the equilibrium stage indicated by the table, CO_2 exists in three forms in both the plasma and the erythrocytes, that is, as physically dissolved CO_2, as carbamino compounds, and as

Table 69-3. Changes in amount and distribution of CO_2 in transition from arterial to venous blood

	Arterial blood	Mixed venous blood	Arteriovenous difference
Total CO_2 in 1 L of blood (mmole)	22.0	23.9	1.9
Total CO_2 in 550 ml of plasma in 1 L of blood (mmole)	14.7	15.7	1.0
As dissolved CO_2	0.7	0.8	0.1
As bicarbonate	13.8	14.7	0.9
As carbamino CO_2	0.2	0.2	0.0
Total CO_2 in 450 ml of red blood cells in 1 L of blood (mmole)	7.3	8.1	0.8
As dissolved CO_2	0.5	0.5	0.0
As bicarbonate	6.0	6.4	0.4
As carbamino CO_2	0.8	1.2	0.4
P_{CO_2} (mm Hg)	41.0	46.0	5.0
pH	7.40	7.37	0.03
Hematocrit (%)	45.0		
Hb saturation (%)	98.0	74.5	23.5
Hb (mmole/L)	9.5		
Buffer slope ($\Delta BHCO_3/\Delta pH$)	−30.1		

bicarbonate ions. Plasma contains the largest amount of CO_2. Of this, most exists in the form of bicarbonate ions, and the ratio of bicarbonate to physically dissolved (molecular) CO_2 is normally about 20 to 1. Part of the physically dissolved CO_2 is hydrated according to the equilibrium reaction $CO_2 + H_2O \rightleftharpoons H_2CO_3$, in which the proportion of molecular CO_2 to H_2CO_3 is approximately 1,000:1. Within the plasma the hydration of CO_2 is not accelerated by the action of carbonic anhydrase as it is within the erythrocyte, where the enzyme occurs. Finally, the amount of CO_2 combined with $-NH_2$ groups of plasma protein to form carbamino protein is so small as to be physiologically negligible.

In the oxygenated erythrocytes of arterial blood the amount of dissolved CO_2 is slightly less than in plasma, since the water content of red blood cells is lower than that of plasma (Table 69-3). However, more CO_2 is present in the carbamino form, since Hb has numerous $-NH_2$ groups that hold CO_2 in the following manner:

$$\underset{\underset{H}{|}}{\overset{\overset{H}{|}}{Hb-N}} + CO_2 \rightleftharpoons \underset{\underset{COO^-}{|}}{\overset{\overset{H}{|}}{Hb-N}} + H^+$$

These are the terminal α-amino groups of the α and β chains of the Hb molecule.[50,72] Molecular O_2 and CO_2 thus are bound at different chemical sites on the Hb molecule. However, by their

binding each influences the reaction of Hb with the other, and the effects have prominent physiologic significance.[6,72]

The amount of CO_2 present as bicarbonate ions in the erythrocyte is considerably less than that carried as bicarbonate in the arterial plasma, the degree of difference depending on the Hb concentration.

Changes within tissue capillaries

The changes produced in the blood in its transition from the arterial state in passing through tissue capillaries will differ with the amount of CO_2 added to each liter of blood. This, in turn, depends on (1) the rate of blood flow per unit of metabolizing tissue and (2) the rate of metabolism of the perfused tissue. For these reasons the CO_2 pressure and composition of venous blood will be different for all tissues. The average or mean composition of blood from all the tissue capillaries is that of the mixed venous blood.

As physically dissolved CO_2 diffuses continuously from the tissues into the capillaries, it must first pass into the blood plasma, but as a result of diffusion within the blood the P_{CO_2} of the plasma and that of the red blood cells are elevated equally. During the dynamic period of transit through the capillary a precise and extremely rapid interplay among physical and chemical factors occurs (Table 69-4), resulting in a stable pattern of CO_2 distribution in the

Table 69-4. Measured reaction rates for CO_2 and O_2*

Reaction or function	Location	Reaction rate ($T_{1/2}$, sec)	Reference
O_2			
Diffusion across RBC membrane to combination with Hb	RBC	0.071	Roughton[21]
CO_2			
Diffusion across RBC membrane	RBC	0.001	Forster[11]
$H_2CO_3 \rightleftarrows H_2O + CO_2$	Plasma (uncatalyzed)	11.5	Forster and Crandall[12]
$H_2CO_3 \rightleftarrows H_2O + CO_2$	RBC (catalyzed)	<0.20	Forster[10]
HCO_3^-/Cl^- shift	Plasma $\leftrightarrow$ RBC	0.16	Dirken and Mook[9]

*Modified from Wagner.[24]

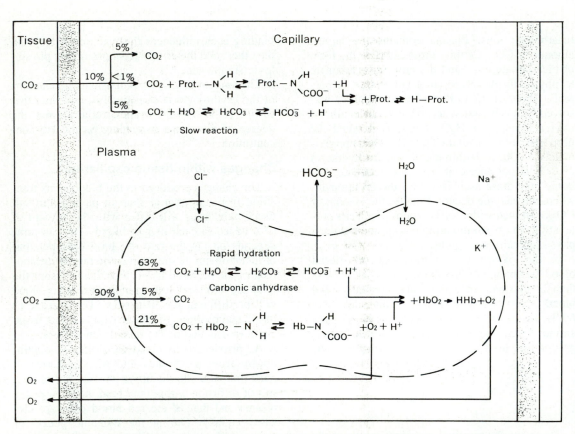

Fig. 69-6. Sequence of reactions involved in exchange of molecular O_2 for molecular CO_2 in tissue capillary as result of partial pressure gradients for these gases between tissue cell and capillary. Values accompanied by percent sign indicate approximate quantitative disposition of CO_2 entering blood from tissues. (Modified from Carlson; from Ruch and Fulton.[74])

mixed venous blood flowing to the lungs (Fig. 69-6).

The diffusion of molecular CO_2 into the capillary blood coincides with the diffusion of slightly greater amounts of O_2 into the tissues and hence with the deoxygenation of Hb. These events have been analyzed theoretically as "physiologic" O_2 and CO_2 dissociation curves to show the probable time course of changes in PO_2, PCO_2 and percentage Hb saturation in capillary blood.[40] Table 69-3 shows that a total of about 2.0 mM CO_2 entered a liter of blood in its passage through the capillaries, elevating PCO_2 from 41 to 46 mm Hg.

The 5 mm Hg rise in capillary PCO_2 corresponds to an increase of only 0.1 mM in physically dissolved CO_2 but is large enough to have considerable physiologic effects on chemical reactions and membrane functions. A similar small quantity of CO_2, about 0.1 mmole, remains in the plasma as bicarbonate ions formed in plasma and buffered by plasma proteins.[20] The balance of the CO_2 taken up by the capillary enters the red blood cell, where additional stages of the intracapillary gas exchange proceed. Thus approximately 1.8/2.0 mM or 90% of the CO_2 taken up by 1 L of blood enters the red blood cells and will depend on intraerythrocytic factors for its buffering and transport. It is in this phase of the rapidly progressing sequence that the role of Hb as a buffer and as the dominant factor in CO_2 transport becomes apparent.

Hemoglobin and carbon dioxide transport: role of hemoglobin and the red blood cell in facilitating formation of bicarbonate ion from molecular carbon dioxide

When Hb becomes deoxygenated in the tissues, it provides two mechanisms for the transport of CO_2. One is the combination of small amounts of CO_2 with the terminal amino groups of polypeptide chains of Hb to form carbaminohemoglobin. Hb aids most extensively in the transport of CO_2 by providing basic groups that accept and neutralize H^+ formed in the equilibrium reaction:

$$Carbonic\ anhydrase$$
$$CO_2 + H_2O \rightleftharpoons H_2CO_3 \rightleftharpoons HCO_3^- + H^+$$
$$Hb^- + H^+ + HCO_3^- \rightleftharpoons HHb + HCO_3^-$$

This acceptance of H^+ by Hb causes the first reaction to proceed to the right as the diffusion of O_2 from and CO_2 into the erythrocyte occurs. It ranks in importance with the function of Hb in O_2 transport.

Carbonic anhydrase and carbon dioxide transport

Meldrun and Roughton discovered that the first (hydration) stage of this reaction is greatly accelerated, nearly 5,000 times, in both directions by a zinc-containing enzyme called carbonic anhydrase.* This enzyme is in high concentration (about 0.1%) in the erythrocytes but is not present in significant amounts in blood plasma. It has similar importance in renal HCO_3^- reabsorption and H^+ excretion,[61] in the secretion of gastric acid, and in the secretion of bicarbonate ions into the pancreatic juice.[61]

The activity of carbonic anhydrase affects the first step of the conversion of molecular CO_2 to bicarbonate in the tissue capillaries and the final step in the evolution of molecular CO_2 from the blood in the lungs. Only these reactions, not the ionic dissociations or the ultimate equilibrium levels, are affected by the enzyme. However, by accelerating the hydration of CO_2 as it enters the red blood cell, this important enzyme makes possible almost the entire sequence of CO_2 uptake within the time, less than 1 sec, spent by the blood in the capillary bed.[21,24,71] On leaving an alveolar capillary the changes are so nearly complete that any alveolar-arterial blood PCO_2 differences are not readily detectable (Chapter 68). Unaided by carbonic anhydrase, these same processes could be expected to take as long as 200 sec for completion.[71] In such a rate-limiting situation, interference with CO_2 elimination would be inevitable. When carbonic anhydrase is inactivated, as can be done by a large number of drugs,[4] the predicted inefficiency of CO_2 elimination can be demonstrated.

Effects of altered oxygenation

The titration curves of reduced and oxygenated Hb illustrate the basis for the ability of Hb to neutralize H^+ formed on the entrance of CO_2 into the blood (Fig. 69-7). For nearly every molecule of CO_2 bound by the blood as HCO_3^- or in carbamino form, one hydrogen ion is liberated.[6] However, because reduced Hb is a weaker acid than oxyhemoglobin, deoxygenation results in basic groups of Hb becoming available for combination with the hydrogen ions. Oxygenation appears to increase the acidity of certain ionizable groups, now called "oxylabile." One such group is considered to exist in each heme, the most likely site being the imidazole group that provides the attachment for the iron.

---

*See references 4, 18, 21, 24, 61, and 63.

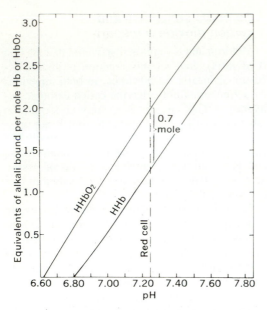

Fig. 69-7. Influence of pH on equivalents of alkali bound per mole of oxygenated or reduced hemoglobin. Titration curves of $HHbO_2$ and HHb are parallel. However, at any pH (e.g., the 7.25 pH within erythrocytes), $HHbO_2$ binds 0.7 equivalent more base than does HHb. Therefore for each mole of O_2 released from $HHbO_2$ in a tissue capillary, 0.7 mole of H^+ formed by entrance of CO_2 can be neutralized by HHb without altering pH. (Modified from Peters and Van Slyke.[20])

Actually the complete deoxygenation of 1 mmole of oxyhemoglobin to free 1 mmole of O_2 results in the neutralization of 0.7 mmole of hydrogen ions without a change in pH (see iso-pH 7.25 in Fig. 69-7).[15] Theoretically, at a tissue or whole-body respiratory exchange ratio of 0.7 (pure fat diet) the entire transport of CO_2 could thus be effected without a change in the pH of the blood as it passes from the arteries through the capillaries. In a more realistic circumstance represented by the brain, where essentially only carbohydrate is metabolized and the respiratory exchange ratio (R) is 1.0, an additional 0.3 mmole of H^+ must be buffered. This is done largely by Hb acting as a protein and involves a change in pH of about 0.02 unit. Under peculiar environmental circumstances such as the breathing of pure O_2 at several atmospheres pressure (Chapter 75) the tissue needs for O_2 may temporarily be met entirely by physically dissolved O_2. Oxyhemoglobin will then not be reduced in the tissue capillaries and will circulate unchanged. Failure to free basic groups of Hb will occur, resulting in less efficient transport of CO_2 from the tissues. The buffer line of venous blood, now fully saturated with O_2, will be the same as for arterial blood. A rise in venous and tissue PCO_2 (and $[H^+]$) will occur, which can be as high as about 5 mm Hg above normal.[52] This is a large rise, being the equivalent of that produced in the brain by breathing 7% CO_2.

Binding of H^+ by Hb and the consequent facilitation of the reaction $H_2CO_3 \rightarrow H^+ + HCO_3^-$ result in a gradient for HCO_3^- from inside the red blood cell to the plasma. The erythrocytic membrane is only slowly permeable to the positively charged Na and K ions, but allows the extremely rapid passage of negative ions. As the fulminating formation of HCO_3^- occurs in the red blood cell, HCO_3^- begins to diffuse along its concentration gradient into the plasma. Development of an appreciable deviation from electrical balance between the interior and exterior of the erythrocyte is prevented by passage of Cl^- ions into the erythrocyte in numbers equal to the outward flow of HCO_3^- ions (Fig. 69-6). This effect is often called the "chloride shift."[20]

Changes within pulmonary capillaries

The sequence of events described for loss of O_2 and uptake of CO_2 in the blood of a tissue capillary is in all respects reversed in the pulmonary capillary.[21] There the diffusion of molecular CO_2 from the blood, and the consequent change in $[H^+]$, facilitates the combination of O_2 with Hb at the same time that oxygenation of the blood renders the oxylabile groups of Hb more acid and speeds the release of CO_2. As in the tissues, the rate of the total gas exchange process is critically important, and in the absence of active carbonic anhydrase in the red blood cells, the uncatalyzed rate of dehydration of H_2CO_3 would be inadequate for normal CO_2 removal, even at rest.[4,21]

The gross differences between the CO_2 and O_2 uptake curves are evident in Fig. 69-8. CO_2 is buffered, although to varying extents, in all body fluids. Except for the reservoir function served by Hb, O_2 in body fluids is present only in physical solution.[39]

The elimination of CO_2 is less likely to become a limiting factor in the gas exchange than is oxygenation, for several reasons:

1. Molecular (uncharged) CO_2 is much more soluble than O_2 in body fluids and therefore is able both to penetrate living membranes more rapidly and to be carried in larger amounts in simple solution in the blood plasma.

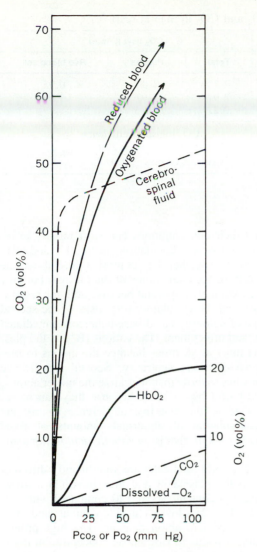

Fig. 69-8. CO_2 uptake curves for blood and cerebrospinal fluid (compared with O_2 uptake by blood). Two lower curves illustrate greater physical solubility of CO_2 as compared with O_2. Upper curves show that at a particular partial pressure, amount of CO_2 combined chemically in blood is much greater than amount of O_2 bound to hemoglobin. Cerebrospinal fluid, having a bicarbonate concentration close to that of arterial blood plasma, contains little protein and therefore has very low buffering capacity. For this reason, slope of CO_2 uptake curve in CSF in normal state is essentially that for physical solubility of CO_2.

2. A specific enzyme (carbonic anhydrase), presiding over the CO_2-bicarbonate relationship, together with the greater diffusion rate of CO_2, affords a massive factor of safety.

3. A number of poisons that are capable of inactivating Hb with regard to the transport of O_2 (CO; methemoglobin-forming substances such as nitrites, chlorates, aniline derivatives, methylene blue, sulfonamides, and ferricyanide; sulfides generating sulfhemoglobin; nitric oxide generating MHb; etc.[5,20]) have no corresponding effect on the transport of CO_2. Chemicals that selectively inactivate carbonic anhydrase such as acetazolamide (Diamox) can be used in limited doses to increase the renal excretion of water and alkali without producing serious disturbances in the respiratory gas exchange.

4. The tissues are capable of binding CO_2 in nongaseous form (as bicarbonate or in carboxyl form) but have no corresponding defense against deficient Po_2.

For such reasons the usual cause of death in cases of inadequate gas exchange is failure of the O_2 supply rather than inadequate removal of CO_2. Under certain circumstances, however, increased Pco_2 or acidosis becomes the lethal factor. One of these is insufficient pulmonary ventilation associated with high inspired Po_2 in individuals with severe depression of the respiratory control mechanisms.

The normal patterns of O_2 and CO_2 transport in the blood and to the tissues summarized in Table 69-5 show that, at rest, O_2 transport is about 23% efficient and the efficiency of CO_2 transport is only about 7%. It is nevertheless the effectiveness of O_2 transport by the circulation that limits physical exertion and not the transport of CO_2.

REGULATION OF HYDROGEN ION CONCENTRATION OF THE BLOOD AND TISSUES

The many factors involved in internal homeostasis provide for maintenance of a $[H^+]$ slightly on the alkaline side of neutrality in the aqueous fluids of body cells.[68] In this internal (extracellular or intracellular) environment occur the enzymatic and biophysical reactions on which metabolism and cell function depend. This desirable acid-base state is accomplished by an interplay among (1) *physicochemical changes in the blood* from which the fluid environment of the cells is derived and (2) *physiologic responses* of pulmonary ventilation, renal excretion of acid and base, and volume flow of blood in actively metabolizing tissues. The end result is to maintain the hydrogen ion concentration of arterial blood serum (pH_s) close to pH 7.40 under normal

Table 69-5. Gas transport efficiency (volume flow of O_2 and CO_2 in man at rest)

Route	O_2 flow (L/min)			CO_2 flow (L/min)		
	Total	HbO$_2$	Dissolved	Total	Plasma	Red blood cell
Arterial	1.09	1.07	0.02	2.75	1.80	0.95
Venous	0.84	0.84	0.005	2.97	1.92	1.05
Tissue	0.25	—	0.25	0.22	—	—

O_2 transport efficiency
$$\frac{\text{Tissue } O_2 \text{ flow}}{\text{Arterial } O_2 \text{ flow}} = \frac{0.25 \text{ L}}{1.09 \text{ L}} = 23\%$$

CO_2 transport efficiency
$$\frac{\text{Tissue } CO_2 \text{ flow}}{\text{Venous } CO_2 \text{ flow}} = \frac{0.22 \text{ L}}{2.97 \text{ L}} = 7\%$$

circumstances and within the range 7.20 to 7.60 under all physiologic conditions that can be borne for more than a few minutes (i.e., steady states). The range in damaging disease is about pH 6.7 to 7.8.

The *physiologic factors* are brought into play by changes in blood pH on either side of the optimum 7.40. They are therefore second lines of defense, activated only when a departure from the normal has already occurred. The first line of defense resides in the blood itself and comprises a set of physicochemical readjustments by which fluctuations in the concentrations of H^+ are minimized by the action of various buffers.[20,21,78]

Anion-cation relationships

The allocation of cations and anions in normal human arterial blood is shown diagrammatically in Fig. 69-9. The distribution is different in plasma and cells, and therefore distribution in whole blood is determined by the proportions of these two components.

In the plasma the anions to be neutralized are mainly those of strong, nonvolatile acids whose concentrations are not subject to rapid fluctuation. They are commonly classed as *fixed acid*. Of the normal total of 150 mEq/L, 110 are accounted for as fixed acid, and of these all but about 5 consist of chloride ion (Cl^-). The remaining 5 mEq (X^-) are referable to nonvolatile metabolic acids such as inorganic phosphate, sulfate, lactate, and ketone bodies, which are normally present only in small amounts. Increase in this fraction is, however, the primary cause of clinical acidoses such as those of diabetes and nephritis and is responsible for the metabolic acidosis of exercise. The remaining anions consist of plasma proteins and bicarbonate (HCO_3^-).

Protein and HCO_3^- are the *buffer* anions, so called because they are anions of weak acids,

and their concentrations can shift rapidly in response to small fluctuations in free H^+ and OH^- present. Together they comprise only about one fourth of the total anions of the plasma, but they are extremely important because both the respiratory and the circulatory responses of the second line of defense in acid-base balance are mediated by and affect them. The cations (B^+) of the plasma must at all times balance the anions to preserve electrical neutrality. Special significance attaches to those that neutralize the buffer anions, the buffer base (BB^+), because they can be regarded as the first line of defense against the accumulation of abnormal amounts of fixed acid or alkali, that is, *metabolic acidosis* or *alkalosis*.

The red blood cells are significantly different from the plasma in that they have (1) a lower total ion content (about 120 compared with 150 mEq/L), (2) a lower chloride content, and (3) a higher protein anion content. The high-protein content is due to Hb, whose concentration in the red blood cells is much higher than that of the proteins in the plasma. In addition, Hb is able to change its acidic properties as it shifts from the oxygenated to the deoxygenated form. The proportion of oxyhemoglobin to reduced Hb therefore is a factor in evaluating the state of acid-base balance.

In whole blood the distribution of ions is determined by the proportion of cells to plasma. In Fig. 69-9 a normal ratio of 45:55 (i.e., a hematocrit value of 0.45) is assumed. An additional factor in buffering the changes in pH of the plasma is the shift of chloride between plasma and cells in conjunction with the transport of CO_2. Such a shift entails the transfer of an equal number of bicarbonate ions in the reverse direction between cells and plasma, with consequent changes in the amounts of gaseous CO_2 to be eliminated by the lungs.

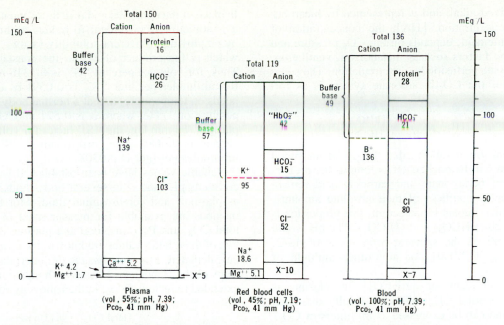

Fig. 69-9. Normal anion-cation balance of human arterial blood. Arranged in milliequivalents per liter for plasma, red blood cells, and whole blood. X^-, Fixed acid other than chloride; B^+, total base. (From Singer: In Altman et al.[26])

pH–carbon dioxide tension–bicarbonate relationship

Among the cations that can be drawn on to minimize or prevent fluctuations in H^+ concentration of the blood or tissues, those that are neutralized by bicarbonate are especially effective. The great importance of bicarbonate as a defense against acidosis lies in the fact that when a fixed acid is introduced and bicarbonate is decomposed, the CO_2 produced can be eliminated promptly by the lungs. As much as 80% of the bicarbonate of the blood can be decomposed before severe fixed acid poisoning results. When, however, the acid added to the blood is carbonic, the end result is an uptake of CO_2 and rise in $[H^+]$. The bicarbonate buffer system is not very effective as a buffer of carbonic acid and the blood proteins (especially Hb) are the buffers of paramount importance for the acidosis produced by CO_2.

The relationships between CO_2, H_2O, Hb, and bicarbonate have already been considered, but thus far the extremely important factor of $[H^+]$ has not been taken into account. The $[H^+]$ of a solution containing both carbonic acid and alkali bicarbonates will depend on their relative proportions. The general relationship is expressed in terms of the *mass law* as follows (Henderson's equation):

$$[H^+] = K' \frac{[HA]}{[BA]}$$

where

$[H^+]$ = Hydrogen ion concentration
$[HA]$ = Concentration of weak acid
$[BA]$ = Concentration of its salts
K' = Dissociation constant of acid (i.e., its tendency to give rise to free hydrogen ions)

By definition, pH is the negative logarithm of $[H^+]$. Hasselbalch converted this equation into the logarithmic form as follows:

$$-\log [H^+] = pH = -\log K' - \log \frac{[HA]}{[BA]}$$

Then by designating $-\log K'$ as pK' and $-\log \frac{[HA]}{[BA]}$ as $+ \log \frac{[BA]}{[HA]}$, he arrived at the now widely used form, the *Henderson-Hasselbalch equation:*

$$pH = pK' + \log \frac{[BA]}{[HA]}$$

This expression is especially useful in physiology to describe quantitatively the interacting influences of factors such as bicarbonate concentration, $[H^+]$, and CO_2 pressure in body fluids such as plasma, cerebrospinal fluid, and urine. For these purposes, $[BA]$ depends on available fixed alkali and is represented by bicarbonate concentration. [HA], the concentration of carbonic acid, depends on the CO_2 pressure and thus on factors such as metabolism, ventilation, and tissue perfusion by the circulation. The equation for the CO_2-bicarbonate system is as follows:

$$pH = pK' + \log \frac{[BHCO_3^-]}{[H_2CO_3]}$$

From this it follows that the pH of the blood plasma will be unchanged as long as the proportions of bicarbonate and carbonic acid remain constant, regardless of their absolute amounts. In normal human blood serum, this proportion is about 20 ($BHCO_3$) to 1 (H_2CO_3). The pH would therefore be the same at proportions of 40/2, 10/0.5, 20.2/1.01, or any other multiple of 20/1.

Measurement of $[BHCO_3]$ in plasma is made as total CO_2 minus physically dissolved CO_2 and can be expressed as milliliters of CO_2/ 100 ml of plasma or as millimoles of CO_2/L of plasma. The denominator $[H_2CO_3]$ represents not the small amount of carbonic acid present in blood but the sum of the carbonic acid and the much larger amount ($1,000\times$) of dissolved molecular CO_2 that is responsible for acid formation and is itself proportional to CO_2 tension. In practice the physically dissolved CO_2 is calculated as the product of PCO_2 and the solubility constant for CO_2 in plasma. Thus, expressed in volumes percent (as measured in the manometric blood gas apparatus):

$$[H_2CO_3]p = \alpha \, \frac{PCO_2}{760} \times 100 = 0.1316 \, \alpha PCO_2$$

where α is the temperature-dependent solubility coefficient for CO_2 expressed in milliliters of CO_2 per milliliter plasma per 760 mm Hg pressure.[20] The factor 100 converts the units to milliliters of CO_2 per 100 ml of plasma.

Should it be desirable to convert values for total CO_2, $[HCO_3^-]$, or $[H_2CO_3]$ measured in milliliters per 100 ml to their equivalents in millimoles per liter, this is accomplished by use of the following expression:

$$mmole/L = \frac{vol\%}{2.226}$$

in which 2.226 is derived from the 22.26 L of gas at STPD, constituting 1 mole of CO_2. Therefore, if the CO_2 concentration is to be expressed in millimoles per liter, the expression becomes as follows:

$$[H_2CO_3]p = \alpha \, \frac{PCO_2}{760} \times \frac{100}{2.226} = 0.0591 \, \alpha PCO_2$$

In place of the constants and α in these equations, it is sometimes convenient to reduce arithmetic by employing a different solubility factor, "S," which is the product of α and the constants.[75] Thus, for volumes percent "S" is 0.1316 α and for millimoles per liter "S" is 0.0591 α. α depends on temperature, and at a normal body temperature of 37° C, where α is 0.521 ml CO_2/ml plasma/760 mm Hg, the "S" factor for volumes percent is 0.0686. The corresponding "S" for millimoles per liter is 0.0308.

Reliable values have been established for the constants pK' and α in several body fluids such as plasma[75] and cerebrospinal fluid. Moreover, methods are available for measurement of pH, total CO_2, and PCO_2 in blood and plasma; if any two of these latter values are known, the third can be calculated. For most usages the interrelationships of the Henderson-Hasselbalch equation are extended to include PCO_2, in the following forms:

$$pH = pK' + \log \frac{[total \, CO_2] - 0.1316 \, \alpha PCO_2}{0.1316 \, \alpha PCO_2}$$

and

$$PCO_2 = \frac{[CO_2]}{0.1316 \, \alpha[10^{(pH-pK')} + 1]}$$

pH-bicarbonate diagram

The manner in which pH, PCO_2, and $[HCO_3^-]$ are related algebraically can also be expressed diagramatically (Figs. 69-10 and 69-11). The pH-carbonate diagram extensively employed by Davenport[7] is in fact a graphic representation of the Henderson-Hasselbalch equation, in which the three-dimensional coordinates express pH, PCO_2, and $[HCO_3^-]$. Fig. 69-10 shows the basic diagram with an indication of the places occupied by a group of normal men at rest, during CO_2 breathing,[55] and during hyperventilation. In this illustration the heavy line labeled *buffer slope* depends on the concentration of Hb as a buffer in whole blood and represents the line of change in the relationship between $[HCO_3^-]$ and pH during acute alterations of PCO_2 above and below normal. The finer lines of constant PCO_2 (PCO_2 isobars) indicate the manner in which adjustment of pH is accomplished through change in $[HCO_3^-]$.

The results of two studies in man are plotted on the diagram. One involves CO_2 administration that thus caused a shift upward and to the left (O = PCO_2 50 mm Hg, $[HCO_3^-]$ 27 mmole/L, and pH 7.34), and the second study involves the simple removal of CO_2 from the arterial blood by voluntary overventilation (Δ = PCO_2 20 mm Hg,

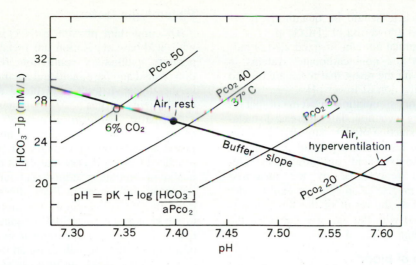

Fig. 69-10. pH-bicarbonate-P_{CO_2} diagram for arterial blood. Solid circle indicates relations of P_{CO_2}, pH, and plasma $[HCO_3^-]$ in resting subjects studied in author's laboratory.[55] Buffer slope, determined largely by hemoglobin concentration, represents pathway for respiratory acidosis, $\bigcirc$, or respiratory alkalosis of hyperventilation, $\triangle$. Deviations of experimentally determined points from buffer line represent changes in its position rather than in slope of buffer lines; these changes are largely related to induced alterations in concentration of hemoglobin, fixed acid, or fixed base.[7]

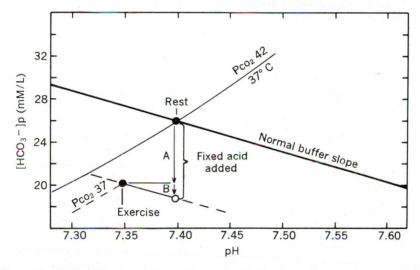

Fig. 69-11. pH-bicarbonate-P_{CO_2} diagram for arterial blood. Solid circles indicate data obtained in normal subjects at rest and in exercise.[57] Amount of "fixed acid" added to blood is represented by sum of *A* (acid combined with bicarbonate to form CO_2 and water) and *B* (acid combined with blood buffers). The latter, which is responsible for pH change, is estimated by assuming titration down buffer slope to original pH. Change in fixed base is estimated in a similar manner.[7]

$[HCO_3^-]$ 22 mmole/L, pH 7.60). Hyperventilation caused some deviation of the pH, P_{CO_2}, $[HCO_3^-]$ relationships above those predicted by the buffer slope, possibly because of changes in hydration.

Fig. 69-11 illustrates, for normal men engaged in moderate muscular work,[57] the manner in

which the addition of fixed acid to blood can be estimated. The entrance of organic acids (lactic and pyruvic) into the blood during exercise is compensated for in the following three ways:

1. The larger amount combines the $BHCO_3$ to cause the formation of CO_2 and H_2O. In the absence of a change in P_{CO_2}, this amount

of fixed acid would be quantitatively reflected as a lowering of $[HCO_3^-]p$.

2. An additional amount of fixed acid is buffered by the nonbicarbonate systems in blood, with the result that these are titrated in the direction of lower pH.

3. Respiratory stimulation in excess of that required to accommodate increased metabolic production of CO_2 may slightly lower Pco_2 and lessen the degree of pH change. Change in fixed acid concentration is therefore not represented by the measured change in $[HCO_3^-]p$, but it can be determined as the result of the three effects mentioned with the aid of a value for the buffer slope of blood.[7]

Buffer slope of blood

The buffer slope of blood is defined as follows:

$$\text{Buffer slope} = \frac{\Delta[HCO_3^-]p}{\Delta pH}$$

This gives a negative slope that is determined predominantly by the concentration of Hb and can be expressed as follows:

$$-\frac{\Delta[HCO_3^-]p}{\Delta pH} = 8.2 + 2.3 \text{ (mmole Hb/L)}$$

In this relationship, 8.2 is the buffer slope attributable to plasma protein and 2.3 is the buffer value of each millimole per liter of Hb.[20] The total buffer value of normal blood containing about 9 mmole Hb/L is therefore close to -29 mmole $\Delta[HCO_3^-]/\Delta pH$.

The addition of fixed acid to the blood in exercise and in other forms of metabolic acidosis is estimated by adding the observed change in $[HCO_3^-]p$ to the action of acid in titrating other blood buffers to greater acidity as follows:

$$\Delta[\text{Fixed acid}] = \Delta[HCO_3^-]p + (\Delta pH \times \text{Buffer slope})$$

Graphically the amount of fixed acid added per liter of plasma can be considered as the vertical distance, at resting pH, between the buffer slope of blood at rest and the buffer slope in a new situation such as diabetic acidosis, ingestion of ammonium chloride, or exercise (Fig. 69-11). It consists of two components (Fig. 69-11), one of which is the amount of fixed acid that is buffered by reaction with bicarbonate to form CO_2 and H_2O. The second component, which is responsible for the pH change, comprises acid combined with the other blood buffers. The method of determining pH change is shown in Fig. 69-11. Changes in fixed base can be similarly estimated from changes in bicarbonate and pH.

Physiologic factors

The important physiologic factors concerned with acid-base regulation are metabolism, temperature, pulmonary ventilation, renal elimination of acid and base, acid and alkaline secretions in the digestive tract, and blood flow through metabolizing tissues.

Normally, *respiratory control mechanisms* cause pulmonary ventilation to be closely adjusted in relation to changes in metabolically produced CO_2. It is very likely that these neuron-mediated ventilatory adjustments are caused by secondary influences of Pco_2 alteration such as those on $[H^+]$ or calcium ionization. According to the Henderson-Hasselbalch equation, diminution in the amount of bicarbonate by introduction of a nonvolatile acid will cause an immediate decrease in plasma pH, an increase in H_2CO_3 content, and a rise in Pco_2. Breathing is then acutely stimulated until the Pco_2 is lowered toward the point at which the $BHCO_3:H_2CO_3$ ratio is restored to 20:1. The actual amount of bicarbonate now may be considerably less than before, but the total cation concentration is unchanged, and the pH will be brought toward normal. Thereafter, as long as $[H^+]$ remains elevated at one of the chemosensitive respiratory control sites, ventilation will remain above normal. If the acid responsible for the decrease in plasma pH is carbonic, the Pco_2 increases first, but the bicarbonate is also increased, the cations being made available by passage of chloride ions into the erythrocytes. Here again breathing is stimulated as Pco_2 rises. Conversely, when the pH of the blood is raised by the introduction of alkali, breathing may be depressed. If CO_2 is lost (as in voluntary hyperventilation, the hyperpnea of high altitudes, or any sustained hyperventilation due to something other than increased Pco_2), there is a *relative* excess of $BHCO_3$ that is eliminated by the kidneys.

The *renal mechanisms* compensate by altering the amounts of ammonium and hydrogen ions secreted and of alkali, chloride, and bicarbonate reabsorbed by the tubular cells. Such compensations are slower both in onset and recovery than the corresponding changes in respiration and the chloride shift between cells and plasma, but they work in the same direction. Abnormal acids or excessive amounts of normal ones usually are eliminated by the kidney together with cations, and the total cation content of the body is thus diminished. The kidney makes an attempt to compensate for this by forming ammonia from some of the amide nitrogen of the blood that ordinarily is made into neutral urea. The ammonia then supplies cations to neutralize some of the excess anions. The effects of this mechanism become detectable only after 1 hr or more and reach a maximum only after several days of acidosis. In alkalosis, ammonia excretion diminishes or ceases and other cations are excreted along with large amounts of bicarbonate.

The *metabolic rate, blood flow through the tissue, and composition of the blood entering the tissue capillaries* establish the tensions of CO_2 and O_2 in a tissue. Blood flow is influenced by general factors such as those responsible for the intensity of sympathetic vasomotor impulse traffic and for blood pressure. These general influences are modified by the effects of arterial gas tensions on arteriolar smooth muscle and by the local effects of metabolism on the O_2 and acid-base environment of the smooth muscle cells in the walls of tissue blood vessels. Increased metabolism tends to lower the Po_2 and raise the Pco_2 of a tissue. Each of these local effects relaxes the smooth muscle of tissue vessels and thus promotes a local increase in blood flow that aids in minimizing deviations from the normal O_2 and acid-base environment. When activity is increased, as in exercise, the cardiovascular system can reinforce the respiratory system in meeting the metabolic requirements for gas exchange.

Despite the combined functional capacities of the lungs, blood, and circulation for meeting the metabolic requirements of the tissues, deviations in O_2 and CO_2 tension do occur in the tissues of normal or ill individuals. These deviations occur when any one of the regulatory functions is handicapped (as by pulmonary insufficiency, anemia, fever, vascular insufficiency, or cardiac abnormality) and the demands of increased tissue metabolism are not met. Such a situation is illustrated by the effects of cardiac insufficiency in Fig. 69-12.

DEVIATIONS FROM NORMAL ACID-BASE BALANCE

Ordinarily the mechanisms described in the preceding section maintain in the plasma of arterial blood a pH of about 7.40, a Pco_2 of about 40 mm Hg, and a plasma bicarbonate concentration of about 25 mmole/L. There are relatively small and transitory normal fluctuations associated with emotional changes in respiration, muscular activity, sleep, the menstrual cycle, and ingestion of foods containing acid or alkali. In a number of diseases, however, the acid-base balance may be shifted severely in either direction, and acidosis or alkalosis ensues (Table 69-6). Pertinent terminology and definitions are summarized in Table 69-7 and in the following paragraphs.

Acidosis is an abnormal condition caused by

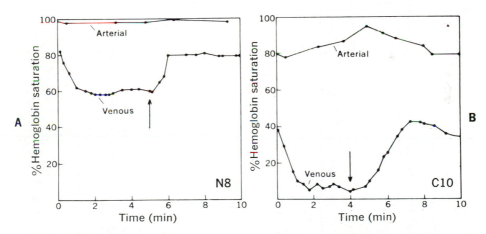

Fig. 69-12. Influence of exercise on arterial and mixed venous oxygenation in normal individual (N8) and in patient (C10) with impaired cardiac function (rheumatic heart disease).[42] Exercise began at 0 time; ended at arrow. Note that in normal subject exercise did not lower arterial O_2 saturation but did somewhat reduce mixed venous %Hb saturation (Chapter 74). Patient had subnormal arterial and venous hemoglobin saturation at rest. However, reduction of venous O_2 saturation by exercise was extreme when ability to increase cardiac output was impaired. Nevertheless, arterial O_2 saturation, which here represents a *result* of pulmonary gas exchange, was actually improved. Hence when lungs are normal, severe tissue anoxia may not be evident from examination of arterial blood alone. (Modified from Donald et al.[42])

Table 69-6. Representative findings in arterial or finger blood in clinical acidoses and alkaloses*

Nature of case	Analytic data				Factors derived from nomogram			Classification of disturbance
	Hematocrit	CO_2 content (mmole/L)	pH	Pco_2 (mm Hg)	Buffer base (mEq/L)	Change from normal mean‡		
						Pco_2 (mm Hg)	Buffer base (mEq/L)	
NaHCO₃ ingestion by normal subject	0.48	29	7.48	48	59	+7.5	+8.5	Primary metabolic alkalosis, slight secondary respiratory acidosis
NH₄Cl ingestion by normal subject	0.42	17	7.25	44§	40	Normal range	−8	Metabolic acidosis
Normal man living at 19,000 ft	0.55	14	7.43	26	43	−14	−8	Primary respiratory alkalosis, slight secondary metabolic acidosis
Acute encephalitis ketosis	0.59	8†	7.59	12	44	−28	−7	Primary respiratory alkalosis, slight secondary metabolic acidosis
CO₂ inhalation by curarized patient	0.50	25	6.89	132	32	+91	−18	Respiratory and metabolic acidosis (both primary)
Emphysema	0.59	31	7.41	62	60	+21	+9	Primary respiratory acidosis, secondary metabolic alkalosis
Diabetic coma	0.49	4.2	6.88	23	18	−17	−32	Primary metabolic acidosis, secondary respiratory alkalosis

*From Singer and Hastings.[78] In the calculation of ΔPco_2 and Δbuffer base, normal values of arterial Pco_2 and buffer base are taken from the nomogram at the measured hematocrit.
†CO_2 content of blood plasma in this case only; in all others the value is for CO_2 content of whole blood.
‡In each the degree of deviation from normal depends on the degree of stress on the buffer systems and the regulatory control mechanisms.
§This value, cited by Singer and Hastings, shows hypercapnia. Metabolic acidosis should show hypocapnia.

Table 69-7. Acid-base factors and respiration*

Term	Definition
Acid	Chemical compound capable of dissociating in solution to form H^+ and negatively charged ions (anions), for example, HCl (strong acid), H_2CO_3 (weak acid)
Base	Chemical compound capable of neutralizing acid or dissociating in solution to form OH^- and positively charged ions (cations), for example, NaOH (strong base), NH_4OH (weak base), $NaHCO_3$ (buffer salt, neutralizes strong acids); this definition avoids undesirable past usage in acid-base literature of "base" as synonymous with "cation," and also more modern but confusing Bronsted definition of base as H^+ acceptor (e.g., anion HCO_3^- would be called "base")
Buffer base	Biologic buffer salts capable of neutralizing strong acids; in blood—appropriate fraction of total cation and equivalent buffer anions, chiefly bicarbonate, hemoglobinate, and proteinate
Acidosis	Abnormal condition caused by accumulation in body of excess of acid or loss from body of base[20]
Alkalosis	Abnormal condition caused by accumulation in body of excess of base or loss from body of acid[20]
Respiratory factor in acidosis or alkalosis	If acid concerned in disturbance is H_2CO_3, acidosis or alkalosis may be called "respiratory"; best index for this factor is CO_2 pressure, P_{CO_2}, of arterial blood, which is normally equal to P_{CO_2} of alveolar air; it can be calculated from plasma pH and total CO_2 by Henderson-Hasselbalch equation or measured directly; peripheral venous P_{CO_2} is less desirable because of variability of arteriovenous difference of 2 to 10 mm Hg or more
Metabolic factor in acidosis or alkalosis	If base or some acid other than H_2CO_3 is concerned in disturbance, acidosis or alkalosis may be called "metabolic"; a satisfactory quantitative index for this factor is whole blood buffer base concentration[78] or plasma bicarbonate concentration at pH 7.4[20]; total CO_2 or bicarbonate concentration is not satisfactory because it also varies with P_{CO_2}, the respiratory factor[20,78]; plasma CO_2-combining power, still widely used, is even less satisfactory because it does not measure directly any variable in blood or plasma[7,20,78]; buffer base can be calculated from pH, total CO_2, hemoglobin, and plasma protein or taken from nomogram
Compensation for acid-base change	In blood, P_{CO_2} and buffer base can be regarded as independent variables sufficient to define state of acid-base balance; pH and total CO_2 or bicarbonate, usually variables determined, are better regarded as dependent variables; primary disturbance in one factor, P_{CO_2} or buffer base, usually results in compensation, one manifestation of which is change in other factor in such a way that pH is returned toward, but not necessarily to, normal range

*Modified from Singer. In Altman et al.[26]

the presence in the body of excessive amounts of acid or by reduction of the amount of alkali to a subnormal level.

Alkalosis is the converse abnormal condition caused by the presence of excessive amounts of alkali or by reduction of the amount of acid to a subnormal level.

Acidoses and alkaloses are of two main types: those in which the primary disturbance is defined by the buffer base content (*metabolic acidosis* or *alkalosis*) and those in which the primary abnormality is in the tension of CO_2 (*respiratory acidosis* or *alkalosis*.)

Metabolic alkalosis may be brought about either by a deficiency of fixed acid (as in the loss of Cl^- during persistent vomiting or continuous loss of gastric juice by fistula or suction drainage) or by the absorption of excessive amounts of alkalinizing salts. In the former

case the quantities of total base and buffer base in the body would be decreased; in the latter they would be increased. The bicarbonate content would change like the buffer base, of which it constitutes a major part (Fig. 69-9). The arterial hydrogen ion concentration would decrease, that is, pH would rise.

Metabolic acidosis can be caused by an excess of fixed acid associated with metabolism (lactic acid in hypoxia and exercise, keto acids in starvation and diabetes, SO_4^{2-} and HPO_4^{2-} in advanced renal deficiency) or by the ingestion of acidifying salts such as NH_4Cl (of which the NH_4 radical is converted to neutral urea in the liver, leaving an excess of Cl^- to be neutralized by the other base), or of $Ca(NO_3)_2$ (of which the NO_3^- is absorbed more readily than the Ca^{2+}, so that more anions than cations enter the circulation). The characteristic pattern of metabolic acidosis is diminution in total base, buffer base, and bicarbonate content of arterial blood along with an increase in hydrogen ion concentration. Acid-

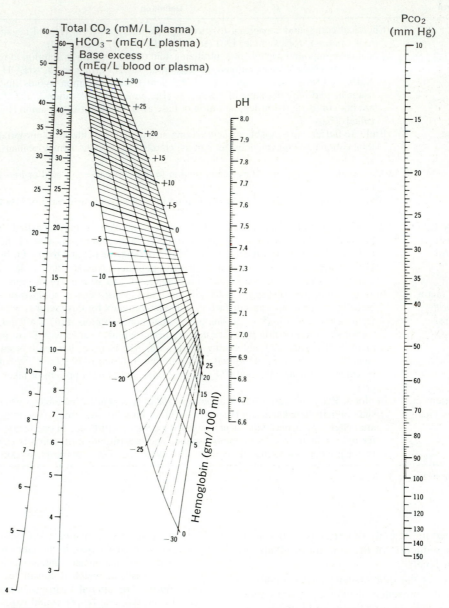

Fig. 69-13. Blood acid-base alignment nomogram. This nomogram, described in detail by its author[76,77] and others,[22] relates principal factors concerned in acid-base composition of blood. It is employed chiefly to determine values for change in fixed acid or fixed base when P_{CO_2}, pH, and Hb concentration have been measured. Changes in fixed acid and fixed base are indicated here as + or − changes in "base excess" in milliequivalents per liter. Values for plasma bicarbonate and total CO_2 can also be estimated, although scales are only approximate. It is pointed out in original description of this nomogram that values presented are for 38° C and when other temperatures are used, appropriate corrections for P_{CO_2} and pH are required. Examples of use of nomogram are presented by its author. These include (1) determination of base excess in pulmonary insufficiency when pH, P_{CO_2}, and Hb concentration are known; (2) determination of P_{CO_2} and fixed acid in diabetic acidosis when pH, total CO_2, and Hb concentration are known; (3) determination of P_{CO_2} and change in fixed acid when pH, plasma bicarbonate, and Hb concentration are known; and (4) other combinations of respiratory and metabolic derangements of acid-base balance.

osis due primarily to withdrawal of secretions high in cation content is encountered in intense diarrhea, particularly in infants, and is due to loss of alkaline pancreatic and intestinal secretions and bile.

Respiratory alkalosis is present when alveolar P_{CO_2} is maintained at a lower than normal level and therefore is produced by an increase in alveolar ventilation referable to anything other than increased arterial P_{CO_2}. Voluntary hyperventilation, anoxemia, hysteria, fever, dyspnea due to congestive heart failure, and irritative or inflammatory lesions involving the brain stem will act in this way. The characteristic effect on the blood is to decrease arterial hydrogen ion concentration, CO_2 tension, and bicarbonate content. The concomitant increase in plasma chloride concentration leads eventually to increased renal excretion of base, which then lowers the buffer base content.

Respiratory acidosis, the term applied to acidosis caused by retention of molecular CO_2, may be produced by inhalation of gas mixtures to which CO_2 has been added, by voluntary breath holding, by pulmonary insufficiency of any type (particularly emphysema), by respiratory obstruction, by depression of the respiratory center, or by paralysis of the respiratory muscles. The characteristic change in the blood is an increase in P_{CO_2}, arterial hydrogen ion concentration, and bicarbonate content. Here the increased renal excretion of chloride will eventually lead to an increase in the buffer base.

It is inevitable that a primary alkalosis, or acidosis, no matter how caused, will elicit physiologic responses, the end result of which is to oppose the original deviation of blood pH from normal. Thus a primarily metabolic alkalosis leads to diminished respiration and therefore to decreased pulmonary elimination of CO_2, which

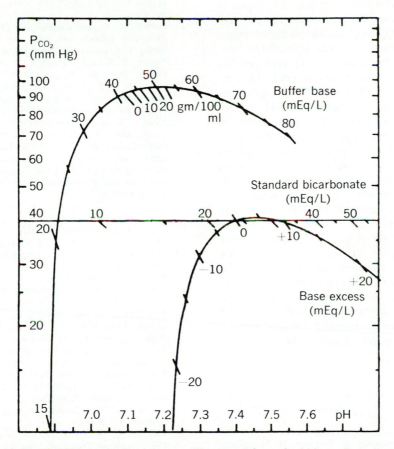

Fig. 69-14. Blood acid-base curve nomogram. The second form of acid-base–P_{CO_2} diagram, from which the foregoing nomogram for oxygenated blood was derived,[76,77] also relates principal factors in acid-base composition of blood. Diagram utilizes graphic plot of relationship of pH to log P_{CO_2} to show changes in base excess and/or buffer base. Normal reference point is found at pH = 7.40 and P_{CO_2} = 40 mm Hg, at which point normal "base excess" = 0. Standard bicarbonate is bicarbonate ion concentration of plasma in milliequivalents per liter at fixed P_{CO_2} of 40 mm Hg and saturated with O_2.

tends to produce a secondary respiratory "acidosis." Conversely, a primary metabolic acidosis tends to cause a secondary respiratory "alkalosis." These changes in P_{CO_2} occur throughout the body and can result in changes in the pH of intracellular and cerebrospinal fluid that are opposite to those in blood (Chapter 71).[58] The respiratory alkalosis resulting from residence at high altitudes leads to increased renal excretion of base, which tends to produce a metabolic acidosis. The change in blood hydrogen ion concentration usually remains in the direction of the primary disturbance (although in diminished degree).

The examples are provided to illustrate that clinical disturbances of acid-base balance are not likely to fall within a single, simple pattern or to be subject to evaluation by any single, simple test. For these reasons, knowledge of only the CO_2 content of blood is inadequate for diagnosis of clinical acid-base derangements. An adequate portrayal of the type and severity of a disturbance of acid-base balance requires information concerning at least three factors: (1) hydrogen ion concentration; (2) bicarbonate or total CO_2 concentration; and (3) CO_2 tension. Additional information of value includes buffer base and Hb concentration or hematocrit. If the blood sample is venous or if arterial blood less than 90% saturated with O_2 is used, precise characterization also requires the percentage of oxyhemoglobin and reduced Hb (but this is unnecessary for ordinary clinical purposes). Clinically, practicable methods for measurement of acid-base factors have become generally available as a consequence of their development for investigative purposes. The direct determination of each of the factors is now convenient except for the buffer base content, but they are so interrelated that the values of all the other factors can be deduced from the hematocrit and any two other measured factors. Nomograms designed for this purpose are shown in Figs. 69-13 and 69-14.

REFERENCES

1. Adamson, J. W., and Finch, C. A.: Hemoglobin function, oxygen affinity, and erythropoietin, Annu. Rev. Physiol. **37:**351, 1975.
2. Bank, A., Rifkind, R. A., and Marks, P. A.: The Thalassemia syndromes. In Surgenor, D. M., editor: The red blood cell, New York, 1975, Academic Press, Inc., vol. 2.
3. Bartels, H., and Baumann, R.: Respiratory function of hemoglobin. In Widdicombe, J. G., editor: Respiratory physiology II. International review of physiology, Baltimore, 1977, University Park Press, vol. 14.
4. Berliner, R. W., and Orloff, J.: Carbonic anhydrase inhibitors, Pharmacol. Rev. **8:**137, 1956.
5. Bodansky, O.: Methemoglobin, and methemoglobin-producing compounds, Pharmacol. Rev. **3:**144, 1951.
6. Bromberg, P. A., and Balcerzak, S. P.: Blood oxygen transport in humans. In Robin, E. D., editor: Extrapulmonary manifestations of respiratory disease, New York, 1978, Marcel Dekker, Inc.
7. Davenport, H. W.: The ABC of acid-base chemistry, ed. 4, Chicago, 1958, University of Chicago Press.
8. Dickens, F., and Neil, E., editors: Oxygen in the animal organism, New York, 1964, Pergamon Press.
9. Dirken, M. N. J., and Mook, H. W.: The rate of gas exchange between blood cells and serum, J. Physiol. (Lond) **73:**394, 1931.
10. Forster, R. E.: Diffusion of gases. In Fenn, W. O., and Rahn, H., editors: Handbook of physiology. Respiration section, Washington, D.C., 1964, American Physiological Society, vol. 1.
11. Forster, R. E.: The rate of CO_2 equilibration between red cells and plasma. In Forster, R. E., et al., editors: CO_2: chemical, biological and physiological aspects, Washington, D.C., NASA SP-188, 1969, Government Printing Office.
12. Forster, R. E., and Crandall, E. D.: Time course of exchanges between red cells and extracellular fluid during CO_2 uptake, J. Appl. Physiol. **38:**710, 1975.
13. Forster, R. E., and Crandall, E. D.: Pulmonary gas exchange, Annu. Rev. Physiol. **38:**69, 1976.
14. Kilmartin, J. V., and Rossi-Bernardi, L.: Interaction of hemoglobin with hydrogen ions, carbon dioxide, and organic phosphates, Physiol. Rev. **53:**836, 1973.
15. Klocke, R. A.: Carbon dioxide transport. In Robin, E. D., editor: Extrapulmonary manifestations of respiratory disease, New York, 1978, Marcel Dekker, Inc.
16. Lambertsen, C. J.: Effects of hyperoxia on organs and their tissues. In Robin, E. D., editor: Extrapulmonary manifestations of respiratory disease, New York, 1978, Marcel Dekker, Inc.
17. Lübbers, D.-W., et al.: Oxygen transport in blood and tissue, Stuttgart, 1968, Georg Thieme Verlag.
18. Maren, T. H.: Carbonic anhydrase: chemistry, physiology and inhibition, Physiol. Rev. **47:**595, 1967.
19. Millikan, G. A.: Muscle hemoglobin, Physiol. Rev. **19:**503, 1939.
20. Peters, J. P., and Van Slyke, D. D.: Quantitative clinical chemistry, methods, Baltimore, 1931, The Williams & Wilkins Co.
21. Roughton, F. J. W.: Transport of oxygen and carbon dioxide. In Fenn, W. O., and Rahn, H., editors: Handbook of physiology. Respiration section, Washington, D.C., 1964, American Physiological Society, vol. 1.
22. Severinghaus, J. W.: Acid-base balance of blood: man. Part I. Sigaard-Andersen alignment nomogram. Part II. Singer-Hastings nomogram. In Altman, P. L., and Dittmer, D. S., editors: Respiration and circulation, Bethesda, Md., 1977, Federation of American Societies for Experimental Biology.
23. Surgenor, D. M., editor: The red blood cell, New York, 1975, Academic Press, Inc., vol. II.
24. Wagner, P. D.: Diffusion and chemical reaction in pulmonary gas exchange, Physiol. Rev. **57:**257, 1977.
25. Adair, G. S.: The hemoglobin system. VI. The oxygen dissociation curve of hemoglobin, J. Biol. Chem. **63:**529, 1925.
26. Altman, P. L., Gibson, J. F., and Wang, C. C.: Hand-

book of respiration, Philadelphia, 1958, W. B. Saunders Co.

27. Bartels, H.: Fetal O_2-dissociation curves, man and various animals. In Altman, P. L., editor: Blood and other body fluids, Bethesda, Md., 1961, Federation of American Societies for Experimental Biology.

28. Bartels, H.: Comparative physiology of oxygen transport in mammals, Lancet **2:**599, 1964.

29. Bartels, H., Hilpert, P., and Riegel, K.: Die O_2-Transportfunktion des Blutes Wahrend der Ersten Lebensmonate von Menschen, Ziegen, und Schafen, Arch. Gesamte Physiol. **271:**169, 1960.

30. Benesch, R., and Benesch, R. E.: The effect of organic phosphates from the human erythrocyte on the allosteric properties of hemoglobin, Biochem. Biophys. Res. Commun. **26:**162, 1967.

31. Benesch, R., Benesch, R. E., and Yu, C. I.: Reciprocal binding of oxygen and diphosphoglycerate by human hemoglobin, Proc. Natl. Acad. Sci. USA **59:**526, 1968.

32. Bock, A. V., Field, H., Jr., and Adair, G. S.: The oxygen and carbon dioxide dissociation curves of human blood, J. Biol. Chem. **59:**353, 1924.

33. Bohr, C., Hasselbalch, K., and Krogh, A.: Ueber Einen in Biologischer Beziehung Wichtigen Einfluss, den die Kohlensäurespannung des Blutes auf Dessen Sauerstoffbindung Ubt, Skand. Arch. Physiol. **16:**402, 1904.

34. Boothby, W., editor: Handbook of respiratory physiology, Washington, D.C., 1954, U.S. Air Force School of Aviation Medicine.

35. Boyes-Watson, J., and Perutz, McF.: X-ray analysis of haemoglobin, Nature **151:**714, 1943.

36. Braunitzer, G., et al.: Die Konstitution des Normalen Adulten Humanhamoglobins, Hoppe-Seylers' Z. Physiol. Chem. **325:**283, 1961.

37. Brown, W. E. L., and Hill, A. V.: The oxygen-dissociation curve of blood and its thermodynamical basis, Proc. R. Soc. Lond. (Biol.) **94:**297, 1923.

38. Caughey, W. S., et al.: Reactions of oxygen with hemoglobin, cytochrome C oxidaze and other hemeproteins, Ann. N.Y. Acad. Sci. **244:**1, 1975.

39. Cherniack, N. S., and Longobardo, G. S.: Oxygen and carbon dioxide gas stores of the body, Physiol. Rev. **50:**196, 1970.

40. Defares, J. G., and Visser, B. F.: On the form of the physiological CO_2 dissociation curve, the physiological O_2 dissociation curve, and the diffusion curves of O_2 and CO_2 along the capillary path, Ann. N.Y. Acad. Sci. **96:**939, 1962.

41. Dill, D. B., et al.: Properties of dog blood, J. Biol. Chem. **95:**143, 1932.

42. Donald, K. W., Bishop, J. M., and Wade, O. L.: A study of minute to minute changes of arterio-venous oxygen content difference, oxygen uptake and cardiac output and rate of achievement of a steady state during exercise in rheumatic heart disease, J. Clin. Invest. **33:**1146, 1954.

43. Drescher, H., and Kunzer, W.: Der Blütfarbstoff des menschlichen Feten, Klin. Wochenschr. **32:**92, 1954.

44. Finch, C. A., and Lenfant, C.: Oxygen transport in man, N. Engl. J. Med. **286:**407, 1972.

45. Granick, S.: Iron and porphyrin metabolism in relation to red blood cell, Ann. N.Y. Acad. Sci. **48:**657, 1947.

46. Handbook of respiratory data in aviation, Washington, D.C., 1944, Division of Medical Sciences, National Research Council.

47. Hurtado, A., and Clark, R. T.: Parameters of human adaptation to altitude. In Benson, O. O., and Strughold, H., editors: Physics and medicine of the atmosphere and space, New York, 1960, John Wiley & Sons, Inc.

48. Jacobson, L. O., et al.: Role of the kidney in erythropoiesis, Nature **179:**633, 1957.

49. Kety, S. S., et al.: The blood flow and oxygen consumption of the human brain in diabetic acidosis and coma, J. Clin. Invest. **27:**500, 1948.

50. Kilmartin, J. V., and Rossi-Bernardi, L.: Inhibition of CO_2 combination and reduction of the Bohr effect in haemoglobin chemically modified at its α-amino groups, Nature **222:**1243, 1969.

51. Kuratowska, Z., Lewortowski, B., and Michalak, E.: Studies on the production of erythropoietin by isolated perfused organs, Blood **18:**527, 1961.

52. Lambertsen, C. J.: Effects of oxygen at high partial pressure. In Fenn, W. O., and Rahn, H., editors: Handbook of physiology. Respiration section. Washington, D.C., 1965, American Physiological Society, vol. 2.

53. Lambertsen, C. J., et al.: Relationship of oxygen tension to hemoglobin oxygen saturation in the arterial blood of normal men, J. Appl. Physiol. **4:**873, 1952.

54. Lambertsen, C. J., et al.: Oxygen toxicity. Effects in man of oxygen inhalation at 1 and 3.5 atmospheres upon blood gas transport, cerebral circulation and cerebral metabolism, J. Appl. Physiol. **5:**471, 1953.

55. Lambertsen, C. J., et al.: Comparison of relationship of respiratory minute volume to Pco_2 and pH of arterial and internal jugular blood in normal man during hyperventilation produced by low concentrations of CO_2 at 1 atmosphere and by O_2 at 3.0 atmospheres, J. Appl. Physiol. **5:**803, 1953.

56. Lambertsen, C. J., et al.: Oxygen-toxicity. Arterial and internal jugular blood gas composition in man during inhalation of air, 100% O_2 and 2% CO_2 in O_2 at 3.5 atmospheres ambient pressure, J. Appl. Physiol. **8:**255, 1955.

57. Lambertsen, C. J., et al.: Respiratory and cerebral circulatory control during exercise at 0.21 and 2.0 atmospheres inspired Po_2, J. Appl. Physiol. **14:**966, 1959.

58. Lambertsen, C. J., et al.: H^+ and Pco_2 as chemical factors in respiratory and cerebral circulatory control, J. Appl. Physiol. **16:**473, 1961.

59. Lenfant, C.: Effect of high F_{Io_2} on measurement of ventilation-perfusion distribution in man at sea level, Ann. N.Y. Acad. Sci. **121:**797, 1965.

60. Lenfant, C., Torrance, J. D., and Reynafarje, C.: Shift in the O_2-Hb dissociation curve at altitude: mechanism and effect, J. Appl. Physiol. **30:**625, 1971.

61. Maren, T. H.: Carbonic anhydrase kinetics and inhibition at 37°: an approach to reaction rates in vivo, J. Pharmacol. Exp. Ther. **139:**129, 1963.

62. Margaria, R.: A mathematical treatment of the blood dissociation curve for oxygen, Clin. Chem. **9:**745, 1963.

63. Meldrun, N. U., and Roughton, F. J. W.: Carbonic anhydrase. Its preparation and properties, J. Physiol. **80:**113, 1933.

64. Mulhausen, R. D.: The affinity of hemoglobin for oxygen, Circulation **42:**195, 1970.

65. Pauling, L., and Coryell, C. D.: The magnetic properties and structure of hemoglobin, oxyhemoglobin and carbon monoxyhemoglobin, Proc. Natl. Acad. Sci. **22:**210, 1936.

66. Perutz, M. F.: X-ray analysis of haemoglobin, Nobel Lecture, Stockholm, December 11, 1962.

67. Perutz, M. F.: Structure and function of hemoglobin, Harvey Lect. **63:**213, 1967-68.

68. Rahn, H., Reeves, R. B., and Howell, B. J.: Hydrogen

ion regulation, temperature, and evolution, Am. Rev. Respir. Dis. **112:**165, 1975.

69. Rapoport, S., and Guest, F. M.: Distribution of acid soluble phosphorus in the blood cells of various vertebrates, J. Biol. Chem. **138:**269, 1941.

70. Riley, R. L., et al.: The relationships of oxygen, carbon dioxide, and hemoglobin in the blood of man: oxyhemoglobin dissociation under various physiological conditions, J. Clin. Invest. **25:**139, 1946.

71. Roughton, F. J. W.: The average time spent by the blood in the human lung capillary and its relation to the rates of CO uptake and elimination in man, Am. J. Physiol. **143:**621, 1945.

72. Roughton, F. J. W.: Some recent work in the interactions of oxygen, carbon dioxide and haemoglobin, Biochem. J. **117:**801, 1970.

73. Roughton, F. J. W., and Forster, R. E.: Relative importance of diffusion and chemical reaction rates in determining rate of exchange of gases in the human lung, with special reference to true diffusing capacity of pulmonary membrane and volume of blood in the lung capillaries, J. Appl. Physiol. **11:**290, 1957.

74. Ruch, T. C., and Fulton, J. F., editors: Medical physiology and biophysics, Philadelphia, 1960, W. B. Saunders Co.

75. Severinghaus, J. W., Stupfel, M., and Bradley, A. F.: Variations of serum carbonic acid pK′ with pH and temperature, J. Appl. Physiol. **9:**197, 1956.

76. Siggaard-Andersen, O.: Blood acid-base alignment nomogram, Scand. J. Clin. Lab. Invest. **15:**211, 1963.

77. Siggaard-Andersen, O.: The acid-base status of the blood, Scand. J. Clin. Lab. Invest. **15**(suppl. 70):1, 1963.

78. Singer, R. B., and Hastings, A. B.: An improved clinical method for the estimation of disturbances of the acid-base balance of human blood, Medicine **27:**223, 1948.

79. Tomita, S., and Riggs, A.: Studies of the interaction of 2,3-diphosphoglycerate and carbon dioxide with hemoglobin from mouse, man and elephant, J. Biol. Chem. **246:**547, 1971.

80. Wyman, J.: Regulation in macromolecules as illustrated by haemoglobin, Q. Rev. Biophys. **1:**35, 1968.

70

CHRISTIAN J. LAMBERTSEN

Neural control of respiration

FUNCTIONAL COMPONENTS OF REGULATION

The central neurons responsible for the motor activity of pulmonary ventilation are remote from contact with either the external atmospheric environment or the metabolic needs of individual tissues. The rhythmic automatic reexcitation function is intrinsically generated, controlled in degree and frequency, and modulated not only by responses to changes in the local chemical environment of the special neurons themselves, but also by remote chemical changes and by reflex influences of the physical act of pulmonary ventilation. Superimposed on this exquisite balance of regulation and demand is the capacity for complete, if temporary, voluntary control.

In this chapter attention will be given to questions concerning the locations of the neuron complexes involved in central respiratory control, the manner of their organization, their mechanisms of autoexcitation, and the nature of the reflex factors responsible for modifying the output of these localized respiratory centers.

CENTRAL MECHANISMS OF RESPIRATORY CONTROL
Voluntary influences on respiration

Voluntary influences on respiration can produce a more powerful respiratory drive than any of the automatic and subconscious regulatory mechanisms. This fact is actually exploited when *maximum voluntary breathing capacity* is used as a test for lung function. Normal values for short periods of voluntary hyperventilation show movement of as much as 150 L air/min as compared with the somewhat lower values associated with the uncontrollable drives of severe muscular exercise.[74]

Conscious control can also completely inhibit automatic mechanisms until accumulation of local and systemic metabolites overpowers the voluntary suppression of respiration.

The cerebral hemispheres contain bilateral areas capable of affecting respiration, in part by modifying the activity of the more basic respiratory mechanisms,* but also in part by bypassing them. In certain well-defined sites, which include the anterior cingulate gyrus, the subcallosal region, the ventral surface of the frontal lobe, the insular region, and the temporal pole,[39] electrical stimulation can produce inhibition of breathing in the form of an expiratory apnea. Increased respiration occurs on stimulation of the piriform cortex.[14] In other portions of the cerebral cortex, stimulation results in crude but rhythmic movements related to the respiratory components of such functions as swallowing and vocalization.

Although they are important in relation to *volitional* influences on respiration and to activities such as talking, singing, whistling, laughing, coughing, sobbing, and sneezing, none of the cortical centers is required to sustain respiration, and there is no indication that the cortical structures concerned with respiration respond specifically with increased activity to administration of CO_2 or to reflex activity. Actually, when the cerebral hemispheres, the cerebellum, and the anterior brain stem are progressively removed in carnivores, no characteristic effects on breathing occur until the anterior border of the pons is reached. Transections at lower levels than this disturb respiration more and more profoundly as the medulla is involved until, following transection at the level of the tip of the *calamus scriptorius,* complete and permanent paralysis of breathing occurs. The centers responsible for the *automatic* spontaneous activation of respiration are therefore located in the pons and upper two thirds of the medulla.

Centers for automatic regulation of respiration

Study of central respiratory control mechanisms has emerged as a prominent interest in physiology. Present studies, superimposed on many decades of work, are aided by exceptional

---

*See references 27, 39, 72, 99, and 135.

advances in electrophysiologic method and computer technology. All indicate that neuronal groups ultimately responsible for originating the spontaneously rhythmic respiratory act are located within the medulla, since respiratory periodicity exists even when the medulla is deprived of connections with higher centers or with afferent nerve fibers from the lungs or peripheral chemoreceptors.*

The concepts that have continued to sharpen involve the following questions: How does the inspiratory discharge develop? In what neuron groups does it develop? How is the inspiratory discharge turned off to allow expiration to occur? How is it modulated by central and by peripheral chemical sensing systems? How is it affected by reflexes from the lungs? These are very old questions whose study has recently begun to provide new concepts.

The early concept of Legallois[109] in 1812 that respiratory rhythmicity originates entirely within the medulla has led to 150 years of meticulous study aimed at determining the locations and basis for rhythmic activation of the central cells responsible for respiratory periodicity. The concept that an autonomous, rhythmic medullary "respiratory center" exists was questioned by Marckwald over 90 years ago; he concluded that respiratory periodicity is imposed on the medulla from without by central and reflex inhibitory influences that interrupt a purely inspiratory medullary drive.[23,27] This view of extramedullary control of rhythmicity, elaborated by Lumsden[112-114] and by Pitts and associates,[27,121] persisted for many years. These and other investigators used discrete electrical stimulation in the neuraxis as well as ablation studies to identify centers of characteristic respiratory response. The evolution of electronic, histochemistry, and microelectrode methods has allowed specifically localized, discrete intracellular recording of phasic electrical activity of neuron groups involved in respiratory control.

Following are some of the functional and anatomic relationships that are now becoming more clearly defined.

Bilaterally paired inspiratory and expiratory "centers"

A respiratory medullary-pontine system of cells phasic in activity with respiration extends from the upper limit of the cervical cord to the pontine pneumotaxic complex (pp. 1765 to 1768).[96] This system is bilaterally symmetrical and contains only two spatial groupings of respiratory cell units—one inspiratory and one expiratory in function. This distribution, in which considerable intermingling of inspiratory and expiratory neurons occurs in the medullary and part of the pontine reticular formation, is not random. Some grouping occurs, which encompasses the smaller volumes of active cells defined by many previous studies in cats, monkeys, and dogs.* The cells involved have no evident special histochemical characteristics.

Inspiratory "center." The main group of cells having inspiratory function is found in the caudal third and dorsolateral portion of the bulbopontine region.[96] Electrical stimulation in locations having concentrations of inspiratory cells can produce maximal inspiratory effort that is sustained as long as the stimulus persists.[21,27]

Expiratory "center." The main group of the expiratory population of neurons is in the middle third of the bulbopontine region, in a ventromedial position.[96] Stimulation of the predominantly expiratory group does not necessarily result in sustained exhalation but can initiate exhalation and inhibit an inspiratory act.[21,27]

Bilaterally paired "pattern generators" for "central inspiratory excitation"

This functional concept extends the original view of intramedullary autorhythmicity. It indicates the necessary function of spontaneous and coordinated rhythmic increase in motor discharge by an inspiratory neuron system† interrupted by intrinsic and/or external influences. The interruption of the inspiratory phase is conceived as occurring through influences of an "off-switch" mechanism, which terminates firing by the population of inspiratory neurons. As inspiration is switched off, an inspiratory inhibition continues through the expiratory phase, delaying the next inhalation. As the inhibitory influence decreases with time, a new inspiration can occur.[77] Expiration is considered passive in eupnea (quiet breathing at rest, in the absence of added chemical stimuli or exercise). Since rhythmic respiration can exist when the brain stem is divided between the pons and medulla[53,94] and rhythmic firing of inspiratory motoneurons occurs even when the respiratory muscles are paralyzed and lung volume is maintained constant, a degree of intrinsic rhythmicity of the central respiratory network

---

*See references 21, 94, 120, 128, and 129.

*See references 21, 36, 39, 49, 55, and 121.
†See references 2, 45, 58, 77, 96, 133, and 137.

must exist, independent of higher centers or inhibitory reflex feedback from the lungs.*

Although the existence of intrinsic autorhythmicity for inspiration is clear, it is not yet known whether the inspiratory motoneurons themselves or other cells acting on them are responsible.

Chemical sensitivity and influences of carbon dioxide. The basic mechanism of CO_2-[H^+] stimulation of respiratory mechanisms is not known, nor are the specific sites of primary chemosensitive action (Chapter 71). However, concepts of CO_2 effects on intrinsic central inspiratory excitation and termination of inspiratory excitation are developing.[2,32,45,77] In the terms used for overall intrinsic respiratory autorhythmicity, CO_2 has been considered to increase the rate of increase in inspiratory motor electrical activity (leading to more rapid initial inspiratory flow) and to increase the threshold of the "inspiratory off-switch" mechanism (allowing deeper inspiration).[77] It is obvious on measurement and by simple observation that chemical stimulus also recruits active and powerful expiratory function.

Bilaterally located pneumotaxic centers in the anterior pons

A functional region concerned with respiratory periodicity is known to occupy definite sites in the rostral few millimeters of the pons.† These regions do not respond to electrical stimulation or to ablation with any characteristic respiratory pattern in an animal with functioning vagal reflexes. However, in the anesthetized, bilaterally vagotomized cat, transection of the neuraxis just caudal to the inferior colliculi causes inspiratory tone to become so prominently exaggerated that the animal may die in sustained inspiratory apnea.

This breath-holding respiration is termed *apneustic* and occurs as sustained inspiratory activity. However, it is not always expressed as a single, long-lasting inspiration. It appears to represent a change in the level of tonic inspiratory outflow on which periodic respirations may be superimposed.[94] It appears likely that the exaggeration of inspiration called apneustic breathing is related, not to release of activity of a specific "apneustic center," as had been proposed,[27] but to loss of the normal function of the pneumotaxic centers in contributing to termination of inspiratory excitation (contribution to inspiratory "off-switch" mechanism).[45,58,96,137] If this is so, the pneumotaxic centers (like the vagal afferents from the lungs) would represent a source of inhibitory influences on the central inspiratory excitation system, permitting an inherent medullary inspiratory rhythmicity to express itself. It is known that the pneumotaxic center is not in itself spontaneously rhythmic and the site of pacemaker activity.[27,39,45] Rhythmicity observed within the pontine pneumotaxic center is now considered secondary to influences of the medullary inspiratory center rather than the reverse.[39,45] The pons therefore has a large influence in eupnea, even though the inspiratory neurons are capable of rhythmic activity.[39] However, the pons is not the site of a respiratory pattern generator and is not necessary for rhythm.

Influence of anesthesia. The sustained inspiratory apnea (apneusis) that occurs in anesthetized vagotomized animals when the pneumotaxic centers are destroyed is now known to revert to rhythmic breathing on recovery from anesthesia,[96,137,138] and animals with pneumotaxic center ablation and bilateral vagotomy remain alive in spontaneous respiration for months.[88] On this basis, it is considered that an intramedullary "inspiratory off-switch" mechanism can operate independently of both pneumotaxic centers and vagal influences, but is functioning via an as yet unknown mechanism and structure interfered with by the anesthetic.

Bilaterally located vagal nuclei

Impulses from the stretch receptors of the lungs are capable of preventing inspiratory spasm (apneustic breathing) in animals deprived of the inspiration-inhibitory function normally provided by pontine pneumotaxic centers. Vagal fibers from the pulmonary stretch receptors penetrate the medulla, reach the solitary tracts, and from there enter the medullary and pontine reticular formation. The degree of crossing is sufficient to permit one intact vagus nerve to prevent the development of inspiratory spasm in animals deprived of the pneumotaxic regions. The intermittent volleys of vagal impulses do not necessarily provide a breath-by-breath suppression of inspiratory tone but may exert inhibitory effects on the brain stem facilitory system, thereby diminishing its effect on the medullary inspiratory center.[21]

Influence of vagal afferents from the lung on the central pattern generator. Although the existence of inhibitoinspiratory influences transmitted to the centers by vagal afferents from

---

*See references 16, 17, 45, 76, 77, 100, and 137.
†See references 17, 21, 27, 45, and 53.

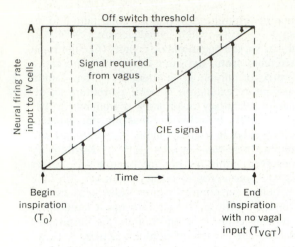

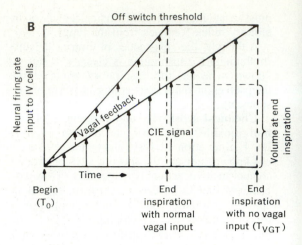

Fig. 70-1. Concept of operation of the inspiratory off-switch. Diagram shows inputs to inhibitory vagal *(IV)* cells. Threshold at which IV cells generate off-switch is shown as constant during a breath. **A,** Central inspiratory excitation *(CIE)* signal increases continuously. It alone will eventually cause IV cells to reach off-switch threshold. Inspiration then terminates at central pattern generator *(CPG)* duration, as seen after vagotomy *(T_{VGT})*. At any time earlier than this a sufficient lung inflation signal can sum with CIE and cause IV cells to reach off-switch threshold early. Volume feedback required to reach off-switch threshold decreases with time from beginning of inspiration. **B,** With vagi intact, lung volume feedback signal sums with CIE, and off-switch threshold is reached earlier. Volume of lungs at any time is roughly proportional to CIE signal. Lung volume attained at inspiratory cutoff with normal vagal input is marked at right of graph. Volume attained without vagal feedback would be proportional to full length of vertical axis. (From Wyman[45]; reproduced, with permission, from the Annual Review of Physiology, Volume 39. © 1977 by Annual Reviews Inc.)

stretch receptors in the bronchi has been known for over 100 years, relations to other components of the central control system and, in particular, the nature of vagal reflex influences on central inspiratory excitation rhythm are now becoming clearer.[6,25,28,42] Vagal reflex inhibition of inspiration by impulses from stretch receptors is evidently not active in normal, quiet breathing.[28,42,77] Impulse generation in the lungs increases progressively with increasing tidal volume, but the effect on the central neuronal net occurs largely in an "all-or-none" fashion. The result is inhibition of central inspiratory excitation only as the bombardment of the centers increases above a certain threshold.[45,77] Vagal afferent activity is therefore one factor in the inspiratory termination or "off-switch" mechanism but not critical to quiet breathing. Deeper breathing appears to require less afferent vagal activity (pulmonary inflation) to terminate inspiration late in the cycle than early in inspiration[45,77] (Fig. 70-1).

Crossover of respiratory tracts

In the central integration of respiratory control functions, it is known that although vagal afferent tracts cross over partially to serve bilateral components of the control system, the efferent respiratory impulses cross over completely in descending from one side of the medulla to the spinal cord. Permanent cessation of breathing occurs on discrete sagittal section of the medulla above the obex.[126] Respiratory neurons continue to fire, but impulse transmission to the respiratory muscles cannot occur.

EFFERENT DISCHARGES TO MUSCLES OF RESPIRATION
Studies of action potentials

The phrenic nerve, containing fibers to the diaphragm, conducts impulses from the inspiratory motor system. Adrian and Bronk[46] found electrical activity in individual fibers of a phrenic nerve to be increased to 20 to 30/sec during a normal inspiration. In related studies of intercostal innervation the volley of impulses related to inspiration occurred predominantly in the nerves to the external intercostal muscles, whereas expiratory activity appeared chiefly in nerve fibers to the internal intercostal muscles.[54] Augmentation of the force of inspiration or expiration is accomplished by the occurrence of impulse

Fig. 70-2. Discharge of motor impulses in nerve twig to external intercostal muscle of cat in three successive inspirations of varying depth (indicated by bottom record, which shows thoracic movements). **A,** Moderate inspiration. **B,** Deep inspiration. **C,** Shallow inspiration. Time in ⅕ sec intervals. At **A** there is short volley of impulses in three fibers; at **B** more neurons are in action, average frequency of discharge is higher and lasts longer; at **C** there is but a single impulse in each of two fibers. (From Bronk and Ferguson.[54])

discharge in additional nerve fibers (recruitment) as well as by increased frequency of firing by individual cells already active. This rate of discharge may increase to 50 to 80/sec during asphyxial stimulation. Quiet exhalation is associated with diminution of electrical activity in nerve fibers to inspiratory muscles, but when respiration is stimulated, active expiration is caused by an increase in the discharge to the muscles of exhalation.[54]

From such studies of action potentials in nerves leading to respiratory muscles it has been learned that pulmonary ventilation is determined by the following factors:

1. The *interval* between successive groups of nerve discharges to the muscles of inspiration (and expiration, if active). (This determines the frequency of breathing, whereas all the others regulate the depth and force of breathing.)
2. The *frequency* of the nerve impulses transmitted by the individual nerve fiber to its motor unit during each inspiratory (or expiratory) discharge.
3. The *duration* of the increased activity in individual nerve fibers.
4. The *number of motor units activated* during each inspiration or expiration. (This applies not only to the total number of muscles that participate in the act but also to the proportion of fibers of the individual muscles.)

In quiet breathing the intervals between the outbursts of nerve impulses producing inspiration are long, the number of muscle fibers that receive impulses is minimal, the number of impulses reaching each contracting fiber at each outburst is low—perhaps only one (Fig. 70-2)—the duration of the outburst is brief, and even in the contracting muscles there are many fibers that are not activated.

When chemical respiratory stimulation is produced, the intervals between the inspiratory outbursts are shortened, more muscles are activated

by each, the number of impulses reaching the muscle fibers at each outburst is greater, and eventually all the fibers of each active muscle are made to contract at each outburst.

In the basal, undriven system, factors related to intrinsic oscillation of the inspiratory motor system are presumably dominant. However, when respiration is driven chemically or by other influences, as in exercise, the inspiratory pacemaker undoubtedly comes more and more under the breath-by-breath influence of the superimposed second lines of control. Therefore in active respiratory stimulation the abrupt termination of the inspiratory discharge and its replacement by expiratory activity are probably due to the combined effects of (1) the inhibitoinspiratory influence from the lungs, (2) the spontaneous decrease in excitability of the inspiratory neurons after they have discharged their volleys of impulses, and (3) the existence of reciprocal inhibitory innervation of inspiratory and expiratory networks.*

The sudden onset of active expiratory activity is in part due to an intrinsic excitation mechanism, released by the abrupt termination of the inhibitoexpiratory influence of the inspiratory discharge. The end of the expiratory discharge then would result from the combined effects of (1) increase in excitability of the inspiratory neurons during the expiratory discharge and (2) release of inspiratory neuronal inhibition induced by activity of the expiratory net. The inspiratory motor neurons then assert their normal dominance and, with the onset of their activity, the expiratory discharge is again brought abruptly to a close.

REFLEX MECHANISMS OF RESPIRATORY CONTROL

The regulation of respiration is only in part attained by sensitivity of central respiratory neu-

*See references 16, 17, 39, 42, 45, 56, and 126.

rons to their local environment and products of their own metabolism. The respiratory centers of the medulla, modulated in rhythm by impulses from sites within the medullar units, are also influenced in rhythm by inhibitory impulses arising in the lungs in vagal receptors activated by the cyclic act of respiration itself. Not only is the automatic function of *rhythm* affected by both central and reflex inhibitory mechanisms, but the control system also can be driven to greatly increased total *activity* both by chemical influences of central origin and by reflex influences from the periphery. In the overall control of respiration, central nervous system (CNS) pathways, patterns of response to central chemical change, and systems of reflex innervation are factors interlocked with one another in a complex system of self-stabilizing loops. Each factor provides a background of effect that can modify the response to the others. In the following discussion, particular attention will be given to reflex influences on the centers for automatic respiratory control. These reflex forms of neurogenic activity, arising outside the CNS in response to stimuli related to respiration, are in many instances able to temporarily override the central, chemical linkage of respiration to the demands of metabolism.

In animals at rest the most important types of reflex influences in respiratory control are (1) those aroused in the lungs by their alternate inflation and deflation and (2) those that originate in the carotid and aortic reflex zones. As long as these are operating, respiration retains its normal character and adaptability, but when either is removed, breathing is modified in important respects.

Other afferent impulses can also affect breathing.[6,26,28,42] Some of these, if involuntary, are probably due to irradiation or overflow of strong central nervous excitations into the respiratory centers (such as the respiratory events associated with emotional disturbances, hysteria, and strong visual, auditory, tactile, or pain sensations).

Identifiable reflexes from the lungs include pulmonary irritant receptors and J-receptors sensitive to some chemical influences.[25,41] In the upper respiratory passages, reflex activations resemble protective devices (such as the suspension of breathing during irritation of the nasal and pharyngeal endings of the trigeminal nerves or the effort at expulsion of an irritant from the air passages by means of a cough or a sneeze).

Reflexes arising from the limbs and elsewhere presumably contribute to the increase in pulmonary ventilation during muscular exercise and combine with chemical stimulation in maintaining nearly constant the P_{CO_2}, P_{O_2}, and pH of arterial blood, although the rate of production of chemical stimuli is increased 10-fold or more (Chapter 74). Afferent impulses capable of affecting respiration more or less intensely can be evoked from receptors situated in the heart and in the pulmonary circulation, but little is known about their physiologic importance.

Respiratory reflexes from pulmonary stretch receptors (Hering-Breuer reflexes)

Reflexes set up by inflation of the lungs with afferent pathways in the vagus nerves were first described by Hering and Breuer in 1868.[92] The respiratory effects transmitted via the vagus nerves were studied by Head,[91] who almost 100 years ago devised a mechanical method of recording active contractions of a slip of the diaphragm in such a way that it was possible to inflate and deflate the lungs without producing passive disturbances in the record. With this sensitive method, he found that if the lungs were made to collapse by allowing air to enter the chest or by sucking air out of the trachea, the diaphragm underwent a prolonged tonic contraction. By sucking air out of the lungs after a quiet inspiration, it was possible to induce an inspiratory movement long before it would have occurred otherwise. All these phenomena were abolished when the vagus nerves were frozen, thus preventing them from carrying the essential afferent impulses.

These results showed that *inflation of the lungs inhibits inspiration and collapse of the lungs stimulates it.* By means of these vagus-mediated influences, the *mechanical* events of pulmonary inflation and collapse are linked to the central, *neurochemical* processes of respiratory control, providing one component in the self-regulation of respiration.

Extensive early and modern studies are described in the Hering-Breuer Centenary Symposium[28] and in recent reviews.[6,25,26,42] The effects of Hering-Breuer stretch receptors as well as lung irritant and J-receptors are shown in Table 70-1. These effects, even in response to volume change, overlap.[42]

The Hering-Breuer reflexes are now known not to be active in normal, quiet breathing. In stimulated breathing, they contribute to the early termination of inspiration, thereby providing for increased respiratory frequency in a central system that otherwise would function only by increases in tidal volume.[45,77] The increase in respiratory frequency in response to CO_2 or exer-

Table 70-1. Respiratory reflexes*

Reflex and stimulus	Receptor site	Afferent pathway	Respiratory effect	
			Ventilatory	Bronchomotor
Pulmonary				
Inflation	Large and small airways	Vagus	Apnea	Dilation
Paradoxical (inflation)	Lungs	Vagus	Inspiration	Unknown
Deflation†	Lungs	Vagus, sympathetic (?)	Hyperpnea	Unknown
Pulmonary chemore-flex† (amidines)	Lungs	Vagus	Apnea, rapid shallow breathing	Constriction
Pulmonary embolism†	Pulmonary vessels	Vagus, sympathetic (?)	Apnea, rapid shallow breathing	Constriction (?)
Pulmonary vascular† (congestion)	Pulmonary vessels	Vagus	Hyperpnea	Unknown
Mechanical irritation	Airways	Vagus, sympathetic (?)	Cough	Constriction
Chemical irritation	Airways, lungs	Vagus, sympathetic (?)	Cough or hyperpnea	Constriction
Cardiovascular				
Chemoreceptor	Carotid bodies	Glossopharyngeal, vagus	Hyperpnea	Constriction
	Aortic bodies			Constriction
Arterial baroreceptor	Carotid sinus	Glossopharyngeal, vagus	Apnea	Dilatation
	Aortic arch			Unknown
Venous baroreceptor	Great veins	Vagus	Hyperpnea (?)	Unknown
Muscle and tendon				
Respiratory	Diaphragm, intercostal	Phrenic, intercostal	Stimulation	Unknown
Limb		—	Hyperpnea	Unknown

*After Widdicombe.[41]
†May be same receptors and afferents.

cise therefore appears to depend on vagal mechanisms.[45,78]

The vagus nerve and respiration

The vagus trunk carries fibers that, when active, evoke at least *nine* respiratory reflexes that originate in thoracic structures.[41] Other reflexes, as outlined by Widdicombe[41] and shown in Table 70-1, can also influence breathing. At least four sets of these nerve fibers are capable of powerfully affecting respiration, these being (1) the fibers mediating the Hering-Breuer reflex, (2) those representing the sensory innervation of the trachea and bronchi, irritation of which causes a cough or at least active expiration, (3) the afferent fibers from baroreceptors in the arch of the aorta, stimulation of which causes inhibition of breathing and fall in blood pressure, and (4) the afferent fibers from the aortic bodies, which on stimulation cause hyperpnea and a rise in blood pressure. The only *efferent* fibers carried by the vagus nerves to respiratory structures below the larynx are those of the motor (constrictor) innervation of the bronchial smooth muscle and secre-

tory glands of the bronchial passages. In the presence of all these fibers and their functions, electrical stimulation of the vagus trunk or central end of the cut vagus cannot provide useful information. What is more surprising is that block of the cervical vagus nerves in the resting human by local anesthetic produces essentially no effect on ventilation.[12]

Chemoreceptors and baroreceptors of the carotid bifurcation and aortic arch

Along with afferent fibers from stretch and other receptors in the heart and lungs, some of which can be activated by pharmacologic agents,[25,41,65] the vagus nerves carry afferent fibers from *baroreceptors* and *chemoreceptors* located in the arch of the aorta.[13,59]

Chemoreceptors and baroreceptors also exist at the bifurcations of the common carotid arteries into external and internal carotid branches. These receptors are functionally equivalent to their counterparts in the aortic arch.[13,14,29]

The afferent fibers from the carotid sinus region ascend, not in the vagus nerves, but in the glossopharyngeal nerves (C_9) (Fig. 70-3). Thus the impulses arising in the two types of receptors, each at the two different locations, enter the medulla and there affect different neural elements. Over these pathways the reflexes aroused by changes in the chemical composition of the arterial blood and by changes in the arterial blood pressure can function in the automatic control of respiration and circulation. The manner in which the effects are mediated and the cell groups acted on are not known.

The anatomic locations of the baroreceptors and chemoreceptors are illustrated in Fig. 70-3, which also shows the characteristic effects of electrical stimulation of fibers leading exclusively either from baroreceptors or from chemoreceptors. It is evident that increased activity of the baroreceptors exerts *inhibitory* influences on the centers controlling respiration and those controlling vasomotor activity. Impulses from the chemoreceptors, on the other hand, have a positive *stimulant,* or *activating,* effect on the medullary respiratory centers. The stimulant effects set up by the chemoreceptors can, if strong enough, extend beyond the medullary centers and even lead to generalized convulsions, the reflex origin of which in some animals is shown by the absence of such convulsions after denervation of the carotid and aortic bodies.[130]

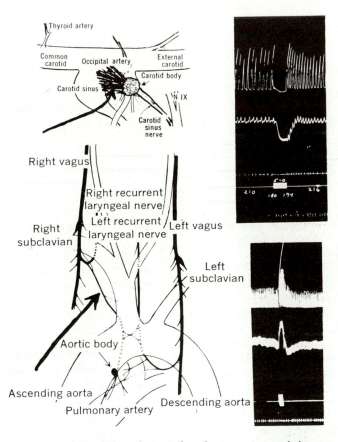

Fig. 70-3. Effects of electrical stimulation of nerves from baroreceptors and chemoreceptors in dog. Upper record: stimulation of fibers from receptors of carotid sinus at position indicated by large arrow (between carotid sinus and carotid body). Tracing shows (from above down) respiration, arterial blood pressure, volume of expired air (in 500 ml), signal, time in 5 sec intervals. Note apnea, fall in blood pressure, and slowing of heart (from 210 to 180 to 174) although both vagodepressor nerves were cut. Lower record: stimulation of fibers from aortic body at point indicated by arrow (before they join other aortic afferents in region of right recurrent laryngeal nerve). Tracing arranged as above (record of volume of expired air omitted here). Note hyperpnea and rise in blood pressure, although vagodepressor nerves were intact. (Carotid reflex zones were denervated.) (From Schmidt.[130])

Respiratory reflexes from baroreceptors

The *baroreceptor end-organs*, morphologically distinct from the chemoreceptors, are not clustered in localized organelles but are spread diffusely in the adventitial coats of the aortic arch and carotid sinuses.[13] In these locations, they are situated so as to be susceptible to their normal stimulus, distortion by expansion of the vessel wall. These "baroreceptors" are, in fact, stretch receptors activated by changes of intravascular pressure. Thus a rise in blood pressure, causing an increase in the frequency of impulses ascending from the baroreceptors, appears to inhibit sympathetic activity, exaggerate vagal cardiac effects, and inhibit respiration (Fig. 70-4). A lowering of blood pressure reduces the degree of inhibitory bombardment. The result of this is a rise in pulse rate, increased sympathetic discharge, and some increase in respiration. Within physiologic limits the baroreceptors appear to be unaffected by hypoxia or chemical products of metabolism, whereas the chemoreceptors are not affected by distortion.[13]

Although baroreceptor reflexes provide an active and important beat-by-beat contribution to the regulation of cardiac and vasomotor activity (Chapter 42), they appear to exert only minor and clinically unimportant influences on normal respiratory control.

Respiratory reflexes from carotid and aortic chemoreceptors

Peripheral chemosensitive cells are localized in discrete structures—the *carotid* and *aortic bodies*. These structures are in turn attached to major arteries that branch to provide the abundant blood supply required for the properties of short lag and rapid rate of response[61,71,73] when chemical changes occur in arterial blood.

The carotid body is more accessible and has been more thoroughly studied than the chemoreceptor tissue of the aortic arch. In experiments that accomplished temporal separation of the influences of the carotid and aortic bodies the aortic bodies have been found to exert prominent stimulant influences on the heart and vasomotor system, whereas the carotid bodies prominently

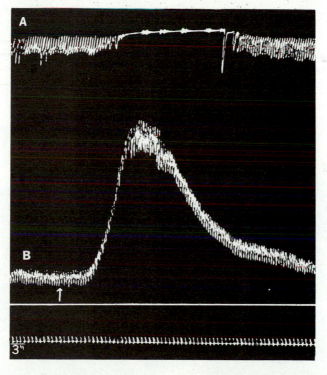

Fig. 70-4. Reflex apnea due to rise in pressure in aorta. Dog: respiratory movements, **A,** are movements of larynx of recipient animal whose head is completely separated from its trunk except for vagodepressor nerves; head is supplied with blood from donor animal. Epinephrine injected into femoral vein of recipient dog; it could not affect movements of head of this animal except through reflexes carried by vagodepressor nerves. Blood pressure, **B,** in femoral artery of recipient dog. (From Heymans and Heymans.[93])

stimulate respiration.[13,60] However, each of these chemoreceptor systems affects both respiration and circulation. Carotid chemoreceptor stimulation induces approximately 7 times the degree of respiratory response induced by aortic body stimulation.[67] The influence of chemical changes at the carotid chemoreceptors is through increases in respiratory frequency and depth, whereas the response to stimulation of aortic chemoreceptors is primarily an increase in frequency alone.[67]

The direct influence of carotid chemoreceptor stimulation is not cardiac stimulation because, when pulmonary ventilation is artificially maintained at a constant level, carotid body stimulation leads to bradycardia and a decrease in cardiac output.[66] For the purposes of this text, emphasis will be given to the carotid body, which has been most extensively studied and reviewed.[9,14,51]

Structural features of carotid body

The carotid chemoreceptor is a discrete, reddish nodule visible to the naked eye after careful dissection of the carotid sinus region; its specific nerve is less evident.[14,81] The body is about 1 to 2 mm in diameter in the adult cat and weighs about 2 mg.[63]

The arterial blood supply of the carotid body differs somewhat in the several species studied. In the dog the arterial supply is often a branch of the occipital artery (Fig. 70-3). As worked out with special care in the cat, arterial blood from the occipitopharyngeal trunk or occipital artery flows through the carotid bodies to emerge in veins that usually drain into the internal jugular vein.[57] The organs are therefore not, as once proposed, blind pouches having only arterial connections to the circulatory system. The exceptional vascularity of this minute organ is shown in Fig. 70-5.

Until the advent of the electron microscope, detailed information concerning the microstructure of these compact sense organs depended largely on the exhaustive studies of DeCastro[68] and others with the light microscope. Most of these initial observations remain valid.[29,51,80]

These microscopic studies indicated that within the carotid body most of the flow of arterial blood enters channels intimately adjacent to cells, while some may pass directly to venules in the form of arteriovenous anastomoses. This observation has prompted suggestions that these arteriovenous connections may function as shunts past the sensitive cells of the carotid body. However, changes in the diameter of parallel arteriovenous connections in a small organ having massive flow relative to other tissues should not appreciably change the characteristics of flow in sinusoidal spaces supplied by the same arterial source.

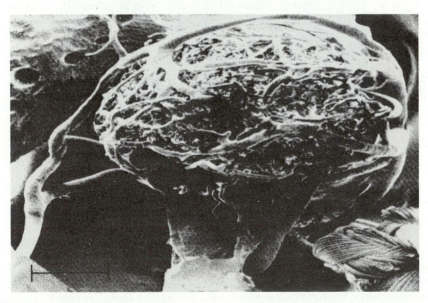

Fig. 70-5. Corrosion casting model of vessels of cat's carotid body as enlarged and demonstrated by scanning electron microscopy. Calibration = 500 μm. (From Seidl[134]; model and micrograph produced by Dr. D. Schäfer.)

Blood flow of carotid body

On the basis of flow of blood per unit of tissue weight the supply of arterial blood to the carotid bodies is very great. In measurements of the volume rate of carotid body blood flow in the cat, it was found that in the normal range of blood pressure an average of 40 mm³ of blood passed through the isolated carotid body each minute (Table 70-2).[63,123] Since the weight of a carotid body in the cat is only about 2 mg, a blood flow of 40 mm³/min, small in the absolute sense, actually represents a blood flow equivalent to about 2,000 ml/100 gm carotid body tissue/min and is thus by far the highest reported for any tissue. It can be compared with the blood flow of about 50 ml/100 gm/min through brain tissue.[102]

Because of its extremely high volume rate of blood flow, the carotid body normally removes only a minute amount of O_2 from the blood flowing through it. Table 70-2 illustrates that at the aforementioned rate of carotid body blood flow only about 0.2 ml O_2 was removed from each 100 ml of blood perfusing the carotid body of the cat.[63,123] Under normal conditions of air breathing, this lowering of blood O_2 content would correspond to a reduction of only about 1% in Hb saturation and a fall of only a few millimeters of mercury Po_2, even at its considerable level of O_2 consumption.[63,86,123] For these reasons the venous blood leaving the carotid body is essentially the same as arterial blood in composition, and the chemoreceptor cells are exposed to levels of Po_2, Pco_2, and [H+] almost identical to those of the blood flowing within the arteries. These cells are therefore more closely concerned than any other regulatory neurons with the interface between the external atmosphere, the alveolar gas, and the cellular environment.

Table 70-2. Carotid body blood flow and arteriovenous oxygen difference*

Condition	Blood flow (mm³/min)	A-V $O_2\Delta$ (ml/100 ml)
Normal blood pressure	40	0.15
Hypotension	9	2.0

*Based on data from Daly et al.[63]

Chemosensitive neural units

Chemosensitive neural units within the carotid body, which were described by DeCastro[68] and more recently examined by electron microscopy techniques,[29,36,51,125] are conceptually illustrated in Fig. 70-6 and include the following cells and fibers.

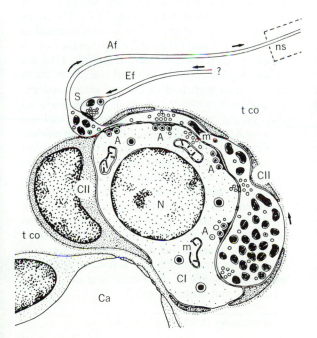

Fig. 70-6. Presumptive diagram of innervation of glomus cells. A type I cell *(CI)* is shown as receiving a highly polymorphous, afferent nerve ending *(Af)* with afferent synaptic complexes *(A)* at level of various aspects of this ending. Clear vesicles are shown both in ending and in type I cell. Diagram indicates that efferent fiber *(Ef)* controls activity of afferent ending at level of axodendritic synapse *(S)*. *t co,* Connective tissue; *Ca,* capillary; *N,* nucleus; *m,* mitochondria; *ns,* sinus nerve. Arrows show proposed direction of circulation of nervous influxes. (Schwann cells have not been shown.) (Fig. 2, article A. Verna, *J. Microscopie,* vol. 16, 1973 pp. 299-308.[139])

1. *Chemoreceptor cells (glomus cells), or type I cells,* are epithelioid cells not clearly neural in origin.[29,80] They are large and contain prominent granules[29,125] and a high concentration of pseudocholinesterase and dopamine.[29]

2. *Afferent nerve fibers* are myelinated and unmyelinated, presumably with cell bodies in the petrosal ganglion of the glossopharyngeal nerve. These fibers degenerate when cut outside the skull but not when cut intracranially.[29] A nerve fiber may originate as branches from up to 10 glomus cells, and 1 glomus cell may have contact with more than one nerve fiber.[29] The form of anatomic and functional connection of afferent fibers and glomus cells is not established, but it is beginning to appear that a plexus of sensory (afferent) nerve endings is held close to the glomus cells and that some form of synaptic mechanism exists (Fig. 70-6).[29,80,125] Extensive summaries of existing information and concepts have been provided by symposia concerning peripheral chemoreceptors.[29,124]

3. *Efferent nerve fibers,* efferent to and ending in the carotid body, are now known to exist,[29,124] but their function has not been established.

4. *Ganglion cells* have been described,[29,80] but they are very few in number and their relationship to other units is not at all certain. The powerful pharmacologic reactivity of the carotid body to ganglion-stimulating drugs such as nicotine and acetylcholine has prompted interest in anatomic identification of a ganglion-like synapse.

5. *Sustentacular (interstitial) cells, or type II cells,* also not of known neural origin, form part of the "terminal net" of interstitial cell fibers and nerve fibers that surrounds and comes into proximity with the glomus cell. This type of cell does not appear to contain granules with possible neurotransmitter function and may have no direct sensory or transmitter role.[29]

It is not now known how or at what sites the chemosensitive system is activated by various stimuli. Concepts are expanding, and most are described in the aforementioned reviews.[29,124] The presumption for many decades has been that the glomus cell is the primary physiologic sensor, but that some stimuli, including certain drugs (p. 1762), may act on sensory nerve endings or other structures within the organ. If the glomus cell is the initiator in normal function, some form of neurochemical synapse is necessary to account for impulse transmission to the endings of afferent nerve fibers. Fig. 70-6 illustrates components of chemoreceptor organization in a concept in which the glomus cell and nerve fiber form a primitive series connection, with the role of the sustentacular cell in chemical and electrical transmission not clear. It is evident that there are at least three possible sites of activation of this unit (the glomus cell, the sustentacular cell, and the nerve ending or fiber itself). Eyzaguirre has stressed that it is highly unlikely that changes in Po_2, Pco_2, and $[H^+]$ uniformly activate the nerve endings themselves, since other nerve fibers are depolarized by hypoxia, hyperpolarized by hypercapnia, and not importantly affected by physiologic changes in extraneuronal $[H^+]$.[81] Emphasis has been placed on the nerve endings by Biscoe[51] in reviewing carotid chemoreceptor function. It is completely likely from information available[29,124] that more than one site, sensor, and biophysical mechanism are involved in activation by different physiologic or pharmacologic stimuli.

Physiologic activation of chemoreceptors

Chemoreceptors of the carotid and aortic bodies respond by an increased rate of electrical discharge to lowered Po_2, elevated Pco_2, increased temperature, and increased $[H^+]$.* In turn, the greater impulse traffic produced by such alterations in acid-base balance and oxygenation acts on the centers to produce stimulation of respiration and circulation (Fig. 70-7). The reflex nature of these effects has been demonstrated in two general ways: (1) by showing that they occur when the corresponding nerves are intact but not when they are cut or inactivated and (2) by eliciting them in perfusion or crossed-circulation experiments in which the exciting agents could act only on the carotid or aortic reflex zone and had no contact with other parts of the reacting animal.

Reversible disruption of carotid chemoreceptor activity has been accomplished in humans by injection of a local anesthetic and irreversible effects by surgical extirpation of the carotid bodies, each removing essentially all respiratory stimulation by hypoxia.[11,90] Extensive study is now being made of (1) the mechanism of action of these stimulant changes on the chemoreceptor cells and (2) the quantitative interactions of similar or opposite changes in these factors at the chemoreceptors and in the centers.[29,124]

---

*See references 13, 48, 52, 61, 79, and 142.

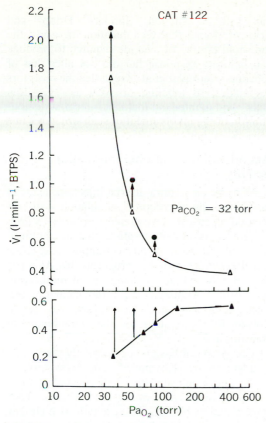

CAT #122

$Pa_{CO_2} = 32$ torr

$\dot{V}_I$ ($l \cdot min^{-1}$, BTPS)

Pa_{O_2} (torr)

Fig. 70-7. Effects of changes in arterial P_{O_2} on ventilation ($\dot{V}_I$) at a maintained arterial P_{CO_2} before (upper panel) and after (lower panel) bilateral section of carotid sinus nerves. Hypoxia increased ventilation before and depressed it after denervation. Arrows indicate amount of depression. (Cat, α-chloralose anesthesia.) (From Lahiri, S.: Introductory remarks: oxygen linked response of carotid chemoreceptors. In Reivich, M., et al., editors: Tissue hypoxia and ischemia. Advances in experimental medicine and biology, New York, 1977, Plenum Publishing Corp., vol. 78, p. 185.)

Decreased P_{O_2} can, by chemoreceptor activation, stimulate respiratory and circulatory centers whose general responsiveness is being depressed by the same lowering of O_2 tension.[9,13,61] Conversely, CO_2 administration can stimulate the respiratory centers even after its action by way of the chemoreceptors is abolished.[13,132] It is now known that the carotid bodies are sensitive to both CO_2 [H^+] and P_{O_2} at the levels normal for arterial blood in humans at sea level.

Early studies involving measurement of the respiratory response to perfusion of the carotid sinus with blood of varied composition[13,132] appear to have underestimated the sensitivity and degree of contribution of the chemoreceptors to respiratory control and relegated the mechanism

to a reserve function instead of considering it as part of the normal regulatory sensory system. This misconception occurred because in deviations from normal respiration at sea level, factors such as acidemia, hypercapnia, and lowered P_{O_2} are often concurrent, both in arterial blood and in the respiratory centers, whereas in the experimental condition of carotid body perfusion with blood of low P_{O_2} or high P_{CO_2} any resulting respiratory stimulation will result in a decrease of the acidity and P_{CO_2} of blood flowing to the respiratory centers. This change in central chemical environment should minimize the effectiveness of bombardment of the centers by impulses generated in the chemoreceptors themselves.

A similar difficulty of interpretation is related to the fact that the chemosensitive cells of the *respiratory centers* in nervous tissue are exposed to levels of CO_2, acid, and O_2 quite different from the levels of these same factors in arterial blood, whereas due to their high blood flow the local environment of the chemosensitive cells in the carotid, and probably the aortic, bodies normally is essentially identical in composition with the arterial blood itself.[63]

Mechanism of chemoreceptor stimulation

Thus far no clear explanation of the manner of chemoreceptor activation has been evolved to satisfactorily explain the actions of the many drugs and physiologic changes that increase chemoreceptor activity.[13,29,105] Surprisingly, biochemical study of carotid body tissue has not shown the presence of appreciable amounts of cytochrome oxidase.[108] It is probable that a prominent factor in the stimulant action of lowered P_{O_2} is the change in [H^+] induced within the carotid and aortic bodies during anaerobic metabolism of the glomus cells.[13,84] This concept has not been substantiated, but how respiratory *stimulation* can be caused by the *lack* of something (P_{O_2}) remains to be discovered. Any such mechanism must be an indirect one.

In experiments involving electrical recording from single chemoreceptor nerve fibers, activity can be elicited by lowered P_{O_2} or elevation of either [H^+] or P_{CO_2}.[13,29,52,104] This indicates that the same receptor and possibly a similar mechanism are involved. The mechanism is unknown but is probably not due simply to an increase in local [H^+] or to depolarization of receptor terminals. Whereas lowered P_{O_2} leads to depolarization of nerve fibers, an increase in P_{CO_2} results in hyperpolarization.[81] This has led to the suggestion that the powerful influences of hypoxia and hypercapnia are indirect, as by release of a transmitter substance,[29,81] which then activates the afferent nerves. However, hypoxia activates both myelinated and nonmyelinated chemoreceptor fibers, whereas endings of myelinated fibers are not prominently affected by acetylcholine (ACh).[81,119]

The carotid bodies are extremely sensitive to temperature change.[50,81,84] However, a rise in temperature, which stimulates the carotid chemoreceptors, leads to increased H^+ activity and hence does not require postulation of the existence of "temperature" receptors.[105]

Increased P_{O_2} suppresses the chemoreceptor activation produced by a rise in P_{CO_2} or $[H^+]$.[81,87,95] This further indicates an interrelationship or common path of stimulation by these factors. Presumably the effect of CO_2 is mediated by way of intracellular change in $[H^+]$ rather than by an action of molecular CO_2 itself. Accumulation of metabolites alone does not appear to be respiratory hypoxia does not further exaggerate chemoreceptor firing.[81]

In addition to activation by a fall in the O_2 tension or a rise in the P_{CO_2} or $[H^+]$ of arterial blood, the carotid and aortic chemoreceptors can be stimulated by other conditions, including administration of drugs or poisons. The conditions and agents capable of chemoreceptor stimulation can be divided into four groups:

1. Those that produce hypoxia in the chemoreceptors, either by interfering with their utilization of O_2 (hypoxia, hypercapnia, acidosis, cyanides, sulfides) or by augmenting their oxidative processes (dinitrophenol, dinitrocresol)

2. Those that exert nicotinic actions, that is, a stimulant action on, followed in higher dosage by depression or paralysis of neurons in the neuraxis and in autonomic ganglia (nicotine, lobeline, coniine, choline and its derivatives such as ACh and therefore agents that inactivate cholinesterase)

3. Drugs that first excite and then depress nerve tissue in general (potassium salts, veratrum alkaloids)

4. Drugs that fit none of the other categories (papaverine, aminophylline, nikethamide, acetaldehyde, CO)[97]

There is reason to believe that the site at which the chemoreceptor response to hypoxia (or acidosis) is initiated is distinct from that affected by nicotinic agents. von Euler et al.,[79] from studies of action potentials, have shown that the response to cyanide can be prevented by previous injection of alkali (ammonia), when response to lobeline

can still be elicited. Similarly Dripps and Dumke[75] found that O_2 inhalation increased the threshold dosage of cyanide required to produce a respiratory response but did not alter that of lobeline. von Euler et al.[79] regarded the receptors responding to lobeline or ACh as "ganglion cells" and pointed out that this is the first demonstration of the involvement of such structures in a sensory excitation.

Acetylcholine and chemoreceptor activity

ACh, by close intra-arterial injection, leads to prominent chemoreceptor stimulation with increased electrical activity and respiratory stimulation.[13] This, and observations that anticholinesterase agents such as physostigmine exaggerate the effect of ACh, led to proposals that ACh is a normal physiologic mediator of chemoreceptor activity.[13,82] Although ACh action cannot be primary, it or other neurotransmitters may be involved in some stage of the intraglomerular transmission.

This proposal has been extensively studied by Eyzaguirre and Koyano[81,82] and Eyzaguirre et al.,[85] who emphasize the high concentration of pseudocholinesterase in the carotid body. They have found by biologic assay a high ACh choline concentration in chemoreceptor tissue and have confirmed the observation that physostigmine increases the activity of the isolated chemoreceptor[82] (Fig. 70-8). They have conducted studies designed to show that fluid washing across one excised, electrically stimulated carotid body can cause activation of a second carotid body[83,85] (Fig. 70-9). ACh is also recoverable from the carotid body of the dog, stimulated during perfusion in vivo, using blood equilibrated with low O_2 and high CO_2 pressures.[116] These responses and the extent of ACh involvement in other neural transmission make it necessary to consider its possible involvement in chemoreceptor activity. However, no evidence thus far indicates a primary role for such a substance.[29,124] It is possible that ACh or another neurotransmitter may be concerned in a chemical transmitter function from one unit to another within the chemoreceptor.

Contributions of chemoreceptor reflexes to respiratory control

Following the discovery of the carotid and aortic chemoreceptors, it became apparent very early that in severe, acute hypoxia the chemoreceptors could sustain respiration reflexly when the centers were no longer responsive to direct

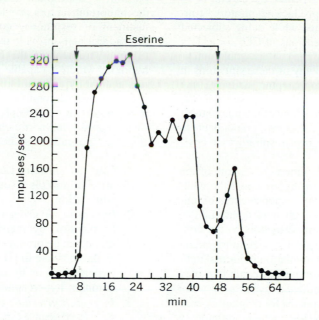

Fig. 70-8. Effects of anticholinesterase agent on chemoreceptor activity in vitro. Activity recorded in whole nerve is shown to increase grossly when physostigmine 10^{-6} wt/vol is directly applied to excised carotid body. Similar response occurs with small doses of ganglion blocking agent hexamethonium, which in higher concentrations proceeds to diminish activity. (From Eyzaguirre and Koyano.[82])

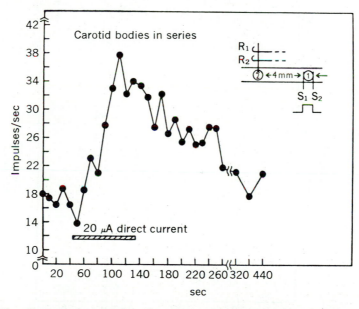

Fig. 70-9. Effect of electrical stimulation of upstream carotid body (glomus 1) on sensory discharge of downstream carotid body (glomus 2). Stimulating electrodes (S_1 and S_2) on glomus 1. Recording electrodes (R_1 and R_2) on carotid nerve of glomus 2. Striped horizontal bar indicates application of 20 μA direct current to glomus 1 for 90 sec. Notice increase in discharge frequency of preparation 2. (From Eyzaguirre et al.[85])

chemical stimulation.[13] This is well shown in perfusion experiments in which the carotid body receptors continue to produce respiratory stimulation for many minutes while being subjected to total O_2 lack, high tensions of CO_2, or a $[H^+]$ as low as pH 6.5. Since the centers would quickly fail when exposed to any of these situations in the absence of chemoreceptor activity, the chemoreceptor reflexes may in one sense be regarded as the *ultimum moriens* of the respiratory-regulating apparatus,[61] the last component to surrender its functions in a respiratory or environmental stress.

The existence of an "emergency" role of the chemoreceptors in severe hypoxia and the well-known lack of respiratory responsiveness of animals and humans to small changes in arterial P_{O_2} (Fig. 71-21) led some investigators to the concept that the chemoreceptors served no important function for respiratory regulation in normal individuals at rest or in exercise but came into play only in severe environmental hypoxia or when the direct central mechanisms for respiratory control were unable to provide adequate pulmonary ventilation.[61]

It is now evident that the chemoreflex mech-

anisms do exert prominent influences on the respiration of the human at rest[7] and are probably the important factors providing fast and fine adjustment of respiratory control.[73,106] However, they additionally provide the increasingly powerful bombardment of the reticular formation in hypoxia, which is capable of sustaining the centers against hypoxic failure. For the general purposes of this discussion on respiratory reflexes, it is important to reemphasize that stimulation of the carotid and aortic chemoreceptors to send impulses centrally may be the result of the following:

1. *Low O_2 tension.* Lowered O_2 tension of arterial blood is a source of chemoreceptor stimulation.

2. *Increased hydrogen ion concentration.* Stimulation by increased $[H^+]$ in arterial blood can be produced regardless of whether the increase in $[H^+]$ is brought about by hypercapnia or by acidemia without concomitant hypercapnia.

3. *Increased arterial P_{CO_2}.* Elevation of arterial P_{CO_2} causes a progressive increase in chemoreceptor activity.[48,104] However, chemoreceptor denervation appears to have

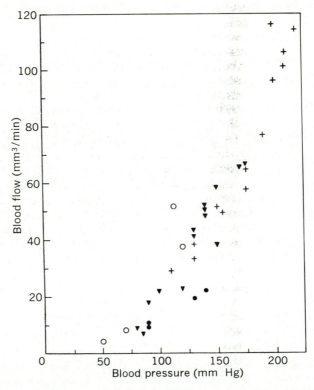

Fig. 70-10. Relationship between mean arterial blood pressure and blood flow through carotid body and superior cervical and nodose ganglia in cat. In two of four experiments (o and •), values for flow are those for carotid body only, following removal of ganglia. (From Daly et al.[63])

a more pronounced proportional effect in reducing the respiratory stimulation caused by low concentrations of inspired CO_2 than the stimulation by high CO_2 tensions.[9,89] The physiologic basis for these paradoxical observations has not been learned, but it must be remembered that elevation of CO_2 tension increases the concentrations of dissolved CO_2 gas and extracellular and intracellular hydrogen ions. The independent effects on chemoreceptor function of these different changes have not been completely studied, in part because changes in P_{CO_2} are not yet separable from changes in intracellular pH. These several alterations produced by CO_2 may eventually prove to have opposite effects on acid-sensitive respiratory neurons.

4. *Interaction of PO_2 and H^+.* A marked exaggeration of carotid body impulse traffic is generated when arterial blood P_{CO_2} is elevated at the same time that PO_2 is lowered.[104,124] This is the usual situation in asphyxia and respiratory failure. Lowered PO_2 alone is characteristic of exposure to high altitude.

5. *Hyperthermia.* Perfusion of the isolated carotid sinus region with fluids at different temperatures shows that chemoreceptor activity increases with local elevation of temperature.[13,81,131] This result is not necessarily specifically related to temperature change; it probably reflects an effect of hyperthermia in altering the acid-base state in the blood[136] or other body fluids.[47]

6. *Hypotension and chemoreceptor ischemia.* Since the chemoreceptors respond to local changes in PO_2, P_{CO_2}, and pH, it is to be expected that they should be stimulated when their own blood supply is reduced. In this circumstance the apparently high metabolic rate of the chemoreceptors[63] should lower the PO_2 and raise the P_{CO_2} and $[H^+]$ within these organs, even though the composition of arterial blood remains normal.[63,107] Thus autostimulation of chemoreceptors is a distinct possibility. Increased impulse traffic in the carotid sinus nerve occurs during hypotension and has actually been found following hemorrhage.[13,98,107,118] In this condition, blood flow through the carotid body of the cat is markedly lowered, resulting in a decrease of the O_2 level in the venous blood from the carotid body[13,63] (Fig. 70-10). The extent to which such a mechanism may contribute to the respiratory stimulation seen in the human during hypotension is not known.

Significance of chemoreceptor reflexes

It is remarkable that, attached to the carotids and aorta of air-breathing animals, there should be a reflex system capable of responding to changes in the gaseous composition of blood as it is delivered from the lungs. This provides the animal with its fastest responding control mechanism (Chapter 71). More surprising is the fact that the system can act by way of each major metabolic index (PO_2, P_{CO_2}, H^+) and can bring about strong stimulation of all the medullary centers and even the entire neuraxis and is responsible, practically alone, for effective protection of the organism against subnormal O_2 tensions in the inhaled air. Finally, the mechanisms are so resistant to adverse circumstances as to be able to keep the inspiratory center neurons actively functioning (through strong excitatory nerve impulses) when they would otherwise lapse into hypoxic inactivity.

INTEGRATION OF RESPIRATORY ACT
Normal state

The neurogenic control of respiration is among the most basic functions of the CNS. From the standpoint of experimental study, this system offers the great advantage that changes in its functional output can be quantitatively measured in terms of work of breathing, electrical activity in neuromotor units, or change in rate, depth, or minute volume of ventilation. As a result of intensive study by many methods, this relatively simple biologic control system is beginning to be understood. Much remains conceptual, and the challenge to relate concept and measurement remains.

It is known that there are neurons capable of rhythmic discharge, and it is presumed by most investigators that this discharge is spontaneous.* However, many questions remain regarding the generation of the primary discharge and its rhythmicity, its site of origin, the site and mechanism of chemical reactivity, the manner in which the activity of these cells is integrated and modified to produce the smoothly increasing efferent volley to muscles of inspiration, and the ultimate biophysical basis for the rather abrupt cessation of inspiratory neuronal discharge on exhalation.

It is not known exactly which mechanisms pro-

*See references 21, 45, and 127 to 129.

duce the graded increase in activity of the respiratory centers in fever, on elevation of blood CO_2 tension, or during the increased metabolic requirements of exercise or how impulses from the peripheral chemoreceptors affect the intensity of the efferent discharges to the muscles of respiration. Each of these influences is prominently involved in the overall process of respiratory control, and there is now abundant evidence for the existence of chemical, central, and reflex feedback influences on an intrinsic oscillating function of groups of central respiratory neurons.

In any concept of the integration of respiratory control functions, the focal point or relay station for the numerous modulators of respiration appears to be the two types of neurons (inspiratory and expiratory) within the reticular formation (Fig. 70-11). Salmoiraghi and von Baumgarten[129] and Burns[56] have proposed that coordinated activity within this neuron network tends to be maintained by a "self-reexciting" mechanism, that the activation of one neuron group provides the inhibition that diminishes the activity of the other, and that the discharge from each group is self-limiting.[129] Spreading, mutual excitation is probable in a system so smoothly coordinated as inspiration and expiration. However, spontaneous initiation of activity is necessary for mutual excitation to occur.

It is known that progressive isolation of the

Fig. 70-11. Components of respiratory control. *Cerebral cortex.* Centers for voluntary control of inspiration and of expiration provide for general utilization of ability of lungs to move air. Pathways between cortex and centers of automatic respiratory regulation are not clearly defined. Automatic and essentially normal respiration continues after removal of cerebral hemispheres, cerebellum, and anterior portion of brain stem. *Pontine pneumotaxic center.* Cells bilaterally located in upper portion of pons inhibit sustained inspiration called "apneusis." Apneusis therefore results in anesthetized animals when inhibitory cells are destroyed. These inhibitory cell groups have been designated as a pneumotaxic center and are considered to function in terminating each inspiration (part of off-switch mechanism). *Medulla: inspiratory and expiratory centers.* Inspiratory and expiratory motoneurons exist but are not necessarily grouped into anatomically discrete inspiratory and expiratory centers in all species. Medullary respiratory neurons appear capable of spontaneous discharge. They are directly or indirectly activated by H^+, perhaps via separate chemosensitive neuron groups. These inspiratory and expiratory cells may have reciprocal influences on each other as well as affecting and being affected by pontine centers. *Chemosensitive regions in medulla.* Structures located somewhere near lateral portions of fourth ventricle respond to induced changes in P_{CO_2} of cerebrospinal fluid.[7,110,111,117] Their relation to medullary inspiratory and expiratory centers is unsettled; they appear to represent not a superficial and unique "central chemoreceptor" sensitive to composition of cerebrospinal fluid, but one of two functionally identifiable central components of response to acid-base change in circulating blood.[106] *Reflex influences: vagal fibers from pulmonary stretch receptors.* Impulses aroused in pulmonary stretch receptors, like pneumotaxic cells, are capable of suppressing normally prominent inspiratory tone and appear to be part of the inspiratory cutoff mechanism (off-switch) in driven respiration. They appear not to be involved in quiet breathing. Although this inhibitory action was once considered a requirement for respiratory automaticity, it is now evident that spontaneous rhythm exists after bilateral vagus section and is therefore a property of isolated central mechanisms for respiratory control. Hering-Breuer reflexes normally serve to aid control of respiratory frequency. *Reflex influences: carotid and aortic chemoreceptors.* Impulses from these peripheral chemosensitive cells ascend via vagus (aortic body) and glossopharyngeal (carotid bodies) nerves. These sense organs appear to be discharging even in normal individuals breathing air at sea level. Impulses from chemoreceptors may influence reticular activating system, shifting level of activity of medullary respiratory neurons. If so, effect of chemoreceptor stimulation would be exerted on cells different from central chemosensitive neurons that respond to changes in central acid-base state. *Reflex influences: carotid and aortic baroreceptors.* Inhibitory impulses from baroreceptors of aortic arch and carotid sinus ascend with chemoreceptor fibers in vagus (aortic baroreceptors) and glossopharyngeal (carotid baroreceptors) nerves. These inhibitory effects of bombardment by increased impulse traffic have little practical significance. They may be brought about by way of stimulant influences on reticular suppressor system. *Integration.* The several centers and reflex factors involved in automatic control of respiration are linked together in an interdependent system in which degree and frequency of spontaneous "oscillating" activity of central respiratory motoneurons can be exaggerated or suppressed by local and distant chemical or physical changes. Changing balance among the many, sometimes simultaneous, excitatory and inhibitory influences on respiratory mechanism determines degree of pulmonary ventilation and its necessary correlation with metabolic requirements.

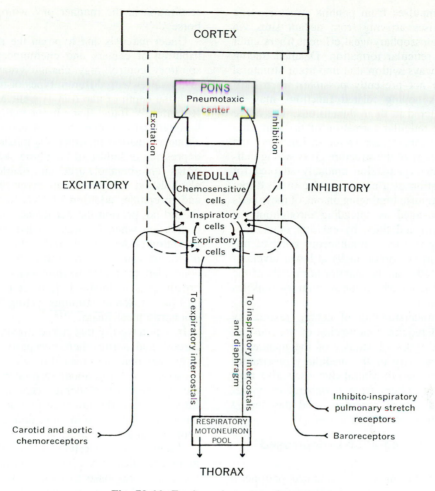

Fig. 70-11. For legend see opposite page.

medullary centers leads to a progressive decrease in the number of spontaneously active inspiratory neurons that can be detected by the probing electrode.[56] This type of observation has led to the assumption that, although the inspiratory and even expiratory "centers" are most likely intrinsically rhythmic, they depend on a heavy, if incidental, electrical traffic in surrounding, nonrespiratory neurons for their natural state of activity.[56]

Inspiratory and expiratory motoneurons are said to respond to increased CO_2 tension.[129] However, it is entirely possible that CO_2 does not produce its characteristic respiratory stimulant effects directly on these neurons of the medullary respiratory center but may act on separate, as yet unidentified, central chemoreceptors (Chapter 71). Respiratory change is produced by perfusing the fourth ventricle and cistern with

cerebrospinal fluid of altered P_{CO_2}.[110,111,117] It is not known how the activation of these unidentified chemosensitive structures affects the electrical activity of the respiratory motor cells within the reticular formation.

Presumably each of the several known nervous influences on respiration (carotid and aortic chemoreceptor reflex, pulmonary stretch reflexes, the baroreceptor reflex from the carotid sinuses and aortic arch, reflexes from proprioceptors in the joints) somehow exerts its effects on the medullary-pontine respiratory neuronal network, which thus functions in the final integration of contradictory excitatory and inhibitory impulses and adjusts the respiratory response to provide the most suitable compromise among them.

A complicating factor in the further exploration of central respiratory control is the manner in which impulses from pontine structures and other impulses arriving from distant sites via vagal and glossopharyngeal efferent fibers enter the diffuse reticular formation. Detailed tracing of the pathways within this important structural complex is not presently possible, since individual cells having related functions may be widely scattered in three dimensions. The nerve cells in the reticular formation of the pons and medulla presumably are *motor* in function, and the morphology of the structure gives strong indication of an organization uniquely adapted for the propagation of a nerve impulse in all directions. An impulse impinging on one of these cells can be envisaged as spreading throughout the entire system and thus, by exciting other neurons, giving rise to a simultaneous burst of impulses to various motor nuclei at lower levels of the system, such as the anterior horn cells of the spinal cord from which the respiratory muscles are innervated.

Further understanding of central respiratory control will require continuation of the present, imaginative types of studies of the manner in which the activity of the medullary respiratory neurons is related to central chemosensitive neurons, afferent fibers from peripheral chemoreceptors and stretch receptors, and descending influences from the higher centers.

Respiratory integration in depressed states

Evolution has provided combined peripheral chemoreflex and central chemosensitive mechanisms of respiratory control, and a state of normal balance between these two interconnected systems has been attained. In normal control, it can be considered that each of these chemical sensing and driving mechanisms is active, sometimes working together and sometimes in opposition, with occasional transient assumption of authority by one over the other.

In one type of abnormal situation, central depression by drugs, hypoxia, or trauma, this normal balance is grossly disturbed and the contribution of the central chemosensitive component of control may be severely diminished or even lost. The peripheral chemoreceptor sense organs are extremely resistant to hypoxia and continue to fire when central neurons can otherwise no longer respond to local chemical stimuli. The greater vulnerability of central structures to disruption by hypoxia is probably a natural consequence of their more complex organization. At any rate, chemoreflex bombardment of centers inactivated by hypoxia or depressant drugs[105,115] can sustain

respiration in the manner of "whipping a tired horse."

Under narcosis and hypoxia the relative contribution of centers and chemoreceptors to respiratory control may change greatly, but the result varies with different types of narcotics and with the depth of narcosis as well. Marshall and Rosenfeld[115] found that administration of phenobarbital and morphine to dogs and cats brought about a situation in which O_2 inhalation caused depression or failure of breathing, whereas other drugs (chlorbutanol, urethane, paraldehyde, and alcohol) did not. In animals given phenobarbital and morphine, addition of CO_2 to the inspired O_2 did not prevent the occurrence of respiratory depression when O_2 was inhaled, but after denervation of the carotids and aorta, O_2 inhalation no longer caused depression of breathing. They concluded that under the influence of these drugs, breathing is maintained "almost if not entirely by the anoxemic stimulus acting through the sinoaortic mechanism."[115]

It is noteworthy that nerve impulses from the carotid and aortic chemoreceptors are not the only ones that can break through a narcotic depression and bring about respiratory activity in the presence of moderate central depression. Impulses from the lower respiratory tract, from the joints of the extremities, and from pain receptors can act similarly.

Whatever its underlying cause, the effect of many narcotics on respiratory control is to reduce the central response to CO_2.[105] Breathing then may be partly maintained through chemoreceptor activation by the lowering of Po_2 and rise in Pco_2 that must accompany a decrease in alveolar ventilation.[101,105] Under such circumstances, inhalation of O_2 will diminish or even stop until the mounting Pco_2 and $[H^+]$ can compensate for the decreased reactivity of the centers to these stimuli. If the degree of narcosis or central damage is severe, respiration may not begin again spontaneously. In such cases, further depression by increasingly high Pco_2 and very low pH will be factors contributing to persistent apnea. When hypoxia exists and O_2 is needed, this interesting chemoreflex suppression of respiration by O_2 should not be feared. The needed oxygenation can be accomplished by using sensible artificial ventilation to prevent the acidosis that would otherwise result from apnea.

OTHER REFLEXES AFFECTING RESPIRATION

Protective reflexes include a group of coordinated, purposeful responses to the inhalation of irritant materials. These reflexes are served by pulmonary irritant fibers[25,42] as well as by innervation of the upper respiratory passages (Fig. 70-12). A *sneeze* is elicited by irritation of the mucous membrane of the nose, a *cough* by irritation of the respiratory passages beyond the nose. The impulses responsible for a sneeze are set up by irritation of sensory receptors of the trigeminal nerves and probably also by stimulation of the olfactory end-organs, whereas those that lead to a cough arise in the pharyngeal distribution of the glossopharyngeal and in the sensory endings of the vagus in the larynx, trachea, and bronchi. The carina at the bifurcation of the trachea is especially sensitive.[122] Both reflexes are characterized by a deep inspiration and a violent expiration; in a cough the vocal cords close completely when the inspiration is completed and do not open until the expiratory movement is well under way, so that the outward movement of air

is an explosive one, with air movement approaching or equal to the speed of sound. It is interesting to note that once the deep inspiration has been made, the violent contraction of the expiratory muscles is harder to control than if the inspiration is voluntarily kept shallow. Perhaps this is another instance of the interaction of the Hering-Breuer reflex and the reciprocal innervation of the inspiratory neurons. The unusually complete reflex inhibition of the inspiratory neurons that follows a deep inspiration may automatically cause the forced expiration because it releases the expiratory neurons from the inhibition that the inspiratory cells ordinarily exert on them.

Irritation of the larynx may cause spasm of the glottis, and irritation of the lower passages may lead to spasm of the bronchial muscle. These effects are likely to follow exposure to violent irritants such as poison gases, smoke, ammonia, sulfur dioxide, or chlorine. Such effects are prob-

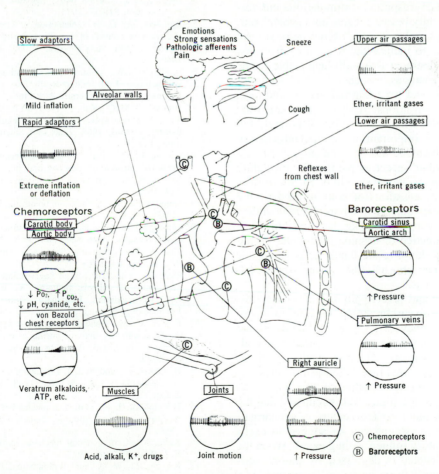

Fig. 70-12. Diagram indicating patterns of effects on respiration of afferent impulses from known baroreceptors and chemoreceptors. (Prepared by D. M. Aviado, Jr.)

ably sometimes due to actual damage to nerve endings and not only to stimulation of specialized pulmonary irritant and J sensory receptors. They do not always disappear as soon as the noxious agent is removed but may require a period of healing.

Reflexes from the joints or muscles appear to be involved in the hyperpnea of muscular exercise.[41,69,70] According to existing evidence, these reflexes, which may not be specific for respiration, arise in the joints and periarticular surfaces, not in the muscles and tendons.

Traction on the tongue has long been known to be capable of setting up a stimulant respiratory reflex.

Reflex stimulant to breathing can also be aroused by *pressure on the chest*. Part of this effect is probably due to the excitoinspiratory Hering-Breuer reflex, aroused by expulsion of air from the lungs by the external pressure. In addition, however, there probably are reflexes from the tissues of the chest wall.[41,140]

Pain causes respiratory stimulation, but this is probably only because it sets up a generalized, subconscious and conscious excitation or arousal that happens to include the respiratory neurons. The same statement is equally applicable to the respiratory changes associated with *emotional disturbances* and with unusual visual, auditory, and tactile sensations. These effects are only complications when respiration is normal, but they may be important under special circumstances. Advantage has been taken of the respiratory stimulant effect of pain in the procedure of spanking a newborn child to make him breathe, and sensations of cold aroused by the infant's sudden change in environment doubtless also play a part in arousing the center to activity.[64] The practice of dashing cold water on the skin and of slapping the hands or pinching the cheeks of one who has lost consciousness rests on the same basis.

Prolonged and severe pain may also cause sufficient increase in breathing to lead to hypocapnia (deficiency of CO_2 in the blood). If the pain is suddenly relieved, as by intravenous administration of an effective analgesic drug, breathing may temporarily diminish because the reflex and conscious influence that had been maintaining it at a high level is removed and the chemical stimulus to the centers is subnormal.

The same reasoning can be applied to excitatory respiratory influences in general. When the respiratory center is depressed by trauma, circulatory disturbances, or drugs, it becomes less responsive to all sorts of stimuli, and a point may

be reached at which it continues to function only because the combination of afferent impulses and chemical stimuli together is just able to arouse the depressed neurons to activity. If now an important group of afferent impulses is suddenly cut off—as by block of pain impulses by analgesic or anesthetic drugs—breathing may be reduced or may stop entirely for a short period of time.

Aspects of the interaction of reflexes and central functions in respiratory control are considered again in Chapters 71 and 74.

REFERENCES

1. Adrian, E. D.: Afferent impulses in the vagus and their effect on respiration, J. Physiol. **79:**332, 1933.
2. Berger, A. J., Mitchell, R. A., and Severinghaus, J. W.: Regulation of respiration, parts I, II, and III, N. Engl. J. Med. **297:**92, 138, 194, 1977.
3. Bianchi, A. L., and Barillot, J. C.: Activity of medullary respiratory neurones during reflexes from the lungs in cats, Respir. Physiol. **25:**335, 1975.
4. Biscoe, T. J.: Carotid body: structure and function, Physiol. Rev. **51:**437, 1971.
5. Bjurstedt, A. G. H.: Interaction of centrogenic and chemoreflex control of breathing during oxygen deficiency at rest, Acta Physiol. Scand. **12:**suppl. 38, 1946.
6. Bradley, G. W.: Control of the breathing pattern. In Widdicombe, J. G., editor: Respiratory physiology II, Baltimore, 1977, University Park Press.
7. Brooks, C. McC., Kao, F. F., and Lloyd, B. B., editors: The cerebrospinal fluid and the regulation of ventilation, Oxford, 1965, Blackwell Scientific Publications, Ltd.
8. Cohen, M. I., et al.: Synaptic connections between medullary inspiratory neurons and phrenic motoneurons as revealed by cross-correlation, Brain Res. **81:** 319, 1974.
9. Dejours, P.: Chemoreflexes in breathing, Physiol. Rev. **42:**335, 1962.
10. Duron, B.: Postural and ventilatory functions of intercostal muscles, Acta Neurobiol. Exp. **33:**355, 1973.
11. Guz, A.: Regulation of respiration in man, Annu. Rev. Physiol. **37:**303, 1975.
12. Guz, A., et al.: The role of vagal inflation reflexes in man and other animals. In Porter, R., editor: Breathing: Hering-Breuer Centenary Symposium, Ciba Foundation Symposium, London, 1970, Churchill.
13. Hesser, C. M.: Central and chemoreflex components in the respiratory activity during acid-base displacements in the blood, Acta Physiol. Scand. **18:**suppl. 64, 1949.
14. Heymans, C., and Neil, E.: Reflexogenic areas of the cardiovascular system, Boston, 1958, Little, Brown & Co.
15. Hildebrandt, J. R.: Gating: a mechanism for selective receptivity in the respiratory center, Fed. Proc. **36:** 2381, 1977.
16. Kalia, M.: Central neural mechanisms of respiration, Fed. Proc. **36:**2365, 1977.
17. Karczewski, W. A., and Widdicombe, J. G., editors: Neural control of breathing, Acta Neurobiol. Exp. **33:**entire volume, 1973.

18. Kim, J. K., and Carpenter, F. G.: Excitation of medullary neurons by chemical agents, Am. J. Physiol. **201:** 1187, 1961.

19. Knowlton, G. C., and Larrabee, M. G.: A unitary analysis of pulmonary volume receptors, Am. J. Physiol. **147:**100, 1946.

20. Larrabee, M. G., and Knowlton, G. C.: Excitation and inhibition of phrenic motoneurons by inflation of the lungs, Am. J. Physiol. **147:**90, 1946.

21. Liljestrand, A.: Neural control of respiration, Physiol. Rev. **38:**691, 1958.

22. MacDonald, D. M., and Mitchell, R. A.: The innervation of glomus cells, ganglion cells and blood vessels in the rat carotid body: a quantitative structural analysis, J. Neurocytol. **4:**177, 1975.

23. Marckwald, M.: The movements of respiration and their innervation in the rabbit, London, 1888, Blackie & Sons. (Translated by Haig.)

24. Mitchell, R. A., and Berger, A. J.: Neural regulation of respiration, Am. Rev. Respir. Dis. **111:**206, 1975.

25. Paintal, A. S.: Vagal sensory receptors and their reflex effects, Physiol. Rev. **53:**159, 1973.

26. Paintal, A. S.: The nature and effects of sensory inputs into the respiratory centers, Fed. Proc. **36:**2428, 1977.

27. Pitts, R. F.: Organization of the respiratory center, Physiol. Rev. **26:**609, 1946.

28. Porter, R., editor: Breathing: Hering-Breuer Centenary Symposium, Ciba Foundation Symposium, London, 1970, Churchill.

29. Purves, M. J., editor: The peripheral arterial chemoreceptors, New York, 1975, Cambridge University Press.

30. Rice, H. V., and Joy, M. S.: Modification of respiratory movements by vagal stimulation, Am. J. Physiol. **149:**24, 1947.

31. Rovainen, C. M.: Neural control of ventilation in the lamprey, Fed. Proc. **36:**2386, 1977.

32. Sorensen, S. C.: The chemical control of ventilation, Acta Physiol. Scand. (suppl.)**361:**entire issue, 1971.

33. Tang, P. C.: Localization of pneumotaxic center in cat, Am. J. Physiol. **172:**645, 1953.

34. Tang, P. C.: Brain stem control of respiratory depth and rate in the cat, Respir. Physiol. **3:**349, 1967.

35. Torrance, R. W.: Arterial chemoreceptors, In Widdicombe, J. G., editor: Respiratory physiology. International review of science, physiology, series one, Baltimore, 1974, University Park Press, p. 247.

36. Torrance, R. W., editor: Arterial chemoreceptors, Oxford, 1968, Blackwell Scientific Publications, Ltd.

37. Trippenbach, T., and Milic-Emili, J.: Vagal contribution to the inspiratory "off-switch" mechanism, Fed. Proc. **36:**2395, 1977.

38. Waldron, I.: Activity patterns in respiratory muscles and in respiratory neurones of the rostral medulla of the cat, J. Physiol. **208:**373, 1970.

39. Wang, S. C., and Ngai, S. H.: General organization of central respiratory mechanisms. In Fenn, W. O., and Rahn, H., editors: Handbook of physiology. Respiration section, Washington, D.C., 1964, American Physiological Society, vol. 1.

40. Wang, S. C., and Ngai, S. H.: Respiration coordinating mechanism of the brain stem: a few controversial points, Ann. N.Y. Acad. Sci. **109:**550, 1963.

41. Widdicombe, J. G.: Respiratory reflexes. In Fenn, W. O., and Rahn, H., editors: Handbook of physiology. Respiration section, Washington, D.C., 1964, American Physiological Society, vol. 1.

42. Widdicombe, J. G.: Reflex control of breathing. In Widdicombe, J. G., editor: Respiratory physiology. International review of science, physiology, series one, Baltimore, 1974, University Park Press, p. 273.

43. Widdicombe, J. G., and Glogowska, M.: Relative roles of irritant, type-J and pulmonary stretch receptors in lung reflexes, Acta Neurobiol. Exp. **33:**21, 1973.

44. Windeo, C. V.: On the mechanism of stimulation of the carotid gland chemoreceptors, Am. J. Physiol. **118:**389, 1937.

45. Wyman, R. J.: Neural generation of the breathing rhythm, Annu. Rev. Physiol. **39:**417, 1977.

46. Adrian, E. D., and Bronk, D. W.: The discharge of impulses in motor nerve fibers. II. The frequency of discharge in reflex and voluntary contractions, J. Physiol. **67:**119, 1929.

47. Alexander, S. C., Workman, R. D., and Lambertsen, C. J.: Hyperthermia, lactic acid infusion, and the composition of arterial blood and cerebrospinal fluid, Am. J. Physiol. **202:**1049, 1962.

48. Bartels, H., and Witzleb, E.: Der Einfluss der Arteriellen CO_2-Druckes auf die Chemoreceptorischen Aktionspotentiale im Carotissinusnerven, Arch. Gesamte Physiol. **262:**466, 1956.

49. Beaton, L. E., and Magoun, H. W.: Localization of the medullary respiratory center in the monkey, Am. J. Physiol. **134:**177, 1941.

50. Bernthal, T., and Weeks, W. F.: Respiratory and vasomotor effects of variations in carotid body temperature, Am. J. Physiol. **127:**94, 1939.

51. Biscoe, T. J.: Carotid body: structure and function, Physiol. Rev. **51:**437, 1971.

52. Biscoe, T. J., Purves, M. J., and Sampson, S. R.: The frequency of nerve impulses in single carotid body chemoreceptor afferent fibres recorded in vivo with intact circulation, J. Physiol. **208:**121, 1970.

53. Breckenridge, C. G., and Moff, H. E.: Pontine and medullary regulation of respiration in the cat, Am. J. Physiol. **160:**385, 1950.

54. Bronk, D. W., and Ferguson, L. K.: The nervous control of intercostal respiration, Am. J. Physiol. **110:** 700, 1935.

55. Brookhart, J. M.: The respiratory effects of localized faradic stimulation of the medulla oblongata, Am. J. Physiol. **129:**709, 1940.

56. Burns, B. D.: The central control of respiratory movements, Br. Med. Bull. **19:**7, 1963.

57. Chungcharoen, D., Daly, M. deB., and Schweitzer, A.: The blood supply of the carotid body in cats, dogs, and rabbits, J. Physiol. **117:**347, 1952.

58. Cohen, M. I., and Feldman, J. L.: Models of respiratory phase-switching, Fed. Proc. **36:**2367, 1977.

59. Comroe, J. H., Jr.: The location and function of the chemoreceptors of the aorta, Am. J. Physiol. **127:** 176, 1939.

60. Comroe, J. H., Jr., and Mortimer, L.: The respiratory and cardiovascular responses of temporally separated aortic and carotid bodies to cyanide, nicotine, phenyldiguanide and serotonin, J. Pharmacol. Exp. Ther. **146:**33, 1964.

61. Comroe, J. H., Jr., and Schmidt, C. F.: The part played by reflexes from the carotid body in the chemical regulation of respiration in the dog, Am. J. Physiol. **121:** 75, 1938.

62. Cropp, G. J. A., and Comroe, J. H., Jr.: Role of mixed venous blood P_{CO_2} in respiratory control, J. Appl. Physiol. **16:**1029, 1961.

63. Daly, M. deB., Lambertsen, C. J., and Schweitzer, A.: Observations of the volume of blood flow and

oxygen utilization of the carotid body in the cat, J. Physiol. **125**:67, 1954.

64. Dawes, G. S.: Oxygen supply and consumption in late fetal life, and the onset of breathing at birth. In Fenn, W. O., and Rahn, H., editors: Handbook of physiology. Respiration section, Washington, D.C., 1965, American Physiological Society, vol. 2.

65. Dawes, G. S., and Comroe, J. H., Jr.: Chemoreflexes from the heart and lungs, Physiol. Rev. **34**:167, 1954.

66. de Burgh Daly, M., and Scott, M. J.: The cardiovascular responses to stimulation of the carotid body chemoreceptors in the dog, J. Physiol. **165**:179, 1963.

67. de Burgh Daly, M., and Ungar, A.: Comparison of the reflex responses elicited by stimulation of the separately perfused carotid and aortic body chemoreceptors in the dog, J. Physiol. **182**:379, 1966.

68. DeCastro, F.: Sur la structure de la synapse dans les chemorecepteurs: leur mecanisme d'excitation et role dans la circulation sanguine locale, Acta Physiol. Scand. **22**:14, 1951.

69. Dejours, P.: La régulation de la ventilation au cours de l'exercise musculaire chez l'homme, J. Physiol. **51**:163, 1959.

70. Dejours, P.: Control of respiration in muscular exercise. In Fenn, W. O., and Rahn, H., editors: Handbook of physiology. Respiration section, Washington, D.C., 1964, American Physiological Society, vol. 1.

71. Dejours, P., et al.: Stimulus oxygene chemoreflexe de la ventilation a basse altitude (50 m.) chez l'homme. I. Au repos, J. Physiol. **49**:115, 1957.

72. Delgado, J. M. R., and Livingston, R. B.: Some respiratory, vascular and thermal responses to stimulation of orbital surface of frontal lobe, J. Neurophysiol. **11**:39, 1948.

73. Downes, J. J., and Lambertsen, C. J.: Dynamic characterics of ventilatory depression in man on abrupt administration of O_2, J. Appl. Physiol. **21**:447, 1966.

74. Dripps, R. D., and Comroe, J. H., Jr.: The respiratory and circulatory response of normal man to inhalation of 7.6 and 10.4 per cent CO_2 with a comparison of the maximal ventilation produced by severe muscular exercise, inhalation of CO_2 and maximal voluntary hyperventilation, Am. J. Physiol. **149**:43, 1947.

75. Dripps, R. D., and Dumke, P. R.: The effect of narcotics on the balance between central and chemoreceptor control of respiration, J. Pharmacol. Exp. Ther. **77**:290, 1943.

76. Eldridge, F. L.: Maintenance of respiration by central neural feedback mechanism, Fed. Proc. **36**:2400, 1977.

77. Euler, C. von: The functional organization of the respiratory phase-switching mechanisms, Fed. Proc. **36**:2375, 1977.

78. Euler, C. von, Herrero, F., and Wexler, I.: Control mechanisms determining rate and depth of respiratory movements, Respir. Physiol. **10**:93, 1970.

79. Euler, U. S. von, Liljestrand, G., and Zotterman, Y.: The excitation mechanism of the chemoreceptors of the carotid body, Scand. Arch. Physiol. **83**:132, 1939.

80. Eyzaguirre, C., and Gallego, A.: An examination of DeCastro's original slides. In Purves, M. J., editor: The peripheral arterial chemoreceptors, New York, 1975, Cambridge University Press.

81. Eyzaguirre, C., and Koyano, H.: Effects of hypoxia, hypercapnia, and pH on the chemoreceptor activity of the carotid body in vitro, J. Physiol. **178**:385, 1965.

82. Eyzaguirre, C., and Koyano, H.: Effects of some pharmacological agents on chemoreceptor discharges, J. Physiol. **178**:410, 1965.

83. Eyzaguirre, C., and Zapata, P.: The release of acetylcholine from carotid body tissues. Further study on the effects of acetylchole and cholinergic blocking agents in the chemosensory discharge, J. Physiol. **195**:589, 1968.

84. Eyzaguirre, C., Baron, M., and Gallego, R.: Intracellular studies of carotid body cells: effects of temperature, "natural" stimuli and chemical substances. In Reivich, M., et al., editors: Tissue hypoxia and ischemia. Advances in Experimental Medicine and Biology, New York, 1977, Plenum Publishing Corp., vol. 78, p. 209.

85. Eyzaguirre, C., Koyano, H., and Taylor, J. R.: Presence of acetylcholine and transmitter release from carotid body chemoreceptors, J. Physiol. **178**:463, 1965.

86. Fay, F. S.: Oxygen consumption of the carotid body, Am. J. Physiol. **218**:518, 1970.

87. Fitzgerald, R. S., and Parks, D. C.: Effect of hypoxia on carotid chemoreceptor responses to carbon dioxide in cats, Respir. Physiol. **12**:218, 1971.

88. Gautier, H., and Bertrand, F.: Respiratory effects of pneumotaxic center lesions and subsequent vagotomy in chronic cats, Respir. Physiol. **23**:71, 1975.

89. Gesell, R., Lapides, J., and Levin, M.: The interaction of central and peripheral chemical control of breathing, Am. J. Physiol. **130**:155, 1940.

90. Guz, A., et al.: Peripheral chemoreceptor block in man, Respir. Physiol. **1**:38, 1966.

91. Head, H.: On the regulation of respiration, J. Physiol. **10**(1):279, 1889.

92. Hering, E., and Breuer, J.: Die Selbststeurung der Athmung durch den Nervus Vagus, Stizungsb Akad. Wiss. **57**:672; **58**:909, 1868.

93. Heymans, J. F., and Heymans, C.: Sur les modifications directes et sur la régulation réflexe de l'activitè du centre respiratoire de la tête isolée du chien, Arch. Int. Pharmacodyn. Ther. **33**:273, 1927.

94. Hoff, H. E., and Breckenridge, C. G.: The medullary origin of respiratory periodicity in the dog, Am. J. Physiol. **158**:157, 1949.

95. Hornbein, T. F., and Roos, A.: Specificity of H ion concentration as a carotid chemoreceptor stimulus, J. Appl. Physiol. **18**:580, 1963.

96. Hugelin, A.: Anatomical organization of bulbopontine respiratory oscillators, Fed. Proc. **36**:2390, 1977.

97. Joels, N., and Neil, E.: Carotid chemoreceptor response to high carbon monoxide tension, J. Physiol. **156**:5P, 1961.

98. Joels, N., and Neil, E.: Simultaneous measurement of carotid body perfusion flow and chemoreceptor impulse activity in the cat, J. Physiol. **184**:8P, 1966.

99. Kaada, B. R., Pribram, K. H., and Epstein, J. A.: Respiratory and vascular responses in monkeys from temporal pole, insula, orbital surface and cingulate gyrus, J. Neurophysiol. **12**:347, 1949.

100. Kalia, M.: Neuroanatomical organization of the respiratory centers, Fed. Proc. **36**:2405, 1977.

101. Kellogg, R. H.: Central chemical regulation of respiration. In Fenn, W. O., and Rahn, H., editors: Handbook of physiology. Respiration section, Washington, D.C., 1964, American Physiological Society, vol. 1.

102. Kety, S. S., and Schmidt, C. F.: The effects of altered arterial tensions of carbon dioxide and oxygen on cerebral blood flow and cerebral oxygen consumption of normal young men, J. Clin. Invest. **27**:484, 1948.

103. Koelle, G. B.: The elimination of enzymatic diffusion artifacts in the histochemical localization of cholines-

terases and a survey of their cellular distributions, J. Pharmacol. Exp. Ther. **103:**153, 1951.

104. Lahiri, S.: Introductory remarks: oxygen linked response of carotid chemoreceptors. In Reivich, M., et al., editors: Tissue hypoxia and ischemia. Advances in experimental medicine and biology, New York, 1977, Plenum Publishing Corp., vol. 78, p. 185.

105. Lambertsen, C. J.: Drugs and respiration, Annu. Rev. Pharmacol. **6:**327, 1966.

106. Lambertsen, C. J., Gelfand, R., and Kemp, R. A.: Dynamic influences of CO_2 in respiratory control. In Brooks, C. McC., Kao, F. F., and Lloyd, B. B., editors: The cerebrospinal fluid and the regulation of ventilation, Oxford, 1965, Blackwell Scientific Publications, Ltd.

107. Landgren, S., and Neil, E.: Chemoreceptor impulse activity following haemorrhage, Acta Physiol. Scand. **23:**158, 1951.

108. Lee, K. D., and Mattenheimer, H.: The biochemistry of the carotid body, Enzymol. Biol. Clin. **4:**199, 1964.

109. Legallois, C. J. J.: Experiences sur le principe de la vie, Paris, 1812, D'Hautel.

110. Leusen, I. R.: Chemosensitivity of the respiratory center. Influence of changes in the H^+ and total buffer concentrations in the cerebral ventricles on respiration, Am. J. Physiol. **176:**45, 1954.

111. Loeschcke, H. H., Koepchen, H. P., and Gertz, K. H.: Über den Einfluss von Wasserstoffionenkonzentration und CO_2-Druck im Liquor cerebrospinalis auf die Atmung, Arch. Gesamte Physiol. **266:**569, 1958.

112. Lumsden, T.: Observations on the respiratory centres in the cat, J. Physiol. **57:**152, 1922-1923.

113. Lumsden, T.: Observations on the respiratory centres, J. Physiol. **57:**354, 1922-1923.

114. Lumsden, T.: The regulation of respiration, J. Physiol. **58:**81, 1923-1924.

115. Marshall, E. K., Jr., and Rosenfeld, M.: Depression of respiration by oxygen, J. Pharmacol. Exp. Ther. **57:**437, 1936.

116. Metz, B.: Release of ACh from the carotid body by hypoxia and hypoxia plus hypercapnia, Respir. Physiol. **6:**386, 1969.

117. Mitchell, R. A., Loeschcke, H. H., Massion, W. H., and Severinghaus, J. W.: Respiratory responses mediated through superficial chemosensitive areas on the medulla, J. Appl. Physiol. **18:**523, 1963.

118. Neil, E., and Joels, N.: The carotid glomus sensory mechanism. In Cunningham, D. J. C., and Lloyd, B. B., editors: The regulation of human respiration, Oxford, 1963, Blackwell Scientific Publications, Ltd.

119. Paintal, A. S.: Mechanism of stimulation of aortic chemoreceptors by natural stimuli and chemical substances, J. Physiol. **189:**63, 1967.

120. Perkins, J. F., Jr.: Respiration, Annu. Rev. Physiol. **22:**245, 1960.

121. Pitts, R. F., Magoun, H. W., and Ranson, S. W.: The origin of respiratory rhythmicity, Am. J. Physiol. **127:**654, 1939.

122. Proctor, D. F.: Physiology of the upper airway. In Fenn, W. O., and Rahn, H., editors: Handbook of physiology. Respiration section, Washington, D.C., 1964, American Physiological Society, vol. 1.

123. Purves, M. J.: The effect of hypoxia, hypercapnia and hypotension upon carotid body blood flow and oxygen consumption in the cat, J. Physiol. **209:**395, 1970.

124. Reivich, M., et al., editors: Tissue hypoxia and ischemia. Advances in experimental medicine and biology, New York, 1977, Plenum Publishing Corp., vol. 78.

125. Ross, L. L.: Electron microscopic observations of the carotid body of the cat, J. Biophys. Biochem. Cytol. **6:**253, 1959.

126. Salmoiraghi, G. C.: Functional organization of brain stem respiratory neurons, Ann. N.Y. Acad. Sci. **109:**571, 1963.

127. Salmoiraghi, G. C., and Burns, B. D.: Localization and patterns of discharge of respiratory neurones in brain-stem of cat, J. Neurophysiol. **23:**2, 1960.

128. Salmoiraghi, G. C., and Burns, B. D.: Notes on mechanism of rhythmic respiration, J. Neurophysiol. **23:**14, 1960.

129. Salmoiraghi, G. C., and von Baumgarten, R.: Intracellular potentials from respiratory neurones in the brain-stem of the cat and the mechanism of rhythmic respiration, J. Neurophysiol. **24:**203, 1961.

130. Schmidt, C. F.: Effect of carotid sinus and carotid body reflexes upon respiration, Anesth. Analg. **19:**261, 1940.

131. Schmidt, C. F.: The respiration. In Bard, P., editor: Medical physiology, ed. 10, St. Louis, 1956, The C. V. Mosby Co.

132. Schmidt, C. F., Dumke, P. L., and Dripps, R. D., Jr.: The part played by carotid body reflexes in the respiratory response of the dog to small changes in the carbon dioxide tension in the arterial blood, Am. J. Physiol. **128:**1, 1939.

133. Sears, T. A.: The respiratory motoneuron and apneusis, Fed. Proc. **36:**2412, 1977.

134. Seidl, E.: On the morphology of the vascular system of the carotid body of cat and rabbit and its relation to the glomus type I cells. In Purves, M. J., editor: The peripheral arterial chemoreceptors, New York, 1975, Cambridge University Press.

135. Spencer, W. G.: The effect produced upon respiration by faradic excitation of the cerebrum in the monkey, dog, cat, and rabbit, Phil. Trans. (Biol.) **185:**609, 1894.

136. Stadie, W. C., Austin, J. H., and Robinson, H. W.: The effect of temperature on the acid-base protein equilibrium and its influence on the CO_2 absorption curve of whole blood, true and separated serum, J. Biol. Chem. **66:**901, 1925.

137. St. John, W. M.: Integration of peripheral and central chemoreceptor stimuli by pontine and medullary respiratory centers, Fed. Proc. **36:**2421, 1977.

138. St. John, W. M., Glasser, R. L., and King, R. A.: Apneustic breathing after vagotomy in cats with chronic pneumotaxic center lesions, Respir. Physiol. **12:**239, 1971.

139. Verna, A.: Terminaisons nerveuses afférentes et efférentes dans le glomus carotidien du lapin, J. Microsc. (Paris) **16:**299, 1973.

140. Whitehead, R. W., and Draper, W. B.: A respiratory reflex originating from the thoracic wall of the dog, Anesthesiology **8:**159, 1947.

141. Winder, C. V.: Pressoreceptor reflexes from the carotid sinus, Am. J. Physiol. **118:**379, 1937.

142. Witzleb, E., Bartels, H., Budde, H., and Mochizucki, M.: Der Einfluss des arteriellen O_2-Drucks auf die chemoreceptorischen Aktionspotentiale im Carotissinusnerven, Arch. Gesamte Physiol. **261:**211, 1955.

71

CHRISTIAN J. LAMBERTSEN

Chemical control of respiration at rest

The neurochemical and related respiratory control mechanisms precisely adjust pulmonary ventilation to the requirements of whole-body metabolism and to alterations of the respiratory environment. This is true not only in the resting state at sea level but even in the extreme conditions of exercise, high altitude, and sleep. Any such versatile system of control must include sensing units as well as machinery for acting on the sensory information received. *In the chemical control of respiration, it is the selective reactivity of chemosensitive neuronal units to induced changes in their own local environment that governs the respiratory influence on the internal acid-base and oxygen environment of the body as a whole.* The sensory neurons involved include peripheral chemosensitive cells of the carotid and aortic bodies as well as those of the brain stem. These receptor neurons are primarily affected by consequences of their own metabolism or by similar influences presented from the external environment by way of the arterial circulation. They are also affected by metabolic products originating in remote structures such as exercising muscles and in addition by the manner in which pulmonary ventilation has modified the blood perfusing the lungs. All this activity is modulated by neural influences that indicate the physical state of the lungs themselves.

The essence of the overall respiratory control system is *feedback* of information concerning the effectiveness of the respiratory response. Since the system has multiple sensors and multiple interrelated influences on them, its reserve, or redundancy, is considerable. The nature of the stimulant factors, the nature of peripheral and central sensory units, the characteristics of the response to chemical stimulation, and the manner in which different chemical factors interact to modify respiratory activity involve some of the most fundamental life processes.

LINKAGE OF STIMULI TO METABOLISM AND ENVIRONMENT

The processes of aerobic metabolism produce carbon dioxide and, through its hydration, HCO_3^- and H^+. In the anaerobic metabolism of severely exercising muscles, fixed acids form that modify H^+, P_{CO_2}, HCO_3^- relationships in blood. Exposure to elevated P_{CO_2} or P_{O_2} or to decreased P_{O_2} in the atmospheric environment modifies oxygen and acid-base composition of blood and tissues, which in turn may induce metabolic effects. The beauty of chemical respiratory control is that, with two sets of sensors (central and peripheral), all these many factors and sometimes overlapping circumstances are smoothly regulated. In fact, the paths of stimulus and regulation inevitably converge so that the respiratory needs in all extremes of activity, metabolism, and environmental stress may ultimately be met through influences of protons (H^+) and oxygen molecules on chemical sites or on sensitive neural membranes.

In considering the interplay between components of the control system and its chemical stimuli, it will be necessary to examine specific components and roles individually. However, this manner of investigation is a handicap to overall understanding of the control system, and it is essential to recognize from the beginning that in any conceivable physiologic situation, multiple sites are being affected by complex combinations of direct and indirect chemical stimulation and that the respiratory activity that emerges is an integrated result of the multiple influences involved. Understanding of central and peripheral components of respiratory control is expanding rapidly through the meticulous efforts of many investigators cited in this chapter. The relative contributions of peripheral and central sensory components of the overall control system are beginning to be defined for the numerous states of

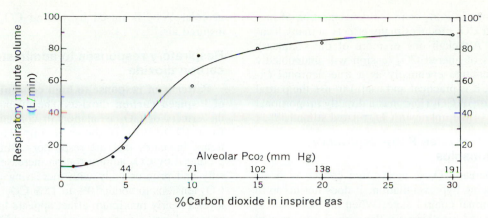

Fig. 71-1. Relationships between respiratory minute volume and inspired CO_2 concentration in humans. Points indicate averages of peak values. Solid circles indicate measurements in normal individuals;[90,236] open circles are measurements in psychoneurotic patients. (From Lambertsen.[149])

Table 71-1. Increases in pulmonary ventilation in man produced by extreme physiologic stress

Procedure	Respiratory volume (L/min)	Remarks
Voluntary hyper-ventilation	166	Average findings in healthy young men[90]
Exercise	110	
Inhalation of 30% CO_2	90	Average in 4 middle-aged psychoneurotic patients[149]
Diabetic acidosis (arterial pH 7.00 or lower)	35	Maximum response to metabolic acidosis in 14 patients[140]
Hypoxemia (4.2% O_2 in N_2)	30	Average in 11 young men[31]

human activity. Despite this progress there is continued uncertainty regarding the sites within the central nervous system (CNS) that are normally responsible for sensing and responding to chemical stimuli. Moreover, the nature of the ultimate excitant or the excitation process for either the central respiratory sensors or the peripheral chemoreceptors remains unknown.

The early and recent historic aspects of studies of respiratory control, as well as the varied interpretations of the available facts, are presented here and in a number of reviews.* Together,

*See references 166, 175, 195, 201, 214, 242, and 249.

these reviews provide exceptionally fine correlation of information and concept. The greatest difficulty in appraising and correlating the voluminous literature has in the past been that imposed by the frequent lack of attention to the manner in which each stimulant factor relates to the others and to the masking influences of secondary physiologic alterations brought about by experimentally induced changes in the level of a single potential stimulus.

Power of chemical influences

The stimulation of respiration by chemical factors can be extremely vigorous. Table 71-1 compares the magnitudes of ventilation produced by CO_2, by hypoxia, and by the acidemia of a disease state with the maximum changes expected in voluntary or exercise hyperpnea.

CARBON DIOXIDE AS STIMULUS INDEX

The most vigorous respiratory stimulation that can be produced in nonexercising individuals is that produced by an increase in the P_{CO_2} of inspired air (Fig. 71-1). This great reactivity to CO_2 led to the early concept that the CO_2 molecule itself was the agent responsible for stimulation of respiratory centers.[26,204] Molecular CO_2 has continued to be implicated over many decades as one of the chemical stimuli in respiratory control,[26,209] almost certainly erroneously.[146]

CO_2 certainly is involved in the chemical regulation of respiration, probably as the dominant factor linking respiration to whole-body and probably to local central chemosensor metabolism in all states. However, as will be described, a change in P_{CO_2} or in the concentration of molecular CO_2 produces many effects at many locations. Although one or more of the secondary effects of altered CO_2 tension will undoubtedly be identified eventually as a true neuronal excitant, it is practical and useful to use the partial pressure of CO_2 (P_{CO_2}) as a directly proportional index of the unknown CO_2-related stimuli.[146]

Normal inverse P_{CO_2}-respiratory relationships

Whereas CO_2 is most often conceived of as producing hyperventilation, it does not do so in the normal control state. When an individual is at rest and breathing air, the normal P_{CO_2}-related chemical control system involves an *inverse relationship* between the level of arterial P_{CO_2} and the magnitude of alveolar ventilation (Fig. 68-6). This means that even though P_{CO_2} is somehow dominant as a *stimulus* to respiration, the normal control situation is one in which the actual role of metabolically produced CO_2 is not to increase but to *prevent a fall in ventilation* below the normal for rest. Metabolism, alveolar ventilation, and P_{CO_2} are necessarily related in the manner described in Fig. 68-6, which in the stable state is based on the statement "metabolic CO_2 production per minute equals alveolar ventilation per minute times the percentage of CO_2 in the alveolar gas." This can be restated as follows:

$$\%\text{Alveolar } CO_2 = \frac{\text{Metabolic } CO_2 \text{ production}}{\text{Alveolar ventilation}}$$

to make it immediately clear that, at any constant level of metabolism (i.e., a stable resting, or even a stable exercising state), doubling alveolar ventilation eventually would halve alveolar P_{CO_2}. It is this lowering of alveolar P_{CO_2} with any slight increase in alveolar ventilation that diminishes the P_{CO_2}-related respiratory stimulus and thus tends to limit the degree of hyperventilation. Conversely, if alveolar ventilation is reduced, alveolar P_{CO_2} will increase. The rise in P_{CO_2} consequent to even slight diminution of alveolar ventilation limits the degree and duration of hypoventilation by increasing the chemical stimulus to respiration. It is therefore evident that in the absence of environmental exposure to CO_2 *the normal "stimulant" role of CO_2 in respiratory control is an inverse one, related to prevention of deviation from the natural or "set" point.* This supportive relationship pertains to all stable environmental or physiologic states and involves principles and mechanisms no different from the direct stimulation of respiration that results when alveolar P_{CO_2} is raised by adding CO_2 to the inspired air.[144]

Respiratory responses to administered carbon dioxide

Patterns of response to high concentrations of inspired carbon dioxide. The magnitude of the effects of CO_2 inhalation in concentrations from 1% to 30% is shown in Fig. 71-1. The figure indicates that the respiratory stimulation produced by CO_2 administration increases markedly and progressively with increasing inspired CO_2 tensions to about 10% to 15% CO_2. At this level a nearly maximum effect appears to be approached, and a further rise in inspired CO_2 concentration to 20% and 30% results in a relatively small additional increase in ventilation, partly due to the interruption of respiratory function by severe convulsive activity.[149] This curve must be appraised from several standpoints, as follows:

1. The ability of CO_2 to stimulate respiration greatly exceeds the maximum respiratory stimulation produced by low inspired O_2 tensions. It was once thought that CO_2 was incapable of driving the respiratory mechanisms to the same degree as do the stimulations of muscular exercise.[90] Actually CO_2 inhalation at high tensions produces a respiratory stimulation considered maximal in the middle-aged psychoneurotic patients studied (Table 71-1)[149] and therefore can be expected to drive respiration to its physical limit in normal young men (Fig. 71-2).

2. CO_2 inhalation, by elevating the P_{CO_2} of alveolar air and body fluids, produces at least two distinct effects on body fluids: it increases the concentration of physically dissolved CO_2 gas and it increases the hydrogen ion concentration. Thus any effect of elevated inspired CO_2 pressures on breathing will be the result of changing dominance of these separate but interdependent alterations.

3. It has been proposed that high concentrations of inspired CO_2 produce respiratory depression by way of a narcotic property of CO_2.[109,142, 234] Such an effect for molecular CO_2 probably does exist and is probably directly related in degree to any increase in P_{CO_2} (concentration of molecular CO_2) above normal. However, a narcotic effect of the CO_2 molecule has not been demonstrated as separate from the disruptive effects of the profound acidosis produced by high CO_2 tensions. Nevertheless, it has been stated that the CNS depression produced by the overall effect of CO_2 (including acidosis) is about 5 times greater than that produced by equivalent tensions of the anesthetic gas nitrous oxide.[188]

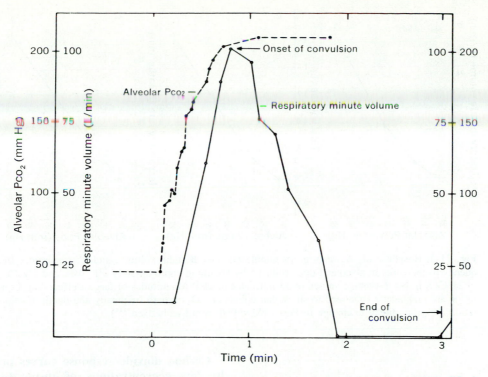

Fig. 71-2. Effects on respiratory minute volume and alveolar P_{CO_2} of inhalation of 30% CO_2 in O_2 (results in one male subject). Alveolar P_{CO_2} reached a plateau of over 200 mm Hg in slightly more than 1 min. In less than 1 min, rapid increase in ventilation was reversed, most probably by abrupt onset of generalized CO_2 convulsions. Apnea was temporarily interrupted at end of seizure but recurred with onset of second seizure. Subsequent substitution of O_2 for CO_2 mixture at 3 min, 40 sec resulted in return of violent hyperventilation before respiration returned to normal level. (From Lambertsen.[149])

4. As will be seen, low concentrations of inspired CO_2 can be tolerated for many days.[75,225] However, as the concentration of CO_2 administered is increased above approximately 7%, both objective and subjective effects become progressively more pronounced, and tolerance is limited to minutes rather than hours. The subject will experience a more rapidly developing and vigorous respiratory stimulation, dyspnea, restlessness, faintness, severe headache, and dulling of consciousness. Unconsciousness inevitably occurs with inspired CO_2 concentrations of 15% and higher. When 15% CO_2 is inhaled, muscular rigidity and tremors occur; with 20% or 30% CO_2, generalized convulsions are produced almost immediately (Fig. 71-2).[149]

5. Because of the rapid occurrence of severe side effects, much of the respiratory data shown in Fig. 71-1 for inspired CO_2 levels above 10% represent *peak, nonequilibrium effects* of CO_2 administration and, as such, may provide only a partial indication of the true capacity for neuronal response to high levels of inspired CO_2 concentration. Fig. 71-2 shows the rates of the changes in respiratory minute volume and alveolar P_{CO_2} when 30% CO_2 in O_2 is administered. In the illustration a rapidly increasing rate of ventilation was abruptly terminated at the onset of generalized convulsions that occurred within 1 min of the onset of CO_2 breathing. Most probably the onset of "apnea" was not the result of respiratory depression by a narcotic action of CO_2 but was related to *mechanical interference with pulmonary ventilation* during the seizure. If so, the S-shaped curve of Fig. 71-1 must grossly underestimate the capacity of the higher levels of CO_2 to produce stimulant effects on the mechanisms of respiratory control.[149] Nevertheless it is inevitable that exposure to the extreme pH change induced by very high P_{CO_2} will prevent neural function and in itself produce apnea.

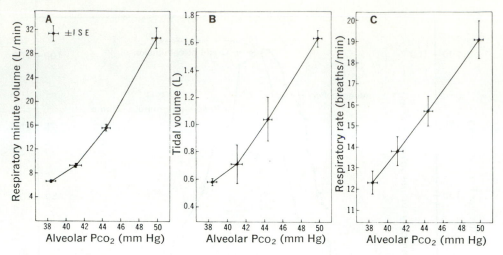

Fig. 71-3. Response of, **A,** respiratory minute volume, **B,** tidal volume, and **C,** respiratory frequency to increases in alveolar P_{CO_2} produced by breathing approximately 2%, 4%, and 6% CO_2 in 21% O_2 in N_2. (Average values in 33 normal subjects.) Magnitudes of largest effects on respiration are graphically equalized to show that effects of CO_2 on both frequency and depth of respiration are proportional to change in P_{CO_2}. (Modified from Lambertsen.[144])

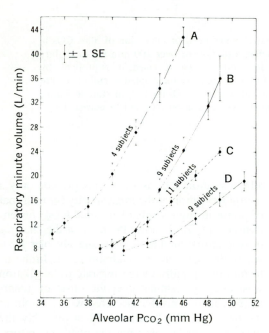

Fig. 71-4. Variability of respiratory response to low concentrations of inspired CO_2. Alveolar P_{CO_2} values determined by using end-tidal values. Same normal subjects and data employed for average curves of Fig. 71-3 are here placed in four groups to show that normal individuals may exhibit different patterns of respiratory response to CO_2. These may range from extreme reactivity of group A to low "sensitivity" shown by subjects of group D. Response of latter group is no greater than that seen in some patients with emphysema or in overdose of a narcotic such a meperidine. (From Lambertsen.[144])

Carbon dioxide response curves produced by low concentrations of inspired carbon dioxide. It is not appropriate to relate respiratory response only to *inspired* concentrations of CO_2; these give an inaccurate indication of the actual change in stimulus level in the blood or brain, since the respiratory response limits the rise in alveolar (arterial) P_{CO_2} produced by an increase in inspired CO_2 concentration. When respiration is instead referred to the levels of alveolar P_{CO_2} corresponding to low concentrations of inspired CO_2 (1% to 6%), it appears that in steady states the respiratory response to increases in *alveolar* P_{CO_2} is progressive and becomes nearly linear for both frequency and depth of breathing (Fig. 71-3). This linearity of response persists up to a tidal volume equal to about one half the vital capacity; above this level further increases in ventilation are produced by a rise in frequency alone.[116] This in turn is self-limited. Part, if not all, of the slight curvature of the alveolar P_{CO_2}-respiration curves in Figs. 71-3 and 71-4 is a result of small errors in estimating mean alveolar P_{CO_2} by direct sampling of the last portion of each exhalation.[144] Such errors are related to incomplete washout of respiratory dead space by alveolar gas and are therefore larger at low than at high tidal volumes. When the P_{CO_2} is measured in *arterial blood* instead of in alveolar air, the P_{CO_2}-respiratory minute volume response curve is essentially linear. This means that no "threshold" for a respiratory response to administered CO_2 can be considered to exist in the

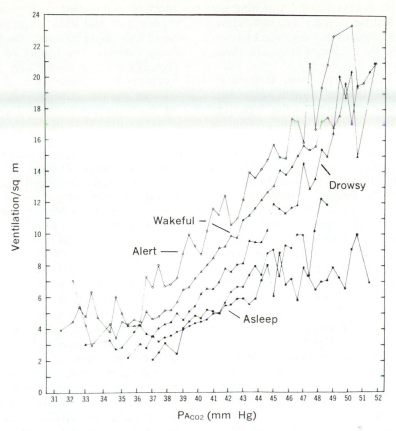

Fig. 71-5. Influence of degree of wakefulness on respiratory response to CO_2. Curves are based on more than 3,000 ventilation-Pco_2 points in normal subjects. State of wakefulness or degree of sleep was determined by analysis of EEG recordings. Curves represent mean findings plotted as differences from data obtained in state of tense alertness. Sleep leads to progressive decrease in reactivity to CO_2. (Modified from Bulow.[8])

normal state. CO_2 is already actively involved in sustaining activity. Thus any rise in inspired CO_2 pressure above that in ambient air will produce an effect on the respiratory drive.

Fig. 71-3 shows that even the smallest elevations of alveolar Pco_2 are associated with an increase both in *depth* and *rate* of respiration. Thus CO_2 affects the interval between bursts of impulses from the respiratory centers as well as the degree of neuromuscular activity associated with each unit of inspiratory discharge.

Absence of response to carbon dioxide. Respiration can continue even in the unusual situations when administered CO_2 fails to induce stimulation of breathing. This situation results when three neurologic lesions are produced simultaneously. These are lesions that destroy the carotid chemoreceptors, tissue including the area postrema in the medulla and rostral diencephalic tissue.[187] Insensitivity to CO_2 occurs as a rare condition in man.

Influence of sleep. When a resting subject falls asleep, alveolar ventilation decreases, the arterial Pco_2, $[HCO_3^-]$ and $[H^+]$ rise, and Po_2 falls.[185,211] These increases in chemical stimulus level, which occur in "non–rapid eye movement" sleep, should increase the impulse generation by both central and peripheral chemosensors. The decrease in ventilation that produced the rise in stimulus level is apparently part of a decrease in reactivity of the respiratory control mechanisms.*

Fig 71-5 shows the progressive development of decreased CO_2 response in the transition from the alert to the sleeping state.[8] The depression, which occurs even in severe hypoxia,[30] is considered to result from diminished bombardment of the reticular formation by the great variety of afferent impulses associated with consciousness.[8,99,129] If true, this would indicate that even

---

*See references 8, 99, 102, 126, 185, 211, 217, and 221.

at rest the waking state is a complex of stimulant influences and that the sleeping state more closely represents the actual basal response of respiratory neurons to the influences of specific chemical stimuli.

Indices of respiratory drive. Respiratory minute volume is extensively used as a composite index of the functional output of the respiratory centers in resting states or when the control mechanisms are driven by ancillary stimuli. In normal individuals exposed to modest stress, minute volume has been a useful index of overall respiratory stimulation or depression. In the presence of pulmonary mechanical limitations imposed by disease or environment (extreme gas density), measures of respiratory work are superior. As studies have become more detailed, it has become desirable to evaluate tidal volume and frequency effects separately. Approaches to this end are the plotting of tidal volume against frequency,[114,116] rates of breath-by-breath changes of tidal volume,[103] and the determination of mean inspiratory flow rate per breath.[193] The respiratory response to administered CO_2 or other stimuli is still sometimes expressed in terms of alveolar ventilation instead of respiratory minute volume, tidal volume, frequency, respiratory work, or inspiratory flow rate. Use of alveolar ventilation is improper if the aim is to designate a value indicative of the total output of the respiratory control system, since the total respiratory act comprises alveolar ventilation *plus* dead space ventilation.[154] Alveolar ventilation represents only a fraction of the total respiratory work.

Fig. 71-4 shows that, in terms of respiratory minute volume, the response to CO_2 inhalation varies greatly even in normal individuals. Certain subjects, indicated by curve *A* in this illustration, appear to have a high sensitivity ($\Delta \dot{V}E/\Delta P_{CO_2}$) to elevated P_{CO_2} and begin to show respiratory stimulation at low levels of CO_2 tension. At the other extreme, indicated by curve *D* in the same illustration, are other normal subjects who have a low sensitivity to elevations of P_{CO_2} and, in addition, have the highest resting levels of alveolar CO_2 pressure.[144] It can hardly be expected that individuals with such varying responses to a single respiratory stimulant will react uniformly to a more complex form of respiratory stimulation such as that associated with exercise.

Relation of respiratory stimulation to physicochemical changes produced by carbon dioxide

The exceptional potency of CO_2 as a respiratory stimulant has from the beginning of respi-

ratory physiology led to study and conjecture to explain how and even where intrinsically produced CO_2 might influence the activity of respiratory neurons. The history of these important investigations, presented here and in several recent reviews,* points to the developing understanding and the remaining uncertainties regarding the qualitative nature and the sites of action of the natural chemical respiratory stimuli.

Two physicochemical characteristics of CO_2 have emerged as being of definite importance in respiratory control. The first is its ability as physically dissolved, uncharged, molecular CO_2 to pass freely and rapidly across all cell membranes and in all directions between the tissue cells, extracellular fluids, and blood.[127] The diffusibility of CO_2 in tissue fluids is approximately 20 times greater than that of O_2. A second characteristic is in the production of a *change in acidity* of body fluids whenever the pressure of CO_2 is altered.

It was ultimately learned that *peripheral chemoreflex as well as central chemosensitive sites are stimulated by CO_2 administration and that at each site either a rise in P_{CO_2} or a rise in $[H^+]$ produces excitation.*

Over many decades of intensive study a major block to understanding has been the well-substantiated observation that the acidemia produced by administration of CO_2 (designated "respiratory acidosis") is associated with a gross increase in respiration, whereas an equivalent acidemia caused by administration of fixed acid such as HCl or lactic acid (designated "metabolic acidosis") results in a lesser degree of respiratory stimulation.[21,28,58] Largely for reasons such as this, it was considered almost to the present time that CO_2 possesses a specific role as a respiratory stimulant or "hormone," unrelated to its ability to alter the pH of blood and body fluids.[28,204]

It was also proposed as early as 1888 that an acid product of metabolism affects the cells of the respiratory centers,[28,104] and since that time, various concepts have repeatedly been put forward proposing respiratory control entirely by actions of the hydrogen ion.[26,151,197,255] There is now no major remaining tendency to implicate CO_2 itself as a specific stimulus.[144,174]

Gradually it became evident that neither CO_2 alone nor the acid reaction of the blood alone could account for the respiratory responses in such states as metabolic (diabetic, renal, NH_4Cl) acidosis and metabolic alkalosis. In metabolic acidosis a gross increase in blood acidity produces only a slight increase in respiration, and this is associated with a lower than normal arte-

---

*See references 166, 175, 195, 214, 242, and 249.

rial Pco_2. In metabolic alkalosis a shift of blood pH in the alkaline direction is accompanied by only slight reduction of ventilation but by an arterial Pco_2 greater than normal.[28,58]

One effort to resolve the failure of change in a single chemical factor in blood to account quantitatively for the respiratory response to acid-base alterations was Gray's derivation of a "multiple-factor" equation in which respiratory effects were quantitatively related in part to changes in pH and in part to changes in Pco_2 of the arterial blood.[21,110]

The second general approach has involved the assumption that in the normoxic state a single factor, usually defined as the hydrogen ion, acts at sites accessible to the arterial blood and in the cells of the respiratory centers to account for the entire chemical control of respiration. This qualitative concept provided the basis for the evolution of the several "reaction (acid) theories" advanced by Winterstein over nearly a 50-year period.[247,252,253,255] These and related current concepts will be discussed in a subsequent part of this chapter.

Ultimate mechanisms of excitation of respiratory chemosensors

Despite the clear correlation of changes in respiration with Pco_2 and with CO_2-induced and other acid-base changes in blood and the fluids of the CNS, it has not yet been feasible to determine the ultimate basis for the effect of any natural or pharmacologic respiratory stimulant directly on neurons concerned with respiratory control. However, in many ways, respiratory regulation is better understood than the processes involved in the firing of other central neuronal circuits whose electrical output is not subject to the quantitative measurement practical for respiration.

Many studies have provided clues to factors that may be involved either secondarily or in addition to acid-base changes. Those that have received most attention are mentioned in the following paragraphs.

An effect of molecular carbon dioxide. CO_2 produces respiratory effects apart from those associated with change in the $[H^+]$ of the blood and CNS fluids, but there is not yet a basis for considering that it induces direct stimulant effects.[146] It has been pointed out that CO_2 has never been shown to produce respiratory stimulation unless the change in Pco_2 somewhere leads to an increase in $[H^+]$.[146] This simple but amazing substance does have *depressant* actions on the CNS[261] that may in part be related to the narcotic effects of the inert gases. However, CO_2 is not entirely inert metabolically. Many of the decarboxylation re-

actions that result in the production of CO_2 from organic acids are now known to be reversible when Pco_2 is increased.[108] CO_2 can be incorporated into amino acids, presumably by way of the citric acid cycle.[68] This "fixation" of molecular CO_2, well known as part of the photosynthetic processes of green plants, is of less certain significance when it occurs within mammalian cells.[260] Although rather high tensions of CO_2 appear to be required for the demonstration of this fixation in vitro, it is actually possible that the availability of CO_2 at adequate pressure is one of the rate-limiting factors in the citric acid cycle of metabolic reactions. No such reentry effect of CO_2 has yet been shown to be related to respiratory control, and it is unlikely that such reactions lead to the dynamic membranal effects involved in respiratory stimulation. However, it does involve a demonstrable reverse link of metabolism to CO_2 itself or to HCO_3^- rather than to the hydrogen derived from its hydration.

Direct or indirect effects of hydrogen ions on respiratory neurons. Next to the electron the hydrogen ion (essentially a proton) is the simplest active particle in the living organism. It is not only a constituent of the biologically universal water but is also reversibly buffered by protein and anions of both intracellular and extracellular fluids. Conceivably the hydrogen ion concentration could affect the activity of respiratory neurons and chemoreceptor cells directly or in a number of indirect ways.

In view of the limited movement of charged particles across living membranes, the influence of pH could be related to physicochemical effects on the electrical state of the cell membrane. Such effects could be direct or be the result of secondary ion flux induced by the change in $[H^+]$. It is possible that change in $[H^+]$ may also modify the rate of energy-yielding reactions within the cell, either by altering the dissociation of the protein enzymes or by entering into cell metabolism as one of the reactants. Despite such possibilities, the evidence for an ultimate role for hydrogen ion concentration in the stimulation of central respiratory neurons or chemoreceptors remains circumstantial. At present, it is not even known whether observed influences of change in $[H^+]$ correlate best with extracellular or intracellular $[H^+]$.

Actions of $[H^+]$ to produce ion flux across cell membranes. Change in $[H^+]$ should alter $[Ca^{2+}]$ and flux in chemosensitive cells, the cells that provide central impulse generation or other components of the integrated control system. When the cerebral ventricles are perfused with fluid containing subnormal amounts of calcium or an excess of potassium, respiration is stimulated.[243] Similarly, direct injection of a solution of potassium chloride or sodium citrate into the medulla, like injections of CO_2-bicarbonate buffers, stimulates respiration.[76,141] All these effects may be related to a relative lowering of Ca^{2+} in the medullary centers. An excess of calcium in the fluid perfusing the ventricles depresses respiration, whereas an absence of potassium in the perfusion fluid has no effect.[243] Such demonstrations of respiratory effects of ions such as Ca^{2+} and K^+ may relate to the ion shifts observed within the CNS of rats exposed to high concentrations of CO_2. In such experiments, Woodbury and colleagues found that acute exposure to high concentrations of inspired CO_2 (12.5%) causes a decrease in "whole-brain" intracellular sodium concentration and an increase of intracellular potassium concentration.[261] Inasmuch as these and other observed changes were accompanied by a *decrease* in brain excitability (as measured by electroshock threshold), it is not clear how such changes may relate to the *increased* activity produced in respiratory neurons by CO_2 administration. Nor is it clear whether these ion fluxes are related to effects of the CO_2 molecule, to extracellular changes in $[H^+]$, or to intracellular changes in $[H^+]$ or $[HCO_3^-]$.

Central sites of CO_2 $[H^+]$-induced stimulation

Respiratory centers. Early in the study of respiratory control the concept evolved of specialized, closely related, and structurally discrete inspiratory and expiratory centers in the medulla (Chapter 70). As direct neurophysiologic recording methods improved, it became evident not only that CO_2 somehow induces excitation of the inspiratory and expiratory neuron groups but that this excitation is expressed in cell groups in centers extending as high as the cerebral cortex.[245] From extensive studies over many decades, it has become clear that the powerful respiratory response to CO_2 is mediated almost entirely through stimulation of structures within the brain with a contribution via peripheral chemoreflex influence on the center. The specific chemical bases for this stimulation and the specific sites of central stimulation are among the most important unsettled questions in respiratory physiology.

Chemoreflex effects of CO_2 $[H^+]$

At normal levels of arterial P_{CO_2} and P_{O_2} there is considerable chemoreceptor impulse traffic over the carotid body component of the sinus nerve. This electrical activity is reduced by O_2 administration[96,97,125] but is not minimal until arterial P_{CO_2} is lowered below about 30 mm Hg.* Conversely, CO_2 administration causes stimulation of the peripheral chemoreceptors, thereby increasing the electrical activity in the chemoreceptor nerve fibers. Both increased CO_2 tension without a rise in arterial $[H^+]$ and increased $[H^+]$ without a rise in P_{CO_2} have been shown to increase chemoreceptor activity in proportion to the change in P_{CO_2} or $[H^+]$.† This indicates either that extracellular *and* intracellular

change in $[H^+]$ can excite the chemoreceptor cells or that the H^+ and HCO_3^- can enter the excitable component of the chemoreceptor with relative ease.

Although the basis for peripheral chemoreceptor activation by H^+ is no better known than that for central neurons, it is highly significant that in single fiber recordings the fibers that transmit the activity generated by acid-base factors appear to be the same as those that respond to lowering of P_{O_2}.[96,128,214] This indicates a common chemosensitive site and further suggests the possibility of a common mechanism of action, not only for H^+ but for decreased O_2 as well. These aspects of chemoreceptor function and the interactions of CO_2 and O_2 effects will be considered later in this chapter.

Although the chemoreceptors of the carotid and aortic bodies are known to react, as sensors, to acid-base changes in the blood, it has proved difficult to establish the degree to which such peripheral chemoreflex stimulation contributes to normal control or to the total respiratory stimulation by CO_2. The quantitative role of the chemoreceptors in producing the respiratory response to hypercapnia has been studied in animals by many methods, including CO_2 administration before and after chemoreceptor denervation* and cold block of the sinus nerves *during* CO_2 breathing.[27,28,106,250] As a result of such studies, it is clear that (1) the chemoreceptors are not at all essential for the stimulation of respiration, (2) resting respiration in the absence of carotid bodies is not appreciably different from normal,[101,183] and (3) when oxygenation is normal, the stimulant effect of CO_2 on the respiratory centers is much greater than the effect mediated via the carotid and aortic bodies.[28,103,250] In early studies the *proportionate* influence of the chemoreceptors appeared to be greater at low levels than at very high levels of CO_2-induced hyperpnea[27,28,106]; it is greatest when hypoxemia is associated with alkalosis.[5] In man the most rapid component of respiratory response to change in P_{O_2}, presumed to be chemoreflex, appears to contribute a nearly constant fraction of the overall ventilatory response at different levels of inspired P_{CO_2}.[89] Such quantitative aspects of respiratory control will be reconsidered later (p. 1792 and 1798).

Influences of carbon dioxide in eupnea

CO_2, the most abundant waste product of nerve cell metabolism, exists within the CNS at

---

*See references 28, 63, 70, 95, 96, 125, 143, 214, and 242.
†See references 96, 124, 128, 133, 166, 214, and 233.

---

*See references 1, 28, 130, 131, and 231.

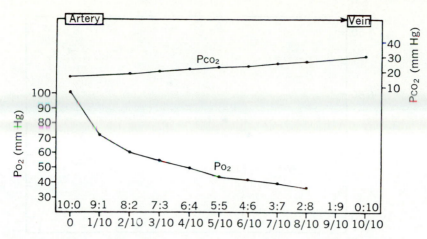

Fig. 71-6. Indirect derivation of P_{CO_2} and P_{O_2} changes as blood flows from arterial to venous end of capillary. Ordinates indicate P_{O_2} and P_{CO_2} in millimeters of mercury, whereas abscissa represents length of any capillary in relative units (tenths). Patterns of gas tension change were determined in vitro by measuring P_{CO_2} and P_{O_2} after mixing arterial and venous blood in proportions indicated on abscissa; for example, 6/10 means that 6 parts of arterial blood were mixed with 4 parts of venous blood. (From Gleichmann et al.[107])

a tension necessarily higher than that in the arterial blood supplying the brain. *Any change in the concentration or pressure of this freely diffusible gas at any site, outside or within any cell, will immediately lead to a proportional change in the hydrogen ion concentration at that site.* The actual level of central P_{CO_2} (and hence $[H^+]$) at any region important to respiratory control will be the result of dynamic equilibria among such factors as (1) the rate of CO_2 production by the respiratory neurons, (2) the characteristics of blood buffers, (3) the volume rate of blood flow through the centers, and (4) the influence of pulmonary ventilation on the acid-base composition of arterial blood.[144,154,220] The P_{CO_2} of the internal jugular venous blood draining the brain is normally about 10 mm Hg higher than that of arterial blood[144,156] and nearly equal to that of cisternal cerebrospinal fluid (CSF).[74] Central neurons should thus be exposed to levels of P_{CO_2} ranging from arterial to venous values, depending on their location with respect to the perfusing stream of blood and their distances from the capillary. The "mean capillary" or "mean brain" level of P_{CO_2} is a mathematical concept rather than a measurable value. It is sometimes useful in studies of respiratory control, and in normal states, it is apparently best indicated as being approximately midway between arterial and venous values (Fig. 71-6).[107] During O_2 breathing this may not be the case if the interference with CO_2 transport shifts the mean P_{CO_2} toward the venous value.

Regardless of the absolute levels of P_{CO_2} and $[H^+]$ in the centers or the number and exact locations of CO_2-reactive structures, the normal dynamic regulation of respiration from moment to moment in *stable states* of rest, exercise, or chronic acid-base disturbances will involve minute, probably unmeasurable, changes in P_{CO_2} and $[H^+]$ at sites within the peripheral and central chemosensors of the respiratory nervous system. Effects in stable states can probably be considered tonic rather than fluctuating.

The changes in $[H^+]$ to be expected in the brain on *sustained alteration of* P_{CO_2} can be seen from the P_{CO_2} dissociation curve of brain in Fig. 71-7,[237] which is not grossly different from that of blood.[213] In acutely changing conditions of CO_2 administration, the beginnings of exercise, or the infusions of fixed acids or alkalies, entirely different patterns of acid-base change in various blood, extracellular fluid, and intercellular sites can be identified.[151]

HYDROGEN ION AS INDEX OF RESPIRATORY STIMULUS
Hydrogen ion concentration as index in stable states of carbon dioxide administration

When CO_2 is administered, the diffusible CO_2 molecule penetrates all membranes and produces an increase in P_{CO_2} in every capillary, tissue, and cell. The hyperpnea of CO_2 breathing can therefore be correlated not only with change in P_{CO_2} in the arterial blood but also to changes at

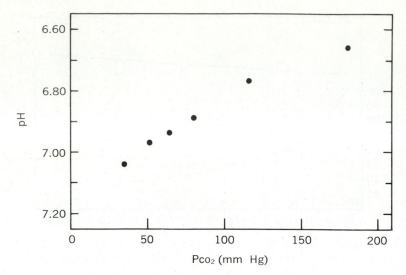

Fig. 71-7. Relation between pH and P_{CO_2} in undiluted brain tissue in vitro. CO_2 tensions were calculated from CO_2 concentrations in gas mixtures used for equilibration. Relationship appears to be slightly curvilinear at lower levels of P_{CO_2}. (Modified from Siesjö.[237])

any other sites such as alveolar gas, brain venous blood, CSF, and mean capillary blood (Fig. 71-8). Each curve in Fig. 71-8 conveniently illustrates the characteristic respiratory response to CO_2, but none of the response curves shown wholly represents the relationship of respiration to absolute stimulus level at the chemosensitive site in all physiologic situations. For example, arterial blood or even alveolar gas is a better index of the absolute stimulus at the highly perfused *peripheral arterial chemoreceptors;* however, mean capillary or venous P_{CO_2} is a closer index of *central stimulus* level than is arterial blood,[144] but is still not pertinent to the exact environment of central chemosensors.

It is important to recognize that CO_2 administration must produce a proportionate increase in [H$^+$] at all sites throughout the body. This means, as shown in Fig. 71-8, that for each P_{CO_2}-respiratory response curve that can be constructed there is a corresponding [H$^+$]-respiratory response curve. Curves are shown for the anatomic regions of interest, and each individual curve describes the *relative* relationships of respiration to change in [H$^+$] at a particular site. Selection of the relationship that best describes the *absolute* stimulus-response pattern requires presently unavailable information about the [H$^+$] at and within the actual central chemosensors.

Derivations of a single pattern for ventilatory response to change in [H$^+$] at a hypothetical site of central chemosensitivity function have been proposed,[67,69,206,207] and these necessarily describe the stable states from which the relationships are derived. However, overall function of a control system includes response to deviations from stability. The [H$^+$] at the control sites in turn depends at least in part on rate of change of environmental influences and on the relationship of blood flow to metabolism. If this relationship is similar in the reactive centers to that observed for the brain as a whole,[156,157] the [H$^+$] of the chemosensitive centers should be close to mean brain capillary or venous levels. In some instances, this appears to be the case.[153] However, if the blood flow through the control centers is high in relation to the average for the brain, the reactive centers should be considerably influenced by acid-base changes of arterial blood even when the venous [H$^+$] and P_{CO_2} are not altered. Since this also appears to happen even in stable states,[212] it is evident that neither arterial blood alone nor mean capillary blood alone nor venous blood alone provides a universally applicable index of central respiratory neuronal acid-base environment, even in the stable state. If, as is now extensively advocated, CSF composition influences the environment of superficial medullary chemosensors, the composition of CSF would represent still another factor modulating sensor [H$^+$].

Influences of fixed acid and fixed base

When a fixed acid such as HCl is infused into the bloodstream, it is partly neutralized by alkali

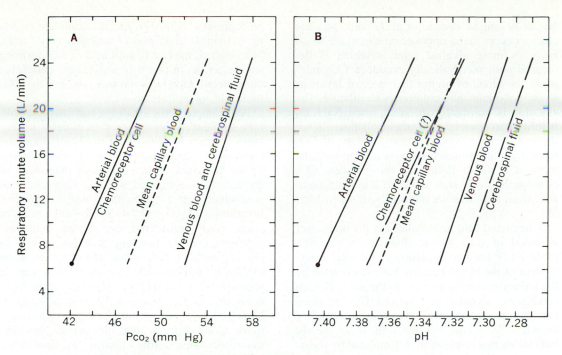

Fig. 71-8. Relationship of respiratory minute volume to P_{CO_2} and pH of various extracellular and intracellular fluids during CO_2 inhalation. Diagram is based only in part on experimental observations and is offered as composite illustration of probable changes in acid-base composition of fluids considered important to respiratory control.

A, Curve relating *arterial* P_{CO_2} and ventilation is regression through actual data obtained in several studies.[144] It ascends from mean value for P_{CO_2} and $\dot{V}_E$ observed during air breathing in 21 subjects in whom resting values for both P_{CO_2} and pH are available.[154,157] Respiratory relationships to brain venous P_{CO_2} were derived by using regression of venous P_{CO_2} on arterial P_{CO_2}[144] and relationships to "mean brain capillary P_{CO_2}" by assuming capillary values to be about midway between arterial and venous values. This is now considered more rational than a previous concept that mean capillary P_{CO_2} was close to venous level.[107,144] P_{CO_2} of CSF is shown as close to that of jugular venous blood. P_{CO_2} of intra- and extracellular fluids of central chemosensitive neurons is not yet deducible, but can differ from cerebrospinal fluid or blood.

B, Curve relating respiration to arterial pH passes through average arterial pH during air breathing in 21 subjects.[154,157] Mean capillary pH is estimated as described for mean capillary P_{CO_2}. Cerebrospinal fluid pH is approximated by calculations based on a fixed bicarbonate concentration of 1.6 mM/L less than that in arterial blood during air breathing and average P_{CO_2} equivalent to that proposed for jugular venous blood. Difference of 1.6 mM/L between arterial and cerebrospinal fluid [HCO_3^-] is derived from data of Bradley and Semple.[74] pH of extra- or intracellular location of central chemosensors is not definable.

that is withdrawn mainly from combination with CO_2 as bicarbonate. The immediate result is to increase the concentration of molecular CO_2 and hence the CO_2 tension of the blood. In an open beaker, CO_2 would be released from the blood. In the closed blood vessels, this cannot happen until the blood reaches the lung. Carry-over of increased P_{CO_2} to the arterial blood leads to a damming back of CO_2 in the tissues, including the respiratory centers and the carotid and aortic chemoreceptors. The infusion of fixed acid thus initially produces respiratory stimulant effects of

the acid itself and secondary influences of the CO_2 released from bicarbonate. The resulting increase in alveolar ventilation causes elimination of CO_2 in the expired air. As a result, the more acid blood now has a lower than normal P_{CO_2}. The acid-base state of the centers is partly restored toward normal by the lowering of P_{CO_2}. However, the acid-base balance of the blood will then have struck an equilibrium at a new level involving metabolic acidosis with a respiratory hypocapnia.

Effects of alkali infusion. If an alkali such as

NaOH is infused instead of acid, corresponding changes occur in the opposite direction; the blood becomes more alkaline, and breathing is depressed until metabolically produced CO_2 piles up in the blood to higher than normal levels as a result of diminished pulmonary ventilation. The situation is transiently different when $NaHCO_3$ is infused to produce a metabolic alkalosis, since the initial effect is a temporary release of CO_2 from the infused bicarbonate on mixing with the more acid blood. Following this transient CO_2 effusion the stable state achieved is indistinguishable from that after infusion of sodium hydroxide.

When fixed acid accumulates in the body over a period of time (as in diabetic or nephritic acidosis), it leads to depletion of the total cation content of the blood because base is excreted by the kidney in combination with the acid. In such conditions, alveolar and arterial CO_2 tensions may be reduced to levels one fourth normal or even less.[137,210] This was one of the clearest early indications that respiration is stimulated by something other than the molecular CO_2 concentration of brain or blood, for otherwise breathing would

be depressed. Although specific stimulant effects by individual abnormal chemical agents cannot be excluded, the fact that all acid or acid-forming substances act in this way justifies the belief that the concentration of hydrogen ion, not the CO_2 molecule, is an important factor.

An example of the small but consistent respiratory stimulation encountered in patients in diabetic acidosis is shown in Fig. 71-9. Breathing tends to increase in direct relation to the severity of the acidemia,[140] but the small magnitude of the effect, as compared with the response to an equivalent degree of acidemia brought about by breathing CO_2 (Fig. 71-9, Table 71-1), was for many years considered one of the major arguments against assigning a dominant role to $[H^+]$ in respiratory control. However, in acute metabolic acidosis the associated hypocapnia should *lower* central $[H^+]$, especially if the fixed acids involved do not readily cross any blood-brain or other barrier interposed between capillary blood and the specific site of the chemosensor mechanism (extra- or intracellular).[60,154,165] Fig. 71-9 shows that when CO_2 tension is held constant in normal subjects the respi-

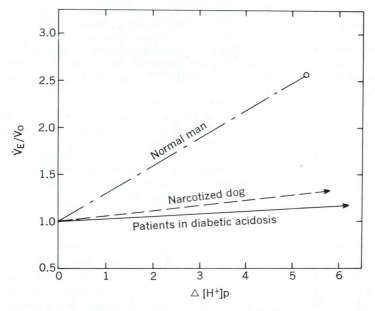

Fig. 71-9. Change in ventilation in response to elevation of $[H^+]$ in blood plasma. ($\dot{V}_E/\dot{V}_0$ is ratio of ventilation per minute during acidosis to ventilation per minute during control state.) In patients with diabetic acidosis, respiratory response to increased acidity of blood is small.[140] In these patients, however, hypocapnia also existed; this should have lowered central $[H^+]$ even though blood $[H^+]$ was elevated. In unnarcotized normal subjects in whom arterial and central P_{CO_2} is held constant while blood $[H^+]$ is altered,[154,179] respiratory response to change in blood $[H^+]$ is about 10 times greater than in diabetic patient with hypocapnia and even 5 times greater than anesthetized dog maintained at constant P_{CO_2}.[88] (From Lambertsen et al.[154])

ratory response to change in blood pH is about 10 times greater than that of the hypocapnic diabetic patients[154] (and considerably greater than that in anesthetized dogs also maintained at iso-P_{CO_2}[88]). The respiratory reactivity of the normal subjects to acute change in blood pH is in agreement with the magnitude of the independent arterial pH influence cited in the multiple-factor equation of Gray.[21,110] Therefore respiration in normal individuals is strongly affected by increases in the [H$^+$] of the blood, even when these are not accompanied by elevation of P_{CO_2} in the blood or brain.[154] In chronic states the passage of charged particles across the blood-brain barrier is probably responsible for the further exaggeration of the respiratory effects of metabolic acidosis.

"Blood-brain" barrier and respiration

It is now clear that blood [H$^+$] and central extra- and intracellular [H$^+$] at certain sites may change in opposite directions in metabolic acidosis and that opposite changes can also occur in metabolic alkalosis (Fig. 71-10).[59,222] This means that the *acid-base environment of the central*

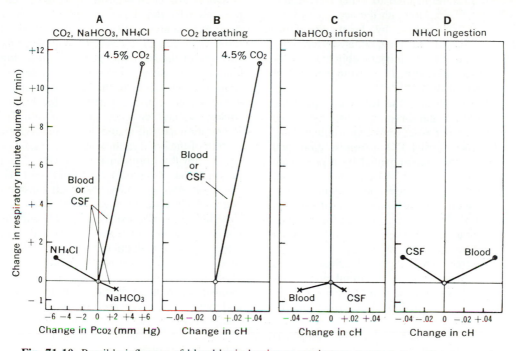

Fig. 71-10. Possible influence of blood-brain barrier on respiratory response to acute exposure to CO_2, fixed acid, and alkali, using calculated values for CSF as indices of changes in brain tissue P_{CO_2} and cH in which cH designates pH change with reversal of direction.*

A, CO_2, NaHCO$_3$, and NH$_4$Cl. Represents observed effects on respiration and arterial acid-base factors in human after breathing CO_2,[157] intravenous injection of sodium bicarbonate,[154] and ingestion of NH$_4$Cl.[163] Similarity of effects on P_{CO_2} of CSF and blood is due to free diffusion of CO_2 across blood-brain and other barriers.

B, CO_2 breathing. Administration of CO_2 leads to rise in [H$^+$] in blood and beyond blood-brain and blood-CSF barrier. pH change in CSF is *calculated* on *assumption* that there is change in P_{CO_2} without change in [HCO$_3^-$].[144,154]

C, NaHCO$_3$ infusion. Respiratory depression associated with *decrease* in acidity of arterial blood leads to *increase* in acidity of CSF due to entry of molecular CO_2 and failure of alkali to cross barrier.

D, NH$_4$Cl ingestion. As for alkali infusion, administration of fixed acid causes pH of blood and central fluids to change in opposite directions. Respiratory stimulation by [H$^+$] change in blood should thus be partly counteracted by diminished stimulus level in central fluids beyond blood-brain barrier.

---

*CSF is used here only as index of changes in fluids of the CNS (cerebrospinal, extracellular, or intracellular), differing in [H$^+$] from the blood because of the influence of blood-brain, blood-CSF or cell membrane barriers to ionic diffusion.

nerve cells critical in respiratory control in many physiologic or disease states may not be altered to the same degree or even in the same direction as is indicated by measurements made only on circulating blood. This interesting and extremely important phenomenon is related to the slowness with which fixed acids and bases cross the blood-brain barrier and the great ease with which molecular CO_2 diffuses through such membranes. This phenomenon of biology was recognized early[127] and has been studied in relation to respiratory control by measuring the differences between blood and CSF pH changes when acids or alkalies were administered to experimental animals.[58,82,166,220,222] Although the accuracy of the early results was limited by the nature of pH methods then available, the qualitative trends determined in animals are unmistakably the same as those summarized in Fig. 71-10 from more recent data obtained in man.[74,154,163]

Fig. 71-10 indicates the following: In *A* the P_{CO_2} of arterial and brain venous blood (and therefore of brain tissue, brain extracellular fluid, and CSF) rises during CO_2 administration, and these changes are accompanied by a prominent respiratory stimulation.[144] The administration of ammonium chloride causes only slight respiratory stimulation with a lowering of blood (and central) P_{CO_2},[163] whereas sodium bicarbonate causes respiration to be decreased with a consequent elevation of P_{CO_2}.[134]

B shows that, when the freely diffusible gas CO_2 is breathed, respiratory stimulation is accompanied by an elevation of $[H^+]$ not only in the blood but presumably also throughout the brain and CSF.[144]

C indicates that when sodium bicarbonate is infused intravenously, the blood becomes more alkaline and respiration decreases slightly. The resulting rise in blood P_{CO_2} should cause an acid shift in all CNS fluids. The degree of acidification of brain tissue by CO_2 depends on the buffer characteristics of the specific brain cells[135,237] (Fig. 71-7).

Finally, in *D*, administration of the acidifying salt NH_4Cl causes an acid shift in the blood as NH_3 is metabolized (e.g., to urea). However, the fixed acid (HCl) does not readily cross the blood-CSF barrier, and the hypocapnia produced by the slight respiratory stimulation should actually result in an alkaline shift within the central neurons protected by a blood-brain barrier. Prolonged administration of CO_2 over 1 hr or more up to 30 days gradually elevates central $[HCO_3^-]$,[62,72,75] diminishing the CO_2-induced central acidification. The initial rate of rise of

$[HCO_3^-]$ is about 1 mEq/L/30 min.[190] The rise in fixed acids in the blood during exercise should not immediately alter the $[HCO_3^-]$ of CSF, since intravenously administered lactic acid does not readily enter the CSF.[61,165] It would therefore not be expected that the fine adjustment of ventilation to metabolism in exercise would result from production of such fixed acids.

Cerebrospinal fluid and superficial medullary chemosensitivity

An important concept of the most recent period is that central chemosensitivity is not a function possessed by the dorsal medullary inspiratory/expiratory centers, but resides in separate structures superficially located on the ventrolateral surfaces of the medulla. The concept of superficial chemosensitivity originated with demonstrations that changes in the acid-base and ionic composition of fluids perfusing the brain ventricles induced changes in respiration.[36,164,166] Respiration is stimulated when the ventricular system of the brain in the dog, cat, or goat is perfused with fluid high in P_{CO_2} or $[H^+]$* and depressed when the fluid is low in P_{CO_2} or $[H^+]$.[36] These observations have been well substantiated and now include localization of two discrete, superficial chemosensitive regions, with maximum functional concentration located approximately 0.2 mm in depth below the medullary surface.[172,175,195] It is considered by the discoverers of these sensitive areas that they represent the structures that sense deviations in the CNS $[H^+]$ and send impulses to the integrating and motor components of the central respiratory control system.[175,195] The key feature of the concept is that, with receptors being at or close to the surface, their activity is influenced by the acid-base state of the adjacent CSF. This effect of CSF was initially considered dominant and sufficient; it is now proposed that the sensitive superficial structures respond to an extracellular fluid influenced in part by the blood and in part by diffusion of bicarbonate ion from the CSF.[175,195]

The experimental work summarized in this chapter is also described in recent extensive reviews,[166,175,195,240] a symposium on CSF and respiratory control,[29] and individual papers.† Whereas the pattern of observations does not establish that a mechanism *responding to* CSF acid-base composition is actually involved in normal respiratory control, the nature of the ob-

---

*See references 98, 164, 166, 167, 180, 196, 197, and 207.
†See references 172, 173, 196, 226, and 235.

servations themselves has theoretical and practical significance in relation to the search for central chemosensitivity and relative influences of arterial blood and central fluids in respiratory control.

Early studies of localized superficial chemosensitivity. During the past 40 years a number of studies have demonstrated respiratory response to drugs applied to superficial regions of the brain stem. Winterstein[59] and Leusen[166] have reviewed the many early studies of central respiratory control relating to the ventricles and CSF. One region of considerable early interest was the floor of the fourth ventricle, in the posterior portion of which is the area postrema.[245] This region is of special interest in respiratory control for two important reasons. First, the reticular formation containing the inspiratory and expiratory neurons lies beneath the floor of the fourth ventricle. Second, the area postrema is a region apparently lacking an important blood-brain barrier.[245]

Respiratory depression has been produced by application of acetylcholine (ACh),[105,197] nicotine, or cocaine[59,196,202,203] to the floor of the fourth ventricle, as well as by cooling this region. Because stimulation of respiration followed application of ACh to the floor of the fourth ventricle, Gesell et al.[105] postulated activation of the underlying medullary respiratory centers.

The present widespread attention to superficial chemosensitivity derives chiefly from extensive efforts to localize the effects of drugs, ions, and electrical stimulation[181,227] on superficial medullary structures. A consistent stimulation of respiration by acid fluids and by ACh applied only to discrete bilateral regions on the ventrolateral surface of the medulla has been demonstrated by Mitchell, Loeschcke, and their co-workers.[175,195] These regions, which apparently contain numerous nerve fibers and nerve endings but not neurons identifiable as chemoreceptors, are increasingly being considered to be the site of central $CO_2[H^+]$ sensitivity.[175,194,196,197] The region is also close to the entry of the glossopharyngeal and vagus nerves, each of which includes afferent fibers from the peripheral carotid and aortic chemoreceptors.[33] The many studies that provide the bases for this localization and the clear demonstrations that respiration is indeed influenced by electrical or chemical stimulation of the designated regions make it evident that these are part of the central respiratory control network. Whether they are the sole central chemoreceptors and whether the proposed susceptibility to $[H^+]$ of CSF is functionally important are not yet known. The studies have greatly increased interest in the roles of CNS fluids other than blood.

Blood-brain barrier and quantitative separation of carbon dioxide and hydrogen ion concentration effects on respiration

Although CO_2 breathing induces greater respiratory stimulation than can be accounted for by change in the $[H^+]$ of blood, several considerations justify abandonment of molecular CO_2 as a direct stimulant factor in chemical respiratory control. Foremost is that change in the $[H^+]$ of the blood can prominently influence respiration in the absence of central[154] or peripheral change in P_{CO_2}.[21,110,163,179] It must also be realized that the powerful respiratory stimulation attributed to P_{CO_2} in the "multiple-factor" concept[110] cannot be related to P_{CO_2} alone, since change in P_{CO_2} will alter intracellular pH everywhere.[144,146] Thus change in P_{CO_2} has never been shown to affect respiration even qualitatively without the P_{CO_2} change being somewhere accompanied by a corresponding change in $[H^+]$ (e.g., surrounding or within central neurons or within chemoreceptor cells).[154]

Regardless of whether CO_2 accomplishes its effects on respiration entirely through acidification of blood and other internal environmental fluids or whether even an event as fundamental as formation of H^+ is only another step toward an unknown ultimate mechanism, it remains uncertain where the clearly demonstrable respiratory responses of changes in blood $[H^+]$ are effected. The peripheral chemoreflex mechanisms are now known to be involved in part, since they respond to $[H^+]$,[28,124] and sudden lowering of P_{CO_2} $[H^+]$ leads to about a 12% fall in ventilation.[65,103] Centrally the blood-CSF barrier does not permit a rapid rate of entrance of fixed acid or base into the CSF.[82,259] However, the blood-brain barrier is not a single structure of uniform function, and certain portions of the barrier may offer less resistance to passage of ions from blood into brain tissue.[82,259] Infused fixed acids or alkalis may thus rapidly leave the circulating blood to cause an extravascular change of $[H^+]$ in localized regions of the CNS concerned with respiratory control even though they cannot be detected in samples of CSF withdrawn from the ventricles or cisterna magna. The area postrema appears to be more readily entered from the blood than is most other CNS tissue by such substances as dyes.[82]

Information concerning the rate of change of acid-base composition of the ventricular and cisternal CSF, as affected across the blood-brain barrier by changes in blood gaseous and ionic composition, is now developing.[62,186,244] Nevertheless, there is as yet no direct information describing the dynamics of permeability of any identifiable central respiratory control structure or compartment except the cisterna magna to fixed acid or base.[151] Until more such information is available, it will be necessary to investigate quantitative aspects of central chemical respiratory control by indirect means.

Stable-state respiratory responses to separate changes in P_{CO_2} and pH

In attempts to separate the quantitative influences of acid-base factors in respiration, change in blood $[H^+]$ has been produced by administration of alkali or acid to normal subjects while employing methods for "alveolar" P_{CO_2} control to hold essentially constant the P_{CO_2} of the blood and brain[88,154] or by utilizing graphic techniques for equalizing arterial P_{CO_2} at different levels of pH.[88,134,179,224] The results obtained in such an experiment in normal men on infusing the normal body constituent sodium bicarbonate to change arterial and internal jugular venous $[H^+]$ at constant levels of P_{CO_2} are summarized in Fig. 71-11.

An aim of this study was to evaluate respiratory effects by measuring changes in the $[H^+]$ of circulating blood and computing changes in fluids beyond the blood-brain barrier. CSF is presumed to have an ionic composition similar to that of other extracellular fluids beyond the blood-brain or cellular barrier. It is used here as an index for calculation of changes in *all* such fluids and not as a source of stimulus per se.

In Fig. 71-11 the changes in CSF or in intra- or extracellular fluid $[H^+]$ that should correspond to the alterations of P_{CO_2} imposed on the jugular venous blood were calculated rather than directly measured. The calculations of change in CSF $[H^+]$ were based on the known near-equality of P_{CO_2} in CSF and internal jugular venous blood[74] and the fact that CSF $[HCO_3^-]$, normally close to that in arterial blood,[74] remains essentially unchanged when large doses of sodium bicarbonate are infused intravenously.[153,222]

Fig. 71-11 indicates the following: In *A* the respiratory stimulation resulting from controlled elevation of P_{CO_2} was partly reversed when sodium bicarbonate was administered to restore blood pH to control levels while holding the blood and brain P_{CO_2} at a nearly constant level above normal.

In *B*, when the acidemia produced by P_{CO_2} elevation was fully reversed by bicarbonate injection, only 45% of the respiratory stimulant action of CO_2 was removed. Therefore approximately 55% of the respiratory response to CO_2 is unrelated to change in the pH of arterial or central venous blood but is somehow related to elevation of P_{CO_2} and is presumed to be related to change in $[H^+]$ beyond the blood-brain barrier. A similar division has been demonstrated in anesthetized dogs.[88]

In *C* the central acidosis produced by elevation of CO_2 tension in regions protected by a blood-brain barrier (here illustrated by CSF pH) was unaltered when sodium bicarbonate was used to reverse the acid change in the circulating blood.

Fig. 71-11, *A* and *C,* shows that correlations exist between 55% of the respiratory effect of CO_2 breathing, the total P_{CO_2} change induced in the blood or the centers, and the total central pH change produced by elevation of CO_2 tension.

Relation to respiratory control

It is evident that in humans there exist (1) a respiratory response to change in blood $[H^+]$ unrelated to change in P_{CO_2} anywhere in the body and (2) a respiratory influence of P_{CO_2} unrelated to change in $[H^+]$ of the blood. As these findings have become more precise in quantitative definition, they have stimulated refinement of theories of normal respiratory control.

While no concept of respiratory control can be simple, major conceptual elements can be descriptively stated in simple terms. The situation with respect to CO_2 stimulation of respiratory minute volume $(\Delta \dot{V}_{E_{CO_2}})$ has been summarized mathematically as follows:

$$\Delta \dot{V}_{E_{CO_2}} = A \Delta[H^+]_{blood} + B\Delta X \tag{1}$$

in which X is an unidentified effect of CO_2 produced somewhere, but distinct from the change in $[H^+]$ of arterial or other circulating blood; A and B represent coefficients of reactivity to the respective stimulus indices.[145,146,151] ΔX has been further illustrated to ensure recognition that it is directly proportional to many different and largely inseparable changes at different loci (Fig. 71-12). These changes, most of which were recognized early in the evolution of respiratory physiology and are still pertinent, include (1) change in the $[CO_2]$ and $[H^+]$ of *all* extravascular, extravascular fluids of the CNS including CSF, (2) change in the $[CO_2]$ and $[H^+]$ within *all* central neurons and central chemosensitive cells, and (3) change in $[CO_2]$ and $[H^+]$ within carotid and aortic chemoreceptors.

Progress has been made with respect to loca-

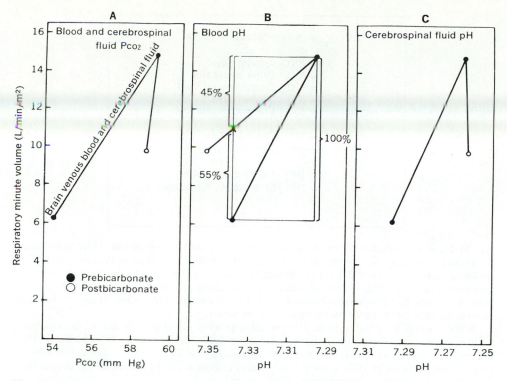

Fig. 71-11. Determination of respiratory effect of change in blood pH at "constant" arterial and central P_{CO_2}.[154]

A, Adjustment of *alveolar* CO_2 concentration from 44 to 50 mm Hg resulted in respiratory stimulation and plotted change of internal jugular venous P_{CO_2} from 54 to about 60 mm Hg ($\bullet—\bullet$). P_{CO_2} of CSF and other CNS intra- and extracellular fluids is presumed to have changed in a similar fashion. Intravenous injection of $NaHCO_3$ while alveolar P_{CO_2} was held constant at 50 mm Hg and brain P_{CO_2} remained essentially unchanged resulted in decrease in respiratory minute volume ($\bullet—\circ$).

B, $\bullet—\bullet$ shows relation of respiration to pH changes produced by elevation of blood P_{CO_2} with its concomitant increases in P_{CO_2} and acidity of blood and brain. This CO_2-induced change is indicated as 100%. Reversal of pH change induced in blood by hypercapnia removed 45% of respiratory response to P_{CO_2} elevation, leaving 55% of respiratory stimulation by CO_2 unaccounted for by changes in blood pH.

C, Relationship of respiration to calculated changes in pH of CSF induced by P_{CO_2} elevation resembles **A.** Intravenous injection of $NaHCO_3$ is presumed to have caused no appreciable immediate change in $[HCO_3^-]$ of CSF. Since central P_{CO_2} was maintained nearly constant, infusion of base would not alter CSF pH. Thus lowering of respiratory drive by alkalinization of blood occurred in regions affected promptly by change in blood $[H^+]$ despite persistent respiratory stimulation and sustained acidosis beyond blood-brain barrier.[154]

tions of the CO_2-related effect, but it has required acceptance that H^+ and not CO_2 itself is the stimulus mediator; this has not derived from true separation (probably impossible), but is based on abstract and instinctive grounds. Regardless of the location and exact basis for respiratory stimulation, the known direct relationship of P_{CO_2} and $[H^+]$ permits the stable-state equation to be rewritten as follows:

$$\Delta \dot{V}_{E_{CO_2}} = a\,\Delta[H^+]_{blood} + b\,\Delta P_{CO_2\ blood} \qquad (2)$$

or, since CO_2 mediates change in $[H^+]$,

$$\cdot\ \Delta \dot{V}_{E_{CO_2}} = \alpha\Delta[H^+]_{blood} + \beta\Delta[H^+]_{extravascular} \qquad (3)$$

Equation 2 resembles in principle the multiple-factor theory,[21,110] whereas equation 3 indicates that the findings illustrated by Fig. 71-11 can be described by a single-stimulus index such as

$$\Delta \dot{V}_{E_{CO_2}} = A \; \Delta[H^+]\text{Blood} \; + \; B \; \Delta X$$

$$
\begin{aligned}
X &= [CO_2] \; \text{Arterial blood} \\
&= [CO_2] \; \text{Other blood site} \\
&= [CO_2] \; \text{Cerebrospinal fluid} \\
&= [CO_2] \; \text{Central cells} \\
&= [CO_2] \; \text{Chemoreceptor cells} \\
\\
&= [H^+] \; \text{Central cells} \\
&= [H^+] \; \text{Chemoreceptor cells} \\
\\
&= [H^+] \; \text{Cerebrospinal fluid} \\
&= [H^+] \; \text{CNS extracellular fluid} \\
&= \; ? \; ? \; ?
\end{aligned}
$$

Fig. 71-12. Relationships between change in pulmonary ventilation and chemical effects produced by inhalation of CO_2. So far the only consistent relationship demonstrated between $\dot{V}_E$ and an acid-base factor has been with $[H^+]$ of circulating blood, and this accounts for only part of respiratory stimulation by CO_2. Figure illustrates that remaining effect of CO_2 is not necessarily an influence of any factor in arterial blood itself. It can be equally well related to any of the other factors and sites listed, including the last one in the series.

While none of the possible bases for residual respiratory effect of CO_2 can yet be separately studied, a tentative simplification may be justified on the basis that change in molecular $[CO_2]$, in circulating blood or centrally, has never been shown to produce respiratory stimulation without concomitant alteration of $[H^+]$ at some site, whereas change in $[H^+]$ in blood and in central fluid does affect respiration in the absence of alterations in P_{CO_2}.

Most important, it should be recognized that while any of the factors listed provides a reasonable stimulus *index*, this does not imply an ultimate biophysical cause-and-effect relationship.

$[H^+]$ acting at different sites. It thus somewhat resembles the qualitative criteria for the reaction theory.[255]

Finally, since the locations of acid-base effects on breathing are not firmly established on a quantitative and generally applicable basis, an overall relationship of $\dot{V}_E$ and $[H^+]$ can also be stated in the following form:

$$\dot{V}_{E_{CO_2}} = 3.45 \; \Delta[H^+]_{\text{effective}} \qquad (4)$$

in which 3.45 is the overall respiratory response in liters per minute to the change in pH produced in man by CO_2 administration, and $\Delta[H^+]_{\text{effective}}$ represents the composite, blended influence of $[H^+]$ change in blood and extracellular or intracellular fluid of the CNS. For the study illustrated in Fig. 71-11:

$$\Delta[H^+]_{\text{effective}} = 0.45 \; \Delta[H^+]_{\text{blood}} + 0.55 \; \Delta[H^+]_{\text{CSF}} \qquad (5)$$

where CSF represents not a specific chemosensor environment but an approximation to CNS fluids beyond the blood-brain barrier.

This expression is limited to $[H^+]$ as a composite single-stimulus index but differs in that the possibility of action on different neuronal groups is not emphasized. It quantitatively describes the total H^+ reactivity in man at rest, but not the individual responses of possibly different component parts of the control system. It should be clear that all these equations are at present intentionally shown as equally applicable to a number of acute acid-base disturbances such as the one from which they are derived.[154] Although none of them offers a final indication of the separate mechanisms of respiratory stimulation, they emphasize the magnitudes and proportions of acid-base changes in blood and other fluids that must be considered in attempts to delineate normal respiratory regulation by products of metabolism.

THEORIES OF RESPIRATORY CONTROL

Evolution of theories of respiratory control. From the earliest queries concerning respiration, theories and concepts of chemical regulation of respiration have evolved progressively as new information has been obtained by experiment. The factors that have been considered as somewhat stimulant to the respiratory mechanisms from the beginning are the same as those recognized as important today and include *lack of O₂*, *excess of CO₂, and increase in acidity of the blood.* During the past 100 years, many individuals have emphasized one or more of these aspects, but even in the present rapid advance of knowledge important questions of true stimulus, location of effect, nature of interaction, and proportionate contribution remain unsettled. The highlights in the history of control concepts are elaborated in several reviews.[26,33,168,247]

Lack of O₂ (p. 1807) was proposed by Rosenthal in 1862[223] as the sustaining drive to the respiratory centers; the effect of hypoxia had been discovered 50 years before by Legallois.[160] After it was learned by Walter in 1877[56] that infusion of acids led to respiratory stimulation, attention was focused on acidity as a stimulant factor, possibly separate from effects of CO₂. In 1885 Miescher-Rusch suggested that the CO₂ of blood was the normal respiratory stimulant, with the lack of O₂ being able to increase the reactivity of the respiratory center.[191] Each of these three early concepts, all based on astute objective observations, still appears to embody an important element of truth, and each has received almost continuous investigative attention up to the present time.

Two aspects of respiratory stimulation confounded early efforts to integrate the studies of potential chemical stimuli. These were the responses to exercise (Chapter 74) and to hypoxia (Chapter 73). It has been known since 1888 that changes in O₂ or CO₂ could not in themselves account for the whole respiratory response to muscular exercise.[104] On this basis, it was presumed that acids produced in the muscles provided the stimulus, but a correlation of some component of respiration with the increase in the acidity of the blood in exercise has only recently been established.[41,158] The stimulation of breathing by hypoxia, once also considered due to acid in the blood or central neurons, was eventually found in 1927 by Heymans and Heymans[120] to be mediated reflexly via peripheral chemoreceptors; the only central effect of hypoxia that could be consistently demonstrated was that of depression rather than stimulation.[5,80,246] This finding provided a new qualitative dimension to the understanding of respiratory regulation and stimulated the studies of the mechanism of chemoreceptor activation and the magnitude of actual contribution of this reflex system.

From the time that the separate central and peripheral chemosensitive components of respiratory control were demonstrated, questions and concepts concerning the mechanisms of overall respiratory control have increased until they now encompass the effects of CO₂, O₂, acidity, and other factors on *two* reactive regions (i.e., central and peripheral reflex). Concepts of normal control by acid-base factors are many and for the most part can be considered to complement rather than to contradict one another. The concepts that have received the most attention or support will be mentioned here. To facilitate comparison, consideration of the multiple relationships of acute and chronic hypoxia to acid-base stimulation of normal respiration will be deferred.

Theories and concepts of acid-base regulation of respiration

Reaction theory. The reaction theory, which refers to the acid or alkaline ''reaction'' of fluids, has been a part of respiratory physiology since 1910, when Winterstein suggested that all resting respiratory stimulations by CO₂ administration, by administration of fixed acids, and by hypoxia were produced by a single factor—an increase in acidity of the blood.[247] Later, when it was learned that acute hypoxia due to a low inspired Po₂ and the hyperventilation associated with it actually led to a *reduction* of arterial acidity, the reaction theory was revised to place the site of acid production and stimulation during hypoxia *within* the cells of the respiratory center.[247,253] This astute conjecture is still quite conceivable in part, but an alternate and experimentally demonstrable real explanation for hypoxic hyperventilation was provided by discovery of the peripheral chemoreceptors and their powerful stimulation by hypoxemia.[120] When the respiration of chemoreceptor-denervated but centrally reactive animals was found to be depressed by hypoxia, the reaction theory was again adjusted to exclude the previously postulated acid-producing effect of hypoxia on the respiratory center and refer it to the peripheral chemoreceptor cells.[247]

From this point on, adaptations of the theory were concentrated on the relatively greater respiratory stimulant action induced by CO₂ rather than on the stimulation by acidification of the blood with fixed acids. The demonstration by Jacobs[127] that molecular CO₂ passes readily across cell membranes to alter pH, whereas fixed acids and alkali do not, provided a basis for explaining the respiratory stimulant effects of CO₂ in terms of ready entry of CO₂ into the centers where it could dissociate to produce hydrogen ions. This concept, not requiring the specific stimulant role repeatedly (and until very recently) proposed for molecular CO₂ itself,[204] was based on the rapid passage of CO₂ across membranes and the realization that central and other neurons could thus be rendered acid by increases in Pco₂ even without a rise in [H⁺] of the circulating blood.[127]

After Leusen[164,167] demonstrated a respiratory response to fluids of varied Pco₂ during perfusion of the cerebral ventricles and Mitchell[196] et al. showed a similar stimulation by change in [H⁺] of the fluid perfusing the fourth ventricle, the reaction theory was further elaborated, as shown in Fig. 71-13.[255,256] As stated, the most recent version included respiratory control by (1) che-

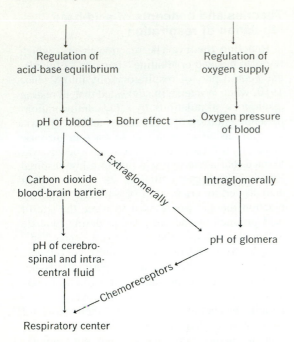

Fig. 71-13. Scheme of most recent version of Winterstein reaction theory of respiratory control. Outline, while qualitative, presents main features of highly rational reaction theory; several proposed locations of P_{CO_2}, pH, and O_2 influence on respiration are all linked through common H^+ mechanism. (From Winterstein.[255])

moreceptor stimulation through an action of hydrogen ions generated *within* chemoreceptor cells during hypoxia; (2) chemoreceptor stimulation by way of an effect of arterial hydrogen ions on the *exterior* of the chemoreceptor cells (the chemoreceptors formerly were considered as providing the dominant respiratory response to acute rise or fall of fixed acid concentration in blood); (3) an effect of fixed acids that only slowly cross the blood-brain barrier to enter the CSF and thereby affect superficial chemosensitive regions; and (4) an effect of CO_2 freely entering the brain and CSF to produce an acid reaction and prompt respiratory stimulation.

The reaction theory is here described in detail for several reasons. Its evolution covers a long and vigorous period of respiratory physiology, and its repeatedly changing character reflects the normal incorporation of important findings, conceptual adjustment, and criticisms resulting from continuing experimentation by many investigators. It is also important to recognize that it is not

a quantitative description of observed responses. Instead it represents several different forms of a purely *qualitative* concept, all of which cite the hydrogen ion as the respiratory stimulus.

Efforts to assign quantitative values to any site of stimulation by hydrogen ions have included those by Winterstein and Gökhan[257] and others[133] who have claimed that as much as 75% of the respiratory stimulation by CO_2 is chemoreflex in origin. This chemoreflex dominance is now considered unlikely.* What is important in the theory is that the one central theme of a role for hydrogen ions in stimulation of respiration has been persistently adhered to through a variety of correct and incorrect extensions of the concept; it is probably as reasonable now as when it was first proposed and certainly should continue to play its present role as promoting a single demonstrable index of central stimulus. Its role is also just as untestable, since any change in hydrogen ion concentration produces in turn further effects that may themselves produce the ultimate electrical changes in the membranes of the respiratory neurons (p. 1783).

Multiple factor theory. While the qualitative reaction theory was evolving, Gray in about 1945 approached the question of respiratory control analytically and quantitatively by deriving descriptive equations to fit the mass of then existing information.[21,110] This effort accurately described the actual numeric data on which it was based. The concept described by Gray avoided qualitative assignment of all respiratory stimulation to unmeasurable changes in hydrogen ion concentration and designated as additive, stimulant factors the $[H^+]$, the P_{CO_2}, and fall in the P_{O_2} of the arterial blood. Values were assigned to the stimulant effects of each of these factors without the necessity of considering the site or ultimate mechanism of action. By dealing only with stable states, a linear relationship of arterial $[H^+]$ and arterial P_{CO_2} to respiration was derived as follows:

$$\text{Ventilation ratio} = 0.22\ H^+ + 0.262\ Pa_{CO_2} - 18.0$$

in which ventilation ratio is the ratio of ventilation in a state of respiratory stimulation or depression to the natural, resting ventilation and H^+ is arterial hydrogen ion concentration in moles per liter $\times 10^{-9}$.[21,110] This concept of multiple-stimulant factors and dominance of arterial blood was not as rigidly conceived as it was stated, since Gray was certainly aware of the in-

---

*See references 65, 89, 103, 130, and 251.

fluence of P_{CO_2} on $[H^+]$ in inaccessible regions.*

At this time, it can be considered that the multiple-factor equation, although a highly useful quantitative description of those situations from which it was derived (CO_2 breathing, acute metabolic acidosis, and metabolic alkalosis), is lacking in qualitative and in general quantitative applicability because it erroneously depends on arterial blood values for all stimulus indices and because it emphasizes more than one index of respiratory stimulation. Aside from these considerations the multiple-factor theory and the qualitative reaction theory are entirely compatible and complementary.

Theory of superficial medullary chemoreceptors affected by cerebrospinal fluid hydrogen ion concentration. This theory is primarily anatomic in basis and has three main components that have developed progressively. They are (1) acceptance that H^+ represents the most appropriate choice of metabolic and CO_2-related stimulus, a premise of essentially all other current concepts; (2) designation of superficial areas on the ventrolateral surface of the medulla, rather than the dorsal medullary respiratory centers, as the primary and only chemosensitive components in central respiratory control, providing chemical drive to the medullary inspiratory-expiratory mechanisms; and (3) definition of the composition of CSF overlying the superficial chemosensitive areas as having a functional influence in their stimulation.

Choice of the superficial location as the central respiratory chemoreceptors derived from their identification as chemosensitive, from clear evidence that topical applications could increase or decrease respiration, and that cooling or local anesthesia could reduce ventilation.†

The initial concept was that surface central respiratory chemoreceptors were exposed directly to CSF and were normally responsive to the acid-base state of CSF itself; this concept was based on the now well-documented evidence of a central respiratory stimulant effect of increased P_{CO_2} or $[H^+]$ in CSF or other fluids used to perfuse the fluids about the medulla.‡

When it was demonstrated that the dynamic characteristics of respiratory response to abrupt alterations of inspired CO_2 did not at all closely follow changes in cisternal CSF,[145,146] it became evident that an overall or "effective" index of the acid-base stimulus at any central receptor site in the dynamic state would have to be one that changed at a rate between the changes in $[H^+]$ cisternal CSF and arterial blood.[145] The superficial medullary chemostat theory was therefore modified to include (1) an influence by the CO_2 tension of circulating blood as well as by the P_{CO_2} and $[H^+]$ of CSF[196] and (2) the assumption that the acid reaction of the blood itself contributes to respiratory stimulation via peripheral chemoreceptors.[196] These adaptations were followed by separate demonstration that the steady-state respiratory response of goats to cisternal perfusion followed neither the acid-base composition of arterial blood nor that of cisternal perfusion fluid but of interpolated values along the electrochemical gradient between the two.[98,206,207]

Extensive additional work has led to refined localization of the superficial chemosensitive tissue, in terms both of area and of depth of localization.* The concept continues to emphasize superficial localization, uniqueness of defined superficial sites as the functional central receptors for H^+ and response to their own local extracellular fluid, and at least partial influence by the composition of adjacent CSF.[67,69,175,195]

Although it is not yet a quantifiable description and it is not known whether the identified structures are unique or in what manner and to what degree the superficial receptors actually influence the medullary centers and respiration even in normal states, the conceptual importance of the identified central chemosensitive areas is large.

Present status of localization and function of medullary chemosensitivity. There is no doubt that perfusion of ventricular spaces with acidic or hypercapnic fluids stimulates respiration. This general perfusion may affect a number of different regions and lead to a greater respiratory response than do more localized, discrete topical applications of the same solutions. Through meticulous and exhaustive effort, qualitative concepts concerning localization of the regions most reactive to drugs and ions have evolved, including an active site in the floor of the fourth ventricle,[105,202,203] the area postrema,[82] the lateral recesses of the fourth ventricle,[177,180] and the ventrolateral surface of the medulla.[171,178,194,196] These sites are all normally exposed to any influences of CSF. The greatest respiratory response is elicited when stimulant

---

*The mathematical description was further extended to include a value for an additive action of another arterial factor, P_{O_2}, to give:

$$VR_{H^+, CO_2, O_2} = 0.22\ H^+ + 0.262\ P_{A_{CO_2}}$$

$$18.0 + \frac{105}{10^{0.038}\ P_{O_2}}$$

† See references 166, 172, 175, 195, and 226.
‡ See references 59, 164, 166, 175, 180, 195, and 196.

---

*See references 67, 69, 175, 195, and 242.

substances are applied to the ventrolateral surfaces of the medulla.

Students of the physiologic significance of superficial chemosensitive regions exposed to CSF have variously considered that:

1. Medullary receptors respond to the composition of CSF itself,[197] are subject to influences of CSF on the extracellular fluid surrounding the chemosensitive neurons,[196] and any superficial central chemosensitive region must be responsive to influences of blood as well as CSF.[151,196]

2. Superficial chemosensitivity represents the medullary CO_2 [H^+] chemoreceptor, the total *medullary* CO_2 sensitivity is located in the bilateral regions on the ventrolateral surfaces of the medulla,[196,197] and at least 40% of the whole-body response to CO_2 is accounted for by these regions.

3. Blood [H^+] changes associated with acute metabolic acidosis or alkalosis produce their respiratory effects by way of peripheral chemoreceptors rather than in the CNS.[132,196,257]

4. In chronic acid-base disturbances the [H^+] of CSF is regulated and maintained constant through transport mechanisms involved in the formation of CSF.[198]

In such studies and considerations there has been a tendency to accept that CSF has a real and dominant influence on the central acid-base regulation of respiration either by direct contact or by ready diffusion of its ions to modify the extracellular fluid environment of subsurface chemosensitive respiratory neurons.[29,196,207] Such assumptions are far from justified for moment-by-moment control and for acute respiratory adaptations. Considerable evidence now exists that (1) the neural structures affected by changes in the acid-base composition of CSF in perfusion studies must lie functionally between the capillary wall and the medullary surface,[146,151,207] with the location along the functional concentration gradient being proposed as two thirds to three fourths of the distance between CSF and blood,[207] (2) CSF overlying the medulla is *not* involved in normal respiratory control, (3) the peripheral chemoreflex component of respiratory stimulation by CO_2 is small, (4) nearly all the respiratory response to acid-base changes in blood is central, whether the disturbance is CO_2-induced or follows fixed acid infusion, and (5) more than one central mechanism sensitive to acid-base change exists.[151]

Although discrete superficial regions capable of affecting respiration are now known to exist,* important questions remain concerning such regions. These questions include (1) whether these are in fact the only chemosensitive regions affected in central respiratory stimulation, (2) whether under normal circumstances such regions are primarily stimulant or damping components of the system of chemical respiratory control, (3) if CSF does have an influence on the local extracellular environment of chemosensitive central structures, to what degree this contributes to the total respi-

ratory activity in various states, and (4) if actual chemosensitive cells can be identified, how such structures relate to the medullary inspiratory and expiratory respiratory neurons. Beyond these questions, nearly all applicable to the peripheral chemoreceptors as well, is the continued uncertainty over whether change in the external acid-base environment of chemosensitive cells, the intracellular environment, or both is the effective basis for neuronal firing.[186a]

"Single function" concept. In addition to the questions of the nature of the ultimate stimulus and the localization of the chemosensitive neurons are questions concerned with the numbers and quantitative contribution of control sites responsive to changes in acid-base composition of blood and body fluids.[151]

Pappenheimer, Fencl, and colleagues[98,206,207] provided extensive information refuting the functional importance of CSF as the dominant environmental fluid affecting even superficial central chemosensitive cells and attempted to determine the relative location of central chemosensitive cells between the boundary extremes of the brain capillary and the ependymal lining of a CSF-containing space. From stable-state measurements in unanesthetized goats, relating alveolar ventilation to values for acid-base factors in blood and cisternal CSF, they deduced that (1) essentially the entire central respiratory response to administration of CO_2, fixed acid, or fixed base could be described as a "single function" of a local extracellular [H^+] affecting sensitive respiratory neurons and (2) the locus for the chemical environment of the respiratory neurons affected is best represented as two thirds to three fourths of the distance along the gradient of [HCO_3^-] between the large cavity containing CSF and the tissue capillary.[207] These derivations have been extended in cats to provide the estimate that the function of central chemosensitivity lies approximately 100 μm from specific contact with actual CSF.[67,69]

This extensive analysis of stable states, also assuming the validity of [H^+] as a single-stimulus index, places essentially all the chemical respiratory control by acid-base changes at a single central system of sensors situated beyond the blood-brain barrier, but does not include an estimation of contribution by peripheral chemoreceptors.

This concept is in general quantitative agreement with the studies of the time courses of changes in respiration relative to changes in the [H^+] of blood and cisternal CSF in the dog in that it emphasizes that unknown central chemosensitive elements are functionally closer to the capillaries than to the surface.[151,207]

---

*See references 169, 175, 176, 195, and 227.

The "single function" concept was derived through trial and error analysis by the use of stable-state measurements relating blood and CSF pH to alveolar ventilation to define the "effective"[154] or single central [H^+] stimulus.[207] Inevitably, this stable-state analysis provides only a single mean stimulus index and a corresponding single mean respiratory response, even where several regions of central chemosensitivity may have been responding with different latencies, at different rates, and to different degrees. The quantitative aspects of the concept of a "single function" depend on the use of alveolar ventilation rather than total minute ventilation (alveolar ventilation plus dead space ventilation) to express the overall output of the respiratory neurons. This use of alveolar ventilation affects computations of both the value of mean or "effective" stimulus and its locus along the gradient between capillary and CSF.

Dynamic measurements of rates of changes in total ventilation and acid-base changes of blood and cisternal CSF further indicate that while useful for stable state description, the "single function" concept is not sufficient, either qualitatively or quantitatively, for description of respiratory control in changing conditions. In addition to the peripheral chemoreceptors, at least two distinct functions of central stimulation by acid-base changes can be demonstrated in man and dog.[65,151] Each must be active in stable and transient alterations of respiration.

Dual center theory. The dual center theory has its basis in descriptions of observed dynamic characteristics of respiratory control. Major deficiencies in theories of chemical respiratory control that delay their evaluation include (1) lack of quantitation, even of the relative contributions of peripheral and central chemosensitive elements, (2) tendency to force observations of ventilatory response into relation with a single fluid or site (e.g., arterial blood or other blood, CSF, brain extracellular fluid) when more than one may be involved, and (3) emphasis on study of stable-state responses, that, if a system incorporates more than one control sensor component or more than one stimulus environment, must in fact mask and blend the response characteristics and influences of individual sensor components and stimulus locations.

The "dual center theory" of chemical respiratory control derived from efforts to examine the integrated peripheral and central control system by temporal separation of its components and their dynamic response characteristics. In such examination of the time courses of changes in

respiration in relation to acid-base changes in fluids related to respiratory control, it became evident that no single fluid environment or dynamic characteristic could adequately describe even responses to acute CO_2 breathing or metabolic acid-base derangement.[151] In administration of carbon dioxide the rates of change of [H^+] of accessible CNS fluids (arterial and brain venous blood and cisternal CSF) were markedly different from each other and not related to rate of change of respiration.[151] When fixed acids or alkalies are introduced into the blood, their extremely slow entrance into the CSF results in very slow changes in CSF bicarbonate concentration[59,61,115,222] with a half time of hours rather than minutes (Fig. 71-14). Since no single recognizable combination of influences by CNS fluids could correlate dynamically throughout the respiratory response, it became sensible to determine the features of the sensor environment by examining the dynamic characteristics of the intact, unanesthetized respiratory control system itself.

Based on dynamic studies using O_2 breathing, respiratory response to CO_2 breathing, or acute infusion of fixed acid or alkali, it has been found that the respiratory stimulation by CO_2 administration involves a *small component* of chemoreflex stimulation and *two large, nearly equal components* of response that differ in time constant, although both are considered to be central[65,103,151] (Fig. 71-15). This led to the proposal of a dual center theory, in which more than one site of central chemosensitivity is required to account for observations of dynamic respiratory responses in acute acid-base alterations.[103,151] One site responds promptly to change in [H^+] produced in blood by infusion of fixed acid or base and thus is related to effects on neurons *unprotected by a blood-brain or other barrier;* the second component is related to change in P_{CO_2} but *not* to change in the [H^+] of the blood and hence may be concerned with alteration of [H^+] *beyond a barrier*. Each is so fast as to be unrelated to changes in the composition of cisternal CSF.[150,151] Thus, in addition to a limited but rapid influence of CO_2 and hydrogen ions on peripheral chemoreceptors, amounting to about 12% of the respiratory response of humans to CO_2, there appears to be a dual central mechanism providing for both prompt and delayed response to acute changes in the acid-base composition of the blood.[151] This "dual center" theory also uses the [H^+] as an index of the respiratory stimulus; however, it is a quantitative functional description and applies equally to the dynamic as well as the stable state. It identifies

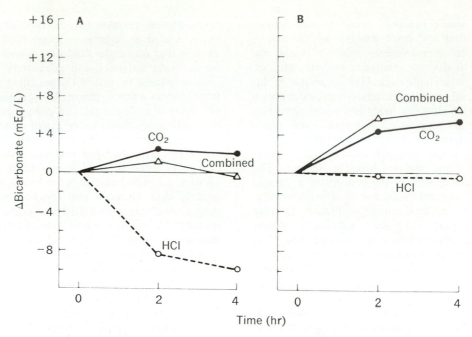

Fig. 71-14. Bicarbonate change in arterial plasma, **A,** and CSF, **B,** in five experimental groups of anesthetized dogs. Mean values in at least six animals in each group ±SEM. Solid circle = respiratory acidosis; open circle = metabolic acidosis; open triangle = respiratory and metabolic acidosis. In animals with HCl or $NaHCO_3$ infusion, zero time values are those obtained just prior to start of infusion. (Modified from Hasan and Kazemi.[115])

the lag, time constant, and proportionate contribution characteristics of the several reactive sites. It is not anatomic and does not identify the location of either of the central components functionally described.

The dual center concept considers that different dynamic characteristics of CO_2 response by central "sites" or by functions of the whole human respiratory system could represent not only differences in the sensors, but also differences in local delivery of fixed acid or CO_2, or even the damping of stimulant effects by partial loss of CO_2 to adjacent tissue fluids or CSF. Response to $[H^+]$ change in blood separate from the effect of CO_2 requires the absence of a barrier to H^+ and presumably represents an extracellular effect. The rapid response of sensors only to CO_2 (beyond a barrier) could imply penetration not of a choroid plexus but of a local medullary blood-brain barrier or even the cell membrane barrier between extra- and intraneuronal fluid. The identification of dual responses emphasizes that attempts to account for all CNS respiratory chemosensitivity in a single mechanism or a single local environment are not likely to succeed.[151]

In normal respiratory control at rest the dual center mechanism is conceived as providing no discernible separation of drive or special separate functions. During normal respiration and during short periods of CO_2 breathing, or the inverse responses associated with eupnea (p. 1776), all sensor input (peripheral and composite central) would be expected to affect respiration together in the same direction, even if at slightly different rates. Also, during sustained respiratory acidosis or alkalosis or in stable states of sustained metabolic alkalosis or acidosis, differences in rapid *dynamic characteristics* will have little significance. However, in these situations the interacting contributions of different central sensor influences of local acid-base change could be critical, varying in degree (and even in direction).

Aspects of the continuing uncertainty concerning actual stimulus mechanism, neural integration, and locations of sensors apply to the "dual center" theory, as to all other concepts cited.

DYNAMIC CHARACTERISTICS OF RESPIRATORY CONTROL SYSTEM

An important difficulty in evaluating concepts of respiratory regulation by chemical stimuli is that stable-state studies usually indicate only proportionality of change in a conceived stim-

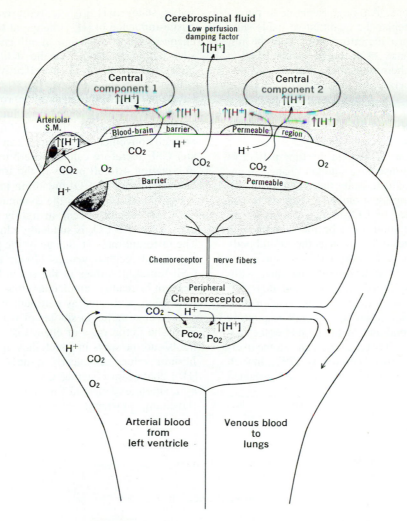

Fig. 71-15. Dual center concept and functional components of respiratory control. Diagram illustrates aspects of respiratory control by H^+, CO_2, and related acid-base factors.

At *all* sites, central and peripheral, molecular CO_2 is shown as capable of freely passing out of blood across blood-brain barriers and into tissue fluids and through cell membranes. In each instance where a rise in $[CO_2]$ occurs, level of $[H^+]$ is thereby raised.

At peripheral chemoreceptors, rise in arterial Pco_2 leads to increased $[H^+]$ both at extra- and at intracellular locations. Rise in *arterial* $[H^+]$ without increase in arterial Pco_2 will immediately raise $[H^+]$; in extracellular fluids of carotid body sites or functions, rate of passage of H^+ into glomus cell or other stimulus site is not now known.

In CNS, two chemosensitive sites or functions illustrate dual center concept. One central function is not protected by blood-brain barrier and hence is readily influenced by charged particles, as by changes in $[H^+]$ or $[HCO_3^-]$ from circulating blood. Second central region is protected by blood-brain barrier and thus is only very slowly affected by changes in fixed acid or bicarbonate. CSF, similarly separated by barrier from blood, is also acutely affected only by freely diffusible molecular CO_2.

During CO_2 administration and in normal breathing it is presumed that peripheral and all central units are activated, since CO_2 will penetrate all tissue sites uniformly. When acid is infused, regions not protected by barrier can be stimulated, whereas regions beyond barrier will not be. If respiratory stimulation of unprotected center leads to hypocapnia, second center (beyond barrier) will experience a *lowering* of acid stimulus level, whereas level of acidity to which unprotected center is exposed is *raised*. Thus counteracting central influences can occur.

CSF is considered to be affected in same manner as other fluids beyond blood-tissue barrier, but to normally act as a damping factor opposing influences of changes in blood, rather than as a primary source of stimulation.

ulus and response. Although superior to purely qualitative concepts, the quantitative stable-state study does not provide information about the lag, the rate of response, or the degree of contribution of a particular chemosensitive location. This difficulty is now being met in part by investigating unstable states.

Dynamic response to change in inspired Pco_2

In the absence of hypoxia the time pattern of respiratory response to a sudden change in inspired Pco_2 should indicate the composite influence of the changing stimulus at each of the several sites affected by CO_2-related stimuli. These rates of change should not be the same at such differently perfused locations as the carotid body and a medullary respiratory chemosensor.

When CO_2 is administered and then withdrawn, the respiration of men* and dogs[65,146] changes less rapidly than does alveolar or arterial Pco_2. By definition, the overall respiratory response of normal men at any time after the sudden initiation of CO_2 breathing can, on the average, be described by equation 4 (p. 1792). In such a description of a point in time, as for a sustained

*See references 102, 103, 146, 216, and 248.

stable state, $\Delta[H^+]_{effective}$ represents the composite influence of $[H^+]$ change at the peripheral chemoreceptor cells and at all chemosensitive sites within the CNS. For the stable-state study illustrated in Fig. 71-11, the effective $\Delta[H^+]$ was defined by equation 5 (p. 1793), with the behavior of CSF itself used as an approximate index of the stable-state changes in a fluid beyond the blood-brain barrier. In the normal dynamic state no such convenient single index can be presumed and CSF therefore cannot be arbitrarily applied in equation 4 to all points in the transition from one stable state of CO_2 breathing to another.

Fig. 71-16 illustrates the dynamic changes in $[H^+]$ that actually occur in lightly anesthetized dogs when 7% CO_2 is suddenly administered.[151] The different rates of change of the $[H^+]$ of arterial blood, central venous blood, and cisternal CSF represent values in the several CNS fluids that can be readily sampled and that have at some time been advocated as pertinent indices of the acid-base stimulus in studies of respiratory control. The initial rates of change of $[H^+]$ in these fluids and the ways in which they approach equilibrium values are clearly quite different. The $[H^+]$ change in cisternal CSF is well described by a single exponential function.[146,151] The arterial change appears to consist of two exponential

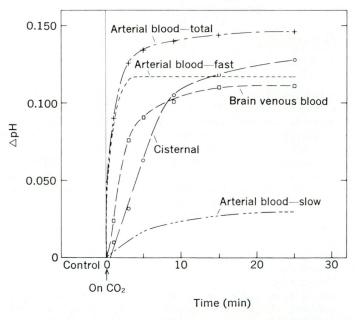

Fig. 71-16. Nature of changes in acidity following abrupt administration of 7% CO_2 in dogs. Figure shows that relatively simple procedure of CO_2 administration produces different temporal patterns of change in acidity of various fluids considered important in respiratory control, with even arterial blood having more than one component of $[H^+]$ change. (See text.) (From Lambertsen et al.[151])

components: a large, fast component of exponential character reflecting the addition of CO_2 to the blood (respiratory acidosis) and a slow, small exponential function representing a gradual increase in arterial fixed acid concentration.[151] The venous changes, which were measured in another series of experiments, occur at a slower rate than arterial $[H^+]$ changes but considerably faster than CSF $[H^+]$ changes. The CSF bicarbonate was not measured, but it is now known that CO_2 administration induces formation of HCO_3^- in CSF (Chapter 52).[62,115,186,244] Fig. 71-

16 therefore shows that not only multiple types of acid-base stimuli but also multiple stimulus locations, multiple lags, multiple time constants, and multiple contributions to stimulation must be taken into account in respiratory control.

Fig. 71-17 compares the respiratory and acid-base changes produced in dogs by sudden administration and removal of CO_2.[146,151] The values for the *on* CO_2 phase are the same as in Fig. 71-16. In this illustration the findings are plotted as percent change from control states in order to directly compare the rates of change de-

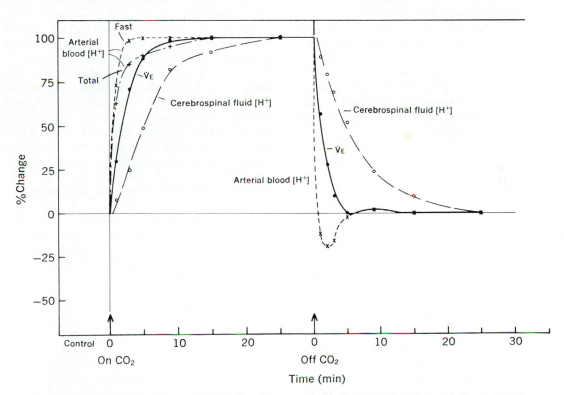

Fig. 71-17. Relative rates of change of ventilation and $[H^+]$ of arterial blood and cisternal CSF in anesthetized dogs. Figures represent changes produced by abrupt administration of 7% CO_2 in 30% O_2 in N_2 and by abrupt return to 30% O_2 in N_2. Curves for changes in arterial and CSF $[H^+]$ in "on CO_2" phase are same as in Fig. 71-16. After first few minutes following abrupt administration of CO_2, ventilation tended to rise more rapidly than did *total* arterial $[H^+]$. This indicates existence of a CO_2-related stimulus that changed faster than did overall $[H^+]$ of arterial blood. Fastest component should be rise in arterial Pco_2 and rise in $[H^+]$ induced by it anywhere. Relatively slow rate of increase in CSF $[H^+]$ detracts from any direct role of this fluid in acute control situations.

When CO_2 breathing is abruptly terminated, respiration returns to control levels more slowly than arterial $[H^+]$ and before change in CSF $[H^+]$ is more than 50% complete. Evidently, "effective" respiratory stimulus is largely represented by acid-base factors that change at rate slower than arterial blood but more rapidly than observed rate of $[H^+]$ change in CSF. Study of this data reveals no proportionate relationship of arterial and CSF $[H^+]$ that could account for temporal pattern of ventilation throughout entire period of CO_2 breathing and recovery. (From Lambertsen et al.[151])

fined in different physiologic units (e.g., liters per minute and hydrogen ions). The figure shows that, when CO_2 is administered, respiration does not closely follow changes in either arterial or cisternal CSF $[H^+]$. The time course of the respiratory change falls between the arterial and CSF values over part of the initial period of CO_2 breathing, and it was once proposed that the stable-state equation might apply to dynamic states of CO_2 breathing as well.[146] However, the relationship does not hold precisely for all points in time, and the apparent tendency for respiration to change at a rate nearly midway between that for arterial blood and CSF $[H^+]$ is considered to be a circumstantial approximation of the composite influences of all stimuli involved.

On close inspection, Fig. 71-17 shows that respiration eventually changes faster than does the *total* $[H^+]$ of arterial blood, indicating that in the study illustrated there was a component of stimulation changing *faster* than did total arterial $[H^+]$ that was compounded of a respiratory and a metabolic acidosis; it was most probably the CO_2-induced component of the acidemia.[151]

Cisternal CSF itself, bathing the medulla, does not appear to be actively involved in inducing respiratory stimulation, since during both the on and the off phases of CO_2 breathing the changes in respiration were essentially complete by the time only 50% of the CSF change had occurred. If CSF had a role, it would be to damp the responses to blood Pco_2 and $[H^+]$ rather than to provide a positive effect.[151]

In other studies the rate of change of brain venous (sagittal sinus) blood was found to be slower than that of respiration when CO_2 was suddenly administered, but it was indistinguishable from the rate of change of respiration when CO_2 administration was suddenly terminated.[151] This result in CO_2 breathing means that mixed venous blood from the brain cannot be expected to provide a satisfactory index of changes in the overall acid-base environment of the chemosensitive respiratory control regions in all states of acid-base derangement. Whenever combined CO_2 acidosis and fixed acid acidosis occur, problems of interpretation concerned with passage of fixed acid across the blood-brain and cell membrane barriers will arise, and no single source of blood alone should be adequate as the sole index of stimulation. Fig. 71-14 shows that CO_2 breathing, with and without associated metabolic acidosis, does lead to altered CSF $[HCO_3^-]$.[62,115,186a] However, the time course of change covers hours rather than the minutes involved in acute respiratory response to CO_2.

Separate effects of carbon dioxide and hydrogen ion concentration in dynamic states

It is possible in stable states to separate the CO_2 stimulation of respiration into one part (45%) related to change in blood acidity and a second part (55%) related to CO_2 or to change in acidity beyond the blood-brain barrier (Fig. 71-11). The nature, the location, and the reactivity of the control mechanisms should be unchanged during the transition from one stable state of acid-base stimulation to another. Hence, if appropriate indices of acid-base stimulus are employed, the basic descriptive equation 3 for stable states should accurately describe each point in time during an acute change in Pco_2 or fixed acid.

Fig. 71-18 compares the effect and time course of respiratory changes induced in dogs whose respiration had become stable on a mixture of about 7% CO_2 in 30% O_2 (1) when CO_2 is suddenly removed, resulting in rapid *lowering of Pco_2 and $[H^+]$* in all body compartments, and (2) when holding alveolar (arterial) CO_2 tension steady, alkali (sodium hydroxide–sodium bicarbonate mixture) is abruptly infused intravenously to mimic the rate and degree of pH change normally associated with cessation of CO_2 breathing. The figure shows that, as in other stable-state studies in man[154] (Fig. 71-11) or dog,[88] a change in blood $[H^+]$ while holding Pco_2 constant across the brain led to less than half the amount of respiratory change produced by the combined effects of CO_2 removal. In this brief situation of iso-Pco_2, no change in CSF bicarbonate or pH should occur (Fig. 71-11).[62,115,186] Quantitative comparison of the *rates of decrease* in respiration induced by alkali infusion and by CO_2 removal requires plotting each as percent change per unit time, as in Fig. 71-19.

Fig. 71-19 shows that the onset and the rate of development of the measured respiratory response to change in $[H^+]$ without corresponding lowering of Pco_2 are as rapid as the different effects derived for *fall* in Pco_2 without corresponding lowering of $[H^+]$ in blood. This indicates that no blood-brain barrier exists for this component of stimulation by H^+.

Relation of dynamic respiratory response to "single function" and "dual control center" concepts

It appears from the previous discussion that there is no complete barrier to a rapid respiratory influence of change in $[H^+]$ in blood. This is in contrast to the lingering assumption that central respiratory sensors are not affected immediately

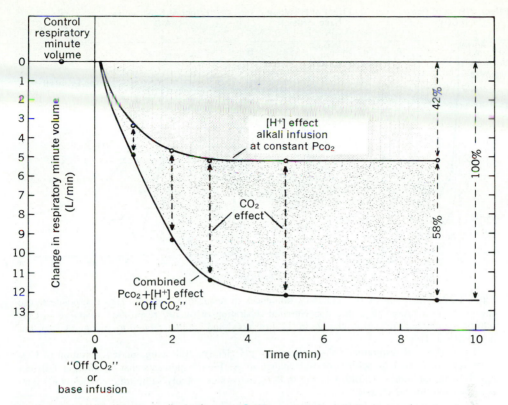

Fig. 71-18. Separation of effects of P_{CO_2} and $[H^+]$ on respiration in dynamic states. Lower curve in composite diagram shows time course of fall in respiration that results on abrupt cessation of CO_2 breathing in anesthetized dogs (combined P_{CO_2} + $[H^+]$ effect or "off CO_2" transient). This comprises one component related to change in blood $[H^+]$ and another component related to change in P_{CO_2}.[88,154] The small contribution of peripheral chemoreceptors is not separately identifiable in such studies in anesthetized dogs.

Time course of respiratory effect of $[H^+]$ change at constant alveolar (arterial) P_{CO_2} is shown by upper curve ($[H^+]$ effect). Difference between the two therefore represents respiratory influence of P_{CO_2}, which is unrelated to change in $[H^+]$ of circulating blood. Proportionate contribution of these components in stable state is essentially as previously reported.[88,154] Note that direct comparison of time courses of these control components requires that each be plotted in terms of percentage of total change. (From Lambertsen et al.[151])

by changes in fixed acid or base, since particles such as HCO_3^-, H^+ or Cl^- cross the blood-brain barrier so slowly that the central effects of metabolic acidosis or alkalosis on respiration should not be expected for many hours.[59,154,222,241] Evidently there is a "center" or functional component of the respiratory control system that is unprotected by a blood-brain barrier to charged particles and freely reached by changes in blood $[H^+]$. The existence of such a rapid component for $[H^+]$ effect provides a basis for analyzing the respiratory effects of acidemia in exercise, the metabolic acidosis of diabetes, and the infusion of fixed acids or alkalies.

Fig. 71-19 also shows that the time course of the respiratory change related to P_{CO_2} but not to blood $[H^+]$ is also rapid (actually in some dogs the P_{CO_2}-related response is faster; in others the $[H^+]$-related response is faster).[151] Because this second (CO_2-related) component is unaffected by acute change in $[H^+]$, it is considered to represent a center beyond the barrier to charged particles.[151] However, following abrupt CO_2 removal, the respiratory influences of even this control component are essentially complete almost before change in the cisternal fluid bathing the medulla has begun. The time responses of each of the reactive components do closely resemble

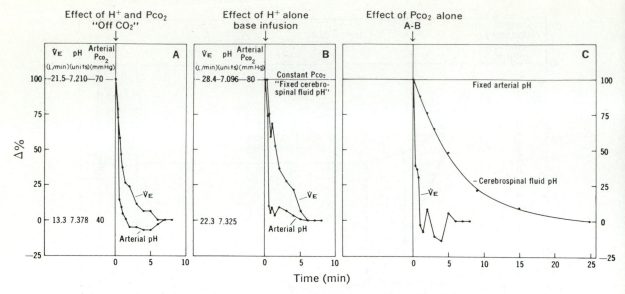

Fig. 71-19. Separation of H^+ and P_{CO_2} influences in respiratory control. This composite figure shows in one representative dog experimental separation of overall respiratory response to CO_2 into one component related to change in $[H^+]$ of blood and another related to change in P_{CO_2} but not to blood $[H^+]$.

Time course of respiratory ($\dot{V}_E$) and arterial pH changes following sudden cessation of CO_2 breathing are plotted as percent of full change in **A.** Figure indicates that ventilation fell less rapidly than did arterial acidity. Change in P_{CO_2} follows essentially same time course as pH (and is responsible for pH change).

B, Rapid decrease in arterial acidity was induced by infusion of alkali while holding P_{CO_2} (and hence CSF pH) constant. The rate of fall of ventilation is almost as rapid as that following abrupt cessation of CO_2 breathing. **B** should indicate predominantly characteristics of blood-brain barrier for passage of HCO_3^- into sites that respond to changes in fixed acid concentration in blood. Any peripheral chemoreflex component will be small and not separable.

Rate of change of $\dot{V}_E$ in **C** was derived as the difference between **B** and **A.** It can be considered to represent rate of entry of uncharged (molecular) CO_2 into reactive regions beyond blood-brain barrier and blood-CSF barrier for charged particles. Reactive center beyond barrier to HCO_3^- and H^+ responds extremely rapidly to CO_2; however, in some dogs it is even exceeded by influence of $[H^+]$ alone. Whether this component of response is in fact more rapid than response to H^+ requires determination of whether **B, C,** or both contain peripheral chemoreflex component. This requires identification and quantitative definition of P_{CO_2} and H^+ contributions in peripheral chemoreceptor effect on respiration.

Curve in **C** for changing pH of CSF is same one illustrated in Fig. 71-17. Note how much more rapid respiratory response is than would be predicted if entry of CO_2 into cisternal CSF or influence of CSF on superficial neurons were significant controlling factor in respiration.

the rate of change of $[H^+]$ or P_{CO_2} in central venous blood after abrupt CO_2 administration; thus it can be considered that each component represents a region in which perfusion with blood is rapid and diffusion distances between capillary and sensor cells are small.

Role of cerebrospinal fluid

Since each of the two components of response to CO_2-induced acid-base change is more rapid than the time course of change in P_{CO_2} or $[H^+]$ of

CSF, cisternal fluid cannot be considered as a local environmental factor directly contributing to stimulation of even superficially located chemosensitive regions in the normal, dynamic control of breathing. Since the reaction of CSF is altered by CO_2 from adjacent tissue and vessels, change in the acid-base composition of CSF must be the *result rather than the cause* of changes in pH or P_{CO_2} at the site of any superficial reactive neurons.[151] In this interpretation the role of CSF would be to provide a slight, passive damping of the rate at which superficially located neurons could respond to the effects of rapid changes induced by way of blood P_{CO_2}.[3,151]

Dynamic adjustments of central nervous system fluids

As with all current concepts of chemical respiratory control, the correlation of chemical change with neurologic and system response information is required. In all these concepts a principal uncertainty has been the rates at which acid-base adjustments occur within the compartments (and ultimately at or in the chemosensitive cells) of the CNS. This uncertainty has been compounded by the recognition that fixed acids or alkalis introduced into the blood enter the CSF extremely slowly.* This extremely slow change in CSF bicarbonate in *isocapnic acidosis* is now to be considered misleading when taken alone with respect to respiratory regulation, since both CO_2 breathing and hypocapnia have been shown to induce changes in $[HCO_3^-]$ of CSF within 1 hr, progressing to prominent further change over 3 to 6 hr[62,115,208] (Chapter 52 and Fig. 71-14). Evidently, P_{CO_2} change can induce subsequent change in central $[HCO_3^-]$. Compared with the rapid entry of molecular CO_2 into all brain compartments and, presumably, all cells, the induced changes in CSF $[H^+]$ and $[HCO_3^-]$ are too slow to be factors in the acute respiratory adjustments of abrupt CO_2 breathing (Fig. 71-17). They have, however, been considered in relation to possible explanations of influences of CSF or extracellular fluid $[H^+]$ in prolonged exposures to hypoxia, hypocapnia, and metabolic acidosis. As described in Chapter 52, changes in cerebral intracellular $[H^+]$ can be distinctly different from those in brain extracellular fluids such as CSF. If the prompt and evidently intrinsic capacity to regulate intracellular $[H^+]$ in the face of hyper- or hypocapnia reflects a capacity of central chemoreceptors as well, the task of actually measuring either stable-state or dynamic responses to the local respiratory stimulus at the sensor site becomes infinitely more complex.

Dynamic components of respiratory control in man

When a stable state of CO_2 breathing is abruptly terminated in normal human subjects or unanesthetized dogs, the fall in respiration is not a single, smooth, exponential function.[65,103] As shown in Fig. 71-20, the decline in tidal volume appears to occur in three identifiable stages, which can be mathematically described by three

*See references 36, 60, 62, 115, and 222.

different exponential functions.[103] Two of these functions have lag times longer than the lung-to-brain circulation time and presumably represent dual central control components. The third and fastest function most probably describes peripheral chemoreflex components and contributes only about 12% to 14% of the total respiratory response to CO_2 breathing. This degree of influence of CO_2 corresponds to most other estimates of chemoreceptor contribution obtained by suddenly changing the inspired O_2 tension.[84,89] Since chemoreceptors respond to change in both P_{CO_2} and $[H^+]$ of arterial blood, it is not yet possible to determine definitely whether this small but rapid unit of stimulation should be ascribed to the P_{CO_2}-related or to the $[H^+]$-related component in the quantitative description of the respiratory control system. However, O_2 can reasonably be expected to produce its actions by indirect, metabolic influence *within* chemoreceptor cell components, and it is also possible that the most rapid influence of acid-base factors on chemoreceptor activity would be by way of CO_2 effects *within* the chemoreceptor cell. Additional influences of $[H^+]$ on chemoreceptors may occur due to metabolic acidosis. If H^+ acts outside the sensor cell, these changes would be as rapid as those mediated by CO_2. If H^+ exerts its influence only inside the chemoreceptor sensor, the rate of its effect would depend on whether there is a barrier to passage of H^+ across the chemoreceptor membrane. No information concerning this question exists, since even the specific site of chemoreceptor stimulation is unknown.

Correlations of mathematical description of acid-base factors

There is a natural interest in the mathematical description of the stimulus-effect characteristics of the respiratory control system. The relation of respiration to various levels of CO_2 and $[H^+]$ in stable states of CO_2 breathing and metabolic derangements of rest was mathematically described by Gray[110] and elaborated by Grodins[23] for exercise. Since then, several further attempts have been made to derive an expression that will take into account the many factors in CO_2 hyperpnea. These analytic efforts include extensive further treatments by Grodins,[112] Grodins et al.,[113] Defares,[83] and Milhorn et al.,[192] and in each case are complex in spite of attention only to the "simple" alteration of inspired P_{CO_2}. Of these efforts, the analysis by Grodins et al.[113] appears to provide the most accurate mathematical description, and it provides a computer-programmed digital basis for appraising concepts

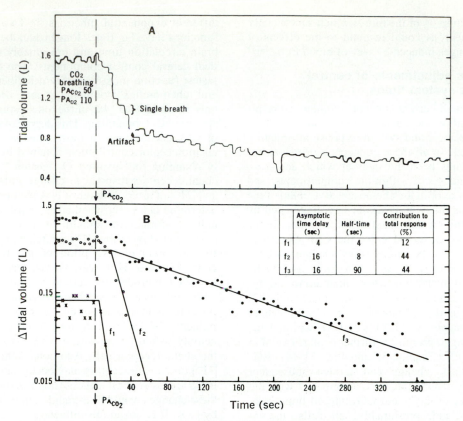

Fig. 71-20. Breath-by-breath changes in ventilation of normal human on abrupt cessation of CO_2 breathing at fixed alveolar Po_2.

A, Example of breath-by-breath decrease in tidal volume on abrupt change from breathing about 6% CO_2 to breathing a gas containing zero Pco_2, shown as tracing of actual recording. Fall of V_T is essentially complete in about 3 min. Decline in tidal volume on abrupt lowering of inspired Pco_2 appears to be a complex of several exponential functions rather than a smoothly decaying, single exponential function.[103]

B shows analytical separation of overall respiratory change into three components. Pattern of these several functions is consistent within same subject from day to day in human and can be identified also in unanesthetized dog. Values for lag, rate of development, and degree of contribution of specific responses are shown in tabular inset. Most rapid component is presumed to be peripheral chemoreflex in origin; others are presumed to be central. (Modified from Gelfand and Lambertsen.[103])

and constants involved in dynamic respiratory regulation as they are experimentally derived. It intentionally includes many aspects of respiratory regulation beyond the component relationships of respiratory response to acid-base changes.

The equations given here describe an intentionally still more fundamental description, encompassing the three evident functional components of CO_2 response in the "dual center concept)) (p. 1797).

Studies such as those illustrated in Figs. 71-11, 71-18, and 71-20 provide the fundamental bases for any model of respiratory control in that they permit direct and relatively simple description of the fully functioning control system in terms of actual measurements and do so in a manner applicable both to the stable state and to dynamic changes produced by CO_2 or fixed acids. The simplest of conceivable models, described previously (p. 1797), presumes the normal regula-

Total response = Chemoreflex + Center with blood-brain barrier + Center with no blood-brain barrier   (6)

$$\Delta \dot{V}_{E_{CO_2,H^+}} = \underset{\text{chemoreceptor}}{A\,\Delta[H^+]_{\text{arterial}}} + B\,\Delta[H^+]_{\text{center 1}} + C\,\Delta[H^+]_{\text{center 2}}$$

$$\Delta \dot{V}_{E_{CO_2,H^+}} = \underset{\text{chemoreceptor}}{0.12a\Delta[H^+]_{\text{arterial}}} + Xb\Delta[H^+]_{\text{center 1}} + (0.88 - X)c\Delta[H^+]_{\text{center 2}} \qquad (7)$$

tion of respiration at rest by centers having dual response characteristics and peripheral chemoreceptors responding to acid-base factors as well as to changes in Po_2.

This basic component descriptive expression is shown in equation 6, which is intended to identify rather than to simplify. The time course of change in $[H^+]$ or Pco_2 at the peripheral chemoreceptor can, because of extremely high blood flow,[79] be considered the same as that characteristic of rate of change in arterial blood. The half-time for acidic activation of the isolated and perfused peripheral chemoreceptor itself is 1 sec or less,[111] which means that the dynamic influence of the chemoreceptors is governed by the rate of change in composition of the arterial blood.[103] Each central function appears to have a time course of change in acid-base stimulus that closely resembles the time course of changes in central (brain) venous blood, even though the changes responsible for neuronal stimulation must in fact be occurring at extravascular sites. For practical purposes, measurements on *arterial blood for the chemoreflex component* and on *jugular venous blood for the two central components should now provide the closest measurable indices of the rates of change of [H$^+$] at the indicated control sites.*

The remaining problem in describing the respiratory response involves assigning values to the coefficients A, B, and C in equation 6. They represent the combined influences of reactivity of each control component and the proportionate response of each to change in stimulus level. If, as appears to be true, the chemoreflex component can be considered to contribute about 12% of the overall respiratory response to change in arterial $[H^+]$, the first term in the equation can be expanded to $0.12a\Delta[H^+]$, where a is the coefficient of overall reactivity of the entire respiratory system to CO_2-induced change in $[H^+]$ and 0.12 is the proportionate influence of this most rapid control component. In normal man, overall reactivity to CO_2 equals $3.45 \text{ L/min}/\Delta[H^+]$.

If it were known that the $[H^+]$ effect on chemoreceptor mechanisms was entirely intracellularly mediated, it would then be possible to assign values to each of the other coefficients in equation 6. This is not known. Therefore it is possible now only to consider that the sum of influences of the dual central functions is represented by the remaining 88% of the respiratory response not ascribed to the fastest component (i.e., chemoreceptor activity).

In its general form the control equation is then written as shown in equation 7, in which X and $(0.88 - X)$ represent the relative contributions of the two centers and b and c represent the coefficients of reactivity to central change in $[H^+]$. In the terms of the "off-CO_2 transient," analysis of Fig. 71-20, a, b, and c can be presumed to be equal to each other, and each to represent the overall reactivity to H^+ change. In normal man, a, b, or c thus should average about $3.45 \text{ L/min}/\Delta[H^+]$. From studies in anesthetized dogs and humans, it appears that the two central responses provide nearly equal contributions to overall respiratory control.[65,103,151]

The relationships shown by equation 7 indicate that in the dynamic regulation of respiration by acid-base factors the response includes a small but very rapid component reflected by changes in arterial blood, and the two slower components are reflected by changes in the CNS. Thus chemical control of respiration by changes in acid-base composition of body fluids is based on more than one-time function. The fastest function, evidently peripheral chemoreflex in origin, represents the *temporal fine adjustment,* superimposed on two somewhat less rapid control functions that are protected from abrupt change by a lower blood flow, by a finite diffusing distance for CO_2 and ions, and by the reactivity of brain vessels to the fluctuations of CO_2 tension in the circulating blood (and possibly by a damping influence of CSF on superficial central receptors).

In normal respiration and when acid-base disturbances are brought about by CO_2 breathing, a central sensor mechanism protected by a blood-brain barrier and a center not thus protected should each be affected essentially equally by the CO_2 level; the dual mechanisms should, in these situations, contribute smoothly together to the overall respiratory stimulation, as though only one site of stimulation existed. This should be the situation at rest via the inverse normal association of stimulus and response already described (p. 1776). However, when a combined fixed acid and CO_2 acidosis or a metabolic acidosis occurs acutely (as in exercise), the two central sites can be expected to respond separately to the different degree, direction, and rates of acute change of $[H^+]$ and Pco_2. Where slowly changing blood and central $[H^+]$ occurs, the sites may respond more in concert.

OXYGEN AND OXYGEN LACK

Hypoxia due to lowered inspired Po_2 can produce a pronounced stimulation of respiration. About 100 years ago, this observation led to the assumption that respiration is normally regulated by effects on the respiratory centers of fluctuations in arterial O_2 content.[28] The status of hypoxia as a conceived central respiratory stimulus underwent a complete transformation following the discovery of a reflex response to hypoxia mediated by chemosensitive cells in the carotid and aortic bodies.[28,117-119] Once the indirect nature of this response was known, it became possible to evaluate (by denervation of the carotid and aortic bodies) the direct effects of hypoxia on the respiratory centers.* This has been studied in patients in whom the nerves from the carotid sinus were transected and the carotid bodies were removed.[122,183] As a result of these studies, the following conclusions can be drawn:

1. In the absence of the carotid chemoreceptors the effect of increasing degrees of hypoxia is a progressive, overall depression of the integrated central mechanisms of respiratory control (Fig. 70-7), leading to complete respiratory failure. It appears that without the carotid chemoreceptors, the aortic receptors alone are not able to sustain hyperventilation in hypoxia.[122]

2. When the chemoreceptors are intact, the depressant effects of even a considerable degree of hypoxia on the respiratory centers can be overcome by the influence of nerve impulses reaching the centers from the chemoreceptors. Through this mechanism of increased reflex stimulation the otherwise depressed centers can sustain respiration against failure or in exposures to low inspired P_{O_2} can drive respiration at greater than normal levels.

3. The degree of reflex respiratory response to hypoxia is related to the change in O_2 tension rather than to the change in O_2 content of arterial blood.[28,70,214,242]

4. The degrees of chemoreceptor activity and reflex respiratory response to hypoxia are magnified when hypoxia occurs in combination with hypercapnia.[77,205,214,242]

5. It is probable that an indirect consequence of O_2 lack, not the level of P_{O_2} itself, causes the electrical discharges.

Peripheral chemoreflex contribution to respiratory stimulation in normal and hypoxic states

The question of whether chemoreflex effects of changes in P_{O_2} in any way contribute to the normal regulation of respiration at sea level was a persistent source of difficulty for several decades. The difficulty stems from problems of interpretation of two apparently divergent but equally well-documented types of observations that are described in continuing series of reviews

of chemoreflex function.* These observations are (1) the relatively small degree of respiratory stimulation induced in normal humans by moderate hypoxia and (2) the progressive increase in chemoreceptor electrical activity with even small degrees of lowering of arterial P_{O_2}.

Relationship of inspired P_{O_2} to ventilation in normal men

Fig. 71-21 shows that the overall ventilatory response to O_2 lack in normal humans is extremely small (often stated as absent[136]) until the inspired O_2 concentration is reduced below about 10%.[91] This degree of oxygenation is equivalent to that produced by air breathing at about 17,000 ft or 0.5 atm. Under these conditions, arterial P_{O_2} is approximately 40 mm Hg and arterial hemoglobin saturation close to 70%. These values, corresponding to those normally found in the *mixed venous* blood of individuals breathing air at sea level, are associated with deterioration of physical performance, visual functions, and mental processes. Although further lowering of arterial P_{O_2} results in more definite respiratory stimulation, the maximum anoxemic hyperventilation observed is relatively small compared with the vigorous stimulation produced by even small changes of P_{CO_2} (Table 71-1).[91]

Effect of change in arterial P_{O_2} on chemoreceptor discharge

To understand the small respiratory response to hypoxia and its relation to normal respiratory control requires information concerning sensor responses to change in P_{O_2} and P_{CO_2}, alone and in combination. For P_{O_2} (1) do the carotid and aortic chemoreceptors respond only when P_{O_2} falls below a low "threshold" or (2) do they exert an increasing influence on the centers from the moment that arterial O_2 tension begins to fall from its normal high level? On the basis of this demonstration of apparent respiratory insensitivity to O_2 lack, it was once considered that, in humans, a "threshold" corresponding to arterial P_{O_2} of about 50 mm Hg exists for chemoreceptor effects on respiration, and hence a chemoreflex response to change in P_{O_2} does not contribute to the normal control of respiration at sea level.[230-232] It is now clear that chemoreflex response to P_{O_2} and P_{CO_2} changes represents an active fine adjustment even in the normal resting state. Studies performed in cats indicate not only the existence of a definite "tonic" chemoreceptor discharge at normal levels of arterial P_{O_2},† but

---

*See references 1, 5, 70, 71, 80, 166, 183, 200, 229, and 246.

*See references 10, 16, 28, 70, 214, 219, and 242.
†See references 63, 95, 125, 214, 219, 242, and 258.

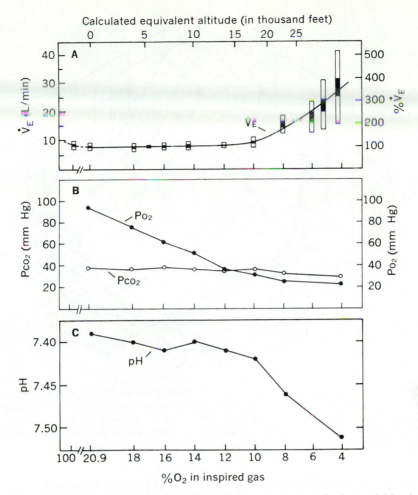

Fig. 71-21. Influence of reduced inspired O_2 partial pressure on respiration and blood composition in man.

　　A, Respiratory response to O_2 lack. Average findings at sea level in normal male subjects, except for subjects breathing 4% O_2, who were schizophrenic patients. Solid bars represent ± 1 SE on each side of mean; open bars, ± 1 SD on each side of mean. Prominent stimulation of respiration began only when O_2 concentrations below 10% were administered. Note that sustained inhalation of 100% O_2 tended not to depress but to stimulate breathing. Similar findings have been obtained in dog.[123]

　　B and **C,** Changes in arterial P_{O_2}, P_{CO_2}, and pH. Note that fall in P_{O_2} tends to be limited by increased ventilation when very low O_2 concentrations are inspired. Since administration of 6% and 4% O_2 was for brief periods only, P_{CO_2} and acidity were not grossly lowered, even by high ventilations.

　　(**A** modified from Dripps and Comroe[91]; **B** and **C** based on data from Lambertsen et al.[152])

also show that electrical activity is not completely abolished even by administration of 100% O_2 at 1 atm.[125] Moreover, the reactivity of the carotid body both in situ[125] and in excised preparations[96,97] is exaggerated when hypoxia and hypercapnia are combined. It is also important to the understanding of chemoreceptor activation mechanisms to note that very high tensions of CO, pre-

sumably by inhibiting cytochrome oxidase, lead to chemoreceptor firing even at normal P_{O_2} and P_{CO_2},[32] Fig. 71-22 illustrates the manner in which electrical impulse activity over the carotid body fibers responds to changes in P_{O_2} and P_{CO_2}. The reduction in impulse traffic when P_{O_2} is raised above normal indicates that the chemoreceptors are actively discharging even at normal

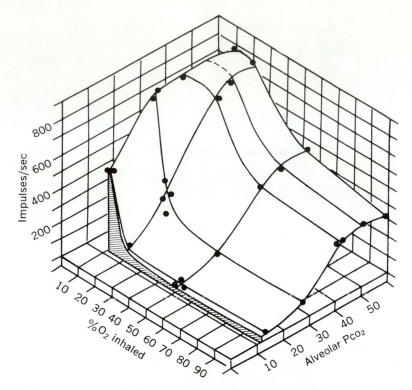

Fig. 71-22. Three-dimensional plot of carotid chemoreceptor activity in vivo as influenced by inhalation of different O_2 concentrations, using variations in ventilation to vary alveolar P_{CO_2}. Note that minimal activity exists at high P_{O_2} (100% O_2) and extremely low P_{CO_2}. Increased P_{CO_2} has little influence at high P_{O_2}. Prominent increases in activity are produced by combined hypoxia and hypercapnia. (From Eyzaguirre and Lewin.[97])

P_{O_2}.[63,95,125] As P_{O_2} falls, the impulse traffic rises progressively in classic "dose-response" fashion without indication of a threshold for activation. Maximal activity of the chemoreceptors is produced by asphyxia (increased P_{CO_2}, decreased P_{O_2}); this combination of changes occurs in breath-holding, in respiratory obstruction, and in suffocation in a closed compartment. The greatest reduction of activity is accomplished by hyperoxygenation in the presence of subnormal P_{CO_2}. In the normal response to low inspired P_{O_2} (as at altitude) the beginnings of an increase in alveolar ventilation lead to a slight lowering of arterial [H^+] and P_{CO_2} (Fig. 71-21).[152] Thus as one peripheral factor tends toward further activation, a second peripheral and central factor tends toward limitation of activity. The respiration that occurs is the composite of these influences, in which one stimulus change masks another. It is most probably this that limits respiratory stimulation by hypoxia to nearly unmeasurable degree.

Magnitude of chemoreceptor influence on respiration. Use of abrupt stimulus change and measure-

ment of dynamic response offers partial resolution of the previously mentioned interactions. Fig. 71-23 shows that when O_2 is abruptly administered to men at controlled alveolar P_{CO_2} at rest, a transient fall in ventilation occurs.[84,85,89] This kind of procedure has also been utilized with one or several inspirations of O_2 in hypoxia or of hypoxic gas in normally oxygenated subjects by Dejours[16,85] and others[93,161] in an "O_2 test" for chemoreceptor activity. When O_2 is administered suddenly and continuously while alveolar P_{CO_2} is held constant, the characteristics shown in Fig. 71-23 of lag (average 3.8 sec), time constant (average 6.5 sec), and percent depression by O_2 (average 12%) are considered to represent the oxygen-related chemoreflex contribution to resting ventilation in man.[89] This is equivalent in magnitude and time course to presumed chemoreflex influences of CO_2.[103] It has similarly been found in cats[130] and rabbits[251] that in the absence of hypoxia the influence of chemoreceptors on respiration is small in comparison with central drives.[130] Evidently a chemoreflex contribution to the ventilation of resting normal men and animals does exist, even at sea level. However, since the effect of 100% O_2 (over 600 mm Hg P_{O_2}) is barely more than 10%,[84,89] the degree to which the normal small fluctuations of P_{O_2} affect breathing must be small indeed. Al-

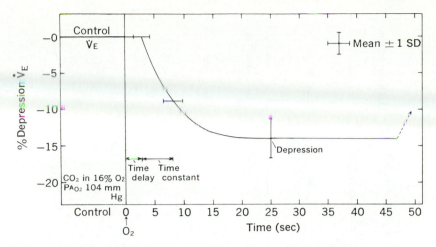

Fig. 71-23. Time course of abrupt depression of respiration by O_2 in normal men. Abrupt increase in inspired P_{O_2} while holding alveolar (arterial) P_{CO_2} constant,[89] or allowing natural change,[84] promptly decreases respiratory minute volume approximately 12%. This rapid component of respiratory control, presumably chemoreflex in origin,[16] begins after lag of only about 4 sec and has time constant of about 7 sec. (From Downes and Lambertsen.[89])

though this component is not the most powerful controlling factor, the promptness and speed of its reaction to deviations of P_{O_2} or P_{CO_2} (p. 1801) should have almost a breath-by-breath influence on ventilation.

A critical link relating the effects of O_2 on chemoreceptor electrical activity to its effects on ventilation has recently been provided by performing these measurements in the cat during transient states of O_2 or CO_2 administration or both.[161,162] Fig. 71-24 shows that the fall in ventilation that occurs on administration of O_2 follows a time course closely resembling the depression of chemoreceptor electrical activity.[161]

Influence of low inspired P_{O_2} on central acid-base stimulus level. As arterial hypoxemia develops and respiration begins to be stimulated, arterial P_{CO_2} will be lowered. This should diminish the chemoreflex stimulant effect of lowered arterial P_{O_2} both at the carotid body and in the centers. This lowered arterial P_{CO_2} and, as hypoxia becomes severe, dilatation of brain vessels, will lead to a lowering of the P_{CO_2} of central cells and their environment. The degree of this influence of hypoxia on central acid-base stimulus level, which is considerable, is shown in Fig. 71-25. In men breathing 8% O_2 a drastic fall of 15 mm Hg in central P_{CO_2} occurs. When any central depressant effect of severe O_2 lack is added to this, it begins to become evident that there are reasons why the respiratory stimulation in acute inspiratory hypoxia is limited. It is likely that the secondary effects of hypoxia, by reducing the level of acid-base stimulus both peripherally and centrally, combine to minimize the overall respiratory stimulant effect of hypoxia in normal man. Thus Fig. 71-21 can be considered descriptive of the

natural, compensated state of hypoxia, in which chemoreceptor activation by lowered P_{O_2} plays only one part.

Influence of carbon dioxide on respiratory response to hypoxia in man. A vigorous respiratory response to even slight degrees of hypoxia occurs when the usual fall in alveolar P_{CO_2} associated with hypoxia is prevented. Fig. 71-26, *A* contrasts such a situation with the natural response to lowered inspired P_{O_2} without control of alveolar P_{CO_2}. The figure shows (1) a progressive fall in P_{CO_2} as alveolar P_{O_2} is lowered (lower curve), (2) the prominent influence of CO_2 tension on the response to hypoxia (a 35 L/min or almost 10-fold increase in respiratory minute volume at 40 mm Hg P_{O_2}) when alveolar P_{CO_2} is prevented from falling below normal, and (3) an apparently smooth increase in respiratory stimulation as alveolar P_{O_2} falls below 100 mm Hg.[176]

The importance of CO_2 tension (or [H^+]) in determining the magnitude of hypoxic respiratory stimulation is evident from the classic study by Nielsen and Smith[205] and from others by Cormack et al.,[77] Loeschcke and Gertz,[176] and Lloyd et al.[169] If the respiratory response to CO_2 is measured at several different O_2 tensions, it becomes evident that the slope of the response to CO_2 becomes steeper as alveolar P_{O_2} is lowered (Fig. 71-26, *B*).[136,169,176,205] The basis for the increased reactivity to CO_2 is not entirely known. Exaggerated chemoreceptor discharge is one explanation, but it still remains possible that the very early and now largely discarded concept of acid formation in the CNS during severe hypoxia may in some degree be a contributing factor.

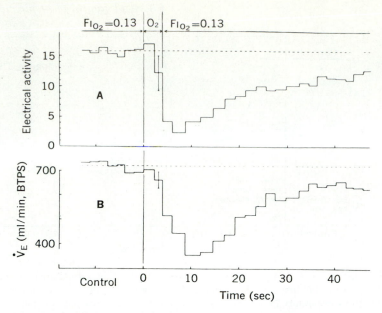

Fig. 71-24. Effect of inhalation of two consecutive breaths of pure O_2 on frequency of chemoreceptor discharge and on ventilation in cat. Mean of five O_2 tests on anesthetized cat. Following control period breathing slightly hypoxic gas mixture (13% O_2 and 87% N_2), cat inhales pure O_2 during two successive respiratory cycles and then returns to breathing hypoxic mixture.

A indicates mean frequency of chemoreceptor impulses measured during each respiratory cycle on some isolated fibers of right Hering nerve. Rapid decrease in impulse frequency occurs during transient O_2 administration.

B shows respiratory minute volume for each respiratory cycle as horizontal lines. Starting with second respiratory cycle following O_2 inhalation, frequency of discharge and ventilation decrease significantly. Degree of decrease in impulse frequency and in ventilation is greater than would be expected for air breathing at sea level.

(From Leitner et al.[161])

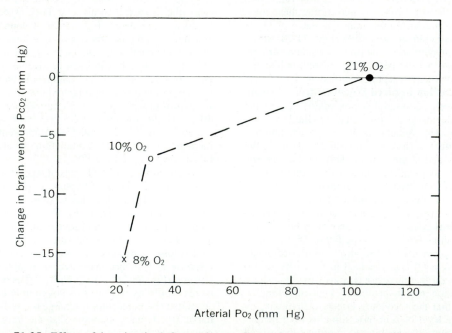

Fig. 71-25. Effect of low inspired O_2 tension on P_{CO_2} of brain venous blood.[139,155] Breathing 10% and 8% O_2 in N_2 at sea level lowers central P_{CO_2} and $[H^+]$ by (1) respiratory stimulation (which lowers arterial P_{CO_2}) and (2) anoxemic dilatation of cerebral vessels (which by increasing blood flow further lowers central P_{CO_2} and increases pH). Hypoxia therefore lowers concentration of central respiratory stimulus at the same time that it induces peripheral chemoreflex stimulation. The real shape of the curve is not known, and there may be small effect on brain P_{CO_2} produced by small degrees of hypoxia.

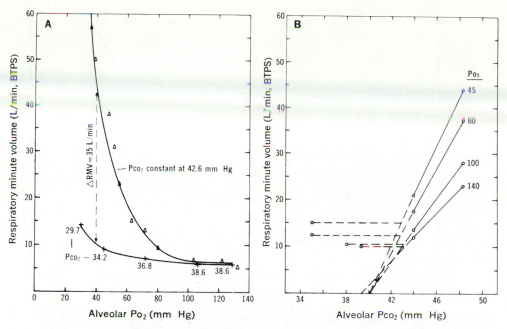

Fig. 71-26. Interaction of P_{O_2} and P_{CO_2} in respiratory control.

A, Influence of P_{CO_2} on respiratory response to O_2 lack. Lower curve shows that progressive lowering of alveolar P_{O_2} produces only slight respiratory stimulation, which, apparently by lowering alveolar and central P_{CO_2}, tends to be self-limiting. Hypocapnia thus may mask anoxic drive. Upper curve, obtained over same range of hypoxia, but *while holding alveolar P_{CO_2} essentially constant,* indicates true capacity of chemoreceptor influence on respiratory control system. Difference between curves, for example, 35 L/min at 40 mm Hg P_{O_2}, represents loss of reactivity of centers resulting from effects of difference in alveolar P_{CO_2} of approximately 8 mm Hg.

B, Influence of P_{O_2} on respiratory response to CO_2. Response to change in alveolar P_{CO_2} is steeper at low than at normal P_{O_2}. Horizontal, dashed lines are employed to suggest that level of ventilation during zero inspired P_{CO_2} will be unchanged by addition of CO_2 until threshold for CO_2 stimulation is reached.[169,176,205]

(Modified from Loeschcke and Gertz.[176])

Respiratory effects of increased inspired P_{O_2}. Respiration of resting man is transiently *decreased* by O_2 administration, but if O_2 breathing is continued for several minutes, ventilation *increases* above air breathing control levels.[153, 157,170] The increase in respiratory minute volume is most likely the result of the central acidosis produced when increased amounts of physically dissolved O_2 diminish liberation of O_2 from hemoglobin and thus interfere with base release in the tissue capillaries. The resulting inefficiency in CO_2 transport leads to a central *hyper*capnia. The slight respiratory stimulation in turn causes an arterial *hypo*capnia that limits the respiratory response.[147,148,157]

If the respiratory *reactivity* to CO_2 is measured during O_2 breathing, it will be found to have decreased below normal.[153] Thus O_2 produces three influences on respiration concurrently: chemore-

flex respiratory inhibition, indirect respiratory stimulation via central accumulation of CO_2, and diminution of the respiratory response to CO_2. The latter two effects increase progressively up to at least 3 atm of inspired P_{O_2}.[147,148] However, the O_2-induced increase in central P_{CO_2} should not exceed about 5 mm Hg. Whether the depression of the respiratory response to CO_2 is related to chemoreceptor inactivation or to direct central effects of high O_2 pressure is not now known. Fig. 71-27 illustrates the effects of O_2 breathing at 1 atm in a stable-state situation. It shows the following:

1. The slight respiratory stimulation that occurs when O_2 is breathed without added CO_2 and the associated central hypercapnia–arterial hypocapnia.

2. A unique pattern of arterial CO_2-ventilation response curves that, without other information,

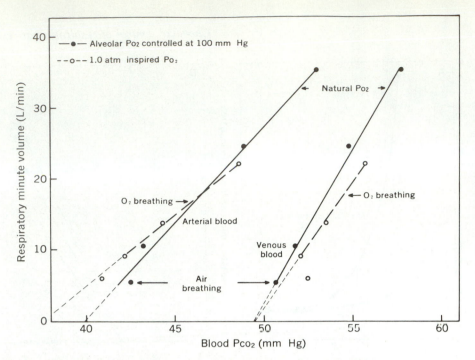

Fig. 71-27. Influence of O_2 inhalation at 1.0 atm on respiratory response to change in P_{CO_2} of arterial and internal jugular venous blood (average values in five normal men). Solid and dashed lines are regression lines. Data for condition of O_2 breathing without added CO_2 were omitted in deriving regression of ventilation on P_{CO_2} during O_2 breathing; this was done to avoid introducing uncertain influences of *subnormal* arterial P_{CO_2} as variables in this study. O_2 decreases slope of respiratory reactivity to CO_2. Peculiar crossing of P_{CO_2}-respiratory response curves occurs but only when respiration is related to arterial P_{CO_2}. When brain venous P_{CO_2} is used as reference for determining respiratory response curves, O_2 breathing curve is seen to be depressed below that for air breathing at all levels of P_{CO_2}. No paradoxical crossing of venous CO_2 response curves occurs. (From Lambertsen et al.[153])

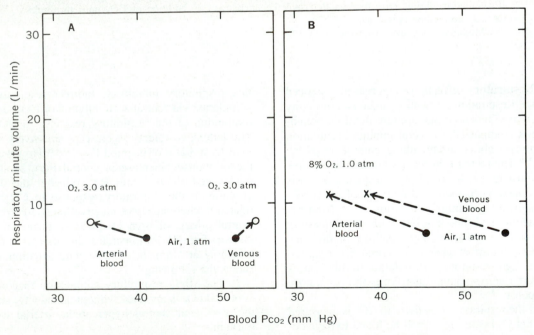

Fig. 71-28. Influences of O_2 and hypoxia on arterial and central P_{CO_2}. Control values relating respiratory to arterial and internal jugular venous P_{CO_2} in air breathing are compared with effects of high inspired P_{O_2}, **A**, and with low inspired P_{O_2}, **B**.

A shows that high O_2 pressures (3.0 atm) raise *central venous* P_{CO_2} while lowering arterial P_{CO_2}. Mild hyperpnea is therefore associated with simultaneous tissue *hyper*capnia and arterial *hypo*capnia. Hyperpnea is evidently related to P_{CO_2}.

B indicates that hypoxemia leads to simultaneous lowering of arterial and central P_{CO_2}. Hyperpnea occurs despite general lowering of P_{CO_2}. (Data from Lambertsen et al.[155,156])

would suggest the unreasonable combination of (a) stimulation of respiration at low P_{CO_2}, (b) no effect of O_2 on respiration at intermediate levels of P_{CO_2}, and (c) depression of breathing at the highest P_{CO_2}. This peculiar pattern indicates that arterial blood alone is no more appropriate as a source of information concerning central respiratory control during O_2 breathing than it is during hypoxia (Fig. 71-25).

3. A consistent relationship of central venous P_{CO_2} to ventilation, indicating the coexistence of an increased central P_{CO_2} (stimulant factor) and a diminished response to CO_2.

These effects of increased P_{O_2} are largely opposite those encountered in hypoxia, where respiratory stimulation is accompanied by *decreased* central acid-base stimulus and *increased* response to CO_2 (Fig. 71-28). Whether these represent the extremes of a physiologic continuum or unrelated circumstances is not yet known.

CENTRAL CIRCULATION AND RESPIRATORY CONTROL

It is probable that extremely minute volumes of nervous tissue provide central respiratory chemosensitivity. The levels of P_{O_2}, P_{CO_2}, and pH in the local environment of even these small components of the respiratory center system depend on maintenance of a fine balance among factors such as pulmonary ventilation, whole-body metabolism, blood gas composition, hemoglobin concentration and buffer state of the blood, volume rate of blood flow through or in proximity to the chemosensitive centers, and metabolism of the respiratory sensors themselves. Of these, all but the last two can be measured in man and other animals. Measurement at the minute sites of sensor function is unlikely. Change in the rate of CO_2 removal from the CNS by flowing blood does affect respiration in animals, even when the composition of arterial blood entering the brain is not changed.[228] The closest approach to determining the influence of central blood flow and metabolism on maintenance of O_2 and acid-base homeostasis in man has been made by studies of these factors in the entire brain or in relatively large divisions of it.[159,189,239] These studies are considered in relation to respiration because they may at least suggest the general characteristics of the factors in gas exchange between the blood and the inaccessible central mechanisms of respiratory control.

Brain metabolism

Reactivity of central neurons, including those involved in respiratory control, depends on intact metabolic state. Determinations of the rate of brain O_2 consumption in humans indicate that, on an average for the entire brain, no detectable changes in its O_2 metabolism occur over a wide range of physiologic stresses, including muscular exercise,[158,239] moderate hypoxia,[139] extreme hyperoxia,[156] and breathing low concentrations of CO_2.[139,239] However, it is quite likely that any changes in neuronal oxidative metabolism occurring within the normal range of functional activity will be small and probably unmeasurable by methods involving mean rates for the entire brain. Appreciable increase in brain O_2 consumption has been shown in generalized convulsions,[239] and diminution has required the use of narcotic drugs[239] or levels of hypoxia close to that which produces unconsciousness.[149] *Such indications that the metabolism of central nerve cells is extremely stable under physiologic conditions suggest that activation of the respiratory neurons by CO_2 may not be associated with an appreciable increase in their metabolism.*

Metabolism–blood flow–activity relationships. There has been a natural tendency, based on teleologic reasoning, to expect that the demands of a rise in neuronal metabolism or activity will somehow be met by a corresponding increase in local circulation. A greater O_2 consumption rate should both lower the P_{O_2} and elevate the P_{CO_2} of the fluids surrounding the active cells. Since anoxemia and hypercapnia have each been shown to dilate cerebral vessels, it has tended to be considered that one or both of these changes in gas tension will adjust local circulation to the greater demands for gas exchange. As an example of this reasoning, if stimulant factors (such as chemoreceptor activation) cause an increased metabolic activity of the respiratory neurons that is not met by a corresponding increase in their blood supply, the cells of the centers might then be stimulated to further activity by accumulation of CO_2 (and perhaps other acids) produced by their own metabolism.

In certain parts of the brain (e.g., the visual cortex) an increase in *functional activity* is known to be accompanied by an increase in blood flow.[159] Whether this is actually related to an increase in *metabolism* ($\downarrow P_{O_2}$, $\uparrow P_{CO_2}$) or represents an incidental vascular action of substances primarily involved in interneuronal transmission and electrical activity rather than with metabolism (e.g., ACh) is not certain. This question relates to respiratory control, since any influence that alters the central neuronal acid-base environment will modify the output of these centers.

Influence of respiratory factors on brain blood flow

Carbon dioxide. Increased CO_2 tension of inspired air has been shown to produce the greatest degree of cerebral vascular dilatation[156,218,239]; conversely, hyperventilation causes diminution in the volume rate of brain blood flow.[138,218] In the normal range of respiratory and cerebral circulatory control, these effects on the smooth muscle cells of brain vessels follow changes in the blood CO_2 tension rather than in the altered arterial pH, since in acute studies in the physiologic range, vascular resistance appears to be unrelated to changes in pH unless the latter are accompanied by altered CO_2 tension[154,158,239] (Table 71-2).

The situation resembles that for respiratory control in that molecular CO_2 itself is probably inert and does not itself affect the brain vessels. Because of the ready passage of molecular CO_2 across cell membranes, it is probable that these powerful cerebral vascular effects of CO_2 are brought about not by a primary action of CO_2 but through CO_2-induced alterations of pH *within* the smooth muscle cells. When, as in diabetic acidosis, a very large increase in the acidity of blood occurs in the presence of hypocapnia, dilatation of brain vessels does occur.[140] However, the vasodilator effect is small and should probably be considered either as being beyond the range of physiologic control mechanisms or as the result of slow penetration of ions through the smooth muscle cell membrane.

Hypoxia. Brain vessels dilate when the arterial Po_2 is acutely lowered[139,159,239] (Table 71-2). This fortunate consequence of hypoxia results in an increased brain blood flow, tending to sustain oxygenation of brain tissue as the level of arterial Po_2 falls. As shown in Fig. 71-29, the hypoxic vessels can still be affected by changes in Pco_2, and the ability of arterial Pco_2 to influence the degree of vasomotor tone persists even at extremely low levels of arterial O_2 tension.[149] Thus the cerebral vasodilatation produced in nor-

Table 71-2. Relationships of brain blood flow in man to changes in composition of arterial and internal jugular venous blood (changes in mean blood pressure minimal)

Experimental condition	Reference	Brain blood flow	Arterial blood		Internal jugular venous blood	
			Pco_2	[H+]	Pco_2	[H+]
A. CO_2 breathing	Kety and Schmidt[139]	↑	↑	↑	↑	↑
B. Hyperventilation	Kety and Schmidt[138]	↓	↓	↓	↓	↓
C. O_2 breathing (normal arterial Pco_2)	Kety and Schmidt[139] Lambertsen et al.[156]	↓	↓	↓	↑	↑
D. O_2 breathing (fixed arterial Pco_2)	Lambertsen[149]	↔	↔	↔	↑	↑
E. $NaHCO_3$ infusion (fixed arterial Pco_2)	Lambertsen et al.[154]	↔	↔	↓	↔	↓
F. Low inspired Po_2 (normal arterial Pco_2)	Kety and Schmidt[139]	↑	↓	↓	↓	↓

In *A* the administration of CO_2 increases the rate of cerebral blood flow with concurrent increases in the Pco_2 and [H+] of both arterial and central venous blood. Conversely in *B*, the reduction of brain blood flow by overbreathing is accompanied by lowered Pco_2 and [H+] in both arterial and venous blood. When, as in these two conditions, all the indicated changes occur in the same direction, it is not possible to ascertain whether blood flow is more subject to deviations in Pco_2 or in [H+]. Nor can it be determined whether changes in composition on the arterial side of the circulation have more influence on brain circulation those those on the venous (tissue?) side.

C to *F* represent a form of chemical dissection in an effort to localize the *site* and *nature* of the powerful effects of CO_2 and O_2 on brain blood flow. *C* shows that when high inspired O_2 pressure reduces brain blood flow, the vasoconstriction occurs in the presence of arterial hypocapnia and despite a central hypercapnia.

In *D*, in which artificial fixation of arterial acid-base composition is employed, O_2 administration no longer causes reduction in brain blood flow. *C* and *D* suggest that the cerebral vasoconstriction of O_2 breathing is an indirect result of the slight hyperventilation (↓Pco_2, ↓[H+]) normally produced by O_2 breathing rather than being due to a direct effect of O_2 on smooth muscle cells. In *D* an elevation of central Pco_2 and [H+] results from interference with hemoglobin reduction, but despite this form of metabolite accumulation, blood flow is not affected. This indicates that the effects of CO_2 on cerebral blood flow are predominantly exerted on the arterial side of the brain circulation.

In *E*, in which the [H+] of blood was changed while both arterial and central Pco_2 were maintained constant, brain blood flow was unaffected. On the basis of these and other findings,[49,140,158] it appears that the powerful effects of CO_2 on brain vessels are *not* mediated by way of the influence of Pco_2 on the [H+] of blood. Brain vascular resistance is therefore responsive to changes in arterial Pco_2, but it cannot be distinguished whether this is an effect of CO_2 itself or (because of the free diffusibility of this metabolite) the result of acid-base changes within the smooth muscle cells.[154]

mal subjects by administration of low O_2 tensions can be overcome by excessive hyperventilation or exaggerated by CO_2 administration (Fig. 71-29). When severe hypocapnia occurs during hypoxia, the result is cerebral vasoconstriction and extreme reduction in brain O_2 tension. Thus unconsciousness can result from excessive ventilation at a low inspired PO_2 that would have sustained consciousness in normocapnia.[149]

Oxygen. When normal men inhale O_2 at sea level or at the increased ambient pressures encountered in diving or hyperoxygenation therapy, there results a distinct constriction of brain vessels[139,156] (Table 71-2). This effect, indicating an increase in smooth muscle tone, has been demonstrated in isolated smooth muscle from a variety of organs.[238] It is usually considered as being due to a direct action of high PO_2, opposite to the dilating effect of hypoxia.[159,239] However, the studies summarized in Table 71-2 indicate that at normal levels of PCO_2 such an effect of O_2 on cerebral vessels is largely an indirect action re-

lated to the O_2-induced respiratory stimulation rather than an effect of O_2 directly on smooth muscle cells. This has considerable pertinence to concepts of central respiratory control.

The slight respiratory stimulation associated with short exposures to high inspired PO_2 appears to be caused by a sequence of related events beginning with the increase in the amount of physically dissolved O_2 in blood. This results in diminished reduction of hemoglobin, decreased availability of base for CO_2 transport, and increased central PCO_2 and pH* (Figs. 71-27 and 71-28). Probably this central acidosis is responsible for the respiratory stimulation that lowers arterial $[H^+]$.[147,157]

When this slight arterial hypocapnia is prevented by use of devices for artificially controlling alveolar PCO_2, no cerebral vasoconstrictor effect of O_2 is seen.[147,148] For this reason the cere-

---

*See references 64, 147, 156, and 157.

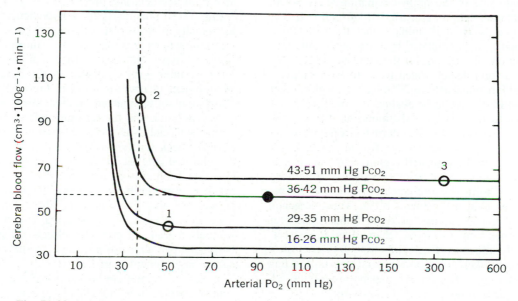

Fig. 71-29. Relationships among arterial PO_2, arterial PCO_2, and rate of brain blood flow (based on 56 measurements in seven normal men). ● represents normal blood flow and arterial gas tensions for air breathing at sea level. Above about 50 mm Hg arterial PO_2, O_2 tension appears to exert no important effect on blood flow. Vasodilator effect of low arterial PO_2 becomes increasingly prominent as arterial PO_2 falls below 50 mm Hg. By varying arterial PCO_2, wide fluctuations in brain blood flow can be produced both in absence and in presence of anoxemia; actually hypocapnia is capable of reversing and hypercapnia of exaggerating cerebral vasodilatation normally associated with anoxemia. Illustration shows that particular rate of cerebral blood flow can be obtained over wide range of variation of arterial PO_2 and PCO_2.

Numbered open circles indicate conditions to be expected in, *1*, hyperventilation of altitude hypoxia; *2*, hypoxia and hypercapnia of insufficient pulmonary ventilation; and *3*, administration of O_2 to patient with pulmonary insufficiency. (Modified from Lambertsen.[147])

bral vasomotor actions of O_2 excess and O_2 lack do not appear to be part of a smoothly continuous physiologic effect of change in P_{O_2}. More likely, high O_2 tensions produce little or no effect on adult vessels that are not hypoxic initially, and hypoxic vasodilatation represents not an active control function, but a pathologic failure of the normal mechanisms for maintenance of vascular tone.[87,147,148] This possibility must be examined in the light of clear demonstration that O_2 administration does cause constriction of the ductus arteriosus in the newborn lamb.[73]

O_2 administration aids respiratory and brain circulatory studies by making it possible to produce central *hypercapnia* concurrently with arterial *hypocapnia*.[147,148] The situation resembles that to be expected if the rate of metabolic CO_2 production by the brain cells were grossly increased. The occurrence of respiratory stimulation in this situation suggests that the tiny regions of respiratory sensor function experience, at least qualitatively, the same acid-base changes measured across the whole brain.

Since hypercapnia limited to the venous or tissue side of the brain circulation leads not to dilatation but to constriction of brain vessels (Table 71-2), it is unlikely that the local brain blood flow is regulated by local metabolite production. To account for the increases in regional circulation demonstrated to occur with exaggerated functional activity,[239] metabolic factors other than P_{CO_2} have to be evaluated. One such metabolic factor is a localized diminution of P_{O_2}. However, even a fall in P_{O_2} during increased metabolism is an unlikely basis for the dilatation of vessels in the activated region, since the P_{O_2} of arterial blood must be reduced nearly 50 mm Hg below normal before cerebral vasodilatation becomes evident[149] (Fig. 71-29). Thus, although local response to increased functional activity (as distinct from metabolic change) is almost certainly a feature of cerebral circulatory control, it is more likely that neurohumoral transmitters link flow to *activity* than that metabolic products adjust flow to the level of *metabolism*.

Blood pressure. The hydrostatic pressure gradient across the vascular bed, along with the mean diameter of these vessels and the viscosity of the blood, determines the volume rate of blood flow through each region of the brain, including the respiratory centers. In many normal physiologic situations, changes in blood pressure are accompanied by changes in the P_{CO_2} and P_{O_2} of the blood. The blood flow through the control centers will therefore be governed by the result of several influences.

INTERACTIONS OF RESPIRATORY AND BRAIN CIRCULATORY CONTROL

Clearly the arterial CO_2 tension provides a powerful link of whole-body oxidative metabolism to the basic mechanisms of respiratory control and the regulation of the brain circulation. Because the medullary vessels in the cat appear to respond to CO_2 in the same qualitative manner as vessels in other parts of the brain,[50,239] it is quite probable that a rise in arterial P_{CO_2} will increase and a fall in arterial P_{CO_2} will decrease the flow of blood through regions of the central respiratory sensors.

In considering the stable state of respiration, it must be recognized that the interacting relationships among respiration, blood P_{CO_2}, and local brain circulation under normal circumstances of air breathing are different from those induced by CO_2 inhalation (p. 1776 and 1816). Normally at rest a trend toward *hypo*ventilation will *increase* arterial P_{CO_2}. Conversely a slight degree of *hyper*ventilation will *decrease* arterial P_{O_2}. The normal situation is one of self-limitation rather than reinforcement and drive.

Fig. 71-30 illustrates, for a hypothetical subject breathing air at rest or in exercise, some of the interactions responsible for limiting the degree of hyperventilation or hypoventilation and for preventing gross changes in the acid-base environment of chemosensors in the CNS. The illustration shows that any reduction of pulmonary ventilation and the resulting arterial hypercapnia will lead to an increased central P_{CO_2} and tend to check the fall in respiration. However, the elevated arterial P_{CO_2}, through its dilating effect on brain vessels, increases the rate of CO_2 removal from the brain, thereby tending to decrease the arteriovenous P_{CO_2} difference and lower central P_{CO_2}. Thus the initial reduction of ventilation leads to opposing effects on central P_{CO_2} and pH, and gross change in the central neural environment is prevented.

At rest the primary event in normal respiratory control appears to be not change in central stimulus level but change in neuronal responsiveness to the sum of many stimuli. Since the level of arterial P_{CO_2} is determined largely *by* respiration, the influence of arterial CO_2 tension in regulating brain circulation would appear to be a consequence of the activity of the respiratory control mechanisms rather than a primary regulatory mechanism in its own right.

In the sequence just proposed, illustrated by Fig. 71-30, the function of brain tissue (and central chemosensor?) oxygenation does not appear

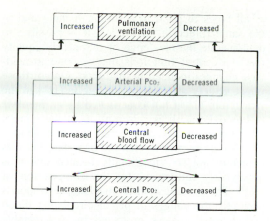

Fig. 71-30. Interactions of ventilation, blood gas levels, central blood flow, and central acid-base environment. This oversimplified diagram shows some of interlocking, self-stabilizing loops involved in central respiratory control. Change can be initiated at any point, resulting in successive readjustment of other factors as new level of dynamic equilibrium is attained. Examples of initial changes include the following: (1) increase in whole-body metabolism at rest may first increase arterial P_{CO_2}; (2) fall in blood pressure may first lower central blood flow; (3) sudden emotional distubance may first increase pulmonary ventilation; (4) use of drugs may stimulate central metabolism and increase central P_{CO_2}. In each case, pattern of related events must be readjusted to new equilibrium, in which the inverse relationship of $[H^+]$ and ventilation is again established.

as an actively regulated process. Except for any tonic chemoreflex influence of O_2 at normal arterial P_{O_2}, oxygenation during rest (and perhaps during exercise) at sea level may indeed be incidental to the more basic respiratory functions of CO_2 elimination and maintenance of central acid-base homeostasis. When inspired P_{O_2} is diminished, as by ascent to regions of low barometric pressure, chemoreflex respiratory adjustment to the requirements for oxygenation supervenes, but considerable reduction of arterial P_{O_2} is required before brain vessels become dilated by O_2 lack[149] and thereby begin to limit the fall of central O_2 tension to aid in sustaining central neural function.

BREATH-HOLDING

The voluntary suppression of respiratory function in the special circumstance of breath-holding becomes of direct practical interest in modern skin diving, as it has been for many centuries in pearl diving and in sea plant harvesting.[215] Analysis and now numerous studies have elaborated

the many interrelated factors that determine maximum duration of voluntary apnea.[81,121,199,215] In simplest terms, it can be considered that each of the chemical stimuli, acting increasingly on the chemical sensor–drive components of peripheral and central respiratory control mechanisms, contributes to a composite respiratory stimulus that ultimately cannot be resisted. At the "breaking point" following breath-holding of air at sea level, arterial P_{O_2} is subnormal while P_{CO_2} and $[H^+]$ are elevated. This means that the true degree of suppression at that stage is large.

Breath-holding can be extended by beginning with a full inspiration, since a larger volume of alveolar gas exists to damp the changes in arterial P_{CO_2}, P_{O_2}, and $[H^+]$ and since the stretch of lung tissue at large lung volumes also stimulates the inhibitoinspiratory reflex from the lungs themselves.

Breath-holding can be extended grossly by prior hyperventilation to reduce blood and tissue P_{CO_2} and $[H^+]$; the time required for metabolic CO_2 production to restore the acid-base state to normal is thus added to what otherwise would have been the breath-holding duration. The hazard in such an approach is that, in the absence of the powerful respiratory drive related to P_{CO_2} and $[H^+]$, the central suppressant influences of progressive hypoxemia can obliterate judgment, competence, and consciousness, causing death in apnea. This form of extension of breath-holding has a peculiar increase in hazard when applied to breath-holding underwater, where confusion or exhaustion can lead to drowning on return to respiration. In water the occurrence of the immersion reflex[78] may contribute to fatality due to excessively long voluntary apnea.

To overcome the inevitable hazard of hypoxia following hyperventilation with air, pure O_2 has been administered to subjects in breath-holding studies. Hyperventilation with pure O_2, minimizing each of the chemical drives, has allowed breath-holding at rest for as long as 15 to 20 min.[199] Termination results from progressive decrease in lung volume and rise in arterial P_{CO_2} $[H^+]$.

SUMMARY

Pulmonary ventilation at rest is precisely adjusted to the level of metabolic activity by interlocked chemical and physical systems; the result is a maintenance of central and arterial P_{CO_2} and $[H^+]$ at levels above atmospheric, which promotes spontaneous firing of specialized respiratory neurons.

It is necessary to conceive a system, active and

unstable, in which electrical impulse generation occurs to hold [H⁺], or proton concentration, *below* a desired level. In the centers, spontaneously oscillating (perhaps mutually exciting) neuronal discharge occurs in the presence of, and possible *maintained by,* a [H⁺], which thus becomes self-limiting and stable. In the peripheral chemoreceptors, spontaneous steady (perhaps random) discharge occurs in the presence of, and possibly maintained by, the level of [H⁺] experienced by the sensor site. The massive firing of these peripheral chemoreceptors as O_2 pressure falls to zero requires consideration that this is *not activation,* but *failure* of membrane function in the face of inadequate oxygenation and that the quieting of chemoreceptor discharge in the presence of O_2 is related to restoration of normal membrane stability. The coincidence of cranial nerve innervation of these "peripheral" sensors has led to the query as to whether, masked in the complexity of the central chemosensor and control system, there is a similar O_2-related sensor mechanism.[103]

The adjustment of ventilation is most probably accomplished by reactivity of chemosensitive respiratory neurons to products of their own metabolism and those of the general metabolism of the body as they spill past the lungs and kidneys to affect neural structures.

Of these metabolic waste products, CO_2 plays a dominant role, since (1) it can exist as a freely diffusible, uncharged, and presumably inert molecule in solution, (2) it can evolve as a gas in the lungs for ready elimination from the body, and (3) it can react with water in proportion to its partial pressure, thus affecting the level of [H⁺] in blood and other body fluids that influence the environment of respiratory neurons. Change in CO_2 pressure can activate peripheral chemoreceptor cells, but whether this is accomplished entirely through an alteration of intracellular [H⁺] is not known. Change in CO_2 pressure can somehow also activate the central respiratory neurons.

Although the effects of CO_2 have not been separated from actions of H⁺, change in [H⁺] has been shown to produce respiratory effects in the absence of Pco_2 alteration. Thus respiration is stimulated by increased [H⁺] in blood independent of change in blood or central Pco_2. The site and mechanism of this effect is not certain, although its magnitude is sufficient to account for almost half the overall respiratory response to breathing CO_2.

When acid-base changes are brought about entirely by alterations of Pco_2, the chemosensors of all types respond in a similar manner and in the same direction. However, it is possible to demonstrate the existence of two nearly equally reactive central functions that respond to acid-base changes induced in the blood. One of these is not acutely affected by change in [H⁺] in the circulating blood but is activated by an increase in blood Pco_2. It is this portion of the control mechanism that is considered to lie beyond a blood-brain barrier to charged particles, but it is evidently not so superficially located that it is appreciably influenced by the composition of CSF. The second central mechanism responds promptly to changes in pH of circulating blood and therefore is not protected by a detectable blood-brain barrier. It is this component that should respond to acidemia in exercise and metabolic acidosis.

Diminished arterial Po_2 results in a chemoreflex respiratory stimulation that is prominently exaggerated by moderate increases in Pco_2 and even by preventing the arterial hypocapnia that is normally an accompaniment of altitude hypoxia.

The basis for peripheral chemoreceptor stimulation by lowered Po_2 is uncertain, but it most likely is an indirect effect resulting from anaerobiosis associated with O_2 lack. Although it has become conventional to consider decreased Po_2 as a "stimulus" to the chemoreceptor cells, it is not reasonable that a lack of O_2 molecules can directly stimulate. Most probably the chemoreceptor activation by decreased Po_2 will prove to have a metabolic basis and a common path with chemoreceptor stimulation by hypercapnia and acidemia. Certainly the end result of a particular degree of chemoreceptor activation, regardless of the exciting cause, should be the production of the same degree of respiratory and circulatory stimulation.

The manner by which peripheral or central chemoreceptor impulses influence the respiratory centers also requires further elaboration. The impulses arriving from the arterial chemoreceptors via fibers in the vagus and glossopharyngeal nerves can exert extremely powerful influences centrally. Through these reflex mechanisms, respiratory centers that are so depressed (as by hypoxia or narcotics) that they no longer respond to a general elevation of CO_2 and [H⁺] do respond to a lowering of the Po_2 in the arterial blood.

The degree of contribution of peripheral chemoreceptors to normal respiratory control at sea level, whether by O_2 tension or CO_2 tension, appears to be small; however, as the most rapidly responding of all the control components, the chemoreceptors may nevertheless provide a fine and rapid adjustment of breathing.

Regardless of the details of its activation, the overall respiratory control mechanism appears to include both *a dual peripheral and a dual central sensing system* capable of detecting changes both in the environment of the central neurons and in the arterial blood that has just been exposed to the external environment through the pulmonary gas exchange process. Either the reflex or central component of this sensing system is capable of driving respiration when the other component is no longer capable of independent function. The centers themselves have dual character with respect to accessibility to acid-base changes in the blood. The peripheral sensing mechanism at least is able to respond to changes in more than one chemical factor. By this series of duplications the total system for respiratory control is capable of maintaining not only the acid-base environment most suitable for central function but also (and perhaps incidentally) the oxygenation of the arterial blood supply to the body tissues.

The many features of the respiratory control system are interlocked via the respiratory gases (in particular CO_2) with other physiological processes, including mechanisms for the regulation of general and local circulation, metabolism, electrolyte and water balance, endocrine secretion, body temperature, and other basic functions that together provide dynamic equilibrium states through sequences of self-stabilizing loops. Deviations produced within one of these systems result in adjustments of each of the others until a new level of equilibrium is reached. As just emphasized, this complex interplay among the many factors concerned with respiratory control appears to have as its primary regulating mechanism the self-oriented response of tiny sensitive masses of central and chemoreflex respiratory neurons to changes in the chemical state of their own local environment. The rest of the body has found ways to be served in the process.

REFERENCES
General reviews

1. Astrom, A.: On the action of combined carbon dioxide excess and oxygen deficiency in the regulation of breathing, Acta Physiol. Scand. **27**:suppl. 98, 1952.
2. Asmussen, E., and Nielsen, M.: Experiments on nervous factors controlling respiration and circulation during exercise employing blocking of the blood flow, Acta Physiol. Scand. **60**:103, 1964.
3. Bainton, C. R.: Canine ventilation after acid-base infusions, exercise, and carotid body denervation, J. Appl. Physiol. **44**(1):28, 1978.
4. Bisgard, G. E., et al.: Cerebrospinal fluid acid-base balance during muscular exercise, J. Appl. Physiol. **45**(1):94, 1978.
5. Bjurstedt, A. G. H.: Interaction of centrogenic and chemoreflex control of breathing during oxygen deficiency at rest, Acta Physiol. Scand. **12**:suppl. 38, 1946.
6. Borison, H. L., Gonsalves, S. F., Montgomery, S. P., and McCarthy, L. E.: Dynamics of respiratory V_I response to isocapnic pH_a forcing in chemodenervated cats, J. Appl. Physiol. **45**(4):502, 1978.
7. Bradley, J. W., Von Euler, C., Marttila, I., and Roos, B.: Transient and steady state effects of CO_2 on mechanisms determining rate and depth of breathing, Acta Physiol. Scand. **92**:341, 1974.
8. Bulow, K.: Respiration and wakefulness in man, Acta Physiol. Scand. **59**:suppl. 209, 1963.
9. Comroe, J. H., Jr.: The hyperpnea of muscular exercise, Physiol. Rev. **24**:319, 1944.
10. Comroe, J. H., Jr.: The peripheral chemoreceptors. In Fenn, W. O., and Rahn, H., editors: Handbook of physiology. Respiration section, Washington, D.C., 1964, American Physiological Society, vol. 1.
11. Comroe, J. H., Jr., Dripps, R. D., Dumke, P. R., and Deming, M.: Oxygen toxicity. The effect of inhalation of high concentrations of oxygen for twenty-four hours on normal men at sea level and at a simulated altitude of 18,000 feet, J.A.M.A. **128**:710, 1945.
12. Comroe, J. H., Jr., et al.: The lung, ed. 2, Chicago, 1975, Year Book Medical Publishers, Inc.
13. Cunningham, D. J. C., and Lloyd, B. B., editors: The regulation of human respiration, Oxford, 1963, Blackwell Scientific Publications, Ltd.
14. Defares, J. G., Derksen, H. E., and Duyff, J. W.: Cerebral blood flow in the regulation of respiration, Acta Physiol. Pharmacol. Neerl. **9**:327, 1960.
15. Dejours, P.: La regulation de la ventilation au cours de l'exercise musculaire chez l'homme, J. Physiol. (Paris) **51**:163, 1959.
16. Dejours, P.: Chemoreflexes in breathing, Physiol. Rev. **42**:335, 1962.
17. Dempsey, J. A.: CO_2 response: stimulus definitions and limitations, Chest **70**:114, 1976.
18. Fencl, V., Miller, T. B., and Pappenheimer, J. R.: Studies on the respiratory response to disturbances of acid-balance base, with deductions concerning the ionic composition of cerebral interstitial fluid, Am. J. Physiol. **210**:459, 1966.
19. Fenn, W. O., and Rahn, H., editors: Handbook of physiology. Respiration section, Washington, D.C., 1964, 1965, American Physiological Society, vols. 1 and 2.
20. Gonzalez, F., Jr., Fordyce, W. E., and Grodins, F. S.: Mechanism of respiratory responses to intravenous $NaHCO_3$, HCl, and KCN, J. Appl. Physiol. **43**:1075, 1977.
21. Gray, J. S.: Pulmonary ventilation and its physiological regulation, American Lecture Series 63, Springfield, Ill., 1950, Charles C Thomas, Publisher.
22. Greco, E. C., Jr., et al.: Respiratory responses to intravenous and intrapulmonary CO_2 in awake dogs, J. Appl. Physiol. **45**(1):109, 1978.
23. Grodins, F. S.: Analysis of factors concerned in the regulation of breathing in exercise, Physiol. Rev. **30**:220, 1950.
24. Grodins, F. S.: Regulation of pulmonary ventilation, Physiologist **7**:319, 1964.
25. Grodins, F. S., et al.: Respiratory responses to CO_2 inhalation. A theoretical study of a nonlinear biological regulator, J. Appl. Physiol. **7**:283, 1954.

26. Haldane, J. S., and Priestly, J. S.: Respiration, London, 1935, Oxford University Press, Ltd.

27. Hesser, C. M.: Central and chemoreflex components in the respiratory activity during acid-base displacements in the blood, Acta Physiol. Scand. **18**:suppl. 64, 1949.

28. Heymans, C., and Neil, E.: Reflexogenic areas of the cardiovascular system, Boston, 1958, Little, Brown & Co.

29. Brooks, C. McC., Kao, F. F., and Lloyd, B. B., editors: The cerebrospinal fluid and the regulation of ventilation, Oxford, 1965, Blackwell Scientific Publications, Ltd.

30. Honda, Y., and Natsui, T.: Effect of sleep on ventilatory response to CO_2 in severe hypoxia, Respir. Physiol. **3**:220, 1967.

31. Horvath, S. M., Dill, D. B., and Corwin, W.: Effects on man of severe oxygen lack, Am. J. Physiol. **138**:659, 1943.

32. Joels, N., and Neil, E.: Carotid chemoreceptor response to high carbon monoxide tension, J. Physiol. **156**:5P, 1961.

33. Kellogg, R. H.: Central chemical regulation of respiration. In Fenn, W. O., and Rahn, H., editors: Handbook of physiology. Respiration section, Washington, D.C., 1964, American Physiological Society, vol. 1.

34. Lahiri, S., and Delaney, R. G.: Stimulus interaction in the responses of carotid body chemoreceptor single afferent fibers, Respir. Physiol. **24**:249, 1975.

35. Lambertsen, C. J.: Drugs and respiration, Annu. Rev. Pharmacol. **6**:327, 1966.

36. Leusen, I.: Aspects of the chemical control of respiration, Verh. K. Vlaam. Acad. Geneeskd. Belg. **24**:576, 1962.

37. Levin, E., Arieff, A., and Kleeman, C. R.: Evidence of different compartments in the brain for extracellular markers, Am. J. Physiol. **221**:1319, 1971.

38. Linton, R., and Cameron, I.: The ventilatory response to intravenous CO_2 loading, Chest **73**:273, 1978.

39. Loeschcke, H. H.: Intracranielle Chemoreceptoren mit Wirkung auf die Atmung, Helvet. Physiol. Pharmacol. Acta **15**:C25, 1957.

40. Martin, B., et al.: Chemical drives to breathe as determinants of exercise ventilation, Chest **73**:283, 1978.

41. Matell, G.: Time-courses of changes in ventilation and arterial gas tensions in man induced by moderate exercise, Acta Physiol. Scand. **58**:suppl. 206, 1963.

42. McDonald, D. M., and Mitchell, R.: The innervation of glomus cells, ganglion cells and blood vessels in the rat carotid body: a quantitative ultrastructural analysis, J. Neurocytol. **4**:177, 1975.

43. Messeter, K., and Siesjo, B. K.: Regulation of the CSF pH in acute and sustained respiratory acidosis, Acta Physiol. Scand. **83**:21, 1971.

44. Monroe, C. B., and Kazemi, H.: Effect of changes in plasma bicarbonate level on CSF bicarbonate in respiratory acidosis, Respir. Physiol. **17**:386, 1973.

45. Olendorf, W. H., and Davson, H.: Brain extracellular space and the sink action of cerebrospinal fluid, Arch. Neurol. **17**:196, 1967.

46. Pelligrino, D. A., and Dempsey, J. A.: Dependence of CSF on plasma bicarbonate during hypocapnia and hypoxemic hypocapnia, Respir. Physiol. **26**:11, 1976.

47. Rahn, H.: Why are pH of 7.4 and PCO2 of 40 normal values for man? (Pourquoi un pH de 7,4 and une Pco2 de 40 mm Hg sont-ils des valeurs normales chez l'homme?), Bull. Europ. Physiopathol. Respir. **12**:5, 1976.

48. Santiago, T. V., and Edelman, N. H.: Mechanism of ventilatory response to carbon monoxide, J. Clin. Invest. **57**:977, 1976.

49. Schieve, J. F., and Wilson, W. P.: The changes in cerebral vascular resistance of man in experimental alkalosis and acidosis, J. Clin. Invest. **32**:33, 1953.

50. Schmidt, C. F., and Pierson, J. C.: The intrinsic regulation of the blood vessels of the medulla oblongata, Am. J. Physiol. **108**:241, 1934.

51. Severinghaus, J. W., Mitchell, R. A., Richardson, B. W., and Singer, M. M.: Respiratory control at high altitude suggesting active transport regulation of CSF pH, J. Appl. Physiol. **18**:1155, 1963.

52. Siesjo, B. K.: The regulation of cerebrospinal fluid pH, Kidney Int. **1**:360, 1972.

53. Stremel, R. W., et al.: Control of ventilation during intravenous CO_2 loading in the awake dog, J. Appl. Physiol. **44**(2): 311, 1978.

54. Swanson, G. D.: The exercise hyperpnea dilemma, Chest **73**:277, 1978.

55. Symposium on carbon dioxide and man, Anesthesiology **21**:585, 1960.

56. Walter, F.: Untersuchungen uber die Wirkung der Sauren auf den thierischen Organismus, Arch. Exp. Pathol. Pharmakol. **7**:148, 1877.

57. Whipp, B. J.: Tenets of the exercise hyperpnea and their degree of corroboration, Chest **73**:274, 1978.

58. Winterstein, H.: Die chemische Steuerung der Atmung, Ergeb. Physiol. **48**:328, 1955.

59. Winterstein, H.: The actions of substances introduced into the cerebrospinal fluid and the problem of intracranial chemoreceptors, Pharmacol. Rev. **13**:71, 1961.

Original papers

60. Alexander, S. C., Gelfand, R., and Lambertsen, C. J.: The pK′ of carbonic acid in cerebrospinal fluid, J. Biol. Chem. **236**:592, 1961.

61. Alexander, S. C., Workman, R. D., and Lambertsen, C. J.: Hyperthermia, lactic acid infusion, and the composition of arterial blood and cerebrospinal fluid, Am. J. Physiol. **202**:1049, 1962.

62. Arieff, A. I., Kerian, A., Massry, S. G., and De Lima, J.: Intracellular pH of brain: alterations in acute respiratory acidosis and alkalosis, Am. J. Physiol. **230**:804, 1976.

63. Bartels, H., and Witzleb, E.: Der Einfluss des Ateriellen CO2-Druckes auf die Chemoreceptorischen Aktionspotentiale im Carotissinusnerven, Arch. Gesamte Physiol. **262**:466, 1956.

64. Behnke, A. R., Johnson, F. S., Poppen, J. R., and Motley, E. P.: The effect of oxygen on man at pressures from 1 to 4 atmospheres, Am. J. Physiol. **110**:565, 1935.

65. Berger, A. J., Krasny, J. A., and Dutton, R. E.: Respiratory recovery from CO_2 breathing in intact and chemodenervated awake dogs, J. Appl. Physiol. **35**:35, 1973.

66. Berger, A. J., Mitchell, R. A., and Severinghaus, J. W.: Regulation of respiration (parts I, II, and III), N. Engl. J. Med. **297**:92, 1977.

67. Berkenbosch, A., de Goede, J., Olievier, C. N., and Quanjer, P. H.: Influence of the CSF bicarbonate concentration on the ventilatory response to CO_2 in relation to the location of the central chemoreceptors, Respir. Physiol. **35**:215, 1978.

68. Berl, S., Takagaki, G., Clarke, D. D., and Waelsch, H.: Carbon dioxide fixation in the brain, J. Biol. Chem. **237**:2570, 1962.

69. Berndt, J., Berger, W., and Mückenhoff, K.: Studies on the central chemosensitive mechanisms of respiration. II. Control of respiration by the extracellular pH in medullary tissue, Arch. Gesamte Physiol. **332**:146, 1972.

70. Biscoe, T. J.: Carotid body: structure and function, Physiol. Rev. **51**:437, 1971.

71. Bisgard, G. E., Forster, H. V., Orr, J. A., Buss, D. D., Rawlings, C. A., and Rasmussen, B.: Hypoventilation in ponies after carotid body denervation, J. Appl. Physiol. **40**:184, 1976.

72. Bleich, H. L., Berkman, P. M., and Schwartz, W. B.: The response of cerebrospinal fluid composition to sustained hypercapnia, J. Clin. Invest. **43**:11, 1964.

73. Born, G. V. R., Dawes, G. S., Mott, J. C., and Rennick, B. R.: The constriction of the ductus arteriosus caused by oxygen and by asphyxia in newborn lambs, J. Physiol. **132**:304, 1956.

74. Bradley, R. D., and Semple, S. J. G.: A comparison of certain acid-base characteristics of arterial blood, jugular venous blood and cerebrospinal fluid in man, and the effect on them of some acute and chronic acid-base disturbances, J. Physiol. **160**:381, 1962.

75. Clark, J. M., Sinclair, R. D., and Welch, B. E.: Rate of acclimatization to chronic hypercapnia in man. In Lambertsen, C. J., editor: Underwater physiology, New York, 1971, Academic Press, Inc.

76. Comroe, J. H., Jr.: The effects of direct chemical and electrical stimulation of the respiratory center in the cat, Am. J. Physiol. **139**:490, 1943.

77. Cormack, R. S., Cunningham, D. J. C., and Gee, J. B. L.: The effect of carbon dioxide on the respiratory response to want of oxygen in man, Q. J. Exp. Physiol. **42**:303, 1957.

78. Daly, M. deB., and James, J. E. A.: Role of the arterial chemoreceptors in the control of the cardiovascular responses to breath-hold diving. In Purves, M. J., editor: The peripheral arterial chemoreceptors, London, 1975, Cambridge University Press.

79. Daly, M. deB., Lambertsen, C. J., and Schweitzer, A.: Observations of the volume of blood flow and oxygen utilization of the carotid body in the cat, J. Physiol. **125**:67, 1954.

80. Davenport, H. W., Brewer, G., Chambers, A. H., and Goldschmidt, S.: The respiratory responses to anoxemia of unanesthetized dogs with chronically denervated aortic and carotid chemoreceptors and their causes, Am. J. Physiol. **148**:406, 1947.

81. Davidson, J. T., et al.: Role of the carotid bodies in breath-holding, N. Engl. J. Med. **290**:819, 1974.

82. Davson, H.: Physiology of the ocular and cerebrospinal fluids, Boston, 1956, Little, Brown & Co.

83. Defares, J. G.: Principles of feedback control and their application to the respiratory control system. In Fenn, W. O., and Rahn, H., editors: Handbook of physiology. Respiration section, Washington, D.C., 1964, American Physiological Society, vol. 1.

84. Dejours, P., Labrousse, Y., Raynaud, J., and Teillac, A.: Stimulus oxygène chémoréflexe de la ventilation à basse altitude (50m), chez l'homme. I. Au repos, J. Physiol. (Paris) **49**:115, 1957.

85. Dejours, P., et al.: Stimulus oxygène de la ventilation au repos et au cours de l'exercise musculaire, à basse altitude (50m), chez l'homme, Rev. Fr. Étud. Clin. Biol. **3**:105, 1958.

86. Dempsey, J. A., et al.: Is brain ECF [H+] an important drive to breathe in man? Chest **73**:251, 1978.

87. Detar, R., and Bohr, D. F.: Oxygen and vascular smooth muscle contraction, Am. J. Physiol. **214**:241, 1968.

88. Domizi, D. B., Perkins, J. F., Jr., and Byrne, J. S.: Ventilatory response to fixed acid evaluated by "Iso-P_{CO_2}" technique, J. Appl. Physiol. **14**:557, 1959.

89. Downes, J. J., and Lambertsen, C. J.: Dynamic characteristics of ventilatory depression in man on abrupt administration of O_2 at 1.0 atm., J. Appl. Physiol. **21**:447, 1966.

90. Dripps, R. D., and Comroe, J. H., Jr.: The respiratory and circulatory response of normal man to inhalation of 7.6 and 10.4 per cent CO_2 with a comparison of the maximal ventilation produced by severe muscular exercise, inhalation of CO_2 and maximal voluntary hyperventilation, Am. J. Physiol. **149**:43, 1947.

91. Dripps, R. D., and Comroe, J. H., Jr.: The effect of the inhalation of high and low oxygen concentrations on respiration, pulse rate, ballistocardiogram and arterial oxygen saturation (oximeter) of normal individuals, Am. J. Physiol. **149**:277, 1947.

92. Edelman, N. H., Chapman, R. W., and Santiago, T. V.: Role of brain blood flow in the control of breathing: effects of flow limitations, Chest **73**:261, 1978.

93. Edelman, N. H., Epstein, P. E., Lahiri, S., and Cherniack, N. S.: Ventilatory responses to transient hypoxia and hypercapnia in man, Respir. Physiol. **17**:302, 1973.

94. Eldridge, F. L.: Central nervous system and chemoreceptor factors in control of breathing, Chest **73**:256, 1978.

95. Euler, U. S., von, Liljestrand, G., and Zotterman, Y.: The excitation mechanism of the chemoreceptors of the carotid body, Acta Physiol. Scand. **83**:132, 1939.

96. Eyzaguirre, C., and Koyano, H.: Effects of hypoxia, hypercapnia, and pH on the chemoreceptor activity of the carotid body in vitro, J. Physiol. **178**:385, 1965.

97. Eyzaguirre, C., and Lewin, J.: Chemoreceptor activity of the carotid body of the cat, J. Physiol. **159**:222, 1961.

98. Fencl, V., Heisey, S. R., Held, D., and Pappenheimer, J. R.: Role of cerebrospinal fluid in the respiratory response to CO_2 as studied in unanesthetized goats. In Brooks, C. M., Kao, F. F., and Lloyd, B. B., editors: The cerebrospinal fluid and the regulation of ventilation, Oxford, 1965, Blackwell Scientific Publications, Ltd.

99. Fink, B. R.: The stimulant effect of wakefulness on respiration: clinical aspects, Br. J. Anaesth. **33**:97, 1961.

100. Fink, B. R., Ngai, S. H., and Hanks, E. C.: The central regulation of respiration during halothane anesthesia, Anesthesiology **23**:200, 1962.

101. Forster, H. V., Bisgard, G. E., Dempsey, J. A., and Orr, J. A.: Role of intracranial [H+] receptor in physiologic regulation of ventilation in ponies, Chest **73**:253, 1978.

102. Fuleihan, F. J. D., et al.: Transient responses to CO_2 breathing of human subjects awake and asleep, J. Appl. Physiol. **18**:239, 1963.

103. Gelfand, R., and Lambertsen, C. J.: Dynamic respiratory response to abrupt change of inspired CO_2 at normal and high P_{O_2}, J. Appl. Physiol. **35**:903, 1973.

104. Geppert, J., and Zuntz, N.: Über die Regulation der Atmung, Arch. Gesamte Physiol. **42**:189, 1888.

105. Gessell, R., Hansen, E. T., and Worzniak, J. J.: Humoral intermediation of nerve cell activation in the central nervous system, Am. J. Physiol. **138**:776, 1943.

106. Gesell, R., Lapides, J., and Levin, M.: The inter-

action of central and peripheral chemical control of breathing, Am. J. Physiol. **130:**155, 1940.

107. Gleichmann, U., et al.: Tissue P_{O_2} and P_{CO_2} of the cerebral cortex, related to blood gas tensions, Acta Physiol. Scand. **55:**127, 1962.

108. Goddard, D. R.: The biological role of carbon dioxide, Anesthesiology **21:**587, 1960.

109. Goodman, L. S., and Gilman, A.: The pharmacological basis of therapeutics, ed. 3, New York, 1965, Macmillan Publishing Co., Inc.

110. Gray, J. S.: The multiple factor theory of the control of respiratory ventilation, Science **103:**739, 1946.

111. Gray, B. A.: On the speed of the carotid chemoreceptor response in relation to the kinetics of CO_2 hydration, Respir. Physiol. **11:**235, 1971.

112. Grodins, F. S.: Control theory and biological systems, New York, 1963, Columbia University Press.

113. Grodins, F. S., Buell, J., and Bart, A. J.: Mathematical analysis and digital simulation of the respiratory control system, J. Appl. Physiol. **22:**260, 1967.

114. Guz, A.: Regulation of respiration in man, Annu. Rev. Physiol. **37:**303, 1975.

115. Hasan, F. M., and Kazemi, H.: Dual contribution theory of regulation of CSF HCO_3^- in respiratory acidosis, J. Appl. Physiol. **40:**559, 1976.

116. Hey, E. N., et al.: Effects of various respiratory stimuli on the depth and frequency of breathing in man, Respir. Physiol. **1:**193, 1966.

117. Heymans, C., and Bouckaert, J. J.: Sinus caroticus and respiratory reflexes. I. Cerebral blood flow and respiration, J. Physiol. **69:**254, 1930.

118. Heymans, C., Bouckaert, J. J., and Regniers, P.: Le sinus carotidien et la zone hemologue cardio-aortigue, Paris, 1933, Gaston Doin & Cie.

119. Heymans, C., and Bouckaert, J. J.: Les chemorecepteurs du sinus carotidien, Ergeb. Physiol. **41:**28, 1939.

120. Heymans, J. F., and Heymans, C.: Sur les modifications directes et sur la régulation réflexe de l'activité du centre respiratoire de la tête isolée du chien, Arch. Int. Pharmacodyn. Ther. **33:**272, 1927.

121. Hill, P. M.: Hyperventilation, breath-holding and alveolar oxygen tensions at the breaking point, Respir. Physiol. **19:**201, 1973.

122. Holton, P., and Wood, J. B.: The effects of bilateral removal of the carotid bodies and denervation of the carotid sinuses in two human subjects, J. Physiol. **181:**365, 1965.

123. Honda, Y., and Kreuzer, F.: P_{O_2}-ventilation response curve with normal pH and P_{CO_2} in the dog, J. Appl. Physiol. **21:**423, 1966.

124. Hornbein, T. F., and Roos, A.: Specificity of H ion concentration as a carotid chemoreceptor stimulus, J. Appl. Physiol. **18:**580, 1963.

125. Hornbein, T. F., Griffo, Z. J., and Roos, A.: Quantitation of chemoreceptor activity: interrelation of hypoxia and hypercapnia, J. Neurophysiol. **24:**561, 1961.

126. Ingvar, D. H., and Bülow, K. B.: Respiratory regulation in sleep, Ann. N.Y. Acad. Sci. **109:**870, 1963.

127. Jacobs, M. H.: Production of intracellular acidity by neutral and alkaline solutions containing carbon dioxide, Am. J. Physiol. **53:**457, 1920.

128. Joels, N., and Neil, E.: The excitation mechanism of the carotid body, Br. Med. Bull. **19:**21, 1963.

129. Jouvet, M.: Neurophysiology of the states of sleep, Physiol. Rev. **47:**117, 1967.

130. Katsaros, B.: Die Rolle der Chemoreceptoren des Carotisgebiets der Narkotisierten Katze für die Antwort der Atmung auf Isolierte Änderung der Wasserstof-

fionen-Konzentration und des CO_2-Drucks des Blutes, Arch. Gesamte Physiol. **282:**157, 1965.

131. Katsaros, B.: Der Effekt der Durchtrennung der Sinusnerven auf die Atmung der Narkotisierten Katze bei konstant gehaltenem arteriellen Druck und seine Abhängigkeit vom CO_2-Druck, Arch. Gesamte Physiol. **282:**179, 1965.

132. Katsaros, B., and Loeschcke, H. H.: Die Rolle der Chemoreceptoren des Carotis und Aortengebiets der Katze für die Antwort der Atmung auf isolierte Änderung der Wasserstoffionenkonzentration im Blut, Arch. Gesamte Physiol. **274:**21, 1961.

133. Katsaros, B., Edwards, A. W. T., Perkins, J. F., Jr., and Loeschcke, H. H.: Antwort der Ventilation auf Perfusion der vom Übrigen Kreislauf Isolierten Chemoreceptoren des Carotisgebiets mit Lösungen, in denen CO_2-Druck und Wasserstoffionen-Konzentration variiert wurden, Arch. Gesamte Physiol. **282:**200, 1965.

134. Katsaros, B., et al.: Wirkung der Bicarbonat-Alkalose auf die Lungenbelüftung beim Menschen Bestimmung der Teilwirkungen von pH and CO_2-Druck auf die Ventilation und Vergleich mit den Ergebnissen bei Acidose, Arch. Gesamte Physiol. **271:**732, 1960.

135. Kazemi, H., and Mithoefer, J. C.: CO_2 dissociation curve of dog brain, Am. J. Physiol. **205:**598, 1963.

136. Kellogg, R. H.: Oxygen and carbon dioxide in the regulation of respiration, Fed. Proc. **36:**1658, 1977.

137. Kellogg, R. H.: Acclimatization to carbon dioxide, Anesthesiology **21:**634, 1960.

138. Kety, S. S., and Schmidt, C. F.: The effects of active and passive hyperventilation on cerebral blood flow, cerebral oxygen consumption, cardiac output and blood pressure of normal young men, J. Clin. Invest. **25:**107, 1946.

139. Kety, S. S., and Schmidt, C. F.: The effects of altered arterial tensions of carbon dioxide and oxygen on cerebral blood flow and cerebral oxygen consumption of normal young men, J. Clin. Invest. **27:**484, 1948.

140. Kety, S. S., Polis, B. D., Nadler, C. S., and Schmidt, C. F.: The blood flow and oxygen consumption of the human brain in diabetic acidosis and coma, J. Clin. Invest. **27:**500, 1948.

141. Kim, J. K., and Carpenter, F. G.: Excitation of medullary neurons by chemical agents, Am. J. Physiol. **201:**1187, 1961.

142. Krogh, A.: The comparative physiology of respiratory mechanisms, Philadelphia, 1959, University of Pennsylvania Press.

143. Lahiri, S.: Introductory remarks: oxygen linked response of carotid chemoreceptors. In Reivich, M., Coburn, R., Lahiri, S., and Chance, B., editors: Tissue hypoxia and ischemia, New York, 1977, Plenum Press.

144. Lambertsen, C. J.: Carbon dioxide and respiration in acid-base homeostasis, Anesthesiology **21:**642, 1960.

145. Lambertsen, C. J.: Chemical control of respiration at rest. In Bard, P., editor: Medical physiology, ed. 11, St. Louis, 1961, The C. V. Mosby Co.

146. Lambertsen, C. J.: Factors in the stimulation of respiration by carbon dioxide. In Cunningham, D. J. C., and Lloyd, B. B., editors: The regulation of human respiration, Oxford, 1963, Blackwell Scientific Publications, Ltd.

147. Lambertsen, C. J.: Effects of hyperoxia on organs and their tissues. In Robin, E., editor: Extrapulmonary manifestations of respiratory disease. Lung biology in health and disease series, New York, 1978, Marcel Dekker, Inc., vol. 8.

148. Lambertsen, C. J.: Effects of oxygen at high partial

pressure. In Fenn, W. O., and Rahn, H., editors: Handbook of physiology. Respiration section, Washington, D.C., 1964, American Physiological Society, vol. 2.

149. Lambertsen, C. J.: Therapeutic gases: oxygen, carbon dioxide and helium. In DiPalma, J. R., editor: Drill's pharmacology in medicine, ed. 4, New York, 1971, McGraw-Hill Book Co.

150. Lambertsen, C. J., and Gelfand, R.: Separation of dynamic response characteristics of H^+-related and CO_2-related components of respiratory control, Fed. Proc. **24:**272, 1965.

151. Lambertsen, C. J., Gelfand, R., and Kemp, R. A.: Dynamic influences of CO_2 in respiratory control. In Brooks, C. M., Kao, F. F., and Lloyd, B. B., editors: The cerebrospinal fluid and the regulation of ventilation, Oxford, 1965, Blackwell Scientific Publications, Ltd.

152. Lambertsen, C. J., Bunce, P. L., Drabkin, D. L., and Schmidt, C. F.: Relationship of oxygen tension to hemoglobin oxygen saturation in the arterial blood of normal men, J. Appl. Physiol. **4:**873, 1952.

153. Lambertsen, C. J., Hall, P., Wollman, H., and Goodman, M. W.: Quantitative interactions of increased Po_2 and Pco_2 upon respiration in man, Ann. N.Y. Acad. Sci. **109:**731, 1963.

154. Lambertsen, C. J., Semple, S. J. G., Smyth, M. G., and Gelfand, R.: H^+ and Pco_2 as chemical factors in respiratory and cerebral circulatory control, J. Appl. Physiol. **16:**473, 1961.

155. Lambertsen, C. J., Wendel, H., Chiodi, H., and Owen, S. G.: Respiratory effects of .08 and .8 atmospheres of inspired Po_2 at a "constant" alveolar Pco_2 of 43 mm. Hg, Fed. Proc. **16:**76, 1957.

156. Lambertsen, C. J., et al.: Oxygen toxicity. Effects in man of oxygen inhalation at 1 and 3.5 atmospheres upon blood gas transport, cerebral circulation and cerebral metabolism, J. Appl. Physiol. **5:**471, 1953.

157. Lambertsen, C. J., et al.: Comparison of relationship of respiratory minute volume to Pco_2 and pH of arterial and internal jugular blood in normal man during hyperventilation produced by low concentrations of CO_2 at 1 atmosphere and by O_2 at 3.0 atmospheres, J. Appl. Physiol. **5:**803, 1953.

158. Lambertsen, C. J., et al.: Respiratory and cerebral circulatory control during exercise at .21 and 2.0 atmospheres inspired Po_2, J. Appl. Physiol. **14:**966, 1959.

159. Lassen, N. A.: Cerebral blood flow and oxygen consumption in man, Physiol. Rev. **39:**183, 1959.

160. Legallois, C. J. J.: Experiences sur le principe de la vie, Paris, 1812, D'Hautel.

161. Leitner, L. M., Pagès, B., Puccinelli, R., and Dejours, P.: Étude simultanée de la ventilation et des décharges des chémorécepteurs du glomus carotidien chez le chat. I. Au cours d'inhalations brèves d'oxygène pur, Arch. Int. Pharmacodyn. **154:**421, 1965.

162. Leitner, L. M., Pagès, B., Puccinelli, R., and Dejours, P.: Étude simultanée de la ventilation et des décharges des chémorécepteurs du glomus carotidien chez le chat. II. Au cours d'inhalations brèves d'anhydride carbonique, Arch. Int. Pharmacodyn. **154:**427, 1965.

163. Lerche, D., Katsaros, B., Lerche, G., and Loeschcke, H. H.: Vergleich der Wirkung verschiedener Acidosen (NH_4Cl, $CaCl_2$, Acetazolamid) auf die Lungenbelüftung beim Menschen, Arch. Gesamte Physiol. **270:**450, 1960.

164. Leusen, I.: Chemosensitivity of the respiratory center. Influences of changes in the H^+ and total buffer con-

centrations in the cerebral ventricles on respiration, Am. J. Physiol. **176:**45, 1954.

165. Leusen, I. R.: Aspects of the chemosensitivity of the respiratory centres. In Cunningham, D. J. C., and Lloyd, B. B., editors: The regulation of human respiration, Oxford, 1963, Blackwell Scientific Publications, Ltd.

166. Leusen, I.: Regulation of cerebrospinal fluid composition with reference to breathing, Physiol. Rev. **52:**1, 1972.

167. Leusen, I. R.: Aspects of the acid-base balance between blood and cerebrospinal fluid. In Brooks, C. M., Kao, F. F., and Lloyd, B. B., editors: The cerebrospinal fluid and the regulation of ventilation, Oxford, 1965, Blackwell Scientific Publications, Ltd.

168. Lloyd, B. B.: The chemical stimulus to breathing, Br. Med. Bull. **19:**10, 1963.

169. Lloyd, B. B., Jukes, M. G. M., and Cunningham, D. J. C.: The relation between alveolar oxygen pressure and the respiratory response to carbon dioxide in man, Q. J. Exp. Physiol. **43:**214, 1958.

170. Loeschcke, G. C.: Spielen für die Ruheatmung des Menschen vom O_2-Druck Abhängige Erregungen der Chemoreceptoren eine Rolle? Arch. Gesamte Physiol. **257:**349, 1953.

171. Loeschcke, H. H.: A concept of central chemical control of ventilation. In Brooks, C. M., Kao, F. F., and Lloyd, B. B., editors: The cerebrospinal fluid and the regulation of ventilation, Oxford, 1965, Blackwell Scientific Publications, Ltd.

172. Loeschcke, H. H.: Der Saure-Basenstatus des Liquor cerebrospinalis und seine Regulation durch die Lungenventilation, Klin. Wochenschr. **50:**581, 1972.

173. Loeschcke, H. H.: Respiratory chemosensitivity in the medulla oblongata, Acta Neurobiol. Exp. **33:**97, 1973.

174. Loeschcke, H. H.: The apparent specificity of CO_2 as a respiratory stimulus, Bull. Physiopathol. Respir. **10:**857, 1974.

175. Loeschcke, H. H.: Central nervous chemoreceptors. In Widdicombe, J. G., editor: Physiology series one: respiratory physiology, Baltimore, 1974, University Park Press, vol. 2.

176. Loeschcke, H. H., and Gertz, K. H.: Einfluss des O_2-Druckes in der Einatmungsluft auf die Atemtatigkeit des Menschen, gepruft Unter Konstanthaltung des alveolaren CO_2-Druckes, Arch. Gesamte Physiol. **267:**460, 1958.

177. Loeschcke, H. H., and Koepchen, H. P.: Versuch Zur Lokalisation des Angriffsortes der Atmungs—Und Kreislaufwirkung von Novocain im Liquor cerebrospinalis, Arch. Gesamte Physiol. **266:**628, 1958.

178. Loeschcke, H. H., and Mitchell, R. A.: Properties and localization of intracranial chemosensitivity. In Cunningham, D. J. C., and Lloyd, B. B., editors: The regulation of human respiration, Oxford, 1963, Blackwell Scientific Publications, Ltd.

179. Loeschcke, H. H., Katsaros, B., and Lerche, D.: Differenzierung der Wirkungen von CO_2-Druck und Wasserstoffionen-Konzentration im Blut auf die Atmung beim Menschen, Arch. Gesamte Physiol. **270:**461, 1960.

180. Loeschcke, H. H., Koepchen, H. P., and Gertz, K. H.: Uber den Einfluss von Wasserstoffionenkonzentration und CO_2-Druck im Liquor cerebrospinalis auf die Atmung, Arch. Gesamte Physiol. **266:**569, 1958.

181. Loeschcke, H. H., De Lattre, J., Schlafke, M. E., and Trouth, C. O.: Effects on respiration and circulation of electrically stimulating the ventral surface of the

medulla oblongata, Respir. Physiol. **10:**184, 1970.

182. Loeschcke, H. H., Katsaros, B., Albers, C., and Michel, C. C.: Uber den Zeitlichen Verlauf von Atemzugvolumen, Atem-Periodendauer, Atemminutenvolumen und Endexspiratorischem CO_2-Druck bei Einatmung von Gasgemischen mit Erhohtem CO_2-Druck, Arch. Gesamte Physiol. **277:**671, 1963.

183. Lugliani, R., Whipp, B. J., Seard, C., and Wasserman, K.: Effect of bilateral carotid-body resection on ventilatory control at rest and during exercise in man, N. Engl. J. Med. **285:**1105, 1971.

184. Macklem, P. T.: Respiratory mechanics, Annu. Rev. Physiol. **40:**157, 1978.

185. Magnussen, G.: Studies on the respiration during sleep, London, 1944, H. K. Lewis.

186. Maren, T. H.: Physiology and chemistry of cerebrospinal fluid, aqueous humor and endolymph in *squalus acanthias,* J. Exp. Zool. **199:**317, 1977.

186a. Maren, T. H.: Effect of varying CO_2 equilibria on rates of HCO_3^- formation in cerebrospinal fluid, J. Appl. Physiol. **47:**471, 1979.

187. Masland, W. S., and Yamamoto, W. S.: Abolition of ventilatory response to inhaled CO_2 by neurological lesions, Am. J. Physiol. **203:**789, 1962.

188. McAleavy, J. C., Way, W. L., Altstatt, A. H., Guadagni, N. P., and Severinghaus, J. W.: The effect of P_{CO_2} on the depth of anesthesia, Anesthesiology **22:**260, 1961.

189. Mchedlishvili, G. I.: Vascular mechanisms pertaining to the intrinsic regulation of the cerebral circulation, Circulation **30:**597, 1964.

190. Michel, C. C.: C.s.f. [HCO_3^-] during respiratory acid-base disturbance, J. Physiol. **170:**66P, 1963.

191. Miescher-Rusch, F.: Bemerkungen zur Lehre von den Athembewegungen, Arch. Anat. Physiol. (Leipzig) **3:**355, 1885.

192. Milhorn, H. T., Jr., Benton, R., Ross, R., and Guyton, A. C.: A mathematical model of the human respiratory control system, Biophys. J. **5:**27, 1965.

193. Milic-Emili, J., and Grunstein, M. M.: Drive and timing components of ventilation, Chest (suppl.) **70:**131, 1976.

194. Mitchell, R. A.: The regulation of respiration in metabolic acidosis and alkalosis. In Brooks, C. M., Kao, F. F., and Lloyd, B. B., editors: The cerebrospinal fluid and the regulation of ventilation, Oxford, 1965, Blackwell Scientific Publications, Ltd.

195. Mitchell, R. A., and Berger, A. J.: Neural regulation of respiration, Am. Rev. Respir. Dis. **111:**206, 1975.

196. Mitchell, R. A., Loeschcke, H. H., Massion, W. H., and Severinghaus, J. W.: Respiratory responses mediated through superficial chemosensitive areas on the medulla, J. Appl. Physiol. **18:**523, 1963.

197. Mitchell, R. A., et al.: Regions of respiratory chemosensitivity on the surface of the medulla, Ann. N.Y. Acad. Sci. **109:**661, 1963.

198. Mitchell, R. A., et al.: Stability of cerebrospinal fluid pH in chronic acid-base disturbances in blood, J. Appl. Physiol. **20:**443, 1965.

199. Mithoefer, J. C.: Breath holding. In Fenn, W. O., and Rahn, H., editors: Handbook of physiology. Respiration section, Washington, D.C., 1964, American Physiological Society, vol. 2.

200. Moyer, C. A., and Beecher, H. K.: Central stimulation of respiration during hypoxia, Am. J. Physiol. **136:**13, 1942.

201. Neff, T. A., and Talmage, D., editors: The 20th Aspen Lung Conference: neuromuscular and chemical control

202. Nicholson, H. C.: Localization of the central respiratory mechanism as studied by local cooling of the surface of the brain stem, Am. J. Physiol. **115:**402, 1936.

203. Nicholson, H. C., and Sobin, S.: Respiratory effects from the application of cocaine, nicotine, and lobeline to the floor of the fourth ventricle, Am. J. Physiol. **123:**776, 1938.

204. Nielsen, M.: Untersuchungen uber die Atemregulation beim Menschen, besonders mit Hinblick auf die Art des chemischen Reizes, Acta Physiol. Scand. **74** (suppl. 10):87, 1936.

205. Nielsen, M., and Smith, H.: Studies on the regulation of respiration in acute hypoxia, Acta Physiol. Scand. **24:**293, 1951.

206. Pappenheimer, J. R.: Cerebral HCO_3^- transport and control of breathing, Fed. Proc. **25:**884, 1966.

207. Pappenheimer, J. R., Fencl, V., Heisey, S. R., and Held, D.: Role of cerebral fluids in control of respiration as studied in unanesthetized goats, Am. J. Physiol. **208:**436, 1965.

208. Pavlin, E. G., and Hornbein, T. F.: Distribution of H^+ and HCO_3^- between CSF and blood during respiratory acidosis in dogs, Am. J. Physiol. **228:**1145, 1975.

209. Perkins, J. F., Jr.: Historical development of respiratory physiology. In Fenn, W. O., and Rahn, H., editors: Handbook of physiology. Respiration section, Washington, D.C., 1964, American Physiological Society, vol. 1.

210. Peters, J. P., and Van Slyke, D. D.: Quantitative clinical chemistry, methods, Baltimore, 1931, The Williams & Wilkins Co.

211. Phillipson, E. A.: Respiratory adaptations in sleep, Annu. Rev. Physiol. **40:**133, 1978.

212. Pierce, E. C., Lambertsen, C. J., Strong, M. J., Alexander, S. C., and Steele, D.: Blood P_{CO_2} and brain oxygenation at reduced ambient pressure, J. Appl. Physiol. **17:**899, 1962.

213. Pontén, U.: Acid-base changes in rat brain tissue during acute respiratory acidosis and baseosis, Acta Physiol. Scand. **68:**152, 1966.

214. Purves, M. J., editor: The peripheral arterial chemoreceptors, London, 1975, Cambridge University Press.

215. Rahn, H., editor: Physiology of breath-hold diving, Washington, D.C., 1965, NAS-NRC Report 1341.

216. Read, D. J. C., Maloney, J. E., and Fowler, K. T.: Reproducibility of transient response to CO_2 inhalation in man, J. Appl. Physiol. **19:**750, 1964.

217. Reed, D. J., and Kellogg, R. H.: Effect of sleep on hypoxic stimulation of breathing at sea level and altitude, J. Physiol. **15:**1130, 1960.

218. Reivich, M.: Arterial P_{CO_2} and cerebral hemodynamics, Am. J. Physiol. **206:**25, 1964.

219. Reivich, M., Coburn, R., Lahiri, S., and Chance, B.: Tissue hypoxia and ischemia, New York, 1977, Plenum Press.

220. Robin, E. D., and Bromberg, P. A.: Claude Bernard's milieu interieur extended: intracellular acid-base relationships, Am. J. Med. **27:**689, 1959.

221. Robin, E. D., Whaley, R. D., Crump, C. H., and Travis, D. M.: Alveolar gas tensions, pulmonary ventilation and blood pH during physiologic sleep in normal subjects, J. Clin. Invest. **37:**981, 1958.

222. Robin, E. D., et al.: Acid-base relations between spinal fluid and arterial blood with special reference to control of ventilation, J. Appl. Physiol. **13:**385, 1958.

223. Rosenthal, I.: Die Athembewegungen und ihre Bezie-

hungen zum Nervus vagus, Berlin, 1862, A. Hirschwald.

224. Saito, K., Honda, Y., and Hasumura, N.: Evaluation of respiratory response to changes in P_{CO_2} and hydrogen ion concentration of arterial blood in rabbits and dogs, Jpn. J. Physiol. **10:**634, 1960.

225. Schaefer, K. E.: Atmung und Saure-Basengleichgewicht bei Langdauerndem Aufenthalt in 3% CO_2, Arch. Gesamte Physiol. **251:**689, 1949.

226. Schlafke, M. E., et al.: Chemosensitive neurons on the ventral medullary surface (Les neurones chemosensibles de la surface ventrale du bulbe), Bull. Physiopathol. Respir. **11:**277, 1975.

227. Schlafke, M. E., See, W. R., and Loeschcke, H. H.: Ventilatory response to alterations of H⁺ ion concentration in small areas of the ventral medulla surface, Respir. Physiol. **10:**198, 1970.

228. Schmidt, C. F.: The influence of cerebral blood-flow on respiration. I. The respiratory responses to changes in cerebral blood-flow, Am. J. Physiol. **34:**202, 1928.

229. Schmidt, C. F.: Carotid sinus reflexes to the respiratory center, Am. J. Physiol. **102:**94, 119, 1932.

230. Schmidt, C. F.: Respiration, Annu. Rev. Physiol. **7:** 231, 1945.

231. Schmidt, C. F.: The respiration. In Bard, P., editor: Medical physiology, ed. 10, St. Louis, 1956, The C. V. Mosby Co.

232. Schmidt, C. F., and Comroe, J. H., Jr.: Functions of the carotid and aortic bodies, Physiol. Rev. **20:**115, 1940.

233. Schmidt, C. F., Comroe, J. H., Jr., and Dripps, R. D.: Carotid body reflexes in the dog, Proc. Soc. Exp. Biol. Med. **42:**31, 1939.

234. Seevers, M. H.: The narcotic properties of CO_2, N.Y. J. Med. **44:**597, 1944.

235. Semple, S. J. G.: Respiration and the cerebrospinal fluid, Br. J. Anaesth. **37:**262, 1965.

236. Shock, N. W., and Soley, M. H.: Effect of oxygen tension of inspired air on the respiratory response of normal subject to carbon dioxide, Am. J. Physiol. **130:**777, 1940.

237. Siesjö, B. K.: The bicarbonate/carbonic acid buffer system of the cerebral cortex of cats, as studied in tissue homogenates, Acta Neurol. Scand. **38:**98, 1962.

238. Smith, D. J., and Vane, J. R.: Effects of oxygen tension on vascular and other smooth muscle, J. Physiol. **186:**284, 1966.

239. Sokoloff, L.: The action of drugs on the cerebral circulation, Pharmacol. Rev. **11:**1, 1959.

240. Sorensen, S. C.: The chemical control of ventilation, Acta Physiol. Scand. **361:**suppl., 1971.

241. Tschirgi, R. D., and Taylor, J. L.: Slowly changing bioelectric potentials associated with the blood-brain barrier, Am. J. Physiol. **195:**7, 1958.

242. Torrance, R. W.: Arterial chemoreceptors. In Widdicombe, J. G., editor: Physiology series one: respiratory physiology, Baltimore, 1974, University Park Press, vol. 2.

243. Verstraeten, J. M.: Influences centrales du calcium et du potassium sur la respiration, Arch. Int. Pharmacodyn. **77:**52, 1948.

244. Vogh, B. P., and Maren, T. H.: Sodium, chloride, and bicarbonate movement from plasma to cerebrospinal fluid in cats, Am. J. Physiol. **228:**673, 1975.

245. Wang, S. C., and Ngai, S. H.: General organization of central respiratory mechanisms. In Fenn, W. O., and Rahn, H., editors: Handbook of physiology, Respiration section, Washington, D.C., 1964, American Physiological Society, vol. 1.

246. Watt, J. G., Dumke, P. R., and Comroe, J. H., Jr.: Effects of inhalation of 100 per cent and 14 per cent oxygen upon respiration of unanesthetized dogs before and after chemoreceptor denervation, Am. J. Physiol. **138:**610, 1943.

247. Weber, H. H., and Loeschcke, H. H.: Hans Winterstein, Ergeb. Physiol. **55:**1, 1964.

248. White, C. S., Humm, J. H., Armstrong, E. D., and Lundren, N. P. V.: Human tolerance to acute exposure to carbon dioxide, J. Aviation Med. **23:**439, 1952.

249. Widdicombe, J. G., editor: Respiratory physiology II. In International review of physiology, Baltimore, 1977, University Park Press, vol. 14.

250. Wiemer, W., Ott, N., and Winterstein, H.: Reflektorische und zentrale Anteile der O_2-Mangel und CO_2-Hyperpnoe des Kaninchens, Z. Biol. **114:**230, 1963.

251. Wiemer, W., Ott, N., and Winterstein, H.: Reflektorische und zentrale Anteile der Hyperpnoe bei HCl-Acidose, Z. Biol. **114:**299, 1964.

252. Winterstein, H.: Die Regulierung der Atmung durch das Blut, Arch. Gesamte Physiol. **138:**167, 1911.

253. Winterstein, H.: Die Reaktionstheorie der Atmungsregulation, Arch. Gesamte Physiol. **187:**293, 1921.

254. Winterstein, H.: Die Narkose, Berlin, 1926, Julius Springer.

255. Winterstein, H.: Chemical control of pulmonary ventilation. III. The "reaction theory" of respiratory control, N. Engl. J. Med. **255:**331, 1956.

256. Winterstein, H.: 50 Jahre Reaktionstheorie der Atmung, Naturwiss Rundsch. **14:**413, 1961.

257. Winterstein, H., and Gökhan, N.: Ammoniumchloride-Acidose und Reactions Theorie der Atmungsregulation, Arch. Int. Pharmacodyn. **93:**212, 1953.

258. Witzleb, E., Bartels, H., Budde, H., and Mochizucki, M.: Der Einfluss des arteriellen O_2-Drucks auf die chemoreceptorischen Aktionspotentiale im Carotissinusnerven, Arch. Gesamte Physiol. **261:**211, 1955.

259. Wolstenholme, G. E. W., and O'Connor, C. M., editors: Ciba Foundation symposium on the cerebrospinal fluid, Boston, 1958, Little, Brown & Co.

260. Wood, H. G.: The fixation of carbon dioxide and the interrelationships of the tricarboxylic acid cycle, Physiol. Rev. **26:**198, 1946.

261. Woodbury, D. M., and Karler, R.: The role of carbon dioxide in the nervous system, Anesthesiology **21:** 686, 1960.

72

CHRISTIAN J. LAMBERTSEN

Dyspnea and abnormal types of respiration

Normal function entails a considerable reserve respiratory capacity for each of the interrelated mechanisms involved in pulmonary ventilation, in the transport of gases by blood, and in the regulation of respiration. This reserve function is sufficient to permit the demands of at least a 10-fold increase in oxidative metabolism to be met by a similar increase in gas exchange between the external environment and the site of reactions within the cell.

In certain abnormal states, gross interference with this gas exchange occurs. The interference may occur abruptly, be readily detectable, and in some situations be completely reversible. However, it is also possible for respiratory insufficiency to develop progressively over many years with a gradual reduction in functional reserve. At first, evidence of respiratory insufficiency may appear only under the stress of exercise, since the reserve may permit adequate adjustment of the gas exchange to the metabolic requirements at rest. Later, as the pathologic changes progress, the reserve may be lost altogether; then the individual will be unable to maintain normal alveolar ventilation even at rest.

When alveolar ventilation becomes inadequate, the resulting changes in blood gas composition are in part responsible for the sensations of respiratory distress. In addition, they contribute to the derangement of the internal oxygen and acid-base environment and to progressive further deteriorations of vital functions that culminate in complete respiratory failure and death. In the following discussion, attention will be given to the manner in which the mechanisms of respiration are altered in insufficiency of respiration and failure of respiration, in producing distressful respiration, and in abnormalities of respiratory rhythm.

RESPIRATORY INSUFFICIENCY

In the normal system for respiratory and circulatory control the balance between chemical and nervous influences on special groups of reactive neurons provides automatic regulation of the acid-base and O_2 levels at the metabolizing cell. This reduplicated and interrelated self-regulation depends on the capacity of the neurons in the control centers to react to deviations from homeostasis of their own local environment. This fundamental requirement of reactivity can be reduced or abolished by all the influences that depress the excitability of any neuron. For convenience, these can be generally considered to include *narcosis, other pharmacologic influences, hypoxia,* and *pathologic processes.* The common result of all these is diminution of the ability of the sensitive system of chemical respiratory control to function, culminating in respiratory failure.[22] This chapter outlines aspects of the pathophysiologic conditions that influence the development and management of respiratory failure. Treatment of respiratory failure[22] and effects of drugs on respiration[54] are discussed in reviews.

Respiratory failure due to narcotic drugs

Depression of respiration by a narcotic drug necessarily leads to a reduction in P_{O_2} and an increase in P_{CO_2} of the arterial blood. Of the several conditions under discussion, this is the one most amenable to treatment and carries the best prognosis because a narcotic depression that does not lead to secondary tissue damage through hypoxia should disappear completely when the drug is metabolized or excreted. Hypoxia, on the other hand, "not only stops the machine, it wrecks the machinery" and deserves classification with pathologic processes rather than the pharmacologic processes involved in narcotic depressions (which secondarily produce hypoxia).

Narcosis and hypoxia

The tendency to regard narcosis as a relatively benign process occurs because it leads not only to diminished functional activity but also to a decreased metabolic requirement for O_2 in the central nervous system (CNS). This diminished demand for O_2 provides a measure of protection against damage to cells

Table 72-1. Effect of narcotic drugs on brain O_2 consumption and oxygenation

Condition	Reference	Brain O_2 consumption (ml O_2/100 gm/min)	Mean art. BP (mm Hg)	Brain blood flow (ml/100 gm/min)	Brain vascular resistance (mm Hg/ml/ 100 gm/min)	Blood gas tensions		
						Art. PCO_2 (mm Hg)	Art. PO_2 (mm Hg)	Jug. ven. PO_2 (mm Hg)*
Normal, conscious subjects	Schmidt[17]	3.3	86	54	1.6	40	95	40
		Changes induced by hypoxia				*Values during hypoxia*		
Hypoxia, 8% O_2	—	−23%	−4 mm	+49%	−36%	34	23	18
		Changes induced by narcosis				*Values during narcosis*		
Thiopental anesthesia	Wechsler et al.[69]	−20%	0	+13%	−30%	54	73	40
Morphine, 60 mg IV	Moyer et al.[60]	−40%	−9 mm	+15%	−25%	49	64	40

*Index of lowest value of *blood* PO_2 to which brain tissue is exposed. When gas mixtures low in PO_2 are inhaled, unconsciousness occurs as internal jugular venous PO_2 falls below about 20 mm Hg. Cellular PO_2 near the venous end of the capillary is lower than that in blood.

by failure of O_2 delivery (Table 72-1). It is possible that the depression of metabolism is primary, the decrease in function the result of the curtailment of energy-yielding reactions.[40,50] There is no doubt that in narcotic poisoning there is a diminution in the metabolic rate of most tissues, including the brain. The degree of this diminution may be judged from the fact that, even with narcotic levels too low to cause complete respiratory failure, the cerebral O_2 consumption of the human may be reduced to almost half the normal value for consciousness. Since the intensity of narcotic processes in general depends on the concentration of the drug, narcotic doses large enough to produce respiratory failure lead to an even greater depression of cerebral metabolism.

The O_2 tension of arterial blood will necessarily be lowered when breathing is depressed. However, the associated narcotic depression of the O_2 utilization elevates or sustains the tissue O_2 pressure and places the narcotized individual in a more favorable position to survive than is the case with one in whom respiratory depression follows a bout of increased neuronal activity or even in one with a normal rate of brain tissue O_2 consumption. The importance of the narcotic depression of O_2 consumption as a protection against *arterial* hypoxemia is shown in Table 72-1. Brain metabolism is reduced at the same time that hypoxia, hypercapnia, and probably the narcosis itself increase brain blood flow by relaxing cerebral vessels. The consequence of these secondary, indirect effects of the narcotic agents is that values for *brain venous PO_2* are not appreciably lowered even in the face of considerable arterial hypoxemia.

The dangers in narcotic poisoning are from what may be called secondary complications rather than the narcotic process itself, which is by definition completely reversible.[71] The most important of these complications are (1) asphyxia from insufficient pulmonary ventilation, (2) pathologic changes (microbial and otherwise) developing over several days in the respiratory tract and lungs, (3) circulatory depression (shock), (4) hypothermia or hyperthermia, (5) renal functional derangements, and (6) consequences of therapeutic measures (acapnia and circulatory depression; pulmonary and cerebral edema or cardiac dilatation from excessive fluid; hypoxia secondary to convulsions produced by stimulant drugs).

Asphyxia

Asphyxia, or the concomitant elevation of arterial PCO_2 and lowering of PO_2, will of course ensue as the respiratory movements become ineffective and, despite the previously mentioned "protective" effect of reduced neuronal metabolism, *the ultimate cause of death in narcosis is hypoxia and acidosis at the cellular level.*

When a degree of hypoxemia and acidosis sufficient to produce damage to cells is superimposed on the existing narcosis, the reversibility of the specific narcotic drug action itself becomes an academic point. Consequently, the first consideration is to bring about an adequate pulmonary gas exchange; this involves provision of an open airway as well as sustainable artificial respiration.

Circulatory depression

Circulatory depression is usually present in narcosis and is attributed to a combination of central vasomotor depression, hypoxemia, and direct narcotic poisoning of blood vessels.[61] The blood supply of the brain at first is maintained or even increased, since the development of hypercapnia leads to a concomitant decrease in cerebral vascular tonus (Table 72-1), but in the presence of severe hypotension coupled with hemo-concentration (which may be present in narcotic poisoning[61]), cerebral blood flow probably would fall and a hypokinetic anoxia would ensue despite the lessened O_2 requirement. Thus support of the circulation by suitable means (blood or blood substitute, vasopressor drugs, cardiotonic drugs) is indicated if severe hypotension develops.

Hypothermia

Hypothermia is likely to be present if the poisoned individual was exposed to a cold environment, since, in addition to diminished O_2 metabolism, the heat-regulating mechanisms are also deranged and the organism becomes poikilothermic. *Hyperthermia* may result from infection.

Renal impairment

Renal impairment in hypotension may lead to oliguria and its consequences. These can be circumvented by intravenous administration of fluids.

The system for proper management of severe narcotic poisoning over the many days of its usual duration has improved along with advances in anesthesiology.* There is no question that it must include oxygenation, ventilation adequate to prevent acidosis, support of the circulation, prevention of infection, adequate hydration and nutrition, and careful nursing. Specific antagonistic drugs against the narcotic analgesics are available.

Respiratory insufficiency due to pulmonary disease

This can be brought about by impediment to the free movement of gases through the respiratory passages and alveoli (as in respiratory obstruction of any type, fluid in the air spaces, or thickening of the alveolar membranes), diminution of the effective ventilating surfaces (by collapse, obstruction, fibrosis, consolidation, adhesions, or other pathologic encroachments), or impairment primarily in the pulmonary circulation (congestion, embolization, obliteration in emphysema, pneumonia, arteriovenous shunts, etc.). In all such cases the impairment, if severe enough to lead to difficulty in eliminating CO_2, will certainly cause difficulty in the absorption of O_2. Therefore hypoxia and all its implications must be suspected whenever CO_2 retention occurs. The great variety of pathologic conditions affecting the lungs requires that students interested in the physiologic basis of pulmonary disease refer to specialized sources of detailed discussion. Since it is in pulmonary dysfunction that respiratory distress is most prominent, one special pathologic condition will be examined here to illustrate the consequences of progressive decrease in efficiency of alveolar ventilation. The condition selected is pulmonary emphysema.[1,33,63]

Pulmonary emphysema

In pulmonary emphysema, a degenerative disorder, several pathologic patterns occur.[1,63] All lead to disruption of alveolar septa and pulmonary fibrosis, resulting in an increase in both residual air and dead space and a decrease in elasticity of the lung. As a result, the efficiency of alveolar ventilation decreases. In addition, there occurs an exaggerated shunting of blood across the lungs (inequality of ventilation and perfusion[65]). Then, at an even greater than normal respiratory minute volume, arterial P_{O_2} falls. In its early stages the lowering of arterial P_{O_2} is not associated with elevation of P_{CO_2}.[27,64] An effect of CO_2 as a respiratory stimulant undoubtedly occurs, however, since normal individuals exposed to comparable hypoxemia would actually show a reduced P_{CO_2}.[9,59] Eventually the ability to accomplish alveolar ventilation will be further diminished. Then the changes in arterial blood gas tensions become extreme, including elevation of arterial CO_2 tension.[27,64]

In Fig. 72-1 the ventilatory state of individuals chronically hypoxemic and hypercapnic due to pulmonary emphysema is compared with the respiratory responses to be expected in a normal subject exposed to a nearly equivalent degree of *asphyxiation,* that is, hypercapnia plus hypoxemia. The data for this comparison were drawn largely from studies by Alexander and co-workers[27] and Richards and colleagues[64] on emphysematous subjects. To compare the state of the emphysematous patient with the respiratory effects of hypoxia and hypercapnia in a normal individual, a subject studied by Loeschcke and Gertz[59] was also used in preparing Fig. 72-1. These two conditions of hypoxia (one environmental and one pathologic) are both contrasted with the respiratory response to simple hypercapnia in normal subjects.[52]

Fig. 72-1 shows that severely emphysematous patients have (1) an elevated arterial P_{CO_2} (i.e., alveolar ventilation is reduced) despite a slightly increased respiratory minute volume and (2) less increase in ventilation per unit rise in P_{CO_2} than does the well-oxygenated normal subject and still less than the normal subject exposed to a comparable degree of hypoxemia (in this case about 45 mm Hg P_{O_2}). The most striking aspect of this figure is the contrast between the low respiratory minute volume (4.5 L/min/m²) of the emphysematous patients, hypoxemic during air breathing, and the approximately 24 L/min/m² ventilation for the hypoxic normal subject at comparable P_{O_2} and P_{CO_2}. In Fig. 72-1, this difference in breathing at "equivalent chemical stimulus levels" is divided into a "hypoxic stimulus component" and a "CO_2-related stimulus component" to indicate the magnitude of the potential respiratory drives to which the emphysematous patient is exposed. That a hypoxic respiratory drive exists in such patients is evident from the further hypercapnia (further reduction in *alveolar ventilation*) seen on administration of 50% O_2.[64] Respiration at the high inspired P_{CO_2} is presumably prevented from a further fall by the exaggerated respiratory acidosis that inevitably must occur as ventilation begins to decrease (Chapter 71).

When the responsiveness of severely hypoxic individuals to CO_2 has become very low, the hypoxic respiratory drive exerts a proportionately greater role in sustaining the inadequate ventilation. It is in such pa-

---

*See references 31, 46, 54, and 61.

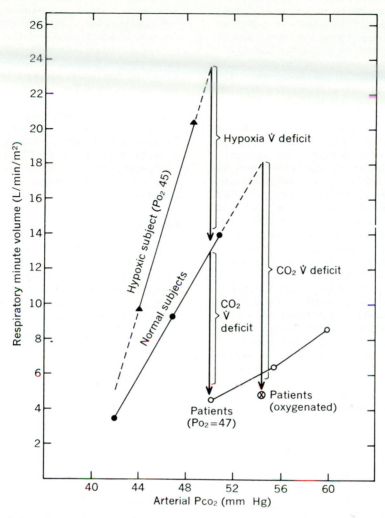

Fig. 72-1. Pulmonary emphysema: deficit in ventilatory response to hypercapnia and hypoxemia. This diagram, constructed from data obtained in several laboratories, compares (1) normal respiratory response to CO_2 in absence of hypoxia or pulmonary disease (●—● representing average of control values in 27 subjects),[27,52,55,58] (2) marked stimulant effect of CO_2 administration in normal subject with arterial Po_2 of 45 mm Hg (▲—▲),[59] (3) elevated Pco_2 and small response to additional hypercapnia in emphysematous patients hypoxic due to disease process (○—○),[27] and (4) influence of improved oxygenation on ventilation and arterial Pco_2 of patients with severe emphysema (⊗).[64] Total ventilation deficit for hypoxic, hypercapnic, emphysematous patients ($Po_2 = 47$ mm Hg) as contrasted with normal individual at comparable arterial Pco_2 and Po_2 is depicted by sum of two vertical lines labeled "hypoxia $\dot{V}$ deficit" and "CO_2 $\dot{V}$ deficit." When such patients are given O_2, removal of hypoxic, chemoreflex drive reduces alveolar ventilation, causing further rise in arterial Pco_2. Ventilation deficit in oxygenated emphysematous patients as compared with well-oxygenated but hypercapnic, normal subjects is indicated by second vertical line (far right) labeled "CO_2 $\dot{V}$ deficit." Considering the ventilatory responses of normal subjects to same levels of Pco_2 and Po_2 seen in patients, it is not surprising that pulmonary insufficiency of emphysema is associated with sensation of dyspnea, especially in exercise. Acclimatization to chronic elevation of CO_2 tension may reduce distress caused by increased Pco_2, but it is not likely that acclimatization to effects of chemoreceptor stimulation ever occurs (Chapter 73).

tients, depressed by central hypoxia and hypercapnia, that administration of O_2 has led to a gross fall in ventilation and unconsciousness, or occasionally even to the so-called *O_2 apnea.*[32] Very likely in such chronically and severely hypoxic persons the slowly adapting chemoreceptors provide not only reflex activation of the respiratory centers but also an "arousal" mechanism for higher functions such as consciousness. Loss of this drive, together with any concurrent depressant effects of Pco_2 elevation, may be the basis for the apparently paradoxical deterioration of central nervous functions that sometimes ensues on relief of the hypoxia. The occurrence of these disturbing effects (respiratory depression and depressed consciousness) of course indicates that the O_2 that caused them was needed. However, unless alveolar ventilation is maintained at a level adequate for the existing metabolism, the now well-oxygenated patient will gradually succumb to the exaggeration of respiratory acidosis. By combining (1) O_2 administration to relieve the dangerous hypoxia and (2) assisted artificial ventilation to accomplish CO_2 elimination, both of the defects in gas exchange can be overcome.

The reduced capacity of emphysematous patients to respond to artificially elevated Pco_2 (Fig. 72-1) was once considered entirely due to progressive desensitization of the respiratory centers to respiratory acidosis. It now appears to be in large part related to a diminished influence on actual ventilation of impulses to the respiratory muscles. Thus, if the change in *work of breathing* in response to elevations of Pco_2 is employed in place of respiratory minute volume as an index of respiratory center activity, emphysematous patients are found to have nearly the same reactivity to a *change* in Pco_2 as do normal subjects.[64] The higher than normal Pco_2 that can occur in advanced disease may nevertheless be related to a form of acclimatization to hypercapnia. Fairly stable levels of arterial Pco_2 as great as 80 mm Hg have occurred in emphysematous patients in association with arterial O_2 tensions of between 30 and 40 mm Hg; such patients are physically and mentally incapacitated. Boutourline-Young and Whittenberger[29] have employed mechanical hyperventilation over a period of several days to restore reactivity to nearly normal levels of Pco_2. The effort was apparently successful, with coincident improvement in ventilation, oxygenation, and mental acuity. Quite apart from the practical implications of such studies is the indication that within a particular individual it is possible to readjust the acid-base environment to which the respiratory neurons react. Very likely the renal adjustments to induced hypocapnia should lead to diminished bicarbonate concentration in the blood and in the central regions readily influenced by blood, and in time even to a reduction of bicarbonate in fluids beyond the blood-brain barrier. Then, on discontinuing the artificial hyperventilation, a particular level of Pco_2 would lead to a higher level of $[H^+]$ at each site. This would represent a form of acute acclimatization to CO_2 (Chapter 73) based on change in central stimulus level.

Respiratory failure due to hypoxia

Regardless of its cause, continued hypoxia of prominent degree leads to progressive damage and to self-perpetuating vicious circles. The pathologic processes that may lead to respiratory failure include involvement of the central neurons (by inflammatory or neoplastic disease, trauma, hemorrhage, etc.), interruption of efferent motor pathways (by similar lesions and diseases, including poliomyelitis), and the peripheral neuromuscular block of myasthenia or curarization. One of the most frequent causes of respiratory difficulties in diseases of the brain is increase in cerebrospinal pressure, which collapses brain vessels and diminishes blood flow to the CNS tissue. Respiratory failure also is the usual immediate cause of death in most toxemias, but respiration typically is the last of central nervous functions to disappear.

APNEA

Apnea simply means absence of breathing but, as generally used, the term implies that the cessation is temporary. Prolonged apnea actually is respiratory failure and is considered elsewhere in this chapter. Theoretically, apnea might be produced by (1) reduction in the stimulus to the respiratory center, the latter having its normal ability to respond; (2) active inhibition of the discharge of impulses by the center, the stimulus being normal or augmented and the center likewise normal; (3) decrease in the ability of the center to react to the stimulus; and (4) any combination of these. In experiments on anesthetized or decerebrate animals, examples of the operation of each of these causes can be obtained without difficulty and most can be demonstrated in humans.

Apnea due to reduction in the stimulus to the respiratory center is seen when a reflex stimulus is suddenly withdrawn or a period of passive hyperventilation of anesthetized animals or humans is terminated.[41] Reduction in the chemical stimulus that is produced by hyperpnea now makes itself manifest, and breathing stops until the chemical stimulus is restored. This form of apnea, which follows the lowering of arterial Pco_2, is apparently not readily demonstrable in conscious individuals, possibly because extrinsic, nonspecific stimulation sustains respiratory activity in the conscious state.[41] A more definite apnea occurs when O_2 is breathed after inhalation of a low O_2 mixture with an inert gas such as nitrogen, helium, or nitrous oxide (Fig. 72-2).

Apnea due to active inhibition of the respiratory neurons can be produced by prolonging the inhibitory Hering-Breuer reflex, as by obstructing the trachea at the end of inspiration. This causes relaxation of the inspiratory muscles and the resulting apnea lasts until the constantly rising CO_2 tension of the blood becomes stronger than the inhibitory nerve influence, for the

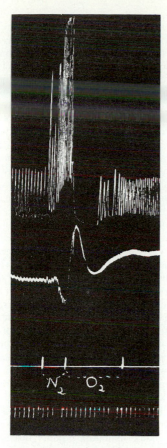

Fig. 72-2. Apnea due to breathing O_2 after nitrogen. Cat: pentobarbital narcosis. Respiration recorded by plethysmograph, blood pressure by mercury manometer connected with femoral artery. Vagodepressor and sinus nerves intact. Nitrogen (undiluted) inhaled between first pair of signals. When dyspnea was at its height, nitrogen was removed and O_2 was supplied through tracheal catheter until third signal mark. Time record shows 5 sec intervals.

latter diminishes only very slowly when the distention of the lungs remains constant.[26]

A similar result is obtained when the carotid sinus pressure receptors are stimulated by a rise in intracarotid pressure. The explanation is also similar; the pressure receptors of the carotid sinuses, like the stretch receptors of the lungs, adapt very slowly[30] and the resumption of breathing is due to the overpowering effect of the increased chemical stimulus.

Apnea due to decreased ability of the respiratory neurons to react to stimuli is produced by depressant drugs, especially when they are given intravenously to produce their effects very rapidly. In all such cases the CO_2 tension of the blood rises above normal and the extent of the rise is a fair measure of the extent to which the reactivity of the neurons has been reduced.

When P_{CO_2} is elevated as a result of hypoventilation, P_{O_2} must be subnormal. If the depression caused by the drug is not too severe, breathing will begin again when the combined effects of these increasing stimuli arouse the depressed neurons. As in the previously mentioned example of pulmonary emphysema (p. 1830), the hypoxic stimulus to the chemoreceptors contributes a larger proportion of the total respiratory drive and administration of pure O_2 in such a situation will further reduce respiration. If the respiratory neurons are so depressed that they no longer react to effects of hypercapnia, cessation of respiration (O_2 apnea) will result.

Combinations of these causes of apnea are frequently encountered. The commonest and most important example is *simultaneous reduction in one form of stimulus and in the ability of the center to react to it*. This occurs in inhalation anesthesia. A similar combination is encountered in *anoxemia* or in O_2 *lack in the central neurons due to cerebral anemia, no matter how produced,* since in anoxemia breathing is driven by reflexes and in cerebral anemia by accumulation of metabolic products in the cells of the center; the normal stimulus is reduced by the hyperpnea and the ability of the cells to react is simultaneously reduced by O_2 lack within them. *Active inhibition of the cells of the center* becomes a serious factor only when it is combined with a diminution in the reactivity of the cells to chemical influences while their reactivity to afferent inhibitory influences is less reduced or actually exaggerated. This composite of effects may, together with actual hypoxia, be the basis for death during prolonged breath holding under water after a preliminary bout of hyperventilation.

DYSPNEA

The term "dyspnea" denotes labored, distressful breathing with exaggerated consciousness of the necessity for increased respiratory effort. It can be severe and incapacitating. By its definition an unconscious patient cannot be considered dyspneic, even if he is hyperventilating. Dyspnea in normal individuals is experienced during exercise that approaches the limit of tolerance. The essential difference between dyspnea and normal breathing at rest is that the unpleasant sensation of breathlessness (i.e., lack of satisfaction of the desire to breathe even by the fullest respiratory efforts) now is present for reasons unrelated to an increased metabolic requirement. It is therefore not always alleviated by rest or by an increase in activity of the respiratory muscles or the nerves supplying them. This probably means that the influence of stimulant factors on the inspiratory neurons does not diminish or disappear as it should at the end of inspiration but persists throughout the respiratory cycle.

Hyperpnea (hyperventilation) denotes an increase in pulmonary ventilation and does not imply distressful subjective sensations. Actually the hyperventilation of normal persons in moderate exercise is usually considered pleasant. Dyspnea, when not due to a form of resiratory paralysis, may involve excessive activity of the respiratory muscles of a type and degree normally reserved for maximal metabolic requirements. This indicates that dyspnea not only is uncomfortable but may in itself increase metabolism through the work of breathing and result in a feedback to further increase the ventilatory requirements.

Causes of dyspnea

Since there are numerous causes of respiratory stimulation in the normal state, there can be no single cause of respiratory distress in disease. In most clinical states, dyspnea is due to multiple factors that vary according to the condition. As in any form of respiratory stimulation, there is a tendency to summation of the various factors involved. The most common example of this is the decrease in exercise tolerance that occurs in latent cardiopulmonary insufficiency. In such patients there is already present an abnormality capable of limiting pulmonary or tissue gas exchange, but it is not severe enough to be felt until the increased metabolic requirements of exercise are superimposed on it. As this abnormality progresses, the added stimulus required to cause dyspnea diminishes until finally no addition is needed to produce strong stimulation of the respiratory centers even at complete rest.

Some of the factors responsible for exaggerating and sustaining the desire to breathe in different conditions are summarized in Table 72-2. They include (1) excitatory nerve impulses of reflex origin but unrelated to the carotid and aortic chemoreceptors, (2) chemical stimulation, whether of central or chemoreceptor reflex origin (3) fever, and (4) increased reactivity of the respiratory centers, whether primary, drug-induced, or of emotional origin.

Dyspnea and excitatory reflexes not related to chemical stimuli

Although the existence of reflex factors in the production of the sensations of dyspnea is likely, it is evident that determination of such influences does not lend itself to experimental study. Therefore at present, it can only be deduced that reflex influences contribute to particular forms of dyspnea for which other bases cannot be determined. Certain neurogenic factors deserve particular attention as possibilities.

Impulses from receptors in limbs. The apparent existence of motion-related impulses, discussed in relation to the hyperpnea of exercise (Chapter 74), deserves mention again as these are potential contributors to the dyspnea of exertion. This form of dyspnea represents an exaggeration by exercise of a preexisting respiratory defect.

Hering-Breuer reflexes. These inhibitory reflexes could, under physiologic conditions, modify hyperpnea originated by some other mechanism but could not of themselves initiate hyperpnea. As long as the receptors responsible for the hyperpnea retain their normal relations to the alveolar walls, they would discharge vigorously only when respiration is being powerfully stimulated.

In abnormal conditions, however, the situation may be different. In *congestive heart failure* the engorged pulmonary capillaries encroach on the air-bearing surfaces. Since the elasticity of the lungs is reduced by the vascular engorgement, the inspiratory discharges may tend to not be inhibited normally, even by a maximal inspiratory effort. In *pneumonia* a corresponding involvement could occur in enough of these receptors to have a similar, although less striking, effect. Similar effects may also be elicited in *asthma*, but here a certain amount of emphysema is also likely to be present. In *emphysema*, it is possible that, due to the breaking of alveolar septa, the inhibitory Hering-Breuer reflex would at the same time have become less effective. This, by necessitating a deeper inspiratory effort to bring about the sensation of satisfaction associated with an adequate inhibitoinspiratory discharge or by making such a sensation impossible, is another possible factor in the situation. It should be understood that not all the Hering-Breuer receptors need be involved to bring about these results but only enough to set up a perceptible change in the balance of factors operating in respiratory control. Naturally, the larger the proportion of receptors involved, the more marked the excitatory influence on the center.

Reflexes from the air passages. Reflexes from the air passages may or may not be normal contributions to respiratory control, but they are included here because they certainly are implicated in the responses to irritation of the deeper air passages.[12,24] If they have any part in the production of dyspnea, it probably lies in their being responsible for active expiratory efforts (coughs) aimed at expelling abnormal material (irritant gases or vapors, foreign particles, secretions, fluid) from the air passages. The normal cough response involves strong activation, first of the inspiratory and then of the expiratory center, and impulses capable of bringing this about are potential contributors to clinical dyspneas associated with the presence of foreign materials in the air passages. Such would be the case in *bronchitis, pneumonia, asthma,* and any condition associated with fluid in the finer air passages and alveoli, such as *congestive heart failure.*

Reflexes from the carotid and aortic baroreceptors. Reflexes from the carotid and aortic baroreceptors, if important at all in producing dyspnea, should

Table 72-2. Dyspnea*

Condition	Alveolar ventilation at rest	Direction of chemical changes in arterial blood			Basis for dyspnea					
					Contribution of reflex factors (other than chemoreceptor)			Contribution of chemical factors		
								Chemoreceptor		Central
		PCO_2	$[H^+]$	PO_2	Hering-Breuer	Baro-receptors	Pain	PO_2	$PCO_2,$ $[H^+]$	$PCO_2,$ $[H^+]$
Metabolic acidosis	↑	↓	↑	↑					+	+
Poliomyelitis	↓	↑	↑	↓				+	+	+
Myasthenia	↓	↑	↑	↓				+	+	+
Emphysema	↓	↑	↑	↓	+			+	+	+
Asthma	↓	↑	↑	↓	+			+	+	+
Pneumonia	±	±	±	↓	+		+	+	±	±
Atelectasis	±	±	±	↓	+			+	±	±
Cardiac decompensation	±	±	±	↓	+	+		+	?	?
Pulmonary embolism	±	±	±	±	+		+			
Anemia	↑	↓	↓	↑						?
Hemorrhage	↑	↓	↓	↑		+				?
Hyperventilation	↑	↓	↓	↑						
Fever	↑	↓	?	↑					?	?
Stimulant drugs	↑	↓	↓	↑						

*Arrows are used to indicate the direction of ventilatory and chemical changes usually observed in the conditions cited. ± indicates inconstant direction. The probable contribution of a particular reflex or chemical factor is shown by +. Factors suspected to be important but for which little or no evidence exists are denoted by ?

The table lists a number of conditions in which dyspnea is a prominent symptom. In most the complete basis for dyspnea is not established. The table suggests that dyspnea may be associated with increased or decreased alveolar ventilation, an increase or decrease in any of the usual indices of chemical stimuli in arterial blood, and may occur even in the absence of known hematogenous or reflex stimuli.

Chemoreceptor and central mechanisms of chemical stimulation are both indicated as potential causes of dyspnea where hypercapnia and acidosis exist. This is done because there is still uncertainty regarding the relative contribution of these mechanisms to respiratory stimulation by acid-base changes. When *arterial hypocapnia* exists in conjunction with a normally reactive CNS circulation (metabolic acidosis, fever, hyperventilation, stimulant drugs), it is presumed that *central* PCO_2 and $[H^+]$ have also been lowered. However, when CNS circulation is diminished (shock) or the buffer capacity of the blood is reduced (anemia), it is considered that *central hypercapnia* and acidosis may exist concurrently with the arterial hypocapnia and alkalosis indicated in the table.

be most likely involved, along with the separate effects of metabolic acidosis, in the shortness of breath or air hunger associated with *hemorrhage, shock,* or *vasomotor collapse* and in the *hypotension produced by vasodilator drugs.* Their full effect, representing complete abolition of the inhibitory influence tonically exerted on the medullary centers, presumably would occur when the systolic pressure falls below a level of about 60 to 70 mm Hg.[49] Further respiratory stimulation on fall in pressure to levels lower than this must be attributed to something else, the most probable cause being accumulation of stimulant factors in the respiratory centers or peripheral chemoreceptors[49] because of inadequate blood flow and fixed acid formation due to tissue ischemia. Positive stimulant effects from the baroreceptors (i.e., effects not due simply to removal of a previously exerted inhibition) have not been proved to exist.

Pain impulses. Pain impulses undoubtedly can stimulate breathing reflexly and they probably play a part in the respiratory difficulties of *pneumonia, surgical shock, atelectasis,* and *pulmonary embolism.* Their influence on the center presumably is a nonspecific form of arousal and can be abolished by means of analgesic drugs. Their part in the total dyspnea must be extremely variable.

Chemical factors in production of dyspnea

Table 72-2 indicates that changes in the chemical composition of the blood and in the acid-base state of the centers should be expected to contribute to respiratory distress in a variety of conditions associated with dyspnea. However, it is not practical in subjective dyspnea to assess the proportion that each of several chemical factors contributes to the sensation of respiratory distress experienced in the various pathologic conditions listed. It is even likely that in any one patient the dominant cause of distress may differ from one time to another. Certainly the relative roles of chemoreceptors and direct central stimulation of the respiratory control system in producing dyspnea can now only be surmised.

On purely qualitative grounds *it appears that the influence of chemical factors in producing respiratory distress has been very much underestimated.* This has resulted from (1) difficulties encountered in the past in measuring the relatively small changes in pH and Pco_2 capable of producing respiratory stimulation, (2) failure to recognize the powerful interaction of factors such as increased Pco_2 and decreased Po_2 (Fig. 71-26), and (3) the common tendency to consider the composition of arterial blood alone (rather than the sometimes opposite changes in arterial blood and the centers) when assessing possible changes in chemical stimulus level (Fig. 71-10).

Subnormal alveolar ventilation not caused by narcotic drugs or other forms of central respiratory depression will result in dyspnea due to respiratory acidosis combined with hypoxemia (bulbar poliomyelitis, emphysema, asthma, and myasthenia gravis). Administration of O_2 to test for the presence of a hypoxic stimulus as a component of dyspnea leads, as ventilation decreases, to further respiratory acidosis and therefore little or no relief of dyspnea. Relief of one form of stimulation thus leads to exaggeration of the level and influence of another. Indirect observations of this type have contributed to the undoubtedly erroneous impression that hypoxemia is not a factor in the severe dyspnea of the conditions just cited.

Acidemia due to the presence of fixed acids in the blood (diabetic acidosis) produces hyperventilation and dyspnea in which the only apparent stimulant factor is increased acidity of the blood.[51] It is even possible that, because of the secondary reduction of arterial and central Pco_2,[51] the dyspnea of an acute metabolic *acidemia* exists in the face of the respiration-diminishing effect of an *alkalosis* in the regions beyond the blood-brain barrier (Chapter 71).

Excessive formation of fixed acids may also contribute to acidemia in severe tissue hypoxia in conditions in which (1) arterial O_2 tension is subnormal (Tables 72-1 and 72-2), (2) arterial Po_2 is normal but tissue blood flow is reduced (shock), and (3) arterial Po_2 is normal but the O_2 capacity of the blood is low (anemia, CO poison-

ing). In each case, tissue Po_2 is low and anaerobic metabolism leads to spilling of lactate and pyruvate into the blood.

Shunting of blood across a part of the lungs without exposure to alveolar gas results in an arterial hypoxemia due to admixture of aerated and mixed venous blood (Chapter 68). Peripheral chemoreflex stimulation and tissue anaerobiosis combine to induce respiratory stimulation. This problem, primary in certain congenital cardiovascular defects such as patent ductus arteriosus, is the cause of acute hypoxia in *atelectasis* and in early stages of lobar pneumonia. Alveolar ventilation is reduced or absent in the affected pulmonary segments and, because of respiratory stimulation (chemoreceptor), is increased in the unaffected alveoli. CO_2 elimination may be adequately accomplished by the functioning lung tissue; the direction of change in Pco_2 and $[H^+]$ is therefore not predictable. With extreme degrees of atelectasis, arterial Pco_2 and $[H^+]$ should rise. O_2 administration in this situation can raise arterial Po_2 only to a limited degree.[53]

Stimulation of the center by inadequate removal of its own metabolic products may be the basis for respiratory stimulation when (1) brain blood flow is subnormal (hypotension, increased CSF pressure) and (2) the buffer capacity of the blood is reduced (anemia, O_2 breathing). In such situations, changes in the gaseous composition of the arterial blood reflect the *influence* of alveolar ventilation and offer no indication of the cause of respiratory stimulation.

Impulses from the higher centers are a highly variable factor. The possible irradiation of impulses in the hyperpnea of exercise has its clinical counterpart in the nervousness, apprehension, and excitement that may be associated with *pain, congestive heart failure, asthma,* and other clinical states.

Dyspnea may be due to the primary increase in excitability of the respiratory centers. A clinical example is the hyperventilation sometimes seen in *encephalitis,*[47] which presumably is due to an irritative pathologic process involving the respiratory neurons themselves or other neurons that relay impulses carried to them from other parts of the neuraxis.

Hyperventilation alkalosis. Physically healthy but emotionally disturbed persons may be subject to attacks of breathlessness that lead to voluntary hyperventilation as the natural reaction to such sensations. This the *hyperventilation syndrome.* In some instances, unconsciousness and convulsions occur but apparently this is exceptional[62] except in epileptics, in whom a major seizure may thus be induced.[66] The motor phenomena are the familiar ones of tetany. The reasons for the different effects of hyperventilation on CNS function are complex and involve the effects of alkalosis and tissue hypoxia. The alkalosis shifts the dissociation curve for oxyhemoglobin to the left (Chapter 69), and this would necessitate a lower Po_2 in the brain tissues to secure the unloading of a given amount of O_2. Hyperventilation during air breathing leads to cerebral hypoxia due to constriction of the blood vessels of the brain. Other possibilities (change in pyruvate oxidation associated with shifts in bicarbonate concentration in the brain[34] and increased rate of hydrolysis of ACh due to alkalosis[42]) have also been suggested.

Interactions of factors in dyspnea

It is evident that dyspnea is associated with many conditions and in any single condition can result from a variety of causes. The dyspnea of *congestive heart failure* is among the most complex[37] and is mentioned to emphasize the necessity for evaluating the total stimulant effect of multiple, interacting factors. In this condition, pulmonary edema can result in interference with diffusion of O_2 from the alveoli into the pulmonary capillaries, even while CO_2 elimination is not appreciably affected. Ultimately, fluid within the alveoli can reduce alveolar ventilation, leading to hypercapnia and respiratory acidosis. The engorgement of lung capillaries and lung tissue may alter function of the Hering-Breuer reflex. Nerve endings in the air passages may be stimulated by the fluid. These factors, together with chemical stimulation of the peripheral chemoreceptors and respiratory centers, may all be potentiated by the increased metabolism resulting from the greater work of breathing and by impulses generated in the cerebral cortex by apprehension and fear. There can hardly be a single treatment for this or any other form of dyspnea due to so many contributing factors. Inability to identify a cause is not reason to doubt the existence or severity of dyspnea.

DISORDERS OF RATE AND RHYTHM

Abnormal respiratory patterns may take the form of acceleration of rate (*tachypnea* or *polypnea*) or changes in rhythm (*periodic breathing*).

Polypnea

Acceleration of respiratory rate without an increase in depth is not a normal response to physiologic stresses. However, it may occur in a variety of pathologic conditions, some of which will be mentioned.

Pain, originating in structures moved by breathing (abdominal wall, pleural membranes) may result in restriction of depth of breathing with a consequent increase in respiratory rate. This is presumably a factor in the production of postoperative atelectasis.

Atelectasis (collapse of pulmonary alveoli) can reduce volume available for inspiration and cause a shunting of unaerated blood, which lowers arterial Po_2. These and other factors sustain an excitatory drive.

During *fever,* depth of respiration is usually increased as well as rate, but during the recovery period, respiratory rate remains high while depth returns to normal or below.[28,56] This rapid, shallow respiration permits elimination of heat by evaporation of water from the respiratory passages and minimizes loss of CO_2. It resembles the normal mechanism for heat loss by furred or feathered animals that do not sweat. The tachypnea and the concomitant increase in cardiac rate may have a common mechanism, but this is not yet established (Chapter 71).

Irritation of the respiratory tract by gases and vapors can cause apnea or a form of tachypnea.

Periodic breathing

Periodic breathing as it occurs clinically is a series of dyspneic respiratory efforts interrupted by periods of no breathing at all. The commonest form encountered clinically is *Cheyne-Stokes respiration,* which is characterized by respiratory efforts that are at first weak but become successively stronger until definite hyperpnea is present. Breathing then gradually diminishes to the point of apnea. Rhythmic waxing and waning in respiratory depth short of complete apnea are more commonly seen and probably depend on the same underlying causes as classic Cheyne-Stokes breathing, which, however, implies distinct apneas.

The cyclic changes in respiratory activity may be accompanied by corresponding changes in consciousness, which is clouded or absent during the apnea, improves as the breathing becomes more active, is nearest normal during the period of increased ventilation, and fades again as the breathing diminishes. The blood pressure also is likely to show cyclic changes, usually (although not always) in the direction of a rise during the increase in breathing and a fall during the decrease; typically, it is highest toward the end of the dyspnea and lowest early in the period of breathing.

In most cases, Cheyne-Stokes breathing is probably a sign of central asphyxia. The course of events may be explained as follows: cortical cells and cells of the respiratory center are depressed by hypoxia and hypercapnia; breathing stops and arterial Po_2 falls (along with rising Pco_2), consciousness is lost, the activity of the vasomotor center is reduced, and blood pressure falls. The chemoreceptors, however, can still respond to the asphyxial changes that now occur in the blood. The resulting buildup of nerve impulses, together with the direct effects of the mounting Pco_2 on the hypoxic centers, are enough to arouse the depressed center and breathing begins. The excitatory reflexes themselves may conceivably indirectly also be a factor in arousing consciousness. The onset of respiration leads to oxygenation of the blood, which relieves the central hypoxia and improves neuronal functions; respiration becomes stronger, consciousness is less clouded, and blood pressure rises because hypoxic depression of both the vasomotor center and heart are relieved and the respiratory movements improve cardiac filling. The improved ventilation leads to an increase in Po_2 and a decrease in Pco_2 of the arterial blood. The rise in Po_2 diminishes any chemoreceptor drive that might have been present. The fall in Pco_2 decreases the chemical stimulus to the abnormal center, and respiration again begins to decrease.

Apparently the occurrence of Cheyne-Stokes respiration depends, in addition to the factors cited, on a prolonged interval between the modification of blood gas composition in the pulmonary capillaries and the arrival of the altered blood in the reactive centers.[57] An experimental analog of this condition has been devised in the dog.[45]

The occurrence of Cheyne-Stokes breathing is an ominous sign in disease, not because it is dangerous in itself, but because it is a sign of a central hypoxia sufficiently severe to depress cortical and subcortical functions below the physiologic level unless abnormal stimuli are brought into action. It is encountered in congestive heart failure, uremia, and disease or injury involving the brain. It may be seen in normal persons at high altitudes and has been described in early aviators and mountain climbers, in the latter while asleep, in the former before consciousness is lost completely during rapid ascent. It can also be produced in normal persons who are able to continue maximal voluntary hyperventilation to the point at which unconsciousness occurs. Finally, it is not uncommon in infants, especially in those born prematurely; in these the apneic periods may be dangerously long, suggesting injury to or imperfection of the respiratory regulating system.

Measures successful in relieving periodic breathing have included administration of O_2 (which should improve central oxygenation, restore normal reactivity to previously hypoxic brain vessels, and aid in stabilizing central Pco_2 and $[H^+]$).

RESUSCITATION

Restoration of respiration after acute respiratory failure requires (1) artificial ventilation of the alveoli to a degree sufficient to oxygenate the arterial blood and (2) the establishment or maintenance of cardiac activity adequate to assure delivery of enough oxygenated blood to the respiratory centers.

Failure of resuscitation results when the initiation of artificial respiration is too long delayed (leading to irreversible medullary or cardiac damage), when the alveoli are not adequately ventilated by the procedure used (allowing progressive hypoxic damage to brain and heart), or when heart action has ceased (or is too weak to sustain the blood pressure required for myocardial and central oxygenation). To these common possibilities must be added the situation in which the primary cause of respiratory failure is a pathologic change in the brain stem (abscess, tumor, cerebral vascular accident) or in the heart (coronary occlusion, ventricular fibrillation). When rhythmic cardiac contraction persists, success in resuscitation by adequate artificial ventilation alone is highly likely. When cardiac output is zero, as in ventricular fibrillation, artificial ventilation alone cannot be of any use.

In recent years, many important advances have been made in the evolution of resuscitation techniques. *Cardiopulmonary resuscitation* is a combined term, which emphasizes the need to consider both respiration and heart action in delivering O_2 to the tissues. Initially the attempts to sustain a degree of cardiac output with a fibrillating or asystolic heart was accomplished by ''openchest'' methods of squeezing the ventricles by hand. In this procedure the compression provides the propulsive force, filling of the heart is passive, and the cardiac valves direct the blood through the lungs and to tissues via normal vascular channels. The open-chest method, which is adapted to the hospital environment, cannot be considered generally useful and has been largely replaced by closed-chest cardiac resuscitation. Either must be begun immediately after acute cardiac standstill has occurred, and each requires artificial ventilation compatible with the cardiac massage. The most universally available and rational method is ''mouth-to-mouth'' ventilation, in which the exhaled air of the rescuer is used as the insufflating air of the victim. Electrical means for overcoming ventricular fibrillation may be needed, and in certain circumstances, continued rhythmic contraction is aided by an electrical cardiac pacemaker.

Nonmechanical methods of artificial respiration
Manual

Each of the several ''push-pull'' methods of manual artificial respiration involves pressure on the chest to expel air and a highly important additional means of causing a flaring of the ribs to expand the chest and draw air into the lungs. None approaches in effectiveness the more rational, oral methods of lung inflation. The now obsolete manual methods are the *back-pressure arm lift,* the *back-pressure hip lift,* and the *chest-pressure arm lift* methods. In each case the small inspiratory phase is accomplished without the aid of diaphragm action. Table 72-3 shows the average results obtained with these techniques in eight anesthetized adults rendered apneic by administration of succinylcholine and compares

Table 72-3. Mean pulmonary ventilation during artificial respiration on eight apneic normal men*†

Method	Gas moved per respiratory cycle (ml)	No. of subjects not ventilated at all
Normal resting tidal volume	540	0
Mouth-to-mouth resuscitation	910	0
Back-pressure arm lift	580	3
Back-pressure hip lift	650	2
Chest-pressure arm lift	450	4

*After Gordon et al.[44]
†Column indicating gas moved applied only to subjects who were actually ventilated by the procedure. The percentage of subjects not ventilated, because of intentionally unrelieved soft tissue obstruction, was as high as 50% with one of the methods and zero with mouth-to-mouth technique.

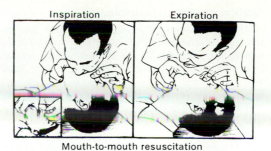

Mouth-to-mouth resuscitation

Fig. 72-3. Expired air inflation. For explanation see text. (From Gordon et al.[44])

these results with those obtained by mouth-to-mouth resuscitation.[43] The table is somewhat misleading in that it shows milliliters per respiratory cycle (tidal volume) in those patients who could be ventilated at all. However, because of respiratory obstruction by soft tissues in the oropharynx, certain of the subjects received no ventilation when one or another of the manual methods was employed.

Expired air inflation

Although a number of methods for artificial respiration not requiring mechanical apparatus have been designed over the years, none is as effective as *expired air inflation,* or mouth-to-mouth ventilation.* Table 72-3 compares results obtained with the several methods in apneic adults. The positive procedure, illustrated by Fig. 72-3, is undoubtedly the most effective of the methods of artificial ventilation not requiring ancillary apparatus.[68] The bases for its effectiveness are several, including (1) the positive nature of the insufflation, by which it is possible to determine that air is actually being delivered to the victim's lungs, (2) the ability to detect obstruction on a breath-by-breath basis, and (3) the free-

dom of the rescuer's hands to relieve soft tissue obstruction by extension of the neck and to prevent gastric distention by air.

The mouth-to-mouth technique is absolutely more certain than any manual method and should unquestionably be selected over a manual method.

Methods of cardiopulmonary resuscitation are being taught to large numbers of the population, with important effect.

Procedure. The victim is placed on his back and the rescuer positions himself at one side of the victim's head.[43] The rescuer puts a thumb in the victim's mouth and grasps the lower jaw between the thumb and fingers of that hand and pulls the lower jaw upward (Fig. 72-3). This pulls the root of the tongue and the epiglottis out of the oropharynx, clearing the airway. The other hand is used to occlude the nostrils. The rescuer then places his mouth over that of the victim, forces his own exhalation smoothly through the victim's airway to inflate his lungs, and then removes his mouth to allow passive exhalation by the victim. This procedure is repeated 12 to 20 times/min, with no need to sustain an inflation. Periodically, pressure over the epigastrium should be applied to aid in removing air from the stomach and thus to prevent gastric distention. Dizziness on the part of the rescuer signifies CO_2 loss by overbreathing and can be corrected by reducing the vigor of ventilation or an occasional brief period of breath holding. The mouth-to-mouth ventilation method can be sustained for several hours.

When used to resuscitate an apneic infant, only minor modifications of the procedure are required. These include more rapid ventilation (about 20/min), application of continuous pressure on the epigastrium with one hand to prevent gastric distention during inflation of lung, and covering of the nose and mouth of the infant with the mouth of the rescuer.

There are times when both cardiac compression and mouth-to-mouth pulmonary ventilation are required. Even this can be carried out by one rescuer who inflates the lungs twice with every 15 compressions of the heart.

*Variations of expired air inflation include methods not requiring equipment (direct mouth-to-mouth and mouth-to-nose) as well as those utilizing minor apparatus (mouth-to-mask, mouth-to-endotracheal tube, mouth-to oropharyngeal airway, or mask-to-mask). Each has its advantages in particular situations.

Carbon dioxide in resuscitation

The mechanisms of respiratory control are extremely sensitive to the effects of CO_2 and normally will regulate breathing so as to maintain the arterial P_{CO_2} and pH characteristics of eupnea, that is, about 40 mm Hg and 7.40. *Under abnormal conditions,* however, this is no longer true, and frequently the sensitivity of the central neurons to chemical stimulation is more or less drastically reduced, sometimes even to the point of abolition.

When ventilation ceases or decreases and metabolism continues, P_{CO_2} and $[H^+]$ must rise. Therefore the usual difficulty in resuscitation is *not* that of *providing* CO_2 to the apneic or subnormally ventilating individual but the *removal* of the CO_2 retained during the period of inadequate alveolar ventilation. Administration of CO_2 with O_2 would, in such individuals, exaggerate a hypercapnia and respiratory acidosis that might already be formidable. For this reason, CO_2 administration is not a normal part of resuscitation.

The toxic effects of CO_2 accumulation in apnea must be considered from the standpoints of (1) possible depressant actions of the CO_2 molecule and (2) derangements resulting from acidification of body fluids. Since CO_2 freely crosses cell membranes, a CO_2-induced increase in blood $[H^+]$ will be accompanied by similar acidification within the cells. For this reason a degree of arterial acidemia produced by CO_2 retention may be more damaging than a metabolic acidemia of the same order of magnitude (since fixed acids do not readily cross the blood-brain barrier and the hypocapnia of metabolic "acidosis" may actually lead to a central "alkalosis" in certain regions). This likely possibility has still not been well evaluated.

Apneic (diffusion) "respiration"

When apnea occurs in an individual who has been breathing air, it will lead to severe hypoxia and death in only a few minutes. However, if the lungs are artificially ventilated *thoroughly* with pure O_2 for 10 to 15 min, most of the alveolar nitrogen will be flushed out and replaced by O_2. If, in this condition, the patient's airway remains connected entirely to a source of pure O_2, apnea may persist for many hours without development of anoxemia.[35] The basis is as follows: During *air breathing* at sea level, alveolar gas pressures in mm Hg can be represented as:

$$Total\ pressure = P_{CO_2} + P_{H_2O} + P_{O_2} + P_{N_2}$$
$$760 = 40 + 47 + 100 + 573$$

Apnea, beginning with these normal alveolar gas tensions, will result in progressive anoxemia, leading rapidly to death. Anoxemia develops because, as O_2 is removed from the alveoli in accordance with the O_2 requirements, the volume of CO_2 that enters the alveoli is considerably smaller due both to the normal respiratory ratio and to the buffering of CO_2 by body tissues. Total alveolar pressure is therefore lowered and gas enters via the trachea to sustain the total pressure. This gas (air) contains only 21% O_2. Thus for each volume of O_2 required, only one fifth of a volume

of O_2 is supplied by the mass movement of air into the lungs, and an exponential decline of alveolar P_{O_2} must occur as the O_2 is displaced by nitrogen. *This leads to death within a period of minutes.*

After nitrogen has been flushed from the lungs by oxygen the initial alveolar gas pressures can be indicated essentially as follows:

$$Total\ pressure = P_{CO_2} + P_{H_2O} + P_{O_2} + P_{N_2}$$
$$760 = 40 + 47 + 673 + 0$$

During apnea, O_2 will again be removed from the alveoli at a rate equal to the metabolic O_2 consumption, but anoxemia and death will be long delayed. For the reasons previously cited (R < 1.0 and buffering of CO_2), CO_2 will enter the alveoli at a slower rate than the O_2 leaves, and total alveolar pressure will tend to fall. With the airway of the patient connected to a source of pure O_2, the pressure difference between the ambient atmosphere and the alveoli will cause a continuous mass flow of O_2 from the external 100% O_2 source. O_2 outflow from the lungs to the blood is almost exactly balanced by O_2 inflow into the lungs. Arterial oxygenation is thus maintained by a high alveolar P_{O_2} that is only slowly encroached on by the rising P_{CO_2}.

The limitations of this amazing situation should be clearly understood. This method is not a substitute for cyclic alveolar ventilation. Diffusion respiration is able to maintain sufficient oxygenation of the blood to forestall acute circulatory failure, but only if the N_2 is first flushed out of the alveoli by O_2. Connecting a source of 100% O_2 to the trachea of a patient who has ceased breathing will not prevent hypoxic death unless his lungs are first flushed thoroughly with pure O_2.

Even in the carefully denitrogenated individual, apnea will eventually produce death as a result of irreversible cellular damage by the extreme acidosis, but a considerable period of time is required, as great as several hours. In a patient maintained on apneic oxygenation with high arterial hemoglobin saturation, the arterial P_{CO_2} was 207 mm Hg after 20 min of apneic oxygenation, and the P_{CO_2} and pH were 314 mm Hg and 6.67, respectively, after 3½ hr.[2] *It is evident that ventilation, not CO_2 administration, is needed in apnea.*[53]

REFERENCES
General reviews

1. Bates, D. V., and Christie, R. V.: Respiratory function in disease, Philadelphia, 1964, W. B. Saunders Co.
2. Comroe, J. H., Jr., and Dripps, R. D.: Artificial respiration, J.A.M.A. **130:**381, 1946.
3. Dill, D. B.: Introduction to symposium on mouth-to-mouth resuscitation (expired air inflation), J.A.M.A. **167:**317, 1958.
4. Haldane, J. S., and Priestley, J. G.: Respiration, London, 1935, Oxford University Press, Ltd.
5. Haldane, J. S., Meakins, J. C., and Priestley, J. G.: The effects of shallow breathing, J. Physiol. **52:**433, 1919.
6. Kafer, E. R., and Leigh, J.: Recurrent respiratory failure associated with the absence of ventilatory response to

hypercapnia and hypoxemia, Am. Rev. Respir. Dis. **106:**100, 1972.

7. Kepron, W., and Cherniack, R. M.: The ventilatory response to hypercapnia and to hypoxemia in chronic obstructive lung disease, Am. Rev. Respir. Dis. **108:** 843, 1973.
8. Lambertsen, C. J., Bond, G., and Jacobson, J. H., II, chairmen: Symposium on hyperbaric oxygenation, Ann. N.Y. Acad. Sci. **117:**647, 1965.
9. Nielsen, M., and Smith, H.: Studies on the regulation of respiration in acute hypoxia, Acta Physiol. Scand. **24:**293, 1951.
10. North, J. B., and Jennett, S.: Abnormal breathing patterns associated with acute brain damage, Arch. Neurol. **31:**388, 1974.
11. Otis, A. B., et al.: Performance as related to composition of alveolar air, Am. J. Physiol. **146:**207, 1946.
12. Paintal, A. S.: Vagal sensory receptors and their reflex effects, Physiol. Rev. **53:**159, 1973.
13. Patterson, J. L, Jr.: Carotid bodies, breath-holding and dyspnea, N. Engl. J. Med. **290:**853, 1974.
14. Rahn, H., et al.: The pressure volume diagram of the thorax and lung, Am. J. Physiol. **146:**161, 1946.
15. Safar, P.: Ventilatory efficacy of mouth-to-mouth artificial respiration, J.A.M.A. **167:**335, 1958.
16. Schmidt, C. F.: Carotid sinus reflexes to the respiratory center, Am. J. Physiol. **102:**119, 1932.
17. Schmidt, C. F.: The cerebral circulation in health and disease, Springfield, Ill., 1950, Charles C Thomas, Publisher.
18. Schwerma, H., et al.: A study of resuscitation from the nearly fatal effects of exposure to carbon monoxide, Occup. Med. **5:**24, 1948.
19. Sokoloff, L.: The action of drugs on the cerebral circulation, Pharmacol. Rev. **11:**1, 1959.
20. Solliday, N. H., et al.: Impaired central chemoreceptor function and chronic hypoventilation many years following poliomyelitis, Respiration **31:**177, 1974.
21. Stroud, M. W., III, et al.: The effects of aminophylline and meperidine alone and in combination on the respiratory response to carbon dioxide inhalation, J. Pharmacol. Exp. Ther. **114:**461, 1955.
22. Symposium on the treatment of respiratory failure, Br. J. Anaesth. **9:**536, 1964.
23. Wechsler, R. L., Kleiss, L. M., and Kety, S. S.: The effects of intravenously administered aminophylline on cerebral circulation and metabolism in man, J. Clin. Invest. **29:**28, 1950.
24. Widdicombe, J. G., editor: Respiratory physiology II. International review of physiology, Baltimore, 1977, University Park Press, vol. 14.
25. Wolff, R. C., and Ward, R. J.: Cardiopulmonary resuscitation, technical documentary report No. SAM-TDR-10-62, Washington, D.C., 1963, USAF School of Aerospace Medicine.

Original papers

26. Adrian, E. D.: Afferent impulses in the vagus and their effect on respiration, J. Physiol. **79:**332, 1933.
27. Alexander, J. K., et al.: Analysis of the respiratory response to carbon dioxide inhalation in varying clinical states of hypercapnia, anoxia, and acid-base derangement, J. Clin. Invest. **34:**511, 1955.
28. Bazett, H. C.: Physiological responses to heat, Physiol. Rev. **7:**531, 1927.
29. Boutourline-Young, H. J., and Whittenberger, J. L.: The use of artificial respiration in pulmonary emphysema accompanied by high carbon dioxide levels, J. Clin. Invest. **30:**838, 1951.
30. Bronk, D. W., and Stella, G.: Afferent impulses from single end organs in the carotid sinus, Proc. Soc. Exp. Biol. Med. **29:**443, 1932.
31. Clemmesen, C.: Treatment of narcotic intoxication. Results and principles of the "Scandinavian" method, especially concerning stimulation, Dan. Med. Bull. **10:**97, 1963.
32. Comroe, J. H., Jr., Bahnson, E. R., and Coates, E. O., Jr.: Mental changes occurring in chronically anoxemic patients during oxygen therapy, J.A.M.A. **143:**1044, 1950.
33. Comroe, J. H., Jr., et al.: The lung, ed. 3, Chicago, 1975, Year Book Medical Publishers, Inc.
34. Craig, F. N.: The effect of carbon dioxide tension on the metabolism of cerebral cortex and medulla oblongata, J. Gen. Physiol. **27:**325, 1944.
35. Draper, W. B., and Whitehead, R. W.: Diffusion respiration in the dog anesthetized with Pentothal sodium, Anesthesiology **5:**262, 1944.
36. Dripps, R. D., and Comroe, J. H., Jr.: The respiratory and circulatory response of normal man to inhalation of 7.6 and 10.4 per cent CO_2 with a comparison of the maximal ventilation produced by severe muscular exercise, inhalation of CO_2 and maximal voluntary hyperventilation, Am. J. Physiol. **149:**43, 1947.
37. Ebert, R. V.: The lung in congestive heart failure, Arch. Intern. Med. **107:**450, 1961.
38. Eger, E. I., and Severinghaus, J. W.: The rate of rise of Pa_{CO_2} in the apneic anesthetized patient, Anesthesiology **22:**419, 1961.
39. Fenn, W. O., Rahn, H., and Otis, A. B.: A theoretical study of the composition of the alveolar air at altitude, Am. J. Physiol. **146:**637, 1946.
40. Field, J., II: Cell respiration and fermentation; a background for the study of narcotic action, Anesthesiology **8:**127, 1947.
41. Fink, B. R.: The stimulant effect of wakefulness on respiration: clinical aspects, Br. J. Anaesth. **33:**97, 1961.
42. Gesell, R., and Hansen, E. T.: Eserine, acetylcholine and nervous integration, Am. J. Physiol. **139:**371, 1943.
43. Gordon, A. S., et al.: Critical survey of manual artificial respiration, J.A.M.A. **147:**1444, 1951.
44. Gordon, A. S., et al.: Mouth-to-mouth versus manual artificial respiration for children and adults, J.A.M.A. **167:**320, 1958.
45. Guyton, A. C., Crowell, J. W., and Moore, J. W.: Basic oscillating mechanism of Cheyne-Stokes breathing, Am. J. Physiol. **187:**395, 1956.
46. Hahn, F.: Analeptics, Pharmacol. Rev. **12:**447, 1960.
47. Harrop, G. A., and Loeb, R. F.: Uncompensated alkalosis in encephalitis, J.A.M.A. **81:**452, 1923.
48. Heller, M. L., Watson, R. T., Jr., and Imredy, D. S.: Apneic oxygenation in man: polarographic arterial oxygen tension study, Anesthesiology **25:**25, 1964.
49. Heymans, C., and Neil, E.: Reflexogenic areas of the cardiovascular system, Boston, 1958, Little, Brown & Co.
50. Jowett, M.: The action of narcotics on brain respiration, J. Physiol. **92:**322, 1938.
51. Kety, S. S., et al.: The blood flow and oxygen consumption of the human brain in diabetic acidosis and coma, J. Clin. Invest. **27:**500, 1948.
52. Lambertsen, C. J.: Carbon dioxide and respiration in acid-base homeostasis, Anesthesiology **21:**642, 1960.
53. Lambertsen, C. J.: Therapeutic gases: oxygen, carbon dioxide, and helium. In DiPalma, J. R., editor: Drill's

pharmacology in medicine, ed. 3, New York, 1965, McGraw-Hill Book Co.

54. Lambertsen, C. J.: Drugs and respiration, Annu. Rev. Pharmacol. **6:**327, 1966.

55. Lambertsen, C. J., et al.: Comparison of relationship of respiratory minute volume to P_{CO_2} and pH of arterial and internal jugular blood in normal man during hyperventilation produced by low concentrations of CO_2 at 1 atmosphere and by O_2 at 3.0 atmospheres, J. Appl. Physiol. **5:**803, 1953.

56. Landis, E. M., et al.: Studies on the effects of baths on man; effects of hot baths on respiration, blood and urine, Am. J. Physiol. **76:**35, 1926.

57. Lange, R. L., and Hecht, H. H.: The mechanism of Cheyne-Stokes respiration, J. Clin. Invest. **41:**42, 1962.

58. Lerche, D., et al.: Vergleich der Wirkung verschiedener Acidosen (NH_4Cl, $CaCl_2$, Acetazolamid) auf die Lungenbelulftung beim Menschen, Arch. Gesamte Physiol. **270:**450, 1960.

59. Loeschcke, H. H., and Gertz, K. H.: Einfluss des O_2-Druckes in der Einatmungsluft auf die Atemtatigkeit des Menschen, gepruft unter Konstanthaltung des alveolaren CO_2-Druckes, Arch. Gesamte Physiol. **267:**460, 1958.

60. Moyer, J. H., et al.: Effect of morphine and n-allylnormorphine on cerebral hemodynamics and oxygen metabolism, Circulation **15:**379, 1957.

61. Nilsson, E.: On treatment of barbiturate poisoning, a modified clinical aspect, Acta Med. Scand. **139**(suppl. 253):1, 1951.

62. O'Donovan, D. K.: The hyperventilation syndrome, Irish J. Med. Sci. series **6:**519, 564, 1943.

63. Richards, D. W.: Pulmonary emphysema: etiologic factors and clinical forms, Ann. Intern. Med. **53:**1105, 1960.

64. Richards, D. W., Fritts, H. W., Jr., and Davis, A. L.: Observations on the control of respiration in emphysema: the effects of oxygen on ventilatory response to CO_2 inhalation, Trans. Assoc. Am. Physicians **71:**142, 1958.

65. Riley, R. L., and Cournand, A.: "Ideal" alveolar air and the analysis of ventilation-perfusion relationships in the lungs, J. Appl. Physiol. **1:**825, 1949.

66. Robinson, L. J.: Electroencephalographic study during epileptic seizures related to hyperventilation, Dis. Nerv. Syst. **5:**87, 1944.

67. Roth, L. W., Whitehead, R. W., and Draper, W. B.: Studies on diffusion respiration. II. Survival of the dog following a prolonged period of respiratory arrest, Anesthesiology **8:**294, 1947.

68. Sadoul, P., et al.: Ventilation par insufflation orale études expérimentales, Ann. Med. Nancy **87:**272, 1962.

69. Wechsler, R. L., et al.: Blood flow and oxygen consumption of the human brain during anesthesia produced by thiopental, Anesthesiology **12:**308, 1951.

70. West, J. B., and Hugh-Jones, P.: Pulsatile gas flow in bronchi caused by the heart beat, J. Appl. Physiol. **16:**697, 1961.

71. Winterstein, H.: Die Narkose, Berlin, 1926, Julius Springer.

CHRISTIAN J. LAMBERTSEN

73

Hypoxia, altitude, and acclimatization

The biologic combustion that provides the energy for all biophysical and chemical mechanisms collectively called "physiologic" depends on adequate O_2 partial pressure at the infinitely numerous sites of combustion. Lack of O_2 (hypoxia) at any or all of these sites may result from a great variety of environmental, toxicologic, and pathologic causes. These encompass acute and chronic disease processes, various forms of accidental suffocation, and acute and chronic exposure to low atmospheric pressures in mountainous regions, as well as stresses of aerospace and undersea operations. Regardless of the large number of mechanisms capable of producing hypoxia, the ultimate defect in each instance is the failure of oxidation *within the tissue cell*. With this in mind, the terminology that has evolved to describe different forms of O_2 lack can be employed for rational description of the circumstances involved.

OXYGEN DEFICIENCY IN DISEASE
Types and causes

The general term *"hypoxia,"* meaning lower than normal O_2, must be expanded to be useful because it does not even indicate location (e.g., atmosphere, arterial blood, or brain cortex) or what is subnormal (e.g., O_2 content, HbO_2 saturation, or Po_2). *Dysoxia*, a term proposed to characterize abnormal *use of O_2 in the tissues*,[29] is likewise general and does not itself define a basis for abnormality. Dysoxia is not synonymous with hypoxia, and in their overall function, they serve the different purposes of describing disruption of oxidation and deficiency of oxygenation, respectively. *Anoxia*, an older term, has the absolute connotation of "without O_2," and remains useful for this purpose.

Classification of the many forms of O_2 deficiency aids in emphasizing both the varied causes and the differences to be expected when in treatment, as with O_2. A useful classification, based on the ways in which hypoxia may be produced, is summarized in Table 73-1 and the following paragraphs.

Hypoxemia

This general term means diminution in the amount of O_2 carried by the blood. Since some forms of cellular anoxia occur despite normal oxygenation of the blood, *hypoxemia* is not synonymous with hypoxia. When the term "hypoxemia" is used without further qualification, it generally refers to the arterial blood, since most clinical considerations involving hypoxemia have to do with this. It then necessarily tends to also involve the entire body, and cardiovascular functions and tissue oxidative systems are affected as a result of the failure of O_2 delivery. From the standpoint of influences on respiration and O_2 delivery it is desirable to divide arterial hypoxemia into two types: hypotonic and isotonic.

Hypotonic hypoxemia. In hypotonic hypoxemia the disturbance is primarily due to a decrease in arterial O_2 tension; arterial O_2 content is correspondingly low.

This situation may be brought about by the following:

1. Interference with the passage of O_2 from the air into the blood in the lungs (mechanical obstruction, drowning, spasm or other occlusion of glottis or bronchi, insufficient respiratory movements, reduction in size or permeability of the aerating surface by congestion, edema, exudates, consolidation, adhesions, collapse, emphysema, or fibrosis)
2. Impairment of the pulmonary circulation (engorgement, admixture of venous and arterial blood due to patent congenital or acquired communications, flow of blood through unventilated areas of the lung)
3. Inhalation of air containing O_2 at subnormal pressure (high altitudes, fire, inade-

Table 73-1. Classification of types of hypoxia

Terminology			Characteristics	Occurrence
Hypoxemia (diminished O_2 in blood)	— Hypotonic		Decreased volume and tension of O_2 in arterial blood; chemoreceptor activation by diminished P_{O_2}; cyanosis when Hb concentration is adequate	Any condition leading to lowered arterial P_{O_2} (respiratory obstruction, decreased alveolar ventilation, drowning, lowered inspired P_{O_2}, asphyxia, transpulmonary shunt, reduced permeability of alveolar membranes)
	— Isotonic		Arterial O_2 content low but P_{O_2} normal; chemoreceptor stimulation not a feature; cyanosis absent; may be skin color of methemoglobinemia, carboxyhemoglobinemia	Anemia, poisoning by CO and by drugs that cause methemoglobinemia (aniline, acetanilid, nitrophenol, chlorates, methylene blue)
Hypokinetic hypoxia (diminished blood flow)	Ischemic	— Local	Subnormal local arterial supply; local cyanosis or pallor; pain	Arterial embolus, thrombosis, spasm, obliteration, damage; high extravascular pressure (e.g., CSF)
		— General	Subnormal arterial supply to entire body; hypotension, pallor, syncope, air hunger, metabolic acidosis	Acute circulatory collapse (spinal anesthesia, depressor drugs, acute cardiac failure)
	Congestive	— Local	Impediment to venous return; edema, capillary engorgement, and cyanosis	Thrombosis, external pressure
		— General	High systemic venous pressure with capillary and venous engorgement; cyanosis	Questionable—perhaps congestive heart failure or polycythemia
Overutilization hypoxia (excessive O_2 requirement)			Demand for O_2 (local or general) increased relative to supply; may be local or general	Convulsions, O_2 debt in exercise, angina, intermittent claudication
Histotoxic hypoxia (failure of metabolism)			Acute depression or inactivation of cellular oxidative systems	Poisoning by cyanide, sulfide, O_2, etc.

quate ventilation of compartments, dilution of the inspired gas with inert gases such as nitrogen, hydrogen, helium, methane, or anesthetic gases such as nitrous oxide and ethylene)

The distinguishing characteristics of hypotonic hypoxemia not due to central depression are dyspnea and respiratory stimulation, cyanosis, and tachycardia. Convulsions may occur if the condition is severe and sudden in onset. The respiratory stimulant phenomena are largely due to chemoreceptor reflexes. If circulatory failure occurs early, cyanosis may be replaced by pallor, and in a concurrent severe anemia, cyanosis is impossible.

Isotonic hypoxemia. In isotonic hypoxemia the arterial O_2 content is reduced, but the O_2 tension remains near the normal level. Such conditions are due primarily to decreased concentration of active hemoglobin (Hb) and therefore are seen in anemia and in poisoning with substances that combine with Hb in such a way as to limit the uptake of O_2 (carboxyhemoglobin or methemoglobin). Some lowering of arterial Po_2 may occur in these conditions because of the greater influence of existing transpulmonary shunts. The distinguishing characteristic is lack of dyspnea or other signs of severe anoxia, a situation caused by the fact that in the presence of nearly normal arterial blood gas tensions the chemoreceptors are not prominently stimulated.[56,69] Cyanosis may be marked if methemoglobinemia is present but is absent in severe poisoning by CO because carboxyhemoglobin is red and the amounts of reduced Hb probably would be below the threshold for cyanosis even in fatal poisoning by this gas.

Hypokinetic hypoxia

This denotes interference with the circulation of blood. It can occur with or without arterial anoxemia. It may even be present during inhalation of pure O_2 at sea level, since the fundamental defect is subnormal flow of Hb molecules through the tissue capillaries. In the *ischemic* form the circulatory abnormality is on the arterial side, and the amount of blood in the capillaries would be diminished. In the *congestive* type the interference with flow is on the venous side and the capillaries are engorged. Either type may be *general,* involving the entire body, or *local,* affecting only certain tissue components, organs, or parts of the vascular bed.

Local ischemic hypoxia results from occlusion, spasm, or interruption of the arterial supply to a given tissue, as in embolism, arteriosclerosis, Raynaud's disease, or traumatic injury. A special type is seen in cerebral ischemia associated with elevated intracranial pressure.

General ischemic hypoxia is encountered in the acute circulatory depressions due to vasomotor paralysis, syncope, peripheral paralysis of vasomotor nerves by block or section, shock, vasodilator drugs, or primary cardiac failure, including the congestive, low-output type.

Local congestive hypoxia is produced by organic obstruction (external pressure, thrombosis) to the venous return from a given tissue. An arteriovenous fistula leading to a local elevation in venous pressure would have a similar effect.

General congestive hypoxia is seen characteristically in congestive heart failure or other conditions in which systemic venous pressure is elevated.

Overutilization hypoxia

Overutilization hypoxia is due primarily to an increase in the demand of living tissues for O_2 to a level so in excess of the available supply as to lead to an acute O_2 deficiency in them. This is a normal consequence of vigorous muscular exercise, leading to diminished O_2 tension in the working muscle, anaerobiosis, and development of an "O_2 debt."A striking but impure example of this relationship is seen in the prolonged period of cerebral depression following convulsions elicited by electrical shock or analeptic drugs or occurring idiopathically in epilepsy. In each case the increased demand for O_2 is compounded by the anoxemia of breath-holding during the violent exercise of a generalized convulsion. The prolonged depression of the brain following a convulsion is associated with a diminution in cerebral O_2 consumption, which indicates a slowly reversible or irreversible derangement of cellular mechanisms.[147]

The most severe and dangerous situations are those in which there is increased functional activity together with restricted blood supply, that is, a combination of hypokinetic and overutilization hypoxias, such as increased cardiac activity in the presence of coronary occlusion. This combination can readily be produced in the laboratory by exercising an arm or leg while its blood supply is cut off by a blood-pressure cuff or tourniquet. It is seen in the muscular weakness and pain associated with exercise in the presence of peripheral arterial disease.

Histotoxic hypoxia

Histotoxic hypoxia is due to interference with the ability of the tissues to use O_2 even though the supply is normal or greater than normal. The underlying defect is an interference with the ability of the cell enzyme systems to use molecular O_2 even though the supply of the gas is normal in quantity and tension. The characteristic example is the union of cyanide or sulfide with the Fe^{2+} of cytochrome oxidase. Another is O_2 toxicity, a condition in which an excessive pressure of O_2 itself inactivates oxidative metabolism.[59,104]

The phenomena of narcosis are not included here because the implications of a narcotic depression of cell oxidations are quite different from those of hypoxia from the standpoints of reversibility, prognosis, management, and probable mechanism.

Effects of generalized hypoxia

Hypoxia localized in a particular volume of tissue is usually due to ischemia, and the effects are the result of local dysfunction with or without pain (coronary occlusion, cerebral thrombosis). The presence of generalized hypoxia is sometimes more difficult to establish. There are so many ways in which hypoxia can develop that it is not surprising that no single pattern of symptoms and signs can be expected and that difficulty may be encountered even in detecting its existence. In addition to the many precipitating causes, the effects of hypoxia will vary with its *severity*, with its *rate of development*, and with its *duration*. These are the features of hypoxia that determine the degree of dysfunction produced and therefore the nature of the observable effects.

Regardless of the dynamics of its development or even the tissue affected, it is probable that the mechanism of the disruptive actions of hypoxia is fundamentally the same in all forms not involving specific poisoning of tissue enzymes. This mechanism entails failure of the processes of energy expenditure involved in maintaining the physicochemical relationships necessary for normal cellular function.

Susceptibility of different types of cells to hypoxia

The susceptibility to hypoxia is often cited as varying widely from one tissue or organ to another. However, in view of the single etiologic event (decreased cellular Po_2) and the specific biochemical effects of most types of hypoxic failure, it is more likely the effect, rather than the basic "sensitivity," that varies.

It is well known that when arterial hypoxemia occurs, the most obvious failure of function associated with this generalized cellular anoxia occurs in the CNS; that is, consciousness is lost at levels of arterial O_2 tension that still support cardiac, renal, and digestive functions. The order in which the different components of the CNS fail follows roughly the same order of "descending depression" seen with increasing dosages of a narcotic drug.[90] The heart is generally considered second to the brain in overall susceptibility to hypoxia, but this is also probably merely an indication of criticality of effect. Actually, little is known regarding susceptibility of various cell types to functional derangement by hypoxia, and it is probable that "sensitivity to hypoxia" of a tissue or organ is more related to relative rates of blood (O_2) flow and O_2 consumption than to differences in susceptibility of common enzyme systems within the cell. The single spinal motoneuron has been found to be remarkably unaffected by hypoxia, maintaining its polarization at extremely low levels of Po_2.[127] Beyond this, it should be recognized that functional relationships among organized groups of nerve cells may be affected by lesser degrees of hypoxia than will seriously modify the activity of individual cells, and the effects on the general organism of cerebral, medullary, or cardiovascular derangement are acutely disruptive and readily apparent. However, hypoxic depression of other vital functions such as insulin secretion, tubular reabsorption of glucose, or gastrointestinal motility is not detected with the same ease as is syncope or acute myocardial failure due to hypoxia.

On the basis of these examples, several aspects of differential susceptibility to hypoxia can be visualized, including the following:

1. The ability of certain cellular reactions and functions to occur at a lower than normal Po_2 can be visualized. Thus *skeletal muscle*, capable of deriving energy for contraction by way of anaerobic processes, can be contrasted with the *brain*, which has practically no functional alternative to aerobic oxidation of glucose.

2. The ability of an organ to sustain a balance between its O_2 supply and O_2 requirement can also be visualized. Again, the brain, with its high rate of O_2 utilization, and the heart, an exercising muscle organ even when the body is at rest, can tolerate less diminution of O_2 supply than can the slowly metabolizing skin.

3. The ability of an organ to carry out its function even when the more susceptible cells of the organ have succumbed to hypoxia can be visualized. In this instance a highly organized system such as the brain, with complex electrophysiologic and chemical interdependence among its components, would be expected to show derangement before an organ such as the liver, which is primarily concerned with synthesis or detoxification. The relative susceptibilities of different parts of the CNS, including the apparent resistance of respiratory and vasomotor centers to hypoxia, may in this manner depend on the nature of the networks and the number of cells involved rather than on the complexity of the chemistry within individual neurons.

Influence of rate of onset of hypoxia

The speed with which O_2 deficiency develops, whether in disease or under environmental stress, has considerable influence on the severity of its effects. Therefore an elderly patient with chronic, severe cardiopulmonary disease may tolerate a degree of hypoxemia that could within several hours cause the death of a young patient with pneumonia. Death can also result from a few minutes of exposure to the extreme O_2 lack of respiratory obstruction, sudden loss of cabin pressure in an aircraft flying at an altitude above 35,000 feet, puncture of a space suit on the lunar surface, or inhalation of an inert gas (such as N_2O). The term *"fulminating anoxia"* has been used to describe the extremely rapid development of unconsciousness (and failure of vital functions) that can take place in situations such as those last mentioned.

By definition, the distinguishing characteristic of fulminating hypoxia is the extremely rapid onset of symptoms; the latent period is a matter of seconds because the O_2-deficient air will quickly approach Po_2 equilibrium with the arterial blood. This process is further accelerated by the fact that the depth and rate of respiration are increased by severe hypotonic hypoxemia. The immediate symptoms are dizziness, dimness of vision, and breathlessness, followed within a few seconds by unconsciousness and collapse. The circulation is usually stimulated at first (via the chemoreceptor reflex) and pulmonary hypertension may occur.[162] If O_2 is not soon supplied, failure of both circulation and respiration will follow. In some cases there is little or no circulatory stimulation.

The extreme rate at which arterial and venous O_2 tensions are lowered in normal men following an abrupt decrease of inspired Po_2 is shown in Fig. 73-1.[74] After a lag of less than 5 sec a precipitous fall in arterial Po_2 occurs. The subjects lose consciousness within another few seconds and, if an O_2-containing gas were not provided, would die in 2 to 4 min. The rate of fall of tissue venous O_2 tension is not as rapid as the decrease in arterial Po_2.

In the "blackout" due to acceleration, as seen characteristically in the excessive gravitational stress that results when a fighter aircraft turns too fast, the episode is referable to cerebral ischemia rather than to extreme arterial hypoxemia, although disturbances of pulmonary circulation and blood distribution within the lungs lead to lowered arterial Po_2.[170] The arterial pressure in positive acceleration is greatly reduced (even to subatmospheric levels), the pulse disappears, vision is impaired or lost ("grayout" or "blackout"), and consciousness is clouded or abolished; the latent period is about 5 sec, this apparently being the time required for the retinal and brain tissues to develop hypoxia of this degree when the cerebral vessels are emptied of blood.

When the cause of a fulminating hypoxia is removed, recovery may be almost as rapid as the onset of symptoms—but only if the exposure was brief. One of the main attractions of nitrous oxide anesthesia for dental extractions has been the rapid and complete recovery expected if the exposure lasts only 1 or 2 min. However, the speed of onset has sometimes been ac-

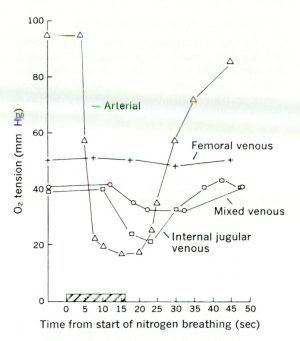

Fig. 73-1. Effect of transient overventilation with 100% nitrogen on O_2 tension of arterial blood and venous blood. Each point represents mean of values obtained from three subjects. (From Ernsting.[74])

celerated by reducing below normal the concentration of O_2 given with the nitrous oxide, even to the point of producing severe hypoxemia. The longer an hypoxia severe enough to produce rapid unconsciousness is allowed to act, the greater are the chances of acute cardiac failure and slowly reversible or irreversible injury to the brain. The latter changes are well known in conjunction with nitrous oxide anesthesia[58] as well as aviation accidents.[167] The heart is particularly vulnerable to hypotonic hypoxemia because the rate of O_2 utilization by the working heart is high and the blood in the coronary veins contains less than 5 vol% O_2 even when the arterial blood has its normal quota of 20 vol%. Therefore there is practically no reserve of O_2 when the arterial content falls sharply.[71,72] If at the same time the work of the heart is increased by an acceleration of rate and a rise in aortic pressure, the danger of acute cardiac failure is further intensified. Conversely, vasomotor collapse with consequent arterial hypotension decreases the work that the heart is required to perform; at very low arterial pressures, some cardiac activity may therefore continue for many minutes after breathing has failed.

Fulminating hypoxia can also be produced by overwhelming absorption of cyanides, sulfides, or CO, but these are special cases involving specific chemical combinations and recovery would not be rapid and complete even if the exposure were terminated immediately after unconsciousness occurred.

Influence of duration of hypoxia

The decrease in functional activity produced by hypoxia depends on derangements that are reversible only if the product of intensity times duration of exposure is small. Otherwise irreversible damage is produced, recovery does not occur when the hypoxia is removed, and permanent morphologic changes may take place. The margin between reversibility and irreversibility may be very narrow.[158] In these respects, hypoxia differs from narcosis, which is fully reversible, has a relatively wide margin of safety, and is therefore much more benign than hypoxia.

The symptoms and effects of acute hypoxia do not disappear immediately when the normal O_2 supply is restored. If the exposure is short, the *patient* successfully treated for atelectasis, the *balloonist or glider pilot* as he returns from a flight at high altitude, the *traveler* who has spent only a few hours on a mountaintop and then returns to a lower altitude, or the *subject of a laboratory experiment* as the O_2 pressure is increased from a severely low level may notice first that lights are brighter and sounds are louder. Muscular power, judgment, and self-criticism return. The longer the exposure lasts, the slower the recovery when normal conditions are restored. Emotional disturbances, nausea, or weakness may continue after the subject has returned to sea level conditions.

Repeated or sustained exposures to reduced O_2 pressures can produce marked deterioration in cerebral functions. Evidently the brain does not recover completely from a brief exposure to severe hypoxia or from a prolonged exposure to moderate hypoxia until many hours (24 to 72)[35] have elapsed, and the more severe the O_2 lack or the longer the exposure the longer the time required for complete recovery. If the exposure is repeated before recovery has taken place, it will have cumulative effects; that is, the derangements produced by today's exposure are superimposed on the remains of yesterday's. The most vulnerable parts of the brain then show signs of damage between exposures.[58,158] Lewy and Drabkin[113] exposed dogs for 5¾ hr daily to a concentration of CO (0.01%) that produced about 20% carboxyhemoglobin, representing an hypoxia too mild to elicit any acute symptoms. After about 3 months the animals showed signs of degenerative changes in the cortex and basal ganglia of the brain, which were confirmed by postmortem examination. The lesions were similar in location and type to those produced by very acute hypoxia. There is evidently a narrow and uncertain margin of safety between completely and promptly reversible neuronal derangements on the one hand and slowly reversible or irreversible changes on the other.

Detection of hypoxia—its presence and degree

Because of the reserves of mental and other functions, it is difficult to recognize the existence of hypoxia in disease, to determine its degree, and to monitor its therapy with O_2. Differences due to the type, cause, severity, and duration of hypoxia are further exaggerated by the presence of other symptoms of the concurrent disease process. For these reasons most of the physiologic signs of hypoxia have limited clinical value.

Dyspnea and *hyperventilation* may be prominent results of chemoreceptor drive, as in hypoxemia caused by atelectasis. However, respiratory stimulation is not a useful guide to hypoxia in the isotonic hypoxemia of CO poisoning (minimal chemoreceptor stimulation) or the hypotonic hypoxemia produced by the reduced ventilation in narcotic poisoning (hypoxia secondary to respiratory depression).

Pulse rate, less well compensated by control mechanisms than respiration, tends to increase progressively when the chemoreceptors are activated by hypoxia. However, this sign is useful only when the hypoxia is one associated with a lowered arterial P_{O_2}.

Cyanosis may be evident as a bluish color change of the skin, nail beds, or mucosae, the degree of coloration in a particular individual being roughly proportional to the unsaturation of the blood. Actually the blue color in cyanosis is due to the *absolute* concentration of reduced or otherwise unoxygenated Hb in the capillaries and not to the relative proportions of reduced and oxygenated Hb. It has been found that about 5 gm reduced Hb/100 ml capillary blood or an O_2 unsaturation of 6.70 ml/100 ml (since 1 gm of Hb can combine with 1.34 ml O_2) is about the threshold level at which cyanosis appears.[118,132] Consequently, an anemic subject who has less than 5 gm Hb/100 ml blood (= an O_2 capacity of about 7 ml/100 ml or less) cannot become cyanotic.

Conversely, a patient with polycythemia and an abnormally high concentration of Hb may show cyanosis even though the degree of Hb unsaturation is less than normal. The cyanotic color of the skin or other surface such as the nail beds or conjunctivae is mainly due to the color of the blood in the capillaries. Since the blood is continually giving up O_2 during its passage through the capillaries, the average unsaturation of the blood in its transit through them may be taken as approximately the mean between the arterial and venous unsaturations.

Cyanosis must for many reasons be considered an imperfect indicator of hypoxia. When present, it reveals only the state of the superficial circulation. In the *absence* of arterial hypoxemia, blue coloration of the skin can be produced by a *sub-*

normal rate of cutaneous blood flow (chilling, peripheral vascular disease, heart disease). Cyanosis increases with the amount of Hb present because at a given O_2 unsaturation there will be more reduced Hb; there may be marked cyanosis in conditions of *polycythemia,* with little or no anoxemia. *Anemia, on the contrary, decreases* the tendency to become cyanotic while increasing the likelihood of hypoxia. Acclimatization reduces the symptoms of hypoxia, but cyanosis remains unchanged. *It is only in conditions in which O_2 lack is due primarily to imperfect oxygenation of the blood, in which the Hb content of the blood remains unaltered, and in which the onset and course are too rapid for acclimatization to occur that a close parallelism exists between cyanosis and symptoms of hypoxia. Such conditions exist in acute hypoxias arising from interference with the gas exchange in the lungs.*

In certain forms of O_2 deficiency there are abnormal discolorations of the skin not related to reduced Hb concentration. *Methemoglobin* is produced by a number of drugs and poisons, among which nitrites, acetanilid, aniline, nitrobenzene, the nitrophenols, sulfanilamide and its congeners, methylene blue, and chlorates are well recognized. The iron is oxidized to the trivalent form and cannot combine reversibly with O_2. Such Hb is dark, and cyanosis appears as a result. This, however, is an isotonic hypoxemia and the subjects may have no chemoreflex respiratory stimulation, although they are deeply cyanosed (in severe isotonic hypoxia the development of a metabolic acidosis due to anaerobic metabolism in the tissues may lead to respiratory stimulation). The color of the skin is due to the methemoglobin, not to reduced Hb.[44] *Hydrogen sulfide* may inactivate Hb, possibly in more than one way. *In poisoning by CO there is no true cyanosis* because carboxyhemoglobin is red, not blue. *In poisoning by cyanide or other agents that produce a histotoxic hypoxia there also is no cyanosis* because the tissues take less O_2 out of the blood than they normally do, and there is therefore less reduced Hb in the capillaries than in the normal state.

Determination of degree of hypoxia by measurement of oxygen in blood

Measurements of oxygenation of *arterial* blood are extremely valuable when the condition involves an arterial hypoxemia, as in situations involving (1) inefficient pulmonary gas exchange (respiratory depression, emphysema) and (2) abnormalities of blood O_2 transport (e.g., CO

poisoning). They are not useful for evaluating the degree of O_2 lack produced by an inadequate rate of blood flow through an organ (coronary insufficiency) or the entire body (cardiac decompensation). For the appraisal of the state of oxygenation of an organ such as the brain, it is sometimes useful to measure the O_2 composition of the brain *venous* blood. This provides an overall index of the combined adequacy of arterial oxygenation (pressure, content, and capacity) and tissue blood flow in meeting the existing level of metabolism. Even this procedure provides only an average for the organ as a whole and cannot identify localized regions of hypoxia within the organ.

Measurements of particular interest for appraising the state of oxygenation are the O_2 partial pressure (P_{O_2}), spectrophotometric measures of oxyhemoglobin and reduced Hb, and the manometrically determined O_2 content and O_2 capacity, which together indicate *percent Hb saturation* (Chapter 69). O_2 capacity can be closely estimated from accurate measurement of Hb concentration, since each gram of Hb can combine with 1.34 ml of O_2. Actually the terms "O_2 capacity" and "percent Hb saturation" are misleading when CO Hb or methemoglobin are present, since these forms are unavailable for combination with O_2.

If measurement of pH is made, the P_{O_2} or O_2 tension can be estimated from the previously mentioned determination of percent Hb saturation by use of Hb dissociation curves at various pH levels.[132,148] When this method is employed, the values of P_{O_2} may be grossly inaccurate in the normal range because of the flatness of the upper portion of the Hb dissociation curve. This flatness of the curve causes a small error in Hb saturation to result in a large error in reading the P_{O_2}. For this reason, measurement of P_{O_2} is now accomplished by a polarographic "O_2 electrode" technique.[109,134] This value, arterial O_2 tension, is necessary for any evaluation of the degree of respiratory insufficiency, and if measurement is made on venous blood from an organ, it provides an approximation of the lowest pressure of O_2 in the tissue capillary.

In the course of determining the state of blood oxygenation, little additional effort is required to estimate the degree of respiratory and metabolic acidosis. The CO_2 content of whole blood is determined in the course of the manometric measurement of O_2 concentration. This, together with pH and O_2 capacity, permits calculation of P_{CO_2} by means of nomograms or by use of the Henderson-Hasselbalch equation, on which the nomograms are based (Chapter 69). P_{CO_2} can also be measured by means of special electrodes.

Failure of symptomatic self-detection of acute hypoxia

When an individual rapidly develops hypoxia, he cannot be counted on to recognize the fact, nature, or degree of his difficulty. As hypoxia develops, the progressive depression of CNS functions affects the powers of introspection, discrimination, logic, and judgment. Confusion and euphoria may be recognized by the patient, but these symptoms are not specific for hypoxia. With increase in the severity of the hypoxia, sensory disturbances develop and include diminished visual and auditory acuity and decreased sensitivity of touch and position sense. Still later, muscular weakness with lack of coordination becomes prominent and, ultimately, unconsciousness occurs. *Ordinarily this sequence of smoothly developing abnormalities will be completely unrecognized by the victim if it occurs during the course of a few minutes or less and does not involve respiratory obstruction or accumulation of CO_2.* Since no distressful sensations are produced, the entire experience is comfortable and even rather pleasant. Recognition of hypoxia may be possible when decreased oxygenation develops slowly. However, by the time it is recognized the individual may lack the ability to aid himself.

VALUE OF OXYGEN ADMINISTRATION IN HYPOXIA

Cellular hypoxia can be relieved only by reducing the demand for O_2 or by increasing the O_2 tension at the cell. Although the second possibility is usually more desirable, the administration of O_2 by inhalation is not uniformly effective in relieving all types of hypoxia at the cel-

lular level. The reasons for this are evident in Table 73-2, in which the degree of improvement in the oxygenation of arterial blood produced by O_2 breathing is shown for six conditions in which death may result from O_2 deficit. These examples include local ischemic hypoxia, depression of the respiratory control mechanisms, inefficiency of alveolar ventilation, defect in the pulmonary circulation, general circulatory collapse, disturbance in O_2 transport, and cellular enzymatic intoxication. In evaluating the effects of O_2 breathing in these conditions, it must be realized that, even without a change in blood flow, cellular P_{O_2} can be elevated by increasing arterial O_2 *content, tension,* or *capacity.* When more than one of these is improved, the gains to the cell are compounded, for example, the use of hyperbaric O_2 and blood transfusion in acute CO poisoning.

Effects of oxygen in relieving hypoxia

In Table 73-2, rounded values are chosen to illustrate (1) the deviations from normal produced by the conditions selected and (2) the influence of O_2 breathing in each form of hypoxia. The alveolar and arterial O_2 values for air and O_2 breathing shown for coronary occlusion, a defect in local circulation, may be considered equivalent to those of normal subjects.

Coronary insufficiency

Administration of 100% O_2 will lower alveolar P_{N_2} to only a few millimeters of mercury within

Table 73-2. Typical initial effects of O_2 administration on alveolar and arterial blood O_2 in different types of hypoxia (values rounded)

Hypoxic condition	Air = 20.9% O_2 = 158 mm Hg P_{O_2}					100% O_2 = 760 mm Hg P_{O_2}					Immediate improvement in arterial oxygenation	
	Alveolar P_{O_2} (mm Hg)	Arterial P_{O_2} (mm Hg)	Arterial Hb saturation (%)	Arterial dissolved O_2 (vol %)	Total arterial O_2 content (vol %)	Alveolar P_{O_2} (mm Hg)	Arterial P_{O_2} (mm Hg)	Arterial Hb saturation (%)	Arterial dissolved O_2 (vol %)	Total arterial O_2 content (vol %)	Increase in O_2 content (%)	Pressure (mm Hg)
Coronary occlusion	100	100	100	0.3	20	670	670	100	2.0	22	10	570
Morphine poisoning	25	25	50	0.07	10	670	670	100	2.0	22	120	645
Emphysema	40	40	75	0.1	15	670	670	100	2.0	22	50	630
Atelectasis	100	40	75	0.1	15	670	45	80	0.13	16	7	5
CO poisoning	100	100	50	0.3	10	670	670	50	2.0	12	20	570
Cyanide poisoning	100	100	100	0.3	20	670	670	100	2.0	22	10	570

the 5 to 10 min of pulmonary washout. Alveolar gas composition (in millimeters of mercury) can then be represented approximately as follows:

$$\text{Total pressure} = P_{CO_2} + P_{H_2O} + P_{O_2}$$
$$760 = 40 + 47 + 673$$

Arterial P_{O_2} will be close to the high P_{O_2} of the alveoli, Hb will become fully saturated, and about 2 vol% of additional O_2 will be forced into physical solution. Because Hb saturation is normally nearly complete during air breathing, most of the improvement in oxygenation of the myocardium is derived from the added volume of physically dissolved O_2 and from the related higher head of O_2 partial pressure. Although the increase in O_2 content is only about 10% above normal, O_2 administration does increase the P_{O_2} of the myocardium surrounding a coronary infarct, reducing the mass of hypoxic tissue.[143] Ionic and nutritional deficits may remain, masking the important benefits of O_2.

Narcotic poisoning

Reduction of alveolar ventilation by any central respiratory depression elevates the P_{CO_2} but lowers the O_2 tension, saturation, and content of arterial blood. When O_2 is administered, alveolar P_{O_2} is raised to the same very high level as in any subject with normal lungs (e.g., the patient with coronary insufficiency). In the presence of central respiratory depression a fall in ventilation and rise in arterial P_{CO_2} may result from the deactivation of the carotid and aortic chemoreceptors by O_2, but alveolar O_2 tension will nevertheless remain high (close to 670 mm Hg).

Atelectasis

The peripheral arterial blood of an individual with atelectasis (or other form of a right-to-left shunt of blood across the lung) is composed of a mixture of aerated blood from normally ventilated lung tissue and mixed venous blood that has not been exposed to alveolar gas. The relative proportions of aerated and nonaerated blood determine the degree of hypoxemia; it may be severe. This form of O_2 deficit is difficult to relieve, even with pure O_2, because improvement in oxygenation can only occur in normal lung tissue where O_2 uptake is already normal. This small gain is partially lost when the hyperoxygenated blood mixes with the shunted, mixed venous blood.[57]

Methemoglobinemia

When the concentration of available active Hb is lowered by severe anemia, by combination with CO, or by conversion to methemoglobin, a subnormal O_2 content of the blood occurs with normal O_2 tension. In the presence of inactive Hb forms, O_2 inhalation will increase the P_{O_2} and physically dissolved O_2 concentration in arterial blood, but it cannot immediately cause much improvement in Hb saturation. Nevertheless, this physically dissolved O_2 is important, and O_2 breathing at the highest practical tension is the means of increasing tissue oxygenation, while the amounts of active Hb are being restored to safe levels. The amount of physically dissolved O_2 can be increased over 2 ml/100 ml blood/atm by O_2 administration at pressures above atmospheric pressure (Chapter 75).

Cyanide poisoning

In cyanide poisoning the initial failure is in oxidation rather than in oxygenation. Because the rate of O_2 utilization by the cells is diminished and there is no primary involvement of pulmonary gas exchange, O_2 transport, or circulation, the levels of O_2 content and tension in the venous blood are higher than normal. It was in the past considered that O_2 administration is useless in this cytotoxic form of hypoxia. This attitude deserves reconsideration, since O_2 administration can reduce the chemoreceptor stimulant action of intravenously injected cyanide[67,75] and at high pressures (greater than 1 atm) does increase tolerance to intravenously administered cyanide.[89] The extreme rapidity of onset remains a practical difficulty in attempts at therapy, even if O_2 at high pressure reduces the degree of poisoning by cyanide.

ALTITUDE, AVIATION, AND SPACE FLIGHT

On ascent to high altitude, the percentage composition of air does not change appreciably, but the P_{O_2} decreases. The effects of reduced barometric pressure, as with other forms of hypoxia, depend on the nature of the exposure. The reduced atmospheric pressure of the high-altitude environment therefore produces effects in the exposed human that differ in the *aviator*, the *transient visitor* to a mountainous region, the *acclimatized mountaineer*, and the *long-term resident* at increased altitude. Actually in each of these individuals the stress that precipitates the physiologic responses (ignoring low environmental temperature) is the lowering of inspired O_2 tension.

The several types of reactions to the same degree of the single stress result from differences in (1) the *rate* at which hypoxia develops, (2) the

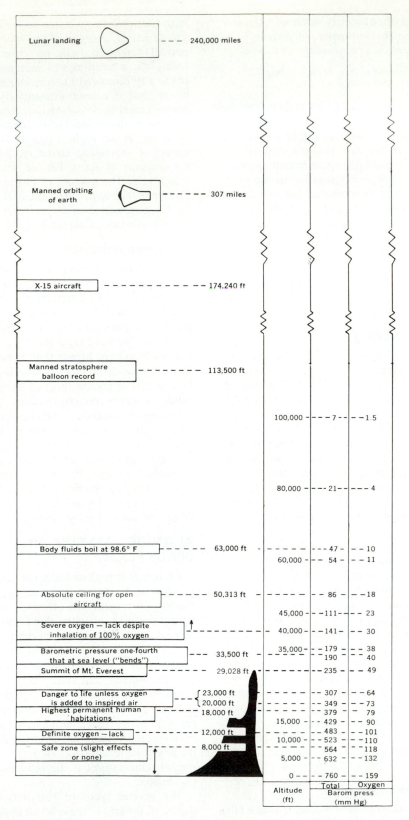

Fig. 73-2. High altitude. Critical levels of reduced barometric and O₂ pressure are shown in relation to significant stages in conquest of upper atmosphere, "near space," and lunar exploration. Scales at right show total air pressure and pressure of O₂ in air at increasing distance from earth's sea surface.

degree of concomitant O_2 requirement imposed by physical exertion, and (3) the *duration* of exposure. The balloonist or unpressurized aviator, the laboratory subject breathing low O_2 mixtures or acutely exposed to decreased ambient pressures in an altitude chamber, and the transient visitor to high altitude have not undergone the secondary changes, collectively called acclimatization, that can increase the ability of normal persons to tolerate an environment of low O_2 tension. On a major climbing expedition, such as that leading to the successful ascents of Mt. Everest,* a planned program of acclimatization at intermediate altitudes precedes the "dash" to the summit. Certainly the permanent resident at high altitude has accomplished the maximal physiologic adaptation to a particular level of inspired Po_2 and may even be subject to deteriorations that can only be considered pathologic.[18,27] Acclimatization and its interesting implications will be described in the succeeding section; attention will be given here to the conditions and reactions involved in acute exposure to decreased ambient pressure.

Pressure-altitude relationships

Important features involved in mankind's conquest of altitude and "near space" are illustrated in Fig. 73-2.

Influence of altitude

The total barometric pressure (mm Hg/cm^2) at a particular altitude is equal to the weight of a 1 cm^2 column of air extending upward to the turbulent upper limits of our atmosphere. Since air is compressible, the decrease in pressure (weight) of the air column that occurs from a 10,000-foot ascent beginning at the floor of the sea of air (sea level) is over 10 times greater than the pressure drop during the same degree of vertical ascent from 60,000 to 70,000 feet (Fig. 73-2). The total atmospheric pressure at the earth's surface averages about 760 mm Hg in those land regions having altitudes close to sea level. Atmospheric pressure is less than 10 mm Hg at an altitude of 100,000 feet. It is essentially zero at the 100- to 200-mile altitude selected for the first manned orbiting of the earth; a hard vacuum characterizes the external environment of a spacecraft engaged in cislunar flight or lunar landing.

Influence of water vapor pressure

Since the O_2 percentage in air does not change appreciably with altitude, the O_2 partial pressure

*See references 37, 51, 70, 86, 135, and 166.

of the atmosphere at any altitude is 20.94% of the existing total pressure of *dry* ambient air. At sea level:

$$\text{Alv. } Po_2 = \text{Total pressure} - (P_{H_2O} + P_{CO_2} + P_{N_2})$$

One of these pressures, alveolar P_{H_2O}, is independent of altitude, being determined entirely by the deep body temperature. Since this temperature is normally maintained close to 37° C, P_{H_2O} or water vapor pressure remains fixed at about 47 mm Hg. It therefore occupies a larger and larger fraction of the total pressure (and alveolar volume) as altitude is increased.

At approximately 63,000 feet the total barometric pressure (or total alveolar gas pressure) is about 47 mm Hg, and water vapor occupies the entire volume of the lungs. At this pressure, body water begins to "boil" at normal body temperature. Animals suddenly exposed to such low ambient pressures become distended by gaseous water[35] and, of course, die of anoxia unless immediately recompressed.[39] The body volume of a dog nearly doubles due to the evolution of water vapor and the expansion of trapped bubbles.[156] However, it was surprising in preparation for manned space flight to learn that exposures for as long as 3½ min to the extreme altitude equivalent to 150,000 feet allowed evidently complete recovery on recompression.[49]

Armstrong[35] and colleagues have seen this phenomenon taking place in a transparent cell through which an animal's blood was passed. No changes were visible until an altitude (simulated) of 55,000 feet was reached, when the film of blood suddenly assumed a grayish appearance, probably because of the liberation of minute bubbles of gas or vapor. At 58,000 feet, microscopic bubbles appeared; at 61,000 feet, these were more numerous; and at 63,000 feet, all the fluid of the blood was suddenly replaced by a mass of bubbles (presumably generated of water vapor). This phenomenon is interesting in relation to extreme, explosive decompression and the time available for recompression. At slower rates of ascent or pressure reduction, death from anoxia would occur at a considerably lower altitude, even if the animals were exposed to breathing 100% O_2. This is because water vapor would displace O_2 from the lungs. The practical importance of recompression after vaporization of body water is limited.

Critical altitudes for oxygenation

Unlike Po_2 in O_2 breathing, alveolar CO_2 pressure does not passively depend on barometric pressure but is determined by the balance between respiration and metabolism. P_{CO_2} falls progressively as the altitude of air breathing increases and the hypoxic stimulus to respiration increases (Table 73-3). At 18,000 feet, when alveolar P_{O_2} has fallen from a normal value of about 100 to below 40 mm Hg, P_{CO_2} has been lowered only to about 30 mm Hg. At this altitude, where barometric pressure is approximately equal to half that at sea level, prolonged exposure of the unacclimatized individual breathing air may lead to death from anoxia.

At an altitude of about 33,700 feet a person breathing pure O_2 will have an alveolar P_{O_2} equal to that during air breathing at sea level (Table 73-3). On ascent to about 44,000 feet the alveolar P_{O_2} falls to the same dangerously subnormal level of 40 mm Hg found with air breathing at 18,000 feet, and hypoxia is pronounced *even though 100% O_2 is being inhaled*. The second critical altitude is reached when the sum of alveolar P_{CO_2} and P_{H_2O} equals total ambient pressure. At this altitude (about 50,000 feet) the lungs will contain no O_2.

These considerations indicate that it is necessary to provide a source of pressure as well as an increased percentage of O_2 in the respired gas when flying at high altitudes or engaged in space activities. For some limited purposes in the evolution of aviation, this pressure was provided only to the lungs by means of a *pressure breathing apparatus* that increased the alveolar gas

pressure but not the pressure surrounding the body. With this method, P_{O_2} was slightly raised but at the expense of interference with circulation through the lungs. More satisfactory approaches now in use involve *pressurized cabins,* such as those employed in military, commercial aviation, and space flight and the pressure suits used in lunar exploration. In each case a local ambient environment is provided, adequate in pressure for breathing the gas supplied. This permits function without hypoxia when the external ambient pressure is low or, as in orbiting missions or extravehicular space operations, zero.

Chronic exposure to pure oxygen at reduced pressures

Although continued breathing of 100% O_2 at 1 atm produces toxic effects within a day (Chapter 75), astronauts in preparation for space flight have breathed pure O_2 for over 50 days at the reduced atmospheric pressure of 0.3 atm. On the lunar surface, pure O_2 was breathed in the pressure suit at about 0.2 atm. At those partial pressures the 100% inspired O_2 does not induce detectable pulmonary toxicity. In fact, in the latter situation the P_{O_2} is essentially identical to that for natural air breathing at sea level.

Respiratory function in low-pressure, pure O_2 breathing is unnatural in that the lungs contain no inert gas vehicle for O_2. The alveolar gas is composed of O_2, CO_2, and water vapor—all rapidly

Table 73-3. Partial pressures of alveolar gases at sea level and at various altitudes when breathing air and when breathing 100% O_2

Alveolar air			Breathing air		Breathing 100% O_2	
O_2 tension (mm Hg)	CO_2 tension (mm Hg)	H_2O vapor tension (mm Hg)	Barometric pressure (mm Hg)	Altitude (ft)	Barometric pressure (mm Hg)	Altitude (ft)
103	40.0	47	760	0	190	33,700
81	37.5	47	632	5,000	166	36,000
61	35.5	47	523	10,000	144	39,500
45	32.5	47	429	15,000	125	42,500
38	31.0	47	380	18,000	116	44,000
35	30.0	47	349	20,000	112	44,800

When the human is breathing pure O_2 at 33,700 feet, the partial pressure of O_2 in the alveoli is the same as the P_{O_2} when breathing air at sea level. Above 34,000 feet the partial pressure of O_2 in the lungs begins to fall below the pressure at sea level, even though 100% O_2 is breathed. At altitudes greater than 40,000 feet the partial pressure of O_2 decreases rapidly and falls below the limit that permits enough O_2 to be absorbed by the blood to maintain the body in a physiologically safe condition. A critical altitude, where the total air pressure drops to 87 mm Hg, is at 50,000 feet. At this pressure the sum of a 47 mm Hg P_{H_2O} and a 40 mm Hg P_{CO_2} is equal to the total alveolar pressure, and O_2 is completely excluded, even though 100% O_2 is available for breathing. (From Flight Surgeon's Manual, USAF,[77] data based on measurements by several authors.)

and completely absorbable by the blood. It was considered a possible hazard of aerospace activity with such a pure O_2 atmosphere that a progressive and diffuse pulmonary atelectasis might develop, leading to fatal hypoxia, especially in the event of respiratory infection.[103] Although this did not prove a problem, concern over fire remained and the spacecraft involved in lunar flight avoided use of pure O_2.[14]

Anoxemia-altitude-performance relations in acute exposures
Time of consciousness at various altitudes or reduced barometric pressures

When an artificially pressured cabin suddenly decompresses to the same pressure as the external environment, the effect on the passengers depends on the altitude at which this loss of cabin pressure occurs. Fig. 73-3 indicates that at 40,000 feet, unconsciousness would be expected to occur in about 20 sec unless O_2 was available and administered. At altitudes above about 55,000 feet, O_2 cannot reach the alveoli and consciousness will be lost within one circulation time.

Performance decrement on acute exposure

Anoxemia-altitude-performance relations are summarized in Fig. 73-4, which is derived from observations made on normal men subjected to simulated altitudes in decompression chambers. The O_2 saturations were determined by the oximeter as well as by direct arterial punctures. Fig. 73-4 indicates normal functions at saturations above 90% arterial O_2 saturation, corresponding with altitudes up to about 8,000 feet. This is the region in which the great majority of the world's population reside. Actually, some physiologic changes develop progressively and can be detected even in this range if the criterion is sensitive enough.

Acceleration of the pulse commonly begins at O_2 pressures equivalent to altitudes below 5,000 feet and appears to be progressive from sea level upward.[66] Deterioration of sensory and cerebral functions becomes more marked with increased altitudes.[121,131] *Cortical involvement* can lead to sleepiness and lassitude or to a sense of comfort, well-being, and self-satisfaction, perhaps associated with outbursts of hilarity or quarrelsomeness; judgment is impaired and a fixity of pur-

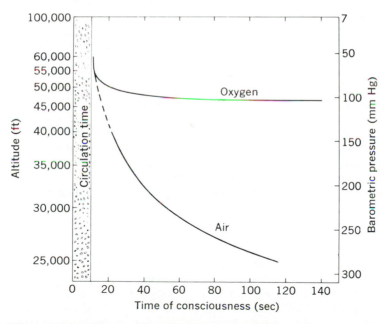

Fig. 73-3. Effect of O_2 breathing on time of consciousness on sudden exposure to reduced ambient pressure at various altitudes. Sudden decompression of a cabin at 35,000 feet results in loss of useful consciousness within about 30 sec. Above about 50,000 feet, even pure O_2 offers no protection, and unconsciousness develops in about one circulation time. (Modified from Flight Surgeon's Manual[77] and based on data from Benzinger[40] and Comfort and Wilson.[55])

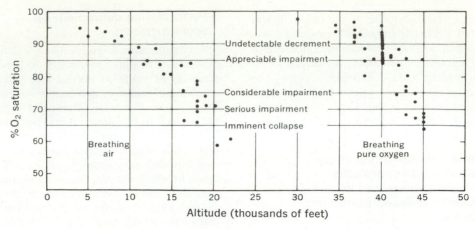

Fig. 73-4. Effect of O_2 breathing on relation of altitude to arterial O_2 saturation and to performance. At left, points represent findings in subjects exposed to simulated altitudes up to about 20,000 feet while breathing air. Fall in O_2 saturation is accompanied by decreased cerebral functional capacity that is difficult to detect but probably exists even at altitudes between 5,000 and 10,000 feet. At right, administration of O_2 prevents diminution of arterial O_2 saturation or performance until altitudes in excess of 30,000 feet are reached. As altitude is further increased, O_2 saturation and performance are impaired despite inhalation of 100% O_2.

pose frequently is seen; pain perception is obtunded early. *Neuromuscular coordination* deteriorates, as shown by the appearance of tremors and clumsiness, slowed reaction time, and greater number of errors in standardized tasks; simple, familiar manipulations, however, may still be well carried out when more specialized ones become difficult or impossible. Visual acuity diminishes early and the sudden brightening of lights in the altitude chamber when O_2 is breathed is a dramatic experience.

These signs and symptoms progress as the acute ascent is continued. They are typically mild, although appreciable over the range of 10,000 to 15,000 feet; the corresponding figures for arterial O_2 saturation and tension are about 87% and 53 mm Hg to 79% and 43 mm Hg, respectively. Above 15,000 feet the deterioration becomes considerable and at 20,000 feet (arterial O_2 saturation 65% to 70%, arterial Po_2 about 40 mm Hg) it becomes incapacitating; that is, the occupants of the balloon or glider, if still conscious, would be incapable of intelligent actions and would be practically blind and deaf and oblivious to the hazard. Above this altitude lapse into complete unconsciousness is likely to occur without warning; the chemoreceptors may keep the breathing going for some time, but circulatory failure is probably the usual cause of death.

Tolerance of unacclimatized individuals to acute hypoxic or altitude exposure

The pattern described for the hypoxia of altitude resembles in all details the equivalent hypoxia that results from breathing gas mixtures low in O_2 concentration at sea level. The factors that determine the ability of a normal human to withstand acute exposure to high altitudes appear to be respiratory response, shift in dissociation curve of oxyhemoglobin, ability of tissues to withstand anoxia, and cardiac reactions.

Respiratory response. Increased depth of breathing brings the arterial Po_2 closer to the ambient Po_2 than before. According to Houston[84] an increase in resting ventilation of only 50% may increase arterial saturation by 10% to 20%. Excessive ventilation, however, leads to arterial alkalosis and can produce cerebral vasoconstriction, even in the presence of a degree of hypoxia that without extreme hypocapnia would cause cerebral vasodilation. Thus hyperventilation exaggerates the cerebral anoxia and hypocapnia.[105,131,137,157] It also somehow contributes to elevation of the blood lactate concentration that accompanies hypoxia.[159] Administration of a carbonic anhydrase inhibitor leads to elevation of arterial Po_2 at altitude, apparently because of respiratory stimulation by retained CO_2.[48]

Shift in dissociation curve of oxyhemoglobin. Due to the hyperventilation the arterial P_{CO_2} is lowered and the blood can take up slightly more O_2 at the prevailing P_{O_2} (Chapter 69). This small gain of hypocapnia is offset by the lower mean tissue P_{O_2} which is then required to bring about the unloading of the extra O_2 in the tissues. However, the increased circulation provides a volume flow of blood sufficient to meet tissue requirements in some organs at a higher mean tissue P_{O_2}. The gain from the greater volume of O_2 carried by more alkaline blood would be greater at saturations below 90% than at higher ones because of the peculiar shapes of the dissociation curves.

Ability of tissues to withstand acute hypoxia. Some individuals become disoriented, confused, or unconscious at Hb saturations as high as 80% (P_{O_2} 44 mm Hg) and most are likely to do so at 70% (P_{O_2} 35 mm Hg), but some persons are still reasonably competent at saturations as low as 60% (P_{O_2} 30 mm Hg).

Although no definitive statements are justified as to the most likely basis for such differences in the resistance of individual brains to hypoxic derangement, at least two factors seem important. The situation resembles that with narcotic drugs and with nitrogen narcosis in that those individuals who have a considerable reserve of mental capacity appear best able to marshal their intellectual resources in states of partial depression. It is also likely that the persons most tolerant to acute hypoxia are those in whom the respiratory, blood pressure, arterial, and especially cerebral vasomotor changes are such that they limit the degree of fall in P_{O_2} of brain tissue. Such a mechanism cannot be construed as an *increased tolerance* of the tissues to hypoxia; rather it entails an ability to minimize the *degree* of central hypoxia. However, the fact that the vasomotor center is capable not only of survival but also of sustained maximal activity when the respiratory center has been paralyzed by acute cerebral ischemia, induced experimentally by high cerebrospinal pressure, indicates that great and unexplained differences in neural susceptibility to hypoxia exist even among the most basic of the control mechanisms in the CNS.

Cardiovascular reactions. From the standpoint of consciousness there are two important cardiovascular responses to hypoxia. These are increased blood pressure (the result of chemoreceptor activation) and dilation of cerebral vessels. The higher head of pressure and the decreased cerebral vascular resistance together result in a greater rate of blood flow through the hypoxic brain. Failure of O_2 utilization results when the O_2 *tension* in the brain capillaries is too low to sustain diffusion of O_2 to the metabolizing cells. This can occur even though the *volume* of O_2 delivered each minute by the blood is greater than that required for metabolism. Similar considerations apply to the heart, in which coronary vasodilation is a prominent mechanism in the support of myocardial oxygenation.

Effects of carbon dioxide on acute tolerance to hypoxia and altitude

In acute hypoxic exposure due to low inspired P_{O_2}, hypocapnia results from chemoreflex stimulation of respiration. This fall in P_{CO_2} may be considerably exaggerated by excitement or other factors, adding the detrimental effects of alkalosis on neuronal function to those of hypoxia. It would appear therefore that the addition of small amounts of CO_2 to the air or O_2 breathed at high altitude should be advantageous (1) in relieving the alkalosis and (2) in improving oxygenation of the brain. This question, which is not a simple one, has received attention from several different standpoints.* It has bearing on the differences in effect and tolerance on exposure to low inspired P_{O_2} alone and exposure to combined hypoxia and hypercapnia.

Breathing air or other nitrogen-containing mixture. When a low concentration of CO_2 is added to air breathed at high altitude or to gas mixtures low in O_2 content, respiration is stimulated, alveolar P_{O_2} increases and symptomatic improvement results.[76,80,81,105] Actually, subjects unconscious due to hyperventilation while breathing 8% O_2 in nitrogen at sea level can be restored to consciousness by adding CO_2 to the 8% O_2 mixture inspired. The basis is an improved delivery of O_2 to the brain, secondary to improvement of brain blood flow as well as to increased arterial P_{O_2} (Fig. 73-5).[20,105]

Breathing pure oxygen. When no inert gas is present in the alveoli, the situation differs considerably from that just described. Hypocapnia during pure O_2 breathing will lead to an increase in alveolar P_{O_2} equal to the fall in P_{CO_2}. Conversely, a rise in alveolar P_{CO_2} brought about by CO_2 administration must lead to a decrease in alveolar O_2 tension exactly equal to the elevation of P_{CO_2}. These facts of arithmetic have led in some studies to the conclusion that no improvement in oxygenation is expected from the addition of CO_2 to O_2 breathed at high altitude.[83,130] This conclusion certainly pertains to the alveolar

---

*See references 20, 76, 80, 81, 83, 91, 130, and 133.

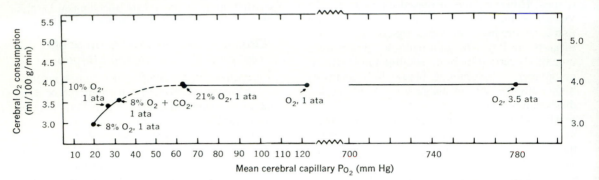

Fig. 73-5. Relations between cerebral O_2 consumption (CMRo$_2$) and mean cerebral capillary Po$_2$. Rate at which brain O_2 is used declines as mean brain capillary Po$_2$ is lowered, and this decline becomes prominent below about 40 mm Hg. Dashed portion of curve points out that no information is yet available to indicate level of Po$_2$ at which decline in CMRo$_2$ begins. O_2 breathing does not alter CMRo$_2$, even to 3.5 ata inspired Po$_2$. When hypocapnia normally accompanying 8% O_2 is prevented by artificially maintaining normal arterial Pco$_2$, mean brain capillary Po$_2$ is maintained by higher blood flow above Po$_2$ associated with 10% O_2 breathing. Preventing hypocapnia during hypoxemia thus minimizes hypoxic depression of brain metabolism.[20] Mean cerebral capillary Po$_2$ is estimated by integration from values for arterial and internal jugular venous O_2 content, capacity, and tension. (Reprinted from Lambertsen[20], by courtesy of Marcel Dekker, Inc.)

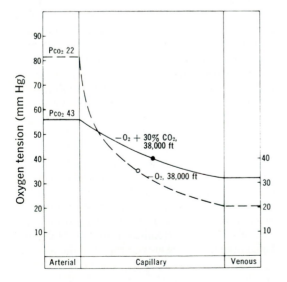

Fig. 73-6. Effect of adding 30% CO_2 to O_2 breathed by hypocapnic subjects at 38,000 feet. (From average data obtained on arterial and internal jugular venous blood in five subjects[133]; ● and ○ indicate calculated mean brain capillary Po$_2$.) When subjects hyperventilate while breathing pure O_2, alveolar Po$_2$ is elevated by an amount equal to fall in alveolar Pco$_2$. Conversely, administration of CO_2 to O_2-breathing subject will lower alveolar Po$_2$ by amount of increase in CO_2 tension. This figure is offered to show that under certain circumstances a fall in *arterial Po$_2$* such as that produced by adding 30% CO_2 to O_2 (solid line) may be associated with restoration of consciousness due to improvement in mean *brain capillary Po$_2$* (●, ○) and in the O_2 tension of brain venous blood.[104] This effect is related to the reversal of cerebral vasoconstriction normally produced by hypocapnia. In such subjects, measurement of arterial O_2 saturation, as with an oximeter, would erroneously indicate better oxygenation of spontaneously hyperventilating, unconscious subjects than of conscious subjects in which cerebral vasoconstriction was prevented.

gas and the arterial blood. However, prevention of extreme hypocapnia should improve *central* oxygenation and well-being at altitudes between about 35,000 and 40,000 feet by improving the blood supply to the brain. This improvement of brain oxygenation has been demonstrated in the human but is not evident when severe hypocapnia has not been allowed to develop.[83,133] There is a narrow range of altitude over which addition of CO_2 should improve central oxygenation. At altitudes below 35,000 feet, arterial Po_2 is adequate, and no appreciable hypoxia should be encountered during O_2 breathing, whereas above 40,000 feet, displacement of alveolar O_2 by added CO_2 should become critical and obviate any advantages of CO_2 as a cerebral vasodilating agent.

These considerations are described to emphasize the interplay among Po_2, Pco_2, $[H^+]$, respiratory control, and brain circulatory control, since all affect competence and even consciousness (p. 1858). They are illustrated in Fig. 73-6, which is based on determinations of arterial and internal jugular venous O_2 tensions in normal men at a simulated altitude of 38,000 feet. The diagram indicates that the rise of arterial Po_2 produced by hyperventilation during O_2 breathing is accompanied by severe reduction in the mean brain capillary Po_2 and the Po_2 of internal jugular venous blood. This form of central hypoxia results from hypocapnic constriction of brain vessels that are detectably dilated by hypoxemia only when arterial Po_2 is reduced to levels of about 50 mm Hg[20,105] (Fig. 71-29). Thus it appears that under certain conditions an artificial increase in arterial Pco_2 during pure O_2 breathing at altitude can both prevent alkalosis and improve the degree of central oxygenation. Any improved tolerance to altitude hypoxia will be accompanied by the increased respiratory response to central and peripheral chemoreceptor stimulation that occurs when the hypocapnia that normally accompanies this form of hypoxemia is prevented from developing (Chapter 71).

ACCLIMATIZATION TO HYPOXIA AND TO HIGH ALTITUDE

Extensive adaptations to prolonged, generalized hypoxia occur in any sustained exposure to subnormal arterial Po_2. Such exposures include not only the ascent to high altitude or lifelong residence at such an altitude, but also many forms of clinical arterial hypoxemia. In some (congenital cardiac right-to-left shunts) an uncorrected disorder may exist from birth; because the lungs are normal, no failure of ventilation or hypercapnia exists, and the conditions tend to

resemble residence at high altitude. In others a slowly developing pulmonary abnormality (as in emphysema) limits alveolar ventilation with eventual addition of hypercapnia and acidosis to the arterial hypoxia; this circumstance has features very different from either short or long exposure to high altitude, the differences chiefly involving (1) capacity for respiratory response and (2) adaptation to the *hyper*capnia in emphysema as opposed to the *hypo*capnia of high altitude.

Finally it must be recognized that individuals at a high altitude, including permanent residents, do suffer acute or chronic cardiorespiratory disease and in such situations have gross exaggeration of hypoxia.

Acclimatization to lowered Po_2 of high altitude

In continued exposure to low barometric pressures, prolonged hypoxia and hypocapnia coexist, and acclimatization to altitude must be considered in relation to effects of lowered Pco_2 as well as those of hypoxia.

Development of tolerance to the low Po_2 of high altitude and the nature of the physiologic adaptations involved should be qualitatively the same for all persons exposed to the same degree and duration of hypoxia. However, individuals should differ greatly in their quantitative capacity for acclimatization, just as they differ in endurance and physical efficiency at sea level. When the character of exposure differs, the pattern of adaptation will also vary, depending on the time available for accomplishing the several steps needed to improve oxygenation at rest and at work. In the section that follows, attention will be given to *altitude intolerance* in the unacclimatized human, to adaptations involved in temporary acclimatization, to adjustments in permanent residents at high altitude, and to failures of adaptation in chronic exposure.

Altitude sickness in the unacclimatized human

When a normal individual accustomed to living at sea level moves at rest by train or automobile to a high altitude over the short period of a day or two, he experiences the usual symptoms of moderate hypoxia, including headache, dizziness, breathlessness, weakness, nausea, sweating, palpitation, dimness of vision, partial deafness, and sleeplessness. Dyspnea and most other symptoms are severely exaggerated on exertion.

This pattern of symptoms on ascent to high altitude has long been termed *mountain sickness* and has been the subject of several famous physiologic studies by pioneers in respiratory physiology.[2,11,54,160] These symptoms will gradually diminish over a period of days or weeks, depending on the altitude. Some of the early effects disappear entirely. As time passes, the individual will become able to carry out considerably more physical activity than was possible on first arriving at the high altitude. The improvement, related in part to an increased capacity for supplying O_2 to the tissues and in part to overcoming the chemical consequences of the hypocapnia produced by excessive breathing, is called *acclimatization*.

The factors involved in this important and complex adaptation are (1) a sustained increase in alveolar ventilation, (2) an adjustment of the acid-base composition of blood and other body fluids, (3) an increase in arterial O_2-carrying capacity, and (4) an increase in cardiac output. If the newcomer partially acclimatized to the high altitude now returns to sea level, he will suffer no symptoms on descent and will soon lose the acclimatization he has acquired.

Programmed acclimatization in major mountain climbing expeditions

A slow increase in altitude, especially when accomplished by the exercise and conditioning of climbing on foot, minimizes the effects of acute mountain sickness. Over the many years of assault on Mt. Everest and other major peaks, climbers have learned to invest many weeks or even months in physiologic adaptation at bases located at successively higher altitudes. By this means and by the judicious use of O_2 in the final stages of ascent, men from sea level regions have succeeded in surmounting Mt. Everest.[28,70,73,86] It now appears that the physiologically almost unbelievable feat of Everest ascent while breathing air has been accomplished,[37,51] although the Po_2 of the inhaled air was only 49 mm Hg (Fig. 73-2) and arterial O_2 saturation was therefore only about 60% even at rest.[135]

By use of a long period of adjustment to increasing altitude, compensatory factors come into play, and the detrimental CNS effects of acute hypoxia and hypocapnia become progressively less marked. At extreme altitudes, some nervous effects inevitably remain, such as decreased patience, lessened ability to concentrate, decreased quality in the performance of tasks involving mental or physical work, and greater liability to fatigue, mental as well as muscular.

Sleep is likely to be disturbed. Pulse rate and blood pressure may become normal again or may remain above normal. Breathing remains deeper than at sea level. Muscular capacity improves, but it is less than normal, partly because of dyspnea on exertion and partly because the oxidations within the muscles are limited. Fatigue results from relatively slight exertion and recovery is unusually slow.

Mountain climbers are a special case in adaptation to hypoxia because they are physically fit, specially trained, and exceptionally motivated, and they take the time required to become acclimatized. Such individuals have repeatedly spent weeks above 22,000 feet, are capable of performing heavy work at 25,000 feet, and have carried heavy loads breathing air above 26,000 feet on Mt. Everest without added O_2.[28,37,51,86]

At these extreme altitudes, pronounced breathlessness and fatigue occur with muscular exertion on the part of the climber, making an ascent of 100 feet/hr an exhausting procedure. The hypoxemic drive alone would make the hyperventilation from exercise greater than it would be in the same individual at sea level. Another factor operating in the same direction is the diminution in blood alkali at high altitudes; because of this, small changes in Pco_2 or fixed acid formation should exert disporportionate effects on pH. The fatigue and muscular weakness of mountain climbers doubtless are due in part to impairment of oxidative processes in the muscles and in part to interference with the neural and cerebral functions concerned with voluntary movement.

The rigorous circumstances of acclimatization to extreme altitude are reported by Pugh in a description of the Himalayan Scientific and Mountaineering Expedition of 1960-1961.[135] On that expedition, one base was established and maintained for 5½ months at 19,000 feet (5,790 m). Several members of the expedition were able to remain continuously at that altitude for periods of 4 to 9 weeks after acclimatization. However, men who had taken nearly 3 weeks to climb to a base at 15,100 feet still could not stay a full day at the 19,000 feet base until they had remained at 15,100 feet for at least an additional 2 weeks. The process of acclimatization must be considered one of adaptation to each existing atmospheric condition. Time is required for each physiologic adjustment, and full acclimatization to a medium altitude represents only partial (and inadequate) acclimatization to a higher altitude.

Certainly there is a limit to the physiologic adjustment to low Po_2 that can be made, regardless of the duration of exposure. It is considered that, although there was no indication of true physical deterioration, 19,000 feet closely approaches this limit and represents an altitude too high to allow full and permanent adaptation to occur in European climbers[135] or Andean natives.[4]

One impressive aspect of acclimatization by vigorous and determined climbers is that, although decrement does occur,[18,27,65a,97] work capacity is sustained at a surprisingly high level even at the low alveolar P_{O_2} and P_{CO_2} associated with extremes of altitude.[97,135] Table 73-4 shows the effect of very high altitude on the resting alveolar gas composition in men acclimatized to 19,000 feet.[135] With these gross deviations of alveolar P_{O_2} and P_{CO_2} from normal, indicating a continuing and marked respiratory stimulation even at rest, the degree of exercise tolerated at altitudes up to 24,400 feet becomes even more remarkable.

In a series of measurements during exercise by acclimatized subjects at 19,000 feet, West et al.[169] studied the influence of exercise on alveolar and arterial oxygenation, using indirect methods adapted to the prevailing conditions. Their findings, at work rates up to 900 kg-m/min (average O_2 consumption 2.07 L/min STPD), indicate that, while alveolar P_{O_2} rose with the relative hyperventilation of exercise, the already low arterial Hb saturation (measured by oximetry) fell to an average of 53%. This corresponded to an arterial P_{O_2} averaging 27 mm Hg and the alveolar-arterial P_{O_2} gradient in these hypoxic exercising individuals averaged 26 mm Hg. On the basis of these findings, it is estimated that a pulmonary diffusing capacity for O_2 less than 80 ml/min/mm Hg would have imposed gross limits on O_2

uptake.[169] The high P_{O_2} gradient across the lungs, high diffusion capacity, competent cardiac response to exercise even in hypoxia, high ventilatory capacity, and prominent increase of blood O_2 capacity (to an average of 26.5 vol%) all contributed to the ability of "plainsmen" to sustain work at altitude. The same processes appear to be the basis for acclimatization of permanent residents in mountainous regions* (Fig. 73-7).

Permanent residents at high altitude

The highest points on the earth's surface at which people permanently reside are found in the Andes and Himalaya mountains at altitudes of about 18,000 feet,[4] where the barometric pressure is almost exactly half that at sea level (Fig. 73-2). Expected values for arterial P_{O_2} and Hb saturation at this altitude would be about 38 mm Hg and 73%. Since the Hb dissociation curves become very steep just below 40 mm Hg (Fig. 69-3), small further increases in altitude lead to relatively large decreases in Hb saturation and O_2 transport to the tissues. Compensation by further increase in polycythemia may actually lead to a rise in blood viscosity that may interfere with the volume flow of blood through organs such as the brain.[146]

The impression that approximately ½ atm represents a limit for permanent residence is supported by the experiences of the 1960-1961 Himalayan expedition[135] and by Dill's finding[4] that Andean natives, resident at 17,500 feet, could not remain at a mining camp at an altitude

---

*See references 11, 18, 24, 27, and 88.

Table 73-4. Effect of acclimatization to increasing altitude on alveolar gases, work capacity, and cardiorespiratory response to work*

Altitude		Barometric pressure (mm Hg)	P_{IO_2} (mm Hg)	Rest		Maximal 5 min exercise			
Feet	Meters			Alv. P_{O_2} (mm Hg)	Alv P_{CO_2} (mm Hg)	Work rate (kg-m/m)	O_2 intake (L/min STPD)	Respiratory minute volume (L/min BTPS)	Heart rate (beats/min)
Sea level	Sea level	750	150	110	38	1,500-1,800	3.40	119.7	192
15,000	4,570	440	—	—	—	1,500	2.58	164.8	159
19,000	5,790	380	69	45	22	900-1,200	2.14	159.1	144
21,000	6,400	344	62	38	21	900-1,050	1.95	168.8	146
24,400	7,430	300	53	34	16	600	1.40	119.8	135
25,700	7,830	288	50	33	14	—	—	—	—

*Data from Pugh.[135]

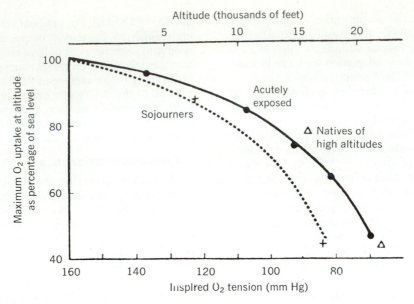

Fig. 73-7. Altitude impairment of muscular work capacity in acute and chronic hypoxia. (From Billings[3]; modified from Cerretelli.[50])

of 19,000 feet for more than a few weeks without disturbances of sleep and appetite.

Permanent residents at altitudes lower than the maximum that permit survival appear able to completely acclimatize and thus develop an amazing tolerance to what is elsewhere considered extreme hypoxia. Some such individuals are then able to carry out mental functions during brief exposure to altitudes of 30,000 feet—a situation in which most of us would be unconscious.[165] Extensive study of fully adapted people in the Andes of South America has been carried out and summarized by Monge[24] and by Hurtado,[11] and current concepts have been summarized by these and other investigators.[114] A particular group, residing in Morococha, Peru, at an altitude of 14,900 feet (4,540 m), has been extensively studied. These native residents, like the Himalayan Sherpas,[18,27,135] have a capacity for physical work and play that is as great or greater than that of individuals adapted to a normal sea level environment. This greater efficiency of individuals acclimatized to moderate altitude is indicated in Table 73-5, which compares native residents of Morococha with physically fit young men of Lima. The table is based on tolerance to treadmill running to exhaustion at 132.4 m/min up an 11% grade. A similar study comparing men acclimatized to altitude with athletes in training at sea level showed

basically the same results,[18] indicating that the chronically hypoxic residents at high altitude (1) ran longer and farther, (2) used less O_2 per unit of work, (3) breathed more air per minute to obtain the required O_2, and (4) had a smaller change in cardiac rate and blood pressure despite the longer exercise. Evidently, exposure to this degree of chronic hypoxia is not limiting.

Failure of acclimatization—chronic mountain sickness

Natives who spend their entire lives at high altitudes become completely acclimatized, but occasionally they suffer from a set of symptoms known as chronic mountain sickness.[11,18,24,25] The symptoms include headache, hoarseness, loss of appetite, weakness and pains in the extremities, visual disturbances, and transient spells of stupor or coma. They occur without obvious precipitating cause in persons who have lived for many years without symptoms at the same altitude. They are likely to disappear spontaneously but return with increasing severity.

Prominent features of the loss of acclimatization include pulmonary hypertension, a diminished degree of hyperventilation, exaggerated arterial hypoxemia, a decreased reactivity to administered CO_2,[11] and a diminished respiratory response to induced hypoxia.[151] As the degree of hyperventilation at altitude decreases, alveolar

Table 73-5. Comparison of response and tolerance to physical activity of subjects in Lima (sea level) and Morococha (4,540 m)*

Subjects	Lima (sea level)		Morococha (4,540 m)	
Number of subjects	10		10	
Age (years)	21		21	
Body height (cm)	164		161	
Body weight (kg)	65		55	
Body surface area (m²)	1.71		1.57	
Exercise tolerance time (min)	34.2		59.4	

Measurements	Rest	Exercise (34.2 min)	Rest	Exercise (59.4 min)
Ventilation (L/min/m² BTPS)	4.6	37.5	5.7	42.2
O_2 consumption (L/min/m² STPD)	0.148	1.333	0.142	1.167
Vent. equiv. (vent./O_2 cons.)	31.2	27.8	40.6	36.1
Pulse (rate/min)	72	183	72	160
Blood pressure, systolic (mm Hg)	116	138	93	97
Blood lactate (mEq/L)	1.35	6.37	1.30	3.19

*Data from Hurtado.[11]

Po_2 and arterial O_2 saturation and tension fall and arterial Pco_2 rises; polycythemia and blood volume are further increased, while general tolerance to hypoxia is lessened.

The cause of chronic mountain sickness is unknown, but there is evidence indicating failure of the normal chemoreceptor influence on respiration.[151] Talbot and Dill,[160] in discussing the possible causes, point out that pathologic changes may occur in the tissues after many years of exposure to subnormal O_2 tensions. Specific suggested contributing possibilities related to the evident diminution of altitude hyperventilation include decreased sensitivity of the respiratory center to chemical stimulation by CO_2[11] and an eventual decrease in sensitivity of the carotid chemoreceptors.[151] It is also possible that the failure is in the renal adjustment of acid-base balance and hence in change in central stimulus level. Treatment requires descent to a lesser altitude.

Composite mechanisms of acclimatization to altitude hypoxia

Studies that have contributed to understanding of acute altitude acclimatization have included expeditions or visits to mountainous regions,* very long-term investigations on native residents,[11,18,24,25] and simulation of the barometric

*See references 4, 58a, 60, 64, 78, 79, 98, 100, 102, 135, and 152.

pressure of altitude in decompression chambers.[65,85] With the complexity of effects of hypoxia, the influences of degree, and the involvement of control systems not fully defined for rest and exercise in the normal sea level environment, the challenge of exploring adaptation to altitude continues its long-standing excitement for respiratory physiologists.

Interrelated components of acclimatization for which explanations are gradually developing include (1) *a sustained respiratory drive with increase in pulmonary ventilation, (2) complex and sustained alterations of arterial and CNS acid-base relationships, (3) a sustained increase in the rate of Hb and erythrocyte formation,* and (4) *increase in cardiac activity and adaptations of brain circulation.* Each factor is concerned with the improvement of cellular oxygenation, each has influence on the other, and, at present, the primary controlling forces are still uncertain.

Respiratory component of acclimatization

The hyperventilation of altitude begins immediately in the newcomer and continues for a lifetime in the resident. This sustained vital response, without which the altitude tolerance of many would be sharply diminished, has more than one physiologic basis, the nature and contributions of which are under active study.[18,26,27]

Several aspects that seem to be well established include the following:

1. A hypoxic chemoreflex stimulation is in-

volved as a primary event in the stimulus to breathing.

2. The chemoreflex effect appears to persist over many years but may progressively diminish during long-sustained residence at high altitude.[151,154]

3. The persistent low O_2–peripheral reflex component becomes partially obscured by a return to primary respiratory control by acid-base factors, central and/or peripheral.

4. Renal adjustments to hypocapnia, by decreasing the concentration of base in body fluids, restore the $[H^+]$ toward normal even in the presence of sustained hypocapnia.

5. Pco_2-related adjustments of $[H^+]$ occur within the CNS,[28] and increase in respiratory reactivity to CO_2 takes place.

Chemoreflex drive. The initial respiratory stimulation on going to a high altitude is clearly related to peripheral chemoreceptor stimulation by lowered arterial Po_2.* In acute exposure to hypoxia the proportionate chemoreflex response diminishes even over several hours as the central contribution to respiration becomes reestablished.[42] During the first hours of exposure to high altitude, hyperpnea can be overcome or prevented by sustained O_2 breathing, which also restores arterial Pco_2 and pH to normal levels[88,136] and abolishes the symptoms of mountain sickness. During continued acclimatization to a high altitude, as renal excretion of base compensates for the initial respiratory alkalosis by restoring arterial pH to normal, the ability of acute administration of O_2 to cause a sustained suppression of respiration decreases.† Eventually, even after full acclimatization of subjects or residents, O_2 administration still does not restore ventilation to sea level values. It produces a small decrease,[53] especially well brought out on abrupt O_2 administration,[50,61,62,120] but this effect is so small that it may be masked in steady-state methods by such factors as an increase in central Pco_2, which obscures the effects of O_2 itself.[108] A chemoreflex effect of O_2 is apparently still prominent in lowlander subjects who have spent several months at 19,000 feet.[123] In stable-state studies, even a small respiratory depression by O_2 is quickly compensated for by the secondary acid-base changes associated with increased Pco_2.

Diminution ("blunting") of hypoxic respiratory drive. Although the chemoreflex drive appears to persist throughout years of exposure to hypoxia, it has also been found that natives of high altitudes continue to be less sensitive to peripheral chemoreflex stimulation by hypoxia than are sea level natives, even after several weeks or possibly even years of staying at sea level.[18,101,155] The diminished respiratory response to hypoxia (generally termed "blunting") seems to eventually return to normal, appearing to exist "unblunted" in migrants who had left high altitude 10 to 25 years before.[18]

Respiratory response to carbon dioxide. Several studies have shown that altitude acclimatization leads to two distinct changes in the respiratory response to administered CO_2.*

One effect is a shift of the CO_2-ventilatory response curve to the left. This shift is sometimes cited as evidence for a change in the threshold to CO_2,† but by itself, without a change in slope of the response, it should indicate a change in central acid-base stimulus level rather than in sensor reactivity.[106] This influence of altitude apparently requires only about 5 to 7 days for near completion.[18,94,124]

A second effect of acclimatization, considered to develop over at least 3 to 5 weeks,[135] has also been reported. This is an actual increase in slope of the CO_2 response curve when alveolar O_2 tension is above normal.[124] Although the magnitude of this response depends on the method used to produce it, it appears that respiratory reactivity to CO_2 does increase during prolonged exposure to altitude.[11,52,95,137] This is to be considered distinct from a hypoxia-induced increase in the slope of the respiratory response to CO_2, which is known to occur in acute exposures to low O_2 mixtures at sea level (Fig. 71-26)[115,128] or to the hypoxia of altitude.[94,123,124,161]

Respiratory acclimatization— consequences of hypoxia and alterations in acid-base balance

Special interest in altitude adaptation continues to be focused on the causes of the continued respiratory drive not related specifically to hypoxic chemoreflex stimulation. On initial exposure to altitude the hypoxic stimulation lowers arterial Pco_2 (and $[H^+]$) and should necessarily decrease the acidity of all body fluids. Nevertheless, a progressive further increase in ventilation occurs, even preceding the compensatory restoration of arterial pH.

---

*See references 18, 26, 27, 50, 52, 120, 136, and 161.
†See references 11, 18, 52, 64, 95, 124, and 135.

*See references 11, 18, 27, 52, 95, 123, 136, 152, and 161.
†See references 11, 115, 135, 152, and 161.

Even after 1 to 3 weeks, subjects acutely acclimatized and still exhibiting hyperventilation even when inspired O_2 is abruptly restored to normal show arterial pH values of from 7.48 to 7.52,[85,152] and no cerebrospinal fluid acidosis. This sustained respiratory stimulation during normoxia, of an as yet unknown cause, is a major component of respiratory acclimatization to altitude.[18,26,27] Eventually, as in permanent residents at high altitude, arterial pH averages a normal 7.39 despite the low P_{CO_2},[11] which persists indefinitely.[99] The "respiratory alkalosis" has been compensated for by a decrease in base in the blood and probably in most tissue fluids as well. The importance of CO_2 tension in the regulation of bicarbonate excretion by the kidney indicates the existence of feedback mechanisms relating to P_{CO_2}, renal elimination of base, and respiration.

Relation to acid-base changes in the central nervous system. Certainly changes in acidity and bicarbonate concentration of blood and brain are induced by the sustained hyperventilation of altitude hypoxia and somehow related to the persistence of hyperventilation when arterial oxygenation is temporarily restored to normal values for residence at 1 atm. With the demonstrations of superficial areas of medullary respiratory chemosensitivity and concepts of influence on them by cerebrospinal and other brain extracellular fluid,[112,117,125] proposals naturally followed that a central component of the sustained hyperventilation response to altitude acclimatization was secondary to an effect mediated via the $[H^+]$ of bulk CSF.[94] Initial observation in humans actually suggested active transport regulation of CSF $[H^+]$ in altitude exposure, with return of a medullary chemoreceptor drive to normal in the face of continued hypocapnia.[152] However, extensive subsequent studies clearly indicate that, following the primary hypoxic, chemoreflex hyperpnea, the secondary additional hyperventilation of altitude acclimatization is contrary to the acid-base changes in CSF.* Actually the increased respiration that persists on temporarily relieving the hypoxic drive occurs in the face of a persistent *decrease of arterial blood and CSF* $[H^+]$, rather than an *increase to or above normal*.† In this situation, neither the $[H^+]$ of CSF nor a physiologic interpolation between it and arterial blood could provide a rational index of a supranormal central respiratory stimulus.

With such a prominent component of altitude hyperventilation not accounted for by recognized forms of chemical stimuli, the customary tendency emerges to credit the increased ventilation not related to hypoxic chemoreflex to a "central sensitization."[18] However, this has no experimental basis, and the search for a change in stimulus and stimulus location continues.[18,26,27] The locations include intra- as well as extracellular fluids.

Relation of intracellular to extracellular $[H^+]$. Although it is not now possible to measure the actual $[H^+]$-related central respiratory stimulus in any situation (p. 1784), the lack of correspondence of altitutde hyperventilation and the $[H^+]$ of CSF or other brain extracellular fluid indicates that such fluids themselves are unlikely sources of the fundamental central respiratory stimulation (Chapter 71). As described in more detail in Chapter 52, in hypocapnia there appear to be P_{CO_2}-related changes in intracellular $[HCO_3^-]$, which result in prompt return of brain cortex *intracellular* pH essentially to normal despite continued subnormal P_{CO_2}, even though CSF $[H^+]$ also remains subnormal.[34,119,119a] The existence of P_{CO_2}-related mechanisms modifying brain intracellular $[H^+]$ and CSF $[HCO_3^-]$ $[H^+]$ is becoming recognized.[34,64,119] These mechanisms, and the intracellular acid-base changes that result, must now be taken into account in exploring respiratory acclimatization to altitude (Chapters 52 and 71). The most difficult aspect will remain that of defining stimulus changes in the as yet not fully defined central respiratory chemosensors (Chapter 71).

Improvement in tissue oxygenation

The secondary respiratory stimulation in the early stages of acclimatization increases arterial O_2 tension. In studies of acutely acclimatized subjects the *percent Hb saturation* at the lowered arterial P_{O_2} and elevated pH corresponds to the values expected from existing Hb dissociation curves.[85] A mechanism considered to contribute to acclimatization to high altitudes is a decrease in the affinity of Hb for O_2 as compared with that found in individuals residing at sea level.[88] This has for many years been known to occur in natives to high altitudes[36] and in sojourners,[96] even when exposure to hypoxia is only of a few days' duration.[111,126] This adaptation is related to a rise in the concentration of 2,3-diphosphoglycerate (DPG), which occurs on exposure to altitude or hypoxia,[68,111] and the accompanying decrease in affinity for O_2 (increase in standard P_{50}) facili-

---

*See references 18, 26, 27, 92, and 94.

†See references 18, 26, 27, 58a, 63, 63a, 64, 78, and 129.

tates release of O_2 in tissues as Po_2 is lowered by metabolic O_2 consumption.

The mechanism of increase in DPG is twofold. In relation to hypoxia itself the larger amounts of deoxyhemoglobin bind existing DPG and H^+. The combination of (1) diminished concentration of free DPG and (2) alkaline shift within the red blood cells stimulates synthesis of DPG.[1,10] Evidently the increase in concentration of DPG that results from exposure to hypoxic states is induced largely by the concomitant alkalemia, since individuals rendered acidotic before exposure to altitude show only a small rise in DPG in red blood cells.[110] This formation occurs within hours[1] and therefore represents one of the first steps in acclimatization. While this mechanism is not as important in acute or chronic adaptation to the hypoxia of altitude as is an increase in pulmonary ventilation,[110] it is advantageous.

In neither acute nor chronic acclimatization to high altitude is the O_2 *content* of arterial blood lowered, despite diminished Po_2 and Hb saturation. Because of the increased O_2 capacity, arterial O_2 content at high altitude was unchanged in the acutely acclimatized subjects and was actually 1.9 vol% greater in the permanent residents at Morococha than in their countrymen who resided at sea level (Table 73-5).

The result of these alterations in the function of O_2 uptake by the blood is that the Po_2 in the average tissue capillary is not lowered as much as the ambient Po_2 is reduced by the decreased barometric pressure at high altitude (Fig. 73-8). This means that although the total gradient of Po_2 from ambient air to the blood in the tissues is considerably diminished, the pressure of O_2 diffusion from the blood to the metabolizing cell tends to be sustained.

Central nervous system blood flow. Two factors in tolerance to extended residence at high altitude relate to blood flow through the brain tissue. These are (1) the possibility, demonstrated in two patients with idiopathic polycythemia,[146] that increased erythrocyte concentration may so increase blood viscosity that brain blood flow is reduced out of proportion to the increase in O_2 content and (2) the question of whether the *vasoconstrictor reaction of cerebral vessels* to low arterial Pco_2 persists indefinitely to hinder brain oxygenation or is lost when hypocapnia persists for a long period of time. This second question is compounded by the first, since brain vascular resistance is greater in polycythemia than with the less viscous, normal blood. Measurements of brain blood flow in recently acclimatized men[142,153] and in permanent high-

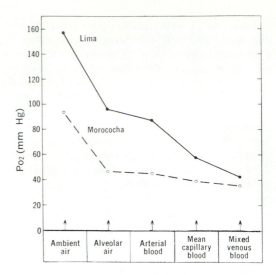

Fig. 73-8. O_2 pressure gradients from ambient air to mixed venous blood in residents of Lima (sea level) and of Morococha (14,900 feet altitude). Despite a nearly 65 mm Hg difference in inspired Po_2, difference in mixed venous Po_2 is less than 10 mm Hg. However, arterial Po_2 to which chemoreceptors are exposed is much lower, and Po_2 in some tissues must be on the verge of producing anoxic dysfunction. (Modified from Hurtado et al.[88])

landers[43] indicate reduction below that of sea level residents.

Hemoglobin concentration, erythropoiesis and polycythemia. Improvement in the efficiency of O_2 transport to the tissues results not only from the effect of hyperventilation on alveolar Po_2 and from a decrease in alveolar-arterial O_2 tension difference, but also in large measure from an increased Hb concentration in the blood. The O_2 capacity increased to 23.5 vol% in subjects acutely acclimatized over a week at 20,000 feet[85] and to 26.5 vol% in acclimatized Himalayan mountaineers at 19,000 feet[135,169]; in the residents of Morococha, O_2 capacity was 28 vol% at 14,000 feet altitude.[11]

The development of hyperhemoglobinemia and polycythemia on exposure to high altitude results from an increased rate of hemopoiesis, with both erythrocytes and Hb being formed more rapidly. There is a corresponding increase in the rate of erythrocyte destruction, but the erythrocyte concentration for equilibrium between formation and destruction is higher than at sea level.[88] The erythropoietic response is a rapid one. In rats exposed for several months to reduced air pressure equivalent to altitudes of

1,000 to 8,000 m, the daily rate of Hb formation was four to five times the normal rate during the first 2 days of hypoxia, regardless of the severity of hypoxia. Following this initial acceleration the rate of formation was a function of the degree of hypoxia.[163]

The stimulus for increased erythropoietic activity is linked to hypoxia. In studies in which parabiotic rats (having an exchange of blood through connecting vessels) were used, exposure of one rat to hypoxia leads to a hemopoietic response in both.[139] In addition, the injection of plasma from native residents of high altitude into inhabitants of sea level regions leads to a reticulocyte response.[122]

It is now known that stimulation of red blood cell and Hb formation during hypoxia is not a direct, local effect of low Po_2 on bone marrow. Rather, it is mediated through a humoral mechanism by a glycoprotein hormone, erythropoietin. The major locus of formation of the erythrocyte-stimulating factor in response to hypoxia is the kidney, evidently with a subsequent stage of activation by a plasma substrate from the liver.[1] Extrarenal sites of hormone formation also appear to be active.*

The effect of erythropoietin appears specific for red blood cell formation, since leukocyte concentration does not increase in hypoxia. However, blood volume is increased, together with red blood cell concentration, and reticulocyte release from the bone marrow is stimulated.[1,11]

An increased rate of erythropoiesis appears also to be present in the hypoxia of chronic pulmonary disease. Nevertheless, in a large number of such patients, gross polycythemia does not develop, since the rate of hemolysis is disproportionately elevated.[141] This "failure of adaptation" has been considered to be due to effects of concurrent infection.[141]

Increased cardiac activity

Tachycardia and increased cardiac output[11,24,28,85] occur in the early stages of adaptation, although the pulse rate has been observed to return to normal after a prolonged stay at 14,000 feet.[2,9,11] In the permanent residents of Morococha the average cardiac output was not elevated, and the resting cardiac rate of 72 was identical with that of the residents of Lima. Both systolic and diastolic blood pressures at rest and in exercise were found to be lower in the permanent residents at high altitude. Thus although aclimatization entails a persistent respiratory

*See references 1, 11, 82, 139, 141, and 163.

simulation and an increased O_2-carrying capacity of the blood, a sustained increase in cardiovascular function does not appear to occur.

Deacclimatization

Following acute acclimatization a return to sea level results in the reversal of all adaptations to hypoxia.[18] On prolonged stay at sea level, residents of Morococha show, in addition to diminished ventilation and fall in Hb concentration, decreases in cardiac rate, pulmonary artery pressure, pulmonary vascular resistance, and systemic vascular resistance. Possibly as a result of the latter, resting cardiac output increased.[38] Those natives of high altitudes who have developed diminished ("blunted") ventilatory response to hypoxia regain their reactivity at sea level, but this may require several years.[18]

Acclimatization to altered carbon dioxide tension

The most difficult aspects of understanding acclimatization to high altitude have involved adaptations to lowered Pco_2 (and H^+). A sustained increase in Pco_2 occurs in chronic pulmonary disease, usually accompanied by hypoxia. Prolonged environmental exposure to elevated Pco_2 is unusual, but has been related conceptually and practically to preparations for multimonth submarine submergence or space flight. In these closed spaces, O_2 deficiency is a less prominent concern than CO_2 accumulation (Chapter 75).

When normal subjects breathe increased concentrations of CO_2 for prolonged periods at sea level, there results a decrease in respiratory response to administered CO_2.[144,145] This is related to a shift of the CO_2 response curve to the right, indicating that a higher alveolar Pco_2 is now required to produce a particular ventilation.[144] This effect parallels a retention of bicarbonate, which means that the shift to the higher Pco_2 is not necessarily associated with a greater degree of acidity at the central or peripheral chemosensitive cells. The elevated arterial Pco_2 of patients with severe pulmonary emphysema complicated by anoxemia also appears to induce a retention of base in the blood.[33] The rise in blood bicarbonate may be paralleled by similar increases in the brain cells, interstitial fluid, and CSF. The CO_2 response curves of emphysematous patients are shifted grossly to the right. In addition, the slope of the ventilatory response of these patients to induced increases in Pco_2 is less than normal.[33] Although this suggests a decreased "sensitivity" to the CO_2 stimulus, the lowered slope is actually considered to be at least in part an artifact related to the greater work required to increase ventilation of the less compliant lungs in severe emphysema[140] (p. 1830).

Increased respiratory response at the natural level of arterial CO_2 is produced by (1) prolonged artificial hyperventilation of normally well-oxygenated subjects,[47] (2) similar artificial overventilation of the anoxic and hypercapnic emphysematous patients,[45] and (3) the natural hyperventilation of normal persons made anoxic by ascent to a high altitude.[88,136] In each instance, renal excretion of base during continued hypocapnia lowers the bicarbonate concentration in the blood and general tissue fluids.[87,88] Change in brain $[H^+]$ toward normal also occurs, evidently due to intrinsic, P_{CO_2}-related mechanisms.[34,119]

Of the many questions raised by such studies, one of considerable importance is whether these altered respiratory responses to CO_2 actually represent changes in the responsiveness of respiratory neurons to their normal chemical stimulus. It is necessary to recognize that it is also possible that the neuron *reactivity* remains unchanged in acclimatization to hypoxia, hypercapnia, or hypocapnia and that the adjustments in respiration are responses of normally reactive respiratory neurons to alterations in local baseline stimulus level.[106]

SUMMARY

Very likely the mechanisms of respiratory regulation at high altitude are the same as those responsible for eupnea at sea level, with differences only in the degree to which individual components participate and the relative importance of different stimulant factors. If so, full understanding of the basis for the higher ventilation at high altitude will be attained only when the interaction of these same factors in respiratory control at sea level becomes more clear.

At present, it appears that at high altitudes the persistence of hyperventilation and hypocapnia is related to a primary sustained chemoreceptor activation by arterial hypoxemia.[61,123,161] The resulting generalized and CNS hypocapnia could be responsible for reestablishing central (intracellular[52]) $[H^+]$, once again tightly superimposing on the active peripheral chemoreflex function the normal inverse control through responsiveness of central respiratory neurons to alterations in their own acid-base environment (Chapter 71).

Whereas at sea level a dominant central effect of CO_2 $[H^+]$ appears to be supported by chemoreflex activity, at high altitude a dominant peripheral chemoreflex effect on respiration seems to be "backed up" by the central effects of CO_2.

If this concept is correct, both a chemoreceptor and a central respiratory control mechanism would be functioning at high altitude just as at sea level, and the administration of O_2 would in each situation produce only a small and transient decrease in ventilation. The existence of a transient respiratory depression produced by O_2 at high altitude has been demonstrated,[50,62,116,120] as has respiratory stimulation by the smallest rise in P_{CO_2} above that normal for the altitude.[161] For these reasons, altitude *acclimatization must be considered to include adaptations both to hypoxia and to low arterial P_{CO_2}*.

Certainly the changes induced by exposure to altitude are complex. However, the tendency to consider permanent residents at high altitude as having adjusted to an "abnormal" environment is no more or no less accurate than to consider individuals residing at sea level to be acclimatized to an abnormally low altitude and hyperoxia.[103] Actually a continuum of natural physiologic adjustments extends from individuals living at sea level through those who reside in Denver (5,300 feet), Mexico City (7,415 feet), and Morococha (14,900 feet) to the people of the highest Andean and Himalayan habitations (about 18,000 feet). Until altitudes are reached at which O_2 lack results in selective elimination or exerts demonstrable pathologic effects, the mechanisms of acclimatization may be viewed as reversible alterations of normal processes in response to change in a natural environment.

REFERENCES
General reviews

1. Adamson, J. W., and Finch, C. A.: Hemoglobin function, oxygen affinity, and erythropoietin, Annu. Rev. Physiol. **37:**351, 1975.
2. Barcroft, J.: The respiratory function of the blood, London, 1925, Cambridge University Press.
3. Billings, C. E.: Atmosphere. In Packer, J. F., and West, V. R., editors: Bioastronautics data book, ed. 2, Washington, D.C., 1973, National Aeronautics and Space Administration.
4. Dill, D. B.: Life, heat, and altitude, Cambridge, 1938, Harvard University Press.
5. Ernsting, J.: Prevention of hypoxia—acceptable compromises, Aviat. Space Environ. Med. **49:**495, 1978.
6. Finch, C. A., and Lenfant, C.: Oxygen transport in man, N. Engl. J. Med. **286:**407, 1972.
7. Forster, H. V., et al.: Evidence of altered regulation of ventilation during exposure to hypoxia, Respir. Physiol. **20:**379, 1974.
8. Frisancho, R. A.: Functional adaptation to high altitude hypoxia, Science **187:**313, 1975.
9. Haldane, J. S., and Priestley, J. G.: Respiration, London, 1935, Oxford University Press, Ltd.
10. Hamasaki, N., Asakura, T., and Minikami, S.: Effect of oxygen tension on glycolysis in human erythrocytes, J. Biochem. (Tokyo) **68:**157, 1970.

11. Hurtado, A.: Animals in high altitudes: resident man. In Dill, D. B., Adolph, E. F., and Wilber, C. G., editors: Handbook of physiology. Adaptation to the environment section, Washington, D.C., 1964, American Physiological Society.

12. Jöbsis, F. F., and Lamanna, J. C.: Kinematic aspects of intracellular redox reaction. In Robin, E. D., editor: Extrapulmonary manifestations of respiratory disease, New York, 1978, Marcel Dekker, Inc.

13. Jacobson, L. O., et al.: Role of the kidney in erythropoiesis, Nature **179:**633, 1957.

14. Johnston, R. S., Dietlein, L. F., and Berry, C. A.: Biomedical results of Apollo, Washington, D.C., 1975, National Aeronautics and Space Administration, Scientific and Technical Information Office.

15. Kazemi, H., et al.: The CSF HCO_3^- increase in hypercapnia relationship to HCO_3^-, glutamate, glutamine and NH_3 in brain, Respir. Physiol. **28:**387, 1976.

16. Kelsen, S. G., et al.: Effect of hypoxia on the pressure developed by inspiratory muscles during airway occlusion, J. Appl. Physiol. **40:**372, 1976.

17. Kuratowska, Z., Lewartowski, B., and Michalak, E.: Studies on the production of erythropoietin by isolated perfused organs, Blood **18:**527, 1961.

18. Lahiri, S.: Physiological responses and adaptations to high altitude. In Robertshaw, D., editor: Environmental physiology II, Baltimore, 1977, University Park Press, vol. 15.

19. Lahiri, S., and Gelfand, R.: Mechanisms of acute ventilatory responses. In Hornbein, T. F., editor: Regulation of breathing. Lung biology in health and disease, New York, Marcel Dekker, Inc. (In press.)

20. Lambertsen, C. J.: Effects of hyperoxia on organs and their tissues. In Robin, E. D., editor: Extrapulmonary manifestations of respiratory disease. Lung biology in health and disease, New York, 1978, Marcel Dekker, Inc., vol. 8.

21. Maker, H. S., and Lehrer, G. M.: The effect of ischemia. In Lajtha, A., editor: Handbook of neurochemistry, New York, 1971, Plenum Press, vol. 6.

22. Maker, H. S., and Nicklas, W. J.: Biochemical responses of body organs to hypoxia and ischemia. In Robin, E. D., editor: Extrapulmonary manifestations of respiratory disease. Lung biology in health and disease, New York, 1978, Marcel Dekker, Inc., vol. 8.

23. Maren, T. H.: Physiology and chemistry of cerebrospinal fluid, aqueous humor and endolymph in *Squalus acanthias*, J. Exp. Zool. **199:**317, 1977.

24. Monge, C.: Aclimatacion en los Andes, Lima, 1960, Facultad de Medicina.

25. Monge, C. M., and Monge, C. C.: High altitude diseases, Springfield, Ill., 1966, Charles C Thomas, Publisher.

26. Neff, T. A., and Talmadge, D., editors: The Twentieth Aspen Lung Conference. Neuromuscular and chemical control of breathing, Chest **73**(suppl.):1, 1978.

27. Porter, R., and Knight, J., editors: High altitude physiology: cardiac and respiratory aspects, Ciba Foundation Symposium, Edinburgh, 1971, Churchill Livingstone.

28. Pugh, L. G. C. E.: Animals in high altitudes: man above 5,000 meters—mountain exploration. In Dill, D. B., Adolph, E. F., and Wilber, C. G., editors: Handbook of physiology. Adaptation to the environment section, Washington, D.C., 1964, American Physiological Society.

29. Robin, E. D., editor: Extrapulmonary manifestations of respiratory disease. Lung biology in health and disease, New York, 1978, Marcel Dekker, Inc., vol. 8.

30. Tenney, S. M., et al.: Suprapontine influences on hypoxic ventilatory control. In Porter, R., and Knight, J., editors: High altitude physiology: cardiac and respiratory aspects, Ciba Foundation Symposium, Edinburgh, 1971, Churchill Livingstone.

31. Weiskopf, R. B., and Gabel, R. A.: Depression of ventilation during hypoxia in man, J. Appl. Physiol. **39:**911, 1975.

32. Wichser, J., and Kazemi, H.: Ammonia and ventilation: site and mechanism of action, Respir. Physiol. **20:**393, 1974.

Original papers

33. Alexander, J. K., et al.: Analysis of the respiratory response to carbon dioxide inhalation in varying clinical states of hypercapnia, anoxia, and acid-base derangement, J. Clin. Invest. **34:**511, 1955.

34. Arieff, A. I., et al.: Intracellular pH of brain: alterations in acute respiratory acidosis and alkalosis, Am. J. Physiol. **230:**804, 1976.

35. Armstrong, H. G.: Principles and practice of aviation medicine, Baltimore, 1939, The Williams & Wilkins Co.

36. Aste-Salazar, H., and Hurtado, A.: The affinity of hemoglobin for oxygen at sea-level and at high altitudes, Am. J. Physiol. **142:**733, 1944.

37. Austrian Everest climb, 1978, New York Times, June 17, 1978, p. 19.

38. Banchero, N., and Cruz, J. C.: Hemodynamic changes in the Andean native after two years at sea level, Aerospace Med. **41:**849, 1970.

39. Bancroft, R. W., and Dunn, J. E.: Experimental animal decompressions to a near vacuum environment, Aerospace Med. **36:**720, 1965.

40. Benzinger, T.: Explosive decompression. In German aviation medicine in World War II, Washington, 1950, U.S. Government Printing Office.

41. Billings, C. E.: Barometric pressure. In Packer, J. F., and West, V. R., editors: Bioastronautics data book, ed. 2, Washington, D.C., 1973, National Aeronautics and Space Administration.

42. Bjurstedt, A. G. H.: Interaction of centrogenic and chemoreflex control of breathing during oxygen deficiency at rest, Acta Physiol. Scand. **12:**suppl. 38, 1946.

43. Blayo, M. C., Marc-Vergnes, J. P., and Pocidalo, J. J.: pH, Pco_2, and Po_2 of cisternal fluid in high altitude natives, Respir. Physiol. **19:**298, 1973.

44. Bodansky, O.: Methemoglobin and methemoglobin-producing compounds, Pharmacol. Rev. **3:**144, 1951.

45. Boutourline-Young, H. J., and Whittenberger, J. L.: The use of artificial respiration in pulmonary emphysema accompanied by high carbon dioxide levels, J. Clin. Invest. **30:**838, 1951.

46. Brown, E. B., Jr.: Changes in brain pH response to CO_2 after prolonged hypoxic hyperventilation, J. Appl. Physiol. **2:**549, 1950.

47. Brown, E. B., Jr., et al.: Changes in response to inhalation of CO_2 before and after 24 hours of hyperventilation in man, J. Appl. Physiol. **1:**333, 1948.

48. Cain, S. M., and Dunn, J. E.: Low doses of acetazolamide to aid accommodation of men to altitude, J. Appl. Physiol. **21:**1195, 1966.

49. Casey, H. W., Bancroft, R. W., and Cooke, J. P.: Residual pathologic changes in the central nervous sys-

tem of a dog following rapid decompression to 1 mm Hg, Aerospace Med. **37:**713, 1966.

50. Cerretelli, P.: Esistenza di una permanente stimolazione ipossica del centro respiratorio in individui acclimatati a quote di 5.000-7.500 m.s.l.m. (Himalaya), Atti Congr. Intern. Med. Aeron. Spaz **2:**866, 1961.

51. Chinese Everest climb, 1960, New York Times, May 27, 1960, p. 9.

52. Chiodi, H.: Respiratory adaptations to chronic high altitude hypoxia, J. Appl. Physiol. **10:**81, 1957.

53. Chiodi, H.: Respiratory adaptation to high altitude. In Cunningham, D. J. C., and Lloyd, B. B., editors: The regulation of human respiration, Oxford, 1963, Blackwell Scientific Publications, Ltd.

54. Christensen, E. H.: Sauerstoffaufnahme und respiratorische funktionen in grossen hohen, Scand. Arch. Physiol. **76:**88, 1937.

55. Comfort, E. C., and Wilson, J. W.: Some factors affecting time consciousness at high altitudes, AF technical report no. 5970, 1949.

56. Comroe, J. H., Jr., and Schmidt, C. F.: The part played by reflexes from the carotid body in the chemical regulation of respiration in the dog, Am. J. Physiol. **121:**75, 1938.

57. Comroe, J. H., Jr., et al.: The lung, ed. 3, Chicago, 1975, Year Book Medical Publishers, Inc.

58. Courville, C. B.: Asphyxia as a consequence of nitrous oxide anesthesia, Medicine **15:**129, 1936.

58a. Crawford, R. D., and Severinghaus, J. W.: CSF pH and ventilatory acclimatization to altitude, J. Appl. Physiol. **45:**275-283, 1978.

59. Davies, H. C., and Davies, R. E.: Biochemical aspects of oxygen poisoning. In Fenn, W. O., and Rahn, H., editors: Handbook of physiology. Respiration section, Washington, D.C., 1965, American Physiological Society, vol. 2.

60. Dejours, P., Kellogg, R. H., and Pace, N.: Regulation of respiration and heart rate response in exercise during altitude acclimatization, J. Appl. Physiol. **18:**10, 1963.

61. Dejours, P., et al.: Stimulus oxygene chemoreflexe de la ventilation à basse altitude (50 m.) chez l'homme. I. Au repos, J. Physiol. (Paris) **49:**115, 1957.

62. Dejours, P., et al.: Étude de la regulation de la ventilation de repos chez l'homme en haute altitude, Rev. Fr. Etud. Clin. Biol. **4:**115, 1959.

63. Dempsey, J. A.: CO_2 response: stimulus definitions and limitations, Chest **70:**114, 1976.

63a. Dempsey, J. A., et al.: Effects of moderate hypoxemia and hypocapnia on CSF [H^+] and ventilation in man, J. Appl. Physiol. **38:**665, 1975.

64. Dempsey, J. A., et al.: Regulation of CSF [HCO_3^-] during long-term hypooxic hypocapnia in man, J. Appl. Physiol. **44:**175, 1978.

65. Dill, D. B., et al.: Work capacity in acute exposures to altitude, J. Appl. Physiol. **21:**1168-1176, 1966.

65a. Dill, D. B., et al.: Work capacity in chronic exposures to altitude, J. Appl. Physiol. **23:**555, 1967.

66. Dripps, R. D., and Comroe, J. H., Jr.: The effect of the inhalation of high and low oxygen concentrations on respiration, pulse rate, ballistocardiogram and arterial oxygen saturation (oximeter) of normal individuals, Am. J. Physiol. **149:**277, 1947.

67. Dripps, R. D., and Dumke, P. R.: The effect of narcotics on the balance between central and chemoreceptor control of respiration, J. Pharmacol. Exp. Ther. **77:**290, 1943.

68. Duhm, J., Deuticke, E., and Gerlach, E.: Metabolism of 2,3-diphosphoglycerate and glycolysis in red blood cells under influence of dipyridamole and inorganic sulfur compounds, Biochim. Biophys. Acta **170:**452, 1968.

69. Duke, H. N., Green, J. H., and Neil, E.: Carotid chemoreceptor impulse activity during inhalation of carbon monoxide mixtures, J. Physiol. **118:**520, 1952.

70. Dyhrenfurth, N. G.: Six to the summit, National Geographic **124:**460, 1963.

71. Eckenhoff, J. E., and Hafkenschiel, J. H.: The effect of nikethamide on coronary blood flow and cardiac oxygen metabolism, J. Pharmacol. Exp. Ther. **91:**362, 1947.

72. Eckenhoff, J. E., et al.: Cardiac oxygen metabolism and control of coronary circulation, Am. J. Physiol. **149:**634, 1947.

73. Eggler, A.: On Lhotse and Mt. Everest, Alpine J. **61:**239, 1956.

74. Ernsting, J.: The effect of brief profound hypoxia upon the arterial and venous oxygen tensions in man, J. Physiol. **169:**292, 1963.

75. Euler, U. S. von, Liljestrand, G., and Zotterman, Y.: Action of lobeline on carotid sinus region, Upsala Lakaref. Forh. **45:**373, 1939.

76. Fenn, W. O., Rahn, H., and Otis, A. B.: A theoretical study of the composition of the alveolar air at altitude, Am. J. Physiol. **146:**637, 1946.

77. Flight surgeon's manual, USAF manual no. 160-5, Washington, D.C., 1954.

78. Forster, H. V., Dempsey, J. A., and Chosy, L. W.: Incomplete compensation of CSF [H^+] in man during acclimatization to high altitude (4,300 m), J. Appl. Physiol. **38:**1067, 1975.

79. Forster, H. V., et al.: CSF [H^+] as a function of changes in alveolar ventilation during deacclimatization from chronic hypoxia, Fed. Proc. **37:**533, 1978.

80. Garasenko, V. M.: The use of carbon dioxide at lowered barometric pressures, Am. Rev. Soviet Med. **2:**119, 1945.

81. Gibbs, F. A., et al.: The value of carbon dioxide in counteracting the effects of low oxygen, J. Aviation Med. **14:**250, 1943.

82. Grant, W. C., and Root, W. S.: Fundamental stimulus for erythropoiesis, Physiol. Rev. **32:**449, 1952.

83. Himwich, H., et al.: Studies on the effects of adding carbon dioxide to oxygen-enriched atmospheres in low pressure chambers. II. The oxygen and carbon dioxide tensions of cerebral blood, J. Aviation Med. **13:**177, 1942.

84. Houston, C. S.: The effect of pulmonary ventilation on anoxemia, Am. J. Physiol. **146:**613, 1946.

85. Houston, C. S., and Riley, R. L.: Respiratory and circulatory changes during acclimatization to high altitude, Am. J. Physiol. **149:**565, 1947.

86. Hunt, J., and Hillary, E.: The conquest of Everest, New York, 1954, E. P. Dutton & Co., Inc. Also in Natl. Geographic **106:**1, 45, July, 1954.

87. Hurtado, A., and Aste-Salazar, H.: Arterial blood gases and acid-base balance at sea level and at high altitudes, J. Appl. Physiol. **1:**304, 1948.

88. Hurtado, A., et al.: Mechanisms of natural acclimatization. Studies on the native resident of Morococha, Peru, at an altitude of 14,900 feet, Technical documentary reort no. SAM-TDR-56-1, Washington, D.C., 1956, USAF School of Aerospace Medicine. Also in Benson, O. O., and Strughold, H., editors: Physics and medicine of the atmosphere and space, New York, 1960, John Wiley & Sons, Inc.

89. Ivanov, K. P.: Effect of increased oxygen pressure on

animals poisoned by potassium cyanide, Farmakol. Toksikol. **22:**468, 1959.

90. Jackson, J. H.: The evolution and dissolution of the nervous system, Br. Med. J. **1:**591, 660, 703, 1884.

91. Johnson, A. E., et al.: Studies on the effects of adding carbon dioxide to oxygen-enriched atmospheres in low pressure chambers, J. Aviation Med. **13:**130, 1942.

92. Kellogg, R. H.: Central chemical regulation of respiration. In Fenn, W. O., and Rahn, H., editors: Handbook of physiology. Respiration section, Washington, D. C., 1964, American Physiological Society, vol. 1.

93. Kellogg, R. H.: Acclimatization to carbon dioxide, Anesthesiology **21:**634, 1960.

94. Kellogg, R. H.: The role of CO_2 in altitude acclimatization. In Cunningham, D. J. C., and Lloyd, B. B., editors: The regulation of human respiration, Oxford, 1963, Blackwell Scientific Publications, Ltd.

95. Kellogg, R. H., et al.: Respiratory response to inspired CO_2 during acclimatization to an altitude of 12,470 feet, J. Appl. Physiol. **11:**65, 1957.

96. Keys, A., Hall, F. G., and Guzman Barron, E. S.: The position of the oxygen dissociation curve of human blood at high altitude, Am. J. Physiol. **115:**292, 1936.

97. Klausen, K., et al.: Effect of high altitude on maximal working capacity, J. Appl. Physiol. **21:**1191, 1966.

98. Kreuzer, F., et al.: Alveolar-arterial oxygen gradient in Andean natives at high altitude, J. Appl. Physiol. **19:**13, 1964.

99. Lahiri, S.: Alveolar gas pressures in man with lifetime hypoxia, Respir. Physiol. **4:**373, 1968.

100. Lahiri, S.: Dynamic aspects of regulation of ventilation in man during acclimatization to high altitude, Respir. Physiol. **16:**245, 1972.

101. Lahiri, S., and Edelman, N. H.: Peripheral chemoreflexes in the regulation of breathing of high altitude natives, Respir. Physiol. **6:**375, 1969.

102. Lahiri, S., et al.: Relative role of environmental and genetic factors in respiratory adaptation to high altitude, Nature **261:**133, 1976.

103. Lambertsen, C. J.: The philosophy of extremes for the gaseous environment of manned, closed ecological systems, Aerospace Med. **34:**291, 1963.

104. Lambertsen, C. J.: Effects of oxygen at high partial pressure. In Fenn, W. O., and Rahn, H., editors: Handbook of physiology. Respiration section, Washington, D.C., 1965, American Physiological Society, vol. 2.

105. Lambertsen, C. J.: Therapeutic gases: oxygen, carbon dioxide, and helium. In DiPalma, J. R., editor: Drill's pharmacology in medicine, ed. 4, New York, 1971, McGraw-Hill Book Co.

106. Lambertsen, C. J.: Drugs and respiration, Annu. Rev. Pharmacol. **6:**327, 1966.

107. Lambertsen, C. J., Gelfand, R., and Kemp, R. A.: Dynamic influences of CO_2 in respiratory control. In Brooks, C. McC., Kao, F. F., and Lloyd, B. B., editors: The cerebrospinal fluid and the regulation of ventilation, Oxford, 1965, Blackwell Scientific Publications, Ltd.

108. Lambertsen, C. J., et al.: Quantitative interactions of increased Po_2 and Pco_2 upon respiration in man, Ann. N.Y. Acad. Sci. **109:**731, 1963.

109. Laver, M. B., and Seifen, A.: Measurement of blood oxygen tension in anesthesia, Anesthesiology **26:**73, 1965.

110. Lenfant, C., Torrance, J. D., and Reynafarje, C.: Shift of the O_2-Hb dissociation curve at altitude: mechanism and effect, J. Appl. Physiol. **30:**625, 1971.

111. Lenfant, C., et al.: Effect of altitude on oxygen binding by hemoglobin and on organic phosphate levels, J. Clin. Invest. **47:**2652, 1968.

112. Leusen, I.: Aspects of the chemical control of respiration, Verh. K. Vlasm. Acad. Geneskd. Belg. **24:**576, 1962.

113. Lewy, F. H., and Drabkin, D. L.: Experimental chronic carbon monoxide poisoning of dogs, Am. J. Med. Sci. **208:**502, 1944.

114. Life at high altitudes. Proceedings of the Fifth Meeting of the Pan American Health Organization, Scientific Publ. 140, Washington, D.C., 1966, World Health Organization.

115. Lloyd, B. B., Jukes, M. G. M., and Cunningham, D. J. C.: The relation between alveolar oxygen pressure and the respiratory response to carbon dioxide in man, Q. J. Exp. Physiol. **43:**214, 1958.

116. Loeschcke, G. C.: Spielen fur die Ruheatmung des Menschen vom O_2-Druck Abhangige Erregungen der Chemoreceptoren eine Rolle? Arch. Gesamte Physiol. **257:**349, 1953.

117. Loeschcke, H. H.: Central nervous chemoreceptors. In Widdicombe, J. G., editor: Physiology series one: respiratory physiology, Baltimore, 1974, University Park Press, vol. 2.

118. Lundsgaard, C., and Van Slyke, D. D.: Cyanosis, Medicine **2:**1, 1923.

119. Maren, T. H.: Bicarbonate formation in cerebrospinal fluid: role in sodium transport and pH regulation, Am. J. Physiol. **222:**885, 1972.

119a. Maren, T. H.: Effect of varying CO_2 equilibria on rates of HCO_3^- formation in cerebrospinal fluid, J. Appl. Physiol. **47:**471, 1979.

120. Margaria, R., and Cerretelli, P.: Physiological aspects of life at extreme altitudes. In Biometerology, London, 1962, Pergamon Press, Ltd.

121. McFarland, R. A.: The effects of altitude on pilot performance. In Hannisdahl, B., and Sem-Jacobsen, C. W., editors: Aviation and space medicine, Oslo, 1969, Universitetsforlaget.

122. Merino, C. F.: The plasma erythropoietic factor in the polycythemia of high altitudes, technical documentary report no. SAM-TDR-56-103, Washington, D.C., 1956, USAF School of Aerospace Medicine.

123. Michel, C. C., and Milledge, J. S.: Respiratory regulation in man during acclimatization to high altitude, J. Physiol. **168:**631, 1963.

124. Milledge, J. S.: Respiratory regulation at 19,000 ft (5,700 m.). In Cunningham, D. J. C., and Lloyd, B. B., editors: The regulation of human respiration, Oxford, 1963, Blackwell Scientific Publications, Ltd.

125. Mitchell, R. A., et al.: Respiratory responses mediated through superficial chemosensitive areas on the medulla, J. Appl. Physiol. **18:**523, 1963.

126. Mulhausen, R. O., Astrup, P., and Mellemgard, K.: Oxygen affinity and acid-base status of human blood during exposure to hypoxia and carbon monoxide, Scand. J. Clin. Lab. Invest. suppl. **103:**9, 1968.

127. Nelson, P. G., and Frank, K.: Intracellularly recorded responses of nerve cells to oxygen deprivation, Am. J. Physiol. **205:**208, 1963.

128. Nielsen, M., and Smith, H.: Studies on the regulation of respiration in acute hypoxia, Acta Physiol. Scand. **24:**293, 1951.

129. Orr, J. A., et al.: Cerebrospinal fluid alkalosis during high altitude sojourn in unanesthetized ponies, Respir. Physiol. **25:**23, 1975.

130. Otis, A. B., Rahn, H., and Chadwick, L. E.: Effects of adding carbon dioxide to inspired oxygen on tolerance

to high altitudes, Proc. Soc. Exp. Biol. Med. **70:**487, 1949.

131. Otis, A. B., et al.: Performance as related to composition of alveolar air, Am. J. Physiol. **146:**207, 1946.

132. Peters, J. P., and Van Slyke, D. D.: Quantitative clinical chemistry methods, Baltimore, 1931, The Williams & Wilkins Co.

133. Pierce, E. C., Jr., et al.: Blood P_{CO_2} and brain oxygenation at reduced ambient pressure, J. Appl. Physiol. **17:**899, 1962.

134. Polgar, G., and Forster, R. E.: Measurement of oxygen tension in unstirred blood with a platinum electrode, J. Appl. Physiol. **15:**706, 1960.

135. Pugh, L. G. C. E.: Physiological and medical aspects of the Himalayan Scientific and Mountaineering Expedition, 1960-1961, Br. Med. J. **2:**621, 1962.

136. Rahn, H., and Otis, A. B.: Man's respiratory response during and after acclimatization to high altitude, Am. J. Physiol. **157:**445, 1949.

137. Rahn, H., et al.: The effects of hypocapnia on performance, J. Aviation Med. **17:**164, 1946.

138. Rahn, H., et al.: Adaptation to high altitude: respiratory response to CO_2 and O_2, J. Appl. Physiol. **6:**158, 1953.

139. Reissmann, K. R.: Studies on the mechanism of erythropoietic stimulation in parabiotic rats during hypoxia, Blood **5:**372, 1950.

140. Richards, D. W., Fritts, H. W., Jr., and Davis, A. L.: Observations on the control of respiration in emphysema: the effects of oxygen on ventilatory response to CO_2 inhalation, Trans. Assoc. Am. Physicians **71:** 142, 1958.

141. Rouing, P. J. E., et al.: Hypoxemia—erythropoieses and hemolysis, Med. Thorac. **19:**26, 1962.

142. Roy, S. B.: Circulatory and ventilatory effects of high altitude acclimatization and deacclimatization of Indian soldiers, Delhi, 1972, General Printing Co.

143. Sayen, J. J., et al.: Studies of coronary disease in the experimental animal. II. Polarographic determinations of local oxygen availability in the dog's left ventricle during coronary occlusion and pure oxygen breathing, J. Clin. Invest. **30:**932, 1951.

144. Schaefer, K. E.: Atmung und Saure-Basengleichgewicht bei Langdauernden Aufenthalt in 3% CO_2, Arch. Gesamte. Physiol. **251:**689, 1949.

145. Schaefer, K. E., et al.: Respiratory acclimatization to carbon dioxide, J. Appl. Physiol. **18:**1071, 1963.

146. Schmidt, C. F.: The cerebral circulation in health and disease, Springfield, Ill., 1950, Charles C Thomas, Publisher.

147. Schmidt, C. F., Kety, S. S., and Pennes, H. H.: The gaseous metabolism of the brain of the monkey, Am. J. Physiol. **143:**33, 1945.

148. Severinghaus, J. W.: Oxyhemoglobin dissociation curve correction for temperature and pH variation in human blood, J. Appl. Physiol. **12:**485, 1958.

149. Severinghaus, J. W., and Carcelen, B. A.: Cerebrospinal fluid in man native to high altitude, J. Appl. Physiol. **19:**319, 1964.

150. Severinghaus, J. W., and Crawford, R. R.: Carotid chemoreceptor role in CSF alkalosis at altitude, Chest **73:**249, 1978.

151. Severinghaus, J. W., Bainton, C. R., and Carcelen, A.: Respiratory insensitivity to hypoxia in chronically hypoxic man, Respir. Physiol. **1:**308, 1966.

152. Severinghaus, J. W., et al.: Respiratory control at high altitude suggesting active transport regulation of CSF pH, J. Appl. Physiol. **18:**1155, 1963.

153. Severinghaus, J. W., et al.: Cerebral blood flow at high altitude, Circ. Res. **19:**274, 1966.

154. Sørensen, S. C., and Severinghaus, J. W.: Respiratory sensitivity to acute hypoxia in man born at sea level living at high altitude, J. Appl. Physiol. **25:**211, 1968.

155. Sørenson, S. C., and Severinghaus, J. W.: Irreversible respiratory insensitivity to acute hypoxia in man born at high altitude, J. Appl. Physiol. **25:**217, 1968.

156. Sprouffske, J. F., Pittman, J. C., and Kaufman, W. C.: Prediction of the final volume of the human body exposed to a vacuum, Aerospace Med. **40:**740, 1969.

157. Sugioka, K., and Davis, D. A.: Hyperventilation with oxygen—a possible cause of cerebral hypoxia, Anesthesiology **21:**135, 1960.

158. Swann, H. G.: The principles of resuscitation, Anesthesiology **14:**126, 1953.

159. Takano, N.: Effect of CO_2 on O_2 transport, O_2 uptake and blood lactate in hypoxia of anesthetized dog, Respir. Physiol. **10:**38, 1970.

160. Talbot, J. H., and Dill, D. B.: Clinical observations at high altitude: observations on six healthy persons living at 17,500 feet and a report of one case of chronic mountain sickness, Am. J. Med. Sci. **192:**626, 1936.

161. Tenney, S. M., Remmers, J. E., and Mithoefer, J. C.: Interaction of CO_2 and hypoxic stimuli on ventilation at high altitude, Q. J. Exp. Physiol. **48:**192, 1963.

162. Thilenius, O. G., et al.: Response of pulmonary circulation of resting unanesthetized dogs to acute hypoxia, Am. J. Physiol. **206:**867, 1964.

163. Tribukait, B.: Der Einfluss Chronischer Hypoxie Entsprechend 1,000-8,000 m. Hohe auf die Erythropoiese der Ratte, Acta Physiol. Scand. **57:**1, 1963.

164. Tribukait, B.: Experimentelle Untersuchungen zue Regulation der Erythropoiese unter Besonderer Berucksichtigung der Bedeutung der Sauerstoffs, Acta Physiol. Scand. **58**(suppl. 208):5, 1963.

165. Velasquez, T.: Tolerance to acute anoxia in high altitude natives, J. Appl. Physiol. **14:**357, 1959.

166. Ullman, J. R.: Americans on Everest, Philadelphia, 1964, J. B. Lippincott Co.

167. Ward, R. L., and Olson, O. C.: Report of severe anoxic anoxia with recovery, J. Aviation Med. **14:**360, 1943.

168. Weiskopf, R. B., and Gabel, R. A.: Depression of ventilation during hypoxia in man, J. Appl. Physiol. **39:** 911, 1975.

169. West, J. B., et al.: Arterial oxygen saturation during exercise at high altitude, J. Appl. Physiol. **17:**617, 1962.

170. Wood, E. H., et al.: Influence of acceleration on pulmonary physiology, Fed. Proc. **22:**1024, 1963.

74

CHRISTIAN J. LAMBERTSEN

Physical, chemical, and nervous interactions in respiratory control

Response of respiration to environmental and metabolic alterations is accomplished by the simultaneous influences of many factors. These influences may represent interactions of different types of chemical effects, a single type of effect acting in a different manner at separate sites, and even differences in the rate of change in the level of a chemical factor at various chemosensitive locations. In hypoxia, simultaneous changes occur in relation to such mechanisms as (1) the concurrent peripheral chemoreflex stimulation and central depression by low P_{O_2}; (2) the simultaneous, similar, and additive effects of lowered P_{O_2} and increased P_{CO_2} on chemoreceptor discharge; (3) the reduction of central P_{CO_2} and $[H^+]$ by the combined results of hypoxic respiratory stimulation and cerebral vasodilation; and (4) the gradual development of metabolic acidosis and its consequences (Chapter 71). These and other interactions differ greatly from one type of hypoxia to another, depending largely on whether the O_2 lack is due to changes in the external gaseous environment or to inadequate pulmonary ventilation. The effects are superimposed on and modify the normal relationships among such functions as general metabolism, pulmonary ventilation, blood gas transport, local metabolism, and circulation.

In addition to the interplay of different factors in the natural resting state, there are situations in which the basis and manner of chemical, physical, and neurologic interactions may be still more complex than is respiratory control in eupnea, hypoxia, or carbon dioxide breathing. Two such situations are altered body temperature and muscular exercise.

BODY TEMPERATURE AND RESPIRATION

Hyperthermia and hypothermia, like changes in barometric pressure or in P_{CO_2}, are physical changes. However, many of the physiologic effects of altered body temperature are undoubtedly related to induced alterations of metabolism and body fluid composition, as well as to localized chemical consequences of the temperature change at discrete locations that include sensory components of the respiratory control system.

Hyperthermia

Body temperature elevation occurs not only in the fever of infectious disease but also in normal exercise and during exposure to environments of increased temperature and humidity. When deep body temperature rises, the respiratory effects appear to be a smoothly continuous extension of those encountered on going from hypothermia to the normal body temperature level. Thus hyperthermia in man at rest or in exercise results in increased pulmonary ventilation.* For this reason, and because whole-body metabolism is not elevated to the same degree as ventilation, both alveolar and arterial P_{CO_2} are diminished below the level for euthermia.[89,110,135]

The fundamental basis for the hyperpnea of hyperthermia has not been established, but several tentative explanations have been considered for this important form of respiratory stimulation. They include (1) a physical effect of increased temperature on the cells of the respiratory centers and/or the peripheral chemoreceptors to exaggerate the reactivity of these respiratory control mechanisms to their normal stimuli,[89] (2) an increase in central stimulus level due to accelerated cellular metabolism, (3) diminished efficiency of the buffering of CO_2 by body fluids,[158] and (4) an increase in activity of hydrogen ions.[62] It is likely that the effect is a modulation of the normal stimulus-sensor function rather than the introduction of a new mechanism. Thus the dif-

---

*See references 75, 89, 135, 147, and 152.

ficulties involved in determining the basis for the respiratory effects of hyperthermia and hypothermia are largely the same as those concerned with understanding the normal mechanisms of respiratory control.

Elevation of carotid body temperature or perfusion of the carotid sinus regions with fluids at a higher than normal temperature does produce prominent respiratory stimulation.[101,157] Certainly, reduced carotid body temperature, exemplified in the extreme by cold block of the carotid chemoreceptor mechanisms, reduces impulse traffic over Hering's nerve.[78,113] The actual contribution of chemoreceptor activity to respiratory regulation at abnormal body temperatures is not known, but it is conceivable that the chemoreflex influences of hypoxia, H^+, and CO_2 may be exaggerated in hyperthermia and reduced in hypothermia.

The "CO_2 response" curves of Fig. 74-1 indicate that a 2° C elevation of body temperature in normal humans produces a gross exaggeration of the respiratory response to elevated alveolar or central venous P_{CO_2}. Since the exaggeration is not due to an altered slope of the CO_2-response curve, "sensitivity" of the control mechanism to change in alveolar P_{CO_2} can be considered to

remain close to normal. Even when alveolar or central nervous system (CNS) P_{CO_2} is held constant, pulmonary ventilation increases approximately 3.9 L/min/m²/Δt °C above normal.[89] Whereas increased body temperature alters the characteristics of a "central impulse generator" in the respiratory control system,[147] the exaggerated respiratory response at a particular level of arterial or central P_{CO_2} does not necessarily imply an increased cellular reactivity to a normal chemical stimulus. The real stimulus at the sensor site could have been changed by hyperthermia. Elevated body temperature, altering as it does the activity of ions in solution in tissue fluids,[155] could increase the functional activity of such potential stimuli as H^+, as suggested in Fig. 74-1. Unfortunately the difficulties involved in measuring the effects of temperature change on $[H^+]$, and especially H^+ activity, make this aspect of respiratory control an elusive subject for study.[62]

Hypothermia

Intentionally lowered deep body temperature has been used to protect vital organs against the ravages of hypoxia during partial or temporary total ischemia.[20] Information concerning effects

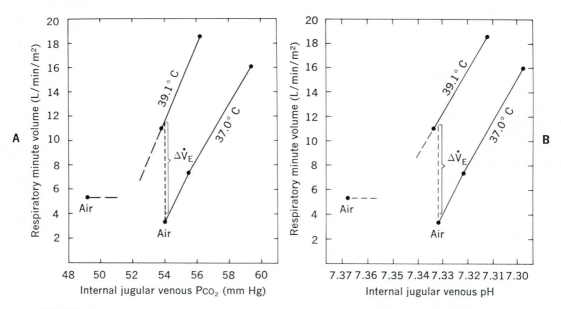

Fig. 74-1. Influence of hyperthermia on respiratory response to CO_2 in normal humans. **A** and **B** represent respiratory response to increased central P_{CO_2} and acidity at two different body temperatures (37.0° and 39.1° C). Vertical lines at normal internal jugular venous P_{CO_2} of 54 mm Hg indicate increase in respiration produced by hyperthermia at a constant level of CNS P_{CO_2}. Note that plot of respiration against pH ignores influence of increased temperature on hydrogen ion activity. It is therefore conceivable that the hyperthermic curve is an extension of the curve for normothermia. (Data from author's laboratory, unpublished observations.)

of hypothermia is summarized in several reviews.* It produces drastic changes in respiration and in the total system for gas exchange between the atmosphere and the tissue cell. The extreme effect of generalized hypothermia is, of course, death, due to respiratory failure or ventricular fibrillation. With more moderate degrees of temperature depression the events related to respiration appear to be as follows:

1. Lowered central temperature somehow reduces the response of the respiratory control mechanisms to their normal stimuli. The mechanisms responsible for this lowered response in hypothermia are probably the same but are no better understood than the opposite changes in the probably identical factors cited above as involved in hyperthermic hyperventilation.

2. The rates of overall O_2 consumption and CO_2 production are diminished.

3. Ventilation falls progressively with the lowering of body temperature and metabolism. Eventually the reduction of ventilation is greater than would be expected from the magnitude of the concomitant decrease in whole-body metabolism. As a result, alveolar P_{CO_2} of the spontaneously breathing, nonshivering animal rises, oxygenation becomes deficient, and hypoxic death supervenes despite the reduced O_2 requirement.[20,119,161] Ventilation ceases before cardiac failure occurs. Substitution of O_2 or artificial ventilation for spontaneous breathing will prevent hypoxia and permit further lowering of deep body temperature; death then results from abnormalities of impulse conduction in the heart that may be related to gain of calcium and loss of potassium by the myocardium.[86]

Extreme degrees of hypothermia are used to preserve excised, unperfused organs for many hours.[144] In vitro and in vivo the effects of decreased temperature on tissue metabolism resemble those of extreme narcosis. Each of these conditions acts to protect the tissue cells against the otherwise more damaging effects of hypoxia. It is still not known whether hypothermia, narcosis, and sustained oxygenation are capable of complementary protective action in preservation of life processes.

RESPIRATORY RESPONSE TO MUSCULAR EXERCISE

A precise match is sustained between ventilation and metabolism as increasing degrees of muscular work present increasing requirements for elimination of CO_2 and uptake of O_2. This match occurs despite the fact that exercise produces both the greatest pulmonary ventilation that can be sustained more than momentarily and the greatest attainable increase in the body's requirement for O_2 (from a resting level of about 250 ml/min to 4 L/min or more). The extensive investigation of the complex and still mysterious response to physiologic stress has been elaborated expertly in reviews that describe observable changes and offer concepts of regulatory processes involved.* These observations and concepts will be summarized and extended here.

Several outstanding features in the hyperpnea of muscular exercise, shown in Fig. 74-2, are firmly established:

1. *O_2 requirement in exercise* is linearly related to the rate at which work is performed.[1,82]

2. *Body temperature* is raised by increased rate of heat production, the rise being about 0.086° C/100 kg-m/min of work load.[134] The rise in body temperature is, within limits, independent of environmental temperature.[148]

3. *Respiratory minute volume* increases smoothly and steadily in direct proportion to the increase in work done and in O_2 consumed; the curve for CO_2 *production* would exactly parallel that for O_2 consumption and would coincide with it if corrected for the existing respiratory quotient.[160a]

4. *Efficiency of the respiratory adjustment,* as judged from the amount of air that must be respired for each liter of O_2 absorbed (known as the *ventilation coefficient*), increases during light and moderate work and decreases during heavy work.[82,160a] The increased efficiency in moderate exertion is probably largely due to an increase in the mean alveolar to pulmonary capillary P_{O_2} gradient secondary to the gross lowering of mixed venous P_{O_2} (Fig. 69-12). It may also be due in part to an increase in the pulmonary capillary area (Chapter 68). As work increases to severe levels a decrease in efficiency (rise in ratio of ventilation/O_2 consumption) coincides with a diminution in overall muscular efficiency, evidenced in a flattening of the curve relating work to O_2 consumption and with steep rises in respiratory minute volume (Fig. 74-2) and in blood lactate level.[1,68,143]

The deviations from the linear relationships of lighter work are obvious characteristics of transition from a level of effort that can be sustained as a steady state (i.e., physiologic) to one that approaches the limit of tolerance and would soon lead to intolerable sensations or to collapse (i.e.,

---

*See references 77, 84, 99, 137, and 161.

*See references 1, 9, 17, 160a, and 164.

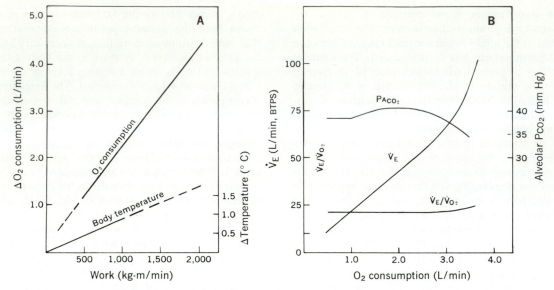

Fig. 74-2. Effects of physical exercise on metabolism and respiration.

A shows pattern of essentially direct increases of metabolic O_2 consumption[1] and deep body temperature[134] as work level is increased.

B indicates changes of respiratory minute volume,[1] alveolar P_{CO_2},[1] and ratio of pulmonary ventilation/unit of O_2 consumption. Respiratory minute volume increases linearly with small or no change in alveolar P_{CO_2} until degree of work leads to O_2 consumption above about 2.5 L/min. Then ventilation rises out of proportion to work and O_2 consumption, resulting in lowering of alveolar P_{CO_2}. (Composite diagrams from several sources.)

pathologic). Part of the indicated ventilatory inefficiency may be referable to the maximal activation of the respiratory muscles themselves, particularly the expiratory group, which must be progressively involved in producing the observed increase in respiratory rate when the depth is already close to maximal.

5. *Increase in tidal volume and frequency components of respiratory minute volume* is progressive, the former rising more steeply than the latter to reach an optimum in moderately heavy work (i.e., at an O_2 consumption of about 2.5 L/min).[82,160a,164] Beyond this point the respiratory adjustment to further increases in work is almost entirely due to acceleration of the rate of breathing.

6. *Limitation of the capacity to increase ventilation* occurs when the ability to increase respiratory frequency without diminishing depth of breathing is reached.[82,160a] Attempts to increase the rate of work beyond this limit must fail, since gross alteration of alveolar ventilation, arterial gaseous composition, and pH will promptly result.

Control of pulmonary ventilation relative to metabolism

Since the ultimate mechanisms and quantitative relationships underlying respiratory control at rest are uncertain, the relation of breathing to metabolism in the more complex state of exercise is not completely understood. A considerable advance in understanding has resulted from improved quantitative study of changes in blood and application of methods for studies of dynamic characteristics of chemical and respiratory changes in exercise.* From such observations the following facts and relationships appear evident:

1. Changes in chemical factors such as H^+, P_{CO_2}, and fixed acids do occur in the arterial and brain venous blood in exercise. However, these changes are not all uniformly progressive from the mildest to the most severe muscular work. Changes in chemical factors such as P_{O_2} and P_{CO_2} may in some individuals and circumstances be in directions opposite to those that normally

*See references 1, 17, 74, 79, 146, 147, 160a, 164, and 167.

Table 74-1. Quantitative relationships of ventilatory, metabolic, thermal, and chemical factors in increasingly strenuous muscular exercise

	Rest	Exercise I	Exercise II
No. of subjects*	13	7	10
Work load (kg-m/min)	0	346	777
O_2 consumption (L/min, STPD)	0.257	1.062	2.036
Respiratory minute volume (L/min/m²)	3.57	13.36	28.15
Immediate change in ventilation†			
Initiation (L/min/m²)	0	+3.51	+6.68
Termination (L/min/m²)	0	−6.23	−13.04
Temperature‡			
Δt (°C)	0	+0.3	+0.7
Respiratory equivalent (L/min/m²)	0	+1.2	+2.8
Arterial P_{CO_2} (mm Hg)	40.5	Δ0	Δ0
Arterial pH (units)	7.405	Δ−0.010	Δ−0.060
Estimated CSF pH§ Δ	0	0	0

All tabulated values for respiratory minute volume are expressed as BTPS.

*Except where otherwise indicated, the subjects for the blood, metabolic, and respiratory measurements in this table are the pooled subjects of two studies offering comparable conditions and technique, as follows: Lambertsen et al.[134] subjects J. E., G. H., R. L., C. M., D. L., E. H., G. O., and J. S.; Asmussen and Nielsen[65] subjects O. W., S. L., M. L., S. S., and J. S. Unpublished pH data and other individual values for the second study were kindly provided by its authors. For determination of the changes induced by exercise, each subject served as his own resting control. Information from the two studies was combined to provide a larger amount of data on arterial blood changes. Values for P_{CO_2} and pH are corrected for the change in body temperature in exercise. As a result, arterial P_{CO_2} and estimated CSF pH are shown to be unchanged in exercise.

†Immediate changes in ventilation are derived from 23 measurements on 6 subjects of Dejours.[16] (See footnote, p. 1885).

‡Regression of change in deep body temperature or work load[134] during stable-state exercise in normal subjects is found to average Δt = +0.086° C/100 kg-m/min. This Δt is related, through work load, to O_2 consumption in the table.

§CSF pH change of zero is estimated on basis that acute change is (1) related to alterations in P_{CO_2}, (2) unrelated to change in blood [HCO_3^-], and (3) not affected by fixed acid level of blood.[131]

produce respiratory stimulation at rest.[1,16,17,110] Thus *arterial* P_{CO_2} may rise slightly or remain essentially unchanged during light and moderate work[1,65,160a,164] (Table 74-2); during extreme exertion, it tends to decrease.* *Arterial and brain venous hydrogen ion concentration* may not be detectably altered in light work (Table 74-1),[122] but as the severity of exercise increases to levels involving O_2 consumptions somewhat greater than 1 L/min, a clear-cut correlation between degree of hyperventilation and increase in blood [H^+] emerges.[134] This fact, unrecognized for many years,[9,110] is evident in Table 74-1. The rise in blood acidity is primarily related to production of *fixed acids*[68,134,143] and is modified in degree by secondary changes related to ventilatory elimination of CO_2. It is not primarily the result of carry-over of metabolically produced CO_2 into arterial blood.

Percent hemoglobin saturation may be unchanged[146] or lowered slightly in exercise,[103,134] but this does not necessarily indicate a lowered arterial P_{O_2}, since the change in hemoglobin saturation is related to the greater acidity of the arterial plasma.[134] The O_2 *tension* in arterial blood appears from most direct measurements not to be lowered appreciably in trained or untrained normal subjects, even during maximal work[65,103,160a,164]; when observed, a fall in arterial P_{O_2} has been small.[136] *Hemoglobin concentration* is increased.*

The effect of O_2 administration on exercising subjects is clearly to reduce the ventilation at a particular level of work.[64,65,124,134] In the presence of the high cardiac output and rapid muscle blood flow in exercise, the high P_{O_2} and the approximately 10% increase in arterial O_2 content produced by O_2 breathing at sea level should significantly reduce the degree of muscle anaerobiosis. Fixed acid levels in blood are reduced by O_2 administration.[64,134] However, it is likely that this represents only a part of the basis for reduction of exercise hyperpnea by O_2.[16,17,65] An additional factor proposed is removal of a chemoreceptor reflex drive, *present even when arterial O_2 tension has not been lowered by exercise.*† This

---

*See references 1, 65, 103, 134, 160a, and 164.

*See references 9, 16, 74, 110, and 146.
†See references 16, 17, 65, 146, 160a, and 164.

latter effect of O_2 in exercise is indicated in Fig. 74-10.[16]

2. Although *increased* PCO_2 *and* $[H^+]$, together *with decreased* PO_2, must necessarily occur in the exercising muscles and hence in mixed venous blood, no receptors have been demonstrated that take advantage of these metabolically induced chemical changes in capillary or venous blood. An active and unsuccessful search for such receptors has included infusion of venous blood into the vessels of the limbs.[9,118] However, when the circulation to exercising limbs is suddenly occluded while holding arterial CO_2 tension constant, the resulting gross increase in the chemical disturbance within the muscles somehow leads to a progressive increase in respiration.[66]

3. *Body temperature* rises progressively with increase in the severity of work and thus with the degree of exercise hyperpnea.[110,134,150] Hyperthermia at rest produces prominent respiratory stimulation,[89] the effect at an artificially fixed arterial PCO_2 of 54 mm Hg approximating 3.9 L/min/m²/Δ °C (Fig. 74-1). However, the rise in temperature accompanying exercise in a cool and dry environment is small (Fig. 74-2), and its maximal effect must therefore represent only a fraction of the total respiratory response to muscular work.[110] Moreover, change in temperature should not be useful as a stimulant factor in the rapid, initial adjustment of respiration to the requirements of metabolism, since the time constant for temperature change is considerably slower than that for most of the increase in exercise ventilation (Fig. 74-3).[148]

The rate of rise in deep body temperature

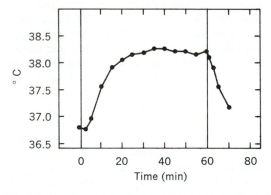

Fig. 74-3. Average rate of change in deep esophageal temperature in normal subject during leg work at 900 kg-m/min. Rise in body temperature is a factor in respiratory control during sustained exercise. (Modified from Nielsen and Nielsen.[148])

varies from one subject to another, even under constant environmental conditions and at a fixed rate of O_2 consumption (heat production), since normal subjects vary in the mass to be heated and in the efficiency of temperature regulation. Rise in temperature and its effects on acid-base balance are probably important factors limiting exercise tolerance.

Components of respiratory regulation in exercise

Most early discussions of respiratory control in exercise have emphasized the inadequacy of any single stimulus, chemical or neurologic, as a basis for the entire phenomenon of exercise hyperpnea. This emphasis has on occasion been so extreme that factors considered individually incapable of accounting for the whole of ventilatory stimulation in exercise tended to be discarded as having no contribution at all.

On the other hand, it is now quite evident that the total respiratory stimulation in exercise cannot be interpreted in terms of direct "dose-response" stimulation of central and peripheral chemoreceptors only by influences of metabolic products.[147,160a,164] The total ventilatory response is brought about not by one stimulus but by the combined and simultaneous influences of several.* Factors that separately appear to have limited importance may together be contributing heavily to the overall response, and the relative importance of different stimuli may change in various degrees and conditions of exercise. Furthermore, it is a serious error to consider exercise to be a single physiologic state when qualitative and quantitative differences, between moderate and severe exercise or in different environmental conditions, may be as great as between rest and moderate work. To this should be added the probability that quantitative features of respiratory control are not necessarily identical in different *individuals,* in different *types* of exercise, and that *conditioning* to exercise ("training") may alter the degree of response to either neurologic or chemical factors. Thus it should not be expected that the data obtained in all studies of exercise will fall into a single clear-cut pattern of quantitative relationships, even though the underlying mechanisms must be identical in all exercise.

From the considerations stated, several conclusions can be reached regarding this powerful and physiologic hyperpnea:

---

*See references 17, 110, 111a, 114, 129, 147, 160a, and 164.

1. *The hyperventilation of exercise cannot be entirely explained solely in terms of any single or composite known chemical change related to metabolism.*
2. *The hyperventilation of exercise is so accurately adjusted to rising metabolic requirements that a chemical sensor mechanism related to products of metabolism must be primarily and intimately concerned in its fine adjustment.*
3. *The preceding conclusions are not mutually exclusive.*

Chemical stimuli. In the many attempts to identify a chemical basis for the clear linkage of ventilation to metabolism, practically every possibility has been suggested. The proposals have included the following:

1. Production by exercise of a specific chemical excitant of unknown identity[17,29,106,160a]
2. So great a sensitivity of the respiratory center to changes in blood [H+] that a hyperpnea of the total observed magnitude could be attributed to changes too small to be detected by existing chemical methods[28]
3. Accumulation of acid inside the center in higher concentration than in the arterial blood[107,168]
4. Production by exercise of an increased sensitivity of the center to its existing normal chemical stimulus[149]
5. An influence of increased body temperature on the metabolism of the respiratory centers[9]
6. The existence of additional peripheral chemoreceptors of unknown location but somehow responsive to the [H+] and P_{CO_2} of mixed venous blood coming from the exercising muscles.[63,147,160a,164]

Although some of these concepts persist, they really are not explanations but only restatements of the problem, each of which supports the basic assumption that the gross and controlled hyperpnea of exercise must be due primarily to the direct and graded action of a chemical or temperature stimulus. Nevertheless, whereas chemical factors cannot be responsible for the entire respiratory drive in work, it is now evident that they are involved in the adjustment of ventilation to metabolism.[17,66,118,134]

Neural factors. In the search for other than chemical factors, particular emphasis has been given to the possibility that the essential cause of hyperpnea of exercise is an increase in the influence of excitatory nerve impulses on the respiratory center. Krogh and Lindhard[125,126] and Nielsen[149] advocated irradiation of motor nerve

impulses into the respiratory center as a partial explanation, particularly for the hyperpnea occurring at the onset of exercise.

Numerous attempts to identify neurologic factors of great potency have been made,* and evidence obtained for the existence of stimulant reflexes from the limbs, activated by active or passive movements.† This proposed neural, peripheral mechanism alone is not capable of explaining either the entire magnitude of the hyperpnea of exercise or its accurate adjustment to the prevailing metabolic requirements.[9,110,160a,164] However, it has become well established that neural factors from the exercising limbs contribute heavily to exercise hyperventilation.‡

Interactions and combined influences

Interest in the combined effects of a number of different forms of respiratory drive has expanded as information concerning the magnitude of respiratory stimulation associated with changes in each of several types of factors has gradually evolved. Although the degree of understanding has not increased greatly over the past several decades,[147,160a,164] improved descriptive information is now accumulating. As this has occurred, it has become less necessary to postulate the existence of an unknown and extremely potent reflex or chemical excitant to the respiratory center or an equally powerful but unknown neurogenic influence. Instead, it has become necessary to seek out the interactions of factors and the variations in different degrees of exercise.

At this point of improved qualitative understanding it is necessary to acknowledge that the individual *quantitative roles* of (1) radiation from higher cortical centers, (2) peripheral neural factors, (3) central chemical stimulation, (4) temperature, or (5) peripheral chemoreflex stimulation have not been firmly established in relation to exercise hyperpnea. Moreover, each has been considered as not being related at all to hyperpnea in some particular exercise situation.§

The subsequent discussion, while recognizing the complexity of and present uncertainties regarding control, will provide and emphasize information concerning *measurable* physiologic changes associated with increasing work. These will be related to the evolution of chemoneurologic concepts of regulation.

---

*See references 83, 104, 105, 112, 147, 160a, and 164.
†See references 17, 83, 112, 160, 160a, and 164.
‡See references 16, 17, 66, 118, 121, 146, 160, and 160a.
§See references 9, 17, 147, 160a, and 164.

Fig. 74-4. Concepts of regulation of alveolar ventilation to metabolic production of CO_2 at rest and in exercise.

A, Concept of limited and positive stimulation of breathing by CO_2 superimposed on stimulation by "neurogenic" drive.

$\dot{V}_A$, Pa_{CO_2} diagram shows iso-CO_2 elimination curves (for 0.25 and 1.5 L/min CO_2 production of rest and exercise). Points on iso 0.25-$\dot{V}_{CO_2}$ line of this figure indicate rest conditions in normoxia and hypoxia. Ventilatory response to CO_2 at rest has been indicated both in normoxia and hypoxia. Points *A* and *B* indicate work conditions in normoxia and hypoxia, and CO_2 response curves during exercise have been drawn for each condition. *b* and *b'* indicate hyperpnea due to nervous drive occurring without any change of Pa_{CO_2} both for normoxia and hypoxia. *a* and *a'* indicate additional hyperpnea due to increased Pa_{CO_2}. *b* and *b'* represent the increase of ventilation observed at onset of exercise in normoxia *(b)* and hypoxia *(b')*.

B, Concept of full positive respiratory response to metabolically produced CO_2, with supplemental stimulation of breathing by neural factors, supplemental stimulation by metabolic acidosis in severe exercise, and final precise adjustment to metabolism by inverse role of CO_2.

At a particular stable level of metabolic CO_2 production, whether at rest or during exercise, alveolar ventilation and alveolar P_{CO_2} are necessarily inversely related (Chapter 68). These fixed relationships are shown by iso-CO_2 production curves for $\dot{V}_{CO_2} = 0.25$ for rest and $\dot{V}_{CO_2} = 2.0$ L/min for exercise (Table 74-1). At rest during air breathing, *rise in alveolar P_{CO_2} results from fall in alveolar ventilation and is not a cause of increased ventilation.* This relationship necessarily pertains also to stable states in exercise. *Stable-state chemical system of control in exercise provides, not only a positive driving stimulus by metabolically produced CO_2, but support of ventilation against its decline.* Diagram reflects this inverse or negative role in exercise so that, in presence of stimuli other than or in addition to *CO_2-generated H^+*, increase in respiration lowers alveolar P_{CO_2}. Rise in alveolar P_{CO_2} thus represents alveolar ventilation subnormal for existing rate of metabolic CO_2 production. Rise in CO_2 checks against further fall in ventilation rather than providing supranormal stimulus.

Point *A* in this figure represents normal $\dot{V}_A$, Pa_{CO_2} relationship at rest, and *A-B*, respiratory stimulation observed during hypercapnia produced by breathing of CO_2.[127a] Point *B*, for ventilation in CO_2 breathing at rest, is shown as being also on $\dot{V}_A$, Pa_{CO_2} curve for metabolic CO_2 production of 2.0 L/min in exercise, with any small increase or decrease in ventilation inducing opposite changes in P_{CO_2}. *A-B* therefore represents full and maximum degree of respiratory stimulant response that could and should be expected in stable-state exercise from effects of metabolically produced CO_2 if no other factors were involved. This full response is considered actually to exist as baseline or primary positive influence of increased metabolism in exercise, on which other factors exert modifying secondary, tertiary, or other effects.

C' represents actual observed (measured) relationship of ventilation and alveolar P_{CO_2} in stable-state exercise (with alveolar P_{CO_2} not changed from its normal value at rest) (Table 74-1). Level of ventilation observed in exercise is higher than that indicated at point *B* as related only to CO_2. The ventilation in excess of that attributable to CO_2 itself is here indicated as related to "neurogenic stimuli" (e.g., of motion). Therefore, in accordance with necessary relationship of $\dot{V}_A$ and Pa_{CO_2} for 2.0 L/min CO_2 production, this additive influence will lower P_{CO_2} and cause stable-state alveolar P_{CO_2} to lie at some real point on iso-$\dot{V}_{CO_2}$ curve, less than the nearly 55 mm Hg of point *B*. It in fact stabilized at *C'* in the human subjects (Table 74-1), with alveolar/arterial P_{CO_2} not detectably different from the normal value at which it is controlled for the resting state.

At *C'*, small changes in ventilation about stable exercising state must induce inverse changes in Pa_{CO_2}; these then in turn counter the ventilatory deviations.

Addition of other forms of positive ventilatory stimuli, such as increase in $[H^+]$ related to metabolic acidosis in severe exercise, must increase ventilation and lower Pa_{CO_2} along iso-$\dot{V}_{CO_2}$ curve, as from *C'* to *D'*. In this new state, small changes in ventilation due to any cause must inversely alter P_{CO_2} along iso-$\dot{V}_{CO_2}$ curve relating ventilation to P_{CO_2} at the existing metabolic rate of CO_2 production. This figure therefore emphasizes the dominant positive role of metabolically produced CO_2, modified by additive lesser positive factors, and ultimate inverse fine adjustment to metabolism by CO_2 itself.[129a] (**A** from Torelli, G., and D'Angelo, E.: The factors affecting ventilation during exercise at sea level and at altitude. In Margaria, R., editor: Exercise at altitude, Amsterdam, 1967, Elsevier Scientific Publ. Co.; **B** from Lambertsen.[129a])

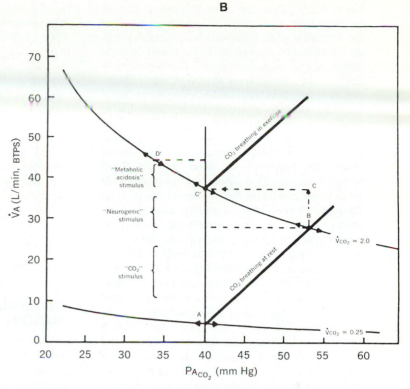

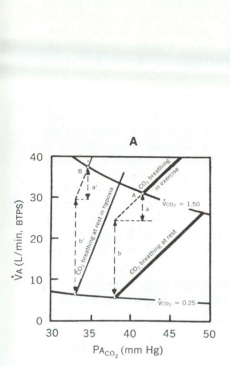

Fig. 74-4. For legend see opposite page.

Additive, positive influences in chemoneurologic regulation

Probably the largest obstacle to understanding has been the persistent attempt to identify a single basis for exercise hyperpnea. If no single neural or chemical influence on respiration can alone account for respiratory drive, then the requirement is to define the manner in which identifiable changes in exercise provide the *full* and *exact* respiratory response together. In simplest interpretation, it becomes necessary to visualize prominent influences of *neural impulses* from various sources (central and peripheral) serving in a *coarse, unregulated* stimulation of respiration in exercise and *not specifically related to metabolic requirements*. Then, to account for the actual accurate adjustment of ventilation to increases and decreases of metabolism, it is necessary to assume that *chemical control* (central and chemoreflex) represents the *fine adjustment* superimposed on the approximate and unregulated drive provided through the barrage of as yet unidentified nerve impulses associated with motion. The chemical stimuli and mechanisms pertinent to the resting state can then be con-

sidered also to be the determining factors of the precise degree of ventilation produced in exercising states (Fig. 74-4, *A*).

Concept of "combined positive and inverse control" of ventilation by chemical factors in exercise

This general concept of positive summation of chemical and neural factors* continues to lack specific and quantitative fit to the variety of responses to exercise. Its extension below includes recognition of the normally inverse relationship of chemical stimulus level and ventilation (Fig. 74-4, *B*) (p. 1776).

As at rest, the ultimate chemical control of breathing most likely functions in air-breathing exercise in the normal *inverse* manner. This means that, when the composite of other, coarse drives is more than required for alveolar ventilation to exactly match the increased CO_2 production and O_2 consumption, a slight *lowering* of arterial P_{CO_2} [H^+] occurs. This should limit and stabilize the degree of respiratory response.

---

*See references 17, 88, 110, 111a, 129, 147, and 160a.

Conversely, when the sum of coarse stimuli to respiration is somewhat *less than required* for a match with metabolism, the beginnings of an increase in arterial P_{CO_2} could lead to further slight *increase* in alveolar ventilation, limiting the elevation of P_{CO_2}. In this conceptual scheme, acid-base of O_2 changes in blood not only provide a major driving stimulus to hyperpnea, as they indeed would in the absence of supplemental factors, but, as stable state is approached, provide a check against excess or deficiency of other less regulated stimulus forms. This limits the occurrence of gross changes in acid-base composition of local fluids to which chemosensitive neural structures respond. Chemical stimuli can, in this concept of combined positive and inverse influences, provide major positive drive as well as the ultimate or *fine adjustment* linking alveolar ventilation to whole-body metabolism, serving as regulators of the degree to which all drives together can affect the respiratory motor activity.

Fig. 74-4 illustrates two related but different concepts concerning summation of neural and metabolically related factors. *A*, from Torreli and D'Angelo,[160a] effectively describes limited and supplementary stimulation of ventilation by CO_2 superimposed on neurogenic drive. This is a concept, elaborated by several laboratories,[88,90,111a,160a] in which a large component of motion-related neural stimulus is *positively supplemented by an additional stimulus* related to the degree of arterial hypercapnia generated by metabolically produced CO_2. The hypothetical example in Fig. 74-4, *A*, indicates such an influence of CO_2 proportionate to a stable-state rise in arterial P_{CO_2} and to the magnitude of respiratory reactivity to change in P_{CO_2} observed in CO_2 breathing at rest. Influences of hypoxia at rest and in exercise are similarly shown in the figure, with a *positive* final contribution by hypercapnia to the overall ventilation of exercise. This concept is quantitatively analyzed later in this chapter.

Fig. 74-4, *B*, is derived from actual data for CO_2 reactivity at rest[127a] and relation of CO_2 production to ventilation in exercise[65,134] (Table 74-1). This figure has the same coordinates and scale as *A*, but extends over a larger range of $\dot{V}_A$, $P_{A_{CO_2}}$ values. The data it describes involve a higher level of work. It illustrates, for air breathing but not for hypoxia, a four-component concept of exercise hyperpnea due to (1) major primary, positive response to metabolically produced CO_2, (2) superimposed coarse contribution of unidentified ''neural'' factors, (3) the ultimate *inverse (negative)* role of CO_2 in

exactly adjusting ventilation to maintain arterial P_{CO_2} at or near resting values, and (4) at high work levels a positive stimulus contribution by increase of $[H^+]$ generated by fixed acid accumulation in the blood, with the inevitable consequent arterial hypocapnia. Since the respiratory control mechanisms appear to be in part protected by the blood-brain barrier from the acute effects of metabolic acidosis, the positive influence of the change in $[H^+]$ due to metabolic acids should be less than the negative influence of the associated hypocapnia, and inverse effects of CO_2 can therefore still be considered as providing the final adjustment to metabolism.[129a]

This pattern of ''combined positive and inverse'' control by CO_2 of respiration should apply to the hypoxic state during exercise as well as to normoxia and hyperoxia. It is qualitatively comparable to the situation during CO_2 administration in any state.

IDENTIFICATION OF SPECIFIC FACTORS IN VENTILATORY CONTROL

Combined chemoneurologic concepts of varied character have been gathering favor, and it has become possible to study certain of the components of respiratory stimulation separately.* It certainly is still to be preferred to simple postulation of the existence of a dominant but unidentified mechanism. Its main weaknesses remain related to the unsettled questions of chemical control of respiration even at rest, the difficulties involved in identifying specific, excitatory nerve impulses in the complex neurogenic patterns of exercise, and the problems involved in quantitative appraisal of any one factor in exercise hyperpnea.

It is in the *quantitative* assessment of potential neurogenic and chemical stimuli that important gains will continue to be made. This is happening with the additional advantages of continuous measurement and control of blood gas tensions.

Study of *stable* resting and exercising states aids in correlating the overall changes in ventilation and possible stimulant factors at more than one level of work. Table 74-1 summarizes some such data obtained from several sources and coordinated into a single group of subjects for comparison. This will be employed in the following appraisal of exercise hyperpnea.

Study of events during the actual *transition* from rest to exercise and back to rest offers a further opportunity to correlate rates of change of

---

*See references 16, 17, 66, 74, 134, 146, 147, 157, 160a, and 164.

ventilation with potential stimulant factors. Fig. 74-5 is an excellent example of a dynamic study performed in normal men.[146] Similarly useful measurements, change in blood acid-base and lactate composition in different degrees of exercise, now exist for "incremental" or step-wise increases from mild to strenuous work.[163]

Dynamic changes

In the examples given in Fig. 74-5 the pulmonary ventilation is seen to rise rapidly after beginning exercise, progressing toward a new level of hyperpnea in the 6 min period of work. On stopping work the return toward the resting level has a similar time course. These changes are now conventionally cited as phase I (initial ventilatory increase at start of exercise), phase II (period of gradual increase to steady state level), and phase III (stable state).[163] The phases are most evident in moderate, sustainable exercise. When work rate is extreme, no stable state can be maintained.

During exercise for 6 min on an ergometer bicycle at 50 rpm and at a work load of 625 kg-m/min, ventilation did not fully match the metabolic production of CO_2 and a rise in P_{CO_2} and $[H^+]$ occurred, returning toward normal resting level as exercise continued toward a stable state. As exercise continued, the changes in blood did not all stabilize; arterial P_{CO_2} began to fall, probably because of a slight further increase in ventilation related to entry of fixed acids into the blood from the tissues. The fall in P_{CO_2} in turn limited the rise in acidity of arterial blood

and limited ventilation. When exercise was discontinued, ventilation decreased considerably before arterial $[H^+]$ began its relatively slow return toward the normal resting level.

When exercise is abruptly terminated, metabolism also decreases. After the abrupt fall in respiratory minute volume immediately on cessation of work, pulmonary ventilation falls more slowly than does O_2 consumption and CO_2 production; the period of relative hyperventilation that then occurs during recovery from exercise leads to a rise in arterial P_{O_2} and a fall in arterial P_{CO_2} below normal.

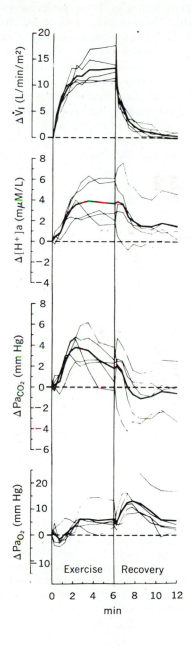

Fig. 74-5. Time courses of changes produced in normal men by exercise (625 kg-m/min) for 6 min and during recovery from exercise. Heavy lines show means of individual time averages (thin lines) determined over successive 30 sec periods (first 3 min of exercise and first 2 min of recovery) and 60 sec periods (remaining phases of exercise and recovery). In plotting time courses for blood chemical changes, due allowance has been made for delays in recordings. Note (1) that much respiratory change occurs before prominent rise in $[H^+]$ or P_{CO_2} develops, (2) that lag in development of acidemia and hypercapnia is about equal to half-time for ventilatory change, (3) that during sustained acidemia rising ventilation gradually lowers arterial P_{CO_2} toward resting level, (4) that acidemia does not rapidly disappear on cessation of exercise, and (5) arterial P_{O_2} is above normal throughout exercise and recovery. The example represents an illustration of ventilatory effect on blood P_{CO_2} at elevated CO_2 production, rather than just an effect of P_{CO_2} on ventilation (p. 1881). (From Matell.[146])

It is clear from Fig. 74-5 that changes in blood gas composition and [H⁺] do occur in exercise, but determination of their relation to the degree and pattern of ventilatory response requires attention to the more coarse underlying drives. One step in that appraisal relates to the evident fact that some component of ventilatory response to exercise changes more rapidly than any of the potential chemical stimuli shown in Fig. 74-5 could arrive at chemosensitive sites, and is dissociated from controls of cardiac output response to exercise.[167]

Factors related to voluntary movement

Afferent impulses from proprioceptors in the limbs. These constitute the most characteristic and best documented neurologic influence in the hyperpnea of exercise. There is abundant experimental evidence* to indicate that active or passive movements of the legs can produce a reflex hyperpnea. The impulses apparently arise in or around the joints. There is still some question regarding whether additional sense organs exist in the muscles or tendons.[16,83] It must, however, be admitted that the respiratory effects thus far

---

*See references 16, 83, 105, 110, 112, and 160a.

attributed to this reflex from the knee alone are small.[83,110] Gray[109] considers that its potency as a factor in the hyperpnea of exercise may have been greatly underestimated because such a reflex hyperventilation diminishes the chemical stimulus when elicited in a resting human or animal, whereas during exercise this would not be the case. Actually alveolar hyperventilation in any situation of stable CO_2 production must lower arterial P_{CO_2}. It should also be noted that, although only a few joints have been investigated, movement of other joints besides the knee gives rise to respiratory stimulation,[16] and if each of the joints involved in exercise were to contribute a small share, the total might well be considerable.

Attempts to identify the exact nature of the receptors and pathways have proceeded along several lines. One approach, employed by Kao (Fig. 74-6),[118] shows that electrical stimulation of the leg muscles of dogs with intact spinal pathways leads to an increase in ventilation. When the lateral spinal columns are cut or the entire cord is transected, the respiratory stimulation does not occur. Some diminution of response also follows cutting of the dorsal columns. In these experiments, metabolic products resulting from the artificial exercise were cleared by cross-

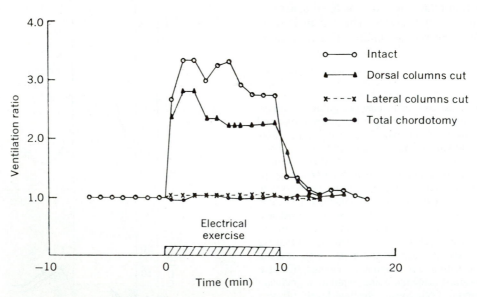

Fig. 74-6. Pathways of neural influences on respiration of dog. Influence of electrical stimulation of muscles (electrical exercise) in dogs in which cross-perfusion of blood from second dog was used to minimize chemical influences on breathing. Increase in ventilation ratio is greatest in animals with intact spinal cords. Total chordotomy abolishes respiratory response, as does section of lateral columns. Lesser degree of diminution occurs when dorsal columns are sectioned. (From Kao.[118])

perfusion with a second dog. A nervous pathway therefore seems to exist.

A second profitable method of study has involved abrupt occlusion of the circulation to and from the exercising limbs to isolate the respiratory centers and peripheral chemoreceptors from effects of chemical products of muscle metabolism.* Several distinct conclusions can be drawn: (1) respiration is stimulated during exercise with occluded circulation, suggesting the existence of nonhumoral factors[66,67]; (2) if an obstructing cuff is released, even after exercise has ended, respiratory stimulation occurs[93]; and (3) if arterial CO_2 tension is maintained constant when occlusion of circulation is effected in exercise, a progressive increase in ventilation occurs.[66] These findings indicate the existence of neural factors in exercise, modifying effects of chemical factors transported by the blood throughout the body.

The question of whether chemical influences within the exercising muscles may modify neural activity, even of mechanoreceptor function, has been raised.[66] Within the expected physiologic range, no respiratory stimulation over a neural pathway has been elicited by inducing local chemical changes in the unexercising limbs.[118]

Irradiation of impulses from higher parts of the neuraxis. Impulses of forebrain origin destined for the exercising muscles must traverse the medullary reticular formation, where they could conceivably affect the respiratory centers. This concept of an irradiation factor does not lend itself to separate study. In one investigation comparing the ventilatory response to electrical stimulation of the leg muscles with that to voluntary leg exercise, no effect attributable to cortical irradiations could be detected.[69] Since the existence of this factor has never been specifically demonstrated, it can not be assigned a magnitude in the quantitative evaluation of factors in exercise hyperpnea. That it could be only a partial explanation of the hyperpnea of exercise is indicated by the findings that passive movements of the legs or stimulation of ventral spinal roots will produce a hyperpnea in anesthetized or decerebrate animals[83]; cortical activity could not be importantly involved in those states.

Composite effects of factors related to muscle activity or motion. Difficulties involved in identifying or separating the *partial* effects of central influences or peripheral mechanisms related to muscle and joint movement in animals and man[9,16,110] do not prevent attempts at estimating that portion of the integrated or *total response* to motor activity that should be unrelated to metabolic or chemical factors. Several important studies have measured in unnarcotized man what is presumed to be the *overall contribution* of neurogenic factors in the hyperpnea of exercise.* The approach, an extension of the earlier efforts of Krogh and Lindhard[125,126] and Nielsen,[148] involves the continuous measurement of respiration to monitor the change in ventilation during the first breaths after beginning or terminating exercise as well as during the approach to a new stable state. These methods have had extensive recent use.

Fig. 74-7 shows the nature of such breath-by-breath changes in respiratory minute volume. As exercise begins, there occurs within the first seconds an abrupt increase in ventilation. After remaining relatively constant for about 30 sec, ventilation then rises progressively toward a stable state. As exercise is abruptly terminated, ventilation falls acutely but only part way to the resting level. The magnitude of the initial decrease in ventilation on cessation of exercise is greater than the initial increase in ventilation as work is begun. Gradual return to a normal resting value then occurs[16]; the slow rate of decrease in ventilation presumably results from the persistence of stimulant factors not related to limb movement.

The abrupt changes in ventilation at the onset and end of exercise have now been observed by many investigators.[17,146,147,160a,164] As can be expected, the character and magnitude of the changes are being found to differ somewhat in various forms of work. Generally, both the initial and the terminal changes in ventilation vary with the severity of the work load and degree of O_2 consumption.[16,17] However, the effects so prominent in abrupt transition from rest to exercise are not routinely seen in abrupt transition from a low level of work to a higher level, or even from ''no-load'' ergometry to the working state.[76,147,160a,163]

Relation of abrupt hyperpnea to work level. Fig. 74-8, *B,* based on the combined studies summarized in Table 74-1, compares reported relationships of initial and terminal abrupt changes in ventilation to work load (here, as in Table 74-1, converted to O_2 consumption).† Fig. 74-8, *A,*

---

*See references 17, 66, 67, 90, 160a, and 164.

---

*See references 16, 17, 92, 147, 160a, and 164.

†In Fig. 74-8 the regressions of immediate changes in ventilation in O_2 consumption differ slightly from those published[16] in that the present regression analysis includes a resting value for each subject, whereas the published regression did not. This evaluation was made possible by P. Dejours, who supplied necessary individual data.

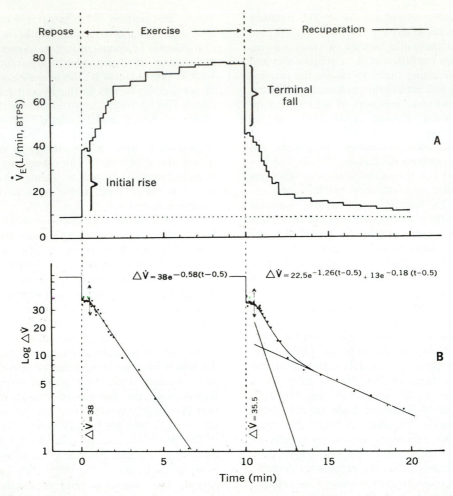

Fig. 74-7. Rapid and slower components of respiratory stimulation in transition from rest to stable state of exercise and return to resting state. Within one breath of beginning exercise, initial rise in ventilation occurs. This is followed by essentially exponential further increase in ventilation. On sudden cessation of work, abrupt terminal fall in ventilation occurs. This is followed by slow return of ventilation toward resting values. Data suggest that this recovery may have more than one component. (From Dejours and Teillac.[91])

indicates the overall respiratory stimulation accompanying exercise in the subjects of Table 74-1. If the sudden *fall* in ventilation is considered a more suitable index of the maximum contribution of "movement-related" factors in stable-state exercise than is the sudden *rise,* the composite influence of such factors is seen to approximate 50% of the total ventilatory response at an O_2 consumption of 2 L/min. If the sudden *rise* at onset of exercise is used as the index of neurogenic influence, this contribution approximates 27% of the total ventilation. The bases for differences in magnitude of sudden change at

beginning and end of exercise are not known. That at the end of exercise may have a "fast" chemical component. As stated previously, these effects represent the *maximum and overall neurogenic response to movement;* breakdown into components such as reflex influences from proprioceptors and any effects of irradiation from higher centers is not yet feasible.

Factors related to metabolism

Carbon dioxide pressure. In the many discussions of respiratory control in exercise, it has been stressed that the minute changes in arterial

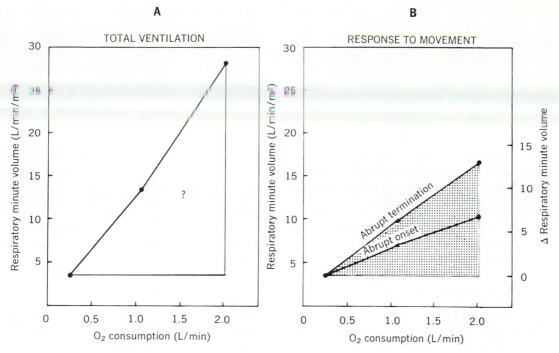

Fig. 74-8. Ventilatory changes during muscular exercise from integration of findings in humans (Table 74-1).

A, Relation between total pulmonary ventilation and rate of O_2 utilization at rest and at two levels of exercise. Question mark indicates long-standing uncertainty regarding not only nature of factors driving respiration but also their proportionate contribution to total response. In subsequent diagrams a quantitative evaluation is made of the degree to which several possible "stimuli" may be involved in producing exercise hyperpnea.

B, Relation between O_2 consumption and immediate respiratory responses to (1) initiation of motion (lower curve) and (2) cessation of motion (upper curve). Data from Table 74-1 were derived for two levels of exercise from regression of individual ventilatory changes on O_2 consumption in six subjects. Curves in this diagram differ slightly from those originally published by Dejours[16] in that resting values were here included to obtain regression desired for present analysis. Lower curve indicates addition of respiratory drive to control state in which other conditions are those normally associated with rest. Upper curve represents effect of removing same type of drive but now from stable state that includes chemical and other features of muscular exercise.

P_{CO_2} observed during muscular activity cannot account for the vigorous hyperpnea of exercise.* However, increased amounts of CO_2 are brought to the lungs in the venous blood, and it must be removed by an alveolar ventilation above normal. An elevation of arterial CO_2 tension produces approximately the same incremental degree of respiratory stimulation, whether the CO_2 is presented as a fixed amount per minute (as from increased metabolism) or as a fixed inspired concentration.[102] Since no appreciable rise in arterial P_{CO_2} or even central P_{CO_2} occurs in exercise, a part of the CO_2 removal by ventilation must be unrelated to the CO_2 stimulus itself. *Despite this recognized situation, it is not contradictory to emphasize that P_{CO_2} is probably the most important factor in providing both primary drive and the ultimate adjustment of pulmonary ventilation to the total increase in metabolism (CO_2 production)* (Fig. 74-4).

Correlation of changes in ventilation and [H^+] in exercise can be found, indicating some small role for fixed acid production as the degree of exercise increases.[134,146] This influence of fixed acid can be considered a positive, added stimulus (Fig. 74-4), summing with neurogenic and posi-

*See references 9, 88, 90, 110, 147, 160a, and 164.

tive effects of CO_2 production. *However, the final role of PCO_2 in adjusting ventilation to metabolism appears to be a negative one*. Thus when the other respiratory drives, including the summation of all neural influences and effects of anaerobiosis and metabolic acidosis, result in a ventilation *excessive* with respect to CO_2 production, hypocapnia occurs, and this counter-effect tends to limit the overall ventilation to the degree required by metabolism. Conversely, if the sum of other respiratory stimuli produces a total ventilatory drive *inadequate* with respect to metabolic formation of CO_2, the alveolar, blood, and central PCO_2 increases; this tendency to hypercapnia then should sustain pulmonary ventilation at a level close to that corresponding to the existing metabolic rate. The critical dual role for CO_2 would be accomplished not only by *producing* much of the entire gross respiratory stimulation proportional to the metabolic rate of exercise but also by *limiting* the degree to which ventilation can deviate from the exact requirements of metabolism. This possibility of an inverse rather than direct correlation of ventilation and blood PCO_2 in exercise has been recognized,[16,110] and its occurrence clearly demonstrated.[134] What is most important now is that this type of inverse relationship, added to a major positive influence of CO_2, strengthens the role of PCO_2 rather than excluding it as a factor linking ventilation to metabolism, since the same inverse pattern characterizes the system of respiratory control at rest (p. 1776).[130]

If PCO_2 is thus considered to provide the fine adjustment of ventilation to the changing rate of metabolic CO_2 production, it is evident that in light exercise, with a variable degree of neural drive and where no appreciable increase in fixed acid occurs, the PCO_2 of arterial blood would rise in some individuals or situations, in others it would fall, and in still others remain unchanged.[134] The direction of initial PCO_2 change would depend on the magnitude of the large but coarse adjustments of respiration to muscular movement and their relation to overall CO_2 production. On the other hand, the nearly uniform tendency to hypocapnia in severe exercise would indicate the existence of a total respiratory stimulation (by neural and metabolic acid factors) that is already excessive with respect to the requirement for elimination of the large amounts of CO_2 produced per minute but not excessive in relation to $[H^+]$. Since a part of the respiratory stimulation in this situation is related to $[H^+]$ derived from anaerobiosis,[134] the hypocapnia should limit increase in $[H^+]$ but should not be expected to restore arterial $[H^+]$ to resting values.

pH and fixed acid formation. Elevation of $[H^+]$ in blood during exercise must be considered to be the result of changes in PCO_2 and changes in the level of fixed acids produced in the anaerobic metabolism that occurs in strenuously working muscles. Thus the pH of arterial blood during exercise should reflect (1) the degree to which disproportion between ventilation and CO_2 production has altered arterial PCO_2 and (2) the rise in acidity resulting from increases in blood lactate and pyruvate. In relative hypoventilation such as encountered in respiratory resistance or pulmonary disease the effects of 1 and 2 can be additive, resulting in gross increases in the acidity of blood.[134,147] Conversely, when ventilation is excessive relative to metabolism, change in pH may be minimized.

The long-standing tendency to disregard acidemia as an important factor in the regulation of breathing in exercise appears to have been related in part to the assumption that a very large change in a product of metabolism would be required to provide the precise link observed between ventilation and metabolism[9,16,110] and in part to a failure to detect a consistent relationship between ventilation and small changes in pH.[9,110] Such a relationship is now well established.[65,74,134,146,160a] Table 74-1 illustrates that the acidity of arterial blood increases only slightly at a rate of O_2 consumption close to 1 L/min.[65,103,134] Even at twice this metabolic rate the rise in acidity that does occur is not great.[65,134] However, the $[H^+]$ changes at the higher level of work, although still too small to account for a large portion of the exercise hyperpnea, are positively correlated with changes in ventilation.[134] In turn, the changes in pH appear related to alterations in the level of fixed acid rather than to the $[H^+]$ changes induced by alterations of PCO_2 in the blood.[16,65,134] This suggests the emergence of $[H^+]$ as a positive chemical stimulant factor linked *not only to total metabolism* but also to the degree of *anaerobic metabolism* in the exercising muscles and having a demonstrable quantitative association with respiratory stimulation at rest.[97,109,131,139,169] Thus a change in acidity produced by fixed acids relates to anaerobiosis and provides a discrete order of control. The influence of PCO_2 on $[H^+]$, reflecting relationships between ventilation and total metabolism, is superimposed overall.

It will not prove easy to define for exercise the kind of quantitative relationships of ventilation and acid-base change descriptive of the resting state.[109,111,131,138] However, attempts such as those of Fig. 74-4 are being made,[17,88,129a,160a] and the data in Table 74-1 are useful in illustrating the problems entailed in such analysis. Among the most important considerations are the following:

1. At rest the respiratory effect of acidosis produced by a rise in P_{CO_2} is more than twice as great as the effect of acidosis produced by fixed acids. Thus a change in blood $[H^+]$ without a corresponding alteration in P_{CO_2} (as in the two degrees of exercise in Table 74-1) produces a positive respiratory response only about 45% as great as would have resulted if the acidosis had been CO_2-induced (p. 1792).[130,131,140] In exercise the changes in arterial P_{CO_2} or $[H^+]$ may be insignificant, or the acidemia may be associated with a *lowered* P_{CO_2}.

2. If arterial CO_2 tension does rise in exercise, the acidosis it produces should add its *dual stimulant effects* to the influence of any separate, metabolic acidosis present. A rise in arterial P_{CO_2}, representing ventilation less than required for the existing CO_2 production, should sustain ventilation against further fall. At rest the contribution of a P_{CO_2}-induced change in acidosis should include a 45% contribution correlated with the change in acidity of the blood and a 55% contribution correlated with the extravascular effects of CO_2 (p. 1792).[130,131]

3. If P_{CO_2} is lowered in severe exercise, a reduction of elements of the central chemical respiratory stimulus should be expected due to decrease of influences of CO_2 beyond the blood-brain barrier. Effects of *decrease* in central or peripheral stimulus level in exercise are not at all well documented.

4. If the respiratory response to alteration in $[H^+]$ or P_{CO_2} is of the same order of magnitude as at rest,[109,131,140] as has been shown by Matell,[146] the contribution of P_{CO_2} and acid-base factors to the magnitude of exercise hyperpnea should be predictable from measurements of arterial P_{CO_2} and $[H^+]$ and use of the relationships for "effective" $[H^+]$ stimulus in stable states at rest (p. 1792).

$$\Delta \dot{V}_{E_{CO_2,H^+}} = 4.07 \ \Delta[H^+]_{effective}$$

$$\Delta \dot{V}_{E_{CO_2,H^+}} = 4.07 \ (0.45 \ \Delta[H^+]_{blood} + 0.55 \ \Delta[H^+]_{CSF})$$

In this equation, 4.07 L/min/$\Delta[H^+]$ is the overall respiratory response to pH change produced by CO_2 administration; CSF is used as an index of the behavior of fluids beyond the blood-brain barrier and hence acutely altered by a change in P_{CO_2} but not by $[H^+]$ change generated by accumulation of metabolic fixed acids[130,131] (p. 1792). These influences would be expected to take the positive form of respiratory drive in exercise and the form of inverse fine adjustment to metabolism (p. 1776).

In Table 74-1 the values for pH and P_{CO_2} are corrected for mild temperature rise in exercise. It is seen that arterial P_{CO_2} was held essentially constant by the balance of normal factors and not altered by either degree of work. According to the *dual center concept*, if arterial P_{CO_2} does not change, this is equivalent to no acute change in $[H^+]$ at the reactive sensor site beyond the blood-brain barrier (p. 1797). In these conditions of exercise, only the chemosensitive mechanisms unprotected by a blood-brain barrier would be expected to be influenced by the very small change in blood $[H^+]$, and the expression for the respiratory effect of acid-base changes at the two levels of exercise in Table 74-1 would be as follows:

Exercise level I:

$$\Delta \dot{V}_{E_{CO_2,H^+}} = 2.0 \ (0.45 \times 0.010 + 0.55 \times 0)$$
$$= 0.8 \ \text{L/min/m}^2$$

Exercise level II:

$$\Delta \dot{V}_{E_{CO_2,H^+}} = 2.0 \ (0.45 \times 0.060 + 0.55 \times 0)$$
$$= 5.4 \ \text{L/min/m}^2$$

These estimated values for contribution of metabolic acidosis to the total exercise hyperpnea are plotted in Fig. 74-9, *B*, using the normal reactivity factor of 2.0 L/min/m²/0.01 ΔpH (equivalent to 4.07 L/min/0.01 ΔpH or 2.73 L/min/mm Hg ΔP_{CO_2}) (p. 131). They indicate that whereas acidemia contributes a negligible respiratory stimulation in the light work (exercise level I), a detectable portion of the respiratory stimulation of more strenuous exercise could be associated with acidemia.[129,134] It should be only this portion that is diminished by hyperoxic reduction of anaerobiosis.[134]

In a subsequent and related estimate, Cunningham[88] used a larger value for the reactivity to acid-base change (4 L/min/mm Hg ΔP_{CO_2}) and also, contrary to the analysis above, assumed that the entire effect of CO_2-induced acidosis at rest should apply to the acidemia of exercise. These differences from the procedure illustrated in Fig. 74-9 resulted in the assignment of a larger fraction of exercise hyperventilation (as much as 60%) to arterial acid-base changes, whereas the appraisal of the data in Table 74-1 and Fig. 74-9, which account for P_{CO_2} and $[H^+]$ influences separately, indicates that the combined influence of deviation from resting values is less than one fourth, even at the heavier work level.[130,131]

5. The "sensitivity" or reactivity to increased

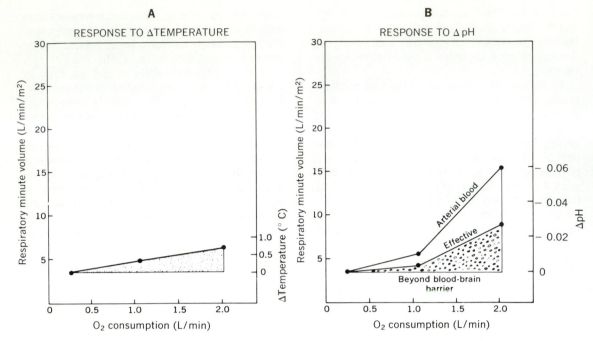

Fig. 74-9. Ventilatory equivalents of changes in temperature and in [H⁺] during exercise.

A, Temperature. Rise in body temperature at various levels of work is 0.09° C/100 kg-m/min.[82,134] This temperature increase is plotted against O_2 consumption values equivalent to imposed increase in work load. Moreover, at fixed central Pco_2 in resting subjects, respiratory equivalent of a 1° C increase in temperature was 3.9 L/min/m²; figure suggests what is probably minimum respiratory effect of small temperature rise in exercise.

B, Hydrogen ion concentration. Changes in hydrogen ion concentration in blood and all chemosensitive regions are brought about by alterations in Pco_2 whereas fixed acid accumulation produces its effects only where no barriers to H⁺ or other charged particles exist.[130,131]

Fixed acids such as lactic acid do not readily cross blood-CSF barrier and thus should not affect regions protected in this manner. However, according to dual center theory (p. 1797), approximately 45% of response to acid-base stimulation at rest is related to chemosensitive regions permeable to changes in fixed acid and fixed base in circulating blood.

In exercise states in Table 74-1, increases in arterial [H⁺] were unaccompanied by elevations of Pco_2. Changes in acidity are plotted in this figure in two ways. Upper curve indicates respiratory response that would have been expected if pH changes had been produced by CO_2 breathing instead of by "metabolic acidosis" of exercise; this curve is undoubtedly *not* representative of H⁺ stimulus in exercise. Horizontal baseline under curves illustrates that, in absence of Pco_2 change, no increase in [H⁺] should be expected in fluids beyond blood-brain barrier. Finally, middle curve, labeled as "effective" change in [H⁺], represents respiratory equivalent of composite influence of [H⁺] at central and peripheral reactive regions (see text). Effective curve, actually based on reactivity in resting state, involves no interaction with neural stimuli and represents positive effect of [H⁺] change due to metabolic acidosis in exercise.

acidity has been proposed to be greater in exercise than at rest.[9,16] No experimental basis exists for such an assumption, and it actually appears that CO_2 reactivity in exercise is similar to that in the resting state.[160a]

Temperature. Rise in temperature cannot be related to the rapid initial or terminal changes in exercise ventilation. The rise of deep body temperature during light stable-state exercise is not great (Table 74-1), but in the first 25 min of heavy work it can amount to as much as 2° C.[82] Even small degrees of hyperthermia at rest result in considerable hyperventilation when "compensatory" hypocapnia is prevented.[89,152] Fig. 74-1 indicates a respiratory stimulant effect of increased body temperature at rest approximating

3.9 L/min/m²/Δt °C at a fixed peripheral and central P_{CO_2}. This factor of progressively increasing temperature could be a component of the similarly increasing ventilation between the abrupt initial rise and the attainment of a stable state. If the effect of temperature change is no greater in exercise than at rest, the quantitative contribution of temperature change to exercise hyperpnea in the subjects referred to in Table 74-1 would be represented by Fig. 74-9, *A*.

Whether the mechanism of this temperature effect is related to changes in the metabolism of respiratory neurons, increase in hydrogen ion activity, or other factors is not known. Although elevation of body temperature should increase H⁺ *activity*, it is not now feasible to determine the exact influence of change in body temperature on the pH of body fluids.[62] This is because pH scales are not absolute and are, to an unknown degree, different for each temperature.[62] For this reason the pH values of Table 74-1 were previously presented for resting body temperature.[129] Regardless of its mechanism, the extent to which temperature change contributes as a drive to ventilation in muscular exercise will depend on the degree of work (heat production), work duration, the mass of the subject to be heated, and the opportunity for heat loss (environmental conditions). Environmental temperatures between 5° and 30° C have little influence on the temperature rise in exercise.[148] However, higher ambient temperatures should lead to a progressive rise in deep body temperature and consequently affect respiration.

Oxygenation of blood and tissues. Since most of the direct evidence not only fails to show an important fall in arterial P_{O_2} during exercise[65,103] but also indicates that a clear-cut rise in P_{O_2} can occur,[74,146] "hypoxic" chemoreflex stimulation cannot be assigned even a tentative quantitative value for air breathing at seal level.

Humans deprived of carotid chemoreceptors show no detectable difference in ventilation from normal in mild and moderate steady-state exercise while breathing air.[142] However, they do fail to further increase exercise ventilation when induced hypoxia is superimposed on the exercise state.[142]

Peripheral chemoreflex contribution becomes a large component in exercise at high altitudes or in hypoxic states.[115,123,145,154,160a] It is included here for two important reasons. The first is that even mild exercise in individuals with disturbed cardiopulmonary function can lead to lowered arterial P_{O_2} and hence to support by exaggeration of chemoreceptor activity. The second is that both abrupt and sustained administration of O_2 at higher than normal tension to exercising subjects does result in a decrease in respiration.*

This decrease in ventilatory response at particular levels of O_2 consumption has been shown to be proportional to work load[64,134] and, within limits, to inspired O_2 tension.[65] The higher the work load, the greater will be the reduction of ventilation when pure O_2 is inhaled; breathing pure O_2 at 1 atm causes a greater decrease in ventilation than does inspiration of 33% O_2 and almost as great a decrease as does pure O_2 at 2 atm ambient pressure.[65,134] Thus O_2 does diminish exercise hyperpnea, even in normal individuals at sea level. Since the extent to which ventilation is lowered by hyperoxia is correlated with the degree of reduction of fixed acid levels and acidity of the blood, at least part of the O_2 action should be related to improved oxygenation in the working muscle.[16,65,134] However, this reduction of muscle anaerobiosis does not account for the entire action of O_2 in reducing exercise ventilation, since sudden substitution of O_2 for the air breathed by exercising subjects results in a reduction in ventilation that is too rapid in onset to be related to an effect in fixed acid production in the muscles. This rapid effect of O_2 appears to occur within one circulation time[16,95,96,124] (Fig. 74-10) and therefore more rapidly than does the lowering of blood lactate levels by O_2.[64,124] It may indeed indicate that normally oxygenated chemoreceptors still contribute about 10% to 15% of the total ventilatory drive of exercise.[65,95,96]

Time-response characteristics of this abrupt effect of O_2 in exercise have not been worked out as clearly as for rest.[98] Moreover, the persistent tendency to cite the effect of O_2 as due to "relief of a hypoxic drive," even when no arterial hypoxia can be demonstrated, is not reasonable. In view of the now well-demonstrated interrelationships of P_{O_2} and H⁺ in producing chemoreceptor stimulation,[22,115] it is possible that the abrupt effect of O_2 actually represents a depression of the chemoreflex activation produced by H⁺ or P_{CO_2} (p. 1808).[98,128] The abrupt effect of O_2 requires appraisal at various levels of work before its quantitative importance can be judged.

Afferent impulses from chemoreceptors in venous side of circulation and in exercising muscles. This possibility is one that has been raised repeatedly† since it was proposed over 90 years ago[61] and continues to be reexplored with

---

*See references 16, 17, 65, 73, 115, 124, and 134.
†See references 17, 63, 108, 156, 165, 166, and 170.

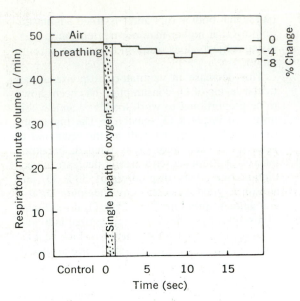

Fig. 74-10. Effect of single inspiration of O_2 on ventilation during stable-state exercise (mean of 20 measurements on each of three subjects). Prompt, transient, small decrease in ventilation produced by one breath of O_2 during exercise has been attributed to removal of chemoreceptor contribution to exercise hyperpnea.[16,65] It is presumed that effect begins with lag shorter than time required for recirculation of blood from limbs and is thus not due to improved oxygenation of working muscles.[16,65] (Based on data from Dejours et al.[95])

each methodologic advance. It is evident that in exercise, because of the increased production of CO_2 by the working muscles, correlations among the rise in mixed venous P_{CO_2} and $[H^+]$, increase in metabolism, and the degree of hyperventilation *must* exist. A necessary coincidence of this type cannot be considered evidence in favor of a hypothetical chemoreflex system responding to changes in venous blood and responsible for most of the respiratory stimulation in exercise.[63] Moreover, as indicated by Fig. 74-4, *B*, it is not necessary to postulate unknown sites or sensors for CO_2 effects. To place complete emphasis on the correlation of exercise hyperpnea with acid-base changes in mixed venous blood, it is in fact necessary to ignore the clearly demonstrated respiratory response to CO_2, to abrupt change in body movement[16,17] (Figs. 74-7 and 74-8), to change in body temperature[89] (Figs. 74-1 and 74-9, *A*), and to central acid-base changes.

Attempts to identify functional chemosensitive elements on the venous side of the circulation have not been successful.* Without demonstration of such sites of stimulant activity, the correlating of chemical changes in venous blood with the ventilatory effects of exercise will be nothing more than a restatement of the well-known fact that both ventilation and CO_2 production are increased by exercise. Available evidence is also against the existence of chemoreceptors or temperature receptors in exercising muscles.

Factors related to circulation

Readjustments in blood supply of the center. Studies in man have indicated that the respiratory stimulant effect of a primary rise in the P_{CO_2} of the arterial blood may be lessened by a concomitant increase in the rate at which the CO_2 produced by the respiratory center's own metabolism is carried away from it.[120,133] This is accomplished by a dilatation of central blood vessels, apparently on the arterial side of the capillary bed.[132,134] Exercise severe enough to produce a lowering of arterial P_{CO_2} would lead to constriction of cerebral arteries and thus limit the removal of the CO_2 produced by the cells of the center. Studies of the behavior of the cerebral circulation and of cerebral venous P_{CO_2} during exercise indicate that whole-brain metabolism is not detectably elevated by the muscular exercise and that the changes in P_{CO_2} and $[H^+]$ in brain venous (internal jugular) blood parallel those in the arterial blood.[122,134] Apparently the slight cerebral vasoconstriction is counterbalanced by the increased head of systemic arterial pressure so that cerebral blood flow remains essentially unchanged (i.e., normal).[122,134] Induced changes in brain blood flow of small degree in animals have shown no significant effects on ventilation.[100] Unfortunately, such studies of the entire brain in man or other animals provide only indirect information regarding events in the respiratory centers themselves.

QUANTITATION OF COMBINED EFFECT OF SEVERAL STIMULI

Of the many possible contributing factors in exercise hyperventilation the preceding analysis and discussion indicates that only four have been demonstrated as distinctly associated with positive influences in exercise. These, of which none has been fully defined as to specific effect, are *increase in metabolically produced CO_2, increase of body temperature, increase in $[H^+]$ of the circulating blood, and movement of myoskeletal components*. The possible influence of each on respiration has been subjected to separate quantitative appraisal, even though not in each case in exercise. Ignoring the fact that the final mechanism of respiratory stimulation is not known for any of these factors, it is at least possible to

---

*See references 16, 63, 71, 87, 118, 159, and 170.

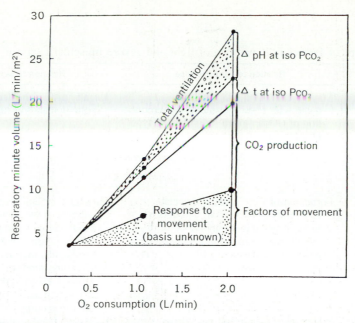

Fig. 74-11. Composite ventilatory response to exercise. Ventilatory equivalents of CO_2 production, "effective" $\Delta[H^+]$ (Fig. 74-9, *B*) and Δtemperature (Fig. 74-9, *A*) and ventilatory change on abrupt initiation of stable-state exercise (Fig. 74-8, *B*) are combined to indicate magnitudes of their probable positive effects in relation to total hyperventilation. For this diagram, no interaction of factors is assumed. Unshaded wedge represents ventilatory change induced by metabolically produced CO_2 and modified by other neural and chemical factors not quantitatively accounted for by three other factors selected and illustrated. It is possible that Δtemperature and ΔpH are related through influence of temperature on hydrogen ion activity. Superimposed on all these positive factors is ultimate adjustment of ventilation, inverse influence of CO_2, providing final relationship of ventilation to metabolism. These relationships are shown to illustrate some questions and observations concerned with identification of nature and quantitative contribution of mechanisms concerned with respiratory stimulation in exercise.

estimate what portion of the total exercise ventilation should be expected to result from their combined effects. Such an estimate is shown in Fig. 74-11, which is derived from the data given in Table 74-1 and the individual "dose-response" curves of Figs. 74-8 and 74-9.

The illustration of combined effects (Fig. 74-11) suggests that even if no greater reactivity to temperature or $[H^+]$ change is presumed for exercise than is found at rest, the combined influence of observed increases in these two "metabolic" factors may account for about one third of the total steady-state ventilatory response at an O_2 consumption of 2 L/min. If this work load produced a lesser degree of metabolic acidosis in better conditioned men, the influence of $[H^+]$ change would also be lessened. It should be reemphasized at this point that the effect of temperature change may eventually prove not to be

distinct from its influence on an acid-base factor such as hydrogen ion *activity*, and therefore not distinct from effects of CO_2 or fixed acid production. A larger part of total ventilation, not more than half at the higher work load, remains attributable to the unknown mechanisms responsible for the relationship of body movements to the hyperpnea of stable-state exercise (Figs. 74-7 and 74-8, *B*). Obviously, to depict this proportionate effect of movement from the data in Table 74-1, it is necessary to make a choice between (1) the increase in respiration that occurs in the sudden transition from rest to exercise and (2) the larger decrease that accompanies abrupt cessation of exercise.[16,17] Unfortunately, it is not possible for the choice between these to be made unequivocally. However, in accordance with the indications of Fig. 74-4 that a large neurogenic role was not required to restore arterial P_{CO_2} to normal, the

Table 74-2. Concept of factors in hyperpnea of light and severe muscular exercise

Factors	Relation to light exercise	Relation to severe exercise
Summation of afferent impulses from proprioceptors in moving limbs and other musculoskeletal structures; influences of higher centers	May be excessive, adequate, or inadequate with respect to metabolic production of CO_2 in given individual or type of exercise	Usually grossly inadequate to meet metabolic demands at high work levels
Fixed acid formation due to muscle anaerobiosis	Formation of fixed acids is detectable but apparently in such small amounts as to be of negligible importance in producing change in blood acidity or respiratory stimulation	Distinct correlation exists between rise in pulmonary ventilation and increase in fixed acid concentration; total ventilatory response is not caused only by fixed acids; increased fixed acid concentration produces positive chemical stimulus effect superimposed on CO_2 production and neurogenic factors related to motion
Change in blood $[H^+]$	Where neurologic respiratory drive is excessive in relation to metabolism, a fall in blood and central $[H^+]$ is brought about by lowered P_{CO_2}, limiting hyperpnea; when neurologic drive is inadequate, elevation of P_{CO_2} leads to rise in blood and central $[H^+]$, sustaining hyperpnea	Excessive fixed acid formation due to muscle anaerobiosis results in rise of blood $[H^+]$ (metabolic acidosis), which adds to increased ventilation; central $[H^+]$ may possibly be lowered by hypocapnia in regions beyond blood-brain barrier, partially limiting ventilation, but raised by H^+ in other chemosensitive regions
Change in P_{CO_2}	Provides both major positive drive and ultimate inverse fine adjustment of ventilation to metabolism. As in all situations, P_{CO_2} of arterial blood reflects relation between metabolism and alveolar ventilation; therefore arterial P_{CO_2} is high when ventilatory drives are low, and P_{CO_2} is low when ventilation is excessive; superimposed on nonchemical stimulation, P_{CO_2} changes provide link to metabolism by sustaining low ventilation and detracting from excessive ventilation; an inverse role of CO_2 thus exists and appears to resemble that at rest; change in blood P_{CO_2} alters blood and central pH concurrently	Provides both positive drive and inverse ultimate fine adjustment of ventilation to metabolism. P_{CO_2} is lowered to less than resting levels as the sum of neurogenic and H^+ ventilatory drives becomes high relative to degree of increased CO_2 production; P_{CO_2} is low because ventilation is high; without this hypocapnia, degree of acidosis in blood would be greater; however, since fixed acids of blood do not readily cross blood-brain barrier, fall in P_{CO_2} should decrease one facet of central stimulation and oppose effect of acidemia elsewhere; again CO_2 probably provides ultimate link to metabolism, as lowering of P_{CO_2} checks further rise of ventilation
Hyperthermia	Effect may be indirect, as by modifying H^+ activity of body fluids; site or mechanism of effect is not known; it is proportional to work rate, work duration, and body mass; effect varies with environmental conditions	Relationship to ventilation is qualitatively similar to that in light exercise
Metabolic production of CO_2	Provides major positive basis for increased ventilation	Provides major positive basis for increased ventilation

values observed for the abrupt change in ventilation on initiating work were employed in constructing Fig. 74-11.

The quantitative allocations of Fig. 74-11 correlate with Fig. 74-4 and represent a considerable evolution of concept from the previous editions of this text.[129,129a] The evolution of the analysis is itself of interest and, being based on quantitative measurement, provides a continuing base for refinement. What is of special interest is that the composite analysis of effects, deduced here from data provided by experiments on Danish,[65] American,[134] and French[16] subjects, does not appear to require effects of the many remaining potential stimulant factors proposed but not plotted in Fig. 74-11, including changes in threshold and/or sensitivity to stimuli, "venous chemoreceptors," chemoreceptor O_2 drive, unidentified chemical excitants, unknown reflex excitants, and interactions of stimulant factors.

Thus there appear to be separately demonstrable forms of stimuli that together are capable of accounting for most of the hyperpnea of exercise. It is likely that the greatest improvement in the understanding of exercise hyperpnea will be achieved by identification and further quantitation of the proportionate contribution of central and peripheral factors involved in the response to movement. Special gains should come from extension of the quantitation and analysis in dynamic states, with special attention to appraisal of the separate, inverse role of CO_2 in final adjustment to metabolism (p. 1776). Meanwhile the data available provide a basis for visualizing at least approximately some of the interactions and summations of the major variables involved in respiratory regulation during the changes in muscular activity involved in work and play. A summary concept of these interactions is presented in Table 74-2.

Breathlessness and exercise

It seems most probable that as long as exercise is not accompanied by unpleasant sensations (such as breathlessness, fatigue, or pain) chemical homeostasis in the neuraxis is being maintained. That is, the positive neurogenic drive supplements the ventilatory response to metabolically produced CO_2 to influence the respiratory centers to the approximate extent required for CO_2 elimination while superimposed fine adjustment by the inverse role of P_{CO_2} prevents gross change in arterial P_{CO_2}, P_{O_2}, or pH. Over the steady-state range the pulmonary ventilation will adjust itself to the elevated metabolic rate, with any tendency to hyper- or hypoventila-

tion being *counteracted* by the resulting changes in the chemical stimulus carried by the blood. The onset of breathlessness signifies that the existing pulmonary ventilation and tissue circulation has not prevented a departure from acid-base homeostasis, either because the metabolic rate has risen beyond the possibility of any compensation or because a stimulus that is not eliminated by the lungs (such as lactic or other fixed acid) has entered the picture. Actually the ideal of perfect homeostasis is approached more closely during light to moderate exercise than it is during vigorous hyperpnea produced by hyperventilation, anoxemia, CO_2 inhalation, or induced metabolic acidosis, in each of which cerebral functions are definitely disturbed. Nevertheless, there will be departures from homeostasis, even during steady-state exercise, in the form of fluctuations in blood sugar[82] and rises in blood lactate,[72,85] blood $[H^+]$,[110] and blood temperature.[82]

INFLUENCE OF HYPOXEMIA ON EXERCISE HYPERPNEA

Lowered environmental P_{O_2} ranks with muscular exercise as a condition in which survival of the individual depends on his ability to increase the amount of air moved in and out of the lungs. As in muscular exercise, the most important respiratory adjustment is increased depth and rate of breathing brought about by neurogenic influences. In acute altitude hypoxemia the initial hyperventilation is due to reflexes from the carotid and aortic bodies, is not associated with a corresponding increase in metabolic rate, and is accompanied by both hypocapnia and a depressant or disorganizing effect on nerve cells in general, including the respiratory center.

When, because of environmental conditions or due to pulmonary insufficiency, an exercising individual is simultaneously subjected to the stress of work and the stress of hypoxemia, the respiratory response to the combined stimulations is greater than that to either alone.* Even after acclimatization the chronic hypoxia of altitude exaggerates exercise hyperpnea[53,94,145,147,160a] (Chapter 73). When, as in rebreathing of the air in a closed space, the lowered inspired P_{O_2} is accompanied by an increase in P_{CO_2}, the hyperpnea of even moderate anoxemia or exercise will be even more severe. Still additional factors may come into play in other forms of severe anoxemia to reinforce the neurogenic and the chemorecep-

---

*See references 65, 111b, 115, 121, 127, 145, 147, 151, 154, and 160a.

tor drive (e.g., an exaggerated metabolic acidosis) which could act both centrally and reflexly, and hyperthermia, which has been described following hypoxia in animals[141] and man,[162] and which also may affect breathing both centrally and reflexly. In a normal man in a normal environment, such circumstances are comparable to the collapse stage of exercise and are therefore to be regarded as pathologic rather than physiologic.

INTERRELATIONS BETWEEN NERVOUS AND CHEMICAL STIMULATION OF RESPIRATORY CENTERS

In each of the several forms of respiratory stimulation thus far described (including CO_2 breathing, anoxemia, exercise, and hyperthermia) there exists a combination of direct chemical and reflex or neurogenic influences on the specialized functions of the respiratory centers. Explanation for the manner of interaction of stimuli to produce the respiratory responses observed in these conditions would therefore have to include at least (1) the manner in which chemical changes in the blood affect central and chemoreflex structures, (2) the nature and mechanism of chemoreflex and other excitatory reflex effects on the respiratory centers, (3) the manner in which inhibitory afferent impulses modify the activities of central respiratory neurons, (4) the ways in which voluntary alterations of ventilation are accomplished, and for exercise (5) the manner in which conscious exercise and movement itself affect the respiratory motor function.

The discussion in these chapters indicates that much is being learned and that much remains to be learned about the interplay of factors involved in the automatic control of breathing, not only at rest but in the great variety of normal and pathologic stresses encountered in life. The history of respiratory physiology has been highlighted by a series of attempts to crowd within the confines of a single, simple, and all-inclusive explanation respiratory phenomena observed in a variety of conditions. This is fundamentally rational because the control mechanisms are the same. However, not all features of these mechanisms are known at any point in time, and discovery of new functions and relationships opens the way for new concepts. As it was learned that chemical stimulation of a respiratory center was not the sole factor in respiratory control in any state, attention was directed to qualitative appraisal of the interplay of numerous chemical and neural influences on the undefined central mechanisms.

When it later became practical to perform quantitative separation of variables, it also became evident that respiratory control, at rest, in exercise, or in environmental stress, depends on a precise balance of numerous factors, the importance and even the nature of which differ from one physiologic condition or degree of stress to another. Increased understanding of this subject depends not only on obtaining improved quantitative data but also on careful and imaginative exploration and interpretation of the now massive amount of information already in existence.

For all the as yet inconceivably intricate and sensitive physiologic mechanisms involved in physical exercise at sea level, the reserves of functional adaptability are still more remarkable. Humans have worked their way up through the thinning natural atmosphere to the summit of Mount Everest,[70,81,153] have demonstrated their capacity for effective work at the extreme pressures and respiratory gas density found 1,600 feet beneath the sea,[37] and have found it possible to work at the low gravity and with the artificial respiratory environment of lunar ambulation.[116] The systems that have provided in advance for adaptation to these environmental extremes deserve admiration even without full understanding.

REFERENCES
General reviews

1. Asmussen, E.: Muscular exercise. In Fenn, W. O., and Rahn, H., editors, Respiration section: Handbook of physiology, Baltimore, 1965, The Williams & Wilkins Co., vol. 2.
2. Beaver, W. L., and Wasserman, K.: Tidal volume and respiratory rate changes at *start* and *end* of exercise, J. Appl. Physiol. **29:**872, 1970.
3. Brodie, B. B., and Shore, P. A.: A concept for a role of serotonin and norepinephrine as chemical mediators in the brain, Ann. N.Y. Acad. Sci. **66:**631, 1957.
4. Burns, B. D., and Salmoiraghi, G. C.: Repetitive firing of respiratory neurons during their burst of activity, J. Neurophysiol. **23:**27, 1960.
5. Casaburi, R., et al.: Ventilatory and gas exchange dynamics in response to sinusoidal work, J. Appl. Physiol. **42:**300, 1977.
6. Churchill, E. D., and Cope, O.: The rapid shallow breathing resulting from pulmonary congestion and edema, J. Exp. Med. **49:**531, 1929.
7. Clark, J. M.: Control of ventilation in man during combined exercise and hypercapnia. (Submitted to J. Appl. Physiol.)
8. Clark, J. M., Sinclair, R. D., and Lenox, J. B.: Metabolic and acid-base responses to combined exercise and hypercapnia in man. (Submitted to J. Appl. Physiol.)
9. Comroe, J. H., Jr.: The hyperpnea of muscular exercise, Physiol. Rev. **24:**319, 1944.
10. Comroe, J. H., Jr.: The effects of direct chemical and

electrical stimulation of the respiratory center in the cat, Am. J. Physiol. **139:**490, 1943.

11. Cooper, D. Y., Emmel, G. L., Kough, R. H., and Lambertsen, C. J.: Effects of CO_2 induced hyperventilation upon the alveolar-arterial Po_2 gradient and the functional respiratory dead space in normal men, Fed. Proc. **12:**28, 1953.

12. Crossland, J.: Chemical transmission in the central nervous system, J. Pharm. Pharmacol. **12:**1, 1960.

13. Daly, I. de B., Ludany, G., Todd, A., and Verney, E. B.: Sensory receptors in the pulmonary vascular bed, Q. J. Exp. Physiol. **27:**123, 1938.

14. Daly, M. de B., Lambertsen, C. J., and Schweitzer, A.: The effects upon the bronchial musculature of altering the oxygen and carbon dioxide tensions of the blood perfusing the brain, J. Physiol. **119:**292, 1953.

15. de Candole, C. A., et al.: The failure of respiration in death by anticholinesterase poisoning, Br. J. Pharmacol. **8:**466, 1953.

16. Dejours, P.: La regulation de la ventilation au cours de l'exercise musculaire chez l'homme, J. Physiol. (Paris) **51:**163, 1959.

17. Dejours, P.: Control of respiration in muscular exercise. In Fenn, W. O., and Rahn, H., editors, Handbook of physiology, respiration section, Washington, D.C., 1964, American Physiological Society, vol. 1.

18. Dempsey, J. A., et al.: Is brain ECF [H^+] an important drive to breathe in man? Chest (suppl.)**73**(2):251, 1978. D., editors: The Twentieth Aspen Lung Conference: Neuromuscular and chemical control of breathing, Chest (suppl.)**73**(2):251, 1978.

19. Douglas, W. W., and Matthews, P. B. C.: Acute tetraethylpyrophosphate poisoning in cats and its modification by atropine or hyoscine, J. Physiol. **116:**202, 1952.

20. Dripps, R. D., editor: The physiology of induced hypothermia, Publ. 451, Washington, D.C., 1956, National Academy of Sciences, National Research Council.

21. Eccles, J. C., Eccles, R. M., and Fatt, P.: Pharmacological investigations of a central synapse operated by acetylcholine, J. Physiol. **131:**154, 1956.

22. Eyzaguirre, C., and Koyano, H.: Effects of hypoxia, hypercapnia, and pH on the chemoreceptor activity of the carotid body in vitro, J. Physiol. **178:**385, 1965.

23. Eyzaguirre, C., Koyano, H., and Taylor, J. R.: Presence of acetylcholine and transmitter release from carotid body chemoreceptors, J. Physiol. **178:**463, 1965.

24. Forster, H. V., Bisgard, G. E., Dempsey, J. A., and Orr, J. A.: Role of intracranial [H^+] receptor in physiologic regulation of ventilation in ponies. In Neff, T. A., and Talmadge, D., editors: The Twentieth Aspen Lung Conference: neuromuscular and chemical control of breathing, Chest (Suppl) **73**(2):253, 1978.

25. Gesell, R., and Hansen, E. T.: Eserine, acetylcholine and nervous integration, Am. J. Physiol. **139:**371, 1943.

26. Gesell, R., Hansen, E. T., and Worzniak, J. J.: Humoral intermediation of nerve cell activation in the central nervous system, Am. J. Physiol. **138:**776, 1943.

27. Giarman, N. J.: Neurohumors in the brain, Yale J. Biol. Med. **32:**73, 1959.

28. Haldane, J. S., and Priestley, J. G.: Respiration, London, 1935, Oxford University Press, Ltd.

29. Henderson, Y.: Adventures in respiration, Baltimore, 1938, The Williams & Wilkins Co.

30. Heymans, C., and Neil, E.: Reflexogenic areas of the cardiovascular system, Boston, 1958, Little, Brown & Co.

31. Holmstedt, B.: Pharmacology of organophosphorus cholinesterase inhibitors, Pharmacol. Rev. **11:**567, 1959.

32. Kelsen, S. J., Altose, M. D., and Cherniack, N. S.: Interaction of lung volume and chemical drive on respiratory muscle EMG and respiratory timing, J. Appl. Physiol. **42:**287, 1977.

33. Klausen, K., Dill, D. B., and Horvath, S. M.: Exercise at ambient and high oxygen pressure at high altitude and at sea level, J. Appl. Physiol. **29**(4):456, 1970.

34. Koelle, G. B.: A new general concept of the neurohumoral functions of acetylcholine and acetylcholinesterase, J. Pharm. Pharmacol. **14:**65, 1962.

35. Koelle, G. B. (editor): Cholinesterases and anticholinesterase agents, Berlin, 1963, Springer Verlag.

36. Krivoy, W. A., Hart, E. R., and Marrazzi, A. S.: Further analysis of the actions of DFP and curare on the respiratory center, J. Pharmacol. Exp. Ther. **103:**351, 1951.

37. Lambertsen, C. J., Gelfand, R., and Clark, J. M., editors: Predictive studies IV. Work capability and physiological effects in He-O_2 excursions to pressures of 400-800-1200-1600 fsw. Institute for environmental medicine report 78-1, Philadelphia, 1978, University of Pennsylvania Press.

38. Leusen, I. R.: Chemosensitivity of the respiratory center. Influence of changes in the H^+ and total buffer concentrations in the cerebral ventricles on respiration, Am. J. Physiol. **176:**45, 1954.

39. Liljestrand, G.: Acetylcholine and respiration, Acta Physiol. Scand. **24:**225, 1951.

40. Loeschcke, H. H., Koepchen, H. P., and Gertz, K. H.: Über den Einfluss von Wasserstoffionenkonzentration und CO_2-Druck im Liquor Cerebrospinalis auf die Atmung, Arch. Gesamte Physiol. **266:**569, 1958.

41. Metz, B.: Brain acetylcholinesterase and a respiratory reflex, Am. J. Physiol. **192:**101, 1958.

42. Metz, B.: The brain ACh-AChE-ChA system in respiratory control, Neurology **11:**37, 1961.

43. Metz, B.: Hypercapnia and acetylcholine release from the cerebral cortex and medulla, J. Physiol. **186:**321, 1966.

44. Mitchell, R. A., Loeschcke, H. H., Massion, W. H., and Severinghaus, J. W.: Respiratory responses mediated through superficial chemosensitive areas on the medulla, J. Appl. Physiol. **18:**523, 1963.

45. Parin, V. V.: Pulmonary hemodynamic reflexes, Am. Rev. Soviet Med. **1:**251, 1944.

46. Parin, V. V.: The role of pulmonary vessels in the reflex control of the blood circulation, Am. J. Med. Sci. **214:**167, 1947.

47. Pearce, D. H., and Milhorn, H. T.: Dynamic and steady-state respiratory responses to bicycle exercise, J. Appl. Physiol. **42:**959, 1977.

48. Raub, W. F., and Yamamoto, W. S.: Absence of blood-brain potential difference fluctuations at the respiratory frequency, J. Appl. Physiol. **21:**1309, 1966.

49. Salmoiraghi, G. C., and Burns, B. D.: Localization and patterns of discharge of respiratory neurons in brain stem of cat, J. Neurophysiol. **23:**2, 1960.

50. Salmoiraghi, G. C., and Burns, B. D.: Notes on mechanism of rhythmic respiration, J. Neurophysiol. **23:**14, 1960.

51. Stewart, W. C.: The effects of sarin and atropine on the respiratory center and neuromuscular junctions of the rat, Can. J. Biochem. Physiol. **37:**651, 1959.

52. Swenson, E. R., and Maren, T. H.: A quantitative

analysis of CO_2 transport at rest and during maximal exercise, Respir. Physiol. **35:**129, 1978.

53. Tenney, S. M., Remmers, G. E., and Mithoefer, J. C.: Hypoxic-hypercapnic interaction at high altitude. In Weihe, W. H., editor: The physiological effects of high altitude, Oxford, 1964, Pergamon Press.

54. Vogel, J. A., Hansen, J. E., and Harris, C. W.: Cardiovascular responses in man during exhaustive work at sea level and high altitude, J. Appl. Physiol. **23**(4):531, 1967.

55. von Euler, C., and Soderberg, U.: Medullary chemosensitive receptors, J. Physiol. **118:**545, 1952.

56. von Euler, U. S., and Pernow, B.: Neurotropic effects of substance P, Acta Physiol. Scand. **36:**265, 1956.

57. Wasserman, K., Van Kessel, A. L., and Burton, G. G.: Interaction of physiological mechanisms during exercise, J. Appl. Physiol. **22:**71, 1967.

58. Wasserman, K., Whipp, B. J., and Castagna, J.: Cardiodynamic hyperpnea: hyperpnea secondary to cardiac output increase, J. Appl. Physiol. **36:**457, 1974.

59. Wigertz, O.: Dynamics of ventilation and heart rate in response to sinusoidal work load in man, J. Appl. Physiol. **29:**208, 1970.

60. Wright, P. G.: An analysis of the central and peripheral components of respiratory failure produced by anticholinesterase poisoning in the rabbit, J. Physiol. **126:**52, 1954.

61. Zuntz, N., and Geppert, J.: Über die Natur der Normalen Atemreize und den Ort ihrer Wirkung, Arch. Gesamte Physiol. **38:**337, 1886.

Original papers

62. Alexander, S. C., Gelfand, R., and Lambertsen, C. J.: The pK' of carbonic acid in cerebrospinal fluid, J. Biol. Chem. **236:**592, 1961.

63. Armstrong, B. W., Hurt, H. H., Jr., Blide, R. W., and Workman, J. M.: The humoral regulation of breathing, Science **133:**1897, 1961.

64. Asmussen, E., and Nielsen, M.: Studies on the regulation of respiration in heavy work, Acta Physiol. Scand. **12:**171, 1946.

65. Asmussen, E., and Nielsen, M.: Pulmonary ventilation and effect of oxygen breathing in heavy exercise, Acta Physiol. Scand. **43:**365, 1958.

66. Asmussen, E., and Nielsen, M.: Experiments on nervous factors controlling respiration and circulation during exercise employing blocking of the blood flow, Acta Physiol. Scand. **60:**103, 1964.

67. Asmussen, E., Christensen, E. H., and Nielsen, M.: Humoral or nervous control of respiration during muscular work? Acta Physiol. Scand. **6:**160, 1943.

68. Asmussen, E., von Dobeln, W., and Nielsen, M.: Blood lactate and oxygen debt after exhaustive work at different oxyten tensions, Acta Physiol. Scand. **15:**57, 1948.

69. Asmussen, E., Nielsen, M., and Wieth-Pedersen, G.: Cortical or reflex control of respiration during muscular work, Acta Physiol. Scand. **6:**168, 1943.

70. Austrian Everest climb, 1978, New York Times, 17 June 1978, p. 19.

71. Aviado, D. M., Jr., et al.: Respiratory and circulatory reflexes from the perfused heart and pulmonary circulation of the dog, Am. J. Physiol. **165:**2, 1951.

72. Bang, O.: The lactate content of the blood during and after muscular exercise in man, Scand. Arch. Physiol. **74**(suppl. 10):51, 1936.

73. Bannister, R. G., and Cunningham, D. J. C.: The effects on the respiration and performance during exercise

of adding oxygen to the inspired air, J. Physiol. **125:**118, 1954.

74. Barr, P. O., et al.: Time courses of blood gas changes provoked by light and moderate exercise in man, Acta Physiol. Scand. **60:**1, 1964.

75. Bazett, H. C.: Physiological responses to heat, Physiol. Rev. **7:**531, 1927.

76. Beaver, W. L., and Wasserman, K.: Transients in ventilation at start and end of exercise, J. Appl. Physiol. **25:**390, 1968.

77. Bering, E. A., Jr., Taren, J. A., McMurrey, J. D., and Bernhard, W. F.: Studies on hypothermia in monkeys. II. The effect of hypothermia on the general physiology and cerebral metabolism of monkeys in the hypothermic state, Surg. Gynecol. Obstet. **102:**134, 1956.

78. Bjurstedt, A. G. H.: Interaction of centrogenic and chemoreflex control of breathing during oxygen deficiency at rest, Acta Physiol. Scand. **12:**suppl. 38, 1946.

79. Bjurstedt, A. G. H., and Wigertz, O.: Dynamics of arterial oxygen tension in response to sinusoidal work load in man, Acta Physiol. Scand. **82:**236, 1971.

80. Bradley, G. W.: Control of the breathing pattern, Int. Rev. Physiol. **14:**185, 1977.

81. Chinese Everest climb, 1960, New York Times, 27 May 1960, p. 9.

82. Christensen, E. H.: Beiträge zur Physiologie schweren körperlicher Arbeit, Arbeitphysiol. **4:**128, 1931, **5:**463, 1932.

83. Comroe, J. H., Jr., and Schmidt, C. F.: Reflexes from the limbs as a factor in the hyperpnea of muscular exercise, Am. J. Physiol. **138:**536, 1943.

84. Cooper, K. E.: Physiology of hypothermia, Br. J. Anaesth. **31:**96, 1959.

85. Courville, C. B.: Asphyxia as a consequence of nitrous oxide anesthesia, Medicine **15:**129, 1936.

86. Covino, B. G., and Hagnauer, A. H.: Electrolytes and pH changes in relation to hypothermic ventricular fibrillation, Circ. Res. **3:**575, 1955.

87. Cropp, G. J. A., and Comroe, J. H.: Role of mixed venous blood P_{CO_2} in respiratory control, J. Appl. Physiol. **16:**1029, 1961.

88. Cunningham, D. J. C.: Some quantitative aspects of the regulation of human respiration in exercise, Br. Med. Bull. **19:**25, 1963.

89. Cunningham, D. J. C., and O'Riordan, J. L. H.: The effect of a rise in the temperature of the body on the respiratory response to carbon dioxide at rest, Q. J. Exp. Physiol. **42:**329, 1957.

90. Dejours, P.: The regulation of breathing during muscular exercise in man. A neurohumoral theory. In Cunningham, D. J. C., and Lloyd, B. B., editors: The regulation of human respiration, Oxford, 1963, Blackwell Scientific Publications, Ltd.

91. Dejours, P., and Teillac, A.: Caractères des variations de la ventilation pulmonaire au cours de l'exercice musculaire dynamique chez l'homme. Étude expérimentale et théorique, Rev. Fr. Etud. Clin. Biol. **8:**439, 1963.

92. Dejours, P., Labrousse, Y., and Teillac, A.: Étude du stimulus ventilatoire proprioceptif mis en jeu par l'activité motrice chez l'homme, Compt. Rend. Acad. Sci. **248:**2129, 1959.

93. Dejours, P., Mithoefer, J. C., and Teillac, A.: Essai de mise en évidence de chemorecepteurs veineux de ventilation, J. Physiol. (Paris) **47:**160, 1955.

94. Dejours, P., Kellogg, R. H., and Pace, N.: Regulation of respiration and heart rate response in exercise

during altitude acclimatization, J. Appl. Physiol. **18:** 10, 1963.

95. Dejours, P., Girard, F., Labrousse, Y., and Raynaud, J.: Stimulus oxygène chemoreflexe de la ventilation à basse altitude (50m) chez l'homme. II. Au cours de l'exercise musculaire, J. Physiol. (Paris) **49:**120, 1957.

96. Dejours, P., et al.: Stimulus oxygène de la ventilation au repos et au cours de l'exercise musculaire, à basse altitude (50m), chez l'homme, Rev. Fr. Etud. Clin. Biol. **3:**105, 1958.

97. Domizi, D. B., Perkins, J. F., Jr., and Byrne, J. S.: Ventilatory response to fixed acid evaluated by "Iso-P_{CO_2}" technique, J. Appl. Physiol. **14:**557, 1959.

98. Downes, J. J., and Lambertsen, C. J.: Dynamic characteristics of ventilatory depression in man on abrupt administration of O_2 at 1.0 atm., J. Appl. Physiol. **21:** 447, 1966.

99. Dundee, J. W., and King, R.: Clinical aspects of induced hypothermia, Br. J. Anaesth. **31:**106, 1959.

100. Edelman, N. H., Chapman, R. W., and Santiago, T. V.: Role of brain blood flow in the control of breathing: effects of flow limitation. Chest **73**(2) Suppl:261, 1978.

101. Eyzaguirre, C., Baron, M., and Gallego, R.: Intracellular studies of carotid body cells: effects of temperature, "natural" stimuli and chemical substances, Adv. Exp. Med. Biol. **78:**209, 1977.

102. Fenn, W. O., and Craig, A. B., Jr.: Effect of CO_2 on respiration using a new method of administering CO_2, J. Appl. Physiol. **18:**1023, 1963.

103. Filley, G. F., Gregoire, F., and Wright, G. W.: Alveolar and arterial oxygen tensions and the significance of the alveolar-arterial oxygen tension difference in normal men, J. Clin. Invest. **33:**517, 1954.

104. Gardner, E.: Physiology of movable joints, Physiol. Rev. **30:**127, 1950.

105. Gardner, E., and Jacobs, J.: Joint reflexes and regulation of respiration during exercise, Am. J. Physiol. **153:** 567, 1948.

106. Geppert, J., and Zuntz, N.: Über die Regulation der Atmung, Pfluegers Arch. **42:**189, 1888.

107. Gesell, R.: The chemical regulation of respiration, Physiol. Rev. **5:**551, 1925.

108. Gonzalez, F., Jr., Fordyce, W. E., and Grodins, F. S.: Mechanism of respiratory responses to intravenous $NaHCO_3$, HCl, and KCN, J. Appl. Physiol. **43**(6): 1075, 1977.

109. Gray, J. S.: The multiple factor theory of the control of respiratory ventilation, Science **103:**739, 1946.

110. Grodins, F. S.: Analysis of factors concerned in the regulation of breathing in exercise, Physiol. Rev. **30:** 220, 1950.

111. Grodins, F. S.: Control theory and biological systems, New York, 1963, Columbia University Press.

111a. Grodins, F. S., Buell, J., and Bart, A. J.: Mathematical analysis and digital simulation of the respiratory control system, J. Appl. Physiol. **22:**260, 1967.

111b. Hansen, J. E., Stelter, G. P., and Vogel, J. A.: Arterial pyruvate, lactate, pH, and P_{CO_2} during work at sea level and high altitude, J. Appl. Physiol. **23:**523, 1967.

112. Harrison, T. R., Harrison, W. G., Jr., Calhoun, J. A., and Marsh, J. P.: Congestive heart failure; mechanism of dyspnea on exertion, Arch. Intern. Med. **50:**690, 1932.

113. Hesser, C. M.: Central and chemoreflex components in the respiratory activity during acid-base displacements in the blood, Acta Physiol. Scand. **18:**suppl. 64, 1949.

114. Hickam, J. B., Pryor, W. W., Page, E. B., and Atwell, R. J.: Respiratory regulation during exercise in unconditioned subjects, J. Clin. Invest. **30:**503, 1951.

115. Hornbein, T. F., and Roos, A.: Effect of mild hypoxia on ventilation during exercise, J. Appl. Physiol. **17:** 239, 1962.

115a. Hornbein, T. F., Griffo, Z. J., and Roos, A.: Quantitation of chemoreceptor activity: interrelation of hypoxia and hypercapnia, J. Neurophysiol. **25:**561, 1961.

116. Johnston, R. S., Dietlein, L. F., and Berry, C. A.: Biomedical results of Apollo, Washington, D.C., 1975, National Aeronautics and Space Administration, Scientific and Technical Information Office.

117. Kao, F. F.: Regulation of respiration during muscular activity, Am. J. Physiol. **185:**145, 1956.

118. Kao, F. F.: An experimental study of the pathways involved in exercise hyperpnea employing cross-circulation techniques. In Cunningham, D. J. C., and Lloyd, B. B., editors: The regulation of human respiration, Oxford, 1963, Blackwell Scientific Publications, Ltd.

119. Kao, F. F., and Schlig, B. B.: Impairment of gas transport and gas exchange in dogs during acute hypothermia, J. Appl. Physiol. **9:**387, 1956.

120. Kety, S. S., and Schmidt, C. F.: The effects of altered arterial tensions of carbon dioxide and oxygen on cerebral blood flow and cerebral oxygen consumption of normal young men, J. Clin. Invest. **27:**484, 1948.

121. Klausen, K., Robinson, S., Michael, E. D., and Myhre, L. G.: Effect of high altitude on maximal working capacity, J. Appl. Physiol. **21:**1191, 1966.

122. Kleinerman, J., and Sancetta, S. M.: Effect of mild steady state exercise on cerebral and general hemodynamics of normal untrained subjects, J. Clin. Invest. **34:**945, 1955.

123. Kollias, J., et al.: Work capacity of long-time residents and newcomers to altitude, J. Appl. Physiol. **24**(6):792, 1968.

124. Kozlowski, S., Rasmussen, B., and Wilkoff, W. G.: The effect of high oxygen tensions on ventilation during severe exercise, Acta Physiol. Scand. **81:**385, 1971.

125. Krogh, A., and Lindhard, J.: The regulation of respiration and circulation during the initial stages of muscular work, J. Physiol. **47:**112, 1913.

126. Krogh, A., and Lindhard, J.: A comparison between voluntary and electrically induced muscular work in man, J. Physiol. **51:**182, 1917.

127. Lahiri, S.: Physiological responses and adaptations to high altitude, Int. Rev. Physiol. **15:**217, 1977.

127a. Lambertsen, C. J.: Carbon dioxide and respiration in acid-base homeostasis, Anesthesiology **21:**642, 1960.

128. Lambertsen, C. J.: Physiological effects of oxygen. In Proceedings of the second underwater physiology symposium. Publ. 1181, Washington, D.C., 1963. National Academy of Sciences, National Research Council.

129. Lambertsen, C. J.: Interactions of physical, chemical and nervous factors in respiratory control. In Bard, P., editor: Medical physiology, ed. 13, St. Louis, 1974, The C. V. Mosby Co.

129a. Lambertsen, C. J.: Concepts of regulation of alveolar ventilation to metabolic production of CO_2 in exercise. Institute for Environmental Medicine Report 78-3, Philadelphia, 1978, University of Pennsylvania Press.

130. Lambertsen, C. J., Gelfand, R., and Kemp, R. A.: Dynamic influences of CO_2 in respiratory control. In Brooks, C. McC., Kao, F. F., and Lloyd, B. B., editors: The cerebrospinal fluid and the regulation of ventilation. Oxford, 1965, Blackwell Scientific Publications, Ltd.

131. Lambertsen, C. J., Semple, S. J. G., Smyth, M. G.,

and Gelfand, R.: H+ and Pco₂ as chemical factors in respiratory and cerebral circulatory control, J. Appl. Physiol. **16**:473, 1961.

132. Lambertsen, C. J., et al.: Oxygen toxicity. Effects in man of oxygen inhalation at 1 and 3.5 atmospheres upon blood gas transport, cerebral circulation and cerebral metabolism, J. Appl. Physiol. **5**:471, 1953.

133. Lambertsen, C. J., et al.: Comparison of relationship of respiratory minute volume to Pco₂ and pH of arterial and internal jugular blood in normal man during hyperventilation produced by low concentrations of CO₂ at 1 atmosphere and by O₂ at 3.0 atmospheres, J. Appl. Physiol. **5**:803, 1953.

134. Lambertsen, C. J., et al.: Respiratory and cerebral circulatory control during exercise at .21 and 2.0 atmospheres inspired Po₂, J. Appl. Physiol. **14**:966, 1959.

135. Landis, E. M., Long, W. L., Dunn, J. W., Jackson, C. L., and Meyer, U.: Studies on the effects of baths on man; effects of hot baths on respiration, blood and urine, Am. J. Physiol. **76**:35, 1926.

136. Lilienthal, J. L., Jr., Riley, R. L., Proemmel, D. D., and Franke, R. E.: An experimental analysis in man of the oxygen pressure gradient from alveolar air to arterial blood during rest and exercise at seal level and at altitude, Am. J. Physiol. **147**:199, 1946.

137. Little, D. M.: Hypothermia, Anesthesiology **20**:842, 1959.

138. Lloyd, B. B.: The chemical stimulus to breathing, Br. Med. Bull. **19**:10, 1963.

139. Loeschcke, H. H., and Katsaros, B.: Die Wirkung von in den Liquor Cerebrospinalis eingebrachtem Ammoniumchlorid auf Atmung und Vasomotorik, Arch. Gesamte Physiol. **270**:147, 1959.

140. Loeschcke, H. H., Katsaros, B., and Lerche, D.: Differenzierung der Wirkungen von CO₂-Druck und Wasserstoffionenkonzentration im Blut auf die Atmung beim Menschen, Arch. Gesamte Physiol. **270**:461, 1960.

141. Lowenbach, H.: Hypoxemia and the temperature of the hypothalamus of the cat, J. Neuropath. Exp. Neurol. **10**:67, 1951.

142. Lugliani, R., Whipp, B. J., Seard, C., and Wasserman, K.: Effect of bilateral carotid-body resection on ventilatory control at rest and during exercise in man, N. Engl. J. Med. **285**(20):1105, 1971.

143. Lundin, G., and Strom, G.: The concentration of blood lactic acid in man during muscular work in relation to the partial pressure of oxygen of the inspired air, Acta Physiol. Scand. **13**:253, 1947.

144. Manax, W. G., et al.: Hypothermia and hyperbaria: simple method for whole organ preservation, J.A.M.A. **192**:755, 1965.

145. Margaria, R., editor: Exercise at altitude, New York, 1967, Excerpta Medica Foundation.

146. Matell, G.: Time-courses of changes in ventilation and arterial gas tensions in man induced by moderate exercise, Acta Physiol. Scand. **58**(suppl. 206):1, 1963.

147. Neff, T. A., and Talmage, D., editors: The Twentieth Aspen Lung Conference: neuromuscular and chemical control of breathing, Chest (suppl.) **73**(2):248, 1978.

148. Nielsen, B., and Nielsen, M.: Body temperature during work at different environmental temperatures, Acta Physiol. Scand. **56**:120, 1962.

149. Nielsen, M.: Untersuchungen über die Atemregulation beim Menschen, besonders mit Hinlick auf die Art des chemischen Reizes, Scand. Arch. Physiol. **74**(suppl. 10):87, 1936.

150. Nielsen, M.: Die Regulation der Korpertemperatur bei Musketarbeit, Scand. Arch. Physiol. **79**:13, 1938.

151. Nielsen, M., and Smith, H.: Studies on the regulation of respiration in acute hypoxia, Acta Physiol. Scand. **24**:293, 1951.

152. Petersen, E. S., and Vejby-Christensen, H.: Effect of body temperature on steady state ventilation and metabolism, Acta Physiol. Scand. **89**:342, 1973.

153. Pugh, L. G. C. E.: Scientific aspects of the expedition to Mount Everest, 1953, Geograph. J., London **120**:183, 1954.

154. Pugh, L. G. C. E., et al.: Muscular exercise at great altitudes, J. Appl. Physiol. **19**(3):431, 1964.

155. Rahn, H., and Baumgardner, F. W.: Temperature and acid-base regulation in fish, Respir. Physiol. **14**:171, 1972.

156. Riley, R. L.: The hyperpnea of exercise. In Cunningham, D. J. C., and Lloyd, B. B., editors: The regulation of human respiration, Oxford, 1963, Blackwell Scientific Publications, Ltd.

157. Schmidt, C. F.: The respiration. In Bard, P., editor: Medical physiology, ed. 10, St. Louis, 1956, The C. V. Mosby Co.

158. Stadie, W. C., Austin, J. H., and Robinson, H. W.: The effect of temperature on the acid-base protein equilibrium and its influence on the CO₂ absorption curve of whole blood, true and separated serum, J. Biol. Chem. **66**:901, 1925.

159. Storey, W. F., and Butler, J.: Evidence that the Pco₂ of mixed venous blood is not a regulator of ventilation during exercise, J. Appl. Physiol. **18**:345, 1963.

160. Torelli, G., and Brandi, G.: Regulation of the ventilation at the beginning of muscular exercise, Int. Z. Angew. Physiol. **19**:134, 1961.

160a. Torelli, G., and D'Angelo, E.: The factors affecting ventilation during exercise at sea level and at altitude. In Margaria, R., editor: Exercise at altitude, Amsterdam, 1967, Elsevier Scientific Publ. Co.

161. Vandam, L. D., and Burnap, T. K.: Hypothermia, N. Engl. J. Med. **261**:546, 595, 1959.

162. Ward, R. L., and Olson, O. C.: Report of severe anoxic anoxia with recovery, J. Aviation Med. **14**:360, 1943.

163. Wasserman, K.: Breathing during exercise, N. Engl. J. Med. **298**:780, 1978.

164. Wasserman, K., and Whipp, B. J.: Exercise physiology in health and disease, Am. Rev. Respir. Dis. **112**:219-249, 1975.

165. Wasserman, K., Whipp, B. J., Casaburi, R., and Beaver, W. L.: Carbon dioxide flow and exercise hyperpnea, cause and effect, Am. Rev. Respir. Dis. **115**:225, 1977.

166. Wasserman, K., et al.: Regulation of arterial Pco₂ during intravenous CO₂ loading, J. Appl. Physiol. **38**(4):651, 1975.

167. Wigertz, O.: Dynamics of ventilation and heart rate in response to sinusoidal work load in man, J. Appl. Physiol. **29**:208, 1970.

168. Winterstein, H.: Die Reaktionstheorie der Atmungsregulation, Arch. Gesamte Physiol. **187**:293, 1921.

169. Winterstein, H.: Die chemische Steuerung der Atmung, Ergeb. Physiol. **48**:328, 1955.

170. Yamamoto, W. S., and Edwards, M. W., Jr.: Homeostasis of carbon dioxide during intravenous infusion of carbon dioxide, J. Appl. Physiol. **15**:807, 1960.

75

CHRISTIAN J. LAMBERTSEN

Effects of excessive pressures of oxygen, nitrogen, helium, carbon dioxide, and carbon monoxide

IMPLICATIONS IN AEROSPACE, UNDERSEA, AND INDUSTRIAL ENVIRONMENTS

Exposure of the normal human to extremes of environment and work not only has demonstrated an amazing capacity for effective performance under physiologic stress but also has opened the way to understanding of failures in disease. This applies not only to the natural environments of high altitude and thermal extremes, but also to the deep undersea and lunar regions reachable only by combining life-support systems and human adaptation. The continued extension of technical activity in industry has further generated physiologic stresses in the working environment.

It is important in these situations to recognize that many of the stresses, and the solutions to relieving them, involve respirable gases.

It is now common in aerospace and undersea activities for artificial atmospheres to be employed to permit useful activity when the breathing of air itself is impractical. These synthetic gaseous environments may have concentrations of O_2 lower or higher than air. There may be no inert gas present, or nitrogen may be replaced by another inert gas such as helium or neon. Total barometric pressure may be subnormal in a spacecraft or extremely high in a submerged diving compartment or lockout submersible craft exposed through an open hatch to high hydrostatic pressure. Hypoxia can occur at either extreme of total barometric pressure whenever P_{O_2} is lowered. O_2 poisoning can result in either case if P_{O_2} is maintained too high and breathed too long; CO_2 intoxication is a risk inherent in any closed-system environment that depends on CO_2 removal systems. Deviations of inert gas concentration or tension from normal introduce several peculiar physical and physiologic problems. An awareness of these effects has enabled humans to adapt and perform useful functions over an extremely wide range of environmental pressures and atmospheric composition.

PHYSIOLOGIC AND TOXIC EFFECTS OF OXYGEN

The vital gas O_2 both sustains life and destroys it, and it is probable that the toxic and metabolic reactions of O_2 are related. Since time is required for serious toxicity to develop, pure O_2 can be breathed safely at pressures up to several atmospheres. However, the duration of this safe exposure period becomes shorter and the pattern of toxicity changes as the partial pressure of O_2 is increased. The physiologic and toxic effects of O_2 are considered in the several symposia and reviews in the references to this chapter.

It is probable that O_2 can poison any living cell or tissue when the dosage and duration of exposure are sufficient.[34,119,133,184] The lining cells of the swim bladder of deep-swimming fish are an amazing exception to this statement. This structure, evidently a derivative of a lung that has lost its respiratory function, can hold O_2 to a pressure of as much as 100 atm[228] and is not poisoned in the process.

In the human the exposure of various cells and tissues is not at all uniform, even at a fixed level of high inspired Po_2.[29,34] For this and other reasons the poisoning of human or animal cells is not uniform in degree or rate of onset, and *O_2 toxicity* tends to express itself most recognizably in one of several forms, including (1) generalized, epileptiform convulsions that occur as a result of CNS toxicity, (2) chemical damage of the respiratory passages that occurs at O_2 pressures below and above 1 atm, and (3) retinal damage (retrolental fibroplasia) in premature infants treated at birth with excessive tensions of O_2 at sea level.[59,120,199]

Other forms of toxicity, probably involving most body cells, should be expected in severe exposure but are difficult to detect and have not been well identified. These are not to be considered important limitations for O_2 use in therapy or environmental exposure. Widespread and repeated use of O_2 in diving and during decompression after diving with nitrogen-containing or helium-containing mixtures would indicate that in the absence of overt manifestations, effects of O_2 are not limiting at rest or in exercise.[29]

Physical effects of oxygen on the lung

O_2 toxicity is related to Po_2 and not to the percentage of O_2 respired. Inhalation of even 100% O_2 will produce no chemical toxicity when the total ambient pressure is so low that alveolar Po_2 during O_2 breathing does not appreciably exceed the normal for air breathing at 1 atm, for example, 100 mm Hg.* This was the situation in the early manned earth-orbital flights, including manned extravehicular activity in orbit[55] and in the "space suits" used in lunar exploration. An idealized example of this is represented by a pure O_2 atmosphere at a reduced total ambient pressure of 187 mm Hg. With an alveolar Pco_2 of 40 mm Hg and $P_{H_2}O$ of 47 mm Hg, the alveolar Po_2 will be 100 mm Hg and nontoxic. As a result of prolonged multiweek exposure of men in O_2-filled altitude chambers at 250 mm Hg total pressure,[135,211] it is clear that the respiratory passages do not suffer chemical O_2 toxicity at this Po_2.

Physical effects of pure O_2 breathing do occur, largely due to the absence of an inert gas in partial pressure equilibrium with the blood and body

fluids. In the absence of nitrogen the gas-containing spaces of the body (e.g., alveoli, paranasal sinuses, middle ear cavity) contain only freely adsorbable O_2, CO_2, and water vapor (Fig. 75-1). Therefore if the communication of these spaces with the O_2 atmosphere is blocked by secretions, spasm, or inflammation, absorption of the gases will occur, leading to complete or partial obliteration of the spaces by distortion or effusion. This diffuse alveolar atelectasis might occur in the lungs, whereas pain, hemorrhage, and exudation might result from diminished pressure in the spaces with more rigid walls. Each of these physical effects occurs in the absence of chemical toxicity. Pulmonary atelectasis does occur in patients with pulmonary disease and occasionally in normal subjects,[194] and each of the problems cited will be grossly aggravated if infection or allergic involvement of the respiratory passages should occur.[161] However, after an initial weeding-out phase, mice have lived for over 7 weeks in a pure O_2 atmosphere.[180] In multiday chamber studies and earth-orbiting situations, obstructions of gas-containing spaces have presented no serious problems in humans breathing pure O_2 at a reduced, nontoxic pressure.

Pulmonary oxygen poisoning in therapy and working environments
Effects at reduced atmospheric pressures and at sea level

Systematic reviews of pulmonary O_2 toxicity in relation to therapeutic and undersea uses of O_2 describe the chemical, physiologic, and pathologic consequences of exposure to increased O_2 pressures.* As for all sites of O_2 toxicity there is a definite relationship between inspired Po_2 and duration of exposure required to produce detectable pulmonary O_2 poisoning.[9] At the relatively low inspired Po_2 of 254 mm Hg, no toxicity has occurred in a period of continuous exposure as long as 30 days.[211] However, when 80% O_2 is breathed at sea level for periods of more than 12 hr, symptoms of irritation of the respiratory passages may be encountered, that is, substernal distress or soreness, coughing, nasal congestion, and sore throat.[34,77,90] Warning of impending severe toxicity is present as a moderately decreased vital capacity at 24 hr.[90] In very few instances the exposure of human subjects at sea level has been as long as 110 hr[65,77,97,98,179]; then the pulmonary consequences have been serious, including persistent severe cough, bronchopneumonia in some subjects, and

---

*See references 90, 157, 159, 161, and 236.

*See references 9, 29, 34, 53, and 54.

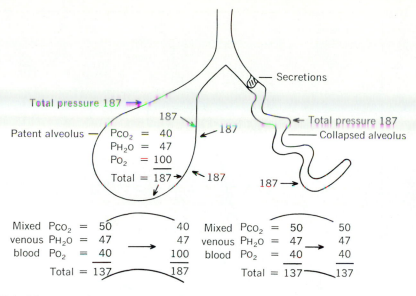

Fig. 75-1. Diagrammatic representation of alveolar and blood gas tensions during prolonged O_2 breathing at nontoxic total ambient pressure of 187 mm Hg. (From Lambertsen.[34,159])

pleural effusion in others. Patients who die after prolonged O_2 therapy for their primary illness show the results of O_2 damage of the lungs.* In one study in humans at 1 atm, brief intermissions (up to 15 min every 3 hr during 1 day) did not significantly diminish the incidence of early pulmonary symptoms.[90] However, a 4 hr daily respite from exposure to O_2 at 1 atm reduces the mortality of mice to one fourth the rate found for uninterrupted exposure,[242] and at increased atmospheric pressures the gains in O_2 tolerance through periodic interruption of exposure are very large (Fig. 75-2).†

These findings indicate that at sea level pure (100%) O_2 can be given continuously without concern over untoward effects for periods up to 12 hr, but if it must be given without appreciable interruption, attention must be paid to the probability that O_2 used for a longer time will irritate the lungs. For many purposes in which O_2 inhalation is indicated for periods involving days at a time, a concentration of 60% may suffice. It is assumed that this pressure of O_2 represents a borderline between absence of toxicity and detectable effects on the lungs.[34] If this does not provide the necessary improvement in arterial oxygenation, higher concentrations should be employed, for intolerable acute hypoxia is much more serious than the early stages of pulmonary

*See references 61, 118, 144, 191, and 205.
†See references 131, 138, 148, 201, and 209.

O_2 toxicity. Brief, periodic interruptions of O_2 administration in ordinary use at least delays the development of pulmonary irritation beyond the times indicated by precisely controlled experiment.

Pulmonary oxygen toxicity at increased atmospheric pressure

The rate of development of pulmonary damage by O_2 is faster at 2 atm inspired P_{O_2} than at 1 atm. Fig. 75-3 shows that although no measurable pulmonary toxicity is expected below a P_{O_2} of 0.5 to 0.6 atm the time required to reach a toxicity level that would induce a 5% decrement in vital capacity becomes progressively shorter as the inspired O_2 pressure increases.[34,85,162] At 2 atm an average of 5% reduction of vital capacity develops in about 6 hr. This initially small effect progresses and is associated with cough and bronchial irritation, which become extreme and dangerous within 10 hr. Recovery occurs over the next 1 to 3 days after return to air breathing.[34,86,162]

Consequences of pulmonary oxygen toxicity

When pulmonary poisoning by O_2 is mild, as after 24 hr at sea level or 4 hr at 2 atm, no limitation of pulmonary gas exchange occurs and recovery is prompt. In fact, the effect is not measurable in some subjects. Continuous exposures

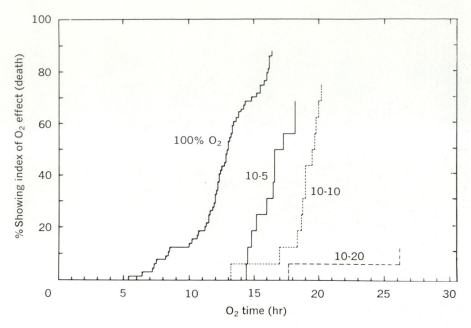

Fig. 75-2. Extension of O_2 tolerance in guinea pigs by interrupting exposure to O_2. Continuous exposure to 100% O_2 at 3.0 ata shows sigmoid dose effect pattern, with each step being death of one animal. The numbers 10-5, 10-10, and 10-20 represent minutes on 100% O_2 alternating with minutes on 7% O_2 in N_2 at 3.0 ata. Systematic, periodic substitution of 7% O_2 in N_2 extends tolerable "O_2 time" (total summed periods of O_2 exposure). (From Hall.[131])

of normal men to 2 atm of O_2 pressure for as long as 9 hr, which cause severe symptoms and prominent reduction of vital capacity, have produced only a minimal decrease in the pulmonary diffusion constant for CO,[207] an increase in the alveolar-arterial O_2 tension difference,[85] and increases in airway resistance and lung compliance.[111] At this stage of poisoning, ventilation and alveolar-arterial gas exchange are not handicapped. However, with more prolonged O_2 exposure, total failure of gas exchange is inevitable, and studies in animals indicate that if O_2 inhalation is continued indefinitely, more severe toxicity does result, progressively leading to loss of pulmonary surfactant, structural changes, pulmonary edema, extravasation of red blood cells into the airways, frothing in the airways, massive atelectasis, CO_2 retention, and acidosis and *death from hypoxia in the presence of and due to the high inspired PO_2*.[146,152] This is a classic example of a pathophysiologic vicious cycle in that when severe pulmonary O_2 poisoning has led to arterial hypoxemia,[169] survival is unlikely because lowering of the inspired Po_2 will aggravate hypoxemia, whereas sustaining a high Po_2 will fur-

ther exaggerate pulmonary damage.[29,34] Death will result in either event.

The cause of the local "irritant" phenomena. Pulmonary damage by O_2 definitely is attributable to dose-related biochemical effects of the abnormally high O_2 tension. Nonspecific effects from the tanks, pipes, regulators, and masks as well as humidity changes have been excluded, not only by proper humidification of the O_2, but also by a study in which individuals breathed room air through the same system used to demonstrate pulmonary irritation and reported no symptoms.[90] The inference that they may be due to the denitrogenation associated with inhalation of pure O_2 is untenable because subjects breathing 100% O_2 continuously for 30 days at an ambient pressure of 254 mm Hg have remained free of symptoms.[139]

Prediction of pulmonary oxygen effects. The importance of balancing O_2 toxicity against benefits of O_2 in therapy and environmental exposures has led to derivation of means for predicting the degree of pulmonary O_2 effect to be expected in exposures to varied pressures and durations of O_2 breathing.[243]

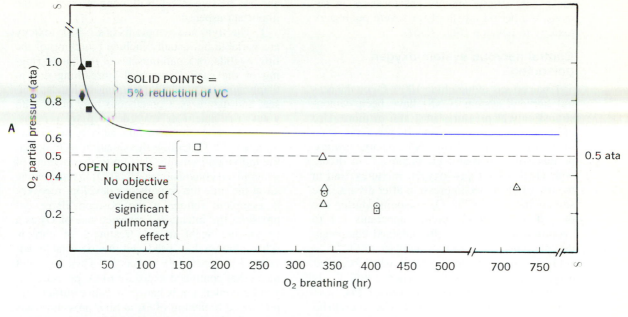

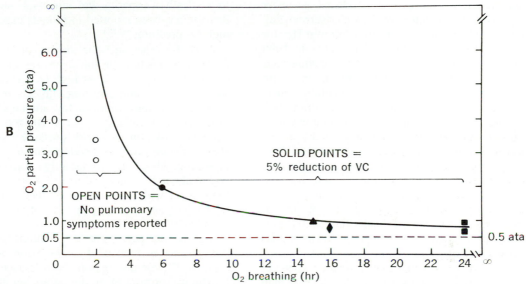

Fig. 75-3. Tolerance to continuous O_2 exposure as indicated by 5% decrease in vital capacity.
   A, Exposure to pressures of 1 atm or less.
   B, Extension of O_2 tolerance curve to pressures above 1 atm.
   Each solid symbol indicates a study in which decreased vital capacity occurred. Open symbols represent studies with no indication of O_2 toxicity. Curves are same rectangular hyperbola. Reference sources for each symbol are provided elsewhere.[85] (Modified from Clark and Lambertsen.[85])

The respiratory tract has been found to be involved at pressures below those producing detectable effects elsewhere, probably because these tissues are exposed to the full tension of the inhaled O_2.

Under the increased barometric pressure of hyperbaric O_2 therapy and diving the latent peri-od for the onset of detectable local irritant phenomena is shortened as the O_2 tension increases, as shown in Figs. 75-3 and 75-5 by the actual findings at 2 atm and the derived predictions of pulmonary tolerance at other high pressures. As will be shown in the following discussion, pure O_2 inhaled at 4 atm (100 feet of seawater) will cause convulsions and unconsciousness in humans in a period too short for severe pulmonary toxicity to develop (Fig. 75-5).

Central nervous system oxygen poisoning

Toxic effects of breathing 100% O_2 on the central nervous system (CNS) have been demonstrated only at pressures of 2 atm or more. Use of O_2 at greater than atmospheric pressures, even to the limits imposed by CNS toxicity, has importance in hyperbaric O_2 therapy,[14] in diving with O_2 and inert gas–oxygen mixtures, and in the use of O_2 in decompression after diving. The serious hazard of CNS O_2 poisoning during use of self-contained O_2 diving apparatus led to evolution of forms of self-contained equipment that permitted air to be used in place of O_2.[177]

O_2 can be toxic to the CNS when breathed in gas mixtures as well as when 100% O_2 is breathed. The effects are related to the P_{O_2} (percentage times absolute pressure). The diver, to be able to breathe, must have his inhaled gas under a pressure equal to that of the water surrounding him, which amounts to 1 atm (760 mm Hg) for each 33 feet of seawater. At the moderate depth of 100 feet, therefore, he would be inhaling a gas under a total pressure of 4 atm, that is, the 1 atm with which he started and three additional ones. Inhalation of pure O_2 under this pressure (3,040 mm Hg), after a latent period of only a few minutes at work, produces violent generalized convulsions together with unconsciousness.[29,34,100] Although the O_2 toxicity itself is reversible if the exposure to O_2 is not extreme,[29,34] unconsciousness is a state having obvious consequences in a diver unattached to the surface. Use of pure O_2 in diving is therefore practically limited to depths of 30 feet or less.[232] Even then it is too hazardous for general use.

O_2 convulsions can also be produced when gas mixtures (O_2-N_2 or O_2-He) are breathed at high pressures. At a diving depth of 1,600 feet (about 50 atm) an O_2 concentration of 0.4% provides an inspired P_{O_2} equivalent to that of air breathing at sea level, and 4% O_2 is equivalent to 100% O_2 at 33 feet of seawater. The range between hypoxia and O_2 toxicity is thus very small at extreme depths.[35,37,212,225]

Practical experience as well as laboratory experimentation has led to considerable understanding of the chemistry and symptoms, but not the functional consequences of CNS O_2 poisoning. It should be assumed that the chemical toxicity is diffuse, that it is progressive, and that the convulsion is an electrical expression of the toxicity

and not the sole effect. Following are some of the important aspects:

1. The signs and symptoms of CNS O_2 toxicity are variable but usually include twitching of the lips, eyelids, or small muscles of the hands; ringing of the ears; "cogwheel" breathing due to diaphragmatic twitching; tingling of hands; nausea; and dizziness. Convulsions may occur with great suddenness or may develop by progressively more severe myoclonic activity over many minutes.[29,34,100] Unlike the situation in hypoxia, the individual remains alert up to the moment of generalized convulsions.[29,155] If O_2 is replaced by air at the time the convulsions develop, recovery is complete following the postconvulsive depression; the intentional repeated use of O_2 as a convulsive agent in the treatment of chronic schizophrenia has produced no detectable adverse (or beneficial) effects.[170] This does not mean that continued exposure to O_2 pressure beyond a convulsion is harmless, since sufficiently prolonged breathing of O_2 at high pressure leads to the death of experimental animals,[4,9] and CNS damage has been reported in rodents exposed to high O_2 pressures.[63,233]

2. As for other sites the toxicity of O_2 to the CNS is a function of the *pressure* and *duration* of exposure to O_2. At high O_2 partial pressures (above 6 atm), resting subjects may develop convulsions within a few minutes of the onset of O_2 breathing.[100] At a 100-foot depth equivalent (4 atm) in a dry pressure chamber, 10% of resting subjects develop symptoms of O_2 poisoning (but not necessarily convulsions) within about 15 min; at 80 feet (3.4 atm), symptoms occur within 25 min; at 60 feet (2.8 atm), no subject developed symptoms of toxicity in 2 hr.[29,156,245]

3. The safe period for O_2 breathing prior to convulsions is reduced by *immersion* and by *exercise*.[100] For this reason the tolerance limits for the use of pure O_2 in diving are set at much shorter periods than indicated by experimental exposures at rest in pressure chambers.[245] In the absence of a full study of O_2 tolerance during work in water, practical limits to such exposure must be applied.[29,192,232]

The bases for accelerated development of O_2 convulsions in exercise and immersion are not known. The effect does not extend to pulmonary O_2 poisoning.

4. O_2 tolerance is also reduced by *CO_2 inhalation*. It is not yet known whether the effect of this factor is additive with exercise in shortening the safe latent period for CNS exposure, but there is no basis for considering that CO_2 exag-

gerates the early stages of pulmonary O_2 toxicity.[9] Actually, on physiologic grounds, it was once considered that the toxic effects of O_2 on the CNS might actually be due to an autointoxication by CO_2.[4] A severe CO_2 retention during O_2 breathing was postulated on the basis that a supply of large amounts of O_2 in physical solution would fully meet the needs of the tissues without the usual reduction of oxyhemoglobin. Failure of Hb reduction would then limit the base available for CO_2 transport, causing CO_2 to be carried at higher than normal P_{CO_2}.[121] It was thought that the resulting tissue hypercapnia or acidosis would be severe and provide the mechanism responsible for the symptoms of O_2 poisoning.[4,121]

It has now been demonstrated that administration of O_2 at a very high pressure can indeed supply enough O_2 in physical solution to meet the demands of tissue metabolism (e.g., at 3.5 atm the O_2 physically dissolved in blood will equal about 6.5 vol%, or more than the normal arteriovenous O_2 difference across most organs.)[29,167,170] Nevertheless, it is now evident from studies in humans that no gross accumulation of CO_2 in the brain occurs during O_2 administration, even up to the time at which convulsions begin.* While the small CO_2 retention that does occur is certainly not the direct basis for O_2 convulsions, inhaled CO_2 has pronounced influences on the latency and the symptom pattern of O_2 toxicity (Table 75-1). The interrelationships of oxygenation and CO_2 transport and the influences of these interactions on respiratory and brain circulatory control in the human are shown in Fig. 75-4.

The ways in which CO_2 might affect O_2 poisoning have been extensively studied. CO_2 does not appear to exaggerate the toxic effect of high O_2 pressures on the metabolism of brain tissue slices[227] or conduction in peripheral nerve.[202] One clearly demonstrable effect of CO_2 is related to dilation of brain blood vessels, increased brain "O_2 flow," and a consequent massive increase in the O_2 tension to which brain tissue is exposed.[170] Table 75-1 shows that the addition of 2% CO_2 to the O_2 breathed at 3.5 atm increased the P_{O_2} of brain venous blood from less than 100 to about 1,000 mm Hg. This important indirect effect of CO_2 on O_2 toxicity should not be confused with the true autointoxication by CO_2 that appears to occur in those divers who show a subnormal ventilatory response to exercise at increased pressure, even when breathing air.[171,175,176] This represents a failure of CO_2 elimination at the lungs.

Other possible effects of CO_2 in O_2 poisoning may be the result of adding the toxic actions of CO_2 itself to the different but specific enzymatic toxicity of O_2. Interactions of increased tensions of CO_2 and O_2 have been studied in small animals over a wide range of pressures.[83,182] In addition to the aforementioned effect of CO_2 in elevating central P_{O_2}, administration of CO_2 can produce profound extracellular and intracellular acidosis and may possibly exert still another action not due to change in $[H^+]$ but related to high concentrations of dissolved molecular CO_2.[182,238] These would be expected only at levels beyond those tolerable in normal human activity.

Comparison of central nervous system and pulmonary oxygen tolerance

This limited information on resting subjects has been analyzed and supplemented by studies at lower pressures, where it was found that no prominent CNS symptoms of O_2 poisoning occurred in subjects exposed to pure O_2 for nearly 12 hr at 2 atm pressure for the purpose of studying pulmonary O_2 toxicity.[34,86] The curves comparing CNS and pulmonary O_2 tolerance in man in Fig. 75-5 were constructed from data obtained by several different laboratories on CNS and pulmonary O_2 tolerance *at rest* at various pressures. In addition to summarizing existing experimental observations, Fig. 75-5 suggests probable limits of tolerance to extreme conditions of pressure and duration that cannot ever be directly experimentally investigated. The legend explains the sources of information and the basis for the CNS and pulmonary tolerance curves.[34,85,162] These curves indicate that (1) CNS symptoms are not likely to be encountered at inspired O_2 pressures below 2 atm, (2) pulmonary toxicity is unlikely at partial pressures below 0.5 atm,[90] (3) at very high pressures of O_2, CNS and pulmonary involvement should be expected to occur immediately, (4) below inspired O_2 pressures of about 2.5 atm the development of pulmonary O_2 toxicity should precede the occurrence of CNS signs, and (5) at O_2 pressures above approximately 2.5 atm, evidence of CNS toxicity should appear before pulmonary changes occur. These relationships, which are now better defined at low than at high pressures, apply to subjects at rest (as in hyperbaric O_2 therapy or in decompression). Since the latent period for CNS poisoning by O_2 is shortened by a number of factors, the

---

*See references 29, 34, 160, and 167 to 171.

Table 75-1. Effects of inhalation of 100% O_2 and 2% CO_2 in O_2 at 3.5 atm on brain blood flow and oxygenation*†

	Cerebral blood flow (ml/100 gm/min)	Cerebral O_2 uptake (ml/100 gm/min)	Cerebral vascular resistance (mm Hg/ml/100 gm/min)	Mean arterial blood pressure (mm Hg)	Pulse rate (per min)
Breathing air at 760 mm Hg	57	3.3‡	1.1	63	66
Breathing O_2 at 2,660 mm Hg	43	3.5‡	1.7	73	58

	Arterial blood						Cerebral venous blood					
	O_2 saturation (%)	O_2 content (vol%)	Po_2 (mm Hg)	CO_2 content (vol%)	Pco_2 (mm Hg)	pH	O_2 saturation (%)	O_2 content (vol%)	Po_2 (mm Hg)	CO_2 content (vol%)	Pco_2 (mm Hg)	pH
Breathing air at 760 mm Hg	96	18.7	91	50	39	7.40	65.2	12.6	38	55.7	50	7.34
Breathing O_2 at 2,660 mm Hg	100	26.0	2,300	46.9	34	7.43	89.3	17.8	75	55.2	53	7.31
Breathing 2% CO_2 in O_2 at 2,660 mm Hg		25.3	2,000	54.5	58	7.27	100	22.3	1,000	57.5	67	7.22

* Based on data from Lambertsen et al.[167,170]

† Average findings in normal and schizophrenic young men.

‡ This is the only change shown in this table that is not statistically significant in comparison with air breathing at 1 atm.

NOTE: This table contains information bearing on the relationship of CO_2 to O_2 toxicity.

Effects of 100% O_2: Measurements of cerebral blood flow and of the composition of arterial and brain venous blood were made in human subjects before and during inhalation of O_2 at a pressure of 3.5 atm (2,660 mm Hg). According to these findings, inhalation of O_2 under a pressure that would have produced convulsions if continued (but had not done so at the time of these observations) leads to constriction of cerebral blood vessels sufficient to produce about a 25% decrease in cerebral blood flow. This constriction is not great enough to elicit an ischemic hypoxia, as is shown by the fact that the Po_2 of cerebral venous blood (75 mm Hg) is still far above its normal level (38 mm Hg). The vascular readjustment, however, is sufficient to slow the blood flow through the brain to a point at which the effluent venous blood is less than 90% saturated with O_2 and therefore still is capable of providing some alkali for combination with CO_2 produced in the brain cells. Thus the Pco_2 of cerebral venous blood is actually only 3 mm Hg higher than it was during the breathing of air at 1 atm. This is the same rise in cerebral venous Pco_2 that is produced by inhalation of about 4% CO_2 in air by normal young men at sea level.[168] Since such inhalations do not cause convulsions, these findings indicate that the manifestations of O_2 toxicity cannot be due to retention of CO_2 in the brain.

The changes in arterial pH, CO_2 content, and CO_2 tension are due to hyperventilation, which may be partly or wholly responsible for the cerebral vasoconstriction.[168] The hyperventilation is most probably the result of the slight central retention of CO_2.[168] This ability of O_2 at high pressure to produce a tissue acidosis and an arterial alkalosis deserves attention as an aid to the localization of the effects of CO_2 and hydrogen ions.[171]

Inhalation of 2% CO_2 with O_2: When patients breathed O_2 with CO_2 added, the most prominent difference from inhalation of pure O_2 was the nearly *1,000 mm Hg rise of brain venous O_2 tension.* This change, reflecting a gross exaggeration of central Po_2, is presumed to have been largely the result of the cerebral vasodilator action of the increase in arterial Pco_2. Because only 3 ml of O_2 was removed from each 100 ml of blood perfusing the brain, Hb in the brain venous blood was 100% saturated.

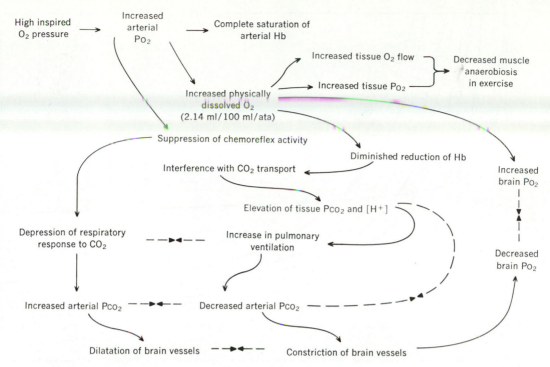

Fig. 75-4. Sequence of acute physiologic effects of O_2 in normal men. As O_2 at pressure of 1 atm or more is administered to normal men, interrelated sequence of changes occurs. Each effect shown in figure has been demonstrated at pressure of 1 ata and, except for chemoreflex suppression, at 3.0 to 3.5 ata.[167,168,170] Each should be considered a physiologic (or pharmacologic) effect rather than a toxic effect and spontaneously reversible on returning to normal levels of inspired P_{O_2}. Sequence includes actions affecting O_2 and CO_2 transport by blood, respiration, smooth muscle of cerebral vascular bed, and degree of anaerobic metabolism in exercising muscle. As O_2 is administered at high partial pressure, initial effects of change in arterial P_{O_2} and content of physically dissolved O_2 lead to readjustments in series of related functions. Direct and indirect effects emerge, often with indirect effects more evident. In stable state of O_2 breathing by normal individual at rest, magnitude of changes from normal are proportional to dose of O_2 (P_{O_2}). As five sets of opposing arrows in figure show, O_2 breathing and its conflicting physiologic effects lead to new state of dynamic balance. In abnormal state, degree and even direction of changes produced by O_2 will depend on balance of physiologic and pathologic factors prior to beginning of O_2 breathing. (From Lambertsen.[29])

curves for rest cannot be applied to work or actual diving.

Effects of high oxygen pressures on the eye

The eye, as a component of the CNS, is affected differently in the adult than in the premature infant. In the adult, O_2 produces a reversible vasoconstriction* similar to the cerebral vasoconstriction produced by O_2 administration.[29,34] At high pressures (e.g., 3 atm), prolonged O_2 breathing also leads to a drastic constriction of

---

*See references 66, 99, 118, 129, 141, 142, 215, and 221.

visual fields with failure of peripheral vision.[68,100] These visual changes are reversible on return to atmospheric levels of inspired P_{O_2} and may provide a sensitive index of CNS O_2 tolerance. If the exposure to high O_2 pressures is longer than that required to produce convulsions, damage and destruction of retinal cells can occur,[195] and blindness is a possible consequence of extreme exposure.[153]

Retrolental fibroplasia. Retrolental fibroplasia is an irreversible pathologic process that was a prominent cause of permanent blindness in premature infants until it was discovered that the vascular changes were the result of O_2 ther-

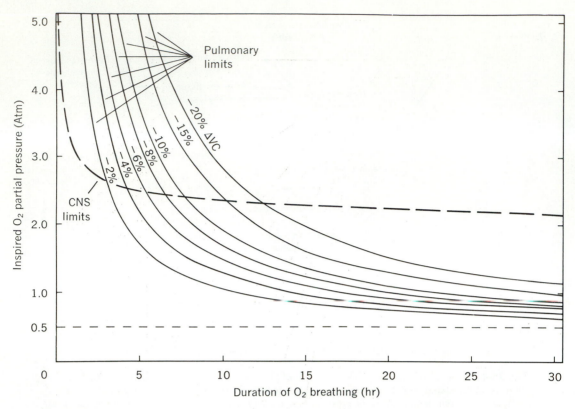

Fig. 75-5. Pulmonary and CNS tolerance of normal men at rest to breathing 100% O_2 at various levels of ambient pressure. Severity of pulmonary O_2 toxicity produced in normal men by various pressure-duration exposures is based on development of indicated change in vital capacity in 50% of subjects. Family of rectangular hyperbolas drawn through experimental findings predicts decrease in vital capacity to be encountered at various tolerable combinations of pressure and duration of exposure.

Construction of pulmonary curves is described elsewhere,[85] as is origin of data showing relationship of pulmonary to CNS O_2 tolerance. (From Lambertsen.[30,85])

apy.[59,120,199] The pathologic changes begin with effects of O_2 on the immature vessels, not the retinal neurons, and have been clearly described by Ashton[58] and Ashton et al.[59] O_2 causes an initial blanching due to retinal vasoconstriction. Under sustained hyperoxic therapy at 1 atm, or even levels of Po_2 not much higher than in atmospheric air, obliteration of retinal vessels occurs, followed by disorganized vascular proliferation, effusion, and finally connective tissue organization with stresses that result in a pulling off of the retina.

Treatment of premature infants with inspired air having a supplemental, high Po_2 has been practiced over many years; it has presumably been responsible for many thousands of cases of blindness, and the problem continues.[81,198] The "normal" Po_2 of the fetus is low, and it might be considered that even the alveolar Po_2 resulting from inhalation of room air may represent a degree of hyperoxia. Certainly, administration of pure O_2 to premature infants for other than emergency lifesaving purposes is to be condemned.

Effects of high oxygen pressures on nerve impulse conduction

Exposure to high partial pressures of O_2 can produce a conduction block of isolated nerve following decreases in conduction velocity, spike amplitude, and increases in rheobase.[92] However, the O_2 pressures required to induce these progressive changes are more than 5 times those that are required to induce pulmonary changes or convulsions.

Effects of oxygen on other tissues

It is likely that O_2 at high pressure is capable of producing toxic effects on any cell or tissue. However, as shown in Fig. 75-6 and Table 75-2, the form of exposure of an organ or tissue to in-

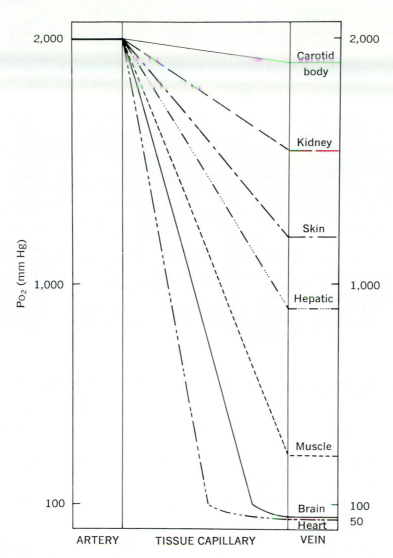

Fig. 75-6. Differences in rate of fall in Po_2 across capillary bed in various tissues during O_2 breathing at 3.5 ata. Curve for brain based on average of actual measurements of arterial and internal jugular venous blood in 16 conscious men.[167,170] Venous values and slopes of change in capillary Po_2 for other organs and tissues calculated from measured arterial values and tables of tissue O_2 consumption and blood flow in man.[62] Fall in Po_2 across tissue capillary is estimated by assuming uniform rate of removal of O_2 as it passes along tissue capillary. Predictions, although only approximate for individual organs, indicate great differences to be expected in oxygenation of various tissues even at single level of arterial Po_2. Predictions indicate that regardless of inspired Po_2, Po_2 in venous blood from carotid chemoreceptor will be no more than 200 mm Hg lower than the 2,000 mm Hg entering arterial O_2 tension.[93] Even within an organ or tissue, there should be local differences in Po_2 due to inequalities of metabolic rate and blood flow. It can be expected that in some tissues a toxic dose of O_2 exists (e.g., kidney) whereas other tissues (e.g., heart) are exposed to a much lower mean level of Po_2. In all tissues a very high Po_2 will be seen for cells influenced by arterial end of capillary, whereas cells or cell constituents at a distance or influenced by Po_2 of venous end of capillary may experience no toxic effects of O_2.[167,170] In pathologic states and under influence of drugs affecting circulation or O_2 metabolism, it should be expected that trends shown by this figure will be grossly altered. (From Lambertsen.[29,29a])

Table 75-2. Effect of O_2 at 3.5 atm on tissue oxygenation

Organ or tissue	Blood flow (ml/100 gm/min)	O_2 use (ml/100 gm/min)	$\Delta(A\text{-}V)O_2$ (ml/100 ml)	Arterial Po_2 (mm Hg) O_2 breathing	Venous Po_2 (mm Hg) Air breathing	O_2 breathing
Heart	84	9.7	11.4	2,000	21	40
Brain* (air/O_2)	57/43	3.3/3.5	6.1/7.9	2,000	40	75
Muscle	3	0.16	6.0	2,000	36	300
Hepatic-portal vein	58	1.96	3.4	2,000	46	900
Skin	13	0.33	2.5	2,000	52	1,190
Kidney	420	6.0	1.4	2,000	63	1,540
Carotid body	2,000	9.0	0.15	2,000	100	1,900

*Experimental data.

NOTE: The table illustrates factors concerned with the fall in Po_2 across a capillary bed as blood changes from the arterial to the venous state. Values for brain are actual experimental observations. For all tissues except the brain and carotid body the average values for tissue blood flow, O_2 utilization, and $\Delta(A\text{-}V)O_2$ are derived from the summary by Bard[62] of data obtained in normal men. Estimates of carotid body oxygenation at normal and high O_2 pressure were derived from measurements of carotid body blood flow in the cat.[93] These data, the normal values for arterial blood during air breathing (arterial O_2 capacity = 20 vol%, arterial Hb saturation = 96.0%, arterial O_2 content = 19.5 vol%, and arterial pH = 7.40), and the conventional assumption of a uniform rate of O_2 loss across the capillary were used to estimate values for venous Po_2 during air breathing. Values for brain blood flow and O_2 consumption during air breathing and O_2 breathing at 3.5 atm were experimentally determined in seven subjects[167]; values for brain $\Delta(A\text{-}V)O_2$ and venous Po_2 were determined in 16 subjects.[167,170] Estimates of venous Po_2 during exposure to high O_2 pressures were made by assuming that an amount of O_2 equal to the air-breathing values for $\Delta(A\text{-}V)O_2$ was removed from arterial blood.

These derived venous Po_2 values illustrate gross differences in tissue oxygenation to be expected in different organs. However, the discrepancy that high O_2 pressure can cause between similarly derived values and actual measured values in brain emphasize the need to improve these estimates for other organs by actual measurement at high inspired Po_2.

creased Po_2 depends on both the metabolic rate and the blood supply of the tissue. Therefore certain tissues such as the carotid chemoreceptor are exposed to a mean Po_2 close to that in arterial blood. Another tissue such as the ventricular myocardium, always at work, removes so much O_2 from each unit of its blood that only a small fraction of tissue near the arterial end of the capillary is exposed to the high Po_2 presented to the whole animal.[29,34,170] If it becomes possible to prevent toxic effects of O_2 on the brain and lung, then effects on other organs such as the kidney may become important.[29,208]

Biochemical mechanisms of oxygen toxicity: effects of oxygen on metabolic function

Although a mixture of glucose and carbon burns faster as O_2 pressure is raised, the rate of brain or whole body O_2 consumption remains constant as inspired Po_2 increases to at least 4 atm[34,167] (Table 75-1). Development of severe toxicity should diminish O_2 consumption, but this is not evident in the brain at the time convulsions develop.[167]

O_2 poisoning is clearly the result of the direct actions of O_2 itself on the chemical processes within the cells. As in hypoxia and with certain drugs, the effects on the CNS are the most obvious because they produce visible consequences. It has been shown that high O_2 pressures interfere with the metabolism of a great variety of cells, from the unicellular paramecium to the highly developed cells of the heart, liver, and brain.* The inhibitory effects on obligate anaerobes, which die in the presence of O_2, is apparently due to lack of the antioxidant catalase in these organisms.[184,204]

The damaging mechanism of O_2 toxicity appears to be related to the concept of indirect effect of O_2 via increased formation of free radicals and inactivation of enzymes by the free radicals.[119] This concept emphasizes that a low-grade degree of O_2 poisoning by free radicals is a normal state during life at sea level.[119]

The discovery of superoxide dismutases[39] has improved understanding of the relationship of the normal biologic role of O_2 to the toxicity of O_2, at least in the chemical and microbiologic systems in which these factors can be studied. The stepwise reduction of molecular O_2 to water in the cell involves 4 electrons, and the normal stages in reduction result in formation of the highly reactive intermediates (superoxide anion

*See references 7, 13, 39, 95, 119, and 133.

radical, hydrogen peroxide, and hydroxyl radical). Scavenging of these potentially damaging substances represents an intrinsic defense against them, without which life could probably not exist. Superoxide dismutases convert O_2^- to $H_2O_2 + O_2$, and the hydrogen peroxide is converted to H_2O + molecular O_2. Together, these steps normally limit the formation of the extremely reactive free radical OH'.

It is presumed that at increased O_2 pressures an increased formation or level of O_2^- is generated, with partial failure of scavenging by dismutases and damage to enzymes, membranes, and other cell components. The nature of the metabolic defect would then depend on the enzymatic inactivations produced, the symptoms on the cells most affected.

O_2 exerts its effects on cellular metabolism by more than one chemical mechanism and at more than one metabolic site.* There is now an abundance of evidence indicating that an excess of O_2 pressure actually interferes with oxidative reactions.† The effect is related to an inactivation of enzymes involved in the early stages of hydrogen transfer. These enzymes include a number of dehydrogenases, especially glyceraldehyde phosphate dehydrogenase and succinic dehydrogenase, in which sulfhydryl (SH) groups are essential for activity.[95,132] A second type of effect is an inhibition of reactions involved in pyruvate oxidation in the tricarboxylic acid cycle,[95,132] and a third relates to interference with formation of high-energy phosphate bonds.[7,132] The inactivation process is not indiscriminate, since many enzymes including cytochrome oxidase, are not affected.[95]

The exact nature of enzyme inactivation by O_2 also may be by more than one mechanism. One proposal is that O_2 excessively drives certain stages in the normal oxidative sequence by mass action, resulting in the removal of vital components of the reaction chain.[95] If this occurs, it is probably limited and readily reversible. This current view relates the chemical toxicity of O_2 to the established concept of free radical formation as a factor in cell damage produced by ionizing radiations and bears on the well-known interaction of O_2 pressure and degree of damage by radiation.[84,127,128,174]

Reversibility. The toxicity of O_2 in its early stages is completely *reversible*. However, continued exposure to high O_2 pressure will result in sustained interference with cell metabolism and ultimately irreversible damage. Therefore

*See references 7, 13, 39, 95, 132, and 133.
†See references 7, 13, 95, 132, and 133.

recovery from severe pulmonary O_2 poisoning requires days,[9] and incomplete recovery from the damage of one acute exposure to high O_2 pressure should result in a diminished tolerance to subsequent exposure.

Prevention of pulmonary and other oxygen toxicity

Prevention of toxicity has been a matter of great practical and theoretical interest. Obviously, if O_2 toxicity is developing, a *lowering of inspired P_{O_2}* (as in substituting a gas mixture containing less O_2) will either prevent toxicity or prolong the latent period. Brief, systematic periodic *interruption* of the O_2 exposure greatly extends O_2 tolerance in small animals (Fig. 75-7) and humans,[138,148,156] suggesting that the rate of recovery from at least some toxic effects of O_2 is more rapid than the rate of their development. *Narcotic drugs* such as phenobarbital can prevent the occurrence of O_2 convulsions, but this finding does not at all imply prevention of cellular O_2 toxicity, which should continue. *Chelating agents,* which remove trace metals from enzyme systems, appear able to prevent depression of metabolism by O_2.[95,133] Finally, certain substances, themselves containing sulfhydryl groups (glutathione, cysteine), can interfere with the inactivation of the enzymatic sulfhydryl group by O_2 toxicity.[95,133]

Hyperbaric oxygenation: therapy with oxygen at high pressures

The degree of oxygenation at the cellular level is determined by (1) the head of O_2 pressure in the arterial blood, (2) the volume flow of O_2 through the tissue, (3) the rate of O_2 metabolism of the cell, (4) the diffusion constant for O_2 in the tissue fluids, and (5) the distance over which the O_2 must diffuse to reach the site of oxidative metabolism. There are numerous pathologic situations in which inadequate whole-body, organ, or cellular oxygenation exists, leading to hypoxic damage or destruction of cells. Table 75-3 illustrates some of these conditions in which the potential use of O_2 administration at pressures even greater than 1 atm ("hyperbaric") is being studied.[5,14,50,207]

Hyperbaric oxygenation for therapeutic purposes in circumstances of inadequate delivery usually involves the use of high inspired P_{O_2} to increase the gradient of O_2 to the metabolizing cell. In CO poisoning, it also speeds CO elimination by improving the ability of O_2 to compete with CO for the reactive site on the Hb molecule.

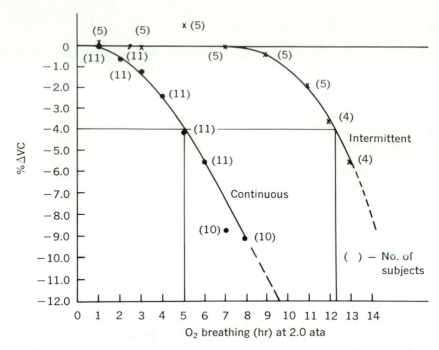

Fig. 75-7. Extension of pulmonary O_2 tolerance in human. When 20 min of 100% O_2 breathing is alternated with 5 min of 10% O_2 breathing, "O_2 time" required to produce specific decrement in vital capacity (e.g., −4%) is more than doubled. O_2 time represents summation of actual time breathing O_2. Figure consists of curve for continuous O_2 exposure[86] and added results of intermittent exposure.[138] (From Hendricks et al.[138])

Table 75-3. Major current applications of intermittent exposures to O_2 at pressures greater than 1 atm*

Condition	Rationale
Pulmonary gas embolism	Compression of bubble size
Cardiac surgery	Oxygenation of tissue
Renal dialysis	Accelerated resolution of bubbles
Arterial gas embolism	
Ruptured lung	
Clinical arterial gas injection	
Decompression sickness	
Diving (spinal cord, inner ear, joints, skin, lungs)	
Aviation bends (joints, lungs, spinal cord)	
Carbon monoxide poisoning	Accelerated dissociation of COHb
	Supply of tissues by physically dissolved O_2
	O_2 diffusion in brain edema
Gas gangrene	Inhibition of toxin production
	Improved oxygenation of affected tissue
Osteomyelitis	Improvement of oxygenation of affected tissue

*Modified from Davis and Hunt.[14]

When O_2 is administered at high ambient pressure, alveolar P_{O_2} is increased in proportion to the elevation of total pressure. For example, at a total pressure of 3 atm absolute inspired P_{O_2} will be:

$$3 \text{ atm} \times 100\% \ O_2 \times 760 \text{ mm Hg} = 2,280 \text{ mm Hg}$$

With no change in respiration the pressure of O_2 in the functioning alveolus will be:

$$\text{Alv. } P_{O_2} = 2,280 \text{ mm Hg} - 40 \text{ mm Hg } P_{CO_2} - 47 \text{ mm Hg } P_{H_2O}$$

O_2 tension in the arterial blood of individuals with normal lungs may, at increased ambient pressures, be several hundred millimeters of mercury less than alveolar P_{O_2} for reasons not yet discovered.[29,87]

The actual benefit to be derived from high O_2 pressures depends on the nature of the hypoxic disorder. Increased ambient pressures are not required when arterial hypoxemia is due entirely to respiratory depression in an individual with normal lungs and circulation. In this instance, alveolar and arterial oxygenation is more than adequate even with O_2 breathing at 1 atm. Since the circulation of blood involves more functions than transport of O_2, an ischemic myocardium or thrombosed cerebral cortex will not necessarily respond to increase in P_{O_2} alone.

A special and unique advantage of hyperbaric oxygenation exists in the treatment of the bends and air embolism. High O_2 pressures are extremely effective, since improved oxygenation is combined with compression of the bubble size as well as accelerated removal of tissue and bubble inert gas, leading to a rapid rate of decrease in the bubble volume and reestablishment of local circulation.

In gas gangrene necrotizing toxin production by the anaerobic organism *Clostridium perfringens* is inhibited by O_2 at high pressure. The organisms are poisoned by oxygen. However, since O_2 cannot reach organisms in tissue that does not have a functioning circulation, hyperbaric oxygenation is only an adjunct to the therapy of gas gangrene.

The major handicap to extending the usefulness of high O_2 therapy to most forms of hypoxia is the toxicity of O_2 (p. 1902). The greater the pressure of O_2 required to relieve local or general hypoxia, the more rapid the development of a toxic effect of O_2 (Fig. 75-5). Toxic effects on normal tissues can occur in the face of inadequate oxygenation of the site of pathology. If a pathologic disturbance is chronic or cannot be acutely relieved, the utility of very high O_2 pressures is small indeed. Considerable gain in procedure for therapy of acute conditions may be expected from the use of moderate elevation of O_2 (to 1.5 or 2 atm absolute pressure) combined with periodic interruption of O_2 administration to permit recovery from the toxic effects of O_2 that occur even at these pressures.[5,29,34,50]

CARBON DIOXIDE

CO_2 is both a metabolic waste product and a key regulator of several fundamental systems. Its roles in life and as a factor in disease are described in several reviews and symposia.[6,28,49]

Intoxication by CO_2 may result from exposure to atmospheres containing high concentrations of CO_2. The gas is increasingly used as a nontoxic preservative of fresh foods and in liquid form as a quick freezing agent. *Exogenous* CO_2 has caused fatalities in brewery workers, food storage workers, and dry ice handlers overcome by accumulation of the gas in closed or poorly ventilated compartments. CO_2 is heavier than air and tends to accumulate in low places. CO_2 disposal is a limiting factor in the duration of manned extra-atmospheric missions, including not only space flight but also submarine, undersea habitat, and bathyscaphe operations and in self-contained diving. CO_2 poisoning can be produced intentionally, as has been done in the administration of 30% CO_2 in the treatment of neuroses.[198]

Intoxication can also result from retention of autogenous CO_2, as in respiratory diseases[49] and overdose of depressant drugs. Although the agent and the cellular effects are the same in each instance, the overt reactions may appear different, depending largely on the rate at which hypercapnia develops or the basis for CO_2 retention (e.g., general anesthesia, pulmonary disease). In each situation the intoxication may result largely from intrinsically produced CO_2, since elevation of the inspired pressure of this gas causes the equivalent of a damming back of the CO_2 produced in the body. The most severe forms of autointoxication by CO_2 occur when O_2 is being administered, since without the O_2, failure in hypoxia would accompany any effects of increasing P_{CO_2}.

Undoubtedly CO_2, capable of freely penetrating cell membranes and increasing intracellular $[H^+]$, can exert toxic effects on any mammalian cell. However, the actions best known are naturally those resulting in distressful symptoms or dynamic signs relating to the CNS, respiration, or the cardiovascular system.

The normal range of P_{CO_2} in the body is quite narrow. Below an arterial P_{CO_2} of about 25 mm

Hg, mental efficiency is impeded and tetany due to decreased calcium ionization is likely to appear. The upper normal limit for arterial blood is probably less than 50 mm Hg.[108] The effects of hypercapnia in humans are properly regarded as those of increased $[H^+]$ and as less damaging than those of hypoxia, but they can be lethal. Although the tissues are equipped with enzyme systems capable of accelerating the transformation of CO_2 into bicarbonate and citrate, these reactions are limited in capacity and rate, and they cannot substitute for pulmonary elimination of CO_2. Since increased P_{CO_2} raises the acidity within the cells to a degree equivalent to that produced in the blood, a respiratory acidosis amounting to an arterial blood pH of 7.30 or less must be taken seriously.

Mechanisms of carbon dioxide toxicity

Elevation of P_{CO_2} in the alveoli, blood, or body fluids is in itself a physical (pressure) phenomenon, and any toxic effects on cells should be referable to one or both of the following reasons: (1) *increased concentration of molecular CO_2* and (2) *increased hydrogen ion concentration.* The toxic actions of these consequences of hypercapnia are not likely to be the same at different levels of P_{CO_2}, even on the same cell. Actions on different cell types further complicate the picture of CO_2 toxicity. Sudden, extreme elevation of CO_2 tension, such as that caused by inhalation of 30% CO_2, produces simultaneous manifestations of *failure and stimulation* in the CNS. Convulsions occur in normal humans, along with unconsciousness, and violent respiratory stimulation is followed by apnea (Chapter 71). The convulsive manifestations of CNS hyperactivity are considered to be the paradoxical result of a disruptive effect of CO_2 $[H^+]$ on the cerebral cortex, releasing subcortical centers from normally powerful inhibitory influences.[238] Thus what may actually be a failure of function is expressed in the form of increased, although disorganized, activity. Biphasic actions of increased $[H^+]$ and differences in effect on various cells should also be expected. Progressive increases in $[H^+]$, known to exert stimulant effects at least on respiration, should eventually so disturb the intracellular acid-base environment that enzymatic and electrochemical processes are adversely affected, resulting in failure of cellular function. The processes involved in CO_2 $[H^+]$ toxicity may involve biophysical membrane phenomena as well as biochemical enzymatic alterations.

Narcosis and metabolic "reentry"

Even in its molecular form, CO_2 may act in more than one way. The possibility of action as an indifferent narcotic, resembling the inert gases, has been mentioned. (CO_2 has approximately the same physical solubility in water as nitrous oxide, and the two gases are about equally soluble in fat.) The inability to separate molecular CO_2 effects from CO_2-induced changes in $[H^+]$ prevents evaluation of CO_2 narcosis.

In addition, CO_2 is known to be more than a waste product of cellular reactions; it can reenter into certain of these processes, its carbon again becoming involved in the synthesis of cell constituents such as *l*-malate.[122,196] Whether at very high pressures active reentry of CO_2 has influences on cellular function or even whether it is a normal feature of metabolism is not known.

Effects of acute exposure to high P_{CO_2}

Abrupt, short-term exposure to CO_2 produces effects that range from barely detectable respiratory stimulation to unconsciousness and convulsions, depending on the partial pressure inspired[30,158] (see Figs. 71-2 and 75-8).

Administration of 1%, 2%, and in most normal individuals, 3% CO_2 in air at sea level produces no recognizable respiratory or other discomfort at rest.[102,158] Distinct respiratory stimulation of increasing degrees is produced by 4%, 5%, and 6% inspired CO_2 (Chapter 71), with an accompanying arousal.[158] Respiratory stimulation to the degree of true dyspnea occurs with 7% and 10% CO_2 in air, and with these concentrations the deterioration of mental competence rapidly becomes evident.[30,102]

At still higher inspired concentrations of CO_2 (15% and 20%), respiratory distress is abrupt and violent in onset and is accompanied by a rapid loss of consciousness and by spasmodic neuromuscular twitching.[30] With a high O_2 concentration provided for safety, 20% to 30% inspired CO_2 produces convulsions that occur within 1 to 3 min after the beginning of the exposure (see Figs. 71-2 and 75-8).[30]

Since the normal human has been exposed continuously to inspired concentrations of 3% for 30 days and 4% for 11 days[88] without signs or severe symptoms of adverse biochemical, physiologic, or endocrine changes, shorter or intermittent exposure to these or lower concentrations cannot be expected to induce adverse changes in the parameters studied. Actually, rhesus monkeys have been exposed to 195 mm Hg P_{CO_2} for as long as 17 hr and successfully restored to normal

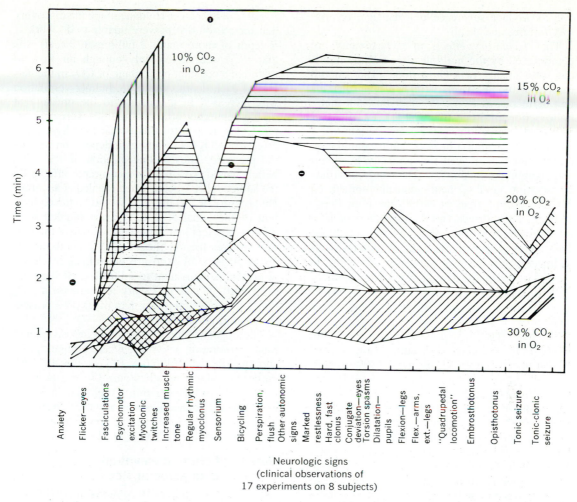

The x-axis labels (Neurologic signs): Anxiety, Flicker—eyes, Fasciculations, Psychomotor excitation, Myoclonic twitches, Increased muscle tone, Regular rhythmic myoclonus, Sensorium, Bicycling, Perspiration, flush, Other autonomic signs, Marked restlessness, Hard, fast clonus, Conjugate deviation—eyes, Torsion spasms, Dilatation—pupils, Flexion—legs, Flex.—arms, ext.—legs, "Quadrupedal locomotion", Embrosthotonus, Opisthotonus, Tonic seizure, Tonic-clonic seizure

Neurologic signs
(clinical observations of
17 experiments on 8 subjects)

Fig. 75-8. Effects of inhalation of high concentrations of CO_2 on brain function in human. Starting with normal air-breathing state, 10%, 15%, 20%, or 30% CO_2 in O_2 suddenly administered leads to progressive changes in neurologic status and consciousness. (From Lambertsen.[30])

by slow reduction of inspired P_{CO_2} to zero.[183] Exercise tolerance will be reduced, since an increase in inspired P_{CO_2} interferes with the elimination of metabolically produced CO_2 and leads to an elevation of arterial hydrogen ion concentration above normal levels for the exercise state in that individual. However, in acute exposures, heavy work has been done at 3% and 4% inspired CO_2, and the maximum exercise for particular subjects has been carried out at 16 and 21 mm Hg inspired P_{CO_2}.[110,185,210,222] Such acute exposures to the simultaneous stresses of increased inspired P_{CO_2} and physical exercise induced no evident residual adverse effects.[222]

Regardless of the several possible actions of CO_2, the response of the human respiratory control mechanism is vigorous and shows no true

depression even at 30% inspired CO_2 (Fig. 71-1). If an "inert gas" depression of the center is indeed occurring, it is being counteracted by progressive, direct, or chemoreflex drives set up by the mounting P_{CO_2} and/or $[H^+]$. The apnea that occurs at the onset of CO_2 convulsions (Fig. 71-2) may actually represent a mechanical interference with breathing due to the sustained period of muscular rigidity rather than a specific respiratory depression.[30] A decision regarding the mechanism of respiratory effects of CO_2 at very high pressures is impossible from the information presently available, but it is clear that, in its effects on the human, CO_2 cannot be regarded as a simple narcotic. It also produces extreme acidosis, and if very high levels of inspired P_{CO_2} are maintained for long periods of time, CO_2 is certain to produce severe cellular damage, respiratory failure, and death.

The *circulatory effects* of CO_2 toxicity are complicated by the multiple sites and the multiple mechanisms of CO_2 action.* The smooth muscle of blood vessels is dilated by a local effect of increased P_{CO_2}. Whether this is a metabolic action of molecular CO_2 or an effect of increased intracellular $[H^+]$ is not certain.[172,206] In addition, elevation of P_{CO_2} causes stimulation of the sympathetic centers, probably by direct and chemoreflex actions.[206] The result is cardiac acceleration and a neurogenically influenced vasoconstriction that, at certain levels of P_{CO_2}, counteract the direct vasodilator action of CO_2 on the renal, skin, and splanchnic vascular beds. This complex overall reaction of blood vessels to hypercapnia is made still less susceptible to predictive evaluation by the fact that increased P_{CO_2} or acidosis interferes with the actions of norepinephrine released at sympathetic nerve endings.[83]

The heart itself is depressed by acidosis[206] but continues to beat even when animals are exposed to concentrations of CO_2 as high as 30% to 40% for many hours. However, as a result of this extreme myocardial acidosis, ionic shifts occur, and when dogs exposed to prolonged hypercapnia (30% to 40%) are suddenly given room air to breathe, ventricular fibrillation and death occur in the majority.[74] This accident does not occur when the hypercapnia is slowly relieved; this presumably allows time for reestablishment of the extracellular-intracellular ionic equilibria required for cardiac function at normal pH.

Prominent *effects on neuroendocrine systems* and actions of some hormones are produced by CO_2.[49,193] These are especially evident in excitation of the sympathoadrenal mechanism, leading to increased liberation of epinephrine and norepinephrine.[193,206] The acidosis limits the effects of these substances and concurrent activation of the parasympathetic system leads to extremely varied results.[193]

Carbon dioxide as an environmental constituent

CO_2 is present in a minute and ineffective concentration in the normal atmosphere (Chapter 68). Exaggerated effects of high CO_2 concentrations and partial pressures are encountered when the atmospheric pressure is raised, as in diving and similar work, since the physicochemical effects of CO_2 are P_{CO_2}-related.

*See references 49, 105, 143, 229, and 234.

Early in his career Haldane made the discovery that the reason divers were unable to do work at a depth of more than 12 fathoms (72 feet) was that they were given only enough air flow to maintain a constant percentage of CO_2 in the helmet at each depth. When the helmet ventilation was increased in proportion to the pressure at diving depth, no further difficulties were encountered. The reason for this is that the physiologic effects of CO_2 (as of any gas) depend on its partial pressure. Thus a concentration of 1% CO_2, which would have barely measurable effects on the subject's well-being when inhaled at an ambient pressure of 760 mm Hg, would be the equivalent of the lethal 50% CO_2 when inhaled by a diver 1,600 feet below the surface where the prevailing pressure on the inhaled gas mixture is about 50 atm. In exposure of animals to He-O_2 at pressures equivalent to 4,000 feet of seawater (122 atm), it was necessary to maintain a CO_2 concentration even lower than the 0.03% present in our normal atmosphere.[181]

The opposite situation exists in aerospace and lunar exploration activities, where ambient pressure may be reduced to 0.5, 0.3, or even 0.2 atm while the subject breathes pure O_2 or a high O_2 mixture.[230] In this situation a high percentage of CO_2 can be breathed, since 5% CO_2 exerts no greater partial pressure at 0.2 atm total ambient pressure than does 1% CO_2 at sea level.

Effects of chronic elevation of inspired or alveolar P_{CO_2}

Chronic elevation of P_{CO_2} in the internal fluids results from pulmonary disease such as emphysema and in normal individuals as a potential consequence of closed-system existence in aerospace and undersea craft and habitations. Acclimatization to moderate elevation of inspired P_{CO_2} can occur[149] with independent adjustments of acid-base, renal, and respiratory functions.

The effects of chronic exposure to increased environmental CO_2 include (1) the initial changes induced at the beginning of the exposure, (2) secondary alterations resulting from the acclimatization processes, and (3) readaptations on restoration of exposure to natural air.

Of the changes studied, the most definite and important alterations have been found in the P_{CO_2} and acid-base characteristics of blood and CSF, the respiration and respiratory responsiveness, the deposition of calcium and carbonate in the skeleton, and the tolerance to physical exercise.[88,217,219,222] In each instance, regardless of the level of CO_2 tension inspired, the specific influences of the CO_2 exposures were completely reversible on termination of the exposure. No temporary or residual pathologic lesions have been produced.

Pco₂ and acid-base changes

On exposure to 1.5%, 3%, or 4% CO_2 in 20% to 21% O_2 in nitrogen, the respiratory changes are qualitatively the same (Chapter 71).[88,219] A rise in alveolar and arterial Pco_2 is necessarily induced by the increased inspired CO_2 pressure. This rise in alveolar-arterial Pco_2 tends to be sustained throughout the chronic exposure, even in studies lasting for as long as 11, 30, or 42 days (Fig. 75-9).[88,219] The rise in arterial Pco_2 induces elevations in Pco_2 in all body tissues, including the brain (as reflected by the increase in Pco_2 of the CSF).[88]

The increase in Pco_2 in itself is harmless. However, the initial Pco_2 elevation effects an immediate acidification of arterial blood and tissue fluids (including CSF) in proportion to the rise in arterial Pco_2.[88,217] Compensatory mechanisms, including retention of base by the kidney and passage of base from blood into brain cells and CSF, lead to a progressive return of acidity toward normal in the blood and other body fluids. In the most precise studies the reestablishment of normal arterial pH occurs over a period of about 5 days in parallel with compensatory elevations of plasma and CSF bicarbonate ion concentration.[88] An earlier study with less exact methods and a very low inspired Pco_2 had indicated that arterial pH compensation was still not complete over 23 days following the beginning of exposure to CO_2.[219] Even with the 30 mm Hg ambient Pco_2 (equivalent to about 4% inspired CO_2), the compensated level of pH in CSF was only about 0.01 lower than normal.[88] Most likely, it was fully compensated *within* the neurons (Chapter 70).

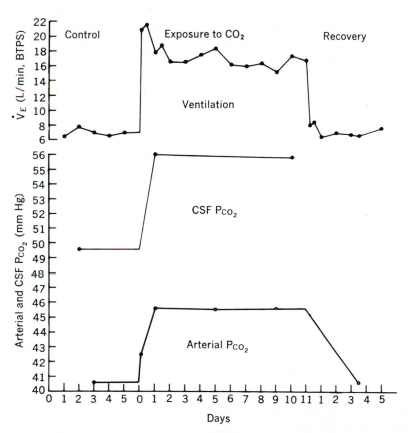

Fig. 75-9. Effect of 11-day exposure to ambient Pco_2 of 30 mm Hg on ventilation, Pa_{CO_2}, and CSF Pco_2 in human. Exposure period was preceded by 5-day control period and was followed by 5-day recovery period. Ambient O_2 concentration was maintained at about 19% to 21% throughout experiment. Ambient Pco_2 was less than 2 mm Hg during control and recovery periods. Data, which represent average results of four subjects, were obtained from Menn et al.[41] (From Clark et al.[88])

Respiration and respiratory reactivity

Respiratory rate, depth, and minute volume begin to increase immediately on exposure to increased inspired CO_2. Within 24 hr the total ventilation stabilizes at the elevated level and remains essentially unchanged until the exposure is terminated.[88]

Respiratory response to an elevation of alveolar or arterial P_{CO_2} is diminished by 8 to 10 L/min by the processes of acclimatization to an increased inspired CO_2 of about 4%.[88] This change is evidently due to bicarbonate elevation and consequent alteration of the influence of a particular level of P_{CO_2} on pH at or in the chemosensitive respiratory neurons. It is not, as has been stated, due to decreased respiratory reactivity of the neural mechanisms to CO_2, since the slope of the P_{CO_2} ventilation response curves is not changed by acclimatization of normal men to CO_2.[88] In disease states with CO_2 retention, such as in emphysema, a decreased reactivity to CO_2 appears to occur,[1] but even this may actually be related to increased work of breathing rather than to decreased respiratory drive. On termination of an exposure to elevated inspired P_{CO_2}, ventilation returns promptly to normal, stabilizing there within less than 1 hr.[88]

Blood electrolyte changes

Changes in calcium storage and mobilization have been described for prolonged exposure (42 days) to 1.5% inspired CO_2.[218] Although these changes were not physiologically significant in degree, they represent a pattern of expected physicochemical change that could become important in prolonged exposures to higher pressures of CO_2. The effects on calcium movement are probably related to the influences of altered acidity on the degree of calcium ionization and reflect a slight increase in deposition of calcium and CO_2 as calcium carbonate in bone, with a new equilibrium established between the calcium in blood and bone.[49,218] On ending the exposure to CO_2 the excess calcium carbonate is released from bone during the postexposure period, the CO_2 being exhaled and the excess calcium excreted in the urine.[217] None of these effects appears sufficient in degree to generate pathologic changes of either a transient or a residual nature. Rarification of bone should not occur in exposures to tolerable increased P_{CO_2}.

Other situations in which toxic concentration of carbon dioxide may be encountered

These can be deduced from two attributes of the gas, that is, it is produced wherever organic matter is oxidized and it is heavier than air. Thus it may be encountered by miners in poorly ventilated passages (''blackdamp''), at the bottoms of wells, in the cellars of buildings destroyed by fire or explosion, etc. The use of solid CO_2 (dry ice) and liquid CO_2 as refrigerants has introduced a new set of risks that are encountered in industry and laboratory. Industries employing large-scale fermentation (brewery) must prevent dangerous accumulations of CO_2.

CARBON MONOXIDE

Carbon monoxide is an insidious poison. It is produced in human metabolism, but at a rate (about 0.42 ml/hr) too low to induce toxic effects.[89] The toxicity of exogenous CO is due to its ability to compete successfully with O_2 for combination with the Fe^{2+} of Hb in the pulmonary capillaries. The immediate result of the competition is an increase in the percentage of carboxyhemoglobin and a corresponding decrease in the percentage of oxyhemoglobin in the arterial blood. Arterial O_2 content is therefore lowered. Since less O_2 is delivered to the tissues by each volume of circulating blood, the tension of O_2 in the metabolizing tissues is reduced, and a severe tissue hypoxia can occur *even though arterial P_{O_2} is not appreciably less than normal.* CO therefore does not produce its lethal effects primarily by a direct action on the tissues. Carbon monoxide can depress tissue oxidation by a direct action on cytochrome enzymes, but the P_{CO} required is about 1,000 times greater than that necessary to produce death in hypoxia by occupation of Hb.

Poisoning by CO is a serious and common condition.[123] The gas is ordinarily formed by incomplete combustion of carbon; it may be evolved from a coal fire that has insufficient draft or from a charcoal brazier. As a major ''feedstock'' of the chemical industry, millions of pounds are produced annually, with the attendant need for protection of workers and safeguards in transport. Considerable amounts of it may be present in mines as a result of explosions or of spontaneous oxidation of coal. The sources of greatest importance to the average physician are the exhaust gases of automobiles and airplanes. The exhaust gas of a gasoline motor contains about 6% CO when the motor is fairly well adjusted, and if the mixture is excessively rich in gasoline, the proportion of CO will be higher than this. The gas obtained as a by-product of the manufacture of coke contains about 10% CO, and natural gas contains almost none. Smoking may provide enough CO to occupy as much as 4% to 7% of the Hb.

The symptoms of monoxide poisoning are the usual symptoms of hypoxia (Chapter 73). One of the outstanding symptoms is headache—a severe, throbbing, frontal pain, worse in the recumbent than in the upright position, accompanied by nausea and associated with decreased mental capacity. This symptom may be due to an increase in CSF pressure consequent to increased permeability of cerebral capillaries, to cerebral vasodilatation, and to cerebral edema resulting from hypoxia. Fainting and collapse occur, and if there is cardiac disease, the heart is likely to be seriously affected. Respiration is not stimulated as much as in most other types of hypoxia because this is an isotonic hypoxemia (Chapter 73) and the receptors of the carotid bodies and aorta are not stimulated even though the quantity of O_2 available to the tissues is greatly reduced.

Theoretically, the human could survive complete occupation of his Hb by CO, if he simultaneously breathed O_2 at very high pressure. Practically, he would then succumb to O_2 poisoning instead of CO-induced hypoxia. Metabolic effects of CO would also express themselves in this situation.

In severe poisoning with marked lowering of tissue Po_2, exaggeration of anaerobic metabolism results in greater formation of lactic and pyruvic acids, combining the effects of hypoxia with those of metabolic acidosis. Even in the terminal stages of poisoning, true cyanosis does not develop; instead the cherry-red pigment of carboxyhemoglobin colors the skin, nail beds, and mucosae bright red.

Acute toxicity

The severity of hypoxia at the cellular level is related to (1) the proportion of the total Hb that is combined with CO instead of O_2 and (2) the influence of carboxyhemoglobin formation on the dissociation characteristics of the Hb still available for O_2 transport. These aspects will be considered separately.

Uptake of carbon monoxide

Estimates of the affinity of CO for Hb range from 200 to 300 times that of O_2 with 210:1 being the ratio usually stated. This means that exposure of a blood sample to a gas mixture in which the concentration of CO is 1/210 that of O_2 should, at the time of equilibrium, result in the blood sample containing equal parts of HbCO and HbO$_2$. Since the normal 100 mm Hg Po_2 of alveolar air in a resting subject at sea level is equivalent to an O_2 percentage of 14.0, the alveolar CO required to convert 50% of the Hb to the carboxy form would be approximately 0.07%. This represents a Pco of about 0.5 mm Hg. Assuming 20 vol% as the O_2 (or CO) capacity of blood and 5 L as the blood volume of a normal subject, saturation of 50% of the total Hb in the subject would require the following:

$$5,000 \text{ ml} \times \frac{20 \text{ ml CO}}{100 \text{ ml blood}} \times 50\% = 500 \text{ ml CO gas}$$

Even assuming full absorption of CO from the volume of air inhaled into the alveoli (total ventilation minus dead space ventilation), the intake of 500 ml CO would necessitate a total alveolar ventilation of 710 L air containing 0.07% CO. At a resting alveolar ventilation of 3.55 L/min, it would take 200 min to inhale and absorb 500 ml CO. Thus in addition to being affected by the *inspired concentration* of CO, the degree of intoxication is determined by the *duration of exposure* and by the *alveolar* ventilation. During *exercise* the rate of uptake of CO from the ambient air would be increased in proportion to the elevation of alveolar ventilation. This, together with the increased O_2 requirement, makes exercise in the presence of CO doubly dangerous. Finally, the rate of CO uptake from air of a given percentage composition should be diminished when the ambient pressure is lowered (as in aviation) and increased when ambient pressure is raised (as in diving). In the latter situation the mass of gas inspired is proportional to the atmospheric pressure in atmospheres— at 4 atm, 4 times as much CO would be inspired per minute as at sea level.[164]

The related influences of concentration, duration of exposure, and degree of physical activity have been studied by Forbes et al.[112] They give the relationships among these factors for normal men as follows:

At rest (ventilation 6 L/min, pulse 70)

$$\%\text{COHb} = 3 \times \%\text{CO inspired} \times t \text{ (min)}$$

Light activity (ventilation 9.5 L/min, pulse 80)

$$\%\text{COHb} = 5 \times \%\text{CO inspired} \times t \text{ (min)}$$

Light work (ventilation 18 L/min, pulse 110)

$$\%\text{COHb} = 8 \times \%\text{CO inspired} \times t \text{ (min)}$$

Heavy work (ventilation 30 L/min, pulse 135)

$$\%\text{COHb} = 11 \times \%\text{CO inspired} \times t \text{ (min)}$$

It is evident that these relationships will not continue to apply when CO uptake is indefinitely prolonged. However, they are useful in describing the early stages of equilibration when the rate of uptake is relatively constant. The basic formula is most applicable to inspired CO concentrations above 0.02%. For 0.01% inspired CO, it holds only to about 7% COHb; at 0.1% inspired CO, it will apply up to formation of 30% COHb and to still higher values of carboxyhemoglobin saturation as the inspired CO concentration is increased.[112] An additional mathematical analysis of CO accumulation and elimination includes consideration of artificial ventilation with O_2 at increased ambient pressure (e.g., 2 to 3 atm).[57]

Formation of carboxyhemoglobin

The severe hazard of CO poisoning is due not only to displacement of O_2 from some Hb molecules, but also to an alteration of O_2 release characteristics in the tissues. When CO combines with Hb, it presumably reacts first with the Hb form that normally enters into the most labile combination with O_2. As a result, the Hb remaining for combination with O_2 comprises more stable Hb forms that release O_2 only at a considerably lowered Po_2. The effect of the presence of carboxyhemoglobin on the dissociation curve for oxyhemoglobin is seen in Fig. 75-10. The illustration shows that in the presence of carboxyhemoglobin the O_2 dissociation curve tends to become less S shaped.[214] The more exponential characteristic indicates that O_2 is more firmly bound to Hb; this means that in the tissues, O_2 is given up at excessively low O_2 tensions. *The relative increase in affinity of Hb for O_2 is the second factor in the hypoxia of CO poisoning.* Its importance as a contributing cause of tissue hypoxia is indicated in Fig. 75-9 by the contrast between the effects of CO poisoning and anemia. In each case the arterial Po_2 is nearly normal. When 60% of the Hb is combined with monoxide, the patient is unconscious and will probably die if the exposure continues. Anemic patients whose Hb content is 40% are in no immediate danger of death from hypoxia. One difference is in the tension of O_2 at which O_2 is unloaded in the tissue capillary. Fig. 75-10 shows that in anemia a given loss of O_2 from the blood occurs at a higher Po_2, which provides a large head of pressure for diffusion to the cells.

The effects and symptoms produced by CO poisoning depend on both the concentration of COHb in the blood and the duration of the resulting hypoxia. Table 75-4 is applicable only in exposures to inhalation of low concentrations of CO. It must be realized that exposure to high inspired CO concentrations will cause such a rapid progression through the sequence that no distress may be noted by the dying victim.

Recovery from carbon monoxide poisoning: treatment with high oxygen pressures

The combination of CO with Hb is freely reversible when the Pco becomes less in the alveolar air than it is in the mixed venous blood. The most rapid elimination of CO is accomplished by (1) reducing inspired CO to zero by removing the victim from contact with CO to ensure a maximal partial pressure gradient for elimination of CO, (2) increasing alveolar ventilation by any appropriate means *not* including CO_2 administration, and (3) employing high inspired O_2 tensions to facilitate the dissociation of CO from Hb. These influences have been studied in animals and humans.[106,178,197]

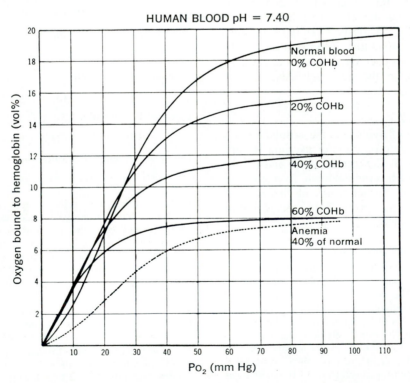

Fig. 75-10. Calculated O_2 dissociation curves of human blood containing various amounts of carboxyhemoglobin, plotting absolute amounts of bound O_2 (O_2 content) rather than percentage of available hemoglobin saturated with O_2. (From Roughton and Darling.[214])

Influence of increased inspired PO₂ and PCO₂

Breathing O₂ at high pressure can drive enough O₂ into physical solution in the arterial blood to sustain life and even consciousness in animals with Hb essentially 100% occupied by CO.[130,178] Although existence based entirely on physically dissolved O₂ requires such high ambient pressures that O₂ toxicity prevents all but brief exposures,[29] increased O₂ pressures are the most rational means for accelerating elimination of CO.[14,106,178,197]

Fig. 75-11 shows curves of CO elimination constructed from findings of Pace et al.[197] in hu-

Table 75-4. Relationships between symptoms of CO poisoning and %COHb in the blood*

Signs and symptoms	%Blood saturation with COHb	
	Short exposure	Long exposure
No prominent symptoms	0-10	0-10
Possibly slight headache, dilatation of cutaneous blood vessels	10-20	
Headache, throbbing in temples	20-30	
Severe headache, weakness, dizziness, dimness of vision, nausea and vomiting, collapse	30-40	20-30
Severe headache, weakness, dizziness, dimness of vision, nausea and vomiting, collapse but with more possibility of collapse and syncope; increased respiration and pulse	40-50	
Syncope, increased respiration and pulse, coma with intermittent convulsions, Cheyne-Stokes respiration	50-60	30-40
Coma with intermittent convulsions, depressed heart action and respiration, possible death	60-70	
Weak pulse and slowed respiration; respiratory failure and death	70-80	40-50

*Symptoms and estimates for short exposure are from Sayers and Davenport.[216] Additional estimates for long exposure are included to emphasize that at increased duration a lesser degree of hypoxia should produce severe effects (see text).

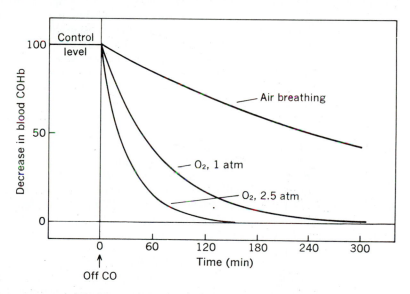

Fig. 75-11. Effect of increased inspired PO₂ on CO elimination in men. Curves represent exponential rates of CO elimination from blood at 0.2, 1.0, and 2.5 atm of inspired PO₂ and illustrate the considerable effect of increased PO₂ in speeding dissociation of carboxyhemoglobin. At 1 atm, O₂ breathing decreased half-time for CO elimination from about 250 min to about 50 min. At 2.5 atm, half-time was further reduced to 20 min. Advantage of breathing O₂ is probably only slightly related to increase in alveolar ventilation, but rather depends on greater driving force for diffusion of O₂ into tissues when arterial PO₂ is high. Increased CO₂ tension further accelerates elimination of CO.[106,136] (Based on data from Pace et al.[197])

man subjects breathing CO prior to administration of various CO-free gases. With spontaneous respiration the rate of fall in the CO content of the blood is essentially exponential, the time constant depending on alveolar ventilation and the affinity of Hb for CO. In Fig. 75-11 the half-time for CO elimination during air breathing at rest is about 250 min.[213] Administration of pure O_2 at sea level not only improves tissue oxygenation but also decreases the duration of serious CO poisoning by a factor of about 5 by competing with CO for combination with Hb. Using O_2 at the therapeutically tolerable pressure of 2.5 atm further improves oxygenation and rate of recovery, with the half-time for CO elimination being further reduced to about 22 min.[197] This principle of hyperbaric oxygenation in CO poisoning is now receiving extensive clinical application.[14,101,154,223] It is important to realize that in treating the existing tissue hypoxia of an acute exposure to CO poisoning (1) preexisting damage may not be reversible, (2) a serious and damaging consequence of CO poisoning is brain edema, which contributes to residual CNS hypoxia even after CO elimination has occurred, and (3) no narcosis or hypothermia exists to reduce O_2 demand and hence protect the brain and other tissues from damage by diminished O_2 supply. Residual damage in untreated CO poisoning is therefore common.

The influence of CO_2 on CO elimination is also prominent but no longer considered an aid to therapy.[106] The addition of 5% to 7% CO_2 to the O_2 breathed at sea level reduces the half-time for CO elimination almost to the theoretical minimum of 12 min.[213] The effect of CO_2 is related both to its stimulation of breathing and to the influence of acidemia in facilitating the release of CO from Hb. Despite the great effectiveness of CO_2 administration in increasing the rate of CO elimination in normal humans and animals,[137,213] it has been pointed out that in CO poisoning administration of CO_2 will exaggerate the acidosis already present as a result of tissue anaerobiosis and excessive fixed acid formation.[220] Moreover, when O_2 is employed at high ambient pressures, CO_2 would accelerate the onset of O_2 convulsions.[4,29]

Residual effects

It is erroneous to assume that unless a victim of CO poisoning dies in the acute stage of hypoxia he has suffered no harm. As in other forms of tissue hypoxia, the extent of damage to cells is proportional to the degree and duration of cellular hypoxia.

Chronic or frequently repeated exposure to low concentrations of CO may result in cumulative effects of repeated hypoxic injury, even though CO itself does not accumulate progressively from day to day.[64]

Patients who survive *acute* intoxication by CO may show evidence of transient or permanent damage to the heart, the CNS, or other organs having a high demand for O_2. In some instances, these effects become evident only after an interval of several days following exposure, and they may ultimately cause death. When death does not result, the patient may be chronically invalided by such disorders as psychosis or severe loss of mental capacity, various types of neurologic damage (including bilateral necrosis of the globus pallidus), and myocardial degeneration with congestive failure.[64] After mild intoxication, recovery may be essentially complete. It is between the severe and the mild forms of CO poisoning that real difficulty in assessing residual effects exists; the methods for evaluating small decrements of organ function and general performance are insensitive, and it cannot be expected that control studies will have been carried out prior to intoxication.

INERT RESPIRATORY GASES

The term ''inert'' is applied to gases that do not appear to enter into fixed chemical combination with other substances in cells. Included in this group are hydrogen, helium, neon, nitrogen, argon, krypton, xenon, and the radioactive gas radon, as well as a variety of general anesthetic agents and industrial gases.

Certainly its designation as an inert gas does not mean that nitrogen is unimportant, for at least three types of effects related to nitrogen have physiologic or pathologic significance. These effects are *nitrogen narcosis, respiratory resistance,* and *production of gas bubble diseases in bends, air embolism,* and *isobaric counterdiffusion.* In addition, the prolonged *absence of nitrogen* in the respired gas may be physiologically important.

Inert gas narcosis

The narcotic effects of inert gases such as nitrogen on cellular function[45] are prominent even though the gases do not enter into chemical metabolic processes. The effects are generally similar in end result to those of narcotic drugs such as morphine, barbiturates, and ethanol. However, the mechanisms are probably not similar, and it is the loss of CNS function that provides the resemblance.

The inert gases differ greatly in narcotic potency, and the partial pressure at which the nitrogen in air produces unconsciousness is extremely high as compared with its oxidation product nitrous oxide. Although administration of an 80% mixture of N_2O with O_2 for dental analgesia will produce unconsciousness, Fig. 75-12 indicates that N_2 is only about $\frac{1}{32}$ as narcotic as N_2O, and it is therefore predictable that to cause the same degree of depression by the nearly 80% nitrogen in air an individual would have to be exposed to a pressure of about 26 atm, equivalent to air breathing at a depth of 900 feet of seawater.[37] Nitrogen and the other narcotic inert gases probably do not have a true threshold for effect but bring about their depressant phenomena by an action that progresses smoothly as the partial pressure of the inert gas in the tissue increases.

Nitrogen and other inert gases at high pressure do not produce merely a euphoric sensation in human beings; they presumably can affect any type of cellular function if the pressure is raised high enough. They can produce actual unconsciousness,[45] prevent the convulsions of electroshock in mice (Table 75-5),[78] impair mentation and motor performance,[69] affect visual reaction time,[140] block conduction in nerve fibers,[78] block synaptic transmission in nerve pathways,[78] slow the metamorphosis of insects,[113] diminish the rate of maturation of plant seedlings,[45] and interfere with the O_2-dependent radiosensitivity of plant, microbial, and tumor cells.[104] These and other

Table 75-5. Characteristics of inert gases at isonarcotic partial pressures*

Gas	Molecular weight	Pressure for equivalent CNS depression (atm)	Concentration in olive oil at isonarcotic pressure (mM/L)
Radon	222.0	—	—
Xenon	131.3	0.51	0.038
Krypton	83.8	1.8	0.034
Argon	39.9	12.6	0.077
Nitrogen	28.0	18.0	0.052
Neon	20.2	—	—
Helium	4.0	163.0	0.107
Hydrogen	2.0	?	?

*Based on data from Carpenter.[79]

NOTE: Equivalent narcotic pressures were determined by effects of the gases on threshold to electroshock seizures in mice. Note that narcotic potency is directly proportional to molecular weight and that at isonarcotic pressures the concentrations of the several gases in olive oil are essentially the same. Neon and radon were not studied.

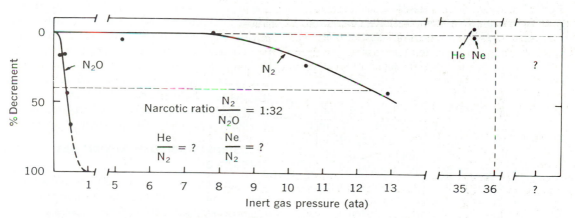

Fig. 75-12. Comparative effects of N_2O, N_2, Ne, and He on mental function (paced arithmetic). Mean values in two subjects. Since previous study had demonstrated that N_2O at inspired pressures to 0.5 ata induces prominent decrement in mental arithmetic skill,[96a] this effect was used as basis for comparison of N_2, Ne, and He. Subjects were remarkably competent during N_2 breathing to nearly 8 ata (about 230 fswg) and showed an $N_2:N_2O$ narcotic ratio of 1:32 while retaining considerable capacity for mental function even to an inspired N_2 pressure of 13 ata (equivalent to 500-foot depth breathing air). Subjects were fully competent, without evident narcosis, during Ne and during He breathing to 37 ata (1,200 fswg). This indicates that any appreciable "narcotic" depression by Ne or He would not now be expected until depths probably more than twice 1,200 feet are reached. (From Lambertsen et al.[37])

influences of inert gases[45] indicate the diverse effects of inert gases that are coming to have considerable importance as our interest in extremes of gaseous environment increases. The effect has extreme practical importance in diving with air or N_2-O_2 mixtures,[232] limiting safe and effective function to depths not greater than about 180 feet of seawater.

It is possible that most, if not all, of the effects of nitrogen on cellular function, both in the CNS and at other sites, are produced by a single biophysical mechanism of action. The mechanism of inert gas narcosis and the important related question concerning the mechanism of action by clinical anesthetics have been the subject of much experiment and thought over many years. The major theories concerning narcosis have been critically reviewed on several occasions,* and in these discussions it has been agreed that many of the biologic effects of nitrogen and the other inert gases can be produced by physical actions on cells, membranes, and solutions. Ionic or covalent chemical bonding with cellular metabolic constituents is not required to cause depression of metabolism or other functional activity. The designation "inert" gases is a reflection of this. Apparently the physical presence of molecules, perhaps concentrated at particular sites on the surface of structures within the cell, can lead to the interferences collectively called narcosis.

A number of different but related theories have been proposed to explain narcosis as the result of physical actions of inert gases. Each theory is based on the ability to demonstrate a correlation between anesthetic or narcotic activity and a physical characteristic of the inert agent.

The Meyer-Overton theory, the best known of the correlations concerned with the *lipid phase* of microcomponents of tissue,[186] considers the narcotic action to be related to the lipoid solubility and oil-water partition coefficients of the inert substances and hence to a concentrating of molecules in undefined cellular lipid elements. However, anesthetic activity is also equally well correlated with such different characteristics as partial (molal-free) energy of the gas molecules,[73,109] lowering of surface tension of water,[51] molecular weight,[45] molecular refraction,[244] boiling point,[244] polarizability of molecules,[45] and Van der Waals' forces.[244]

A more recent concept, which relates narcosis to the *water phase* of cells, but one still necessarily correlated with lipid solubility is the proposal that narcosis in general and by inert gases

in particular is the result of a formation of gas hydrate "microcrystals" or the development of ordered water molecules around each inert gas molecule, which then stabilize charged side chains of chemical components of nerve cells.[188,200]

In the face of the many proposals for a mechanism of narcosis, it is important to recognize that most of the properties found to correlate well with anesthetic action are not in fact independent of each other,[109] and that high correlation does not imply mechanism. Quite possibly the common factor will prove to be the distribution of gas molecules at water-lipid interfaces such as the surfaces of mitochondria and in cell membranes, with resultant interferences in ionic movement.

Alteration of metabolic processes by inert gases

Carpenter[79] has shown that other indifferent narcotics, including nitrous oxide gas at high partial pressure, depress pyruvate oxidation by enzymatic reactions in the citric acid cycle. Enzymes responsible for a number of stages in the oxidation of carbohydrate fragments, including pyruvate, are present largely in mitochondria, and the mitochondrial fraction also represents the greatest concentration of phospholipid in the CNS. It is still not certain whether these additional coincidences are factors in the narcotic action of nitrogen.[79]

Whether or not the mechanism of inert gas narcosis will prove to be one of these many possibilities, it is a real phenomenon and one that has practical importance at high air pressures. The prominent effects are evidently on higher centers and synapses, since in classic studies with frog nerve-muscle preparations[125] neither nitrogen nor argon had adverse effects on nerve threshold, neuromuscular junction, or muscle.

Relation of narcosis to inert gas breathed

Comparisons of narcotic effects have been made in animals and humans. *Xenon* produces a marked depressant effect with a decrease in the ability of the muscle to generate tension, but it requires about 100 atm of xenon to demonstrate a slight depressant effect on nerve excitability.[125] In the intact animal, inert gas narcosis is more readily demonstrable.

The *nitrogen* in compressed air produced detectable subjective effects even at pressures of about 3 atm.[80] These effects are not great at this low pressure, and it is possible at 3 to 4 atm to

---

*See references 73, 78, 188, 237, and 244.

perform intricate, timed, experimental procedures such as measurement of brain blood flow[167] and to perform surgical procedures.[5,50] In controlled studies, nitrogen at pressures up to 3.9 atm has little effect on psychomotor performance.[115] Apparently objective evidence of a performance decrement can be obtained at air pressures of 4 to 5 atm,[115,140,151] but the decrement in *performance* appears to be extremely small both at 5 atm of air pressure[140] and in other narcotic states where *subjective* effects are prominent.[114]

Apparently high O_2 *pressures* somehow add to the overall narcotic influence of nitrogen,[115] but whether through a central elevation of P_{CO_2}[115] or by additional contribution of an "inert gas narcosis" effect of O_2 itself[160] is not known. It is now well established that a rise in alveolar P_{CO_2}, as when increased airway resistance at high ambient pressures interferes with alveolar ventilation, leads to an increase in the degree of narcosis associated with air breathing.[30,176] Practical trials over the years has shown that competent, purposeful individuals can work accurately and effectively at air pressures up to 4 to 5 atm. This should not be taken to mean that any individual will perform any procedure without difficulty under a pressure of 0.80×5 atm of nitrogen. Judgment, comprehension, awareness, and skill will hardly be improved by exposure to this environment.[114] As pressure is increased, as P_{CO_2} rises,[75,140,176] or as a gas having an increased level of O_2 pressure as well as increased nitrogen pressure is breathed,[140] the degree of narcosis and the associated detrimental effects will increase.

Helium is used extensively in deep undersea work because of its essentially undetectable narcotic effect and its relatively low density (airway resistance) as compared with argon, nitrogen, and neon at the pressures involved (Fig. 75-12). It has required 163 atm of *helium* to produce suppression of electroshock seizures in mice,[79] and since *hydrogen* is more soluble in lipid than is helium, it is not at all certain that higher pressures of hydrogen would be needed. Neither of these gases has yet produced detectable narcosis in humans. In saturation exposures of mice to 122 atm of gas containing 0.3% O_2 and 99.7% helium the animals remained active and conscious.[181] However, at 135 to 149 atm, exposure to helium caused death, possibly due to the effect of hydrostatic forces rather than to a narcotic action of helium itself.[173,187]

Men have remained conscious and competent during exposure to He-O_2 breathing at pressures as great as 2,000 feet of seawater,* and in a 6-day saturation exposure of men to a He-O_2 environment at approximately 37 atm (equivalent to 1,200 feet of seawater), no prominent subjective or objective mental, psychomotor, or physical impairment was produced by the helium.[37] In the same study, no gross narcotic effect was found during periods of neon breathing.

The heavier inert gases do produce narcosis, the depressant activity being proportional to the molecular weight of the gas. Thus *argon* produces about twice as great a central depression as nitrogen at the same partial pressure.[45] *Xenon* is capable of producing unconsciousness and light surgical anesthesia even at 0.8 atm.[45,91] For these reasons the inert gases that might be useful in extending diving are essentially limited to hydrogen, helium, neon, nitrogen, and argon. In some instances the usefulness is as a respiratory diluent for O_2, in others as a special component of the surrounding atmosphere.

Gas lesion diseases

Serious "gas lesion" diseases of several types occur in occupations involving exposure to high ambient pressures and in clinical circumstances. Recognized conditions, each of which can be lethal, include *decompression sickness, isobaric inert gas counterdiffusion, arterial and venous gas embolism*, and *pneumothorax*.

Decompression sickness (bends, caisson disease)

Decompression sickness (often called bends because of a tendency to produce pain in or about the joints) is a condition involving formation of bubbles in the blood and body tissues by escape of excess gases from physical solution. It is a problem in diving, aviation, space operations, caisson work, and the use of pressure chambers in medicine. For decompression sickness to occur, a situation must develop in which the partial pressure of inert gas in some location in the tissues is in excess of the ambient environmental pressure, thereby allowing evolution of gas and the formation of the free gas phase. This gas phase may, in severe situations, be thousands of bubbles carried in venous blood, with some filtering by the lungs.

The ways in which the situation can develop are many and are not limited to diving. The nitrogen in the atmospheric air at sea level (about 573 mm Hg) is in partial pressure equilibrium with the gaseous nitrogen dissolved in the tissues. When a mountain climber makes an ascent, the

*See references 35, 37, 76, 116, 117, 212, and 225.

lowered ambient and alveolar partial pressure of nitrogen allows the gradual escape of nitrogen from body fluids until a new equilibrium is established at the lower inspired P_{N_2}. This loss of nitrogen by way of the lungs produces no symptoms or harmful effects. Conversely, when a diver breathes air at many atmospheres pressure, the high P_{N_2} will cause additional nitrogen to dissolve in his body fluids. On programmed, slow return to the surface the excess nitrogen will diffuse from the tissues into the blood and be eliminated by the lungs. Except for the previously mentioned phenomenon of nitrogen narcosis, the uptake and subsequent slow elimination of excess nitrogen by the diver produce no detectable effects. When a diver ascends too rapidly for the excess nitrogen in the tissues to escape by diffusion from the cells into the tissue capillaries, transport in the blood to the lung capillaries, and elimination from the blood into the alveoli, the fall in ambient pressure results in the local escape of nitrogen from solution and the formation of bubbles. This is *decompression sickness,* a painful, potentially permanently crippling, and sometimes lethal phenomenon. It can be produced by any respirable inert gas usable at high pressures in diving.

Decompression sickness can also occur when an aviator, equilibrated for a lifetime, years, or days to the nitrogen pressure at sea level, ascends to the lowered ambient pressure of a high altitude too rapidly for gradual escape of nitrogen to occur. An aviator ascending from sea level to an altitude of about 33,000 feet is being subjected to exactly the same degree of decompression as a diver coming to the surface after working at a depth of 100 feet in seawater; in the former case, decompression occurs from the pressure of 1 atm to one fourth that value, whereas in the latter it occurs from a total pressure of 4 atm to 1.

The caissonier and river tunnel worker are exposed to the increased air pressure used to counterbalance hydrostatic pressure and prevent water or mud from entering the excavation or caisson.

Inert gas uptake. The time required to reach equilibrium is the same regardless of the pressure of the exposure. During air breathing at sea level the inspired P_{N_2} is about 0.8 atm. As diving depth increases, the partial pressure of each gas inspired increases in direct proportion to the increased ambient pressure. Every 33 feet of seawater (about 34 feet of fresh water) adds another atmosphere of pressure. Total or "absolute" pressure is therefore equal to the initial atmosphere of air pressure prior to descent, plus the number of additional atmospheres added as water pressure. This can be expressed as follows:

$$\text{Absolute pressure (atm)} = 1 + \frac{\text{Depth (ft)}}{33}$$

If air is breathed at 99 feet (4 atm), the partial pressure of inert gas (nitrogen) inspired is therefore about 0.8×4 or 3.2 atm. With initiation of air breathing at 4 atm, there will be a P_{N_2} gradient of 3.2 atm $-$ 0.8 atm or about 2.4 atm, resulting in nitrogen uptake in the lungs, its distribution throughout the body, and its diffusion into all body fluids.

The rates of *uptake* and equilibration of inert gas in various tissue fluids on exposure of the individual to a high ambient pressure are extremely different. The blood in the pulmonary capillaries equilibrates with nitrogen or other inert gas almost immediately, but because the solubility of nitrogen in blood is low, the amount carried to the tissues per minute by the circulating blood is small. Tissues having a large blood flow are saturated more rapidly than poorly perfused tissues. In addition, the rate of rise in inert gas pressure in a tissue depends on the solubility of the gas in the tissue components. The solubility of nitrogen in fat is about 5 times greater than its solubility in water.[67,79] Therefore the capacity of fatty structures for nitrogen is large, and the volume (and time) required for their saturation is great. Since fatty tissues are also rather poorly vascularized, near-equilibrium with respect to P_{N_2} may require many hours longer than to saturate an aqueous fluid. For example, an organ such as the kidney, with a high blood flow and little fat, should equilibrate in less than 1 hr. Thus the rates of gas uptake (and elimination) vary greatly from one tissue or tissue component to another.

The desaturation of various discrete tissues follows essentially the same time course as for saturation. This means (1) that after a short exposure to increased air pressure, rapid ascent will not produce bends in the "slow tissues," since these have not had time to acquire much additional nitrogen and (2) when exposure has occurred nearly to the point of P_{N_2} equilibrium with all tissues, extremely long periods of decompression may be required to allow nitrogen to escape from the poorly perfused tissues.

The rates of inert gas uptake and elimination from the whole body are described by a complex curve comprising the summation of an infinite number of exponential rates of elimination. For particular sites in the tissues the rates of uptake and elimination differ for each inert gas. For convenience and to aid in computing safe decompression schedules, it is now customary to designate arbitrary "fast" or "slow" tissues by assigning values to the exponential curves describing clearance of gas from hypothetical tissues. The arbitrariness is in large part due to the inability to perform appropriate measurements of inert gas exchange at the many thousands of sites involved. Such curves, as shown in Fig. 75-13, do not describe inert gas elimination from particular organs or gross masses of tissue, but are mathematical counterparts of the patterns of inert gas exchange to be expected in a variety of diffusely scattered, sometimes microscopic, sites.

Studies in humans of actual rates of elimination of inert gases such as nitrogen, helium, and krypton indicate that despite differences in diffusion characteristics the rates of elimination of the inert gases from the most rapidly perfused tissues are indistinguishable.[147] This indicates that except in the least well-perfused tissues, the development of bends is related to limitations of local perfusion of blood through tissue capillaries, not to diffusion of gas through the tissue fluids.[147,190,224] However, the patterns of inert gas exchange in the poorly perfused tissues have not been studied, and on grounds of logic alone, it is evident that limitations of diffusion must also contribute to the development of bends.[162] These aspects of inert gas exchange have been of great theoretical and practical importance in the evolution of undersea activity.*

Symptoms of decompression sickness. The

---

*See references 31, 33, 35, 44, and 232.

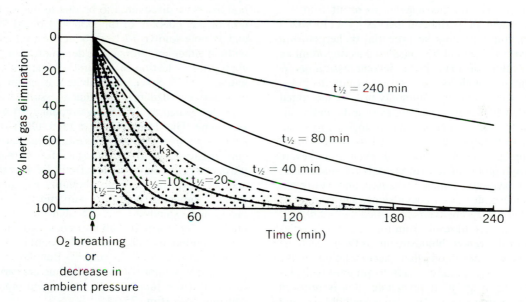

Fig. 75-13. Elimination of inert gas following exposure to increased ambient pressures.[162] When O_2 is breathed, or when individual returns toward surface after period of breathing an inert gas–O_2 mixture at increased ambient pressure, inert gas dissolved in cells and body fluids is eliminated. Rate of inert gas elimination from various gross and microscopic units of body tissue varies, depending largely on rate of perfusion of tissue with blood. Thus, organelle such as carotid body, with extremely high blood flow per unit weight of tissue, should be cleared of inert gas extremely rapidly. In contrast, scarified lesion with low perfusion rate should have slow rate of inert gas elimination, even if located in organ with high average blood flow. Figure shows several exponential curves that represent elimination of inert gas from series of hypothetical "tissues" ranging from half-time of 5 min (which would resemble cranial CSF) to slowest rate (240 min half-time) that has been detected by studies of bends induction in humans. Shaded portion, bounded by curve labeled k_3,[147] defines area where direct determinations have been made. Unshaded region of diagram is more pertinent to bends in deep or prolonged diving. It shows inverse images of uptake curves of the same "tissues." Use of different rates of inert gas exchange to compute decompression schedules for air, He-O_2, or other gas mixtures is based on procedure originated by Haldane,[72] which recently has been extensively facilitated by computer methods. "Fast" tissues are limiting following short dives, and tissues with slow inert gas exchange limit rate of decompression after dives of long duration. Following saturation dive (e.g., for durations of 24 hr or more), rate of ascent is limited by elimination of gas from body tissue with slowest perfusion rate.

signs and symptoms resulting from bubble formation depend on the locations and size of the bubbles. The variety of symptom patterns is therefore large, ranging from mild and fleeting localized joint pain or itching of the skin to blindness, deafness, disrupted vestibular function, paralysis, coma, and shock.[18,19,150,232] The symptoms may occur during or even several hours after decompression.

Joint pains, usually in the shoulder, elbow, ankle, or knee, are not due to gas in the joint cavity but in periarticular tissues or vessels supplying them.[82] Pain, which can become excruciating, is probably due primarily to interference with local blood supply. Gas can be demonstrated roentgenologically in the periarticular tissues,[56,82] in tendon sheaths, and in other extravascular locations.[56]

Cutaneous decompression sickness in the form of rash and pruritus appears to result from involvement of peripheral nerves. The localized skin eruptions may be urticarial or herpetiform in type and should be considered as important an indication of inadequate decompression as is arthralgia, with its more severe symptoms.

CNS *bends* most commonly involve the spinal cord, with severity ranging from mild and localized muscle weakness and paresthesias, to widespread paralysis and sensory loss, to complete paraplegia. Less common but even more severe are symptoms of intracranial CNS disturbances. Such symptoms include nausea and vomiting, dizziness, deafness, convulsions, and coma.

Pulmonary involvement, called the "chokes," is considered to result from formation of bubbles *within* the venous bloodstream; as these reach the lungs via the blood, their accumulation in the pulmonary capillaries leads to progressively increasing dyspnea. If reversible, this is relieved by recompression to decrease the bubble size and by hyperbaric O_2 therapy. However, since the formation of large gas bubbles in the marrow has forced marrow into the circulation, pulmonary fat embolism is another possible cause of severe respiratory disturbance in bends.

Massive collapse may result from gross inadequacy of decompression as occurs in accidental "blow-up," without programmed decompression. In severe cases the rate of progression from first twinge to coma may be only minutes, and at such times, decompression sickness may involve both intravascular and extravascular evolution of gas as bubbles. When massive numbers of bubbles do appear in the circulating blood, they can obstruct small vessels within vital tissues or even lead to frothing in the ventricles. Untreated

occlusion of vessels in the CNS can lead to permanent paralysis, sensory defects, or respiratory failure and death. If treatment is prompt and adequate, necessarily including both compression and hyperbaric O_2, full function can often be restored even to victims of severe bends.

Isobaric inert gas counterdiffusion

A newly recognized and potentially lethal gas lesion disease results from gas bubble formation in tissue and blood *without decompression*.[166] Isobaric failure of normal inert gas exchange occurs in two forms: superficial and deep tissue.

The *superficial* counterdiffusion form occurs at stable increased ambient pressures as itching and development of cutaneous gas lesions when one inert gas–O_2 mixture is breathed (e.g., N_2-O_2, or air) while the individual is surrounded by another (e.g., helium)[70,126,166] This phenomenon has been discovered to be due to the formation of gas bubbles in the skin and blood[126,166] and is now known to be producible at 1 atm, without either compression or decompression, by appropriate selection of respiratory and ambient gases.[145,165,166] Since the bubble formation occurs without change in ambient pressure, it has been designated the isobaric gas counterdiffusion syndrome. Gas lesions appear to result from a stable state of localized supersaturation that occurs because one diffusible gas, for example, nitrogen, is lost more slowly from the skin than another gas, for example, helium, is gained. Bubble formation begins in superficial dermal layers and extends to the subcutaneous tissues; bubbles then form in subcutaneous capillaries, and a continuous lethal gas embolization of the right heart and lungs occurs.[165] Bubbles break through the pulmonary capillaries and are carried to the vascular supply of all tissues and organs, causing death (Fig. 75-14).

A peculiar feature of this syndrome is the development of severe and incapacitating vestibular dysfunction during inert gas counterdiffusion, as though the counterdiffusion were able to take place between the middle ear and inner ear through the round window, with disruptive bubble formation in the fluids of the inner ear.[166]

Deep tissue counterdiffusion inert gas supersaturation occurs in the isobaric state at any ambient pressure when a more rapidly exchanged gas is breathed following prolonged respiration of a more slowly exchanged gas (Fig. 75-15). In this case the ambient and respired gases are not different. Rather, it is the sequencing of respiratory gases, unrelated to ambient environment, that induces the *transient* supersaturation. In

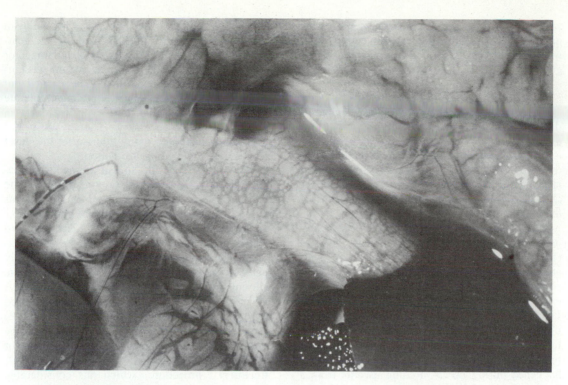

Fig. 75-14. Occurrence of massive amount of gas in vena cava of pig killed by N_2O/He isobaric counterdiffusion at 1 ata. (From Lambertsen et al.[36])

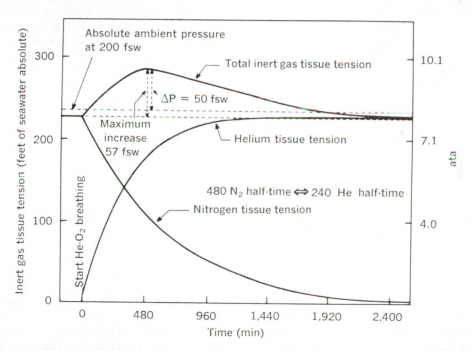

Fig. 75-15. Deep tissue form of isobaric inert gas counterdiffusion. Diagrammatic representation of time courses of changes in He, N_2, and sum of He + N_2 pressures in hypothetical tissue or cellular fluid. Tissue selected has half-time for N_2 elimination of 480 min, compared with 240 min for He uptake. At unchanged ambient pressure (isobaric state) equivalent to depth of 200 feet of seawater, abrupt change from N_2 to He breathing (with natural Po_2) leads to more rapid uptake of He than elimination of N_2. Result is transient inert gas supersaturation (sum of P_{He} + P_{N_2} exceeds ambient pressure), which may be several atmospheres in magnitude. (From Lambertsen and Idicula.[166])

such situations at increased ambient pressures, gas bubbles appear in the mixed venous blood when helium is substituted for respired nitrogen.[94]

Arterial and venous gas embolism

The term "air (or gas) embolism" denotes a distinct entity that is clearly due to abrupt entrance of air into the circulation. It is encountered in several situations in which the lungs are overexpanded, stretched, and torn by intrapulmonary pressures above the ambient level and in clinical accidents. Following are discussions of these situations.

Ascent of a diver from one depth to a shallower depth while holding his breath. Expansion of the air within the chest results in overstretching of lung tissue, tearing of lung tissue, entrance of air into pulmonary veins, and generalized air embolism of the arterial circulation. Unconsciousness is usually immediate due to blockage of the cerebral circulation or to interference with cardiac action by the presence of air in the left cardiac chambers. This accident has happened during breath holding ascents from a depth of only 15 feet. It cannot occur when the individual without breathing apparatus takes a breath at the surface, dives, and returns to the surface still holding his breath. It is not a necessary consequence of ascent if pulmonary air is continuously vented, since "free ascent" (ascent without diving apparatus) has been practiced now for many years.

Explosive decompression of an aircraft cockpit, spacecraft, or aerospace pressure suit. Sudden reduction of the extrathoracic pressure may cause overdistention, stretching, and laceration of the lung as the intrapulmonary gas expands. The results and the only effective treatment are the same as for gas embolism due to lung rupture in divers.

Clinical gas embolism. Air enters the venous circulation by way of traumatized veins in cardiac surgery or neurosurgery, in renal dialysis, or in insufflation of the oviduct. In such situations involving the right heart and lungs, tolerance to air embolization is greater than in arterial gas embolism. However, large amounts of gas interfere with cardiac pumping of blood. Clinical gas embolism of the arterial circulation also occurs and blocks coronary, brain, and other vital circulatory beds.

Treatment of gas lesion diseases

Each form of gas lesion disease, whether due to decompression, isobaric counterdiffusion, or accidental clinical or occupational gas embolism requires therapy with pressure and O_2.[232] Treatment is carried out to prevent residual damage or extension of the bends, and in accidental gas embolism treatment should ideally be carried out within the few minutes that the brain can tolerate ischemia but should be done even if a delay in pressurization is necessary. Therapy is useful even when instituted many hours after the onset of symptoms of bubble formation.

The only effective treatment of gas lesion diseases is compression and O_2 breathing. In the current approach to treatment of bends and gas embolism, these measures are combined (i.e., as hyperbaric O_2 therapy).[14,232] The standard methods are based on recompression of the patient to reduce bubble size; however, their highly physiologic bases also depend on the immediate administration of pure O_2 to (1) improve oxygenation of tissues whose circulation is impaired by intravascular or extravascular bubbles and (2) speed the *rate of decrease* of bubble size by establishing a maximum diffusion gradient for elimination of nitrogen.[124] This modern "O_2 treatment" procedure, which has considerable advantage over the earlier, 2- to 3-day therapy with compressed air, employs pressures to 3 atm and is limited by O_2 toxicity rather than by subsequent bends liability. It has usefulness in treatment of air embolism as well as bends, although higher pressure therapy to produce immediate reduction of bubble size must be held in reserve.[189] The effects of O_2 are prompt, and the duration of therapy required is in most situations limited to 1 or 2 hr instead of days. The air-breathing attendants enjoy the benefits of the reduced pressure and duration of compressed air exposure. When no recompression chamber is immediately available, pure O_2 administration at 1 atm should be carried out until recompression becomes possible.[232]

Diving methods in relation to development of gas lesion (bubble) diseases

Three basic types of diving are practiced, including (1) pure O_2 diving at relatively shallow depths, (2) diving for limited periods to any appropriate depth using O_2–inert gas mixtures such as air, other O_2-N_2 mixtures (Fig. 75-16), or He-O_2 mixtures, and (3) saturation diving with an O_2–inert gas mixture. The risk of bends and the decompression procedures for these types of diving are different.

In "*O_2 diving*," depth and time are limited by O_2 poisoning, but this is not pertinent to decom-

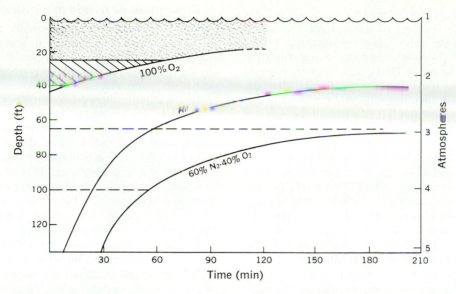

Fig. 75-16. Comparison of O_2, air, and "mixed-gas" diving. O_2, Stippled and crosshatched areas above upper curve show range of diving depth and duration in which no symptoms of O_2 toxicity were encountered during breathing of pure O_2 while engaged in sustained underwater exercise.[232] *U.S. Navy Diving Manual* recommends that when 100% O_2 is used in diving, limits shown by stippled area should not be exceeded. *Air and N_2-O_2 gas mixtures,* Two lower curves together show influence of reduced inspired inert gas concentration on diving depths allowable without need for stepwise decompression on ascent ("no-decompression" diving). Middle curve illustrates depth-duration relationships when diving is done with *air* as breathing medium. As an example, gains in safe "no-decompression" diving are shown when *mixture* of 40% O_2 with 60% N_2 is breathed in an "open-circuit" or demand apparatus. This mixture, at depth of slightly more than 66 feet, has P_{N_2} nearly the same as air at 35 feet; hence this gas mixture can be used for long diving durations at depths of 66 feet or less without need for decompression stops. At 100 feet (approximately 4 atm pressure), O_2 partial pressure in gas containing 40% O_2 is same as P_{O_2} of pure O_2 breathed at 20 feet (1 atm). At this inspired P_{O_2}, no symptoms of O_2 toxicity were found even during work.[232] *Thus limitations normally imposed by O_2 toxicity or bends are both reduced by use of compromise gas mixture,* and it is possible to select gas mixture best suited for particular diving situation. Further, use of He in place of N_2 offers additional advantages in "mixed-gas" diving. It is important to recognize that use of any inert gas–O_2 mixture in rebreathing system presents hazard of hypoxia, unconsciousness, and death should flow of gas be inadequate or cease. In an "open-circuit" system, in which each breath is drawn from the gas cylinder, this hazard does not occur without definite warning of difficulty in inspiration.

pression sickness, since the body actually *loses* inert gas during the dive rather than gains it. The O_2 diver can return rapidly to the surface without stepwise decompression, since O_2 itself is consumed in the tissues and tissue O_2 continuously falls parallel with the decreasing ambient pressure at a rate related to the circulation time.

In *saturation diving* the diver remains for so long at the increased pressure that the tissues come into equilibrium with the partial pressure of the inert gas (N_2, Ne, He) tension inspired.[71,241] The near equilibration of the least well-perfused tissues may require 12 to 24 hr.[239] Once the stay at pressure has led to saturation of

all tissues the rate of decompression will be the same regardless of how long thereafter the exposure is sustained. This approach to multiday diving has been demonstrated in the sea at increasing pressures up to 1,000 feet and in the laboratory at pressures reaching 2,000 feet of seawater.* The rate of decompression (return to the surface) depends on the fraction of O_2 tolerable in the respired gas and approximates 12 to 24 hr for each 100 feet of ascent. This is equivalent to ascending at the rate required for elimination of gases from the tissue having the slowest

---

*See references 35, 38, 96, 117, 212, 225, and 226.

gas exchange. As shown in Fig. 75-17, a 2-day stay at a pressure equivalent to 1,200 feet of seawater requires up to 6 days for surfacing.[37,163]

Limited duration excursion diving from the surface with air or He-O$_2$ gas mixtures presents the greatest difficulty in estimating efficient decompression requirements. After a short period at high pressure the fastest perfused tissues have taken up the most inert gas. Therefore in the initial phase of ascent from a dive of short duration, it is these "fast" tissues that limit the rate at which surfacing can be accomplished. Even though return to the surface is relatively rapid in the first stages, inert gas continues to be taken up by the most slowly perfused tissues even after ascent has begun, and in the latter phase, it is the requirement for inert gas elimination from the slowly perfused tissues that limits the rate of ascent. The pattern of ascent in such a situation is therefore not linear but roughly exponential. After a 1 hr exposure to helium to a depth of 400 feet, it will require approximately 18 hr to return to atmospheric pressure. It is the avoidance of the repeated requirement for such long periods of decompression that provides the gain in saturation diving.

Prevention of bends in air diving

When air is the breathing medium, prevention of bends is based on limiting diving depth and duration and careful control of the rate of return to sea level. By such means, it is possible to accomplish some short and relatively shallow air diving without prolonged decompression.

Limiting diving depth. Since the tissues can be moderately supersaturated with inert gas without formation of bubbles to a degree that produces symptoms,[23] it is possible to spend unlimited time at diving depths less than 35 feet without need for slow decompression when ultimately returning to the surface (Fig. 75-16). In aviation, rapid ascent rarely causes bends until altitudes of 30,000 feet or more are reached.

Limiting diving duration. Table 75-6 indicates empirically determined durations at which dives breathing air can be carried out without the need for subsequent, stepwise decompression stops.[232] Ascent can then be accomplished at a uniform rate of 60 feet/min or less. When diving is restricted to these depths and durations, the uptake of nitrogen is limited and the incidence of bends is reduced to a low, almost zero, level.[232]

Using established safe rates of decompression.

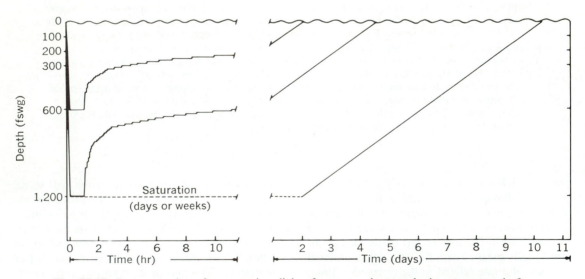

Fig. 75-17. Decompression after excursion diving from normal atmospheric pressure and after "saturation diving" at a constant high pressure. Diagram shows, for actual and probable future saturation exposures at increasing depths, the slow, linear, and identical rates of return to surface that must follow prolonged exposures to high inert gas pressures. When all body tissues have become saturated with inert gas, further exposure cannot result in further uptake of inert gas. Therefore requirement for decompression will not increase, and decompression will be the same for a 1-month diving duration as for a 1-day exposure. In these situations, rate of return to surface is governed by elimination of inert gas from tissues having slowest gas exchange.

When a dive is of long duration, bends are prevented by eliminating nitrogen during a succession of waiting periods at stops that are precisely defined for each dive.[232]

Extending diving depth and duration

Major extension of useful diving, decreasing the requirement for decompression, and lessening the risk of bends have been accomplished by employing one or more procedures having physiologic advantage in limiting the uptake of inert gas or facilitating its elimination during decompression.[162] The more important of these include the following.

Use of helium-oxygen mixtures. Substitution of helium as the inert gas in place of nitrogen eliminates inert gas narcosis even at considerable depths[45,107] but does not prevent bends. It does reduce the time required for decompression, making possible the performance of deeper and longer dives. Helium is only about half as soluble as nitrogen in body tissues, while at the same time it diffuses about twice as rapidly as nitrogen through tissue fluids. Consequently, the amount of helium taken up in the diver's body in a given time and at a given pressure is considerably less

than the amount of nitrogen absorbed under the same conditions. The rate at which the excess helium is eliminated from the body when decompression takes place also appears to be considerably more rapid than is the rate of nitrogen elimination after exposure to a N_2-O_2 atmosphere. This is of evident importance in preventing bends despite the observations that most of the inert gas elimination from "fast tissues" in humans is perfusion limited.[147] Hydrogen has been similarly used in diving,[246] but the advantages of H_2-O_2 mixtures are in part cancelled by the explosion hazard that exists except at extreme concentrations.[246] Helium is so much less dense than nitrogen that a He-O_2 atmosphere can be moved within the respiratory passages with less effort. This is important in maintaining adequate alveolar ventilation and preventing CO_2 accumulation during exertion at great depth. Fig. 75-18 shows that as the density of respired gas increases with the pressure at which it is breathed, the maximal breathing capacity is progressively diminished.[37] Despite this, men exposed to a respiratory gas density equivalent to that to be expected on breathing helium at 5,000 feet of seawater have effectively maintained respiration functions at rest and in moderate exercise.[35,37]

Oxygen decompression. When pure O_2 is substituted for the air breathed at a particular ambient pressure (e.g., 40 feet of seawater), the alveolar P_{N_2} rapidly approaches zero, a maximal gradient between the pressures of nitrogen in the tissues and the external atmosphere is attained, the rate of nitrogen elimination becomes maximum, and decompression time can be shortened. The large P_{N_2} gradient does not result in bubble formation because the total or hydrostatic pressure remains high. The depth at which this important procedure can be instituted is limited by the risk of O_2 toxicity. Fig. 75-16 illustrates approximate limits for O_2 tolerance in exercise. Since in the resting subject O_2 tolerance is much greater than during exercise, breathing pure O_2 at rest as a highly practical aid to decompression has become a standard feature of long, deep dives with He-O_2 mixtures.[37] The physiologic principles involved are also extremely valuable in other diving situations such as decompression after air diving or after diving with any inert gas–O_2 mixture. Decompression procedures that fully utilize the principles of O_2 decompression are gradually being established.

Use of multiple inert gas mixtures. Since gases in solution behave independently and should escape from the tissues as though each were the only one present and should exert inde-

Table 75-6. "No-decompression" table for air breathing at increased ambient pressure

Depth in seawater (ft)	Diving limits for "no decompression" (min)
0	—
10	—
15	—
20	—
25	—
30	—
35	310
40	200
50	100
60	60
70	50
80	40
90	30
100	25
110	20
120	15
130	10

NOTE: "No decompression" diving limits are the diving times that allow return to the surface without need for decompression stops. Rate of ascent should not exceed 60 feet/min. Note that at depths less than 35 feet even extreme diving duration does not require slow decompression.

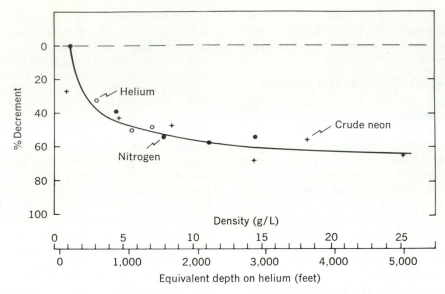

Fig. 75-18. Influence of increased gas density on human pulmonary function. Respiration at rest is comfortable and nonlimiting to inspired gas densities equivalent to breathing He-O$_2$ at 5,000 feet of seawater. (From Lambertsen et al.[37])

pendent tendencies to bubble formation in the supersaturated state, it was conceived that advantage might accrue from simultaneous use of more than one inert gas in a breathing mixture.[156] This concept has been subjected to experimental and mathematical evaluation,[162,235,240] with relatively slight practical advantage being found thus far. The major difficulty is that any bubble formation or cavitation in the presence of a high *total* tissue gas pressure leads to rapid growth of the bubble by entry of all gases. Therefore although bends liability should theoretically be lessened, in practice the incidence and severity of bends are not diminished by multiple gas mixtures. This suggests that bubble formation is a usual event during decompression, even when no symptoms are produced.

Purposeful variation of the composition of inspired gas. During ascent, at diving depth, and during early stages of decompression the partial pressure gradient of any inert gas determines the amount that enters or leaves the tissues. At any stage in an exposure the length of decompression required to prevent bends depends on the amount of each inert gas that has accumulated. Thus it has been proposed that systematic alternation of inert gases in the breathing mixture (e.g., He-O$_2$, Ne-O$_2$, N$_2$-O$_2$) will limit the accumulation of any one gas but cannot limit the total inert gas pressure to less than the in-

spired level. This situation therefore superficially resembles the simultaneous use of several inert gases (e.g., N$_2$-Ne-He-O$_2$), as just described. It differs in that, if the order of gas administration is proper, a limitation of total inert gas uptake could result. Although this does not grossly minimize the decompression requirement, the physiologic trend is in the right direction.

Alternation of high and low inspired oxygen tensions. To minimize the uptake and facilitate elimination of inert gas the concentration of O$_2$ inspired should be maximal for each depth and condition during diving and decompression. In view of the toxicity of excessive pressure and duration of exposure to O$_2$ itself and of the influences of factors such as exercise in lowering O$_2$ tolerance, considerable compromise is necessary in practical situations to avoid pulmonary or CNS O$_2$ toxicity (p. 1906). This compromise is accomplished by planned alternation of gases high and low in O$_2$ tension.[156] Safe periods of relatively high O$_2$ pressure aid in the elimination of inert gas, whereas shorter periods of low inspired Po$_2$ provide at least partial reestablishment of O$_2$ tolerance. This approach has markedly extended the tolerance of guinea pigs to O$_2$ breathing at 3 atm[131,148,156] and is now part of the procedure for pure O$_2$ therapy of bends.[124,232] It has become important in facilitating decompression after deep He-O$_2$ diving.

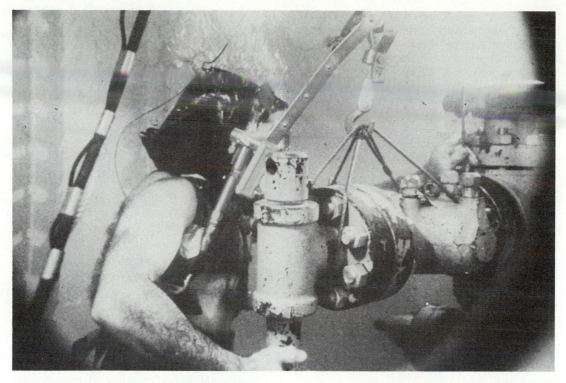

Fig. 75-19. Diver-subject performing practical underwater work task in undersea simulator chamber at pressure equivalent of 1,600 feet of seawater, breathing He with O_2. (From Lambertsen et al.[35])

Combining approaches. The methods cited have been under consideration for a long time, but they have only slowly begun to be incorporated into diving operations. The recent increase in interest in deep, prolonged diving makes consideration of each necessary. The several possibilities are not mutually exclusive. Under special circumstances, it is possible to employ the physiologic advantages of more than one established method with consequent improvement in diving depth and duration, decreased decompression time, or reduced incidence of bends. The most obvious combination of procedures is the simultaneous substitution of helium for nitrogen, increased O_2 percentage to lower helium partial pressure, alternation of high and low inspired Po_2 during most of the decompression, employment of pure O_2 breathing (O_2 decompression) during the shallower decompression stages, and alternation of air and O_2 breathing in the phase of decompression at less than 2 atm. Without awareness of the limits for these still unstandardized, physiologic techniques, great hazard of permanent disability is involved in their use, and more routine approaches[232] should continue to be employed by those individuals not involved in this work or investigation.

LIMITS OF MANNED UNDERSEA ACTIVITY

In the short interval since publication of the previous edition of this text, extensive advances in manned undersea activity have derived from laboratory experiment and open-sea application. The situation is remarkably similar to the parallel developments of physiology and applications in space flight.

At present, humans work effectively 1,000 feet under the sea for weeks on end without decompressing to atmospheric pressure. Stepwise exposures to helium in chambers have established that the human is capable of full mental and physiologic activity under these conditions.[37]

It has also been found that the human is physiologically competent and can accomplish detailed and strenuous practical work in water at hydrostatic pressures equivalent to 1,600 feet of seawater[35] (Fig. 75-19).

Exposure to helium pressures equal to 2,000

feet of seawater has produced no residual effects, but when combined with rapid compression, exposure to less than half that extreme pressure produces definite symptoms. The physical effect of hydrostatic pressure itself, known to incapacitate small animals,[173] becomes a progressively more limiting stress on the human as depth and speed of compression are increased.* The influences of hydrostatic pressure are largely on neural and neuromuscular functions, with tremor, confusion, weakness, and electroencephalographic changes.† Although these were initially considered effects of helium,[60] this seems clearly not to be the case.

The hydrostatic pressure effects become prominent and may be limiting at depths greater than 1,600 feet.[35,117,212] Nevertheless, the same decade that saw the ascent of Mount Everest and exploration of the lunar surface also saw all of the continental shelves and a thousand feet beyond opened to direct activity by humans.

---

*See references 35, 37, 60, 76, 117, and 212.
†See references 35, 60, 76, 117, and 212.

REFERENCES
General reviews

1. Alexander, J. K., et al.: Analysis of respiratory response to carbon dioxide inhalation in varying clinical states of hypercapnia, anoxia and acid-base derangement, J. Clin. Invest. **34:**511, 1955
2. Armstrong, H. G.: Principles and practice of aviation medicine, Baltimore, 1939, The Williams & Wilkins Co.
3. Atwater, J., and Vaughan, R.: Room at the bottom of the sea, Sat. Evening Post, Sept. 5, 1964.
4. Bean, J. W.: Effects of oxygen at increased pressure, Physiol. Rev. **25:**1, 1945.
5. Boerema, I., Brummelkamp, W. H., and Meijne, N. G., editors: Clinical application of hyperbaric oxygen, Amsterdam, 1964, Elsevier Publishing Co.
6. Carbon Dioxide Symposium, Anesthesiology **21:**585, 1960.
7. Chance, B., and Boveris, A.: Hyperoxia and hydroperoxide metabolism. In Robin, E. D., editor: Extrapulmonary manifestations of respiratory disease. Lung biology in health and disease, New York, 1978, Marcel Dekker, vol. 8.
8. Chance, B., and Jamieson, D.: Control of energy-linked reactions in mitochondria by hyperbaric oxygen. In Hyperbaric medicine, Publ. 1404, Washington, D.C., 1966, National Academy of Sciences, National Research Council.
9. Clark, J. M., and Lambertsen, C. J.: Pulmonary oxygen toxicity: a review, Pharmacol. Rev. **23:**37, 1971.
10. Cousteau, J. Y.: At home in the sea, Natl. Geographic **125:**465, 1964.
11. Damato, M. J., et al.: Rapid decompression hazards after prolonged exposure to 50 per cent oxygen—50 per cent nitrogen atmosphere, Aerospace Med. **34:**1037, 1963.
12. D'Aoust, B. G., et al.: Venous gas bubbles: produc-

tion by transient, deep isobaric counterdiffusion of helium against nitrogen, Science **197:**889, 1977.
13. Davies, H. C., and Davies, R. E.: Biochemical aspects of oxygen poisoning. In Fenn, W. O., and Rahn, H., editors: Handbook of physiology. Respiration section, Washington, D.C., 1965, American Physiological Society, vol. 2.
14. Davis, J. C., and Hunt, T. K., editors: Hyperbaric oxygen therapy, Bethesda, Md., 1977, Undersea Medical Society.
15. Dickens, F., and Neil, E., editors: Oxygen in the animal organism, Oxford, 1964, Pergamon Press, Ltd.
16. Fenn, W. O., et al.: Recovery from oxygen poisoning in Drosophila, Am. J. Physiol. **213:**663, 1967.
17. Fridovich, I.: The biology of oxygen radicals, Science **201:**875, 1978.
18. Fulton, J. F., editor: Decompression sickness: caisson sickness, diver's and flier's bends and related syndromes, Philadelphia, 1951, W. B. Saunders Co.
19. Fundamentals of hyperbaric medicine, Publ. 1298, Washington, D.C., 1966, National Academy of Sciences, National Research Council.
20. Gerschman, R., Gilbert, D. L., and Caccamise, D.: Effect of various substances on survival times of mice exposed to different high oxygen tensions, Am. J. Physiol. **192:**563, 1958.
21. Gottlieb, S. F.: Effect of hyperbaric oxygen on microorganisms, Annu. Rev. Microbiol. **25:**111, 1971.
22. Hallenbeck, J. M., and Sokoloff, L.: Blood flow studies during spinal cord–damaging decompression sickness in dogs. In Shilling, C. W., and Beckett, M. W., editors: Underwater physiology. VI. Proceedings of the Sixth Symposium on Underwater Physiology, Bethesda, Md., 1978, FASEB.
23. Harvey, E. N.: Bubble formation. In Proceedings of the Underwater Physiology Symposium, Publ. 377, Washington, D.C., 1955, National Academy of Sciences, National Research Council.
24. Haugaard, N.: The toxic action of oxygen on metabolism and the role of trace metals. In Dickens, F., and Neil, E., editors: Oxygen in the animal organism, Oxford, 1964, Pergamon Press, Ltd.
25. Hesser, C. M., Adolfson, J., and Fagraeus, L.: Role of CO_2 in compressed-air narcosis, Aerospace Med. **42:**163, 1971.
26. Ivanov, K. P.: Effect of increased oxygen pressure on animals poisoned by potassium cyanide, Farmakol. Toksikol. **22:**468, 1959.
27. Kelley, M. A., and Kazemi, H.: Role of ammonia as a buffer in the central nervous system, Respir. Physiol. **22:**345, 1974.
28. Klocke, R. A.: Carbon dioxide transport. In Robin, E. D., editor: Extrapulmonary manifestations of respiratory disease. Lung biology in health and disease, New York, 1978, Marcel Dekker, vol. 8.
29. Lambertsen, C. J.: Effects of oxygen at high partial pressure. In Fenn, W. O., and Rahn, H., editors: Handbook of physiology. Respiration section, Washington, D.C., 1965, American Physiological Society, vol. 2.
29a. Lambertsen, C. J.: Medical implications of high oxygen pressures, Tr. Coll. Physicians Phila. **33:**1, 1965.
30. Lambertsen, C. J.: Therapeutic gases: oxygen, carbon dioxide and helium. In DiPalma, J. R., editor: Drill's pharmacology in medicine, ed. 4, New York, 1971, McGraw-Hill Book Co.
31. Lambertsen, C. J., editor: Underwater physiology. Proceedings of the Fourth Symposium on Underwater Physiology, New York, 1971, Academic Press, Inc.

32. Lambertsen, C. J., editor: Multiday exposure of men to high nitrogen pressure and increased airway resistance at natural inspired oxygen tension: a 14-day continuous exposure to 5.2% O_2 in N_2 at 4.0 atmospheres absolute pressure, Aerospace Med. **44:**821, 1973.

33. Lambertsen, C. J., editor: Underwater physiology. V. Proceedings of the Fifth Symposium on Underwater Physiology, Bethesda, Md., 1976, FASEB.

34. Lambertsen, C. J.: Effects of hyperoxia on organs and their tissues. In Robin, E. D., editor: Extrapulmonary manifestations of respiratory disease. Lung biology in health and disease, New York, 1978, Marcel Dekker, vol. 8.

35. Lambertsen, C. J., Gelfand, R., and Clark, J. M., editors: Predictive studies. IV. Work capability and physiological effects in He-O_2 excursions to pressures of 400-800-1200-1600 fsw, Institute for Environmental Medicine Report 78-1, Philadelphia, 1978, University of Pennsylvania Press.

36. Lambertsen, C. J., O'Neil, S. R., and Long, M. L., editors: The human factor in North Sea operational diving. Proceedings of a symposium organized by the Institute for Environmental Medicine, University of Pennsylvania, and the United Kingdom Diving Medical Advisory Committee, London, 1976 and Allentown, Pa., 1978, Air Products and Chemicals.

37. Lambertsen, C. J., et al.: Human tolerance to He, Ne, and N_2 at respiratory gas densities equivalent to He-O_2 breathing at depths to 1200, 2000, 3000, 4000, and 5000 feet of sea water (Predictive studies III), Aviat. Space Environ. Med. **48:**843, 1977.

38. Link, E. A., and Stenuit, R.: Outpost under the sea, Natl. Geographic **127:**530, 1965.

39. McCord, J. M., and Fridovich, I.: The biology and pathology of oxygen radicals, Ann. Intern. Med. **89:**122, 1978.

40. Meduna, L. J.: Alteration of neurotic pattern by use of carbon dioxide inhalations, J. Nerv. Ment. Dis. **108:**373, 1948.

41. Menn, S. J., Sinclair, R. D., and Welch, B. E.: Effect of inspired P_{CO_2} up to 30 mm Hg on response of normal man to exercise, J. Appl. Physiol. **28:**663, 1970.

42. Michel, E. L., Langeven, R. W., and Gell, C. F.: Effects of continuous human exposure to oxygen tension of 418 mm Hg for 168 hours, Aerospace Med. **31:**138, 1960.

43. Pratt, P. C.: Pathology of pulmonary oxygen toxicity. Part 2, Am. Rev. Respir. Dis. **110**(6):51, 1974.

44. Proceedings of the Second Underwater Physiology Symposium, Publ. 1181, Washington, D.C., 1963, National Academy of Sciences, National Research Council.

45. Rinfret, A. P., and Doebbler, G. F.: Physiological and biochemical effects and applications. In Cook, G. A., editor: Argon, helium and the rare gases, New York, 1961, John Wiley & Sons, Inc., vol. 2.

46. Roth, E. M.: Selection of space cabin atmospheres: oxygen toxicity, NASA Technical Note TN-2008, Washington, D.C., 1963.

47. Santiago, T. V., and Edelman, N. H.: Mechanism of the ventilatory response to carbon monoxide, J. Clin. Invest. **57:**977, 1976.

47a. Shilling, C. W., and Beckett, M. W., editors: Underwater physiology. VI. Proceedings of the Sixth Symposium on Underwater Physiology, Bethesda, Md., 1978, FASEB.

48. Singer, M., et al.: Oxygen toxicity in man. A prospective study in patients after open-heart surgery, N. Engl. J. Med. **283:**1473, 1970.

49. Symposium on Carbon Dioxide, Bull. Eur. Physiol. Respir. **12:**Nos. 1 and 2, 1976.

50. Symposium on hyperbaric oxygenation, Ann. N.Y. Acad. Sci. **117:**647, 1965.

51. Traube, J.: Theorie der Osmose und Narkose, Arch. Gesamte Physiol. **105:**541, 1904.

52. Williams, C. M., and Beecher, H. K.: Sensitivity of Drosophila to poisoning by oxygen, Am. J. Physiol. **140:**566, 1943.

53. Winter, P., and Smith, G.: The toxicity of oxygen, Anesthesiology **37:**210, 1972.

54. Wolfe, W. G., and De Vries, W. C.: Oxygen toxicity, Annu. Rev. Med. **26:**203, 1975.

Original papers

55. America's 6,000 mile walk in space, Natl. Geographic **128:**440, 1965.

56. Adler, H. F.: Dysbarism, aeromedical reviews, Technical Documentary Report No. SAM-TDR-1-64, Washington, D.C., 1964, USAF School of Aerospace Medicine.

57. Allen, T. H., and Allard, R. W.: Fundamental parameters influencing the accumulation and elimination of carbon monoxide by adult human beings, Report 261, Sept., 1961, U.S. Army Medical Research and Nutrition Laboratory.

58. Ashton, N.: Discussion. In Dickens, F., and Neil, E., editors: Oxygen in the animal organism, Oxford, 1964, Pergamon Press, Ltd.

59. Ashton, N. B., Ward, B., and Serpell, G.: Role of oxygen in the genesis of retrolental fibroplasia, Br. J. Ophthalmol. **37:**513, 1953.

60. Bachrach, A. J., and Bennett, P. B.: The high pressure nervous syndrome during human deep saturation and excursion diving, Forsvarsmedicin **9:**490, 1973.

61. Barber, R. E., Lee, J., and Hamilton, W. K.: Oxygen toxicity in man—a prospective study in patients with irreversible brain damage, N. Engl. J. Med. **283:**1478, 1970.

62. Bard, P.: Blood supply of special regions. In Bard, P., editor: Medical physiology, ed. 11, St. Louis, 1961, The C. V. Mosby Co.

63. Bean, J. W.: Problems of oxygen toxicity. In Boerema, I., Brummelkamp, W. H., and Meijne, N. G., editors: Clinical application of hyperbaric oxygen, Amsterdam, 1964, Elsevier Publishing Co.

64. Beck, H. G., and Suter, G. M.: Role of carbon monoxide in the causation of myocardial disease, J.A.M.A. **110:**1982, 1938.

65. Becker-Freyseng, H., and Clamann, H. G.: Zur Frage der Sauerstoffvergiftung, Klin. Wochenschr. **18:**1382, 1939.

66. Beehler, C. C.: Oxygen and the eye, Survey Ophthalmol. **9:**549, 1964.

67. Behnke, A. R., and Yarbrough, O. D.: Respiratory resistance, oil-water solubility and mental effects of argon, compared with helium and nitrogen, Am. J. Physiol. **126:**409, 1939.

68. Behnke, A. R., Forbes, H. S., and Motley, E. P.: Circulatory and visual effects of oxygen at 3 atmospheres pressure, Am. J. Physiol. **114:**436, 1936

69. Behnke, A. R., Thomson, R. M., and Motley, E. P.: The physiological effects from breathing air at four atmospheres pressure, Am. J. Physiol. **112:**554, 1935.

70. Blenkarn, G. D., et al.: Urticaria following the sequential breathing of various inert gases at a constant ambient pressure of 7 ATA: a possible manifestation of gas-induced osmosis, Aerospace Med. **42:**141, 1971.

71. Bond, G. F.: Medical problems of multi-day saturation diving in open water. In Lambertsen, C. J., editor: Underwater physiology. Proceedings of the Third Symposium on Underwater Physiology, Baltimore, 1967, The Williams & Wilkins Co.

72. Boycott, A. E., Damant, G. C., and Haldane, J. S.: The prevention of compressed air illness, J. Hyg. **8:** 342, 1908.

73. Brink, F., and Posternak, J. M.: Thermodynamic analysis of relative effectiveness of narcotics, J. Cell. Comp. Physiol. **32:**211, 1948.

74. Brown, E. B., and Miller, F.: Ventricular fibrillation following a rapid fall in alveolar carbon dioxide concentration, Am. J. Physiol. **169:**56, 1952.

75. Buhlmann, A. A.: Respiratory resistance with hyperbaric gas mixtures. In Proceedings of the second underwater physiology symposium, Publ. 1181, Washington, D.C., 1963, National Academy of Sciences, National Research Council.

76. Broussolle, B., et al.: Respiratory function during a simulated saturation dive to 51 ata (500 meters) with a helium-oxygen mixture. In Lambertsen, C. J., editor: Underwater physiology. V. Proceedings of the Fifth Symposium on Underwater Physiology, Bethesda, Md., 1976, FASEB.

77. Caldwell, P. R. B., et al.: Changes in lung volume, diffusing capacity and blood gases in men breathing oxygen, J. Appl. Physiol. **21:**1477, 1966.

78. Carpenter, F. G.: Anesthetic action of inert and unreactive gases on intact animals and isolated tissues, Am. J. Physiol. **178:**505, 1954.

79. Carpenter, F. G.: Inert gas narcosis. In Proceedings of the Underwater Physiology Symposium, Publ. 377, Washington, D.C., 1955, National Academy of Sciences, National Research Council.

80. Case, E. M., and Haldane, J. B. S.: Human physiology under high pressures: the effects of nitrogen, carbon dioxide and cold, J. Hyg. **41:**225, 1941.

81. Castren, J. A.: Das Auge Beim Frühgeborenen Unter Besonderer Berücksichtigung Der Retrolentalen Fibroplasie, Klin. Wochenschr. **37:**165, 1959.

82. Catchpole, H. R., and Gersh, I.: Pathogenetic factors and pathological consequences of decompression sickness, Physiol. Rev. **27:**360, 1947.

83. Chapin, J. L.: Anticonvulsant threshold of CO_2 in oxygen under high pressure, Proc. Soc. Exp. Biol. Med. **90:**663, 1955.

84. Churchill-Davidson, I., Sanger, C., and Thomlinson, R. H.: Oxygenation in radiotherapy. II. Clinical application, Br. J. Radiol. **30:**406, 1957.

85. Clark, J. M., and Lambertsen, C. J.: Pulmonary oxygen tolerance and the rate of development of pulmonary oxygen toxicity in man. In Proceedings of the Third Symposium on Underwater Physiology, Baltimore, 1967, The Williams & Wilkins Co.

86. Clark, J. M., and Lambertsen, C. J.: Rate of development of pulmonary O_2 toxicity in man during O_2 breathing at 2.0 ata, J. Appl. Physiol. **30:**739, 1971.

87. Clark, J. M., and Lambertsen, C. J.: Alveolar-arterial O_2 differences in man at 0.2, 1.0, 2.0 and 3.5 atm inspired Po_2, J. Appl. Physiol. **30:**753, 1971.

88. Clark, J. M., Sinclair, R. D., and Welch, B. E.: Rate of acclimatization to chronic hypercapnia in man. In Lambertsen, C. J., editor: Underwater physiology, New York, 1971, Academic Press, Inc.

89. Coburn, R. F., Blakemore, W. S., and Forster, R. E.: Endogenous carbon monoxide production in man, J. Clin. Invest. **42:**1172, 1963.

90. Comroe, J. H., Jr., et al.: Oxygen toxicity. The effect of inhalation of high concentrations of oxygen for twenty-four hours on normal men at sea level and at a simulated altitude of 18,000 feet, J.A.M.A. **128:**710, 1945.

91. Cullen, S. C., and Gross, E. G.: Anesthetic properties of xenon in animals and human beings with additional observations on krypton, Science **113:**580, 1951.

92. Cymerman, A., and Gottlieb, S. F.: Effects of increased oxygen tensions on bioelectric properties of frog sciatic nerve, Aerospace Med. **41:**36, 1970.

93. Daly, M. deB., Lambertsen, C. J., and Schweitzer, A.: Observations of the volume of blood flow and oxygen utilization of the carotid body in the cat, J. Physiol. **125:**67, 1954.

94. D'Aoust, B. G., et al.: Venous gas bubbles: production by transient deep isobaric counterdiffusion of helium against nitrogen, Science **197:**889, 1977.

95. Dickens, F.: The toxic effect of oxygen on nervous tissue. In Elliott, K. A. C., editor: Neurochemistry, Springfield, Ill., 1955, Charles C Thomas, Publisher.

96. Dickson, J. G., and MacInnis, J. B.: Confluence of physiological, environmental and engineering factors in prolonged diving at extreme depths. In Proceedings of the Third Symposium on Underwater Physiology, Baltimore, 1967, The Williams & Wilkins Co.

96a. Dickson, J. G., Lambertsen, C. J., and Cassils, J. G.: Quantitation of performance decrements in narcotized man. In Lambertsen, C. J., editor: Underwater physiology. Proceedings of the Fourth Symposium on Underwater Physiology, New York, 1971, Academic Press, Inc., p. 449.

97. Doležal, V.: Some humoral changes in man produced by continuous oxygen inhalation at normal barometric pressure, Riv. Med. Aeronaut. **25:**219, 1962.

98. Doležal, V.: The effect of long lasting oxygen inhalation upon respiratory parameters in man, Physiol. Bohemoslov. **11:**145, 1962.

99. Dollery, C. T., et al.: High oxygen pressure and the retinal blood vessels, Lancet **2:**291, 1964.

100. Donald, K. W.: Oxygen poisoning in man, Br. Med. J. **1:**667, 1947.

101. Douglas, T. A., et al.: Carbon monoxide poisoning. In Boerema, I., Brummelkamp, W. H., and Meijne, N. G., editors: Clinical application of hyperbaric oxygen, Amsterdam, 1964, Elsevier Publishing Co.

102. Dripps, R. D., and Comroe, J. H., Jr.: The respiratory and circulatory response of normal man to inhalation of 7.6 and 10.4 percent CO_2 with a comparison of the maximal ventilation produced by severe muscular exercise, inhalation of CO_2 and maximal voluntary hyperventilation, Am. J. Physiol. **149:**43, 1947.

103. DuBois, A. B.: Oxygen toxicity, Anesthesiology **23:** 473, 1962.

104. Ebert, M., Hornsey, S., and Howard, A.: Effect of radiosensitivity of inert gases, Nature **181:**613, 1958.

105. Eckenhoff, J. E., chairman: Symposium on carbon dioxide and man, Anesthesiology **21:**585, 1960.

106. End, E., and Long, C. W.: Oxygen under pressure in carbon monoxide poisoning. I. Effect on dogs and guinea pigs, J. Indust. Hyg. Toxicol. **24:**302, 1942.

107. Featherstone, R. M., and Muehlbaecher, C. A.: The current role of inert gases in the search for anesthesia mechanisms, Pharmacol. Rev. **15:**97, 1963.

108. Fenn, W. O., Rahn, H., and Otis, A. B.: A theoretical study of the composition of the alveolar air at altitude, Am. J. Physiol. **146:**637, 1946.

109. Ferguson, J.: Use of chemical potential as indices of

toxicity, Proc. R. Soc. Lond. (Biol.) **127**:387, 1939.

110. Finkelstein, S., Elliott, J. E., and Luft, U. C.: The effects of breathing low concentrations of CO_2 on exercise tolerance. Presented at the Thirty-ninth Annual Scientific Meeting of the Aerospace Medical Association, 1968.

111. Fisher, A. B., et al.: Effect of oxygen at two atmospheres upon the pulmonary mechanics of normal man, J. Appl. Physiol. **24**:529, 1968.

112. Forbes, W. H., Sargent, F., and Roughton, F. J. W.: The rate of carbon monoxide uptake by normal men, Am. J. Physiol. **143**:594, 1945.

113. Frankel, J., and Schneiderman, H. A.: The effects of nitrogen, helium, argon, and sulfur hexafluoride on the development of insects, J. Cell. Comp. Physiol. **52**:431, 1958.

114. Frankenhaeuser, M., and Jaerpe, G.: Subjective intoxication induced by nitrous oxide in various concentrations, Scand. J. Psychol. **3**:171, 1962.

115. Frankenhaeuser, M., Graff-Lonnevig, V., and Hesser, C. M.: Effects on psychomotor functions of different nitrogen-oxygen gas mixtures at increased ambient pressures, Acta Physiol. Scand. **59**:400, 1963.

116. Fructus, X. R.: Report of 1700-foot dive. Presented at the Fifth Symposium on Underwater Physiology, August, 1972.

117. Fructus, X., et al.: Postponing the "high pressure nervous syndrome" to 1640 feet and beyond. In Lambertsen, C. J., editor: Underwater physiology. V. Proceedings of the Fifth Symposium on Underwater Physiology, Bethesda, Md., 1976, FASEB.

118. Fuson, R. L., et al.: Clinical hyperbaric oxygenation with severe oxygen toxicity, N. Engl. J. Med. **273**:415, 1965.

119. Gerschman, R.: Oxygen effects in biological systems. In Dickens, F., and Neil, E., editors: Oxygen in the animal organism, Oxford, 1964, Pergamon Press, Ltd.

120. Gerschman, R., et al.: Effect of high oxygen concentrations on eyes of newborn mice, Am. J. Physiol. **179**:115, 1954.

121. Gesell, R.: On the chemical regulation of respiration: regulation of respiration with special reference to the metabolism of the respiratory center and the coordination of the dual function of hemoglobin, Am. J. Physiol. **66**:5, 1923.

122. Goddard, D. R.: The biological role of carbon dioxide, Anesthesiology **21**:587, 1960.

123. Goldsmith, J. R., and Landaw, S. A.: Carbon monoxide and human health, Science **162**:1352, 1968.

124. Goodman, M. W.: Decompression sickness treated with compression to 2-6 atmospheres absolute, Aerospace Med. **35**:1204, 1964.

125. Gottlieb, S. F., and Weatherly, J. M.: Physiological effects of the noble gases on frog sciatic nerve and gastrocnemius muscle, Am. J. Physiol. **208**:407, 1965.

126. Graves, D. J., et al.: Bubble formation resulting from counterdiffusion supersaturation: a possible explanation for inert gas "urticaria" and vertigo, Phys. Med. Biol. **18**:256, 1973.

127. Gray, L. H.: Oxygenation in radiotherapy. I. Radiobiological consideration, Br. J. Radiol. **30**:403, 1957.

128. Gray, L. H.: Radiobiologic basis of oxygen as a modifying factor in radiation therapy, Am. J. Roentgenol. **85**:803, 1961.

129. Haddad, H. M., and Leopold, I. H.: Effect of hyperbaric oxygenation of microcirculation: use in therapy of retinal vascular disorders, Invest. Ophthalmol. **4**:1141, 1965.

130. Haldane, J. S.: The relation of the action of carbonic oxide to oxygen tension, J. Physiol. **18**:201, 1895.

131. Hall, D. A.: The influence of the systematic fluctuation of Po_2 upon the nature and rate of the development of oxygen toxicity in guinea pigs. Masters thesis, University of Pennsylvania, 1967.

132. Haugaard, N.: Poisoning of cellular reactions by oxygen, Ann. N.Y. Acad. Sci. **117**:736, 1965.

133. Haugaard, N.: Cellular mechanisms of oxygen toxicity, Physiol. Rev. **48**:311, 1968.

134. Helvey, W. M.: A problem of man and milieu: prolonged exposure to pure oxygen, Fed. Proc. **22**:1057, 1963.

135. Helvey, W. M., et al.: Effects of prolonged exposure to pure oxygen on human performance, Report 393-1 (NASA Contr. NASr-92), Washington, D.C., 1962, Republic Aviation Corp.

136. Henderson, Y., and Haggard, H. W.: The elimination of carbon monoxide from the blood after a dangerous degree of asphyxiation, and a therapy for accelerating the elimination, J. Pharmacol. Exp. Ther. **16**:11, 1920.

137. Henderson, Y., and Haggard, H. W.: Noxious gases and the principles of respiration influencing their action, New York, 1927, Chemical Catalog Co.

138. Hendricks, P. L., et al.: Extension of pulmonary O_2 tolerance in man at 2 ATA by intermittant O_2 exposure, J. Appl. Physiol. **42**:593, 1977.

139. Herlocher, J. E., et al.: Physiologic response to increased oxygen partial pressure. I. Clinical observations, Aerospace Med. **35**:613, 1964.

140. Hesser, C. M.: Measurement of inert gas narcosis in man. In Proceedings of the Second Underwater Physiology Symposium, Publ. 1181, Washington, D.C., 1963, National Academy of Sciences, National Research Council.

141. Hickam, J. B., and Frayser, R.: Studies of the retinal circulation in man, Circulation **33**:302, 1966.

142. Hickam, J. B., Sieker, H. O., and Frayser, R.: Studies of retinal circulation and A-V oxygen difference in man, Trans. Am. Clin. Climatol. Assoc. **71**:34, 1959.

143. Horwitz, L. D., Bishop, V. S., and Stone, H. L.: Effects of hypercapnia on the cardiovascular system of conscious dogs, J. Appl. Physiol. **25**:346, 1968.

144. Hyde, R. W., and Rawson, A. J.: Unintentional oxygen pneumonitis—response to therapy, Ann. Intern Med. **71**:517, 1969.

145. Idicula, J., et al.: Bubble formation resulting from the steady counterdiffusion of two inert gases. In Lambertsen, C. J., editor: Underwater physiology. V. Proceedings of the Fifth Symposium on Underwater Physiology, Bethesda, Md., 1976, FASEB.

146. Jamieson, D., and van den Brenk, H. A.: Pulmonary damage due to high pressure oxygen breathing in rats. V. Changes in the surface active lung alveolar lining, Aust. J. Exp. Biol. Med. Sci. **42**:483, 1964.

147. Jones, H. B.: Respiratory system: nitrogen elimination. In Glasser, O., editor: Medical physics, Chicago, 1950, Year Book Medical Publishers, Inc., vol. 2.

148. Kaufman, B. D., Owen, S. G., and Lambertsen, C. J.: Effects of brief interruptions of pure oxygen breathing upon central nervous system tolerance to oxygen, Fed. Proc. **15**:107, 1956.

149. Kellogg, R. H.: Acclimatization to carbon dioxide, Anesthesiology **21**:634, 1960.

150. Kern, J. D.: The etiology and pathological physiology of decompression sickness, Report 345, Groton, Conn., Dec., 1960, U.S. Naval Medical Research Laboratory.

151. Kiessling, R. J., and Maag, C. H.: Performance impairment as a function of nitrogen narcosis, Research Report 3-60, Washington, D.C., 1960, U.S. Navy Experimental Diving Unit.

152. Kistler, G. S., Caldwell, P. R. B., and Weibel, E. R.: Development of fine structural damage to alveolar and capillary lining cells in oxygen-poisoned rat lungs, J. Cell Biol. **32:**605, 1967.

153. Kobayashi, T., and Murakami, S.: Blindness of an adult caused by oxygen, J.A.M.A. **219:**741, 1972.

154. Kokame, G. M., and Shuler, S. E.: Carbon monoxide poisoning—treatment by hyperbaric oxygenation, Arch. Surg. **96:**211, 1968.

155. Lambertsen, C. J.: Problems of shallow water diving; report based on experiences of operational swimmers of the Office of Strategic Services, Occup. Med. **3:**230, 1947.

156. Lambertsen, C. J.: Respiratory and circulatory actions of high oxygen pressure. In Proceedings of the Underwater Physiology Symposium, Publ. 377, Washington, D.C., 1955, National Academy of Sciences, National Research Council.

157. Lambertsen, C. J.: From submarines to satellites, Circ. Res. **6:**405, 1958.

158. Lambertsen, C. J.: Carbon dioxide and respiration in acid-base homeostasis, Anesthesiology **21:**642, 1960.

159. Lambertsen, C. J.: The philosophy of extremes for the gaseous environment of manned, closed ecological systems, Aerospace Med. **34:**291, 1963.

160. Lambertsen, C. J.: Physiological effects of oxygen. In Proceedings of the Second Underwater Physiology Symposium, Publ. 1181, Washington, D.C., 1963, National Academy of Sciences, National Research Council.

161. Lambertsen, C. J.: Physiological interactions and gaseous environment in manned exploration of space, Fed. Proc. **22:**1046, 1963.

162. Lambertsen, C. J.: Basic requirements for improving diving depth and decompression tolerance. In Proceedings of the Third Symposium on Underwater Physiology, Baltimore, 1967, The Williams & Wilkins Co.

163. Lambertsen, C. J., editor: Effects of high ambient pressures of nitrogen, neon, and helium on respiratory, neurophysiological and performance functions. Report of the Institute for Environmental Medicine, Philadelphia, 1972, University of Pennsylvania Medical Center.

164. Lambertsen, C. J., and Bascom, W.: The permissible level of carbon monoxide in inspired air, Reference 54-3, 1953, Scripps Institution of Oceanography.

165. Lambertsen, C. J., and Idicula, J.: Cutaneous gas lesions and continuous, lethal gas embolization in animals due to isobaric inert gas counterdiffusion, Fed. Proc. **33:**455, 1974.

166. Lambertsen, C. J., and Idicula, J.: A new gas lesion syndrome in man, induced by "isobaric gas counterdiffusion," J. Appl. Physiol. **39:**434, 1975.

167. Lambertsen, C. J., et al.: Oxygen toxicity. Effects in man of oxygen inhalation at 1 and 3.5 atmospheres upon blood gas transport, cerebral circulation and cerebral metabolism, J. Appl. Physiol. **5:**471, 1953.

168. Lambertsen, C. J., et al.: Comparison of relationship of respiratory minute volume to Pco_2 and pH of arterial and internal jugular blood in normal man during hyperventilation produced by low concentrations of CO_2 at 1 atmosphere and by O_2 at 3.0 atmospheres, J. Appl. Physiol. **5:**803, 1953.

169. Lambertsen, C. J., et al.: Oxygen toxicity. Effects of oxygen breathing at increased ambient pressure upon Pco_2 of subcutaneous gas depots in men, dogs, rabbits and cats, J. Appl. Physiol. **6:**358, 1953.

170. Lambertsen, C. J., et al.: Oxygen toxicity. Arterial and internal jugular blood gas composition in man during inhalation of air, 100% O_2 and 2% CO_2 in O_2 at 3.5 atmospheres ambient pressure, J. Appl. Physiol. **8:**255, 1955.

171. Lambertsen, C. J., et al.: Respiratory and cerebral circulatory control during exercise at .21 and 2.0 atmospheres inspired Po_2, J. Appl. Physiol. **14:**966, 1959.

172. Lambertsen, C. J., et al.: H^+ and Pco_2 as chemical factors in respiratory and cerebral circulatory control, J. Appl. Physiol. **16:**473, 1961.

173. Landau, J. V.: Hydrostatic effects on cellular function. In Lambertsen, C. J., editor: Underwater physiology. Proceedings of the Fourth Symposium on Underwater Physiology, New York, 1971, Academic Press, Inc.

174. Langham, W. H., Brooks, P. M., and Grahn, D.: Radiation biology and space environment parameters in manned spacecraft design and operations, Aerospace Med. **36:**1, 1965.

175. Lanphier, E. H.: Nitrogen-oxygen mixture physiology: end-tidal gas sampling system, carbon dioxide regulation in divers, carbon dioxide sensitivity tests, Research Report 2-56 (Project NS185-005, Sub. No. 5), Washington, D.C., 1955, Experimental Diving Unit, U.S. Naval Gun Factory.

176. Lanphier, E. H.: Influence of increased ambient pressure upon alveolar ventilation. In Proceedings of the Second Underwater Physiology Symposium, Publ. 1181, Washington, D.C., 1963, National Academy of Sciences, National Research Council.

177. Larson, H. E.: A history of self-contained diving and underwater swimming, Publ. 469, Washington, D.C., 1959, National Academy of Sciences, National Research Council.

178. Lawson, D. D., McAllister, R. A., and Smith, G.: Treatment of acute experimental carbon-monoxide poisoning with oxygen under pressure, Lancet **1:**800, 1961.

179. Lee, W. L., Jr., Caldwell, P. B., and Schildkraut, H. S.: Changes of lung volume, diffusion capacity, and blood gases in oxygen toxicity in humans, Fed. Proc. **22:**395, 1963.

180. MacHattie, L., and Rahn, H.: Survival of mice in absence of inert gas, Proc. Soc. Exp. Biol. Med. **104:**772, 1960.

181. MacInnis, J. B., Dickson, J. G., and Lambertsen, C. J.: Exposure of mice to a helium-oxygen atmosphere at pressures to 122 atmospheres (4000 feet of sea water), J. Appl. Physiol. **22:**694, 1967.

182. Marshall, J. R., and Lambertsen, C. J.: Interactions of increased Po_2 and Pco_2 effects in producing convulsions and death in mice, J. Appl. Physiol. **16:**1, 1961.

183. Mattsson, J. L., and Stinson, J. M.: Tolerance of rhesus monkeys to Pco_2 of 195 mm Hg at 0.5 atmosphere total pressure, Aerospace Med. **41:**1051, 1970.

184. McAllister, T. A., et al.: Inhibitory effects of hyperbaric oxygen on bacteria and fungi, Lancet **2:**1040, 1963.

185. Menn, S. J., Sinclair, R. D., and Welch, B. E.: Response of normal man to graded exercise in progressive elevations of CO_2, SAM-TR68-116. Brooks Air Force Base, Texas, 1968, USAF School of Aerospace Medicine.

186. Meyer, H.: Zur Theorie der Alkoholnarkose, Arch. Exp. Pathol. Pharmakol. **42:**109, 1899.

187. Miller, K. W., et al.: Animals at very high pressures of helium and neon, Science **157:**97, 1967.

188. Miller, S. L.: A theory of gaseous anesthetics, Proc. Natl. Acad. Sci. **47:**1515, 1961.

189. Modern aspects of treatment of decompression sickness. Symposium on Undersea-aerospace Medicine at the First Annual Meeting of the Undersea Medical Society, Aerospace Med. **39:**1055, 1968.

190. Morales, M. F., and Smith, R. E.: The physiological factors which govern inert gas exchange, Bull. Math. Biophys. **7:**99, 1945.

191. Nash, G., Blennerhassett, J. B., and Pontoppidan, H.: Pulmonary lesions associated with oxygen therapy and artificial ventilation, N. Engl. J. Med. **276:**368, 1967.

192. National Oceanic and Atmospheric Administration, Manned Undersea Science and Technology Office: NOAA diving manual: diving for science and technology, Washington, D.C., 1975, Government Printing Office.

193. Nielsen, M.: Untersuchungen Uber die Atemregulation Beim Menschen. Skand. Arch. Physiol. **74**(suppl. 10): 83, 1936.

194. Nobrega, F. T., Turaids, T., and Gallagher, T. J.: Production of absorptional atelectasis in high oxygen environment at a simulated altitude of 27,000 feet, NAEC-ACEL-528, Philadelphia, 1965, U.S. Naval Air Engineering Center, Aerospace Crew Equipment Laboratory.

195. Noell, W. K.: Recovery of rabbit retina from hyperoxia. In Hyperbaric medicine, Publ. 1404, Washington, D.C., 1966, National Academy of Sciences, National Research Council.

196. Ochoa, S.: Enzymatic mechanisms of carbon dioxide fixation. In Sumner, J. B., and Myrback, K., editors: The enzymes, New York, 1952, Academic Press, Inc., vol. 2.

197. Pace, N., Strajman, E., and Walker, E. L.: Acceleration of carbon monoxide elimination in man by high pressure oxygen, Science **111:**652, 1950.

198. Patz, A.: Retrolental fibroplasia, Pediatr. Clin. North Am., Feb.: 239, 1958.

199. Patz, A., et al.: Oxygen studies in retrolental fibroplasia, Am. J. Ophthalmol. **36:**1511, 1953.

200. Pauling, L.: A molecular theory of general anesthesia, Science **134:**15, 1961.

201. Penrod, K. E.: Effect of intermittent nitrogen exposures on tolerance to oxygen at high pressures, Am. J. Physiol. **186:**149, 1956.

202. Perot, P. L., and Stein, S. N.: Conduction block in peripheral nerve produced by oxygen at high pressure, Fed. Proc. **15:**144, 1956.

203. Peters, J. P., and Van Slyke, D. D.: Quantitative clinical chemistry, Baltimore, 1931, The Williams & Wilkins Co.

204. Porter, J. R.: Bacterial chemistry and physiology, New York, 1946, John Wiley & Sons, Inc.

205. Pratt, P. C.: The reaction of the human lung to enriched oxygen atmosphere, Ann. N.Y. Acad. Sci. **121:**809, 1965.

206. Price, H. L.: Effects of carbon dioxide on the cardiovascular system, Anesthesiology **21:**652, 1960.

207. Puy, R. J. M., et al.: Alterations in the pulmonary capillary bed during early O_2 toxicity in man, J. Appl. Physiol. **24:**537, 1968.

208. Rennie, D. W., and Knox, F. G.: Effect of O_2 at high

209. Richmond, V., and D'Aoust, B. G.: Effects of intermittent hyperbaric oxygen on guinea pig lung elastin and collagen, J. Appl. Physiol. **41:**295, 1976.

210. Rizzo, A., et al.: Influence d'une atmosphere de CO_2 à 4% sur le comportement metabolique à l'exercice d'hommes jeunes, Bull. Eur. Pathophysiol. Respir. **12:** 209, 1976.

211. Robertson, W. G., et al.: Physiologic response to increased oxygen partial pressure. II. Respiratory studies, Aerospace Med. **35:**618, 1964.

211a. Rostain, J. C., and Naquet, R.: Le syndrome nerveux des hautes pressions: Caracteristiques et evolution en fonction de divers modes de compression, Rev. Electroencephalogr. Neurophysiol. Clin. **4:**107, 1974.

212. Rostain, J. C., and Naquet, R.: Human neurophysiologic data obtained from two simulated heliox dives to a depth of 610 meters. In Shilling, C. W., and Beckett, M. W., editors: Underwater physiology. VI. Proceedings of the Sixth Symposium on Underwater Physiology, Bethesda, Md., 1978, FASEB.

213. Roughton, F. J. W.: The average time spent by the blood in the human lung capillary and its relation to the rates of CO uptake and elimination in man, Am. J. Physiol. **143:**621, 1945.

214. Roughton, F. J. W., and Darling, R. C.: The effect of carbon monoxide on the oxyhemoglobin dissociation curve, Am. J. Physiol. **141:**17, 1944.

215. Saltzman, H. A., et al.: Retinal vascular response to hyperbaric oxygenation, J.A.M.A. **191:**290, 1965.

216. Sayers, P. R., and Davenport, S. J.: Review of carbon monoxide poisoning, Public Health Bull. No. 195, Washington, D.C., 1930.

217. Schaefer, K. E., Nichols, G., Jr., and Carey, C. R.: Calcium phosphorus metabolism in man during acclimatization to carbon dioxide, J. Appl. Physiol. **18:** 1079, 1963.

218. Schaefer, K. E., Nichols, G., Jr., and Carey, C. R.: Acid-base balance and blood and urine electrolytes of man during acclimatization to CO_2, J. Appl. Physiol. **19:**48, 1964.

219. Schaefer, K. E., et al.: Respiratory acclimatization to carbon dioxide, J. Appl. Physiol. **18:**1071, 1963.

220. Schwerma, H., et al.: A study of resuscitation from the nearly fatal effects of exposure to carbon monoxide, Occup. Med. **5:**24, 1948.

221. Sieker, H. O., and Hickam, J. B.: Normal and impaired retinal vascular reactivity, Circulation **7:**79, 1953.

222. Sinclair, R. D., Clark, J. M., and Welch, B. E.: Comparison of physiological responses of normal man to exercise in air and in acute and chronic hypercapnia. In Lambertsen, C. J., editor: Underwater physiology, New York, 1971, Academic Press, Inc.

223. Smith, G.: Carbon monoxide poisoning, Ann. N.Y. Acad. Sci. **117:**684, 1965.

224. Smith R. E., and Morales, M. F.: On the theory of blood-tissue exchanges: fundamental equation, Bull. Math. Biophys. **6:**125, 1944.

225. Spaur, W. H.: 1600 foot dive. In Spalsbury, D., editor: The working diver—1974, Washington, D.C., 1974, The Marine Technology Society.

226. Spaur, W. H., et al.: Dyspnea in divers at 49.5 ata: mechanical, not chemical in origin, Undersea Biomed. Res. **4:**183, 1977.

227. Stadie, W. C., Riggs, B. C., and Haugaard, N.: Oxy-

gen poisoning. III. The effect of high oxygen pressures upon the metabolism of the brain, J. Biol. Chem. **160:** 191, 1945.

228. Steen, J. B.: Oxygen secretion in the swim-bladder. In Dickens, F., and Neil, E., editors: Oxygen in the animal organism, Oxford, 1964, Pergamon Press, Ltd.

229. Suutarinen, T.: Cardiovascular response to changes in arterial carbon dioxide tension, Acta Physiol. Scand. **67:**266, 1966.

230. Symposium on the respiratory physiology in manned spacecraft, Fed. Proc. **22:**1022, 1963.

231. Third Symposium on Underwater Physiology, Washington, D.C., 1966, National Academy of Sciences, National Research Council.

232. U.S. Navy diving manual, Washington, D.C., 1979, Department of the Navy.

233. Van den Brenk, H. A. S., and Jamieson, D.: Potentiation by anaesthetics of brain damage due to breathing high-pressure oxygen in mammals, Nature **194:**777, 1962.

234. Wada, J., and Iwa, T., editors: Proceedings of the Fourth International Congress on Hyperbaric Medicine, Tokyo, 1970, Igaku Shoin, Ltd.

235. Webster, A. P.: Some theoretical aspects of the use of multiple gas mixtures for deep-sea diving. In Proceedings of the Underwater Physiology Symposium, Publ. 377, Washington, D.C., 1955, National Academy of Sciences, National Research Council.

236. Welch, B. E., Morgan, T. E., and Ulvedal, F.: Observations in the SAM two-man space cabin simulator, Aerospace Med. **32:**583, 1961.

237. Winterstein, H.: Die Narkose, Berlin, 1926, Julius Springer.

238. Woodbury, D. M., and Karler, R.: The role of carbon dioxide in the nervous system, Anesthesiology **21:** 686, 1960.

239. Workman, R. D.: Calculation of decompression schedules for nitrogen-oxygen and helium-oxygen dives, Research report 6-65, Washington, D.C., May, 1965, U.S. Navy Experimental Diving Unit.

240. Workman, R. D.: Standard decompression procedures and their modification in preventing the bends, Ann. N.Y. Acad. Sci. **117:**834, 1965.

241. Workman, R. D., Bond, G. F., and Mazzone, W. F.: Prolonged exposure of animals to pressurized normal and synthetic atmospheres, Report 374, Washington, D.C., 1962, U.S. Navy Medical Laboratory.

242. Wright, R. A., et al.: Risk of mortality in interrupted exposure to 100% O_2: role of air vs. lowered Po_2, Am. J. Physiol. **210:**1015, 1966.

243. Wright, W. B.: Use of the University of Pennsylvania [Institute for Environmental Medicine] procedure for calculation of cumulative pulmonary oxygen toxicity, Report 2-72, Washington, D.C., 1972, U. S. Navy Experimental Diving Unit.

244. Wulf, R. J., and Featherstone, R. M.: A correlation of Van Der Waals constants with anesthetic potency, Anesthesiology **18:**97, 1957.

245. Yarbrough, O. D., et al.: Symptoms of oxygen poisoning and limits of tolerance at rest and at work, Project X-337 (Sub. No. 62, Report 1), 1947, Washington, D.C., U.S. Naval Experimental Diving Unit.

246. Zetterstrom, A.: Deep-sea diving with synthetic gas mixtures, Mil. Surgeon **103:**104, 1948.

XV

THE PHYSIOLOGY OF DEVELOPMENT AND AGING

76

JOHN D. BIGGERS

Fetal and neonatal physiology

NORMAL DEVELOPMENT[6,45]

The human zygote is a single cell that arises by the fusion of a sperm and an ovum. This process is called fertilization. The zygote weighs approximately 1.25 μg, and by successive multiplication and division, this cell may give rise to a newborn infant weighing 3 to 4 kg and containing approximately $2^{42} \sim 4$ trillion cells. Eventually this infant grows into an adult human being who weighs, on the average, approximately 50 kg if female and 70 kg if male. Fertilization completes a process begun earlier during meiosis in both the male and female, which results in a new sample of the human gene pool. This sample may contain the information necessary for the development of the next individual life. Not all samples of the gene pool that are produced are compatible with normal development, and these samples account to a considerable extent for the normal high level of early spontaneous abortion.[20]

Embryologists conventionally divide the prenatal period into several stages (Table 76-1). The human zygote, as in all mammals, is formed in the ampullar region of the oviduct. It soon begins to move down the oviduct to the uterus, a journey that lasts about 3 days.[13] During this time the developing preimplantation embryo is "free living" in the sense that it can be flushed from the genital tract without disrupting any fixed morphologic connections with its mother. Throughout the tubal journey the embryo undergoes a series of mitoses called the cleavage divisions. At first the individual cells are easily recognized, but after about 16 cells are produced the identity of the separate blastomeres is lost. This process is called *compaction*, and it is associated with the development of tight junctions between the outer *trophoblast* cells.[15] The result is the formation of the first tissue in the new multicellular organism. This tissue is epithelial in nature, and it is called the *trophectoderm*. The trophectoderm provides the machinery for producing the extracellular fluid that results in the development of the blastocyst (Fig. 76-1). This fluid has at least two primary functions. One is to provide the milieu intérieur for the cells of the inner cell mass, which will eventually develop into the fetus, and the other is to cause the expansion of the blastocyst that is necessary for implantation. A somewhat later function of the trophectoderm is to invade the maternal epithelium during implantation—a process that leads to the development of the placenta. Implantation begins in women approximately 6 to 7 days after fertilization. Up to this time it is customary to speak of the *preimplantation stages of pregnancy*.

The invasion of the endometrium by the embryo is so complete that it becomes buried beneath the lining epithelium of the uterus (Fig. 76-2). This type of implantation is called interstitial implantation. By this time, further regionalization of the embryo has occurred; thus the embryo consists of the three primary germ layers—the ectoderm, entoderm, and mesoderm—and also the amnion, amniotic cavity, and primitive yolk sac. The trophectoderm surrounds the entire embryo but becomes differentiated into two parts: the inner *cytotrophoblast* and the outer *syncytiotrophoblast*. At this stage of development a *conceptus* has formed, consisting of embryonic and extraembryonic parts. The endometrium of the pregnant uterus is called the *decidua* and eventually forms three principle regions: the *decidua basalis,* the *decidua capsularis,* and the *decidua parietalis* (Figs. 76-3, *A, B,* and *C*). During this period the human embryo is still very small and can conduct its vital processes through the syncytiotrophoblast and cytotrophoblast by local diffusion and active transport.

Soon after, the germ layers begin to differentiate into the organ systems, and at the same time the extraembryonic parts and their associated fluids begin to increase rapidly in size. Dif-

1947

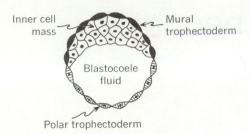

Fig. 76-1. Schematic diagram of mammalian blastocyst. This stage consists of three primary compartments: inner cell mass, trophectoderm, and blastocoele fluid. Trophectoderm is simple squamous epithelium consisting of trophoblast cells bound together by tight junctions. Two regions of trophectoderm are recognized: mural trophectoderm overlying inner cell mass and remaining polar trophectoderm.

Table 76-1. Human developmental stages

Period	Time after conception	Stages	Approximate conceptual age (days)
Preimplantation	1-55 days	Cleavage	1-3
		Blastocyst	4-6
Embryo		Gastrula	7-8
		Neurula	21-28
		Period of embryo completion	29-56
Fetus	56 days to term (267 days)	First fetal	56-70
		Second fetal	70-140
		Third fetal	140-term

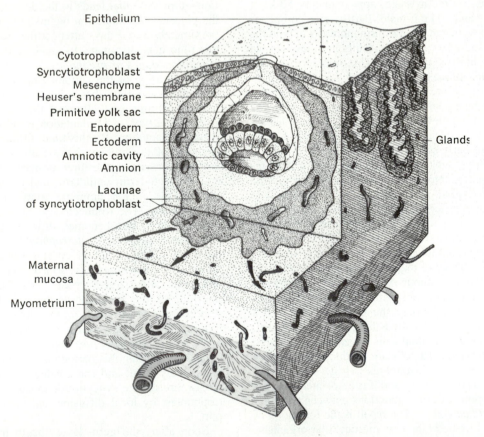

Fig. 76-2. Schematic diagram showing position of early human embryo at end of implantation. Embryo has become buried in endometrium, and it is separated from maternal decidual tissue by syncytiotrophoblast and cytotrophoblast. (From Tuchmann-Duplessis et al.[96])

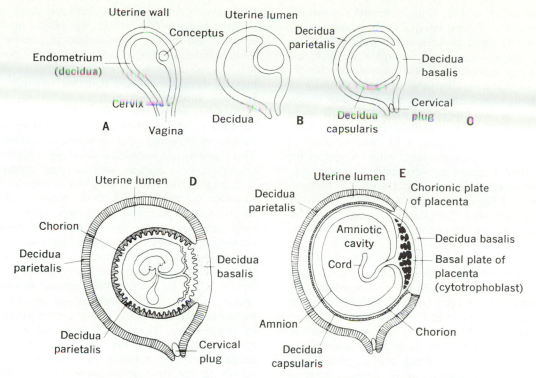

Fig. 76-3. Relationships between conceptus and decidua at several stages of human pregnancy. **A** to **C**, Development of three principal regions of decidua: *decidua basalis, decidua capsularis,* and *decidua parietalis.* **D,** Shows 5½-week conceptus with villi distributed all over *decidua basalis* and *decidua capsularis.* **E,** Approximately 10-week conceptus where villi have been lost from *decidua capsularis* and are confined to *decidua basalis* that has developed into placenta. (From Steven and Morriss.[94])

fusion and active transport mechanisms quickly become insufficient to subserve the exchanges needed by the fetus. Thus by about the twenty-first day after fertilization a fetal circulation begins to function.[6] For the first 12 weeks the entire conceptus is covered by *villi* through which exchanges with the mother take place. These villi are found on both the decidua capsularis and the decidua basalis (Fig. 76-3, *D*). By 10 weeks, however, the villi on the decidua capsularis become rarefied, and by 12 weeks they are confined to one pole of the embryonic vesicle and form the placenta (Fig. 76-3, *E*). This structure is a highly specialized transitory organ peculiar to pregnancy through which most of the exchanges occur between the mother and the fetus. The rest of the embryonic vesicle is translucent and is called the *chorion laeve.* It, too, may be the site of exchange between mother and fetus, but only for special functions.

The human fetus is supported by the placenta for approximately 28 weeks until birth, when the fetus is expelled from the mother into the external world. Then the alimentary, respiratory, and renal systems of the fetus replace the placenta. Under natural conditions the newborn infant is not suddenly and totally cut off from maternal support, since it is fed by milk produced by the mother. Lactation may last many months, but in some societies, such as in the United States, this maternal function is frequently terminated at birth and replaced by artificial substitutes.[53] The adaptation to a physiologically independent life occurs gradually after birth, but most major changes are complete within 1 month. From birth to 1 month of age is arbitrarily defined as the *neonatal period.*

PROBLEMS IN FETAL AND NEONATAL PHYSIOLOGY

Two fundamental concepts have played major roles in shaping our current physiologic ideas: Schoenheimer's theory of the dynamic state of body constituents[83] and Cannon's theory of homeostasis.[22] The dynamic state and the maintenance of a constant internal environment involve relatively rapid processes. All multicellular organisms, however, pass through a life cycle that involves morphologic and physiologic change. This cycle arises because of the mechanism of forming a new sample of the gene pool by the fusion of two unicellular stages at fertilization. The return to the multicellular form is a relatively slow process compared with those involved in homeostasis. Nevertheless, within a species the patterns of change are very stable, a concept embodied in Waddington's theory of homeorrhesis.[100] The developmental changes are usually sufficiently slow that effective homeostatic regulation operates throughout, although set points may change.

The overall growth and final size of an individual and the development of the structure and function of the different organ systems are ultimately under genetic controls determined by the sample of genes from the gene pool that are acquired at fertilization. The interpretation of the information carried by these genes is under epigenetic control, involving the sequential transcription and translation of constellations of genes. Maternal genetic and epigenetic controls, however, begin to operate in an oocyte prior to fertilization as part of the mechanisms that ensure the continuity of life between generations. In the words of E. V. Wilson, "Embryogenesis begins in oogenesis." Thus the preimplantation phase of development is programmed by the maternal genome, since parthenogenetic development to the blastocyst stage can be induced without the involvement of a sperm.[40] A primary question in the physiology of development concerns the nature of the genetic and epigenetic controls. Since these processes result in large increases in the size of the organism, another major question in the physiology of development concerns the transport of materials into the embryo for growth and metabolism and the elimination of waste products.

Mammalian development is complicated by *viviparity*—the bearing of live young. In general the control of embryonic and neonatal life involves processes occurring in four periods, which are separated by three relatively abrupt events. The periods are oogenesis, preimplantation development, fetal development, and neonatal development. The first abrupt event is fertilization, which is the normal trigger to activate the mature ovum that is poised for development.[14] The second and third abrupt events are implantation and parturition, respectively. These latter events and the period between them are singular because they utilize physiologic mechanisms involving communication between mother and fetus. Implantation is a control point that is used in some species to synchronize embryonic development and maternal functions.[12] In many species, such as some carnivores and marine mammals, the embryo develops to the blastocyst stage and then stops. After a time, development resumes and implantation occurs. This phenomenon is called *delayed implantation,* or *embryonic diapause,* and in some species, such as the river otter, it may last as long as 11 months. The mink, in which the natural period of arrested development is 49 days, is particularly interesting, since the resumption of development in this species is triggered by a maternal signal. Thus the period of arrest can be artificially shortened by increasing the daily amounts of light received by the pregnant females. In some species, including the human, there appears to be no maternal control of development at the time of implantation. However, a hormone is produced by the cells of the trophectoderm called human chorionic gonadotropin (HCG), which passes to the mother, suppresses the ovarian cycle, and establishes the pattern of maternal ovarian function characteristic of pregnancy. HCG is the first hormone to be produced by the embryo. Birth is triggered by corticosteroids that are produced by the fetal adrenal cortex under the control of fetal ACTH. Thus anencephaly, a congenital abnormality in which the fetus lacks many structures in the head, including the pituitary gland, is associated with prolonged pregnancy. During the period between implantation and birth the conceptus secretes progesterone, which maintains the pregnancy after the corpus luteum ceases to function 3 months after conception. The conceptus secretes human chorionic somatomammotropin (human placental lactogen), which has a glucose-sparing effect on the maternal metabolism, thus favoring the transport of glucose to the fetus. This hormone may also cause the growth of the maternal mammary glands before birth.

Another set of physiologic questions is involved with the transient changes that are associated with birth, particularly the replacement of the placental circulatory system with the pulmonary circulatory system. Up to the time of birth the fetus develops in a protected, very stable environment controlled by homeostatic mechanisms operating within the mother. After birth the individual suddenly encounters a new and relatively unstable environment. A further set of questions therefore concerns the mechanisms mobilized in the neonate to maintain the constancy of its own internal environment.

QUANTITATIVE ASPECTS OF FETAL AND NEONATAL GROWTH[78]

The prenatal growth of the fetus depends on many factors such as parity, sex, multiple births, country, and socioeconomic class (Fig. 76-4).

Furthermore, there are important changes in the velocity of growth (Fig. 76-5). Up to the twentieth week the rate of growth is relatively slow. Thereafter it accelerates and reaches a maximum at about the thirty-fourth week of prenatal age. The rate of growth then declines until birth, after which time it again increases to reach a peak at about 8 weeks of postnatal age.

The increase in fetal mass that occurs during

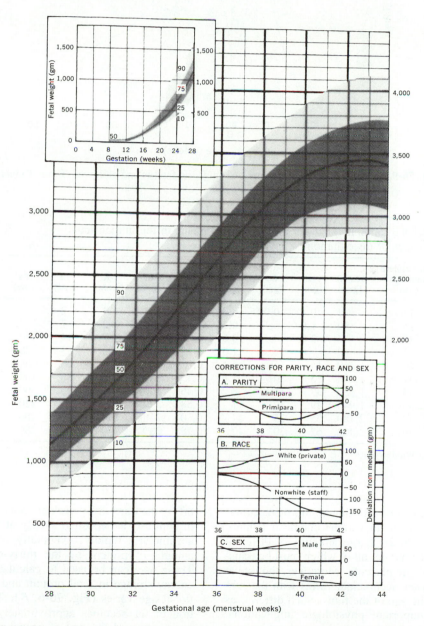

Fig. 76-4. Mean fetal weight and 50th (median), 10th, 25th, 75th, and 90th percentiles of fetal weight in grams throughout pregnancy and correction factors for parity, race (socioeconomic factors), and sex. (From Brenner et al.[21])

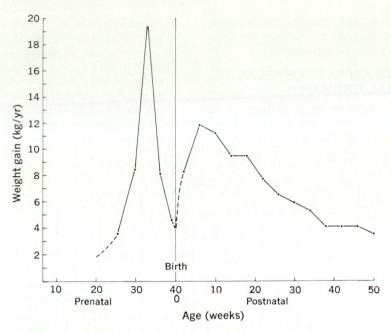

Fig. 76-5. Velocity of growth in weight of singleton fetuses and children. (From Tanner.[95])

Table 76-2. Total amounts of water and several elements in the human fetus at different stages of development*

Body weight (gm)	Approximate fetal age (wk)	Water (gm)	N (gm)	Ca (gm)	P (gm)	Mg (gm)	Na (mM)	K (mM)	Cl (mM)	Fe (mg)	Cu (gm)	Zn (gm)
30	13	27	0.4	0.09	0.09	0.003	3.6	1.4	2.4	—	—	—
100	15	89	1.0	0.3	0.2	0.01	9	2.6	7	5.1	—	—
200	17	177	2.8	0.7	0.6	0.03	20	7.9	14	10	0.7	2.6
500	23	440	7.0	2.2	1.5	0.10	49	22	33	28	2.4	9.4
1,000	26	860	14	6.0	3.4	0.22	90	11	66	64	3.5	16
1,500	31	1,270	25	10	5.6	0.35	125	60	96	100	5.6	25
2,000	33	1,620	37	15	8.2	0.46	160	84	120	160	8.0	35
2,500	35	1,940	49	20	11	0.58	200	110	130	220	10	43
3,000	38	2,180	55	25	14	0.70	240	130	150	260	12	50
3,500	40	2,400	62	30	17	0.78	280	150	160	280	14	53

*Modified from Widdowson.[102]

development can be partitioned into the accumulations attributable to various chemical components (Table 76-2). Large amounts of different materials accumulate, all of which must be transported across the placenta. Moreover, the rate at which materials accumulate varies with the substance and the age of the fetus. Such differences may have important physiologic implications. Fig. 76-6, *A* compares the amounts of protein and fat in the human fetus at different ages. Protein accumulates in the early fetus and then levels off by the thirty-fifth week at about 300 gm. Fat, in contrast, is deposited slowly at first but then accumulates rapidly, eventually overtaking the amount of protein by the thirty-eighth week. These data can be used to calculate the rate of energy storage in fetal protein and fat as gestational age varies (Fig. 76-6, *B*). The contribution of fat becomes approximately three times greater toward the end of pregnancy. Much of this fat is deposited subcutaneously rather than as deep body fat.

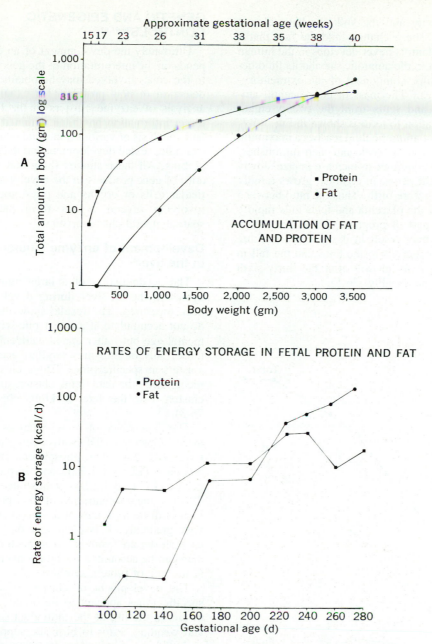

Fig. 76-6. A, Total amount of fat and protein in human fetus from 15th to 40th week of gestation. **B,** Rate of energy storage from fat and protein in human fetus from 100th to 280th day of gestation. (From Southgate and Hey.[92])

Another important change in the chemical composition of the fetus during development is due to a decrease in the relative amount of water.[92] The young fetus has as much as 95% water, but the proportion of solids increases as development proceeds. Both intracellular and extracellular water are involved in the changes. Thus the ratio of extracellular to intracellular water is 4 at 10 weeks of age, 1.8 at 30 weeks, and unity near term.

The conceptus can be partitioned into the embryonic (fetal), fluid, and placental components. All three components increase in size during early pregnancy. Two fluid-filled cavities appear early in human development: the amnionic sac and the chorionic sac. By 9 weeks, however, the chorionic cavity and the yolk sac have almost disappeared. These changes and the fact that a significant allantois never develops in the human species result in the amniotic sac and its fluid becoming a major component of the extraembryonic structures.[6] The mass of the placenta, amniotic fluid, and fetus from the twenty-eighth week of gestation to term is shown in Fig. 76-7. The amount of amniotic fluid reaches a maximum by the thirty-fourth week and then diminishes. The placenta continues to increase in size slowly until birth. The fetus, in contrast, grows rapidly until shortly before birth. Thus the ratio between the weight of the placenta and fetus falls rapidly in the latter part of pregnancy. The diminution of this ratio may result in the placenta limiting transport of materials to the fetus and the fall in the velocity of its growth after the thirty-sixth week of pregnancy (Fig. 76-5).

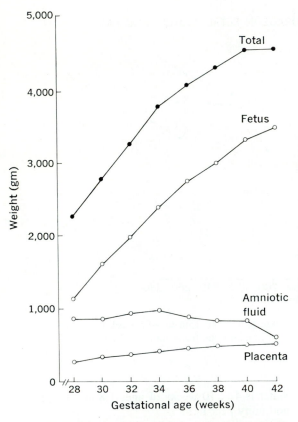

Fig. 76-7. Total weight of human conceptus and weights of fetus, amniotic fluid, and placenta from 28th to 44th week of conception. Weights of placenta and fetus are from Gruenwald[43] and weight of amniotic fluid has been estimated from mean volumes of amniotic fluid observed by Queenan et al.[77]

GENETIC AND EPIGENETIC CONTROLS[39,41,61]

Ultimately the development of an embryo depends on the interpretation of the genome. A clue to the genes involved may be obtained from the detection of new proteins, for example, in the analysis of enzyme clusters in the liver. More direct information has been obtained by the genetic analysis of various abnormal conditions, for example, sexual development and the fetal hemoglobins. All these studies show that the appearance of gene products in the development of particular cells or organs and physiologic systems involving several organs form patterns with spatiotemporal characteristics.

Development of enzyme clusters in the liver[42]

The accumulation of a large number of enzymes in the rat liver during development has been measured. The results show that enzymes do not accumulate at a linear rate with developmental age but rather appear suddenly. Furthermore, sets of enzymes tend to accumulate in clusters at specific times. Three clusters can be recognized: the late fetal cluster, the neonatal cluster, and the late suckling cluster (Table 76-3).

The appearance of representatives of each of these clusters—UDPG-glycogen-glucosyltransferase (EC 2.4.1.11), phosphoenolpyruvate carboxylase (EC 4.1.1.32), and tryptophan oxygenase (EC 1.13.1.12)—is illustrated in Fig. 76-8. Since the concentration of total protein in the liver at different stages of development increases very gradually, it is clear that the enzymes in each cluster are synthesized at much greater rates than can be accounted for by the overall increase in the rate of protein synthesis.

The development of clusters of enzymes at specific times appears to be related to epochal events in the life of the individual such as birth and weaning. Prior to birth the supply of nutrients and the elimination of waste products are mediated by the placenta. After birth the infant takes over these functions. An important function of the liver is the elimination of nitrogen in the form of urea, and the urea cycle enzymes from carbamylphosphate synthetase (EC 2.7.2.5) through arginase (EC 3.5.3.1) emerge in the late fetal cluster. A major breakdown product of hemoglobin is bilirubin; this substance is eliminated from the fetus unchanged via the placenta and then the maternal liver, where it is first conjugated with glucuronide (p. 1132). After birth, bilirubin is eliminated by the liver in the infant

recessive trait.[52] These individuals develop normal gonads and normal male internal secondary genital organs, but externally the genital organs, derived from the urogenital sinus, are female. The patients are deficient in the enzyme 5α-reductase, which converts testosterone to 5α-dihydrotestosterone. It is now known that the wolffian ducts can respond to testosterone directly, whereas the urogenital sinus cannot unless the circulating testosterone is first converted to 5α-dihydrotestosterone.

The sexual development of a male fetus thus depends on at least five genes located on both X and Y chromosomes and an autosome. This

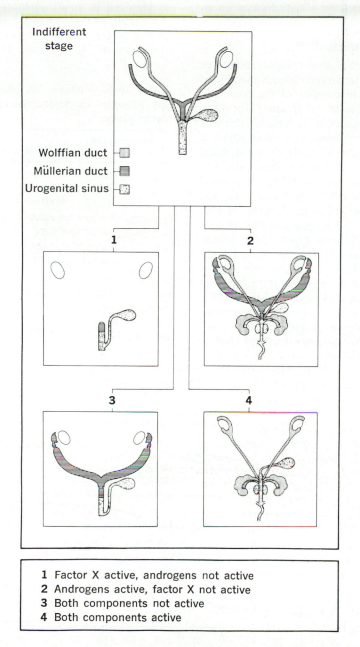

1	Factor X active, androgens not active
2	Androgens active, factor X not active
3	Both components not active
4	Both components active

Fig. 76-9. Possible routes of development of genital tract from indifferent stage. Indifferent stage possesses both müllerian and wolffian ducts. Normal females do not produce androgens nor müllerian inhibitory factor (MIF) and follow route *3*. Normal males produce both androgens and MIF and follow route *4*. Routes *1* and *2* are pathologic pathways of development. (From Neumann et al.[71])

constellation of genes, however, exerts its effects in sequence. In the human male fetus the testes develop by 40 days of gestation under the control of the H-Y antigen involving three gene products. The wolffian duct continues to develop, mediated by a gene on the X chromosome, and the urogenital sinus develops by about 70 days of gestation, mediated by a gene located on an autosome.

Hemoglobin synthesis[73,103]

Hemoglobins are a class of complex molecules with a high affinity for oxygen found in red blood cells. Each molecule consists of the prosthetic group heme bound to an apoprotein. The variation among the hemoglobins is due to differences among the apoprotein moieties, each of which is a tetramer that consists of two pairs of identical globin chains. Seven different globins exist normally during development, and they are denoted α, ζ, ϵ, G_γ, A_γ, β, and δ, respectively. G_γ and A_γ differ only in the amino acid at position 136; G_γ contains glycine and A_γ contains alanine. The abundance of the different types during human development is shown in Fig. 76-10. α-Globin appears very early in development, and its synthesis reaches a plateau by the fifth gestational week, where it remains throughout the life of the individual. ϵ-Globin and ζ-globin appear very early but disappear by the eighth week of pregnancy. γ-Globin also appears very early, and its

synthesis rises to a plateau that persists until shortly before birth. The synthesis of this globin then falls and reaches a low level by the eighteenth week after birth. β-Globin is synthesized in small amounts in fetal life. Shortly before birth the rate of synthesis rises and reaches a plateau by the eighteenth week after birth. δ-Globin is synthesized in small amounts only after birth. As a result of these changes, the types of hemoglobin present in the red blood cells alter with development (Table 76-6). Three types of em-

Table 76-6. Globin chain composition of the hemoglobins produced during human development*

Developmental stage	Hemoglobin	α-Like chain	β-Like chain
Embryonic	Gower 1†	ζ_2	ϵ_2
	Gower 2	α_2	ϵ_2
	Portland 1	ζ_2	γ_2
Fetal	F	α_2	G_{γ_2}
	F	α_2	A_{γ_2}
	A	α_2	β_2
Adult	A	α_2	β_2
	A_2	α_2	δ_2

*From Wood.[103]
†The composition of Hb Gower 1 may be ϵ_4; the composition of $\zeta_2\epsilon_2$ is based on the recent suggestion of Huehns and Farooqui.[51]

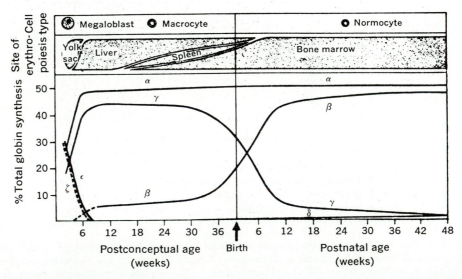

Fig. 76-10. Changes in erythroid cell type, site of erythropoiesis, and globin synthesis during human development. (From Wood.[103])

bryonic hemoglobin have been found up to the eighth week of development. Hb F exists in two forms, $\alpha_2 G_{\gamma_2}$ and $\alpha_2 A_{\gamma_2}$, in the ratio of $3:1$ and is found in abundance throughout fetal life. A small amount of Hb A is also found during this period. After birth the predominant Hb is $\alpha_2\beta_2$, with small amounts of $\alpha_2\delta_2$ and Hb F. The major change at birth is the replacement of $\alpha_2\gamma_2$ by $\alpha_2\beta_2$. This phenomenon is called *the switch*.[72]

The formation of red blood cells occurs in several organs (Fig. 76-10). Very early on in development, erythropoiesis is in the yolk sac. By the sixth week the process ceases in this organ and is taken over by the fetal liver. At about 15 weeks of development the spleen and bone marrow begin to manufacture red blood cells. Shortly after birth, erythropoiesis ceases in the liver and spleen, and red blood cells are produced solely by the bone marrow. There is no evidence that the different organs or clones of erythroid cells produce specific globins because the different species of Hb are present together in single red blood cells.

The synthesis of the different types of globin chains is under genetic control. By the use of somatic cell hybrids, it has been shown that the genes coding for α- and β-globin are on separate chromosomes. The gene for β-globin is part of a gene complex present in chromosome 11 that consists of two γ-globin genes, a δ-globin gene, and a β-globin gene.[33] The complex also contains a genetic locus that influences the production of Hb F in the adult. The mechanisms controlling the switch that occurs at birth are not understood, but they presumably involve interactions within this gene complex.

TRANSPORT ACROSS THE MATERNAL-EMBRYONIC INTERFACE
Early mammalian embryo

Prior to the formation of tight junctions between the outer cells of the 16-cell stage, all cells of the embryo are exposed to the environment provided by the oviduct or uterus. Exchanges between these cells and their environment occur through the plasma membranes by mechanisms common to all cells—diffusion, facilitated diffusion, active transport, and pinocytosis. Once the outer cells become intimately associated a simple epithelium develops that permits vectorial exchanges between the extracellular fluid of the embryo and its environment. The mechanisms involved in this new capability are probably not fundamentally different from mechanisms present in other epithelia.

Little is known about the transport processes between the mother and embryo during the period between implantation and the onset of placental circulation, other than the organization of the structures involved. Materials pass through a barrier consisting of the endothelium of the maternal capillary and its basement membrane and a bilaminar epithelial structure that develops from the trophectoderm.

Placental transport[30]

The main route of exchange between mother and fetus occurs across a thin cellular barrier that separates the maternal placental circulation from the fetal *placental circulation*. In primates the anatomic vascular unit is called a *lobule*, and about 200 of these units are present in the fully developed placenta.[26] The structure of a lobule is shown in Fig. 76-11. A main stem villus branches into secondary and tertiary stems, which in turn give rise to the terminal villi. The accompanying fetal artery and vein subdivide in a parallel fashion. The major cotyledons are anchored to the basal plate of the placenta by a ring of villous stems, while free villi are closely packed at the periphery. The center of the cotyledon is relatively free of villi, and it is in this region that a maternal spiral artery discharges its blood to perfuse the intervillous space. The maternal veins that drain the intervillous space leave via the basal plate. The arrangement of the blood vessels in a terminal villus is shown in Fig. 76-12. The capillaries of the terminal villi are vascularized from a previous network of fine anastomosing blood vessels called the paravascular net. Ultrastructural studies have demonstrated that the mature human placenta is of the hemomonochorial type (Fig. 76-13). In many areas the maternal and fetal bloods are separated by the endothelial cells and basement membrane of the fetal capillary, the basement membrane of the trophoblast, and the syncytiotrophoblast. In earlier stages of development, another layer of cells called the cytotrophoblast (Langhan's cells) is present everywhere between the basement membrane of the trophoblast and the syncytiotrophoblast. In the mature placenta the cytotrophoblast persists only in patches. The barrier between the fetal and maternal blood is therefore cellular, and consequently the physiologic principles involved in the analysis of placental transport are those characteristic of epithelial transcellular transport.

Unfortunately, relatively little is known about the physiologic nature of this epithelial barrier, such as its electrical resistance and its leakiness. The fact that the syncytiotrophoblast is a continuous cellular structure with no junctions implies

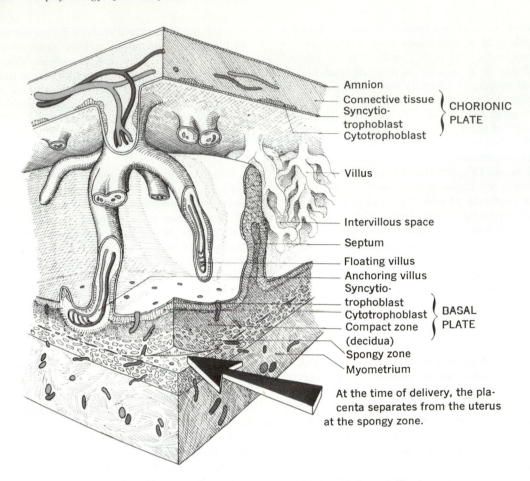

Amnion
Connective tissue
Syncytio-
trophoblast
Cytotrophoblast
} CHORIONIC
PLATE

Villus

Intervillous space

Septum

Floating villus
Anchoring villus
Syncytio-
trophoblast
Cytotrophoblast } BASAL
Compact zone PLATE
(decidua)
Spongy zone
Myometrium

At the time of delivery, the placenta separates from the uterus at the spongy zone.

Fig. 76-11. Structure of mature placenta. (From Tuchmann-Duplessis et al.[96])

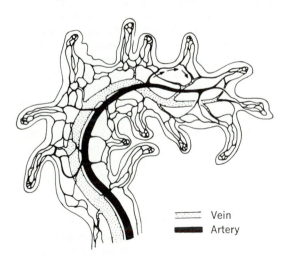

Vein
Artery

Fig. 76-12. Schematic drawing to show arrangement of fetal blood vessels in human placenta. Capillaries of villi are vascularized exclusively from previllous network of fine, anastomosing blood vessels, the paravascular net. This arrangement may enable short-circuiting of blood away from villous capillaries, as indicated by arrows. (From Bøe.[18])

that there is a very high electrical resistance across the membrane. However, it is also known that cells can cross from the fetal to the maternal circulation, suggesting that a considerable degree of leakiness exists.[9,30] The passage of cells from fetus to mother can be the cause of very serious conditions. Fetal trophoblast cells enter the maternal circulation in all pregnancies. Most women, however, are able to destroy these cells, but in the rare cases in which this destruction does not occur a rapidly fatal malignancy—choriocarcinoma—results. Many individuals carry a surface antigen (D antigen) on their erythrocytes. This antigen is inherited as a dominant factor. Homozygotes or heterozygotes carrying the antigen are called Rh+ and homozygotes without the antigen are called Rh−. If an Rh− mother carries an Rh+ fetus, the fetal red blood cells may enter the maternal circulation and stimulate the production of anti-D antibodies. If these antibodies then pass back into the fetal circulation, they destroy fetal erythrocytes and

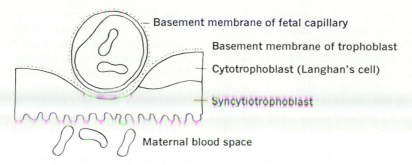

- Basement membrane of fetal capillary

Basement membrane of trophoblast

Cytotrophoblast (Langhan's cell)

Syncytiotrophoblast

Maternal blood space

Fig. 76-13. Structure of human hemomonochorial placenta. Fetal and maternal blood systems are separated by cellular syncytiotrophoblast, basement membrane of trophoblast, basement membrane, and endothelium of fetal capillary. (From Steven.[93])

cause a serious syndrome called *erythroblastosis fetalis*. The mechanisms involved in the passage of cells across the placenta have naturally attracted considerable attention. Early clinical observations on erythroblastosis fetalis indicated that the leakage of erythrocytes tended to be from fetus to mother.[24] By progressively raising the pressure in the umbilical vein in the guinea pig, which also has a hemomonochorial placenta, an increase in the transfer from fetus to mother, first of water, then of plasma proteins, and then of blood cells, occurred.[31] The transfers disappeared on returning the umbilical vein pressure to normal. Thus it has been suggested that leakage of cells from fetus to mother is caused by transient increases in pressure in the venous placental circulation.

A rise in maternal intravillous pressure would tend to collapse the fetal villous capillaries and not favor passage of cells. Moreover the leakage of fetal cells into the mother need only be small in number to elicit the immunologic response or malignancy. For these reasons, it will therefore be assumed that during most of the pregnancy the placenta is an effective epithelial barrier in which well-known transcellular transport mechanisms occur.

Comparison of the concentrations of several substances in maternal cord blood shows that they may be about equal, lower in the fetus, or higher in the fetus (Table 76-7). The presence of a substance in the fetus does not necessarily imply it was acquired from the mother. Many substances are synthesized in the fetus from precursor substances. For example, phospholipids are released from maternal plasma proteins and hydrolyzed in the placenta. The products are transferred to the fetus and resynthesized into phospholipids in the fetal liver. The concentra-

Table 76-7. Relative levels of some constituents of maternal and cord blood*

	About equal	Lower in fetus	Higher in fetus
Amino acids			+
Urea	+		
Uric acid	+		
Creatinine	+		
Inorganic phosphorus			+
Free fatty acids		+	
Cholesterol		+	
Glucose		+	
Lactic acid			+
Calcium			+
Magnesium	+		
Chloride	+		
Sodium	+		
Potassium	+		
Iron			+
Vitamins			
Fat-soluble		+	
Water-soluble			+
Chorionic gonadotropin		+	
Placental lactogen		+	
Growth hormone			+

*From Dancis and Schneider.[30]

tion of growth hormone in cord blood is higher than maternal blood. The growth hormone in the fetus is synthesized by the fetal anterior pituitary gland from precursor amino acids and is not derived from the maternal anterior pituitary gland. Thus anencephalic monsters have low levels of growth hormone.

The mechanisms involved in the transport of many substances across the placenta are bidirectional. The majority, however, are associated with a net transfer from mother to fetus. Under normal circumstances, only the waste products of fetal metabolism are associated with a net transfer from fetus to mother. These substances are carbon dioxide and the nitrogenous products urea, uric acid, creatinine, and bilirubin.

All types of transport mechanisms are involved in the transfer of substances across the placenta. Some, such as oxygen and many lipid-soluble, nonpolar substances, cross by simple diffusion down a concentration gradient. Other water-soluble, polar molecules such as glucose cannot cross the syncytiotrophoblast and require carrier-mediated mechanisms. Still other substances such as amino acids are actively transported.

The direction in which some substances move across the placenta may be controlled by chemical modification. For example, the nonpolar bilirubin molecule passes freely across the placenta in either direction. In contrast, the polar conjugate bilirubin glucuronide cannot cross the placenta. This difference in the properties of the two compounds is important in the elimination of bilirubin from the fetus. Fig. 76-14 shows the antepartum excretion of bilirubin. In the mother the bilirubin is transported in the blood by serum albumin to the liver, where it undergoes glucuronidation and is excreted in the bile. The fetal liver, however, does not possess the enzyme UDP-glucuronyltransferase, which is necessary for the glucuronidation of bilirubin. Bilirubin produced by the breakdown of fetal hemoglobins is thus transported to the placenta bound to serum albumin, where it is dissociated to diffuse down a concentration gradient to the maternal blood. The bilirubin is then carried to the maternal liver, where it is conjugated and excreted. Thus no bilirubin accumulates in the fetus, even if it is affected with congenital hemolytic disease. On the other hand, severe jaundice of the mother also causes jaundice of the fetus. At birth the UDP-

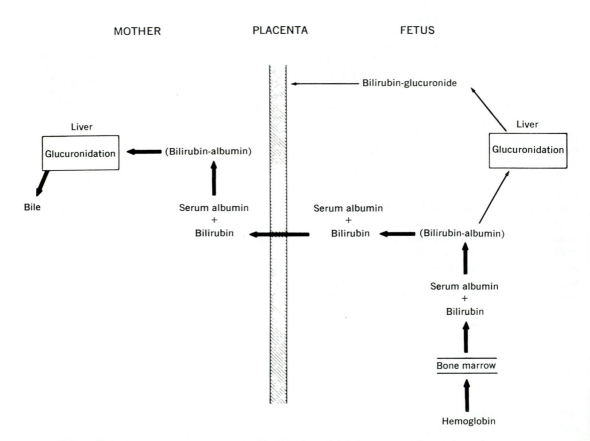

MOTHER   PLACENTA   FETUS

Fig. 76-14. Antepartum excretion of bilirubin. Fetal bilirubin is transferred from fetal serum albumin through placenta to maternal circulation, then conjugated by maternal liver and excreted into bile (thick arrows). Fetal synthesis of poorly transferred glucuronides is suppressed (thin arrows). (From Dancis.[29])

glucuronyltransferase is synthesized in the fetal liver as part of the neonatal cluster, and the newborn assumes its own excretion of bilirubin.[42]

The passage of hormones across the placenta is of particular importance. In general, maternal polypeptide hormones and thyroxine pass so slowly across the placenta that they do not significantly affect the fetus. The only protein hormone that crosses the barrier is chorionic gonadotropin. This hormone, however, is synthesized by the trophoblast cells of the fetus and delivered immediately to the mother. Unconjugated estrogens can readily pass across the placenta in either direction. Their polar conjugates, the sulfates and glucuronates, however, are very restricted in their passage across the placenta. Sulfatases are present in the placenta, which cleave estrogen sulfates to facilitate estrogen transfer. Progesterone can rapidly cross the human placenta.

Two of the major substances transported into the fetus are O_2 and water. O_2 is necessary for the provision of energy for the large amount of work that occurs in the growing fetus. The O_2 is required in a relatively uninterrupted supply, since the O_2 stores of the fetus are very small. For example, a fetus near term weighing 3 kg needs about 18 ml O_2/min. Since the O_2 stores are only about 36 ml, the reserve can only last about 2 min. Water is by far the most abundant molecular species in the fetus, and a large amount also accumulates in the amniotic sac (Fig. 76-7). The amount is in excess of that produced metabolically by the fetus; thus a net transfer of water from mother to fetus occurs. Near term the total flux of water into the fetus is approximately 30 ml/day.

Despite these relatively large total fluxes, the flux per unit area of maternal-fetal interface is relatively small because of the very large surface area for exchange in the human placenta. At term the mean surface area of the fetal vessels in the villi is about 12 m².[3] Thus the fluxes of O_2 and water across the near-term human placenta are only $0.15\,\mu l \cdot cm^{-2} \cdot min^{-1}$ and $0.17\,nl \cdot cm^{-2} \cdot cm^{-2} \cdot min^{-1}$, respectively. Other substances are accumulated at a much smaller rate, and hence the fluxes will be even less. Since the placental barrier is very permeable to O_2 and water, the gradients necessary to move these substances into the fetus are very small.

Transport of oxygen and carbon dioxide

O_2 is a nonpolar molecule that can readily pass through the placental barrier by simple diffusion. Its behavior can therefore be expressed in terms of Fick's law of diffusion.[30] The equation for this law, when applied to the entire placenta, takes the following form:

$$\frac{dm}{dt} = -\frac{DF(C_1 - C_2)}{b}$$

where

dm/dt = Amount of gas transferred per unit time across placenta
D = Diffusion coefficient multiplied by solubility coefficient of O_2 in water
F = Area of exchange
b = Thickness of placenta
$(C_1 - C_2)$ = Partial pressure gradient between maternal and fetal circulations

The term "DF/b" has the dimensions of a permeability coefficient and is called the *diffusion capacity*. To estimate the diffusion capacity, it is necessary to measure $(C_1 - C_2)$. Unfortunately, it has been technically impossible to sample intervillous maternal blood and fetal capillary blood for analysis. Fig. 76-15 shows estimates of the P_{O_2} in the uterine artery, uterine vein, umbilical artery, and umbilical vein. The P_{O_2} in the uterine vein will be higher than that in the maternal intravillous space, since it receives O_2-rich blood via arteriovenous shunts. The P_{O_2} in the umbilical vein may also be lower than that in the fetal venous capillary, since it may contain a contribution of O_2-poor blood via shunts in the fetus. Thus estimates of $(C_1 - C_2)$ based on observed differences in the P_{O_2} in the uterine and umbilical veins may be excessively high. The analysis is also complicated by the fact that the cells of the placenta also consume 10% to 30% of the O_2 supplied to the organ. More recently, indirect studies have suggested that the P_{O_2} gradient across the placental barrier is much lower than formerly believed. A large P_{O_2} gradient is not required, however, because of the very extensive maternal-fetal interface available for exchange.

The P_{O_2} difference across the placental barrier is caused by only about 2% of the total O_2 in the blood on either side. Under normal conditions, approximately 0.3 ml O_2 is dissolved in 100 ml plasma, whereas approximately 15.8 ml O_2 is carried by 100 ml of arterial blood. The remaining O_2 is bound to the hemoglobin in the maternal and fetal red blood cells. The amount of O_2 bound in red blood cells depends on the allosteric properties of the hemoglobin molecule, since the capacity of hemoglobin to bind with O_2 is modified by the presence of 2,3-diphosphoglycerate (DPG) and CO_2 and protons (Bohr effect).[58] DPG lowers the O_2 affinity of Hb, as

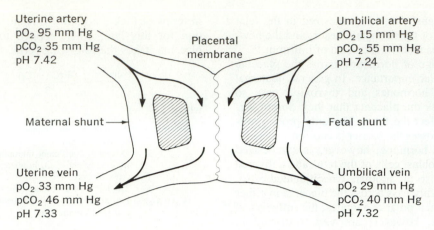

Fig. 76-15. Schematic representation of placental gas exchange. "Shunts" represent blood flows that are not exposed to membrane for exchange. Uterine and umbilical veins contain mixture of blood from shunts and blood exposed to exchange at membrane. (From Dancis and Schneider.[30])

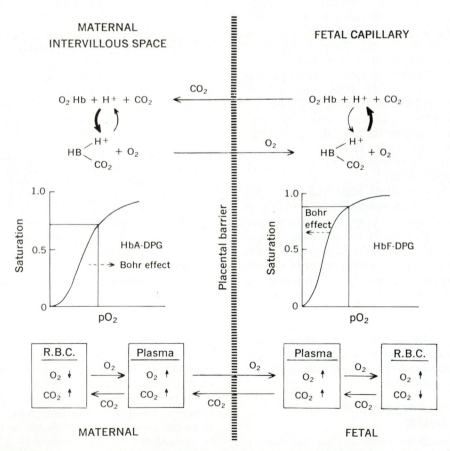

Fig. 76-16. Factors regulating transport of O_2 and CO_2 between mother and fetus. (For explanation, see text.)

do increasing concentrations of CO_2 and protons. The effect of increasing the concentrations of these substances is to displace the O_2 dissociation curve to the right.

It has been known for many years that fetal blood has a higher O_2 affinity than maternal blood. In contrast, purified Hb F has a lower O_2 affinity than Hb A. The reversal of affinity observed when the hemoglobin is within the red blood cell is due to the differential effects of DPG on Hb A and Hb F.[8] DPG binds more strongly to Hb A than Hb F and therefore has a greater effect on maternal blood in displacing the O_2 dissociation curve to the right. Thus for a given Po_2, fetal blood binds more O_2 than maternal blood (Fig. 76-16).

Fig. 76-15 also shows estimates of the Pco_2 and pH in the uterine artery, uterine vein, umbilical artery, and umbilical vein. The same technical difficulties as those described for O_2 apply to the determination of the Pco_2 and pH gradients across the placenta. CO_2 is carried in the blood in three forms, which are in equilibrium. These are in physical solution ($\sim 8\%$), bicarbonate ($\sim 62\%$), and bound to hemoglobin ($\sim 30\%$). CO_2 freely crosses the placental barrier, its diffusion constant being 20 times higher than that of O_2.

The factors and interactions that control the kinetics of gas exchange across the placenta may now be summarized (Fig. 76-16). An O_2 gradient exists between the intervillous maternal plasma and fetal plasma. Similarly, a gradient in CO_2 exists between the fetal plasma and the maternal plasma. As a result, O_2 diffuses passively from mother to fetus and CO_2 diffuses passively from fetus to mother. The increase in concentration of protons and in the Pco_2 in the maternal plasma influences the maternal red blood cells and causes the release of O_2 into the maternal plasma, thereby augmenting the gradient across the placental barrier. These changes promote an increase in the amount of O_2 and a fall in the amounts of CO_2 and protons in the fetal plasma. As a result, O_2 binding by the Hb F with a high O_2 affinity is favored, with the release of CO_2 and protons into the plasma. These effects further augment the gradients across the placenta, which facilitate the transport of O_2 from mother to fetus and CO_2 from fetus to mother.

Although fetal blood has a higher O_2 affinity than adult blood, it is not necessary for fetal survival. Thus fetal survival is unaffected in mothers with Hb Rainier, whose blood has a greater affinity for O_2 than that of their fetuses. Furthermore, fetal blood can be almost totally replaced with adult blood by intrauterine transfusions without destroying the fetus. Thus the primary forces governing gas exchange across the placental barrier are the O_2, CO_2, and proton gradients arising from fetal metabolism. The differential binding of DPG by maternal and fetal hemoglobins reinforces the O_2 gradient and makes the system more efficient.

Transport of water[85]

There are two regions where water can be exchanged between mother and fetus: the placenta and the chorion laeve. Tracer studies with labeled water have shown that the placenta is very permeable to water; the chorion laeve is also very permeable but to a lesser extent. Most of the very large exchange demonstrated in these experiments, however, is due to one-for-one diffusional exchange of molecules. These measurements do not reflect the net transfer of water across the two membranes.

There is no evidence that active transport of water occurs across the wall of the conceptus. Thus the net accumulation of water in the conceptus must be the result of gradients of the chemical potential of water across the placenta between the maternal compartment and the compartments in the conceptus. The flux of water in or out of the conceptus is given by the product of this gradient and the hydraulic conductivity of either the placenta or the chorion laeve. Relatively little is known about the quantitative aspects of these processes, particularly in early pregnancy. There is evidence that the main exchange occurs across the placenta and that transport across the chorion laeve is considerably less. The physical principles involved can be illustrated by considering the transfer of water across the near-term placenta.

The factors determining the transfer of water across the placenta are related by the following expression:

$$J_v = Lp(\Delta p - \Sigma \sigma_i \Delta \pi_i)$$

where

J_v = Net volume flux ($cm^3 \cdot cm^{-2} \cdot min^{-1}$)

Lp = Hydraulic conductivity of barrier

$\Delta p = (p_m - p_f)$, where p_m and p_f are hydrostatic pressures in maternal intervillous space and fetal placental capillary, respectively

$\Delta \pi = (\pi_{mi} - \pi_{fi})$, where π_{mi} and π_{fi} are osmotic pressures of solute i in maternal intervillous space and fetal placental capillary, respectively

σ_i = Reflection coefficient of placental barrier to solute i (for impermeable solutes, $\sigma_i = 0$; for fully permeable solutes, $\sigma_i = 1$; and for partially permeable solutes, $0 < \sigma_i < 1$)

An estimate of the gradient of chemical activity of water across the near-term human placental barrier is therefore given by $\frac{J_v}{L_p}$. J_v has been shown to be approximately $0.17 \; nl \cdot cm^{-2} \cdot min^{-1}$. Although estimates of L_p are not available for the human placenta, it is known that it is greater than that of the chorion laeve, which is approximately $0.015 \; nl \cdot cm^{-2} \cdot sec^{-1} \cdot cm \; H_2O^{-1}$. An upper estimate of $(\Delta p - \Delta \pi)$ can therefore be calculated to give a value of approximately 0.14 mm Hg. Thus the driving force needed to move water into the fetus is extremely small.

Experimentally, it has been demonstrated in the guinea pig that water moves across the placenta in response to increased gradients of hydrostatic pressure.[31] Artificially induced osmotic gradients also cause the movement of water across the placenta in the human and several other species. Although Δp is only slightly greater than $\Delta \pi$, the differences between the hydraulic pressures in the maternal intervillous space and fetal placental capillary, respectively, and the osmotic pressures in the same sites could be large. However, attempts to demonstrate a gradient in hydraulic pressure and osmotic pressure in the placenta of humans and animals have failed. These results imply that the necessary gradient in the chemical activity of water across the placental barrier arises from very small gradients in hydraulic pressure and osmotic pressure, which are too small to detect by present techniques. There are clearly selective advantages to the small hydraulic gradient across the placenta. If $p_m \gg p_f$, the villi would be compressed and the blood flow through the fetal capillaries retarded. As yet it is impossible to distinguish whether the driving force that moves water from mother to fetus is only a small hydraulic gradient, only a small osmotic gradient, or a combination of both.

INTERNAL TRANSPORT—THE FETAL CIRCULATION[32,70,80]

William Harvey, the discoverer of the adult circulation, also studied the circulation in the fetus, and he recognized that two major differences exist between them. The differences are that the two fetal ventricles pump in parallel rather than in series, as in the adult, and also there are vascular shunts in the fetus not found in the adult.

Fig. 76-17 shows a diagram of the fetal circulation. It is convenient to begin its description in the descending aorta. A large vessel, the um-

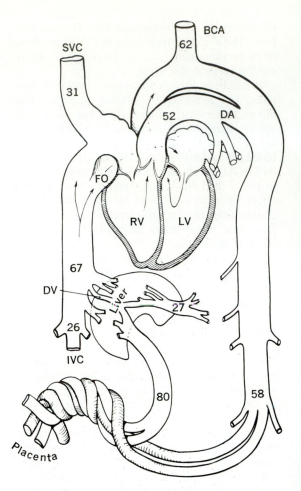

Fig. 76-17. Diagram of circulation in mature fetal lamb. Numerals indicate mean O_2 saturation (%) in great vessels of 6 lambs. *RV,* Right ventricle; *LV,* left ventricle; *SVC,* superior vena cava; *IVC,* inferior vena cava; *BCA,* brachiocephalic artery; *FO,* foramen ovale; *DA,* ductus arteriosus; *DV,* ductus venosus. (From Born et al.[19])

bilical artery, leaves the aorta and delivers blood to the placenta. After passage through the placenta the blood emerges in the umbilical vein. This large vessel carries blood to the fetal liver, where it splits into two channels. The larger channel is the *ductus venosus,* a fetal shunt that bypasses the fetal hepatic circulation and delivers blood directly into the inferior vena cava. The smaller channel feeds into the hepatic microcirculation, and this blood eventually enters the inferior vena cava via the hepatic veins. The inferior vena cava carries the blood to the right heart. On entering the heart the blood encounters a ridge called the *crista dividens,* which splits the

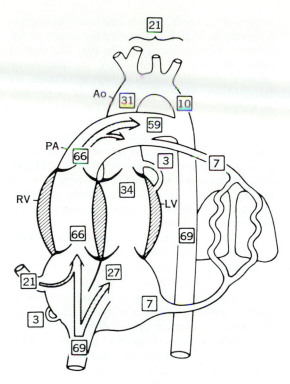

Fig. 76-18. Percentages of combined ventricular output ejected by left and right ventricles and distributed to major vascular channels in fetus. *RV,* Right ventricle; *LV,* left ventricle; *PA,* pulmonary artery; *Ao,* aorta. (From Rudolph and Heymann.[80])

flow into two main streams. The largest stream passes into the left atrium by a fetal shunt called the *foramen ovale,* whereas the smaller stream enters the right atrium. The blood returning from the head region in the superior vena cava flows entirely into the right atrium, as does the returning coronary blood. The blood returning from the lungs in the pulmonary vein discharges into the left ventricle, after which it passes from the left atrium to the left ventricle and is expelled through the aorta. The blood in the right atrium flows into the right ventricle and is discharged through the pulmonary artery. This vessel soon splits into two channels. The largest channel is a fetal shunt, the *ductus arteriosus,* which carries a stream of blood to the aorta. The smaller channel conveys blood to the lungs. Before the aorta is joined by the ductus arteriosus the vessels supplying the coronary circulation and head region leave the aortic arch.

Very few studies have been made on the quantitative aspects of the fetal circulation in human beings, and the main source of information has

therefore been obtained on fetal sheep. In the adult, both right and left ventricles pump in series, and the output of the two sides of the heart is normally equal. In contrast, in the fetus the right and left ventricles pump in parallel, and the outputs of the two sides may be very different. Moreover, the outputs of the two sides contribute jointly to the blood flowing to many organs and the placenta. Thus the cardiac output of the fetal heart is taken to be the combined output of both ventricles. As may be expected, the combined cardiac output increases as the fetus grows and develops. The quantitative aspects of the passage of blood through the fetal circulation of the sheep has been clarified by the use of radionuclide-labeled microspheres. At term in the sheep the percentage of the combined cardiac output is distributed between the organs as follows: placenta, 41%; lungs, 7%; kidneys, 2%; myocardium, 3%; brain, 3%; gastrointestinal tract, 6%; and fetal body, 38%. The results demonstrate clearly the large amount of blood that is pumped through the placental circulation. The blood returning to the right and left atria is made up of the following percentage contributions: inferior vena cava, 69%; superior vena cava, 21%; coronary circulation, 3%; pulmonary vein, 7% (Fig. 76-18); 27% of the total cardiac output that returns via the inferior vena cava is shunted into the left atrium to be joined by 7% from the pulmonary vein. This blood, representing 34% of the total cardiac output, is eventually discharged from the left ventricle in the aorta. The remainder of the flow in the inferior vena cava passes into the right atrium and is joined by 21% and 3% from the superior vena cava and coronary circulation. This blood, representing 66% of the total cardiac output, is eventually discharged into the pulmonary artery. Thus the combined cardiac output is about one third from the left ventricle and two thirds from the right ventricle. The blood leaving the left ventricle, in terms of the total cardiac output, is partitioned into 3% to the coronary circulation, 21% to the head, and 10% to the descending aorta. Similarly, the blood leaving the right ventricle is partitioned into 7% to the lungs and 59% shunted through the ductus arteriosus. The latter is joined by the blood from the aortic arch to give 69% to the lower regions of the body.

The three fetal vascular shunts—the ductus venosus, foramen ovale, and ductus arteriosus—can have profound effects on the concentrations of blood constituents. This fact is amply demonstrated by an examination of the O_2 concentration at several points in the fetal circulation. Fig.

76-17 shows estimates of the O_2 saturation in the great vessels of the fetal lamb. O_2-rich blood (80% saturated) leaves the placenta, and some of it is shunted past the liver to mix with the O_2-poor blood (26% saturated) in the inferior vena cava. This mixing of the two streams results in blood reaching the heart via the inferior vena cava that is 67% saturated. About 40% of this blood is shunted into the left atrium and is slightly diluted by the blood returning from the fetal lungs. As a result, an O_2-rich blood (62% saturated) is discharged into the brachiocephalic artery, a large proportion of which passes to the circulation in the head region. The remaining 60% of the O_2-rich blood from the inferior vena cava passes into the right atrium and mixes with the O_2-poor blood (31% saturated) returning via the superior vena cava. Consequently, the blood discharged into the pulmonary artery has an O_2 saturation of 52%. A large part of this blood then mixes with the O_2-rich blood in the aorta to give an O_2 saturation of 58% in the descending aorta. Clearly the combined functions of the ductus venosus, foramen ovale, and ductus arteriosus ensure that high levels of O_2 are supplied to the head region by preventing dilution of the blood from the placenta with the venous blood from the cranial regions.

The fetal circulation changes quantitatively during development both absolutely and relatively. The pulmonary circulation provides an important example, which has been studied extensively in fetal lambs during the final two trimesters of pregnancy (Fig. 76-19). The flow of blood through the pulmonary circulation increases eightfold and can be accounted for largely by a corresponding decrease in the pulmonary vascular resistance. This increase in flow through the lungs, however, involves only a doubling of the drain on the total cardiac output because of the simultaneous growth and increased pumping capacity of the fetal heart. Nevertheless, the relatively small differentials in flow and peripheral vascular resistances result in a 75% increase in pulmonary arterial pressure during this period of gestation.

The total pulmonary vascular resistance is the sum of the individual conductances of all the contributing arterioles in the system. The total pulmonary resistance is thus the harmonic mean of the individual resistances of each arteriole in the fetal pulmonary vascular system. Thus if the individual resistances are relatively constant, a large increase in the number of arterioles during development will reduce the total pulmonary resistance. In fetal lambs the number of arterioles increases nearly 40-fold in the second half of pregnancy.[79] This increase in arterioles is due partly to the overall growth of the lung and partly to an increase in the density of arterioles per unit volume of lung tissue.

The fetal lung arterioles have a unique struc-

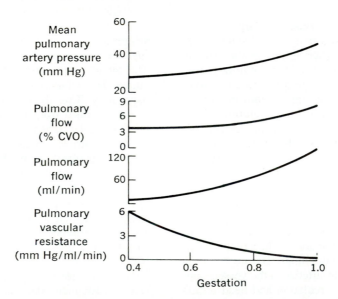

Fig. 76-19. Diagrammatic representation of changes in mean pulmonary arterial pressure, proportion of combined ventricular output (CVO) distributed to lungs, actual pulmonary blood flow, and calculated pulmonary vascular resistance in fetal lambs during gestational development from 0.4 to 1.0 (full-term) gestation. (From Rudolph.[79])

ture in that the muscular medial coat is very thick. This smooth muscle is also very sensitive to the local P_{O_2} and pH (Fig. 76-20). If the P_{O_2} falls, the arterioles undergo intense vasoconstriction, thereby increasing the pulmonary vascular resistance. This response contributes in part to the reaction of the fetus to hypoxia. The vaso constriction of the arterioles in the lungs results in a rise of blood pressure in the pulmonary ar-

tery, favoring the flow of blood through the ductus arteriosus. The heart rate also increases, possibly through a rise in pressure in the right atrium. This change favors the flow of blood to the left side of the heart through the foramen ovale. Simultaneously, vasoconstriction occurs in the limbs, while vasodilation occurs in the cerebral and coronary circulations. Thus the response to hypoxia favors increased blood flow through the placenta, the source of additional O_2, and the maintenance of the O_2 supply to the fetal brain and heart.

AMNIOTIC FLUID[67,85,101]

The technique called amniocentesis allows samples of amniotic fluid to be obtained by the sixteenth week of gestation. The development of this capability has diagnostic importance and has led to a renewed interest in the composition and formation of amniotic fluid.

The volume of amniotic fluid 8 weeks after conception is about 15 ml. The volume increases at an accelerating rate until about the twentieth week, reaching a volume of about 450 ml. Thereafter the net rate of production of amniotic fluid declines and becomes zero by the thirty-fourth week. These values are very approximate indications of the changes in volume of amniotic fluid during pregnancy, since large variation is observed at all times (Fig. 76-21). Recently, dilution methods for determining amniotic fluid volume have been developed. For example, a

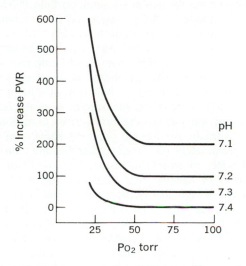

Fig. 76-20. Percentage increases in newborn calves of pulmonary vascular resistance (PVR) associated with changes in pH. (From Rudolph and Yuan.[81])

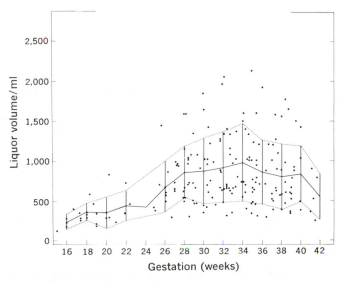

Fig. 76-21. Volume of amniotic fluid from 16th to 42nd week of pregnancy demonstrating large variability between individuals. Nevertheless, volume tends to peak at 34th week and rapidly fall as term approaches. (From Queenan et al.[77])

known dose of a solution of *p*-aminohippurate (PAH) is injected into the amniotic sac. After 20 to 30 min to allow for mixing, a sample of amniotic fluid is withdrawn and the concentration of PAH determined. The volume of amniotic fluid can then be calculated from the volume of PAH solution injected and the dilution that occurs after mixing. This technique allows serial determinations of the volume of amniotic fluid in a patient. Serial determinations in patients from 16 weeks to term have shown the occurrence of large, rapid fluctuations in amniotic fluid volume. Thus the amniotic fluid is in a very dynamic state.

The compositions of amniotic fluid during the various trimesters are shown in Table 76-8. Amniotic fluid is mainly water. During the first and second trimesters the composition of amniotic fluid is very similar to that of maternal and fetal serum, with the exception of total protein, which is 4% to 5% of that in serum. These observations suggest that amniotic fluid in early pregnancy is a dialysate of maternal and/or fetal fluids. During the third trimester the total solute concentration in the amniotic fluid falls, and the concentrations of urea, uric acid, and creatinine increase considerably. A specific compound that is of diagnostic interest is α-fetoprotein,[74,88] the dominant protein species in serum for part of fetal life. The concentration of this protein in serum in late fetal life falls, and this is reflected in the amniotic fluid. If the fetus has defects of the neural tube, for example, anencephaly, the concentration of α-fetoprotein even at term may be several times greater than the concentration at the twenty-eighth week of pregnancy.

The production and composition of amniotic fluid depends on several pathways that control the ingress and egress of water and solutes into the amniotic cavity (Fig. 76-22). These pathways are the several epithelial structures that line the amniotic cavity—the epithelia lining the placental and reflexed membranes (chorioamnion), the umbilical cord and fetal skin, and the three fetal organs that have direct communications with the amniotic cavity—gastrointestinal tract, urinary tract, and respiratory tract. Removal of the fetus in the rhesus monkey does not prevent the formation of amniotic fluid.[10] Thus the epithelia of the placenta and chorioamnion are able to form amniotic fluid alone. Many of the cells lining the amnion are separated by intercellular channels that open directly into the amniotic cavity. The membrane is therefore very permeable and allows the bulk flow of water from one side to the other. Measurements of the bulk flow of water across the term amnion in response to hydraulic or osmotic gradients show it to be approximately 100 times greater than the transfer of water by simple diffusion. Prior to 17 to 20 weeks of age the fetal skin consists of a few layers of cells that lie between a basal layer and a superficial layer called the *periderm*. After this time the periderm disappears, and the superficial cell layers become keratinized. Before keratinization occurs, the fetal skin is permeable to water, sodium, urea, and creatinine. By the twenty-fifth week, when keratinization is complete, the skin is no longer a pathway for exchange. The presence of vernix caseosa, lanugo hair, and epithelial cells in the fetal stomach and intestine indicates that the fetus swallows during fetal life. If a radiopaque dye is injected into the amniotic fluid of a woman at term, it enters the

Table 76-8. Compositions of amniotic fluid in early and late pregnancy and the full-term maternal and fetal serum*

Fluid	Total osmotic pressure (mosmols)	Na (mM)	Cl (mM)	K (mM)	NPN (mg/ 100 ml)	Urea (mg/ 100 ml)	Uric acid (mg/ 100 ml)	Creatinine (mg/ 100 ml)	Total protein (gm/ 100 ml)	Water content (%)
Amniotic fluid, first and second trimester	283	134	110	4.2	24	25	3.2	1.23	0.28	98.7
Amniotic fluid, third trimester	262	126	105	4.0	27	34	5.6	2.17	0.26	98.8
Maternal serum, full-term	289	137	105	3.6	22	21	—	1.55	6.5	91.6
Fetal serum, full-term	290	140	106	4.5	23	25	3.6	1.02	5.5	—

*Modified from Seeds.[85]

fetal stomach and becomes concentrated, presumably because most of the water is absorbed by the fetus.[66] Thus the fetal gastrointestinal tract is a pathway for the removal of amniotic fluid. The amount of amniotic fluid swallowed per day is large, particularly in late pregnancy.[76] At 16 weeks the fetus swallows 7 ml · hr[-1]; at 20 to 21 weeks, 16 ml · hr[-1]; at 28 weeks, 120 ml · hr[-1]. Thus at the end of pregnancy the fetus may swallow approximately half of the amniotic fluid in a 24 hr period. The fetal lung produces a fluid that fills the alveoli. Some of this respiratory fluid enters the amniotic fluid, since surface-active lecithins produced by the lungs are found in the fluid (p. 1972). The amount of respiratory fluid produced in human fetuses is unknown. Urine is present in the fetal bladder from 11 weeks of age until term, indicating continual renal activity in fetal life. Micturition occurs into the amniotic sac. An ultrasonic technique has been used to estimate the rate of urine formation in human fetuses from the twenty-fifth to the forty-second week.[97] At 25 weeks of age, urine is produced at 3 to 5 ml · hr[-1]. The rate of urine production increases linearly until the thirty-ninth week, when 26 ml · hr[-1] is produced. After the fortieth week, the rate of urine production drops rapidly.

The amount of amniotic fluid present at a particular time will be the result of the ingress and egress of water and solutes through the different pathways for exchange. Some of these provide for continuous transfers throughout fetal life, such as the chorioamnion, placenta, umbilical cord, and fetal respiratory tract; some provide for exchange for a restricted period of fetal life, such as the fetal skin; others are phasic in function, such as fetal swallowing and micturition. The relative importance of these pathways for exchange at different times of development is still poorly understood and can only be described in general terms.[101] For example, the 20-week-old human fetus produces amniotic fluid at the rate of 16 ml · day[-1]. Urine is excreted into the amniotic sac at the rate of 50 ml · day[-1], and the fetus swallows about 16 ml · day[-1]. Thus an excess of 18 ml · day[-1] is contributed by the urine, which is probably reabsorbed through the chorioamnion into the maternal tissue. The major molecular species involved in this process is water, and ultimately these molecules are derived from the mother. A circulation of water molecules therefore occurs, passing from mother to fetus through the placenta, into the amniotic sac via the urine, where it is either recirculated through the fetus via the fetal gastrointestinal tract or passes back to the mother via the chorioamnion. Presumably solutes will also move through similar pathways.

In late pregnancy the excretion of large amounts of hypotonic urine by the fetus signifi-

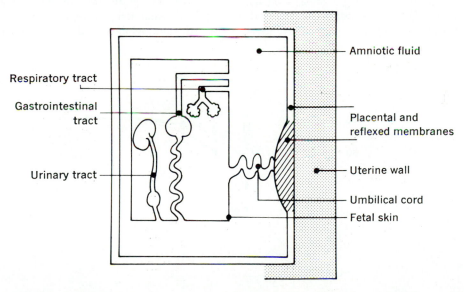

Fig. 76-22. Pathways involved in formation of amniotic fluid. Placenta, umbilical cord, and chorioamnion (reflexed membranes) exchange materials continuously. Fetal skin exchanges materials up to time of keratinization. Kidney and gastrointestinal tract exchange materials phasically. (From Wallenburg.[101])

cantly affects the composition of amniotic fluid. Moreover, no fluid enters or leaves the amniotic cavity through the fetal skin. These developmental changes may account for the fall in osmolality and in the concentration of sodium ions in amniotic fluid as pregnancy advances and also the increased concentrations of urea, uric acid, and creatinine.

The control of amniotic fluid formation may be under hormonal control. Recently, it has been shown that prolactin, which is present in amniotic fluid, can increase the permeability of the chorioamnion to water. The injection of prolactin into the amniotic sac of rhesus monkeys in the third trimester significantly reduces the volume of amniotic fluid for 24 hr.[54]

MATURATION OF THE FETUS FOR POSTNATAL LIFE[63]

The embryonic foregut is the forerunner of the digestive tract from the mouth to far along the small intestine. In addition, it gives rise to many other derivatives, chief of which are the respiratory tract, liver, pancreas, thyroid, parathyroid, and thymus. Shortly before birth, several of these organs undergo functional changes that facilitate the transfer of the infant from an intrauterine, dependent existence to independent life. The coordination of these changes depends on the secretion of fetal corticosteroids.

Lungs[36,82]

Birth puts severe physiologic stresses on the newborn infant. One stress results from the sudden cut off of the O_2 supply from the placenta at a time when the infant requires a sudden acceleration of energy production to sustain the respiratory movements that begin immediately after delivery and to maintain body temperature. The O_2 needed to produce this energy necessitates the sudden onset of efficient lung function.

Major structural changes occur in the lungs as the time of birth is approached. These changes are reflected functionally by alterations of the pressure-volume curve, which relates the pressure required to inflate the lungs to the proportionate changes in volume. Fig. 76-23 shows these curves on days 25 to 31 in the rabbit fetal lung. Clearly the pressure required to expand the fetal lung diminishes as the time of birth approaches. The physical properties that control the expansion of the lung are the structural components of the lung tissue and the surface tension of the fluid within the lung. The lower the surface tension of the fetal pulmonary fluid, the more easily the lung expands. The presence of a surface-active agent called *surfactant* in the tracheal fluid was first reported in sheep.[1] Since then, surfactant has been found in the fetal lungs of all species that have been studied. The chemical structure of surfactant is not completely known. However, it consists of a phospholipid prosthetic group attached to an apoprotein. The phospholipid is a disaturated lecithin, predominantly dipalmitoyllecithin, and it is this moiety that has the surface-active properties. The biosynthesis of the prosthetic group of surfactant involves the production of two substances—a specific saturated 1,2-diglyceride, particularly 1,2-dipalmitoyldiglyceride, and CDP-choline;

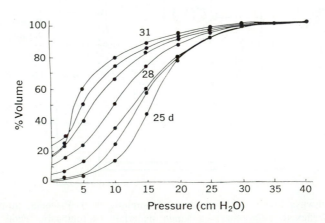

Fig. 76-23. Mean pressure-volume curves in fetal rabbits at different stages of development. Pressure required to expand lungs decreases as term is approached. (From Kotas and Avery.[60])

these substances then interact to give a dipalmitoyllecithin (Fig. 76-24). CDP-choline occupies a central role in the synthesis of several phospholipids, but the nature of the products depends on the molecule with which it interacts. Thus the biosynthesis of the surface-active lecithin involves both a specific and a nonspecific pathway. The suggestion has been made that the interaction of CDP-choline with the 1,2-diglyceride, catalyzed by choline phosphotransferase, is the rate-limiting step in the biosynthesis of surfactant. This theory, however, assumes that the supply of the specific saturated 1,2-diglyceride is more than adequate, a fact that has yet to be demonstrated. Presumably, specific biosynthetic pathways develop in the type II pneumocytes, providing them with the ability to synthetize surfactant.

Considerable evidence has accumulated in recent years, demonstrating that substances in the fetal pulmonary fluid, including surfactant, can pass into the amniotic fluid. Thus estimates of the concentration of lecithin in amniotic fluid can be used as an indicator of surfactant production in the fetal lungs. Fig. 76-25 shows that the concentration of lecithin in the amniotic fluid of human fetuses rises rapidly after the twenty-fifth week of gestation and reaches a maximal level at about the thirty-sixth week of gestation. This rise in the lecithin content of human amniotic fluid shortly before birth is the basis of the lecithin/sphingomyelin ratio test for monitoring fetal development and well-being. Sphingomyelin is a phospholipid whose concentration in amniotic fluid does not change as expected time of birth nears, and therefore it provides a baseline with which to compare the change in concentration of lecithin. A ratio greater than 2 is normal, whereas a ratio less than 2 indicates that surfactant production is below normal and that the fetus may be a victim of *hyaline membrane disease*.

The increased production of surfactant toward the end of pregnancy is probably stimulated by rising circulating levels of corticosteroids produced by the fetal adrenal gland.[63] Fig. 76-25 shows the close correspondence between the concentration of lecithin in the amniotic fluid and the concentration of corticosteroids in the umbilical

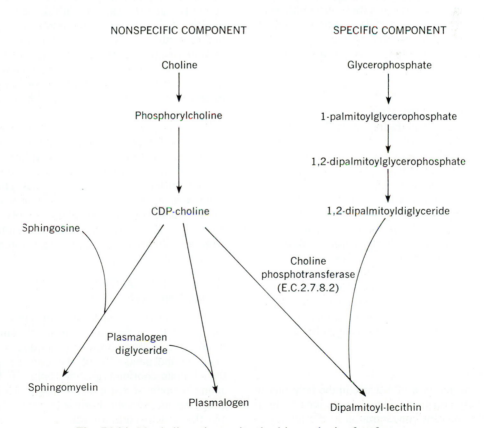

Fig. 76-24. Metabolic pathways involved in synthesis of surfactant.

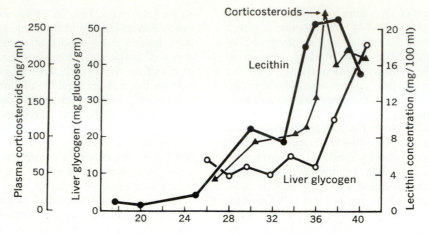

Fig. 76-25. Time courses in human pregnancy of concentrations of corticosteroids in umbilical cord plasma, lecithin in amniotic fluid, and glycogen in fetal liver. (From Liggins.[64])

cord plasma in the human as pregnancy draws to an end. Experimentally, it has been demonstrated in rabbits that the injection of 9α-fluoroprednisolone into the fetuses on the twenty-fourth day of pregnancy increases the concentration of lecithin in the fetal lungs.[60] The administration of this synthetic corticosteroid to fetal rabbits also shifts the pressure-volume curve of the fetal lung to the left.

Liver[63]

Shortly before birth a rapid increase occurs in the glycogen concentration in the human fetal liver (Fig. 76-25). This increase is also observed in the fetal rat liver and is associated with the development of the late fetal cluster of enzymes (Table 76-3). Studies on newborn rats have shown that corticosteroids can induce the liver enzymes required for glycogen synthesis and gluconeogenesis.[42] Thus the rise in liver glycogen that occurs in rats shortly before birth is prevented by decapitation of the fetus, which eliminates the production of ACTH by the fetal anterior pituitary gland, necessary to stimulate production of fetal corticosteroids. The administration of corticosteroids to these decapitated fetuses causes the accumulation of glycogen by the fetal liver.[57]

Pancreas[63]

The concentration of glucose in the fetal blood is relatively constant, since it is derived from a relatively constant concentration in the maternal blood. For most of fetal life the response of the β cells in the fetal pancreas to short-term increases in circulating glucose levels is absent or weak. Even near birth the production of insulin by the fetal pancreas in response to hyperglycemia is sluggish, like an adult diabetic. The injection of ACTH into fetal lambs several days before birth increases the sensitivity of the β cells and stimulates the development of the adult type of response.[64] The fetal corticoids may regulate maturation of the fetal islets of Langerhans in anticipation of the need to regulate blood glucose after birth.

The fetal pancreas, in contrast, is responsive to sustained hyperglycemia. Insulin secretion is stimulated as well as β cell hyperplasia. Under such conditions, which occur naturally if the mother suffers from diabetes, fetal growth is stimulated by the combined effect of insulin on tissue anabolism and the excess supply of glucose. This syndrome results in the unnaturally large and abnormal fetuses that develop in cases of maternal diabetes.[98]

CONCEPTUS-MOTHER COMMUNICATION[62,98]

Hormones produced by the conceptus that influence maternal functions can be classified into three groups: first, hormones produced by the placenta independently of the embryo and fetus—human chorionic gonadotropin (HCG), human chorionic somatomammotropin (HCS), and progesterone; second, hormones produced solely by the fetus—corticosteroids; and third, hormones that can only be produced by the joint action of the fetus and placenta—estrogens. The system that synthesizes fetal estrogens is called the *fetoplacental unit*.

Human chorionic gonadotropin

HCG is the first hormone to be produced by the conceptus during development. It is synthesized by the syncytiotrophoblast. The hormone has now been purified from pregnancy urine, placental chorionic tissue, and tumors of the trophoblast. The molecule is a glycoprotein and consists of two nonidentical subunits, both containing amino acid and carbohydrate residues.[49] The α subunit has a molecular weight of 14,000, contains 92 amino acid residues, and is identical to the α subunit of FSH, LH, and TSH. The β subunit has a molecular weight of 24,000 and contains 145 amino acid residues, and it is this subunit that determines the biologic specificity of the hormone. Despite its close chemical relation with the three hormones produced by the anterior pituitary gland, its mode of secretion is very different, since it is not stored in granules but is immediately released into the maternal blood as it is synthesized.[35]

By the use of biologic and immunologic assays, it is known that HCG is secreted into the maternal circulation by the trophoblast cells at or shortly after implantation.[7] The concentration in the blood rises to a peak between the seventh and twelfth weeks of pregnancy, and then it falls to about 10% to 20% of the peak value by the fifteenth week. The blood concentration of HCG remains at this low level for the remainder of pregnancy. It has been estimated that at the peak concentration, HCG is secreted at a rate of 0.5 to 1.0×10^6 IU $\cdot$ day^{-1} and at a rate of 0.08 to 0.12×10^6 IU $\cdot$ day^{-1} in late pregnancy.[62] About 90% of the hormone is metabolized in the body, and the remaining 10% is excreted in the urine. The detection of HCG soon after implantation is the basis of the test for pregnancy.

The blood concentration of HCG is elevated throughout the period when a functioning corpus luteum is essential for the maintenance of pregnancy.[68] Bilateral ovariectomy of women at the seventh week of pregnancy results in abortion within 3 to 7 days, whereas a similar operation at the tenth week may produce delayed abortion in some cases or may not terminate the pregnancy.[27] By this time the conceptus has begun to synthesize steroids, particularly progesterone and estrogens in amounts sufficient to take over the function of the corpus luteum. Thus by the time the fetus is 3 months old the secretions of the conceptus maintain the pregnancy.

Human chorionic somatomammotropin[62]

HCS is produced by the trophoblast, and most of it passes into the maternal circulation. The molecule is a protein with a molecular weight between 19,000 and 30,000. The hormone is first secreted by the sixth week after conception, and the concentration in maternal blood rises throughout pregnancy to reach a plateau by the thirty-sixth week. By this time, it is secreted at a rate of 0.5 to 3 gm $\cdot$ day^{-1}. The rise in blood concentration closely parallels the rise in placental weight, suggesting that the rate of synthesis of the hormone per unit weight of placenta remains constant throughout pregnancy.[87] The primary function of this hormone is to ensure that the supply of glucose, the chief energy source of the fetus, is maintained even when the intake of food by the mother is restricted.

Progesterone[62,84]

The conceptus is the main source of progesterone 3 months after conception. This hormone passes into the maternal circulation in ever-increasing amounts; thus the blood concentration continually rises, reaching about 150 ng $\cdot$ ml^{-1}. Biosynthesis of progesterone occurs primarily in the placenta, since neither fetal death nor ligation of the umbilical cord diminishes its production.[23] By term the placenta produces about 250 mg $\cdot$ day^{-1} progesterone; 70% passes into the maternal circulation and 30% into the fetal circulation.[84]

The primary precursor of progesterone is cholesterol. The placenta, however, is unable to synthesize cholesterol from acetate. Thus the production of progesterone depends on a nonplacental source of cholesterol. This precursor can be exclusively provided by the mother, since progesterone continues to be produced after fetal death. Cholesterol is converted to pregnenolone by the side chain–splitting enzyme, and the pregnenolone is converted to progesterone by Δ^5-3β-ol dehydrogenase (EC 1.1.1.145) and Δ^5-isomerase.

Corticosteroids[62]

The two major corticosteroids circulating in fetal blood are the active cortisol and the inactive cortisone. They are present in a ratio of 1:10.[50] The concentration of cortisol in fetal blood is one third that in the mother. Cortisol is produced by the fetal adrenal cortex under the control of fetal ACTH. Although fetal corticoids are implicated in the initiation of parturition, their role is not fully understood. Delivery of the placenta in women occurs about 1 week after fetal death, showing that the initiation of parturition does not absolutely depend on the secretion of fetal corticoids.[28] This observation is difficult to reconcile with the prolonged pregnancy associated with anencephaly, in which the fetal pituitary gland is missing and the fetal adrenal gland is hypoplastic.

The metabolism of corticosteroids in the fetus is very different from that found in the adult. In the adult, cortisol is converted into excretion products such as tetrahydrocortisol by A-ring reductase activity. In the conceptus, the metabolism is dominated by two enzymes—11β-hydroxysteroid dehydrogenase (EC 1.1.146) in the placenta,[75] which converts cortisol to cortisone, and 21-hydroxysteroid sulfokinase in the fetal adrenal and liver, which conjugates corticosteroids with sulfate.[59] The source of cortisol that is converted to cortisone in the placenta could be maternal or fetal, but the relative contribution of each is unknown. The corticosteroid sulfates can pass from fetus to mother intact and are finally excreted in the maternal urine. Thus the detection of these compounds in maternal urine may form the basis of a test of fetal adrenal gland function.

Fetoplacental unit and estrogen biosynthesis[62,84]

Estrogens begin to appear in maternal blood at about the hundredth day of pregnancy. The total concentration rise parallels the growth of the placenta and near term reaches a concentration of about 175 ng · ml⁻¹. The major circulating estrogens are estradiol 17β, estrone, and estriol. The most abundant is estriol, and near term as much as 50 mg · day⁻¹ of this compound is excreted in the maternal urine.[11] The source of all three estrogens is the fetoplacental unit, and little comes from the maternal ovaries and adrenal cortex. Ovariectomy and bilateral adrenalectomy of women in pregnancy do not reduce the levels of circulating estrogens.[46] Interruption

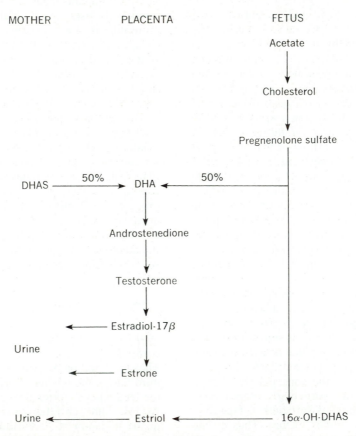

Fig. 76-26. Metabolic pathways in fetoplacental unit for synthesis of estrogens. There are two routes, one for synthesis of estradiol-17β and estrone and one for synthesis of estriol.

of the fetal circulation, however, causes an immediate large fall in the maternal blood levels of all three estrogens.[23] Thus the placenta cannot synthesize estrogens unless the fetus is present.

The human placenta lacks the enzymes 17α-hydroxylase (EC 1.14.99.9) and desmolase.[84] Thus the C_{21} steroids pregnenolone and progesterone synthesized by the placenta cannot serve as precursors for androgen and estrogen production by the placenta. In contrast, the placenta readily converts neutral C_{19} steroids, such as testosterone and androstenedione, to estrogens by a very effective aromatization system. To synthesize estrogens the placenta depends on an exogenous supply of C_{19} precursors. The observation that women carrying anencephalic monsters excrete low levels of estrogens suggested that the fetal adrenal cortex is a source of these C_{19} steroids.[37] The fetal adrenal cortex is now known to produce large quantities of dehydroepiandrosterone (DHA). This steroid circulates in fetal blood conjugated to sulfate (DHAS).[25] DHAS is probably synthesized in the fetal zone of the adrenal cortex, a specialized region that regresses after birth. The region lacks 3β-hydroxysteroid dehydrogenase (EC 1.1.1.145), which would convert DHA to androstenedione and testosterone. This zone and the fetal liver, also, have high sulfuryltransferase activity, which tends to conjugate both neutral and phenolic steroids to sulfate. These conjugates are biologically inactive, and this system may provide a mechanism for the transport of large quantities of steroids in the fetus in an inactive form.

Dehydroepiandrosterone sulfate is utilized in the fetoplacental unit via two pathways (Fig. 76-26). One pathway involves passage to the placenta, where a sulfatase hydrolyzes the conjugate to yield the free steroid. This compound is converted to androstenedione and testosterone, which are then aromatized to estradiol 17β and estrone. These estrogens enter the maternal circulation and are excreted. The other route involves the conversion of DHAS to 16α-hydroxydehydroepiandrosterone by 16α-hydroxylating enzymes in the fetal liver. 16α-OH-DHAS then passes to the placenta, where it is hydrolyzed to yield the free steroid, which is aromatized to estriol. The estriol enters the maternal circulation and is excreted. The 16α-hydroxylating enzymes are very active in the fetus, which results in the concentration of 16α-OH-DHAS in fetal plasma being two to three times greater than DHAS.

A major quantitative difference exists between these two pathways.[90] The DHAS that is used for the synthesis of estradiol 17β and estrone is derived equally from the mother and fetus. In contrast, nearly all the DHAS that is converted to estriol is derived from the fetus, with little maternal contribution. This difference accounts for the fact that in women carrying anencephalic monsters the ratio of estrone and estradiol 17β to estriol in the maternal urine is relatively high. Monitoring the urinary secretion of estriol in pregnancy is used clinically to assess the viability of the fetus. A sudden premature fall in the estriol concentration in the mother's urine may indicate fetal and/or placental failure.

RESPIRATORY SYSTEMS OF THE FETUS AND NEONATE[16,47,69,70]

Only recently has it been accepted that movements characteristic of breathing occur in the fetus. Formerly the observation that respiratory movements occur in exteriorized fetuses in animals was regarded as an artifact. Extensive studies on sheep have now shown that the movements are real. The respiratory movements in human fetuses have recently been studied in situ by ultrasonic techniques.

Breathing movements in human fetuses can be detected as early as 11 weeks after conception, although they are easier to observe at 13 to 14 weeks. During this early period the amplitude of the movements is irregular compared to the amplitude of the movements observed near term (Fig. 76-27). Fetal breathing is also episodic, in contrast to the continuous pattern seen after the first postnatal breath is taken. The periods of intermittent respiration last 30 to 70 min, and they occupy 50% to 90% of the time. The frequency of the periods also varies diurnally, being less frequent in the morning and higher in the evening. The effect is particularly large when mothers have starved overnight and may be associated with hypoglycemia. The intravenous injection of glucose into a mother starved overnight causes a rapid increase in the respiratory activity of her fetus.

Fetal breathing does not involve the flow of air into and out of the respiratory passages, since during this period the air passages are filled with fluid. The observations that a radiopaque dye injected into the amnion does not enter the fetal lungs and that surfactant appears in the amniotic fluid suggest that lung fluid tends to leave the respiratory passages. It is not clear to what extent fetal respiratory movements influence the movement of lung fluids.

After the first breath, air moves in and out of the lungs continuously. Different patterns of respiration are seen in the human newborn infant.

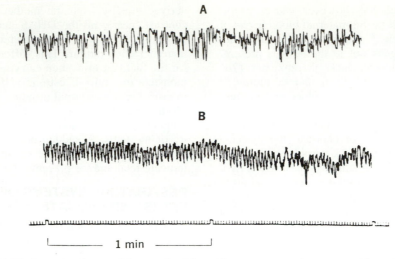

Fig. 76-27. Records of normal human fetal breathing movements in utero at 14 weeks' *(a)* and 40 weeks' gestation *(b)*. (From Boddy et al.[17])

For example, in sleep, regular and irregular breathing is observed. Regular breathing is associated with nonrapid eye movement in sleep, whereas irregular, more rapid breathing is accompanied by rapid eye movements, movement of the limbs, and mouthing. Sudden movements accompanied by deep sighs occur more frequently in nonrapid eye movement sleep.

Two major, equally critical, physiologic changes occur in the fetal respiratory system at birth. One change is the removal of the fluids in the respiratory pathways so that exchange between gaseous inspired air and the blood can be rapidly established. The second change is from prenatal episodic breathing to postnatal continuous breathing. In natural childbirth, fluid in the lungs is partially expelled through the mouth by a rise in intrathoracic pressure. The mechanism is denied infants delivered by cesarean section. Fluid may also be eliminated from the respiratory passages by absorption by the fetus. Whatever the mechanisms, the postnatal gaseous exchange function of the lungs is established relatively rapidly, perhaps less than 15 min after birth. The change from episodic to continuous breathing is not well understood. Although tactile and gravitational stimuli and cold exposure enhance respiration after birth, they are not absolutely necessary for the initiation of continuous postnatal breathing.

CHANGES IN THE FETAL CIRCULATION AT BIRTH[32,69]

At birth the placental circulation is replaced by the pulmonary circulation. The change in-volves a switch from the parallel system (fetal) of pumping by the left and right ventricles to the series system (adult) of pumping (Fig. 76-28). At the same time the umbilical circulation is cut off, and the ductus venosus closes. The switch in the pattern of blood flow through the heart involves closure of the foramen ovale and the ductus arteriosus. These changes, however, are not immediate, and therefore a transitional type of circulation is recognized, which occurs between the fetal and adult types in neonatal and postnatal life (Fig. 76-28). During this transitional period the bypasses may open up periodically, and the flow of blood through them may reverse.

The functional closure of the ductus arteriosus and foramen ovale at delivery is almost entirely due to a fall of resistance in the pulmonary vascular bed. As a result, a large rapid increase in the flow of blood through the lungs occurs with a sudden increase in the venous return to the left atrium. Studies on fetal sheep have shown that the flow of blood through the lungs may increase 5-fold to 10-fold. Also, the pulmonary arterial pressure falls rapidly during the first few minutes after birth (Fig. 76-29). The fall in pressure continues less precipitously for about 36 hr after birth.

These functional changes brought about by the elimination of the placental circulation and the rapid expansion of the pulmonary circulation result in the pressure in the left atrium exceeding the pressure in the right atrium. As a result, the foramen ovale tends to close, reducing and finally cutting off the flow of blood from the infe-

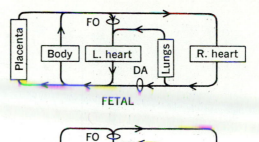

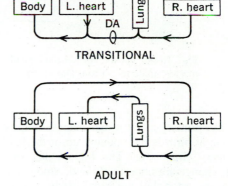

Fig. 76-28. Diagrams of circulation during fetal, transitional, and adult stages showing transition from parallel to series system of pumping through heart. (From Born et al.[19])

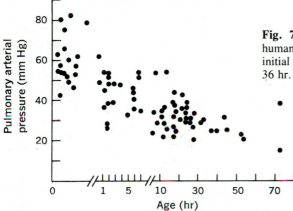

Fig. 76-29. Pulmonary arterial pressure in normal human infants at different times after birth showing initial rapid fall in first hour and slower fall for next 36 hr. (From Dawes.[32])

rior vena cava into the left atrium. Major alterations occur in the O_2 saturation of the blood in the right and left sides of the heart as the result of these changes in flow. The Po_2 in the right atrium falls because of the elimination of the placental circulation. As a result, the Po_2 of blood leaving the right ventricle also falls. In contrast, the Po_2 of blood in the left atrium rises and consequently the Po_2 of blood flowing through the left ventricle into the aorta. Also, the Po_2 entering the coronary circulation increases rapidly.

The flow of blood through the ductus arteriosus and the Po_2 of this blood are complex during the transition period. The increase of pressure on the left side of the circulation above that in the right side results in a reversal of flow through the ductus arteriosus. Consequently, O_2-rich blood from the aorta enters the lumen of the ductus arteriosus. There is extensive evidence that the muscular wall of the ductus arteriosus is very sensitive to Po_2 near term. High concentrations of O_2 cause strong contraction of the wall of the ductus arteriosus. Thus the reversal of flow of blood through the ductus arteriosus may lead to its temporary closure until the Po_2 drops, whereupon more O_2-rich blood enters it again to promote further closure. Thus an intermittent flow of blood may occur in the vessel during the period of transition after birth.

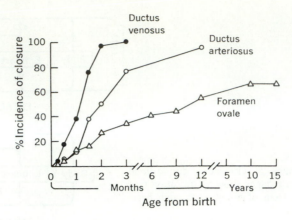

Fig. 76-30. Closure of special vascular channels of fetus after birth as judged by human postmortem material. (From Dawes.[32])

The primary event in the changes in the circulation at birth is the opening up of the fetal pulmonary circulation. The factors that stimulate this change are imperfectly understood. The pulmonary arterioles are particularly sensitive to Po_2 (Fig. 76-20). At high Po_2 the vessels are maximally dilated. The rapid rise in alveolar and pulmonary venous Po_2 that occurs after the first breath is believed to be the major mechanism causing the rapid fall in pulmonary vascular resistance.

The transitory stage between the fetal and adult types of circulation occurs because the foramen ovale and ductus arteriosus are only closed functionally. They are thus able to reopen if conditions controlling their patency changes. Irreversible anatomic closure depends on morphologic changes, and these are slow to reach completion and may even not occur at all (Fig. 76-30). The ductus venosus is permanently closed in almost all individuals 3 months after birth. The ductus arteriosus is closed in most individuals by 1 year. The foramen ovale closes irreversibly very slowly and only in about 60% of individuals.

TEMPERATURE REGULATION
AFTER BIRTH[5,48,89]

Birth not only involves pulmonary and circulatory adaptations in the newborn but also the assumption of functions by the gastrointestinal tract and the renal system. There is in addition another unique change—adaptation to an ambient temperature well below body temperature.

At delivery the warm infant, with wet skin, is suddenly exposed to a drop in ambient temperature. Heat is immediately lost by radiant, convective, and evaporative routes. The rate of heat loss in the first few minutes has been estimated at $0.2 \text{ kcal} \cdot \text{kg}^{-1} \cdot \text{min}^{-1}$,[2] which results in a fall in skin temperature of about $0.3°$ C $\cdot$ min^{-1}. The sudden cooling of the skin may be an important stimulus for the first breath.

The thermal controlling mechanisms are partly functional at birth, and the infant is able to maintain body temperature above ambient temperature. Fig. 76-31 shows the heat production by naked newborn infants at different ambient temperatures. If the temperature falls below $32.5°$ C, the infant responds by generating heat. If the ambient temperature rises above $33.5°$ C, thermal balance is maintained by sweating. The interval between $32.5°$ C and $33.5°$ C is called the thermoneutral zone, which is defined as the range of ambient temperatures within which the metabolic rate is at a minimum and temperature regulation is achieved by nonevaporative physical processes alone. As the ambient temperature falls, the amount of heat produced is matched by the heat dissipated (Fig. 76-31).

The smaller the infant, the less developed is its thermal controlling mechanism, and heat is lost more rapidly. Consequently, the operative environmental temperature needed to ensure thermal neutrality is higher for smaller infants and must be maintained for longer periods (Table 76-9). Normal infants rapidly develop thermoregulatory mechanisms, compared with premature infants.

Newborn infants respond to lowered ambient temperature by increased muscular movement and only in a minor way by shivering. Yet these responses cannot account for all the heat produced in response to cold. The extra heat is gen-

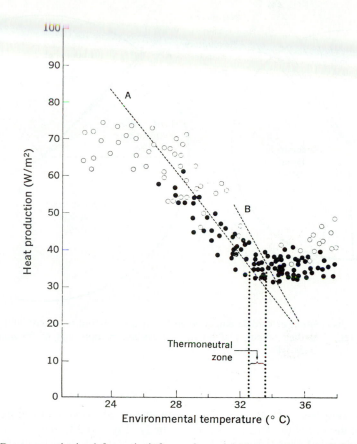

Fig. 76-31. Data were obtained from six infants who weighed approximately 2.5 kg when 7 to 11 days old, lying naked on a mattress in draught-free surroundings of uniform temperature and moderate humidity. Superimposed lines *(A* and *B)* show amount of heat normally dissipated by radiation and convection evaporation (in the absence of sweating) when skin blood flow is at minimum (line *A*) and maximum (line *B*). Open circles relate to data obtained during periods of restlessness or activity. Zone of minimum heat production extends from 32.5° to about 36.5° C, but thermal balance is maintained by sweat loss above a temperature of 33.5° C, and, by international agreement, term "thermoneutral" is now restricted to zone from 32.5° to 33.5° C. (From Hey.[48])

Table 76-9. Mean temperature needed to provide thermal neutrality for a healthy infant nursed naked in draught-free surroundings of uniform temperature and moderate humidity after birth*

Birth weight (kg)	Operative environmental temperature			
	35° C	34° C	33° C	32° C
1.0	For 10 days	After 10 days	After 3 wk	After 5 wk
1.5	—	For 10 days	After 10 days	After 4 wk
2.0	—	For 2 days	After 2 days	After 3 wk
>2.5	—	For 2 days	For 2 days	After 2 days

*From Hey.[48]

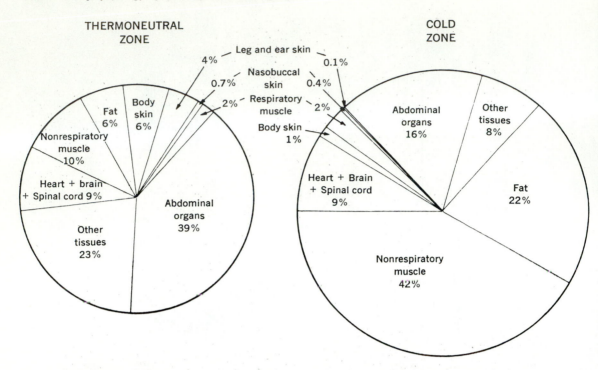

Fig. 76-32. Redistribution of cardiac output when lambs are exposed to severe cold. Total areas of circles represent cardiac output, and sectional areas are proportional to fraction of cardiac output received by tissues. (From Alexander.[5])

erated by the unique properties of "brown fat," which is abundant in the newborn. This fat in the newborn infant is deposited in five sites[4]: (1) a mass lying in a thin diamond-shaped sheath between the scapulae, (2) smaller masses around the blood vessels of the neck, (3) large deposits in the axillae, (4) moderately large masses in the mediastinum between the esophagus and the trachea and deposits around the internal mammary vessels, and (5) a large mass around the kidneys and adrenal glands. The brown fat cell is different in structure from the white fat cell. The lipid is distributed in a multilocular fashion, and a large number of mitochondria with prominent cristae lie in close apposition to the lipid droplets.[91] The injection of norepinephrine produces a large increase in heat production in newborn animals but not in adults. It is now known that the thermogenic response of brown fat is sympathetically mediated. The molecular basis of the thermogenic response of brown fat is not completely understood; one possibility is that Na^+, K^+-ATPase is stimulated by norepinephrine with the formation of ADP and the stimulation of respiration.[38]

Exposure of newborn animals to cold causes a major redistribution of blood circulating in the body. Fig. 76-32 compares the distributions of the cardiac output of lambs in the thermoneutral zone and in the cold zone. Blood flow through the nonrespiratory muscles and brown fat is greatly enhanced in lambs exposed to the cold and is greatly reduced in the abdominal organs and extremities. Blood flow through the heart, central nervous system, and respiratory muscles is not changed.

As development proceeds, the thermogenic brown fat becomes less important as a source of heat and the cells become transformed into white fat cells.[32,91]

REFERENCES

1. Adams, F. H., and Fujiwara, T.: Surfactant in fetal lamb tracheal fluid. J. Pediatr. **63**:537, 1963.
2. Adamsons, K., and Jowell, M. E.: Thermal homeostasis in the fetus and newborn, Anesthesiology **26**: 531, 1965.
3. Aherne, W.: Morphometry. In Gruenwald, P., editor: The placenta, Baltimore, 1975, University Park Press.
4. Aherne, W., and Hull, D.: Brown adipose tissue and heat production in the newborn infant, J. Pathol. **91**: 223, 1966.
5. Alexander, G.: Body temperature control in mammalian young, Br. Med. Bull. **31**:62, 1975.

6. Arey, L. B. A.: Developmental anatomy, ed. 7 (rev.), Philadelphia, 1974, W. B. Saunders Co.
7. Aschheim, S., and Zondek, B.: Hypophysenvorderlappenhormon und Ovarialhormon im Harn von Schwangeren, Klin. Wochenschr. **6:**1322, 1927.
8. Bauer, C., Ludwig, I., and Ludwig, M.: Different effects of 2,3-diphosphoglycerate and adenosine triphosphate on the oxygen affinity of adult and foetal human haemoglobin, Life Sci. **7:**1339, 1968.
9. Beer, A. E., and Billingham, R. E.: The immunobiology of mammalian reproduction, Englewood Cliffs, N.J., 1976, Prentice-Hall, Inc.
10. Behrman, R. E., Parer, J. T., and De Lannoy, C. W.: Placental growth and the formation of amniotic fluid, Nature **214:**678, 1967.
11. Beling, C. G.: Estrogens. In Fuchs, F., Klopper, A., editors: Endocrinology of pregnancy, New York, 1971, Harper & Row, Publishers.
12. Biggers, J. D.: Methods for studying the blastocyst. In Yoshinaga, K., Meyer, R. K., and Greep, R. O., editors: Implantation of the ovum, Cambridge, Mass., 1976, Harvard University Press, pp. 27-42.
13. Biggers, J. D.: Fertilization and blastocyst formation. In Alexander, N., editor: Symposium on animal models for contraceptive and fertility research, Harper & Row, Publishers. (In press.)
14. Biggers, J. D., and Borland, R. M.: Physiological aspects of growth and development of the preimplantation mammalian embryo, Annu. Rev. Physiol. **38:**95, 1976.
15. Biggers, J. D., and Powers, R. D.: Na$^+$ transport and swelling of the mammalian blastocyst: effect of amiloride. In Cuthbert, A. W., Fanelli, G. M., and Scriabine, A., editors: Amiloride and epithelial sodium transport, Baltimore, 1979, Urban and Schwarzenberg.
16. Boddy, K., and Dawes, G. S.: Fetal breathing, Br. Med. Bull. **31:**3, 1975.
17. Boddy, K., Dawes, G. S., and Robinson, J. S.: In Gluck, L., editor: Modern perinatal medicine, Chicago, 1974, Year Book Medical Publishers, Inc.
18. Bøe, F.: Studies on the human placenta, Acta Obstet. Gynecol. Scand. **48:**159, 1969.
19. Born, G. V. R., Dawes, G. S., Mott, J. C., and Widdicombe, J. G.: Changes in the heart and lungs at birth, Cold Spring Harbor Symp. Quant. Biol. **19:**102, 1954.
20. Boué, J. G., and Boué, A.: Chromosome anomalies in early spontaneous abortion, Curr. Topics Pathol. **62:**193, 1976.
21. Brenner, W. E., Edelman, D. A., and Hendricks, C. H.: A standard of fetal growth for the United States of America, Am. J. Obstet. Gynecol. **126:**555, 1976.
21a. Burns, R. K.: Urogenital system. In Willier, B. H., Weiss, P. A., and Hamburger, V., editors: Analysis of development, Philadelphia, 1955, W. B. Saunders Co.
22. Cannon, W. B.: Wisdom of the body, New York, 1932, W. W. Norton.
23. Cassmer, O.: Hormone production of the isolated human placenta, Acta Endocrinol. (Suppl.) **45:**1, 1959.
24. Cohen, F., and Zuelzer, W. W.: The transplancental passage of maternal erythrocytes into the fetus, Am. J. Obstet. Gynecol. **93:**566, 1965.
25. Colás, A., and Heinrichs, L.: Pettenkofer chromogens in the maternal and fetal circulations: anencephalic pregnancies, cesarean sections, and tentative identification of 3β, 17β-dihydroxyandrost-5-en-16-one in umbilical cord blood, Steroids **5:**753, 1964.
26. Crawford, J. M.: Vascular anatomy of the human placenta, Am. J. Obstet. Gynecol. **84:**1543, 1962.
27. Csapo, A.: The four direct regulatory factors of myometrial function, In Wolstenholme, G. E. W., and Knight, J., editors: Progesterone: its regulatory effect on the myometrium, London, 1969, Churchill Livingstone.
28. Csapo, A. I., and Wood, C.: The endocrine control of the initiation of labor in the human. In James, V. H. T., editor: Recent advances in endocrinology, ed. 8, London, 1968, Churchill Livingstone.
29. Dancis, J.: Feto-maternal interaction. In Avery, G. B., editor: Neonatology, Philadelphia, 1975, J. B. Lippincott Co.
30. Dancis, J., and Schneider, H.: Physiology: transfer and barrier function. In Gruenwald, P., editor: The placenta, Baltimore, 1975, University Park Press, ch. 6.
31. Dancis, J., Brenner, M., and Money, W. L.: Some factors affecting the permeability of the guinea pig placenta, Am. J. Obstet. Gynecol. **84:**570, 1962.
32. Dawes, G. S.: Foetal and neonatal physiology, Chicago, 1968, Year Book Medical Publishers, Inc.
33. Deisseroth, A., et al.: Chromosomal localization of human beta globin gene on human chromosome 11 in somatic cell hybrids, Proc. Natl. Acad. Sci. USA **75:**1456, 1978.
34. Donahoe, P. K., Ito, Y., Price, J. M., and Hendren, W. H.: Studies of müllerian inhibiting substance. IV. Müllerian inhibiting activity in fetal, newborn, and prepubertal testes, Biol. Reprod. **16:**238, 1977.
35. Dreskin, R. B., Spicer, S. S., and Greene, W. B.: Ultrastructural localization of chorionic gonadotropin in human term placenta, J. Histochem. Cytochem. **18:**862, 1970.
36. Farrell, P. M.: Fetal lung development and the influence of glucocorticoids on pulmonary surfactant, J. Steroid Biochem. **8:**463, 1977.
37. Frandsen, V. A., and Stakemann, G.: The site of production of oestrogenic hormones in human pregnancy. Hormone excretion in pregnancy with anencephalic foetus, Acta Endocrinol. Metab. **38:**383, 1961.
38. Girardier, L., Seydoux, J., and Clausen, T.: Membrane potential of brown adipose tissue, J. Gen. Physiol. **52:**925, 1968.
39. Goodwin, B. C.: Analytical physiology of cells and developing organisms, New York, 1976, Academic Press, Inc.
40. Graham, C. F.: The production of parthogenetic mammalian embryos, Biol. Rev. **49:**393, 1974.
41. Graham, C. F., and Wareing, P. F.: The developmental biology of plants and animals, Philadelphia, 1976, W. B. Saunders Co.
42. Greengard, O.: Enzymic differentiation in mammalian tissues, Essay Biochem. **7:**159, 1971.
43. Gruenwald, P.: Chronic fetal distress and placental insufficiency, Biol. Neonat. **5:**215, 1963.
44. Hales, J. R. S.: MTP international review of science: Physiology series one. In Robertshaw, D., editor: Environmental physiology, London, 1973, Butterworth & Co. (Publishers), Ltd.
45. Hamilton, W. J., and Mossman, H. W.: Human embryology, ed. 4, Cambridge, Mass., 1972, Heffer and Sons.
46. Harkness, R. A., et al.: Studies of urinary steroid excretion by an adrenalectomized woman during and after pregnancy, Acta Endocrinol. Metab. **52:**409, 1966.
47. Hathorn, M. K. S.: Analysis of the rhythm of infantile breathing, Br. Med. Bull. **31:**8, 1975.
48. Hey, E.: Thermal neutrality, Br. Med. Bull. **31:**69, 1975.
49. Hilgenfeldt, U., Wolfgang, E. M., and Brossmer, R.:

Circular dichroism of human chorionic gonadotropin, Hoppe Seylers Z. Physiol. Chem. **355:**1051, 1974.

50. Hillman, D. A., and Giroud, C. J.: Plasma cortisone and cortisol levels at birth and during the neonatal period, J. Clin. Endocrinol. Metab. **25:**243, 1965.

51. Huehns, E. R., and Farooqui, A. M.: Oxygen dissociation properties of human embryonic red cells, Nature **254:**335, 1975.

52. Imperoeto-McQuinley, J., Guerrerol, L., Gautier, T., and Peterson, R. E.: Steroid 5,α-reductase deficiency in man: an inherited form of male pseudohermaphrodism, Science **186:**1213, 1974.

53. Jelliffe, D. B.: Community and sociopolitical considerations of breast-feeding, Ciba Found. Symp. **45** (new series):231, 1976.

54. Josimovich, J. B., Merisko, K., and Boccella, L.: Amniotic prolactin control of amniotic and fetal extracellular fluid water and electrolytes in the rhesus monkey, Endocrinology **100:**564, 1977.

55. Josso, N.: Evolution of the müllerian inhibiting activity of the human testes, Biol. Neonate **20:**368, 1972.

56. Jost, A.: Sexual differentiation. In Beard, R. W., and Nathanielsz, P. W., editors: Fetal physiology and medicine, Philadelphia, 1976, W. B. Saunders Co.

57. Jost, A., and Jacquot, R.: Recherches sur le contrôle hormonal de la charge en glycogène du foie foetal du lapin et du rat, C. R. Acad. Sci. Paris, **239:**98, 1954.

58. Kilmartin, J. C.: Interaction of haemoglobin with protons, CO_2 and 2,3-diphosphoglycerate, Br. Med. Bull. **32:**209, 1976.

59. Klein, G. P., et al.: Identification and possible origin of C-21 steroid sulfates within the foetoplacental unit. In Pecile, A., and Finzi, C., editors: The foetoplacental unit, Amsterdam, 1969, Excerpta Medica.

60. Kotas, R. V., and Avery, M. E.: Accelerated appearance of pulmonary surfactant in the fetal rabbit, J. Appl. Physiol. **30:**358, 1971.

61. Lash, J., and Whittaker, J. R.: Concepts of development, Stamford, Conn., 1974, Sinauer Associates.

62. Liggins, G. C.: Endocrinology of the foeto-maternal unit. In Shearman, R. P., editor: Human reproductive physiology, Oxford, 1972, Blackwell Scientific Publications, ch. 4.

63. Liggins, G. C.: The drive to fetal growth. In Beard, R. W., and Nathanielsz, P. W., editors: Fetal physiology and medicine, Philadelphia, 1976, W. B. Saunders Co., pp. 254-270.

64. Liggins, G. C., and Rees, L.: The effect of adrenal stimulation on the insulin response to glucose in the fetal lamb, N.Z. Med. J. **81:**486, 1975.

65. Lyon, M. F., and Hawkes, S. G.: X-linked gene for testicular feminization in the mouse, Nature **227:**1217, 1970.

66. McLain, C. R.: Amniographic studies of the gastrointestinal motility of the human fetus, Am. J. Obstet. Gynecol. **86:**1079, 1963.

67. Nadler, H. L.: Biochemical and cytologic components of amniotic fluid. In Barnes, A. C., and Seeds, A. E., editors: The water metabolism of the fetus, Springfield, Ill., 1972, Charles C Thomas, Publisher.

68. Neill, J. D., and Knobil, E.: On the nature of the initial luteotropic stimulus of pregnancy in the rhesus monkey, Endocrinology **90:**34, 1972.

69. Nelson, N. M.: Respiration and circulation after birth. In Smith, C. A., and Nelson, N. M., editors: The physiology of the newborn infant, ed. 4, Springfield, Ill., 1976, Charles C Thomas, Publisher, ch. 4.

70. Nelson, N. M.: Respiration and circulation before birth.

In Smith, C. A., and Nelson, N. M., editors: The physiology of the newborn infant, ed. 4, Springfield, Ill., 1976, Charles C Thomas, Publisher, ch. 3.

71. Neuman, F., Elger, W., Steinbeck, H., and Gräf, K. J.: The role of androgens in sexual differentiation of mammals. In Reinboth, R., editor: Intersexuality in the animal kingdom, Heidelberg, 1975, Springer-Verlag.

72. Nienhuis, A. W.: Hemoglobin switching: a new experimental model, J. Lab. Clin. Med. **91:**857, 1978.

73. Nienhuis, A. W., and Benz, E. J.: Regulation of hemoglobin synthesis during the development of the red cell, New Engl. J. Med. **297:**1318, 1977.

74. Nørgaard-Pedersen, B., Jørgensen, P. I., and Trolle, D.: Alpha-fetoprotein concentration in amniotic fluid during the last trimester in normal pregnancies and in pregnancies with severe fetal abnormalities, Acta Obstet. Gynecol. Scand. **55:**59, 1976.

75. Osinski, P. A.: Steroid 11β-ol-dehydrogenase in human placenta, Nature **187:**777, 1960.

76. Pritchard, J. A.: Deglutition by normal and anencephalic fetuses, Obstet. Gynecol. **25:**289, 1965.

77. Queenan, J. T., Thomson, W., Whitfield, C. R., and Shah, S. I.: Amniotic fluid volumes in normal pregnancies, Am. J. Obstet. Gynecol. **114:**34, 1972.

78. Roberts, D. F., and Thomson, A. M.: The biology of human fetal growth, New York, 1976, Halsted Press.

79. Rudolph, A. M.: Fetal and neonatal pulmonary circulation, Am. Rev. Res. Dis. (Suppl.) **115:**11, 1977.

80. Rudolph, A. M., and Heymann, M. A.: Fetal and neonatal circulation and respiration, Annu. Rev. Physiol. **36:**187, 1974.

81. Rudolph, A. M., and Yuan, S.: Response of the pulmonary vasculature to hypoxia and H^+ ion concentration changes, J. Clin. Invest. **45:**399, 1966.

82. Scarpelli, E. M.: The surfactant system of the lung, Int. Anesthesiol. Clin. **15**(4):19, 1977.

83. Schoenheimer, R.: The dynamic state of body constituents, Cambridge, Mass., 1940, Harvard University Press.

84. Schubert, K., and Schade, K.: Placental steroid hormones, J. Ster. Biochem. **8:**359, 1977.

85. Seeds, A. E.: Amniotic fluid. In Barnes, A. C., and Seeds, A. E., editors: The water metabolism of the fetus, Springfield, Ill., 1972, Charles C Thomas, Publisher, chap. 3.

86. Seeds, A. E.: Mechanisms of intrauterine water transfer in pregnancy. In Barnes, A. C., and Seeds, A. E., editors: The water metabolism of the fetus, Springfield, Ill., 1972, Charles C Thomas, Publisher, chap. 2.

87. Selenkow, H. A., Saxena, B. N., Dana, C. L., and Emerson, K.: Measurement and pathophysiologic significance of human placental lactogen. In Pecile, A., and Finzi, C., editors: The foetoplacental unit, Amsterdam, 1969, Excerpta Medica.

88. Sell, S., and Becker, F. F.: Alpha-fetoprotein, J. Natl. Canc. Inst. **60:**19, 1978.

89. Sinclair, J. C.: Metabolic state and temperature control. In Smith, C. A., and Nelson, N. M., editors: The physiology of the newborn infant, ed. 4, Springfield, Ill., 1976, Charles C Thomas, Publisher, chap. 7.

90. Sitteri, P. K., and MacDonald, P. C.: Placental estrogen biosynthesis during human pregnancy, J. Clin. Endocrinol. Metab. **26:**751, 1966.

91. Smith, R. E., and Horwitz, B. A.: Brown fat and thermogenesis, Physiol. Rev. **49:**330, 1969.

92. Southgate, D. A. T., and Hey, E. N.: Chemical and biochemical development of the human fetus. In

Roberts, D. F., and Thomson, A. M., editors: The biology of human fetal growth, New York, 1976, Halsted Press.

93. Steven, D. H.: Anatomy of the placental barrier. In Steven, D. H., editor: Comparative placentation, London, 1975, Academic Press, Ltd., chap. 2.

94. Steven, D. H., and Morriss, G.: Development of the foetal membranes. In Steven, D. H., editor: Comparative placentation, London, 1975, Academic Press, Ltd., chap. 3.

95. Tanner, J. M.: The regulation of human growth, Child Dev. **34:**817, 1963.

96. Tuchmann-Duplessis, H., David, G., and Haegel, P.: Illustrated human embryology, vol. 1. Embryogenesis, Paris, 1972, Masson et Cie, Editeurs.

97. Van Otterlo, L. C., Wladimiroff, J. W., and Wallenburg, H. C. S.: Relationship between fetal urine production and amniotic fluid volume in normal pregnancy and pregnancy complicated by diabetes, Br. J. Obstet. Gynaecol. **84:**205, 1977.

98. Villee, D. B.: Human endocrinology. A developmental approach, Philadelphia, 1975, W. B. Saunders Co.

99. Wachtel, S. S.: H-Y antigen and the genetics of sex determination, Science **198:**797, 1977.

100. Waddington, C. H.: The basic ideas of biology. In Towards a theoretical biology. I. Prolegomena, Edinburgh, 1968, Edinburgh University Press, pp. 1-41.

101. Wallenburg, H. C. S.: The amniotic fluid. I. Water and electrolyte homeostasis, J. Perinat. Med. **5:**193, 1977.

102. Widdowson, E. M.: Growth and composition of the fetus and newborn. In Assali, N. S., editor: The biology of gestation, vol. 2, New York, 1968, Academic Press, Inc.

103. Wood, W. G.: Haemoglobin synthesis during human fetal development, Br. Med. Bull. **32:**282, 1976.

77

PAOLA S. TIMIRAS

Physiology of aging

Most textbooks in human and animal physiology focus on the mature organism as their model of optimal integrated function of all body systems and as a standard against which to measure degrees of physiologic or pathologic deviation. The preceding chapter traced the numerous physiologic events and changes that occur in body systems and organs in the course of growth and development and brought us to the threshold of maturity. The purpose of this chapter is to follow the changes that take place in the late years of life, that is, during senescence, when aging processes become superimposed on normal functions and gradually transform them. In this sense the aging process can be viewed as an integral aspect of the continuous development of the organism. The concept of Minot[39] that aging is essentially a decline in growth with time and represents the price one pays for differentiation has been enlarged by Weiss,[68] who regards aging itself as differentiation, thus inseparable from developmental processes. Indeed from a physiologic standpoint, all changes occurring throughout the life span reflect a prior state or foreshadow events to follow.

Although the pathology of aging has received considerable study in terms of combating specific diseases associated with old age, the physiology of aging has attracted less interest, partly because of the difficulty of isolating "normal" from "abnormal" aging processes. Furthermore, although old age in humans has been conventionally accepted as the stage of the life span that starts at 65 years—an age chosen on the basis of social security legislation, retirement policies, and actuarial statistics rather than biologic criteria—and terminates with death, it is extremely difficult to circumscribe its temporal boundaries in physiologic terms; thus aging is commonly described as having its onset at some indeterminate point following maturity. In contrast with other periods of the life span that are clearly marked by specific physiologic events (e.g., menarche at puberty), aging has so far defied any attempt to establish objective landmarks that would precisely characterize its earlier stages.

It is undeniable that older persons show certain common traits with respect to the age of onset and the rate and manifestations of aging processes; nevertheless, still essentially unanswered are the fundamental questions of which physiologic events herald the beginning of old age, how the rate of aging is modified by genetic and environmental factors, and, most importantly, what the nature of the aging process is and what the mechanisms are by which aging affects living organisms at molecular, cellular, and organismic levels. Lacking these answers, an explanation of the aging process has been extremely difficult to construct; in the absence of a satisfactory explanation, a plethora of theories has been advanced, each representing the subjective view of a specific investigator.

In this chapter, discussion will center around aspects of aging of interest to physiologists: a definition of aging that emphasizes the decline in functional competence with the passage of time, a brief survey of the differential timetables of physiologic aging in humans, and the major theories of aging. I will also attempt to integrate some of the current information within the concept of aging as part of a physiologic ontogenetic program that is regulated by neuroendocrine control systems essential for homeostasis and survival.

DEFINITIONS OF PHYSIOLOGIC AGING

As indicated previously, biologists exploring aging phenomena have long held the view that growth, development and aging could be considered as stages in a continuum of physiologic events having their onset at fertilization and their termination at death. Defined as "the sum total of all changes that occur in a living organism with the passage of time," this "global appreciation" of the biologic history of the individual suggests that aging may best be understood by studying development. Although the concept of aging as an integral part of development, but differentiated from it in time, has opened many avenues for investigations of both theoretical and practical interest, it still remains to be fortified by substantive evidence. Tempting as it may be for a physiologist to visualize aging as an orderly unfolding of precisely timed events from birth to death, such a view leads us to think of the aging process as a series of "frames" successively flashed on a screen and disregards the fact that chronologic and physiologic age rarely coincide because of the continuing interplay of genetic and environmental factors. A more accurate picture of aging must recognize the decrement in physiologic competence that occurs at various levels from molecular to organismic throughout the life span and that contributes to the progressively decreasing capacity of the organism to maintain its viability. Thus in man and higher vertebrates, aging takes the form of a morphologic, functional, and biochemical involution, always regressive and often silent, that affects most of the organs and results in a gradual decline in performance. Taking cognizance of such decrement the previous definition of aging has been extended to include "the sum total of all changes that occur in a living organism with the passage of time and lead to functional impairment and death." One of the principal expressions of functional competence is the capacity of the organism to adapt to an ever-changing environment, and this adaptability is progressively impaired with time. Therefore the definition of aging as functional decrement has been further modified to mean a "decreasing ability to survive stress." However, as will be emphasized in the following section, the decline in functional competence with aging is not uniform but shows considerable variability among individuals and species in both the rate and the magnitude of age-related changes in cells, tissues, and organs. Furthermore, deteriorative changes in one such element or structure do not always signal the aging of the whole organism. Indeed, localized cell aging and death is a phenomenon that is also associated with organogenesis[52] and structures that emerge, age, and cease to function are not uncommon in embryologic development (e.g., yolk sac, allantois, placenta) and in adult tissues and systems (e.g., hematopoietic system, skin, mucosae).

Another aspect of the relationship between physiologic decrement and aging is that the decline in function with age proceeds with linear precision, whereas the risk of mortality undergoes a sharp exponential increase.[51] People rarely die of "old age"; omitting fatal accidents, death generally occurs when the capacity of the body to withstand a specific challenge is surpassed. Further, stress may not be in itself sufficient to challenge the body's tolerance, but the functional impairments of old age may increase the frequency with which such stresses are experienced. In other words, the loss of visual acuity, auditory perception, and muscle strength, for example, may predispose the elderly to accident or disease, which, in association with other physiologic decrements, overtaxes their endurance.

Because of these complexities, systematic longitudinal studies of aging in humans are relatively rare, and adequate cross-sectional studies are also lacking. Therefore for information on physiologic aging we rely primarily on comparative data secured in animals whose life is shorter than our own but who may differ in important respects. It is also to be recognized that biologic research in aging has received its major impetus in recent years from the growth of molecular biology and consequently tends to be predisposed in this direction. Thus the definition presented previously has been modified to emphasize "the sum total of cellular and/or molecular aging" as the cause of aging and death in lower as well as higher organisms including humans. Proponents of this view approach aging studies in terms of the progressive occurrence of somatic mutations, alterations in genetic coding, defects in information readout, and other manifestations of molecular and cellular deterioration as discussed briefly in the following section (see also p. 1993).

Maximum life span in humans. Among mammals, man is the longest-living species, and his maximum life span potential (i.e., the intrinsic biologic survival depending only on chronologic time and independent from accidents and diseases) is related to a variety of physiologic factors, of which the most important appears to be the higher ratio of brain size to body size.[23] Accurate records of human longevity, however, are as difficult to collect as those of animals; not only are they subject to unavoidable recording errors, but it is well known that beyond a certain age, elderly individuals show a tendency to exaggerate their age. Claims of extreme longevity abound in particular districts. For example, investigators have frequently reported the celebrated longevity of some inhabitants of Siberia and the Caucasus, such as the Abkhasians, who claim to be 130 years old and over. The existence of individuals with very advanced old age but few debilitating illnesses has also been reported in some Andean villages of Ecuador and some valleys of the Hunza state in the Karakorum mountains. Although there is no question that many of the subjects interviewed were very old, even centenarians, their exact age and their proportion in the total population have been questioned. Records supported by proper documentary evidence have fixed the highest human age at 120 years. Apart from examination of the observed records, attempts have been made to predict the maximum human life span from statistical analysis of life tables. In these terms the actuarial probability of an individual exceeding the age of 150 years is extremely small. Indeed, the number of persons reaching even 100 years is too limited to permit statistically significant estimates of the duration of the life span beyond 90 years of age.

DIFFERENTIAL AGING IN HUMANS

The body's diversity and the complexity of its integrative regulations, discussed repeatedly in previous chapters, are also reflected in the differential patterns that characterize the aging of discrete functions. Thus whereas in certain functions the regulation of the organism remains quite efficient until the age of 80 or 90 years, in others, functional decrements appear at an early age. However, recognition of this fact does not permit us to classify the onset of aging changes with any precision. Because aging is a continuous but slow process, we observe its effects only when they have progressed sufficiently to induce alterations that can be identified or validated by available testing methods. An illustrative example is atherosclerosis (Chapter 00), the consequences of which become manifest in middle and old age even though the atherosclerotic lesion may start early in infancy.[27,29] Thus although "aging charts" to establish a profile of physiologic aging comparable to the "growth charts" employed for evaluating development in childhood would be extremely valuable, the compilation of the necessary data is hampered by the lack of systematic studies in healthy and aged populations.

Regulation of body temperature, continuance of blood sugar levels, and maintenance of acid-base balance represent functions that, having a number of alternate control mechanisms, remain relatively stable into old age. This is an observation, however, based on measurements taken under basal conditions. When the same tests are made after increased physiologic demand (e.g., sustained muscular exercise or environmental changes), the efficiency with which the organism is capable of maintaining levels within normal limits or the rapidity with which these levels return to normal demonstrate marked differences between young and old. For example, when thermoregulatory competence was tested by comparing the speed of body temperature recovery following a 3 min immersion in ice water, marked differences were observed among adult, middle-aged, and old rats (Fig. 77-1): the drop

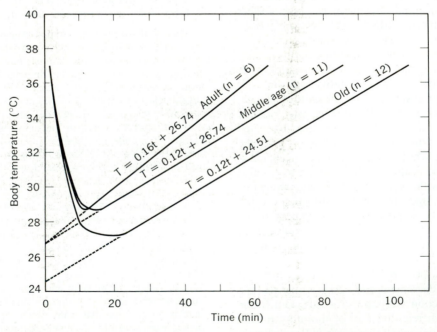

Fig. 77-1. Age-related differences in body temperature changes following 3 min whole-body ice water immersion.[53] Average body temperature plotted against time for three age groups. *T*, Body temperature; *t*, time. Rats were immersed in ice water at 0 time and removed 3 min later. Temperature loss and recovery profile for three age groups were calculated by method of least squares. *n*, Number of animals. (Reprinted from Segall and Timiras: Fed. Proc. **34**:83, 1975.)

in body temperature was greater in older animals than in either the middle-aged or adult rats, and the slope of the recovery curve indicated that the younger animals recovered significantly faster than the middle-aged and old animals, and that even though the recovery rate is similar between the latter two groups, body temperature of the old animals remained lower than that of the other two age groups throughout the test period.[53] Therefore placing a system under stress not only brings to light age differences not otherwise detectable, but also clearly demonstrates the declining ability of the aging organism to withstand or respond adequately to stress.

Other functions of the body begin to age relatively early in adult life and then decline very rapidly. For example, the speed of accommodation of the eye begins to diminish in the young adult and by age 50 years, on the average, has reached a minimum; similarly, dark adaptation and critical flicker frequency—two other easily quantifiable aspects of ocular function—decline at an almost linear rate after the 20th year of life (Fig. 77-2). Aging of auditory function has been shown to begin at adolescence and to continue steadily thereafter, culminating around 50 years of age. In evaluating the effects of aging on these and other sensory functions, one must question whether it is simply the availability of sensitive testing methods for these functions that accounts for the disparity in their aging timetable compared to that of other functions or whether these differences are real. It should be noted, for example, that great variability has been reported both among individuals and among different populations; data indicate that auditory senescence occurs at a considerable later age in some African tribes living in relatively noise-free environments than in populations of Europe and the United States.[15,28,48]

The locomotor system represents another system in which early aging phenomena have been well demonstrated. Irrespective of the muscle group considered, the peak of muscular strength occurs between 20 and 30 years of age; the decline is continuous thereafter, accelerating progressively with age, although the rate of aging may vary depending on the muscle group considered and the type of test used for measurement of muscle function (Fig. 77-3). The progressive senescence of the skeletal muscle, involving biochemical changes in the muscular fibers as well as alterations in the myoneural junctions, goes pari passu with a gradual modification of the skeleton, characterized by a steady diminution in the density of the long bones and vertebrae and inflammatory and regressive changes in the cartilage and articulations. Whereas many of the age-related changes identified in muscle are common to other tissues of the body, those occurring in muscle are particularly prominent because of the proportionally high distribution of muscle throughout the body and its relationship to lean body mass. Inasmuch as 95% of the body's potassium is in the living cell and 60% is in muscle alone, the decline with age in total potassium content[26,47] may be the consequence of the diminution of body weight and the muscular involution that accompany old age.[14,41] This observation has been confirmed by other methods used to assess lean body mass; for example, basal metabolic rate (BMR) also declines with age in healthy subjects who show no signs of thyroid disorder, one of the most frequent causes of alterations in BMR (Chapter 55). A further correlation has emerged from studies conducted in both the young rat and the human infant that relate lean body mass to the quantity of creatinine excreted in the urine in any 24 hr period,[40] but whether creatinuria can be considered a valid index of muscle mass in the adult and elderly has been questioned.[46,50] Further, the diminution in cellular functional mass with age is often associated with an increase in the mass of connective tissue and in fat deposition and replacement of specific, competent cells with nonspecific, less differentiated cellular elements. In general, it can be said that as growth is represented by an incremental increase in protoplasmic mass, aging can be viewed as a progressive decrease in this element. Whether the onset and magnitude of this reversal can be evaluated by measuring total body potassium, lean body mass, or protein metabolism, the onset of aging thus established remains a subject of active interest.

The aging of other functions has been charted by several authors; of these, the renal and cardiorespiratory functions offer representative examples, and their decline with aging has been diagrammatically summarized by Shock[57] (Fig. 77-4). The decline in kidney function represents one of the irreversible involutionary processes that accompanies advancing age. Parallel with a progressive reduction in the number of functioning nephrons, a progressive decline occurs in the glomerular filtration and tubular excretion rates, as well as in the maximum tubular excretion (Tm-Diodrast) and resorption capacities (Tm-glucose) and in the renal blood flow. The significant decrease in renal blood flow is often consequent to the arteriosclerotic involvement of the renal blood vessels, but that does not entirely ex-

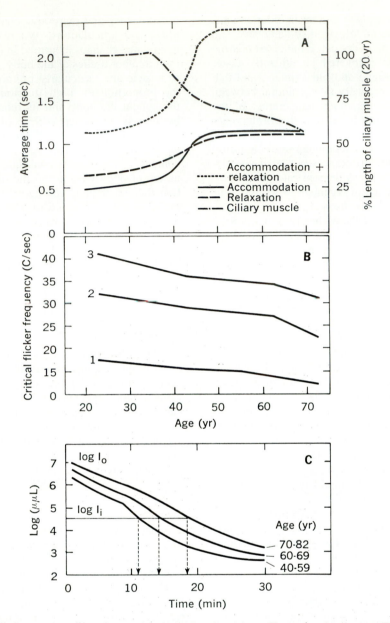

Fig. 77-2. Some examples of age-related changes in vision. (For more details, see Timiras.[12]) **A,** Speed of accommodation and relaxation. With age, time of accommodation and relaxation for near and far vision progressively increases (reaching a maximum at about 50 years) as a consequence of hardening of lens and atrophy (shortening) of ciliary muscle. **B,** Critical flicker frequency for three levels of stimulus brightness. Intermittent light stimuli delivered to retina are perceived as flicker, except when rate of delivery (in cycles per second) reaches critical value, when flicker disappears and stimuli are perceived as continuous light source. Critical flicker frequency increases with brightness of stimulus but decreases with age at all levels of brightness, probably as consequence of senile alterations in retina, pupil, and lens. **C,** Dark adaptation time. Values are expressed in terms of logarithm of brightness in $\mu\mu L$ (L = lambert, unit of brightness) on ordinate, measured against time on abscissa; each curve represents different age group. Threshold of illumination (i.e., minimal brightness perceived as light) rises with age, and data presented indicate this difference in three age groups. Curves expressing rate of dark adaptation do not vary significantly among three groups, but because of progressively lowering sensitivity to light (as indicated by progressively higher threshold) of aging eye, time necessary after exposure to sunlight (log I_0) to reach given level of dark adaptation (log I_1) is increased with age. (From Timiras, P. S.: Developmental physiology and aging, New York, Copyright © 1972, Macmillan Publishing Co., Inc.)

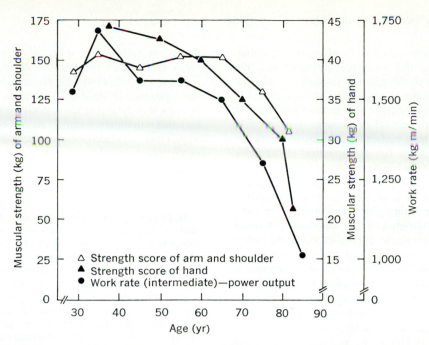

Fig. 77-3. Changes with age in muscular strength and maximum work rate. Test used to assess changes in strength of arms and shoulders with age is carried out in subject who is seated and asked to pull, with maximum effort, different levers that are arranged to demand exertion of these muscles.[58] Composite strength scores are calculated by summating highest values recorded (in kilograms) for each pull. Muscular strength did not change significantly from age 20 to age 65 years, and even in 70- and 80-year-old subjects, reduction in muscle strength (although significant) was less marked when allowance was made for lower body weight of this age group. Conversely, however, muscular strength as measured in muscle of hand with Smedley dynamometer appears to undergo progressive decline already significant at 50 years of age.[18] Thus, decline in strength varies with muscular group considered. Maximum work rate is assessed by having subject, recumbent in bed, crank drive sprocket wheel with maximum effort, over 10 to 15 sec periods at four different "load" settings. Since duration of cranking exercise is short, involvement of cardiovascular and pulmonary responses is minimal. Work rate is measured in terms of power output expressed in kilogram-meters per minute; only values for intermediate loads are presented here. With all load rates, power output is seen to decline at 40 years of age and to progressively decrease thereafter, even when values are corrected for lower body weights in 70- and 80-year-old subjects. It has been suggested that reduced coordination, involving both muscular and nervous function, is most likely cause for decline in power output with age.[58] (From Timiras, P. S.: Developmental physiology and aging, New York, Copyright © 1972, Macmillan Publishing Co., Inc.)

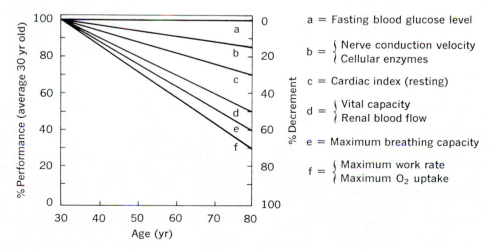

Fig. 77-4. Age-related decrements in physiologic performance. Schematic representation of decline in selected functional capacity and physiologic measurements with age in humans.[57] (From Timiras, P. S.: Developmental physiology and aging, New York, Copyright © 1972, Macmillan Publishing Co., Inc.)

plain the functional changes of the organ; rather, it is believed that significant involutional processes occur in the glomerulus and the renal tubule and that these, together with advancing circulatory impairment, are responsible for the progressive functional decrement.

As mentioned earlier, responses to stress frequently reveal the failing capacities of the aging individual; comparison of the cardiorespiratory responses of young and old subjects during sustained muscular exercise illustrates this point. The increased muscular requirements for extra oxygen and nutrients and for elimination of metabolic waste products during exercise are generally met by the significant increase (up to 30-fold) in blood flow, a response that is known to depend on the activation of several local and systemic mechanisms, among which the increase in cardiac output is predominant. In the young adult, cardiac output during exercise may increase from 5 L/min to as high as 35 L/min (Chapter 40). In the aged, however, cardiac output is already significantly reduced as compared to that of the young adult and cannot effectively respond to the additional demands imposed by the contracting muscles. Thus in the elderly exercise causes a marked acceleration in cardiac rate and a rise in blood pressure, which although greater than in the young, are not sufficient to compensate for the inadequate increase in cardiac output. The failure to satisfy the circulatory demands of the exercising muscle imposes severe limits on the amount of muscular work that the elderly can perform.

Some of the fundamental functions of the lung are also impaired with age. For example, the amount of oxygen that the blood takes up from the lung and transports to the tissues falls substantially with advancing years, from approximately 4 L/min at 20 years to 1.5 L/min at 75 years. Thus to maintain the increased activity of exercising muscles the older individual must move in excess of 50% more air in and out of the lungs. No doubt the decline in oxygen consumption partially reflects the reduction in cardiac output, for less blood flows through the lungs in the older person at any given time; the magnitude of the difference in oxygen uptake between young and old indicates that marked changes take place in the lung tissue as well.

The decline in respiratory function also reflects a loss in mechanical efficiency. With age, less air turns over in the lungs and the size of the dead air space is proportionately greater, although the total lung volume remains unchanged. Vital capacity is decreased with age, and this is re-flected in a 40% decline in maximum breathing capacity between the ages of 20 and 80 years. Since the older person expels approximately the same quantity of air at each breath as the younger person, it is clear that his capacity is less because he cannot maintain a rapid rate of breathing; this impairment has been interpreted as an expression of the general decline in neuromuscular capacity reflecting deficits in both muscle strength and the central nervous system.

Immunology and aging. The immune system, in view of its protective action against invading foreign substances and the changes it undergoes with age, is another system that has been strongly implicated in the etiopathogenesis of aging processes. In the last decade, numerous arguments have been advanced in favor of the concept that aging is a generalized, mild, and prolonged type of autoimmune phenomenon resulting from the disruption of the mechanisms that preside at the acquisition and maintenance of self-recognition and tolerance.[3,13,20,35] Indeed, alterations in all facets of immune responses (i.e., tolerance, immunity, and hypersensitivity) have been related to diseases of senescence (e.g., autoimmune diseases, amyloidosis, and increased susceptibility to infectious disease). In general the immunologic inadequacies of the aged reflect both intrinsic deficiencies in the immunocompetent cell populations (e.g., defect in T cell proliferative activity) and extrinsic deficiencies (e.g., decreased opsonic levels) in the environment of the cells.[35] In addition, it has been suggested that suppressor T cells (i.e., thymus-dependent cells with immunosuppressive properties) increase in number and/or activity with age and that this increase plays a major role in the age-related decline of the immune potential.[56] Whereas the response to exogenous antigens decreases with age, the reverse occurs with respect to endogenous antigens and leads to an increase in the incidence and severity of the so-called autoimmune diseases. This increase has been ascribed either to the emergence of new antigenic stimuli that evade self-censorship or to the loss of immunologic memory and faulty self-recognition, or to the enhanced antigenicity resulting from tissue alterations induced by mutations, viruses, and other factors. Changes in antigenicity due to slow viruses and related to the declining function of immunocompetent cells with aging have also been interpreted as responsible for the increase in neoplastic diseases in later life.[63]

Immunologic competence is an integral component of physiologic homeostasis, and changes in this competence with age will lead inevitably to decrements in overall adaptation and survival. Therefore whether immunologic responses represent the cause, as purported by the proponents of the immunologic theories of aging, or the consequence, among others, of aging, this theory provides a useful framework within which to test the relationship between some of the diseases of old age and the decline in physiologic competence.

Furthermore, the marked influence exerted by endocrine and metabolic factors on immunologic responses provides the physiologist both with a model for testing functional adequacy at different ages and with tools (e.g., hormones) for modifying the aging process.

For a more complete description of the changes with aging in all functions of the body, the reader is referred to specialized textbooks and reviews on the subject.[1-13] Some observations on the aging of the nervous and endocrine systems are presented in the next section in relation to the hypothesis that aging and death are governed by one or more "biologic clocks" that eventually "run down."

THEORIES OF AGING

An initial exposure to the literature on the theories of aging will most probably leave an overall impression of confusion, not only because of our general lack of knowledge in this area, but also because of the ambiguous way in which aging theories are sometimes presented and the uncertainty of their relation to one another. The ambiguity almost always revolves around the claim of which factor among multiple factors is the "cause" of aging. It is possible that different theories of aging, some of which will be briefly summarized here, each focusing on a different cause, are all correct. Living in a quiet, rural environment may help one to stay in a fit state longer. Reengineering some of our genes may also prolong our vigor and life. Aging is environmental or genetic, depending on how one wants to look at it.

Of the numerous theories that have been and continue to be advanced with respect to the causative factors in aging, the following may be taken as representative of the major work in the field thus far: (1) fundamental views of cellular wear and tear; (2) mechanicochemical explanations of structural and functional changes in nervous, endocrine, vascular, and immunologic systems and connective tissue; (3) concepts of progressively increasing autointoxication with age; (4) general developmental theories relating senescence to morphogenesis and cessation of growth; (5) loss of proliferative capacity within a finite life span; and (6) genetic hypotheses regarding the accumulation of somatic mutations and the loss of biologic information. Those related to physiology are summarized here; for others, see general reviews (p. 1997).

Cellular and molecular theories

Theories of cytologic aging suggest that the body is victimized by the additive effects of insults and injuries to individual cells or cellular macromolecules. The "insults" have been variously viewed as ionizing radiation; mechanical and thermal alterations[5,72]; uncontrolled chemical reactions leading to the liberation of "free radicals" affecting membranes, DNA, and collagen[31,45]; rebounding imbalances in which a domino effect takes place from a single physiologic malfunction[6,59,60]; and a progressive glut of insoluble elements such as lipofuscin (or age pigment), noticed particularly in the nondividing cells of the nervous and muscular systems and characterized by progressive increase with age.[61]

According to other theories that view aging as caused primarily by intrinsic factors, proteins and nucleic acids become irreversibly immobilized with age as a result of the formation of strong bonds (cross-linking) between molecules and the consequent creation of a frozen metabolic pool with no space for active molecules.[16] Genetic "errors," including the incorrect insertion of an amino acid into a protein, have also been postulated as fundamental to the deterioration of the older organism; if the mistaken information is inserted into a protein important for protein synthesis—as may be the case for tRNA synthetase—then the error, as it increases exponentially throughout the life span, culminates in an "error catastrophe" that disrupts the functioning of the cell and leads to its death.[43, 44]

Finally, nuclear studies have focused on alterations with aging in the transcription of genetic events from DNA into mRNA,[67] in the order of DNA nucleotides, in DNA repair and redundancy of critical genes,[38] in differences with development and aging in codon modulation in the production of tRNA species and tRNA synthetases,[62] or in a sequential activation or suppression of genes depending on intrinsic and extrinsic factors.[32] Experiments based on transplants of nuclei and cytoplasm from old to young cells and vice versa show interesting results that may help us to distinguish between nuclear and cytoplasmic contribution to in vitro cellular senescence. Although evidence is still too ambiguous to permit a definitive conclusion at this time, the available results nonetheless suggest that nuclear factors and not cytoplasmic factors limit the proliferative capacity of cultured normal human cells.[70]

Neuroendocrine regulation of aging

Another theory put forward to explain the cause of aging suggests that "a program for aging" genetically encoded in the brain may be expressed by a precisely programmed timetable of growth, maturation, and aging.[6,25] From specific centers in the brain the controlling signal would be relayed to the peripheral tissues and organs by hormonal and neural (somatic and autonomic) stimulation or inhibition. Evidence in support of such a program, possibly analogous to similar programs regulating other "timed" physiologic interactions (i.e., biologic clocks) is still very tentative, and its nature and localization remain to be clearly identified. Nevertheless, because of the central role of the brain and endocrine glands in the control of all bodily functions, it appears reasonable to formulate, at least as a working hypothesis, the concept that such a clock does exist, that it is located in some discrete area of the brain, and that it is influenced by environmental factors. Some of the data in support of this theory derive from observations of the aging of the nervous and endocrine systems and from experiments that show that alterations in neuroendocrine interrelations may modify the course of growth, maturation, and aging.

Early studies of the aging brain reported a decrease in size associated with a loss of neurons. Subsequent studies, however, failed to recognize a progressive decrease in neuronal number under conditions of "normal" aging (i.e., not associated with neurologic and extensive vascular pathology) in several animal species including man. Rather an increase in glial cell number (gliosis) was observed in some brain areas and interpreted as a compensatory reaction to the declining functional capacity of the neurons.[19] Indeed, neurons, even though unchanged in number, undergo clear-cut and progressive structural and neurochemical damage characterized, among other changes, by accumulation of lipofuscin, loss of Nissl substance, presence of neurofibrillary tangles, and a striking loss of dendrites.[65] The loss of dendrites and consequently of synaptic connections in aging may be related to concomitant alterations in neurotransmission. Despite several differences with respect to animal species, brain region, and neurotransmitter system, the overall picture of the aging brain appears to be one of neurotransmission imbalance. In view of the differential regional distribution of neurotransmitters, preferential decrements may occur in some neurotransmitters while others remain unaffected. In this case, excess or deficit of a specific neurotransmitter would result not only in functional alterations ascribable to a single loss or decrease (e.g., the decrease in dopamine in the corpus striatum of individuals with Parkinson's disease, Chapter 29), but, more importantly, imbalances among several neurotrans-

mitters. We know that normal brain function depends on the delicate tuning of inhibitory and excitatory impulses and that a disequilibrium of this balance such as may occur during aging will have severe and widespread functional repercussions, involving somatic, visceral, and endocrinologic functions. This balance could become so unstable in old age that deviations would be induced even by minor shifts in neurotransmission.[64,66]

Assessment of psychologic age in humans. The pathophysiologic alterations in the old brain briefly described in the preceding section are usually associated with behavioral and intellectual changes.[2] Indeed, any adequate assessment of biologic age in humans must simultaneously consider psychologic as well as physiologic aspects. The decrease of mental capacities with advancing age has been studied both clinically and experimentally, and the data thus far collected have shown that the overall decline in psychologic competence and the severity of behavioral impairment with age show considerable individual variability. The nature and degree of the individual differences depend on intrinsic and extrinsic factors, and besides genetic and environmental factors, one of the most important is the ability of the cerebral circulation to provide the brain with the adequate oxygenation crucial for normal metabolism and function (Chapter 44). Inasmuch as with aging the cerebral circulation is progressively impaired by the atherosclerotic involvement of its arteries and the blood flow and oxygen supply correspondingly diminish, some of the behavioral and intellectual deficits observed in old people may be secondary to circulatory changes and hypoxia rather than to aging of the nerve cells. In addition, the rate of decline varies considerably from one function to another in a given individual: those functions related to experience and utilization of acquired knowledge resist the assault fairly well, whereas those functions that depend on the capacity for adjustment to new situations show significant deterioration with advancing age. Some of the most frequent manifestations of changes in brain function with aging include the following: slowing of sensori-motor responses due to both peripheral and central alterations (e.g., impairment of receptor and effector organs, lengthening of conduction time, loss and impairment of synaptic activity); preferential loss of short-term memory (e.g., increased susceptibility to interference in immediate memory); alterations in EEG and sleep patterns (probably due to a slowing down of brain metabolism consequent to hypoxia); increased frequency of communication disorders (e.g., changes in speech and voice, aphasia); decreased efficiency of autonomic control (e.g., decreased responsiveness of baroreceptor adjustments to vasomotor changes); and increased incidence of psychiatric disorders (e.g., from depression to dementia). In the preceding section the relationship of monoamines to hypothalamic and pituitary hormones was underlined. It is important to point out that alterations in monoamine metabolism have also been implicated in some of the neurologic and psychiatric manifestations of senescence just listed. For example, disturbances in monoamine metabolism have been linked to such age-related phenomena as graying of the hair (L-dopa is a melanin precursor), slowness in initiating and executing voluntary movements, defects and hesitancy in speech, increased incidence of extrapyramidal behavioral disorders, disruption of sleep patterns, and impairment in thinking, concentration, and alterness, and recovery from stress. In view of the central role of the brain in regulating numerous bodily functions, deteriorative changes occurring with age and their effects on performance represent a subject of interest and active research not only for physiologists but also for psychologists and social scientists; this is particularly true with respect to the progressive decline in adaptive capacity, for the assets and liabilities of old age, viewed from any vantage point, must be assessed with respect to the demands of the environment in which the elderly live.

As for the brain, many investigators have attempted to delineate changes in endocrine function identifiable as having a causative role in aging. However, the resulting information has been for the most part circumstantial, inasmuch as cessation of endocrine function or clear-cut alterations in hormone levels and metabolism could not be detected except for the arrest of the ovarian function at menopause. A typical example of the dichotomy between signs of endocrine insufficiency and unaltered hormone levels is represented by the case of the thyroid gland. Whereas its crucial role during early development on body growth and brain maturation and its influence during adulthood on basal metabolic rate and brain excitability are undisputed (Chapter 71), its involvement in aging processes is uncertain. Several investigators have related some of the features of senescence (e.g., decreased metabolic rate, diminished motor activity, impaired muscular strength, dryness of skin, sparseness of hair) to those of the hypothyroid state even though comparison of thyroid activity as a function of age in adult individuals has not revealed profound differences and the administration of thyroid hormones has not rejuvenated older persons.[65] Thus the hypothesis of hypothyroidism as a concomitant of aging was discarded for some time. However, more recent work seems to indicate that even if thyroid function is unchanged, the sensitivity of the target tissues to thyroid hormones decreases with old age. According to evidence based on age-related observations of thyroid and hypophyseal interrelations, responsiveness of the thyroid gland to thyroid-stimulating hormone also decreases with

old age.[42] In addition, other studies have suggested the existence of a pituitary factor, the release of which increases with aging and is responsible for the decreased responsiveness to thyroid hormones of tissues from old rats.[24] The hypothesis that aging induces alterations in the ability of animals and tissues to respond to hormonal signals rather than alterations in the secretory activity of the endocrine glands is supported by the demonstration that the number and/or affinity of receptors for steroid hormones change with development and aging.[49]

From what we know of neuroendocrine relations in the adult (Chapter 34) and the changes in the nervous and endocrine systems with aging, the following sequence of events relating neurotransmission to neuroendocrine balance may be postulated (Fig. 77-5). During development, neurotransmitters in several cortical and subcortical centers progressively mature[34] and begin to exert a regulatory action on the neurosecretory cells of the hypothalamus, reaching stability during adulthood.[36,37,71] Coincident with the maturation of these hypothalamic neurosecretory systems the releasing and inhibiting hormones they produce trigger the anterior pituitary gland to synthesize and secrete its tropic hormones, which, in turn, stimulate the target endocrine glands to secrete their respective hormones. These hormones, especially thyroid, adrenal, and gonadal hormones, reach the target effector cells where they may act as "gene triggers," causing target tissues to undergo developmental changes. The activity of each of the control systems—cortex, hypothalamus, pituitary, and other endocrine glands—and the positive and negative feedbacks that regulate them may be impaired with aging. If, as has been postulated, the monoaminergic neurons in the higher brain centers represent the "pacemakers" that regulate the biologic clock governing development and aging,[6,25,50-54] then it follows that any slowing down of the activity of these neurons with aging would lead to alterations in neurotransmitter activity. Such a slowing down would, in turn, have repercussions in neural, endocrine, muscular, secretory, and motor functions as variously manifested in the aged; examples are, involution of reproductive organs, loss of fertility, diminished muscular strength, lessened ability to recover from stress, and impairment of secretory, cardiovascular, and respiratory activity (Fig. 77-5).

The possibilities that programmed changes in the activities of genes are not confined only to fertilization but may continue at later ages and

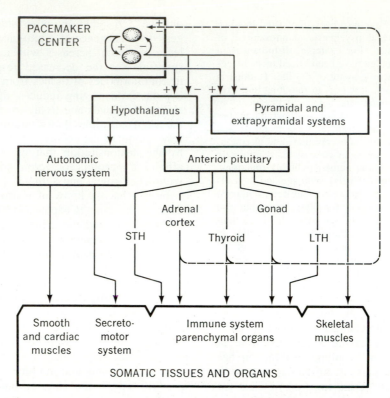

Fig. 77-5. Schematic drawing of CNS-endocrine theory of programmed aging. Specific "pacemaker" centers (25) composed of groups of excitatory (+) or inhibitory (−) neurons, which feed back to each other by direct or indirect neural processes, create stimulatory (+) or repressive (−) influences on brain areas such as hypothalamus (which controls autonomic nervous and neuroendocrine systems) and pyramidal and extrapyramidal nervous systems (which control skeletal muscle tone and movement). These influences then affect structure and function of musculature (smooth, cardiac, and striated), exocrine glands, immune system, and parenchymal organs. With aging, it is hypothesized that inhibitory neural influences are programmed to become predominant,[64] creating loss in muscle tone and function, alterations or decline in elaboration of many secretions, cardiovascular impairment, immunologic suppression, parenchymal organs—and therefore homeostatic—performance loss, and many other physiologic deficits associated with senescence. Dashed line represents hypothetical feedback influences of thyroid and steroid hormones on regulatory centers of brain.

that these changes are triggered by hormonal influences are well documented in insects and amphibians.[4,7,12] If we assume that such programming operates in mammals as well and is regulated by neural and endocrine systems, then interventions at the level of these systems may influence aging as well as growth and development. Speculative as this hypothesis may appear, it has the advantage of providing a practical handle with which to investigate aging processes. There are now various means of altering monoamine levels and metabolism in the brain, and it is rational to expect that these alterations affect several functions throughout the life span. Old and new experiments seem to support this point of view. Thus dietary and pharmacologic manipulations of brain monoamines, such as administration of diets low in total calories or in a single amino acid (e.g., tryptophan, the precursor of serotonin) or those containing a monoamine inhibitor (e.g., parachlorophenylalanine) have been shown to influence growth and development in rats and to delay age-related decrements in homeostatic and reproductive functions as well as to prolong the life span.[42,53-55] Similarly, prolonged administration of L-dopa, a precursor of dopamine, significantly extends the life span in mice,[22] and pharmacologic (e.g., administration of iproniazide, a monoamine oxidase inhibitor), hormonal, and electrical stimulation reinitiate estrous cyclicity in old rats and restore some of their physiologic competence.[6,21]

From the few considerations on the physiology of aging presented here, the guidelines for future studies in this area remain broadly drawn. It is in the conviction that studies of aging in higher animals and particularly in humans will benefit from an integrative physiologic approach that we have emphasized new avenues of investigation, even though some of them are still little explored.

RESEARCH PERSPECTIVES

Of the theories of aging briefly mentioned in the preceding section, none have so far been completely substantiated. Yet in recent years the impressive gain in longevity (in terms of the greater number of individuals expected to reach advanced age) and the perspective of even more dramatic increases in the near future (on the basis of the current increases, it has been calculated that, by 1985, the population of individuals 70 years of age and over will show a dramatic rise of 11% for men and 13.7% for women) make this area a subject of urgent concern for its medical and socioeconomic implications. From a medical point of view a better understanding of functional changes with aging becomes necessary not only to provide appropriate treatment for the ill, but also to devise means to improve physiologic competence for the healthy.

As physiologists become aware of the possibility of modifying the life span of laboratory animals by a variety of experimental (e.g., nutritional, endocrine, and pharmacologic) manipulations, they may also use these experimental results as models for new approaches to lengthen the life span in humans as well. Several areas of research that have the potential of altering aging and death are now in the early stages of development; although their application may appear far away, their potential contributions to progress not only in gerontology but also in physiology are worthy of consideration. Techniques of deep hypothermia[17] and those of total body washout and replacement by blood substitutes have already been developed and experimentally tested in humans and animals.[30,33] Mammalian cells and tissues, including early stage wholemouse embryos perfused with various cryoprotective solutions, have been reversibly frozen to temperatures as low as that of liquid helium and then successfully restored to normal function.[69] The not-too-distant future will undoubtedly see the marriage of deep hypothermia and cryogenic biology for the reversible induction of long-term solid-state suspended animation. These techniques, along with the development of the transplantation of natural organs or the implantation of their synthetic counterparts, coupled with breakthroughs in the science of resuscitation (which may lead to a considerable extension of the duration of reversible clinical death),[73] could rapidly lead to a marked increase in the length and quality of the human life span. Perhaps it is befitting that such technology appears as mankind takes its first steps on extraterrestrial soil.

REFERENCES
General reviews

1. Andrew, W.: The anatomy of aging in man and animals, New York, 1971, Grune & Stratton, Inc.
2. Birren, J. E., and Schaie, K. W., editors: Handbook of the psychology of aging, Cincinnati, 1976, Van Nostrand Reinhold Co.
3. Burch, P. R. J.: Growth, disease and ageing, Toronto, Canada, 1969, University of Toronto Press.
4. Comfort, A.: Aging, the biology of senescence, New York, 1964, Holt, Rinehart and Winston, Inc.
5. Curtis, H. J.: Biological mechanisms of aging, Springfield, Ill., 1966, Charles C Thomas, Publisher.
6. Everitt, A. V., and Burgess, J. A., editors: Hypothalamus, pituitary and aging, Springfield, Ill., 1976, Charles C Thomas, Publisher.
7. Finch, D. E., and Hayflick, L., editors: Handbook of the biology of aging, New York, 1977, Van Nostrand Reinhold Co.
8. Goldman, R., and Rockstein, M.: The physiology and pathology of human aging, New York, 1975, Academic Press, Inc.
9. Korenchevsky, V.: Physiological and pathological aging, New York, 1961, Hafner Publishing Co.
10. Strehler, B. L.: Time, cells and aging, ed. 2, New York, 1977, Academic Press, Inc.
11. Thorbecke, G. J., editor: Biology of aging and development, New York, 1975, Plenum Press.
12. Timiras, P. S.: Developmental physiology and aging, New York, 1972, Macmillan Publishing Co., Inc.
13. Walford, R. L.: The immunologic theory of aging, Copenhagen, 1969, Munksgaard.

Original papers

14. Allen, T. H., Anderson, E. C., and Langham, W. H.: Total body potassium and gross body composition in relation to age, J. Gerontol. **15**:348, 1960.
15. Bergman, M.: Changes in hearing with age, Gerontologist **11**:148, 1971.
16. Bjorksten, J.: The crosslinkage theory of aging, J. Am. Geriatr. Soc. **16**:408, 1968.
17. Black, P. R., Van Devanter, S., and Cohn, L. H.: Effects of hypothermia on systemic and organ system metabolism and function, J. Surg. Res. **20**:49, 1976.
18. Bourlière, F.: Principes et méthodes de mesure de l'âge biologique chez l'homme, Bull. Mem. Soc. Anthropol. Paris **4**:561, 1963.
19. Brizzee, K. R., Sherwood, N., and Timiras, P. S.: A comparison of cell populations at various depth levels in cerebral cortex of young adult and aged Long-Evans rats, J. Gerontol. **23**:289, 1968.
20. Burnet, F. M.: An immunological approach to aging, Lancet **2**:358, 1970.
21. Clemens, J. A., Amenomori, Y., Jenkins, T., and Meites, J.: Effects of hypothalamic stimulation, hormones, and drugs on ovarian function in old female rats, Proc. Soc. Exp. Biol. Med. **132**:561, 1969.
22. Cotzias, G. C., et al.: Levodopa, fertility and longevity, Science **197**:549, 1977.
23. Cutler, R. G.: Evolution of longevity in primates, J. Hum. Evol. **5**:169, 1976.
24. Denckla, W. D.: Role of the pituitary and thyroid glands in the decline of minimal O_2 consumption with age, Q. Rev. Biol. **51**:49, 1976.

25. Finch, C. E.: Physiological changes of aging in mammals, Q. Rev. Biol. **51**:49, 1976.

26. Forbes, G. B., and Reina, J. C.: Adult lean body mass declines with age: some longitudinal observations, Metabolism **19**:653, 1970.

27. Friedman, G. M.: Atherosclerosis and the pediatrician. In Winick, M., editor: Childhood obesity, New York, 1975, John Wiley & Sons, Inc.

28. Glorig, A., and Nixon, J.: Distribution of hearing loss in various populations, Ann. Otol. Rhinol. Laryngol. **69**:497, 1960.

29. Glueck, C. J., Fallat, R. W., and Tsang, R.: A pediatric approach to atherosclerosis prevention. In Winick, M., editor: Childhood obesity, New York, 1975, John Wiley & Sons, Inc.

30. Haff, R. C., Klebanoff, G., Brown, B. G., and Koreski, W. R.: Asanguineous hypothermic perfusion as a means of total organism preservation, J. Surg. Res. **19**:13, 1975.

31. Harman, D.: Free radical theory of aging, Triangle **12**:153, 1973.

32. Kanungo, M. S.: A model for ageing, J. Theor. Biol. **53**:253, 1975.

33. Klebanoff, G., et al.: Resuscitation of a patient in stage IV hepatic coma using total body washout, J. Surg. Res. **13**:159, 1972.

34. Loizou, L. A.: The postnatal ontogeny of monoamine-containing neurons in the central nervous system of the albino rat, Brain Res. **40**:395, 1972.

35. Makinodan, T.: Immunity and aging. In Finch, D. E., and Hayflick, L., editors: Handbook of the biology of aging, New York, 1977, Litton Educational Publishing, Inc.

36. Martini, L., and Meites, J.: Neurochemical aspects of hypothalamic function, New York, 1970, Academic Press, Inc.

37. McCann, S. M., et al.: The role of monoamines in the control of gonadotropin and prolactin secretion. In Saxena, B. B., Beling, C. G., and Gandy, H. M., editors: Gonadotropins, New York, 1972, Wiley-Interscience.

38. Medvedev, Z.: Possible role of repeated nucleotide sequences in DNA in the evolution of life spans of differentiated cells, Nature **237**:453, 1972.

39. Minot, C.: The problems of age, growth and death, London, 1908, G. P. Putnam's Sons.

40. Neumaster, T. D., and Ring, G. C.: Creatinine excretion and its relation to whole body potassium and muscle mass in inbred rats, J. Gerontol. **20**:379, 1965.

41. Novak, L.: Aging, total body potassium, fat-free mass, and cell mass in males and females between ages 18 and 85 years, J. Gerontol. **27**:438, 1972.

42. Ooka, H., Segall, P. E., and Timiras, P. S.: Neural and endocrine development after chronic tryptophan deficiency in rats. II. Pituitary-thyroid axis, Mech. Ageing Dev. **7**:19, 1978.

43. Orgel, L. E.: The maintenance of the accuracy of protein synthesis and its relevance to aging, Proc. Natl. Acad. Sci. USA **49**:517, 1963.

44. Orgel, L. E.: Aging of clones of mammalian cells, Nature **243**:441, 1973.

45. Packer, L., Deamer, D. W., and Heath, R. L.: Regulation and deterioration of structures in membranes. In Strehler, B. L., editor: Advances in gerontological research, New York, 1967, Academic Press, Inc., vol. 2.

46. Parot, S.: Les méthodes de mesure de la masse métabolique active et de la masse grasse: principes et études

47. Pierson, R. N., Jr., Lin, D. H. Y., and Phillips, R. A.: Total-body potassium in health: effects of age, sex, height, and fat, Am. J. Physiol. **226**:206, 1974.

48. Rosen, S., et al.: Presbycusis study of a relatively noise-free population in the Sudan, Trans. Am. Otol. Soc. **50**:135, 1962.

49. Roth, G. S.: Reduced glucocorticoid binding site concentration in cortical neuronal perikarya from senescent rats, Brain Res. **107**:345, 1976.

50. Rowe, J. W., et al.: The effect of age on creatinine clearance in men: a cross-sectional and longitudinal study, J. Gerontol. **31**:155, 1976.

51. Sacher, G. A., and Trucco, E.: The stochastic theory of mortality, Ann. N.Y. Acad. Sci. **96**:985, 1962.

52. Saunders, J. W., Jr.: Death in embryonic systems. Science **154**:604, 1966.

53. Segall, P. E., and Timiras, P. S.: Age-related changes in thermoregulatory capacity of tryptophan-deficient rats, Fed. Proc. **34**:83, 1975.

54. Segall, P. E., and Timiras, P. S.: Patho-physiological findings in chronic tryptophan restriction: a model for delayed growth and aging in the rat, Mech. Ageing Dev. **5**(2):109, 1976.

55. Segall, P. E., Ooka, H., Rose, K. and Timiras, P. S.: Neural and endocrine development after chronic tryptophan deficiency in rats: I. Brain monoamine and pituitary responses, Mech. Ageing Dev. **7**:1, 1978.

56. Segre, D., and Segre, M.: Humoral immunity in aged mice. II. Increased suppressor T cell activity in immunologically deficient old mice, J. Immunol. **116**:735, 1976.

57. Shock, N. W.: Energy metabolism, caloric intake and physical activity of the aging. In Carlson, L. A., editor: Nutrition in old age, Tenth Symposium of the Swedish Nutrition Foundation, Uppsala, Sweden, 1972, Almqvist & Wiksell.

58. Shock, N. W., and Norris, A. H.: Neuromuscular coordination as a factor in age changes in muscular exercise. In Jokl, E., and Brunner, D., editors: Physical activity and aging, Basel, 1970, S. Kargerag, vol. 4.

59. Sinex, F. M.: Aging and the lability of irreplaceable molecules. II. Amide groups of collagen, J. Gerontol. **15**:15, 1960.

60. Sinex, F. M.: The mutation theory of aging. In Rockstein, M., editor: Theoretical aspects of aging, New York, 1974, Academic Press, Inc.

61. Strehler, B. L., Mark, D. D., Mildvan, A. S. and Gee, M. V.: Rate and magnitude of age pigment accumulation in the human myocardium, J. Gerontol. **14**:430, 1959.

62. Strehler, B., et al.: Codon-restriction theory of aging and development, J. Theor. Biol. **33**:429, 1971.

63. Teller, M. N.: Age changes and immune resistance to cancer, Adv. Geront. Res. **4**:25, 1972.

64. Timiras, P. S.: Neurophysiological factors in aging: recent advances, Jerusalem, Israel, June, 1975, Tenth International Congress of Gerontology, vol. 1.

65. Timiras, P. S., and Bignami, A.: Pathophysiology of the aging brain. In Elias, M. F., Eleftheriou, B. E., and Elias, P. K., editors: Progress in biology, Bar Harbor, 1976, Ear, Inc.

66. Timiras, P. S., Hudson, D. B., and Oklund, S.: Changes in central nervous system free amino acids with development and aging. In Ford, D. H., editor: Neurobiological aspects of maturation and aging, New York, 1973, Elsevier Scientific Publishing Co., vol. 40.

critiques. In Bourliere, F., editor: Progres en gerontologie, Paris, 1969, Flammarion et Cie.

67. von Hahn, H. P.: Structural and functional changes in nucleoprotein during the ageing of the cell, Gerontologia **16:**116, 1970.

68. Weiss, P.: Aging, a corollary of development. In Shock, N. W., editor: Perspectives in experimental gerontology, Springfield, Ill., 1966, Charles C Thomas, Publisher.

69. Whittingham, D. G., Leibo, S. P., and Mazur, P.: Survival of mouse embryos frozen to −196° and −269° C, Science **178:**411, 1972.

70. Wright, W. E., and Hayflick, L.: Contributions of cytoplasmic factors to *in vitro* cellular senescence. In Thorbecke, G. J., editor: Biology of aging and development, New York, 1975, Plenum Press.

71. Wurtman, R. J.: Brain catecholamines and the control of secretion from the anterior pituitary gland. In Meites, J., editor: Hypophysiotropic hormones on the hypothalamus: assay and chemistry, Baltimore, 1970, The Williams & Wilkins Co.

72. Upton, A. C.: Ionizing radiation and the aging process, a review, J. Gerontol. **12:**306, 1957.

73. Zimmermann, V., and Hossmann, K.-A.: Resuscitation of the monkey brain after one hour's complete ischemia. II. Brain water and electrolytes, Brain Res. **85:**1, 1975.

Index*

A

Ablation
   of anterior lobe, results of, 845-846
   of pyramis and paramedian lobule, effects of, 846
   of tuber vermis and ansiform lobule, results of, 846
Absolute refractory period, 68
Absorption
   of carbohydrates in alimentary canal, 1264-1271; *see also*
      Carbohydrates, absorption of, in alimentary canal
   of fluids and nutrients by alimentary canal, *1255-1286;*
      see also Alimentary canal, absorptive function
      of
   intestinal, methods for investigation of, 1258-1259
   of lipids in alimentary canal, 1278-1286; *see also* Lipids,
      absorption of, in alimentary canal
   of monosaccharides, 1267-1270
   of proteins in alimentary canal, 1271-1277; *see also* Pro-
      tein(s), absorption of, in alimentary canal
   of water and electrolytes in alimentary canal, 1259-1264;
      see also Water and electrolytes, absorption of, in
      alimentary canal
Acceleration, effects of, on normal circulation, 1043-1044
Accelerometers, linear, otolith organs as, 821-823
Accession rate in fluid formation, 1226
Acclimatization, 1853, 1859-1868
   to altered carbon dioxide tension, 1867-1868
   to altitude hypoxia, mechanisms of, 1863-1867
      deacclimatization as, 1867
      improvement in oxygenation, 1865-1867
      increased cardiac activity as, 1867
      respiratory component of, 1863-1864
      respiratory consequences of alterations in acid-base bal-
        ance as, 1864-1865
   factors involved in, 1860
   failure of, 1862-1863
   to lowered partial pressure of oxygen of high altitude,
      1859-1863
   programmed, in major mountain climbing expeditions,
      1860-1861
   respiratory component of, 1863-1864
   thermoregulatory adjustments in, 1452-1455
Accommodation
   changes in, with age, 488-489, 491-492
   ciliary muscle and, 491-492
   definition of, 491
   to excitation, 48, *49,* 70
   measurement of, 488
   nerve, sensory receptor adaptation and, 335-336
   speed of, reduced, in elderly, 1989
   voltage clamp parameters in explanation of, 70

Acetazolamide
   effect of
      on bicarbonate transfer in cerebrospinal fluid, 1231
      on rate of aqueous formation, 484
   for hydrocephalus, 1239
   for Ménière's disease, 1248
Acetylcholine (ACh)
   and acetylcholinesterase, formation of complex between,
      162
   in ACTH release, 1574
   and chemoreceptor activity, 1762, *1763*
   as cholinergic transmitter, 902-903, 905
   distribution of, in central nervous system, 207
   effect of, on heart rate, 982
   enzymatic hydrolysis of, 161-163
   enzymatic synthesis and hydrolysis of, 160-163
   as ganglionic transmitter agent, 906
   in gastric secretion, 1299, 1300
   in hydrochloric acid secretion, 1300
   increased sensitivity of denervated mammalian skeletal
      muscle to, 176
   influence of
      on vascular smooth muscle, 136
      on visceral smooth muscle performance, 132
   as muscarinic receptor, 907
   as neurochemical mediator of salivary secretion, 1292
   in neuromuscular transmission of skeletal muscle, 152
   as neurotransmitter
      in basal ganglia, 810
      in skeletal muscle excitation, 88
   in production of depolarization associated with gastric peri-
      stalsis, 1334
   in production of miniature potentials in smooth muscle,
      138-139
   receptors for, 203
   in reduction of intraocular pressure, 1243
   release of, by nerve impulse, decline in, in neuromuscular
      fatigue, 173
   sensitivity to, changes in, following preganglionic nerve
      section, 207
   spontaneous and activated release of, from motor nerve
      terminals, 163-165
   spontaneous release of, and release evoked by nerve im-
      pulse, relation between, 165-167
   synthesis, distribution, and storage of, 160-161
   transmission of, postsynaptic response of muscle cell to,
      167-172
   as transmitter agent, 203-204, 210*t*
      in central cholinergic transmission, 207-209
      at neuromuscular junction, 155-157
Acetylcholinesterase (AChE)
   and acetylcholine, formation of complex between, 162
   in central cholinergic transmission, 207
   in cholinergic axons of smooth muscle, 141, 143, *144*

---

*Page numbers in *italics* indicate illustrations; those followed
by *t* indicate tables; those followed by *n* indicate footnotes.